CHILTON®

GENERAL MOTORS
DIAGNOSTIC SERVICE
2006 EDITION

THOMSON

DELMAR LEARNING

Australia • Canada • Mexico • Singapore • Spain • United Kingdom • United States

CHILTON®

GENERAL MOTORS
DIAGNOSTIC SERVICE
2006 Edition

**Vice President,
Technology Professional Business Unit:**
Gregory L. Clayton

**Publisher,
Professional Business Unit:**
David Koontz

Production Director:
Mary Ellen Black

Marketing Director:
Beth A. Lutz

Marketing Specialist:
Brian McGrath

Marketing Coordinator:
Marissa Mariella

Marketing Assistant:
Jennifer Stall

Sr. Production Editor:
Elizabeth Hough

Editorial Assistant:
Christine Wade

Editors:
Dennis Bailey
Timothy A. Crain

Publishing Coordinator:
Paula Baillie

Cover Design:
Melinda Possinger

ISBN: 1-4180-2120-2

NOTICE TO THE READER

TABLE OF CONTENTS

1 - INTRODUCTION AND APPLICATION

OBD II VEHICLE APPLICATIONS 1-2
NOTES & CAUTIONS 1-7
PRELIMINARY DIAGNOSTICS 1-8
DIAGNOSTIC TOOLS & CIRCUIT TESTING. 1-11
EFFECTIVE DIAGNOSTICS 1-12

2 - IDENTIFYING THE PROBLEM

INTRODUCTION 2-2
WHERE TO BEGIN 2-3

3 - DIAGNOSTIC TROUBLE CODES

HOW TO USE THIS SECTION 3-3
GM OBD II SYSTEMS 3-4
GM OBD II SYSTEM TERMINOLOGY. 3-8
GM OBD II MONITORS 3-21

4 - COMPONENT TESTING

GM CARS(EXC. CADILLAC) 4-2
CADILLAC 4-173
GM TRUCKS & VANS 4-297

5 - SYMPTOM DIAGNOSIS (NO CODES)

WHAT TO DO WHEN THERE ARE NO DTCS 5-2
SYMPTOM DIAGNOSIS TESTS 5-3
INTERMITTENT FAULT TESTS. 5-11
OTHER DIAGNOSIS AND TESTING 5-12

6 - PID CHARTS

BUICK CENTURY 6-6
LACROSSE 6-17
LESABRE 6-23
PARK AVENUE 6-23
REGAL 6-23
RENDEVOUS. 6-38
RIVIERA 6-23
ROADMASTER 6-42
SKYLARK 6-47
TERRAZA 6-55
CADILLAC CATERA 6-59
CTS. 6-63
DEVILLE. 6-73
ELDORADO 6-73
ESCALADE, ESCALADE EXT, ESCALADE ESV 6-86
FLEETWOOD, FLEETWOOD BROUGHAM 6-95
SEVILLE 6-73
STS. 6-99
CHEVROLET BERETTA, CORSICA 6-105
CAMARO 6-110
CAPRICE 6-136
CAVALIER. 6-144
COBALT 6-148
CORVETTE 6-154
IMPALA. 6-173
LUMINA & MONTE CARLO 6-178
LUMINA 6-195
MALIBU 6-200
MONTE CARLO 6-212
UPLA NDER 6-220
C/K, M/L & S/T SERIES VEHICLES. 6-224
ASTRO & SAFARI 6-224
BLAZER JIMMY, SONMA, S10 PICKUP 6-249
ENVOY & TRAILBLAZER UTILITY VEHICLE 6-254
EQUINOX UTILITY VEHICLE 6-257
SSR 6-261
VENTURE. 6-265
OLDSMOBILE ACHIEVA 6-268
ALERO 6-276
AURORA 6-285
CUTLASS 6-301
CUTLASS CIERA 6-307

CUTLASS CIERA, CRUISER 6-312
CUTLASS SUPREME 6-313
EIGHTY-EIGHT, NINETY EIGHT 6-320
INTRIGUE 6-326
LSS, REGENCY 6-334
SILHOUETTE 6-338
PONTIAC AZTEC 6-341
BONNEVILLE 6-344
FIREBIRD 6-352
G6 6-378
GRAND AM 6-382
GRAND PRIX 6-401
GTO 6-426
MONTANA SV6 6-432
MONTANA 6-436
SUNFIRE 6-439
TRANS SPORT 6-436
SATURN L100 4-DOOR SEDAN 6-443
L200 4-DOOR SEDAN 6-443
L300 4-DOOR SEDAN 6-446
LS 4-DOOR SEDAN 6-443
LS1 4-DOOR SEDAN 6-443
LS2 4-DOOR SEDAN 6-443
LW1 4-DOOR WAGON 6-443
LW2 4-DOOR WAGON 6-443
LW200 4-DOOR WAGON 6-443
LW300 4-DOOR WAGON 6-446
RELAY 6-449
SC1 2-DOOR COUPE 6-453
SC2 2-DOOR COUPE 6-453
SL 4-DOOR SEDAN 6-453
SL1 4-DOOR SEDAN 6-453
SL2 4-DOOR SEDAN 6-453
SW1 4-DOOR WAGON 6-453
SW2 4-DOOR WAGON 6-453
VUE 4-DOOR UTILITY 6-443

USING THIS INFORMATION

Organization

To find where a particular model section or procedure is located, look in the Table of Contents. Main topics are listed with the page number on which they may be found. Following the main topics is a listing of all of the subjects within the section and their page numbers.

Manufacturer and Model Coverage

This product covers 1995-2005 General Motors models that are produced in sufficient quantities to warrant coverage, and which have technical content available from the vehicle manufacturers before our publication date. Although this information is as complete as possible at the time of publication, some manufacturers may make changes which cannot be included here. While striving for total accuracy, the publisher cannot assume responsibility for any errors, changes, or omissions that may occur in the compilation of this data.

Part Numbers & Special Tools

Part numbers and special tools are recommended by the publisher and vehicle manufacturer to perform specific jobs. Before substituting any part or tool for the one recommended, you must be completely satisfied that neither your personal safety, nor the performance of the vehicle will be endangered.

ACKNOWLEDGEMENT

Portions of materials contained herein have been reprinted with permission from General Motors Corporation, Service and Parts Operations under License Agreement #0510757. The publisher would like to express appreciation to General Motors Corporation for its assistance in producing this publication. No further reproduction or distribution of the material in this manual is allowed without the expressed written permission of the publisher.

PRECAUTIONS

Before servicing any vehicle, please be sure to read all of the following precautions, which deal with personal safety, prevention of component damage, and important points to take into consideration when servicing a motor vehicle:

• Always wear safety glasses or goggles when drilling, cutting, grinding or prying.

• Steel-toed work shoes should be worn when working with heavy parts. Pockets should not be used for carrying tools. A slip or fall can drive a screwdriver into your body.

• Work surfaces, including tools and the floor should be kept clean of grease, oil or other slippery material.

• When working around moving parts, don't wear loose clothing. Long hair should be tied back under a hat or cap, or in a hair net.

• Always use tools only for the purpose for which they were designed. Never pry with a screwdriver.

• Keep a fire extinguisher and first aid kit handy.

• Always properly support the vehicle with approved stands or lift.

• Always have adequate ventilation when working with chemicals or hazardous material.

• Carbon monoxide is colorless, odorless and dangerous. If it is necessary to operate the engine with vehicle in a closed area such as a garage, always use an exhaust collector to vent the exhaust gases outside the closed area.

• When draining coolant, keep in mind that small children and some pets are attracted by ethylene glycol antifreeze, and are quite likely to drink any left in an open container, or in puddles on the ground. This will prove fatal in sufficient quantity. Always drain the coolant into a sealable container.

• To avoid personal injury, do not remove the coolant pressure relief cap while the engine is operating or hot. The cooling system is under pressure; steam and hot liquid can come out forcefully when the cap is loosened slightly. Failure to follow these instructions may result in personal injury. The coolant must be recovered in a suitable, clean container for reuse. If the coolant is contaminated it must be recycled or disposed of correctly.

• When carrying out maintenance on the starting system be aware that heavy gauge leads are connected directly to the battery. Make sure the protective caps are in place when maintenance is completed. Failure to follow these instructions may result in personal injury.

• Do not remove any part of the engine emission control system. Operating the engine without the engine emission control system will reduce fuel economy and engine ventilation. This will weaken engine performance and shorten engine life. It is also a violation of Federal law.

• Due to environmental concerns, when the air conditioning system is drained, the refrigerant must be collected using refrigerant recovery/recycling equipment. Federal law requires that refrigerant be recovered into appropriate recovery equipment and the process be conducted by qualified technicians who have been certified by an approved organization, such as MACS, ASI, etc. Use of a recovery machine dedicated to the appropriate refrigerant is necessary to reduce the possibility of oil and refrigerant incompatibility concerns. Refer to the instructions provided by the equipment manufacturer when removing refrigerant from or charging the air conditioning system.

• Always disconnect the battery ground when working on or around the electrical system.

• Batteries contain sulfuric acid. Avoid contact with skin, eyes, or clothing. Also, shield your eyes when working near batteries to protect against possible splashing of the acid solution. In case of acid contact with skin or eyes, flush immediately with water for a minimum of 15 minutes and get prompt medical attention. If acid is swallowed, call a physician immediately. Failure to follow these instructions may result in personal injury.

• Batteries normally produce explosive gases. Therefore, do not allow flames, sparks or lighted substances to come near the battery. When charging or working near a battery, always shield your face and protect your eyes. Always provide ventilation. Failure to follow these instructions may result in personal injury.

• When lifting a battery, excessive pressure on the end walls could cause acid to spew through the vent caps, resulting in personal injury, damage to the vehicle or battery. Lift with a battery carrier or with your hands on opposite corners. Failure to follow these instructions may result in personal injury.

• Observe all applicable safety precautions when working around fuel. Whenever servicing the fuel system, always work in a well-ventilated area. Do not allow fuel spray or vapors to come in contact with a spark, open flame, or excessive heat (a hot drop light, for example). Keep a dry chemical fire extinguisher near the work area. Always keep fuel in a container specifically designed for fuel storage; also, always properly seal fuel containers to avoid the possibility of fire or explosion. Do not smoke or carry lighted tobacco or open flame of any type when working on or near any fuel-related components.

• Fuel injection systems often remain pressurized, even after the engine has been turned OFF. The fuel system pressure must be relieved before disconnecting any fuel lines. Failure to do so may result in fire and/or personal injury.

• The evaporative emissions system contains fuel vapor and condensed fuel vapor. Although not present in large quantities, it still presents the danger of explosion or fire. Disconnect the battery ground cable from the battery to minimize the possibility of an electrical spark occurring, possibly causing a fire or explosion if fuel vapor or liquid fuel is present in the area. Failure to follow these instructions can result in personal injury.

• The EPA warns that prolonged contact with used engine oil may cause a number of skin disorders, including cancer! You should make every effort to minimize your exposure to used engine oil. Protective gloves should be worn when changing oil. Wash your hands and any other exposed skin areas as soon as possible after exposure to used engine oil. Soap and water, or waterless hand cleaner should be used.

• Some vehicles are equipped with an air bag system, often referred to as a Supplemental Restraint System (SRS) or Supplemental Inflatable Restraint (SIR) system. The system must be disabled before performing service on or around system components, steering column, instrument panel components, wiring and sensors. Failure to follow safety and disabling procedures could result in accidental air bag deployment, possible personal injury and unnecessary system repairs.

• Always wear safety goggles when working with, or around, the air bag system. When carrying a non-deployed air bag, be sure the bag and trim cover are pointed away from your body. When placing a non-deployed air bag on a work surface, always face the bag and trim cover upward, away from the surface. This will reduce the motion of the module if it is accidentally deployed.

• Electronic modules are sensitive to electrical charges. The ABS module can be damaged if exposed to these charges.

• Brake pads and shoes may contain asbestos, which has been determined to be a cancer-causing agent. Never clean brake surfaces with compressed air. Avoid inhaling brake dust. Clean all brake surfaces with a commercially available brake cleaning fluid.

• When replacing brake pads, shoes, discs or drums, replace them as complete axle sets.

• When servicing drum brakes, disassemble and assemble one side at a time, leaving the remaining side intact for reference.

• Brake fluid often contains polyglycol ethers and polyglycols. Avoid contact with the eyes and wash your hands thoroughly after handling brake fluid. If you do get brake fluid in your eyes, flush your eyes with clean, running water for 15 minutes. If eye irritation persists, or if you have taken brake fluid internally, immediately seek medical assistance.

• Clean, high quality brake fluid from a sealed container is essential to the safe and proper operation of the brake system. You should always buy the correct type of brake fluid for your vehicle. If the brake fluid becomes contaminated, completely flush the system with new fluid. Never reuse any brake fluid. Any brake fluid that is removed from the system should be discarded. Also, do not allow any brake fluid to come in contact with a painted or plastic surface; it will damage the paint.

• Never operate the engine without the proper amount and type of engine oil; doing so will result in severe engine damage.

• Timing belt maintenance is extremely important! Many models utilize an interference-type, non-freewheeling engine. If the timing belt breaks, the valves in the cylinder head may strike the pistons, causing potentially serious (also time-consuming and expensive) engine damage.

• Disconnecting the negative battery cable on some vehicles may interfere with the functions of the on-board computer system(s) and may require the computer to undergo a relearning process once the negative battery cable is reconnected.

• Steering and suspension fasteners are critical parts because they affect performance of vital components and systems and their failure can result in major service expense. They must be replaced with the same grade or part number or an equivalent part if replacement is necessary. Do not use a replacement part of lesser quality or substitute design. Torque values must be used as specified during reassembly to ensure proper retention of these parts.

OBD II VEHICLE APPLICATIONS
     Cars...Page 1-2
     SUVs, Trucks & Vans ....Page 1-4

NOTES & CAUTIONS
     Notes & Cautions ....Page 1-7

PRELIMINARY DIAGNOSTICS
     History of OBD Systems ....Page 1-8
     OBD I System Diagnostics ....Page 1-8
     Changes In Diagnostic Routines ....Page 1-8
     OBD II System Overview ....Page 1-10
     Common Terminology ....Page 1-11

DIAGNOSTIC TOOLS & CIRCUIT TESTING
     Hand Tools & Meter Operation ....Page 1-11
     Scan Tools ....Page 1-11
     Malfunction Indicator Lamp ....Page 1-12
     Electronic Controls ....Page 1-12
     Electricity & Electrical Circuits ....Page 1-12
     Circuit Testing Tools ....Page 1-12

EFFECTIVE DIAGNOSTICS
     Getting Started ....Page 1-12

General Motors OBD II Vehicle Coverage

<u>GM CAR APPLICATIONS</u>

A-Car Body Codes

1996 Century, Ciera SL & Wagon
Engine: 2.2L I4 .... VIN 4
Engine: 3.1L V6 .... VIN M
2005-06 Cobalt
Engine: 2.0L I4 .... VIN F
Engine: 2.2L I4 .... VIN F

B-Car Body Codes

1996 Caprice, Impala SS, Roadmaster & Wagon

Engine: 4.3L V8 (Caprice only) .... VIN W
Engine: 5.7L V8 .... VIN P

C-Car Body Codes (CW - Park Avenue, CU - Park Avenue Ultra)

1994-96 Ninety-Eight, Park Avenue

Engine: 3.8L V6 .... VIN L
Engine: 3.8L V6 .... VIN K
Engine: 3.8L V6 Super Charged .... VIN 1
1996-2005 Park Avenue
Engine: 3.8L V6 .... VIN K
Engine: 3.8L V6 Super Charged .... VIN 1

D-Car Body Codes

1996 Fleetwood, Fleetwood Brougham, 2003-06 CTS, 2005-06 STS

Engine: 5.7L V8 (Fleetwood, 1996) .... VIN P
Engine: 2.6L V6 (CTS, 2003-04) .... VIN M
Engine: 2.8L V6 (CTS, 2005) .... VIN T
Engine: 3.2L V6 (CTS, 2003-04) .... VIN N
Engine: 3.6L V6 (CTS, 2004-06; STS, 2005-06) .... VIN 7
Engine: 4.6L V8 (STS, 2005-06) .... VIN A
Engine: 5.7L V8 (CTS, 2004) .... VIN P
Engine: 5.7L V8 (CTS, 2005-06) .... VIN S

E-Car Body Codes

1996-2002 Eldorado

Engine: 4.6L V8 .... VIN Y
Engine: 4.6L V8 .... VIN 9

F-Car Body Codes

1995-2002 Camaro, Firebird

Engine: 3.8L V6 .... VIN K
Engine: 5.7L V8 .... VIN G
Engine: 5.7L V8 (1997) .... VIN P

G-Car Body Codes

1996-02 Aurora

Engine: 3.5L V6 .... VIN H
Engine: 4.0L V8 .... VIN C
1996-99 Riviera
Engine: 3.8L V6 .... VIN K
Engine: 3.8L V6 Super Charged .... VIN 1

H-Car Body Codes

1994-2005 Bonneville, Eighty-Eight, LeSabre & LSS

Engine: 3.8L V6 .... VIN K
Engine: 3.8L V6 .... VIN L
Engine: 3.8L V6 Super Charged .... VIN 1
Engine: 4.6L V8 .... VIN I

J-Car Body Codes

1996-2005 Cavalier, Sunfire

Engine: 2.2L I4 .... VIN 4, F
Engine: 2.2L CNG .... VIN 4
Engine: 2.4L I4 .... VIN T

K-Car Body Codes

1996-2005 Deville

Engine: 4.6L V8 .... VIN Y

1996-99 Concours, d'Elegance (Deville)

Engine: 4.6L V8 .... VIN 9

1996-2004 Seville (KS - SLS, KY - STS)

Engine: 4.6L V8 .... VIN 9
Engine: 4.6L V8 .... VIN Y

L-Car Body Codes

1996 Beretta, Corsica

Engine: 2.2L I4 .... VIN 4
Engine: 3.1L V6 .... VIN M

L/N Car Body Codes

1997-2003 Malibu, Cutlass

Engine: 2.4L I4 .... VIN T
Engine: 3.1L V6 .... VIN J
Engine: 3.1L V6 .... VIN M

M-Car Body Codes

1996-2001 Geo Metro

Engine: 1.0L I3 .... VIN 6
Engine: 1.3L I4 .... VIN 9

N-Car Body Codes

1996-2005 Achieva, Alero, Grand Am, Skylark

Engine: 2.2L I4 .... VIN F
Engine: 2.4L I4 .... VIN T
Engine: 3.1L V6 .... VIN M
Engine: 3.4L V6 .... VIN E

2004-2005 Classic

Engine: 2.2L I4 .... VIN F

2000-2005 Malibu

Engine: 3.4L V6 .... VIN N

S-Car Body Codes

1996-2002 Geo Prism

Engine: 1.6L I4 .... VIN 6
Engine: 1.8L I4 .... VIN 8

2003-2005 Aveo, Vibe

Engine: 1.5L SOHC .... VIN Y

Engine: 1.5L DOHC .... VIN V
Engine: 1.6L DOHC .... VIN 6
Engine: 1.8L SOHC .... VIN 8
Engine: 1.8L DOHC .... VIN 3, L

V Car Body Codes

2005-2006 GTO
Engine: 3.5L V6 .... VIN U

V/R Car Body Codes

1997-2001 Catera
Engine: 3.0L V6 .... VIN R

W-Car Body Codes

1994-2005 Allure, Century, Cutlass Supreme, Grand Prix, Impala, Intrigue, LaCrosse, Lumina, Monte Carlo, Regal
Engine: 2.2L I4 (1996) .... VIN 4
Engine: 3.1L V6 High Output (2000-05) .... VIN J
Engine: 3.1L V6 .... VIN M
Engine: 3.4L V6 (1994-97) .... VIN X
Engine: 3.4L V6 (2000-05 Impala) .... VIN E
Engine: 3.5L V6 (1999-2002 Intrigue) .... VIN H
Engine: 3.6L V6 (2005 Allure, LaCrosse) .... VIN 7
Engine: 3.8L V6 .... VIN L, 2, 4
Engine: 3.8L V6 High Output .... VIN K, I
Engine: 3.8L V6 Supercharged .... VIN 1
Engine: 5.3L V8 .... VIN C

Y-Car Body Codes

1994-2006 Corvette
Engine: 5.7L V8 (1994-96) .... VIN P, J
Engine: 5.7L V8 (1997-2002) .... VIN G, S
Engine: 5.7L V8 (1996) .... VIN 5
Engine: 6.0L V8 (2005-06) .... VIN U
Engine: 7.0L V8 (2006) .... VIN Y, E

2004-05 XLR
Engine: 4.6L V8 .... VIN A

Z-Car Body Codes

1996-2003 Saturn, 3D, & Wagon
Engine: 1.9L I4 .... VIN 7, 8
Engine: 2.2L I4 .... VIN F
Engine: 3.0L V7 .... VIN R

2004-06 G6, Malibu
Engine: 2.2L I4 .... VIN F
Engine: 3.5L V6 .... VIN 8
Engine: 3.9L V6 .... VIN 1

GM SUV, TRUCK & VAN APPLICATIONS

A/B Body Codes

2002-05 Rendezvous AWD, FWD Utility (A-2WD, B-4WD)
Engine: 3.4L V6 .... VIN E
Engine: 3.6L V6 .... VIN 7

C/K Body Codes

1996-2005 C/K Series, Escalade, Sierra, Silverado, Suburban, Tahoe, Yukon 2 & 4-Door Utility
Engine: 4.3L V6 .... VIN W

Engine: 4.8L V8 .... VIN V
Engine: 5.0L V8 .... VIN M
Engine: 5.3L V8 .... VIN T
Engine: 5.3L V8 Flexible Fuel .... VIN Z
Engine: 5.7L V8 (GAS/CNG) .... VIN K
Engine: 5.7L V8 .... VIN R
Engine: 6.0L V8 .... VIN N, U
Engine: 6.5L V8 Diesel (1996-2000) .... VIN F
Engine: 6.5L V8 Turbo Diesel (1997-98) .... VIN P
Engine: 6.5L V8 Turbo Diesel (1997-2000) .... VIN S
Engine: 6.6L V8 Turbo Diesel (2001-02) .... VIN 1, J
Engine: 7.2L V8 Diesel .... VIN C
Engine: 7.4L V8 .... VIN J
Engine: 8.1L V8 .... VIN E, G

E/J Body Codes

1996-2004 Geo Tracker Utility (,I) & Utility (E)
Engine: 1.6L I4 .... VIN 6
Engine: 2.0L I4 .... VIN C
Engine: 2.5L V6 .... VIN 1

2004-05 SRX
Engine: 3.6L V6.. VIN 7
Engine: 4.6L V8 .... VIN A

G Body Codes

1996-2003 Step Van, Express, Rally, Savana Van, Cargo

Engine: 4.3L V6 .... VIN W
Engine: 5.0L V8 .... VIN M
Engine: 5.7L V8 (Gas/CNG) .... VIN K
Engine: 5.7L V8 .... VIN R
Engine: 6.5L V8 Diesel .... VIN F
Engine: 7.4L V8 .... VIN N
Engine: 8.1L V8 .... VIN G

G/H Body Codes

2003-05 Express
Engine: 4.3L V6.. VIN X
Engine: 4.8L V8 .... VIN V
Engine: 5.3L V8 .... VIN T
Engine: 6.0L V8 .... VIN U

L Body Codes

2005-2006 Equinox

Engine: 3.4L V6.. VIN F

M/L Body Codes

1996-2003 Astro, Safari Van Passenger & Cargo

Engine: 4.3L V6 .... VIN W

S/T Body Codes

1995-2006 Blazer, Bravada, Colorado, Envoy, Jimmy, S-10, Sonoma, Trailblazer
Engine: 2.2L I4 .... VIN 4
Engine: 2.2L I4 Flexible Fuel .... VIN 5
Engine: 2.8L I4 .... VIN 6
Engine: 3.5L I5 .... VIN 8
Engine: 4.2L V6 .... VIN S
Engine: 4.3L V6 .... VIN W, X

Engine: 4.8L V8 .... VIN V
Engine: 5.3L V8 .... VIN P
Engine: 6.0L V8 .... VIN N, U

S/T Roadster

2003-2005 SSR

Engine: 4.8l V8 .... VIN V
Engine: 5.3L V8 .... VIN P
Engine: 6.0L V8 .... VIN H

U Body Codes

1994-2005 Lumina APV, Montana, Montana SV6, Silhouette, Terraza, Trans Sport, Uplander, Venture

Engine: 3.4L V6 HO .... VIN E
Engine: 3.5L V6 .... VIN L
Engine: 3.8L V6 .... VIN L

NOTES & CAUTIONS

Before servicing any vehicle, please be sure to read all of the following precautions, which deal with personal safety, prevention of component damage, and important points to take into consideration when servicing a motor vehicle:

- Observe all applicable safety precautions when working around fuel. Whenever servicing the fuel system, always work in a well-ventilated area. Do NOT allow fuel spray or vapors to come in contact with a spark, open flame, or excessive heat (a hot drop light, for example). Keep a dry chemical fire extinguisher near the work area. Always keep fuel in a container specifically designed for fuel storage; also, always properly seal fuel containers to avoid the possibility of fire or explosion. Refer to the additional fuel system precautions later in this section.

- Fuel injection systems often remain pressurized, even after the engine has been turned **OFF**. The fuel system pressure must be relieved before disconnecting any fuel lines. Failure to do so may result in fire and/or personal injury.

- Brake fluid often contains Polyglycol Ethers and Polyglycols. Avoid contact with the eyes and wash your hands thoroughly after handling brake fluid. If you do get brake fluid in your eyes, flush your eyes with clean, running water for 15 minutes. If eye irritation persists, or if you have taken brake fluid internally, IMMEDIATELY seek medical assistance.

- The EPA warns that prolonged contact with used engine oil may cause a number of skin disorders, including cancer. You should make every effort to minimize your exposure to used engine oil. Protective gloves should be worn when changing oil. Wash your hands and any other exposed skin areas as soon as possible after exposure to used engine oil. Soap and water, or waterless hand cleaner should be used.

- The air bag system must be disabled (negative battery cable disconnected and/or air bag system main fuse removed) for at least 30 seconds before performing service on or around system components, steering column, instrument panel components, wiring and sensors. Failure to follow safety and disabling procedures could result in accidental air bag deployment, possible personal injury and unnecessary system repairs.

- Always wear safety goggles when working with, or around, the air bag system. When carrying a non-deployed air bag, be sure the bag and trim cover are pointed away from your body. When placing a non-deployed air bag on a work surface, always face the bag and trim cover upward, away from the surface. This will reduce the motion of the module if it is accidentally deployed. Refer to the additional air bag system precautions later in this section.

- Disconnecting the negative battery cable on some vehicles may interfere with the functions of the on-board computer system(s) and may require the computer to undergo a relearning process once the negative battery cable is reconnected.

- It is critically important to observe all instructions regarding ground disconnects, ignition switch positions, etc., in each diagnostic routine provided. Ignoring these instructions can result in false readings, damage to electronic components or circuits, or personal injury.

Preliminary Diagnostics

HISTORY OF OBD SYSTEMS

Starting in 1978, several vehicle manufacturers introduced a new type of control for several vehicle systems and computer control of engine management systems. These computer-controlled systems included programs to test for problems in the engine mechanical area, electrical fault identification and tests to help diagnose the computer control system. Early attempts at diagnosis involved expensive and specialized diagnostic testers that hooked up externally to the computer in series with the wiring connector and monitored the input/output operations of the computer.

By early 1980, vehicle manufacturers had designed systems in which the onboard computer incorporated programs to monitor selected components, and to store a trouble code in its memory that could be retrieved at a later time. These trouble codes identified failure conditions that could be used to refer a technician to diagnostic repair charts or test procedures to help pinpoint the problem area.

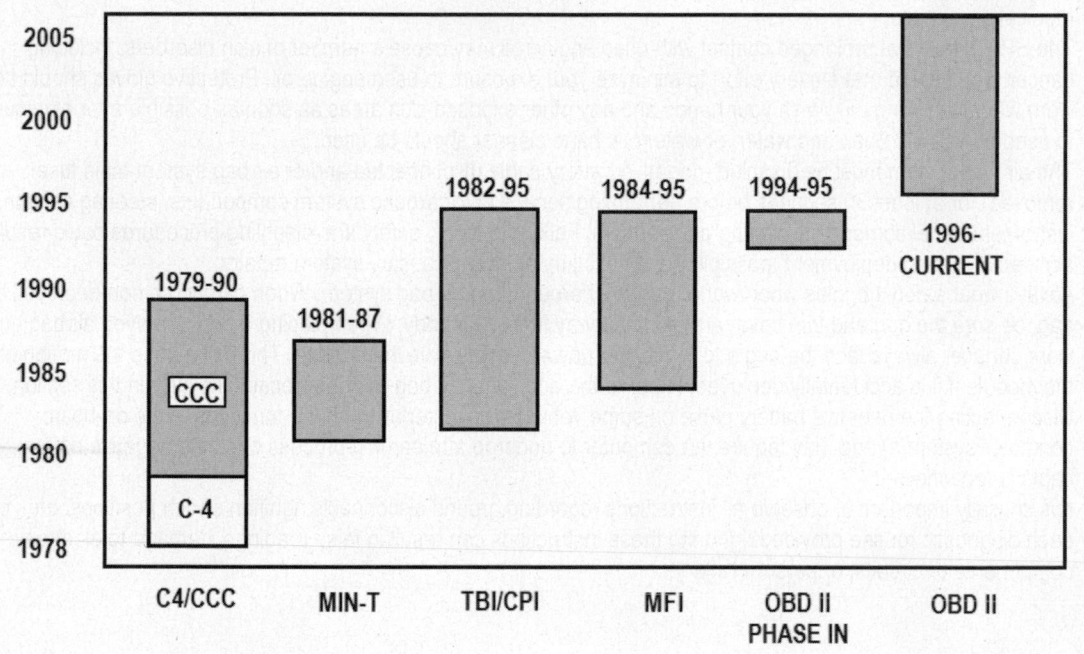

Evolution of GM Computerized Engine Controls

OBD I SYSTEM DIAGNOSTICS

One of the most important things to understand about the automotive repair industry is the fact that you have to continually learn new systems and new diagnostic routines (the test procedures designed to isolate a problem on a vehicle system). For OBD I and II systems, a diagnostic routine can be defined as a procedure (a series of steps) that you follow to find the cause of a problem, make a repair and then verify the problem is fixed.

CHANGES IN DIAGNOSTIC ROUTINES

In some cases, a new Engine Control system may be similar to an earlier system, but it can have more indepth control of vehicle emissions, input and output devices and it may include a diagnostic "monitor" embedded in the engine controller designed to run a thorough set of emission control system tests.

OBD I System Diagnostics

One of the most important things to understand about the automotive repair industry is the fact that you have to continually learn new systems and new diagnostic routines (the test procedures designed to isolate a problem on a vehicle system). For OBD I and II systems, a diagnostic routine can be defined as a procedure (a series of steps) that you follow to find the cause of a problem, make a repair and then verify the problem is fixed.

The OBD I Diagnostic Flowchart on this page can be used to find the cause of problems related to Engine Control system trouble codes or driveability symptoms detected on OBD I systems. It includes a step-by-step procedure to use to repair these systems. To compare this flowchart with the one used on OBD II systems, refer to the next page.

The steps in this flow chart should be followed as described below (from top to bottom).

- Do the Pre-Computer Checks.
- Check for any trouble codes stored in memory.
- Read the trouble codes - If trouble codes are set, record them and then clear the codes.
- Start the vehicle and see if the trouble code(s) reset. If they do, then use the correct trouble code repair chart to make the repair.
- If the codes do not reset, than the problem may be intermittent in nature. In this case, refer to the test steps used to find the cause of an intermittent fault (wiggle test).
- In no trouble codes are found at the initial check, then determine if a driveability symptom is present. If so, then refer to the approriate driveability symptom repair chart to make the repair. If the first symptom chart does not isolate the cause of the condition, then go on to another driveability symptom and follow that procedure to conclusion.

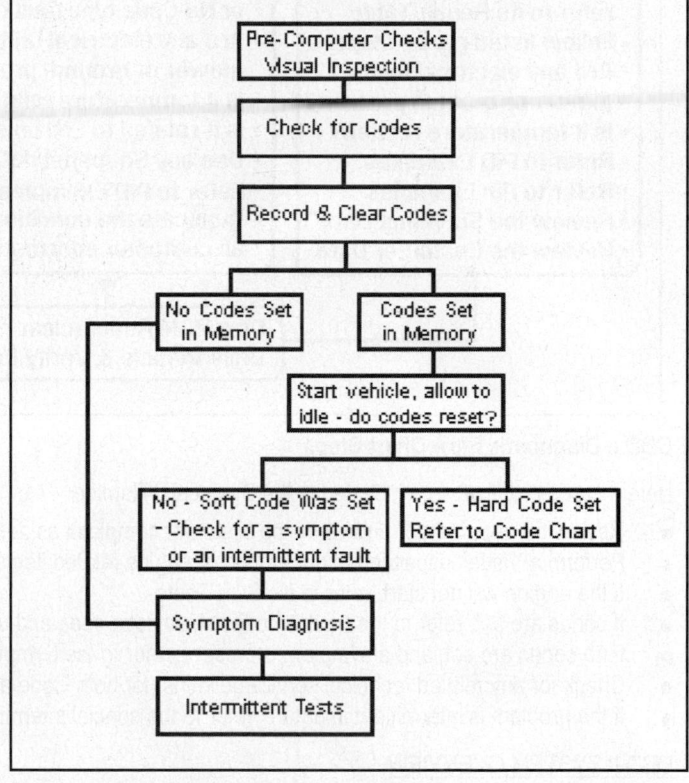

- If the problem is intermittent in nature, then refer to the special intermittent tests. Follow all available intermittent tests to determine the cause of this type of fault (usually an electrical connection problem).

OBD II System Diagnostics

The diagnostic approach used in OBD II systems is more complex than that of the one for OBD I systems. This complexity will effect how you approach diagnosing the vehicle. On an OBD II system, the onboard diagnostics will identify sensor faults (i.e., open, shorted or grounded circuits) as well as those that lose calibration. Another new test that arrived with OBD II is the rationality test (a test that checks whether the value for one input makes rational sense when compared against other sensor input values). The changes plus the use of OBD II Monitors have dramatically changed OBD II diagnostics.

The use of a repeatable test routine can help you quickly get to the root cause of a customer complaint, save diagnostic time and result in a higher percentage of properly repaired vehicles. You can use this Diagnostic Flow Chart to keep on track as you diagnose an Engine Control problem or a base engine fault on vehicles with OBD II.

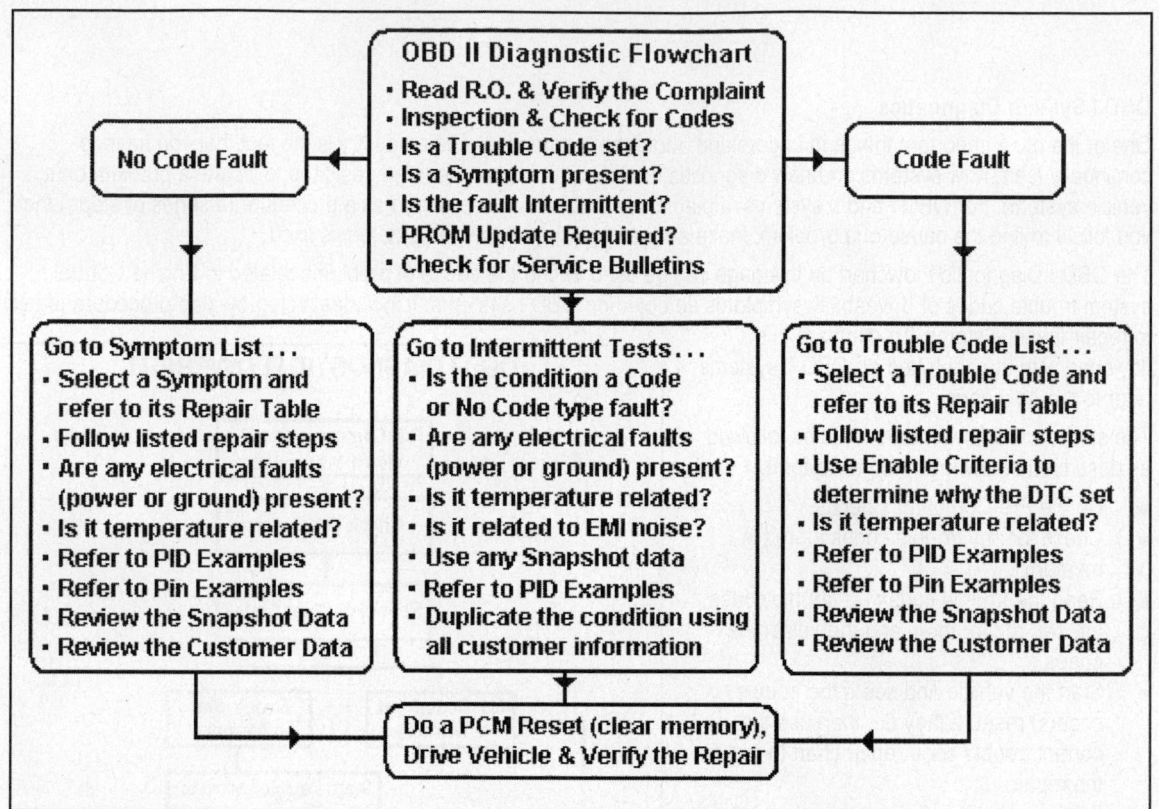

OBD II Diagnostic Flow Chart Steps

Here are some of the steps included in the Diagnostic Routine:

- Review the repair order and verify the customer complaint as described
- Perform a Visual Inspection of underhood or engine related items
- If the engine will not start, refer to No Start Tests
- If codes are set, refer to the trouble code list, select a code and use the repair chart
- If no codes are set, and a symptom is present, refer to the Symptom List
- Check for any related technical service bulletins (for both Code and No Code Faults)
- If the problem is intermittent in nature, refer to the special Intermittent Tests

OBD II SYSTEM OVERVIEW

The OBD II system was developed as a step toward compliance with California and Federal regulations that set standards for vehicle emission control monitoring for all automotive manufacturers. The primary goal of this system is to detect when the degradation or failure of a component or system will cause emissions to rise by 50%. Every manufacturer must meet OBD II standards by the 1996 model year. Some manufacturers began programs that were OBD II mandated as early as 1992, but most manufacturers began an OBD II phase-in period starting in 1994.

The changes to On-Board Diagnostics influenced by this new program include:

- Common Diagnostic Connector
- Expanded Malfunction Indicator Light Operation
- Common Trouble Code and Diagnostic Language
- Common Diagnostic Procedures
- New Emissions-Related Procedures, Logic and Sensors
- Expanded Emissions-Related Monitoring

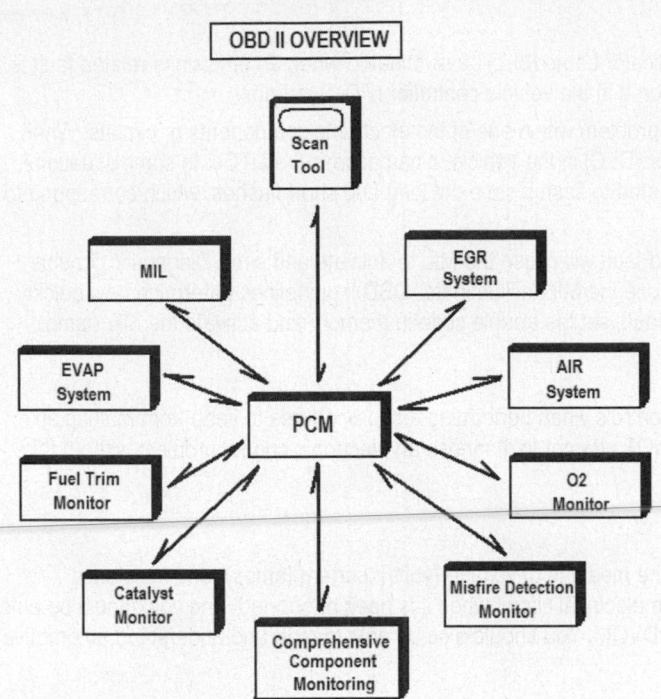

OBD II OVERVIEW

COMMON TERMINOLOGY

OBD II introduces common terms, connectors, diagnostic language and new emissions-related monitoring procedures. The most important benefit of OBD II is that all vehicles will have a common data output system with a common connector. This allows equipment Scan Tool manufacturers to read data from every vehicle and pull codes with common names and similar descriptions of fault conditions. In the future, emissions testing will require the use of an OBD II certifiable Scan Tool.

Diagnostic Tools & Circuit Testing

HAND TOOLS & METER OPERATION

To effectively use this or any diagnostic information, you should have a solid understanding of how to operate required tools and test equipment.

SCAN TOOLS

Domestic vehicle manufacturers designed their computers to have an accessible data line where a diagnostic tester could retrieve data on sensors and the status of operation for components.

These testers became known in the automotive repair industry as "Scan Tools" because they scanned the data on the computers and provided information for the technician.

The Scan Tool is your basic tool link into the on-board electronic control system of the vehicle. Scan Tools are

SCAN TOOL
1. DLC Cable Connection
2. SAE 16/19 Pin Adapter

equipped with, or have separate software cards, for each OEM needed to be diagnosed. In this case, always secure a scan tool that has the latest OEM-specific diagnostic software included. Spend some time in the scan tool user's manual to ensure you know how to properly operate the tool and how to select the necessary programs required for full and proper diagnostics.

The manufacturer may specify the use of a specific scan tool with its diagnostic processes. However, there are aftermarket scan tools, when equipped with the right software that can provide proper diagnosis as well.

MALFUNCTION INDICATOR LAMP

Emission regulations require that a Malfunction Indicator Lamp (MIL) be illuminated when an emissions related fault is detected and that a Diagnostic Trouble Code be stored in the vehicle controller (PCM) memory.

When the MIL is illuminated, it is an indication of a problem within one of the electronic components or circuits. When the scan tool is attached to the Data Link Connector (DLC) in the vehicle, it can access the DTCs. In some situations, without the use of a scan tool, the MIL can be activated to flash a series of long and short flashes, which correspond to the numbering of the DTC.

OBD II guidelines define *when* an emissions-related fault will cause the MIL to activate and set a Diagnostic Trouble Code (DTC). There are some DTCs that will not cause the MIL to illuminate. OBD II guidelines determine how quickly the onboard diagnostics must be able to identify a fault, set the trouble code in memory and activate the MIL (lamp).

ELECTRONIC CONTROLS

You should have a basic knowledge of electronic controls when performing test procedures to keep from making an incorrect diagnosis or damaging components. Do NOT attempt to diagnose an electronic control problem without this basic knowledge!

ELECTRICITY & ELECTRICAL CIRCUITS

You should understand basic electricity and know the meaning of voltage (volts), current (amps), and resistance (ohms). You should understand what happens in an electrical circuit when it is open or shorted, and you should be able to identify an open circuit or shorted circuit using a DVOM. You should also be able to read and understand automotive electrical wiring diagrams and schematics.

CIRCUIT TESTING TOOLS

You should know when to use and when NOT to use a 12-volt test light during diagnosis of electronic controls (Do NOT use this tester unless specifically instructed to do so by a test procedure). Instead of using a 12-volt test light, you should use a DVOM or Lab Scope with a breakout box whenever a diagnostic procedure calls for a measurement at a PCM connector or component wiring harness.

Effective Diagnostics

GETTING STARTED

If you are reasonably certain that the problem is related to a particular electronic control system, the first step is to check for any stored trouble codes in that controller.

On vehicles with more than one vehicle controller (i.e., PCM, BCM, MIC, TCM, etc.), if you are unsure whether the problem is Powertrain related, start by checking for codes in the other controllers to determine if the problem is related to another vehicle system.

If there are no codes set, and you are certain which Powertrain subsystem has a problem, you can start by checking one of the subsystems. The subsystems include the Charging, Cooling, Fuel, Ignition and Speed Control systems.

If a wiring problem is found during testing, you will need to refer to wiring diagrams in the appropriate information resource. Using a wiring schematic can help you determine:

- Wiring circuits, circuit numbers, and wire colors
- Electrical component connector and component relationships within a circuit
- Power, ground, and splice locations within a circuit
- Related circuits connected into the circuit you are reviewing

Once you decide how to repair the vehicle, in addition to performing the repair, it is a good idea to clear any trouble codes that were set and to verify they do not reset.

To verify a repair, you should confirm that the Check Engine Light is operational and goes out after the 4-second key-on bulb check. Then, you need to duplicate the conditions present when the customer complaint occurred or when a trouble code set; these are the actual code conditions that caused a code to set. The individual code conditions and possible causes are included in **Section 3**. You can use this information to find out how to drive a vehicle for problem verification.

Contents

INTRODUCTION

System Control Modules ....Page 2-2

Powertrain Subsystems ....Page 2-2

WHERE TO BEGIN

Six-Step Procedure ....Page 2-3

Base Engine Tests ....Page 2-4

Ignition System Tests – Distributor ....Page 2-6

Ignition System Tests – Distributorless..Page 2-6

Symptom Diagnosis ....Page 2-7

Accessing Components & Circuits ....Page 2-7

INTRODUCTION

System Control Modules

Before attempting diagnosis of the Electronic Engine Control system, familiarize yourself with the basics of how the system is designed to operate. It consists of a central processing unit: Powertrain Control Module (PCM), Engine Control Module (ECM), Transmission Control Module (TCM) and/or the Body Control Module (BCM). These units are the "heart" of the electronic control systems on the vehicle. In some cases, these units are integral with one another, and on some applications, they are separate. As you get deeper into actual diagnostic testing, you will find out which units are used on the vehicle you are testing.

The PCM is a digital computer that contains a microprocessor. The PCM receives input signals from various sensors and switches that are referred to as PCM inputs. Based on these inputs, the PCM adjusts various engine and vehicle operations through devices that are referred to as PCM outputs. Examples of the input and output devices are shown in the graphic below.

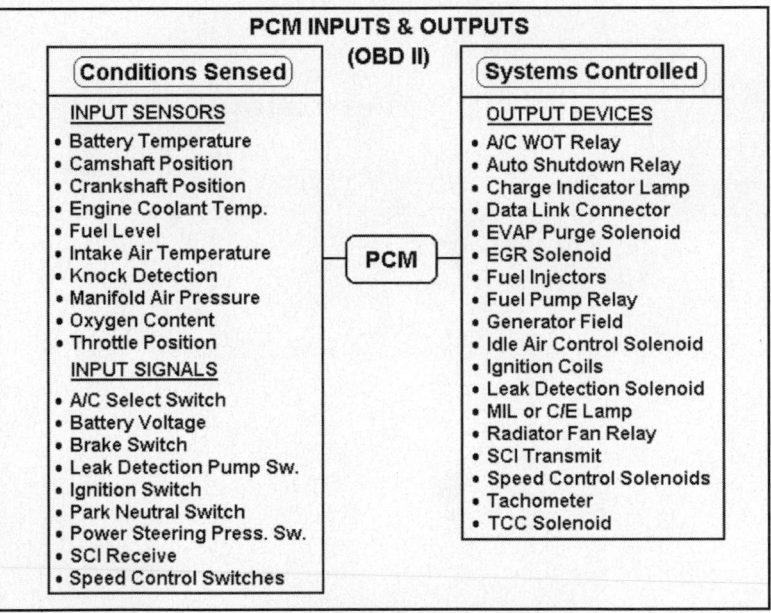

Input & Output Device Graphic (Example)

Powertrain Subsystems

A key to the diagnosis of the PCM and its subsystems is to determine which subsystems are on a vehicle. Examples of typical subsystems appear below:

- Cranking & Charging System
- Emission Control Systems
- Engine Cooling System
- Engine Air/Fuel Controls
- Exhaust System
- Ignition System
- Speed Control System
- Transaxle Controls

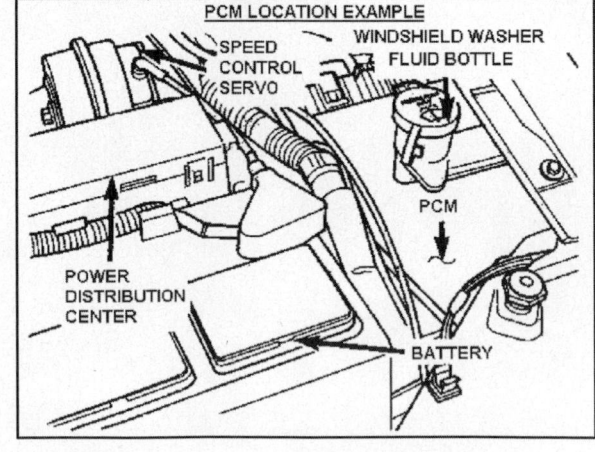

WHERE TO BEGIN

Diagnosis of engine performance or drivability problems on a vehicle with an onboard computer requires that you have a logical plan on how to approach the problem. The "Six Step Test Procedure" is designed to provide a uniform approach to repair any problems that occur in one or more of the vehicle subsystems.

The diagnostic flow built into this test procedure has been field-tested for several years at dealerships - *it is the starting point when a repair is required!*

It should be noted that a commonly overlooked part of the "Problem Resolution" step is to check for any related Technical Service Bulletins.

Six-Step Test Procedure

The steps outlined on this page were defined to help you determine how to perform a proper diagnosis. Refer to the flow chart that outlines the Six Step Test Procedure on the previous page as needed. The recommended steps include:

1. VERIFY THE COMPLAINT &
CHECK FOR TSBS

To verify the customer complaint, the technician should understand the normal operation of the system. Conduct a thorough visual and operational inspection, review the service history, detect unusual sounds or odors, and gather diagnostic trouble code (DTC) information resources to achieve an effective repair.

SIX STEP TROUBLESHOOTING PROCEDURE

- VERIFICATION OF THE COMPLAINT & CHECK FOR ANY RELATED TSBs
- CHECK FOR TROUBLE CODES
- SYMPTOM OR TROUBLE CODE TESTS
- PROBLEM RESOLUTION & REPAIR
- PCM RESET
- REPAIR VERIFICATION

This check should include videos, newsletters, and any other information in the form of TSBs or Dealer Service Bulletins. Analyze the complaint and then use the recommended Six Step Test Procedure. Utilize the wiring diagrams and theory of operation articles. Combine your own knowledge with efficient use of the available service information.

Verify the cause of any related symptoms that may or may not be supported by one or more trouble codes. There are various checks that can be performed to Engine Controls that will help verify the cause of a related symptom. This step helps to lead you in an organized diagnostic approach.

2. CHECK FOR TROUBLE CODES OR SYMPTOMS

Determine if the problem is a Code or a No Code Fault. Then refer to the appropriate published service diagnostic information to make the repair.

3. PROBLEM RESOLUTION & REPAIR

Once the problem component or circuit has been properly identified and verified using published diagnostic procedures, make any needed repairs or replacement to restore the vehicle to proper working order. If the condition has set a DTC, follow the designated repair chart to make an effective repair. If there is not a DTC set, but you can determine specific symptoms that are evident during the failure, select the symptom from the symptom tables and follow the diagnostic paths or suggestions to complete the repair or refer to the applicable component or system in service information.

4. SYMPTOM OR TROUBLE CODE TESTS

If the vehicle does not set a DTC and has only intermittent operating failures or concerns, to resolve an intermittent fault, perform the following steps:

- Observe trouble codes, DTC modes and freeze frame data.
- Evaluate the symptoms and conditions described by the customer.
- Use a check sheet to identify the circuit or electrical system component.
- Many Aftermarket Scan Tools and Lab Scopes have data capturing features.

5. PCM RESET

It is a good idea, prior to tracing any faults, to clear the DTCs, attempt to replicate the condition and see if the same DTC resets. Also, once any repairs are made, it will be necessary to clear the DTC(s) – PCM Reset – to ensure the repair has totally resolved the problem. For procedures on PCM Reset, see the DIAGNOSTIC TROUBLE CODES section.

6. REPAIR VERIFICATION

Once a repair is completed, the next step is to verify the vehicle operates properly and that the original symptom was corrected. Verification Tests, related to specific DTC diagnostic steps, can be used to verify a repair.

Base Engine Tests

To determine that an engine is mechanically sound, certain tests need to be performed to verify that the correct A/F mixture enters the engine, is compressed, ignited, burnt, and then discharged out of the exhaust system. These tests can be used to help determine the mechanical condition of the engine.

To diagnose an engine-related complaint, compare the results of the Compression, Cylinder Balance, Engine Cylinder Leakage (not included) and Engine Vacuum Tests.

ENGINE COMPRESSION TEST

The Engine Compression Test is used to determine if each cylinder is contributing its equal share of power. The compression readings of all the cylinders are recorded and then compared to each other and to the manufacturer's specification (if available).

Cylinders that have low compression readings have lost their ability to seal. It this type of problem exists, the location of the compression leak must be identified. The leak can be in any of these areas: piston, head gasket, spark plugs, and exhaust or intake valves.

The results of this test can be used to determine the overall condition of the engine and to identify any problem cylinders as well as the most likely cause of the problem.

CAUTION: *Prior to starting this procedure, set the parking brake, place the gear selector in P/N and block the drive wheels for safety. The battery must be fully charged.*

COMPRESSION TEST PROCEDURE

1. Allow the engine to run until it is fully warmed up.

2. Remove the spark plugs and disable the Ignition system and the Fuel system for safety. Disconnecting the CKP sensor harness connector will disable both fuel and ignition (except on NGC vehicles).

3. Carefully block the throttle to the wide-open position.

4. Insert the compression gauge into the cylinder and tighten it firmly by hand.

5. Use a remote starter switch or ignition key and crank the engine for 3-5 complete engine cycles. If the test is interrupted for any reason, release the gauge pressure and retest. Repeat this test procedure on all cylinders and record the readings.

The lowest cylinder compression reading should not be less than 70% of the highest cylinder compression reading and no cylinder should read less than 100 psi.

EVALUATING THE TEST RESULTS

To determine why an individual cylinder has a low compression reading, insert a small amount of engine oil (3 squirts) into the suspect cylinder. Reinstall the compression gauge and retest the cylinder and record the reading. Review the explanations below.

Reading is higher - If the reading is higher at this point, oil inserted into the cylinder helped to seal the piston rings against the cylinder walls. Look for worn piston rings.

Reading did not change - If the reading didn't change, the most likely cause of the low cylinder compression reading is the head gasket or valves.

Low readings on companion cylinders - If low compression readings were recorded from cylinders located next to each other, the most likely cause is a blown head gasket.

Readings are higher than normal - If the compression readings are higher than normal, excessive carbon may have collected on the pistons and in the exhaust areas. One way to remove the carbon is with an approved brand of "Top Engine Cleaner."

Note: *Always clean spark plug threads and seat with a spark plug thread chaser and seat cleaning tool prior to reinstallation. Use anti-seize compound on aluminum heads.*

ENGINE VACUUM TESTS

An engine vacuum test can be used to determine if each cylinder is contributing an equal share of power. Engine vacuum, defined as any pressure lower than atmospheric pressure, is produced in each cylinder during the intake stroke. If each cylinder produces an equal amount of vacuum, the measured vacuum in the intake manifold will be even during engine cranking, at idle speed, and at off-idle speeds.

Engine vacuum is measured with a vacuum gauge calibrated to show the difference between engine vacuum (the lack of pressure in the intake manifold) and atmospheric pressure. Vacuum gauge measurements are usually shown in inches of Mercury (in. Hg).

Note: *In the tests described in this article, connect the vacuum gauge to an intake manifold vacuum source at a point below the throttle plate on the throttle body.*

ENGINE CRANKING VACUUM TEST PROCEDURE

The Engine Cranking Vacuum Test can be used to verify that low engine vacuum is not the cause of a No Start, Hard Start, Starts and Dies or Rough Idle condition (symptom).

The vacuum gauge needle fluctuations that occur during engine cranking are indications of individual cylinder problems. If a cylinder produces less than normal engine vacuum, the needle will respond by fluctuating between a steady high reading (from normal cylinders) and a lower reading (from the faulty cylinder). If more than one cylinder has a low vacuum reading, the needle will fluctuate very rapidly.

1. Prior to starting this test, set the parking brake, place the gearshift in P/N and block the drive wheels for safety. Then block the PCV valve and disable the idle air control device.

2. Disable the fuel and/or ignition system to prevent the vehicle from starting during the test (while it is cranking).

3. Close the throttle plate and connect a vacuum gauge to an intake manifold vacuum source. Crank the engine for three seconds (do this step at least twice).

The test results will vary due to engine design characteristics, the type of PCV valve and the position of the AIS or IAC motor and throttle plate. However, the engine vacuum should be steady between 1.0-4.0 in. Hg during normal cranking.

ENGINE RUNNING VACUUM TEST PROCEDURE

1. Allow the engine to run until fully warmed up. Connect a vacuum gauge to a clean intake manifold source. Connect a tachometer or Scan Tool to read engine speed.

2. Start the engine and let the idle speed stabilize. Raise the engine speed rapidly to just over 2000 rpm. Repeat the test (3) times. Compare the idle and cruise readings.

EVALUATING THE TEST RESULTS

If the engine wear is even, the gauge should read over 16 in. Hg and be steady. Test results can vary due to engine design and the altitude above or below sea level.

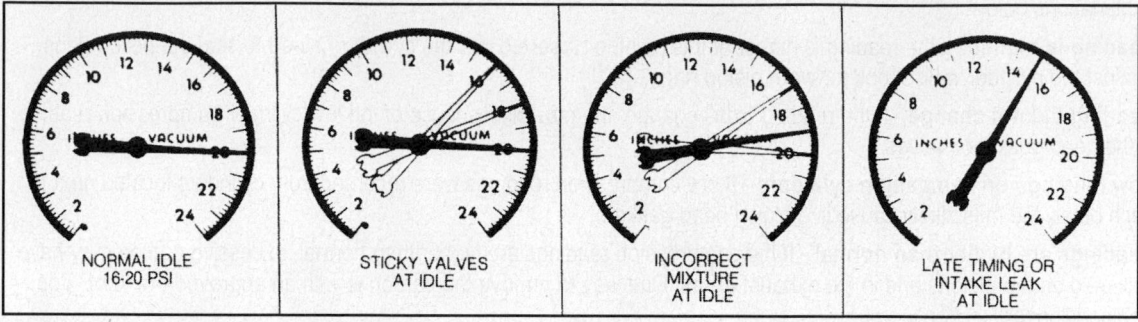

Engine Running Vacuum Test Graphic

Ignition System Tests - Distributor

This next section gives an overview of ignition tests (with examples) for a Distributor Ignition System.

PRELIMINARY INSPECTION

1. Perform these checks prior to connecting the Engine Analyzer:

2. Check the battery condition (verify that it can sustain a cranking voltage of 9.6v).

3. Inspect the ignition coil for signs of damage or carbon tracking at the coil tower.

4. Remove the coil wire and check for signs of corrosion on the wire or tower.

5. Test the coil wire resistance with a DVOM (it should be less than 7 k/ohm per foot).

6. Connect a *low* output spark tester to the coil wire and engine ground. Verify that the ignition coil can sustain adequate spark output while cranking for 3-6 seconds.

7. Connect the Engine Analyzer to the Ignition System, and choose Parade display. Run the engine at 2000 RPM, and note the display patterns, looking for any abnormalities.

Ignition System Tests - Distributorless

Perform the following checks prior to connecting the Engine Analyzer:

1. Check the battery condition (verify that it can sustain a cranking voltage of 9.6v).

2. Inspect the ignition coils for signs of damage or carbon tracking at the coil towers.

3. Remove the secondary ignition wires and check for signs of corrosion.

4. Test the plug wire resistance with a DVOM (specification varies from 15-30 k/ohm).

5. Connect a *low* output spark tester to a plug wire and to engine ground. Verify that the ignition coil can sustain adequate spark output for 3-6 seconds.

SECONDARY IGNITION SYSTEM SCOPE PATTERNS (V6 ENGINE)

1. Connect the Engine Analyzer to the ignition system.

2. Turn the scope selector to view the "Parade Display" of the ignition secondary.

3. Start the engine in Park or Neutral and slowly increase the engine speed from idle to 2000 rpm.

4. Compare actual display to the secondary ignition system examples below.

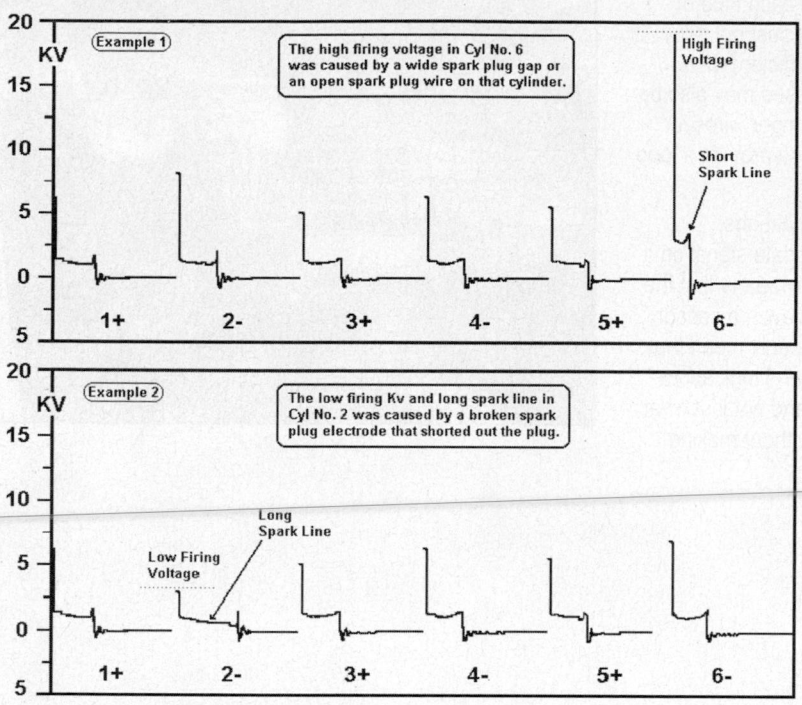

Example 1

20 KV
15
10
5
0
5

The high firing voltage in Cyl No. 6 was caused by a wide spark plug gap or an open spark plug wire on that cylinder.

High Firing Voltage

Short Spark Line

1+   2-   3+   4-   5+   6-

Example 2

20 KV
15
10
5
0
5

The low firing Kv and long spark line in Cyl No. 2 was caused by a broken spark plug electrode that shorted out the plug.

Long Spark Line

Low Firing Voltage

1+   2-   3+   4-   5+   6-

Symptom Diagnosis

To determine whether vehicle problems are identified by a set Diagnostic Trouble Code, you will first have to connect a proper scan tool to the Data Link Connector and retrieve any set codes. See DIAGNOSTIC TROUBLE CODES section for information on retrieving and reading codes.

If no codes are set, the problem must be diagnosed using only vehicle operating symptoms. A complete set of "No Code" symptoms is found in the SYMPTOM DIAGNOSIS (NO CODES) section.

Do NOT attempt to diagnose driveability symptoms without having a logical plan to use to determine which engine control system is the cause of the symptom - this plan should include a way to determine which systems do NOT have a problem! Remember, there are 2 kinds of NO CODE conditions:

- Symptom diagnosis, in which a continuous problem exists, but no DTC is set as a result. Therefore, only the operating symptoms of the vehicle can be used to pinpoint the root cause of the problem.
- Intermittent problem diagnosis, in which the problem does not occur all the time and does not set any DTCs.

Both of these NO CODE conditions are covered in the SYMPTOM DIAGNOSIS (NO CODES) section.

Accessing Components & Circuits

Every vehicle and every diagnostic situation is different. It is a good idea to first determine the best diagnostic path to follow using flow charts, wiring diagrams, TSBs, etc. Part of choosing steps is to determine how time-consuming and effective each step will be. It may be easy to access a component or circuit in one vehicle, but difficult in another. Many circuits are integrated into a large harness and are difficult to test. Many components are inaccessible without disassembly of unrelated systems.

In the graphic, you will note that the protective covers have been removed from the PCM connectors, and any circuit can be easily identified and back probed. In other cases, PCM access is difficult, and it may be easier to access circuits at the component side of the harness.

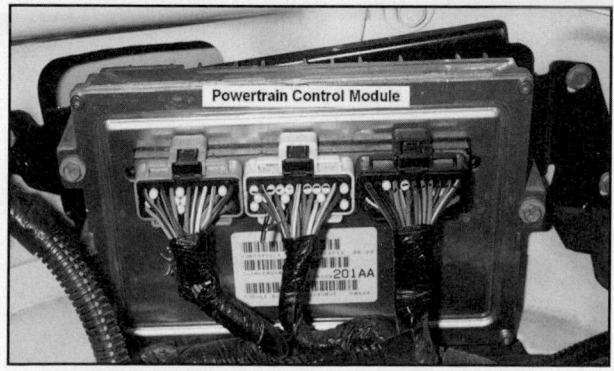

Powertrain Control Module

Another important point to remember is that any circuit or component controlled by a relay or fused circuit can be monitored from the appropriate fuse box.

There is generally more than one of each type of relay or fuse. Therefore, swapping a suspect relay from another system may be more efficient than testing the relay itself. Relays and fuses may also be removed and replaced with fused jumper wires for testing circuits. Jumper wires can also provide a loop for inductive amperage tests.

Choosing the easiest way has its limitations, however. Remember that an appropriate signal on a PCM controlled circuit at an actuator means that the signal at the PCM is also good. However, a sensor signal at the sensor does not necessarily mean that the PCM is receiving the same signal. Think about the direction flow through a circuit, and not just what signal is appropriate, to save time without making costly assumptions.

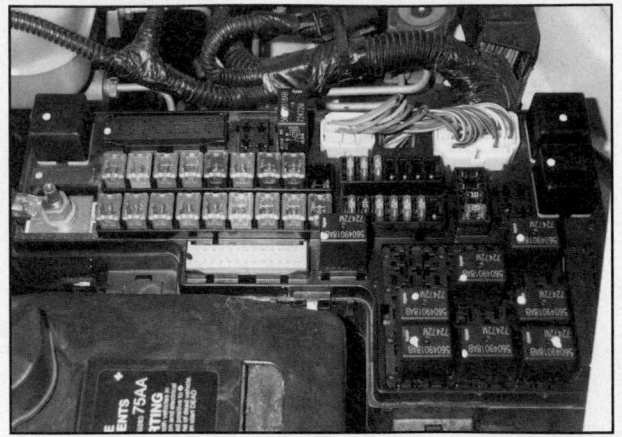

GENERAL MOTORS OBD II CONTENTS

How To Use This Section

Introduction ....Page 3-3
PCM Diagnostics ....Page 3-3
Reference Information ....Page 3-3

GM OBD II Systems

Introduction ....Page 3-4
OBD I Systems ....Page 3-4
Purpose of Onboard Diagnostics ....Page 3-4
Powertrain Control Module ....Page 3-5
OBD II Monitor Tests ....Page 3-5
Scan Tool Test Modes ....Page 3-5
Diagnostic Executive ....Page 3-6
GM Repair Information ....Page 3-6
Customer Snapshot Mode ....Page 3-6
Passive & Active Diagnostic Tests ....Page 3-7
UART Serial Data ....Page 3-7
Class II Serial Data ....Page 3-7

GM OBD II System Terminology

Enable Criteria ....Page 3-8
OBD II Warmup Cycle ....Page 3-8
Similar Operating conditions ....Page 3-8
Failure Records ....Page 3-8
OBD II Trip Definition ....Page 3-9
Flash EEPROM ....Page 3-9
I/M Readiness Status ....Page 3-9
Malfunction Indicator Lamp ....Page 3-10
Data Link Connector ....Page 3-12
Diagnostic Trouble Codes ....Page 3-13
Diagnostic Trouble Code Display ....Page 3-14
Standard Corporate Protocol ....Page 3-15
OBD II Monitor Software ....Page 3-15
Cylinder Bank Identification ....Page 3-15
Misfire Diagnostics ....Page 3-16
Fuel Trim Diagnostics ....Page 3-16
Heated Oxygen Sensor ....Page 3-17
Freeze Frame Data ....Page 3-19
Failure Records ....Page 3-19
OBD II Drive Cycle ....Page 3-20
OBD II Drive Cycle Graphic ....Page 3-20

GM OBD II Monitors

Comprehensive Component Monitor ....Page 3-21
Catalyst Monitor..Page 3-22
ECT Sensor Monitor...Page 3-23
EGR System Monitor..Page 3-24
EVAP System Monitor..Page 3-25
Fuel System Monitor ....Page 3-27
Idle Air Control System Monitor ....Page 3-28
Misfire Monitor ....Page 3-29
Oxygen Sensor Monitor ....Page 3-31
Oxygen Sensor Heater Monitor...Page 3-32
Secondary AIR Monitor ....Page 3-33
Non-Monitored Systems & Components ....Page 3-34

General Motors OBD II Trouble Codes

OBD II Trouble Code Lists - P0xxx Codes ....Page 3-35
OBD II Trouble Code Lists - P1xxx Codes ....Page 3-330
OBD II Trouble Code Lists - U1xxx Codes ....Page 3-465
OBD II Trouble Code Lists - B1xxx Codes ....Page 3-472

GEO OBD II Trouble Codes (1996-97)

OBD II Trouble Code Lists - P0xxx Codes ....Page 3-473
OBD II Trouble Code Lists - P1xxx Codes ....Page 3-489

Chevrolet Geo OBD II Trouble Codes (1998-2004)

OBD II Trouble Code Lists - P0xxx Codes ....Page 3-491
OBD II Trouble Code Lists - P1xxx Codes ....Page 3-523

Pontiac Vibe OBD II Trouble Codes (2003-05)

OBD II Trouble Code Lists - P0xxx Codes ....Page 3-530
OBD II Trouble Code Lists - P1xxx Codes ....Page 3-537

How To Use This Section

Introduction

The General Motors (GM) OBD II Section contains information grouped into the three main categories described below:

Vehicle Coverage - This information identifies GM vehicles equipped with OBD II Systems by year, body and engine. Each vehicle has its own Vehicle Identification Number (VIN). This information can be used to look up OBD II P0xxx and P1xxx type codes and their individual descriptions. GM began phase-in of the OBD II System in 1995 on 115,000 vehicles spread across several models. All GM vehicles were equipped with OBD II Systems from 1996-2002.

PCM Diagnostics - This information includes an explanation of how the OBD II system incorporates changes to the PCM diagnostics that include new emission control monitoring and control of the Malfunction Indicator Lamp. The articles in this section explain how certain test Conditions: or enable criteria must be met before an OBD II Monitor will run a diagnostic test. An explanation of how each OBD II Monitor runs a particular test is also included in this section.

Reference Information - This information includes OBD II Code Descriptions in a two-column table format. The codes are split into two categories: a P0xxx series (SAE defined) and a P1xxx series (OEM defined). The first column in each row contains code numbers and trip designations that identify the type of monitor test (CCM, EGR, etc.) and whether the code is a One Trip **(1T)** or Two-Trip **(2T)** type of code. The next column contains the trouble code conditions. Special notes are shown in *italics* at the bottom of the condition explanation.

To use this information, first look up the year, make, model, and in some cases, the specific engine application. Then read the code conditions containing the *enable criteria* that explain why a code set. *Note: Trouble code repair charts are not included in this manual.*

Example of Trouble Code Information

DTC	Trouble Code Title, Conditions & Possible Causes
P0131 1T O2S, MIL: Yes 1999, 2000, 2001, 2002, 2003, 2004, 2005, 2004, 2005 C/K Series Truck, G Series Van, S/T Series Pickup and Blazer, Envoy Escalade & TrailBlazer 4.8L VIN V, 5.3L VIN P, 5.3L VIN T, 5.3L VIN Z, 6.0L VIN N, 6.0L VIN U, 8.1L VIN G Transmissions: All	**HO2S-11 (Bank 1 Sensor 1) Circuit Low Input Conditions:** DTC P0101, P0102, P0103, P0106, P0107, P0108, P0112, P0113, P0116, P0117, P0118, P0120, P0121, P0122, P0123, P0169, P0178, P0179, P0200, P0220, P0300, P0442, P0446, P0452, P0453, P0455, P0496, P1125, P1258, P1514, P1515, P1516, P1518, P2108 and P2135 not set, engine started, engine running in closed loop, system voltage from 10-18v, Fuel Alcohol content less than 90%, fuel level over 10%, TP angle from 3-70% more than the idle value, then during the Lean Test period, the PCM detected the HO2S signal was less than 200 mv for 165 seconds or with engine runtime over 30 seconds, and during the Power Enrichment test period, the PCM detected the HO2S signal was less than 400 mv for 10 seconds. **Possible Causes** ● Engine misfire condition present (look for P0300 series codes) ● Fuel system too lean (possible low fuel pressure, water in fuel) ● HO2S signal circuit is shorted to the sensor or chassis ground ● HO2S is damaged (i.e., cracked) or air reference hole clogged ● PCM has failed
P0131 2T O2S, MIL: Yes 1996, 1997, 1998, 1999, 2000, 2001, 2002, 2003, 2004, 2005, 2004, 2005 Deville, Eldorado & Seville 4.6L VIN 9, 4.6L VIN Y Transmissions: A/T	**HO2S-11 (Bank 1 Sensor 1) Circuit Low Input Conditions:** DTC P0101-P0103, P0106-P0108, P0112-P0113, P0116-P0118, P0121-P0123, P0125, P0128, P0200, P0300, P0410-P0446, P0452-P0453, P1258, P1415-P1416 and P1441 not set, AIR, Catalyst and EGR Flow tests off, system voltage 8-18v, no injectors disabled, TP angle from 3-25%, A/F ratio at 14.5-14.8:1, engine running in closed loop for 3 seconds, and the PCM detected the HO2S signal was less than 75 mv for 5 seconds while in closed loop, or it was less than 575 mv for 8 seconds while operating in Power Enrichment mode. **Possible Causes** ● Air leaks in the exhaust system, intake manifold, vacuum lines ● Engine misfire condition present (look for P0300 series codes) ● Fuel system too lean (possible low fuel pressure, water in fuel) ● HO2S signal circuit is shorted to the sensor or chassis ground ● HO2S is damaged (i.e., cracked) or air reference hole clogged ● PCM has failed

GM OBD II Systems

INTRODUCTION

The California Air Resources Board (CARB) began regulating On-Board Diagnostic (OBD) systems for vehicles sold in California beginning with the 1988 model year. The initial requirements, known as OBD I, required the identification of the likely area of a fault with regard to the fuel metering system, EGR system, emission-related components and the PCM. Implementation of this new vehicle emission controls monitoring regulation was done in several phases.

OBD I Systems

A malfunction indicator lamp (MIL) labeled *Check Engine Lamp* or *Service Engine Soon* was required to illuminate and alert the driver of a fault, and the need to service the emission controls. A diagnostic trouble code (DTC) was required to assist in identifying the system or component associated with the fault. If the fault that caused the MIL goes away, the MIL will go out and the code associated with the fault will disappear after a predetermined number of ignition cycles.

Following extensive research, CARB determined by the time an Emission System component failed and caused the MIL to illuminate, that the vehicle could have emitted excess emissions over a long period of time. CARB also concluded that semi-annual or annual tailpipe tests were not catching enough of the vehicles with Emission Control systems operating at less than normal efficiency.

To take advantage of improvements in vehicle manufacturer adaptive and failsafe strategies, CARB developed new requirements designed to monitor the performance of Emission Control components, as well as to detect circuit and component hard faults. The new diagnostics were designed to operate under normal driving conditions, and the results of its tests would be viewable without any special equipment.

DIFFERENCES BETWEEN OBD I & OBD II

As with OBD I, if the PCM detects an emissions-related fault on an OBD II system, the MIL is activated and a code is set. However, that is where the similarity between the systems ends. OBD II procedures that define emissions component and system tests, code clearing and drive cycles are more comprehensive than tests in the OBD I system.

Purpose of Onboard Diagnostics

The purpose of Onboard Diagnostics is to provide optimum control of the engine and transmission while meeting the objectives of the OBD II regulations. At the center of this system is a PCM connected to input and output devices via a wiring harness. The PCM receives information from sensors and switches, performs calculations based on data stored in memory and controls various output devices.

Powertrain Control Module

PCM Hardware & Software

As with earlier GM systems, the PCM is divided into two main parts: the system hardware and software. The system hardware includes:

- All related actuators, relays, and solenoids
- All related sensors and switches
- All interconnecting wires, connectors and terminals
- The Power Control Module (PCM)

System software includes programs that contain strategies used by the PCM to control engine system outputs based on related inputs. These are the strategies used to control the operation of the engine, electronic transmission, idle speed, fuel delivery control, and backup circuitry that is used if a major failure occurs inside the PCM.

PCM "Learning Capability"

The PCM includes some "learning" capability that allows it to make corrections for minor variations to improve driveability. If power to the PCM is disconnected, the "learning" process resets and begins all over again. Most vehicles will relearn operating parameters in a short driving period. Most Scan Tools have a special test mode that can be used on some 1989 and newer models to allow the "relearn" steps.

OBD II Monitor Tests

An OBD II Monitor is a PCM controlled (diagnostic) test run on one or more of the Emission Control systems to determine if a component or system is operating properly. As discussed, some tests either run once per trip or run continuously. The OBD II Monitor tests include:

- Catalyst & Fuel System Monitors
- EGR & EVAP Monitors
- Misfire Monitor
- Oxygen Sensor & Oxygen Sensor Heater Monitors
- Secondary AIR Monitor

Diagnostic Test Modes

The "test mode" messages available on a Scan Tool are listed below:

- Mode $01: Used to display Powertrain Data (PID data)
- Mode $02: Used to display any stored Freeze Frame data
- Mode $03: Used to request any trouble codes stored in memory
- Mode $04: Used to request that any trouble codes be cleared
- Mode $05: Used to monitor the Oxygen sensor test results
- Mode $06: Used to monitor Non-Continuous Monitor test results
- Mode $07: Used to monitor the Continuous Monitor test results
- Mode $08: Used to request control of a special test (EVAP Leak)
- Mode $09: Used to request vehicle information (INFO MENU)

Diagnostic Executive

The Diagnostic Executive is a unique segment of software in the PCM designed to coordinate and prioritize the diagnostic procedures as well as define the protocol for recording and displaying their results. The main responsibilities of the Diagnostic Executive are listed below:

- Command the MIL either On or Off
- Log and clear a diagnostic trouble code (DTC)
- Record the Freeze Frame data for the first emission related DTC
- Control any non-emission related Service Lamps
- Control the operating conditions buffer in the Failure Records
- Display current status information on each diagnostic function
- Update and display the current I/M Readiness Status "flags"

A key function of the Diagnostic Executive is to record codes and turn on the MIL when an emission related fault is detected. It can also turn off the MIL if the Conditions: go away that caused the code to set.

The PCM continually tests the operation of certain engine control functions. This diagnostic capability is complimented by the diagnostic procedures contained in repair manuals. The language of communicating the source of the fault detected occurs through a set of trouble codes. When a fault is detected by the PCM, a code is set and the MIL is illuminated for all emission related trouble codes.

GM Repair Information

GM repair charts are organized by section in their GM publications. In most cases Section 'A' includes the PCM power and ground circuit tests, No MIL, No DLC Communication, No Start Tests and the Code Repair Tables. Section 'B' includes Driveability Symptoms. Section 'C' includes Component checks of various PCM systems (i.e., Emission Controls, Fuel, Speed Control and Ignition systems). Diagnosis of the OBD II system on GM vehicles starts with an OBD System Check.

Customer Snapshot Mode

On Cadillac and Saturn models, the PCM contains a feature that allows it to record a snapshot or movie. The feature only works when the vehicle diagnostic tests are off. To activate this feature, press "OFF" and "FRONT DEFROST" on the dash controls simultaneously.

The next step is to retrieve the data captured with this feature with a Scan Tool. The data will appear as a DTC Snapshot with the trouble code designation of P0001. Be aware that any previous Customer Snapshot data will be overwritten when another snapshot is initiated.

Passive & Active Diagnostic Tests

A passive test is a diagnostic test that simply monitors a vehicle component or system. Conversely, an active test actually takes some sort of action when performing diagnostic functions, often in response to a failed passive test. For example, the EGR active test will force the EGR valve open during closed throttle deceleration and may not force the EGR valve closed during a steady state mode. Either action should result in a change in manifold pressure on the MAP sensor.

<u>Intrusive Diagnostic Tests</u>

An intrusive test refers to a test controlled by the PCM as part of OBD II that may have an effect on vehicle performance or emission levels.

UART Serial Data

There are two methods of data transmission used on OBD II systems. One method is the Universally Asynchronous Receiving/Transmitting (UART) protocol. UART is an interface device that allows a controller to send and receive serial data. Serial data refers to information that is transferred in a linear fashion (over a single line) one bit at a time. A data bus describes the electronic pathway through which serial data travels. UART receives serial data, converts the data to parallel format, and then places it on the data bus for use by the vehicle computer. UART can also convert data in parallel format to serial format, and then transmit the converted data to the Scan Tool.

Class II Serial Data

OBD II regulations require that all vehicle manufacturers establish a common communications system called Class II data. Each bit of information in Class II data can have one or two lengths, long or short. This feature allows a Scan Tool to access communication data from any make or model vehicle sold with Class II data capability.

This feature allows vehicle wiring to carry multiple signals over a single wire. Messages carried on Class II data streams can be prioritized. If two messages attempt to establish communication on the data line at the same time, only the message with the higher priority will continue. The message with the lower priority must wait.

Class II Data Example Graphic

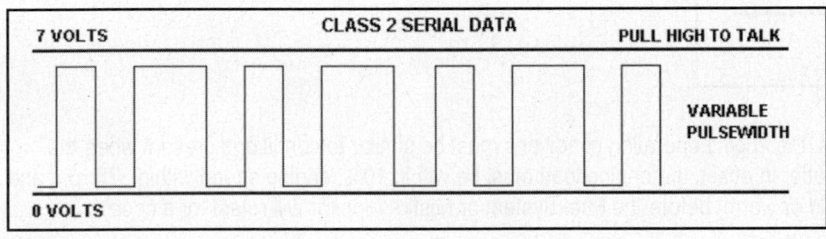

GM OBD II System Terminology

Enable Criteria

The term "enable criteria" is used to describe the *exact Conditions: necessary for a diagnostic test to run*. Each Monitor has a specific set of Conditions: that must be met before the diagnostic test is run. In this manual, the enable criteria for each trouble code is listed in the column to the right of the DTC and vehicle/engine identification column under the heading "***Trouble Code Title & Conditions***" in bold lettering. Enable criteria is different for each Monitor test and its related code. It may include, but is not limited to the following items:

- Air Conditioning (A/C) on
- BARO, ECT, IAT, MAP, TP and Vehicle Speed Sensors
- EVAP Canister Purge Enabled or Disabled
- Engine Speed (RPM) and Engine Load
- Short Term and Long Term Fuel Trim

OBD II Warmup Cycle

Once the MIL is off, a stored trouble code will remain in memory until 40 warmup cycles are completed without the fault occurring.

A warmup cycle is defined as a trip that includes a change in engine temperature of at least 40ºF and where it reaches 160ºF.

OBD II Warmup Graphic

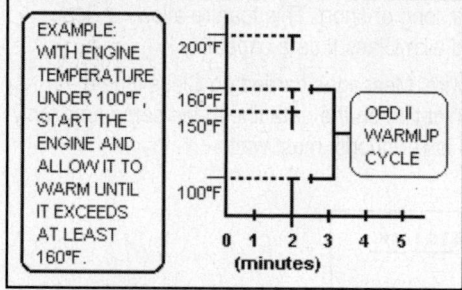

Similar Operating conditions

For Fuel System and Misfire codes, the engine operating conditions must be similar to conditions present when the fault was first detected to clear a code. In effect, the engine load must be within 10%, engine speed within 375 rpm and the engine temperature similar (cold or warm) before the Fuel System or Misfire Monitor will retest for a code.

Failure Records

Failure Records are records in the PCM that contain the engine operating conditions present when a trouble code is stored and the MIL is illuminated. On GM vehicles, the PCM can store multiple records (i.e., the Aurora 4.0L V8 engine can store up to three records while other vehicles can store up to five records) and can update the records at any time. As each record is updated, the first record is dropped and the new failure event is recorded. These records are an enhancement of the OBD II Freeze Frame capture feature, but they can store data for any fault in memory (not just data for a MIL fault)

OBD II Trip Definition

An OBD II "trip" is official when all the enable criteria for a given test (Monitor) are met. Because enable criteria vary from one Monitor to another, the definition of a trip varies as well. The trip requirements (criteria) can include seemingly unrelated items such as driving style, the length of the trip, and ambient temperature. A minimum requirement for a trip includes one key cycle, and in most cases, the engine must run for a period of time before the test is enabled.

Vehicle tests vary in length - some are performed only once per trip and some are performed continuously. The Catalyst, EGR, EVAP and Oxygen Sensor tests are performed once per trip. The Component Monitor, Fuel and Misfire tests are performed continuously. An OBD II trip is defined as a "key on-drive-the-vehicle-key-off" cycle in which the vehicle is operated in a manner that satisfies the criteria for a test.

Flash EEPROM

GM vehicles with OBD II systems (including 1991-95 Saturn models) use a flash erasable programmable read only memory device to make running changes.

In most cases, a computer is used to download the latest changes (listed by VIN code) into a PC or the Scan Tool. The service bay PC or Scan Tool is then connected to the vehicle so that it can verify the current PROM calibration to determine if the PROM needs updating.

I/M Readiness Status

The Scan Tool can identify the **"Flags" or I/M Readiness Status**. A flag ON for a system means that the test has been run. A flag OFF for a system means that the test has not been run. If a vehicle comes in with a problem, the technician should first look at the "Flags" screen on the Scan Tool to see if all flags were set to ON. If the EGR flag is OFF, there is a possibility that the EGR system has a fault or that the EGR system tests have not have been run (the EGR system may be okay or it may have a problem). The technician needs to drive the vehicle under the trip Conditions: and get the flag set to ON before proceeding with testing the EGR system. It should be noted that OBD II trips are different for vehicles with different body codes and engines.

If power to the PCM is removed (by removing a fuse or disconnecting the battery) the Inspection & Maintenance (I/M) Flags will be reset to "off". In effect, the vehicle must be driven under specific Conditions: until all flags are set to "on".

If the power is removed, a Scan Tool can be used to do a "quick relearn step" which resets the Fuel Trim and Idle Speed to the default settings. These steps are part of the **MISC, FUNCTIONAL or SPECIAL menus** found on many Aftermarket Scan Tools.

Malfunction Indicator Lamp

The Malfunction Indicator Lamp (MIL) looks similar to the lamp used on earlier vehicles (i.e., the "Check Engine" or "Service Engine Soon" lamp). However, on OBD II systems, the MIL is activated under a strict set of guidelines that dictate that the lamp must be turned on when the PCM detects an emissions related fault that could impact the vehicle tailpipe emissions or Evaporative Loss system.

The Malfunction Indicator Lamp (MIL) lamp is mounted in the instrument panel. It provides several functions as described below:

- To inform the driver that a fault affecting vehicle emission levels has occurred and to bring the vehicle in for service immediately.
- To check the bulb and related circuit, the MIL will is turned on with the key in the "on" position with the engine off (KOEO). Once the engine is started, the MIL should go out.
- If the MIL remains on with the engine running, or if a fault is suspected due to an emissions related problem, an OBD System Check of the PCM diagnostics should be performed. The tests contained in this diagnostic procedure will help find any faults that might not be detected using other diagnostic routines.

MIL Circuit Graphic

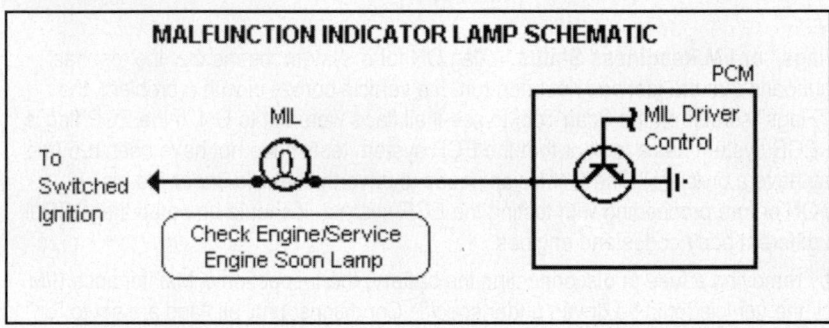

MIL On/Off Guidelines

If an emission-related fault is detected, the Diagnostic Executive will activate the MIL and allow it to remain "on" until the system or component passes the same test for three consecutive trips without the fault reoccurring.

MIL On or Flashing - Fuel System or Misfire Conditions:

If a Fuel system problem or misfire condition is present that could damage the catalyst, the MIL will flash once per second. The MIL will continue to flash until the vehicle is outside of the speed and load conditions that could cause possible catalyst damage. It will stop flashing and remain on once these conditions: are no longer present.

Malfunction Indicator Lamp (Continued)

Once the MIL is on, the Diagnostic Executive will turn the MIL off after three consecutive trips when "test passed" is recorded for the trouble code that originally caused the MIL to be activated.

If this situation occurs while the MIL is off the DTC that set when the emission-related fault occurred will be stored by the PCM in the Freeze Frame and Failure Records. The DTC will remain in memory until 40 warmup cycles (without a new fault) have been completed.

If the MIL was set due to either a Fuel system or Misfire related fault, there are other requirements to meet before the code can be cleared. Once all the requirements for these types of faults are met, the onboard diagnostics can validate that the emissions fault that caused the MIL to be activated has been corrected.

The additional requirements for a Fuel system or Misfire fault are:

- Diagnostic tests that are passed must occur within 375 rpm of the engine speed data stored at the time that the last test failed
- The engine must be at ±10% of the engine load that was stored at the time the last test failed
- Engine operating conditions must be similar to the conditions present (warmed up or warming up) when the last test failed

Intermittent MIL "On" Conditions:

If the PCM detects a fault and then the fault goes away, the MIL will remain on until after three trips are completed without the same fault reoccurring. This type of MIL "on" condition could appear to be an intermittent fault, but most OBD II faults are not intermittent in nature.

An example of an intermittent MIL condition is described next. If the customer were to leave the fuel filler cap loose or off, and the vehicle was driven under the correct code conditions (meeting all enable criteria for an EVAP large leak trouble code), the PCM would turn the MIL on the second time it ran the EVAP Monitor and failed the test. The MIL would remain on until the vehicle was refueled and the customer tightened the fuel cap properly or noticed it was off. Once the vehicle was driven under the correct EVAP large leak code conditions, the EVAP Monitor would run the test. If the EVAP test passed for three consecutive trips, the PCM would turn off the MIL (this is true for some vehicles depending upon the model year).

However, in this case, the related trouble code would remain in memory until 40 OBD II warmup cycles occur (80 OBD II warmup cycles for Fuel system and Misfire faults) without the fault reoccurring. If the vehicle was brought in for service and a PCM Reset step was done, the MIL would then go out and the codes would clear.

Data Link Connector

OBD II Systems use a standardized test connector called the Data Link Connector (DLC). It is located beneath the instrument panel somewhere between the left end of the instrument panel and 12 inches (300 mm) past the vehicle centerline.

The DLC is located out of the sight of vehicle passengers, but should be easily viewable by a technician from a kneeling position outside the vehicle. The DLC is rectangular in design, capable of accommodating up to 16 terminals and has keying features to allow for easy connection. The DCL and Scan Tool connector have latching features that ensure the Scan Tool connector will remain mated when properly connected. Some common uses of the Scan Tool are:

- To identify and clear any stored diagnostic trouble codes
- To read the serial data stream information (e.g., PID data)
- To perform Enhanced Diagnostic Tests (bi-directional Scan Tool)

Data Link Connector Graphic

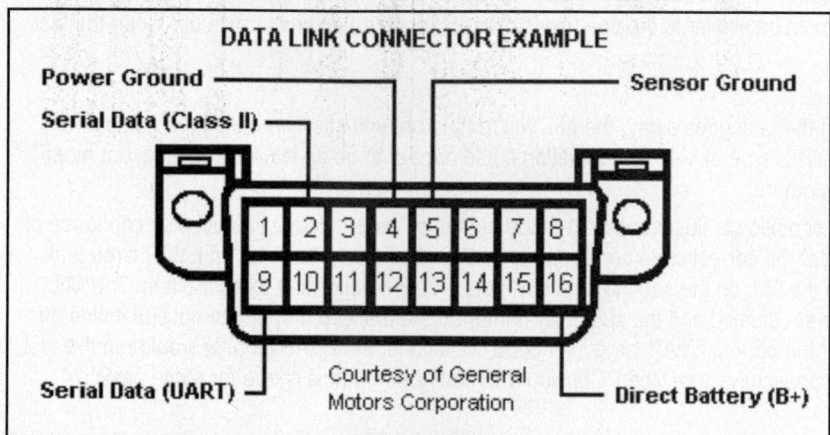

DLC Pin Assignment Table

An example of the GM DLC Pin Assignments is shown below.

Cavity	Pin Assignment	Cavity	Pin Assignment
1	Secondary UART 8192 Baud	9	Primary UART
2	Class II or J1850 Bus + L-Line	10	J1850 Bus- Line (2-wire)
3	Ride Control Diagnostic	11	EVO or MSVA Steering
4	Chassis Ground	12	ABS or CCM Enable
5	Signal Ground	13	SIR Diagnostic Enable
6	PCM/VCM Diagnostic Enable	14	E & C Bus
7	ISO 9141 (K-Line) Bus	15	ISO 9141 (L-Line) Bus
8	Keyless Entry or MRD Enable	16	Fused Battery Power

Diagnostic Trouble Codes

Each OBD II Diagnostic Trouble Code (DTC) is directly related to a particular Monitor Test. The Diagnostic Management System sets codes based on the failure of the tests during a trip (or trips). Certain tests must fail during two consecutive trips before the DTC is set.

Failure Records

GM vehicles can store up to 5 Failure Records on some models. Each record is for a different code. It is also possible that there will not be Failure Records for every code if multiple codes are set. The four types of codes and their related characteristics are shown below:

Type 'A' Codes

- Type 'A' Codes are emissions-related
- Request the MIL "on" the first trip a fault is detected
- Store a History Code on first trip a fault is detected
- Store Freeze Frame data the first trip a fault is detected (if empty)
- Store a Failure Record
- Update the Failure Record each time a Monitor Test fails

Type 'B' Codes

- Type 'B' Codes are emissions-related
- Are "armed" or pending after one trip when a fault is detected
- Are "disarmed" after 1-Trip if the second consecutive trip passes
- Request the MIL "on" the 2nd consecutive trip if the fault occurs
- Set a History Code on the 2nd consecutive trip if the fault occurs
- Store Freeze Frame Data the 1st trip that a fault is detected
- Store a Failure Record the first trip a fault is detected
- Update a Failure Record the first time the test fails each key cycle

Type 'C' Codes (changed to Type C1 in mid-1997)

- Type 'C' Codes are non-emissions related
- Request the SES (lamp) or DIC the first time the fault is detected
- Store a History Code the first time the fault is detected
- Do not store a Freeze Frame Record
- Store a Failure Record the first time the test fails each key
- Update a Failure Record the first time a test fails each key cycle

Type 'D' Codes (changed to Type C0 in mid-1997)

- Type 'D' Codes are non-emission related
- Do not request the PCM or other controller to turn on a lamp
- Store a History Code on the first trip that a fault is detected
- Do not store a Freeze Frame Record
- Store a Failure Record if a non-emission related test fails
- Update the Failure Record each time a non-emissions test fails

Diagnostic Trouble Code Display

The Scan Tool can display up to five DTC options of enhanced code data on GM Vehicles They are DTC Info, Specific DTC, Freeze Frame, Fail Records (not used on all applications) and Clear Info.

<u>DTC Info Mode</u>

This mode is used to search for a specific type of stored DTC information. There are seven different selections. The repair charts may instruct the technician to test for a DTC in a particular manner.

- History
- MIL Request
- Last Test Fail
- Test Fail Since Codes Cleared
- Not Run Since Codes Cleared
- Fail This Ignition
- DTC Status

<u>Specific DTC Mode</u>

In this Mode, the PCM checks the status of individual diagnostic tests by their DTC number. This selection can be accessed if a DTC has passed, failed or a combination of both. Some of the individual DTC tests that are available on the Scan Tool are shown below:

- Failed Last Test
- Failed Since Clear
- Failed This Ignition
- History DTC
- MIL Requested
- Not Run Since Clear
- Not Run This Ignition
- Test Ran and Passed

<u>Conditions: to Clear Diagnostic Trouble Codes</u>

Here are 3 methods for clearing codes from the PCM memory:

- Use the Scan Tool (also clears Freeze Frame & Failure Records)
- If battery power to the PCM is removed (battery cable or PCM fuse), all current data (DTC, Freeze Frame, Fail Records, Statistical Filters and I/M Readiness Flags) will be cleared
- If the fault that caused a DTC to set has been corrected, the PCM will begin to count warmup cycles. Once it has counted 40 warmup cycles with no further faults detected, the DTC is cleared

Standard Corporate Protocol

On vehicles equipped with OBD II, a Standard Corporate Protocol (SCP) communication language is used to exchange bi-directional messages between stand-alone modules and devices. With this type of system, two or more messages can be sent over one circuit.

OBD II Monitor Software

The Diagnostic Executive contains software designed to allow the PCM to organize and prioritize the Main Monitor tests and procedures, and to record and display test results and diagnostic trouble codes.

The functions controlled by this software include:

- To control the diagnostic system so that the vehicle will continue to operate in a normal manner during testing.
- To ensure that all of the OBD II Monitors run during the first two sample periods of the Federal Test Procedure.
- To ensure that all OBD II Monitors and their related tests are sequenced so that required inputs (enable criteria) for a particular Monitor are present prior to running that particular Monitor.
- To sequence the running of the Monitors to eliminate the possibility of different Monitor tests interfering with each other or upsetting normal vehicle operation.
- To provide a Scan Tool interface by coordinating the operation of special tests or data requests.

Cylinder Bank Identification

Engine sensors are identified for each engine cylinder bank by SAE regulations as explained below.

Bank - A specific group of engine cylinders that share a common control sensor (e.g., Bank 1 identifies the location of Cyl 1 while Bank 2 identifies the cylinders on the opposite bank).

An example of the cylinder bank configuration for a Chevrolet Lumina with FWD and a 3.4L V6 (VIN X) transverse mounted engine is shown in the Graphic to the right.

Cylinder Bank Identification Graphic

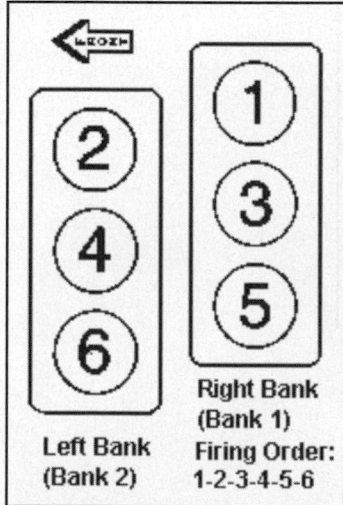

Misfire Diagnostics

The Misfire Diagnostics represents a special type of trouble code diagnostics. Each time a misfire is detected, the engine load, engine speed and coolant temperature at that moment are recorded and the last reported set of Conditions: are stored in Freeze Frame when the key is turned off. During the next few key on/off cycles, the stored Conditions: are used as a reference for similar conditions.

If an emissions threatening misfire occurs for two consecutive trips, the PCM treats the fault as a normal Type B trouble code (it turns on the MIL and stores a code). However if a misfire is detected on two non-consecutive trips, the stored Conditions: are compared with the current conditions. If a misfire occurs on a second non-consecutive trip under the similar Conditions: listed below, the MIL is illuminated:

- The engine load conditions are within 10% of the previous failure
- Engine speed is within 375 rpm of the previous failure
- Engine temperature is in the same range as the previous failure

Misfire Diagnostics

The Diagnostic Executive has the capability of alerting the driver of potentially damaging levels of engine misfire that could damage the catalyst. If this type of condition is detected, the MIL is commanded to flash on/off once per second during the actual misfire condition.

Fuel Trim Diagnostics

In order to meet OBD II regulations, fuel trim information is displayed on a Scan Tool in percentages. This is different from the way fuel trim has been traditionally displayed on a Scan Tool. Short term and long term fuel trim functions within OBD II are similar to past usage, only their measurement units will differ. The fault detection logic and MIL operation is the same as that described for Misfire Diagnosis.

Fuel Trim Conversion Graphic

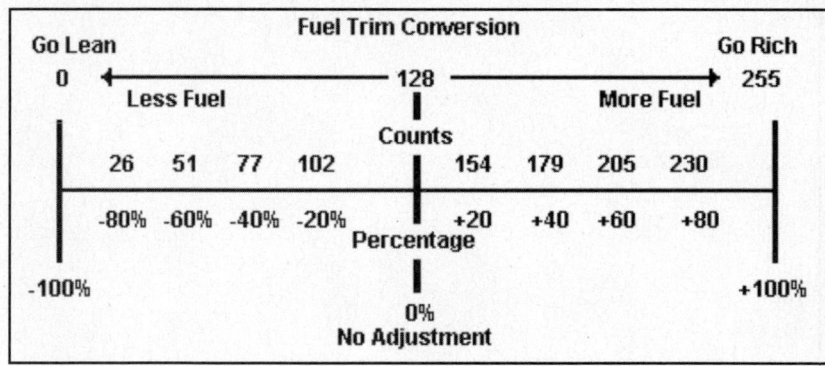

Heated Oxygen Sensor

The Heated Oxygen Sensor (HO2S) detects the presence of oxygen in the exhaust and produces a variable voltage according to the amount of oxygen detected. The HO2S outputs a voltage from 0-1v. A value less than 0.4v indicates lean A/F ratio and a value over 0.6v indicates a rich A/F ratio.

HO2S-11, HOS-12, HO2S-13 Locations - Example 1

Throughout the manual, there are references to Heated Oxygen Sensors identified with acronyms HO2S-11, HO2S-12 and HO2S-13. In addition, the sensors are identified as either cylinder Bank 1 or cylinder Bank 2. It should be understood that Bank 1 always contains engine cylinder number 1 (**Cyl 1**).

HO2S Location Graphic

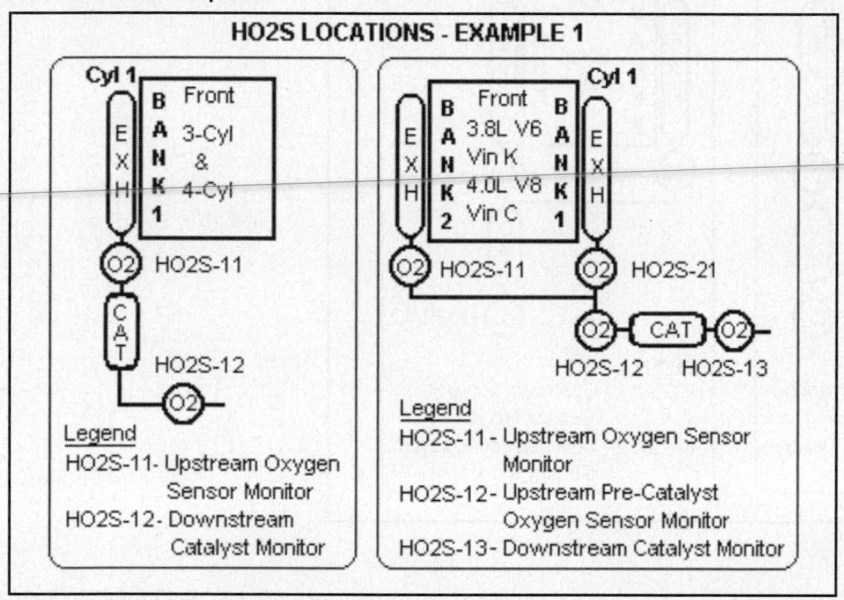

Explanation

In the V6 and V8 examples in the Graphic above, HO2S-11 refers to the upstream oxygen sensor and HO2S-12 refers to the downstream oxygen sensor. The downstream oxygen sensor or third oxygen is referred to as HO2S-13. The upstream HO2S-11 (or HO2S-12 on the V6 and V8 engines as noted) signal is used with the Oxygen Sensor Monitor test function. The HO2S-12 (or HO2S-13 as noted) signal is used with the Catalyst Monitor test function.

Heated Oxygen Sensor (Continued)

HO2S-11, HOS2 Locations - Example 2

Throughout the manual, there are references to Heated Oxygen Sensors identified with acronyms HO2S-11, HO2S-12 and HO2S-13. In addition, the sensors are identified as either cylinder Bank 1 or cylinder Bank 2. It should be understood that Bank 1 always contains engine cylinder number 1 (**Cyl 1**).

HO2S Location Graphic

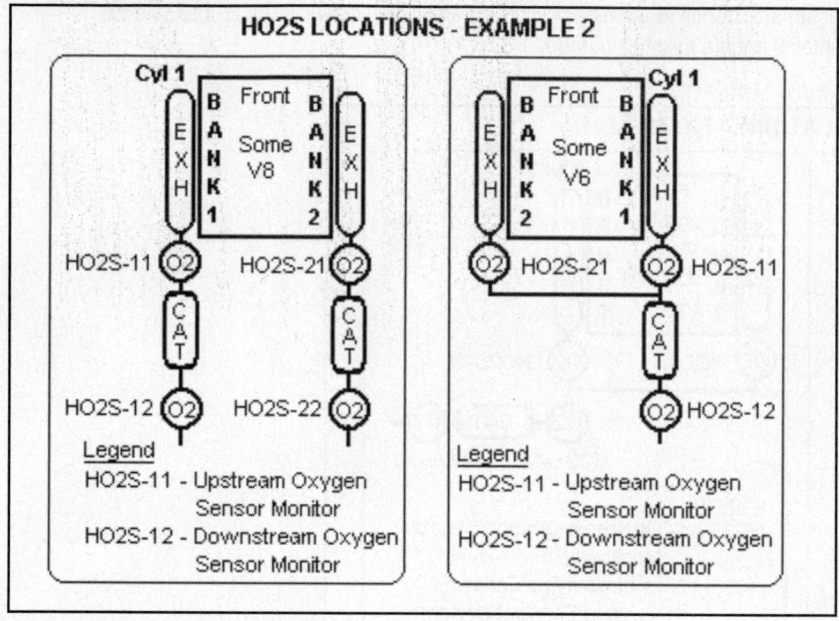

Explanation

In both of the pictures in Graphic above, HO2S-11 refers to the upstream oxygen sensor while HO2S-12 refers to the downstream oxygen sensor.

In these examples, the upstream HO2S-11 signal is used with the Oxygen Sensor Monitor test function. The downstream HO2S-12 signal is used with the Catalyst Monitor test function. HO2S location information is very important when attempting to identify the correct oxygen sensor as it relates to a trouble code repair chart.

Freeze Frame Data

Freeze Frame is an element of the Diagnostic Executive that stores engine operating conditions at the moment an emission-related fault is stored in memory (when the MIL is commanded on). This data can be used to help identify the cause of an emissions-related fault.

Regulations related to the OBD II System require that certain engine-operating conditions be captured and stored whenever the MIL is illuminated. The data captured is called Freeze Frame data. This data can be thought of as a single record of a certain set of operating conditions. Whenever the MIL is turned on, the corresponding record of operating conditions is recorded to the Freeze Frame buffer.

The Freeze Frame data can only be overwritten by data associated with a Fuel Trim or Misfire fault because data from these faults takes priority over data associated with any other type of fault. The Freeze Frame data will not be erased unless the associated History DTC is cleared.

Failure Records

Failure records store data about the operating conditions when the code was stored and the MIL was illuminated. The PCM can store multiple records and can also update these records at any time. Some vehicles can store up to 5 failure records. However, the Aurora with the Northstar 4.0L engine will only store up to 3 failure records.

The current engine operating conditions are recorded in the Failure Records buffer each time a test fails. As each record is updated, the first record is dropped and the new failure event is recorded.

The operating conditions for a diagnostic test that failed may include one or more of the following engine operating parameters:

- Air Fuel Ratio
- Airflow Rate
- Barometric Pressure
- Engine Load
- Engine Coolant Temperature
- Engine Speed
- Fuel Trim
- Injector Base Pulsewidth
- Manifold Absolute Pressure
- Open or Closed Loop Status
- Throttle Position Angle
- Vehicle Speed

OBD II Drive Cycle

General Motors OBD II systems implement the usual Strategy Based Diagnostic procedures built into the Powertrain Control Module (PCM) and Transmission Control Module (TCM). The first step in diagnosis of a problem on an OBD II system is to identify it as either a trouble code fault (Code Fault) or a driveability symptom (No Code Fault). The OBD II Drive Cycle is used to verify any repair to the system.

Strategy Based Diagnostics

Strategy Based Diagnostics can be used to repair all Electrical and Electronic systems on GM vehicles. The diagnostic approach in this methodology can also be used to solve problems in an OBD II system.

OBD II Drive Cycle Procedure

The main intention of the OBD II Drive Cycle is to run the OBD II Main Monitors in order to determine the status of the Inspection & Maintenance (I/M) Readiness Tests. A cold engine startup (e.g., ambient air temperature of 40-100ºF) is a necessary step in preparation to run a complete OBD II Drive Cycle. In most cases the engine coolant temperature must be below 122ºF.

OBD II Drive Pattern

The drive pattern shown below can be used to help solve:

- Trouble Code Faults - Refer to the Code List (in this section) or look in electronic media or repair manuals for a code repair chart.
- Driveability Symptoms & Intermittent Faults - Refer to the special repair instructions under No Code Faults in other repair manuals.

GM OBD II Drive Cycle Graphic

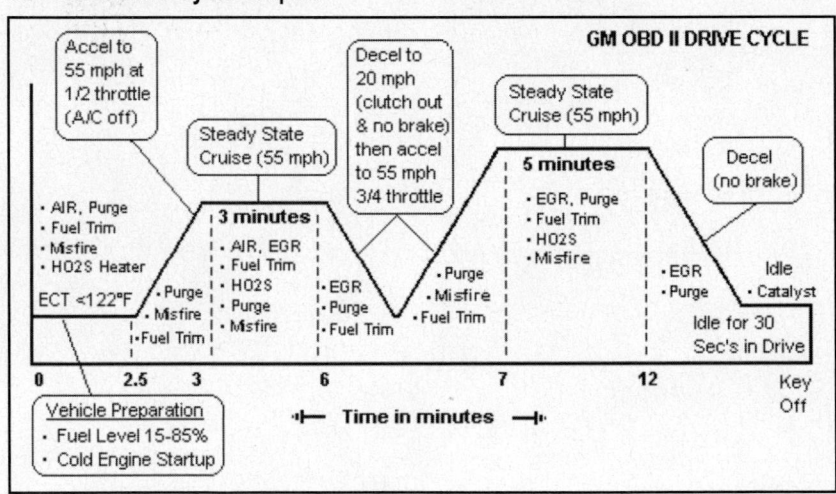

GM OBD II Monitors

Comprehensive Component Monitor

OBD II regulations require that the PCM monitor all components and systems that could cause tailpipe emissions to exceed 1.5 times the FTP Standard. The Comprehensive Component Monitor (CCM) is designed to continuously monitor components for a short-to-ground, open circuit or short-to-power condition. It also conducts rationality checks of *input devices* and functionality checks of *output devices*.

Input Devices and the CCM

Devices that provide input signals are monitored for circuit continuity and out-of-range values including performance checks. Performance checks are defined as checks that indicate a fault when the signal from a sensor does not seem reasonable. An example would be a high TP sensor input at low engine load (e.g., with a low MAP input).

PCM Input Devices that are monitored by the CCM include:

- Camshaft and Crankshaft Position Sensors
- ECT, IAT, MAF, MAP, TFT and TP Sensors
- Knock Sensor and Vehicle Speed Sensor

Output Devices and the CCM

The PCM uses an Output Device Monitor to "watch" the voltage level change on a controlled device as it is switched from "on" and "off". The eyeballs in the Graphic below represent the point at which the PCM monitors the output command signal for a change in voltage.

PCM (controlled) Output Devices monitored by the CCM include:

- Air Conditioning, Cooling Fan and Fuel Pump Relays
- EVAP Purge and EVAP Vent Control Solenoids
- Cruise Control Vacuum and Vent Solenoids
- Electronic Transaxle Controls
- Idle Air Control Motor and Malfunction Indicator Lamp

Output Device Monitor Graphic

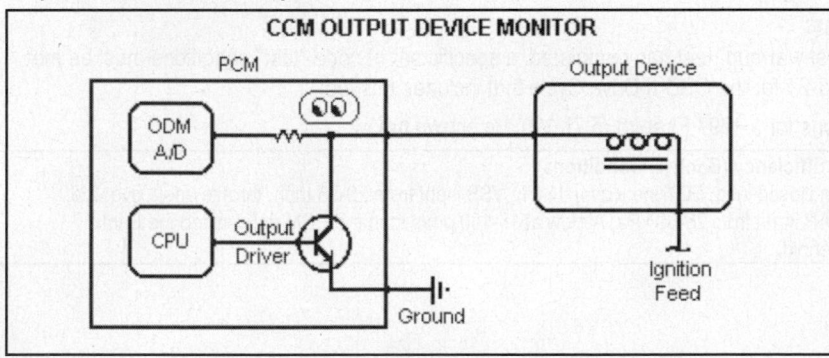

Catalyst Monitor

EPA regulations require that the onboard diagnostics must monitor the catalyst once per trip. The catalyst is considered degraded if the hydrocarbon levels at the tailpipe exceed the FTP standards.

The OBD II Catalyst Monitor measures oxygen storage capacity. To test the catalyst efficiency, oxygen sensors are installed before and after the Three-Way Catalyst. Voltage variations between the oxygen sensors allow the PCM to determine the performance of the catalyst.

The Catalyst Monitor test is based on the correlation conversion efficiency and oxygen storage capacity of the catalyst. An efficient catalyst will show a relatively flat output voltage on the post-catalyst heated oxygen sensor. If the catalyst is degraded, the signal from the post-catalyst HO2S will show increased activity (refer to the Graphic).

Catalyst Monitor Graphic

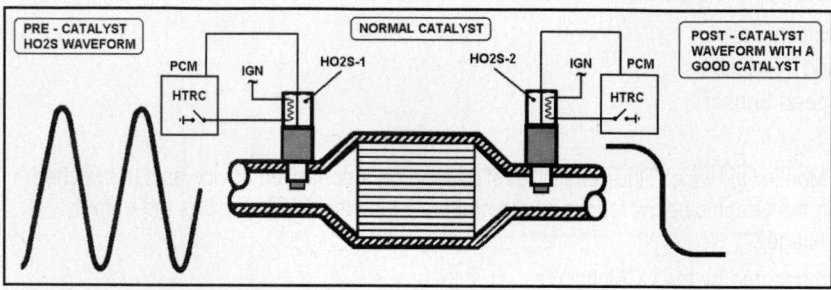

Catalyst Degradation and Failure

When a catalyst DTC sets, remember that the converter may have internal or external damage. A few items to check are shown below:

- Check for any exhaust leaks (they can hide a degraded catalyst)
- Check for contaminated fuel and for miss-matched O2 sensors
- Check for the use of alternate fuels and for any aftermarket parts

Catalyst Monitor Test Conditions:

Once the PCM detects the "catalyst warmup" test has completed, a specific set of code "test" conditions must be met to run the Monitor. Refer to Page 3-21 for the OBD II Drive Cycle that includes this test.

The Catalyst trouble code conditions for a 1997 Firebird (5.7L V8) are shown below.

DTC P0420 - Catalyst System Low Efficiency (Bank 1) Conditions:
No PCM or TCM codes set, system in closed loop, ECT input over 124°F, VSS input from 20-75 mph, throttle angle over 2%, engine speed less than 3500 rpm, MAP input from 25-80 kPa, Airflow at 15-100 gm/s then the PCM determined the Bank 1 Catalyst had degraded below its threshold.

ECT Sensor Monitor

The ECT Monitor is an on-board diagnostic designed to test the ECT Sensor input for out-of-range conditions. It is also used to detect how long it takes for the engine coolant to reach closed loop temperature.

ECT Sensor Wiring Graphic

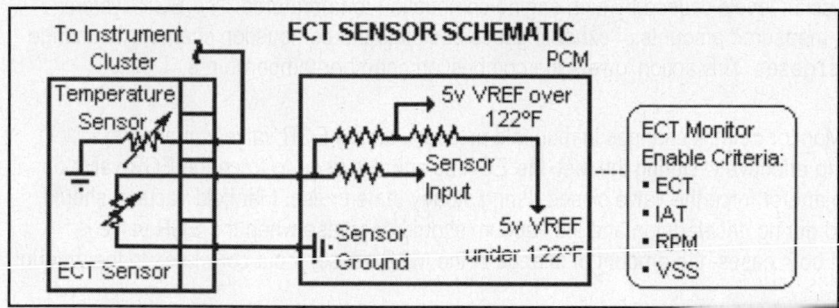

ECT Closed Loop Enable Test

The Closed Loop Enable Test measures the amount of engine run time required for the ECT sensor input to reach closed loop threshold.

ECT Sensor Out-of-Range Test

The ECT sensor out-of-range test monitors the temperature reading from the ECT sensor approximately every 100 milliseconds. For a fixed interval of time during the test, the PCM counts the number of ECT sensor inputs outside of its expected range. If the number of ECT sensor inputs in a high or low range exceeds a stored threshold, the ECT Sensor Monitor determines the sensor has failed high or low.

If a relatively small number of samples fall into the high range, DTC P1115 is set (ECT Sensor Intermittent High Input). If the samples fall into the low range, DTC P1114 is set (ECT Sensor Intermittent Low Input). This intermittent test is not required by the OBD II legislation. Refer to Page 3-21 for the OBD II Drive Cycle that includes this test.

ECT Monitor Test Information

The ECT Monitor Test runs all the time except for these conditions:

* Right after startup for a short period of time due to poor linear initial accuracy of the sensor at extremely cold temperatures
* When vehicle speed is zero mph (a fault could occur during vehicle service such as cooling system flushing)

The intermittent code test criteria on a 1997 Seville 4.6L V8 include the items listed below.

DTC P1114 - ECT Sensor Circuit Intermittent Low Input Conditions:
No IAT sensor codes set, IAT input less than 158°F, engine runtime over 10 seconds, then the PCM detected an intermittent ECT input that was less than 0.25v (271°F), condition met for 1 second. Refer to the correct code repair chart in other manuals or electronic media as needed.

EGR System Monitor

Government regulations require that the onboard diagnostics must monitor the EGR system for abnormally high or low EGR flow rates. This system is considered to have failed if a fault is detected that could change the EGR flow rate to a level below the FTP standards.

The EGR system lowers NOx emission levels caused by high engine combustion temperatures. The EGR system accomplishes this task by feeding measured amounts of exhaust gases back into the combustion chamber to change the A/F ratio and dilute the exhaust gases. This action lowers the combustion chamber temperatures.

EGR System Monitor Tests

In most cases, the EGR System Monitor detects changes in manifold pressure during EGR valve actuation to determine if the system is operating effectively. During the test, the EGR solenoid is used to force the EGR valve open during closed throttle deceleration and/or force the valve closed during steady state cruise. Manifold vacuum should increase when the valve is opened during deceleration and the vacuum should decrease when the EGR valve is closed during cruise conditions. In both cases, the amount of change in the MAP sensor input correlates to the amount of EGR flow through the valve.

EGR System Graphic

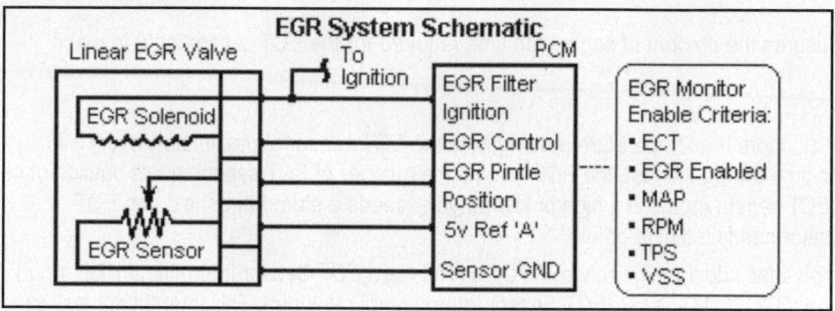

EGR Monitor Test Conditions:

The PCM runs the EGR Monitor once its code test criteria are met.

Refer to Page 3-21 for the OBD II Drive Cycle that includes this test. The test conditions for a 1998 Cavalier with a 2.2L I4 engine include:

DTC P0404 - EGR Flow Insufficient Conditions:

ECT sensor input over 176°F, engine speed 900-1700 rpm (engine speed 100-2200 on M/T or 900-1800 on A/T on VIN T), VSS over 20 mph, BARO input over 60 kPa, MAP input 13-32 kPa, IAC change below 4-5 counts, throttle angle under 1%, then the PCM detected the Decel Exponentially Weighted Moving Average was more than +2 kPa (it should be below -3). Refer to the correct code repair chart in other manuals or electronic media.

EVAP System Monitor

The EVAP System Monitor is a PCM diagnostic run once per trip that monitors the Evaporative Emissions Control System in order to detect a loss of system integrity or leaks of more than 0.040" in the system.

The GM vehicles included in this section are equipped with an EVAP Purge System or an Enhanced EVAP System (1997-2002 models).

EVAP Purge System

This system includes an EVAP canister (with vapor lines), a purge solenoid (with purge lines), a vacuum switch, fuel tank and cap and related pipes and hoses. The vacuum switch is used to detect when the system is purging. It senses the flow of vacuum from the engine through the purge valve. It is closed when the system is purging.

EVAP Purge System Graphic

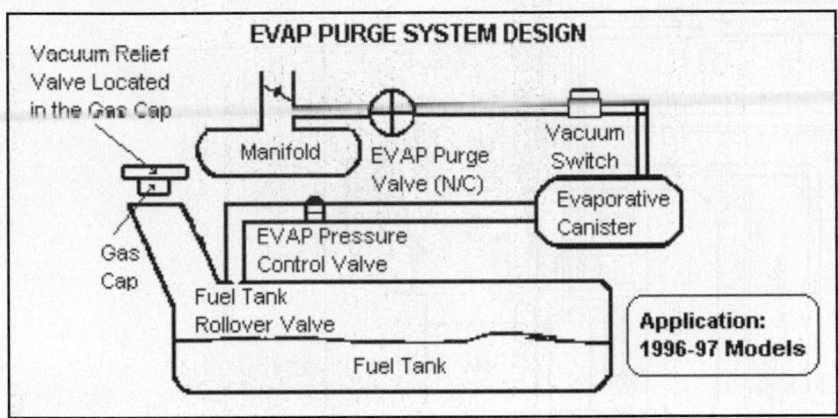

EVAP Purge Monitor Test Conditions:

The PCM runs the EVAP Monitor once its code test criteria are met. Refer to Page 3-21 for the OBD II Drive Cycle that includes this test. The test conditions for a 1997 Camaro with a 3.8L V6 engine include:

P0441 (1T) - EVAP System No Flow During Purge Conditions:
No PCM codes set, IAT input at 50-158°F, ECT input under 237°F, the difference between ECT and IAT under 18°F, BARO input over 73 kPa, engine speed at 900-5000 rpm, throttle angle at 2.5-40%, Purge PWM over 85%, then the PCM detected that the EVAP vacuum switch was closed for 2 seconds with the Purge solenoid enabled.

EVAP Purge System Faults

If the charcoal canister fresh air vent is clogged, and the EVAP purge valve is closed, source vacuum will be trapped in the purge hose from the vacuum switch to the canister. This action will indicate the purge valve is open or the switch is faulty during the test. Be sure to check for restrictions or leaks in the purge hose or switch on this system.

EVAP System Monitor (Continued)

Enhanced EVAP System

The Enhanced EVAP System includes an EVAP canister, fuel tank vapor pressure (FTVP) sensor, purge and vent solenoids, a fuel level sensor, a service port and the fuel cap. The FTVP sensor is used as part of a diagnostic strategy that is based on applying vacuum to the system and then monitoring the amount of vacuum decay over time.

Once the test criteria are met, source vacuum is used to draw a small amount of vacuum on the entire system by mechanically sealing off the designed vent path. After a calibrated amount of vacuum has been achieved, the vacuum source is turned off (system is closed). Leaks are detected by testing the amount of vacuum decay over time.

Enhanced EVAP System Graphic

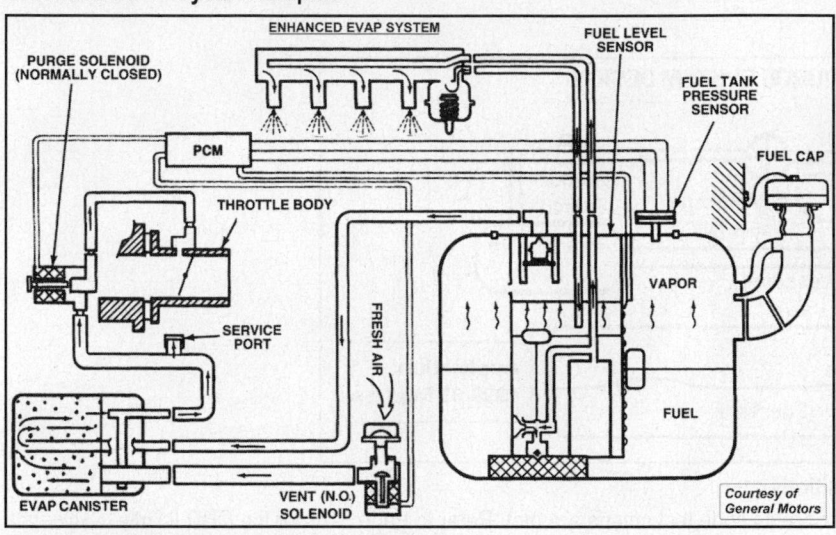

EVAP Enhanced Monitor Test Conditions:

The PCM runs the EVAP Monitor once its code test criteria are met. Refer to Page 3-21 for the OBD II Drive Cycle that includes this test. The test conditions for a 2000 Camaro with a 5.7L V8 engine include:

P0440 (2T) - EVAP System Performance Conditions:
DTC P0107, P0108, P0112, P0113, P0117, P0118, P0125, P0420, P0430, P0500, P0502, P0503, P0562, P0563, P1120, P1220 and P1221 not set and no HO2S codes set, system voltage 9-16v, ECT and IAT inputs from 39-86°F and within 19°F of each other at startup, BARO input over 75 kPa, fuel level from 12.5-87.5%, then the PCM detected the EVAP system was unable to develop a vacuum level greater than a predetermined value.

Fuel System Monitor

The Fuel System Monitor is a PCM diagnostic that continuously monitors the Fuel system to verify its ability to provide compliance with OBD II regulations. This diagnostic must fail if it detects a fault that could cause tailpipe emissions to exceed the FTP standards.

Long and Short Fuel Trim

The Fuel System Monitor is designed to monitor the averages of Long Term fuel trim (LONGFT) and Short Term Fuel Trim (SHRTFT). If the PCM determines that the fuel trim values reach and stay at their limits for too long a period of time, a fault is indicated and a code is set.

This Monitor compares the average of LONGFT values with SHRTFT values to the rich and lean limits (*calibrated fail thresholds*) of the test. A Pass is recorded if either value is correct with the system controlling fuel authority. If both values fall into the failure thresholds, a failure condition is exists. In if occurs a code to set and the engine conditions at that time are recorded in Freeze Frame and in the Failure Records.

The Fuel System Monitor also conducts an *intrusive test* to detect if a rich condition is being caused by excessive vapors from the canister.

Fuel System Monitor Test Conditions:

The PCM runs the Fuel System Monitor once its test criteria are met. Refer to Page 3-21 for the OBD II Drive Cycle that includes this test. The code test criteria for a 1998 Rivera with a 3.8L V6 engine include:

P0172 (2T) - Fuel Trim System Rich (Bank 1) Conditions:
No CKP, CMP, ECT, EGR, EST, EVAP, IAT, Injector, MAF, MAP, Misfire, TP or VSS codes set, ECT input from 68-230°F, IAT from 0-158°F, BARO input more than 70 kPa, MAP input from 15-85 kPa, throttle angle below 90%, VSS under 82 mph, MAF input from 3-150 gm/s, engine speed from 600-5000 rpm, then the PCM detected the Bank 1 LONGFT was at or near the maximum authority of -20%, and the SHRTFT was at or near maximum authority of -20%, all conditions met in the appropriate Fuel Trim Cells.

Fuel System Faults

A Fuel system fault and related DTC may be triggered by a whole list of engine Conditions: or faults. To diagnose the cause of a Fuel system fault, use all available information including any trouble codes stored in memory, a rich or lean condition at the tailpipe or any engine or vehicle driveability symptoms. Check the list of components below:

- A defective catalytic converter
- A leaking or weak fuel pump
- Any faults in the ECT, IAT, MAF, MAP or pre-catalyst HO2S
- Any faults in the Cooling, EGR, Fuel or Ignition Systems
- Worn engine mechanical components

Idle Air Control System Monitor

The PCM controls the air entering into the engine with an Idle Air Control valve. To increase the idle speed, the PCM commands the pintle inside the valve away from the throttle body seat. This action allows more air to bypass through the throttle blade. To decrease the engine speed, the PCM commands the pintle towards the seat. This action reduces the amount of air bypassing the throttle blade.

IAC System Graphic

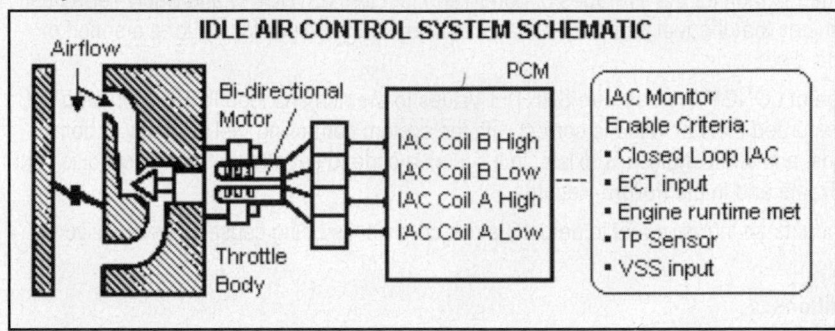

IAC Passive Test

The Passive test monitors the measured intake airflow to determine if the IAC system is delivering a calibrated amount of throttle bypass air. The test is enabled at idle with the IAC valve operating normally. The passive test criteria for a 2000 Camaro with a 5.7L V8 engine include:

P0506 (2T) - Idle Speed Low Conditions:
DTC P0101, P0102, P0103, P0107, P0108, P0112, P0113, P0117, P0118, P0121, P0122, P0123, P0125, P0171, P0172, P0174, P0175, P0220, P0300, P0401, P0404, P0405, P0443, P0500, P1414 and P1441 not set, system voltage 9-17v, engine runtime 60 seconds, ECT input over 140°F, IAT input more than -14°F, BARO input over 65 kPa, throttle angle under 1%, VSS input under 1 mph, then the PCM detected the Actual idle speed was more than 75 rpm lower than Desired idle speed for 5 seconds.

IAC Intrusive Test

If the PCM determines that an Intrusive test is needed, this test is performed at cruise speed while changing the IAC valve position. The IAC Monitor measures the amount of change in the intake air rate or MAP input to detect normal operation, or if a fault is present. The intrusive test criteria for a 1997 Saturn with a 1.9L I4 engine includes:

P1508 (2T) - Idle Air Control Valve Stuck Open Conditions:
DTC P0506 or P0507 not set, VSS input at 20-44 mph at steady cruise, Airflow input from 4-10 gm/s, then the PCM detected that the IAC failed the IAC Intrusive Test (the IAC valve may be stuck in open position).

Misfire Monitor

The Misfire Monitor is a PCM diagnostic required by OBD regulations that continuously monitors the engine for misfires under all engine positive load and speed conditions. The test operates on the principle that crankshaft rotational velocity fluctuates as each engine cylinder contributes its power. When a misfire occurs, the crankshaft speed will slow down temporarily.

The PCM monitors crankshaft rotational velocity using a crankshaft position (CKP) sensor. The CMP sensor is for cylinder identification.

Misfire Definition

Misfire is defined as lack of combustion in a cylinder due to weak compression, lack of adequate ignition spark, poor fuel metering, or any other engine mechanical, fuel or ignition system fault.

Catalyst Damaging Diagnostic

This diagnostic operates when the level of misfire is sufficient to cause catalyst damage under the current operating conditions. It is designed to detect a misfire within 200-1000 crankshaft revolutions. If a Type 1-Trip code is set (or a 2-Trip code with misfire relief), the PCM will command the MIL to flash. If a catalyst-damaging misfire is no longer present, the MIL will remain on steady instead of flashing.

Emission-Threatening Diagnostic

The Misfire test is designed to detect levels of misfire sufficient to result in emissions levels that exceed 1.5 times the FTP standard. It is designed to detect misfire within 1000-4000 crankshaft revolutions. The PCM *arms* the Misfire trouble code (2-Trip code) on the first trip. If the same misfire is detected on two consecutive trips, the PCM will store a History Code and illuminate the MIL. The PCM will also set this type of code and illuminate the MIL if it detects a misfire condition on non-consecutive trip if the misfire occurs under the same operating conditions (within 375 rpm of engine speed and 20% of engine load under similar engine coolant temperature) in the last 80 trips.

Rough Road Detection

Driving over rough roads can cause engine speed variations that are similar to an engine Misfire condition and may cause false detection of a misfire. OBD II regulations allow the misfire detection test to be disabled for a short time while driving on rough road conditions.

A rough road can cause torque to be applied to the drive wheels and drive train. This torque can temporarily and intermittently decrease engine speed, and thereby cause the Misfire Monitor to incorrectly detect a misfire condition.

Misfire Monitor (Continued)

Misfire Counters

Whenever a misfire is detected, the Misfire Monitor counts the misfire and notes the crankshaft position at the time that the misfire occurs. These misfire counters consist of a file on each engine cylinder.

A Current and History misfire counter is maintained for each cylinder. Misfire Current Counters (Misfire Current Cyl 1-8) indicate the number of firing events out of the last 200 crankshaft revolutions that were misfires. Misfire Current Counters display real time data without a misfire code stored. Misfire History Counters (Misfire History Cyl 1-8) indicate the total number of cylinder firing events that were misfires. Misfire History counters display zero (0) until the Misfire Monitor has detected a failure and DTC P0300 is set.

Once the DTC P0300 sets, the Misfire History counters are updated every 200 revolutions of the crankshaft. If the Misfire Monitor reports a failure, the Diagnostic Executive reviews all of the Misfire counters before reporting a trouble code (to report the most current data).

Engine Misfire with Misfire Relief

GM introduced full-range misfire detection in 1997. This system senses misfire under all positive speed and load conditions up to vehicle "redline." Misfire is not monitored under negative speed and load conditions (deceleration). Misfire detection was only enabled under speed and load conditions to meet FTP standards before 1997.

Misfire Monitor Test Graphic

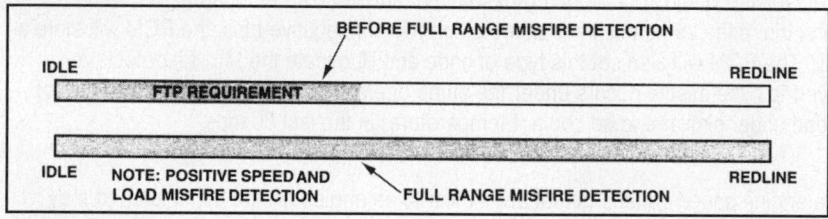

Misfire Monitor Test Conditions:

The PCM runs the Misfire Monitor once its test criteria are met. Refer to Page 3-21 for the OBD II Drive Cycle that includes this test. The code test criteria for a 1999 G-Van with a 3.1L V6 engine include:

DTC P0300 (1T or 2T) - Engine Misfire Detected Conditions:

No ABS, CKP, CMP, MAF, TP or VSS codes set, fuel level over 10%, system voltage from 11-16v, engine speed from 450-5000 rpm, throttle position steady within 2% for 100 ms, if the ECT input is less than 20°F at startup, misfire testing is delayed until the ECT input exceeds 70°F, if the ECT input is more than 20°F at startup, misfire detection is delayed for 5 seconds, then the PCM detected a deceleration in crankshaft speed characteristic of an engine misfire.

Oxygen Sensor Monitor

OBD II regulations require that the Oxygen Sensor Monitor test the front and rear oxygen sensors for faults or deterioration that could cause tailpipe emissions to exceed 1.5 times the FTP standard. The Oxygen Sensor Monitor continuously tests the voltage output and response rate of the front and rear oxygen sensor (see the Graphic).

Fuel Control HO2S Operation

The main function of the front oxygen sensor (O2S or HO2S) is to provide the PCM with exhaust stream information to allow proper fuel control. Once it reaches operating temperature, the sensor generates a voltage that is inversely proportional to the amount of oxygen in the exhaust gases. Once the system is in closed loop, the PCM uses the signal from the oxygen sensor to adjust the injector pulsewidth to maintain an A/F ratio that controls tailpipe emissions and driveability.

HO2S Response Time Graphic

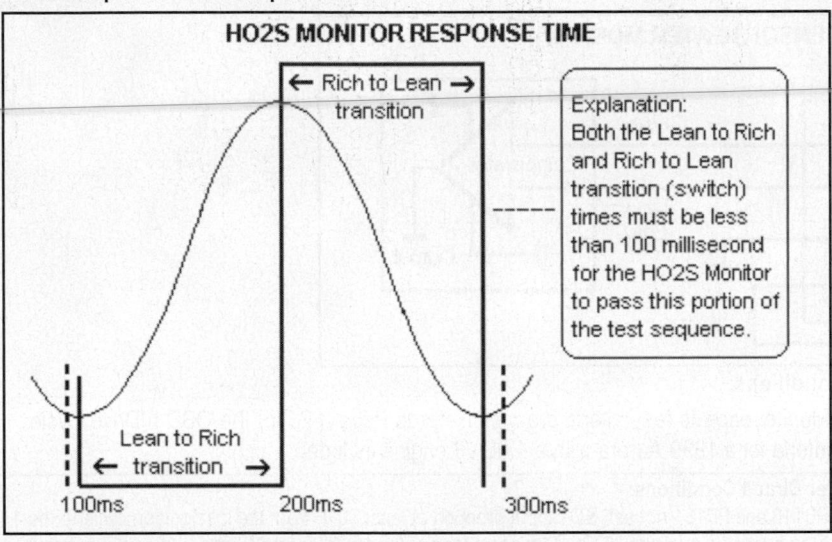

HO2S MONITOR RESPONSE TIME

← Rich to Lean → transition

Explanation:
Both the Lean to Rich and Rich to Lean transition (switch) times must be less than 100 millisecond for the HO2S Monitor to pass this portion of the test sequence.

Lean to Rich
← transition →

100ms   200ms   300ms

Oxygen Sensor Monitor Test Conditions:

The PCM runs the HO2S Monitor once its test criteria are met. Refer to Page 3-21 for the OBD II Drive Cycle that includes this test. The code test criteria for a 1998 Venture with a 3.4L V6 engine include:

P0132 (2T) - HO2S-11 (B1 S1) Circuit High Input Conditions:
No AIR, ECT, EGR, EVAP, Injector, MAF, MAP or TP sensor codes set, A/F ratio from 14.5-14.8, throttle angle from 3-40%, Air Pump off, system voltage 9-18v, then the PCM detected that the HO2S-11 signal remained above 977 mv in closed loop, or remained above 200 mv in Decel Fuel Shutoff, either condition met for 10 seconds.
• Refer to the correct code repair chart in other manuals or media.

Oxygen Sensor Heater Monitor

OBD II regulations require the Oxygen Sensor Heater Monitor to test the operation of the front and rear oxygen sensor heaters. This type of diagnostic is conducted right after a cold engine startup occurs.

<u>HO2S Heater Test</u>

The HO2S Heater Test measures the time required for the HO2S to become active after a cold engine startup. The time required for the oxygen sensor to become active is compared to a calibrated fault threshold in the PCM to determine the capability of the HO2S heater.

If the heater circuit has deteriorated or failed, the amount of time required for the HO2S to become active will increase. If the engine is started warm, or if a heater failure occurs during normal operation, other HO2S tests are used to detect a fault in the heater or its circuit.

Heated Oxygen Sensor Graphic

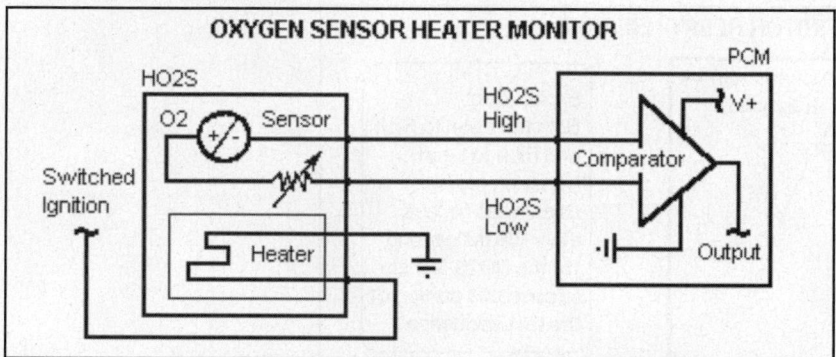

Oxygen Sensor Monitor Test Conditions:

The PCM runs the HO2S Heater Monitor once its test criteria are met. Refer to Page 3-21 for the OBD II Drive Cycle that includes this test. The code criteria for a 1999 Aurora with a 4.0L V8 engine include:

P0135 (2T) - HO2S-11 (B1 S1) Heater Circuit Conditions:
DTC P0101, P0102, P0103, P0117, P0118 and P0134 not set, ECT input dropped at least 50°F from the last ignition cycle to the start of this ignition cycle, system voltage over 11v, average HO2S-11 bias voltage between 352 and 546 mv, engine did not stall during the test, and then the PCM monitored the HO2S-11 signal to determine when it became either 151 mv less than or 151 mv more than average Bank 1 HO2S-11 bias voltage for 100 milliseconds. If it takes too long, the PCM will set DTC P0135.

<u>Heated Oxygen Sensor Faults</u>

When an HO2S related code is set, check the items listed below.

*   An exhaust system leak "masking" the HO2S signal output
*   Moisture or oil contamination inside the HO2S connector

Secondary AIR Monitor

Some GM vehicles are equipped with a Secondary Air Injection (AIR) system. OBD II regulations require that the presence of secondary airflow in the exhaust stream be monitored along with the functional monitoring of the AIR pump and its related switching valves.

Secondary AIR System Graphic

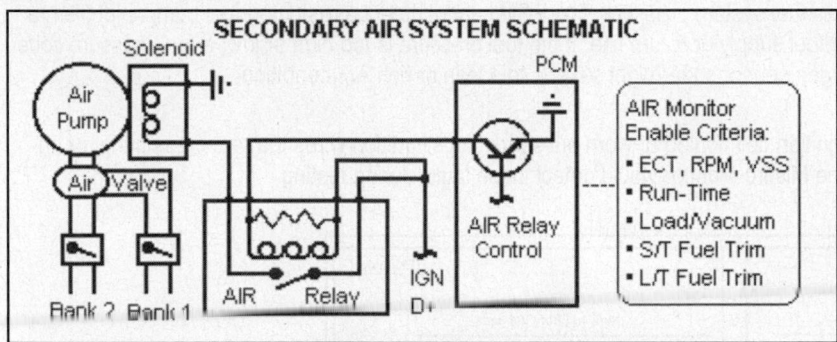

Secondary AIR Monitor Passive Test

When the Secondary AIR system is activated, excess air flows into the exhaust causing the HO2S-11 signal to drop. Then the PCM monitors the HO2S-11 signal and/or SHRTFT value for a response.

A Passive Test monitors the HO2S-11 signal after startup and also prior to closed loop operation. The AIR pump is normally enabled at this point to clean-up exhaust emissions. The HO2S-11 signal should be approximated 0-200mv if the AIR pump is delivering additional air to the exhaust during this period. The AIR Monitor will indicate a Pass if HO2S-11 indicates lean prior to closed loop operation. The Monitor also looks for the HO2S-11 to toggle as the AIR pump is disabled.

Secondary AIR Monitor Active Test

If a Passive Test fails or is inconclusive, the Active Test is started. The AIR pump is turned on during closed loop operation, and the Monitor will indicate Pass or Fail based on the HO2S-11 and SHRTFT values. A low HO2S-11 signal and increase in SHRTFT values indicates the AIR system is operating as designed.

Secondary AIR Monitor Test Conditions:

The PCM runs the AIR System Monitor once its test criteria are met. Refer to Page 3-21 for the OBD II Drive Cycle that includes this test. The code criteria for a 1997 Catera with a 3.0L V6 engine include:

P0410 (2) - Secondary AIR System Constant Flow Detected Condition
No PCM codes set, running in closed loop, EVAP Purge is disabled, IAT sensor input more than 41.5°F, ECT input over 65.8°F, VSS at 0 mph, then the PCM detected that SHRTFT exceeded a value of -25% for 20 seconds.

Non-Monitored Systems & Components

The PCM and CCM cannot monitor the Base Engine systems and components for faults or conditions that might cause a driveability problem. This situation can cause some confusing diagnostic situations when a Base Engine problem is present in this system.

Fuel Pressure

The fuel pressure regulator controls fuel system pressure. The PCM cannot detect a restricted fuel pump inlet filter, a dirty in-line fuel filter, or a pinched fuel supply or return line. If the fuel pressure is too high or low, a fuel pressure code would not set, but a misfire or oxygen sensor code might set due to a lean or rich A/F condition.

Ignition System Secondary

The PCM cannot detect a faulty ignition coil, fouled or worn out spark plugs, ignition wires that are cross firing, or an open spark plug wire. However, the Misfire Monitor would detect these faults during testing.

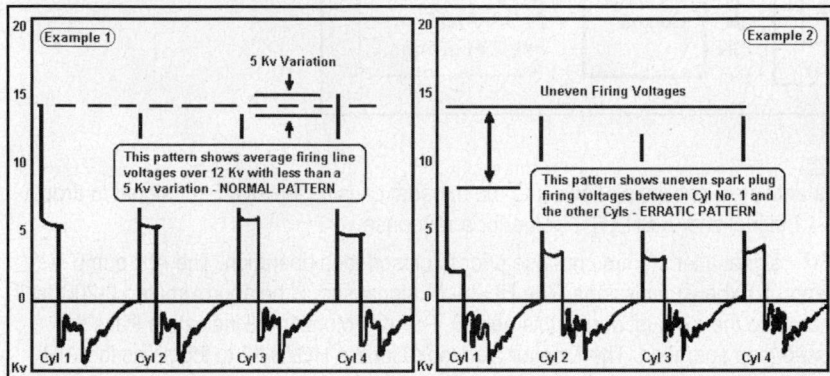

Engine Analyzer Testing the Ignition System

Engine Compression

The PCM cannot detect uneven, low, or high engine cylinder compression. However, a fault in one of these areas could cause the Oxygen Sensor or Misfire Monitor to fail during testing.

Exhaust System

The PCM cannot detect a restriction or leak in the Exhaust system. However, a fault in one of these areas could cause the EGR System, Fuel System, HO2S or Misfire Monitor to fail during testing.

Fuel Injector Mechanical Fault

The PCM cannot detect if a fuel injector is restricted, stuck open or closed. However, a fault in one of these areas could result in a rich or lean condition and cause the Fuel System or HO2S Monitor to fail.

GM Reference Information

OBD II TROUBLE CODE LIST

To use this information, first read and record all codes in memory along with Freeze Frame data. *If a PCM Reset function is done prior to recording this data, all codes and freeze frame data are lost!*

Look up the appropriate trouble code in the list on the following pages. The left hand column includes the code number, the number of trips to set the code (e.g., **1T or 2T**), the year, model description and type of OBD II Monitor that failed (e.g., **CCM or O2S**). This data can be used to determine how to drive a vehicle after a repair in order to validate the repair has been completed.

The **(N/MIL)** designator in the left hand column indicates the trouble code does not turn on the Malfunction Indicator Lamp or MIL. The **(STS Lamp)** indicator in the left column indicates a code that turns on the Service Transmission Soon lamp. This code may or may not turn "on" the MIL.

OBD II Trouble Code List (P0xxx Codes)

DTC	Trouble Code Title, Conditions & Possible Causes
P0005 **1T CCM, MIL: Yes** 2003, 2004, 2005, 2004, 2005 C/K Series Trucks & G Series Vans 6.0L VIN U CNG engine Transmissions: All	**Camshaft Phasing System Malfunction Conditions:** Engine started; system voltage from 6-18v, and the PCM detected the Actual and Commanded state of the High Pressure Lock-Off (HPL) solenoid did not match for more than one second. **Possible Causes** ● AF lock-off relay circuit to the HPL solenoid is open ● HPL solenoid control circuit is open, shorted or grounded ● HPL solenoid is damaged or it has failed ● PCM has failed
P00013 **2T CCM, MIL: Yes** 2002, 2003, 2004, 2005, 2004, 2005 Envoy & TrailBlazer 4.2L VIN S engine Transmissions: A/T	**Camshaft Position Actuator Circuit Malfunction Conditions:** Engine started, system voltage over 10.0v and the PCM detected an unexpected voltage condition on the Camshaft Position actuator high control or low reference circuit for 250 ms. Note: A 50% duty cycle is used to maintain a steady retard angle. **Possible Causes** ● CMP actuator high control circuit is open or shorted to ground ● CMP actuator low reference circuit is open ● CMP actuator is damaged or has failed ● PCM has failed
P00014 **2T CCM, MIL: Yes** 2002, 2003, 2004, 2005, 2004, 2005 Envoy & TrailBlazer 4.2L VIN S engine Transmissions: A/T	**Camshaft Phasing System Malfunction Conditions:** Engine started, vehicle driven with the Cam Phaser active, and the PCM detected the difference between the Actual and Desired Cam Phase angle was over 1.5 degrees with the Cam Phaser steady for 20 seconds (50% duty cycle signal achieves a steady retard angle). **Possible Causes** ● CKP or CMP sensor loose/damaged (causes signal variation) ● Contamination or debris interfering with the actuator operation ● Timing chain and gear assembly has excessive free-play ● TSB 01-06-04-052 contains a repair procedure for this code
P00016 **2T CCM, MIL: Yes** 2003, 2004, 2005, 2004, 2005 C/K Series Trucks 6.6L VIN 1 Diesel engine Transmissions: All	**Camshaft Position Sensor Circuit Malfunction Conditions:** Engine cranking or running, and the PCM detected the CMP sensor pulses received by the ECM did not equal 3 or 0 for two crankshaft revolutions. The ECM monitors the CKP and CMP sensor signals to determine if they are synchronized. An error is detected if both of these signals are not observed within a narrow window of time. **Possible Causes** ● CMP sensor is damaged causing a variance in sensor signal ● Distributor installed one tooth off in advance or retard position ● Distributor hold-down bolt loose or the rotor is loose on shaft ● Excessive free play in the timing chain and gear assembly
P0016 **1T CCM, MIL: Yes** 2003, 2004, 2005, 2004, 2005 Astro & Safari Vans, S & T Blazer & Pickup 4.3L VIN X Transmissions: All	**Camshaft Position Actuator Circuit Malfunction Conditions:** Engine started; system voltage over 10.0v and the PCM detected the CMP sensor was not at the correct position relative to position of the CKP sensor pulse signal. The PCM supplies a 12v reference and a low reference circuit to the CKP and CMP sensors. The CKP sensor sends a signal to the PCM with each crankshaft revolution. The CMP sensor sends a signal to the PCM during each camshaft revolution. Use the Camshaft Timing Offset test to test this sensor. **Possible Causes** ● CMP sensor is damaged causing a variance in sensor signal ● Distributor installed one tooth off in advance or retard position ● Distributor hold-down bolt loose or the rotor is loose on shaft ● Excessive free play in the timing chain and gear assembly

OBD II Trouble Code List (P0xxx Codes)

DTC	Trouble Code Title, Conditions & Possible Causes
P0030 **2T CCM, MIL: Yes** 2001, 2002, 2003, 2004, 2005, 2004, 2005 Montana, Rendezvous, Silhouette, Venture 3.4L VIN E engine Transmissions: All	**HO2S-11 (Bank 1 Sensor 1) Heater Circuit Conditions:** Engine started; system voltage from 9-18v, and the PCM detected the HO2S-11 heater current was more than 1.25 amps for 20 seconds. The PCM controls the HO2S-11 heater low control circuit with a low side driver, and tests the current draw through the driver. **Possible Causes** ● HO2S low side control circuit is shorted to system power (B+) ● HO2S low side control circuit driver is shorted inside the PCM ● HO2S heater has failed (it may be shorted internally) ● PCM has failed
P0030 **2T CCM, MIL: Yes** 2003, 2004, 2005, 2004, 2005 CTS (Cadillac) 2.6L V6 VIN M & 3.2L V6 VIN N engines Transmissions: All	**HO2S-11 (Bank 1 Sensor 1) Heater Low Side Open Circuit** **Conditions:** Engine started and the PCM detected that the HO2S-11 heater low control circuit signal was between 2.6v to 4.6v with the heater commanded off, indicating an open HO2S heater circuit condition. **Possible Causes** ● HO2S heater low side control circuit open or shorted to ground ● HO2S heater power circuit is open (check the PRE O2 fuse) ● HO2S heater is damaged or it has failed ● PCM has failed
P0030 **2T CCM, MIL: Yes** 2000, 2001, 2002, 2003, 2004, 2005, 2004, 2005 DeVille, Eldorado, Seville 4.6L VIN 9, 4.6L VIN Y Transmissions: A/T	**HO2S-11 (Bank 1 Sensor 1) Heater Circuit Malfunction** **Conditions:** Key on, system voltage from 9-18v, and the PCM detected an unexpected voltage on the high side driver circuit of the HO2S-11. **Possible Causes** ● HO2S low side circuit is shorted to ground or to the high side ● HO2S high side circuit is shorted to ground or to the low side ● HO2S heater is damaged or it has failed ● PCM has failed
P0030 **2T CCM, MIL: Yes** 2001, 2002, 2003, 2004, 2005, 2004, 2005 Alero, Aurora, Century, Grand Am, Impala, Intrigue, Lumina, Malibu, Monte Carlo 3.1L VIN J, 3.4L VIN E, 3.5L V6 VIN H engines Transmissions: All	**HO2S-11 (Bank 1 Sensor 1) Heater Circuit Fault Conditions:** Engine started, system voltage from 9-18v, and the PCM detected the heater low control circuit current was more than the capacity of the PCM internal driver for over 20 seconds. **Possible Causes** ● HO2S low control circuit is shorted to system power (B+) ● HO2S low control circuit driver is shorted inside the PCM ● HO2S is damaged or it has failed ● PCM has failed
P0030 **2T CCM, MIL: Yes** 2001, 2002, 2003, 2004, 2005, 2004, 2005 Aurora 4.0L V8 VIN C engine Transmissions: A/T	**HO2S-11 (Bank 1 Sensor 1) Heater Circuit Fault Conditions:** Engine started; system voltage from 9-18v, and the PCM detected an incorrect voltage on the Bank1 HO2S heater high control circuit. **Possible Causes** ● HO2S-11 high control circuit is shorted to ground ● HO2S-11 assembly is damaged or it has failed ● HO2S-21 assembly is damaged or it has failed ● Ignition 1 voltage circuit may be shorted to ground ● PCM has failed
P0031 **2T CCM, MIL: Yes** 2003, 2004, 2005, 2004, 2005 CTS (Cadillac) 2.6L V6 VIN M & 3.2L V6 VIN N engines Transmissions: All	**HO2S-11 (Bank 1 Sensor 1) Heater Low Side Short Circuit Conditions:** Engine started and the PCM detected that the HO2S-11 heater low control circuit signal was less than 2.6v with the heater commanded off, indicating a shorted HO2S heater circuit condition. **Possible Causes** ● HO2S heater low side control circuit is shorted to ground ● HO2S heater low side circuit is shorted to signal or ground ● HO2S heater is damaged or it has failed ● PCM has failed
P0032 **2T CCM, MIL: Yes** 2003, 2004, 2005, 2004, 2005 CTS (Cadillac) 2.6L V6 VIN M & 3.2L V6 VIN N engines Transmissions: All	**HO2S-11 (Bank 1 Sensor 1) Heater Low Side Short To B+ Conditions:** Engine started and the PCM detected that the HO2S-11 heater low control circuit signal was more than 4.6v with the heater commanded off, indicating the HO2S heater low side circuit is shorted to power. **Possible Causes** ● HO2S heater low side control circuit is shorted to system power ● HO2S heater is damaged or it has failed ● PCM has failed

OBD II Trouble Code List (P0xxx Codes)

DTC	Trouble Code Title, Conditions & Possible Causes
P0036 **2T CCM, MIL: Yes** 2003, 2004, 2005, 2004, 2005 CTS (Cadillac) 2.6L V6 VIN M & 3.2L V6 VIN N engines Transmissions: All	**HO2S-12 (Bank 1 Sensor 2) Heater Low Side Open Circuit Conditions:** Engine started and the PCM detected that the HO2S-12 heater low control circuit signal was between 2.6v to 4.6v with the heater commanded off, indicating an open HO2S heater circuit condition. **Possible Causes** ● HO2S heater low side control circuit open or shorted to ground ● HO2S heater power circuit is open (check the PRE O2 fuse) ● HO2S heater is damaged or it has failed ● PCM has failed
P0036 **2T CCM, MIL: Yes** 2000, 2001, 2002, 2003, 2004, 2005, 2004, 2005 DeVille, Eldorado, Seville 4.6L VIN 9, 4.6L VIN Y Transmissions: A/T	**HO2S-12 (Bank 1 Sensor 2) Heater Circuit Fault Conditions:** Engine running, system voltage from 9-18v, and the PCM detected an incorrect value at the low side driver control circuit of the HO2S. **Possible Causes** ● HO2S low side circuit is shorted to ground or to the high side ● HO2S high side circuit is shorted to ground or to the low side ● HO2S heater is damaged or it has failed ● PCM has failed
P0036 **2T CCM, MIL: Yes** 2001, 2002, 2003, 2004, 2005, 2004, 2005 Aurora 4.0L V8 VIN C engine Transmissions: A/T	**HO2S-21 (Bank 2 Sensor 1) Heater Circuit Fault Conditions:** Engine started; system voltage from 9-18v, and the PCM detected an incorrect voltage on the Bank 2 HO2S heater high control circuit. **Possible Causes** ● HO2S-21 connector is damaged, open or shorted ● HO2S-21 high control circuit is open or shorted to power ● HO2S-21 assembly is damaged or it has failed ● HO2S heater power circuit is open (test the OXY SEN fuse) ● PCM has failed
P0037 **2T CCM, MIL: Yes** 2003, 2004, 2005, 2004, 2005 CTS (Cadillac) 2.6L V6 VIN M & 3.2L V6 VIN N engines Transmissions: All	**HO2S-12 (Bank 1 Sensor 2) Heater Low Side Short Circuit Conditions:** Engine started and the PCM detected that the HO2S-12 heater low control circuit signal was less than 2.6v with the heater commanded off, indicating a shorted HO2S heater circuit condition. **Possible Causes** ● HO2S heater low side control circuit is shorted to ground ● HO2S heater low side circuit is shorted to signal or ground ● HO2S heater is damaged or it has failed ● PCM has failed
P0038 **2T CCM, MIL: Yes** 2003, 2004, 2005, 2004, 2005 CTS (Cadillac) 2.6L V6 VIN M & 3.2L V6 VIN N engines Transmissions: All	**HO2S-12 (Bank 1 Sensor 2) Heater Low Side Short To B+ Conditions:** Engine started and the PCM detected that the HO2S-12 heater low control circuit signal was more than 4.6v with the heater commanded off, indicating the HO2S heater low side circuit is shorted to power. **Possible Causes** ● HO2S heater low side control circuit is shorted to system power ● HO2S heater is damaged or it has failed ● PCM has failed
P0050 **2T CCM, MIL: Yes** 2003, 2004, 2005, 2004, 2005 CTS (Cadillac) 2.6L V6 VIN M & 3.2L V6 VIN N engines Transmissions: All	**HO2S-21 (Bank 2 Sensor 1) Heater Low Side Open Circuit Conditions:** Engine started and the PCM detected that the HO2S-21 heater low control circuit signal was between 2.6v to 4.6v with the heater commanded off (i.e., an open circuit). **Possible Causes** ● HO2S heater low side control circuit is open shorted to ground ● HO2S heater power circuit is open (check the POST O2 fuse) ● HO2S heater is damaged or it has failed ● PCM has failed
P0051 **2T CCM, MIL: Yes** 2003, 2004, 2005, 2004, 2005 CTS (Cadillac) 2.6L V6 VIN M & 3.2L V6 VIN N engines Transmissions: All	**HO2S-21 (Bank 2 Sensor 1) Heater Low Side Short Circuit Conditions:** Engine started and the PCM detected the HO2S-21 heater low control circuit was less than 2.6v with the heater commanded off. **Possible Causes** ● HO2S heater low side control circuit is shorted to ground ● HO2S heater low side circuit is shorted to signal or ground ● HO2S heater is damaged or it has failed ● PCM has failed
P0052 **2T CCM, MIL: Yes** 2003, 2004, 2005, 2004, 2005 CTS (Cadillac) 2.6L V6 VIN M & 3.2L V6 VIN N engines Transmissions: All	**HO2S-21 (Bank 2 Sensor 1) Heater Low Side Short To B+ Conditions:** Engine started and the PCM detected that the HO2S-21 heater low control circuit signal was more than 4.6v with the heater commanded off, indicating the HO2S heater low side circuit is shorted to power. **Possible Causes** ● HO2S heater low side control circuit is shorted to system power ● HO2S heater is damaged or it has failed ● PCM has failed

OBD II Trouble Code List (P0xxx Codes)

DTC	Trouble Code Title, Conditions & Possible Causes
P0056 **2T CCM, MIL: Yes** 2003, 2004, 2005, 2004, 2005 CTS (Cadillac) 2.6L V6 VIN M & 3.2L V6 VIN N engines Transmissions: All	**HO2S-22 (Bank 2 Sensor 2) Heater Low Side Open Circuit Conditions:** Engine started and the PCM detected that the HO2S-22 heater low control circuit signal was between 2.6v to 4.6v with the heater commanded off, indicating an open HO2S heater circuit condition. **Possible Causes** ● HO2S heater low side control circuit open or shorted to ground ● HO2S heater power circuit is open (check the POST O2 fuse) ● HO2S heater is damaged or it has failed ● PCM has failed
P0057 **2T CCM, MIL: Yes** 2003, 2004, 2005, 2004, 2005 CTS (Cadillac) 2.6L V6 VIN M & 3.2L V6 VIN N engines Transmissions: All	**HO2S-22 (Bank 2 Sensor 2) Heater Low Side Short Circuit Conditions:** Engine started and the PCM detected that the HO2S-22 heater low control circuit signal was less than 2.6v with the heater commanded off, indicating a shorted HO2S heater circuit condition. **Possible Causes** ● HO2S heater low side control circuit is shorted to ground ● HO2S heater low side circuit is shorted to the signal circuit ● HO2S heater is damaged or it has failed ● PCM has failed
P0058 **2T CCM, MIL: Yes** 2003, 2004, 2005, 2004, 2005 CTS (Cadillac) 2.6L V6 VIN M & 3.2L V6 VIN N engines Transmissions: All	**HO2S-22 (Bank 2 Sensor 2) Heater Low Side Short To B+ Conditions:** Engine started and the PCM detected that the HO2S-21 heater low control circuit signal was more than 4.6v with the heater commanded off, indicating the HO2S heater low side circuit is shorted to power. **Possible Causes** ● HO2S heater low side control circuit is shorted to system power ● HO2S heater is damaged or it has failed ● PCM has failed
P0087 **1T CCM, MIL: Yes** 2001, 2002, 2003, 2004, 2005, 2004, 2005 C/K Series Trucks 6.6L VIN 1 engine Transmissions: A/T	**Fuel Rail Pressure Sensor Circuit Low Input Conditions:** DTC P0192, P0193 or P0641 not set, Key on or engine running; and the PCM detected the actual FRP sensor signal was less than 0.0 MPa at 0-400 rpm, or the actual FRP was less than 22.5 MPa at more than 600 rpm. The ECM monitors the fuel rail pressure (FRP) using the FRP sensor. If the sensor indicates a pressure less than the commanded rail pressure plus a possible transitional overshoot, the ECM will set DTC P0087 for fuel rail pressure too low. **Possible Causes** ● Check the engine oil for fuel contamination ● Fuel filter is clogged or restricted ● Fuel supply lines between fuel tank and injector pump clogged ● Fuel rail pressure sensor signal circuit is shorted to ground ● Fuel rail pressure sensor is out-of-calibration or it has failed ● ECM has failed
P0088 **1T CCM, MIL: Yes** 2001, 2002, 2003, 2004, 2005, 2004, 2005 C/K Series Trucks 6.6L VIN 1 engine Transmissions: A/T	**Fuel Rail Pressure Sensor Circuit High Input Conditions:** DTC P0192, P0193 or P0641 not set, Key on or engine running; and the PCM detected the actual FRP sensor signal was more than 167 MPa. The ECM monitors the fuel rail pressure (FRP) using the FRP sensor. If the sensor indicates a pressure less than the commanded rail pressure plus a possible transitional overshoot, the ECM will set DTC P0088 for fuel rail pressure too high. **Possible Causes** ● FRP pressure sensor signal circuit is open ● FRP sensor signal circuit has a high resistance condition ● FRP sensor is damaged or it has failed ● FRP regulator is damaged or it has failed ● ECM has failed
P0089 **1T CCM, MIL: Yes** 2001, 2002, 2003, 2004, 2005, 2004, 2005 C/K Series Trucks 6.6L VIN 1 engine Transmissions: A/T	**Fuel Rail Pressure Sensor Signal Range/Performance Conditions:** DTC P0192, P0193 or P0641 not set, key on, and the PCM detected the difference between the Actual and Desired FRP sensor was over 20 MPa with Commanded fuel pump flow was 100 mm³/second or less. The ECM uses the commanded fuel pump flow to determine a desired fuel rail pressure. The actual fuel pressure is monitored using the FRP sensor. If the PCM detects the FRP sensor pressure is 20 MPa more than desired value, it will set DTC P0089. **Possible Causes** ● FRP pressure sensor signal circuit is shorted to ground ● FRP sensor is damaged or it has failed ● FRP regulator is damaged or it has failed ● ECM has failed

OBD II Trouble Code List (P0xxx Codes)

DTC	Trouble Code Title, Conditions & Possible Causes
P0090 **1T CCM, MIL: Yes** 2001, 2002, 2003, 2004, 2005, 2004, 2005 C/K Series Trucks 6.6L VIN 1 engine Transmissions: A/T	**Fuel Rail Pressure Regulator Circuit Malfunction Conditions:** DTC P0192, P0193 or P0641 not set, key on, and the PCM detected the difference between the Actual and Desired FRP sensor was over 20 MPa with Commanded fuel pump flow was 100 mm³/second or less. The ECM supplies power and ground to the fuel rail pressure (FRP) regulator. The PCM monitors the circuit current to detect when it is out of its normal range. **Possible Causes** ● Fuel rail pressure regulator control circuit is open or grounded ● Fuel rail pressure regulator supply circuit is open or grounded ● Fuel rail pressure regulator is damaged or it has failed ● ECM has failed
P0100 **1T CCM, MIL: Yes** 1996, 1997 Camaro, Caprice, Corvette, Firebird, Fleetwood, Impala, Roadmaster 4.3L V6 VIN W, 5.7L V8 VIN 5, 5.7L V8 VIN P Transmissions: All	**MAF Sensor Circuit Insufficient Activity Conditions:** DTC P0101, P0102 and P0103 not set, engine speed over 500 rpm, system voltage over 11v, and the PCM detected the MAF sensor was less than a preset minimum voltage amount for 1 second. The PCM uses Speed Density fuel calculation when this code sets. **Possible Causes** ● MAF sensor signal circuit is open or shorted to ground ● MAF sensor power circuit is open ● MAF sensor is damaged (it may have dirt on it) or has failed ● PCM has failed
P0100 **1T CCM, MIL: Yes** 1997, 1998, 1999, 2000, 2001 Catera 3.0L V6 VIN R engine Transmissions: A/T	**MAF Sensor Signal Range/Performance Conditions:** DTC P1120 and P1220 not set, system voltage more than 11.0v, engine speed over 500 rpm, and the PCM detected the MAF signal was less than 1.38 g/sec, or that it was out of a calculated range with the engine running for over 100 ms. The ranges are as follows: 84 g/s at 1000 rpm 111 g/s at 2000 rpm 140 g/s at 3000 rpm 180 g/sec at 4000 rpm 222 g/s at 5000 rpm and 251 g/s at 6000 rpm. **Possible Causes** ● MAF sensor signal circuit is open, shorted to ground or to B+ ● MAF sensor VREF circuit is open ● MAF sensor signal circuit shorted to ground at another device ● MAF sensor is damaged (it may be dirt or contaminated) ● PCM has failed
P0100 **1T CCM, MIL: Yes** 1995 Aurora, Bonneville, Century, Ciera, LeSabre, LSS, Lumina Van, Park Avenue, Riviera, 88' 3.1L VIN M, 3.4L VIN X, 3.8L VIN 1, VIN K, VIN L Transmissions: All	**MAF Sensor Signal Range/Performance Conditions:** DTC P0107, P0108, P0112, P0113, P0122 and P0123 not set, engine started, system voltage over 11v, and the PCM did not detect a MAF signal for 2 seconds. **Possible Causes** ● Base engine vacuum leak, PCV valve leaking or stuck open ● Engine oil dipstick missing or not fully seated ● MAF sensor element (wire) is contaminated or dirty ● MAF sensor signal or ground circuit fault or sensor has failed ● PCM has failed
P0101 **2T CCM, MIL: Yes** 1996, 1997, 1998 Achieva, Beretta, Century, Ciera, Corsica, Cutlass Supreme, Grand Am, Grand Prix, Intrigue, Lumina, Malibu, Monte Carlo, Skylark 3.1L VIN M, 3.4L VIN X, 3.8L VIN 1, 3.8L VIN K Transmissions: All	**MAF Sensor Signal Range/Performance Conditions:** DTC P0106, P0107, P0122, P0123, P1106, P1107, P1121 and P1122 not set, engine started, system voltage from 9-16v, EGR duty cycle below 50%, EGR pintle position less than 50%, TP angle under 50%, and the PCM detected the actual MAF sensor signal was significantly more or less than the MAF sensor predicted value. **Possible Causes** ● MAF sensor or MAP sensor ground circuit has high resistance ● MAF minimum airflow rate to low at idle or during deceleration ● MAF sensor interference (i.e., electrical noise from the ignition) ● MAP sensor source vacuum line is loose, or it may be leaking ● PCM has failed
P0101 **2T CCM, MIL: Yes** 1999, 2000, 2001, 2002, 2003, 2004, 2005, 2004, 2005 Aurora, Bonneville, Grand Prix, Century, Cutlass Supreme, Grand Prix, Intrigue, LSS, LeSabre, Lumina, Monte Carlo, 88', 98', Regal, Park Avenue 3.1L VIN J, 3.1L VIN M, 3.4L VIN E, 3.5L V6 VIN H, 3.8L VIN 1, 3.8L VIN K, 4.0L V8 VIN C engines Transmissions: All	**MAF Sensor Signal Range/Performance Conditions:** DTC P0102, P0103, P0106, P0107, P0108, P0121, P0122, P0123, P0401, P0404, P0405, P0440, P0442, P0446, P1404 or P1441 not set, engine started, system voltage at 9-18v, TP angle under 15% (± 5%), MAP signal under 80 kPa (± 5 kPa), Traction Control off, Purge command below 50%, and the PCM detected the MAF signal did not agree with calculated MAF value for a period of 5 to 40 seconds. **Possible Causes** ● Base engine vacuum leak, PCV valve leaking or stuck open ● Engine oil dipstick missing or not fully seated ● MAF sensor signal or ground circuit fault or sensor has failed ● PCM has failed

OBD II Trouble Code List (P0xxx Codes)

DTC	Trouble Code Title, Conditions & Possible Causes
P0101 **1T CCM, MIL: Yes** 1996, 1997 Camaro, Caprice, Corvette, Firebird, Fleetwood, Impala, Roadmaster, 4.3L V6 VIN W, 5.7L V8 VIN 5, 5.7L V8 VIN P Transmissions: All	**MAF Sensor Signal Range/Performance Conditions:** DTC P0100, P0102, P0103, P0107, P0108, P0122 and P0123 not set, engine speed over 500 rpm, and the PCM detected the Actual signal did not match the Calculate MAF sensor signal for 1 second. **Possible Causes** ● Base engine vacuum leak, PCV valve leaking or stuck open ● Engine oil dipstick missing or not fully seated ● MAF sensor signal or ground circuit fault or sensor has failed ● PCM has failed ● TSB 61-65-50 contains a repair procedure for this code
P0101 **2T CCM, MIL: Yes** 1999, 2000, 2001 Catera 3.0L V6 VIN R engine Transmissions: A/T	**MAF Sensor Signal Range/Performance Conditions:** DTC P0102, P0103, P0106, P0107, P0108, P0121, P0122, P0123, P0401, P0404, P0405, P0440, P0442, P0446, P1404 or P1441 not set, engine started, system voltage at 11-18v, TP angle under 45% (± 3), MAP signal below 59 kPa (± 4 kPa), Purge command under 90%, Traction Control off, and the PCM detected the Actual MAF sensor was not within a preset range of the calculated MAF value. **Possible Causes** ● Excessive deposits on the throttle plate or in the throttle bore ● MAF sensor power or ground circuit has high resistance ● MAF sensor is damaged or it has failed ● MAP sensor is "skewed" (causes an invalid BARO signal) ● Verify the oil dipstick is fully seated and oil filler cap is secure ● PCM has failed
P0101 **2T CCM, MIL: Yes** 1996, 1997, 1998, 1999, 2000, 2001, 2002, 2003, 2004, 2005, 2004, 2005 Montana, Silhouette, Venture, Rendezvous 3.1L V6 VIN M, 3.1L V6 VIN J, 3.4L V6 VIN E Transmissions: All	**MAF Sensor Signal Range/Performance Conditions:** DTC P0102, P0103, P0107, P0108, P0121, P0122, P0123, P0401, P0403, P0404, P0405, P0440, P0442, P0443, P0446, P0449, P1404 and P1441 not set, engine started, MAP sensor less than 63 kPa (± 3 kPa), TP angle less than 25% (± 1.5%), and the PCM detected the Actual MAF frequency value was not within a predetermined range of the calculated MAF value for 10 seconds. **Possible Causes** ● Base engine vacuum leak, PCV valve leaking or stuck open ● Engine oil dipstick missing or not fully seated ● MAF sensor element (wire) is contaminated or dirty ● MAF sensor signal or ground circuit has a high resistance ● MAF sensor is damaged or it has failed ● PCM has failed
P0101 **2T CCM, MIL: Yes** 1996, 1997, 1998, 1999, 2000, 2001, 2002, 2003, 2004, 2005, 2004, 2005 C/K Cab & Chassis, C/K Series Truck, G Series Van, L/M Vans 4.3L V6 VIN W, 4.3L V6 VIN X, 5.0L VIN M, 5.7L VIN K, 5.7L VIN R, 7.4L VIN J engines Transmissions: All	**MAF Sensor Signal Range/Performance Conditions:** DTC P0102-P0103, P0106, P0107, P0108, P0112, P0113, P0120, P0121, P0122, P0123, P0220, P0442, P0443, P0446, P0449, P0455, P0496, P1111, P1112, P1120, P1122, P1220, P1221 and P2135 not set, engine cranking or running, TP angle less than 95% (± 5%), MAP sensor less than 80 kPa (± 3 kPa) for 1.5 seconds, system voltage from 11-18v, and the PCM detected the Actual MAF sensor value was not within a predetermined range of the Calculated MAF value for 4 seconds. **Possible Causes** ● Base engine vacuum leak, PCV valve leaking or stuck open ● Engine oil dipstick missing or not fully seated ● MAF sensor element (wire) is contaminated or it has failed ● MAF sensor signal circuit or ground circuit has high resistance ● PCM has failed
P0101 **2T CCM, MIL: Yes** 1999, 2000, 2001, 2002, 2003, 2004, 2005, 2004, 2005 C/K Series Truck, G Series Van, Envoy, Escalade, TrailBlazer 4.8L VIN V, 5.3L VIN P, 5.3L VIN T, 5.3L VIN Z, 6.0L VIN N, 6.0L VIN U, 6.6L VIN 1, 8.1L VIN G Transmissions: All	**MAF Sensor Signal Range/Performance Conditions:** DTC P0102, P0103, P0106, P0107, P0108, P0120, P0121, P0122, P0123, P0220, P0442, P0443, P0446, P0449, P0455, P0496, P1404 and P2135 not set, engine cranking or running, system voltage from 11-18v, TP angle under 95% (± 5%), MAP sensor over 17 kPa (± 3 kPa), all conditions met for 1.5 seconds, and the PCM detected the Actual MAF sensor frequency was not within a predetermined range of the Calculated MAF value for 4 seconds. **Possible Causes** ● Base engine vacuum leak, PCV valve leaking or stuck open ● Engine oil dipstick missing or not fully seated ● MAF sensor element (wire) is contaminated or it has failed ● MAF sensor signal circuit or ground circuit has high resistance ● PCM has failed

OBD II Trouble Code List (P0xxx Codes)

DTC	Trouble Code Title, Conditions & Possible Causes
P0101 **2T CCM, MIL: Yes** 2003, 2004, 2005, 2004, 2005 CTS (Cadillac) 2.6L V6 VIN M & 3.2L V6 VIN N engines engine Transmissions: All	**MAF Sensor Signal Range/Performance Conditions:** Engine started; engine runtime over 500 ms to allow the MAF sensor hot film element to heat up, more than 20 crankshaft revolutions detected, system voltage over 10v, and the PCM detected the Actual MAF sensor signal was not within a predetermined range of the Predicted MAF sensor value for more than 2 seconds. **Possible Causes** ● Base engine vacuum leak, PCV valve leaking or stuck open ● Engine oil dipstick missing or not fully seated ● MAF sensor element (wire) is contaminated or dirty ● MAF sensor signal or ground circuit fault or sensor has failed ● PCM has failed
P0101 **2T CCM, MIL: Yes** 1996, 1997, 1998, 1999, 2000, 2001, 2002, 2003, 2004, 2005, 2004, 2005 DeVille, Eldorado, Seville 4.0L VIN 9, 4.6L VIN Y engine Transmissions: A/T	**MAF Sensor Signal Range/Performance Conditions:** DTC P0102, P0103, P0106, P0107, P0108, P0121, P0122, P0123, P0401, P0404, P0405, P0440, P0442, P0446, P1404 and P1441 not set, engine started, EVAP Purge less than 90%, total EGR pintle position change less than 90%, TP angle under 48% (± 3%), MAP sensor less than 59 kPa (± 4 kPa), Traction Control not active, and the PCM detected the actual MAF sensor frequency signal was not within a predetermined range of the calculated MAF sensor value. **Possible Causes** ● Base engine vacuum leak, PCV valve leaking or stuck open ● Engine oil dipstick missing or not fully seated ● MAF sensor element (wire) is contaminated or dirty ● MAF sensor signal or ground circuit fault or sensor has failed ● PCM has failed
P0101 **2T CCM, MIL: Yes** 1995, 1996, 1997, 1998, 1999, 2000, 2001, 2002 Camaro & Firebird 3.8L V6 VIN K engine Transmissions: All	**MAF Sensor Signal Range/Performance Conditions:** DTC P0102-P0103, P0107-P0108, P0401-P0405, P0440-P0449, P0606, P1106-1107, P1404, P1441, P1514-P1517 not set, and P1120, P1220 not set as a combination, and P1518 not set in combination with any of these codes: P1120, P1125, P1220, P1221, P1271-P1276 and P1280-P1286, engine running, MAP signal less than 80 kPa (± 3 kPa), TP angle less than 5% and steady, all conditions met for 2 seconds, and the PCM detected the MAF signal was not within a preset range of the calculated MAF for 20 seconds. **Possible Causes** ● Base engine vacuum leak, PCV valve leaking or stuck open ● Engine oil dipstick missing or not fully seated ● MAF sensor element (wire) is contaminated or it has failed ● MAF sensor signal circuit or ground circuit has high resistance ● PCM has failed
P0101 **2T CCM, MIL: Yes** 1998, 1999, 2000, 2001, 2002 Camaro & Firebird 5.7L VIN G engine Transmissions: All	**MAF Sensor Signal Range/Performance Conditions:** DTC P0102, P0103, P0107, P0108, P0112, P0113, P0121-P0123, P1111, P1112, P1120-P1122, P1220 and P1221 not set, engine started, system voltage from 11-18v, TP angle under 95% (± 5%), MAP signal over 17 kPa (± 3 kPa), conditions met for 1.5 seconds, and the PCM detected the MAF sensor frequency was 50% more or less than the Speed Density internal calculation for 5 seconds. **Possible Causes** ● Base engine vacuum leak, PCV valve leaking or stuck open ● Engine oil dipstick missing or not fully seated ● MAF sensor element (wire) is contaminated or it has failed ● MAF sensor signal circuit or ground circuit has high resistance ● PCM has failed
P0101 **2T CCM, MIL: Yes** 1996, 1997, 1998, 1999, 2001, 2002, 2003, 2004, 2005, 2004, 2005 Aurora 4.0L V8 VIN C Transmissions: A/T	**MAF Sensor Signal Range/Performance Conditions:** DTC P0102, P0103, P0106, P0107, P0108, P0121- P0123, P0560 or P1108 not set, P0401 not active, engine started, engine running with the EVAP Purge below 99.6%, EGR rescaled pintle less than 240, TP angle less than 50%, Throttle Control "off", 100 ms Delta MAP below 5 kPa, and the PCM detected a large variation between the Actual and Calculated MAF. **Possible Causes** ● Base engine vacuum leak, PCV valve leaking or stuck open ● Engine oil dipstick missing or not fully seated ● MAF sensor element (wire) is contaminated or dirty ● MAF sensor signal or ground circuit has a high resistance ● MAF sensor is damaged or it has failed ● PCM has failed

OBD II Trouble Code List (P0xxx Codes)

DTC	Trouble Code Title, Conditions & Possible Causes
P0101 **2T CCM, MIL: Yes** 2001, 2002 Aurora 3.5L V6 VIN H engine Transmissions: A/T	**MAF Sensor Signal Range/Performance Conditions:** DTC P0102, P0103, P0106, P0107, P0108, P0121- P0123, P0560 or P1108 not set, engine started, engine speed over 250 rpm, system voltage from 9-18v, change in MAP signal less than 3 kPa, TP angle less than 50% with any change less than 1.5%, Purge command less than 100%, EGR valve position signal less than 50% for 5 seconds, and the PCM detected the Actual MAF frequency was not within a Calculated MAF frequency for 5 seconds. **Possible Causes** ● Base engine vacuum leak, PCV valve leaking or stuck open ● Engine oil dipstick missing or not fully seated ● MAF sensor element (wire) is contaminated or dirty ● MAF sensor signal or ground circuit has a high resistance ● MAF sensor is damaged or it has failed ● PCM has failed
P0101 **2T CCM, MIL: Yes** 1996, 1997, 1998, 1999, 2000, 2001, 2002, 2003, 2004, 2005, 2004, 2005 S/T Blazer & S/T Pickup 4.3L VIN W, 4.3L VIN X engines Transmissions: All	**MAF Sensor Signal Range/Performance Conditions:** DTC P0102-P0103, P0106-P0108, P0112-P0113, P0121-P0123, P0335-P0336, P0401, P0440-P0449, P1106-P1107, P1111, P1112, P1121-P1122 and P1441 not set, system voltage 11-18v, TP angle less than 95% (± 5%), MAP signal less than 80 kPa (± 3 kPa), all conditions met for 1-3 seconds, and the PCM detected the actual MAF frequency was not within a preset range of the calculated MAF value, condition met for more than 4.0 seconds. **Possible Causes** ● Base engine vacuum leak, PCV valve leaking or stuck open ● Engine oil dipstick missing or not fully seated ● MAF sensor element (wire) is contaminated or dirty ● MAF sensor signal or ground circuit has a high resistance ● MAF sensor is damaged or it has failed ● PCM has failed
P0101 **2T CCM, MIL: Yes** 1997, 1998 Catera 3.0L V6 VIN R engine Transmissions: A/T	**MAF Sensor Signal Range/Performance Conditions:** DTC P0100, P0120, P0505, P0506 and P0507 not set, engine started, engine speed over 520 rpm, ECT sensor more than 176°F, and the PCM detected the TP sensor load signal indicate a value of from 2.0-3.3 ms for over 7 seconds. **Possible Causes** ● Base engine vacuum leak, PCV valve leaking or stuck open ● Engine oil dipstick missing or not fully seated ● MAF sensor element (wire) is contaminated or dirty ● MAF sensor signal or ground circuit has a high resistance ● MAF sensor is damaged or it has failed ● PCM has failed
P0101 **2T CCM, MIL: Yes** 1997, 1998, 1999, 2000, 2001, 2002, 2003, 2004, 2005 Corvette 5.7L VIN G, 5.7L VIN S engines Transmissions: All	**MAF Sensor Signal Range/Performance Conditions:** DTC P0102, P0103, P0107, P0108, P0112, P0113, P1120, P1220, P1221 and P1441 not set, engine started, engine running with the system voltage from 11-18v, TP angle less than 95% (± 5%), MAP signal more than 17 kPa (± 3 kPa), conditions met for 1-4 seconds, and the PCM detected the MAF sensor was not within a preset range of the its calculated frequency, condition met for 0.5 seconds. The MAF sensor is an airflow meter that measures the amount of air entering the engine. The PCM uses the MAF sensor to provide the correct fuel delivery for all engine speeds and loads. A small quantity of air entering the engine indicates an idle condition or deceleration while large quantity of air indicates acceleration or high engine load. **Possible Causes** ● Base engine vacuum leak, PCV valve leaking or stuck open ● Engine oil dipstick missing or not fully seated ● MAF sensor element (wire) is contaminated or dirty ● MAF sensor signal or ground circuit fault or sensor has failed ● PCM has failed
P0102 **2T CCM, MIL: Yes** 1996, 1997 Beretta, Century, Ciera, Corsica, Cutlass, Malibu, Grand Prix, Regal, Cutlass Supreme, Monte Carlo, Lumina APV & Lumina 3.1L VIN M, 3.4L VIN E, 3.4L VIN X, 3.8L VIN 1, 3.8L VIN K engines Transmissions: All	**MAF Sensor Circuit Low Frequency Conditions:** Engine started, system voltage from 10-18v, IAC motor more than 2 counts, and the PCM detected the MAF sensor frequency was less than 1,200 Hz, condition met for 1 second. **Possible Causes** ● MAF sensor element hot wire contaminated or the sensor failed ● MAF sensor signal shorted to ground or ground circuit problem ● MAF sensor wiring routed close to ignition wires, generator, solenoids or electric motors ● PCM has failed

OBD II Trouble Code List (P0xxx Codes)

DTC	Trouble Code Title, Conditions & Possible Causes
P0102 **2T CCM, MIL: Yes** 1998, 1999, 2000, 2001, 2002, 2003, 2004, 2005 Achieva, Alero, Century, Cutlass Ciera, Grand Prix, Regal, Lumina, Malibu, '88 & Monte Carlo 3.1L VIN J, 3.1L VIN M, 3.4L VIN E, 3.5L V6 VIN H, 3.8L VIN 1, 3.8L VIN K Transmissions: All	**MAF Sensor Circuit Low Frequency Conditions:** Engine started, system voltage from 10-18v, IAC motor more than 2 counts, and the PCM detected the MAF sensor frequency was less than 1,200 Hz, condition met for 1 second. **Possible Causes** • MAF sensor element hot wire contaminated or the sensor failed • MAF sensor signal shorted to ground or ground circuit problem • MAF sensor wiring routed close to the ignition wires, generator, solenoids or electric motors • PCM has failed
P0102 **2T CCM, MIL: Yes** 1997, 1998, 1999, 2000 Cutlass, Malibu, Montana, Silhouette, Venture 3.1L VIN J, 3.1L VIN M, 3.4L VIN E engines Transmissions: All	**MAF Sensor Circuit Low Frequency Conditions:** Engine started IAC position over 2 counts, system voltage from 8-18v, engine running for 0.5 seconds under these conditions, and the PCM detected the MAF sensor frequency was less than 10 Hz for over 500 ms. The MAF sensor is an airflow meter that measures the amount of air that enters the engine. The PCM uses the MAF sensor to provide the correct fuel delivery for all engine speeds and loads. A small quantity of air entering the engine indicates a deceleration or idle condition. A large quantity of air entering the engine indicates an acceleration or high load condition. **Possible Causes** • MAF sensor element hot wire contaminated or the sensor failed • MAF sensor signal shorted to ground or ground circuit problem • MAF sensor wiring routed close to the ignition wires, generator, solenoids or electric motors • PCM has failed
P0102 **1T CCM, MIL: Yes** 1996, 1997 Camaro, Caprice, Corvette, Firebird, Fleetwood 4.3L VIN W, 5.7L VIN 5, 5.7L VIN P engines Transmissions: All	**MAF Sensor Circuit Low Frequency Conditions:** Engine started engine speed less than 500 rpm, system voltage over 11v, and the PCM detected that the MAF sensor signal was less than 3.9 g/sec for 1 second during the test. **Possible Causes** • MAF sensor element hot wire contaminated or the sensor failed • MAF sensor signal shorted to ground or ground circuit problem • MAF sensor wiring routed close to the ignition wires, generator, solenoids or electric motors • PCM has failed
P0102 **2T CCM, MIL: Yes** 2001, 2002, 2003, 2004, 2005 Impala, Malibu, Montana, Monte Carlo, Silhouette, Venture, Rendezvous 3.1L VIN J, 3.1L VIN M, 3.4L VIN E engines Transmissions: All	**MAF Sensor Circuit Low Frequency Conditions:** Engine started, IAC position over 2 counts, system voltage from 8-18v, engine running for 0.5 seconds under these conditions, and the PCM detected the MAF sensor frequency was less than 1,200 Hz for over 500 ms. The MAF sensor is an airflow meter that measures the amount of air that enters the engine. The PCM uses the MAF sensor to provide the correct fuel delivery for all engine speeds and loads. A small quantity of air entering the engine indicates a deceleration or idle condition. A large quantity of air entering the engine indicates an acceleration or high load condition. **Possible Causes** • MAF sensor element hot wire contaminated or the sensor failed • MAF sensor signal shorted to ground or ground circuit problem • MAF sensor wiring routed close to ignition wires, generator, solenoids or electric motors • PCM has failed
P0102 **2T CCM, MIL: Yes** 1996, 1997, 1998, 1999, 2000, 2001, 2002, 2003, 2004, 2005 C/K Cab & Chassis, C/K Series Truck, G Series Van, L/M Vans 4.3L VIN W, 4.3L VIN X, 5.0L VIN M, 5.7L VIN K, 5.7L VIN R Transmissions: All	**MAF Sensor Circuit Low Frequency Conditions:** Engine started; engine runtime over 3 seconds, system voltage over 8v, and the PCM detected the MAF sensor frequency was 1300 Hz or less for 1.2 seconds. The MAF sensor is an airflow meter that measures how much air enters the engine. The PCM uses the MAF sensor signal to provide the correct fuel delivery for all engine speeds and loads. A small quantity of air entering the engine indicates a deceleration or idle condition. A large quantity of air entering the engine indicates an acceleration or high load condition. **Possible Causes** • MAF sensor element hot wire contaminated or the sensor failed • MAF sensor signal shorted to ground or ground circuit problem • MAF sensor wiring routed close to the ignition wires, generator, solenoids or electric motors (this causes it to pick up EMI/RFI) • PCM has failed • TSB 76-65-04 contains a repair procedure for this code

OBD II Trouble Code List (P0xxx Codes)

DTC	Trouble Code Title, Conditions & Possible Causes
P0102 **2T CCM, MIL: Yes** 1999, 2000, 2001, 2002, 2003, 2004, 2005 C/K Series Truck, G Series Van, Envoy, Escalade, TrailBlazer 4.8L VIN V, 5.3L VIN P, 5.3L VIN T, 5.3L VIN Z, 6.0L VIN N, 6.0L VIN U, 6.6L VIN 1, 8.1L VIN G Transmissions: All	**MAF Sensor Circuit Low Frequency Conditions:** Engine started; engine runtime over 3 seconds, system voltage over 8v, and the PCM detected the MAF sensor frequency was less than 1,200 Hz or less for 1.2 seconds. The MAF sensor is an airflow meter that measures how much air enters the engine. The PCM uses the MAF sensor signal to provide the correct fuel delivery for all engine speeds and loads. A small quantity of air entering the engine indicates a deceleration or idle condition. A large quantity of air entering the engine indicates acceleration or high load condition. **Possible Causes** ● MAF sensor element hot wire contaminated or the sensor failed ● MAF sensor signal shorted to ground or ground circuit problem ● MAF sensor wiring routed close to the ignition wires, generator, solenoids or electric motors (this causes it to pick up EMI/RFI) ● PCM has failed ● TSB 76-65-04 contains a repair procedure for this code
P0102 **2T CCM, MIL: Yes** 1996, 1997, 1998, 1999, 2000 C/K Series Truck, G Series Van, L & M Series Vans 7.4L VIN J engine Transmissions: All	**MAF Sensor Circuit Low Frequency Conditions:** Engine started; engine runtime over 3 seconds, system voltage over 8v, MAF sensor less than 60 Hz, TP angle less than 89%, and the PCM detected the MAF sensor signal was below 2 Hz for 2 seconds. **Possible Causes** ● MAF sensor element hot wire contaminated or the sensor failed ● MAF sensor signal shorted to ground or ground circuit problem ● MAF sensor wiring routed close to the ignition wires, generator, solenoids or electric motors (this causes it to pick up EMI/RFI) ● PCM has failed
P0102 **2T CCM, MIL: Yes** 1996, 1997, 1998, 1999, 2000, 2001, 2002, 2003, 2004, 2005 DeVille, Eldorado, Seville 4.6L VIN 9, 4.6L VIN Y engines Transmissions: A/T	**MAF Sensor Circuit Low Frequency Conditions:** DTC P0106, P0107, P0108, P0121, P0122, P0123, P0560, P0562 and P0563 not set, engine started, engine speed over 400 rpm, system voltage over 10.5v, and the PCM detected the MAF sensor signal was less than 1135 Hz for 3 seconds during the CCM test. **Possible Causes** ● MAF sensor element hot wire contaminated or the sensor failed ● MAF sensor signal shorted to ground or ground circuit problem ● MAF sensor wiring routed close to the ignition wires, generator, solenoids or electric motors (this causes it to pick up EMI/RFI) ● PCM has failed ● TSB 71-65-26 contains a repair procedure for this code
P0102 **2T CCM, MIL: Yes** 2003, 2004, 2005 CTS (Cadillac) 2.6L V6 VIN M & 3.2L V6 VIN N engines engine Transmissions: All	**MAF Sensor Circuit Low Input Conditions:** Engine started; engine runtime over 500 ms to allow the MAF sensor hot film element to heat up, more than 20 crankshaft revolutions detected, system voltage over 10v, and the PCM detected the MAF sensor signal was less than 0.42v for over 2 seconds, or the MAF sensor was below a calculated value using the TP and RPM inputs. **Possible Causes** ● MAF sensor element hot wire contaminated or the sensor failed ● MAF sensor signal shorted to ground or ground circuit problem ● MAF sensor wiring routed close to the ignition wires or motors ● PCM has failed
P0102 **2T CCM, MIL: Yes** 1995, 1996, 1997, 1998, 1999, 2000, 2001, 2002 Camaro & Firebird 3.8L VIN K engine Transmissions: All	**MAF Sensor Circuit Low Frequency Conditions:** Engine cranking or running for 500 ms, system voltage over 8v, and the PCM detected the MAF sensor was less than 1,200 Hz for 12 seconds. This sensor measures how much air enters the engine. **Possible Causes** ● MAF sensor element hot wire contaminated or the sensor failed ● MAF sensor signal shorted to ground or ground circuit problem ● MAF sensor wiring routed close to the ignition wires, generator, solenoids or electric motors (this causes it to pick up EMI/RFI) ● PCM has failed
P0102 **2T CCM, MIL: Yes** 1997, 1998, 1999, 2000, 2001 Corvette, Camaro & Firebird 5.7L VIN G, 5.7L VIN S Transmissions: All	**MAF Sensor Circuit Low Frequency Conditions:** Engine started; engine runtime over 3 seconds, engine speed more than 400 rpm, system voltage 9-16v, and the PCM detected the MAF sensor frequency was less than 10 Hz for 1.2 seconds. **Possible Causes** ● Engine oil dipstick missing or not fully seated ● MAF sensor element hot wire contaminated or the sensor failed ● MAF sensor signal shorted to ground or ground circuit problem ● MAF sensor wiring routed close to the ignition wires or motors ● PCM has failed

OBD II Trouble Code List (P0xxx Codes)

DTC	Trouble Code Title, Conditions & Possible Causes
P0102 **2T CCM, MIL: Yes** 2002, 2003, 2004, 2005 Corvette, Camaro & Firebird 5.7L VIN G, 5.7L VIN S Transmissions: All	**MAF Sensor Circuit Low Frequency Conditions:** Engine started engine speed over 400 rpm for 3 seconds, system voltage 9-16v, and the PCM detected the MAF sensor frequency was less than 1,300 Hz for 1.2 seconds. **Possible Causes** ● Engine oil dipstick missing or not fully seated ● MAF sensor element hot wire contaminated or the sensor failed ● MAF sensor signal shorted to ground or ground circuit problem ● MAF sensor wiring routed close to the ignition wires or motors ● PCM has failed
P0102 **2T CCM, MIL: Yes** 1996, 1997, 1998, 1999, 2001, 2002, 2003, 2004, 2005 Aurora 4.0L V6 VIN C engine Transmissions: A/T	**MAF Sensor Circuit Low Frequency Conditions:** Engine started; system voltage over 10.5v and the PCM detected the MAF sensor frequency was less than 1,135 Hz. The MAF sensor is an airflow meter that measures the amount of air that enters the engine. The PCM uses the MAF sensor to provide the correct fuel delivery for all engine speeds and loads. **Possible Causes** ● MAF sensor element hot wire contaminated or the sensor failed ● MAF sensor signal shorted to ground or ground circuit problem ● MAF sensor wiring routed close to ignition wires or generator ● PCM has failed
P0102 **2T CCM, MIL: Yes** 2001, 2002 Aurora 3.5L V6 VIN H engine Transmissions: A/T	**MAF Sensor Circuit Low Frequency Conditions:** Engine started, system voltage over 10.5v, and the PCM detected the MAF sensor frequency was less than 1,200 Hz during the CCM test period. **Possible Causes** ● Base idle speed set to low, oil dipstick missing or not seated ● MAF sensor element hot wire contaminated or the sensor failed ● MAF sensor signal shorted to ground or ground circuit problem ● MAF sensor wiring routed close to ignition wires or generator ● PCM has failed
P0102 **2T CCM, MIL: Yes** 1996, 1997, 1998, 1999, 2000 S/T Blazer & S/T Pickup 4.3L VIN W, 4.3L VIN X Transmissions: All	**MAF Sensor Circuit Low Frequency Conditions:** Engine started engine speed over 400 rpm, system voltage over 8v, MAF sensor frequency stable for over 1 second, and the PCM/VCM detected the MAF sensor indicated less than 10 Hz for one second. The MAF sensor is an airflow meter that measures how much air enters the engine. It is used to provide the correct fuel delivery for all engine speeds and loads. A small quantity of air entering the engine indicates a deceleration or idle condition. A large quantity of air entering the engine indicates an acceleration or high load condition. **Possible Causes** ● Base idle speed is set to low, oil dipstick missing or not seated ● MAF sensor element hot wire contaminated or the sensor failed ● MAF sensor signal shorted to ground or ground circuit problem ● MAF sensor wiring routed close to ignition wires, generator, solenoids or electric motors ● PCM has failed ● TSB 76-65-04 contains a repair procedure for this code
P0102 **2T CCM, MIL: Yes** 2001, 2002, 2003, 2004, 2005 S/T Blazer & S/T Pickup 4.3L VIN W, 4.3L VIN X Transmissions: All	**MAF Sensor Circuit Low Frequency Conditions:** Engine started, system voltage over 8v, MAF sensor signal frequency stable for 1 second, and the PCM/VCM detected the MAF sensor indicated less than 1300 Hz for 1 second. The MAF sensor is an airflow meter that measures how much air enters the engine. It is used to provide the correct fuel delivery for all engine speeds and loads. **Possible Causes** ● Engine oil dipstick missing or not fully seated ● MAF sensor element hot wire contaminated or the sensor failed ● MAF sensor signal shorted to ground or ground circuit problem ● MAF sensor wiring routed close to ignition wires, generator, solenoids or electric motors ● PCM has failed
P0103 **2T CCM, MIL: Yes** 1996, 1997 Beretta, Century, Ciera, Corsica, Cutlass, Malibu, Grand Prix, Regal, Cutlass Supreme, Monte Carlo, Lumina APV & Lumina 3.1L VIN M, 3.4L VIN E, 3.4L VIN X, 3.8L VIN 1, 3.8L VIN K engines Transmissions: All	**MAF Sensor Circuit High Frequency Conditions:** Engine started, system voltage from 10-18v, IAC motor more than 2 counts, and the PCM detected the MAF sensor frequency was more than 11,500 Hz for 1 second. **Possible Causes** ● MAF sensor power circuit has a high resistance condition ● MAF sensor is contaminated, dirty or it has failed ● MAF sensor wiring routed too close to the Generator or to close to ignition wires (this causes a EMI/REFI high frequency signal) ● Water enters the air intake system reaches the MAF sensor, cools it, and causes it to indicate excessive airflow (check AIR system) ● PCM has failed

OBD II Trouble Code List (P0xxx Codes)

DTC	Trouble Code Title, Conditions & Possible Causes
P0103 **2T CCM, MIL: Yes** 1996, 1997 Camaro, Caprice, Corvette, Firebird, Fleetwood 4.3L VIN W, 5.7L VIN 5, 5.7L VIN P engines Transmissions: All	**MAF Sensor Circuit High Frequency Conditions:** Engine started engine running for 1 second, system voltage over 8v, engine conditions stable for 2 seconds, and the PCM detected the MAF sensor signal was more than 14,000 Hz for 1 second. **Possible Causes** ● MAF sensor power circuit has a high resistance condition ● MAF sensor is contaminated, dirty or it has failed ● MAF sensor wiring routed too close to the Generator or to close to ignition wires (this causes a EMI/REFI high frequency signal) ● Water enters the air intake system reaches the MAF sensor, cools it, and causes it to indicate excessive airflow (check AIR system) ● PCM has failed
P0103 **2T CCM, MIL: Yes** 1997, 1998, 1999, 2000, 2001, 2002, 2003, 2004, 2005 Alero, Achieva, Cutlass, Malibu, Montana, Silhouette, Venture 3.1L VIN J, 3.1L VIN M, 3.4L VIN E engines Transmissions: All	**MAF Sensor Circuit High Frequency Conditions:** Engine started; system voltage over 8v, IAC position more than 2 counts, and the PCM detected the MAF sensor indicated more than 11,500 Hz for over 500 milliseconds. **Possible Causes** ● MAF sensor power circuit has a high resistance condition ● MAF sensor is contaminated, dirty or it has failed ● MAF sensor wiring routed close to Generator or ignition wires ● Water enters the air intake system reaches the MAF sensor, cools it, and causes it to indicate excessive airflow (check AIR system) ● PCM has failed
P0103 **2T CCM, MIL: Yes** 1998, 1999, 2000, 2001, 2002, 2003, 2004, 2005 Century, LeSabre, Park Avenue, Regal 3.1L VIN J, 3.1L VIN M, 3.4L VIN E, 3.5L V6 VIN H, 3.8L VIN 1, 3.8L VIN K Transmissions: All	**MAF Sensor Circuit High Frequency Conditions:** Engine cranking or running, IAC position over 5 counts, TP angle under 50%, conditions met for 0.5 seconds plus 400 3X reference periods (133 crankshaft revolutions), and the PCM detected the MAF frequency was more than 11,500 Hz for 12 seconds. The MAF sensor is an airflow meter that measures how much airs enters the engine. This sensor is used to provide the correct fuel delivery for all engine speeds and loads. A small quantity of air entering the engine indicates a deceleration or idle condition. A large quantity of air entering the engine indicates an acceleration or high load condition. **Possible Causes** ● MAF sensor power circuit has a high resistance condition ● MAF sensor is contaminated, dirty or it has failed ● MAF sensor wiring routed close to Generator or ignition wires ● Water enters the air intake system reaches the MAF sensor, cools it, and causes it to indicate excessive airflow (check AIR system) ● PCM has failed
P0103 **2T CCM, MIL: Yes** 1996, 1997, 1998, 1999, 2000, 2001, 2002, 2003, 2004, 2005 C/K Cab & Chassis, C/K Series Truck, G Series Vans & L/M Vans 4.3L VIN W, 4.3L VIN X, 5.0L VIN M, 5.7L VIN K, 5.7L VIN R engines Transmissions: All	**MAF Sensor Circuit High Frequency Conditions:** Engine started; engine runtime over 3 seconds, engine speed over 400 rpm, and the PCM detected the MAF sensor frequency was 1100-13500 Hz or more for 1.2 seconds. The MAF sensor is an airflow meter that measures how much air enters the engine. The PCM uses the MAF sensor to provide the correct fuel delivery for all engine speeds and loads. A small quantity of air entering the engine indicates a deceleration or idle condition. A large quantity of air entering the engine indicates an acceleration or high load condition. **Possible Causes** ● MAF sensor power circuit has a high resistance condition ● MAF sensor is contaminated, dirty or it has failed ● MAF sensor wiring routed close to Generator or ignition wires ● Water enters the air intake system reaches the MAF sensor, cools it, and causes it to indicate excessive airflow (check AIR system) ● PCM has failed
P0103 **2T CCM, MIL: Yes** 1999, 2000, 2001, 2002, 2003, 2004, 2005 C/K Series Truck, G Series Van, Envoy, Escalade, TrailBlazer 4.8L VIN V, 5.3L VIN P, 5.3L VIN T, 5.3L VIN Z, 6.0L VIN N, 6.0L VIN U, 6.6L VIN 1, 8.1L VIN G Transmissions: All	**MAF Sensor Circuit High Frequency Conditions:** Engine started; engine runtime over 3 seconds and the PCM detected the MAF sensor frequency was more than 13,500 Hz for over 1.2 seconds. The MAF sensor is an airflow meter that measures how much air enters the engine. The PCM uses the MAF sensor to provide the correct fuel delivery for all engine speeds and loads. A small quantity of air entering the engine indicates a deceleration or idle condition. A large quantity of air entering the engine indicates an acceleration or high load condition. **Possible Causes** ● MAF sensor power circuit has a high resistance condition ● MAF sensor is contaminated, dirty or it has failed ● MAF sensor wiring routed close to Generator or ignition wires ● Water enters the air intake system reaches the MAF sensor, cools it, and causes it to indicate excessive airflow (check AIR system) ● PCM has failed

OBD II Trouble Code List (P0xxx Codes)

DTC	Trouble Code Title, Conditions & Possible Causes
P0103 **2T CCM, MIL: Yes** 1996, 1997, 1998, 1999, 2000, 2001, 2002, 2003, 2004, 2005 C/K Series Truck, G Series Van, L & M Series Vans 7.4L VIN J engine Transmissions: All	**MAF Sensor Circuit High Frequency Conditions:** Engine started; engine runtime 3 seconds, system voltage over 11v, engine vacuum less than 90 kPa, and the PCM detected the MAF sensor was over 11,000 Hz for 2 seconds. **Possible Causes** ● MAF sensor power circuit has a high resistance condition ● MAF sensor is contaminated, dirty or it has failed ● MAF sensor wiring routed close to Generator or ignition wires ● Water enters the air intake system reaches the MAF sensor, cools it, and causes it to indicate excessive airflow (check AIR system) ● PCM has failed
P0103 **2T CCM, MIL: Yes** 1996, 1997, 1998, 1999, 2000, 2001, 2002, 2003, 2004, 2005 DeVille, Eldorado, Seville 4.6L VIN 9, 4.6L VIN Y Transmissions: All	**MAF Sensor Circuit High Frequency Conditions:** DTC P0106, P0107, P0108, P0121, P0122, P0123, P0560, P0562 or P0563 not set, engine speed over 400 rpm, system voltage over 10.5v, and the PCM detected the MAF sensor was over 11,000 Hz for 3 seconds. **Possible Causes** ● MAF sensor power circuit has a high resistance condition ● MAF sensor is contaminated, dirty or it has failed ● MAF sensor wiring routed too close to the Generator or to close to ignition wires (this causes a EMI/REFI high frequency signal) ● Water enters the air intake system reaches the MAF sensor, cools it, and causes it to indicate excessive airflow (check AIR system) ● PCM has failed
P0103 **2T CCM, MIL: Yes** 2003, 2004, 2005 CTS (Cadillac) 2.6L V6 VIN M & 3.2L V6 VIN N engines Transmissions: All	**MAF Sensor Circuit High Input Conditions:** Engine started; engine runtime over 500 ms, more than 20 crankshaft revolutions detected, system voltage over 10v, and the PCM detected the MAF sensor signal was more than 4.88v for over 2 seconds, or the MAF sensor was above a calculated value stored in its memory. **Possible Causes** ● MAF sensor element hot wire contaminated or the sensor failed ● MAF sensor signal circuit is open or shorted to system power ● MAF sensor wiring routed close to the ignition wires, generator, solenoids or electric motors (this causes it to pick up EMI/RFI) ● PCM has failed
P0103 **2T CCM, MIL: Yes** 1995, 1996, 1997, 1998, 1999, 2000, 2001, 2002 Camaro & Firebird 3.8L VIN K Transmissions: All	**MAF Sensor Circuit High Frequency Conditions:** Engine runtime over 3 seconds, engine speed more than 300 rpm, system voltage over 8v, and the PCM detected the MAF sensor frequency was more than 11,500 Hz for over 12 seconds. The range of this sensor is 2000 (idle) to 10000 Hz (high load). **Possible Causes** ● MAF sensor power circuit has a high resistance condition ● MAF sensor is contaminated, dirty or it has failed ● MAF sensor wiring routed too close to the Generator or to close to ignition wires (this causes a EMI/REFI high frequency signal) ● Water enters the air intake system reaches the MAF sensor, cools it, and causes it to indicate excessive airflow (check AIR system) ● PCM has failed
P0103 **2T CCM, MIL: Yes** 1997, 1998, 1999, 2000, 2001 Corvette, Camaro & Firebird 5.7L VIN G, 5.7L VIN S Transmissions: All	**MAF Sensor Circuit High Frequency Conditions:** Engine started engine speed more than 400 rpm, system voltage over 8.0v, and the PCM detected the MAF frequency indicated more than 14,000 Hz for over 1.2 seconds. **Possible Causes** ● MAF sensor power circuit has a high resistance condition ● MAF sensor is contaminated, dirty or it has failed ● MAF sensor wiring routed too close to the Generator or to close to ignition wires (this causes a EMI/REFI high frequency signal) ● Water enters the air intake system reaches the MAF sensor, cools it, and causes it to indicate excessive airflow (check AIR system) ● PCM has failed ● TSB 99-06-04-007 contains a repair procedure for this code
P0103 **2T CCM, MIL: Yes** 2002, 2003, 2004, 2005 Corvette, Camaro & Firebird 5.7L VIN G, 5.7L VIN S Transmissions: All	**MAF Sensor Circuit High Frequency Conditions:** Engine started engine speed more than 400 rpm, system voltage over 8.0v, and the PCM detected the MAF frequency indicated more than 13,500 Hz for over 1.2 seconds. **Possible Causes** ● MAF sensor power circuit has a high resistance condition ● MAF sensor is contaminated, dirty or it has failed ● MAF sensor wiring routed too close to the Generator or to close to ignition wires (this causes a EMI/REFI high frequency signal) ● Water enters the air intake system reaches the MAF sensor, cools it, and causes it to indicate excessive airflow (check AIR system) ● PCM has failed

OBD II Trouble Code List (P0xxx Codes)

DTC	Trouble Code Title, Conditions & Possible Causes
P0103 **2T CCM, MIL: Yes** 1996, 1997, 1998, 1999, 2001, 2002, 2003, 2004, 2005 Aurora 4.0L V8 VIN C engine Transmissions: A/T	**MAF Sensor Circuit High Frequency Conditions:** Engine started; system voltage over 10.5v, TP angle less than 50 degrees, and the PCM detected the MAF sensor signal was more than 11,000 Hz. The range of the MAF sensor is 2000 (idle) to 10,000 Hz (high load). The MAF sensor is an airflow meter that measures the amount of air entering the engine. It is used to determine the correct fuel delivery at all engine speeds and loads. **Possible Causes** ● MAF sensor power circuit has a high resistance condition ● MAF sensor is contaminated, dirty or it has failed ● MAF sensor wiring routed too close to the Generator or to close to ignition wires (this causes a EMI/REFI high frequency signal) ● Water enters the air intake system reaches the MAF sensor, cools it, and causes it to indicate excessive airflow (check AIR system) ● PCM has failed
P0103 **2T CCM, MIL: Yes** 2001, 2002 Aurora 3.5L V6 VIN H engine Transmissions: A/T	**MAF Sensor Circuit High Frequency Conditions:** Engine started; system voltage over 10.5v and the PCM detected the MAF sensor frequency was more than 11,500 Hz. The MAF sensor is an airflow meter that measures how much air enters the engine. It is used to provide the correct fuel delivery for all engine speeds and loads. A small quantity of air entering the engine indicates a deceleration or idle condition. A large quantity of air entering the engine indicates an acceleration or high load condition. **Possible Causes** ● MAF sensor power circuit has a high resistance condition ● MAF sensor is contaminated, dirty or it has failed ● MAF sensor wiring routed too close to the Generator or to close to ignition wires (this causes a EMI/REFI high frequency signal) ● Water enters the air intake system reaches the MAF sensor, cools it, and causes it to indicate excessive airflow (check AIR system) ● PCM has failed
P0103 **2T CCM, MIL: Yes** 1996, 1997, 1998, 1999, 2000, 2001, 2002, 2003, 2004, 2005 S/T Blazer & S/T Pickup 4.3L VIN W, 4.3L VIN X Transmissions: All	**MAF Sensor Circuit High Frequency Conditions:** Engine started, system voltage over 8v, MAF sensor signal stable for 1 second, and the PCM detected the MAF sensor was over 12000 Hz for 1 second. The MAF sensor is an airflow meter that measures how much air enters the engine. The PCM uses the MAF sensor to provide the correct fuel delivery for all engine speeds and loads. A small quantity of air entering the engine indicates a deceleration or idle condition. A large quantity of air entering the engine indicates an acceleration or high load condition. **Possible Causes** ● MAF sensor power circuit has a high resistance condition ● MAF sensor is contaminated, dirty or it has failed ● MAF sensor wiring routed too close to the Generator or to close to ignition wires (this causes a EMI/REFI high frequency signal) ● Water enters the air intake system reaches the MAF sensor, cools it, and causes it to indicate excessive airflow (check AIR system) ● PCM has failed
P0105 **2T CCM, MIL: Yes** 1996, 1997, 1998, 1999 Aurora 4.0L V8 VIN C Transmissions: A/T	**MAF Sensor Signal Insufficient Activity Conditions:** DTC P0106, P0107, P0108, P0122, P0123 and P1108 not set, MAP sensor signal at least 21.5 kPa, engine vacuum at least 11.8 kPa, and the PCM detected the MAP signal changed less than 4 kPa in the one second following a 3.2 degree change in throttle position, and the MAP signal was not within 17.3 kPa of the expected MAP reading. This failure must occur five times before this code will set. The MAF sensor is an airflow meter that measures how much air enters the engine. The PCM uses the MAF sensor to provide the correct fuel delivery for all engine speeds and loads. **Possible Causes** ● MAP sensor signal or ground circuit has high resistance ● MAP sensor is contaminated (moisture or icing problems) ● MAP sensor is damaged, sticking or has failed
P0105 **1T CCM, MIL: Yes** 1996, 1997, 1998, 1999 DeVille, Eldorado, Seville 4.6L VIN 9, 4.6L VIN Y Transmissions: All	**MAP Sensor Insufficient Activity Conditions:** DTC P0106-P0108, P0122, P0123 and P1108 not set, MAP sensor signal at least 21.5 kPa, engine vacuum at least 11.8 kPa, and the PCM detected the MAP signal changed less than 4 kPa in 1 second following a 3.2 degree change in TP angle, and the Actual MAP was not within 17.3 kPa of Expected MAP (the fault must occur 5 times). **Possible Causes** ● MAP sensor signal or ground circuit has high resistance ● MAP sensor is contaminated (moisture or icing problems) ● MAP sensor is damaged, sticking or has failed

OBD II Trouble Code List (P0xxx Codes)

DTC	Trouble Code Title, Conditions & Possible Causes
P0105 **2T CCM, MIL: Yes** 1999, 2000, 2001, 2002, 2003, 2004, 2005 Cavalier & Sunfire 2.2L VIN 4, 2.4L VIN T Transmissions: All	**MAP Sensor Signal Range/Performance Conditions:** DTC P0107-P0108, P0117-P0118, P0122-P0123, P0125, P0128, P0130-P0132, P0171-P0172, P0201-P0204, P0300-P0304, P0336, P0440-P0446, P0452-P0453, P0502, P0506, P0507, P1441, P1680, P1682 and P1860 not set, engine speed from 600-6375 rpm for 40 seconds, TCC stable within 2.5%, IAC valve stable within 5 counts for 1.5 seconds, and the PCM detected the MAP or TP signal was out of its expected range for 14-16 seconds. **Possible Causes** ● MAP sensor signal or ground circuit has high resistance ● MAP sensor is contaminated (moisture or icing problems) ● MAP sensor is damaged, sticking or has failed
P0105 **2T CCM, MIL: Yes** 1999, 2000, 2001, 2002, 2003, 2004, 2005 S/T Blazer & S/T Pickup, Envoy & TrailBlazer 2.2L VIN 4, 2.2L VIN 5, 2.2L VIN F, 2.2L VIN H, 4.2L VIN S	**MAP Sensor Signal Range/Performance Conditions:** DTC P0013, P0014, P0107, P0108, P0116, P0117, P0118, P0122, P0123, P0125, P0128, P0130, P0131, P0132, P0171, P0172, P0201-P0204, P0300, P0301-P0304, P0335, P0336, P0351-P0356, P0440, P0442, P0446, P0452, P0453, P0482, P0502, P0506, P0507, P0601, P0602, P0604, P0606, P1441, P1621 and P1860 not set, engine started, engine speed at 600-6375 rpm (± 50 rpm) for 40 seconds, IAC position stable (± 5%), TP angle stable (± 2%), engine speed stable (+ 50 rpm), TCC command stable (± 2.5%), and the PCM detected the MAP sensor was out-of-range for 10-14 seconds. **Possible Causes** ● MAP sensor signal or ground circuit has high resistance ● MAP sensor is contaminated (moisture or icing problems) ● MAP sensor is damaged, sticking or has failed
P0106 **2T CCM, MIL: Yes** 1996, 1997, 1998 Achieva, Beretta, Cavalier, Ciera, Corsica, Century, Grand Am, Malibu, Skylark, Sunfire 2.2L VIN 4, 2.4L VIN T Transmissions: A/T	**MAP Sensor Signal Range/Performance Conditions:** No CKP, CMP, ECT, EGR, EVAP, Fuel Trim, HO2S, MAP, PCM Memory, TP or VSS codes set, engine started, engine speed over 600 rpm, MAP sensor at idle over 60 kPa, TP angle under 50%, then after the throttle angle was changed by over 12%, the PCM detected the MAP sensor signal did not correlated to the amount of expected change in the TP angle. Cruise & Idle Test Initial MAP sensor at idle under 65 kPa, TP angle under 50%, then the TP angle changed over 12% and PCM detected the MAP sensor did not change as expected during the test. **Possible Causes** ● MAP sensor is contaminated, dirty, skewed or has failed ● MAP sensor vacuum line is loose, restricted or contains "ice" ● PCM has failed
P0106 **2T CCM, MIL: Yes** 1996, 1997, 1998, 1999 Achieva, Beretta, Cavalier, Ciera, Corsica, Century, Grand Am, Malibu, Skylark, Skyhawk, Sunfire 2.2L VIN 4, 2.4L VIN T Transmissions: M/T	**MAP Sensor Signal Range/Performance Conditions:** No CKP, CMP, ECT, EGR, EVAP, Fuel Trim, HO2S, MAP, PCM Memory, TP or VSS codes set, engine started, engine speed over 900 rpm, MAP sensor at idle over 60 kPa, TP angle under 50%, then after the throttle angle was changed by over 12%, the PCM detected the MAP sensor signal did not correlated to the amount of expected change in the TP angle. Cruise & Idle Test Initial MAP sensor at idle under 65 kPa, TP angle under 50%, then the TP angle changed over 12% and PCM detected the MAP sensor did not change as expected during the test. **Possible Causes** ● MAP sensor seal is missing or damage, intake manifold leaks ● MAP sensor is contaminated, dirty, skewed or has failed ● MAP sensor vacuum line is loose, restricted or contains "ice" ● PCM has failed
P0106 **2T CCM, MIL: Yes** 1996, 1997, 1998, 1999 Achieva, Beretta, Ciera, Corsica, Century, Grand Am, Malibu, Skylark 3.1L VIN M engine Transmissions: All	**MAP Sensor Signal Range/Performance Conditions:** DTC P0107 and P0108 not set, engine started, engine running with the engine speed, EGR flow rate, throttle angle and IAC counts all steady, and the PCM detected the Actual MAP sensor and Expected Map value were too far apart. The MAP sensor responds to pressure changes in the intake manifold. The pressure changes occur based on the engine load. **Possible Causes** ● MAP sensor seal is missing or damage, intake manifold leaks ● MAP sensor is contaminated, dirty, skewed or has failed ● MAP sensor vacuum line is loose, restricted or contains "ice" ● PCM has failed
P0106 **1T CCM, MIL: Yes** 1996, 1997 Camaro, Caprice, Corvette, Firebird, Fleetwood 4.3L VIN W, 5.7L VIN 5, 5.7L VIN P Transmissions: All	**MAP Sensor Signal Range/Performance Conditions:** DTC P0107, P0108, P0122 and P0123 not set, engine started, engine running with any speed change less than 100 rpm, throttle angle change under 5%, ASR/TCS not in a Traction Event, EGR position change under 25%, A/C Clutch, Brake Switch, IAC counts and TR sensor signals all constant for 500 ms, and the PCM detected too large a change in the MAP sensor without a previous large change in throttle angle and speed for 4.5 seconds. **Possible Causes** ● MAP sensor seal is missing or damage, intake manifold leaks ● MAP sensor is contaminated, dirty, skewed or has failed ● MAP sensor vacuum line is loose, restricted or contains "ice" ● PCM has failed

OBD II Trouble Code List (P0xxx Codes)

DTC	Trouble Code Title, Conditions & Possible Causes
P0106 **2T CCM, MIL: Yes** 1996, 1997, 1998, 1999, 2000, 2001, 2002, 2003, 2004, 2005 C/K Cab & Chassis, C/K Series Truck, G Series Vans & L/M Series Vans 4.3L VIN W, 4.3L VIN X, 5.0L VIN M, 5.7L VIN R, 7.4L VIN J engines Transmissions: All	**MAP Sensor Signal Range/Performance Conditions:** DTC P0101, P0102, P0103, P0107, P0108, P0121, P0122, P0123, P0401, P0404, P0405, P0410, P0440, P0442, P0443 and P0446 not set, engine started, engine running at 400-5000 rpm with any change less than 125 rpm, PTO and Traction Control inactive, any change in IAC position less than 10 g/sec, any change in EGR position less than 20%, A/C clutch, power steering, Clutch and Brake switch signals all constant, and the PCM detected the MAP sensor signal was not within a predicted range for 2 seconds. The MAP sensor responds to pressure changes in the intake manifold that occur based on the amount of engine load. **Possible Causes** ● MAP sensor seal is missing or damage, intake manifold leaks ● MAP sensor is contaminated, dirty, skewed or has failed ● MAP sensor vacuum line is loose, restricted or contains "ice" ● PCM has failed
P0106 **2T CCM, MIL: Yes** 1999, 2000, 2001, 2002, 2003, 2004, 2005 C/K Series Truck, G Series Van, Envoy, Escalade, TrailBlazer 4.8L VIN V, 5.3L VIN P, 5.3L VIN T, 5.3L VIN Z, 6.0L VIN N, 6.0L VIN U, 8.1L VIN G engines Transmissions: All	**MAP Sensor Signal Range/Performance Conditions:** DTC P0101, P0102, P0103, P0107, P0108, P0120, P0121, P0122, P0123, P0220, P0442, P0443, P0446, P0455, P1125, P1514, P1515, P1516, P1518, P2108, P2120, P2121, P2125, P2126, P2130, P2131and P2135 not set, engine speed from 400-5000 rpm (± 125 rpm), PTO and Traction Control inactive, A/C Clutch, Brake, Clutch and Power Steering switch signals all constant, and the PCM detected the MAP sensor signal was out of range for 2 seconds. **Possible Causes** ● MAP sensor seal is missing or damage, intake manifold leaks ● MAP sensor is contaminated, dirty, skewed or has failed ● MAP sensor vacuum line is loose, restricted or contains "ice" ● PCM has failed
P0106 **2T CCM, MIL: Yes** 1996, 1997, 1998, 1999, 2000, 2001, 2002, 2003, 2004, 2005 DeVille, Eldorado, Seville 4.6L VIN 9, 4.6L VIN Y Transmissions: A/T	**MAP Sensor Signal Range/Performance Conditions:** DTC P0122 and P0123 not set, engine started, engine speed from 1000-4000 rpm and steady, throttle angle, IAC valve, A/C clutch status and TCC status all steady, EGR position steady, brakes not applied, and the PCM detected the Actual MAP sensor value did not correlate with to the Predicted MAP sensor value for 5 seconds. The MAP sensor responds to pressure changes in the intake manifold that occur based on the engine load. **Possible Causes** ● MAP sensor seal is missing or damage, intake manifold leaks ● MAP sensor is contaminated, dirty, skewed or has failed ● MAP sensor vacuum line is loose, restricted or contains "ice" ● PCM has failed
P0106 **2T CCM, MIL: Yes** 1995, 1996, 1997, 1998, 1999, 2000, 2001, 2002 Camaro & Firebird 3.8L VIN K engines Transmissions: All	**MAP Sensor Signal Range/Performance Conditions:** Engine started A/C Clutch, Brake and Clutch switch status, EGR flow rate, engine speed, throttle angle and IAC counts all steady, and the PCM detected the MAP sensor changed too much without a corresponding change in engine load during 200 3X REF periods. **Possible Causes** ● MAP sensor seal is missing or damage, intake manifold leaks ● MAP sensor is contaminated, dirty, skewed or has failed ● MAP sensor vacuum line is loose, restricted or contains "ice" ● PCM has failed
P0106 **2T CCM, MIL: Yes** 2001, 2002 Camaro & Firebird 5.7L VIN G engine Transmissions: All	**MAP Sensor Signal Range/Performance Conditions:** DTC P0101-P0103, P0107, P0108, P0121-P0123 or P0440-P0446 not set, engine speed from 400-5000 rpm with any change less than 125 rpm, Traction Control "off", change in IAC position less than 10 g/sec, A/C clutch, Power Steering, Brake and Clutch switch signals stable for 1 second, and the PCM detected the MAP sensor was out of range for 2 seconds. **Possible Causes** ● MAP sensor seal is missing or damage, intake manifold leaks ● MAP sensor is contaminated, dirty, skewed or has failed ● MAP sensor vacuum line is loose, restricted or contains "ice" ● PCM has failed
P0106 **2T CCM, MIL: Yes** 1996, 1997, 1998, 1999, 2001, 2002, 2003, 2004, 2005 Aurora 3.5L V6 VIN H, 4.0L V8 VIN C engines Transmissions: A/T	**MAP Sensor Signal Range/Performance Conditions:** DTC P0121, P0122 and P0123 not set, engine speed from 1000-4000 rpm, TP angle stable (± 4%), IAC position stable (± 15 counts), engine speed stable (± 125 rpm), EGR position stable (± 10%), A/C request stable, Traction Control "off", Engine Over-Temperature not active, Power Steering load stable, brake pedal switch position stable, conditions met for 3 seconds, and the PCM detected the MAP sensor signal was out of range for 2 seconds. **Possible Causes** ● MAP sensor seal is missing or damage, intake manifold leaks ● MAP sensor is contaminated, dirty, skewed or has failed ● MAP sensor vacuum line is loose, restricted or contains "ice" ● PCM has failed

OBD II Trouble Code List (P0xxx Codes)

DTC	Trouble Code Title, Conditions & Possible Causes
P0106 **2T CCM, MIL: Yes** 1995, 1996, 1997, 1998, 1999, 2000, 2001, 2002, 2003, 2004, 2005 S/T Blazer & S/T Pickup 4.3L VIN W, 4.3L VIN X Transmissions: All	**MAP Sensor Signal Range/Performance Conditions:** DTC P0101-P0103, P0107, P0108, P0121-P0123, P0401, P0404, P0405, P0410, P0440-P0443 and P0446 not set, engine speed from 400-5000 rpm with any change less than 125 rpm, Traction Control inactive, change in IAC position less than 10 counts, A/C Clutch, Power Steering, Clutch and Brake signals all stable for 1 second, and the PCM detected the MAP sensor value was out of its normal range during the CCM test. The MAP sensor responds to pressure changes in the intake manifold that occur based on the engine load. **Possible Causes** ● MAP sensor seal is missing or damage, intake manifold leaks ● MAP sensor is contaminated, dirty, skewed or has failed ● MAP sensor vacuum line is loose, restricted or contains "ice" ● PCM has failed
P0106 **2T CCM, MIL: Yes** 1996, 1997 Grand Prix, Lumina, Monte Carlo 3.4L VIN X engine Transmissions: All	**MAP Sensor Signal Range/Performance Conditions:** DTC P0107 and P0108 not set, engine started, engine running with the engine speed, EGR flow rate, throttle angle and IAC counts all steady, and the PCM detected the Actual MAP sensor and Expected Map value were too far apart for 5 seconds and 200 3X REF periods. The MAP sensor responds to pressure changes in the intake manifold. **Possible Causes** ● MAP sensor seal is missing or damage, intake manifold leaks ● MAP sensor is contaminated, dirty, skewed or has failed ● MAP sensor vacuum line is loose, restricted or contains "ice" ● PCM has failed
P0106 **2T CCM, MIL: Yes** 2001, 2002, 2003, 2004, 2005 Corvette 5.7L VIN G, 5.7L VIN S Transmissions: All	**MAP Sensor Signal Range/Performance Conditions:** DTC P0101- P0103, P0107, P0108, P0440, P0442, P0443, P0446, P1120, P1125, P1220, P1221, P1275, P1276, P1280, P1281, P1285, P1286, P1514-P1517 and P1518 not set, engine speed from 400-5000 rpm (± 125 rpm), Traction Control off, IAC stable (± 10%), A/C clutch, power steering, clutch and brake switch signals all stable, and the PCM detected the MAP sensor was out of its normal range for 2 seconds. The MAP sensor responds to pressure changes in the intake manifold that occur based on the engine load **Possible Causes** ● MAP sensor seal is missing or damage, intake manifold leaks ● MAP sensor is contaminated, dirty, skewed or has failed ● MAP sensor vacuum line is loose, restricted or contains "ice" ● PCM has failed
P0107 **2T CCM, MIL: Yes** 1996, 1997, 1998, 1999, 2000, 2001, 2002, 2003, 2004, 2005 Achieva, Alero, Beretta, Century, Ciera, Cavalier, Envoy, Grand Am, Sunfire Skyhawk, TrailBlazer, S/T Blazer & S/T Pickup 2.2L VIN 4, 2.2L VIN 5, 2.2L VIN F, 2.2L VIN H, 2.4L VIN T, 4.2L VIN S Transmissions: All	**MAP Sensor Circuit Low Input Conditions:** DTC P0121, P0122 and P0123 not set, engine started, TP angle at 0% with the engine speed less than 1000 rpm, or TP angle over 15% with the engine speed more than 1000 rpm, and the PCM detected the MAP sensor indicated less than 0.20v (Scan Tool reads less than 11.8 kPa) for 6.25. The MAP sensor responds to pressure changes in the intake manifold that occur based on the engine load. **Possible Causes** ● MAP sensor signal circuit shorted to sensor or chassis ground ● MAP sensor power circuit open between the sensor and PCM ● MAP sensor is damaged or has failed ● PCM has failed
P0107 **2T CCM, MIL: Yes** 1996, 1997, 1998, 1999 Achieva, Beretta, Ciera, Corsica, Century, Grand Am, Grand Prix, Malibu, Skylark 3.1L VIN M Transmissions: All	**MAP Sensor Circuit Low Input Conditions:** DTC P0121, P012 and P0123 not set, engine started, system voltage over 9v, TP angle from 0-6% with engine speed under 1000 rpm or more than 10% with engine speed over 1000 rpm, and the PCM detected the MAP sensor was below 0.10v during the CCM test period. **Possible Causes** ● MAP sensor signal circuit shorted to sensor or chassis ground ● MAP sensor power circuit open between the sensor and PCM ● MAP sensor is damaged or has failed ● PCM has failed
P0107 **2T CCM, MIL: Yes** 1996, 1997, 1998, 1999, 2000, 2001, 2002, 2003, 2004, 2005 Aurora, Century, LeSabre, Park Avenue, Regal 3.1L VIN J, 3.1L VIN M, 3.4L VIN X, 3.5L V6 VIN H, 3.8L VIN 1, 3.8L VIN K, 4.0L V8 VIN C engines Transmissions: All	**MAP Sensor Circuit Low Input Conditions:** DTC P0121, P0122 and P0123 not set, system voltage at 8-18v, TP angle over 0% with engine speed under 1000 rpm or more than 10% with engine speed above 1000 rpm, and the PCM detected the MAP sensor was less than 0.10v (Scan Tool reads 12 kPa) for 3 seconds. **Possible Causes** ● MAP sensor signal circuit shorted to sensor ground ● MAP sensor power circuit open between the sensor and PCM ● MAP sensor is damaged or has failed ● PCM has failed

OBD II Trouble Code List (P0xxx Codes)

DTC	Trouble Code Title, Conditions & Possible Causes
P0107 **1T CCM, MIL: Yes** 1996, 1997 Camaro, Caprice, Corvette, Firebird, Fleetwood 4.3L VIN W, 5.7L VIN 5, 5.7L VIN P Transmissions: All	**MAP Sensor Circuit Low Input Conditions:** DTC P0122 and P0123 not set, engine started, throttle angle over 15%, and the PCM detected the MAP sensor signal was less than 0.24v for 4.5 seconds during the CCM test. **Possible Causes** ● MAP sensor signal circuit shorted to sensor or chassis ground ● MAP sensor power circuit open between the sensor and PCM ● MAP sensor is damaged or has failed ● PCM has failed
P0107 **2T CCM, MIL: Yes** 1996, 1997, 1998, 1999, 2000, 2001, 2002, 2003, 2004, 2005 Alero, Cutlass, Impala, Malibu, Montana, Venture Silhouette, Rendezvous 3.1L VIN J, 3.1L VIN M, 3.4L VIN E Transmissions: All	**MAP Sensor Circuit Low Input Conditions:** DTC P0121, P0122 and P0123 not set, system voltage at 8-18v, TP angle over 0% with engine speed under 1000 rpm or more than 10% with engine speed above 1000 rpm, and the PCM detected the MAP sensor was less than 0.10v (Scan Tool reads 12 kPa) for 3 seconds. The PCM supplies the MAP sensor with a 5v reference and a ground circuit. **Possible Causes** ● MAP sensor signal circuit shorted to sensor ground ● MAP sensor power circuit open between the sensor and PCM ● MAP sensor is damaged or has failed ● PCM has failed
P0107 **2T CCM, MIL: Yes** 1999, 2000, 2001, 2002, 2003, 2004, 2005 C/K Series Truck, G Vans Envoy & TrailBlazer 4.8L VIN V, 5.3L VIN P, 5.3L VIN T, 5.3L VIN Z, 6.0L VIN N, 6.0L VIN U, 8.1L VIN G Transmissions: All	**MAP Sensor Circuit Low Input Conditions:** DTC P0120, P0121, P0122, P0123, P0220, P1125, P1514, P1515, P1516, P1518, P2108, P2120, P2121, P2125, P2126, P2130, P2131 and P2135 not set, TP angle at 0% with engine speed under 800 rpm, or TP angle over 12.5% with engine speed over 800 rpm, and PCM detected the MAP sensor was under 0.10v for 4 seconds. **Possible Causes** ● MAP sensor signal circuit shorted to sensor or chassis ground ● MAP sensor power circuit open between the sensor and PCM ● MAP sensor is damaged or has failed ● PCM has failed
P0107 **2T CCM, MIL: Yes** 1996, 1997, 1998, 1999, 2000, 2001, 2002, 2003, 2004, 2005 C/K Cab & Chassis, C/K Series Truck, G Series Vans & L/M Series Vans 4.3L VIN W, 4.3L VIN X, 5.0L VIN M, 5.7L VIN K, 5.7L VIN R, 7.4L VIN J Transmissions: All	**MAP Sensor Circuit Low Input Conditions:** DTC P0121, P0122 and P0123 not set, engine running, system voltage 10-18v, TP angle over 0% with engine speed under 800 rpm, or TP angle over 12.5% with engine speed over 800 rpm, and the PCM detected the MAP sensor was less than 0.10v for 2 seconds. **Possible Causes** ● MAP sensor signal circuit shorted to sensor or chassis ground ● MAP sensor power circuit open between the sensor and PCM ● MAP sensor is damaged or has failed ● PCM has failed
P0107 **2T CCM, MIL: Yes** 1996, 1997, 1998, 1999, 2000, 2001, 2002, 2003, 2004, 2005 DeVille, Eldorado, Seville 4.6L VIN 9, 4.6L VIN Y Transmissions: All	**MAP Sensor Circuit Low Input Conditions:** DTC P0121-P0123 not set, TP angle at 0% with engine speed under 800 rpm or over 12.5% with engine speed over 800 rpm, and the PCM detected the MAP sensor was under 0.10v. **Possible Causes** ● MAP sensor signal circuit shorted to sensor or chassis ground ● MAP sensor power circuit open between the sensor and PCM ● MAP sensor is damaged or has failed ● PCM has failed
P0107 **2T CCM, MIL: Yes** 1995, 1996, 1997, 1998, 1999, 2000, 2001, 2002 Camaro & Firebird Camaro & Firebird 3.8L VIN K Transmissions: All	**MAP Sensor Circuit Low Input Conditions:** DTC P0606, P1120, P1125, P1220-P1221, P1271-P1276, P1280-P1286 and P1514-P1518 not set, TP angle at 0% with the engine speed under 1000 rpm or over 10% with the engine speed over 1000 rpm, and the PCM detected the MAP sensor was under 0.1v for 7 seconds. **Possible Causes** ● MAP sensor signal circuit shorted to sensor or chassis ground ● MAP sensor power circuit open between the sensor and PCM ● MAP sensor is damaged or has failed ● PCM has failed
P0107 **2T CCM, MIL: Yes** 1998, 1999, 2000, 2001, 2002 Camaro & Firebird 5.7L VIN G Transmissions: All	**MAP Sensor Circuit Low Input Conditions:** DTC P0121, P0122 and P0123 not set, TP angle 0% with engine speed under 800 rpm or over 12.5% with engine speed above 800 rpm, and the PCM detected the MAP sensor was under 0.10v for 4 seconds. **Possible Causes** ● MAP sensor signal circuit shorted to sensor or chassis ground ● MAP sensor power circuit open between the sensor and PCM ● MAP sensor is damaged or has failed ● PCM has failed

OBD II Trouble Code List (P0xxx Codes)

DTC	Trouble Code Title, Conditions & Possible Causes
P0107 **2T CCM, MIL: Yes** 1996, 1997, 1998, 1999, 2001, 2002, 2003, 2004, 2005 Aurora 3.5L V6 VIN H, 4.0L V8 VIN C engine Transmissions: A/T	**MAP Sensor Circuit Low Input Conditions:** DTC P0121, P0122 and P0123 not set, gear selector not in P/N, TP angle over 0% with the engine speed below 1000 rpm, or the TP angle over 9% with engine speed over 1000 rpm, and the PCM detected the MAP sensor was less than 0.10v for 10 seconds. **Possible Causes** ● MAP sensor signal circuit shorted to sensor or chassis ground ● MAP sensor power circuit open between the sensor and PCM ● MAP sensor is damaged or has failed ● PCM has failed
P0107 **2T CCM, MIL: Yes** 1995, 1996, 1997, 1998, 1999, 2000, 2001, 2002, 2003, 2004, 2005 S/T Blazer & S/T Pickup 4.3L VIN W, 4.3L VIN X Transmissions: All	**MAP Sensor Circuit Low Input Conditions:** DTC P0121-P0123 not set, TP angle over 0% with engine speed below 800 rpm, or TP angle over 12.5% with engine speed over 800 rpm, and the PCM detected the MAP sensor was less than 0.10v. **Possible Causes** ● MAP sensor signal circuit shorted to sensor or chassis ground ● MAP sensor power circuit open between the sensor and PCM ● MAP sensor is damaged or has failed ● PCM has failed
P0107 **2T CCM, MIL: Yes** 1996, 1997 Grand Prix, Lumina & Monte Carlo 3.4L VIN X engine Transmissions: All	**MAP Sensor Circuit High Input Conditions:** DTC P0121, P0122 and P0123 not set, engine runtime from 1-2 minutes (depends upon the ECT sensor at startup), system voltage from 10-18v, TP angle from 1-2% with engine speed under 1000 rpm, TP angle over 6% with engine speed over 1000 rpm, and the PCM detected the MAP sensor was less than 0.10v for 200 3X reference periods (67 crankshaft revolutions). **Possible Causes** ● MAP sensor signal circuit shorted to sensor or chassis ground ● MAP sensor power circuit open between the sensor and PCM ● MAP sensor is damaged or has failed ● PCM has failed
P0107 **2T CCM, MIL: Yes** 1997, 1998, 1999, 2002, 2001, 2002, 2003, 2004, 2005 Corvette 5.7L VIN G, 5.7L VIN S Transmissions: All	**MAP Sensor Circuit Low Input Conditions:** DTC P1120, P1125, P1220, P1221, P1275, P1276, P1280, P1281, P1285, P1286, P1514, P1515, P1516, P1517 and P1518 not set, engine started, TP angle at 0% with engine speed under 800 rpm, or TP angle over 12.5% with engine speed more than 800 rpm, and the PCM detected the MAP sensor was less than 0.10v for 4 seconds. **Possible Causes** ● MAP sensor signal circuit shorted to sensor or chassis ground ● MAP sensor power circuit open between the sensor and PCM ● MAP sensor is damaged or has failed ● PCM has failed
P0108 **2T CCM, MIL: Yes** 1996, 1997, 1998, 1999, 2000, 2001, 2002 Achieva, Alero, Beretta, Century, Ciera, Cavalier, Envoy, Grand Am, Sunfire Skyhawk, TrailBlazer, S/T Blazer & S/T Pickup 2.2L VIN 4, 2.2L VIN 5, 2.2L VIN F, 2.2L VIN H, 2.4L VIN T, 4.2L VIN S Transmissions: All	**MAP Sensor Circuit High Input Conditions:** DTC P0122 and P0123 not set, engine runtime 20-40 seconds, throttle angle under 12%, VSS signal under 1 mph, and the PCM detected the MAP sensor was more than 4.20v for 1.25 seconds. **Possible Causes** ● MAP sensor signal circuit is open or it is shorted to VREF ● MAP sensor ground circuit open between sensor and the PCM ● MAP sensor is damaged or has failed ● PCM has failed ● This code can set due to a backfire or engine cranking too long
P0108 **2T CCM, MIL: Yes** 1996, 1997, 1998, 1999 Achieva, Beretta, Ciera, Corsica, Century, Grand Am, Grand Prix, Malibu, Skylark 3.1L VIN M Transmissions: All	**MAP Sensor Circuit High Input Conditions:** DTC P0121-P0123 not set, engine runtime at 1-2 minutes, system voltage 10-18v, TP angle under 2% with engine speed under 3000 rpm, or TP angle over 30% with engine speed over 3000 rpm, VSS under 1 mph, and PCM detected the MAP sensor was over 3.8-4.2v. **Possible Causes** ● MAP sensor signal circuit is open or it is shorted to VREF ● MAP sensor ground circuit open between sensor and the PCM ● MAP sensor is damaged or has failed ● This code can set due to a backfire or engine cranking too long ● PCM has failed

OBD II Trouble Code List (P0xxx Codes)

DTC	Trouble Code Title, Conditions & Possible Causes
P0108 **2T CCM, MIL: Yes** 1996, 1997, 1998, 1999, 2000, 2001, 2002, 2003, 2004, 2005 Aurora, Century, LeSabre, Park Avenue, Regal 3.1L VIN J, 3.1L VIN M, 3.4L VIN X, 3.5L V6 VIN H, 3.8L VIN 1, 3.8L VIN K, 4.0L V8 VIN C Transmissions: All	**MAP Sensor Circuit High Input Conditions:** DTC P0121, P0122 or P0123 not set, engine runtime 1-2 minutes, VSS signal less than 1 mph, TP angle below 2% with engine speed under 1500 rpm or above 10% with engine speed over 1500 rpm, and the PCM detected the MAP sensor was over 4.3v for 3 seconds. The PCM supplies the MAP sensor with a 5v reference circuit, a ground circuit and in turn, the MAP sensor sends the PCM a MAP sensor signal that is relative to pressure changes in the manifold. The MAP sensor voltage should be low with low MAP (idle speed or deceleration), or a high voltage with high MAP (KOEO or at WOT). **Possible Causes** ● MAP sensor signal circuit is open or it is shorted to VREF ● MAP sensor ground circuit open between sensor and the PCM ● MAP sensor is damaged or has failed ● PCM has failed
P0108 **1T CCM, MIL: Yes** 1996, 1997 Caprice, Camaro, Corvette & Firebird, Fleetwood, Brougham 4.3L VIN W, 5.7L VIN 5, 5.7L VIN P Transmissions: All	**MAP Sensor Circuit High Input Conditions:** DTC P0121, P0122 and P0123 not set, throttle angle less than 5% with engine speed less than 1000 rpm or throttle angle less than 18% with engine speed more than 1000 rpm, and the PCM detected the MAP sensor was more than 4.30v for 4 seconds. **Possible Causes** ● MAP sensor signal circuit is open or it is shorted to VREF ● MAP sensor ground circuit open between sensor and the PCM ● MAP sensor is damaged or has failed ● PCM has failed
P0108 **2T CCM, MIL: Yes** 1996, 1997, 1998, 1999, 2000, 2001, 2002, 2003, 2004, 2005 Alero, Cutlass, Impala, Malibu, Montana, Venture Silhouette, Rendezvous 3.1L VIN J, 3.1L VIN M, 3.4L VIN E Transmissions: All	**MAP Sensor Circuit High Input Conditions:** DTC P0121-P0123 not set, engine runtime 1-2 minutes (depends on the ECT sensor at startup), TP angle below 2% with engine speed under 3000 rpm or TP angle over 30% with engine speed over 3000 rpm, and the PCM detected the MAP sensor was over 4.30v for 3 seconds. The MAP sensor signal is relative to the pressure changes in the manifold. The MAP signal voltage is with low MAP (idle speed or deceleration), and a high voltage with high MAP (KOEO & WOT). **Possible Causes** ● MAP sensor signal circuit is open or it is shorted to VREF ● MAP sensor ground circuit open between sensor and the PCM ● MAP sensor is damaged or has failed ● PCM has failed ● This code can set due to a backfire or engine cranking too long
P0108 **2T CCM, MIL: Yes** 1996, 1997, 1998, 1999, 2000, 2001, 2002, 2003, 2004, 2005 C/K Cab & Chassis, C/K Series Truck, G Series 4.3L VIN W, 4.3L VIN X, 5.0L VIN M, 5.7L VIN K, 5.7L VIN R, 7.4L VIN J Transmissions: All	**MAP Sensor Circuit High Input Conditions:** DTC P012, P0122 and P0123 not set, engine started, system voltage 10-18v, TP angle over 0% with the engine speed under 1200 rpm (under 600 rpm on 7.4L VIN J), or TP angle less than 20% with engine speed over 1200 rpm (over 600 rpm on 7.4L VIN J), and the PCM detected the MAP sensor was more than 4.40v for 2 seconds. **Possible Causes** ● MAP sensor signal circuit is open or it is shorted to VREF ● MAP sensor ground circuit open between sensor and the PCM ● MAP sensor is damaged or has failed ● This code can set due to a backfire or engine cranking too long ● PCM has failed
P0108 **2T CCM, MIL: Yes** 1999, 2000, 2001, 2002, 2003, 2004, 2005 C/K Series Truck, G Vans Envoy & TrailBlazer 4.8L VIN V, 5.3L VIN P, 5.3L VIN T, 5.3L VIN Z, 6.0L VIN N, 6.0L VIN U, 8.1L VIN G Transmissions: All	**MAP Sensor Circuit High Input Conditions:** DTC P0120, P0121, P0122, P0123, P0220, P1125, P1514, P1515, P1516, P1518, P2108, P2120, P2121, P2125, P2126, P2130, P2131 and P2135 not set, engine started, TP angle under 1% with engine speed under 1200 rpm, or TP angle under 20% with engine speed over 1200 rpm, and the PCM detected the MAP sensor was over 4.90v for 4 seconds. **Possible Causes** ● MAP sensor signal circuit is open or it is shorted to VREF ● MAP sensor ground circuit open between sensor and the PCM ● MAP sensor is damaged or has failed ● PCM has failed
P0108 **2T CCM, MIL: Yes** 1996, 1997, 1998, 1999, 2000, 2001, 2002, 2003, 2004, 2005 DeVille, Eldorado, Seville 4.6L VIN 9, 4.6L VIN Y Transmissions: All	**MAP Sensor Circuit High Input Conditions:** DTC P0121-P0123 not set, TP angle at 0% with engine speed under 1200 rpm or TP angle over 20% with engine speed over 1200 rpm, and the PCM detected the MAP sensor was over 4.9v for 4 seconds. **Possible Causes** ● MAP sensor signal circuit is open or it is shorted to VREF ● MAP sensor ground circuit open between sensor and the PCM ● MAP sensor is damaged or has failed ● PCM has failed

OBD II Trouble Code List (P0xxx Codes)

DTC	Trouble Code Title, Conditions & Possible Causes
P0108 **2T CCM, MIL: Yes** 1995, 1996, 1997, 1998, 1999, 2000, 2001, 2002 Camaro & Firebird 3.8L VIN K engine Transmissions: All	**MAP Sensor Circuit High Input Conditions:** DTC P0606, P1120, P1125, P1220, P1221, P1271-P1276, P1280-P1286 and P1514-P1517 not set, engine runtime 2 minutes with startup ECT input below -22ºF to 1 second over 86ºF, TP angle under 0.5% with engine speed under 900 rpm, and PCM detected the MAP sensor was over 4.3v for 7 seconds. This code can set due to a backfire or engine cranking too long. **Possible Causes** ● MAP sensor signal circuit is open or it is shorted to VREF ● MAP sensor ground circuit open between sensor and the PCM ● MAP sensor is damaged or has failed ● PCM has failed
P0108 **2T CCM, MIL: Yes** 1998, 1999, 2000, 2001, 2002 Camaro & Firebird 5.7L VIN G engine Transmissions: All	**MAP Sensor Circuit High Input Conditions:** DTC P0121-P0123 not set, engine runtime determined by the ECT sensor at startup (it ranges from 4 minutes at below -22ºF to under 30 seconds at above 86ºF), TP angle 0% with engine speed under 1200 rpm or less than 12.5% with engine speed over 1200 rpm, and the PCM detected the MAP sensor was under 0.10v for 4 seconds. **Possible Causes** ● MAP sensor signal circuit is open or it is shorted to VREF ● MAP sensor ground circuit open between sensor and the PCM ● MAP sensor is damaged or has failed ● PCM has failed ● This code can set due to a backfire or engine cranking too long
P0108 **2T CCM, MIL: Yes** 2001, 2002 Aurora 3.5L V6 VIN H engine Transmissions: A/T	**MAP Sensor Circuit High Input Conditions:** DTC P0121, P0122, P0123 not set, engine runtime from 1-2 minutes (depends on ECT sensor signal at startup), TP angle under 2% with the engine speed under 3000 rpm, and PCM detected the MAP sensor signal indicated more than 4.20v for 10 seconds. **Possible Causes** ● MAP sensor signal circuit is open or it is shorted to VREF ● MAP sensor ground circuit open between sensor and the PCM ● MAP sensor is damaged or has failed ● PCM has failed
P0108 **2T CCM, MIL: Yes** 1996, 1997, 1998, 1999, 2001, 2002, 2003, 2004, 2005 Aurora 4.0L V8 VIN C engine Transmissions: A/T	**MAP Sensor Circuit High Input Conditions:** DTC P0122 and P0123 not set, engine started, TP angle less near 0% with the engine speed under 1000 rpm, or the TP angle less than 98% with the engine speed more than 1000 rpm, and the PCM detected the MAP sensor signal was more than 4.50 for 2 seconds. **Possible Causes** ● MAP sensor signal circuit is open or it is shorted to VREF ● MAP sensor ground circuit open between sensor and the PCM ● MAP sensor is damaged or has failed ● PCM has failed ● This code can set due to a backfire or engine cranking too long
P0108 **2T CCM, MIL: Yes** 1995, 1996, 1997, 1998, 1999, 2000, 2001, 2002, 2003, 2004, 2005 S/T Blazer & S/T Pickup 4.3L VIN W, 4.3L VIN X Transmissions: All	**MAP Sensor Circuit High Input Conditions:** DTC P0121-P0123 not set, TP angle at 0% with engine speed under 1200 rpm, or under 20% and engine speed over 1200 rpm, and the PCM detected the MAP sensor was over 4.9v. **Possible Causes** ● MAP sensor signal circuit is open or it is shorted to VREF ● MAP sensor ground circuit open between sensor and the PCM ● MAP sensor is damaged or has failed ● PCM has failed
P0108 **2T CCM, MIL: Yes** 1996, 1997 Grand Prix, Lumina, Monte Carlo 3.4L VIN X engine Transmissions: All	**MAP Sensor Circuit High Input Conditions:** DTC P0121, P0122 and P0123 not set, engine runtime from 1-2 minutes, system voltage at 10-18v, TP angle under 2% with engine speed under 900 rpm, and the PCM detected the MAP sensor was more than 4.2v for 200 3X reference periods (67 crankshaft revolutions). **Possible Causes** ● MAP sensor signal circuit is open or it is shorted to VREF ● MAP sensor ground circuit open between sensor and the PCM ● MAP sensor is damaged or has failed ● PCM has failed
P0108 **2T CCM, MIL: Yes** 1997, 1998, 1999, 2002, 2001, 2002, 2003, 2004, 2005 Corvette 5.7L VIN G, 5.7L VIN S Transmissions: All Corvette	**MAP Sensor Circuit High Input Conditions:** DTC P1120, P1125, P1220, P1221, P1275, P1276, P1280-P1286, P1514, P1515-P1517 and P1518 not set, TP angle below 5% with engine speed less than 1000 rpm or over 18% with the speed over 1000 rpm, and the PCM detected the MAP sensor over 4.90v for 4 seconds. **Possible Causes** ● MAP sensor signal circuit is open or it is shorted to VREF ● MAP sensor ground circuit open between sensor and the PCM ● MAP sensor is damaged or has failed ● PCM has failed ● This code can set due to a backfire or engine cranking too long

OBD II Trouble Code List (P0xxx Codes)

DTC	Trouble Code Title, Conditions & Possible Causes
P0110 **2T CCM, MIL: Yes** 1997, 1998, 1999, 2000 Catera 3.0L V6 VIN R engine Transmissions: A/T	**IAT Sensor Circuit Malfunction Conditions:** Engine started; engine runtime over 3 minutes and the PCM detected the IAT sensor was more than 282ºF, or the IAT sensor signal was less than -40ºF for 10 seconds at idle speed. **Possible Causes** ● IAT sensor signal circuit is open between the sensor and PCM ● IAT sensor signal circuit is shorted to sensor or chassis ground ● IAT sensor signal circuit is shorted to VREF ● IAT ground circuit is open between the sensor and the PCM ● IAT sensor is damaged or has failed ● PCM has failed
P0111 **2T CCM, MIL: Yes** 1996, 1997, 1998, 1999 Aurora, DeVille, Eldorado, Seville 4.0L V8 VIN C, 4.6L VIN 9, 4.6L VIN Y Transmissions: All	**IAT Sensor Signal Range/Performance Conditions:** Engine started; engine runtime for over 2 seconds, and the PCM detected the IAT sensor signal changed 0.30v in 250 milliseconds. **Possible Causes** ● IAT sensor signal circuit is open (intermittent fault) ● IAT sensor signal circuit is shorted to ground (intermittent fault) ● IAT sensor is contaminated, damaged or has failed ● PCM has failed
P0112 **2T CCM, MIL: Yes** 1996, 1997, 1998, 1999, 2000, 2001, 2002, 2003, 2004, 2005 Achieva, Alero, Beretta, Century, Ciera, Cavalier, Envoy, Grand Am, Sunfire Skyhawk, TrailBlazer, S/T Blazer & S/T Pickup 2.2L VIN 4, 2.2L VIN 5, 2.2L VIN F, 2.2L VIN H, 2.4L VIN T, 4.2L VIN S Transmissions: All	**IAT Sensor Circuit Low Input Conditions:** DTC P0117, P0118, P0125, P0502 and P0503 not set, engine started, engine runtime over 320 seconds, VSS more than 15 mph, and the PCM detected the IAT sensor indicated more than 262ºF for 3.25 seconds. The IAT sensor is a variable resistor that includes an IAT signal circuit and a low reference circuit to measure the temperature of the air entering the engine. The PCM supplies the sensor with a 5v signal circuit and a low reference ground circuit. When the IAT sensor is cold, its resistance is high. When the air temperature increases, its resistance decreases. **Possible Causes** ● IAT sensor signal circuit is shorted to sensor or chassis ground ● IAT sensor is damaged or has failed (it may be shorted) ● PCM has failed
P0112 **2T CCM, MIL: Yes** 1996, 1997, 1998, 1999 Achieva, Beretta, Ciera, Corsica, Century, Grand Am, Grand Prix, Malibu, Skylark 3.1L VIN M, 3.4L VIN X Transmissions: All	**IAT Sensor Circuit Low Input Conditions:** DTC P0100, P0102, P0103, P0117, P0118, P0502 and P0503 not set, engine started, engine runtime over 10 seconds, VSS more than 25 mph, and the PCM detected the IAT sensor indicated more than 275ºF for 18-20 seconds during the CCM test. **Possible Causes** ● IAT sensor signal circuit is shorted to sensor or chassis ground ● IAT sensor is damaged or has failed (it may be shorted) ● PCM has failed ● TSB 02-06-03-005 contains a repair procedure for this code
P0112 **1T CCM, MIL: Yes** 1995 Century, Cutlass Ciera, Cutlass Supreme, Grand Prix, Lumina, Monte Carlo Park Avenue Regal, Riviera, 88' & 98' 3.1L VIN M, 3.4L VIN X, 3.8L VIN 1, 3.8L VIN K, 3.8L VIN L engines Transmissions: All	**Intake Air Temperature Sensor Circuit Low Temperature Conditions:** Engine started, MAF sensor less than 12 g/sec, ECT sensor over 140ºF, VSS less than 25 mph, and the PCM detected the IAT sensor indicated less than -29ºF for 5 seconds. **Possible Causes** ● IAT sensor signal circuit is open between the sensor and PCM ● IAT sensor signal circuit is shorted to VREF or system power ● IAT sensor is damaged or has failed (it may be open) ● PCM has failed
P0112 **2T CCM, MIL: Yes** 1996, 1997, 1998, 1999, 2000, 2001, 2002, 2003, 2004, 2005 Aurora, Century, LeSabre, Park Avenue, Regal 3.1L VIN J, 3.1L VIN M, 3.4L VIN X, 3.5L V6 VIN H, 3.8L VIN 1, 3.8L VIN K, 4.0L V8 VIN C Transmissions: All	**IAT Sensor Circuit Low Input Conditions:** DTC P0101, P0102, P0103, P0116, P0117, P0118, P0125, P0128, P0502 and P0503 not set, engine started, engine runtime over 10 seconds, VSS more than 25 mph, and the PCM detected the IAT sensor was less than 0.10v (Scan Tool reads over 253ºF) for 20 seconds. The IAT sensor is a variable resistor that includes an IAT signal circuit and a low reference circuit to measure the temperature of the air entering the engine. The PCM supplies the sensor with a 5v signal circuit and a low reference ground circuit. When the IAT sensor is cold, its resistance is high. When the air temperature increases, its resistance decreases. With high sensor resistance, the IAT sensor signal voltage is high. With lower sensor resistance, the IAT sensor signal voltage should be a lower voltage. **Possible Causes** ● IAT sensor signal circuit is shorted to sensor or chassis ground ● IAT sensor is damaged or has failed (it may be shorted) ● PCM has failed

OBD II Trouble Code List (P0xxx Codes)

DTC	Trouble Code Title, Conditions & Possible Causes
P0112 **1T CCM, MIL: Yes** 1996, 1997 Camaro, Caprice, Corvette, Firebird, Fleetwood 4.3L VIN W, 5.7L VIN 5, 5.7L VIN P Transmissions: All	**IAT Sensor Circuit Low Input Conditions:** DTC P0100, P0102, P0103, P0117, P0118, P0502 and P0503 not set, engine started, engine runtime over 30 seconds, VSS more than 25 mph, and the PCM detected the IAT sensor indicated more than 302°F for 4.5 seconds. The IAT sensor is a variable resistor with both an IAT signal and a low reference circuit. It measures the temperature of the air entering the engine. When the IAT sensor is cold, its resistance is high. When the air temperature increases, its resistance decreases. With high sensor resistance, the IAT sensor voltage is high. With lower sensor resistance, the IAT sensor signal voltage should be a lower voltage. **Possible Causes** ● IAT sensor signal circuit is shorted to sensor or chassis ground ● IAT sensor is damaged or has failed (it may be shorted) ● PCM has failed
P0112 **2T CCM, MIL: Yes** 1996, 1997, 1998, 1999, 2000, 2001, 2002, 2003, 2004, 2005 Alero, Cutlass, Impala, Malibu, Montana, Venture Silhouette, Rendezvous 3.1L VIN J, 3.1L VIN M, 3.4L VIN E engines Transmissions: All	**IAT Sensor Circuit Low Input Conditions:** DTC P0101, P0102, P0103, P0116, P0117, P0118, P0125, P0128, P0502 and P0503 not set, engine started, engine runtime over 10 seconds, VSS more than 25 mph, and the PCM detected the IAT sensor was more than 253-275°F for 20 seconds in the CCM test. **Possible Causes** ● IAT sensor signal circuit is shorted to sensor or chassis ground ● IAT sensor is damaged or has failed (it may be shorted) ● PCM has failed ● TSB 02-06-03-005 contains a repair procedure for this code
P0112 **2T CCM, MIL: Yes** 1996, 1997, 1998, 1999, 2000, 2001, 2002, 2003, 2004, 2005 C/K Cab & Chassis, C/K Series Truck, G Series 4.3L VIN W, 4.3L VIN X, 5.0L VIN M, 5.7L VIN K, 5.7L VIN R, 7.4L VIN J Transmissions: All	**IAT Sensor Circuit Low Input Conditions:** DTC P0522 and P0523 not set, engine started, engine runtime over 45 seconds, VSS more than 2-40 mph, and the PCM detected the IAT sensor was more than 262-282°F for 1 second. The IAT sensor is a variable resistor that includes a signal circuit and low reference circuit to measure the temperature of the air entering the engine. **Possible Causes** ● IAT sensor signal circuit is shorted to sensor or chassis ground ● IAT sensor is damaged or has failed (it may be shorted) ● PCM has failed
P0112 **2T CCM, MIL: Yes** 1999, 2000, 2001, 2002, 2003, 2004, 2005 C/K Series Truck, G Vans Envoy & TrailBlazer 4.8L VIN V, 5.3L VIN P, 5.3L VIN T, 5.3L VIN Z, 6.0L VIN N, 6.0L VIN U, 8.1L VIN G Transmissions: All	**IAT Sensor Circuit Low Input Conditions:** DTC P0502 and P0503 not set, engine runtime more than 45 seconds, VSS more than 25 mph, and the PCM detected the IAT sensor was more than 262°F for 5 seconds. The IAT sensor is a variable resistor that includes a signal circuit and a low reference circuit to measure the temperature of the air entering the engine. **Possible Causes** ● IAT sensor signal circuit is shorted to sensor or chassis ground ● IAT sensor is damaged or has failed (it may be shorted) ● PCM has failed
P0112 **2T CCM, MIL: Yes** 1996, 1997, 1998, 1999, 2000, 2001, 2002, 2003, 2004, 2005 C/K Series Truck, G Vans 6.5L VIN F, 6.5L VIN S, 6.6L VIN 1 Diesel engines Transmissions: All	**IAT Sensor Circuit Low Input Conditions:** Key on or engine running, ECT sensor less than 109°F, and the PCM detected the IAT sensor was more than 304°F for 2 seconds. The IAT sensor is a variable resistor that has both an IAT signal and a low reference circuit to measure the temperature of incoming air. **Possible Causes** ● IAT sensor signal circuit is shorted to sensor or chassis ground ● IAT sensor is damaged or has failed (it may be shorted) ● PCM has failed
P0112 **2T CCM, MIL: Yes** 2003, 2004, 2005 CTS (Cadillac) 2.6L V6 VIN M & 3.2L V6 VIN N engines Transmissions: All	**IAT Sensor Circuit Low Input Conditions:** Engine started; engine runtime over 1 minute and 20 seconds, engine at idle speed for over 10 seconds, and the PCM detected the IAT sensor was more than 282°F for 2 seconds during the CCM test. **Possible Causes** ● IAT sensor signal circuit is shorted to sensor or chassis ground ● IAT sensor is damaged or has failed (it may be shorted) ● PCM has failed
P0112 **2T CCM, MIL: Yes** 1996, 1997, 1998, 1999, 2000, 2001, 2002, 2003, 2004, 2005 DeVille, Eldorado, Seville 4.6L VIN 9, 4.6L VIN Y Transmissions: A/T	**IAT Sensor Circuit Low Input Conditions:** DTC P0101, P0102, P0103, P0116, P0117, P0118, P0125, P0128, P0502 and P0503 not set, engine started, engine runtime over 10 seconds, ECT sensor less than 212°F, VSS over 25 mph, and the PCM detected the IAT sensor was less than 0.08v (Scan Tool reads over 278°F) for 20 seconds during the CCM continuous test. **Possible Causes** ● IAT sensor signal circuit is shorted to sensor or chassis ground ● IAT sensor is damaged or has failed (it may be shorted) ● PCM has failed

OBD II Trouble Code List (P0xxx Codes)

DTC	Trouble Code Title, Conditions & Possible Causes
P0112 **2T CCM, MIL: Yes** 1995, 1996, 1997, 1998, 1999, 2000, 2001, 2002 Camaro & Firebird 3.8L VIN K, 5.7L VIN G Transmissions: All	**IAT Sensor Circuit Low Input Conditions:** DTC P0101, P0102, P0103, P0116, P0117, P0118, P0125, P0128, P0500, P0502 and P0503 not set, engine started, engine runtime from 10-45 seconds, VSS more than 25 mph, and the PCM detected the IAT sensor was from 253-282°F for 3-5 seconds in the CCM test. **Possible Causes** ● IAT sensor signal circuit is shorted to sensor or chassis ground ● IAT sensor is damaged or has failed (it may be shorted) ● PCM has failed
P0112 **2T CCM, MIL: Yes** 2001, 2002 Aurora 3.5L V6 VIN H engine Transmissions: A/T	**IAT Sensor Circuit Low Input Conditions:** DTC P0101, P0102, P0103, P0116, P0117, P0118, P0125, P0128, P0502 and P0503 not set, engine started, engine runtime over 10 seconds, VSS more than 25 mph, and the PCM detected the IAT sensor was less than 0.10v (Scan Tool reads more than 282°F). **Possible Causes** ● IAT sensor signal circuit is shorted to sensor or chassis ground ● IAT sensor is damaged or has failed (it may be shorted) ● PCM has failed
P0112 **2T CCM, MIL: Yes** 1996, 1997, 1998, 1999, 2001, 2002, 2003, 2004, 2005 Aurora 4.0L V8 VIN C Transmissions: A/T	**IAT Sensor Circuit Low Input Conditions:** DTC P0101, P0102, P0103, P0116, P0117, P0118, P0125, P0128, P0502 and P0503 not set, engine runtime over 10 seconds, ECT sensor below 212°F, VSS over 9 mph, and PCM detected the IAT sensor was under 0.10v (Scan Tool reads over 282°F) for 1 second. **Possible Causes** ● IAT sensor signal circuit is shorted to sensor or chassis ground ● IAT sensor is damaged or has failed (it may be shorted) ● PCM has failed
P0112 **2T CCM, MIL: Yes** 1995, 1996, 1997, 1998, 1999, 2000, 2001, 2002, 2003, 2004, 2005 S/T Blazer & S/T Pickup 4.3L VIN W, 4.3L VIN X Transmissions: All	**IAT Sensor Circuit Low Input Conditions:** DTC P0502 and P0503 not set, engine started; engine runtime over 45 seconds, VSS over 25 mph, and the PCM detected the IAT sensor indicated more than 262°F for over 5 seconds. **Possible Causes** ● IAT sensor connector is damaged or shorted ● IAT sensor signal circuit is shorted to sensor or chassis ground ● IAT sensor is damaged or has failed (it may be shorted) ● PCM has failed
P0112 **2T CCM, MIL: Yes** 1996, 1997 Grand Prix, Lumina, Monte Carlo 3.4L VIN X Transmissions: All	**IAT Sensor Circuit Low Input Conditions:** DTC P0100, P0102, P0103, P0117, P0118, P0502 and P0503 not set, engine running, ECT sensor more than 140°F, MAF sensor less than 12 g/sec, VSS less than 35 mph, and the PCM detected the IAT sensor was less than 0.10v (Scan Tool reads over 262°F). **Possible Causes** ● IAT sensor connector is damaged or shorted ● IAT sensor signal circuit is shorted to sensor or chassis ground ● IAT sensor is damaged or has failed (it may be shorted) ● PCM has failed
P0112 **2T CCM, MIL: Yes** 1997, 1998, 1999, 2002, 2001, 2002, 2003, 2004, 2005 Corvette 5.7L VIN G, 5.7L VIN S Transmissions: All	**IAT Sensor Circuit Low Input Conditions:** DTC P0101, P0102, P0103, P0117, P0118, P0125, P0500, P0502, P0503 and P1258 not set, engine started, engine runtime over 45 seconds, VSS more than 25 mph, and the PCM detected the IAT sensor was more than 262°F for 5 seconds during the CCM test. **Possible Causes** ● IAT sensor signal circuit is shorted to sensor or chassis ground ● IAT sensor is damaged or has failed (it may be shorted) ● PCM has failed
P0113 **2T CCM, MIL: Yes** 1996, 1997, 1998, 1999, 2000, 2001, 2002, 2003, 2004, 2005 Achieva, Alero, Beretta, Century, Ciera, Cavalier, Envoy, Grand Am, Sunfire Skyhawk, TrailBlazer, S/T Blazer & S/T Pickup 2.2L VIN 4, 2.2L VIN 5, 2.2L VIN F, 2.2L VIN H, 2.4L VIN T, 4.2L VIN S Transmissions: All	**IAT Sensor Circuit High Input Conditions:** DTC P0117, P0118, P0125, P0502 or P0503 are not set, engine runtime over 320 seconds, VSS less than 15 mph, and the PCM detected the IAT sensor was less than -38°F for 3.25 seconds. **Possible Causes** ● IAT sensor signal circuit is open between the sensor and PCM ● IAT sensor signal circuit is shorted to VREF or system power ● IAT sensor is damaged or has failed (it may be open) ● PCM has failed ● TSB 02-06-03-005 contains a repair procedure for this code

OBD II Trouble Code List (P0xxx Codes)

DTC	Trouble Code Title, Conditions & Possible Causes
P0113 **2T CCM, MIL: Yes** 1996, 1997, 1998, 1999 Achieva, Beretta, Ciera, Corsica, Century, Grand Am, Grand Prix, Malibu, Skylark 3.1L VIN M Transmissions: All	**IAT Sensor Circuit High Input Conditions:** DTC P0100, P0102, P0103, P0117, P0118, P0502 and P0503 not set, engine started, ECT sensor more than 140ºF, MAF sensor less than 12 g/sec, VSS less than 35 mph, and the PCM detected the IAT sensor was less than -35ºF for 20 seconds during the CCM test. **Possible Causes** ● IAT sensor signal circuit is open between the sensor and PCM ● IAT sensor signal circuit is shorted to VREF or system power ● IAT sensor is damaged or has failed (it may be open) ● PCM has failed ● TSB 02-06-03-005 contains a repair procedure for this code
P0113 **1T CCM, MIL: Yes** 1995 Century, Cutlass Ciera, Cutlass Supreme, Grand Prix, Lumina, Monte Carlo Park Avenue Regal, Riviera, 88' & 98' 3.1L VIN M, 3.4L VIN X, 3.8L VIN 1, 3.8L VIN K, 3.8L VIN L engines Transmissions: All	**Intake Air Temperature Sensor Circuit High Temperature Conditions:** Engine started vehicle drive at over 35 mph and the PCM detected the IAT sensor was more than 284ºF for 5 seconds. The IAT sensor is a variable resistor. The PCM connects to the IAT signal with a signal and low reference circuit to measure the temperature of air that enters the engine. When the IAT sensor is cold, its resistance is high. When the air temperature increases, its resistance decreases. **Possible Causes** ● IAT sensor signal circuit is shorted to sensor or chassis ground ● IAT sensor is damaged or has failed (it may be shorted) ● PCM has failed
P0113 **2T CCM, MIL: Yes** 1996, 1997, 1998, 1999, 2000, 2001, 2002, 2003, 2004, 2005 Aurora, Century, LeSabre, Park Avenue, Regal 3.1L VIN J, 3.1L VIN M, 3.4L VIN X, 3.5L V6 VIN H, 3.8L VIN 1, 3.8L VIN K, 4.0L V8 VIN C Transmissions: All	**IAT Sensor Circuit High Input Conditions:** DTC P0116, P0117, P0118, P0125, P0128, P0502, or P0503 not set, engine runtime over 3 minutes, ECT sensor more than 140ºF, MAF sensor less than 8 g/sec, VSS under 35 mph, and the PCM detected the IAT sensor was less than -38ºF for 3 seconds. The IAT sensor is a variable resistor that is used to measure the temperature of the air entering the engine. When the IAT sensor is cold, its resistance is high. With high sensor resistance, the IAT sensor voltage is high. With lower sensor resistance, the IAT sensor signal will be a lower voltage. **Possible Causes** ● IAT sensor signal circuit is open between the sensor and PCM ● IAT sensor signal circuit is shorted to VREF or system power ● IAT sensor is damaged or has failed (it may be open) ● PCM has failed
P0113 **2T CCM, MIL: Yes** 1996, 1997 Camaro, Corvette, Caprice, Firebird & Fleetwood 4.3L VIN W, 5.7L VIN 5, 5.7L VIN P Transmissions: All	**IAT Sensor Circuit High Input Conditions:** DTC P0101-P0103, P0117, P0118, P0125, P0500-P0503, P1258 not set, engine started, engine runtime over 100 seconds, ECT sensor over 32ºF, VSS below 7 mph, MAF sensor less than 15 g/sec, and the PCM detected the IAT sensor was below -31ºF for 4.5 seconds. **Possible Causes** ● IAT sensor signal circuit is open between the sensor and PCM ● IAT sensor signal circuit is shorted to VREF or system power ● IAT sensor is damaged or has failed (it may be open) ● PCM has failed
P0113 **2T CCM, MIL: Yes** 1996, 1997, 1998, 1999, 2000, 2001, 2002, 2003, 2004, 2005 Alero, Cutlass, Impala, Malibu, Montana, Venture Silhouette, Rendezvous 3.1L VIN J, 3.1L VIN M, 3.4L VIN E engines Transmissions: All	**IAT Sensor Circuit High Input Conditions:** DTC P0116-P0118, P0125, P0128, P0502 and P0503 not set, engine started, engine runtime over 180 seconds, ECT sensor more than 140ºF, MAF sensor less than 12 g/sec, VSS less than 35 mph, and the PCM detected the IAT sensor indicated less than -38ºF for a period of 3-20 seconds. The IAT sensor is a variable resistor that includes an IAT signal circuit and a low reference circuit to measure the temperature of the air entering the engine. When the IAT sensor is cold, its resistance is high. When the air temperature increases, its resistance decreases. With high sensor resistance, the IAT sensor signal voltage is high. With lower sensor resistance, the IAT sensor signal voltage should be a lower voltage. **Possible Causes** ● IAT sensor signal circuit is open between the sensor and PCM ● IAT sensor signal circuit is shorted to VREF or system power ● IAT sensor is damaged or has failed (it may be open) ● PCM has failed ● TSB 02-06-03-005 contains a repair procedure for this code
P0113 **2T CCM, MIL: Yes** 1996, 1997, 1998, 1999, 2000, 2001, 2002, 2003, 2004, 2005 C/K Cab & Chassis, C/K Series Truck, G Series 4.3L VIN W, 4.3L VIN X, 5.0L VIN M, 5.7L VIN K, 5.7L VIN R, 7.4L VIN J Transmissions: All	**IAT Sensor Circuit High Input Conditions:** DTC P0101, P0102, P0103, P0116, P0117, P0118, P0125, P0128, P0502 and P0503 not set, engine runtime over 120 seconds, ECT sensor more than 140ºF (more than 32ºF on 7.4L VIN J), MAF sensor less than 15 g/sec, VSS less than 7 mph, and the PCM detected the IAT sensor was less than -36ºF for 1 second. **Possible Causes** ● IAT sensor signal circuit is open between the sensor and PCM ● IAT sensor signal circuit is shorted to VREF or system power ● IAT sensor is damaged or has failed (it may be open) ● PCM has failed

OBD II Trouble Code List (P0xxx Codes)

DTC	Trouble Code Title, Conditions & Possible Causes
P0113 **2T CCM, MIL: Yes** 1999, 2000, 2001, 2002, 2003, 2004, 2005 C/K Series Truck, G Vans Envoy & TrailBlazer 4.8L VIN V, 5.3L VIN P, 5.3L VIN T, 5.3L VIN Z, 6.0L VIN N, 6.0L VIN U, 8.1L VIN G Transmissions: All	**IAT Sensor Circuit High Input Conditions:** DTC P0101, P0102, P0103, P0116, P0117, P0118, P0121, P0122, P0123, P0125, P0128, P0502 and P0503 not set, engine runtime over 120 seconds, ECT sensor more than 140°F, MAF sensor less than 15 g/sec, VSS less than 7 mph, and the PCM detected the IAT sensor indicated less than −36°F for 5 seconds during the CCM test. **Possible Causes** ● IAT sensor signal circuit is open between the sensor and PCM ● IAT sensor signal circuit is shorted to VREF (5v)// ● IAT sensor is damaged or has failed (it may be open) ● PCM has failed
P0113 **2T CCM, MIL: Yes** 1996, 1997, 1998, 1999, 2000, 2001, 2002, 2003, 2004, 2005 1996, 1997, 1998, 1999, 2000, 2001, 2002, 2003, 2004, 2005 C/K Series Truck, G Vans 6.5L VIN F, 6.5L VIN S, 6.6L VIN 1 Diesel engines Transmissions: All	**IAT Sensor Circuit High Input Conditions:** Engine started; engine runtime over 8 minutes and the PCM detected the IAT sensor indicated less than or equal to −40°F for 2 seconds during the CCM test. **Possible Causes** ● IAT sensor signal circuit is open between the sensor and PCM ● IAT sensor signal circuit is shorted to VREF or system power ● IAT sensor is damaged or has failed (it may be open) ● PCM has failed
P0113 **2T CCM, MIL: Yes** 2003, 2004, 2005 CTS (Cadillac) 2.6L V6 VIN M & 3.2L V6 VIN N engines Transmissions: All	**IAT Sensor Circuit High Input Conditions:** Engine started; engine runtime over 1 minute and 20 seconds, engine at idle speed for over 10 seconds, and the PCM detected the IAT sensor was less than −38°F for 2 seconds. **Possible Causes** ● IAT sensor signal circuit is open between the sensor and PCM ● IAT sensor signal circuit is shorted to VREF or system power ● IAT sensor is damaged or has failed (it may be open) ● PCM has failed
P0113 **2T CCM, MIL: Yes** 1996, 1997, 1998, 1999, 2000, 2001, 2002 DeVille, Eldorado, Seville 4.6L VIN 9, 4.6L VIN Y Transmissions: All	**IAT Sensor Circuit High Input Conditions:** DTC P0101, P0102, P0103, P0116, P0117, P0118, P0125, P0128, P0502 and P0503 not set, engine started, engine runtime over 3 minutes, MAF sensor less than 12 g/sec, ECT sensor more than 32°F, VSS less than 35 mph for 5 seconds, and the PCM detected the IAT sensor was more than 4.8 volts (Scan Tool read under −33°F) for 1 second during the test. **Possible Causes** ● IAT sensor signal circuit is open between the sensor and PCM ● IAT sensor signal circuit is shorted to VREF or system power ● IAT sensor is damaged or has failed (it may be open) ● PCM has failed
P0113 **2T CCM, MIL: Yes** 1995, 1996, 1997, 1998, 1999, 2000, 2001, 2002 Camaro & Firebird 3.8L VIN K, 5.7L VIN G Transmissions: All	**IAT Sensor Circuit High Input Conditions:** DTC P0101, P0102, P0103, P0116, P0117, P0118, P0125, P0128, P0500, P0502 and P0503 not set, engine started, engine runtime over 3 minutes, MAF sensor less than 8-15 g/sec, VSS less than 5-7 mph, ECT sensor more than 140°F, and the PCM detected the IAT sensor indicated less than −38°F for 3-5 seconds in the CCM test. **Possible Causes** ● IAT sensor signal circuit is open between the sensor and PCM ● IAT sensor signal circuit is shorted to VREF or system power ● IAT sensor is damaged or has failed (it may be open) ● PCM has failed
P0113 **2T CCM, MIL: Yes** 2001, 2002 Aurora 3.5L V6 VIN H engine Transmissions: A/T	**IAT Sensor Circuit High Input Conditions:** DTC P0101-P0103, P0116, P0117, P0118, P0125, P0128, P0502 and P0503 not set, engine runtime over 3 minutes, ECT sensor more than 140°F, VSS less than 35 mph, and the PCM detected the IAT sensor indicated less than −38°F for 20 seconds during testing. **Possible Causes** ● IAT sensor signal circuit is open between the sensor and PCM ● IAT sensor signal circuit is shorted to VREF or system power ● IAT sensor is damaged or has failed (it may be open) ● PCM has failed
P0113 **2T CCM, MIL: Yes** 1996, 1997, 1998, 1999, 2001, 2002, 2003, 2004, 2005 Aurora 4.0L V8 VIN C engine Transmissions: A/T	**IAT Sensor Circuit High Input Conditions:** DTC P0101-P0103, P0121-P0123 not set, engine runtime over 20 seconds, VSS from 7-50 mph for 5 seconds, ECT sensor over 32°F, IAT sensor below 81°F, MAF sensor under 60 g/sec, and the PCM detected the IAT sensor was over 4.90v (Scan Tool reads −35°F). **Possible Causes** ● IAT sensor signal circuit is open between the sensor and PCM ● IAT sensor signal circuit is shorted to VREF or system power ● IAT sensor is damaged or has failed (it may be open) ● PCM has failed

OBD II Trouble Code List (P0xxx Codes)

DTC	Trouble Code Title, Conditions & Possible Causes
P0113 **2T CCM, MIL: Yes** 1995, 1996, 1997, 1998, 1999, 2000, 2001, 2002, 2003, 2004, 2005 S/T Blazer & S/T Pickup 4.3L VIN W, 4.3L VIN X Transmissions: All	**IAT Sensor Circuit High Input Conditions:** DTC P0101-P0103, P0116-P0118, P0125, P0128, P0502 and P0503 not set, engine runtime over 2 minutes, ECT sensor more than 140ºF, MAF sensor under 15 g/sec, VSS under 7 mph, and the PCM detected the IAT sensor was less than -38ºF for 5 seconds. **Possible Causes** ● IAT sensor signal circuit is open between the sensor and PCM ● IAT sensor signal circuit is shorted to VREF or system power ● IAT sensor is damaged or has failed (it may be open) ● PCM has failed
P0113 **2T CCM, MIL: Yes** 1997, 1998, 1999, 2000, 2001 Catera 3.0L V6 VIN R engine Transmissions: A/T	**ECT Sensor Circuit High Input Conditions:** Key on or engine running; and the PCM detected the ECT sensor signal indicated more than 282ºF, or it indicated less than -40ºF. **Possible Causes** ● ECT sensor signal circuit is open between the sensor and PCM ● ECT sensor signal circuit shorted to chassis or sensor ground ● ECT sensor is damaged or has failed (it may be open) ● PCM has failed
P0113 **2T CCM, MIL: Yes** 1996, 1997 Grand Prix, Lumina, Monte Carlo 3.4L VIN X Transmissions: A/T	**IAT Sensor Circuit High Input Conditions:** DTC P0100, P0102, P0103, P0117, P0118, P0502 and P0503 not set, engine running, ECT sensor more than 140ºF, MAF sensor less than 12 g/sec, VSS less than 35 mph, and the PCM detected the IAT sensor was less than -27ºF during the CCM continuous test. **Possible Causes** ● IAT sensor signal circuit is open between the sensor and PCM ● IAT sensor signal circuit is shorted to VREF or system power ● IAT sensor is damaged or has failed (it may be open) ● PCM has failed
P0113 **2T CCM, MIL: Yes** 1997, 1998, 1999, 2002, 2001, 2002, 2003, 2004, 2005 Corvette 5.7L VIN G, 5.7L VIN S Transmissions: All	**IAT Sensor Circuit High Input Conditions:** DTC P0101-P0103, P0117, P0118, P0125, P0500, P0502, P0503 and P1258 not set, engine runtime over 2 minutes, ECT sensor over 140ºF, MAF sensor less than 15 g/sec, VSS less than 7 mph, and the PCM detected the IAT sensor was less than -36ºF for 5 seconds. **Possible Causes** ● IAT sensor signal circuit is open between the sensor and PCM ● IAT sensor signal circuit is shorted to VREF or system power ● IAT sensor is damaged or has failed (it may be open) ● PCM has failed
P0116 **2T CCM, MIL: Yes** 2002, 2003, 2004, 2005 Aurora, Aztec, Bravada, Envoy, Impala, Malibu, Montana, Silhouette, Rendezvous, Venture 3.1L VIN J, 3.1L VIN M, 3.4L VIN E, 3.5L V6 VIN H, 4.0L V8 VIN C Transmissions: All	**ECT Sensor Signal Range/Performance Conditions:** DTC P0112, P0113, P0117, P0118, P0125, P0128, P0601, P0602, P0604, P0606, P1621 and P1683 not set, minimum soak time over 8 hours, key on, IAT sensor more than 59ºF, and the PCM detected the difference between the ECT sensor and IAT sensor values was more than 59ºF. If the vehicle soak time is from 8-10 hours, the ECT and IAT sensors should be with 10ºF of each other at initial key on. **Possible Causes** ● ECT sensor circuit has an intermittent high resistance condition ● ECT sensor circuit has an intermittent grounded condition ● ECT sensor is out of calibration or "skewed" high ● IAT sensor is out of calibration or it is "skewed" high or low ● TSB 01-06-04-052 contains a repair procedure for this code
P0116 **2T CCM, MIL: Yes** 1996, 1997, 1998, 1999 DeVille, Eldorado, Seville 4.6L VIN 9, 4.6L VIN Y Transmissions: A/T	**ECT Sensor Signal Range/Performance Conditions:** DTC P0117 and P0118 not set, engine started, engine running for 2 seconds, PCM command to shift in the ECT pull-up resistors "on" for 2 seconds (this shift occurs at 122ºF), and then the PCM detected the ECT sensor changed 0.30v or more within one second. **Possible Causes** ● ECT sensor circuit has an intermittent high resistance condition ● ECT sensor circuit has an intermittent grounded condition ● ECT sensor is out of calibration or it is "skewed"
P0116 **2T CCM, MIL: Yes** 2002, 2003, 2004, 2005 Alero, Bonneville, Century, Intrigue, Regal, LeSabre, Park Avenue 3.1L VIN J, 3.4L VIN E, 3.5L V6 VIN H, 3.8L VIN K, 3.8L VIN 1 Transmissions: All	**ECT Sensor Signal Range/Performance Conditions:** DTC P0112, P0113, P0117, P0118, P0125, P0128, P0601, P0602, P1621 or P1683 not set, minimum soak time of 8 hours, engine started, and the PCM detected the difference between the ECT and IAT sensors was over 180ºF, or the time spent cranking the engine without it starting was more than 5 seconds, with a temperature difference between the ECT sensor and the IAT sensor of more than 27ºF, or the difference between the ECT and IAT sensors was over 27ºF, after the vehicle was driven for 5 minutes at over 15 mph. If the IAT sensor temperature decreases more than 12.6ºF, a block heater is detected and the test is aborted. If the temperature does not decrease, then a block heater is not detected and this code is set. **Possible Causes** ● ECT sensor circuit has an intermittent high resistance condition ● ECT sensor circuit has an intermittent grounded condition ● ECT sensor is out of calibration or "skewed"

OBD II Trouble Code List (P0xxx Codes)

DTC	Trouble Code Title, Conditions & Possible Causes
P0116 **2T CCM, MIL: Yes** 2002, 2003, 2004, 2005 C/K Series Truck, G Vans Envoy & TrailBlazer 4.3L VIN W, 4.8L VIN V, 5.0L VIN M, 5.3L VIN P, 5.3L VIN T, 5.3L VIN Z, 5.7L VIN R, 6.0L VIN U, 6.6L VIN 1, 8.1L VIN G Transmissions: All	**ECT Sensor Signal Range/Performance Conditions:** DTC P0112, P0113, P0117, P0118, P0125, P0128, P0601, P0602, P1621 or P1683 not set, minimum soak time of 8 hours, key on, the difference between the ECT and IAT sensors over 27-36ºF, engine started, VSS over 15 mph for 5 minutes. If the IAT sensor decreases more than 12.6ºF, a block heater is indicated and the test is aborted. If the IAT sensor does not decrease and the PCM detects difference between the ECT and IAT sensor signals at startup is more than 252ºF, and the engine is cranked for 10 seconds, this code will set. **Possible Causes** ● ECT sensor circuit has an intermittent high resistance condition ● ECT sensor circuit has an intermittent grounded condition ● ECT sensor is out of calibration or "skewed"
P0116 **2T CCM, MIL: Yes** 2003, 2004, 2005 CTS (Cadillac) 2.6L V6 VIN M & 3.2L V6 VIN N engines Transmissions: All	**ECT Sensor Signal Range/Performance Conditions:** DTC P0112, P0113, P0117, P0118, P0125, P0128, P0601, P0602, P0603, P0604 and P0606 not set, engine runtime over 8 minutes, engine speed over 400 rpm, MAF value equal to a calculated value, and the PCM detected a temperature difference of more than 36ºF between the actual ECT sensor signal and the predicted ECT signal. **Possible Causes** ● IAT sensor signal circuit is open between the sensor and PCM ● IAT sensor signal circuit is shorted to VREF or system power ● IAT sensor is damaged or has failed (it may be open) ● PCM has failed
P0116 **2T CCM, MIL: Yes** 2002, 2003, 2004, 2005 Aurora, DeVille, Eldorado, Seville 4.0L V8 VIN C, 4.6L VIN 9, 4.6L VIN Y engines Transmissions: A/T	**ECT Sensor Signal Range/Performance Conditions:** DTC P0112, P0113, P0117, P0118, P0125, P0128, P0601, P0602, P1621 and P1683 not set, minimum soak time over 8 hours, engine started. If the PCM detects a temperature difference of over 36ºF between the ECT and the IAT sensors, the vehicle must be driven for 5 minutes at over 15 mph. If the IAT sensor decreases more than 12.6ºF, then a block heater is indicated and the test is aborted. If the IAT sensor temperature does not decrease, a block heater is not present and this code is set. If the difference between the ECT and IAT sensors at startup is over 252ºF, and the time spent cranking the engine without starting is more than 10 seconds, then this code will set. **Possible Causes** ● ECT sensor circuit has an intermittent high resistance condition ● ECT sensor circuit has an intermittent grounded condition ● ECT sensor is out of calibration or it is "skewed"
P0116 **2T CCM, MIL: Yes** 1996, 1997, 1998, 1999 Aurora 4.0L V8 VIN C Transmissions: A/T	**ECT Sensor Signal Range/Performance Conditions:** Key on or engine runtime more than 2 seconds, then two seconds after the shift command occurred (at a temperature of 122ºF) on the ECT pullup resistor, and then the PCM detected the ECT sensor signal changed more than 0.30v in less than one second. **Possible Causes** ● ECT sensor circuit has an intermittent high resistance condition ● ECT sensor circuit has an intermittent grounded condition ● ECT sensor is out of calibration or "skewed" ● ECT sensor is damaged or it has failed
P0116 **2T CCM, MIL: Yes** 2002, 2003, 2004, 2005 Camaro, Corvette & Firebird 3.8L VIN K, 5.7L VIN G, 5.7L VIN S engines Transmissions: All	**ECT Sensor Signal Range/Performance Conditions:** DTC P0112, P0113, P0117, P0118, P0125, P0128, P0601, P0602, P1621 or P1683 not set, minimum soak time of 8 hours, key on, the difference between the ECT and IAT sensor over 27-36ºF, engine started, VSS over 15 mph for 5 minutes. If the IAT sensor decreases more than 12.6ºF, a block heater is indicated and the test is aborted. If the IAT sensor does not decrease and the PCM detects difference between the ECT and IAT sensor signals at startup is more than 252ºF, and the engine is cranked for 10 seconds, this code will set. **Possible Causes** ● ECT sensor circuit has an intermittent high resistance condition ● ECT sensor circuit has an intermittent grounded condition ● ECT sensor is out of calibration or "skewed"
P0116 **2T CCM, MIL: Yes** 2002, 2003, 2004, 2005 Alero, Cavalier, Envoy, Envoy XL, Sunfire & TrailBlazer 2.2L VIN F, 4.2L VIN S Transmissions: All Alero, Envoy, Envoy XL, Sunfire	**ECT Sensor Signal Range/Performance Conditions:** DTC P0112, P0113, P0117, P0118, P0125, P0128, P0601, P0602, P0604, P0606, P1621 or P1683 not set, minimum soak time of 8 hours, key on, IAT sensor more than 59ºF, and the PCM detected the difference between the ECT and IAT sensor signals was more than 59ºF during the CCM test. If the vehicle soak time is from 8-10 hours, the ECT and IAT sensors should be with 10ºF of each other. **Possible Causes** ● ECT sensor circuit has an intermittent high resistance condition ● ECT sensor circuit has an intermittent grounded condition ● ECT sensor is out of calibration or "skewed" high ● IAT sensor is out of calibration or it is "skewed" high or low

OBD II Trouble Code List (P0xxx Codes)

DTC	Trouble Code Title, Conditions & Possible Causes
P0116 **2T CCM, MIL: Yes** 1997, 1998, 1999, 2000, 2001 Catera 3.0L V6 VIN R engine Transmissions: A/T	**ECT Sensor Circuit Malfunction Conditions:** Engine started and the PCM detected the difference between the Actual ECT sensor and the Calculated ECT sensor signal was more than 22°F during the CCM continuous test. **Possible Causes** ● ECT sensor circuit has an intermittent high resistance condition ● ECT sensor circuit has an intermittent grounded condition ● ECT sensor is out of calibration or "skewed"
P0116 **2T CCM, MIL: No** 1995 Corvette 5.7L V8 VIN P engine Transmissions: All	**ECT Sensor Signal Range/Performance Conditions:** ECT and IAT sensors more than -13°F and within 11°F but less than 90°F at startup, engine started, and the PCM detected the ECT sensor did not reach 138°F after 22 minutes of sustained engine operation. The IAT sensor is a variable resistor that measures the temperature of the air entering the engine. The IAT is connected to the PCM with a 5v signal and a low reference ground circuit. When the IAT sensor is cold, its resistance is high. As the air temperature increases, its resistance decreases. With high sensor resistance, the IAT sensor signal voltage is high. With lower sensor resistance, the IAT sensor signal voltage should be a lower voltage. **Possible Causes** ● ECT sensor circuit has an intermittent high resistance condition ● ECT sensor circuit has an intermittent grounded condition ● ECT sensor is out of calibration or "skewed"
P0117 **2T CCM, MIL: Yes** 1996, 1997, 1998, 1999, 2000, 2001, 2002, 2003, 2004, 2005 Achieva, Alero, Beretta, Century, Ciera, Cavalier, Envoy, Grand Am, Sunfire Skyhawk, TrailBlazer, S/T Blazer & S/T Pickup 2.2L VIN 4, 2.2L VIN 5, 2.2L VIN F, 2.2L VIN H, 2.4L VIN T, 4.2L VIN S Transmissions: All	**ECT Sensor Circuit Low Input Conditions:** Engine started; engine runtime over 128 seconds, and the PCM detected the ECT sensor was more than 280°F for 6.25 seconds. The ECT sensor is a variable resistor that includes a signal and low reference circuit to measure the temperature of the engine coolant. **Possible Causes** ● ECT sensor connector is damaged or shorted ● ECT sensor signal circuit shorted to chassis or sensor ground ● ECT sensor is damaged or has failed (it may be shorted) ● PCM has failed
P0117 **1T CCM, MIL: Yes** 1995 Century, Cutlass Ciera, Cutlass Supreme, Grand Prix, Lumina, Monte Carlo Park Avenue Regal, Riviera, 88' & 98' 3.1L VIN M, 3.4L VIN X, 3.8L VIN 1, 3.8L VIN K, 3.8L VIN L engines Transmissions: All	**ECT Sensor Circuit Low Temperature Conditions:** Engine runtime over 3 seconds and the PCM detected the ECT sensor was less than -38°F for 400 ms. The ECT sensor is a variable resistor that measures the engine coolant temperature. The ECT sensor connects to the PCM with a 5v signal and low reference ground circuit. When the ECT sensor is cold, its resistance is high. As the coolant temperature increases, its resistance decreases. **Possible Causes** ● ECT sensor signal circuit is open between the sensor and PCM ● ECT sensor signal circuit is shorted to VREF or system power ● ECT sensor is damaged or has failed (it may be open) ● PCM has failed
P0117 **2T CCM, MIL: Yes** 1996, 1997, 1998, 1999 Achieva, Beretta, Ciera, Corsica, Century, Grand Am, Grand Prix, Malibu, Skylark 3.1L VIN M engine Transmissions: All	**ECT Sensor Circuit Low Input Conditions:** Engine started; engine runtime from 3 to 15 seconds and the PCM detected the ECT sensor indicated more than 283°F for 25 seconds. The ECT sensor is a variable resistor that includes an ECT signal circuit and a low reference circuit to measure the temperature of the engine coolant. If the ECT sensor is cold, the ECT sensor resistance is high. As the temperature increases, the ECT sensor resistance decreases. **Possible Causes** ● ECT sensor signal circuit shorted to sensor or chassis ground ● ECT sensor is damaged or has failed (it may be shorted) ● PCM has failed
P0117 **2T CCM, MIL: Yes** 1996, 1997, 1998, 1999, 2000, 2001, 2002, 2003, 2004, 2005 Aurora, Century, LeSabre, Park Avenue, Regal 3.1L VIN J, 3.1L VIN M, 3.4L VIN X, 3.5L V6 VIN H, 3.8L VIN 1, 3.8L VIN K, 4.0L V8 VIN C Transmissions: All	**ECT Sensor Circuit Low Input Conditions:** Engine started, engine runtime over 15 seconds and the PCM detected the ECT sensor was more than 282°F for 15 seconds. The ECT sensor is a variable resistor that measures the engine coolant temperature. The ECT sensor connects to the PCM with a 5v signal and low reference ground circuit. When the ECT sensor is cold, its resistance is high. As the coolant temperature increases, its resistance decreases. **Possible Causes** ● ECT sensor signal circuit shorted to sensor or chassis ground ● ECT sensor is damaged or has failed (it may be shorted) ● PCM has failed

OBD II Trouble Code List (P0xxx Codes)

DTC	Trouble Code Title, Conditions & Possible Causes
P0117 **2T CCM, MIL: Yes** 1996, 1997 Camaro, Corvette, Caprice, Firebird & Fleetwood 4.3L VIN W, 5.7L VIN 5, 5.7L VIN P engines Transmissions: All	**ECT Sensor Circuit Low Input Conditions:** Engine started; engine runtime over 10 seconds (or less than 10 seconds if the IAT sensor is less than 122ºF at startup), and the PCM detected the ECT sensor indicated more than 282ºF for 20 seconds during the CCM test period. **Possible Causes** ● ECT sensor signal circuit shorted to sensor or chassis ground ● ECT sensor is damaged or has failed (it may be shorted) ● PCM has failed
P0117 **2T CCM, MIL: Yes** 1996, 1997, 1998, 1999, 2000, 2001, 2002, 2003, 2004, 2005 Alero, Cutlass, Impala, Malibu, Montana, Silhouette, Venture, Rendezvous 3.1L VIN J, 3.1L VIN M, 3.4L VIN E engines Transmissions: All	**ECT Sensor Circuit Low Input Conditions:** Engine started; engine runtime over 3 seconds and the PCM detected the ECT sensor was less than 0.10v (Scan Tool reads over 283ºF) for 15-25 seconds. The PCM supplies the ECT sensor with a 5v signal and a low reference ground circuit. When the ECT sensor is cold, its resistance is high. As the engine coolant temperature increases, its resistance decreases. With high sensor resistance, the ECT sensor signal voltage is high. With lower sensor resistance, the ECT sensor signal voltage should be a lower voltage. **Possible Causes** ● ECT sensor signal circuit shorted to sensor or chassis ground ● ECT sensor is damaged or has failed (it may be shorted) ● PCM has failed
P0117 **2T CCM, MIL: Yes** 1996, 1997, 1998, 1999, 2000, 2001, 2002, 2003, 2004, 2005 C/K Cab & Chassis, C/K Series Truck, G Series 4.3L VIN W, 4.3L VIN X, 5.0L VIN M, 5.7L VIN K, 5.7L VIN R, 7.4L VIN J Transmissions: All	**ECT Sensor Circuit Low Input Conditions:** Engine started; engine runtime over 10 seconds or with the engine runtime under 10 seconds and the IAT sensor signal less than 122ºF, the PCM detected the ECT sensor was more than 282ºF for 20 seconds. When the coolant is cold, sensor resistance is high, and as it warms, the sensor resistance decreases. With high sensor resistance, the signal voltage is high. **Possible Causes** ● ECT sensor signal circuit shorted to sensor or chassis ground ● ECT sensor is damaged or has failed (it may be shorted) ● PCM has failed
P0117 **2T CCM, MIL: Yes** 1999, 2000, 2001, 2002, 2003, 2004, 2005 C/K Series Truck, G Vans Envoy & TrailBlazer 4.3L VIN W, 4.8L VIN V, 5.0L VIN M, 5.3L VIN P, 5.3L VIN T, 5.3L VIN Z, 5.7L VIN R, 6.0L VIN U, 6.6L VIN 1, 8.1L VIN G Transmissions: All	**ECT Sensor Circuit Low Input Conditions:** Engine started; engine runtime less than 10 seconds and IAT sensor signal less than 122ºF or engine runtime over 10 seconds, and the PCM detected the ECT sensor was less than 0.10v (Scan Tool reads over 280ºF) for 20 seconds. **Possible Causes** ● ECT sensor connector is damaged or shorted ● ECT sensor signal circuit shorted to sensor ground ● ECT sensor is damaged or has failed (it may be shorted) ● PCM has failed
P0117 **2T CCM, MIL: Yes** 1996, 1997, 1998, 1999, 2000, 2001, 2002, 2003, 2004, 2005 C/K Series Truck, G Vans 6.5L VIN F, 6.5L VIN S, 6.6L VIN 1 Diesel engines Transmissions: All	**ECT Sensor Circuit Low Input Conditions:** Key on or engine running; and the PCM detected the ECT sensor signal indicated more than 303ºF for 2 seconds in the CCM test. **Possible Causes** ● ECT sensor connector is damaged or shorted ● ECT sensor signal circuit shorted to sensor or chassis ground ● ECT sensor is damaged or has failed (it may be shorted) ● PCM has failed
P0117 **2T CCM, MIL: Yes** 1996, 1997, 1998, 1999, 2000, 2001, 2002, 2003, 2004, 2005 DeVille, Eldorado, Seville 4.6L VIN 9, 4.6L VIN Y Transmissions: A/T	**ECT Sensor Circuit Low Input Conditions:** DTC P0112, P0113, P1111 and P1112 not set, engine started, engine runtime over 10 seconds, IAT sensor less than 158ºF, and the PCM detected the ECT sensor was less than 0.08v (Scan Tool indicates more than 278ºF) for less than 1 second during the CCM test. **Possible Causes** ● ECT sensor connector is damaged or shorted ● ECT sensor signal circuit shorted to sensor or chassis ground ● ECT sensor is damaged or has failed (it may be shorted) ● PCM has failed
P0117 **2T CCM, MIL: Yes** 2003, 2004, 2005 CTS (Cadillac) 2.6L V6 VIN M & 3.2L V6 VIN N engines Transmissions: All	**ECT Sensor Circuit Low Input Conditions:** Engine cranking or running, and the PCM detected the ECT sensor signal was more than 280ºF for more than 300 ms during the test. **Possible Causes** ● ECT sensor connector is damaged or shorted ● ECT sensor signal circuit shorted to sensor ground ● ECT sensor is damaged or has failed (it may be shorted) ● PCM has failed

OBD II Trouble Code List (P0xxx Codes)

DTC	Trouble Code Title, Conditions & Possible Causes
P0117 **2T CCM, MIL: Yes** 1995, 1996, 1997, 1998, 1999, 2000, 2001, 2002, 2003, 2004, 2005 Camaro, Corvette & Firebird 3.8L VIN K, 5.7L VIN G, 5.7L VIN S engines Transmissions: All	**ECT Sensor Circuit Low Input Conditions:** Engine started; engine runtime under 10 seconds (or more than 10 seconds if the ECT sensor is less than 122°F at startup), and the PCM detected the ECT sensor was more than 282°F for 20 seconds. **Possible Causes** • ECT sensor connector is damaged or shorted • ECT sensor signal circuit shorted to sensor or chassis ground • ECT sensor is damaged or has failed (it may be shorted) • PCM has failed
P0117 **2T CCM, MIL: Yes** 1996, 1997, 1998, 1999, 2001, 2002, 2003, 2004, 2005 Aurora 4.0L V8 VIN C engine Transmissions: A/T	**ECT Sensor Circuit Low Input Conditions:** DTC P0112, P0113, P111 and P1112 not set, engine started, engine running with the TP angle from 5-35 degrees for 3-4 minutes, IAT sensor less than 158°F, and the PCM detected the ECT sensor indicated more than 300°F. When the ECT sensor is cold, its resistance is high. As the engine coolant temperature increases, its resistance decreases. With high sensor resistance, the ECT sensor signal voltage is high. With lower sensor resistance, the ECT sensor signal voltage should be a lower voltage. **Possible Causes** • ECT sensor signal circuit shorted to sensor or chassis ground • ECT sensor is damaged or has failed (it may be shorted) • PCM has failed
P0117 **2T CCM, MIL: Yes** 2001, 2002 Aurora 3.5L V6 VIN H engine Transmissions: A/T	**ECT Sensor Circuit Low Input Conditions:** Engine started; engine runtime over 15 seconds and the PCM detected the ECT sensor indicated more than 282°F. The ECT sensor is a variable resistor that includes an ECT signal circuit and a low reference circuit to measure the temperature of the engine coolant. **Possible Causes** • ECT sensor signal circuit shorted to sensor or chassis ground • ECT sensor is damaged or has failed (it may be shorted) • PCM has failed
P0117 **2T CCM, MIL: Yes** 1995, 1996, 1997, 1998, 1999, 2000, 2001, 2002, 2003, 2004, 2005 S/T Blazer & S/T Pickups 4.3L VIN W, 4.3L VIN X Transmissions: All	**ECT Sensor Circuit Low Input Conditions:** Engine started; engine runtime over 10 seconds, and the PCM detected the ECT sensor was less than 0.10v (Scan Tool reads over 282°F) for 20 seconds. The ECT sensor is a variable resistor with a signal and low reference circuit used to measure engine coolant temperature. **Possible Causes** • ECT sensor signal circuit shorted to sensor or chassis ground • ECT sensor is damaged or has failed (it may be shorted) • PCM has failed
P0117 **2T CCM, MIL: Yes** 1996, 1997 Grand Prix, Lumina, Monte Carlo 3.4L V6 VIN X engine Transmissions: All	**ECT Sensor Circuit Low Input Conditions:** Engine runtime from 3 to 15 seconds and the PCM detected the ECT sensor was less than 0.10v (Scan Tool read over 237°F) for 10 seconds during the CCM continuous period. **Possible Causes** • ECT sensor connector is damaged or shorted • ECT sensor signal circuit shorted to sensor or chassis ground • ECT sensor is damaged or has failed (it may be shorted) • PCM has failed
P0117 **2T CCM, MIL: No** 1995 Corvette 5.7L V8 VIN P engine Transmissions: All	**ECT Sensor Circuit Low Input Conditions:** Key on or engine running; and the PCM detected the ECT sensor signal was less than 0.10v for 2 seconds (Scan Tool reads over 282°F) during the CCM continuous test. **Possible Causes** • ECT sensor signal circuit shorted to sensor or chassis ground • ECT sensor is damaged or has failed (it may be shorted) • PCM has failed
P0118 **2T CCM, MIL: Yes** 1996, 1997, 1998, 1999, 2000, 2001, 2002, 2003, 2004, 2005 Achieva, Alero, Beretta, Century, Ciera, Cavalier, Envoy, Grand Am, Sunfire Skyhawk, TrailBlazer, S/T Blazer & S/T Pickup 2.2L VIN 4, 2.2L VIN 5, 2.2L VIN F, 2.2L VIN H, 2.4L VIN T, 4.2L VIN S Transmissions: All	**ECT Sensor Circuit High Input Conditions:** Engine started; engine runtime over 60 seconds and the PCM detected the ECT sensor indicated less than -38°F for 6.25 seconds. The ECT sensor is a variable resistor that includes an ECT signal circuit and a low reference circuit to measure the temperature of the engine coolant. The PCM supplies the ECT sensor with a 5v signal circuit and a low reference ground circuit. When the ECT sensor is cold, its resistance is high. As the engine coolant temperature increases, its resistance decreases. With high sensor resistance, the ECT sensor signal voltage is high. With lower sensor resistance, the ECT sensor signal voltage should be a lower voltage. **Possible Causes** • ECT sensor signal circuit is open between the sensor and PCM • ECT sensor signal circuit is shorted to VREF or system power • ECT sensor is damaged or has failed (it may be open) • PCM has failed

OBD II Trouble Code List (P0xxx Codes)

DTC	Trouble Code Title, Conditions & Possible Causes
P0118 **1T CCM, MIL: Yes** 1995 Century, Cutlass Ciera, Cutlass Supreme, Grand Prix, Lumina, Monte Carlo Park Avenue Regal, Riviera, 88' & 98' 3.1L VIN M, 3.4L VIN X, 3.8L VIN 1, 3.8L VIN K, 3.8L VIN L engines Transmissions: All	**ECT Sensor Circuit High Temperature Conditions:** Engine runtime over 3 seconds and the PCM detected the ECT sensor was more than 284°F for 400 ms. The ECT sensor is a variable resistor that includes an ECT signal and a low reference circuit to measure the temperature of the engine coolant. **Possible Causes** ● ECT sensor signal circuit shorted to sensor or chassis ground ● ECT sensor is damaged or has failed (it may be shorted) ● PCM has failed
P0118 **2T CCM, MIL: Yes** 1996, 1997, 1998, 1999 Achieva, Beretta, Ciera, Corsica, Century, Grand Am, Grand Prix, Malibu, Skylark 3.1L VIN M engine Transmissions: All	**ECT Sensor Circuit High Input Conditions:** Engine started; engine runtime from 3-15 seconds and the PCM detected the ECT sensor indicated less than -37°F for 3 seconds. The ECT sensor is a variable resistor that includes an ECT signal circuit and a low reference circuit to measure the temperature of the engine coolant. When the ECT sensor is cold, its resistance is high. As the coolant temperature increases, its resistance decreases. **Possible Causes** ● ECT sensor signal circuit is open between the sensor and PCM ● ECT sensor signal circuit is shorted to VREF or system power ● ECT sensor is damaged or has failed (it may be open) ● PCM has failed
P0118 **2T CCM, MIL: Yes** 1996, 1997, 1998, 1999, 2000, 2001, 2002, 2003, 2004, 2005 Aurora, Century, LeSabre, Park Avenue, Regal 3.1L VIN J, 3.1L VIN M, 3.4L VIN X, 3.5L V6 VIN H, 3.8L VIN 1, 3.8L VIN K, 4.0L V8 VIN C Transmissions: All	**ECT Sensor Circuit High Input Conditions:** Engine started; engine runtime over 15 seconds and the PCM detected the ECT sensor indicated less than -36°F for 24 seconds. The ECT sensor is a variable resistor that includes an ECT signal circuit and a low reference circuit to measure the temperature of the engine coolant. The PCM supplies the ECT sensor with a 5v signal circuit and a low reference ground circuit. When the ECT sensor is cold, its resistance is high, and as the temperature increases, its resistance decreases. With high sensor resistance, the ECT sensor signal voltage is high, and it is lower with a lower sensor resistance. **Possible Causes** ● ECT sensor signal circuit is open between the sensor and PCM ● ECT sensor signal circuit is shorted to VREF or system power ● ECT sensor is damaged or has failed (it may be open) ● PCM has failed
P0118 **1T CCM, MIL: Yes** 1996, 1997 Camaro, Corvette, Caprice, Firebird & Fleetwood 4.3L VIN W, 5.7L VIN 5, 5.7L VIN P engines Transmissions: All	**ECT Sensor Circuit High Input Conditions:** Engine started; engine runtime over 60 seconds (or less than 60 seconds with IAT sensor more than 32°F at startup), and the PCM detected the ECT sensor indicated less than -38°F for 25 seconds. **Possible Causes** ● ECT sensor signal circuit is open between the sensor and PCM ● ECT sensor signal circuit is shorted to VREF or system power ● ECT sensor is damaged or has failed (it may be open) ● PCM has failed
P0118 **2T CCM, MIL: Yes** 1996, 1997, 1998, 1999, 2000, 2001, 2002, 2003, 2004, 2005 Alero, Cutlass, Impala, Malibu, Montana, Silhouette, Venture, Rendezvous 3.1L VIN J, 3.1L VIN M, 3.4L VIN E engines Transmissions: All	**ECT Sensor Circuit High Input Conditions:** Engine started and the PCM detected the ECT sensor indicated less than -36°F for a period of 15-25 seconds during the CCM test period. **Possible Causes** ● ECT sensor signal circuit is open between the sensor and PCM ● ECT sensor signal circuit is shorted to VREF or system power ● ECT sensor is damaged or has failed (it may be open) ● PCM has failed
P0118 **2T CCM, MIL: Yes** 1996, 1997, 1998, 1999, 2000, 2001, 2002, 2003, 2004, 2005 C/K Cab & Chassis, C/K Series Truck, G Series 4.3L VIN W, 4.3L VIN X, 5.0L VIN M, 5.7L VIN K, 5.7L VIN R, 7.4L VIN J Transmissions: All	**ECT Sensor Circuit High Input Conditions:** Engine started; engine runtime over 60 seconds, or less than 60 seconds with the IAT sensor more than 32°F at startup, and the PCM detected the ECT sensor was less than -38°F for 20 seconds. **Possible Causes** ● ECT sensor signal circuit is open between the sensor and PCM ● ECT sensor signal circuit is shorted to VREF or system power ● ECT sensor is damaged or has failed (it may be open) ● PCM has failed

OBD II Trouble Code List (P0xxx Codes)

DTC	Trouble Code Title, Conditions & Possible Causes
P0118 **2T CCM, MIL: Yes** 1999, 2000, 2001, 2002, 2003, 2004, 2005 C/K Series Truck, G Vans Envoy & TrailBlazer 4.3L VIN W, 4.8L VIN V, 5.0L VIN M, 5.3L VIN P, 5.3L VIN T, 5.3L VIN Z, 5.7L VIN R, 6.0L VIN U, 6.6L VIN 1, 8.1L VIN G Transmissions: All	**ECT Sensor Circuit High Input Conditions:** Engine started; engine runtime over 10 seconds, or engine runtime under 10 seconds with the IAT sensor more than 32ºF, and the PCM detected the ECT sensor was over 4.90v (Scan Tool reads less than -36ºF) for 20 seconds during the CCM test period. **Possible Causes** • ECT sensor connector is damaged, loose or open • ECT sensor signal circuit is open between the sensor and PCM • ECT sensor signal circuit is shorted to VREF • ECT sensor is damaged or has failed (it may be open) • PCM has failed
P0118 **2T CCM, MIL: Yes** 1996, 1997, 1998, 1999, 2000, 2001, 2002, 2003, 2004, 2005 C/K Series Truck, G Vans 6.5L VIN F, 6.5L VIN S, 6.6L VIN 1 Diesel engines Transmissions: All	**ECT Sensor Circuit High Input Conditions:** Engine started; engine runtime over 8 minutes and the PCM detected the ECT sensor indicated less than -22ºF for 2 seconds. The ECT sensor is a variable resistor that includes an ECT signal and low reference circuit to measure the temperature of the engine coolant. **Possible Causes** • ECT sensor signal circuit is open between the sensor and PCM • ECT sensor signal circuit is shorted to VREF or system power • ECT sensor is damaged or has failed (it may be open) • PCM has failed
P0118 **2T CCM, MIL: Yes** 1996, 1997, 1998, 1999, 2000, 2001, 2002, 2003, 2004, 2005 DeVille, Eldorado, Seville 4.6L VIN 9, 4.6L VIN Y Transmissions: A/T	**ECT Sensor Circuit High Input Conditions:** DTC P0112 and P0113 not set, engine runtime over 3 seconds, IAT sensor over 19ºF, and the PCM detected the ECT sensor indicated more than 4.9v (Scan Tool reads below -35ºF) for 1 second during the CCM test period. **Possible Causes** • ECT sensor signal circuit is open between the sensor and PCM • ECT sensor signal circuit is shorted to VREF or system power • ECT sensor is damaged or has failed (it may be open) • PCM has failed • TSB 71-65-25 contains a repair procedure for this code
P0118 **2T CCM, MIL: Yes** 2003, 2004, 2005 CTS (Cadillac) 2.6L V6 VIN M & 3.2L V6 VIN N engines Transmissions: All	**ECT Sensor Circuit High Input Conditions:** Engine cranking or running, and the PCM detected the ECT sensor signal was less than -38ºF for more than 300 ms during the test. **Possible Causes** • ECT sensor connector is damaged, loose or open • ECT sensor signal circuit is open between the sensor and PCM • ECT sensor signal circuit is shorted to VREF • ECT sensor is damaged or has failed (it may be open) • PCM has failed
P0118 **2T CCM, MIL: Yes** 1995, 1996, 1997, 1998, 1999, 2000, 2001, 2002, 2003, 2004, 2005 Camaro, Corvette & Firebird 3.8L VIN K, 5.7L VIN G, 5.7L VIN S engines Transmissions: All	**ECT Sensor Circuit High Input Conditions:** Engine runtime 15-60 seconds (or less than 60 seconds with the ECT sensor more than 32ºF at startup), and the PCM detected the ECT sensor indicated less than -36ºF for 20 seconds. **Possible Causes** • ECT sensor connector is damaged, loose or open • ECT sensor signal circuit is open between the sensor and PCM • ECT sensor signal circuit is shorted to VREF or system power • ECT sensor is damaged or has failed (it may be open) • PCM has failed
P0118 **2T CCM, MIL: Yes** 1996, 1997, 1998, 1999, 2001, 2002, 2003, 2004, 2005 Aurora 4.0L V8 VIN C engine Transmissions: A/T	**ECT Sensor Circuit High Input Conditions:** DTC P0112, P0113, P0122, P0123 P1111 and P1112 not set, engine started, engine running with TP angle over 7 degrees for at least 3.5 minutes, IAT sensor signal more than 19ºF, and the PCM detected the ECT sensor indicated less than -36ºF for 10 seconds. **Possible Causes** • ECT sensor signal circuit is open between the sensor and PCM • ECT sensor signal circuit is shorted to VREF or system power • ECT sensor is damaged or has failed (it may be open) • PCM has failed
P0118 **2T CCM, MIL: Yes** 2001, 2002 Aurora 3.5L V6 VIN H engine Transmissions: All	**ECT Sensor Circuit High Input Conditions:** Engine started; engine runtime over 3 seconds and the PCM detected the ECT sensor indicated less than -36ºF during the test. **Possible Causes** • ECT sensor signal circuit is open between the sensor and PCM • ECT sensor signal circuit is shorted to VREF or system power • ECT sensor is damaged or has failed (it may be open) • PCM has failed

OBD II Trouble Code List (P0xxx Codes)

DTC	Trouble Code Title, Conditions & Possible Causes
P0118 **2T CCM, MIL: Yes** 1995, 1996, 1997, 1998, 1999, 2000, 2001, 2002, 2003, 2004, 2005 S/T Blazer & S/T Pickup 4.3L VIN W, 4.3L VIN X Transmissions: All	**ECT Sensor Circuit High Input Conditions:** Engine started; engine runtime over 60 seconds and the PCM detected the ECT sensor indicated less than -36° for 20 seconds. The ECT sensor is a variable resistor that includes an ECT signal circuit and a low reference circuit to measure the engine coolant temperature. **Possible Causes** ● ECT sensor signal circuit is open between the sensor and PCM ● ECT sensor signal circuit is shorted to VREF or system power ● ECT sensor is damaged or has failed (it may be open) ● PCM has failed
P0118 **2T CCM, MIL: Yes** 1996, 1997 Grand Prix, Lumina, Monte Carlo 3.4L V6 VIN X engine Transmissions: All	**ECT Sensor Circuit High Input Conditions:** Engine started; engine runtime from 3-15 seconds and the PCM detected the ECT sensor indicated less than -27°F for 10 seconds. **Possible Causes** ● ECT sensor signal circuit is open between the sensor and PCM ● ECT sensor signal circuit is shorted to VREF or system power ● ECT sensor is damaged or has failed (it may be open) ● PCM has failed
P0118 **2T CCM, MIL: No** 1995 Corvette 5.7L V8 VIN P engine Transmissions: All	**ECT Sensor Circuit High Input Conditions:** Key on or engine running; and the PCM detected the ECT sensor was more than 4.8v for 30 seconds. The ECT sensor is a variable resistor that includes an ECT signal circuit and a low reference circuit to measure the temperature of the engine coolant. The PCM supplies the ECT sensor with a 5v signal circuit and a low reference ground circuit. When the ECT sensor is cold, its resistance is high. With high sensor resistance, the ECT sensor signal voltage is high. With lower sensor resistance, the ECT sensor signal is lower **Possible Causes** ● ECT sensor signal circuit is open between the sensor and PCM ● ECT sensor signal circuit is shorted to VREF or system power ● ECT sensor is damaged or has failed (it may be open) ● PCM has failed
P0120 **1T CCM, MIL: Yes** 2003, 2004, 2005 Avalanche, C/K Trucks, Envoy, Escalade, G Vans, TrailBlazer 4.8L VIN V, 5.3L VIN P, 5.3L VIN T, 5.3L VIN Z, 6.0L VIN N, 6.0L VIN U, 8.1L VIN G Transmissions: All	**TP Sensor 1 Signal Range/Performance Conditions:** DTC P1510 and P2108 not set, engine cranking or running, system voltage over 5.23v, and the PCM detected the TP sensor 1 signal was less than 0.37v or more than 4.51v for 1 second. If the TAC module detects an internal condition, several TAC system codes can be set due to the many redundant tests run continuously on this system. Locating and repairing one individual condition may correct more than one code. **Possible Causes** ● TAC connector contaminated with water (causes other codes) ● TP Sensor 1 low reference circuit is shorted to ground ● TP Sensor 1 signal circuit is open or shorted to ground ● APP Sensor signal or VREF circuit is open or shorted to power ● APP Sensor ground circuit is open or shorted to system power ● APP Sensor is damaged or it has failed (it may be cracked) ● APP Sensor assembly is damaged or it has failed ● TAC assembly is damaged or it has failed ● TSB 03-04-06-034 contains a repair procedure for this code
P0120 **2T CCM, MIL: Yes** 1996, 1997, 1998, 1999 Aurora, DeVille, Eldorado & Seville 4.0L V8 VIN C, 4.6L VIN 9, 4.6L VIN Y engines Transmissions: All	**TP Sensor Signal Range/Performance Conditions:** DTC P0107 and P0108 not set, engine started, MAP sensor more than 21.8 kPa and not within 12 kPa of BARO sensor, engine speed over 500 rpm, Engine Metal Over-Temperature and Traction Control "off", and the PCM detected the MAP sensor changed less than 3 kPa in a 210 ms after a 5 degree increase in throttle position, or the MAP sensor changed less than 2.2 kPa in a 210 ms period after a 4.7 degree decrease in the throttle position during the test. **Possible Causes** ● Misfire trouble code (DTP P0300 through P0308) is present ● TP sensor signal circuit is shorted to ground (intermittent fault) ● TP sensor signal circuit is open (an intermittent fault) ● TP sensor is damaged or has failed (it may be cracked)
P0120 **2T CCM, MIL: Yes** 1997, 1998 Catera 3.0L V6 VIN R engine Transmissions: All	**TP Sensor Signal Range/Performance Conditions:** DTC P0107 and P0108 not set, engine started, and the PCM detected the TP angle was over 90% or less than 3.9% during cranking for 50 ms or for 2 second with the engine cranking. **Possible Causes** ● TP sensor signal circuit is open or grounded (intermittent fault) ● TP sensor is damaged or has failed (it may be cracked) ● TP sensor VREF circuit is shorted to ground on the VREF circuit that connects to the Fuel Tank Pressure sensor ● TP sensor VREF circuit is open on the VREF circuit that connects to the Fuel Tank Pressure sensor

OBD II Trouble Code List (P0xxx Codes)

DTC	Trouble Code Title, Conditions & Possible Causes
P0121 **2T CCM, MIL: Yes** 1996, 1997, 1998 Achieva, Beretta, Ciera, Cavalier, Century, Corsica Grand Am, Malibu, S/T Blazer, S/T Pickup, Skylark 2.2L VIN 4, 2.4L VIN T Transmissions: A/T	**TP Sensor Signal Range/Performance Conditions:** DTC P0106, P0107, P0108, P0171, P0172, P0200, P0300, P0301, P0302, P0303, P0304, P0325, P0335, P0341, P0342, P0404, P0405, P0440, P0442, P0452, P0453, P0502, P0506, P0507, P0601, P0602, P1441 not set, engine started, ECT sensor more than 68ºF, TP angle stable (± 2%), MAP sensor less than 45 kPa, and the PCM detected the TP angle was more than 2% at 0 rpm, or more 10% at 800 rpm, or more than 20% at 1600 rpm, or more than 25% at 2400 rpm, or more than 30% at 3200 rpm, or more than 35% at 4000 rpm, or more than 35% at 4800 rpm, or more than 40% at 5600 rpm, or more than 40% at 6400 rpm. **Possible Causes** ● Misfire trouble code (DTP P0300 through P0308) is present ● TP sensor signal circuit is shorted to ground (intermittent fault) ● TP sensor signal circuit is open (an intermittent fault) ● TP sensor is damaged or has failed (it may be cracked)
P0121 **2T CCM, MIL: Yes** 1996, 1997, 1998 Achieva, Beretta, Ciera, Cavalier, Century, Corsica Grand Am, Malibu, S/T Blazer, S/T Pickup, Skylark 2.2L VIN 4, 2.4L VIN T Transmissions: M/T	**TP Sensor Signal Range/Performance Conditions:** DTC P0106, P0107, P0108, P0171, P0172, P0200, P0300, P0301, P0302, P0303, P0304, P0325, P0335, P0341, P0342, P0404, P0405, P0440, P0442, P0452, P0453, P0502, P0506, P0507, P0601, P0602, P1441 not set, engine started, ECT sensor more than 68ºF, TP angle stable (± 2%), MAP sensor less than 30 kPa, and the PCM detected the TP angle was more than 2% at 0 rpm, or more than 10% at 800 rpm, or more than 20% at 1600 rpm, or more than 25% at 2400 rpm, or more than 30% at 3200 rpm, or more than 35% at 4000 rpm, or more than 35% at 4800 rpm, or more than 40% at 5600 rpm or more than 40% at 6400 rpm. **Possible Causes** ● TP sensor signal circuit is shorted to ground (intermittent fault) ● TP sensor signal circuit is open (an intermittent fault) ● TP sensor VREF circuit is open or shorted (intermittent fault) ● TP sensor is damaged or it failed (may be cracked or sticking)
P0121 **2T CCM, MIL: Yes** 1996, 1997, 1998, 1999 Achieva, Beretta, Ciera, Corsica, Century, Grand Am, Grand Prix, Lumina, Malibu, Monte Carlo, Skylark 3.1L VIN M, 3.4L VIN X Transmissions: All	**TP Sensor Signal Range/Performance Conditions:** DTC P0107, P0108, P0122 and P0123 not set, engine started, MAP sensor less than 55 kPa and steady for 5 seconds, ECT sensor more than 167ºF, and the PCM detected the TP sensor was more than the Predicted value, or with the MAP sensor more than 70 kPa, the TP sensor was less than the Predicted value for 10 seconds. **Possible Causes** ● TP sensor signal circuit open or shorted to ground (intermittent) ● TP sensor VREF circuit is open or shorted (intermittent fault) ● TP sensor is damaged or it failed (may be cracked or sticking) ● PCM has failed
P0121 **2T CCM, MIL: Yes** 1996, 1997, 1998, 1999, 2000, 2001, 2002, 2003, 2004, 2005 Aurora, Century, LeSabre, Park Avenue, Regal 3.1L VIN J, 3.1L VIN M, 3.4L VIN X, 3.5L V6 VIN H, 3.8L VIN 1, 3.8L VIN K, 4.0L V8 VIN C Transmissions: All	**TP Sensor Signal Range/Performance Conditions:** DTC P0106, P0107, P0108, P0122, P0123, P1106, P1107, P1121, or P1122 not set, engine runtime over 2 minutes, ECT sensor over 167ºF, MAP sensor less than 50 kPa (skewed low test), or more than 70 kPa (skewed high test), and PCM detected the TP sensor was more than the expected value with the MAP sensor under 50 kPa, or less than the expected value with the MAP sensor over 70 kPa for 10 seconds. The TP sensor signal is used to determine the throttle plate angle for various engine controls. **Possible Causes** ● TP sensor signal circuit open or shorted to ground (intermittent) ● TP sensor VREF circuit is open or shorted (intermittent fault) ● TP sensor is damaged or it failed (may be cracked or sticking) ● PCM has failed
P0121 **2T CCM, MIL: Yes** 1996, 1997 Camaro, Corvette, Caprice, Firebird & Fleetwood 4.3L VIN W, 5.7L VIN 5, 5.7L VIN P engines Transmissions: All	**TP Sensor Signal Range/Performance Conditions:** DTC P0107, P0108, P0122 and P0123 not set, engine started, MAP sensor under 60 kPa with TP angle steady, and the PCM detected the Actual TP angle was under the Predicted TP angle for 38 sec's. **Possible Causes** ● TP sensor signal circuit open or shorted to ground (intermittent) ● TP sensor VREF circuit is open or shorted (intermittent fault) ● TP sensor is damaged or it failed (may be cracked or sticking)
P0121 **2T CCM, MIL: Yes** 1996, 1997, 1998, 1999, 2000, 2001, 2002, 2003, 2004, 2005 Alero, Cutlass, Impala, Malibu, Montana, Silhouette, Venture, Rendezvous 3.1L VIN J, 3.1L VIN M, 3.4L VIN E Transmissions: All	**TP Sensor Signal Range/Performance Conditions:** DTC P0106, P0107, P0108, P0122, P0123, P1106, P1107, P1121 and P1122 not set, engine runtime over 2 minutes, ECT sensor more than 167ºF, MAP sensor less than 50 kPa for TP sensor skewed "high" test, or MAP sensor more than 70 kPa for TP sensor skewed "low" test, MAP sensor steady for 5 seconds, and the PCM detected the TP sensor was too high with the MAP sensor below 50 kPa or it was too low with MAP sensor over 70 kPa for 10 seconds. **Possible Causes** ● TP sensor signal circuit is shorted to ground (intermittent fault) ● TP sensor signal circuit is open (an intermittent fault) ● TP sensor VREF circuit is open or shorted (intermittent fault) ● TP sensor is damaged or it failed (may be cracked or sticking)

OBD II Trouble Code List (P0xxx Codes)

DTC	Trouble Code Title, Conditions & Possible Causes
P0121 **2T CCM, MIL: Yes** 1996, 1997, 1998, 1999, 2000, 2001, 2002 C/K Cab & Chassis, C/K Series Truck, G Series 4.3L VIN W, 4.3L VIN X, 5.0L VIN M, 5.7L VIN K, 5.7L VIN R, 7.4L VIN J Transmissions: All	**TP Sensor Signal Range/Performance Conditions:** DTC P0101, P0102, P0103, P0106, P0107, P0108, P0122, P0123, P0506, P0507 are not set, engine runtime over 2 minutes, ECT sensor more than 158ºF, MAP sensor under 43 kPa for the TP sensor skewed "high" test, or the MAP sensor over 67 kPa for the TP sensor skewed "low" test, MAP and TP sensor steady for two seconds, and the PCM detected the TP sensor was more than the predicted value with the MAP less than 43 kPa, or it was less than a predicted value with the MAP sensor above 67 kPa for 1 seconds. **Possible Causes** • TP sensor signal circuit open or shorted to ground (intermittent) • TP sensor VREF circuit is open or shorted (intermittent fault) • TP sensor is damaged or it failed (may be cracked or sticking) • TSB 76-65-04 contains a repair procedure for this code
P0121 **2T CCM, MIL: Yes** 2003, 2004, 2005 C/K Series Truck, G Vans & L/M Series Vans 4.3L VIN W, 4.3L VIN X Transmissions: All	**TP Sensor Signal Range/Performance Conditions:** DTC P0101, P0102, P0103, P0106, P0107, P0108, P0122, P0123, P0506 and P0507 not set, engine runtime 2 minutes, ECT sensor over 158ºF, MAP sensor under 43 kPa for the TP sensor "skewed" high test, or the MAP sensor above 67 kPa for TP sensor "skewed" low test, MAP sensor steady and TP sensor steady (± 1.5%) for 2 seconds, and the PCM detected the TP sensor was more than the predicted value with the MAP under 43 kPa, or it was below a predicted value with MAP sensor more than 67 kPa for 1 second. **Possible Causes** • TP sensor signal circuit open or shorted to ground (intermittent) • TP sensor VREF circuit is open or shorted (intermittent fault) • TP sensor is damaged or it failed (may be cracked or sticking) • TSB 76-65-04 contains a repair procedure for this code
P0121 **2T CCM, MIL: Yes** 1996, 1997, 1998, 1999, 2000, 2001, 2002, 2003, 2004, 2005 C/K Series Truck, G Vans 6.5L VIN F, 6.5L VIN S, 6.6L VIN 1 Diesel engines Transmissions: All	**Accelerator Pedal Position Sensor 1 Performance Conditions:** Engine speed over 300 rpm, system voltage over 8.0v, and the PCM detected a difference of over 230 mv between the APP 1 and APP 2 signals, a difference between the APP 1 and APP 3 signals of over 500 mv for 2 seconds. STS lamp is "on" with multiple APP faults. **Possible Causes** • APP sensor circuit open or shorted to ground (intermittent) • APP sensor VREF circuit is open or shorted (intermittent fault) • APP sensor is damaged or it failed (may be cracked or sticking)
P0121 **2T CCM, MIL: Yes** 1999, 2000, 2001, 2002 Avalanche, C/K Trucks, Envoy, Escalade, G Vans & TrailBlazer 4.8L VIN V, 5.3L VIN P, 5.3L VIN T, 5.3L VIN Z, 6.0L VIN N, 6.0L VIN U, 8.1L VIN G Transmissions: All	**TP Sensor Signal Range/Performance Conditions:** DTC P0106, P0107, P0108, P0122, P0123, P1106, P1107, P1121 and P1122 not set, engine runtime over 2 minutes, ECT sensor more than 140º F, MAP sensor under 55 kPa for the TP sensor skewed "high" test, or the MAP sensor over 65 kPa for the TP sensor skewed "low" test, then with the MAP sensor steady for one second, the PCM detected the TP sensor was more than the predicted value with the MAP less than 55 kPa, or it was less than a predicted value with the MAP sensor above 65 kPa for 1 seconds. **Possible Causes** • TP sensor signal circuit open or shorted to ground (intermittent) • TP sensor is damaged or it failed (may be cracked or sticking) • TSB 76-65-04 contains a repair procedure for this code
P0121 **2T CCM, MIL: Yes** 1996, 1997, 1998, 1999, 2000, 2001, 2002, 2003, 2004, 2005 DeVille, Eldorado, Seville 4.6L VIN 9, 4.6L VIN Y Transmissions: All	**TP Sensor Circuit Insufficient Activity Conditions:** DTC P0106, P0107, P0108, P0122, P0123, P0506 and P0507 not set, engine runtime over 30 seconds, ECT sensor more than 32ºF, IAC position from 10-160 counts, MAP sensor less than 55 kPa for a "skewed" high test, or more than 64 kPa for a "skewed" low test, and the PCM detected the TP sensor signal was more than the predicted value when the MAP sensor pressure less than 55 kPa, or the PCM detected the TP sensor voltage was less than the predicted value when the MAP sensor pressure was more than 64 kPa. **Possible Causes** • TP sensor signal circuit open or shorted to ground (intermittent) • TP sensor VREF circuit is open or shorted (intermittent fault) • TP sensor is damaged or it failed (may be cracked or sticking)
P0121 **2T CCM, MIL: Yes** 2003, 2004, 2005 CTS (Cadillac) 2.6L V6 VIN M & 3.2L V6 VIN N engines Transmissions: All	**TP Sensor 1-2 Correlation Error Conditions:** Engine started, TP sensor voltage between 1.7-4.6v, and the PCM detected the TP Sensor 1 disagreed more than 6% with the TP Sensor 2, or the TP Sensor 1 disagreed more than 9% with the predicted value for 280 ms. When the throttle plate is in the closed position, the TP Sensor 1 signal voltage is near the low reference and increases as the throttle plate is opened. TP Sensor 2 signal voltage at closed throttle is near the 5v VREF. It decreases as the throttle plate opens. **Possible Causes** • TP sensor connector is damaged, open or shorted • TP Sensor low reference circuit is open or has high resistance • TP Sensor 1 signal is open, shorted to ground or to 5v VREF • TP Sensor 1 VREF circuit is open or shorted to ground • Throttle body assembly is damaged or it has failed • PCM has failed

OBD II Trouble Code List (P0xxx Codes)

DTC	Trouble Code Title, Conditions & Possible Causes
P0121 **2T CCM, MIL: Yes** 1995, 1996, 1997, 1998, 1999, 2000 Camaro, Corvette & Firebird 3.8L VIN K, 5.7L VIN G, 5.7L VIN S engines Transmissions: All	**TP Sensor Circuit Insufficient Activity Conditions:** DTC P0107, P0108, P0122, P0123, P1122P and 1123 not set, engine started, engine runtime over 2 minutes, ECT sensor more than 158°F, MAP sensor less than 40 kPa, or with the MAP more than 67 kPa and steady, and the PCM detected the predicted TP angle did not match the Actual throttle angle for 20-40 seconds. **Possible Causes** ● TP sensor signal circuit open or shorted to ground (intermittent) ● TP sensor VREF circuit is open or shorted (intermittent fault) ● TP sensor is damaged or it failed (may be cracked or sticking)
P0121 **2T CCM, MIL: Yes** 2003, 2004, 2005 G Series Van, Van Cargo 4.8L VIN V, 5.3L VIN T, 6.0L VIN U engines Transmissions: All	**TP Sensor Signal Range/Performance Conditions:** DTC P0106, P0107, P0108, P0122 and P0123 not set, engine started, engine runtime over 2 minutes, ECT sensor more than 140° F, MAP sensor under 55 kPa for the TP sensor skewed "high" test, or the MAP sensor over 65 kPa for the TP sensor skewed "low" test, then with the MAP sensor steady for one second, the PCM detected the TP sensor was more than the predicted value with the MAP less than 55 kPa, or it was less than a predicted value with the MAP sensor above 65 kPa for 1 seconds. **Possible Causes** ● TP sensor signal circuit open or shorted to ground (intermittent) ● TP sensor VREF circuit is open or shorted (intermittent fault) ● TP sensor is damaged or it failed (may be cracked or sticking) ● TSB 76-65-04 contains a repair procedure for this code
P0121 **2T CCM, MIL: Yes** 1996, 1997, 1998, 1999, 2000, 2001, 2002 Camaro, Corvette & Firebird 5.7L VIN G engine Transmissions: All	**TP Sensor Signal Range/Performance Conditions:** DTC P0606, P1517 and P1518 not set, key in the crank or run mode, system voltage over 5.23v, and the PCM detected the APP 1 disagreed with the APP 2 sensor by over 10.5%, or the APP 1 sensor disagreed with APP 3 sensor by over 13% for 1 second. **Possible Causes** ● TP sensor signal circuit is shorted to ground (intermittent fault) ● TP sensor signal circuit is open (an intermittent fault) ● TP sensor VREF circuit is open or shorted (intermittent fault) ● TP sensor is damaged or it failed (may be cracked or sticking)
P0121 **2T CCM, MIL: Yes** 2001, 2002, 2003, 2004, 2005 Aurora 3.5L V6 VIN H engine Transmissions: A/T	**TP Sensor Signal Range/Performance Conditions:** DTC P0106, P0107, P0108, P0122 and P0123 not set, engine runtime over 2 minutes, MAP sensor less than 50 kPa for the TP sensor skewed "high'" test, or MAP sensor more than 70 kPa for the TP sensor skewed "low" test, MAP sensor steady for 5 seconds, and the PCM detected the TP sensor was more than a predicted value with the MAP sensor less than 50 kPa, or the TP sensor was less than a predicted value with the MAP over 70 kPa in the test. **Possible Causes** ● TP sensor signal circuit is shorted to ground (intermittent fault) ● TP sensor signal circuit is open (an intermittent fault) ● TP sensor VREF circuit is open or shorted (intermittent fault) ● TP sensor is damaged or it failed (may be cracked or sticking)
P0121 **2T CCM, MIL: Yes** 1996, 1997, 1998, 1999, 2001, 2002, 2003, 2004, 2005 Aurora 4.0L V8 VIN C Transmissions: A/T	**TP Sensor Signal Range/Performance Conditions:** DTC P0106, P0107, P0108, P0122, P0123, P0506, P0507, P1106, P1107, P1121, or P1122 are not set, engine started, engine runtime over 30 seconds, ECT sensor more than 32°F, MAP sensor less than 55 kPa for the TP sensor skewed "high'" test, or MAP sensor more than 64 kPa for the TP sensor skewed "low" test, MAP sensor steady for 5 seconds, and the PCM detected the TP sensor was more than a predicted value with the MAP sensor less than 55 kPa, or it was less than a predicted value with the MAP over 64 kPa. **Possible Causes** ● TP sensor signal circuit is shorted to ground (intermittent fault) ● TP sensor signal circuit is open (an intermittent fault) ● TP sensor VREF circuit is open or shorted (intermittent fault) ● TP sensor is damaged or it failed (may be cracked or sticking)
P0121 **2T CCM, MIL: Yes** 1995, 1996, 1997, 1998, 1999, 2000, 2001, 2002, 2003, 2004, 2005 S/T Blazer & S/T Pickup 4.3L VIN W, 4.3L VIN X Transmissions: All	**TP Sensor Signal Range/Performance Conditions:** DTC P0101, P0102, P0103, P0106, P0106, P0107, P0108, P0122, P0123, P0506, P0507, P1106, P1107, P1121 and P1122 not set, engine started, engine runtime over 2 minutes, ECT sensor more than 158°F, MAP sensor less than 43 kPa for TP sensor skewed high test, or MAP sensor more than 67 kPa for the TP sensor skewed low test, then with the TP angle steady (± 1.5%) for 2 seconds, the PCM detected the TP sensor was more than a preset value with the MAP sensor less than 43 kPa, or the TP sensor was less than a preset value with the MAP more than 67 kPa for 1 second. **Possible Causes** ● TP sensor signal circuit is shorted to ground (intermittent fault) ● TP sensor signal circuit is open (an intermittent fault) ● TP sensor VREF circuit is open or shorted (intermittent fault) ● TP sensor is damaged or it failed (may be cracked or sticking) ● TSB 76-65-04 contains a repair procedure for this code

OBD II Trouble Code List (P0xxx Codes)

DTC	Trouble Code Title, Conditions & Possible Causes
P0122 **2T CCM, MIL: Yes** 1996, 1997, 1998, 1999, 2000, 2001, 2002, 2003, 2004, 2005 Achieva, Alero, Beretta, Century, Ciera, Cavalier, Envoy, Grand Am, Sunfire Skyhawk, TrailBlazer, S/T Blazer & S/T Pickup 2.2L VIN 4, 2.2L VIN 5, 2.2L VIN F, 2.2L VIN H, 2.4L VIN T Transmissions: All	**TP Sensor Circuit Low Input Conditions:** Key on or engine running; and the PCM detected the TP sensor indicated less than 0.10v for 2 seconds. The PCM uses the TP sensor signal to determine the throttle plate angle for various engine controls. The TP sensor is a potentiometer type sensor with a 5v reference circuit, ground circuit and varying signal circuit (0-5v). **Possible Causes** ● TP sensor signal circuit is shorted to sensor ground ● TP sensor signal circuit is open (except 1996-1998 models) ● TP sensor VREF circuit is open between sensor and PCM ● TP sensor is damaged or it failed (it may be shorted) ● PCM is damaged or has failed
P0122 **2T CCM, MIL: Yes** 1996, 1997, 1998, 1999 Achieva, Beretta, Ciera, Corsica, Century, Grand Am, Grand Prix, Lumina, Malibu, Monte Carlo 3.1L VIN M, 3.4L VIN X Transmissions: All	**TP Sensor Circuit Low Input Conditions:** Key on or engine running; and the PCM detected the TP sensor signal was less than 0.16v for over 1 second. The PCM uses the TP sensor signal to determine the throttle plate angle for various engine controls. The TP sensor is a potentiometer type sensor with a 5v reference circuit, ground circuit and varying signal circuit (0-5v). **Possible Causes** ● TP sensor signal circuit is shorted to sensor ground ● TP sensor signal circuit is open (except 1996-1998 models) ● TP sensor VREF circuit is open between sensor and PCM ● TP sensor is damaged or it failed (it may be shorted) ● PCM is damaged or has failed
P0122 **1T CCM, MIL: Yes** 1995 Century, Cutlass Ciera, Cutlass Supreme, Grand Prix, Lumina, Monte Carlo Park Avenue Regal, Riviera, 88' & 98' 3.1L VIN M, 3.4L VIN X, 3.8L VIN 1, 3.8L VIN K, 3.8L VIN L engines Transmissions: All	**TP Sensor Circuit Low Input Conditions:** Key on or engine running; and the PCM detected the TP sensor signal was less than 0.16v for 4-10 seconds. The PCM uses the TP sensor signal to determine the throttle plate angle for various engine controls. The TP sensor is a potentiometer type sensor with a 5v reference circuit, ground circuit and a varying signal circuit (0-5v). **Possible Causes** ● TP sensor signal circuit is shorted to sensor ground ● TP sensor signal circuit is open ● TP sensor VREF circuit is open between sensor and PCM ● TP sensor is damaged or it failed (it may be shorted) ● PCM is damaged or has failed
P0122 **2T CCM, MIL: Yes** 1996, 1997, 1998, 1999, 2000, 2001, 2002, 2003, 2004, 2005 Aurora, Century, LeSabre, Park Avenue, Regal 3.1L VIN J, 3.1L VIN M, 3.4L VIN X, 3.5L V6 VIN H, 3.8L VIN 1, 3.8L VIN K, 4.0L V8 VIN C Transmissions: All	**TP Sensor Circuit Low Input Conditions:** DTC P1635 and P1639 not set Key on or engine running; and the PCM detected the TP sensor was less than 0.10v for 1 second. The PCM uses the TP sensor signal to determine the throttle plate angle for various engine controls. It is a potentiometer type sensor with a 5v reference circuit, ground circuit and varying signal circuit (0-5v). **Possible Causes** ● TP sensor signal circuit is shorted to sensor ground ● TP sensor signal circuit is open (except 1996-1998 models) ● TP sensor VREF circuit is open between sensor and PCM ● TP sensor is damaged or it failed (it may be shorted) ● PCM is damaged or has failed
P0122 **1T CCM, MIL: Yes** 1996, 1997 Camaro, Corvette, Caprice, Firebird & Fleetwood 4.3L VIN W, 5.7L VIN 5, 5.7L VIN P engine Transmissions: All	**TP Sensor Circuit Low Input Conditions:** Key on or engine running; and the PCM detected the TP sensor signal was less than 0.20v for over 1 second. The PCM uses the TP sensor signal to determine the throttle plate angle for various engine controls. The TP sensor is a potentiometer type sensor with a 5v reference circuit, ground circuit and varying signal circuit (0-5v). **Possible Causes** ● TP sensor signal circuit is shorted to sensor ground ● TP sensor VREF circuit is open between sensor and PCM ● TP sensor is damaged or it failed (it may be shorted) ● PCM is damaged or has failed ● TSB 61-65-52 contains a repair procedure for this code
P0122 **1T CCM, MIL: Yes** 1996, 1997, 1998 Alero, Cutlass, Impala, Malibu, Montana, Silhouette, Venture 3.1L VIN J, 3.1L VIN M, 3.4L VIN E Transmissions: All	**TP Sensor Circuit Low Input Conditions:** Key on or engine running; and the PCM detected the TP sensor signal was less than 0.1v for over 1 second. The TP sensor is a potentiometer type sensor that connects to the PCM with a 5v reference circuit, ground circuit and varying signal circuit (0-5v). **Possible Causes** ● TP sensor signal circuit is shorted to sensor ground ● TP sensor VREF circuit is open between sensor and PCM ● TP sensor is damaged or it failed (it may be shorted) ● PCM is damaged or has failed

OBD II Trouble Code List (P0xxx Codes)

DTC	Trouble Code Title, Conditions & Possible Causes
P0122 **1T CCM, MIL: Yes** 1999, 2000, 2001, 2002, 2003, 2004, 2005 Alero, Aztek, Cutlass, Impala, Malibu, Montana, Silhouette, Venture, Rendezvous 3.1L VIN J, 3.1L VIN M, 3.4L VIN E Transmissions: All	**TP Sensor Circuit Low Input Conditions:** Key on or engine running; and the PCM detected the TP sensor signal was less than 0.1v for over 1 second. The PCM uses the TP sensor signal to determine the throttle plate angle for various engine controls. The TP sensor output is an analog signal that varies from 0-5v. **Possible Causes** ● TP sensor signal circuit is shorted to sensor ground ● TP sensor signal circuit is open ● TP sensor VREF circuit is open between sensor and PCM ● TP sensor is damaged or it failed (it may be shorted) ● PCM is damaged or has failed
P0122 **1T CCM, MIL: Yes** 1996, 1997, 1998, 1999, 2000, 2001, 2002 C/K Cab & Chassis, C/K Series Truck, G Series 4.3L VIN W, 4.3L VIN X, 6.0L VIN M, 5.7L VIN K, 5.7L VIN R, 7.4L VIN J Transmissions: All	**TP Sensor Circuit Low Input Conditions:** Key on or engine running; and the PCM detected the TP sensor indicated less than 0.10v for 1 second. The PCM uses the TP sensor signal to determine the throttle plate angle for various engine controls. The TP sensor output is an analog signal that varies from 0-5v. **Possible Causes** ● TP sensor signal circuit is shorted to sensor ground ● TP sensor VREF circuit is open between sensor and PCM ● TP sensor is damaged or it failed (it may be shorted) ● PCM is damaged or has failed
P0122 **1T CCM, MIL: Yes** 2003, 2004, 2005 C/K Series truck G Series Van, L/M Series Van 4.3L VIN W, 4.3L VIN X Transmissions: All	**TP Sensor Circuit Low Input Conditions:** Key on or engine running; and the PCM detected the TP sensor indicated less than 0.10v for 1 second. The PCM uses the TP sensor signal to determine the throttle plate angle for various engine controls. The TP sensor output is an analog signal that varies from 0-5v. **Possible Causes** ● TP sensor signal circuit is shorted to sensor ground ● TP sensor VREF circuit is open between sensor and PCM ● TP sensor is damaged or it failed (it may be shorted) ● PCM is damaged or has failed
P0122 **2T CCM, MIL: Yes** 1999, 2000, 2001, 2002 Avalanche, C/K Series Truck, Envoy, Escalade, G Vans, TrailBlazer 4.8L VIN V, 5.3L VIN P, 5.3L VIN T, 5.3L VIN Z, 6.0L VIN N, 6.0L VIN U, 8.1L VIN G Transmissions: All	**TP Sensor Circuit Low Input Conditions:** Key on or engine running; and the PCM detected the TP sensor indicated less than 0.25v for 1 second. The PCM uses the TP sensor signal to determine the throttle plate angle for various engine controls. The TP sensor output is an analog signal that varies from 0-5v. **Possible Causes** ● TP sensor signal circuit is shorted to sensor ground ● TP sensor signal circuit is open (except 1996-1998 models) ● TP sensor VREF circuit is open between sensor and PCM ● TP sensor is damaged or it failed (it may be shorted) ● PCM is damaged or has failed
P0122 **2T CCM, MIL: Yes** 1996, 1997, 1998, 1999, 2000, 2001, 2002 C/K Series Truck, G Vans 6.5L VIN F, 6.5L VIN S, 6.6L VIN 1 Diesel engines Transmissions: All	**Accelerator Pedal Position Sensor 1 Circuit Low Input Conditions:** Key on or engine running; and the PCM detected the APP 1 sensor signal was less than 250 mv for 2 seconds. The PCM will turn "on" the STS lamp if multiple APP sensor faults are set. **Possible Causes** ● APP sensor signal circuit shorted to chassis or sensor ground ● APP sensor is damaged or it failed (it may be shorted) ● PCM is damaged or has failed
P0122 **2T CCM, MIL: Yes** 1996, 1997, 1998, 1999, 2000, 2001, 2002, 2003, 2004, 2005 DeVille, Eldorado, Seville 4.6L VIN 9, 4.6L VIN Y Transmissions: All	**TP Sensor Circuit Low Input Conditions:** Engine started, IAC position counts from 10-110, and the PCM detected the TP sensor was less than 0.25v for 1 second. Note: The PCM disables TCC, calculates a TP sensor value based on MAP and RPM signals and disables 4th gear when this code is set. **Possible Causes** ● TP sensor signal circuit is shorted to ground or open (except 1996-1998 models) ● TP sensor VREF circuit is open between sensor and PCM ● TP sensor is damaged or it failed (it may be shorted) ● PCM is damaged or has failed
P0122 **2T CCM, MIL: Yes** 2003, 2004, 2005 CTS (Cadillac) 2.6L V6 VIN M & 3.2L V6 VIN N engines Transmissions: All	**TP Sensor 1 Circuit Low Input Conditions:** Engine started; system voltage over 10v, and PCM detected the TP Sensor 1 signal was less than 0.195v for 140 ms. The TP Sensor 1 signal is used to determine the throttle plate angle. **Possible Causes** ● APP2 Sensor 5v VREF circuit is shorted to ground ● ECM power ground circuit is open (a high resistance condition) ● TP Sensor low reference circuit is open or has high resistance ● TP Sensor 1 signal is open, shorted to ground or low reference ● TP Sensor 1 VREF circuit is open or shorted to ground or to B+ ● Throttle body assembly is damaged or it has failed

OBD II Trouble Code List (P0xxx Codes)

DTC	Trouble Code Title, Conditions & Possible Causes
P0122 **2T CCM, MIL: Yes** 1995, 1996, 1997, 1998, 1999, 2000, 2001, 2002 Camaro, Corvette & Firebird 3.8L VIN K, 5.7L VIN G, 5.7L VIN S engines Transmissions: All	**TP Sensor Circuit Low Input Conditions:** Key on or engine running; and the PCM detected the TP sensor signal was less than 0.20v for 10 seconds. The PCM uses the TP sensor signal to determine the throttle plate angle for various engine controls. The TP sensor is a potentiometer type sensor with a 5v reference circuit, ground circuit and varying signal circuit (0-5v). **Possible Causes** ● TP sensor signal circuit is shorted to sensor ground ● TP sensor signal circuit is open (except 1996-1998 models) ● TP sensor VREF circuit is open between sensor and PCM ● TP sensor is damaged or it failed (it may be shorted) ● PCM is damaged or has failed
P0122 **2T CCM, MIL: Yes** 1996, 1997, 1998, 1999, 2001, 2002, 2003, 2004, 2005 Aurora 3.5L V6 VIN H, 4.0L V8 VIN C engines Transmissions: A/T	**TP Sensor Circuit Low Input Conditions:** Engine speed 600 rpm or greater, IAC position at 10-110 counts, and the PCM detected the TP signal was less than 0.10v. Note: The PCM disables TCC, calculates TP angle value based on MAP and RPM signals and disables 4th gear when this code sets. **Possible Causes** ● TP sensor signal circuit is shorted to sensor ground ● TP sensor signal circuit is open (except 1996-1998 models) ● TP sensor VREF circuit is open between sensor and PCM ● TP sensor is damaged or it failed (it may be shorted) ● PCM is damaged or has failed
P0122 **2T CCM, MIL: Yes** 2003, 2004, 2005 G Series Van, Cargo Van 4.8L VIN V, 5.3L VIN T, 6.0L VIN U Transmissions: All	**TP Sensor Circuit Low Input Conditions:** Key on or engine running; and the PCM detected the TP sensor indicated less than 0.25v for 1 second. The PCM uses the TP sensor signal to determine the throttle plate angle for various engine controls. The TP sensor is a potentiometer type sensor with a 5v reference circuit, ground circuit and varying signal circuit (0-5.0v). **Possible Causes** ● TP sensor signal circuit is shorted to sensor ground ● TP sensor signal circuit is open (except 1996-1998 models) ● TP sensor VREF circuit is open between sensor and PCM ● TP sensor is damaged or it failed (it may be shorted) ● PCM is damaged or has failed
P0122 **2T CCM, MIL: Yes** 1995, 1996, 1997, 1998, 1999, 2000, 2001, 2002, 2003, 2004, 2005 S/T Blazer & S/T Pickup 4.3L VIN W, 4.3L VIN X Transmissions: All	**TP Sensor Circuit Low Input Conditions:** Key on or engine running; and the PCM/VCM detected the TP sensor signal was less than 0.10v, conditions met for 1 second. **Possible Causes** ● TP sensor signal circuit is shorted to sensor ground ● TP sensor signal circuit is open (except 1995-1998 models) ● TP sensor VREF circuit is open between sensor and PCM ● TP sensor is damaged or it failed (it may be shorted) ● PCM is damaged or has failed ● TSB 81-65-39 contains a repair procedure for this code
P0122 **2T CCM, MIL: Yes** 2002, 2003, 2004, 2005 Envoy & TrailBlazer 4.2L VIN S engine Transmissions: All	**TP Sensor Circuit Low Input Conditions:** DTC P1635 not set, engine cranking or running, system voltage over 5.23v, and the PCM detected the TP Sensor 1 and TP Sensor 2 signals were out of their calibrated range. **Possible Causes** ● TP sensor 1 or 2 circuit is shorted to chassis or sensor ground ● TP sensor VREF circuit is open between sensor and PCM ● TP sensor is damaged or it failed (it may be shorted) ● PCM is damaged or has failed
P0123 **1T CCM, MIL: Yes** 1995 Century, Cutlass Ciera, Cutlass Supreme, Grand Prix, Lumina, Monte Carlo Park Avenue Regal, Riviera, 88' & 98' 3.1L VIN M, 3.4L VIN X, 3.8L VIN 1, 3.8L VIN K, 3.8L VIN L engines Transmissions: All	**TP Sensor Circuit High Input Conditions:** DTC P0102 and P0103 not set, engine started, and the PCM detected the TP sensor was above 4.80v, or with MAF sensor below 15 g/sec, the TP sensor was over 1.1v for 5 seconds. **Possible Causes** ● TP sensor signal circuit is shorted to VREF or system power ● TP sensor ground circuit is open between the sensor and PCM ● TP sensor is damaged or it failed (it may be open) ● PCM is damaged or has failed

OBD II Trouble Code List (P0xxx Codes)

DTC	Trouble Code Title, Conditions & Possible Causes
P0123 **2T CCM, MIL: Yes** 1996, 1997, 1998, 1999, 2000, 2001, 2002, 2003, 2004, 2005 Achieva, Alero, Beretta, Century, Ciera, Cavalier, Grand Am, S/T Blazer & Pickup, Sunfire, Skylark 2.2L VIN 4, 2.2L VIN 5, 2.2L VIN F, 2.2L VIN H, 2.4L VIN T Transmissions: All	**TP Sensor Circuit High Input Conditions:** DTC P0106 and P0107 not set, engine started, MAP sensor less than 60 kPa, engine speed less than 1500 rpm, and the PCM detected the TP sensor signal was more than 4.80v for 2 seconds. **Possible Causes** ● TP sensor signal circuit is shorted to VREF or system power ● TP sensor ground circuit is open between the sensor and PCM ● TP sensor is damaged or it failed (it may be open) ● PCM is damaged or has failed
P0123 **2T CCM, MIL: Yes** 1996, 1997, 1998, 1999 3.1L VIN 4, 3.4L VIN X Achieva, Beretta, Ciera, Corsica, Century, Grand Am, Grand Prix, Lumina, Malibu, Monte Carlo Transmissions: All	**TP Sensor Circuit High Input Conditions:** Key on or engine running; and the PCM detected the TP sensor signal was more than 4.90v for over 1 second. **Possible Causes** ● TP sensor signal circuit is shorted to VREF or system power ● TP sensor ground circuit is open between the sensor and PCM ● TP sensor is damaged or it failed (it may be open) ● PCM is damaged or has failed
P0123 **2T CCM, MIL: Yes** 1996, 1997, 1998, 1999, 2000, 2001, 2002 Achieva, Cavalier, Grand Am, Malibu, S/T Pickup, Skylark 2.4L VIN T Transmissions: All	**TP Sensor Circuit High Input Conditions:** DTC P0106 and P0107 not set, engine started, MAP sensor less than 60 kPa, engine speed less than 1500 rpm, and the PCM detected the TP sensor indicated over 3.90v for 2 seconds. **Possible Causes** ● TP sensor signal circuit is shorted to VREF or system power ● TP sensor ground circuit is open between the sensor and PCM ● TP sensor is damaged or it failed (it may be open) ● PCM is damaged or has failed
P0123 **2T CCM, MIL: Yes** 1996, 1997, 1998, 1999, 2000, 2001, 2002, 2003, 2004, 2005 Aurora, Century, LeSabre, Park Avenue, Regal 3.1L VIN J, 3.1L VIN M, 3.4L VIN X, 3.5L V6 VIN H, 3.8L VIN 1, 3.8L VIN K, 4.0L V8 VIN C Transmissions: All	**TP Sensor Circuit High Input Conditions:** Engine started engine speed over 600 rpm, TP sensor more than 1.0v, calculated airflow less than 17 g/sec, and the PCM detected the TP sensor signal was more than 4.90v for over 5 seconds. The TP sensor is a potentiometer type sensor with a 5v reference circuit, ground circuit and varying signal circuit (0-5v). **Possible Causes** ● TP sensor signal circuit is shorted to VREF or system power ● TP sensor ground circuit is open between the sensor and PCM ● TP sensor is damaged or it failed (it may be open) ● PCM is damaged or has failed
P0123 **2T CCM, MIL: Yes** 1996, 1997 Camaro, Corvette, Caprice, Firebird & Fleetwood 4.3L VIN W, 5.7L VIN 5, 5.7L VIN P engine Transmissions: All	**TP Sensor Circuit High Input Conditions:** Key on or engine running; and the PCM detected the TP sensor signal was over 4.80v for 1 second during the CCM test period. **Possible Causes** ● TP sensor signal circuit is shorted to VREF or system power ● TP sensor ground circuit is open between the sensor and PCM ● TP sensor is damaged or it failed (it may be open) ● PCM is damaged or has failed
P0123 **2T CCM, MIL: Yes** 1996, 1997, 1998, 1999, 2000, 2001, 2002, 2003, 2004, 2005 Alero, Aztek, Cutlass, Impala, Malibu, Montana, Silhouette, Venture, Rendezvous 3.1L VIN J, 3.1L VIN M, 3.4L VIN E Transmissions: All	**TP Sensor Circuit High Input Conditions:** Key on or engine running; and the PCM detected the TP sensor signal was more than 4.90v for 1 second. Rotation of the TP sensor rotor from closed throttle position to the wide open throttle (WOT) position provides the PCM with a signal voltage from below 1.0v to over 4.0v. **Possible Causes** ● TP sensor signal circuit is shorted to VREF or system power ● TP sensor ground circuit is open between the sensor and PCM ● TP sensor is damaged or it failed (it may be open) ● PCM is damaged or has failed
P0123 **2T CCM, MIL: Yes** 2003, 2004, 2005 C/K Series Truck, G Series Van, L/M Vans 4.3L VIN W, 4.3L VIN X Transmissions: All	**TP Sensor Circuit High Input Conditions:** DTC P1635 and P1639 not set, Key on or engine running; and the PCM detected the TP sensor signal was more than 4.70-4.90v for over 1 second during the CCM test. **Possible Causes** ● TP sensor signal circuit is shorted to VREF or system power ● TP sensor ground circuit is open between the sensor and PCM ● TP sensor is damaged or it failed (it may be open) ● PCM is damaged or has failed

OBD II Trouble Code List (P0xxx Codes)

DTC	Trouble Code Title, Conditions & Possible Causes
P0123 **2T CCM, MIL: Yes** 1996, 1997, 1998, 1999, 2000, 2001, 2002 C/K Cab & Chassis, C/K Series Truck, G Series 4.3L VIN W, 4.3L VIN X, 5.0L VIN M, 5.7L VIN K, 5.7L VIN R, 7.4L VIN J Transmissions: All	**TP Sensor Circuit High Input Conditions:** DTC P1635 and P1639 not set, Key on or engine running; and the PCM detected the TP sensor signal was more than 4.70-4.90v for over 1 second during the CCM test. **Possible Causes** • TP sensor signal circuit is shorted to VREF or system power • TP sensor ground circuit is open between the sensor and PCM • TP sensor is damaged or it failed (it may be open) • PCM is damaged or has failed
P0123 **2T CCM, MIL: Yes** 1999, 2000, 2001, 2002, 2003, 2004, 2005 Avalanche, C/K Series Truck, Envoy, Escalade, G Vans, TrailBlazer 4.8L VIN V, 5.3L VIN P, 5.3L VIN T, 5.3L VIN Z, 6.0L VIN N, 6.0L VIN U, 8.1L VIN G Transmissions: All	**TP Sensor Circuit High Input Conditions:** DTC P0641 and P0651 not set, Key on or engine running; and the PCM detected the TP sensor was more than 4.90v for 1 second during the CCM test period. **Possible Causes** • TP sensor signal circuit is shorted to VREF or system power • TP sensor ground circuit is open between the sensor and PCM • TP sensor is damaged or it failed (it may be open) • PCM is damaged or has failed
P0123 **2T CCM, MIL: Yes** 1996, 1997, 1998, 1999, 2000, 2001, 2002 C/K Series Truck, G Vans 6.5L VIN F, 6.5L VIN S, 6.6L VIN 1 Diesel engines Transmissions: All	**Accelerator Pedal Position Sensor 1 High Input Conditions:** Key on or engine running; and the PCM detected the APP 1 sensor input was more than 4.75v, condition met for 2 seconds. Note: The PCM will turn "on" the STS lamp if multiple APP sensor faults are detected during the CCM Rationality test and/or CCM test. **Possible Causes** • APP sensor signal circuit is shorted to VREF or system power • APP sensor ground circuit open between the sensor and PCM • APP sensor is damaged or it failed (it may be open) • PCM is damaged or has failed
P0123 **2T CCM, MIL: Yes** 1996, 1997, 1998, 1999, 2000, 2001, 2002, 2003, 2004, 2005 DeVille, Eldorado, Seville 4.6L VIN 9, 4.6L VIN Y Transmissions: A/T	**TP Sensor Circuit High Input Conditions:** Engine started IAC position from 10-110 counts, and the PCM detected the TP sensor was more than 4.80v for over 1 second during the CCM test period. **Possible Causes** • TP sensor connector is damaged, loose or open • TP sensor signal circuit is shorted to VREF or system power • TP sensor ground circuit is open between the sensor and PCM • TP sensor is damaged or it failed (it may be open) • PCM is damaged or has failed • TSB 71-65-25 contains a repair procedure for this code
P0123 **2T CCM, MIL: Yes** 2003, 2004, 2005 CTS (Cadillac) 2.6L V6 VIN M & 3.2L V6 VIN N engines Transmissions: All	**TP Sensor 1 Circuit High Input Conditions:** Key on or engine running, system voltage over 10v, and PCM detected the TP Sensor 1 indicated over 4.60v for 140 ms. This sensor has a VREF, low reference and signal circuit. **Possible Causes** • APP2 Sensor 5v VREF circuit is shorted to system power (B+) • ECM power ground circuit is open (a high resistance condition) • Throttle body assembly is damaged or it has failed • TP Sensor low reference circuit is open or has high resistance • TP Sensor 1 signal is shorted to 5v VREF, or to APP2 5v VREF • PCM has failed
P0123 **2T CCM, MIL: Yes** 1995, 1996, 1997, 1998, 1999, 2000, 2001, 2002 Camaro, Corvette & Firebird 3.8L VIN K, 5.7L VIN G, 5.7L VIN S Transmissions: All	**TP Sensor Circuit High Input Conditions:** Key on or engine running; and the PCM detected the TP sensor was more than 4.75v for 10 seconds. The TP sensor is a potentiometer with reference, ground circuit and signal circuit. **Possible Causes** • TP sensor signal circuit is shorted to VREF or system power • TP sensor ground circuit is open between the sensor and PCM • TP sensor is damaged or it failed (it may be open) • PCM is damaged or has failed
P0123 **2T CCM, MIL: Yes** 1996, 1997, 1998, 1999, 2001, 2002, 2003, 2004, 2005 Aurora 3.5L V6 VIN H, 4.0L V8 VIN C engines Transmissions: All	**TP Sensor Circuit High Input Conditions:** Engine started engine speed from 25-3000 rpm, IAC position from 10-110 counts, and the PCM detected the TP sensor signal was more than 4.96v during the CCM test period. **Possible Causes** • TP sensor signal circuit is shorted to VREF or system power • TP sensor ground circuit is open between the sensor and PCM • TP sensor is damaged or it failed (it may be open) • PCM is damaged or has failed

OBD II Trouble Code List (P0xxx Codes)

DTC	Trouble Code Title, Conditions & Possible Causes
P0123 **2T CCM, MIL: Yes** 2003, 2004, 2005 C/K Series Truck, G Series Van, Cargo Van 4.8L VIN V, 5.3L VIN T, 6.0L VIN U engines Transmissions: All	**TP Sensor Circuit High Input Conditions:** DTC P0641 and P0651 not set, Key on or engine running; and the PCM detected the TP sensor was more than 4.90v for 1 second during the CCM test period. **Possible Causes** ● TP sensor signal circuit is shorted to VREF or system power ● TP sensor ground circuit is open between the sensor and PCM ● TP sensor is damaged or it failed (it may be open) ● PCM is damaged or has failed
P0123 **2T CCM, MIL: Yes** 2002, 2003, 2004, 2005 Envoy & TrailBlazer 4.2L VIN S engine Transmissions: All	**TP Sensor Circuit High Input Conditions:** DTC P1635 not set, engine cranking or running, system voltage over 5.23v, and the PCM detected the TP Sensor 1 and TP Sensor 2 signals were out of their calibrated range. **Possible Causes** ● TP sensor 1 or 2 circuit is shorted to VREF or system power ● TP sensor ground circuit is open between the sensor and PCM ● TP sensor is damaged or it failed (it may be open internally) ● PCM is damaged or has failed
P0123 **2T CCM, MIL: Yes** 1995, 1996, 1997, 1998, 1999, 2000, 2001, 2002 S/T Blazer & S/T Pickup 4.3L VIN W, 4.3L VIN X Transmissions: All	**TP Sensor Circuit High Input Conditions:** DTC P1635 and P1639 not set, then Key on or engine running; and the PCM detected the TP sensor signal indicated more than 4.90v for over 1 second during the CCM test period. **Possible Causes** ● TP sensor connector is damaged, loose or shorted ● TP sensor signal circuit is shorted to VREF or system power ● TP sensor ground circuit is open between the sensor and PCM ● TP sensor is damaged or it failed (it may be open) ● PCM is damaged or has failed
P0123 **2T CCM, MIL: Yes** 2003, 2004, 2005 S/T Blazer & S/T Pickup 4.3L VIN W, 4.3L VIN X Transmissions: All	**TP Sensor Circuit High Input Conditions:** DTC P0641 and P0651 not set, then Key on or engine running; and the PCM detected the TP sensor signal indicated more than 4.90v for over 1 second during the CCM test period. **Possible Causes** ● TP sensor connector is damaged, loose or shorted ● TP sensor signal circuit is shorted to VREF or system power ● TP sensor ground circuit is open between the sensor and PCM ● TP sensor is damaged or it failed (it may be open) ● PCM is damaged or has failed
P0125 **2T CCM, MIL: Yes** 1996, 1997, 1998, 1999, 2000, 2001, 2002, 2003, 2004, 2005 Achieva, Alero, Beretta, Century, Ciera, Cavalier, Grand Am, S/T Blazer & Pickup, Sunfire, Skylark 2.2L VIN 4, 2.2L VIN 5, 2.2L VIN F, 2.2L VIN H Transmissions: All	**ECT Excessive Time To Enter Closed Loop Conditions:** DTC P0105, P0107, P0108, P0112-P0118, P0122, P0123, P0130-P0132, P0171, P0172, P0201-P0204, P0300, P0325, P0336, P0420, P0440-P0446, P0452, P0453, P0480, P0502, P0503, P0506 and P1441 not set, engine runtime from 30 seconds to 20 minutes, minimum IAT sensor more than 27°F, ECT sensor less than 149°F at startup, average MAF over 20 g/sec, vehicle driven more than 1.5 miles at over 25 mph, and the PCM determined after enough airflow had entered the ECT sensor did not rise to 104°F after 30 seconds. The ECT sensor monitors the temperature of the engine coolant. This input is used by the PCM for various engine controls and as enabling criteria for some engine diagnostics. The airflow into the engine is accumulated and used to determine if the vehicle has been driven within conditions that would allow the engine coolant to heat up normally to closed loop temperature. If the temperature does not increase normally or does not reach closed loop temperature, the diagnostics that use engine coolant temperature as enabling criteria may not run when expected. This code test will only run once per key cycle once the enabling conditions are met. If the PCM detects the calibrated amount of airflow and runtime are met, and the engine coolant does not reach Closed Loop temperature, DTC P0125 is set. **Possible Causes** ● Check the operation of the thermostat (it may be stuck open) ● Coolant level is too low, or the coolant mixture is incorrect ● ECT sensor signal circuit has a high resistance condition ● ECT sensor is damaged or it has failed
P0125 **2T CCM, MIL: Yes** 1996, 1997 Camaro, Corvette, Caprice, Firebird & Fleetwood 4.3L VIN W, 5.7L VIN 5, 5.7L VIN P engine Transmissions: All	**ECT Excessive Time To Enter Closed Loop Conditions:** DTC P0112, P0113, P0117 or P0118 not set, ECT and IAT sensors more than 50°F at startup, engine started, VSS over 5 mph while in closed loop, and the PCM detected the ECT sensor signal did not reach 140°F after 7 minutes during the test period. This trouble code test will only run once per ignition cycle once the enabling conditions are met. If the PCM detects the calibrated amount of airflow and run time are met and the engine coolant does not reach Closed Loop temperature, the PCM will set this trouble code (DTC P0125). **Possible Causes** ● Check the operation of the thermostat (it may be stuck open) ● Coolant level is too low, or the coolant mixture is incorrect ● ECT sensor signal circuit has a high resistance condition ● ECT sensor is damaged or it has failed

OBD II Trouble Code List (P0xxx Codes)

DTC	Trouble Code Title, Conditions & Possible Causes
P0125 **2T CCM, MIL: Yes** 1996, 1997, 1998, 1999, 2000, 2001, 2002, 2003, 2004, 2005 Achieva, Beretta, Ciera, Corsica, Century, Grand Am, Grand Prix, Lumina, Malibu, Monte Carlo 3.1L VIN M, 3.4L VIN X Transmissions: All 3.1L VIN M, 3.4L VIN X Transmissions: All	**ECT Excessive Time To Enter Closed Loop Conditions:** DTC P0112, P0113, P0117 and P0118 not set, startup ECT sensor more than 19°F, minimum IAT sensor more than 19°F, then while in: Region 1 Conditions: Startup ECT sensor more than 50°F and IAT sensor more than 50°F, then with engine runtime over 127 seconds, the engine did not reach a closed loop temperature of 64°F after a calibrated amount of total airflow and maximum idle time of more than 95 seconds. Region 2 Conditions: Startup ECT sensor between 20-50°F and IAT sensor more than 20°F, then with engine runtime over 20 seconds, the engine did not reach closed loop temperature of 64°F after a calibrated amount of total airflow and maximum idle time of more than 210 seconds. Region 3 Conditions: Startup ECT from -20°F to 40°F and IAT sensor more than 20°F, then with engine runtime over 439 seconds, the engine did not reach closed loop temperature of 64°F after a calibrated amount of total airflow and maximum idle time of more than 329 seconds. **Possible Causes** • Check the operation of the thermostat (it may be stuck open) • Coolant level is too low, or the coolant mixture is incorrect • ECT sensor signal circuit has a high resistance condition • ECT sensor is damaged or it has failed
P0125 **2T CCM, MIL: Yes** 1996, 1997, 1998, 1999, 2000, 2001, 2002, 2003, 2004, 2005 Aurora, Century, LeSabre, Park Avenue, Regal 3.1L VIN J, 3.1L VIN M, 3.4L VIN X, 3.5L V6 VIN H, 3.8L VIN 1, 3.8L VIN K, 4.0L V8 VIN C Transmissions: All	**ECT Excessive Time To Enter Closed Loop Conditions:** DTC P0110, P0111, P0112, P0113, P0116, P0117, P0118, P0128, P1114, or P1115 not set, startup ECT sensor from -40°F to 104°F, minimum IAT sensor more than 19°F, then while in: Region 1 Conditions: IAT sensor more than 50°F and the startup ECT sensor above 14°F for VIN K or above 50°F for VIN 1, then with engine runtime over 6 seconds (VIN K) or over 150 seconds (VIN 1), the engine did not reach a closed loop temperature of 64°F (VIN K) or 68°F for VIN 1 after a calibrated amount of total airflow and maximum idle time of more than 5 seconds (VIN K) or more than 125 seconds for VIN 1. Region 2 Conditions: IAT sensor more than 20°F and startup ECT sensor between 20-50°F, then with engine runtime over 154 seconds (VIN K) or over 250 seconds (VIN 1), the engine did not reach closed loop temperature of 64°F (VIN K) or 68°F for VIN 1 after a calibrated amount of total airflow and maximum idle time of more than 120 seconds (VIN K) or more than 200 seconds for VIN 1 was exceeded. Region 3 Conditions: IAT sensor more than 20°F and startup ECT sensor from -40-20°F, then with engine runtime over 287 seconds (VIN K) or over 375 seconds (VIN 1), the engine did not reach closed loop temperature of 64°F (VIN K) or 68°F for VIN 1 after a calibrated amount of total airflow and maximum idle time of more than 255 seconds (VIN K) or more than 300 seconds for VIN 1 was exceeded. **Possible Causes** • Check the operation of the thermostat (it may be stuck open) • Coolant level is too low, or the coolant mixture is incorrect • ECT sensor signal circuit has a high resistance condition • ECT sensor is damaged or it has failed
P0125 **2T CCM, MIL: Yes** 1996, 1997, 1998, 1999, 2000, 2001, 2002, 2003, 2004, 2005 C/K Cab & Chassis, C/K Series Truck, G Series 4.3L VIN W, 4.3L VIN X, 5.0L VIN M, 5.7L VIN K, 5.7L VIN R, 7.4L VIN J Transmissions: All	**ECT Excessive Time To Enter Closed Loop Conditions:** DTC P0101, P0102, P0103, P0112, P0113, P0116, P0117, P0118, P0500, P0502, P0503, P1111, P1112, P1114 and P1115 not set, and with the engine started with the ECT sensor from -31°F to 104°F, IAT sensor more than -40°F, accumulated airflow since startup more than 9000 grams and idle time under 360 seconds, and the PCM detected the ECT sensor was less than 68°F after 8 minutes (Test 1), or with the engine was started with the ECT sensor more than 20°F, accumulated airflow since startup more than 5500 grams with idle time less than 225 seconds, and the PCM detected the ECT sensor was less than 68°F after 2 minutes (Test 2), or the engine was started with the ECT sensor more than 50°F, accumulated airflow since startup more than 2000 grams and idle time less than 90 seconds, and the PCM detected the ECT sensor was less than 68°F after 2 minutes (Test 3). **Possible Causes** • Check the operation of the thermostat (it may be stuck open) • Coolant level is too low, or the coolant mixture is incorrect • ECT sensor signal circuit has a high resistance condition • ECT sensor is damaged or it has failed

OBD II Trouble Code List (P0xxx Codes)

DTC	Trouble Code Title, Conditions & Possible Causes
P0125 **2T CCM, MIL: Yes** 2002, 2003, 2004, 2005 Montana, Silhouette, Venture, Rendezvous 3.1L VIN J, 3.4L VIN E Transmissions: All	**Engine Coolant Temperature Insufficient For Closed Loop Operation Conditions:** DTC P0112, P0113, P0116, P0117 and P0118 not set, engine runtime from 120-440 seconds, minimum air temperature more than 19°F, ECT sensor more than -40°F and less than 50°F at startup, accumulated mass airflow (MAF) from 1252-5670 g/sec, and the PCM detected the ECT sensor signal did not reach a minimum closed loop temperature value of at least 59°F during the test period. **Possible Causes** ● Check the operation of the thermostat (it may be stuck open) ● Coolant level is too low, or the coolant mixture is incorrect ● ECT sensor signal circuit has a high resistance condition ● ECT sensor is damaged or it has failed
P0125 **2T CCM, MIL: Yes** 1999, 2000, 2001, 2002, 2003, 2004, 2005 C/K Cab & Chassis, C Series Vans & Cargo Van, Envoy & TrailBlazer 4.8L VIN V, 5.3L VIN P, 5.3L VIN T, 5.3L VIN Z, 6.0L VIN N, 6.0L VIN U, 8.1L VIN G Transmissions: All	**ECT Excessive Time To Enter Closed Loop Conditions:** DTC P0101, P0102, P0103, P0112, P0113, P0116, P0117, P0118, P0500, P0502 and P0503 not set, engine started, ECT sensor from 33-83°F at startup, IAT sensor from 19-131°F, MAF sensor from 24-75 g/sec with the average more than 12 g/sec, engine runtime from 120-3200 seconds, vehicle speed over 5 mph for 0.5 miles, and the PCM detected the engine coolant temperature did not reach 93°F with the calibrated amount of engine runtime and engine airflow met. **Possible Causes** ● Check the operation of the thermostat (it may be stuck open) ● Coolant level is too low, or the coolant mixture is incorrect ● ECT sensor signal circuit has a high resistance condition ● ECT sensor is damaged or it has failed
P0125 **2T CCM, MIL: Yes** 1996, 1997, 1998, 1999, 2000, 2001, 2002, 2003, 2004, 2005 C/K Series Truck, G Vans 6.5L VIN F, 6.5L VIN S, 6.6L VIN 1 Diesel engines Transmissions: All	**Coolant Temperature Insufficient For Stable Operation Conditions:** P0112, P0113, P0117 and P0118 not set, ECT sensor less than 133°F, IAT sensor more than 20°F at startup, engine started, engine runtime over 5 minutes, and the PCM detected the amount of fuel consumed since startup was over 1,000, cu. mm with the total time since startup under 450 seconds, or the PCM detected the amount of fuel consumed since startup was over 468,120 cu. mm with the total time since startup less than 225 seconds. **Possible Causes** ● Check the operation of the thermostat (it may be stuck open) ● Coolant level is too low, or the coolant mixture is incorrect ● ECT sensor signal circuit has a high resistance condition ● ECT sensor is damaged or it has failed
P0125 **2T CCM, MIL: Yes** 1996, 1997, 1998, 1999, 2000, 2001, 2002, 2003, 2004, 2005 DeVille, Eldorado, Seville 4.6L VIN 9, 4.6L VIN Y Transmissions: All	**ECT Excessive Time To Enter Closed Loop Conditions:** DTC P0110, P0111, P0112, P0113, P0116, P0117, P0118, P1114 and P1115 not set, engine started, ECT sensor from -40°F to 104°F at startup, IAT sensor more than 19°F, vehicle speed over 1 mph, and the PCM detected that it took an excessive amount of time before the ECT reached the Closed Loop operation (50°F) threshold. The amount of time ranges between 2-13 minutes, depending on the ECT sensor value at startup and the amount of air cycled through the engine since the engine started. This input is used by the PCM for various engine controls and as enabling criteria for some engine diagnostics. **Possible Causes** ● Check the operation of the thermostat (it may be stuck open) ● Coolant level is too low, or the coolant mixture is incorrect ● ECT sensor signal circuit has a high resistance condition ● ECT sensor is damaged or it has failed
P0125 **2T CCM, MIL: Yes** 2003, 2004, 2005 CTS (Cadillac) 2.6L V6 VIN M & 3.2L V6 VIN N engines Transmissions: All	**ECT Excessive Time To Enter Closed Loop Conditions:** DTC P0101, P0102, P0103, P0112, P0113, P0116, P0117 and P0118 not set, engine runtime more than 5 minutes, and the PCM detected an excessive amount of time passed before the ECT sensor reached the Closed Loop operating temperature. **Possible Causes** ● Check the operation of the thermostat (it may be stuck open) ● Coolant level is too low, or the coolant mixture is incorrect ● ECT sensor signal circuit has a high resistance condition ● ECT sensor is damaged or it has failed
P0125 **2T CCM, MIL: Yes** 1995, 1996, 1997, 1998, 1999, 2000, 2001, 2002 Camaro & Firebird 3.8L VIN K Transmissions: All	**ECT Excessive Time To Enter Closed Loop Conditions:** DTC P0110-P0113, P0116-P0118, P0128, P1114 and P1115 not set, engine started, minimum IAT sensor more than 19°F, ECT sensor from -40°F to 104°F at startup, and the PCM detected the engine runtime exceeded 5 minutes before the ECT exceeded 41°F after a calibrated amount of airflow was exceeded at idle speed. **Possible Causes** ● Check the operation of the thermostat (it may be stuck open) ● Coolant level is too low, or the coolant mixture is incorrect ● ECT sensor signal circuit has a high resistance condition ● ECT sensor is damaged or it has failed

OBD II Trouble Code List (P0xxx Codes)

DTC	Trouble Code Title, Conditions & Possible Causes
P0125 **2T CCM, MIL: Yes** 1997, 1998, 1999, 2000, 2001, 2002, 2003, 2004, 2005 Camaro, Corvette & Firebird 5.7L VIN G, 5.7L VIN S Transmissions: All	**ECT Excessive Time To Enter Closed Loop Conditions:** DTC P0110-P0113, P0116-P0118, P0128, P1114 and P1115 not set, engine runtime from 2 to 22 minutes, ECT sensor below 83ºF at startup, IAT sensor from 19-131ºF, MAF sensor from 15-75 g/sec with the average over 14 g/sec, VSS over 3 mph for 1.5 miles, and the PCM detected the engine temperature did not reach 93ºF. **Possible Causes** ● Check the operation of the thermostat (it may be stuck open) ● Coolant level is too low, or the coolant mixture is incorrect ● ECT sensor signal circuit has a high resistance condition ● ECT sensor is damaged or it has failed
P0125 **2T CCM, MIL: Yes** 1996, 1997, 1998, 1999, 2001, 2002, 2003, 2004, 2005 Aurora 3.5L V6 VIN H, 4.0L V8 VIN C engines Transmissions: All	**ECT Excessive Time To Enter Closed Loop Conditions:** DTC P0110, P0111, P0112, P0113, P0116, P0117, P0118, P1114 and P1115 not set, ECT sensor from -40ºF to 122ºF at startup, IAT sensor more than 19ºF, engine runtime from 2-22 minutes, vehicle speed over 1 mph, and the PCM detected an excessive amount of time passed before the ECT sensor signal reached the Closed Loop mode 50ºF threshold. The engine runtime ranges from 2-13 minutes, depending on the ECT at startup, and the amount of air cycled through the engine since startup. **Possible Causes** ● Check the operation of the thermostat (it may be stuck open) ● Coolant level is too low, or the coolant mixture is incorrect ● ECT sensor signal circuit has a high resistance condition ● ECT sensor is damaged or it has failed
P0125 **2T CCM, MIL: Yes** 1996, 1997, 1998, 1999 Achieva, Beretta, Ciera, Corsica, Century, Grand Am, Grand Prix, Lumina, Malibu, Monte Carlo 3.1L VIN M, 3.4L VIN X Transmissions: All	**Excessive Time To Closed Loop Fuel Control Conditions:** DTC P0112, P0113, P0116, P0117, P0118, P1111, P1112, P1114 and P1115 not set, minimum air temperature more than 19ºF, ECT sensor less than 122ºF at startup, then under these conditions: Region 1 - Automatic Transmission Startup ECT sensor over 50ºF, IAT sensor over 50ºF, and the PCM detected the engine required 127 seconds to reach a closed loop temperature of 64ºF after exceeding the calibrated amount of airflow and a maximum idle time of less than 95 seconds during the test. Region 2 - Automatic Transmission Startup ECT sensor from 20-50ºF, IAT sensor more than 20ºF, and the PCM detected the engine required over 280 seconds to reach a closed loop temperature of 64ºF after exceeding a calibrated amount of airflow and a maximum idle time of less than 210 seconds. Region 3 - Automatic Transmission Startup ECT sensor from -40-20ºF, IAT sensor more than 20ºF, and the PCM detected the engine required over 439 seconds to reach a closed loop temperature of 64ºF after exceeding a calibrated amount of airflow and a maximum idle time of less than 329 seconds. **Possible Causes** ● Check the operation of the thermostat (it may be stuck open) ● Coolant level is too low, or the coolant mixture is incorrect ● ECT sensor signal circuit has a high resistance condition ● ECT sensor is damaged or it has failed
P0125 **2T CCM, MIL: Yes** 1999, 2000, 2001, 2002, 2003, 2004, 2005 Alero, Aztek, Cutlass, Impala, Malibu, Montana, Silhouette, Venture, Rendezvous 3.1L VIN J, 3.1L VIN M, 3.4L VIN E Transmissions: All	**Excessive Time To Closed Loop Fuel Control Conditions:** DTC P0112, P0113, P0116, P0117, P0118, P1111, P1112, P1114 and P1115 not set, engine runtime from 16 seconds to 50 seconds, ECT sensor more than -40ºF and less than 50ºF at startup, HO2S signal varying indicating it is hot enough to operate, IAT sensor more than 50ºF during testing, and the PCM detected the ECT sensor signal did not reach a closed loop temperature value of at least 71.6ºF after 4 minutes of engine operation. **Possible Causes** ● Check the operation of the thermostat (it may be stuck open) ● Coolant level is too low, or the coolant mixture is incorrect ● ECT sensor signal circuit has a high resistance condition ● ECT sensor is damaged or it has failed
P0125 **2T CCM, MIL: Yes** 1995, 1996, 1997, 1998, 1999, 2000, 2001, 2002 S/T Blazer & S/T Pickup 4.3L VIN W, 4.3L VIN X Transmissions: All	**Excessive Time To Closed Loop Fuel Control Conditions:** DTC P0101, P0102, P0103, P0110, P0111, P0112, P0113, P0116, P0117, P0118, P0125, P0500, P0501, P0503, P1114 and P1115 not set, IAT sensor above 19ºF at startup, ECT sensor below 104ºF at startup, engine started, engine runtime from 2-17 minutes, VSS over 5 mph, average MAF reading above 15 g/sec, and the PCM detected the ECT sensor did not reach 104ºF after a calibrated amount of time (depends on the temperature at startup). **Possible Causes** ● Check the operation of the thermostat (it may be stuck open) ● Coolant level is too low, or the coolant mixture is incorrect ● ECT sensor signal circuit has a high resistance condition ● ECT sensor is damaged or it has failed

OBD II Trouble Code List (P0xxx Codes)

DTC	Trouble Code Title, Conditions & Possible Causes
P0125 **2T CCM, MIL: Yes** 2003, 2004, 2005 S/T Blazer & S/T Pickup 4.3L VIN W, 4.3L VIN X Transmissions: All	**Excessive Time To Closed Loop Fuel Control Conditions:** DTC P0101, P0102, P0103, P0112, P0113, P0116, P0117, P0118, P0500 P0502 and P0503 not set, IAT sensor from 19-1341°F, ECT sensor from -33°F to 83°F at startup, engine started, engine runtime from 120-2100 seconds, VSS over 5 mph for 0.5 miles, MAF sensor from 15-75 g/sec with the average MAF at 12 g/sec, and the PCM detected the ECT sensor did not reach 93°F after a calibrated amount of airflow and engine runtime had expired. **Possible Causes** ● Check the operation of the thermostat (it may be stuck open) ● Coolant level is too low, or the coolant mixture is incorrect ● ECT sensor signal circuit has a high resistance condition ● ECT sensor is damaged or it has failed
P0125 **2T CCM, MIL: Yes** 2002, 2003, 2004, 2005 Envoy & TrailBlazer 4.2L VIN S engine Transmissions: All	**ECT Excessive Time To Enter Closed Loop Conditions:** DTC P0105-P0108, P0112-P0118, P0122, P0123, P0130-P0134, P0171, P0172, P0201-P0206, P0300-P0306, P0325, P0335, P0336, P0420, P0440-P0480, P0502, P0503, P0601-P0606, P1120, P1220, P1221-P1275, P1280, P1281, P1441, P1481-P1484, P1512-P1515, P1621, P1635 and P1639 not set, engine runtime from 30 seconds to 20 minutes, ECT sensor less than 104°F at startup IAT sensor more than 19°F, MAF sensor more than 20 g/sec, vehicle speed over 25 mph, and the PCM detected the engine did not reach 104°F. **Possible Causes** ● Check the operation of the thermostat (it may be stuck open) ● Coolant level is too low, or the coolant mixture is incorrect ● ECT sensor signal circuit has a high resistance condition ● ECT sensor is damaged or it has failed
P0125 **2T CCM, MIL: Yes** 1995 Regal 3.8L VIN L engine Transmissions: All	**ECT Excessive Time To Enter Closed Loop Conditions:** DTC P0117 and P0118 not set, IAT sensor indicating a cold engine at startup, vehicle speed from 5-10 mph for 4-9 minutes, and the PCM detected the ECT sensor did not exceed 54°F. **Possible Causes** ● Check the operation of the thermostat (it may be stuck open) ● Coolant level is too low, or the coolant mixture is incorrect ● ECT sensor signal circuit has a high resistance condition ● ECT sensor is damaged or it has failed
P0125 **2T CCM, MIL: Yes** 2002, 2003, 2004, 2005 Intrigue, Regal 3.8L VIN 1, 3.8L VIN K Transmissions: All	**Engine Coolant Temperature Insufficient For Closed Loop Operation Conditions:** DTC P0101, P0102, P0103, P0112, P0113, P0116, P0117, P0118, P0502 and P0503 not set, engine runtime from 200-1500 seconds, minimum air temperature over 19°F, ECT sensor more than -38°F and less than 14°F at startup, vehicle speed over 15 mph for more than 0.5 miles, accumulated mass airflow (MAF) more than 5 g/sec, and the PCM detected the ECT sensor did not reach a minimum closed loop temperature value of at least 14°F in the test. The airflow into the engine is accumulated and used to determine if the vehicle has been driven within conditions that would allow the engine coolant to heat up normally to the closed loop temperature. If the coolant temperature does not increase normally or does not reach the closed loop temperature, the diagnostics that use engine coolant temperature as enabling criteria, may not run when expected. **Possible Causes** ● Check the operation of the thermostat (it may be stuck open) ● Coolant level is too low, or the coolant mixture is incorrect ● ECT sensor signal circuit has a high resistance condition ● ECT sensor is damaged or it has failed
P0128 **2T CCM, MIL: Yes** 2000, 2001, 2002, 2003, 2004, 2005 Bonneville, Century, Intrigue, LeSabre, Lumina, Park Avenue 3.1L VIN J, 3.1L VIN M, 3.5L V6 VIN H, 3.8L VIN 1, 3.8L VIN K engines Transmissions: All	**ECT Sensor Below Thermostat Regulating Temperature Conditions:** DTC P0100, P0101, P0103, P0110, P0111-P0113, P0116-P0118, P0125, P0500, P0502, P0503, P1114, or P1115 not set, ECT sensor within a range of -40°F to 172°F and IAT sensor over 19°F at startup, engine runtime from 2-20 minutes, MAF sensor more than 15 g/sec, vehicle driven to a speed over 15 mph for 1 mile, and the PCM detected the engine temperature did not reach 158°F-170°F after the time period completed. The PCM uses the ECT sensor for various engine controls and as enabling criteria for engine diagnostics. The airflow into the engine is accumulated and used to determine if the vehicle has been driven within conditions that would allow the engine coolant to warm to closed loop temperature. If the temperature does not increase normally or does not reach closed loop temperature, tests that use the ECT sensor signal as enabling criteria may not run as expected. This test runs once per key cycle after enabling criteria are met. The PCM sets P0125 if it detects the correct airflow and run time is met, and the ECT sensor does not reach a closed loop value. **Possible Causes** ● Check the operation of the thermostat (it may be stuck open) ● Coolant level is too low, or the coolant mixture is incorrect ● ECT sensor signal circuit has a high resistance condition ● ECT sensor is damaged or it has failed

OBD II Trouble Code List (P0xxx Codes)

DTC	Trouble Code Title, Conditions & Possible Causes
P0128 **2T CCM, MIL: Yes** 1996, 1997, 1998, 1999, 2000, 2001, 2002, 2003, 2004, 2005 Alero, Aztek, Cutlass, Impala, Malibu, Montana, Silhouette, Venture, Rendezvous 3.1L VIN J, 3.1L VIN M, 3.4L VIN E Transmissions: All	**ECT Sensor Below Thermostat Regulating Temperature Conditions:** DTC P0112, P0113, P0116, P0117, P0118, P1111, P1112, P1114 or P1115 not set, ECT sensor from -40°F to 172°F and IAT sensor more than 19°F at started, engine started, engine runtime from 2 to 30 minutes, vehicle driven to over 15 mph for 1 mile, average MAF reading more than 15 g/sec, and the PCM detected the time it took tool long for the ECT sensor to reach 170°F (one test per key cycle). **Possible Causes** ● Check the operation of the thermostat (it may be stuck open) ● Coolant level is too low, or the coolant mixture is incorrect ● ECT sensor signal circuit has a high resistance condition ● ECT sensor is damaged or it has failed
P0128 **2T CCM, MIL: Yes** 2000, 2001, 2002, 2003, 2004, 2005 C/K Cab & Chassis, C/K Series Trucks & G Vans 4.3L VIN W, 4.3L VIN X, 5.0L VIN M, 5.7L VIN K, 5.7L VIN R, 7.4L VIN J Transmissions: All 4.3L VIN W, 4.3L VIN X, 5.0L VIN M, 7.4L VIN J Transmissions: All	**ECT Sensor Below Thermostat Regulating Temperature Conditions:** DTC P0101, P0102, P0103, P0112, P0113, P0116, P0117, P0118, P0125, P0500, P0502, P0503, P1111, P1112, P1114 and P1115 not set, ECT sensor from -40°F to 158°F, IAT sensor 19°F or more, engine runtime from 2-22 minutes, VSS over 5 mph for 1.5 miles, MAF average reading over 23 g/sec, and the PCM detected the time to reach closed loop temperature had been exceeded. **Possible Causes** ● Check the operation of the thermostat (it may be stuck open) ● Coolant level is too low, or the coolant mixture is incorrect ● ECT sensor signal circuit has a high resistance condition ● ECT sensor is damaged or it has failed
P0128 **2T CCM, MIL: Yes** 1999, 2000, 2001, 2002, 2003, 2004, 2005 Avalanche, C/K Trucks, G Vans, Envoy, TrailBlazer 4.8L VIN V, 5.3L VIN P, 5.3L VIN T, 5.3L VIN Z, 6.0L VIN N, 6.0L VIN U, 6.6L VIN 1, 8.1L VIN G Transmissions: All	**ECT Excessive Time To Enter Closed Loop Conditions:** DTC P0101, P0102, P0103, P0112, P0113, P0116, P0117, P0118, P0500, P0502 and P0503 not set, started, ECT sensor less than 49°F at startup, IAT sensor from 19-131°F, MAF sensor from 24-75 g/sec with the average more than 12 g/sec, engine runtime from 120-3200 seconds, vehicle speed over 5 mph for 1.5 miles, this test has not run previously on current ignition cycle, and the PCM detected the engine coolant temperature did not reach 167°F after the calibrated amount of engine runtime and engine airflow met. **Possible Causes** ● Check the operation of the thermostat (it may be stuck open) ● Coolant level is too low, or the coolant mixture is incorrect ● ECT sensor signal circuit has a high resistance condition ● ECT sensor is damaged or it has failed
P0128 **2T CCM, MIL: Yes** 2003, 2004, 2005 C/K Series Trucks & Vans 4.3L VIN W, 4.3L VIN X Transmissions: All	**ECT Sensor Below Thermostat Regulating Temperature Conditions:** DTC P0101, P0102, P0103, P0112, P0113, P0116, P0117, P0118, P0500 P0502 and P0503 not set, IAT sensor from 19-1341°F, ECT sensor less than 169°F at startup, engine started, engine runtime from 120-2100 seconds, VSS over 5 mph for more than 0.5 miles, MAF sensor from 15-75 g/sec with the average MAF at 12 g/sec, and the PCM detected the ECT sensor did not reach 178°F after the calibrated amount of airflow and engine runtime expired **Possible Causes** ● Check the operation of the thermostat (it may be stuck open) ● ECT sensor is damaged or it has failed ● Inspect for low coolant level or an incorrect coolant mixture ● PCM has failed
P0128 **2T CCM, MIL: Yes** 2002, 2003, 2004, 2005 Aurora, DeVille, Eldorado & Deville 4.0L V8 VIN C, 4.6L VIN 9, 4.6L VIN Y engines Transmissions: All	**ECT Sensor Below Thermostat Regulating Temperature Conditions:** DTC P0101, P0102, P0103, P0112, P0113, P0116, P0117, P0118, P0125, P0500, P0501-P0503, P1111, P1112, P1114, P1115 not set, ECT sensor from -40 to 158°F at startup, IAT sensor over 19°F, engine runtime from 2-22 minutes, average MAF sensor more than 23 g/sec, VSS over 5 mph for 1.5 miles, and the PCM detected it took too long for the ECT sensor to reach a set temperature. **Possible Causes** ● Check the operation of the thermostat (it may be stuck open) ● ECT sensor is damaged or it has failed ● Inspect for low coolant level or an incorrect coolant mixture ● PCM has failed
P0128 **2T CCM, MIL: Yes** 2003, 2004, 2005 CTS (Cadillac) 2.6L V6 VIN M & 3.2L V6 VIN N engines Transmissions: All	**ECT Sensor Below Thermostat Regulating Temperature Conditions:** DTC P0101, P0102, P0103, P0112, P0113, P0116, P0117 and P0118 not set, engine block heater has not been detected or delayed for 20 seconds, ignition off time over 15 minutes, ECT sensor less than 122°F at startup engine speed more than 960 rpm, ambient air temperature between -38°F to 122°F, VSS over 9 mph, MAF sensor more than 23 g/sec, and the PCM detected the time for the engine coolant to reach a set temperature was exceeded. **Possible Causes** ● Check the operation of the thermostat (it may be stuck open) ● Coolant level is too low, or the coolant mixture is incorrect ● ECT sensor signal circuit has a high resistance condition ● ECT sensor is damaged or it has failed

OBD II Trouble Code List (P0xxx Codes)

DTC	Trouble Code Title, Conditions & Possible Causes
P0128 **2T CCM, MIL: Yes** 2001, 2002, 2003, 2004, 2005 Camaro, Corvette & Firebird 3.8L VIN K, 5.7L VIN G, 5.7L VIN S engines Transmissions: All	**ECT Sensor Below Thermostat Regulating Temperature Conditions:** DTC P0100-P0103, P0110-P0113, P0116-P0118, P0125, P0500, P0501-P0503, P1111-P1115 not set, ECT sensor less than 158°F at startup, IAT sensor from 19-131°F, engine runtime at 2-22 minutes, MAF sensor at 15-75 g/sec (avg. 14 g/sec), VSS over 3 mph for 1.5 miles, and PCM detected the engine coolant did not reach 167°F. **Possible Causes** ● Check the operation of the thermostat (it may be stuck open) ● Coolant level is too low, or the coolant mixture is incorrect ● ECT sensor signal circuit has a high resistance condition ● ECT sensor is damaged or it has failed
P0128 **2T CCM, MIL: Yes** 2000, 2001, 2002, 2003, 2004, 2005 Alero, Century, Cavalier, Grand Am, Sunfire, Envoy & TrailBlazer, S/T Blazer & S/T Pickup 2.2L VIN 4, 2.2L VIN 5, 2.2L VIN F, 2.2L VIN H, 2.4L VIN T, 4.2L VIN S Transmissions: All	**ECT Sensor Below Thermostat Regulating Temperature Conditions:** DTC P0105-P0108, P0112-P0118, P0122, P0123, P0130-P0132, P0171, P0172, P0201-P0204, P0300, P0325, P0336, P0420, P0440-P0453, P0480, P0502, P0503, P0506 and P1441 not set, engine runtime from 30 seconds to 20 minutes, IAT sensor more than 19°F and ECT sensor less than 158°F at startup, average MAF more than 20 g/sec, vehicle driven at over 25 mph for 1.5 miles, and after enough airflow had entered the engine, the PCM detected the engine temperature did not rise to 167°F after another 30 seconds. **Possible Causes** ● Check the operation of the thermostat (it may be stuck open) ● Coolant level is too low, or the coolant mixture is incorrect ● ECT sensor signal circuit has a high resistance condition ● ECT sensor is damaged or it has failed
P0128 **2T CCM, MIL: Yes** 2001, 2002, 2003, 2004, 2005 S/T Blazer & S/T Pickup 4.3L VIN W, 4.3L VIN X Transmissions: All S/T Blazer, S/T Pickup	**ECT Sensor Below Thermostat Regulating Temperature Conditions:** DTC P0100-P0103, P0110-P0113, P0116-P0118, P0125, P0500, P0501, P0503, P1114 and P1115 not set, ECT sensor under 178°F and IAT sensor over 19°F at startup, engine runtime 2-17 minutes, MAF sensor over 13 g/sec, VSS over 5 mph for 1 mile, and the PCM detected the engine did not reach normal temperature during testing. **Possible Causes** ● Check the operation of the thermostat (it may be stuck open) ● ECT sensor is damaged or it has failed ● Inspect for low coolant level or an incorrect coolant mixture ● PCM has failed!
P0128 **2T CCM, MIL: Yes** 2002, 2003, 2004, 2005 Intrigue, Regal 3.8L VIN 1, 3.8L VIN K Transmissions: All	**ECT Sensor Below Thermostat Regulating Temperature Conditions:** DTC P0101, P0102, P0103, P0112, P0113, P0116, P0117, P0118, P0502 and P0503 not set, engine runtime over 4 minutes, minimum air temperature over 19°F, ECT sensor between -38°F and 176°F at startup, VSS over 15 mph for 3 miles, MAF sensor over 20 g/sec, and the PCM detected the ECT sensor did not reach a calibrated engine coolant temperature of at least 176°F during the test period. **Possible Causes** ● Check the operation of the thermostat (it may be stuck open) ● Coolant level is too low, or the coolant mixture is incorrect ● ECT sensor is damaged or it has failed
P0130 **2T O2S, MIL: Yes** 2000, 2001, 2002, 2003, 2004, 2005 Aztek, Malibu, Montana, Silhouette, Venture, Rendezvous 3.1L VIN J, 3.1L VIN M, 3.4L VIN E engines Transmissions: All	**HO2S-11 (Bank 1 Sensor 1) Closed Loop Performance Conditions:** DTC P0101-P0103, P0107, P0108, P0112, P0113, P0116-P0118, P0121-P0123, P0125, P0128, P0201-P0206, P0410, P0440, P0442, P0443, P0446, P0449 and P1441 not set, engine started, engine speed 550-3000 rpm for 4 minutes, MAF sensor from 8-35 g/sec, TP angle from 3-35%, ECT sensor more than 158°F, and the PCM detected the HO2S signal had an improper voltage amplitude. The HO2S is used for fuel control and post-catalyst monitoring. This sensor compares the oxygen content of the surrounding air with the oxygen content of the exhaust stream. At initial startup, the PCM operates in open loop mode, ignoring the HO2S signal when calculating the air/fuel ratio. The PCM supplies the HO2S with a reference (or bias) voltage of about 450 mv. The HO2S generates a voltage within a range of 0-1000 mv that fluctuates above and below the bias voltage once in closed loop. A high HO2S voltage indicates a rich fuel mixture. A low HO2S voltage indicates a lean mixture. Heating elements in the HO2S shorten the time required for the sensor to reach normal temperature, and an accurate voltage signal. **Possible Causes** ● Air leaks in the exhaust system, intake manifold, vacuum lines ● EVAP Purge system malfunction or charcoal canister problems ● HO2S signal circuit is open between the sensor and the PCM ● HO2S signal circuit is shorted to sensor or chassis ground ● HO2S is damaged, contaminated or air reference hole clogged ● PCM has failed

OBD II Trouble Code List (P0xxx Codes)

DTC	Trouble Code Title, Conditions & Possible Causes
P0130 **2T O2S, MIL: Yes** 2000, 2001, 2002, 2003, 2004, 2005 Bonneville, Century, Intrigue, LeSabre, Lumina, Park Avenue 3.1L VIN J, 3.1L VIN M, 3.5L V6 VIN H, 3.8L VIN 1, 3.8L VIN K engines Transmissions: All	**HO2S-11 (Bank 1 Sensor 1) Circuit Closed Loop Performance Conditions:** DTC P0101-P0103, P0107, P0108, P0112, P0113, P0116-P0118, P0121, P0122, P0123, P0125, P0128, P0201-P0206, P0410, P0440, P0442, P0443, P0446, P0449 and P1441 not set, engine started, system voltage from 9-18v, engine speed from 550-3000 rpm for 4 minutes, ECT sensor more than 167ºF, MAF sensor from 8-35 g/sec, TP angle from 3-35%, and the PCM detected the HO2S signal amplitude was less than the calibrated minimum amplitude. At startup, the PCM operates in open loop mode, ignoring the HO2S signal when calculating the air/fuel ratio. The PCM supplies the HO2S with a bias voltage of 450 mv. **Possible Causes** ● Air leaks in the exhaust system, intake manifold, vacuum lines ● EVAP Purge system malfunction or charcoal canister problems ● HO2S signal circuit is open between the sensor and the PCM ● HO2S signal circuit is shorted to sensor or chassis ground ● HO2S is damaged, contaminated or air reference hole clogged ● PCM has failed
P0130 **2T O2S, MIL: Yes** 2003, 2004, 2005 CTS (Cadillac) 2.6L V6 VIN M & 3.2L V6 VIN N engines Transmissions: All	**HO2S-11 (Bank 1 Sensor 1) Circuit Closed Loop Conditions:** DTC P0443, P0444 and P0445 not set, engine started, TP angle from 5-35%, system voltage over 10.5v, MAF sensor at 10-35 g/sec, Calculated Converter Temperature under 1,472ºF, and the PCM detected the HO2S-11 signal was 60-400 mv while the HO2S-12 signal was over 499 mv, or the HO2S-11 signal was 600-1080 mv while the HO2S-12 signal was less than 104 mv for 10 seconds. **Possible Causes** ● EVAP system leaks or restrictions, charcoal canister problems ● EVAP system leaks or restrictions, charcoal canister problems ● HO2S air reference hole plugged (check for dirt on the outside) ● HO2S signal circuit is shorted to sensor or chassis ground ● HO2S is damaged or contaminated due to improper fuel usage ● PCM has failed
P0130 **2T O2S, MIL: Yes** 2000, 2001, 2002, 2003, 2004, 2005 Aurora, DeVille, Eldorado & Seville 4.0L V8 VIN C, 4.6L VIN 9, 4.6L VIN Y engines Transmissions: All	**HO2S-11 (Bank 1 Sensor 1) Circuit Closed Loop Performance Conditions:** DTC P0101-P0103, P0106-P0108, P0116-P0118, P0121-P0123, P0125, P0131-P0134, P0151-P0154, P0201-P0208, P1133, P1134, P1153 and P1154 not set, engine speed at 500-5000 rpm in closed loop for 2 minutes, ECT sensor over 176ºF, HO2S temperature over 1112ºF, Catalyst, DFCO, Traction Control and P/E tests "off", MAF sensor from 3-30 g/sec for 5 seconds, and the PCM detected the HO2S signal was fixed at 350-500 mv for 16 seconds. **Possible Causes** ● EVAP system leaks or restrictions, charcoal canister problems ● Fuel system restricted or fuel pressure regulator has failed ● HO2S air reference hole plugged (check for dirt on the outside) ● HO2S signal circuit shorted to low reference or chassis ground ● HO2S is damaged or contaminated due to improper fuel usage ● PCM has failed
P0130 **2T O2S, MIL: Yes** 2000, 2001, 2002 Camaro & Firebird 3.8L VIN K engine Transmissions: All	**HO2S-11 (Bank 1 Sensor 1) Circuit Closed Loop Performance Conditions:** DTC P0101-P0103, P0107, P0108, P0112-P0113, P0117-P0118, P0121-P0123, P0125-P0128, P0201-P0206, P0410, P0440, P0442, P0446, P0449 and P1441 not set, engine runtime 4 minutes, ECT sensor more than 122ºF, engine speed from 1000-3000 rpm, airflow from 13-30 g/sec, APP sensor indicated angle from 5-40%, and the PCM detected an active HO2S with an improper voltage amplitude. **Possible Causes** ● Air leaks in the exhaust system, intake manifold, vacuum lines ● EVAP Purge system malfunction or charcoal canister problems ● HO2S signal circuit is open between the sensor and the PCM ● HO2S signal circuit is shorted to sensor or chassis ground ● HO2S is damaged, contaminated or air reference hole clogged ● PCM has failed
P0130 **2T O2S, MIL: Yes** 2000, 2001, 2002, 2003, 2004, 2005 Alero, Century, Cavalier, Grand Am, Sunfire, S/T Blazer & S/T Pickup 2.2L VIN 4, 2.2L VIN 5, 2.2L VIN F, 2.2L VIN H, 2.4L VIN T engines Transmissions: All	**HO2S-11 (Bank 1 Sensor 1) Circuit Closed Loop Performance Conditions:** DTC P0107-P0108, P0117-P0118, P0122-P0123, P0131-P0134, P0201- P0204, P0300-P0304 P0325, P0336, P0440-P0446, P0502, P0601, P0602, P1441 and P1621 not set, engine started, TP angle at 10-40%, ECT sensor over 149ºF, engine speed from 1200-3400 rpm for 3 minutes, and the PCM detected the engine did not enter closed loop for over 5 seconds under these test conditions. **Possible Causes** ● Air leaks in the exhaust system, intake manifold, vacuum lines ● EVAP Purge system malfunction or charcoal canister problems ● HO2S signal circuit is open between the sensor and the PCM ● HO2S signal circuit is shorted to sensor or chassis ground ● HO2S is damaged, contaminated or air reference hole clogged ● PCM has failed

OBD II Trouble Code List (P0xxx Codes)

DTC	Trouble Code Title, Conditions & Possible Causes
P0130 **2T O2S, MIL: Yes** 2002, 2003, 2004, 2005 Envoy & TrailBlazer 4.2L VIN S engine Transmissions: All	**HO2S-11 (Bank 1 Sensor 1) Circuit Closed Loop Performance Conditions:** DTC P0105, P0107, P0108, P0116-P0118, P0122, P0123, P0125, P0131-P0134, P0201-P0206, P0300-P0306, P0335, P0336, P0351-P0356, P0601-P0606, P1120, P1133, P1220, P1221, P1271, P1275, P1280, P1484, P1512-P1516, P1635, P1639 and P1681 not set, engine started, engine speed from 1200-3400 rpm for over 100 seconds, TP indicated angle from 15-50%, ECT sensor more than 176°F, APP sensor indicated angle over 1.2% for 5 seconds, and the PCM detected an open loop condition for 7.5 out of 12 seconds. The HO2S is used for fuel control and post-catalyst monitoring. This sensor compares the oxygen content of the surrounding air with the oxygen content of the exhaust stream. At initial startup, the PCM operates in open loop mode, ignoring the HO2S signal when calculating the air/fuel ratio. The PCM supplies the HO2S with a reference (or bias) voltage of about 450 mv. The HO2S generates a voltage within a range of 0-1000 mv that fluctuates above and below the bias voltage once in closed loop. A high HO2S voltage indicates a rich fuel mixture. A low HO2S voltage indicates a lean mixture. Heating elements in the HO2S shorten the time required for the sensor to reach normal temperature, and an accurate voltage signal. **Possible Causes** ● Air leaks in the exhaust system, intake manifold, vacuum lines ● EVAP Purge system malfunction or charcoal canister problems ● HO2S signal circuit is open between the sensor and the PCM ● HO2S signal circuit is shorted to sensor or chassis ground ● HO2S is damaged, contaminated or air reference hole clogged ● PCM has failed
P0130 **2T O2S, MIL: Yes** 1997, 1998, 1999, 2000, 2001 Catera 3.0L V6 VIN R engine Transmissions: A/T	**HO2S-11 (Bank 1 Sensor 1) Closed Loop Performance Conditions:** DTC P0135, P0139, P0140 and P0141 not set, engine speed from 1400-2400 rpm, engine load at 8-12%, MAF sensor at 18-22 g/sec, throttle angle at 6-10%, and the PCM detected the HO2S Lambda signal was over 4.78v, or with the front HO2S signal from 0.9-1v for 4 seconds, the rear HO2S signal was above 650 mv or below 300 mv, or with the front HO2S signal over 1.2v, the rear HO2S signal was more than 650 mv, or with the front HO2S Lambda signal less than 0.80v, the rear HO2S signal was less than 300 mv in the test. **Possible Causes** ● Air leaks in the exhaust system, intake manifold, vacuum lines ● EVAP Purge system malfunction or charcoal canister problems ● HO2S signal circuit is open between the sensor and the PCM ● HO2S signal circuit is shorted to sensor or chassis ground ● HO2S is damaged, contaminated or air reference hole clogged ● PCM has failed
P0131 **2T CCM, MIL: Yes** 1996, 1997, 1998, 1999, 2000, 2001, 2002, 2003, 2004, 2005 Achieva, Alero, Beretta, Century, Cavalier, Ciera, Century, Corsica, Envoy, Grand Am, Malibu, S/T Blazer & Pickup, Skylark, Sunfire, TrailBlazer 2.2L VIN 4, 2.2L VIN 5, 2.2L VIN F, 2.2L VIN H, 4.2L VIN S Transmissions: All	**HO2S-11 (Bank 1 Sensor 1) Circuit Low Input Conditions:** DTC P0105-P0108, P0112-P0113, P0117-P0118, P0122-P0123, P0171, P0201-P0204, P0300, P0335, P0440-P0446, P0506, P0507, P0601-P0602 or P1441 not set, engine started, engine running with the system voltage over 10.0v, ECT sensor more than 158°F, fuel level over 10%, MAP sensor more than 25 kPa, TP angle from 8-50% for 4 seconds in closed loop, and the PCM detected the HO2S signal was less than 52 mv for 125 seconds. **Possible Causes** ● Air leaks in the exhaust system, intake manifold, vacuum lines ● Engine misfire condition present (look for P0300 series codes) ● Fuel system too lean (possible low fuel pressure, water in fuel) ● HO2S signal circuit is shorted to the sensor or chassis ground ● HO2S is damaged (i.e., cracked) or air reference hole clogged ● MAP sensor is skewed indicating a false high vacuum condition - disconnect MAP sensor and check the HO2S for a lean signal ● PCM has failed
P0131 **2T CCM, MIL: Yes** 1995 Century, Cutlass Ciera, Cutlass Supreme, Grand Prix, Lumina, Monte Carlo Park Avenue Regal, Riviera, 88' & 98' 3.1L VIN M, 3.4L VIN X, 3.8L VIN 1, 3.8L VIN K, 3.8L VIN L engines Transmissions: All	**HO2S-11 (Bank 1 Sensor 1) Circuit Low Input Conditions:** DTC P0101-P0103, P0106-P0108, P0112-P0113, P0116-P0118, P0121-P0123, P0125, P0128, P0201-P0206, P0410, P0440, P0442, P0443, P0446, P0449 and P1441 not set, engine started, TP angle from 5-40%, and the PCM detected the HO2S signal was less than 175 mv for 45 seconds, or was less than 600 mv for 55 seconds while operating in P/E mode during the test. **Possible Causes** ● Air leaks in the exhaust system, intake manifold, vacuum lines ● Engine misfire condition present (look for P0300 series codes) ● Fuel system too lean (possible low fuel pressure, water in fuel) ● HO2S signal circuit is shorted to the sensor or chassis ground ● HO2S is damaged (i.e., cracked) or air reference hole clogged ● PCM has failed

OBD II Trouble Code List (P0xxx Codes)

DTC	Trouble Code Title, Conditions & Possible Causes
P0131 **2T CCM, MIL: Yes** 1996, 1997, 1998, 1999, 2000, 2001, 2002 Achieva, Beretta, Ciera, Corsica, Century, Grand Am, Grand Prix, Malibu, Monte Carlo, Skylark 3.1L VIN M engine Transmissions: All	**HO2S-11 (Bank 1 Sensor 1) Circuit Low Input Conditions:** DTC P0101-P0103, P0107, P0108, P0112, P0113, P0116, P0117, P0118, P0121-P0123, P0125, P0128, P0201-P0206, P0410, P0440, P0442, P0443, P0446, P0449 and P1441 not set, engine started, engine running with the TP angle from 3-35%, A/F ratio from 12.5-16.5:1, system voltage over 10.0v and the PCM detected the HO2S signal was less than 175 mv for 30 seconds, or with P/E Mode active, the HO2S signal was less than 600 mv for 10 seconds. **Possible Causes** ● Air leaks in the exhaust system, intake manifold, vacuum lines ● Engine misfire condition present (look for P0300 series codes) ● Fuel system too lean (possible low fuel pressure, water in fuel) ● HO2S signal circuit is shorted to the sensor or chassis ground ● HO2S is damaged (i.e., cracked) or air reference hole clogged ● PCM has failed
P0131 **2T CCM, MIL: Yes** 1996, 1997, 1998, 1999, 2000, 2001, 2002, 2003, 2004, 2005 Aurora, Bonneville, Century, Cutlass Supreme, Grand Prix, Intrigue, LeSabre, LSS, Lumina, Monte Carlo, 88', 98', Regal, Park Avenue 3.1L VIN J, 3.1L VIN M, 3.4L VIN X, 3.5L V6 VIN H, 3.8L VIN 1, 3.8L VIN K, 4.0L V8 VIN C engines Transmissions: All	**HO2S-11 (Bank 1 Sensor 1) Circuit Low Input Conditions:** DTC P0101-P0103, P0106-P0108, P0112-P0113, P0116-P0118, P0121-P0123, P0125, P0128, P0201-P0206, P0410, P0440, P0442, P0443, P0446, P0449 and P1441 not set, engine started, TP angle from 5-40%, and the PCM detected the HO2S signal was less than 175 mv for 45 seconds, or less than 600 mv for 55 seconds while operating in P/E mode during the test. The HO2S is used for fuel control and post-catalyst monitoring. This sensor compares the oxygen content of the surrounding air with the oxygen content of the exhaust stream. At initial startup, the PCM operates in open loop mode, ignoring the HO2S signal when calculating the air/fuel ratio. The PCM supplies the HO2S with a reference (or bias) voltage of about 450 mv. The HO2S generates a voltage within a range of 0-1000 mv that fluctuates above and below the bias voltage once in closed loop. A high HO2S voltage indicates a rich fuel mixture. A low HO2S voltage indicates a lean mixture. Heating elements in the HO2S shorten the time required for the sensor to reach normal temperature, and an accurate voltage signal. **Possible Causes** ● Air leaks in the exhaust system, intake manifold, vacuum lines ● Engine misfire condition present (look for P0300 series codes) ● Fuel system too lean (possible low fuel pressure, water in fuel) ● HO2S signal circuit is shorted to the sensor or chassis ground ● HO2S is damaged (i.e., cracked) or air reference hole clogged ● PCM has failed
P0131 **2T CCM, MIL: Yes** 1996, 1997 Camaro, Corvette, Caprice, Firebird & Fleetwood 4.3L VIN W, 5.7L VIN 5, 5.7L VIN P engine Transmissions: All	**HO2S-11 (Bank 1 Sensor 1) Circuit Low Input Conditions:** DTC P0101-P0103, P0106-P0108, P0112-P0113, P0116-P0118, P0121-P0123, P0125, P0128, P0200, P0300, P0410-P0446, P0452-P0453, P1258, P1415-P1416 and P1441 not set engine started, engine running in closed loop, ECT sensor more than 118°F, TP angle from 3-20%, Fuel Trim control enabled, and the PCM detected the HO2S signal was less than 200 mv for 30 seconds. The HO2S is used for fuel control and post-catalyst monitoring. This sensor compares the oxygen content of the surrounding air with the oxygen content of the exhaust stream. At initial startup, the PCM operates in open loop mode, ignoring the HO2S signal when calculating the air/fuel ratio. The PCM supplies the HO2S with a reference (or bias) voltage of about 450 mv. **Possible Causes** ● Air leaks in the exhaust system, intake manifold, vacuum lines ● Engine misfire condition present (look for P0300 series codes) ● Fuel system too lean (possible low fuel pressure, water in fuel) ● HO2S signal circuit is shorted to the sensor or chassis ground ● HO2S is damaged (i.e., cracked) or air reference hole clogged ● PCM has failed
P0131 **2T CCM, MIL: Yes** 1996, 1997, 1998, 1999, 2000, 2001, 2002, 2003, 2004, 2005 Alero, Aztek, Cutlass, Impala, Malibu, Montana, Silhouette, Venture, Rendezvous 3.1L VIN J, 3.1L VIN M, 3.4L VIN E Transmissions: All	**HO2S-11 (Bank 1 Sensor 1) Circuit Low Input Conditions:** DTC P0101, P0102, P0103, P0107, P0108, P0112, P0113, P0116, P0117, P0118, P0121, P0122, P0123, P0125, P0128, P0201, P0202, P0203, P0204, P0205, P0206, P0410, P0440, P0442, P0443, P0446, P0449 and P1441 not set, A/F ratio at 13.0-16.5:1, TP angle from 3-40%, Air Pump "off", and the PCM detected the HO2S signal was less than 175 mv in closed loop, or with the P/E mode active, the HO2S signal was under 600 mv for 15 seconds. The HO2S is used for fuel control and post-catalyst monitoring. This sensor compares the oxygen content of the surrounding air with the oxygen content of the exhaust stream. At startup, the PCM operates in open loop mode, ignoring the HO2S signal when calculating the air/fuel ratio. The HO2S voltage range is from 0-1000 mv as it fluctuates above and below 450 mv. **Possible Causes** ● Air leaks in the exhaust system, intake manifold, vacuum lines ● Engine misfire condition present (look for P0300 series codes) ● Fuel system too lean (possible low fuel pressure, water in fuel) ● HO2S signal circuit is shorted to the sensor or chassis ground ● HO2S is damaged (i.e., cracked) or air reference hole clogged ● PCM has failed

OBD II Trouble Code List (P0xxx Codes)

DTC	Trouble Code Title, Conditions & Possible Causes
P0131 **2T CCM, MIL: Yes** 1996, 1997, 1998, 1999, 2000, 2001, 2002, 2003, 2004, 2005 C/K Cab & Chassis, C/K Series Trucks & G Vans 4.3L VIN W, 4.3L VIN X, 5.0L VIN M, 5.7L VIN K, 5.7L VIN R, 7.4L VIN J Transmissions: All	**HO2S-11 (Bank 1 Sensor 1) Circuit Low Input Conditions:** DTC P0101-P0103, P0106-P0108, P0112, P0113, P0116-P0118, P0121-P0123, P0200, P0300, P0401, P0404, P0405, P0440, P0442, P0446, P0452, P0453, P1120, P1125, P1220, P1221, P1258, P1404, P1441, P1514, P1515, P1516, P1517 and P1518 not set, engine started, engine running in closed loop, fuel level over 10%, system voltage over 10v, TP angle from 8-50%, or on models with TAC, the APP sensor indicated angle from 3-70%, MAP sensor more than 25 kPa, Intrusive and Scan Tool tests both off, then with the Lean Test enabled, the PCM detected the HO2S signal was less than 200 mv for 50 seconds or during the P/E Mode test, the PCM detected the HO2S signal was less than 360 mv for 10 seconds. **Possible Causes** ● Air leaks in the exhaust system, intake manifold, vacuum lines ● Engine misfire condition present (look for P0300 series codes) ● Fuel system too lean (possible low fuel pressure, water in fuel) ● HO2S signal circuit is shorted to the sensor or chassis ground ● HO2S is damaged (i.e., cracked) or air reference hole clogged ● PCM has failed
P0131 **2T CCM, MIL: Yes** 1999, 2000, 2001, 2002, 2003, 2004, 2005 Avalanche, C/K Trucks, G Vans, Envoy, TrailBlazer 4.8L VIN V, 5.3L VIN P, 5.3L VIN T, 5.3L VIN Z, 6.0L VIN N, 6.0L VIN U, 6.6L VIN 1, 8.1L VIN G Transmissions: All	**HO2S-11 (Bank 1 Sensor 1) Circuit Low Input Conditions:** DTC P0101, P0102, P0103, P0106, P0107, P0108, P0112, P0113, P0116, P0117, P0118, P0120, P0121, P0122, P0123, P0160, P0170, P0179, P0200, P0220, P0300, P0442, P0446, P0452, P0453, P0455, P0496, P1125, P1258, P1514, P1515, P1516, P1518, P2108 and P2135 not set, engine started, engine running in closed loop, system voltage from 10-18v, Fuel Alcohol content less than 90%, fuel level over 10%, TP angle from 3-70% more than the idle value, then during the Lean Test period, the PCM detected the HO2S signal was less than 200 mv for 165 seconds or with engine runtime over 30 seconds, and during the Power Enrichment test, the PCM detected the HO2S signal was less than 400 mv for 10 seconds. **Possible Causes** ● Engine misfire condition present (look for P0300 series codes) ● Fuel system too lean (possible low fuel pressure, water in fuel) ● HO2S signal circuit is shorted to the sensor or chassis ground ● HO2S is damaged (i.e., cracked) or air reference hole clogged ● PCM has failed
P0131 **2T CCM, MIL: Yes** 2003, 2004, 2005 CTS (Cadillac) 2.6L V6 VIN M & 3.2L V6 VIN N engines Transmissions: All	**HO2S-11 (Bank 1 Sensor 1) Circuit Low Input Conditions:** DTC P0443, P0444 and P0445 not set, engine started, TP angle from 5-35%, system voltage over 10.5v, MAF sensor at 10-35 g/sec, Calculated Converter Temperature under 1,472°F, and the PCM detected the HO2S-11 signal was less than 40 mv for 10 seconds. **Possible Causes** ● EVAP system leaks or restrictions, charcoal canister problems ● Fuel system too lean (possible low fuel pressure, water in fuel) ● HO2S signal circuit is shorted to sensor or chassis ground ● HO2S is damaged (air hole plugged) or it is contaminated due to improper fuel usage ● PCM has failed
P0131 **2T CCM, MIL: Yes** 1996, 1997, 1998, 1999, 2000, 2001, 2002, 2003, 2004, 2005 DeVille, Eldorado, Seville 4.6L VIN 9, 4.6L VIN Y Transmissions: All	**HO2S-11 (Bank 1 Sensor 1) Circuit Low Input Conditions:** DTC P0101-P0103, P0106-P0108, P0112-P0113, P0116-P0118, P0121-P0123, P0125, P0128, P0200, P0300, P0410-P0446, P0452-P0453, P1258, P1415-P1416 and P1441 not set, engine running in closed loop for 3 seconds, system voltage 8-18v, no injectors disabled, TP angle from 3-25%, A/F ratio at 14.5-14.8:1, and the PCM detected the HO2S signal was less than 75 mv for 5 seconds while in closed loop, or it was less than 575 mv for 8 seconds while operating in Power Enrichment mode. **Possible Causes** ● Air leaks in the exhaust system, intake manifold, vacuum lines ● Engine misfire condition present (look for P0300 series codes) ● Fuel system too lean (possible low fuel pressure, water in fuel) ● HO2S signal circuit is shorted to the sensor or chassis ground ● HO2S is damaged (i.e., cracked) or air reference hole clogged ● PCM has failed
P0131 **2T CCM, MIL: Yes** 1998, 1999, 2000, 2001, 2002 Camaro & Firebird 5.7L VIN G engine Transmissions: All	**HO2S-11 (Bank 1 Sensor 1) Circuit Low Input Conditions:** DTC P0101-P0103, P0106-P0108, P0112-P0113, P0116-P0118, P0121-P0123, P0125, P0128, P0200, P0300, P0410-P0446, P0452-P0453, P1258, P1415, P1416 and P1441 not set, engine running in closed loop, Intrusive and Scan Tool tests off, A/F ratio at 14.5-14.8:1, TP angle from 3-70%, Lean Test enabled, and the PCM detected the HO2S input was under 200 mv for 165 seconds or with the P/E test enabled, it was less than 360 mv for 10 seconds. **Possible Causes** ● Air leaks in the exhaust system, intake manifold, vacuum lines ● Engine misfire condition present (look for P0300 series codes) ● Fuel system too lean (possible low fuel pressure, water in fuel) ● HO2S signal circuit is shorted to the sensor or chassis ground ● HO2S is damaged (i.e., cracked) or the air reference hole clogged ● PCM has failed

OBD II Trouble Code List (P0xxx Codes)

DTC	Trouble Code Title, Conditions & Possible Causes
P0131 **2T CCM, MIL: Yes** 1995, 1996, 1997, 1998, 1999, 2000, 2001, 2002 Camaro & Firebird 3.8L VIN K engine Transmissions: All	**HO2S-11 (Bank 1 Sensor 1) Circuit Low Input Conditions:** DTC P0101-P0103, P0107-P0108, P0112-P0113, P0117-P0118, P0121-P0125, P0128, P0201-P0206, P0410, P0440-P0449 and P1441 not set, running in closed loop for 4 minutes, A/F ratio from 14.5-14.8:1, TP angle from 3-40%, or on models with TAC, the APP sensor indicated angle from 5-40%, ECT sensor more than 122°F, and the PCM detected the HO2S signal was less than 175 mv, or it was less than 600 mv while in P/E mode for up to 15 seconds. **Possible Causes** • Air leaks in the exhaust system, intake manifold, vacuum lines • Engine misfire condition present (look for P0300 series codes) • Fuel system too lean (possible low fuel pressure, water in fuel) • HO2S signal circuit is shorted to the sensor or chassis ground • HO2S is damaged (i.e., cracked) or air reference hole clogged • PCM has failed
P0131 **2T CCM, MIL: Yes** 1996, 1997, 1998, 1999, 2001, 2002, 2003, 2004, 2005 Aurora 3.5L V6 VIN H, 4.0L V8 VIN C engine Transmissions: A/T	**HO2S-11 (Bank 1 Sensor 1) Circuit Low Input Conditions:** DTC P0101-P0103, P0107-P0108, P0117-P0118, P0121-P0123, P0132, and P0134, not set, engine started, engine speed over 800 rpm, TP angle from 5.4-25 degrees, ECT sensor more than 179°F, MAP sensor more than 32 kPa, DTC P0151 test passed, and the PCM detected the HO2S signal was under 356 mv for 45 seconds. **Possible Causes** • Air leaks in the exhaust system, intake manifold, vacuum lines • Engine misfire condition present (look for P0300 series codes) • Fuel system too lean (possible low fuel pressure, water in fuel) • HO2S signal circuit is shorted to the sensor or chassis ground • HO2S is damaged (i.e., cracked) or air reference hole clogged • PCM has failed
P0131 **2T CCM, MIL: Yes** 1995, 1996, 1997, 1998, 1999, 2000, 2001, 2002, 2003, 2004, 2005 S/T Blazer & S/T Pickup 4.3L VIN W, 4.3L VIN X Transmissions: All	**HO2S-11 (Bank 1 Sensor 1) Circuit Low Input Conditions:** DTC P0101-P0103, P0106-P0108, P0112, P0113, P0116-P0118, P0121-P0123, P0200, P0300, P0440-P0446, P0452, P0453 and P0496 not set, engine running in closed loop, Scan Tool tests off, fuel level over 10%, Lean Test enabled, TP angle from 3-70% for 5 seconds, and the PCM detected the HO2S signal was under 200 mv for 165 seconds or with the Power Enrichment test enabled for 2 seconds, the HO2S signal was under 400 mv for 10 seconds. **Possible Causes** • Engine misfire condition present (look for P0300 series codes) • Fuel system too lean (possible low fuel pressure, water in fuel) • HO2S signal circuit is shorted to the sensor or chassis ground • HO2S is damaged (i.e., cracked) or air reference hole clogged • PCM has failed
P0131 **2T CCM, MIL: Yes** 1997, 1998, 1999, 2000, 2001 Catera 3.0L V6 VIN R engine Transmissions: All	**HO2S-11 (Bank 1 Sensor 1) Circuit Low Input Conditions:** DTC P0300-P0306, P0171, P0172, P0174, P0175, P0440, P0442, P0443, P0446, P0455 and P0460 are not set, engine started, engine speed 1400-2400 rpm, engine load from 8-12%, MAF sensor from 18-22 g/sec, TP angle from 6-10%, EVAP high loading not active, AIR test "off", and the PCM detected the HO2S signal was too low. **Possible Causes** • Air leaks in the exhaust system, intake manifold, vacuum lines • Engine misfire condition present (look for P0300 series codes) • Fuel system too lean (possible low fuel pressure, water in fuel) • HO2S signal circuit is shorted to the sensor or chassis ground • HO2S is damaged (i.e., cracked) or air reference hole clogged • PCM has failed
P0131 **2T CCM, MIL: Yes** 1996, 1997 Grand Prix, Lumina, Monte Carlo 3.4L VIN X Transmissions: All	**HO2S-11 (Bank 1 Sensor 1) Circuit Low Input Conditions:** DTC P0101, P0102, P0103, P0107, P0108, P0112, P0113, P0116, P0117, P0118, P0121, P0122, P0123, P0125, P0128, P0201, P0202, P0203, P0204, P0205, P0206, P0410, P0440, P0442, P0443, P0446, P0449 and P1441 not set, engine started, engine running with the TP angle from 3-40%, A/F ratio from 14.5-14.8:1, Air Pump "off", and the PCM detected the HO2S signal was less than 175 mv for 45 seconds, or with the P/E Mode active, that the HO2S signal was less than 600 mv for 55 seconds. **Possible Causes** • Air leaks in the exhaust system, intake manifold, vacuum lines • Engine misfire condition present (look for P0300 series codes) • Fuel system too lean (possible low fuel pressure, water in fuel) • HO2S signal circuit is shorted to the sensor or chassis ground • HO2S is damaged (i.e., cracked) or air reference hole clogged • PCM has failed

OBD II Trouble Code List (P0xxx Codes)

DTC	Trouble Code Title, Conditions & Possible Causes
P0131 **2T CCM, MIL: Yes** 1995 Corvette 5.7L VIN P engine Transmissions: All	**HO2S-11 (Bank 1 Sensor 1) Circuit Low Input Conditions:** DTC P0116, P0117, P0118, P0135, P0171, P0172, P0174, P0175, P1114 and P1115 not set, engine started, engine running in closed loop, TP angle from 2-10%, and the PCM detected the HO2S signal was less than 141 mv for 102 seconds. **Possible Causes** ● Air leaks in the exhaust system, intake manifold, vacuum lines ● Engine misfire condition present (look for P0300 series codes) ● Fuel system too lean (possible low fuel pressure, water in fuel) ● HO2S signal circuit is shorted to the sensor or chassis ground ● HO2S is damaged (i.e., cracked) or air reference hole clogged ● PCM has failed
P0131 **2T CCM, MIL: Yes** 1997, 1998, 1999, 2000, 2001, 2002, 2003, 2004, 2005 Corvette 5.7L VIN G, 5.7L VIN S Transmissions: All	**HO2S-11 (Bank 1 Sensor 1) Circuit Low Input Conditions:** DTC P0101-P0103, P0106-P0108, P0112, P0113, P0116, P0117, P0118, P0200, P0300, P0410, P0440-P0446, P0452-P0453, P1120, P1125, P1220, P1221, P1258, P1415-P1416, P1441, P1514-1518 not set, Scan Tool and Intrusive tests all "off", fuel level over 10%, TP angle from 3-70%, then with the Lean Test enabled, the PCM detected the HO2S signal was less than 200 mv for 165 seconds or with Power Enrichment mode active for 1 second, the PCM detected the HO2S signal was less than 360 mv for 10 seconds. The HO2S is used for fuel control and post-catalyst monitoring. This sensor compares the oxygen content of the surrounding air with the oxygen content of the exhaust stream. At initial startup, the PCM operates in open loop mode, ignoring the HO2S signal when calculating the air/fuel ratio. The PCM supplies the HO2S with a reference (or bias) voltage of about 450 mv. The HO2S generates a voltage within a range of 0-1000 mv that fluctuates above and below the bias voltage once in closed loop. A high HO2S voltage indicates a rich fuel mixture. A low HO2S voltage indicates a lean mixture. Heating elements in the HO2S shorten the time required for the sensor to reach normal temperature, and an accurate voltage signal. **Possible Causes** ● Air leaks in the exhaust system, intake manifold, vacuum lines ● Engine misfire condition present (look for P0300 series codes) ● Fuel system too lean (possible low fuel pressure, water in fuel) ● HO2S signal circuit is shorted to the sensor or chassis ground ● HO2S is damaged (i.e., cracked) or air reference hole clogged ● PCM has failed
P0132 **2T CCM, MIL: Yes** 1996, 1997, 1998, 1999, 2000, 2001, 2002, 2003, 2004, 2005 Achieva, Alero, Beretta, Century, Cavalier, Ciera, Century, Corsica, Envoy, Grand Am, Malibu, S/T Blazer & Pickup, Skylark, Sunfire, TrailBlazer 2.2L VIN 4, 2.2L VIN 5, 2.2L VIN F, 2.2L VIN H, 4.2L VIN S Transmissions: All	**HO2S-11 (Bank 1 Sensor 1) Circuit High Input Conditions:** DTC P0105, P0107, P0108, P0112, P0113, P0117, P0118, P0122, P0123, P0125, P0201-P0204, P0300, P0301-P0304, P0336, P0440, P0446, P0452, P0453, P0506, P0507, P0601, P0602, P1441 and P1621 not set, engine started, engine runtime in closed loop, system voltage over 10.0v, MAP sensor over 20 kPa, ECT sensor over 158ºF, fuel level over 10%, TP angle from 8-50%, and the PCM detected the HO2S signal was more than 946 mv for 50 seconds, or more than 1042 mv during Decel Fuel Cutoff mode for 2.5 seconds. **Possible Causes** ● Fuel system is too rich (fuel pressure too high, fuel pressure regulator leaking, or one or more fuel injectors sticking/leaking) ● HO2S element is silicon, water or fuel contaminated ● HO2S signal circuit is shorted to system power (B+) ● HO2S signal tracking (water intrusion) in the connector causing a short between the HO2S signal and heater power circuits ● PCM has failed
P0132 **2T CCM, MIL: Yes** 1995 Century, Cutlass Ciera, Cutlass Supreme, Grand Prix, Lumina, Monte Carlo Park Avenue Regal, Riviera, 88' & 98' 3.1L VIN M, 3.4L VIN X, 3.8L VIN 1, 3.8L VIN K, 3.8L VIN L Transmissions: All	**HO2S-11 (Bank 1 Sensor 1) Circuit High Input Conditions:** DTC P0101-P0103, P0106-P0108, P0112-P0118, P0121-P0123, P0125, P0128, P0201-P0206, P0410 and P0440 not set, engine started, engine running with the TP angle from 2-40% in closed loop, system voltage over 10.0v and the PCM detected the HO2S signal was more than 900 mv for 2-4 minutes. **Possible Causes** ● Fuel system is too rich (fuel pressure too high, fuel pressure regulator leaking, or one or more fuel injectors sticking/leaking) ● HO2S element is silicon, water or fuel contaminated ● HO2S signal tracking (water intrusion) in the connector causing a short between the HO2S signal and heater power circuits ● PCM has failed

OBD II Trouble Code List (P0xxx Codes)

DTC	Trouble Code Title, Conditions & Possible Causes
P0132 **2T CCM, MIL: Yes** 1996, 1997, 1998, 1999, 2000, 2001, 2002 1996, 1997, 1998, 1999 Achieva, Beretta, Ciera, Corsica, Century, Grand Am, Malibu, Skylark 3.1L VIN M engine Transmissions: All	**HO2S-11 (Bank 1 Sensor 1) Circuit High Input Conditions:** DTC P0101, P0102, P0103, P0107, P0108, P0112, P0113, P0116, P0117, P0118, P0121, P0122, P0123, P0125, P0128, P0201-P0204, P0205, P0206, P0410, P0440, P0442, P0443, P0446, P0449, or P1441 not set, TP angle from 3-35%, A/F ratio at 12-16:5:1 in closed loop, and the PCM detected the HO2S signal was more than 975 mv for 45 seconds, or with Decel Fuel Shutoff active, the HO2S signal was more than 200 mv for 55 seconds. The HO2S is used for fuel control and post-catalyst monitoring. This sensor compares the oxygen content of the surrounding air with the oxygen content of the exhaust stream. At initial startup, the PCM operates in open loop mode, ignoring the HO2S signal when calculating the air/fuel ratio. The PCM supplies the HO2S with a reference (or bias) voltage of about 450 mv. The HO2S generates a voltage within a range of 0-1000 mv that fluctuates above and below the bias voltage once in closed loop. A high HO2S voltage indicates a rich fuel mixture. A low HO2S voltage indicates a lean mixture. Heating elements in the HO2S shorten the time required for the sensor to reach normal temperature, and an accurate voltage signal. **Possible Causes** ● Fuel system is too rich (fuel pressure too high, fuel pressure regulator leaking, or one or more fuel injectors sticking/leaking) ● HO2S element is silicon, water or fuel contaminated ● HO2S signal circuit is shorted to system power (B+) ● HO2S signal tracking (water intrusion) in the connector causing a short between the HO2S signal and heater power circuits ● PCM has failed
P0132 **2T CCM, MIL: Yes** 1996, 1997 Camaro, Caprice, Corvette, Firebird, Fleetwood 4.3L VIN W, 5.7L VIN 5, 5.7L VIN P engines Transmissions: All	**HO2S-11 (Bank 1 Sensor 1) Circuit High Input Conditions:** DTC P0101-P0103, P0112-P0118, P0125, P0200, P0335, P0336, P0351-P0358, P0371 and P1372 not set, engine started, AIR and Catalyst Tests inactive, TP angle from 2-70% in closed loop, and the PCM detected the HO2S signal was above 774 mv for 30 seconds. **Possible Causes** ● Fuel system is too rich (fuel pressure too high, fuel pressure regulator leaking, or one or more fuel injectors sticking/leaking) ● HO2S element is silicon, water or fuel contaminated ● HO2S signal tracking (water intrusion) in the connector causing a short between the HO2S signal and heater power circuits ● PCM has failed
P0132 **2T CCM, MIL: Yes** 1996, 1997, 1998, 1999, 2000, 2001, 2002, 2003, 2004, 2005 Alero, Aztek, Cutlass, Impala, Malibu, Montana, Silhouette, Venture, Rendezvous 3.1L VIN J, 3.1L VIN M, 3.4L VIN E engines Transmissions: All	**HO2S-11 (Bank 1 Sensor 1) Circuit High Input Conditions:** DTC P0101, P0102, P0103, P0107, P0108, P0112, P0113, P0116, P0117, P0118, P0121, P0122, P0123, P0125, P0128, P0201, P0202, P0203, P0204, P0205, P0206, P0410, P0440, P0442, P0443, P0446, P0449 and P1441 not set, A/F ratio at 12.0-16.5:1, TP angle from 3-35%, Air Pump "off", and the PCM detected the HO2S signal was more than 975 mv for 45 seconds, or the HO2S signal was more than 200 mv while in Decel Fuel Cutoff mode. **Possible Causes** ● Fuel system is too rich (fuel pressure too high, fuel pressure regulator leaking, or one or more fuel injectors sticking/leaking) ● HO2S element is silicon, water or fuel contaminated ● HO2S signal circuit is shorted to system power (B+) ● HO2S signal tracking (water intrusion) in the connector causing a short between the HO2S signal and heater power circuits ● PCM has failed
P0132 **2T CCM, MIL: Yes** 1996, 1997, 1998, 1999, 2000, 2001, 2002, 2003, 2004, 2005 Aurora, Bonneville, Century, LSS, Cutlass Supreme, Grand Prix, Intrigue, LeSabre, Monte Carlo, Park Avenue, Regal, 88', 98' 3.1L VIN J, 3.1L VIN M, 3.4L VIN X, 3.5L V6 VIN H, 3.8L VIN 1, 3.8L VIN K, 4.0L V8 VIN C Transmissions: All	**HO2S-11 (Bank 1 Sensor 1) Circuit High Input Conditions:** DTC P0101-P0103, P0106-P0108, P0112-P0118, P0121-P0123, P0125, P0128, P0201-P0206, P0410, P0440-P0449 and P1441 not set, engine started, engine running with the TP angle from 5-40% in closed loop, and the PCM detected the HO2S signal was more than 975 mv, or it was more than 200 mv for 55 seconds in DFCO mode. **Possible Causes** ● Fuel system is too rich (fuel pressure too high, fuel pressure regulator leaking, or one or more fuel injectors sticking/leaking) ● HO2S element is silicon, water or fuel contaminated ● HO2S signal tracking (water intrusion) in the connector causing a short between the HO2S signal and heater power circuits ● PCM has failed

OBD II Trouble Code List (P0xxx Codes)

DTC	Trouble Code Title, Conditions & Possible Causes
P0132 **2T CCM, MIL: Yes** 1996, 1997, 1998, 1999, 2000, 2001, 2002, 2003, 2004, 2005 C/K Cab & Chassis, C/K Series Pickup, G vans, M/L Series Vans 4.3L VIN W, 4.3L VIN X, 5.0L VIN M, 5.7L VIN K, 5.7L VIN R, 7.4L VIN J Transmissions: All	**HO2S-11 (Bank 1 Sensor 1) Circuit High Input Conditions:** DTC P0101-P0103, P0106-P0108, P0112, P0113, P0116-P0118, P0121-P0123, P0200, P0300, P0401, P0404, P0405, P0440, P0442, P0446, P0452, P0453, P1120, P1125, P1220, P1221, P1258, P1404, P1441, P1514-P1517 and P1518 not set, engine started, fuel level over 10%, Intrusive and Scan Tool Tests inactive, then with the Rich Test enabled, A/F ratio from 14.5-14.7:1, TP angle from 3.5-70% for 5 seconds, or for vehicles with TAC, with the TP indicated angle from 3-70%, the PCM detected the HO2S signal was more than 775 mv for 165 seconds or with Decel Fuel Cutoff active and the time since the test started over 1 second, the PCM detected the HO2S signal was more than 540 mv for 5 seconds. **Possible Causes** ● Fuel system is too rich (fuel pressure too high, fuel pressure regulator leaking, or one or more fuel injectors sticking/leaking) ● HO2S element is silicon, water or fuel contaminated ● HO2S signal tracking (water intrusion) in the connector causing a short between the HO2S signal and heater power circuits ● PCM has failed
P0132 **2T CCM, MIL: Yes** 1999, 2000, 2001, 2002, 2003, 2004, 2005 C/K Series Truck, G Series Van, Envoy, Escalade, TrailBlazer 4.8L VIN V, 5.3L VIN P, 5.3L VIN T, 5.3L VIN Z, 6.0L VIN N, 6.0L VIN U, 8.1L VIN G Transmissions: All	**HO2S-11 (Bank 1 Sensor 1) Circuit High Input Conditions:** DTC P0101, P0102, P0103, P0106, P0107, P0108, P0112, P0113, P0116, P0117, P0118, P0120, P0121, P0122, P0123, P0169, P0178, P0179, P0200, P0220, P0300, P0442, P0446, P0452, P0453, P0455, P0496, P1125, P1258, P1514, P1515, P1516, P1518, P2108 and P2135 not set, engine started, engine running in closed loop, system voltage from 10-18v, Fuel Alcohol content less than 90%, fuel level over 10%, TP angle from 3-70% more than the idle value, then during the Rich Test period, the PCM detected the HO2S signal was more than 900 mv for 165 seconds or with engine runtime over 30 seconds, and during the Decel Fuel Cutoff test, the PCM detected the HO2S signal was less than 250 mv for 5 seconds. **Possible Causes** ● Fuel system is too rich (fuel pressure too high, fuel pressure regulator leaking, or one or more fuel injectors sticking/leaking) ● HO2S element is silicon, water or fuel contaminated ● HO2S signal tracking (water intrusion) in the connector causing a short between the HO2S signal and heater power circuits ● PCM has failed
P0132 **2T CCM, MIL: Yes** 2003, 2004, 2005 CTS (Cadillac) 2.6L V6 VIN M & 3.2L V6 VIN N engines Transmissions: All	**HO2S-11 (Bank 1 Sensor 1) Circuit High Input Conditions:** Engine started, system voltage over 10.5v, Converter Temperature less than 1,472ºF, and the PCM detected the HO2S-11 signal was more than 1,080 mv for over 5 seconds. **Possible Causes** ● Air leaks in the exhaust system, intake manifold, vacuum lines ● Fuel system too rich (fuel pressure too high, fuel pressure regulator leaking, or one or more fuel injectors sticking/leaking) ● HO2S element is silicon, water or fuel contaminated ● HO2S signal tracking (water intrusion) in the connector causing a short between the HO2S signal and heater power circuits ● HO2S low reference circuit is open or shorted to power (B+) ● PCM has failed
P0132 **2T CCM, MIL: Yes** 1995, 1996, 1997, 1998, 1999, 2000, 2001, 2002 Camaro & Firebird 3.8L VIN K engine Transmissions: All	**HO2S-11 (Bank 1 Sensor 1) Circuit High Input Conditions:** DTC P0101, P0102, P0103, P0107, P0108, P0112, P0113, P0117, P0118, P0121, P0122, P0123, P0125, P0128, P0201-P0206, P0410, P0440, P0442, P0443, P0446, P0449, or P1441not set, engine started, system voltage over 10.0v, engine running in closed loop for 4 minutes, ECT sensor more than 122ºF, A/F ratio from 14.5-14.8:1, TP angle from 3-40% or on models with TAC, the APP sensor indicated angle from 5-40%, and the PCM detected the HO2S signal was more 975 mv, or it was more than 200 mv in Decel Fuel Cutoff mode for 15 seconds. The HO2S is used for fuel control and post-catalyst monitoring. The PCM supplies the HO2S with a reference (or bias) voltage of about 450 mv. The HO2S generates a voltage within a range of 0-1000 mv that fluctuates above and below bias voltage once in closed loop. **Possible Causes** ● Fuel system is too rich (fuel pressure too high, fuel pressure regulator leaking, or one or more fuel injectors sticking/leaking) ● HO2S element is silicon, water or fuel contaminated ● HO2S signal circuit is shorted to system power (B+) ● HO2S signal tracking (water intrusion) in the connector causing a short between the HO2S signal and heater power circuits ● PCM has failed

OBD II Trouble Code List (P0xxx Codes)

DTC	Trouble Code Title, Conditions & Possible Causes
P0132 **2T CCM, MIL: Yes** 1996, 1997, 1998, 1999, 2000, 2001, 2002, 2003, 2004, 2005 DeVille, Eldorado, Seville 4.6L VIN 9, 4.6L VIN Y Transmissions: All	**HO2S-11 (Bank 1 Sensor 1) Circuit High Input Conditions:** DTC P0101-P0103, P0106-P0108, P0112-P0113, P0116-P0118, P0121-P0123, P0125, P0128, P0200, P0300, P0410-P0446, P0452-P0453, P1258, P1415-P1416 and P1441 not set, AIR, Catalyst and EGR Flow tests "off", A/F ratio at 14.4-14.9:1, no injectors disabled, TP angle from 3-25% for 3 seconds, and the PCM detected the HO2S signal was more than 900 mv for 50 seconds, or that it was more than 200 mv during Decel Fuel Cutoff mode. **Possible Causes** ● Fuel system is too rich (fuel pressure too high, fuel pressure regulator leaking, or one or more fuel injectors sticking/leaking) ● HO2S element is silicon, water or fuel contaminated ● HO2S signal tracking (water intrusion) in the connector causing a short between the HO2S signal and heater power circuits ● PCM has failed
P0132 **2T CCM, MIL: Yes** 1998, 1999, 2000, 2001, 2002 Camaro & Firebird 5.7L VIN G engines Transmissions: All	**HO2S-11 (Bank 1 Sensor 1) Circuit High Input Conditions:** DTC P0101, P0102, P0103, P0106, P0107, P0108, P0112-P0118, P0121, P0122, P0123, P0125, P0128, P0200, P0300, P0410, P0440-P0446, P0452, P0453, P1258, P1415, P1416 and P1441 not set, engine started, system voltage over 10.0v, fuel level over 10%, Intrusive and Scan Tool tests "off", A/F ratio 14.5-14.7:1, then with the Rich Test enabled and the TP angle from 3-70%, the PCM detected the HO2S signal was more than 775 mv for 165 seconds or with DFCO mode enabled for 1 second, the PCM detected the HO2S signal was less than 540 mv for 5 seconds. **Possible Causes** ● Fuel system is too rich (fuel pressure too high, fuel pressure regulator leaking, or one or more fuel injectors sticking/leaking) ● HO2S element is silicon, water or fuel contaminated ● HO2S signal tracking (water intrusion) in the connector causing a short between the HO2S signal and heater power circuits ● PCM has failed
P0132 **2T CCM, MIL: Yes** 1996, 1997, 1998, 1999, 2001, 2002, 2003, 2004, 2005 Aurora 3.5L V6 VIN H, 4.0L V8 VIN C engines Transmissions: A/T	**HO2S-11 (Bank 1 Sensor 1) Circuit High Input Conditions:** DTC P0101, P0102, P0103, P0106, P0107, P0108, P0112, P0113, P0116, P0117, P0118, P0121, P0122, P0123, P0125, P0201-P0208, P0410, P0412, P0418, P0419, P0440, P0442, P0443, P0446, P0449, P1415, P1416 and P1441 are not set, engine started, TP angle from 3-40 degrees, ECT sensor more than 179°F, A/F ratio from 14.4-14.9:1, AIR, Catalyst and EGR Tests all inactive, system voltage over 10.0v, MAP sensor more than 32 kPa, and the PCM detected the HO2S signal was more than 900 mv in closed loop for 3 seconds, or that it was more than 200 mv in DFCO mode. **Possible Causes** ● Fuel system is too rich (fuel pressure too high, fuel pressure regulator leaking, or one or more fuel injectors sticking/leaking) ● HO2S element is silicon, water or fuel contaminated ● HO2S signal tracking (water intrusion) in the connector causing a short between the HO2S signal and heater power circuits ● PCM has failed
P0132 **2T CCM, MIL: Yes** 1995, 1996, 1997, 1998, 1999, 2000, 2001, 2002, 2003, 2004, 2005 S/T Blazer & S/T Pickup 4.3L VIN W, 4.3L VIN X Transmissions: All	**HO2S-11 (Bank 1 Sensor 1) Circuit High Input Conditions:** DTC P0101, P0102, P0103, P0106, P0107, P0108, P0112, P0113, P0116, P0117, P0118, P0121, P0122, P0123, P0200, P0300, P0440, P0442, P0446, P0452, P0453, P0455, P0496 and P1441 not set, engine started, system voltage over 10.0v, fuel level over 10%, Intrusive and Scan Tool tests "off", then with the Rich Test enabled, A/F ratio from 14.5-14.7:1 and the TP angle at 3-70% for 5 seconds, the PCM detected the HO2S signal was more than 900 mv for 165 seconds or the HO2S signal was more than 250 mv for 5 seconds while operating in Decel Fuel Cutoff mode. The HO2S is used for fuel control and post-catalyst monitoring. This sensor compares the oxygen content of the surrounding air with the oxygen content of the exhaust stream. At initial startup, the PCM operates in open loop mode, ignoring the HO2S signal when calculating the air/fuel ratio. **Possible Causes** ● Fuel system is too rich (fuel pressure too high, fuel pressure regulator leaking, or one or more fuel injectors sticking/leaking) ● HO2S element is silicon, water or fuel contaminated ● HO2S signal circuit is shorted to system power (B+) ● HO2S signal tracking (water intrusion) in the connector causing a short between the HO2S signal and heater power circuits ● PCM has failed

OBD II Trouble Code List (P0xxx Codes)

DTC	Trouble Code Title, Conditions & Possible Causes
P0132 **2T CCM, MIL: Yes** 1997, 1998, 1999, 2000, 2001 Catera 3.0L V6 VIN R engine Transmissions: All	**HO2S-11 (Bank 1 Sensor 1) Circuit High Input Conditions:** DTC P0137 and P0138 not set, engine started, system voltage over 10.0v, engine speed from 1400-2400 rpm, ECT sensor more than 158°F, engine load from 8-12%, MAF sensor 18-22 g/sec, TP angle at 6-10%, HO2S-12 test finished, and the PCM detected too much difference between the front Lambda HO2S and rear HO2S signals. **Possible Causes** ● Fuel system is too rich (fuel pressure too high, fuel pressure regulator leaking, or one or more fuel injectors sticking/leaking) ● HO2S element is silicon, water or fuel contaminated ● HO2S signal circuit is shorted to system power (B+) ● HO2S signal tracking (water intrusion) in the connector causing a short between the HO2S signal and heater power circuits ● PCM has failed
P0132 **2T CCM, MIL: Yes** 1996, 1997 Grand Prix, Lumina, Monte Carlo 3.4L VIN X engine Transmissions: All	**HO2S-11 (Bank 1 Sensor 1) Circuit High Input Conditions:** DTC P0101, P0102, P0103, P0107, P0108, P0112, P0113, P0116, P0117, P0118, P0121, P0122, P0123, P0125, P0128, P0201, P0202, P0203, P0204, P0205, P0206, P0410, P0440, P0442, P0443, P0446, P0449, or P1441 not set, engine started, TP angle from 3-35%, AIR pump inactive, system voltage over 10.0v, A/F ratio at 12.5-16.5:1 in closed loop, and the PCM detected the HO2S signal was more than 975 mv for 45 seconds or with Decel Fuel Shutoff enabled, the HO2S signal was over 110 mv for 55 seconds. **Possible Causes** ● Fuel system is too rich (fuel pressure too high, fuel pressure regulator leaking, or one or more fuel injectors sticking/leaking) ● HO2S element is silicon, water or fuel contaminated ● HO2S signal tracking (water intrusion) in the connector causing a short between the HO2S signal and heater power circuits ● PCM has failed
P0132 **2T CCM, MIL: Yes** 1995 Corvette 5.7L VIN P engine Transmissions: All Corvette	**HO2S-11 (Bank 1 Sensor 1) Circuit High Input Conditions:** DTC P0116-P0118, P0135, P0171, P0172, P0174, P0175, P1114 and P1115 not set, engine started, system voltage over 10.0v, engine running in closed loop, TP angle from 2-10%, and the PCM detected the HO2S signal was more than 853 mv for 102 seconds. **Possible Causes** ● Fuel system too rich (fuel pressure too high, fuel pressure regulator leaking, or one or more fuel injectors sticking/leaking) ● HO2S element is silicon, water or fuel contaminated ● HO2S signal circuit is shorted to system power (B+) ● HO2S signal tracking (water intrusion) in the connector causing a short between the HO2S signal and heater power circuits ● PCM has failed
P0132 **2T CCM, MIL: Yes** 1997, 1998, 1999, 2000, 2001, 2002, 2003, 2004, 2005 Corvette 5.7L VIN G, 5.7L VIN S Transmissions: All	**HO2S-11 (Bank 1 Sensor 1) Circuit High Input Conditions:** DTC P0101-P0103, P0106-P0108, P0112, P0113, P0116, P0117, P0118, P0200, P0300, P0410, P0440, P0442, P0446, P0452, P0453, P1120, P1125, P1220, P1221, P1258, P1415, P1416, P1441, P1514, P1515, P1516, P1517, or P1518 not set, engine started, TP angle from 3-70% in closed loop, AIR and Catalyst Tests "off", then with the Rich Test enabled, the PCM detected the HO2S signal was above 775 mv for 165 seconds or with Decel Fuel Cutoff Test active for 1 second, the PCM detected the HO2S signal was over 540 mv for 5 seconds. **Possible Causes** ● Fuel system too rich (fuel pressure too high, fuel pressure regulator leaking, or one or more fuel injectors sticking/leaking) ● HO2S element is silicon, water or fuel contaminated ● HO2S signal circuit is shorted to system power (B+) ● HO2S signal tracking (water intrusion) in the connector causing a short between the HO2S signal and heater power circuits ● PCM has failed
1996, 1997, 1998, 1999, 2000, 2001, 2002, 2003, 2004, 2005 Achieva, Alero, Beretta, Century, Cavalier, Ciera, Century, Corsica, Envoy, Grand Am, Malibu, S/T Blazer & Pickup, Skylark, Sunfire, TrailBlazer 2.2L VIN 4, 2.2L VIN 5, 2.2L VIN F, 2.2L VIN H, 4.2L VIN S Transmissions: All	**HO2S-11 (Bank 1 Sensor 1) Slow Response Conditions:** DTC P0105, P0107, P0108, P0112, P0113, P0117, P0118, P0122, P0123, P0171, P0201-P0206, P0300, P0301-P0306, P0335, P0440, P0442, P0446, P0506, P0507, P0601, P0602 and P1441 not set, engine started, system voltage over 10.0v, ECT sensor over 158°F, fuel level over 10%, engine speed from 1600-2450 rpm for 10 second in closed loop, TP angle from 9-18%, Purge command over 36%, and the PCM detected the HO2S rich-to-lean response time was over 1119 ms, or the lean-to-rich response time was over 760 ms. **Possible Causes** ● Exhaust leak present in the exhaust manifold or exhaust pipes ● Fuel system is too rich (fuel pressure too high, fuel pressure regulator leaking, or one or more fuel injectors sticking/leaking) ● HO2S element is silicon, water or fuel contaminated or it failed ● TP sensor element broken (can cause false acceleration event) ● PCM has failed

OBD II Trouble Code List (P0xxx Codes)

DTC	Trouble Code Title, Conditions & Possible Causes
P0133 **2T O2S, MIL: Yes** 1995 Century, Cutlass Ciera, Cutlass Supreme, Grand Prix, Lumina, Monte Carlo Park Avenue Regal, Riviera, 88' & 98' 3.1L VIN M, 3.4L VIN X, 3.8L VIN 1, 3.8L VIN K, 3.8L VIN L Transmissions: All	**HO2S-11 (Bank 1 Sensor 1) Slow Response Conditions:** DTC P0116, P0117, P0118, P0131, P0132, P0135, P0171 P0172, P0174, P0175, P0420, P1115, P1133, P1153 and P1158 not set, engine started, engine speed from 1000-3000 rpm in closed loop, ECT sensor more than 122ºF, MAF sensor from 13-32 g/s, and the PCM detected the average HO2S rich-lean or lean-rich response time was more than 135-145 ms. **Possible Causes** ● Exhaust leak present in the exhaust manifold or exhaust pipes ● Fuel system is too rich (fuel pressure too high, fuel pressure regulator leaking, or one or more fuel injectors sticking/leaking) ● HO2S element is silicon, water or fuel contaminated or it failed ● TP sensor element broken (can cause false acceleration event) ● PCM has failed
P0133 **2T O2S, MIL: Yes** 1996, 1997, 1998, 1999, 2000, 2001, 2002 Achieva, Beretta, Ciera, Corsica, Century, Grand Am, Grand Prix, Lumina, Malibu, Monte Carlo, Skylark 3.1L VIN M, 3.4L VIN X Transmissions: All	**HO2S-11 (Bank 1 Sensor 1) Slow Response Conditions:** DTC P0101-P0103, P0105, P0107, P0108, P0112, P0113, P0117, P0118, P0121-P0123, P0171, P0201-P0204, P0300, P0301-P0304, P0335, P0440, P0442, P0446, P0506, P0507, P0601, P0602 and P1441 not set, engine started, engine speed from 1000-3000 rpm in closed loop, TP angle from 3-40%, system voltage over 10.0v, ECT sensor more than 122ºF, MAF sensor from 10-30 g/sec, Air Pump "off", and the PCM detected the HO2S rich-lean average transition response time was longer than 105-148 ms. The HO2S is used for fuel control and post-catalyst monitoring. This sensor compares the oxygen content of the surrounding air with the oxygen content of the exhaust stream. At initial startup, the PCM operates in open loop mode, ignoring the HO2S signal when calculating the air/fuel ratio. The PCM supplies the HO2S with a reference (or bias) voltage of about 450 mv. The HO2S generates a voltage within a range of 0-1000 mv that fluctuates above and below the bias voltage once in closed loop. A high HO2S voltage indicates a rich fuel mixture. A low HO2S voltage indicates a lean mixture. **Possible Causes** ● Exhaust leak present in the exhaust manifold or exhaust pipes ● Fuel system is too rich (fuel pressure too high, fuel pressure regulator leaking, or one or more fuel injectors sticking/leaking) ● HO2S element is silicon, water or fuel contaminated or it failed ● TP sensor element broken (can cause false acceleration event) ● PCM has failed
P0133 **2T O2S, MIL: Yes** 1996, 1997 Camaro, Corvette, Caprice, Firebird & Fleetwood 4.3L VIN W, 5.7L VIN 5, 5.7L VIN P Transmissions: All	**HO2S-11 (Bank 1 Sensor 1) Slow Response Conditions:** DTC P0101-P0103, P0112-P0115, P0125, P0131-P0135, P0171-P0175, P0410, P1114-P1115 and P1133 not set, engine started, engine speed from 1000-1700 rpm in closed loop for 2 minutes, MAF sensor from 12-28 g/sec, Purge command less than 90%, and the PCM detected the average HO2S lean-rich or rich-lean response time was more than 110-150 ms for over 100 seconds. **Possible Causes** ● Exhaust leak present in the exhaust manifold or exhaust pipes ● Fuel system is too rich (fuel pressure too high, fuel pressure regulator leaking, or one or more fuel injectors sticking/leaking) ● HO2S element is silicon, water or fuel contaminated or it failed ● TP sensor element broken (can cause false acceleration event) ● PCM has failed
P0133 **2T O2S, MIL: Yes** 1996, 1997, 1998, 1999, 2000, 2001, 2002, 2003, 2004, 2005 Alero, Aztek, Cutlass, Impala, Malibu, Montana, Silhouette, Venture, Rendezvous 3.1L VIN J, 3.1L VIN M, 3.4L VIN E engines Transmissions: All	**HO2S-11 (Bank 1 Sensor 1) Slow Response Conditions:** DTC P0101, P0102, P0103, P0107, P0108, P0112, P0113, P0116, P0117, P0118, P0121, P0122, P0123, P0125, P0128, P0201, P0202, P0203, P0204, P0205, P0206, P0410, P0440, P0442, P0443, P0446, P0449 and P1441 not set, engine started, engine speed from 1000-3000 rpm in closed loop, A/F ratio from 14.5-14.8, system voltage over 10.0v, ECT sensor more than 122ºF, MAF sensor from 10-30 g/sec, gear selector not in Reverse or P/N, Air Pump "off", and the PCM detected the HO2S lean to rich average response time was more than 94 milliseconds, or the average rich to lean response time was more than 105 ms. **Possible Causes** ● Exhaust leak present in the exhaust manifold or exhaust pipes ● Fuel system is too rich (fuel pressure too high, fuel pressure regulator leaking, or one or more fuel injectors sticking/leaking) ● HO2S element is silicon, water or fuel contaminated or it failed ● TP sensor element broken (can cause false acceleration event) ● PCM has failed

OBD II Trouble Code List (P0xxx Codes)

DTC	Trouble Code Title, Conditions & Possible Causes
P0133 **2T O2S, MIL: Yes** 1996, 1997, 1998, 1999, 2000, 2001, 2002, 2003, 2004, 2005 Bonneville, Century, Cutlass Supreme, 88'/98', Grand Prix, Intrigue, LeSabre, LSS, Lumina, Monte Carlo, Regal, Park Avenue 3.1L VIN J, 3.1L VIN M, 3.4L VIN X, 3.5L V6 VIN H, 3.8L VIN 1, 3.8L VIN K, 4.0L V8 VIN C engines Transmissions: All	**HO2S-11 (Bank 1 Sensor 1) Slow Response Conditions:** DTC P0101, P0102, P0103, P0106, P0107, P0108, P0112, P0113, P0116, P0117, P0118, P0121, P0122, P0123, P0125, P0128, P0201 P0202, P0203, P0204, P0205, P0206, P0410, P0440, P0442, P0443, P0446, P0449 and P1441 not set, engine started, engine speed from 1000-3000 rpm in closed loop for 1 minute, ECT sensor more than 122°F, MAF sensor from 13-32 g/s, and the PCM detected the average HO2S rich-lean response time was over 145 ms or the average lean-rich response time was over 135 ms. **Possible Causes** ● Exhaust leak present in the exhaust manifold or exhaust pipes ● Fuel system is too rich (fuel pressure too high, fuel pressure regulator leaking, or one or more fuel injectors sticking/leaking) ● HO2S element is silicon, water or fuel contaminated or it failed ● TP sensor element broken (can cause false acceleration event) ● PCM has failed
P0133 **2T O2S, MIL: Yes** 1996, 1997, 1998, 1999, 2000, 2001, 2002, 2003, 2004, 2005 C/K Cab & Chassis, C/K Series Truck, G Series Van, M/L Series Vans 4.3L VIN W, 4.3L VIN X, 5.0L VIN M, 5.7L VIN K, 5.7L VIN R, 7.4L VIN J Transmissions: All	**HO2S-11 (Bank 1 Sensor 1) Slow Response Conditions:** DTC P0101-P0103, P0106-P0108, P0112-P0113, P0116-P0118, P0121 P0123, P0131-P0135, P0151-P0155, P0200, P0300, P0401, P0404-P0405, P0440-P0446, P0452-P0453, P1120, P1125, P1220-P1221, P1258, P1404, P1441, P1514 and P1518 not set, engine started, engine speed from 1200-3000 rpm for 2 minutes in closed loop, ECT sensor more than 149°F, Purge command over 1%, MAF sensor from 23-50 g/sec, TP angle over 5% or for models with TAC, TP angle more than 5% higher than the idle value, fuel level over 10%, Scan Tool and Intrusive tests all off, conditions met for 100 seconds, and the PCM detected the HO2S rich-to-lean or the lean-to-rich response time was more than a calibrated value. **Possible Causes** ● Exhaust leak present in the exhaust manifold or exhaust pipes ● Fuel system is too rich (fuel pressure too high, fuel pressure regulator leaking, or one or more fuel injectors sticking/leaking) ● HO2S element is silicon, water or fuel contaminated or it failed ● TP sensor element broken (can cause false acceleration event) ● PCM has failed
P0133 **2T O2S, MIL: Yes** 1999, 2000, 2001, 2002, 2003, 2004, 2005 C/K Series Truck, G Series Van, Envoy, Escalade, TrailBlazer 4.8L VIN V, 5.3L VIN P, 5.3L VIN T, 5.3L VIN Z, 6.0L VIN N, 6.0L VIN U, 8.1L VIN G Transmissions: All	**HO2S-11 (Bank 1 Sensor 1) Slow Response Conditions:** DTC P0101, P0102, P0103, P0106, P0107, P0108, P0112, P0113, P0116, P0117, P0118, P0120, P0121, P0122, P0123, P0169, P0178, P0179, P0200, P0220, P0300, P0442, P0446, P0452, P0453, P0455, P0496, P1125, P1258, P1514, P1515, P1516, P1518, P2108 and P2135 not set, engine runtime in closed loop over 160 seconds, ECT sensor over 149°F, engine speed from 1200-3000 rpm, system voltage from 10-18v, Fuel Alcohol content less than 90%, fuel level over 10%, TP indicated angle more than 5% higher than the idle value, and the PCM detected the HO2S signal rich to lean or lean to rich average response time was more than a calibrated value during the test. **Possible Causes** ● Fuel system is too rich (fuel pressure too high, fuel pressure regulator leaking, or one or more fuel injectors sticking/leaking) ● HO2S element is silicon, water or fuel contaminated ● HO2S signal tracking (water intrusion) in the connector causing a short between the HO2S signal and heater power circuits ● PCM has failed
P0133 **2T O2S, MIL: Yes** 1996, 1997, 1998, 1999, 2000, 2001, 2002, 2003, 2004, 2005 DeVille, Eldorado, Seville 4.6L VIN 9, 4.6L VIN Y Transmissions: A/T	**HO2S-11 (Bank 1 Sensor 1) Slow Response Conditions:** DTC P0101-P0103, P0106-P0108, P0112-P0113, P0116-P0118, P0121-P0123, P0125, P0128, P0200, P0300, P0410-P0446, P0452-P0453, P1258, P1415-P1416 and P1441 not set, engine started, AIR, Catalyst, EGR and Intrusive Tests all "off", engine speed from 1200-2800 rpm for 202 seconds in closed loop, TP angle over 3%, ECT sensor more than 167°F, MAF sensor from 15-35 g/sec, gear selector not in Reverse, Park or Neutral, and the PCM detected the HO2S lean-rich or the rich-lean response time was over 200 ms, or the HO2S signal was fixed at 325-625 mv during the HO2S test. **Possible Causes** ● Exhaust leak present in the exhaust manifold or exhaust pipes ● Fuel system is too rich (fuel pressure too high, fuel pressure regulator leaking, or one or more fuel injectors sticking/leaking) ● HO2S element is silicon, water or fuel contaminated or it failed ● TP sensor element broken (can cause false acceleration event) ● PCM has failed

OBD II Trouble Code List (P0xxx Codes)

DTC	Trouble Code Title, Conditions & Possible Causes
P0133 **2T O2S, MIL: Yes** 2003, 2004, 2005 CTS (Cadillac) 2.6L V6 VIN M & 3.2L V6 VIN N engines Transmissions: All	**HO2S-11 (Bank 1 Sensor 1) Slow Response Conditions:** DTC P0030, P0031, P0032, P0036, P0037, P0038, P0101, P0102, P0103, P0112, P0113, P0116, P0117, P0118, P0125, P0128, P0135, P0141, P0171, P0172, P0174, P0175, P0300, P0301-P0306, P0443, P0444 and P0445 not set, engine speed 1520-3000 rpm while running in closed loop, Calculated Converter Temperature more than 662ºF, volumetric efficiency from 30-70%, Purge solenoid commanded off for 10 seconds, and the PCM detected the HO2S-11 signal took longer than 1.3 seconds to switch, after 21 valid switching cycles were detected (test runs once per ignition cycle). **Possible Causes** ● EVAP system leaks or restrictions, charcoal canister problems ● Exhaust system is leaking or severely restricted ● HO2S air reference hole plugged (check for dirt on the outside) ● HO2S signal circuit is open, or the low reference circuit is open ● HO2S is damaged or contaminated due to improper fuel usage ● PCM has failed
P0133 **2T O2S, MIL: Yes** 1995, 1996, 1997, 1998, 1999, 2000, 2001, 2002 Camaro & Firebird 3.8L VIN K engine Transmissions: All	**HO2S-11 (Bank 1 Sensor 1) Slow Response Conditions:** DTC P0101-P0103, P0107-P0108, P0112-P0113, P0117-P0118, P0121-P0123, P0125, P0128, P0201-P0206, P0300, P0410, P0440-P0449 and P1441 not set, engine runtime over 4 minutes, ECT sensor more than 122ºF, engine speed from 1000-3000 rpm in closed loop, MAF sensor from 13-30 g/sec, and the PCM detected the HO2S lean-rich response time was more than 63 ms, or the rich-lean response time was more than 190 ms. **Possible Causes** ● Exhaust leak present in the exhaust manifold or exhaust pipes ● Fuel system is too rich (fuel pressure too high, fuel pressure regulator leaking, or one or more fuel injectors sticking/leaking) ● HO2S element is silicon, water or fuel contaminated or it failed ● TP sensor element broken (can cause false acceleration event) ● PCM has failed
P0133 **2T O2S, MIL: Yes** 1998, 1999, 2000, 2001, 2002 Camaro & Firebird 5.7L VIN G Transmissions: All	**HO2S-11 (Bank 1 Sensor 1) Slow Response Conditions:** DTC P0101-P0103, P0106-P0108, P0112-P0118, P0121-P0123, P0125, P0128, P-131-P0135, P0151-P0155, P0200, P0300, P0410, P0440-P0446, P0452-P0453, P1258, P1415-P1416 and P1441 not set, engine started, engine speed 1000-2300 rpm for 60 seconds, system voltage over 10.0v, fuel level over 10%, ECT sensor more than 122ºF, AIR, EGR and Catalyst tests all "off", MAF sensor from 20-50 g/sec, Purge command over 0%, and the PCM detected the HO2S lean to rich or rich to lean response time was over 250 ms. **Possible Causes** ● Exhaust leak present in the exhaust manifold or exhaust pipes ● Fuel system is too rich (fuel pressure too high, fuel pressure regulator leaking, or one or more fuel injectors sticking/leaking) ● HO2S element is silicon, water or fuel contaminated or it failed ● TP sensor element broken (can cause false acceleration event) ● PCM has failed
P0133 **2T O2S, MIL: Yes** 1996, 1997, 1998, 1999, 2001, 2002, 2003, 2004, 2005 Aurora 3.5L V6 VIN H, 4.0L V8 VIN C Transmissions: A/T	**HO2S-11 (Bank 1 Sensor 1) Slow Response Conditions:** DTC P0101, P0102, P0103, P0106, P0107, P0108, P0112, P0113, P0116, P0117, P0118, P0121, P0122, P0123, P0125, P0128, P0131, P0132, P0135, P0151, P0152, P0201-P0208, P0300, P0410, P0410, P0418, P0419, P0440, P0442, P0443, P0446, P0449, P1133, P1415, P1416, or P1441 not set, engine started, engine speed from 1200-2300 rpm in closed loop for 200 seconds, ECT sensor over 154ºF, TP angle over 3%, MAF sensor from 15-35 g/sec, Catalyst, AIR and EGR tests off for 3 seconds, TR signal not in Reverse or P/N, system voltage over 10.0v and PCM detected the HO2S lean-rich or rich-lean response time was over 219 ms. **Possible Causes** ● Exhaust leak present in the exhaust manifold or exhaust pipes ● Fuel system is too rich (fuel pressure too high, fuel pressure regulator leaking, or one or more fuel injectors sticking/leaking) ● HO2S element is silicon, water or fuel contaminated or it failed ● TP sensor element broken (can cause false acceleration event) ● PCM has failed

OBD II Trouble Code List (P0xxx Codes)

DTC	Trouble Code Title, Conditions & Possible Causes
P0133 **2T O2S, MIL: Yes** 1995, 1996, 1997, 1998, 1999, 2000, 2001, 2002, 2003, 2004, 2005 S/T Blazer & S/T Pickup 4.3L VIN W, 4.3L VIN X Transmissions: All	**HO2S-11 (Bank 1 Sensor 1) Slow Response Conditions:** DTC P0101, P0102, P0103, P0106, P0107, P0108, P0112, P0113, P0116, P0117, P0118, P0121, P0122, P0123, P0131, P0132, P0134, P0135, P0151, P0152, P0154, P0155, P0200, P0300, P0440, P0442, P0446, P0452, P0453, P0496 and P1441 not set, engine started, engine speed at 1200-3000 rpm for 160 seconds in closed loop, ECT sensor over 140°F, MAF sensor from 18-55 g/sec, TP angle over 5%, system voltage over 10.0v, fuel level over 10%, Intrusive and Scan Tool tests off, Purge command over 1%, and the PCM detected the HO2S rich-to-lean or lean-to-rich response average time was more than a calibrated value. **Possible Causes** ● Exhaust leak present in the exhaust manifold or exhaust pipes ● Fuel system is too rich (fuel pressure too high, fuel pressure regulator leaking, or one or more fuel injectors sticking/leaking) ● HO2S element is silicon, water or fuel contaminated or it failed ● TP sensor element broken (can cause false acceleration event) ● PCM has failed
P0133 **2T O2S, MIL: Yes** 1997, 1998 Catera 3.0L V6 VIN R engine Transmissions: A/T	**HO2S-11 (Bank 1 Sensor 1) Slow Response Conditions:** DTC P0137 and P0138 not set, engine running in closed loop for 1 minute, system voltage over 10.0v, and the PCM detected the average HO2S response time was over 3.3 seconds. **Possible Causes** ● Exhaust leak present in the exhaust manifold or exhaust pipes ● Fuel system is too rich (fuel pressure too high, fuel pressure regulator leaking, or one or more fuel injectors sticking/leaking) ● HO2S element is silicon, water or fuel contaminated or it failed ● TP sensor element broken (can cause false acceleration event) ● PCM has failed
P0133 **2T O2S, MIL: Yes** 1997, 1998, 1999, 2000, 2001, 2002, 2003, 2004, 2005 Corvette 5.7L VIN G, 5.7L VIN S Transmissions: All	**HO2S-11 (Bank 1 Sensor 1) Slow Response Conditions:** DTC P0101-P0103, P0106-P0108, P0112-P0118, P0131-P0135, P0151-P0155, P0200, P0300, P0410, P0440-P0446, P0452, P0453, P1120, P1125, P1220, P1221, P1258, P1415, P1416, P1441, P1514-P1518 not set, engine started, engine speed from 1000-2300 rpm in closed loop for 2 minutes, ECT sensor more than 122°F, system voltage over 10.0v, fuel level over 10%, MAF sensor from 18-50 g/sec, Purge command over 0%, TP indicated angle 5% more than the idle value, and the PCM detected the HO2S lean-rich or the rich-lean average response time was over 250 ms for over 1 minute. The PCM supplies the HO2S with a reference (or bias) voltage of about 450 mv. The HO2S generates a voltage in a range of 0-1000 mv that fluctuates above and below the bias voltage once in closed loop. **Possible Causes** ● Exhaust leak present in the exhaust manifold or exhaust pipes ● Fuel system is too rich (fuel pressure too high, fuel pressure regulator leaking, or one or more fuel injectors sticking/leaking) ● HO2S element is silicon, water or fuel contaminated or it failed ● TP sensor element broken (can cause false acceleration event) ● PCM has failed
P0133 **1T O2S, MIL: NO** 1995 Corvette 5.7L VIN P Transmissions: All	**HO2S-11 (Bank 1 Sensor 1) Slow Response Conditions:** DTC P0116, P0117, P0118, P0135, P0171, P0172, P0174, P0175, P1114 and P1115 not set, engine started, system voltage over 10.0v, engine running in closed loop, TP angle from 2-10%, engine running in closed loop for 1 minute, and the PCM detected the HO2S lean-rich or rich-lean response time was more than 100 ms. **Possible Causes** ● Exhaust leak present in the exhaust manifold or exhaust pipes ● Fuel system is too rich (fuel pressure too high, fuel pressure regulator leaking, or one or more fuel injectors sticking/leaking) ● HO2S element is silicon, water or fuel contaminated or it failed ● TP sensor element broken (can cause false acceleration event) ● PCM has failed
P0133 **2T O2S, MIL: Yes** 1996, 1997, 1998, 1999, 2000, 2001, 2002, 2003, 2004, 2005 Achieva, Alero, Beretta, Century, Cavalier, Ciera, Century, Corsica, Envoy, Grand Am, Malibu, S/T Blazer & Pickup, Skylark, Sunfire, TrailBlazer 2.2L VIN 4, 2.2L VIN 5, 2.2L VIN F, 2.2L VIN H, 4.2L VIN S Transmissions: All	**HO2S-11 (Bank 1 Sensor 1) Insufficient Activity Conditions:** DTC P0105-P0108, P0112, P0113, P0117, P0118, P0122, P0123, P0171, P0201-P0206, P0300-P0306, P0335, P0440, P0442, P0446, P0506, P0507, P0601, P0602 and P1441 not set, engine runtime over 30 seconds, ECT sensor above 158°F, system voltage over 10v, fuel level over 10%, calculated airflow above 3 g/sec, TP angle from 8-56%, MAP sensor over 25 kPa, and the PCM detected the HO2S signal was fixed from 399-499 mv for 125 seconds. **Possible Causes** ● HO2S heater is damaged or has failed ● HO2S signal or ground circuit has a high resistance condition ● HO2S signal circuit is open or shorted to system power (B+) ● HO2S has failed (i.e., it is silicon, water or fuel contaminated) ● PCM has failed

OBD II Trouble Code List (P0xxx Codes)

DTC	Trouble Code Title, Conditions & Possible Causes
P0134 **1T O2S, MIL: No** 1995 Century, Cutlass Ciera, Cutlass Supreme, Grand Prix, Lumina, Monte Carlo Park Avenue Regal, Riviera, 88' & 98' 3.1L VIN M, 3.4L VIN X, 3.8L VIN 1, 3.8L VIN K, 3.8L VIN L Transmissions: All	**HO2S-11 (Bank 1 Sensor 1) Insufficient Activity Conditions:** DTC P0116, P0117, P0118, P0131, P0132, P0135, P0171 P0172, P0174, P0175, P0420, P1115, P1133, P1153 and P1158 not set, engine started, engine runtime over 200 seconds, ECT sensor more than 110ºF, TP angle over 3%, and the PCM detected the HO2S signal was fixed between 350-557 mv for 30 seconds. **Possible Causes** ● HO2S heater is damaged or has failed ● HO2S signal or ground circuit has a high resistance condition ● HO2S signal circuit is open or shorted to system power (B+) ● HO2S has failed (i.e., it is silicon, water or fuel contaminated) ● PCM has failed
P0134 **2T O2S, MIL: Yes** 1996, 1997, 1998, 1999, 2000, 2001, 2002 Achieva, Beretta, Ciera, Corsica, Century, Grand Am, Grand Prix, Lumina, Malibu, Monte Carlo, Skylark 3.1L VIN M, 3.4L VIN X Transmissions: All	**HO2S-11 (Bank 1 Sensor 1) Insufficient Activity Conditions:** DTC P0101, P0102, P0103, P0107, P0108, P0112, P0113, P0116, P0117, P0118, P0121, P0122, P0123, P0125, P0128, P0201, P0202, P0203, P0204, P0205, P0206, P0410, P0440, P0442, P0443, P0446, P0449, or P1441 not set, engine started, system voltage over 10.0v, engine speed from 1000-3000 rpm for 4 minutes in closed loop, fuel level over 10%, ECT sensor more than 122ºF, TP angle from 3-40%, gear selector not in Reverse or P/N, MAF sensor from 10-30 g/sec, Air Pump "off", and the PCM detected the HO2S signal was fixed between 350-550 mv for 30 seconds. **Possible Causes** ● HO2S heater is damaged or has failed ● HO2S signal or ground circuit has a high resistance condition ● HO2S signal circuit is open or shorted to system power (B+) ● HO2S has failed (i.e., it is silicon, water or fuel contaminated) ● PCM has failed
P0134 **1T O2S, MIL: Yes** 1996, 1997 Camaro, Corvette, Caprice, Firebird & Fleetwood 4.3L VIN W, 5.7L VIN 5, 5.7L VIN P Transmissions: All	**HO2S-11 (Bank 1 Sensor 1) Insufficient Activity Conditions:** DTC P0100, P0102, P0103, P0107, P0108, P0112, P0113, P0117, P0118, P0125, P0200, P0372 and P1371 not set, engine started, engine speed from 1000-3000 rpm in closed loop for 4 minutes, AIR and Catalyst Tests "off", ECT sensor more than 118ºF, and the PCM detected the HO2S signal remained at 352-552 mv for 30 seconds. **Possible Causes** ● HO2S heater is damaged or has failed ● HO2S signal or ground circuit has a high resistance condition ● HO2S signal circuit is open or shorted to system power (B+) ● HO2S has failed (i.e., it is silicon, water or fuel contaminated) ● PCM has failed
P0134 **2T O2S, MIL: Yes** 1996, 1997, 1998, 1999, 2000, 2001, 2002, 2003, 2004, 2005 Alero, Aztek, Cutlass, Impala, Malibu, Montana, Silhouette, Venture, Rendezvous 3.1L VIN J, 3.1L VIN M, 3.4L VIN E Transmissions: All	**HO2S-11 (Bank 1 Sensor 1) Insufficient Activity Conditions:** DTC P0101, P0102, P0103, P0106, P0107, P0108, P0112, P0113, P0116, P0117, P0118, P0121, P0122, P0123, P0125, P0128, P0131, P0132, P0135, P0151, P0152, P0201-P0208, P0300, P0410, P0410, P0418, P0419, P0440, P0442, P0443, P0446, P0449, P1133, P1415, P1416, or P1441 not set, engine started, engine runtime over 200 seconds, system voltage over 10.0v, ECT sensor more than 122ºF, and the PCM detected the HO2S signal was fixed between 408-512 mv for over 29 seconds. **Possible Causes** ● HO2S heater is damaged or has failed ● HO2S signal or ground circuit has a high resistance condition ● HO2S signal circuit is open or shorted to system power (B+) ● HO2S has failed (i.e., it is silicon, water or fuel contaminated) ● PCM has failed
P0134 **2T O2S, MIL: Yes** 1996, 1997, 1998, 1999, 2000, 2001, 2002, 2003, 2004, 2005 Bonneville, Century, Cutlass Supreme, 88'/98', Grand Prix, Intrigue, LSS, LeSabre, Lumina, Monte Carlo, Regal-Park Avenue 3.1L VIN J, 3.1L VIN M, 3.4L VIN X, 3.5L V6 VIN H, 3.8L VIN 1, 3.8L VIN K, 4.0L V8 VIN C Transmissions: All	**HO2S-11 (Bank 1 Sensor 1) Insufficient Activity Conditions:** DTC P0101, P0102, P0103, P0106, P0107, P0108, P0112, P0113, P0116, P0117, P0118, P0121, P0122, P0123, P0125, P0128, P0201 P0202, P0203, P0204, P0205, P0206, P0410, P0440, P0442, P0443, P0446, P0449 and P1441 not set, engine started, engine runtime over 200 seconds, and the PCM detected the HO2S signal was fixed between 400-500 mv for over 30 seconds. **Possible Causes** ● HO2S heater is damaged or has failed ● HO2S signal or ground circuit has a high resistance condition ● HO2S signal circuit is open or shorted to system power (B+) ● HO2S has failed (i.e., it is silicon, water or fuel contaminated) ● PCM has failed

OBD II Trouble Code List (P0xxx Codes)

DTC	Trouble Code Title, Conditions & Possible Causes
P0134 **2T O2S, MIL: Yes** 1996, 1997, 1998, 1999, 2000, 2001, 2002, 2003, 2004, 2005 C/K Cab & Chassis, C/K Series Truck, G Series Van, M/L Series Vans 4.3L VIN W, 4.3L VIN X, 5.0L VIN M, 5.7L VIN K, 5.7L VIN R, 7.4L VIN J Transmissions: All	**HO2S-11 (Bank 1 Sensor 1) Insufficient Activity Conditions:** DTC P0101-P0103, P0106-P0108, P0112, P0113, P0116-P0118, P0121-P0123, P0200, P0300, P0401, P0404-P0405, P0440, P0442, P0446, P0452, P0453, P1120, P1125, P1220, P1221, P1258, P1404, P1441, P1514, P1515, P1516, P1517 and P1518 not set, engine started, system voltage over 10.0v, Scan Tool and Intrusive tests "off", engine runtime over 409 seconds, and the PCM detected the HO2S signal remained 350-550 mv for 60 seconds. **Possible Causes** ● HO2S heater is damaged or has failed ● HO2S signal or ground circuit has a high resistance condition ● HO2S signal circuit is open or shorted to system power (B+) ● HO2S has failed (i.e., it is silicon, water or fuel contaminated) ● PCM has failed
P0134 **2T O2S, MIL: Yes** 1999, 2000, 2001, 2002, 2003, 2004, 2005 C/K Series Truck, G Series Van, Envoy, Escalade, TrailBlazer 4.8L VIN V, 5.3L VIN P, 5.3L VIN T, 5.3L VIN Z, 6.0L VIN N, 6.0L VIN U, 8.1L VIN G Transmissions: All	**HO2S-11 (Bank 1 Sensor 1) Insufficient Activity Conditions:** DTC P0101-P0103, P0106-P0108, P0112, P0113, P0116-P0118, P0120, P0169, P0178, P0179, P0200, P0220, P0121-P0123, P0300, P0442, P0446, P0452, P0453, P0455, P0496, P1125, P1258, P1514, P1515, P1516, P1518, P2108 and P2135 not set, engine runtime over 300 seconds, system voltage from 10-18v, Fuel Alcohol content less than 90%, and the PCM detected the HO2S signal remained between 350-550 mv for 60 seconds during the test. **Possible Causes** ● HO2S heater is damaged or has failed ● HO2S signal or ground circuit has a high resistance condition ● HO2S signal circuit is open or shorted to system power (B+) ● HO2S has failed (i.e., it is silicon, water or fuel contaminated) ● PCM has failed
P0134 **2T O2S, MIL: Yes** 1996, 1997, 1998, 1999, 2000, 2001, 2002, 2003, 2004, 2005 DeVille, Eldorado, Seville 4.6L VIN 9, 4.6L VIN Y Transmissions: All	**HO2S-11 (Bank 1 Sensor 1) Insufficient Activity Conditions:** DTC P0101-P0103, P0106-P0108, P0112-P0113, P0116-P0118, P0121-P0123, P0125, P0128, P0200, P0300, P0410-P0446, P0452-P0453, P1258, P1415-P1416 and P1441 not set, engine runtime 200 seconds in closed loop, AIR, Catalyst, EGR and Intrusive Tests "off", and the PCM detected the HO2S signal was fixed within a range of 400-500 mv for 1 minute. **Possible Causes** ● HO2S heater is damaged or has failed ● HO2S signal or ground circuit has a high resistance condition ● HO2S signal circuit is open or shorted to system power (B+) ● HO2S has failed (i.e., it is silicon, water or fuel contaminated) ● PCM has failed
P0134 **2T O2S, MIL: Yes** 2003, 2004, 2005 CTS (Cadillac) 2.6L V6 VIN M & 3.2L V6 VIN N engines Transmissions: All	**HO2S-11 (Bank 1 Sensor 1) Insufficient Activity Conditions:** Engine started; system voltage over 10.5v, Calculated Converter temperature below 1,472°F, and the PCM detected the HO2S-11 signal was fixed from 400-597 mv for 7 seconds, or the HO2S signals remained above 200 mv during Decel Fuel Cutoff for more than 3 seconds. **Possible Causes** ● EVAP system leaks or restrictions, charcoal canister problems ● Exhaust system is leaking or severely restricted ● HO2S air reference hole plugged (check for dirt on the outside) ● HO2S signal circuit is open, or the low reference circuit is open ● HO2S is damaged or contaminated due to improper fuel usage ● PCM has failed
P0134 **2T O2S, MIL: Yes** 1995, 1996, 1997, 1998, 1999, 2000, 2001, 2002 Camaro & Firebird 3.8L VIN K engine Transmissions: All	**HO2S-11 (Bank 1 Sensor 1) Insufficient Activity Conditions:** DTC P0101-P0103, P0107, P0108, P0112, P0113, P0116-P0118, P0121-P0123, P0125, P0128, P0201-P0206, P0410, P0440- P0443, P0446, P0449 and P1441 not set, engine runtime over 4 minutes, system voltage over 10.0v and the PCM detected the HO2S signal was fixed between 400-500 mv for 60 seconds. **Possible Causes** ● HO2S heater is damaged or has failed ● HO2S signal or ground circuit has a high resistance condition ● HO2S signal circuit is open or shorted to system power (B+) ● HO2S has failed (i.e., it is silicon, water or fuel contaminated) ● PCM has failed
P0134 **2T O2S, MIL: Yes** 1998, 1999, 2000, 2001, 2002 Camaro & Firebird 5.7L VIN G engine Transmissions: All	**HO2S-11 (Bank 1 Sensor 1) Insufficient Activity Conditions:** DTC P0101-P0103, P0106-P0108, P0112-P0118, P0121-P0123, P0125, P0128, P0200, P0300, P0410, P0440-P0446, P0452, P0453, P1258, P1415-P1416 and P1441 not set, engine runtime over 409 seconds, Scan Tool and Intrusive tests "off", system voltage over 10.0v and the PCM detected the HO2S signal was fixed from 300-550 mv for 60 seconds. **Possible Causes** ● HO2S heater is damaged or has failed ● HO2S signal or ground circuit has a high resistance condition ● HO2S signal circuit is open or shorted to system power (B+) ● HO2S has failed (i.e., it is silicon, water or fuel contaminated) ● PCM has failed

OBD II Trouble Code List (P0xxx Codes)

DTC	Trouble Code Title, Conditions & Possible Causes
P0134 **2T O2S, MIL: Yes** 1996, 1997, 1998, 1999, 2001, 2002, 2003, 2004, 2005 Aurora 3.5L V6 VIN H, 4.0L V8 VIN C engines Transmissions: A/T	**HO2S-11 (Bank 1 Sensor 1) Insufficient Activity Conditions:** DTC P0101-P0103, P0106-P0108, P0112, P0113, P0116-P0118, P0121-P0123, P0125, P0201-P0208, P0410, P0412, P0418, P0419, P0440, P0442, P0443, P0446, P0449, P1133, P1415, P1416, or P1441 not set, engine runtime over 200 seconds, system voltage over 10.0v, ECT sensor above 163°F, throttle position switch open, AIR, Catalyst and EGR tests all "off", and the PCM detected the HO2S signal was between 400 to 500 mv for 1 minute. **Possible Causes** • HO2S heater is damaged or has failed • HO2S signal or ground circuit has a high resistance condition • HO2S signal circuit is open or shorted to system power (B+) • HO2S has failed (i.e., it is silicon, water or fuel contaminated) • PCM has failed
P0134 **2T O2S, MIL: Yes** 1995, 1996, 1997, 1998, 1999, 2000, 2001, 2002, 2003, 2004, 2005 S/T Blazer & S/T Pickup 4.3L VIN W, 4.3L VIN X Transmissions: All	**HO2S-11 (Bank 1 Sensor 1) Insufficient Activity Conditions:** DTC P0101, P0102, P0103, P0106, P0107, P0108, P0112, P0113, P0116, P0117, P0118, P0121, P0122, P0123, P0200, P0300, P0440, P0442, P0446, P0452, P0453, P0455, P0496 and P1441 not set, engine started, engine runtime over 409 seconds, system voltage over 10.0v, Intrusive and Scan Tool tests off, and the PCM detected the HO2S signal was fixed from 350-550 mv for 1 minute. **Possible Causes** • HO2S heater is damaged or has failed • HO2S signal or ground circuit has a high resistance condition • HO2S signal circuit is open or shorted to system power (B+) • HO2S has failed (i.e., it is silicon, water or fuel contaminated) • PCM has failed
P0134 **2T O2S, MIL: Yes** 1997, 1998, 1999, 2000, 2001 Catera 3.0L V6 VIN R engine Transmissions: A/T	**HO2S-11 (Bank 1 Sensor 1) Insufficient Activity Conditions:** DTC P0130 and P0150 not set, engine started, engine speed from 1400-2400 rpm for 25 seconds, system voltage over 10.0v, engine load from 8-12%, MAF sensor from 18-22 g/sec, TP angle from 6-10%, and the PCM detected the difference in HO2S signal during a 10 ms period was more than the allowable limit. **Possible Causes** • HO2S heater is damaged or has failed • HO2S signal or ground circuit has a high resistance condition • HO2S signal circuit is open or shorted to system power (B+) • HO2S has failed (i.e., it is silicon, water or fuel contaminated) • PCM has failed
P0134 **2T O2S, MIL: Yes** 1995 Corvette 5.7L VIN P Transmissions: All	**HO2S-11 (Bank 1 Sensor 1) Insufficient Activity Conditions:** DTC P0116, P0117, P0118, P0135, P0171, P0172, P0174, P0175, P1114 and P1115 not set, engine started, engine running in closed loop enabled, TP angle from 2-10%, and the PCM detected the HO2S signal was fixed from 356-497 mv for 102 seconds. **Possible Causes** • HO2S heater is damaged or has failed • HO2S signal or ground circuit has a high resistance condition • HO2S signal circuit is open or shorted to system power (B+) • HO2S has failed (i.e., it is silicon, water or fuel contaminated) • PCM has failed
P0134 **2T O2S, MIL: Yes** 1997, 1998, 1999, 2000, 2001, 2002, 2003, 2004, 2005 Corvette 5.7L VIN G, 5.7L VIN S Transmissions: All	**HO2S-11 (Bank 1 Sensor 1) Insufficient Activity Conditions:** DTC P0101-P0103, P0106-P0108, P0112-P0118, P0200, P0300, P0410, P0440-P0446, P0452, P0453, P1120, P1125, P1220, P1221, P1258, P1415, P1416, P1441, P1514 to P1518 not set, engine runtime over 409 seconds, Scan Tool tests "off", system voltage over 10.0v, and the PCM detected the HO2S signal was fixed from 350-550 mv for over 1 minute. **Possible Causes** • HO2S heater is damaged or has failed • HO2S signal or ground circuit has a high resistance condition • HO2S signal circuit is open or shorted to system power (B+) • HO2S has failed (i.e., it is silicon, water or fuel contaminated) • PCM has failed
P0135 **2T O2S HTR, MIL: Yes** 1996, 1997, 1998, 1999 Achieva, Beretta, Ciera, Corsica, Century, Grand Am, Grand Prix, Lumina, Malibu, Monte Carlo 3.1L VIN M, 3.4L VIN X Transmissions: All	**HO2S-11 (Bank 1 Sensor 1) Heater Circuit Malfunction Conditions:** DTC P0101-P0103, P0106-P0108, P0112, P0113, P0116-P0118, P0121-P0123, P0125, P0128, P0201-P0206, P0410, P0440, P0442-P0446, P0449 and P1441 not set, ECT and IAT sensors under 95°F and within 43°F at startup, system voltage over 10v, MAF sensor below 20 g/sec, Air Pump "off", and the PCM detected the HO2S signal was within 150 mv of bias voltage for too long (depends upon the engine temperature and average airflow after startup). **Possible Causes** • HO2S heater ground circuit is open or has high resistance • HO2S heater power circuit is open (test O2S fuse in fuse block) • HO2S heater element is damaged or has failed • PCM has failed

OBD II Trouble Code List (P0xxx Codes)

DTC	Trouble Code Title, Conditions & Possible Causes
P0135 **2T O2S HTR, MIL: Yes** 1996, 1997 Camaro, Corvette, Caprice, Firebird & Fleetwood 4.3L VIN W, 5.7L VIN 5, 5.7L VIN P Transmissions: All	**HO2S-11 (Bank 1 Sensor 1) Heater Circuit Malfunction Conditions:** DTC P0101, P0102, P0103, P0112, P0113, P0117, P0118, P0125, P0132, P0133, P0134 and P0200 not set, ECT and IAT sensor less than 122°F and within 5°F at startup, engine started, system voltage over 10.0v, AIR and Catalyst tests "off", MAF sensor less than 40 g/sec, TP angle less than 20%, and the PCM detected the HO2S signal remained from 300-700 mv for too long a period of time. **Possible Causes** ● HO2S heater ground circuit is open or has high resistance ● HO2S heater power circuit is open (test O2S fuse in fuse block) ● HO2S heater element is damaged or has failed ● PCM has failed
P0135 **2T O2S HTR, MIL: Yes** 1996, 1997, 1998, 1999, 2000, 2001 Alero, Aztek, Cutlass, Impala, Malibu, Montana, Silhouette, Venture, Rendezvous 3.1L VIN J, 3.1L VIN M, 3.4L VIN E Transmissions: All	**HO2S-11 (Bank 1 Sensor 1) Heater Circuit Malfunction Conditions:** DTC P0101-P0103, P0107, P0108, P0112, P0113, P0116-P0118, P0121, P0122, P0123, P0125, P0128, P0201, P0202, P0203, P0204, P0205, P0206, P0410, P0440, P0442, P0443, P0446, P0449 and P1441 not set, ECT and IAT sensor less than 95°F and within 42°F at startup, engine runtime 200 seconds, system voltage over 10.0v and the PCM detected the HO2S signal was within 150 mv of bias voltage for too long (depends on ECT/MAF at startup). **Possible Causes** ● HO2S heater ground circuit is open or it has high resistance ● HO2S heater power circuit is open (test O2S fuse in fuse block) ● HO2S heater element is damaged or it has failed ● PCM has failed
P0135 **2T O2S HTR, MIL: Yes** 2002, 2003, 2004, 2005 Aztek, Impala, Malibu, Montana, Silhouette, Venture, Rendezvous 3.1L VIN J, 3.4L VIN E Transmissions: All	**HO2S-11 (Bank 1 Sensor 1) Heater Circuit Malfunction Conditions:** DTC P0101, P0102, P0103, P0107, P0108, P0112, P0113, P0116, P0117, P0118, P0121, P0122, P0123, P0125, P0128, P0201-P0206, P0410, P0440, P0442, P0443, P0446, P0449 and P1441 not set, engine runtime over 4 minutes, engine speed from 650-2500 rpm, ECT sensor more than 158°F, system voltage from 9-18.0v, MAF sensor from 4-26 g/sec, and the PCM detected the HO2S-11 heater current was less than 0.25 amps or more than 0.90 amps. **Possible Causes** ● HO2S heater ground circuit is open or it has high resistance ● HO2S heater power circuit is open (test O2S fuse in fuse block) ● HO2S heater element is damaged or it has failed ● PCM has failed
P0135 **2T O2S HTR, MIL: Yes** 1996, 1997, 1998, 1999, 2000, 2001, 2002, 2003, 2004, 2005 Bonneville, Century, Cutlass Supreme, 88'/98', Grand Prix, Intrigue, LSS, LeSabre, Lumina, Monte Carlo, Regal-Park Avenue 3.1L VIN J, 3.1L VIN M, 3.4L VIN X, 3.5L V6 VIN H, 3.8L VIN 1, 3.8L VIN K, 4.0L V8 VIN C Transmissions: All	**HO2S-11 (Bank 1 Sensor 1) Heater Circuit Malfunction Conditions:** DTC P0101-P0103, P0106-P0108, P0112, P0113, P0116-P0118, P0121-P0123, P0125, P0128, P0201-P0206, P0410, P0440, P0442, P0443, P0446, P0449 and P1441 not set, IAT and ECT sensors with 11°F of each other at startup, MAF sensor less than 17-20 g/sec, HO2S signal within 100 mv of bias voltage at startup, and the PCM detected the HO2S signal remained within 150 mv of bias voltage (450 mv) for 50-80 seconds (depends on ECT and MAF at startup). **Possible Causes** ● HO2S heater ground circuit is open or has high resistance ● HO2S heater power circuit is open (test O2S fuse in fuse block) ● HO2S heater element is damaged or has failed ● PCM has failed
P0135 **2T O2S HTR, MIL: Yes** 1996, 1997, 1998, 1999, 2000, 2001, 2002, 2003, 2004, 2005 C/K Cab & Chassis, C/K Series Truck, G Series Van, M/L Series Vans 4.3L VIN W, 4.3L VIN X, 5.0L VIN M, 5.7L VIN K, 5.7L VIN R, 7.4L VIN J Transmissions: All	**HO2S-11 (Bank 1 Sensor 1) Heater Circuit Malfunction Conditions:** DTC P0101-P0103, P0106-P0108, P0112, P0113, P0116-P0118, P0121-P0123, P0131, P0132, P0134, P0137, P0138, P0140, P0151, P0152, P0154, P0157, P0158, P0160, P0200, P0300, P0401, P0404, P0405, P0440, P0442, P0446, P0452, P0453, P1120, P1125, P1220, P1221, P1258, P1404, P1441, P1514, P1515, P1516, P1517 and P1518 not set, engine started, ECT and IAT sensors less than 122°F and within 14.5°F at startup, HO2S signal from 425-475 mv right after startup, Intrusive and Scan Tool tests off, MAF sensor less than 25 g/sec, and the PCM detected the HO2S signal remained within 150 mv of startup HO2S signal for a predetermined amount of time based on the ECT and airflow inputs. **Possible Causes** ● HO2S heater ground circuit is open or has high resistance ● HO2S heater power circuit is open (test O2A fuse in fuse block) ● HO2S heater element is damaged or has failed ● PCM has failed

OBD II Trouble Code List (P0xxx Codes)

DTC	Trouble Code Title, Conditions & Possible Causes
P0135 **2T O2S HTR, MIL: Yes** 1999, 2000, 2001, 2002, 2003, 2004, 2005 C/K Series Truck, G Series Van, Envoy, Escalade, TrailBlazer 4.8L VIN V, 5.3L VIN P, 5.3L VIN T, 5.3L VIN Z, 6.0L VIN N, 6.0L VIN U, 8.1L VIN G Transmissions: All	**HO2S-11 (Bank 1 Sensor 1) Heater Circuit Malfunction Conditions:** DTC P0101-P0103, P0106-P0108, P0112, P0113, P0116-P0118, P0120, P0121, P0122, P0123, P0169, P0178, P0179, P0200, P0220, P0300, P0442, P0446, P0452, P0453, P0455, P0496, P1125, P1258, P1514, P1515, P1516, P1518, P2108 and P2135 not set, engine runtime over 120 seconds, engine speed from 500-3000 rpm, system voltage from 10-18v, ECT sensor over 122ºF, MAF sensor from 3-40 g/sec, Fuel Alcohol content less than 90%, and the PCM detected the HO2S heater current was less than 0.25 amps, or more than 3.125 amps (more than 1.375 amps on 4.8L V8). **Possible Causes** ● HO2S heater low control circuit is open or shorted to ground ● HO2S heater circuit is open or it is shorted to ground ● HO2S heater power circuit is open (test O2A fuse in fuse block) ● HO2S heater element is damaged or has failed ● PCM has failed
P0135 **2T O2S HTR, MIL: Yes** 1996, 1997, 1998, 1999 DeVille, Eldorado, Seville 4.6L VIN 9, 4.6L VIN Y Transmissions: All	**HO2S-11 (Bank 1 Sensor 1) Heater Circuit Malfunction Conditions:** DTC P0101-P0103, P0117, P0118, and P0134 not set, ECT sensor change more than 50ºF since the last ignition cycle, engine started, system voltage over 11.0v, average HO2S bias voltage from 352-546 mv, engine did not stall, and the PCM detected the HO2S signal was fixed within 150 mv of the bias voltage for too long (depends on ECT and MAF at startup). **Possible Causes** ● HO2S heater ground circuit is open or has high resistance ● HO2S heater power circuit is open (test O2S fuse in fuse block) ● HO2S heater element is damaged or has failed ● PCM has failed
P0135 **2T O2S HTR, MIL: Yes** 2000, 2001, 2002, 2003, 2004, 2005 DeVille, Eldorado, Seville 4.6L VIN 9, 4.6L VIN Y Transmissions: All	**HO2S-11 (Bank 1 Sensor 1) Heater Circuit Malfunction Conditions:** DTC P0030, P1031, or P1032 not set, engine started, engine speed from 500-3000 rpm for 3 minutes, system voltage is steady (± 1v), MAF sensor from 4-30 g/sec, HO2S Over Temperature Control "off", and the PCM detected the HO2S heater current was out-of-range. **Possible Causes** ● HO2S heater high side driver circuit has high resistance (open) ● HO2S heater low side driver circuit has high resistance (open) ● HO2S heater power circuit is open (test O2S fuse in fuse block) ● HO2S heater element is damaged or has failed ● PCM has failed
P0135 **2T O2S HTR, MIL: Yes** 2003, 2004, 2005 CTS (Cadillac) 2.6L V6 VIN M & 3.2L V6 VIN N engines Transmissions: All	**HO2S-11 (Bank 1 Sensor 1) Heater Circuit Malfunction Conditions:** DTC P0030, P0031, P0032, P0036, P0037, P0038, P0050, P0051, P0052, P0056, P0057, P0058, P0157, P0158, P0160 and P0161 not set, engine started, Calculated Converter temperature from 626-1,112ºF, system voltage over 10.5v, system voltage from 10-16v, and the PCM detected the HO2S-11 heater current was out-of-range with the heater enabled. **Possible Causes** ● HO2S heater high side driver circuit has high resistance (open) ● HO2S heater low side driver circuit has high resistance (open) ● HO2S heater power circuit is open (test O2S fuse in fuse block) ● HO2S heater element is damaged or has failed ● PCM has failed
P0135 **2T O2S HTR, MIL: Yes** 1995, 1996, 1997, 1998, 1999, 2000, 2001, 2002 Camaro & Firebird 3.8L VIN K engine Transmissions: All	**HO2S-11 (Bank 1 Sensor 1) Heater Circuit Conditions:** DTC P0101-P0103, P0107, P0108, P0112, P0113, P0117, P0118, P0121-P0123, P0125, P0128, P0201-P0206, P0410, P0440, P0442-P0449 and P1441 not set, ECT and IAT sensors less than 95ºF at startup and within 11ºF at startup, engine started, system voltage over 10.0v, MAF sensor less than 20 g/sec, and the PCM detected the HO2S signal was within 150 mv of bias voltage for a period of 42 to 120 seconds - depends on ECT at startup). **Possible Causes** ● HO2S heater ground circuit is open or has high resistance ● HO2S heater power circuit is open (test O2S fuse in fuse block) ● HO2S heater element is damaged or has failed ● PCM has failed
P0135 **2T O2S HTR, MIL: Yes** 1998, 1999, 2000, 2001, 2002 Camaro & Firebird 5.7L VIN G engine Transmissions: All	**HO2S-11 (Bank 1 Sensor 1) Heater Circuit Malfunction Conditions:** DTC P0101-P0103, P0106-P0108, P0112-P0118, P0121-P0123, P0125, P0128, P0131-P0138, P0140, P0151-P0158, P0160, P0200, P0300, P0410, P0440-P0446, P0452, P0453, P1258, P1415, P1416 or P1441 not set, ECT and IAT sensors under 122ºF and within 14.5ºF and HO2S 425-475 mv at startup, Scan Tool and Intrusive tests off, MAF sensor under 18 g/sec, and the PCM detected the HO2S Signal was fixed between 300 mv and 700 mv for too long (from 42 seconds to 2 minutes - depends on ECT and MAF inputs). **Possible Causes** ● HO2S heater ground circuit is open or has high resistance ● HO2S heater power circuit is open (test O2S fuse in fuse block) ● HO2S heater element is damaged or has failed ● PCM has failed

OBD II Trouble Code List (P0xxx Codes)

DTC	Trouble Code Title, Conditions & Possible Causes
P0135 **2T O2S HTR, MIL: Yes** 1996, 1997, 1998, 1999, 2001, 2002, 2003, 2004, 2005 Aurora 3.5L V6 VIN H, 4.0L V8 VIN C engines Transmissions: All	**HO2S-11 (Bank 1 Sensor 1) Heater Circuit Malfunction Conditions:** DTC P0101-P0103, P0117, P0118 and P0134 not set, ECT sensor at least 50°F less than the last key cycle, average HO2S-11 bias voltage from 352-546 mv, engine did not stall during the test, and the PCM detected the HO2S-11 signal was either 151 mv less than or 151 mv more than the average HO2S-11 bias voltage for 100 ms. **Possible Causes** ● HO2S heater ground circuit is open or has high resistance ● HO2S heater power circuit is open (test O2S fuse in fuse block) ● HO2S heater element is damaged or has failed ● PCM has failed
P0135 **2T O2S HTR, MIL: Yes** 2002, 2003, 2004, 2005 Alero, Cavalier, Grand Am S/T Blazer & S/T Pickup 2.2L VIN 4, 2.2L VIN 5, 2.2L VIN F, 4.2L VIN S Transmissions: All	**HO2S-11 (Bank 1 Sensor 1) Heater Circuit Malfunction Conditions:** DTC P0105-P0108, P0112-P0118, P0122, P0123, P0171, P0172, P0201-P0206, P0300, P0336, P0351-P0356, P0440, P0446, P0452, P0453, P0461, P0506, P0507, P0601, P0602, P1220, P1221, P1441, P1635, P1639 and P1681 not set, engine runtime 1 minute, ECT sensor over 158°F, Calculated Airflow less than 16 g/sec, fuel level over 10%, system voltage over 10.0v and the PCM detected the HO2S heater current was not within range for over 200 seconds. **Possible Causes** ● HO2S heater ground circuit is open or has high resistance ● HO2S heater power circuit is open (test O2S fuse in fuse block) ● HO2S heater element is damaged or has failed ● PCM has failed
P0135 **2T O2S HTR, MIL: Yes** 1995, 1996, 1997, 1998, 1999, 2000, 2001, 2002, 2003, 2004, 2005 S/T Blazer & S/T Pickup 4.3L VIN W, 4.3L VIN X Transmissions: All	**HO2S-11 (Bank 1 Sensor 1) Heater Circuit Malfunction Conditions:** DTC P0101, P0102, P0103, P0106, P0107, P0108, P0112, P0113, P0116, P0117, P0118, P0121-P0123, P0200, P0300, P0440, P0452, P0453, P0455, P0496 and P1441 not set, engine running in closed loop, ECT and IAT sensors less than 122°F and within 14.5°F at startup, Intrusive and Scan Tool tests off, and the PCM detected the HO2S signal was fixed within 150 mv of the bias voltage (450 mv) for too long (depends on ECT & MAF at startup). **Possible Causes** ● HO2S heater ground circuit is open or has high resistance ● HO2S heater power circuit is open (test O2S fuse in fuse block) ● HO2S heater element is damaged or has failed ● PCM has failed
P0135 **2T O2S HTR, MIL: Yes** 1997, 1998, 1999, 2000, 2001 Catera 3.0L V6 VIN R engine Transmissions: A/T	**HO2S-11 (Bank 1 Sensor 1) Heater Circuit Malfunction Conditions:** Engine started; system voltage over 10.0v, HO2S heater "on", and the PCM detected the HO2S heater circuit was over 3.6v, or less than 2.3v for 50 ms, or System Readiness Monitor was not detected 20 seconds after a startup that was 4 minutes from previous startup. **Possible Causes** ● HO2S heater ground circuit is open or has high resistance ● HO2S heater power circuit is open (test O2S fuse in fuse block) ● HO2S heater element is damaged or has failed ● PCM has failed
P0135 **2T O2S HTR, MIL: Yes** 1995 Corvette 5.7L VIN P engine Transmissions: All	**HO2S-11 (Bank 1 Sensor 1) Heater Circuit Malfunction Conditions:** DTC P0116, P0117, P0118, P0135, P0171, P0172, P0174, P0175, P1114 and P1115 not set, engine started, ECT and IAT sensors less than 122°F and within 11°F at startup, TP angle less than 16%, and the PCM detected the HO2S signal was 300-700 mv for from 49-200 seconds (depends on the ECT sensor value at engine startup). **Possible Causes** ● HO2S heater ground circuit is open or has high resistance ● HO2S heater power circuit is open (test O2S fuse in fuse block) ● HO2S heater element is damaged or has failed ● PCM has failed
P0135 **2T O2S HTR, MIL: Yes** 1997, 1998, 1999, 2000, 2001, 2002, 2003, 2004, 2005 Corvette 5.7L VIN G, 5.7L VIN S Transmissions: All	**HO2S-11 (Bank 1 Sensor 1) Heater Circuit Malfunction Conditions:** DTC P0101-P0103, P0106-P0108, P0112-P0118, P0131-P0138, P0140, P0151-P0158, P0160, P0200, P0300, P0410, P0440, P0442-P0446, P0452, P0453, P1120, P1125, P1220-P1222, P1258, P1415, P1416, P1441, P1514-P1518 not set, HO2S signal from 425-475 mv at startup, ECT and IAT sensors under 122°F and within 14.5°F at startup, MAF sensor under 21 g/sec, system voltage from 9-18v, AIR and Catalyst Tests off, and the PCM detected the HO2S signal remained within 150 mv of bias voltage (450 mv) for too long a period of time. **Possible Causes** ● HO2S heater ground circuit is open or has high resistance ● HO2S heater power circuit is open (test O2S fuse in fuse block) ● HO2S heater element is damaged or has failed ● PCM has failed

OBD II Trouble Code List (P0xxx Codes)

DTC	Trouble Code Title, Conditions & Possible Causes
P0136 **2T CCM, MIL: Yes** 2003, 2004, 2005 CTS (Cadillac) 2.6L V6 VIN M & 3.2L V6 VIN N engines Transmissions: All	**HO2S-12 (Bank 1 Sensor 2) Circuit Closed Loop Conditions:** DTC P0443, P0444 and P0445 not set, engine started, TP angle from 5-35%, system voltage over 10.5v, MAF sensor at 10-35 g/sec, Calculated Converter Temperature under 1,472°F, and the PCM and detected the HO2S-11 signal was 60-400 mv while the HO2S-12 signal was over 499 mv, or it detected the HO2S-11 signal was 600-1080 mv while the HO2S-12 signal was less than 104 mv for 10 seconds during the CCM test period. **Possible Causes** ● EVAP system leaks or restrictions, charcoal canister problems ● EVAP system leaks or restrictions, charcoal canister problems ● HO2S air reference hole plugged (check for dirt on the outside) ● HO2S signal circuit is shorted to sensor or chassis ground ● HO2S is damaged or contaminated due to improper fuel usage ● PCM has failed
P0136 **2T CCM, MIL: Yes** 1997, 1998, 1999, 2000, 2001 Catera 3.0L V6 VIN R engine Transmissions: A/T	**HO2S-12 (Bank 1 Sensor 2) Circuit Malfunction Conditions:** DTC P0135, P0139, P0140 and P0141 not set, engine speed from 1400-2400 rpm, HO2S-12 signal varying, engine load from 8-12%, MAF sensor from 18-22 g/sec, throttle angle from 6-10%, AIR System test "off", engine runtime over 10 minutes, low fuel not indicated, TWC temperature less than 1472°F, and the PCM detected the HO2S signal was less than 40 mv. **Possible Causes** ● Air leaks in the exhaust system, intake manifold, vacuum lines ● EVAP Purge system malfunction or charcoal canister problems ● HO2S signal circuit is open between the sensor and the PCM ● HO2S signal circuit is shorted to sensor or chassis ground ● HO2S is damaged, contaminated or air reference hole clogged ● PCM has failed
P0137 **2T CCM, MIL: Yes** 1996, 1997, 1998, 1999, 2000, 2001, 2002, 2003, 2004, 2005 Achieva, Alero, Beretta, Century, Cavalier, Ciera, Century, Corsica, Envoy, Grand Am, Malibu, S/T Blazer & Pickup, Skylark, Sunfire, TrailBlazer 2.2L VIN 4, 2.2L VIN 5, 2.2L VIN F, 2.2L VIN H, 4.2L VIN S Transmissions: All	**HO2S-12 (Bank 1 Sensor 2) Circuit Low Input Conditions:** DTC P0105-P0108, P0112-P0113, P0117-P0118, P0122-P0123, P0171, P0201-P0204, P0300, P0335, P0440-P0446, P0506, P0507, P0601-P0602 or P1441 not set, engine started, engine running with the system voltage over 10.0v, ECT sensor more than 158°F, fuel level over 10%, MAP sensor more than 25 kPa, TP angle from 8-50% for 4 seconds in closed loop, and the PCM detected the HO2S signal was less than 43 mv for 150 seconds during testing. **Possible Causes** ● Air leaks in the exhaust system, intake manifold, vacuum lines ● Engine misfire condition present (look for P0300 series codes) ● Fuel system too lean (possible low fuel pressure, water in fuel) ● HO2S signal circuit is shorted to the sensor or chassis ground ● HO2S is damaged (i.e., cracked) or air reference hole clogged ● MAP sensor is skewed indicating a false high vacuum condition - disconnect MAP sensor and check the HO2S for a lean signal ● PCM has failed
P0137 **2T CCM, MIL: Yes** 1996, 1997, 1998, 1999, 2000, 2001, 2002 Achieva, Beretta, Ciera, Century, Corsica, Grand Am, Grand Prix, Lumina, Malibu & Monte Carlo 3.1L VIN M engine Transmissions: All	**HO2S-12 (Bank 1 Sensor 2) Circuit Low Input Conditions:** DTC P0101-P0103, P0107, P0108, P0112, P0113, P0116, P0117, P0118, P0121-P0123, P0125, P0128, P0201-P0206, P0410, P0440, P0442, P0443, P0446, P0449 and P1441 not set, engine started, engine running with the TP angle from 3-35%, A/F ratio from 12.5-16.5:1, system voltage over 10.0v and the PCM detected the HO2S signal was less than 175 mv for 30 seconds, or with P/E Mode active, the HO2S signal was less than 600 mv for 10 seconds. The HO2S is used for fuel control and post-catalyst monitoring. This sensor compares the oxygen content of the surrounding air with the oxygen content of the exhaust stream. At initial startup, the PCM operates in open loop mode, ignoring the HO2S signal when calculating the air/fuel ratio. The PCM supplies the HO2S with a reference (or bias) voltage of about 450 mv. The HO2S generates a voltage within a range of 0-1000 mv that fluctuates above and below the bias voltage once in closed loop. A high HO2S voltage indicates a rich fuel mixture. A low HO2S voltage indicates a lean mixture. Heating elements in the HO2S shorten the time required for the sensor to reach normal temperature, and an accurate voltage signal. **Possible Causes** ● Air leaks in the exhaust system, intake manifold, vacuum lines ● Engine misfire condition present (look for P0300 series codes) ● Fuel system too lean (possible low fuel pressure, water in fuel) ● HO2S signal circuit is shorted to sensor or chassis ground ● HO2S is damaged (i.e., cracked) or air reference hole clogged ● PCM has failed

OBD II Trouble Code List (P0xxx Codes)

DTC	Trouble Code Title, Conditions & Possible Causes
P0137 **2T CCM, MIL: Yes** 1996, 1997, 1998, 1999, 2000, 2001, 2002, 2003, 2004, 2005 Bonneville, Century, Cutlass Supreme, 88'/98', Grand Prix, Intrigue, LSS, LeSabre, Lumina, Monte Carlo, Regal-Park Avenue 3.1L VIN J, 3.1L VIN M, 3.4L VIN X, 3.5L V6 VIN H, 3.8L VIN 1, 3.8L VIN K, 4.0L V8 VIN C engines Transmissions: All	**HO2S-12 (Bank 1 Sensor 2) Circuit Low Input Conditions:** DTC P0101-P0103, P0106-P0108, P0112-P0113, P0116-P0118, P0121-P0123, P0125, P0128, P0201-P0206, P0410, P0440, P0442, P0443, P0446, P0449 and P1441 not set, engine started, TP angle from 5-40%, and the PCM detected the HO2S signal was less than 30 mv for 45 seconds, or was less than 600 mv for 100 ms while operating in P/E mode during the test. **Possible Causes** ● Air leaks in the exhaust system, intake manifold, vacuum lines ● Engine misfire condition present (look for P0300 series codes) ● Fuel system too lean (possible low fuel pressure, water in fuel) ● HO2S signal circuit is shorted to the sensor or chassis ground ● HO2S is damaged (i.e., cracked) or air reference hole clogged ● PCM has failed
P0137 **2T CCM, MIL: Yes** 1996, 1997 Camaro, Corvette, Caprice, Firebird & Fleetwood 4.3L VIN W, 5.7L VIN 5, 5.7L VIN P engine Transmissions: All	**HO2S-12 (Bank 1 Sensor 2) Circuit Low Input Conditions:** DTC P0101, P0102, P0103, P0112, P0113, P0117, P0118, P0125, P0200, P0335, P0336, P0351-P0358, P0372 and P1371 not set, engine started, system voltage over 9v, AIR and Catalyst Tests inactive, TP angle 2-70%, running in closed loop, and the PCM detected the HO2S signal was below 39 mv, condition met for 38 seconds. **Possible Causes** ● Air leaks in the exhaust system, intake manifold, vacuum lines ● Engine misfire condition present (look for P0300 series codes) ● Fuel system too lean (possible low fuel pressure, water in fuel) ● HO2S signal circuit is shorted to the sensor or chassis ground ● HO2S is damaged (i.e., cracked) or air reference hole clogged ● PCM has failed
P0137 **2T CCM, MIL: Yes** 1996, 1997, 1998, 1999, 2000, 2001, 2002, 2003, 2004, 2005 Alero, Aztek, Cutlass, Impala, Malibu, Montana, Silhouette, Venture, Rendezvous 3.1L VIN J, 3.1L VIN M, 3.4L VIN E engines Transmissions: All	**HO2S-12 (Bank 1 Sensor 2) Circuit Low Input Conditions:** DTC P0101, P0102, P0103, P0107, P0108, P0112, P0113, P0116, P0117, P0118, P0121, P0122, P0123, P0125, P0128, P0201, P0202, P0203, P0204, P0205, P0206, P0410, P0440, P0442, P0443, P0446, P0449 and P1441 not set, A/F ratio at 13.0-16.5:1, TP angle from 3-40%, Air Pump "off", and the PCM detected the HO2S signal was less than 10 mv in closed loop, or with the P/E mode active, the HO2S signal was under 600 mv for 15 seconds. The HO2S is used for fuel control and post-catalyst monitoring. This sensor compares the oxygen content of the surrounding air with the oxygen content of the exhaust stream. At initial startup, the PCM operates in open loop mode, ignoring the HO2S signal when calculating the air/fuel ratio. The PCM supplies the HO2S with a reference (or bias) voltage of about 450 mv. The HO2S generates a voltage within a range of 0-1000 mv that fluctuates above and below the bias voltage once in closed loop. A high HO2S voltage indicates a rich fuel mixture. A low HO2S voltage indicates a lean mixture. Heating elements in the HO2S shorten the time required for the sensor to reach normal temperature, and an accurate voltage signal. **Possible Causes** ● Air leaks in the exhaust system, intake manifold, vacuum lines ● Engine misfire condition present (look for P0300 series codes) ● Fuel system too lean (possible low fuel pressure, water in fuel) ● HO2S signal circuit is shorted to the sensor or chassis ground ● HO2S is damaged (i.e., cracked) or air reference hole clogged ● PCM has failed
P0137 **2T CCM, MIL: Yes** 1999, 2000, 2001, 2002, 2003, 2004, 2005 C/K Series Truck, G Series Van, Envoy, Escalade, TrailBlazer 4.8L VIN V, 5.3L VIN P, 5.3L VIN T, 6.0L VIN Z, 6.0L VIN N, 6.0L VIN U, 8.1L VIN G engines Transmissions: All	**HO2S-12 (Bank 1 Sensor 2) Circuit Low Input Conditions:** DTC P0101, P0102, P0103, P0106, P0107, P0108, P0112, P0113, P0116, P0117, P0118, P0120, P0121, P0122, P0123, P0169, P0178, P0179, P0200, P0220, P0300, P0442, P0446, P0452, P0453, P0455, P0496, P1125, P1258, P1514, P1515, P1516, P1518, P2108 and P2135 not set, engine started, engine running in closed loop, system voltage from 10-18v, Fuel Alcohol content less than 90%, fuel level over 10%, TP angle from 3-70% over the idle value, then during the Lean Test, the PCM detected the HO2S signal was less than 80 mv for 200 seconds or with engine runtime over 30 seconds during the Power Enrichment test, the PCM detected the HO2S signal was less than 490 mv for 10 seconds. **Possible Causes** ● Air leaks in the exhaust system, intake manifold, vacuum lines ● Engine misfire condition present (look for P0300 series codes) ● Fuel system too lean (possible low fuel pressure, water in fuel) ● HO2S signal circuit is shorted to the sensor or chassis ground ● HO2S is damaged (i.e., cracked) or air reference hole clogged ● PCM has failed

OBD II Trouble Code List (P0xxx Codes)

DTC	Trouble Code Title, Conditions & Possible Causes
P0137 **2T CCM, MIL: Yes** 1996, 1997, 1998, 1999, 2000, 2001, 2002, 2003, 2004, 2005 C/K Cab & Chassis, C/K Series Truck, G Series Van, M/L Series Vans 4.3L VIN W, 4.3L VIN X, 5.0L VIN M, 5.7L VIN K, 5.7L VIN R, 7.4L VIN J Transmissions: All	**HO2S-12 (Bank 1 Sensor 2) Circuit Low Input Conditions:** DTC P0101-P0103, P0106-P0108, P0112, P0113, P0116-P0118, P0121-P0123, P0200, P0300, P0401, P0404, P0405, P0440, P0442, P0446, P0452, P0453, P1120, P1125, P1220, P1221, P1258, P1404, P1441, P1514, P1515, P1516, P1517 and P1518 not set, engine started, engine running in closed loop, fuel level over 10%, system voltage over 10v, TP angle from 8-50% or on models with TAC, the APP sensor indicated angle from 3-70%, MAP sensor more than 25 kPa, Intrusive and Scan Tool Tests "off", then with the Lean Test enabled, the PCM detected the HO2S signal was less than 26 mv for 110 seconds or during the P/E Mode test, the HO2S signal was less than 399 mv for 40 seconds. The HO2S compares the oxygen content of the surrounding air with the oxygen content of the exhaust stream. The PCM supplies the HO2S with a reference (or bias) voltage of 450 mv. The HO2S generates a voltage within a range of 0-1000 mv that fluctuates above and below the bias voltage once in closed loop. Heating elements shorten the time required to reach normal temperature and to provide an accurate voltage signal. **Possible Causes** ● Air leaks in the exhaust system, intake manifold, vacuum lines ● Engine misfire condition present (look for P0300 series codes) ● Fuel system too lean (possible low fuel pressure, water in fuel) ● HO2S signal circuit is shorted to the sensor or chassis ground ● HO2S is damaged (i.e., cracked) or air reference hole clogged ● PCM has failed ● TSB 81-65-37 contains a repair procedure for this code
P0137 **2T CCM, MIL: Yes** 1996, 1997, 1998, 1999, 2000, 2001, 2002, 2003, 2004, 2005 DeVille, Eldorado, Seville 4.6L VIN 9, 4.6L VIN Y Transmissions: All	**HO2S-12 (Bank 1 Sensor 2) Circuit Low Input Conditions:** DTC P0101-P0103, P0106-P0108, P0112-P0113, P0116-P0118, P0121-P0123, P0125, P0128, P0200, P0300, P0410-P0446, P0452-P0453, P1258, P1415-P1416 and P1441 not set, AIR, Catalyst and EGR Flow tests off, system voltage 8-18v, no injectors disabled, TP angle from 3-25%, A/F ratio at 14.5-14.8:1, engine running in closed loop for 3 seconds, and the PCM detected the HO2S signal was less than 9 mv for 5 seconds while in closed loop, or that it was less than 575 mv for 8 seconds while operating in Power Enrichment mode. **Possible Causes** ● Air leaks in the exhaust system, intake manifold, vacuum lines ● Engine misfire condition present (look for P0300 series codes) ● Fuel system too lean (possible low fuel pressure, water in fuel) ● HO2S signal circuit is shorted to the sensor or chassis ground ● HO2S is damaged (i.e., cracked) or air reference hole clogged ● PCM has failed
P0137 **2T CCM, MIL: Yes** 2003, 2004, 2005 CTS (Cadillac) 2.6L V6 VIN M & 3.2L V6 VIN N engines Transmissions: All	**HO2S-12 (Bank 1 Sensor 2) Circuit Low Input Conditions:** DTC P0030, P0031, P0032, P0036, P0037, P0038, P0135, P0155, P0171-P0175, P0300, P0301-P0306, P0341, P0342, P0343, P0443, P0444 and P0445 not set, engine started, TP angle from 5-35%, system voltage over 10.5v, MAF sensor more than 9.72 g/sec, Calculated Converter Temperature less than 1,472°F, and the PCM detected the HO2S-12 signal was less than 40 mv for 680 seconds. **Possible Causes** ● EVAP system leaks or restrictions, charcoal canister problems ● Fuel system too lean (possible low fuel pressure, water in fuel) ● HO2S air reference hole plugged (check for dirt on the outside) ● HO2S signal circuit is shorted to sensor or chassis ground ● HO2S is damaged or contaminated due to improper fuel usage ● PCM has failed
P0137 **2T CCM, MIL: Yes** 1995, 1996, 1997, 1998, 1999, 2000, 2001, 2002 Camaro & Firebird 3.8L VIN K engines Transmissions: All	**HO2S-12 (Bank 1 Sensor 2) Circuit Low Input Conditions:** DTC P0101-P0103, P0107-P0108, P0112-P0113, P0117-P0118, P0121-P0125, P0128, P0201-P0206, P0410, P0440-P0449 and P1441 not set, running in closed loop for 4 minutes, A/F ratio from 14.5-14.8:1, TP angle from 3-40%, or on models with TAC, the APP sensor indicated angle from 5-40%, ECT sensor more than 122°F, and the PCM detected the HO2S signal was less than 60 mv, or it was less than 600 mv while in P/E mode for up to 15 seconds. **Possible Causes** ● Air leaks in the exhaust system, intake manifold, vacuum lines ● Engine misfire condition present (look for P0300 series codes) ● Fuel system too lean (possible low fuel pressure, water in fuel) ● HO2S signal circuit is shorted to the sensor or chassis ground ● HO2S is damaged (i.e., cracked) or air reference hole clogged ● PCM has failed

OBD II Trouble Code List (P0xxx Codes)

DTC	Trouble Code Title, Conditions & Possible Causes
P0137 **2T CCM, MIL: Yes** 1998, 1999, 2000, 2001, 2002 Camaro & Firebird 5.7L VIN G engine Transmissions: All	**HO2S-12 (Bank 1 Sensor 2) Circuit Low Input Conditions** DTC P0101-P0103, P0106-P0108, P0112-P0113, P0116-P0118, P0121-P0123, P0125, P0128, P0200, P0300, P0410-P0446, P0452-P0453, P1258, P1415-P1416 and P1441 not set, engine started, engine running in closed loop, Intrusive and Scan Tool tests "off", A/F ratio at 14.5-14.8:1, TP angle from 3-70%, then with the Lean Test enabled, the PCM detected the HO2S signal was less than 80 mv for 165 seconds or with P/E test enabled for 1 second, the PCM detected the HO2S signal was less than 420 mv for 400 seconds. **Possible Causes** • Air leaks in the exhaust system, intake manifold, vacuum lines • Engine misfire condition present (look for P0300 series codes) • Fuel system too lean (possible low fuel pressure, water in fuel) • HO2S signal circuit is shorted to the sensor or chassis ground • HO2S is damaged (i.e., cracked) or air reference hole clogged • PCM has failed
P0137 **2T CCM, MIL: Yes** 2001, 2002, 2003, 2004, 2005 Aurora 3.5L V6 VIN H, 4.0L V8 VIN C engines Transmissions: A/T	**HO2S-12 (Bank 1 Sensor 2) Circuit Low Input Conditions:** DTC P0101-P0103, P0107-P0108, P0117-P0118, P0121-P0123, P0132, and P0134, not set, engine started, engine speed over 800 rpm with the A/F ratio from 14.4-14.9:1, TP angle from 3-25 degrees, ECT sensor more than 179°F, AIR, Catalyst and EGR tests "off", and the PCM detected the HO2S signal was less than 9 mv for 5 seconds while in closed loop, or that it was less than 575 mv for 8 seconds while operating in Power Enrichment mode. **Possible Causes** • Air leaks in the exhaust system, intake manifold, vacuum lines • Engine misfire condition present (look for P0300 series codes) • Fuel system too lean (possible low fuel pressure, water in fuel) • HO2S signal circuit is shorted to the sensor or chassis ground • HO2S is damaged (i.e., cracked) or air reference hole clogged • PCM has failed
P0137 **2T CCM, MIL: Yes** 1995, 1996, 1997, 1998, 1999, 2000, 2001, 2002, 2003, 2004, 2005 S/T Blazer & S/T Pickup 4.3L VIN W, 4.3L VIN X Transmissions: All	**HO2S-11 (Bank 1 Sensor 1) Circuit Low Input Conditions:** DTC P0101-P0103, P0106-P0108, P0112, P0113, P0116-P0118, P0121-P0123, P0200, P0300, P0440-P0446, P0452, P0453, P0455, P0496 and P1441 not set, engine started, engine running in closed loop, fuel level over 10%, Scan Tool Intrusive tests off, then with the Lean Test enabled, TP angle from 3-70% for 5 seconds, the PCM detected the HO2S signal was less than 80 mv for 200 seconds or with the Power Enrichment mode active for 2 seconds, the PCM detected the HO2S signal was less than 490 mv for 10 seconds. The HO2S is used for fuel control and post-catalyst monitoring. This sensor compares the oxygen content of the surrounding air with the oxygen content of the exhaust stream. At initial startup, the PCM operates in open loop mode, ignoring the HO2S signal when calculating the air/fuel ratio. The PCM supplies the HO2S with a reference (or bias) voltage of about 450 mv. The HO2S generates a voltage within a range of 0-1000 mv that fluctuates above and below the bias voltage once in closed loop. A high HO2S voltage indicates a rich fuel mixture. A low HO2S voltage indicates a lean mixture. Heating elements in the HO2S shorten the time required for the sensor to reach normal temperature, and an accurate voltage signal. **Possible Causes** • Air leaks in the exhaust system, intake manifold, vacuum lines • Engine misfire condition present (look for P0300 series codes) • Fuel system too lean (possible low fuel pressure, water in fuel) • HO2S signal circuit is shorted to the sensor or chassis ground • HO2S is damaged (i.e., cracked) or air reference hole clogged • PCM has failed
P0137 **2T CCM, MIL: Yes** 1997, 1998, 1999, 2000, 2001 Catera 3.0L V6 VIN R engine Transmissions: A/T	**HO2S-12 (Bank 1 Sensor 2) Circuit Low Input Conditions:** DTC P0300-P0306, P0171, P0172, P0174, P0175, P0440, P0442, P0443, P0446, P0455 and P0460 are not set, engine started, engine speed 1400-2400 rpm, engine load from 8-12%, MAF sensor from 18-22 g/sec, TP angle from 6-10%, EVAP high loading not active, AIR test "off", and the PCM detected the HO2S signal was less than 148 mv for 200 ms during the CCM test of the HO2S circuit. **Possible Causes** • Air leaks in the exhaust system, intake manifold, vacuum lines • Engine misfire condition present (look for P0300 series codes) • Fuel system too lean (possible low fuel pressure, water in fuel) • HO2S signal circuit is shorted to the sensor or chassis ground • HO2S is damaged (i.e., cracked) or air reference hole clogged • PCM has failed

OBD II Trouble Code List (P0xxx Codes)

DTC	Trouble Code Title, Conditions & Possible Causes
P0137 **2T CCM, MIL: Yes** 1996, 1997 Grand Prix, Lumina, Monte Carlo 3.4L VIN X engine Transmissions: All	**HO2S-12 (Bank 1 Sensor 2) Circuit Low Input Conditions:** DTC P0101, P0102, P0103, P0107, P0108, P0112, P0113, P0116, P0117, P0118, P0121, P0122, P0123, P0125, P0128, P0201, P0202, P0203, P0204, P0205, P0206, P0410, P0440, P0442, P0443, P0446, P0449 and P1441 not set, engine started, engine running with the TP angle from 3-40%, A/F ratio from 14.5-14.8:1, Air Pump "off", and the PCM detected the HO2S signal was less than 75 mv for 100 ms, or with P/E Mode enabled, that the HO2S signal was less than 600 mv for 100 ms during the CCM test. **Possible Causes** ● Air leaks in the exhaust system, intake manifold, vacuum lines ● Engine misfire condition present (look for P0300 series codes) ● Fuel system too lean (possible low fuel pressure, water in fuel) ● HO2S signal circuit is shorted to the sensor or chassis ground ● HO2S is damaged (i.e., cracked) or air reference hole clogged ● PCM has failed
P0137 **2T CCM, MIL: Yes** 1997, 1998, 1999, 2000, 2001, 2002, 2003, 2004, 2005 Corvette 5.7L VIN G, 5.7L VIN S Transmissions: All	**HO2S-12 (Bank 1 Sensor 2) Circuit Low Input Conditions:** DTC P0101-P0103, P0106-P0108, P0112-P0118, P0200, P0300, P0410, P0440-P0446, P0452-P0453, P1120, P1125, P1220, P1221, P1258, P1415-P1416, P1441, P1514-P1518 not set, Scan Tool and Intrusive tests all "off", fuel level over 10%, TP angle from 3-70%, then with the Lean Test enabled, the PCM detected the HO2S signal was less than 80 mv for 400 seconds or with Power Enrichment mode active for 1 second, the PCM detected the HO2S signal was less than 420 mv for 10 seconds. The HO2S is used for fuel control and post-catalyst monitoring. At initial startup, the PCM operates in open loop mode, ignoring the HO2S signal when calculating the air/fuel ratio. The PCM supplies the HO2S with a reference (or bias) voltage of about 450 mv. The HO2S generates a voltage within a range of 0-1000 mv that fluctuates above and below the bias voltage once in closed loop. Heating elements in the HO2S shorten the time required for the sensor to reach normal temperature, and an accurate voltage signal. **Possible Causes** ● Air leaks in the exhaust system, intake manifold, vacuum lines ● Engine misfire condition present (look for P0300 series codes) ● Fuel system too lean (possible low fuel pressure, water in fuel) ● HO2S signal circuit is shorted to the sensor or chassis ground ● HO2S is damaged (i.e., cracked) or air reference hole clogged ● PCM has failed
P0138 **2T CCM, MIL: Yes** 1996, 1997, 1998, 1999, 2000, 2001, 2002 Achieva, Beretta, Ciera, Century, Corsica, Grand Am, Grand Prix, Lumina, Malibu, Monte Carlo 3.1L VIN M Transmissions: All	**HO2S-12 (Bank 1 Sensor 2) Circuit High Input Conditions:** DTC P0101-P0103, P0107, P0108, P0112, P0113, P0116-P0118, P0121-P0123, P0125, P0128, P0201-P0206, P0410, P0440, P0442, P0443, P0446, P0449, or P1441 not set, engine started, ECT sensor more than 149°F, TP angle from 3-40%, A/F ratio at 14.5-14:8:1 in closed loop, and the PCM detected the HO2S signal was more than 975 mv for 10 seconds, or with Decel Fuel Shutoff active, the HO2S signal was more than 200 mv for 10 seconds. The HO2S is used for fuel control and post-catalyst monitoring. This sensor compares the oxygen content of the surrounding air with the oxygen content of the exhaust stream. Heating elements in the HO2S shorten the time required for the sensor to reach normal temperature, and an accurate voltage signal. **Possible Causes** ● Fuel system is too rich (fuel pressure too high, fuel pressure regulator leaking, or one or more fuel injectors sticking/leaking) ● HO2S element is silicon, water or fuel contaminated ● HO2S signal circuit is shorted to system power (B+) ● HO2S signal tracking (water intrusion) in the connector causing a short between the HO2S signal and heater power circuits ● PCM has failed
P0138 **2T CCM, MIL: Yes** 1996, 1997, 1998, 1999, 2000, 2001, 2002, 2003, 2004, 2005 Achieva, Alero, Beretta, Century, Cavalier, Ciera, Century, Corsica, Envoy, Grand Am, Malibu, S/T Blazer & Pickup, Skylark, Sunfire, TrailBlazer 2.2L VIN 4, 2.2L VIN 5, 2.2L VIN F, 2.2L VIN H, 4.2L VIN S Transmissions: All	**HO2S-12 (Bank 1 Sensor 2) Circuit High Input Conditions:** DTC P0105, P0107, P0108, P0112, P0113, P0117, P0118, P0122, P0123, P0125, P0201-P0204, P0300, P0301-P0304, P0336, P0440, P0446, P0452, P0453, P0506, P0507, P0601, P0602, P1441, or P1621 not set, engine started, engine runtime over 10 seconds, system voltage over 10v, ECT sensor more than 158°F, fuel level over 10v, TP angle from 8-50% for 3.8 seconds, running in closed loop, MAP sensor greater than 20 kPa, and the PCM detected the HO2S signal was more than 1042 mv for 50-75 seconds. **Possible Causes** ● Fuel system is too rich (fuel pressure too high, fuel pressure regulator leaking, or one or more fuel injectors sticking/leaking) ● HO2S element is silicon, water or fuel contaminated ● HO2S signal circuit is shorted to system power (B+) ● HO2S signal tracking (water intrusion) in the connector causing a short between the HO2S signal and heater power circuits ● PCM has failed

OBD II Trouble Code List (P0xxx Codes)

DTC	Trouble Code Title, Conditions & Possible Causes
P0138 **2T CCM, MIL: Yes** 1996, 1997 Camaro, Corvette, Caprice, Firebird & Fleetwood 4.3L VIN W, 5.7L VIN 5, 5.7L VIN P engines Transmissions: All	**HO2S-12 (Bank 1 Sensor 2) Circuit High Input Conditions:** DTC P0100-P0103, P0107, P0108, P0117, P0118, P0125, P0200, P0372 and P1371 not set, engine running in closed loop, A/F ratio from 14.6-14.8:1, TP angle from 2-70%, Catalyst and AIR Tests "off", and the PCM detected the HO2S signal was above 930 mv for 38 seconds. **Possible Causes** ● Fuel system is too rich (fuel pressure too high, fuel pressure regulator leaking, or one or more fuel injectors sticking/leaking) ● HO2S element is silicon, water or fuel contaminated ● HO2S signal circuit is shorted to system power (B+) ● HO2S signal tracking (water intrusion) in the connector causing a short between the HO2S signal and heater power circuits ● PCM has failed
P0138 **2T CCM, MIL: Yes** 1996, 1997, 1998, 1999, 2000, 2001, 2002, 2003, 2004, 2005 Alero, Aztek, Cutlass, Impala, Malibu, Montana, Silhouette, Venture, Rendezvous 3.1L VIN J, 3.1L VIN M, 3.4L VIN E Transmissions: All	**HO2S-12 (Bank 1 Sensor 2) Circuit High Input Conditions:** DTC P0101-P0103, P0107, P0108, P0112, P0113, P0116-P0118, P0121-P0125, P0128, P0201-P0206, P0410, P0440, P0442, P0443, P0446, P0449 and P1441 not set, A/F ratio at 12.0-16.5:1, TP angle from 3-35%, Air Pump "off", and the PCM detected the HO2S signal was over 975 mv for 45 seconds or it was more than 200 mv during Decel Fuel Cutoff mode. **Possible Causes** ● Fuel system is too rich (fuel pressure too high, fuel pressure regulator leaking, or one or more fuel injectors sticking/leaking) ● HO2S element is silicon, water or fuel contaminated ● HO2S signal circuit is shorted to system power (B+) ● HO2S signal tracking (water intrusion) in the connector causing a short between the HO2S signal and heater power circuits ● PCM has failed
P0138 **2T CCM, MIL: Yes** 1996, 1997, 1998, 1999, 2000, 2001, 2002, 2003, 2004, 2005 C/K Cab & Chassis, C/K Series Truck, G Series Van, M/L Series Vans 4.3L VIN W, 4.3L VIN X, 5.0L VIN M, 5.7L VIN K, 5.7L VIN R, 7.4L VIN J Transmissions: All	**HO2S-12 (Bank 1 Sensor 2) Circuit High Input Conditions:** DTC P0101-P0103, P0106-P0108, P0112, P0113, P0116-P0118, P0121-P0123, P0200, P0300, P0401, P0404, P0405, P0440, P0442, P0446, P0452, P0453, P1120, P1125, P1220, P1221, P1258, P1404, P1441, P1514-P1517 and P1518 not set, engine started, fuel level over 10%, Intrusive and Scan Tool Tests "off", Rich Test enabled, A/F ratio from 14.5-14.7:1, TP angle from 3.5-70% for 5 seconds, or for vehicles with TAC, with the TP indicated angle from 3-70%, the PCM detected the HO2S signal was more than 930 mv for 200 seconds or while in DFCO mode, the PCM detected the HO2S signal was above 480 mv for 5 seconds. **Possible Causes** ● Fuel system is too rich (fuel pressure too high, fuel pressure regulator leaking, or one or more fuel injectors sticking/leaking) ● HO2S element is silicon, water or fuel contaminated ● HO2S signal circuit is shorted to system power (B+) ● HO2S signal tracking (water intrusion) in the connector causing a short between the HO2S signal and heater power circuits ● PCM has failed
P0138 **2T CCM, MIL: Yes** 1996, 1997, 1998, 1999, 2000, 2001, 2002, 2003, 2004, 2005 Bonneville, Century, Cutlass Supreme, 88'/98', Grand Prix, Intrigue, LSS, LeSabre, Lumina, Monte Carlo, Regal-Park Avenue 3.1L VIN J, 3.1L VIN M, 3.4L VIN X, 3.5L V6 VIN H, 3.8L VIN 1, 3.8L VIN K Transmissions: All	**HO2S-12 (Bank 1 Sensor 2) Circuit High Input Conditions:** DTC P0101-P0103, P0106-P0108, P0112-P0118, P0121-P0123, P0125, P0128, P0201-P0206, P0410, P0440-P0449 and P1441 not set, engine running in closed loop, TP angle from 3-40%, and the PCM detected the HO2S signal was more than 975 mv for 55 seconds, or that it was more than 200 mv while operating in DFCO mode. **Possible Causes** ● Fuel system is too rich (fuel pressure too high, fuel pressure regulator leaking, or one or more fuel injectors sticking/leaking) ● HO2S element is silicon, water or fuel contaminated ● HO2S signal tracking (water intrusion) in the connector causing a short between the HO2S signal and heater power circuits ● PCM has failed
P0138 **2T CCM, MIL: Yes** 1999, 2000, 2001, 2002, 2003, 2004, 2005 C/K Series Truck, G Series Van, Envoy, Escalade, TrailBlazer 4.8L VIN V, 5.3L VIN P, 5.3L VIN T, 5.3L VIN Z, 6.0L VIN N, 6.0L VIN U, 8.1L VIN G Transmissions: All	**HO2S-12 (Bank 1 Sensor 2) Circuit High Input Conditions:** DTC P0101, P0102, P0103, P0106, P0107, P0108, P0112, P0113, P0116, P0117, P0118, P0120, P0121, P0122, P0123, P0169, P0178, P0179, P0200, P0220, P0300, P0442, P0446, P0452, P0453, P0455, P0496, P1125, P1258, P1514, P1515, P1516, P1518, P2108 and P2135 not set, engine running in closed loop, Fuel Alcohol content less than 90%, fuel level over 10%, TP angle from 3-70% over the idle value, system voltage from 10-18v, Rich Test enabled, the PCM detected the HO2S signal was above 950 mv for 200 seconds or while operating in Decel Fuel Cutoff mode, the HO2S signal was more than 250 mv for 5 seconds. **Possible Causes** ● Fuel system is too rich (fuel pressure too high, fuel pressure regulator leaking, or one or more fuel injectors sticking/leaking) ● HO2S element is silicon, water or fuel contaminated ● HO2S signal tracking in connector (short between the HO2S signal and power circuits) ● PCM has failed

OBD II Trouble Code List (P0xxx Codes)

DTC	Trouble Code Title, Conditions & Possible Causes
P0138 **2T CCM, MIL: Yes** 1996, 1997, 1998, 1999, 2000, 2001, 2002, 2003, 2004, 2005 DeVille, Eldorado, Seville 4.6L VIN 9, 4.6L VIN Y Transmissions: A/T	**HO2S-12 (Bank 1 Sensor 2) Circuit High Input Conditions:** DTC P0101-P0103, P0106-P0108, P0112-P0113, P0116-P0118, P0121-P0123, P0125, P0128, P0200, P0300, P0410-P0446, P0452-P0453, P1258, P1415-P1416 and P1441 not set, AIR, Catalyst and EGR Flow tests "off", A/F ratio at 14.4-14.9:1, no injectors disabled, TP angle from 3-25% for 5 seconds, and the PCM detected the HO2S signal was more than 950 mv for 50 seconds, or that it was more than 200 mv during Decel Fuel Cutoff mode. **Possible Causes** • Fuel system is too rich (fuel pressure too high, fuel pressure regulator leaking, or one or more fuel injectors sticking/leaking) • HO2S element is silicon, water or fuel contaminated • HO2S signal tracking (water intrusion) in the connector causing a short between the HO2S signal and heater power circuits • PCM has failed
P0138 **2T CCM, MIL: Yes** 2003, 2004, 2005 CTS (Cadillac) 2.6L V6 VIN M & 3.2L V6 VIN N engines Transmissions: All	**HO2S-12 (Bank 1 Sensor 2) Insufficient Activity Conditions:** Engine started; system voltage over 10.5v, Calculated Converter Temperature less than 1,472°F, and the PCM detected the HO2S-12 signal remained above 1080 mv for 5 seconds, or it detected the HO2S-12 signal remained over 200 mv during DFCO for 3 seconds **Possible Causes** • EVAP system leaks or restrictions, charcoal canister problems • Exhaust system is leaking or severely restricted • HO2S air reference hole plugged (check for dirt on the outside) • HO2S signal circuit is open, or the low reference circuit is open • HO2S is damaged or contaminated due to improper fuel usage • PCM has failed
P0138 **2T CCM, MIL: Yes** 1995, 1996, 1997, 1998, 1999, 2000, 2001, 2002 Camaro & Firebird 3.8L VIN K engine Transmissions: All	**HO2S-12 (Bank 1 Sensor 2) Circuit High Input Conditions:** DTC P0101, P0102, P0103, P0107, P0108, P0112, P0113, P0117, P0118, P0121, P0122, P0123, P0125, P0128, P0201-P0206, P0410, P0440, P0442, P0443, P0446, P0449, or P1441 not set, engine running in closed loop for 4 minutes, ECT sensor over 122°F, system voltage over 10v, TP angle at 3-40% (on models with TAC, APP sensor angle at 5-40%), and PCM detected the HO2S signal was over 999 mv, or over 200 mv in DFCO for 15 seconds. **Possible Causes** • Fuel system too rich (fuel pressure too high, regulator leaking) • HO2S element is silicon, water or fuel contaminated • HO2S signal tracking (water intrusion) in the connector causing a short between the HO2S signal and heater power circuits • PCM has failed
P0138 **2T CCM, MIL: Yes** 1998, 1999, 2000, 2001, 2002 Camaro & Firebird 5.7L VIN G engine Transmissions: All	**HO2S-12 (Bank 1 Sensor 2) Circuit High Input Conditions:** DTC P0101, P0102, P0103, P0106, P0107, P0108, P0112-P0118, P0121, P0122, P0123, P0125, P0128, P0200, P0300, P0410, P0440-P0446, P0452, P0453, P1258, P1415, P1416 and P1441 not set, engine started, system voltage over 10.0v, fuel level over 10%, Intrusive and Scan Tool tests "off", A/F ratio 14.5-14.7:1, then with the Rich Test enabled and the TP angle from 3-70%, the PCM detected the HO2S signal was more than 930 mv for 200 seconds or with DFCO mode enabled for 1 second, the PCM detected the HO2S signal was more than 480 mv for 5 seconds. **Possible Causes** • Fuel system is too rich (fuel pressure too high, fuel pressure regulator leaking, or one or more fuel injectors sticking/leaking) • HO2S element is silicon, water or fuel contaminated • HO2S signal circuit is shorted to system power (B+) • HO2S signal tracking (water intrusion) in the connector causing a short between the HO2S signal and heater power circuits • PCM has failed
P0138 **2T CCM, MIL: Yes** 1996, 1997, 1998, 1999, 2001, 2002, 2003, 2004, 2005 Aurora 3.5L V6 VIN H, 4.0L V8 VIN C engines Transmissions: All	**HO2S-12 (Bank 1 Sensor 2) Circuit High Input Conditions:** DTC P0101-P0103, P0106-P0108, P0112, P0113, P0116-P0118, P0121-P0123, P0125, P0201-P0208, P0410, P0412, P0418, P0419, P0440, P0442, P0443, P0446, P0449, P1415, P1416, or P1441 are not set, engine started, TP angle from 3-40 degrees, ECT sensor more than 179°F, A/F ratio from 14.4-14.9:1, AIR, Catalyst and EGR Tests all "off", system voltage over 10.0v, MAP sensor over 32 kPa, and the PCM detected the HO2S signal was more than 950 mv in closed loop for 50 seconds, or it was more than 200 mv while in DFCO mode. **Possible Causes** • Fuel system is too rich (fuel pressure too high, fuel pressure regulator leaking, or one or more fuel injectors sticking/leaking) • HO2S element is silicon, water or fuel contaminated • HO2S signal circuit is shorted to system power (B+) • HO2S signal tracking (water intrusion) in the connector causing a short between the HO2S signal and heater power circuits • PCM has failed

OBD II Trouble Code List (P0xxx Codes)

DTC	Trouble Code Title, Conditions & Possible Causes
P0138 **2T CCM, MIL: Yes** 1995, 1996, 1997, 1998, 1999, 2000, 2001, 2002, 2003, 2004, 2005 S/T Blazer & S/T Pickup 4.3L VIN W, 4.3L VIN X Transmissions: All	**HO2S-11 (Bank 2 Sensor 2) Circuit High Input Conditions:** DTC P0101, P0102, P0103, P0106, P0107, P0108, P0112, P0113, P0116, P0117, P0118, P0121, P0122, P0123, P0200, P0300, P0440, P0442, P0446, P0452, P0453, P0455, P0496 and P1441 not set, engine started, engine running in closed loop, system voltage over 10.0v, fuel level over 10%, Scan Tool tests all off, TP angle from 3-70% for 5 seconds, then with the Rich Test enabled, the PCM detected the HO2S signal was more than 950 mv for 200 seconds or while operating in Decel Fuel Cutoff mode, the PCM detected the HO2S signal was more than 250 mv for 5 seconds. **Possible Causes** ● Fuel system is too rich (fuel pressure too high, fuel pressure regulator leaking, or one or more fuel injectors sticking/leaking) ● HO2S element is silicon, water or fuel contaminated ● HO2S signal circuit is shorted to system power (B+) ● HO2S signal tracking (water intrusion) in the connector causing a short between the HO2S signal and heater power circuits ● PCM has failed
P0138 **2T CCM, MIL: Yes** 1997, 1998, 1999, 2000, 2001 Catera 3.0L V6 VIN R engine Transmissions: All	**HO2S-12 (Bank 1 Sensor 2) Circuit High Input Conditions.** DTC P0137 not set, engine started, system voltage over 11.0v, Catalytic Converter temperature less than 1472°F, and the PCM detected the HO2S signal was more than 1080 mv for over 5.1 seconds during the CCM test. **Possible Causes** ● Fuel system is too rich (fuel pressure too high, fuel pressure regulator leaking, or one or more fuel injectors sticking/leaking) ● HO2S element is silicon, water or fuel contaminated ● HO2S signal circuit is shorted to system power (B+) ● HO2S signal tracking (water intrusion) in the connector causing a short between the HO2S signal and heater power circuits ● PCM has failed
P0138 **2T CCM, MIL: Yes** 1996, 1997 Grand Prix, Lumina, Monte Carlo 3.4L VIN X Transmissions: All	**HO2S-12 (Bank 1 Sensor 2) Circuit High Input Conditions:** DTC P0101-P0103, P0107, P0108, P0112, P0113, P0116-P0118, P0121-P0123, P0125, P0128, P0201-P0206, P0410, P0440, P0442, P0443, P0446, P0449, or P1441 not set, engine started, TP angle from 3-35%, AIR Pump "off", system voltage over 10.0v, A/F ratio from 13.0-16.0:1 in closed loop, and the PCM detected the HO2S signal was above 975 mv for 45 seconds or with Decel Fuel Shutoff enabled, it was over 110 mv for 55 seconds. **Possible Causes** ● Fuel system is too rich (fuel pressure too high, fuel pressure regulator leaking, or one or more fuel injectors sticking/leaking) ● HO2S element is silicon, water or fuel contaminated ● HO2S signal circuit is shorted to system power (B+) ● HO2S signal tracking (water intrusion) in the connector causing a short between the HO2S signal and heater power circuits ● PCM has failed
P0138 **2T CCM, MIL: Yes** 1997, 1998, 1999, 2000, 2001, 2002, 2003, 2004, 2005 Corvette 5.7L VIN G, 5.7L VIN S Transmissions: All	**HO2S-12 (Bank 1 Sensor 2) Circuit High Input Conditions:** DTC P0101, P0102, P0103, P0106, P0107, P0108, P0112, P0113, P0116, P0117, P0118, P0200, P0300, P0410, P0440, P0442, P0446, P0452, P0453, P1120, P1125, P1220, P1221, P1258, P1415, P1416, P1441, P1514, P1515, P1516, P1517, or P1518 not set, engine started, TP indicated angle from 3-70% in closed loop, AIR and Catalyst Tests "off", then with the Rich Test enabled, the PCM detected the HO2S signal was more than 930 mv for 200 seconds or with Decel Fuel Cutoff Test active for over 1 second, the PCM detected the HO2S signal was over 480 mv for 5 seconds. The HO2S is used for fuel control and post-catalyst monitoring. This sensor compares the oxygen content of the surrounding air with the oxygen content of the exhaust stream. At initial startup, the PCM operates in open loop mode, ignoring the HO2S signal when calculating the air/fuel ratio. The PCM supplies the HO2S with a reference (or bias) voltage of about 450 mv. The HO2S generates a voltage within a range of 0-1000 mv that fluctuates above and below the bias voltage once in closed loop. A high HO2S voltage indicates a rich fuel mixture. A low HO2S voltage indicates a lean mixture. Heating elements in the HO2S shorten the time required for the sensor to reach normal temperature, and an accurate voltage signal. **Possible Causes** ● Fuel system is too rich (fuel pressure too high, fuel pressure regulator leaking, or one or more fuel injectors sticking/leaking) ● HO2S element is silicon, water or fuel contaminated ● HO2S signal circuit is shorted to system power (B+) ● HO2S signal tracking (water intrusion) in the connector causing a short between the HO2S signal and heater power circuits ● PCM has failed

OBD II Trouble Code List (P0xxx Codes)

DTC	Trouble Code Title, Conditions & Possible Causes
P0138 **2T O2S, MIL: Yes** 1996, 1997 DeVille, Eldorado, Seville 4.6L VIN 9, 4.6L VIN Y Transmissions: All	**HO2S-12 (Bank 1 Sensor 2) Slow Response Conditions:** DTC P0117, P0118, P0121, P0122, P0123, P0131, P0132, P0133, P0134, P0135, P0151, P0152, P0153, P0154, P0155, P0171, P0172, P0174, P0175, P0300, P133, P1134, P1139, P1153 and P1154 not set, engine started, Low Coolant Level fault not present, engine speed from 1200-2800 rpm for 202 seconds in closed loop, ECT sensor more than 167°F, gear selector not indicating Reverse, Park or Neutral, and the PCM detected the HO2S lean-rich or rich-lean response time was more than 150 ms during the CCM test. **Possible Causes** ● Exhaust leak present in the exhaust manifold or exhaust pipes ● Fuel system is too rich (fuel pressure too high, fuel pressure regulator leaking, or one or more fuel injectors sticking/leaking) ● HO2S element is silicon, water or fuel contaminated ● TP sensor element broken (can cause false acceleration event) ● PCM has failed
P0138 **2T O2S, MIL: Yes** 1997, 1998, 1999, 2000, 2001 Catera 3.0L V6 VIN R engine Transmissions: A/T	**HO2S-12 (Bank 1 Sensor 2) Slow Response Conditions:** DTC P0131, P0132, P0133, P0134, P0135 and P0141 not set, engine started, engine running with the HO2S-12 indicating closed loop operation, system voltage over 11v, and the PCM detected the HO2S signal was fixed at over 600 mv or fixed at less than 600 mv or the HO2S signal was more than 200 mv during Decel Fuel Cutoff. **Possible Causes** ● Exhaust leak present in the exhaust manifold or exhaust pipes ● Fuel system is too rich (fuel pressure too high, fuel pressure regulator leaking, or one or more fuel injectors sticking/leaking) ● HO2S element is silicon, water or fuel contaminated ● TP sensor element broken (can cause false acceleration event) ● PCM has failed
P0140 **2T O2S, MIL: Yes** 1996, 1997, 1998, 1999, 2000, 2001, 2002, 2003, 2004, 2005 Achieva, Alero, Beretta, Century, Cavalier, Ciera, Century, Corsica, Envoy, Grand Am, Malibu, S/T Blazer & Pickup, Skylark, Sunfire, TrailBlazer 2.2L VIN 4, 2.2L VIN 5, 2.2L VIN F, 2.2L VIN H, 4.2L VIN S Transmissions: All	**HO2S-12 (Bank 1 Sensor 2) Insufficient Activity Conditions:** DTC P0105, P0107, P0108, P0112, P0113, P0117, P0118, P0122, P0123, P0171, P0201-P0204, P0300, P0301-P0304, P0335, P0440, P0442, P0446, P0506, P0507, P0601, P0602 and P1441 not set, engine started, engine runtime over 30 seconds, ECT sensor more than 158°F, system voltage over 10v, fuel level over 10%, calculated airflow more than 3 g/sec, TP angle from 8-56%, MAP sensor more than 25 kPa, and the PCM detected the HO2S- signal was fixed between 425-460 mv for 125 seconds during the CCM test. **Possible Causes** ● Exhaust leak present in the exhaust manifold or exhaust pipes ● HO2S signal or ground circuit has a high resistance condition ● HO2S element is silicon, water or fuel contaminated ● PCM has failed ● TSB 02-06-04-011 contains a repair procedure for this code
P0140 **1T O2S, MIL: No** 1995 Bonneville, Century, Cutlass Supreme, 88'/98', Grand Prix, Intrigue, LSS, LeSabre, Lumina, Monte Carlo, Regal-Park Avenue 3.1L VIN M, 3.4L VIN X, 3.8L VIN 1, 3.8L VIN K, 3.8L VIN L Transmissions: All	**HO2S-12 (Bank 1 Sensor 2) Insufficient Activity Conditions:** DTC P0116, P0117, P0118, P0131, P0132, P0135, P0171 P0172, P0174, P0175, P0420, P1115, P1133, P1153 and P1158 not set, engine started, engine runtime over 200 seconds, ECT sensor more than 110°F, TP angle over 3%, and the PCM detected the HO2S signal was fixed between 350-560 mv for 120 seconds. **Possible Causes** ● Exhaust leak present in the exhaust manifold or exhaust pipes ● HO2S signal or ground circuit has a high resistance condition ● HO2S element is silicon, water or fuel contaminated ● PCM has failed
P0140 **2T O2S, MIL: Yes** 1996, 1997, 1998, 1999, 2000, 2001, 2002 Achieva, Beretta, Ciera, Century, Corsica, Grand Am, Grand Prix, Lumina, Malibu & Monte Carlo 3.1L VIN M, 3.4L VIN X Transmissions: All	**HO2S-12 (Bank 1 Sensor 2) Insufficient Activity Conditions:** DTC P0101, P0102, P0103, P0107, P0108, P0112, P0113, P0116, P0117, P0118, P0121, P0122, P0123, P0125, P0128, P0201, P0202, P0203, P0204, P0205, P0206, P0410, P0440, P0442, P0443, P0446, P0449, or P1441 not set, engine started, system voltage over 10.0v, engine speed from 1000-3000 rpm for 4 minutes in closed loop, fuel level over 10%, ECT sensor more than 122°F, TP angle from 3-40%, gear selector not in Reverse or P/N, MAF sensor from 10-30 g/sec, Air Pump "off", and the PCM detected the HO2S signal was fixed between 412-499 mv for 30-90 seconds. **Possible Causes** ● Exhaust leak present in the exhaust manifold or exhaust pipes ● HO2S signal or ground circuit has a high resistance condition ● HO2S element is silicon, water or fuel contaminated ● PCM has failed

OBD II Trouble Code List (P0xxx Codes)

DTC	Trouble Code Title, Conditions & Possible Causes
P0140 **1T O2S, MIL: Yes** 1996, 1997 Camaro, Corvette, Caprice, Firebird & Fleetwood 4.3L VIN W, 5.7L VIN 5, 5.7L VIN P engines Transmissions: All	**HO2S-12 (Bank 1 Sensor 2) Insufficient Activity Conditions:** DTC P0100-P0103, P0107, P0108, P0112, P0113, P0117, P0118, P0125, P0200, P0372 and P1371 not set; engine speed from 1000-3000 rpm in closed loop for 4 minutes, AIR and Catalyst Tests "off", ECT sensor over 118°F, and the PCM detected the HO2S signal was fixed between 391-491 mv for 68 seconds during the CCM test of the Oxygen sensor circuit. **Possible Causes** ● Exhaust leak present in the exhaust manifold or exhaust pipes ● HO2S signal or ground circuit has a high resistance condition ● HO2S element is silicon, water or fuel contaminated ● PCM has failed
P0140 **2T O2S, MIL: Yes** 1996, 1997, 1998, 1999, 2000, 2001, 2002, 2003, 2004, 2005 Alero, Aztek, Cutlass, Impala, Malibu, Montana, Silhouette, Venture, Rendezvous 3.1L VIN J, 3.1L VIN M, 3.4l VIN E engines Transmissions: All	**HO2S-12 (Bank 1 Sensor 2) Insufficient Activity Conditions:** DTC P0101-P0103, P0106-P0108, P0112, P0113, P0116-P0118, P0121-P0123, P0125, P0128, P0131, P0132, P0135, P0151, P0152, P0201-P0208, P0300, P0410, P0410, P0418, P0419, P0440, P0442, P0443, P0446, P0449, P1133, P1415, P1416, or P1441 not set, engine runtime over 200 seconds, system voltage over 10.0v, ECT sensor over 122°F, and the PCM detected the HO2S signal was fixed from 412-499 mv for more than 29 seconds. **Possible Causes** ● Exhaust leak present in the exhaust manifold or exhaust pipes ● HO2S signal or ground circuit has a high resistance condition ● HO2S element is silicon, water or fuel contaminated ● PCM has failed
P0140 **2T O2S, MIL: Yes** 1996, 1997, 1998, 1999, 2000, 2001, 2002, 2003, 2004, 2005 Bonneville, Century, Cutlass Supreme, 88'/98', Grand Prix, Intrigue, LSS, LeSabre, Lumina, Monte Carlo, Regal-Park Avenue 3.1L VIN J, 3.1L VIN M, 3.4L VIN X, 3.5L V6 VIN H, 3.8L VIN 1, 3.8L VIN K, 4.0L V8 VIN C Transmissions: All	**HO2S-12 (Bank 1 Sensor 2) Insufficient Activity Conditions:** DTC P0101, P0102, P0103, P0106, P0107, P0108, P0112, P0113, P0116, P0117, P0118, P0121, P0122, P0123, P0125, P0128, P0201, P0202, P0203, P0204, P0205, P0206, P0410, P0440, P0442, P0443, P0446, P0449 and P1441 not set, engine started, engine runtime over 200 seconds, and the PCM detected the HO2S signal was fixed between 400-500 mv for over 240 seconds. **Possible Causes** ● Exhaust leak present in the exhaust manifold or exhaust pipes ● HO2S signal or ground circuit has a high resistance condition ● HO2S element is silicon, water or fuel contaminated ● PCM has failed
P0140 **2T O2S, MIL: Yes** 1996, 1997, 1998, 1999, 2000, 2001, 2002, 2003, 2004, 2005 C/K Cab & Chassis, C/K Series Truck, G Series Van, M/L Series Vans 4.3L VIN W, 4.3L VIN X, 5.0L VIN M, 5.7L VIN K, 5.7L VIN R, 7.4L VIN J Transmissions: All	**HO2S-12 (Bank 1 Sensor 2) Insufficient Activity Conditions:** DTC P0101-P0103, P0106-P0108, P0112, P0113, P0116-P0118, P0121-P0123, P0200, P0300, P0401, P0404, P0405, P0440, P0442, P0446, P0452, P0453, P1120, P1125, P1220, P1221, P1258, P1404, P1441, P1514-P1517 and P1518 not set, engine runtime over 409 seconds, system voltage over 10.0v, Intrusive and Scan Tool tests "off", TP angle over 5%, and the PCM detected the HO2S signal was fixed from 410-490 mv for 150 seconds. **Possible Causes** ● Exhaust leak present in the exhaust manifold or exhaust pipes ● HO2S signal or ground circuit has a high resistance condition ● HO2S element is silicon, water or fuel contaminated ● PCM has failed
P0140 **2T O2S, MIL: Yes** 1999, 2000, 2001, 2002, 2003, 2004, 2005 C/K Series Truck, G Series Van, Envoy, Escalade, TrailBlazer 4.8L VIN V, 5.3L VIN P, 5.3L VIN T, 5.3L VIN Z, 6.0L VIN N, 8.0L VIN U, 8.1L VIN G Transmissions: All	**HO2S-12 (Bank 1 Sensor 2) Insufficient Activity Conditions:** DTC P0101, P0102, P0103, P0106, P0107, P0108, P0112, P0113, P0116, P0117, P0118, P0120, P0121, P0122, P0123, P0169, P0178, P0179, P0200, P0220, P0300, P0442, P0446, P0452, P0453, P0455, P0496, P1125, P1258, P1514, P1515, P1516, P1518, P2108 and P2135 not set, engine runtime over 300 seconds, system voltage from 10-18v, Fuel Alcohol content less than 90%, then after a TP indicated angle change of over 5% within one second six times, the PCM detected the HO2S signal was fixed from 410-490 mv for 150 seconds. **Possible Causes** ● HO2S signal or ground circuit has a high resistance condition ● HO2S signal circuit is open or shorted to system power (B+) ● HO2S has failed (i.e., it is silicon, water or fuel contaminated) ● PCM has failed
P0140 **2T O2S, MIL: Yes** 1996, 1997, 1998, 1999, 2000, 2001, 2002, 2003, 2004, 2005 DeVille, Eldorado, Seville 4.6L VIN 9, 4.6L VIN Y Transmissions: All	**HO2S-12 (Bank 1 Sensor 2) Insufficient Activity Conditions:** DTC P0101-P0103, P0106-P0108, P0112-P0113, P0116-P0118, P0121-P0123, P0125, P0128, P0200, P0300, P0410, P0446-P0453, P1258, P1415, P1416 and P1441 not set, engine runtime over 200 seconds in closed loop, AIR, Catalyst, EGR and Intrusive Tests "off", TP angle from 3-25%, and the PCM detected the HO2S signal was fixed from 400-500 mv. **Possible Causes** ● Exhaust leak present in the exhaust manifold or exhaust pipes ● HO2S signal or ground circuit has a high resistance condition ● HO2S element is silicon, water or fuel contaminated ● PCM has failed

OBD II Trouble Code List (P0xxx Codes)

DTC	Trouble Code Title, Conditions & Possible Causes
P0140 **2T O2S, MIL: Yes** 2003, 2004, 2005 CTS (Cadillac) 2.6L V6 VIN M & 3.2L V6 VIN N engines Transmissions: All	**HO2S-12 (Bank 1 Sensor 2) Insufficient Activity Conditions:** Engine started; system voltage over 10.5v, Calculated Converter Temperature less than 1,472°F, and the PCM did not detect any activity from the HO2S-12, and it detected the HO2S-12 signal remained between 400-520 mv for more than 600 seconds. **Possible Causes** • EVAP system leaks or restrictions, charcoal canister problems • Exhaust system is leaking or severely restricted • HO2S air reference hole plugged (check for dirt on the outside) • HO2S signal circuit is open, or the low reference circuit is open • HO2S is damaged or contaminated due to improper fuel usage • PCM has failed
P0140 **2T O2S, MIL: Yes** 1995, 1996, 1997, 1998, 1999, 2000, 2001, 2002 Camaro & Firebird 3.8L VIN K engine Transmissions: All	**HO2S-12 (Bank 1 Sensor 2) Insufficient Activity Conditions:** DTC P0101-P0103, P0107, P0108, P0112, P0113, P0117, P0118, P0121-P0123, P0125, P0128, P0201-P0206, P0410, P0440, P0442, P0443, P0446, P0449 and P1441 not set, engine started, engine runtime over 409 seconds, Scan Tool and Intrusive tests all "off", system voltage over 10.0v, TP angle or APP indicated angle changed over 1.5% six times, and the PCM detected the HO2S signal was fixed from 410-490 mv for 150 seconds. **Possible Causes** • Exhaust leak present in the exhaust manifold or exhaust pipes • HO2S signal or ground circuit has a high resistance condition • HO2S element is silicon, water or fuel contaminated • PCM has failed
P0140 **2T O2S, MIL: Yes** 1998, 1999, 2000, 2001, 2002 Camaro & Firebird 5.7L VIN G engine Transmissions: All	**HO2S-12 (Bank 1 Sensor 2) Insufficient Activity Conditions:** DTC P0101-P0103, P0106-P0108, P0112-P0118, P0121-P0123, P0125, P0128, P0200, P0300, P0410, P0440-P0446, P0452, P0453, P1258, P1415-P1416 and P1441 not set, engine started, engine runtime over 409 seconds, Scan Tool and Intrusive tests "off", system voltage over 10.0v, TP angle or APP indicated angle changed over 1.5% six times, and the PCM detected the HO2S signal was fixed from 412-490 mv for 1 minute (test failed 6 times). **Possible Causes** • Exhaust leak present in the exhaust manifold or exhaust pipes • HO2S signal or ground circuit has a high resistance condition • HO2S element is silicon, water or fuel contaminated • PCM has failed
P0140 **2T O2S, MIL: Yes** 1996, 1997, 1998, 1999, 2001, 2002, 2003, 2004, 2005 Aurora 3.5L V6 VIN H, 4.0L V8 VIN C engines Transmissions: All	**HO2S-12 (Bank 1 Sensor 2) Insufficient Activity Conditions:** DTC P0101, P0102, P0103, P0106-P0108, P0112, P0113, P0116-P0118, P0121-P0123, P0125, P0201-P0208, P0410, P0412, P0418, P0419, P0440, P0442, P0443, P0446, P0449, P1133, P1415, P1416, or P1441 not set, engine runtime over 200 seconds, system voltage over 10.0v, ECT sensor more than 163°F, TP Switch open, Catalyst AIR, and EGR tests "off", and the PCM detected the HO2S signal was fixed between 400 to 500 mv for 80 seconds. **Possible Causes** • Exhaust leak present in the exhaust manifold or exhaust pipes • HO2S signal or ground circuit has a high resistance condition • HO2S element is silicon, water or fuel contaminated • PCM has failed
P0140 **2T O2S, MIL: Yes** 1995, 1996, 1997, 1998, 1999, 2000, 2001, 2002, 2003, 2004, 2005 S/T Blazer & S/T Pickup 4.3L VIN W, 4.3L VIN X Transmissions: All	**HO2S-12 (Bank 1 Sensor 2) Insufficient Activity Conditions:** DTC P0101-P0103, P0106-P0108, P0112, P0113, P0116-P0118, P0121-P0123, P0200, P0300, P0440, P0442, P0446, P0452, P0453, P0496 and P1441 not set, engine runtime over 409 seconds, TP angle change over 1.5% 6 times, system voltage over 10.0v, Intrusive tests "off", and the PCM detected the HO2S signal was fixed at 410-490 mv for 150 seconds. **Possible Causes** • Exhaust leak present in the exhaust manifold or exhaust pipes • HO2S signal or ground circuit has a high resistance condition • HO2S element is silicon, water or fuel contaminated • PCM has failed
P0140 **2T O2S, MIL: Yes** 1997, 1998, 1999, 2000, 2001 Catera 3.0L V6 VIN R engine Transmissions: All	**HO2S-12 (Bank 1 Sensor 2) Insufficient Activity Conditions:** DTC P0130 and P0150 not set, engine speed from 1400-2400 rpm in closed loop for 25 seconds, system voltage over 10.0v, engine load at 8-12%, MAF sensor at 18-22 g/sec, TP angle has changed over 1.5% at least 6 times, and the PCM detected the HO2S signal remained fixed between 401-519 mv for more than 60 seconds. **Possible Causes** • Exhaust leak present in the exhaust manifold or exhaust pipes • HO2S signal or ground circuit has a high resistance condition • HO2S element is silicon, water or fuel contaminated • PCM has failed

OBD II Trouble Code List (P0xxx Codes)

DTC	Trouble Code Title, Conditions & Possible Causes
P0140 **2T O2S, MIL: Yes** 1997, 1998, 1999, 2000, 2001, 2002, 2003, 2004, 2005 Corvette 5.7L VIN G, 5.7L VIN S Transmissions: All	**HO2S-12 (Bank 1 Sensor 2) Insufficient Activity Conditions:** DTC P0101-P0103, P0106-P0108, P0112-P0118, P0200, P0300, P0410, P0440-P0446, P0452, P0453, P1120, P1125, P1220, P1221, P1258, P1415, P1416, P1441, P1514 to P1518 not set, engine runtime 70 seconds in closed loop, Intrusive and Scan Tool tests "off", system voltage over 10.0v and the PCM detected the HO2S signal was fixed at 409-489 mv. **Possible Causes** ● Exhaust leak present in the exhaust manifold or exhaust pipes ● HO2S signal or ground circuit has a high resistance condition ● HO2S element is silicon, water or fuel contaminated ● PCM has failed
P0141 **2T O2S HTR, MIL: Yes** 2002, 2003, 2004, 2005 Alero, Cavalier, Grand AM and S/T Pickup, TrailBlazer 2.2L VIN F, 2.2L VIN H, 4.2L VIN S Transmissions: All	**HO2S-12 (Bank 1 Sensor 2) Heater Circuit Malfunction Conditions:** DTC P0105-P0108, P0112-P0118, P0122, P0123, P0171, P0172, P0201-P0206, P0300, P0336, P0351-P0356, P0440, P0446, P0452, P0453, P0461, P0506, P0507, P0601, P0602, P1220, P1221, P1441, P1635, P1639 and P1681 not set, engine runtime over 1 minute, ECT sensor more than 158°F, system voltage over 10.0v, Calculated Airflow under 16 g/sec, fuel level over 10%, and the PCM detected the HO2S heater current was not within range for 200 seconds during the HO2S Heater Monitor test. **Possible Causes** ● HO2S heater ground circuit is open or has high resistance ● HO2S heater power circuit is open (test O2S fuse in fuse block) ● HO2S heater element is damaged or has failed ● PCM has failed
P0141 **2T O2S HTR, MIL: Yes** 1996, 1997, 1998, 1999, 2000, 2001, 2002 Achieva, Alero, Beretta, Century, Cavalier, Ciera, Century, Corsica, Envoy, Grand Am, Malibu, S/T Blazer & Pickup, Skylark, Sunfire, TrailBlazer 2.2L VIN 4, 2.2L VIN 5, 2.4L VIN T Transmissions: All	**HO2S-12 (Bank 1 Sensor 2) Heater Circuit Malfunction Conditions:** DTC P0105-P0108, P0112-P0118, P0122, P0123, P0171, P0201-P0204, P0301-P0304, P0335, P0440-P0446, P0506, P0507, P0601, P0602 and P1441 not set, ECT and IAT sensors below 113°F and within 45°F at startup, fuel level over 10%, HO2S signal from 395-495 mv, and the PCM detected the HO2S signal did not go below 300 mv or above 600 mv within 110-255 seconds after a cold startup with the average airflow less than 20 g/sec. **Possible Causes** ● HO2S heater ground circuit is open or has high resistance ● HO2S heater power circuit is open (test O2S fuse in fuse block) ● HO2S heater element is damaged or has failed ● PCM has failed ● TSB 01-06-04-003 contains a repair procedure for this code
P0141 **2T O2S HTR, MIL: Yes** 1996, 1997, 1998, 1999, 2000, 2001, 2002 Achieva, Beretta, Ciera, Century, Corsica, Grand Am, Grand Prix, Lumina, Malibu, Monte Carlo 3.1L VIN M, 3.4L VIN X Transmissions: All	**HO2S-12 (Bank 1 Sensor 2) Heater Circuit Malfunction Conditions:** DTC P0101, P0102, P0103, P0106, P0107, P0108, P0112, P0113, P0116, P0117, P0118, P0121, P0122, P0123, P0125, P0128, P0201, P0202, P0203, P0204, P0205, P0206, P0410, P0440, P0442, P0443, P0446, P0449 and P1441 not set, ECT and IAT sensors less than 95°F and within 43°F at startup, engine started, system voltage over 10v, MAF sensor less than 20 g/sec, Air Pump "off", and the PCM detected the HO2S signal was within 150 mv of bias voltage (450 mv) for up to 2.5 minutes (depends on ECT and MAF input at startup). **Possible Causes** ● HO2S heater ground circuit is open or has high resistance ● HO2S heater power circuit is open (test O2S fuse in fuse block) ● HO2S heater element is damaged or has failed ● PCM has failed
P0141 **2T O2S HTR, MIL: Yes** 1996, 1997 Camaro, Corvette, Caprice, Firebird & Fleetwood 4.3L VIN W, 5.7L VIN 5, 5.7L VIN P engines Transmissions: All	**HO2S-12 (Bank 1 Sensor 2) Heater Circuit Malfunction Conditions:** DTC P0101-P0103, P0112, P0113, P0117, P0118, P0125, P0132-P0134 and P0200 not set, ECT and IAT sensors less than 122°F and within 5°F at startup, engine started, system voltage over 10.0v, AIR and Catalyst tests "off", MAF sensor less than 40 g/sec, TP angle under 20%, and the PCM detected the HO2S signal was fixed at 300-700 mv too long a time. **Possible Causes** ● HO2S heater ground circuit is open or has high resistance ● HO2S heater power circuit is open (test O2S fuse in fuse block) ● HO2S heater element is damaged or has failed ● PCM has failed
P0141 **2T O2S HTR, MIL: Yes** 1996, 1997, 1998, 1999, 2000, 2001 Alero, Aztek, Cutlass, Impala, Malibu, Montana, Silhouette, Venture, Rendezvous 3.1L VIN J, 3.1L VIN M, 3.4L VIN E engines Transmissions: All	**HO2S-12 (Bank 1 Sensor 2) Heater Circuit Malfunction Conditions:** DTC P0101-P0103, P0107, P0108, P0112, P0113, P0116-P0118, P0121-P0123, P0125, P0128, P0201-P0206, P0410, P0440-P0449 and P1441 not set, ECT and IAT sensors less than 95°F and within 42°F at startup, engine runtime over 200 seconds, system voltage over 10.0v and the PCM detected the HO2S signal remained within 150 mv of the bias voltage. **Possible Causes** ● HO2S heater ground circuit is open or has high resistance ● HO2S heater power circuit is open (test O2S fuse in fuse block) ● HO2S heater element is damaged or has failed ● PCM has failed

OBD II Trouble Code List (P0xxx Codes)

DTC	Trouble Code Title, Conditions & Possible Causes
P0141 **2T O2S HTR, MIL: Yes** 2002, 2003, 2004, 2005 Alero, Aztek, Cutlass, Impala, Malibu, Montana, Silhouette, Venture, Rendezvous 3.1L VIN J, 3.1L VIN M, 3.4L VIN E engines Transmissions: All	**HO2S-12 (Bank 1 Sensor 2) Heater Circuit Malfunction Conditions:** DTC P0101, P0102, P0103, P0107, P0108, P0112, P0113, P0116, P0117, P0118, P0121, P0122, P0123, P0125, P0128, P0201-P0206, P0410, P0440, P0442, P0443, P0446, P0449, P1441 not set, ECT sensor and IAT sensor more than 95ºF at engine startup, engine runtime over 200 seconds, system voltage from 9-18.0v, and the PCM detected the HO2S signal was fixed within 74 mv of the bias voltage (450 mv) for 2 minutes (depends on ECT at startup). **Possible Causes** ● HO2S assembly connector is damaged, open or shorted ● HO2S heater ground circuit is open or it has high resistance ● HO2S heater power circuit is open (test O2S fuse in fuse block) ● HO2S heater element is damaged or has failed ● PCM has failed
P0141 **2T O2S HTR, MIL: Yes** 1996, 1997, 1998, 1999, 2000, 2001, 2002, 2003, 2004, 2005 Bonneville, Century, Cutlass Supreme, 88'/98', Grand Prix, Intrigue, LSS, LeSabre, Lumina, Monte Carlo, Regal-Park Avenue 3.1L VIN J, 3.1L VIN M, 3.4L VIN X, 3.5L V6 VIN H, 3.8L VIN 1, 3.8L VIN K, 4.0L V8 VIN C Transmissions: All	**HO2S-12 (Bank 1 Sensor 2) Heater Circuit Malfunction Conditions:** DTC P0101-P0103, P0106-P0108, P0112, P0113, P0116-P0118, P0121-P0123, P0125, P0128, P0201-P0206, P0410, P0440, P0442, P0443, P0446, P0449 and P1441 not set, IAT and ECT sensors with 11ºF of each other at startup, MAF sensor less than 17-20 g/sec, HO2S signal within 100 mv of bias voltage at startup, and the PCM detected the HO2S signal remained within 150 mv of bias voltage (450 mv) for 50-80 seconds (depends on the ECT/MAF at startup). **Possible Causes** ● HO2S heater ground circuit is open or has high resistance ● HO2S heater power circuit is open (test O2 fuse in fuse block) ● HO2S heater element is damaged or has failed ● PCM has failed
P0141 **2T O2S HTR, MIL: Yes** 1996, 1997, 1998, 1999, 2000, 2001, 2002, 2003, 2004, 2005 C/K Cab & Chassis, C/K Series Truck, G Series Van, M/L Series Vans 4.3L VIN W, 4.3L VIN X, 5.0L VIN M, 5.7L VIN K, 5.7L VIN R, 7.4L VIN J Transmissions: All	**HO2S-12 (Bank 1 Sensor 2) Heater Circuit Malfunction Conditions:** DTC P0101-P0103, P0106-P0108, P0112, P0113, P0116-P0118, P0121-P0123, P0131, P0132, P0134, P0137, P0138, P0140, P0200, P0300, P0401, P0404, P0405, P0440, P0442, P0446, P0452, P0453, P1120, P1125, P1220, P1221, P1258, P1404, P1441, P1514, P1515, P1516, P1517 and P1518 not set, ECT and IAT sensors less than 122ºF and with 14.5ºF at startup, engine started, HO2S signal from 425-475 mv right after startup, Intrusive and Scan Tool tests "off", MAF sensor less than 25 g/sec, and the PCM detected the HO2S signal was fixed within 150 mv of startup HO2S signal for too long (depends on ECT/MAF at startup). **Possible Causes** ● HO2S heater ground circuit is open or has high resistance ● HO2S heater power circuit is open (test O2S fuse in fuse block) ● HO2S heater element is damaged or has failed ● PCM has failed ● TSB 00-06-04-006 contains a repair procedure for this code
P0141 **2T O2S HTR, MIL: Yes** 1999, 2000, 2001, 2002, 2003, 2004, 2005 C/K Series Truck, G Series Van, Envoy, Escalade, TrailBlazer 4.8L VIN V, 5.3L VIN P, 5.3L VIN T, 5.3L VIN Z, 6.0L VIN N, 6.0L VIN U, 8.1L VIN G Transmissions: All	**HO2S-12 (Bank 1 Sensor 2) Heater Circuit Malfunction Conditions:** DTC P0101, P0102, P0103, P0106, P0107, P0108, P0112, P0113, P0116, P0117, P0118, P0120, P0121, P0122, P0123, P0169, P0178, P0179, P0200, P0220, P0300, P0442, P0446, P0452, P0453, P0455, P0496, P1125, P1258, P1514, P1515, P1516, P1518, P2108 and P2135 not set, engine runtime over 120 seconds, engine speed from 500-3000 rpm, system voltage from 10-18v, ECT sensor more than 122ºF, MAF sensor signal from 3-40 g/sec, Fuel Alcohol content less than 90%, and the PCM detected the HO2S heater current was less than 0.25 amps, or more than 3.125 amps (more than 1.375 amps on 4.8L V8). **Possible Causes** ● HO2S heater low control circuit is open or shorted to ground ● HO2S heater circuit is open or it is shorted to ground ● HO2S heater power circuit is open (test O2A fuse in fuse block) ● HO2S heater element is damaged or has failed ● PCM has failed
P0141 **2T O2S HTR, MIL: Yes** 1996, 1997, 1998, 1999, 2000, 2001, 2002, 2003, 2004, 2005 DeVille, Eldorado, Seville 4.6L VIN 9, 4.6L VIN Y Transmissions: A/T	**HO2S-12 (Bank 1 Sensor 2) Heater Circuit Malfunction Conditions:** DTC P0101, P0102, P0103, P0117, P0118, and P0134 not set, ECT sensor at least 50ºF less than the last ignition cycle, engine started, system voltage over 11.0v, average HO2S bias voltage from 352-546 mv, engine did not stall, and the PCM detected the HO2S signal was fixed within 150 mv of the bias voltage for too long (depends on ECT/MAF at startup). **Possible Causes** ● HO2S heater ground circuit is open or has high resistance ● HO2S heater power circuit is open (test O2S fuse in fuse block) ● HO2S heater element is damaged or has failed ● PCM has failed

OBD II Trouble Code List (P0xxx Codes)

DTC	Trouble Code Title, Conditions & Possible Causes
P0141 **2T O2S HTR, MIL: Yes** 2003, 2004, 2005 CTS (Cadillac) 2.6L V6 VIN M & 3.2L V6 VIN N engines Transmissions: All	**HO2S-12 (Bank 1 Sensor 2) Heater Circuit Malfunction Conditions:** DTC P0030, P0031, P0032, P0036, P0037, P0038, P0050, P0051, P0052, P0056, P0057, P0058, P0157-P0158, P0160-P0161 not set, engine started, system voltage from 10-16v, Converter Temperature from 626-1112°F, and the PCM detected the HO2S-12 heater current was not within the expected range with it commanded "on". **Possible Causes** ● HO2S heater high side driver circuit has high resistance (open) ● HO2S heater low side driver circuit has high resistance (open) ● HO2S heater power circuit is open (test O2S fuse in fuse block) ● HO2S heater element is damaged or has failed ● PCM has failed
P0141 **2T O2S HTR, MIL: Yes** 1995, 1996, 1997, 1998, 1999, 2000, 2001, 2002 Camaro & Firebird 3.8L VIN K engine Transmissions: All	**HO2S-12 (Bank 1 Sensor 2) Heater Circuit Conditions:** DTC P0101-P0103, P0107, P0108, P0112, P0113, P0117, P0118, P0121-P0123, P0125, P0128, P0201-P0206, P0410, P0440, P0442-P0449 and P1441 not set, ECT and IAT sensors less than 95°F at startup and within 11°F at startup, engine started, system voltage over 10.0v, MAF sensor less than 20 g/sec, and the PCM detected the HO2S signal was within 150 mv of bias voltage for from 42 seconds to 2 minutes (depends on the ECT/MAF signals at startup). **Possible Causes** ● HO2S heater ground circuit is open or has high resistance ● HO2S heater power circuit is open (test O2S fuse in fuse block) ● HO2S heater element is damaged or has failed ● PCM has failed
P0141 **2T O2S HTR, MIL: Yes** 1998, 1999, 2000, 2001, 2002 Camaro & Firebird 5.7L VIN G engine Transmissions: All	**HO2S-12 (Bank 1 Sensor 2) Heater Circuit Malfunction Conditions:** DTC P0101-P0103, P0106-P0108, P0112-P0118, P0121-P0123, P0125, P0128, P0131-P0138, P0140, P0151-P0158, P0160, P0200, P0300, P0410, P0440-P0446, P0452, P0453, P1258, P1415, P1416 or P1441 not set, engine running in closed loop, ECT and IAT sensors under 122°F and within 14.5°F at startup, Intrusive Tests "off", MAF sensor under 18 g/sec, and the PCM detected the HO2S signal was fixed from 300-700 mv for 42-120 seconds (depends startup ECT and MAP sensor inputs at engine startup). **Possible Causes** ● HO2S heater ground circuit is open or has high resistance ● HO2S heater power circuit is open (test O2S fuse in fuse block) ● HO2S heater element is damaged or has failed ● PCM has failed
P0141 **2T O2S HTR, MIL: Yes** 1995, 1996, 1997, 1998, 1999, 2000, 2001, 2002, 2003, 2004, 2005 S/T Blazer & S/T Pickup 4.3L VIN W, 4.3L VIN X Transmissions: All	**HO2S-12 (Bank 1 Sensor 2) Heater Circuit Malfunction Conditions:** DTC P0101-P0103, P0106-P0108, P0112, P0113, P0116-P0118, P0121-P0123, P0200, P0300, P0440, P0442, P0452-P0453, P0455, P0496 and P1441 not set, ECT and IAT sensors under 122°F and within 14.5°F at startup, engine running in closed loop, Scan Tool tests off, and the PCM detected the HO2S signal was within 150 mv of bias voltage (450 mv) for too long (depends on ECT at startup). **Possible Causes** ● HO2S heater ground circuit is open or has high resistance ● HO2S heater power circuit is open (test O2S fuse in fuse block) ● HO2S heater element is damaged or has failed ● PCM has failed
P0141 **2T O2S HTR, MIL: Yes** 2001, 2002, 2003, 2004, 2005 Aurora 4.0L V8 VIN C engine Transmissions: A/T	**HO2S-12 (Bank 1 Sensor 2) Heater Circuit Malfunction Conditions:** DTC P0137, P0138 and P0140 not set, cold engine startup finished, system voltage from 11-18v, average MAF less than 30 g/sec, HO2S-12 signal from 360-540 mv at startup, and the PCM detected the HO2S remained within 150 mv of the bias voltage for too long. **Possible Causes** ● HO2S heater low control circuit is open or has high resistance ● HO2S heater power circuit is open (test the O2S SEN fuse) ● HO2S heater element is damaged or has failed ● PCM has failed
P0141 **2T O2S HTR, MIL: Yes** 1997, 1998, 1999, 2000, 2001 Catera 3.0L V6 VIN R engine Transmissions: A/T	**HO2S-12 (Bank 1 Sensor 2) Heater Circuit Malfunction Conditions:** DTC P0136, P0138, P0139 and P0140 not set, engine started, system voltage over 11.0v, HO2S heater "on", Catalytic Converter temperature from 626-928°F, and the PCM detected the HO2S heater circuit resistance was out of normal range for 15 seconds during the CCM test of the Oxygen Sensor circuit. **Possible Causes** ● HO2S heater ground circuit is open or has high resistance ● HO2S heater power circuit is open (test O2S fuse in fuse block) ● HO2S heater element is damaged or has failed ● PCM has failed

OBD II Trouble Code List (P0xxx Codes)

DTC	Trouble Code Title, Conditions & Possible Causes
P0141 **2T O2S HTR, MIL: Yes** 1997, 1998, 1999, 2000, 2001, 2002, 2003, 2004, 2005 Corvette 5.7L VIN G, 5.7L VIN S Transmissions: All	**HO2S-12 (Bank 1 Sensor 2) Heater Circuit Malfunction Conditions:** DTC P0101, P0102, P0103, P0106, P0107, P0108, P0112, P0113, P0116, P0117, P0118, P0131, P0132, P0134, P0137, P0138, P0140, P0151, P0152, P0154, P0157, P0158, P0160, P0200, P0300, P0410, P0440, P0442, P0446, P0452, P0453, P1120, P1125, P1220, P1221, P1221, P1258, P1415, P1416, P1441, P1514, P1515, P1516, P1517 and P1518 not set, HO2S signal from 425-475 mv at startup, ECT and IAT sensors less than 122°F and within 14.5°F at startup, MAF sensor less than 21 g/sec, system voltage 9-18v, AIR and Catalyst Tests off, and the PCM detected the HO2S signal remained within 150 mv of the bias voltage for too long a period (depends on the ECT/MAF sensor signals at startup). **Possible Causes** ● HO2S heater ground circuit is open or has high resistance ● HO2S heater power circuit is open (test O2S fuse in fuse block) ● HO2S heater element is damaged or has failed ● PCM has failed
P0143 **2T CCM, MIL: Yes** 1996, 1997, 1998, 1999, 2000 C/K Cab & Chassis, C/K Series Truck, G Series Van, M/L Series Vans 4.3L VIN W, 5.0L VIN M, 5.7L VIN K, 5.7L VIN R, 7.4L VIN J Transmissions: All	**HO2S-13 (Bank 1 Sensor 3) Circuit Low Input Conditions:** DTC P0101-P0103, P0106-P0108, P0112, P0113, P0116-P0118, P0121-P0123, P0131-P0138, P0140, P0151-P0158, P0160, P0200, P0300, P0401, P0404, P0405, P0440, P0442, P0446, P0452, P0453, P1120, P1125, P1220, P1221, P1258, P1404, P1441, P1514-P1517 and P1518 not set, engine started, engine running in closed loop, A/F ratio from 14.5-14.8:1, TP angle from 3.5-99%, Catalyst and EGR Tests all "off" for over 5 seconds, system voltage over 10.0v, then with the Lean Test enabled, the PCM detected the HO2S-13 signal was less than 26 mv for 110 seconds or with the Power Enrichment Lean Test enabled and Fuel Cutoff Mode "off", time since the test started over 1 second, the PCM detected the HO2S-13 signal was under 399 mv for 40 seconds. **Possible Causes** ● Air leaks in the exhaust system, intake manifold, vacuum lines ● Engine misfire condition present (look for P0300 series codes) ● Fuel system too lean (possible low fuel pressure, water in fuel) ● HO2S signal circuit is shorted to the sensor or chassis ground ● HO2S is damaged (i.e., cracked) or air reference hole clogged ● PCM has failed
P0143 **2T CCM, MIL: Yes** 1996, 1997, 1998, 1999 DeVille, Eldorado, Seville 4.6L VIN 9, 4.6L VIN Y Transmissions: A/T	**HO2S-13 (Bank 1 Sensor 3) Circuit Low Input Conditions:** DTC P0101-P0103, P0107, P0108, P0118, P0121, P0122, P0123, P0131-P0134, P0151-P0154, P0300, P1133, P1134, P1153 and P1154 not set, engine started, ECT sensor more than 167°F, engine speed over 800 rpm, HO2S-13 signal varying while in closed loop, TP angle from 5-25%, MAP sensor more than 32 kPa, and the PCM detected the HO2S-13 signal was less than 0.49 mv for 2 minutes. **Possible Causes** ● Air leaks in the exhaust system, intake manifold, vacuum lines ● Engine misfire condition present (look for P0300 series codes) ● Fuel system too lean (possible low fuel pressure, water in fuel) ● HO2S signal circuit is shorted to the sensor or chassis ground ● HO2S is damaged (i.e., cracked) or air reference hole clogged ● PCM has failed
P0143 **2T CCM, MIL: Yes** 1995, 1996, 1997, 1998, 1999 Camaro & Firebird 3.8L VIN K engine Transmissions: All	**HO2S-13 (Bank 1 Sensor 3) Circuit Low Input Conditions:** DTC P0101-P0103, P0107-P0108, P0112-P0113, P0117-P0118, P0121-P0125, P0128, P0201-P0206, P0410, P0440-P0449 and P1441 not set, engine running in closed loop, A/F ratio from 14.5-14.8:1, TP angle from 3-40%, ECT sensor over 122°F, and the PCM detected the HO2S signal was under 175 mv or it was under 600 mv during P/E mode for 15 seconds. **Possible Causes** ● Air leaks in the exhaust system, intake manifold, vacuum lines ● Engine misfire condition present (look for P0300 series codes) ● Fuel system too lean (possible low fuel pressure, water in fuel) ● HO2S signal circuit is shorted to the sensor or chassis ground ● HO2S is damaged (i.e., cracked) or air reference hole clogged ● PCM has failed
P0143 **2T CCM, MIL: Yes** 1996, 1997, 1998, 1999 Aurora 4.0L V8 VIN C engine Transmissions: A/T	**HO2S-13 (Bank 1 Sensor 3) Circuit Low Input Conditions:** DTC P0101-P0103, P0107-P0108, P0117-P0118, P0121-P0123, P0132, and P0134, not set, engine started, engine speed over 800 rpm, TP angle from 5.4-25 degrees, ECT sensor more than 179°F, MAP sensor more than 32 kPa for over 3 seconds and the PCM detected the HO2S signal was less than 490 mv for 100 seconds. **Possible Causes** ● Air leaks in the exhaust system, intake manifold, vacuum lines ● Engine misfire condition present (look for P0300 series codes) ● Fuel system too lean (possible low fuel pressure, water in fuel) ● HO2S signal circuit is shorted to the sensor or chassis ground ● HO2S is damaged (i.e., cracked) or air reference hole clogged ● PCM has failed

OBD II Trouble Code List (P0xxx Codes)

DTC	Trouble Code Title, Conditions & Possible Causes
P0143 **2T CCM, MIL: Yes** 1995, 1996, 1997, 1998, 1999, 2000 S/T Blazer & S/T Pickup 4.3L VIN W, 4.3L VIN X Transmissions: All	**HO2S-13 (Bank 1 Sensor 3) Circuit Low Input Conditions:** DTC P0101-P0103, P0106-P0108, P0112, P0113, P0116-P0118, P0121-P0123, P0200, P0300, P0440-P0446, P0452, P0453 and P1441 not set, engine started, fuel level over 10%, Scan Tool Intrusive Tests off, then with the Lean Test enabled, A/F ratio from 14.5-14.7:1, TP angle from 3-70% for 5 seconds, the PCM detected the HO2S signal was less than 86 mv for 50 seconds or with Power Enrichment mode active for 2 seconds, the PCM detected the HO2S signal was less than 598 mv for 30 seconds. **Possible Causes** ● Air leaks in the exhaust system, intake manifold, vacuum lines ● Engine misfire condition present (look for P0300 series codes) ● Fuel system too lean (possible low fuel pressure, water in fuel) ● HO2S signal circuit is shorted to the sensor or chassis ground ● HO2S is damaged (i.e., cracked) or air reference hole clogged ● PCM has failed
P0144 **2T CCM, MIL: Yes** 1996, 1997, 1998, 1999, 2000 C/K Cab & Chassis, C/K Series Truck, G Series Van, M/L Series Vans 4.3L VIN W, 5.0L VIN M, 5.7L VIN R, 7.4L VIN J Transmissions: Alls	**HO2S-13 (Bank 1 Sensor 3) Circuit High Input Conditions:** DTC P0101-P0103, P0106-P0108, P0112, P0113, P0116-P0118, P0121-P0123, P0131-P0138, P0140, P0151, P0158, P0160, P0200, P0300, P0401, P0404, P0405, P0440, P0442, P0446, P0452, P0453, P1120, P1125, P1220, P1221, P1258, P1404, P1441, P1514-P1517 and P1518 not set, engine started, A/F ratio from 14.5-14.8:1, TP angle from 0-50%, Catalyst and EGR Tests off, and the PCM detected the HO2S-13 signal was more than 994 mv for 110 seconds or with Fuel Cutoff Mode enabled for 2 seconds, the PCM detected the HO2S-13 signal was over 469 mv for 40 seconds. **Possible Causes** ● Fuel system is too rich (fuel pressure too high, fuel pressure regulator leaking, or one or more fuel injectors sticking/leaking) ● HO2S element is silicon, water or fuel contaminated ● HO2S signal moisture tracking in the connector to signal circuit ● PCM has failed
P0144 **2T CCM, MIL: Yes** 1996, 1997, 1998, 1999 DeVille, Eldorado, Seville 4.6L VIN 9, 4.6L VIN Y Transmissions: All	**HO2S-13 (Bank 1 Sensor 3) Circuit High Input Conditions:** DTC P0101-P0103, P0107, P0108, P0118, P0121, P0122, P0123, P0131-P0134, P0151-P0154, P0300, P1133, P1134, P1153 and P1154 not set, ECT sensor from 167-179°F, engine speed over 800 rpm, MAP sensor more than 32 kPa, TP angle 5-25%, closed loop enabled and HO2S ready for 3 seconds and the PCM detected the HO2S signal was 534 mv or more for 100 out of 120 seconds. **Possible Causes** ● Fuel system is too rich (fuel pressure too high, fuel pressure regulator leaking, or one or more fuel injectors sticking/leaking) ● HO2S element is silicon, water or fuel contaminated ● HO2S signal moisture tracking in the connector to signal circuit ● PCM has failed
P0144 **2T CCM, MIL: Yes** 1995, 1996, 1997, 1998, 1999 Camaro & Firebird 3.8L VIN K engine Transmissions: All	**HO2S-13 (Bank 1 Sensor 3) Circuit High Input Conditions:** P0121-P0125, P0201-P0206, P0410, P0440-P0449 and P1441 not set, engine running in closed loop, TP angle at 3-40%, ECT sensor over 122°F, and the PCM detected the HO2S input was over 999 mv, or while in Decel Fuel mode, it was over 200 mv for 2 minutes. **Possible Causes** ● Fuel system is too rich (fuel pressure too high, fuel pressure regulator leaking, or one or more fuel injectors sticking/leaking) ● HO2S element is silicon, water or fuel contaminated ● HO2S signal moisture tracking in the connector to signal circuit ● PCM has failed
P0144 **2T CCM, MIL: Yes** 1996, 1997, 1998, 1999 Aurora 4.0L V8 VIN C engine Transmissions: A/T	**HO2S-13 (Bank 1 Sensor 3) Circuit High Input Conditions:** DTC P0101, P0102, P0103, P0107, P0108, P0117, P0118, P0121, P0122, P0123, P0131, P0132, P0133, P0134, P0151, P0152, P0153, P0154, P0300, P1133, P1134, P1153 and P1154 not set, speed over 800 rpm in closed loop, ECT sensor more than 179°F, HO2S ready to test, engine MAP sensor over 32 kPa, TP angle from 5.4-25 degrees for 3 seconds, and the PCM detected the HO2S signal was less than 900 mv for 100 out of 120 seconds. **Possible Causes** ● Fuel system is too rich (fuel pressure too high, fuel pressure regulator leaking, or one or more fuel injectors sticking/leaking) ● HO2S element is silicon, water or fuel contaminated ● HO2S signal moisture tracking in the connector to signal circuit ● PCM has failed

OBD II Trouble Code List (P0xxx Codes)

DTC	Trouble Code Title, Conditions & Possible Causes
P0144 **2T CCM, MIL: Yes** 1995, 1996, 1997, 1998, 1999, 2000 S/T Blazer & S/T Pickup 4.3L VIN W, 4.3L VIN X Transmissions: All	**HO2S-13 (Bank 1 Sensor 3) Circuit High Input Conditions:** DTC P0101-P0103, P0106-P0108, P0112, P0113, P0116-P0118, P0121-P0123, P0200, P0300, P0440-P0446, P0452, P0453 and P1441 not set, engine started, A/F ratio from 14.5-14.8:1, TP angle from 0-50%, Scan Tool Intrusive Tests off for 5 seconds, then with the Rich Test enabled, the PCM/VCM detected the HO2S was more than 994 mv for 110 seconds or with Decel Fuel Cutoff Rich Test enabled for 2 seconds, the PCM detected the HO2S was more than 468 mv for 40 seconds. **Possible Causes** ● Fuel system rich (high fuel pressure, fuel pressure regulator leaking or leaking injector) ● HO2S element is silicon, water, fuel contaminated or shorted to power in the connector ● HO2S signal circuit is shorted to system power (B+) ● PCM has failed
P0146 **2T O2S, MIL: Yes** 1996, 1997, 1998, 1999, 2000 C/K Cab & Chassis, C/K Series Truck, G Series Van, M/L Series Vans 4.3L VIN W, 5.0L VIN M, 5.7L VIN K, 5.7L VIN R, 7.4L VIN J Transmissions: All	**HO2S-13 (Bank 1 Sensor 3) Insufficient Activity Conditions:** DTC P0101-P0103, P0106-P0108, P0112, P0113, P0116-P0118, P0121-P0123, P0200, P0300, P0440-P0446, P0452, P0453 and P1441 not set, engine speed at 1100-3000 rpm for 2 minutes, ECT sensor over 137ºF, A/F ratio from 14.5-14.8:1, TP angle from 0-50%, MAF sensor over 13 g/sec, Device Controls and Scan Tool Tests "off" for 5 seconds, DTC P0147 not testing, and the PCM detected the HO2S signal was from 399-473 mv for 100 seconds. **Possible Causes** ● HO2S heater is damaged or has failed ● HO2S signal or ground circuit has a high resistance condition ● HO2S signal circuit is open or shorted to system power (B+) ● HO2S has failed (i.e., it is silicon, water or fuel contaminated) ● PCM has failed
P0146 **2T O2S, MIL: Yes** 1996, 1997, 1998, 1999 DeVille, Eldorado, Seville 4.6L VIN 9, 4.6L VIN Y Transmissions: All	**HO2S-13 (Bank 1 Sensor 3) Insufficient Activity Conditions:** DTC P0118, P0118, P0122 and P0123 not set, engine speed from 800-3000 rpm, ECT more than 167ºF, TP angle from 2-81.6%, Throttle Position Switch open, and the PCM detected the HO2S input was steady at 307-609 mv for 250 of 300 seconds (HO2S signal is toggling). **Possible Causes** ● HO2S heater is damaged or has failed ● HO2S signal or ground circuit has a high resistance condition ● HO2S signal circuit is open or shorted to system power (B+) ● HO2S has failed (i.e., it is silicon, water or fuel contaminated) ● PCM has failed
P0146 **2T O2S, MIL: Yes** 1995, 1996, 1997, 1998, 1999 Camaro & Firebird 3.8L VIN K engine Transmissions: All	**HO2S-13 (Bank 1 Sensor 3) Insufficient Activity Conditions:** DTC P0101-P0103, P0107, P0108, P0112, P0113, P0117, P0118, P0121-P0123, P0125, P0171-P0175, P0300, P0336, P0401, P0440, P0442, P0446 and P0449 not set, engine runtime over 4 minutes, and the PCM detected the HO2S signal was fixed from 425-475 mv. **Possible Causes** ● HO2S heater is damaged or has failed ● HO2S signal or ground circuit has a high resistance condition ● HO2S signal circuit is open or shorted to system power (B+) ● HO2S has failed (i.e., it is silicon, water or fuel contaminated) ● PCM has failed
P0146 **2T O2S, MIL: Yes** 1996, 1997, 1998, 1999 Aurora 4.0L V8 VIN C engine Transmissions: A/T	**HO2S-13 (Bank 1 Sensor 3) Insufficient Activity Conditions:** DTC P0117, P0118, P0122 and P0123 not set, engine started, engine speed from 800-3000 rpm, ECT sensor more than 167ºF, TP angle from 2-81.6%, Calculated Throttle Position Switch open, engine speed 800-3000 rpm, and the PCM detected the HO2S signal was between 307-609 mv for 250 out of 300 seconds (i.e., the HO2S signal not toggling high-low). **Possible Causes** ● HO2S heater is damaged or has failed ● HO2S signal or ground circuit has a high resistance condition ● HO2S signal circuit is open or shorted to system power (B+) ● HO2S has failed (i.e., it is silicon, water or fuel contaminated) ● PCM has failed
P0146 **2T O2S, MIL: Yes** 1995, 1996, 1997, 1998, 1999, 2000 S/T Blazer & S/T Pickup 4.3L VIN W, 4.3L VIN X Transmissions: All	**HO2S-13 Insufficient Activity (Bank 1 Sensor 3) Conditions:** DTC P0101-P0103, P0106-P0108, P0112, P0113, P0116-P0118, P0121-P0123, P0200, P0300, P0440-P0446, P0452, P0453 and P1441 not set, engine runtime over 2 minutes, ECT sensor over 137ºF, MAF sensor over 13 g/sec, Device Controls Tests "off", O2 Temperature Test True, the PCM detected the HO2S signal was 399-473 mv for 2 minutes. **Possible Causes** ● HO2S heater is damaged or has failed ● HO2S signal or ground circuit has a high resistance condition ● HO2S signal circuit is open or shorted to system power (B+) ● HO2S has failed (i.e., it is silicon, water or fuel contaminated) ● PCM has failed

OBD II Trouble Code List (P0xxx Codes)

DTC	Trouble Code Title, Conditions & Possible Causes
P0147 **2T HTR O2S, MIL: Yes** 1996, 1997, 1998, 1999, 2000 C/K Cab & Chassis, C/K Series Truck, G Series Van, M/L Series Vans 4.3L VIN W, 5.0L VIN M, 5.7L VIN K, 5.7L VIN R, 7.4L VIN J Transmissions: All	**HO2S-13 (Bank 1 Sensor 3) Heater Circuit Malfunction Conditions:** DTC P0101-P0103, P0106-P0108, P0112, P0113, P0116-P0118, P0121-P0123, P0200, P0300, P0440-P0446, P0452, P0453 and P1441 not set, ECT and IAT sensors less than 91°F and within 9°F at startup, engine runtime over 2 seconds, system voltage over 10.0v, MAF sensor below 27-35 g/sec, and the PCM detected the HO2S signal remained within 150 mv of the startup HO2S signal voltage for 245-270 seconds after a cold engine startup. **Possible Causes** ● HO2S heater ground circuit is open, or HO2S heater power circuit is open (O2S fuse) ● HO2S heater element has failed, or the PCM has failed ● TSB 01-06-04-004 contains a repair procedure for this code
P0147 **2T HTR O2S, MIL: Yes** 1996, 1997, 1998, 1999 DeVille, Eldorado, Seville 4.6L VIN 9, 4.6L VIN Y Transmissions: A/T	**HO2S-13 (Bank 1 Sensor 3) Heater Circuit Conditions:** DTC P0101, P0102, P0103, P0117, P0118 and P0146 not set, ECT has dropped at least 50°F from the end of last key cycle to the current key cycle, engine did not stall, average HO2S bias voltage from 352-546 mv, and the PCM detected the HO2S signal required to long a time (over 100 ms) to change from 151 mv above or below the HO2S bias voltage. **Possible Causes** ● HO2S heater ground circuit is open, or HO2S heater power circuit is open (O2S fuse) ● HO2S heater element is damaged or has failed ● PCM has failed
P0147 **2T HTR O2S, MIL: Yes** 1995, 1996, 1997, 1998, 1999 Camaro & Firebird 3.8L VIN K engine Transmissions: All	**HO2S-13 (Bank 1 Sensor 3) Heater Circuit Conditions:** DTC P0101-P0103, P0107, P0108, P0112, P0113, P0117, P0118, P0121-P0123, P0125, P0171, P0172, P0174, P0175, P0300, P0336, P0401, P0440, P0442, P0446 and P0449 not set, engine started, ECT and IAT sensor less than 95°F and within 11°F at startup, engine running with the average MAF for sample period less than 20 g/sec, and the PCM detected the HO2S signal remained within ±150 mv of the bias voltage (about 450 mv) for too long a period of time (depends on the ECT/MAF sensor signals at startup). **Possible Causes** ● HO2S heater ground circuit is open or has high resistance ● HO2S heater power circuit is open (test O2S fuse in fuse block) ● HO2S heater element is damaged or has failed ● PCM has failed
P0147 **2T HTR O2S, MIL: Yes** 1996, 1997, 1998, 1999 Aurora 4.0L V8 VIN C engine Transmissions: A/T	**HO2S-13 (Bank 1 Sensor 3) Heater Circuit Conditions:** DTC P0101, P0102, P0103, P0117, P0118 and P0146 not set, then with the ECT sensor 50°F lower than it was after the last key cycle, and with the average HO2S-13 bias voltage at 352-546 mv, engine did not stall, the PCM detected the HO2S-13 signal was either 151 mv less than or 151 mv more than the average HO2S-13 bias voltage for over 100 milliseconds. **Possible Causes** ● HO2S heater ground circuit is open or has high resistance ● HO2S heater power circuit is open (test O2S fuse in fuse block) ● HO2S heater element is damaged or has failed ● PCM has failed
P0147 **2T HTR O2S, MIL: Yes** 1995, 1996, 1997, 1998, 1999, 2000 S/T Blazer & S/T Pickup 4.3L VIN W, 4.3L VIN X Transmissions: All	**HO2S-13 (Bank 1 Sensor 3) Heater Circuit Conditions:** DTC P0101-P0103, P0106-P0108, P0112, P0113, P0116-P0118, P0121-P0123, P0200, P0300, P0440-P0446, P0452, P0453 and P1441 not set, engine started, ECT and IAT sensors less than 91°F and within 9°F at startup, engine runtime over 2 seconds, system voltage over 10.0v, MAF sensor under 27 g/sec, then the PCM detected the HO2S-13 signal remained within 150 mv of the startup HO2S bias voltage for 245 seconds after startup. **Possible Causes** ● HO2S heater ground circuit is open or has high resistance ● HO2S heater power circuit is open (test O2S fuse in fuse block) ● HO2S heater element is damaged or has failed ● PCM has failed
P0148 **2T CCM, MIL: Yes** 2003, 2004, 2005 C/K Series Truck, G Series Van, Cargo Van 6.0L VIN U CNG engine Transmissions: All	**A/F Enable Circuit Malfunction Conditions:** Key on or engine running, system voltage from 6-18v, and the PCM detected the Actual and Commanded state of the AF Fuel Enable circuit did not match for over two seconds during the test. The PCM opens the AF enable circuit when operating on CNG. When the AF enable circuit is open, the fuel injector control module (FICM) operates the CNG injectors based upon PCM fuel injector control pulse width signals. The PCM grounds the AF enable circuit when gasoline operation is desired. The switchover from one fuel to the other is always performed in an orderly, sequential manner. Since some injectors are in the middle of injecting the previous fuel, the FICM will wait until that cylinders fuel delivery is complete and then will switch over in sequential firing order to complete the operation. **Possible Causes** ● AF enable circuit is open, shorted to ground ● AF enable circuit is shorted to system power (B+) ● FICM connector is damaged, open or shorted ● FICM assembly had failed, or the PCM has failed

OBD II Trouble Code List (P0xxx Codes)

DTC	Trouble Code Title, Conditions & Possible Causes
P0150 **2T O2S, MIL: Yes** 2003, 2004, 2005 CTS (Cadillac) 2.6L V6 VIN M & 3.2L V6 VIN N engines Transmissions: All	**HO2S-21 (Bank 2 Sensor 1) Circuit Closed Loop Conditions:** DTC P0443, P0444 and P0445 not set, engine started, TP angle from 5-35%, system voltage over 10.5v, MAF sensor at 10-35 g/sec, Calculated Converter Temperature under 1,472°F, and the PCM detected the HO2S-21 signal was 60-400 mv while the HO2S-22 signal was over 499 mv, or the HO2S-21 signal was 600-1080 mv while the HO2S-22 signal was less than 104 mv for 10 seconds. **Possible Causes** ● EVAP system leaks or restrictions, charcoal canister problems ● EVAP system leaks or restrictions, charcoal canister problems ● HO2S air reference hole plugged (check for dirt on the outside) ● HO2S signal circuit is shorted to sensor or chassis ground ● HO2S is damaged or contaminated due to improper fuel usage ● PCM has failed
P0150 **2T O2S, MIL: Yes** 2000, 2001, 2002, 2003, 2004, 2005 Body Codes: E, G, K 4.0L V8 VIN C, 4.6L VIN 9, 4.6L VIN Y Transmissions: All	**HO2S-21 (Bank 2 Sensor 1) Closed Loop Performance Conditions:** DTC P0101-P0103, P0106-P0108, P0116-P0118, P0121-P0123, P0125, P0131, P0132, P0133, P0134, P0151, P0152, P0153, P0154, P0201-P0208, P1133, P1134, P1153 and P1154 not set, engine speed from 500-5000 rpm in closed loop for 2 minutes, ECT sensor more than 176°F, HO2S heater temperature over 1112°F, Catalyst, DFCO, Traction Control and Power Enrichment mode "off", MAF sensor from 3-30 g/sec for 5 seconds, and the PCM detected the HO2S signal remained fixed from 350-500 mv for 16 seconds. **Possible Causes** ● Air leaks in the exhaust system, intake manifold, vacuum lines ● EVAP Purge system malfunction or charcoal canister problems ● HO2S signal circuit is open between the sensor and the PCM ● HO2S signal circuit is shorted to sensor or chassis ground ● HO2S is damaged, contaminated or air reference hole clogged ● PCM has failed
P0150 **2T O2S, MIL: Yes** 2000, 2001, 2002 Camaro & Firebird 3.8L VIN K engine Transmissions: All	**HO2S-21 (Bank 2 Sensor 1) Closed Loop Performance Conditions:** DTC P0101-P0103, P0107, P0108, P0112-P0113, P0117-P0118, P0121-P0123, P0125, P0128, P0201-P0206, P0410, P0440, P0442, P0446, P0449 and P1441 not set; engine runtime 4 minutes, engine speed at 1000-3000 rpm, ECT sensor over 122°F, airflow from 13-30 g/sec, APP sensor from 5-40%, and the PCM detected the voltage amplitude was wrong. **Possible Causes** ● Air leaks in the exhaust system, intake manifold, vacuum lines ● EVAP Purge system malfunction or charcoal canister problems ● HO2S signal circuit is open between the sensor and the PCM ● HO2S signal circuit is shorted to sensor or chassis ground ● HO2S is damaged, contaminated or air reference hole clogged ● PCM has failed
P0150 **2T O2S, MIL: Yes** 1997, 1998, 1999, 2000, 2001 Catera 3.0L V6 VIN R engine Transmissions: A/T	**HO2S-21 (Bank 2 Sensor 1) Closed Loop Performance Conditions:** DTC P0135, P0139, P0140 and P0141 not set, engine speed from 1400-2400 rpm, engine load at 8-12%, MAF sensor at 18-22 g/sec, TP angle at 6-10%, and the PCM detected the HO2S-21 signal exceeded 4.78v, or with HO2S-21 from 0.9-1v for 4 seconds, the HO2S-22 signal was over 650 mv or below 300 mv, or with HO2S-21 over 1.2v, the HO2S-22 was over 650 mv, or with HO2S-21 less than 0.80v, the HO2S-22 was less than 300 mv during testing. **Possible Causes** ● Air leaks in the exhaust system, intake manifold, vacuum lines ● EVAP Purge system malfunction or charcoal canister problems ● HO2S signal circuit is open between the sensor and the PCM ● HO2S signal circuit is shorted to sensor or chassis ground ● HO2S is damaged, contaminated or air reference hole clogged ● PCM has failed
P0151 **2T CCM, MIL: Yes** 1996, 1997, 1998, 1999, 2000, 2001, 2002, 2003, 2004, 2005 C/K Cab & Chassis, C/K Series Truck, G Series Van, M/L Series Vans 4.3L VIN W, 4.3L VIN X, 5.0L VIN M, 5.7L VIN K, 5.7L VIN R, 7.4L VIN J Transmissions: All	**HO2S-21 (Bank 2 Sensor 1) Circuit Low Input Conditions:** DTC P0101-P0103, P0106-P0108, P0112, P0113, P0116-P0118, P0121-P0123, P0200, P0300, P0401, P0404, P0405, P0440, P0442, P0446, P0452, P0453, P1120, P1125, P1220, P1221, P1258, P1404, P1441, P1514, P1515, P1516, P1517 and P1518 not set, engine started, engine running in closed loop, fuel level over 10%, system voltage over 10v, TP angle from 8-50% or on models with TAC, the APP sensor indicated angle from 3-70%, MAP sensor more than 25 kPa, Intrusive and Scan Tool Tests "off", then with the Lean Test enabled, the PCM detected the HO2S signal was less than 20 mv for 50 seconds or during the P/E Mode test, the PCM detected the HO2S signal was less than 360 mv for 10 seconds. **Possible Causes** ● Air leaks in the exhaust system, intake manifold, vacuum lines ● Engine misfire condition present (look for P0300 series codes) ● Fuel system too lean (possible low fuel pressure, water in fuel) ● HO2S signal circuit is shorted to the sensor or chassis ground ● HO2S is damaged (i.e., cracked) or air reference hole clogged ● PCM has failed

OBD II Trouble Code List (P0xxx Codes)

DTC	Trouble Code Title, Conditions & Possible Causes
P0151 **2T CCM, MIL: Yes** 1996, 1997 Camaro, Corvette, Caprice, Firebird & Fleetwood 4.3L VIN W, 5.7L VIN 5, 5.7L VIN P engines Transmissions: All	**HO2S-21 (Bank 2 Sensor 1) Circuit Low Input Conditions:** DTC P0101-P0103, P0106-P0108, P0112-P0113, P0116-P0118, P0121-P0123, P0125, P0128, P0200, P0300, P0410-P0446, P0452-P0453, P1258, P1415-P1416 and P1441 not set engine running in closed loop, ECT sensor over 118°F, TP angle at 3-20%, and the PCM detected the HO2S signal was below 200 mv for 30 seconds. **Possible Causes** ● Air leaks in the exhaust system, intake manifold, vacuum lines ● Engine misfire condition present (look for P0300 series codes) ● Fuel system too lean (possible low fuel pressure, water in fuel) ● HO2S signal circuit is shorted to the sensor or chassis ground ● HO2S is damaged (i.e., cracked) or air reference hole clogged ● PCM has failed
P0151 **2T CCM, MIL: Yes** 1999, 2000, 2001, 2002, 2003, 2004, 2005 O/K Series Truck, G Series Van, Envoy, Escalade, TrailBlazer 4.8L VIN V, 5.3L VIN P, 5.3L VIN T, 5.3L VIN Z, 6.0L VIN N, 6.0L VIN U, 8.1L VIN G Transmissions: All	**HO2S-21 (Bank 2 Sensor 1) Circuit Low Input Conditions:** DTC P0101-P0103, P0106-P0108, P0112, P0113, P0116-P0118, P0120, P0121-P0123, P0169, P0178, P0179, P0200, P0220, P0300, P0442, P0446, P0452, P0453, P0455, P0496, P1125, P1258, P1514, P1515, P1516, P1518, P2108 and P2135 not set, engine running in closed loop, system voltage from 10-18v, Fuel Alcohol content under 90%, fuel level over 10%, TP angle from 3-70% over the idle value, Lean Test enabled, the PCM detected the HO2S signal was below 200 mv for 2 minutes or with engine runtime over 30 seconds during the P/E test period, the PCM detected the HO2S input was under 400 mv for 10 seconds. **Possible Causes** ● Air leaks in the exhaust system, intake manifold, vacuum lines ● Engine misfire condition present (look for P0300 series codes) ● Fuel system too lean (possible low fuel pressure, water in fuel) ● HO2S signal circuit is shorted to the sensor or chassis ground ● HO2S is damaged (i.e., cracked) or air reference hole clogged ● PCM has failed
P0151 **2T CCM, MIL: Yes** 1996, 1997, 1998, 1999, 2000, 2001, 2002, 2003, 2004, 2005 DeVille, Eldorado, Seville 4.6L VIN 9, 4.6L VIN Y Transmissions: All	**HO2S-21 (Bank 2 Sensor 1) Circuit Low Input Conditions:** DTC P0101-P0103, P0106-P0108, P0112-P0113, P0116-P0118, P0121-P0123, P0125, P0128, P0200, P0300, P0410-P0446, P0452-P0453, P1258, P1415-P1416 and P1441 not set, engine running in closed loop for 3 seconds, AIR, Catalyst and EGR Flow tests "off", system voltage 8-18v, TP angle from 3-25%, no injectors disabled, and the PCM detected the HO2S signal was below 75 mv for 5 seconds, or below 575 mv for 8 seconds in the P/E test. **Possible Causes** ● Air leaks in the exhaust system, intake manifold, vacuum lines ● Engine misfire condition present (look for P0300 series codes) ● Fuel system too lean (possible low fuel pressure, water in fuel) ● HO2S signal circuit is shorted to the sensor or chassis ground ● HO2S is damaged (i.e., cracked) or air reference hole clogged ● PCM has failed
P0151 **2T CCM, MIL: Yes** 2003, 2004, 2005 CTS (Cadillac) 2.6L V6 VIN M & 3.2L V6 VIN N engines Transmissions: All	**HO2S-21 (Bank 2 Sensor 1) Circuit Low Input Conditions:** DTC P0443, P0444 and P0445 not set, engine started, TP angle from 5-35%, system voltage over 10.5v, MAF sensor at 10-35 g/sec, Calculated Converter Temperature under 1,472°F, and the PCM detected the HO2S-21 signal was less than 40 mv for 10 seconds. **Possible Causes** ● EVAP system leaks or restrictions, charcoal canister problems ● Fuel system too lean (possible low fuel pressure, water in fuel) ● HO2S air reference hole plugged (check for dirt on the outside) ● HO2S signal circuit is shorted to sensor or chassis ground ● HO2S is damaged or contaminated due to improper fuel usage ● PCM has failed
P0151 **2T CCM, MIL: Yes** 1995, 1996, 1997, 1998, 1999, 2000, 2001, 2002 Camaro & Firebird 3.8L VIN K engine Transmissions: All	**HO2S-21 (Bank 2 Sensor 1) Circuit Low Input Conditions:** DTC P0101-P0103, P0107-P0108, P0112-P0113, P0117-P0118, P0121-P0125, P0128, P0201-P0206, P0410, P0440-P0449 and P1441 not set, running in closed loop for 4 minutes, A/F ratio from 14.5-14.8:1, TP angle from 3-40%, or on models with TAC, the APP sensor indicated angle from 5-40%, ECT sensor more than 122°F, and the PCM detected the HO2S signal was less than 175 mv, or it was less than 600 mv while in P/E mode for up to 15 seconds. **Possible Causes** ● Air leaks in the exhaust system, intake manifold, vacuum lines ● Fuel system too lean (possible low fuel pressure, water in fuel) ● HO2S signal circuit is shorted to the sensor or chassis ground ● HO2S is damaged (i.e., cracked) or air reference hole clogged ● PCM has failed

OBD II Trouble Code List (P0xxx Codes)

DTC	Trouble Code Title, Conditions & Possible Causes
P0151 **2T CCM, MIL: Yes** 1998, 1999, 2000, 2001, 2002 Camaro & Firebird 5.7L VIN G engine Transmissions: All	**HO2S-21 (Bank 2 Sensor 1) Circuit Low Input Conditions:** DTC P0101-P0103, P0106-P0108, P0112-P0113, P0116-P0118, P0121-P0123, P0125, P0128, P0200, P0300, P0410-P0446, P0452-P0453, P1258, P1415-P1416 and P1441 not set, engine started, engine running in closed loop, Intrusive and Scan Tool tests "off", A/F ratio at 14.5-14.8:1, TP angle from 3-70%, then with the Lean Test enabled, the PCM detected the HO2S signal was less than 200 mv for 165 seconds or with P/E test enabled for 1 second, the PCM detected the HO2S signal was less than 360 mv for 10 seconds. **Possible Causes** ● Air leaks in the exhaust system, intake manifold, vacuum lines ● Engine misfire condition present (look for P0300 series codes) ● Fuel system too lean (possible low fuel pressure, water in fuel) ● HO2S signal circuit is shorted to the sensor or chassis ground ● HO2S is damaged (i.e., cracked) or air reference hole clogged ● PCM has failed
P0151 **2T CCM, MIL: Yes** 1996, 1997, 1998, 1999 Aurora 4.0L V8 VIN C engine Transmissions: A/T	**HO2S-21 (Bank 2 Sensor 1) Circuit Low Input Conditions:** DTC P0101-P0103, P0107-P0108, P0117-P0118, P0121-P0123, P0132, and P0134, not set, engine started, engine speed over 800 rpm, TP angle from 5.4-25 degrees, ECT sensor over 179°F, MAP sensor above 32 kPa, DTC P0151 test passed, and the PCM detected the HO2S signal was under 356 mv for 45 seconds or less than 575 mv during P/E mode for 5 seconds. **Possible Causes** ● Air leaks in the exhaust system, intake manifold, vacuum lines ● Fuel system too lean (possible low fuel pressure, water in fuel) ● HO2S signal circuit is shorted to the sensor or chassis ground ● HO2S is damaged (i.e., cracked) or air reference hole clogged ● PCM has failed
P0151 **2T CCM, MIL: Yes** 2001, 2002, 2003, 2004, 2005 Aurora 4.0L V8 VIN C engine Transmissions: All	**HO2S-21 (Bank 2 Sensor 1) Circuit Low Input Conditions:** DTC P0101-P0103, P0106-P0108, P0112, P0113, P0116-P0118, P0121-P0123, P0125, P0128, P0201-P0208, P0410, P0412, P0418, P0419, P0440-P0449, P1415, P1416, P1441 not set, engine started, AIR and Catalyst Tests off, TP angle 3-25% in closed loop, and the PCM detected the HO2S signal was below 75 mv for 5 seconds. **Possible Causes** ● Air leaks in the exhaust system, intake manifold, vacuum lines ● Engine misfire condition present (look for P0300 series codes) ● Fuel system too lean (possible low fuel pressure, water in fuel) ● HO2S signal circuit is shorted to the sensor or chassis ground ● HO2S is damaged (i.e., cracked) or air reference hole clogged ● PCM has failed
P0151 **2T CCM, MIL: Yes** 1995, 1996, 1997, 1998, 1999, 2000, 2001, 2002, 2003, 2004, 2005 S/T Blazer & S/T Pickup 4.3L VIN W, 4.3L VIN X Transmissions: All	**HO2S-21 (Bank 2 Sensor 1) Circuit Low Input Conditions:** DTC P0101-P0103, P0106-P0108, P0112, P0113, P0116-P0118, P0121-P0123, P0200, P0300, P0440, P0442, P0446, P0452, P0453, P0496 and P1441 not set, engine started, engine running in closed loop, fuel level over 10%, Scan Tool tests off, then with the Lean Test enabled, TP angle from 3-70% for 5 seconds, the PCM detected the HO2S signal was less than 200 mv for 165 seconds or with Power Enrichment mode active for 2 seconds, the PCM detected the HO2S signal was less than 400 mv for 10 seconds. **Possible Causes** ● Air leaks in the exhaust system, intake manifold, vacuum lines ● Engine misfire condition present (look for P0300 series codes) ● Fuel system too lean (possible low fuel pressure, water in fuel) ● HO2S signal circuit is shorted to the sensor or chassis ground ● HO2S is damaged (i.e., cracked) or air reference hole clogged ● PCM has failed
P0151 **2T CCM, MIL: Yes** 1997, 1998, 1999, 2000, 2001 Catera 3.0L V6 VIN R engine Transmissions: All	**HO2S-21 (Bank 2 Sensor 1) Circuit Low Input Conditions:** DTC P0300-P0306, P0171, P0172, P0174, P0175, P0440, P0442, P0443, P0446, P0455 and P0460 are not set, engine started, engine speed 1400-2400 rpm, engine load from 8-12%, MAF sensor from 18-22 g/sec, TP angle from 6-10%, EVAP high loading not active, AIR test "off", and the PCM detected the HO2S signal was too low. **Possible Causes** ● Air leaks in the exhaust system, intake manifold, vacuum lines ● Engine misfire condition present (look for P0300 series codes) ● Fuel system too lean (possible low fuel pressure, water in fuel) ● HO2S signal circuit is shorted to the sensor or chassis ground ● HO2S is damaged (i.e., cracked) or air reference hole clogged ● PCM has failed

OBD II Trouble Code List (P0xxx Codes)

DTC	Trouble Code Title, Conditions & Possible Causes
P0151 **2T CCM, MIL: Yes** 1995 Corvette 5.7L VIN P Transmissions: All	**HO2S-21 (Bank 2 Sensor 1) Circuit Low Input Conditions:** DTC P0116, P0117, P0118, P0135, P0171, P0172, P0174, P0175, P1114 and P1115 not set, engine started, engine running in closed loop, TP angle from 2-10%, and the PCM detected the HO2S signal was less than 141 mv for 102 seconds. **Possible Causes** • Air leaks in the exhaust system, intake manifold, vacuum lines • Engine misfire condition present (look for P0300 series codes) • Fuel system too lean (possible low fuel pressure, water in fuel) • HO2S signal circuit is shorted to the sensor or chassis ground • HO2S is damaged (i.e., cracked) or air reference hole clogged • PCM has failed
P0151 **2T CCM, MIL: Yes** 1997, 1998, 1999, 2000, 2001, 2002, 2003, 2004, 2005 Corvette 5.7L VIN C, 5.7L VIN 3 Transmissions: All	**HO2S-21 (Bank 2 Sensor 1) Circuit Low Input Conditions:** DTC P0101-P0103, P0106-P0108, P0112-P0118, P0200, P0300, P0410, P0440-P0446, P0452-P0453, P1120, P1125, P1220, P1221, P1258, P1415-P1416, P1441, P1514-1518 not set, Scan Tool and Intrusive tests all "off", fuel level over 10%, TP angle from 3-70%, then with the Lean Test enabled, the PCM detected the HO2S signal was less than 200 mv for 165 seconds or with Power Enrichment mode active for 1 second, the PCM detected the HO2S signal was less than 360 mv for 10 seconds. **Possible Causes** • Air leaks in the exhaust system, intake manifold, vacuum lines • Engine misfire condition present (look for P0300 series codes) • Fuel system too lean (possible low fuel pressure, water in fuel) • HO2S signal circuit is shorted to the sensor or chassis ground • HO2S is damaged (i.e., cracked) or air reference hole clogged • PCM has failed
P0152 **1T CCM, MIL: Yes** 1996, 1997 Camaro, Corvette, Caprice, Firebird & Fleetwood 4.3L VIN W, 5.7L VIN 5, 5.7L VIN P engines Transmissions: All	**HO2S-21 (Bank 2 Sensor 1) Circuit High Input Conditions:** DTC P0101-P0103, P0112-P0118, P0125, P0200, P0335, P0336, P0351-P0358, P0371 and P1372 not set, engine started, AIR and Catalyst Tests inactive, TP angle from 2-70% in closed loop, and the PCM detected the HO2S signal was above 774 mv for 30 seconds. **Possible Causes** • Fuel system is too rich (fuel pressure too high, fuel pressure regulator leaking, or one or more fuel injectors sticking/leaking) • HO2S element is silicon, water or fuel contaminated • HO2S signal tracking in the connector causing a short to power • PCM has failed
P0152 **2T CCM, MIL: Yes** 1996, 1997, 1998, 1999, 2000, 2001, 2002, 2003, 2004, 2005 C/K Cab & Chassis, C/K Series Truck, G Series Van, M/L Series Vans 4.3L VIN W, 4.3L VIN X, 5.0L VIN M, 5.7L VIN K, 5.7L VIN R, 7.4L VIN J Transmissions: All	**HO2S-21 (Bank 2 Sensor 1) Circuit High Input Conditions:** DTC P0101-P0103, P0106-P0108, P0112, P0113, P0116-P0118, P0121-P0123, P0200, P0300, P0401, P0404, P0405, P0440, P0442, P0446, P0452, P0453, P1120, P1125, P1220, P1221, P1258, P1404, P1441, P1514-P1517 and P1518 not set, engine started, fuel level over 10%, Intrusive Tests all off, then with the Rich Test enabled, A/F ratio from 14.5-14.7:1, TP angle from 3.5-70% for 5 seconds (TP indicated angle at 3-70% on vehicles with TAC), the PCM detected the HO2S signal was over 775 mv for 165 seconds or it was more than 540 mv with DFCO enabled for over 5 seconds. **Possible Causes** • Fuel system is too rich (fuel pressure too high, fuel pressure regulator leaking, or one or more fuel injectors sticking/leaking) • HO2S element is silicon, water or fuel contaminated • HO2S signal tracking in the connector causing a short to power • PCM has failed
P0152 **2T CCM, MIL: Yes** 1999, 2000, 2001, 2002, 2003, 2004, 2005 C/K Series Truck, G Series Van, Envoy, Escalade, TrailBlazer 4.8L VIN V, 5.3L VIN P, 5.3L VIN T, 5.3L VIN Z, 6.0L VIN N, 6.0L VIN U, 8.1L VIN G Transmissions: All	**HO2S-21 (Bank 2 Sensor 1) Circuit High Input Conditions:** DTC P0101, P0102, P0103, P0106, P0107, P0108, P0112, P0113, P0116, P0117, P0118, P0120, P0121, P0122, P0123, P0169, P0178, P0179, P0200, P0220, P0300, P0442, P0446, P0452, P0453, P0455, P0496, P1125, P1258, P1514, P1515, P1516, P1518, P2108 and P2135 not set, engine started, engine running in closed loop, system voltage from 10-18v, Fuel Alcohol content less than 90%, fuel level over 10%, TP angle from 3-70% more than the idle value, then during the Rich Test period, the PCM detected the HO2S signal was more than 900 mv for 165 seconds or with engine runtime over 30 seconds, and during the Decel Fuel Cutoff test, the PCM detected the HO2S signal was less than 250 mv for 5 seconds. **Possible Causes** • Fuel system is too rich (fuel pressure too high, fuel pressure regulator leaking, or one or more fuel injectors sticking/leaking) • HO2S element is silicon, water or fuel contaminated • HO2S signal tracking in the connector causing a short to power • PCM has failed

OBD II Trouble Code List (P0xxx Codes)

DTC	Trouble Code Title, Conditions & Possible Causes
P0152 **2T CCM, MIL: Yes** 2003, 2004, 2005 CTS (Cadillac) 2.6L V6 VIN M & 3.2L V6 VIN N engines Transmissions: All	**HO2S-21 (Bank 2 Sensor 1) Circuit High Input Conditions:** Engine started, system voltage over 10.5v, Converter Temperature less than 1,472°F, and the PCM detected the HO2S-21 signal was more than 1,080 mv for over 5 seconds. **Possible Causes** ● Air leaks in the exhaust system, intake manifold, vacuum lines ● Fuel system too rich (fuel pressure too high, fuel pressure regulator leaking, or one or more fuel injectors sticking/leaking) ● HO2S element is silicon, water or fuel contaminated ● HO2S signal tracking (water intrusion) in the connector causing a short between the HO2S signal and heater power circuits ● HO2S low reference circuit is open or shorted to power (B+) ● PCM has failed
P0152 **2T CCM, MIL: Yes** 1996, 1997, 1998, 1999, 2000, 2001, 2002, 2003, 2004, 2005 DeVille, Eldorado, Seville 4.6L VIN 9, 4.6L VIN Y Transmissions: A/T	**HO2S-21 (Bank 2 Sensor 1) Circuit High Input Conditions:** DTC P0101-P0103, P0106-P0108, P0112-P0113, P0116-P0118, P0121-P0123, P0125, P0128, P0200, P0300, P0410-P0446, P0452-P0453, P1258, P1415-P1416 and P1441 not set, AIR, Catalyst and EGR Flow tests "off", A/F ratio at 14.4-14.9:1, no injectors disabled, TP angle from 3-25% for 3 seconds, and the PCM detected the HO2S signal was more than 900 mv for 50 seconds, or that it was more than 200 mv during Decel Fuel Cutoff mode. **Possible Causes** ● Fuel system is too rich (fuel pressure too high, fuel pressure regulator leaking, or one or more fuel injectors sticking/leaking) ● HO2S element is silicon, water or fuel contaminated ● HO2S signal circuit is shorted to system power (B+) ● HO2S signal tracking (water intrusion) in the connector causing a short between the HO2S signal and heater power circuits ● PCM has failed
P0152 **2T CCM, MIL: Yes** 1995, 1996, 1997, 1998, 1999, 2000, 2001, 2002 Camaro & Firebird 3.8L VIN K engine Transmissions: All	**HO2S-21 (Bank 2 Sensor 1) Circuit High Input Conditions:** DTC P0101, P0102, P0103, P0107, P0108, P0112, P0113, P0117, P0118, P0121, P0122, P0123, P0125, P0128, P0201-P0206, P0410, P0440, P0442, P0443, P0446, P0449, or P1441 not set, engine started, system voltage over 10.0v, engine running in closed loop for 4 minutes, ECT sensor more than 122°F, A/F ratio from 14.5-14.8:1, TP angle from 3-40% or on models with TAC, the APP sensor angle from 5-40%, and the PCM detected the HO2S signal was over 975 mv, or it was more than 200 mv in Decel Fuel Cutoff mode for 15 seconds. **Possible Causes** ● Fuel system rich (high fuel pressure, fuel pressure regulator leaking, or leaking injector) ● HO2S element is silicon, water or fuel contaminated ● HO2S signal circuit is shorted to system power (B+) ● HO2S signal tracking (water intrusion) in the connector causing a short between the HO2S signal and heater power circuits ● PCM has failed
P0152 **2T CCM, MIL: Yes** 1998, 1999, 2000, 2001, 2002 Camaro & Firebird 5.7L VIN G engine Transmissions: All	**HO2S-21 (Bank 2 Sensor 1) Circuit High Input Conditions:** DTC P0101, P0102, P0103, P0106, P0107, P0108, P0112-P0118, P0121, P0122, P0123, P0125, P0128, P0200, P0300, P0410, P0440-P0446, P0452, P0453, P1258, P1415, P1416 and P1441 not set, engine started, system voltage over 10.0v, fuel level over 10%, Intrusive and Scan Tool tests "off", A/F ratio 14.5-14.7:1, Rich Test enabled, TP angle from 3-70%, the PCM detected the HO2S signal was over 775 mv for 2 minutes, or with DFCO mode enabled for 1 second, the PCM detected the HO2S signal was less than 540 mv for 5 seconds. **Possible Causes** ● Fuel system rich (high fuel pressure, fuel pressure regulator leaking, or leaking injector) ● HO2S element is silicon, water or fuel contaminated ● HO2S signal circuit is shorted to system power (B+) ● HO2S signal tracking (water intrusion) in the connector causing a short between the HO2S signal and heater power circuits ● PCM has failed
P0152 **2T CCM, MIL: Yes** 1996, 1997, 1998, 1999 Aurora 4.0L V8 VIN C engine Transmissions: A/T	**HO2S-21 (Bank 2 Sensor 1) Circuit High Input Conditions:** DTC P0101, P0102, P0103, P0106, P0107, P0108, P0112, P0113, P0116, P0117, P0118, P0121, P0122, P0123, P0125, P0201-P0208, P0410, P0412, P0418, P0419, P0440, P0442, P0443, P0446, P0449, P1415, P1416 and P1441 are not set, engine started, TP angle from 3-40 degrees, ECT sensor more than 179°F, A/F ratio from 14.4-14.9:1, AIR, Catalyst and EGR Tests all inactive, system voltage over 10.0v, MAP sensor more than 32 kPa, and the PCM detected the HO2S signal was more than 900 mv in closed loop for 3 seconds, or that it was more than 200 mv in DFCO mode. **Possible Causes** ● Fuel system rich (high fuel pressure, fuel pressure regulator leaking, or leaking injector) ● HO2S element is silicon, water or fuel contaminated ● HO2S signal circuit is shorted to system power (B+) ● HO2S signal tracking in the connector causing a short to power ● PCM has failed

OBD II Trouble Code List (P0xxx Codes)

DTC	Trouble Code Title, Conditions & Possible Causes
P0152 **2T CCM, MIL: Yes** 2001, 2002, 2003, 2004, 2005 Aurora 4.0L V8 VIN C engine Transmissions: A/T	**HO2S-21 (Bank 2 Sensor 1) Circuit High Input Conditions:** DTC P0101-P0103, P0106-P0108, P0112, P0113, P0116-P0118, P0121-P0123, P0125, P0128, P0201-P0208, P0410, P0412, P0418, P0419, P0440-P0449, P1415, P1416, P1441 not set, engine started, AIR and Catalyst Tests off, TP angle 3-25% in closed loop, and the PCM detected the HO2S signal was over 774 mv for 30 seconds. **Possible Causes** ● Fuel system is too rich (fuel pressure too high, fuel pressure regulator leaking, or one or more fuel injectors sticking/leaking) ● HO2S element is silicon, water or fuel contaminated ● HO2S signal tracking in the connector causing a short to power ● PCM has failed
P0152 **2T CCM, MIL: Yes** 1995, 1996, 1997, 1998, 1999, 2000, 2001, 2002, 2003, 2004, 2005 O/T Blazer & S/T Pickup 4.3L VIN W, 4.3L VIN X Transmissions: All	**HO2S-21 (Bank 2 Sensor 1) Circuit High Input Conditions:** DTC P0101, P0102, P0103, P0106-P0108, P0112, P0113, P0116, P0117, P0118, P0121-P0123, P0200, P0300, P0440-P0446, P0452, P0453, P0455, P0496 and P1441 not set, engine started, engine running in closed loop, system voltage over 10.0v, fuel level over 10%, Scan Tool tests off, TP angle at 3-70% for 5 seconds, then with the Rich Test enabled, the PCM detected the HO2S signal was over 900 mv for 165 seconds or the HO2S signal was more than 250 mv for 5 seconds while operating in Decel Fuel Cutoff mode. **Possible Causes** ● Fuel system is too rich (fuel pressure too high, fuel pressure regulator leaking, or one or more fuel injectors sticking/leaking) ● HO2S element is silicon, water or fuel contaminated ● HO2S signal tracking (water intrusion) in the connector causing a short between the HO2S signal and heater power circuits ● PCM has failed
P0152 **2T CCM, MIL: Yes** 1997, 1998, 1999, 2000, 2001 Catera 3.0L V6 VIN R engine Transmissions: A/T	**HO2S-21 (Bank 2 Sensor 1) Circuit High Input Conditions:** DTC P0137 and P0138 not set, engine started, system voltage over 10.0v, engine speed from 1400-2400 rpm, ECT sensor more than 158°F, engine load from 8-12%, MAF sensor 18-22 g/sec, TP angle at 6-10%, HO2S-12 test finished, and the PCM detected too much difference between the front Lambda HO2S and rear HO2S signals. **Possible Causes** ● Fuel system is too rich (fuel pressure too high, fuel pressure regulator leaking, or one or more fuel injectors sticking/leaking) ● HO2S element is silicon, water or fuel contaminated ● HO2S signal circuit is shorted to system power (B+) ● HO2S signal tracking (water intrusion) in the connector causing a short between the HO2S signal and heater power circuits ● PCM has failed
P0152 **2T CCM, MIL: Yes** 1995 Corvette 5.7L VIN P Transmissions: All	**HO2S-21 (Bank 2 Sensor 1) Circuit High Input Conditions:** DTC P0116, P0117, P0118, P0135, P0171, P0172, P0174, P0175, P1114 and P1115 not set, engine started, system voltage over 10.0v, engine running in closed loop, TP angle from 2-10%, and the PCM detected the HO2S signal was over 853 mv for 102 seconds. **Possible Causes** ● Fuel system is too rich (fuel pressure too high, fuel pressure regulator leaking, or one or more fuel injectors sticking/leaking) ● HO2S element is silicon, water or fuel contaminated ● HO2S signal circuit is shorted to system power (B+) ● HO2S signal tracking (water intrusion) in the connector causing a short between the HO2S signal and heater power circuits ● PCM has failed
P0152 **2T CCM, MIL: Yes** 1997, 1998, 1999, 2000, 2001, 2002, 2003, 2004, 2005 Corvette 5.7L VIN G, 5.7L VIN S Transmissions: All	**HO2S-21 (Bank 2 Sensor 1) Circuit High Input Conditions:** DTC P0101, P0102, P0103, P0106, P0107, P0108, P0112, P0113, P0116, P0117, P0118, P0200, P0300, P0410, P0440, P0442, P0446, P0452, P0453, P1120, P1125, P1220, P1221, P1258, P1415, P1416, P1441, P1514, P1515, P1516, P1517, or P1518 not set, engine started, TP indicated angle from 3-70% in closed loop, AIR and Catalyst Tests "off", then with the Rich Test enabled, the PCM detected the HO2S signal was more than 775 mv for 165 seconds or with Decel Fuel Cutoff Test active for 1 second, the PCM detected the HO2S signal was over 540 mv for 5 seconds. **Possible Causes** ● Fuel system is too rich (fuel pressure too high, fuel pressure regulator leaking, or one or more fuel injectors sticking/leaking) ● HO2S element is silicon, water or fuel contaminated ● HO2S signal circuit is shorted to system power (B+) ● HO2S signal tracking (water intrusion) in the connector causing a short between the HO2S signal and heater power circuits ● PCM has failed

OBD II Trouble Code List (P0xxx Codes)

DTC	Trouble Code Title, Conditions & Possible Causes
P0153 **1T O2S, MIL: Yes** 1996, 1997 Camaro, Corvette, Caprice, Firebird & Fleetwood 4.3L VIN W, 5.7L VIN 5, 5.7L VIN P engines Transmissions: All	**HO2S-21 (Bank 2 Sensor 1) Slow Response Conditions:** DTC P0101-P0103, P0112-P0115, P0125, P0131-P0135, P0171-P0175, P0410, P1114-P1115 and P1133 not set, engine speed from 1000-1700 rpm in closed loop for 2 minutes, MAF sensor from 12-28 g/sec, Purge command below 90%, and the PCM detected the average HO2S lean-rich or rich-lean response time was over 110-150 ms for 100 seconds. **Possible Causes** ● Exhaust leak present in the exhaust manifold or exhaust pipes ● Fuel system rich (high fuel pressure, fuel pressure regulator leaking, or leaking injectors)HO2S element is silicon, water or fuel contaminated or it failed ● TP sensor element broken (can cause false acceleration event) ● PCM has failed
P0153 **2T O2S, MIL: Yes** 1996, 1997, 1998, 1999, 2000, 2001, 2002, 2003, 2004, 2005 C/K Cab & Chassis, C/K Series Truck, G Series Van, M/L Series Vans 4.3L VIN W, 4.3L VIN X, 5.0L VIN M, 5.7L VIN K, 5.7L VIN R, 7.4L VIN J Transmissions: All	**HO2S-21 (Bank 2 Sensor 1) Slow Response Conditions:** DTC P0101-P0103, P0106-P0108, P0112-P0113, P0116-P0118, P0121-P0123, P0131-135, P0151-P0155, P0200, P0300, P0401, P0404-P0405, P0440-P0446, P0452-P0453, P1120, P1125, P1220-P1221, P1258, P1404, P1441, P1514 and P1518 not set, engine started, engine speed from 1200-3000 rpm for 2 minutes in closed loop, ECT sensor more than 149°F, Purge command over 1%, MAF sensor from 23-50 g/sec, TP angle over 5% or for models with TAC, TP angle more than 5% higher than the idle value, fuel level over 10%, Scan Tool and Intrusive tests "off", conditions met for 100 seconds, and the PCM detected the HO2S rich-to-lean or the lean-to-rich response time was more than a calibrated value. **Possible Causes** ● Exhaust leak present in the exhaust manifold or exhaust pipes ● Fuel system rich (high fuel pressure, fuel pressure regulator leaking, or leaking injectors)HO2S element is silicon, water or fuel contaminated or it failed ● TP sensor element broken (can cause false acceleration event) ● PCM has failed
P0153 **2T O2S, MIL: Yes** 1999, 2000, 2001, 2002, 2003, 2004, 2005 C/K Series Truck, G Series Van, Envoy, Escalade, TrailBlazer 4.8L VIN V, 5.3L VIN P, 5.3L VIN T, 5.3L VIN Z, 6.0L VIN N, 6.0L VIN U, 8.1L VIN G Transmissions: All	**HO2S-21 (Bank 2 Sensor 1) Slow Response Conditions:** DTC P0101, P0102, P0103, P0106, P0107, P0108, P0112, P0113, P0116, P0117, P0118, P0120, P0121, P0122, P0123, P0169, P0178, P0179, P0200, P0220, P0300, P0442, P0446, P0452, P0453, P0455, P0496, P1125, P1258, P1514, P1515, P1516, P1518, P2108 and P2135 not set, engine runtime in closed loop over 160 seconds, ECT sensor over 149°F, engine speed from 1200-3000 rpm, system voltage from 10-18v, Fuel Alcohol content less than 90% on TAC equipped models, fuel level over 10%, TP indicated angle more than 5% higher than the idle value, and the PCM detected the HO2S signal rich to lean or lean to rich average response time was more than a calibrated value during the test. **Possible Causes** ● Fuel system rich (high fuel pressure, fuel pressure regulator leaking, or leaking injectors) ● HO2S element is silicon, water or fuel contaminated ● HO2S signal tracking (water intrusion) in the connector causing a short between the HO2S signal and heater power circuits ● PCM has failed
P0153 **2T O2S, MIL: Yes** 1996, 1997, 1998, 1999, 2000, 2001, 2002, 2003, 2004, 2005 DeVille, Eldorado, Seville 4.6L VIN 9, 4.6L VIN Y Transmissions: A/T	**HO2S-21 (Bank 2 Sensor 1) Slow Response Conditions:** DTC P0101-P0103, P0106-P0108, P0112-P0113, P0116-P0118, P0121-P0123, P0125, P0128, P0200, P0300, P0410-P0446, P0452-P0453, P1258, P1415-P1416 and P1441 not set, engine speed from 1200-2800 rpm for 202 seconds in closed loop, AIR, Catalyst, EGR and Intrusive Tests "off", TP angle over 3%, ECT sensor over 167°F, MAF sensor from 15-35 g/sec, gear position not indicating Reverse, Park or Neutral, and the PCM detected the HO2S lean-rich or rich-lean response time was over 200 ms, or the HO2S was fixed at 325-625 mv. **Possible Causes** ● Exhaust leak present in the exhaust manifold or exhaust pipes ● Fuel system rich (high fuel pressure, fuel pressure regulator leaking, or leaking injectors) ● HO2S element is silicon, water or fuel contaminated or it failed ● TP sensor element broken (can cause false acceleration event) ● PCM has failed
P0153 **2T O2S, MIL: Yes** 2003, 2004, 2005 CTS (Cadillac) 2.6L V6 VIN M & 3.2L V6 VIN N engines Transmissions: All	**HO2S-21 (Bank 2 Sensor 1) Slow Response Conditions:** DTC P0030, P0031, P0032, P0036, P0037, P0038, P0101, P0102, P0103, P0112, P0113, P0116, P0117, P0118, P0125, P0128, P0135, P0141, P0171, P0172, P0174, P0175, P0300, P0301-P0306, P0443, P0444 and P0445 not set, engine speed 1520-3000 rpm while running in closed loop, Calculated Converter Temperature more than 662°F, volumetric efficiency from 30-70%, Purge solenoid commanded off for 10 seconds, and the PCM detected the HO2S-21 signal took longer than 1.3 seconds to switch, after 21 valid switching cycles were detected (test runs once per ignition cycle). **Possible Causes** ● EVAP system leaks or restrictions, charcoal canister problems ● Exhaust system is leaking or severely restricted ● HO2S signal circuit is open, or the low reference circuit is open ● HO2S is damaged or contaminated due to improper fuel usage ● PCM has failed

OBD II Trouble Code List (P0xxx Codes)

DTC	Trouble Code Title, Conditions & Possible Causes
P0153 **2T O2S, MIL: Yes** 1998, 1999, 2000, 2001, 2002 Camaro & Firebird 5.7L VIN G engine Transmissions: All	**HO2S-21 (Bank 2 Sensor 1) Slow Response Conditions:** DTC P0101-P0103, P0106-P0108, P0112, P0113, P0116-P0118, P0121-P0123, P0125, P0128, P-131-P0135, P0151-P0155, P0200, P0300, P0410, P0440-P0446, P0452-P0453, P1258, P1415-P1416 and P1441 not set, engine started, engine speed 1000-2300 rpm for 60 seconds, system voltage over 10.0v, fuel level over 10%, ECT sensor more than 122ºF, AIR, EGR and Catalyst tests all "off", MAF sensor from 20-50 g/sec, Purge command over 0%, and the PCM detected the HO2S lean to rich or rich to lean response time was over 250 ms. **Possible Causes** ● Exhaust leak present in the exhaust manifold or exhaust pipes ● Fuel system is too rich (fuel pressure too high, fuel pressure regulator leaking, or one or more fuel injectors sticking/leaking) ● HO2S element is silicon, water or fuel contaminated or it failed ● TP sensor element broken (can cause false acceleration event) ● PCM has failed
P0153 **2T O2S, MIL: Yes** 1995, 1996, 1997, 1998, 1999, 2000, 2001, 2002 Camaro & Firebird 3.8L VIN K engine Transmissions: All	**HO2S-21 (Bank 2 Sensor 1) Slow Response Conditions:** DTC P0101-P0103, P0107-P0108, P0112, P0113, P0117-P0118, P0121-P0123, P0125, P0128, P0201-P0206, P0300, P0410, P0440-P0449 and P1441 not set, engine runtime over 4 minutes, ECT sensor more than 122ºF, engine speed from 1000-3000 rpm in closed loop, MAF sensor from 13-30 g/sec, and the PCM detected the HO2S lean-rich response time was more than 63 ms, or the rich-lean response time was more than 190 ms. **Possible Causes** ● Exhaust leak present in the exhaust manifold or exhaust pipes ● Fuel system is too rich (fuel pressure too high, fuel pressure regulator leaking, or one or more fuel injectors sticking/leaking) ● HO2S element is silicon, water or fuel contaminated or it failed ● TP sensor element broken (can cause false acceleration event) ● PCM has failed
P0153 **2T O2S, MIL: Yes** 1996, 1997, 1998, 1999 Aurora 4.0L V8 VIN C engine Transmissions: A/T	**HO2S-21 (Bank 2 Sensor 1) Slow Response Conditions:** DTC P0101, P0102, P0103, P0106, P0107, P0108, P0112, P0113, P0116, P0117, P0118, P0121, P0122, P0123, P0125, P0128, P0131, P0132, P0135, P0151, P0152, P0201-P0208, P0300, P0410, P0410, P0418, P0419, P0440, P0442, P0443, P0446, P0449, P1133, P1415, P1416, or P1441 not set, engine started, engine speed from 1200-2300 rpm in closed loop for 200 seconds, gear selector not in Reverse or P/N, ECT sensor more than 154ºF, TP angle over 3%, MAF sensor from 15-35 g/sec, system voltage over 10.0v, AIR, Catalyst and EGR Tests all "off" for 3 seconds, and the PCM detected the average HO2S lean-rich or rich-lean response time was more than 219 ms. **Possible Causes** ● Exhaust leak present in the exhaust manifold or exhaust pipes ● Fuel system is too rich (fuel pressure too high, fuel pressure regulator leaking, or one or more fuel injectors sticking/leaking) ● HO2S element is silicon, water or fuel contaminated or it failed ● TP sensor element broken (can cause false acceleration event) ● PCM has failed
P0153 **2T O2S, MIL: Yes** 2001, 2002, 2003, 2004, 2005 Aurora 4.0L V8 VIN C engine Transmissions: A/T	**HO2S-21 (Bank 2 Sensor 1) Slow Response Conditions:** DTC P0101-P0103, P0106-P0108, P0112, P0113, P0116-P0118, P0121-P0123, P0125, P0128, P0201-P0208, P0410, P0412, P0418, P0419, P0440-P0449, P1415, P1416, P1441 not set, engine started, AIR and Catalyst Tests off, TP angle 3-25% in closed loop, and the PCM detected the HO2S signal was over 774 mv for 30 seconds. **Possible Causes** ● Exhaust leak present in the exhaust manifold or exhaust pipes ● Fuel system too rich (fuel pressure too high, injector(s) leaking) ● HO2S element is silicon, water or fuel contaminated or it failed ● TP sensor element broken (can cause false acceleration event) ● PCM has failed
P0153 **2T O2S, MIL: Yes** 1995, 1996, 1997, 1998, 1999, 2000, 2001, 2002, 2003, 2004, 2005 S/T Blazer & S/T Pickup 4.3L VIN W, 4.3L VIN X Transmissions: All	**HO2S-21 (Bank 2 Sensor 1) Slow Response Conditions:** DTC P0101-P0103, P0106-P0108, P0116-P0118, P0121-P0123, P0131, P0132, P0134, P0135, P0151, P0152, P0154, P0155, P0200, P0300, P0440, P0442, P0446, P0452, P0453, P0455, P0496 and P1441 not set, engine speed from 1300-3000 rpm for 2 minutes in closed loop, ECT sensor over 135ºF, MAF sensor from 18-45 g/sec, TP angle over 5%, system voltage over 10.0v, fuel level over 10%, Intrusive Tests off, Purge command over 1%, and the PCM detected the HO2S rich-to-lean or lean-to-rich response average time was more than a calibrated value in memory. **Possible Causes** ● Exhaust leak present in the exhaust manifold or exhaust pipes ● Fuel system too rich (fuel pressure too high, injector(s) leaking) ● HO2S element is silicon, water or fuel contaminated or it failed ● TP sensor element broken (can cause false acceleration event) ● PCM has failed

OBD II Trouble Code List (P0xxx Codes)

DTC	Trouble Code Title, Conditions & Possible Causes
P0153 **2T O2S, MIL: Yes** 1997, 1998 Catera 3.0L V6 VIN R engine Transmissions: A/T	**HO2S-21 (Bank 2 Sensor 1) Slow Response Conditions:** DTC P0137 and P0138 not set, engine running in closed loop for 1 minute, system voltage over 10.0v, and the PCM detected the average HO2S response time was over 3.3 seconds. **Possible Causes** ● Exhaust leak present in the exhaust manifold or exhaust pipes ● Fuel system too rich (fuel pressure too high, injector(s) leaking) ● HO2S element is silicon, water or fuel contaminated or it failed ● TP sensor element broken (can cause false acceleration event) ● PCM has failed
P0153 **2T O2S, MIL: Yes** 1997, 1998, 1999, 2000, 2001, 2002, 2003, 2004, 2005 Corvette 5.7L VIN G, 5.7L VIN S Transmissions: All	**HO2S-21 (Bank 2 Sensor 1) Slow Response Conditions:** DTC P0101-P0103, P0106-P0108, P0112-P0118, P0131-P0135, P0151-P0155, P0200, P0300, P0410, P0440-P0446, P0452, P0453, P1120, P1125, P1220, P1221, P1258, P1415, P1416, P1441, P1514-P1518 not set, engine speed from 1000-2300 rpm in closed loop for 2 minutes, ECT sensor over 122ºF, MAF sensor at 18-50 g/sec, system voltage over 10.0v, fuel level over 10%, TP angle 5% over idle value, Purge command over 0%, and PCM detected the HO2S lean-rich or the rich-lean response time was over 250 ms. **Possible Causes** ● Exhaust leak present in the exhaust manifold or exhaust pipes ● Fuel system too rich (fuel pressure too high, injector(s) leaking) ● HO2S element is silicon, water or fuel contaminated or it failed ● TP sensor element broken (can cause false acceleration event) ● PCM has failed
P0153 **2T O2S, MIL: Yes** 1995 Corvette 5.7L VIN P Transmissions: All	**HO2S-21 (Bank 2 Sensor 1) Slow Response Conditions:** DTC P0116-P0118, P0135, P0171, P0172, P0174, P0175, P1114 and P1115 not set, engine runtime in closed loop over 1 minute, system voltage over 10.0v, TP angle from 2-10%, and the PCM detected the average HO2S lean-rich or rich-lean response time was over 100 ms. **Possible Causes** ● Exhaust leak present in the exhaust manifold or exhaust pipes ● Fuel system is too rich (fuel pressure too high, fuel pressure regulator leaking, or one or more fuel injectors sticking/leaking) ● HO2S element is silicon, water or fuel contaminated or it failed ● TP sensor element broken (can cause false acceleration event) ● PCM has failed
P0153 **1T O2S, MIL: Yes** 1996, 1997 Camaro, Corvette, Caprice, Firebird & Fleetwood 4.3L VIN W, 5.7L VIN 5, 5.7L VIN P engines Transmissions: All	**HO2S-21 (Bank 2 Sensor 1) Insufficient Activity Conditions:** DTC P0100-P0103, P0107, P0108, P0112-P0118, P0125, P0200, P0372 and P1371 not set; engine speed at 1000-3000 rpm for 4 minutes, AIR and Catalyst Tests "off", ECT sensor over 118ºF, and the PCM detected the HO2S signal was fixed from 352-552 mv for 30 seconds. **Possible Causes** ● HO2S heater is damaged or has failed ● HO2S signal or ground circuit has a high resistance condition ● HO2S signal circuit is open or shorted to system power (B+) ● HO2S has failed (i.e., it is silicon, water or fuel contaminated) ● PCM has failed
P0153 **2T O2S, MIL: Yes** 1996, 1997, 1998, 1999, 2000, 2001, 2002, 2003, 2004, 2005 C/K Cab & Chassis, C/K Series Truck, G Series Van, M/L Series Vans 4.3L VIN W, 4.3L VIN X, 5.0L VIN M, 5.7L VIN K, 5.7L VIN R, 7.4L VIN J Transmissions: All	**HO2S-21 (Bank 2 Sensor 1) Insufficient Activity Conditions:** DTC P0101-P0103, P0106-P0108, P0112, P0116-P0118, P0121-P0123, P0200, P0300, P0401, P0404-P0405, P0440, P0442, P0446, P0452, P0453, P1120, P1125, P1220, P1221, P1258, P1404, P1441, P1514, P1515, P1516, P1517 and P1518 not set, engine runtime over 409 seconds, system voltage over 10.0v, Scan Tool and Intrusive tests "off", and the PCM detected the HO2S signal remained between 350-550 mv for 60 seconds. **Possible Causes** ● HO2S heater is damaged or has failed ● HO2S signal or ground circuit has a high resistance condition ● HO2S signal circuit is open or shorted to system power (B+) ● HO2S has failed (i.e., it is silicon, water or fuel contaminated) ● PCM has failed
P0153 **2T O2S, MIL: Yes** 1999, 2000, 2001, 2002, 2003, 2004, 2005 C/K Series Truck, G Series Van, Envoy, Escalade, TrailBlazer 4.8L VIN V, 5.3L VIN P, 5.3L VIN T, 5.3L VIN Z, 6.0L VIN N, 6.0L VIN U, 8.1L VIN G Transmissions: All	**HO2S-21 (Bank 2 Sensor 1) Insufficient Activity Conditions:** DTC P0101-P0103, P0106-P0108, P0112, P0113, P0116-P0118, P0120, P0121-P0123, P0169, P0178, P0179, P0200, P0220, P0300, P0442, P0446, P0452, P0453, P0455, P0496, P1125, P1258, P1514, P1515, P1516, P1518, P2108 and P2135 not set, engine runtime over 300 seconds, system voltage from 10-18v, Fuel Alcohol content less than 90%, and the PCM detected the HO2S signal remained fixed between 350-550 mv for 60 seconds. **Possible Causes** ● HO2S heater is damaged or has failed ● HO2S signal or ground circuit has a high resistance condition ● HO2S signal circuit is open or shorted to system power (B+) ● HO2S has failed (i.e., it is silicon, water or fuel contaminated) ● PCM has failed

OBD II Trouble Code List (P0xxx Codes)

DTC	Trouble Code Title, Conditions & Possible Causes
P0154 **2T O2S, MIL: Yes** 1996, 1997, 1998, 1999, 2000, 2001, 2002, 2003, 2004, 2005 DeVille, Eldorado, Seville 4.6L VIN 9, 4.6L VIN Y Transmissions: A/T	**HO2S-21 (Bank 2 Sensor 1) Insufficient Activity Conditions:** DTC P0101-P0103, P0106-P0108, P0112-P0113, P0116-P0118, P0121-P0123, P0125, P0128, P0200, P0300, P0410-P0446, P0452-P0453, P1258, P1415-P1416 and P1441 not set, engine started, engine runtime over 200 seconds in closed loop, AIR, Catalyst, EGR and Intrusive Tests all "off", system voltage over 10.0v and the PCM detected the HO2S signal remained fixed within a range of 400-500 mv for more than 1 minute during the CCM test. **Possible Causes** ● HO2S heater is damaged or has failed ● HO2S signal or ground circuit has a high resistance condition ● HO2S signal circuit is open or shorted to system power (B+) ● HO2S has failed (i.e., it is silicon, water or fuel contaminated) ● PCM has failed
P0154 **2T O2S, MIL: Yes** 2003, 2004, 2005 CTS (Cadillac) 2.6L V6 VIN M & 3.2L V6 VIN N engines Transmissions: All	**HO2S-21 (Bank 2 Sensor 1) Insufficient Activity Conditions:** Engine started; system voltage over 10.5v, Calculated Converter temperature below 1,472°F, and the PCM detected the HO2S-21 signal was fixed from 400-597 mv for 7 seconds, or it detected the HO2S-21 and HO2S-22 signals remained above 200 mv during DFCO mode. **Possible Causes** ● EVAP system leaks or restrictions, charcoal canister problems ● Exhaust system is leaking or severely restricted ● HO2S signal circuit is open, or the low reference circuit is open ● HO2S is damaged or contaminated due to improper fuel usage ● PCM has failed
P0154 **2T O2S, MIL: Yes** 1998, 1999, 2000, 2001, 2002 Camaro & Firebird 5.7L VIN G engine Transmissions: All	**HO2S-21 (Bank 2 Sensor 1) Insufficient Activity Conditions:** DTC P0101-P0103, P0106-P0108, P0112-P0118, P0121-P0123, P0125, P0128, P0200, P0300, P0410, P0440-P0446, P0452, P0453, P1258, P1415-P1416 and P1441 not set, engine runtime over 409 seconds, Scan Tool and Intrusive tests "off", system voltage over 10.0v and the PCM detected the HO2S signal was fixed from 300-550 mv for 60 seconds. **Possible Causes** ● HO2S heater is damaged or has failed ● HO2S signal or ground circuit has a high resistance condition ● HO2S signal circuit is open or shorted to system power (B+) ● HO2S has failed (i.e., it is silicon, water or fuel contaminated) ● PCM has failed
P0154 **2T O2S, MIL: Yes** 1995, 1996, 1997, 1998, 1999, 2000, 2001, 2002 Camaro & Firebird 3.8L VIN K engine Transmissions: All	**HO2S-21 (Bank 2 Sensor 1) Insufficient Activity Conditions:** DTC P0101-P0103, P0107, P0108, P0112, P0113, P0116-P0118, P0121-P0123, P0125, P0128, P0201-P0206, P0410, P0440-P0449 and P1441 not set; engine runtime over 4 minutes and the PCM detected the HO2S signal was fixed at 400-500 mv for 60 seconds. **Possible Causes** ● HO2S heater is damaged or has failed ● HO2S signal or ground circuit has a high resistance condition ● HO2S signal circuit is open or shorted to system power (B+) ● HO2S has failed (i.e., it is silicon, water or fuel contaminated) ● PCM has failed
P0154 **2T O2S, MIL: Yes** 1996, 1997, 1998, 1999 Aurora 4.0L V8 VIN C engine Transmissions: All	**HO2S-21 (Bank 2 Sensor 1) Insufficient Activity Conditions:** DTC P0101-P0103, P0106-P0108, P0112, P0113, P0116-P0118, P0121-P0123, P0125, P0201-P0208, P0410, P0412, P0418, P0419, P0440, P0442-P0446, P0449, P1133, P1415, P1416, or P1441 not set, engine runtime over 200 seconds, system voltage over 10.0v, ECT sensor over 163°F, Calculated throttle position switch open, AIR, Catalyst and EGR tests all "off", and the PCM detected the HO2S signal remained from 400 to 500 mv for 1 minute. **Possible Causes** ● HO2S signal or ground circuit has a high resistance condition ● HO2S signal circuit is open or shorted to system power (B+) ● HO2S has failed (i.e., it is silicon, water or fuel contaminated) ● PCM has failed
P0154 **2T O2S, MIL: Yes** 1995, 1996, 1997, 1998, 1999, 2000, 2001, 2002, 2003, 2004, 2005 S/T Blazer & S/T Pickup 4.3L VIN W, 4.3L VIN X Transmissions: All	**HO2S-21 (Bank 2 Sensor 1) Insufficient Activity Conditions:** DTC P0101-P0103, P0106-P0108, P0112, P0113, P0116-P0118, P0121-P0123, P0200, P0300, P0440, P0442, P0446, P0452, P0453, P0455, P0496 and P1441 not set, engine started, engine runtime over 409 seconds, system voltage over 10.0v, Scan Tool tests all off, and the PCM detected the HO2S signal remained within a 350-550 mv range for 1 minute. **Possible Causes** ● HO2S heater is damaged or has failed ● HO2S signal or ground circuit has a high resistance condition ● HO2S signal circuit is open or shorted to system power (B+) ● HO2S has failed (i.e., it is silicon, water or fuel contaminated) ● PCM has failed ● TSB 81-65-39 contains a repair procedure for this code

OBD II Trouble Code List (P0xxx Codes)

DTC	Trouble Code Title, Conditions & Possible Causes
P0154 **2T O2S, MIL: Yes** 1997, 1998, 1999, 2000, 2001 Catera 3.0L V6 VIN R engine Transmissions: All	**HO2S-21 (Bank 2 Sensor 1) Insufficient Activity Conditions:** DTC P0130 and P0150 not set, engine started, engine speed from 1400-2400 rpm for 25 seconds, system voltage over 10.0v, engine load from 8-12%, MAF sensor from 18-22 g/sec, TP angle from 6-10%, and the PCM detected the difference in HO2S signal during a 10 ms period was more than the allowable limit. **Possible Causes** • HO2S heater is damaged or has failed • HO2S signal or ground circuit has a high resistance condition • HO2S signal circuit is open or shorted to system power (B+) • HO2S has failed (i.e., it is silicon, water or fuel contaminated) • PCM has failed
P0154 **2T O2S, MIL: Yes** 1995 Corvette 5.7L VIN P Transmissions: All	**HO2S-21 (Bank 2 Sensor 1) Insufficient Activity Conditions:** DTC P0116, P0117, P0118, P0135, P0171, P0172, P0174, P0175, P1114 and P1115 not set, engine started, engine running in closed loop, TP angle from 2-10%, and the PCM detected the HO2S signal was fixed from 356-497 mv for 102 seconds during the CCM test. **Possible Causes** • HO2S heater is damaged or has failed • HO2S signal or ground circuit has a high resistance condition • HO2S signal circuit is open or shorted to system power (B+) • HO2S has failed (i.e., it is silicon, water or fuel contaminated) • PCM has failed
P0154 **2T O2S, MIL: Yes** 1997, 1998, 1999, 2000, 2001, 2002, 2003, 2004, 2005 Corvette 5.7L VIN G, 5.7L VIN S Transmissions: All	**HO2S-21 (Bank 2 Sensor 1) Insufficient Activity Conditions:** DTC P0101-P0103, P0106-P0108, P0112-P0118, P0200, P0300, P0410, P0440-P0446, P0452, P0453, P1120, P1125, P1220, P1221, P1258, P1415, P1416, P1441, P1514 to P1518 not set, engine runtime over 409 seconds in closed loop, system voltage over 10.0v, Intrusive and Scan Tool Tests off, and the PCM detected the HO2S signal remained fixed from 350-550 mv for over 1 minute. **Possible Causes** • HO2S heater is damaged or has failed • HO2S signal or ground circuit has a high resistance condition • HO2S signal circuit is open or shorted to system power (B+) • HO2S has failed (i.e., it is silicon, water or fuel contaminated) • PCM has failed
P0155 **2T O2S HTR, MIL: Yes** 1996, 1997 Camaro, Corvette, Caprice, Firebird & Fleetwood 4.3L VIN W, 5.7L VIN 5, 5.7L VIN P engines Transmissions: All	**HO2S-21 (Bank 2 Sensor 1) Heater Circuit Malfunction Conditions:** DTC P0101, P0102, P0103, P0112, P0113, P0117, P0118, P0125, P0132, P0133, P0134 and P0200 not set, ECT and IAT sensor less than 122ºF and within 5ºF at startup, engine started, system voltage over 10.0v, AIR and Catalyst tests "off", MAF sensor less than 40 g/sec, TP angle less than 20%, and the PCM detected the HO2S signal remained from 300-700 mv for too long a period of time. **Possible Causes** • HO2S heater ground circuit is open or has high resistance • HO2S heater power circuit is open (test O2S fuse in fuse block) • HO2S heater element is damaged or has failed • PCM has failed
P0155 **2T O2S HTR, MIL: Yes** 1996, 1997, 1998, 1999, 2000, 2001, 2002, 2003, 2004, 2005 DeVille, Eldorado, Seville 4.6L VIN 9, 4.6L VIN Y Transmissions: A/T	**HO2S-21 (Bank 2 Sensor 1) Heater Circuit Malfunction Conditions:** DTC P0030, P1031, or P1032 not set, engine started, engine speed from 500-3000 rpm for 3 minutes, system voltage is steady (± 1v), MAF sensor from 4-30 g/sec, HO2S Over Temperature Control "off", and the PCM detected the HO2S heater current was out-of-range. **Possible Causes** • HO2S heater high side driver circuit has high resistance (open) • HO2S heater low side driver circuit has high resistance (open) • HO2S heater power circuit is open (test O2S fuse in fuse block) • HO2S heater element is damaged or has failed • PCM has failed
P0155 **2T O2S HTR, MIL: Yes** 2000, 2001, 2002, 2003, 2004, 2005 DeVille, Eldorado, Seville 4.6L VIN 9, 4.6L VIN Y Transmissions: A/T	**HO2S-21 (Bank 2 Sensor 1) Heater Circuit Malfunction Conditions:** DTC P0030, P1031, or P1032 not set, engine started, engine speed from 500-3000 rpm for 3 minutes, system voltage is steady (± 1v), MAF sensor from 4-30 g/sec, HO2S Over Temperature Control "off", and the PCM detected the HO2S heater current was out-of-range. **Possible Causes** • HO2S heater high side driver circuit has high resistance (open) • HO2S heater low side driver circuit has high resistance (open) • HO2S heater power circuit is open (test O2S fuse in fuse block) • HO2S heater element is damaged or has failed • PCM has failed

OBD II Trouble Code List (P0xxx Codes)

DTC	Trouble Code Title, Conditions & Possible Causes
P0155 **2T O2S HTR, MIL: Yes** 1996, 1997, 1998, 1999, 2000, 2001, 2002, 2003, 2004, 2005 C/K Cab & Chassis, C/K Series Truck, G Series Van, M/L Series Vans 4.3L VIN W, 4.3L VIN X, 5.0L VIN M, 5.7L VIN K, 5.7L VIN R, 7.4L VIN J Transmissions: All	**HO2S-21 (Bank 2 Sensor 1) Heater Circuit Malfunction Conditions:** DTC P0101-P0103, P0106-P0108, P0112, P0113, P0116-P0118, P0121-P0123, P0131, P0132, P0134, P0137, P0138, P0140, P0151, P0152, P0154, P0157, P0158, P0160, P0200, P0300, P0401, P0404, P0405, P0440, P0442, P0446, P0452, P0453, P1120, P1125, P1220, P1221, P1258, P1404, P1441, P1514, P1515, P1516, P1517 and P1518 not set, ECT and IAT sensors less than 122ºF and with 14.5ºF at startup, engine started, engine running in closed loop right after startup, Intrusive and Scan Tool tests "off", MAF sensor less than 25 g/sec, and the PCM detected the HO2S signal remained within 150 mv of startup HO2S signal for a predetermined amount of time based on ECT and airflow signals. **Possible Causes** ● HO2S heater ground circuit is open or has high resistance ● HO2S heater power circuit is open (test O2S fuse in fuse block) ● HO2S heater element is damaged or has failed ● PCM has failed
P0155 **2T O2S HTR, MIL: Yes** 1999, 2000, 2001, 2002, 2003, 2004, 2005 C/K Series Truck, G Series Van, Envoy, Escalade, TrailBlazer 4.8L VIN V, 5.3L VIN P, 5.3L VIN T, 5.3L VIN Z, 6.0L VIN N, 6.0L VIN U, 8.1L VIN G Transmissions: All	**HO2S-21 (Bank 2 Sensor 1) Heater Circuit Malfunction Conditions:** DTC P0101-P0103, P0106-P0108, P0112, P0113, P0116-P0118, P0120, P0121-P0123, P0169, P0178, P0179, P0200, P0220, P0300, P0442, P0446, P0452, P0453, P0155, P0490, P1125, P1258, P1514, P1515, P1516, P1518, P2108 and P2135 not set, engine speed from 500-3000 rpm for 120 seconds, system voltage at 10-18v, ECT sensor over 122ºF, MAF sensor at 3-40 g/sec, Fuel Alcohol content below 90%, and the PCM detected the HO2S heater current was below 0.25 amps or over 3.125 amps (more than 1.375 amps on 4.8L V8). **Possible Causes** ● HO2S heater low control circuit is open or shorted to ground ● HO2S heater circuit is open or it is shorted to ground ● HO2S heater power circuit is open (test O2A fuse in fuse block) ● HO2S heater element is damaged or has failed ● PCM has failed
P0155 **2T O2S HTR, MIL: Yes** 1998, 1999, 2000, 2001, 2002 Camaro & Firebird 5.7L VIN G engine Transmissions: All	**HO2S-21 (Bank 2 Sensor 1) Heater Circuit Malfunction Conditions:** DTC P0101-P0103, P0106-P0108, P0112-P0118, P0121-P0123, P0125, P0128, P0131-P0138, P0140, P0151-P0158, P0160, P0200, P0300, P0410, P0440-P0446, P0452, P0453, P1258, P1415, P1416 or P1441 not set, engine started, ECT and IAT sensors under 122ºF and within 14.5ºF, HO2S sensor at 425-475 mv at startup, Scan Tool and Intrusive tests off, MAF sensor under 18 g/sec, and the PCM detected the HO2S Signal was fixed from 300 to 700 mv for too long (from 42 seconds to 2 minutes - depends on ECT and MAF inputs). **Possible Causes** ● HO2S heater ground circuit is open or has high resistance ● HO2S heater power circuit is open (test O2S fuse in fuse block) ● HO2S heater element is damaged or has failed ● PCM has failed
P0155 **2T O2S HTR, MIL: Yes** 1995, 1996, 1997, 1998, 1999, 2000, 2001, 2002 Camaro & Firebird 3.8L VIN K engine Transmissions: All	**HO2S-21 (Bank 2 Sensor 1) Heater Circuit Conditions:** DTC P0101-P0103, P0107, P0108, P0112, P0113, P0117, P0118, P0121-P0123, P0125, P0128, P0201-P0206, P0410, P0440, P0442-P0449 and P1441 not set, engine started, ECT and IAT sensors less than 95ºF at startup and within 11ºF at startup, system voltage over 10.0v, MAF sensor below 20 g/sec, and the PCM detected the HO2S signal was within 150 mv of bias voltage for too long (from 42 seconds to 2 minutes - depends on ECT at startup). **Possible Causes** ● HO2S heater ground circuit is open or has high resistance ● HO2S heater power circuit is open (test O2S fuse in fuse block) ● HO2S heater element is damaged or has failed ● PCM has failed
P0155 **2T O2S HTR, MIL: Yes** 1996, 1997, 1998, 1999 Aurora 4.0L V8 VIN C engine Transmissions: A/T	**HO2S-21 (Bank 2 Sensor 1) Heater Circuit Malfunction Conditions:** DTC P0101-P0103, P0117, P0118 and P0134 not set, ECT sensor change over 50ºF from last ignition cycle, engine did not stall, HO2S-11 bias voltage from 352-546 mv, and the PCM detected it took to long for the HO2S signal to reach 151 mv over or under the bias voltage. **Possible Causes** ● HO2S heater ground circuit is open or has high resistance ● HO2S heater power circuit is open (test O2S fuse in fuse block) ● HO2S heater element is damaged or has failed ● PCM has failed
P0155 **2T O2S HTR, MIL: Yes** 2001, 2002, 2003, 2004, 2005 Aurora 4.0L V8 VIN C engine Transmissions: All	**HO2S-21 (Bank 2 Sensor 1) Heater Circuit Malfunction Conditions:** DTC P0030, P1031, or P1032 not set, engine speed from 500-3000 rpm for 3 minutes, system voltage steady (± 1v), MAF sensor from 4-30 g/sec, HO2S Over Temperature Control "off", and the PCM detected the HO2S heater current was out-of-range during the test. **Possible Causes** ● HO2S heater high side or low side driver circuit has high resistance (open) ● HO2S heater power circuit is open (test the O2S SEN fuse) ● HO2S heater element is damaged or has failed ● PCM has failed

OBD II Trouble Code List (P0xxx Codes)

DTC	Trouble Code Title, Conditions & Possible Causes
P0155 **2T O2S HTR, MIL: Yes** 1995, 1996, 1997, 1998, 1999, 2000, 2001, 2002, 2003, 2004, 2005 S/T Blazer & S/T Pickup 4.3L VIN W, 4.3L VIN X Transmissions: All	**HO2S-21 (Bank 2 Sensor 1) Heater Circuit Malfunction Conditions:** DTC P0101, P0102, P0103, P0106, P0107, P0108, P0112, P0116, P0117, P0118, P0121, P0122, P0123, P0200, P0300, P0440, P0442, P0446, P0453, P0455, P0496 and P1441 not set, ECT and IAT sensors less than 122ºF and within 14.5ºF at startup, engine started, engine running in closed loop at startup, Scan Tool tests off, and the PCM detected the HO2S signal was within 150 mv of bias voltage (450 mv) for too long (depends on ECT and MAF inputs). **Possible Causes** ● HO2S heater ground circuit is open or has high resistance ● HO2S heater power circuit is open (test O2S fuse in fuse block) ● HO2S heater element is damaged or has failed ● PCM has failed
P0155 **2T O2S HTR, MIL: Yes** 1997, 1998, 1999, 2000, 2001 Catera 3.0L V6 VIN R engine Transmissions: All	**HO2S-21 (Bank 2 Sensor 1) Heater Circuit Malfunction Conditions:** Engine started; system voltage over 10.0v, HO2S heater "on", and the PCM detected the HO2S heater circuit indicated more than 3.6v, or less than 2.3v for 50 ms, or that the System Readiness Monitor was not detected after 20 seconds of operation on a startup that occurred at least 4 minutes after the previous engine startup. **Possible Causes** ● HO2S heater ground circuit is open or has high resistance ● HO2S heater power circuit is open (test O2S fuse in fuse block) ● HO2S heater element is damaged or has failed ● PCM has failed
P0155 **2T O2S HTR, MIL: Yes** 1995 Corvette 5.7L VIN P Transmissions: All	**HO2S-21 (Bank 2 Sensor 1) Heater Circuit Malfunction Conditions:** DTC P0116, P0117, P0118, P0135, P0171, P0172, P0174, P0175, P1114 and P1115 not set, engine started, ECT and IAT sensors less than 122ºF and within 11ºF at startup, TP angle under 16%, and the PCM detected the HO2S signal was 300-700 mv for 200 seconds (ECT sensor under -40ºF at startup) or over 48 seconds (ECT sensor over 148ºF at startup). **Possible Causes** ● HO2S heater ground circuit is open or has high resistance ● HO2S heater power circuit is open (test O2S fuse in fuse block) ● HO2S heater element is damaged or has failed ● PCM has failed
P0155 **2T O2S HTR, MIL: Yes** 1997, 1998, 1999, 2000, 2001, 2002, 2003, 2004, 2005 Corvette 5.7L VIN G, 5.7L VIN S Transmissions: All	**HO2S-21 (Bank 2 Sensor 1) Heater Circuit Malfunction Conditions:** DTC P0101-P0103, P0106-P0108, P0112-P0118, P0131-P0138, P0140, P0151-P0158, P0160, P0200, P0300, P0410, P0440, P0442-P0446, P0452, P0453, P1120, P1125, P1220-P1222, P1258, P1415, P1416, P1441, P1514-P1518 not set, HO2S signal from 425-475 mv at startup, ECT and IAT sensors less than 122ºF and within 14.5ºF at startup, MAF sensor less than 21 g/sec, system voltage 9-18v, AIR and Catalyst Tests off, and the PCM detected the HO2S signal remained within 150 mv of the bias voltage for too long a period (depends on the ECT/MAF sensor signals at startup). **Possible Causes** ● HO2S heater ground circuit is open or has high resistance ● HO2S heater power circuit is open (test O2S fuse in fuse block) ● HO2S heater element is damaged or has failed ● PCM has failed
P0156 **2T CCM, MIL: Yes** 1997, 1998, 1999, 2000, 2001 Catera 3.0L V6 VIN R engine Transmissions: A/T	**HO2S-22 (Bank 2 Sensor 2) Circuit Malfunction Conditions:** DTC P0135, P0139, P0140 and P0141 not set, engine speed from 1400-2400 rpm, HO2S-12 signal is operational, engine load from 8-12%, MAF sensor from 18-22 g/sec, throttle angle from 6-10%, AIR System test "off", engine runtime over 10 minutes, low fuel not "on", TWC temperature less than 1472ºF, and the PCM detected the HO2S signal was below 40 mv. **Possible Causes** ● Air leaks in the exhaust system, intake manifold, vacuum lines ● EVAP Purge system malfunction or charcoal canister problems ● HO2S signal circuit is open between the sensor and the PCM ● HO2S signal circuit is shorted to sensor or chassis ground ● HO2S is damaged, contaminated or air reference hole clogged ● PCM has failed
P0157 **2T CCM, MIL: Yes** 1996, 1997 Camaro, Corvette, Caprice, Firebird & Fleetwood 4.3L VIN W, 5.7L VIN 5, 5.7L VIN P engines Transmissions: All	**HO2S-22 (Bank 2 Sensor 2) Circuit Low Input Conditions:** DTC P0101-P0103, P0112, P0113, P0117, P0118, P0125, P0200, P0335, P0336, P0351-P0358, P0372 and P1371 not set, engine running in closed loop, TP angle 2-70%, AIR and Catalyst Tests "off", and the PCM detected the HO2S input was below 39 mv for 38 seconds. **Possible Causes** ● Air leaks in the exhaust system, intake manifold, vacuum lines ● Engine misfire condition present (look for P0300 series codes) ● Fuel system too lean (possible low fuel pressure, water in fuel) ● HO2S signal circuit is shorted to the sensor or chassis ground ● HO2S is damaged (i.e., cracked) or air reference hole clogged ● PCM has failed

OBD II Trouble Code List (P0xxx Codes)

DTC	Trouble Code Title, Conditions & Possible Causes
P0157 **2T CCM, MIL: Yes** 1996, 1997, 1998, 1999, 2000, 2001, 2002, 2003, 2004, 2005 Body Codes: C, G, K, P 5.0L VIN M, 5.7L VIN K, 5.7L VIN R, 7.4L VIN J Transmissions: All	**HO2S-22 (Bank 2 Sensor 2) Circuit Low Input Conditions:** DTC P0101-P0103, P0106-P0108, P0112, P0113, P0116-P0118, P0121-P0123, P0200, P0300, P0401, P0404, P0405, P0440, P0442, P0446, P0452, P0453, P1120, P1125, P1220, P1221, P1258, P1404, P1441, P1514, P1515, P1516, P1517 and P1518 not set, engine started, engine running in closed loop, fuel level over 10%, system voltage over 10v, TP angle from 8-50% or on models with TAC, the APP sensor indicated angle from 3-70%, MAP sensor more than 25 kPa, Intrusive and Scan Tool Tests "off", then with the Lean Test enabled, the PCM detected the HO2S signal was less than 26 mv for 110 seconds or during the P/E Mode test, the PCM detected the HO2S signal was less than 399 mv for 40 seconds. **Possible Causes** ● Air leaks in the exhaust system, intake manifold, vacuum lines ● Engine misfire condition present (look for P0300 series codes) ● Fuel system too lean (possible low fuel pressure, water in fuel) ● HO2S signal circuit is shorted to the sensor or chassis ground ● HO2S is damaged (i.e., cracked) or air reference hole clogged ● PCM has failed
P0157 **2T CCM, MIL: Yes** 1999, 2000, 2001, 2002, 2003, 2004, 2005 C/K Series Truck, G Series Van, Envoy, Escalade, TrailBlazer 4.8L VIN V, 5.3L VIN P, 5.3L VIN T, 5.3L VIN Z, 6.0L VIN N, 6.0L VIN U, 8.1L VIN G Transmissions: All	**HO2S-22 (Bank 2 Sensor 2) Circuit Low Input Conditions:** DTC P0101-P0103, P0106-P0108, P0112, P0113, P0116-P0118, P0120, P0121-P0123, P0169, P0178, P0179, P0200, P0220, P0300, P0442, P0446, P0452-P0496, P1125, P1258, P1514, P1515, P1516, P1518, P2108 and P2135 not set, engine started, engine running in closed loop, system voltage from 10-18v, Fuel Alcohol content less than 90%, fuel level over 10%, TP angle from 3-70% over the idle value, Lean Test enabled, the PCM detected the HO2S signal was below 80 mv for 200 seconds or with engine runtime over 30 seconds, and during the P/E test, the PCM detected the HO2S signal was below 490 mv for 10 seconds. **Possible Causes** ● Air leaks in the exhaust system, intake manifold, vacuum lines ● Engine misfire condition present (look for P0300 series codes) ● Fuel system too lean (possible low fuel pressure, water in fuel) ● HO2S signal circuit is shorted to the sensor or chassis ground ● HO2S is damaged (i.e., cracked) or air reference hole clogged ● PCM has failed
P0157 **2T CCM, MIL: Yes** 2003, 2004, 2005 CTS (Cadillac) 2.6L V6 VIN M & 3.2L V6 VIN N engines Transmissions: All	**HO2S-22 (Bank 2 Sensor 2) Circuit Low Input Conditions:** DTC P0030-P0038, P0135, P0155, P0171-P0175, P0300, P0301-P0306, P0341, P0342, P0343, P0443, P0444 and P0445 not set, engine started, TP angle from 5-35%, system voltage over 10.5v, MAF sensor more than 9.72 g/sec, Calculated Converter temperature under 1,472°F, and the PCM detected the HO2S-22 input was below 40 mv for 680 seconds. **Possible Causes** ● EVAP system leaks or restrictions, charcoal canister problems ● Fuel system too lean (possible low fuel pressure, water in fuel) ● HO2S air reference hole plugged (check for dirt on the outside) ● HO2S signal circuit is shorted to sensor or chassis ground ● HO2S is damaged or contaminated due to improper fuel usage ● PCM has failed
P0157 **2T CCM, MIL: Yes** 1998, 1999, 2000, 2001, 2002, 2003, 2004, 2005 Camaro, Corvette & Firebird 5.7L VIN G engine Transmissions: All	**HO2S-22 (Bank 2 Sensor 2) Circuit Low Input Conditions:** DTC P0101-P0103, P0106-P0108, P0112-P0113, P0116-P0118, P0121-P0123, P0125, P0128, P0200, P0300, P0410-P0446, P0452-P0453, P1258, P1415-P1416 and P1441 not set, engine started, engine running in closed loop, Intrusive and Scan Tool tests "off", A/F ratio at 14.5-14.8:1, TP angle from 3-70%, then with the Lean Test enabled, the PCM detected the HO2S signal was less than 80 mv for 165 seconds or with P/E test enabled for 1 second, the PCM detected the HO2S signal was less than 420 mv for 400 seconds. **Possible Causes** ● Air leaks in the exhaust system, intake manifold, vacuum lines ● Engine misfire condition present (look for P0300 series codes) ● Fuel system too lean (possible low fuel pressure, water in fuel) ● HO2S signal circuit is shorted to the sensor or chassis ground ● HO2S is damaged (i.e., cracked) or air reference hole clogged ● PCM has failed
P0157 **2T CCM, MIL: Yes** 1996, 1997 Camaro, Corvette, Caprice, Firebird & Fleetwood 4.3L VIN W, 5.7L VIN 5, 5.7L VIN P engines Transmissions: All Fleetwood	**HO2S-22 (Bank 2 Sensor 2) Circuit High Input Conditions:** DTC P0100, P0102, P0103, P0107, P0108, P0117, P0118, P0125, P0200, P0372 and P1371 not set, engine started, engine running in closed loop, TP angle from 2-70%, Catalyst & AIR Tests "off", and the PCM detected the HO2S signal was more than 930 mv for 38 seconds. **Possible Causes** ● Fuel system rich (high fuel pressure, fuel pressure regulator leaking or injector leaking) ● HO2S element is silicon, water or fuel contaminated ● HO2S signal circuit is shorted to system power (B+) ● HO2S signal tracking (water intrusion) in the connector causing a short between the HO2S signal and heater power circuits ● PCM has failed

OBD II Trouble Code List (P0xxx Codes)

DTC	Trouble Code Title, Conditions & Possible Causes
P0158 **2T CCM, MIL: Yes** 1996, 1997, 1998, 1999, 2000, 2001, 2002, 2003, 2004, 2005 Body Codes: C, G, K, P 5.0L VIN M, 5.7L VIN K, 5.7L VIN R, 7.4L VIN J Transmissions: All	**HO2S-22 (Bank 2 Sensor 2) Circuit High Input Conditions:** DTC P0101-P0103, P0106-P0108, P0112, P0113, P0116-P0118, P0121-P0123, P0200, P0300, P0401, P0404, P0405, P0440, P0442, P0446, P0452, P0453, P1120, P1125, P1220, P1221, P1258, P1404, P1441, P1514-P1517 and P1518 not set, engine started, fuel level over 10%, Intrusive and Scan Tool Tests all "off", then with the Rich Test enabled, A/F ratio from 14.5-14.7:1, TP angle from 3.5-70% for 5 seconds, or for vehicles with TAC, with the TP indicated angle from 3-70%, the PCM detected the HO2S signal was more than 930 mv for 200 seconds or with Decel Fuel Cutoff active and the time since the test started over 1 second, the PCM detected the HO2S signal was more than 480 mv for 5 seconds. **Possible Causes** ● Fuel system rich (high fuel pressure, fuel pressure regulator leaking, or injector sticking) ● HO2S element is silicon, water or fuel contaminated ● HO2S signal circuit is shorted to system power (B+) ● HO2S signal tracking (water intrusion) in the connector causing a short between the HO2S signal and heater power circuits ● PCM has failed
P0158 **2T CCM, MIL: Yes** 1999, 2000, 2001, 2002, 2003, 2004, 2005 C/K Series Truck, G Series Van, Envoy, Escalade, TrailBlazer 4.8L VIN V, 5.3L VIN P, 5.3L VIN T, 5.3L VIN Z, 6.0L VIN N, 6.0L VIN U, 8.1L VIN G Transmissions: All	**HO2S-22 (Bank 2 Sensor 2) Circuit High Input Conditions:** DTC P0101, P0102, P0103, P0106, P0107, P0108, P0112, P0113, P0116, P0117, P0118, P0120, P0121, P0122, P0123, P0169, P0178, P0179, P0200, P0220, P0300, P0442, P0446, P0452, P0453, P0455, P0496, P1125, P1258, P1514, P1515, P1516, P1518, P2108 and P2135 not set, engine started, engine running in closed loop, system voltage from 10-18v, Fuel Alcohol content less than 90%, fuel level over 10%, TP angle from 3-70% more than the idle value, then during the Rich Test, the PCM detected the HO2S signal was more than 950 mv for 200 seconds or with engine runtime over 30 seconds, and during the Decel Fuel Cutoff test, the PCM detected the HO2S signal was less than 250 mv for 5 seconds. **Possible Causes** ● Fuel system rich (high fuel pressure, fuel pressure regulator leaking, or injector sticking) ● HO2S element is silicon, water or fuel contaminated ● HO2S signal tracking (water intrusion) in the connector causing a short between the HO2S signal and heater power circuits ● PCM has failed
P0158 **2T CCM, MIL: Yes** 2003, 2004, 2005 CTS (Cadillac) 2.6L V6 VIN M & 3.2L V6 VIN N engines Transmissions: All	**HO2S-22 (Bank 2 Sensor 2) Insufficient Activity Conditions:** Engine started; system voltage over 10.5v, Calculated Converter Temperature less than 1,472ºF, and the PCM detected the HO2S-12 signal remained above 1080 mv for 5 seconds, or it detected the HO2S-22 signal remained over 200 mv during DFCO for 3 seconds **Possible Causes** ● EVAP system leaks or restrictions, charcoal canister problems ● Exhaust system is leaking or severely restricted ● HO2S air reference hole plugged (check for dirt on the outside) ● HO2S signal circuit is open, or the low reference circuit is open ● HO2S is damaged or contaminated due to improper fuel usage ● PCM has failed
P0158 **2T CCM, MIL: Yes** 1998, 1999, 2000, 2001, 2002 Camaro & Firebird 5.7L VIN G engine Transmissions: All	**HO2S-22 (Bank 2 Sensor 2) Circuit High Input Conditions:** DTC P0101-P0103, P0106-P0108, P0112-P0118, P0121-P0123, P0125, P0128, P0200, P0300, P0410, P0440-P0453, P1258, P1415, P1416 and P1441 not set, system voltage over 10.0v, fuel level over 10%, Intrusive and Scan Tool tests "off", Rich Test enabled, TP angle from 3-70%, the PCM detected the HO2S signal was over 930 mv for 200 seconds, or with DFCO enabled, that the HO2S signal was over 480 mv for 5 seconds. **Possible Causes** ● Fuel system rich (high fuel pressure, fuel pressure regulator leaking, or injector sticking) ● HO2S element is silicon, water or fuel contaminated ● HO2S signal circuit is shorted to system power (B+) ● HO2S signal tracking (water intrusion) in the connector causing a short between the HO2S signal and heater power circuits ● PCM has failed
P0158 **2T CCM, MIL: Yes** 1997, 1998, 1999, 2000, 2001, 2002, 2003, 2004, 2005 Corvette 5.7L VIN G, 5.7L VIN S Transmissions: All	**HO2S-22 (Bank 2 Sensor 2) Circuit High Input Conditions:** DTC P0101-P0103, P0106-P0108, P0112, P0113, P0116-P0118, P0200, P0300, P0410, P0440, P0442, P0446, P0452, P0453, P1120, P1125, P1220, P1221, P1258, P1415, P1416, P1441, P1514, P1515, P1516, P1517, or P1518 not set, engine started, TP angle at 3-70% in closed loop, AIR and Catalyst Tests "off", Rich Test enabled, the PCM detected the HO2S signal was over 930 mv for 200 seconds or it was over 480 mv for 5 seconds in DFCO mode. **Possible Causes** ● Fuel system rich (high fuel pressure, fuel pressure regulator leaking, or injector sticking) ● HO2S element is silicon, water or fuel contaminated ● HO2S signal circuit is shorted to system power (B+) ● HO2S signal tracking (water intrusion) in the connector causing a short between the HO2S signal and heater power circuits ● PCM has failed

OBD II Trouble Code List (P0xxx Codes)

DTC	Trouble Code Title, Conditions & Possible Causes
P0158 **2T CCM, MIL: Yes** 1995 Corvette 5.7L VIN P Transmissions: All	**HO2S-22 (Bank 2 Sensor 2) Circuit High Input Conditions:** Engine started engine running in closed loop, TP angle from 2-20%, and the PCM detected the HO2S signal was more than 924 mv. **Possible Causes** ● Fuel system is too rich (fuel pressure too high, fuel pressure regulator leaking, or one or more fuel injectors sticking/leaking) ● HO2S element is silicon, water or fuel contaminated ● HO2S signal circuit is shorted to system power (B+) ● HO2S signal tracking (water intrusion) in the connector causing a short between the HO2S signal and heater power circuits ● PCM has failed
P0158 **2T CCM, MIL: Yes** 1997, 1998, 1999, 2000, 2001 Catera 3.0L V6 VIN R engine Transmissions: All	**HO2S-22 (Bank 2 Sensor 2) Circuit High Input Conditions:** DTC P0157 not set, engine started, Catalytic Converter temperature below 1472°F, system voltage over 11v, and the PCM detected the HO2S signal was over 1080 mv for 5.1 seconds. **Possible Causes** ● Fuel system is too rich (fuel pressure too high, fuel pressure regulator leaking, or one or more fuel injectors sticking/leaking) ● HO2S element is silicon, water or fuel contaminated ● HO2S signal circuit is shorted to system power (B+) ● HO2S signal tracking (water intrusion) in the connector causing a short between the HO2S signal and heater power circuits ● PCM has failed
P0159 **2T CCM, MIL: Yes** 1997, 1998, 1999, 2000, 2001 Catera 3.0L V6 VIN R engine Transmissions: All	**HO2S-22 (Bank 2 Sensor 2) Slow Response Conditions:** DTC P0151, P0152, P0153, P0154, P0155 and P0161 not set, engine started, engine running with the HO2S-12 indicating closed loop operation, system voltage over 11v, and the PCM detected the HO2S signal was fixed at over 600 mv or fixed at less than 600 mv or the HO2S signal was more than 200 mv during Decel Fuel Cutoff. **Possible Causes** ● Exhaust leak present in the exhaust manifold or exhaust pipes ● Fuel system is too rich (fuel pressure too high, fuel pressure regulator leaking, or one or more fuel injectors sticking/leaking) ● HO2S element is silicon, water or fuel contaminated ● TP sensor element broken (can cause false acceleration event)
P0160 **2T O2S, MIL: Yes** 1996, 1997 Camaro, Corvette, Caprice, Firebird & Fleetwood 4.3L VIN W, 5.7L VIN 5, 5.7L VIN P engines Transmissions: All	**HO2S-22 (Bank 2 Sensor 2) Insufficient Activity Conditions:** DTC P0100, P0102, P0103, P0107, P0108, P0112, P0113, P0117, P0118, P0125, P0200, P0372 and P1371 not set, engine speed from 1000-3000 rpm in closed loop for 4 minutes, AIR and Catalyst Tests "off", ECT sensor more than 118°F, and the PCM detected the HO2S signal was fixed between 391-491 mv for 68 seconds during the CCM test period. **Possible Causes** ● Exhaust leak present in the exhaust manifold or exhaust pipes ● HO2S signal or ground circuit has a high resistance condition ● HO2S element is silicon, water or fuel contaminated
P0160 **2T O2S, MIL: Yes** 1996, 1997, 1998, 1999, 2000, 2001, 2002, 2003, 2004, 2005 Body Codes: C, G, K, P 5.0L VIN M, 5.7L VIN K, 5.7L VIN R, 7.4L VIN J Transmissions: All	**HO2S-22 (Bank 2 Sensor 2) Insufficient Activity Conditions:** DTC P0101-P0103, P0106-P0108, P0112, P0113, P0116-P0118, P0121-P0123, P0200, P0300, P0401, P0404, P0405, P0440, P0442, P0446, P0452-P0453, P1120, P1125, P1220-P1221, P1258, P1404, P1441, P1514-P1517 and P1518 not set, engine runtime over 409 seconds, system voltage over 10.0v, Intrusive tests all off, TP angle change over 5%, and the PCM detected the HO2S signal was 410-490 mv for 150 seconds. **Possible Causes** ● Exhaust leak present in the exhaust manifold or exhaust pipes ● HO2S signal or ground circuit has a high resistance condition ● HO2S element is silicon, water or fuel contaminated
P0160 **2T O2S, MIL: Yes** 2003, 2004, 2005 CTS (Cadillac) 2.6L V6 VIN M & 3.2L V6 VIN N engines Transmissions: All	**HO2S-22 (Bank 2 Sensor 2) Insufficient Activity Conditions:** Engine started; system voltage over 10.5v, Calculated Converter Temperature less than 1,472°F, and the PCM did not detect any activity from the HO2S-22, or it detected the HO2S-22 signal remained between 400-520 mv for more than 600 seconds. **Possible Causes** ● EVAP system leaks or restrictions, charcoal canister problems ● Exhaust system is leaking or severely restricted ● HO2S air reference hole plugged (check for dirt on the outside) ● HO2S signal circuit is open, or the low reference circuit is open ● HO2S is damaged or contaminated due to improper fuel usage ● PCM has failed

OBD II Trouble Code List (P0xxx Codes)

DTC	Trouble Code Title, Conditions & Possible Causes
P0160 **2T O2S, MIL: Yes** 1999, 2000, 2001, 2002, 2003, 2004, 2005 C/K Series Truck, G Series Van, Envoy, Escalade, TrailBlazer 4.8L VIN V, 5.3L VIN P, 5.3L VIN T, 5.3L VIN Z, 6.0L VIN N, 6.0L VIN U, 8.1L VIN G Transmissions: All	**HO2S-22 (Bank 2 Sensor 2) Insufficient Activity Conditions:** DTC P0101, P0102, P0103, P0106, P0107, P0108, P0112, P0113, P0116, P0117, P0118, P0120, P0121, P0122, P0123, P0169, P0178, P0179, P0200, P0220, P0300, P0442, P0446, P0452, P0453, P0455, P0496, P1125, P1258, P1514, P1515, P1516, P1518, P2108 and P2135 not set, engine runtime over 300 seconds, system voltage from 10-18v, Fuel Alcohol content less than 90%, then after the TP indicated angle changed more than 5% within one seconds six times on models with a TAC system, the PCM detected the HO2S signal remained between 410-490 mv for 150 seconds. **Possible Causes** ● HO2S heater is damaged or it has failed ● HO2S signal or ground circuit has a high resistance condition ● HO2S has failed (i.e., it is silicon, water or fuel contaminated) ● PCM has failed
P0160 **2T O2S, MIL: Yes** 1998, 1999, 2000, 2001, 2002 Camaro & Firebird 5.7L VIN G engine Transmissions: All	**HO2S-22 (Bank 2 Sensor 2) Insufficient Activity Conditions:** DTC P0101-P0103, P0106-P0108, P0112-P0118, P0121-P0123, P0125, P0128, P0200, P0300, P0410, P0440-P0446, P0452, P0453, P1258, P1415-P1416 and P1441 not set, engine started, engine runtime over 409 seconds, Scan Tool and Intrusive tests "off", system voltage over 10.0v, TP angle or APP indicated angle changed over 1.5% six times, and the PCM detected the HO2S signal was fixed from 412-490 mv for 1 minute (test failed 6 times). **Possible Causes** ● Exhaust leak present in the exhaust manifold or exhaust pipes ● HO2S signal or ground circuit has a high resistance condition ● HO2S element is silicon, water or fuel contaminated ● PCM has failed
P0160 **2T O2S, MIL: Yes** 1997, 1998, 1999, 2000, 2001 Catera 3.0L V6 VIN R engine Transmissions: A/T	**HO2S-22 (Bank 2 Sensor 2) Insufficient Activity Conditions:** DTC P0130 and P0150 not set, engine started, engine speed from 1400-2400 rpm in closed loop for 25 seconds, system voltage over 10.0v, engine load from 8-12%, MAF sensor from 18-22 g/sec, TP angle changed at least 1.5% at least six times, and the PCM detected the HO2S signal was fixed at 401-519 mv for 60 seconds. **Possible Causes** ● Exhaust leak present in the exhaust manifold or exhaust pipes ● HO2S signal or ground circuit has a high resistance condition ● HO2S element is silicon, water or fuel contaminated ● PCM has failed
P0160 **2T O2S, MIL: Yes** 1997, 1998, 1999, 2000, 2001, 2002, 2003, 2004, 2005 Corvette 5.7L VIN G, 5.7L VIN S Transmissions: All	**HO2S-22 (Bank 2 Sensor 2) Insufficient Activity Conditions:** DTC P0101-P0103, P0106-P0108, P0112-P0118, P0200, P0300, P0410, P0440-P0446, P0452, P0453, P1120, P1125, P1220, P1221, P1258, P1415, P1416, P1441, P1514 to P1518 not set, engine runtime 70 seconds while in closed loop, system voltage over 10.0v, Intrusive and Scan Tool tests off, and the PCM detected the HO2S signal was fixed from 409-489 mv for 1 minute. **Possible Causes** ● Exhaust leak present in the exhaust manifold or exhaust pipes ● HO2S signal or ground circuit has a high resistance condition ● HO2S element is silicon, water or fuel contaminated ● PCM has failed
P0160 **2T O2S, MIL: No** 1995 Corvette 5.7L VIN P Transmissions: All	**HO2S-22 (Bank 2 Sensor 2) Insufficient Activity Conditions:** Engine started engine running in closed loop, TP angle from 2-10%, and the PCM detected the HO2S signal was fixed from 356-497 mv for 102 seconds during the CCM test of the Oxygen sensor circuit. **Possible Causes** ● Exhaust leak present in the exhaust manifold or exhaust pipes ● HO2S signal or ground circuit has a high resistance condition ● HO2S element is silicon, water or fuel contaminated
P0161 **2T O2S HTR, MIL: Yes** 1996, 1997 Camaro, Corvette, Caprice, Firebird & Fleetwood 4.3L VIN W, 5.7L VIN 5, 5.7L VIN P engines Transmissions: All	**HO2S-22 (Bank 2 Sensor 2) Heater Circuit Malfunction Conditions:** DTC P0101-P0103, P0112, P0113, P0117, P0118, P0125, P0132, P0133, P0134 and P0200 not set, ECT and IAT sensors less than 122°F and within 5°F at startup, engine started, system voltage over 10.0v, AIR and Catalyst tests off, MAF sensor less than 40 g/sec, TP angle under 20%, and the PCM detected the HO2S signal was fixed from 300-700 mv too long (depends on ECT/MAF at startup). **Possible Causes** ● HO2S heater ground circuit is open or has high resistance ● HO2S heater power circuit is open (test O2S fuse in fuse block) ● HO2S heater element is damaged or has failed ● PCM has failed

OBD II Trouble Code List (P0xxx Codes)

DTC	Trouble Code Title, Conditions & Possible Causes
P0161 **2T O2S HTR, MIL: Yes** 1996, 1997, 1998, 1999, 2000, 2001, 2002, 2003, 2004, 2005 Body Codes: C, G, K, P 5.0L VIN M, 5.7L VIN K, 5.7L VIN R, 7.4L VIN J Transmissions: All	**HO2S-22 (Bank 2 Sensor 2) Heater Circuit Malfunction Conditions:** DTC P0101-P0103, P0106-P0108, P0112, P0113, P0116-P0118, P0121-P0123, P0151, P0152, P0154, P0157, P0158, P0160, P0200, P0300, P0401, P0404, P0405, P0440, P0442, P0446, P0452, P0453, P1120, P1125, P1220, P1221, P1258, P1404, P1441, P1514, P1515, P1516, P1517 and P1518 not set, ECT and IAT sensors less than 122°F and with 14.5°F at startup, engine running, HO2S signal from 425-475 mv right after startup, Intrusive and Scan Tool tests "off", MAF sensor below 25 g/sec, and the PCM detected the HO2S signal was within 150 mv of startup HO2S signal for too long (depends on ECT/MAF signals at startup). **Possible Causes** ● HO2S heater ground circuit is open or has high resistance ● HO2S heater power circuit is open (test O2S fuse in fuse block) ● HO2S heater element is damaged or has failed ● PCM has failed ● TSB 00-06-04-006 contains a repair procedure for this code
P0161 **2T O2S HTR, MIL: Yes** 1999, 2000, 2001, 2002, 2003, 2004, 2005 C/K Series Truck, G Series Van, Envoy, Escalade, TrailBlazer 4.8L VIN V, 5.3L VIN P, 5.3L VIN T, 5.3L VIN Z, 6.0L VIN N, 6.0L VIN U, 8.1L VIN G Transmissions: All	**HO2S-22 (Bank 2 Sensor 2) Heater Circuit Malfunction Conditions:** DTC P0101, P0102, P0103, P0106, P0107, P0108, P0112, P0113, P0116, P0117, P0118, P0120, P0121, P0122, P0123, P0169, P0178, P0179, P0200, P0220, P0300, P0442, P0446, P0452, P0453, P0455, P0496, P1125, P1258, P1514, P1515, P1516, P1518, P2108 and P2135 not set, engine runtime 2 minutes, engine speed from 500-3000 rpm, system voltage from 10-18v, ECT sensor more than 122°F, MAF sensor from 3-40 g/sec, Fuel Alcohol content less than 90%, and the PCM detected the HO2S heater current was less than 0.25 amps, or over 3.125 amps (over 1.375 amps on 4.8L). **Possible Causes** ● HO2S heater low control circuit is open or shorted to ground ● HO2S heater circuit is open or it is shorted to ground ● HO2S heater power circuit is open (test O2A fuse in fuse block) ● HO2S heater element is damaged or has failed ● PCM has failed
P0161 **2T O2S HTR, MIL: Yes** 2003, 2004, 2005 CTS (Cadillac) 2.6L V6 VIN M & 3.2L V6 VIN N engines Transmissions: All	**HO2S-22 (Bank 2 Sensor 2) Heater Circuit Malfunction Conditions:** DTC P0030, P0031, P0032, P0036, P0037, P0038, P0050, P0051, P0052, P0056, P0057, P0058, P0157, P0158, P0160 and P0161 not set, engine started, Calculated Converter Temperature from 626-1,112°F, system voltage over 10.5v, system voltage from 10.5-16v, and the PCM detected the HO2S-22 heater current was not within the expected range with the appropriate heater commanded "on". **Possible Causes** ● HO2S heater high side driver circuit has high resistance (open) ● HO2S heater low side driver circuit has high resistance (open) ● HO2S heater power circuit is open (test O2S fuse in fuse block) ● HO2S heater element is damaged or has failed ● PCM has failed
P0161 **2T O2S HTR, MIL: Yes** 1998, 1999, 2000, 2001, 2002 Camaro & Firebird 5.7L VIN G engine Transmissions: All	**HO2S-22 (Bank 2 Sensor 2) Heater Circuit Malfunction Conditions:** DTC P0101-P0103, P0106-P0108, P0112-P0118, P0121-P0123, P0125, P0128, P0131-P0138, P0140, P0151-P0158, P0160, P0200, P0300, P0410, P0440-P0446, P0452, P0453, P1258, P1415, P1416 or P1441 not set, ECT and IAT sensors under 122°F and within 14.5°F at startup, engine running, HO2S from 425-475 mv, Scan Tool and Intrusive tests "off", MAF sensor under 18 g/sec, and the PCM detected the HO2S signal was fixed at 300-700 mv for 42 seconds to 2 minutes. **Possible Causes** ● HO2S heater ground circuit is open or has high resistance ● HO2S heater power circuit is open (test O2S fuse in fuse block) ● HO2S heater element is damaged or has failed ● PCM has failed
P0161 **2T O2S HTR, MIL: Yes** 1997, 1998, 1999, 2000, 2001 Catera 3.0L V6 VIN R engine Transmissions: A/T	**HO2S-22 (Bank 2 Sensor 2) Heater Circuit Malfunction Conditions:** DTC P0136, P0138, P0139 and P0140 not set, engine started, system voltage over 11.0v, HO2S heater "on", Catalytic Converter temperature from 626-928°F, and the PCM detected the HO2S heater circuit resistance was out of normal range for 15 seconds during the CCM test of the Oxygen Sensor circuit. **Possible Causes** ● HO2S heater ground circuit is open or has high resistance ● HO2S heater power circuit is open (test O2S fuse in fuse block) ● HO2S heater element is damaged or has failed ● PCM has failed

OBD II Trouble Code List (P0xxx Codes)

DTC	Trouble Code Title, Conditions & Possible Causes
P0161 **2T O2S HTR, MIL: Yes** 1997, 1998, 1999, 2000, 2001, 2002, 2003, 2004, 2005 Corvette 5.7L VIN G, 5.7L VIN S Transmissions: All	**HO2S-22 (Bank 2 Sensor 2) Heater Circuit Malfunction Conditions:** DTC P0101-P0103, P0106-P0108, P0112, P0113, P0116, P0117, P0118, P0151, P0152, P0154, P0157, P0158, P0160, P0200, P0300, P0410, P0440, P0442, P0446, P0452, P0453, P1120, P1125, P1220, P1221, P1221, P1258, P1415, P1416, P1441, P1514, P1515, P1516, P1517 and P1518 not set, HO2S signal from 425-475 mv at startup, ECT and IAT sensors less than 122°F and within 14.5°F at startup, MAF sensor less than 21 g/sec, system voltage 9-18v, AIR and Catalyst Tests off, and the PCM detected the HO2S signal remained within 150 mv of the bias voltage for too long. **Possible Causes** ● HO2S heater ground circuit is open or has high resistance ● HO2S heater power circuit is open (test O2S fuse in fuse block) ● HO2S heater element is damaged or has failed ● PCM has failed
P0161 **2T O2S HTR, MIL: Yes** 1995 Corvette 5.7L VIN P Transmissions: All	**HO2S-22 (Bank 2 Sensor 2) Heater Circuit Malfunction Conditions:** Engine started ECT and IAT sensors less than 122°F and within 11°F at startup, engine running, TP angle less than 16%, and the PCM detected the HO2S signal was fixed from 300-700 mv for 200 seconds with an ECT sensor at -40°F, or it was fixed from 300-700 mv for over 48 seconds with an ECT sensor more than 154°F. **Possible Causes** ● HO2S heater ground circuit is open or has high resistance ● HO2S heater power circuit is open (test O2S fuse in fuse block) ● HO2S heater element is damaged or has failed ● PCM has failed
P0168 **1T CCM, MIL: No** 2003, 2004, 2005 Body Codes: C, K 6.6L VIN 1 Transmissions: All	**Fuel Temperature Sensor Signal Range/Performance Conditions:** Key on or engine running; and the PCM detected the Fuel Temperature sensor was under 0.10v (Scan Tool reads over 252°F). **Possible Causes** ● Inspect fuel cooler in front of fuel tank for debris restricting the airflow or damage to the cooling fins, and fuel lines for damage. ● Fuel temperature sensor connector is damaged or shorted ● Fuel temperature sensor signal circuit is shorted to ground ● Fuel temperature sensor is damaged or it has failed ● ECM has failed
P0169 **1T CCM, MIL: No** 2000, 2001, 2002 S/T Blazer & S/T Pickup 2.2L VIN 5 CNG engine Transmissions: All	**Fuel Composition Sensor Circuit Malfunction Conditions:** Engine started, ECT and IAT sensors within less than 8°F at startup, and after the PCM monitored the Fuel Composition Sensor (FCS) frequency during the first 30 seconds of runtime, it detected the fuel composition was too high for 25 out of a 30 seconds. **Fuel Composition Sensor Overview** The FCS supplies the PCM with the alcohol content and alcohol percentage, information is supplied to the PCM. If the PCM detects that the alcohol content is too high for a given ambient temperature, it will set this code. This code can set if the fuel is contaminated, and causing the calculated alcohol content to be higher than expected. The FCS frequency at 32°F or higher is 138 Hz at 88% Ethanol. **Possible Causes** ● FCS is contaminated, damaged or skewed ● Fuel is contaminated due to a higher than normal alcohol level ● Ethanol concentration is higher than 85% (E85) ● Methanol fuel is not recommended, and can cause this code
P0169 **1T CCM, MIL: No** 2001, 2002, 2003, 2004, 2005 Body Codes: C, G, K 6.0L VIN N, 6.0 VIN U Transmissions: All	**Fuel Composition Sensor Circuit Malfunction Conditions:** Engine started ECT and IAT sensors within less than 8°F at startup, and the PCM detected the Fuel Composition Sensor (FCS) signal was above the given threshold for the ambient temperature for 25 seconds out of a 30 second interval. **Fuel Composition Sensor Overview** The PCM uses data from the IAT and ECT sensors to determine ambient temperature at cold engine startup. Alcohol content (in percentage) information is supplied to the PCM from the Fuel Composition sensor. This data can be to diagnose symptom related complaints such as hard starting in cold ambient temperatures. The PCM monitors the fuel frequency during the first 30 seconds of run time. When the PCM detects alcohol content that is too high for a given ambient temperature, it sets this trouble code (DTC P0169). The FCS frequency at 32°F or higher is 138 Hz at 88% Ethanol. **Possible Causes** ● FCS is contaminated, damaged or skewed ● Fuel is contaminated due to a higher than normal alcohol level ● Ethanol concentration is higher than 85% (E85) ● Methanol fuel is not recommended, and can cause this code

OBD II Trouble Code List (P0xxx Codes)

DTC	Trouble Code Title, Conditions & Possible Causes
P0171 **2T FUEL, MIL: Yes** 1996, 1997, 1998, 1999, 2000, 2001, 2002, 2003, 2004, 2005 Achieva, Alero, Beretta, Century, Cavalier, Ciera, Century, Corsica, Envoy, Grand Am, Malibu, S/T Blazer & Pickup, Skylark, Sunfire, TrailBlazer 2.2L VIN 4, 2.2L VIN 5, 2.2L VIN F, 2.2L VIN H, 4.2L VIN S Transmissions: All	**Fuel System Lean (Bank 1) Conditions:** DTC P0105, P0107, P0108, P0112-P0118, P0122, P0123, P0201-P0204, P0300-P0304, P0335, P0440-P0446, P0506, P0507, P0601, P0602 and P1441 not set, engine started, vehicle driven at less than 72 mph at 550-3400 rpm, BARO sensor over 72 kPa, ECT sensor from 140-239°F, IAT sensor from -13°F to 239°F, MAP sensor more than 26 kPa, fuel ethanol percentage under 88%, and the PCM detected the Long Term fuel trim was more than +23%. **Possible Causes** ● Air leaks in intake manifold, exhaust pipes or exhaust manifold ● Fuel control sensor is out of calibration (ECT, IAT or MAP) ● Low fuel pressure (fuel filter clogged, pressure regulator failure) ● One or more injectors restricted or pressure regulator has failed ● HO2S element is contaminated, deteriorated or has failed ● Vacuum hose is disconnected, broken, leaking or loose
P0171 **2T FUEL, MIL: Yes** 1996, 1997, 1998, 1999, 2000, 2001, 2002 Body Codes: A, L, N, W 3.1L VIN M, 3.4L VIN X Transmissions: All	**Fuel System Lean (Bank 1) Conditions:** DTC P0101-P0103, P0107, P0108, P0121-P0123, P0130-P0138, P0140, P0141, P0201-P0206 P0300, P0401, P0403, P0404, P0405, P0410, P0440, P0442, P0446, P0506, P0507, P1404 and P1441 not set, engine started, vehicle speed under 82 mph at 600-4000 rpm, BARO sensor more than 70 kPa, ECT sensor from 68-230°F, IAT sensor from 0-158°F, MAF sensor from 3-150 gm/s, MAP sensor from 15-85 kPa, TP angle less than 90%, fuel level over 10%, and the PCM detected the Long Term fuel trim value was over +20%. **Possible Causes** ● Air leaks in intake manifold, exhaust pipes or exhaust manifold ● Low fuel pressure (fuel filter clogged, pressure regulator failure) ● Fuel control sensor is out of calibration (ECT, IAT or MAF) ● One or more injectors restricted or pressure regulator has failed ● HO2S element is contaminated, deteriorated or has failed ● Vacuum hose is disconnected, broken, leaking or loose
P0171 **1T FUEL, MIL: Yes** 1995 Bonneville, Century, Cutlass Supreme, 88'/98', Grand Prix, Intrigue, LSS, LeSabre, Lumina, Monte Carlo, Regal-Park Avenue 3.1L VIN M, 3.4L VIN X, 3.8L VIN 1, 3.8L VIN K, 3.8L VIN L Transmissions: All	**Fuel Trim System Lean (Bank 1) Conditions:** DTC P0101, P0117, P0118, P0122, P0123, P0134, P0501 and P0502 not set, engine started, vehicle driven at under 70 mph at 1000-4000 rpm, ECT signal from 176-230°F, Purge command less than 80%, TP angle less than 91% and steady, and the PCM detected the Long Term fuel trim count was over 158, or the Short Term count was 180 for 25 seconds. **Possible Causes** ● Air leaks in intake manifold, exhaust pipes or exhaust manifold ● Fuel control sensor is out of calibration (ECT, IAT or MAF) ● Low fuel pressure (fuel filter clogged, pressure regulator failure) ● One or more injectors restricted or pressure regulator has failed ● HO2S element is contaminated, deteriorated or has failed ● Vacuum hose is disconnected, broken, leaking or loose
P0171 **1T FUEL, MIL: Yes** 1996, 1997 Camaro, Corvette, Caprice, Firebird & Fleetwood 4.3L VIN W, 5.7L VIN 5, 5.7L VIN P engines Transmissions: All	**Fuel Trim System Lean (Bank 1) Conditions:** DTC P0100-P0108, P0112-P0123, P0131-P0141, P0152-P0155, P0200, P0300, P0372, P0441, P0443, P0500, P0601, P1133, P1151, P1351, P1361, P1371, P1414, P1416 and P1441 not set, engine speed at 500-4000 rpm, VSS under 75 mph, BARO sensor over 70 kPa, ECT sensor at 120-239°F, IAT sensor at -22 to 302°F, MAF sensor from 5-100 gm/s, MAP sensor at 26-90 kPa, TP angle under 90%, Fuel Trim enabled, Purge solenoid "off", and the PCM detected the Long Term fuel trim was above 20% for 10 seconds. **Possible Causes** ● Air leaks in intake manifold, exhaust pipes or exhaust manifold ● Fuel control sensor is out of calibration (ECT, IAT or MAF) ● Fuel component fault (fuel filter, fuel injector, low fuel pressure) ● HO2S element is contaminated, deteriorated or has failed
P0171 **2T FUEL, MIL: Yes** 1996, 1997, 1998, 1999, 2000, 2001, 2002, 2003, 2004, 2005 Alero, Aztek, Cutlass, Impala, Malibu, Montana, Silhouette, Venture, Rendezvous 3.1L VIN J, 3.1L VIN M, 3.4L VIN E engines Transmissions: All	**Fuel Trim System Lean (Bank 1) Conditions:** DTC P0101, P0103, P0103, P0107, P0108, P0121, P0122, P0123, P0130-P0141, P0201-P0206, P0300, P0401-P0405, P0410, P0440-P0446, P0506, P0507, P1404 and P1441 not set, engine speed from 550-4000 rpm, ECT sensor from 68-230°F, IAT sensor from 64-158°F, MAF sensor at 2.8-150 gm/s, BARO sensor over 70 kPa, MAP sensor at 15-105 kPa, VSS under 82 mph, fuel level over 10%, and the PCM detected the Short Term fuel trim was more than 20% for 6 seconds. **Possible Causes** ● Air leaks in intake manifold, exhaust pipes or exhaust manifold ● Fuel control sensor is out of calibration (ECT, IAT or MAF) ● Fuel component fault (fuel filter, fuel injector, low fuel pressure) ● HO2S element is contaminated, deteriorated or has failed

OBD II Trouble Code List (P0xxx Codes)

DTC	Trouble Code Title, Conditions & Possible Causes
P0171 **2T FUEL, MIL: Yes** 1996, 1997, 1998, 1999, 2000, 2001, 2002, 2003, 2004, 2005 Bonneville, Century, Cutlass Supreme, Grand Prix, Intrigue, LeSabre, LSS, Lumina, Monte Carlo, Regal, Park Avenue, 88' & 98' 3.1L VIN J, 3.1L VIN M, 3.4L VIN X, 3.5L V6 VIN H, 3.8L VIN 1, 3.8L VIN K, 4.0L V8 VIN C Transmissions: All	**Fuel Trim System Lean (Bank 1) Conditions:** DTC P0101-P0103, P0107, P0108, P0121-P0123, P0130-P0138, P0140, P0141, P0201-P0206, P0300, P0401-P0405, P0410, P0440-P0446, P0506, P0507, P1404 or P1441 not set, engine started, vehicle driven at less than 82 mph at 550-4000 rpm, BARO sensor over 70 kPa (10.1 psi), ECT sensor from 68-230ºF, IAT sensor from 64-158ºF, MAF sensor from 2.8-150 gm/s, MAP sensor from 16-105 kPa (2.6-15.2 psi), fuel level over 10%, and the PCM detected the Long Term fuel trim value was more than +20% for 6 seconds. The PCM controls the air/fuel metering system to good overall fuel economy, driveability, and emission control. During open loop, the PCM determines fuel delivery based on sensor signals, without the oxygen sensor input. During closed loop, the PCM adds oxygen sensor inputs to calculate Short and Long term fuel trim fuel delivery adjustments. Short Term fuel trim values change rapidly in response to changes in the oxygen sensor signals. Long Term fuel trim makes coarse adjustments in order to maintain an air/fuel ratio of 14.7:1. **Possible Causes** ● Air leaks in intake manifold, exhaust pipes or exhaust manifold ● Fuel control sensor is out of calibration (ECT, IAT or MAF) ● Low fuel pressure (fuel filter clogged, pressure regulator failure) ● One or more injectors restricted or pressure regulator has failed ● HO2S element is contaminated, deteriorated or has failed ● Vacuum hose is disconnected, broken, leaking or loose
P0171 **2T FUEL, MIL: Yes** 1996, 1997, 1998, 1999, 2000, 2001, 2002, 2003, 2004, 2005 C/K Cab & Chassis, C/K Series Truck, G Series Van, M/L Series Vans 4.3L VIN W, 4.3L VIN X, 5.0L VIN M, 5.7L VIN K, 5.7L VIN R, 7.4L VIN J Transmissions: All	**Fuel Trim System Lean (Bank 1) Conditions:** DTC P0101, P0102, P0103, P0108, P0135, P0137, P0141, P0200, P0300, P0410, P0420, P0430, P0440, P0442, P0443, P0446, P0449, P0506, P0507 and P1441 not set, engine started, vehicle driven at less than 85 mph at 400-3000 rpm, BARO sensor more than 74 kPa (10.7 psi), ECT sensor from 167-239ºF, IAT sensor from 4-194ºF, MAF sensor from 5-90 gm/s, MAP sensor from 26-90 kPa (3.7-13 psi), TP angle less than 90%, fuel level over 10%, and the PCM detected the Long Term fuel trim value was more than +23% for 6 seconds (i.e., a lean A/F mixture existed). **Possible Causes** ● Air leaks in intake manifold, exhaust pipes or exhaust manifold ● Fuel control sensor is out of calibration (ECT, IAT or MAF) ● Low fuel pressure (fuel filter clogged, pressure regulator failure) ● One or more injectors restricted or pressure regulator has failed ● HO2S element is contaminated, deteriorated or has failed ● Vacuum hose is disconnected, broken, leaking or loose
P0171 **2T FUEL, MIL: Yes** 1999, 2000, 2001, 2002, 2003, 2004, 2005 C/K Series Truck, G Series Van, Envoy, Escalade, TrailBlazer 4.8L VIN V, 5.3L VIN P, 5.3L VIN T, 5.3L VIN Z, 6.0L VIN N, 6.0L VIN U, 8.1L VIN G Transmissions: All	**Fuel Trim System Lean (Bank 1) Conditions:** DTC P0101-P0103, P0108, P0135, P0137, P0141, P0200, P0300, P0410, P0420, P0430, P0440, P0442, P0443, P0446, P0449, P0506, P0507 and P1441 not set, engine started, ECT sensor from 167-239ºF, IAT sensor from 4-194ºF, engine speed from 400-3000 rpm, BARO sensor over 74 kPa, MAF sensor from 5-90 gm/s, TP angle less than 90%, VSS less than 85 mph, and the PCM detected the Long Term fuel trim value was more than 23% for 6 seconds (i.e., indicating that a lean A/F mixture was present). **Possible Causes** ● Air leaks in intake manifold, exhaust pipes or exhaust manifold ● Fuel control sensor is out of calibration (ECT, IAT or MAF) ● Low fuel pressure (fuel filter clogged, pressure regulator failure) ● One or more injectors restricted or pressure regulator has failed ● HO2S element is contaminated, deteriorated or has failed ● Vacuum hose is disconnected, broken, leaking or loose
P0171 **2T FUEL, MIL: Yes** 2003, 2004, 2005 C/K Series Truck, G Series Van, Cargo Van 6.0L VIN U CNG engine Transmissions: All	**Fuel Trim System Lean (Bank 1) Conditions:** DTC P0101, P0102, P0103, P0106, P0107, P0108, P0112, P0113, P0116-P0118, P0121-P0123, P0125, P0200, P0300, P0327, P0332, P0335, P0336, P0351-P0358, P0401, P0402, P0403, P0443, P0446, P0449, P0496, P0502, P0503, P1020, or P1258 not set, engine started, ECT sensor from 167-239ºF, IAT sensor from 4-194ºF, engine speed 400-3000 rpm, BARO sensor over 74 kPa, MAF sensor from 5-90 gm/s, TP angle under 90%, VSS less than 85 mph, and the PCM detected the Long Term fuel trim value was over 28% for 6 seconds (indicating a lean A/F mixture was present). **Possible Causes** ● Air leaks in intake manifold, exhaust pipes or exhaust manifold ● Fuel control sensor is out of calibration (ECT, IAT or MAF) ● Low fuel pressure (fuel filter clogged, pressure regulator failure) ● One or more injectors restricted or pressure regulator has failed ● HO2S element is contaminated, deteriorated or has failed ● Vacuum hose is disconnected, broken, leaking or loose

OBD II Trouble Code List (P0xxx Codes)

DTC	Trouble Code Title, Conditions & Possible Causes
P0171 **2T FUEL, MIL: Yes** 1996, 1997, 1998, 1999, 2000, 2001, 2002, 2003, 2004, 2005 DeVille, Eldorado, Seville 4.6L VIN 9, 4.6L VIN Y Transmissions: A/T	**Fuel Trim System Lean (Bank 1) Conditions:** DTC P0101-P0103, P0106-P0108, P0112-P0113, P0116-P0118, P0121-P0123, P0125, P0128, P0200, P0300, P0410-P0446, P0452-P0453, P1258, P1415-P1416 and P1441 not set, vehicle speed under 82 mph at 500-4000 rpm in closed loop, BARO sensor over 75 kPa, ECT sensor from 68-239ºF, IAT sensor from -4ºF to 212ºF, MAF sensor from 3-60 g/sec, EGR Test "off", MAP sensor from 15-85 kPa, TP angle under 65%, and the PCM detected the Long Term fuel trim value was more than +23 for 6 seconds. **Possible Causes** ● Air leaks in intake manifold, exhaust pipes or exhaust manifold ● Fuel control sensor is out of calibration (ECT, IAT or MAF) ● Low fuel pressure (fuel filter clogged, pressure regulator failure) ● One or more injectors restricted or pressure regulator has failed ● HO2S element is contaminated, deteriorated or has failed ● Vacuum hose is disconnected, broken, leaking or loose
P0171 **2T FUEL, MIL: Yes** 1995, 1996, 1997, 1998, 1999, 2000, 2001, 2002 Camaro & Firebird 3.8L VIN K engine Transmissions: All	**Fuel Trim System Lean (Bank 1) Conditions:** DTC P0101-P0103, P0108, P0135, P0137, P0141, P0200, P0300, P0410, P0420, P0430, P0440-P0449, P0506, P0507 and P1441 not set, engine started, vehicle driven at less than 86 mph at 400-0000 rpm, BARO sensor more than 74 kPa (10.7 psi), ECT sensor from 167-239ºF, IAT sensor from 4-194ºF, MAF sensor from 5-90 gm/s, MAP sensor from 26-90 kPa (3.7-13 psi), TP angle less than 90%, fuel level over 10%, and the PCM detected the Long Term fuel trim value was over +23% for 6 seconds during the Fuel Trim test. **Possible Causes** ● Air leaks in intake manifold, exhaust pipes or exhaust manifold ● Fuel control sensor is out of calibration (ECT, IAT or MAF) ● Low fuel pressure (fuel filter clogged, pressure regulator failure) ● One or more injectors restricted or pressure regulator has failed ● HO2S element is contaminated, deteriorated or has failed ● Vacuum hose is disconnected, broken, leaking or loose
P0171 **2T FUEL, MIL: Yes** 1998, 1999, 2000, 2001, 2002 Camaro & Firebird 5.7L VIN G engine Transmissions: All	**Fuel Trim System Lean (Bank 1) Conditions:** DTC P0101-P0103, P0107-P0108, P0112-P0113, P0117-P0118, P0121-P0125, P0200, P0300, P0335-P0336, P0351-P0358, P0405, P0410, P0412, P0418, P0443, P0500-P0503, P1258, P1404, P1415 and P1416 not set, engine running, VSS under 85 mph at 400-3000 rpm, BARO sensor over 74 kPa, ECT sensor from 122-239ºF, IAT sensor from -4ºF-194ºF, MAF sensor from 3-150 gm/s, MAP sensor from 26-90 kPa, TP angle below 90% and steady, Fuel Level over 10%, and the PCM detected the LT fuel trim was over +23% for 6 seconds. **Possible Causes** ● Air leaks in intake manifold, exhaust pipes or exhaust manifold ● Fuel control sensor is out of calibration (ECT, IAT or MAF) ● Low fuel pressure (fuel filter clogged, pressure regulator failure) ● One or more injectors restricted or pressure regulator has failed ● HO2S element is contaminated, deteriorated or has failed ● Vacuum hose is disconnected, broken, leaking or loose
P0171 **2T FUEL, MIL: Yes** 1996, 1997, 1998, 1999, 2001, 2002, 2003, 2004, 2005 Aurora 3.5L V6 VIN H, 4.0L V8 VIN C engines Transmissions: All	**Fuel Trim System Lean (Bank 1) Conditions:** DTC P0101-P0103, P0106-P0108, P0121, P0122, P0131-P0138, P0140-P0146, P0151, P0154, P0562, P0563, P0201-P0208, P0300, P0325, P0327, P0340, P0401, P0506, P0507, P1121, P1133, P1153, P1154, P1441, P1508 and P1509 not set, engine started, vehicle driven at less than 70 mph at 400-3000 rpm, BARO sensor more than 70.8 kPa, ECT sensor from 184-221ºF, IAT sensor from 40ºF to 218ºF, MAF sensor from 3-200 gm/s, MAP sensor from 27-103 kPa, EVAP purge tank not full, TP angle less than 19.8%, Fuel Trim enabled, and the PCM detected the Short Term fuel trim value was over +23% for 6 seconds. **Possible Causes** ● Air leaks in intake manifold, exhaust pipes or exhaust manifold ● Fuel control sensor is out of calibration (ECT, IAT or MAF) ● Low fuel pressure (fuel filter clogged, pressure regulator failure) ● One or more injectors restricted or pressure regulator has failed ● HO2S element is contaminated, deteriorated or has failed ● Vacuum hose is disconnected, broken, leaking or loose
P0171 **2T FUEL, MIL: Yes** 2003, 2004, 2005 CTS (Cadillac) 2.6L V6 VIN M & 3.2L V6 VIN N engines Transmissions: All	**Fuel Trim System Lean (Bank 1) Conditions:** DTC P0101-P0103, P0130-P0138, P0140 and P0141 not set, engine running in closed loop, Fuel Level over 10% and the PCM detected the Long Term fuel trim in Cruise/Acceleration was over +23%, or the Long Term fuel trim Idle/Deceleration parameter was over 8.5%. **Possible Causes** ● Air leaks in intake manifold, exhaust pipes or exhaust manifold ● Fuel control sensor is out of calibration (ECT, IAT or MAF) ● Low fuel pressure (fuel filter clogged, pressure regulator failure) ● One or more injectors restricted or pressure regulator has failed ● HO2S element is contaminated, deteriorated or has failed ● Vacuum hose is disconnected, broken, leaking or loose

OBD II Trouble Code List (P0xxx Codes)

DTC	Trouble Code Title, Conditions & Possible Causes
P0171 **2T FUEL, MIL: Yes** 1995, 1996, 1997, 1998, 1999, 2000, 2001, 2002, 2003, 2004, 2005 S/T Blazer & S/T Pickup 4.3L VIN W, 4.3L VIN X Transmissions: All	**Fuel System Lean (Bank 1) Conditions:** DTC P0101-P0103, P0108, P0135, P0137, P0141, P0200, P0300, P0410, P0420, P0430, P0440-P0449, P0506, P0507 and P1441 not set, engine speed from 400-3000 rpm, VSS under 85 mph, BARO sensor over 74 kPa, ECT sensor from 167-239°F, IAT sensor from 4-194°F, MAF sensor from 5-90 gm/s, MAP sensor from 26-90 kPa (3.7-13 psi), Fuel Level over 10%, TP angle under 90%, and the PCM detected the Long Term fuel trim was over 23% for 6 seconds. **Possible Causes** ● Air leaks in intake manifold, exhaust pipes or exhaust manifold ● Fuel control sensor is out of calibration (ECT, IAT or MAF) ● Low fuel pressure (fuel filter clogged, pressure regulator failure) ● One or more injectors restricted or pressure regulator has failed ● HO2S element is contaminated, deteriorated or has failed ● Vacuum hose is disconnected, broken, leaking or loose
P0171 **2T FUEL, MIL: Yes** 1997, 1998, 1999, 2000, 2001, 2002, 2003, 2004, 2005 Corvette 5.7L VIN G, 5.7L VIN S Transmissions: All	**Fuel Trim System Lean (Bank 1) Conditions:** DTC P0101-P0103, P0108, P0135, P0137, P0141, P0200, P0300, P0410, P0420, P0430, P0440-P0449, P0506, P0507 or P1441 not set, engine started, vehicle driven at less than 85 mph at an engine speed 500-3000 rpm, BARO sensor over 74 kPa (10.7 psi), ECT sensor from 167-239°F, IAT sensor from 4-194°F, MAF sensor from 5-90 gm/s, MAP sensor from 26-90 kPa, TP angle less than 90%, fuel level over 10%, and the PCM detected the Long Term fuel trim cell value was over +23% for 6 seconds during the Fuel Trim test. If a lean condition is detected, Fuel Trim values will be more than 0%. If a rich condition is detected, Fuel Trim values will be less than 0%. **Possible Causes** ● Air leaks in intake manifold, exhaust pipes or exhaust manifold ● Fuel control sensor is out of calibration (ECT, IAT or MAF) ● Low fuel pressure (fuel filter clogged, pressure regulator failure) ● One or more injectors restricted or pressure regulator has failed ● HO2S element is contaminated, deteriorated or has failed ● Vacuum hose is disconnected, broken, leaking or loose
P0171 **2T FUEL, MIL: Yes** 1995 Corvette 5.7L VIN P Transmissions: All	**Fuel Trim System Lean (Bank 1) Conditions:** DTC P0116, P0117, P0118, P0135, P0171, P0172, P0174, P0175, P1114 and P1115 not set, engine started, vehicle driven at less than 75 mph at 500-4000 rpm with Fuel Trim enabled, BARO sensor over 70 kPa, ECT sensor from 140-239°F, IAT sensor from -22 to 140°F, MAF from 5-100 gm/s, MAP sensor at 20-90 kPa, Purge command under 75%, TP angle under 70%, and the PCM detected the LTFT was counts were 155, or the STFT counts were 138 for 10 seconds. **Possible Causes** ● Air leaks in intake manifold, exhaust pipes or exhaust manifold ● Fuel control sensor is out of calibration (ECT, IAT or MAF) ● One or more injectors restricted or pressure regulator has failed ● O2S/HO2S element is contaminated, deteriorated or has failed ● Vacuum hose is disconnected, broken, leaking or loose
P0171 **2T FUEL, MIL: Yes** 1997, 1998, 1999, 2000, 2001 Catera 3.0L V6 VIN R engine Transmissions: A/T	**Fuel Trim System Lean (Bank 1) Conditions:** Engine started, vehicle driven at steady cruise speed in closed loop, Fuel Trim enabled, and the PCM detected the Long Term fuel trim was over 23% with Short Term fuel trim over 25%. **Possible Causes** ● Air leaks in intake manifold, exhaust pipes or exhaust manifold ● Engine oil cap loose or missing, oil dip stick not fully seated ● Fuel control sensor is out of calibration (ECT, IAT or MAF) ● Low fuel pressure (fuel filter clogged, pressure regulator failure) ● One or more injectors restricted or pressure regulator has failed ● HO2S element is contaminated, deteriorated or has failed ● Vacuum hose is disconnected, broken, leaking or loose
P0172 **2T FUEL, MIL: Yes** 1996, 1997, 1998, 1999, 2000, 2001, 2002, 2003, 2004, 2005 Achieva, Alero, Beretta, Century, Cavalier, Ciera, Century, Corsica, Envoy, Grand Am, Malibu, S/T Blazer & Pickup, Skylark, Sunfire, TrailBlazer 2.2L VIN 4, 2.2L VIN 5, 2.2L VIN F, 2.2L VIN H, 4.2L VIN S Transmissions: All	**Fuel System Rich (Bank 1) Conditions:** DTC P0105-P0108, P0112-P0118, P0122, P0123, P0201-P0204, P0300-P0304, P0335, P0440-P0446, P0506, P0507, P0601, P0602 and P1441 not set, engine speed at 550-3400 rpm, VSS under 72 mph, BARO sensor over 72 kPa, ECT sensor from 140-239°F, IAT sensor from -13°F to 239°F, MAP sensor over 26 kPa, fuel ethanol % less than 88, and the PCM detected the LTFT was below -42%. **Possible Causes** ● Base engine "mechanical" fault affecting one or more cylinders ● EVAP system component has failed or canister fuel saturated ● Fuel control sensor is out of calibration (i.e., ECT, IAT or MAP) ● Fuel system supplying too much fuel at idle speed or at cruise ● Fuel injector(s) is leaking or stuck partially open (one or more) ● HO2S is contaminated, deteriorated or it has failed

OBD II Trouble Code List (P0xxx Codes)

DTC	Trouble Code Title, Conditions & Possible Causes
P0172 **2T FUEL, MIL: Yes** 1996, 1997, 1998, 1999, 2000, 2001, 2002, 2003, 2004, 2005 Achieva, Beretta, Ciera, Century, Corsica, Grand Am, Grand Prix, Lumina, Malibu & Monte Carol 3.1L VIN M, 3.4L VIN X Transmissions: All	**Fuel System Rich (Bank 1) Conditions:** DTC P0101-P0103, P0107, P0108, P0121-P0123, P0130-P0138, P0140, P0141, P0201-P0206, P0300, P0401-P0404, P0405, P0410, P0440, P0442, P0446, P0506, P0507, P1404 and P1441 not set, engine started, vehicle driven at less than 82 mph at 600-4000 rpm, BARO sensor more than 70 kPa, ECT sensor from 68-230°F, IAT sensor from 0-158°F, MAF sensor from 3-150 gm/s, MAP sensor from 15-85 kPa, TP angle less than 90%, fuel level over 10%, and the PCM detected the Long Term fuel trim was less than -13%. **Possible Causes** ● Base engine "mechanical" fault affecting one or more cylinders ● EVAP system component has failed or canister fuel saturated ● Fuel control sensor is out of calibration (i.e., ECT, IAT or MAF) ● Fuel delivery system supplying too much fuel during cruise or idle periods (e.g., faulty fuel pump, or faulty pressure regulator) ● Fuel injector(s) is leaking or stuck partially open (one or more) ● HO2S is contaminated, deteriorated or it has failed
P0172 **2T FUEL, MIL: Yes** 1995 Bonneville, Century, Cutlass Supreme, 88'/98', Grand Prix, Intrigue, LSS, LeSabre, Lumina, Monte Carlo, Regal-Park Avenue 3.1L VIN M, 3.4L VIN X, 3.8L VIN 1, 3.8L VIN K, 3.8L VIN L Transmissions: All	**Fuel Trim System Rich (Bank 1) Conditions:** DTC P0101, P0117, P0118, P0122, P0123, P0134, P0501 and P0502 not set, engine started, vehicle speed less than 70 mph at 1000-4000 rpm, ECT signal from 176-230°F, Purge command below 80%, TP angle under 91% and steady, and the PCM detected the Long Term fuel trim was under 100, or the Short Term count was below 94 for 25 seconds. **Possible Causes** ● Base engine "mechanical" fault affecting one or more cylinders ● EVAP system component has failed or canister fuel saturated ● Fuel control sensor is out of calibration (i.e., ECT, IAT or MAF) ● Fuel delivery system supplying too much fuel during cruise or idle periods (e.g., faulty fuel pump, or faulty pressure regulator) ● Fuel injector(s) is leaking or stuck partially open (one or more) ● HO2S is contaminated, deteriorated or it has failed
P0172 **2T FUEL, MIL: Yes** 1996, 1997, 1998, 1999, 2000, 2001, 2002, 2003, 2004, 2005 Alero, Aztek, Cutlass, Impala, Malibu, Montana, Silhouette, Venture, Rendezvous 3.1L VIN J, 3.1L VIN M, 3.4L VIN E engines Transmissions: All	**Fuel Trim System Rich (Bank 1) Conditions:** DTC P0101-P0103, P0107, P0108, P0121- P0123, P0130-P0135, P0137, P0138, P0140, P0141, P0201-P0206, P0300, P0401-P0405, P0410, P0440-P0446, P0506, P0507, P1404 or P1441 not set, engine started, vehicle driven at less than 82 mph at 550-4000 rpm, BARO sensor more than 70 kPa, ECT sensor from 68-230°F, IAT sensor from 64-158°F, MAF sensor from 2.8-150 gm/s, MAP sensor from 15-105 kPa, fuel level over 10%, and the PCM detected the Long Term fuel trim value was below -13% for 40 seconds. During open loop, the PCM determines fuel delivery based on sensor signals, without the oxygen sensor input. During closed loop, the PCM adds oxygen sensor inputs to calculate Short and Long term fuel trim fuel delivery adjustments. If the oxygen sensors indicate a lean condition, the fuel trim values will be above 0 percent. If the oxygen sensors indicate a rich condition, the fuel trim values will be below 0 percent. Short Term fuel trim values change rapidly in response to HO2S signals. Long Term fuel trim makes its adjustments to maintain an A/F ratio of 14.7:1. **Possible Causes** ● Base engine "mechanical" fault affecting one or more cylinders ● Excess fuel vapors in crankcase (the oil needs to be changed) ● EVAP system component has failed or canister fuel saturated ● Fuel control sensor is out of calibration (i.e., ECT, IAT or MAF) ● Fuel delivery system supplying too much fuel during cruise or idle periods (e.g., faulty fuel pump, or faulty pressure regulator) ● Fuel injector(s) is leaking or stuck partially open (one or more) ● HO2S is contaminated, deteriorated or it has failed
P0172 **1T FUEL, MIL: Yes** 1996, 1997 Camaro, Corvette, Caprice, Firebird & Fleetwood 4.3L VIN W, 5.7L VIN 5, 5.7L VIN P engines Transmissions: All	**Fuel Trim System Rich (Bank 1) Conditions:** DTC P0100-P0108, P0112-P0123, P0131-P0141, P0152-P0155, P0200, P0300, P0372, P0441, P0443, P0500, P0601, P1133, P1151, P1351, P1361, P1371, P1414, P1416 and P1441 not set, engine started, vehicle driven at less than 75 mph at 500-4000 rpm, BARO sensor over 70 kPa, ECT sensor from 120-239°F, IAT sensor from -22°F to 302°F, MAF sensor from 5-100 gm/s, MAP sensor from 26-90 kPa, TP angle less than 90%, Fuel Trim enabled, EVAP purge not enabled, and the PCM detected the Long Term fuel trim value was less than -13 for 10 seconds. **Possible Causes** ● Base engine "mechanical" fault affecting one or more cylinders ● EVAP system component has failed or canister fuel saturated ● Fuel control sensor is out of calibration (i.e., ECT, IAT or MAF) ● Fuel delivery system supplying too much fuel during cruise or idle periods (e.g., faulty fuel pump, or faulty pressure regulator) ● Fuel injector(s) is leaking or stuck partially open (one or more) ● HO2S is contaminated, deteriorated or it has failed

OBD II Trouble Code List (P0xxx Codes)

DTC	Trouble Code Title, Conditions & Possible Causes
P0172 **2T FUEL, MIL: Yes** 1996, 1997, 1998, 1999, 2000, 2001, 2002, 2003, 2004, 2005 Aurora, Bonneville, Century, Cutlass Supreme, Grand Prix, Intrigue, LeSabre, LSS, Lumina, Monte Carlo, 88', Regal, Park Avenue & 98' 3.1L VIN J, 3.1L VIN M, 3.4L VIN X, 3.5L V6 VIN H, 3.8L VIN 1, 3.8L VIN K, 4.0L V8 VIN C Transmissions: All	**Fuel Trim System Rich (Bank 1) Conditions:** DTC P0101-P0103, P0107, P0108, P0121-P0123, P0130-P0138, P0140, P0141, P0201-P0206, P0300, P0401-P0405, P0410, P0440-P0446, P0506, P0507, P1404 or P1441 not set, engine started, vehicle driven at less than 82 mph at 550-4000 rpm, BARO sensor more than 70 kPa (10.1 psi), ECT sensor from 68-230°F, IAT sensor from 64-158°F, MAF sensor from 2.8-150 gm/s, MAP sensor from 16-105 kPa, fuel level over 10%, and the PCM detected the Long Term fuel trim value was less than -13% for 6 seconds. **Possible Causes** ● Base engine "mechanical" fault affecting one or more cylinders ● Excess fuel vapors in crankcase (the oil needs to be changed) ● EVAP system component has failed or canister fuel saturated ● Fuel control sensor is out of calibration (i.e., ECT, IAT or MAF) ● Fuel delivery system supplying too much fuel during cruise or idle periods (e.g., faulty fuel pump, or faulty pressure regulator) ● Fuel injector(s) is leaking or stuck partially open (one or more) ● HO2S is contaminated, deteriorated or it has failed
P0172 **2T FUEL, MIL: Yes** 1996, 1997, 1998, 1999, 2000, 2001, 2002, 2003, 2004, 2005 C/K Cab & Chassis, C/K Series Truck, G Series Van, M/L Series Vans 4.3L VIN W, 4.3L VIN X, 5.0L VIN M, 5.7L VIN K, 5.7L VIN R, 7.4L VIN J Transmissions: All	**Fuel Trim System Rich (Bank 1) Conditions:** DTC P0101-P0103, P0108, P0135, P0137, P0141, P0200, P0300, P0410, P0420, P0440, P0442, P0443, P0446, P0449, P0506, P0507 and P1441 not set, engine started, vehicle driven at less than 85 mph at 400-3000 rpm, BARO sensor more than 74 kPa (10.7 psi), ECT sensor from 167-239°F, IAT sensor from 4-194°F, MAF sensor from 5-90 gm/s, MAP sensor from 26-90 kPa, TP angle less than 90%, fuel level over 10%, and the PCM detected the Long Term fuel trim was less than -13% for 40 seconds. **Possible Causes** ● Base engine "mechanical" fault affecting one or more cylinders ● EVAP system component has failed or canister fuel saturated ● Fuel control sensor is out of calibration (i.e., ECT, IAT or MAF) ● Fuel delivery system supplying too much fuel during cruise or idle periods (e.g., faulty fuel pump, or faulty pressure regulator) ● Fuel injector(s) is leaking or stuck partially open (one or more) ● HO2S is contaminated, deteriorated or it has failed ● TSB 81-65-37 contains a repair procedure for this code
P0172 **2T FUEL, MIL: Yes** 1999, 2000, 2001, 2002, 2003, 2004, 2005 C/K Series Truck, G Series Van, Envoy, Escalade, TrailBlazer 4.8L VIN V, 5.3L VIN P, 5.3L VIN T, 5.3L VIN Z, 6.0L VIN N, 6.0L VIN U, 8.1L VIN G Transmissions: All	**Fuel Trim System Rich (Bank 1) Conditions:** DTC P0101-P0103, P0108, P0135, P0137, P0141, P0200, P0300, P0410, P0420, P0430, P0440, P0442, P0443, P0446, P0449, P0506, P0507 and P1441 not set, engine started, ECT sensor from 167-239°F, IAT sensor from 4-194°F, engine speed from 400-3000 rpm, BARO sensor over 74 kPa, MAF sensor from 5-90 gm/s, TP angle less than 90%, VSS less than 85 mph, and the PCM detected the Long Term fuel trim value was less than -13% for 6 seconds (i.e., indicating that a rich A/F mixture was present). **Possible Causes** ● Base engine "mechanical" fault affecting one or more cylinders ● Excess fuel vapors in crankcase (the oil needs to be changed) ● EVAP system component has failed or canister fuel saturated ● Fuel control sensor is out of calibration (i.e., ECT, IAT or MAF) ● Fuel delivery system supplying too much fuel during cruise or idle periods (e.g., faulty fuel pump, or faulty pressure regulator) ● Fuel injector(s) is leaking or stuck partially open (one or more) ● HO2S is contaminated, deteriorated or it has failed
P0172 **2T FUEL, MIL: Yes** 2003, 2004, 2005 C/K Series Truck, G Series Van, Cargo Van 6.0L VIN U CNG engine Transmissions: All	**Fuel Trim System Rich (Bank 1) Conditions:** DTC P0101, P0102, P0103, P0106, P0107, P0108, P0112, P0113, P0116-P0118, P0121-P0123, P0125, P0200, P0300, P0327, P0332, P0335, P0336, P0351-P0358, P0401, P0402, P0403, P0443, P0446, P0449, P0496, P0502, P0503, P1020, or P1258 not set, engine started, ECT sensor from 167-239°F, IAT sensor from 4-194°F, engine speed 400-3000 rpm, BARO sensor over 74 kPa, MAF sensor from 5-90 gm/s, TP angle under 90%, VSS under 85 mph, and the PCM detected the Long Term fuel trim was less than -15% for 40 seconds (indicating a lean A/F mixture was present). **Possible Causes** ● Base engine "mechanical" fault affecting one or more cylinders ● Excess fuel vapors in crankcase (the oil needs to be changed) ● EVAP system component has failed or canister fuel saturated ● Fuel control sensor is out of calibration (i.e., ECT, IAT or MAF) ● Fuel delivery system supplying too much fuel during cruise or idle periods (e.g., faulty fuel pump, or faulty pressure regulator) ● Fuel injector(s) is leaking or stuck partially open (one or more) ● HO2S is contaminated, deteriorated or it has failed

OBD II Trouble Code List (P0xxx Codes)

DTC	Trouble Code Title, Conditions & Possible Causes
P0172 **2T FUEL, MIL: Yes** 1996, 1997, 1998, 1999, 2001, 2002, 2003, 2004, 2005 Aurora 3.5L V6 VIN H, 4.0L V8 VIN C engines Transmissions: A/T	**Fuel Trim System Rich (Bank 1) Conditions:** DTC P0101-P0103, P0106-P0108, P0121, P0122, P0131-P0138, P0140-P0146, P0151, P0154, P0562, P0563, P0201-P0208, P0300, P0325, P0327, P0340, P0401, P0506, P0507, P1121, P1133, P1153, P1154, P1441, P1508 and P1509 not set, engine started, vehicle speed less than 70 mph at 400-3000 rpm, BARO sensor over 70.8 kPa, ECT sensor from 184-221°F, IAT sensor from 40°F to 218°F, MAF sensor from 3-200 gm/s, MAP sensor from 27-103 kPa, EVAP purge tank not full, TP angle below 19.8%, Fuel Trim enabled, and the PCM detected the Short Term fuel trim value was over +23% for 6 seconds during the test. **Possible Causes** • Base engine "mechanical" fault affecting one or more cylinders • Excess fuel vapors in crankcase (the oil needs to be changed) • EVAP system component has failed or canister fuel saturated • Fuel control sensor is out of calibration (i.e., ECT, IAT or MAF) • Fuel delivery system supplying too much fuel during cruise or idle periods (e.g., faulty fuel pump, or faulty pressure regulator) • Fuel injector(s) is leaking or stuck partially open (one or more) • HO2S is contaminated, deteriorated or it has failed
P0172 **2T FUEL, MIL: Yes** 2003, 2004, 2005 CTS (Cadillac) 2.6L V6 VIN M & 3.2L V6 VIN N engines Transmissions: All	**Fuel Trim System Rich (Bank 1) Conditions:** DTC P0101, P0102, P0103, P0130, P0131, P0132, P0133, P0134, P0135, P0137, P0138, P0140 and P0141 not set, engine running in closed loop, fuel level over 10%, and the PCM detected the Long Term fuel trim Cruise/Acceleration parameter was over -23% or the Long Term fuel trim Idle/Deceleration parameter was over -8.5%. **Possible Causes** • Base engine "mechanical" fault affecting one or more cylinders • Excess fuel vapors in crankcase (the oil needs to be changed) • EVAP system component has failed or canister fuel saturated • Fuel control sensor is out of calibration (i.e., ECT, IAT or MAF) • Fuel delivery system supplying too much fuel during cruise or idle periods (e.g., faulty fuel pump, or faulty pressure regulator) • Fuel injector(s) is leaking or stuck partially open (one or more) • HO2S is contaminated, deteriorated or it has failed
P0172 **2T FUEL, MIL: Yes** 1998, 1999, 2000, 2001, 2002 Camaro & Firebird 5.7L VIN G engine Transmissions: All	**Fuel Trim System Rich (Bank 1) Conditions:** DTC P0101-P0103, P0107-P0108, P0112-P0113, P0116-P0118, P0121-P0125, P0200, P0300, P0335-P0336, P0351-P0358, P0405, P0410, P0412, P0418, P0443, P0500-P0503, P1258, P1404, P1415 and P1416 not set, engine started, vehicle driven at less than 85 mph at 400-3000 rpm, BARO sensor more than 74 kPa, ECT sensor from 122-239°F, IAT sensor from -4°F-194°F, MAF sensor airflow from 3-150 gm/s, MAP sensor from 26-90 kPa, TP angle less than 90% and steady, fuel level over 10%, and the PCM detected the Long Term fuel trim value was less than -13% for 6 seconds. **Possible Causes** • Base engine "mechanical" fault affecting one or more cylinders • Excess fuel vapors in crankcase (the oil needs to be changed) • EVAP system component has failed or canister fuel saturated • Fuel control sensor is out of calibration (i.e., ECT, IAT or MAF) • Fuel delivery system supplying too much fuel during cruise or idle periods (e.g., faulty fuel pump, or faulty pressure regulator) • Fuel injector(s) is leaking or stuck partially open (one or more) • HO2S is contaminated, deteriorated or it has failed
P0172 **2T FUEL, MIL: Yes** 1996, 1997, 1998, 1999, 2000, 2001, 2002, 2003, 2004, 2005 DeVille, Eldorado, Seville 4.6L VIN 9, 4.6L VIN Y Transmissions: All	**Fuel Trim System Rich (Bank 1) Conditions:** DTC P0101-P0103, P0106-P0108, P0112-P0113, P0116-P0118, P0121-P0123, P0125, P0128, P0200, P0300, P0410-P0446, P0452-P0453, P1258, P1415-P1416 and P1441 not set, engine started, vehicle driven at less than 82 mph at 500-4000 rpm, BARO sensor more than 75 kPa, ECT sensor from 68-239°F, IAT sensor from -4°F to 212°F, MAF sensor from 3-60 g/sec, EGR Flow Test "off", MAP sensor from 15-85 kPa, TP angle less than 65%, and the PCM detected the Long Term fuel trim was below -13% for 40 seconds. **Possible Causes** • Base engine "mechanical" fault affecting one or more cylinders • EVAP system component has failed or canister fuel saturated • Fuel control sensor is out of calibration (i.e., ECT, IAT or MAF) • Fuel delivery system supplying too much fuel during cruise or idle periods (e.g., faulty fuel pump, or faulty pressure regulator) • Fuel injector(s) is leaking or stuck partially open (one or more) • HO2S is contaminated, deteriorated or it has failed

OBD II Trouble Code List (P0xxx Codes)

DTC	Trouble Code Title, Conditions & Possible Causes
P0172 **2T FUEL, MIL: Yes** 1995, 1996, 1997, 1998, 1999, 2000, 2001, 2002 Camaro & Firebird 3.8L VIN K engine Transmissions: All	**Fuel Trim System Rich (Bank 1) Conditions:** DTC P0101-P0103, P0108, P0135, P0137, P0141, P0200, P0300, P0410, P0420, P0430, P0440-P0449, P0506, P0507 and P1441 not set, engine started, vehicle driven at less than 85 mph at 400-3000 rpm, BARO sensor more than 74 kPa (10.7 psi), ECT sensor from 167-239°F, IAT sensor from 4-194°F, MAF sensor from 5-90 gm/s, MAP sensor from 26-90 kPa (3.7-13 psi), TP angle less than 90%, fuel level over 10%, and the PCM detected the Long Term fuel trim value was less than -13% for 40 seconds during the Fuel Trim test. **Possible Causes** ● Base engine "mechanical" fault affecting one or more cylinders ● Excess fuel vapors in crankcase (the oil needs to be changed) ● EVAP system component has failed or canister fuel saturated ● Fuel control sensor is out of calibration (i.e., ECT, IAT or MAF) ● Fuel delivery system supplying too much fuel during cruise or idle periods (e.g., faulty fuel pump, or faulty pressure regulator) ● Fuel injector(s) is leaking or stuck partially open (one or more) ● HO2S is contaminated, deteriorated or it has failed
P0172 **2T FUEL, MIL: Yes** 1996, 1997, 1998, 1999, 2001, 2002, 2003, 2004, 2005 Aurora 3.5L V6 VIN H, 4.0L V8 VIN C engines Transmissions: A/T	**Fuel Trim System Rich (Bank 1) Conditions:** DTC P0101-P0103, P0106-P0108, P0121, P0122, P0131-P0138, P0140-P0146, P0151, P0154, P0562, P0563, P0201-P0208, P0300, P0325, P0327, P0340, P0401, P0506, P0507, P1121, P1133, P1153, P1154, P1441, P1508 and P1509 not set, engine started, vehicle driven at less than 70 mph at 400-3000 rpm, BARO sensor more than 70.8 kPa, ECT sensor from 184-221°F, IAT sensor from 40°F to 218°F, MAF sensor from 3-200 gm/s, MAP sensor from 27-103 kPa, EVAP purge tank not full, TP angle less than 19.8%, Fuel Trim enabled, and the PCM detected the Short Term fuel trim value was less than -13% for 40 seconds during the Fuel Trim test. **Possible Causes** ● Base engine "mechanical" fault affecting one or more cylinders ● Excess fuel vapors in crankcase (the oil needs to be changed) ● EVAP system component has failed or canister fuel saturated ● Fuel control sensor is out of calibration (i.e., ECT, IAT or MAF) ● Fuel delivery system supplying too much fuel during cruise or idle periods (e.g., faulty fuel pump, or faulty pressure regulator) ● Fuel injector(s) is leaking or stuck partially open (one or more) ● HO2S is contaminated, deteriorated or it has failed
P0172 **2T FUEL, MIL: Yes** 1997, 1998, 1999, 2000, 2001 Catera 3.0L V6 VIN R engine Transmissions: All	**Fuel Trim System Rich (Bank 1) Conditions:** Engine started, vehicle driven at a steady cruise speed in closed loop, Fuel Trim enabled, and the PCM detected the Fuel Trim long term values were greater than 23% and the Short Term fuel trim values were greater than 25% during the Fuel System Monitor test. **Possible Causes** ● Base engine "mechanical" fault affecting one or more cylinders ● Excess fuel vapors in crankcase (the oil needs to be changed) ● EVAP system component has failed or canister fuel saturated ● Fuel control sensor is out of calibration (i.e., ECT, IAT or MAF) ● Fuel delivery system supplying too much fuel during cruise or idle periods (e.g., faulty fuel pump, or faulty pressure regulator) ● Fuel injector(s) is leaking or stuck partially open (one or more) ● HO2S is contaminated, deteriorated or it has failed
P0172 **2T FUEL, MIL: Yes** 1995, 1996, 1997, 1998, 1999, 2000, 2001, 2002, 2003, 2004, 2005 S/T Blazer & S/T Pickup 4.3L VIN W, 4.3L VIN X Transmissions: All	**Fuel System Rich (Bank 1) Conditions:** DTC P0101, P0102, P0103, P0106, P0107, P0108, P0112, P0113, P0117, P0118, P0125, P0135, P0137, P0141, P0200, P0300, P0410, P0420, P0430, P0440, P0442, P0452, P0453, P0455, P0449, P0506, P0507 and P1441 not set, engine started, vehicle driven at less than 85 mph at an engine speed of 400-3000 rpm, BARO sensor over 74 kPa (10.7 psi), ECT sensor from 167-239°F, IAT sensor from 4-194°F, MAF sensor from 5-90 gm/s, MAP sensor from 26-90 kPa (3.7-13 psi), fuel level over 10%, TP angle less than 90%, and the PCM detected the average Long Term fuel trim value was less than -13 for 40 seconds (i.e., a rich A/F condition existed). **Possible Causes** ● Base engine "mechanical" fault affecting one or more cylinders ● Excess fuel vapors in crankcase (the oil needs to be changed) ● EVAP system component has failed or canister fuel saturated ● Fuel control sensor is out of calibration (i.e., ECT, IAT or MAF) ● Fuel delivery system supplying too much fuel during cruise or idle periods (e.g., faulty fuel pump, or faulty pressure regulator) ● Fuel injector(s) is leaking or stuck partially open (one or more) ● HO2S is contaminated, deteriorated or it has failed ● TSB 76-65-04 contains a repair procedure for this code

OBD II Trouble Code List (P0xxx Codes)

DTC	Trouble Code Title, Conditions & Possible Causes
P0172 **2T FUEL, MIL: No** 1995 Corvette 5.7L VIN P Transmissions: All	**Fuel Trim System Rich (Bank 1) Conditions:** DTC P0116, P0117, P0118, P0135, P0171, P0172, P0174, P0175, P1114 and P1115 not set, engine speed from 500-4000 rpm, VSS less than 75 mph, Fuel Trim enabled, BARO sensor over 70 kPa, ECT sensor from 140-239ºF, IAT sensor from -22 to 140ºF, MAF from 5-100 gm/s, MAP sensor from 20-90 kPa, Purge command under 75%, TP angle under 70%, and the PCM detected the LTFT counts were under 118, or the STFT counts were 12 for 10 seconds. **Possible Causes** ● Base engine "mechanical" fault affecting one or more cylinders ● Excess fuel vapors in crankcase (the oil needs to be changed) ● EVAP system component has failed or canister fuel saturated ● Fuel control sensor is out of calibration (i.e., ECT, IAT or MAF) ● Fuel delivery system supplying too much fuel during cruise or idle periods (e.g., faulty fuel pump, or faulty pressure regulator) ● Fuel injector(s) is leaking or stuck partially open (one or more) ● HO2S is contaminated, deteriorated or it has failed
P0172 **2T FUEL, MIL: Yes** 1997, 1998, 1999, 2000, 2001, 2002, 2003, 2004, 2005 Corvette 5.7L VIN G, 5.7L VIN S Transmissions: All	**Fuel Trim System Rich (Bank 1) Conditions:** DTC P0101-P0103, P0108, P0135, P0137, P0141, P0200, P0300, P0410, P0420, P0430, P0440-P0449, P0506, P0507 or P1441 not set, engine started, vehicle driven at less than 85 mph at an engine speed 500-3000 rpm, BARO sensor over 74 kPa (10.7 psi), ECT sensor from 167-239ºF, IAT sensor from 4-194ºF, MAF sensor from 5-90 gm/s, MAP sensor from 26-90 kPa, TP angle less than 90%, fuel level over 10%, and the PCM detected the Long Term fuel trim cell value was over -13% for 40 seconds during the Fuel Trim test. **Possible Causes** ● Base engine "mechanical" fault affecting one or more cylinders ● Excess fuel vapors in crankcase (the oil needs to be changed) ● EVAP system component has failed or canister fuel saturated ● Fuel control sensor is out of calibration (i.e., ECT, IAT or MAF) ● Fuel delivery system supplying too much fuel during cruise or idle periods (e.g., faulty fuel pump, or faulty pressure regulator) ● Fuel injector(s) is leaking or stuck partially open (one or more) ● HO2S is contaminated, deteriorated or it has failed
P0174 **1T FUEL, MIL: Yes** 1996, 1997 Camaro, Corvette, Caprice, Firebird & Fleetwood 4.3L VIN W, 5.7L VIN 5, 5.7L VIN P engines Transmissions: All	**Fuel Trim System Lean (Bank 2) Conditions:** DTC P0100-P0108, P0112-P0123, P0131-P0141, P0152-P0155, P0200, P0300, P0372, P0441, P0443, P0500, P0601, P1133, P1151, P1351, P1361, P1371, P1414, P1416 and P1441 not set, engine started, vehicle driven at less than 75 mph at 500-4000 rpm, BARO sensor over 70 kPa, ECT sensor from 120-239ºF, IAT sensor from -22ºF to 302ºF, MAF sensor from 5-100 gm/s, MAP sensor from 26-90 kPa, TP angle less than 90%, Fuel Trim enabled, EVAP purge not enabled, and the PCM detected the Long Term fuel trim value was more than +20 for 10 seconds during the Fuel Trim test. **Possible Causes** ● Air leaks in intake manifold, exhaust pipes or exhaust manifold ● Fuel control sensor is out of calibration (ECT, IAT or MAF) ● Low fuel pressure (fuel filter clogged, pressure regulator failure) ● One or more injectors restricted or pressure regulator has failed ● HO2S element is contaminated, deteriorated or has failed ● Vacuum hose is disconnected, broken, leaking or loose
P0174 **2T FUEL, MIL: Yes** 1996, 1997, 1998, 1999, 2000, 2001, 2002, 2003, 2004, 2005 C/K Cab & Chassis, C/K Series Truck, G Series Van, M/L Series Vans 4.3L VIN W, 4.3L VIN X, 5.0L VIN M, 5.7L VIN K, 5.7L VIN R, 7.4L VIN J Transmissions: All	**Fuel Trim System Lean (Bank 2) Conditions:** DTC P0101, P0102, P0103, P0108, P0135, P0137, P0141, P0200, P0300, P0410, P0420, P0430, P0440, P0442, P0443, P0446, P0449, P0506, P0507 and P1441 not set, vehicle driven at less than 85 mph at 400-3000 rpm, BARO sensor over 74 kPa, ECT sensor from 167-239ºF, IAT sensor from 4-194ºF, MAF sensor at 5-90 gm/s, MAP sensor from 26-90 kPa (3.7-13 psi), TP angle below 90%, fuel level over 10%, and the PCM detected the Long Term fuel trim was over +23% for 6 seconds (i.e., a lean A/F mixture existed). **Possible Causes** ● Air leaks in intake manifold, exhaust pipes or exhaust manifold ● Fuel control sensor is out of calibration (ECT, IAT or MAF) ● Low fuel pressure (fuel filter clogged, pressure regulator failure) ● One or more injectors restricted or pressure regulator has failed ● HO2S element is contaminated, deteriorated or has failed ● Vacuum hose is disconnected, broken, leaking or loose

OBD II Trouble Code List (P0xxx Codes)

DTC	Trouble Code Title, Conditions & Possible Causes
P0174 **2T FUEL, MIL: Yes** 1999, 2000, 2001, 2002, 2003, 2004, 2005 C/K Series Truck, G Series Van, Envoy, Escalade, TrailBlazer 4.8L VIN V, 5.3L VIN P, 5.3L VIN T, 5.3L VIN Z, 6.0L VIN N, 6.0L VIN U, 8.1L VIN G Transmissions: All	**Fuel Trim System Lean (Bank 2) Conditions:** DTC P0101-P0103, P0108, P0135, P0137, P0141, P0200, P0300, P0410, P0420, P0430, P0440, P0442, P0443, P0446, P0449, P0506, P0507 and P1441 not set, engine started, ECT sensor from 167-239ºF, IAT sensor from 4-194ºF, engine speed from 400-3000 rpm, BARO sensor over 74 kPa, MAF sensor from 5-90 gm/s, TP angle less than 90%, VSS less than 85 mph, and the PCM detected the Long Term fuel trim value was more than 23% for 6 seconds (i.e., indicating that a lean A/F mixture was present). **Possible Causes** ● Air leaks in intake manifold, exhaust pipes or exhaust manifold ● Fuel control sensor is out of calibration (ECT, IAT or MAF) ● Low fuel pressure (fuel filter clogged, pressure regulator failure) ● One or more injectors restricted or pressure regulator has failed ● HO2S element is contaminated, deteriorated or has failed ● Vacuum hose is disconnected, broken, leaking or loose
P0174 **2T FUEL, MIL: Yes** 2003, 2004, 2005 C/K Series Truck, G Series Van, Cargo Van 6.0L VIN U CNG engine Transmissions: All	**Fuel Trim System Lean (Bank 2) Conditions:** DTC P0101-P0103, P0106-P0108, P0112, P0113, P0116-P0118, P0121-P0123, P0125, P0200, P0300, P0327, P0332, P0335, P0336, P0351-P0358, P0401, P0402, P0403, P0443, P0446, P0449, P0496, P0502, P0503, P1020, or P1258 not set, engine started, ECT sensor from 167-239ºF, IAT sensor from 4-194ºF, engine speed 400-3000 rpm, BARO sensor over 74 kPa, MAF sensor from 5-90 gm/s, TP angle under 90%, VSS less than 85 mph, and the PCM detected the LT fuel trim value was over 28% for 6 seconds (e.g., a lean A/F mixture). **Possible Causes** ● Air leaks in intake manifold, exhaust pipes or exhaust manifold ● Fuel control sensor is out of calibration (ECT, IAT or MAF) ● Low fuel pressure (fuel filter clogged, pressure regulator failure) ● One or more injectors restricted or pressure regulator has failed ● HO2S element is contaminated, deteriorated or has failed ● Vacuum hose is disconnected, broken, leaking or loose
P0174 **2T FUEL, MIL: Yes** 2003, 2004, 2005 CTS (Cadillac) 2.6L V6 VIN M & 3.2L V6 VIN N engines Transmissions: All	**Fuel Trim System Lean (Bank 2) Conditions:** DTC P0101-P0103, P0130-P0138, P0140 and P0141 not set, engine running in closed loop, fuel level over 10%, and the PCM detected the LT fuel trim Cruise/Acceleration parameter was more than +23% or the LT fuel trim Idle/Deceleration parameter was more than 8.5%. **Possible Causes** ● Air leaks in intake manifold, exhaust pipes or exhaust manifold ● Fuel control sensor is out of calibration (ECT, IAT or MAF) ● Low fuel pressure (fuel filter clogged, pressure regulator failure) ● One or more injectors restricted or pressure regulator has failed ● HO2S element is contaminated, deteriorated or has failed ● Vacuum hose is disconnected, broken, leaking or loose
P0174 **2T FUEL, MIL: Yes** 1996, 1997, 1998, 1999, 2000, 2001, 2002, 2003, 2004, 2005 DeVille, Eldorado, Seville 4.6L VIN 9, 4.6L VIN Y Transmissions: All	**Fuel Trim System Lean (Bank 2) Conditions:** DTC P0101-P0103, P0106-P0108, P0112-P0113, P0116-P0118, P0121-P0123, P0125, P0128, P0200, P0300, P0410-P0446, P0452-P0453, P1258, P1415-P1416 and P1441 not set, engine started, vehicle driven at less than 82 mph at 500-4000 rpm in closed loop, BARO sensor over 75 kPa, ECT sensor from 68-239ºF, IAT sensor from -4ºF to 212ºF, MAF sensor from 3-60 g/sec, EGR Flow Test "off", MAP sensor from 15-85 kPa, TP angle below 65%, and the PCM detected the LT fuel trim was more than +23 for 6 seconds during the test. **Possible Causes** ● Air leaks in intake manifold, exhaust pipes or exhaust manifold ● Fuel control sensor is out of calibration (ECT, IAT or MAF) ● Low fuel pressure (fuel filter clogged, pressure regulator failure) ● One or more injectors restricted or pressure regulator has failed ● HO2S element is contaminated, deteriorated or has failed ● Vacuum hose is disconnected, broken, leaking or loose
P0174 **2T FUEL, MIL: Yes** 1998, 1999, 2000, 2001, 2002 Camaro & Firebird 5.7L VIN G engine Transmissions: All	**Fuel Trim System Lean (Bank 2) Conditions:** DTC P0101-P0103, P0107-P0108, P0112-P0113, P0117-P0118, P0121-P0125, P0200, P0300, P0335-P0336, P0351-P0358, P0405, P0410, P0412, P0418, P0443, P0500-P0503, P1258, P1404, P1415 and P1416 not set, vehicle speed less than 85 mph at 400-3000 rpm, BARO sensor over 74 kPa, ECT sensor from 122-239ºF, IAT sensor from -4ºF-194ºF, MAF sensor from 3-150 gm/s, MAP sensor from 26-90 kPa, TP angle below 90% and steady, fuel level over 10%, and the PCM detected the LT fuel trim was more than +23% for 6 seconds. **Possible Causes** ● Air leaks in intake manifold, exhaust pipes or exhaust manifold ● Fuel control sensor is out of calibration (ECT, IAT or MAF) ● Low fuel pressure (fuel filter clogged, pressure regulator failure) ● One or more injectors restricted or pressure regulator has failed ● HO2S element is contaminated, deteriorated or has failed ● Vacuum hose is disconnected, broken, leaking or loose

OBD II Trouble Code List (P0xxx Codes)

DTC	Trouble Code Title, Conditions & Possible Causes
P0174 **2T FUEL, MIL: Yes** 1995, 1996, 1997, 1998, 1999, 2000, 2001, 2002 Camaro & Firebird 3.8L VIN K engine Transmissions: All	**Fuel Trim System Lean (Bank 2) Conditions:** DTC P0101-P0103, P0108, P0135, P0137, P0141, P0200, P0300, P0410, P0420, P0430, P0440-P0449, P0506, P0507 and P1441 not set, vehicle speed less than 85 mph at 400-3000 rpm, BARO sensor over 74 kPa, ECT sensor from 167-239°F, IAT sensor from 4-194°F, MAF sensor from 5-90 gm/s, MAP sensor from 26-90 kPa (3.7-13 psi), TP angle below 90%, fuel level over 10%, and the PCM detected the LT fuel trim value was +23% for 6 seconds. **Possible Causes** ● Air leaks in intake manifold, exhaust pipes or exhaust manifold ● Fuel control sensor is out of calibration (ECT, IAT or MAF) ● Low fuel pressure (fuel filter clogged, pressure regulator failure) ● One or more injectors restricted or pressure regulator has failed ● HO2S element is contaminated, deteriorated or has failed ● Vacuum hose is disconnected, broken, leaking or loose
P0174 **2T FUEL, MIL: Yes** 1996, 1997, 1998, 1999, 2001, 2002, 2003, 2004, 2005 Aurora 3.5L V6 VIN H, 4.0L V8 VIN C engines Transmissions: A/T	**Fuel Trim System Lean (Bank 2) Conditions:** DTC P0101-P0103, P0106-P0108, P0121, P0122, P0131-P0138, P0140-P0146, P0151, P0154, P0562, P0563, P0201-P0208, P0300, P0325, P0327, P0340, P0401, P0506, P0507, P1121, P1133, P1153, P1154, P1441, P1508 and P1509 not set, VSS under 70 mph at 400-3000 rpm, BARO sensor over 70.8 kPa, ECT sensor at 184-221°F, IAT sensor at 40-218°F, MAF sensor at 3-200 gm/s, MAP sensor at 27-103 kPa, Purge Tank not full, TP angle below 19.8%, and the PCM detected the ST fuel trim was over +23% for 6 seconds during the test. **Possible Causes** ● Air leaks in intake manifold, exhaust pipes or exhaust manifold ● Fuel control sensor is out of calibration (ECT, IAT or MAF) ● Low fuel pressure (fuel filter clogged, pressure regulator failure) ● One or more injectors restricted or pressure regulator has failed ● HO2S element is contaminated, deteriorated or has failed ● Vacuum hose is disconnected, broken, leaking or loose
P0174 **2T FUEL, MIL: Yes** 1997, 1998, 1999, 2000, 2001 Catera 3.0L V6 VIN R engine Transmissions: A/T	**Fuel Trim System Lean (Bank 2) Conditions:** Engine started, vehicle driven at a steady cruise speed in closed loop, Fuel Trim enabled, and the PCM detected the Fuel Trim long term values were greater than 23% and the Short Term fuel trim values were greater than 25% during the Fuel System Monitor test. **Possible Causes** ● Air leaks in intake manifold, exhaust pipes or exhaust manifold ● Engine oil cap loose or missing, oil dip stick not fully seated ● Fuel control sensor is out of calibration (ECT, IAT or MAF) ● Low fuel pressure (fuel filter clogged, pressure regulator failure) ● One or more injectors restricted or pressure regulator has failed ● HO2S element is contaminated, deteriorated or has failed ● Vacuum hose is disconnected, broken, leaking or loose
P0174 **2T FUEL, MIL: Yes** 1995, 1996, 1997, 1998, 1999, 2000, 2001, 2002, 2003, 2004, 2005 S/T Blazer & S/T Pickup 4.3L VIN W, 4.3L VIN X Transmissions: All	**Fuel System Lean (Bank 2) Conditions:** DTC P0101, P0102, P0103, P0106, P0107, P0108, P0112, P0113, P0117, P0118, P0135, P0137, P0141, P0200, P0300, P0420, P0430, P0440, P0442, P0446, P0455, P0449, P0496, P0506, P0507 and P1441 not set, engine started, vehicle driven at less than 85 mph at 400-3000 rpm, BARO sensor more than 74 kPa (10.7 psi), ECT sensor from 167-239°F, IAT sensor from 4-194°F, MAF sensor from 5-90 gm/s, MAP sensor from 26-90 kPa (3.7-13 psi), fuel level over 10%, TP angle less than 90%, and the PCM detected the Long Term fuel trim value was more than 23% for 6 seconds (i.e., a lean A/F condition existed during the test). **Possible Causes** ● Air leaks in intake manifold, exhaust pipes or exhaust manifold ● Fuel control sensor is out of calibration (ECT, IAT or MAF) ● Low fuel pressure (fuel filter clogged, pressure regulator failure) ● One or more injectors restricted or pressure regulator has failed ● HO2S element is contaminated, deteriorated or has failed ● Vacuum hose is disconnected, broken, leaking or loose
P0174 **2T FUEL, MIL: Yes** 1995 Corvette 5.7L VIN P Transmissions: All	**Fuel Trim System Lean (Bank 2) Conditions:** DTC P0116, P0117, P0118, P0135, P0171, P0172, P0174, P0175, P1114 and P1115 not set, engine started, vehicle driven at less than 75 mph at 500-4000 rpm with Fuel Trim enabled, BARO sensor more than 70 kPa, ECT sensor from 140-239°F, IAT sensor from -22 to 140°F, MAF from 5-100 gm/s, MAP sensor from 20-90 kPa, Purge command under 75%, TP angle less than 70%, and the PCM detected the Long Term fuel trim counts was more than 155, or the Short Term fuel trim count was 138 for 10 seconds. **Possible Causes** ● Air leaks in intake manifold, exhaust pipes or exhaust manifold ● Fuel control sensor is out of calibration (ECT, IAT or MAF) ● Low fuel pressure (fuel filter clogged, pressure regulator failure) ● One or more injectors restricted or pressure regulator has failed ● HO2S element is contaminated, deteriorated or has failed ● Vacuum hose is disconnected, broken, leaking or loose

OBD II Trouble Code List (P0xxx Codes)

DTC	Trouble Code Title, Conditions & Possible Causes
P0174 **2T FUEL, MIL: Yes** 1997, 1998, 1999, 2000, 2001, 2002, 2003, 2004, 2005 Corvette 5.7L VIN G, 5.7L VIN S Transmissions: All	**Fuel Trim System Lean (Bank 2) Conditions:** DTC P0101-P0103, P0108, P0135, P0137, P0141, P0200, P0300, P0410, P0420, P0430, P0440-P0449, P0506, P0507 or P1441 not set, VSS below 85 mph at 500-3000 rpm, BARO sensor over 74 kPa, ECT sensor from 167-239°F, IAT sensor from 4-194°F, MAF sensor from 5-90 gm/s, MAP sensor from 26-90 kPa, TP angle below 90%, fuel level over 10%, and the PCM detected the Long Term fuel trim was more than +23% for 6 seconds during the test. **Possible Causes** ● Air leaks in intake manifold, exhaust pipes or exhaust manifold ● Low fuel pressure (fuel filter clogged, pressure regulator failure) ● One or more injectors restricted or pressure regulator has failed ● HO2S element is contaminated, deteriorated or has failed ● Vacuum hose is disconnected, broken, leaking or loose
P0175 **2T FUEL, MIL: Yes** 1996, 1997 Camaro, Corvette, Caprice, Firebird & Fleetwood 4.3L VIN W, 5.7L VIN 5, 5.7L VIN P engines Transmissions: All	**Fuel Trim System Rich (Bank 2) Conditions:** DTC P0100-P0108, P0112-P0123, P0131-P0141, P0152-P0155, P0200, P0300, P0372, P0441, P0443, P0500, P0601, P1133, P1151, P1351, P1361, P1371, P1414, P1416 and P1441 not set, VSS under 75 mph at 500-4000 rpm, BARO sensor over 70 kPa, ECT sensor at 120-239°F, IAT sensor at -22°F to 302°F, MAF sensor at 5-100 gm/s, MAP sensor at 26-90 kPa, TP angle less than 90%, Purge "off", and the PCM detected the LT fuel trim was -13%. **Possible Causes** ● Base engine "mechanical" fault affecting one or more cylinders ● EVAP system component has failed or canister fuel saturated ● Fuel delivery system supplying too much fuel at cruise or idle periods (faulty regulator) ● Fuel injector(s) is leaking or stuck partially open (one or more) ● HO2S is contaminated, deteriorated or it has failed
P0175 **2T FUEL, MIL: Yes** 1996, 1997, 1998, 1999, 2000, 2001, 2002, 2003, 2004, 2005 C/K Cab & Chassis, C/K Series Truck, G Series Van, M/L Series Vans 4.3L VIN W, 4.3L VIN X, 5.0L VIN M, 5.7L VIN K, 5.7L VIN R, 7.4L VIN J Transmissions: All	**Fuel Trim System Rich (Bank 2) Conditions:** DTC P0101, P0102, P0103, P0108, P0135, P0137, P0141, P0200, P0300, P0410, P0420, P0430, P0440, P0442, P0443, P0446, P0449, P0506, P0507 and P1441 not set, engine started, vehicle speed less than 85 mph at 400-3000 rpm, BARO sensor over 74 kPa, ECT sensor from 167-239°F, IAT sensor from 4-194°F, MAF sensor from 5-90 gm/s, MAP sensor from 26-90 kPa (3.7-13 psi), TP angle under 90%, fuel level over 10%, and the PCM detected the LT fuel trim was less than -13% for 40 seconds (i.e., a rich A/F mixture existed). **Possible Causes** ● Base engine "mechanical" fault affecting one or more cylinders ● Excess fuel vapors in crankcase (the oil needs to be changed) ● EVAP system component has failed or canister fuel saturated ● Fuel control sensor is out of calibration (i.e., ECT, IAT or MAF) ● Fuel delivery system supplying too much fuel at cruise or idle periods (faulty regulator) ● Fuel injector(s) is leaking or stuck partially open (one or more) ● HO2S is contaminated, deteriorated or it has failed ● TSB 76-65-04 contains a repair procedure for this code
P0175 **2T FUEL, MIL: Yes** 1999, 2000, 2001, 2002, 2003, 2004, 2005 C/K Series Truck, G Series Van, Envoy, Escalade, TrailBlazer 4.8L VIN V, 5.3L VIN P, 5.3L VIN T, 5.3L VIN Z, 6.0L VIN N, 6.0L VIN U, 8.1L VIN G Transmissions: All	**Fuel Trim System Rich (Bank 2) Conditions:** DTC P0101-P0103, P0108, P0135, P0137, P0141, P0200, P0300, P0410, P0420, P0430, P0440, P0442, P0443, P0446, P0449, P0506, P0507 and P1441 not set, engine speed from 400-3000 rpm, ECT sensor from 167-239°F, IAT sensor from 4-194°F, BARO sensor over 74 kPa, MAF sensor from 5-90 gm/s, TP angle under 90%, VSS below 85 mph, and the PCM detected the LT fuel trim was less than -13% for 6 seconds (i.e., a possible rich A/F mixture). **Possible Causes** ● Air leaks in intake manifold, exhaust pipes or exhaust manifold ● Fuel control sensor is out of calibration (ECT, IAT or MAF) ● Low fuel pressure (fuel filter clogged, pressure regulator failure) ● One or more injectors restricted or pressure regulator has failed ● HO2S element is contaminated, deteriorated or has failed ● Vacuum hose is disconnected, broken, leaking or loose
P0175 **2T FUEL, MIL: Yes** 2003, 2004, 2005 C/K Trucks & G Vans 6.0L VIN U CNG engine Transmissions: All	**Fuel Trim System Rich (Bank 2) Conditions:** DTC P0101-P0103, P0106-P0108, P0112, P0113, P0116-P0118, P0121-P0123, P0125, P0200, P0300, P0327, P0332, P0335, P0336, P0351-P0358, P0401-P0403, P0443, P0446, P0449, P0496, P0502, P0503, P1020 and P1258 not set, engine speed 400-3000 rpm, ECT sensor from 167-239°F, IAT sensor from 4-194°F, BARO sensor over 74 kPa, MAF sensor from 5-90 gm/s, TP angle under 90%, VSS under 85 mph, and the PCM detected the LT fuel trim was less than -15% for 40 seconds (indicating a lean A/F mixture was present). **Possible Causes** ● Base engine "mechanical" fault affecting one or more cylinders ● Excess fuel vapors in crankcase (the oil needs to be changed) ● EVAP system component has failed or canister fuel saturated ● Fuel delivery system supplying too much fuel at cruise or idle periods (faulty regulator) ● Fuel injector(s) is leaking or stuck partially open (one or more) ● HO2S is contaminated, deteriorated or it has failed

OBD II Trouble Code List (P0xxx Codes)

DTC	Trouble Code Title, Conditions & Possible Causes
P0175 **2T FUEL, MIL: Yes** 2003, 2004, 2005 CTS (Cadillac) 2.6L V6 VIN M & 3.2L V6 VIN N engines Transmissions: All	**Fuel Trim System Rich (Bank 2) Conditions:** DTC P0101, P0102, P0103, P0130, P0131, P0132, P0133, P0134, P0135, P0137, P0138, P0140 and P0141 not set, engine running in closed loop, fuel level over 10%, and the PCM detected the Long Term fuel trim Cruise/Acceleration parameter was over -23% or the Long Term fuel trim Idle/Deceleration parameter was over -8.5%. **Possible Causes** ● Base engine "mechanical" fault affecting one or more cylinders ● Excess fuel vapors in crankcase (the oil needs to be changed) ● EVAP system component has failed or canister fuel saturated ● Fuel control sensor is out of calibration (i.e., ECT, IAT or MAF) ● Fuel delivery system supplying too much fuel during cruise or idle periods (e.g., faulty fuel pump, or faulty pressure regulator) ● Fuel injector(s) is leaking or stuck partially open (one or more) ● HO2S is contaminated, deteriorated or it has failed
P0175 **2T FUEL, MIL: Yes** 1996, 1997, 1998, 1999, 2000, 2001, 2002, 2003, 2004, 2005 DeVille, Eldorado, Seville 4.6L VIN 9, 4.6L VIN Y Transmissions: All	**Fuel Trim System Rich (Bank 2) Conditions:** DTC P0101-P0103, P0106-P0108, P0112-P0113, P0116-P0118, P0121 P0123, P0125, P0128, P0200, P0300, P0410-P0446, P0452-P0453, P1258, P1415-P1416 and P1441 not set, engine started, vehicle driven at less than 82 mph at 500-4000 rpm, BARO sensor more than 75 kPa, ECT sensor from 68-239°F, IAT sensor from -4°F to 212°F, MAF sensor from 3-60 g/sec, EGR Flow Test "off", MAP sensor from 15-85 kPa, TP angle less than 65%, and the PCM detected the Long Term fuel trim was less than -13 for 40 seconds. **Possible Causes** ● Base engine "mechanical" fault affecting one or more cylinders ● Excess fuel vapors in crankcase (the oil needs to be changed) ● EVAP system component has failed or canister fuel saturated ● Fuel control sensor is out of calibration (i.e., ECT, IAT or MAF) ● Fuel delivery system supplying too much fuel during cruise or idle periods (e.g., faulty fuel pump, or faulty pressure regulator) ● Fuel injector(s) is leaking or stuck partially open (one or more) ● HO2S is contaminated, deteriorated or it has failed
P0175 **2T FUEL, MIL: Yes** 1995, 1996, 1997, 1998, 1999, 2000, 2001, 2002 Camaro & Firebird 3.8L VIN K engine Transmissions: All	**Fuel Trim System Rich (Bank 2) Conditions:** DTC P0101-P0103, P0108, P0135, P0137, P0141, P0200, P0300, P0410, P0420, P0430, P0440-P0449, P0506, P0507 and P1441 not set, engine started, vehicle driven at less than 85 mph at 400-3000 rpm, BARO sensor more than 74 kPa (10.7 psi), ECT sensor from 167-239°F, IAT sensor from 4-194°F, MAF sensor from 5-90 gm/s, MAP sensor from 26-90 kPa (3.7-13 psi), TP angle less than 90%, fuel level over 10%, and the PCM detected the Long Term fuel trim value was less than -13% for 40 seconds during the Fuel Trim test. **Possible Causes** ● Base engine "mechanical" fault affecting one or more cylinders ● Excess fuel vapors in crankcase (the oil needs to be changed) ● EVAP system component has failed or canister fuel saturated ● Fuel control sensor is out of calibration (i.e., ECT, IAT or MAF) ● Fuel delivery system supplying too much fuel during cruise or idle periods (e.g., faulty fuel pump, or faulty pressure regulator) ● Fuel injector(s) is leaking or stuck partially open (one or more) ● HO2S is contaminated, deteriorated or it has failed
P0175 **2T FUEL, MIL: Yes** 1998, 1999, 2000, 2001, 2002 Camaro & Firebird 5.7L VIN G engine Transmissions: All	**Fuel Trim System Rich (Bank 2) Conditions:** DTC P0101-P0103, P0107-P0108, P0112-P0113, P0116-P0118, P0121-P0125, P0200, P0300, P0335-P0336, P0351-P0358, P0405, P0410, P0412, P0418, P0443, P0500-P0503, P1258, P1404, P1415 and P1416 not set, engine started, vehicle driven at less than 85 mph at 400-3000 rpm, BARO sensor more than 74 kPa, ECT sensor from 122-239°F, IAT sensor from -4°F-194°F, MAF sensor airflow from 3-150 gm/s, MAP sensor from 26-90 kPa, TP angle less than 90% and steady, fuel level over 10%, and the PCM detected the Long Term fuel trim value was less than -13% for 6 seconds. **Possible Causes** ● Base engine "mechanical" fault affecting one or more cylinders ● Excess fuel vapors in crankcase (the oil needs to be changed) ● EVAP system component has failed or canister fuel saturated ● Fuel control sensor is out of calibration (i.e., ECT, IAT or MAF) ● Fuel delivery system supplying too much fuel during cruise or idle periods (e.g., faulty fuel pump, or faulty pressure regulator) ● Fuel injector(s) is leaking or stuck partially open (one or more) ● HO2S is contaminated, deteriorated or it has failed

OBD II Trouble Code List (P0xxx Codes)

DTC	Trouble Code Title, Conditions & Possible Causes
P0175 **2T FUEL, MIL: Yes** 1996, 1997, 1998, 1999, 2001, 2002, 2003, 2004, 2005 Aurora 3.5L V6 VIN H, 4.0L V8 VIN C engines Transmissions: A/T	**Fuel Trim System Rich (Bank 2) Conditions:** DTC P0101-P0103, P0106-P0108, P0121, P0122, P0131-P0138, P0140-P0146, P0151, P0154, P0562, P0563, P0201-P0208, P0300, P0325, P0327, P0340, P0401, P0506, P0507, P1121, P1133, P1153, P1154, P1441, P1508 and P1509 not set, engine started, vehicle driven at less than 70 mph at 400-3000 rpm, BARO sensor more than 70.8 kPa, ECT sensor from 184-221ºF, IAT sensor from 40ºF to 218ºF, MAF sensor from 3-200 gm/s, MAP sensor from 27-103 kPa, EVAP purge tank not full, TP angle less than 19.8%, and the PCM detected the Short Term fuel trim value was less than -13% for 40 seconds during the Fuel Trim test. **Possible Causes** ● Base engine "mechanical" fault affecting one or more cylinders ● Excess fuel vapors in crankcase (the oil needs to be changed) ● EVAP system component has failed or canister fuel saturated ● Fuel control sensor is out of calibration (i.e., ECT, IAT or MAF) ● Fuel delivery system supplying too much fuel during cruise or idle periods (e.g., faulty fuel pump, or faulty pressure regulator) ● Fuel injector(s) is leaking or stuck partially open (one or more) ● HO2S is contaminated, deteriorated or it has failed
P0175 **2T FUEL, MIL: Yes** 1997, 1998, 1999, 2000, 2001 Catera 3.0L V6 VIN R engine Transmissions: A/T	**Fuel Trim System Rich (Bank 2) Conditions:** Engine started, vehicle driven at a steady cruise speed in closed loop, Fuel Trim enabled, and the PCM detected the Fuel Trim long term values were greater than 23% and the Short Term fuel trim values were greater than 25% during the Fuel System Monitor test. **Possible Causes** ● Base engine "mechanical" fault affecting one or more cylinders ● Excess fuel vapors in crankcase (the oil needs to be changed) ● EVAP system component has failed or canister fuel saturated ● Fuel delivery system supplying too much fuel during cruise or idle periods (e.g., faulty fuel pump, or faulty pressure regulator) ● Fuel injector(s) is leaking or stuck partially open (one or more) ● HO2S is contaminated, deteriorated or it has failed
P0175 **2T FUEL, MIL: Yes** 1995, 1996, 1997, 1998, 1999, 2000, 2001, 2002, 2003, 2004, 2005 S/T Blazer & S/T Pickup 4.3L VIN W, 4.3L VIN X Transmissions: All	**Fuel System Rich (Bank 2) Conditions:** DTC P0101, P0102, P0103, P0106, P0107, P0108, P0112, P0113, P0117, P0118, P0125, P0135, P0137, P0141, P0200, P0300, P0410, P0420, P0430, P0440-P0449, P0506, P0507 and P1441 not set, engine started, vehicle driven at less than 85 mph at 400-3000 rpm, BARO sensor more than 74 kPa (10.7 psi), ECT sensor from 167-239ºF, IAT sensor from 4-194ºF, MAF sensor from 5-90 gm/s, MAP sensor from 26-90 kPa, fuel level over 10%, TP angle less than 90%, and the PCM detected the average Long Term fuel trim value was less than -13 for 40 seconds (i.e., a rich condition). **Possible Causes** ● Base engine "mechanical" fault affecting one or more cylinders ● Excess fuel vapors in crankcase (the oil needs to be changed) ● EVAP system component has failed or canister fuel saturated ● Fuel control sensor is out of calibration (i.e., ECT, IAT or MAF) ● Fuel delivery system supplying too much fuel during cruise or idle periods (e.g., faulty fuel pump, or faulty pressure regulator) ● Fuel injector(s) is leaking or stuck partially open (one or more) ● HO2S is contaminated, deteriorated or it has failed ● TSB 76-65-04 contains a repair procedure for this code
P0175 **2T FUEL, MIL: Yes** 1995 Corvette 5.7L VIN P Transmissions: All	**Fuel Trim System Rich (Bank 2) Conditions:** DTC P0116-P0118, P0135, P0174, P0175, P1114 and P1115 not set, engine started, vehicle driven at less than 75 mph at 500-4000 rpm with Fuel Trim enabled, BARO sensor more than 70 kPa, ECT sensor from 140-239ºF, IAT sensor from -22 to 140ºF, MAF from 5-100 gm/s, MAP sensor from 20-90 kPa, Purge command under 75%, TP angle less than 70%, and the PCM detected the LONGFT was under 118, or the SHRTFT count was 112 for 10 seconds. **Possible Causes** ● Base engine "mechanical" fault affecting one or more cylinders ● Excess fuel vapors in crankcase (the oil needs to be changed) ● EVAP system component has failed or canister fuel saturated ● Fuel control sensor is out of calibration (i.e., ECT, IAT or MAF) ● Fuel delivery system supplying too much fuel during cruise or idle periods (e.g., faulty fuel pump, or faulty pressure regulator) ● Fuel injector(s) is leaking or stuck partially open (one or more) ● HO2S is contaminated, deteriorated or it has failed

OBD II Trouble Code List (P0xxx Codes)

DTC	Trouble Code Title, Conditions & Possible Causes
P0175 **2T FUEL, MIL: Yes** 1997, 1998, 1999, 2000, 2001, 2002, 2003, 2004, 2005 Corvette 5.7L VIN G, 5.7L VIN S Transmissions: All	**Fuel Trim System Rich (Bank 2) Conditions:** DTC P0101-P0103, P0108, P0135, P0137, P0141, P0200, P0300, P0410, P0420, P0430, P0440-P0449, P0506, P0507 or P1441 not set, engine started, vehicle driven at less than 85 mph at an engine speed 500-3000 rpm, BARO sensor over 74 kPa (10.7 psi), ECT sensor from 167-239°F, IAT sensor from 4-194°F, MAF sensor from 5-90 gm/s, MAP sensor from 26-90 kPa, TP angle less than 90%, fuel level over 10%, and the PCM detected the Long Term fuel trim cell value was over -13% for 40 seconds during the Fuel Trim test. **Possible Causes** ● Base engine "mechanical" fault affecting one or more cylinders ● Excess fuel vapors in crankcase (the oil needs to be changed) ● EVAP system component has failed or canister fuel saturated ● Fuel delivery system supplying too much fuel in cruise or idle period (faulty regulator) ● Fuel injector(s) is leaking or stuck partially open (one or more) ● HO2S is contaminated, deteriorated or it has failed
P0178 **2T CCM, MIL: Yes** 2000, 2001, 2002, 2003, 2004, 2005 C/K Series Truck, G Van, S/T Series Pickup 2.2L VIN 5, 6.0L VIN N, 6.0L VIN U Transmissions: All	**Fuel Composition Sensor Low Frequency Conditions:** Engine started; engine runtime over 30 seconds, system voltage over 10.9v, and the PCM detected the Fuel Composition Sensor (FCS) signal frequency was less than 45 Hz for 12.5 seconds. The FCS signal frequency at 32°F or higher is 138 Hz at 88% Ethanol. **Possible Causes** ● FCS signal circuit is open or shorted to ground ● FCS power circuit is open (check power from the ECMI fuse) ● FCS ground circuit is open between the sensor and the PCM ● FCS is damaged or has failed ● PCM is damaged
P0179 **2T CCM, MIL: Yes** 2000, 2001, 2002, 2003, 2004, 2005 C/K Series Truck, G Van, S/T Series Pickup 2.2L VIN 5, 6.0L VIN N, 6.0L VIN U Transmissions: All	**Fuel Composition Sensor High Frequency Conditions:** Engine started; engine runtime over 30 seconds, system voltage over 10.9v, and the PCM detected the Fuel Composition Sensor (FCS) signal frequency was more than 155 Hz for 12.5 seconds. The FCS signal frequency at 32°F or higher is 138 Hz at 88% Ethanol. **Possible Causes** ● FCS signal circuit is shorted to system power ● FCS is damaged or has failed ● PCM is damaged
P0181 **1T CCM, MIL: Yes** 1999, 2000, 2001, 2002, 2003, 2004, 2005 C/K Series Truck, G Vans 6.6L VIN 1 Diesel engine Transmissions: All	**Fuel Tank Temperature Sensor Signal Range/Performance Conditions:** Key off for 10 hours, DTC P0112, P0113, P0182, P0183, P0500 and P1683 not set, Fuel Tank Temperature (FTT) and ECT sensor within 18°F, IAT sensor over 59°F at startup, vehicle driven to over 15 mph, and the PCM detected the FTT signal dropped less than 10°F after a period of 400 seconds. The PCM supplies the sensor with a 5v signal and a low reference circuit. If the fuel temperature sensor is cold, its resistance is high. The FTT signal voltage remains near the signal voltage cold and decreases as the sensor warms. The PCM monitors the FTT signal circuit to calculate the temperature of fuel entering the engine **Possible Causes** ● FTT sensor signal circuit has a high resistance fault ● FTT sensor is damaged, skewed or it has failed ● PCM has failed
P0181 **2T CCM, MIL: Yes** 1996, 1997, 1998, 1999, 2000 C/K Series Truck, G Vans 6.5L VIN F, 6.5L VIN S Diesel engines Transmissions: All	**Fuel Temperature Sensor Circuit Low Input Conditions:** Key on or engine running; and the PCM detected the Optical Fuel Temperature Sensor (OFT) indicated more than 215°F for 2 seconds during the CCM test. **Possible Causes** ● FTS signal circuit is shorted to chassis or sensor ground ● FTS is damaged, skewed or it has failed ● PCM has failed
P0182 **1T CCM, MIL: Yes** 1999, 2000, 2001, 2002, 2003, 2004, 2005 C/K Series Truck, G Vans 6.6L VIN 1 Diesel engine Transmissions: All	**Fuel Tank Temperature Sensor Circuit Low Input Conditions:** Key on or engine running; and the PCM detected the Fuel Tank Temperature (FTT) sensor indicated more than 248°F for 2 seconds. **Possible Causes** ● FTT sensor signal circuit is shorted to chassis or sensor ground ● FTT sensor is damaged, skewed or it has failed ● PCM has failed
P0182 **1T CCM, MIL: Yes** 1999, 2000, 2001, 2002, 2003, 2004, 2005 C/K Series Truck 6.0L VIN U CNG engine Transmissions: All	**Fuel Tank Temperature Sensor Circuit Low Input Conditions:** DTC P1207 not set, key on or engine running, IAT sensor more than -31°F, and the PCM detected the Fuel Tank Temperature (FTT) sensor was under 0.1v (Scan Tool reads over 248°F) for 5 seconds. **Possible Causes** ● FTT sensor signal circuit is shorted to chassis or sensor ground ● FTT sensor is damaged, skewed or it has failed ● PCM has failed

OBD II Trouble Code List (P0xxx Codes)

DTC	Trouble Code Title, Conditions & Possible Causes
P0182 **1T CCM, MIL: No** 2003, 2004, 2005 G Series Van, Car Van 6.0L VIN U CNG engine Transmissions: All	**Fuel Tank Temperature Sensor Circuit Low Input Conditions:** Key on or engine running, IAT sensor more than -31ºF, and the PCM detected the Fuel Tank Temperature (FTT) sensor indicated less than 0.05v (Scan Tool reads over 248ºF) for 5 seconds. **Possible Causes** ● FTT sensor signal circuit is shorted to chassis or sensor ground ● FTT sensor is damaged, skewed or it has failed ● PCM has failed
P0183 **1T CCM, MIL: Yes** 2003, 2004, 2005 C/K Series Truck 6.0L VIN U CNG engine Transmissions: All	**Fuel Tank Temperature Sensor Circuit High Input Conditions:** DTC P0112, P0113 and P1207 not set, engine started, IAT sensor over 14ºF, and the PCM detected the Fuel Tank Temperature (FTT) sensor was more than 4.95v (Scan Tool reads -22ºF) for 5 seconds. **Possible Causes** ● FTT sensor signal circuit or ground circuit is open ● FTT sensor signal circuit is shorted to VREF ● FTT sensor is damaged, skewed or it has failed ● PCM has failed
P0183 **1T CCM, MIL: No** 2003, 2004, 2005 G Series Van, Cargo Van 6.0L VIN U CNG engine Transmissions: All	**Fuel Tank Temperature Sensor Circuit High Input Conditions:** Key on or engine running, IAT sensor more than -31ºF, and the PCM detected the Fuel Tank Temperature (FTT) sensor was over 4.95v (Scan Tool reads less than -22ºF) for 5 seconds. **Possible Causes** ● FTT sensor signal circuit or ground circuit is open ● FTT sensor signal circuit is shorted to VREF ● FTT sensor is damaged, skewed or it has failed ● PCM has failed
P0183 **2T CCM, MIL: Yes** 1996, 1997, 1998, 1999, 2000, 2001, 2002 C/K Series Truck, G Series Van, Cargo Van 6.5L VIN F, 6.5L VIN S Diesel engines Transmissions: All	**Fuel Temperature Sensor Circuit High Input Conditions:** Engine started; engine runtime over 8 minutes and the PCM detected the Optical Fuel Temperature Sensor (OFT) indicated less than 64ºF for 2 seconds during the CCM test. **Possible Causes** ● FTS signal circuit is open between the sensor and the PCM ● FTS ground circuit is open between the sensor and the PCM ● FTS is damaged, skewed or it has failed ● PCM has failed
P0183 **1T CCM, MIL: No** 1999, 2000, 2001, 2002 C/K Series Truck, G Series Van, Van Cargo 6.6L VIN 1 Diesel engine Transmissions: All	**Fuel Tank Temperature Sensor Circuit High Input Conditions:** Engine started; engine runtime over 8 minutes and the PCM detected the Fuel Tank Temperature (FTT) sensor indicated less than -22ºF for 2 seconds during the CCM test. **Possible Causes** ● FTT sensor signal circuit or ground circuit is open ● FTT sensor signal circuit is shorted to VREF ● FTT sensor is damaged, skewed or it has failed ● PCM has failed
P0187 **2T CCM, MIL: Yes** 1999, 2000, 2001, 2002, 2003, 2004, 2005 C/K Series Truck 6.0L VIN U CNG engine Transmissions: All	**Fuel Rail Temperature Sensor Circuit Low Input Conditions:** Key on or engine running; and the PCM detected the Fuel Rail Temperature (FRT) sensor was less than 0.10v for 5 seconds. The FRT sensor is a variable resistor that measures the temperature of the fuel in the CNG fuel rail. The fuel injector control module (FICM) supplies 5v to the FRT signal circuit and supplies a ground to the low reference circuit. The FICM monitors the FRT sensor signal and communicates data to the PCM by a dedicated PWM circuit. This code sets when the FRT signal is below normal operating range. **Possible Causes** ● FRT sensor signal circuit shorted to chassis or sensor ground ● FRT sensor is damaged, skewed or it has failed ● PCM has failed
P0187 **2T CCM, MIL: Yes** 2003, 2004, 2005 G Series Van, Cargo Van 6.0L VIN U CNG engine Transmissions: All	**Fuel Rail Temperature Sensor Circuit Low Input Conditions:** Key on or engine running, IAT sensor more than -31ºF, and the PCM detected the Fuel Rail Temperature (FRT) sensor was less than 0.05v for 5 seconds. The FRT sensor is a variable resistor that measures the temperature of the fuel in the CNG fuel rail. The fuel injector control module (FICM) supplies 5v to the FRT signal circuit and supplies a ground to the low reference circuit. The FICM monitors the FRT sensor signal and communicates data to the PCM by a dedicated PWM circuit. This code sets when the PCM detects that the FRT signal is less than the normal operating range. **Possible Causes** ● FRT sensor signal circuit shorted to chassis or sensor ground ● FRT sensor is damaged, skewed or it has failed ● PCM has failed

OBD II Trouble Code List (P0xxx Codes)

DTC	Trouble Code Title, Conditions & Possible Causes
P0188 **2T CCM, MIL: Yes** 1999, 2000, 2001, 2002, 2003, 2004, 2005 C/K Series Trucks 6.0L VIN U CNG engine Transmissions: All	**Fuel Rail Temperature Sensor Circuit High Input Conditions:** DTC P0112, P0113 and P1207 not set key on or engine running, IAT sensor more than 14°F, and the PCM detected the Fuel Rail Temperature (FRT) sensor was more than 4.95v for 2 seconds. Note: The FRT sensor should read approximately 2.20v at 86°F. **Possible Causes** ● FRT sensor signal circuit or ground circuit is open ● FRT sensor signal circuit is shorted to VREF ● FRT sensor is damaged, skewed or it has failed ● PCM has failed
P0188 **2T CCM, MIL: Yes** 2003, 2004, 2005 G Series Van, Cargo Van 6.0L VIN U CNG engine Transmissions: All	**Fuel Rail Temperature Sensor Circuit High Input Conditions:** Key on or engine running, IAT sensor more than -31°F, and the PCM detected the Fuel Rail Temperature (FRT) sensor indicated was more than 4.95v for 5 seconds. The FRT sensor should read approximately 2.20v at 86°F. The FRT sensor is a variable resistor that measures the temperature of the fuel in the CNG fuel rail. The fuel injector control module (FICM) supplies 5v to the FRT signal circuit and supplies a ground to the low reference circuit. The FICM monitors the FRT sensor signal and communicates data to the PCM by a dedicated PWM circuit. This code sets when the PCM detects that the FRT signal is less than the normal operating range **Possible Causes** ● FRT sensor signal circuit or ground circuit is open ● FRT sensor signal circuit is shorted to VREF ● FRT sensor is damaged, skewed or it has failed ● PCM has failed
P0191 **2T CCM, MIL: Yes** 1999, 2000, 2001, 2002, 2003, 2004, 2005 C/K Series Truck 6.0L VIN U CNG engine Transmissions: All	**Fuel Rail Pressure Sensor Circuit Low Input Conditions:** DTC P0005, P0192, P0193, P0336, P1207, P1432, P1433 and P2665 not set, then after engine startup with the engine speed over 100 rpm, ECT sensor more than 68°F, MAF sensor less than 200 g/sec, and the PCM detected the FRP sensor was less than 206 kPa (30 psi) or more than 620 kPa (90 psi) for 3 seconds. **Possible Causes** ● FRP Sensor 5v VREF circuit has a high resistance condition ● FRP sensor signal circuit has a high resistance condition ● FRP sensor is damaged or it has failed ● PCM has failed
P0191 **1T CCM, MIL: Yes** 2003, 2004, 2005 G Series Van, Cargo Van 6.0L VIN U CNG engine Transmissions: All	**Fuel Rail Pressure Sensor Circuit Low Input Conditions:** DTC P0005, P0192, P0193, P0336, P1207, P1432, P1433 and P2665 not set, then after engine startup with the engine speed over 600 rpm, ECT sensor more than 68°F, MAF sensor less than 175 g/sec, CNG fuel tank pressure more than 750 psi, and the PCM detected the FRP sensor was less than 120 psi or more than 220 psi for 5 seconds. The fuel injector control module (FICM) supplies 5v on the FRP sensor reference voltage circuit. The FICM also supplies a ground circuit and a signal circuit to the FRP sensor. When the fuel rail pressure is normal, the FRP signal voltage rises to near 2.5v. If the fuel rail pressure increases, the FRP signal voltage increases. The FICM monitors the FRP sensor and communicates the data to the PCM by a dedicated pulse width modulated (PWM) circuit. **Possible Causes** ● FRP Sensor 5v VREF circuit has a high resistance condition ● FRP sensor signal circuit has a high resistance condition ● FRP sensor is damaged or it has failed ● PCM has failed
P0192 **1T CCM, MIL: No** 1999, 2000, 2001, 2002, 2003, 2004, 2005 C/K Series Truck, G Series Van, Van Cargo 6.6L VIN 1 Diesel engine Transmissions: All	**Fuel Rail Pressure Sensor Circuit Low Input Conditions:** DTC P0005, P0192, P0193, P0336, P1207, P1432, P1433 and P2665 not set, key on or engine running; and the PCM detected the Fuel Rail Pressure sensor was less than 1.2 MPa. **Possible Causes** ● FRP sensor 5-volt power circuit is open or shorted to ground ● FRP Sensor signal circuit is shorted to ground ● FRP Sensor is damaged or has failed ● PCM has failed
P0192 **1T CCM, MIL: No** 1999, 2000, 2001, 2002, 2003, 2004, 2005 C/K Series Truck 6.0L VIN U CNG engine Transmissions: All	**Fuel Rail Pressure Sensor Circuit Low Input Conditions:** DTC P0191 not set, engine started, engine speed over 100 rpm, the PCM detected the FRP sensor was less than 0.10v for 2.5 seconds. The fuel injector control module (FICM) supplies 5v on the FRP sensor reference voltage circuit. The FICM also supplies a ground circuit and a signal circuit to the FRP sensor. When the fuel rail pressure is normal, the FRP signal rises to near 2.5v. As the fuel rail pressure increases, the FRP signal voltage increases. The FICM monitors the FRP sensor and communicates the data to the PCM by a discrete PWM circuit. **Possible Causes** ● FRP sensor 5-volt power circuit is open or shorted to ground ● FRP Sensor signal circuit is shorted to ground ● FRP Sensor is damaged or has failed ● PCM has failed

OBD II Trouble Code List (P0xxx Codes)

DTC	Trouble Code Title, Conditions & Possible Causes
P0192 **2T CCM, MIL: Yes** 2003, 2004, 2005 G Series Van, Cargo Van 6.0L VIN U Transmissions: All	**Fuel Rail Pressure Sensor Circuit Low Input Conditions:** Key on or engine running; and the PCM detected the FRP sensor was less than 0.45v for 5 seconds. The fuel rail pressure (FRP) sensor is a pressure sensor. The fuel injector control module (FICM) supplies 5v on the FRP sensor reference voltage circuit. The FICM also supplies a ground circuit and a signal circuit to the FRP sensor. When the fuel rail pressure is normal, the FRP signal voltage rises to near 2.5v. If the fuel rail pressure increases, the FRP signal voltage increases. The FICM monitors the FRP sensor and communicates the data to the PCM by a dedicated pulse width modulated (PWM) circuit. **Possible Causes** ● FRP sensor 5-volt power circuit is open or shorted to ground ● FRP Sensor signal circuit is shorted to ground ● FRP Sensor is damaged or has failed ● PCM has failed
P0193 **1T CCM, MIL: Yes** 1999, 2000, 2001, 2002, 2003, 2004, 2005 C/K Series Truck G Series Van, Cargo Van 6.6L VIN 1 Diesel engine Transmissions: All	**Fuel Rail Pressure Sensor Circuit High Input Conditions:** DTC P1635 and P1639 not set, key on or engine running system not in Power-Down mode, and the PCM detected the Fuel Rail Pressure (FRP) sensor was more than 75 MPa during the CCM test. **Possible Causes** ● FRP sensor signal circuit shorted to VREF or system power ● FRP sensor is damaged or has failed ● PCM has failed
P0193 **2T CCM, MIL: Yes** 1999, 2000, 2001, 2002, 2003, 2004, 2005 C/K Series Truck 6.0L VIN U CNG engine Transmissions: All	**Fuel Rail Pressure Sensor Circuit High Input Conditions:** DTC P0336 not set, and after the engine started, engine speed over 100 rpm, the PCM detected the Fuel Rail Pressure (FRP) sensor was more than 4.95v for 10 seconds (the fault is continuous). **Possible Causes** ● FRP sensor signal circuit is open between sensor and the PCM ● FRP Sensor ground circuit is open between sensor and PCM ● FRP sensor signal circuit is shorted to VREF or system power ● FRP Sensor is damaged or has failed ● PCM has failed
P0193 **2T CCM, MIL: Yes** 2003, 2004, 2005 G Series Van, Cargo Van 6.0L VIN U CNG engine Transmissions: All	**Fuel Rail Pressure Sensor Circuit High Input Conditions:** Key on or engine running; and the PCM detected the Fuel Rail Pressure (FRP) sensor was more than 4.95v for 5 seconds. **Possible Causes** ● FRP sensor signal circuit is open between sensor and the PCM ● FRP Sensor ground circuit is open between sensor and PCM ● FRP sensor signal circuit is shorted to VREF or system power ● FRP Sensor is damaged or has failed ● PCM has failed
P0200 **2T CCM, MIL: Yes** 1996, 1997, 1998, 1999 Achieva, Beretta, Ciera, Cavalier, Century, Corsica, Grand Am, Malibu, Skylark, Sunfire 2.2L VIN 4, 2.4L VIN T Transmissions: All	**Fuel Injector Circuit Malfunction Conditions:** Engine started; system voltage over 9v, and the PCM detected the injector current level was under 1 amp or too high for 7 seconds. Note: Drive the vehicle at cruise speed. Record the misfire current counters for review to detect if more than one cylinder is misfiring. **Possible Causes** ● Fuel injector power circuit (B+) is open (check the power fuse) ● Fuel injector control circuit is open between injector and PCM ● Fuel injector control circuit is grounded between injector and PCM ● Fuel injector is damaged or has failed ● PCM is damaged
1996, 1997, 1998, 1999, 2000, 2001, 2002, 2003, 2004, 2005 Camaro, Caprice, Corvette, Firebird, Fleetwood 4.3L VIN W, 5.7L VIN 5, 5.7L VIN P, 5.7L VIN G, 5.7L VIN S engines Transmissions: All	**Fuel Injector Circuit Malfunction Conditions:** Engine started; system voltage over 10.0v and the PCM detected an unexpected voltage on one or more of the Fuel Injector driver circuits for 5 seconds. Note: Drive the vehicle at off-idle speeds and have an assistant monitor the misfire current counters. Observe if more than one cylinder is misfiring. This may not be apparent until after a repair is completed. If an injector fuse is open for one cylinder bank, the Scan Tool may only display 2 or 3 cylinders misfiring. **Possible Causes** ● Fuel injector power circuit (B+) is open (check the power fuse) ● Fuel injector control circuit is open between injector and PCM ● Fuel injector control circuit is grounded between injector and PCM ● Fuel injector is damaged or has failed ● PCM is damaged

OBD II Trouble Code List (P0xxx Codes)

DTC	Trouble Code Title, Conditions & Possible Causes
P0200 **2T CCM, MIL: Yes** 1995 Regal, LeSabre 3.8L VIN L engine Transmissions: All	**Fuel Injector Circuit Malfunction Conditions:** Engine started; engine speed over 400 rpm, system voltage over 10.0v and the PCM detected an unexpected voltage on the one of the Fuel Injector driver circuits for 5 seconds. **Possible Causes** • Fuel injector power circuit is open (test Fuse 7 in Relay Center) • Fuel injector control circuit is open between injector and PCM • Fuel injector control circuit is grounded between injector and PCM • Fuel injector is damaged or has failed • PCM is damaged
P0200 **2T CCM, MIL: Yes** 1996, 1997, 1998, 1999, 2000, 2001, 2002, 2003, 2004, 2005 C/K Cab & Chassis, C/K Series Truck, G Series Vans & L/M Series Vans 4.3L VIN W, 4.3L VIN X, 5.0L VIN M, 5.7L VIN R, 7.4L VIN J engines Transmissions: All	**Fuel Injector Circuit Malfunction Conditions:** Engine started; engine speed over 400 rpm, system voltage 6-18v, and the PCM detected an unexpected voltage on one or more of the Fuel Injector driver circuits for 5 seconds. Drive the vehicle at off-idle speeds and monitor the misfire current counters. Observe if more than one cylinder is misfiring. Repeat this step after a repair is done to confirm the repair. **Possible Causes** • Fuel injector power circuit is open (test ECM fuse in fuse block) • Fuel injector control circuit is open between injector and PCM • Fuel injector control circuit is grounded between injector and PCM • Fuel injector is damaged or has failed • PCM is damaged
P0200 **2T CCM, MIL: Yes** 1999, 2000, 2001, 2002, 2003, 2004, 2005 C/K Series Truck, G Series Van, Envoy, Escalade, TrailBlazer 4.8L VIN V, 5.3L VIN P, 5.3L VIN T, 5.3L VIN Z, 6.0L VIN N, 6.0L VIN U CNG engine, 8.1L VIN G engines Transmissions: All	**Fuel Injector Circuit Malfunction Conditions:** Engine started; engine speed over 400 rpm, system voltage 6-18v, and the PCM detected an unexpected voltage on one or more of the Fuel Injector driver circuits for 5 seconds. Drive the vehicle at off-idle speeds and monitor the misfire current counters. Observe if more than one cylinder is misfiring. This may not be apparent until after a repair is completed. If an injector fuse is open on one cylinder bank, the Scan Tool may only display 2 or 3 cylinders as misfiring. **Possible Causes** • Fuel injector control circuit is open between injector and PCM • Fuel injector control circuit is grounded between injector and PCM • Fuel injector power circuit is open (test INJ A, B in fuse block) • Fuel injector is damaged or has failed • PCM is damaged
P0200 **2T CCM, MIL: Yes** 2001, 2002, 2003, 2004, 2005 S/T Blazer & S/T Pickup 4.3L VIN W, 4.3L VIN X Transmissions: All	**Fuel Injector Circuit Malfunction Conditions:** Engine started; system voltage at 6-18v, and the PCM detected an unexpected voltage on one of the fuel injector circuits for 5 seconds. **Possible Causes** • Fuel injector power circuit is open (test the power fuse) • Fuel injector control circuit is open or shorted to ground • Fuel injector is damaged or has failed • PCM is damaged
P0200 **2T CCM, MIL: Yes** 2003, 2004, 2005 G Series Van, Cargo Van 6.0L VIN U CNG engine CNG engine Transmissions: All	**Fuel Injector Circuit Malfunction Conditions:** Engine started; system voltage from 6-18v and the PCM detected an unexpected voltage on one of the Fuel Injector driver circuits for 5 seconds. Drive the vehicle at off-idle speeds and monitor the misfire current counters to determine if more than one cylinder is misfiring. This may not be apparent until after a repair is completed. **Possible Causes** • Fuel injector control circuit is open between injector and PCM • Fuel injector control circuit is grounded between injector and PCM • Fuel injector power circuit is open (test INJ A, B in fuse block) • Fuel injector is damaged or has failed • PCM is damaged
P0200 **2T CCM, MIL: Yes** 1996, 1997, 1998, 1999, 2000 S/T Blazer & S/T Pickup 2.2L VIN 4, 2.2L VIN 5, 2.4L VIN T engines Transmissions: All	**Fuel Injector Circuit Malfunction Conditions:** Engine started; system voltage more than 9v, and the PCM detected that the injector current level was less than 1 amp or it was continuously high for 7 seconds. Drive the vehicle at cruise speed. Record the misfire current counters to detect the misfiring cylinders. **Possible Causes** • Fuel injector power circuit is open (test the power fuse) • Fuel injector control circuit is open between injector and PCM • Fuel injector control circuit is grounded between injector and PCM • Fuel injector is damaged or has failed • PCM is damaged

OBD II Trouble Code List (P0xxx Codes)

DTC	Trouble Code Title, Conditions & Possible Causes
P0201 **2T CCM, MIL: Yes** 1998, 1999, 2000, 2001, 2002, 2003, 2004, 2005 Aurora, Bonneville, Century, Cutlass Supreme, Grand Prix, Intrigue, LeSabre, LSS, Lumina, Monte Carlo, 88', 98', Regal, Park Avenue 3.1L VIN J, 3.1L VIN M, 3.5L VIN H, 3.8L VIN 1, 3.8L VIN K, 4.0L VIN C Transmissions: All	**Fuel Injector 1 Control Circuit Malfunction Conditions:** Engine started; engine speed over 400 rpm, system voltage over 10.0v and the PCM detected an unexpected voltage condition on the Fuel Injector driver circuit for Cylinder 1 for 5 seconds. Note: Drive the vehicle at cruise speed. Record the misfire current counters for review to detect if more than one cylinder is misfiring. **Possible Causes** ● Injector 1 power circuit (B+) is open (check the power fuse) ● Injector 1 control circuit is open between injector and PCM ● Injector 1 control circuit is grounded between injector and PCM ● Injector 1 is damaged or it has failed ● PCM is damaged
P0202 **2T CCM, MIL: Yes** 1998, 1999, 2000, 2001, 2002, 2003, 2004, 2005 Aurora, Bonneville, Century, Cutlass Supreme, Grand Prix, Intrigue, LeSabre, LSS, Lumina, Monte Carlo, 88', 98', Regal, Park Avenue 3.1L VIN J, 3.1L VIN M, 3.5L VIN H, 3.8L VIN 1, 3.8L VIN K, 4.0L VIN C Transmissions: All	**Fuel Injector 2 Control Circuit Malfunction Conditions:** Engine started; engine speed over 400 rpm, system voltage over 10.0v and the PCM detected an unexpected voltage condition on the Fuel Injector driver circuit for Cylinder 2 for 5 seconds. Note: Drive the vehicle at cruise speed. Record the misfire current counters for review to detect if more than one cylinder is misfiring. **Possible Causes** ● Injector 2 power circuit (B+) is open (check the power fuse) ● Injector 2 control circuit is open between injector and PCM ● Injector 2 control circuit is grounded between injector and PCM ● Injector 2 is damaged or it has failed ● PCM is damaged
P0203 **2T CCM, MIL: Yes** 1998, 1999, 2000, 2001, 2002, 2003, 2004, 2005 Aurora, Bonneville, Century, Cutlass Supreme, Grand Prix, Intrigue, LeSabre, LSS, Lumina, Monte Carlo, 88', 98', Regal, Park Avenue 3.1L VIN J, 3.1L VIN M, 3.5L VIN H, 3.8L VIN 1, 3.8L VIN K, 4.0L VIN C Transmissions: All	**Fuel Injector 3 Control Circuit Malfunction Conditions:** Engine started; engine speed over 400 rpm, system voltage over 10.0v and the PCM detected an unexpected voltage condition on the Fuel Injector driver circuit for Cylinder 3 for 5 seconds. Note: Drive the vehicle at cruise speed. Record the misfire current counters for review to detect if more than one cylinder is misfiring. **Possible Causes** ● Injector 3 power circuit (B+) is open (check the power fuse) ● Injector 3 control circuit is open between injector and PCM ● Injector 3 control circuit is grounded between injector and PCM ● Injector 3 is damaged or it has failed ● PCM is damaged
P0204 **2T CCM, MIL: Yes** 1998, 1999, 2000, 2001, 2002, 2003, 2004, 2005 Aurora, Bonneville, Century, Cutlass Supreme, Grand Prix, Intrigue, LeSabre, LSS, Lumina, Monte Carlo, 88', 98', Regal, Park Avenue 3.1L VIN J, 3.1L VIN M, 3.5L VIN H, 3.8L VIN 1, 3.8L VIN K, 4.0L VIN C Transmissions: All	**Fuel Injector 4 Control Circuit Malfunction Conditions:** Engine started; engine speed over 400 rpm, system voltage over 10.0v and the PCM detected an unexpected voltage condition on the Fuel Injector driver circuit for Cylinder 4 for 5 seconds. Note: Drive the vehicle at cruise speed. Record the misfire current counters for review to detect if more than one cylinder is misfiring. **Possible Causes** ● Injector 4 power circuit (B+) is open (check the power fuse) ● Injector 4 control circuit is open between injector and PCM ● Injector 4 control circuit is grounded between injector and PCM ● Injector 4 is damaged or it has failed ● PCM is damaged
P0205 **2T CCM, MIL: Yes** 1998, 1999, 2000, 2001, 2002, 2003, 2004, 2005 Aurora, Bonneville, Century, Cutlass Supreme, Grand Prix, Intrigue, LeSabre, LSS, Lumina, Monte Carlo, 88', 98', Regal, Park Avenue 3.1L VIN J, 3.1L VIN M, 3.5L VIN H, 3.8L VIN 1, 3.8L VIN K, 4.0L VIN C	**Fuel Injector 5 Control Circuit Malfunction Conditions:** Engine started; engine speed over 400 rpm, system voltage over 10.0v and the PCM detected an unexpected voltage condition on the Fuel Injector driver circuit for Cylinder 5 for 5 seconds. Note: Drive the vehicle at cruise speed. Record the misfire current counters for review to detect if more than one cylinder is misfiring. **Possible Causes** ● Injector 5 power circuit (B+) is open (check the power fuse) ● Injector 5 control circuit is open between injector and PCM ● Injector 5 control circuit is grounded between injector and PCM ● Injector 5 is damaged or it has failed ● PCM is damaged

OBD II Trouble Code List (P0xxx Codes)

DTC	Trouble Code Title, Conditions & Possible Causes
P0206 **2T CCM, MIL: Yes** 1998, 1999, 2000, 2001, 2002, 2003, 2004, 2005 Aurora, Bonneville, Century, Cutlass Supreme, Grand Prix, Intrigue, LeSabre, LSS, Lumina, Monte Carlo, 88', 98', Regal, Park Avenue 3.1L VIN J, 3.1L VIN M, 3.5L VIN H, 3.8L VIN 1, 3.8L VIN K, 4.0L VIN C Transmissions: All	**Fuel Injector 6 Control Circuit Malfunction Conditions:** Engine started; engine speed over 400 rpm, system voltage over 10.0v and the PCM detected an unexpected voltage condition on the Fuel Injector driver circuit for Cylinder 6 for 5 seconds. Note: Drive the vehicle at cruise speed. Record the misfire current counters for review to detect if more than one cylinder is misfiring. **Possible Causes** • Injector 6 power circuit (B+) is open (check the power fuse) • Injector 6 control circuit is open between injector and PCM • Injector 6 control circuit is grounded between injector and PCM • Injector 6 is damaged or it has failed • PCM is damaged
P0207 **2T CCM, MIL: Yes** 1998, 1999, 2000, 2001, 2002, 2003, 2004, 2005 Aurora 4.0L V8 VIN C engine Transmissions: A/T	**Fuel Injector 7 Control Circuit Malfunction Conditions:** Engine started; engine speed over 400 rpm, system voltage over 10.0v and the PCM detected an unexpected voltage condition on the Fuel Injector driver circuit for Cylinder 7 for 5 seconds. Note: Drive the vehicle at cruise speed. Record the misfire current counters for review to detect if more than one cylinder is misfiring. **Possible Causes** • Injector 7 power circuit (B+) is open (check the power fuse) • Injector 7 control circuit is open between injector and PCM • Injector 7 control circuit is grounded between injector and PCM • Injector 7 is damaged or it has failed • PCM is damaged
P0208 **2T CCM, MIL: Yes** 1998, 1999, 2000, 2001, 2002, 2003, 2004, 2005 Aurora 4.0L V8 VIN C Transmissions: A/T	**Fuel Injector 8 Control Circuit Malfunction Conditions:** Engine started; engine speed over 400 rpm, system voltage over 10.0v and the PCM detected an unexpected voltage condition on the Fuel Injector driver circuit for Cylinder 8 for 5 seconds. Note: Drive the vehicle at cruise speed. Record the misfire current counters for review to detect if more than one cylinder is misfiring. **Possible Causes** • Injector 8 power circuit (B+) is open (check the power fuse) • Injector 8 control circuit is open between injector and PCM • Injector 8 control circuit is grounded between injector and PCM • Injector 8 is damaged or it has failed • PCM is damaged
P0201 **1T CCM, MIL: Yes** 1995, 1996, 1997, 1998, 1999, 2000, 2001, 2002 Camaro & Firebird 3.8L VIN K Transmissions: All	**Fuel Injector 1 Control Circuit Malfunction Conditions:** Engine started; system voltage over 10.0v and the PCM detected an unexpected voltage condition on the Fuel Injector 1 circuit for 5 seconds. Drive the vehicle at cruise speed and record the misfire current counters to determine if more than one cylinder is misfiring. **Possible Causes** • Injector 1 power circuit (B+) is open (check the power fuse) • Injector 1 control circuit is open between injector and PCM • Injector 1 control circuit is grounded between injector and PCM • Injector 1 is damaged or it has failed • PCM is damaged
P0202 **1T CCM, MIL: Yes** 1995, 1996, 1997, 1998, 1999, 2000, 2001, 2002 Camaro & Firebird 3.8L VIN K engine Transmissions: All	**Fuel Injector 2 Control Circuit Malfunction Conditions:** Engine started; system voltage over 10.0v and the PCM detected an unexpected voltage condition on the Fuel Injector 2 circuit for 5 seconds. Drive the vehicle at cruise speed and record the misfire current counters to determine if more than one cylinder is misfiring. **Possible Causes** • Injector 2 power circuit (B+) is open (check the power fuse) • Injector 2 control circuit is open between injector and PCM • Injector 2 control circuit is grounded between injector and PCM • Injector 2 is damaged or it has failed • PCM is damaged
P0203 **1T CCM, MIL: Yes** 1995, 1996, 1997, 1998, 1999, 2000, 2001, 2002 Camaro & Firebird 3.8L VIN K engine Transmissions: All	**Fuel Injector 3 Control Circuit Malfunction Conditions:** Engine started; system voltage over 10.0v and the PCM detected an unexpected voltage condition on the Fuel Injector 3 circuit for 5 seconds. Drive the vehicle at cruise speed and record the misfire current counters to determine if more than one cylinder is misfiring. **Possible Causes** • Injector 3 power circuit (B+) is open (check the power fuse) • Injector 3 control circuit is open between injector and PCM • Injector 3 control circuit is grounded between injector and PCM • Injector 3 is damaged or it has failed • PCM is damaged

OBD II Trouble Code List (P0xxx Codes)

DTC	Trouble Code Title, Conditions & Possible Causes
P0204 **1T CCM, MIL: Yes** 1995, 1996, 1997, 1998, 1999, 2000, 2001, 2002 Camaro & Firebird 3.8L VIN K engine Transmissions: All	**Fuel Injector 4 Control Circuit Malfunction Conditions:** Engine started; system voltage over 10.0v and the PCM detected an unexpected voltage condition on the Fuel Injector 4 circuit for 5 seconds. Drive the vehicle at cruise speed and record the misfire current counters to determine if more than one cylinder is misfiring. **Possible Causes** ● Injector 4 power circuit (B+) is open (check the power fuse) ● Injector 4 control circuit is open between injector and PCM ● Injector 4 control circuit is grounded between injector and PCM ● Injector 4 is damaged or it has failed ● PCM is damaged
P0205 **1T CCM, MIL: Yes** 1995, 1996, 1997, 1998, 1999, 2000, 2001, 2002 Camaro & Firebird 3.8L VIN K engine Transmissions: All	**Fuel Injector 5 Control Circuit Malfunction Conditions:** Engine started; system voltage over 10.0v and the PCM detected an unexpected voltage condition on the Fuel Injector 5 circuit for 5 seconds. Drive the vehicle at cruise speed and record the misfire current counters to determine if more than one cylinder is misfiring. **Possible Causes** ● Injector 5 power circuit (B+) is open (check the power fuse) ● Injector 5 control circuit is open between injector and PCM ● Injector 5 control circuit is grounded between injector and PCM ● Injector 5 is damaged or it has failed ● PCM is damaged
P0206 **1T CCM, MIL: Yes** 1995, 1996, 1997, 1998, 1999, 2000, 2001, 2002 Camaro & Firebird 3.8L VIN K engine Transmissions: All	**Fuel Injector 6 Control Circuit Malfunction Conditions:** Engine started; system voltage over 10.0v and the PCM detected an unexpected voltage condition on the Fuel Injector 6 circuit for 5 seconds. Drive the vehicle at cruise speed and record the misfire current counters to determine if more than one cylinder is misfiring. **Possible Causes** ● Injector 6 power circuit (B+) is open (check the power fuse) ● Injector 6 control circuit is open between injector and PCM ● Injector 6 control circuit is grounded between injector and PCM ● Injector 6 is damaged or it has failed ● PCM is damaged
P0201 **1T CCM, MIL: Yes** 2003, 2004, 2005 CTS (Cadillac) 2.6L VIN M, 3.2L VIN N Transmissions: All	**Fuel Injector 1 Control Circuit Malfunction Conditions:** Engine started; engine speed more than 40 rpm; system voltage from 8-18v, and the PCM detected an unexpected open condition on the Fuel Injector 1 driver circuit for 30 seconds. **Possible Causes** ● Injector 1 control circuit is open (check for connector problem) ● Injector 1 power circuit is open (test INJ/Coil fuse in fuse block) ● Injector 1 is damaged or has failed ● PCM is damaged
P0202 **1T CCM, MIL: Yes** 2003, 2004, 2005 CTS (Cadillac) 2.6L VIN M, 3.2L VIN N Transmissions: All	**Fuel Injector 2 Control Circuit Malfunction Conditions:** Engine started; engine speed more than 40 rpm; system voltage from 8-18v, and the PCM detected an unexpected open condition on the Fuel Injector 2 driver circuit for 30 seconds. **Possible Causes** ● Injector 2 control circuit is open (check for connector problem) ● Injector 2 power circuit is open (test INJ/Coil fuse in fuse block) ● Injector 2 is damaged or has failed ● PCM is damaged
P0203 **1T CCM, MIL: Yes** 2003, 2004, 2005 CTS (Cadillac) 2.6L VIN M, 3.2L VIN N Transmissions: All	**Fuel Injector 3 Control Circuit Malfunction Conditions:** Engine started; engine speed more than 40 rpm; system voltage from 8-18v, and the PCM detected an unexpected open condition on the Fuel Injector 3 driver circuit for 30 seconds. **Possible Causes** ● Injector 3 control circuit is open (check for connector problem) ● Injector 3 power circuit is open (test INJ/Coil fuse in fuse block) ● Injector 3 is damaged or has failed ● PCM is damaged
P0204 **1T CCM, MIL: Yes** 2003, 2004, 2005 CTS (Cadillac) 2.6L VIN M, 3.2L VIN N Transmissions: All	**Fuel Injector 4 Control Circuit Malfunction Conditions:** Engine started; engine speed more than 40 rpm; system voltage from 8-18v, and the PCM detected an unexpected open condition on the Fuel Injector 4 driver circuit for 30 seconds. **Possible Causes** ● Injector 4 control circuit is open (check for connector problem) ● Injector 4 power circuit is open (test INJ/Coil fuse in fuse block) ● Injector 4 is damaged or has failed ● PCM is damaged

OBD II Trouble Code List (P0xxx Codes)

DTC	Trouble Code Title, Conditions & Possible Causes
P0205 **1T CCM, MIL: Yes** 2003, 2004, 2005 CTS (Cadillac) 2.6L VIN M, 3.2L VIN N Transmissions: All	**Fuel Injector 5 Control Circuit Malfunction Conditions:** Engine started; engine speed more than 40 rpm; system voltage from 8-18v, and the PCM detected an unexpected circuit condition on the Fuel Injector 5 driver circuit for 30 seconds. **Possible Causes** ● Injector 5 control circuit is open (check for connector problem) ● Injector 5 power circuit is open (test INJ/Coil fuse in fuse block) ● Injector 5 is damaged or has failed ● PCM is damaged
P0206 **1T CCM, MIL: Yes** 2003, 2004, 2005 CTS (Cadillac) 2.6L VIN M, 3.2L VIN N Transmissions: All	**Fuel Injector 6 Control Circuit Malfunction Conditions:** Engine started; engine speed more than 40 rpm; system voltage from 8-18v, and the PCM detected an unexpected open condition on the Fuel Injector 6 driver circuit for 30 seconds. **Possible Causes** ● Injector 6 control circuit is open (check for connector problem) ● Injector 6 power circuit is open (test INJ/Coil fuse in fuse block) ● Injector 6 is damaged or has failed ● PCM is damaged
P0201 **1T CCM, MIL: Yes** 1996, 1997, 1998, 1999, 2000, 2001, 2002, 2003, 2004, 2005 DeVille, Eldorado, Seville 4.6L VIN 9, 4.6L VIN Y Transmissions: All	**Fuel Injector 1 Control Circuit Malfunction Conditions:** DTC P0560 and P1376, engine started, system voltage over 10.0v and the PCM detected an unexpected voltage on the Fuel Injector 1 driver circuit for 5 seconds. Drive at cruise speed and record the Misfire current counters to detect the injector(s) with a problem. **Possible Causes** ● Injector 1 control circuit is open or grounded between the injector and the PCM ● Injector 1 power circuit is open (test INJ fuse in fuse block) ● Injector 1 is damaged or has failed ● PCM is damaged
P0202 **1T CCM, MIL: Yes** 1996, 1997, 1998, 1999, 2000, 2001, 2002, 2003, 2004, 2005 DeVille, Eldorado, Seville 4.6L VIN 9, 4.6L VIN Y Transmissions: All	**Fuel Injector 2 Control Circuit Malfunction Conditions:** DTC P0560 and P1376, engine started, system voltage over 10.0v and the PCM detected an unexpected voltage on the Fuel Injector 2 driver circuit for 5 seconds. Drive at cruise speed and record the Misfire current counters to detect the injector(s) with a problem. **Possible Causes** ● Injector 2 control circuit is open or grounded between the injector and the PCM ● Injector 2 power circuit is open (test INJ fuse in fuse block) ● Injector 2 is damaged or has failed ● PCM is damaged
P0203 **1T CCM, MIL: Yes** 1996, 1997, 1998, 1999, 2000, 2001, 2002, 2003, 2004, 2005 DeVille, Eldorado, Seville 4.6L VIN 9, 4.6L VIN Y Transmissions: All	**Fuel Injector 3 Control Circuit Malfunction Conditions:** DTC P0560 and P1376, engine started, system voltage over 10.0v and the PCM detected an unexpected voltage on the Fuel Injector 3 driver circuit for 5 seconds. Drive at cruise speed and record the Misfire current counters to detect the injector(s) with a problem. **Possible Causes** ● Injector 3 control circuit is open or grounded between the injector and the PCM ● Injector 3 power circuit is open (test INJ fuse in fuse block) ● Injector 3 is damaged or has failed ● PCM is damaged
P0204 **1T CCM, MIL: Yes** 1996, 1997, 1998, 1999, 2000, 2001, 2002, 2003, 2004, 2005 DeVille, Eldorado, Seville 4.6L VIN 9, 4.6L VIN Y Transmissions: All	**Fuel Injector 4 Control Circuit Malfunction Conditions:** DTC P0560 and P1376, engine started, system voltage over 10.0v and the PCM detected an unexpected voltage on the Fuel Injector 4 driver circuit for 5 seconds. **Possible Causes** ● Injector 4 control circuit is open or grounded between the injector and the PCM ● Injector 4 power circuit is open (test INJ fuse in fuse block) ● Injector 4 is damaged or has failed ● PCM is damaged
P0205 **1T CCM, MIL: Yes** 1996, 1997, 1998, 1999, 2000, 2001, 2002, 2003, 2004, 2005 DeVille, Eldorado, Seville 4.6L VIN 9, 4.6L VIN Y Transmissions: All	**Fuel Injector 5 Control Circuit Malfunction Conditions:** DTC P0560 and P1376, engine started, system voltage over 10.0v and the PCM detected an unexpected voltage on the Fuel Injector 5 driver circuit for 5 seconds. **Possible Causes** ● Injector 5 control circuit is open or grounded between the injector and the PCM ● Injector 5 power circuit is open (test INJ fuse in fuse block) ● Injector 5 is damaged or has failed ● PCM is damaged
P0206 **1T CCM, MIL: Yes** 1996, 1997, 1998, 1999, 2000, 2001, 2002, 2003, 2004, 2005 DeVille, Eldorado, Seville 4.6L VIN 9, 4.6L VIN Y Transmissions: All	**Fuel Injector 6 Control Circuit Malfunction Conditions:** DTC P0560 and P1376, engine started, system voltage over 10.0v and the PCM detected an unexpected voltage on the Fuel Injector 6 driver circuit for 5 seconds. **Possible Causes** ● Injector 6 control circuit is open or grounded between the injector and the PCM ● Injector 6 power circuit is open (test INJ fuse in fuse block) ● Injector 6 is damaged or has failed ● PCM is damaged

OBD II Trouble Code List (P0xxx Codes)

DTC	Trouble Code Title, Conditions & Possible Causes
P0207 **1T CCM, MIL: Yes** 1996, 1997, 1998, 1999, 2000, 2001, 2002, 2003, 2004, 2005 DeVille, Eldorado, Seville 4.6L VIN 9, 4.6L VIN Y Transmissions: All	**Fuel Injector 7 Control Circuit Malfunction Conditions:** DTC P0560 and P1376, engine started, system voltage over 10.0v and the PCM detected an unexpected voltage on the Fuel Injector 7 driver circuit for 5 seconds. **Possible Causes** ● Injector 7 control circuit is open or grounded between the injector and the PCM ● Injector 7 power circuit is open (test INJ fuse in fuse block) ● Injector 7 is damaged or has failed ● PCM is damaged
P0208 **1T CCM, MIL: Yes** 1996, 1997, 1998, 1999, 2000, 2001, 2002, 2003, 2004, 2005 DeVille, Eldorado, Seville 4.6L VIN 9, 4.6L VIN Y Transmissions: All	**Fuel Injector 8 Control Circuit Malfunction Conditions:** DTC P0560 and P1376, engine started, system voltage over 10.0v and the PCM detected an unexpected voltage on the Fuel Injector 8 driver circuit for 5 seconds. **Possible Causes** ● Injector 8 control circuit is open or grounded between the injector and PCM ● Injector 8 power circuit is open (test INJ fuse in fuse block) ● Injector 8 is damaged or has failed ● PCM is damaged
P0201 **1T CCM, MIL: Yes** 1997, 1998, 1999, 2000, 2001 Catera 3.0L V6 VIN R engine Transmissions: A/T	**Fuel Injector 1 Control Circuit Malfunction Conditions:** Engine started; engine speed more than 40 rpm; system voltage from 7.5-15v, and the PCM detected an unexpected voltage on the Cylinder 1 Injector driver circuit for 5 seconds. **Possible Causes** ● Injector 1 control circuit is open or grounded between the Injector and the PCM ● Injector 1 power circuit is open (test for B+ from the main relay) ● Injector 1 is damaged or has failed ● PCM is damaged
P0202 **1T CCM, MIL: Yes** 1997, 1998, 1999, 2000, 2001 Catera 3.0L V6 VIN R engine Transmissions: A/T	**Fuel Injector 2 Control Circuit Malfunction Conditions:** Engine started; engine speed more than 40 rpm; system voltage from 7.5-15v, and the PCM detected an unexpected voltage on the Cylinder 2 Injector driver circuit for 5 seconds. **Possible Causes** ● Injector 2 control circuit is open or grounded between the Injector and the PCM ● Injector 2 power circuit is open (test for B+ from the main relay) ● Injector 2 is damaged or has failed ● PCM is damaged
P0203 **1T CCM, MIL: Yes** 1997, 1998, 1999, 2000, 2001 Catera 3.0L V6 VIN R engine Transmissions: A/T	**Fuel Injector 3 Control Circuit Malfunction Conditions:** Engine started; engine speed more than 40 rpm; system voltage from 7.5-15v, and the PCM detected an unexpected voltage on the Cylinder 3 Injector driver circuit for 5 seconds. **Possible Causes** ● Injector 3 control circuit is open or grounded between the Injector and the PCM ● Injector 3 power circuit is open (test for B+ from the main relay) ● Injector 3 is damaged or has failed ● PCM is damaged
P0204 **1T CCM, MIL: Yes** 1997, 1998, 1999, 2000, 2001 Catera 3.0L V6 VIN R engine Transmissions: A/T	**Fuel Injector 4 Control Circuit Malfunction Conditions:** Engine started; engine speed more than 40 rpm; system voltage from 7.5-15v, and the PCM detected an unexpected voltage on the Cylinder 4 Injector driver circuit for 5 seconds. **Possible Causes** ● Injector 4 control circuit is open or grounded between the Injector and the PCM ● Injector 4 power circuit is open (test for B+ from the main relay) ● Injector 4 is damaged or has failed ● PCM is damaged
P0205 **1T CCM, MIL: Yes** 1997, 1998, 1999, 2000, 2001 Catera 3.0L V6 VIN R engine Transmissions: A/T	**Fuel Injector 5 Control Circuit Malfunction Conditions:** Engine started; engine speed more than 40 rpm; system voltage from 7.5-15v, and the PCM detected an unexpected voltage on the Cylinder 5 Injector driver circuit for 5 seconds. **Possible Causes** ● Injector 5 control circuit is open or grounded between the Injector and the PCM ● Injector 5 power circuit is open (test for B+ from the main relay) ● Injector 5 is damaged or has failed ● PCM is damaged
P0206 **1T CCM, MIL: Yes** 1997, 1998, 1999, 2000, 2001 Catera 3.0L V6 VIN R engine Transmissions: A/T	**Fuel Injector 6 Control Circuit Malfunction Conditions:** Engine started; engine speed more than 40 rpm; system voltage from 7.5-15v, and the PCM detected an unexpected voltage on the Cylinder 6 Injector driver circuit for 5 seconds. **Possible Causes** ● Injector 6 control circuit is open or grounded between the Injector and the PCM ● Injector 6 power circuit is open (test for B+ from the main relay) ● Injector 6 is damaged or has failed ● PCM is damaged

OBD II Trouble Code List (P0xxx Codes)

DTC	Trouble Code Title, Conditions & Possible Causes
P0201 **2T CCM, MIL: Yes** 2000, 2001, 2002, 2003, 2004, 2005 Cavalier, Envoy, Grand Am, TrailBlazer, Sunfire, S/T Blazer & S/T Pickup 2.2L VIN 4, 2.2L VIN 5, 2.2L VIN F, 2.2L VIN H, 2.4L VIN T, 4.2L VIN S Transmissions: All	**Fuel Injector 1 Control Circuit Malfunction Conditions:** Engine started; system voltage over 10.0v and the PCM detected an unexpected voltage on the Injector 1 driver circuit. Note: Drive the vehicle at cruise speed. Record the misfire current counters for review to detect if more than one cylinder is misfiring. **Possible Causes** ● Injector 1 control circuit is open or grounded between the injector and the PCM ● Injector 1 power circuit is open (test INJ fuse in fuse block) ● Injector 1 is damaged or has failed ● PCM is damaged
P0202 **2T CCM, MIL: Yes** 2000, 2001, 2002, 2003, 2004, 2005 Cavalier, Envoy, Grand Am, TrailBlazer, Sunfire, S/T Blazer & S/T Pickup 2.2L VIN 4, 2.2L VIN 5, 2.2L VIN F, 2.2L VIN H, 2.4L VIN T, 4.2L VIN S Transmissions: All	**Fuel Injector 2 Control Circuit Malfunction Conditions:** Engine started; system voltage over 10.0v and the PCM detected an unexpected voltage on the Injector 2 driver circuit. Note: Drive the vehicle at cruise speed. Record the misfire current counters for review to detect if more than one cylinder is misfiring. **Possible Causes** ● Injector 2 control circuit is open or grounded between the injector and the PCM ● Injector 2 power circuit is open (test INJ fuse in fuse block) ● Injector 2 is damaged or has failed ● PCM is damaged
P0203 **2T CCM, MIL: Yes** 2000, 2001, 2002, 2003, 2004, 2005 Cavalier, Envoy, Grand Am, TrailBlazer, Sunfire, S/T Blazer & S/T Pickup 2.2L VIN 4, 2.2L VIN 5, 2.2L VIN F, 2.2L VIN H, 2.4L VIN T, 4.2L VIN S Transmissions: All	**Fuel Injector 3 Control Circuit Malfunction Conditions:** Engine started; system voltage over 10.0v and the PCM detected an unexpected voltage on the Injector 3 driver circuit. Note: Drive the vehicle at cruise speed. Record the misfire current counters for review to detect if more than one cylinder is misfiring. **Possible Causes** ● Injector 3 control circuit is open or grounded between the injector and the PCM ● Injector 3 power circuit is open (test INJ fuse in fuse block) ● Injector 3 is damaged or has failed ● PCM is damaged
P0204 **2T CCM, MIL: Yes** 2000, 2001, 2002, 2003, 2004, 2005 Cavalier, Envoy, Grand Am, TrailBlazer, Sunfire, S/T Blazer & S/T Pickup 2.2L VIN 4, 2.2L VIN 5, 2.2L VIN F, 2.2L VIN H, 2.4L VIN T, 4.2L VIN S Transmissions: All	**Fuel Injector 4 Control Circuit Malfunction Conditions:** Engine started; system voltage over 10.0v and the PCM detected an unexpected voltage on the Injector 4 driver circuit. Note: Drive the vehicle at cruise speed. Record the misfire current counters for review to detect if more than one cylinder is misfiring. **Possible Causes** ● Injector 4 control circuit is open or grounded between the injector and the PCM ● Injector 4 power circuit is open (test INJ fuse in fuse block) ● Injector 4 is damaged or has failed ● PCM is damaged
P0205 **2T CCM, MIL: Yes** 2002, 2003, 2004, 2005 Envoy & TrailBlazer 4.2L VIN S engine Transmissions: All	**Fuel Injector 5 Control Circuit Malfunction Conditions:** Engine started; system voltage over 10.0v and the PCM detected an unexpected voltage on the Injector 5 driver circuit. Record the misfire current counters for review to detect the fault. **Possible Causes** ● Injector 5 control circuit is open or grounded between the injector and the PCM ● Injector 5 power circuit is open (test INJ fuse in fuse block) ● Injector 5 is damaged or has failed ● PCM is damaged d
P0206 **2T CCM, MIL: Yes** 2002, 2003, 2004, 2005 Envoy & TrailBlazer 4.2L VIN S engine Transmissions: All	**Fuel Injector 6 Control Circuit Malfunction Conditions:** Engine started; system voltage over 10.0v and the PCM detected an unexpected voltage on the Injector 6 driver circuit. Record the misfire current counters for review to detect the fault. **Possible Causes** ● Injector 6 control circuit is open or grounded between the injector and the PCM ● Injector 6 power circuit is open (test INJ fuse in fuse block) ● Injector 6 is damaged or has failed ● PCM is damaged
P0201 **2T CCM, MIL: Yes** 1998, 1999, 2000, 2001, 2002, 2003, 2004, 2005 Achieva, Alero, Century, Cutlass, Cutlass Supreme Grand Prix, Impala, Monte Carlo, Intrigue, Lumina, Malibu, Montana, Venture Silhouette, Rendezvous 3.1L VIN J, 3.1L VIN M, 3.4L VIN E engines Transmissions: All	**Fuel Injector 1 Control Circuit Malfunction Conditions:** Engine started; system voltage over 10.0v and the PCM detected an unexpected voltage on the Fuel Injector 1 driver circuit for 30 seconds. Note: Drive the vehicle at cruise speed. Record the misfire current counters to detect if more than one cylinder is misfiring. **Possible Causes** ● Injector 1 control circuit is open between injector and PCM ● Injector 1 control circuit is grounded between injector and PCM ● Injector 1 power circuit is open (test INJ fuse in fuse block) ● Injector 1 is damaged or it has failed ● PCM is damaged

OBD II Trouble Code List (P0xxx Codes)

DTC	Trouble Code Title, Conditions & Possible Causes
P0202 **2T CCM, MIL: Yes** 1998, 1999, 2000, 2001, 2002, 2003, 2004, 2005 Achieva, Alero, Century, Cutlass, Cutlass Supreme Grand Prix, Impala, Monte Carlo, Intrigue, Lumina, Malibu, Montana, Venture Silhouette, Rendezvous 3.1L VIN J, 3.1L VIN M, 3.4L VIN E engines Transmissions: All	**Fuel Injector 2 Control Circuit Malfunction Conditions:** Engine started; system voltage over 10.0v and the PCM detected an unexpected voltage on the Fuel Injector 2 driver circuit for 30 seconds. Note: Drive the vehicle at cruise speed. Record the misfire current counters to detect if more than one cylinder is misfiring. **Possible Causes** ● Injector 2 control circuit is open between injector and PCM ● Injector 2 control circuit is grounded between injector and PCM ● Injector 2 power circuit is open (test INJ fuse in fuse block) ● Injector 2 is damaged or it has failed ● PCM is damaged
P0203 **2T CCM, MIL: Yes** 1998, 1999, 2000, 2001, 2002, 2003, 2004, 2005 Achieva, Alero, Century, Cutlass, Cutlass Supreme Grand Prix, Impala, Monte Carlo, Intrigue, Lumina, Malibu, Montana, Venture Silhouette, Rendezvous 3.1L VIN J, 3.1L VIN M, 3.4L VIN E engines Transmissions: All	**Fuel Injector 3 Control Circuit Malfunction Conditions:** Engine started; system voltage over 10.0v and the PCM detected an unexpected voltage on the Fuel Injector 3 driver circuit for 30 seconds. Note: Drive the vehicle at cruise speed. Record the misfire current counters to detect if more than one cylinder is misfiring. **Possible Causes** ● Injector 3 control circuit is open between injector and PCM ● Injector 3 control circuit is grounded between injector and PCM ● Injector 3 power circuit is open (test INJ fuse in fuse block) ● Injector 3 is damaged or it has failed ● PCM is damaged
P0204 **2T CCM, MIL: Yes** 1998, 1999, 2000, 2001, 2002, 2003, 2004, 2005 Achieva, Alero, Century, Cutlass, Cutlass Supreme Grand Prix, Impala, Monte Carlo, Intrigue, Lumina, Malibu, Montana, Venture Silhouette, Rendezvous 3.1L VIN J, 3.1L VIN M, 3.4L VIN E engines Transmissions: All	**Fuel Injector 4 Control Circuit Malfunction Conditions:** Engine started; system voltage over 10.0v and the PCM detected an unexpected voltage on the Fuel Injector 4 driver circuit for 30 seconds. Note: Drive the vehicle at cruise speed. Record the misfire current counters to detect if more than one cylinder is misfiring **Possible Causes** ● Injector 4 control circuit is open between injector and PCM ● Injector 4 control circuit is grounded between injector and PCM ● Injector 4 power circuit is open (test INJ fuse in fuse block) ● Injector 4 is damaged or it has failed ● PCM is damaged
P0205 **2T CCM, MIL: Yes** 1998, 1999, 2000, 2001, 2002, 2003, 2004, 2005 Achieva, Alero, Century, Cutlass, Cutlass Supreme Grand Prix, Impala, Monte Carlo, Intrigue, Lumina, Malibu, Montana, Venture Silhouette, Rendezvous 3.1L VIN J, 3.1L VIN M, 3.4L VIN E engines Transmissions: All	**Fuel Injector 5 Control Circuit Malfunction Conditions:** Engine started; system voltage over 10.0v and the PCM detected an unexpected voltage on the Fuel Injector 5 driver circuit for 30 seconds. Note: Drive the vehicle at cruise speed. Record the misfire current counters to detect if more than one cylinder is misfiring **Possible Causes** ● Injector 5 control circuit is open between injector and PCM ● Injector 5 control circuit is grounded between injector and PCM ● Injector 5 power circuit is open (test INJ fuse in fuse block) ● Injector 5 is damaged or it has failed ● PCM is damaged
P0206 **2T CCM, MIL: Yes** 1998, 1999, 2000, 2001, 2002, 2003, 2004, 2005 Achieva, Alero, Century, Cutlass, Cutlass Supreme Grand Prix, Impala, Monte Carlo, Intrigue, Lumina, Malibu, Montana, Venture Silhouette, Rendezvous 3.1L VIN J, 3.1L VIN M, 3.4L VIN E engines Transmissions: All	**Fuel Injector 6 Control Circuit Malfunction Conditions:** Engine started; system voltage over 10.0v and the PCM detected an unexpected voltage on the Fuel Injector 6 driver circuit for 30 seconds. Note: Drive the vehicle at cruise speed. Record the misfire current counters to detect if more than one cylinder is misfiring **Possible Causes** ● Injector 6 control circuit is open between injector and PCM ● Injector 6 control circuit is grounded between injector and PCM ● Injector 6 power circuit is open (test INJ fuse in fuse block) ● Injector 6 is damaged or it has failed ● PCM is damaged

OBD II Trouble Code List (P0xxx Codes)

DTC	Trouble Code Title, Conditions & Possible Causes
P0201 **1T CCM, MIL: Yes** 2001, 2002, 2003, 2004, 2005 C/K Series Truck 6.0L VIN U CNG engine Transmissions: All	**Fuel Injector 1 Control Circuit Malfunction Conditions:** U1800 and U2104 not set; engine started; system voltage from 6-18v, and the PCM detected and incorrect current level on the Fuel Injector 1 control circuit. The fuel injection control module (FICM) supplies power to each fuel injector via the ignition voltage circuit. The FICM energizes each injector by grounding the control circuit between the FICM and the injector. The FICM monitors the status of the ignition voltage and fuel injector command circuits. **Possible Causes** ● Fuel injector 1 control circuit is open, shorted to ground or shorted to power ● Fuel injector 1 is damaged or it has failed, or the connector is damaged ● FICM (module) has failed
P0202 **1T CCM, MIL: Yes** 2001, 2002, 2003, 2004, 2005 C/K Series Truck 6.0L VIN U CNG engine Transmissions: All	**Fuel Injector 2 Control Circuit Malfunction Conditions:** U1800 and U2104 not set; engine started; system voltage from 6-18v, and the PCM detected and incorrect current level on the Fuel Injector 2 control circuit. The fuel injection control module (FICM) supplies power to each fuel injector via the ignition voltage circuit. The FICM energizes each injector by grounding the control circuit between the FICM and the injector. The FICM monitors the status of the ignition voltage and fuel injector command circuits. **Possible Causes** ● Fuel injector 2 control circuit is open, shorted to ground or shorted to power ● Fuel injector 2 is damaged or it has failed, or the connector is damaged ● FICM (module) has failed
P0203 **1T CCM, MIL: Yes** 2001, 2002, 2003, 2004, 2005 C/K Series Truck 6.0L VIN U CNG engine Transmissions: All	**Fuel Injector 3 Control Circuit Malfunction Conditions:** U1800 and U2104 not set; engine started; system voltage from 6-18v, and the PCM detected and incorrect current level on the Fuel Injector 3 control circuit. The fuel injection control module (FICM) supplies power to each fuel injector via the ignition voltage circuit. The FICM energizes each injector by grounding the control circuit between the FICM and the injector. The FICM monitors the status of the ignition voltage and fuel injector command circuits. **Possible Causes** ● Fuel injector 3 control circuit is open, shorted to ground or shorted to power ● Fuel injector 3 is damaged or it has failed, or the connector is damaged ● FICM (module) has failed
P0204 **1T CCM, MIL: Yes** 2001, 2002, 2003, 2004, 2005 C/K Series Truck 6.0L VIN U CNG engine Transmissions: All	**Fuel Injector 4 Control Circuit Malfunction Conditions:** U1800 and U2104 not set; engine started; system voltage from 6-18v, and the PCM detected and incorrect current level on the Fuel Injector 4 control circuit. The fuel injection control module (FICM) supplies power to each fuel injector via the ignition voltage circuit. The FICM energizes each injector by grounding the control circuit between the FICM and the injector. **Possible Causes** ● Fuel injector 4 control circuit is open, shorted to ground or shorted to power ● Fuel injector 4 is damaged or it has failed, or the connector is damaged ● FICM (module) has failed
P0205 **1T CCM, MIL: Yes** 2001, 2002, 2003, 2004, 2005 C/K Series Truck 6.0L VIN U CNG engine Transmissions: All	**Fuel Injector 5 Control Circuit Malfunction Conditions:** U1800 and U2104 not set; engine started; system voltage from 6-18v, and the PCM detected and incorrect current level on the Fuel Injector 5 control circuit. The fuel injection control module (FICM) supplies power to each fuel injector via the ignition voltage circuit. The FICM energizes each injector by grounding the control circuit between the FICM and the injector. **Possible Causes** ● Fuel injector 5 control circuit is open, shorted to ground or shorted to power ● Fuel injector 5 is damaged or it has failed, or the connector is damaged ● FICM (module) has failed
P0206 **1T CCM, MIL: Yes** 2001, 2002, 2003, 2004, 2005 C/K Series Truck 6.0L VIN U CNG engine Transmissions: All	**Fuel Injector 6 Control Circuit Malfunction Conditions:** U1800 and U2104 not set; engine started; system voltage from 6-18v, and the PCM detected and incorrect current level on the Fuel Injector 6 control circuit. The fuel injection control module (FICM) supplies power to each fuel injector via the ignition voltage circuit. The FICM energizes each injector by grounding the control circuit between the FICM and the injector. **Possible Causes** ● Fuel injector 6 control circuit is open, shorted to ground or shorted to power ● Fuel injector 6 is damaged or it has failed, or the connector is damaged ● FICM (module) has failed
P0207 **1T CCM, MIL: Yes** 2001, 2002, 2003, 2004, 2005 C/K Series Truck 6.0L VIN U CNG engine Transmissions: All	**Fuel Injector 7 Control Circuit Malfunction Conditions:** U1800 and U2104 not set; engine started; system voltage from 6-18v, and the PCM detected and incorrect current level on the Fuel Injector 7 control circuit. The fuel injection control module (FICM) supplies power to each fuel injector via the ignition voltage circuit. The FICM energizes each injector by grounding the control circuit between the FICM and the injector. The FICM monitors the status of the ignition voltage and fuel injector command circuits. **Possible Causes** ● Fuel injector 7 control circuit is open, shorted to ground or shorted to power ● Fuel injector 7 is damaged or it has failed, or the connector is damaged ● FICM (module) has failed

OBD II Trouble Code List (P0xxx Codes)

DTC	Trouble Code Title, Conditions & Possible Causes
P0208 **1T CCM, MIL: Yes** 2001, 2002, 2003, 2004, 2005 C/K Series Truck 6.0L VIN U CNG engine Transmissions: All	**Fuel Injector 8 Control Circuit Malfunction Conditions:** U1800 and U2104 not set; engine started; system voltage from 6-18v, and the PCM detected and incorrect current level on the Fuel Injector 8 control circuit. The fuel injection control module (FICM) supplies power to each fuel injector via the ignition voltage circuit. The FICM energizes each injector by grounding the control circuit between the FICM and the injector. The FICM monitors the status of the ignition voltage and fuel injector command circuits. **Possible Causes** ● Fuel injector 8 connector is damaged, loose or shorted ● Fuel injector 8 control circuit is open, shorted to ground or shorted to power ● Fuel injector 8 is damaged or it has failed ● FICM (module) has failed
P0215 **2T CCM, MIL: No** 1996, 1997, 1998, 1999, 2000, 2001, 2002 C/K Series Truck, G Series Van, Van Cargo 6.5L VIN F, 6.5L VIN S, 6.5L VIN Y Transmissions: All	**Engine Shutoff Solenoid Control Circuit Malfunction Conditions:** The PCM requested ESO "on" (Test 1), and the PCM detected the ESO control circuit voltage was less than 8v (Test 1), or with ESO requested "off" (Test 2), the PCM detected the ESO control circuit voltage was over 8v for 2 seconds. Note: The engine will not start if the ESO circuit to the PCM is open. **Possible Causes** ● ESC control circuit is open or shorted to ground ● ESO power circuit is open (test ECM fuse in fuse relay center) ● PCM has failed ● TSB 87-63-04 contains a repair procedure for this code
P0216 **2T CCM, MIL: Yes** 1996, 1997, 1998, 1999, 2000, 2001, 2002 C/K Series Truck, G Series Van, Van Cargo 6.5L VIN F, 6.5L VIN S Transmissions: All	**Injection Timing Control Circuit Conditions:** DTC P0251, P0335 and P0370 not set; engine started, engine speed stable (± 56 rpm) for 5 seconds, and the PCM detected over a 5 degree change in Actual and Desired Injector timing for 2 seconds. **Possible Causes** ● Injector Timing 1 high or low circuit is open or shorted to ground ● Injector Timing 2 high or low circuit is open or shorted to ground ● CKP sensor signal circuit has high resistance
P0218 **1T CCM, MIL: No** 1997, 1998, 1999, 2000, 2001, 2002, 2003, 2004, 2005 Alero, Cutlass, Malibu, U/X Van, Rendezvous 2.2L VIN F, 3.1L VIN J, 3.1L VIN M, 3.4L VIN E Transmissions: A/T	**Transmission Fluid Over-Temperature (4T40-E, 4T45-E) Conditions:** DTC P0711, P0712 and P0713 not set, engine running, and the PCM detected the TFT sensor was more than 266°F for 10 minutes. **Possible Causes** ● ATF is low, contaminated, burnt or dirty ● Engine cooling system has an airflow restriction ● Transmission cooling system has an airflow restriction ● Transmission cooler lines are bent, damaged or restricted ● Transmission internal failure (i.e., low line pressure, TCC fault)
P0218 **1T CCM, MIL: No** 1997, 1998, 1999, 2000, 2001, 2002, 2003, 2004, 2005 C/K Cab & Chassis, C/K Series Truck, G Series Vans & L/M Series Vans 4.3L VIN W, 4.8L VIN V, 5.0L VIN M, 5.3L VIN T, 5.3L VIN Z, 5.7L VIN K, 5.7L VIN R, 6.0L VIN N, 6.0L VIN U CNG engine, 6.6L VIN 1 Diesel engine, 7.4L VIN J, 8.1L VIN G Transmissions: A/T	**Transmission Fluid Over-Temperature (4L60-E, 4L80-E) Conditions:** DTC P0711, P0712 and P0713 not set, key on for 5 seconds or engine running; and the PCM detected the Transmission Fluid Temperature (TFT sensor indicated more than 266°F for 10 minutes. **Possible Causes** ● ATF is low, contaminated, burnt or dirty ● Customer driving habits (i.e., excessive trailer towing) ● Engine cooling system has an airflow restriction ● Transmission cooling system has an airflow restriction ● Transmission cooler lines are bent, damaged or restricted ● Transmission internal failure (i.e., low line pressure, TCC fault)
P0218 **1T CCM, MIL: No** 1997, 1998, 1999, 2000, 2001, 2002, 2003, 2004, 2005 Aurora, Century, LeSabre, Park Avenue, Regal 3.1L VIN J, 3.1L VIN M, 3.4L VIN X, 3.5L VIN H, 3.8L VIN 1, 3.8L VIN K Transmissions: A/T	**Transmission Fluid Over-Temperature (4T60-E, 4T65-E) Conditions:** DTC P0711, P0712 and P0713 not set, engine started, and the PCM detected the TFT sensor indicated more than 266°F for 10 minutes. **Possible Causes** ● ATF is low, contaminated, burnt or dirty ● Engine cooling system has an airflow restriction ● Transmission cooling system has an airflow restriction ● Transmission cooler lines are bent, damaged or restricted ● Transmission internal failure (i.e., low line pressure, TCC fault)

OBD II Trouble Code List (P0xxx Codes)

DTC	Trouble Code Title, Conditions & Possible Causes
P0218 **1T CCM, MIL: No** 1996, 1997, 1998, 1999, 2000, 2001, 2002, 2003, 2004, 2005 Aurora, DeVille, Eldorado, Seville 3.5L VIN H, 4.0L VIN C, 4.6L VIN 9, 4.6L VIN Y Transmissions: A/T	**Transaxle Fluid Over-Temperature (4L65E, 4L80-E) Conditions:** DTC P0116-P0118, P0711, P0712 or P0713 not set, engine started, and the PCM detected the TFT sensor was over 291°F for 15 seconds or over 270°F for 10 minutes. **Possible Causes** • ATF is low, contaminated, burnt or dirty • Engine cooling system has an airflow restriction • Transmission cooling system has an airflow restriction • Transmission cooler lines are bent, damaged or restricted • Transmission internal failure (i.e., low line pressure, TCC fault)
P0218 **1T CCM, MIL: No** 2002, 2003, 2004, 2005 Escalade, S/T Blazer & S/T Pickup 4.2L VIN S, 5.3L VIN P Transmissions: A/T	**Transmission Fluid Over-Temperature (4L60-E, 4L65-E) Conditions:** DTC P0711, P0712 and P0713 not set, engine started, and the PCM detected the TFT sensor was more than 266°F for 5 seconds. **Possible Causes** • Check for any Trans Fluid, Trans Hot, Idle Engine messages displayed • ATF is low, contaminated, burnt or dirty • Engine or transmission cooling system has an airflow restriction • Transmission cooler lines are bent, damaged or restricted • Transmission internal failure (i.e., low line pressure, TCC fault)
P0218 **1T CCM, MIL: No** 1997, 1998, 1999, 2000, 2001 Catera 3.0L V6 VIN R engine Transmissions: A/T	**Transaxle Fluid Over-Temperature (4L30-E) Conditions:** Engine started; and the PCM detected the TFT sensor was over 329°F for 19.9 seconds. **Possible Causes** • ATF is low, contaminated, burnt or dirty • Engine or transmission cooling system has an airflow restriction • Transmission cooler lines are bent, damaged or restricted • Transmission internal failure (i.e., low line pressure, TCC fault)
P0218 **1T CCM, MIL: No** 2003, 2004, 2005 CTS (Cadillac) 3.2L VIN N Transmissions: A/T	**Transmission Fluid Over-Temperature (5L40-E) Conditions:** Key on, system voltage 8-18v, TFT sensor -38 to 300°F (5 seconds), and PCM detected the TFT sensor was over 270°F for 10 minutes. **Possible Causes** • ATF is low, contaminated, burnt or dirty • Transmission cooling system has an airflow restriction • Transmission cooler lines are bent, damaged or restricted • Transmission internal failure (i.e., low line pressure, TCC fault)
P0219 **1T CCM, MIL: No** 1996, 1997, 1998, 1999, 2000, 2001, 2002 C/K Series Truck, G Van 6.5L VIN F, 6.5L VIN S Transmissions: All	**Engine Overspeed Condition Conditions:** Engine started; and the PCM detected five (5) Engine Shutoff (ESO) cycles occurred with a resulting engine speed change from 800-1200 rpm occurred continuously. **Possible Causes** • Fuel injection pump may need to be replaced • Fuel injection pump timing adjustment needs to be performed • PCM has failed
P0219 **1T CCM, MIL: No** 2003, 2004, 2005 CTS (Cadillac) 2.6L VIN M, 3.2L VIN N Transmissions: All	**Engine Overspeed Condition Conditions:** Engine started; vehicle driven and the PCM detected that the engine speed exceeded 7,000 rpm. This code is used to indicate the engine speed exceeded the fuel cutoff limit in gear. **Possible Causes** • This code may set if the transmission is placed in a low gear while driving down a steep grade or traveling at a high speed • Operate the vehicle under the Freeze Frame/Failure Records.
P0219 **2T CCM, MIL: Yes** 1996, 1997, 1998, 1999, 2000, 2001, 2002 C/K Series Truck, G Series Van, Van Cargo 6.5L VIN F, 6.5L VIN S Transmissions: All	**Accelerator Pedal Position Sensor 2 Circuit Malfunction Conditions:** Key on or engine running; and the PCM detected the reference voltage for APP Sensor 2 was less than 4.80v for 2 seconds. **Possible Causes** • APP VREF signal is open between the sensor and the PCM • APP VREF signal is grounded between the sensor and PCM • APP sensor is damaged or has failed • PCM is damaged
P0220 **1T CCM, MIL: No** 2003, 2004, 2005 Escalade, C/K Truck, G Series Van, Van Cargo 4.8L VIN V, 5.3L VIN P, 5.3L VIN T, 5.3L VIN Z, 6.0L VIN N, 6.0L VIN U CNG engine, 8.1L VIN G Transmissions: All	**Throttle Position Sensor 2 Circuit Malfunction Conditions:** DTP P1518 and P2108 not set, engine cranking or running; system voltage over 5.23v, and the PCM detected the TP Sensor 2 signal was less than 0.28v or more than 4.60v for one second. The PCM provides the TP sensor with a 5v, low reference and signal circuit. The signal is low at closed throttle and higher as the throttle opens. **Possible Causes** • TP Sensor 2 signal circuit is open or shorted to ground • TP Sensor 2 VREF (5v) circuit is open, or TP Sensor 2 ground circuit is open • TP Sensor 2 is damaged or has failed • PCM is damaged • TSB 03-04-06-034 contains a repair procedure for this code

OBD II Trouble Code List (P0xxx Codes)

DTC	Trouble Code Title, Conditions & Possible Causes
P0221 **1T CCM, MIL: No** 1996, 1997, 1998, 1999, 2000, 2001, 2002 C/K Series Truck, G Series Van, Van Cargo 6.5L VIN F, 6.5L VIN S Transmissions: All	**Accelerator Pedal Position Sensor 2 Range/Performance Conditions:** No APP sensor codes set, system voltage over 6.4v, engine started, and the PCM detected the difference between the APP-2 and APP-3 signals was over 500 mv, or the difference between the APP-2 and APP-1 signals was over 230 mv for 2 seconds during the CCM test. **Possible Causes** ● APP2 signal circuit is open or shorted to ground ● APP2 sensor is binding, damaged or has failed ● PCM is damaged
P0221 **2T CCM, MIL: Yes** 2003, 2004, 2005 CTS (Cadillac) 2.6L VIN M, 3.2L VIN N Transmissions: All	**Throttle Position Sensor 1-2 Range Correlation Conditions:** Engine started; system voltage over 10.0v, TP Sensor voltage more than 0.17v and less than 4.60v, engine speed more than 1,320 rpm, and the PCM detected that the TP Sensor 1 disagreed by more than 6% from the TP sensor 2 value, or that the TP Sensor 2 disagreed by more than 9% from the predicted value, either condition met for 280 ms. The throttle position (TP) Sensors 1 and 2 are located within the throttle body assembly. The TP sensors share a common 5-volt reference circuit and a common low reference circuit. The 5-volt reference circuit is also shared with accelerator pedal position (APP) sensor 2. The 5-volt reference voltage is supplied on 2 separate ECM terminals, but the terminals are connected internally to the same voltage supply. Each TP sensor has an individual signal circuit, which provides the ECM with a signal voltage proportional to throttle the plate movement. When the throttle plate is in the closed position, the TP sensor 1 signal voltage is near the low reference and increases as the throttle plate is opened. The TP Sensor 2 signal voltage at closed throttle should be near the 5 volt reference and decrease as the throttle plate is opened. The ECM compares the signal of the TP sensors to the MAF sensor when the engine is running to determine if the sensor readings are correct. The ECM also compares the signal of the TP sensor 1 and TP sensor 2 through the entire range. **Possible Causes** ● TP Sensor 1 circuit is open, shorted to ground or has a high resistance condition ● Throttle body assembly is damaged, or it has failed ● ECM is damaged
P0222 **1T CCM, MIL: No** 1996, 1997, 1998, 1999, 2000, 2001, 2002 C/K Series Truck, G Series Van, Van Cargo 6.5L VIN F, 6.5L VIN S Transmissions: All	**Accelerator Pedal Position Sensor 2 Low Input Conditions:** Key on or engine running; and the PCM detected the APP-2 sensor voltage was less than 250 mv for 2 seconds during the CCM test. Note: The APP2 Sensor should read near 4.5v at idle speed. **Possible Causes** ● APP2 signal circuit is shorted to chassis or sensor ground ● APP2 VREF circuit has a high resistance condition (low VREF) ● APP2 sensor is damaged or has failed ● PCM is damaged
P0222 **2T CCM, MIL: Yes** 2003, 2004, 2005 CTS (Cadillac) 2.6L VIN M, 3.2L VIN N Transmissions: All	**Throttle Position Sensor 2 Circuit Low Input Conditions:** Engine started; engine speed below 1,310 rpm, system voltage over 10v, and the PCM detected the TP Sensor 2 signal was less than 0.195v for over 140 ms. The 5v VREF circuit to the TP sensors is shared with the Accelerator Pedal Position Sensor 2 (APP2). **Possible Causes** ● APP2 signal circuit is shorted to chassis or sensor ground ● TP1 Sensor signal is open, shorted to ground or low reference ● TP2 Sensor signal circuit is shorted to sensor ground ● TP Sensor 5v VREF circuit is open or shorted to ground ● PCM is damaged (check the power ground circuit resistance)
P0223 **1T CCM, MIL: No** 1996, 1997, 1998, 1999, 2000, 2001, 2002 C/K Series Truck, G Series Van, Van Cargo 6.5L VIN F, 6.5L VIN S Transmissions: All	**Accelerator Pedal Position Sensor 2 High Input Conditions:** Key on or engine running; and the PCM detected that the APP-2 sensor voltage was more than 4.75v, condition met for 2 seconds. **Possible Causes** ● APP2 sensor signal circuit is open between sensor and PCM ● APP2 sensor signal circuit is shorted to sensor VREF ● APP2 sensor ground circuit is open between sensor and PCM ● APP sensor is damaged or has failed ● PCM is damaged
P0223 **2T CCM, MIL: Yes** 2003, 2004, 2005 CTS (Cadillac) 2.6L VIN M, 3.2L VIN N Transmissions: All	**Throttle Position Sensor 2 Circuit High Input Conditions:** Engine started; engine speed below 1,310 rpm, system voltage over 10v, and the PCM detected the TP Sensor 2 (TP2) signal was more than 4.60v for over 140 ms. **Possible Causes** ● APP Sensor 2 signal circuit is shorted to the 5v VREF circuit ● TP2 Sensor signal circuit is shorted to the 5v VREF circuit ● TP Sensor 5v VREF circuit is open or shorted to ground ● PCM is damaged (check for an open power ground circuit)

OBD II Trouble Code List (P0xxx Codes)

DTC	Trouble Code Title, Conditions & Possible Causes
P0225 **1T CCM, MIL: No** 1996, 1997, 1998, 1999, 2000, 2001, 2002 C/K Series Truck, G Series Van, Van Cargo 6.5L VIN F, 6.5L VIN S Transmissions: All	**Accelerator Pedal Position Sensor 3 Circuit Malfunction Conditions:** Key on or engine running; and the PCM detected the reference voltage for APP Sensor 3 was less than 4.80v for 2 seconds. **Possible Causes** ● APP VREF signal is open or grounded between the sensor and the PCM ● APP VREF signal is grounded between the sensor and PCM ● APP sensor is damaged or has failed ● PCM is damaged
P0226 **1T CCM, MIL: No** 1996, 1997, 1998, 1999, 2000, 2001, 2002 C/K Series Truck, G Series Van, Van Cargo 6.5L VIN F, 6.5L VIN S Transmissions: All	**Accelerator Pedal Position Sensor 3 Range/Performance Conditions:** No APP sensor codes set, system voltage over 6.4v, engine started, and the PCM detected the difference between the APP-3 and APP-1 signals was more than 230 mv, or the difference between the APP-3 and APP-2 signals was more than 500 mv for 2 seconds. **Possible Causes** ● APP3 signal circuit is open or shorted to ground ● APP3 sensor is binding, damaged or has failed ● PCM is damaged
P0227 **1T CCM, MIL: No** 1996, 1997, 1998, 1999, 2000, 2001, 2002 C/K Series Truck, G Series Van, Van Cargo 6.5L VIN F, 6.5L VIN S Transmissions: All	**Accelerator Pedal Position Sensor 3 Low Input Conditions:** Key on or engine running; and the PCM detected the APP-3 sensor voltage was less than 250 mv for 2 seconds. Note: The APP3 Sensor should read near 4.5v at idle speed. **Possible Causes** ● APP3 signal circuit is shorted to chassis or sensor ground ● APP3 VREF circuit has a high resistance condition (low VREF) ● APP3 sensor is damaged or has failed ● PCM is damaged
P0228 **1T CCM, MIL: No** 1996, 1997, 1998, 1999, 2000, 2001, 2002 C/K Series Truck, G Series Van, Van Cargo 6.5L VIN F, 6.5L VIN S Transmissions: All	**Accelerator Pedal Position Sensor 3 High Input Conditions:** Key on; and the PCM detected the APP-3 sensor was more than 4.75v for 2 seconds. **Possible Causes** ● APP3 sensor signal circuit is open between sensor and PCM ● APP3 sensor signal circuit is shorted to VREF or system power ● APP3 sensor ground circuit is open between sensor and PCM ● APP sensor is damaged or has failed ● PCM is damaged
P0230 **2T CCM, MIL: Yes** 1999, 2001, 2002, 2003, 2004, 2005 C/K Series Truck, G Series Van, Envoy, Escalade, TrailBlazer 4.8L VIN V, 5.3L VIN P, 5.3L VIN T, 5.3L VIN Z, 6.0L VIN N, 6.0L VIN U, 8.1L VIN G engines Transmissions: All	**Fuel Pump Relay Control Circuit Malfunction Conditions:** Engine started; engine speed more than 400 rpm, system voltage from 6-18v, and the PCM detected the Actual state and the Commanded state of the Fuel Pump control circuit did not match for 2.5 seconds. **Possible Causes** ● Fuel pump relay control circuit is open or shorted to ground ● Fuel pump relay power circuit is open (PCM Fuse B fuse block) ● Fuel pump relay is damaged or it has failed ● PCM is damaged
P0230 **2T CCM, MIL: Yes** 1998, 1999, 2000, 2001, 2002, 2003, 2004, 2005 Aurora, Bonneville, Grand Prix, Century, Cutlass Supreme, Grand Prix, Intrigue, LSS, LeSabre, Lumina, Monte Carlo, 88', 98', Regal, Park Avenue 3.1L VIN J, 3.1L VIN M, 3.4L VIN E engines, 3.5L VIN H, 3.8L VIN 1, 3.8L VIN K, 4.0L VIN C Transmissions: All	**Fuel Pump Control Circuit Malfunction Conditions:** Engine started; engine speed over 400 rpm, system voltage over 10.0v and the PCM detected that the Actual and Commanded state of the Fuel Pump driver control circuit did not match for 2.5 seconds. **Possible Causes** ● Fuel pump relay power circuit is open (test B+ from fuse box) ● Fuel pump control circuit is open or shorted to ground ● Fuel pump control circuit is shorted to system power ● PCM has failed ● TSB 00-06-04-023 contains a repair procedure for this code
P0230 **1T CCM, MIL: No** 2002, 2003, 2004, 2005 Aztek, Montana, Silhouette, Venture, Rendezvous 3.1L VIN J, 3.4L VIN E Transmissions: All	**Fuel Pump Control Circuit Malfunction Conditions:** Key on or engine running; system voltage from 9-18.0v, and the PCM detected an unexpected voltage condition on the Fuel Pump relay driver circuit for less than 1 second. **Possible Causes** ● Fuel pump relay power circuit is open (test B+ from at fuse box) ● Fuel pump control circuit is open or shorted to ground ● Fuel pump control circuit is shorted to system power ● PCM has failed

OBD II Trouble Code List (P0xxx Codes)

DTC	Trouble Code Title, Conditions & Possible Causes
P0230 **2T CCM, MIL: Yes** 1998, 1999, 2000, 2001, 2002, 2003, 2004, 2005 C/K Series Truck, G Series Van, Van Cargo, L/M Series Van 4.3L VIN W, 4.3L VIN X, 5.0L VIN M, 5.7L VIN R, 7.4L VIN J engines Transmissions: All	**Fuel Pump Control Circuit Malfunction Conditions:** Engine started; system voltage from 6-18v, and the PCM detected that the Commanded state and the Actual state of the Fuel Pump control circuit did not match for 2-5 continuous seconds. **Possible Causes** ● Fuel pump relay power circuit is open (test B+ from fuse box) ● Fuel pump control circuit is open or shorted to ground ● Fuel pump control circuit is shorted to system power ● PCM has failed
P0230 **1T CCM, MIL: No** 2000, 2001, 2002, 2003, 2004, 2005 CTS, DeVille, Eldorado, Seville 4.6L VIN 9, 4.6L VIN Y Transmissions: All	**Fuel Pump Control Circuit Malfunction Conditions:** Engine started; system voltage over 10.0v and the PCM detected an unexpected voltage condition on the Fuel Pump control circuit. **Possible Causes** ● Fuel pump relay power circuit is open (test B+ from fuse box) ● Fuel pump control circuit is open or shorted to ground ● Fuel pump control circuit is shorted to system power ● PCM has failed
P0230 **2T CCM, MIL: Yes** 1997, 1998, 1999, 2000, 2001 Catera 3.0L V6 VIN R engine Transmissions: A/T	**Fuel Pump Control Circuit Malfunction Conditions:** Engine started; engine speed over 400 rpm, system voltage over 10.0v and the PCM detected an unexpected voltage condition on the fuel pump control circuit after power-down. **Possible Causes** ● Fuel pump relay power circuit is open (test B+ from main relay) ● Fuel pump control circuit is open, shorted to ground or to power ● PCM has failed
P0230 **1T CCM, MIL: No** 1997, 1998, 1999, 2000, 2001, 2002, 2003, 2004, 2005 Corvette 5.7L VIN G, 5.7L VIN S Transmissions: All	**Fuel Pump Control Circuit Malfunction Conditions:** Engine started; system voltage over 10.0v and the PCM detected the Commanded state and Actual state of the Fuel Pump control circuit did not match for 2-5 continuous seconds. **Possible Causes** ● Fuel pump relay power circuit is open (test B+ from fuse box) ● Fuel pump control circuit is open or shorted to ground ● Fuel pump control circuit is shorted to system power ● PCM has failed
P0231 **1T CCM, MIL: No** 1996 CTS, DeVille, Eldorado & Seville 4.6L VIN 9, 4.6L VIN Y Transmissions: A/T	**Fuel Pump Feedback Circuit Low Input Conditions:** Engine started; and the PCM detected the difference between system and the feedback voltage was 2v for 1.3 seconds. **Possible Causes** ● Fuel pump relay power circuit is open (test B+ from fuse center) ● Fuel pump control circuit is open, shorted to ground or to power ● PCM has failed ● TSB 71-65-29 contains a repair procedure for this code
P0231 **1T CCM, MIL: No** 1996 Aurora 4.0L VIN C engine Transmissions: A/T	**Fuel Pump Feedback Circuit Low Input Conditions:** Engine started; and the PCM detected the difference between system and the feedback voltage was 2v for 1.3 seconds. **Possible Causes** ● Fuel pump relay power circuit is open (test B+ from fuse center) ● Fuel pump control circuit is open, shorted to ground or shorted to system power ● PCM has failed
P0231 **2T CCM, MIL: Yes** 1996, 1997, 1998, 1999, 2000, 2001, 2002 C/K Series Truck, G Series Van, Van Cargo 6.5L VIN F, 6.5L VIN S Transmissions: All	**Fuel Pump Feedback Circuit Low Input Conditions:** Key on or engine running; Fuel Lift Pump commanded "on", and the PCM detected the difference between system voltage and fuel pump feedback signal was more than 4v for 2 seconds (continuous test). **Possible Causes** ● Fuel pump feedback Circuit 120 is open or shorted to ground ● Fuel pump relay power circuit is open, or the relay is damaged ● Fuel pump relay control circuit is open or shorted to power ● PCM is damaged
P0231 **1T CCM, MIL: No** 1997, 1998, 1999 CTS, DeVille, Eldorado & Seville 4.0L VIN C, 4.6L VIN 9, 4.6L VIN Y Transmissions: A/T	**Fuel Pump Feedback Circuit Low Input Conditions:** Engine started; and the PCM detected the difference between system and the feedback voltage was 2v for 1.3 seconds. **Possible Causes** ● Fuel pump relay power circuit is open (test B+ from fuse center) ● Fuel pump control circuit is open or shorted to ground ● Fuel pump control circuit is shorted to system power ● PCM has failed

OBD II Trouble Code List (P0xxx Codes)

DTC	Trouble Code Title, Conditions & Possible Causes
P0232 **1T CCM, MIL: No** 1996 DeVille, Eldorado, Seville 4.6L VIN 9, 4.6L VIN Y Transmissions: A/T	**Fuel Pump Feedback Circuit High Input Conditions:** DTC P0117, P0118 and P0232 not set, key on, no reference pulses received, ECT sensor over 122ºF, and PCM detected the Fuel Pump feedback input was over 7.0v for 6 seconds. **Possible Causes** ● Fuel pump relay is sticking (closed), damaged or has failed ● Fuel pump ground circuit has a high resistance condition ● Fuel pump feedback circuit is shorted to system power ● PCM has failed ● TSB 71-65-29 contains a repair procedure for this code
P0232 **1T CCM, MIL: No** 1996 Aurora 4.0L VIN C engine Transmissions: A/T	**Fuel Pump Feedback Circuit High Input Conditions:** DTC P0117, P0118 and P0232 not set, key on, no reference pulses received, ECT sensor over 122ºF, and PCM detected the Fuel Pump Feedback signal was over 7.0v for 6 seconds. **Possible Causes** ● Fuel pump relay is sticking (closed), damaged or has failed ● Fuel pump ground circuit has a high resistance condition ● Fuel pump feedback circuit is shorted to system power ● PCM has failed
P0232 **1T CCM, MIL: No** 1997, 1998, 1999 Aurora, DeVille, Eldorado & Seville 4.0L VIN C, 4.6L VIN 9, 4.6L VIN Y engines Transmissions: A/T	**Fuel Pump Feedback Circuit High Input Conditions:** DTC P0117, P0118 and P0232 not set, key on, no reference pulses received, ECT sensor over 122ºF, and PCM detected the Fuel Pump Feedback signal was over 7.0v for 6 seconds. **Possible Causes** ● Fuel pump relay is sticking (closed), damaged or has failed ● Fuel pump ground circuit has a high resistance condition ● Fuel pump feedback circuit is shorted to system power ● PCM has failed
P0234 **2T CCM, MIL: Yes** 2001, 2002, 2003, 2004, 2005 C/K Series Trucks 6.6L VIN 1 6.6L VIN 1 Diesel engine Transmissions: All	**Turbocharger Boost Circuit Range/Performance Conditions:** DTC P0238 not set, engine speed over 500 rpm, less than 2700 rpm on RPO NF2, NF4, fuel quality less than 70 mm³, total fuel quality burned less than 2000 mm³, and the PCM detected the measured Boost Pressure sensor was above the expected range by 35 kPa or more for 12 seconds. This sensor responds to pressure changes in the intake manifold created by the turbocharger along with changes in the accelerator pedal position (APP) and engine speed. The ECM uses this data to assist in diagnosis of the BARO sensor, and to provide engine overboost protection. The Boost sensor has a 5-volt reference circuit, a low reference circuit, and a Boost sensor signal circuit that connect to the ECM. **Possible Causes** ● Boost sensor signal, VREF or ground circuit connection faults ● Boost sensor is damaged or has failed ● Pressure hose from the charged air tube to the Wastegate actuator is disconnected or ruptured ● Wastegate or Turbocharger is damaged or has failed
P0236 **2T CCM, MIL: Yes** 1996, 1997, 1998, 1999, 2000, 2001, 2002 C/K Series Truck, G Series Van, Van Cargo 6.5L VIN F, 6.5L VIN S Transmissions: All	**Turbocharger Boost System Performance Conditions:** Engine started; engine speed over 2400 rpm (Test 1), and the PCM detected the Actual Boost Pressure sensor was higher than, lower than or equal to 20 kPa of the Desired value, or with engine speed at 1800-2400 (Test 2), the PCM detected the Actual Boost Pressure was higher than, lower than or equal to 20 kPa of the Desired value. **Possible Causes** ● Engine vacuum line is loose, open or pinched, or misrouted ● Engine source vacuum restricted where vacuum line connects ● Turbocharger solenoid vent filter or vent orifice is restricted (the vacuum to the solenoid should fluctuate in normal operation)
P0236 **2T CCM, MIL: Yes** 2001, 2002, 2003, 2004, 2005 C/K Series Trucks 6.6L VIN 1 6.6L VIN 1 Diesel engine Transmissions: All	**Turbocharger Boost Circuit High Input Conditions:** DTC P0237, P2227, P2228 and P2229 not set, engine speed over 500 rpm (500-2700 rpm on RPO NF2, NF4), ECT sensor from 140-158ºF, total fuel quality burned under 2000 mm³, and the PCM detected the measure Boost Pressure sensor was below the expected range by more than 35 kPa for 12 seconds. **Possible Causes** ● Air cleaner or intake tube is severely restricted ● Boost sensor vacuum hose is disconnected, loose or clogged ● Boost sensor is damaged or has failed ● Charge air pipes or intake manifold leaking ● EGR throttle valve is stuck closed ● Turbocharger is damaged or has failed

OBD II Trouble Code List (P0xxx Codes)

DTC	Trouble Code Title, Conditions & Possible Causes
P0237 **2T CCM, MIL: Yes** 1996, 1997, 1998, 1999, 2000, 2001, 2002, 2003, 2004, 2005 C/K Series Truck, G Series Van, Van Cargo 6.5L VIN F, 6.5L VIN S Transmissions: All	**Turbocharger Boost Circuit Low Input Conditions:** Key on or engine running; and the PCM detected the Boost Pressure sensor was less than 0.10v (Scan Tool reads less than 40 kPa) for 2 seconds during the CCM continuous test. **Possible Causes** ● MAP sensor signal circuit shorted to sensor or chassis ground ● MAP sensor VREF circuit open between the sensor and PCM ● MAP sensor is damaged or has failed ● PCM has failed
P0237 **2T CCM, MIL: Yes** 2001, 2002, 2003, 2004, 2005 C/K Series Trucks 6.6L VIN 1 6.6L VIN 1 Diesel engine Transmissions: All	**Boost Sensor Circuit Low Input Conditions:** DTC P1635 not set, key on or engine running and the PCM detected the Boost sensor signal was less than 38 kPa for 2 seconds. The Boost sensor responds to pressure changes in the intake manifold. This pressure is created by the turbocharger and changes to the accelerator pedal position (APP) and engine speed. The ECM uses this information to provide engine Overboost protection. The boost sensor has a 5-volt reference circuit, a low reference circuit, and a signal circuit. The ECM supplies 5 volts to the boost sensor on a 5-volt reference circuit, and provides a ground on a low reference circuit. The boost sensor provides a voltage signal to the ECM on a signal circuit relative to the pressure changes. The ECM monitors the boost sensor signal for voltage outside of the normal range. The ECM sets this code if it detects a boost sensor signal that is too low. **Possible Causes** ● Boost sensor signal circuit shorted to sensor or chassis ground ● Boost sensor VREF circuit open between the sensor and PCM ● Boost sensor is damaged or has failed ● PCM has failed
P0238 **2T CCM, MIL: Yes** 2001, 2002, 2003, 2004, 2005 C/K Series Trucks 6.6L VIN 1 Diesel engine Transmissions: All	**Boost Sensor Circuit High Input Conditions:** DTC P1635 not set, key on or engine running and the PCM detected the Boost sensor signal was more than 4.80v (Scan Tool reads more than 254 kPa) for 2 seconds. The Boost sensor responds to pressure changes in the intake manifold. This pressure is created by the turbocharger and changes to the accelerator pedal position (APP) and engine speed. The ECM uses this data to provide engine Overboost protection. The boost sensor has a 5-volt reference circuit, a low reference circuit, and a signal circuit. The ECM supplies 5 volts to the boost sensor on a 5-volt reference circuit, and provides a ground on a low reference circuit. The boost sensor provides a voltage signal to the ECM on a signal circuit relative to the pressure changes. The ECM monitors the boost sensor signal for voltage outside of the normal range. The ECM sets this code if it detects a boost sensor signal that is too low. **Possible Causes** ● Boost sensor signal circuit is open or it is shorted to VREF ● Boost sensor ground circuit open between sensor and the PCM ● Boost sensor is damaged or has failed ● PCM has failed
P0238 **2T CCM, MIL: Yes** 1996, 1997, 1998, 1999, 2000, 2001, 2002, 2003, 2004, 2005 C/K Series Truck, G Series Van, Van Cargo 6.5L VIN F, 6.5L VIN S Transmissions: All	**Turbocharger Boost Circuit High Input Conditions:** Engine started; engine speed less than 3506 rpm, and the PCM detected the Turbocharger Boost Pressure sensor was more than, or was equal to 4.80v (202 kPa) for 2 seconds. **Possible Causes** ● MAP sensor signal circuit is open or it is shorted to VREF ● MAP sensor ground circuit open between sensor and the PCM ● MAP sensor is damaged or has failed ● PCM has failed
P0243 **1T CCM, MIL: No** 1996, 1997, 1998, 1999, 2000, 2001, 2002, 2003, 2004, 2005 Bonneville, Grand Prix, LSS, Regal, Riviera 3.8L VIN 1 SC engine Transmissions: All	**Supercharger Boost Solenoid Control Circuit Malfunction Conditions:** Key on or engine running; and the PCM detected an unexpected voltage on the Boost Control Solenoid control circuit for 30 seconds. The PCM uses an output driver module (ODM) to enable several current-driven devices that are needed to control various engine and transaxle functions. Each ODM can control several output device functions. **Possible Causes** ● Boost solenoid circuit is open between the solenoid and PCM ● Boost solenoid circuit is shorted to ground ● Boost solenoid power circuit is open to system power ● Boost solenoid is damaged or has failed ● PCM has failed

OBD II Trouble Code List (P0xxx Codes)

DTC	Trouble Code Title, Conditions & Possible Causes
P0251 **1T CCM, MIL: Yes** 1996, 1997, 1998, 1999, 2000, 2001, 2002 C/K Series Truck, G Van 6.5L VIN F, 6.5L VIN S Transmissions: All	**Injection Pump Camshaft System Malfunction Conditions:** Engine started; and the PCM detected that (8) consecutive Camshaft signals were missing, or that an average of eight (8) Camshaft signals were missing for 32 Cylinder 1 events. **Possible Causes** ● CMP sensor signal circuit is open or shorted (intermittent fault) ● Vehicle driven during hard vehicle maneuvers while low on fuel, or run until out of fuel ● PCM has failed
P0261 **2T CCM, MIL: Yes** 2003, 2004, 2005 CTS (Cadillac) 2.6L VIN M, 3.2L VIN N Transmissions: All	**Fuel Injector 1 Control Circuit Low Input Conditions:** Engine started; system voltage from 8-18v, and the PCM detected a low voltage condition on the Fuel Injector 1 circuit for 30 seconds. **Possible Causes** ● Fuel injector 1 connector is damaged or shorted ● Fuel injector 1 power circuit is open (check the INJ/Coil fuse) ● Fuel injector 1 control circuit is shorted to ground ● Fuel injector 1 is damaged or it has failed ● PCM is damaged
P0262 **2T CCM, MIL: Yes** 2003, 2004, 2005 CTS (Cadillac) 2.6L VIN M, 3.2L VIN N Transmissions: All	**Fuel Injector 1 Control Circuit High Input Conditions:** Engine started; system voltage from 8-18v, and the PCM detected a high voltage condition on the Fuel Injector 1 circuit for 30 seconds. **Possible Causes** ● Fuel injector 1 connector is damaged or shorted ● Fuel injector 1 control circuit is shorted to system power (B+) ● Fuel injector 1 is damaged or it has failed ● PCM is damaged
P0263 **1T CCM, MIL: No** 1996, 1997, 1998, 1999, 2000, 2001, 2002 C/K Series Truck, G Series Van, Van Cargo 6.5L VIN F, 6.5L VIN S Transmissions: All	**Cylinder 1 Balance System Malfunction Conditions:** Engine at hot idle for 90 seconds, cylinder fault detected in Misfire test, and PCM detected the Cylinder 1 fuel reduction level was too high for 2 seconds (use Scan Tool snap shot). **Possible Causes** ● Injector 1 fuel nozzle dirty, damaged or restricted ● Fuel feed lines between the tank and fuel pump are restricted ● Fuel filter id contaminated, dirty or clogged
P0264 **2T CCM, MIL: Yes** 2003, 2004, 2005 CTS (Cadillac) 2.6L VIN M, 3.2L VIN N Transmissions: All	**Fuel Injector 2 Control Circuit Low Input Conditions:** Engine started; system voltage from 8-18v, and the PCM detected a low voltage condition on the Fuel Injector 2 circuit for 30 seconds. **Possible Causes** ● Fuel injector 2 connector is damaged or shorted ● Fuel injector 2 power circuit is open (check the INJ/Coil fuse) ● Fuel injector 2 control circuit is shorted to ground ● Fuel injector 2 is damaged or it has failed ● PCM is damaged
P0265 **2T CCM, MIL: Yes** 2003, 2004, 2005 CTS (Cadillac) 2.6L VIN M, 3.2L VIN N Transmissions: All	**Fuel Injector 2 Control Circuit High Input Conditions:** Engine started; system voltage from 8-18v, and the PCM detected a high voltage condition on the Fuel Injector 2 circuit for 30 seconds. **Possible Causes** ● Fuel injector 2 connector is damaged or shorted ● Fuel injector 2 control circuit is shorted to system power (B+) ● Fuel injector 2 is damaged or it has failed ● PCM is damaged
P0267 **2T CCM, MIL: Yes** 2003, 2004, 2005 CTS (Cadillac) 2.6L VIN M, 3.2L VIN N Transmissions: All	**Fuel Injector 3 Control Circuit Low Input Conditions:** Engine started; system voltage from 8-18v, and the PCM detected a low voltage condition on the Fuel Injector 3 circuit for 30 seconds. **Possible Causes** ● Fuel injector 3 connector is damaged or shorted ● Fuel injector 3 power circuit is open (check the INJ/Coil fuse) ● Fuel injector 3 control circuit is shorted to ground ● Fuel injector 3 is damaged or it has failed ● PCM is damaged
P0268 **2T CCM, MIL: Yes** 2003, 2004, 2005 CTS (Cadillac) 2.6L VIN M, 3.2L VIN N Transmissions: All	**Fuel Injector 3 Control Circuit High Input Conditions:** Engine started; system voltage from 8-18v, and the PCM detected a high voltage condition on the Fuel Injector 3 circuit for 30 seconds. **Possible Causes** ● Fuel injector 3 connector is damaged or shorted ● Fuel injector 3 control circuit is shorted to system power (B+) ● Fuel injector 3 is damaged or it has failed ● PCM is damaged

OBD II Trouble Code List (P0xxx Codes)

DTC	Trouble Code Title, Conditions & Possible Causes
P0263 **1T CCM, MIL: No** 1996, 1997, 1998, 1999, 2000, 2001, 2002 C/K Series Truck, G Van 6.5L VIN F, 6.5L VIN S Transmissions: All	**Cylinder 2 Balance System Malfunction Conditions:** Engine at hot idle for 90 seconds, cylinder fault detected in Misfire test, and PCM detected the Cylinder 2 fuel reduction level was too high for 2 seconds (use Scan Tool snap shot). **Possible Causes** ● Injector 2 fuel nozzle dirty, damaged or restricted ● Fuel feed lines between the tank and fuel pump are restricted ● Fuel filter id contaminated, dirty or clogged
P0269 **1T CCM, MIL: No** 1996, 1997, 1998, 1999, 2000, 2001, 2002 C/K Series Truck, G Series Van, Van Cargo 6.5L VIN F, 6.5L VIN S Transmissions: All	**Cylinder 3 Balance System Malfunction Conditions:** Engine at hot idle for 90 seconds, cylinder fault detected in Misfire test, and the PCM detected the Cylinder 3 fuel reduction level was too high for 2 seconds (use Scan Tool snap shot to locate the fault). **Possible Causes** ● Injector 3 fuel nozzle dirty, damaged or restricted ● Fuel feed lines between the tank and fuel pump are restricted ● Fuel filter id contaminated, dirty or clogged
P0270 **2T CCM, MIL: Yes** 2003, 2004, 2005 CTS (Cadillac) 2.6L VIN M, 3.2L VIN N Transmissions: All	**Fuel Injector 4 Control Circuit Low Input Conditions:** Engine started; system voltage from 8-18v, and the PCM detected a low voltage condition on the Fuel Injector 4 circuit for 30 seconds. **Possible Causes** ● Fuel injector 4 connector is damaged or shorted ● Fuel injector 4 power circuit is open (check the INJ/Coil fuse) ● Fuel injector 4 control circuit is shorted to ground ● Fuel injector 4 is damaged or it has failed ● PCM is damaged
P0271 **2T CCM, MIL: Yes** 2003, 2004, 2005 CTS (Cadillac) 2.6L VIN M, 3.2L VIN N Transmissions: All	**Fuel Injector 4 Control Circuit High Input Conditions:** Engine started; system voltage from 8-18v, and the PCM detected a high voltage condition on the Fuel Injector 4 circuit for 30 seconds. **Possible Causes** ● Fuel injector 4 connector is damaged or shorted ● Fuel injector 4 control circuit is shorted to system power (B+) ● Fuel injector 4 is damaged or it has failed ● PCM is damaged
P0272 **1T CCM, MIL: No** 1996, 1997, 1998, 1999, 2000, 2001, 2002 C/K Series Truck, G Series Van, Van Cargo 6.5L VIN F, 6.5L VIN S Transmissions: All	**Cylinder 4 Balance System Malfunction Conditions:** Engine at hot idle for 90 seconds, cylinder fault detected in Misfire test, and PCM detected the Cylinder 4 fuel reduction level was too high for 2 seconds (use Scan Tool snap shot). **Possible Causes** ● Injector 4 fuel nozzle dirty, damaged or restricted ● Fuel feed lines between the tank and fuel pump are restricted ● Fuel filter id contaminated, dirty or clogged
P0273 **2T CCM, MIL: Yes** 2003, 2004, 2005 CTS (Cadillac) 2.6L VIN M, 3.2L VIN N Transmissions: All	**Fuel Injector 5 Control Circuit Low Input Conditions:** Engine started; system voltage from 8-18v, and the PCM detected a low voltage condition on the Fuel Injector 5 circuit for 30 seconds. **Possible Causes** ● Fuel injector 5 connector is damaged or shorted ● Fuel injector 5 power circuit is open (check the INJ/Coil fuse) ● Fuel injector 5 control circuit is shorted to ground ● Fuel injector 5 is damaged or it has failed ● PCM is damaged
P0274 **2T CCM, MIL: Yes** 2003, 2004, 2005 CTS (Cadillac) 2.6L VIN M, 3.2L VIN N Transmissions: All	**Fuel Injector 5 Control Circuit High Input Conditions:** Engine started; system voltage from 8-18v, and the PCM detected a high voltage condition on the Fuel Injector 5 circuit for 30 seconds. **Possible Causes** ● Fuel injector 5 connector is damaged or shorted ● Fuel injector 5 control circuit is shorted to system power (B+) ● Fuel injector 5 is damaged or it has failed ● PCM is damaged
P0275 **1T CCM, MIL: No** 1996, 1997, 1998, 1999, 2000, 2001, 2002 C/K Series Truck, G Series Van, Van Cargo 6.5L VIN F, 6.5L VIN S Transmissions: All	**Cylinder 5 Balance System Malfunction Conditions:** Engine at hot idle for 90 seconds, cylinder fault detected in Misfire test, and PCM detected the Cylinder 5 fuel reduction level was too high for 2 seconds (use Scan Tool snap shot). **Possible Causes** ● Injector 5 fuel nozzle dirty, damaged or restricted ● Fuel feed lines between the tank and fuel pump are restricted ● Fuel filter id contaminated, dirty or clogged

OBD II Trouble Code List (P0xxx Codes)

DTC	Trouble Code Title, Conditions & Possible Causes
P0276 **2T CCM, MIL: Yes** 2003, 2004, 2005 CTS (Cadillac) 2.6L VIN M, 3.2L VIN N Transmissions: All	**Fuel Injector 6 Control Circuit Low Input Conditions:** Engine started; system voltage from 8-18v, and the PCM detected a low voltage condition on the Fuel Injector 5 circuit for 30 seconds. **Possible Causes** ● Fuel injector 6 connector is damaged or shorted ● Fuel injector 6 power circuit is open (check the INJ/Coil fuse) ● Fuel injector 6 control circuit is shorted to ground ● Fuel injector 6 is damaged or it has failed ● PCM is damaged
P0277 **2T CCM, MIL: Yes** 2003, 2004, 2005 CTS (Cadillac) 2.6L VIN M, 3.2L VIN N Transmissions: All	**Fuel Injector 6 Control Circuit High Input Conditions:** Engine started; system voltage from 8-18v, and the PCM detected a high voltage condition on the Fuel Injector 5 circuit for 30 seconds. **Possible Causes** ● Fuel injector 6 connector is damaged or shorted ● Fuel injector 6 control circuit is shorted to system power (B+) ● Fuel injector 6 is damaged or it has failed ● PCM is damaged
P0278 **1T CCM, MIL: No** 1996, 1997, 1998, 1999, 2000, 2001, 2002 C/K Series Truck, G Series Van, Van Cargo 6.5L VIN F, 6.5L VIN S Transmissions: All	**Cylinder 6 Balance System Malfunction Conditions:** Engine at hot idle for 90 seconds, cylinder fault detected in Misfire test, and the PCM detected the Cylinder 6 fuel reduction level was too high for 2 seconds (use Scan Tool snap shot to locate the fault). **Possible Causes** ● Injector 6 fuel nozzle dirty, damaged or restricted ● Fuel feed lines between the tank and fuel pump are restricted ● Fuel filter id contaminated, dirty or clogged
P0281 **1T CCM, MIL: No** 1996, 1997, 1998, 1999, 2000, 2001, 2002 C/K Series Truck, G Series Van, Van Cargo 6.5L VIN F, 6.5L VIN S Transmissions: All	**Cylinder 7 Balance System Malfunction Conditions:** Engine at hot idle for 90 seconds, cylinder fault detected in Misfire test, and the PCM detected the Cylinder 7 fuel reduction level was too high for 2 seconds (use Scan Tool snap shot to locate the fault). **Possible Causes** ● Injector 7 fuel nozzle dirty, damaged or restricted ● Fuel feed lines between the tank and fuel pump are restricted ● Fuel filter id contaminated, dirty or clogged
P0284 **1T CCM, MIL: No** 1996, 1997, 1998, 1999, 2000, 2001, 2002 C/K Series Truck, G Series Van, Van Cargo 6.5L VIN F, 6.5L VIN S Transmissions: All	**Cylinder 8 Balance System Malfunction Conditions:** Engine at hot idle for 90 seconds, cylinder fault detected in Misfire test, and the PCM detected the Cylinder 8 fuel reduction level was too high for 2 seconds (use Scan Tool snap shot to locate the fault). **Possible Causes** ● Injector 8 fuel nozzle dirty, damaged or restricted ● Fuel feed lines between the tank and fuel pump are restricted ● Fuel filter id contaminated, dirty or clogged
P0300 **2T MISFIRE, MIL: Yes** 1996, 1997, 1998, 1999, 2000, 2001, 2002, 2003, 2004, 2005 Achieva, Alero, Beretta, Century, Ciera, Cavalier, Envoy, Grand Am, Sunfire Skyhawk, TrailBlazer, S/T Blazer & S/T Pickup 2.2L VIN 4, 2.2L VIN 5, 2.2L VIN F, 2.2L VIN H, 2.4L VIN T Transmissions: All	**Multiple Engine Misfire Detected Conditions:** DTC P0105, P0107, P0112, P0113, P0117, P0118, P0122, P0123, P0125, P0128, P0130, P0131, P0132, P0133, P0134, P0137, P0138, P0140, P0171, P0172, P0325, P0336, P0502, P0503, P0506, P0507, P0601, P0602, P1133, P1137, P1138, P1171, 1336 and P1621 not set, engine started, engine speed from 469-5900 rpm, system voltage over 10.0v, ECT sensor from 20-254°F, TP angle change less than 8% and has not decreased more than 1.5% within a 1 second period, and the PCM detected a crankshaft speed variation in one or more cylinders characteristic of a misfire. *Note: If the misfire is severe, the MIL will flash on/off on the 1st trip!* **Possible Causes** ● Base engine mechanical fault that affects one or more cylinders ● Fuel metering fault that affects more than one cylinder ● Fuel pressure too low or too high, fuel supply contaminated ● EVAP system problem or the EVAP canister is fuel saturated ● EGR valve is stuck open or PCV system has a vacuum leak ● IC control circuit is shorted to ground (an intermittent fault) ● Ignition system fault (a coil) that affects more than one cylinder ● TSB 87-65-08 contains a repair procedure for this code ● TSB 03-06-04-030 contains a repair procedure for this code ● TSB 03-06-04-055 contains a repair procedure for this code

OBD II Trouble Code List (P0xxx Codes)

DTC	Trouble Code Title, Conditions & Possible Causes
P0301 **2T MISFIRE, MIL: Yes** 1996, 1997, 1998, 1999, 2000, 2001, 2002, 2003, 2004, 2005 Achieva, Alero, Beretta, Century, Ciera, Cavalier, Envoy, Grand Am, Sunfire Skyhawk, TrailBlazer, S/T Blazer & S/T Pickup 2.2L VIN 4, 2.2L VIN 5, 2.2L VIN F, 2.2L VIN H, 2.4L VIN T, 4.2L VIN S Transmissions: All	**Cylinder 1 Misfire Detected Conditions:** DTC P0013, P0014, P0105, P0107, P0112, P0113, P0117, P0118, P0122, P0123, P0125, P0128, P0130-P0140, P0171-P0172, P0325, P0336, P0502-P0503, P0506-P0507, P0601, P0602, P1133, P1137, P1138, P1171, 1336 and P1621 not set, engine speed from 469-5900 rpm, DFCO and EGR Test off, system voltage over 10.0v, ECT sensor from 20-254°F, TP angle stable and the PCM detected a crankshaft speed variation characteristic of a misfire in Cylinder 1. *Note: If the misfire is severe, the MIL will flash on/off on one trip!* **Possible Causes** ● Air leak in the intake manifold, or in the EGR or PCV system ● Base engine mechanical fault that affects only Cylinder 1 ● Fuel delivery component fault that affects only Cylinder 1 (i.e., a contaminated, dirty or sticking fuel injector) ● Ignition system problem (coil or plug) that affects Cylinder 1 ● TSB 87-65-08 contains a repair procedure for this code
P0302 **2T MISFIRE, MIL: Yes** 1996, 1997, 1998, 1999, 2000, 2001, 2002, 2003, 2004, 2005 Achieva, Alero, Beretta, Century, Ciera, Cavalier, Envoy, Grand Am, Sunfire Skyhawk, TrailBlazer, S/T Blazer & S/T Pickup 2.2L VIN 4, 2.2L VIN 5, 2.2L VIN F, 2.2L VIN H, 2.4L VIN T, 4.2L VIN S Transmissions: All	**Cylinder 2 Misfire Detected Conditions:** DTC P0013, P0014, P0105, P0107, P0112, P0113, P0117, P0118, P0122, P0123, P0125, P0128, P0130-P0140, P0171-P0172, P0325, P0336, P0502-P0503, P0506-P0507, P0601, P0602, P1133, P1137, P1138, P1171, 1336 and P1621 not set, engine speed from 469-5900 rpm, DFCO and EGR Test off, system voltage over 10.0v, ECT sensor from 20-254°F, TP angle stable and the PCM detected a crankshaft speed variation characteristic of a misfire in Cylinder 2. *Note: If the misfire is severe, the MIL will flash on/off on one trip!* **Possible Causes** ● Air leak in the intake manifold, or in the EGR or PCV system ● Base engine mechanical fault that affects only Cylinder 2 ● Fuel delivery component fault that affects only Cylinder 2 (i.e., a contaminated, dirty or sticking fuel injector) ● Ignition system problem (coil or plug) that affects Cylinder 2 ● TSB 87-65-08 contains a repair procedure for this code
P0303 **2T MISFIRE, MIL: Yes** 1996, 1997, 1998, 1999, 2000, 2001, 2002, 2003, 2004, 2005 Achieva, Alero, Beretta, Century, Ciera, Cavalier, Envoy, Grand Am, Sunfire Skyhawk, TrailBlazer, S/T Blazer & S/T Pickup 2.2L VIN 4, 2.2L VIN 5, 2.2L VIN F, 2.2L VIN H, 2.4L VIN T, 4.2L VIN S Transmissions: All	**Cylinder 3 Misfire Detected Conditions:** DTC P0013, P0014, P0105, P0107, P0112, P0113, P0117, P0118, P0122, P0123, P0125, P0128, P0130-P0140, P0171-P0172, P0325, P0336, P0502-P0503, P0506-P0507, P0601, P0602, P1133, P1137, P1138, P1171, 1336 and P1621 not set, engine speed from 469-5900 rpm, DFCO and EGR Test off, system voltage over 10.0v, ECT sensor from 20-254°F, TP angle stable and the PCM detected a crankshaft speed variation characteristic of a misfire in Cylinder 3. *Note: If the misfire is severe, the MIL will flash on/off on one trip!* **Possible Causes** ● Air leak in the intake manifold, or in the EGR or PCV system ● Base engine mechanical fault that affects only Cylinder 3 ● Fuel delivery component fault that affects only Cylinder 3 (i.e., a contaminated, dirty or sticking fuel injector) ● Ignition system problem (coil or plug) that affects Cylinder 3 ● TSB 87-65-08 contains a repair procedure for this code
P0304 **2T MISFIRE, MIL: Yes** 1996, 1997, 1998, 1999, 2000, 2001, 2002, 2003, 2004, 2005 Achieva, Alero, Beretta, Century, Ciera, Cavalier, Envoy, Grand Am, Sunfire Skyhawk, TrailBlazer, S/T Blazer & S/T Pickup 2.2L VIN 4, 2.2L VIN 5, 2.2L VIN F, 2.2L VIN H, 2.4L VIN T, 4.2L VIN S Transmissions: All	**Cylinder 4 Misfire Detected Conditions:** DTC P0013, P0014, P0105, P0107, P0112, P0113, P0117, P0118, P0122, P0123, P0125, P0128, P0130-P0140, P0171-P0172, P0325, P0336, P0502-P0503, P0506-P0507, P0601, P0602, P1133, P1137, P1138, P1171, 1336 and P1621 not set, engine speed from 469-5900 rpm, DFCO and EGR Test off, system voltage over 10.0v, ECT sensor from 20-254°F, TP angle stable and the PCM detected a crankshaft speed variation characteristic of a misfire in Cylinder 4. *Note: If the misfire is severe, the MIL will flash on/off on one trip!* **Possible Causes** ● Air leak in the intake manifold, or in the EGR or PCV system ● Base engine mechanical fault that affects only Cylinder 4 ● Fuel delivery component fault that affects only Cylinder 4 (i.e., a contaminated, dirty or sticking fuel injector) ● Ignition system problem (coil or plug) that affects one cylinder ● TSB 87-65-08 contains a repair procedure for this code

OBD II Trouble Code List (P0xxx Codes)

DTC	Trouble Code Title, Conditions & Possible Causes
P0300 **2T MISFIRE, MIL: Yes** 2002, 2003, 2004, 2005 Envoy & TrailBlazer 4.2L VIN S Transmissions: All	**Multiple Engine Misfire Detected Conditions:** DTC P0013, P0014, P0105, P0107, P0108, P0117, P0118, P0122, P0123, P0125, P0128, P0217, P0218, P0336, P0340, P0341, P0365, P0366, P0502, P0503, P1114, P1115, P1121, P1122, P1336 and P1345 not set, engine started, system voltage over 10.0v, ECT sensor from 19-266ºF, TP angle steady, A/C Clutch status not changing, EGR test off, Decel Fuel Cutoff off, fuel level over 10 %, and the PCM detected variation in the crankshaft speed in two or more cylinders characteristic of a misfire condition. *Note: If the misfire is severe, the MIL will flash on/off on the 1st trip!* **Possible Causes** ● Base engine mechanical fault that affects two or more cylinders ● Fuel metering fault that affects more two or more cylinders ● Fuel pressure too low or too high, fuel supply contaminated ● EVAP system problem or the EVAP canister is fuel saturated ● EGR valve is stuck open or the PCV system has a vacuum leak ● IC control circuit is shorted to ground (an intermittent fault) ● Ignition system fault (a coil) that affects more than one cylinder ● TSB 03-06-04-055 contains a repair procedure for this code
P0305 **2T MISFIRE, MIL: Yes** 2002, 2003, 2004, 2005 Envoy & TrailBlazer 4.2L VIN S Transmissions: All	**Cylinder 5 Misfire Detected Conditions:** DTC P0013, P0014, P0105, P0107, P0108, P0117, P0118, P0122, P0123, P0125, P0128, P0217, P0218, P0336, P0340, P0341, P0365, P0366, P0502, P0503, P1114, P1115, P1121, P1122, P1336, or P1345 not set, engine started, system voltage over 10.0v, ECT sensor from 19-266ºF, TP angle steady, A/C Clutch status not changing, EGR Test and DFCO inactive, fuel level over 10 %, and the PCM detected a crankshaft speed variation in one cylinder characteristic of a misfire condition during the Misfire Monitor Test. *Note: If the misfire is severe, the MIL will flash on/off on the 1st trip!* **Possible Causes** ● Air leak in the intake manifold, or in the EGR or PCV system ● Base engine mechanical fault that affects only one cylinder ● Fuel delivery component fault that affects only one cylinder (i.e., a contaminated, dirty or sticking fuel injector) ● Ignition system problem (coil or plug) that affects one cylinder
P0306 **2T MISFIRE, MIL: Yes** 2002, 2003, 2004, 2005 Envoy & TrailBlazer 4.2L VIN S Transmissions: All	**Cylinder 6 Misfire Detected Conditions:** DTC P0013, P0014, P0105, P0107, P0108, P0117, P0118, P0122, P0123, P0125, P0128, P0217, P0218, P0336, P0340, P0341, P0365, P0366, P0502, P0503, P1114, P1115, P1121, P1122, P1336, or P1345 not set, engine started, system voltage over 10.0v, ECT sensor from 19-266ºF, TP angle steady, A/C Clutch status not changing, EGR Test and DFCO inactive, fuel level over 10 %, and the PCM detected a crankshaft speed variation in one cylinder characteristic of a misfire condition during the Misfire Monitor Test. *Note: If the misfire is severe, the MIL will flash on/off on the 1st trip!* **Possible Causes** ● Air leak in the intake manifold, or in the EGR or PCV system ● Base engine mechanical fault that affects only one cylinder ● Fuel delivery component fault that affects only one cylinder (i.e., a contaminated, dirty or sticking fuel injector) ● Ignition system problem (coil or plug) that affects one cylinder
P0300 **2T MISFIRE, MIL: Yes** 1996, 1997, 1998, 1999, 2000, 2001, 2002, 2003, 2004, 2005 Achieva, Alero, Beretta, Bonneville, Century, Corsica, Cutlass Supreme, Grand Am, Grand Prix, Intrigue, LeSabre, LSS, Lumina, Monte Carlo, Park Avenue, Regal, Skylark, 88' & 98' 3.1L VIN J, 3.1L VIN M, 3.4L VIN X, 3.5L VIN H, 3.8L VIN 1, 3.8L VIN K Transmissions: All	**Multiple Engine Misfire Detected Conditions:** DTC P0101, P0102, P0103, P0107, P0108, P0116-P0118, P0121-P0123, P0125, P0336, P0341, P0502-P0503, P1106-P1107, P1114, P1115, P1121, P1122, P1336, 1351, P1361, P1362 and P1374 not set, engine speed from 525-6600 rpm, ECT sensor at 21-255ºF, TP angle steady, system voltage over 10.0v and the PCM detected a crankshaft speed variation in 2 or more cylinders characteristic of a misfire. If the misfire is severe, the MIL will flash on/off on 1st trip! **Possible Causes** ● Base engine mechanical fault that affects one or more cylinders ● Fuel metering fault (high fuel pressure or fuel contaminated) ● EVAP system problem or the EVAP canister is fuel saturated ● EGR valve is stuck open or the PCV system has a vacuum leak ● Ignition system fault (a coil) that affects more than one cylinder ● MAF sensor contamination (it can cause a very lean condition) ● TSB 99-06-04-005B contains a repair procedure for this code ● TSB 03-06-04-030 contains a repair procedure for this code

OBD II Trouble Code List (P0xxx Codes)

DTC	Trouble Code Title, Conditions & Possible Causes
P0301 **2T MISFIRE, MIL: Yes** 1996, 1997, 1998, 1999, 2000, 2001, 2002, 2003, 2004, 2005 Achieva, Alero, Beretta, Bonneville, Century, Corsica, Cutlass Supreme, Grand Am, Grand Prix, Intrigue, LeSabre, LSS, Lumina, Monte Carlo, Park Avenue, Regal, Skylark, 88' & 98' 3.1L VIN J, 3.1L VIN M, 3.4L VIN X, 3.5L VIN H, 3.8L VIN 1, 3.8L VIN K Transmissions: All	**Cylinder 1 Misfire Detected Conditions:** DTC P0101, P0102, P0103, P0107, P0108, P0116-P0118, P0121-P0123, P0125, P0336, P0341, P0502-P0503, P1106-P1107, P1114, P1115, P1121, P1122, P1336, 1351, P1361, P1362 and P1374 not set, engine speed from 525-6600 rpm, ECT sensor from 21-255ºF, TP angle steady, system voltage over 10.0v and the PCM detected a crankshaft speed variation in Cylinder 1 characteristic of a misfire. *Note: If the misfire is severe, the MIL will flash on/off on the 1st trip!* **Possible Causes** ● Air leak in the intake manifold, or in the EGR or PCV system ● Base engine mechanical fault that affects only Cylinder 1 ● Fuel delivery component fault that affects only Cylinder 1 (i.e., a contaminated, dirty or sticking fuel injector) ● Ignition system problem (coil, plug) that affects only Cylinder 1
P0302 **2T MISFIRE, MIL: Yes** 1996, 1997, 1998, 1999, 2000, 2001, 2002, 2003, 2004, 2005 Achieva, Alero, Beretta, Bonneville, Century, Corsica, Cutlass Supreme, Grand Am, Grand Prix, Intrigue, LeSabre, LSS, Lumina, Monte Carlo, Park Avenue, Regal, Skylark, 88' & 98' 3.1L VIN J, 3.1L VIN M, 3.4L VIN X, 3.5L VIN H, 3.8L VIN 1, 3.8L VIN K Transmissions: All	**Cylinder 2 Misfire Detected Conditions:** DTC P0101, P0102, P0103, P0107, P0108, P0116-P0118, P0121-P0123, P0125, P0336, P0341, P0502-P0503, P1106-P1107, P1114, P1115, P1121, P1122, P1336, 1351, P1361, P1362 and P1374 not set, engine speed from 525-6600 rpm, ECT sensor from 21-255ºF, TP angle steady, system voltage over 10.0v and the PCM detected a crankshaft speed variation in Cylinder 2 characteristic of a misfire. *Note: If the misfire is severe, the MIL will flash on/off on the 1st trip!* **Possible Causes** ● Air leak in the intake manifold, or in the EGR or PCV system ● Base engine mechanical fault that affects only Cylinder 2 ● Fuel delivery component fault that affects only Cylinder 2 (i.e., a contaminated, dirty or sticking fuel injector) ● Ignition system problem (coil, plug) that affects only Cylinder 2
P0303 **2T MISFIRE, MIL: Yes** 1996, 1997, 1998, 1999, 2000, 2001, 2002, 2003, 2004, 2005 Achieva, Alero, Beretta, Bonneville, Century, Corsica, Cutlass Supreme, Grand Am, Grand Prix, Intrigue, LeSabre, LSS, Lumina, Monte Carlo, Park Avenue, Regal, Skylark, 88' & 98' 3.1L VIN J, 3.1L VIN M, 3.4L VIN X, 3.5L VIN H, 3.8L VIN 1, 3.8L VIN K Transmissions: All	**Cylinder 3 Misfire Detected Conditions:** DTC P0101, P0102, P0103, P0107, P0108, P0116-P0118, P0121-P0123, P0125, P0336, P0341, P0502-P0503, P1106-P1107, P1114, P1115, P1121, P1122, P1336, 1351, P1361, P1362 and P1374 not set, engine speed from 525-6600 rpm, ECT sensor from 21-255ºF, TP angle steady, system voltage over 10.0v and the PCM detected a crankshaft speed variation in Cylinder 3 characteristic of a misfire. *Note: If the misfire is severe, the MIL will flash on/off on the 1st trip!* **Possible Causes** ● Air leak in the intake manifold, or in the EGR or PCV system ● Base engine mechanical fault that affects only Cylinder 3 ● Fuel delivery component fault that affects only Cylinder 3 (i.e., a contaminated, dirty or sticking fuel injector) ● Ignition system problem (coil, plug) that affects only Cylinder 3
P0304 **2T MISFIRE, MIL: Yes** 1996, 1997, 1998, 1999, 2000, 2001, 2002, 2003, 2004, 2005 Achieva, Alero, Beretta, Bonneville, Century, Corsica, Cutlass Supreme, Grand Am, Grand Prix, Intrigue, LeSabre, LSS, Lumina, Monte Carlo, Park Avenue, Regal, Skylark, 88' & 98' 3.1L VIN J, 3.1L VIN M, 3.4L VIN X, 3.5L VIN H, 3.8L VIN 1, 3.8L VIN K Transmissions: All	**Cylinder 4 Misfire Detected Conditions:** DTC P0101, P0102, P0103, P0107, P0108, P0116-P0118, P0121-P0123, P0125, P0336, P0341, P0502-P0503, P1106-P1107, P1114, P1115, P1121, P1122, P1336, 1351, P1361, P1362 and P1374 not set, engine speed from 525-6600 rpm, ECT sensor from 21-255ºF, TP angle steady, system voltage over 10.0v and the PCM detected a crankshaft speed variation in Cylinder 4 characteristic of a misfire. *Note: If the misfire is severe, the MIL will flash on/off on the 1st trip!* **Possible Causes** ● Air leak in the intake manifold, or in the EGR or PCV system ● Base engine mechanical fault that affects only Cylinder 4 ● Fuel delivery component fault that affects only Cylinder 4 (i.e., a contaminated, dirty or sticking fuel injector) ● Ignition system problem (coil, plug) that affects only Cylinder 4

OBD II Trouble Code List (P0xxx Codes)

DTC	Trouble Code Title, Conditions & Possible Causes
P0305 **2T MISFIRE, MIL: Yes** 1996, 1997, 1998, 1999, 2000, 2001, 2002, 2003, 2004, 2005 Achieva, Alero, Beretta, Bonneville, Century, Corsica, Cutlass Supreme, Grand Am, Grand Prix, Intrigue, LeSabre, LSS, Lumina, Monte Carlo, Park Avenue, Regal, Skylark, 88' & 98' 3.1L VIN J, 3.1L VIN M, 3.4L VIN X, 3.5L VIN H, 3.8L VIN 1, 3.8L VIN K Transmissions: All	**Cylinder 5 Misfire Detected Conditions:** DTC P0101, P0102, P0103, P0107, P0108, P0116-P0118, P0121-P0123, P0125, P0336, P0341, P0502-P0503, P1106-P1107, P1114, P1115, P1121, P1122, P1336, 1351, P1361, P1362 and P1374 not set, engine speed from 525-6600 rpm, ECT sensor from 21-255ºF, TP angle steady, system voltage over 10.0v and the PCM detected a crankshaft speed variation in Cylinder 5 characteristic of a misfire. *Note: If the misfire is severe, the MIL will flash on/off on the 1st trip!* **Possible Causes** ● Air leak in the intake manifold, or in the EGR or PCV system ● Base engine mechanical fault that affects only Cylinder 5 ● Fuel delivery component fault that affects only Cylinder 5 (i.e., a contaminated, dirty or sticking fuel injector) ● Ignition system problem (coil, plug) that affects only Cylinder 5
P0306 **2T MISFIRE, MIL: Yes** 1996, 1997, 1998, 1999, 2000, 2001, 2002, 2003, 2004, 2005 Achieva, Alero, Beretta, Bonneville, Century, Corsica, Cutlass Supreme, Grand Am, Grand Prix, Intrigue, LeSabre, LSS, Lumina, Monte Carlo, Park Avenue, Regal, Skylark, 88' & 98' 3.1L VIN J, 3.1L VIN M, 3.4L VIN X, 3.5L VIN H, 3.8L VIN 1, 3.8L VIN K Transmissions: All	**Cylinder 6 Misfire Detected Conditions:** DTC P0101, P0102, P0103, P0107, P0108, P0116-P0118, P0121-P0123, P0125, P0336, P0341, P0502-P0503, P1106-P1107, P1114, P1115, P1121, P1122, P1336, 1351, P1361, P1362 and P1374 not set, engine speed from 525-6600 rpm, ECT sensor from 21-255ºF, TP angle steady, system voltage over 10.0v and the PCM detected a crankshaft speed variation in Cylinder 6 characteristic of a misfire. *Note: If the misfire is severe, the MIL will flash on/off on the 1st trip!* **Possible Causes** ● Air leak in the intake manifold, or in the EGR or PCV system ● Base engine mechanical fault that affects only Cylinder 6 ● Fuel delivery component fault that affects only Cylinder 6 (i.e., a contaminated, dirty or sticking fuel injector) ● Ignition system problem (coil, plug) that affects only Cylinder 6
P0300 **2T MISFIRE, MIL: Yes** 1996, 1997 Camaro, Caprice, Corvette, Firebird, Fleetwood 4.3L VIN W, 5.7L VIN 5, 5.7L VIN P Transmissions: All	**Multiple Engine Misfire Detected Conditions:** DTC P0101-P0103, P0117, P0118, P0122, P0123, P0335, P0336, P0500, P0502, P0641 and P0742 not set, engine started, engine speed from 425-3000 rpm, system voltage over 10.0v, ECT sensor from 19-266ºF, AIR, Traction Control and DFCO all "off", TP angle steady (±1%), transmission not shifting, A/C status steady, ABS signals not exceeding the Rough Road thresholds, and the PCM detected a crankshaft speed variation in one or more cylinders characteristic of a misfire condition during the Misfire Monitor test. *Note: If the misfire is severe, the MIL will flash on/off on the 1st trip!* **Possible Causes** ● Base engine mechanical fault that affects one or more cylinders ● Fuel metering fault that affects more than one cylinder ● Fuel pressure too low or too high, fuel supply contaminated ● EVAP system problem or the EVAP canister is fuel saturated ● EGR valve is stuck open or the PCV system has a vacuum leak ● IC control circuit is shorted to ground (an intermittent fault) ● Ignition system fault (a coil) that affects more than one cylinder ● MAF sensor contamination (it can cause a very lean condition)
P0301 **2T MISFIRE, MIL: Yes** 1996, 1997 Camaro, Caprice, Corvette, Firebird, Fleetwood 4.3L VIN W, 5.7L VIN 5, 5.7L VIN P Transmissions: All	**Cylinder 1 Misfire Detected Conditions:** DTC P0101, P0102, P0103, P0117, P0118, P0122, P0123, P0335, P0336, P0500, P0502, P0641 and P0742 not set, engine started, engine speed from 425-3000 rpm, system voltage over 10.0v, ECT sensor from 19-266ºF, AIR, Traction Control and DFCO all "off", TP angle steady (±1%), transmission not shifting, A/C status steady, signals for the ABS module not exceeding the Rough Road thresholds, and the PCM detected a crankshaft speed variation in one cylinder characteristic of a misfire condition during the Misfire Monitor test. *Note: If the misfire is severe, the MIL will flash on/off on the 1st trip!* **Possible Causes** ● Air leak in the intake manifold, or in the EGR or PCV system ● Base engine mechanical fault that affects only one cylinder ● Fuel component fault that affects only one cylinder (i.e., a dirty or sticking fuel injector) ● Ignition system problem (coil or plug) that affects one cylinder

OBD II Trouble Code List (P0xxx Codes)

DTC	Trouble Code Title, Conditions & Possible Causes
P0302 **2T MISFIRE, MIL: Yes** 1996, 1997 Camaro, Caprice, Corvette, Firebird, Fleetwood 4.3L VIN W, 5.7L VIN 5, 5.7L VIN P Transmissions: All	**Cylinder 2 Misfire Detected Conditions:** DTC P0101, P0102, P0103, P0117, P0118, P0122, P0123, P0335, P0336, P0500, P0502, P0641 and P0742 not set, engine started, engine speed from 425-3000 rpm, system voltage over 10.0v, ECT sensor from 19-266°F, AIR, Traction Control and DFCO all "of", TP angle steady (±1%), transmission not shifting, A/C status steady, signals for the ABS module not exceeding the Rough Road thresholds, and the PCM detected a crankshaft speed variation in one cylinder characteristic of a misfire condition during the Misfire Monitor test. *Note: If the misfire is severe, the MIL will flash on/off on the 1st trip!* **Possible Causes** ● Air leak in the intake manifold, or in the EGR or PCV system ● Base engine mechanical fault that affects only one cylinder ● Fuel component fault that affects only one cylinder (i.e., a dirty or sticking fuel injector) ● Ignition system problem (coil or plug) that affects one cylinder
P0303 **2T MISFIRE, MIL: Yes** 1996, 1997 Camaro, Caprice, Corvette, Firebird, Fleetwood 4.3L VIN W, 5.7L VIN 5, 5.7L VIN P Transmissions: All	**Cylinder 3 Misfire Detected Conditions:** DTC P0101, P0102, P0103, P0117, P0118, P0122, P0123, P0335, P0336, P0500, P0502, P0641 and P0742 not set, engine started, engine speed from 425-3000 rpm, system voltage over 10.0v, ECT sensor from 19-266°F, AIR, Traction Control and DFCO all "of", TP angle steady (±1%), transmission not shifting, A/C status steady, signals for the ABS module not exceeding the Rough Road thresholds, and the PCM detected a crankshaft speed variation in one cylinder characteristic of a misfire condition during the Misfire Monitor test. *Note: If the misfire is severe, the MIL will flash on/off on the 1st trip!* **Possible Causes** ● Air leak in the intake manifold, or in the EGR or PCV system ● Base engine mechanical fault that affects only one cylinder ● Fuel component fault that affects only one cylinder (i.e., a dirty or sticking fuel injector) ● Ignition system problem (coil or plug) that affects one cylinder
P0304 **2T MISFIRE, MIL: Yes** 1996, 1997 Camaro, Caprice, Corvette, Firebird, Fleetwood 4.3L VIN W, 5.7L VIN 5, 5.7L VIN P Transmissions: All	**Cylinder 4 Misfire Detected Conditions:** DTC P0101, P0102, P0103, P0117, P0118, P0122, P0123, P0335, P0336, P0500, P0502, P0641 and P0742 not set, engine started, engine speed from 425-3000 rpm, system voltage over 10.0v, ECT sensor from 19-266°F, AIR, Traction Control and DFCO all "of", TP angle steady (±1%), transmission not shifting, A/C status steady, signals for the ABS module not exceeding the Rough Road thresholds, and the PCM detected a crankshaft speed variation in one cylinder characteristic of a misfire condition during the Misfire Monitor test. *Note: If the misfire is severe, the MIL will flash on/off on the 1st trip!* **Possible Causes** ● Air leak in the intake manifold, or in the EGR or PCV system ● Base engine mechanical fault that affects only one cylinder ● Fuel component fault that affects only one cylinder (i.e., a dirty or sticking fuel injector) ● Ignition system problem (coil or plug) that affects one cylinder
P0305 **2T MISFIRE, MIL: Yes** 1996, 1997 Camaro, Caprice, Corvette, Firebird, Fleetwood 4.3L VIN W, 5.7L VIN 5, 5.7L VIN P Transmissions: All	**Cylinder 5 Misfire Detected Conditions:** DTC P010, P0102, P0103, P0117, P0118, P0122, P0123, P0335, P0336, P0500, P0502, P0641 and P0742 not set, engine started, engine speed from 425-3000 rpm, system voltage over 10.0v, ECT sensor from 19-266°F, AIR, Traction Control and DFCO all "of", TP angle steady (±1%), transmission not shifting, A/C status steady, signals for the ABS module not exceeding the Rough Road thresholds, and the PCM detected a crankshaft speed variation in one cylinder characteristic of a misfire condition during the Misfire Monitor test. *Note: If the misfire is severe, the MIL will flash on/off on the 1st trip!* **Possible Causes** ● Air leak in the intake manifold, or in the EGR or PCV system ● Base engine mechanical fault that affects only one cylinder ● Fuel component fault that affects only one cylinder (i.e., a dirty or sticking fuel injector) ● Ignition system problem (coil or plug) that affects one cylinder
P0306 **2T MISFIRE, MIL: Yes** 1996, 1997 Camaro, Caprice, Corvette, Firebird, Fleetwood 4.3L VIN W, 5.7L VIN 5, 5.7L VIN P Transmissions: All	**Cylinder 6 Misfire Detected Conditions:** DTC P0101, P0102, P0103, P0117, P0118, P0122, P0123, P0335, P0336, P0500, P0502, P0641 and P0742 not set, engine started, engine speed from 425-3000 rpm, system voltage over 10.0v, ECT sensor from 19-266°F, AIR, Traction Control and DFCO all "of", TP angle steady (±1%), transmission not shifting, A/C status steady, signals for the ABS module not exceeding the Rough Road thresholds, and the PCM detected a crankshaft speed variation in one cylinder characteristic of a misfire condition during the Misfire Monitor test. *Note: If the misfire is severe, the MIL will flash on/off on the 1st trip!* **Possible Causes** ● Air leak in the intake manifold, or in the EGR or PCV system ● Base engine mechanical fault that affects only one cylinder ● Fuel component fault that affects only one cylinder (i.e., a dirty or sticking fuel injector) ● Ignition system problem (coil or plug) that affects one cylinder

OBD II Trouble Code List (P0xxx Codes)

DTC	Trouble Code Title, Conditions & Possible Causes
P0307 **2T MISFIRE, MIL: Yes** 1996, 1997 Camaro, Caprice, Corvette, Firebird, Fleetwood 5.7L VIN 5, 5.7L VIN P Transmissions: All	**Cylinder 7 Misfire Detected Conditions:** DTC P0101, P0102, P0103, P0117, P0118, P0122, P0123, P0335, P0336, P0500, P0502, P0641 and P0742 not set, engine started, engine speed from 425-3000 rpm, system voltage over 10.0v, ECT sensor from 19-266°F, AIR, Traction Control and DFCO all "of", TP angle steady (±1%), transmission not shifting, A/C status steady, signals for the ABS module not exceeding the Rough Road thresholds, and the PCM detected a crankshaft speed variation in one cylinder characteristic of a misfire condition during the Misfire Monitor test. *Note: If the misfire is severe, the MIL will flash on/off on the 1st trip!* **Possible Causes** ● Air leak in the intake manifold, or in the EGR or PCV system ● Base engine mechanical fault that affects only one cylinder ● Fuel component fault that affects only one cylinder (i.e., a dirty or sticking fuel injector) ● Ignition system problem (coil or plug) that affects one cylinder
P0308 **2T MISFIRE, MIL: Yes** 1996, 1997 Camaro, Caprice, Corvette, Firebird, Fleetwood 5.7L VIN 5, 5.7L VIN P Transmissions: All	**Cylinder 8 Misfire Detected Conditions:** DTC P0101, P0102, P0103, P0117, P0118, P0122, P0123, P0335, P0336, P0500, P0502, P0641 and P0742 not set, engine started, engine speed from 425-3000 rpm, system voltage over 10.0v, ECT sensor from 19-266°F, AIR, Traction Control and DFCO all "of", TP angle steady (±1%), transmission not shifting, A/C status steady, signals for the ABS module not exceeding the Rough Road thresholds, and the PCM detected a crankshaft speed variation in one cylinder characteristic of a misfire condition during the Misfire Monitor test. *Note: If the misfire is severe, the MIL will flash on/off on the 1st trip!* **Possible Causes** ● Air leak in the intake manifold, or in the EGR or PCV system ● Base engine mechanical fault that affects only one cylinder ● Fuel component fault that affects only one cylinder (i.e., a dirty or sticking fuel injector) ● Ignition system problem (coil or plug) that affects one cylinder
P0300 **2T MISFIRE, MIL: Yes** 1995 Century, Cutlass Ciera, Cutlass Supreme, Grand Prix, Lumina, Monte Carlo Park Avenue Regal, Riviera, 88' & 98' 3.1L VIN M, 3.4L VIN X, 3.8L VIN 1, 3.8L VIN K, 3.8L VIN L Transmissions: All	**Multiple Cylinder Misfire Detected Conditions:** DTC P0101, P0117, P0118, P0122, P0123 and P0502 not set, ECT sensor from -13°F to 248°F, engine speed from 525-4775 rpm, TP angle steady, and the PCM detected a crankshaft speed variation characteristic of an engine misfire condition two or more cylinders. *Note: If the misfire is severe, the MIL will flash on/off on the 1st trip!* **Possible Causes** ● Base engine mechanical fault that affects one or more cylinders ● Fuel metering fault that affects more than one cylinder ● Fuel pressure too low or too high, fuel supply contaminated ● EVAP system problem or the EVAP canister is fuel saturated ● EGR valve is stuck open or the PCV system has a vacuum leak ● IC control circuit is shorted to ground (an intermittent fault) ● Ignition system fault (a coil) that affects more than one cylinder ● MAF sensor contamination (it can cause a very lean condition)
P0300 **2T MISFIRE, MIL: Yes** 1996, 1997, 1998, 1999, 2000, 2001, 2002, 2003, 2004, 2005 Aurora, Bonneville, Century, Cutlass, Intrigue, LeSabre, LeSabre, LSS, Lumina, Monte Carlo, 88', Park Avenue, Regal, 98' 3.1L VIN J, 3.1L VIN M, 3.4L VIN X, 3.5L VIN H, 3.8L VIN 1, 3.8L VIN K, 4.0L VIN C Transmissions: All	**Multiple Cylinder Misfire Detected Conditions:** DTC P0101-P0103, P0107, P0108, P0116-P0118, P0121-P0123, P0125, P0336, P0341, P0502, P0503, P1106, P1107, P1114, P1115, P1121, P1122, P1336 and P1374 not set, engine speed from 550-5850 rpm, ECT sensor from 21-248°F, TP angle steady, and the PCM detected a crankshaft speed variation characteristic of a misfire condition in two or more cylinders. *Note: If the misfire is severe, the MIL will flash on/off on the 1st trip!* **Possible Causes** ● Base engine mechanical fault that affects one or more cylinders ● Fuel metering fault that affects more than one cylinder ● Fuel pressure too low or too high, fuel supply contaminated ● EVAP system problem or the EVAP canister is fuel saturated ● EGR valve is stuck open or the PCV system has a vacuum leak ● IC control circuit is shorted to ground (an intermittent fault) ● Ignition system fault (a coil) that affects more than one cylinder ● MAF sensor contamination (it can cause a very lean condition) ● TSB 77-65-30 contains a repair procedure for this code ● TSB 03-06-04-030 contains a repair procedure for this code

OBD II Trouble Code List (P0xxx Codes)

DTC	Trouble Code Title, Conditions & Possible Causes
P0301 **2T MISFIRE, MIL: Yes** 1996, 1997, 1998, 1999, 2000, 2001, 2002, 2003, 2004, 2005 Aurora, Bonneville, Century, Cutlass, Intrigue, LeSabre, LeSabre, LSS, Lumina, Monte Carlo, 88', Park Avenue, Regal, 98' 3.1L VIN J, 3.1L VIN M, 3.4L VIN X, 3.5L VIN H, 3.8L VIN 1, 3.8L VIN K, 4.0L VIN C Transmissions: All	**Cylinder 1 Misfire Detected Conditions:** DTC P0101-P0103, P0107, P0108, P0116-P0118, P0121-P0123, P0125, P0336, P0341, P0502-0503, P1106, P1107, P1114, P1115, P1121, P1122, P1336 and P1374 not set; engine speed from 550-5850 rpm, ECT sensor at 21-248°F, TP angle steady, and the PCM detected a crankshaft speed variation in Cylinder 1 characteristic of a misfire. *Note: If the misfire is severe, the MIL will flash on/off on 1st trip!* **Possible Causes** ● Air leak in the intake manifold, or in the EGR or PCV system ● Base engine mechanical fault that affects only Cylinder 1 ● Fuel component fault on Cylinder 1 (e.g., restricted fuel injector) ● Ignition system problem (coil, plug) that affects only Cylinder 1
P0302 **2T MISFIRE, MIL: Yes** 1996, 1997, 1998, 1999, 2000, 2001, 2002, 2003, 2004, 2005 Aurora, Bonneville, Century, Cutlass, Intrigue, LeSabre, LeSabre, LSS, Lumina, Monte Carlo, 88', Park Avenue, Regal, 98' 3.1L VIN J, 3.1L VIN M, 3.4L VIN X, 3.5L VIN H, 3.8L VIN 1, 3.8L VIN K, 4.0L VIN C Transmissions: All	**Cylinder 2 Misfire Detected Conditions:** DTC P0101-P0103, P0107, P0108, P0116-P0118, P0121-P0123, P0125, P0336, P0341, P0502-0503, P1106, P1107, P1114, P1115, P1121, P1122, P1336 and P1374 not set; engine speed from 550-5850 rpm, ECT sensor at 21-248°F, TP angle steady, and the PCM detected a crankshaft speed variation in Cylinder 2 characteristic of a misfire. If the misfire is severe, the MIL will flash on/off on 1st trip! **Possible Causes** ● Air leak in the intake manifold, or in the EGR or PCV system ● Base engine mechanical fault that affects only Cylinder 2 ● Fuel component fault on Cylinder 2 (e.g., restricted fuel injector) ● Ignition system problem (coil, plug) that affects only Cylinder 2
P0303 **2T MISFIRE, MIL: Yes** 1996, 1997, 1998, 1999, 2000, 2001, 2002, 2003, 2004, 2005 Aurora, Bonneville, Century, Cutlass, Intrigue, LeSabre, LeSabre, LSS, Lumina, Monte Carlo, 88', Park Avenue, Regal, 98' 3.1L VIN J, 3.1L VIN M, 3.4L VIN X, 3.5L VIN H, 3.8L VIN 1, 3.8L VIN K, 4.0L VIN C Transmissions: All	**Cylinder 3 Misfire Detected Conditions:** DTC P0101-P0103, P0107, P0108, P0116-P0118, P0121-P0123, P0125, P0336, P0341, P0502-0503, P1106, P1107, P1114, P1115, P1121, P1122, P1336 and P1374 not set; engine speed from 550-5850 rpm, ECT sensor at 21-248°F, TP angle steady, and the PCM detected a crankshaft speed variation in Cylinder 3 characteristic of a misfire. If the misfire is severe, the MIL will flash on/off on 1st trip! **Possible Causes** ● Air leak in the intake manifold, or in the EGR or PCV system ● Base engine mechanical fault that affects only Cylinder 3 ● Fuel component fault on Cylinder 3 (e.g., restricted fuel injector) ● Ignition system problem (coil, plug) that affects only Cylinder 3
P0304 **2T MISFIRE, MIL: Yes** 1996, 1997, 1998, 1999, 2000, 2001, 2002, 2003, 2004, 2005 Aurora, Bonneville, Century, Cutlass, Intrigue, LeSabre, LeSabre, LSS, Lumina, Monte Carlo, 88', Park Avenue, Regal, 98' 3.1L VIN J, 3.1L VIN M, 3.4L VIN X, 3.5L VIN H, 3.8L VIN 1, 3.8L VIN K, 4.0L VIN C	**Cylinder 4 Misfire Detected Conditions:** DTC P0101-P0103, P0107, P0108, P0116-P0118, P0121-P0123, P0125, P0336, P0341, P0502-0503, P1106, P1107, P1114, P1115, P1121, P1122, P1336 and P1374 not set; engine speed from 550-5850 rpm, ECT sensor at 21-248°F, TP angle steady, and the PCM detected a crankshaft speed variation in Cylinder 4 characteristic of a misfire. If the misfire is severe, the MIL will flash on/off on 1st trip! **Possible Causes** ● Air leak in the intake manifold, or in the EGR or PCV system ● Base engine mechanical fault that affects only Cylinder 4 ● Fuel component fault on Cylinder 4 (e.g., restricted fuel injector) ● Ignition system problem (coil, plug) that affects only Cylinder 4
P0305 **2T MISFIRE, MIL: Yes** 1996, 1997, 1998, 1999, 2000, 2001, 2002, 2003, 2004, 2005 Aurora, Bonneville, Century, Cutlass, Intrigue, LeSabre, LeSabre, LSS, Lumina, Monte Carlo, 88', Park Avenue, Regal, 98' 3.1L VIN J, 3.1L VIN M, 3.4L VIN X, 3.5L VIN H, 3.8L VIN 1, 3.8L VIN K, 4.0L VIN C	**Cylinder 5 Misfire Detected Conditions:** DTC P0101-P0103, P0107, P0108, P0116-P0118, P0121-P0123, P0125, P0336, P0341, P0502-0503, P1106, P1107, P1114, P1115, P1121, P1122, P1336 and P1374 not set; engine speed from 550-5850 rpm, ECT sensor at 21-248°F, TP angle steady, and the PCM detected a crankshaft speed variation in Cylinder 5 characteristic of a misfire. If the misfire is severe, the MIL will flash on/off on 1st trip! **Possible Causes** ● Air leak in the intake manifold, or in the EGR or PCV system ● Base engine mechanical fault that affects only Cylinder 5 ● Fuel component fault on Cylinder 1 (e.g., restricted fuel injector) ● Ignition system problem (coil, plug) that affects only Cylinder 5

OBD II Trouble Code List (P0xxx Codes)

DTC	Trouble Code Title, Conditions & Possible Causes
P0306 **2T MISFIRE, MIL: Yes** 1996, 1997, 1998, 1999, 2000, 2001, 2002, 2003, 2004, 2005 Aurora, Bonneville, Century, Cutlass, Intrigue, LeSabre, LeSabre, LSS, Lumina, Monte Carlo, 88', Park Avenue, Regal, 98' 3.1L VIN J, 3.1L VIN M, 3.4L VIN X, 3.5L VIN H, 3.8L VIN 1, 3.8L VIN K, 4.0L VIN C Transmissions: All	**Cylinder 6 Misfire Detected Conditions:** DTC P0101-P0103, P0107, P0108, P0116-P0118, P0121-P0123, P0125, P0336, P0341, P0502-0503, P1106, P1107, P1114, P1115, P1121, P1122, P1336 and P1374 not set; engine speed from 550-5850 rpm, ECT sensor at 21-248ºF, TP angle steady, and the PCM detected a crankshaft speed variation in Cylinder 6 characteristic of a misfire. If the misfire is severe, the MIL will flash on/off on 1st trip! **Possible Causes** ● Air leak in the intake manifold, or in the EGR or PCV system ● Base engine mechanical fault that affects only Cylinder 6 ● Fuel component fault on Cylinder 1 (e.g., restricted fuel injector) ● Ignition system problem (coil, plug) that affects only Cylinder 6
P0307 **2T MISFIRE, MIL: Yes** 1996, 1997, 1998, 1999, 2001, 2002, 2003, 2004, 2005 Aurora 4.0L VIN C engine Transmissions: A/T	**Cylinder 7 Misfire Detected Conditions:** DTC P0101-P0103, P0107, P0108, P0116-P0118, P0121-P0123, P0125, P0336, P0341, P0502, P0503, P1106, P1107, P1114, P1115, P1121, P1122, P1336 and P1374 not set, engine started, engine speed from 550-5050 rpm, ECT sensor from 21-248ºF, TP angle steady, and the PCM detected a crankshaft speed variation in Cylinder 7 characteristic of a misfire condition. *Note: If the misfire is severe, the MIL will flash on/off on the 1st trip!* **Possible Causes** ● Air leak in the intake manifold, or in the EGR or PCV system ● Base engine mechanical fault that affects only Cylinder 7 ● Fuel component fault that affects only Cylinder 7 (i.e., a dirty or sticking fuel injector) ● Ignition system problem (coil, plug) that affects only Cylinder 7
P0308 **2T MISFIRE, MIL: Yes** 1996, 1997, 1998, 1999, 2001, 2002, 2003, 2004, 2005 Aurora 4.0L VIN C engine Transmissions: A/T	**Cylinder 8 Misfire Detected Conditions:** DTC P0101-P0103, P0107, P0108, P0116-P0118, P0121-P0123, P0125, P0336, P0341, P0502, P0503, P1106, P1107, P1114, P1115, P1121, P1122, P1336 and P1374 not set, engine started, engine speed from 550-5850 rpm, ECT sensor from 21-248ºF, TP angle steady, and the PCM detected a crankshaft speed variation in Cylinder 8 characteristic of a misfire condition. *Note: If the misfire is severe, the MIL will flash on/off on the 1st trip!* **Possible Causes** ● Air leak in the intake manifold, or in the EGR or PCV system ● Base engine mechanical fault that affects only Cylinder 8 ● Fuel component fault that affects only Cylinder 8 (i.e., a dirty or sticking fuel injector) ● Ignition system problem (coil, plug) that affects only Cylinder 8
P0300 **2T MISFIRE, MIL: Yes** 1996, 1997, 1998, 1999, 2000, 2001, 2002, 2003, 2004, 2005 Cab & Chassis, C/K Series Truck, G Series Van, Van Cargo, L/M Series Vans (Astro), S/T Blazer & S/T Pickup 4.3L VIN W, 4.3L VIN X, 5.0L VIN M, 5.7L VIN K, 5.7L VIN R, 7.4L VIN J Transmissions: All	**Multiple Cylinder Misfire Detected Conditions:** DTC P0101-P0103, P0116-P0118, P0125, P0128, P0335, P0336, P0341, P0343, P0502, P0503, P1114, P1115, P1120, P1220, P1221 or P1336 not set, engine speed from 450-5000 rpm, system voltage over 10.0v, ECT sensor from 19-266ºF, fuel level over 10%, TP angle steady (± 1%), ABS, AIR, Traction Control and DFCO all "off", transmission not shifting, A/C status steady, ABS signals less than the rough road thresholds, and the PCM detected a crankshaft speed variation characteristic of a misfire in more than one cylinder. *Note: If the misfire is severe, the MIL will flash on/off on the 1st trip!* **Possible Causes** ● Base engine mechanical fault that affects one or more cylinders ● Fuel delivery component fault that affects more than 1 cylinder ● EVAP system problem or the EVAP canister is fuel saturated ● EGR valve is stuck open or PCV system has a vacuum leak ● Ignition system fault (a coil) that affects more than one cylinder ● TSB 03-06-04-030 contains a repair procedure for this code ● TSB 03-06-04-041 contains a repair procedure for this code
P0301 **2T MISFIRE, MIL: Yes** 1996, 1997, 1998, 1999, 2000, 2001, 2002, 2003, 2004, 2005 Cab & Chassis, C/K Series Truck, G Series Van, Van Cargo, L/M Series Vans (Astro), S/T Blazer & S/T Pickup 4.3L VIN W, 4.3L VIN X, 5.0L VIN M, 5.7L VIN K, 5.7L VIN R, 7.4L VIN J Transmissions: All	**Cylinder 1 Misfire Detected Conditions:** DTC P0101-P0103, P0116-P0118, P0125, P0128, P0335, P0336, P0341, P0343, P0502, P0503, P1114, P1115, P1120, P1220, P1221 or P1336 not set, engine speed from 450-5001 rpm, system voltage over 10.0v, ECT sensor from 19-266ºF, fuel level over 10%, TP angle steady (±1%), ABS, AIR, Traction Control and DFCO all "off", transmission not shifting, A/C status steady, ABS signals less than the rough road thresholds, and the PCM detected a crankshaft speed variation in one cylinder characteristic of a misfire condition. *Note: If the misfire is severe, the MIL will flash on/off on the 1st trip!* **Possible Causes** ● Air leak in the intake manifold, or in the EGR or PCV system ● Base engine mechanical fault that affects only one cylinder ● Fuel delivery component fault that affects only one cylinder (i.e., a contaminated, dirty or sticking fuel injector) ● Ignition system problem (coil or plug) that affects one cylinder ● TSB 00-06-04-024 contains a repair procedure for this code

OBD II Trouble Code List (P0xxx Codes)

DTC	Trouble Code Title, Conditions & Possible Causes
P0302 **2T MISFIRE, MIL: Yes** 1996, 1997, 1998, 1999, 2000, 2001, 2002, 2003, 2004, 2005 Cab & Chassis, C/K Series Truck, G Series Van, Van Cargo, L/M Series Vans (Astro), S/T Blazer & S/T Pickup 4.3L VIN W, 4.3L VIN X, 5.0L VIN M, 5.7L VIN K, 5.7L VIN R, 7.4L VIN J Transmissions: All	**Cylinder 2 Misfire Detected Conditions:** DTC P0101-P0103, P0116-P0118, P0125, P0128, P0335, P0336, P0341, P0343, P0502, P0503, P1114, P1115, P1120, P1220, P1221 or P1336 not set, engine speed from 450-5001 rpm, system voltage over 10.0v, ECT sensor from 19-266°F, fuel level over 10%, TP angle steady (±1%), ABS, AIR, Traction Control and DFCO all "off", transmission not shifting, A/C status steady, ABS signals less than the rough road thresholds, and the PCM detected a crankshaft speed variation in one cylinder characteristic of a misfire condition. *Note: If the misfire is severe, the MIL will flash on/off on the 1st trip!* **Possible Causes** ● Air leak in the intake manifold, or in the EGR or PCV system ● Base engine mechanical fault that affects only one cylinder ● Fuel delivery component fault that affects only one cylinder (i.e., a contaminated, dirty or sticking fuel injector) ● Ignition system problem (coil or plug) that affects one cylinder ● TSB 00-06-04-024 contains a repair procedure for this code
P0303 **2T MISFIRE, MIL: Yes** 1996, 1997, 1998, 1999, 2000, 2001, 2002, 2003, 2004, 2005 Cab & Chassis, C/K Series Truck, G Series Van, Van Cargo, L/M Series Vans (Astro), S/T Blazer & S/T Pickup 4.3L VIN W, 4.3L VIN X, 5.0L VIN M, 5.7L VIN K, 5.7L VIN R, 7.4L VIN J Transmissions: All	**Cylinder 3 Misfire Detected Conditions:** DTC P0101-P0103, P0116-P0118, P0125, P0128, P0335, P0336, P0341, P0343, P0502, P0503, P1114, P1115, P1120, P1220, P1221 or P1336 not set, engine speed from 450-5001 rpm, system voltage over 10.0v, ECT sensor from 19-266°F, fuel level over 10%, TP angle steady (±1%), ABS, AIR, Traction Control and DFCO all "off", transmission not shifting, A/C status steady, ABS signals less than the rough road thresholds, and the PCM detected a crankshaft speed variation in one cylinder characteristic of a misfire condition. *Note: If the misfire is severe, the MIL will flash on/off on the 1st trip!* **Possible Causes** ● Air leak in the intake manifold, or in the EGR or PCV system ● Base engine mechanical fault that affects only one cylinder ● Fuel delivery component fault that affects only one cylinder (i.e., a contaminated, dirty or sticking fuel injector) ● Ignition system problem (coil or plug) that affects one cylinder ● TSB 00-06-04-024 contains a repair procedure for this code
P0304 **2T MISFIRE, MIL: Yes** 1996, 1997, 1998, 1999, 2000, 2001, 2002, 2003, 2004, 2005 Cab & Chassis, C/K Series Truck, G Series Van, Van Cargo, L/M Series Vans (Astro), S/T Blazer & S/T Pickup 4.3L VIN W, 4.3L VIN X, 5.0L VIN M, 5.7L VIN K, 5.7L VIN R, 7.4L VIN J Transmissions: All	**Cylinder 4 Misfire Detected Conditions:** DTC P0101-P0103, P0116-P0118, P0125, P0128, P0335, P0336, P0341, P0343, P0502, P0503, P1114, P1115, P1120, P1220, P1221 or P1336 not set, engine speed from 450-5001 rpm, system voltage over 10.0v, ECT sensor from 19-266°F, fuel level over 10%, TP angle steady (±1%), ABS, AIR, Traction Control and DFCO all "off", transmission not shifting, A/C status steady, ABS signals less than the rough road thresholds, and the PCM detected a crankshaft speed variation in one cylinder characteristic of a misfire condition. *Note: If the misfire is severe, the MIL will flash on/off on the 1st trip!* **Possible Causes** ● Air leak in the intake manifold, or in the EGR or PCV system ● Base engine mechanical fault that affects only one cylinder ● Fuel delivery component fault that affects only one cylinder (i.e., a contaminated, dirty or sticking fuel injector) ● Ignition system problem (coil or plug) that affects one cylinder ● TSB 00-06-04-024 contains a repair procedure for this code
P0305 **2T MISFIRE, MIL: Yes** 1996, 1997, 1998, 1999, 2000, 2001, 2002, 2003, 2004, 2005 Cab & Chassis, C/K Series Truck, G Series Van, Van Cargo, L/M Series Vans (Astro), S/T Blazer & S/T Pickup 4.3L VIN W, 4.3L VIN X, 5.0L VIN M, 5.7L VIN K, 5.7L VIN R, 7.4L VIN J Transmissions: All	**Cylinder 5 Misfire Detected Conditions:** DTC P0101-P0103, P0116-P0118, P0125, P0128, P0335, P0336, P0341, P0343, P0502, P0503, P1114, P1115, P1120, P1220, P1221 or P1336 not set, engine speed from 450-5001 rpm, system voltage over 10.0v, ECT sensor from 19-266°F, fuel level over 10%, TP angle steady (±1%), ABS, AIR, Traction Control and DFCO all "off", transmission not shifting, A/C status steady, ABS signals less than the rough road thresholds, and the PCM detected a crankshaft speed variation in one cylinder characteristic of a misfire condition. *Note: If the misfire is severe, the MIL will flash on/off on the 1st trip!* **Possible Causes** ● Air leak in the intake manifold, or in the EGR or PCV system ● Base engine mechanical fault that affects only one cylinder ● Fuel delivery component fault that affects only one cylinder (i.e., a contaminated, dirty or sticking fuel injector) ● Ignition system problem (coil or plug) that affects one cylinder ● TSB 00-06-04-024 contains a repair procedure for this code

OBD II Trouble Code List (P0xxx Codes)

DTC	Trouble Code Title, Conditions & Possible Causes
P0306 **2T MISFIRE, MIL: Yes** 1996, 1997, 1998, 1999, 2000, 2001, 2002, 2003, 2004, 2005 Cab & Chassis, C/K Series Truck, G Series Van, Van Cargo, L/M Series Vans (Astro), S/T Blazer & S/T Pickup 4.3L VIN W, 4.3L VIN X, 5.0L VIN M, 5.7L VIN K, 5.7L VIN R, 7.4L VIN J Transmissions: All	**Cylinder 6 Misfire Detected Conditions:** DTC P0101-P0103, P0116-P0118, P0125, P0128, P0335, P0336, P0341, P0343, P0502, P0503, P1114, P1115, P1120, P1220, P1221 or P1336 not set, engine speed from 450-5001 rpm, system voltage over 10.0v, ECT sensor from 19-266°F, fuel level over 10%, TP angle steady (±1%), ABS, AIR, Traction Control and DFCO all "off", transmission not shifting, A/C status steady, ABS signals less than the rough road thresholds, and the PCM detected a crankshaft speed variation in one cylinder characteristic of a misfire condition. *Note: If the misfire is severe, the MIL will flash on/off on the 1st trip!* **Possible Causes** ● Air leak in the intake manifold, or in the EGR or PCV system ● Base engine mechanical fault that affects only one cylinder ● Fuel delivery component fault that affects only one cylinder (i.e., a contaminated, dirty or sticking fuel injector) ● Ignition system problem (coil or plug) that affects one cylinder ● TSB 00-06-04-024 contains a repair procedure for this code
P0307 **2T MISFIRE, MIL: Yes** 1996, 1997, 1998, 1999, 2000, 2001, 2002, 2003, 2004, 2005 Cab & Chassis, C/K Series Truck, G Series Van & Van Cargo 5.0L VIN M, 5.7L VIN K, 5.7L VIN R, 7.4L VIN J Transmissions: All	**Cylinder 7 Misfire Detected Conditions:** DTC P0101-P0103, P0116-P0118, P0125, P0128, P0335, P0336, P0341, P0343, P0502, P0503, P1114, P1115, P1120, P1220, P1221 or P1336 not set, engine speed from 450-5001 rpm, system voltage over 10.0v, ECT sensor from 19-266°F, fuel level over 10%, TP angle steady (±1%), ABS, AIR, Traction Control and DFCO all "off", transmission not shifting, A/C status steady, ABS signals less than the rough road thresholds, and the PCM detected a crankshaft speed variation in one cylinder characteristic of a misfire condition. *Note: If the misfire is severe, the MIL will flash on/off on the 1st trip!* **Possible Causes** ● Air leak in the intake manifold, or in the EGR or PCV system ● Base engine mechanical fault that affects only one cylinder ● Fuel delivery component fault that affects only one cylinder (i.e., a contaminated, dirty or sticking fuel injector) ● Ignition system problem (coil or plug) that affects one cylinder ● TSB 00-06-04-024 contains a repair procedure for this code
308 **2T MISFIRE, MIL: Yes** 1996, 1997, 1998, 1999, 2000, 2001, 2002, 2003, 2004, 2005 Cab & Chassis, C/K Series Truck, G Series Van & Van Cargo 5.0L VIN M, 5.7L VIN K, 5.7L VIN R, 7.4L VIN J Transmissions: All	**Cylinder 8 Misfire Detected Conditions:** DTC P0101-P0103, P0116-P0118, P0125, P0128, P0335, P0336, P0341, P0343, P0502, P0503, P1114, P1115, P1120, P1220, P1221 or P1336 not set, engine speed from 450-5001 rpm, system voltage over 10.0v, ECT sensor from 19-266°F, fuel level over 10%, TP angle steady (±1%), ABS, AIR, Traction Control and DFCO all "off", transmission not shifting, A/C status steady, ABS signals less than the rough road thresholds, and the PCM detected a crankshaft speed variation in one cylinder characteristic of a misfire condition. *Note: If the misfire is severe, the MIL will flash on/off on the 1st trip!* **Possible Causes** ● Air leak in the intake manifold, or in the EGR or PCV system ● Base engine mechanical fault that affects only one cylinder ● Fuel delivery component fault that affects only one cylinder (i.e., a contaminated, dirty or sticking fuel injector) ● Ignition system problem (coil or plug) that affects one cylinder ● TSB 00-06-04-024 contains a repair procedure for this code
P0300 **2T MISFIRE, MIL: Yes** 1999, 2001, 2002, 2003, 2004, 2005 C/K Series Truck, G Series Van, Envoy, Escalade, TrailBlazer 4.8L VIN V, 5.3L VIN P, 5.3L VIN T, 5.3L VIN Z, 6.0L VIN N, 6.0L VIN U CNG engine, 8.1L VIN G engines Transmissions: All	**Multiple Cylinder Misfire Detected Conditions:** DTC P0101-P0103, P0106-P0108, P0116-P0118, P0121-P0123, P0125, P0128, P0220, P0315, P0335, P0336, P0341-P0343, P0502, P0503, P1114, P1115, P1120, P1258 are not set, engine speed from 450-5000 rpm, system voltage from 10-18v, ECT sensor from 19-266°F, Fuel Level over 10%, TP angle steady (± 1%), ABS, DFCO and Traction Control off, transmission shift and A/C steady, no ABS Rough Road signals, and the PCM detected a crankshaft speed variation characteristic of a misfire in two or more cylinders. *Note: If the misfire is severe, the MIL will flash on/off on the 1st trip!* **Possible Causes** ● Base engine mechanical fault that affects one or more cylinders ● EGR valve is stuck open or PCV system has a vacuum leak ● EVAP system problem or the EVAP canister is fuel saturated ● Fuel metering fault that affects more than one cylinder ● Fuel pressure too low or too high, fuel supply contaminated ● IC control circuit is shorted to ground (an intermittent fault) ● Ignition system fault (a coil) that affects more than one cylinder ● MAF sensor contamination (it can cause a very lean condition) ● TSB 03-06-04-030 contains a repair procedure for this code

OBD II Trouble Code List (P0xxx Codes)

DTC	Trouble Code Title, Conditions & Possible Causes
P0300 **2T MISFIRE, MIL: Yes** 2003, 2004, 2005 G Series Van, Van Cargo 6.0L VIN U CNG engine Transmissions: All	**Multiple Cylinder Misfire Detected Conditions:** DTC P0101, P0102, P0103, P0106, P0107, P0108, P0117, P0112, P0113, P0116, P0117, P0118, P0121, P0122, P0123, P0125, P0128, P0315, P0335, P0336, P0341, P0342, P0343, P0502, P0503, P1114, P1115, P1121, P1122 and P1258 not set, engine speed from 450-5200 rpm; if the ECT sensor is less than 20ºF at startup, misfire detection is delayed until the ECT is more than 70ºF; and if the ECT sensor is more than 20ºF at startup, misfire detection delayed for 5 seconds; system voltage from 11-18v, ECT sensor less than 266ºF, Fuel Level over 10%, TP angle steady (± 1%), A/C clutch steady, and the PCM detected a crankshaft speed variation characteristic of a misfire in two or more cylinders. *Note: If the misfire condition is severe, the MIL will flash on/off on the 1st trip!* **Possible Causes** ● Base engine mechanical fault that affects one or more cylinders ● EGR valve is stuck open or PCV system has a vacuum leak ● EVAP system problem or the EVAP canister is fuel saturated ● Fuel metering fault that affects more than one cylinder ● Fuel pressure too low or too high, fuel supply contaminated ● Ignition system fault (a coil) that affects more than one cylinder ● MAF sensor contamination (it can cause a very lean condition)
P0301 **2T MISFIRE, MIL: Yes** 1999, 2001, 2002, 2003, 2004, 2005 C/K Series Truck, G Series Van, Envoy, Escalade, TrailBlazer 4.8L VIN V, 5.3L VIN P, 5.3L VIN T, 5.3L VIN Z, 6.0L VIN N, 6.0L VIN U CNG engine, 8.1L VIN G engines Transmissions: All	**Cylinder 1 Misfire Detected Conditions:** DTC P0101-P0103, P0106-P0108, P0116-P0118, P0121-P0123, P0125, P0128, P0220, P0315, P0335, P0336, P0341-P0343, P0502, P0503, P1114, P1115, P1120, P1258 are not set, engine speed from 450-5000 rpm, system voltage from 10-18v, ECT sensor from 19-266ºF, Fuel Level over 10%, TP angle steady (± 1%), ABS, DFCO and Traction Control "off", A/C and transmission steady, ABS Rough Road signal not present, and the PCM detected a crankshaft speed variation characteristic of a misfire condition in Cylinder 1. *Note: If the misfire is severe, the MIL will flash on/off on the 1st trip!* **Possible Causes** ● Base engine mechanical problem that affects only Cylinder 1 ● Fuel delivery component fault (injector) that affects Cylinder 1 ● Ignition system problem (coil, plug) that affects only Cylinder 1
P0302 **2T MISFIRE, MIL: Yes** 1999, 2001, 2002, 2003, 2004, 2005 C/K Series Truck, G Series Van, Envoy, Escalade, TrailBlazer 4.8L VIN V, 5.3L VIN P, 5.3L VIN T, 5.3L VIN Z, 6.0L VIN N, 6.0L VIN U CNG engine, 8.1L VIN G engines Transmissions: All	**Cylinder 2 Misfire Detected Conditions:** DTC P0101-P0103, P0106-P0108, P0116-P0118, P0121-P0123, P0125, P0128, P0220, P0315, P0335, P0336, P0341-P0343, P0502, P0503, P1114, P1115, P1120, P1258 are not set, engine speed from 450-5000 rpm, system voltage from 10-18v, ECT sensor from 19-266ºF, Fuel Level over 10%, TP angle steady (± 1%), ABS, DFCO and Traction Control "off", A/C and transmission steady, ABS Rough Road signal not present, and the PCM detected a crankshaft speed variation characteristic of a misfire condition in Cylinder 2. *Note: If the misfire is severe, the MIL will flash on/off on the 1st trip!* **Possible Causes** ● Base engine mechanical problem that affects only Cylinder 2 ● Fuel delivery component fault (injector) that affects Cylinder 2 ● Ignition system problem (coil, plug) that affects only Cylinder 2
P0303 **2T MISFIRE, MIL: Yes** 1999, 2001, 2002, 2003, 2004, 2005 C/K Series Truck, G Series Van, Envoy, Escalade, TrailBlazer 4.8L VIN V, 5.3L VIN P, 5.3L VIN T, 5.3L VIN Z, 6.0L VIN N, 6.0L VIN U CNG engine, 8.1L VIN G engines Transmissions: All	**Cylinder 3 Misfire Detected Conditions:** DTC P0101-P0103, P0106-P0108, P0116-P0118, P0121-P0123, P0125, P0128, P0220, P0315, P0335, P0336, P0341-P0343, P0502, P0503, P1114, P1115, P1120, P1258 are not set, engine speed from 450-5000 rpm, system voltage from 10-18v, ECT sensor from 19-266ºF, Fuel Level over 10%, TP angle steady (± 1%), ABS, DFCO and Traction Control "off", A/C and transmission steady, ABS Rough Road signal not present, and the PCM detected a crankshaft speed variation characteristic of a misfire condition in Cylinder 3. *Note: If the misfire is severe, the MIL will flash on/off on the 1st trip!* **Possible Causes** ● Base engine mechanical problem that affects only Cylinder 3 ● Fuel delivery component fault (injector) that affects Cylinder 3 ● Ignition system problem (coil, plug) that affects only Cylinder 3
P0304 **2T MISFIRE, MIL: Yes** 1999, 2001, 2002, 2003, 2004, 2005 C/K Series Truck, G Series Van, Envoy, Escalade, TrailBlazer 4.8L VIN V, 5.3L VIN P, 5.3L VIN T, 5.3L VIN Z, 6.0L VIN N, 6.0L VIN U CNG engine, 8.1L VIN G engines Transmissions: All	**Cylinder 4 Misfire Detected Conditions:** DTC P0101-P0103, P0106-P0108, P0116-P0118, P0121-P0123, P0125, P0128, P0220, P0315, P0335, P0336, P0341-P0343, P0502, P0503, P1114, P1115, P1120, P1258 are not set, engine speed from 450-5000 rpm, system voltage from 10-18v, ECT sensor from 19-266ºF, Fuel Level over 10%, TP angle steady (± 1%), ABS, DFCO and Traction Control "off", A/C and transmission steady, ABS Rough Road signal not present, and the PCM detected a crankshaft speed variation characteristic of a misfire condition in Cylinder 4. *Note: If the misfire is severe, the MIL will flash on/off on the 1st trip!* **Possible Causes** ● Base engine mechanical problem that affects only Cylinder 4 ● Fuel delivery component fault (injector) that affects Cylinder 4 ● Ignition system problem (coil, plug) that affects only Cylinder 4

OBD II Trouble Code List (P0xxx Codes)

DTC	Trouble Code Title, Conditions & Possible Causes
P0305 **2T MISFIRE, MIL: Yes** 1999, 2001, 2002, 2003, 2004, 2005 C/K Series Truck, G Series Van, Envoy, Escalade, TrailBlazer 4.8L VIN V, 5.3L VIN P, 5.3L VIN T, 5.3L VIN Z, 6.0L VIN N, 6.0L VIN U CNG engine, 8.1L VIN G engines Transmissions: All	**Cylinder 5 Misfire Detected Conditions:** DTC P0101-P0103, P0106-P0108, P0116-P0118, P0121-P0123, P0125, P0128, P0220, P0315, P0335, P0336, P0341-P0343, P0502, P0503, P1114, P1115, P1120, P1258 are not set, engine speed from 450-5000 rpm, system voltage from 10-18v, ECT sensor from 19-266ºF, Fuel Level over 10%, TP angle steady (± 1%), ABS, DFCO and Traction Control "off", A/C and transmission steady, ABS Rough Road signal not present, and the PCM detected a crankshaft speed variation characteristic of a misfire condition in Cylinder 5. *Note: If the misfire is severe, the MIL will flash on/off on the 1st trip!* **Possible Causes** <ul><li>Base engine mechanical problem that affects only Cylinder 5</li><li>Fuel delivery component fault (injector) that affects Cylinder 5</li><li>Ignition system problem (coil, plug) that affects only Cylinder 5</li></ul>
P0306 **2T MISFIRE, MIL: Yes** 1999, 2001, 2002, 2003, 2004, 2005 C/K Series Truck, G Series Van, Envoy, Escalade, TrailBlazer 4.8L VIN V, 5.3L VIN P, 5.3L VIN T, 5.3L VIN Z, 6.0L VIN N, 6.0L VIN U CNG engine, 8.1L VIN G engines Transmissions: All	**Cylinder 6 Misfire Detected Conditions:** DTC P0101-P0103, P0106-P0108, P0116-P0118, P0121-P0123, P0125, P0128, P0220, P0315, P0335, P0336, P0341-P0343, P0502, P0503, P1114, P1115, P1120, P1258 are not set, engine speed from 450-5000 rpm, system voltage from 10-18v, ECT sensor from 19-266ºF, Fuel Level over 10%, TP angle steady (± 1%), ABS, DFCO and Traction Control "off", A/C and transmission steady, ABS Rough Road signal not present, and the PCM detected a crankshaft speed variation characteristic of a misfire condition in Cylinder 6. *Note: If the misfire is severe, the MIL will flash on/off on the 1st trip!* **Possible Causes** <ul><li>Base engine mechanical problem that affects only Cylinder 6</li><li>Fuel component fault that affects only Cylinder 6 (i.e., a contaminated, dirty or sticking fuel injector)</li><li>Ignition system problem (coil, plug) that affects only Cylinder 6</li></ul>
P0307 **2T MISFIRE, MIL: Yes** 1999, 2001, 2002, 2003, 2004, 2005 C/K Series Truck, G Series Van, Envoy, Escalade, TrailBlazer 4.8L VIN V, 5.3L VIN P, 5.3L VIN T, 5.3L VIN Z, 6.0L VIN N, 6.0L VIN U CNG engine, 8.1L VIN G engines Transmissions: All	**Cylinder 7 Misfire Detected Conditions:** DTC P0101-P0103, P0106-P0108, P0116-P0118, P0121-P0123, P0125, P0128, P0220, P0315, P0335, P0336, P0341-P0343, P0502, P0503, P1114, P1115, P1120, P1258 are not set, engine speed from 450-5000 rpm, system voltage from 10-18v, ECT sensor from 19-266ºF, Fuel Level over 10%, TP angle steady (± 1%), ABS, DFCO and Traction Control "off", A/C and transmission steady, ABS Rough Road signal not present, and the PCM detected a crankshaft speed variation characteristic of a misfire condition in Cylinder 7. *Note: If the misfire is severe, the MIL will flash on/off on the 1st trip!* **Possible Causes** <ul><li>Base engine mechanical problem that affects only Cylinder 7</li><li>Fuel component fault that affects only Cylinder 7 (i.e., a dirty or sticking fuel injector)</li><li>Ignition system problem (coil, plug) that affects only Cylinder 7</li></ul>
P0308 **2T MISFIRE, MIL: Yes** 1999, 2001, 2002, 2003, 2004, 2005 C/K Series Truck, G Series Van, Envoy, Escalade, TrailBlazer 4.8L VIN V, 5.3L VIN P, 5.3L VIN T, 5.3L VIN Z, 6.0L VIN N, 6.0L VIN U CNG engine, 8.1L VIN G engines Transmissions: All	**Cylinder 8 Misfire Detected Conditions:** DTC P0101-P0103, P0106-P0108, P0116-P0118, P0121-P0123, P0125, P0128, P0220, P0315, P0335, P0336, P0341-P0343, P0502, P0503, P1114, P1115, P1120, P1258 are not set, engine speed from 450-5000 rpm, system voltage from 10-18v, ECT sensor from 19-266ºF, Fuel Level over 10%, TP angle steady (± 1%), ABS, DFCO and Traction Control "off", A/C and transmission steady, ABS Rough Road signal not present, and the PCM detected a crankshaft speed variation characteristic of a misfire condition in Cylinder 8. *Note: If the misfire is severe, the MIL will flash on/off on the 1st trip!* **Possible Causes** <ul><li>Base engine mechanical problem that affects only Cylinder 8</li><li>Fuel component fault that affects only Cylinder 8 (i.e., a dirty or sticking fuel injector)</li><li>Ignition system problem (coil, plug) that affects only Cylinder 8</li></ul>
P0300 **2T MISFIRE, MIL: Yes** 1996, 1997, 1998, 1999, 2000, 2001, 2002, 2003, 2004, 2005 C/K Series Truck, G Series Van, Van Cargo 6.5L VIN F, 6.5L VIN S, 6.6L VIN 1 Diesel engine Transmissions: All	**Multiple Cylinder Misfire Detected Conditions:** Engine started; engine at idle over 90 seconds, ECT sensor more than 132ºF, and the PCM detected the fuel adjustment was more than a calibrated amount due to a misfire in one or more cylinders. *Note: The Misfire Monitor identifies a cylinder that needs more fuel.* **Possible Causes** <ul><li>Base engine mechanical fault that affects one or more cylinders</li><li>Fuel injection (high pressure) nozzles or fuel lines leaking</li><li>Fuel injection nozzles stuck closed, dirty or restricted</li></ul>
P0301 **2T MISFIRE, MIL: Yes** 1996, 1997, 1998, 1999, 2000, 2001, 2002, 2003, 2004, 2005 C/K Series Truck, G Series Van, Van Cargo 6.5L VIN F, 6.5L VIN S, 6.6L VIN 1 Diesel engine Transmissions: All	**Cylinder 1 Misfire Detected Conditions:** Engine started; engine at idle over 90 seconds, ECT sensor more than 132ºF, and the PCM detected the fuel adjustment was more than a calibrated amount due to a misfire in Cylinder Number 1. *Note: The Misfire Monitor identifies a cylinder that needs more fuel.* **Possible Causes** <ul><li>Base engine mechanical fault that affects only Cylinder 1</li><li>Fuel injection (high pressure) nozzle leaking only on Cylinder 1</li><li>Fuel injection nozzle stuck closed or dirty affecting Cylinder 1</li></ul>

OBD II Trouble Code List (P0xxx Codes)

DTC	Trouble Code Title, Conditions & Possible Causes
P0302 **2T MISFIRE, MIL: Yes** 1996, 1997, 1998, 1999, 2000, 2001, 2002, 2003, 2004, 2005 C/K Series Truck, G Series Van, Van Cargo 6.5L VIN F, 6.5L VIN S, 6.6L VIN 1 Diesel engine Transmissions: All	**Cylinder 2 Misfire Detected Conditions:** Engine started; engine at idle over 90 seconds, ECT sensor more than 132ºF, and the PCM detected the fuel adjustment was more than a calibrated amount due to a misfire in Cylinder Number 2. *Note: The Misfire Monitor identifies a cylinder that needs more fuel.* **Possible Causes** ● Base engine mechanical fault that affects only Cylinder 2 ● Fuel injection (high pressure) nozzle leaking only on Cylinder 2 ● Fuel injection nozzle stuck closed or dirty affecting Cylinder 2
P0303 **2T MISFIRE, MIL: Yes** 1996, 1997, 1998, 1999, 2000, 2001, 2002, 2003, 2004, 2005 C/K Series Truck, G Series Van, Van Cargo 6.5L VIN F, 6.5L VIN S, 6.6L VIN 1 Diesel engine Transmissions: All	**Cylinder 3 Misfire Detected Conditions:** Engine started; engine at idle over 90 seconds, ECT sensor more than 132ºF, and the PCM detected the fuel adjustment was more than a calibrated amount due to a misfire in Cylinder Number 3. *Note: The Misfire Monitor identifies a cylinder that needs more fuel.* **Possible Causes** ● Base engine mechanical fault that affects only Cylinder 3 ● Fuel injection (high pressure) nozzle leaking only on Cylinder 3 ● Fuel injection nozzle stuck closed or dirty affecting Cylinder 3
P0304 **2T MISFIRE, MIL: Yes** 1996, 1997, 1998, 1999, 2000, 2001, 2002, 2003, 2004, 2005 C/K Series Truck, G Series Van, Van Cargo 6.5L VIN F, 6.5L VIN S, 6.6L VIN 1 Diesel engine Transmissions: All	**Cylinder 4 Misfire Detected Conditions:** Engine started; engine at idle over 90 seconds, ECT sensor more than 132ºF, and the PCM detected the fuel adjustment was more than a calibrated amount due to a misfire in Cylinder Number 4. *Note: The Misfire Monitor identifies a cylinder that needs more fuel.* **Possible Causes** ● Base engine mechanical fault that affects only Cylinder 4 ● Fuel injection (high pressure) nozzle leaking only on Cylinder 4 ● Fuel injection nozzle stuck closed or dirty affecting Cylinder 4
P0305 **2T MISFIRE, MIL: Yes** 1996, 1997, 1998, 1999, 2000, 2001, 2002, 2003, 2004, 2005 C/K Series Truck, G Series Van, Van Cargo 6.5L VIN F, 6.5L VIN S, 6.6L VIN 1 Diesel engine Transmissions: All	**Cylinder 5 Misfire Detected Conditions:** Engine started; engine at idle over 90 seconds, ECT sensor more than 132ºF, and the PCM detected the fuel adjustment was more than a calibrated amount due to a misfire in Cylinder Number 5. *Note: The Misfire Monitor identifies a cylinder that needs more fuel.* **Possible Causes** ● Base engine mechanical fault that affects only Cylinder 5 ● Fuel injection (high pressure) nozzle leaking only on Cylinder 5 ● Fuel injection nozzle stuck closed or dirty affecting Cylinder 5
P0306 **2T MISFIRE, MIL: Yes** 1996, 1997, 1998, 1999, 2000, 2001, 2002, 2003, 2004, 2005 C/K Series Truck, G Series Van, Van Cargo 6.5L VIN F, 6.5L VIN S, 6.6L VIN 1 Diesel engine Transmissions: All	**Cylinder 6 Misfire Detected Conditions:** Engine started; engine at idle over 90 seconds, ECT sensor more than 132ºF, and the PCM detected the fuel adjustment was more than a calibrated amount due to a misfire in Cylinder Number 6. *Note: The Misfire Monitor identifies a cylinder that needs more fuel.* **Possible Causes** ● Base engine mechanical fault that affects only Cylinder 6 ● Fuel injection (high pressure) nozzle leaking only on Cylinder 6 ● Fuel injection nozzle stuck closed or dirty affecting Cylinder 6
P0307 **2T MISFIRE, MIL: Yes** 1996, 1997, 1998, 1999, 2000, 2001, 2002, 2003, 2004, 2005 C/K Series Truck, G Series Van, Van Cargo 6.5L VIN F, 6.5L VIN S, 6.6L VIN 1 Diesel engine Transmissions: All	**Cylinder 7 Misfire Detected Conditions:** Engine started; engine at idle over 90 seconds, ECT sensor more than 132ºF, and the PCM detected the fuel adjustment was more than a calibrated amount due to a misfire in Cylinder Number 7. Note: The Misfire Monitor identifies a cylinder that needs additional fuel. **Possible Causes** ● Base engine mechanical fault that affects only Cylinder 7 ● Fuel injection (high pressure) nozzle leaking only on Cylinder 7 ● Fuel injection nozzle stuck closed or dirty affecting Cylinder 7
P0308 **2T MISFIRE, MIL: Yes** 1996, 1997, 1998, 1999, 2000, 2001, 2002, 2003, 2004, 2005 C/K Series Truck, G Series Van, Van Cargo 6.5L VIN F, 6.5L VIN S, 6.6L VIN 1 Diesel engine Transmissions: All	**Cylinder 8 Misfire Detected Conditions:** Engine started; engine at idle over 90 seconds, ECT sensor more than 132ºF, and the PCM detected the fuel adjustment was more than a calibrated amount due to a misfire in Cylinder Number 8. Note: The Misfire Monitor identifies a cylinder that needs additional fuel. **Possible Causes** ● Base engine mechanical fault that affects only Cylinder 8 ● Fuel injection (high pressure) nozzle leaking only on Cylinder 8 ● Fuel injection nozzle stuck closed or dirty affecting Cylinder 8

OBD II Trouble Code List (P0xxx Codes)

DTC	Trouble Code Title, Conditions & Possible Causes
P0300 **2T MISFIRE, MIL: Yes** 1996, 1997, 1998, 1999, 2000, 2001, 2002, 2003, 2004, 2005 DeVille, Eldorado, Seville 4.6L VIN 9, 4.6L VIN Y Transmissions: A/T	**Engine Misfire Detected Conditions:** DTC P0101-P0103, P0106-P0108, P0112-P0113, P0116-P0118, P0121-P0123, P0125, P0128, P0200, P0300, P0410-P0446, P0452-P0453, P1258, P1415-P1416 and P1441 not set, engine speed from 500-5850 rpm, ECT sensor from 19-248ºF, A/C Clutch and TP angle steady, EGR Test not active, DFCO and Torque Control "off", fuel level over 10%, and the PCM detected a crankshaft speed variation in two or more cylinders characteristic of a misfire condition. *Note: If the misfire is severe, the MIL will flash on/off on the 1st trip!* **Possible Causes** ● Base engine mechanical or fuel metering fault that affects one or more cylinders ● EGR valve is stuck open or the PCV system has a vacuum leak ● EVAP system problem or the EVAP canister is fuel saturated ● Ignition system fault (a coil) that affects more than one cylinder
P0301 **2T MISFIRE, MIL: Yes** 1996, 1997, 1998, 1999, 2000, 2001, 2002, 2003, 2004, 2005 DeVille, Eldorado, Seville 4.6L VIN 9, 4.6L VIN Y Transmissions: A/T	**Cylinder 1 Misfire Detected Conditions:** DTC P0101, P0102, P0103, P0106-P0108, P0112-P0113, P0116-P0118, P0121-P0123, P0125, P0128, P0200, P0300, P0410-P0446, P0452-P0453, P1258, P1415-P1416 and P1441 not set, engine speed from 500-5850 rpm, ECT sensor from 19-248ºF, A/C Clutch and TP angle both steady, EGR Test inactive, Torque Control and DFCO "off", fuel level over 10%, and the PCM detected a crankshaft speed variation in Cylinder 1 characteristic of a misfire condition. *Note: If the misfire is severe, the MIL will flash on/off on the 1st trip!* **Possible Causes** ● Base engine mechanical fault that affects only Cylinder 1 ● Fuel component fault that affects only Cylinder 4 (i.e., a dirty or sticking fuel injector) ● Ignition system problem (coil, plug) that affects only Cylinder 1
P0302 **2T MISFIRE, MIL: Yes** 1996, 1997, 1998, 1999, 2000, 2001, 2002, 2003, 2004, 2005 DeVille, Eldorado, Seville 4.6L VIN 9, 4.6L VIN Y Transmissions: A/T	**Cylinder 2 Misfire Detected Conditions:** DTC P0101, P0102, P0103, P0106-P0108, P0112-P0113, P0116-P0118, P0121-P0123, P0125, P0128, P0200, P0300, P0410-P0446, P0452-P0453, P1258, P1415-P1416 and P1441 not set, engine speed from 500-5850 rpm, ECT sensor from 19-248ºF, A/C Clutch and TP angle both steady, EGR Test inactive, Torque Control and DFCO "off", fuel level over 10%, and the PCM detected a crankshaft speed variation in Cylinder 2 characteristic of a misfire condition. *Note: If the misfire is severe, the MIL will flash on/off on the 1st trip!* **Possible Causes** ● Base engine mechanical fault that affects only Cylinder 2 ● Fuel component fault that affects only Cylinder 4 (i.e., a dirty or sticking fuel injector) ● Ignition system problem (coil, plug) that affects only Cylinder 2
P0302 **2T MISFIRE, MIL: Yes** 1996, 1997, 1998, 1999, 2000, 2001, 2002, 2003, 2004, 2005 DeVille, Eldorado, Seville 4.6L VIN 9, 4.6L VIN Y Transmissions: A/T	**Cylinder 3 Misfire Detected Conditions:** DTC P0101, P0102, P0103, P0106-P0108, P0112-P0113, P0116-P0118, P0121-P0123, P0125, P0128, P0200, P0300, P0410-P0446, P0452-P0453, P1258, P1415-P1416 and P1441 not set, engine speed from 500-5850 rpm, ECT sensor from 19-248ºF, A/C Clutch and TP angle both steady, EGR Test inactive, Torque Control and DFCO "off", fuel level over 10%, and the PCM detected a crankshaft speed variation in Cylinder 3 characteristic of a misfire condition. *Note: If the misfire is severe, the MIL will flash on/off on the 1st trip!* **Possible Causes** ● Base engine mechanical fault that affects only Cylinder 3 ● Fuel component fault that affects only Cylinder 4 (i.e., a dirty or sticking fuel injector) ● Ignition system problem (coil, plug) that affects only Cylinder 3
P0304 **2T MISFIRE, MIL: Yes** 1996, 1997, 1998, 1999, 2000, 2001, 2002, 2003, 2004, 2005 DeVille, Eldorado, Seville 4.6L VIN 9, 4.6L VIN Y Transmissions: A/T	**Cylinder 4 Misfire Detected Conditions:** DTC P0101, P0102, P0103, P0106-P0108, P0112-P0113, P0116-P0118, P0121-P0123, P0125, P0128, P0200, P0300, P0410-P0446, P0452-P0453, P1258, P1415-P1416 and P1441 not set, engine speed from 500-5850 rpm, ECT sensor from 19-248ºF, A/C Clutch and TP angle both steady, EGR Test inactive, Torque Control and DFCO "off", fuel level over 10%, and the PCM detected a crankshaft speed variation in Cylinder 4 characteristic of a misfire condition. *Note: If the misfire is severe, the MIL will flash on/off on the 1st trip!* **Possible Causes** ● Base engine mechanical fault that affects only Cylinder 4 ● Fuel component fault that affects only Cylinder 4 (i.e., a dirty or sticking fuel injector) ● Ignition system problem (coil, plug) that affects only Cylinder 4
P0305 **2T MISFIRE, MIL: Yes** 1996, 1997, 1998, 1999, 2000, 2001, 2002, 2003, 2004, 2005 DeVille, Eldorado, Seville 4.6L VIN 9, 4.6L VIN Y Transmissions: A/T e	**Cylinder 5 Misfire Detected Conditions:** DTC P0101, P0102, P0103, P0106-P0108, P0112-P0113, P0116-P0118, P0121-P0123, P0125, P0128, P0200, P0300, P0410-P0446, P0452-P0453, P1258, P1415-P1416 and P1441 not set, engine speed from 500-5850 rpm, ECT sensor from 19-248ºF, A/C Clutch and TP angle both steady, EGR Test inactive, Torque Control and DFCO "off", fuel level over 10%, and the PCM detected a crankshaft speed variation in Cylinder 5 characteristic of a misfire condition. *Note: If the misfire is severe, the MIL will flash on/off on the 1st trip!* **Possible Causes** ● Base engine mechanical fault that affects only Cylinder 5 ● Fuel component fault that affects only Cylinder 4 (i.e., a dirty or sticking fuel injector) ● Ignition system problem (coil, plug) that affects only Cylinder 5

OBD II Trouble Code List (P0xxx Codes)

DTC	Trouble Code Title, Conditions & Possible Causes
P0306 **2T MISFIRE, MIL: Yes** 1996, 1997, 1998, 1999, 2000, 2001, 2002, 2003, 2004, 2005 DeVille, Eldorado, Seville 4.6L VIN 9, 4.6L VIN Y Transmissions: A/T	**Cylinder 6 Misfire Detected Conditions:** DTC P0101, P0102, P0103, P0106-P0108, P0112-P0113, P0116-P0118, P0121-P0123, P0125, P0128, P0200, P0300, P0410-P0446, P0452-P0453, P1258, P1415-P1416 and P1441 not set, engine speed from 500-5850 rpm, ECT sensor from 19-248°F, A/C Clutch and TP angle both steady, EGR Test inactive, Torque Control and DFCO "off", fuel level over 10%, and the PCM detected a crankshaft speed variation in Cylinder 6 characteristic of a misfire condition. *Note: If the misfire is severe, the MIL will flash on/off on the 1st trip!* **Possible Causes** ● Base engine mechanical fault that affects only Cylinder 6 ● Fuel component fault that affects only Cylinder 4 (i.e., a dirty or sticking fuel injector) ● Ignition system problem (coil, plug) that affects only Cylinder 6
P0307 **2T MISFIRE, MIL: Yes** 1996, 1997, 1998, 1999, 2000, 2001, 2002, 2003, 2004, 2005 DeVille, Eldorado, Seville 4.6L VIN 9, 4.6L VIN Y Transmissions: A/T e	**Cylinder 7 Misfire Detected Conditions:** DTC P0101, P0102, P0103, P0106-P0108, P0112-P0113, P0116-P0118, P0121-P0123, P0125, P0128, P0200, P0300, P0410-P0446, P0452-P0453, P1258, P1415-P1416 and P1441 not set, engine speed from 500-5850 rpm, ECT sensor from 19-248°F, A/C Clutch and TP angle both steady, EGR Test inactive, Torque Control and DFCO "off", fuel level over 10%, and the PCM detected a crankshaft speed variation in Cylinder 7 characteristic of a misfire condition. *Note: If the misfire is severe, the MIL will flash on/off on the 1st trip!* **Possible Causes** ● Base engine mechanical fault that affects only Cylinder 7 ● Fuel component fault that affects only Cylinder 4 (i.e., a dirty or sticking fuel injector) ● Ignition system problem (coil, plug) that affects only Cylinder 7
P0308 **2T MISFIRE, MIL: Yes** 1996, 1997, 1998, 1999, 2000, 2001, 2002, 2003, 2004, 2005 DeVille, Eldorado, Seville 4.6L VIN 9, 4.6L VIN Y Transmissions: A/T	**Cylinder 8 Misfire Detected Conditions:** DTC P0101, P0102, P0103, P0106-P0108, P0112-P0113, P0116-P0118, P0121-P0123, P0125, P0128, P0200, P0300, P0410-P0446, P0452-P0453, P1258, P1415-P1416 and P1441 not set, engine speed from 500-5850 rpm, ECT sensor from 19-248°F, A/C Clutch and TP angle both steady, EGR Test inactive, Torque Control and DFCO "off", fuel level over 10%, and the PCM detected a crankshaft speed variation in Cylinder 8 characteristic of a misfire condition. *Note: If the misfire is severe, the MIL will flash on/off on the 1st trip!* **Possible Causes** ● Base engine mechanical fault that affects only Cylinder 8 ● Fuel component fault that affects only Cylinder 4 (i.e., a dirty or sticking fuel injector) ● Ignition system problem (coil, plug) that affects only Cylinder 8
P0300 **2T MISFIRE, MIL: Yes** 2003, 2004, 2005 CTS (Cadillac) 2.6L VIN M, 3.2L VIN N Transmissions: All	**Multiple Cylinder Misfire Detected Conditions:** DTC P0121-P0123, P0221-P0223, P0335, P0336, P0337, P0338, P0341, P0342, P0343, P0440, P0442-P0444 and P0445 not set, engine speed from 520-6520 rpm, IAT sensor more than 17°F, ABS/TC, Traction Control and Fuel Cutoff all "off", no Rough Road signals detected, and the PCM detected a crankshaft speed variation in two or more cylinders characteristic of a misfire. *Note: If the misfire is severe, the MIL will flash on/off on 1st trip!* **Possible Causes** ● Base engine mechanical fault that affects two or more cylinders ● Check for leaks in the throttle body, in the vacuum hoses, and in the PCV valve/hoses ● Check ECM power grounds (verify they are clean and secure) ● Fuel metering problem that affects two or more cylinders ● Ignition system fault (coil, plug) that affects 2 or more cylinders
P0301 **2T MISFIRE, MIL: Yes** 2003, 2004, 2005 CTS (Cadillac) 2.6L VIN M, 3.2L VIN N Transmissions: All	**Cylinder 1 Misfire Detected Conditions:** DTC P0121-P0123, P0221-P0223, P0335, P0336, P0337, P0338, P0341, P0342, P0343, P0440, P0442-P0444 and P0445 not set, engine speed from 520-6520 rpm, IAT sensor more than 17°F, ABS/TC, Traction Control and Fuel Cutoff all "off", no Rough Road signals detected, and the PCM detected a crankshaft speed variation in Cylinder 1 characteristic of a misfire. *Note: If the misfire is severe, the MIL will flash on/off on the 1st trip!* **Possible Causes** ● Base engine mechanical fault that affects only Cylinder 1 ● Check throttle body and intake manifold for any vacuum leaks ● Fuel metering problem that affects only Cylinder 1 ● Ignition system fault (coil, plug) that affects only Cylinder 1
P0302 **2T MISFIRE, MIL: Yes** 2003, 2004, 2005 CTS (Cadillac) 2.6L VIN M, 3.2L VIN N Transmissions: All	**Cylinder 2 Misfire Detected Conditions:** DTC P0121-P0123, P0221-P0223, P0335, P0336, P0337, P0338, P0341, P0342, P0343, P0440, P0442-P0444 and P0445 not set, engine speed from 520-6520 rpm, IAT sensor more than 17°F, ABS/TC, Traction Control and Fuel Cutoff all "off", no Rough Road signals detected, and the PCM detected a crankshaft speed variation in Cylinder 2 characteristic of a misfire. *Note: If the misfire is severe, the MIL will flash on/off on the 1st trip!* **Possible Causes** ● Base engine mechanical fault that affects only Cylinder 2 ● Check throttle body and intake manifold for any vacuum leaks ● Fuel metering problem that affects only Cylinder 2 ● Ignition system fault (coil, plug) that affects only Cylinder 2

OBD II Trouble Code List (P0xxx Codes)

DTC	Trouble Code Title, Conditions & Possible Causes
P0303 **2T MISFIRE, MIL: Yes** 2003, 2004, 2005 CTS (Cadillac) 2.6L VIN M, 3.2L VIN N Transmissions: All	**Cylinder 3 Misfire Detected Conditions:** DTC P0121-P0123, P0221-P0223, P0335, P0336, P0337, P0338, P0341, P0342, P0343, P0440, P0442-P0444 and P0445 not set, engine speed from 520-6520 rpm, IAT sensor more than 17°F, ABS/TC, Traction Control and Fuel Cutoff all "off", no Rough Road signals detected, and the PCM detected a crankshaft speed variation in Cylinder 3 characteristic of a misfire. *Note: If the misfire is severe, the MIL will flash on/off on the 1st trip!* **Possible Causes** ● Base engine mechanical fault that affects only Cylinder 3 ● Check throttle body and intake manifold for any vacuum leaks ● Fuel metering problem that affects only Cylinder 3 ● Ignition system fault (coil, plug) that affects only Cylinder 3
P0304 **2T MISFIRE, MIL: Yes** 2003, 2004, 2005 CTS (Cadillac) 2.6L VIN M, 3.2L VIN N Transmissions: All	**Cylinder 4 Misfire Detected Conditions:** DTC P0121-P0123, P0221-P0223, P0335, P0336, P0337, P0338, P0341, P0342, P0343, P0440, P0442-P0444 and P0445 not set, engine speed from 520-6520 rpm, IAT sensor more than 17°F, ABS/TC, Traction Control and Fuel Cutoff all "off", no Rough Road signals detected, and the PCM detected a crankshaft speed variation in Cylinder 4 characteristic of a misfire. *Note: If the misfire is severe, the MIL will flash on/off on the 1st trip!* **Possible Causes** ● Base engine mechanical fault that affects only Cylinder 4 ● Check throttle body and intake manifold for any vacuum leaks ● Fuel metering problem that affects only Cylinder 4 ● Ignition system fault (coil, plug) that affects only Cylinder 4
P0305z **2T MISFIRE, MIL: Yes** 2003, 2004, 2005 CTS (Cadillac) 2.6L VIN M, 3.2L VIN N Transmissions: All	**Cylinder 5 Misfire Detected Conditions:** DTC P0121-P0123, P0221-P0223, P0335, P0336, P0337, P0338, P0341, P0342, P0343, P0440, P0442-P0444 and P0445 not set, engine speed from 520-6520 rpm, IAT sensor more than 17°F, ABS/TC, Traction Control and Fuel Cutoff all "off", no Rough Road signals detected, and the PCM detected a crankshaft speed variation in Cylinder 5 characteristic of a misfire. *Note: If the misfire is severe, the MIL will flash on/off on the 1st trip!* **Possible Causes** ● Base engine mechanical fault that affects only Cylinder 5 ● Check throttle body and intake manifold for any vacuum leaks ● Fuel metering problem that affects only Cylinder 5 ● Ignition system fault (coil, plug) that affects only Cylinder 5
P0306 **2T MISFIRE, MIL: Yes** 2003, 2004, 2005 CTS (Cadillac) 2.6L VIN M, 3.2L VIN N Transmissions: All	**Cylinder 6 Misfire Detected Conditions:** DTC P0121-P0123, P0221-P0223, P0335, P0336, P0337, P0338, P0341, P0342, P0343, P0440, P0442-P0444 and P0445 not set, engine speed from 520-6520 rpm, IAT sensor more than 17°F, ABS/TC, Traction Control and Fuel Cutoff all "off", no Rough Road signals detected, and the PCM detected a crankshaft speed variation in Cylinder 6 characteristic of a misfire. *Note: If the misfire is severe, the MIL will flash on/off on the 1st trip!* **Possible Causes** ● Base engine mechanical fault that affects only Cylinder 6 ● Check throttle body and intake manifold for any vacuum leaks ● Fuel metering problem that affects only Cylinder 6 ● Ignition system fault (coil, plug) that affects only Cylinder 6
P0300 **2T MISFIRE, MIL: Yes** 1997, 1998, 1999, 2000, 2001 Catera 3.0L V6 VIN R engine Transmissions: A/T	**Multiple Cylinder Misfire Detected Conditions:** DTC P0100, P0115, P0116, P0335, P1120 and P1220 not set; engine speed from 520-6250 rpm, IAT sensor over 17°F, Torque Control, ABS/TC and Fuel Cutoff all "off", EVAP Test off, and the PCM detected a crankshaft speed variation in two or more cylinders characteristic of a misfire condition. *Note: If the misfire is severe, the MIL will flash on/off on the 1st trip!* **Possible Causes** ● Base engine mechanical fault or fuel metering problem that affects more than one cylinder ● EVAP system problem or the EVAP canister is fuel saturated ● EGR valve is stuck open or the PCV system has a vacuum leak ● IC control circuit is shorted to ground (an intermittent fault) ● Ignition system fault (a coil) that affects more than one cylinder ● MAF sensor contamination (it can cause a very lean condition)
P0301 **2T MISFIRE, MIL: Yes** 1997, 1998, 1999, 2000, 2001 Catera 3.0L V6 VIN R engine Transmissions: A/T	**Cylinder 1 Misfire Detected Conditions:** DTC P0100, P0115, P0116, P0335, P1120 and P1220 not set; engine speed from 520-6250 rpm, IAT sensor over 17°F, Torque Control, ABS/TC and Fuel Cutoff all "off", EVAP Test off, and the PCM detected a crankshaft speed variation in Cylinder 1 characteristic of a misfire condition. *Note: If the misfire is severe, the MIL will flash on/off on the 1st trip!* **Possible Causes** ● Base engine mechanical fault that affects only Cylinder 1 ● Fuel delivery component fault that affects only Cylinder 1 (i.e., a contaminated, dirty or sticking fuel injector) ● Ignition system problem (coil, plug) that affects only Cylinder 1

OBD II Trouble Code List (P0xxx Codes)

DTC	Trouble Code Title, Conditions & Possible Causes
P0302 **2T MISFIRE, MIL: Yes** 1997, 1998, 1999, 2000, 2001 Catera 3.0L V6 VIN R engine Transmissions: A/T	**Cylinder 2 Misfire Detected Conditions:** DTC P0100, P0115, P0116, P0335, P1120 and P1220 not set; engine speed from 520-6250 rpm, IAT sensor over 17°F, Torque Control, ABS/TC and Fuel Cutoff all "off", EVAP Test "off", and the PCM detected a crankshaft speed variation in Cylinder 2 characteristic of a misfire condition. *Note: If the misfire is severe, the MIL will flash on/off on the 1st trip!* **Possible Causes** ● Base engine mechanical fault that affects only Cylinder 2 ● Fuel delivery component fault that affects only Cylinder 2 (i.e., a contaminated, dirty or sticking fuel injector) ● Ignition system problem (coil, plug) that affects only Cylinder 2
P0303 **2T MISFIRE, MIL: Yes** 1997, 1998, 1999, 2000, 2001 Catera 3.0L V6 VIN R engine Transmissions: A/T	**Cylinder 3 Misfire Detected Conditions:** DTC P0100, P0115, P0116, P0335, P1120 and P1220 not set; engine speed from 520-6250 rpm, IAT sensor over 17°F, Torque Control, ABS/TC and Fuel Cutoff all "off", EVAP Test "off", and the PCM detected a crankshaft speed variation in Cylinder 3 characteristic of a misfire condition. *Note: If the misfire is severe, the MIL will flash on/off on the 1st trip!* **Possible Causes** ● Base engine mechanical fault that affects only Cylinder 3 ● Fuel component fault that affects only Cylinder 3 (i.e., a dirty or sticking fuel injector)Ignition system problem (coil, plug) that affects only Cylinder 3
P0304 **2T MISFIRE, MIL: Yes** 1997, 1998, 1999, 2000, 2001 Catera 3.0L V6 VIN R engine Transmissions: A/T	**Cylinder 4 Misfire Detected Conditions:** DTC P0100, P0115, P0116, P0335, P1120 and P1220 not set; engine speed from 520-6250 rpm, IAT sensor over 17°F, Torque Control, ABS/TC and Fuel Cutoff all "off", EVAP Test "off", and the PCM detected a crankshaft speed variation in Cylinder 4 characteristic of a misfire condition. *Note: If the misfire is severe, the MIL will flash on/off on the 1st trip!* **Possible Causes** ● Base engine mechanical fault that affects only Cylinder 4 ● Fuel component fault that affects only Cylinder 4 (i.e., a dirty or sticking fuel injector) ● Ignition system problem (coil, plug) that affects only Cylinder 4
P0305 **2T MISFIRE, MIL: Yes** 1997, 1998, 1999, 2000, 2001 Catera 3.0L V6 VIN R engine Transmissions: A/T	**Cylinder 5 Misfire Detected Conditions:** DTC P0100, P0115, P0116, P0335, P1120 and P1220 not set; engine speed from 520-6250 rpm, IAT sensor over 17°F, Torque Control, ABS/TC and Fuel Cutoff all "off", EVAP Test "off", and the PCM detected a crankshaft speed variation in Cylinder 5 characteristic of a misfire condition. *Note: If the misfire is severe, the MIL will flash on/off on the 1st trip!* **Possible Causes** ● Base engine mechanical fault that affects only Cylinder 5 ● Fuel component fault that affects only Cylinder 5 (i.e., a dirty or sticking fuel injector) ● Ignition system problem (coil, plug) that affects only Cylinder 5
P0306 **2T MISFIRE, MIL: Yes** 1997, 1998, 1999, 2000, 2001 Catera 3.0L V6 VIN R engine Transmissions: A/T	**Cylinder 6 Misfire Detected Conditions:** DTC P0100, P0115, P0116, P0335, P1120 and P1220 not set; engine speed from 520-6250 rpm, IAT sensor over 17°F, Torque Control, ABS/TC and Fuel Cutoff all "off", EVAP Test "off", and the PCM detected a crankshaft speed variation in Cylinder 6 characteristic of a misfire condition. *Note: If the misfire is severe, the MIL will flash on/off on the 1st trip!* **Possible Causes** ● Base engine mechanical fault that affects only Cylinder 6 ● Fuel component fault that affects only Cylinder 6 (i.e., a dirty or sticking fuel injector) ● Ignition system problem (coil, plug) that affects only Cylinder 6
P0300 **2T MISFIRE, MIL: Yes** 1997, 1998, 1999, 2000, 2001, 2002, 2003, 2004, 2005 Camaro, Corvette & Firebird 5.7L VIN G, 5.7L VIN S Transmissions: All	**Multiple Engine Misfire Detected Conditions:** DTC P0101-P0103, P0116-P0118, P0121-P0123, P0125, P0335, P0336, P0341-P0343, P0500-P0503 and P1258 not set, engine speed from 450-3000 rpm, system voltage over 10.0v, ECT sensor from 19-230°F, fuel level over 10%, TP angle steady (± 1%), ABS and Traction Control inactive, ABS signal not indicating rough road thresholds, transmission not shifting, A/C clutch stable, AIR Test and DFCO "off", and the PCM detected a crankshaft speed variation in two or more cylinders characteristic of a misfire condition. *Note: If the misfire is severe, the MIL will flash on/off on the 1st trip!* **Possible Causes** ● Base engine mechanical fault that affects one or more cylinders ● Fuel metering fault that affects more than one cylinder ● Fuel pressure too low or too high, fuel supply contaminated ● EVAP system problem or the EVAP canister is fuel saturated ● EGR valve is stuck open or the PCV system has a vacuum leak ● IC control circuit is shorted to ground (an intermittent fault) ● Ignition system fault (a coil) that affects more than one cylinder ● MAF sensor contamination (it can cause a very lean condition) ● TSB 03-06-04-030 contains a repair procedure for this code

OBD II Trouble Code List (P0xxx Codes)

DTC	Trouble Code Title, Conditions & Possible Causes
P0301 **2T MISFIRE, MIL: Yes** 1997, 1998, 1999, 2000, 2001, 2002, 2003, 2004, 2005 Camaro, Corvette & Firebird 5.7L VIN G, 5.7L VIN S Transmissions: All	**Cylinder 1 Misfire Detected Conditions:** DTC P0101, P0102, P0103, P0116, P0117, P0118, P0121-P0123, P0125, P0335, P0336, P0341-P0343, P0500-P0503 and P1258 not set, engine speed from 450-3000 rpm, system voltage over 10.0v, ECT sensor from 19-230ºF, fuel level over 10%, TP angle steady (± 1%), ABS and Traction Control inactive, ABS signal not indicating rough road thresholds, transmission not shifting, A/C clutch stable, AIR Test and DFCO "off", and the PCM detected a crankshaft speed variation on Cylinder 1 characteristic of a misfire condition. *Note: If the misfire is severe, the MIL will flash on/off on the 1st trip!* **Possible Causes** ● Air leak in the intake manifold, or in the EGR or PCV system ● Base engine mechanical fault that affects only Cylinder 1 ● Fuel delivery component fault that affects only Cylinder 1 (i.e., a contaminated, dirty or sticking fuel injector) ● Ignition system fault (coil or plug) that affects only Cylinder 1
P0302 **2T MISFIRE, MIL: Yes** 1997, 1998, 1999, 2000, 2001, 2002, 2003, 2004, 2005 Camaro, Corvette & Firebird 5.7L VIN G, 5.7L VIN S Transmissions: All	**Cylinder 2 Misfire Detected Conditions:** DTC P0101, P0102, P0103, P0116, P0117, P0118, P0121-P0123, P0125, P0335, P0336, P0341-P0343, P0500-P0503 and P1258 not set, engine speed from 450-3000 rpm, system voltage over 10.0v, ECT sensor from 10-200ºF, fuel level over 10%, TP angle steady (± 1%), ABS and Traction Control inactive, ABS signal not indicating rough road thresholds, transmission not shifting, A/C clutch stable, AIR Test and DFCO "off", and the PCM detected a crankshaft speed variation on Cylinder 2 characteristic of a misfire condition. *Note: If the misfire is severe, the MIL will flash on/off on the 1st trip!* **Possible Causes** ● Air leak in the intake manifold, or in the EGR or PCV system ● Base engine mechanical fault that affects only Cylinder 2 ● Fuel component fault that affects only Cylinder 2 (i.e., a dirty or sticking fuel injector) ● Ignition system fault (coil or plug) that affects only Cylinder 2
P0303 **2T MISFIRE, MIL: Yes** 1997, 1998, 1999, 2000, 2001, 2002, 2003, 2004, 2005 Camaro, Corvette & Firebird 5.7L VIN G, 5.7L VIN S Transmissions: All	**Cylinder 3 Misfire Detected Conditions:** DTC P0101, P0102, P0103, P0116, P0117, P0118, P0121-P0123, P0125, P0335, P0336, P0341-P0343, P0500-P0503 and P1258 not set, engine speed from 450-3000 rpm, system voltage over 10.0v, ECT sensor from 19-230ºF, fuel level over 10%, TP angle steady (± 1%), ABS and Traction Control inactive, ABS signal not indicating rough road thresholds, transmission not shifting, A/C clutch stable, AIR Test and DFCO "off", and the PCM detected a crankshaft speed variation on Cylinder 3 characteristic of a misfire condition. *Note: If the misfire is severe, the MIL will flash on/off on the 1st trip!* **Possible Causes** ● Air leak in the intake manifold, or in the EGR or PCV system ● Base engine mechanical fault that affects only Cylinder 3 ● Fuel component fault that affects only Cylinder 3 (i.e., a dirty or sticking fuel injector) ● Ignition system fault (coil or plug) that affects only Cylinder 3
P0304 **2T MISFIRE, MIL: Yes** 1997, 1998, 1999, 2000, 2001, 2002, 2003, 2004, 2005 Camaro, Corvette & Firebird 5.7L VIN G, 5.7L VIN S Transmissions: All	**Cylinder 4 Misfire Detected Conditions:** DTC P0101, P0102, P0103, P0116, P0117, P0118, P0121-P0123, P0125, P0335, P0336, P0341-P0343, P0500-P0503 and P1258 not set, engine speed from 450-3000 rpm, system voltage over 10.0v, ECT sensor from 19-230ºF, fuel level over 10%, TP angle steady (± 1%), ABS and Traction Control inactive, ABS signal not indicating rough road thresholds, transmission not shifting, A/C clutch stable, AIR Test and DFCO "off", and the PCM detected a crankshaft speed variation on Cylinder 4 characteristic of a misfire condition. *Note: If the misfire is severe, the MIL will flash on/off on the 1st trip!* **Possible Causes** ● Air leak in the intake manifold, or in the EGR or PCV system ● Base engine mechanical fault that affects only Cylinder 4 ● Fuel component fault that affects only Cylinder 4 (i.e., a dirty or sticking fuel injector) ● Ignition system fault (coil or plug) that affects only Cylinder 4
P0305 **2T MISFIRE, MIL: Yes** 1997, 1998, 1999, 2000, 2001, 2002, 2003, 2004, 2005 Camaro, Corvette & Firebird 5.7L VIN G, 5.7L VIN S Transmissions: All	**Cylinder 5 Misfire Detected Conditions:** DTC P0101, P0102, P0103, P0116, P0117, P0118, P0121-P0123, P0125, P0335, P0336, P0341-P0343, P0500-P0503 and P1258 not set, engine speed from 450-3000 rpm, system voltage over 10.0v, ECT sensor from 19-230ºF, fuel level over 10%, TP angle steady (± 1%), ABS and Traction Control inactive, ABS signal not indicating rough road thresholds, transmission not shifting, A/C clutch stable, AIR Test and DFCO "off", and the PCM detected a crankshaft speed variation on Cylinder 5 characteristic of a misfire condition. *Note: If the misfire is severe, the MIL will flash on/off on the 1st trip!* **Possible Causes** ● Air leak in the intake manifold, or in the EGR or PCV system ● Base engine mechanical fault that affects only Cylinder 5 ● Fuel component fault that affects only Cylinder 5 (i.e., a dirty or sticking fuel injector) ● Ignition system fault (coil or plug) that affects only Cylinder 5

OBD II Trouble Code List (P0xxx Codes)

DTC	Trouble Code Title, Conditions & Possible Causes
P0306 **2T MISFIRE, MIL: Yes** 1997, 1998, 1999, 2000, 2001, 2002, 2003, 2004, 2005 Camaro, Corvette & Firebird 5.7L VIN G, 5.7L VIN S Transmissions: All	**Cylinder 6 Misfire Detected Conditions:** DTC P0101, P0102, P0103, P0116, P0117, P0118, P0121-P0123, P0125, P0335, P0336, P0341-P0343, P0500-P0503 and P1258 not set, engine speed from 450-3000 rpm, system voltage over 10.0v, ECT sensor from 19-230ºF, fuel level over 10%, TP angle steady (± 1%), ABS and Traction Control inactive, ABS signal not indicating rough road thresholds, transmission not shifting, A/C clutch stable, AIR Test and DFCO "off", and the PCM detected a crankshaft speed variation on Cylinder 6 characteristic of a misfire condition. *Note: If the misfire is severe, the MIL will flash on/off on the 1st trip!* **Possible Causes** ● Air leak in the intake manifold, or in the EGR or PCV system ● Base engine mechanical fault that affects only Cylinder 6 ● Fuel component fault that affects only Cylinder 6 (i.e., a dirty or sticking fuel injector) ● Ignition system fault (coil or plug) that affects only Cylinder 6
P0307 **2T MISFIRE, MIL: Yes** 1997, 1998, 1999, 2000, 2001, 2002, 2003, 2004, 2005 Camaro, Corvette & Firebird 5.7L VIN G, 5.7L VIN S Transmissions: All	**Cylinder 7 Misfire Detected Conditions:** DTC P0101, P0102, P0103, P0116, P0117, P0118, P0121-P0123, P0125, P0335, P0336, P0341-P0343, P0500-P0503 and P1258 not set, engine speed from 450-3000 rpm, system voltage over 10.0v, ECT sensor from 19-230ºF, fuel level over 10%, TP angle steady (± 1%), ABS and Traction Control inactive, ABS signal not indicating rough road thresholds, transmission not shifting, A/C clutch stable, AIR Test and DFCO "off", and the PCM detected a crankshaft speed variation on Cylinder 7 characteristic of a misfire condition. *Note: If the misfire is severe, the MIL will flash on/off on the 1st trip!* **Possible Causes** ● Air leak in the intake manifold, or in the EGR or PCV system ● Base engine mechanical fault that affects only Cylinder 7 ● Fuel component fault that affects only Cylinder 7 (i.e., a dirty or sticking fuel injector) ● Ignition system fault (coil or plug) that affects only Cylinder 7
P0308 **2T MISFIRE, MIL: Yes** 1997, 1998, 1999, 2000, 2001, 2002, 2003, 2004, 2005 Camaro, Corvette & Firebird 5.7L VIN G, 5.7L VIN S Transmissions: All	**Cylinder 8 Misfire Detected Conditions:** DTC P0101, P0102, P0103, P0116, P0117, P0118, P0121-P0123, P0125, P0335, P0336, P0341-P0343, P0500-P0503 and P1258 not set, engine speed from 450-3000 rpm, system voltage over 10.0v, ECT sensor from 19-230ºF, fuel level over 10%, TP angle steady (± 1%), ABS and Traction Control inactive, ABS signal not indicating rough road thresholds, transmission not shifting, A/C clutch stable, AIR Test and DFCO "off", and the PCM detected a crankshaft speed variation on Cylinder 8 characteristic of a misfire condition. *Note: If the misfire is severe, the MIL will flash on/off on the 1st trip!* **Possible Causes** ● Air leak in the intake manifold, or in the EGR or PCV system ● Base engine mechanical fault that affects only Cylinder 8 ● Fuel component fault that affects only Cylinder 8 (i.e., a dirty or sticking fuel injector) ● Ignition system fault (coil or plug) that affects only Cylinder 8
P0300 **2T MISFIRE, MIL: Yes** 1996, 1997, 1998, 1999, 2001, 2002 Aurora 3.5L VIN H, 4.0L VIN G Transmissions: All	**Multiple Cylinder Misfire Detected Conditions:** DTC P0101-P0103, P0107, P0108, P0112-P0118, P0122, P0123, P0340 and P0502 not set, engine runtime 5 seconds, engine speed from 500-5850 rpm, ECT sensor from 19-248ºF, TP angle steady with any change less than 1%, Traction Control and DFCO inactive, EGR Flow Test "off", A/C clutch stable, rough road indication not present, and the PCM detected a crankshaft speed variation in more than one cylinder characteristic of a misfire condition. *Note: If the misfire is severe, the MIL will flash on/off on the 1st trip!* **Possible Causes** ● Base engine mechanical or fuel metering fault that affects one or more cylinders ● EVAP system problem or the EVAP canister is fuel saturated ● EGR valve is stuck open or the PCV system has a vacuum leak ● Ignition system fault (a coil) that affects more than one cylinder ● MAF sensor contamination (it can cause a very lean condition)
P0300 **2T MISFIRE, MIL: Yes** 1995, 1996, 1997, 1998, 1999, 2000, 2001, 2002 Camaro & Firebird 3.8L VIN K Transmissions: All	**Multiple Engine Misfire Detected Conditions:** DTC P0101-P0103, P0107, P0108, P0117, P0118, P0121-P0123, P0125, P0336, P0341, P0502-P0503, P1106-P1107, P1114, P1115, P1121, P1122, P1336 and P1374 not set, engine speed from 550-5850 rpm, ECT sensor from 21-248ºF, system voltage over 10.0v, TP angle steady, and the PCM detected a crankshaft speed change in more than one cylinder characteristic of a misfire condition. *Note: If the misfire is severe, the MIL will flash on/off on the 1st trip!* **Possible Causes** ● Base engine mechanical or fuel metering fault that affects one or more cylinders ● Fuel pressure too low or too high, fuel supply contaminated ● EVAP system problem or the EVAP canister is fuel saturated ● EGR valve is stuck open or the PCV system has a vacuum leak ● Ignition system fault (a coil) that affects more than one cylinder ● MAF sensor contamination (it can cause a very lean condition) ● TSB 03-06-04-030 contains a repair procedure for this code

OBD II Trouble Code List (P0xxx Codes)

DTC	Trouble Code Title, Conditions & Possible Causes
P0300 **2T MISFIRE, MIL: Yes** 1996, 1997, 1998, 1999, 2000, 2001, 2002, 2003, 2004, 2005 Alero, Aztek, Cutlass, Impala, Malibu, Montana, Venture Silhouette, Rendezvous 3.1L VIN J, 3.1L VIN M, 3.4L VIN E engines Transmissions: All	**Multiple Cylinder Misfire Detected Conditions:** DTC P0101-P0103, P0107, P0108, P0116-P0118, P0121, P0122, P0123, P0125, P0336, P0341, P0502, P0503, P1106, P1107, P1114, P1115, P1121, P1122, P1336, P1351, P1352, P1361, P1362 and P1374 not set, engine speed from 525-5900 rpm, ECT sensor from 21-255°F, system voltage over 10.0v, TP angle steady, and the PCM detected a crankshaft speed variation characteristic of a misfire condition in two or more cylinders. *Note: If the misfire is severe, the MIL will flash on/off on the 1st trip!* **Possible Causes** ● Base engine mechanical or fuel metering fault that affects one or more cylinders ● EVAP system problem or the EVAP canister is fuel saturated ● EGR valve is stuck open or the PCV system has a vacuum leak ● IC control circuit is shorted to ground (an intermittent fault) ● Ignition system fault (a coil) that affects more than one cylinder ● MAF sensor contamination (it can cause a very lean condition) ● TSB 99-06-04-005B contains a repair procedure for this code ● TSB 03-06-04-030 contains a repair procedure for this code
P0301 **2T MISFIRE, MIl · Yes** 1996, 1997, 1998, 1999, 2000, 2001, 2002, 2003, 2004, 2005 Alero, Aztek, Cutlass, Impala, Malibu, Montana, Venture Silhouette, Rendezvous 3.1L VIN J, 3.1L VIN M, 3.4L VIN E engines Transmissions: All	**Cylinder 1 Misfire Detected Conditions:** DTC P0101-P0103, P0107, P0108, P0116-P0118, P0121-P0123, P0125, P0336, P0341, P0502, P0503, P1106, P1107, P1114, P1115, P1121, P1122, P1336, P1351, P1352, P1361, P1362 and P1374 not set, engine speed from 525-5900 rpm, ECT sensor from 21-255°F, system voltage over 10.0v, TP angle steady, and the PCM detected a crankshaft speed variation characteristic of a misfire in Cylinder 1. *Note: If the misfire is severe, the MIL will flash on/off on the 1st trip!* **Possible Causes** ● Air leak in the intake manifold, or in the EGR or PCV system ● Base engine mechanical fault that affects only Cylinder 1 ● Fuel component fault that affects only Cylinder 1 (i.e., a dirty or sticking fuel injector) ● Ignition system problem (coil, plug) that affects only Cylinder 1
P0302 **2T MISFIRE, MIL: Yes** 1996, 1997, 1998, 1999, 2000, 2001, 2002, 2003, 2004, 2005 Alero, Aztek, Cutlass, Impala, Malibu, Montana, Venture Silhouette, Rendezvous 3.1L VIN J, 3.1L VIN M, 3.4L VIN E engines Transmissions: All	**Cylinder 2 Misfire Detected Conditions:** DTC P0101-P0103, P0107, P0108, P0116-P0118, P0121-P0123, P0125, P0336, P0341, P0502, P0503, P1106, P1107, P1114, P1115, P1121, P1122, P1336, P1351, P1352, P1361, P1362 and P1374 not set, engine speed from 525-5900 rpm, ECT sensor from 21-255°F, system voltage over 10.0v, TP angle steady, and the PCM detected a crankshaft speed variation characteristic of a misfire in Cylinder 2. *Note: If the misfire is severe, the MIL will flash on/off on the 1st trip!* **Possible Causes** ● Air leak in the intake manifold, or in the EGR or PCV system ● Base engine mechanical fault that affects only Cylinder 2 ● Fuel component fault that affects only Cylinder 2 (i.e., a dirty or sticking fuel injector) ● Ignition system problem (coil, plug) that affects only Cylinder 2
P0303 **2T MISFIRE, MIL: Yes** 1996, 1997, 1998, 1999, 2000, 2001, 2002, 2003, 2004, 2005 Alero, Aztek, Cutlass, Impala, Malibu, Montana, Venture Silhouette, Rendezvous 3.1L VIN J, 3.1L VIN M, 3.4L VIN E engines Transmissions: All	**Cylinder 3 Misfire Detected Conditions:** DTC P0101-P0103, P0107, P0108, P0116-P0118, P0121-P0123, P0125, P0336, P0341, P0502, P0503, P1106, P1107, P1114, P1115, P1121, P1122, P1336, P1351, P1352, P1361, P1362 and P1374 not set, engine speed from 525-5900 rpm, ECT sensor from 21-255°F, system voltage over 10.0v, TP angle steady, and the PCM detected a crankshaft speed variation characteristic of a misfire in Cylinder 3. *Note: If the misfire is severe, the MIL will flash on/off on the 1st trip!* **Possible Causes** ● Air leak in the intake manifold, or in the EGR or PCV system ● Base engine mechanical fault that affects only Cylinder 3 ● Fuel component fault that affects only Cylinder 3 (i.e., a dirty or sticking fuel injector) ● Ignition system problem (coil, plug) that affects only Cylinder 3
P0304 **2T MISFIRE, MIL: Yes** 1996, 1997, 1998, 1999, 2000, 2001, 2002, 2003, 2004, 2005 Alero, Aztek, Cutlass, Impala, Malibu, Montana, Venture Silhouette, Rendezvous 3.1L VIN J, 3.1L VIN M, 3.4L VIN E engines Transmissions: All	**Cylinder 4 Misfire Detected Conditions:** DTC P0101-P0103, P0107, P0108, P0116-P0118, P0121-P0123, P0125, P0336, P0341, P0502, P0503, P1106, P1107, P1114, P1115, P1121, P1122, P1336, P1351, P1352, P1361, P1362 and P1374 not set, engine speed from 525-5900 rpm, ECT sensor from 21-255°F, system voltage over 10.0v, TP angle steady, and the PCM detected a crankshaft speed variation characteristic of a misfire in Cylinder 4. *Note: If the misfire is severe, the MIL will flash on/off on the 1st trip!* **Possible Causes** ● Air leak in the intake manifold, or in the EGR or PCV system ● Base engine mechanical fault that affects only Cylinder 4 ● Fuel component fault that affects only Cylinder 4 (i.e., a dirty or sticking fuel injector) ● Ignition system problem (coil, plug) that affects only Cylinder 4

OBD II Trouble Code List (P0xxx Codes)

DTC	Trouble Code Title, Conditions & Possible Causes
P0305 **2T MISFIRE, MIL: Yes** 1996, 1997, 1998, 1999, 2000, 2001, 2002, 2003, 2004, 2005 Alero, Aztek, Cutlass, Impala, Malibu, Montana, Venture Silhouette, Rendezvous 3.1L VIN J, 3.1L VIN M, 3.4L VIN E engines Transmissions: All	**Cylinder 5 Misfire Detected Conditions:** DTC P0101-P0103, P0107, P0108, P0116-P0118, P0121-P0123, P0125, P0336, P0341, P0502, P0503, P1106, P1107, P1114, P1115, P1121, P1122, P1336, P1351, P1352, P1361, P1362 and P1374 not set, engine speed from 525-5900 rpm, ECT sensor from 21-255°F, system voltage over 10.0v, TP angle steady, and the PCM detected a crankshaft speed variation characteristic of a misfire in Cylinder 5. *Note: If the misfire is severe, the MIL will flash on/off on the 1st trip!* **Possible Causes** ● Air leak in the intake manifold, or in the EGR or PCV system ● Base engine mechanical fault that affects only Cylinder 5 ● Fuel component fault that affects only Cylinder 5 (i.e., a dirty or sticking fuel injector) ● Ignition system problem (coil, plug) that affects only Cylinder 5
P0306 **2T MISFIRE, MIL: Yes** 1996, 1997, 1998, 1999, 2000, 2001, 2002, 2003, 2004, 2005 Alero, Aztek, Cutlass, Impala, Malibu, Montana, Venture Silhouette, Rendezvous 3.1L VIN J, 3.1L VIN M, 3.4L VIN E engines Transmissions: All	**Cylinder 6 Misfire Detected Conditions:** DTC P0101-P0103, P0107, P0108, P0116-P0118, P0121-P0123, P0125, P0336, P0341, P0502, P0503, P1106, P1107, P1114, P1115, P1121, P1122, P1336, P1351, P1352, P1361, P1362 and P1374 not set, engine speed from 525-5900 rpm, ECT sensor from 21-255°F, system voltage over 10.0v, TP angle steady, and the PCM detected a crankshaft speed variation characteristic of a misfire in Cylinder 6. *Note: If the misfire is severe, the MIL will flash on/off on the 1st trip!* **Possible Causes** ● Air leak in the intake manifold, or in the EGR or PCV system ● Base engine mechanical fault that affects only Cylinder 6 ● Fuel component fault that affects only Cylinder 6 (i.e., a dirty or sticking fuel injector) ● Ignition system problem (coil, plug) that affects only Cylinder 6
P0313 **2T MISFIRE, MIL: Yes** 2003, 2004, 2005 CTS (Cadillac) 2.6L VIN M, 3.2L VIN N Transmissions: All	**Misfire Detected, Lean Air Fuel Condition Present Conditions:** DTC P0171, P0174, P0300 or P0301-P0306 are not set, engine speed from 520-6520 rpm, fuel level signal less than 0.79 gallons, and the PCM detected a lean condition or a misfire condition for 10 seconds. **Possible Causes** ● Verify that there is an adequate fuel supply in the fuel tank ● If this code is set and a misfire or lean condition is not currently present, the vehicle has run low on fuel and caused this code.
P0315 **1T CCM, MIL: Yes** 2003, 2004, 2005 Cab & Chassis, C/K Series Truck, G Series Van, L/M Series Van, S/T Blazer & S/T Pickup 4.3L VIN W, 4.3L VIN X, 4.8L VIN V, 5.3L VIN T, 5.3L V8 VIN Z, 6.0L V8 VIN N, 6.0L V8 VIN U CNG, 8.1L VIN G engines Transmissions: All	**Crankshaft Position Sensor Variation Not Learned Conditions:** DTC P0335, P0336, P0341, P0342 and P0343 not set, engine started, ECT sensor more than 149°F, and the PCM determined the CKP sensor variation values were not stored in memory. The CKP System variation "learning" feature is used to calculate reference period errors caused by slight tolerance variations in the crankshaft and the CKP sensor. The calculated error allows the PCM to accurately compensate for reference period variations. The PCM stores CKP variation values after a learn procedure is done. **Possible Causes** ● CKP sensor signal circuit has an interference condition (EMI) ● Crankshaft main bearings worn or reluctor wheel is damaged ● Crankshaft run-out is excessive or the crankshaft is damaged ● ECT sensor not within the conditions for running the code test ● Ignition switch is on, but the battery has insufficient voltage ● PCM power disconnected with key on (erases learned values) ● Debris that passes between the CKP sensor and reluctor wheel
P0315 **1T MISFIRE, MIL: Yes** 2003, 2004, 2005 Envoy & TrailBlazer 5.3L VIN P Transmissions: All	**Crankshaft Position Sensor Variation Not Learned Conditions:** DTC P0335, P0336, P0341, P0342 and P0343 not set, engine started, ECT sensor over 149°F, and the PCM determined the CKP sensor variation values were not stored in memory. The CKP "learn" feature is used to calculate reference period errors caused by slight tolerance variations in the crankshaft and the CKP sensor. This calculated error allows compensation for reference period variations. The CKP variation values are stored after a learn procedure is done. **Possible Causes** ● CKP sensor signal circuit has an interference condition (EMI) ● Crankshaft main bearings worn or reluctor wheel is damaged ● Crankshaft run-out is excessive or the crankshaft is damaged ● ECT sensor not within the conditions for running the code test ● Ignition switch is on, but the battery has insufficient voltage ● PCM power disconnected with key on (erases learned values) ● Debris that passes between the CKP sensor and reluctor wheel

OBD II Trouble Code List (P0xxx Codes)

DTC	Trouble Code Title, Conditions & Possible Causes
P0318 **1T MISFIRE, MIL: No** 2003, 2004, 2005 CTS (Cadillac) 2.6L VIN M, 3.2L VIN N Transmissions: All	**Misfire Detected, Lean Air Fuel Condition Present Conditions:** Engine started; DTC P0300 set with MIL requested "on", and the PCM detected it lost communication with the EBCM for 10 seconds. **Possible Causes** ● Perform the Diagnostic Circuit Check for the PCM ● Perform the Diagnostic Circuit Check for the EBCM ● Check the serial data connections to the EBCM and the PCM
P0321 **1T CCM, MIL: No** 1995 Century, Cutlass Ciera, Cutlass Supreme, Grand Prix, Lumina, Monte Carlo Park Avenue Regal, Riviera, 88' & 98' 3.1L VIN M, 3.4L VIN X, 3.8L VIN 1, 3.8L VIN K, 3.8L VIN L Transmissions: All	**IC Module Spark Reference Circuit Malfunction Conditions:** Engine started; engine speed less than 1200 rpm, and the PCM detected normal fuel control pulses without detecting any spark reference pulses during the CCM test period. **Possible Causes** ● Spark reference signal circuit is open ● Spark reference signal circuit is shorted to ground ● IC module is damaged or has failed ● PCM has failed
P0321 **1T CCM, MIL: No** 1996, 1997, 1998, 1999 Aurora, DeVille, Eldorado & Seville 4.0L VIN C, 4.6L VIN 9, 4.6L VIN Y Transmissions: All	**ICM 4X Reference Circuit Malfunction Conditions:** DTC P0340 and P1376 not set, engine started, CMP sensor signal detected in the last 4 seconds, and the PCM did not detect any 4X fuel control pulses for 4 seconds in the test. **Possible Causes** ● Fuel control signal circuit is open ● Fuel control signal circuit is shorted to ground ● IC module is damaged or has failed ● PCM has failed ● TSB 71-65-69 contains a repair procedure for this code
P0321 **1T CCM, MIL: No** 1996, 1997 Camaro, Caprice, Corvette, Firebird, Fleetwood 4.3L VIN W, 5.7L VIN 5, 5.7L VIN P Transmissions: All	**DI Ignition Low-resolution Circuit Malfunction Conditions:** Engine started; and the PCM detected an interruption in DI optical distributor low-resolution signal or in the high-resolution signals. This code can be caused by a misfire condition. **Possible Causes** ● DI low-resolution circuit is open, shorted to ground or to power ● DI high-resolution circuit is open, shorted to ground or to power ● IC module is damaged or has failed ● PCM has failed
P0324 **1T CCM, MIL: No** 2003, 2004, 2005 CTS (Cadillac) 2.6L VIN M, 3.2L VIN N Transmissions: All	**Knock Sensor Circuit Malfunction Conditions:** Engine started; engine speed from 1,000-5520 rpm, ECT sensor over 104°F, volumetric efficiency more than 40%, and the PCM detected a fault the KS diagnostic circuit that did not allow proper diagnosis of the KS system. The KS system is used to detect any engine detonation. The PCM retards the spark timing based on the signals from the KS. The KS produces an AC voltage proportional to the amount of knock to the PCM. **Possible Causes** ● Clear the codes and then recheck for the same code to reset. ● PCM has failed
P0325 **1T CCM, MIL: No** 1995 Century, Cutlass Ciera, Cutlass Supreme, Grand Prix, Lumina, Monte Carlo Park Avenue Regal, Riviera, 88' & 98' 3.1L VIN M, 3.4L VIN X, 3.8L VIN L, 3.8L VIN 1	**Knock Sensor Circuit Malfunction Conditions:** Engine started; and the PCM detected the Knock Sensor signal was below 0.8v or above 2.0v for 20 seconds during the CCM test. **Possible Causes** ● Knock sensor signal circuit is open, shorted to ground or power ● Knock sensor ground circuit is open (i.e., not mounted properly) ● Knock sensor is damaged or has failed ● PCM has failed
P0325 **2T CCM, MIL: Yes** 1996, 1997, 1998, 1999, 2000, 2001, 2002, 2003, 2004, 2005 Achieva, Alero, Beretta, Century, Ciera, Cavalier, Envoy, Grand Am, Sunfire Skyhawk, TrailBlazer, S/T Blazer & S/T Pickup 2.2L VIN 4, 2.2L VIN 5, 2.2L VIN H, 2.4L VIN T Transmissions: All	**Knock Sensor Circuit Malfunction Conditions:** DTC P0122 and P0123 not set, engine started, vehicle driven at an engine speed of 1600-6400 rpm for 20 seconds, ECT sensor over 131°F, MAP sensor more than 60 kPa, and the PCM detected the KS sensor signal variation was out of normal range for 15 seconds. **Possible Causes** ● Knock sensor signal circuit is open, shorted to ground or power ● Knock sensor ground circuit is open (not mounted properly) ● Knock sensor is damaged or has failed ● On modules with an integrated sensor, clear the codes and retest for codes. If the same code resets, the PCM has failed.

OBD II Trouble Code List (P0xxx Codes)

DTC	Trouble Code Title, Conditions & Possible Causes
P0325 **2T CCM, MIL: Yes** 1996, 1997, 1998, 1999, 2000, 2001, 2002 Achieva, Beretta, Ciera, Corsica, Century, Grand Am, Malibu & Skylark 3.1L VIN M Transmissions: All	**Knock Sensor Circuit Malfunction Conditions:** DTC P0101-P0103, P0116-P0118, P0121-P0123, P0125, P0336, P0341, P0502, P0503, P1114, P1115, P1121, P1122 and P1336 not set, engine speed from 1000-5000 rpm, ECT sensor more than 153°F, TP angle from 10-15%, Calculated engine load over 45%, spark retard less than 15 degrees, and the PCM detected an unexpected voltage condition on the Knock Sensor circuit that prevented diagnosis of the KS circuit for 30 seconds. **Possible Causes** ● Knock sensor pins are bent, broken or miss-aligned in the PCM ● On modules with an integrated sensor, clear the codes and retest for codes. If the same code resets, the PCM has failed.
P0325 **2T CCM, MIL: Yes** 1996, 1997, 1998, 1999, 2000, 2001, 2002, 2003, 2004, 2005 Aurora, Bonneville, Century, Cutlass Supreme, Grand Prix, Intrigue, LeSabre, LSS, Lumina, Monte Carlo, 88', Regal, Park Avenue & 98' 3.1L VIN J, 3.1L VIN M, 3.4L VIN X, 3.5L VIN H, 3.8L VIN 1, 3.8L VIN K Transmissions: All	**Knock Sensor Circuit Malfunction (Bank 1) Conditions:** DTC P0101, P0102, P0103, P0116, P0117, P0118, P0121, P0122, P0123, P0125, P0128, P0336, P0341, P0502, P0503, P1114, P1115, P1121 and P1336 are not set, engine speed from 1000-4000 rpm for 30 seconds, system voltage over 10.0v, ECT sensor more than 140°F, TP angle from 3-15%, engine load from 20-45%, spark retard less than 15 degrees, and the PCM detected an unexpected voltage condition on the Knock sensor circuit. **Possible Causes** ● Knock sensor signal circuit is open, shorted to ground or power ● Knock sensor ground circuit is open (i.e., not mounted properly) ● Knock sensor is damaged or has failed ● On modules with an integrated sensor, clear the codes and retest for codes. If the same code resets, the PCM has failed.
P0325 **2T CCM, MIL: Yes** 1996, 1997, 1998, 1999, 2000, 2001, 2002, 2003, 2004, 2005 Cab & Chassis, C/K Series Truck, G Series Van, L/M Series Van 4.3L VIN W, 4.3L VIN X, 5.0L VIN M, 5.7L VIN K, 5.7L VIN R, 7.4L VIN J Transmissions: All	**Knock Sensor Circuit Malfunction Conditions:** DTC P0327 not set, engine started, engine runtime from 10 seconds to 2 minutes, system voltage over 10.0v and the PCM detected an unexpected voltage condition for a period of 5-25 seconds on the diagnostic circuit used during diagnosis of the Knock sensor. **Possible Causes** ● Knock sensor signal circuit is open, shorted to ground or power ● Knock sensor ground circuit is open (i.e., not mounted properly) ● Knock sensor is damaged or has failed ● On modules with an integrated sensor, clear the codes and retest for codes. If the same code resets, the PCM has failed.
P0325 **2T CCM, MIL: Yes** 1999, 2000, 2001, 2002, 2003, 2004, 2005 C/K Series Truck, G Series Van, Envoy, Escalade, TrailBlazer 4.8L VIN V, 5.3L VIN P, 5.3L VIN T, 5.3L VIN Z, 6.0L VIN N, 6.0L VIN U CNG engine, 8.1L VIN G Transmissions: All	**Knock Sensor Circuit Malfunction Conditions:** Engine started; engine runtime more than 10 seconds, system voltage over 10.0v and the PCM detected an unexpected voltage condition for 12 seconds on the Knock sensor circuit. **Possible Causes** ● Knock sensor signal circuit is open, shorted to ground or power ● Knock sensor ground circuit is open (i.e., not mounted properly) ● Knock sensor is damaged or has failed ● On modules with an integrated sensor, clear the codes and retest for codes. If the same code resets, the PCM has failed.
P0325 **1T CCM, MIL: No** 1996, 1997, 1998, 1999, 2000, 2001, 2002, 2003, 2004, 2005 DeVille, Eldorado, Seville 4.6L VIN 9, 4.6L VIN Y Transmissions: All	**Knock Sensor Module Circuit Malfunction Conditions:** Engine started; engine speed from 600-3000 rpm, system voltage over 10.0v and the PCM detected an unexpected voltage condition on the Knock Sensor circuit used during testing. **Possible Causes** ● Knock sensor signal circuit is open, shorted to ground or power ● Knock sensor ground circuit is open (i.e., not mounted properly) ● Knock sensor is damaged or has failed ● On modules with an integrated sensor, clear the codes and retest for codes. If the same code resets, the PCM has failed.
P0325 **2T CCM, MIL: Yes** 1996, 1997, 1998, 1999, 2001, 2002, 2003, 2004, 2005 Aurora 3.5L VIN H, 4.0L VIN C Transmissions: A/T	**Knock Sensor Module Circuit Malfunction Conditions:** DTC P0101, P0102, P0103, P0121, P0122, P0123, P0335, P0340, P0385, P0502, P0503, P1114, P1115, P1121, P1122 and P1336 not set; engine started, system voltage over 10.0v and the PCM detected the Knock sensor signal that indicated a knock event longer than 99.99 ms was detected for 3 continuous seconds. **Possible Causes** ● Knock sensor signal circuit is open, shorted to ground or power ● Knock sensor ground circuit is open (i.e., not mounted properly) ● Knock sensor is damaged or has failed ● On modules with an integrated sensor, clear the codes and retest for codes. If the same code resets, the PCM has failed.

OBD II Trouble Code List (P0xxx Codes)

DTC	Trouble Code Title, Conditions & Possible Causes
P0325 **2T CCM, MIL: Yes** 1996, 1997, 1998, 1999, 2000, 2001, 2002, 2003, 2004, 2005 Alero, Cutlass, Impala, Malibu, Montana, Venture Silhouette, Rendezvous 3.1L VIN J, 3.1L VIN M, 3.4L VIN E engines Transmissions: All	**Knock Sensor Circuit Malfunction Conditions:** DTC P0101, P0102, P0103, P0116, P0117, P0118, P0121, P0122, P0123, P0125, P0336, P0341, P0502, P0503, P1114, P1115, P1121, P1122 and P1336 are not set, engine speed from 1000-5000 rpm for 30 seconds, TP sensor over 15%, engine load over 45%, ECT sensor over 140°F, spark retard less than 15 degrees, and the PCM detected an unexpected voltage condition on the Knock Sensor circuit used by the PCM to test the sensor. **Possible Causes** ● Knock sensor signal circuit is open, shorted to ground or power ● Knock sensor ground circuit is open (i.e., not mounted properly) ● Knock sensor is damaged or has failed ● On modules with an integrated sensor, clear the codes and retest for codes. If the same code resets, the PCM has failed.
P0325 **2T CCM, MIL: Yes** 1995, 1996, 1997, 1998, 1999, 2000, 2001, 2002 Camaro & Firebird 3.8L VIN K engine Transmissions: All	**Knock Sensor Range/Performance Conditions:** DTC P0101-P0103, P0116-P0118, P0121-P0123, P0125, P0128, P0336, P0341, P0502, P0503, P1114, P1115, P1121 and P1336 not set, engine started, engine speed from 1000-2500 rpm for over 30 seconds, ECT sensor more than 140°F, TP angle over 10%, engine load over 40%, spark retard less than 15 degrees, and the PCM detected an unexpected voltage condition on the diagnostic circuit used by the PCM to diagnose the Knock sensor. **Possible Causes** ● Knock sensor signal circuit is open, shorted to ground or power ● Knock sensor ground circuit is open (i.e., not mounted properly) ● Knock sensor is damaged or has failed ● On modules with an integrated sensor, clear the codes and retest for codes. If the same code resets, the PCM has failed.
P0325 **2T CCM, MIL: Yes** 1997, 1998, 1999, 2000, 2001, 2002, 2003, 2004, 2005 Camaro, Corvette & Firebird 5.7L VIN G, 5.7L VIN S Transmissions: All	**Knock Sensor Range/Performance Conditions:** DTC P0117, P0118, P0121, P0122, P0123, P0125, P1114, P1115, P112 and P1122 not set, engine started, engine runtime over 10 seconds, system voltage over 10.0v and the PCM detected an unexpected voltage condition on the diagnostic circuit used during testing. **Possible Causes** ● Knock sensor signal circuit is open, shorted to ground or power ● Knock sensor ground circuit is open (i.e., not mounted properly) ● Knock sensor is damaged or has failed ● On modules with an integrated sensor, clear the codes and retest for codes. If the same code resets, the PCM has failed.
P0325 **2T CCM, MIL: Yes** 1995, 1996, 1997, 1998, 1999, 2000, 2001, 2002, 2003, 2004, 2005 S/T Blazer & S/T Pickup 4.3L VIN W, 4.3L VIN X Transmissions: All	**Knock Sensor Circuit Malfunction Conditions:** DTC P0327 not set, engine started, engine runtime from 10 seconds to 2 minutes, system voltage over 10.0v and the PCM detected an unexpected voltage condition for a period of 10 seconds on the diagnostic circuit used during diagnosis of the Knock sensor. **Possible Causes** ● Knock sensor signal circuit is open, shorted to ground or power ● Knock sensor ground circuit is open (i.e., not mounted properly) ● Knock sensor is damaged or has failed ● On modules with an integrated sensor, clear the codes and retest for codes. If the same code resets, the PCM has failed.
P0325 **2T CCM, MIL: Yes** 1997, 1998, 1999, 2000, 2001 Catera 3.0L V6 VIN R engine Transmissions: A/T	**Knock Sensor Circuit Malfunction (Bank 1) Conditions:** Engine started; engine runtime at idle 5 seconds (to allow the PCM to learn the minimum noise level), ECT sensor more than 104°F, then driven to an engine speed of over 2,000 rpm, and the PCM did not detect any difference between the noise level at idle and the noise level at 2500-3000 rpm for 5 seconds during the CCM test period. **Possible Causes** ● Knock sensor signal circuit is open, shorted to ground or power ● Knock sensor ground circuit is open (i.e., not mounted properly) ● Knock sensor is damaged or has failed ● PCM has failed
P0325 **2T CCM, MIL: Yes** 1996, 1997 Grand Prix, Lumina, Monte Carlo 3.4L VIN X Transmissions: All	**Knock Sensor Circuit Malfunction Conditions:** DTC P0100, P0102, P0103, P0117, P0118, P0340, P0502 and P0503 not set, engine started, engine speed from 1000-5000 rpm, ECT sensor more than 153°F, TP angle from 10-15%, engine load more than 45%, spark retard less than 15°, and the PCM detected an unexpected voltage condition on the Knock sensor circuit. **Possible Causes** ● Knock sensor pins are bent, broken or miss-aligned in the PCM ● On modules with an integrated sensor, clear the codes and retest for codes. If the same code resets, the PCM has failed.

OBD II Trouble Code List (P0xxx Codes)

DTC	Trouble Code Title, Conditions & Possible Causes
P0326 **2T CCM, MIL: Yes** 2002, 2003, 2004, 2005 Envoy & TrailBlazer 4.2L VIN S Transmissions: All	**Knock Sensor Excessive Spark Retard Conditions:** DTC P0117, P0118, P0122, P0123, P0327, or P0332 are not set, engine speed from 2000-6400 rpm, ECT sensor over 158°F, MAP sensor over 60 kPa, and the PCM detected an unexpected voltage condition on KS diagnostic circuit. **Possible Causes** ● Clear the codes and retest for this same code. If it resets, the PCM has failed and needs to be replaced.
P0326 **1T CCM, MIL: No** 1996, 1997 Achieva, Beretta, Ciera, Corsica, Century, Grand Am, Malibu & Skylark 3.1L VIN M Transmissions: All	**Knock Sensor Noise Channel High Input Conditions:** DTC P0117, P0118, P0122, P0123, P0502 and P0503 not set, engine started, engine speed from 2200-2900 rpm for 10 seconds, ECT sensor more than 149°F, throttle angle over 10%, and the PCM detected the KS noise channel voltage was more than 4.80v. **Possible Causes** ● Knock sensor circuit is open between the sensor and the PCM ● Knock sensor signal wire routed close to the spark plug wires (causes EMI/RFI) ● Knock sensor is damaged or has failed ● PCM has failed
P0326 **1T CCM, MIL: No** 1996, 1997 Bonneville, Century, Cutlass Supreme, Grand Prix, LeSabre, LSS, Lumina, Monte Carlo, 88', 98', Park Avenue, Regal 3.4L VIN X, 3.8L VIN K, 3.8L VIN 1 Transmissions: All	**Knock Sensor Noise Channel High Input Conditions:** DTC P0117, P0118, P0122, P0123, P0502 and P0503 not set, engine started, engine speed from 2500-2900 rpm for 10 seconds, ECT sensor more than 149°F, throttle angle over 1.5%, and the PCM detected the KS noise channel voltage was more than 4.80v. **Possible Causes** ● Knock sensor circuit is open between the sensor and the PCM ● Knock sensor signal wire is routed to close to the spark plug wires or any other possible cause of EMI/RFI under the hood ● Knock sensor is damaged or has failed ● PCM has failed
P0326 **1T CCM, MIL: No** 1996, 1997, 1998, 1999, 2000, 2001, 2002, 2003, 2004, 2005 Aurora, DeVille, Eldorado, Seville 4.0L VIN C, 4.6L VIN 9, 4.6L VIN Y engines Transmissions: All	**Knock Sensor Excessive Spark Retard Conditions:** DTC P0325 and P0327 not set, engine speed over 600 rpm, Knock Detection enabled and Octane Learn/Lock-In not done, and the PCM detected a knock signal with the PCM spark retard command for a given load and speed more than a calibrated amount for 3.5 seconds. **Possible Causes** ● Enable Service Spark Command with Scan Tool - adjust timing ● Check engine for possible source of an engine knock condition ● This vehicle requires the use of premium fuel - verify its use ● Knock sensor is damaged or has failed ● PCM has failed
P0326 **1T CCM, MIL: No** 1995, 1996, 1997 Camaro & Firebird 3.8L VIN K engine Transmissions: All	**Knock Sensor Noise Channel High Input Conditions:** DTC P0117, P0118, P0122, P0123, P0502 and P0503 not set engine started, engine speed from 2500-2900 rpm for 10 seconds, ECT sensor more than 149°F, throttle angle over 1.5%, and the PCM detected the Knock Sensor noise channel voltage indicated more than 5.0v. **Possible Causes** ● Knock sensor circuit is open between the sensor and the PCM ● Knock sensor signal wire routed close to the spark plug wires (causes EMI/RFI) ● Knock sensor is damaged or has failed ● PCM has failed
P0326 **1T CCM, MIL: No** 1995 S/T Blazer & S/T Pickup 4.3L VIN W engine Transmissions: All	**Knock Sensor Range/Performance Conditions:** Engine runtime over 2 minutes, ECT sensor over 140°F, engine speed over 2200 rpm, TP angle over 6%, and the PCM detected the KS AC sensor was less than a minimum value. **Possible Causes** ● Knock sensor circuit is open between the sensor and the PCM ● Knock sensor signal wire routed close to the spark plug wires (causes EMI/RFI) ● Knock sensor is damaged or has failed ● PCM has failed
P0327 **1T CCM, MIL: No** 1996, 1997, 1998, 1999, 2000 Achieva, Beretta, Ciera, Corsica, Century, Grand Am, Malibu & Skylark 3.1L VIN M Transmissions: All	**Knock Sensor Circuit Malfunction Conditions:** DTC P0101-P0103, P0116-P0118, P0121-P0123, P0125, P0336, P0341, P0502, P0503, P1114, P1115, P1121, P1122 and P1336 not set, engine speed from 1000-5000 rpm, system voltage over 10.0v, ECT sensor more than 140°F, TP angle from 10-15%, engine load more than 45%, maximum spark retard less than 15 degrees, and the PCM detected the Knock sensor signal was within a calculated average voltage range for 10 seconds. **Possible Causes** ● Knock sensor signal circuit is shorted to ground ● Knock sensor is damaged or has failed ● PCM has failed ● TSB 02-06-04-045 contains a repair procedure for this code

OBD II Trouble Code List (P0xxx Codes)

DTC	Trouble Code Title, Conditions & Possible Causes
P0327 **2T CCM, MIL: Yes** 1996, 1997, 1998, 1999, 2000, 2001, 2002, 2003, 2004, 2005 Aurora, Century, LeSabre, Park Avenue, Regal 3.1L VIN J, 3.1L VIN M, 3.4L VIN X, 3.5L VIN H, 3.8L VIN 1, 3.8L VIN K Transmissions: All	**Knock Sensor Circuit Malfunction Conditions:** DTC P0101-P0103, P0116-P0118, P0121-P0125, P0128, P0336, P0341, P0502-P0503, P1114-P1115, P1121-P1122 and P1336 not set, engine speed from 1000-3000 rpm for 30 seconds, system voltage over 10.0v, TP angle from 10-15%, engine load above 45, ECT sensor more than 140ºF, maximum spark retard less than 15 degrees, and the PCM detected the Knock sensor signal was within an assigned average range for 30 seconds. **Possible Causes** ● Knock sensor signal circuit is open, shorted to ground or power ● Knock sensor ground circuit is open (check for proper torque) ● Knock sensor is damaged or has failed ● PCM has failed
P0327 **2T CCM, MIL: Yes** 1996, 1997 Camaro, Caprice, Corvette, Firebird, Fleetwood 4.3L VIN W, 5.7L VIN 5, 5.7L VIN P Transmissions: All	**Knock Sensor Circuit Low Input (Bank 1) Conditions:** DTC P0117, P0118, P0122, P0123, P0125, P1114 and P1115 not set, engine started, engine runtime over 20 seconds, ECT sensor more than 176ºF, engine speed from 1100-2500 rpm, TP angle more than 25%, and the PCM detected an unexpected low frequency on the Knock Sensor circuit for over 3 seconds during the CCM test. **Possible Causes** ● Knock sensor signal circuit is shorted to ground ● Knock sensor is damaged or has failed ● PCM has failed
P0327 **1T CCM, MIL: No** 1996, 1997, 1998 C/K Series Truck, G Series Van, Van Cargo, L, M, P, S, T 4.3L VIN W, 4.3L VIN X, 5.0L VIN M, 5.7L VIN K, 5.7L VIN R, 7.4L VIN J Transmissions: All	**Knock Sensor Noise Channel Low Input Conditions:** DTC P0117, P0118, P0121 and P0122 not set, engine started, engine speed from 500-900 rpm for 2 minutes, ECT sensor more than 140ºF, then with the Knock Sensor updated completed, TP angle less than 5.9%, engine speed from 2000-3000 rpm with the spark timing retard less than 0 degrees, the PCM detected that the ESC noise channel signal voltage indicated less than 500 mv. **Possible Causes** ● Knock sensor signal circuit is open, shorted to ground or power ● Knock sensor ground circuit is open (check for proper torque) ● Knock sensor is damaged or has failed ● PCM has failed
P0327 **2T CCM, MIL: Yes** 1999, 2000, 2001, 2002, 2003, 2004, 2005 C/K Series Truck, G Series Van, Van Cargo, L, M, P, S, T 4.3L VIN W, 4.3L VIN X, 5.0L VIN M, 5.7L VIN K, 5.7L VIN R, 7.4L VIN J Transmissions: All	**Knock Sensor Circuit Low Input (Bank 1) Conditions:** DTC P0117, P0118, P0121, P0122, P0123, P0125, P1114, P1115, P1121, P1122 and P1258 not set, engine speed from 475-975 for 10 seconds, ECT sensor over 140ºF, system voltage over 10.0v, minimum noise level learned, then with the engine speed from 1500-3000 rpm for 10 seconds, the MAP sensor less than 49 kPa, TP angle over 0%, and the PCM detected the Knock sensor was within an assigned average range for 9 seconds. **Possible Causes** ● Knock sensor signal circuit is open, shorted to ground or power ● Knock sensor ground circuit is open (check for proper torque) ● Knock sensor is damaged or has failed ● PCM has failed
P0327 **2T CCM, MIL: Yes** 1999, 2000, 2001, 2002, 2003, 2004, 2005 C/K Series Truck, G Series Van, Envoy, Escalade, TrailBlazer 4.8L VIN V, 5.3L VIN P, 5.3L VIN T, 5.3L VIN Z, 6.0L VIN U CNG engine, 6.0L VIN N, 8.1L VIN G Transmissions: All	**Knock Sensor Circuit Low Input (Bank 1) Conditions:** DTC P0117, P0118 and P0125 not set, engine runtime 10 seconds, minimum noise level learned with the engine speed from 475-975 rpm, then with the engine speed from 1500-3000, ECT sensor more than 140ºF, MAP sensor under 49 kPa, TP angle over 0%, system voltage over 10.0v, the PCM detected the Knock Sensor signal was within a calculated voltage range or no signal existed for 9 seconds. **Possible Causes** ● Knock sensor signal circuit is open, shorted to ground or power ● Knock sensor ground circuit is open (check for proper torque) ● Knock sensor is damaged or it has failed ● PCM has failed
P0327 **2T CCM, MIL: Yes** 1996, 1997, 1998, 1999, 2000, 2001, 2002, 2003, 2004, 2005 Alero, Cutlass, Impala, Malibu, Montana, Venture Silhouette, Rendezvous 3.1L VIN J, 3.1L VIN M, 3.4L VIN E engines Transmissions: All	**Knock Sensor Circuit Malfunction Conditions:** DTC P0101, P0102, P0103, P0116, P0117, P0118, P0121, P0122, P0123, P0125, P0336, P0341, P0502, P0503, P1114, P1115, P1121, P1122 and P1336 not set, engine started, engine speed 1000-5000 rpm for over 30 seconds, ECT sensor over 140ºF, engine load over 45%, TP angle over 1.5%, maximum spark retard less than 15 degrees, and the PCM detected the Knock Sensor signal was within the average voltage range for over 10 seconds. **Possible Causes** ● Knock sensor signal circuit is open, shorted to ground or power ● Knock sensor ground circuit is open (check for proper torque) ● Knock sensor is damaged or has failed ● PCM has failed

OBD II Trouble Code List (P0xxx Codes)

DTC	Trouble Code Title, Conditions & Possible Causes
P0327 **2T CCM, MIL: Yes** 2003, 2004, 2005 CTS (Cadillac) 2.6L VIN M, 3.2L VIN N Transmissions: All	**Knock Sensor 1 Circuit Low Input (Bank 1) Conditions:** Engine started; engine speed more than 2000 rpm, volumetric efficiency more than 40%; and the PCM detected the KS signal was too low, or that the KS signal was missing. The KS system is used to detect any engine detonation, or spark knock. The ECM will retard the spark timing based on the signals from the KS. The KS produces an AC voltage proportional to the amount of knock to the PCM. An operating engine produces a normal amount of engine mechanical vibration, or noise. The knock sensors will produce an AC voltage signal from this noise. When the engine is operating, the ECM will learn the minimum and maximum amplitude of the noise produced. **Possible Causes** ● Knock sensor connector is damaged or shorted ● Knock sensor signal circuit is shorted to ground ● Knock sensor is damaged or it has failed ● PCM has failed
P0327 **2T CCM, MIL: Yes** 1996, 1997, 1998, 1999, 2000, 2001, 2002, 2003, 2004, 2005 Aurora, DeVille, Eldorado & Seville 4.0L VIN C, 4.6L VIN 9, 4.6L VIN Y engines Transmissions: A/T	**Knock Sensor Circuit Low Input Conditions:** Engine started; engine speed over 3000 rpm, ECT sensor more than 104ºF, TP angle over 5%, system voltage over 10.0v and the PCM detected the Knock sensor was below its minimum calibrated range. **Possible Causes** ● Knock sensor signal circuit is open, shorted to ground or power ● Knock sensor ground circuit is open (check for proper torque) ● Knock sensor is damaged or has failed ● PCM has failed
P0327 **2T CCM, MIL: Yes** 1995, 1996, 1997, 1998, 1999, 2000, 2001, 2002 Camaro & Firebird 3.8L VIN K engine Transmissions: All	**Knock Sensor Circuit Malfunction (Bank 1) Conditions:** DTC P0101-P0103, P0116-P0118, P0121-P0123, P0128, P0336, P0341, P0502, P0503, P1114-P1122 and P1336 not set, engine started, engine runtime 30 seconds at 1000-2500 rpm, ECT sensor more than 140ºF, TP angle over 10%, engine load over 40%, and the PCM detected the Knock Sensor signal remained with the assigned voltage range for 30 seconds. **Possible Causes** ● Knock sensor signal circuit is open, shorted to ground or power ● Knock sensor ground circuit is open (check for proper torque) ● Knock sensor is damaged or has failed ● PCM has failed
P0327 **2T CCM, MIL: Yes** 1997, 1998, 1999, 2000, 2001, 2002, 2003, 2004, 2005 Camaro, Corvette & Firebird 5.7L VIN G, 5.7L VIN S Transmissions: All	**Knock Sensor Circuit Malfunction (Bank 1) Conditions:** DTC P0117, P0118, P0121-P0123, P0125, P1114, P1115, P1120 and P1122 not set, engine started, engine speed from 475-975 for 10 seconds, system voltage from 10-18v, ECT sensor more than 140ºF, minimum noise level learned, then with the engine speed from 1500-3000 rpm, MAP sensor less than 49 kPa, TP angle more than 0% for 10 seconds, the PCM detected the KS signal was less than or more than the expected amount for 9 seconds. **Possible Causes** ● Knock sensor signal circuit is open, shorted to ground or power ● Knock sensor ground circuit is open (check for proper torque) ● Knock sensor is damaged or has failed ● PCM has failed
P0327 **2T CCM, MIL: Yes** 2002, 2003, 2004, 2005 Envoy & TrailBlazer 4.2L VIN S Transmissions: All	**Knock Sensor Circuit Low Input (Bank 1) Conditions:** DTC P0117, P0118, P0122 and P0123 not set, engine runtime over 20 seconds, minimum noise level learned, engine speed from 2,000-6,400 rpm, ECT sensor more than 158ºF, system voltage over 10.0v, MAP sensor more than 60 kPa, TP angle over 0%, and the PCM detected the Knock sensor signal was not within an assigned average range or that the Knock sensor was missing during the test. **Possible Causes** ● Knock sensor signal circuit is open, shorted to ground or power ● Knock sensor ground circuit is open (check for proper torque) ● Knock sensor is damaged or has failed ● PCM has failed
P0327 **1T CCM, MIL: No** 1996, 1997, 1998, 1999, 2000, 2001, 2002 Century, Grand Prix, Intrigue, Lumina, Monte Carlo, Regal 3.8L VIN K engine Transmissions: All	**Knock Sensor Circuit Malfunction Conditions:** DTC P0101-P0103, P0116-P0118, P0121-P0123, P0125, P0128, P0336, P0341, P0502, P0503, P1114, P1115, P1121, P1122 and P1336 not set, engine started, engine runtime 30 seconds at 1000-2500 rpm, ECT sensor more than 140ºF, TP angle over 10%, engine load over 40%, and the PCM detected the Knock Sensor signal was within a predetermined voltage range for 10 seconds during the test. **Possible Causes** ● Knock sensor signal circuit is open, shorted to ground or power ● Knock sensor ground circuit is open (check for proper torque) ● Knock sensor is damaged or has failed ● PCM has failed

OBD II Trouble Code List (P0xxx Codes)

DTC	Trouble Code Title, Conditions & Possible Causes
P0327 **1T CCM, MIL: No** 1996, 1997 Grand Prix, Lumina, Monte Carlo 3.4L VIN X Transmissions: All	**Knock Sensor Circuit Malfunction Conditions:** DTC P0117, P0118, P0121, P0122, P0502 and P0503 not set, engine started, engine speed from 2600-2900 rpm, ECT sensor more than 149ºF, TP angle over 1%, and the PCM detected the Knock Sensor signal was less than 0.40v during the CCM test. **Possible Causes** ● Knock sensor signal circuit is shorted to ground ● Knock sensor is damaged or has failed ● PCM has failed
P0328 **2T CCM, MIL: Yes** 2003, 2004, 2005 CTS (Cadillac) 2.6L VIN M, 3.2L VIN N Transmissions: All	**Knock Sensor 2 Circuit Low Input (Bank 2) Conditions:** Engine started; engine speed more than 2000 rpm, volumetric efficiency more than 40%; and the PCM detected the KS signal was too low, or that the KS signal was missing. The KS system is used to detect any engine detonation, or spark knock. The ECM will retard the spark timing based on the signals from the KS. The KS produces an AC voltage proportional to the amount of knock to the PCM. An operating engine produces a normal amount of engine mechanical vibration, or noise. The knock sensors will produce an AC voltage signal from this noise. When the engine is operating, the ECM will learn the minimum and maximum amplitude of the noise produced. **Possible Causes** ● Knock Sensor 2 connector is damaged or open ● Knock Sensor 2 circuit is shorted to ground ● Knock Sensor 2 is damaged or it has failed ● PCM has failed
P0330 **1T CCM, MIL: No** 1997, 1998, 1999, 2000, 2001 Catera 3.0L V6 VIN R engine Transmissions: A/T	**Knock Sensor Circuit Malfunction (Bank 2) Conditions:** Engine runtime at idle 5 seconds (to learn the minimum noise level), ECT sensor more than 104ºF, vehicle driven at engine speed of over 2,000 rpm, and the PCM did not detect any change between the KS noise level at idle and noise level at 2500-3000 rpm for 5 seconds. **Possible Causes** ● Knock sensor signal circuit is open, shorted to ground or power ● Knock sensor ground circuit is open (i.e., not mounted properly) ● Knock sensor is damaged or has failed ● PCM has failed
P0332 **1T CCM, MIL: No** 1996, 1997 Camaro, Caprice, Corvette, Firebird, Fleetwood 4.3L VIN W, 5.7L VIN 5, 5.7L VIN P Transmissions: All	**Knock Sensor Circuit Low Input (Bank 2) Conditions:** DTC P0117, P0118, P0122, P0123, P0125, P1114 and P1115 not set, engine started, engine runtime over 20 seconds, ECT sensor more than 176ºF, engine speed from 1100-2500 rpm, TP angle more than 25%, and the PCM detected an unexpected low frequency on the Knock Sensor circuit for over 3 seconds during the CCM test. **Possible Causes** ● Knock sensor signal circuit is shorted to ground ● Knock sensor is damaged or has failed ● PCM has failed
P0332 **2T CCM, MIL: Yes** 1998, 1999, 2000, 2001, 2002, 2003, 2004, 2005 Aurora, Bonneville, Century, Cutlass Supreme, Grand Prix, Intrigue, LeSabre, LSS, Lumina, Monte Carlo, 88', 98', Regal, Park Avenue 3.8L VIN 1, 3.8L VIN K Transmissions: All	**Knock Sensor Circuit Malfunction (Bank 2) Conditions:** DTC P0101-P0103, P0116-P0118, P0121-P0123, P0125, P0128, P0336, P0341, P0502, P0503, P1114, P1115, P1121 and P1336 not set, engine started, engine speed at 1000-4000 rpm for 30 seconds, system voltage over 10.0v, ECT sensor over 140ºF, TP angle from 3-15%, engine load 20-45%, spark retard less than 15 degrees, and the PCM detected an invalid voltage on the Knock sensor circuit. **Possible Causes** ● Knock sensor signal circuit is open, shorted to ground or power ● Knock sensor ground circuit is open (i.e., not mounted properly) ● Knock sensor is damaged or has failed ● On modules with an integrated sensor, clear the codes and retest for codes. If the same code resets, the PCM has failed.
P0332 **2T CCM, MIL: Yes** 1999, 2000, 2001, 2002, 2003, 2004, 2005 C/K Series Truck, G Series Van, Van Cargo, L/M Series Van 5.0L VIN M, 7.4L VIN J Transmissions: All	**Knock Sensor Circuit Low Input (Bank 2) Conditions:** DTC P0117-P0118, P0121-P0125, P1114-1115, P1121, P1122 and P1258 not set, engine started, engine speed from 475-975 for 10 seconds, ECT sensor more than 140ºF, system voltage over 10.0v, minimum noise level learned, then with the engine speed from 1500-3000 rpm for 10 seconds, MAP sensor less than 49 kPa, TP angle over 0%, the PCM detected the Knock Sensor signal was within an assigned average range for 9 seconds. **Possible Causes** ● Knock sensor signal circuit is open, shorted to ground or power ● Knock sensor ground circuit is open (check for proper torque) ● Knock sensor is damaged or has failed ● PCM has failed

OBD II Trouble Code List (P0xxx Codes)

DTC	Trouble Code Title, Conditions & Possible Causes
P0332 **2T CCM, MIL: Yes** 1999, 2000, 2001, 2002, 2003, 2004, 2005 C/K Series Truck, G Series Van, Envoy, Escalade, TrailBlazer 4.8L VIN V, 5.3L VIN P, 5.3L VIN T, 5.3L VIN Z, 6.0L VIN N, 6.0L VIN U CNG engine, 8.1L VIN G Transmissions: All	**Knock Sensor Circuit Low Input (Bank 2) Conditions:** DTC P0117, P0118 and P0125 not set, engine runtime 10 seconds, minimum noise level learned with the engine speed from 475-975 rpm, then with the engine speed from 1500-3000, ECT sensor more than 140ºF, MAP sensor under 49 kPa, TP angle over 0%, system voltage over 10.0v, the PCM detected the Knock Sensor signal was within a calculated voltage range or no signal existed for 9 seconds. **Possible Causes** ● Knock sensor signal circuit is open, shorted to ground or power ● Knock sensor ground circuit is open (check for proper torque) ● Knock sensor is damaged or it has failed ● PCM has failed
P0332 **2T CCM, MIL: Yes** 2003, 2004, 2005 CTS (Cadillac) 2.6L VIN M, 3.2L VIN N Transmissions: All	**Knock Sensor 1 Circuit Low Input (Bank 1) Conditions:** Engine started; engine speed more than 2000 rpm, volumetric efficiency more than 40%; and the PCM detected the KS signal was too low, or that the KS signal was missing. The KS system is used to detect any engine detonation, or spark knock. The ECM will retard the spark timing based on the signals from the KS. The KS produces an AC voltage proportional to the amount of knock to the PCM. An operating engine produces a normal amount of engine mechanical vibration, or noise. The knock sensors will produce an AC voltage signal from this noise. When the engine is operating, the ECM will learn the minimum and maximum amplitude of the noise produced. **Possible Causes** ● Knock Sensor 1 connector is damaged, open or shorted ● Knock Sensor 1 signal circuit is shorted to ground ● Knock Sensor 1 is damaged, open or shorted to ground ● PCM has failed
P0333 **2T CCM, MIL: Yes** 2003, 2004, 2005 CTS (Cadillac) 2.6L VIN M, 3.2L VIN N Transmissions: All	**Knock Sensor 2 Circuit High Input (Bank 2) Conditions:** Engine started; engine speed more than 2000 rpm, volumetric efficiency more than 40%, and the PCM detected the KS signal was higher than the normal operating range. The KS system is used to detect any engine detonation, or spark knock. The ECM will retard the spark timing based on the signals from the KS. The KS produces an AC voltage proportional to the amount of knock to the PCM. An operating engine produces a normal amount of engine mechanical vibration, or noise. The knock sensors will produce an AC voltage signal from this noise. When the engine is operating, the ECM will learn the minimum and maximum amplitude of the noise produced. **Possible Causes** ● Knock Sensor 2 connector is damaged or shorted ● Knock Sensor 2 circuit is shorted to system power ● Knock Sensor 2 is damaged or it has failed ● PCM has failed
P0332 **2T CCM, MIL: Yes** 1997, 1998, 1999, 2000, 2001, 2002, 2003, 2004, 2005 Camaro, Corvette & Firebird 5.7L VIN G, 5.7L VIN S Transmissions: All	**Knock Sensor Circuit Malfunction (Bank 2) Conditions:** DTC P0117, P0118, P0121-P0123, P0125, P1114, P1115, P1120 and P1122 not set, engine started, engine speed from 475-975 for 10 seconds, system voltage from 10-18v, ECT sensor more than 140ºF, minimum noise level learned, then with the engine speed from 1500-3000 rpm, MAP sensor less than 49 kPa, TP angle more than 0% for 10 seconds, the PCM detected the KS signal was less than or more than the expected amount for 9 seconds. **Possible Causes** ● Knock sensor signal circuit is open, shorted to ground or power ● Knock sensor ground circuit is open (check for proper torque) ● Knock sensor is damaged or has failed ● PCM has failed ● TSB 02-06-04-023A contains a repair procedure for this code
P0332 **2T CCM, MIL: Yes** 1998, 1999, 2000, 2001, 2002 Camaro & Firebird, Grand Prix 3.8L VIN K engine Transmissions: All	**Knock Sensor Circuit Malfunction (Bank 2) Conditions:** DTC P0101-P0103, P0116-P0118, P0121-P0123, P0128, P0336, P0341, P0502, P0503, P1114-P1122 and P1336 not set, engine started, engine runtime 30 seconds at 1000-2500 rpm, ECT sensor more than 140ºF, TP angle over 10%, engine load over 40%, and the PCM detected the Knock Sensor signal remained with the assigned voltage range for 30 seconds during the CCM test. **Possible Causes** ● Knock sensor signal circuit is open, shorted to ground or power ● Knock sensor ground circuit is open (check for proper torque) ● Knock sensor is damaged or has failed ● PCM has failed

OBD II Trouble Code List (P0xxx Codes)

DTC	Trouble Code Title, Conditions & Possible Causes
P0332 **2T CCM, MIL: Yes** 2002, 2003, 2004, 2005 Envoy & TrailBlazer 4.2L VIN S Transmissions: All	**Knock Sensor Circuit Low Input (Bank 2) Conditions:** DTC P0117, P0118, P0122 and P0123 not set, engine runtime over 20 seconds, minimum noise level learned, engine speed from 2,000-6,400 rpm, ECT sensor more than 158°F, system voltage over 10.0v, MAP sensor more than 60 kPa, TP angle over 0%, and the PCM detected the Knock sensor signal was not within an assigned average range or that the Knock sensor signal was missing during the test. **Possible Causes** ● Knock sensor signal circuit is open, shorted to ground or power ● Knock sensor ground circuit is open (check for proper torque) ● Knock sensor is damaged or has failed ● PCM has failed
P0335 **1T CCM, MIL: Yes** 1996, 1997, 1998, 1999 Achieva, Beretta, Ciera, Cavalier, Corsica, Grand Am, Century, Malibu, Skylark, Sunfire, S/T Blazer & O/T Pickup 2.2L VIN 4, 2.2L VIN 5, 2.4L VIN T engines Transmissions: All	**Crankshaft Position Sensor Circuit Malfunction Conditions:** DTC P0341 not set, engine started; and the PCM detected the CKP resynchronization counter incremented to more than 15 within 4 minutes and 15 seconds during the CCM test due an unexpected loss of the Medium Resolution 7X signal to the ICM. **Possible Causes** ● CKP sensor signal (+) circuit is open or shorted to ground ● CKP sensor signal (-) circuit is open or shorted to ground ● CKP sensor is damaged or has failed ● ICM is damaged or has failed ● PCM has failed
P0335 **1T CCM, MIL: No** 1996, 1997 Camaro, Caprice, Corvette, Firebird, Fleetwood 4.3L VIN W, 5.7L VIN 5, 5.7L VIN P Transmissions: All	**DI Crankshaft Position Sensor Circuit Malfunction Conditions:** Engine speed from 500-4000 rpm, and the PCM detected the CKP low-resolution angle was not within the normal deviation range of -21° to +51°, or it did not detect any CKP signals. **Possible Causes** ● CKP sensor connection problems, it is damaged or has failed ● Crankshaft reluctor wheel is damaged or improper installation ● Distributor mechanical fault (i.e., too much distributor backlash) ● Engine front cover is damaged ● Engine timing gear is worn excessively or the gear is worn out ● PCM has failed
P0335 **2T CCM, MIL: Yes** 1999, 2000, 2001, 2002, 2003, 2004, 2005 C/K Series Truck, G Series Van, Van Cargo, L/M Series Van 4.3L VIN W, 4.3L VIN X, 5.0L VIN M, 7.4L VIN J Transmissions: All	**CKP Sensor Circuit Malfunction Conditions:** DTC P0101, P0102, P0103, P0341, P0342 and P0343 not set; engine cranking, CMP signal varying, MAF sensor more than 3 g/sec, and the PCM did not detect any signals from the CKP sensor for less than 8 seconds during the CCM test period. **Possible Causes** ● CKP sensor signal circuit is open or shorted to ground ● CKP sensor ground (low reference) circuit is open ● CKP sensor power circuit is open between sensor and the PCM ● Crankshaft reluctor wheel is damaged or improper installation ● PCM has failed
P0335 **1T CCM, MIL: Yes** 1999, 2000, 2001, 2002, 2003, 2004, 2005 C/K Series Truck, G Series Van, Envoy, Escalade, TrailBlazer 4.8L VIN V, 5.3L VIN P, 5.3L VIN T, 5.3L VIN Z, 6.0L VIN N, 6.0L VIN U CNG engine, 8.1L VIN G Transmissions: All	**CKP Sensor Circuit Malfunction Conditions:** DTC P0101-P0103, P0341, P0342 and P0343 not set, engine cranking, CMP signal incrementing, MAF sensor over 3 g/sec, and the PCM did not detect any CKP sensor signals for 8 seconds. **Possible Causes** ● CKP sensor signal circuit is open or shorted to ground ● CKP sensor ground (low reference) circuit is open ● CKP sensor power circuit is open between sensor and the PCM ● Crankshaft reluctor wheel is damaged or improper installation ● CKP sensor contacting the reluctor wheel or it has failed ● PCM has failed
P0335 **1T CCM, MIL: Yes** 1996, 1997, 1998, 1999, 2000, 2001, 2002, 2003, 2004, 2005 C/K Series Truck, G Series Van, Van Cargo 6.5L VIN F, 6.5L VIN S, 6.5L VIN Y Transmissions: All	**CKP Sensor Signal Range/Performance Conditions:** Engine cranking, engine speed under 300 rpm, and the PCM detected that 8 consecutive CKP sensor (Hall Effect 4X) signals were missing for 8 Cylinder 1 events; or engine running, the PCM detected 8 consecutive CKP signals were missing for 32 Cylinder 1 events. **Possible Causes** ● CKP sensor signal circuit is open or shorted to ground ● CKP sensor VREF circuit is open between the sensor and PCM ● CKP sensor ground circuit is open ● CKP sensor is damaged or it failed (check crankshaft sprocket) ● PCM has failed

OBD II Trouble Code List (P0xxx Codes)

DTC	Trouble Code Title, Conditions & Possible Causes
P0335 **1T CCM, MIL: No** 2001, 2002, 2003, 2004, 2005 C/K Series Trucks 6.6L VIN 1 Diesel engine Transmissions: All	**CKP Sensor Signal Range/Performance Conditions:** Engine cranking or engine running, CMP sensor signals detected, and the PCM did not detect any CKP sensor (Hall Effect 57-1) signals for less than 8 seconds in the CCM test. **Possible Causes** • CKP sensor signal circuit is open or shorted to ground • CKP sensor VREF circuit is open between the sensor and PCM • CKP sensor ground circuit is open • CKP sensor is damaged or it failed (check crankshaft sprocket) • PCM has failed
P0335 **1T CCM, MIL: No** 2000, 2001, 2002, 2003, 2004, 2005 Aurora, DeVille, Eldorado, Seville 3.5L VIN H, 4.0L VIN C, 4.6L VIN 9, 4.6L VIN Y Transmissions: All	**CKP Sensor 'A' Circuit Malfunction Conditions:** Engine cranking or running; and the PCM did not detect any CKP Sensor 'A' signals during the CCM test. **Possible Causes** • CKP sensor signal (+) circuit is open or shorted to ground • CKP sensor signal (-) circuit is open or shorted to ground • CKP sensor is damaged or has failed • PCM has failed
P0335 **2T CCM, MIL: Yes** 2003, 2004, 2005 CTS (Cadillac) 2.6L VIN M, 3.2L VIN N Transmissions: All	**Crankshaft Position Sensor Circuit Malfunction Conditions:** DTC P0341, P0342 and P0343 not set, engine cranking or running; and the PCM did not detect any signals from the CKP sensor. **Possible Causes** • CKP sensor connector is damaged, open or shorted • CKP sensor positive (+) circuit or (-) circuit is open or shorted to ground • CKP sensor is physically damaged or it is improperly installed • CKP sensor has failed • Electromagnetic interference in the CKP sensor circuits • Excessive air gap between the CKP sensor and reluctor ring • Foreign material lodged between CKP sensor and reluctor ring • PCM has failed
P0335 **2T CCM, MIL: Yes** 1997, 1998, 1999, 2000, 2001, 2002, 2003, 2004, 2005 Camaro, Corvette & Firebird 5.7L VIN G, 5.7L VIN S Transmissions: All	**Crankshaft Position Sensor Circuit Malfunction Conditions:** DTC P0101, P0102, P0103, P0341, P0342 and P0343 not set, engine cranking, CMP sensor signals transitioning, MAF sensor more than 3 g/sec, and the PCM did not detect any signals from the CKP sensor (Hall Effect) for up 4-8 seconds during the CCM test. **Possible Causes** • CKP sensor signal circuit is open or shorted to ground • CKP sensor VREF circuit is open between the sensor and PCM • CKP sensor ground (Low Reference) circuit is open • CKP sensor is damaged or it failed (check crankshaft reluctor) • PCM has failed
P0335 **2T CCM, MIL: Yes** 2001, 2002, 2003, 2004, 2005 S/T Blazer & S/T Pickup 4.3L VIN W, 4.3L VIN X Transmissions: All	**Crankshaft Position Sensor Circuit Malfunction Conditions:** DTC P0101, P0102, P0103 and P0341 not set, engine cranking, CMP sensor signal transitioning, MAF sensor over 3 g/sec or MAF sensor more than 5 g/sec in "run" mode, and the PCM did not detect any CKP sensor signals for 3 seconds during the CCM test period. **Possible Causes** • CKP sensor signal circuit is open or shorted to ground • CKP sensor VREF circuit is open between the sensor and PCM • CKP sensor ground (Low Reference) circuit is open • CKP sensor is damaged or it failed (check crankshaft reluctor) • PCM has failed • Vehicle may be have been driven until it was run completely out of fuel
P0335 **2T CCM, MIL: Yes** 2002, 2003, 2004, 2005 Envoy & TrailBlazer 4.2L VIN S Transmissions: All	**Crankshaft Position Sensor Circuit Malfunction Conditions:** DTC P0562 not set, engine started; system voltage less than 18v, and the PCM did not detect any CKP sensor signals. **Possible Causes** • CKP sensor signal (+) circuit or (-) circuit is open or shorted to ground • CKP sensor is damaged or has failed • PCM has failed
P0335 **1T CCM, MIL: No** 1997, 1998, 1999, 2000, 2001 Catera 3.0L V6 VIN R engine Transmissions: All	**Crankshaft Position Sensor Circuit Malfunction Conditions:** Engine cranking or running; CMP sensor signals transitioning, and the PCM did not detect any CKP sensor pulses during the CCM test. **Possible Causes** • CKP sensor signal (+) circuit or (-) circuit is open or shorted to ground • CKP sensor is damaged or failed (check the reluctor wheel) • PCM has failed

OBD II Trouble Code List (P0xxx Codes)

DTC	Trouble Code Title, Conditions & Possible Causes
P0336 **2T CCM, MIL: Yes** 1996, 1997, 1998, 1999, 2000, 2001, 2002 ; Achieva, Beretta, Ciera, Corsica, Century, Grand Am, Malibu, Skylark 3.1L VIN M Transmissions: All	**Crankshaft Reference 24X Circuit Malfunction Conditions:** Engine started; 3X signals detected for 3 seconds, and the PCM detected the ratio of Crankshaft Sensor 'A' signals (24X to 3X REF Hall Effect pulses) was incorrect in the test. **Possible Causes** ● CKP sensor signal circuit is open or shorted to ground ● CKP sensor VREF circuit is open between the sensor and PCM ● CKP sensor ground (Low Reference) circuit is open ● CKP sensor is damaged or it failed (check crankshaft reluctor) ● CKP sensor wiring routed close to spark plug wires (EMI/RFI) ● PCM has failed
P0336 **1T CCM, MIL: No** 1996, 1997 Camaro, Caprice, Corvette, Firebird, Fleetwood, Roadmaster 4.3L VIN W, 5.7L VIN 5, 5.7L VIN P Transmissions: All	**DI Crankshaft Position Sensor Range/Performance Conditions:** Engine started; engine speed from 500-4000 rpm, and the PCM detected a deviation of from -10° to +7° on the DI CKP low-resolution signal while it was in the -21° to +51° deviation range during the test. **Possible Causes** ● CKP sensor connection problems, it is damaged or has failed ● Crankshaft reluctor wheel is damaged or improper installation ● Distributor mechanical fault (i.e., too much distributor backlash) ● Engine front cover damaged or timing gear is worn excessively ● PCM has failed
P0336 **2T CCM, MIL: Yes** 2000, 2001, 2002, 2003, 2004, 2005 Alero, Century, Grand Prix, Impala, Monte Carlo, Intrigue, Lumina, Malibu, Montana, Silhouette, Rendezvous, Venture 3.1L VIN J, 3.1L VIN M, 3.4L VIN E engines Transmissions: All	**Crankshaft Reference 24X Circuit Malfunction Conditions:** Engine started; 3X signals detected for 3 seconds, and the PCM detected an invalid ratio of 24X to 3X CKP REF pulses. The circuit uses 2 different types of crankshaft position (CKP) sensors. The CKP Sensor 'A' connects directly to the PCM through the 12v VREF, Medium Resolution engine speed signal and the low reference circuits. The CKP Sensor 'B' connects directly to the ignition control (IC) module via the CKP 'B' signal and low reference circuits. **Possible Causes** ● CKP sensor signal circuit is open or shorted to ground ● CKP sensor ground (Low Reference) circuit is open ● CKP sensor is damaged or it failed (check crankshaft reluctor) ● PCM has failed
P0336 **2T CCM, MIL: Yes** 1996, 1997, 1998, 1999, 2000, 2001, 2002, 2003, 2004, 2005 Aurora, Bonneville, Century, Cutlass Supreme, Grand Prix, Intrigue, LeSabre, LSS, Lumina, Monte Carlo, 88', Park Avenue, Regal, 98' 3.1L VIN J, 3.5L VIN H, 3.8L VIN 1, 3.8L VIN K Transmissions: All	**Crankshaft Reference 18X Circuit Malfunction Conditions:** Engine started; 3X REF signals detected, and the PCM did not detect any 18X pulses, or the ratio of 18X REF pulses to 3X REF pulses did not equal 6:1, or ratio of 3X REF pulses to CMP pulses equaled 6:1, conditions met for 290 of 300 samples. The Crankshaft Position Sensor (CKP) circuit uses two types of CKP sensors. CKP Sensor 'B' is connected directly to the ignition control module (ICM), while CKP sensor 'A' connects directly to the PCM. **Possible Causes** ● CKP sensor signal circuit is open or shorted to ground ● CKP sensor VREF circuit is open, or the ground (Low Reference) circuit is open ● CKP sensor is damaged or it failed (check the crankshaft reluctor) ● CKP sensor wiring routed close to spark plug wires (EMI/RFI) ● PCM has failed
P0336 **2T CCM, MIL: Yes** 1996, 1997, 1998, 1999, 2000, 2001, 2002, 2003, 2004, 2005 C/K Series Truck, G Van, Van Cargo & L/M Van 4.3L VIN W, 4.3L VIN X, 5.0L VIN M, 5.7L VIN K, 5.7L VIN R, 7.4L VIN J Transmissions: All	**Crankshaft Reference Sensor Range/Performance Conditions:** Engine cranking or engine running, CMP sensor signals detected (more than 4), and the PCM detected an invalid Crankshaft Position sensor signal occurred for over 3 seconds. **Possible Causes** ● CKP sensor signal circuit or ground circuit has a high resistance condition ● Crankshaft reluctor wheel is damaged or improper installation ● CKP sensor contacting reluctor, or crankshaft turns backwards if clutch out (stalls - M/T) ● PCM has failed ● Vehicle is running out of fuel
P0336 **2T CCM, MIL: Yes** 1999, 2000, 2001, 2002, 2003, 2004, 2005 C/K Series Truck, G Series Van, Envoy, Escalade, TrailBlazer 4.8L VIN V, 5.3L VIN P, 5.3L VIN T, 5.3L VIN Z, 6.0L VIN N, 6.0L VIN U CNG engine, 6.6L VIN 1 Diesel engine, 8.1L VIN G Transmissions: All	**Crankshaft Reference Sensor Signal Range/Performance Conditions:** Engine cranking or engine running, CMP sensor signals detected, and the PCM detected the CKP sensor signal was out-of-range for a period of less than 2 seconds during the CCM continuous test. **Possible Causes** ● CKP sensor connector is damaged, loose or shorted ● CKP sensor signal circuit is open or shorted (intermittent fault) ● CKP sensor contacting the reluctor wheel or it has failed ● CKP sensor is damaged or it has failed ● PCM has failed

OBD II Trouble Code List (P0xxx Codes)

DTC	Trouble Code Title, Conditions & Possible Causes
P0336 **2T CCM, MIL: Yes** 2003, 2004, 2005 CTS (Cadillac) 2.6L VIN M, 3.2L VIN N Transmissions: All	**Crankshaft Position Sensor Signal Range/Performance Conditions:** DTC P0341, P0342 and P0343 not set; engine cranking or running; and the PCM did not detect any CKP sensor reference sync gap. The CKP sensor signal indicates the crankshaft speed and position. The CKP sensor works in conjunction with the 58-rib reluctor ring attached to the crankshaft. This is a 60-2 ring with two ribs missing. The ECM synchronizes the crankshaft position by the reference gap that is created by the two missing ribs. **Possible Causes** ● CKP sensor wires routed close to other wiring or components ● CKP sensor wires routed close to after-market add-on devices ● CKP sensor wires routed close to solenoids, relays, and motors ● CKP sensor is physically damaged or it is improperly installed ● CKP sensor is damaged or it has failed ● Electromagnetic interference in CKP sensor circuit (see above) ● Excessive air gap between the CKP sensor and reluctor ring ● Foreign material lodged between CKP sensor and reluctor ring. ● PCM has failed
P0336 **2T CCM, MIL: Yes** 2000, 2001, 2002, 2003, 2004, 2005 Aurora, DeVille, Eldorado, Seville 3.5L VIN H, 4.0L VIN C, 4.6L VIN 9, 4.6L VIN Y Transmissions: A/T	**Crankshaft Position Sensor 'A' Range/Performance Conditions:** DTC P0385 not set, engine cranking or running; and the PCM detected an unexpected interruption of the CKP Sensor 'A' signal. **Possible Causes** ● Check for signs of a loose connection at the harness connector ● CKP sensor signal (+) circuit is open or shorted to ground ● CKP sensor signal (-) circuit is open or shorted to ground ● CKP sensor is damaged or has failed (an intermittent fault) ● PCM has failed
P0336 **2T CCM, MIL: Yes** 1995, 1996, 1997, 1998, 1999, 2000, 2001, 2002 Camaro & Firebird, Grand Prix 3.8L VIN K engine Transmissions: All	**Crankshaft Reference 18X Range/Performance Conditions:** Engine started; with 3X REF signals detected, and the PCM did not detect any 18X reference pulses for one engine cycle (i.e., 720 degrees of crankshaft rotation) during the CCM test. **Possible Causes** ● CKP sensor signal circuit is open or shorted to ground ● CKP sensor VREF circuit is open between the sensor and PCM ● CKP sensor ground (Low Reference) circuit is open ● CKP sensor is damaged or it failed (check crankshaft reluctor) ● Check for CKP sensor wiring routed to close to the plug wires ● PCM has failed
P0336 **2T CCM, MIL: Yes** 1997, 1998, 1999, 2000, 2001, 2002, 2003, 2004, 2005 Camaro, Corvette & Firebird 5.7L VIN G, 5.7L VIN S Transmissions: All	**Crankshaft Position Sensor Range/Performance Conditions:** Engine cranking or running; and the PCM detected the CKP sensor signal was out-of-range for 2 seconds during the CCM test. **Possible Causes** ● CKP sensor signal circuit has a high resistance condition ● CKP sensor ground (Low Reference) circuit has high resistance ● CKP sensor is damaged or it failed (check crankshaft reluctor) ● PCM has failed ● Vehicle have been driven while very low on fuel
P0336 **2T CCM, MIL: Yes** 2000, 2001, 2002, 2003, 2004, 2005 Alero, Cavalier, Century, Grand Am, Malibu, Skylark, Sunfire, S/T Blazer & S/T Pickup 2.2L VIN 4, 2.2L VIN 5, 2.2L VIN F, 2.2L VIN H, 2.4L VIN T engines Transmissions: All	**Crankshaft Position Sensor Range/Performance Conditions:** DTC P0341 not set; engine started, and the PCM detected the Crankshaft Position resynchronization counter changed (counted) over 15 times in a 4 minutes and 15 seconds. **Possible Causes** ● CKP sensor signal circuit is open or shorted to ground ● CKP sensor VREF circuit is open between the sensor and PCM ● CKP sensor ground (Low Reference) circuit is open ● CKP sensor is damaged or it failed (check crankshaft reluctor) ● Check for CKP sensor wiring routed to close to the plug wires ● PCM has failed
P0336 **2T CCM, MIL: Yes** 1996, 1997, 1998, 1999 Alero, Cutlass, Malibu, Montana, Silhouette, Venture, Rendezvous 3.1L VIN J, 3.1L VIN M, 3.4L VIN E engines Transmissions: All	**Crankshaft Reference 24X Circuit Malfunction Conditions:** Engine started; 3X signals detected for 3 seconds, and the PCM detected an invalid ratio of 24X to 3X CKP REF pulses. The circuit uses 2 different types of crankshaft position (CKP) sensors. The CKP Sensor 'A' connects directly to the PCM through the 12v VREF, Medium Resolution engine speed signal and low reference circuits. **Possible Causes** ● CKP sensor signal circuit is open or shorted to ground ● CKP sensor ground (Low Reference) circuit is open ● CKP sensor is damaged or it failed (check crankshaft reluctor) ● Check for CKP sensor wiring routed to close to the plug wires ● PCM has failed

OBD II Trouble Code List (P0xxx Codes)

DTC	Trouble Code Title, Conditions & Possible Causes
P0336 **1T CCM, MIL: No** 1997, 1998 Catera 3.0L V6 VIN R engine Transmissions: A/T	**Crankshaft Position Sensor Range/Performance Conditions:** Engine cranking or running; and the PCM detected less than 58 CKP sensor teeth before it counted a signature tooth occurrence. **Possible Causes** ● CKP sensor signal (+) circuit is open or shorted to ground ● CKP sensor signal (-) circuit is open or shorted to ground ● CKP sensor is damaged or failed (check the reluctor wheel) ● CKP reluctor wheel clearance (gap) is out of specification ● CKP sensor wiring may be routed close to the spark plug wires or any other possible cause of EMI/RFI under the hood ● PCM has failed
P0336 **1T CCM, MIL: No** 1996, 1997 Grand Prix, Lumina, Monte Carlo 0.4L VIN E, 3.4L VIN X Transmissions: All	**Crankshaft Reference 24X Circuit Malfunction Conditions:** Engine started; 3X signals detected for 3 seconds, and the PCM detected the ratio of 24X to 3X REF pulses did not equal 8, or the ratio of 3X REF pulses to CMP pulses equaled 6. **Possible Causes** ● CKP sensor signal circuit is open or shorted to ground ● CKP sensor VREF circuit is open between the sensor and PCM ● CKP sensor ground (Low Reference) circuit is open ● CKP sensor is damaged or it failed (check crankshaft reluctor) ● CKP sensor wiring routed close to spark plug wires (EMI/RFI) ● PCM has failed ● TSB 61-65-33 contains a repair procedure for this code
P0337 **1T CCM, MIL: No** 1996, 1997, 1998 C/K Cab & Chassis, C/K Series Truck, G Van, Van Cargo, L/M Series Van, S/T Blazer & S/T Pickup 4.3L VIN W, 4.3L VIN X, 5.0L VIN M, 5.7L VIN K, 5.7L VIN R engines Transmissions: All	**CKP Sensor Circuit Low Frequency Conditions:** Engine started; engine speed less than 4000 rpm, MAF sensor more than 5 g/sec, and the PCM detected the CKP reference signal duty cycle was less than 50%, or the ratio of High REF to Low REF pulses was less than 0.18 during the CCM test. **Possible Causes** ● CKP sensor signal or ground circuit is open (intermittent fault) ● CKP reluctor wheel is chipped, damaged or the wrong part ● CKP sensor is not aligned properly to the reluctor wheel ● CKP sensor face or tip is covered with metallic particles ● Crankshaft has excessive end play ● PCM has failed
P0337 **2T CCM, MIL: Yes** 1999, 2000 C/K Cab & Chassis, C/K Series Truck, G Series Van, Van Cargo, L, M, P, S, T 4.3L VIN W, 4.3L VIN X, 5.0L VIN M, 5.7L VIN K, 5.7L VIN R, 7.4L VIN J Transmissions: All	**CKP Sensor Circuit Low Duty Cycle Conditions:** Engine started; engine speed less than 4000 rpm, MAF sensor more than 5 g/sec, and the PCM detected the CKP sensor signal (duty cycle) was less than a calibrated percentage during the CCM test. **Possible Causes** ● CKP reluctor wheel is chipped, damaged or the wrong part ● CKP sensor is not aligned properly to the reluctor wheel ● CKP sensor face or tip is covered with metallic particles ● Crankshaft has excessive end play ● PCM has failed
P0337 **2T CCM, MIL: Yes** 2003, 2004, 2005 CTS (Cadillac) 2.6L VIN M, 3.2L VIN N Transmissions: All	**Crankshaft Position Sensor Circuit Low Frequency Conditions:** DTC P0341, P0342 and P0343 not set, engine cranking or running; and the PCM detected less than 58 CKP sensor reference signals. The CKP sensor works in conjunction with the 58-rib reluctor ring attached to the crankshaft. This is a 60-2 ring with two ribs missing. The ECM synchronizes crankshaft position by the reference gap created by the 2 missing ribs. **Possible Causes** ● CKP sensor wires routed close to other wiring or components ● CKP sensor positive (+) signal circuit is open ● CKP sensor is damaged or it has failed ● Excessive air gap between CKP sensor and reluctor ring, or material lodged in the ring ● PCM has failed
P0337 **1T CCM, MIL: No** 1995 S/T Blazer & S/T Pickup 4.3L VIN W engine Transmissions: All	**CKP Sensor Circuit Low Frequency Conditions:** Engine started; engine speed less than 2000 rpm, MAF sensor more than 5 g/sec, and the PCM detected the CKP reference signal duty cycle was less than 50%, or the ratio of High REF to Low REF pulses was less than 0.18 during the CCM test period. **Possible Causes** ● CKP sensor signal, ground or VREF circuit (intermittent fault) ● CKP reluctor wheel is chipped, damaged or the wrong part ● CKP sensor is not aligned properly to the reluctor wheel ● CKP sensor face or tip is covered with metallic particles ● Crankshaft has excessive end play ● PCM has failed

OBD II Trouble Code List (P0xxx Codes)

DTC	Trouble Code Title, Conditions & Possible Causes
P0338 **1T CCM, MIL: No** 1995 S/T Blazer & S/T Pickup 4.3L VIN W engine Transmissions: All	**CKP Sensor Circuit High Frequency Conditions:** Engine started; engine speed less than 2000 rpm, MAF sensor more than 5 g/sec, and the PCM detected the CKP reference signal duty cycle was over 50%, or the ratio of High REF to Low REF pulses was more than 5.66 during the CCM test. **Possible Causes** ● CKP reluctor wheel is chipped, damaged or the wrong part ● CKP sensor is not aligned properly to the reluctor wheel ● CKP sensor face or tip is covered with metallic particles ● Crankshaft has excessive end play ● PCM has failed
P0338 **1T CCM, MIL: No** 1996, 1997, 1998 C/K Cab & Chassis, C/K Series Truck, G Series Van, L/M Series Van 4.3L VIN W, 4.3L VIN X, 5.0L VIN M, 5.7L VIN K, 5.7L VIN R engines Transmissions: All	**CKP Sensor Circuit High Frequency Conditions:** Engine started; engine speed less than 4000 rpm, MAF sensor more than 5 g/sec, and the PCM detected the CKP reference signal duty cycle was more than 50%, or the ratio of High REF to Low REF pulses was more than 5.66 during the CCM test. **Possible Causes** ● CKP reluctor wheel is chipped, damaged or the wrong part ● CKP sensor is not aligned properly to the reluctor wheel ● CKP sensor face or tip is covered with metallic particles ● Crankshaft has excessive end play ● PCM has failed ● TSB 00-06-04-014 contains a repair procedure for this code
P0338 **1T CCM, MIL: No** 1999, 2000 C/K Cab & Chassis, C/K Series Truck, G Series Van, L/M Series Van 4.3L VIN W, 4.3L VIN X, 5.0L VIN M, 5.7L VIN K, 5.7L VIN R engines Transmissions: All	**CKP Sensor Circuit High Duty Cycle Conditions:** Engine started; engine speed less than 4000 rpm, MAF sensor more than 5 g/sec, and the PCM detected the CKP sensor signal (duty cycle) was more than a calibrated percentage. **Possible Causes** ● CKP reluctor wheel is chipped, damaged or the wrong part ● CKP sensor is not aligned properly to the reluctor wheel ● CKP sensor face or tip is covered with metallic particles ● Crankshaft has excessive end play ● PCM has failed ● TSB 00-06-04-014 contains a repair procedure for this code
P0338 **2T CCM, MIL: Yes** 2003, 2004, 2005 CTS (Cadillac) 2.6L VIN M, 3.2L VIN N Transmissions: All	**Crankshaft Position Sensor Circuit High Frequency Conditions:** DTC P0341, P0342 and P0343 not set, engine cranking or running; and the PCM detected more than 58 CKP sensor reference signals. The CKP sensor works in conjunction with the 58-rib reluctor ring attached to the crankshaft. This is a 60-2 design ring in which two ribs are missing. The ECM synchronizes the crankshaft position by the reference gap that is created by the two missing ribs. **Possible Causes** ● CKP sensor wires routed close to other wiring or components ● CKP sensor resistance out of specification (700-1100 ohms) ● CKP sensor is damaged or it has failed ● Excessive air gap between the CKP sensor and reluctor ring ● Foreign material lodged between CKP sensor and reluctor ring. ● PCM has failed
P0339 **2T CCM, MIL: Yes** 1996, 1997, 1998, 1999, 2000 C/K Cab & Chassis, C/K Series Truck, G Series Van, L/M Series Van 4.3L VIN W, 4.3L VIN X, 5.0L VIN M, 5.7L VIN R, 7.4L VIN J engines Transmissions: All	**CKP Sensor Circuit Malfunction (Intermittent) Conditions:** Engine started; MAF sensor 5 g/sec or more, and the PCM detected that the measured change in engine speed was 1000 rpm or more during a 125 ms period, or the measured change in engine speed was 0 rpm over a period of 2-3 seconds. **Possible Causes** ● CKP sensor signal, ground or VREF circuit fault (intermittent) ● CKP reluctor wheel is chipped, damaged or the wrong part ● CKP sensor is not aligned properly to the reluctor wheel ● CKP sensor face or tip is covered with metallic particles ● Crankshaft has excessive end play ● PCM has failed
P0339 **2T CCM, MIL: Yes** 1996, 1997, 1998, 1999, 2000 C/K Cab & Chassis, C/K Truck, G Van, L/M Van, S/T Blazer & S/T Truck 4.3L VIN W, 4.3L VIN X, 5.0L VIN M, 5.7L VIN K, 5.7L VIN R, 7.4L VIN J Transmissions: All	**CMP Sensor Circuit Malfunction Conditions:** Engine started; and the PCM did not detect any CMP sensor (Hall Effect) pulses at least once during 2 complete crankshaft revolutions during the CCM test. **Possible Causes** ● CMP sensor signal circuit is open or shorted to ground ● CMP sensor VREF circuit is open between sensor and PCM ● CMP sensor ground (Low Reference) circuit is open ● CMP sensor is damaged or it failed (check the reluctor wheel) ● PCM has failed ● TSB 02-06-04-043 contains a repair procedure for this code

OBD II Trouble Code List (P0xxx Codes)

DTC	Trouble Code Title, Conditions & Possible Causes
P0340 **1T CCM, MIL: Yes** 2001, 2002, 2003, 2004, 2005 C/K Series Trucks 6.6L VIN 1 Diesel engine Transmissions: All	**CMP Sensor Circuit Malfunction Conditions:** Engine cranking (engine speed more than 50 rpm), and the PCM did not detect any CMP sensor pulses at least once for two (2) seconds. The hall effect camshaft position (CMP) sensor produces three (3) On-Off pulses for each revolution of the camshaft. The CMP output is pulsewidth encoded. The ECM uses the crankshaft position (CKP) and CMP pulses to determine engine speed and position. Note: The engine will not start without this signal. **Possible Causes** ● CMP sensor signal circuit is open, shorted to ground or VREF ● CMP sensor VREF circuit is open between sensor and PCM ● CMP sensor low reference circuit is open from sensor to PCM ● CMP sensor is damaged or it failed ● PCM has failed
P0340 **1T CCM, MIL: Yes** 1996, 1997, 1998, 1999 Aurora, DeVille, Eldorado & Deville 4.0L VIN C, 4.6L VIN 9, 4.6L VIN Y engines Transmissions: A/T	**CMP Sensor Circuit Malfunction Conditions:** DTC P0322 not set, engine started, engine speed below 1600 rpm, and the PCM did not detect any CMP sensor pulses for 5.3 seconds. Note: The resistance of the CMP sensor is 800-1200 ohms at 68°F. **Possible Causes** ● CMP sensor signal (+) or (-) circuit is open or shorted to ground ● CMP sensor is damaged or it failed (check the reluctor wheel) ● PCM has failed
P0340 **2T CCM, MIL: Yes** 2000, 2001, 2002, 2003, 2004, 2005 Aurora, DeVille, Eldorado & Seville 4.0L VIN C, 4.6L VIN 9, 4.6L VIN Y engines Transmissions: A/T	**CMP Sensor Circuit Malfunction Conditions:** DTC P0615 not set, engine started, engine speed above 400 rpm, and the PCM did not detect any CMP sensor pulses during the test. **Possible Causes** ● CMP sensor signal circuit is open, shorted to ground or power ● CMP sensor VREF circuit is open between sensor and PCM ● CMP sensor low reference (ground) circuit is open ● CMP sensor is damaged or it failed (check the reluctor wheel) ● PCM has failed ● TSB 03-06-04-013 contains a repair procedure for this code
P0340 **2T CCM, MIL: Yes** 2001, 2002, 2003, 2004, 2005 Aurora 3.5L VIN H engine Transmissions: A/T	**CMP Sensor Circuit Malfunction Conditions:** DTC P0335 not set, engine started, engine speed above 400 rpm, and the PCM did not detect any CMP sensor pulses. Note: This code can set if the camshaft timing is not correct, or if an attempt is made to start the vehicle in a gear position other than in P/N. **Possible Causes** ● CMP sensor signal circuit is open, shorted to ground or power ● CMP sensor VREF circuit is open between sensor and PCM ● CMP sensor ground (Low Reference) circuit is open ● CMP sensor is damaged or it failed (check the reluctor wheel) ● PCM has failed
P0340 **2T CCM, MIL: Yes** 2000, 2001, 2002, 2003, 2004, 2005 Alero, Cavalier, Grand Am, Sunfire, S/T Pickup, Envoy & TrailBlazer 2.2L VIN 4, 2.2L VIN 5, 2.4L VIN T engines, 2.2L VIN H, 4.2L VIN S Transmissions: All	**CMP Sensor Circuit Malfunction Conditions:** Engine started; and the PCM detected the CMP sensor (Hall Effect) Active Counter did not increment (i.e., no change detected in the CMP sensor activity for 30 crankshaft revolutions). **Possible Causes** ● CMP sensor signal circuit is open, shorted to ground or shorted to VREF between the sensor and the PCM ● CMP sensor VREF circuit is open between sensor and PCM ● CMP sensor ground circuit or "shielded" ground circuit is open ● CMP sensor is cracked or damaged (check the reluctor wheel) ● PCM has failed
P0340 **2T CCM, MIL: Yes** 1995 S/T Blazer & S/T Pickup 4.3L VIN W engine Transmissions: All	**CMP Sensor Circuit Malfunction Conditions:** Engine started; and the PCM did not detect a CMP pulse for each crankshaft revolution. **Possible Causes** ● CMP sensor signal circuit is open, shorted to ground or VREF ● CMP sensor VREF circuit is open between sensor and PCM ● CMP sensor ground circuit or "shielded" ground circuit is open ● CMP sensor is cracked or damaged (check the reluctor wheel) ● PCM has failed
P0340 **2T CCM, MIL: Yes** 1997, 1998, 1999, 2000, 2001 Catera 3.0L V6 VIN R engine Transmissions: All	**CMP Sensor Circuit Malfunction Conditions:** Engine started; and the PCM did not detect any CMP sensor pulses for five engine cycles. **Possible Causes** ● CMP sensor signal circuit open, shorted to ground or to VREF ● CMP sensor 5v VREF or ground (Low REF) circuit is open ● CMP sensor is cracked or damaged (check the reluctor wheel) ● PCM has failed

OBD II Trouble Code List (P0xxx Codes)

DTC	Trouble Code Title, Conditions & Possible Causes
P0341 **2T CCM, MIL: Yes** 1995 Century, Ciera, Cutlass Supreme, Grand Prix, 88', Lumina, Monte Carlo Park Avenue-Regal-Riviera, 98' 3.1L VIN M, 3.4L VIN X, 3.8L VIN 1, 3.8L VIN K, 3.8L VIN L Transmissions: All	**CMP Sensor Signal Range/Performance Conditions:** Engine started; and PCM detected the ratio of CKP sensor pulses to CMP pulses did not equal 6:1 at least 10 times in 1engine revolution. **Possible Causes** ● CMP sensor signal circuit is open, shorted to ground or shorted to VREF between the sensor and the PCM (intermittent fault) ● CMP sensor signal wire is routed to close to the Generator, spark plug wires or any other possible cause of EMI/RFI ● CMP sensor is cracked or damaged (check the reluctor wheel) ● PCM has failed
P0341 **2T CCM, MIL: Yes** 1996, 1997, 1998, 1999, 2000, 2001, 2002, 2003, 2004, 2005 Achieva, Alero, Beretta, Century, Ciera, Cavalier, Envoy, Grand Am, Sunfire Skyhawk, TrailBlazer, S/T Blazer & S/T Pickup 2.2L VIN 4, 2.2L VIN 5, 2.4L VIN T engines, 2.2L VIN H, 4.2L VIN S Transmissions: All	**CMP Sensor Signal Range/Performance Conditions:** Engine started; and the PCM detected more than 15 CMP sensor resynchronizations during a 4 minute 16 second period. **Possible Causes** ● CMP sensor signal circuit is open, shorted to ground or VREF ● CMP sensor signal wire is routed to close to the Generator, spark plug wires or any other possible cause of EMI/RFI under the hood (check for high power receivers causing interference) ● CMP sensor "shield" ground circuit is open (intermittent fault) ● CMP sensor is cracked or damaged (check the reluctor wheel) ● PCM has failed
P0341 **2T CCM, MIL: Yes** 1996, 1997, 1998, 1999, 2000, 2001, 2002, 2003, 2004, 2005 Aurora, Century, LeSabre, Park Avenue, Regal 3.1L VIN J, 3.1L VIN M, 3.5L VIN H, 3.8L VIN 1, 3.8L VIN K engines Transmissions: All	**CMP Sensor Signal Range/Performance Conditions:** Engine started; with 3X REF pulses received, and the PCM did not detect the CMP sensor pulse during every engine revolution during the CCM test. The PCM compares the number of 24X and 3X pulses to the CMP sensor pulses to check for an error. **Possible Causes** ● CMP sensor signal circuit is open, shorted to ground or VREF ● CMP sensor is cracked or damaged (check the reluctor wheel) ● PCM has failed ● TSB 02-06-04-008 contains a repair procedure for this code
P0341 **2T CCM, MIL: Yes** 1996, 1997, 1998, 1999, 2000, 2001, 2002, 2003, 2004, 2005 Alero, Achieva, Aztek, Cutlass, Impala, Malibu, Montana, Venture Silhouette, Rendezvous 3.1L VIN J, 3.1L VIN M, 3.4L VIN E, 3.8L VIN K Transmissions: All	**CMP Sensor Signal Range/Performance Conditions:** Engine started; with 3X signals received, and the PCM did not detect a CMP sensor pulse for each engine revolution. **Possible Causes** ● CMP sensor signal circuit is open, shorted to ground or shorted to VREF between the sensor and the PCM (intermittent fault) ● CMP sensor signal wire is routed to close to the Generator, spark plug wires or any other possible cause of EMI/RFI ● CMP sensor is cracked, damaged or has failed ● PCM has failed ● TSB 02-06-04-008 contains a repair procedure for this code
P0341 **2T CCM, MIL: Yes** 1996, 1997, 1998, 1999, 2000, 2001, 2002, 2003, 2004, 2005 C/K Series, Truck, G Series Van, L/M Van 4.3L VIN W, 4.3L VIN X, 5.0L VIN M, 5.7L VIN K, 5.7L VIN R, 7.4L VIN J Transmissions: All	**CMP Sensor Signal Range/Performance Conditions:** Engine started; at less than 4000 rpm, and the PCM detected that the CKP sensor pulses and CMP sensor pulses did not match correctly for each engine revolution during the test. **Possible Causes** ● CMP sensor signal circuit is open, shorted to ground or VREF ● CMP sensor is cracked or damaged (check the reluctor wheel) ● CMP sensor signal wire is routed to close to the Generator, spark plug wires or any other possible cause of EMI/RFI ● Reluctor wheel is damaged or the sensor is touching the wheel ● PCM has failed
P0341 **2T CCM, MIL: Yes** 1999, 2000, 2001, 2002, 2003, 2004, 2005 C/K Series Truck, G Series Van, Envoy, Escalade, TrailBlazer 4.8L VIN V, 5.3L VIN P, 5.3L VIN T, 5.3L VIN Z, 6.0L VIN N, 6.0L VIN U CNG engine, 8.1L VIN G Transmissions: All	**CMP Sensor Signal Range/Performance Conditions:** Engine speed less than 4000 rpm, and the PCM detected the CKP sensor pulses and CMP sensor pulses did not match during each engine revolution. The CMP sensor works with the 1X reluctor wheel on the camshaft. The PCM provides a 12v VREF to the CMP sensor as well as low reference and signal circuits. As the camshaft rotates, the reluctor wheel interrupts a magnetic field produced by a magnet in the sensor to produce the CMP signal. **Possible Causes** ● CMP sensor signal circuit is open, shorted to ground or VREF ● CMP sensor is cracked or damaged (check the reluctor wheel) ● CMP sensor wiring routed to close to Generator or plug wires ● Reluctor wheel is damaged or the sensor is touching the wheel ● PCM has failed

OBD II Trouble Code List (P0xxx Codes)

DTC	Trouble Code Title, Conditions & Possible Causes
P0341 **2T CCM, MIL: Yes** 2001, 2002, 2003, 2004, 2005 C/K Series Trucks 6.6L VIN 1 Diesel engine Transmissions: All	**CMP Sensor Signal Range/Performance Conditions:** Engine cranking (engine speed more than 50 rpm), and the ECM detected the CMP sensor pulses were out-of-range for two seconds. The hall effect camshaft position (CMP) sensor produces three (3) On-Off pulses for each revolution of the camshaft. The CMP output is pulsewidth encoded. The ECM uses the crankshaft position (CKP) and CMP pulses to determine the engine speed and position. Note: The engine will not start without this signal. **Possible Causes** ● CMP sensor signal circuit is open, shorted to ground or VREF ● CMP sensor VREF circuit is open between sensor and PCM ● CMP sensor low reference circuit is open from sensor to PCM ● CMP sensor is damaged or it failed ● PCM has failed
P0341 **2T CCM, MIL: Yes** 1996, 1997, 1998, 1999, 2000, 2001, 2002 DeVille, Eldorado, Seville 4.0L VIN C, 4.6L VIN 9, 4.6L VIN Y Transmissions: A/T	**CMP Sensor Signal Range/Performance Conditions:** Engine started; at less than 2000 rpm, and the PCM detected incorrect correlation between the CKP and CMP signals. **Possible Causes** ● CMP sensor signal circuit is open, shorted to ground or VREF ● CMP sensor is cracked or damaged (check the reluctor wheel) ● CMP sensor signal wire is routed to close to the Generator, spark plug wires or any other possible cause of EMI/RFI ● Reluctor wheel is damaged or the sensor is touching the wheel ● PCM has failed
P0341 **2T CCM, MIL: Yes** 2003, 2004, 2005 CTS (Cadillac) 2.6L VIN M, 3.2L VIN N Transmissions: All	**Camshaft Position Sensor Signal Range/Performance Conditions:** Engine started; and the PCM the CMP sensor signal was inconsistent for eight (8) or more crankshaft revolutions. The CMP sensor is a hall effect switching device that works in conjunction with a single tooth reluctor wheel used to determine the position of the Bank 2 Exhaust Camshaft. The ECM expects the CMP sensor signal to be low (0v), as the single tooth in the reluctor wheel passes the sensor, and high (12v), during the remainder of the reluctor wheel rotation. The ECM supplies a 12v pullup voltage on the CMP sensor signal circuit. The ECM expects to see one transition from high to low every two revolutions of the crankshaft. This signal, when combined with the CKP sensor signal, enables the ECM to properly synchronize ignition timing, fuel delivery and knock control. As long as the CKP signal is available, the engine can start and run. The ECM will default to a non-sequential fuel injector mode even if there is no CMP sensor signal. **Possible Causes** ● CMP sensor wires routed close to other wiring or components ● Camshaft reluctor wheel damage, incorrect sensor installation ● CMP sensor is contacting the reluctor wheel ● CMP sensor is damaged (cracked), or it has failed ● Electromagnetic interference in CMP sensor circuit (due to the sensor wires routed to close to ignition cables or motors) ● Excessive air gap between reluctor wheel and sensor magnet ● Foreign material lodged between sensor and the reluctor wheel ● PCM has failed
P0341 **2T CCM, MIL: Yes** 2003, 2004, 2005 DeVille, Eldorado, Seville 4.0L VIN G, 4.6L VIN 9, 4.6L VIN Y engines Transmissions: All	**CMP Sensor Signal Range/Performance Conditions:** Engine started; at less than 2000 rpm, and the PCM detected incorrect correlation between the CKP and CMP signals. **Possible Causes** ● CMP sensor signal circuit is open, shorted to ground or VREF ● CMP sensor is cracked or damaged (check the reluctor wheel) ● CMP sensor signal wire is routed to close to the Generator, spark plug wires or any other possible cause of EMI/RFI ● Reluctor wheel is damaged or the sensor is touching the wheel ● PCM has failed ● TSB 03-06-04-013 contains a repair procedure for this code
P0341 **2T CCM, MIL: Yes** 1995, 1996, 1997, 1998, 1999, 2000, 2001, 2002 Camaro & Firebird 3.8L VIN K engine Transmissions: All	**CMP Sensor Signal Range/Performance Conditions:** Engine started; engine running with 3X signals received, and the PCM did not detect a CMP sensor pulse for each engine revolution. **Possible Causes** ● CMP sensor signal circuit is open, shorted to ground or VREF ● CMP sensor signal wire is routed to close to the Generator, spark plug wires or any other possible cause of EMI/RFI ● CMP sensor is cracked, damaged or has failed ● PCM has failed ● TSB 02-06-04-008 contains a repair procedure for this code

OBD II Trouble Code List (P0xxx Codes)

DTC	Trouble Code Title, Conditions & Possible Causes
P0341 **2T CCM, MIL: Yes** 1997, 1998, 1999, 2000, 2001, 2002, 2003, 2004, 2005 Camaro, Corvette & Firebird 5.7L VIN G, 5.7L VIN S Transmissions: All	**CMP Sensor Signal Range/Performance Conditions:** Engine started; at less than 4000 rpm, and the PCM detected incorrect correlation between the CKP and CMP signals. **Possible Causes** ● CMP sensor signal circuit is open, shorted to ground or VREF ● CMP sensor signal wire is routed to close to the Generator, spark plug wires or any other possible cause of EMI/RFI ● CMP sensor is cracked, damaged or has failed ● PCM has failed ● TSB 02-06-04-008 contains a repair procedure for this code
P0341 **2T CCM, MIL: Yes** 1995, 1996, 1997, 1998, 1999, 2000, 2001, 2002, 2003, 2004, 2005 S/T Blazer & S/T Pickup 4.3L VIN W, 4.3L VIN X Transmissions: All	**CMP Sensor Signal Range/Performance Conditions:** Engine started; and the PCM did not detect a CMP sensor reference pulse at least once for each engine revolution. **Possible Causes** ● CMP sensor signal circuit is open, shorted to ground or VREF ● CMP sensor signal wire is routed to close to the Generator, spark plug wires or any other possible cause of EMI/RFI ● CMP sensor is cracked, damaged or has failed ● PCM has failed
P0341 **2T CCM, MIL: Yes** 1997, 1998, 1999, 2000, 2001 Catera 3.0L V6 VIN R engine Transmissions: A/T	**CMP Sensor Signal Range/Performance Conditions:** Engine starting; and the PCM determined the CMP sensor signals were not rational when compared to the CKP sensor signals. **Possible Causes** ● CMP sensor signal circuit is open, shorted to ground or shorted to VREF between the sensor and the PCM (intermittent fault) ● CMP sensor signal wire is routed to close to the Generator, spark plug wires or any other possible cause of EMI/RFI ● CMP sensor is cracked, damaged or has failed ● PCM has failed
P0341 **2T CCM, MIL: Yes** 1996, 1997 Cutlass Supreme, Grand Prix, Lumina, Monte Carlo 3.4L VIN X engine Transmissions: All	**CMP Sensor Signal Range/Performance Conditions:** Engine started; with 3X signals received, and the PCM did not detect any CMP sensor signals, or it detected that the ratio of 3X to CMP sensor pulses did not equal 6:1, or the ratio of 3X REF pulses to 18X REF pulses equaled 6:1 for 290 of 300 events. **Possible Causes** ● CMP sensor signal circuit is open, shorted to ground or VREF ● CMP sensor signal wire is routed to close to the Generator, spark plug wires or any other possible cause of EMI/RFI ● CMP sensor is cracked or damaged (check the reluctor wheel) ● PCM has failed ● TSB 02-06-04-008 contains a repair procedure for this code
P0342 **2T CCM, MIL: Yes** 1995 Century, Cutlass Ciera, Cutlass Supreme, Grand Prix, Lumina, Monte Carlo Park Avenue Regal, Riviera, 88' & 98' 3.1L VIN M, 3.4L VIN X, 3.8L VIN 1, 3.8L VIN K, 3.8L VIN L Transmissions: All	**CMP Sensor Circuit Low Input Conditions:** Engine started; and the PCM did not detect any CMP sensor signals for 5 seconds during the CCM test **Possible Causes** ● CMP sensor signal circuit is open, shorted to ground or VREF ● CMP sensor signal wire is routed to close to the Generator, spark plug wires or any other possible cause of EMI/RFI ● CMP sensor is cracked or damaged (check the reluctor wheel) ● PCM has failed
P0342 **2T CCM, MIL: Yes** 1996, 1997, 1998, 1999 Achieva, Beretta, Cavalier, Ciera, Corsica, S/T Blazer & S/T Pickup 2.2L VIN 4, 2.2L VIN 5, 2.4L VIN T engines Transmissions: All	**CMP Sensor Circuit Low Input Conditions:** Engine started; and the PCM determined the CMP Sensor Activity Counter did not increment (i.e., it did not count up or increase). **Possible Causes** ● CMP sensor signal circuit is open, shorted to ground or VREF ● CMP sensor ground circuit (low reference) is open ● CMP sensor VREF circuit is open between sensor and PCM ● CMP sensor is cracked, damaged or has failed ● PCM has failed

OBD II Trouble Code List (P0xxx Codes)

DTC	Trouble Code Title, Conditions & Possible Causes
P0342 **2T CCM, MIL: Yes** 1998, 1999, 2000, 2001, 2002, 2003, 2004, 2005 C/K Cab & Chassis, C/K Series Truck, Envoy & TrailBlazer 4.8L VIN V, 5.3L VIN P, 5.3L VIN T, 5.3L VIN Z, 6.0L VIN N, 6.0L VIN U CNG engine, 8.1L VIN G Transmissions: All	**CMP Sensor Circuit Low Input Conditions:** Engine started; at less than 4000 rpm, and the PCM detected the CMP sensor signal was in a low state (when the signal should have been in a high state) for 1.5 seconds in the test. **Possible Causes** ● Camshaft reluctor wheel is damaged or foreign material present ● CMP sensor signal circuit is open, shorted to ground or VREF ● CMP sensor is contacting the reluctor wheel or is damaged ● PCM has failed
P0342 **2T CCM, MIL: Yes** 2003, 2004, 2005 CTS (Cadillac) 2.6L VIN M, 3.2L VIN N Transmissions: All	**Camshaft Position Sensor Circuit Low Input Conditions:** Engine started; and the PCM detected the CMP sensor signal was continuously low for (8) or more crankshaft revolutions. The CMP sensor is a hall effect switching device that works in conjunction with a single tooth reluctor wheel to determine the position of the Bank 2 Exhaust Camshaft. The ECM expects the CMP sensor signal to be low (0v), as the single tooth in the reluctor wheel passes the sensor, and high (12v), during the remainder of the reluctor wheel rotation. The CMP sensor signal is connected to a 12v pullup voltage at the ECM. The ECM expects to see one transition from high to low for each crankshaft revolution. **Possible Causes** ● CMP sensor connector is damaged or open ● CMP sensor signal circuit is shorted to ground ● CMP sensor signal circuit is open or it has high resistance ● CMP sensor reluctor wheel is damaged or loose ● PCM has failed
P0342 **2T CCM, MIL: Yes** 1997, 1998, 1999, 2000, 2001, 2002, 2003, 2004, 2005 Camaro, Corvette & Firebird 5.7L VIN G, 5.7L VIN S Transmissions: All	**CMP Sensor Circuit Low Input Conditions:** Engine speed less than 4000 rpm and the PCM detected the CMP sensor signal was in a low voltage state when it should have been in a high voltage state for 1.5 seconds. **Possible Causes** ● CMP sensor signal circuit is open, shorted to ground or VREF ● Camshaft reluctor wheel is damaged or foreign material present ● CMP sensor is contacting the reluctor wheel or is damaged ● PCM has failed
P0343 **2T CCM, MIL: Yes** 1998, 1999, 2000, 2001, 2002, 2003, 2004, 2005 C/K Cab & Chassis, C/K Series Truck, G Van, Envoy & TrailBlazer 4.8L VIN V, 5.3L VIN P, 5.3L VIN T, 5.3L VIN Z, 6.0L VIN N, 6.0L VIN U CNG engine, 8.1L VIN G Transmissions: All	**CMP Sensor Circuit High Input Conditions:** Engine started; engine speed less than 4000 rpm and the PCM detected the CMP sensor signal was stuck high (when the signal should have been in a low state) for 1.5 seconds in the CCM test. **Possible Causes** ● CMP sensor connector is damaged, loose or shorted ● CMP sensor low reference circuit is open or shorted to VREF ● Camshaft reluctor wheel is damaged or foreign material present ● CMP sensor is contacting the reluctor wheel or is damaged ● PCM has failed
P0343 **2T CCM, MIL: Yes** 2003, 2004, 2005 CTS (Cadillac) 2.6L VIN M, 3.2L VIN N Transmissions: All	**Camshaft Position Sensor Circuit Low Input Conditions:** Engine started; and the PCM detected the CMP sensor signal was continuously high for (8) or more crankshaft revolutions. The CMP sensor is a hall effect unit working in conjunction with a single tooth reluctor wheel to determine the position of the Bank 2 Exhaust Camshaft. The ECM expects the CMP sensor signal to be 0v, as the single tooth in the reluctor wheel passes the sensor and 12v during the remainder of the reluctor wheel rotation. **Possible Causes** ● CMP sensor connector is damaged or shorted ● CMP sensor signal circuit is open ● CMP sensor signal circuit is shorted to system power ● CMP sensor is damaged or it has failed ● CMP sensor reluctor wheel is damaged or loose ● PCM has failed
P0343 **2T CCM, MIL: Yes** 1997, 1998, 1999, 2000, 2001, 2002, 2003, 2004, 2005 Camaro, Corvette & Firebird 5.7L VIN G, 5.7L VIN S Transmissions: All	**CMP Sensor Circuit High Input Conditions:** Engine started; at less than 4000 rpm, and the PCM detected the CMP sensor signal was in a high voltage state when it should have been in a low voltage state for 1.5 seconds. **Possible Causes** ● CMP sensor signal circuit is open, shorted to ground or shorted to VREF ● Camshaft reluctor wheel is damaged or foreign material present ● CMP sensor is contacting the reluctor wheel or is damaged ● PCM has failed

OBD II Trouble Code List (P0xxx Codes)

DTC	Trouble Code Title, Conditions & Possible Causes
P0351 **2T CCM, MIL: Yes** 2001, 2002, 2003, 2004, 2005 L/M Series Van, S/T Blazer & S/T Pickup 4.3L VIN W, 4.3L VIN X Transmissions: All	**Ignition Coil Control Circuit Malfunction Conditions:** Engine cranking, ignition control enabled with the engine speed less than 250 rpm, and the PCM detected the ignition control voltage was not within the specified range of 0.04-4.9 volts. The CKP sensor signal provides a timing input to the ICM so that it can determine the spark timing for each cylinder. Each timing pulse received from the PCM allows the ICM to energize the coil. A high ignition voltage is induced in the coil secondary winding by the primary coil winding. The distributor switches high voltage to each spark plug. If the PCM detects an unusually high or low voltage on this circuit, P0351 is set. **Possible Causes** ● IC circuit is open, shorted to ground or shorted to power (B+) ● IC ground (Low Reference) circuit is open ● IC power circuit is open (check the ECM fuse in U/H fuse block) ● Ignition coil is damaged or has failed ● PCM has failed
P0351 **2T CCM, MIL: Yes** 1999, 2000, 2001, 2002, 2003, 2004, 2005 C/K Series Truck, G Series Van, Envoy, Escalade, TrailBlazer 4.8L VIN V, 5.3L VIN P, 5.3L VIN T, 5.3L VIN Z, 6.0L VIN N, 6.0L VIN U CNG engine, 8.1L VIN G Transmissions: All	**Ignition Coil 1 Control Circuit Malfunction Conditions:** Engine started; and the PCM detected an unexpected voltage condition on the Coil Near Plug Ignition Control (IC) 1 circuit for less than one second during the CCM test period. **Possible Causes** ● IC circuit is open, shorted to ground or shorted to power (B+) ● IC ground (Low REF) circuit or Module ground circuit is open ● IC power circuit is open (check the INJ fuse in U/H fuse block) ● Ignition Coil 1 is damaged or it has failed ● PCM has failed
P0352 **2T CCM, MIL: Yes** 1999, 2000, 2001, 2002, 2003, 2004, 2005 C/K Series Truck, G Series Van, Envoy, Escalade, TrailBlazer 4.8L VIN V, 5.3L VIN P, 5.3L VIN T, 5.3L VIN Z, 6.0L VIN N, 6.0L VIN U CNG engine, 8.1L VIN G Transmissions: All	**Ignition Coil 2 Control Circuit Malfunction Conditions:** Engine started; and the PCM detected an unexpected voltage condition on the Coil Near Plug Ignition Control (IC) 2 circuit for less than one second during the CCM test period. **Possible Causes** ● IC circuit is open, shorted to ground or shorted to power (B+) ● IC ground (Low REF) circuit or Module ground circuit is open ● IC power circuit is open (check the INJ fuse in U/H fuse block) ● Ignition Coil 2 is damaged or it has failed ● PCM has failed
P0353 **2T CCM, MIL: Yes** 1999, 2000, 2001, 2002, 2003, 2004, 2005 C/K Series Truck, G Series Van, Envoy, Escalade, TrailBlazer 4.8L VIN V, 5.3L VIN P, 5.3L VIN T, 5.3L VIN Z, 6.0L VIN N, 6.0L VIN U CNG engine, 8.1L VIN G Transmissions: All	**Ignition Coil 3 Control Circuit Malfunction Conditions:** Engine started; and the PCM detected an unexpected voltage condition on the Coil Near Plug Ignition Control (IC) 3 circuit for less than one second during the CCM test period. **Possible Causes** ● IC circuit is open, shorted to ground or shorted to power (B+) ● IC ground (Low REF) circuit or Module ground circuit is open ● IC power circuit is open (check the INJ fuse in U/H fuse block) ● Ignition Coil 3 is damaged or it has failed ● PCM has failed
P0354 **2T CCM, MIL: Yes** 1999, 2000, 2001, 2002, 2003, 2004, 2005 C/K Series Truck, G Series Van, Envoy, Escalade, TrailBlazer 4.8L VIN V, 5.3L VIN P, 5.3L VIN T, 5.3L VIN Z, 6.0L VIN N, 6.0L VIN U CNG engine, 8.1L VIN G Transmissions: All	**Ignition Coil 4 Control Circuit Malfunction Conditions:** Engine started; and the PCM detected an unexpected voltage condition on the Coil Near Plug Ignition Control (IC) 4 circuit for less than one second during the CCM test period. **Possible Causes** ● IC circuit is open, shorted to ground or shorted to power (B+) ● IC ground (Low REF) circuit or Module ground circuit is open ● IC power circuit is open (check the INJ fuse in U/H fuse block) ● Ignition Coil 4 is damaged or it has failed ● PCM has failed

OBD II Trouble Code List (P0xxx Codes)

DTC	Trouble Code Title, Conditions & Possible Causes
P0355 **2T CCM, MIL: Yes** 1999, 2000, 2001, 2002, 2003, 2004, 2005 C/K Series Truck, G Series Van, Envoy, Escalade, TrailBlazer 4.8L VIN V, 5.3L VIN P, 5.3L VIN T, 5.3L VIN Z, 6.0L VIN N, 6.0L VIN U CNG engine, 8.1L VIN G Transmissions: All	**Ignition Coil 5 Control Circuit Malfunction Conditions:** Engine started; and the PCM detected an unexpected voltage condition on the Coil Near Plug Ignition Control (IC) 5 circuit for less than one second during the CCM test period. **Possible Causes** ● IC circuit is open, shorted to ground or shorted to power (B+) ● IC ground (Low REF) circuit or Module ground circuit is open ● IC power circuit is open (check the INJ fuse in U/H fuse block) ● Ignition Coil 5 is damaged or it has failed ● PCM has failed
P0356 **2T CCM, MIL: Yes** 1999, 2000, 2001, 2002, 2003, 2004, 2005 C/K Series Truck, G Series Van, Envoy, Escalade, TrailBlazer 4.8L VIN V, 5.3L VIN P, 5.3L VIN T, 5.3L VIN Z, 6.0L VIN N, 6.0L VIN U CNG engine, 8.1L VIN G Transmissions: All	**Ignition Coil 6 Control Circuit Malfunction Conditions:** Engine started; and the PCM detected an unexpected voltage condition on the Coil Near Plug Ignition Control (IC) 6 circuit for less than one second during the CCM test period. **Possible Causes** ● IC circuit is open, shorted to ground or shorted to power (B+) ● IC ground (Low REF) circuit or Module ground circuit is open ● IC power circuit is open (check the INJ fuse in U/H fuse block) ● Ignition Coil 6 is damaged or it has failed ● PCM has failed
P0357 **2T CCM, MIL: Yes** 1999, 2000, 2001, 2002, 2003, 2004, 2005 C/K Series Truck, G Series Van, Envoy, Escalade, TrailBlazer 4.8L VIN V, 5.3L VIN P, 5.3L VIN T, 5.3L VIN Z, 6.0L VIN N, 6.0L VIN U CNG engine, 8.1L VIN G Transmissions: All	**Ignition Coil 7 Control Circuit Malfunction Conditions:** Engine started; and the PCM detected an unexpected voltage condition on the Coil Near Plug Ignition Control (IC) 7 circuit for less than one second during the CCM test period. **Possible Causes** ● IC circuit is open, shorted to ground or shorted to power (B+) ● IC ground (Low REF) circuit or Module ground circuit is open ● IC power circuit is open (check the INJ fuse in U/H fuse block) ● Ignition Coil 7 is damaged or it has failed ● PCM has failed
P0358 **2T CCM, MIL: Yes** 1999, 2000, 2001, 2002, 2003, 2004, 2005 C/K Series Truck, G Series Van, Envoy, Escalade, TrailBlazer 4.8L VIN V, 5.3L VIN P, 5.3L VIN T, 5.3L VIN Z, 6.0L VIN N, 6.0L VIN U CNG engine, 8.1L VIN G	**Ignition Coil 8 Control Circuit Malfunction Conditions:** Engine started; and the PCM detected an unexpected voltage condition on the Coil Near Plug Ignition Control (IC) 8 circuit for less than one second during the CCM test period. **Possible Causes** ● IC circuit is open, shorted to ground or shorted to power (B+) ● IC ground (Low REF) circuit or Module ground circuit is open ● IC power circuit is open (check the INJ fuse in U/H fuse block) ● Ignition Coil 8 is damaged or it has failed ● PCM has failed
P0351 **2T CCM, MIL: Yes** 2000, 2001, 2002, 2003, 2004, 2005 DeVille, Eldorado, Seville 4.0L VIN C, 4.6L VIN 9, 4.6L VIN Y engines Transmissions: All	**Ignition Coil 1 Control Circuit Malfunction Conditions:** Engine started; engine speed over 400 rpm and the PCM detected an unexpected low voltage condition on Coil On Plug (COP) Ignition Control 1 circuit for less than one second. **Possible Causes** ● IC circuit is shorted to ground between the coil and the PCM ● Ignition Coil (COP) 1 is damaged or has failed ● PCM has failed
P0352 **2T CCM, MIL: Yes** 2000, 2001, 2002, 2003, 2004, 2005 DeVille, Eldorado, Seville 4.0L VIN C, 4.6L VIN 9, 4.6L VIN Y engines Transmissions: All	**Ignition Coil 2 Control Circuit Malfunction Conditions:** Engine started; engine speed over 400 rpm and the PCM detected an unexpected low voltage condition on Coil On Plug (COP) Ignition Control 2 circuit for less than one second. **Possible Causes** ● IC circuit is shorted to ground between the coil and the PCM ● Ignition Coil (COP) 2 is damaged or has failed ● PCM has failed
P0353 **2T CCM, MIL: Yes** 2000, 2001, 2002, 2003, 2004, 2005 DeVille, Eldorado, Seville 4.0L VIN C, 4.6L VIN 9, 4.6L VIN Y engines Transmissions: All	**Ignition Coil 3 Control Circuit Malfunction Conditions:** Engine started; engine speed over 400 rpm and the PCM detected an unexpected low voltage condition on Coil On Plug (COP) Ignition Control 3 circuit for less than one second. **Possible Causes** ● IC circuit is shorted to ground between the coil and the PCM ● Ignition Coil (COP) 3 is damaged or has failed ● PCM has failed

OBD II Trouble Code List (P0xxx Codes)

DTC	Trouble Code Title, Conditions & Possible Causes
P0354 **2T CCM, MIL: Yes** 2000, 2001, 2002, 2003, 2004, 2005 DeVille, Eldorado, Seville 4.0L VIN C, 4.6L VIN 9, 4.6L VIN Y engines Transmissions: All	**Ignition Coil 4 Control Circuit Malfunction Conditions:** Engine started; engine speed over 400 rpm and the PCM detected an unexpected low voltage condition on Coil On Plug (COP) Ignition Control 4 circuit for less than one second. **Possible Causes** ● IC circuit is shorted to ground between the coil and the PCM ● Ignition Coil (COP) 4 is damaged or has failed ● PCM has failed
P0355 **2T CCM, MIL: Yes** 2000, 2001, 2002, 2003, 2004, 2005 DeVille, Eldorado, Seville 4.0L VIN C, 4.6L VIN 9, 4.6L VIN Y engines Transmissions: All	**Ignition Coil 5 Control Circuit Malfunction Conditions:** Engine started; engine speed over 400 rpm and the PCM detected an unexpected low voltage condition on Coil On Plug (COP) Ignition Control 5 circuit for less than one second. **Possible Causes** ● IC circuit is shorted to ground between the coil and the PCM ● Ignition Coil (COP) 5 is damaged or has failed ● PCM has failed
P0356 **2T CCM, MIL: Yes** 2000, 2001, 2002, 2003, 2004, 2005 DeVille, Eldorado, Seville 4.0L VIN C, 4.6L VIN 9, 4.6L VIN Y engines Transmissions: All	**Ignition Coil 6 Control Circuit Malfunction Conditions:** Engine started; engine speed over 400 rpm and the PCM detected an unexpected low voltage condition on Coil On Plug (COP) Ignition Control circuit for less than one second. **Possible Causes** ● IC circuit is shorted to ground between the coil and the PCM ● Ignition Coil (COP) 6 is damaged or has failed ● PCM has failed
P0357 **2T CCM, MIL: Yes** 2000, 2001, 2002, 2003, 2004, 2005 DeVille, Eldorado, Seville 4.0L VIN C, 4.6L VIN 9, 4.6L VIN Y engines Transmissions: All	**Ignition Coil 7 Control Circuit Malfunction Conditions:** Engine started; engine speed over 400 rpm and the PCM detected an unexpected low voltage condition on Coil On Plug (COP) Ignition Control 7 circuit for less than one second. **Possible Causes** ● IC circuit is shorted to ground between the coil and the PCM ● Ignition Coil (COP) 7 is damaged or has failed ● PCM has failed
P0358 **2T CCM, MIL: Yes** 2000, 2001, 2002, 2003, 2004, 2005 DeVille, Eldorado, Seville 4.0L VIN C, 4.6L VIN 9, 4.6L VIN Y engines Transmissions: All	**Ignition Coil 8 Control Circuit Malfunction Conditions:** Engine started; engine speed over 400 rpm and the PCM detected an unexpected low voltage condition on Coil On Plug (COP) Ignition Control 8 circuit for less than one second. **Possible Causes** ● IC circuit is shorted to ground between the coil and the PCM ● Ignition Coil (COP) 8 is damaged or has failed ● PCM has failed
P0351 **2T CCM, MIL: Yes** 1997, 1998, 1999, 2000, 2001, 2002, 2003, 2004, 2005 Camaro, Corvette & Firebird 5.7L VIN G, 5.7L VIN S Transmissions: All	**Ignition Coil 1 Control Circuit Malfunction Conditions:** Engine started; and the PCM detected an unexpected low or high voltage condition on Coil On Plug (COP) Ignition Control circuit for less than one second during the CCM test. *Note: Watch the Scan Tool Misfire Counters to identify the fault.* **Possible Causes** ● IC circuit is open, shorted to ground or shorted to power ● Ignition coil (COP) is damaged or has failed ● PCM has failed
P0352 **2T CCM, MIL: Yes** 1997, 1998, 1999, 2000, 2001, 2002, 2003, 2004, 2005 Camaro, Corvette & Firebird 5.7L VIN G, 5.7L VIN S Transmissions: All	**Ignition Coil 2 Control Circuit Malfunction Conditions:** Engine started; and the PCM detected an unexpected low or high voltage condition on Coil On Plug (COP) Ignition Control circuit for less than one second during the CCM test. *Note: Watch the Scan Tool Misfire Counters to identify the fault.* **Possible Causes** ● IC circuit is open, shorted to ground or shorted to power ● Ignition coil (COP) is damaged or has failed ● PCM has failed
P0353 **2T CCM, MIL: Yes** 1997, 1998, 1999, 2000, 2001, 2002, 2003, 2004, 2005 Camaro, Corvette & Firebird 5.7L VIN G, 5.7L VIN S Transmissions: All	**Ignition Coil 3 Control Circuit Malfunction Conditions:** Engine started; and the PCM detected an unexpected low or high voltage condition on Coil On Plug (COP) Ignition Control circuit for less than one second during the CCM test. *Note: Watch the Scan Tool Misfire Counters to identify the fault.* **Possible Causes** ● IC circuit is open, shorted to ground or shorted to power between the coil and the PCM ● Ignition coil (COP) is damaged or has failed ● PCM has failed
P0354 **2T CCM, MIL: Yes** 1997, 1998, 1999, 2000, 2001, 2002, 2003, 2004, 2005 Camaro, Corvette & Firebird 5.7L VIN G, 5.7L VIN S	**Ignition Coil 4 Control Circuit Malfunction Conditions:** Engine started; and the PCM detected an unexpected low or high voltage condition on Coil On Plug (COP) Ignition Control circuit for less than one second during the CCM test. *Note: Watch the Scan Tool Misfire Counters to identify the fault.* **Possible Causes** ● IC circuit is open, shorted to ground or shorted to power between the coil and the PCM ● Ignition coil (COP) is damaged or has failed, or the PCM has failed

OBD II Trouble Code List (P0xxx Codes)

DTC	Trouble Code Title, Conditions & Possible Causes
P0355 **2T CCM, MIL: Yes** 1997, 1998, 1999, 2000, 2001, 2002, 2003, 2004, 2005 Camaro, Corvette & Firebird 5.7L VIN G, 5.7L VIN S Transmissions: All	**Ignition Coil 5 Control Circuit Malfunction Conditions:** Engine started; and the PCM detected an unexpected low or high voltage condition on Coil On Plug (COP) Ignition Control circuit for less than one second during the CCM test. *Note: Watch the Scan Tool Misfire Counters to identify the fault.* **Possible Causes** ● IC circuit is open, shorted to ground or shorted to power between the coil and the PCM ● Ignition coil (COP) is damaged or has failed ● PCM has failed
P0356 **2T CCM, MIL: Yes** 1997, 1998, 1999, 2000, 2001, 2002, 2003, 2004, 2005 Camaro, Corvette & Firebird 5.7L VIN G, 5.7L VIN S Transmissions: All	**Ignition Coil 6 Control Circuit Malfunction Conditions:** Engine started; and the PCM detected an unexpected low or high voltage condition on Coil On Plug (COP) Ignition Control circuit for less than one second during the CCM test. *Note: Watch the Scan Tool Misfire Counters to identify the fault.* **Possible Causes** ● IC circuit is open, shorted to ground or shorted to power between the coil and the PCM ● Ignition coil (COP) is damaged or has failed ● PCM has failed
P0357 **2T CCM, MIL: Yes** 1997, 1998, 1999, 2000, 2001, 2002, 2003, 2004, 2005 Camaro, Corvette & Firebird 5.7L VIN G, 5.7L VIN S Transmissions: All	**Ignition Coil 7 Control Circuit Malfunction Conditions:** Engine started; and the PCM detected an unexpected low or high voltage condition on Coil On Plug (COP) Ignition Control circuit for less than one second during the CCM test. *Note: Watch the Scan Tool Misfire Counters to identify the fault.* **Possible Causes** ● IC circuit is open, shorted to ground or shorted to power between the coil and the PCM ● Ignition coil (COP) is damaged or has failed ● PCM has failed
P0358 **2T CCM, MIL: Yes** 1997, 1998, 1999, 2000, 2001, 2002, 2003, 2004, 2005 Camaro, Corvette & Firebird 5.7L VIN G, 5.7L VIN S Transmissions: All	**Ignition Coil 8 Control Circuit Malfunction Conditions:** Engine started; and the PCM detected an unexpected low or high voltage condition on Coil On Plug (COP) Ignition Control circuit for less than one second during the CCM test. *Note: Watch the Scan Tool Misfire Counters to identify the fault.* **Possible Causes** ● IC circuit is open, shorted to ground or shorted to power between the coil and the PCM ● Ignition coil (COP) is damaged or has failed ● PCM has failed
P0370 **1T CCM, MIL: Yes** 1996, 1997, 1998, 1999, 2000, 2001, 2002 C/K Series Truck, G Series Van, Van Cargo 6.5L VIN F, 6.5L VIN S, 6.5L VIN Y Diesel engine Transmissions: All	**Timing Reference High-Resolution System Performance Conditions:** Engine started; and the PCM detected several High Resolution signals were missing for every 8 CMP Reference signals. **Possible Causes** ● Optical fuel temperature sensor circuit has a problem ● PCM has failed ● Vehicle driven in an aggressive manner while low on fuel
P0370 **1T CCM, MIL: Yes** 2001, 2002, 2003, 2004, 2005 C/K Series Trucks 6.6L VIN 1 Diesel engine Transmissions: All	**Engine Speed Signal Circuit Malfunction Conditions:** Engine cranking or running; and the FICM did not receive a crank signal, or it detected an invalid crank signal from the ECM. The CKP sensor signal is replicated and sent to the FICM as an engine speed signal. If the Engine speed signal is lost, the engine will not start. The ECM replicates the signal received from the CKP sensor. This signal is sent to the FICM through the engine speed signal circuit. The FICM uses this replicated signal to generate injection current and control the recharge of the fuel injection high voltage circuits. The FICM has full control of the fuel injectors during cranking. The only input the FICM uses at this time is the engine speed signal from the ECM. The FICM monitors the signal along with the injection request signals from the ECM after the engine is running. **Possible Causes** ● Engine speed signal circuit is open, shorted to ground or VREF ● FICM has failed ● PCM has failed
P0371 **1T CCM, MIL: Yes** 1996, 1997, 1998, 1999 Aurora, DeVille, Eldorado & Seville 4.0L VIN C, 4.6L VIN 9, 4.6L VIN Y Transmissions: A/T	**IC 24X Signal Circuit Too Many Pulses Conditions:** DTC P1376 not set, engine running with at least one CMP pulse received in last 250 ms, engine speed 496-3500 rpm, 7 or more CMP edges since the ignition sequence started, eight 4X pulses received between CMP pulses, and the PCM detected over 49 24X pulses between CMP pulses, or that the last 24X variance was 49 along with 49 current 24X pulses. **Possible Causes** ● Battery charger connected to the vehicle with engine running ● CMP sensor 24X signal wire is routed to close to the Generator, spark plug wires or any other possible cause of EMI/RFI ● Reluctor wheel is damaged or the sensor is touching the wheel ● PCM has failed ● TSB 71-65-69 contains a repair procedure for this code

OBD II Trouble Code List (P0xxx Codes)

DTC	Trouble Code Title, Conditions & Possible Causes
P0372 **1T CCM, MIL: Yes** 1996, 1997 Camaro, Caprice, Corvette, Firebird, Fleetwood 4.3L VIN W, 5.7L VIN 5, 5.7L VIN P Transmissions: All	**DI High Resolution Signal Circuit Malfunction Conditions:** Engine started; and the PCM detected DI Low Resolution signals present without detecting any DI High Resolution signals. **Possible Causes** ● CKP High Resolution signal circuit is open or shorted to ground ● CKP High Resolution signal circuit is shorted to system power ● Ignition control module is damaged or has failed ● PCM has failed
P0372 **1T CCM, MIL: Yes** 1996, 1997, 1998, 1999 Aurora, DeVille, Eldorado & Seville 4.0L VIN C, 4.6L VIN 9, 4.6L VIN Y engines Transmissions: A/T	**IC 24X Signal Circuit Too Few Pulses Conditions:** DTC P1376 not set, engine started with at least one CMP pulse received in last 250 ms, engine speed 496-3500 rpm, 7 or more CMP edges since the ignition sequence started, eight 4X pulses received between CMP pulses, and the PCM detected less than 47 24X pulses between CMP pulses, or it received 47 24X pulses twice in succession. **Possible Causes** ● CMP sensor 24X signal wire is open or shorted to ground (fault may be intermittent) ● CMP sensor or the Ignition Control Module is damaged or failed ● PCM has failed ● TSB 71-65-69 contains a repair procedure for this code
P0374 **1T CCM, MIL: Yes** 2001, 2002, 2003, 2004, 2005 C/K Series Trucks 6.6L VIN 1 Diesel engine Transmissions: All	**Engine Speed Signal Circuit Malfunction Conditions:** Engine cranking or running; and the FICM did not receive a crank signal, or it detected an invalid crank signal from the ECM. The CKP sensor signal is replicated and sent to the FICM as an engine speed signal. If the Engine speed signal is lost, the engine will not start. **Possible Causes** ● Engine speed signal circuit is open, shorted to ground or VREF ● FICM has failed ● PCM has failed
P0380 **2T CCM, MIL: Yes** 1996, 1997, 1998, 1999, 2000, 2001, 2002 C/K Series Truck, G Series Van, Van Cargo 6.5L VIN F, 6.5L VIN S, 6.5L VIN Y engines Transmissions: All	**Glow Plug Range/Performance Conditions:** Engine cranking, Glow Plugs "on", and the PCM detected glow plug voltage was below 4.0v, or with the plugs "off", it detected the glow plug voltage was over 4.0v, or with the Glow Plugs commanded "on", over a 2.0v difference between system voltage and glow plug voltage. *Note: Use the Scan Tool Glow Plug command to test the relay.* **Possible Causes** ● Glow plug relay control circuit is open, shorted to power, or the ground circuit is open ● Glow plug relay is stuck in "on" position causing constant power ● Glow plug relay is damaged or has failed ● PCM has failed
P0380 **1T CCM, MIL: No** 2001, 2002, 2003, 2004, 2005 C/K Series Trucks 6.6L VIN 1 Diesel engine Transmissions: All	**Glow Plug Range/Performance Conditions:** Engine cranking, Glow [lugs "on", and the PCM detected the glow plug voltage was less than 4.0v, or with the Glow plugs "off", the PCM detected the Glow plug voltage was over 4.0v or not between 5v and 6.2v (California models). The Scan Tool Glow can help to test the relay. **Possible Causes** ● Glow plug relay control circuit is open, shorted to power, or the ground circuit is open ● Glow plug relay is stuck in "on" position causing constant power ● Glow plug relay is damaged or has failed ● ECM has failed
P0385 **1T CCM, MIL: Yes** 2000, 2001, 2002, 2003, 2004, 2005 DeVille, Eldorado, Seville 4.0L VIN C, 4.6L VIN 9, 4.6L VIN Y engines Transmissions: All	**Crankshaft Position Sensor 'B' Circuit Malfunction Conditions:** Engine started; MAF sensor more than 2.5 g/sec, and the PCM detected an interruption of the CKP Sensor 'B' signal. **Possible Causes** ● Start attempt performed with a defective starter relay or starter, or while not in P/N ● CKP Sensor 'B' signal circuit is open or shorted to ground ● CKP Sensor 'B' VREF circuit is shorted to ground (a No Start) ● CKP Sensor 'B' is damaged or has failed
P0386 **1T CCM, MIL: Yes** 2000, 2001, 2002, 2003, 2004, 2005 DeVille, Eldorado, Seville 4.0L VIN C, 4.6L VIN 9, 4.6L VIN Y engines Transmissions: A/T	**Crankshaft Position Sensor 'B' Performance Conditions:** DTC P0335 not set, engine cranking or running (over 20 rpm), no CMP faults present, then with the timing in Time 'B' mode, the PCM detected an interruption of the CKP 'B' signals. **Possible Causes** ● CKP Sensor 'B' signal circuit is open (an intermittent fault) ● CKP Sensor 'B' signal circuit is shorted to ground (intermittent) ● CKP Sensor 'B' VREF circuit is open (an intermittent fault) ● CKP Sensor 'B' ground circuit is open (an intermittent fault) ● CKP Sensor 'B' is damaged or has failed

OBD II Trouble Code List (P0xxx Codes)

DTC	Trouble Code Title, Conditions & Possible Causes
P0400 **1T EGR, MIL: Yes** 1996, 1997 Camaro, Caprice, Corvette, Firebird, Fleetwood, Roadmaster 4.3L VIN W, 5.7L VIN 5, 5.7L VIN P engines Transmissions: All	**EGR System Performance Conditions:** DTC P0107, P0108, P0112, P0113, P0117, P0118, P0122, P0123, P0601, P0718 and P0724 not set, engine started, vehicle driven to a speed of over 14 mph at an engine speed of 500-900 rpm, BARO sensor more than 95 kPa, ECT sensor more than 81ºF, MAP sensor less than 24 kPa, IAC counts steady, A/C Clutch, Brake switch and TCC command all steady, followed by a deceleration period with the throttle closed, DFCO inactive, and the PCM detected the change in the MAP sensor signal was less than a calculated value during the EGR System Monitor flow test; or after the vehicle driven to over 19 mph at 900-2500 rpm, TP angle from 6-19% and steady, the PCM detected too little change after the EGR solenoid was cycled on and off. **Possible Causes** ● Base engine problem (e.g., a severely restricted exhaust) ● EGR vacuum hoses damaged, loose or routed incorrectly ● EGR passages or intake passages clogged or restricted ● EGR solenoid valve is clogged (carbon), damaged or has failed
P0401 **2T EGR, MIL: Yes** 1996, 1997, 1998 Achieva, Beretta, Ciera, Cavalier, Century, Corsica, Grand Am, Malibu, Skylark, Sunfire, S/T Blazer & S/T Pickup 2.2L VIN 4, 2.4L VIN T Transmissions: All	**Insufficient EGR Flow Detected Conditions:** DTC P0106, P0107, P0108, P0112, P0113, P0117, P0118, P0121, P0122, P0123, P0200, P0300, P0301-P0304, P0335, P0502, P0506, P0507 and P1441 not set, engine started, vehicle driven to over 20 mph at an engine speed of 900-1800 rpm, BARO sensor more than 72 kPa, ECT sensor more than 167ºF, MAP sensor from 12-32 kPa, IAC change less than 4-5 counts, TP angle under 1%, and the PCM detected the Decel Exponentially Weighted Moving Average was more than +2 kPa (a normal value is less than -3 kPa) because of a restriction somewhere in the EGR system. **Possible Causes** ● EGR valve pipe assembly is damaged or leaking ● EGR vacuum hoses damaged, loose or routed incorrectly ● EGR passages or intake passages clogged or restricted ● EGR solenoid valve is clogged (carbon), damaged or has failed ● Base engine problem (e.g., a severely restricted exhaust)
P0401 **1T EGR, MIL: Yes** 1996, 1997, 1998 Achieva, Beretta, Cavalier, Grand Am, Skylark, Sunfire, S/T Blazer & S/T Pickup 2.2L VIN 4, 2.4L VIN T Transmissions: M/T	**Insufficient EGR Flow Detected Conditions:** DTC P0106, P0107, P0108, P0112, P0113, P0117, P0118, P0121, P0122, P0123, P0200, P0300, P0301-P0304, P0335, P0502, P0506, P0507 and P1441 not set, engine started, vehicle driven to over 20 mph at an engine speed of 1100-2200 rpm, BARO sensor more than 72 kPa, ECT sensor more than 176ºF, MAP sensor from 12-32 kPa, IAC change less than 4-5 counts, TP angle under 1%, and the PCM detected the Decel Exponentially Weighted Moving Average was more than +2 kPa (a normal value is less than -3 kPa) because of a restriction somewhere in the EGR system. **Possible Causes** ● Base engine problem (e.g., a severely restricted exhaust) ● EGR valve pipe assembly is damaged or leaking ● EGR vacuum hoses damaged, loose or routed incorrectly ● EGR passages or intake passages clogged or restricted ● EGR solenoid valve is clogged (carbon), damaged or has failed
P0401 **1T EGR, MIL: Yes** 1996, 1997, 1998, 1999, 2000, 2001, 2002, 2003, 2004, 2005 Aurora, Bonneville, Century, Cutlass Supreme, Grand Prix, Intrigue, LeSabre, LSS, Lumina, Monte Carlo, 88', Regal, Park Avenue & 98' 3.1L VIN J, 3.1L VIN M, 3.5L VIN H, 3.8L VIN 1, 3.8L VIN K engines Transmissions: All	**Insufficient EGR Flow Detected Conditions:** DTC P0101, P0102, P0103, P0107, P0108, P0112, P0113, P0116, P0117, P0118, P0121, P0122, P0123, P0403, P0404, P0405, P0502, P0503, P0506, P0507, P0641, P0651, P1374, P1404 not set, engine started, ECT sensor more than 167ºF, IAT sensor from 32-212ºF, system voltage 11-18v, BARO sensor more than 74 kPa, IAC steady (± 5 counts), A/C Clutch and TR signals stable, then vehicle driven to over 50 mph at an engine speed of 1050-1300 rpm, MAP sensor from 15-70 kPa, followed by a deceleration period with the TP angle less than 1.3%, and the PCM detected the amount of MAP sensor change monitored with the valve open and then closed during deceleration indicated insufficient EGR flow. The EGR flow test is enabled by the PCM during deceleration. A change from 0 to a value over +0 in the Desired EGR and Actual EGR Position PID will appear on a Scan Tool. The PCM allows one EGR flow test in each key cycle. To verify a repair, the PCM will allow up to 12 EGR flow test counts during the first key cycle after codes are cleared. From 9-12 EGR flow tests are enough to detect adequate EGR flow. **Possible Causes** ● Base engine problem (e.g., a severely restricted exhaust) ● EGR vacuum hoses damaged, loose or routed incorrectly ● EGR passages or intake passages clogged or restricted ● EGR solenoid valve is clogged (carbon), damaged or has failed ● TSB 87-65-22 contains a repair procedure for this code

OBD II Trouble Code List (P0xxx Codes)

DTC	Trouble Code Title, Conditions & Possible Causes
P0401 **1T EGR, MIL: Yes** 1995 Century, Cutlass Ciera, Cutlass Supreme, Grand Prix, Lumina, Monte Carlo Park Avenue Regal, Riviera, 88' & 98' 3.1L VIN M, 3.4L VIN X, 3.8L VIN 1, 3.8L VIN K, 3.8L VIN L engines Transmissions: All	**Insufficient EGR Flow Detected Conditions:** DTC P0122, P0123, P0501 and P1406 not set, engine started, vehicle driven to over 25 mph at an engine speed of 900-1100 rpm, Brake switch "off", ECT sensor more than 185ºF, followed by a deceleration period with the TP angle less than 1%, and the PCM detected the MAP changes monitored during deceleration with the EGR valve opened and closed indicated insufficient EGR flow. **Possible Causes** ● Base engine problem (e.g., a severely restricted exhaust) ● EGR vacuum hoses damaged, loose or routed incorrectly ● EGR passages or intake passages clogged or restricted ● EGR solenoid valve is clogged (carbon), damaged or has failed
P0401 **1T EGR, MIL: Yes** 1996, 1997, 1998, 1999, 2000, 2001 Achieva, Alero, Century, Cutlass, Cutlass Supreme Grand Prix, Impala, Monte Carlo, Intrigue, Lumina, Malibu, Montana, Silhouette & Venture 3.1L VIN J, 3.1L VIN M, 3.4L VIN E engines Transmissions: All	**Insufficient EGR Flow Detected Conditions:** DTC P0101-P0103, P0107, P0108, P0112, P0113, P0116-P0118, P0121-P0123, P0201-P0206, P0300, P0336, P0403, P0404, P0502, P0503, P0506, P0507, P1106, P1107, P1111, P1112, P1114, P1115, P1121, P1122, P1374 and P1404 not set, engine started, vehicle driven to over 50 mph at an engine speed of 1050-1400 rpm, system voltage over 10.0v, BARO sensor more than 70 kPa, ECT sensor over 167ºF, IAT sensor over 176ºF, A/C Clutch, IAC position and A/T TR signals stable, MAP sensor from 15-70 kPa, then during a deceleration period from over 30 mph with the TP angle less than 1%, the PCM detected the MAP changes during testing indicated insufficient EGR flow. Note that during the test that the Desired EGR and EGR Position PID will change from 0 to a positive number (+ 0). **Possible Causes** ● Base engine problem (e.g., a severely restricted exhaust), or any other problem that causes the engine to run poorly ● EGR passages or intake passages clogged or restricted ● EGR pipe is clogged, dirty or otherwise restricted ● EGR vacuum hoses damaged, loose or routed incorrectly ● EGR solenoid valve is clogged (carbon), damaged or has failed ● MAP sensor is dirty, damaged or it is "skewed"
P0401 **1T EGR, MIL: Yes** 2002, 2003, 2004, 2005 Alero, Aztek, Century, Grand Am, Grand Prix, Impala, Monte Carlo, Intrigue, Malibu, Montana, Silhouette, Rendezvous, Venture 3.1L VIN J, 3.1L VIN M, 3.4L VIN E engines Transmissions: All	**Insufficient EGR Flow Detected Conditions:** DTC P0101, P0102, P0103, P0107, P0108, P0112, P0113, P0116, P0117, P0118, P0121, P0122, P0123, P0125, P0201, P0202, P0203, P0204, P0205, P0206, P0300, P0336, P0403, P0404, P0405, P0502, P0503, P0506, P0507, P1106, P1107, P1111, P1112, P1114, P1115, P1121, P1122, P1374 and P1404 not set, engine runtime up to 3 minutes, engine speed of 1050-1300 rpm, VSS over 35 mph, system voltage from 11-18v, BARO sensor over 74 kPa, ECT sensor more than 167ºF, IAT sensor from 32-212ºF, IAC counts stable (± 5 counts), A/C Clutch and current gear stable, gear selector not in Park or Neutral, Decel Fuel Cutoff and Power Enrichment not active, MAP sensor from 15-70 kPa, then during a deceleration period from over 30 mph with the TP angle less than 1%, the PCM detected the MAP changes during testing indicated insufficient EGR flow. Note that during the test that the Desired EGR and EGR Position PID will change from 0 to a positive number (+ 0). **Possible Causes** ● Base engine problem (e.g., a severely restricted exhaust), or any other problem that causes the engine to run poorly ● EGR passages or intake passages clogged or restricted ● EGR pipe is clogged, dirty or otherwise restricted ● EGR vacuum hoses damaged, loose or routed incorrectly ● EGR solenoid valve is clogged (carbon), damaged or has failed ● MAP sensor is dirty, damaged or it is "skewed"
P0401 **2T EGR, MIL: Yes** 1996, 1997, 1998, 1999, 2000, 2001, 2002 C/K Cab & Chassis, C/K Series Truck, Escalade, G Series Van, L/M Van, S/T Blazer & S/T Pickup C/K Truck, Escalade, G Series Van, Cargo Van, L, M, P, S, T 4.3L VIN W, 4.3L VIN X, 4.8L VIN V, 5.0L VIN M, 5.3L VIN T, 5.7L VIN K, 5.7L VIN R, 6.0L VIN U, 7.4L VIN J, 8.1L VIN G Transmissions: All	**Insufficient EGR Flow Detected Conditions:** DTC P0106-P0108, P0112-P0118, P0121-P0123, P0300, P0404, P0405, P0502-P0507, P1111-P1112, P1120-P1125, P1220, P1221, P1404, P1514-P1518, P1635 and P1639 not set, engine started, vehicle driven to 25-70 mph at 725-2000 rpm, BARO sensor over 70 kPa, ECT sensor from 140-244ºF, IAT sensor from 37-167ºF, altitude compensated MAP from 10-60 kPa (2-7 psi) with any change less than 0.8 kPa, IAC position and A/C Clutch both steady, followed by a deceleration period with the TP angle under 1.2% and Deceleration Fuel Cutoff off, and the PCM detected the change in the MAP sensor was less than a calculated value during the EGR flow test for 2 seconds. The PCM allows one EGR flow test in each key cycle. To verify a repair, the PCM will allow up to 12 EGR flow test counts during the first key cycle after codes are cleared. From 9-12 EGR flow tests are enough to detect adequate EGR flow. **Possible Causes** ● Base engine problem (e.g., a severely restricted exhaust) ● EGR vacuum hoses damaged, loose or routed incorrectly ● EGR passages or intake passages clogged or restricted ● EGR solenoid valve is clogged (carbon), damaged or has failed

OBD II Trouble Code List (P0xxx Codes)

DTC	Trouble Code Title, Conditions & Possible Causes
P0401 **2T EGR, MIL: Yes** 2003, 2004, 2005 C/K Series Truck, G Series Van 8.1L VIN G engine Transmissions: All	**Insufficient EGR Flow Detected Conditions:** DTC P0106-P0108, P0112-P0118, P0121-P0123, P0300, P0404, P0405, P0502-P0507, P1111-P1112, P1120-P1125, P1220, P1221, P1404, P1514-P1518, P1635 and P1639 not set, engine started, vehicle driven to 25-70 mph at 725-2000 rpm, BARO sensor over 70 kPa, ECT sensor from 140-244ºF, IAT sensor from 37-167ºF, altitude compensated MAP from 10-60 kPa (2-7 psi) with any change less than 0.8 kPa, IAC position and A/C Clutch both steady, followed by a deceleration period with the TP angle under 1.2% and Deceleration Fuel Cutoff off, and the PCM detected the change in the MAP sensor was less than a calculated value during the EGR flow test for 2 seconds. The PCM allows one EGR flow test in each key cycle. To verify a repair, the PCM allows up to 12 EGR flow test counts during the first key cycle after codes are cleared (from 9-12 EGR flow tests are needed) **Possible Causes** • Base engine problem (e.g., a severely restricted exhaust) • EGR vacuum hoses damaged, loose or routed incorrectly • EGR passages or intake passages clogged or restricted • EGR solenoid valve is clogged (carbon), damaged or has failed
P0401 **2T EGR, MIL: Yes** 2001, 2002, 2003, 2004, 2005 C/K Series Truck 6.6L VIN 1 Diesel engine Transmissions: All	**Insufficient EGR Flow Detected Conditions:** DTC P0101, P0102, P0103, P0106, P0107, P0108, P0112, P0113, P0116, P0117, P0118, P0405, P0406, P0489, P0490, P0500, P0651, P2142, P2144, and P2145 not set, engine runtime over 5 seconds, ECM not operating in Reduced Engine Power mode, system voltage from 11-18v, BARO sensor over 72 kPa, engine speed from 610-820 rpm (± 50 rpm) for 3 seconds, ECT sensor from 140-212ºF, IAT sensor more than 32ºF, Calculated Fuel Rate from 3-20 mm³, Power Take Off not enabled, APP indicated angle less than 1%, vehicle speed less than 0.25 mph, conditions met for 3 seconds, and the ECM detected a calibrated difference between the Expected MAF sensor rate and Actual MAF sensor rate during EGR system operation, or the ECM did not detect any difference between the Expected EGR Vacuum sensor signal and the Actual EGR Vacuum sensor signal for over 4 seconds. **Possible Causes** • EGR ports are clogged or restricted • EGR throttle valve vacuum control solenoid is leaking/restricted • EGR throttle valve is binding or sticking, or it is damaged • EGR throttle valve vacuum diaphragm is damaged or has failed • Vacuum hose from vacuum pump to EGR throttle valve diaphragm leaking or restricted • Exhaust system is restricted, or exhaust system was modified
P0401 **2T EGR, MIL: Yes** 1996, 1997, 1998, 1999, 2000, 2001, 2002, 2003, 2004, 2005 DeVille, Eldorado, Seville 4.6L VIN 9, 4.6L VIN Y Transmissions: A/T	**Insufficient EGR Flow Detected Conditions:** DTC P0101-P0103, P0106-P0108, P0112-P0113, P0116-P0118, P0121-P0123, P0125, P0128, P0200, P0300, P0410-P0446, P0452-P0453, P1258, P1415-P1416 and P1441 not set, engine started, vehicle driven to over 12 mph at an engine speed of 1000-1500 rpm, system voltage over 10.0v, BARO sensor over 70 kPa, ECT sensor from 176-230ºF, A/C Clutch, EVAP Purge, IAC position, TCC and A/T TR position all steady, then during a deceleration period with the TP angle less than 2%, the PCM detected the MAP sensor changes monitored during the EGR Flow Test indicated insufficient EGR flow. **Possible Causes** • Base engine problem or any other problem that causes the engine to run poorly • EGR passages or intake passages clogged or restricted • EGR pipe is clogged, dirty or otherwise restricted • EGR vacuum hoses damaged, loose or routed incorrectly • EGR solenoid valve is clogged (carbon), damaged or has failed • MAP sensor is dirty, damaged or it is "skewed" • TSB 71-65-26 contains a repair procedure for this code
P0401 **2T EGR, MIL: Yes** 1995, 1996, 1997, 1998, 1999, 2000, 2001, 2002 Camaro & Firebird 3.8L VIN K engine Transmissions: All	**Insufficient EGR Flow Detected Conditions:** DTC P0101, P0102, P0103, P0107, P0108, P0112, P0113, P0116, P0117, P0118, P0121, P0122, P0123, P0201-P0206, P0300, P0336, P0403, P0404, P0405, P0502, P0503, P0506, P0507, P0604, P0606, P1106, P1107, P1112, P1114, P1115, P1120, P1121, P1122, P1125, P1220, P1221, P1374, P1404, P1514, P1515, P1516, P1517 and P1518 not set, engine started, BARO sensor more than 70 kPa, system voltage from 11-18v, ECT sensor more than 167ºF, IAT sensor less than 176ºF, A/C Clutch, IAC position and A/T TR sensor signal all stable, then during a deceleration period from more than 50 mph to a speed over 25 mph at 900-1400 rpm with the TP angle under 1.1%, the PCM detected that changes monitored in the MAP sensor signal during the EGR Flow Test indicated insufficient EGR flow. **Possible Causes** • Base engine problem or any other problem that causes the engine to run poorly • EGR passages or intake passages clogged or restricted • EGR pipe is clogged, dirty or otherwise restricted • EGR vacuum hoses damaged, loose or routed incorrectly • EGR solenoid valve is clogged (carbon), damaged or has failed • MAP sensor is dirty, damaged or it is "skewed"

OBD II Trouble Code List (P0xxx Codes)

DTC	Trouble Code Title, Conditions & Possible Causes
P0401 **1T EGR, MIL: Yes** 1997, 1998, 1999, 2000 Camaro & Firebird 5.7L VIN G Transmissions: All	**Insufficient EGR Flow Detected Conditions:** DTC P0101, P0102, P0103, P0107, P0108, P0112, P0113, P0117, P0118, P0121, P0122, P0123, P0405, P0500, P0502, P0503, P0562, P0563 and P1404 not set, engine started, ECT sensor from 140-243°F, IAT sensor less than 140°F, engine vacuum from 60-83 kPa (BARO value minus MAP value), system voltage from 11-18v, vehicle driven at 42-70 mph at a speed of 800-2000 rpm, A/C Clutch and IAC position all stable, DFCO not enabled, then during a deceleration period from over 40 mph with the TP angle under 1.1%, the PCM detected that the change monitored in the MAP sensor during the EGR Flow Test indicated insufficient EGR flow. Note that the Desired EGR and EGR Position PID will change from zero (0) to a calibrated value above zero (0) under these test conditions. **Possible Causes** ● Base engine problem (e.g., a severely restricted exhaust), or any other problem that causes the engine to run poorly ● EGR passages or intake passages clogged or restricted ● EGR pipe is clogged, dirty or otherwise restricted ● EGR vacuum hoses damaged, loose or routed incorrectly ● EGR solenoid valve is clogged (carbon), damaged or has failed ● MAP sensor is dirty, damaged or it is "skewed"
P0401 **1T EGR, MIL: Yes** 1996, 1997 Aurora 4.0L VIN C engine Transmissions: A/T	**Insufficient EGR Flow Detected Conditions:** DTC P0101-P0103, P0106-P0108, P0112, P0113, P0116-P0118, P0121-P0123, P0201-P0208, P0300, P0335, P0336, P0385, P0386, P0405, P0449, P0502, P0503, P0506 and P0507 not set, engine started, vehicle driven to 20-70 mph at an engine speed of 700-1200 rpm, system voltage over 10.0v, BARO sensor more than 72 kPa, ECT sensor from 176-230°F, EGR pintle within 5% its command, throttle angle 1 degree or less, A/C Clutch state, EVAP solenoid, Transaxle Gear and TCC state all stable for 300 ms, then with the MAP sensor stable (± 1 kPa) and from 24.7-44 kPa with the engine speed at 700-1300 rpm, EGR Pintle position at 0.0v with EGR off, the PCM recorded the maximum RPM, minimum and maximum MAP sensor signals and the EGR pintle position at 0.00v (EGR off). At this point the EGR valve is turned "on" and held open for 600 ms and then turned off. The change in MAP during this test is compared to the expected change in MAP. If the difference between expected and actual MAP change is more than 4 kPa, this trouble code is set. **Possible Causes** ● Base engine problem (e.g., a severely restricted exhaust), or any other problem that causes the engine to run poorly ● EGR passages or intake passages clogged or restricted ● EGR pipe is clogged, dirty or otherwise restricted ● EGR vacuum hoses damaged, loose or routed incorrectly ● EGR solenoid valve is clogged (carbon), damaged or has failed ● MAP sensor is dirty, damaged or it is "skewed" ● TSB 71-65-26 contains a repair procedure for this code
P0401 **1T EGR, MIL: Yes** 1998, 1999, 2001, 2002, 2003, 2004, 2005 Aurora 4.0L VIN C engine Transmissions: A/T	**Insufficient EGR Flow Detected Conditions:** DTC P0101-P0103, P0106-P0108, P0112, P0113, P0116-P0118, P0121-P0123, P0201-P0208, P0300, P0335, P0336, P0385, P0386, P0405, P0449, P0502, P0503, P0506 and P0507 not set, engine started, vehicle driven to 20-70 mph at an engine speed of 700-1200 rpm, system voltage over 10.0v, BARO sensor more than 72 kPa, ECT sensor from 176-230°F, EGR pintle within 5% its command, throttle angle 1 degree or less, A/C Clutch state, EVAP solenoid, Transaxle Gear and TCC state all stable for 300 ms, then with the MAP sensor stable (± 1 kPa) and from 24.7-44 kPa with the engine speed at 700-1300 rpm, EGR Pintle position at 0.0v with EGR off, the PCM recorded the maximum RPM, minimum and maximum MAP sensor signals and the EGR pintle position at 0.00v (EGR off). At this point the EGR valve is turned "on" and held open for 600 ms and then turned off. The change in MAP during this test is compared to the expected change in MAP. If the difference between expected and actual MAP change is more than 4 kPa, this trouble code is set. **Possible Causes** ● Base engine problem (e.g., a severely restricted exhaust), or any other problem that causes the engine to run poorly ● EGR passages or intake passages clogged or restricted ● EGR pipe is clogged, dirty or otherwise restricted ● EGR vacuum hoses damaged, loose or routed incorrectly ● EGR solenoid valve is clogged (carbon), damaged or has failed ● MAP sensor is dirty, damaged or it is "skewed"

OBD II Trouble Code List (P0xxx Codes)

DTC	Trouble Code Title, Conditions & Possible Causes
P0401 **1T EGR, MIL: Yes** 1996, 1997 Grand Prix, Lumina, Monte Carlo 3.4L VIN X engine Transmissions: All	**Insufficient EGR Flow Detected Conditions:** DTC P0106, P0107, P0108, P0112, P0113, P0117, P0118, P0121, P0122, P0123, P0200, P0300, P0301-P0304, P0335, P0502, P0506, P0507 and P1441 not set, engine started, vehicle driven to over 35 mph at an engine speed of 1275-1450 rpm, ECT sensor more than 167°F, system voltage over 10.0v, BARO sensor more than 70 kPa, MAP sensor from 15-75 kPa and steady, IAC position steady, A/C Clutch and A/T TR signals stable, followed by a deceleration period with the TP angle less than 1%, and the PCM detected that the MAP sensor indicated insufficient EGR flow. *Note: The vehicle will need to be driven to about 50 mph and then allowed to decelerate. The PCM will enable the EGR flow test during deceleration. The change from 0 to over 0 in the Desired EGR and Actual EGR Position PID can be viewed on a Scan Tool.* **Possible Causes** ● Base engine problem (e.g., a severely restricted exhaust) ● EGR vacuum hoses damaged, loose or routed incorrectly ● EGR passages or intake passages clogged or restricted ● EGR solenoid valve is clogged (carbon), damaged or has failed
P0403 **1T CCM, MIL: Yes** 1996, 1997 Camaro, Caprice, Corvette, Firebird, Fleetwood, Roadmaster 4.3L VIN W, 5.7L VIN 5, 5.7L VIN P engines Transmissions: All	**EGR Solenoid Control Circuit Malfunction Conditions:** Engine started; engine speed over 600 rpm, ECT sensor more than 180°F, and the PCM detected the commanded state of the EGR solenoid driver did not match the Actual state of the EGR solenoid driver for 25.5 seconds during the CCM test. **Possible Causes** ● EGR solenoid control circuit open, shorted to ground or power ● EGR solenoid B+ circuit is open (test Actuator fuse in U/H box) ● EGR solenoid is damaged or has failed ● PCM has failed
P0403 **2T CCM, MIL: Yes** 1996, 1997, 1998, 1999, 2000, 2001, 2002, 2003, 2004, 2005 Achieva, Alero, Aztek, Century, Cutlass, Cutlass Supreme, Grand Prix, Impala, Monte Carlo, Intrigue, Lumina, Malibu, Montana, Silhouette, Rendezvous, Venture 3.1L VIN J, 3.1L VIN M, 3.4L VIN E engines Transmissions: All	**EGR Solenoid Control Circuit Malfunction Conditions:** Engine started; system voltage from 9-18.0v, and the PCM detected an unexpected voltage condition on the EGR Solenoid high or low control circuit for 20 seconds in the CCM test. **Possible Causes** ● EGR solenoid control circuit is open, shorted to ground or B+ ● EGR solenoid high control (VREF) circuit is open ● EGR solenoid is damaged or has failed ● PCM has failed ● TSB 02-06-04-053 contains a repair procedure for this code
P0403 **1T CCM, MIL: Yes** 1998, 1999, 2000, 2001, 2002, 2003, 2004, 2005 Aurora, Bonneville, Century, Cutlass Supreme, Grand Prix, Intrigue, LeSabre, LSS, Lumina, Monte Carlo, 88', Regal, Park Avenue & 98' 3.1L VIN J, 3.1L VIN M, 3.5L VIN H, 3.8L VIN 1, 3.8L VIN K Transmissions: All	**EGR Solenoid Control Circuit Malfunction Conditions:** Engine started; system voltage over 10.0v and the PCM detected an unexpected voltage condition on the EGR valve control circuit for over 20 seconds. The PCM controls the EGR valve with a solid-state device called a driver. The driver supplies the EGR solenoid with 12 volts that is pulsewidth modulated (PWM) signal through the EGR solenoid high control circuit. A ground path is provided by the PCM through the EGR solenoid low control circuit. If the PCM detects the driver has detected a circuit failure an EGR circuit, it sets this code. **Possible Causes** ● EGR solenoid control circuit is open or shorted to ground ● EGR solenoid control circuit is shorted to system power (B+) ● EGR solenoid high control (VREF) circuit is open ● EGR solenoid is damaged or has failed ● PCM has failed
P0403 **2T CCM, MIL: Yes** 1998, 1999, 2000, 2001, 2002, 2003, 2004, 2005 DeVille, Eldorado, Seville 4.6L VIN 9, 4.6L VIN Y Transmissions: A/T	**Exhaust Gas Recirculation Solenoid Circuit Malfunction Conditions:** Engine running, system voltage 11-16v, and the PCM detected an unexpected voltage condition on the EGR ignition, EGR solenoid control circuit or the EGR solenoid ground circuit for 10 seconds. **Possible Causes** ● EGR solenoid control circuit is open or shorted to ground ● EGR solenoid control circuit is shorted to system power (B+) ● EGR solenoid high control (VREF) circuit is open ● EGR solenoid is damaged or has failed ● PCM has failed

OBD II Trouble Code List (P0xxx Codes)

DTC	Trouble Code Title, Conditions & Possible Causes
P0403 **2T CCM, MIL: Yes** 1998, 1999, 2000, 2001, 2002 Camaro & Firebird 3.8L VIN K engine Transmissions: All	**EGR Solenoid Control Circuit Malfunction Conditions:** Engine started; system voltage over 10.0v and the PCM detected an unexpected voltage condition on the EGR solenoid control circuit for 20 seconds during the CCM test. **Possible Causes** • EGR solenoid control circuit is open or shorted to ground • EGR solenoid control circuit is shorted to system power (B+) • EGR solenoid high control (VREF) circuit is open • EGR solenoid is damaged or has failed • PCM has failed
P0403 **2T CCM, MIL: Yes** 1998, 1999 Achieva, Beretta, Ciera, Corsica, Century, Grand Am, Malibu & Skylark 3.1L VIN M Transmissions: All	**EGR Solenoid Control Circuit Malfunction Conditions:** DTC P0101, P0102, P0103, P0107, P0108, P0112, P0113, P0117, P0118, P0201-P0206, P0335, P0502, P0503, P0506 and P0507 not set, engine running, system voltage over 10.0v, Desired EGR position commanded to 0%, and the PCM detected the EGR feedback signal was 0.40v more than the EGR Closed Valve pintle position for 20 seconds in the test. **Possible Causes** • EGR solenoid control circuit is open or shorted to ground • EGR solenoid high control (VREF) circuit is open • EGR solenoid is damaged or has failed • PCM has failed
P0403 **2T CCM, MIL: Yes** 2000, 2001, 2002, 2003, 2004, 2005 Alero, Century, Grand Am, Malibu, Monte Carlo 3.1L VIN J Transmissions: All	**EGR Solenoid Control Circuit Malfunction Conditions:** Engine started; system voltage over 10.0v and the PCM detected an unexpected voltage condition on the EGR Solenoid control circuit for 20 seconds in the test. **Possible Causes** • EGR solenoid control circuit is open or shorted to ground • EGR solenoid high control (VREF) circuit is open • EGR solenoid is damaged or has failed • PCM has failed
P0404 **1T CCM, MIL: Yes** 1996, 1997, 1998, 1999, 2000, 2001, 2002, 2003, 2004, 2005 Achieva, Alero, Aztek, Century, Cutlass, Cutlass Supreme Grand Prix, Impala, Monte Carlo, Intrigue, Lumina, Malibu, Montana, Silhouette, Rendezvous, Venture 3.1L VIN J, 3.1L VIN M, 3.4L VIN E engines Transmissions: All	**EGR Open Position Signal Range/Performance Conditions:** Engine started; system voltage over 11.0v, EGR solenoid enabled, EGR flow test not active, and the PCM detected the difference between the Actual EGR position and the Desired EGR position was more than 15% for over 20 seconds. The PCM will disable the EGR command if the startup ECT sensor value is less than 41°F, and will not enable the EGR solenoid until the ECT is more than 167°F. **Possible Causes** • EGR sensor signal circuit is open, shorted to ground or power • EGR sensor ground or VREF circuit is open • EGR valve seat contains debris or carbon (inspect and clean) • EGR valve pintle contains debris or carbon (inspect and clean) • EGR valve is contaminated, clogged, damaged or has failed • PCM has failed
P0404 **1T CCM, MIL: Yes** 1998, 1999, 2000, 2001, 2002, 2003, 2004, 2005 Aurora, Bonneville, Century, Cutlass Supreme, Grand Prix, Intrigue, LeSabre, LSS, Lumina, Monte Carlo, 88', Regal, Park Avenue & 98' 3.1L VIN J, 3.1L VIN M, 3.5L VIN H, 3.8L VIN 1, 3.8L VIN K Transmissions: All	**EGR Open Position Signal Range/Performance Conditions:** Engine started; system voltage over 10.0v, EGR valve commanded open, and the PCM detected the difference between the Actual EGR position and the Desired EGR position was more than 15% for 20 seconds during the CCM test period. **Possible Causes** • EGR sensor signal circuit is open, shorted to ground or power • EGR sensor ground or VREF circuit is open • EGR valve seat or pintle contains debris or plugged with carbon • PCM has failed
P0404 **2T CCM, MIL: Yes** 1998, 1999, 2000, 2001, 2002, 2003, 2004, 2005 Cab & Chassis, C/K truck G Van, S/T Blazer, pickup 4.3L VIN W, 4.3L VIN X, 5.0L VIN M, 5.7L VIN K, 5.7L VIN R, 7.4L VIN J Transmissions: All	**EGR Open Position Signal Range/Performance Conditions:** Engine started; commanded EGR position over 0%, EGR flow test "off", Desired EGR position change less than 20%, then after the vehicle was driven, the PCM detected the difference between the Actual EGR position and the Desired EGR position was more than 10% for 13 seconds during the CCM test. **Possible Causes** • EGR sensor signal circuit is open, shorted to ground or power • EGR sensor ground or VREF circuit is open • EGR valve seat or valve pintle contains debris or carbon (inspect and clean) • EGR valve is contaminated, clogged, damaged or has failed • PCM has failed • TSB 01-06-04-043 contains a repair procedure for this code

OBD II Trouble Code List (P0xxx Codes)

DTC	Trouble Code Title, Conditions & Possible Causes
P0404 **2T CCM, MIL: Yes** 1999, 2000, 2001, 2002, 2003, 2004, 2005 C/K Truck, Escalade, G Series Van, Cargo Van 4.8L VIN V, 5.3L VIN T, 6.0L VIN N, 6.0L VIN U, 8.1L VIN G engines Transmissions: All	**EGR Open Position Signal Range/Performance Conditions:** Engine started; commanded EGR position over 0%, EGR flow test inactive, Desired EGR position did not change more than 20%, then after the vehicle was driven, the PCM detected the difference between the Actual EGR position and the Desired EGR position was more than 10% for over 13 seconds during the test. **Possible Causes** • EGR sensor signal circuit is open, shorted to ground or power • EGR sensor ground or VREF circuit is open • EGR valve seat or pintle contains debris or carbon (inspect and clean) • EGR valve is contaminated, clogged, damaged or has failed • PCM has failed
P0404 **1T CCM, MIL: Yes** 1996, 1997, 1998 C/K Truck, G Series Van 6.5L VIN F, 6.5L VIN S, 6.5L VIN Y Diesel engine Transmissions: All	**Exhaust Gas Recirculation System Malfunction Conditions:** DTC P0405 and P0406 not set, engine speed over 506 rpm, EVAP Vent Solenoid commanded "on", conditions met for 2 seconds, and the PCM detected the difference between the ambient air pressure and EGR pressure was more than 15 kPa for 2 seconds. **Possible Causes** • EGR system vacuum hose disconnected, leaking or pinched • EGR valve vacuum hose is disconnected, damaged or leaking • EGR valve vacuum hose routing is not correct • EGR valve is leaking (a small leak is present)
P0404 **2T CCM, MIL: Yes** 2001, 2002, 2003, 2004, 2005 C/K Series Truck 6.6L VIN 1 Diesel engine Transmissions: All	**Exhaust Gas Recirculation System Malfunction Conditions:** DTC P0101, P0102, P0103, P0106, P0107, P0108, P0112, P0113, P0116, P0117, P0118, P0405, P0406, P0489, P0490, P0500, P0651, P2142, P2144, and P2145 not set, engine runtime over 5 seconds, PCM not operating in Reduced Engine Power mode, system voltage from 11-18v, BARO sensor over 72 kPa, engine speed from 610-820 rpm (± 50 rpm) for 3 seconds, ECT sensor from 140-212°F, IAT sensor more than 32°F, Calculated Fuel Rate from 3-20 mm³, PTO "off", APP indicated angle less than 1%, vehicle speed less than 0.25 mph, conditions met for 3 seconds, and the PCM detected a lower than expected EGR Vacuum sensor signal along with a higher than expected MAF sensor rate for 8 seconds. **Possible Causes** • Air intake pipe is clogged or damaged • EGR system vacuum lines are clogged or leaking • EGR throttle valve throttle solenoid is damaged or it had failed • EGR valve vacuum sensor is damaged or it has failed • EGR valve vent solenoid is damaged or it has failed
P0404 **2T CCM, MIL: Yes** 1997, 1998, 1999, 2000, 2001, 2002, 2003, 2004, 2005 DeVille, Eldorado, Seville 4.6L VIN 9, 4.6L VIN Y Transmissions: A/T	**EGR Open Position Signal Range/Performance Conditions:** DTC P0101, P0102, P0103, P0107, P0108, P0112, P0113, P0117, P0118, P0201-P0206, P0335, P0502, P0503, P0506 and P0507 not set, engine started, system voltage over 10.0v, EGR system commanded "off", and the PCM detected the Actual EGR position was 20% less than the Desired EGR position in the test. **Possible Causes** • EGR sensor signal circuit is open, shorted to ground or power • EGR sensor ground circuit or the VREF circuit is open • EGR valve seat or pintle contains debris or carbon (inspect and clean) • EGR valve is contaminated, clogged, damaged or has failed • PCM has failed
P0404 **2T CCM, MIL: Yes** 1998, 1999, 2000, 2001, 2002 Camaro & Firebird 3.8L VIN K engine Transmissions: All	**EGR Open Position Signal Range/Performance Conditions:** Engine started; system voltage over 10.0v, EGR solenoid "on", and the PCM detected more than a 15% difference between the Actual and Desired EGR Position signal for 25 seconds. **Possible Causes** • EGR sensor signal circuit is open, shorted to ground or power • EGR sensor ground or VREF circuit is open • EGR valve seat or pintle contains debris or carbon (inspect and clean) • EGR valve is contaminated, clogged, damaged or has failed • PCM has failed
P0404 **2T CCM, MIL: Yes** 1997, 1998, 1999, 2000 Camaro & Firebird 5.7L VIN G engine Transmissions: All	**EGR Open Position Signal Range/Performance Conditions:** Engine started; system voltage over 11.7v, Desired EGR position over 0%, and the PCM detected more than a 20% difference between the Actual and Desired EGR position with the EGR failure counter more than a calibrated value. The ability of the PCM to control the EGR valve is verified with this test. The PCM sets DTC P0404 if the error is too great. **Possible Causes** • EGR sensor signal circuit is open, shorted to ground or power • EGR sensor ground or VREF circuit is open • EGR valve seat or pintle contains debris or carbon (inspect and clean) • EGR valve is contaminated, clogged, damaged or has failed • PCM has failed

OBD II Trouble Code List (P0xxx Codes)

DTC	Trouble Code Title, Conditions & Possible Causes
P0404 **2T CCM, MIL: Yes** 1997, 1998, 1999, 2001, 2002, 2003, 2004, 2005 Aurora 4.0L VIN C engine Transmissions: A/T	**EGR Open Position Signal Range/Performance Conditions:** Engine started; system voltage over 10.0v, MAP sensor more than 49.7 kPa, TFT sensor more than 193.4ºF, EGR solenoid commanded "on", and the PCM detected the difference between the Actual EGR position and the Desired EGR position was more than 8% for more than 10 seconds during the CCM test. **Possible Causes** ● EGR sensor signal circuit is open, shorted to ground or power ● EGR sensor ground or VREF circuit is open ● EGR valve seat contains debris or carbon (inspect and clean) ● EGR valve pintle contains debris or carbon (inspect and clean) ● EGR valve is contaminated, clogged, damaged or has failed ● PCM has failed
P0404 **2T CCM, MIL: Yes** 1997, 1998 Achieva, Beretta, Ciera, Cavalier, Century, Corsica, Grand Am, Malibu, Skylark, Sunfire, S/T Blazer & S/T Pickup 2.2L VIN 4, 2.4L VIN T Transmissions: All	**EGR Open Position Signal Range/Performance Conditions:** DTC P0106, P0107, P0108, P0112, P0113, P0117, P0118, P0121, P0122, P0123, P0200, P0300, P0301-P0304, P0335, P0502, P0505, P0507 and P1441 not set, engine started, commanded EGR position more than 0%, system voltage more than 11.7v, then after the vehicle was driven, the PCM detected the difference between the Actual EGR position and the Desired EGR position was more than 10% for 11 seconds during the CCM test. **Possible Causes** ● EGR sensor signal circuit is open, shorted to ground or power ● EGR sensor ground or VREF circuit is open ● EGR valve seat contains debris or carbon (inspect and clean) ● EGR valve pintle contains debris or carbon (inspect and clean) ● EGR valve is contaminated, clogged, damaged or has failed ● PCM has failed
P0405 **1T CCM, MIL: Yes** 1996, 1997, 1998, 1999, 2000, 2001, 2002, 2003, 2004, 2005 Achieva, Alero, Aztek, Century, Cutlass, Cutlass Supreme Grand Prix, Impala, Monte Carlo, Intrigue, Lumina, Malibu, Montana, Silhouette, Rendezvous, Venture 3.1L VIN J, 3.1L VIN M, 3.4L VIN E engines Transmissions: All	**EGR Sensor Circuit Low Input Conditions:** Engine started; system voltage over 10.0v and the PCM detected the EGR Pintle Position sensor was less than 0.11v for 2 seconds. The PCM is connected to the sensor with a 5v VREF, low reference and EGR valve position signal circuit to determine the EGR valve position. This code is set if the EGR sensor voltage is pulled too low. **Possible Causes** ● EGR position sensor signal circuit is open or shorted to ground ● EGR position sensor VREF circuit is open or shorted to ground ● EGR position sensor is damaged or has failed ● PCM has failed
P0405 **2T CCM, MIL: Yes** 1998, 1999, 2000, 2001, 2002, 2003, 2004, 2005 Aurora, Bonneville, Century, Cutlass Supreme, Grand Prix, Intrigue, LeSabre, LSS, Lumina, Monte Carlo, 88', Regal, Park Avenue & 98' 3.1L VIN J, 3.1L VIN M, 3.5L VIN H, 3.8L VIN 1, 3.8L VIN K Transmissions: All	**EGR Position Sensor Circuit Low Input Conditions:** Key on or engine running; system voltage at 10-16v, and the PCM detected that the Actual EGR (feedback) sensor signal was less than 0.14v, conditions et for over 20 seconds during the test period. **Possible Causes** ● EGR position sensor signal circuit is open ● EGR position sensor signal circuit is shorted to ground ● EGR position sensor VREF circuit is open or shorted to ground ● EGR position sensor is damaged or has failed ● PCM has failed ● TSB 99-06-04-45 contains a repair procedure for this code
P0405 **2T CCM, MIL: Yes** 1998, 1999, 2000, 2001, 2002, 2003, 2004, 2005 C/K Cab & Chassis, C/K Series Truck, G Series Van, L/M Series Van, S/T Blazer & S/T Truck 4.3L VIN W, 4.3L VIN X, 5.0L VIN M, 5.7L VIN K, 5.7L VIN R, 7.4L VIN J Transmissions: All	**EGR Position Sensor Circuit Low Input Conditions:** Engine started; system voltage over 10.0v, EGR Position sensor VREF stable from 4-5v, and the PCM detected the EGR Position signal was less than 0.14v for a period of 5 seconds. **Possible Causes** ● EGR position sensor signal circuit is open ● EGR position sensor signal circuit is shorted to ground ● EGR position sensor VREF circuit is open or shorted to ground ● EGR position sensor is damaged or has failed ● PCM has failed ● TSB 01-06-04-043 contains a repair procedure for this code

OBD II Trouble Code List (P0xxx Codes)

DTC	Trouble Code Title, Conditions & Possible Causes
P0405 **2T CCM, MIL: Yes** 1996, 1997, 1998, 1999 C/K Truck, G Series Van 6.5L VIN F, 6.5L VIN S, 6.5L VIN Y Diesel engine Transmissions: All	**EGR Control BARO Sensor Circuit Low Input Conditions:** Engine started; and the PCM detected the EGR sensor was less than or equal to 0.24v (15 kPa) for 2 seconds. *Note: This sensor monitors the vacuum level to the EGR valve.* **Possible Causes** ● EGR BARO sensor signal circuit is open ● EGR BARO sensor signal circuit is shorted to ground ● EGR BARO sensor VREF circuit is open or shorted to ground ● EGR BARO sensor is damaged or has failed ● PCM has failed
P0405 **2T CCM, MIL: Yes** 1999, 2000, 2001, 2002, 2003, 2004, 2005 C/K Truck, Escalade, G Series Van, Cargo Van 4.8L VIN V, 5.3L VIN T, 6.0L VIN N, 6.0L VIN U, 8.1L VIN G engines Transmissions: All	**EGR Position Sensor Circuit Low Input Conditions:** Engine started; system voltage over 10.0v, EGR Position sensor VREF stable from 4-5v, and the PCM detected the EGR Position signal was less than 0.14v for a period of 5 seconds. **Possible Causes** ● EGR position sensor signal circuit is open ● EGR position sensor signal circuit is shorted to ground ● EGR position sensor VREF circuit is open or shorted to ground ● EGR position sensor is damaged or has failed ● PCM has failed
P0405 **2T CCM, MIL: Yes** 2001, 2002, 2003, 2004, 2005 C/K Series Truck 6.6L VIN 1 Diesel engine Transmissions: All	**EGR Vacuum Sensor Circuit Low Input Conditions:** DTC P0101, P0489 and P0651 not set, engine started, system voltage from 11-18v, and the PCM detected the EGR Vacuum sensor was less than 19 kPa for 5 seconds. The EGR Vacuum sensor is used to monitor the amount of vacuum is available to the EGR valve. A low reference, 5v VREF and a Vacuum sensor signal circuit connect this sensor to the PCM. **Possible Causes** ● EGR vacuum sensor connector is damaged or shorted ● EGR vacuum sensor VREF circuit is open or shorted to ground ● EGR vacuum sensor signal circuit is shorted to ground ● EGR vacuum sensor is damaged or it has failed ● ECM has failed
P0405 **2T CCM, MIL: Yes** 1997, 1998, 1999, 2000, 2001, 2002, 2003, 2004, 2005 DeVille, Eldorado, Seville 4.6L VIN 9, 4.6L VIN Y Transmissions: A/T	**EGR Position Sensor Circuit Low Input Conditions:** Key on or engine running; system voltage over 10.0v and the PCM detected the EGR Pintle Position sensor voltage was less than 0.11v for two seconds during the CCM test. **Possible Causes** ● EGR position sensor signal circuit is open ● EGR position sensor signal circuit is shorted to ground ● EGR position sensor VREF circuit is open or shorted to ground ● EGR position sensor is damaged or has failed ● PCM has failed
P0405 **2T CCM, MIL: Yes** 1998, 1999, 2000, 2001, 2002 Camaro & Firebird 3.8L VIN K engine Transmissions: All	**EGR Sensor Circuit Low Input Conditions:** Engine started; system voltage over 10.0v and the PCM detected the EGR signal was less than 0.14v for 20 seconds. **Possible Causes** ● EGR position sensor signal circuit is open or shorted to ground ● EGR position sensor VREF circuit is open or shorted to ground ● EGR position sensor is damaged or has failed ● PCM has failed
P0405 **2T CCM, MIL: Yes** 1997, 1998, 1999, 2000 Camaro & Firebird 5.7L VIN G engine Transmissions: All	**EGR Sensor Circuit Low Input Conditions:** Engine started; system voltage over 10.0v and the PCM detected the EGR Feedback signal was less than 0.14v for 10 seconds. An EGR system is used to lower Oxides of Nitrogen (NOx) emission levels. This function is accomplished by feeding small amounts of exhaust gas back into the combustion chamber. High combustion temperatures cause NOx. Combustion temperatures are reduced when the A/F mixture is diluted with exhaust gases. The EGR valve is designed to accurately supply exhaust gases to the engine without the use of intake manifold vacuum. The EGR valve controls the exhaust flow into the intake manifold from the exhaust manifold through an orifice with a PCM controlled pintle. The PCM controls the pintle position using inputs from the TP sensor, the MAP sensor and the ECT sensor. The PCM commands the EGR valve to control the amount of exhaust gas recirculation for all engine operating conditions. This action can be monitored on a Scan Tool as the Desired EGR Position. The PCM monitors the position of the EGR valve through a feedback signal. A voltage signal that represents the EGR valve pintle position is sent to the PCM from the EGR valve. The Actual EGR Position should always be near the Desired EGR Position. **Possible Causes** ● EGR position sensor signal circuit is open or shorted to ground ● EGR position sensor VREF circuit is open or shorted to ground ● EGR position sensor is damaged or has failed ● PCM has failed

OBD II Trouble Code List (P0xxx Codes)

DTC	Trouble Code Title, Conditions & Possible Causes
P0405 **2T CCM, MIL: Yes** 1997 Aurora 4.0L VIN C engine Transmissions: A/T	**EGR Sensor Circuit Low Input Conditions:** Engine started; system voltage over 10.0v and the PCM detected the EGR Pintle Position sensor signal was less than 0.11v for more than 2 seconds during the CCM test. **Possible Causes** • EGR position sensor signal circuit is open or shorted to ground • EGR position sensor VREF circuit is open or shorted to ground • EGR position sensor is damaged or has failed • PCM has failed • TSB 71-65-28 contains a repair procedure for this code
P0405 **2T CCM, MIL: Yes** 1998, 1999, 2001, 2002, 2003, 2004, 2005 Aurora 4.0L VIN C engine Transmissions: A/T	**EGR Sensor Circuit Low Input Conditions:** Engine started; system voltage over 10.0v and the PCM detected the EGR Pintle Position sensor was less than 0.11v for 2 seconds. **Possible Causes** • EGR position sensor signal circuit is open or shorted to ground • EGR position sensor VREF circuit is open or shorted to ground • EGR position sensor is damaged or has failed • PCM has failed
P0405 **2T CCM, MIL: Yes** 1997, 1998 Achieva, Beretta, Ciera, Cavalier, Century, Corsica, Grand Am, Malibu, Skylark, Sunfire, S/T Blazer & S/T Pickup 2.2L VIN 4, 2.4L VIN T Transmissions: All	**EGR Position Sensor Circuit Low Input Conditions:** DTC P0106, P0107, P0108, P0112, P0113, P0117, P0118, P0121, P0122, P0123, P0200, P0300, P0301-P0304, P0335, P0502, P0505, P0507 and P1441 not set, engine started, system voltage more than 11.7v, and the PCM detected the EGR Position signal was less than 0.11v for a period of 25 seconds. The PCM monitors the EGR valve position sensor. The PCM is connected to the sensor with a 5v VREF, low reference and EGR valve position signal circuit in order to determine the EGR valve position. If the EGR valve position sensor voltage is pulled too low, the PCM sets DTC P0405. **Possible Causes** • EGR position sensor signal circuit is open • EGR position sensor signal circuit is shorted to ground • EGR position sensor VREF circuit is open or shorted to ground • EGR position sensor is damaged or has failed • PCM has failed
P0406 **2T CCM, MIL: Yes** 1996, 1997, 1998, 1999 C/K Truck, G Series Van 6.5L VIN F, 6.5L VIN S, 6.5L VIN Y Diesel engine Transmissions: All	**EGR Control BARO Sensor Circuit Low Input Conditions:** Engine started; and the PCM detected the EGR sensor was more than or equal to 3.96v (85 kPa) for 2 seconds. *Note: This sensor monitors the vacuum level to the EGR valve.* **Possible Causes** • EGR BARO sensor signal circuit is shorted to VREF or power • EGR BARO sensor ground circuit is open • EGR BARO sensor is damaged or has failed • PCM has failed
P0406 **2T CCM, MIL: Yes** 2001, 2002, 2003, 2004, 2005 C/K Series Truck 6.6L VIN 1 Diesel engine Transmissions: All	**EGR Vacuum Sensor Circuit High Input Conditions:** DTC P0651 not set, engine started, system voltage from 11-18v, and the ECM detected the EGR Vacuum sensor was more than 158 kPa for 5 seconds during the test. The EGR Vacuum sensor is used to monitor the amount of vacuum available to the EGR valve. A low reference, 5v VREF and a sensor signal circuit connect to the ECM. **Possible Causes** • EGR vacuum sensor connector is damaged or open • EGR vacuum sensor ground circuit is open • EGR vacuum sensor signal circuit is open • EGR vacuum sensor signal circuit is shorted to VREF • EGR vacuum sensor is damaged or it has failed • ECM has failed
P0410 **2T AIR, MIL: Yes** 1996, 1997 Camaro, Caprice, Corvette, Firebird, Fleetwood, Roadmaster 4.3L VIN W, 5.7L VIN 5, 5.7L VIN P engines Transmissions: All	**Secondary Air System Performance Conditions:** DTC P0107, P0108, P0116, P0117, P0133, P0135, P0153, P0155, P0171, P0172, P0174 and P0175 not set, ECT sensor less than 167°F at startup, vehicle driven with the engine speed over 600 rpm, IAT sensor over 50°F, engine load less than 6.6%, MAF sensor less than 26 g/sec, TP angle less than 1%, and the PCM detected the HO2S signal was more than 222 mv for 19 seconds, or with the ECT sensor more than 167°F at startup, it detected the HO2S signal was more than 700 mv for 19 seconds during the AIR Monitor test. **Possible Causes** • AIR solenoid power circuit is open (check the U/H No. 7 fuse) • AIR pump is damaged or has failed (inspect air pump for water) • AIR system check valves and/or pipes are damaged or leaking • AIR check valve (in hose to vacuum bleed valve) stuck closed • TSB 63-65-02 contains a repair procedure for this code

OBD II Trouble Code List (P0xxx Codes)

DTC	Trouble Code Title, Conditions & Possible Causes
P0410 **2T AIR, MIL: Yes** 1996, 1997, 1998, 1999, 2000, 2001, 2002 Montana, Silhouette, Venture, Rendezvous, S/T Blazer & S/T Pickup 2.2L VIN 5, 2.2L VIN H, 3.4L VIN E, 3.5L VIN H, 3.8L VIN K Transmissions: All	**Secondary AIR System Performance Conditions:** DTC P0101, P0102, P0103, P0107, P0108, P0112, P0113, P0117, P0118, P0121-P0123, P0171, P0172, P0131-P0141, P0300, P0412, P0418, P0442, P0443 and P1441 not set, engine runtime over 2 seconds, then after the HO2S-11 signal changed less than 6 mv or was went over 1.1v during the 2.5 seconds the Air pump was "on", the PCM detected the HO2S signal was over 300 mv (Passive Test) or over 150 mv during the Active Test for too long, and the SHRTFT value did not increase a calibrated amount on 3 consecutive tests. **Possible Causes** ● AIR solenoid power circuit is open (check the U/H IGN fuse) ● AIR pump is damaged or has failed (inspect air pump for water) ● AIR system check valves and/or pipes are damaged or leaking ● AIR system hoses or lines are damaged, pinched or kinked ● Base engine problem (e.g., excessive exhaust back pressure) ● PCM has failed
P0410 **2T AIR, MIL: Yes** 2003, 2004, 2005 S/T Series Pickup 2.2L VIN H Transmissions: All	**Secondary AIR System Performance Conditions:** DTC P0105, P0107, P0108, P0112, P0113, P0122, P0123, P0130, P0131, P0132, P0133, P0134, P0171, P0172, P0300, P0301-P0304, P0341, P0506, P0507, P0601, P0602 not set, engine speed over 1500 rpm (over 1000 rpm for M/T), engine runtime over 200 seconds, system voltage over 11v, vehicle operating in Fuel Trim cell 16 or 17, TP angle change less than 5%, ECT sensor from 41-230°F, IAT sensor from 32-302°F, MAP sensor less than 30 kPa, injector pulsewidth more than 650 ms, Air system "on" for 20 seconds, and the PCM detected the HO2S signal was less than 50 mv for 1 second during the 3 second active test, and the failure occurred for 3 consecutive active tests. **Possible Causes** ● AIR solenoid power circuit is open (check the U/H IGN fuse) ● AIR pump is damaged or has failed (inspect air pump for water) ● AIR system check valves and/or pipes are damaged or leaking ● AIR system hoses or lines are damaged, pinched or kinked ● Base engine problem (e.g., excessive exhaust back pressure) ● PCM has failed
P0410 **2T AIR, MIL: Yes** 2000, 2001, 2002 Aurora, Bonneville, Century, Cutlass Supreme, Grand Prix, Intrigue, LeSabre, LSS, Lumina, Monte Carlo, 88', Regal, Park Avenue & 98' 3.5L VIN H, 3.8L VIN 1, 3.8L VIN K, 4.0L VIN C Transmissions: All	**Secondary AIR System Performance Conditions:** DTC P0030, P0036, P0101-P0103, P0112, P0113, P0116-P0118, P0121-P0123, P0125, P0131-P0138, P0140, P0141, P0171, P0172, P0300, P0440, P0442, P0446, P0449, P0502-P0503, P1111-P1115, P1121-P1122, P1133, P1134, P1380, P1381 and P1441 not set, engine started, vehicle driven to over 25 mph at over 1200 rpm for 10 seconds, BARO sensor over 75 kPa, ECT sensor from 41-230°F, MAF sensor at 2-25 g/sec, IAT sensor from 41-158°F, A/F Ratio over 13.0:1, engine load from 0-50%, and the PCM detected the HO2S signal was too high and the SHRTFT did not increase a calibrated amount with the AIR system "on" (test must fail 3 times). **Possible Causes** ● AIR solenoid power circuit is open (check the U/H IGN fuse) ● AIR pump is damaged or has failed (inspect air pump for water) ● AIR system check valves and/or pipes are damaged or leaking ● AIR system hoses or lines are damaged, pinched or kinked ● Base engine problem (e.g., excessive exhaust back pressure) ● PCM has failed ● TSB 99-06-04-048 contains a repair procedure for this code
P0410 **2T AIR, MIL: Yes** 1996, 1997, 1998, 1999, 2000, 2001, 2002 C/K Cab & Chassis, C/K Truck, G Series Van, Van Cargo & L/M Series Van 4.3L VIN W, 5.0L VIN M, 5.7L VIN K, 5.7L VIN R, 7.4L VIN J Transmissions: All	**Secondary Air System Performance Conditions:** DTC P0101, P0102, P0103, P0106, P0107, P0108, P0112, P0113, P0117, P0118, P0121, P0122, P0123, P0131, P0132, P0133, P0134, P0135, P0137, P0138, P0140, P0141, P0151, P0152, P0153, P0154, P0155, P0157, P0158, P0160, P0161, P0171, P0172, P0174, P0175, P0200, P0300, P0335, P0336, P0351-P0358, P0442, P0443, P0446 and P0449 not set, engine started, vehicle driven to over 25 mph at over 900 rpm for 2 seconds, engine load less than 33%, MAF sensor less than 28 g/sec, system voltage over 10.0v, A/F ratio at 13.125:1, ECT sensor from 36-230°F, IAT sensor from 68-212°F, DFCO and Power Enrichment "off", operating in Fuel Trim cells 4, 5 or 6, and the PCM detected the HO2S signal remained over 222 mv for 4 seconds, and the Short Term fuel trim did not increase by 6% for each bank with the Air Pump "on". **Possible Causes** ● AIR solenoid power circuit is open (check the U/H IGN fuse) ● AIR pump is damaged or has failed (inspect air pump for water) ● AIR system check valves and/or pipes are damaged or leaking ● AIR system hoses or lines are damaged, pinched or kinked ● Base engine problem (e.g., excessive exhaust back pressure) ● PCM has failed

OBD II Trouble Code List (P0xxx Codes)

DTC	Trouble Code Title, Conditions & Possible Causes
P0410 **2T AIR, MIL: Yes** 1997, 1998, 1999, 2000, 2001, 2002 Camaro, Corvette & Firebird 5.7L VIN G, 5.7L VIN S Transmissions: All	**Secondary AIR System Performance Conditions:** DTC P0101-P0103, P0107, P0108, P0112-P0118, P0125, P0171-P0175, P0200, P0300, P0335, P0336, P0351-P0358, P0440, P0442-P0449, P1120, P1220, P1221, P1258 and P1441 not set, fuel level at 12.5-87.5%, engine started, ECT sensor from 14-230ºF, IAT sensor from 14-212ºF, MAF sensor less than 22 g/sec, fuel level from 12.5-87%, A/F ratio at 13.125:1, Catalyst Temperature, DFCO and P/E Modes all "off", VSS over 25 mph, engine speed over 850 rpm in closed loop for 15 seconds, operating in Fuel Trim cells 1, 2, 4, 5 or 6, the PCM detected the HO2S signal was over 222 mv for 1 second, or the SHRTFT did not change with the AIR pump enabled. **Possible Causes** • AIR solenoid power circuit is open (check the U/H IGN fuse) • AIR pump is damaged or has failed (inspect air pump for water) • AIR system check valves and/or pipes are damaged or leaking • AIR system hoses or lines are damaged, pinched or kinked • Base engine problem (e.g., excessive exhaust back pressure) • PCM has failed
P0410 **2T AIR, MIL: Yes** 2000, 2001, 2002, 2003, 2004, 2005 DeVille, Eldorado, Seville 4.6L VIN 9, 4.6L VIN Y Transmissions: A/T	**Secondary AIR System Performance Conditions:** DTC P0102, P0103 P0106-P0108, P0112, P0113, P0116-P0118, P0121-P0123, P0131-P0138, P0140, P0141, P0151-P0158, P0161, P0200, P0300, P0335, P0336, P0351-P0358, P0506-0507, P1133, P1134, P1138 and P1171 not set, engine started, vehicle driven to over 25 mph with TP angle steady, engine load less than 80%, system voltage over 10.0v, ECT sensor from 41-230ºF, A/F ratio more than 12.5:1, MAF sensor less than 35 g/sec, DFCEO and Power Enrichment "off", then with the Air Pumps operating, the PCM detected the HO2S signals were over 250 mv for 20 seconds, or the HO2S signals were over 200 mv for 7 seconds after a hot restart, or with the AIR pumps off, the HO2S signals were below 740 mv for 20 seconds or for 7 seconds after a hot startup. Then the PCM runs the Active Test with the engine speed over 600 rpm at steady throttle, engine load less than 80%, MAF sensor under 35 g/sec in closed loop, Purge enabled, ECT sensor over 154ºF, and the PCM detected the HO2S signals were over 250 mv. **Possible Causes** • AIR solenoid power circuit is open (check the U/H IGN fuse) • AIR pump is damaged or has failed (inspect air pump for water) • AIR system check valves and/or pipes are damaged or leaking • AIR system hoses or lines are damaged, pinched or kinked • Base engine problem (e.g., excessive exhaust back pressure) • PCM has failed • TSB 02-06-04-024 contains a repair procedure for this code • TSB 03-06-04-013 contains a repair procedure for this code
P0410 **2T AIR, MIL: Yes** 2001, 2002, 2003, 2004, 2005 Aurora 4.0L VIN C engine Transmissions: All	**Secondary AIR System Performance Conditions:** DTC P0101, P0102, P0103, P0106, P0107, P0108, P0112, P0113, P0116, P0117, P0118, P0121, P0122, P0123, P0130, P0131, P0132, P0133, P0134, P0135, P0137, P0138, P0140, P0141, P0150, P0151, P0152, P0153, P0154, P0155, P0201, P0202, P0203, P0204, P0205, P0206, P0207, P0208, P0300, P0335, P0336, P0351, P0352, P0353, P0354, P0355, P0356, P0357, P0358, P0412, P0418, P0506, P0507, P1133 and P1134 not set, engine runtime over 30 seconds at over 600 rpm with the throttle steady, BARO sensor over 75 kPa, ECT sensor from 41-227ºF, MAF sensor under 35 g/sec, IAT sensor from 41-162ºF, A/F Ratio over 12.5:1, engine load less than 80%, and the PCM detected the HO2S signal was over 470 mv for 20 seconds (AIR pump "on"), or HO2S signal did not toggle over 600 mv for 25 seconds (AIR pump "off"). **Possible Causes** • AIR pump or relay power circuit is open (check the IGN 1 fuse) • AIR pump is damaged or has failed (inspect air pump for water) • AIR system check valves and/or pipes are damaged or leaking • AIR system hoses or lines are damaged, pinched or kinked • Base engine problem (e.g., excessive exhaust back pressure) • PCM has failed

OBD II Trouble Code List (P0xxx Codes)

DTC	Trouble Code Title, Conditions & Possible Causes
P0410 **2T AIR, MIL: Yes** 2003, 2004, 2005 Camaro, Corvette & Firebird 5.7L VIN G, 5.7L VIN S Transmissions: All	**Secondary AIR System Performance Conditions:** DTC P0101, P0102, P0103, P0107, P0108, P0112, P0113, P0116, P0117, P0118, P0125, P0128, P0131, P0132, P0133, P0134, P0135, P0137, P0138, P0140, P0141, P0151, P0152, P0153, P0154, P0155, P0157, P0158, P0160, P0161, P0171, P0172, P0174, P0175, P0200, P0300, P0335, P0336, P0351, P0352, P0353, P0354, P0355, P0356, P0357, P0358, P0455, P0442, P0443, P0446, P0449, P1120, P1133, P1134, P1153 and P1154 not set, fuel level from 12.5-87.5%, engine runtime over 30 seconds, ECT sensor from 14-230°F, IAT sensor from 14-212°F, MAF sensor less than 23 g/sec, fuel level from 12.5-87%, engine load less than 40%, A/F ratio at 14.7:1, Catalyst Over Temperature, DFCO and P/E Modes "off", vehicle driven at over 25 mph at over 850 rpm while in closed loop for 15 seconds (operating in Fuel Trim cells 1, 2, 4, 5 or 6), and the PCM detected the HO2S signal was over 222 mv for 1.5 seconds, or the Short Term fuel trim did not change with the AIR pump commanded on. If the PCM detects the HO2S signals for both banks did not respond correctly during testing, DTC P0410 sets. If only one HO2S responds, the PCM sets P1415 or P1416. **Possible Causes** ● AIR solenoid supply voltage circuit is open or shorted to ground ● AIR pump is damaged or has failed (inspect air pump for water) ● AIR pump relay is damaged or it has failed ● AIR system check valves and/or pipes are damaged or leaking ● AIR system hoses or lines are damaged, pinched or kinked ● AIR solenoid vacuum hose to shutoff valve leaking or blocked ● AIR shutoff valve is restricted or blocked ● Base engine problem (e.g., excessive exhaust back pressure) ● PCM has failed
P0410 **2T AIR, MIL: Yes** 1996, 1997, 1998, 1999, 2000, 2001 S/T Blazer & S/T Pickup 4.3L VIN W, 4.3L VIN X Transmissions: All	**Secondary Air System Performance Conditions:** DTC P0101-P0103, P0106-P0108, P0112, P0113, P0117, P0118, P0121-P0123, P0131-P0138, P0140, P0141, P0151-P0158, P0160, P0161, P0171-P0172, P0174-P0175, P0200, P0300, P0335, P0336, P0351-P0358, P0442, P0443, P0446 and P0449 not set, engine started, vehicle driven to over 25 mph at an engine speed over 500 rpm for 20 seconds, system voltage over 10.0v, ECT sensor from 176-230°F, MAF sensor less than 25 g/sec, IAT sensor more than 36°F, engine load sensor less than 34%, engine load under 34%, A/F ratio at 14.7:1, DFCO, Catalyst Over Temperature and P/E Modes inactive, Short Term fuel trim from 124-132, and the PCM detected the HO2S signal was over 222 mv for 1.3 seconds, or there was too little change in Short Term fuel trim with the AIR pump "on". **Possible Causes** ● AIR solenoid power circuit is open (check the U/H IGN fuse) ● AIR pump is damaged or has failed (inspect air pump for water) ● AIR system check valves and/or pipes are damaged or leaking ● AIR system hoses or lines are damaged, pinched or kinked ● Base engine problem (e.g., excessive exhaust back pressure) ● PCM has failed
P0410 **2T AIR, MIL: Yes** 1997, 1998, 1999, 2000, 2001 Catera 3.0L V6 VIN R engine Transmissions: A/T	**Secondary AIR System Constant Flow Detected Conditions:** DTC P0110, P0115, P0116, P0130, P0131, P0132, P0134, P0150, P0151, P0152, P0154, P0171, P0172, P0174, P0175, P0300-P0306, P0412, P0418, P0440, P0442, P0443, P0446 and P0455 not set, engine running in closed loop, vehicle stopped, ECT sensor from 65-221°F, IAT sensor from 41-104°F, AIR pump enabled, and the PCM detected reduced Secondary airflow or no Secondary airflow occurred with the air pump "on" during the test. **Possible Causes** ● AIR pump is damaged or has failed (inspect air pump for water) ● AIR relay power circuit is open (check the 50A fuse in ECM/H) ● AIR relay control circuit is open between the relay and the PCM ● AIR system check valves and/or pipes are damaged or leaking ● AIR system hoses or lines are damaged, pinched or kinked ● PCM has failed
P0410 **2T AIR, MIL: Yes** 1996 Cutlass Supreme, Grand Prix, Lumina, Monte Carlo 3.4L VIN X engine Transmissions: All	**Secondary Air System Performance Conditions:** DTC P0107, P0108, P0116, P0117, P0118, P0133, P0135, P0153, P0155, P0171, P0172, P0174 and P0175 not set, ECT sensor over 167°F at startup, engine speed over 600 rpm, IAT sensor over 50°F, engine load below 6.6%, MAF sensor under 26 g/sec, TP angle below 1%, and the PCM detected the HO2S signal was over 222 mv for 19 seconds, or with ECT sensor more than 167°F at startup, the HO2S signal was more than 700 mv for 19 seconds. **Possible Causes** ● AIR solenoid power circuit is open (check the U/H No. 7 fuse) ● AIR pump is damaged or has failed (inspect air pump for water) ● AIR system check valves and/or pipes are damaged or leaking ● AIR check valve (in hose to vacuum bleed valve) stuck closed ● PCM has failed ● TSB 67-65-22 contains a repair procedure for this code

OBD II Trouble Code List (P0xxx Codes)

DTC	Trouble Code Title, Conditions & Possible Causes
P0411 **2T CCM, MIL: Yes** 1997, 1998, 1999, 2000, 2001 Catera 3.0L V6 VIN R engine Transmissions: A/T	**Secondary Air System Malfunction Conditions:** DTC P0110, P0115, P0116, P0130, P0131, P0132, P0134, P0150, P0151, P0152, P0154, P0171, P0172, P0174, P0175, P0300-P0306, P0412, P0418, P0440, P0442, P0443, P0446 and P0455 not set, ECT sensor from 65.8°F-221°F, IAT sensor from 41°F-104°F, engine at idle speed, AIR system not active, VSS sensor at 0 mph, and the PCM detected reduced Secondary airflow. **Possible Causes** ● AIR solenoid control circuit is open, shorted to ground or B+ ● AIR solenoid power circuit is open (test power from main relay) ● AIR solenoid is damaged or has failed, or the PCM has failed ● TSB 00-06-04-032 contains a repair procedure for this code
P0412 **2T CCM, MIL: Yes** 1996, 1997 Camaro, Caprice, Corvette, Firebird, Fleetwood, Roadmaster 4.3L VIN W, 5.7L VIN 5, 5.7L VIN P engines Transmissions: All	**Secondary Air System Control Circuit Malfunction Conditions:** Engine started; system voltage over 10.0v and the PCM detected the Actual and Commanded state of the AIR solenoid driver did not match for 5 seconds during the test. **Possible Causes** ● AIR solenoid control circuit is open, shorted to ground or B+ ● AIR solenoid power circuit is open (test power from IGN fuse) ● AIR solenoid is damaged or has failed, or PCM has failed
P0412 **2T CCM, MIL: Yes** 1996, 1997, 1998, 1999, 2000, 2001, 2002 Regal, Montana, Silhouette, Venture 3.4L VIN E, 3.5L VIN H, 3.8L VIN K engines Transmissions: All	**Secondary Air System Solenoid Circuit Malfunction Conditions:** Engine started; and the PCM detected the HO2S signal changed less than 600 mv or remained above 225 mv for 1.8 seconds with the AIR pump enabled for 3 seconds during the CCM test period. **Possible Causes** ● AIR solenoid control circuit is open, shorted to ground or B+ ● AIR solenoid power circuit is open (test power from IGN1 fuse) ● AIR solenoid is damaged or has failed ● PCM has failed
P0412 **2T CCM, MIL: Yes** 2000, 2001, 2002 Aurora, Bonneville, Century, Cutlass Supreme, Grand Prix, Intrigue, LeSabre, LSS, Lumina, Monte Carlo, 88', Regal, Park Avenue & 98' 3.5L VIN H, 3.8L VIN 1, 3.8L VIN K, 4.0L VIN C Transmissions: All	**Secondary Air System Control Circuit Malfunction Conditions:** Engine started; system voltage over 10v, AIR solenoid commanded "on" and "off", and the PCM detected an unexpected voltage condition on the AIR solenoid control circuit for 30 seconds during the CCM test period. **Possible Causes** ● AIR solenoid control circuit is open, shorted to ground or B+ ● AIR solenoid power circuit is open (test power from IGN fuse) ● AIR solenoid is damaged or has failed ● PCM has failed
P0412 **2T CCM, MIL: Yes** 2003, 2004, 2005 Aurora 4.0L VIN C engine Transmissions: All	**Secondary Air Vacuum Solenoid Circuit Malfunction Conditions:** Engine started; system voltage at 8-16v, AIR solenoid commanded "on" and "off", and the PCM detected an unexpected voltage on the AIR vacuum solenoid circuit for 10 seconds. **Possible Causes** ● AIR solenoid control circuit is open, shorted to ground or B+ ● AIR solenoid power circuit is open (test power from IGN fuse) ● AIR solenoid is damaged or has failed ● PCM has failed
P0412 **2T CCM, MIL: Yes** 2000, 2001, 2002, 2003, 2004, 2005 Aurora, DeVille, Eldorado & Seville 4.0L VIN C engine, 4.6L VIN 9, 4.6L VIN Y Transmissions: All	**Secondary Air System Solenoid Circuit Fault (Bank 1) Conditions:** Engine started; system voltage over 10.0v and the PCM detected an unexpected voltage on the AIR solenoid circuit for 10 seconds. **Possible Causes** ● AIR solenoid control circuit is open, shorted to ground or B+ ● AIR solenoid power circuit is open (test power from IGN1 fuse) ● AIR solenoid is damaged or has failed ● PCM has failed
P0412 **2T CCM, MIL: Yes** 1997, 1998, 1999, 2000, 2001, 2002, 2003, 2004, 2005 Camaro, Corvette & Firebird 5.7L VIN G, 5.7L VIN S Transmissions: All	**Secondary Air System Solenoid Control Circuit Malfunction Conditions:** Engine started; system voltage over 10.0v and the PCM detected that the Actual and Commanded state of the AIR solenoid driver did not match for 5 seconds during the test. **Possible Causes** ● AIR solenoid control circuit is open, shorted to ground or power ● AIR solenoid control circuit is shorted to system power (B+) ● AIR solenoid power circuit is open (test power from IGN1 fuse) ● AIR solenoid is damaged or has failed ● PCM has failed

OBD II Trouble Code List (P0xxx Codes)

DTC	Trouble Code Title, Conditions & Possible Causes
P0412 **2T CCM, MIL: Yes** 1997, 1998, 1999, 2000, 2001 Catera 3.0L V6 VIN R engine Transmissions: A/T	**Secondary AIR Solenoid Valve Control Circuit Malfunction Conditions:** Engine started; engine running for 300 milliseconds, system voltage over 10.0v and the PCM detected an unexpected voltage condition on the AIR solenoid control circuit during the test. **Possible Causes** ● AIR solenoid control circuit is open, shorted to ground or power ● AIR solenoid control circuit is shorted to system power (B+) ● AIR solenoid power circuit is open (test power from IGN fuse) ● AIR solenoid is damaged or has failed ● PCM has failed
P0418 **2T CCM, MIL: Yes** 2000, 2001, 2002 Bonneville, Century, Grand Prix, Intrigue, LeSabre, Monte Carlo, Regal & Park Avenue 3.5L VIN H, 3.8L VIN 1, 3.8L VIN K engines Transmissions: All	**Secondary Air System Pump Relay Control Circuit Malfunction Conditions:** Engine started; system voltage from 8-16v, AIR relay commanded "on", and the PCM detected an unexpected voltage condition on the AIR relay control circuit for 30 seconds. **Possible Causes** ● AIR relay control circuit is open, shorted to ground or power ● AIR relay control circuit is shorted to system power (B+) ● AIR relay power circuit is open (test power from IGN fuse) ● AIR relay is damaged or has failed ● PCM has failed
P0418 **2T CCM, MIL: Yes** 2001, 2002, 2003, 2004, 2005 Aurora 4.0L VIN C engine Transmissions: A/T	**Secondary Air System Pump Relay Circuit Malfunction Conditions:** Engine started; system voltage from 8-16v, AIR relay commanded "on", and the PCM detected an unexpected voltage condition on the AIR relay control circuit for 10 seconds. **Possible Causes** ● AIR relay control circuit is open, shorted to ground or power ● AIR relay control circuit is shorted to system power (B+) ● AIR relay power circuit is open (test power from IGN fuse) ● AIR relay is damaged or has failed ● PCM has failed
P0418 **2T CCM, MIL: Yes** 1996, 1997, 1998, 1999, 2000, 2001, 2002, 2003, 2004, 2005 C/K Cab & Chassis, C/K Truck, G Series Van, Van Cargo & L/M Series Van 4.3L VIN W, 5.0L VIN M, 5.7L VIN R engine Transmissions: All	**Secondary Air System Pump Relay Control Circuit Malfunction (Bank 1) Conditions:** Engine started; system voltage over 10.0v and the PCM detected the Actual and Commanded state of the AIR Pump Relay driver did not match for 5 seconds in the test. **Possible Causes** ● AIR relay control circuit is open, shorted to ground or power ● AIR relay control circuit is shorted to system power (B+) ● AIR relay power circuit is open (test power from IGN fuse) ● AIR relay is damaged or has failed ● PCM has failed
P0418 **2T CCM, MIL: Yes** 2000, 2001, 2002, 2003, 2004, 2005 Aurora, DeVille, Eldorado & Seville 4.0L VIN C engine, 4.6L VIN 9, 4.6L VIN Y Transmissions: All	**Secondary Air Pump Relay Control Circuit Malfunction (Bank 2) Conditions:** Engine started; system voltage over 10.0v and the PCM detected an unexpected voltage condition on the AIR relay control circuit for Bank 2 for 10 seconds during the CCM test. **Possible Causes** ● AIR relay control circuit is open, shorted to ground or power ● AIR relay control circuit is shorted to system power (B+) ● AIR relay power circuit is open (test power from IGN fuse) ● AIR relay is damaged or has failed ● PCM has failed
P0418 **2T CCM, MIL: Yes** 1997, 1998, 1999, 2000, 2001, 2002, 2003, 2004, 2005 Camaro, Corvette & Firebird 5.7L VIN G, 5.7L VIN S Transmissions: All	**Secondary Air System Pump Relay Control Circuit Malfunction (Bank 1) Conditions:** Engine started; system voltage over 10.0v and the PCM detected the Actual and Commanded state of the Secondary Air Pump Relay driver did not match for 5 seconds. **Possible Causes** ● AIR relay control circuit is open, shorted to ground or power ● AIR relay control circuit is shorted to system power (B+) ● AIR relay power circuit is open (test power from IGN fuse) ● AIR relay is damaged or has failed ● PCM has failed
P0418 **2T CCM, MIL: Yes** 2001 S/T Pickup 4.3L VIN W engine Transmissions: All	**Secondary Air System Pump Relay Circuit Malfunction (Bank 1) Conditions:** Engine started; system voltage over 10.0v and the PCM detected the Actual and Commanded state of the Secondary AIR Pump Relay driver did not match for 5 seconds. **Possible Causes** ● AIR relay control circuit is open, shorted to ground or power ● AIR relay control circuit is shorted to system power (B+) ● AIR relay power circuit is open (test power from IGN fuse) ● AIR relay is damaged or has failed ● PCM has failed

OBD II Trouble Code List (P0xxx Codes)

DTC	Trouble Code Title, Conditions & Possible Causes
P0418 **2T CCM, MIL: Yes** 1996, 1997, 1998, 1999, 2000, 2001, 2002 Regal, Montana, Venture, Silhouette, Rendezvous 3.4L VIN E, 3.5L VIN H, 3.8L VIN K engines Transmissions: All	**Secondary Air System Pump Relay Control Circuit Malfunction (Bank 1) Conditions:** Engine started; system voltage over 10.0v, AIR relay commanded on and off, and the PCM detected an unexpected voltage condition on the relay control circuit for 30 seconds. **Possible Causes** ● AIR relay control circuit is open, shorted to ground or power ● AIR relay control circuit is shorted to system power (B+) ● AIR relay power circuit is open (test power from IGN fuse) ● AIR relay is damaged or has failed ● PCM has failed
P0418 **2T CCM, MIL: Yes** 1997, 1998, 1999, 2000, 2001 Catera 3.0L V6 VIN R engine Transmissions: All	**Secondary Air Pump Relay Control Circuit Malfunction Conditions:** Engine started; system voltage over 10.0v and the PCM detected an unexpected voltage on the AIR Solenoid Relay control circuit. **Possible Causes** ● AIR relay control circuit is open, shorted to ground or power ● AIR relay control circuit is shorted to system power (B+) ● AIR relay power circuit is open (test power from main relay) ● AIR relay is damaged or has failed ● PCM has failed
P0419 **2T CCM, MIL: Yes** 2000 Aurora, DeVille, Eldorado & Seville 4.0L VIN C engine, 4.6L VIN 9, 4.6L VIN Y Transmissions: A/T	**Secondary Air Pump Relay Control Circuit Malfunction (Bank 2) Conditions:** Engine started; system voltage over 10.0v and the PCM detected an unexpected voltage on the AIR relay circuit for 10 seconds. **Possible Causes** ● AIR relay control circuit is open, shorted to ground or shorted to system power (B+) ● AIR relay power circuit is open (test power from IGN fuse) ● AIR relay is damaged or has failed ● PCM has failed
P0420 **1T CAT, MIL: Yes** 1996, 1997, 1998, 1999, 2000, 2001, 2002, 2003, 2004, 2005 Achieva, Beretta, Ciera, Cavalier, Century, Corsica, Grand Am, Malibu, Skylark, Sunfire, S/T Blazer & S/T Pickup 2.2L VIN 4, 2.2L VIN 5, 2.2L VIN F, 2.2L VIN H, 2.4L VIN T, 4.2L VIN S Transmissions: All	**Catalyst System Low Efficiency Conditions:** DTC P0105-P0108, P0112-P0118, P0122, P0123, P0131-P0138, P0140, P0141, P0171, P0172, P0201-P0204, P0300, P0301-P0304, P0325, P0336, P0340, P0341, P0440-P0453, P0480, P0502-P0503, P0506, P0507, P0601, P0602, P1133, P1336, P1441 and P1621 not set, engine runtime over 530 seconds since last TP change, engine speed over 1000 rpm for 35 seconds, BARO sensor over 72 kPa, ECT sensor from 156-257°F, IAT sensor from -4°F to 176°F, VSS over 3 mph, then back to Desired idle speed (± 150 rpm), TP angle at 0%, Predicted Catalyst Temperature from 934-1382°F, SHRTFT from -28 to 28%, and the PCM detected the Catalyst was degraded. **Possible Causes** ● Air leaks at the exhaust manifold or in the exhaust pipes ● Base engine problems (i.e., high engine oil or coolant usage) ● Catalytic converter is damaged, contaminated or has failed ● Continuous engine misfire conditions, or weak or low coil output ● Front HO2S or rear HO2S is contaminated with fuel or moisture ● Rear HO2S is loose in the mounting hole (check it for a leak) ● Front HO2S older (aged) than the rear HO2S (HO2S-12 is lazy)
P0420 **1T CAT, MIL: Yes** 1996, 1997, 1998, 1999, 2000, 2001, 2002, 2003, 2004, 2005 Aurora, Bonneville, Century, Cutlass Supreme, Grand Prix, Intrigue, LeSabre, LSS, Lumina, Monte Carlo, 88', Regal, Park Avenue & 98' 3.1L VIN J, 3.1L VIN M, 3.5L VIN H, 3.8L VIN 1, 3.8L VIN K Transmissions: All	**Catalyst System Bank 1 Low Efficiency Conditions:** DTC P0030, P0101-P0103, P0107, P0108, P0112, P0113, P0116-P0118, P0121-P0123, P0128, P0130-P0138, P0140, P0141, P0171, P0172, P0201-P0206, P0300, P0336, P0341, P0404, P0405, P0410, P0440, P0442, P0443, P0502, P0503, P0506, P0507, P1133, P1134, P1351, P1352, P1361, P1362 and P1441 not set, engine runtime over 10 minutes, system voltage over 10.0v, BARO sensor more than 75 kPa, ECT sensor from 169-255°F, IAT sensor from -4°F to 212°F, engine running in closed loop, and the PCM detected the catalyst oxygen storage capacity had degraded. Test Instructions: To activate the test, return to idle and place vehicle in Drive (depress the clutch pedal for manual transmission vehicles). Then within 60 seconds, the A/F ratio will go below 14.1 for up to 8 seconds (and may go to above 15.3 for up to 10 seconds). Use a Scan Tool to monitor DTC P0420 to determine if the current trip passes or fails. The catalytic catalyst promotes a chemical reaction that oxidizes the amount of HC and CO in the exhaust gas to convert them into water vapor and CO2. It also reduces NOx by converting it to nitrogen. The converter has the ability to store excess oxygen and then release it. **Possible Causes** ● Air leaks at the exhaust manifold or in the exhaust pipes ● Base engine problems (i.e., high engine oil or coolant usage) ● Catalytic converter is damaged, contaminated or has failed ● Continuous engine misfire conditions, or weak or low coil output ● Front HO2S or rear HO2S is contaminated with fuel or moisture ● Rear HO2S is loose in the mounting hole (check it for a leak) ● Front HO2S older (aged) than the rear HO2S (HO2S-12 is lazy)

OBD II Trouble Code List (P0xxx Codes)

DTC	Trouble Code Title, Conditions & Possible Causes
P0420 **1T CAT, MIL: Yes** 1996, 1997 Camaro, Caprice, Corvette, Firebird, Fleetwood, Roadmaster 4.3L VIN W, 5.7L VIN 5, 5.7L VIN P engines Transmissions: All	**Catalyst System Low Efficiency (Bank 1) Conditions:** DTC P0101-P0103, P0117, P0118, P0122, P0123, P0335, P0336, P0171-P0175, P0500, P0502, P0641 and P0742 not set, vehicle driven to a speed of 20-75 mph at below 3500 rpm in closed loop, ECT sensor more than 124°F, MAP sensor from 25-80 kPa, MAF sensor from 15-100 g/sec, TP angle over 2%, and the PCM detected the Bank 1 Catalyst was degraded. **Possible Causes** ● Air leaks at the exhaust manifold or in the exhaust pipes ● Base engine problems (i.e., high engine oil or coolant usage) ● Catalytic converter is damaged, contaminated or has failed ● Continuous engine misfire conditions, or weak or low coil output ● Front HO2S or rear HO2S is contaminated with fuel or moisture ● Rear HO2S is loose in the mounting hole (check it for a leak) ● Front HO2S older (aged) than the rear HO2S (HO2S-12 is lazy)
P0420 **2T CAT, MIL: Yes** 1996, 1997, 1998, 1999, 2000, 2001, 2002, 2003, 2004, 2005 CK Cab & Chassis, C/K Truck, G Series Van, L/M Van, S/T Series Truck 4.3L VIN W, 4.3L VIN X, 5.0L VIN M, 5.7L VIN K, 5.7L VIN R, 7.4L VIN J Transmissions: All	**Catalyst System Low Efficiency (Bank 1) Conditions:** DTC P0101-P0103, P0106-P0108, P0112, P0113, P0117, P0118, P0125, P0131, P0132-P0138, P0140, P0141, P0151-P0158, P0160, P0161, P0171-P0175, P0200, P0300, P0325, P0327, P0335, P0000, P0041, P0310, P0051-P0338, P0443-P0449, P0502, P0503, P0506, P0507, P1120, P1125, P1133, P1134, P1153, P1154, P1220, P1221, P1275, P1276, P1280-P1286, P1441, P1514, P1518 not set, engine started, engine runtime over 6 minutes, engine speed over 1000 rpm for 32-40 since last idle period, BARO sensor over 72 kPa, ECT sensor more than 167°F, IAT sensor over 16°F, MAF sensor from 15-50 g/sec, Catalyst Temperature over 840°F, engine running at idle speed for under 2 minutes with the Actual idle speed within 100-125 rpm of the Desired idle speed, then vehicle driven to 22-85 mph in closed loop, TP angle over 2% on 7.4L V8, Long Term fuel trim stable, any change in engine load less than 10%, and the PCM detected the Bank 1 Catalyst was degraded. **Possible Causes** ● Air leaks at the exhaust manifold or in the exhaust pipes ● Base engine problems (i.e., high engine oil or coolant usage) ● Catalytic converter is damaged, contaminated or has failed ● Continuous engine misfire conditions, or weak or low coil output ● Front HO2S or rear HO2S is contaminated with fuel or moisture ● Rear HO2S is loose in the mounting hole (check it for a leak) ● Front HO2S older (aged) than the rear HO2S (HO2S-12 is lazy) ● TSB 81-65-37 contains a repair procedure for this code
P0420 **2T CAT, MIL: Yes** 1996, 1997, 1998, 1999, 2000, 2001, 2002, 2003, 2004, 2005 Achieva, Alero, Century, Cutlass, Cutlass Supreme Grand Prix, Impala, Monte Carlo, Intrigue, Lumina, Malibu, Montana, Silhouette, Rendezvous, Venture 3.1L VIN J, 3.1L VIN M, 3.4L VIN E engines Transmissions: All	**Catalyst System Low Efficiency (Bank 1) Conditions:** DTC P0030, P0101-P0103, P0107, P0108, P0112, P0113, P0116-P0118, P0121-P0123, P0128, P0130-P0138, P0140, P0141, P0171, P0172, P0201-P0206, P0300, P0336, P0341, P0404-P0405, P0410, P0440, P0442, P0443, P0502-P0503, P0506-P0507, P1133, P1134, P1351, P1352, P1361, P1362 and P1441 not set, engine runtime over 10 minutes, system voltage over 10.0v, BARO sensor over 75 kPa, ECT sensor at 169-255°F, IAT sensor at -4°F to 212°F, MAF sensor at 12-32 g/sec, Catalyst Temperature at 788-1202°F for 3-4 minutes, engine speed at 1000-3000 rpm, engine load below 63%, VSS at 30-75 mph, and the PCM detected the Catalyst was degraded. **Possible Causes** ● Air leaks at the exhaust manifold or in the exhaust pipes ● Base engine problems (i.e., high engine oil or coolant usage) ● Catalytic converter is damaged, contaminated or has failed ● Continuous engine misfire conditions, or weak or low coil output ● Front HO2S or rear HO2S is contaminated with fuel or moisture ● Rear HO2S is loose in the mounting hole (check it for a leak) ● Front HO2S older (aged) than the rear HO2S (HO2S-12 is lazy)
P0420 **2T CAT, MIL: Yes** 1999, 2000, 2001, 2002, 2003, 2004, 2005 C/K Cab & Chassis, C/K Truck, Envoy, Escalade, G Series Van, TrailBlazer 4.8L VIN V, 5.3L VIN P, 5.3L VIN T, 6.0L VIN N, 6.0L VIN U, 8.1L VIN G Transmissions: All	**Catalyst System Low Efficiency (Bank 1) Conditions:** DTC P0101-P0103, P0106-P0108, P0112, P0113, P0117, P0118, P0120, P0121-123, P0125, P0128, P0131-P0138, P0140, P0141, P0171-P0172, P0177-P0179, P0200, P0220, P0300, P0325, P0327, P0332, P0335, P0336, P0341-P0343, P0351-P0358, P0442-P0446, P0452-P0453, P0455, P0496, P0502-P0503, P1125, P1133, P1153, P1258, P1514- P1518, P2108 or P2135 not set, engine started, ECT sensor from 158-248°F, BARO sensor over 74 kPa, IAT sensor at 5-185°F, vehicle driven in closed loop at cruise speed for 40-45 seconds, and the PCM detected that the Oxygen storage capability of the Catalyst was degraded. **Possible Causes** ● Air leaks at the exhaust manifold or in the exhaust pipes ● Base engine problems (i.e., high engine oil or coolant usage) ● Catalytic converter is damaged, contaminated or has failed ● Continuous engine misfire conditions, or weak or low coil output ● Front HO2S or rear HO2S is contaminated with fuel or moisture ● Rear HO2S is loose in the mounting hole (check it for a leak) ● TSB 81-65-37 contains a repair procedure for this code

OBD II Trouble Code List (P0xxx Codes)

DTC	Trouble Code Title, Conditions & Possible Causes
P0420 **1T CAT, MIL: Yes** 1996, 1997, 1998, 1999, 2000, 2001 DeVille, Eldorado, Seville 4.6L VIN 9, 4.6L VIN Y Transmissions: A/T	**Catalyst System Low Efficiency (Bank 1) Conditions:** DTC P0101-P0103, P0106-P0108, P0112-P0113, P0116-P0118, P0121-P0123, P0125-P0128, P0171, P0172, P0174, P0175, P0200, P0300, P0410-P0446, P0452-P0453, P1258, P1415-P1416 and P1441 not set, engine started, vehicle driven to over 800 rpm for 46 seconds in closed loop (Actual idle speed within 100-125 rpm of Desired speed), engine runtime over 7.5 minutes, BARO over 75 kPa, ECT sensor from 160-248ºF, IAT sensor more than 21ºF, Catalyst Temperature from 788-1202ºF, AIR tests this key cycle less than 12, and the PCM detected the catalyst efficiency had degraded. **Possible Causes** • Air leaks at the exhaust manifold or in the exhaust pipes • Base engine problems (i.e., high engine oil or coolant usage) • Catalytic converter is damaged, contaminated or has failed • Continuous engine misfire conditions, or weak or low coil output • Front HO2S or rear HO2S is contaminated with fuel or moisture • Rear HO2S is loose in the mounting hole (check it for a leak)
P0420 **1T CAT, MIL: Yes** 2002, 2003, 2004, 2005 DeVille, Eldorado, Seville 4.6L VIN 9, 4.6L VIN Y Transmissions: A/T	**Catalyst System Low Efficiency (Bank 1) Conditions:** DTC P0101-P0103, P0106-P0108, P0112-P0113, P0116-P0118, P0121-P0123, P0125-P0128, P0171, P0172, P0174, P0175, P0200, P0300, P0410-P0446, P0452-P0453, P1258, P1415-P1416 and P1441 not set, engine started, vehicle driven to over 800 rpm for 46 seconds in closed loop (Actual idle speed within 100-125 rpm of Desired speed), engine runtime over 7.5 minutes, BARO over 75 kPa, ECT sensor from 160-248ºF, IAT sensor more than 21ºF, Catalyst Temperature from 788-1202ºF, AIR tests this key cycle less than 12, and the PCM detected the catalyst efficiency had degraded. **Possible Causes** • Air leaks at the exhaust manifold or in the exhaust pipes • Base engine problems (i.e., high engine oil or coolant usage) • Catalytic converter is damaged, contaminated or has failed • Continuous engine misfire conditions, or weak or low coil output • Front HO2S or rear HO2S is contaminated with fuel or moisture • Rear HO2S is loose in the mounting hole (check it for a leak) • TSB 03-06-04-028 contains a repair procedure for this code
P0420 **2T CAT, MIL: Yes** 2003, 2004, 2005 CTS (Cadillac) 2.6L VIN M, 3.2L VIN N Transmissions: All	**Catalyst System Low Efficiency (Bank 1) Conditions:** DTC P0030-P0038, P0050-P0058, P0101-P0103, P0121-P0123, P0130-P0138, P0140, P0141, P0150-P0158, P0160, P0161, P0221, P0222, P0223, P0300, P0301-P0306, P0313, P0440, P0442-P0448, P0449, P0496, P0638, P2096, P2097, P2098, P2099, P2100, P2101, P2105, P2107, P2119, P2122-P2123, P2127-P2128, P2135, P2138 and P2176 not set, engine speed at 200-3000 rpm in closed loop, engine load from 25-30%, TWC Temperature at 968-1454ºF, and the PCM detected the Catalyst efficiency was degraded. **Possible Causes** • Air leaks at the exhaust manifold or in the exhaust pipes • Base engine problems (i.e., high engine oil or coolant usage) • Catalytic converter is damaged, contaminated or has failed • Continuous engine misfire conditions, or weak or low coil output • Front HO2S or rear HO2S is contaminated with fuel or moisture • Rear HO2S is loose in the mounting hole (check it for a leak)
P0420 **2T CAT, MIL: Yes** 1995, 1996, 1997, 1998, 1999, 2000, 2001, 2002 Camaro & Firebird 3.8L VIN K engine Transmissions: All ; Camaro, Firebird	**Catalyst System Low Efficiency (Bank 1) Conditions:** DTC P0101-P0103, P0107, P0108, P0112-P0118, P0125, P0130 to P0140, P0141, P0150-P0155, P0171-P0175, P0201-P0206, P0300, P0325, P0336, P0341, P0401-P0405, P0440-P0449, P0502-P0507, P1120, P1125, P1133-P1134, P1153-P1154, P1220, P1221, P1336, P1352, P1361, P1362, P1441 or P1523 not set, engine runtime 10 minutes, engine speed from 1500-2000 rpm, then back to idle, ECT sensor at 158-255ºF, IAT sensor at -4ºF to 212ºF, BARO sensor over 75 kPa, and the PCM detected the Catalyst was degraded. **Possible Causes** • Air leaks at the exhaust manifold or in the exhaust pipes • Base engine problems (i.e., high engine oil or coolant usage) • Catalytic converter is damaged, contaminated or has failed • Continuous engine misfire conditions, or weak or low coil output • Front HO2S or rear HO2S is contaminated with fuel or moisture • Rear HO2S is loose in the mounting hole (check it for a leak) • Front HO2S older (aged) than the rear HO2S (HO2S-12 is lazy)

OBD II Trouble Code List (P0xxx Codes)

DTC	Trouble Code Title, Conditions & Possible Causes
P0420 **1T CAT, MIL: Yes** 1997, 1998, 1999, 2000, 2001, 2002, 2003, 2004, 2005 Camaro, Corvette & Firebird 5.7L VIN G, 5.7L VIN S Transmissions: All	**Catalyst System Low Efficiency (Bank 1) Conditions:** DTC P0101-P0103, P0106-P0108, P0112-P0118, P0128, P0131-P0137, P0140, P0141, P0151-P0158, P0160, P0161, P0171-P0175, P0200, P0300, P0335, P0336, P0341-P0343, P0351-P0358, P0410, P0440, P0502-P0503, P0506-P0507, P0606, P1120, P1133, P1134, P1153, P1154, P1220, P1336, P1415, P1416 and P1441 not set, engine speed over 850 rpm for 230 seconds since last idle, VSS under 85 mph, BARO sensor over 75 kPa, ECT sensor over 167°F, IAT sensor at 19-167°F, MAF sensor at 14-40 g/sec, MAP sensor from 25-80 kPa, and the PCM detected the Catalyst was degraded. **Possible Causes** ● Air leaks at the exhaust manifold or in the exhaust pipes ● Base engine problems (i.e., high engine oil or coolant usage) ● Catalytic converter is damaged, contaminated or has failed ● Continuous engine misfire conditions, or weak or low coil output ● Front HO2S or rear HO2S is contaminated with fuel or moisture ● Rear HO2S is loose in the mounting hole (check it for a leak) ● Front HO2S older (aged) than the rear HO2S (HO2S-12 is lazy)
P0420 **1T CAT, MIL: Yes** 1996, 1997, 1998, 1999, 2001, 2002, 2003, 2004, 2005 Aurora 4.0L VIN C engine Transmissions: A/T	**Catalyst System Low Efficiency (Bank 1) Conditions:** DTC P0101-P0103, P0105-P0108, P0116-P0118, P0120-P0125, P0131-P0147, P0151-P0155, P0171-P0175, P0300, P0322, P0371-P0372, P0440-P0446, P0502-P0507, P1108, P1114-P1115, P1133-P1139-P1140, P1153-P1154, P1320, P1350, P1370, P1441, P1508 and P1509 not set, engine runtime over 7.5 minutes, engine speed over 800 rpm since last idle for 46 seconds, idle speed within 100-125 rpm of Desired idle, BARO sensor over 75 kPa, ECT sensor at 160-248°F, IAT sensor at 21-392°F, Predicted Catalyst Temperature from 788-1202°F, and the PCM detected the Catalyst was degraded. **Possible Causes** ● Air leaks at the exhaust manifold or in the exhaust pipes ● Base engine problems (i.e., high engine oil or coolant usage) ● Catalytic converter is damaged, contaminated or has failed ● Continuous engine misfire conditions, or weak or low coil output ● Front HO2S or rear HO2S is contaminated with fuel or moisture ● Rear HO2S is loose in the mounting hole (check it for a leak) ● Front HO2S older (aged) than the rear HO2S (HO2S-12 is lazy)
P0420 **1T CAT, MIL: Yes** 1995 S/T Blazer & S/T Pickup 4.3L VIN W Transmissions: All	**Catalyst System Low Efficiency (Bank 1) Conditions:** DTC P0106-P0108, P0117, P0118, P0121-P0123, P0125, P0131-P0138, P0140, P0141, P0171-P0172, P0200, P0300-P0306, P0500, P0502 and P0503 not set, engine speed below 4900 in closed loop, BARO sensor more than 75 kPa, ECT sensor more than 167°F, IAT sensor from -15°F to 167°F, Calculated engine load stable, A/F ratio at 14.7:1, and the PCM detected the catalyst was degraded. **Possible Causes** ● Air leaks at the exhaust manifold or in the exhaust pipes ● Base engine problems (i.e., high engine oil or coolant usage) ● Catalytic converter is damaged, contaminated or has failed ● Continuous engine misfire conditions, or weak or low coil output ● Front HO2S or rear HO2S is contaminated with fuel or moisture ● Rear HO2S is loose in the mounting hole (check it for a leak) ● Front HO2S older (aged) than the rear HO2S (HO2S-12 is lazy)
P0420 **1T CAT, MIL: Yes** 1997, 1998, 1999, 2000, 2001 Catera 3.0L V6 VIN R engine Transmissions: All	**Catalyst System Low Efficiency (Bank 1) Conditions:** DTC P0100, P0131-P0141, P0150-P0156, P0158-P0161, P0171, P0172, P0174, P0175, P0300, P0301-P0306, P0440-P0455, P1120 and P1220 not set, engine started, engine speed from 1400-2200 rpm, engine load from 8-12%, Predicted TWC temperature over 716°F, and the PCM detected the Bank 1 Catalyst was degraded. **Possible Causes** ● Air leaks at the exhaust manifold or in the exhaust pipes ● Base engine problems (i.e., high engine oil or coolant usage) ● Catalytic converter is damaged, contaminated or has failed ● Continuous engine misfire conditions, or weak or low coil output ● Front HO2S or rear HO2S is contaminated with fuel or moisture ● Rear HO2S is loose in the mounting hole (check it for a leak) ● Front HO2S older (aged) than the rear HO2S (HO2S-12 is lazy)

OBD II Trouble Code List (P0xxx Codes)

DTC	Trouble Code Title, Conditions & Possible Causes
P0420 **1T CAT, MIL: Yes** 1996, 1997 Grand Prix, Lumina, Monte Carlo 3.4L VIN X engine Transmissions: All	**Catalyst System Bank 1 Low Efficiency Conditions:** DTC P0101-P0103, P0106-P0108, P0112, P0113, P0116-P0118, P0121-P0123, P0201-P0206, P0300, P0336, P0403, P0404, P0405, P0502, P0503, P0506, P0507, P1106-P1107, P1111-P1112, P1114, P1115, P1121, P1122, P1374 and P1404 not set, engine runtime over 10 minutes, system voltage over 10.0v, BARO sensor above 75 kPa, ECT sensor from 169-255°F, IAT sensor from -4°F to 212°F, and the PCM determined the catalyst oxygen storage capacity had degraded. *Note: This test can run up to 18 times during one ignition cycle.* **Possible Causes** ● Air leaks at the exhaust manifold or in the exhaust pipes ● Base engine problems (i.e., high engine oil or coolant usage) ● Catalytic converter is damaged, contaminated or has failed ● Continuous engine misfire conditions, or weak or low coil output ● Front HO2S or rear HO2S is contaminated with fuel or moisture ● Rear HO2S is loose in the mounting hole (check it for a leak) ● Front HO2S older (aged) than the rear HO2S (HO2S-12 is lazy)
P0420 **1T CAT, MIL: Yes** 1995 Corvette 5.7L VIN P engine Transmissions: All	**Catalyst System Low Efficiency (Bank 1) Conditions:** DTC P0116, P0117, P0118, P0131-P0138, P0140, P0151-P0160, P0171-P0175, P114, P1115, P1153 and P1155 not set, engine started, VSS from 25-68 mph for 2 minutes, and then back to idle speed, ECT sensor over 185°F, MAF sensor from 14-32 g/sec, and the PCM detected the Catalyst was degraded. **Possible Causes** ● Air leaks at the exhaust manifold or in the exhaust pipes ● Base engine problems (i.e., high engine oil or coolant usage) ● Catalytic converter is damaged, contaminated or has failed ● Continuous engine misfire conditions, or weak or low coil output ● Front HO2S or rear HO2S is contaminated with fuel or moisture ● Rear HO2S is loose in the mounting hole (check it for a leak)
P0430 **1T CAT, MIL: Yes** 1996, 1997 Camaro, Caprice, Corvette, Firebird, Fleetwood, Roadmaster 4.3L VIN W, 5.7L VIN 5, 5.7L VIN P engines Transmissions: All	**Catalyst System Low Efficiency (Bank 2) Conditions:** DTC P0101-P0103, P0117, P0118, P0122, P0123, P0335, P0336, P0171-P0175, P0500, P0502, P0641 and P0742 not set, engine speed under 3500 at 20-75 mph in closed loop, ECT sensor over 124°F, MAP sensor at 25-80 kPa, MAF sensor at 15-100 g/sec, TP angle over 2%, and the PCM detected the Catalyst was degraded. **Possible Causes** ● Air leaks at the exhaust manifold or in the exhaust pipes ● Base engine problems (i.e., high engine oil or coolant usage) ● Catalytic converter is damaged, contaminated or has failed ● Front HO2S or rear HO2S is contaminated with fuel or moisture ● Rear HO2S is loose in the mounting hole (check it for a leak)
P0430 **1T CAT, MIL: Yes** 1996, 1997, 1998, 1999, 2000, 2001, 2002, 2003, 2004, 2005 C/K Cab & Chassis, C/K Truck, G Series Van, L/M Series Vans, S/T Pickup 4.3L VIN W, 4.3L VIN X, 5.0L VIN M, 5.7L VIN K, 5.7L VIN R, 7.4L VIN J Transmissions: All	**Catalyst System Low Efficiency (Bank 2) Conditions:** DTC P0101-P0103, P0106-P0108, P0112-P0118, P0125, P0131, P0132-P0138, P0140, P0141, P0151-P0158, P0160, P0161, P0171-P0175, P0200, P0300, P0325, P0327, P0335, 336, P0341, P0343, P0351-P0358, P0443-P0449, P0502, P0503, P0506, P0507, P1120, P1125, P1133, P1134, P1153, P1154, P1220, P1221, P1275, P1276, P1280-P1286, P1441, P1514-P1518 not set, engine runtime over 6 minutes, ECT sensor over 167°F, BARO sensor over 72 kPa, IAT sensor over 16°F, MAF sensor from 15-50 g/sec, Catalyst Temperature over 840°F, engine running at idle speed for 2 minutes with Actual idle speed within 100-125 rpm of the Desired idle speed, vehicle driven to 22-85 mph, less than a 10% change in engine load, fuel trim stable, and the PCM detected the Catalyst was degraded. **Possible Causes** ● Air leaks at the exhaust manifold or in the exhaust pipes ● Base engine problems (i.e., high engine oil or coolant usage) ● Catalytic converter is damaged, contaminated or has failed ● Continuous engine misfire conditions, or weak or low coil output ● Front HO2S or rear HO2S is contaminated with fuel or moisture ● Rear HO2S is loose in the mounting hole (check it for a leak)

OBD II Trouble Code List (P0xxx Codes)

DTC	Trouble Code Title, Conditions & Possible Causes
P0430 **2T CAT, MIL: Yes** 1999, 2000, 2001, 2002, 2003, 2004, 2005 C/K Truck, Escalade, G Series Van, Cargo Van, Envoy & TrailBlazer 4.8L VIN V, 5.3L VIN P, 5.3L VIN T, 6.0L VIN N, 6.0L VIN U, 8.1L VIN G Transmissions: All	**Catalyst System Low Efficiency (Bank 2) Conditions:** DTC P0101-P0103, P0106-P0108, P0112, P0113, P0117, P0118, P0120, P0121-123, P0125, P0128, P0151-P0155, P0157, P0158, P0160-P0161, P0174, P0175, P0177, P0178, P0179, P0200, P0220, P0300, P0325, P0327, P0332, P0335, 336, P0341-P0343, P0351-P0358, P0442, P0443, P0446, P0452, P0453, P0455, P0496, P0502, P0503, P1125, P1133, P1153, P1258, P1514, P1516, P1518 and P2108, P2135 not set, vehicle driven at 25-75 mph for 40-45 seconds in closed loop, BARO sensor over 74 kPa, ECT sensor at 158-248ºF, IAT sensor at 5-185ºF, and the PCM detected the Catalyst had degraded to below a calibrated threshold. **Possible Causes** • Air leaks at the exhaust manifold or in the exhaust pipes • Base engine problems (i.e., high engine oil or coolant usage) • Catalytic converter is damaged, contaminated or has failed • Continuous engine misfire conditions, or weak or low coil output • Front HO2S or rear HO2S is contaminated with fuel or moisture • Rear HO2S is loose in the mounting hole (check it for a leak) • TSB 81-65-37 contains a repair procedure for this code
P0430 **2T CAT, MIL: Yes** 2003, 2004, 2005 CTS (Cadillac) 2.6L VIN M, 3.2L VIN N Transmissions: All	**Catalyst System Low Efficiency (Bank 2) Conditions:** DTC P0030, P0031, P0032, P0036, P0037, P0038, P0050, P0051, P0052, P0056, P0057, P0058, P0101, P0102, P0103, P0121, P0122, P0123, P0130, P0131, P0132, P0133, P0134, P0135, P0136, P0137, P0138, P0140, P0141, P0150, P0151, P0152, P0153, P0154, P0155, P0156, P0157, P0158, P0160, P0161, P0221, P0222, P0223, P0300, P0301-P0306, P0313, P0440, P0442, P0443, P0444, P0445, P0446, P0447, P0448, P0449, P0496, P0638, P2096, P2097, P2098, P2099, P2100, P2101, P2105, P2107, P2119, P2122, P2123, P2127, P2128, P2135, P2138 and P2176 not set, engine speed at 1,200-3,000 rpm in closed loop, engine load from 25-30%, TWC Temperature from 968-1454ºF, and the PCM detected the Bank 2 Catalyst efficiency had degraded. **Possible Causes** • Air leaks at the exhaust manifold or in the exhaust pipes • Base engine problems (i.e., high engine oil or coolant usage) • Catalytic converter is damaged, contaminated or has failed • Continuous engine misfire conditions, or weak or low coil output • Front HO2S or rear HO2S is contaminated with fuel or moisture • Rear HO2S is loose in the mounting hole (check it for a leak)
P0430 **2T CAT, MIL: Yes** 1997, 1998, 1999, 2000, 2001, 2002, 2003, 2004, 2005 Camaro, Corvette & Firebird 5.7L VIN G, 5.7L VIN S Transmissions: All	**Catalyst System Low Efficiency (Bank 2) Conditions:** DTC P0101-P0103, P0107, P0108, P0112-P0118, P0121-P0123, P0125, P0171-P0175, P0200, P0230, P0300, P0325-P0327, P0332-P0336, P0341-P0343, P0351-P0358, P0401-P0405, P0410, P0412, P0418, P0440-P0449, P0500, P0704, P0801-0803, P1258, P1336, P1404, P1415, P1416, P1441, no HO2S codes set, engine started, engine speed over 1000 rpm for a period of 37-44 seconds, BARO sensor more than 75 kPa, ECT sensor from 167-248ºF, IAT sensor from 64-176ºF, MAF sensor from 12-32 g/sec, and the PCM detected the Bank 2 Catalyst was degraded below a calibrated level. **Possible Causes** • Air leaks at the exhaust manifold or in the exhaust pipes • Base engine problems (i.e., high engine oil or coolant usage) • Catalytic converter is damaged, contaminated or has failed • Continuous engine misfire conditions, or weak or low coil output • Front HO2S or rear HO2S is contaminated with fuel or moisture • Rear HO2S is loose in the mounting hole (check it for a leak) • Front HO2S older (aged) than the rear HO2S (HO2S-12 is lazy)
P0430 **2T CAT, MIL: Yes** 1997, 1998, 1999, 2000, 2001 Catera 3.0L V6 VIN R engine Transmissions: A/T	**Catalyst System Low Efficiency (Bank 2) Conditions:** DTC P0100, P0131-P0141, P0150-P0156, P0158-P0161, P0171, .P0172, P0174, P0175, P0300, P0301-P0306, P0440-P0455, P1120 and P1220 not set, engine started, engine speed from 1400-2200 rpm, engine load from 8-12%, Predicted TWC temperature over 716ºF, and the PCM detected the Bank 2 Catalyst was degraded. **Possible Causes** • Air leaks at the exhaust manifold or in the exhaust pipes • Base engine problems (i.e., high engine oil or coolant usage) • Catalytic converter is damaged, contaminated or has failed • Continuous engine misfire conditions, or weak or low coil output • Front HO2S or rear HO2S is contaminated with fuel or moisture • Rear HO2S is loose in the mounting hole (check it for a leak) • Front HO2S older (aged) than the rear HO2S (HO2S-12 is lazy)

OBD II Trouble Code List (P0xxx Codes)

DTC	Trouble Code Title, Conditions & Possible Causes
P0440 **2T EVAP, MIL: Yes** 1996, 1997, 1998, 1999, 2000, 2001, 2002, 2003, 2004, 2005 Aurora, Bonneville, Century, Cutlass Supreme, Grand Prix, Impala, Intrigue, LeSabre, LSS, Lumina, Monte Carlo, Park Avenue, Regal, 88' & 98' 3.1L VIN J, 3.1L VIN M, 3.4L VIN E, 3.5L VIN H, 3.8L VIN 1, 3.8L VIN K Transmissions: All	**EVAP System No Flow During Purge Conditions:** DTC P0107, P0108, P0112, P0113, P0116, P0117, P0118, P0121, P0122, P0123, P0125, P0443, P0449, P0452, P0453, P1106, P1107, P1112, P1114, P1115, P1121 and P1122 not set, engine started, system voltage over 10.0v, ECT and IAT sensors from 39-86°F and within 16°F at startup, vehicle driven at a steady speed of less than 75 mph, BARO more than 75 kPa, fuel level from 15-85%, and the PCM detected the EVAP system was unable to achieve or maintain enough vacuum during the EVAP test. *Note: An Ultrasonic Leak Detector can be used to help detect leaks in the EVAP system.* **Possible Causes** ● Charcoal canister is loaded with fuel or moisture ● Fuel filler cap is loose, cross-threaded, damaged or wrong part ● Fuel tank, fuel filler neck or fuel sending unit 'O' ring is leaking ● Fuel tank pressure sensor is damaged, disconnected or it failed ● Fuel tank vapor line(s) is clogged, damaged or disconnected ● Purge valve vapor line is clogged, damaged, or disconnected ● Purge or vent solenoid power circuit is open (check the fuse) ● PCM has failed
P0440 **2T EVAP, MIL: Yes** 1998, 1999, 2000, 2001, 2002, 2003, 2004, 2005 C/K Cab & Chassis, C/K Truck, G Series Van, Van Cargo, L/M Series Van, S/T Blazer & S/T Pickup 4.3L VIN W, 4.3L VIN X, 5.0L VIN M, 5.7L VIN K, 5.7L VIN R, 7.4L VIN J Transmissions: All	**EVAP System No Flow During Purge Conditions:** DTC P0107, P0108, P0112, P0113, P0115, P0116-P0118, P0121, P0122, P0123, P0125, P0443, P0449, P0452, P0453, P1106, P1107, P1112, P1114, P1115, P1121 and P1122 not set, engine started, ECT and IAT sensors from 39-86°F and within 16°F at startup, vehicle driven at a steady speed less than 72 mph, system voltage over 10.0v, BARO sensor more than 75 kPa, fuel level from 15-85%, and the PCM detected the EVAP system was unable to achieve and maintain vacuum during the EVAP flow and leak test. **Possible Causes** ● Charcoal canister is loaded with fuel or moisture ● Fuel filler cap is loose, cross-threaded, damaged or wrong part ● Fuel tank, fuel filler neck or fuel sending unit 'O' ring is leaking ● Fuel tank pressure sensor is damaged, disconnected or it failed ● Fuel tank vapor line(s) is clogged, damaged or disconnected ● Purge valve vapor line is clogged, damaged, or disconnected ● Purge or vent solenoid power circuit is open (check the fuse) ● PCM has failed
P0440 **2T EVAP, MIL: Yes** 2003, 2004, 2005 CTS (Cadillac) 2.6L VIN M, 3.2L VIN N Transmissions: All	**EVAP System Large Leak Detected Conditions:** DTC P0112, P0113, P0116, P0117, P0118, P0121, P0122, P0123, P0125, P0443, P0444, P0445, P0447, P0448, P0449, P0451, P0452 and P0453 not set, engine running at idle in closed loop, BARO sensor over 75 kPa, system voltage from 10-18v, fuel level 15-85%, ECT and IAT within 16°F at startup, IAT sensor from 39-149°F, VSS at 0 mph, and the PCM detected the EVAP system was not able to achieve or maintain a vacuum during the diagnostic test period. **Possible Causes** ● Fuel filler cap is loose, cross-threaded, damaged or missing ● Fuel tank, fuel filler neck or fuel sending unit 'O' ring is leaking ● Fuel tank vapor line(s) is clogged, damaged or disconnected ● Fuel tank pressure sensor low reference circuit is open ● Fuel tank pressure sensor is damaged or it has failed ● EVAP charcoal canister is clogged or loaded with fuel or water ● EVAP purge or EVAP vent valve is damaged or it has failed ● PCM has failed
P0440 **1T EVAP, MIL: Yes** 1998, 1999, 2000, 2001, 2002, 2003, 2004, 2005 DeVille, Eldorado, Seville 4.6L VIN 9, 4.6L VIN Y Transmissions: A/T	**EVAP System Malfunction Conditions:** No ECT, EVAP Vent Control, EVAP Purge Control, Fuel Tank Level Sensor, Fuel Tank Pressure Sensor, IAT, MAP and TP Sensor codes set, engine started, ECT and IAT sensors from 35.6°F-86°F and within 57.5°F at startup, system voltage over 10.0v, fuel level from 15-85% for 3 seconds (to prevent fuel slosh in the tank from causing the fuel level to vary outside its limits), BARO sensor more than 72 kPa, and the PCM detected that the EVAP system was unable to achieve and maintain vacuum during the EVAP leak test. **Possible Causes** ● Charcoal canister is loaded with fuel or moisture ● Fuel filler cap is loose, cross-threaded, damaged or wrong part ● Fuel tank, fuel filler neck or fuel sending unit 'O' ring is leaking ● Fuel tank pressure sensor is damaged, disconnected or it failed ● Fuel tank vapor line(s) is clogged, damaged or disconnected ● Purge valve vapor line is clogged, damaged, or disconnected ● Purge or vent solenoid power circuit is open (check the fuse) ● PCM has failed

OBD II Trouble Code List (P0xxx Codes)

DTC	Trouble Code Title, Conditions & Possible Causes
P0440 **1T EVAP, MIL: Yes** 1998, 1999, 2000, 2001, 2002 Camaro & Firebird 3.8L VIN K engine Transmissions: All	**EVAP System Malfunction Conditions:** DTC P0107, 0108, P0112, P0113, P0116-P0118, P0125, P0443, P0449, P0452, P0453, P1106, P1107, P1112, P1114, P1120 and P1221 not set, engine started, ECT and IAT sensors from 39-86ºF with within 16ºF at startup, vehicle driven at a steady speed of less than 75 mph, BARO sensor over 75 kPa, fuel level from 15-85%, and the PCM detected the EVAP system was unable to achieve and maintain vacuum in the EVAP flow and leak test. **Possible Causes** • Charcoal canister is loaded with fuel or moisture • Fuel filler cap is loose, cross-threaded, damaged or wrong part • Fuel tank, fuel filler neck or fuel sending unit 'O' ring is leaking • Fuel tank pressure sensor is damaged, disconnected or it failed • Fuel tank vapor line(s) is clogged, damaged or disconnected • Purge valve vapor line is clogged, damaged, or disconnected • Purge or vent solenoid power circuit is open (check the fuse) • PCM has failed
P0440 **1T EVAP, MII · Yes** 1999, 2000, 2001, 2002, 2003, 2004, 2005 Camaro, Corvette & Firebird 5.7L VIN G, 5.7L VIN S Transmissions: All	**EVAP System Malfunction Conditions:** DTO P0107, P0108, P0112, P0113, P0116-P0118, P0121-P0123, P0125, P0131-P0141, P0443, P0449, P0452, P0453, P1111-P1115, P1121 and P1122 not set, engine started, vehicle driven at a steady speed of less than 75 mph, system voltage over 10.0v, ECT and IAT sensors from 39-86ºF and within 16ºF at startup, BARO sensor more than 75 kPa, fuel level from 15-85%, and the PCM detected the EVAP system was unable to achieve and maintain proper vacuum. *Note: This trouble code may not report a first failed test. A first fail of this code may show a Scan Tool status of "Not Run". Read the EVAP Test Result to determine the pass/fail for this ignition cycle.* **Possible Causes** • Charcoal canister is loaded with fuel or moisture • Fuel filler cap is loose, cross-threaded, damaged or wrong part • Fuel tank, fuel filler neck or fuel sending unit 'O' ring is leaking • Fuel tank pressure sensor is damaged, disconnected or it failed • Fuel tank vapor line(s) is clogged, damaged or disconnected • Purge valve vapor line is clogged, damaged, or disconnected • Purge or vent solenoid power circuit is open (check the fuse) • PCM has failed
P0440 **2T EVAP, MIL: Yes** 1998, 1999, 2001, 2002, 2003, 2004, 2005 Aurora 4.0L VIN C engine Transmissions: A/T	**EVAP System Malfunction Conditions:** DTC P0107-P0108, P0112-P0118, P0122-P0125, P0131-P0135, P0141, P0151-P0154, P0452, P0453, P0503, P1111-P1115, P1133 and P1153 not set, engine started, ECT and IAT sensors from 35.6-86ºF and within 57.5ºF at engine startup, BARO sensor more than 72 kPa, fuel level from 10-89.9% for 3.2 seconds (to prevent fuel slosh in the tank from causing the fuel level to vary outside its limits), and the PCM detected the EVAP system was unable to achieve and maintain vacuum during the EVAP flow and leak test. **Possible Causes** • Charcoal canister is loaded with fuel or moisture • Fuel filler cap is loose, cross-threaded, damaged or wrong part • Fuel tank, fuel filler neck or fuel sending unit 'O' ring is leaking • Fuel tank pressure sensor is damaged, disconnected or it failed • Fuel tank vapor line(s) is clogged, damaged or disconnected • Purge valve vapor line is clogged, damaged, or disconnected • Purge or vent solenoid power circuit is open (check the fuse) • PCM has failed
P0440 **2T EVAP, MIL: Yes** 1996, 1997, 1998, 1999, 2000, 2001, 2002, 2003, 2004, 2005 Achieva, Beretta, Ciera, Cavalier, Century, Corsica, Envoy, Grand Am, Malibu, Skylark, Sunfire, S/T Blazer & S/T Pickup & TrailBlazer 2.2L VIN 4, 2.2L VIN 5, 2.4L VIN T, 2.2L VIN H, 4.2L VIN S Transmissions: All	**EVAP System Malfunction Conditions:** DTC P0107, P0108, P0112, P0113, P0117, P0118, P0122, P0123, P0125, P0452 and P0453 not set, engine started, ECT and IAT sensors from 39-86ºF and within 16ºF at startup, vehicle driven at a steady speed of less than 75 mph, BARO sensor more than 75 kPa, fuel level from 15-85%, and the PCM detected the EVAP System was unable to achieve and maintain vacuum during the EVAP test. **Possible Causes** • Charcoal canister is loaded with fuel or moisture • Fuel filler cap is loose, cross-threaded, damaged or wrong part • Fuel tank, fuel filler neck or fuel sending unit 'O' ring is leaking • Fuel tank pressure sensor is damaged, disconnected or it failed • Fuel tank vapor line(s) is clogged, damaged or disconnected • Purge valve vapor line is clogged, damaged, or disconnected • Purge or vent solenoid power circuit is open (check the fuse) • PCM has failed • TSB 02-06-04-044 contains a repair procedure for this code

OBD II Trouble Code List (P0xxx Codes)

DTC	Trouble Code Title, Conditions & Possible Causes
P0440 **2T EVAP, MIL: Yes** 1997, 1998, 1999, 2000, 2001 Catera 3.0L V6 VIN R engine Transmissions: A/T	**EVAP System Malfunction Conditions:** DTC P0100, P0115, P0116, P0130-P0161, P0300-P0306, P0443, P0446, P0450, P0500, P0506, P0507, P0560, P1120 and P1220 not set, engine started, ECT sensor from 17-167°F and IAT sensor from 17°F-122°F at startup, engine runtime over 17 minutes, engine at idle speed in closed loop, system voltage over 10.0v, engine load under 30%, MAF sensor less than 7.5 g/sec, VSS at 0 mph, AIR system inactive, Short Term greater than -5%, and the PCM detected the fuel tank pressure was below 3.2" Hg (-6 mm Hg) with the Purge solenoid open, or the EVAP vapor pressure did not decay, or it decayed too slowly with the Vent solenoid open, or the fuel tank pressure was below -5.6" H2O (-10 mm Hg) for 15 seconds. **Possible Causes** • Charcoal canister is loaded with fuel or moisture • Fuel filler cap is loose, cross-threaded, damaged or wrong part • Fuel tank, fuel filler neck or fuel sending unit 'O' ring is leaking • Fuel tank pressure sensor is damaged, disconnected or it failed • Fuel tank vapor line(s) is clogged, damaged or disconnected • Purge valve vapor line is clogged, damaged, or disconnected • Purge or vent solenoid power circuit is open (check the fuse) • PCM has failed
P0441 **2T EVAP, MIL: Yes** 1996, 1997 Body Codes: A, C, G, H, L, N, W Achieva, Century, Ciera, Beretta, Corsica, Cutlass, Grand Prix, Regal, Monte Carlo, Malibu & Lumina 3.1L VIN M, 3.4L VIN X, 3.8L VIN K, 3.8L VIN 1 Transmissions: All	**EVAP System No Flow During Purge Conditions:** DTC P0106, P0107, P0108, P0112, P0113, P0121, P0122, P0123, P1641 and P1642 not set, engine started, engine speed from 650-5000 rpm, ECT sensor from 156-239°F, IAT sensor less than 158°F, ECT and IAT sensors within 18°F, TP angle from 2.5-40%, Purge command over 75-85%, and the PCM detected the EVAP Vacuum Switch was closed (EVAP switch circuit reads 12v) for 4 seconds. **Possible Causes** • Charcoal canister is loaded with fuel or moisture • EVAP switch is damaged, disconnected or it failed • Fuel filler cap is loose, cross-threaded, damaged or wrong part • Fuel tank vapor line(s) is clogged, damaged or disconnected • Purge valve vapor line is clogged, damaged, or disconnected • Purge solenoid is damaged or sticking (it may be stuck closed) • Purge solenoid power circuit is open (check the fuse) • PCM has failed
P0441 **2T EVAP, MIL: Yes** 1996, 1997 Camaro, Caprice, Corvette, Firebird, Fleetwood, Roadmaster 4.3L VIN W, 5.7L VIN 5, 5.7L VIN P engines Transmissions: All	**EVAP System No Flow During Purge Conditions:** DTC P0100, P0101, P0102, P0103, P0106, P0323, P0325, P0327, P0335, P0336, P0372, P0400, P0403, P0410-P0412, P0420-P0430, P0440, P0443, P0500, P0502, P0503, P0506, P0507, P0530, P0537, P0719, P0758, P1415, P1642 and P1652 not set, vehicle driven at a steady speed of 800-3000 rpm, BARO sensor more than 77 kPa, ECT sensor less than 239°F, IAT sensor more than 37°F, MAP sensor from 33-77 kPa, Purge command over 60%, TP angle from 2-50%, and the PCM detected a low voltage signal on the EVAP vacuum switch circuit with EVAP Purge commanded "on" for 5 seconds during the EVAP flow test. **Possible Causes** • Charcoal canister is loaded with fuel or moisture • EVAP switch is damaged, disconnected or it failed • Fuel filler cap is loose, cross-threaded, damaged or wrong part • Fuel tank vapor line(s) is clogged, damaged or disconnected • Purge valve vapor line is clogged, damaged, or disconnected • Purge solenoid is damaged or sticking (it may be stuck closed) • Purge solenoid power circuit is open (check the fuse) • PCM has failed
P0441 **2T EVAP, MIL: Yes** 1996, 1997 C/K Cab & Chassis, C/K Truck, G Series Van, Van Cargo, L/M Series Van, S/T Blazer & Pickup 4.3L VIN W, 4.3L VIN X, 5.0L VIN M, 5.7L VIN R, 7.4L VIN J engines Transmissions: All	**EVAP System No Flow During Purge Conditions:** DTC P0106, P0107, P0108, P0112, P0113, P0121, P0122, P0123, P0401 and P0403 not set, engine started, engine speed from 800-3000 rpm, ECT sensor less than 226°F, IAT sensor less than 198°F with the startup IAT sensor over 0°F, BARO sensor more than 75 kPa, change in ECT and IAT sensor 198°F or less, Purge command 90% or more, MAP sensor from 20-80 kPa, TP angle from 5-60%, and the PCM detected the vacuum switch indicated it was closed with the Purge solenoid enabled during the EVAP Monitor flow test. **Possible Causes** • Charcoal canister is loaded with fuel or moisture • EVAP switch is damaged, disconnected or it failed • Fuel filler cap is loose, cross-threaded, damaged or wrong part • Fuel tank vapor line(s) is clogged, damaged or disconnected • Purge valve vapor line is clogged, damaged, or disconnected • Purge solenoid is damaged or sticking (it may be stuck closed) • Purge solenoid power circuit is open (check the fuse) • PCM has failed

OBD II Trouble Code List (P0xxx Codes)

DTC	Trouble Code Title, Conditions & Possible Causes
P0441 **2T EVAP, MIL: Yes** 1996, 1997 DeVille, Eldorado, Seville 4.6L VIN 9, 4.6L VIN Y Transmissions: All	**EVAP System No Flow During Purge Conditions:** DTC P0106-P0108, P0112, P0113, P0121, P0122, P0123, P1641 and P1642 not set, engine speed from 500-6375 rpm, BARO more than 72 kPa, ECT sensor 50°F less than previous key off and less than 257°F, IAT sensor from -13°F to 211°F, MAP sensor from 20-60 kPa, TP angle from 0-81.6°, Purge command from 60-85%, and the PCM detected the EVAP Vacuum Switch was closed for 9 seconds or did not open for 2 seconds during the flow test. **Possible Causes** ● Charcoal canister is loaded with fuel or moisture ● EVAP switch is damaged, disconnected or it failed ● Fuel filler cap is loose, cross-threaded, damaged or wrong part ● Fuel tank vapor line(s) is clogged, damaged or disconnected ● Purge valve vapor line is clogged, damaged, or disconnected ● Purge solenoid is damaged or sticking (it may be stuck closed) ● Purge solenoid power circuit is open (check the fuse) ● PCM has failed
P0441 **2T EVAP, MIL: Yes** 1995, 1996, 1997 Camaro & Firebird 3.8L VIN K engine Transmissions: All	**EVAP System No Flow During Purge Conditions:** DTC P0106, P0107, P0108, P0112, P0113, P0121, P0122, P0123, P1641 and P1642 not set, engine speed from 900-5000 rpm, BARO sensor more than 73 kPa, ECT sensor less than 237°F, IAT sensor from 50-158°F, difference between ECT and IAT sensor less than 18°F, TP angle from 2.5-40%, Purge command more than 85%, and the PCM detected the EVAP vacuum switch indicated it was closed for 2 seconds with Purge solenoid enabled. **Possible Causes** ● Charcoal canister is loaded with fuel or moisture ● EVAP switch is damaged, disconnected or it failed ● Fuel filler cap is loose, cross-threaded, damaged or wrong part ● Fuel tank vapor line(s) is clogged, damaged or disconnected ● Purge valve vapor line is clogged, damaged, or disconnected ● Purge solenoid is damaged or sticking (it may be stuck closed) ● Purge solenoid power circuit is open (check the fuse) ● PCM has failed
P0441 **2T EVAP, MIL: Yes** 1996, 1997 Aurora 4.0L VIN C engine Transmissions: All	**EVAP System No Flow During Purge Conditions:** DTC P0101, P0102, P0103, P0107, P0108, P0122, P0123, P0401, P0506, P0507, P1441, P1508, P1509 and P1645 not set, IAT sensor more than -13°F at key on, IAT sensor remains below 211°F, ECT sensor at least 50°F below the previous key off reading, engine started, ECT sensor less than 257°F, MAP sensor from 20.2-60 kPa, TP angle at 0-81.6 degrees, engine speed from 375-6375 rpm, BARO sensor over 72 kPa, engine vacuum over 12 kPa, purge duty cycle over 85%, and the PCM detected the EVAP Vacuum switch was closed for 9 seconds or it did not open for 2 seconds at any time during the EVAP flow test. **Possible Causes** ● Charcoal canister is loaded with fuel or moisture ● EVAP switch is damaged, disconnected or it failed ● Fuel filler cap is loose, cross-threaded, damaged or wrong part ● Fuel tank vapor line(s) is clogged, damaged or disconnected ● Purge valve vapor line is clogged, damaged, or disconnected ● Purge solenoid is damaged or sticking (it may be stuck closed) ● Purge solenoid power circuit is open (check the power fuse) ● PCM has failed
P0441 **2T EVAP, MIL: Yes** 1995 S/T Series Pickup 4.3L VIN W engine Transmissions: All	**EVAP System Continuous Open Purge Flow Conditions:** DTC P0106, P0107, P0108, P0112, P0113, P0121, P0122, P0123, P1641 and P1642 not set, engine started, BARO sensor more than 75 kPa, ECT sensor less than 226°F, ECT and IAT sensor within 198°F or less, MAP sensor from 20-80 kPa, TP angle from 5-60%, Purge command more than 90%, and the PCM detected the EVAP Vacuum Switch indicated it was open for 4 seconds during the test. **Possible Causes** ● Charcoal canister is loaded with fuel or moisture ● EVAP switch is damaged, disconnected or it failed ● Fuel filler cap is loose, cross-threaded, damaged or wrong part ● Fuel tank vapor line(s) is clogged, damaged or disconnected ● Purge valve vapor line is clogged, damaged, or disconnected ● Purge solenoid is damaged or sticking (it may be stuck closed) ● Purge solenoid power circuit is open (check the fuse) ● PCM has failed

OBD II Trouble Code List (P0xxx Codes)

DTC	Trouble Code Title, Conditions & Possible Causes
P0441 **2T EVAP, MIL: Yes** 1997, 1998 Catera 3.0L V6 VIN R engine Transmissions: A/T	**EVAP System Incorrect Purge Flow Conditions:** DTC P0100, P0115, P0116, P0130-P0161, P0300-P0306, P0443, P0446, P0450, P0500, P0506, P0507, P0560, P1120 and P1220 not set, engine started, system voltage over 10.0v, engine at idle speed in closed loop, engine runtime over 16 minutes, ECT sensor at startup from 17°F-212°F, IAT sensor more than 17°F, engine load less than 2.7%, MAF sensor less than 7.5 g/sec, fuel tank pressure under 4" H2O, VSS at 0 mph, AIR Test inactive, SHRTFT at start of test over -5%, and the PCM the EVAP system failed the flow test. **Possible Causes** ● Charcoal canister is loaded with fuel or moisture ● EVAP switch is damaged, disconnected or it failed ● Fuel filler cap is loose, cross-threaded, damaged or wrong part ● Fuel tank vapor line(s) is clogged, damaged or disconnected ● Purge valve vapor line is clogged, damaged, or disconnected ● Purge solenoid is damaged or sticking (it may be stuck closed) ● Purge solenoid power circuit is open (check the fuse) ● PCM has failed
P0441 **2T EVAP, MIL: Yes** 1997, 1998 Corvette 5.7L VIN G engine Transmissions: All	**EVAP System No Flow During Purge Conditions:** DTC P0107-P0108, P0112-P0113, P0117-P0118, P0125, P0443, P1120, P1220-P1221 and P1441 not set, DTC P0440 test passed, engine started, engine speed from 800-1000 rpm, IAT sensor more than 37°F at startup, ECT and IAT sensors within 19°F at startup, ECT sensor less than 239°F during testing, BARO sensor more than 75 kPa, TP angle from 8-60%, MAP sensor more than 15 kPa, Purge command over 90%, and the PCM detected the EVAP Vacuum switch indicated low with Purge enabled for 5 seconds. **Possible Causes** ● Charcoal canister is loaded with fuel or moisture ● EVAP switch is damaged, disconnected or it failed ● Fuel filler cap is loose, cross-threaded, damaged or wrong part ● Fuel tank vapor line(s) is clogged, damaged or disconnected ● Purge valve vapor line is clogged, damaged, or disconnected ● Purge solenoid is damaged or sticking (it may be stuck closed) ● Purge solenoid power circuit is open (check the fuse) ● PCM has failed
P0442 **2T EVAP, MIL: Yes** 1996, 1997, 1998, 1999, 2000, 2001, 2002, 2003, 2004, 2005 Aztek, Montana, Silhouette, Rendezvous & Venture 3.1L VIN J, 3.1L VIN M, 3.4L VIN E engines Transmissions: All	**EVAP System Small Leak (0.040") Detected Conditions:** DTC P0107, P0108, P0112, P0113, P0116-P0118, P0121, P0122, P0123, P0125, P0440, P0443, P0449, P0452, P0453, P1106, P1107, P1112, P1114, P1115, P1121 and P1122 not set, engine started, ECT and IAT sensors from 39-86°F and within 16°F, system voltage over 10.0v, BARO sensor more than 75 kPa, vehicle driven to a steady speed of less than 75 mph, fuel level from 15-80%, and the PCM detected the EVAP system achieved proper vacuum, but a vacuum decay condition was detected during the EVAP leak test. **Possible Causes** ● Charcoal canister is loaded with fuel or moisture ● Fuel filler cap is loose, cross-threaded, damaged or wrong part ● Fuel tank, fuel filler neck or fuel sending unit 'O' ring is leaking ● Fuel tank pressure sensor is damaged, disconnected or it failed ● Fuel tank vapor line(s) is clogged, damaged or disconnected ● Purge valve vapor line is clogged, damaged, or disconnected ● Purge solenoid or Vent solenoid has a small leaking (sticking) ● PCM has failed
P0442 **2T EVAP, MIL: Yes** 1998, 1999, 2000, 2001, 2002, 2003, 2004, 2005 C/K Cab & Chassis, C/K Series Truck, G Series Van & Van Cargo, L/M Series Van, S/T Blazer & S/T Pickup 4.3L VIN W, 4.3L VIN X, 5.0L VIN M, 5.7L VIN K, 5.7L VIN R Transmissions: All	**EVAP System Small Leak (0.040") Detected Conditions:** DTC P0107, P0108, P0112, P0113, P0116, P0117, P0118, P0125, P0440, P0443, P0455, P0449, P0452, P0453, P1111, P1112, P1114, P1115, P1120, P1220 and P1221 not set, ECT and IAT sensors from 39-86°F and within 16°F at startup, engine started, vehicle driven at less than 75 mph, system voltage over 10.0v, BARO sensor over 75 kPa, fuel level from 15-85%, DTC P0125 not active, and the PCM detected the EVAP system was able to achieve proper vacuum, but that a vacuum decay condition was detected. **Possible Causes** ● Charcoal canister is loaded with fuel or moisture ● Fuel filler cap is loose, cross-threaded, damaged or wrong part ● Fuel tank, fuel filler neck or fuel sending unit 'O' ring is leaking ● Fuel tank pressure sensor is damaged, disconnected or it failed ● Fuel tank vapor line(s) is clogged, damaged or disconnected ● Purge valve vapor line is clogged, damaged, or disconnected ● Purge solenoid or Vent solenoid has a small leaking (sticking) ● PCM has failed

OBD II Trouble Code List (P0xxx Codes)

DTC	Trouble Code Title, Conditions & Possible Causes
P0442 **2T EVAP, MIL: Yes** 1996, 1997, 1998, 1999, 2000, 2001, 2002, 2003, 2004, 2005 Aurora, Bonneville, Century, Cutlass Supreme, Grand Prix, Impala, Intrigue, LeSabre, LSS, Lumina, Monte Carlo, Park Avenue, Regal, 88' & 98' 3.1L VIN J, 3.1L VIN M, 3.4L VIN E, 3.5L VIN H, 3.8L VIN 1, 3.8L VIN K Transmissions: All	**EVAP System Small Leak (0.040") Detected Conditions:** DTC P0107, P0108, P0112, P0113, P0116, P0117, P0118, P0121-P0123, P0125, P0440-P0449, P0452, P0453, P1106-P1107, P1112, P1114, P1115, P1121 and P1122 not set, engine started, ECT and IAT sensors from 39-86°F and within 16°F at start, system voltage over 10.0v, vehicle speed less than 75 mph and steady, BARO sensor over 75 kPa, fuel level from 15-85%, and the PCM detected the EVAP system achieved proper vacuum, but a vacuum decay was detected. The PCM monitors the FTP sensor to determine the vacuum decay rate. At an appropriate time, the PCM turns the EVAP Purge valve "on" and the EVAP vent solenoid "off" to allow the engine to draw a vacuum on the system. After a calibrated time (vacuum level), the PCM turns the Purge solenoid off, to seal the system, and monitors the FTP sensor to determine the amount of vacuum decay in the EVAP system. **Possible Causes** ● Charcoal canister is loaded with fuel or moisture ● Fuel filler cap is loose, cross-threaded, damaged or wrong part ● Fuel tank, fuel filler neck or fuel sending unit 'O' ring is leaking ● Fuel tank pressure sensor is damaged, disconnected or it failed ● Fuel tank vapor line(s) is clogged, damaged or disconnected ● Purge valve vapor line is clogged, damaged, or disconnected ● Purge solenoid or Vent solenoid has a small leaking (sticking) ● PCM has failed
P0442 **2T EVAP, MIL: Yes** 1999, 2000, 2001, 2002, 2003, 2004, 2005 C/K Truck, Escalade, G Series Van, Cargo Van, Envoy & TrailBlazer 4.8L VIN V, 5.3L VIN P, 5.3L VIN T, 5.3L VIN Z, 6.0L VIN N, 6.0L VIN U, 8.1L VIN G engines Transmissions: All	**EVAP System Small Leak (0.040") Detected Conditions:** DTC P0100, P0101-P0103, P0106-P0108, P0112, P0113, P0116, P0117, P0118, P0125, P0335, P0336, P0443, P0446, P0449, P0452, P0453, P0455, P0496, P0500, P0502, P1106, P1107 and P1683 not set, engine runtime over 600 seconds, ECT and IAT sensors from 39-86°F and within 15°F at startup, vehicle driven over 3 miles this trip, BARO sensor over 74 kPa, fuel level from 15-85%, P0455 ran and passed, and the PCM detected a pressure change in the EVAP system that was less than a calibrated value. **Possible Causes** ● Charcoal canister is loaded with fuel or moisture ● Fuel filler cap is loose, cross-threaded, damaged or wrong part ● Fuel tank, fuel filler neck or fuel sending unit 'O' ring is leaking ● Fuel tank pressure sensor is damaged, disconnected or it failed ● Fuel tank or purge valve vapor lines clogged or disconnected ● Purge solenoid or Vent solenoid has a small leaking (sticking) ● PCM has failed
P0442 **2T EVAP, MIL: Yes** 2003, 2004, 2005 CTS (Cadillac) 2.6L VIN M, 3.2L VIN N Transmissions: All	**EVAP System Small Leak (0.040") Detected Conditions:** DTC P0112, P0113, P0116, P0117, P0118, P0121, P0122, P0123, P0125, P0443, P0444, P0445, P0447, P0448, P0449, P0451, P0452 and P0453 not set, engine running at idle in closed loop, BARO sensor over 75 kPa, system voltage from 10-18v, fuel level from 15-85%, ECT and IAT within 16°F at startup, IAT sensor from 39-149°F, VSS at 0 mph, and the PCM detected the EVAP system did not achieve a vacuum, but a vacuum decay was detected. **Possible Causes** ● Fuel filler cap is loose, cross-threaded, damaged or wrong part ● Fuel tank, fuel filler neck or fuel sending unit 'O' ring is leaking ● Fuel tank vapor line(s) is clogged, damaged or disconnected ● Fuel tank pressure sensor low reference circuit is open ● Fuel tank pressure sensor is damaged or it has failed ● EVAP charcoal canister is clogged or loaded with fuel or water ● EVAP purge or EVAP vent valve is damaged or it has failed ● PCM has failed
P0442 **2T EVAP, MIL: Yes** 1998, 1999, 2000, 2001, 2002, 2003, 2004, 2005 DeVille, Eldorado, Seville 4.6L VIN 9, 4.6L VIN Y Transmissions: A/T	**EVAP System Small Leak (0.040") Detected Conditions:** DTC P0107, P0108, P0112, P0113, P0115, P0116-P0118, P0121, P0122, P0123, P0125, P0443, P0449, P0452, P0453, P1106, P1107, P1112, P1114, P1115, P1121 and P1122 not set, engine started, ECT and IAT sensors from 39-86°F and within 16°F at startup, vehicle driven at a steady speed of less than 72 mph, and the PCM detected the EVAP system achieved proper vacuum, but a vacuum decay condition was detected in the EVAP leak test. *Note: These are 1T tests, yet they act like a 2T test under certain conditions. If the Monitor detects the system passes or that a code cleared, the test must fail twice to set this code.* **Possible Causes** ● Charcoal canister is loaded with fuel or moisture ● Fuel filler cap is loose, cross-threaded, damaged or wrong part ● Fuel tank, fuel filler neck or fuel sending unit 'O' ring is leaking ● Fuel tank pressure sensor is damaged, disconnected or it failed ● Fuel tank vapor line(s) is clogged, damaged or disconnected ● Purge valve vapor line is clogged, damaged, or disconnected ● Purge solenoid or Vent solenoid has a small leaking (sticking) ● PCM has failed

OBD II Trouble Code List (P0xxx Codes)

DTC	Trouble Code Title, Conditions & Possible Causes
P0442 **2T EVAP, MIL: Yes** 1998, 1999, 2000, 2001, 2002 Camaro & Firebird 3.8L VIN K engine Transmissions: All	**EVAP System Small Leak (0.040") Detected Conditions:** DTC P0107, P0108, P0112, P0113, P0116-P0118, P0125, P0443, P0449, P0452, P0453, P1106, P1107, P1112, P1114, P1120 and P1221 not set, engine started, ECT and IAT sensors from 39-86°F and within 16°F at startup, BARO sensor more than 75 kPa, fuel level from 15-85%, and the PCM detected the EVAP system could achieve vacuum, or it detected the system achieved proper vacuum, but a vacuum decay condition was detected in the test. **Possible Causes** ● Charcoal canister is loaded with fuel or moisture ● Fuel filler cap is loose, cross-threaded, damaged or wrong part ● Fuel tank, fuel filler neck or fuel sending unit 'O' ring is leaking ● Fuel tank pressure sensor is damaged, disconnected or it failed ● Fuel tank vapor line(s) is clogged, damaged or disconnected ● Purge valve vapor line is clogged, damaged, or disconnected ● Purge solenoid or Vent solenoid has a small leaking (sticking) ● PCM has failed
P0442 **2T EVAP, MIL: Yes** 1998, 1999, 2000, 2001, 2002, 2003, 2004, 2005 Camaro, Corvette & Firebird 5.7L VIN G, 5.7L VIN S Transmissions: All	**EVAP System Small Leak (0.040") Detected Conditions:** DTC P0107, P0108, P0112-P0118, P0121-P0123, P0125, P0443, P0449, P0452, P0453, P1111-P1115, P1121 and P1122 not set, engine started, ECT and IAT sensors from 39-86°F and within 16°F at startup, BARO sensor over 75 kPa, system voltage over 10.0v, vehicle speed less than 75 mph and steady, fuel level at 15-85%, and the PCM detected the EVAP system achieved proper vacuum, but a vacuum decay condition was detected in the test. **Possible Causes** ● Charcoal canister is loaded with fuel or moisture ● Fuel filler cap is loose, cross-threaded, damaged or wrong part ● Fuel tank, fuel filler neck or fuel sending unit 'O' ring is leaking ● Fuel tank pressure sensor is damaged, disconnected or it failed ● Fuel tank vapor line(s) is clogged, damaged or disconnected ● Purge valve vapor line is clogged, damaged, or disconnected ● Purge solenoid or Vent solenoid has a small leaking (sticking) ● PCM has failed
P0442 **1T EVAP, MIL: Yes** 1997, 1998, 1999, 2000, 2001, 2002, 2003, 2004, 2005 Camaro, Corvette & Firebird 5.7L VIN G, 5.7L VIN S Transmissions: All	**EVAP System Small Leak (0.040") Detected Conditions:** DTC P0107, P0108, P0112, P0113, P0116-P0118, P0125, P0131-P0141, P0151-P0161, P0443, P0449, P0452, P0453, P1106, P1107, P1111, P1112, P1114, P1115, P1120, P1220 and P1221 not set, engine started, ECT and IAT sensors from 39-86°F and within 16°F at startup, BARO sensor more than 75 kPa, vehicle driven to a steady speed of less than 70 mph, TP angle less than 75%, fuel level from 15-85%, and the PCM detected the EVAP system achieved proper vacuum, but a vacuum decay condition was detected during the EVAP leak test. Note: A first fail of this code will show a Scan Tool status as "Not Run". **Possible Causes** ● Charcoal canister is loaded with fuel or moisture ● Fuel filler cap is loose, cross-threaded, damaged or wrong part ● Fuel tank, fuel filler neck or fuel sending unit 'O' ring is leaking ● Fuel tank pressure sensor is damaged, disconnected or it failed ● Fuel tank vapor line(s) is clogged, damaged or disconnected ● Purge valve vapor line is clogged, damaged, or disconnected ● Purge solenoid or Vent solenoid has a small leaking (sticking) ● PCM has failed
P0442 **2T EVAP, MIL: Yes** 1998, 1999, 2001, 2002, 2003, 2004, 2005 Aurora 4.0L VIN C engine Transmissions: A/T	**EVAP System Small Leak (0.040") Detected Conditions:** DTC P0107, P0108, P0112, P0113, P0117, P0118, P0122, P0123, P0125, P0131, P0132, P0133, P0134, P0135, P0141, P0151, P0152, P0153, P0154, P0443, P0449, P0452, P0453, P1106, P1107, 1112, P1114, P1115, P1121 and P1122 are not set, engine started, ECT and IAT sensors from 39-86°F and within 16°F at startup, system voltage over 10.0v, BARO sensor more than 75 kPa, vehicle driven to a speed of less than 75 mph, fuel level from 15-85 for more than 3.2 seconds (this time limit is required to prevent fuel slosh in the tank from causing the fuel level indication to vary outside the fuel level limits), and the PCM detected the vacuum level in the EVAP system decayed at too rapid of a rate in the test. **Possible Causes** ● Charcoal canister is loaded with fuel or moisture ● Fuel filler cap is loose, cross-threaded, damaged or wrong part ● Fuel tank, fuel filler neck or fuel sending unit 'O' ring is leaking ● Fuel tank pressure sensor is damaged, disconnected or it failed ● Fuel tank vapor line(s) is clogged, damaged or disconnected ● Purge valve vapor line is clogged, damaged, or disconnected ● Purge solenoid or Vent solenoid has a small leaking (sticking) ● PCM has failed

OBD II Trouble Code List (P0xxx Codes)

DTC	Trouble Code Title, Conditions & Possible Causes
P0442 **2T EVAP, MIL: Yes** 1996, 1997, 1998, 1999, 2000, 2001, 2002, 2003, 2004, 2005 Achieva, Beretta, Ciera, Cavalier, Century, Corsica, Envoy, Grand Am, Malibu, Skylark, Sunfire, S/T Blazer & S/T Pickup & TrailBlazer 2.2L VIN 4, 2.2L VIN 5, 2.4L VIN T, 2.2L VIN H, 4.2L VIN S Transmissions: All	**EVAP System Small Leak (0.040") Detected Conditions:** DTC P0107, P0108, P0112, P0113, P0116, P0117, P0118, P0122, P0123, P0452, or P0453 not set, engine started, ECT and IAT sensors from 38-86°F and within 16°F at startup, system voltage over 10.0v, BARO sensor more than 75 kPa, vehicle driven to a steady speed of less than 75 mph, fuel level from 15-85%, and the PCM detected the EVAP system achieved proper vacuum, but a vacuum decay condition was detected in the EVAP leak test. **Possible Causes** • Charcoal canister is loaded with fuel or moisture • Fuel filler cap is loose, cross-threaded, damaged or wrong part • Fuel tank, fuel filler neck or fuel sending unit 'O' ring is leaking • Fuel tank pressure sensor is damaged, disconnected or it failed • Fuel tank vapor line(s) is clogged, damaged or disconnected • Purge valve vapor line is clogged, damaged, or disconnected • Purge solenoid or Vent solenoid has a small leaking (sticking) • PCM has failed • TSB 01-06-04-007A contains a repair procedure for this code
P0442 **2T EVAP, MIL: Yes** 1997, 1998, 1999, 2000, 2001 Catera 3.0L V6 VIN R engine Transmissions: A/T	**EVAP System Small Leak (0.040") Detected Conditions:** DTC P0100, P0115, P0116, P0130-P0141, P0150-P0161, P0300, P0301-P0306, P0443, P0446, P0450, P0500, P0506, P0507, P0560, P1120 and P1220 not set, engine started, ECT sensor at startup from 17°F-212°F, IAT sensor at startup from 17°F-167°F, engine running at idle speed in closed loop, engine load less than 30%, MAF sensor less than 7.5 g/sec, system voltage over 10.0v, VSS sensor at 0 mph, engine runtime over 17 minutes, Secondary AIR System and Test inactive, Short Term fuel trim at the start of the test more than -5%, and the PCM detected the vacuum in the EVAP system decayed at too fast a rate during the EVAP leak test. **Possible Causes** • Charcoal canister is loaded with fuel or moisture • Fuel filler cap is loose, cross-threaded, damaged or wrong part • Fuel tank, fuel filler neck or fuel sending unit 'O' ring is leaking • Fuel tank pressure sensor is damaged, disconnected or it failed • Fuel tank vapor line(s) is clogged, damaged or disconnected • Purge valve vapor line is clogged, damaged, or disconnected • Purge solenoid or Vent solenoid has a small leaking (sticking) • PCM has failed
P0443 **2T CCM, MIL: Yes** 1996, 1997 Camaro, Caprice, Corvette, Firebird, Fleetwood, Roadmaster 4.3L VIN W, 5.7L VIN 5, 5.7L VIN P engines Transmissions: All	**EVAP Purge Solenoid Control Circuit Malfunction Conditions:** Engine started; system voltage over 10.0v and the PCM detected the Actual and Commanded state of the Purge Solenoid driver control circuit did not match for 2.5 seconds. **Possible Causes** • Purge solenoid control circuit is open or shorted to ground • Purge solenoid control circuit is shorted to system power (B+) • Purge solenoid power circuit is open (test ACTUATORS fuse) • Purge solenoid is damaged or has failed • PCM has failed
P0443 **2T CCM, MIL: Yes** 1996, 1997, 1998, 1999, 2000, 2001, 2002, 2003, 2004, 2005 Aztek, Montana, Silhouette, Rendezvous & Venture 3.1L VIN J, 3.1L VIN M, 3.4L VIN E engines Transmissions: All	**EVAP Purge Solenoid Control Circuit Malfunction Conditions:** Engine started; system voltage over 10.0v and the PCM detected an unexpected voltage condition on the Purge Solenoid control circuit for over 30 seconds during the CCM test. **Possible Causes** • Purge solenoid control circuit is open or shorted to ground • Purge solenoid control circuit is shorted to system power (B+) • Purge solenoid power circuit is open (check the IGN1 UH fuse) • Purge solenoid is damaged or has failed • PCM has failed
P0443 **2T CCM, MIL: Yes** 1996, 1997, 1998, 1999, 2000, 2001, 2002, 2003, 2004, 2005 Cab & Chassis, C/K Truck, G Series Van, L/M Series Van, S/T Blazer & Pickup 4.3L VIN W, 4.3L VIN X, 5.0L VIN M, 5.7L VIN R Transmissions: All	**EVAP Purge Solenoid Control Circuit Malfunction Conditions:** Engine started; system voltage from 6-18v, and the PCM detected the Actual and Commanded state of the EVAP Purge solenoid driver control circuit did not match for over 5 seconds during the CCM test. **Possible Causes** • Purge solenoid control circuit is open or shorted to ground • Purge solenoid control circuit is shorted to system power (B+) • Purge solenoid power circuit is open (test the ENG1 fuse) • Purge solenoid is damaged or has failed • PCM has failed

OBD II Trouble Code List (P0xxx Codes)

DTC	Trouble Code Title, Conditions & Possible Causes
P0443 **2T CCM, MIL: Yes** 1999, 2000, 2001, 2002, 2003, 2004, 2005 Aurora, Bonneville, Century, Cutlass Supreme, Grand Prix, Impala, Intrigue, LeSabre, LSS, Lumina, Monte Carlo, Park Avenue, 88' & 98' 3.1L VIN J, 3.1L VIN M, 3.4L VIN E engines, 3.5L VIN H, 3.8L VIN 1, 3.8L VIN K engines Transmissions: All	**EVAP Purge Solenoid Control Circuit Malfunction Conditions:** Engine started; system voltage over 10.0v and the PCM detected the Actual state and the Commanded state of the Purge Solenoid driver control circuit did not match for 30 seconds. An ignition voltage is supplied directly to the EVAP canister purge solenoid valve. The EVAP canister purge solenoid is driven by a pulse width modulated (PWM) signal. The Scan Tool displays the amount of signal on-time as a percentage. The PCM monitors the status of the solenoid driver. The PCM controls the EVAP canister purge valve on-time by grounding the control circuit via an internal switch called a driver. If the PCM detects an incorrect voltage for the commanded state of the driver, it will set this trouble code (P0443). **Possible Causes** ● Purge solenoid control circuit is open or shorted to ground ● Purge solenoid control circuit is shorted to system power (B+) ● Purge solenoid power circuit is open (test the IGN1 fuse) ● Purge solenoid is damaged or has failed ● PCM has failed
P0443 **2T CCM, MIL: Yes** 1999, 2000, 2001, 2002, 2003, 2004, 2005 C/K Truck, Escalade, G Series Van, Cargo Van, Envoy & TrailBlazer 4.8L VIN V, 5.3L VIN P, 5.3L VIN T, 5.3L VIN Z, 6.0L VIN N, 6.0L VIN U, 8.1L VIN G engines Transmissions: All	**EVAP Purge Solenoid Control Circuit Malfunction Conditions:** Engine started; system voltage at 6-18v, and the PCM detected the Actual and Commanded state of the EVAP Purge solenoid driver control circuit did not match for over 5 seconds during the CCM test period. **Possible Causes** ● Purge solenoid control circuit is open or shorted to ground ● Purge solenoid control circuit is shorted to system power (B+) ● Purge solenoid power circuit is open (test the ENG1 fuse) ● Purge solenoid is damaged or has failed ● PCM has failed
P0443 **1T CCM, MIL: Yes** 2003, 2004, 2005 CTS (Cadillac) 2.6L VIN M, 3.2L VIN N Transmissions: All	**EVAP Purge Solenoid Control Circuit Malfunction Conditions:** Engine started; engine speed over 400 rpm, system voltage from 8-18v, and the PCM detected 3.5v on the EVAP purge valve control circuit with the control driver commanded "off" for over 50 seconds during the CCM test period. **Possible Causes** ● Purge solenoid connector is damaged or open ● Purge solenoid control circuit is open ● Purge solenoid power circuit is open (test the Manifold fuse) ● Purge solenoid is damaged or it has failed ● PCM has failed
P0443 **2T CCM, MIL: Yes** 2000, 2001, 2002, 2003, 2004, 2005 Aurora, DeVille, Eldorado & Seville 4.0L VIN C engine, 4.6L VIN 9, 4.6L VIN Y Transmissions: A/T	**EVAP Solenoid No. 1 Control Circuit Malfunction Conditions:** Engine started; system voltage over 10.0v and the PCM detected an unexpected voltage condition on the Purge Solenoid Valve No. 1 driver control circuit for over 10 seconds. **Possible Causes** ● Purge Solenoid 1 control circuit is open or shorted to ground ● Purge Solenoid 1 control circuit is shorted to system power ● Purge Solenoid 1 power circuit is open (test RUN/CRANK fuse) ● Purge Solenoid 1 is damaged or has failed ● PCM has failed
P0443 **2T CCM, MIL: Yes** 1998, 1999, 2000, 2001, 2002, 2003, 2004, 2005 Camaro, Corvette & Firebird 3.8L VIN K, 5.7L VIN G, 5.7L VIN S Transmissions: All	**EVAP Purge Solenoid Control Circuit Malfunction Conditions:** Engine started; system voltage over 10.0v and the PCM detected the Commanded and Actual state of the Purge Solenoid driver control circuit did not match for over 5 seconds. **Possible Causes** ● Purge solenoid control circuit is open or shorted to ground ● Purge solenoid control circuit is shorted to system power (B+) ● Purge solenoid power circuit is open (test ENG CTRL fuse) ● Purge solenoid is damaged or has failed ● PCM has failed
P0443 **2T CCM, MIL: Yes** 1997, 1998, 1999, 2000, 2001 Catera 3.0L V6 VIN R engine Transmissions: A/T	**EVAP Purge Valve Control Circuit Malfunction Conditions:** Engine started; system voltage over 10.0v and the PCM detected an unexpected voltage condition on the Purge Solenoid control circuit under these operating conditions. **Possible Causes** ● Purge solenoid control circuit is open or shorted to ground ● Purge solenoid control circuit is shorted to system power (B+) ● Purge solenoid power circuit is open (test power from the relay) ● Purge solenoid is damaged or has failed ● PCM has failed

OBD II Trouble Code List (P0xxx Codes)

DTC	Trouble Code Title, Conditions & Possible Causes
P0444 **1T CCM, MIL: Yes** 2003, 2004, 2005 CTS (Cadillac) 2.6L VIN M, 3.2L VIN N Transmissions: All	**EVAP Purge Solenoid Control Circuit Low Input Conditions:** Engine started; engine speed over 400 rpm, system voltage from 8-18v, and the PCM detected 0v on the EVAP purge valve control circuit with the control driver commanded "off" for over 50 seconds. **Possible Causes** ● Purge solenoid connector is damaged or shorted ● Purge solenoid control circuit is shorted to ground ● PCM has failed
P0445 **1T CCM, MIL: Yes** 2003, 2004, 2005 CTS (Cadillac) 2.6L VIN M, 3.2L VIN N Transmissions: All	**EVAP Purge Solenoid Control Circuit High Input Conditions:** Engine started; engine speed over 400 rpm, system voltage from 8-18v, and the PCM detected 12v on the EVAP purge valve control circuit with the control driver commanded "on" for over 50 seconds. **Possible Causes** ● Purge solenoid connector is damaged or shorted ● Purge solenoid control circuit is shorted to system power (B+) ● Purge solenoid valve is damaged or it has failed ● PCM has failed
P0446 **2T EVAP, MIL: Yes** 1996, 1997, 1998, 1999, 2000, 2001, 2002, 2003, 2004, 2005 Aztek, Montana, Venture, Silhouette & Rendezvous 3.1L VIN J, 3.1L VIN M, 3.4L VIN E engines Transmissions: All	**EVAP Vent System Performance Conditions:** DTC P0107, P0108, P0112, P0113, P0116-P0118, P0121-P0123, P0125, P0440, P0442, P0443, P0449, P0452, P0453, P1106, P1107, P1112, P1114, P1115, P1121 and P1122 not set, engine started, system voltage over 10.0v, fuel level from 15-85%, BARO sensor over 75 kPa, ECT and IAT sensors from 39-86°F and within 16°F at startup, VSS under 75 mph, and the PCM detected the fuel tank pressure was less than -10" H2O for 30 seconds. **Possible Causes** ● EVAP vent fresh air hose is clogged, kinked or restricted ● EVAP Vent solenoid is contaminated, damaged or has failed ● FTP sensor is out-of-calibration, damaged or "skewed" ● PCM has failed
P0446 **2T EVAP, MIL: Yes** 1996, 1997, 1998, 1999, 2000, 2001, 2002, 2003, 2004, 2005 Aurora, Bonneville, Century, Cutlass Supreme, Grand Prix, Impala, Intrigue, LeSabre, LSS, Lumina, Monte Carlo, Park Avenue, Regal, 88' & 98' 3.1L VIN J, 3.1L VIN M, 3.4L VIN E engines, 3.5L VIN H, 3.8L VIN 1, 3.8L VIN K engines Transmissions: All	**EVAP Canister Purge Vent Blocked Conditions:** DTC P0107, P0108, P0112, P0113, P0116-P0118, P0121-P0125, P0440-P0453, P1106-P1107, P1111-P1115, P1121 and P0122 not set, engine started, ECT and IAT sensors from 39-86°F and within 16°F at startup, vehicle driven to a seed less than 75 mph, BARO sensor more than 75 kPa, fuel level from 15-85%, and the PCM detected the FTP sensor was less than -10" H2O for 30 seconds. The PCM tests the EVAP system for a restricted or blocked EVAP vent path. The PCM commands the EVAP canister purge solenoid open and the EVAP canister vent solenoid closed to allow vacuum to be applied to the EVAP system. Once a calibrated vacuum level is reached, the PCM commands the purge solenoid closed and the vent solenoid open, and monitors the FTP sensor for a decrease in vacuum. It expects it to read near 0 inches H2O in a calibrated time. **Possible Causes** ● EVAP vent fresh air hose is clogged, kinked or restricted ● EVAP Vent solenoid is contaminated, damaged or has failed ● FTP sensor is out-of-calibration, damaged or "skewed" ● PCM has failed
P0446 **2T EVAP, MIL: Yes** 1996, 1997, 1998, 1999, 2000, 2001, 2002, 2003, 2004, 2005 C/K Cab & Chassis, C/K Series Truck, G Series Van & Van Cargo, L/M Series Van, S/T Blazer & Pickup 4.3L VIN W, 4.3L VIN X, 5.0L VIN M, 5.7L VIN K, 5.7L VIN R engines Transmissions: All	**EVAP Vent System Performance Conditions:** DTC P0106, P0107, P0108, P0112, P0113, P0116-P0118, P0125, P0440-P0453, P1111-P1115, P1120, P1220 and P1221 not set, engine started, ECT and IAT sensors from 39-86°F and within 16°F at startup, BARO over 75 kPa, fuel level from 15-85%, vehicle driven to a speed of less than 75 mph, and the PCM detected the fuel tank pressure sensor indicated less than -10 inches H2O for 20 seconds. **Possible Causes** ● EVAP vent fresh air hose is clogged, kinked or restricted ● EVAP Vent solenoid is contaminated, damaged or has failed ● EVAP Canister plugged or severely restricted ● Fuel Cap or EVAP Service Port leaking ● Fuel vapor lines or purge lines damaged or leaking ● FTP sensor is out-of-calibration, damaged or "skewed" ● PCM has failed ● TSB 02-06-04-037 contains a repair procedure for this code

OBD II Trouble Code List (P0xxx Codes)

DTC	Trouble Code Title, Conditions & Possible Causes
P0446 **2T EVAP, MIL: Yes** 1999, 2000, 2001, 2002, 2003, 2004, 2005 C/K Truck, Escalade, G Series Van, Cargo Van, Envoy & TrailBlazer 4.8L VIN V, 5.3L VIN P, 5.3L VIN T, 5.3L VIN Z, 6.0L VIN N, 6.0L VIN U, 8.1L VIN G engines Transmissions: All	**EVAP Vent System Performance Conditions:** DTC P0106, P0107, P0108, P0112, P0113, P0116, P0117, P0118, P0120, P0121, P0122, P0123, P0125, P0131, P0132, P0133, P0134, P0135, P0137, P0138, P0140, P0141, P0147, P0151, P0152, P0153, P0154, P0155, P0157, P0158, P0160, P0161, P0167, P0220, P0442, P0443, P0449, P0452, P0453, P0455, P0502, P0503, P1111, P1112, P1114, P1115, P1120 are not set, engine started, ECT and IAT sensors from 39-86°F and within 16°F at startup, BARO sensor over 75 kPa, system voltage from 10-18v, fuel level from 15-85%, and the PCM detected the fuel tank pressure sensor was less than -10 inches H2O for 30 seconds. **Possible Causes** ● EVAP vent fresh air hose is clogged, kinked or restricted ● EVAP Vent solenoid is contaminated, damaged or has failed ● EVAP Canister plugged or severely restricted ● Fuel Cap or EVAP Service Port leaking ● Fuel vapor lines or purge lines damaged or leaking ● FTP sensor is out-of-calibration, damaged or "skewed" ● PCM has failed ● TSB 02-06-04-037 contains a repair procedure for this code
P0446 **1T EVAP, MIL: Yes** 2003, 2004, 2005 CTS (Cadillac) 2.6L VIN M, 3.2L VIN N Transmissions: All	**EVAP Vent Solenoid Vent Path Restricted Conditions:** DTC P0112, P0113, P0116-P0118, P0121-P0123, P0125, P0440, P0442, P0443, P0444, P0445, P0447, P0448, P0449, P0451, P0452, P0453 and P0496 are not set, engine started, BARO sensor over 75 kPa, system voltage from 10-18v, fuel level from 15-85%, ECT sensor from 39-149°F, IAT sensor from 39-167°F, ECT and IAT sensors within 16°F at startup, VSS at 0 mph, and the PCM detected the FTP sensor was less than -10 H2O for 30 seconds. **Possible Causes** ● Charcoal canister fresh air inlet is blocked or restricted ● EVAP vent fresh air hose is clogged, kinked or restricted ● EVAP vent solenoid is damaged or it has failed ● FTP sensor is out-of-calibration, damaged or "skewed" ● PCM has failed
P0446 **2T EVAP, MIL: Yes** 1998, 1999, 2000, 2001, 2002, 2003, 2004, 2005 DeVille, Eldorado, Seville 4.6L VIN 9, 4.6L VIN Y Transmissions: A/T	**EVAP Vent System Performance Conditions:** DTC P0107, P0108, P0112, P0113, P0116, P0117, P0118, P0121, P0122, P0123, P0125, P0440, P0442, P0443, P0449, P0452, P0453, P1106. P1107, P1111, P1112, P1114, P1115, P1121, or P1122 not set, engine started, ECT and IAT sensors from 39-86°F and within 16°F at startup, BARO sensor more than 72 kPa, FLI sensor indicating 15-85%, vehicle driven to a speed of less than 75 mph, and the PCM detected the fuel tank pressure sensor was less than -10 inches H2O for 20 seconds during the Vent System test. **Possible Causes** ● EVAP vent fresh air hose is clogged, kinked or restricted ● EVAP Vent solenoid is contaminated, damaged or has failed ● FTP sensor is out-of-calibration, damaged or "skewed" ● PCM has failed
P0446 **1T EVAP, MIL: Yes** 1998, 1999, 2000, 2001, 2002, 2003, 2004, 2005 Camaro, Corvette & Firebird 5.7L VIN G, 5.7L VIN S Transmissions: All	**EVAP Vent System Performance Conditions:** DTC P0107. P0108, P0112, P0113, P0116, P0117, P0118, P0121, P0122, P0123, P0125, P0440, P0442, P0443, P0449, P0452, P0453, P1111, P1112, P1114, P1115, P1121, and P1122 not set, engine started, ECT and IAT sensors from 39-86°F and within 16°F at startup, system voltage over 10.0v, BARO sensor over 75 kPa, fuel level from 15-85%, vehicle driven to a speed of less than 75 mph, and the PCM detected the fuel tank pressure (from the FTP sensor signal) was less than -10" H2O for up to 30 seconds. *Note: This trouble code does not report a first failed test. A first fail of this code will show the Scan Tool status as "Not Run".* **Possible Causes** ● EVAP vent fresh air hose is clogged, kinked or restricted ● EVAP Vent solenoid is contaminated, damaged or has failed ● FTP sensor is out-of-calibration, damaged or "skewed" ● PCM has failed
P0446 **2T EVAP, MIL: Yes** 1996, 1997, 1998, 1999, 2000, 2001, 2002 Camaro & Firebird 3.8L VIN K engine Transmissions: All	**EVAP Vent System Performance Conditions:** DTC P0107, P0108, P0112, P0113, P0116-P0118, P0125, P0443, P0449, P0452, P0453, P1106, P1107, P1112, P1114, P1120 and P1221 not set, engine started, ECT and IAT sensors from 39-86°F and within 16°F at startup, BARO sensor over 75 kPa, fuel level from 15-85%, vehicle driven at a steady 75 mph, and the PCM detected the fuel tank pressure was less than -10" H2O for 30 seconds. **Possible Causes** ● EVAP vent fresh air hose is clogged, kinked or restricted ● EVAP Vent solenoid is contaminated, damaged or has failed ● FTP sensor is out-of-calibration, damaged or "skewed" ● PCM has failed

OBD II Trouble Code List (P0xxx Codes)

DTC	Trouble Code Title, Conditions & Possible Causes
P0446 1T EVAP, MIL: Yes 1998, 1999, 2001, 2002, 2003, 2004, 2005 Aurora 4.0L VIN C engine Transmissions: A/T	**EVAP Vent System Performance Conditions:** DTC P0107, P0108, P0112, P0113, P0117, P0118, P0122, P0123, P0125, P0131-P0135, P0141, P0151-P0154, P0452, P0453, P0503, P1111-P1115, P1133 and P1153 not set, engine started, ECT and IAT sensors from 35-86°F and within 16°F at startup, BARO sensor over 72 kPa, fuel level from 10-90% for at least 3.2 seconds (this time limit is needed to prevent fuel sloshing in the tank from causing the fuel level to indicate an incorrect value), and the PCM detected the fuel tank pressure was less than -10" H2O for 30 seconds. A first fail of this trouble code will have a Scan Tool status as Not Run. **Possible Causes** ● EVAP vent fresh air hose is clogged, kinked or restricted ● EVAP Vent solenoid is contaminated, damaged or has failed ● FTP sensor is out-of-calibration, damaged or "skewed" ● PCM has failed
P0446 2T EVAP, MIL: Yes 1996, 1997, 1998, 1999, 2000, 2001, 2002, 2003, 2004, 2005 Achieva, Beretta, Ciera, Cavalier, Century, Corsica, Envoy, Grand Am, Malibu, Skylark, Sunfire, S/T Blazer & S/T Pickup & TrailBlazer 2.2L VIN 4, 2.2L VIN 5, 2.4L VIN T, 2.2L VIN H, 4.2L VIN S engines Transmissions: All	**EVAP Vent System Performance Conditions:** DTC P0106-P0108, P0112, P0113, P0117, P0118, P0122, P0123, P0131-P0134, P0452, P0453 and P1133 not set, ECT and IAT sensors from 39-86°F and within 12°F at startup, system voltage over 10.0v, BARO sensor over 75 kPa, fuel level from 15-85%, EVAP Purge solenoid operating at a 50% PWM with 65 seconds of startup, vehicle speed less than 75 mph, and the PCM detected the fuel tank pressure was less than -10" H2O for 30 seconds. **Possible Causes** ● EVAP vent fresh air hose is clogged, kinked or restricted ● EVAP Vent solenoid is contaminated, damaged or has failed ● EVAP Canister plugged or severely restricted ● Fuel Cap or EVAP Service Port leaking ● Fuel vapor lines or purge lines damaged or leaking ● FTP sensor is out-of-calibration, damaged or "skewed" ● PCM has failed ● TSB 61-65-31A contains a repair procedure for this code
P0446 2T EVAP, MIL: Yes 1997, 1998, 1999, 2000, 2001 Catera 3.0L V6 VIN R engine Transmissions: A/T	**EVAP Purge Vent Valve Circuit Malfunction Conditions:** Engine started; system voltage over 10.0v and the PCM detected an unexpected voltage on the Purge Vent Valve control circuit. **Possible Causes** ● Vent solenoid control circuit is open or shorted to ground ● Vent solenoid control circuit is shorted to system power (B+) ● Vent solenoid power circuit is open (test power from the relay) ● Vent solenoid is damaged or has failed ● PCM has failed
P0447 2T CCM, MIL: Yes 2003, 2004, 2005 CTS (Cadillac) 2.6L VIN M, 3.2L VIN N Transmissions: All	**EVAP Vent Solenoid Control Circuit Low Input Conditions:** Engine started; engine speed over 400 rpm, system voltage from 8-18v, and the PCM detected 0v on the EVAP vent valve control circuit with the control driver commanded "off" for over 50 seconds. **Possible Causes** ● Vent solenoid connector is damaged or shorted ● Vent solenoid control circuit is shorted to ground ● PCM has failed
P0448 2T CCM, MIL: Yes 2003, 2004, 2005 CTS (Cadillac) 2.6L VIN M, 3.2L VIN N Transmissions: All	**EVAP Vent Solenoid Control Circuit High Input Conditions:** Engine started; engine speed over 400 rpm, system voltage from 8-18v, and the PCM detected 12v on the EVAP vent valve control circuit with the control driver commanded "on" for over 50 seconds. **Possible Causes** ● Vent solenoid connector is damaged or open ● Vent solenoid control circuit is shorted to power ● Vent solenoid is damaged or it has failed ● PCM has failed
P0449 2T CCM, MIL: Yes 1999, 2000, 2001, 2002, 2003, 2004, 2005 Aztek, Montana, Venture, Silhouette & Rendezvous 3.4L VIN E engine Transmissions: All ; Montana, Silhouette, Venture, Rendezvous	**EVAP Vent Solenoid Control Circuit Malfunction Conditions:** Engine started; system voltage from 6-18.0v, and the PCM detected the Actual and the Commanded state of the EVAP Vent Solenoid driver control circuit did not match for over 5 seconds during the CCM test period. **Possible Causes** ● Vent solenoid control circuit is open, shorted to ground or B+ ● Vent solenoid power circuit is open (test the ENG CTRL fuse) ● Vent solenoid is damaged or has failed ● PCM has failed

OBD II Trouble Code List (P0xxx Codes)

DTC	Trouble Code Title, Conditions & Possible Causes
P0449 **2T CCM, MIL: Yes** 1999, 2000, 2001, 2002, 2003, 2004, 2005 Aurora, Bonneville, Century, Cutlass Supreme, Grand Prix, Impala, Intrigue, LeSabre, LSS, Lumina, Monte Carlo, Park Avenue, Regal, 88' & 98' 3.1L VIN J, 3.1L VIN M, 3.4L VIN E engines, 3.5L VIN H, 3.8L VIN 1, 3.8L VIN K Transmissions: All	**EVAP Vent Solenoid Control Circuit Malfunction Conditions:** Engine started; system voltage over 10.0v and the PCM detected the Actual and Commanded state of the Vent Solenoid driver control circuit did not match for 30 seconds. **Possible Causes** ● Vent solenoid control circuit is open or shorted to ground ● Vent solenoid control circuit is shorted to system power (B+) ● Vent solenoid power circuit is open (test the VENT SOL fuse) ● Vent solenoid is damaged or has failed ● PCM has failed
P0449 **2T CCM, MIL: Yes** 1998, 1999, 2000, 2001, 2002, 2003, 2004, 2005 C/K Cab & Chassis, C/K Truck, G Van, L/M Series Van, S/T Blazer & Pickup 4.3L VIN W, 4.3L VIN X, 5.0L VIN M, 5.7L VIN R, 7.4L VIN J engines Transmissions: All	**EVAP Vent Solenoid Control Circuit Malfunction Conditions:** Engine started; system voltage from 6-18v, and the PCM detected the Actual and Commanded state of the Vent Solenoid driver control circuit did not match for over 5 seconds. **Possible Causes** ● Vent solenoid control circuit is open or shorted to ground ● Vent solenoid control circuit is shorted to system power (B+) ● Vent solenoid power circuit is open (test the ENG1 fuse) ● Vent solenoid is damaged or has failed ● PCM has failed
P0449 **2T CCM, MIL: Yes** 1999, 2000, 2001, 2002, 2003, 2004, 2005 C/K Truck, Escalade, G Series Van, Cargo Van, Envoy & TrailBlazer 4.8L VIN V, 5.3L VIN P, 5.3L VIN T, 5.3L VIN Z, 6.0L VIN N, 6.0L VIN U, 8.1L VIN G engines Transmissions: All	**EVAP Vent Solenoid Control Circuit Malfunction Conditions:** Engine started; system voltage from 6-18v, and the PCM detected the Actual and Commanded state of the Vent Solenoid driver control circuit did not match for over 5 seconds during the CCM test period. **Possible Causes** ● Vent solenoid control circuit is open or shorted to ground ● Vent solenoid control circuit is shorted to system power (B+) ● Vent solenoid power circuit is open (test the ENG1 fuse) ● Vent solenoid is damaged or has failed ● PCM has failed
P0449 **2T CCM, MIL: Yes** 2000, 2001, 2002, 2003, 2004, 2005 Aurora, DeVille, Eldorado & Seville 4.0L VIN C engine, 4.6L VIN 9, 4.6L VIN Y Transmissions: All	**EVAP Vent Solenoid Control Circuit Malfunction Conditions:** Engine started; system voltage over 10.0v and the PCM detected the Commanded and Actual state of the Vent Solenoid driver control circuit did not match for over 5 seconds. **Possible Causes** ● Vent solenoid control circuit is open or shorted to ground ● Vent solenoid control circuit is shorted to system power (B+) ● Vent solenoid power circuit is open (test the VENT SOL fuse) ● Vent solenoid is damaged or has failed ● PCM has failed
P0449 **1T CCM, MIL: Yes** 2003, 2004, 2005 CTS (Cadillac) 2.6L VIN M, 3.2L VIN N Transmissions: All	**EVAP Vent Solenoid Control Circuit Malfunction Conditions:** Engine started; engine speed over 400 rpm, system voltage from 8-18v, and the PCM detected 3.5v on the EVAP vent valve control circuit with the control driver commanded "off" for over 50 seconds. An ignition voltage is supplied to the EVAP canister vent valve. The ECM grounds the EVAP canister vent valve control circuit to close the valve by means of an internal switch called a driver. The Scan Tool displays the commanded state of the EVAP canister vent valve as on or off. The control module monitors the status of the driver **Possible Causes** ● Vent solenoid connector is damaged or open ● Vent solenoid control circuit is open ● Vent solenoid power circuit is open (test the Manifold fuse) ● Vent solenoid is damaged or it has failed ● PCM has failed

OBD II Trouble Code List (P0xxx Codes)

DTC	Trouble Code Title, Conditions & Possible Causes
P0449 **2T CCM, MIL: Yes** 1999, 2000, 2001, 2002, 2003, 2004, 2005 Camaro, Corvette & Firebird 3.8L VIN K, 5.7L VIN G, 5.7L VIN S Transmissions: All	**EVAP Vent Solenoid Control Circuit Malfunction Conditions:** Engine started; system voltage over 10.0v and the PCM detected the Commanded and Actual state of the Vent Solenoid driver control circuit did not match for over 5 seconds. **Possible Causes** ● Vent solenoid control circuit is open, shorted to ground or B+ ● Vent solenoid power circuit is open (test the ENG CTRL fuse) ● Vent solenoid is damaged or has failed ● PCM has failed
P0450 **2T CCM, MIL: Yes** 1997, 1998, 1999, 2000, 2001 Catera 3.0L V6 VIN R engine Transmissions: A/T	**Fuel Tank Pressure Sensor Circuit Malfunction Conditions:** Engine started; and the PCM detected that pressure in the fuel tank was more than 12.05" H2O (22.50 mm Hg) for 5 seconds, or with the ECT sensor less than 91°F at startup, engine runtime at idle speed from 2-10 seconds, the PCM detected the FTP sensor indicated more than -11.80" H2O for 5 seconds, or that the pressure in the fuel tank was more than 3.76" H2O for 3 seconds during the test. **Possible Causes** ● FTP sensor signal circuit is open, shorted to ground or to VREF ● FPT sensor VREF is open or shorted to ground (is P0100 set?) ● FTP sensor is contaminated, damaged, skewed or has failed ● PCM has failed
P0451 **2T CCM, MIL: Yes** 2003, 2004, 2005 CTS (Cadillac) 2.6L VIN M, 3.2L VIN N Transmissions: All	**Fuel Tank Pressure Sensor Signal Range/Performance Conditions:** Engine started; ECT sensor less than 95°F, system voltage from 8-18v, and the PCM detected the FTP sensor was less than 0.78v, or over 2.94v within 12 seconds of startup. **Possible Causes** ● FTP sensor signal circuit is open, shorted to ground or to VREF ● FPT sensor VREF is open or shorted to ground ● FTP sensor is contaminated, damaged, skewed or has failed ● PCM has failed
P0452 **2T CCM, MIL: Yes** 1998, 1999, 2000, 2001, 2002, 2003, 2004, 2005 Aurora, Bonneville, Century, Cutlass Supreme, Grand Prix, Intrigue, LeSabre, LSS, Lumina, Monte Carlo, 88', Regal, Park Avenue & 98' 3.1L VIN J, 3.1L VIN M, 3.4L VIN E, 3.5L VIN H, 3.8L VIN 1, 3.8L VIN K Transmissions: All	**Fuel Tank Pressure Sensor Circuit Low Input Conditions:** Key on or engine running; and the PCM detected an unexpected low voltage condition (less than 0.10v) on the Fuel Tank Pressure sensor signal circuit for 5 seconds. The FTP sensor measures the difference between the air pressure and vacuum in the EVAP system, and the outside air pressure. The PCM supplies a 5v VREF and a low reference circuit to the FTP sensor. The FTP sensor signal varies depending on EVAP system pressure or vacuum. **Possible Causes** ● FTP sensor signal circuit is open or shorted to ground ● FTP sensor VREF circuit is open or shorted to ground ● FTP sensor is damaged or has failed ● PCM has failed
P0452 **2T CCM, MIL: Yes** 1998, 1999, 2000, 2001, 2002, 2003, 2004, 2005 C/K Truck, G Series Van & Cargo Van, L/M Vans, S/T Blazer & S/T Pickup 4.3L VIN W, 4.3L VIN X, 4.8L VIN V, 5.0L VIN M, 5.3L VIN P, 5.3L VIN T, 5.3L VIN Z, 5.7L VIN R, 6.0L VIN N, 6.0L VIN U, 8.1L VIN G engines Transmissions: All	**Fuel Tank Pressure Sensor Circuit Low Input Conditions:** Key on or engine running; and the PCM detected the Fuel Tank Pressure (FTP) sensor circuit was less than 0.10v for 5 seconds during the CCM test period. **Possible Causes** ● FTP sensor connector is damaged or shorted ● FTP sensor signal circuit is open or shorted to ground ● FTP sensor VREF circuit is open or shorted to ground ● FTP sensor is damaged or has failed ● PCM has failed
P0452 **2T CCM, MIL: Yes** 2003, 2004, 2005 CTS (Cadillac) 2.6L VIN M, 3.2L VIN N Transmissions: All	**Fuel Tank Pressure Sensor Circuit Low Input Conditions:** Key on or engine running; and the PCM detected the FTP sensor signal was less than 0.10 for 50 seconds during the CCM test. **Possible Causes** ● FTP sensor connector is damaged or shorted ● FTP sensor signal circuit is shorted to ground ● FTP sensor is damaged or it has failed ● PCM has failed

OBD II Trouble Code List (P0xxx Codes)

DTC	Trouble Code Title, Conditions & Possible Causes
P0452 **2T CCM, MIL: Yes** 1998, 1999, 2000, 2001, 2002, 2003, 2004, 2005 DeVille, Eldorado, Seville 4.6L VIN 9, 4.6L VIN Y Transmissions: All	**Fuel Tank Pressure Sensor Circuit Low Input Conditions:** Key on or engine running; and the PCM detected the Fuel Tank Pressure (FTP) sensor signal was less than 0.1v for 5 seconds. **Possible Causes** ● FTP sensor signal circuit is open or shorted to ground ● FTP sensor VREF circuit is open or shorted to ground ● FTP sensor is damaged or has failed ● PCM has failed
P0452 **2T CCM, MIL: Yes** 1999, 2000, 2001, 2002, 2003, 2004, 2005 Camaro, Corvette & Firebird 5.7L VIN G, 5.7L VIN S Transmissions: All	**Fuel Tank Pressure Sensor Circuit Low Input Conditions:** Key on or engine running; and the PCM detected the Fuel Tank Pressure (FTP) sensor circuit was less than 0.1v for 5 seconds. **Possible Causes** ● FTP sensor signal circuit is open or shorted to ground ● FTP sensor VREF circuit is open or shorted to ground ● FTP sensor is damaged or has failed ● PCM has failed
P0452 **2T CCM, MIL: Yes** 1998, 1999, 2000, 2001, 2002 Camaro & Firebird 3.8L VIN K engine Transmissions: All	**Fuel Tank Pressure Sensor Circuit Low Input Conditions:** Key on or engine running; and the PCM detected an unexpected "low" voltage condition (less than 0.10v) on the Fuel Tank Pressure sensor signal circuit for 5 seconds in the CCM test. **Possible Causes** ● FTP sensor signal circuit is open or shorted to ground ● FTP sensor VREF circuit is open or shorted to ground ● FTP sensor is damaged or has failed ● PCM has failed
P0452 **2T CCM, MIL: Yes** 1998, 1999, 2001, 2002, 2003, 2004, 2005 Aurora 4.0L VIN C engine Transmissions: A/T	**Fuel Tank Pressure Sensor Circuit Low Input Conditions:** Key on or engine running; and the PCM detected an unexpected "low" voltage condition (less than 0.10v) on the Fuel Tank Pressure sensor signal circuit for 5 seconds in the CCM test. **Possible Causes** ● FTP sensor signal circuit is open or shorted to ground ● FTP sensor VREF circuit is open or shorted to ground ● FTP sensor is damaged or has failed ● PCM has failed
P0452 **2T CCM, MIL: Yes** 1998, 1999, 2000, 2001, 2002, 2003, 2004, 2005 Achieva, Beretta, Ciera, Cavalier, Century, Corsica, Grand Am, Malibu, Skylark, Sunfire, S/T Blazer & S/T Pickup 2.2L VIN 4, 2.2L VIN 5, 2.4L VIN T, 2.2L VIN H, 4.2L VIN S engines Transmissions: All	**Fuel Tank Pressure Sensor Circuit Low Input Conditions:** Key on or engine running; and the PCM detected an unexpected "low" voltage condition (less than 0.10v) on the Fuel Tank Pressure sensor signal circuit for 5 seconds in the CCM test. **Possible Causes** ● FTP sensor signal circuit is open or shorted to ground ● FTP sensor VREF circuit is open or shorted to ground ● FTP sensor is damaged or has failed ● PCM has failed
P0453 **2T CCM, MIL: Yes** 1998, 1999, 2000, 2001, 2002, 2003, 2004, 2005 Aurora, Bonneville, Century, Cutlass Supreme, Grand Prix, Intrigue, LeSabre, LSS, Lumina, Monte Carlo, 88', Regal, Park Avenue & 98' 3.1L VIN J, 3.1L VIN M, 3.4L VIN E engines, 3.5L VIN H, 3.8L VIN 1, 3.8L VIN K engines Transmissions: All	**Fuel Tank Pressure Sensor Circuit High Input Conditions:** Key on or engine running; and the PCM detected the Fuel Tank Pressure (FTP) sensor signal was over 4.90v for 5 seconds during the CCM test period. The FTP sensor measures the difference between the air pressure and vacuum in the EVAP system. The PCM supplies a 5v VREF and a low reference circuit to the FTP sensor. The FTP sensor signal varies depending on EVAP system pressure or vacuum. **Possible Causes** ● FTP sensor signal circuit is shorted to VREF or system power ● FTP sensor ground circuit is open between sensor and PCM ● FTP sensor is damaged or has failed ● PCM has failed

OBD II Trouble Code List (P0xxx Codes)

DTC	Trouble Code Title, Conditions & Possible Causes
P0453 **2T CCM, MIL: Yes** 1998, 1999, 2000, 2001, 2002, 2003, 2004, 2005 C/K Cab & Chassis, C/K Truck, G Van, L/M Van, S/T Blazer & S/T Pickup 4.3L VIN W, 4.3L VIN X, 4.8L VIN V, 5.0L VIN M, 5.3L VIN P, 5.3L VIN T, 5.3L VIN Z, 5.7L VIN K, 5.7L VIN R, 6.0L VIN N, 6.0L VIN U, 8.1L VIN G Transmissions: All	**Fuel Tank Pressure Sensor Circuit High Input Conditions:** Key on or engine running; and the PCM detected the Fuel Tank Pressure (FTP) sensor circuit was more than 4.90v for 5 seconds during the CCM test period. **Possible Causes** ● FTP sensor connector is damaged, loose or open ● FTP sensor signal circuit is shorted to VREF (5v) ● FTP sensor ground circuit is open between sensor and PCM ● FTP sensor is damaged or has failed ● PCM has failed
P0453 **2T CCM, MIL: Yes** 2003, 2004, 2005 CTS (Cadillac) 2.6L VIN M, 3.2L VIN N Transmissions: All	**Fuel Tank Pressure Sensor Circuit High Input Conditions:** Key on or engine running; and the PCM detected the FTP sensor signal was less than 4,900.10 for 50 seconds during the CCM test. **Possible Causes** ● FTP sensor connector is damaged or open ● FTP sensor signal circuit is open or shorted to 5v VREF circuit ● FTP sensor 5v VREF circuit shorted to low reference circuit ● FTP sensor is damaged or it has failed ● PCM has failed
P0453 **2T CCM, MIL: Yes** 1998, 1999, 2000, 2001, 2002, 2003, 2004, 2005 DeVille, Eldorado, Seville 4.6L VIN 9, 4.6L VIN Y Transmissions: All	**Fuel Tank Pressure Sensor Circuit High Input Conditions:** Key on or engine running; and the PCM detected an unexpected high voltage condition (more than 4.90v) on the Fuel Tank Pressure sensor signal circuit for 5 seconds in the test. **Possible Causes** ● FTP sensor signal circuit is shorted to VREF or system power ● FTP sensor ground circuit is open between sensor and PCM ● FTP sensor is damaged or has failed ● PCM has failed
P0453 **2T CCM, MIL: Yes** 1999, 2000, 2001, 2002, 2003, 2004, 2005 Camaro, Corvette & Firebird 5.7L VIN G, 5.7L VIN S Transmissions: All	**Fuel Tank Pressure Sensor Circuit High Input Conditions:** Key on or engine running; and the PCM detected an unexpected "high" voltage condition (more than 4.90v) on the Fuel Tank Pressure sensor signal circuit for 5 seconds in the test. **Possible Causes** ● FTP sensor signal circuit is shorted to VREF or system power ● FTP sensor ground circuit is open between sensor and PCM ● FTP sensor is damaged or has failed ● PCM has failed
P0453 **2T CCM, MIL: Yes** 1998, 1999, 2000, 2001, 2002 Camaro & Firebird 3.8L VIN K engine Transmissions: All	**Fuel Tank Pressure Sensor Circuit High Input Conditions:** Key on or engine running; and the PCM detected an unexpected "high" voltage condition (more than 4.90v) on the Fuel Tank Pressure sensor signal circuit for 5 seconds in the test. **Possible Causes** ● FTP sensor signal circuit is shorted to VREF or system power ● FTP sensor ground circuit is open between sensor and PCM ● FTP sensor is damaged or has failed ● PCM has failed
P0453 **2T CCM, MIL: Yes** 1998, 1999, 2001, 2002, 2003, 2004, 2005 Aurora 4.0L VIN C engine Transmissions: A/T	**Fuel Tank Pressure Sensor Circuit High Input Conditions:** Key on or engine running; and the PCM detected an unexpected "high" voltage condition (more than 4.90v) on the Fuel Tank Pressure sensor signal circuit for 5 seconds in the test. **Possible Causes** ● FTP sensor signal circuit is open or shorted to VREF or system power ● FTP sensor is damaged or has failed ● PCM has failed
P0453 **2T CCM, MIL: Yes** 1998, 1999, 2000, 2001, 2002, 2003, 2004, 2005 Cavalier, Grand Am, Malibu, Skylark, Sunfire, Envoy & TrailBlazer, S/T Blazer & S/T Pickup 2.2L VIN 4, 2.2L VIN 5, 2.4L VIN T, 2.2L VIN H, 4.2L VIN S Transmissions: All	**Fuel Tank Pressure Sensor Circuit High Input Conditions:** Key on or engine running; and the PCM detected an unexpected "high" voltage condition (more than 4.90v) on the Fuel Tank Pressure sensor signal circuit for 5 seconds in the test. **Possible Causes** ● FTP sensor signal circuit is shorted to VREF or system power ● FTP sensor ground circuit is open between sensor and PCM ● FTP sensor is damaged or has failed ● PCM has failed

OBD II Trouble Code List (P0xxx Codes)

DTC	Trouble Code Title, Conditions & Possible Causes
P0455 **2T EVAP, MIL: Yes** 1997, 1998, 1999, 2000, 2001 Catera 3.0L V6 VIN R engine Transmissions: A/T	**EVAP System Large Leak (0.080") Detected Conditions:** DTC P0100, P0115, P0116, P0130-P0161, P0300-P0306, P0443, P0446, P0450, P0500-P0507, P0560, P1120 and P1220 not set, engine started, ECT and IAT sensors from 17-167°F at startup, engine load less than 30%, engine runtime over 17 minutes, VSS at 0 mph, MAF sensor less than 7.5 g/sec, FTP signal less than 4" H2O, Air pump off, Short Term fuel trim over -5%, and the PCM detected it was unable to achieve a vacuum in the EVAP system, or it took too long to achieve proper vacuum in the EVAP system. **Possible Causes** ● Fuel filler cap is very loose, missing or the wrong part ● Fuel tank, fuel filler neck or fuel sending unit 'O' ring is leaking ● Fuel tank pressure sensor is damaged, disconnected or it failed ● Fuel tank vapor line(s) is clogged, damaged or disconnected ● Purge valve vapor line is clogged, damaged, or disconnected ● Purge solenoid is not opening (it may be damaged or sticking) ● Vent solenoid is not closing (it may be damaged or sticking) ● PCM has failed
P0455 **2T EVAP, MIL: Yes** 2003, 2004, 2005 C/K Series Truck, G Series Van & Cargo Van, L/M Series Van, Escalade, Envoy & TrailBlazer, S/T Series Blazer & S/T Pickup 4.3L VIN W, 4.3L VIN X, 4.8L VIN V, 5.3L VIN P, 5.3L VIN T, 5.3L VIN Z, 6.0L VIN N, 6.0L VIN U, 8.1L VIN G engines Transmissions: All	**EVAP System Large Leak (0.080") Detected Conditions:** DTC P0106-P0108, P0112, P0113, P0116-P0118, P0120-P0123, P0125, P0131-P0138, P0140, P0141, P0147, P0151-P0158, P0160, P0161, P0167, P0220, P0442-P0443, P0449, P0452-P0453, P0455, P0502, P0503, P1111, P1112, P1114, P1115, P1120 not set, engine started, ECT and IAT sensors from 39-167°F and within 16°F at startup, system voltage from 10-18v, BARO sensor more than 75 kPa, Fuel Level from 15-85%, and the PCM detected it was unable to achieve or maintain vacuum during the EVAP system. The PCM monitors the FTP sensor signal to determine the EVAP system vacuum level. Once conditions are correct, the PCM commands the Purge valve open and the EVAP vent valve closed to allow engine vacuum to enter the system. After a calibrated time or vacuum level, the PCM commands the Purge valve closed to seal the system, and monitors the FTP sensor to determine the EVAP system vacuum level. If the system is unable to achieve the correct vacuum level, or the vacuum level decreases too rapidly, the PCM will set this code. **Possible Causes** ● Fuel filler cap is very loose, missing or the wrong part ● Fuel tank, fuel filler neck or fuel sending unit 'O' ring is leaking ● Fuel tank pressure sensor is damaged, disconnected or it failed ● Fuel tank vapor line(s) is clogged, damaged or disconnected ● Purge valve vapor line is clogged, damaged, or disconnected ● Purge solenoid is not opening (it may be damaged or sticking) ● Vent solenoid is not closing (it may be damaged or sticking) ● PCM has failed
P0461 **1T CCM, MIL: No** 1998, 1999, 2000 C/K Series Truck, G Series Van & Cargo Van, L/M Series Van, Escalade, Envoy & TrailBlazer, S/T Series Blazer & S/T Pickup 4.3L VIN W, 4.3L VIN X, 4.8L VIN V, 5.3L VIN P, 5.3L VIN T, 5.3L VIN Z, 6.0L VIN N, 6.0L VIN U, 8.1L VIN G engines Transmissions: All	**Fuel Level Sensor Range/Performance Conditions:** Engine started; fuel level indicating from 15-85%, Fuel Tank Level Slosh Test and Fuel Level Main Test completed, then vehicle driven for a distance of 200 miles, and the PCM did not detect a change (decrease) in the fuel level under these conditions. **Possible Causes** ● Fuel level sensor signal circuit is open or shorted to ground between the fuel level sensor and the PCM connector ● Fuel level sensor ground circuit is open or has high resistance ● Fuel level sender is damaged, binding or not aligned properly ● Instrument cluster or the PCM has failed
P0461 **1T CCM, MIL: No** 1998, 1999, 2000, 2001, 2002, 2003, 2004, 2005 Aurora, DeVille, Eldorado & Seville 4.0L VIN C engine, 4.6L VIN 9, 4.6L VIN Y Transmissions: A/T	**Fuel Level Sensor Signal Range/Performance Conditions:** DTC P0502 and P0503 not set, engine started, vehicle driven at a speed of over 3 mph for a distance of over 37.5 miles, and the PCM detected a Fuel Level sensor signal that indicated the fuel volume did not change under these conditions. **Possible Causes** ● Fuel level sensor signal circuit is open or shorted to ground between the fuel level sensor and the PCM connector ● Fuel level sensor ground circuit is open or has high resistance ● Fuel level sender is damaged, binding or not aligned properly ● Instrument cluster or the PCM has failed

OBD II Trouble Code List (P0xxx Codes)

DTC	Trouble Code Title, Conditions & Possible Causes
P0461 **1T CCM, MIL: No** 1997, 1998, 1999, 2000 Camaro, Corvette & Firebird 3.8L VIN K, 5.7L VIN G, 5.7L VIN G Transmissions: All	**Fuel Level Sensor Signal Range/Performance Conditions:** Engine started; Fuel Level Sensor indicating over 9.1 gallons (34.4 liters) of fuel present, then the vehicle driven for 150 accumulated miles, and the PCM detected the fuel tank level did not decrease by at least 3/4 gallon (3 liters) under these conditions. **Possible Causes** ● Fuel level sensor signal circuit is open or shorted to ground between the fuel level sensor and the PCM connector ● Fuel level sensor ground circuit is open or has high resistance ● Fuel level sender is damaged, binding or not aligned properly ● Instrument cluster or the PCM has failed
P0461 **1T CCM, MIL: No** 1996, 1997, 1998, 1999, 2000 Achieva, Beretta, Ciera, Cavalier, Century, Corsica, Grand Am, Malibu, Skylark, Sunfire, S/T Blazer & S/T Pickup 2.2L VIN 4, 2.4L VIN T Transmissions: All	**Fuel Level Sensor Range/Performance Conditions:** Engine started; and the PCM detected Fuel Level Sensor indicated the fuel level did not change more than 1.6% (4 counts) after the vehicle was driven over 120 miles during the test. *Note: DTC P1601 may be along with this particular trouble code.* **Possible Causes** ● If DTC U1016 sets along with this trouble code, diagnose the cause of this code (a serial data fault) prior to testing for P0460. ● Fuel level sensor is damaged, out-of-range or "skewed" ● Instrument cluster or the PCM has failed ● TSB 01-06-04-026 contains a repair procedure for this code
P0461 **1T CCM, MIL: No** 1996, 1997, 1998, 1999, 2000 S/T Blazer & S/T Pickup 4.3L VIN W, 4.3L VIN X Transmissions: All	**Fuel Level Sensor Range/Performance Conditions:** Engine started; fuel level indicating from 15-85%, Fuel Tank Level Slosh Test and Fuel Level Main Test completed, then vehicle driven for a distance of 200 miles, and the PCM did not detect a change (decrease) in the fuel level under these conditions. **Possible Causes** ● Fuel level sensor signal circuit is open or shorted to ground between the fuel level sensor and the PCM connector ● Fuel level sensor ground circuit is open or has high resistance ● Fuel level sender is damaged, binding or not aligned properly ● Instrument cluster or the PCM has failed
P0462 **1T CCM, MIL: No** 1998, 1999, 2000 Century, Grand Prix, Impala, Intrigue, Lumina, Monte Carlo, Regal 3.1L VIN M, 3.8L VIN K, 3.8L VIN 1 engines Transmissions: All	**Fuel Level Sensor Low Input Conditions:** Engine started; system voltage over 10.0v and the PCM was notified over the Class 2 Bus that the IPC had detected a Fuel Level sensor signal that was less than 0.40v for 10-20 seconds. **Possible Causes** ● Fuel level sensor signal circuit is shorted to ground between the fuel level sensor and the PCM connector ● Fuel level sender is damaged, binding or not aligned properly ● Instrument cluster or the PCM has failed
P0462 **1T CCM, MIL: No** 1998, 1999, 2000 C/K Truck, Escalade, G Van & Cargo Van, L/M Van, S/T Blazer & Pickup, Envoy & TrailBlazer 4.3L VIN W, 4.3L VIN X, 4.8L VIN V, 5.0L VIN M, 5.3L VIN T, 5.3L VIN Z, 5.7L VIN K, 5.7L VIN R, 6.0L VIN N, 6.0L VIN U, 8.1L VIN G engines Transmissions: All	**Fuel Level Sensor Low Input Conditions:** Engine started; system voltage over 10.0v and the PCM detected the Fuel Level sensor signal was less than 0.39v for 20 seconds under these conditions during the CCM test. **Possible Causes** ● Fuel level sensor signal circuit is shorted to ground between the fuel level sensor and the PCM connector ● Fuel level sender is damaged, binding or not aligned properly ● PCM has failed
P0462 **1T CCM, MIL: No** 1998, 1999, 2000, 2001, 2002, 2003, 2004, 2005 Aurora, DeVille, Eldorado & Seville 4.0L VIN C, 4.6L VIN 9, 4.6L VIN Y engines Transmissions: All	**Fuel Level Sensor Low Input Conditions:** DTC P0601 not set; engine running, system voltage over 10.0v and the PCM detected the Fuel Level sensor signal indicated less than 0.50v for over 10 seconds during the CCM test. **Possible Causes** ● Fuel level sensor signal circuit is shorted to ground between the fuel level sensor and the PCM connector ● Fuel level sender is damaged, binding or not aligned properly ● PCM has failed

OBD II Trouble Code List (P0xxx Codes)

DTC	Trouble Code Title, Conditions & Possible Causes
P0462 **1T CCM, MIL: No** 1997, 1998, 1999, 2000 Camaro, Corvette & Firebird 3.8L VIN K, 5.7L VIN G, 5.7L VIN G engines Transmissions: All	**Fuel Level Sensor Low Input Conditions:** DTC P0601 not set; engine running, system voltage over 10.0v and the PCM detected the Fuel Level sensor signal indicated less than 0.39v for over 6 minutes during the CCM test. **Possible Causes** ● Fuel level sensor signal circuit is shorted to ground between the fuel level sensor and the PCM connector ● Fuel level sender is damaged, binding or not aligned properly ● Instrument cluster or the PCM has failed ● TSB 01-06-04-025 contains a repair procedure for this code
P0462 **1T CCM, MIL: No** 1998, 1999, 2000 Achieva, Century, Envoy, Grand Am, Malibu, S/T Blazer & Pickup, Skylark, Sunfire & TrailBlazer 2.2L VIN 4, 2.2L VIN 5, 2.4L VIN T Transmissions: All	**Fuel Level Sensor Low Input Conditions:** Engine started; system voltage over 10.0v and the PCM detected the Fuel Level sensor signal indicated less than 0.40v for over 10 seconds during the CCM test. **Possible Causes** ● Fuel level sensor signal circuit is shorted to ground between the fuel level sensor and the PCM connector ● Fuel level sender is damaged, binding or not aligned properly ● PCM has failed
P0462 **1T CCM, MIL: No** 1996, 1997, 1998, 1999, 2000, 2001, 2002, 2003, 2004, 2005 Montana, Silhouette, Venture, Rendezvous 3.4L VIN E engine Transmissions: All	**Fuel Level Sensor Circuit Low Input Conditions:** Engine started; system voltage over 10.0v and the PCM detected the Fuel Level sensor signal was less than 0.4v (less than 3%) for 10 seconds during the CCM continuous test. **Possible Causes** ● Fuel level sensor signal circuit is shorted to ground between the fuel level sensor and the PCM connector ● Fuel level sender is damaged, binding or not aligned properly ● PCM has failed
P0462 **1T CCM, MIL: No** 1998, 1999, 2000 S/T Blazer & S/T Pickup 4.3L VIN W, 4.3L VIN X Transmissions: All	**Fuel Level Sensor Circuit Low Input Conditions:** Engine started; system voltage over 10.0v and the PCM detected the Fuel Level sensor signal was less than 0.39v for 10 seconds under these conditions. **Possible Causes** ● Fuel level sensor signal circuit is shorted to ground between the fuel level sensor and the PCM connector ● Fuel level sender is damaged, binding or not aligned properly ● Instrument cluster or the PCM has failed
P0463 **1T CCM, MIL: No** 1998, 1999, 2000 Century, Grand Prix, Impala, Intrigue, Lumina, Monte Carlo, Regal 3.1L VIN M, 3.8L VIN K, 3.8L VIN 1 engines Transmissions: All	**Fuel Level Sensor High Input Conditions:** Engine started; system voltage over 10.0v and the PCM detected the Fuel Level sensor signal was more than 3.0v for 10 seconds under these conditions during the CCM test. **Possible Causes** ● Fuel level sensor signal circuit is shorted to system power ● Fuel level sensor ground circuit is open ● Fuel level sender is damaged, binding or not aligned properly ● PCM has failed
P0463 **1T CCM, MIL: No** 1998, 1999, 2000 Achieva, Beretta, Ciera, Cavalier, Century, Corsica, Grand Am, Malibu, Skylark, Sunfire, S/T Blazer & S/T Pickup 2.2L VIN 4, 2.2L VIN 5, 2.4L VIN T Transmissions: All	**Fuel Level Sensor High Input Conditions:** Engine started; system voltage over 10.0v and the PCM detected the Fuel Level sensor signal was more than 3.0v for 10 seconds under these conditions. **Possible Causes** ● Fuel level sensor signal circuit is shorted to system power ● Fuel level sensor ground circuit is open ● Fuel level sender is damaged, binding or not aligned properly ● PCM has failed
P0463 **1T CCM, MIL: No** 1998, 1999, 2000, 2001, 2002, 2003, 2004, 2005 Aurora, DeVille, Eldorado & Seville 4.0L VIN C engine, 4.6L VIN 9, 4.6L VIN Y Transmissions: A/T	**Fuel Level Sensor Circuit High Input Conditions:** Engine started; system voltage over 10.0v, vehicle speed over 5 mph, and the PCM detected the Fuel Level sensor signal was more than 3.0v for 10 seconds under these conditions. **Possible Causes** ● Fuel level sensor signal circuit is shorted to system power ● Fuel level sensor ground circuit is open ● Fuel level sender is damaged, binding or not aligned properly ● PCM has failed

OBD II Trouble Code List (P0xxx Codes)

DTC	Trouble Code Title, Conditions & Possible Causes
P0463 **1T CCM, MIL: No** 1997, 1998, 1999, 2000, 2001, 2002, 2003, 2004, 2005 Camaro, Corvette & Firebird 3.8L VIN K, 5.7L VIN G, 5.7L VIN G Transmissions: All	**Fuel Level Sensor High Input Conditions:** Engine started; system voltage over 10.0v and the PCM detected the Fuel Level sensor signal indicated more than 2.90v for 6 minutes under these conditions during the CCM test. **Possible Causes** ● Fuel level sensor signal circuit is shorted to system power ● Fuel level sensor ground circuit is open ● Fuel level sender is damaged, binding or not aligned properly ● PCM has failed
P0463 **1T CCM, MIL: No** 1998, 1999, 2000, 2001, 2002, 2003, 2004, 2005 C/K Series Truck, G Series Van, S/T Series Pickup and Blazer, Envoy Escalade & TrailBlazer 4.0L VIN W, 4.3L VIN X, 4.8L VIN V, 5.0L VIN M, 5.3L VIN T, 5.3L VIN Z, 5.7L VIN K, 5.7L VIN R, 6.0L VIN N, 6.0L VIN U, 8.1L VIN G engines Transmissions: All	**Fuel Level Sensor High Input Conditions:** Engine started; system voltage over 10.0v and the PCM detected the Fuel Level sensor signal indicated more than 2.9-3.0v for 10-20 seconds under these conditions during the CCM test. **Possible Causes** ● Fuel level sensor signal circuit is shorted to system power ● Fuel level sensor ground circuit is open ● Fuel level sender is damaged, binding or not aligned properly ● PCM has failed
P0463 **1T CCM, MIL: No** 1996, 1997, 1998, 1999, 2000, 2001, 2002, 2003, 2004, 2005 Montana, Silhouette, Venture, Rendezvous 3.4L VIN E engine Transmissions: All	**Fuel Level Sensor Circuit High Input Conditions:** Engine started; system voltage over 10.0v and the PCM detected the Fuel Level sensor signal was more than 3.0v (more than 98%) for 10 seconds during the CCM continuous test. **Possible Causes** ● Fuel level sensor signal circuit is shorted to system power ● Fuel level sensor ground circuit is open ● Fuel level sender is damaged, binding or not aligned properly ● PCM has failed
P0463 **1T CCM, MIL: No** 1996, 1997, 1998, 1999, 2000, 2001, 2002, 2003, 2004, 2005 S/T Blazer & S/T Pickup 4.3L VIN W, 4.3L VIN X Transmissions: All	**Fuel Level Sensor Signal Range/Performance Conditions:** Engine started; system voltage over 10.0v and the PCM detected the Fuel Level sensor indicated more than 2.90v for 20 seconds under these conditions. Note: The PCM uses a "fuel remaining value" of 40% as a default for the EVAP test when this trouble code is set. **Possible Causes** ● Fuel level sensor signal circuit is shorted to system power ● Fuel level sensor ground circuit is open ● Fuel level sender is damaged, binding or not aligned properly ● PCM has failed
P0480 **2T CCM, MIL: Yes** 1996, 1997, 1998, 1999, 2000, 2001, 2002, 2003, 2004, 2005 Impala, Monte Carlo, Montana, Silhouette, Venture & Rendezvous 3.4L VIN E engine Transmissions: All	**Cooling Fan Relay 1 Control Circuit Malfunction Conditions:** Key on or engine running; system voltage from 9-18v, and the PCM detected the Actual state and Commanded state of the Fan Relay 1 control circuit (Low Speed Fan) did not match for 30 seconds. **Possible Causes** ● Fan control relay control circuit is open or shorted to ground ● Fan control relay control circuit is shorted to system power ● Fan control relay power circuit is open (check Cool Fan 1 fuse) ● Fan control relay is damaged or has failed ● PCM has failed
P0480 **2T CCM, MIL: Yes** 1998, 1999, 2000, 2001, 2002, 2003, 2004, 2005 Century, Grand Prix, Impala, Intrigue, Lumina, Monte Carlo, Regal 3.1L VIN M, 3.8L VIN K, 3.8L VIN 1 engines Transmissions: All	**Cooling Fan Relay 1 Control Circuit Malfunction Conditions:** Key on or engine running; system voltage over 10.0v and the PCM detected the Actual state and Commanded state of the Fan Relay 1 control circuit (Low Speed Fan) did not match for 30 seconds. **Possible Causes** ● Fan control relay control circuit is open or shorted to ground ● Fan control relay control circuit is shorted to system power ● Fan control relay power circuit is open (check Cool Fan 1 fuse) ● Fan control relay is damaged or has failed ● PCM has failed

OBD II Trouble Code List (P0xxx Codes)

DTC	Trouble Code Title, Conditions & Possible Causes
P0480 **2T CCM, MIL: Yes** 1998, 1999, 2000, 2001, 2002, 2003, 2004, 2005 Eldorado, CTS (Cadillac), DeVille, Eldorado, Seville 3.2L VIN N, 4.0L VIN C, 4.6L VIN 9, 4.6L VIN Y Transmissions: All	**Cooling Fan Relay 1 Control Circuit Malfunction Conditions:** Engine started; system voltage over 10.0v and the PCM detected an invalid voltage on the Cooling Fan Relay 1 circuit for 10 seconds. **Possible Causes** ● Fan control relay control circuit is open or shorted to ground ● Fan control relay control circuit is shorted to system power ● Fan control relay power circuit is open (check Cool Fan 1 fuse) ● Fan control relay is damaged or has failed ● PCM has failed
P0480 **2T CCM, MIL: Yes** 1997, 1998, 1999, 2000, 2001, 2002, 2003, 2004, 2005 Camaro, Corvette & Firebird 3.8L VIN K, 5.7L VIN G, 5.7L VIN G engines Transmissions: All	**Cooling Fan Relay 1 Control Circuit Malfunction Conditions:** Engine started; system voltage over 10.0v and the PCM detected the Commanded state and Actual state of the Fan Relay 1 control circuit (Low Speed Fan) did not match for 5 seconds. **Possible Causes** ● Fan control relay control circuit is open or shorted to ground ● Fan control relay control circuit is shorted to system power ● Fan control relay power circuit is open (check Cool Fan 1 fuse) ● Fan control relay is damaged or has failed ● PCM has failed
P0480 **2T CCM, MIL: Yes** 1997, 1998, 1999, 2000, 2001, 2002, 2003, 2004, 2005 Alero, Achieva, Cavalier, Century, Grand Am, Malibu, Skylark, Sunfire 2.2L VIN 4, 2.4L VIN T Transmissions: All	**Cooling Fan Relay 1 Control Circuit Malfunction Conditions:** Engine started, system voltage over 10.0v and the PCM detected the Commanded state and Actual state of the Fan Relay 1 control circuit (Low Speed Fan) did not match for 6 seconds. **Possible Causes** ● Fan control relay control circuit is open or shorted to ground ● Fan control relay control circuit is shorted to system power ● Fan control relay power circuit is open (check Cool Fan 1 fuse) ● Fan control relay is damaged or has failed ● PCM has failed
P0480 **2T CCM, MIL: Yes** 1996, 1997, 1998, 1999, 2000, 2001, 2002, 2003, 2004, 2005 Impala, Monte Carlo, Montana, Silhouette, Venture, Rendezvous 3.4L VIN E engine Transmissions: All	**Cooling Fan Relay 2 Control Circuit Malfunction Conditions:** Engine started, system voltage from 9-18v, and the PCM detected the Commanded state and Actual state of the Fan Relay 2 control circuit (High Speed Fan) did not match for 30 seconds. **Possible Causes** ● Fan control relay control circuit is open or shorted to ground ● Fan control relay control circuit is shorted to system power ● Fan control relay power circuit is open or shorted to ground ● Fan control relay is damaged or has failed ● PCM has failed
P0481 **2T CCM, MIL: Yes** 1999, 2000, 2001, 2002, 2003, 2004, 2005 Century, Grand Prix, Impala, Intrigue, Lumina, Monte Carlo, Regal 3.1L VIN M, 3.8L VIN K, 3.8L VIN 1 engines Transmissions: All	**Cooling Fan Relay 2 Control Circuit Malfunction Conditions:** Key on or engine running; system voltage from 9-18v, and the PCM detected the Actual state and Commanded state of the Fan Relay 2 control circuit (High Speed Fan) did not match for 30 seconds. **Possible Causes** ● Fan control relay control circuit is open or shorted to ground ● Fan control relay control circuit is shorted to system power ● Fan control relay power circuit is open (check Cool Fan 2 fuse) ● Fan control relay is damaged or has failed ● PCM has failed
P0481 **2T CCM, MIL: Yes** 1998, 1999, 2000, 2001, 2002, 2003, 2004, 2005 Aurora, CTS (Cadillac), DeVille, Eldorado, Seville 3.2L VIN N, 4.0L VIN C, 4.6L VIN 9, 4.6L VIN Y Transmissions: All	**Cooling Fan Relay 2 Control Circuit Malfunction Conditions:** Engine started; system voltage over 10.0v and the PCM detected an invalid voltage on the Cooling Fan Relay 1 circuit for 10 seconds. **Possible Causes** ● Fan control relay control circuit is open or shorted to ground ● Fan control relay control circuit is shorted to system power ● Fan control relay power circuit is open (check Cool Fan 2 fuse) ● Fan control relay is damaged or has failed ● PCM has failed
P0481 **2T CCM, MIL: Yes** 1997, 1998, 1999, 2000, 2001, 2002, 2003, 2004, 2005 Camaro, Corvette & Firebird 3.8L VIN K, 5.7L VIN G, 5.7L VIN G engines Transmissions: All	**Cooling Fan Relay 2-3 Control Circuit Malfunction Conditions:** Engine started; system voltage over 10.0v and the PCM detected the Actual state and the Commanded state of the Cooling Fan Relay 2-3 control circuit did not match for 10 seconds. **Possible Causes** ● Fan control relay control circuit is open or shorted to ground ● Fan control relay control circuit is shorted to system power ● Fan control relay power circuit is open or shorted to ground ● Fan control relay is damaged or has failed ● PCM has failed

OBD II Trouble Code List (P0xxx Codes)

DTC	Trouble Code Title, Conditions & Possible Causes
P0489 **2T CCM, MIL: Yes** 2001, 2002, 2003, 2004, 2005 C/K Series Truck 6.6L VIN 1 Diesel engine Transmissions: All	**EGR Valve Vacuum Solenoid Circuit Low Input Conditions:** Engine started; engine runtime over 0.5 seconds, system voltage from 11-18v, EGR solenoid command less than 71%, and the ECM detected an unexpected low voltage condition on the EGR Valve Solenoid control circuit for 2 seconds. The EGR valve is vacuum operated. A belt-driven vacuum pump is used to supply vacuum for the EGR Valve Control system. The EGR valve vacuum control solenoid and the EGR valve vacuum vent solenoid work together to control the position of the EGR valve. The PCM controls the EGR valve vacuum control solenoid with a PWM control signal. **Possible Causes** ● EGR valve solenoid connector is damaged or shorted ● EGR valve solenoid control circuit is open or shorted to ground ● EGR valve solenoid is damaged or it has failed ● ECM has failed
P0490 **2T CCM, MIL: Yes** 2001, 2002, 2003, 2004, 2005 C/K Series Truck 0.0L VIN 1 Diesel engine Transmissions: All	**EGR Valve Vacuum Solenoid Circuit High Input Conditions:** Engine started; engine runtime over 0.5 seconds, system voltage from 11-18v, EGR solenoid command signal more than 10%, and the ECM detected an unexpected high voltage condition on the EGR Valve Solenoid control circuit for 2 seconds. The EGR valve is vacuum operated. A belt-driven vacuum pump is used to supply vacuum for the EGR Valve Control system. The EGR valve vacuum control solenoid and the EGR valve vacuum vent solenoid work together to control the position of the EGR valve. The PCM controls the EGR valve vacuum control solenoid with a PWM control signal. **Possible Causes** ● EGR valve solenoid connector is damaged or open ● EGR valve solenoid control circuit is open or shorted to power ● EGR valve solenoid is damaged or it has failed ● ECM has failed
P0496 **2T EVAP, MIL: Yes** 2003, 2004, 2005 C/K Truck, Escalade, G Van, Cargo Van, M/L Van S/T Blazer & S/T Pickup 4.3L VIN W, 4.3L VIN X, 4.8L VIN V, 5.3L VIN P, 5.3L VIN T, 5.3L VIN Z, 6.0L VIN N, 6.0L VIN U, 8.1L VIN G engines Transmissions: All	**EVAP Canister Purge System High Purge Flow Conditions:** DTC P0106-P0108, P0112, P0113, P0116-P0118, P0120-P0123, P0125, P0131-P0138, P0140, P0141, P0147, P0151-P0158, P0160, P0161, P0167, P0220, P0442-P0443, P0449, P0452-P0453, P0455, P0502, P0503, P1111, P1112, P1114, P1115, P1120 not set, engine started, ECT and IAT sensors from 39-86°F and within 16°F at startup, system voltage from 10-18v, BARO sensor more than 75 kPa, fuel level at 15-85%, and the PCM detected a continuous open purge flow condition in the system (FTP less than -11 H2O). This diagnostic test is designed to test for undesired intake manifold vacuum flow to the EVAP system. During this test, the PCM seals the EVAP system by commanding the EVAP Purge valve closed and the EVAP canister vent valve closed. The PCM monitors the FTP sensor signal in order to determine if a vacuum is being drawn on the EVAP system. If vacuum in the EVAP system is more than a predetermined value within a certain time, this code is set. **Possible Causes** ● EVAP charcoal canister is damaged or restricted ● EVAP purge pipe is damaged or restricted ● FTP sensor is damaged or it has failed ● Purge solenoid is damaged (it may be sticking) ● Purge solenoid valve has failed
P0496 **2T EVAP, MIL Yes** 2003, 2004, 2005 CTS (Cadillac) 2.6L VIN M, 3.2L VIN N Transmissions: All	**EVAP System Vacuum During Non-Purge Condition Conditions:** DTC P0112, P0113, P0116-P0118, P0121, P0122, P0123, P0125, P0440, P0442, P0443, P0444, P0445, P0447, P0448, P0449, P0451, P0452 and P0453 not set, engine started, BARO sensor over 75 kPa, system voltage from 10-18v, fuel level from 15-85%, ECT sensor from 39-149°F and IAT sensor from 39-167°F, ECT and IAT sensors within 16°F at startup, VSS at 0 mph, and the PCM detected vacuum in the system during a non-purge condition. **Possible Causes** ● EVAP purge valve is damaged or it has failed ● Use the Service Bay Test to diagnose the cause of this code
P0500 **2T CCM, MIL: Yes** 1996, 1997, 1998, 1999, 2000 Cab & Chassis, C/K Truck, G Series Van, L/M Van, S/T Blazer & Pickup 4.3L VIN W, 4.3L VIN X, 4.8L VIN V, 5.0L VIN M, 5.3L VIN T, 5.7L VIN K, 5.7L VIN R, 6.0L VIN U, 7.4L VIN J engines Transmissions: A/T	**Vehicle Speed Sensor Circuit Malfunction Conditions:** DTC P0106, P0107 and P0108 not set, engine started, vehicle driven at a speed of 1400-4000 rpm, ECT sensor more than 140°F, MAP sensor less than 20 kPa, TP angle less than 3%, and the PCM detected the VSS indicated less than 1 mph during deceleration. Note: The VSS frequency is 40 cycles per output shaft revolution. **Possible Causes** ● Output shaft rotor is chipped or damaged ● Output shaft rotor is not aligned properly with the VSS unit ● VSS tip contains debris or metal shavings (an intermittent fault) ● VSS positive (+) signal circuit is open or shorted to ground ● VSS negative (-) signal circuit is open or shorted to ground ● VSS is damaged or has failed

OBD II Trouble Code List (P0xxx Codes)

DTC	Trouble Code Title, Conditions & Possible Causes
P0500 **2T CCM, MIL: Yes** 2001, 2002, 2003, 2004, 2005 Cab & Chassis, C/K Truck, G Series Van, L/M Van, S/T Blazer & Pickup 4.3L VIN W, 4.3L VIN X, 4.8L VIN V, 5.3L VIN T, 5.3L VIN Z, 6.0L VIN N, 6.0L VIN U, 6.5L VIN F, 6.6L VIN 1 Diesel engine, 8.1L VIN G engine Transmissions: M/T	**Vehicle Speed Sensor Circuit Malfunction Conditions:** DTC P0106, P0107, P0108, P0335, P0336, P1120, P1125, P1128, P1220, P1221, P1514, P1515, P1516, P1517 and P1518 not set, engine started, vehicle driven at a speed over 1000 rpm, ECT sensor more than 95ºF, MAP sensor from 40-100 kPa (Turbo Boost Pressure from 40-100 kPa on Diesel) TP angle from 5-95%, and the PCM did not detect any VSS signals for from 50-100 seconds. **Possible Causes** ● Output shaft rotor is chipped or damaged ● Output shaft rotor is not aligned properly with the VSS unit ● VSS tip contains debris or metal shavings (an intermittent fault) ● VSS positive (+) signal circuit is open or shorted to ground ● VSS negative (-) signal circuit is open or shorted to ground ● VSS is damaged or has failed
P0500 **2T CCM, MIL: Yes** 1996, 1997 Camaro, Caprice, Corvette, Firebird, Fleetwood, Roadmaster 4.3L VIN W, 5.7L VIN 5, 5.7L VIN P engines Transmissions: All	**Vehicle Speed Sensor Circuit Malfunction Conditions:** DTC P0106, P0107, P0108, P0117, P0118, P0122, P0123, P0506, P0507, P0508 and P1509 not set, engine started, vehicle driven to a speed over 1000 rpm, ECT more than 40ºF, throttle angle over 5%, MAP sensor more than 40 kPa with the A/C "off" or more than 45 kPa with the A/C "on", engine load and acceleration indicate the vehicle is in gear, conditions met for 25 seconds, and the PCM detected the VSS signal was less than 3 mph. **Possible Causes** ● Output shaft rotor is chipped or damaged ● Output shaft rotor is not aligned properly with the VSS unit ● VSS tip contains debris or metal shavings (an intermittent fault) ● VSS positive (+) signal circuit is open or shorted to ground ● VSS negative (-) signal circuit is open or shorted to ground ● VSS is damaged or has failed
P0500 **2T CCM, MIL: Yes** 1997, 1998, 1999, 2000 Camaro, Corvette & Firebird 5.7L VIN G, 5.7L VIN S Transmissions: All	**Vehicle Speed Sensor Circuit Malfunction Conditions:** DTC P0107, P0108, P0117, P0118, P0121, P0122, P0123, P0125, P0506 and P0507 not set, vehicle driven to a speed over 1000 rpm, ECT sensor more than 95ºF, MAP sensor from 40-100 kPa with the A/C "off" or between 45-100 kPa with the A/C "on", TP angle from 5-100%, and the PCM detected the VSS indicated zero (0) mph for 100 seconds in the test. **Possible Causes** ● Output shaft rotor is chipped or damaged ● Output shaft rotor is not aligned properly with the VSS unit ● VSS tip contains debris or metal shavings (an intermittent fault) ● VSS positive (+) signal circuit is open or shorted to ground ● VSS negative (-) signal circuit is open or shorted to ground ● VSS is damaged or has failed
P0500 **2T CCM, MIL: Yes** 1996, 1997, 1998, 1999, 2000 Camaro & Firebird 3.8L VIN K engines Transmissions: All	**Vehicle Speed Sensor Circuit Malfunction Conditions:** Engine started; vehicle driven at an engine speed and load that indicate the vehicle is moving in gear for over 2 seconds, and the PCM detected the VSS signal indicated a speed of less than 3 mph for over 20 seconds under these conditions in the CCM test. **Possible Causes** ● Output shaft rotor is chipped or damaged ● Output shaft rotor is not aligned properly with the VSS unit ● VSS tip contains debris or metal shavings (an intermittent fault) ● VSS positive (+) signal circuit is open or shorted to ground ● VSS negative (-) signal circuit is open or shorted to ground ● VSS is damaged or has failed
P0500 **2T CCM, MIL: Yes** 2003, 2004, 2005 CTS (Cadillac) 2.6L VIN M, 3.2L VIN N Transmissions: All	**Vehicle Speed Sensor Range/Performance (5L40-E) Conditions:** DTC U2106 not set, engine speed at 1,800-2,200 rpm, DFCO active, ECT sensor over 149ºF, TCM commanded gear is 4th gear, and the PCM detected the VSS signal was less than 3 mph for 50 seconds. **Possible Causes** ● Perform a Diagnostic Circuit Check for the PCM ● If DTC U2100, U2104, U2105 or U2106 is set, repair them first ● If DTC P0722 or P0723 are set, repair these trouble codes first
P0500 **2T CCM, MIL: Yes** 1997, 1998, 1999, 2000, 2001 Catera 3.0L V6 VIN R engine Transmissions: All	**Vehicle Speed Sensor Range/Performance Conditions:** Vehicle driven at a speed of 1800-2200 rpm, ECT sensor over 149ºF, then with Decel Fuel Cutoff enabled, the PCM detected the VSS signal was less than 12 mph for one second. **Possible Causes** ● L/R WSS positive (+) signal circuit is open or shorted to ground ● L/R WSS positive (-) signal circuit is open or shorted to ground ● L/R WSS is damaged or it failed (check for WSS trouble codes) ● VSS signal from the EBTCM to the PCM is missing (the same circuit used to send the signal to the Theft Deterrent Module)

OBD II Trouble Code List (P0xxx Codes)

DTC	Trouble Code Title, Conditions & Possible Causes
P0501 **1T CCM, MIL: Yes** 1995 Century, Cutlass Ciera, Cutlass Supreme, Grand Prix, Lumina, Monte Carlo Park Avenue Regal, Riviera, 88' & 98' 3.1L VIN M, 3.4L VIN X, 3.8L VIN 1, 3.8L VIN K, 3.8L VIN L Transmissions: All	**Vehicle Speed Sensor Range/Performance Conditions:** Engine started; vehicle driven at cruise speed for over 1 minute, then with the brakes not applied, the PCM detected the VSS signal went from more than 18 mph to 0 mph in less than 2 seconds during the CCM test period. **Possible Causes** ● Output shaft rotor is not aligned properly with the VSS unit ● VSS tip contains debris or metal shavings (an intermittent fault) ● VSS positive (+) signal circuit is open or shorted to ground ● VSS negative (-) signal circuit is open or shorted to ground ● VSS is damaged or has failed
P0501 **2T CCM, MIL: Yes** 1997, 1998 Catera 3.0L V6 VIN R engine Transmissions: A/T	**Vehicle Speed Signal Range/Performance Conditions:** Vehicle driven at cruise speed under light to medium load conditions for 10 minutes, and the PCM detected the VSS signal was less than 3 mph for 5 seconds during the CCM test. **Possible Causes** ● L/R WSS positive (+) signal circuit is open or shorted to ground ● L/R WSS positive (-) signal circuit is open or shorted to ground ● L/R WSS is damaged or it failed (check for WSS trouble codes) ● VSS signal from the EBTCM to the PCM is missing (the same circuit used to send the signal to the Theft Deterrent Module)
P0501 **1T CCM, MIL: No** 1996, 1997, 1998, 1999, 2000 C/K Truck, G Series Van 6.5L VIN F, 6.5L VIN S, 6.5L VIN Y Diesel engine Transmissions: All	**Vehicle Speed Sensor Range/Performance (4L60-E) Conditions:** Engine started; vehicle driven at a steady cruise speed at more than 20 mph, 4WD Low not selected, and the PCM detected the VSS Buffer Calculated speed was less than half of the Transmission Calculated speed, or the Calculated speed was more than the Transmission shaft speed by over 20 mph during the CCM test. **Possible Causes** ● Output shaft rotor is chipped or damaged ● Output shaft rotor is not aligned properly with the VSS unit ● VSS tip contains debris or metal shavings (an intermittent fault) ● VSS positive (+) signal circuit is open or shorted to ground ● VSS negative (-) signal circuit is open or shorted to ground ● VSS buffer is damaged, or one of the buffer circuits has failed ● VSS is damaged or has failed
P0502 **2T CCM, MIL: Yes** 1996, 1997, 1998, 1999, 2000 Achieva, Beretta, Ciera, Cavalier, Century, Corsica, Grand Am, Malibu, Skylark, Sunfire, S/T Blazer & S/T Pickup 2.2L VIN 4, 2.2L VIN 5, 2.4L VIN T Transmissions: A/T	**Vehicle Speed Sensor Low Input (3T40, 4T40-E, 4T45-E) Conditions:** DTC P0106, P0107, P0108, P0122, P0123, P0716, P0717 and P1810 not set, engine started, vehicle driven to a speed of 1600-4800 rpm, MAP sensor from 50-80 50 kPa, Gear position in Drive, TP angle above 12%, and the PCM detected the VSS signal was less than 2 mph for 5 seconds. *Note: The Torque Converter Clutch is disabled with this code set* **Possible Causes** ● Output shaft rotor is chipped or damaged ● Output shaft rotor is not aligned properly with the VSS unit ● VSS tip contains debris or metal shavings (an intermittent fault) ● VSS positive (+) signal circuit is open or shorted to ground ● VSS negative (-) signal circuit is open or shorted to ground ● VSS buffer is damaged, or one of the buffer circuits has failed ● VSS is damaged or has failed ● TSB 71-65-49 contains a repair procedure for this code
P0502 **2T CCM, MIL: Yes** 2001, 2002 Achieva, Beretta, Ciera, Cavalier, Century, Corsica, Grand Am, Malibu, Skylark, Sunfire, S/T Blazer & S/T Pickup 2.2L VIN 4, 2.2L VIN 5, 2.4L VIN T engines Transmissions: A/T	**Vehicle Speed Sensor Low Input (4T40-E, 4T45-E, 4L60-E) Conditions:** DTC P0106, P0107, P0108, P0122, P0123, P0716, P0717 and P1810 not set, engine started, vehicle driven to a speed over 1500 rpm, MAP sensor from 0-105 kPa, Input speed more than 1500 rpm, engine torque more than 50 lb ft., TP angle over 12%, and the PCM detected the OSS signal indicated less than 150 rpm for 3 seconds. **Possible Causes** ● Output shaft rotor is chipped or damaged ● OSS tip contains debris or metal shavings (an intermittent fault) ● OSS positive (+) signal circuit is open or shorted to ground ● OSS negative (-) signal circuit is open or shorted to ground ● OSS is damaged or has failed

OBD II Trouble Code List (P0xxx Codes)

DTC	Trouble Code Title, Conditions & Possible Causes
P0502 **2T CCM, MIL: Yes** 2001, 2002 Cavalier, Grand Am, Sunfire, S/T Blazer & S/T Pickup 2.2L VIN 4, 2.2L VIN 5, 2.4L VIN T engines Transmissions: M/T	**Vehicle Speed Sensor Low Input (M86/M94 Getrag) Conditions:** DTC P0106, P0107, P0108, P0122 and P0123 not set, engine started, vehicle driven to a speed of 1799-3200 rpm, MAP sensor from 70-80 kPa, TP angle from 0-1%, and the PCM detected the VSS signal indicated less than 3 mph for 5 seconds during the test. **Possible Causes** • Transaxle shaft rotor is chipped or damaged • VSS tip contains debris or metal shavings (an intermittent fault) • VSS positive (+) signal circuit is open or shorted to ground • VSS negative (-) signal circuit is open or shorted to ground • VSS is damaged or has failed
P0502 **1T CCM, MIL: No** 1995 Century, Cutlass Ciera, Cutlass Supreme, Grand Prix, Lumina, Monte Carlo Park Avenue Regal, Riviera, 88' & 98' 3.1L VIN M, 3.4L VIN X, 3.8L VIN 1, 3.8L VIN K, 3.8L VIN L engines Transmissions: All	**Vehicle Speed Sensor Circuit Low Input (4T40-E) Conditions:** DTC P0705 not set, vehicle driven to a speed over 3000 rpm, transmission not in P/N, conditions met for 4 seconds, and the PCM detected the VSS indicated less than 3 mph. **Possible Causes** • Transmission shaft rotor is chipped or damaged • VSS tip contains debris or metal shavings (an intermittent fault) • VSS positive (+) signal circuit is open or shorted to ground • VSS negative (-) signal circuit is open or shorted to ground • VSS is damaged or has failed
P0502 **2T CCM, MIL: Yes** 1996, 1997, 1998, 1999, 2000, 2001, 2002, 2003, 2004, 2005 Aurora, Bonneville, Century, Cutlass Supreme, Grand Prix, Intrigue, LeSabre, LSS, Lumina, Monte Carlo, 88', Regal, Park Avenue & 98' 3.1L VIN J, 3.1L VIN M, 3.5L VIN H, 3.8L VIN 1, 3.8L VIN K engines Transmissions: All	**VSS Circuit Low Input (4T40-E, 4T45-E, 4T65-E) Conditions:** DTC P0106, P0107, P0108, P1106, P1107, P0121, P0122, P0123, P1121 and P1122 not set, engine started, Input Shaft Speed signal over 1500 rpm, MAP sensor from 12-15 kPa, transaxle not in P/N, TP angle over 12%, engine torque more than 25-150 lb ft., and the PCM detected the OSS signal was less than 150 rpm for 3 seconds. **Possible Causes** • Output shaft rotor is chipped or damaged • OSS tip contains debris or metal shavings (an intermittent fault) • OSS positive (+) signal circuit is open or shorted to ground • OSS negative (-) signal circuit is open or shorted to ground • OSS is damaged or has failed
P0502 **2T CCM, MIL: Yes** 1996, 1997, 1998, 1999, 2000, 2001, 2002, 2003, 2004, 2005 Impala, Monte Carlo, Montana, Silhouette, Venture, Rendezvous 3.4L VIN E engine Transmissions: All	**VSS Circuit Low Input (4T65-E) Conditions:** DTC P0107, P0108, P0121, P0122, P0123, P0716, P0717 and P1810 not set, engine started, ISS signal more than 1500 rpm, TP angle over 12%, MAP sensor from 0-150 kPa, engine torque from 40-300 ft-lbs, gearshift not in P/N, and the PCM detected the OSS sensor was less than 150 rpm for 2.5 seconds during the CCM test. **Possible Causes** • Output shaft rotor is chipped or damaged • OSS tip contains debris or metal shavings (an intermittent fault) • OSS positive (+) signal circuit is open or shorted to ground • OSS negative (-) signal circuit is open or shorted to ground • OSS is damaged or has failed
P0502 **2T CCM, MIL: Yes** 1996, 1997, 1998, 1999, 2000, 2001, 2002, 2003, 2004, 2005 C/K Truck, Escalade, G Van, Cargo Van, L/M Van, S/T Blazer & Pickup 4.3L VIN W, 4.3L VIN X, 4.8L VIN V, 5.0L VIN M, 5.3L VIN T, 5.3L VIN Z, 5.7L VIN K, 5.7L VIN R, 6.0L VIN N, 6.0L VIN U, 6.6L VIN 1 Diesel engine, 7.4L VIN J, 8.1L VIN G Transmissions: A/T	**VSS Circuit Low Input (4L60-E, 4L80-E) Conditions:** DTC P0107, P0108, P0122, P0123 and P1810 not set, engine started, vehicle driven with the engine speed from 3000-4800 rpm, engine vacuum from 0-150 kPa, TP angle over 20%, engine torque from 40-400 lb ft., gear selector not in P/N, and the PCM detected the Output Shaft Speed sensor was less than 150 rpm for 3 seconds. **Possible Causes** • Output shaft rotor is chipped or damaged • OSS tip contains debris or metal shavings (an intermittent fault) • OSS positive (+) signal circuit is open or shorted to ground • OSS negative (-) signal circuit is open or shorted to ground • OSS is damaged or has failed (an intermittent fault)

OBD II Trouble Code List (P0xxx Codes)

DTC	Trouble Code Title, Conditions & Possible Causes
P0502 **2T CCM, MIL: Yes** 2001, 2002, 2003, 2004, 2005 C/K Truck, G Series Van 6.5L VIN F, 6.6L VIN 1 Diesel engines Transmissions: A/T	**Vehicle Speed Sensor Low Input (4L60-E) Conditions:** DTC P0107, P0108, P0122, P0123 and P1810 not set, engine speed over 3000 rpm, TP angle over 12%, not in P/N, engine vacuum from 0-15 kPa, engine torque from 40-400 lb ft., and the PCM detected the OSS signal was under 150 rpm for three seconds. **Possible Causes** ● Output shaft rotor is chipped or damaged ● Output shaft rotor is not aligned properly with the VSS unit ● VSS tip contains debris or metal shavings (an intermittent fault) ● VSS (+) or (-) signal circuit is open or shorted to ground ● VSS buffer is damaged, or one of the buffer circuits has failed ● VSS is damaged or has failed
P0502 **2T CCM, MIL: Yes** 1996, 1997, 1998, 1999, 2000 Aurora, DeVille, Eldorado & Seville 4.0L VIN C, 4.6L VIN 9, 4.0L VIN Y engines Transmissions: A/T	**VSS Circuit Low Input (4T80-E) Conditions:** DTC P0121-P0123, P0503, P0716, P0717, P1820, P1822, P1823 and P1825 not set, engine started, ISS from 1000-5000 rpm, not in P/N, system voltage over 10.0v, TP angle over 12%, engine runtime 5 seconds, DFCP not active, engine torque from 44-291 lb ft., and the PCM detected the OSS signal was under 372 rpm for 3 seconds. **Possible Causes** ● Output shaft rotor is chipped or damaged ● OSS tip contains debris or metal shavings (an intermittent fault) ● OSS positive (+) signal circuit is open or shorted to ground ● OSS negative (-) signal circuit is open or shorted to ground ● OSS is damaged or has failed (an intermittent fault)
P0502 **2T CCM, MIL: Yes** 2001, 2002, 2003, 2004, 2005 Aurora, DeVille, Eldorado & Seville 4.0L VIN C, 4.6L VIN 9, 4.6L VIN Y engines Transmissions: A/T	**VSS Circuit Low Input (4T80-E) Conditions:** DTC P0121-P0123, P0503, P0716, P0717, P1820, P1822, P1823 and P1825 not set, engine started, ISS from 1000-5000 rpm, not in P/N, system voltage over 10.0v, TP angle over 12%, engine runtime 5 seconds, DFCP not active, engine torque from 44-291 lb ft., and the PCM detected the OSS signal was under 372 rpm for 3 seconds. **Possible Causes** ● Output shaft rotor is chipped or damaged ● OSS tip contains debris or metal shavings (an intermittent fault) ● OSS positive (+) signal circuit is open or shorted to ground ● OSS negative (-) signal circuit is open or shorted to ground ● OSS is damaged or has failed (an intermittent fault) ● TSB 03-06-04-013 contains a repair procedure for this code
P0502 **2T CCM, MIL: Yes** 1995, 1996, 1997, 1998, 1999, 2000, 2001, 2002 Camaro & Firebird 3.8L VIN K engines Transmissions: All	**VSS Circuit Low Input (4L60-E) Conditions:** DTC P0107, P0108, P0121, P0122 and P0123 not set, engine started, not in P/N, vehicle driven to a speed over 2900 rpm, TP angle over 15%, engine torque 40-150 lb ft., and the PCM detected the Output Shaft Speed signal was less than 150 rpm for 3 seconds. **Possible Causes** ● Output shaft rotor is chipped or damaged ● OSS tip contains debris or metal shavings (an intermittent fault) ● OSS positive (+) signal circuit is open or shorted to ground ● OSS negative (-) signal circuit is open or shorted to ground ● OSS is damaged or has failed (an intermittent fault)
P0502 **1T CCM, MIL: Yes** 1996, 1996, 1998, 1999, 2000 Achieva, Alero, Beretta, Cavalier, Sunfire, Grand Am, S/T Blazer & Pickup 2.2L VIN 4, 2.2L VIN 5, 2.2L VIN F, 2.4L VIN T Transmissions: M/T	**Vehicle Speed Sensor Low Input (M5/MJ1) Conditions:** DTC P0107, P0108, P0122 and P0123 not set, engine started, vehicle driven to a speed of 1700-3600 rpm, engine vacuum from 70-80 kPa, TP angle from 0-1%, and the PCM detected the VSS signal indicated less than 2 mph for 5 seconds during the CCM test. **Possible Causes** ● Transmission shaft rotor is chipped or damaged ● VSS tip contains debris or metal shavings (an intermittent fault) ● VSS positive (+) signal circuit is open or shorted to ground ● VSS negative (-) signal circuit is open or shorted to ground ● VSS is damaged or has failed
P0502 **2T CCM, MIL: Yes** 2002, 2003, 2004, 2005 Envoy & TrailBlazer 4.2L VIN S, 5.3L VIN P Transmissions: A/T	**Vehicle Speed Sensor Low Input (4L60-E, 4L65-E) Conditions:** DTC P0107, P0108, P0122, P0123 and P1810 not set, engine started, engine torque from 40-400 ft lb., engine vacuum from 0-105 kPa, Input speed more than 3,000 rpm, TP angle over 12%, gear selector not in Park or Neutral, and the PCM detected the transmission output speed was less than 150 rpm for 3 seconds. **Possible Causes** ● Output shaft, AWD, 2WD or 4WD rotor is chipped or damaged ● VSS tip contains debris or metal shavings (an intermittent fault) ● VSS positive (+) signal circuit is open or shorted to ground ● VSS negative (-) signal circuit is open or shorted to ground ● VSS is damaged or has failed

OBD II Trouble Code List (P0xxx Codes)

DTC	Trouble Code Title, Conditions & Possible Causes
P0503 **2T CCM, MIL: Yes** 1996, 1997, 1998, 1999, 2000, 2001, 2002, 2003, 2004, 2005 Aurora, Bonneville, Century, Cutlass Supreme, Grand Prix, Intrigue, LeSabre, LSS, Lumina, Monte Carlo, 88', Regal, Park Avenue & 98' 3.1L VIN J, 3.1L VIN M, 3.5L VIN H, 3.8L VIN 1, 3.8L VIN K engines Transmissions: All	**VSS Circuit Malfunction (4T40-E, 4T45-E, 4T65-E, 4L80-E) Conditions:** DTC P0502 and P1810 not set, engine running, at least 6 seconds have passed since the last gear change, Decel Fuel Cutoff inactive, Transmission output shaft speed did not increase over 250 rpm for 2 seconds, and the PCM detected the OSS signal dropped more than 1500 rpm in 2 seconds during the CCM test. **Possible Causes** ● OSS assembly connector is damaged, loose or shorted ● Output shaft rotor is chipped or damaged (intermittent fault) ● OSS tip contains debris or metal shavings (an intermittent fault) ● OSS (+) signal circuit is open or shorted to ground (intermittent) ● OSS (-) signal circuit is open or shorted to ground (intermittent) ● OSS is damaged or has failed (an intermittent fault)
P0503 **2T CCM, MIL: Yes** 1996, 1997, 1998, 1999, 2000, 2001, 2002, 2003, 2004, 2005 Impala, Monte Carlo, Montana, Silhouette, Venture, Rendezvous 3.4L VIN E engine Transmissions: All	**VSS Circuit Malfunction (4T65-E) Conditions:** Engine speed started, engine running, Fuel Cutoff inactive, not in Park or Neutral, time since last gear range change more than 6 seconds, Transmission Output Shaft speed rise more than 250 rpm in 2 seconds, and the PCM detected a drop in the Output Shaft Speed sensor signal of over 1500 rpm within 3 seconds. **Possible Causes** ● Output shaft rotor is chipped or damaged (intermittent fault) ● OSS tip contains debris or metal shavings (intermittent fault) ● OSS (+) signal circuit is open or shorted to ground (intermittent) ● OSS (-) signal circuit is open or shorted to ground (intermittent) ● OSS is damaged or has failed (an intermittent fault)
P0503 **2T CCM, MIL: Yes** 1996, 1997, 1998, 1999, 2000, 2001, 2002, 2003, 2004, 2005 C/K Truck, Escalade, G Van, Cargo Van, M/L Van S/T Blazer & S/T Pickup 4.3L VIN W, 4.3L VIN X, 4.8L VIN V, 5.0L VIN M, 5.3L VIN T, 5.3L VIN Z, 5.7L VIN K, 5.7L VIN R, 6.0L VIN N, 6.0L VIN U, 7.4L VIN J, 8.1L VIN G Transmissions: A/T	**Title: VSS Circuit Malfunction (4L60-E, 4L80-E)** **Trouble Code Conditions** DTC P1810 not set, engine running, 6 seconds have passed since the gear change or change in 4WD Switch status, Transmission output shaft speed did not increase over 600 rpm for 2 seconds, and the PCM detected the Output Shaft Speed decreased over 300 rpm for three seconds. **Possible Causes** ● Output shaft rotor is chipped or damaged (intermittent fault) ● OSS tip contains debris or metal shavings (an intermittent fault) ● OSS (+) signal circuit is open or shorted to ground (intermittent) ● OSS (-) signal circuit is open or shorted to ground (intermittent) ● OSS is damaged or has failed (an intermittent fault)
P0503 **2T CCM, MIL: Yes** 2001, 2002, 2003, 2004, 2005 C/K Truck, G Series Van 6.5L VIN F, 6.6L VIN 1 Diesel engine Transmissions: A/T	**VSS Circuit Malfunction (4L60-E) Conditions:** DTC P1810 not set, engine started, 6 seconds elapsed since last gear change or change in 4WD Switch status, output shaft speed no more than 600 rpm for 2 seconds, and the PCM detected the Output Shaft Speed decreased over 300 rpm for three seconds. **Possible Causes** ● Output shaft rotor is chipped or damaged (intermittent fault) ● OSS tip contains debris or metal shavings (an intermittent fault) ● OSS (+) signal circuit is open or shorted to ground (intermittent) ● OSS (-) signal circuit is open or shorted to ground (intermittent) ● OSS is damaged or has failed (an intermittent fault)
P0503 **2T CCM, MIL: Yes** 1996, 1997, 1998, 1999, 2000 Aurora, DeVille, Eldorado & Seville 4.0L VIN C, 4.6L VIN 9, 4.6L VIN Y engines Transmissions: All	**Vehicle Speed Sensor Circuit Malfunction (4T80-E) Conditions:** DTC P0502, P0716, P0717, P1810, P1820, P1822, P1823, P1825 and P1843 not set, engine speed over 500 rpm for 5 seconds, Fuel Cutoff not active, not in P/N, time since last gearshift change over 6 seconds, OSS signal more than 1484 rpm and did increase over 250 rpm in the last 2 seconds, and the PCM detected the OSS signal dropped by more than 1100-1500 rpm for 1 second during the test. **Possible Causes** ● Output shaft rotor is chipped or damaged (intermittent fault) ● OSS tip contains debris or metal shavings (intermittent fault) ● OSS (+) signal circuit is open or shorted to ground (intermittent) ● OSS (-) signal circuit is open or shorted to ground (intermittent) ● OSS is damaged or has failed (an intermittent fault) ● TSB 99-07-30-015A contains a repair procedure for this code

OBD II Trouble Code List (P0xxx Codes)

DTC	Trouble Code Title, Conditions & Possible Causes
P0503 **2T CCM, MIL: Yes** 2001, 2002, 2003, 2004, 2005 Aurora, DeVille, Eldorado & Seville 4.0L VIN C, 4.6L VIN 9, 4.6L VIN Y engines Transmissions: A/T	**Vehicle Speed Sensor Signal Intermittent (4T80-E) Conditions:** DTC P0716, P0717, P1820, P1822, P1823, P1825 and P1843 not set, engine runtime over 5 seconds, not in P/N, ISS signal stable (± 500 rpm) for 5 seconds, OSS over 1244 rpm for 2 seconds, system voltage over 10.0v, TP angle over 12%, Decel Fuel Cutoff off, and the PCM detected the OSS signal dropped 1,089 rpm for 1 second. **Possible Causes** ● Output shaft rotor is chipped or damaged ● OSS tip contains debris or metal shavings (an intermittent fault) ● OSS positive (+) signal circuit is open or shorted to ground ● OSS negative (-) signal circuit is open or shorted to ground ● OSS is damaged or has failed (an intermittent fault) ● TSB 03-06-04-013 contains a repair procedure for this code
P0503 **2T CCM, MIL: Yes** 1995, 1996, 1997, 1998, 1999, 2000, 2001, 2002 Camaro & Firebird 3.8L VIN K engines Transmissions: All	**Vehicle Speed Sensor Signal Intermittent (4L60-E) Conditions:** DTC P1810 not set, engine running, time since last gear change over 6 seconds, not in P/N, and the PCM detected a drop in the VSS output signal of more than 1800 rpm, or with vehicle in P/N, the VSS output speed dropped over 1300 rpm for 3 seconds. **Possible Causes** ● Transmission shaft rotor is chipped or damaged (intermittent) ● VSS tip contains debris or metal shavings (an intermittent fault) ● VSS (+) or (-) signal circuit is open or shorted to ground ● VSS is damaged or has failed (an intermittent fault)
P0503 **1T CCM, MIL: Yes** 1996, 1996, 1998, 1999, 2000, 2001, 2002, 2003, 2004, 2005 Achieva, Alero, Beretta, Cavalier, Century, Grand Am, Skylark, Sunfire, S/T Blazer & S/T Pickup 2.2L VIN 4, 2.2L VIN 5, 2.2L VIN F, 2.4L VIN T Transmissions: M/T	**Vehicle Speed Sensor Signal Intermittent (4T40, 4T45-E) Conditions:** DTC P0107, P0108, P0122, P0123, P0716, P07171 and P1810 not set, engine started, vehicle driven with the last gear shift change more than 6 seconds earlier, the OSS signal did not change more than 500 rpm for 5 seconds, and the PCM detected the OSS signal decreased (dropped) by over 1200 rpm within 3 seconds. **Possible Causes** ● Transmission shaft rotor is chipped or damaged ● OSS tip contains debris or metal shavings (an intermittent fault) ● OSS positive (+) or (-) signal circuit open or shorted to ground ● OSS is damaged or has failed
P0503 **1T CCM, MIL: Yes** 2002, 2003, 2004, 2005 Envoy & TrailBlazer 4.2L VIN S, 5.3L VIN P Transmissions: A/T	**VSS Circuit Intermittent Sensor (4L60-E, 4L65-E) Conditions:** DTC P1810 not set, engine started, time since the last gearshift change and last 4WD low state change over 6 seconds, VSS signal increase not more than 600 rpm within the last 2 seconds, and the PCM detected the VSS signal decreased by more than 1300 rpm for 3 seconds while not in P/N, or the VSS signal decreased by over 8,129 rpm for 409 seconds with the gearshift not in Park or Neutral. **Possible Causes** ● Output shaft, AWD, 2WD or 4WD rotor debris or it is damaged ● VSS tip contains debris or metal shavings (an intermittent fault) ● VSS (+) signal circuit is open or shorted to ground (intermittent) ● VSS (-) signal circuit is open or shorted to ground (intermittent) ● VSS is damaged or has failed (an intermittent fault)

OBD II Trouble Code List (P0xxx Codes)

DTC	Trouble Code Title, Conditions & Possible Causes
P0505 **2T CCM, MIL: Yes** 1997, 1998 Catera 3.0L VIN R engine Transmissions: All	**Idle Air Control Valve Circuit Malfunction Conditions:** Engine started, and the PCM detected an unexpected voltage condition on the IAC motor control circuit during the CCM test period **Possible Causes** ● IAC motor open circuit is open, shorted to ground or to power ● IAC motor closed circuit is open, shorted to ground or to power ● IAC motor power circuit is open (test for power from main relay) ● IAC motor is damaged or has failed ● PCM has failed
P0506 **2T CCM, MIL: Yes** 1996, 1997, 1998, 1999, 2000, 2001, 2002, 2003, 2004, 2005 Achieva, Alero, Beretta, Cavalier, Ciera, Century, Envoy, Grand Am, Sunfire Skyhawk, S/T Blazer, S/T Pickup & TrailBlazer 2.2L VIN 4, 2.2L VIN F, 2.4L VIN T, 2.2L VIN H & 4.2L VIN S engines Transmissions: All	**Idle Speed Too Low Conditions:** DTC P0105, P0107, P0108, P0112, P0113, P0117, P0118, P0122, P0123, P0125, P0128, P0130-P0134, P0171, P0172, P0201-P0204, P0300, P0301-P0304, P0336, P0440, P0442, P0446, P0452, P0453, P0502, P0503, P1133 and P1441 not set, engine started, engine runtime over 20 seconds, system voltage over 10.0v, ECT sensor more than 104ºF, BARO sensor more than 72 kPa, and the PCM detected the Actual idle speed was 60 rpm less than Desired idle speed for 13 seconds with the IAC position over 145 counts. **Possible Causes** ● Air inlet duct is collapsed, loose or air filter element is clogged ● Base engine problem (i.e., compression or misfire condition) ● Idle air inlet passage or throttle bore is dirty or full of deposits ● IAC solenoid control circuit has a high resistance condition ● IAC valve is damaged or has failed ● MAF sensor is dirty, out-of-calibration or it is "skewed" ● Throttle plate, throttle shaft or linkage is damaged or sticking ● PCM has failed
P0506 **2T CCM, MIL: Yes** 1996, 1997, 1998, 1999, 2000, 2001, 2002, 2003, 2004, 2005 Achieva, Alero, Beretta, Cavalier, Grand Am, Sunfire, Skyhawk, S/T Blazer & S/T Pickup 2.2L VIN 4, 2.2L VIN 5, 2.2L VIN F, 2.4L VIN T Transmissions: M/T	**Idle Speed Too Low Conditions:** DTC P0105, P0107, P0108, P0112, P0113, P0117, P0118, P0122, P0123, P0125, P0128, P0130-P0134, P0171, P0172, P0201-P0204, P0300, P0301-P0304, P0336, P0440, P0442, P0446, P0452-P0453, P0502, P0503, P1133 and P1441 not set, engine runtime over 20 seconds, system voltage over 10.0v, ECT sensor over 104ºF, BARO sensor over 72 kPa, and the PCM detected the Actual speed was 100 rpm less than Desired idle speed for 13 seconds (Scan Tool shows IAC over 145). **Possible Causes** ● Air inlet duct is collapsed, loose or air filter element is clogged ● Base engine problem (i.e., compression or misfire condition) ● Idle air inlet passage or throttle bore is dirty or full of deposits ● IAC solenoid control circuit has a high resistance condition ● IAC valve is damaged or has failed ● MAP sensor is dirty, out-of-calibration or "skewed" ● Throttle plate, throttle shaft or linkage is damaged or sticking ● PCM has failed
P0506 **2T CCM, MIL: Yes** 1996, 1997, 1998, 1999, 2000, 2001, 2002, 2003, 2004, 2005 Alero, Aztek, Cutlass, Impala, Malibu, Montana, Rendezvous, Silhouette & Venture 3.1L VIN J, 3.1L VIN M, 3.4L VIN E engines Transmissions: All	**Idle Speed Too Low Conditions:** DTC P0102, P0103, P0107, P0108, P0121-P0123, P0300, P0301-P0306, P0401-P0405, P0440, P0442, P0446, P0502, P0503, P1404 and P1441 not set, engine started, engine runtime over 2 minutes, system voltage over 10.0v, vehicle speed less than 3 mph, ECT sensor more than 158ºF, IAT sensor more than 5ºF, TP angle less than 1%, BARO sensor more than 65 kPa, and the PCM detected that Actual idle speed was more than 100 rpm lower than Desired idle speed for 15 seconds. The IAC valve, mounted on the throttle body, is used to control the engine idle speed. The IAC valve pintle moves in and out of an idle air passage bore to control airflow past the throttle plate. The IAC valve consists of a movable pintle, driven by a gear attached to an electric motor called a stepper motor. The stepper motor is capable of highly accurate rotation (called steps). The stepper motor has two separate windings called coils. Each coil is supplied current by two circuits from the PCM. Each time the coil polarity is changed, the stepper motor moves one step. The PCM uses a preset number of counts to determine IAC pintle position. **Possible Causes** ● Air inlet duct is collapsed, loose or air filter element is clogged ● Base engine problem (i.e., compression or misfire condition) ● Idle air inlet passage or throttle bore is dirty or full of deposits ● IAC solenoid control circuit has a high resistance condition ● IAC valve is damaged or has failed ● MAF sensor is dirty, out-of-calibration or it is "skewed" ● Throttle plate, throttle shaft or linkage is damaged or sticking ● PCM has failed

OBD II Trouble Code List (P0xxx Codes)

DTC	Trouble Code Title, Conditions & Possible Causes
P0506 **2T CCM, MIL: Yes** 1996, 1997 Camaro, Caprice, Corvette, Firebird, Fleetwood & Roadmaster 4.3L VIN W, 5.7L VIN 5, 5.7L VIN P Transmissions: All	**Idle Speed Too Low Conditions:** DTC P0107, P0108, P0122, P0123, P0174, P0175, P0300 and P0325 not set, engine started, engine runtime 25 seconds, vehicle speed less than 1 mph, BARO sensor more than 77 kPa, ECT sensor more than 140°F, IAT sensor more than 14°F, TP angle at 0%, system voltage over 10.0v, then vehicle driven to 20-77 mph, MAF sensor from 13-35 g/sec, TP sensor and RPM steady for 5 seconds, idle speed too low, then with the Active test enabled, the PCM detected a 6 g/sec change in MAF and DTC P0506 is set. If the change in MAF is less than 6 g/sec, it will set DTC P1508. The IAC valve, mounted on the throttle body, is used to control the engine idle speed. The IAC valve pintle moves in and out of an idle air passage bore to control airflow past the throttle plate. The IAC valve consists of a movable pintle, driven by a gear attached to an electric motor called a stepper motor. The stepper motor is capable of highly accurate rotation (called steps). **Possible Causes** • Air inlet duct is collapsed, loose or air filter element is clogged • Base engine problem (i.e., compression or misfire condition) • Idle air inlet passage or throttle bore is dirty or full of deposits • IAC solenoid control circuit has a high resistance condition • IAC valve is damaged or has failed • MAF sensor is dirty, out-of-calibration or it is "skewed" • Throttle plate, throttle shaft or linkage is damaged or sticking • PCM has failed
P0506 **2T CCM, MIL: Yes** 1996, 1997, 1998, 1999, 2000, 2001, 2002, 2003, 2004, 2005 Impala, Monte Carlo, Montana, Silhouette, Venture, Rendezvous 3.4L VIN E engine Transmissions: All	**Idle Speed Too Low Conditions:** DTC P0101-P0103, P0107-P0108, P0112-P0113, P0116-P0118, P0121-P0123, P0171-P0172, P0201-P0206, P0300, P0401-P0405 P0443, P1121, P1404 and P1441 not set, engine runtime over 2 minutes, ECT sensor over 158°F, IAT sensor over 4°F, BARO sensor over 70 kPa, system voltage over 10.0v, TPS angle under 1.5%, VSS less than 3 mph, and the PCM detected the Actual speed was 100 rpm less than the Desired speed for 8 seconds. **Possible Causes** • Air inlet duct is collapsed, loose or air filter element is clogged • Base engine problem (i.e., compression or misfire condition) • Idle air inlet passage or throttle bore is dirty or full of deposits • IAC valve is damaged or has failed • MAF sensor is dirty, out-of-calibration or it is "skewed" • Throttle plate, throttle shaft or linkage is damaged or sticking • PCM has failed
P0506 **2T CCM, MIL: Yes** 1996, 1997, 1998, 1999, 2000, 2001, 2002, 2003, 2004, 2005 Aurora, Bonneville, Century, Cutlass Supreme, Grand Prix, Intrigue, LeSabre, LSS, Lumina, Monte Carlo, 88', Regal, Park Avenue & 98' 3.1L VIN J, 3.1L VIN M, 3.5L VIN H, 3.8L VIN 1, 3.8L VIN K engines Transmissions: All	**Idle Speed Too Low Conditions:** DTC P0102, P0103, P0107, P0108, P0121-P0123, P0300, P0301-P0306, P0401-P0405, P0440, P0442, P0446, P0502, P0503, P1404 and P1441 not set, engine started, engine runtime over 2 minutes, system voltage over 10.0v, BARO sensor over 65-70 kPa, ECT sensor more than 158°F, IAT sensor more than 4°F, TP angle at 0%, and the PCM detected the Actual idle speed was more than 100 rpm (175 on VIN 1) lower than the Desired idle speed for 8 seconds. *Note: The PCM performs this test 5 consecutive times per key cycle.* **Possible Causes** • Air inlet duct is collapsed, loose or air filter element is clogged • Base engine problem (i.e., compression or misfire condition) • Idle air inlet passage or throttle bore is dirty or full of deposits • IAC solenoid control circuit has a high resistance condition • IAC valve is damaged or has failed • MAF sensor is dirty, out-of-calibration or it is "skewed" • Throttle plate, throttle shaft or linkage is damaged or sticking • PCM has failed
P0506 **2T CCM, MIL: Yes** 1996, 1997, 1998, 1999, 2000, 2001, 2002, 2003, 2004, 2005 C/K Cab & Chassis, C/K Series Truck, L/M Vans, S/T Blazer & S/T Pickup 4.3L VIN W, 4.3L VIN X, 5.0L VIN M, 5.7L VIN K, 5.7L VIN R, 7.4L VIN J Transmissions: All	**Idle Speed Too Low Conditions:** DTC P0101-P0103, P0106-P0108, P0112, P0113, P0116-P0118, P0121-P0123, P0125, P0128, P0171-P0175, P0200, P0300, P0440, P0442, P0443, P0446, P0449, P1111-P1115, P1121, P1122, P1380, P1381 and P1441 not set, engine runtime over 60 seconds, ECT sensor from 140-241°F, IAT sensor over 14°F, TP angle under 0.7%, BARO sensor over 65 kPa, VSS less than 1 mph, system voltage over 10.0v, and the PCM detected the Actual speed was 100 rpm below the Desired speed with a MAF sensor change of under 3 g/sec. **Possible Causes** • Air inlet duct is collapsed, loose or air filter element is clogged • Base engine problem (i.e., compression or misfire condition) • IAC solenoid control circuit has a high resistance condition • IAC valve is damaged or has failed • Idle air inlet passage or throttle bore is dirty or full of deposits • MAF sensor is dirty, out-of-calibration or it is "skewed" • Throttle plate, throttle shaft or linkage is damaged or sticking • PCM has failed

OBD II Trouble Code List (P0xxx Codes)

DTC	Trouble Code Title, Conditions & Possible Causes
P0506 **2T CCM, MIL: Yes** 1999, 2000, 2001, 2002, 2003, 2004, 2005 C/K Truck, Escalade, G Van, S/T Blazer & Pickup 4.8L VIN V, 5.3L VIN P, 5.3L VIN T, 5.3L VIN Z, 6.0L VIN N, 6.0L VIN U, 8.1L VIN G Transmissions: All	**Idle Speed Too Low Conditions:** DTC P0101-P0103, P0106-P0108, P0112, P0113, P0116-P0118, P0120, P0121-P0123, P0125, P0128, P0171-P0175, P0200, P0220, P0300, P0442, P0443, P0452, P0453, P0455, P0496, P0500-P0503 and P2135 not set, engine runtime over 60 seconds, ECT sensor over 140ºF, IAT sensor over 14ºF, TP angle under 0.7%, APP Sensor 1 at 0% on vehicles with TAC, BARO sensor over 65 kPa, VSS below 1 mph, system voltage at 9-18v, and the PCM detected the Actual speed was 100 rpm below the Desired speed for 5 seconds. **Possible Causes** ● Air inlet duct is collapsed, loose or air filter element is clogged ● Base engine problem (i.e., compression or misfire condition) ● IAC solenoid control circuit has a high resistance condition ● IAC valve is damaged or has failed ● Idle air inlet passage or throttle bore is dirty or full of deposits ● MAF sensor is dirty, out-of-calibration or it is "skewed" ● PCM has failed
P0506 **1T CCM, MIL: Yes** 2003, 2004, 2005 CTS (Cadillac) 2.6L VIN M, 3.2L VIN N Transmissions: All	**Idle Speed Too Low Conditions:** DTC P0112, P0113, P0116-P0118, P0121-P0123, P0221, P0222, P0223, P0440, P0442, P0443, P0444, P0445, P0446, P0447, P0448, P0449, P0451, P0452, P0453 and P0500 not set, engine started, ECT sensor over 140ºF, IAT sensor over 32ºF, engine load below 40%, EVAP purge not enabled and EVAP test not running, VSS at 0 mph, and the PCM detected that the Actual idle speed was less than the Desired idle speed by over 100 rpm. **Possible Causes** ● Base engine problem (i.e., compression or misfire condition) ● TAC motor connector is damaged, loose or shorted ● Throttle valves not in their at rest position ● Throttle valves binding open or are binding closed ● Throttle valves moving open or closed without spring pressure ● Throttle body assembly is damaged or it has failed ● PCM has failed
P0506 **2T CCM, MIL: Yes** 1997, 1998, 1999, 2000, 2001 Camaro, Corvette & Firebird 5.7L VIN G, 5.7L VIN S Transmissions: All	**Idle Speed Too Low Conditions:** DTC P0101-P0103, P0106-P0108, P0112-P0118, P0121-P0123, P0125, P0128, P0171-P0175, P0200, P0300, P0440-P0449, P1111-P1115, P1121, P1122, P1380, P1381 and P1441 not set, engine started, engine runtime over 1 minute, ECT sensor from 140-241ºF, IAT sensor more than 14ºF, TP angle less than 0.7%, BARO sensor more than 65 kPa, VSS sensor less than 1 mph, system voltage over 10.0v, and the PCM detected the Actual idle speed was more than 100 rpm less than the Desired idle speed for 8 seconds. **Possible Causes** ● Air inlet duct is collapsed, loose or air filter element is clogged ● Base engine problem (i.e., compression or misfire condition) ● IAC valve is damaged or has failed ● Idle air inlet passage or throttle bore is dirty or full of deposits ● MAF sensor is dirty, out-of-calibration or it is "skewed" ● Throttle plate, throttle shaft or linkage is damaged or sticking ● PCM has failed
P0506 **2T CCM, MIL: Yes** 1996, 1997, 1998, 1999, 2000, 2001, 2002, 2003, 2004, 2005 DeVille, Eldorado, Seville 4.6L VIN 9, 4.6L VIN Y Transmissions: A/T	**Idle Speed Too Low Conditions:** DTC P0101-P0103, P0106-P0108, P0112, P0113, P0116-P0118, P0121, P0122, P0123, P0125, P0201-P0208, P0335, P0336, P0340, P0341, P0385, P0386, P0401, P0403, P0404, P0405, P0440, P0442, P0443, P0446, P0449, P1106, P1107, P1111, P1112, P1114, P1115, P1121, P1122, P1336, P1372, P1404 and P1441 not set, engine started, engine runtime over 2 minutes, system voltage over 10.0v, BARO sensor more than 85 kPa, ECT sensor more than 122ºF, IAT sensor over -13ºF, and the PCM detected the Actual idle speed was more than 75 rpm lower than the Desired idle for 15 seconds. The IAC valve, mounted on the throttle body, is used to control the engine idle speed. The IAC valve pintle moves in and out of an idle air passage bore to control airflow past the throttle plate. The IAC valve consists of a movable pintle, driven by a gear attached to an electric motor called a stepper motor. The stepper motor is capable of highly accurate rotation (called steps). The stepper motor has two separate windings called coils. Each coil is supplied current by two circuits from the PCM. Each time the coil polarity is changed, the stepper motor moves one step. **Possible Causes** ● Air inlet duct is collapsed, loose or air filter element is clogged ● Base engine problem (i.e., compression or misfire condition) ● Idle air inlet passage or throttle bore is dirty or full of deposits ● IAC solenoid control circuit has a high resistance condition ● IAC valve is damaged or has failed ● MAF sensor is dirty, out-of-calibration or it is "skewed" ● Throttle plate, throttle shaft or linkage is damaged or sticking ● PCM has failed

OBD II Trouble Code List (P0xxx Codes)

DTC	Trouble Code Title, Conditions & Possible Causes
P0506 **2T CCM, MIL: Yes** 1995, 1996, 1997, 1998, 1999, 2000, 2001, 2002 Camaro & Firebird 3.8L VIN K engine Transmissions: All	**Idle Speed Too Low Conditions:** DTC P0101-P0103, P0107, P0108, P0112-P0118, P0171-P0175, P0200-P0206, P0300, P0336, P0401-P0405, P0502, P0503, P1112-P1115, P1120, P1220, P1221, P1374, P1380 and P1381 not set, engine runtime over 2 minutes, ECT and IAT sensors more than -40°F, BARO sensor over 65 kPa, system voltage 9-18v, VSS sensor under 3 mph, for models with TAC, APP sensor indicated angle less than 25%, and the PCM detected the Actual idle speed was 150 rpm lower than the Desired idle speed for 15 seconds. **Possible Causes** ● Air inlet duct is collapsed, loose or air filter element is clogged ● Base engine problem (i.e., compression or misfire condition) ● Idle air inlet passage or throttle bore is dirty or full of deposits ● IAC valve is damaged or has failed ● MAF sensor is dirty, out-of-calibration or it is "skewed" ● Throttle plate, throttle shaft or linkage is damaged or sticking ● PCM has failed
P0506 **2T CCM, MIL: Yes** 1996, 1997, 1998, 1999, 2000, 2001, 2002, 2003, 2004, 2005 Aurora 4.0L VIN C engine Transmissions: A/T	**Idle Speed Too Low Conditions:** DTC P0101-P0103, P0106-P0108, P0112-P0118, P0121 P0123, P0201-P0208 and P0502 not set, DTC P1370, P1371, P1406 and P1441 not set, engine runtime over 40 seconds, system voltage over 10.0v, BARO sensor more than 65 kPa, ECT sensor from -4 to 232°F, MAF sensor from 2-35 g/sec, Actual idle speed less than Desired idle speed by at least 80 rpm, TP angle less than 1%, Calculated TP switch closed, vehicle not moving, no change in the Transmission gear state or TCC for 3 seconds, EGR system test inactive, and the PCM detected the Actual idle speed was more than 80 rpm below the Desired idle for 15 seconds. *Note: The PCM performs this test 3 consecutive times per key cycle.* **Possible Causes** ● Air inlet duct is collapsed, loose or air filter element is clogged ● Base engine problem (i.e., compression or misfire condition) ● Idle air inlet passage or throttle bore is dirty or full of deposits ● IAC valve is damaged or has failed ● MAF sensor is dirty, out-of-calibration or it is "skewed" ● Throttle plate, throttle shaft or linkage is damaged or sticking ● PCM has failed
P0506 **2T CCM, MIL: Yes** 1995 S/T Blazer & S/T Pickup 4.3L VIN W Transmissions: All	**Idle Speed Too Low Conditions:** DTC P0106, P0107, P0108, P0117, P0118, P0121, P0122, P0123, P0500, P0502 and P0503 not set, engine runtime over 1 minute, ECT sensor from 140-241°F, IAT sensor over 14°F, BARO sensor over 65 kPa, TP angle below 0.7%, system voltage over 10.0v, VSS under 1 mph, and the PCM detected the Actual idle speed was 75 rpm less than the Desired idle speed for 3 seconds. **Possible Causes** ● Air inlet duct is collapsed, loose or air filter element is clogged ● Base engine problem (i.e., compression or misfire condition) ● Idle air inlet passage or throttle bore is dirty or full of deposits ● IAC valve is damaged or has failed ● MAP sensor is dirty, out-of-calibration or "skewed" ● Throttle plate, throttle shaft or linkage is damaged or sticking ● PCM has failed
P0506 **2T CCM, MIL: Yes** 1997, 1998, 1999, 2000, 2001 Catera 3.0L VIN R engine Transmissions: A/T	**Idle Speed Too Low Conditions:** DTC P0110, P0115, P0116, P0411, P0440, P0442, P0443, P0446, P0455, P0500, P1120, P1220 not set, engine started, engine running with the engine load less than 2%, ECT sensor more than 176°F, IAT sensor more than 32°F, AIR and EVAP tests inactive, VSS sensor at 0 mph, and the PCM detected the Actual idle speed was 100 rpm less than the Desired idle speed for 5 seconds. **Possible Causes** ● Air inlet duct is collapsed, loose or air filter element is clogged ● Base engine problem (i.e., compression or misfire condition) ● Idle air inlet passage or throttle bore is dirty or full of deposits ● IAC valve is damaged or has failed ● MAF sensor is dirty, out-of-calibration or it is "skewed" ● Throttle plate, throttle shaft or linkage is damaged or sticking ● PCM has failed

OBD II Trouble Code List (P0xxx Codes)

DTC	Trouble Code Title, Conditions & Possible Causes
P0506 **2T CCM, MIL: Yes** 1996, 1997 Grand Prix, Lumina, Monte Carlo 3.4L VIN X Transmissions: All	**Idle Speed Too Low Conditions:** DTC P0106, P0107, P0108, P0112, P0113, P0117, P0118, P0121, P0122, P0123, P0200, P0300, P0301-P0304, P0335, P0502, P0506, P0507 and P1441 not set, engine started, BARO sensor over 65 kPa, vehicle speed less than 3 mph, ECT sensor over 158ºF, system voltage over 10.0v, engine runtime over 2 minutes, IAT sensor more than 5ºF, TP angle less than 1%, and the PCM detected the Actual idle speed was more than 100 rpm lower than the Desired idle speed for 15 seconds in the CCM Rationality test. **Possible Causes** ● Air inlet duct is collapsed, loose or air filter element is clogged ● Base engine problem (i.e., compression or misfire condition) ● Idle air inlet passage or throttle bore is dirty or full of deposits ● IAC solenoid control circuit has a high resistance condition ● IAC valve is damaged or has failed ● MAF sensor is dirty, out-of-calibration or it is "skewed" ● Throttle plate, throttle shaft or linkage is damaged or sticking ● PCM has failed
P0506 **2T CCM, MIL: Yes** 2002, 2003, 2004, 2005 Corvette 5.7L VIN G, 5.7L VIN S Transmissions: All	**Idle Speed Too Low Conditions:** DTC P0107, P0108, P0112, P0113, P0117, P0118, P0125, P0171, P0172, P0200, P0300, P0336, P0440, P0442, P0446, P0452, P0453, P0502, P0503, P1120, P1220, P1221, P1514, P1515, P1516, P1635 and P1639 not set, engine runtime over 2 seconds, ECT sensor more than -40ºF, IAT sensor more than -40ºF, BARO sensor more than 65 kPa, system voltage from 6-18v, VSS less than 3 mph, and the PCM detected the Actual idle speed was more than 105 rpm less than the Desired idle speed for 15 seconds. **Possible Causes** ● Air inlet duct is collapsed, loose or air filter element is clogged ● Base engine problem (i.e., compression or misfire condition) ● IAC valve is damaged or has failed ● Idle air inlet passage or throttle bore is dirty or full of deposits ● MAF sensor is dirty, out-of-calibration or it is "skewed" ● Throttle plate, throttle shaft or linkage is damaged or sticking ● PCM has failed
P0507 **2T CCM, MIL: Yes** 1996, 1997, 1998, 1999, 2000, 2001, 2002, 2003, 2004, 2005 Achieva, Alero, Beretta, Cavalier, Ciera, Century, Envoy, Grand Am, Sunfire Skyhawk, S/T Blazer, S/T Pickup & TrailBlazer 2.2L VIN 4, 2.2L VIN 5, 2.2L VIN F, 2.2L VIN H, 2.4L VIN T, 4.2L VIN S Transmissions: All	**Idle Speed Too High Conditions:** DTC P0105, P0107, P0108, P0112, P0113, P0117, P0118, P0122, P0123, P0125, P0128, P0130-P0134, P0171, P0172, P0201-P0204, P0300, P0301-P0304, P0336, P0440, P0442, P0446, P0452, P0453, P0502, P0503, P1133 and P1441 not set, engine started, engine runtime over 20 seconds, system voltage over 10.0v, ECT sensor more than 104ºF, BARO sensor more than 72 kPa, and the PCM detected the Actual idle speed was 60 rpm more than Desired idle speed for 13 seconds with the IAC position under 2 counts. **Possible Causes** ● Engine vacuum leaks, PCM valve is leaking or the wrong valve ● Idle air inlet passage or throttle bore is dirty or full of deposits ● IAC valve is damaged or has failed ● MAF sensor is dirty, "skewed" or installed improperly ● Throttle plate, throttle shaft or linkage is damaged or sticking ● TP sensor is out-of-range or "skewed" high ● PCM has failed
P0507 **2T CCM, MIL: Yes** 1996, 1997, 1998, 1999, 2000, 2001, 2002, 2003, 2004, 2005 Achieva, Alero, Beretta, Cavalier, Ciera, Century, Envoy, Grand Am, Sunfire Skyhawk, S/T Blazer, S/T Pickup & TrailBlazer 2.2L VIN 4, 2.2L VIN F, 2.4L VIN T Transmissions: M/T	**Idle Speed Too High Conditions:** DTC P0105, P0107, P0108, P0112, P0113, P0117, P0118, P0122, P0123, P0125, P0128, P0130-P0134, P0171, P0172, P0201-P0204, P0300, P0301-P0304, P0336, P0440, P0442, P0446, P0452, P0453, P0502, P0503, P1133 and P1441 not set, engine started, engine runtime over 20 seconds, system voltage over 10.0v, ECT sensor more than 104ºF, BARO sensor more than 72 kPa, and the PCM detected the Actual idle speed was 100 rpm more than Desired idle speed for 19 seconds with the IAC position under 2 counts. The IAC valve, mounted on the throttle body, is used to control the engine idle speed. The IAC valve pintle moves in and out of an idle air passage bore to control airflow past the throttle plate. The IAC valve consists of a movable pintle, driven by a gear attached to an electric motor called a stepper motor. Each coil is supplied current by two circuits within the PCM. Each time the PCM changes the coil polarity, the motor moves one step. The PCM uses a predetermined number of counts to calculate IAC pintle position. **Possible Causes** ● Engine vacuum leaks, PCM valve is leaking or the wrong valve ● Idle air inlet passage or throttle bore is dirty or full of deposits ● IAC valve is damaged or has failed ● MAF sensor is dirty, "skewed" or installed improperly ● Throttle plate, throttle shaft or linkage is damaged or sticking ● TP sensor is out-of-range or "skewed" high ● PCM has failed

OBD II Trouble Code List (P0xxx Codes)

DTC	Trouble Code Title, Conditions & Possible Causes
P0507 **2T CCM, MIL: Yes** 1996, 1997, 1998, 1999, 2000, 2001, 2002, 2003, 2004, 2005 Alero, Aztek, Cutlass, Impala, Malibu, Montana, Rendezvous, Silhouette & Venture 3.1L VIN J, 3.1L VIN M, 3.4L VIN E engines Transmissions: All	**Idle Speed Too High Conditions:** DTC P0102, P0103, P0107, P0108, P0121-P0123, P0300, P0301-P0306, P0401-P0405, P0440, P0442, P0446, P0502, P0503, P1404 and P1441 not set, engine started, engine runtime over 2 minutes, system voltage over 10.0v, vehicle speed less than 3 mph, ECT sensor more than 158°F, IAT sensor more than 5°F, TP angle less than 1%, BARO sensor more than 65 kPa, and the PCM detected that Actual idle speed was more than 75 rpm higher than Desired idle speed for 15 seconds. The IAC valve, mounted on the throttle body, is used to control the engine idle speed. The IAC valve pintle moves in and out of an idle air passage bore to control airflow past the throttle plate. The IAC valve consists of a movable pintle, driven by a gear attached to an electric motor called a stepper motor. The stepper motor is capable of highly accurate rotation (called steps). The stepper motor has two separate windings called coils. Each coil is supplied current by two circuits from the PCM. Each time the coil changes polarity, the stepper motor moves one step. The PCM uses a predetermined number of counts to calculate IAC pintle position. **Possible Causes** ● Engine vacuum leaks, PCM valve is leaking or the wrong valve ● Idle air inlet passage or throttle bore is dirty or full of deposits ● IAC valve is damaged or has failed ● MAF sensor is dirty, "skewed" or installed improperly ● Throttle plate, throttle shaft or linkage is damaged or sticking ● PCM has failed
P0507 **2T CCM, MIL: Yes** 1996, 1997 Camaro, Caprice, Corvette, Firebird, Fleetwood & Roadmaster 4.3L VIN W, 5.7L VIN 5, 5.7L VIN P Transmissions: All	**Idle Speed Too High Conditions:** DTC P0107, P0108, P0122, P0123, P0174, P0175, P0300 and P0325 not set, engine started, engine runtime 25 seconds, vehicle speed less than 1 mph, BARO sensor more than 77 kPa, ECT sensor more than 140°F, IAT sensor more than 14°F, TP angle at 0%, system voltage over 10.0v, then vehicle driven to 20-77 mph, MAF sensor from 13-35 g/sec, engine speed and TP sensor steady for 5 seconds, then with the Active test enabled, the PCM detected a 6 g/sec change in MAF and DTC P0507 is set. If the change in MAF is more than 6 g/sec, it will set DTC P1509 under these conditions. **Possible Causes** ● Engine vacuum leaks, PCM valve is leaking or the wrong valve ● Idle air inlet passage or throttle bore is dirty or full of deposits ● IAC valve is damaged or has failed ● MAF sensor is dirty, "skewed" or installed improperly ● Throttle plate, throttle shaft or linkage is damaged or sticking ● PCM has failed
P0507 **2T CCM, MIL: Yes** 1996, 1997, 1998, 1999, 2000, 2001, 2002, 2003, 2004, 2005 Impala, Monte Carlo, Montana, Silhouette, Venture, Rendezvous 3.4L VIN E engine Transmissions: All	**Idle Speed Too High Conditions:** DTC P0101-P0103, P0107-P0108, P0112-P0113, P0116-P0118, P0121-P0123, P0171-P0172, P0201-P0206, P0300, P0401-P0405 P0443, P1121, P1404 and P1441 not set, engine runtime 2 minutes, ECT sensor over 158°F, IAT sensor over 4°F, BARO sensor over 70 kPa, system voltage over 10.0v, TPS angle under 1.5%, VSS less than 3 mph, and the PCM detected the Actual speed was 150 rpm more than the Desired speed for 8 seconds. **Possible Causes** ● Engine vacuum leaks, PCM valve is leaking or the wrong valve ● Idle air inlet passage or throttle bore is dirty or full of deposits ● IAC valve is damaged or has failed ● MAF sensor is dirty, "skewed" or installed improperly ● Throttle plate, throttle shaft or linkage is damaged or sticking ● PCM has failed
P0507 **2T CCM, MIL: Yes** 1996, 1997, 1998, 1999, 2000, 2001, 2002, 2003, 2004, 2005 Aurora, Bonneville, Century, Cutlass Supreme, Grand Prix, Intrigue, LeSabre, LSS, Lumina, Monte Carlo, 88', Regal, Park Avenue & 98' 3.1L VIN J, 3.1L VIN M, 3.5L VIN H Transmissions: All	**Idle Speed Too High Conditions:** DTC P0102, P0103, P0107, P0108, P0121-P0123, P0300, P0301-P0306, P0401-P0405, P0440, P0442, P0446, P0502, P0503, P1404 and P1441 not set, engine started, engine runtime over 2 minutes, system voltage over 10.0v, BARO sensor over 65-70 kPa, ECT sensor more than 158°F, IAT sensor more than 4°F, TP angle at 0%, and the PCM detected the Actual idle speed was more than 100 rpm higher than the Desired idle speed for 8 seconds. The PCM performs this test 5 consecutive times per key cycle. **Possible Causes** ● Engine vacuum leaks, PCM valve is leaking or the wrong valve ● Idle air inlet passage or throttle bore is dirty or full of deposits ● IAC valve is damaged or has failed ● MAF sensor is dirty, "skewed" or installed improperly ● Throttle plate, throttle shaft or linkage is damaged or sticking ● PCM has failed

OBD II Trouble Code List (P0xxx Codes)

DTC	Trouble Code Title, Conditions & Possible Causes
P0507 **2T CCM, MIL: Yes** 1996, 1997, 1998, 1999, 2000, 2001, 2002, 2003, 2004, 2005 Aurora, Bonneville, Century, Cutlass Supreme, Grand Prix, Intrigue, LeSabre, LSS, Lumina, Monte Carlo, 88', Regal, Park Avenue & 98' 3.8L VIN 1, 3.8L VIN K Transmissions: All	**Idle Speed Too High Conditions:** DTC P0102, P0103, P0107, P0108, P0121-P0123, P0300, P0301-P0306, P0401-P0405, P0440, P0442, P0446, P0502, P0503, P1404 and P1441 not set, engine started, engine runtime over 2 minutes, system voltage over 10.0v, BARO sensor over 65-70 kPa, ECT sensor more than 158ºF, IAT sensor more than 4ºF, TP angle at 0%, and the PCM detected the Actual idle speed was more than 100 rpm (180 rpm on VIN 1) higher than the Desired idle speed for 8 seconds. This test is performed 5 consecutive times each key cycle. **Possible Causes** ● Engine vacuum leaks, PCM valve is leaking or the wrong valve ● Idle air inlet passage or throttle bore is dirty or full of deposits ● IAC valve is damaged or has failed ● MAF sensor is dirty, "skewed" or installed improperly ● Throttle plate, throttle shaft or linkage is damaged or sticking ● PCM has failed
P0507 **2T CCM, MIL: Yes** 1996, 1997, 1998, 1999, 2000, 2001, 2002, 2003, 2004, 2005 C/K Cab & Chassis, C/K Series Truck, L/M Vans, S/T Blazer & S/T Pickup 4.3L VIN W, 4.3L VIN X, 5.0L VIN M, 5.7L VIN K, 5.7L VIN R, 7.4L VIN J Transmissions: A/T	**Idle Speed Too High Conditions:** DTC P0101-P0103, P0106-P0108, P0112, P0113, P0116-P0118, P0121-P0123, P0125, P0128, P0171-P0172, P0174-P0175, P0200, P0300, P0440, P0442, P0443, P0446, P0449, P1111, P1112, P1114, P1115, P1121, P1122, P1380, P1381 and P1441 not set, engine started, engine runtime over 60 seconds, ECT sensor from 140-241ºF, IAT sensor more than 14ºF, TP angle less than 0.7%, BARO sensor more than 65 kPa, VSS sensor less than 1 mph, system voltage over 10.0v, and the PCM detected the Actual idle speed was more than 200 rpm higher than the Desired idle speed for 6 seconds during the CCM test. **Possible Causes** ● Engine vacuum leaks, PCM valve is leaking or the wrong valve ● IAC valve is damaged or has failed ● Idle air inlet passage or throttle bore is dirty or full of deposits ● MAF sensor is dirty, "skewed" or installed improperly ● Throttle plate, throttle shaft or linkage is damaged or sticking ● TP sensor is out-of-range or "skewed" high ● PCM has failed
P0507 **2T CCM, MIL: Yes** 1999, 2000, 2001, 2002, 2003, 2004, 2005 C/K Truck, Escalade, G Van & Cargo Van, Envoy & TrailBlazer 4.8L VIN V, 5.3L VIN P, 5.3L VIN T, 5.3L VIN Z, 6.0L VIN N, 6.0L VIN U, 8.1L VIN G Transmissions: All	**Idle Speed Too High Conditions:** DTC P0101-P0103, P0106-P0108, P0112, P0113, P0116-P0118, P0120, P0121-P0123, P0125, P0128, P0171, P0172, P0174, P0175, P0200, P0220, P0300, P0442, P0443, P0452, P0453, P0455, P0496, P0500, P0502, P0503 and P2135 not set, engine runtime over 60 seconds, ECT sensor over 140ºF, IAT sensor over 14ºF, TP angle less than 0.7%, APP Sensor 1 at 0% on vehicles with TAC system, BARO sensor over 65 kPa, VSS sensor less than 1 mph, system voltage 9-18v, and the PCM detected the Actual idle speed was 200 rpm above the Desired idle speed for 5 seconds. **Possible Causes** ● Engine vacuum leaks, PCM valve is leaking or the wrong valve ● Idle air inlet passage or throttle bore is dirty or full of deposits ● IAC valve is damaged or has failed ● MAF sensor is dirty, "skewed" or installed improperly ● Throttle plate, throttle shaft or linkage is damaged or sticking ● TP sensor is out-of-range or "skewed" high ● PCM has failed
P0507 **2T CCM, MIL: Yes** 1996, 1997, 1998, 1999, 2000, 2001, 2002, 2003, 2004, 2005 DeVille, Eldorado, Seville 4.6L VIN 9, 4.6L VIN Y Transmissions: A/T	**Idle Speed Too High Conditions:** DTC P0101-P0103, P0106-P0108, P0112, P0113, P0116-P0118, P0121, P0122, P0123, P0125, P0201-P0208, P0335, P0336, P0340, P0341, P0385, P0386, P0401, P0403, P0404, P0405, P0440, P0442, P0443, P0446, P0449, P1106, P1107, P1111, P1112, P1114, P1115, P1121, P1122, P1336, P1372, P1404 and P1441 not set, engine started, engine runtime over 2 minutes, system voltage over 10.0v, BARO sensor more than 85 kPa, ECT sensor more than 122ºF, IAT sensor over -13ºF, and the PCM detected the Actual idle speed was more than 75 rpm higher than the Desired idle for 15 seconds during the CCM test. **Possible Causes** ● Engine vacuum leaks, PCM valve is leaking or the wrong valve ● Idle air inlet passage or throttle bore is dirty or full of deposits ● IAC valve is damaged or has failed ● MAF sensor is dirty, "skewed" or installed improperly ● Throttle plate, throttle shaft or linkage is damaged or sticking ● TP sensor is out-of-range or "skewed" high ● PCM has failed

OBD II Trouble Code List (P0xxx Codes)

DTC	Trouble Code Title, Conditions & Possible Causes
P0507 **2T CCM, MIL: Yes** 1995, 1996, 1997, 1998, 1999, 2000, 2001, 2002 Camaro & Firebird 3.8L VIN K engines Transmissions: All	**Idle Speed Too High Conditions:** DTC P0101-P0103, P0107, P0108, P0112-P0118, P0171-P0175, P0200-P0206, P0300, P0336, P0401-P0405, P0502, P0503, P1112-P1115, P1120, P1220, P1221, P1374, P1380 and P1381 not set, engine runtime over 2 minutes, ECT and IAT sensors more than -40°F, BARO sensor over 65 kPa, system voltage 9-18v, VSS sensor under 3 mph, for models with TAC, APP sensor indicated angle less than 25%, and the PCM detected the Actual idle speed was 175 rpm more than the Desired idle speed for 15 seconds. **Possible Causes** ● Engine vacuum leaks, PCM valve is leaking or the wrong valve ● Idle air inlet passage or throttle bore is dirty or full of deposits ● IAC valve is damaged or has failed ● MAF sensor is dirty, "skewed" or installed improperly ● Throttle plate, throttle shaft or linkage is damaged or sticking ● PCM has failed
P0507 **2T CCM, MIL: Yes** 1997, 1998, 1999, 2000, 2001 Camaro, Corvette & Firebird 5.7L VIN G, 5.7L VIN S Transmissions: All	**Idle Speed Too High Conditions:** DTC P0101-P0103, P0106-P0108, P0112-P0118, P0121-P0123, P0125, P0128, P0171-P0175, P0200, P0300, P0440-P0449, P1111-P1115, P1121, P1122, P1000, P1381 and P1441 not set, engine started, engine runtime over 1 minute, ECT sensor from 140-241°F, IAT sensor more than 14°F, TP angle less than 0.7%, BARO sensor more than 65 kPa, VSS sensor less than 1 mph, system voltage over 10.0v, and the PCM detected the Actual idle speed was more than 100 rpm more than the Desired idle speed for 8 seconds. **Possible Causes** ● Engine vacuum leaks, PCM valve is leaking or the wrong valve ● Idle air inlet passage or throttle bore is dirty or full of deposits ● IAC valve is damaged or has failed ● MAF sensor is dirty, "skewed" or installed improperly ● Throttle plate, throttle shaft or linkage is damaged or sticking ● TP sensor is out-of-range or "skewed" high ● PCM has failed
P0507 **2T CCM, MIL: Yes** 1996, 1997, 1998, 1999, 2000, 2001, 2002, 2003, 2004, 2005 Aurora 4.0L VIN C engine Transmissions: A/T	**Idle Speed Too High Conditions:** DTC P0101-P0103, P0106-P0108, P0112-P0118, P0121-P0123, P0201-P0208 and P0502 not set, DTC P1370, P1371, P1406 and P1441 not set, engine runtime over 40 seconds, system voltage over 10.0v, BARO sensor more than 65 kPa, ECT sensor from -4 to 232°F, MAF sensor from 2-35 g/sec, Actual idle speed less than Desired idle speed by at least 80 rpm, TP angle less than 1%, Calculated TP switch closed, vehicle not moving, no change in the Transmission gear state or TCC for 3 seconds, EGR system test inactive, and the PCM detected the Actual idle speed was more than 80 rpm higher than the Desired idle for 15 seconds. The PCM performs this test 3 consecutive times per key cycle. The IAC valve consists of a movable pintle, driven by a gear attached to an electric motor called a stepper motor that is capable of highly accurate rotation. The motor has two separate windings called coils. Each coil is supplied current by two circuits from the PCM. Each time the coil changes polarity, the motor moves one step. The PCM uses a preset number of counts in order to calculate IAC pintle position. **Possible Causes** ● Engine vacuum leaks, PCM valve is leaking or the wrong valve ● Idle air inlet passage or throttle bore is dirty or full of deposits ● IAC valve is damaged or has failed ● MAF sensor is dirty, "skewed" or installed improperly ● Throttle plate, throttle shaft or linkage is damaged or sticking ● PCM has failed
P0507 **2T CCM, MIL: Yes** 1995 S/T Blazer & S/T Pickup 4.3L VIN W engine Transmissions: All	**Idle Speed Too High Conditions:** DTC P0106, P0107, P0108, P0117, P0118, P0121, P0122, P0123, P0500, P0502 and P0503 not set, engine started, engine runtime over 1 minute, ECT sensor from 140-241°F, IAT sensor more than 14°F, BARO sensor more than 65 kPa, TP angle less than 0.7%, VSS sensor under 1 mph, system voltage over 10.0v, and the PCM detected the Actual idle speed was 200 rpm more than the Desired idle speed for 3 seconds during the IAC Passive Test. **Possible Causes** ● Engine vacuum leaks, PCM valve is leaking or the wrong valve ● Idle air inlet passage or throttle bore is dirty or full of deposits ● IAC valve is damaged or has failed ● MAF sensor is dirty, "skewed" or installed improperly ● Throttle plate, throttle shaft or linkage is damaged or sticking ● PCM has failed

OBD II Trouble Code List (P0xxx Codes)

DTC	Trouble Code Title, Conditions & Possible Causes
P0507 **2T CCM, MIL: Yes** 1997, 1998, 1999, 2000, 2001 Catera 3.0L VIN R engine Transmissions: A/T	**Idle Speed Too High Conditions:** DTC P0110, P0115, P0116, P0411, P0440, P0442, P0443, P0446, P0455, P0500, P1120, P1220 not set, engine running with the engine load less than 2%, ECT sensor over 176ºF, IAT sensor over 32ºF, AIR and EVAP tests inactive, VSS sensor at 0 mph, and the PCM detected the Actual idle speed was 200 rpm more than the Desired idle speed for 5 seconds. **Possible Causes** • Engine vacuum leaks, PCM valve is leaking or the wrong valve • Idle air inlet passage or throttle bore is dirty or full of deposits • IAC valve is damaged or has failed • MAF sensor is dirty, "skewed" or installed improperly • Throttle plate, throttle shaft or linkage is damaged or sticking • TP sensor is out-of-range or "skewed" high • PCM has failed
P0507 **2T CCM, MIL: Yes** 2003, 2004, 2005 CTS (Cadillac) 2.6L VIN M, 3.2L VIN N Transmissions: All	**Idle Speed Too High Conditions:** DTC P0112, P0113, P0116, P0117, P0118, P0121, P0122, P0123, P0221, P0222, P0223, P0440, P0442, P0443, P0444, P0445, P0446, P0447, P0448, P0449, P0451, P0452, P0453 and P0500 not set, engine started, ECT sensor over 140ºF, IAT sensor over 32ºF, engine load below 40%, EVAP purge inactive and EVAP test not running, VSS at 0 mph, and the PCM detected that the Actual idle speed exceeded the Desired idle speed by over 200 rpm. **Possible Causes** • Engine vacuum leaks, PCM valve is leaking or the wrong valve • Base engine problem (i.e., compression or misfire condition) • TAC motor connector is damaged, loose or shorted • Throttle valves not in their at rest position • Throttle valves binding open or are binding closed • Throttle valves moving open or closed without spring pressure • Throttle body assembly is damaged or it has failed • PCM has failed
P0507 **2T CCM, MIL: Yes** 2002, 2003, 2004, 2005 Corvette 5.7L VIN G, 5.7L VIN S Transmissions: All	**Idle Speed Too High Conditions:** DTC P0107, P0108, P0112, P0113, P0117, P0118, P0125, P0171, P0172, P0200, P0300, P0336, P0440, P0442, P0446, P0452, P0453, P0502, P0503, P1120, P1220, P1221, P1514, P1515, P1516, P1635 and P1639 not set, engine runtime over 2 seconds, ECT sensor more than -40ºF, IAT sensor more than -40ºF, BARO sensor more than 65 kPa, system voltage from 6-18v, VSS less than 3 mph, and the PCM detected the Actual idle speed was more than 100 rpm more than the Desired idle speed for 15 seconds. **Possible Causes** • Air inlet duct is collapsed, loose or air filter element is clogged • Base engine problem (i.e., compression or misfire condition) • IAC valve is damaged or has failed • Idle air inlet passage or throttle bore is dirty or full of deposits • MAF sensor is dirty, out-of-calibration or it is "skewed" • Throttle plate, throttle shaft or linkage is damaged or sticking • PCM has failed
P0507 **2T CCM, MIL: Yes** 1996, 1997 Grand Prix, Lumina, Monte Carlo 3.4L VIN X Transmissions: All	**Idle Speed Too High Conditions:** DTC P0106-P0108, P0112, P0113, P0117, P0118, P0121-P0123, P0200, P0300-P0304, P0335, P0502, P0506, P0507 and P1441 not set, engine started, engine runtime over 2 minutes, BARO sensor over 65 kPa, VSS under 3 mph, ECT sensor over 158ºF, system voltage over 10.0v, IAT sensor over 5ºF, TP angle under 1%, and the PCM detected the Actual idle speed was more than 75 rpm higher than the Desired idle speed for 15 seconds. **Possible Causes** • Engine vacuum leaks, PCM valve is leaking or the wrong valve • Idle air inlet passage or throttle bore is dirty or full of deposits • IAC valve is damaged or has failed • MAF sensor is dirty, "skewed" or installed improperly • Throttle plate, throttle shaft or linkage is damaged or sticking • PCM has failed
P0522 **1T CCM, MIL: No** 1999, 2000, 2001, 2002, 2003, 2004, 2005 Aurora, Bonneville, C/K Truck, Corvette, LeSabre 3.5L VIN H, 3.8L VIN K, 3.8L VIN 1, 4.0L VIN C, 5.7L VIN G, 5.7L VIN S, 6.0L VIN N Transmissions: All	**Engine Oil Pressure Sensor Circuit Low Input Conditions:** DTC P1635 not set, engine started, and the PCM detected the Engine Oil Pressure (EOP) signal was less than 0.48v for 9 seconds. The sensor range is 0.5v (0 psi) to 4.5v (128 psi). **Possible Causes** • Engine oil level it too low • EOP sensor signal circuit is open or shorted to ground • EOP sensor VREF circuit is open • EOP sensor is damaged or has failed • Instrument Cluster or PCM has failed

OBD II Trouble Code List (P0xxx Codes)

DTC	Trouble Code Title, Conditions & Possible Causes
P0523 **1T CCM, MIL: No** 1999, 2000, 2001, 2002, 2003, 2004, 2005 Aurora, Bonneville, C/K Truck, Corvette, LeSabre 3.5L VIN H, 3.8L VIN K, 3.8L VIN 1, 4.0L VIN C, 5.7L VIN G, 5.7L VIN S, 6.0L VIN N Transmissions: All	**Engine Oil Pressure Sensor Circuit High Input Conditions:** DTC P1635 not set, engine started, and the PCM detected the Engine Oil Pressure (EOP) signal was more than 4.60v for 9 seconds. The sensor range is 0.5v (0 psi) to 4.5v (128 psi). **Possible Causes** • EOP sensor signal circuit is shorted to VREF or system power • EOP sensor ground circuit is open • EOP sensor is damaged or has failed • Instrument Cluster or PCM has failed
P0530 **1T CCM, MIL: No** 1996, 1997, 1998, 1999, 2000 Achieva, Alero, Beretta, Cavalier, Ciera, Century, Envoy, Grand Am, Sunfire Skylark, S/T Blazer, S/T Pickup & TrailBlazer 2.2L VIN 4, 2.4L VIN T Transmissions: All	**A/C Refrigerant Pressure Sensor Circuit Malfunction Conditions:** DTC P0112 and P0113 not set, engine started, and the PCM detected the A/C Refrigerant Pressure (ACP) sensor was more than 4.92v (453 psi) with the A/C requested or more than 3.98v (363 psi) with the A/C not requested; or it detected the ACP sensor was less than 0.10v (0 psi) for 15 seconds during the CCM test. **Possible Causes** • ACP sensor signal circuit is open or shorted to ground • ACP sensor signal circuit is shorted to VREF or system power • ACP sensor ground circuit is open • ACP sensor is damaged or has failed • PCM has failed
P0530 **1T CCM, MIL: No** 1996, 1997, 1998, 1999, 2000, 2001, 2002, 2003, 2004, 2005 Achieva, Alero, Century, Cutlass Supreme, Grand Prix, Impala, Intrigue, Lumina, Malibu, Monte Carlo, Montana, Venture, Rendezvous, Silhouette 3.1L VIN J, 3.1L VIN M, 3.4L VIN E, 3.5L VIN H, 3.8L VIN 1, 3.8L VIN K Transmissions: All	**A/C Refrigerant Pressure Sensor Circuit Malfunction Conditions:** Engine started, engine running with A/C requested "on", and the PCM detected the A/C Pressure sensor was less than 0.10v (9 psi) or it was more than 4.94v (488 kPa) for 20 seconds during the test. **Possible Causes** • ACP sensor signal circuit is open or shorted to ground • ACP sensor signal circuit is shorted to VREF or system power • ACP sensor ground circuit is open • ACP sensor is damaged or has failed • PCM has failed • TSB 61-65-61 contains a repair procedure for this code
P0530 **1T CCM, MIL: No** 1996, 1997 Camaro, Caprice, Corvette, Firebird, Fleetwood & Roadmaster 4.3L VIN W, 5.7L VIN 5, 5.7L VIN P Transmissions: All	**A/C Refrigerant Pressure Sensor Circuit Malfunction Conditions:** Engine running with A/C requested "on", and the PCM detected the A/C Pressure sensor was less than 0.10v (9 psi) or it was more than 4.88v (448 kPa) for 5 seconds during the test. **Possible Causes** • ACP sensor signal circuit is open or shorted to ground • ACP sensor signal circuit is shorted to VREF or system power • ACP sensor ground circuit is open • ACP sensor is damaged or has failed • PCM has failed
P0530 **1T CCM, MIL: No** 1995, 1996, 1997, 1998, 1999, 2000, 2003, 2004, 2005 Camaro, Corvette & Firebird 3.8L VIN K, 5.7L VIN G, 5.7L VIN S Transmissions: All	**A/C Refrigerant Pressure Sensor Circuit Malfunction Conditions:** Engine running with A/C requested "on", and the PCM detected the A/C Pressure sensor was less than 0.10v (9 psi) or it was more than 4.88v (448 kPa) for 5 seconds during the test. **Possible Causes** • ACP sensor signal circuit is open or shorted to ground • ACP sensor signal circuit is shorted to VREF or system power • ACP sensor ground circuit is open • ACP sensor is damaged or has failed • PCM has failed
P0530 **1T CCM, MIL: No** 1996, 1997 Grand Prix, Lumina, Monte Carlo3.4L VIN X Transmissions: All	**A/C Refrigerant Pressure Sensor Circuit Malfunction Conditions:** Engine running with A/C requested "on", and the PCM detected the A/C Pressure sensor was less than 0.10v (9 psi) or more than 4.90v (472 kPa) for 20 seconds during the test. **Possible Causes** • ACP sensor signal circuit is open or shorted to ground • ACP sensor signal circuit is shorted to VREF or system power • ACP sensor ground circuit is open • ACP sensor is damaged or has failed • PCM has failed

OBD II Trouble Code List (P0xxx Codes)

DTC	Trouble Code Title, Conditions & Possible Causes
P0531 **1T CCM, MIL: No** 1996, 1997 Camaro, Caprice, Corvette, Firebird, Fleetwood & Roadmaster 4.3L VIN W, 5.7L VIN 5, 5.7L VIN P Transmissions: All	**A/C Refrigerant Pressure Sensor Range/Performance Conditions:** Engine started, engine running with A/C requested "on", and the PCM detected the Air Conditioning Pressure (ACP) sensor signal did not change at least 4 psi after the A/C clutch was cycled "on" for 10 seconds, or after it has been cycled "off" for 10 seconds. **Possible Causes** ● A/C status signal circuit is open or shorted to ground ● A/C status signal circuit is shorted to VREF or system power ● ACP sensor is damaged (sticking) or has failed ● PCM has failed
P0532 **1T CCM, MIL: No** 2000, 2001, 2002, 2003, 2004, 2005 DeVille, Eldorado, Seville 4.6L VIN 9, 4.6L VIN Y Transmissions: A/T	**A/C Refrigerant Pressure Sensor Circuit Low Input Conditions:** Engine started, engine running with A/C requested "on", and the PCM detected the ACP sensor was less than 0.10v for 3 seconds. **Possible Causes** ● ACP sensor signal circuit is shorted to ground ● ACP sensor is damaged or has failed ● PCM has failed
P0533 **1T CCM, MIL: No** 2000, 2001, 2002, 2003, 2004, 2005 DeVille, Eldorado, Seville 4.6L VIN 9, 4.6L VIN Y Transmissions: A/T	**A/C Refrigerant Pressure Sensor Circuit High Input Conditions:** Engine started, engine running with A/C requested "on", and the PCM detected the ACP sensor was more than 4.90v for 3 seconds. **Possible Causes** ● ACP sensor signal circuit is open between the sensor and PCM ● ACP sensor ground circuit is open ● ACP sensor is damaged or has failed ● PCM has failed
P0540 **1T CCM, MIL: No** 2001, 2002, 2003, 2004, 2005 C/K Series Truck 6.6L VIN 1 Diesel engine Transmissions: All	**Intake Air Heater Circuit Malfunction Conditions:** Key on, engine off for over 3 seconds, ECT sensor less than 121°F, system voltage from 10-18v, or engine running with the IAT sensor signal less than 73°, and the ECM detected the Heater line voltage was more than 8.1v with the Heater relay off, or the Heater line signal was from 3.8-8.1v with the key off, or the Heater line voltage was at least 0.5v below system voltage with the ignition on, or the Heater line signal was below 3.8v with the relay on, or the reference line voltage was low with the relay off. The ECM uses an Intake Air Heater (IAH) to warm incoming air for proper cylinder combustion. **Possible Causes** ● IAT relay connector is damaged, loose or shorted ● IAT heater connector is damaged, loose or shorted ● IAT assembly is damaged, or it has failed ● IAH relay control circuit is open, shorted to ground or power ● ECM has failed
P0540 **1T CCM, MIL: No** 2001, 2002, 2003, 2004, 2005 C/K Series Truck 6.6L VIN 1 Diesel engine Transmissions: All	**Intake Air Heater Diagnostic Circuit Malfunction Conditions:** Key on, engine off for over 3 seconds, ECT sensor less than 121°F, system voltage from 10-18v, or engine running with the IAT sensor signal less than 73°, and the ECM detected the Heater line voltage was between 3.8-8.1v with the IAH relay "off", or the Heater line voltage was less than 0.5v below the battery voltage value with the IAH relay "on". The ECM uses an intake air heater (IAH) to warm the incoming air for proper cylinder combustion. The ECM grounds the control coil of the IAH relay to energize the heater during cold operation. The ECM sends a bias voltage on Diagnostic Circuits 1 and 2. The ECM sets this code if the voltage does not go low with the relay off, or if the voltage did not go high with the relay "on". **Possible Causes** ● IAT relay connector is damaged, loose or shorted ● IAT heater connector is damaged, loose or shorted ● IAT assembly is damaged, or it has failed ● IAH relay control circuit is open, shorted to ground or power ● Intake heater diagnostic circuit 1 or circuit 2 is open or shorted ● ECM has failed
P0550 **1T CCM, MIL: No** 1996, 1997, 1998, 1999 Aurora, DeVille, Eldorado, Seville 4.0L VIN C, 4.6L VIN 9, 4.6L VIN Y engines Transmissions: All	**Power Steering Switch Circuit Low Input Conditions:** Engine started, vehicle driven to a speed of over 50 mph, and the PCM detected the PSP switch circuit was open (0v) for 60 seconds. **Possible Causes** ● PSP switch signal circuit is open or shorted to ground ● PSP switch power circuit is open (test CRUISE fuse in R/CTR) ● ACP sensor is damaged (sticking open) or has failed ● PCM has failed

OBD II Trouble Code List (P0xxx Codes)

DTC	Trouble Code Title, Conditions & Possible Causes
P0560 **1T CCM, MIL: No** 1996, 1997, 1998, 1999, 2000, 2001, 2002, 2003, 2004, 2005 Achieve, Alero, Aurora, Bonneville, Century, Cutlass Supreme, Grand Prix, Impala, Intrigue, LeSabre, LSS, Lumina, Monte Carlo, Park Avenue, Regal, 88' & 98' 3.1L VIN J, 3.1L VIN M, 3.4L VIN E, 3.5L VIN H Transmissions: All	**System Voltage Malfunction Conditions:** Engine started, engine speed over 650 rpm, and the PCM detected the system voltage was less than 9v, or that it was more than 18v for 25 seconds during the CCM test. **Possible Causes** ● Check for high resistance at the battery connections or at the Underhood Fuse Block power circuit connection to the PCM ● Check the drive belt for excessive wear and the proper tension ● Check the condition of the battery and the Generator output ● Vehicle may have been used to jump-start another vehicle
P0560 **1T CCM, MIL: No** 1996, 1997, 1998, 1999, 2000, 2001, 2002, 2003, 2004, 2005 C/K Cab & Chassis, C/K Series Truck, L/M Vans, S/T Blazer & S/T Pickup 4.3L VIN W, 4.3L VIN X, 5.0L VIN M, 5.7L VIN R, 7.4L VIN J, 6.5L VIN F, 6.5L VIN S engines Transmissions: M/T	**System Voltage Malfunction Conditions:** Engine started, engine speed from 1400-1550 rpm, and the PCM detected the system voltage was less than 10.5v at a maximum TFT sensor signal of 305°F, or that it was less than 6.7v at a minimum TFT sensor signal of -40°F, or that it was more than 19v at any time. **Possible Causes** ● Check for high resistance at the battery connections or at the Underhood Fuse Block power circuit connection to the PCM ● Check the drive belt for excessive wear and the proper tension ● Check the condition of the battery and the Generator output ● Vehicle may have been used to jump-start another vehicle
P0560 **1T CCM, MIL: No** 2003, 2004, 2005 CTS (Cadillac) 2.6L VIN M, 3.2L VIN N Transmissions: All	**System Voltage Malfunction Conditions:** Engine started, engine speed more than 650 rpm, system voltage below 2.5v, and the PCM detected the system voltage was out of its normal operating range for 30 seconds. **Possible Causes** ● Check for high resistance at battery connections or the U/H Fuse Block power circuit ● Check the drive belt for excessive wear and the proper tension ● Check the condition of the battery and the Generator output
P0560 **1T CCM, MIL: No** 1996, 1997, 1998, 1999, 2001, 2002, 2003, 2004, 2005 Aurora, DeVille, Eldorado, Seville 4.0L VIN C, 4.6L VIN 9, 4.6L VIN Y engine Transmissions: All	**System Voltage Malfunction Conditions:** Engine started, engine speed over 1,500 rpm, and the PCM detected the system voltage was less than 11.0v for 2 seconds. **Possible Causes** ● Check for high resistance at the battery connections or at the Underhood Fuse Block power circuit connection to the PCM ● Check the drive belt for excessive wear and the proper tension ● Check the condition of the battery and the Generator output ● Vehicle may have been used to jump-start another vehicle
P0560 **1T CCM, MIL: No** 1995, 1996, 1997, 1998, 1999, 2000, 2001, 2002 Camaro & Firebird 3.8L VIN K engine Transmissions: All	**System Voltage Malfunction Conditions:** Engine started, engine running and the PCM detected the system voltage was less than 10v or more than 16v for over 2 seconds. **Possible Causes** ● Check for high resistance at the battery connections or at the GAGES fuse power circuit connection to the PCM ● Check the drive belt for excessive wear and the proper tension ● Check the condition of the battery and the Generator output ● Vehicle may have been used to jump-start another vehicle
P0560 **1T CCM, MIL: No** 1996, 1997, 1998, 1999, 2000, 2001, 2002, 2003, 2004, 2005 Achieva, Alero, Grand Am, Skylark, S/T Pickup 2.2L VIN 4, 2.4L VIN T Transmissions: All	**System Voltage Malfunction Conditions:** Engine started, engine speed over 1300 rpm, and the PCM detected the system voltage was less than 10v, or it detected the system voltage was over 16-17.0v for over 2 seconds. **Possible Causes** ● Check for high resistance at battery connections or at the U/H Fuse Block power circuit ● Check the drive belt for excessive wear and the proper tension ● Check the condition of the battery and the Generator output ● Vehicle may have been used to jump-start another vehicle
P0560 **1T CCM, MIL: No** 1997, 1998, 1999, 2000, 2001 Catera 3.0L VIN R engine Transmissions: A/T	**System Voltage Malfunction Conditions:** Engine started, vehicle moving, and the PCM detected the system voltage was less than 9v, or that it was more than 16v for 200 ms during the CCM test. **Possible Causes** ● Check for high resistance at battery connections or the main relay circuit power circuit ● Check the drive belt for excessive wear and the proper tension ● Check the condition of the battery and the Generator output ● Vehicle may have been used to jump-start another vehicle

OBD II Trouble Code List (P0xxx Codes)

DTC	Trouble Code Title, Conditions & Possible Causes
P0562 **1T CCM, MIL: No** 1996, 1997, 1998, 1999, 2000 Achieva, Alero, Beretta, Cavalier, Ciera, Century, Grand Am, Sunfire, S/T Blazer & Pickup, Skyhawk 2.2L VIN 4, 2.4L VIN T Transmissions: All	**System Voltage Too Low Conditions:** Engine started, vehicle driven to a speed of over 5 mph at an engine speed over 1200-1300 rpm, and the PCM detected the system voltage was less than 10v for 240 seconds. **Possible Causes** ● Check for high resistance at the battery connections or at the PCM IGN or PCM BAT fuse circuit connection to the PCM ● Check the drive belt for excessive wear and the proper tension ● Check the condition of the battery and the Generator output ● Vehicle may have been used to jump-start another vehicle
P0562 **1T CCM, MIL: No** 1996, 1997 Camaro, Caprice, Corvette, Firebird, Fleetwood & Roadmaster 4.3L VIN W, 5.7L VIN 5, 5.7L VIN P Transmissions: All	**System Voltage Too Low Conditions:** Engine started, and the PCM detected the system voltage was less than 8.0v for 6 seconds. **Possible Causes** ● Check for high resistance at the battery connections or at the PCM IGN or PCM BAT fuse circuit connection to the PCM ● Check the drive belt for excessive wear and the proper tension ● Check the condition of the battery and the Generator output ● Vehicle may have been used to jump-start another vehicle
P0562 **1T CCM, MIL: No** 2003, 2004, 2005 CTS Cadillac 2.6L VIN M, 3.2L VIN N Transmissions: All	**System Voltage Circuit Low Input Conditions:** Engine started, engine speed more than 1500 rpm, and the PCM detected the system voltage was less than 10v for 5 seconds. **Possible Causes** ● Check for high resistance at battery connections or the U/H Fuse Block power circuit ● Check the drive belt for excessive wear and the proper tension ● Check the condition of the battery and the Generator output ● Vehicle may have been used to jump-start another vehicle
P0562 **1T CCM, MIL: No** 2000, 2001, 2002, 2003, 2004, 2005 DeVille, Eldorado, Seville 4.0L VIN C, 4.6L VIN 9, 4.6L VIN Y engine Transmissions: A/T	**System Voltage Too Low Conditions:** Engine speed over 1,500 rpm, system voltage from 9.5-18v, and the PCM detected the system voltage was under 10.0v for 5 seconds. **Possible Causes** ● Check for high resistance at battery connections or the U/H Fuse Block power circuit ● Check the drive belt for excessive wear and the proper tension ● Check the condition of the battery and the Generator output ● Vehicle may have been used to jump-start another vehicle
P0562 **1T CCM, MIL: No** 1997, 1998, 1999, 2000, 2001, 2002, 2003, 2004, 2005 C/K Truck, Camaro, Corvette & Firebird 5.7L VIN G, 5.7L VIN S, 6.0L VIN N, 6.0L VIN U Transmissions: All	**System Voltage Too Low Conditions:** Engine started, vehicle driven to a speed of over 5 mph at an engine speed over 1000 rpm, and the PCM detected the system voltage was less than 8.0v for 5 seconds during the test. **Possible Causes** ● Check for high resistance at battery connections or the MINI fuse power circuit ● Check the drive belt for excessive wear and the proper tension ● Check the condition of the battery and the Generator output ● Vehicle may have been used to jump-start another vehicle
P0563 **1T CCM, MIL: No** 1997, 1998, 1999, 2000, 2001, 2002, 2003, 2004, 2005 C/K Truck, Camaro, Corvette, Firebird, G Van, S/T Blazer & S/T Pickup 4.2L VIN S, 4.3L VIN X, 4.3L VIN W, 4.8L VIN V, 5.3L VIN P, 5.3L VIN T, 5.7L VIN G, 5.7L VIN S, 6.0L VIN N, 6.0L VIN U, 8.1L VIN G engines Transmissions: All	**System Voltage Too High Conditions:** Engine started, vehicle driven to a speed of over 5 mph at an engine speed over 1000 rpm, and the PCM detected the system voltage was more than 18.0v for 5 seconds during the CCM test. **Possible Causes** ● Check the condition of the battery (it may be worn out) ● Test the operation of the Generator (it may have failed) ● Vehicle may have been used to jump-start another vehicle
P0563 **1T CCM, MIL: No** 1997, 1998, 1999, 2000, 2001, 2002, 2003, 2004, 2005 Achieva, Alero, Beretta, Cavalier, Ciera, Century, Grand Am, Sunfire, S/T Blazer & Pickup, Skyhawk 2.2L VIN 4, 2.4L VIN T Transmissions: All	**System Voltage Too High Conditions:** Engine started, vehicle driven to a speed of over 5 mph, and the PCM detected the system voltage was more than 17.0v **Possible Causes** ● Check the condition of the battery (it may be worn out) ● Test the operation of the Generator (it may be overcharging) ● Vehicle may have been used to jump-start another vehicle

OBD II Trouble Code List (P0xxx Codes)

DTC	Trouble Code Title, Conditions & Possible Causes
P0563 **1T CCM, MIL: No** 1997, 1998, 1999, 2000, 2001, 2002, 2003, 2004, 2005 C/K Truck, Camaro, Corvette, Firebird, G Van, S/T Blazer & S/T Pickup 4.2L VIN S, 4.3L VIN X, 4.3L VIN W, 4.8L VIN V, 5.3L VIN P, 5.3L VIN T, 5.7L VIN G, 5.7L VIN S, 6.0L VIN N, 6.0L VIN U, 8.1L VIN G engines Transmissions: All	**System Voltage Too High Conditions:** Engine started, and the PCM detected the system voltage was more than 17.0v for 6 seconds during the CCM test. **Possible Causes** ● Check the condition of the battery (it may be worn out) ● Test the operation of the Generator (it may be overcharging) ● Vehicle may have been used to jump-start another vehicle
P0563 **1T CCM, MIL: No** 2003, 2004, 2005 CTS (Cadillac) 2.6L VIN M, 3.2L VIN N Transmissions: All	**System Voltage Circuit High Input Conditions:** Engine speed over 1500 rpm and the PCM detected the voltage was over 16v for 5 seconds. **Possible Causes** ● Check the condition of the battery (it may be worn out) ● Test the operation of the Generator (it may be overcharging) ● Vehicle may have been used to jump-start another vehicle
P0563 **1T CCM, MIL: No** 1997, 1998, 1999, 2000, 2001, 2002, 2003, 2004, 2005 DeVille, Eldorado, Seville 4.0L VIN C, 4.6L VIN 9, 4.6L VIN Y engines Transmissions: All	**System Voltage Too High Conditions:** Engine speed more than 1,500 rpm, system voltage from 9.5-16v, and the PCM detected the system voltage was more than 6.0v for 1 second in the CCM continuous test. **Possible Causes** ● Check the condition of the battery (it may be worn out) ● Test the operation of the Generator (it may be overcharging) ● Vehicle may have been used to jump-start another vehicle
P0567 **1T CCM, MIL: No** 1997, 1998, 1999, 2000, 2001, 2002, 2003, 2004, 2005 C/K Truck, G Series Van 6.5L VIN F, 6.5L VIN S, 6.5L VIN Y engines Transmissions: All	**Cruise Control Resume Switch Circuit Malfunction Conditions:** Engine started; and the PCM detected high voltage on the Cruise Resume Switch circuit with the switch off, or with the switch on; it detected the Resume Switch was "on" for 25 seconds. **Possible Causes** ● C/C resume switch is shorted to VREF or system power ● C/C resume switch is stuck in the "on" position or has failed ● PCM has failed
P0567 **1T CCM, MIL: No** 1997, 1998, 1999, 2000, 2001, 2002, 2003, 2004, 2005 Camaro, Corvette & Firebird 5.7L VIN G, 5.7L VIN S Transmissions: All	**Cruise Control Resume Switch Circuit Malfunction Conditions:** Engine started, engine running and the PCM or TAC module (with Cruise enabled) detected the Resume/Acceleration switch indicated "on" for 90 seconds during the CCM test. **Possible Causes** ● C/C resume switch is shorted to VREF or system power ● C/C resume switch is stuck in the "on" position or has failed ● PCM has failed
P0568 **1T CCM, MIL: No** 1997, 1998, 1999, 2000, 2001, 2002, 2003, 2004, 2005 C/K Truck, G Series Van 6.5L VIN F, 6.5L VIN S, 6.5L VIN Y engines Transmissions: All	**Cruise Control Set Switch Circuit Malfunction Conditions:** Engine started; and the PCM detected high voltage on the Set Switch circuit with the Cruise Switch "off", or with Cruise Switch "on", the Set Switch indicated "on" for over 25 seconds. **Possible Causes** ● Set/Coast switch circuit is shorted to VREF or system power ● Set/Coast switch is stuck in "on" position or it has failed ● PCM has failed
P0568 **1T CCM, MIL: No** 1997, 1998, 1999, 2000, 2001, 2002, 2003, 2004, 2005 Camaro, Corvette & Firebird 5.7L VIN G, 5.7L VIN S Transmissions: All	**Cruise Control Set/Coast Switch Circuit Malfunction Conditions:** Engine started, engine running and the PCM or TAC module (with Cruise enabled) detected the Set/Coast switch indicated "on" for 90 seconds during the CCM test. **Possible Causes** ● C/C Set/Coast switch is shorted to VREF or system power ● C/C Set/Coast switch is stuck in the "on" position or has failed ● PCM has failed
P0571 **1T CCM, MIL: No** 1997, 1998, 1999, 2000, 2001, 2002, 2003, 2004, 2005 C/K Truck, G Series Van 6.5L VIN F, 6.5L VIN S, 6.5L VIN Y engines Transmissions: All	**Cruise/Brake Switch Circuit Malfunction Conditions:** Engine started, and the PCM detected the TCC and Stop/Lamp/Cruise Brake Switch signals were different for 10 consecutive minutes, or the TCC and Cruise Brake Switch signals did not cycle open and closed for 6 brake events on one trip. **Possible Causes** ● Cruise brake switch signal circuit is shorted to ground ● Cruise brake switch B+ circuit is open (test STOP/BRAKE fuse) ● Cruise brake switch is damaged (closed) or has failed

OBD II Trouble Code List (P0xxx Codes)

DTC	Trouble Code Title, Conditions & Possible Causes
P0571 **1T CCM, MIL: No** 1997, 1998, 1999, 2000, 2001, 2002, 2003, 2004, 2005 Camaro, Corvette & Firebird 5.7L VIN G, 5.7L VIN S Transmissions: All	**Cruise Control Brake Switch Circuit Malfunction Conditions:** Engine started, vehicle driven to a speed over 30 mph at an engine speed over 700 rpm (the test is disabled below 10 mph), and the PCM detected the a voltage signal on the Cruise Control Brake Switch circuit (closed) when the Brake Switch circuit should be open. **Possible Causes** ● Cruise brake switch signal circuit is shorted to ground ● Cruise brake switch B+ circuit is open (test MINI Relay & fuse) ● Cruise brake switch is damaged (closed) or has failed
P0571 **1T CCM, MIL: No** 1999, 2000, 2001 Catera 3.0L VIN R engine Transmissions: A/T	**Cruise/Brake Switch Circuit Malfunction Conditions:** Engine started, and the PCM detected the Cruise release signal did not match the stop lamp signal at least 20 times. **Possible Causes** ● Cruise brake switch signal circuit is shorted to ground ● Cruise brake switch B+ circuit is open (test Fuses in fuse block) ● Cruise brake switch is damaged (closed) or has failed
P0574 **1T CCM, MIL: No** 2000, 2001, 2002, 2003, 2004, 2005 DeVille, Eldorado, Seville 4.6L VIN 9, 4.6L VIN Y Transmissions: All	**Vehicle Speed Too High - Cruise Control Disabled Conditions:** Engine running at cruise speed with the Cruise Control system enabled and engaged, and the PCM detected the vehicle speed sensor indicated more than 110 mph (176 km/h). **Possible Causes** ● Severe wheel spin condition due to icy, wet/slippery conditions ● Vehicle operated at too high a rate of speed ● PCM has failed
P0600 **1T PCM, MIL: Yes** 1996 Cavalier, Century, Ciera, Beretta, Corsica, Grand Am, Sunfire, S/T Pickup 2.2L VIN 4, 2.4L VIN T Transmissions: All	**Loss Of Serial Communication Link Conditions:** Key on or engine running; and the PCM did not detected that it could not communicate properly with the two sides of its internal controller. **Possible Causes** ● The PCM must be replaced to correct the problem. A new PCM must be programmed with the correct software/calibration.
P0601 **1T PCM, MIL: Yes** 1996, 1997, 1998, 1999, 2000, 2001, 2002, 2003, 2004, 2005 All makes & models Transmissions: All	**Control Module ROM Malfunction Conditions:** Key in crank or the run position, and the PCM detected more than 3 incorrect checksums during its initial self-test. The PCM uses an EEPROM to store software and calibration data. The PCM uses a checksum to verify the integrity of the information. At the time of programming, the PCM calculates a checksum and stores the value in the EEPROM. The PCM retrieves this data, performs a checksum test to compare the key "on" value to the value stored in EEPROM. If these two values do not match at key "on", it sets DTC P0601. **Possible Causes** ● The PCM must be replaced to correct this problem. A new PCM must be programmed with the correct software/calibration. ● TSB 67-65-23 contains a repair procedure for this code
P0602 **1T PCM, MIL: Yes** 1996, 1997, 1998, 1999, 2000, 2001, 2002, 2003, 2004, 2005 All makes & models Transmissions: All	**Control Module Not Programmed Conditions:** Key on, and the PCM detected it did not have the correct program to operate or that the EEPROM had been programmed incorrectly. **Possible Causes** ● Reprogram the PCM with the correct software and calibration. If this step does not correct the problem, the PCM must be replaced and programmed with the correct software/calibration.
P0603 **1T PCM, MIL: Yes** 1996, 1997, 1998, 1999, 2000, 2001, 2002, 2003, 2004, 2005 All makes & models Transmissions: All	**Control Module Long Term Memory Reset Conditions:** DTC P0604 not set, and then with the key on, the PCM detected the calculated checksum that did not match the previous checksum. **Possible Causes** ● An interruption to the PCM main power and/or ground circuits ● Check the PCM power and ground circuits and make repairs as necessary. Clear the codes and recheck. If it resets, the PCM must be replaced and programmed with the correct software.
P0604 **1T PCM, MIL: Yes** 1996, 1997, 1998, 1999, 2000, 2001, 2002, 2003, 2004, 2005 All makes & models Transmissions: All	**Control Module Random Access Memory Failure Conditions:** Key on for 5 seconds, and the PCM detected the internal data test of its RAM failed. The PCM copies the program information stored in the RAM. This allows the PCM to work with, and make any updates to this data. The PCM checks for problems in all areas of the RAM. **Possible Causes** ● The PCM must be replaced to correct this problem. A new PCM must be programmed with the correct software/calibration.

OBD II Trouble Code List (P0xxx Codes)

DTC	Trouble Code Title, Conditions & Possible Causes
P0605 **1T PCM, MIL: Yes** 1998, 1999, 2000, 2001, 2002, 2003, 2004, 2005 C/K Cab & Chassis, C/K Truck, G Van, Cargo Van, M/L Series Van, S/T Blazer & S/T Pickup 4.3L VIN W, 4.3L VIN X, 5.0L VIN M, 5.7L VIN K, 5.7L VIN R, 6.6L VIN 1, 7.4L VIN J engines Transmissions: All	**Control Module Programming Read Only Memory Conditions:** Key on, and the PCM detected the data checksum did not match the expected value, or that it was unable to read its flash memory data. **Possible Causes** ● The PCM must be replaced to correct this problem. A new PCM must be programmed with the correct software/calibration.
P0605 **1T PCM, MIL: Yes** 1998, 1999, 2000, 2001, 2002, 2003, 2004, 2005 Cab & Chassis, C/K Truck, Escalade, G Van, & Cargo Van, L/M Van, S/T Blazer & S/T Pickup 4.8L VIN V, 5.3L VIN T, 5.3L VIN Z, 6.0L VIN N, 6.0L VIN U, 8.1L VIN G Transmissions: All	**Control Module Programming Read Only Memory Conditions:** Key on, and the PCM detected the data checksum did not match the expected value, or that it was unable to read its flash memory data. **Possible Causes** ● The PCM must be replaced to correct this problem. A new PCM must be programmed with the correct software/calibration.
P0606 **1T PCM, MIL: Yes** 1996, 1997, 1998, 1999, 2000, 2001, 2002, 2003, 2004, 2005 All makes & models Transmissions: All	**Control Module Internal Performance Conditions:** DTC P0601 and P0604 not set, key on, and the PCM determined that an internal performance problem existed within its controller. **Possible Causes** ● The PCM must be replaced to correct this problem. A new PCM must be programmed with the correct software/calibration.
P0607 **1T PCM, MIL: Yes** 2001, 2002, 2003, 2004, 2005 All makes & models Transmissions: All	**Control Module Performance Conditions:** Key on or engine running; then after the initial PCM power up sequence, the PCM detected an internal performance problem. **Possible Causes** ● The PCM must be replaced to correct this problem. A new PCM must be programmed with the correct software/calibration.
P0608 **1T PCM, MIL: Yes** 1997, 1998, 1999, 2000, 2001, 2002, 2003, 2004, 2005 C/K Truck, Camaro, Corvette & Firebird 5.7L VIN G, 5.7L VIN S, 6.0L VIN N, 6.0L VIN U Transmissions: All	**Vehicle Speed Output Circuit Malfunction Conditions:** Engine started, engine speed over 600 rpm, and the PCM detected the Actual and Commanded state of the VSS output circuit did not match for 5 seconds. The PCM creates the VSS output signal by causing the circuit to pulse to ground, and monitoring the operation. **Possible Causes** ● VSS output signal circuit is open, shorted to ground or to power ● VSS output signal problem related to the Instrument Cluster or the Electronic Suspension Control Module (internal problem) ● PCM has failed
P0610 **1T PCM, MIL: Yes** 2003, 2004, 2005 CTS (Cadillac) 2.6L VIN M, 3.2L VIN N Transmissions: All	**Control Module Performance Conditions:** Key on, and the engine control module (ECM) detected it was not programmed for the correct transmission application. **Possible Causes** ● The ECM must be reprogrammed to correct this problem. Once this step is done, recheck the code to verify the repair is done.
P0611 **2T CCM, MIL: Yes** 2001, 2002, 2003, 2004, 2005 C/K Series Truck 6.6L VIN 1 Diesel engine Transmissions: All	**Fuel Injection Control Module Performance Conditions:** Key on, U1800 and U2104 not set, and the FICM detected an internal fault. The fuel injection control module (FICM) performs internal circuit checks on the FICM microprocessor, the status of the monitoring module, and status of the FICM A/D conversion module. If the FICM senses a problem in the FICM circuits, the FICM will send an error message to the ECM, and it will set DTC P0611 **Possible Causes** ● FICM is damaged
P0611 **1T CCM, MIL: Yes** 1999, 2000, 2001, 2002, 2003, 2004, 2005 C/K Truck, G Series Van 6.0L VIN U CNG engine Transmissions: All	**Control Module Performance Conditions:** The Fuel Injector Control Module (FICM) monitors the CNG FTP sensor, FTT sensor, and the FRT sensor signals. The FTP, FTT, and FRT sensor values, and diagnostic data, is communicated to the PCM by two pulsewidth modulated (PWM) circuits. This code sets if the PCM detects an incorrect PWM signal from the FICM. **Possible Causes** ● FICM is damaged

OBD II Trouble Code List (P0xxx Codes)

DTC	Trouble Code Title, Conditions & Possible Causes
P0612 **1T CCM, MIL: Yes** 2001, 2002, 2003, 2004, 2005 C/K Series Truck 6.6L VIN 1 Diesel engine Transmissions: All	**Ignition Relay Control Circuit Malfunction Conditions:** Key on, and the ECM detected the feedback voltage did not match the output state. The ECM monitors the condition of the ignition relay control circuit. If the ECM senses excessive voltage on the feedback circuit, it will set DTC P0612. **Possible Causes** ● Ignition relay circuit is shorted to system power ● Ignition relay is damaged or it has failed ● ECM has failed
P0615 **1T CCM, MIL: No** 2000, 2001, 2002, 2003, 2004, 2005 Aurora, CTS (Cadillac), DeVille, Eldorado, Seville, C/K Series Truck 2.6L VIN M, 3.2L VIN N, 3.5L VIN H, 4.0L VIN C, 4.6L VIN 9, 4.6L VIN Y, 6.0L VIN N, 6.0L VIN U Transmissions: All	**Starter Relay Control Circuit Malfunction Conditions:** Engine cranking, system voltage over 10.0v, and the PCM detected an unexpected voltage condition on the Starter Relay control circuit for two seconds during the CCM test. *Note: This code can set if a condition exists that prevents cranking.* **Possible Causes** ● Starter relay control circuit is open or shorted to ground ● Starter relay control circuit is shorted to system power (B+) ● Starter relay is damaged or has failed ● PCM has failed
P0616 **1T CCM, MIL: No** 2003, 2004, 2005 CTS (Cadillac) 2.6L VIN M, 3.2L VIN N Transmissions: All	**Starter Relay Control Circuit Low Input Conditions:** Engine started, system voltage from 8-18v, and the PCM detected the voltage on the Starter Relay control circuit did not match the commanded state for at least two seconds. **Possible Causes** ● Starter relay connector is damaged or shorted to ground ● Starter relay control circuit is shorted to ground ● PCM has failed
P0617 **1T CCM, MIL: No** 2003, 2004, 2005 CTS (Cadillac) 2.6L VIN M, 3.2L VIN N Transmissions: All	**Starter Relay Control Circuit Low Input Conditions:** Engine started, system voltage from 8-18v, and the PCM detected the voltage on the Starter Relay control circuit did not match the commanded state for at least two seconds. **Possible Causes** ● Starter relay connector is damaged or shorted to power ● Starter relay control circuit is shorted to system power ● PCM has failed
P0620 **1T CCM, MIL: No** 1999, 2000, 2001, 2002, 2003, 2004, 2005 Bonneville, Century, Grand Prix, Impala, Intrigue, LeSabre, Lumina, Monte Carlo, Regal, Park Avenue, Montana, Rendezvous, Silhouette, Venture 3.1L VIN J, 3.1L VIN M, 3.4L VIN E, 3.5L VIN H, 3.8L VIN 1, 3.8L VIN K Transmissions: All	**Generator Signal Range/Performance Conditions:** Engine started, the Voltage Telltale lamp is "on", or less than 1000 rpm for the low duty cycle test, or more than 1000 rpm for high duty cycle test, and the PCM detected the 'L' terminal voltage was low with the Generator commanded "on", or the 'F' terminal PWM was less than 5% with the engine speed below 2500 rpm for 30 seconds. Note: Refer to the Freeze Frame Records for additional information. **Possible Causes** ● Generator 'L' terminal circuit is open, shorted to ground or B+ ● Generator 'F' terminal circuit is open, shorted to ground or B+ ● PCM has failed
P0621 **1T CCM, MIL: No** 2000, 2001, 2002, 2003, 2004, 2005 Aurora, DeVille, Eldorado, Seville 4.0L VIN C, 4.6L VIN 9, 4.6L VIN Y Transmissions: All	**Generator 'L' Terminal Circuit Malfunction Conditions:** DTC P0335, P0340 and P0622 not set, key and engine off, and the PCM detected the Generator 'L' Terminal voltage was high for 5 seconds; or with the engine running, the PCM detected the Generator 'L' Terminal voltage was low for over 15 seconds. **Possible Causes** ● Generator 'L' terminal circuit is shorted to system power (B+) ● Generator is damaged or has failed ● PCM has failed
P0621 **1T CCM, MIL: No** 2000, 2001, 2002, 2003, 2004, 2005 S/T Pickup 2.2L VIN 5 CNG engine Transmissions: All	**Generator 'L' Terminal Circuit Malfunction Conditions:** Key on, and the PCM detected the 'L' Terminal was Active with the key on, or the PCM detected the 'L' Terminal was Inactive with the engine running, condition met for 6 seconds. **Possible Causes** ● Generator 'L' terminal circuit is open or shorted to power (B+) ● Generator 'F' terminal circuit is shorted to ground ● Generator is damaged or has failed (no output) ● PCM has failed

OBD II Trouble Code List (P0xxx Codes)

DTC	Trouble Code Title, Conditions & Possible Causes
P0622 **1T CCM, MIL: No** 2000, 2001, 2002, 2003, 2004, 2005 DeVille, Eldorado, Seville 4.0L VIN C, 4.6L VIN 9, 4.6L VIN Y Transmissions: All	**Generator 'F' Terminal Circuit Malfunction Conditions:** DTC P0335, P0340 and P0621 not set, key on and engine off, and the PCM detected the Generator PWM signal was more than 65% for 5 seconds, or with the engine speed under 3000 rpm, it detected the Generator PWM signal was less than 5% for 15 seconds. **Possible Causes** • Generator 'L' terminal circuit is shorted to ground • Generator 'F' terminal is open or shorted to ground • Generator is damaged or has failed • PCM has failed
P0625 **1T CCM, MIL: No** 2003, 2004, 2005 CTS (Cadillac) 2.6L VIN M, 3.2L VIN N Transmissions: All	**Generator Control Circuit Low Input Conditions:** DTC P0335, P0336, P0337, P0341, P0342, P0343 and P0626 not set, engine started, engine speed less than 3,000 rpm, Generator not commanded "off" by the Scan Tool, and the PCM detected the Generator PWM signal was less than 5% for 15 seconds. **Possible Causes** • Battery positive cable is open or has a high resistance condition • Generator connector is damaged or shorted to power • Generator control circuit is shorted to ground • PCM has failed
P0626 **1T CCM, MIL: No** 2003, 2004, 2005 CTS (Cadillac) 2.6L VIN M, 3.2L VIN N Transmissions: All	**Generator Control Circuit High Input Conditions:** DTC P0335, P0336, P0337, P0341, P0342, P0343 and P0625 not set; key on and engine off, and the PCM detected the Generator PWM signal was more than 5% for 15 seconds. **Possible Causes** • Generator connector is damaged or shorted to system power • Generator control circuit is shorted to system power (B+) • PCM has failed
P0628 **1T CCM, MIL: No** 2003, 2004, 2005 CTS (Cadillac) 2.6L VIN M, 3.2L VIN N Transmissions: All	**Fuel Pump Control Circuit Low Input Conditions:** Engine started, engine speed over 40 rpm, system voltage from 8-18v, and the PCM detected an unexpected low voltage on the Fuel Pump control circuit for 1 second. The ECM provides system voltage to the coil side of the fuel pump relay. When the ignition switch is turned "on", the ECM energizes the fuel pump relay, which applies power to the fuel pump. The ECM applies a low current signal of about 7v to the fuel pump relay control circuit with the key on, engine off. The ECM uses this signal to detect the integrity of the fuel pump relay control circuit. The ECM enables the relay as long as the engine is cranking or running. **Possible Causes** • Fuel pump relay connector is damaged or shorted • Fuel pump relay control circuit is shorted to ground • Fuel pump relay is damaged or it has failed • PCM is damaged
P0629 **1T CCM, MIL: No** 2003, 2004, 2005 CTS (Cadillac) 2.6L VIN M, 3.2L VIN N Transmissions: All	**Fuel Pump Control Circuit High Input Conditions:** Engine started, engine speed over 40 rpm, system voltage from 8-18v, and the PCM detected an unexpected high voltage on the Fuel Pump control circuit for 1 second. The ECM provides system voltage to the coil side of the fuel pump relay. When the ignition switch is turned "on", the ECM energizes the fuel pump relay, which applies power to the fuel pump. The ECM applies a low current signal of about 7v to the fuel pump relay control circuit with the key on, engine off. The ECM uses this signal to detect the integrity of the fuel pump relay control circuit. The ECM enables the relay as long as the engine is cranking or running. **Possible Causes** • Fuel pump relay connector is damaged or open • Fuel pump relay control circuit is open or shorted to power (B+) • Fuel pump relay is damaged or it has failed • PCM is damaged
P0638 **1T CCM, MIL: Yes** 2003, 2004, 2005 CTS (Cadillac) 2.6L VIN M, 3.2L VIN N Transmissions: All	**TAC Motor System Malfunction Conditions:** Engine started, system voltage over 10v, IAT sensor over 41ºF, ECT sensor from 41-212ºF, and the PCM detected the required PWM signal to move the throttle blade was more than a predetermined value. The ECM opens the throttle blades by controlling the throttle valve motor with a pulsewidth-modulated (PWM) signal. The ECM reverses the polarity on the throttle valve motor control circuits to close the throttle blades. The ECM increases the pulsewidth as necessary to open the throttle blades. The ECM monitors the throttle position (TP) sensors 1 and 2 to determine the actual blade position. **Possible Causes** • TAC motor control circuit is open, shorted to ground or shorted to system power (B+) • Throttle valves not in their rest position • Throttle valves binding open or binding closed • Throttle valves moving open or closed without spring pressure • Throttle body assembly is damaged or it has failed • PCM is damaged

OBD II Trouble Code List (P0xxx Codes)

DTC	Trouble Code Title, Conditions & Possible Causes
P0641 **2T CCM, MIL: Yes** 2003, 2004, 2005 Cab & Chassis, C/K Truck, Escalade, G Van, & Cargo Van, L/M Van, S/T Blazer & S/T Pickup 4.3L VIN W, 4.3L VIN X, 4.8L VIN V, 5.3L VIN P, 5.3L VIN T, 5.3L VIN Z, 6.0L VIN U, 8.1L VIN G Transmissions: All	**5-Volt Reference 1 Circuit Malfunction Conditions:** Key on or engine running; and the PCM detected the 5v Reference circuit was out of tolerance for 10 seconds. The 5v VREF 1 circuit from the PCM is used to provide power to the Engine Oil Pressure (EOP), Manifold Air Pressure (MAP) sensor and the Throttle Position (TP) sensor on these vehicle applications. **Possible Causes** ● 5v VREF circuit is shorted to chassis or sensor ground ● 5v VREF circuit to MAP or TP sensor circuit is shorted to (B+) ● PCM has failed
P0641 **2T CCM, MIL: Yes** 2001, 2002, 2003, 2004, 2005 C/K Series Truck 6.6L VIN 1 engine Transmissions: All	**5-Volt Reference 1 Circuit Malfunction Conditions:** Key on or engine running; and the PCM detected the 5v Reference circuit was out of tolerance for 10 seconds. The 5v VREF 1 circuit is used to provide power to the Accelerator Pedal Position (APP1) sensor, Accelerator Pedal Position 2 (APP3) sensor, Engine Oil Pressure (EOP) and Fuel Rail Pressure (FPR) sensor. **Possible Causes** ● 5v VREF circuit to APP1 or APP3 is shorted to ground or shorted to system power ● FRP or EOP sensor is shorted to ground or shorted to system power ● PCM has failed
P0641 **2T CCM, MIL: Yes** 2002, 2003, 2004, 2005 C/K Truck, Escalade 6.0L VIN N engine Transmissions: All	**5-Volt Reference 1 Circuit Malfunction Conditions:** Key on or engine running; and the PCM detected the 5v Reference circuit was out of tolerance for 10 seconds. The 5v VREF 1 circuit is used to provide power to the Engine Oil Pressure (EOP) sensor and Manifold Air Pressure (MAP) sensor on these applications. **Possible Causes** ● 5v VREF circuit to EOP sensor or MAP sensor is shorted to ground or it is shorted to system power ● PCM has failed
P0645 **1T CCM, MIL: No** 1997, 1998, 1999, 2000, 2001, 2002, 2003, 2004, 2005 Camaro, Corvette & Firebird 5.7L VIN G, 5.7L VIN S Transmissions: All	**Air Conditioning Clutch Relay Control Circuit Malfunction Conditions:** Engine started, engine speed over 400 rpm, system voltage over 10.0v, and the PCM detected the Actual and Commanded state of the A/C Relay Control circuit did not match for 5 seconds. **Possible Causes** ● A/C relay control circuit is open, shorted to ground or to power ● A/C relay power circuit is open (test the A/C CRUISE mini fuse) ● A/C relay is damaged or has failed ● PCM has failed
P0650 **1T CCM, MIL: No** 1999, 2000, 2001, 2002, 2003, 2004, 2005 Bonneville, Century, Grand Prix, Impala, Intrigue, LeSabre, Lumina, Monte Carlo, Regal, Park Avenue, Montana, Rendezvous, Silhouette, Venture 3.1L VIN J, 3.1L VIN M, 3.4L VIN E, 3.5L VIN H, 3.8L VIN 1, 3.8L VIN K Transmissions: All	**Malfunction Indicator Lamp Circuit Malfunction Conditions:** Engine started, system voltage over 10.0v, and the PCM detected an unexpected voltage condition on the ODM 'D' Output 1 circuit that controls the Malfunction Indicator Lamp (MIL) for over 30 seconds. **Possible Causes** ● MIL control circuit is open or shorted to ground ● MIL control circuit is shorted to system power ● MIL control power circuit is open in the Instrument Cluster ● MIL (the lamp) is damaged or has failed ● PCM has failed
P0650 **1T CCM, MIL: No** 2003, 2004, 2005 C/K Series Truck, L/M Van, S/T Blazer & S/T Pickup 4.3L VIN W, 4.3L VIN X Transmissions: All	**Malfunction Indicator Lamp Circuit Malfunction Conditions:** Engine started, system voltage over 10.0v, and the PCM detected the Actual state and Commanded state of the Malfunction Indicator Lamp (MIL) circuit did not match for 30 seconds during the test. **Possible Causes** ● MIL control circuit is open or shorted to ground ● MIL control circuit is shorted to system power ● MIL control power circuit is open in the Instrument Cluster ● MIL (the lamp) is damaged or has failed ● Instrument Cluster or the PCM has failed

OBD II Trouble Code List (P0xxx Codes)

DTC	Trouble Code Title, Conditions & Possible Causes
P0650 **1T CCM, MIL: No** 2000, 2001, 2002, 2003, 2004, 2005 C/K Series Truck, Escalade, G Van & Cargo Van, Envoy & TrailBlazer 4.8L VIN V, 5.3L VIN P, 5.3L VIN T, 5.3L VIN Z, 6.0L VIN N, 6.0L VIN U, 8.1L VIN G engines Transmissions: All	**Malfunction Indicator Lamp Circuit Malfunction Conditions:** Engine started, system voltage over 10.0v, and the PCM detected the Actual and Commanded state of the Malfunction Indicator Lamp (MIL) driver did not match. **Possible Causes** ● MIL control circuit is open or shorted to ground ● MIL control circuit is shorted to system power ● MIL control power circuit is open in the Instrument Cluster ● MIL (the lamp) is damaged or has failed ● Instrument Cluster or the PCM has failed
P0650 **1T CCM, MIL: No** 2001, 2002, 2003, 2004, 2005 C/K Series Truck 6.6L VIN 1 Diesel engine Transmissions: All	**Malfunction Indicator Lamp Circuit Malfunction Conditions:** Engine started, system voltage from 10-18v, and the PCM detected an unexpected voltage on the MIL control circuit for 30 seconds. **Possible Causes** ● MIL control circuit is open or shorted to ground ● MIL control circuit is shorted to system power ● MIL control power circuit is open in the Instrument Cluster ● MIL (the lamp) is damaged or has failed ● Instrument Cluster or the PCM has failed
P0650 **1T CCM, MIL: No** 2003, 2004, 2005 CTS (Cadillac) 2.6L VIN M, 3.2L VIN N Transmissions: All	**Malfunction Indicator Lamp Circuit Malfunction Conditions:** Engine started, system voltage from 8-18v, and the PCM detected an unexpected voltage condition on the MIL control circuit for 10 seconds during the CCM continuous test. **Possible Causes** ● MIL control circuit is open or shorted to ground ● MIL control circuit is shorted to system power ● MIL control power circuit is open in the Instrument Cluster ● MIL (lamp) is damaged or has failed ● Instrument Cluster or the PCM has failed
P0650 **1T CCM, MIL: No** 2000, 2001, 2002, 2003, 2004, 2005 Aurora, DeVille, Eldorado, Seville 4.0L VIN C, 4.6L VIN 9, 4.6L VIN Y engines Transmissions: A/T	**Malfunction Indicator Lamp Circuit Malfunction Conditions:** Engine started, system voltage over 10.0v, and the PCM detected an unexpected voltage condition on the MIL control circuit for 10 seconds during the CCM continuous test. **Possible Causes** ● MIL control circuit is open, shorted to ground or shorted to system power ● MIL control power circuit is open in the Instrument Cluster ● MIL (the lamp) is damaged or has failed ● Instrument Cluster or the PCM has failed
P0650 **1T CCM, MIL: No** 1997, 1998, 1999, 2000, 2001, 2002, 2003, 2004, 2005 Camaro, Corvette & Firebird 5.7L VIN G, 5.7L VIN S Transmissions: All	**Malfunction Indicator Lamp Circuit Malfunction Conditions:** Engine started, system voltage over 10.0v, and the PCM detected the Commanded state and Actual state of the MIL control driver circuit did not match for 5-30 seconds in the test. **Possible Causes** ● MIL control circuit is open, shorted to ground or shorted to system power ● MIL control power circuit is open in the Instrument Cluster ● MIL (the lamp) is damaged or has failed ● Instrument Cluster or the PCM has failed
P0650 **1T CCM, MIL: No** 1999, 2000, 2001 Catera 3.0L VIN R engine Transmissions: A/T	**Malfunction Indicator Lamp Circuit Malfunction Conditions:** Engine started, system voltage over 10.0v, and the PCM detected an unexpected voltage condition on the MIL Control Circuit during the CCM test. **Possible Causes** ● MIL control circuit is open, shorted to ground or shorted to system power ● MIL control power circuit is open (check I/P Cluster IGN fuse) ● MIL (the lamp) is damaged or has failed ● PCM has failed
P0651 **1T CCM, MIL: No** 2003, 2004, 2005 Cab & Chassis, C/K Truck, Escalade, G Van, & Cargo Van, L/M Van, S/T Blazer & S/T Pickup 4.3L VIN W, 4.3L VIN X, 4.8L VIN V, 5.7L VIN P, 5.3L VIN T, 5.3L VIN Z, 6.0L VIN N, 6.0L VIN U, 8.1L VIN G Transmissions: All	**5-Volt Reference 2 Circuit Malfunction Conditions:** Key on or engine running; and the PCM detected the 5v Reference circuit was out of tolerance for 10 seconds. The 5v VREF 2 circuit is used to provide power to the Fuel Tank Pressure (FTP) sensor on this vehicle application. **Possible Causes** ● 5v VREF circuit is shorted to chassis or sensor ground ● 5v VREF circuit to MAP or TP sensor circuit is shorted to (B+) ● PCM has failed

OBD II Trouble Code List (P0xxx Codes)

DTC	Trouble Code Title, Conditions & Possible Causes
P0651 **1T CCM, MIL: No** 2001, 2002, 2003, 2004, 2005 C/K Series Truck 6.6L VIN 1 Diesel engine Transmissions: All	**5-Volt Reference 2 Circuit Malfunction Conditions:** Engine started; and the PCM detected the 5v Reference circuit was out of tolerance for 10 seconds. The 5v VREF 1 circuit is used to provide power to the Accelerator Pedal Position (APP2) sensor, BARO Sensor, Boost Pressure sensor, and the EGR Vacuum sensor. **Possible Causes** ● 5v VREF circuit to APP2 or APP2, FRP or EOP sensor is shorted to ground or shorted to VREF ● ECM has failed
P0651 **1T CCM, MIL: No** 2002, 2003, 2004, 2005 C/K Truck, Escalade 6.0L VIN N engine Transmissions: All	**5-Volt Reference 2 Circuit Malfunction Conditions:** Key on or engine running; and the PCM detected the 5v Reference circuit was out of tolerance for 10 seconds. The 5v VREF 2 circuit is used to provide power to the Air Conditioning Pressure (ACP) sensor and the Fuel Tank Pressure (FTP) sensor. **Possible Causes** ● 5v VREF circuit to ACP sensor or FTP sensor is shorted to ground or it is shorted to system power ● ECM has failed
P0654 **1T CCM, MIL: No** 1997, 1998, 1999, 2000, 2001, 2002, 2003, 2004, 2005 C/K Truck, Camaro, Corvette & Firebird 3.8L VIN F, 5.7L VIN G, 6.0L VIN N engines Transmissions: All	**Engine Speed Output Control Circuit Malfunction Conditions:** Engine started, engine speed over 600 rpm, and the PCM detected the Commanded and Actual state of the Engine Speed Output circuit did not match for 10 seconds in the test. **Possible Causes** ● Engine speed circuit is open or shorted to ground ● Engine speed is shorted to system power ● Instrument Cluster has failed ● PCM has failed
P0656 **1T CCM, MIL: No** 1997, 1998, 1999, 2000, 2001, 2002, 2003, 2004, 2005 Camaro & Firebird 5.7L VIN G engine Transmissions: All; Camaro, Firebird	**Fuel Gauge Output Control Circuit Malfunction Conditions:** Engine started, engine speed over 600 rpm, and the PCM detected the Actual and Commanded state of the Fuel Gauge Output Control circuit did not match for 10 seconds. **Possible Causes** ● Fuel gauge circuit is open or shorted to ground ● Fuel gauge control circuit is shorted to system power ● Instrument Cluster has failed ● PCM has failed
P0656 **1T CCM, MIL: No** 1999, 2000, 2001, 2002, 2003, 2004, 2005 Malibu, Montana, Silhouette, Venture 3.4L VIN E engine Transmissions: All	**Fuel Gauge Output Control Circuit Malfunction Conditions:** Engine started, engine speed over 600 rpm, and the PCM detected the Actual and Commanded state of the Fuel Gauge Output Control circuit did not match for 10 seconds. **Possible Causes** ● Fuel gauge circuit is open or shorted to ground ● Fuel gauge control circuit is shorted to system power ● Instrument Cluster has failed ● PCM has failed
P0660 **1T CCM, MIL: No** 2003, 2004, 2005 CTS (Cadillac) 2.6L VIN M, 3.2L VIN N Transmissions: All	**IMT Solenoid Control Circuit Malfunction Conditions:** Engine started, system voltage from 8-18v, IMT solenoid control state changed from open to closed position, and the PCM detected the IMT solenoid control circuit voltage was from 2.4-4.6v with the solenoid commanded "off". The IMT valve plenum is an H type configuration containing the IMT valve assembly. During low speed, high load conditions the IMT valve is closed creating two separate runner paths in the incoming air ducts which aids in increasing torque. At engine speeds over 3,200 rpm, the IMT valve opens creating a path between the two intake air duct runners allowing equalization of the intake air pressure which aids in increasing horsepower. The IMT valve is controlled by an IMT vacuum solenoid. Manifold vacuum is supplied to the IMT solenoid. A 12v supply from the Manifold fuse to the IMT solenoid (N.C. to vacuum) is applied when the ignition key turned on. **Possible Causes** ● IMT solenoid control circuit is open or shorted to ground ● IMT solenoid power circuit is open (check the Manifold fuse) ● IMT solenoid is damaged or it has failed ● PCM has failed
P0661 **1T CCM, MIL: No** 2003, 2004, 2005 CTS (Cadillac) 2.6L VIN M, 3.2L VIN N Transmissions: All	**IMT Solenoid Control Circuit Low Input Conditions:** Engine started, system voltage at 8-18v, IMT solenoid control state changed from open to closed position, and the PCM detected the voltage on the IMT control circuit was near 0v with the solenoid "off". The ECM applies a pullup voltage of 2.6-4.6v on the control circuit to differentiate between an open, short to ground, or a short to voltage. **Possible Causes** ● IMT solenoid IMT solenoid control circuit is open or shorted to ground ● IMT solenoid power circuit is open (check the Manifold fuse) ● IMT solenoid is damaged or it has failed ● PCM has failed

OBD II Trouble Code List (P0xxx Codes)

DTC	Trouble Code Title, Conditions & Possible Causes
P0662 **1T CCM, MIL: No** 2003, 2004, 2005 CTS (Cadillac) 2.6L VIN M, 3.2L VIN N Transmissions: All	**IMT Solenoid Control Circuit High Input Conditions:** Engine started, system voltage at 8-18v, IMT solenoid control state changed from open to closed position, and the PCM detected the voltage on the IMT control circuit was 12v with the solenoid "on". The ECM applies a pullup voltage of 2.6-4.6v on the control circuit to differentiate between an open, short to ground, or a short to voltage. **Possible Causes** ● IMT solenoid control connector is damaged, open or shorted ● IMT solenoid control circuit is open or shorted to system power ● IMT solenoid is damaged or it has failed ● PCM has failed
P0700 **1T CCM, MIL: Yes** 1999, 2000, 2001, 2002, 2003, 2004, 2005 C/K Series Truck, G Van 4.8L VIN V, 5.3L VIN T, 5.3L VIN Z, 6.0L VIN N, 0.0L VIN U, 6.6L VIN 1, 8.1L VIN G engines Transmissions: A/T	**Malfunction Indicator Lamp Circuit Malfunction (TCM) Conditions:** Key on or engine running; and the PCM received a signal from the TCM requesting that the Malfunction Indicator Lamp be illuminated. **Possible Causes** ● MIL control circuit is shorted to ground ● Check the TCM for any trouble codes in memory that are responsible for the request to turn on the MIL ● TCM has failed
P0700 **1T CCM, MIL: Yes** 2003, 2004, 2005 CTS (Cadillac) 2.6L VIN M, 3.2L VIN N Transmissions: A/T	**Malfunction Indicator Lamp Circuit Malfunction (TCM) Conditions:** Key on or engine running; and the PCM received a signal from the TCM requesting that the Malfunction Indicator Lamp be illuminated. **Possible Causes** ● MIL control circuit is shorted to ground ● Check the TCM for any trouble codes in memory that are responsible for the request to turn on the MIL ● TCM has failed
P0703 **1T CCM, MIL: No** 1995 Century, Cutlass Ciera, Cutlass Supreme, Grand Prix, Lumina, Monte Carlo Park Avenue Regal, Riviera, 88' & 98' 3.1L VIN M, 3.4L VIN X, 3.8L VIN 1, 3.8L VIN K, 3.8L VIN L engines Transmissions: A/T	**TCC Brake Sensor Circuit Malfunction Conditions:** DTC P0501 and P0502 not set, engine started, vehicle driven to over 35 mph for 10 seconds and back to 0 mph; this event repeated at least 5 times, and the PCM did not detect any change in the Brake status. **Possible Causes** ● TCC brake switch signal circuit is open or shorted to ground ● TCC brake switch B+ circuit is open (check the FAN ALT fuse) ● TCC brake switch is out of adjustment or damaged ● PCM has failed
P0703 **1T CCM, MIL: No** 1997, 1998, 1999, 2000, 2001 Catera 3.0L VIN R engine Transmissions: A/T	**TCC Brake Sensor Circuit Malfunction (4L30-E) Conditions:** U2100 and U2105 not set, engine started, vehicle driven with the Transmission output speed above 192 rpm and the TCM detected the Brake switch signal did not change for 20 minutes. **Possible Causes** ● TCC brake switch signal circuit is open or shorted to ground ● TCC brake switch B+ circuit is open (check the FAN ALT fuse) ● TCC brake switch is out of adjustment or damaged ● TCM has failed, or serial data link to the PCM has failed
P0704 **2T CCM, MIL: Yes** 1996, 1997, 1998, 1999, 2000, 2001, 2002 C/K Cab & Chassis, C/K Series Truck, L/M Vans, S/T Blazer & S/T Pickup 4.3L VIN W, 4.3L VIN X, 4.8L VIN V, 5.0L VIN M, 5.7L VIN K, 5.7L VIN R, 6.0L VIN U, 7.4L VIN J Transmissions: M/T	**Clutch Switch Circuit Malfunction Conditions:** DTC P0500, P0502 and P0503 not set, engine started, vehicle driven to a speed of over 40 mph and back to 0 mph 7 times, and the PCM did not detect any change in the Clutch switch status. **Possible Causes** ● Clutch switch circuit is open, shorted to ground or to power ● Clutch switch power circuit is open (test the ENG IGN fuse) ● Clutch switch is out of adjustment, damaged or has failed ● PCM has failed

OBD II Trouble Code List (P0xxx Codes)

DTC	Trouble Code Title, Conditions & Possible Causes
P0704 **2T CCM, MIL: Yes** 1997, 1998, 1999, 2000, 2001, 2002, 2003, 2004, 2005 Camaro, Corvette & Firebird 3.8L VIN K, 5.7L VIN G, 5.7L VIN S engines Transmissions: M/T	**Clutch Switch Circuit Malfunction (M49/MM6) Conditions:** DTC P0500, P0502 and P0503 not set, engine started, engine load and vehicle acceleration indicate vehicle is in gear and moving, and the PCM detected a VSS signal indicating the vehicle went from 0 to over 24 mph and then back to 0 mph within 2 seconds without the PCM detecting a change in the Clutch Anticipate switch circuit. *Note: This fault must occur 7 times before PCM will set this code.* **Possible Causes** ● Clutch switch circuit is open, shorted to ground or to power ● Clutch switch power circuit is open (test the ENG IGN fuse) ● Clutch switch is out of adjustment, damaged or has failed ● PCM has failed
P0705 **1T CCM, MIL: No** 1995 Century, Cutlass Ciera, Cutlass Supreme, Grand Prix, Lumina, Monte Carlo Park Avenue Regal, Riviera, 88' & 98' 3.1L VIN M, 3.4L VIN X, 3.8L VIN 1, 3.8L VIN K, 3.8L VIN L engines Transmissions: A/T	**Transaxle Range Switch (PRNDL) Circuit Malfunction Conditions:** Engine started, vehicle not moving, and the PCM detected a PRNDL Switch signal indicating a gear position other than P/N for 130 ms for 3 consecutive startups for 10 seconds. **Possible Causes** ● TR switch circuit is open or shorted to ground ● TR switch circuit is shorted to VREF or system power ● TR switch is out of adjustment ● TR switch is damaged or has failed ● PCM has failed
P0705 **1T CCM, MIL: No** 1996, 1997, 1998, 1999, 2000 Achieva, Alero, Beretta, Cavalier, Ciera, Century, Envoy, Grand Am, Sunfire Skyhawk, S/T Blazer, S/T Pickup & TrailBlazer 2.2L VIN 4, 2.2L VIN 5, 2.4L VIN T, 3.1L VIN J, 3.1L VIN M engines Transmissions: A/T	**Transaxle Range Switch (PRNDL) Circuit Malfunction Conditions:** Engine started, and the PCM detected an invalid combination of Transaxle Range (TR) Switch signals for 10 seconds. The TR switch, part of the PNP unit, is a multi-signal switch that sends a signal to the PCM to indicate a gear selection. The TR switch uses 4 discrete circuits to pull four (4) PCM voltages low in various combinations to indicate each gear range (LO = grounded, HI = open circuit). The 4 states displayed represents the encoder P, A, B, and C signals in sequence **Possible Causes** ● TR switch circuit(s) is open or shorted to ground ● TR switch circuit(s) is shorted to system power (B+) ● TR switch is damaged or out of adjustment ● PCM has failed ● TSB 71-65-37 contains a repair procedure for this code
P0705 **1T CCM, MIL: No** 1996, 1997, 1998, 1999 Aurora, Bonneville, Century, Cutlass Supreme, Grand Prix, Intrigue, LeSabre, LSS, Lumina, Monte Carlo, 88', Regal, Park Avenue & 98' 3.1L VIN J, 3.1L VIN M, 3.5L VIN H, 3.8L VIN 1, 3.8L VIN K engines Transmissions: A/T	**Transaxle Range Switch (PRNDL) Circuit Malfunction Conditions:** Engine started, and the PCM detected the Transaxle Range (TR) Switch indicated an invalid combination for 10 seconds. The TR Switch is part of the Transaxle PNP switch mounted on the transaxle manual shaft. The four inputs from this switch indicate to the PCM which position is selected by the Transaxle selector lever. This information is used for EGR control, EVAP purge, IAC valve control and Ignition Timing. The combination of the (4) TR switch states determines the PCM commanded shift pattern. The input voltage level at the PCM is high (B+) when the transaxle range switch is open and low when the switch is closed to ground. These "states" are represented as High, and Low. These parameters represent transaxle range switch Parity, A, B, and C signals respectively **Possible Causes** ● TR switch circuit(s) is open, shorted to ground or shorted to B+ ● TR switch is damaged or out of adjustment ● PCM has failed
P0705 **1T CCM, MIL: No** 1996, 1997, 1998, 1999, 2000, 2001, 2002, 2003, 2004, 2005 Aurora, DeVille, Eldorado, Seville 4.0L VIN C, 4.6L VIN 9, 4.6L VIN Y engines Transmissions: A/T	**Transaxle Range Switch (PRNDL) Circuit Malfunction Conditions:** Engine started, and the PCM detected an invalid combination of Transaxle Range (TR) Switch signals for 10 seconds. **Possible Causes** ● TR switch circuit(s) is open or shorted to ground ● TR switch circuit(s) is shorted to system power (B+) ● TR switch is damaged or out of adjustment ● PCM has failed
P0705 **1T CCM, MIL: No** 1997, 1998, 1999 Camaro, Corvette & Firebird 5.7L VIN G, 5.7L VIN S Transmissions: A/T	**Transaxle Range Switch Circuit Malfunction Conditions:** Engine started, system voltage over 10.0v, and the PCM detected an invalid combination of Transaxle Range (TR) switch signals for 10 seconds. **Possible Causes** ● TR switch circuit(s) is open or shorted to ground ● TR switch circuit(s) is shorted to system power (B+) ● TR switch is damaged or out of adjustment ● PCM has failed

OBD II Trouble Code List (P0xxx Codes)

DTC	Trouble Code Title, Conditions & Possible Causes
P0705 **1T CCM, MIL: No** 2002, 2003, 2004, 2005 Envoy & TrailBlazer 4.2L VIN S engine Transmissions: A/T	**Transaxle Range Switch (PRNDL) Circuit Malfunction Conditions:** Engine started, engine runtime over 5 seconds, vehicle speed more than 5 mph, and the PCM detected the Transaxle Range (TR) Switch signals were invalid for 5 seconds. The TR switch, part of the PNP unit, is a multi-signal switch that sends a signal to the PCM to indicate a gear selection. The TR switch uses 4 discrete circuits to pull four (4) PCM voltages low in various combinations to indicate each gear range (LO = grounded, HI = open circuit). The 4 states displayed represent the encoder P, A, B, and C signals in sequence. **Possible Causes** • TR switch circuit(s) is open or shorted to ground • TR switch circuit(s) is shorted to system power (B+) • TR switch is damaged or out of adjustment • PCM has failed
P0705 **1T CCM, MIL: No** 1997, 1998, 1999, 2000, 2001 Catera 3.0L VIN R engine Transmissions: A/T	**Transaxle Range Switch Circuit Malfunction (4L30-E) Conditions:** Engine started, system voltage over 10.0v, and the PCM detected an invalid TR switch position for over 20 seconds. The A/T manual shaft shift position switch is attached to the left side of the transmission. This switch provides the TCM and console gear position indicator which both the gear the driver selects. Ignition voltage is supplied to the switch on terminal 'D'. Depending on the position of the Gear Selector lever, ignition voltage connects to various combinations of terminals A, B, C, and G of the A/T manual shaft shift position switch. The TCM decodes these combinations in order to determine gear selector lever position. *Note: The Sport Mode Lamp flashes when this trouble code is set.* **Possible Causes** • TR switch circuit(s) is open or shorted to ground • TR switch circuit(s) is shorted to system power (B+) • TR switch is damaged or out of adjustment • PCM has failed
P0706 **1T CCM, MIL: No** 1996, 1997, 1998, 1999 Bonneville, Century, Cutlass Supreme, Grand Prix, Intrigue, LeSabre, LSS, Lumina, Monte Carlo, 88', Regal, Park Avenue & 98' 3.1L VIN J, 3.1L VIN M, 3.8L VIN 1, 3.8L VIN K Transmissions: A/T	**Transaxle Range Switch (PRNDL) Performance Conditions:** Engine cranking; and the PCM detected a Transaxle Range switch sensor other than Park or Neutral during startup during 2 startups. **Possible Causes** • TR switch circuit is or shorted to ground • TR switch circuit is shorted to another PRNDL circuit • TR switch is damaged or out of adjustment • PCM has failed
P0706 **1T CCM, MIL: No** 2002, 2003, 2004, 2005 C/K Truck, Escalade, G Van & Cargo Van, Envoy & TrailBlazer 5.3L VIN T, 5.7L VIN P, 5.7L VIN Z, 6.0L VIN N, 8.1L VIN G Transmissions: All	**A/T TR Switch Circuit Malfunction (4L60-E, 4L65-E) Conditions:** DTC P0121, P0122, P0123, P1120, P1220 or P1221 not set, engine started; system voltage from 6-18v, and the PCM detected a Drive or a Reverse position signal at startup; or with the TP angle over 5%, VSS over 20 mph, it detected a Park or Neutral signal for 20 seconds. The TR switch is part of the Park Neutral position and Back Up lamp switch assembly mounted on the transmission manual shaft. The TR switch is a multi-signal switch. The PCM supplies ignition voltage to the TR switch on 4 signal circuits (A, B, C, and P). Each gear selector lever position grounds one or more switch circuits. The PCM compares the voltage combinations on the signal circuits to determine the gear range selected by the driver. Switch input to the PCM is represented as HI and Low. HI indicates an ignition voltage signal. Low indicates a zero voltage signal. The four switch parameters are A, B, C and Parity. **Possible Causes** • TR switch circuit(s) is shorted to the VREF circuit • TR switch circuit(s) is shorted to another switch circuit • TR switch is damaged or out of adjustment • PCM has failed
P0706 **1T CCM, MIL: No** 1997, 1998, 1999, 2000, 2001, 2002, 2003, 2004, 2005 Camaro, Corvette & Firebird 5.7L VIN G, 5.7L VIN S Transmissions: A/T	**Transaxle Range Switch (PRNDL) Performance Conditions:** Engine started, system voltage over 10.0v, and the PCM detected the Transaxle Range switch signal indicated a range other than P/N during startup. *Note: This test must fail 2 out of 4 consecutive tests in order to set this trouble code.* **Possible Causes** • TR switch circuit is or shorted to ground • TR switch circuit is shorted to another PRNDL circuit • TR switch is damaged or out of adjustment • PCM has failed

OBD II Trouble Code List (P0xxx Codes)

DTC	Trouble Code Title, Conditions & Possible Causes
P0706 **1T CCM, MIL: No** 1997, 1998, 1999, 2000, 2001 Catera 3.0L VIN R engine Transmissions: A/T	**Transaxle Range Switch Performance (4L30-E) Conditions:** DTC P0120 not set, engine started, engine speed from 500-3008 rpm with the TP angle more than 34%, and the PCM detected a 'P' or 'N' select signal; or with the transmission output speed at more than 8160 RPM, the PCM detected a Reverse position signal. **Possible Causes** ● TR switch circuit is or shorted to ground, or it is shorted to another PRNDL circuit ● TR switch is damaged or out of adjustment ● PCM has failed
P0711 **1T CCM, MIL: No** 1997, 1998, 1999, 2000, 2001, 2002, 2003, 2004, 2005 Achieva, Alero, Cavalier, Ciera, Century, Grand Am, Sunfire Skyhawk, S/T Blazer, S/T Pickup 2.2L VIN 4, 2.2L VIN 5, 2.4L VIN T, 3.1L VIN J, 3.1L VIN M engines Transmissions: A/T	**TFT Sensor Range/Performance (4T40-E/4T45-E) Conditions:** DTC P0502, P0503, P0716 and P0717 not set, engine started, engine runtime over 5 minutes, system voltage over 10.0v, ECT sensor more than 158ºF and changed more than 90ºF since startup, startup TFT sensor from -40ºF to 70ºF, TFT sensor from -36ºF to 304ºF, VSS over 5 mph with the TCC slip speed more than 300 rpm for 6 minutes and 49 seconds cumulative in the current key cycle, and the PCM detected the TFT sensor did not change more than 2ºF for 20 seconds after startup; or the TFT sensor changed more than 36ºF in 200 ms (fault occurred 14 times in 7 seconds). **Possible Causes** ● ATF fluid level too low or the fluid is contaminated or burnt ● TFT sensor is contaminated, damaged or "skewed" ● PCM has failed
P0711 **1T CCM, MIL: No** 1999, 2000, 2001, 2002, 2003, 2004, 2005 Aurora, Bonneville, Century, Cutlass Supreme, Grand Prix, Intrigue, LeSabre, LSS, Lumina, Monte Carlo, 88', Regal, Park Avenue, 98', Montana, Rendezvous, Silhouette, Venture 3.1L VIN J, 3.1L VIN M, 3.4L VIN E, 3.5L VIN H, 3.8L VIN 1, 3.8L VIN K Transmissions: A/T	**TFT Sensor Signal Range/Performance (4T65-E) Conditions:** DTC P0502, P0503 and P1870 not set, engine runtime over 5 minutes, system voltage over 10.0v, ECT sensor more than 158°F and changed by 90°F since startup, TFT sensor from 0.2-4.92v at startup, startup TFT sensor from -40 to 69°F, VSS more than 5 mph for 409 seconds cumulative, TCC slip speed over 80 rpm for 4-9 seconds cumulative, and the PCM detected the TFT sensor did not change more than 2ºF after 409 seconds cumulative since startup; or the TFT sensor changed more than 36ºF in 200 ms (fault detected 14 times within a 7 second period). **Possible Causes** ● ATF fluid level too low or the fluid is contaminated or burnt ● TFT sensor is contaminated, damaged or "skewed" ● PCM has failed
P0711 **1T CCM, MIL: No** 1996, 1997, 1998, 1999, 2000, 2001, 2002, 2003, 2004, 2005 Camaro, Corvette & Firebird, C/K Series truck, G Van, L/M Van, S/T Blazer & S/T Pickup 4.3L VIN W, 4.3L VIN X, 4.8L VIN V, 5.0L VIN M, 5.3L VIN T, 5.3L VIN Z, 5.7L VIN K, 5.7L VIN R, 5.7L VIN G, 5.7L VIN S, 6.0L VIN N, 6.0L VIN U, 6.6L VIN 1, 7.4L VIN J, 8.1L VIN G Transmissions: A/T	**TFT Sensor Range/Performance (4L60-E, 4L80-E) Conditions:** DTC P0502, P0503, P0894 or P1870 not set, engine started, engine runtime over 409 seconds, system voltage over 10.0v, ECT sensor more than 158ºF and a change of 90ºF since startup, TFT sensor from -40 to 70ºF at startup, then TFT sensor from -36 to 304ºF, vehicle driven to over 5 mph for 409 seconds cumulative, TCC slip speed over 120 rpm cumulative in the current key cycle, and the PCM detected the TFT sensor did not change more than 2.7ºF for 409 seconds after startup; or the TFT sensor changed more than 36ºF within 200 milliseconds (fault occurred 14 times in 7 seconds). **Possible Causes** ● ATF fluid level too low or the fluid is contaminated or burnt ● TFT sensor is contaminated, damaged or "skewed" ● PCM has failed
P0711 **1T CCM, MIL: No** 1996, 1997, 1998, 1999, 2000, 2001, 2002, 2003, 2004, 2005 Aurora, DeVille, Eldorado, Seville 4.0L VIN C, 4.6L VIN 9, 4.6L VIN Y engines Transmissions: A/T	**TFT Sensor Signal Range/Performance (4L80-E) Conditions:** DTC P0116-P0118, P0502, P0503, P0716, P0717, P1114 and P1115 not set, engine started, TFT sensor at startup -36ºF to 289ºF, TFT sensor from -35ºF to 300ºF, P0711 test not passed this key cycle, ECT sensor over 158ºF with a change of over 90ºF since startup, TFT sensor at startup from -36ºF to 70ºF, VSS over 5 mph for 15 minutes, TCC slip speed over 120 rpm for 15 minutes (Condition 1), and the PCM detected the TFT sensor was below 4ºF for 120 seconds; or ECT sensor over 158ºF with a change of over 90ºF since startup, TFT sensor at startup from 264-289ºF, VSS over 5 mph for 15 minutes, TCC slip speed over 120 rpm for 15 minutes (Condition 2), and the PCM detected a change in the TFT sensor of under 4ºF for 50 seconds; or after the PCM detected the TFT sensor changed over 36ºF in 200 ms (Condition 3). *Note: This test must fail 14 times within 7 seconds to set this code.* **Possible Causes** ● ATF fluid level too low or the fluid is contaminated or burnt ● TFT sensor is contaminated, damaged or "skewed" ● PCM has failed

OBD II Trouble Code List (P0xxx Codes)

DTC	Trouble Code Title, Conditions & Possible Causes
P0711 **1T CCM, MIL: No** 2003, 2004, 2005 CTS (Cadillac) 2.6L VIN M, 3.2L VIN N Transmissions: A/T	**TFT Sensor Signal Range/Performance (5L40-E) Conditions:** DTC P0711, P0716, P0717, P0722, P0723 and P1792 not set, engine runtime over 5 seconds, system voltage from 10-18v, TFT sensor from -38 to +300°F, VSS over 5 mph, for Conditions 1 and 2 below, the ECT sensor more than 158°F and changed by at least 90°F since startup, and the TCC slip speed was 120 rpm or more for 10 minutes, then during Test 1: TFT sensor signal at startup from -40 to +70°F, VSS over 5 mph for 15 minutes, and the PCM detected the TFT sensor changed over 4°F since startup within 1 minute, 40 seconds; or during Condition 2: TFT sensor signal at 264-302°F, VSS over 5 mph for 10 minutes, and the PCM detected the TFT sensor changed over 4°F within 1 minute, 40 seconds; or during Condition 3: The PCM detected the TFT sensor signal changed by 36°F or more within 250 milliseconds (14 times within 7 seconds). **Possible Causes** • ATF fluid level too low or the fluid is contaminated or burnt • TFT sensor is contaminated, damaged or "skewed" • PCM has failed
P0711 **1T CCM, MIL: No** 2002, 2003, 2004, 2005 Envoy & TrailBlazer 4.2L VIN S engine Transmissions: A/T	**TFT Sensor Signal Range/Performance (4L60-E) Conditions:** DTC P0502, P0503 and P1810 not set, engine runtime over 409 seconds, system voltage from 10-18v, ECT sensor over 158°F with a change of over 90°F since startup, startup TFT sensor from -40°F to 70°F, VSS over 5 mph, TCC slip speed more than 120 rpm for 409 seconds cumulative in the current key cycle, and the PCM detected the TFT sensor did not change over 4°F for 409 ms after startup or the TFT sensor changed over 36°F in 200 ms. **Possible Causes** • ATF fluid level too low or the fluid is contaminated or burnt • TFT sensor is contaminated, damaged or "skewed" • PCM has failed
P0711 **1T CCM, MIL: No** 2002, 2003, 2004, 2005 Envoy & TrailBlazer 5.3L VIN P Transmissions: A/T	**TFT Sensor Signal Range/Performance (4L65-E) Conditions:** DTC P0502, P0503 and P0894 not set, engine runtime over 409 seconds, system voltage from 10-18v, ECT sensor more than 151°F with a change of over 97°F since startup, VSS over 5 mph, TFT sensor from -40°F to 70°F at startup, TCC slip speed more than 120 rpm for 409 seconds cumulative during the current key cycle, and the PCM detected the TFT sensor did not change over 2.7°F for 409 ms after startup or the TFT sensor changed more than 36°F within 200 ms (test must fail at least 14 times within 7 seconds). **Possible Causes** • ATF fluid level too low or the fluid is contaminated or burnt • TFT sensor is contaminated, damaged or "skewed" • PCM has failed
P0712 **1T CCM, MIL: No** 1995 Century, Cutlass Ciera, Cutlass Supreme, Grand Prix, Lumina, Monte Carlo Park Avenue Regal, Riviera, 88' & 98' 3.1L VIN M, 3.4L VIN X, 3.8L VIN 1, 3.8L VIN K, 3.8L VIN L engines Transmissions: A/T	**TFT Sensor Circuit High Input Conditions:** Engine started, engine runtime over 3 minutes, and the PCM detected a TFT sensor signal indicating a transmission fluid temperature of less than -40°F (high voltage signal) for 1.6 seconds. **Possible Causes** • TFT sensor signal circuit is open between the sensor and PCM • TFT sensor signal circuit is shorted to VREF or system power • TFT sensor is damaged or has failed (it may be open) • PCM has failed
P0712 **1T CCM, MIL: No** 1996, 1997, 1998, 1999, 2000, 2001, 2002, 2003, 2004, 2005 Aurora, Bonneville, Century, Cutlass Supreme, Grand Prix, Intrigue, LeSabre, LSS, Lumina, Monte Carlo, 88', Regal, Park Avenue, 98', Montana, Rendezvous, Silhouette, Venture 3.1L VIN J, 3.1L VIN M, 3.4L VIN E, 3.5L VIN H, 3.8L VIN 1, 3.8L VIN K Transmissions: A/T	**TFT Sensor Circuit Low Input Conditions:** DTC P0560 not set, engine started, engine running and the PCM detected the TFT sensor was more than 298°F for 10 seconds. **Possible Causes** • TFT sensor signal circuit is shorted to sensor or chassis ground • TFT sensor is damaged or has failed (it may be shorted) • PCM has failed

OBD II Trouble Code List (P0xxx Codes)

DTC	Trouble Code Title, Conditions & Possible Causes
P0712 **1T CCM, MIL: No** 1996, 1997, 1998, 1999, 2000, 2001, 2002, 2003, 2004, 2005 Achieva, Alero, Beretta, Cavalier, Century, Ciera, Corsica, Grand Am, Sunfire Skyhawk, S/T Blazer & S/T Pickup 2.2L VIN 4, 2.2L VIN 5, 2.4L VIN T, 3.1L VIN J, 3.1L VIN M engines Transmissions: A/T	**TFT Sensor Circuit Low Input (4T40-E/4T45-E) Conditions:** Engine started, system voltage over 10.0v, and the PCM detected the TFT sensor was less than 0.20v for 10 seconds. **Possible Causes** • TFT sensor signal circuit is shorted to sensor or chassis ground • TFT sensor is damaged or has failed (it may be shorted) • PCM has failed
P0712 **1T CCM, MIL: No** 2003, 2004, 2005 CTS (Cadillac) 2.6L VIN M, 3.2L VIN N Transmissions: A/T	**TFT Sensor Circuit Low Input (5L40-E) Conditions:** Engine runtime over 5 seconds, system voltage from 8-18v, and the PCM detected the TFT sensor signal indicated less than 0.10v (Scan Tool reads over 302°F) for over 10 seconds. **Possible Causes** • TFT sensor signal circuit is shorted to ground • TFT sensor is damaged • TCM has failed
P0712 **1T CCM, MIL: No** 1996, 1997, 1998, 1999, 2000, 2001, 2002, 2003, 2004, 2005 Aurora, DeVille, Eldorado, Seville 4.0L VIN C, 4.6L VIN 9, 4.6L VIN Y engines Transmissions: A/T	**TFT Sensor Circuit Low Input (4L80-E) Conditions:** Engine started, engine runtime over 5 seconds, system voltage over 10.0v, and the PCM detected the TFT sensor was more than 298°F (a signal of 0.8v or less) for 10 seconds during the CCM test. **Possible Causes** • TFT sensor connector is damaged or shorted • TFT sensor signal circuit is shorted to sensor or chassis ground • TFT sensor is damaged or has failed (it may be shorted) • PCM has failed
P0712 **1T CCM, MIL: No** 1996, 1997, 1998, 1999, 2000, 2001, 2002, 2003, 2004, 2005 Cab & Chassis, C/K truck, G Van, M/L Series Van, S/T Blazer & S/T Pickup 4.3L VIN W, 4.3L VIN X, 4.8L VIN V, 5.0L VIN M, 5.3L VIN T, 5.3L VIN Z, 5.7L VIN K, 5.7L VIN R, 5.7L VIN G, 5.7L VIN S, 6.0L VIN N, 6.0L VIN U, 6.6L VIN 1, 7.4L VIN J, 8.1L VIN G engines Transmissions: A/T	**TFT Sensor Circuit Low Input (4L60-E, 4L80-E) Conditions:** Engine started, system voltage over 10.0v, and the PCM detected the TFT sensor was less than 0.25v for 10 seconds. Note: A Scan Tool does not indicate the default TFT sensor reading. **Possible Causes** • TFT sensor signal circuit is shorted to sensor or chassis ground • TFT sensor is damaged or has failed (it may be shorted) • PCM has failed
P0712 **1T CCM, MIL: No** 1996, 1997, 1998, 1999, 2000, 2001, 2002, 2003, 2004, 2005 Camaro, Corvette & Firebird 3.8L VIN K, 5.7L VIN G, 5.7L VIN S engines Transmissions: A/T	**TFT Sensor Circuit Low Input Conditions:** DTC P0560 not set, engine started, engine running and the PCM detected the TFT sensor was more than 298°F for 10 seconds. **Possible Causes** • TFT sensor connector is damaged or shorted • TFT sensor signal circuit is shorted to sensor or chassis ground • TFT sensor is damaged or has failed (it may be shorted) • PCM has failed
P0712 **1T CCM, MIL: No** 2002, 2003, 2004, 2005 Envoy & TrailBlazer 4.2L VIN S, 5.3L VIN P Transmissions: A/T	**TFT Sensor Circuit Low Input (4L60-E/4L65-E) Conditions:** Engine running; and the PCM detected the TFT sensor was less than 0.25v for 10 seconds. **Possible Causes** • TFT sensor signal circuit is shorted to sensor or chassis ground • TFT sensor is damaged or has failed (it may be shorted) • PCM has failed
P0712 **1T CCM, MIL: No** 1997, 1998, 1999, 2000, 2001 Catera 3.0L VIN R engine Transmissions: A/T	**TFT Sensor Circuit Low Input (4L30-E) Conditions:** Engine started, and the PCM detected the TFT sensor was less than 0.98v for 20 seconds during the CCM test. *Note: The Sport Mode Lamp flashes when this trouble code is set.* **Possible Causes** • TFT sensor signal circuit is shorted to sensor or chassis ground • TFT sensor is damaged or has failed (it may be shorted) • PCM has failed

OBD II Trouble Code List (P0xxx Codes)

DTC	Trouble Code Title, Conditions & Possible Causes
P0713 **1T CCM, MIL: No** 1995 Century, Cutlass Ciera, Cutlass Supreme, Grand Prix, Lumina, Monte Carlo Park Avenue Regal, Riviera, 88' & 98' 3.1L VIN M, 3.4L VIN X, 3.8L VIN 1, 3.8L VIN K, 3.8L VIN L engines Transmissions: A/T	**TFT Sensor Low Input Conditions:** Engine runtime over 10 seconds; and the PCM detected a TFT sensor that indicated a transmission fluid temperature of over 292ºF (Low sensor) for longer than 1.6 seconds. **Possible Causes** ● TFT sensor connector is damaged or shorted ● TFT sensor signal circuit is shorted to sensor or chassis ground ● TFT sensor is damaged or has failed (it may be shorted) ● PCM has failed
P0713 **1T CCM, MIL: No** 1996, 1997, 1998, 1999, 2000, 2001, 2002, 2003, 2004, 2005 Achieva, Alero, Beretta, Cavalier, Ciera, Century, Corsica, Grand Am, Sunfire Skyhawk, S/T Blazer & S/T Pickup 2.2L VIN 4, 2.2L VIN 5, 2.4L VIN T, 3.1L VIN J, 3.1L VIN M engines Transmissions: A/T	**TFT Sensor Circuit High Input (4T40-E/4T45-E) Conditions:** Key on or engine running; system voltage 8-18v, and the PCM detected that the TFT sensor was more than 4.92v for 409 seconds. **Possible Causes** ● TFT sensor connector is damaged, loose or open ● TFT sensor signal circuit is open between the sensor and PCM ● TFT sensor signal circuit is shorted to VREF or system power ● TFT sensor is damaged or has failed (it may be open) ● PCM has failed
P0713 **1T CCM, MIL: No** 1999, 2000, 2001, 2002, 2003, 2004, 2005 Aurora, Bonneville, Century, Cutlass Supreme, Grand Prix, Intrigue, LeSabre, LSS, Lumina, Monte Carlo, 88', Regal, Park Avenue, 98', Montana, Rendezvous, Silhouette, Venture 3.1L VIN J, 3.1L VIN M, 3.4L VIN E, 3.5L VIN H, 3.8L VIN 1, 3.8L VIN K Transmissions: A/T	**TFT Sensor Circuit High Input (4T65-E) Conditions:** DTC P0117, P0118 and P0560 not set, engine started, and the PCM detected the TFT sensor was less than -33ºF (a voltage of 4.92v or higher) for 10 seconds during the CCM test. **Possible Causes** ● TFT sensor signal circuit is open between the sensor and PCM ● TFT sensor signal circuit is shorted to VREF or system power ● TFT sensor is damaged or has failed (it may be open) ● PCM has failed ● TSB 02-07-30-15 contains a repair procedure for this code
P0713 **1T CCM, MIL: No** 1996, 1997, 1998, 1999, 2000, 2001, 2002, 2003, 2004, 2005 Cab & Chassis, C/K truck, G Van, M/L Series Van, S/T Blazer & S/T Pickup 4.3L VIN W, 4.3L VIN X, 4.8L VIN V, 5.0L VIN M, 5.3L VIN T, 5.3L VIN Z, 5.7L VIN K, 5.7L VIN R, 5.7L VIN G, 5.7L VIN S, 6.0L VIN N, 6.0L VIN U, 6.6L VIN 1, 7.4L VIN J, 8.1L VIN G engines Transmissions: A/T	**TFT Sensor Circuit High Input (4L60-E. 4L80-E) Conditions:** Engine started; system voltage more than 10.0v, and the PCM detected the TFT sensor was more than 4.92v for 6-8 minutes. Note: A Scan Tool does not indicate the default TFT sensor reading. **Possible Causes** ● TFT sensor signal circuit is open between the sensor and PCM ● TFT sensor signal circuit is shorted to VREF or system power ● TFT sensor is damaged or has failed (it may be open) ● PCM has failed
P0713 **1T CCM, MIL: No** 2003, 2004, 2005 CTS (Cadillac) 2.6L VIN M, 3.2L VIN N Transmissions: A/T	**TFT Sensor Circuit High Input (5L40-E) Conditions:** DTC P0716, P0717, P0722 and P0723 not set, engine runtime over 5 seconds, system voltage from 8-18v, OSS signal more than 200 rpm, TCC slip speed more than 120 rpm for 3 minutes, 20 seconds, and the PCM detected the TFT sensor signal indicated more than 4.90v (Scan Tool reads less than -38ºF) for over 25 seconds. The TFT sensor is part of the A/T internal wiring harness assembly. The TFT sensor is a thermistor (i.e., a resistor that changes value as the temperature changes). **Possible Causes** ● TFT sensor connector is damaged or open (intermittent fault) ● TFT sensor signal circuit is open or the ground circuit is open ● TFT sensor is damaged or the TCM has failed

OBD II Trouble Code List (P0xxx Codes)

DTC	Trouble Code Title, Conditions & Possible Causes
P0713 **1T CCM, MIL: No** 1996, 1997, 1998, 1999, 2000, 2001, 2002, 2003, 2004, 2005 Aurora, DeVille, Eldorado, Seville 4.0L VIN C, 4.6L VIN 9, 4.6L VIN Y engines Transmissions: A/T	**TFT Sensor Signal High Input (4L80-E) Conditions:** DTC P0502, P0503, P0716 and P0717 not set, engine started, system voltage over 10.0v, ECT sensor indicating above -13ºF, OSS signal more than 738 rpm for 5 minutes, TCC slip speed over 49 rpm for 6 minutes and 40 seconds, Fuel Cutoff inactive, and the PCM detected the TFT sensor was less than -40ºF (a voltage of over 5.04v) for six seconds during the CCM test. **Possible Causes** ● TFT sensor signal circuit is open between the sensor and PCM ● TFT sensor signal circuit is shorted to VREF or system power ● TFT sensor is damaged or has failed (it may be open) ● PCM has failed
P0713 **1T CCM, MIL: No** 1996, 1997, 1998, 1999, 2000, 2001, 2002, 2003, 2004, 2005 Camaro, Corvette & Firebird 3.8L VIN K, 5.7L VIN G, 5.7L VIN S Transmissions: A/T	**TFT Sensor Circuit High Input Conditions:** DTC P0117, P0118 and P0560 not set, engine started, and the PCM detected the TFT sensor was less than -33ºF (a voltage of over 4.92v) for 10 seconds during the CCM test. **Possible Causes** ● TFT sensor connector is damaged, loose or open ● TFT sensor signal circuit is open between the sensor and PCM ● TFT sensor signal circuit is shorted to VREF or system power ● TFT sensor is damaged or has failed (it may be open) ● PCM has failed
P0713 **1T CCM, MIL: No** 2002, 2003, 2004, 2005 Envoy & TrailBlazer 4.2L VIN S, 5.3L VIN P Transmissions: A/T	**TFT Sensor Circuit High Input (4L60-E/4L65-E) Conditions:** Key on or engine running; system voltage 8-18v, and the PCM detected that the TFT sensor was more than 4.92v for 409 seconds. **Possible Causes** ● TFT sensor connector is damaged, loose or open ● TFT sensor signal circuit is open between the sensor and PCM ● TFT sensor signal circuit is shorted to VREF or system power ● TFT sensor is damaged or has failed (it may be open) ● PCM has failed
P0713 **1T CCM, MIL: No** 1997, 1998, 1999, 2000, 2001 Catera 3.0L VIN R engine Transmissions: A/T	**TFT Sensor Circuit High Input (4L30-E) Conditions:** Engine started, and the PCM detected the TFT sensor was more than 4.30v for 20 seconds during the CCM test. *Note: The Sport Mode Lamp flashes when this trouble code is set.* **Possible Causes** ● TFT sensor signal circuit is open between the sensor and PCM ● TFT sensor signal circuit is shorted to VREF or system power ● TFT sensor is damaged or has failed (it may be open) ● PCM has failed
P0716 **2T CCM, MIL: Yes** 1997, 1998, 1999, 2000, 2001, 2002, 2003, 2004, 2005 Achieva, Alero, Cavalier, Century, Grand Am, Malibu, Skylark, Sunfire, S/T Pickup, Montana, Silhouette & Venture 2.2L VIN 4, 2.2L VIN 5, 2.4L VIN T, 3.1L VIN J, 3.1L VIN M, 3.4L VIN E Transmissions: A/T	**A/T Input Speed Sensor Circuit Malfunction (4T40/4T45-E) Conditions:** DTC P0121, P0122, P0123, P0502, P0503, P0717, P0751, P0752 and P0758 not set, system voltage from 8-18v, engine running, throttle position over 8%, VSS sensor over 5 mph, and the PCM detected the Sensor speed changed by over 300 rpm in 800 ms. **Possible Causes** ● ISS positive (+) circuit is open, shorted to ground or to power ● ISS negative (-) circuit is open, shorted to ground or to power ● ISS is damaged or has failed ● PCM has failed ● TSB 02-07-30-022A contains a repair procedure for this code
P0716 **2T CCM, MIL: Yes** 1999, 2000, 2001, 2002, 2003, 2004, 2005 Aurora, Bonneville, Century, Cutlass Supreme, Grand Prix, Intrigue, LeSabre, LSS, Lumina, Monte Carlo, 88', Regal, Park Avenue, 98', Montana, Rendezvous, Silhouette, Venture 3.1L VIN J, 3.1L VIN M, 3.4L VIN E, 3.5L VIN H, 3.8L VIN 1, 3.8L VIN K Transmissions: A/T	**A/T Input Speed Sensor Circuit Malfunction (4T65-E) Conditions:** DTC P0121, P0122, P0123, P0502, P0503, P0717, P0751, P0752, P0753, P0756, P0757 and P0758 not set, DTC P0717 test passed this key cycle, engine started, engine speed over 500 rpm for 5 seconds, Fuel Cutoff inactive, TP angle more than 14%, VSS over 5 mph, and the PCM detected the Input Shaft Sensor speed changed by more than 1300 rpm within 800 ms during the CCM test. **Possible Causes** ● ISS positive (+) circuit is open, shorted to ground or to power ● ISS negative (-) circuit is open, shorted to ground or to power ● ISS is damaged or has failed ● PCM has failed ● TSB 02-07-30-022A contains a repair procedure for this code

OBD II Trouble Code List (P0xxx Codes)

DTC	Trouble Code Title, Conditions & Possible Causes
P0716 **2T CCM, MIL: Yes** 1996, 1997, 1998, 1999, 2000, 2001, 2002, 2003, 2004, 2005 Cab & Chassis, C/K truck, G Van, M/L Series Van, S/T Blazer & S/T Pickup 4.8L VIN V, 5.0L VIN M, 5.3L VIN T, 5.3L VIN Z, 5.7L VIN K, 5.7L VIN R, 6.0L VIN N, 6.0L VIN U, 6.6L VIN 1, 7.4L VIN J, 8.1L VIN G Transmissions: A/T	**A/T Input Shaft Speed Sensor Circuit Malfunction (4L80-E) Conditions:** DTC P0121-P0123, P0502, P0503, P0717, P0751 and P0753 not set, engine running for 7 seconds while not in P/N, TP sensor over 10%, and the PCM detected the Input Shaft Speed sensor signal varied by more than 1300 rpm for a period of time over 5 seconds. **Possible Causes** ● ISS positive (+) circuit is open, shorted to ground or to power ● ISS negative (-) circuit is open, shorted to ground or to power ● ISS terminals are corroded or damaged (check for moisture) ● ISS signal wires routed too close to the Ignition system cables ● PCM has failed
P0716 **2T CCM, MIL: Yes** 2003, 2004, 2005 CTS (Cadillac) 2.6L VIN M, 3.2L VIN N Transmissions: A/T	**A/T Input Shaft Speed Sensor Circuit Malfunction (5L40-E) Conditions:** DTC P0717, P0722, P0723, P0752, P1791, P1795, P1842 and P1843 not set, engine runtime over 5 seconds, system voltage from 8-18v, gear range is D5, VSS over 10 mph, ICC sensor over 1,050 rpm for 2 seconds, accelerator pedal position over 12%, ISS signal stable for 2 seconds, then the ISS signal increased more than 500 rpm for 2 seconds, and the PCM detected the ISS sensor signal dropped more than 1,000 rpm for 4 seconds. **Possible Causes** ● ISS positive (+) or negative (-) circuit is open, shorted to ground or to power ● ISS is damaged - check for ISS rotor damage, excessive air gap between the reverse clutch input housing assembly (401) rotor and the ISS or Incorrect alignment between the ISS and the reverse clutch input housing assembly (401) rotor. ● TCM has failed
P0716 **2T CCM, MIL: Yes** 1996, 1997, 1998, 1999, 2000, 2001, 2002, 2003, 2004, 2005 Aurora, Century, Deville, Eldorado, Impala, Intrigue, Lumina, Monte Carlo, Regal, Seville 3.4L VIN E, 3.8L VIN K, 3.8L VIN 1, 4.0L VIN C, 4.6L VIN 9, 4.6L VIN Y Transmissions: A/T	**A/T Input Speed Sensor Circuit Malfunction (4L80-E) Conditions:** DTC P0502, P0503, P0717, P0751, P0752, P1820, P1822, P1823, P1825, P1842 and P1843 not set, engine started, engine runtime over 5 seconds, Fuel Cutoff inactive, gearshift not in P/N, VSS over 10 mph, and the PCM detected the ISS indicated less than 51 rpm for 6 seconds during the CCM test. **Possible Causes** ● ISS positive (+) or negative (-) circuit is open, shorted to ground or to power ● ISS is damaged or has failed ● PCM has failed ● TSB 77-71-72 contains a repair procedure for this code
P0717 **2T CCM, MIL: Yes** 1996, 1997, 1998, 1999, 2000, 2001, 2002, 2003, 2004, 2005 Cab & Chassis, C/K truck, G Van, M/L Series Van, S/T Blazer & S/T Pickup 4.8L VIN V, 5.0L VIN M, 5.3L VIN T, 5.3L VIN Z, 5.7L VIN K, 5.7L VIN R, 6.0L VIN N, 6.0L VIN U, 6.6L VIN 1, 7.4L VIN J, 8.1L VIN G engines Transmissions: A/T	**A/T Input Shaft Speed Sensor Low Input (4L80-E) Conditions:** DTC P0502 and P1810 not set, engine started, system voltage over 10.0v, Transmission not in Park or Neutral, vehicle driven to a speed of over 20 mph, Fuel Cutoff mode inactive, and the PCM detected the Input Shaft Speed sensor was less than 100 rpm for 5 seconds. **Possible Causes** ● ISS positive (+) circuit is open, shorted to ground or to power ● ISS negative (-) circuit is open, shorted to ground or to power ● ISS is damaged or has failed ● PCM has failed ● TSB 00-07-30-015 contains a repair procedure for this code
P0717 **2T CCM, MIL: Yes** 1999, 2000, 2001, 2002, 2003, 2004, 2005 Aurora, Bonneville, Century, Cutlass Supreme, Grand Prix, Intrigue, LeSabre, LSS, Lumina, Monte Carlo, 88', Regal, Park Avenue, 98', Montana, Rendezvous, Silhouette, Venture 3.1L VIN J, 3.1L VIN M, 3.4L VIN E, 3.5L VIN H, 3.8L VIN 1, 3.8L VIN K Transmissions: A/T	**A/T Input Speed Sensor Circuit Low Input (4T65-E) Conditions:** DTC P0502, P0503 and P1810 not set, engine started, TFP manual valve position switch not indicating Park or Neutral, system voltage over 10.0v, vehicle driven to a speed of over 5 mph, Fuel Cutoff inactive, and the PCM detected the Input Shaft Speed sensor signal was less than 50 rpm for over 5 seconds during the CCM test. **Possible Causes** ● ISS positive (+) circuit is open, shorted to ground or to power ● ISS negative (-) circuit is open, shorted to ground or to power ● ISS is damaged or has failed ● PCM has failed ● TSB 02-07-30-022A contains a repair procedure for this code

OBD II Trouble Code List (P0xxx Codes)

DTC	Trouble Code Title, Conditions & Possible Causes
P0717 **2T CCM, MIL: Yes** 2003, 2004, 2005 CTS (Cadillac) 2.6L VIN M, 3.2L VIN N Transmissions: A/T	**A/T Input Shaft Speed Sensor Circuit Low Input (5L40-E) Conditions:** DTC P0717, P0722, P0723, P0752, P1791, P1795, P1842 and P1843 not set, engine runtime over 5 seconds, system voltage from 8-18v, transmission not in Park or Neutral, VSS over 10 mph, and the PCM detected the A/T input shaft speed (ISS) sensor signal was less than 100 rpm for 5 seconds during the CCM continuous test. **Possible Causes** ● ISS positive (+) circuit is shorted to ground ● ISS negative (-) circuit is shorted to ground ● ISS is damaged - check for ISS rotor damage, excessive air gap between the reverse clutch input housing assembly (401) rotor and the ISS or Incorrect alignment between the ISS and the reverse clutch input housing assembly (401) rotor. ● TCM has failed
P0717 **2T CCM, MIL: Yes** 1996, 1997, 1998, 1999, 2000, 2001, 2002, 2003, 2004, 2005 Aurora, Century, Deville, Eldorado, Impala, Intrigue, Lumina, Monte Carlo, Regal, Seville 3.4L VIN E, 3.8L VIN K, 3.8L VIN 1, 4.0L VIN C, 4.6L VIN 9, 4.6L VIN Y Transmissions: A/T	**A/T Input Shaft Speed Sensor Low Input (4T65-E, 4L80-E) Conditions:** DTC P0502, P0503m P1820, P1822, P1823 and P1825 not set, engine started, system voltage over 10.0v, Transmission not in Park or Neutral, vehicle driven to a speed of over 10 mph, Fuel Cutoff inactive, VSS over 10 mph, and the PCM detected the Input Shaft Speed sensor indicated less than 51 rpm for six seconds. **Possible Causes** ● ISS positive (+) circuit is open, shorted to ground or to power ● ISS negative (-) circuit is open, shorted to ground or to power ● ISS is damaged or has failed ● PCM has failed ● TSB 77-71-72 contains a repair procedure for this code
P0717 **2T CCM, MIL: Yes** 1997, 1998, 1999, 2000, 2001, 2002, 2003, 2004, 2005 Cavalier, Sunfire, Montana, Silhouette & Venture 2.2L VIN 4, 2.4L VIN T, 3.4L VIN E Transmissions: A/T	**A/T Input Shaft Speed Sensor Low Input (4T40/4T45-E) Conditions:** DTC P0502, P0503 and P1820 not set; engine started, system voltage over 10.0v, TFP manual valve position indicated vehicle was not in P/N, vehicle driven to over 5 mph, and the PCM detected the Input Shaft Speed signal was less than 100 RPM for 5 seconds. **Possible Causes** ● ISS positive (+) circuit is open, shorted to ground or to power ● ISS negative (-) circuit is open, shorted to ground or to power ● ISS is damaged or has failed ● PCM has failed ● TSB 02-07-30-022A contains a repair procedure for this code
P0719 **1T CCM, MIL: No** 1996, 1997, 1998, 1999, 2000, 2001, 2002, 2003, 2004, 2005 Achieva, Alero, Beretta, Cavalier, Ciera, Century, Envoy, Grand Am, Sunfire Skyhawk, S/T Blazer, S/T Pickup & TrailBlazer 2.2L VIN 4, 2.2L VIN 5, 2.4L VIN T, 3.1L VIN J, 3.1L VIN M engines Transmissions: A/T	**TCC Brake Switch Circuit Low Input (3T40, 4T40/4T45-E) Conditions:** DTC P0502 not set, engine started, vehicle speed less than 5 mph, then the vehicle speed was 5-20 mph for 3 seconds, followed by a period with the speed over 20 mph for 6-8 seconds, and the PCM detected the Brake Switch circuit indicated (0) volts for 15 minutes. **Possible Causes** ● TCC brake switch circuit is open or shorted to ground ● TCC brake switch power circuit is open (test ABS or ERL fuse) ● TCC brake switch is out of adjustment or damaged ● PCM has failed
P0719 **1T CCM, MIL: No** 1999, 2000, 2001, 2002, 2003, 2004, 2005 Aurora, Bonneville, Century, Cutlass Supreme, Grand Prix, Intrigue, LeSabre, LSS, Lumina, Monte Carlo, 88', Regal, Park Avenue, 98', Montana, Rendezvous, Silhouette, Venture 3.1L VIN J, 3.1L VIN M, 3.4L VIN E, 3.5L VIN H, 3.8L VIN 1, 3.8L VIN K Transmissions: A/T	**TCC Brake Switch Circuit Low Input (4T60-E, 4T65-E) Conditions:** DTC P0502 and P0503 not set, engine started, vehicle speed less than 5 mph, then vehicle speed was from 5-20 mph for 3 seconds, then after the speed more than 20 mph for 6 seconds, the PCM detected the Brake Switch status was open for 5 minutes under these conditions. The TCC brake switch indicates brake pedal status to the PCM (i.e., when the brake pedal is applied or released). The N.C. switch supplies battery voltage signal to the PCM. Applying the brake pedal opens the TCC brake switch, interrupting voltage to the PCM. When the PCM detects 0 volts at the TCC brake switch circuit, the PCM turns OFF the TCC PWM solenoid valve. **Possible Causes** ● TCC brake switch circuit is open or shorted to ground ● TCC brake switch power circuit is open (check ABS PCM fuse) ● TCC brake switch is out of adjustment or damaged ● PCM has failed

OBD II Trouble Code List (P0xxx Codes)

DTC	Trouble Code Title, Conditions & Possible Causes
P0719 **1T CCM, MIL: No** 1996, 1997, 1998 Bonneville, Century, Cutlass Supreme, Grand Prix, Intrigue, LeSabre, LSS, Lumina, Monte Carlo, 88', Park Avenue, Regal & 98' 3.1L VIN M, 3.8L VIN 1, 3.8L VIN K engines Transmissions: A/T	**TCC Brake Switch Circuit Low Input (4T60-E, 4T65-E) Conditions:** DTC P0502 not set, engine started, vehicle speed less than 5 mph, then the vehicle speed was 5-20 mph for 3 seconds, followed by a period with the speed over 20 mph for 6 seconds, and the PCM detected the Brake Switch circuit indicated (0) volts for 15 minutes. **Possible Causes** ● TCC brake switch circuit is open or shorted to ground ● TCC brake switch power circuit is open (check the CRUISE, FAN ALT or other related fuse in underhood junction block) ● TCC brake switch is out of adjustment or damaged ● PCM has failed
P0719 **1T CCM, MIL: No** 1996, 1997, 1998, 1999, 2000, 2001, 2002, 2003, 2004, 2005 Cab & Chassis, C/K truck, G Van, M/L Series Van, S/T Blazer & S/T Pickup 4.3L VIN W, 4.3L VIN X, 4.8L VIN V, 5.0L VIN M, 5.3L VIN T, 5.3L VIN Z, 5.7L VIN K, 5.7L VIN R, 5.7L VIN G, 5.7L VIN S, 6.0L VIN N, 6.0L VIN U, 6.6L VIN 1, 7.4L VIN J, 8.1L VIN G engines Transmissions: A/T	**TCC Brake Switch Circuit Low Input (4L60-E, 4L80-E) Conditions:** DTC P0502 and P0503 not set, engine started, vehicle speed less than 5 mph, then the vehicle speed from 5-20 mph for 4 seconds, followed by a period with the vehicle speed over 20 mph for 6-8 seconds, and the PCM detected the Brake Switch circuit indicated zero (0) volts for 15 minutes without changing for 2 seconds. Note that this series of events must occur (8) times for this code to set. **Possible Causes** ● TCC brake switch circuit is open or shorted to ground ● TCC brake switch power circuit is open (check the CRUISE, FAN ALT or other related fuse in underhood junction block) ● TCC brake switch is out of adjustment or damaged ● PCM has failed
P0719 **1T CCM, MIL: No** 2003, 2004, 2005 CTS (Cadillac) 2.6L VIN M, 3.2L VIN N Transmissions: A/T	**TCC Brake Switch Circuit Low Input (5L40-E) Conditions:** DTC P0722 and P0723 not set, P0724 not pending, system voltage from 8-18v, engine started, and the PCM detected the TCC brake switch signal was at 0v for 15 minutes, and the following conditions occurred (8) times without a status change on the TCC brake input; VSS less than 5 mph, then the VSS from 20-5 mph for a period of from 3-7 seconds, then the VSS signal remained over for 6 seconds. The TCC brake switch is used to indicate the status of the brake pedal, applied or released, to the TCM. The N.C. TCC brake switch supplies 12v signal to the TCM. Applying the brake pedal opens the switch, interrupting voltage to the TCM. Releasing the brake pedal resumes voltage to the TCM. When the TCM senses 0v at the TCC brake switch, the TCM de-energizes the TCC solenoid valve. **Possible Causes** ● TCC brake switch circuit is open or shorted to ground ● TCC brake switch power circuit is open (test the TCC/ET fuse) ● TCC brake switch is out of adjustment or damaged ● PCM has failed
P0719 **1T CCM, MIL: No** 1996, 1997, 1998, 1999, 2000, 2001, 2002, 2003, 2004, 2005 Aurora, DeVille, Eldorado, Seville 4.0L VIN C, 4.6L VIN 9, 4.6L VIN Y engines Transmissions: A/T	**TCC Brake Switch Circuit Low Input (4L80-E) Conditions:** DTC P0502, P0503 not set, P0719 test not passed this key cycle, engine started, vehicle speed less than 5 mph, then the vehicle speed was 5-20 mph for 3 seconds, followed by a period with the vehicle speed over 20 mph for 8 seconds, and the PCM detected the Brake Switch circuit indicated zero (0) volts for 15 minutes. **Possible Causes** ● TCC brake switch circuit is open or shorted to ground ● TCC brake switch power circuit is open (check the CRUISE, FAN ALT or other related fuse in underhood junction block) ● TCC brake switch is out of adjustment or damaged ● PCM has failed
P0719 **1T CCM, MIL: No** 1995, 1996, 1997, 1998, 1999, 2000, 2001, 2002, 2003, 2004, 2005 Camaro, Corvette & Firebird 3.8L VIN K, 5.7L VIN G, 5.7L VIN S Transmissions: A/T	**TCC Brake Switch Circuit Low Input (4L60-E) Conditions:** DTC P0502 not set, engine started, vehicle speed less than 5 mph, then the vehicle speed was 5-20 mph for 3 seconds, followed by a period with the speed over 20 mph for 6-8 seconds, and the PCM detected the Brake Switch circuit indicated (0) volts for 15 minutes. **Possible Causes** ● TCC brake switch circuit is open or shorted to ground ● TCC brake switch power circuit is open (check the PCM IGN, ENG GEN or other related fuse in underhood junction block) ● TCC brake switch is out of adjustment or damaged ● PCM has failed

OBD II Trouble Code List (P0xxx Codes)

DTC	Trouble Code Title, Conditions & Possible Causes
P0719 **1T CCM, MIL: No** 2002, 2003, 2004, 2005 Envoy & TrailBlazer 4.2L VIN S, 5.3L VIN P Transmissions: A/T	**TCC Brake Switch Circuit Low Input (4L60-E/4L65-E) Conditions:** DTC P0502, P0503 and P0719 not set, Key on or engine running; Brake switch status is open and DTC P0719 has not passed, and the PCM detected the Brake switch or circuit indicated open (0v) for 15 minutes without changing for 2 seconds, and the following events occurred (7) times: vehicle speed less than 5 mph, then the vehicle speed from 5-20 mph for 4 seconds; and then the vehicle speed was more than 20 mph for 6 seconds during the test. **Possible Causes** ● TCC brake switch circuit is open or shorted to ground ● TCC brake switch power circuit is open (test ABS or ERL fuse) ● TCC brake switch is out of adjustment or damaged ● PCM has failed
P0722 **2T CCM, MIL: Yes** 1996, 1997, 1998, 1999, 2000, 2001, 2002, 2003, 2004, 2005 C/K Truck & G Van 6.5L VIN F, 6.5L VIN S, 6.6L VIN 1 Diesel engines Transmissions: A/T	**Output Shaft Speed Sensor Circuit Low Input (4L80-E) Conditions:** DTC P0106-P0108 and DTC P1810 not set, vehicle driven with the APP angle over 10% while not in P/N at an engine speed less than 3800 rpm for 7 seconds, ISS signal over 1500 rpm, and the PCM detected the OSS signal indicated less than 300 rpm for 3 seconds. **Possible Causes** ● OSS positive (+) circuit is open or shorted to ground ● OSS negative (-) circuit is open or shorted to ground ● OSS terminals are corroded or damaged (check for moisture) ● PCM has failed
P0722 **2T CCM, MIL: Yes** 2003, 2004, 2005 CTS (Cadillac) 2.6L VIN M, 3.2L VIN N Transmissions: A/T	**Output Shaft Speed Sensor Circuit Low Input (5L40-E) Conditions:** DTC P0716, P0717, P0723, P1791, P1795, P1815, p1818, P1822, P1823, P1825 and P1826 not set, engine runtime over 5 seconds, system voltage from 8-18v, transmission not in P/N, APP angle over 12%, engine torque from 52-258 lb ft., engine torque default values not used, ISS signal from 1,500-6,000 rpm, and the PCM detected the OSS signal indicated less than 100 rpm for 3 seconds. **Possible Causes** ● OSS positive (+) circuit is open or shorted to ground ● OSS negative (-) circuit is open or shorted to ground ● OSS terminals are corroded or damaged (check for moisture) ● OSS unit is damaged or it has failed ● PCM has failed
P0722 **2T CCM, MIL: Yes** 1997, 1998, 1999, 2000, 2001 Catera 3.0L VIN R engine Transmissions: A/T	**Output Shaft Speed Sensor Circuit Low Input (4L30-E) Conditions:** Engine started, vehicle driven to a speed of over 3008 rpm, Transmission not in Park, Reverse or Neutral, and the PCM detected the Output Shaft Speed sensor indicated zero (0) rpm. *Note: The Sport Mode Lamp flashes when this trouble code is set.* **Possible Causes** ● OSS positive (+) circuit is open, shorted to ground or to power ● OSS negative (-) circuit is open, shorted to ground or to power ● OSS is damaged or has failed ● PCM has failed
P0723 **2T CCM, MIL: Yes** 1996, 1997, 1998, 1999, 2000, 2001, 2002, 2003, 2004, 2005 C/K Truck & G Van 6.5L VIN F, 6.5L VIN S, 6.6L VIN 1 Diesel engines Transmissions: A/T	**Output Shaft Speed Sensor Circuit High Input (4L80-E) Conditions:** DTC P1810 and DTC P1810 not set, engine running with no TFP Value Position switch change for 7 seconds, vehicle not in 4WD Low position, and the PCM detected the OSS signal decreased more than 1000 rpm while driven in Drive position for over 3.5 seconds. **Possible Causes** ● OSS positive (+) circuit is open or shorted to system power ● OSS negative (-) circuit is open or shorted to system power ● OSS terminals are corroded or damaged (check for moisture) ● PCM has failed
P0723 **2T CCM, MIL: Yes** 2003, 2004, 2005 CTS (Cadillac) 2.6L VIN M, 3.2L VIN N Transmissions: A/T	**Output Shaft Speed Sensor Circuit High Input (5L40-E) Conditions:** DTC P0717, P0722, P0723, P0752, P1791, P1795, P1842 and P1843 not set, engine runtime over 5 seconds, system voltage from 8-18v, transmission not in P/N, VSS over 10 mph, OSS sensor over 1,400 rpm for 2 seconds with no change for 6 seconds, then the ISS signal stable for 2 seconds with the transmission output shaft speed change less than 500 rpm for 2 seconds, and the PCM detected the A/T OSS sensor signal dropped more than 1,300 rpm for 4 seconds. **Possible Causes** ● OSS positive (+) circuit is open or shorted to ground ● OSS negative (-) circuit is open or shorted to ground ● OSS terminals are corroded or damaged (check for moisture) ● OSS unit is damaged or it has failed ● PCM has failed

OBD II Trouble Code List (P0xxx Codes)

DTC	Trouble Code Title, Conditions & Possible Causes
P0724 **1T CCM, MIL: No** 1996, 1997, 1998, 1999, 2000, 2001, 2002, 2003, 2004, 2005 Achieva, Alero, Beretta, Cavalier, Ciera, Century, Corsica, Grand Am, Sunfire Skyhawk, S/T Blazer & S/T Pickup 2.2L VIN 4, 2.2L VIN 5, 2.4L VIN T, 3.1L VIN J, 3.1L VIN M engines Transmissions: A/T	**Brake Switch Circuit High Input (3T40, 4T40-E, 4T45-E) Conditions:** DTC P0502 and P0503 not set, vehicle speed more than 20 mph for 6 seconds, then the speed decreased to 5 mph for 2-6 seconds, followed by a period with the vehicle speed less than 5 mph, and the PCM detected the Brake switch circuit was 12-14v at least 7 times. **Possible Causes** ● Check for ABS trouble codes (they can cause this code to set) ● TCC brake switch circuit is shorted to system power ● TCC brake switch is out of adjustment or damaged
P0724 **1T CCM, MIL: No** 1996, 1997, 1998, 1999, 2000, 2001, 2002, 2003, 2004, 2005 Aurora, Bonneville, Century, Cutlass Supreme, Grand Prix, Intrigue, LeSabre, LSS, Lumina, Monte Carlo, 88', Regal, Park Avenue, 98', Montana, Rendezvous, Silhouette, Venture 3.1L VIN J, 3.1L VIN M, 3.4L VIN E, 3.5L VIN H, 3.8L VIN 1, 3.8L VIN K Transmissions: A/T	**TCC Brake Switch Circuit High Input (4T65-E) Conditions:** DTC P0502 and P0503 not set, vehicle speed more than 20 mph for 6 seconds, then the speed decreased to 5 mph for 2-6 seconds, followed by a period with the vehicle speed less than 5 mph, and the PCM detected the Brake switch circuit was 12-14v at least 7 times. **Possible Causes** ● Check for ABS trouble codes (they can cause this code to set) ● TCC brake switch circuit is shorted to system power ● TCC brake switch is out of adjustment or damaged ● PCM has failed
P0724 **1T CCM, MIL: No** 1996, 1997, 1998, 1999, 2000, 2001, 2002, 2003, 2004, 2005 Cab & Chassis, C/K truck, G Van, M/L Series Van, S/T Blazer & S/T Pickup 4.3L VIN W, 4.3L VIN X, 4.8L VIN V, 5.0L VIN M, 5.3L VIN T, 5.3L VIN Z, 5.7L VIN K, 5.7L VIN R, 5.7L VIN G, 5.7L VIN S, 6.0L VIN N, 6.0L VIN U, 6.6L VIN 1, 7.4L VIN J, 8.1L VIN G engines Transmissions: A/T	**TCC Brake Switch High Input (4L60-E, 4L80-E) Conditions:** DTC P0502, P0503 and P0724 not set, engine started, then VSS signal more than 20 mph for 6 seconds, then the VSS decreased to 5 mph for 2-6 seconds, followed by a period with the vehicle speed less than 5 mph, and the PCM detected the Brake switch circuit indicated from 12-14v at least 7 times (i.e., no change in the signal). **Possible Causes** ● Check for ABS trouble codes (they can cause this code to set) ● TCC brake switch circuit is shorted to system power ● TCC brake switch is out of adjustment or damaged ● PCM has failed
P0724 **1T CCM, MIL: No** 2003, 2004, 2005 CTS (Cadillac) 2.6L VIN M, 3.2L VIN N Transmissions: A/T	**TCC Brake Switch Circuit Low Input (5L40-E) Conditions:** DTC P0722 and P0723 not set, P0724 not pending, system voltage from 8-18v, engine started, and the PCM detected the TCC brake switch signal was at 0v for 15 minutes, and the following conditions occurred (8) times without a status change on the TCC brake input: VSS more than 20 mph for 6 seconds, then VSS between 20 mph and 5 mph for a period from 3-7 seconds, then the VSS less than 5 mph. The TCC brake switch is used to indicate the status of the brake pedal, applied or released, to the TCM. The TCC brake switch (N.C.) supplies a 12v signal to the TCM. **Possible Causes** ● TCC brake switch circuit is open or shorted to ground ● TCC brake switch power circuit is open (test the TCC/ET fuse) ● TCC brake switch is out of adjustment or damaged ● PCM has failed
P0724 **1T CCM, MIL: No** 1996, 1997, 1998, 1999, 2000, 2001, 2002, 2003, 2004, 2005 DeVille, Eldorado, Seville 4.0L VIN C, 4.6L VIN 9, 4.6L VIN Y engines Transmissions: A/T	**TCC Brake Switch Circuit Low Input (4L80-E) Conditions:** DTC P0502, P0503 not set, P0719 test not passed this key cycle, engine started, vehicle speed more than 20 mph for 6 seconds, then the speed decreased to 5 mph for 2-6 seconds, followed by a period with the vehicle speed less than 5 mph, and the PCM detected the Brake switch circuit was 12-14v at least 7 times. **Possible Causes** ● Check for ABS trouble codes (they can cause this code to set) ● TCC brake switch circuit is shorted to system power ● TCC brake switch is out of adjustment or damaged ● PCM has failed

OBD II Trouble Code List (P0xxx Codes)

DTC	Trouble Code Title, Conditions & Possible Causes
P0724 **1T CCM, MIL: No** 1995, 1996, 1997, 1998, 1999, 2000, 2001, 2002, 2003, 2004, 2005 Camaro, Corvette & Firebird 3.8L VIN K, 5.7L VIN G, 5.7L VIN S engines Transmissions: A/T	**TCC Brake Switch Circuit High Input (4L60-E) Conditions:** DTC P0502 and P0503 not set, vehicle speed more than 20 mph for 6 seconds, then the speed decreased to 5 mph for 2-6 seconds, followed by a period with the vehicle speed less than 5 mph, and the PCM detected the Brake switch circuit was 12-14v at least 7 times. **Possible Causes** ● Check for ABS trouble codes (they can cause this code to set) ● TCC brake switch circuit is shorted to system power ● TCC brake switch is out of adjustment or damaged ● PCM has failed
P0724 **1T CCM, MIL: No** 2002, 2003, 2004, 2005 Envoy & TrailBlazer 4.2L VIN S, 5.3L VIN P Transmissions: A/T	**Brake Switch Circuit High Input (4L60-E, 4L65-E) Conditions:** DTC P0502 and P0503 not set, Key on or engine running; and the PCM detected the Brake switch circuit indicated closed (12v) without changing for two seconds after the following events occur (8) times: vehicle speed was more than 20 mph for 6 seconds; then the vehicle speed was between 5-20 mph for 4 seconds; and then the vehicle speed was less than 5 mph during the CCM test period. **Possible Causes** ● Check for ABS trouble codes (they can cause this code to set) ● TCC brake switch circuit is shorted to system power ● TCC brake switch is out of adjustment or damaged
P0725 **2T CCM, MIL: Yes** 1997, 1998, 1999, 2000, 2001 Catera 3.0L VIN R engine Transmissions: A/T	**Engine Speed Signal Circuit Low Input (4L30-E) Conditions:** Engine started, Transmission not in Park, Reverse or Neutral, TP angle more than 12%, Transmission output speed more than 1024 rpm, and the PCM detected the Engine Speed Sensor signal into the TCM was less than 122 rpm for 4 seconds. Engine speed data is sent to the TCM by the PCM through the CAN Network or CAN. If this network fails, other codes will set before P0725 is set. *Note: The Sport Mode Lamp flashes when this code is set.* **Possible Causes** ● Engine speed sensor circuit is open, shorted to ground or to B+ ● ECM or TCM has failed
P0725 **2T CCM, MIL: Yes** 2003, 2004, 2005 CTS (Cadillac) 2.6L VIN M, 3.2L VIN N Transmissions: A/T	**TCC Brake Switch Circuit Low Input (5L40-E) Conditions:** Engine started, system voltage from 8-18v, and the PCM detected the engine speed was invalid for 2 seconds. Engine speed data is sent to the TCM by the PCM via the controller area network (CAN). Two circuits are used to communicate CAN data between the PCM and TCM. A fault in the CAN will not cause DTC P0727 to set by itself. If a CAN fault occurs, other codes will set before DTC P0727 is set. **Possible Causes** ● TCC brake switch circuit is open or shorted to ground ● TCC brake switch power circuit is open (test the TCC/ET fuse) ● TCC brake switch is out of adjustment or damaged ● PCM has failed
P0727 **2T CCM, MIL: Yes** 1999, 2000, 2001 Catera 3.0L VIN R engine Transmissions: A/T	**Engine Speed Signal Circuit Malfunction (4L30-E) Conditions:** Engine started, and the PCM detected the engine speed signal was too low, too high, or the speed signal was unreliable for 4 seconds. Engine speed information is sent to the TCM by the PCM through the Controller Area Network or CAN. Two circuits are used to communicate CAN data between the ECM and TCM. If the CAN (network) fails, other trouble codes will set before DTC P0727 is set. *Note: The Sport Mode Lamp flashes when this trouble code is set.* **Possible Causes** ● Check for CAN network codes (i.e., U2100, 2104, 2105, 2108) ● ECM or TCM has failed
P0730 **1T CCM, MIL: No** 1996, 1997, 1998, 1999, 2000, 2001, 2002, 2003, 2004, 2005 Aurora, Bonneville, Century, Cutlass Supreme, Grand Prix, Intrigue, LeSabre, LSS, Lumina, Monte Carlo, Regal, Park Avenue, 88' & 98', Montana, Silhouette, Venture, Rendezvous3.1L VIN J, 3.1L VIN M, 3.4L VIN E, 3.5L VIN H, 3.8L VIN 1, 3.8L VIN K Transmissions: A/T	**Incorrect Gear Ratio (4T65-E) Conditions:** DTC P0121, P0122, P0123, P0502, P0503, P0716, P0717 and P1810 not set, engine started, vehicle driven to over 7 mph, Fuel Cutoff inactive, Transmission not in Park or Neutral, time since last gear select lever change over 6 seconds, TP angle over 14%, TFT sensor more than 68°F, engine torque from 50-300 ft lbs, and the PCM detected one of these conditions occurred for 7 seconds: - The gear ratio was more than 2.97:1 or it was 1.62:1 to 2.33:1 - The gear ratio was 1.05:1 to 1.52:1 or it was 0.75:1 to 0.95:1 **Possible Causes** ● ATF level is too low, or the fluid is burnt or contaminated ● ISS or OSS signal circuit has an intermittent fault condition ● Inspect for debris in the transmission pan or internal damaged ● Possible vehicle overloading, exceeding the trailer towing limit, or towing in overdrive events occurred (discuss with customer) ● TSB 02-07-30-022A contains a repair procedure for this code

OBD II Trouble Code List (P0xxx Codes)

DTC	Trouble Code Title, Conditions & Possible Causes
P0730 **1T CCM, MIL: No** 1996, 1997, 1998, 1999 Aurora, DeVille, Eldorado, Seville 4.0L VIN C, 4.6L VIN 9, 4.6L VIN Y engines Transmissions: A/T	**Incorrect Gear Ratio (4L80-E) Conditions:** DTC P0101-P0103, P0121-P0123, P0502, P0503, P0716, P0717, P0753, P0758 and P1810 not set, engine started, Fuel Cutoff inactive, Transmission not in Park or Neutral, time since last gear select lever change over 3 seconds, TP angle over 15.7 degrees, VSS over 7 mph, TFT sensor over 68°F, delivered torque more than 100 ft lbs, and the PCM detected at one or more steady state adaptive cells were at the maximum and one of the conditions listed below was met for four seconds: - The 1st gear ratio was less than 2.87:1 or greater than 3.11:1 - The 2nd gear ratio was less than 1.54:1 or greater than 1.71:1 - The 3rd gear ratio was less than 0.95:1 or greater than 1.05:1 - Fourth gear ratio is less than 0.65:1 or greater than 0.71:1 - Reverse gear ratio is less than 2.02:1 or greater than 2.23:1 **Possible Causes** ● ATF level is too low, or the fluid is burnt or contaminated ● ISS or OSS signal circuit has an intermittent fault condition ● Inspect for debris in the transmission pan or internal damaged ● Possible vehicle overloading, exceeding the trailer towing limit, or towing in overdrive events occurred (discuss with customer)
P0730 **1T CCM, MIL: No** 1996, 1997, 1998, 1999, 2000, 2001, 2002, 2003, 2004, 2005 Cab & Chassis, C/K truck, G Van, M/L Series Van, S/T Blazer & S/T Pickup 5.0L VIN M, 5.3L VIN T, 5.3L VIN Z, 5.7L VIN K, 5.7L VIN R, 5.7L VIN G, 5.7L VIN S, 6.0L VIN N, 6.0L VIN U, 6.6L VIN 1, 7.4L VIN J, 8.1L VIN G Transmissions: A/T	**Incorrect Gear Ratio (4L80-E) Conditions:** DTC P0106-P0108, P0121-P0123, P0502, P0503, P0716, P0717, P1810 and P1875 not set, engine started, vehicle driven to a speed over 4 mph, TP angle over 15%, TFT sensor signal over 68°F, delivered torque more than 400 ft lbs, and the PCM detected one or more of the following gear ratio conditions existed for four seconds: - The gear ratio was more than 2.50: 1 or less than 2.43:1 - The gear ratio was more than 1.50:1 or less than 1.44:1 - The gear ratio was more than 1.03:1 or less than 0.25:1 - The gear ratio was more than 2.12:1 or less than 2.04:1 **Possible Causes** ● ATF level is too low, or the fluid is burnt or contaminated ● ISS or OSS signal circuit has an intermittent fault condition ● Inspect for debris in the transmission pan or internal damaged ● Possible vehicle overloading, exceeding the trailer towing limit, or towing in overdrive events occurred (discuss with customer)
P0730 **2T CCM, MIL: Yes** 1997, 1998, 1999, 2000, 2001 Catera 3.0L VIN R engine Transmissions: A/T	**Incorrect Gear Ratio 4L30-E) Conditions:** Engine started, Transmission not in Park, Neutral or Reverse, or running in Safety mode, enable conditions for DTC P0757 fail Case 1 not met, engine speed more than 3488 rpm, then after the TCM detected the absolute slippage calculated with 1st gear "on", output speed exceeding 352 rpm for 4 seconds, with 2nd gear "on", output speed exceeded 576 rpm for 4 seconds, or with 3rd gear "on", output speed exceeding 806 rpm for 4 seconds (Condition 1); or engine running, OSS speed less than 2976 rpm in 3rd gear (Condition 2), and the TCM detected the OSS sensor speed exceeded 806 rpm for 4 seconds. The Sport Mode Lamp will flash if this trouble code is set. **Possible Causes** ● ATF level is too low, or the fluid is burnt or contaminated ● ISS or OSS signal circuit has an intermittent fault condition ● Inspect for debris in transmission pan or a sticking shift valve ● Possible vehicle overloading, exceeding the trailer towing limit, or towing in overdrive events occurred (discuss with customer)
P0731 **1T CCM, MIL: No** 2000, 2001, 2002, 2003, 2004, 2005 Aurora, DeVille, Eldorado, Seville 4.0L VIN C, 4.6L VIN 9, 4.6L VIN Y engines Transmissions: A/T	**Incorrect 1st Gear Ratio (4L80-E) Conditions:** DTC P0121-P0123, P0502, P0503, P0716, P0717, P1820, P1822, P1823 and P1825 not set, engine started, vehicle speed over 3 mph, Fuel Cutoff inactive, Transmission not in P/N, time since last gear select lever change over 6 seconds, TP angle over 10%, TFT sensor from 68°F-271°F, delivered torque from 30-100 ft lbs, and the PCM commanded 1st gear "on" and detected the gear ratio was greater than 3.11:1; the gear ratio was 2.23:1 to 2.87:1; the gear ratio was 1.71:1 to 2.02:1; the gear ratio was 1.05:1 to 1.54:1; the gear ratio was 0.71:1 to 0.95:1; or the gear ratio was less 0.65: 1. **Possible Causes** ● ATF level is too low, or the fluid is burnt or contaminated ● ISS or OSS signal circuit has an intermittent fault condition ● Inspect for debris in transmission pan or a sticking shift valve ● Possible vehicle overloading, exceeding the trailer towing limit, or towing in overdrive events occurred (discuss with customer) ● TSB 01-07-30-009B contains a repair procedure for this code

OBD II Trouble Code List (P0xxx Codes)

DTC	Trouble Code Title, Conditions & Possible Causes
P0731 **1T CCM, MIL: No** 2003, 2004, 2005 CTS (Cadillac) 2.6L VIN M, 3.2L VIN N Transmissions: A/T	**Incorrect 1st Gear Ratio (5L40-E) Conditions:** DTC P0716, P0717, P0722, P0723, P1791, P1795, P1815, P1818, P1822, P1823, P1825 and P1826 not set, engine runtime over 5 seconds, TFT sensor from 68-266ºF, Transmission not in P/N, time since most recent gear change over 6 seconds, engine torque from 41-258 lb ft., APP angle over 10%, gear ratio between 0.61:1-4.14:1, gear ratio is not Reverse: 3.1:1 to 2.95:1, VSS more than 3 mph, then with 1st gear commanded "on", and the PCM detected the actual gear ratio was not from 3.76:1 to 3.33:1; gear ratio not within 2.44:1 to 2.10:1; the gear ratio was not within 1.76:1 to 1.44:1; the gear ratio was not within 1.10:1 to 0.90:1; or the gear ratio was not within 0.83:1 to 0.68:1. If the TCM detects an invalid 1st gear ratio, indicating too much slip or drag in the transmission, it sets this code. **Possible Causes** • ATF level is too low, or the fluid is burnt or contaminated • 1-2 shift valve stuck in the release position • 1-2 shift solenoid valve assembly mechanically stuck on • 1-2 shift control valve stuck in the release position • 2-3 shift solenoid valve mechanically stuck off or leaking • 2-3 shift control valve stuck or 2-3 shift valve stuck • 2nd clutch backing plate retainer ring out of position/damaged • 2nd clutch plate assembly slipping, dragging or damaged • 2nd clutch apply plate seized or damaged • 2nd clutch piston cracked or worn, 2nd clutch spring damaged • Control valve accumulator assembly is damaged or has failed • Visually inspect the transmission cooling lines for signs of leaks
P0732 **1T CCM, MIL: No** 2000, 2001, 2002, 2003, 2004, 2005 Aurora, DeVille, Eldorado, Seville 4.0L VIN C, 4.6L VIN 9, 4.6L VIN Y engines Transmissions: A/T	**Incorrect 2nd Gear Ratio (4L80-E) Conditions:** DTC P0121-P0123, P0502, P0503, P0716, P0717, P1820, P1822, P1823 and P1825 not set, engine started, vehicle speed over 3 mph, Fuel Cutoff inactive, Transmission not in P/N, time since last gear select lever change over 6 seconds, TP angle over 10%, TFT sensor from 68ºF-271ºF, delivered torque from 30-100 ft lbs, 2nd gear commanded "on", and the PCM detected the gear ratio was more than 3.11:1; or the gear ratio was 2.23:1 to 2.87:1; or the gear ratio was 1.71:1 to 2.02:1; or the gear ratio was 1.05:1 to 1.54:1; or gear ratio was 0.71:1 to 0.95:1;or gear ratio was less than 0.65: 1. **Possible Causes** • ATF level is too low, or the fluid is burnt or contaminated • ISS or OSS signal circuit has an intermittent fault condition • Inspect for debris in transmission pan or a sticking shift valve • Possible vehicle overloading, exceeding the trailer towing limit, or towing in overdrive events occurred (discuss with customer) • TSB 01-07-30-009B contains a repair procedure for this code
P0732 **1T CCM, MIL: No** 2003, 2004, 2005 CTS (Cadillac) 2.6L VIN M, 3.2L VIN N Transmissions: A/T	**Incorrect 2nd Gear Ratio (5L40-E) Conditions:** DTC P0716, P0717, P0722, P0723, P1791, P1795, P1815, P1818, P1822, P1823, P1825 and P1826 not set, engine runtime over 5 seconds, TFT sensor from 68-266ºF, Transmission not in P/N, time since most recent gear change over 6 seconds, engine torque from 41-258 lb ft., APP angle over 10%, gear ratio between 0.61:1-4.14:1, gear ratio is not Reverse: 3.1:1 to 2.95:1, VSS more than 3 mph, then with 2nd gear commanded "on", and the PCM detected the actual gear ratio was not from 3.76:1 to 3.33:1; gear ratio not within 2.44:1 to 2.10:1; the gear ratio was not within 1.76:1 to 1.44:1; the gear ratio was not within 1.10:1 to 0.90:1; or the gear ratio was not within 0.83:1 to 0.68:1. If the TCM detects an invalid 2nd gear ratio, indicating too much slip or drag in the transmission, it sets this code. **Possible Causes** • ATF level is too low, or the fluid is burnt or contaminated • A/T control valve body and accumulator for 2nd clutch leaking • 1-2 shift solenoid mechanically stuck off or leaking • 1-2 shift control valve stuck • 2nd clutch backing plate retainer ring out of position/damaged • Second clutch plate assembly slipping, dragging or damaged • Second clutch apply plate seized or damaged • Transmission case center support fluid passage sleeve leaking • Transmission has internal damage (may require replacement)

OBD II Trouble Code List (P0xxx Codes)

DTC	Trouble Code Title, Conditions & Possible Causes
P0733 **1T CCM, MIL: No** 2000, 2001, 2002, 2003, 2004, 2005 Aurora, DeVille, Eldorado, Seville 4.0L VIN C, 4.6L VIN 9, 4.6L VIN Y engines Transmissions: A/T	**Incorrect 3rd Gear Ratio (4L80-E) Conditions:** DTC P0121-P0123, P0502, P0503, P0716, P0717, P1820, P1822, P1823 or P1825 not set, engine started, vehicle speed over 3 mph, Fuel Cutoff inactive, Transmission not in P/N, time since last gear select lever change over 6 seconds, TP angle over 10%, TFT sensor from 68°F-271°F, delivered torque from 30-100 ft lbs, and the PCM commanded 3rd gear "on" and received one of these conditions: - The gear ratio was greater than 3.11:1, or it was 2.23:1 to 2.87:1 - The gear ratio was 1.71:1 to 2.02:1, or it was 1.05:1 to 1.54:1 - The gear ratio was 0.71:1 to 0.95:1, or it was less 0.65: 1 **Possible Causes** ● ATF level is too low, or the fluid is burnt or contaminated ● ISS or OSS signal circuit has an intermittent fault condition ● Inspect for debris in transmission pan or a sticking shift valve ● Possible vehicle overloading, exceeding the trailer towing limit, or towing in overdrive events occurred (discuss with customer)
P0733 **1T CCM, MIL: No** 2003, 2004, 2005 CTS (Cadillac) 2.6L VIN M, 3.2L VIN N Transmissions: A/T	**Incorrect 3rd Gear Ratio (5L40-E) Conditions:** DTC P0716, P0717, P0722, P0723, P1791, P1795, P1815, P1818, P1822, P1823, P1825 and P1826 not set, engine runtime over 5 seconds, TFT sensor from 68-266°F, Transmission not in P/N, time since most recent gear change over 6 seconds, engine torque from 41-258 lb ft., APP angle over 10%, gear ratio between 0.61:1-4.14:1, gear ratio is not Reverse: 3.1:1 to 2.95:1, VSS more than 3 mph, then with 3rd gear commanded "on", and the PCM detected the actual gear ratio was not from 3.76:1 to 3.33:1; gear ratio not within 2.44:1 to 2.10:1; the gear ratio was not within 1.76:1 to 1.44:1; the gear ratio was not within 1.10:1 to 0.90:1; or the gear ratio was not within 0.83:1 to 0.68:1. If the TCM detects an invalid 3rd gear ratio, indicating too much slip or drag in the transmission, it sets this code. **Possible Causes** ● ATF level is too low, or the fluid is burnt or contaminated ● 2-3 shift solenoid valve assembly mechanically stuck on ● 2-3 shift control valve stuck ● 2-3 shift valve stuck ● Intermediate clutch housing retainer ring out of position ● Intermediate clutch plates slipping, dragging or damaged ● Intermediate clutch apply plate seized or damaged ● Intermediate clutch piston assembly cracked, worn or damaged ● Intermediate clutch spring assembly damaged ● Overdrive clutch housing cracked or fluid feed hole blocked ● Intermediate sprag clutch assembly for the intermediate clutch sprag assembly not holding or damaged ● Transmission has internal damage (may require replacement)
P0733 **1T CCM, MIL: No** 2003, 2004, 2005 CTS (Cadillac) 2.6L VIN M, 3.2L VIN N Transmissions: A/T	**Incorrect 4th Gear Ratio (5L40-E) Conditions:** DTC P0716, P0717, P0722, P0723, P1791, P1795, P1815, P1818, P1822, P1823, P1825 and P1826 not set, engine runtime over 5 seconds, TFT sensor from 68-266°F, Transmission not in P/N, time since most recent gear change over 6 seconds, engine torque from 41-258 lb ft., APP angle over 10%, gear ratio between 0.61:1-4.14:1, gear ratio is not Reverse: 3.1:1 to 2.95:1, VSS more than 3 mph, then with 4th gear commanded "on", and the PCM detected the actual gear ratio was not from 3.76:1 to 3.33:1; gear ratio not within 2.44:1 to 2.10:1; the gear ratio was not within 1.76:1 to 1.44:1; the gear ratio was not within 1.10:1 to 0.90:1; or the gear ratio was not within 0.83:1 to 0.68:1. If the TCM detects an invalid 4th gear ratio, indicating too much slip or drag in the transmission, it sets this code. **Possible Causes** ● ATF level is too low, or the fluid is burnt or contaminated ● 1-2 shift solenoid valve assembly mechanically stuck on ● 3-4 shift valve stuck ● Direct clutch backing plate retainer ring out of position ● Direct clutch plate assembly slipping, dragging or damaged ● Direct clutch apply plate seized or damaged ● Direct clutch piston assembly cracked or damaged ● Direct clutch and release clutch spring damaged ● Direct clutch housing ball check valve leaking ● Direct clutch drum and shaft assembly for the direct clutch input and hub assembly damaged ● Fluid pump unit for reverse clutch housing fluid seal ring leaking ● Transmission has internal damage (may require replacement)

OBD II Trouble Code List (P0xxx Codes)

DTC	Trouble Code Title, Conditions & Possible Causes
P0734 **1T CCM, MIL: No** 2000, 2001, 2002, 2003, 2004, 2005 Aurora, DeVille, Eldorado, Seville 4.0L VIN C, 4.6L VIN 9, 4.6L VIN Y engines Transmissions: A/T	**Incorrect 4th Gear Ratio (4L80-E) Conditions:** DTC P0121-P0123, P0502, P0503, P0716, P0717, P1820, P1822, P1823 and P1825 not set, engine started, VSS over 3 mph, Fuel Cutoff inactive, Transmission not in P/N, time since last gear shift change 6 seconds, TP angle over 10%, TFT sensor at 68°F-271°F, delivered torque from 30-100 ft lbs, then with 4th gear commanded "on", the PCM detected the gear ratio was more than 3.11:1, or 2.23:1 to 2.87:1; or gear ratio was 1.71:1 to 2.02:1, or it was 1.05:1 to 1.54:1; or gear ratio was 0.71:1 to 0.95:1, or less than 0.65: 1 **Possible Causes** ● ATF level is too low, or the fluid is burnt or contaminated ● ISS or OSS signal circuit has an intermittent fault condition ● Inspect for debris in transmission pan or a sticking shift valve ● Possible vehicle overloading, exceeding the trailer towing limit, or towing in overdrive events occurred (discuss with customer)
P0735 **1T CCM, MIL: No** 2003, 2004, 2005 CTS (Cadillac) 2.6L VIN M, 3.2L VIN N Transmissions: A/T	**Incorrect 5th Gear Ratio (5L40-E) Conditions:** DTC P0716, P0717, P0722, P0723, P1791, P1795, P1815, P1818, P1822, P1823, P1825 and P1826 not set, engine runtime over 5 seconds, TFT sensor from 68-266°F, Transmission not in P/N, time since most recent gear change over 6 seconds, engine torque from 41-258 lb ft., APP angle over 10%, gear ratio between 0.61:1-4.14:1, gear ratio is not Reverse: 3.1:1 to 2.95:1, VSS more than 3 mph, then with 5th gear commanded "on", and the PCM detected the actual gear ratio was not from 3.76:1 to 3.33:1; gear ratio not within 2.44:1 to 2.10:1; the gear ratio was not within 1.76:1 to 1.44:1; the gear ratio was not within 1.10:1 to 0.90:1; or the gear ratio was not within 0.83:1 to 0.68:1. If the TCM detects an invalid 5th gear ratio, indicating too much slip or drag in the transmission, it sets this code. **Possible Causes** ● ATF level is too low, or the fluid is burnt or contaminated ● 4-5 shift solenoid valve assembly mechanically stuck on ● 4-5 shift control valve stuck ● 4-5 shift valve stuck ● Control valve body ball check valve #12 stuck or missing ● Overdrive clutch accumulator leaking ● Overdrive clutch backing plate retainer ring out of position ● Overdrive clutch plates slipping, dragging or damaged ● Overdrive clutch spacer cracked or damaged ● Overdrive piston cracked, rolled, worn or damaged ● Overdrive clutch spring damaged ● Transmission has internal damage (may require replacement)
P0740 **1T CCM, MIL: Yes** 1995 Century, Cutlass Ciera, Cutlass Supreme, Grand Prix, Lumina, Monte Carlo Park Avenue Regal, Riviera, 88' & 98' 3.1L VIN M, 3.4L VIN X, 3.8L VIN 1, 3.8L VIN K, 3.8L VIN L engines Transmissions: A/T	**Torque Converter Clutch Not Engaging Conditions:** Engine started, vehicle driven in 3rd or 4th gear, TCC commanded "on", TP angle over 10%, and the PCM detected a vehicle or engine speed signal indicating no TCC lockup occurred for 10 seconds. **Possible Causes** ● TCC solenoid control circuit is open, shorted to ground or to B+ ● TCC solenoid power circuit is open (test the TRANS fuse) ● TCC solenoid is damaged or has failed ● PCM has failed
P0740 **2T CCM, MIL: Yes** 1996, 1997, 1998, 1999, 2000, 2001, 2002, 2003, 2004, 2005 Cab & Chassis, C/K truck, G Van, M/L Series Van, S/T Blazer & S/T Pickup 4.3L VIN W, 4.3L VIN X, 4.8L VIN V, 5.0L VIN M, 5.3L VIN T, 5.7L VIN R, 6.0L VIN N, 6.0L VIN U, 7.4L VIN J, 8.1L VIN G Transmissions: A/T	**TCC Solenoid Circuit Malfunction (4L60-E, 4L65-E) Conditions:** Engine started, system voltage over 10.0v, Fuel Cutoff inactive, and the PCM detected the TCC feedback voltage was high with the TCC Solenoid commanded "on", or it was "low" with the TCC Solenoid commanded "off" for 5 seconds. **Possible Causes** ● TCC solenoid control circuit is open, shorted to ground or to B+ ● TCC solenoid power circuit is open (test the TRANS fuse) ● TCC solenoid is damaged or has failed ● PCM has failed ● TSB 01-07-30-002C contains a repair procedure for this code

OBD II Trouble Code List (P0xxx Codes)

DTC	Trouble Code Title, Conditions & Possible Causes
P0740 **2T CCM, MIL: Yes** 1998, 1999, 2000, 2001, 2002, 2003, 2004, 2005 Camaro, Corvette & Firebird 3.8L VIN K, 5.7L VIN G, 5.7L VIN S engines Transmissions: A/T	**TCC Solenoid Circuit Malfunction (4L60-E) Conditions:** Engine started, Fuel Cutoff "off", and the PCM detected the TCC voltage was high with the TCC Solenoid commanded "on", or it was "low" with it commanded "off" for 5 seconds. **Possible Causes** ● TCC solenoid control circuit is open, shorted to ground or to B+ ● TCC solenoid power circuit is open (test the TRANS fuse) ● TCC solenoid is damaged or has failed ● PCM has failed
P0740 **2T CCM, MIL: Yes** 2002, 2003, 2004, 2005 Envoy & TrailBlazer 4.2L VIN S, 5.3L VIN P Transmissions: A/T	**TCC Solenoid Circuit Malfunction (4L60-E, 4L65-E) Conditions:** Engine started, system voltage from 10-18v, Decel Fuel Cutoff not active, and the PCM detected the TCC solenoid feedback voltage was high with the TCC Solenoid commanded "on", or it was low with the TCC Solenoid commanded "off" for 5 seconds. **Possible Causes** ● TCC solenoid control circuit is open, shorted to ground or to B+ ● TCC solenoid power circuit is open (test the IGN 0 fuse) ● TCC solenoid is damaged or has failed ● PCM has failed
P0741 **1T CCM, MIL: No** 1999, 2000, 2001, 2002, 2003, 2004, 2005 Aurora, Bonneville, Century, Cutlass Supreme, Grand Prix, Intrigue, LeSabre, LSS, Lumina, Monte Carlo, 88', Regal, Park Avenue, 98', Montana, Rendezvous, Silhouette, Venture 3.1L VIN J, 3.1L VIN M, 3.4L VIN E, 3.5L VIN H, 3.8L VIN 1, 3.8L VIN K Transmissions: A/T	**TCC System Stuck Off - Mechanical (4T65-E) Conditions:** DTC P0121-P0123, P0502, P0503, P0716, P0717, P0742, P1820, P1860 and P1887 not set, engine started, Fuel Cutoff inactive, Transmission gear range was D2, D3 or D4, time since last gear select lever change more than 6 seconds, TFT sensor from 68-266°F, TP angle from 4-35%, TCC PWM solenoid commanded "on" for over 500 ms, TCC commanded to maximum apply pressure, and the PCM detected the TCC slip speed was more than 180 rpm twice during a 7 second period during this key cycle. **Possible Causes** ● ATF level is too low, or the fluid is burnt or contaminated ● Inspect transmission lines to radiator for bends or restrictions ● Oil pressure screen is clogged or debris in the oil pan ● TCC control valve is stuck "off" due to sediment or binding ● TCC regulator valve is stuck "off" due to sediment or binding ● TCC solenoid valve O-ring or turbine shaft seals leaking or cut ● TSB 00-07-30-007A contains a repair procedure for this code
P0741 **2T CCM, MIL: Yes** 1996, 1997, 1998, 1999, 2000, 2001, 2002, 2003, 2004, 2005 Aurora, DeVille, Eldorado, Seville, C/K Series Truck 4.0L VIN C, 4.6L VIN 9, 4.6L VIN Y, 6.0L VIN N, 6.0L VIN U, 6.6L VIN 1, 8.1L VIN G engines Transmissions: A/T	**TCC System Stuck Off - Mechanical (4L80-E) Conditions:** DTC P0121, P0122, P0123, P0502, P0503, P0716, P0717, P0742, P1820, P1822, P1823, P1825 and P1860 not set, engine started, Fuel Cutoff inactive, time since last gear lever change more than 6 seconds, IMS indicates D2, D3 or D4, Transmission gear ratio indicates 2nd, 3rd or 4th gear, TFT sensor from 68°F-302°F, TP angle from 10-99%, engine torque from 32-159 ft lbs, TCC solenoid commanded "on", and the PCM detected the TCC slip speed was more than 125-130 rpm for a given torque for 20 seconds. **Possible Causes** ● ATF level is too low, or the fluid is burnt or contaminated ● Inspect transmission lines to radiator for bends or restrictions ● Oil pressure screen is clogged or debris in the oil pan ● TCC control valve is stuck "off" due to sediment or binding ● TCC regulator valve is stuck "off" due to sediment or binding ● TCC feed valve is stuck "off" due to sediment or binding ● TCC solenoid valve O-ring or turbine shaft seals leaking or cut
P0741 **2T CCM, MIL: Yes** 2002, 2003, 2004, 2005 Envoy, Escalade & TrailBlazer 5.3L VIN P, 5.3L VIN T, 6.0L VIN N engines Transmissions: A/T	**TCC System Mechanically Stuck Off (4L60-E, 4L65-E) Conditions:** DTC P0502, P0503, P0740, P0742, P0753, P1120, P1220 and P1810 not set, engine runtime over 5 seconds, not in Fuel Cutoff mode, TFT sensor from 68-302°F, TP angle from 20-99%, speed ratio from 0.89-1.02, gear range is D2, D3 or D4 with no gear change for over 6 seconds, then with the TCC commanded "on" at over 75% for 5 seconds, the PCM detected the TCC slip speed was over 130 rpm for 20 seconds (fault detected 3 times). The TCC solenoid valve is a N.O. exhaust valve used with the TCC PWM solenoid to control fluid acting on the converter clutch apply valve. When the TCC solenoid is grounded, the valve stops converter signal oil from exhausting. This causes converter signal oil pressure to increase and move the converter clutch apply valve against spring force to the apply position. In this position, release fluid is open to an exhaust port and converter feed fluid fills the apply circuit. The converter feed fluid applies the TCC. **Possible Causes** ● Converter clutch apply valve stuck in "off" (release) position ● Misaligned or damaged valve body gasket ● Restricted apply valve passage ● TCC PWM valve exhaust orifice in damaged or it has failed ● TCC solenoid valve mechanically stuck in "off" position

OBD II Trouble Code List (P0xxx Codes)

DTC	Trouble Code Title, Conditions & Possible Causes
P0741 **2T CCM, MIL: Yes** 2003, 2004, 2005 CTS (Cadillac) 2.6L VIN M, 3.2L VIN N Transmissions: A/T	**TCC System Stuck Off - Mechanical Problem (5L40-E) Conditions:** DTC P0716, P0717, P0722, P0723, P0742, P1779, P1791, P1795, P1815, P1818, P1820, P1822, P1823, P1825, P1826, P1866 and P1867 not set, engine runtime over 5 seconds, system voltage from 8-18v, IMS range is D2, D3 or D4, time since last gear select lever change more than 6 seconds, engine torque from 41-258 lb ft., TFT sensor from 68-266ºF, APP angle from 12-90%, gear ratio within the following ranges: 1.56:1 to 1.64:1 (3rd gear), 0.98:1 to 1.03:1 (4th gear), 0.73:1 to 0.77:1 (5th gear), TCC apply pressure greater than 150 kPa for more than 5 seconds, TCC duty cycle over 80% for 5 seconds, then with the TCC commanded "on", the PCM detected the TCC slip speed was 150-250 rpm for more than 8 seconds. The TCM controls the torque converter clutch (TCC) solenoid valve pulsewidth modulation (PWM). The solenoid directs the hydraulic fluid for TCC apply and release. When the TCC is applied, the engine is coupled directly to the transmission through the TCC. The TCC slip speed should be near 0. If the TCM detects high torque converter slip with the TCC "on", DTC P0741 is set. **Possible Causes** ● ATF level is too low, or the fluid is burnt or contaminated ● Inspect the forward clutch input housing assembly for the input shaft fluid seal ring leaking or worn ● Inspect the rear control valve body assembly for the control valve body ball check valve #2 stuck or missing ● TCC control valve stuck in the off position ● TCC enable valve stuck in the off position ● TCC PWM solenoid valve mechanically stuck off or leaking ● TCC regulator apply valve stuck
P0741 **2T CCM, MIL: Yes** 1997, 1998, 1999, 2000, 2001, 2002, 2003, 2004, 2005 Cavalier, Malibu, Sunfire & S/T Series Pickup 2.2L VIN 4, 2.2L VIN 5, 2.4L VIN T, 3.1L VIN J, 3.1L VIN M engines Transmissions: A/T	**TCC System Mechanically Stuck Off (4T40, 4T45-E) Conditions:** DTC P0121-P0123, P0502, P0503, P0742, P1810 and P1887 not set, engine started, TFT sensor from 70-266ºF, time since last gear select lever change 6 seconds, TFP manual valve position switch indicating D2, D3 or D4, TP angle from 8-75%, commanded gear is 2nd, 3rd or 4th, TCC is "on" for 3 seconds, and the PCM detected the TCC slip speed was over 250 rpm for 8 seconds twice in 1 trip. **Possible Causes** ● ATF level is too low, or the fluid is burnt or contaminated ● Inspect transmission lines to radiator for bends or restrictions ● Oil pressure screen is clogged or debris in the oil pan ● TCC control valve is stuck "off" due to sediment or binding ● TCC regulator valve is stuck "off" due to sediment or binding ● TCC feed valve is stuck "off" due to sediment or binding ● TCC solenoid valve O-ring or turbine shaft seals leaking or cut ● TSB 02-07-30-021 contains a repair procedure for this code
P0741 **2T CCM, MIL: Yes** 2002, 2003, 2004, 2005 Envoy & TrailBlazer 4.2L VIN S engine Transmissions: A/T	**TCC System Mechanically Stuck Off (4L60-E) Conditions:** DTC P0502, P0503, P0740, P0742, P0753, P0758, P1120, P1220, P1810 and P1860 not set, engine runtime over 5 seconds, not in Fuel Cutoff mode, TP angle from 20-99%, gear ratio from 0.89-1.02, TFT sensor from 68-300ºF, gear range is D2, D3 or D4 with no gear change for over 6 seconds, then with the TCC commanded "on" to a duty cycle rate over 75% for 5 seconds, the PCM detected the TCC slip speed was over 130 rpm for 20 seconds. The TCC solenoid valve is a N.O. exhaust valve used with the TCC PWM solenoid to control fluid acting on the converter clutch apply valve. The TCC solenoid valve attaches to the transmission case assembly extending into the pump cover. When the TCC solenoid is grounded, the valve stops converter signal oil from exhausting. This causes converter signal oil pressure to increase and move the converter clutch apply valve against spring force to the apply position. In this position, release fluid is open to an exhaust port and converter feed fluid fills the apply circuit. The converter feed fluid applies the TCC. **Possible Causes** ● TCC PWM valve exhaust orifice in damaged or it has failed ● Converter clutch apply valve stuck in "off" (release) position ● Misaligned or damaged valve body gasket ● Restricted apply valve passage ● TCC solenoid valve mechanically stuck in "off" position

OBD II Trouble Code List (P0xxx Codes)

DTC	Trouble Code Title, Conditions & Possible Causes
P0742 **1T CCM, MIL: No** 1996, 1997, 1998, 1999, 2000, 2001, 2002, 2003, 2004, 2005 Achieva, Alero, Beretta, Cavalier, Ciera, Century, Corsica, Grand Am, Sunfire Skyhawk, S/T Blazer & S/T Pickup 2.2L VIN 4, 2.2L VIN 5, 2.4L VIN T, 3.1L VIN J, 3.1L VIN M engines Transmissions: A/T	**TCC System Stuck On - Mechanical (4T40-E, 4T45-E) Conditions:** DTC P0121-P0123 and P1887 not set, engine started, vehicle speed from 35-75 mph, TP angle over 8%, time since last gear select lever change over 6 seconds, TCC commanded "off", speed ratio from 0.9-1.375, TCC slip speed from -20 to +20 rpm for 10 seconds, and the PCM detected the TCC release switch was closed (i.e., release fluid is not present) for 6 seconds (fault occurs 4 times on one trip). **Possible Causes** ● Transmission seal(s) leaking or damaged ● TCC release switch contains debris or sediment ● TCC release switch contacts are damaged or switch has failed ● TCC release switch circuit is shorted to ground
P0742 **2T CCM, MIL: Yes** 1999, 2000, 2001, 2002, 2003, 2004, 2005 Aurora, Bonneville, Century, Cutlass Supreme, Grand Prix, Intrigue, LeSabre, LSS, Lumina, Monte Carlo, 88', Regal, Park Avenue, 98', Montana, Rendezvous, Silhouette, Venture 3.1L VIN J, 3.1L VIN M, 3.4L VIN E, 3.5L VIN H, 3.8L VIN 1, 3.8L VIN K Transmissions: A/T	**Torque Converter Clutch Circuit Stuck On (4T65-E) Conditions:** DTC P0121, P0122, P0123, P1860 and P1887 not set, engine started, Fuel Cutoff inactive, engine torque from 70-200 ft lbs, TP position from 5-45%, TFT sensor from 68-266°F, time since last gear range change more than 6 seconds, speed ratio less than 7.0, TCC commanded "off", slip speed from -20 to 25 for 8 seconds, and the PCM detected the TCC release switch was closed 6 times for 4 seconds each time during the current key cycle. **Possible Causes** ● TCC PWM solenoid valve for the fluid exhaust is restricted ● TCC regulated apply valve is stuck in the TCC apply position ● TCC control valve is stuck in the TCC apply position ● TCC feed limit valve is stuck (i.e., the TCC feed limit pressure, and the TCC release pressure to be low or nonexistent) ● Pressure regulator valve is stuck ● TCC fluid circuits leaking or abnormally low/high line pressure ● TCC release switch circuit is shorted to ground ● TSB 00-07-30-007A contains a repair procedure for this code
P0742 **2T CCM, MIL: Yes** 1996, 1997, 1998, 1999, 2000, 2001, 2002, 2003, 2004, 2005 Cab & Chassis, C/K truck, G Van, M/L Series Van, S/T Blazer & S/T Pickup 4.3L VIN W, 4.3L VIN X, 4.8L VIN V, 5.0L VIN M, 5.3L VIN T, 5.3L VIN Z, 5.7L VIN R, 5.7L VIN G, 5.7L VIN S, 6.0L VIN N, 6.0L VIN U, 6.6L VIN 1, 7.4L VIN J, 8.1L VIN G Transmissions: A/T	**TCC System Stuck Off - Mechanical (4L60-E) Conditions:** DTC P0122, P0123, P0502, P0503, P0740, P1810 and P1860 not set, engine speed from 1000-3000 rpm, TP angle from 17-45%, ECT sensor from 68-266°F, Fuel Cutoff "off", engine vacuum from 0-105 kPa, engine torque from 50-400 ft lbs, commanded gear not in 1st, speed ratio from 0.64-1.35, gear range is D4, no gear range change for 5 seconds, TCC commanded "off", vehicle speed from 15-50 mph, and the PCM detected the TCC slip speed was -20 to +20 rpm for over 5 seconds (the fault must occur twice for this code set). **Possible Causes** ● Exhaust orifice in the TCC solenoid valve is clogged ● Converter clutch apply valve stuck in the "on" (apply) position ● Valve body gasket is damaged or misaligned ● Release pass is clogged or restricted ● Transmission cooler line is bent or restricted
P0742 **2T CCM, MIL: Yes** 2002, 2003, 2004, 2005 Envoy, Escalade & TrailBlazer 5.3L VIN P, 6.0L VIN N Transmissions: A/T	**TCC System Mechanically Stuck Off (4L60-E, 4L65-E) Conditions:** DTC P0120, P0220, P0502, P0503, P0740, P0742, P0753, P0758, P1810 and P1860 not set, engine runtime over 6 seconds, not in Fuel Cutoff mode, TP angle from 17-45%, engine torque at 50-400 lb ft., engine vacuum 0-105 kPa (0-15 psi), speed ratio at 0.64-1.35, TFT sensor from 68-266°F, gear range is D4 with no gear change for over 6 seconds, engine speed from 1,000-3,000 rpm, vehicle speed from 15-50 mph, then with the TCC commanded "off", the PCM detected the TCC slip speed was -20 to +20 rpm for 5 seconds (fault occurs twice during one trip). The TCC solenoid valve is a normally open (N.O.) exhaust valve that is used with the TCC PWM solenoid to control the fluid that acts on the converter clutch apply valve. The TCC solenoid valve attaches to the transmission case assembly extending into the pump cover. When the TCC solenoid is grounded, the valve stops converter signal oil from exhausting. This causes converter signal oil pressure to increase and move the converter clutch apply valve against spring force to the apply position. In this position, release fluid is open to an exhaust port and converter feed fluid fills the apply circuit. The converter feed fluid applies the TCC. **Possible Causes** ● Apply valve passage is restricted ● Converter clutch apply valve stuck in "off" (release) position ● Misaligned or damaged valve body gasket ● TCC PWM valve exhaust orifice in damaged or it has failed ● TCC solenoid valve is mechanically stuck in the "off" position

OBD II Trouble Code List (P0xxx Codes)

DTC	Trouble Code Title, Conditions & Possible Causes
P0742 **2T CCM, MIL: Yes** 2003, 2004, 2005 CTS (Cadillac) 2.6L VIN M, 3.2L VIN N Transmissions: A/T	**TCC System Mechanical Stuck Off (5L40-E) Conditions:** DTC P0716, P0717, P0722, P0723, P0741, P1791, P1795, P1815, P1779, P1818, P1820, P1822, P1823, P1825, P1826, P1866 and P1867 not set, engine runtime over 5 seconds, system voltage from 8-18v, IMS range is D5, engine torque from 59-258 lb ft., TFT sensor from 68-266ºF, not in 1st gear, gear ratio from 0.73:1 to 2.27:1, VSS over 9 mph, APP angle at 15-90%, then after the TCM commanded the TCC "off", it detected the TCC slip speed was from -20 to +20 rpm for over 5 seconds. The TCM controls the torque converter clutch (TCC) solenoid valve pulsewidth modulation (PWM). The solenoid directs the hydraulic fluid for TCC apply and release. **Possible Causes** ● ATF level is too low, or the fluid is burnt or contaminated ● Rapid fluctuation in the fluid line pressure may set this code ● TCC PWM solenoid valve may be stuck "on" due to sediment ● TCC regulator apply valve may be stuck due to sediment
P0742 **2T CCM, MIL: Yes** 1996, 1997, 1998, 1999, 2000, 2001, 2002, 2003, 2004, 2005 Aurora, DeVille, Eldorado, Seville 4.0L VIN C, 4.6L VIN 9, 4.6L VIN Y engines Transmissions: A/T	**TCC System Stuck On - Mechanical (4L80-E) Conditions:** DTC P0121-P0123, P0502, P0503, P0716, P0717, P0741, P1820, P1822, P1823, P1825 and P1860 not set, engine started, Fuel Cutoff inactive, Transmission gear ratio indicates 2nd, 3rd or 4th gear, IMS indicates D4, TFT sensor from 68ºF-271ºF, TP angle from 14-90%, engine torque from 114-217 ft-lbs, TCC commanded off, VSS sensor from 10-80 mph, and the PCM detected that the TCC slip speed was from -20 to +35 for 4.5 seconds (fault must be detected twice during the same vehicle trip). **Possible Causes** ● TCC PWM solenoid valve is stuck "on" due to sediment ● TCC converter feed limit valve is stuck "on" due to sediment ● TCC fluid circuits leaking or abnormally low/high line pressure
P0742 **2T CCM, MIL: Yes** 1998, 1999, 2000, 2001, 2002, 2003, 2004, 2005 Camaro, Corvette & Firebird 3.8L VIN K, 5.7L VIN G, 5.7L VIN S engines Transmissions: A/T	**TCC System Mechanically Stuck Off (4L60-E) Conditions:** DTC P0122, P0123, P0502, P0503, P0740, P1120, P1220, P1810 and P1860 not set, engine runtime over 5 seconds, system voltage over 10.0v, not in Fuel Cutoff mode, engine vacuum from 0-105 kPa, engine torque from 50-400 ft lbs, commanded gear not 1st, speed ratio from 0.64-1.35, gear range is D4 with no gear range change for over 5 seconds, vehicle speed from 15-50 mph, TCC commanded "off", and the PCM detected the TCC slip speed was -20 to +20 rpm for 5 seconds (fault must occur twice in one trip to set this code). **Possible Causes** ● Converter clutch apply valve stuck in the "on" (apply) position ● Exhaust orifice in the TCC solenoid valve is clogged ● Release pass is clogged or restricted ● Transmission cooler line is bent or restricted ● Valve body gasket is damaged or misaligned
P0742 **1T CCM, MIL: No** 2003, 2004, 2005 Envoy & TrailBlazer 4.2L VIN S engine Transmissions: A/T	**TCC System Mechanically Stuck Off (4L60-E) Conditions:** DTC P0502, P0503, P0740, P1120, P1220, P1810 and P1860 not set, engine runtime over 5 seconds, engine speed from 1,000-3,000 rpm, Fuel Cutoff not active, vehicle speed from 15-50 mph, TP angle from 17-45%, gear ratio from 0.64-1.35, engine vacuum from 0-105 kPa, engine torque from 50-400 lb ft., TFT sensor from 68-266ºF, commanded gear is not 1st gear, gear range is D4 with no gear change for over 5 seconds, TCC commanded "off", and the PCM detected the TCC slip speed was from -20 to +20 rpm for 5 seconds (fault must be detected at least twice during the same trip). **Possible Causes** ● Clogged exhaust orifice in the TCC solenoid valve ● Converter clutch apply valve stuck in the apply position ● Misaligned or damaged valve body gasket ● Restricted release passage ● Restricted transmission cooler line
P0742 **2T CCM, MIL: Yes** 1999, 2000, 2001 Catera 3.0L VIN R engine Transmissions: A/T	**TCC System Mechanically Stuck Off (4L30-E) Conditions:** DTC P0120, P0121, P0122, P0123, P0722, P0725, P0727 and P0743 not set, engine speed from 15-74 mph, TP angle from 18-100%, engine torque from 68-271 ft lbs., commanded gear not 1st, gear range is D4, TCC commanded "off", and the PCM detected the TCC slip speed was from -20 to +64 rpm for 3 seconds. **Possible Causes** ● Exhaust orifice in the TCC solenoid valve is clogged ● Converter clutch apply valve stuck in the apply position ● Valve body gasket is damaged or misaligned ● Release pass is clogged or restricted ● Sport Mode Lamp flashes when this trouble code is set ● Transmission cooler line is bent or restricted

OBD II Trouble Code List (P0xxx Codes)

DTC	Trouble Code Title, Conditions & Possible Causes
P0743 **2T CCM, MIL: Yes** 1999, 2000, 2001 Catera 3.0L VIN R engine Transmissions: A/T	**TCC Solenoid Circuit Malfunction (4L30-E) Conditions:** Engine started, vehicle driven with TCC commanded "off", TCM high side driver closed, and the TCM detected less than 2.45v on the control circuit, or the TCM detected from 2.45-3.53 volts on the control circuit, or the TCM detected more than 4.02 volts, any condition met for 0.18 seconds. *Note: The TCM sends DTC data to the PCM on the CAN network.* **Possible Causes** ● TCC solenoid control circuit is open, shorted to ground or shorted to system power ● TCC solenoid is damaged or has failed ● TCM has failed
P0748 **1T CCM, MIL: No** 1997, 1998, 1999, 2000, 2001, 2002, 2003, 2004, 2005 Aurora, Bonneville, Century, Cutlass Supreme, Grand Prix, Intrigue, LeSabre, LSS, Lumina, Monte Carlo, 88', Regal, Park Avenue, 98', Montana, Rendezvous, Silhouette, Venture 3.1L VIN J, 3.1L VIN M, 3.4L VIN E, 3.5L VIN H, 3.8L VIN 1, 3.8L VIN K Transmissions: A/T	**Pressure Control Solenoid Circuit Malfunction (4T65-E) Conditions:** Engine started; system voltage over 11v with the TFT sensor less than -40ºF or over 13.0v with the TFT sensor more than 304ºF, Pressure Control solenoid commanded "on", and the PCM detected the solenoid duty cycle was outside of its normal operating range of 0.5-95% for 200 milliseconds. The PC solenoid valve controls transmission line pressure based on current flow through its windings. The PCM determines desired line pressure based on throttle position and other inputs. The PCM then varies the duty cycle on the high side of the PC solenoid valve to control current flow to the solenoid. Current is controlled from about 0.02 amps for maximum line pressure to 1.1 amps for minimum line pressure. The PCM monitors the actual current to the solenoid. **Possible Causes** ● PC solenoid high side circuit is open, shorted to ground or B+ ● PC solenoid low side circuit is open, shorted to ground or to B+ ● PC solenoid high or low side driver is damaged or has failed ● TSB 00-07-30-002B contains a repair procedure for this code
P0748 **1T CCM, MIL: No** 1996, 1997, 1998, 1999, 2000, 2001, 2002, 2003, 2004, 2005 Cab & Chassis, C/K truck, G Van, M/L Series Van, S/T Blazer & S/T Pickup 4.3L VIN W, 4.3L VIN X, 5.0L VIN M, 5.7L VIN R, 5.7L VIN G, 5.7L VIN S, 7.4L VIN J engines Transmissions: A/T	**Pressure Control Solenoid Circuit Malfunction (4L60-E, 4L65-E) Conditions:** Engine started, system voltage over 10.0v, and the PCM detected the Pressure Control solenoid duty cycle reached its high limit of around 95%, or it reached its low limit of around 0%. **Possible Causes** ● PC solenoid high side circuit is open, shorted to ground or B+ ● PC solenoid low side circuit is open, shorted to ground or to B+ ● PC solenoid high or low side driver is damaged or has failed
P0748 **1T CCM, MIL: No** 2003, 2004, 2005 CTS (Cadillac) 2.6L VIN M, 3.2L VIN N Transmissions: A/T	**Pressure Control Solenoid Circuit Malfunction (5T40-E) Conditions:** Engine started, system voltage from 8-18v, and the PCM detected an unexpected voltage condition on the Pressure Control solenoid feedback circuit for 700 ms during the CCM continuous test period. The pressure control (PC) solenoid valve regulates actuator feed fluid passing through the solenoid into torque signal pressure. The TCM compares various signals to determine the appropriate pressure for a given load. **Possible Causes** ● PC solenoid control circuit is open or shorted to ground ● Pressure control solenoid is damaged or it has failed ● SES lamp displays on the driver information center (DIC) ● TCM has failed
P0748 **1T CCM, MIL: No** 1996, 1997, 1998, 1999, 2000, 2001, 2002, 2003, 2004, 2005 Aurora, DeVille, Eldorado, Seville 4.0L VIN C, 4.6L VIN 9, 4.6L VIN Y engines Transmissions: A/T	**Pressure Control Solenoid Circuit Malfunction (4L80-E) Conditions:** Engine started, system voltage over 10.0v, and the PCM detected an unexpected voltage condition on the Pressure Control solenoid feedback circuit for 100 ms during the CCM continuous test period. **Possible Causes** ● PC solenoid connector is damaged, open or shorted ● PC solenoid control circuit is open or shorted to ground ● Pressure control solenoid is damaged or it has failed ● PCM has failed
P0748 **1T CCM, MIL: No** 1995, 1996, 1997, 1998, 1999, 2000, 2001, 2002, 2003, 2004, 2005 Camaro, Corvette & Firebird 3.8L VIN K, 5.7L VIN G, 5.7L VIN S Transmissions: A/T	**Pressure Control Solenoid Circuit Malfunction (4L60-E) Conditions:** Engine started, system voltage over 10.0v, and the PCM detected an unexpected voltage condition on the Pressure Control (PC) solenoid circuit for 100 ms during the CCM test. **Possible Causes** ● PC solenoid high side circuit is open, shorted to ground or B+ ● PC solenoid low side circuit is open, shorted to ground or to B+ ● PC solenoid high or low side driver is damaged or has failed

OBD II Trouble Code List (P0xxx Codes)

DTC	Trouble Code Title, Conditions & Possible Causes
P0748 **1T CCM, MIL: No** 1997, 1998, 1999, 2000, 2001, 2002, 2003, 2004, 2005 Achieva, Alero, Cavalier, Ciera, Century, Grand Am, Sunfire Skyhawk, S/T Blazer & S/T Pickup 2.2L VIN 4, 2.2L VIN 5, 2.4L VIN T, 3.1L VIN J, 3.1L VIN M engines Transmissions: A/T	**Pressure Control Solenoid Circuit Malfunction (4T40-E, 4T45-E) Conditions:** Engine started, system voltage over 10.0v, and the PCM detected the Pressure Control (PC) PWM solenoid signal was close to 95%, or it was close to 0% during the CCM test. **Possible Causes** ● PC solenoid high side circuit is open, shorted to ground or B+ ● PC solenoid low side circuit is open, shorted to ground or to B+ ● PC solenoid high or low side driver is damaged or has failed
P0748 **1T CCM, MIL: No** 1999, 2000, 2001, 2002, 2003, 2004, 2005 Envoy, Escalade & TrailBlazer 4.2L VIN S, 4.8L VIN V, 5.3L VIN P, 5.3L VIN T, 5.3L VIN Z, 6.0L VIN N, 6.0L VIN U, 6.6L VIN 1, 8.1L VIN G engines Transmissions: A/T	**Pressure Control Solenoid Circuit Malfunction (4L60-E, 4L65-E, 4L80-E) Conditions:** Engine runtime over 7 seconds, system voltage over 11v at -40ºF, or over 12v at over 302ºF, and the PCM detected the Pressure Control solenoid PWM signal was over 95%, or less than 0.5% for 700 ms. **Possible Causes** ● PC solenoid connector is damaged, open or shorted ● PCS high side circuit is open, shorted to ground or to low side ● PCS low side circuit is open, shorted to ground or to high side ● PCS assembly is damaged or it has failed ● PCM had failed (low or high side driver is damaged)
P0748 **2T CCM, MIL: Yes** 1999, 2000, 2001 Catera 3.0L VIN R engine Transmissions: A/T	**Pressure Control Solenoid Circuit Malfunction (4L30-E) Conditions:** Engine started, vehicle driven with the TCC commanded "off", TCM Power Relay enabled, high side driver closed, then after the TCM commanded the Pressure Control solenoid current, it detected an unexpected voltage condition on the PC solenoid circuit for 54 ms. **Possible Causes** ● PC solenoid high side circuit is open, shorted to ground or B+ ● PC solenoid low side circuit is open, shorted to ground or to B+ ● PC solenoid high or low side driver is damaged or has failed ● Sport Mode Lamp flashes when this trouble code is set.
P0751 **2T CCM, MIL: Yes** 1997, 1998, 1999, 2000, 2001, 2002, 2003, 2004, 2005 Achieva, Alero, Cavalier, Ciera, Century, Grand Am, Sunfire Skyhawk, S/T Blazer & S/T Pickup 2.2L VIN 4, 2.2L VIN 5, 2.4L VIN T, 3.1L VIN J, 3.1L VIN M engines Transmissions: A/T	**A/T 1-2 Shift Solenoid - No 1st or 4th Gear (4T40-E/4T45-E) Conditions:** DTC P0121-P0123, P0502, P0503, P0716, P0717, P0753, P0758 and P1810 not set, engine started, TFT sensor from 68-233ºF, TFP manual valve switch indicated not in Park, Neutral or Reverse, TP angle over 8%, vehicle driven to over 5 mph, engine torque more than 20 lb ft., Transmission Output speed more than 150 rpm with the Input speed 150-6000 rpm, then with 1st Gear commanded "on", the PCM detected a gear ratio of 1.54:1-1.71:1 that equaled 2nd gear for 4 seconds; or with 4th Gear commanded "on", the gear ratio was 0.91:1-1.07:1 and equaled 3rd gear for 4 seconds in the test. **Possible Causes** ● ATF is burnt or contaminated, or the level is incorrect ● Transmission has an internal damage to the torque converter ● Shift solenoid valve seals are damaged or leaking ● Transmission has failed
P0751 **2T CCM, MIL: Yes** 1999, 2000, 2001, 2002, 2003, 2004, 2005 Aurora, Bonneville, Century, Cutlass Supreme, Grand Prix, Intrigue, LeSabre, LSS, Lumina, Monte Carlo, 88', Regal, Park Avenue, 98', Montana, Rendezvous, Silhouette, Venture 3.1L VIN J, 3.1L VIN M, 3.4L VIN E, 3.5L VIN H, 3.8L VIN 1, 3.8L VIN K Transmissions: A/T	**A/T 1-2 Shift Solenoid - No 1st or 4th Gear (4T65-E) Conditions:** DTC P0121-P0123, P0502, P0503, P0716, P0717, P1820, P1822, P1823, P1825, P1842, P1843, P1845 and P1847, Engine started, Transmission not in Park, Neutral or Reverse, TFT sensor from 68-266ºF, TP angle over 10%, VSS over 5 mph, engine torque from 20-200 lb ft., ISS signal from 150-8000 rpm, OSS signal more than 300 rpm, last gear range change over 1 second, then after the PCM commanded 1st gear and the gear ratio indicates 2nd gear (1.52:1-1.62:1) for 1 second, or the PCM commanded 4th gear and the gear ratio indicated 3rd gear (0.95:1-1.05:1) for 1 second. **Possible Causes** ● ATF is burnt or contaminated ● Transmission has an internal malfunction. ● Shift solenoid valve seals are damaged ● Transmission has failed

OBD II Trouble Code List (P0xxx Codes)

DTC	Trouble Code Title, Conditions & Possible Causes
P0751 **1T CCM, MIL: Yes** 1996, 1997, 1998, 1999, 2000, 2001, 2002, 2003, 2004, 2005 Cab & Chassis, C/K truck, G Van, M/L Series Van, S/T Blazer & S/T Pickup 4.3L VIN W, 4.3L VIN X, 5.0L VIN M, 5.7L VIN R, 5.7L VIN G, 5.7L VIN K, 5.7L VIN S, 7.4L VIN J Transmissions: A/T	**A/T 1-2 Shift Solenoid - No 1st or 4th Gear (4L60-E/4L80-E) Conditions:** DTC P0122, P0123, P0502, P0503, P0740, P0742, P0753, P0758, P0785, P1810 and P1860 not set, engine started, vehicle driven to over 5 mph, Fuel Cutoff not active, TP angle from 10-35% (± 7%), gear range or TFP manual valve position switch D4 with no change for 6 seconds, TFT sensor from 68-266°F, engine torque 80-400 lb ft., and the PCM detected the commanded gear equaled 1st Gear and the ratio equaled 2nd Gear; or commanded gear equaled 4th Gear with TCC locked and the ratio equaled 3rd Gear for 4 seconds. **Possible Causes** ● ATF is burnt or contaminated, or the level is incorrect ● Transmission has an internal damage to the torque converter ● Shift solenoid valve seals are damaged or leaking ● Transmission has failed
P0751 **2T CCM, MIL: Yes** 1999, 2000, 2001, 2002, 2003, 2004, 2005 Envoy, Escalade, O/K Series Truck, G Van 4.8L VIN V, 5.3L VIN T, 5.3L VIN Z, 6.0L VIN N, 6.0L VIN U, 6.6L VIN 1, 8.1L VIN G engines Transmissions: A/T	**A/T 1-2 Shift Solenoid - No 1st Or 4th Gear (4L60-E, 4L65-E, 4L80-E) Conditions:** DTC P0122, P0123, P0502, P0503, P0740, P0742, P0753, P0758, P0785, P1810 and P1860 not set, vehicle driven to over 5 mph, Fuel Cutoff "off", TP angle over 10%, TFT sensor at 68-266°F, gear range is D4, engine torque from 50-400 lb ft., output speed over 150 rpm, Transfer Case ratio in 4WD Low from 0.9-1.2, or in 4WD High at 2.6-2.85; running in 1st gear for 2 seconds, and the PCM detected the gear ratio was 1.2-1.825 for 500 ms; or while running in 4th Gear for 1 second, the estimated gear ratio was 0.95-1.15 for 6 seconds. **Possible Causes** ● ATF is burnt or contaminated, or the level is incorrect ● Transmission has an internal damage to the torque converter ● Shift solenoid valve seals are damaged or leaking ● Transmission has failed
P0751 **1T CCM, MIL: Yes** 1996, 1997, 1998, 1999, 2000, 2001, 2002, 2003, 2004, 2005 C/K Truck & G Van 6.5L VIN F, 6.5L VIN S, 6.5L VIN Y Diesel engine Transmissions: A/T	**A/T 1-2 Shift Solenoid - No 1st or 4th Gear (4L80-E) Conditions:** DTC P0122, P0123, P0502, P0503, P0740, P0742, P0753, P0758, P0785, P1810 and P1860 not set, engine speed below 3750 rpm, VSS over 3 mph, APP angle or TP angle over 10%, TFT sensor at 68-266°F, engine torque at 80-475 lb ft., and the PCM detected the commanded gear equaled 1st Gear with a speed ratio equal to 2nd Gear for 2 seconds; or commanded gear equaled 4th Gear (TCC "on") with a ratio equal to 3rd Gear for 3 seconds. **Possible Causes** ● ATF is burnt or contaminated, or the level is incorrect ● Transmission has an internal damage to the torque converter ● Shift solenoid valve seals are damaged or leaking ● Transmission has internal damage (it may need to be replaced)
P0751 **2T CCM, MIL: Yes** 2003, 2004, 2005 CTS (Cadillac) 2.6L VIN M, 3.2L VIN N Transmissions: A/T	**A/T 1-2 Shift Solenoid - No 1st or 4th Gear (5L40-E) Conditions:** DTC P0716, P0717, P0722, P0723, P0742, P1779, P1791, P1795, P1815, P1818, P1820, P1822, P1823, P1825, P1826, P1842, P1843, P1845, P1847, P1864 and P1865 not set, engine runtime over 5 seconds, system voltage from 8-18v, IMS range is D2, D3, D4 or D5, TFT sensor from 68-266°F, APP angle over 10%, ISS speed from 200-6,800 rpm, OSS speed over 100 rpm, then after the TCM commanded 1st gear with the engine torque from 30-258 lb ft., and the resulting gear ratio was 2.16:1 to 2.27:1 for 1.9 seconds, then it commanded 4th or 5th gear with engine torque from 27-258 lb ft., with the resulting gear ratio from 1.56:1 to 1.64: 1 for 5 seconds. **Possible Causes** ● 1-2 shift solenoid is mechanically stuck "on" ● 1-2 shift solenoid valve O-ring worn or damaged ● 1-2 shift valve is stuck in release position ● 1-2 shift solenoid is damaged or it has failed
P0751 **2T CCM, MIL: Yes** 1996, 1997, 1998, 1999, 2000, 2001, 2002, 2003, 2004, 2005 Aurora, DeVille, Eldorado, Seville 4.0L VIN C, 4.6L VIN 9, 4.6L VIN Y engines Transmissions: A/T	**A/T 1-2 Shift Solenoid - No 1st or 4th Gear (4L80-E) Conditions:** DTC P0121-P0123, P0502, P0503, P0716, P0717, P1820, P1822, P1823, P1825, P1842, P1843, P1845 and P1847, Engine started, IMS indicates D1, D2, D3 or D4, TFT sensor from 68-271°F, TP angle over 7.5%, VSS over 5 mph, engine torque from 59-291 lb ft., then with 1st Gear commanded "on", the PCM detected the gear ratio was 2nd Gear (1.54:1-1.71:1) for 1.5 seconds, or with 4th Gear commanded "on", the PCM detected the gear ratio was 3rd Gear (0.95:1-1.05:1) for 4 seconds during the PCM test. **Possible Causes** ● A/T 1-2 Shift Solenoid valve is stuck ● A/T 1-2 Shift Valve is stick "off" ● ATF is burnt or contaminated ● A/T fluid circuits are plugged or restricted ● Transmission has failed

OBD II Trouble Code List (P0xxx Codes)

DTC	Trouble Code Title, Conditions & Possible Causes
P0751 **1T CCM, MIL: Yes** 1996, 1997, 1998, 1999, 2000, 2001, 2002, 2003, 2004, 2005 Camaro, Corvette & Firebird 3.8L VIN F, 5.7L VIN G, 5.7L VIN S engines Transmissions: A/T	**A/T 1-2 Shift Solenoid - No 1st or 4th Gear (4L60-E) Conditions:** DTC P0122, P0123, P0502, P0503, P0740, P0742, P0753, P0758, P0785, P1810 and P1860 not set, engine started, vehicle driven to over 5 mph, Fuel Cutoff inactive, TP angle more than 9%, TFT sensor from 68-266ºF, gear range is D4, D3, D2 or D1, engine torque was 50-400 lb ft., transmission output speed was more than 150 rpm, then with 1st Gear "on" for 2 seconds, the PCM detected the engine speed was more than 2.44 times the TCC slip speed with an estimated gear ratio of 1.2-1.85 for 2 seconds; or with 4th Gear "on" for 1 second, the PCM detected the engine speed was 2.44 times the TCC slip speed with an estimated gear ratio of 0.95-1.15 for 6 seconds during the CCM test. **Possible Causes** • ATF is burnt or contaminated, or the level is incorrect • Transmission has an internal damage to the torque converter • Shift solenoid valve seals are damaged or leaking • Transmission is damaged or it has failed
P0751 **2T CCM, MIL: Yes** 2002, 2003, 2004, 2005 Envoy & TrailBlazer 4.2L VIN S, 5.3L VIN P Transmissions: A/T	**A/T 1-2 Shift Solenoid Malfunction (4L60-E, 4L65-E) Conditions:** DTC P0502, P0503, P0740, P0742, P0753, P0758, P1120, P1220, P1810 and P1860 not set, engine runtime over 6 seconds, engine speed from 1,000-3,000 rpm, vehicle speed from 15-50 mph, TP angle from 15-45%, speed ratio from 0.64-1.35, TFT sensor from 68-266ºF, engine vacuum from 0-105 kPa, engine torque from 50-400 lb ft., time since last gearshift change over 5 seconds, transmission output speed over 150 rpm, transfer case ratio in 4WD Low was 0.9 to 1.2, or the transfer case ratio in 4WD High was 2.6 to 2.85, then after the PCM commanded 1st gear "on" for 2 seconds, and the estimated gear ratio from 1.2-1.825 (Condition 1), the PCM detected the torque converter efficiency was greater than or equal to 0.35 for 500 ms; or after the PCM commanded 4th gear "on" for 1 second (Condition 2), it detected the gear ratio was 0.95-1.15 for 6 seconds. **Possible Causes** • ATF is burnt or contaminated • Transmission has an internal malfunction • Shift solenoid valve seals are damaged • Transmission is damaged or it has failed
P0751 **2T CCM, MIL: Yes** 1999, 2000, 2001 Catera 3.0L VIN R engine Transmissions: A/T	**A/T 1-2 Shift Solenoid Stuck Off - Mechanical (4L30-E) Conditions:** DTC P0705, P0722, P0743, P0753, P0758, P1600, P1625 and P1890 not set, engine started, gear shift lever in D4, TP angle more than 30.8% with the engine speed from 3104-3904 rpm, 2nd Gear commanded "on" for 700 ms with the transmission speed over 992 rpm, engine torque more than 68 ft-lb; and the PCM detected the TCC slip speed was from 1400-2000 rpm for 0.9 seconds; or with the TP angle over 24.9%, engine speed from 1216-2016 rpm; 3rd Gear commanded "on" for 700 ms, Transmission speed more than 2016 rpm, engine torque over 68 lb ft., the PCM detected the TCC slip speed was from -1000 to -100 rpm for 3 seconds. **Possible Causes** • ATF is burnt or contaminated, or the level is incorrect • Transmission has an internal damage to the torque converter • Shift solenoid valve seals are damaged or leaking • Sport Mode Lamp flashes when this trouble code is set • Transmission has internal damage (it may need to be replaced)
P0752 **2T CCM, MIL: Yes** 1999, 2000, 2001, 2002, 2003, 2004, 2005 Aurora, Bonneville, Century, Cutlass Supreme, Grand Prix, Intrigue, LeSabre, LSS, Lumina, Monte Carlo, 88', Regal, Park Avenue, 98', Montana, Rendezvous, Silhouette, Venture 3.1L VIN J, 3.1L VIN M, 3.4L VIN E, 3.5L VIN H, 3.8L VIN 1, 3.8L VIN K Transmissions: A/T	**A/T 1-2 Shift Solenoid- No 2nd Or 3rd Gear (4T65-E) Conditions:** DTC P0121-P0123, P0502, P0503, P0716, P0717, P1820, P1822, P1823, P1825, P1842, P1843, P1845 and P1847, engine running, Transmission not in P/N or Reverse, TFT sensor from 68-266ºF, TP angle over 10%, VSS over 5 mph, engine torque from 20-200 lb ft., ISS signal from 150-8000 rpm, OSS signal over 300 rpm, 2nd Gear commanded "on" with last gear change over 1 second, and the PCM detected 1st gear (gear ratio 2.87:1-2.97:1) for 1 second, or with 3rd Gear commanded "on", the PCM detected 4th Gear (0.65:1-0.75:1). **Possible Causes** • ATF is burnt or contaminated • Transmission has plugged or restricted fluid circuits • Shift solenoid valve seals are leaking or damaged • Transmission has failed

OBD II Trouble Code List (P0xxx Codes)

DTC	Trouble Code Title, Conditions & Possible Causes
P0752 **1T CCM, MIL: Yes** 1996, 1997, 1998, 1999, 2000, 2001, 2002, 2003, 2004, 2005 Cab & Chassis, C/K truck, G Van, M/L Series Van, S/T Blazer & S/T Pickup 4.3L VIN W, 4.3L VIN X, 5.0L VIN M, 5.7L VIN R, 5.7L VIN G, 5.7L VIN K, 5.7L VIN S, 7.4L VIN J Transmissions: A/T	**A/T 1-2 Shift Solenoid - No 2nd Or 3rd Gear (4L80-E) Conditions:** DTC P0101-P0103, P0106-P0108, P0121-P0123, P0502, P0503, P0716, P0717, P0742, P0753, P0758, P0785, P1810, P1860 and P1870 not set, engine runtime over 5 seconds, vehicle driven to over 7 mph, Fuel Cutoff inactive, TP angle over10%, gear range or TFP manual valve position switch is D4 with no change for 6 seconds, TFT sensor from 68-266ºF, engine torque from 80-400 lb ft., then with 2nd Gear commanded "on", the PCM detected the gear ratio equaled 1st Gear for 2.25 seconds (fault detected 5 times on 1 trip). **Possible Causes** ● ATF is burnt or contaminated, or the level is incorrect ● Transmission has an internal damage to the torque converter ● Shift solenoid valve seals are damaged or leaking ● Transmission has failed
P0752 **1T CCM, MIL: Yes** 1996, 1997, 1998, 1999, 2000, 2001, 2002, 2003, 2004, 2005 O/K Truck & G Van 6.5L VIN F, 6.5L VIN S, 6.5L VIN Y Diesel engine Transmissions: A/T	**A/T 1-2 Shift Solenoid - No 2nd Or 3rd Gear (4L80-E) Conditions:** DTC P0122, P0123, P0502, P0503, P0740, P0742, P0753, P0758, P0785, P1810 and P1860 not set, engine started, vehicle driven to over 3 mph at an engine speed less than 3750 rpm, APP or TP angle more than 10%, TFT sensor from 68-266ºF, engine torque from 80-476 lb ft., then with 2nd Gear commanded "on", the PCM detected the estimated gear ratio was 1 Gear for 2.25 seconds (fault must occur 5 times in one key cycle to set the code). **Possible Causes** ● ATF is burnt or contaminated, or the level is incorrect ● Transmission has an internal damage to the torque converter ● Shift solenoid valve seals are damaged or leaking ● Transmission has failed
P0752 **2T CCM, MIL: Yes** 1999, 2000, 2001, 2002, 2003, 2004, 2005 C/K Truck, Escalade, L/M Van & G Van 4.8L VIN V, 5.3L VIN T, 5.3L VIN Z, 6.0L VIN N, 6.0L VIN U, 6.6L VIN 1, 8.1L VIN G engines Transmissions: A/T	**A/T 1-2 Shift Solenoid - No 2nd Or 3rd Gear (4L60-E, 4L80-E) Conditions:** DTC P0122, P0123, P0502, P0503, P0740, P0742, P0753, P0758, P0785, P1810 and P1860 not set, vehicle driven to over 5 mph, Fuel Cutoff inactive, TP angle more than 10%, gear range is D4, TFT sensor from 68-266ºF, engine torque from 50-400 lb ft., transmission output speed more than 150 rpm, Transfer Case low ratio in 4WD Low at 0.9-1.2 or in 4WD High at 2.6-2.85; engine torque from 25-650 lb ft., then with 2nd Gear commanded "on" for 1 second, the PCM detected the estimated gear ratio was 3.0-3.3 for 2 seconds; or with 3rd Gear commanded "on" for 1 second, the gear ratio was 0.65-0.95 for 3 seconds. **Possible Causes** ● ATF is burnt or contaminated, or the level is incorrect ● Transmission has an internal damage to the torque converter ● Shift solenoid valve seals are damaged or leaking ● Transmission has failed
P0752 **2T CCM, MIL: Yes** 2000, 2001, 2002, 2003, 2004, 2005 Aurora, DeVille, Eldorado, Seville 4.0L VIN C, 4.6L VIN 9, 4.6L VIN Y engine Transmissions: A/T	**A/T 1-2 Shift Solenoid - No 2nd or 3rd Gear (4L80-E) Conditions:** DTC P0121-P0123, P0502, P0503, P0716, P0717, P1820, P1822, P1823, P1825, P1842, P1843, P1845 and P1847, engine runtime over 5 seconds, IMS indicates D1, D2, D3 or D4, TFT sensor from 68-271ºF, TP angle over 7.5%, VSS over 5 mph, engine torque from 59-291 lb ft., then with 2nd Gear commanded "on", the PCM detected the gear ratio was 1st Gear (speed ratio of 2.87:1-3.11:1) for 2 seconds, or with 3rd Gear commanded "on", the PCM detected the gear ratio was 4th Gear (speed ratio of 0.65:1-0.71:1) for 4 seconds. **Possible Causes** ● ATF is burnt or contaminated ● Transmission has plugged or restricted fluid circuits ● Shift solenoid valve seals are leaking or damaged ● Transmission has failed
P0752 **2T CCM, MIL: Yes** 2003, 2004, 2005 CTS (Cadillac) 2.6L VIN M, 3.2L VIN N Transmissions: A/T	**A/T 1-2 Shift Solenoid - No 2nd or 3rd Gear (5L40-E) Conditions:** DTC P0716, P0717, P0722, P0723, P0742, P1779, P1791, P1795, P1815, P1818, P1820, P1822, P1823, P1825, P1826, P1842, P1843, P1845, P1847, P1864 and P1865 not set, engine runtime over 5 seconds, system voltage from 8-18v, IMS range is D2, D3, D4 or D5, TFT sensor from 68-266ºF, APP angle over 10%, ISS speed from 200-6,800 rpm, OSS speed over 100 rpm, then after the TCM commanded 2nd gear with the engine torque from 24-258 lb ft., and the resulting gear ratio was 3.33:1 to 3.50:1 for 2 seconds, then it commanded 3rd gear with engine torque from 24-258 lb ft., with the resulting gear ratio from 0.98:1 to 1.03: 1 for 5 seconds. **Possible Causes** ● 1-2 shift solenoid is mechanically stuck "off" ● 1-2 shift solenoid valve O-ring worn or damaged ● 1-2 shift valve is stuck in applied position ● 1-2 shift solenoid is damaged or it has failed

OBD II Trouble Code List (P0xxx Codes)

DTC	Trouble Code Title, Conditions & Possible Causes
P0752 **2T CCM, MIL: Yes** 1995, 1996, 1997, 1998, 1999, 2000, 2001, 2002, 2003, 2004, 2005 Camaro, Corvette & Firebird 3.8L VIN F, 5.7L VIN G, 5.7L VIN S engines Transmissions: A/T	**A/T 1-2 Shift Solenoid - No 2nd or 3rd Gear (4L60-E) Conditions:** DTC P0101-P0103, P0107, P0108, P0122, P0123, P0502, P0503, P0740, P1810 and P1860 not set, engine speed at 1000-3000 rpm, VSS at 20-75 mph, Fuel Cutoff "off", TP angle from 13-99%, TFT sensor at 68-266°F, engine torque at 50-400 lb ft., vacuum at 0-15 kPa, gear range is D4 with no gear change for 6 seconds, transmission speed ratio at 0.95-1.7, not in 1st Gear, TCC off and the PCM detected the TCC slip speed was -20 to +58 for 3.8 seconds. **Possible Causes** ● ATF is burnt or contaminated, or the level is incorrect ● Transmission has an internal damage to the torque converter ● Shift solenoid valve seals are damaged or leaking ● Transmission has failed
P0752 **1T CCM, MIL: Yes** 2002, 2003, 2004, 2005 Envoy & TrailBlazer 4.2L VIN S engine Transmissions: A/T	**A/T 1-2 Shift Solenoid- No 2nd or 3rd Gear (4L60-E) Conditions:** DTC P0502, P0503, P0740, P0742, P0753, P0758, P0785, P1120, P1220, P1810 and P1860 not set, engine runtime over 6 seconds, system voltage from 10-18v, gear range is D4, TP angle over 10%, not in Fuel Cutoff mode, speed ratio from 0.64-1.35, TFT sensor from 68-266°F, engine vacuum from 0-105 kPa, engine torque from 50-400 lb ft., time since last gearshift change over 5 seconds, transmission output speed over 150 rpm, Transfer case ratio in 4WD Low at 0.9 to 1.2, or Transfer Case ratio in 4WD High at 2.6 to 2.85, then after the PCM commanded 2nd gear "on" for 1 second, engine torque at 25-400 lb ft., TCC efficiency more than 0.5, it detected the gear ratio was 3.0-3.3 for 2 seconds; or after the PCM commanded third gear "on" for 1 second, engine torque at 50-400 lb ft., TCC efficiency equal to 0.5, it detected the estimated gear ratio was from 0.65-0.9 for 3 seconds. **Possible Causes** ● ATF is burnt or contaminated ● Transmission has an internal malfunction or it has failed ● Shift solenoid valve seals are damaged
P0752 **1T CCM, MIL: Yes** 2003, 2004, 2005 Envoy & TrailBlazer 5.3L VIN P Transmissions: A/T	**A/T 1-2 Shift Solenoid- No 2nd or 3rd Gear (4L65-E) Conditions:** DTC P0120, P0220, P0502, P0503, P0740, P0742, P0753, P0758, P0785, P1810 and P1860 not set, engine runtime over 5 seconds, gear range is D4, TP angle over10%, Fuel Cutoff "off", TFT sensor from 68-266°F, system voltage is 10-18v, transmission output speed over 150 rpm, Transfer Case ratio in 4WD Low from 0.9 to 1.2, or Transfer Case ratio in 4WD High from 2.6 to 2.85, then during Condition 1: PCM commanded 2nd gear "on" for 1 second, estimated gear ratio at 3.0-3.3, engine torque from 25-400 lb ft., conditions met for 2 seconds; then during Condition 2: PCM commanded 3rd gear on for 1 second, estimated gear ratio from 0.65-0.9, engine torque at 50-500 lb ft., conditions met for 3 seconds. **Possible Causes** ● ATF is burnt or contaminated ● Shift solenoid valve seals are damaged ● Transmission has internal damaged or it has failed
P0752 **2T CCM, MIL: Yes** 1999, 2000, 2001 Catera 3.0L VIN R engine Transmissions: A/T	**A/T 1-2, 3-4 Shift Solenoid Stuck On - Mechanical (4L30-E) Conditions:** DTC P0705, P0722, P0743, P0753, P0758, P1600, P1625 and P1890 not set, engine started, gear selector in D4, TP angle over 41.1%, engine speed from 2400-3008 rpm, 1st Gear commanded "on" for 1.5 seconds, Transmission output speed over 704 rpm, engine torque more than 68 ft-lb, and the PCM detected the TCC slip speed was from -711 to 350 rpm for 600 ms (fault detected twice); or with the TP angle over 15%, engine speed from 696-4992 rpm, 4th Gear commanded "on" for over 500 ms, Transmission speed over 1696 rpm, engine torque over 68 ft-lb with TCC commanded "on", the PCM detected the TCC slip speed was from 666 to 1000 rpm for 5 seconds (fault detected twice during one trip). **Possible Causes** ● ATF is burnt or contaminated ● Shift solenoid valve seals are leaking or damaged ● Sport Mode Lamp flashes when this trouble code is set ● Transmission has plugged or restricted fluid circuits ● Transmission has internal damaged or it has failed
P0753 **2T CCM, MIL: Yes** 1997, 1998, 1999, 2000, 2001, 2002, 2003, 2004, 2005 Achieva, Alero, Beretta, Cavalier, Ciera, Century, Envoy, Grand Am, Sunfire Skyhawk, S/T Blazer, S/T Pickup & TrailBlazer, U, X 2.2L VIN 4, 2.2L VIN 5, 2.4L VIN T, 3.1L VIN J, 3.1L VIN M, 3.4L VIN E Transmissions: A/T	**A/T 1-2 Shift Solenoid Circuit Malfunction (4T40-3/4T45-E) Conditions:** Engine started, Fuel Cutoff inactive, system voltage over 10.0v, and the PCM detected an unexpected voltage condition on the 1-2 Shift Solenoid control circuit during the CCM test. **Possible Causes** ● 1-2 shift solenoid control circuit is open or shorted to ground ● 1-2 shift solenoid control circuit is shorted to system power ● 1-2 shift solenoid is damaged or has failed ● PCM has failed ● TSB 02-07-30-022A contains a repair procedure for this code

OBD II Trouble Code List (P0xxx Codes)

DTC	Trouble Code Title, Conditions & Possible Causes
P0753 **2T CCM, MIL: Yes** 1996, 1997, 1998, 1999, 2000, 2001, 2002, 2003, 2004, 2005 C/K Truck & G Van 6.5L VIN F, 6.5L VIN S, 6.5L VIN Y, 6.6L VIN 1 Transmissions: A/T	**A/T 1-2 Shift Solenoid Circuit Malfunction (4L80-E) Conditions:** Engine started, Fuel Cutoff inactive, system voltage over 10.0v, and the PCM detected an unexpected voltage condition on the 1-2 Shift Solenoid control circuit during the CCM test. **Possible Causes** • 1-2 shift solenoid control circuit is open or shorted to ground • 1-2 shift solenoid control circuit is shorted to system power • 1-2 shift solenoid is damaged or has failed • PCM has failed
P0753 **2T CCM, MIL: Yes** 1996, 1997, 1998, 1999, 2000, 2001, 2002, 2003, 2004, 2005 Aurora, Bonneville, Century, Cutlass Supreme, Grand Prix, Intrigue, LeSabre, LSS, Lumina, Monte Carlo, 88', Regal, Park Avenue, 98, Montana, Rendezvous, Silhouette, Venture 3.1L VIN J, 3.1L VIN M, 3.4L VIN E, 3.5L VIN H, 3.8L VIN 1, 3.8L VIN K Transmissions: A/T	**A/T 1-2/3 Shift Solenoid Circuit Malfunction (4T65-E) Conditions:** Engine started, Fuel Cutoff inactive, system voltage over 10.0v, and the PCM detected an unexpected voltage condition on the 1-2/3 Shift Solenoid control circuit during the CCM test. **Possible Causes** • 1-2/3 shift solenoid control circuit is open or shorted to ground • 1-2/3 shift solenoid control circuit is shorted to system power • 1-2/3 shift solenoid is damaged or has failed • PCM has failed • TSB 02-07-30-022A contains a repair procedure for this code
P0753 **2T CCM, MIL: Yes** 1996, 1997, 1998, 1999, 2000, 2001, 2002, 2003, 2004, 2005 Cab & Chassis, C/K truck, G Van, M/L Series Van, S/T Blazer & S/T Pickup 4.3L VIN W, 4.3L VIN X, 5.0L VIN M, 5.7L VIN G, 5.7L VIN K, 5.7L VIN R, 5.7L VIN S, 7.4L VIN J, 8.1L VIN G engines Transmissions: A/T	**A/T 1-2 Shift Solenoid Circuit Malfunction (4L60-E, 4L80-E) Conditions:** Engine started, Fuel Cutoff inactive, system voltage over 10.0v, and the PCM detected an unexpected voltage condition on the 1-2 Shift Solenoid control circuit during the CCM test. **Possible Causes** • 1-2 shift solenoid control circuit is open or shorted to ground • 1-2 shift solenoid control circuit is shorted to system power • 1-2 shift solenoid is damaged or has failed • PCM has failed • TSB 57-65-08A contains a repair procedure for this code
P0753 **2T CCM, MIL: Yes** 1999, 2000, 2001, 2002, 2003, 2004, 2005 Envoy, Escalade, C/K Truck, G Van, L/M Van 4.8L VIN V, 5.3L VIN T, 5.3L VIN Z, 6.0L VIN N, 6.0L VIN U engines Transmissions: A/T	**A/T 1-2 Shift Solenoid Circuit Malfunction (4L60-E, 4L80-E) Conditions:** Engine started, Fuel Cutoff inactive, system voltage over 10.0v, and the PCM detected an unexpected voltage condition on the 1-2 Shift Solenoid control circuit during the CCM continuous test. **Possible Causes** • 1-2 shift solenoid control circuit is open or shorted to ground • 1-2 shift solenoid control circuit is shorted to system power • 1-2 shift solenoid is damaged or has failed • PCM has failed • TSB 01-07-30-002C contains a repair procedure for this code
P0753 **2T CCM, MIL: Yes** 1995, 1996, 1997, 1998, 1999, 2000, 2001, 2002, 2003, 2004, 2005 Camaro, Corvette & Firebird 3.8L VIN F, 5.7L VIN G, 5.7L VIN S engines Transmissions: A/T	**A/T 1-2 Shift Solenoid Circuit Malfunction (4L60-E) Conditions:** Engine started, Fuel Cutoff inactive, system voltage over 10.0v, and the PCM detected an unexpected voltage condition on the 1-2 Shift Solenoid control circuit during the CCM test. **Possible Causes** • 1-2 shift solenoid control circuit is open or shorted to ground • 1-2 shift solenoid control circuit is shorted to system power • 1-2 shift solenoid is damaged or has failed • PCM has failed
P0753 **2T CCM, MIL: Yes** 2001, 2002 Aurora 3.5L VIN H engine Transmissions: A/T	**A/T 1-2 Shift Solenoid Circuit Malfunction (4L65-E) Conditions:** Engine started, Fuel Cutoff inactive, system voltage over 10.0v, and the PCM detected an unexpected voltage condition on the 1-2 Shift Solenoid control circuit during the CCM test. **Possible Causes** • 1-2 shift solenoid control circuit is open or shorted to ground • 1-2 shift solenoid control circuit is shorted to system power • 1-2 shift solenoid is damaged or has failed • PCM has failed

OBD II Trouble Code List (P0xxx Codes)

DTC	Trouble Code Title, Conditions & Possible Causes
P0753 **2T CCM, MIL: Yes** 2002, 2003, 2004, 2005 Envoy & TrailBlazer 4.2L VIN S, 5.3L VIN P Transmissions: A/T	**A/T 1-2 Shift Solenoid Circuit Malfunction (4L60-E, 4L65-E) Conditions:** Engine runtime over 5 seconds, system voltage at 10-18v, engine not in Fuel Cutoff mode, and the PCM detected an unexpected voltage condition on the A/T 1-2 Shift solenoid circuit. **Possible Causes** ● 1-2 shift solenoid connector is damaged, loose or shorted ● 1-2 shift solenoid control circuit is open or shorted to ground ● 1-2 shift solenoid control circuit is shorted to system power ● 1-2 shift solenoid is damaged or has failed ● PCM has failed
P0753 **2T CCM, MIL: Yes** 1999, 2000, 2001 Catera 3.0L VIN R engine Transmissions: A/T	**A/T 1-2 Shift Solenoid Circuit Malfunction (4L30-E) Conditions:** Engine started, TCM high side driver "on", and the PCM detected the feedback voltage was less than 2.45v with the 1-2/3-4 shift solenoid commanded "off"; or the feedback voltage was from 2.45-3.53v with the 2-3 shift solenoid commanded "off", or the feedback voltage was over 4.0v for 180 ms with 2-3 shift solenoid commanded "on". The Sport Mode Lamp flashes when the PCM sets this code. **Possible Causes** ● 1-2/3-4 shift solenoid control circuit open or shorted to ground ● 1-2/3-4 shift solenoid control circuit is shorted to system power ● 1-2/3-4 shift solenoid is damaged or has failed ● 2-3 shift solenoid control circuit is open or shorted to ground ● 2-3 shift solenoid is damaged or has failed ● PCM has failed
P0755 **1T CCM, MIL: No** 1995 Century, Cutlass Ciera, Cutlass Supreme, Grand Prix, Lumina, Monte Carlo Park Avenue Regal, Riviera, 88' & 98' 3.1L VIN M, 3.4L VIN X, 3.8L VIN 1, 3.8L VIN K, 3.8L VIN L engines Transmissions: A/T	**A/T Shift Solenoid 'B' Range/Performance Conditions:** DTC P0122, P0123, P0501 and P0705 not set, engine started, vehicle driven to a speed of over 5 mph, TP sensor from 3-10%, and the PCM detected the transmission was operating in the wrong gear (the correct gear is determined when the engine speed divided by the vehicle speed does not equal 1st Gear) for 2-5 seconds. **Possible Causes** ● ATF is burnt or contaminated, or the level is incorrect ● SSB valve seals are damaged or leaking ● SSB is damaged or has failed ● Transmission has failed
P0756 **1T CCM, MIL: No** 1997, 1998, 1999, 2000, 2001, 2002, 2003, 2004, 2005 Alero, Cavalier, Ciera, Century, Grand Am, Sunfire Skyhawk, S/T Blazer & S/T Pickup 2.2L VIN 4, 2.2L VIN 5, 2.4L VIN T, 3.1L VIN J, 3.1L VIN M engines Transmissions: A/T	**A/T 2-3 Shift Solenoid - No 2nd Or 3rd Gear (4T40-E/4T45-E) Conditions:** DTC P0121-P0123, P0502, P0503, P0716, P0717, P0753, P0758, P1644 and P1810 not set, engine started, VSS over 5 mph, TFT sensor from 66-233°F, TFP manual valve switch indicating not in Park, Neutral or Reverse, TP angle more than 8%, engine torque more than 2- lb ft., Transmission Input speed 150-3000 rpm, Output speed more than 150 rpm, TCC slip speed over 200 rpm, then with 1st Gear commanded "on", the PCM detected the gear ratio equaled 4th Gear (0.61-0.72:1) for 4 seconds; or with 2nd Gear commanded "on", the PCM detected the gear ratio equaled 3rd Gear (0.91-1.07:1) for 4 seconds during the test. **Possible Causes** ● ATF is burnt or contaminated ● Transmission has plugged or restricted fluid circuits ● Shift solenoid valve seals are leaking or damaged ● Transmission has failed
P0756 **1T CCM, MIL: Yes** 1999, 2000, 2001, 2002, 2003, 2004, 2005 Aurora, Bonneville, Century, Cutlass, Grand Prix, Intrigue, LeSabre, LSS, Lumina, Monte Carlo, 88', Regal, Park Avenue, 98', Montana, Silhouette, Venture 3.1L VIN J, 3.1L VIN M, 3.4L VIN E, 3.5L VIN H, 3.8L VIN 1, 3.8L VIN K Transmissions: A/T	**A/T 2-3 Shift Solenoid- No 2nd Or 3rd Gear (4T65-E) Conditions:** DTC P0121-P0123, P0502, P0503, P0716, P0717, P1820, P1822, P1823, P1825, P1842, P1843, P1845 and P1847, Engine started, Transmission not in Park, Neutral or Reverse, time since last gear range change over 1 second, TFT sensor from 68-266°F, TP angle over 10%, VSS over 5 mph, engine torque from 20-200 lb ft., ISS signal from 150-8000 rpm, OSS signal over 300 rpm, then with 1st Gear commanded "on", the PCM detected the gear ratio indicated 4th gear (0.65:1-0.75:1) for 1 second, or with 2nd Gear commanded "on", the PCM detected the gear ratio indicated 3rd Gear (0.95:1-0.05:1) for 1 second during the CCM test. **Possible Causes** ● ATF is burnt or contaminated ● Transmission has plugged or restricted fluid circuits ● Shift solenoid valve seals are leaking or damaged ● Transmission has failed ● TSB 02-07-30-13 contains a repair procedure for this code

OBD II Trouble Code List (P0xxx Codes)

DTC	Trouble Code Title, Conditions & Possible Causes
P0756 **1T CCM, MIL: Yes** 1996, 1997, 1998, 1999, 2000, 2001, 2002, 2003, 2004, 2005 Cab & Chassis, C/K truck, G Van, M/L Series Van, S/T Blazer & S/T Pickup 4.3L VIN W, 4.3L VIN X, 5.0L VIN M, 5.7L VIN R, 5.7L VIN G, 5.7L VIN K, 5.7L VIN S, 7.4L VIN J Transmissions: A/T	**2-3 Shift Solenoid - No 2nd Or 3rd Gear (4L80-E) Conditions:** DTC P0122, P0123, P0502, P0503, P0740, P0742, P0753, P0758, P0785, P1810 and P1860 not set, engine started, Fuel Cutoff inactive, vehicle speed over 5 mph, TP angle from 10-35% (± 7%), gear range is D4, TFT sensor from 68-266°F, engine torque from 80-650 lb ft., then with 1st Gear commanded "on", the PCM detected the gear ratio indicated 4th Gear for 2.5 seconds; or with 2nd Gear commanded "on", the PCM detected the gear ratio indicated 3rd Gear for 2.7 seconds during the CCM test. **Possible Causes** ● ATF is burnt or contaminated ● Transmission has plugged or restricted fluid circuits ● Shift solenoid valve seals are leaking or damaged ● Transmission has failed
P0756 **2T CCM, MIL: Yes** 1999, 2000, 2001, 2002, 2003, 2004, 2005 C/K Truck, Escalade, L/M Van & G Van 4.8L VIN V, 5.3L VIN T, 5.3L VIN Z, 6.0L VIN N, 6.0L VIN U, 8.1L VIN G Transmissions: A/T	**2-3 Shift Solenoid - No 2nd Or 3rd Gear (4L60-E, 4L65-E) Conditions:** DTC P0122, P0123, P0502, P0503, P0740, P0742, P0753, P0758, P0785, P1810 and P1860 not set, engine started, system voltage over 10.0v, vehicle speed over 5 mph, TP angle over 10%, gear range is D4, TFT sensor from 68-266°F, engine torque from 50-400 lb ft, transmission output shaft speed more than 150 rpm, Fuel Cutoff inactive, then with 1st Gear commanded "on", the PCM detected the gear ratio indicated 4th Gear for 2.5 seconds; or with 2nd Gear commanded "on" for 1 second, the PCM detected the estimate gear ratio was 0.9-1.2 for 2 seconds during the CCM test **Possible Causes** ● ATF is burnt or contaminated ● Transmission has plugged or restricted fluid circuits ● Shift solenoid valve seals are leaking or damaged ● Transmission has failed ● TSB 01-07-30-036A contains a repair procedure for this code
P0756 **2T CCM, MIL: Yes** 1996, 1997, 1998, 1999, 2000, 2001, 2002, 2003, 2004, 2005 C/K Truck & G Van 6.5L VIN F, 6.5L VIN S, 6.5L VIN Y, 6.6L VIN 1 Diesel engines Transmissions: A/T	**2-3 Shift Solenoid - No 2nd or 3rd Gear (4L80-E) Conditions:** DTC P0712, P0713, P0716, P0717, P0722, P0723, P0753, P0758, and P1810 not set, engine started, engine speed from 475-3750 rpm, system voltage over 10.0v, vehicle speed over 3 mph, APP angle more than 10%, TFT sensor from 68-266°F, engine torque from 80-475 lb ft., then with 1st Gear commanded "on", the PCM detected he gear ratio equaled 4th Gear for 2.75 seconds; or with 2nd Gear commanded "on", the PCM detected the gear ratio equaled 3rd Gear for 2.75 at least twice; or with 3rd Gear commanded "on", the PCM detected the gear ratio equaled 2nd Gear for 3.25 seconds at least 7 times during the test. **Possible Causes** ● ATF is burnt or contaminated ● Transmission has plugged or restricted fluid circuits ● Shift solenoid valve seals are leaking or damaged ● Transmission has failed
P0756 **2T CCM, MIL: Yes** 2003, 2004, 2005 CTS (Cadillac) 2.6L VIN M, 3.2L VIN N Transmissions: A/T	**A/T 2-3 Shift Solenoid - No 2nd or 3rd Gear (5L40-E) Conditions:** DTC P0716, P0717, P0722, P0723, P0742, P1779, P1791, P1795, P1815, P1818, P1820, P1822, P1823, P1825, P1826, P1842, P1843, P1845, P1847, P1864 and P1865 not set, engine runtime over 5 seconds, system voltage from 8-18v, IMS range is D2, D3, D4 or D5, TFT sensor from 68-266°F, APP angle over 10%, ISS speed from 200-6,800 rpm, OSS speed over 100 rpm, then after the TCM commanded 1st gear with the OSS speed over 200 rpm and engine torque from 30-258 lb ft., and the resulting gear ratio was 0.73:1 to 0.77:1 for 4 seconds, then after the PCM commanded 2nd gear on with the engine torque from 27-258 lb ft., the resulting gear ratio was from 1.56:1 to 1.64: 1 for 2 seconds. **Possible Causes** ● 2-3 shift solenoid is mechanically stuck "off" ● 2-3 shift solenoid valve O-ring worn or damaged ● 2-3 shift valve is stuck in the applied position ● 2-3 shift solenoid is damaged or it has failed
P0756 **1T CCM, MIL: Yes** 1998, 1999, 2000, 2001, 2002, 2003, 2004, 2005 Aurora, DeVille, Eldorado, Seville 4.0L VIN C, 4.6L VIN 9, 4.6L VIN Y engines Transmissions: A/T	**A/T 2-3 Shift Solenoid - No 2nd or 3rd Gear (4L80-E) Conditions:** DTC P0121-P0123, P0502, P0503, P0716, P0717, P1820, P1822, P1823, P1825, P1842, P1843, P1845 and P1847 not set, vehicle driven to over 5 mph, TP angle more than 10%, Fuel Cutoff inactive, IMS indicates D1, D2, D3 or D4, TFT sensor from 68°F-271°F, engine torque from 59-291 lb ft., then with 1st Gear commanded "on", the PCM detected the gear ratio indicated 4th Gear (0.65:1-0.71:1) for 1 second; or with 2nd Gear commanded "on", the PCM detected the gear ratio indicated 3rd gear (0.95:1-1.05:1) for 500 ms during the test. **Possible Causes** ● ATF is burnt or contaminated ● Transmission has plugged or restricted fluid circuits ● Shift solenoid valve seals are leaking or damaged ● Transmission has failed

OBD II Trouble Code List (P0xxx Codes)

DTC	Trouble Code Title, Conditions & Possible Causes
P0756 **1T CCM, MIL: Yes** 1995, 1996, 1997, 1998, 1999, 2000, 2001, 2002, 2003, 2004, 2005 Camaro, Corvette & Firebird 3.8L VIN F, 5.7L VIN G, 5.7L VIN S engines Transmissions: A/T	**A/T 2-3 Shift Solenoid - No 2nd or 3rd Gear (4L60-E) Conditions:** DTC P0101-P0103, P0107, P0108, P0122, P0123, P0502, P0503, P0740, P0742, P0753, P0758, P0785, P1810, P1860 and P1870 not set, engine started, vehicle driven to a speed of over 5 mph, system voltage over 10.0v, Fuel Cutoff inactive, TP angle more than 9%, TFT sensor from 68-266ºF, gear range is D4, no gear change for 6 seconds, engine torque from 50-400 lb ft., engine vacuum from 0-15 kPa, A/T output speed over 150 rpm, gear range is D4, D3, D2 or D1, then with 1st Gear commanded "on", engine speed more than the TCC slip speed and the Output speed over 300 rpm, the PCM detected the gear ratio was 0.0-0.895 for 1 second; or with 2nd Gear commanded "on", engine speed 5.26 times more than the TCC slip speed, the PCM detected the gear ratio was 0.9-1.2 for 2 seconds. **Possible Causes** ● ATF is burnt or contaminated ● Transmission has plugged or restricted fluid circuits ● Shift solenoid valve seals are leaking or damaged ● Transmission has failed
P0756 **1T CCM, MIL: Yes** 2002, 2003, 2004, 2005 Envoy & TrailBlazer 4.2L VIN S engine Transmissions: A/T	**A/T 2-3 Shift Solenoid Malfunction (4L60-E) Conditions:** DTC P0502, P0503, P0740, P0742, P0753, P0758, P1120, P1220, P0785, P1810 and P1860 not set, engine runtime over 5 seconds, engine not in Fuel Cutoff mode, gear range is D4, TP angle over 10%, TFT sensor from 68-266ºF, system voltage from 10-18v, engine torque from 50-400 lb ft., transmission output speed more than 150 rpm, then during Condition 1: PCM commands 1st gear on for 2 seconds, estimated gear ratio 0 to 1, engine torque from 50-400 lb ft., conditions met for 1 second, then during Condition 2: PCM commands 2nd gear on for 1 second, TCC efficiency more than or equal to 0.5, estimated gear ratio from 0.9 to 1.2, engine torque from 50-400 lb ft., conditions met for 2 seconds. **Possible Causes** ● ATF is burnt or contaminated ● Transmission has an internal malfunction ● Shift solenoid valve seals are damaged ● Transmission has failed
P0756 **1T CCM, MIL: Yes** 2003, 2004, 2005 Envoy & TrailBlazer 5.3L VIN P engine Transmissions: A/T	**A/T 2-3 Shift Solenoid Malfunction (4L65-E) Conditions:** DTC P0120, P0220, P0502, P0503, P0740, P0742, P0753, P0758, P0785, P1810 and P1860 not set, engine speed from 1,000-3,000 rpm for 5 seconds, gear range is D4, not in Fuel Cutoff mode, TP angle over 10%, TFT sensor from 68-266ºF, system voltage from 10-18v, engine torque from 50-400 lb ft., transmission output speed over 150 rpm, Transfer Case ratio in 4WD Low at 0.9 to 1.2, or Transfer Case ratio in 4WD High at 2.6 to 2.85, then during Condition 1: PCM commands 1st gear on for 2 seconds, estimated gear ratio 0 to 1.4,conditions met for 1 second, then during Condition 2: PCM commands 2nd gear on for 1 second, estimated gear ratio at 0.9 to 1.2, all conditions met for 2 seconds. **Possible Causes** ● ATF is burnt or contaminated ● Transmission has an internal malfunction ● Shift solenoid valve seals are damaged ● Transmission has failed
P0756 **2T CCM, MIL: Yes** 1999, 2000, 2001 Catera 3.0L VIN R engine Transmissions: A/T	**2-3 Shift Solenoid - No 1st Or 2nd Gear (4L30-E) Conditions:** DTC P0705, P0722, P0743, P0753, P0758, P1600, P1625 and P1890 not set, engine speed from 2400-3008 rpm, Transmission in D4, TP angle over 41.1%, 1st Gear commanded "on" for 1.5 seconds, Transmission output speed over 704 rpm, engine torque more than 68 lb ft., and the PCM detected the TCC slip speed was from -3000 to +100 rpm for 1.5 seconds (twice); or with the TP angle over 41.8%, engine speed from 2592-3008 rpm, 2nd Gear commanded "on" for over 700 ms, Transmission output speed more than 1408 rpm, engine torque more than 68 lb ft., and the PCM detected the TCC slip speed was from -500 to 400 rpm for 1.5 seconds (fault must occur twice for this trouble code to set). *Note: The Sport Mode Lamp flashes when this trouble code is set.* **Possible Causes** ● ATF is burnt or contaminated ● Transmission has plugged or restricted fluid circuits ● Shift solenoid valve seals are leaking or damaged ● Transmission has failed

OBD II Trouble Code List (P0xxx Codes)

DTC	Trouble Code Title, Conditions & Possible Causes
P0756 **2T CCM, MIL: Yes** 1999, 2000, 2001, 2002, 2003, 2004, 2005 Aurora, Bonneville, Century, Cutlass Supreme, Grand Prix, Intrigue, LeSabre, LSS, Lumina, Monte Carlo, 88', Regal, Park Avenue, 98', Montana, Rendezvous, Silhouette, Venture 3.1L VIN J, 3.1L VIN M, 3.4L VIN E, 3.5L VIN H, 3.8L VIN 1, 3.8L VIN K Transmissions: A/T	**2-3 Shift Solenoid - No 3rd Or 4th Gear (4T65-E) Conditions:** DTC P0121-P0123, P0502, P0503, P0716, P0717, P0730, P0753, P0758, P1810, P1814 and P1860 not set, engine started, vehicle speed over 5 mph, Fuel Cutoff inactive, gearshift not in Park, Neutral or Reverse, time since last gear range change 1 second, TP angle over 10%, TFT sensor at 68-264ºF, engine torque from 20-200 lb ft., ISS sensor at 150-8000 rpm, OSS more than 300 rpm, then with 3rd Gear commanded "on", the PCM detected the gear ratio equaled 2nd gear (1.52:1 to 1.62:1) for 1 second; or with 4th Gear commanded "on", the PCM detected the gear ratio equaled 1st gear (2.87:1 to 2.97:1) for 1 second. **Possible Causes** ● ATF is burnt or contaminated ● Transmission has plugged or restricted fluid circuits ● Shift solenoid valve seals are leaking or damaged ● Transmission has failed ● TSB 02-07-30-13 contains a repair procedure for this code
P0757 **1T CCM, MIL: No** 2001, 2002, 2003, 2004, 2005 C/K Truck & G Van, L/M Van, S/T Blazer & Pickup 4.3L VIN W, 4.3L VIN X, 5.7L VIN R, 5.7L VIN G, 5.7L VIN K, 5.7L VIN S, 7.4L VIN J engines Transmissions: A/T	**2-3 Shift Solenoid - No 3rd Or 4th Gear (4T60 E, 4T65 E) Conditions:** DTC P0121-P0123, P0502, P0503, P0716, P0717, P0730, P0753, P0758, P1810, P1814 and P1860 not set, VSS over 5 mph, Fuel Cutoff "off", Transmission not in Park, Neutral or Reverse, time since last gear range change over 1 second, TP angle over 10%, TFT sensor from 68-264ºF, engine torque from 20-200 lb ft., ISS sensor from 150-8000 rpm, OSS sensor more than 300 rpm, then with 3rd Gear commanded "on" for 1 second, the PCM detected the estimated gear ratio 1.6-1.8 and the engine torque was from 50-500 lb ft; or with 4th Gear commanded "on" for two seconds, the PCM detected the estimated gear ratio equaled 1.8-3.3 and the engine torque was from 50-400 lb ft for 2 seconds during the CCM test. **Possible Causes** ● ATF is burnt or contaminated ● Transmission has plugged or restricted fluid circuits ● Shift solenoid valve seals are leaking or damaged, or the Transmission has failed ● TSB 01-07-30-038 contains a repair procedure for this code
P0757 **1T CCM, MIL: No** 1998, 1999, 2000, 2001, 2002, 2003, 2004, 2005 Aurora, DeVille, Eldorado, Seville 4.0L VIN C, 4.6L VIN 9, 4.6L VIN Y, 6.0L VIN N, 6.0L VIN U, 6.6L VIN 1, 8.1L VIN G engines Transmissions: A/T	**2-3 Shift Solenoid - No 3rd Or 4th Gear (4L60-E, 4L80-E) Conditions:** DTC P0121-P0123, P0502, P0503, P0716, P0717, P1820, P1822, P1823, P1825, P1842, P1843, P1845 and P1847 not set, VSS over 5 mph, system voltage over 10.0v, Fuel Cutoff "off", IMS indicates D1, D2, D3 or D4, TFT sensor from 68ºF-271ºF, TP angle over 10%, engine torque from 59-291 lb ft., then with 3rd Gear commanded "on", the PCM detected the gear ratio equaled 2nd Gear (1.54:1-1.71:1) for 3 seconds; or with 4th Gear commanded "on", the PCM detected the gear ratio equaled 1st Gear (2.87:1-3.11:1) for 2 seconds. **Possible Causes** ● ATF is burnt or contaminated ● Transmission has plugged or restricted fluid circuits ● Shift solenoid valve seals are leaking or damaged, or the Transmission has failed
P0757 **1T CCM, MIL: No** 2003, 2004, 2005 CTS (Cadillac) 2.6L VIN M, 3.2L VIN N Transmissions: A/T	**A/T 2-3 Shift Solenoid - No 3rd Or 4th Gear (5L40-E) Conditions:** DTC P0716, P0717, P0722, P0723, P0742, P1779, P1791, P1795, P1815, P1818, P1820, P1822, P1823, P1825, P1826, P1842, P1843, P1845, P1847, P1864 and P1865 not set, engine runtime over 5 seconds, system voltage from 8-18v, IMS range is D2, D3, D4 or D5, TFT sensor from 68-266ºF, APP angle over 10%, ISS speed from 200-6,800 rpm, OSS speed over 100 rpm, then after the TCM commanded 1st gear with the OSS speed over 200 rpm and engine torque from 30-258 lb ft., and the resulting gear ratio was 0.73:1 to 0.77:1 for 4 seconds, then after the PCM commanded 2nd gear on with the engine torque from 27-258 lb ft., the resulting gear ratio was from 1.56:1 to 1.64: 1 for 2 seconds. **Possible Causes** ● 2-3 shift solenoid is mechanically stuck "on" ● 2-3 shift solenoid valve O-ring worn or damaged ● 2-3 shift valve is stuck in the released position, or the solenoid is damaged or has failed
P0757 **1T CCM, MIL: No** 2002, 2003, 2004, 2005 Envoy & TrailBlazer 4.2L VIN S, 5.3L VIN T Transmissions: A/T	**Title: A/T 2-3 Shift Solenoid - No 3rd Or 4th Gear (4L60-E)** **Trouble Code Conditions** DTC P0502, P0503, P0740, P0742, P0753, P0758, P1120, P1220, P1810 and P1860 not set, engine runtime over 5 seconds, Fuel Cutoff "off", gear range is D4, TP angle over 10%, TFT sensor from 68-266ºF, system voltage from 10-18v, engine torque from 0-400 lb ft., transmission output speed over 150 rpm, and during Condition 1: PCM commanded 3rd gear "on" for 1 second, estimated gear ratio from 1.6-1.8, engine torque at 50-400 lb ft., conditions met for 2 seconds, then during Condition 2: PCM commands 4th gear on for 1 second, estimated gear ratio at 1.8-3.3, engine torque at 0-400 lb ft., conditions met for 2 seconds. **Possible Causes** ● ATF is burnt or contaminated ● Transmission has an internal malfunction ● Shift solenoid valve seals are damaged, or the Transmission has failed

OBD II Trouble Code List (P0xxx Codes)

DTC	Trouble Code Title, Conditions & Possible Causes
P0757 **1T CCM, MIL: Yes** 2003, 2004, 2005 Envoy & TrailBlazer 5.3L VIN P engine Transmissions: A/T	**A/T 2-3 Shift Solenoid - No 3rd Or 4th Gear (4L65-E) Conditions:** DTC P0120, P0220, P0502, P0503, P0740, P0742, P0753, P0758, P0785, P1810 and P1860 not set, engine runtime over 5 seconds, engine not in Fuel Cutoff mode, gear range is D4, TP angle less than 10%, TFT sensor from 68-266ºF, system voltage from 10-18v, engine torque from 0-400 lb ft., transmission output speed over 150 rpm, Transfer Case ratio in 4WD Low at 0.9 to 1.2, or the Transfer Case ratio in 4WD High at 2.6 to 2.85, then during Condition 1: PCM commands 3rd gear on for 1 second, estimated gear ratio at 1.6-1.8, engine torque from 50-500 lb ft., conditions met for 2 seconds, then during Condition 2: PCM commands 4th gear on for 1 second, estimated gear ratio from 1.8-3.3, engine torque from 50-400 lb ft., conditions met for 2 seconds. **Possible Causes** ● ATF is burnt or contaminated ● Transmission has an internal malfunction ● Shift solenoid valve seals are damaged ● Transmission has failed
P0757 **2T CCM, MIL: Yes** 1999, 2000, 2001 Catera 3.0L VIN R engine Transmissions: A/T	**2-3 Shift Solenoid Valve - No 3rd or 4th Gear (4L30-E) Conditions:** DTC P0705, P0722, P0743, P0753, P0758, P1600, P1625 and P1890 not set, engine started, gearshift in Drive, TP angle over 26.4%, engine speed from 3104-4192 rpm, 3rd Gear commanded "on" for over 700 ms, Transmission output speed over 1,088 rpm, engine torque over 68 lb ft., and the PCM detected the TCC slip speed was 1200-1800 rpm, or with the TP angle over 13.8%, engine speed from 3008-6016 rpm, 4th Gear commanded "on" for 500 ms, Transmission output speed over 1216 rpm, engine torque over 0 lb ft., the PCM detected the TCC slip speed was 2000-6000 rpm for 1.3 seconds (twice). The Sport Mode Lamp flashes if this code is set. **Possible Causes** ● ATF is burnt or contaminated ● Shift solenoid valve seals are leaking or damaged ● Transmission has restricted fluid circuits or it has failed
P0758 **1T CCM, MIL: Yes** 1997, 1998, 1999, 2000, 2001, 2002, 2003, 2004, 2005 Achieva, Alero, Cavalier, Ciera, Century, Grand Am, Sunfire Skyhawk, S/T Blazer & S/T Pickup 2.2L VIN 4, 2.2L VIN 5, 2.4L VIN T, 3.1L VIN J, 3.1L VIN M, 3.4L VIN E Transmissions: A/T	**A/T 2-3 Shift Solenoid Circuit (4T40-E/4T45-E) Conditions:** P0560 not set, engine started, engine speed over 500 rpm for 5 seconds, and the PCM detected an unexpected voltage condition on the 2-3 Shift Solenoid Control circuit for 5 seconds. **Possible Causes** ● 2-3 shift solenoid control circuit is open or shorted to ground ● 2-3 shift solenoid control circuit is shorted to system power ● 2-3 shift solenoid power circuit is open (check the ERLS fuse) ● 2-3 shift solenoid is damaged or has failed ● PCM has failed ● TSB 02-07-30-022A contains a repair procedure for this code
P0758 **1T CCM, MIL: Yes** 1996, 1997, 1998, 1999, 2000, 2001, 2002, 2003, 2004, 2005 Aurora, Bonneville, Century, Cutlass Supreme, Grand Prix, Intrigue, LeSabre, LSS, Lumina, Monte Carlo, 88', Regal, Park Avenue, 98' 3.1L VIN J, 3.1L VIN M, 3.4L VIN E, 3.5L VIN H, 3.8L VIN 1, 3.8L VIN K Transmissions: A/T	**A/T 2-3 Shift Solenoid Circuit Malfunction (4T65-E) Conditions:** DTC P0560 not set, engine started, engine speed over 500 rpm for 5 seconds, 2-3 Shift Solenoid commanded "on" and then "off", and the PCM detected an unexpected voltage condition on the 2-3 Shift Solenoid Control circuit for 5 seconds during the CCM test. **Possible Causes** ● 2-3 shift solenoid control circuit is open or shorted to ground ● 2-3 shift solenoid control circuit is shorted to system power ● 2-3 shift solenoid power circuit is open (test TRANS SOL fuse) ● 2-3 shift solenoid is damaged or has failed ● PCM has failed ● TSB 02-07-30-022A contains a repair procedure for this code
P0758 **1T CCM, MIL: Yes** 1999, 2000, 2001, 2002, 2003, 2004, 2005 C/K Truck, Escalade, L/M Van & G Van, Cargo Van 4.8L VIN V, 5.3L VIN T, 5.3L VIN Z, 6.0L VIN U, 6.0L VIN N, 8.1L VIN G Transmissions: A/T	**A/T 2-3 Shift Solenoid Circuit Malfunction (4L60-E, 4L80-E) Conditions:** Engine started, engine speed over 450 rpm for 5 seconds, system voltage over 10.0v, and the PCM detected an unexpected voltage condition on the 2-3 Shift Solenoid control circuit for 5 seconds. **Possible Causes** ● 2-3 shift solenoid control circuit is open or shorted to ground ● 2-3 shift solenoid control circuit is shorted to system power ● 2-3 shift solenoid power circuit is open (test the TRANS fuse) ● 2-3 shift solenoid is damaged or has failed ● PCM has failed ● TSB 01-07-30-002C contains a repair procedure for this code

OBD II Trouble Code List (P0xxx Codes)

DTC	Trouble Code Title, Conditions & Possible Causes
P0758 **1T CCM, MIL: Yes** 1996, 1997, 1998, 1999, 2000, 2001, 2002 Cab & Chassis, C/K truck, G Van, M/L Series Van, S/T Blazer & S/T Pickup 4.3L VIN W, 4.3L VIN X, 5.0L VIN M, 5.7L VIN R, 5.7L VIN G, 5.7L VIN K, 5.7L VIN S, 7.4L VIN J Transmissions: A/T	**A/T 2-3 Shift Solenoid Circuit Malfunction (4L80-E) Conditions:** Engine started, engine speed over 450 rpm for 7 seconds, system voltage over 10.0v, and the PCM detected an unexpected voltage condition on the 2-3 Shift Solenoid control circuit for 4-5 seconds. **Possible Causes** • 2-3 shift solenoid control circuit is open or shorted to ground • 2-3 shift solenoid control circuit is shorted to system power • 2-3 shift solenoid power circuit is open (test the TRANS fuse) • 2-3 shift solenoid is damaged or has failed • PCM has failed • TSB 57-65-08A contains a repair procedure for this code
P0758 **1T CCM, MIL: Yes** 2003, 2004, 2005 C/K truck, G Van, M/L Series Van, S/T Blazer & S/T Pickup 4.3L VIN W, 4.3L VIN X Transmissions: A/T	**A/T 2-3 Shift Solenoid Circuit Malfunction (4L80-E) Conditions:** Engine speed over 450 rpm for 7 seconds, system voltage over 10.0v, and the PCM detected an unexpected voltage on the 2-3 Shift Solenoid control circuit for 4-5 seconds. **Possible Causes** • 2-3 shift solenoid control circuit is open, shorted to ground or shorted to system power • 2-3 shift solenoid power circuit is open (test the TRANS fuse) • 2-3 shift solenoid has failed, or the PCM has failed
P0758 **1T CCM, MIL: Yes** 1996, 1997, 1998, 1999, 2000, 2001, 2002, 2003, 2004, 2005 C/K Truck & G Van 6.5L VIN F, 6.5L VIN S, 6.5L VIN Y, 6.6L VIN 1 Diesel engines Transmissions: A/T	**A/T 2-3 Shift Solenoid Circuit Malfunction (4L80-E) Conditions:** Engine speed over 450 rpm for 7 seconds, system voltage over 10.0v, and the PCM detected an unexpected voltage on the 2-3 Shift Solenoid control circuit for 4-5 seconds. **Possible Causes** • 2-3 shift solenoid control circuit is open, shorted to ground or shorted to system power • 2-3 shift solenoid power circuit is open (test the TRANS fuse) • 2-3 shift solenoid is damaged or has failed • PCM has failed
P0758 **1T CCM, MIL: Yes** 1995, 1996, 1997, 1998, 1999, 2000, 2001, 2002, 2003, 2004, 2005 Camaro, Corvette & Firebird 3.8L VIN F, 5.7L VIN G, 5.7L VIN S engines Transmissions: A/T	**A/T 2-3 Shift Solenoid Circuit Malfunction (4L60-E) Conditions:** Engine speed over 450 rpm for 5 seconds, system voltage over 10.0v, and the PCM detected an unexpected voltage on the 2-3 Shift Solenoid control circuit for 4-5 seconds. **Possible Causes** • 2-3 shift solenoid control circuit is open or shorted to ground • 2-3 shift solenoid control circuit is shorted to system power • 2-3 shift solenoid power circuit is open (test the TRANS fuse) • 2-3 shift solenoid is damaged or has failed • PCM has failed
P0758 **1T CCM, MIL: Yes** 1995, 1996, 1997, 1998, 1999, 2000, 2001, 2002, 2003, 2004, 2005 Camaro, Corvette & Firebird 3.8L VIN F, 5.7L VIN G, 5.7L VIN S engines Transmissions: A/T	**A/T 2-3 Shift Solenoid Circuit Malfunction (4L60-E) Conditions:** P0560 not set, engine speed over 500 rpm for 5 seconds, and the PCM detected an unexpected voltage condition on the 2-3 Shift Solenoid Control circuit for 5 seconds. **Possible Causes** • 2-3 shift solenoid control circuit is open or shorted to ground • 2-3 shift solenoid control circuit is shorted to system power • 2-3 shift solenoid power circuit is open (check ENG CTRL fuse) • 2-3 shift solenoid is damaged or has failed • PCM has failed
P0758 **1T CCM, MIL: Yes** 2002, 2003, 2004, 2005 Envoy & TrailBlazer 4.2L VIN S, 5.3L VIN P Transmissions: A/T	**A/T 2-3 Shift Solenoid Circuit Malfunction (4L60-E, 4L65-E) Conditions:** Engine runtime over 5 seconds, system voltage from 10-18v, engine not in Fuel Cutoff mode, and the PCM detected an unexpected voltage condition on the A/T 2-3 Shift solenoid circuit. **Possible Causes** • 2-3 shift solenoid control circuit is open, shorted to ground or shorted to system power • 2-3 shift solenoid is damaged or has failed • PCM has failed
P0758 **1T CCM, MIL: Yes** 1997, 1998, 1999, 2000, 2001 Catera 3.0L VIN R engine Transmissions: A/T	**A/T 2-3 Shift Solenoid Circuit Malfunction (4L30-E) Conditions:** Engine started, TCM power relay enabled, then with the 2-3 Shift Solenoid commanded "off", the PCM detected the feedback voltage was less than 2.45v; or with the 2-3 Shift Solenoid commanded "off", the PCM detected the feedback voltage was from 2.45-3.53v; or with the 2-3 Shift solenoid commanded "on", the PCM detected the feedback voltage was more than 4.02v for 180 ms. *Note: The Sport Mode Lamp flashes when this trouble code is set.* **Possible Causes** • 2-3 shift solenoid control circuit is open, shorted to ground or shorted to system power • 2-3 shift solenoid power circuit is open (check B+ at main case) • 2-3 shift solenoid is damaged or has failed • TCM has failed

OBD II Trouble Code List (P0xxx Codes)

DTC	Trouble Code Title, Conditions & Possible Causes
P0761 **2T CCM, MIL: Yes** 2003, 2004, 2005 CTS (Cadillac) 2.6L VIN M, 3.2L VIN N Transmissions: A/T	**A/T 4-5 Shift Solenoid - No 4th Or 5th Gear (5L40-E) Conditions:** DTC P0716, P0717, P0722, P0723, P0742, P1779, P1791, P1795, P1815, P1818, P1820, P1822, P1823, P1825, P1826, P1842, P1843, P1845, P1847, P1864 and P1865 not set, engine runtime over 5 seconds, system voltage from 8-18v, IMS range is D2, D3, D4 or D5, TFT sensor from 68-266ºF, APP angle over 10%, ISS speed from 200-6,800 rpm, OSS speed over 100 rpm, then after the TCM commanded 4th gear with engine torque from 27-258 lb ft., it detected the resulting gear ratio was 0.73:1 to 0.77:1 for 4 seconds. **Possible Causes** ● 4-5 shift solenoid is mechanically stuck "off" ● 4-5 shift solenoid valve O-ring worn or damaged ● 4-5 shift valve is stuck in the applied position ● 4-5 shift solenoid is damaged or it has failed
P0762 **2T CCM, MIL: Yes** 2003, 2004, 2005 CTS (Cadillac) 2.6L VIN M, 3.2L VIN N Transmissions: A/T	**A/T 4-5 Shift Solenoid - No 4th Or 5th Gear (5L40-E) Conditions:** DTC P0716, P0717, P0722, P0723, P0742, P1779, P1791, P1795, P1815, P1818, P1820, P1822, P1823, P1825, P1826, P1842, P1843, P1845, P1847, P1864 and P1865 not set, engine runtime over 5 seconds, system voltage from 8-18v, IMS range is D2, D3, D4 or D5, TFT sensor from 68-266ºF, APP angle over 10%, ISS speed from 200-6,800 rpm, OSS speed over 100 rpm, then after the TCM commanded 2nd or 3rd gear for 4 seconds, then it commanded 5th gear with engine torque from 27-258 lb ft., it detected the resulting gear ratio was 0.98:1 to 1.03:1 for 3.5 seconds. **Possible Causes** ● 4-5 shift solenoid is mechanically stuck "on" ● 4-5 shift solenoid valve O-ring worn or damaged ● 4-5 shift valve is stuck in the released position ● 4-5 shift solenoid is damaged or it has failed
P0785 **1T CCM, MIL: Yes** 1997, 1998, 1999, 2000, 2001, 2002, 2003, 2004, 2005 Cab & Chassis, C/K truck, G Van, M/L Series Van, S/T Blazer & S/T Pickup 4.3L VIN W, 4.3L VIN X, 5.0L VIN M, 5.7L VIN R, 5.7L VIN G, 5.7L VIN K, 5.7L VIN S, 7.4L VIN J Transmissions: A/T	**A/T 3-2 Shift Solenoid Circuit Malfunction (4L80-E) Conditions:** Engine started, engine speed over 450 rpm for 7 seconds, system voltage over 10.0v, and the PCM detected an unexpected voltage condition on the 3-2 Shift Solenoid control circuit for 4-5 seconds. **Possible Causes** ● 3-2 shift solenoid control circuit is open or shorted to ground ● 3-2 shift solenoid control circuit is shorted to system power ● 3-2 shift solenoid power circuit is open (test the TRANS fuse) ● 3-2 shift solenoid is damaged or has failed ● PCM has failed
P0785 **1T CCM, MIL: Yes** 1997, 1998, 1999, 2000, 2001, 2002, 2003, 2004, 2005 C/K Truck & G Van 6.5L VIN F, 6.5L VIN S, 6.5L VIN Y Diesel engine Transmissions: A/T	**A/T 3-2 Shift Solenoid Circuit Malfunction (4L80-E) Conditions:** Engine speed over 450 rpm for 7 seconds, system voltage over 10.0v, and the PCM detected an unexpected voltage on the 3-2 Shift Solenoid control circuit for 4-5 seconds. **Possible Causes** ● 3-2 shift solenoid control circuit is open or shorted to ground ● 3-2 shift solenoid control circuit is shorted to system power ● 3-2 shift solenoid power circuit is open (check the TRANS fuse) ● 3-2 shift solenoid is damaged or has failed ● PCM has failed
P0785 **1T CCM, MIL: Yes** 1997, 1998, 1999, 2000, 2001, 2002, 2003, 2004, 2005 C/K Truck, Escalade, L/M Van, G Van & Cargo Van 4.8L VIN V, 5.3L VIN T, 5.3L VIN Z, 6.0L VIN N, 6.0L VIN U engines Transmissions: A/T	**A/T 3-2 Shift Solenoid Circuit Malfunction (4L60-E) Conditions:** Engine started, engine speed over 450 rpm for 5 seconds, system voltage over 10.0v, and the PCM detected an unexpected voltage condition on the 3-2 Shift Solenoid control circuit for 4-5 seconds. **Possible Causes** ● 3-2 shift solenoid control circuit is open, shorted to ground or shorted to system power ● 3-2 shift solenoid power circuit is open (check the TRANS fuse) ● 3-2 shift solenoid is damaged or has failed ● PCM has failed ● TSB 01-07-30-002C contains a repair procedure for this code
P0785 **1T CCM, MIL: Yes** 1995, 1996, 1997, 1998, 1999, 2000, 2001, 2002, 2003, 2004, 2005 Camaro, Corvette & Firebird 3.8L VIN F, 5.7L VIN G, 5.7L VIN S engines Transmissions: A/T	**A/T 2-3 Shift Solenoid Circuit Malfunction (4L60-E) Conditions:** P0560 not set, engine started, engine speed over 500 rpm for 5 seconds, and the PCM detected an unexpected voltage on the 2-3 Shift Solenoid Control circuit for 5 seconds. **Possible Causes** ● 2-3 shift solenoid control circuit is open or shorted to ground ● 2-3 shift solenoid control circuit is shorted to system power ● 2-3 shift solenoid power circuit is open (check ENG CTRL fuse) ● 2-3 shift solenoid is damaged or has failed ● PCM has failed

OBD II Trouble Code List (P0xxx Codes)

DTC	Trouble Code Title, Conditions & Possible Causes
P0785 **2T CCM, MIL: Yes** 2002, 2003, 2004, 2005 Envoy & TrailBlazer 4.2L VIN S, 5.3L VIN P Transmissions: A/T	**A/T 3-2 Solenoid Circuit Malfunction (4L60-E, 4L65-E) Conditions:** Engine started, system voltage from 10-18v, Decel Fuel Cutoff not active, and the PCM detected the A/T 3-2 solenoid feedback voltage was high with the 3-2 Solenoid commanded "on", or it was low with the 3-2 Solenoid commanded "off" for 5 seconds. **Possible Causes** • 3-2 solenoid control circuit is open, shorted to ground or to B+ • 3-2 solenoid power circuit is open (test the IGN 0 fuse) • 3-2 solenoid is damaged or has failed • PCM has failed
P0801 **1T CCM, MIL: Yes** 1997, 1998, 1999, 2000, 2001, 2002, 2003, 2004, 2005 Camaro, Corvette & Firebird 5.7L VIN G, 5.7L VIN S Transmissions: M/T	**Reverse Inhibit Solenoid Circuit Malfunction (MM6) Conditions:** Engine started, system voltage over 10.0v, and the PCM detected an unexpected voltage condition on the Reverse Inhibit Solenoid control circuit for 5 seconds during the CCM test. **Possible Causes** • Reverse inhibit solenoid circuit is open, shorted to ground or shorted to system power • Reverse inhibit solenoid B+ circuit open (test ENG SENS fuse) • Reverse inhibit solenoid is damaged or has failed • PCM has failed
P0802 **1T CCM, MIL: No** 2001, 2002, 2003, 2004, 2005 C/K Series Truck 6.6L VIN 1 Diesel engine Transmissions: All	**Malfunction Indicator Lamp Circuit Malfunction Conditions:** Engine started and the PCM detected an incorrect voltage on the TCM MIL request circuit. **Possible Causes** • TCM MIL Request circuit is open, shorted to ground or to B+ • ECM has failed • TCM has failed
P0803 **1T CCM, MIL: No** 1997, 1998, 1999, 2000, 2001, 2002, 2003, 2004, 2005 Camaro, Corvette & Firebird 5.7L VIN G, 5.7L VIN S Transmissions: M/T	**Skip Shift Solenoid Circuit Malfunction (MM6) Conditions:** Engine started, system voltage over 10.0v, and the PCM detected an unexpected voltage condition on the Skip Shift Solenoid control circuit for 5 seconds during the CCM test. **Possible Causes** • Skip shift solenoid circuit is open, shorted to ground or shorted to system power • Skip shift solenoid B+ circuit open (test ENG SENS fuse) • Skip shift solenoid is damaged or has failed • PCM has failed
P0804 **1T CCM, MIL: No** 1997, 1998, 1999, 2000, 2001, 2002, 2003, 2004, 2005 Camaro, Corvette & Firebird 5.7L VIN G, 5.7L VIN S Transmissions: M/T	**Skip Shift Lamp Control Circuit Malfunction (MM6) Conditions:** Engine started, system voltage over 10.0v, and the PCM detected an unexpected voltage condition on the Skip Shift Lamp control circuit for 5 seconds during the CCM test. **Possible Causes** • Skip shift lamp control (Class 2) circuit is open, shorted to ground or to system power • Skip shift lamp power circuit is open (test Instrument Cluster) • Skip shift lamp has failed (in the Instrument Cluster) • PCM has failed
P0850 **1T CCM, MIL: No** 2003, 2004, 2005 CTS (Cadillac) 2.6L VIN M, 3.2L VIN N Transmissions: A/T	**Title: Manual Shift Position Switch Circuit Malfunction (5L40-E)** **Trouble Code Conditions** Engine started, system voltage from 8-18v, and the PCM detected the P/N switch indicated low (0v) for 100 seconds while the TCM reported the gear selector in Reverse gear or a forward range gear, or the P/N switch indicated high (12v) for 200 seconds while the TCM reported the gear selector indicated Park or Neutral position. **Possible Causes** • Manual shift position switch connector is damaged or shorted • PNP switch signal circuit is open or shorted to ground • Manual shift position switch is damaged or it has failed • PCM has failed
P0850 **1T CCM, MIL: No** 2002, 2003, 2004, 2005 C/K Truck, G Van, Envoy & TrailBlazer 5.3L VIN P, 5.3L VIN T, 6.0 VIN N, 6.0L VIN U, 6.6L VIN 1 engines Transmissions: A/T	**Title: A/T TCC PWM Solenoid Malfunction (4L60-E, 4L65-E)** **Trouble Code Conditions** DTC P0502, P0503, P0740, P0742, P0753, P0758, P1120, P1220, P1810 and P1860 not set, engine speed from 1,500-3,000 rpm for 6 seconds, VSS from 30-82 mph, TP angle at 20-99%, speed ratio from 0.64-1.35, TFT sensor from 68-302°F, engine vacuum from 0-105 kPa, engine torque from 50-400 lb ft., not in 1st gear, gear range is D4, Shift Solenoid diagnostic counters = 0, TCC command at 40% duty cycle, TCC slip speed from 130-800 rpm for 7 seconds, then during Condition 1: With the TCC slip speed at 130-800 rpm for 7 seconds, and the PCM commanded maximum line pressure and prevented the freeze shift adapts from being updated, or during Condition 2: or with the TCC slip speed at 130-800 rpm for 7 seconds, the PCM commanded the TCC off for 1.5 seconds, or during Condition 3: with the TCC slip speed at 130-800 rpm for 7 seconds, and the current fail counter incremented. **Possible Causes** • ATF is burnt or contaminated • Transmission has an internal malfunction, or the TCC PWM solenoid seal is damaged • TCC PWM solenoid has failed, or the Transmission has failed

OBD II Trouble Code List (P1xxx Codes)

DTC	Trouble Code Title, Conditions & Possible Causes
DTC P1020 **1T CCM, MIL: Yes** 2003, 2004, 2005 C/K Truck 6.0L VIN U CNG engine Transmissions: All	**Camshaft Phasing System Malfunction Conditions:** DTC P0148, P0611 and P1209 not set; engine operating on CNG, PCM not in Fuel Shutoff or Decel Fuel Shutoff modes, system voltage from 6-18v, and the FICM (module) detected an invalid fuel injector PWM signal. The PCM controls the fuel delivery and determines the fuel system operation on Bi-Fuel (KL6) vehicles. The FICM receives the 8 fuel injector PWM signals from the PCM, and generates duplicate PWM signals to operate the CNG injectors. **Possible Causes** ● AF fuel mode relay circuit (odd bank, even bank) is open ● AF enable circuit is open or shorted to ground ● FICM is damaged or it has failed
DTC P1021 **1T CCM, MIL: Yes** 2003, 2004, 2005 C/K Truck 6.0L VIN U CNG engine Transmissions: All	**CNG Fuel Injector 1 Circuit Malfunction Conditions:** Engine started, system voltage from 6-18v, and the FICM detected an incorrect voltage condition on the circuit that controls the CNG Fuel Injector 1 operation during the CCM test. **Possible Causes** ● CNG fuel injector 1 control circuit is open or shorted to ground ● CNG fuel injector 1 is damaged or it has failed ● FICM has failed
DTC P1022 **1T CCM, MIL: Yes** 2003, 2004, 2005 C/K Truck 6.0L VIN U CNG engine Transmissions: All	**CNG Fuel Injector 2 Circuit Malfunction Conditions:** Engine started, system voltage from 6-18v, and the FICM detected an incorrect voltage condition on the circuit that controls the CNG Fuel Injector 2 operation during the CCM test. **Possible Causes** ● CNG fuel injector 2 control circuit is open or shorted to ground ● CNG fuel injector 2 is damaged or it has failed ● FICM has failed
DTC P1023 **1T CCM, MIL: Yes** 2003, 2004, 2005 C/K Truck 6.0L VIN U CNG engine Transmissions: All	**CNG Fuel Injector 3 Circuit Malfunction Conditions:** Engine started, system voltage from 6-18v, and the FICM detected an incorrect voltage condition on the circuit that controls the CNG Fuel Injector 3 operation during the CCM test. **Possible Causes** ● CNG fuel injector 3 control circuit is open or shorted to ground ● CNG fuel injector 3 is damaged or it has failed ● FICM has failed
DTC P1024 **1T CCM, MIL: Yes** 2003, 2004, 2005 C/K Truck 6.0L VIN U CNG engine Transmissions: All	**CNG Fuel Injector 4 Circuit Malfunction Conditions:** Engine started, system voltage from 6-18v, and the FICM detected an incorrect voltage condition on the circuit that controls the CNG Fuel Injector 4 operation during the CCM test. **Possible Causes** ● CNG fuel injector 4 control circuit is open or shorted to ground ● CNG fuel injector 4 is damaged or it has failed ● FICM has failed
DTC P1025 **1T CCM, MIL: Yes** 2003, 2004, 2005 C/K Truck 6.0L VIN U CNG engine Transmissions: All	**CNG Fuel Injector 5 Circuit Malfunction Conditions:** Engine started, system voltage from 6-18v, and the FICM detected an incorrect voltage condition on the circuit that controls the CNG Fuel Injector 5 operation during the CCM test. **Possible Causes** ● CNG fuel injector 5 control circuit is open or shorted to ground ● CNG fuel injector 5 is damaged or it has failed ● FICM has failed
DTC P1026 **1T CCM, MIL: Yes** 2003, 2004, 2005 C/K Truck 6.0L VIN U CNG engine Transmissions: All	**CNG Fuel Injector 6 Circuit Malfunction Conditions:** Engine started, system voltage from 6-18v, and the FICM detected an incorrect voltage condition on the circuit that controls the CNG Fuel Injector 6 operation during the CCM test. **Possible Causes** ● CNG fuel injector 6 control circuit is open or shorted to ground ● CNG fuel injector 6 is damaged or it has failed ● FICM has failed
DTC P1027 **1T CCM, MIL: Yes** 2003, 2004, 2005 C/K Truck 6.0L VIN U CNG engine Transmissions: All	**CNG Fuel Injector 7 Circuit Malfunction Conditions:** Engine started, system voltage from 6-18v, and the FICM detected an incorrect voltage condition on the circuit that controls the CNG Fuel Injector 7 operation during the CCM test. **Possible Causes** ● CNG fuel injector 7 control circuit is open or shorted to ground ● CNG fuel injector 7 is damaged or it has failed ● FICM has failed
DTC P1028 **1T CCM, MIL: Yes** 2003, 2004, 2005 C/K Truck 6.0L VIN U CNG engine Transmissions: All	**CNG Fuel Injector 8 Circuit Malfunction Conditions:** Engine started, system voltage from 6-18v, and the FICM detected an incorrect voltage condition on the circuit that controls the CNG Fuel Injector 8 operation during the CCM test. **Possible Causes** ● CNG fuel injector 8 control circuit is open or shorted to ground ● CNG fuel injector 8 is damaged or it has failed ● FICM has failed

OBD II Trouble Code List (P1xxx Codes)

DTC	Trouble Code Title, Conditions & Possible Causes
DTC P1031 **2T CCM, MIL: Yes** 2000, 2001, 2002, 2003, 2004, 2005 Aurora, DeVille, Eldorado & Seville 4.0L VIN C, 4.6L VIN 9 & 4.6L VIN Y engines Transmissions: All	**HO2S Heater Current Monitor Control (Bank 1 Sensor 1) Conditions:** Key on or engine running; system voltage over 10.0v, and the PCM detected an unexpected voltage condition on the HO2S-11 Heater Current Monitor Driver circuit. The HO2S heater is a device used to reduce the time the sensor takes to go active. The PCM controls the HO2S heaters using three separate drivers. The PCM turns "on" the heaters using a high side and low side driver to help the sensor warmup. The PCM uses a third driver to test the oxygen sensor heaters. This driver contains a current monitoring circuit that is used in a diagnostic sequence with the other drivers to indicate the condition of the heaters. **Possible Causes** ● HO2S heater current monitor low or high circuit is open ● HO2S heater current monitor low or high circuit shorted to B+ ● HO2S heater is damaged or has failed ● PCM has failed
DTC P1032 **2T CCM, MIL: Yes** 2000, 2001, 2002, 2003, 2004, 2005 Aurora, DeVille, Eldorado & Seville 4.0L VIN C, 4.6L VIN 9 & 4.6L VIN Y engines Transmissions: All	**HO2S Heater Current Monitor Control (Bank 2 Sensor 1) Conditions:** Key on or engine running; system voltage over 10.0v, and the PCM detected an unexpected voltage condition on the HO2S-21 Heater Current Monitor Driver circuit. The HO2S heater is a device used to reduce the time the sensor takes to go active. The PCM controls the HO2S heaters using three separate drivers. With the engine running, the PCM turns "on" the heaters using a high side and low side driver to help the sensor warmup. The PCM uses a third driver to test the oxygen sensor heaters. This driver contains a current monitoring circuit used in a diagnostic sequence with the other drivers to indicate the condition of the heaters. **Possible Causes** ● HO2S heater current monitor low or high circuit is open ● HO2S heater current monitor low or high circuit shorted to B+ ● HO2S heater is damaged or has failed ● PCM has failed
DTC P1093 **1T CCM, MIL: Yes** 2001, 2002, 2003, 2004, 2005 C/K Truck 6.6L VIN 1 Diesel engine Transmissions: All	**Fuel Rail Pressure Range/Performance Conditions:** DTC P0192, P0193 and P0641 not set; key on or engine running, and the ECM detected the difference between the Commanded fuel pressure and Actual fuel pressure was more than 20 MPa, or the FRP regulator fuel flow was more than 15,000 mm³ at 800 rpm, or that it was more than 38,000 mm³ at 2,000 rpm. The ECM uses the fuel rail pressure (FRP) sensor to detect the fuel pressure to the fuel injectors. This value is compared to the calculated target fuel pressure as determined by the ECM. The ECM adjusts the FRP by modulating the duty cycle of the control driver of the fuel pressure regulator. Injector pulse duration is determined by detecting the measured rail pressure and the target injection fuel to each cylinder. **Possible Causes** ● CNG fuel injector 8 connector is damaged, open or shorted ● CNG fuel injector 8 control circuit is open or shorted to ground ● CNG fuel injector 8 is damaged or it has failed ● FICM has failed
DTC P1094 **1T CCM, MIL: Yes** 2001, 2002, 2003, 2004, 2005 C/K Truck 6.6L VIN 1 Diesel engine Transmissions: All	**Fuel Rail Pressure Solenoid Circuit Malfunction Conditions:** DTC P0192, P0193 and P0641 not set; engine started, and the ECM detected the difference between the FRP sensor and commanded fuel injector fuel flow was more than 30 MPa. The ECM uses the fuel rail pressure sensor to detect the fuel injector fuel pressure. This value is compared to a calculated target fuel pressure in the ECM. The ECM adjusts the FRP by modulating the duty cycle of the fuel pressure regulator control driver. The pulse duration is determined by detecting the measured rail pressure and target injection fuel to each cylinder. **Possible Causes** ● Check the engine oil for fuel contamination ● Fuel pressure regulator solenoid circuit is shorted to ground ● If you have to prime the fuel system, inspect for a restricted fuel supply line or a fuel supply line air leak ● ECM has failed
DTC P1106 **1T CCM, MIL: No** 1996, 1997, 1998, 1999, 2000, 2001, 2002, 2003, 2004, 2005 Aurora, Bonneville, Century, Cutlass Supreme, Grand Prix, Intrigue, LeSabre, LSS, Lumina, Monte Carlo, 88', Regal, Park Avenue & 98' 3.1L VIN J, 3.1L VIN M, 3.4L VIN E, 3.5L VIN H, 3.8L VIN 1, 3.8L VIN K Transmissions: All	**MAP Sensor Circuit Intermittent High Input Conditions:** DTC P0121, P0122 and P0123 not set, engine started, engine runtime from 1-2 minutes (depends on the ECT sensor at startup), system voltage over 10.0v, TP angle under 2% with engine speed less than 3000 rpm, or TP angle under 30% with engine speed more than 3000 rpm, and the PCM detected an intermittent high voltage (over 4.20v) condition on the MAP sensor circuit. The MAP sensor responds to pressure changes in the intake manifold that occur based on the engine load. The PCM is connected to the MAP sensor by a 5v VREF, low reference ground and MAP sensor signal circuit. **Possible Causes** ● MAP sensor signal circuit is shorted to VREF (intermittent fault) ● MAP sensor ground circuit is open (intermittent fault) ● MAP sensor is damaged or has failed ● PCM has failed

OBD II Trouble Code List (P1xxx Codes)

DTC	Trouble Code Title, Conditions & Possible Causes
DTC P1106 **1T CCM, MIL: No** 1996, 1997, 1998, 1999, 2000, 2001, 2002, 2003, 2004, 2005 Cab & Chassis, C/K Truck, G Van, L/M Van, S/T Blazer & S/T Pickup 4.3L VIN W, 4.3L VIN X, 5.0L VIN M, 5.7L VIN R, 7.4L VIN J engines Transmissions: All	**MAP Sensor Circuit Intermittent High Input Conditions:** DTC P0121, P0122 and P0123 not set, engine started, system voltage over 10.0v, TP angle less than 0.4% with engine speed below 1200 rpm, or TP angle less than 20% with engine speed over 1200 rpm, and the PCM/VCM detected an unexpected "high" voltage (over 4.40v) on the MAP sensor signal circuit for over 1 second. **Possible Causes** • MAP sensor signal circuit is shorted to VREF (intermittent fault) • MAP sensor ground circuit is open (intermittent fault) • MAP sensor is damaged or has failed • PCM has failed
DTC P1106 **1T CCM, MIL: No** 2003, 2004, 2005 C/K Truck, Envoy, G Van, Escalade & TrailBlazer 4.8L VIN V, 5.3L VIN P, 5.3L VIN T, 5.3L VIN Z, 6.0L VIN N, 6.0L VIN U, 8.1L VIN G engines Transmissions: All	**MAP Sensor Circuit Intermittent High Input Conditions:** DTC P0120, P0220, P1125, P1514, P1515, P1516, P1518, P2108, P2120, P2121, P2125, P2126, P2130, P2131, P2135 are not set, engine started, TP sensor less than 1% with the engine speed below 1200 rpm, or TP angle less than 20% with engine speed over 1200 rpm, and the PCM detected an unexpected high voltage (over 4.90v) on the MAP sensor signal circuit for more than 6 seconds. **Possible Causes** • MAP sensor signal circuit is shorted to VREF (intermittent fault) • MAP sensor ground circuit is open (intermittent fault) • MAP sensor is damaged or has failed • PCM has failed
DTC P1106 **1T CCM, MIL: No** 1996, 1997, 1998, 1999, 2000, 2001, 2002, 2003, 2004, 2005 Aurora, DeVille, Eldorado & Seville 4.0L VIN C, 4.6L VIN 9 & 4.6L VIN Y engines Transmissions: All	**MAP Sensor Circuit Intermittent High Input Conditions:** DTC P0121, P0122 and DTC P0123 not set, engine runtime from 1-2 minutes (depends on the ECT sensor at startup), TP angle below 89% with engine speed under 1000 rpm or TP angle below 97% with engine speed above 1000 rpm, and the PCM detected an intermittent high voltage condition (over 4.90v) on the MAP sensor signal circuit for under 1 second. **Possible Causes** • MAP sensor signal circuit is shorted to VREF (intermittent fault) • MAP sensor ground circuit is open (intermittent fault) • MAP sensor is damaged or has failed • PCM has failed
DTC P1106 **1T CCM, MIL: No** 1996, 1997, 1998, 1999, 2000, 2001, 2002, 2003, 2004, 2005 Aztek, Montana, Venture, Silhouette & Rendezvous 3.4L VIN E engine Transmissions: All	**MAP Sensor Circuit Intermittent High Input Conditions:** DTC P0121, P0122 and P0123 not set, engine started, engine runtime from 1-2 minutes (depends on the ECT sensor at startup), system voltage over 10.0v, TP angle under 2% with engine speed less than 3000 rpm, or TP angle under 30% with engine speed more than 3000 rpm, and the PCM detected an intermittent high voltage (over 4.20v) condition on the MAP sensor circuit. The MAP sensor responds to pressure changes in the intake manifold that occur based on the engine load. The PCM is connected to the MAP sensor by a 5v VREF, low reference ground and MAP sensor signal circuit. **Possible Causes** • MAP sensor signal circuit is shorted to VREF (intermittent fault) • MAP sensor ground circuit is open (intermittent fault) • MAP sensor is damaged or has failed (intermittent fault) • PCM has failed
DTC P1106 **1T CCM, MIL: No** 1995, 1996, 1997, 1998, 1999, 2000, 2001, 2002 Camaro & Firebird 3.8L VIN K engine Transmissions: All	**MAP Sensor Circuit Intermittent High Input Conditions:** DTC P0606, P1120, P1125, P1220, P1221, P1271-P1276, P1280, P1281-P1286 or P1514-P1517 not set, engine runtime determined by ECT input at startup (up to 2 minutes at less than -22ºF to less than 12 seconds at more than 86ºF), TP angle under 0.5% with engine speed under 900 rpm, then PCM detected an intermittent high voltage condition (over 4.20v or 90 kPa) on the MAP sensor signal circuit for 5 seconds during the CCM test. **Possible Causes** • MAP sensor signal circuit is shorted to VREF (intermittent fault) • MAP sensor ground circuit is open (intermittent fault) • MAP sensor is damaged or has failed (intermittent fault) • PCM has failed
DTC P1106 **1T CCM, MIL: No** 1995 S/T Blazer & S/T Pickup 4.3L VIN W engine Transmissions: All	**MAP Sensor Circuit Intermittent High Input Conditions:** DTC P0121, P0122 and P0123 not set, engine started, system voltage over 10.0v, TP angle less than 0.4% with engine speed under 575 rpm, or TP angle less than 20% with engine speed over 1200 rpm, and the PCM detected an intermittent high voltage condition on the MAP sensor signal circuit for 5 seconds. **Possible Causes** • MAP sensor signal circuit is shorted to VREF (intermittent fault) • MAP sensor ground circuit is open (intermittent fault) • MAP sensor is damaged or has failed (intermittent fault) • PCM has failed

OBD II Trouble Code List (P1xxx Codes)

DTC	Trouble Code Title, Conditions & Possible Causes
DTC P1107 **1T CCM, MIL: No** 1996, 1997, 1998, 1999, 2000, 2001, 2002, 2003, 2004, 2005 Aurora, Bonneville, Century, Cutlass Supreme, Grand Prix, Intrigue, LeSabre, LSS, Lumina, Monte Carlo, 88', Regal, Park Avenue & 98' 3.1L VIN J, 3.1L VIN M, 3.4L VIN E, 3.5L VIN H, 3.8L VIN 1, 3.8L VIN K Transmissions: All	**MAP Sensor Circuit Intermittent Low Input Conditions:** DTC P0121, P0122 and P0123 not set, engine started, system voltage over 10.0v, engine runtime from 1-2 minutes (depends on ECT sensor at startup), TP angle above 0% with engine speed less than 1000 rpm, or TP angle above 10% with engine speed more than 1000 rpm, and the PCM detected an intermittent low voltage condition (under 0.1v) on the MAP sensor signal circuit. The MAP sensor responds to pressure changes in the intake manifold that occur based on the engine load. The PCM is connected to the MAP sensor by a 5v VREF, low reference ground and MAP signal circuit. **Possible Causes** ● MAP sensor signal circuit shorted to ground (intermittent fault) ● MAP sensor VREF circuit is open (intermittent fault) ● MAP sensor is damaged or has failed (intermittent fault) ● PCM has failed
DTC P1107 **1T CCM, MIL: No** 1996, 1997 Camaro, Caprice, Corvette, Firebird, Fleetwood & Roadmaster 4.3L VIN W, 5.7L VIN 5, 5.7L VIN P engines Transmissions: All	**MAP Sensor Circuit Intermittent Low Input Conditions:** DTC P0121, P0122 and P0123 not set, engine started, system voltage over 10.0v, and the PCM detected an intermittent low voltage condition (under 0.1v) on the MAP sensor signal circuit for 300 ms during the CCM test. **Possible Causes** ● MAP sensor signal circuit shorted to ground (intermittent fault) ● MAP sensor VREF circuit is open (intermittent fault) ● MAP sensor is damaged or has failed (intermittent fault) ● PCM has failed
DTC P1107 **1T CCM, MIL: No** 1996, 1997, 1998, 1999, 2000, 2001, 2002, 2003, 2004, 2005 Aztek, Montana, Venture, Silhouette & Rendezvous 3.4L VIN E engine Transmissions: All	**MAP Sensor Circuit Intermittent Low Input Conditions:** DTC P0121, P0122 and P0123 not set, engine started, system voltage over 10.0v, engine runtime from 1-2 minutes (depends on ECT sensor at startup), TP angle above 0% with engine speed less than 1000 rpm, or TP angle above 10% with engine speed more than 1000 rpm, and the PCM detected an intermittent low voltage condition (under 0.1v) on the MAP sensor signal circuit. The MAP sensor responds to pressure changes in the intake manifold that occur based on the engine load. The PCM is connected to the MAP sensor by a 5v VREF, low reference ground and MAP signal circuit. **Possible Causes** ● MAP sensor signal circuit shorted to ground (intermittent fault) ● MAP sensor VREF circuit is open (intermittent fault) ● MAP sensor is damaged or has failed (intermittent fault) ● PCM has failed
DTC P1107 **1T CCM, MIL: No** 1996, 1997, 1998, 1999, 2000, 2001, 2002, 2003, 2004, 2005 Cab & Chassis, C/K Truck, G Van, L/M Van, S/T Blazer & S/T Pickup 4.3L VIN W, 4.3L VIN X, 5.0L VIN M, 5.7L VIN R, 7.4L VIN J engines Transmissions: All	**MAP Sensor Circuit Intermittent Low Input Conditions:** DTC P0121, P0122 and P0123 not set, engine started, system voltage over 10.0v, TP angle at 0% with engine speed below 800 rpm or TP angle less than 12.5% with engine speed over 800 rpm, and the PCM/VCM detected an unexpected "Low" voltage (under 0.40v) on the MAP sensor signal circuit for over 1 second. **Possible Causes** ● MAP sensor signal circuit shorted to ground (intermittent fault) ● MAP sensor VREF circuit is open (intermittent fault) ● MAP sensor is damaged or has failed (intermittent fault) ● PCM has failed
DTC P1107 **1T CCM, MIL: No** 2003, 2004, 2005 C/K Truck, Envoy, G Van, Escalade & TrailBlazer 4.8L VIN V, 5.3L VIN P, 5.3L VIN T, 5.3L VIN Z, 6.0L VIN N, 6.0L VIN U, 8.1L VIN G engines Transmissions: All	**MAP Sensor Circuit Intermittent Low Input Conditions:** DTC P0120, P0220, P1125, P1514, P1515, P1516, P1518, P2108, P2120, P2121, P2125, P2126, P2130, P2131, P2135 are not set, TP sensor at 0% with the engine speed below 800 rpm, or TP angle more than 12.5% with engine speed over 800 rpm, and the PCM detected an unexpected low voltage (below 0.10v) on the MAP sensor signal circuit for 6 seconds. **Possible Causes** ● MAP sensor signal circuit is shorted to VREF (intermittent fault) ● MAP sensor ground circuit is open (intermittent fault) ● MAP sensor is damaged or has failed ● PCM has failed
DTC P1107 **1T CCM, MIL: No** 1996, 1997, 1998, 1999, 2000, 2001, 2002, 2003, 2004, 2005 Aurora, DeVille, Eldorado & Seville 4.0L VIN C, 4.6L VIN 9 & 4.6L VIN Y engines Transmissions: All	**MAP Sensor Circuit Intermittent Low Input Conditions:** DTC P0121, P0122 and DTC P0123 not set, engine started, TP angle more than 10% with engine speed over 1000 rpm or TP angle more than 0% with engine speed under 1000 rpm, and the PCM detected an intermittent low voltage condition (under 0.25v) on the MAP sensor signal circuit for 1 second. **Possible Causes** ● MAP sensor signal circuit shorted to ground (intermittent fault) ● MAP sensor VREF circuit is open (intermittent fault) ● MAP sensor is damaged or has failed (intermittent fault) ● PCM has failed

OBD II Trouble Code List (P1xxx Codes)

DTC	Trouble Code Title, Conditions & Possible Causes
DTC P1107 **1T CCM, MIL: No** 1995, 1996, 1997, 1998, 1999, 2000, 2001, 2002 Camaro & Firebird 3.8L VIN K engine Transmissions: All	**MAP Sensor Circuit Intermittent Low Input Conditions:** DTC P0606, P1120, P1125, P1220, P1221, P1271-P1276, P1280, P1281-P1286, P1514 and P1517 not set, engine started, engine runtime 1-2 minutes (depends on the ECT sensor at startup), TP angle over 0% with engine speed under1000 rpm or TP angle over 10% with engine speed over 1000 rpm, and the PCM detected an intermittent low voltage condition (under 0.10v or 2 kPa) for more than 5 seconds during the CCM test. **Possible Causes** ● MAP sensor signal circuit shorted to ground (intermittent fault) ● MAP sensor VREF circuit is open (intermittent fault) ● MAP sensor is damaged or has failed (intermittent fault) ● PCM has failed
DTC P1107 **1T CCM, MIL: No** 1995 S/T Blazer & S/T Pickup 4.3L VIN W engine Transmissions: All	**MAP Sensor Circuit Intermittent Low Input Conditions:** DTC P0121, P0122 and P0123 not set, engine started, system voltage over 10.0v, TP angle less than 0.4% with engine speed under 575 rpm, or TP angle less than 20% with engine speed over 1200 rpm, and the PCM detected an intermittent low voltage condition (<0.10v) on the MAP sensor signal circuit for 5 seconds. **Possible Causes** ● MAP sensor signal circuit shorted to ground (intermittent fault) ● MAP sensor VREF circuit is open (intermittent fault) ● MAP sensor is damaged or has failed (intermittent fault) ● PCM has failed
DTC P1107 **1T CCM, MIL: No** 1996, 1997, 1998, 1999 Aurora, DeVille, Eldorado & Seville 4.0L VIN C, 4.6L VIN 9 & 4.6L VIN Y engines Transmissions: All	**BARO to MAP Sensor Input Comparison Too High Conditions:** DTC P0107, P0108, P0122 and P0123 not set, Engine started; Calculated Throttle Switch indicating at idle, TP angle less than 18 degrees, BARO sensor more than 75 kPa, and the PCM detected the difference between MAP and Calculated BARO was 11 kPa or less for 15 seconds. *Note: The difference in these values should be over 11 kPa at idle.* **Possible Causes** ● Base engine (anything that can cause low engine vacuum) ● MAP sensor signal, ground or VREF circuit problems ● MAP sensor is contaminated, out-of-calibration or "skewed" ● MAP sensor source vacuum line or manifold port is clogged ● PCM has failed
DTC P1111 **1T CCM, MIL: No** 1996, 1997, 1998, 1999, 2000, 2001, 2002, 2003, 2004, 2005 Aurora, Bonneville, Century, Cutlass Supreme, Grand Prix, Intrigue, LeSabre, LSS, Lumina, Monte Carlo, 88', Regal, Park Avenue & 98' 3.1L VIN J, 3.1L VIN M, 3.4L VIN E, 3.5L VIN H, 3.8L VIN 1, 3.8L VIN K Transmissions: All	**IAT Sensor Circuit Intermittent High Input Conditions:** DTC P0101, P0102, P0103, P0116, P0117, P0118, P0502 and P0503 not set, engine started, engine runtime over 3 minutes, vehicle speed less than 35 mph, ECT sensor more than 140°F, MAF sensor less than 12 g/sec, and the PCM detected an intermittent high voltage condition (Scan Tool reads -36°F) on the IAT sensor signal circuit for 3 seconds. The IAT sensor is a variable resistor that includes an IAT signal circuit and a low reference circuit to measure the temperature of the air that enters the engine. The PCM connects to the IAT sensor with a 5v signal and a low reference ground circuit. When this sensor is cold, its resistance is high. As the temperature of the air increases, its resistance decreases. With high sensor resistance, the IAT sensor signal voltage is high. With lower sensor resistance, the IAT sensor signal voltage will be a lower value. **Possible Causes** ● IAT sensor signal circuit is open (intermittent fault) ● IAT sensor ground circuit is open (intermittent fault) ● IAT sensor is damaged (an intermittent "open" condition) ● PCM has failed ● TSB 02-06-03-005 contains a repair procedure for this code
DTC P1111 **1T CCM, MIL: No** 1996, 1997, 1998, 1999, 2000, 2001, 2002, 2003, 2004, 2005 Aztek, Montana, Venture, Silhouette & Rendezvous 3.4L VIN E engine Transmissions: All	**IAT Sensor Circuit Intermittent High Input Conditions:** DTC P0101, P0102, P0103, P0116, P0117, P0118, P0502 and P0503 not set, engine started, engine runtime over 3 minutes, MAF sensor less than 12 g/sec, VSS less than 35 mph, ECT sensor more than 140°F, and the PCM detected an intermittent high voltage condition (Scan Tool reads -35°F) on the IAT sensor signal circuit for 3-5 seconds. The IAT sensor is a variable resistor that includes an IAT signal circuit and a low reference circuit to measure the temperature of the air entering the engine. The PCM connects to the IAT sensor with a 5v signal and a low reference ground circuit. When the IAT sensor is cold, its resistance is high. As air temperature increases, its resistance decreases. With high sensor resistance, the IAT sensor signal voltage is high. With lower sensor resistance, the IAT sensor signal voltage should be lower. **Possible Causes** ● IAT sensor signal circuit is open (intermittent fault) ● IAT sensor ground circuit is open (intermittent fault) ● IAT sensor is damaged (an intermittent "open" condition) ● PCM has failed

OBD II Trouble Code List (P1xxx Codes)

DTC	Trouble Code Title, Conditions & Possible Causes
DTC P1111 **1T CCM, MIL: No** 1996, 1997 Camaro, Caprice, Corvette, Firebird, Fleetwood & Roadmaster 4.3L VIN W, 5.7L VIN 5, 5.7L VIN P engines Transmissions: All	**IAT Sensor Circuit Intermittent High Input Conditions:** DTC P0100, P0102, P0103, P0117, P0118, P0500, P0502 and P0503 not set, engine started, engine runtime over 100 seconds, vehicle speed less than 7 mph, ECT sensor more than 32°F, MAF sensor less than 15 g/sec, and the PCM detected an unexpected voltage condition (Scan Tool reads -31°F) on the IAT sensor circuit for 1 second during the test. **Possible Causes** • IAT sensor signal circuit is open (intermittent fault) • IAT sensor ground circuit is open (intermittent fault) • IAT sensor is damaged (an intermittent "open" condition) • PCM has failed
DTC P1111 **1T CCM, MIL: No** 1996, 1997, 1998, 1999, 2000, 2001, 2002, 2003, 2004, 2005 Cab & Chassis, C/K Truck, G Van, L/M Van, S/T Blazer & S/T Pickup 4.3L VIN W, 4.3L VIN X, 5.0L VIN M, 5.7L VIN K, 5.7L VIN R, 7.4L VIN J engines Transmissions: All	**IAT Sensor Circuit Intermittent High Input Conditions:** DTC P0101, P0102, P0103, P0116, P0117, P0118, P0125, P0128, P0502, P0503, P1114 and P1115 not set, engine started, engine runtime over 120 seconds, ECT sensor more than 140°F, VSS less than 7 mph, MAF input less than 15 g/sec, and the PCM detected an intermittent high voltage condition (over 4.90v) on the IAT sensor signal circuit for 1 second during the CCM test. **Possible Causes** • IAT sensor signal circuit is open (intermittent fault) • IAT sensor ground circuit is open (intermittent fault) • IAT sensor is damaged (an intermittent "open" condition) • PCM has failed
DTC P1111 **1T CCM, MIL: No** 1999, 2000, 2001, 2002, 2003, 2004, 2005 C/K Truck, Envoy, G Van, Escalade & TrailBlazer 4.8L VIN V, 5.3L VIN P, 5.3L VIN T, 5.3L VIN Z, 6.0L VIN N, 6.0L VIN U, 8.1L VIN G engines Transmissions: All	**IAT Sensor Circuit Intermittent High Input Conditions:** DTC P0101, P0102, P0103 and P0113 not set, engine runtime over 120 seconds, ECT sensor more than 140°F, VSS less than 7 mph, MAF input less than 15 g/sec, and the PCM detected an intermittent high voltage condition (Scan Tool reads below -36°F) on the IAT sensor signal circuit during the CCM test period. **Possible Causes** • IAT sensor signal circuit is open (intermittent fault) • IAT sensor ground circuit is open (intermittent fault) • IAT sensor is damaged (an intermittent "open" condition) • PCM has failed
DTC P1111 **1T CCM, MIL: No** 1996, 1997, 1998, 1999, 2000, 2001, 2002, 2003, 2004, 2005 Aurora, DeVille, Eldorado & Seville 4.0L VIN C, 4.6L VIN 9 & 4.6L VIN Y engines Transmissions: All ; Aurora, Deville, Eldorado, Seville	**IAT Sensor Circuit Intermittent High Input Conditions:** DTC P0101, P0102, P0103, P0116, P0117, P0118, P0125, P0128, P0502, P0503, P1114 and P1115 not set, engine started, engine runtime 5 seconds, ECT sensor more than 140°F, MAF sensor less than 12-60 g/sec, VSS less than 50 mph, and the PCM detected an intermittent high voltage condition (Scan Tool reads -30°F) on the IAT sensor signal circuit for 1 second during the CCM test. **Possible Causes** • IAT sensor signal circuit is open (intermittent fault) • IAT sensor ground circuit is open (intermittent fault) • IAT sensor is damaged (an intermittent "open" condition) • PCM has failed
DTC P1111 **1T CCM, MIL: No** 1995, 1996, 1997, 1998, 1999, 2000 Camaro & Firebird 3.8L VIN K engine Transmissions: All	**IAT Sensor Circuit Intermittent High Input Conditions:** DTC P0101-P0103, P0116-P0118, P0125, P0128, P0502, P0503, P1114 and P1115 not set, engine started, engine runtime over 3 minutes, VSS under 35 mph, MAF sensor less than 12 g/sec, ECT sensor more than 140°F, and the PCM detected an intermittent high voltage (Scan Tool reads -29°F) on the IAT sensor signal circuit for 5 seconds during the CCM test. **Possible Causes** • IAT sensor signal circuit is open (intermittent fault) • IAT sensor ground circuit is open (intermittent fault) • IAT sensor is damaged (an intermittent "open" condition) • PCM has failed
DTC P1111 **1T CCM, MIL: No** 1997, 1998, 1999, 2000, 2001, 2002, 2003, 2004, 2005 Camaro, Corvette & Firebird 5.7L VIN G, 5.7L VIN S Transmissions: All	**IAT Sensor Circuit Intermittent High Input Conditions:** DTC P0101-P0103, P0116-P0118, P0125, P0128, P0502, P0503, P1114 and P1115 not set, engine started, engine runtime over 3 minutes, VSS under 35 mph, MAF sensor less than 15 g/sec, ECT sensor more than 140°F, and the PCM detected an intermittent high voltage condition (Scan Tool reads -38°F) on the IAT sensor signal circuit for 5 seconds during the CCM test. **Possible Causes** • IAT sensor signal circuit is open (intermittent fault) • IAT sensor ground circuit is open (intermittent fault) • IAT sensor is damaged (an intermittent "open" condition) • PCM has failed

OBD II Trouble Code List (P1xxx Codes)

DTC	Trouble Code Title, Conditions & Possible Causes
DTC P1111 **1T CCM, MIL: No** 1995 S/T Blazer & S/T Pickup 4.3L VIN W engine Transmissions: All	**IAT Sensor Circuit Intermittent High Input Conditions:** DTC P0117, P0118, P0502 and P0503 not set, engine started, engine runtime over 2 minutes, ECT sensor more than 185ºF, VSS less than 1 mph, and the PCM detected an intermittent high voltage condition (over 4.90v) on the IAT sensor signal circuit for 1 second. **Possible Causes** ● IAT sensor signal circuit is open (intermittent fault) ● IAT sensor ground circuit is open (intermittent fault) ● IAT sensor is damaged (an intermittent "open" condition) ● PCM has failed
DTC P1111 **1T CCM, MIL: No** 1996, 1997, 1998, 1999, 2000, 2001, 2002, 2003, 2004, 2005 Aurora, Bonneville, Century, Cutlass Supreme, Grand Prix, Intrigue, LeSabre, LSS, Lumina, Monte Carlo, 88', Regal, Park Avenue & 98' 3.1L VIN J, 3.1L VIN M, 3.4L VIN E, 3.5L VIN H, 3.8L VIN 1, 3.8L VIN K Transmissions: All	**IAT Sensor Circuit Intermittent Low Input Conditions:** DTC P0101, P0102, P0103, P0116-P0118, P0502 and P0503 not set, engine started, engine runtime over 3 minutes, vehicle driven at a speed of more than 25 mph, ECT sensor more than 140ºF, MAF sensor less than 12 g/sec, and the PCM detected an intermittent low voltage condition (Scan Tool reads more than 253ºF) on the IAT sensor signal circuit for 3 seconds. The IAT sensor is a variable resistor that includes an IAT signal circuit and a low reference circuit to measure the temperature of the air entering the engine. The PCM connects to the IAT sensor with a 5v signal and a low reference ground circuit. When the IAT sensor is cold, its resistance is high. As air temperature increases, its resistance decreases. With high sensor resistance, the IAT sensor signal voltage is high. With lower sensor resistance, the IAT sensor signal voltage should be lower. **Possible Causes** ● IAT sensor signal circuit is shorted to ground (intermittent fault) ● IAT sensor is damaged (an intermittent "shorted" condition) ● PCM has failed ● TSB 02-06-03-005 contains a repair procedure for this code
DTC P1112 **1T CCM, MIL: No** 1996, 1997 Camaro, Caprice, Corvette, Firebird, Fleetwood & Roadmaster 4.3L VIN W, 5.7L VIN 5, 5.7L VIN P engines Transmissions: All	**IAT Sensor Circuit Intermittent Low Input Conditions:** DTC P0100, P0102, P0103, P0117, P0118, P0500, P0502 and P0503 not set, engine runtime over 100 seconds, vehicle speed less than 7 mph, ECT sensor more than 32ºF, MAF sensor less than 15 g/sec, and the PCM detected an intermittent low voltage condition (Scan Tool reads 282ºF) on the IAT sensor circuit for 1 second. **Possible Causes** ● IAT sensor signal circuit is shorted to ground (intermittent fault) ● IAT sensor is damaged (an intermittent "shorted" condition) ● PCM has failed
DTC P1112 **1T CCM, MIL: No** 1996, 1997, 1998, 1999, 2000, 2001, 2002, 2003, 2004, 2005 Aztek, Montana, Venture, Silhouette & Rendezvous 3.4L VIN E engine Transmissions: All	**IAT Sensor Circuit Intermittent Low Input Conditions:** DTC P0101, P0102, P0103, P0116, P0117, P0118, P0502 and P0503 not set, engine runtime over 10 seconds, MAF sensor less than 12 g/sec, VSS less than 25 mph, ECT sensor more than 140ºF, and the PCM detected an intermittent low voltage condition (Scan Tool reads 253ºF) on the IAT sensor signal circuit for 20 seconds. **Possible Causes** ● IAT sensor signal circuit is shorted to ground (intermittent fault) ● IAT sensor is damaged (an intermittent "shorted" condition) ● PCM has failed
DTC P1112 **1T CCM, MIL: No** 1996, 1997, 1998, 1999, 2000, 2001, 2002, 2003, 2004, 2005 Cab & Chassis, C/K Truck, G Van, L/M Van, S/T Blazer & S/T Pickup 4.3L VIN W engine 4.3L VIN X, 5.0L VIN M, 5.7L VIN K, 5.7L VIN R, 7.4L VIN J engines Transmissions: All	**IAT Sensor Circuit Intermittent Low Input Conditions:** DTC P0101, P0102, P0103, P0116, P0117, P0118, P0125, P0128, P0502, P0503, P1114 and P1115 not set, engine started, engine runtime over 120 seconds, ECT sensor more than 140ºF, VSS less than 7 mph, MAF input less than 15 g/sec, and the PCM detected an intermittent low voltage condition (Scan Tool reads 282ºF) on the IAT sensor signal circuit for 6 seconds during the CCM test. **Possible Causes** ● IAT sensor signal circuit is shorted to ground (intermittent fault) ● IAT sensor is damaged (an intermittent "shorted" condition) ● PCM has failed
DTC P1112 **1T CCM, MIL: No** 1999, 2000, 2001, 2002, 2003, 2004, 2005 C/K Truck, Envoy, G Van, Escalade & TrailBlazer 4.8L VIN V, 5.3L VIN P, 5.3L VIN T, 5.3L VIN Z, 6.0L VIN N, 6.0L VIN U, 8.1L VIN G engines Transmissions: All	**IAT Sensor Circuit Intermittent Low Input Conditions:** DTC P0112, P0500, P0502 and P0503 not set, engine runtime over 45 seconds, ECT sensor less than 257ºF, VSS more than 25 mph, and the PCM detected an intermittent low voltage condition (Scan Tool reads over 262ºF) on the IAT sensor signal circuit. **Possible Causes** ● IAT sensor signal circuit is open (intermittent fault) ● IAT sensor ground circuit is open (intermittent fault) ● IAT sensor is damaged (an intermittent "open" condition) ● PCM has failed

OBD II Trouble Code List (P1xxx Codes)

DTC	Trouble Code Title, Conditions & Possible Causes
DTC P1112 **1T CCM, MIL: No** 1996, 1997, 1998, 1999, 2000, 2001, 2002, 2003, 2004, 2005 Aurora, DeVille, Eldorado & Seville 4.0L VIN C, 4.6L VIN 9 & 4.6L VIN Y engines Transmissions: All	**IAT Sensor Circuit Intermittent Low Input Conditions:** DTC P0101, P0102, P0103, P0116, P0117, P0118, P0125, P0128, P0502, P0503, P1114 and P1115 not set, engine runtime 5 seconds, ECT sensor more than 140ºF, MAF sensor less than 12-60 g/sec, VSS less than 50 mph, and the PCM detected an intermittent low voltage condition (Scan Tool reads 275ºF) on the IAT sensor signal circuit for 1 second. **Possible Causes** ● IAT sensor signal circuit is shorted to ground (intermittent fault) ● IAT sensor is damaged (an intermittent "shorted" condition) ● PCM has failed
DTC P1112 **1T CCM, MIL: No** 1995, 1996, 1997, 1998, 1999, 2000 Camaro & Firebird 3.8L VIN K engine Transmissions: All	**IAT Sensor Circuit Intermittent Low Input Conditions:** DTC P0101-P0103, P0116-P0118, P0125, P0128, P0502, P0503, P1114 and P1115 not set, engine runtime over 3 minutes, VSS under 35 mph, MAF sensor less than 12 g/sec, ECT sensor more than 140ºF, and the PCM detected an intermittent low voltage condition (Scan Tool reads 274ºF) on the IAT sensor signal circuit for 5 seconds during the CCM test. **Possible Causes** ● IAT sensor signal circuit is shorted to ground (intermittent fault) ● IAT sensor is damaged (an intermittent "shorted" condition) ● PCM has failed
DTC P1112 **1T CCM, MIL: No** 1997, 1998, 1999, 2000, 2001, 2002, 2003, 2004, 2005 Camaro, Corvette & Firebird 5.7L VIN G, 5.7L VIN S Transmissions: All	**IAT Sensor Circuit Intermittent Low Input Conditions:** DTC P0101-P0103, P0116-P0118, P0125, P0128, P0502, P0503, P1114 and P1115 not set, engine runtime over 3 minutes, VSS under 35 mph, MAF sensor less than 15 g/sec, ECT sensor more than 140ºF, and the PCM detected an intermittent low voltage condition (Scan Tool reads 282ºF) on the IAT sensor signal circuit for 6 seconds during the CCM test. **Possible Causes** ● IAT sensor signal circuit is shorted to ground (intermittent fault) ● IAT sensor is damaged (an intermittent "shorted" condition) ● PCM has failed
DTC P1112 **1T CCM, MIL: No** 1995 S/T Blazer & S/T Pickup 4.3L VIN W engine Transmissions: All	**IAT Sensor Circuit Intermittent Low Input Conditions:** DTC P0117, P0118, P0502 and P0503 not set, engine runtime over 2 minutes, ECT sensor more than 185ºF, VSS less than 1 mph, and the PCM detected an intermittent low voltage on (Scan Tool reads 282ºF) on the IAT sensor signal circuit for 1 second during the CCM test. **Possible Causes** ● IAT sensor signal circuit is shorted to ground (intermittent fault) ● IAT sensor is damaged (an intermittent "shorted" condition) ● PCM has failed
DTC P1112 **1T CCM, MIL: No** 1999, 2000, 2001 Catera 3.0L VIN R engine Transmissions: A/T	**Intake Plenum Switchover Solenoid Control Circuit Malfunction Conditions:** Engine started; system voltage over 10.0v and the PCM detected an unexpected voltage condition on the Intake Plenum Switchover solenoid control circuit during the CCM test. **Possible Causes** ● Intake plenum switchover solenoid control circuit is open, shorted to ground or to B+ ● Intake plenum switchover solenoid power circuit is open (check for power circuit from the power relay on Circuit 128) ● Intake plenum switchover solenoid is damaged or has failed ● PCM has failed
DTC P1113 **1T CCM, MIL: No** 1999, 2000, 2001 Catera 3.0L VIN R engine Transmissions: A/T	**Intake Resonance Switchover Valve Circuit Malfunction Conditions:** Engine started; system voltage over 10.0v and the PCM detected an unexpected voltage condition on the Intake Resonance Switchover Valve control circuit during the CCM test. **Possible Causes** ● Intake resonance switchover solenoid control circuit is open, shorted to ground or to B+ ● Intake resonance switchover solenoid power circuit is open (check for power circuit from the power relay on Circuit 128) ● Intake resonance switchover solenoid is damaged or has failed ● PCM has failed
DTC P1114 **1T CCM, MIL: No** 1996, 1997, 1998, 1999, 2000, 2001, 2002, 2003, 2004, 2005 Aurora, Bonneville, Century, Cutlass Supreme, Grand Prix, Intrigue, LeSabre, LSS, Lumina, Monte Carlo, 88', Regal, Park Avenue & 98' 3.1L VIN J, 3.1L VIN M, 3.4L VIN E, 3.5L VIN H, 3.8L VIN 1, 3.8L VIN K Transmissions: All	**ECT Sensor Circuit Intermittent Low Input Conditions:** Engine runtime over 5 seconds; system voltage over 10.0v, and the PCM detected an intermittent low voltage (Scan Tool reads more than 282ºF) on the ECT sensor signal circuit for 5 seconds. The ECT sensor is a variable resistor that connects to the PCM with an ECT signal and a low reference circuit to measure the temperature of the engine coolant. When the ECT sensor is cold, its resistance is high. As the temperature of the engine coolant increases, its resistance decreases. With high sensor resistance, the ECT sensor signal voltage is high. With lower sensor resistance, the ECT sensor signal voltage should be a lower voltage. **Possible Causes** ● ECT sensor signal circuit shorted to ground (intermittent fault) ● ECT sensor has failed (possible intermittent shorted condition) ● PCM has failed

OBD II Trouble Code List (P1xxx Codes)

DTC	Trouble Code Title, Conditions & Possible Causes
DTC P1114 **1T CCM, MIL: No** 1996, 1997 Camaro, Caprice, Corvette, Firebird, Fleetwood & Roadmaster 4.3L VIN W, 5.7L VIN 5, 5.7L VIN P engines Transmissions: All	**ECT Sensor Circuit Intermittent Low Input Conditions:** Engine started, engine runtime over 20 seconds, system voltage over 10.0v, and the PCM detected an intermittent low voltage condition (Scan Tool reads over 304°F) on the ECT sensor signal circuit for 5-10 seconds during the CCM test. **Possible Causes** • ECT sensor signal circuit shorted to ground (intermittent fault) • ECT sensor has failed (possible intermittent shorted condition) • PCM has failed
DTC P1114 **1T CCM, MIL: No** 1996, 1997, 1998, 1999, 2000, 2001, 2002, 2003, 2004, 2005 Aztek, Montana, Venture, Silhouette & Rendezvous 3.4L VIN E engine Transmissions: All	**ECT Sensor Circuit Intermittent Low Input Conditions:** Engine started, engine runtime over 5 seconds, and the PCM detected an intermittent low voltage condition (Scan Tool reads over 282°F) on the ECT sensor signal circuit for 5 seconds. The ECT sensor is a variable resistor that includes an ECT signal and a low reference circuit to measure the temperature of the engine coolant. **Possible Causes** • ECT sensor signal circuit shorted to ground (intermittent fault) • ECT sensor has failed (possible intermittent shorted condition) • PCM has failed
DTC P1114 **1T CCM, MIL: No** 1996, 1997, 1998, 1999, 2000, 2001, 2002, 2003, 2004, 2005 Cab & Chassis, C/K Truck, G Van, L/M Van, S/T Blazer & S/T Pickup 4.3L VIN W, 4.3L VIN X, 5.0L VIN M, 5.7L VIN K, 5.7L VIN R, 7.4L VIN J engines Transmissions: All	**ECT Sensor Circuit Intermittent Low Input Conditions:** Engine started, engine runtime over 10 seconds, system voltage over 10.0v, and the PCM detected an intermittent low voltage condition (Scan Tool reads over 282°F) on the ECT sensor signal circuit for 1 second out of a 20 second period during the CCM test. **Possible Causes** • ECT sensor signal circuit shorted to ground (intermittent fault) • ECT sensor has failed (possible intermittent shorted condition) • PCM has failed
DTC P1114 **1T CCM, MIL: No** 1999, 2000, 2001, 2002, 2003, 2004, 2005 C/K Truck, Envoy, G Van, Escalade & TrailBlazer 4.8L VIN V, 5.3L VIN P, 5.3L VIN T, 5.3L VIN Z, 6.0L VIN N, 6.0L VIN U, 8.1L VIN G engines Transmissions: All	**ECT Sensor Circuit Intermittent Low Input Conditions:** Engine started, engine runtime over 10 seconds, and the PCM detected an intermittent low voltage condition (Scan Tool reads over 280°F) on the ECT sensor signal circuit during the CCM test period. **Possible Causes** • ECT sensor signal circuit shorted to ground (intermittent fault) • ECT sensor has failed (possible intermittent shorted condition) • PCM has failed
DTC P1114 **1T CCM, MIL: No** 1996, 1997, 1998, 1999, 2000, 2001, 2002, 2003, 2004, 2005 Aurora, DeVille, Eldorado & Seville 4.0L VIN C, 4.6L VIN 9 & 4.6L VIN Y engines Transmissions: All	**ECT Sensor Circuit Intermittent Low Input Conditions:** DTC P0112, P0113, P1111 and P1112 not set, engine started, IAT sensor less than 158°F, engine runtime over 10 seconds, and the PCM detected an intermittent low voltage condition on the ECT sensor signal circuit (Scan Tool reads over 300°F) for 10 seconds. **Possible Causes** • ECT sensor signal circuit shorted to ground (intermittent fault) • ECT sensor has failed (possible intermittent shorted condition) • PCM has failed
DTC P1114 **1T CCM, MIL: No** 1997, 1998, 1999, 2000, 2001, 2002, 2003, 2004, 2005 Camaro, Corvette & Firebird 5.7L VIN G, 5.7L VIN S Transmissions: All	**ECT Sensor Circuit Intermittent Low Input Conditions:** Engine started, engine runtime over 10 seconds, system voltage over 10.0v, and the PCM detected an intermittent "low" voltage condition (Scan Tool reads over 282°F) for 1 second out of a 20 second period during the CCM test. **Possible Causes** • ECT sensor signal circuit shorted to ground (intermittent fault) • ECT sensor has failed (possible intermittent shorted condition) • PCM has failed
DTC P1114 **1T CCM, MIL: No** 1995, 1996, 1997, 1998, 1999, 2000, 2001, 2002 Camaro & Firebird 3.8L VIN K engine Transmissions: All	**ECT Sensor Circuit Intermittent Low Input Conditions:** Engine started, engine runtime over 5 seconds, and the PCM detected an intermittent low voltage condition on the ECT sensor signal circuit (Scan Tool reads over 284°F) for 5 seconds. **Possible Causes** • ECT sensor signal circuit shorted to ground (intermittent fault) • ECT sensor has failed (possible intermittent shorted condition) • PCM has failed

GM REFERENCE INFORMATION

OBD II Trouble Code List (P1xxx Codes)

DTC	Trouble Code Title, Conditions & Possible Causes
DTC P1114 **1T CCM, MIL: No** 1995 S/T Blazer & S/T Pickup 4.3L VIN W engine Transmissions: All	**ECT Sensor Circuit Intermittent Low Input Conditions:** Engine started; and the PCM detected an intermittent low voltage condition (Scan Tool reads over 282ºF) on the ECT sensor signal circuit for 5 seconds during the CCM test. **Possible Causes** ● ECT sensor signal circuit shorted to ground (intermittent fault) ● ECT sensor has failed (possible intermittent shorted condition) ● PCM has failed
DTC P1114 **1T CCM, MIL: No** 1996, 1997 Grand Prix, Lumina & Monte Carlo Engines: 3.4L VIN X Transmissions: All	**ECT Sensor Circuit Intermittent Low Input Conditions:** Engine started, engine runtime over 15 seconds, system voltage over 10.0v, and the PCM detected an intermittent low voltage condition (Scan Tool reads over 284ºF) on the ECT sensor signal circuit for 10 seconds during the CCM test. **Possible Causes** ● ECT sensor signal circuit shorted to ground (intermittent fault) ● ECT sensor has failed (possible intermittent shorted condition) ● PCM has failed
DTC P1114 **1T CCM, MIL: No** 1995 Corvette 5.7L VIN P engine Transmissions: All	**ECT Sensor Circuit Intermittent Low Input Conditions:** Engine started; and the PCM detected an intermittent low voltage condition (Scan Tool reads over 282ºF) on the ECT sensor signal circuit for 2 seconds during the CCM test. **Possible Causes** ● ECT sensor signal circuit shorted to ground (intermittent fault) ● ECT sensor has failed (possible intermittent shorted condition) ● PCM has failed
DTC P1115 **1T CCM, MIL: No** 1996, 1997, 1998, 1999, 2000, 2001, 2002, 2003, 2004, 2005 Aurora, Bonneville, Century, Cutlass Supreme, Grand Prix, Intrigue, LeSabre, LSS, Lumina, Monte Carlo, 88', Regal, Park Avenue & 98' 3.1L VIN J, 3.1L VIN M, 3.4L VIN E, 3.5L VIN H, 3.8L VIN 1, 3.8L VIN K Transmissions: All	**ECT Sensor Circuit Intermittent High Input Conditions:** Engine runtime over 5 seconds; system voltage over 10.0v and the PCM detected an intermittent high voltage (Scan Tool reads less than -36ºF) on the ECT sensor signal circuit for 5 seconds. The ECT sensor is a variable resistor that includes an ECT signal circuit and a low reference circuit to measure the temperature of the engine coolant. The PCM supplies the ECT sensor with a 5v signal circuit and a low reference ground circuit. When the ECT sensor is cold, its resistance is high. As the engine coolant temperature increases, its resistance decreases. With high sensor resistance, the ECT sensor signal voltage is high. With lower sensor resistance, the ECT sensor signal voltage should be lower. **Possible Causes** ● ECT sensor signal circuit is open between the sensor and PCM ● ECT sensor ground circuit is open between sensor and PCM ● ECT sensor has failed (possible intermittent "open" condition) ● PCM has failed
DTC P1115 **1T CCM, MIL: No** 1996, 1997 Camaro, Caprice, Corvette, Firebird, Fleetwood & Roadmaster 4.3L VIN W, 5.7L VIN 5, 5.7L VIN P engines Transmissions: All	**ECT Sensor Circuit Intermittent High Input Conditions:** Engine started; system voltage over 10.0v and the PCM detected an intermittent high voltage condition (Scan Tool reads below -40ºF) on the ECT sensor signal circuit for 3 seconds. **Possible Causes** ● ECT sensor signal circuit is open between the sensor and PCM ● ECT sensor ground circuit is open between sensor and PCM ● ECT sensor has failed (possible intermittent "open" condition) ● PCM has failed
DTC P1115 **1T CCM, MIL: No** 1996, 1997, 1998, 1999, 2000, 2001, 2002, 2003, 2004, 2005 Aztek, Montana, Venture, Silhouette & Rendezvous 3.4L VIN E engine Transmissions: All	**ECT Sensor Circuit Intermittent High Input Conditions:** Engine runtime over 5 seconds and the PCM detected an intermittent high voltage condition on the ECT sensor signal circuit (Scan Tool reads below -36ºF) for 5 seconds. When the ECT sensor is cold, its resistance is high. As the engine coolant temperature increases, its resistance decreases. With high sensor resistance, the ECT sensor signal voltage is high. With lower sensor resistance, the ECT sensor signal voltage should be lower. **Possible Causes** ● ECT sensor signal circuit is open between the sensor and PCM ● ECT sensor ground circuit is open between sensor and PCM ● ECT sensor has failed (possible intermittent "open" condition) ● PCM has failed
DTC P1115 **1T CCM, MIL: No** 1996, 1997, 1998, 1999, 2000, 2001, 2002, 2003, 2004, 2005 Cab & Chassis, C/K Truck, G Van, L/M Van, S/T Blazer & S/T Pickup 4.3L VIN W, 4.3L VIN X, 5.0L VIN M, 5.7L VIN K, 5.7L VIN R, 7.4L VIN J Transmissions: All	**ECT Sensor Circuit Intermittent High Input Conditions:** Engine runtime over 10 seconds, system voltage over 10.0v, and the PCM detected an intermittent high voltage condition (Scan Tool reads below -31ºF) on the ECT sensor signal circuit for 1 second out of a 20 second period during the CCM test. **Possible Causes** ● ECT sensor signal circuit is open between the sensor and PCM ● ECT sensor ground circuit is open between sensor and PCM ● ECT sensor has failed (possible intermittent "open" condition) ● PCM has failed

OBD II Trouble Code List (P1xxx Codes)

DTC	Trouble Code Title, Conditions & Possible Causes
DTC P1115 **1T CCM, MIL: No** 1999, 2000, 2001, 2002, 2003, 2004, 2005 C/K Truck, Envoy, G Van, Escalade & TrailBlazer 4.8L VIN V, 5.3L VIN P, 5.3L VIN T, 5.3L VIN Z, 6.0L VIN N, 6.0L VIN U, 8.1L VIN G engines Transmissions: All	**ECT Sensor Circuit Intermittent High Input Conditions:** Engine runtime over 10 seconds and the PCM detected an intermittent low voltage condition (Scan Tool reads under -36ºF) on the ECT sensor signal circuit during the test period. **Possible Causes** ● ECT sensor signal circuit is open between the sensor and PCM ● ECT sensor ground circuit is open between sensor and PCM ● ECT sensor has failed (possible intermittent "open" condition) ● PCM has failed
DTC P1115 **1T CCM, MIL: No** 1996, 1997, 1998, 1999, 2000, 2001, 2002, 2003, 2004, 2005 Aurora, DeVille, Eldorado & Seville 4.0L VIN C, 4.6L VIN 9 & 4.6L VIN Y engines Transmissions: All	**ECT Sensor Circuit Intermittent High Input Conditions:** DTC P0112, P0113, P1111 and P1112 not set, engine started, IAT sensor more than 19ºF, engine runtime over 60 seconds, and the PCM detected an intermittent high voltage condition on the ECT sensor signal circuit (Scan Tool reads below -36ºF) for 10 seconds. **Possible Causes** ● ECT sensor signal circuit is open between the sensor and PCM ● ECT sensor ground circuit is open between sensor and PCM ● ECT sensor has failed (possible intermittent "open" condition) ● PCM has failed
DTC P1115 **1T CCM, MIL: No** 1997, 1998, 1999, 2000, 2001, 2002, 2003, 2004, 2005 Camaro, Corvette & Firebird 5.7L VIN G, 5.7L VIN S Transmissions: All	**ECT Sensor Circuit Intermittent High Input Conditions:** Engine started, engine runtime over 60 seconds, system voltage over 10.0v, and the PCM detected an intermittent "high" voltage condition (Scan Tool reads below -31ºF) for 1 second out of a 20 second period during the CCM test. **Possible Causes** ● ECT sensor signal circuit is open between the sensor and PCM ● ECT sensor ground circuit is open between sensor and PCM ● ECT sensor has failed (possible intermittent "open" condition) ● PCM has failed
DTC P1115 **1T CCM, MIL: No** 1995, 1996, 1997, 1998, 1999, 2000, 2001, 2002 Camaro & Firebird 3.8L VIN K engine Transmissions: All	**ECT Sensor Circuit Intermittent High Input Conditions:** Engine runtime over 5 seconds and the PCM detected an intermittent high voltage condition on the ECT sensor signal circuit (Scan Tool reads less than -31ºF) for 3 seconds. When the ECT sensor is cold, its resistance is high. As the engine coolant temperature increases, its resistance decreases. With high sensor resistance, the ECT sensor signal voltage is high. With lower sensor resistance, the ECT sensor signal voltage should be lower. **Possible Causes** ● ECT sensor signal circuit is open between the sensor and PCM ● ECT sensor ground circuit is open between sensor and PCM ● ECT sensor has failed (possible intermittent "open" condition) ● PCM has failed
DTC P1115 **1T CCM, MIL: No** 1995 S/T Blazer & S/T Pickup 4.3L VIN W engine Transmissions: All	**ECT Sensor Circuit Intermittent High Input Conditions:** Engine started, and the PCM detected an intermittent high voltage condition (Scan Tool reads below -31ºF) on the ECT sensor signal circuit for 5 seconds. When the ECT sensor is cold, its resistance is high. With high sensor resistance, the ECT sensor voltage is high. With lower sensor resistance, the signal voltage should be lower. **Possible Causes** ● ECT sensor signal circuit is open between the sensor and PCM ● ECT sensor ground circuit is open between sensor and PCM ● ECT sensor has failed (possible intermittent "open" condition) ● PCM has failed
DTC P1115 **1T CCM, MIL: No** 1996, 1997 Grand Prix, Lumina & Monte Carlo Engines: 3.4L VIN X Transmissions: All	**ECT Sensor Circuit Intermittent High Input Conditions:** Engine started; system voltage over 10.0v and the PCM detected an intermittent high voltage (Scan Tool reads below -27ºF) on the ECT sensor signal circuit for 10 seconds. When the ECT sensor is cold, its resistance is high. As the engine coolant temperature increases, its resistance decreases. With high sensor resistance, the ECT sensor signal voltage is high. With lower sensor resistance, the ECT sensor signal voltage should be lower. **Possible Causes** ● ECT sensor signal circuit is open between the sensor and PCM ● ECT sensor ground circuit is open between sensor and PCM ● ECT sensor has failed (possible intermittent "open" condition) ● PCM has failed

OBD II Trouble Code List (P1xxx Codes)

DTC	Trouble Code Title, Conditions & Possible Causes
DTC P1115 **1T CCM, MIL: No** 1995 Corvette 5.7L VIN P engine Transmissions: All	**ECT Sensor Circuit Intermittent High Input Conditions:** Engine started; and the PCM detected an intermittent high voltage condition (Scan Tool reads less than -31ºF) on the ECT sensor signal circuit for 5 seconds. When the ECT sensor is cold, its resistance is high. As the engine coolant temperature increases, its resistance decreases. With high sensor resistance, the ECT sensor signal voltage is high. With lower sensor resistance, the ECT sensor signal voltage should be lower. **Possible Causes** ● ECT sensor signal circuit is open between the sensor and PCM ● ECT sensor ground circuit is open between sensor and PCM ● ECT sensor has failed (possible intermittent "open" condition) ● PCM has failed
DTC P1120 **1T CCM, MIL: Yes** 1999, 2000, 2001, 2002 Camaro & Firebird 3.8L VIN K engine Transmissions: All	**TP Sensor 1 Circuit Malfunction Conditions:** DTC P0606, P1517 and P1518 not set, key in crank or run mode, ETC serial data circuit operational, and the PCM detected the TP Sensor 1 signal was less than 0.34v or more than 4.40v. If the TAC module detects a condition within the ETC system, the PCM may receives a serial data message with more than one ETC related code set because many redundant tests run continuously on this system. Be aware that locating and repairing an individual condition may correct more than one code when reviewing "captured" code data. **Possible Causes** ● TP1 sensor signal circuit is open, shorted to ground or to power ● TP1 sensor low reference circuit has high resistance condition ● TP1 sensor is damaged or has failed
DTC P1120 **1T CCM, MIL: Yes** 2002, 2003, 2004, 2005 C/K Truck, G Van, S/T Blazer & S/T Pickup 4.8L VIN V, 5.3L VIN P, 5.3L VIN T, 6.0L VIN N, 6.0L VIN U, 7.4L VIN J, 8.1L VIN G engines Transmissions: All	**TP Sensor 1 Circuit Malfunction Conditions:** DTC P1517 and P1518 not set, key in crank or run position, system voltage over 5.23v, and the PCM detected the TP1 signal was less than 0.13v or more than 4.87v for 1 second during the CCM test. **Possible Causes** ● TP1 sensor signal circuit is open, shorted to ground or to power ● TP1 sensor VREF circuit is open or shorted to ground or to B+ ● TP1 sensor ground circuit has a high resistance condition ● TP1 sensor is damaged or has failed
DTC P1120 **1T CCM, MIL: Yes** 1999, 2000, 2001 Catera 3.0L VIN R engine Transmissions: A/T	**TP Sensor 1 Circuit Malfunction Conditions:** Engine started; system voltage over 10.0v and the PCM detected the TP Sensor 1 signal was less than 0.195v or more than 4.60v, or the difference between the TP1 and TP2 signals was more than 13% for 140 ms during the CCM test. **Possible Causes** ● TP1 sensor signal circuit is open, shorted to ground or to power ● TP1 sensor low reference circuit has high resistance condition ● TP1 sensor is damaged or has failed
DTC P1120 **1T CCM, MIL: Yes** 2002, 2003, 2004, 2005 S/T Blazer & S/T Pickup 4.2L VIN S engine Transmissions: All	**TP Sensor 1 Circuit Malfunction Conditions:** DTC P1635 not set, key in crank or run mode, system voltage over 5.23v, Electronic Throttle Control serial data operational, and the PCM detected the TP sensor (1) input was less than 0.13v or over 4.87v for 1 second during the CCM test. TP Sensors 1 and 2 are located in the throttle body assembly. Each sensor has a 5v VREF, low reference and signal circuit. These circuits provide the PCM with a signal proportional to throttle plate movement. The TP Sensor 1 signal at closed throttle is near the 4.90v and decreases as the throttle plate opens. The TP Sensor 2 signal at closed throttle is near the 0.10v and increases as the throttle plate opens. When the TP sensor 1 voltage is not within the predicted range, this code sets. **Possible Causes** ● TP1 sensor signal circuit is open, shorted to ground or to power ● TP1 sensor signal circuit is shorted to VREF or system power ● TP1 sensor low reference circuit has high resistance condition ● TP sensor VREF circuit is open or shorted to ground ● TP1 sensor is damaged or has failed
DTC P1120 **1T CCM, MIL: Yes** 1997, 1998, 1999, 2000, 2001, 2002, 2003, 2004, 2005 Corvette 5.7L VIN G, 5.7L VIN S Transmissions: All	**TP Sensor 1 Circuit Malfunction Conditions:** DTC P0606, P1517 and P1518 not set, key in crank or run mode, system voltage over 5.23v, Electronic Throttle Control serial data operational, and the PCM detected the TP sensor (1) input was less than 0.13v or over 4.87v for 1 second during the CCM test. **Possible Causes** ● TP1 sensor signal circuit is open, shorted to ground or to power ● TP1 sensor low reference circuit has high resistance condition ● TP1 sensor is damaged or has failed

OBD II Trouble Code List (P1xxx Codes)

DTC	Trouble Code Title, Conditions & Possible Causes
DTC P1121 **1T CCM, MIL: No** 1996, 1997, 1998, 1999, 2000, 2001, 2002, 2003, 2004, 2005 Aurora, Bonneville, Century, Cutlass Supreme, Grand Prix, Intrigue, LeSabre, LSS, Lumina, Monte Carlo, 88', Regal, Park Avenue & 98' 3.1L VIN J, 3.1L VIN M, 3.4L VIN E, 3.5L VIN H, 3.8L VIN 1, 3.8L VIN K Transmissions: All	**TP Sensor Circuit Intermittent High Input Conditions:** Key on or engine running; and the PCM detected an intermittent high voltage condition (more than 4.90v) on the TP sensor signal circuit for 1 second out of a 20 second period. The TP sensor signal is used by the PCM to determine the throttle plate angle for various engine management systems. Rotation of the TP sensor rotor from the closed throttle position to the WOT position provides the PCM with a signal voltage from less than 1.0v too more than 4.0v through the TP sensor signal circuit. If the PCM detects an intermittent excessively high voltage signal, it will set this code. **Possible Causes** ● TP sensor connector is damaged, loose or open ● TP sensor signal circuit is shorted to VREF (intermittent fault) ● TP sensor ground circuit is open (an intermittent fault) ● TP sensor is damaged or has failed (possible intermittent open) ● PCM has failed
DTC P1121 **1T CCM, MIL: No** 1996, 1997 Camaro, Caprice, Corvette, Firebird, Fleetwood & Roadmaster 4.3L VIN W, 5.7L VIN 5, 5.7L VIN P engines Transmissions: All	**TP Sensor Circuit Intermittent High Input Conditions:** Engine started, engine runtime over 4 seconds, system voltage over 10.0v, and the PCM detected an intermittent high voltage condition (more than 4.90v) on the TP sensor signal circuit for 3 seconds. **Possible Causes** ● TP sensor signal circuit is shorted to VREF (intermittent fault) ● TP sensor ground circuit is open (an intermittent fault) ● TP sensor is damaged or has failed (possible intermittent open) ● PCM has failed
DTC P1121 **1T CCM, MIL: No** 1996, 1997, 1998, 1999, 2000, 2001, 2002, 2003, 2004, 2005 Aztek, Montana, Venture, Silhouette & Rendezvous 3.4L VIN E engine Transmissions: All	**TP Sensor Circuit Intermittent High Input Conditions:** Key on or engine running; and the PCM detected an intermittent high voltage condition on the TP sensor signal circuit (more than 4.90v) for 1 second during the CCM test. **Possible Causes** ● TP sensor connector is damaged, loose or open ● TP sensor signal circuit is shorted to VREF (intermittent fault) ● TP sensor ground circuit is open (an intermittent fault) ● TP sensor is damaged or has failed (possible intermittent open) ● PCM has failed
DTC P1121 **1T CCM, MIL: No** 1996, 1997, 1998, 1999, 2000, 2001, 2002, 2003, 2004, 2005 Cab & Chassis, C/K Truck, G Van, L/M Van, S/T Blazer & S/T Pickup 4.3L VIN W, 4.3L VIN X, 5.0L VIN M, 5.7L VIN K, 5.7L VIN R, 6.0L VIN N, 7.4L VIN J, 8.1L VIN G Transmissions: All	**TP Sensor Circuit High Input (Intermittent) Conditions:** DTC P1639 and P1639 not set, engine started, engine runtime over 5 seconds, system voltage over 10.0v, and the PCM detected an intermittent high voltage condition (more than 4.70-4.90v) on the TP sensor signal circuit for 1 second during the CCM test. **Possible Causes** ● TP sensor signal circuit is shorted to VREF (intermittent fault) ● TP sensor ground circuit is open (an intermittent fault) ● TP sensor is damaged or has failed (possible intermittent open) ● PCM has failed
DTC P1121 **1T CCM, MIL: No** 1996, 1997, 1998, 1999, 2000, 2001, 2002, 2003, 2004, 2005 Aurora, DeVille, Eldorado & Seville 4.0L VIN C, 4.6L VIN 9 & 4.6L VIN Y engines Transmissions: All	**TP Sensor Circuit High Input (Intermittent) Conditions:** Key on or engine running; and the PCM detected an intermittent high voltage condition on the TP sensor signal circuit (more than 4.80v) for 1 second. The TP sensor signal is used by the PCM to determine the throttle plate angle to manage the engine controls. Rotation of the TP sensor rotor from the closed throttle position to the wide open throttle (WOT) position provides the PCM with a signal of less than 1.0v to over 4.0v via the TP sensor signal circuit. If the PCM detects an intermittent excessively high voltage signal, it will set this code. **Possible Causes** ● TP sensor signal circuit is shorted to VREF (intermittent fault) ● TP sensor ground circuit is open (an intermittent fault) ● TP sensor is damaged or has failed (possible intermittent open) ● PCM has failed
DTC P1121 **1T CCM, MIL: No** 1995, 1996, 1997, 1998 Camaro, Firebird, Grand Prix, Lumina, Monte Carlo, Regal 3.4L VIN X, 3.8L VIN K Transmissions: All	**TP Sensor Circuit High Input (Intermittent) Conditions:** Engine started, engine runtime over 3 seconds, and the PCM detected an intermittent high voltage condition (more than 4.90v) on the TP sensor signal circuit for more than one second. **Possible Causes** ● TP sensor signal circuit is shorted to VREF (intermittent fault) ● TP sensor ground circuit is open (an intermittent fault) ● TP sensor is damaged or has failed (possible intermittent open) ● PCM has failed

OBD II Trouble Code List (P1xxx Codes)

DTC	Trouble Code Title, Conditions & Possible Causes
DTC P1121 **1T CCM, MIL: No** 1999, 2000, 2001, 2002 Camaro & Firebird Engines: 5.7L VIN G Transmissions: All	**TP Sensor Circuit Intermittent High Input Conditions:** Key on or engine running; and the PCM detected an intermittent high voltage condition on the TP sensor signal circuit (more than 4.80v) for a 10 second period during the CCM test. **Possible Causes** • TP sensor signal circuit is shorted to VREF (intermittent fault) • TP sensor ground circuit is open (an intermittent fault) • TP sensor is damaged or has failed (possible intermittent open) • PCM has failed
DTC P1121 **1T CCM, MIL: No** 1995 S/T Blazer & S/T Pickup 4.3L VIN W engine Transmissions: All	**TP Sensor Circuit Intermittent High Input Conditions:** Key on or engine running; and the PCM detected an intermittent high voltage condition (more than 4.70v) on the TP sensor signal circuit for 10 seconds during the CCM test. **Possible Causes** • TP sensor signal circuit is shorted to VREF (intermittent fault) • TP sensor ground circuit is open (an intermittent fault) • TP sensor is damaged or has failed (possible intermittent open) • PCM has failed
DTC P1122 **1T CCM, MIL: No** 1996, 1997 Camaro, Caprice, Corvette, Firebird, Fleetwood & Roadmaster 4.3L VIN W, 5.7L VIN 5, 5.7L VIN P engines Transmissions: All	**TP Sensor Circuit Intermittent Low Input Conditions:** Engine started, engine runtime over 4 seconds, system voltage over 10.0v, and the PCM detected an intermittent low voltage condition (less than 0.20v) on the TP sensor signal circuit for 1 second. **Possible Causes** • TP sensor signal circuit is shorted to ground (intermittent fault) • TP sensor VREF circuit is open or grounded (intermittent fault) • TP sensor is damaged or has failed (possible intermittent short) • PCM has failed
DTC P1122 **1T CCM, MIL: No** 2001, 2002 C/K Truck, G Van, S/T Blazer & S/T Pickup 4.8L VIN V, 5.3L VIN T, 5.3L VIN Z, 6.0L VIN N, 6.0L VIN U engines Transmissions: All	**TP Sensor Circuit Intermittent High Input Conditions:** P0641 and P0651 not set, key on or engine running, system voltage over 10.0v, and the PCM detected an intermittent high voltage condition (more than 4.90v) on the TP sensor signal for 1 second. **Possible Causes** • TP sensor signal circuit is shorted to VREF (intermittent fault) • TP sensor ground circuit is open (an intermittent fault) • TP sensor is damaged or has failed (possible intermittent open) • PCM has failed
DTC P1122 **1T CCM, MIL: No** 2003, 2004, 2005 G/H Series Van, L/M Van, S/T Blazer & S/T Pickup 4.3L VIN X, 4.8L VIN V, 5.3L VIN T, 6.0L VIN U Transmissions: All	**TP Sensor Circuit Intermittent High Input Conditions:** P0641 and P0651 not set, key on or engine running, system voltage over 10.0v, and the PCM detected an intermittent high voltage condition (more than 4.90v) on the TP sensor signal for 1 second. **Possible Causes** • TP sensor signal circuit is shorted to VREF (intermittent fault) • TP sensor ground circuit is open (an intermittent fault) • TP sensor is damaged or has failed (possible intermittent open) • PCM has failed
DTC P1122 **1T CCM, MIL: No** 1996, 1997, 1998, 1999, 2000, 2001, 2002, 2003, 2004, 2005 Aurora, Bonneville, Century, Cutlass Supreme, Grand Prix, Intrigue, LeSabre, LSS, Lumina, Monte Carlo, 88', Regal, Park Avenue & 98' 3.1L VIN J, 3.1L VIN M, 3.4L VIN E, 3.5L VIN H, 3.8L VIN 1, 3.8L VIN K Transmissions: All	**TP Sensor Circuit Intermittent High Input Conditions:** Key on or engine running; and the PCM detected an intermittent high voltage condition (less than 0.10v) on the TP sensor signal circuit for 500 ms. Rotation of the TP sensor rotor from the closed throttle position to the WOT position provides the PCM with a signal voltage from less than 1.0v too more than 4.0v through the TP sensor signal circuit. If the PCM detects an intermittent excessively high voltage signal, it will set DTC P1122. **Possible Causes** • TP sensor signal circuit is shorted to ground (intermittent fault) • TP sensor VREF circuit is open or grounded (intermittent fault) • TP sensor is damaged or has failed (possible intermittent short) • PCM has failed
DTC P1122 **1T CCM, MIL: No** 1996, 1997, 1998, 1999, 2000, 2001, 2002, 2003, 2004, 2005 Aztek, Montana, Venture, Silhouette & Rendezvous 3.4L VIN E engine Transmissions: All	**TP Sensor Circuit Intermittent Low Input Conditions:** Key on or engine running; and the PCM detected an intermittent low voltage (less than 0.10v) on the TP sensor signal for 1 second. **Possible Causes** • TP sensor connector is damaged or shorted • TP sensor signal circuit is shorted to ground (intermittent fault) • TP sensor VREF circuit is open or grounded (intermittent fault) • TP sensor is damaged or has failed (possible intermittent short) • PCM has failed

OBD II Trouble Code List (P1xxx Codes)

DTC	Trouble Code Title, Conditions & Possible Causes
DTC P1122 **1T CCM, MIL: No** 1996, 1997, 1998, 1999, 2000, 2001, 2002 Cab & Chassis, C/K Truck, G Van, L/M Van, S/T Blazer & S/T Pickup 4.3L VIN W, 4.3L VIN X, 4.8L VIN V, 5.3L VIN P, 5.3L VIN T, 5.0L VIN M, 5.7L VIN K, 5.7L VIN R, 6.0L VIN N, 6.0L VIN U, 7.4L VIN J, 8.1L VIN G Transmissions: All	**TP Sensor Circuit Intermittent Low Input Conditions:** DTC P1639 and P1639 not set, engine started, engine runtime over 5 seconds, system voltage over 10.0v, and the PCM detected an intermittent low voltage condition (less than 0.15v) on the TP sensor signal circuit for 1 second during the CCM test. **Possible Causes** ● TP sensor signal circuit is shorted to ground (intermittent fault) ● TP sensor VREF circuit is open or grounded (intermittent fault) ● TP sensor is damaged or has failed (possible intermittent short) ● PCM has failed
DTC P1122 **1T CCM, MIL: No** 1996, 1997, 1998, 1999, 2000, 2001, 2002, 2003, 2004, 2005 Aurora, DeVille, Eldorado & Seville 4.0L VIN C, 4.6L VIN 9 & 4.6L VIN Y engines Transmissions: All	**TP Sensor Circuit Intermittent Low Input Conditions:** Key on or engine running; and the PCM detected an intermittent low voltage condition on the TP sensor signal circuit (less than 0.16v) for 1 second during the test. **Possible Causes** ● TP sensor signal circuit is shorted to ground (intermittent fault) ● TP sensor VREF circuit is open or grounded (intermittent fault) ● TP sensor is damaged or has failed (possible intermittent short) ● PCM has failed
DTC P1122 **1T CCM, MIL: No** 1995, 1996, 1997, 1998 Camaro & Firebird 3.8L VIN K engine Transmissions: All	**TP Sensor Circuit Intermittent Low Input Conditions:** Key on or engine running; and the PCM detected an intermittent low voltage condition on the TP sensor signal circuit (less than 0.16v) for a 10 second period during the CCM test. **Possible Causes** ● TP sensor signal circuit is shorted to ground (intermittent fault) ● TP sensor VREF circuit is open or grounded (intermittent fault) ● TP sensor is damaged or has failed (possible intermittent short) ● PCM has failed
DTC P1122 **1T CCM, MIL: No** 1999, 2000, 2001, 2002 Camaro & Firebird Engines: 5.7L VIN G Transmissions: All	**TP Sensor Circuit Intermittent Low Input Conditions:** Engine started and the PCM detected an intermittent high voltage condition on the TP sensor signal circuit (less than 0.20v) for 1 second out of a 20 second period during the CCM test. **Possible Causes** ● TP sensor signal circuit is shorted to ground (intermittent fault) ● TP sensor VREF circuit is open or grounded (intermittent fault) ● TP sensor is damaged or has failed (possible intermittent short) ● PCM has failed
DTC P1122 **1T CCM, MIL: No** 2003, 2004, 2005 G/H Series Van, L/M Van, S/T Blazer & S/T Pickup 4.3L VIN X, 4.8L VIN V, 5.3L VIN T, 6.0L VIN U Transmissions: All	**TP Sensor Circuit Intermittent Low Input Conditions:** P0641 and P0651 not set, engine started, system voltage over 10.0v and the PCM detected an intermittent low voltage condition (less than 0.15v) on the TP sensor signal for 1 second. **Possible Causes** ● TP sensor signal circuit is shorted to ground (intermittent fault) ● TP sensor VREF circuit is open or grounded (intermittent fault) ● TP sensor is damaged or has failed (possible intermittent short) ● PCM has failed
DTC P1122 **1T CCM, MIL: No** 1995 S/T Blazer & S/T Pickup 4.3L VIN W engine Transmissions: All	**TP Sensor Circuit Intermittent Low Input Conditions:** Key on or engine running; and the PCM detected an intermittent low voltage condition (less than 0.15v) on the TP sensor signal circuit for 10 seconds. The PCM uses the TP sensor to determine the throttle plate angle for various engine management systems. Rotation of the TP sensor rotor from the closed throttle position to the WOT position provides the PCM with a signal voltage from less than 1.0v too more than 4.0v through the TP sensor signal circuit. If the PCM detects an intermittent excessively high voltage signal, it will set DTC P1122. **Possible Causes** ● TP sensor signal circuit is shorted to VREF (intermittent fault) ● TP sensor ground circuit is open (an intermittent fault) ● TP sensor is damaged or has failed (possible intermittent open) ● PCM has failed
DTC P1122 **1T CCM, MIL: No** 1996, 1997 Grand Prix, Lumina & Monte Carlo Engines: 3.4L VIN X Transmissions: All	**TP Sensor Circuit Intermittent Low Input Conditions:** Engine started, system voltage over 10.0v and the PCM detected an intermittent low voltage condition (less than 0.10v) on the TP sensor signal circuit for 1 second during the CCM test. **Possible Causes** ● TP sensor signal circuit is shorted to ground (intermittent fault) ● TP sensor VREF circuit is open or grounded (intermittent fault) ● TP sensor is damaged or has failed (possible intermittent short) ● PCM has failed

OBD II Trouble Code List (P1xxx Codes)

DTC	Trouble Code Title, Conditions & Possible Causes
DTC P1125 **1T CCM, MIL: Yes** 2000, 2001, 2002 C/K Series Truck 4.8L VIN V, 5.3L VIN T, 5.3L VIN Z, 6.0L VIN N, 6.0L VIN U engines, 7.4L VIN J, 8.1L VIN G Transmissions: All	**Accelerator Pedal Position System Malfunction Conditions:** DTC P1517 and P1518 not set, key in crank or run position, system voltage over 5.23v, ETC serial data operational, and the PCM detected that two or more APP sensor signals were out of range, or all three APP sensor signals disagreed, or one APP sensor signal was out of range while the other two APP sensors disagreed. The App assembly contains three APP sensors. Three separate signals, grounds, and 5v VREF circuits are used to connect the APP assembly to the TAC Module. If one APP code is set, redundant APP systems allow the TAC system to continue operating normally. **Possible Causes** ● APP sensor connector contaminated with oil or moisture (this condition can cause signal "tracking" between the circuits) ● When the TAC module detects a fault within the TAC system, it may set more than one code. This is due to the redundant tests run continuously on this system. Locating and repairing a single fault may correct more than one code. Keep this in mind as you review the Captured Info available with a Scan Tool. ● Refer to the related TAC System trouble codes
DTC P1125 **1T CCM, MIL: Yes** 2003, 2004, 2005 C/K Truck, Envoy, G Van, Escalade & TrailBlazer 4.8L VIN V, 5.3L VIN P, 5.3L VIN T, 5.3L VIN Z, 6.0L VIN N, 6.0L VIN U, 8.1L VIN G engines Transmissions: All	**Accelerator Position Sensor 1 and 2 Circuit Malfunction Conditions:** DTC P1508, P2108 and U0107 not set, key in crank or run position, system voltage over 5.23v, valid communication between PCM and the TAC Module established, and the PCM detected the difference between the APP1 and APP2 sensors was over a predicted value. **Possible Causes** ● APP sensor assembly is damaged, loose or shorted ● APP1 sensor signal circuit shorted to the APP2 sensor circuit ● APP 1 or APP2 sensor signal circuit has high resistance ● APP sensor assembly is damaged or TAC module has failed
DTC P1125 **1T CCM, MIL: Yes** 2001, 2002, 2003, 2004, 2005 C/K Truck 6.6L VIN 1 Diesel engine Transmissions: All	**Accelerator Pedal Position Sensor Circuit Malfunction Conditions:** DTC P0641 not set, engine cranking or running, system voltage from 7-16v, and the ECM detected a condition where more than one APP sensor trouble codes were detected for over one seconds. The APP sensor is made up of 3 individual sensors in a single housing. The ECM uses the APP sensor to detect the amount of acceleration or deceleration desired to drive the vehicle via fuel injector control. **Possible Causes** ● If DTC P0641 is set, repair the cause of this trouble code first. ● If DTC P0641 is not set, refer to the other APP sensor codes that are set to repair the cause of this trouble code.
DTC P1125 **1T CCM, MIL: Yes** 1996, 1997, 1998, 1999, 2000, 2001, 2002 C/K Truck, G Van 6.5L VIN F, 6.5L VIN S, 6.5L VIN Y, 6.6L VIN 1 Diesel engines Transmissions: All	**Accelerator Pedal Position System Malfunction Conditions:** Key on or engine running; system voltage over 10.0vv, and the PCM detected an unexpected voltage condition on the Accelerator Pedal Position (APP) sensor circuit with more than one fault in memory. The APP sensor is mounted on the accelerator pedal control unit. The sensor is made up of 3 individual sensors in a single housing. It has separate signal, low reference, and 5-volt reference circuits that are used to interface the APP sensor signals with the PCM. Each sensor has a unique function to determine pedal position. The PCM uses the APP sensor to determine the amount of acceleration or deceleration required for proper fuel injector control. This code sets if the PCM detects more than one problem in the APP sensor **Possible Causes** ● APP sensor connector contaminated with oil or moisture (this condition can cause signal "tracking" between the circuits) ● When the TAC module detects a fault in this system, it may set more than one code (due to redundant tests running constantly) ● Refer to the related TAC system trouble codes
DTC P1125 **1T CCM, MIL: Yes** 1997, 1998, 1999, 2000, 2001, 2002, 2003, 2004, 2005 Corvette 5.7L VIN G, 5.7L VIN S Transmissions: All ; Corvette	**TP Sensor 1 Circuit Malfunction Conditions:** DTC P0606, P1517 and P1518 not set, key in crank or run position, system voltage over 5.23v, ETC serial data operational, and the PCM detected that two or more APP sensor signals were out of range, or all three APP sensor signals disagreed, or one APP sensor signal was out of range while the other two APP sensors disagreed. If only one APP sensor code is set, redundant APP systems allow the TAC system to continue operating normally. One or two APP sensor codes set will not cause the Reduced Engine Power message to be displayed. However, if two or more codes are set that relate to more than one sensor, this code will set and the Reduced Engine Power message will be displayed. **Possible Causes** ● APP sensor connector contaminated with oil or moisture (this condition can cause signal "tracking" between the circuits) ● When the TAC module detects a fault within the TAC system, it may set more than one code. This is due to the redundant tests run continuously on this system. Locating and repairing a single fault may correct more than one code. Keep this in mind as you review the Captured Info available with a Scan Tool. ● Refer to the related TAC system trouble codes

OBD II Trouble Code List (P1xxx Codes)

DTC	Trouble Code Title, Conditions & Possible Causes
DTC P1133 **2T O2S, MIL: Yes** 1996, 1997, 1998, 1999, 2000, 2001, 2002, 2003, 2004, 2005 Achieva, Alero, Beretta, Cavalier, Ciera, Century, Corsica, Grand Am, Sunfire, Skyhawk, S/T Blazer & S/T Pickup 2.2L VIN 4, 2.2L VIN 5, 2.2L VIN F, 2.2L VIN H, 2.4L VIN T engines Transmissions: All	**HO2S-11 (Bank 1 Sensor 1) Insufficient Switching Conditions:** DTC P0105-P0108, P0112-P0118, P0122, P0123, P0171, P0201-P0204, P0300-P0304, P0335, P0440-P0446, P0506, P0507, P0601, P0602 and P1441 are not set, engine speed from 1600-2450, ECT sensor over 158ºF, TP angle from 9-18%, Purge command over 36%, Purge Learn Memory over 0.86, fuel level over 10%, and the PCM detected the HO2S signal voltage switched from rich-lean or lean-rich less than 20 times in a 100 ms monitoring period. **Possible Causes** • Air leaks present in the exhaust manifold or the exhaust pipes • Fuel pressure is too high (i.e. causing a rich air fuel mixture) • HO2S may be contaminated (due to improper fuel or silicone) • HO2S signal high or low reference circuit has high resistance • HO2S heater element has failed, or the heater circuit is open • PCM has failed
DTC P1133 **2T O2S, MIL: Yes** 1996, 1997 Camaro, Caprice, Corvette, Firebird, Fleetwood & Roadmaster 4.3L VIN W, 5.7L VIN 5, 5.7L VIN P engines Transmissions: All	**HO2S-11 (Bank 1 Sensor 1) Insufficient Switching Conditions:** DTC P0100, P0102, P0103, P0117, P0118, P0123, P0200, P0372 and P1371 not set, engine speed from 1000-1700 rpm in closed loop, MAF sensor from 12-28 g/sec, and the PCM detected less than 20 rich-to-lean or less than 30 lean-to-rich HO2S switch counts. **Possible Causes** • Air leaks present in the exhaust manifold or the exhaust pipes • Fuel pressure is too high (i.e., causing a rich air fuel mixture) • HO2S may be contaminated (due to improper fuel or silicone) • HO2S signal high or low reference circuit has high resistance • HO2S heater element has failed, or the heater circuit is open • PCM has failed
DTC P1133 **2T O2S, MIL: Yes** 1996, 1997, 1998, 1999, 2000, 2001, 2002, 2003, 2004, 2005 Aurora, Bonneville, Century, Cutlass Supreme, Grand Prix, Intrigue, LeSabre, LSS, Lumina, Monte Carlo, 88', Regal, Park Avenue & 98' 3.1L VIN J, 3.1L VIN M, 3.4L VIN E, 3.5L VIN H, 3.8L VIN 1, 3.8L VIN K Transmissions: All	**HO2S-11 (Bank 1 Sensor 1) Insufficient Switching Conditions:** DTC P0101-P0103, P0107, P0108, P0112, P0113, P0116-P0118, P0121-P0123, P0125, P0128, P0201-P0206, P0410, P0440, P0442, P0443, P0446, P0449 and P1441 not set, engine started, engine runtime over 3.3 minutes, engine speed from 1300-3000 rpm, ECT sensor more than 158ºF, not in P/N, MAF sensor from 13-29 g/sec, Air Pump inactive while running in closed loop, TP angle over 2%, and the PCM detected the HO2S signal voltage switched from rich-lean or lean-rich less than 20 times within a 100 ms test period. **Possible Causes** • Air leaks present in the exhaust manifold or the exhaust pipes • Fuel pressure is too high (i.e., causing a rich air fuel mixture) • HO2S may be contaminated (due to improper fuel or silicone) • HO2S signal high or low reference circuit has high resistance • HO2S heater element has failed, or the heater circuit is open • PCM has failed
DTC P1133 **2T O2S, MIL: Yes** 1996, 1997, 1998, 1999, 2000, 2001, 2002, 2003, 2004, 2005 Aztek, Montana, Venture, Silhouette & Rendezvous 3.4L VIN E engine Transmissions: All	**HO2S-11 (Bank 1 Sensor 1) Insufficient Switching Conditions:** DTC P0101-P0103, P0107, P0108, P0112-P0118, P0121-P0123, P0125, P0128, P0201-P0206, P0410, P0440-P0449 and P1441 not set, engine speed from 1000-3000 rpm for 3 minutes in closed loop, system voltage over 10.0v, ECT sensor over 167ºF, MAF sensor from 10-30 g/sec, gear selector not in Park or Neutral, and the PCM detected the HO2S signal voltage switched from rich-lean or lean-rich less than 20 times within a 100 ms period. **Possible Causes** • Air leaks present in the exhaust manifold or the exhaust pipes • Fuel pressure is too high (i.e., causing a rich air fuel mixture) • HO2S may be contaminated (due to improper fuel or silicone) • HO2S signal high or low reference circuit has high resistance • HO2S heater element has failed, or the heater circuit is open • PCM has failed
DTC P1133 **2T O2S, MIL: Yes** 1996, 1997, 1998, 1999, 2000, 2001, 2002, 2003, 2004, 2005 Aurora, DeVille, Eldorado & Seville 4.0L VIN C, 4.6L VIN 9 & 4.6L VIN Y engines Transmissions: All	**HO2S-11 (Bank 1 Sensor 1) Insufficient Switching Conditions:** DTC P0101-P0103, P0106-P0108, P0112, P0113, P0117, P0118, P0121-P0123, P0125, P0131-P0135, P0151-P0155, P0201-P0208, P0300, P0410, P0440, P0442, P0443, P0446, P0449, P1415, P1416 and P1441 not set, engine speed from 1200-2800 rpm, gear selector in Drive, system voltage over 10.0v, ECT sensor over 122ºF, MAF sensor from 15-35 g/sec, Purge command over 20%, TP angle over 3%, A/F ratio at 14.5-14.8:1 for 3 minutes, and the PCM detected less than (5) lean-to-rich or five rich-to-lean switch counts on the HO2S signal within a 100 second period, or the signal remained within a 325-625 mv operating range. **Possible Causes** • Air leaks present in the exhaust manifold or the exhaust pipes • Fuel pressure is too high (i.e., causing a rich air fuel mixture) • HO2S may be contaminated (due to improper fuel or silicone) • HO2S signal high or low reference circuit has high resistance • HO2S heater element has failed, or the heater circuit is open • PCM has failed

OBD II Trouble Code List (P1xxx Codes)

DTC	Trouble Code Title, Conditions & Possible Causes
DTC P1133 **2T O2S, MIL: Yes** 1997, 1998, 1999, 2000, 2001, 2002, 2003, 2004, 2005 Camaro, Corvette & Firebird 5.7L VIN G, 5.7L VIN S Transmissions: All	**HO2S-11 (Bank 1 Sensor 1) Insufficient Switching Conditions:** DTC P0101-P0103, P0106-P0108, P0112-P0118, P0131-P0135, P0151-P0155, P0200, P0300, P0410, P0440-P0446, P0452, P0453, P1120, P1125, P1220, P1221, P1258, P1415, P1416, P1441, P1514-P1518 not set, engine started, engine speed from 1000-2300 rpm for 160 seconds in closed loop, system voltage over 10.0v, fuel level over 10%, ECT sensor more than 122ºF, Purge command over 0%, MAF sensor from 18-50 g/sec, TP indicated angle 5% over the idle value for 60 seconds, and the PCM less than 10 lean-to-rich or rich-lean switch counts on the HO2S signal circuit during the test. **Possible Causes** • Air leaks present in the exhaust manifold or the exhaust pipes • Fuel pressure is too high (i.e., causing a rich air fuel mixture) • HO2S may be contaminated (due to improper fuel or silicone) • HO2S signal high or low reference circuit has high resistance • HO2S heater element has failed, or the heater circuit is open • PCM has failed
DTC P1133 **2T O2S, MIL: Yes** 1996, 1997, 1998, 1999, 2000, 2001, 2002, 2003, 2004, 2005 Cab & Chassis, C/K Truck, G Van, L/M Van, S/T Blazer & S/T Pickup 4.3L VIN W, 4.3L VIN X, 5.0L VIN M, 5.7L VIN K, 5.7L VIN R, 7.4L VIN J Transmissions: All	**HO2S-11 (Bank 1 Sensor 1) Insufficient Switching Conditions:** DTC P0101-P0103, P0106-P0108, P0112-P0118, P0131-P0133, P0131 P0135, P0151 P0155, P0200, P0300, P0401-P0405, P0440-P0446, P0452, P0453, P1120, P1125, P1220, P1221, P1258, P1404, P1441 and P01514, and P1518 not set, engine started, engine speed from 1200-3000 rpm for over 3 minutes in closed loop, system voltage over 10.0v, ECT sensor more than 149ºF, fuel level over 10%, Purge command over 1%, MAF sensor from 23-50 g/sec, TP angle 5% over the idle value on models with TAC, Intrusive and Scan Tool tests inactive for 100 seconds, and the PCM detected the number of rich-to-lean or lean-to-rich HO2S signal transitions in a 100 second sample period were below a calibrated value. **Possible Causes** • Air leaks present in the exhaust manifold or the exhaust pipes • Fuel pressure is too high (i.e., causing a rich air fuel mixture) • HO2S may be contaminated (due to improper fuel or silicone) • HO2S signal high or low reference circuit has high resistance • HO2S heater element has failed, or the heater circuit is open • PCM has failed
DTC P1133 **2T O2S, MIL: Yes** 1995 S/T Blazer & S/T Pickup 4.3L VIN W engine Transmissions: All	**HO2S-11 (Bank 1 Sensor 1) Insufficient Switching Conditions:** DTC P0106, P0107, P0108, P0112, P0113, P0116, P0117, P0118, P0121, P0122, P0123, P0135, P0171, P0172 and P0401, P1441 and P1441 not set, engine running in closed loop for 3 minutes, ECT sensor over 135ºF, MAF sensor from 18-45 g/sec, TP angle at 10-20%, Purge command over 1%, conditions met for 100 seconds, and the PCM detected less than 30 rich-to-lean or 30 lean-to-rich counts on the HO2S signal during a 100 second period. **Possible Causes** • Air leaks present in the exhaust manifold or the exhaust pipes • Fuel pressure is too high (i.e., causing a rich air fuel mixture) • HO2S may be contaminated (due to improper fuel or silicone) • HO2S signal high or low reference circuit has high resistance • HO2S heater element has failed, or the heater circuit is open • PCM has failed
DTC P1133 **2T O2S, MIL: Yes** 2002, 2003, 2004, 2005 S/T Blazer & S/T Pickup 4.2L VIN S engine Transmissions: All	**HO2S-11 (Bank 1 Sensor 1) Insufficient Switching Conditions:** DTC P0105, P0107, P0108, P0112, P0113, P0117, P0118, P0122, P0123, P0201-P0206, P0300, P0336, P0440, P0446, P0452, P0453, P0507, P0601, P0602, P0604, P0606, P1120, P1220, P1221, P1271, P1275, P1280, P1484, P1512, P1514, P1515, P1516, P1621, P1635, P1639, P1680 and P1681 not set, engine started, engine speed from 1450-1900 for 200 seconds in closed loop, ECT sensor more than 158ºF, TP angle from 25-30%, Purge command over 10%, fuel level over 10%, conditions met for more than 30 seconds, and the PCM detected the HO2S rich-to-lean or lean-to-rich counts were less than 2 during the test. **Possible Causes** • Air leaks present in the exhaust manifold or the exhaust pipes • Fuel pressure is too high (i.e., causing a rich air fuel mixture) • HO2S may be contaminated (due to improper fuel or silicone) • HO2S signal high or low reference circuit has high resistance • HO2S heater element has failed, or the heater circuit is open • MAP sensor or TP sensor is out-of-calibration or "skewed" • PCM has failed

OBD II Trouble Code List (P1xxx Codes)

DTC	Trouble Code Title, Conditions & Possible Causes
DTC P1133 **2T O2S, MIL: Yes** 1999, 2000, 2001, 2002, 2003, 2004, 2005 C/K Truck, Envoy, G Van, Escalade & TrailBlazer 4.8L VIN V, 5.3L VIN P, 5.3L VIN T, 5.3L VIN Z, 6.0L VIN N, 6.0L VIN U, 8.1L VIN G engines Transmissions: All	**HO2S-11 (Bank 1 Sensor 1) Insufficient Switching Conditions:** DTC P0101, P0102, P0103, P0106, P0107, P0108, P0112, P0113, P0116, P0117, P0118, P0120, P0131, P0132, P0134, P0135, P0151, P0152, P0154, P0155, P0169, P0178, P0179, P0200, P0220, P0300, P0442, P0446, P0452, P0453, P0455, P0496, P1125, P1258, P1514, P1515, P1516, P1518, P2108 and P2135 not set, engine runtime over 160 seconds, engine speed at 1200-3000 rpm in closed loop, system voltage from 10-18v, ECT sensor more than 149°F, fuel level over 10%, Purge command over 1%, MAF sensor from 23-50 g/sec, TP indicated angle more than 5% above the idle value on models with TAC, Fuel Alcohol content less than 90%, and the PCM detected the number of rich-to-lean or lean-to-rich HO2S signal transitions were less than a calibrated value. **Possible Causes** • Air leaks present in the exhaust manifold or the exhaust pipes • Fuel pressure is too high (i.e., causing a rich air fuel mixture) • HO2S may be contaminated (due to improper fuel or silicone) • HO2S signal high or low reference circuit has high resistance • HO2S heater element has failed, or the heater circuit is open • PCM has failed
DTC P1133 **2T O2S, MIL: Yes** 1995, 1996, 1997, 1998, 1999, 2000, 2001, 2002 Camaro & Firebird 3.8L VIN K engine Transmissions: All	**HO2S-11 (Bank 1 Sensor 1) Insufficient Switching Conditions:** DTC P0101-P0103, P0107, P0108, P0112, P0113, P0117, P0118, P0121-P0123, P0125, P0128, P0201-P0206, P0300, P0410, P0440, P0442, P0443, P0446, P0449 and P1441 not set, DTC P0101-P0103, P0107-P0108, P0112-P0113, P0117-P0118, P0121-P0123, P0125, P0128, P0201-P0206, P0300, P0410, P0440-P0449 and P1441 not set, engine speed from 1000-3000 rpm in closed loop for 4 minutes, ECT sensor more than 122°F, MAF sensor from 13-30 g/sec, and the PCM detected less than 55 rich-to-lean or lean-to-rich switch counts on the HO2S signal, or the HO2S signal was fixed from 300-600 mv over a period of 2 minutes. **Possible Causes** • Air leaks present in the exhaust manifold or the exhaust pipes • Fuel pressure is too high (i.e., causing a rich air fuel mixture) • HO2S may be contaminated (due to improper fuel or silicone) • HO2S signal high or low reference circuit has high resistance • HO2S heater element has failed, or the heater circuit is open • PCM has failed
DTC P1133 **2T O2S, MIL: Yes** 1996, 1997 Grand Prix, Lumina & Monte Carlo Engines: 3.4L VIN X Transmissions: All	**HO2S-11 (Bank 1 Sensor 1) Insufficient Switching Conditions:** DTC P0101-P0103, P0107, P0108, P0112, P0113, P0116-P0118, P0121-P0123, P0125, P0201-P0206, P0441 and P1441 not set, engine speed from 1000-3000 rpm for 100 seconds in closed loop, ECT sensor over 122°F, MAF sensor from 10-30 g/sec, and the PCM detected less than 40 rich-to-lean or 40 lean-to-rich HO2S switch counts. **Possible Causes** • Air leaks present in the exhaust manifold or the exhaust pipes • Fuel pressure is too high (i.e., causing a rich air fuel mixture) • HO2S may be contaminated (due to improper fuel or silicone) • HO2S signal high or low reference circuit has high resistance • HO2S heater element has failed, or the heater circuit is open • PCM has failed
DTC P1133 **2T O2S, MIL: Yes** 1995 Corvette 5.7L VIN P engine Transmissions: All	**HO2S-11 (Bank 1 Sensor 1) Insufficient Switching Conditions:** DTC P0116, P0117, P0118, P0171, P0172, P0174, P0175, P1114 and P1115 not set, engine running in closed loop for 100 seconds, ECT and IAT sensors less than 122°F and within 11°F at startup, TP angle from 2-10%, and the PCM detected less than 48 lean-to-rich or 48 rich-to-lean switch counts on the HO2S circuit during a 100 second period. **Possible Causes** • Air leaks present in the exhaust manifold or the exhaust pipes • Fuel pressure is too high (i.e., causing a rich air fuel mixture) • HO2S may be contaminated (due to improper fuel or silicone) • HO2S signal high or low reference circuit has high resistance • HO2S heater element has failed, or the heater circuit is open • PCM has failed

OBD II Trouble Code List (P1xxx Codes)

DTC	Trouble Code Title, Conditions & Possible Causes
DTC P1134 **2T O2S, MIL: Yes** 1996, 1997, 1998, 1999, 2000, 2001, 2002, 2003, 2004, 2005 Aurora, Bonneville, Century, Cutlass Supreme, Grand Prix, Intrigue, LeSabre, LSS, Lumina, Monte Carlo, 88', Regal, Park Avenue & 98' 3.1L VIN J, 3.1L VIN M, 3.4L VIN E, 3.5L VIN H, 3.8L VIN 1, 3.8L VIN K Transmissions: All	**HO2S-11 (Bank 1 Sensor 1) Transition Time Ratio Conditions:** DTC P0101-P0103, P0107, P0108, P0112, P0113, P0116-P0118, P0121-P0123, P0125, P0128, P0201-P0206, P0410, P0440, P0442, P0443, P0446, P0449 and P1441 not set, engine started, engine runtime over 3.3 minutes, engine speed from 1300-3000 rpm, ECT sensor more than 169°F, not in P/N, MAF sensor from 13-29 g/sec, Air Pump inactive while running in closed loop, TP angle over 2%, and the PCM detected the average transition time ratio of the HO2S were not within a range of 0.4-4.5 during a 100 second test period. The HO2S is used for fuel control and for post catalyst monitoring. **Possible Causes** • Air leaks present in the exhaust manifold or the exhaust pipes • HO2S may be contaminated (due to improper fuel or silicone) • HO2S signal low reference circuit has high resistance • HO2S heater element has failed, or the heater circuit is open • PCM has failed
DTC P1134 **2T O2S, MIL: Yes** 1996, 1997 Camaro, Caprice, Corvette, Firebird, Fleetwood & Roadmaster 4.3L VIN W, 5.7L VIN 5, 5.7L VIN P engines Transmissions: All	**HO2S-11 (Bank 1 Sensor 1) Transition Time Ratio Conditions:** DTC P0100, P0102, P0103, P0117, P0118, P0123, P0200, P0372 and P1371 not set, engine started, engine speed from 1000-1700 rpm in closed loop, MAF sensor from 12-28 g/sec, and the PCM detected the ratio of average response times from the HO2S was less than 0.3 or more than 3.0 during a 100-second sampling period. **Possible Causes** • Air leaks present in the exhaust manifold or the exhaust pipes • HO2S may be contaminated (due to improper fuel or silicone) • HO2S signal low reference circuit has high resistance • HO2S heater element has failed, or the heater circuit is open • PCM has failed
DTC P1134 **2T O2S, MIL: Yes** 1996, 1997, 1998, 1999, 2000, 2001, 2002, 2003, 2004, 2005 Aztek, Montana, Venture, Silhouette & Rendezvous 3.4L VIN E engine Transmissions: All	**HO2S-11 (Bank 1 Sensor 1) Transition Time Ratio Conditions:** DTC P0101-P0103, P0107, P0108, P0112-P0118, P0121-P0123, P0125, P0128, P0201-P0206, P0410, P0440-P0449 and P1441 not set, engine speed from 1000-3000 rpm for 3 minutes in closed loop, system voltage over 10.0v, ECT sensor over 167°F, MAF sensor from 10-30 g/sec, gear selector not in Park or Neutral, and the PCM detected the transition time ratio was less than 0.4 or more than 4.5 during a 100 second monitoring period. **Possible Causes** • Air leaks present in the exhaust manifold or the exhaust pipes • Fuel pressure is too high (i.e., causing a rich air fuel mixture) • HO2S may be contaminated (due to improper fuel or silicone) • HO2S signal high or low reference circuit has high resistance • HO2S heater element has failed, or the heater circuit is open • PCM has failed
DTC P1134 **2T O2S, MIL: Yes** 1996, 1997, 1998, 1999, 2000, 2001, 2002, 2003, 2004, 2005 Cab & Chassis, C/K Truck, G Van, L/M Van, S/T Blazer & S/T Pickup 4.3L VIN W, 4.3L VIN X, 4.8L VIN V, 5.0L VIN M, 5.3L VIN T, 5.7L VIN K, 5.7L VIN R, 6.0L VIN U, 7.4L VIN J, 8.1L VIN G Transmissions: All	**HO2S-11 (Bank 1 Sensor 1) Transition Time Ratio Conditions:** DTC P0101-P0103, P0106-P0108, P0112-P0118, P0121-P0123, P0131-P0135, P0151-P0155, P0200, P0300, P0401-P0405, P0440-P0446, P0452, P0453, P1120, P1125, P1220, P1221, P1258, P1404, P1441and P01514, and P1518 not set, engine speed from 1200-3000 rpm for over 3 minutes in closed loop, system voltage over 10.0v, ECT sensor over 149°F, fuel level over 10%, Purge command over 1%, MAF sensor from 23-50 g/sec, TP angle at 5% over idle value on models with TAC, Intrusive and Scan Tool tests "off" for 100 seconds, and the PCM detected the HO2S time ratio value was not within the calibrated range. **Possible Causes** • Air leaks present in the exhaust manifold or the exhaust pipes • HO2S may be contaminated (due to improper fuel or silicone) • HO2S signal low reference circuit has high resistance • HO2S heater element has failed, or the heater circuit is open • PCM has failed
DTC P1134 **2T O2S, MIL: Yes** 1996, 1997, 1998, 1999, 2000, 2001, 2002, 2003, 2004, 2005 Aurora, DeVille, Eldorado & Seville 4.0L VIN C, 4.6L VIN 9 & 4.6L VIN Y engines Transmissions: All	**HO2S-11 (Bank 1 Sensor 1) Transition Time Ratio Conditions:** DTC P0101-P0103, P0106-P0108, P0112, P0113, P0117, P0118, P0121-P0123, P0125, P0131-P0135, P0151-P0155, P0201-P0208, P0300, P0410, P0440-0442, P0443, P0446, P0449, P1415, P1416 and P1441 not set, DTC P1258 not testing, engine speed from 1200-2800 rpm, system voltage over 10.0v, gear selector in Drive, ECT sensor more than 122°F, MAF sensor from 15-35 g/sec, Purge command over 20%, TP angle more than 3%, A/F ratio at 14.5-14.8:1 for 3 minutes in closed loop, and the PCM detected the ratio of average transition times was not between 0.50 and 3.90 or the HO2S voltage was from 325-635 mv. **Possible Causes** • Air leaks present in the exhaust manifold or the exhaust pipes • HO2S may be contaminated (due to improper fuel or silicone) • HO2S signal low reference circuit has high resistance • HO2S heater element has failed, or the heater circuit is open • PCM has failed

OBD II Trouble Code List (P1xxx Codes)

DTC	Trouble Code Title, Conditions & Possible Causes
DTC P1134 **2T O2S, MIL: Yes** 1995, 1996, 1997, 1998, 1999, 2000, 2001, 2002 Camaro & Firebird 3.8L VIN K engine Transmissions: All	**HO2S-11 (Bank 1 Sensor 1) Transition Time Ratio Conditions:** DTC P0101-P0103, P0107, P0108, P0112, P0113, P0117, P0118, P0121-P0123, P0125, P0128, P0201-P0206, P0300, P0410, P0440, P0442, P0443, P0446, P0449 and P1441 not set, DTC P0101-P0103, P0107-P0108, P0112-P0113, P0117-P0118, P0121-P0123, P0125, P0128, P0201-P0206, P0300, P0410, P0440-P0449 and P1441 not set, engine speed from 1000-3000 rpm in closed loop for over 4 minutes, ECT sensor more than 122ºF, MAF sensor from 13-30 g/sec, and the PCM detected the HO2S transition time ratio was less than 0.6 or was more than 3.3 during a 2 minute sample period during the test. **Possible Causes** • Air leaks present in the exhaust manifold or the exhaust pipes • Fuel pressure is too high (i.e., causing a rich air fuel mixture) • HO2S may be contaminated (due to improper fuel or silicone) • HO2S signal high or low reference circuit has high resistance • HO2S heater element has failed, or the heater circuit is open • PCM has failed
DTC P1134 **2T O2S, MIL: Yes** 1997, 1998, 1999, 2000, 2001, 2002, 2003, 2004, 2005 Camaro, Corvette & Firebird 5.7L VIN G, 5.7L VIN S Transmissions: All	**HO2S-11 (Bank 1 Sensor 1) Transition Time Ratio Conditions:** DTC P0101-P0103, P0106-P0108, P0112-P0118, P0131-P0135, P0151-P0155, P0200, P0300, P0410, P0440-P0446, P0452, P0453, P1120, P1125, P1220, P1221, P1258, P1415, P1416, P1441, P1514 and P1518 not set, engine speed at 1000-2300 rpm for 160 seconds in closed loop, system voltage over 10.0v, fuel level over 10%, ECT sensor over 122ºF, Purge over 0%, MAF sensor from 18-50 g/sec, TP angle 5% above idle value for 60 seconds (TAC models), and the PCM detected the HO2S transition time ratio value was out-of-range. **Possible Causes** • Air leaks present in the exhaust manifold or the exhaust pipes • HO2S may be contaminated (due to improper fuel or silicone) • HO2S signal low reference circuit has high resistance • HO2S heater element has failed, or the heater circuit is open • PCM has failed
DTC P1134 **2T O2S, MIL: Yes** 1996, 1997 Grand Prix, Lumina & Monte Carlo Engines: 3.4L VIN X Transmissions: All	**HO2S-11 (Bank 1 Sensor 1) Transition Time Ratio Conditions:** DTC P0101-P0103, P0107, P0108, P0112, P0113, P0116-P0118, P0121-P0123, P0125, P0201-P0206, P0441 and P1441 not set, engine speed from 1000-3000 rpm in closed loop for 100 seconds, ECT sensor over 122ºF, MAF sensor at 10-30 g/sec, and the PCM detected the ratio of average HO2S response times was under 0.3 or over 3.0 in a 100 second period. **Possible Causes** • Air leaks present in the exhaust manifold or the exhaust pipes • Fuel pressure is too high (i.e., causing a rich air fuel mixture) • HO2S may be contaminated (due to improper fuel or silicone) • HO2S signal high or low reference circuit has high resistance • HO2S heater element has failed, or the heater circuit is open • PCM has failed
DTC P1134 **2T O2S, MIL: Yes** 2002, 2003, 2004, 2005 S/T Blazer & S/T Pickup 4.2L VIN S engine Transmissions: All	**HO2S-11 (Bank 1 Sensor 1) Transition Time Ratio Conditions:** DTC P0105, P0107, P0108, P0112, P0113, P0117, P0118, P0122, P0123, P0201-P0206, P0300, P0336, P0440, P0446, P0452, P0453, P0507, P0601-P0606, P1120-P1221, P1271, P1275, P1280, P1484, P1512-P1516, P1621, P1635, P1639, P1680 and P1681 not set, engine speed from 1,000-2,500 rpm for 200 seconds in closed loop, ECT sensor over 158ºF, TP angle from 5-30%, Purge command over 10%, fuel level over 10%, MAP sensor from 24-104 kPa, Calculated Airflow over 16 g/sec for 60 seconds, and the PCM detected that the HO2S-11 rich-to-lean or lean-to-rich transition time ratio was not within a calibrated range. **Possible Causes** • Air leaks present in the exhaust manifold or the exhaust pipes • Fuel pressure is too high (i.e., causing a rich air fuel mixture) • HO2S may be contaminated (due to improper fuel or silicone) • HO2S signal high or low reference circuit has high resistance • HO2S heater element has failed, or the heater circuit is open • MAP sensor or TP sensor is out-of-calibration or "skewed" • PCM has failed
DTC P1137 **2T O2S, MIL: Yes** 2000, 2001, 2002, 2003, 2004, 2005 Alero, Cavalier, Century, Envoy, Grand Am, Sunfire, S/T Blazer, S/T Pickup & TrailBlazer 2.2L VIN 4, 2.2L VIN 5, 2.4L VIN T, 2.2L VIN H, 4.2L VIN S engines Transmissions: All	**HO2S-12 Low Voltage During Power Enrichment Conditions:** DTC P0105-P0108, P0112-P0118, P0122, P0123, P0169, P0171, P0172, P0201-P0204, P0300-P0304, P0336, P0440, P0446, P0452, P0453, P0506, P0507, P0601, P0602 and P1441 not set, engine runtime over 10 seconds, ECT sensor over 158ºF, Power Enrichment active, fuel ethanol composition below 88%, fuel level over 10%, and the PCM detected the HO2S signal was above 700 mv while the rear HO2S was less than 399 mv for 9.5 seconds. **Possible Causes** • Air leaks present in exhaust manifold, exhaust pipes, or in the HO2S mounting location • HO2S may be contaminated (due to improper fuel or silicone) • PCM has failed

OBD II Trouble Code List (P1xxx Codes)

DTC	Trouble Code Title, Conditions & Possible Causes
DTC P1137 **2T O2S, MIL: Yes** 2002, 2003, 2004, 2005 S/T Blazer & S/T Pickup 4.2L VIN S engine Transmissions: All	**HO2S-12 Low Voltage During Power Enrichment Conditions:** DTC P0105, P0107, P0108, P0112, P0113, P0117, P0118, P0122, P0123, P0201-P0206, P0300, P0336, P0440, P0446, P0452, P0453, P0507, P0601, P0602, P0604, P0606, P1120, P1220, P1221, P1271, P1275, P1280, P1484, P1512-P1516, P1621, P1635, P1639, P1680 and P1681 not set, engine runtime over 10 seconds in closed loop, fuel level over 10%, ECT sensor over 158ºF, Power Enrichment mode "off", and the PCM detected the HO2S-11signal was more than 700 mv while the HO2S-12 signal was less than 399 mv for 9.5 seconds. **Possible Causes** ● Air leaks present in the exhaust manifold, exhaust pipes or the HO2S mounting location ● Air leaks present at the Catalytic Converter flange area ● HO2S may be contaminated (due to improper fuel or silicone) ● PCM has failed
DTC P1137 **2T O2S, MIL: Yes** 2000, 2001, 2002, 2003, 2004 2005 Alero, Cavalier, Century, Envoy, Grand Am, Sunfire, S/T Blazer, S/T Pickup & TrailBlazer 2.2L VIN 4, 2.2L VIN 5, 2.4L VIN T, 2.2L VIN H, 4.2L VIN S engines Transmissions: All	**HO2S-12 High Voltage During Decel Fuel Cutoff Conditions:** DTC P0105-P0108, P0112-P0118, P0122, P0123, P0169, P0171, P0172, P0201-P0204, P0300-P0304, P0336, P0440-P0453, P0506, P0507, P0601, P0602 and P1441 not set, engine runtime over 10 seconds, ECT sensor over 167ºF, DFCO mode active, fuel level over 10%, and the PCM detected the HO2S signal was over 618 mv for 5 out of 11 seconds. **Possible Causes** ● Fuel injector(s) damaged or leaking ● Fuel pressure is too high (i.e., causing a rich air fuel mixture) ● Fuel pressure regulator is damaged or leaking ● HO2S may be contaminated (due to improper fuel or silicone) ● PCM has failed
DTC P1138 **2T O2S, MIL: Yes** 2002, 2003, 2004, 2005 Envoy & TrailBlazer 4.2L VIN S engine Transmissions: All	**HO2S-12 High Voltage During Decel Fuel Cutoff Conditions:** DTC P0105, P0107, P0108, P0112, P0113, P0117, P0118, P0122, P0123, P0201-P0206, P0300, P0336, P0440, P0446, P0452, P0453, P0507, P0601, P0602, P0604, P0606, P1120, P1220, P1221, P1271, P1275, P1280, P1484, P1512, P1514, P1515, P1516, P1621, P1635, P1639, P1680 and P1681 not set, engine runtime over 10 seconds in closed loop, fuel level over 10%, ECT sensor over 158ºF, Decel Fuel Cutoff "off", and the PCM detected the HO2S-11signal was more than 648 mv for 10 seconds while operating in Decel Fuel Cutoff mode. **Possible Causes** ● Fuel injector(s) damaged or leaking ● Fuel pressure is too high (i.e., causing a rich air fuel mixture) ● Fuel pressure regulator is damaged or leaking ● HO2S may be contaminated (due to improper fuel or silicone) ● PCM has failed
DTC P1139 **2T O2S, MIL: Yes** 1996, 1997 Aurora, DeVille, Eldorado & Seville 4.0L VIN C, 4.6L VIN 9 & 4.6L VIN Y engines Transmissions: All	**HO2S-12 (Bank 1 Sensor 2) Insufficient Switching Conditions:** DTC P0117, P0118, P0121, P0122, P0123, P0131-P0141, P0151-P0151, P0300, P1133, P1134, P1153 and P1154 not set, Engine started; test not completed this key cycle, Low Coolant Level "off", and the PCM detected the number of rich-to-lean or lean-to-rich rear HO2S transitions reached 20 during a 120 second period during the HO2S Monitor test. **Possible Causes** ● Fuel pressure is too high (i.e., causing a rich air fuel mixture) ● HO2S may be contaminated (due to improper fuel or silicone) ● HO2S signal high or low reference circuit has high resistance ● PCM has failed
DTC P1140 **2T O2S, MIL: Yes** 1996, 1997 Aurora, DeVille, Eldorado & Seville 4.0L VIN C, 4.6L VIN 9 & 4.6L VIN Y engines Transmissions: All	**HO2S-12 (Bank 1 Sensor 2) Transition Time Ratio Conditions:** DTC P0117, P0118, P0121, P0122, P0123, P0131-P0141, P0151-P0151, P0300, P1133, P1134, P1153, P1154 and P1258 not set, Engine started; test not completed this key cycle, Engine Metal Over Temperature "off", and the PCM detected the HO2S-12 RL/LR transition time ratio was over 64 counts and the LR/RL transition time ratio was over 48 counts. **Possible Causes** ● Fuel pressure is too high (i.e., causing a rich air fuel mixture) ● HO2S may be contaminated (due to improper fuel or silicone) ● HO2S signal high or low reference circuit has high resistance ● PCM has failed
DTC P1141 **2T CCM, MIL: Yes** 1997, 1998, 1999, 2000, 2001 Catera 3.0L VIN R engine Transmissions: A/T	**HO2S-12 (Bank 1 Sensor 2) Heater Circuit Malfunction Conditions:** Engine started; system voltage over 10.0v, HO2S heater enabled, and the PCM detected an unexpected voltage condition on the HO2S-12 heater circuit during the CCM test. **Possible Causes** ● Air leaks present in the exhaust manifold or the exhaust pipes ● If DTC P0135, P0141, P0155, P0161, P1141 and P1161 are set, check Fuse 43 located in the ECM housing for a problem ● HO2S heater control circuit is open or shorted to ground ● HO2S heater power circuit is open (test power circuit to HO2S) ● HO2S is damaged or has failed

OBD II Trouble Code List (P1xxx Codes)

DTC	Trouble Code Title, Conditions & Possible Causes
DTC P1153 **2T O2S, MIL: Yes** 1996, 1997 Camaro, Caprice, Corvette, Firebird, Fleetwood & Roadmaster 4.3L VIN W, 5.7L VIN 5, 5.7L VIN P engines Transmissions: All	**HO2S-21 (Bank 2 Sensor 1) Insufficient Switching Conditions:** DTC P0100, P0102, P0103, P0117, P0118, P0123, P0200, P0372 and P1371 not set, engine started, engine speed from 1000-1700 rpm in closed loop, MAF sensor from 12-28 g/sec, and the PCM detected less than 20 rich-to-lean or less than 30 lean-to-rich switch counts on the HO2S signal during the HO2S Monitor test. **Possible Causes** ● Air leaks present in the exhaust manifold or the exhaust pipes ● Fuel pressure is too high (i.e., causing a rich air fuel mixture) ● HO2S may be contaminated (due to improper fuel or silicone) ● HO2S signal high or low reference circuit has high resistance ● HO2S heater element has failed, or the heater circuit is open ● PCM has failed
DTC P1153 **2T O2S, MIL: Yes** 1996, 1997, 1998, 1999, 2000, 2001, 2002, 2003, 2004, 2005 Cab & Chassis, C/K Truck, G Van, L/M Van, S/T Blazer & S/T Pickup 4.3L VIN W, 4.3L VIN X, 5.0L VIN M, 5.7L VIN K, 5.7L VIN R, 7.4L VIN J Transmissions: All	**HO2S-21 (Bank 2 Sensor 1) Insufficient Switching Conditions:** DTC P0101-P0103, P0106-P0108, P0112-P0118, P0121-P0123, P0131-P0135, P0151-P0155, P0200, P0300, P0401-P0405, P0440-P0446, P0452, P0453, P1120, P1125, P1220, P1221, P1258, P1404, P1441 and P01514, and P1518 not set, engine speed from 1200-3000 rpm for over 3 minutes in closed loop, system voltage over 10.0v, ECT sensor over 149ºF, fuel level over 10%, Purge command over 1%, MAF sensor from 23-50 g/sec, TP indicated angle at 5% over the idle value (TAC models), Intrusive and Scan Tool tests "off" for 100 seconds, and the PCM detected the number of rich-to-lean or lean-to-rich HO2S signal transitions during a 100 second sample period were less than a calibrated value. **Possible Causes** ● Air leaks present in the exhaust manifold or the exhaust pipes ● Fuel pressure is too high (i.e., causing a rich air fuel mixture) ● HO2S may be contaminated (due to improper fuel or silicone) ● HO2S signal high or low reference circuit has high resistance ● HO2S heater element has failed, or the heater circuit is open ● MAP sensor or TP sensor is out-of-calibration or "skewed" ● PCM has failed
DTC P1153 **2T O2S, MIL: Yes** 1999, 2000, 2001, 2002, 2003, 2004, 2005 C/K Truck, Escalade, G/H Van, Envoy & TrailBlazer 4.8L VIN V, 5.3L VIN P, 5.3L VIN T, 5.3L VIN Z, 6.0L VIN N, 6.0L VIN U, 8.1L VIN G engines Transmissions: All	**HO2S-21 (Bank 2 Sensor 1) Insufficient Switching Conditions:** DTC P0101, P0102, P0103, P0106, P0107, P0108, P0112, P0113, P0116, P0117, P0118, P0120, P0131, P0132, P0134, P0135, P0151, P0152, P0154-P0155, P0169, P0178-P0179, P0200, P0220, P0300, P0442, P0446, P0452-P0453, P0455-P0496, P1125, P1258, P1404, P1514, P1515, P1516, P1518, P2108 and P2135 not set, engine runtime over 160 seconds, engine speed at 1200-3000 rpm in closed loop, system voltage from 10-18v, ECT sensor more than 149ºF, Fuel Level over 10%, Purge command over 1%, MAF sensor from 23-50 g/sec, TP indicated angle more than 5% above the idle value on models with TAC, Fuel Alcohol content less than 90%, and the PCM detected the number of rich-to-lean or lean-to-rich HO2S signal transitions were less than a calibrated value. **Possible Causes** ● Air leaks present in the exhaust manifold or the exhaust pipes ● Fuel pressure is too high (i.e., causing a rich air fuel mixture) ● HO2S may be contaminated (due to improper fuel or silicone) ● HO2S signal high or low reference circuit has high resistance ● HO2S heater element has failed, or the heater circuit is open ● PCM has failed
DTC P1153 **2T O2S, MIL: Yes** 1996, 1997, 1998, 1999, 2000, 2001, 2002, 2003, 2004, 2005 Aurora, DeVille, Eldorado & Seville 4.0L VIN C, 4.6L VIN 9 & 4.6L VIN Y engines Transmissions: All	**HO2S-21 (Bank 2 Sensor 1) Insufficient Switching Conditions:** DTC P0101-P0103, P0106-P0108, P0112, P0113, P0117, P0118, P0121-P0123, P0125, P0131, P0132, P0135, P0151, P0152, P0155, P0201-P0208, P0300, P0410, P0440, P0442, P0443, P0446, P0449, P1415, P1416 and P1441 not set, engine started, engine speed from 1200-2800 rpm, system voltage over 10.0v, not in Park, Neutral or Reverse position, ECT sensor more than 122ºF, MAF sensor from 15-35 g/sec, Purge command over 20%, TP angle more than 3%, A/F ratio at 14.5-14.8:1 for 3 minutes in closed loop, and the PCM detected less than 5 lean-to-rich or 5 rich-to-lean switch counts on the HO2S signal within a 100 second period, and that the HO2S signal remained within a 325-625 mv operating range. **Possible Causes** ● Air leaks present in the exhaust manifold or the exhaust pipes ● Fuel pressure is too high (i.e., causing a rich air fuel mixture) ● HO2S may be contaminated (due to improper fuel or silicone) ● HO2S signal high or low reference circuit has high resistance ● HO2S heater element has failed, or the heater circuit is open ● MAP sensor or TP sensor is out-of-calibration or "skewed" ● PCM has failed

OBD II Trouble Code List (P1xxx Codes)

DTC	Trouble Code Title, Conditions & Possible Causes
DTC P1153 **2T O2S, MIL: Yes** 1997, 1998, 1999, 2000, 2001, 2002, 2003, 2004, 2005 Camaro, Corvette & Firebird 5.7L VIN G, 5.7L VIN S Transmissions: All	**HO2S-21 (Bank 2 Sensor 1) Insufficient Switching Conditions:** DTC P0101-P0103, P0106-P0108, P0112-P0118, P0131-P0135, P0151-P0155, P0200, P0300, P0410, P0440-P0446, P0452, P0453, P1120, P1125, P1220, P1221, P1258, P1415, P1416, P1441, P1514-P1518 not set, engine started, engine speed from 1000-2300 rpm for 160 seconds in closed loop, system voltage over 10.0v, fuel level over 10%, ECT sensor more than 122°F, Purge command over 0%, MAF sensor from 18-50 g/sec, TP indicated angle 5% over the idle value for 60 seconds, and the PCM less than 10 lean-to-rich or rich-lean switch counts on the HO2S signal circuit during the test. **Possible Causes** ● Air leaks present in the exhaust manifold or the exhaust pipes ● Fuel pressure is too high (i.e., causing a rich air fuel mixture) ● HO2S may be contaminated (due to improper fuel or silicone) ● HO2S signal high or low reference circuit has high resistance ● HO2S heater element has failed, or the heater circuit is open ● PCM has failed
DTC P1153 **2T O2S, MIL: Yes** 1995 S/T Blazer & S/T Pickup 4.3L VIN W engine Transmissions: All	**HO2S-21 (Bank 2 Sensor 1) Insufficient Switching Conditions:** DTC P0106-P0108, P0112, P0113, P0116-P0118, P0121-P0123, P0135, P0171, P0172 and P0401, P1441 and P1441 not set, engine runtime over 3 minutes, ECT sensor over 135°F, MAF sensor from 18-45 g/sec, TP angle at 10-20%, Purge over 1% for 100 seconds, and the PCM detected less than 30 rich-to-lean or lean-to-rich counts on the HO2S signal. **Possible Causes** ● Air leaks present in the exhaust manifold or the exhaust pipes ● Fuel pressure is too high (i.e., causing a rich air fuel mixture) ● HO2S may be contaminated (due to improper fuel or silicone) ● HO2S signal high or low reference circuit has high resistance ● HO2S heater element has failed, or the heater circuit is open ● PCM has failed
DTC P1153 **2T O2S, MIL: Yes** 1995, 1996, 1997, 1998, 1999, 2000, 2001, 2002 Camaro & Firebird 3.8L VIN K engine Transmissions: All	**HO2S-21 (Bank 2 Sensor 1) Insufficient Switching Conditions:** DTC P0101-P0103, P0107, P0108, P0112, P0113, P0117, P0118, P0121-P0123, P0125, P0128, P0201-P0206, P0300, P0410, P0440, P0442, P0443, P0446, P0449 and P1441 not set, DTC P0101-P0103, P0107-P0108, P0112-P0113, P0117-P0118, P0121-P0123, P0125, P0128, P0201-P0206, P0300, P0410, P0440-P0449 and P1441 not set, engine speed from 1000-3000 rpm in closed loop for over 4 minutes, ECT sensor over 122°F, MAF sensor from 13-30 g/sec, and the PCM detected less than 55 rich-to-lean or lean-to-rich switch counts on the HO2S signal, or the HO2S signal was fixed from 300-600 mv over a period of 2 minutes. **Possible Causes** ● Air leaks present in the exhaust manifold or the exhaust pipes ● Fuel pressure is too high (i.e., causing a rich air fuel mixture) ● HO2S may be contaminated (due to improper fuel or silicone) ● HO2S signal high or low reference circuit has high resistance ● HO2S heater element has failed, or the heater circuit is open ● PCM has failed
DTC P1153 **2T O2S, MIL: Yes** 1996, 1997 Grand Prix, Lumina & Monte Carlo Engines: 3.4L VIN X Transmissions: All	**HO2S-21 (Bank 1 Sensor 1) Insufficient Switching Conditions:** DTC P0101-P0103, P0107, P0108, P0112, P0113, P0116-P0118, P0121-P0123, P0125, P0201-P0206, P0441 and P1441 not set, engine speed at 1000-3000 rpm for 100 seconds in closed loop, ECT sensor over 122°F, MAF sensor from 10-30 g/sec, and the PCM detected less than 40 rich-to-lean or 40 lean-to-rich switch counts on the HO2S signal during testing. **Possible Causes** ● Air leaks present in the exhaust manifold or the exhaust pipes ● Fuel pressure is too high (i.e., causing a rich air fuel mixture) ● HO2S may be contaminated (due to improper fuel or silicone) ● HO2S signal high or low reference circuit has high resistance ● HO2S heater element has failed, or the heater circuit is open ● PCM has failed
DTC P1153 **1T O2S, MIL: No** 1995 Corvette 5.7L VIN P engine Transmissions: All	**HO2S-21 (Bank 2 Sensor 1) Insufficient Switching Conditions:** DTC P0116-P0118, P0171, P0172, P0174, P0175, P1114 and P1115 not set, engine running in closed loop for 100 seconds, ECT and IAT sensors less than 122°F and within 11°F at startup, TP angle from 2-10%, and the PCM detected less than 48 lean-to-rich or 48 rich-to-lean switch counts on the HO2S signal during a 100 second period during testing. **Possible Causes** ● Air leaks present in the exhaust manifold or the exhaust pipes ● Fuel pressure is too high (i.e., causing a rich air fuel mixture) ● HO2S may be contaminated (due to improper fuel or silicone) ● HO2S signal high or low reference circuit has high resistance ● HO2S heater element has failed, or the heater circuit is open ● PCM has failed

OBD II Trouble Code List (P1xxx Codes)

DTC	Trouble Code Title, Conditions & Possible Causes
DTC P1154 **2T O2S, MIL: Yes** 1996, 1997, 1998, 1999, 2000, 2001, 2002, 2003, 2004, 2005 Cab & Chassis, C/K Truck, G Van, L/M Van, S/T Blazer & S/T Pickup 4.3L VIN W, 4.3L VIN X, 4.8L VIN V, 5.0L VIN M, 5.3L VIN T, 5.7L VIN K, 5.7L VIN R, 6.0L VIN U, 7.4L VIN J, 8.1L VIN G Transmissions: All	**HO2S-21 (Bank 2 Sensor 1) Transition Time Ratio Conditions:** DTC P0101-P0103, P0106-P0108, P0112-P0118, P0121-P0123, P0131-P0135, P0151-P0155, P0200, P0300, P0401-P0405, P0440-P0446, P0452, P0453, P1120, P1125, P1220, P1221, P1258, P1404, P1441and P01514, and P1518 not set, engine speed from 1200-3000 rpm for over 3 minutes in closed loop, system voltage over 10.0v, ECT sensor over 149°F, fuel level over 10%, Purge command over 1%, MAF sensor from 23-50 g/sec, TP angle at 5% over idle value on models with TAC, Intrusive and Scan Tool tests "off" for 100 seconds, and the PCM detected the HO2S time ratio value was not within the calibrated range. **Possible Causes** ● Air leaks present in the exhaust manifold or the exhaust pipes ● HO2S may be contaminated (due to improper fuel or silicone) ● HO2S signal low reference circuit has high resistance ● HO2S heater element has failed, or the heater circuit is open ● PCM has failed
DTC P1154 **2T O2S, MIL: Yes** 1996, 1997 Camaro, Caprice, Corvette, Firebird, Fleetwood & Roadmaster 4.3L VIN W, 5.7L VIN 5, 5.7L VIN P engines Transmissions: All	**HO2S-21 (Bank 2 Sensor 1) Transition Time Ratio Conditions:** DTC P0100-P0103, P0117, P0118, P0123, P0200, P0372 and P1371 not set; engine speed at 1000-1700 rpm in closed loop, MAF sensor from 12-28 g/sec, and the PCM detected the average HO2S response time ratio was under 0.3 or over 3.0 for 100 seconds. **Possible Causes** ● Air leaks present in the exhaust manifold or the exhaust pipes ● HO2S may be contaminated (due to improper fuel or silicone) ● HO2S signal low reference circuit has high resistance ● HO2S heater element has failed, or the heater circuit is open ● PCM has failed
DTC P1154 **2T O2S, MIL: Yes** 1996, 1997, 1998, 1999, 2000, 2001, 2002, 2003, 2004, 2005 Aurora, DeVille, Eldorado & Seville 4.0L VIN C, 4.6L VIN 9 & 4.6L VIN Y engines Transmissions: All	**HO2S-21 (Bank 2 Sensor 1) Transition Time Ratio Conditions:** DTC P0101-P0103, P0106-P0108, P0112, P0113, P0117, P0118, P0121-P0123, P0125, P0131-P0135, P0151-P0155, P0201-P0208, P0300, P0410, P0440-0442, P0443, P0446, P0449, P1415, P1416 and P1441 not set, DTC P1258 not testing, engine speed from 1200-2800 rpm, system voltage over 10.0v, gear selector in Drive, ECT sensor more than 122°F, MAF sensor from 15-35 g/sec, Purge command over 20%, TP angle more than 3%, A/F ratio at 14.5-14.8:1 for 3 minutes in closed loop, and the PCM detected the ratio of average transition times was not between 0.50 and 3.90 or the HO2S voltage was from 325-635 mv. **Possible Causes** ● Air leaks present in the exhaust manifold or the exhaust pipes ● HO2S may be contaminated (due to improper fuel or silicone) ● HO2S signal low reference circuit has high resistance ● HO2S heater element has failed, or the heater circuit is open ● PCM has failed
DTC P1154 **2T O2S, MIL: Yes** 1997, 1998, 1999, 2000, 2001, 2002, 2003, 2004, 2005 Camaro, Corvette & Firebird 5.7L VIN G, 5.7L VIN S Transmissions: All	**HO2S-21 (Bank 2 Sensor 1) Transition Time Ratio Conditions:** DTC P0101-P0103, P0106-P0108, P0112-P0118, P0131-P0135, P0151-P0155, P0200, P0300, P0410, P0440-P0446, P0452, P0453, P1120, P1125, P1220, P1221, P1258, P1415, P1416, P1441, P1514-P1518 not set, engine speed from 1000-2300 rpm for 160 seconds in closed loop, system voltage over 10.0v, fuel level over 10%, ECT sensor over 122°F, Purge over 0%, MAF sensor from 18-50 g/sec, TP angle over 5% over the idle value for 1 minute (TAC models), and the PCM detected the HO2S transition time ratio was out of range. **Possible Causes** ● Air leaks present in the exhaust manifold or the exhaust pipes ● HO2S may be contaminated (due to improper fuel or silicone) ● HO2S signal low reference circuit has high resistance ● HO2S heater element has failed, or the heater circuit is open ● PCM has failed
DTC P1154 **2T O2S, MIL: Yes** 1995, 1996, 1997, 1998, 1999, 2000, 2001, 2002 Camaro & Firebird 3.8L VIN K engine Transmissions: All	**HO2S-21 (Bank 2 Sensor 1) Transition Time Ratio Conditions:** DTC P0101-P0103, P0107, P0108, P0112, P0113, P0117, P0118, P0121-P0123, P0125, P0128, P0201-P0206, P0300, P0410, P0440, P0442, P0443, P0446, P0449 and P1441 not set, DTC P0101-P0103, P0107-P0108, P0112-P0113, P0117-P0118, P0121-P0123, P0125, P0128, P0201-P0206, P0300, P0410, P0440-P0449 and P1441 not set, engine speed from 1000-3000 rpm in closed loop for over 4 minutes, ECT sensor over 122°F, MAF sensor at 13-30 g/sec, and the PCM detected the HO2S transition time ratio was less than 0.6 or was more than 3.3 during a 2 minute sample period. **Possible Causes** ● Air leaks present in the exhaust manifold or the exhaust pipes ● Fuel pressure is too high (i.e., causing a rich air fuel mixture) ● HO2S may be contaminated (due to improper fuel or silicone) ● HO2S signal high or low reference circuit has high resistance ● HO2S heater element has failed, or the heater circuit is open ● PCM has failed

OBD II Trouble Code List (P1xxx Codes)

DTC	Trouble Code Title, Conditions & Possible Causes
DTC P1154 **1T O2S, MIL: No** 1995 Corvette 5.7L VIN P engine Transmissions: All	**HO2S-22 (Bank 2 Sensor 2) Signal Shifted Rich Conditions:** DTC P0116-P0118, P0171, P0172, P0174, P0175, P1114 and P1115 not set, engine running in closed loop for 100 seconds, ECT and IAT sensors below 122ºF and within 11ºF at startup, TP angle from 2-10%, and the PCM detected he majority of the rear HO2S signals were from 729-923 mv during a 129 second test period. **Possible Causes** ● Fuel injector(s) or fuel pressure regulator damaged or leaking ● Fuel pressure is too high (i.e., causing a rich air fuel mixture) ● HO2S may be contaminated (due to improper fuel or silicone) ● HO2S heater element has failed, or the heater circuit is open ● PCM has failed
DTC P1161 **2T CCM, MIL: Yes** 1997, 1998, 1999, 2000, 2001 Catera 3.0L VIN R engine Transmissions: All	**HO2S-22 (Bank 2 Sensor 2) Heater Circuit Malfunction Conditions:** Engine started; system voltage over 10.0v, HO2S heater enabled, and the PCM detected an unexpected voltage condition on the rear HO2S heater circuit during the CCM test. **Possible Causes** ● Air leaks present in the exhaust manifold or the exhaust pipes ● If DTC P0135, P0141, P0155, P0161, P1111 and P1141 are set, check Fuse 43 located in the ECM housing for a problem ● HO2S heater control circuit is open or shorted to ground ● HO2S heater power circuit is open (test power circuit to HO2S) ● HO2S is damaged or has failed
DTC P1171 **2T FUEL, MIL: Yes** 1996, 1997 Camaro, Caprice, Corvette, Firebird, Fleetwood & Roadmaster 4.3L VIN W, 5.7L VIN 5, 5.7L VIN P engines Transmissions: All	**Fuel System Lean During Acceleration Conditions:** DTC P0131-141, P0151-P0155, P1133 and P1153 not set, engine started, engine runtime over 20 seconds in closed loop, ECT sensor more than 70ºF, Decel Fuel Cutoff inactive, Power Enrichment mode enabled, and the PCM detected the HO2S was lean for 3 seconds. The PCM can identify if the fuel delivery system can supply enough fuel during heavy acceleration (Power Enrichment Mode). When this mode is requested during closed loop operation, the PCM provides extra fuel to the engine, and the PCM should detect a rich condition. If it does not detect a rich condition at this time, it will set this code. **Possible Causes** ● Fuel filter clogged or restricted ● Fuel injectors dirty or restricted ● Fuel level too low during heavy acceleration events ● Fuel pressure regulator has failed (not supplying enough fuel) ● Water or alcohol in fuel (low HO2S signal during acceleration)
DTC P1171 **2T FUEL, MIL: Yes** 1996, 1997, 1998, 1999, 2000, 2001, 2002, 2003, 2004, 2005 Achieva, Alero, Beretta, Cavalier, Ciera, Century, Grand Am, Sunfire Skyhawk, S/T Blazer & S/T Pickup 2.2L VIN 4, 2.2L VIN 5, 2.2L VIN H, 2.4L VIN T Transmissions: All	**Fuel System Lean During Acceleration Conditions:** DTC P0131, P0132, P0133, P0134 and P1133 not set, engine started, engine runtime over 20 seconds in closed loop, ECT sensor more than 70ºF, Power Enrichment mode enabled, and the PCM detected the HO2S signal was less than 300 mv for 5 seconds. The PCM can identify if the fuel delivery system can supply adequate fuel during heavy acceleration (Power Enrichment Mode). When this mode is requested during closed loop operation, the PCM provides extra fuel to the engine, and checks for an overly rich condition. If it a rich condition is not detected at this time, it will set this code. **Possible Causes** ● Fuel filter clogged or restricted ● Fuel injectors dirty or restricted ● Fuel level too low during heavy acceleration events ● Fuel pressure regulator has failed (not supplying enough fuel) ● Water or alcohol in fuel (low HO2S signal during acceleration)
DTC P1171 **2T FUEL, MIL: Yes** 2002, 2003, 2004, 2005 Envoy & TrailBlazer 4.2L VIN S engine Transmissions: All	**Fuel System Lean During Acceleration Conditions:** DTC P0122, P0123, P0131, P0132, P0133 and P0134 not set, engine started, engine runtime over 20 seconds in closed loop, ECT sensor more than 68ºF, Power Enrichment mode active, and the PCM detected the HO2S signal was less than 300 mv for 5 seconds. The PCM can identify if the fuel delivery system can supply adequate fuel during heavy acceleration (Power Enrichment Mode). When this mode is requested during closed loop operation, the PCM provides extra fuel to the engine, and checks for an overly rich condition. If a rich condition is not detected, it will set this code. **Possible Causes** ● Fuel filter clogged or restricted ● Fuel injectors dirty or restricted ● Fuel level too low during heavy acceleration events ● Fuel pressure regulator has failed (not supplying enough fuel) ● Water or alcohol in fuel (low HO2S signal during acceleration)

OBD II Trouble Code List (P1xxx Codes)

DTC	Trouble Code Title, Conditions & Possible Causes
DTC P1172 **1T CCM, MIL: No** 2003, 2004, 2005 CTS (Cadillac) 2.6L VIN M, 3.2L VIN N Transmissions: All; Cadillac CTS models	**Secondary Or Primary Fuel Level Sensor Malfunction Conditions:** Engine runtime over 2 minutes; and the PCM detected a difference of less than a 1-gallon between the Fuel Level 1 and Fuel Level 2 sending unit signal. This test checks for a restriction in the Fuel system between the secondary and primary of the fuel tank. The PCM sets this code if it detects the fuel level, secondary side of tank, signal appears to be stuck based on a lack of signal variation occurs during normal vehicle operation. **Possible Causes** ● Fuel level sending unit circuit has high resistance condition ● Primary or Secondary fuel tank module is damaged or failed ● PCM has failed
DTC P1172 **1T CCM, MIL: No** 2000, 2001, 2002, 2003, 2004, 2005 C/K Series Truck, Escalade & G Series Van Engines: 4.8L VIN V, 5.3L VIN T, 5.3L VIN Z, 6.0L VIN U, 6.0L VIN N Transmissions: All	**Secondary Fuel Pump Insufficient/No Fuel Flow Conditions:** DTC P0461, P0462, P0463, P1431, P1432 and P1433 not set, engine started, vehicle not moving, primary fuel level less than 25 L (6.6 gallons), secondary fuel level between 3-10 L (0.7-2.6 gallons), conditions met for 20 seconds before secondary pump commanded "on" for 2 minutes, and the PCM detected the change in the primary and secondary fuel level sensors was less than 4 liters (1.06 gallon). The secondary fuel pump, located in the rear fuel tank, is powered by a secondary fuel pump relay. Fuel is transferred from the rear fuel tank to the front fuel tank to ensure all of the usable fuel volume is available to the primary fuel pump. Secondary fuel pump relay supply voltage is received from the primary fuel pump relay when the primary fuel pump is "on". If the PCM commands the secondary fuel pump "on", and it does not detect a predetermined change in both the front and rear fuel level sensors, it will set this trouble code. **Possible Causes** ● Fuel tank level is too low (must be within 25-75% to test relay) ● Secondary F/P relay power circuit is open (test the ENG1 fuse) ● Secondary F/P relay control circuit is open, shorted to ground or shorted to power ● Secondary F/P relay is damaged or has failed ● PCM has failed
DTC P1172 **1T CCM, MIL: No** 2001, 2002, 2003, 2004, 2005 C/K Truck 6.6L VIN 1 Diesel engine Transmissions: All	**Secondary Fuel Pump Insufficient/No Fuel Flow Conditions:** DTC P0461, P0462, P0463, P0500, P2066, P2067 and P2068 not set, engine started, VSS at zero mph, Primary fuel level less than 25 liters (6.6 gallons), Secondary fuel level between 3-10 liters (0.8-2.6 gallons), conditions met for 20 seconds before the fuel transfer pump was commanded on, then with the fuel transfer pump commanded on for 120 seconds, and the ECM detected a primary fuel level increase and a secondary fuel level decrease of less than 4 liters (1.06 gallons) each. **Possible Causes** ● Fuel transfer relay control circuit is open or shorted to ground ● Fuel pump relay ground circuit is open ● Fuel pump relay power circuit is open (check the ECB B fuse) ● Fuel pump relay or the Fuel transfer pump is damaged or has failed ● ECM has failed
DTC P1187 **1T CCM, MIL: No** 1996, 1997 Corvette 5.7L VIN 5, 5.7L VIN P, 5.7L VIN G Transmissions: All	**Engine Oil Temperature Sensor Low Input Conditions:** Engine runtime over 30 minutes; and the PCM detected the Engine Oil Temperature (EOT) sensor was more than 306°F for 4.5 seconds. The ECT and EOT sensors should be within a few degrees after sitting overnight. Refer to the Failure Records for additional information. **Possible Causes** ● EOT sensor signal circuit is shorted to ground ● EOT sensor is damaged or it has failed (possible short circuit) ● PCM has failed
DTC P1188 **1T CCM, MIL: No** 1996, 1997 Corvette 5.7L VIN 5, 5.7L VIN P, 5.7L VIN G Transmissions: All	**Engine Oil Pressure Sensor High Input Conditions:** Engine running over 30 minutes, and the PCM detected the Engine Oil Pressure (EOP) sensor indicated less than -23°F for 4.5 seconds. The ECT and EOP sensors should be within a few degrees after sitting overnight. Note: Refer to information in the Failure Records. **Possible Causes** ● EOP sensor signal circuit is open, or the EOP sensor ground circuit is open ● EOP sensor is damaged or it has failed (possible open circuit) ● PCM has failed
DTC P1189 **1T CCM, MIL: No** 1998, 1999, 2000, 2001, 2002, 2003, 2004, 2005 DeVille, Eldorado, Seville 4.6L VIN 9, 4.6L VIN Y Transmissions: All	**Engine Oil Pressure Switch Circuit Malfunction Conditions:** DTC P0116, P0117, P0118 and P1114 not set, starter request not activated this key cycle, ECT sensor at last key off over 176°F and current ECT sensor signal 18°F less than the signal at last key off, or with the time between last key off and current key on cycle over 4 minutes, and the PCM detected a high voltage on the EOT switch circuit for 5 seconds. **Possible Causes** ● EOT switch circuit is open or the EOT switch ground circuit is open ● Engine oil temperature switch is damaged or has failed ● PCM has failed ● TSB 87-65-24 contains a repair procedure for this code

OBD II Trouble Code List (P1xxx Codes)

DTC	Trouble Code Title, Conditions & Possible Causes
DTC P1189 **1T CCM, MIL: Yes** 2000, 2001, 2002, 2003, 2004, 2005 Alero, Aztek, Century, Grand Am, Impala, Regal, Lumina, Monte Carlo, Montana, Silhouette, Rendezvous, Venture 3.1L VIN J, 3.4L VIN E, 3.8L VIN K engines Transmissions: All	**Engine Oil Pressure Switch Circuit Malfunction Conditions:** DTC P0117, P0118, P1111 and P1114 codes set, engine started, ECT sensor less than 50ºF at last key off, and the PCM detected an open EOP switch circuit for 10 seconds. Check the Failure Records. **Possible Causes** • EOP switch circuit is open between the sensor and the PCM • Engine oil pressure switch is damaged (possible open circuit) • PCM has failed
DTC P1200 **1T CCM, MIL: Yes** 1995 Models: Century Engines: 3.1L VIN M Transmissions: All	**Fuel Injector Circuit Malfunction Conditions:** Engine started; and the PCM detected an unexpected voltage condition on one or more of the fuel injector driver circuits for 5 seconds during the CCM test. **Possible Causes** • Fuel injector control circuit is open, shorted to ground or the power circuit is open • Fuel injector is damaged or has failed • PCM has failed
DTC P1200 **1T CCM, MIL: Yes** 1995 Beretta, Camaro, Ciera, Corsica, Century, Regal, Firebird, Riviera & Lumina APV 3.4L VIN E, 3.4L VIN X, 3.8L VIN 1, 3.8L VIN K Transmissions: All	**Fuel Injector Circuit Malfunction Conditions:** Engine started; and the PCM detected an unexpected voltage condition on one or more of the fuel injector driver circuits for 5 seconds during the CCM test. **Possible Causes** • Fuel injector control circuit is open or shorted to ground • Fuel injector power circuit is open (check the ECM fuse) • Fuel injector is damaged or has failed • PCM has failed
DTC P1200 **1T CCM, MIL: Yes** 1996 Beretta, Camaro, Ciera, Corsica, Century, Regal, Firebird, Riviera, Lumina APV & Trans Sport APV 3.1L VIN M, 3.4L VIN E, 3.4L VIN X, 3.8L VIN 1, 3.8L VIN K engines Transmissions: All	**Fuel Injector Circuit Malfunction Conditions:** Engine started; and the PCM detected an unexpected voltage condition on one or more of the fuel injector driver circuits for 5 seconds during the CCM test. **Possible Causes** • Fuel injector control circuit is open or shorted to ground • Fuel injector power circuit is open (check the ECM fuse) • Fuel injector is damaged or has failed • PCM has failed
DTC P1201 **1T CCM, MIL: Yes** 1999, 2000, 2001, 2002 Cavalier 2.2L VIN 4 CNG engine Transmissions: All	**Gas Mass Sensor Range/Performance Conditions:** No A/F ECU, ECT, O2S or GMS codes set, engine sped more than 500 rpm while operating on alternate fuel, system voltage over 10v, Fuel Pressure Sensor indicated the CNG fuel level over 25%, ECT sensor from 176-221ºF, AF ECU not operating in Decel Fuel Cutoff or Speed Limiting mode, and the PCM detected the difference between the Desired Gas Flow and the Actual Gas Flow was more than 45% for over 5 seconds. The Gas Mass Sensor (GMS) and the Mixture Control Valve (MCV) are contained in one non-serviceable unit. The GMS monitors the mass and flow of the gaseous fuel entering the engine and converts this data into a GMS ACTUAL Gas Flow signal circuit frequency. The alternative fuels engine control module commands fuel flow by supplying the MCV with a frequency signal. The Desired Gas Flow signal frequency varies from 1050 Hz (0.40 g/sec) at idle, to 5000 Hz (18 g/sec) at wide open throttle. The AF ECM converts the ACTUAL Gas Flow signal circuit frequency into a g/sec value. During low fuel flow rates the GMS sensor Actual Gas circuit will produce a low frequency signal of around 700 Hz (0.21 g/sec). During high fuel flow rates (i.e., at WOT road load) the GMS sensor will produce a high frequency signal of around 2700 Hz (16.65 g/sec). The AF ECM monitors and compares the Actual Gas flow rate to a calculated gas flow rate. The calculated gas flow rate is based upon a calculated Mass Air Flow rate. This trouble code is used to indicate that the Actual Gas flow rate does not match a calculated gas flow rate. The vehicle switches to gasoline operation when this code sets. **Possible Causes** • Air filter or air inlet duct is clogged or restricted • Air leaks in the exhaust pipes or manifold, or the HO2S location • Base engine problem (i.e., engine compression, vacuum leak) • CNG system fuel pressure is incorrect • PCV system problem (i.e., PCV vacuum hose or valve leaking) • EVAP canister purge system malfunction • Gasoline contaminated engine oil • Fuel pressure regulator damage or leaking • PCM has failed

OBD II Trouble Code List (P1xxx Codes)

DTC	Trouble Code Title, Conditions & Possible Causes
DTC P1201 **1T CCM, MIL: Yes** 1999, 2000, 2001, 2002 C/K Truck, G Van 6.0L VIN U CNG engine Transmissions: All	**Gas Mass Sensor Range/Performance Conditions:** No A/F ECU, ECT, O2S or GMS codes set, engine is operating on alternative fuel, engine speed more than 500 rpm, system voltage over 10v, Fuel Pressure Sensor indicated the CNG fuel level over 25%, ECT sensor from 176-221ºF, AF ECU not operating in Decel Fuel Cutoff or Speed Limiting mode, and the PCM detected the difference between the Desired Gas Flow and the Actual Gas Flow was more than 45% for over 5 seconds. The Gas Mass Sensor (GMS) and the Mixture Control Valve (MCV) are contained in one non-serviceable assembly. The GMS/MCV is supplied with ignition voltage and ground. The GMS monitors the mass and flow of the gaseous fuel entering the engine and converts this information into a GMS Actual Gas Flow signal frequency. The AF ECU commands fuel flow by supplying the MCV with a frequency signal. The Desired Gas Flow signal frequency varies from 1050 Hz (0.40 g/sec) at idle, to around 5000 Hz (18 g/sec) at wide open throttle. The AF ECU converts the Actual Gas Flow signal circuit frequency to a gram per second value. During low fuel flow rates (such as idle speed) the GMS sensor produces a low frequency signal of around 700 Hz (0.21 g/sec). During high fuel flow rates (e.g., wide open throttle at road load) the GMS sensor produces a high frequency signal of around 2700 Hz (16.65 g/sec). This code is used to indicate the Actual Gas flow rate is not equal to the Desired Gas flow rate. **Possible Causes** • Air filter or air inlet duct is clogged or restricted • Air leaks in the exhaust pipes or manifold, or the HO2S location • Base engine problem (i.e., engine compression, vacuum leak) • CNG system fuel pressure is incorrect • PCV system problem (i.e., PCV vacuum hose or valve leaking) • EVAP canister purge system malfunction • Gasoline contaminated engine oil • Fuel pressure regulator damage or leaking • PCM has failed
DTC P1202 **1T CCM, MIL: Yes** 1999, 2000, 2001, 2002 Cavalier, C/K Series Truck, G Series Van 2.2L VIN 4, 6.0L VIN U CNG engines Transmissions: All	**Gas Mass Flow Sensor Low Frequency Conditions:** Engine cranking or running (engine speed from 25-500 rpm) while on alternative fuel, system voltage over 10v, and the AF ECU detected the GMS Actual Gas Flow signal frequency was less than 250 Hz for 500 ms; or engine speed over 500 rpm for 10 seconds on alternative fuel, system voltage over 10v, the AF ECU detected the GMS Actual Gas Flow frequency was less than 500 Hz for 500 ms. The vehicle switches to gasoline when this code is set. **Possible Causes** • GMS/MCV power circuit is open (test the CNG fuse underhood) • GMS/MCV power ground circuit is open or has high resistance • GMS/MCV motor power or ground circuit is open or grounded • MCV valve is damaged or has failed or the GMS has failed
DTC P1203 **1T CCM, MIL: Yes** 1999, 2000, 2001, 2002 Cavalier, C/K Series Truck, G Series Van 2.2L VIN 4, 6.0L VIN U Transmissions: All	**Gas Mass Flow Sensor High Frequency Conditions:** Engine cranking or running (engine speed from 25-500 rpm) while on alternative fuel, system voltage over 8v, then with engine speed over 500 while operating on alternative fuel for 10 seconds, and the AF ECU detected the GMS Actual Gas Flow signal frequency was over 2900 Hz for 500 ms continuously. The vehicle switches to gasoline fuel when the code is set. **Possible Causes** • Actual Gas Flow signal circuit routed near secondary ignition components or near high current components (motors, relays) • High power transceivers operating in or near the vehicle • GMS/MCV motor power or ground circuit has high resistance • MCV valve is damaged or has failed or the GMS has failed
DTC P1204 **1T CCM, MIL: Yes** 2003, 2004, 2005 Cavalier, C/K Series Truck, G Series Van 2.2L VIN 4, 6.0L VIN U CNG engines Transmissions: All	**Engine Will Not Start In CNG Mode Conditions:** DTC P0005, P0148, P0191, P0192, P0193, P0611, P1020, P1021-P1028, P1209, P2146 and P2665 not set, Engine started, system voltage from 6-18v, CNG operation is not inhibited, and the FICM detected the cranking time to start in CNG mode was 8 seconds. The PCM controls fuel delivery, and on KL6 vehicles, determines which fuel system the engine is operating on. The PCM controls the low-pressure lock-off (LPL) solenoid, and the high-pressure lock-off (HPL) solenoid. The PCM commands the HPL solenoid open for 1 second at key on, to prime the CNG system. The PCM commands the HPL and LPL open with the engine cranking or running on CNG. **Possible Causes** • Diagnose the other related codes, then recheck for this code • FRP sensor parameter is out of range

OBD II Trouble Code List (P1xxx Codes)

DTC	Trouble Code Title, Conditions & Possible Causes
DTC P1207 **1T CCM, MIL: Yes** 2003, 2004, 2005 C/K Truck, G Series Van 6.0L VIN U CNG engine Transmissions: All	**FICM PWM Signal To PCM Circuit Malfunction Conditions:** Key on or engine running; and the PCM did not detect a Fuel Rail Pressure sensor PWM output signal for 2-5 seconds. The Fuel Injector Control Module (FICM) monitors the CNG FTP sensor, the FTT sensor, and fuel rail temperature sensor. The FTP, FTT, and FRT sensor signal and FICM diagnostic status is communicated to the PCM by two PWM circuits. **Possible Causes** ● FICM sensor PWM signal circuit is open or shorted to ground ● FICM is damaged or it has failed ● PCM has failed
DTC P1208 **1T CCM, MIL: Yes** 2003, 2004, 2005 G Series Van, Cargo Van 6.0L VIN U CNG engine Transmissions: All	**Fuel Rail Pressure Signal To PCM Circuit Malfunction Conditions:** Key on or engine running; and the PCM did not receive the Fuel Rail Temperature sensor signal for five seconds. The Fuel Injector Control Module (FICM) monitors the fuel rail temperature (FRT) sensor voltage, and sends a signal to the PCM via a PWM circuit. This code sets if the FRP output signal is not received by the PCM. **Possible Causes** ● FICM sensor PWM signal circuit is open or shorted to ground ● FICM is damaged or it has failed ● PCM has failed
DTC P1208 **1T CCM, MIL: Yes** 2003, 2004, 2005 Cavalier, C/K Series Truck, G Series Van 2.2L VIN 4, 6.0L VIN U CNG engines Transmissions: All	**FICM Diagnostic Status Signal Circuit Malfunction Conditions:** Engine cranking or engine running, system voltage from 6-18v, and the PCM did not receive the FICM diagnostic status signal for at least two seconds. The fuel injector control module (FICM) monitors the CNG fuel tank pressure (FTP) sensor, the fuel tank temperature (FTT) sensor, and the fuel rail temperature (FRT) sensor. The FTP, FTT, and FRT sensor values, and FICM diagnostic status, is communicated to the PCM by two PWM circuits. **Possible Causes** ● FICM diagnostic status circuit is open or shorted to ground ● FICM is damaged or it has failed ● PCM has failed
DTC P1214 **2T CCM, MIL: Yes** 1996, 1997, 1998, 1999, 2000, 2001, 2002 C/K Truck, G Van 6.5L VIN F, 6.5L VIN S, 6.5L VIN Y, 6.6L VIN 1 Diesel engines Transmissions: All	**Injection Pump Timing Offset Conditions:** Key on or engine running; and the PCM detected the Timing Offset was more than 2.46 degrees or less than -2.46 degrees. Note: This code may set due to an intermittent fault. **Possible Causes** ● The PCM will only run the test if the TDC offset adjustment time set procedure has been activated. The vehicle will most likely not be brought in with this code in memory. ● CKP sensor incorrectly installed or mounting tab is broken ● Fuel injection pump is damaged or has failed (a new pump will need to be "timed" using the correct pump timing procedure)
DTC P1215 **1T CCM, MIL: Yes** 1999, 2000, 2001, 2002 Cavalier 2.2L VIN 4 CNG engine Transmissions: All	**High Pressure Or Low Pressure Lockout, MIL, NGO Circuit Malfunction Conditions:** Engine started and the AF ECU detected an unexpected voltage on the Natural Gas Output Enable, High Pressure Lockout, Low Pressure Lockout or MIL control circuit for 1 second. **Possible Causes** ● Natural Gas Output Enable, High Pressure Lockout, Low Pressure Lockout or MIL control circuit is open, shorted to ground or shorted to system power ● Alternative Fuels ECU is damaged or it has failed
DTC P1215 **1T CCM, MIL: Yes** 1999, 2000, 2001, 2002 C/K Truck 6.0L VIN U CNG engine Transmissions: All	**High Pressure Or Low Pressure Lockout, MIL, NGO Circuit Malfunction Conditions:** Key on or engine running; and the AF ECU detected an unexpected voltage condition on the Natural Gas Output Enable, High Pressure Lockout, Low Pressure Lockout or MIL control circuit for 1 second. The Alternative Fuels (AF) ECM contains output driver modules that provide control circuits for operating solenoids, relays, telltales, and other devices. **Possible Causes** ● Natural Gas Output Enable, High Pressure Lockout, Low Pressure Lockout or MIL control circuit is open, shorted to ground or shorted to system power ● Alternative Fuels ECU is damaged or has failed
DTC P1216 **1T CCM, MIL: No** 1996, 1997, 1998, 1999, 2000, 2001, 2002 C/K Truck, G Series Van 6.5L VIN F, 6.5L VIN S, 6.5L VIN Y, 6.6L VIN 1 Diesel engines Transmissions: All	**Fuel Solenoid Response Time Too Short Conditions:** Engine started, engine speed over 506 rpm, ECT sensor more than 34ºF, system voltage over 10.0v, requested fuel rate more than 0.0 mm, and the PCM detected the injection pump closure time was less than 0.75 ms during the CCM test. **Possible Causes** ● Intermittent P0251, P0370, and P1216 codes may be caused by air entering the fuel system when fuel levels get below 1/8 of a tank during hard acceleration or turning maneuvers. Codes P0251, P0370, and P1216 may set if the vehicle has run out of fuel. Customer driving habits should be checked to determine if the vehicle has been performing in these manners. If the vehicle has performed in these conditions, bleed the fuel system of all air and then perform a complete vehicle test drive. ● Injection pump timing circuit is open, shorted to ground or to B+ ● TSB 77-63-06A contains a repair procedure for this code

OBD II Trouble Code List (P1xxx Codes)

DTC	Trouble Code Title, Conditions & Possible Causes
DTC P1217 **1T CCM, MIL: No** 1996, 1997, 1998, 1999, 2000, 2001, 2002 C/K Truck, G Van 6.5L VIN F, 6.5L VIN S, 6.5L VIN Y, 6.6L VIN 1 Diesel engines Transmissions: All	**Fuel Solenoid Response Time Too Long Conditions:** Engine started, engine speed over 506 rpm, ECT sensor more than 34ºF, requested fuel rate more than 0.0 mm, and the PCM detected the injection pump closure time was more than 2.5 ms in the test. The injection pump delivers fuel to the individual cylinders by opening and closing a solenoid control fuel valve. The PCM monitors the amount of time the fuel solenoid valve takes to physically close after commanded to close. Any closure time that is out of range is seen as a fault. This response time is measured in milliseconds **Possible Causes** ● Fuel solenoid driver power circuit is open (test FUEL SOL fuse) ● Fuel solenoid high circuit is open or shorted to ground ● Fuel solenoid low circuit is open ● Fuel solenoid is damaged or has failed (solenoid may be weak) ● Fuel solenoid driver has failed
DTC P1218 **2T CCM, MIL: Yes** 1996, 1997, 1998, 1999, 2000, 2001, 2002 C/K Truck, G Van 6.5L VIN F, 6.5L VIN S, 6.5L VIN Y, 6.6L VIN 1 Diesel engines Transmissions: All	**Injection Pump Calibration Circuit Malfunction Conditions:** Key on or engine running; and the PCM did not detect the correct current valid resistor value, or it determined that it was unable to read the current resistor value during the CCM test. **Possible Causes** ● Fuel solenoid driver connections have a problem ● Fuel solenoid is damaged or has failed ● Fuel injection pump is damaged or has failed (a new pump will need to be "timed" using the correct pump timing procedure)
DTC P1220 **1T CCM, MIL: Yes** 1999, 2000, 2001, 2002, 2003, 2004, 2005 C/K Truck, Envoy, G Van, Escalade & TrailBlazer 4.8L VIN V, 5.3L VIN P, 5.3L VIN T, 6.0L VIN U, 7.4L VIN J, 8.1L VIN G Transmissions: All	**TP Sensor 2 Circuit Malfunction Conditions:** DTC P1517 and P1518 not set, key in crank or run position, system voltage over 5.23v, and the PCM detected the TP2 signal was less than 0.13v or more than 4.87v for 1 second during the CCM test. **Possible Causes** ● TP2 sensor signal circuit is open, shorted to ground or to power ● TP2 sensor VREF circuit is open, shorted to ground or shorted to system power (B+) ● TP2 sensor ground circuit has a high resistance condition ● TP2 sensor is damaged or has failed
DTC P1220 **1T CCM, MIL: Yes** 1999, 2000, 2001 Catera 3.0L VIN R engine Transmissions: All	**TP Sensor 2 Circuit Malfunction Conditions:** Key on or engine running; and the PCM detected the TP sensor 2 signal was below 0.156v or more than 4.80v, or the difference between TP2 and TP1 signal was over 13% for 140 ms. **Possible Causes** ● TP2 sensor signal circuit is open, shorted to ground or to power ● TP2 sensor VREF circuit is open, shorted to ground or shorted to system power (B+) ● TP2 sensor ground circuit has a high resistance condition ● TP2 sensor is damaged or has failed
DTC P1220 **1T CCM, MIL: Yes** 2002, 2003, 2004, 2005 Envoy & TrailBlazer 4.2L VIN S engine Transmissions: All	**TP Sensor 2 Circuit Malfunction Conditions:** DTC P1635 not set, key in crank or run mode, system voltage over 5.23v, Electronic Throttle Control serial data present, and the PCM detected the TP Sensor 2 was less than 0.13v or more than 4.70v. **Possible Causes** ● TP2 sensor signal circuit is open, shorted to ground or to power ● TP2 sensor VREF circuit is open, shorted to ground or shorted to system power (B+) ● TP2 sensor ground circuit has a high resistance condition ● TP2 sensor is damaged or has failed
DTC P1220 **1T CCM, MIL: Yes** 1997, 1998, 1999, 2000, 2001, 2002, 2003, 2004, 2005 Corvette 5.7L VIN G, 5.7L VIN S Transmissions: All	**TP Sensor 2 Circuit Malfunction Conditions:** DTC P0606, P1517 and P1518 not set, key in crank or run mode, system voltage over 5.23v, Electronic Throttle Control serial data operating, and the PCM detected the TP Sensor 2 signal was less than 0.13v or more than 4.87v during the CCM test. **Possible Causes** ● TP2 sensor signal circuit is open, shorted to ground or to power ● TP2 sensor VREF circuit is open or shorted to ground ● TP2 sensor VREF circuit is shorted to system power (B+) ● TP2 sensor ground circuit has a high resistance condition ● TP2 sensor is damaged or has failed

OBD II Trouble Code List (P1xxx Codes)

DTC	Trouble Code Title, Conditions & Possible Causes
DTC P1221 **1T CCM, MIL: Yes** 1999, 2000, 2001, 2002, 2003, 2004, 2005 C/K Truck, Envoy, G Van, Escalade & TrailBlazer 4.8L VIN V, 5.3L VIN P, 5.3L VIN T, 6.0L VIN U, 7.4L VIN J, 8.1L VIN G Transmissions: All	**TP Sensor 2 Signal Correlation Conditions:** DTC P1517 and P1518 not set, key in crank or run position TP Sensor 1 (TP1) and TP Sensor 2 (TP2) more than 15% for 140 ms, and the PCM detected the TP2 signal disagreed with the TP1 signal by more than 7.5% for 1 second. The TP sensor has two separate signal, ground, and 5 volt reference circuits that are used to connect the TP sensor to the TAC module. These sensors have opposite functionality. The TP1 voltage increases from below 1.0v at 0% throttle to above 3.5v at 100% throttle opening. The TP2 voltage decreases from around 3.8v at 0 percent throttle to below 1.0v at 100% throttle opening. The TP1 signal circuit is pulled up to 5.0v and the TP2 signal circuit is pulled to ground in the TAC module. **Possible Causes** ● TP2 sensor connector is contaminated, dirty or contains water ● TP2 sensor signal, ground or VREF circuit has high resistance ● TP2 sensor VREF circuit has a high resistance condition ● TP2 sensor ground circuit has a high resistance condition ● TP2 sensor is damaged or has failed ● TAC controller or the throttle body is damaged or has failed ● TSB 02-06-04-005 contains a repair procedure for this code
DTC P1221 **1T CCM, MIL: Yes** 1997, 1998, 1999, 2000, 2001, 2002, 2003, 2004, 2005 Corvette 5.7L VIN G, 5.7L VIN S Transmissions: All	**TP Sensor 1-2 Signal Correlation Conditions:** DTC P0606, P1517 and P1518 not set, key in crank or run mode, ETC serial data normal, and the PCM detected the TP1 disagreed with the TP2 input by over 7.5% for 1 second. **Possible Causes** ● TP1, 2 sensor signal circuit has a high resistance condition ● TP1, 2 sensor VREF circuit has a high resistance condition ● TP1, 2 sensor ground circuit has a high resistance condition ● TP1, 2 sensor is damaged or has failed ● TAC controller or the throttle body is damaged or has failed ● TSB 02-06-04-005 contains a repair procedure for this code
DTC P1221 **1T CCM, MIL: Yes** 2002, 2003, 2004, 2005 S/T Blazer & S/T Pickup 4.2L VIN S engine Transmissions: All	**TP Sensor 2 Circuit Malfunction Conditions:** DTC P1120, P1220, P1635 and P1639 not set, key on, system voltage over 5.23v, and the PCM detected the difference between the TP Sensor 1 and 2 circuits was out-of-range. **Possible Causes** ● TP1, 2 sensor signal circuit has a high resistance condition ● TP1, 2 sensor VREF circuit has a high resistance condition ● TP1, 2 sensor ground circuit has a high resistance condition ● TP1, 2 sensor is damaged or has failed ● TAC controller or the throttle body is damaged or has failed
DTC P1221 **1T CCM, MIL: No** 1995 S/T Blazer & S/T Pickup 4.3L VIN W engine Transmissions: All	**Fuel Pump Secondary System Circuit Low Input Conditions:** Key on or engine running; and the PCM detected the voltage at the Fuel Pump Signal circuit (BK14) was less than 2v for 1.5 seconds since the last ignition reference pulse was detected. **Possible Causes** ● Fuel pump relay power circuit is open (test power at Fuse 9) ● Fuel pump relay control circuit is open or shorted to ground ● Fuel pump relay signal circuit is open ● Fuel pump relay is damaged or has failed ● PCM has failed
DTC P1222 **1T CCM, MIL: No** 1996, 1997 Camaro, Caprice, Corvette, Firebird, Fleetwood & Roadmaster 4.3L VIN W, 5.7L VIN 5, 5.7L VIN P engines Transmissions: All	**Injector Control Circuit Malfunction (Intermittent) Conditions:** Engine started, system voltage over 10.0v, and the PCM detected an unexpected voltage value on an injector driver circuit for 500 ms. **Possible Causes** ● Fuel injector control circuit is open (an intermittent fault) ● Fuel injector control circuit shorted to ground (intermittent fault) ● Fuel injector power circuit is open (test the INJ 1, INJ 2 fuses) ● Fuel injector is damaged or has failed ● PCM has failed
DTC P1223 **1T CCM, MIL: No** 2001, 2002, 2003, 2004, 2005 C/K Truck 6.6L VIN 1 Diesel engine Transmissions: All	**Fuel Injector 1 Control Circuit Low Input Conditions:** Key on or engine running system voltage from 6-18v, and the PCM detected the Command and Actual state of the Injector 1 control circuit did not match for over 2 seconds in the test. **Possible Causes** ● FICM (module) connector is damaged, open or shorted ● FICM (module) is damaged or it has failed ● Fuel injector 1 control circuit is open, shorted to ground or to B+ ● Fuel injector 1 is damaged or it has failed ● PCM has failed

OBD II Trouble Code List (P1xxx Codes)

DTC	Trouble Code Title, Conditions & Possible Causes
DTC P1226 **1T CCM, MIL: No** 2001, 2002, 2003, 2004, 2005 C/K Truck 6.6L VIN 1 Diesel engine Transmissions: All	**Fuel Injector 2 Control Circuit Low Input Conditions:** Key on or engine running system voltage from 6-18v, and the PCM detected the Command and Actual state of the Injector 2 control circuit did not match for over 2 seconds in the test. **Possible Causes** ● FICM (module) connector is damaged, open or shorted ● FICM (module) is damaged or it has failed ● Fuel injector 2 control circuit is open, shorted to ground or to B+ ● Fuel injector 2 is damaged or it has failed ● PCM has failed
DTC P1226 **1T CCM, MIL: No** 2001, 2002, 2003, 2004, 2005 C/K Truck 6.6L VIN 1 Diesel engine Transmissions: All	**Fuel Injector 3 Control Circuit Low Input Conditions:** Key on or engine running system voltage from 6-18v, and the PCM detected the Command and Actual state of the Injector 3 control circuit did not match for over 2 seconds in the test. **Possible Causes** ● FICM (module) connector is damaged, open or shorted ● FICM (module) is damaged or it has failed ● Fuel injector 3 control circuit is open, shorted to ground or to B+ ● Fuel injector 3 is damaged or it has failed ● PCM has failed
DTC P1232 **1T CCM, MIL: No** 2001, 2002, 2003, 2004, 2005 C/K Truck 6.6L VIN 1 Diesel engine Transmissions: All	**Fuel Injector 4 Control Circuit Low Input Conditions:** Key on or engine running system voltage from 6-18v, and the PCM detected the Command and Actual state of the Injector 4 control circuit did not match for over 2 seconds in the test. **Possible Causes** ● FICM (module) connector is damaged, open or shorted ● FICM (module) is damaged or it has failed ● Fuel injector 4 control circuit is open, shorted to ground or to B+ ● Fuel injector 4 is damaged or it has failed ● PCM has failed
DTC P1235 **1T CCM, MIL: No** 2001, 2002, 2003, 2004, 2005 C/K Truck 6.6L VIN 1 Diesel engine Transmissions: All	**Fuel Injector 5 Control Circuit Low Input Conditions:** Key on or engine running system voltage from 6-18v, and the PCM detected the Command and Actual state of the Injector 5 control circuit did not match for over 2 seconds in the test. **Possible Causes** ● FICM (module) connector is damaged, open or shorted ● FICM (module) is damaged or it has failed ● Fuel injector 5 control circuit is open, shorted to ground or to B+ ● Fuel injector 5 is damaged or it has failed ● PCM has failed
DTC P1238 **1T CCM, MIL: No** 2001, 2002, 2003, 2004, 2005 C/K Truck 6.6L VIN 1 Diesel engine Transmissions: All	**Fuel Injector 6 Control Circuit Low Input Conditions:** Key on or engine running system voltage from 6-18v, and the PCM detected the Command and Actual state of the Injector 6 control circuit did not match for over 2 seconds in the test. **Possible Causes** ● FICM (module) connector is damaged, open or shorted ● FICM (module) is damaged or it has failed ● Fuel injector 6 control circuit is open, shorted to ground or to B+ ● Fuel injector 6 is damaged or it has failed ● PCM has failed
DTC P1241 **1T CCM, MIL: No** 2001, 2002, 2003, 2004, 2005 C/K Truck 6.6L VIN 1 Diesel engine Transmissions: All	**Fuel Injector 7 Control Circuit Low Input Conditions:** Key on or engine running system voltage from 6-18v, and the PCM detected the Command and Actual state of the Injector 7 control circuit did not match for over 2 seconds in the test. **Possible Causes** ● FICM (module) connector is damaged, open or shorted ● FICM (module) is damaged or it has failed ● Fuel injector 7 control circuit is open, shorted to ground or to B+ ● Fuel injector 7 is damaged or it has failed ● PCM has failed
DTC P1244 **1T CCM, MIL: No** 2001, 2002, 2003, 2004, 2005 C/K Truck 6.6L VIN 1 Diesel engine Transmissions: All	**Fuel Injector 8 Control Circuit Low Input Conditions:** Key on or engine running system voltage from 6-18v, and the PCM detected the Command and Actual state of the Injector 8 control circuit did not match for over 2 seconds in the test. **Possible Causes** ● FICM (module) connector is damaged, open or shorted ● FICM (module) is damaged or it has failed ● Fuel injector 8 control circuit is open, shorted to ground or to B+ ● Fuel injector 8 is damaged or it has failed ● PCM has failed

OBD II Trouble Code List (P1xxx Codes)

DTC	Trouble Code Title, Conditions & Possible Causes
DTC P1257 **2T CCM, MIL: Yes** 1996, 1997, 1998 Bonneville, LSS, Park Avenue, Regal & Riviera 3.8L VIN 1 SC engine Transmissions: All	**Supercharger System Overboost Malfunction Conditions:** Engine started, IAT sensor more than 14ºF, and the PCM detected the engine torque was above a maximum threshold value (varies by the commanded gear and rpm) for 6 seconds. **Possible Causes** ● Boost control solenoid control circuit shorted to ground ● Supercharger bypass valve is sticking or binding (mechanical) ● Supercharger bypass actuator sticking or binding (mechanical)
DTC P1258 **2T CCM, MIL: Yes** 1996, 1997,1998, 1999, 2000, 2001, 2002, 2003, 2004, 2005 Aurora, DeVille, Eldorado & Seville 4.0L VIN C, 4.6L VIN 9 & 4.6L VIN Y engines Transmissions: A/T	**Engine Metal Over-Temperature Protection Conditions:** DTC P0117, P0118 and P0125 not set, Engine started; and the PCM detected the ECT sensor indicated more than 270ºF for 10 seconds. The PCM uses the ECT sensor to detect if the engine is running too hot (when the ECT sensor is over 268ºF). If the engine is overheated, DTC P1258 will set. The PCM will disable 2 groups of four cylinders by turning off the fuel injectors. By switching between the two groups of cylinders, the PCM is able to reduce the temperature of the coolant and lower the temperature. **Possible Causes** ● Engine cooling system problem (test the cooling fan operation) ● Engine coolant level is low or the mixture is not correct ● ECT sensor is out-of-calibration or "skewed"
DTC P1260 **1T CCM, MIL: No** 2000 Bonneville, LSS, 88', 98', Park Avenue, Regal & Riviera 3.8L VIN 1, 3.8L VIN K Transmissions: All	**Fuel Pump Speed Relay Control Circuit Malfunctions Conditions:** Key on, and the PCM detected an improper voltage on the Fuel Pump Speed Control PWM circuit for 30 seconds during the test. Output driver modules are used by the PCM to turn on many of the current-driven devices that are needed to control various engine and transaxle functions. Each ODM can control up to 7 separate outputs by applying ground to the device that the PCM commands "on". When this code is set (P1260), this indicates an invalid voltage level has been detected on ODM 'A' Output 3, which controls the fuel pump speed PWM signal to the fuel pump control module. **Possible Causes** ● FP speed relay control circuit is open or shorted to ground ● FP speed relay control circuit is shorted to system power ● FP speed relay power circuit is open (test the FP relay fuse) ● FP speed relay is damaged or has failed ● PCM has failed
DTC P1261 **1T CCM, MIL: Yes** 2001, 2002, 2003, 2004, 2005 C/K Truck 6.6L VIN 1 Diesel engine Transmissions: All	**Fuel Pump Relay Circuit Malfunction (Injectors 1, 4, 6 or 7) Conditions:** Key on or engine running; and the PCM detected an unexpected voltage condition on one or more fuel injector circuits. The AF ECM (Alternative Fuels) monitors the status of each driver. If a circuit fault is detected on a fuel injector circuit for engine cylinders 1, 4, 6 or 7, DTC P0201, P0204, P0206, P0207 will set, along with P1261. **Possible Causes** ● One or more injector control circuits open or shorted to ground ● One or more injector power circuits open or injector is damaged ● Injector relay power or ground circuit is open, or relay has failed ● FICM (module) has failed ● ECM (module) has failed
DTC P1262 **1T CCM, MIL: Yes** 2001, 2002, 2003, 2004, 2005 C/K Truck 6.6L VIN 1 Diesel engine Transmissions: All	**Fuel Pump Relay Circuit Malfunction (Injectors 2, 3, 5 or 8) Conditions:** Key on or engine running; and the PCM detected an unexpected voltage condition on one or more fuel injector circuits. The AF ECM (Alternative Fuels) monitors the status of each driver. If a circuit fault is detected on a fuel injector circuit for engine cylinders 2, 3, 5 or 8, then DTC P0201, P0204, P0206, P0207 will set, along with P1262. **Possible Causes** ● One or more injector control circuits open or shorted to ground ● One or more injector power circuits open or injector is damaged ● Injector relay power or ground circuit is open, or relay has failed ● FICM (module) has failed ● ECM (module) has failed
DTC P1270 **1T CCM, MIL: No** 2001, 2002, 2003, 2004, 2005 C/K Truck 6.6L VIN 1 Diesel engine Transmissions: All	**Fuel Pump Relay Circuit Malfunction (Injectors 2, 3, 5 or 8) Conditions:** Engine cranking or running, system voltage from 7-16v, and the ECM detected that it could not process the Accelerator Pedal Position (APP) sensor analog data into digital data. **Possible Causes** ● ECM is damaged or it has failed. Clear the codes and then retest for this same trouble code. If it resets, the ECM must be replaced and reprogrammed to repair this code.

OBD II Trouble Code List (P1xxx Codes)

DTC	Trouble Code Title, Conditions & Possible Causes
DTC P1271 **1T CCM, MIL: Yes** 1999, 2000, 2001, 2002 Camaro & Firebird 3.8L VIN K engine Transmissions: All	**Accelerator Pedal Position Sensor 1-2 Correlation Conditions:** DTC P0606, P1275, P1280, P1517, P1518, P1635, P1639, P1776 and P1781 not set, key in crank or run position, ETC serial data operating, and the PCM detected the APP1 and APP 2 disagreed by over 10.5%, or the APP1 circuit was shorted to the APP2 circuit. The APP sensors are located in a throttle actuator control (TAC) module mounted to the accelerator pedal bracket. There are 3 individual APP sensors within the TAC module, each with three circuits. The APP sensor signal voltages increase when the accelerator pedal is depressed. **Possible Causes** ● When the TAC module detects a fault in the ETC system, the PCM receives a message on the serial data line. More than one related code might set because of the many redundant tests run continuously on this system. Locating and repairing a single condition may correct more than one code. Be aware of this when reviewing captured DTC Info on the Scan Tool. ● Refer to any other related TAC system trouble codes. If there are no other TAC system codes, the APP Sensor is defective.
DTC P1271 **1T CCM, MIL: Yes** 2002, 2003, 2004, 2005 Envoy & TrailBlazer 4.2L VIN S engine Transmissions: All	**Accelerator Pedal Position Sensor 1-2 Correlation Conditions:** DTC P1275, P1280, P1535 and P1639 not set, engine started, accelerator pedal leaving the idle position, and the PCM detected the APP1 and APP 2 signal voltage was more than a predetermined limit stored in memory. The APP sensors are located in a TAC module mounted to the accelerator pedal bracket. There are three individual APP sensors within the TAC module. Three separate signal circuits, low reference circuits, and 5v reference circuits are used to determine the accelerator pedal position. The APP sensor signal voltages increase when the accelerator pedal is depressed. **Possible Causes** ● When the TAC module detects a fault in the ETC system, the PCM receives a message on the serial data line. More than one related code might set because of the many redundant tests run continuously on this system. Locating and repairing a single condition may correct more than one code. Be aware of this when reviewing captured DTC Info on the Scan Tool. ● Refer to any other related TAC system trouble codes. If there are no other TAC system codes, the APP Sensor is defective.
DTC P1271 **1T CCM, MIL: Yes** 1999, 2000, 2001 Catera 3.0L VIN R engine Transmissions: A/T	**Accelerator Pedal Position Sensor 1-2 Correlation Conditions:** Engine started; system voltage over 10.0v, ETC serial data operating, the accelerator pedal is leaving idle position, the PCM detected the voltage difference between the APP 1 and APP 2 sensors was not correct when leaving the idle position. The APP1 voltage increases as the throttle opens, from about 0.4 volt at closed throttle to above 3.5 volts at WOT. The APP2 voltage also increases as the throttle opens, but at a slightly different rate. The APP2 voltage increases from about 0.20v at closed throttle to about 1.8v at WOT. The APP1 is the main control of the system. The APP2 signal is used for comparison of the APP1 signal. The ECM constantly monitors these two sensors and if the sensors are not within a calibrated value, a diagnostic trouble code (DTC) will set. **Possible Causes** ● APP sensor mounting bolts loose or damaged ● APP1, 2 sensor signal circuit open, shorted to ground or to B+ ● APP1, 2 sensor VREF circuit is open or shorted to ground ● APP1, 2 sensor VREF circuit is shorted to system power (B+) ● APP1, 2 sensor ground circuit has a high resistance condition ● APP1, 2 sensor is damaged or has failed
DTC P1272 **1T CCM, MIL: No** 1999, 2000, 2001, 2002 Camaro & Firebird 3.8L VIN K engine Transmissions: All	**Accelerator Pedal Position Sensor 2-3 Correlation Conditions:** DTC P0606, P1517, P1518, P1281 or P1286 not set, key in crank or run position, and the PCM detected the APP sensor 1 disagreed with the APP sensor 3 signal by more than 10.5%, or the APP sensor 1 was shorted to the APP sensor 3 signal. The APP sensors are located within the throttle actuator control (TAC) module that is mounted to the accelerator pedal bracket. There are 3 individual APP sensors within the TAC module. Three separate signal circuits, low reference circuits, and 5-volt reference circuits are used in order to determine the accelerator pedal position. The APP sensor signal voltages increase when the accelerator pedal is depressed. **Possible Causes** ● When the TAC module detects a fault in the ETC system, the PCM receives a message on the serial data line. More than one related code might set because of the many redundant tests run continuously on this system. Locating and repairing a single condition may correct more than one code. Be aware of this when reviewing captured DTC Info on the Scan Tool. ● Refer to any other related TAC system trouble codes. If there are no other TAC system codes, the APP Sensor has failed.

OBD II Trouble Code List (P1xxx Codes)

DTC	Trouble Code Title, Conditions & Possible Causes
DTC P1273 **1T CCM, MIL: No** 1999, 2000, 2001, 2002 Camaro & Firebird 3.8L VIN K engine Transmissions: All	**Accelerator Pedal Position Sensor 1-3 Correlation Conditions:** DTC P0606, P1517, P1518, P1276 or P1286 not set, key in crank or run position, and the PCM detected the APP sensor 2 disagreed with the APP sensor 3 signal by more than 10.5%, or the APP sensor 2 was shorted to the APP sensor 3 signal. The APP sensors are located within the throttle actuator control (TAC) module that is mounted to the accelerator pedal bracket. There are 3 individual APP sensors within the TAC module. Three separate signal circuits, low reference circuits, and 5-volt reference circuits are used in order to determine the accelerator pedal position. The APP sensor signal voltages increase when the accelerator pedal is depressed. *Note: Refer to the information in the Failure Records as needed.* **Possible Causes** ● When the TAC module detects a fault in the ETC system, the PCM receives a message on the data bus. More than 1 trouble code might set due to the many redundant tests that run. ● Refer to any other related TAC system trouble codes. If there are no other TAC system codes, the APP Sensor has failed.
DTC P1275 **1T CCM, MIL: No** 1999, 2000, 2001, 2002, 2003, 2004, 2005 C/K Truck, Envoy, G Van, Escalade, TrailBlazer & Corvette 4.8L VIN V, 5.3L VIN P, 5.3L VIN T, 5.7L VIN G, 5.7L VIN S, 6.0L VIN U, 7.4L VIN J engines Transmissions: All	**Accelerator Pedal Position Sensor 1 Circuit Malfunction Conditions:** DTC P0601, P0602, P0606, P1517 and P1518 not set, key in crank or run position, system voltage over 5.23v, and the PCM detected the APP1 sensor signal voltage ranged between 0.25v and 4.22v for less than 1 second during the test. **Possible Causes** ● APP1 sensor connector is contaminated, oily or contains water ● APP1 sensor signal, ground or VREF circuit high resistance ● APP1 sensor VREF circuit is open, shorted to ground or to B+ ● APP1 sensor signal or ground circuit has high resistance ● APP1 sensor is damaged or has failed ● TAC module is damaged or has failed
DTC P1275 **1T CCM, MIL: No** 1999, 2000, 2001, 2002 Camaro & Firebird 3.8L VIN K engine Transmissions: All; Camaro, Firebird	**Accelerator Pedal Position Sensor 1 Circuit Malfunction Conditions:** DTC P0606, P1517 and P1518 not set, key in run or crank mode, system voltage over 5.23v, and the PCM detected the APP Sensor 1 signal was less than 0.68v or more than 4.50v for 1 second. Note: Refer to the information in the Failure Records as needed. **Possible Causes** ● When the TAC module detects a fault in this system, the PCM receives a message on the data bus. One or more related codes might set because of the many redundant tests that run. ● Refer to any other related TAC system trouble codes. If there are no other TAC system codes, the APP Sensor has failed.
DTC P1275 **1T CCM, MIL: Yes** 2002, 2003, 2004, 2005 Envoy & TrailBlazer 4.2L VIN S engine Transmissions: All	**Accelerator Pedal Position Sensor 1 Circuit Malfunction Conditions:** DTC P1635 not set, Key on or engine running; system voltage over 5.23v, and the PCM detected the APP1 signal voltage was less than 0.13v or more than 4.87v. The APP sensors are located within the accelerator pedal assembly. These sensors provide the PCM with a signal voltage proportional to accelerator pedal movement. The APP Sensor 1 signal voltage at rest position is near the low reference and increases as the pedal is actuated. The APP Sensor 2 signal voltage at rest position is near the 5-volt reference and decreases as the pedal is actuated. When APP sensor 1 signal voltage is not within the predicted range, the PCM will set this trouble code (DTC P1275). **Possible Causes** ● APP1 sensor connector is contaminated, oily or contains water ● APP1 sensor signal, ground or VREF circuit high resistance ● APP1 sensor VREF circuit is open, shorted to ground or to B+ ● APP1 sensor signal or ground circuit has high resistance ● APP1 sensor is damaged or has failed ● TAC module is damaged or has failed
DTC P1276 **1T CCM, MIL: No** 1999, 2000, 2001, 2002, 2003, 2004, 2005 C/K Truck, Envoy, G Van, Escalade, TrailBlazer & Corvette 4.8L VIN V, 5.3L VIN P, 5.3L VIN T, 5.7L VIN G, 5.7L VIN S, 6.0L VIN U, 7.4L VIN J engines Transmissions: All	**Accelerator Pedal Position Sensor 1 Range/Performance Conditions:** DTC P0606, P1517 and P1518 not set, key in crank or run position, system voltage over 5.23v, and the PCM detected the APP Sensor 1 and the APP Sensor 2 signals disagreed by more than 10%, or the APP Sensor 1 and APP Sensor 3 signals disagreed by over 13%. *Note: Refer to the information in the Failure Records as needed.* **Possible Causes** ● APP1 sensor connector is contaminated, oily or contains water ● APP1 sensor signal circuit is open or shorted to ground ● APP1 sensor signal circuit is shorted to VREF or system power ● APP1 sensor ground circuit is open or has high resistance ● APP1 sensor VREF circuit is open or shorted to ground ● APP1 sensor is damaged or has failed

OBD II Trouble Code List (P1xxx Codes)

DTC	Trouble Code Title, Conditions & Possible Causes
DTC P1280 **1T CCM, MIL: No** 1999, 2000, 2001, 2002, 2003, 2004, 2005 C/K Truck, Envoy, G Van, Escalade, TrailBlazer & Corvette 4.8L VIN V, 5.3L VIN P, 5.3L VIN T, 5.7L VIN G, 5.7L VIN S, 6.0L VIN U, 7.4L VIN J, 8.1L VIN G Transmissions: All	**Accelerator Pedal Position Sensor 2 Circuit Malfunction Conditions:** DTC P0601, P0602, P0606, P1517 and P1518 not set, key in crank or run position, system voltage over 5.23v, and the PCM detected the TP2 signal was less than 0.83v, or it was more than 4.81v for 1 second during the test. **Possible Causes** ● APP2 sensor connector is contaminated, oily or contains water ● APP2 sensor signal, ground or VREF circuit high resistance ● APP2 sensor VREF circuit is open, shorted to ground or to B+ ● APP2 sensor signal or ground circuit has high resistance ● APP2 sensor is damaged or has failed ● TAC module is damaged or has failed
DTC P1280 **1T CCM, MIL: Yes** 2002, 2003, 2004, 2005 Envoy & TrailBlazer 4.2L VIN S engine Transmissions: All	**Accelerator Pedal Position Sensor 2 Circuit Malfunction Conditions:** DTC P1635 not set, Key on or engine running; system voltage over 5.23v, and the PCM detected the APP2 signal voltage was less than 0.13v or more than 4.87v. The APP sensors are located within the accelerator pedal assembly. These sensors provide the PCM with a signal voltage proportional to accelerator pedal movement. The APP Sensor 1 signal voltage at rest position is near the low reference and increases as the pedal is actuated. The APP Sensor 2 signal voltage at rest position is near the 5-volt reference and decreases as the pedal is actuated. When APP Sensor 2 signal voltage is not within the predicted range, the PCM will set this trouble code (DTC P1280). **Possible Causes** ● APP2 sensor connector is contaminated, oily or contains water ● APP2 sensor signal, ground or VREF circuit high resistance ● APP2 sensor VREF circuit is open, shorted to ground or to B+ ● APP2 sensor signal or ground circuit has high resistance ● APP2 sensor is damaged or has failed ● TAC module is damaged or has failed
DTC P1281 **1T CCM, MIL: No** 1999, 2000, 2001, 2002, 2003, 2004, 2005 C/K Truck, G Van & Corvette 4.8L VIN V, 5.3L VIN T, 5.7L VIN G, 5.7L VIN S, 6.0L VIN U, 7.4L VIN J, 8.1L VIN G engines Transmissions: All	**Accelerator Pedal Position Sensor 2 Range/Performance Conditions:** DTC P1517 and P1518 not set, key in crank or run position, system voltage over 5.23v, and the PCM detected the APP sensor 2 signal disagreed with APP Sensor 1 by over 10.5% or the APP Sensor 2 disagreed with the APP Sensor 3 by over 13% for under 1 second. The APP sensor is mounted on the accelerator pedal assembly. The assembly contains three APP sensors in a single housing. Three separate signal, low-reference and 5-volt reference circuits connect the APP sensor unit to the throttle actuator control (TAC) module. Each of the three APP sensors has a unique functionality **Possible Causes** ● TP2 sensor connector is contaminated, oily or contains water ● TP2 sensor signal, ground or VREF circuit has high resistance ● TP2 sensor VREF circuit is open, shorted to ground or to B+ ● TP2 sensor signal or ground circuit has high resistance ● TP2 sensor is damaged or has failed ● TAC module is damaged or has failed
DTC P1285 **1T CCM, MIL: No** 1999, 2000, 2001, 2002, 2003, 2004, 2005 C/K Truck, G Van & Corvette 4.8L VIN V, 5.3L VIN T, 5.7L VIN G, 5.7L VIN S, 6.0L VIN U, 7.4L VIN J, 8.1L VIN G engines Transmissions: All	**Accelerator Pedal Position Sensor 3 Circuit Malfunction Conditions:** DTC P0606, P1517, or P1518 not set, key in crank or run mode, system voltage over 5.23v, and the PCM detected the APP sensor 3 signal was less than 1.63v, or more than 4.28v for under 1 second. **Possible Causes** ● APP3 sensor connector is contaminated, oily or contains water ● APP3 sensor signal, ground or VREF circuit high resistance ● APP3 sensor VREF circuit is open, shorted to ground or to B+ ● APP3 sensor signal or ground circuit has high resistance ● APP3 sensor or the TAC module is damaged or has failed
DTC P1286 **1T CCM, MIL: No** 1999, 2000, 2001, 2002, 2003, 2004, 2005 C/K Truck, G Van & Corvette 4.8L VIN V, 5.3L VIN T, 5.7L VIN G, 5.7L VIN S, 6.0L VIN U, 7.4L VIN J, 8.1L VIN G engines Transmissions: All	**Accelerator Pedal Position Sensor 3 Range/Performance Conditions:** DTC P0606, P1517 and P1518 not set, key in run or crank mode, system voltage over 5.23v, and the PCM detected the APP Sensor 3 disagreed with the APP sensor 1 by over 13%, or the APP Sensor 3 signal disagreed with APP Sensor 2 by over 13% for 1 second. The APP sensor is mounted on the accelerator pedal assembly. The assembly contains three APP sensors in a single housing. Three separate signal, low-reference and 5-volt reference circuits connect the APP sensor unit to the throttle actuator control (TAC) module. **Possible Causes** ● APP3 sensor connector is contaminated, oily or moisture ● APP3 sensor signal, ground or VREF circuit high resistance ● APP3 sensor signal or ground circuit has high resistance ● APP3 sensor or the TAC module is damaged or has failed

OBD II Trouble Code List (P1xxx Codes)

DTC	Trouble Code Title, Conditions & Possible Causes
DTC P1320 **1T CCM, MIL: No** 1996, 1997, 1998, 1999 Aurora, DeVille, Eldorado & Seville 4.0L VIN C, 4.6L VIN 9 & 4.6L VIN Y engines Transmissions: All	**ICM 4X Reference Circuit No Pulses (Intermittent) Conditions:** DTC P0232 and P1376 not set, engine started, engine speed over 568 rpm, and the PCM detected an interruption of the 4X Fuel Control pulses for 400 ms during the CCM test. **Possible Causes** ● Ignition control (4X) circuit is open, shorted to ground or to B+ ● Ignition module power circuit is open (test power from DS fuse) ● Ignition module is damaged or has failed ● PCM has failed ● TSB 71-65-69 contains a repair procedure for this code
DTC P1323 **1T CCM, MIL: Yes** 1996, 1997, 1998, 1999 Aurora, DeVille, Eldorado & Seville 4.0L VIN C, 4.6L VIN 9 & 4.6L VIN Y engines Transmissions: All	**ICM 24X Reference Circuit Low Frequency Conditions:** DTC P0340 and P1376 not set, engine started, engine speed over 496 rpm, at least 7 CMP sensor edges received this key cycle and at least one CMP REF pulse received in the last 4 seconds, and the PCM did not detect any 24X REF pulses (24X low for 4 seconds). **Possible Causes** ● Ignition control (24X) circuit is open, shorted to ground or to B+ ● Ignition control (24X) circuit is shorted to another module circuit ● Ignition module or the PCM is damaged or has failed ● TSB 71-65-69 contains a repair procedure for this code
DTC P1336 **2T CCM, MIL: Yes** 1997, 1998, 1999 Alero, Bonneville, Cavalier, Ciera, Century, Grand Am, Cutlass Supreme, Grand Prix, Intrigue, LeSabre, LSS, Lumina, Monte Carlo, Regal, Park Avenue, 88', 98', Skylark, Sunfire 2.2L VIN 4, 2.4L VIN T, 3.1L VIN M, 3.4L VIN E, 3.4L VIN X, 3.8L VIN 1, 3.8L VIN K engines Transmissions: All	**CKP Sensor System Variation Not Learned Conditions:** DTC P0336, P0341 and P1374 not set, engine started, ECT sensor more than 158°F, and the PCM did not detect any CKP variation values. Perform the "relearn" procedure after a CKP related repair. The Crankshaft Position System variation "learn" feature is used to calculate reference period errors caused by slight tolerance variations in the crankshaft, and the CKP sensor(s). The calculated error allows the PCM to accurately compensate for reference period variations to enhance the Misfire Detection capability of the system. **Possible Causes** ● Perform the Crankshaft System variation "learn" procedure ● If the Scan Tool indicates that DTC P1336 ran and passed, the CKP system variation "learn" procedure is complete. If not, look for other codes. If no codes are set, repeat the test procedure.
DTC P1336 **2T CCM, MIL: Yes** 2000, 2001, 2002, 2003, 2004, 2005 All makes & models All engines Transmissions: All	**CKP Sensor System Variation Not Learned Conditions:** DTC P0336, P0341 and P1374 not set, engine started, ECT sensor more than 158°F, and the PCM did not detect any CKP variation values. The Crankshaft Position system variation-learning feature is used to calculate reference period errors caused by slight tolerance variations in the crankshaft, and the CKP sensor(s). The calculated error allows the PCM to accurately compensate for reference period variations to enhance the Misfire Detection capability of the system. **Possible Causes** ● Set the parking brake and block the drive wheels for safety. ● Verify the hood is closed. ● Read the trouble codes. If a code is set, refer to that code. ● Start the engine. Allow engine temperature to reach at least 158°F (70°C). Then key off. ● Select Crankshaft Position Variation Learn procedure on Scan Tool & start the vehicle. ● Apply the brake pedal firmly and verify the selector is in Park. ● Increase accelerator pedal position until fuel cutoff is reached at the test rpm (e.g., 5150). Quickly release the accelerator pedal after fuel cutoff is reached. The CKP system variation compensating values are learned when the engine speed (rpm) decreases back to idle speed and the procedure terminates. ● Read the trouble codes and recheck for DTC P1336. ● If DTC P1336 runs and passes, the CKP system variation "learn" procedure is complete. If not, look for other codes. If no codes are set, repeat the test procedure.
DTC P1336 **2T CCM, MIL: Yes** 1998, 1999 C/K Truck, S, T 4.3L VIN W, 4.3L VIN X, 5.0L VIN M, 5.7L VIN K, 5.7L VIN R engines Transmissions: All	**CKP Sensor System Variation Not Learned Conditions:** DTC P0336, P0341 and P1374 not set, engine started, ECT sensor more than 158°F, and the PCM did not detect any CKP variation values. Perform the "relearn" procedure after a CKP related repair. The Crankshaft Position System variation "learn" feature is used to calculate reference period errors caused by slight tolerance variations in the crankshaft, and the CKP sensor(s). The calculated error allows the PCM to accurately compensate for reference period variations to enhance the Misfire Detection capability of the system. **Possible Causes** ● Perform the Crankshaft System variation "learn" procedure ● If the Scan Tool indicates that DTC P1336 ran and passed, the CKP system variation "learn" procedure is complete. If not, look for other codes. If no codes are set, repeat the test procedure.

OBD II Trouble Code List (P1xxx Codes)

DTC	Trouble Code Title, Conditions & Possible Causes
DTC P1345 **1T CCM, MIL: Yes** 1996, 1997, 1998, 1999, 2000, 2001, 2002 Cab & Chassis, C/K Truck, G Van, L/M Van, S/T Blazer & S/T Pickup 4.3L VIN W, 4.3L VIN X, 5.0L VIN M, 5.7L VIN K, 5.7L VIN R, 7.4L VIN J Transmissions: All	**Camshaft To Crankshaft Position Correlation Malfunction Conditions:** Engine started; and the PCM detected the CMP pulses missing at the relative position of the CKP pulses or the CMP signal was 15 degrees out of phase with CKP falling edge (7.4L V8). **Possible Causes** ● CMP sensor is loose causing a variation in the sensor signal ● Excessive free play in the timing chain and gear assembly ● Incorrectly installed distributor - 1 tooth off in either advance or retard positions ● Distributor rotor is loose on the distributor shaft, or hold down bolt is loose or missing ● Perform the Camshaft retard offset test procedure ● Ignition module has failed ● TSB 61-65-60A contains a repair procedure for this code
DTC P1345 **2T CCM, MIL: Yes** 1995 S/T Blazer & S/T Pickup 4.3L VIN W engine Transmissions: All	**Camshaft To Crankshaft Position Correlation Conditions:** Engine started; and the PCM did not detect the CMP sensor pulse at the correct relative position to the CKP sensor pulse. **Possible Causes** ● CMP sensor is loose causing a variation in the sensor signal ● CMP sensor with a low duty cycle can cause this code to set ● Excessive free play in the timing chain and gear assembly ● Incorrectly installed distributor - 1 tooth off in either advance or retard positions ● Distributor rotor is loose on the distributor shaft, or hold down bolt is loose or missing ● Perform the Camshaft retard offset test procedure
DTC P1345 **2T CCM, MIL: Yes** 2002, 2003, 2004, 2005 Envoy & TrailBlazer 4.2L VIN S engine Transmissions: All	**Variable Cam Phaser Circuit Malfunction Conditions:** Engine running at idle and the PCM detected a Variable Cam Phaser command of (0) for 16 seconds. The PCM uses the CKP and CMP signals to monitor the correlation between the crankshaft and camshaft positions. This code sets if the deviation exceeds a certain value. **Possible Causes** ● Camshaft position actuator solenoid high control circuit shorted to voltage or stuck open ● Engine rebuilt incorrectly or harmonic balancer not tightened ● The PCM disables the ability to command the Variable Cam Phaser on the Scan Tool to prevent engine damage from excessive valve overlap during engine operation.
DTC P1350 **1T CCM, MIL: Yes** 1995 Century, Cutlass Ciera, Cutlass Supreme, Grand Prix, Lumina, Monte Carlo Park Avenue Regal, Riviera, 88' & 98' 3.1L VIN M, 3.4L VIN X, 3.8L VIN 1, 3.8L VIN K, 3.8L VIN L engines Transmissions: All	**Bypass Line Monitor Circuit Malfunction Conditions:** Engine cranking; and the PCM detected an unexpected voltage on the IC Control Circuit 423 to the IC Module. *Note: The engine will start and run in Bypass mode in this situation.* **Possible Causes** ● IC signal circuit is open ● IC module has failed
DTC P1350 **2T CCM, MIL: Yes** 1996, 1997 Century, Cutlass Ciera, Cutlass Supreme, Grand Prix, Lumina, Monte Carlo Park Avenue, Regal, Riviera, 88' & 98' 3.1L VIN M, 3.4L VIN X, 3.8L VIN 1, 3.8L VIN K Transmissions: All	**Bypass Line Monitor Circuit Malfunction Conditions:** Engine cranking at less than 450 rpm, and the PCM detected Ignition Control (IC) pulses present with Bypass mode (module timing control of spark advance) enabled during the test. **Possible Causes** ● IC signal circuit is open ● IC module has failed
DTC P1350 **1T CCM, MIL: Yes** 1995, 1996, 1997 Camaro & Firebird 3.8L VIN K engine Transmissions: All	**Bypass Line Monitor Circuit Malfunction Conditions:** Engine cranking, and the PCM detected IC pulses with the Bypass timing control circuit "on". **Possible Causes** ● Bypass circuit is open, shorted to ground or shorted to power ● IC signal circuit is open, shorted to ground or shorted to power ● IC module has failed
DTC P1350 **1T CCM, MIL: Yes** 1996 DeVille, Eldorado, Seville 4.6L VIN 9, 4.6L VIN Y Transmissions: A/T	**Ignition Control (IC) Circuit Malfunction Conditions:** DTC P0322 and P1376 not set, engine cranking with at least one 4X REF pulse detected, and the PCM detected Ignition Control (IC) pulses with the engine cranking, or it detected at least one 4X REF pulse while it did not detect any IC pulses with the engine running. **Possible Causes** ● Bypass circuit or the IC Signal circuit is open, shorted to ground or shorted to power ● IC module has failed ● TSB 71-65-69 contains a repair procedure for this code

OBD II Trouble Code List (P1xxx Codes)

DTC	Trouble Code Title, Conditions & Possible Causes
DTC P1350 **1T CCM, MIL: Yes** 1996 Aurora Engines: 4.0L VIN C Transmissions: All	**Ignition Control (IC) Circuit Malfunction Conditions:** DTC P0322 and P1376 not set, engine cranking with at least one 4X REF pulse detected, and the PCM detected Ignition Control (IC) pulses with the engine cranking, or it detected at least one 4X REF pulse while it did not detect any IC pulses with the engine running. **Possible Causes** ● Bypass circuit is open, shorted to ground or shorted to power ● IC signal circuit is open, shorted to ground or shorted to power ● IC module has failed ● TSB 71-65-28 contains a repair procedure for this code
DTC P1350 **1T CCM, MIL: Yes** 1997, 1998, 1999 Aurora, DeVille, Eldorado & Seville 4.0L VIN C, 4.6L VIN 9 & 4.6L VIN Y engines Transmissions: All	**Ignition Control (IC) Circuit Malfunction Conditions:** DTC P0322 and P1376 not set, engine cranking with at least one 4X REF pulse detected, and the PCM detected Ignition Control (IC) pulses with the engine cranking, or it detected at least one 4X REF pulse while it did not detect any IC pulses with the engine running. **Possible Causes** ● Bypass circuit is open, shorted to ground or shorted to power ● IC signal circuit is open, shorted to ground or shorted to power ● IC module has failed
DTC P1350 **1T CCM, MIL: Yes** 1996, 1997 Lumina APV, Silhouette & Trans Sport 3.4L VIN E engine Transmissions: All	**Bypass Line Monitor Circuit Malfunction Conditions:** Engine cranking, and the PCM detected Ignition Control (IC) pulses present with the Bypass timing control circuit "on" during the test. **Possible Causes** ● Bypass circuit is open, shorted to ground or shorted to power ● IC signal circuit is open, shorted to ground or shorted to power ● IC module has failed
DTC P1351 **2T CCM, MIL: Yes** 1998, 1999, 2000, 2001, 2002, 2003, 2004, 2005 Aurora, Bonneville, Century, Cutlass Supreme, Grand Prix, Intrigue, LeSabre, LSS, Lumina, Monte Carlo, 88', Regal, Park Avenue & 98' 3.1L VIN J, 3.1L VIN M, 3.4L VIN E, 3.4L VIN X, 3.5L VIN H, 3.8L VIN 1, 3.8L VIN K engines Transmissions: All	**ICM Ignition Control Circuit High Input Conditions:** Engine started, engine speed over 600 rpm, and the PCM detected an unexpected open (high voltage) condition on the ICM control circuit for over 300 3X reference periods during 100 crankshaft revolutions. The ICM sends 3X signals to the PCM and controls the timing advance during engine cranking. The timing advance changes to PCM control after the following actions occur: the PCM receives the second 3X signal, the PCM applies 5v to the IC timing signal circuit and then the timing advance switches to PCM control. **Possible Causes** ● ICM control circuit is open ● ICM low reference (ground) circuit has high resistance condition ● ICM is damaged or has failed ● PCM has failed
DTC P1351 **1T CCM, MIL: Yes** 1996, 1997, 1998, 1999, 2000 Cab & Chassis, C/K Truck, G Van, L/M Van, S/T Blazer & S/T Pickup 4.3L VIN W, 4.3L VIN X, 5.0L VIN M, 5.7L VIN K, 5.7L VIN R, 7.4L VIN J Transmissions: All	**ICM Ignition Control Circuit High Input Conditions:** Engine started, engine speed less than 250 rpm, ignition control "on", system voltage over 4.90v, and the PCM detected the ignition control circuit indicated over 4.90v during the test. **Possible Causes** ● ICM control circuit is open, shorted to ground or shorted to B+ ● ICM power circuit is open (check power from the ECM1 fuse) ● ICM is damaged or has failed ● PCM has failed
DTC P1351 **1T CCM, MIL: Yes** 1996, 1997 Camaro, Caprice, Corvette, Firebird, Fleetwood & Roadmaster 4.3L VIN W, 5.7L VIN 5, 5.7L VIN P engines Transmissions: All	**ICM Ignition Control Circuit High Input Conditions:** Engine started; and the PCM detected an intermittent high voltage condition (over 4.60v) on the IC circuit during the CCM test. **Possible Causes** ● Bypass circuit is open, shorted to ground or shorted to power ● IC signal circuit is open, shorted to ground or shorted to power ● IC module has failed
DTC P1351 **1T CCM, MIL: Yes** 1998, 1999, 2000, 2001, 2002 Camaro & Firebird 3.8L VIN K engine Transmissions: All	**ICM Ignition Control Circuit High Input Conditions:** Engine speed over 600 rpm, and the PCM detected an unexpected voltage condition on the ICM Control circuit for at least 300 3X reference periods (100 crankshaft revolutions). **Possible Causes** ● IC signal circuit is open ● IC module has failed ● PCM has failed

OBD II Trouble Code List (P1xxx Codes)

DTC	Trouble Code Title, Conditions & Possible Causes
DTC P1351 **1T CCM, MIL: Yes** 1998, 1999, 2000, 2001, 2002, 2003, 2004, 2005 Aztek, Montana, Venture, Silhouette & Rendezvous 3.4L VIN E engine Transmissions: All	**Ignition Coil Timing Circuit High Input Conditions:** Engine started, engine speed over 600 rpm, and the PCM detected an unexpected open (high voltage) condition on the ICM Coil Timing control circuit for 300 3X reference periods during 100 crankshaft revolutions. The ignition control module (ICM) has independent power and ground circuits that connect to the PCM. They are the IC timing signal, IC timing control signal, low-resolution engine speed signal and the low reference (ground) signal. The ICM sends 3X signals to the PCM. The ICM controls the timing advance when the engine is cranking. The PCM controls the timing advance once it receives a second 3X signal. Then it applies 5v to the IC timing signal circuit so the timing advance can switch to PCM control. **Possible Causes** ● IC signal circuit is open, shorted to ground or shorted to power ● IC module has failed
DTC P1351 **1T CCM, MIL: Yes** 1995 S/T Blazer & S/T Pickup 4.3L VIN W engine Transmissions: All	**ICM Control Circuit High Or Pulses Detected When Open Conditions:** Engine started, engine speed less than 250 rpm, ICM ignition control "on", and the PCM detected the ICM control circuit was over 4.90v. The ICM provides the PCM/VCM with spark timing input with the CKP sensor signals. Once the PCM/VCM calculates the desired amount of spark timing, it sends a timing control signal to the ignition coil on the ignition control (IC) circuit. Each timing pulse received by the ICM triggers the ICM to fire the ignition coil at the correct time. The IC circuit toggles between 0.5-4.5v during normal operation. **Possible Causes** ● IC signal circuit is open ● IC module has failed ● PCM has failed
DTC P1352 **2T CCM, MIL: Yes** 1998, 1999, 2000, 2001, 2002, 2003, 2004, 2005 Aurora, Bonneville, Century, Cutlass Supreme, Grand Prix, Intrigue, LeSabre, LSS, Lumina, Monte Carlo, 88', Regal, Park Avenue & 98' 3.1L VIN J, 3.1L VIN M, 3.4L VIN E, 3.4L VIN X, 3.5L VIN H, 3.8L VIN 1, 3.8L VIN K engines Transmissions: All	**Ignition Control Bypass Circuit High Input Conditions:** Engine started, engine speed over 600 rpm, and the PCM detected an unexpected high voltage condition on the ICM signal circuit for 300 3X reference periods (100 crankshaft revolutions). The ICM has independent power and ground circuits that connect to the PCM. They are the IC timing signal, IC timing control signal, low-resolution engine speed signal and low reference (ground) signal. The ICM sends 3X signals to the PCM. The IC module controls the timing advance during engine cranking. Control of the spark timing advance changes to PCM control after the PCM receives the second 3X signal. The PCM applies a 5v signal to the IC timing signal circuit and the timing advance switches to PCM control. **Possible Causes** ● ICM timing signal circuit is open ● ICM unit is damaged or it has failed ● PCM has failed
DTC P1352 **2T CCM, MIL: Yes** 1998, 1999, 2000, 2001, 2002, 2003, 2004, 2005 Aztek, Montana, Venture, Silhouette & Rendezvous 3.4L VIN E engine Transmissions: All	**Ignition Signal Circuit High Input Conditions:** Engine started; and the PCM detected an unexpected high voltage condition on the ICM signal circuit for at least 300 3X reference periods (100 crankshaft revolutions). The ICM has independent power and ground circuits that connect to the PCM. They are the IC timing signal, IC timing control signal, low-resolution engine speed signal and low reference (ground) signal. The ICM sends 3X signals to the PCM. The IC module controls the timing advance during engine cranking. Control of the spark timing advance changes to PCM control after the PCM receives the second 3X signal. The PCM applies 50volts to the IC timing signal circuit in order to switch the timing advance to PCM control of spark timing. **Possible Causes** ● IC signal circuit is open ● IC module has failed ● PCM has failed
DTC P1352 **2T CCM, MIL: Yes** 1998, 1999, 2000, 2001, 2002 Camaro & Firebird 3.8L VIN K engine Transmissions: All	**ICM Signal Circuit High Input Conditions:** Engine started and the PCM detected an unexpected voltage condition on the ICM signal circuit for at least 300 3X reference periods (100 crankshaft revolutions). The ICM has independent power and ground circuits that connect to the PCM. They are the IC timing signal, IC timing control signal, low-resolution engine speed signal and low reference (ground) signal. The ICM sends 3X signals to the PCM. The IC module controls the timing advance during engine cranking. Control of the spark timing advance changes to PCM control after the PCM receives the second 3X signal. The PCM applies a 5v signal to the IC timing signal circuit and the timing advance switches to PCM control. **Possible Causes** ● ICM signal circuit is open ● ICM unit is damaged or it has failed ● PCM has failed

OBD II Trouble Code List (P1xxx Codes)

DTC	Trouble Code Title, Conditions & Possible Causes
DTC P1359 **2T CCM, MIL: Yes** 2000, 2001, 2002, 2003, 2004, 2005 Aurora, DeVille, Eldorado & Seville 4.0L VIN C, 4.6L VIN 9 & 4.6L VIN Y engines Transmissions: A/T	**Ignition Coil Group 1 Control Circuit Malfunction Conditions:** Key on or engine running; then during PCM power-up mode, the PCM detected an open circuit or short to power condition on the IC control circuit for Ignition Coil Group 1 (cylinders 1, 4, 6 and 7) for less than 1 second during the CCM test. There are two separate ignition module assemblies, one for each bank of cylinders. Each assembly contains an ignition control (IC) module and four ignition coils. Each IC module has these circuits: Ignition 1 voltage, ground, IC timing low reference, and IC circuits for cylinders 1-8. Each ignition coil connects directly to its spark plug via a short boot. *Note: Monitor the Misfire Current Cyl # / Misfire History Cyl # displays on the Scan Tool for additional information for this code.* **Possible Causes** ● IC control circuit for Coil Group 1 is open or shorted to power ● ICM power or ground circuit is open or has high resistance ● ICM is damaged or has failed ● PCM has failed
DTC P1360 **2T CCM, MIL: Yes** 2000, 2001, 2002, 2003, 2004, 2005 Aurora, DeVille, Eldorado & Seville 4.0L VIN C, 4.6L VIN 9 & 4.6L VIN Y engines Transmissions: All	**Ignition Coil Group 2 Control Circuit Malfunction Conditions:** Engine started; then during PCM power-up mode, the PCM detected an open circuit or short to power condition on the IC control circuit for Ignition Coil Group 2 (cylinders 2, 3, 5 and 8) for less than 1 second during the CCM test. There are two separate ignition module assemblies, one for each bank of cylinders. Each assembly contains an ignition control (IC) module and four ignition coils. Each IC module has these circuits: Ignition 1 voltage, ground, IC timing low reference, and IC circuits for cylinders 1-8. Each ignition coil connects directly to its spark plug via a short boot. The IC circuits carry timing pulses from the PCM to the IC modules that trigger the coils to fire the spark plugs in the correct sequence. *Note: Monitor the Misfire Current Cyl # / Misfire History Cyl # displays on the Scan Tool for more help.* **Possible Causes** ● IC control circuit for Coil Group 2 is open or shorted to power ● ICM power or ground circuit is open or has high resistance ● ICM is damaged or has failed ● PCM has failed
DTC P1361 **2T CCM, MIL: Yes** 1996, 1997, 1998, 1999, 2000, 2001, 2002, 2003, 2004, 2005 Aurora, Bonneville, Century, Cutlass Supreme, Grand Prix, Intrigue, LeSabre, LSS, Lumina, Monte Carlo, 88', Regal, Park Avenue & 98' 3.1L VIN J, 3.1L VIN M, 3.4L VIN E, 3.4L VIN X, 3.5L VIN H, 3.8L VIN 1, 3.8L VIN K engines Transmissions: All	**ICM Ignition Control Circuit Low Input Conditions:** DTC P1351 not set, engine speed over 600 rpm, and the PCM did not detect any IC pulses after the IC spark advance mode was commanded "on" for 300 3X reference periods (100 crankshaft revolutions). The ICM has independent power and ground circuits that connect the module to the PCM. The PCM is connected to the ICM with these circuits: IC timing signal, IC timing control, low-resolution engine speed signal and low reference signal. The ICM sends 3X signals to the PCM. The ICM controls the timing advance during engine cranking. The PCM controls the timing advance after it receives the second 3X signal. The PCM applies a 5v signal to the IC timing signal circuit and timing advance switches to PCM control. **Possible Causes** ● IC timing control circuit is shorted to ground or shorted to power ● IC module has failed ● PCM has failed
DTC P1361 **2T CCM, MIL: Yes** 1995 Century, Cutlass Ciera, Cutlass Supreme, Grand Prix, Lumina, Monte Carlo Park Avenue, Regal, Riviera, 88' & 98' 3.1L VIN M, 3.4L VIN X, 3.8L VIN 1, 3.8L VIN K engines, 3.8L VIN L Transmissions: All	**ICM Ignition Control Not Toggling Conditions:** Engine cranking and the PCM detected an unexpected voltage on IC Control Circuit 423, or an unexpected voltage on the IC Bypass Circuit 424; or an unexpected voltage condition. **Possible Causes** ● IC timing control circuit is open or shorted to ground ● IC timing control circuit is shorted to system power ● IC bypass circuit is open or shorted to ground ● IC module has failed ● PCM has failed
DTC P1361 **2T CCM, MIL: Yes** 1996, 1997 Camaro, Caprice, Corvette, Firebird, Fleetwood & Roadmaster 4.3L VIN W, 5.7L VIN 5, 5.7L VIN P engines Transmissions: All	**Ignition Control Module Low Input Conditions:** Engine cranking, and the PCM detected 84 revolutions of the crankshaft occurred without detecting any Ignition Control circuit pulses during the CCM test. **Possible Causes** ● IC timing control circuit is shorted to ground or shorted to power ● IC module has failed ● PCM has failed

OBD II Trouble Code List (P1xxx Codes)

DTC	Trouble Code Title, Conditions & Possible Causes
DTC P1361 **2T CCM, MIL: Yes** 1998, 1999, 2000, 2001, 2002, 2003, 2004, 2005 Aztek, Montana, Venture, Silhouette & Rendezvous 3.4L VIN E engine Transmissions: All	**ICM Ignition Control Circuit Low Input Conditions:** DTC P1351 not set, engine speed over 600 rpm, and the PCM did not detect any IC pulses with IC mode advance enabled for 300 3X reference periods (100 crankshaft revolutions). **Possible Causes** ● IC timing signal circuit or control circuit is shorted to ground or shorted to power ● IC module has failed, or the PCM has failed
DTC P1361 **2T CCM, MIL: Yes** 1996, 1997, 1998, 1999, 2000 Cab & Chassis, C/K Truck, G Van, L/M Van, S/T Blazer & S/T Pickup 4.3L VIN W, 4.3L VIN X, 5.0L VIN M, 5.7L VIN K, 5.7L VIN R, 7.4L VIN J Transmissions: All	**ICM Ignition Control Circuit Low Input Conditions:** Engine started, engine speed less than 250 rpm, ignition control active, and the PCM/VCM detected the Ignition Control (IC) circuit indicate less than 0.04v during the CCM test. **Possible Causes** ● IC timing control circuit is open or shorted to ground ● IC timing control circuit is shorted to system power ● IC bypass circuit is open or shorted to ground ● IC module has failed ● PCM has failed
DTC P1361 **2T CCM, MIL: Yes** 1995, 1996, 1997, 1998, 1999, 2000, 2001, 2002 Camaro & Firebird 3.8L VIN K engine Transmissions: All	**ICM Ignition Control Circuit Low Input Conditions:** Engine started, and the PCM did not detect any ICM ignition control (IC) pulses after the IC mode spark advance was commanded "on". These circuits include the Ignition Control (IC) timing signal, IC timing control signal, low-resolution engine speed signal, medium resolution engine signal, camshaft position signal and the low reference (ground) signal. Both the CMP sensor and CKP sensor signals are input directly to the ICM. The ICM sends 3X signals to the PCM, and controls the timing advance during engine cranking. The timing advance changes to PCM control after it receives the second 3X signal. At this point, the PCM applies 5 volts to the ignition control (IC) timing signal circuit. **Possible Causes** ● IC timing signal circuit is shorted to ground or shorted to power ● IC timing control circuit is shorted to ground or shorted to power ● IC module has failed ● PCM has failed
DTC P1361 **1T CCM, MIL: Yes** 1995 S/T Blazer & S/T Pickup 4.3L VIN W engine Transmissions: All	**ICM Ignition Control Circuit Low Input Conditions:** Engine started, engine speed less than 250 rpm, ignition control active, and the PCM/VCM detected the ICM ignition control circuit was less than 0.04v. Once the PCM/VCM calculates the desired amount of spark timing, it sends a timing control signal to the ignition coil on the ignition control (IC) circuit. Each timing pulse received by the ICM triggers the ICM to fire the ignition coil at the correct time. The IC circuit toggles between 0.5-4.5v in normal operation. **Possible Causes** ● IC timing control circuit is open or shorted to ground ● IC module is damaged or it has failed ● PCM has failed
DTC P1362 **2T CCM, MIL: Yes** 1998, 1999, 2000, 2001, 2002, 2003, 2004, 2005 Body Codes: B, C, F, G, H, L, N, U, W, X 3.1L VIN J, 3.1L VIN M, 3.4L VIN E, 3.4L VIN X, 3.5L VIN H, 3.8L VIN 1, 3.8L VIN K engines Transmissions: All	**ICM Control Circuit High Input Conditions:** Engine started; and the PCM detected an intermittent high voltage condition on the IC timing signal circuit for 300 3X reference periods (100 crankshaft revolutions). The ICM has independent power and ground circuits that connect it to the PCM. Both the CMP sensor and CKP sensor signals are input directly to the ICM. The ICM sends 3X signals to the PCM, and controls the timing advance during engine cranking. The timing advance changes to PCM control after the PCM receives the second 3X signal. At this point, the PCM applies a 5v signal to the to the ignition control (IC) timing signal circuit. **Possible Causes** ● IC timing signal is shorted to system power ● IC timing control circuit and IC timing signal circuits are shorted ● IC module is damaged or it has failed ● PCM has failed
DTC P1370 **2T CCM, MIL: Yes** 1996, 1997, 1998, 1999 Aurora, DeVille, Eldorado & Seville 4.0L VIN C, 4.6L VIN 9 & 4.6L VIN Y engines Transmissions: A/T	**IC Module 4X Reference Circuit Too Many Pulses Conditions:** DTC P1376 not set, engine speed over 496 rpm, at least 7 CMP pulses received since key "on", and one CMP pulse received in last 250 ms, 48 24X pulses received between CMP pulses, and the PCM detected more than eight (8) 4X REF pulses between CMP pulses. **Possible Causes** ● Battery charger connected to the vehicle with engine running ● Fuel control 4X circuit is open or shorted (intermittent fault) ● IC 24X signal wire routed close to Generator, spark plug wires or EMI/RFI interference ● High power transmitters mounted to close (e.g., mobile radio) ● IC module is damaged or has failed ● TSB 71-65-69 contains a repair procedure for this code

OBD II Trouble Code List (P1xxx Codes)

DTC	Trouble Code Title, Conditions & Possible Causes
DTC P1371 **1T CCM, MIL: Yes** 1996, 1997 Camaro, Caprice, Corvette, Firebird, Fleetwood & Roadmaster 4.3L VIN W, 5.7L VIN 5, 5.7L VIN P engines Transmissions: All	**DI Ignition Low-resolution Circuit Malfunction Conditions:** Engine cranking or running; and the PCM detected a number of DI high-resolution pulses occurred without detecting any DI low-resolution pulses. **Possible Causes** ● DI ignition low-resolution signal circuit is open ● DI ignition low-resolution signal circuit shorted to ground or B+ ● DI ignition control module is damaged or has failed ● PCM has failed
DTC P1371 **1T CCM, MIL: Yes** 1996, 1997, 1998, 1999 Aurora, DeVille, Eldorado & Seville 4.0L VIN C, 4.6L VIN 9 & 4.6L VIN Y engines Transmissions: A/T	**IC Module 4X Reference Circuit Too Few Pulses Conditions:** DTC P0322 and P1376 not set, engine started, engine speed over 496 rpm, at least 7 CMP pulses received since key "on", and one CMP pulse received in last 250 ms, 48 24X pulses received between CMP pulses, and the PCM detected less than eight (8) 4X REF pulses between CMP pulses during the CCM test. **Possible Causes** ● Battery charger connected to the vehicle with engine running ● Fuel control 4X circuit is open or shorted (intermittent fault) ● IC 24X signal wire is routed to close to the Generator, spark plug wires or any other possible cause of EMI/RFI interference ● High power transmitters mounted to close (e.g., mobile radio) ● IC module is damaged or has failed ● TSB 71-65-69 contains a repair procedure for this code
DTC P1372 **2T CCM, MIL: Yes** 2000, 2001, 2002, 2003, 2004, 2005 Aurora, DeVille, Eldorado & Seville 4.0L VIN C, 4.6L VIN 9 & 4.6L VIN Y engines Transmissions: A/T	**Crankshaft Position Sensor A-B Correlation Conditions:** Engine cranking or running, PCM operating in Angle Based mode to decode the engine position, and the PCM detected an interruption of the CKP Sensor 'A' or CKP Sensor 'B' signal during the CCM test. The PCM uses two CKP sensors (CKP 'A' and CKP 'B') to determine crankshaft position. The PCM supplies ignition voltage and ground for each sensor. During engine rotation, a slotted ring, machined into the crankshaft, causes the sensors to return a series of ON and OFF pulses to the PCM. The PCM uses two basic methods of decoding the engine position: Angle Based and Time Based using either CKP 'A' or CKP 'B' sensor signals. The PCM uses the angle-based method during normal operation by receiving signals from both CKP sensors. These signals are used to determine an initial crankshaft position, and to generate MEDRES (24X reference) and LORES (4X reference) signals. Once the initial crank position is determined, the PCM continuously monitors both sensors for valid signal inputs. As long as both signals are present, the PCM uses angle-based mode. When either CKP signal is lost, the PCM will compare the MEDRES signal to the CMP sensor signal. If the PCM detects a valid CMP signal, and the MEDRES to CMP signal correlation is correct, the PCM determines that CKP Sensor 'A' failed. However, if the MEDRES to CMP correlation is incorrect, the PCM determines the CKP Sensor 'B' failed. If, while in the angle-based mode, the PCM detects an intermittent loss of either CKP signal, DTC P1372 will set. **Possible Causes** ● CKP Sensor 'B' signal circuit is open, shorted to ground or shorted to system power (an intermittent fault may be present) ● CKP Sensor 'B' is damaged or has failed (an intermittent fault) ● ICM is damaged or has failed (an intermittent fault)
DTC P1374 **2T CCM, MIL: Yes** 1996, 1997, 1998, 1999, 2000, 2001, 2002, 2003, 2004, 2005 Achieva, Alero, Beretta, Century, Corsica, Cutlass, Cutlass Supreme Grand Prix, Impala, Monte Carlo, Intrigue, Lumina, Malibu, Montana, Rendezvous, Silhouette & Venture 3.1L VIN J, 3.1L VIN M, 3.4L VIN E, 3.4L VIN X Transmissions: All	**CKP Sensor High To Low Resolution Frequency Correlation Conditions:** Engine started, engine running with 24X signals received, and the PCM detected the ratio of 24X REF pulses to 3X pulses did not equal 8, or the ratio of 24X pulses to CMP pulses equaled 48 for 10 seconds. The ICM (module) produces a 3X reference signal. The ICM calculates the 3X reference signal by dividing the CKP sensor 7X pulses by two (2) with the engine running with CKP synchronizing pulses received. The PCM uses the 3X reference signal to calculate the engine speed. It uses the CKP sensor signal at engine speeds above 1,600 rpm. The PCM also uses these pulses to initiate injector pulses. The PCM compares the 3X reference pulses to the 24X CKP pulses and the CMP sensor pulses. The engine can start and run using only the 24X CKP and CMP sensor signals. The PCM sets P1374 if it detects an invalid number of pulses occurred on the low-resolution engine speed circuit. **Possible Causes** ● IC 3X signal wire is routed to close to the Generator, spark plug wires or any other possible cause of EMI/RFI interference ● IC 3X signal circuit is open, shorted to ground or shorted to B+ ● IC module is damaged or has failed ● PCM has failed

OBD II Trouble Code List (P1xxx Codes)

DTC	Trouble Code Title, Conditions & Possible Causes
DTC P1374 **2T CCM, MIL: Yes** 1996, 1997, 1998, 1999, 2000, 2001, 2002, 2003, 2004, 2005 Bonneville, Century, Cutlass Supreme, Grand Prix, LeSabre, LSS, Lumina, Monte Carlo, 88', Regal, Park Avenue & 98' 3.8L VIN 1, 3.8L VIN K Transmissions: All	**IC Module 3X Reference Circuit Malfunction Conditions:** Engine started, engine runtime over 3 seconds, and the PCM detected the ratio of 18X pulses received equaled 36:1, or the ratio of 18X REF pulses to 3X pulses did not equal 6:1 for 30 seconds. **Possible Causes** ● IC 3X signal wire is routed to close to the Generator, spark plug wires or any other possible cause of EMI/RFI interference ● IC 3X signal circuit is open, shorted to ground or shorted to B+ ● IC module is damaged or has failed ● PCM has failed ● TSB 99-06-04-45 contains a repair procedure for this code
DTC P1374 **2T CCM, MIL: Yes** 1995 Camaro & Firebird 3.8L VIN K engine Transmissions: All	**IC Module 3X Reference Circuit Malfunction Conditions:** Engine runtime over 3 seconds, and the PCM detected the ratio of 18X pulses received equaled 36:1, and the ratio of 18X REF pulses to 3X pulses did not equal 6:1 for 30 seconds. **Possible Causes** ● IC 3X signal wire routed close to Generator, spark plug wires or EMI/RFI interference ● IC 3X signal circuit is open, shorted to ground or shorted to B+ ● IC module is damaged or has failed ● PCM has failed
DTC P1375 **1T CCM, MIL: Yes** 1996, 1997, 1998, 1999 Aurora, DeVille, Eldorado & Seville 4.0L VIN C, 4.6L VIN 9 & 4.6L VIN Y engines Transmissions: A/T	**IC Module 24X Reference Circuit High Input Conditions:** DTC P1376 not set, engine started, engine speed over 496 rpm, at least 7 CMP pulses detected since key "on", and the PCM detected the 24X signal was high, and that it did not receive any 24X REF pulses during the last eight (8) 4X REF pulses during the CCM test. **Possible Causes** ● CKP Sensor 'A' signal circuit is open or shorted to ground ● Crank Sensor 'A' is damaged or has failed ● Crank Sensor 'B' low circuit is shorted to CKP 'A' high circuit ● Crank Sensor 'B' is damaged or has failed ● IC module is damaged or has failed ● PCM had failed
DTC P1376 **1T CCM, MIL: Yes** 1996, 1997, 1998, 1999 Aurora, DeVille, Eldorado & Seville 4.0L VIN C, 4.6L VIN 9 & 4.6L VIN Y engines Transmissions: A/T	**IC Module Ignition Ground Circuit Malfunction Conditions:** DTC P0560 not set, engine started, and the PCM detected the ICM REF Low signal was over 1.48v, or the ICM REF Low signal was under -1.01v, either condition met for 1 second. **Possible Causes** ● IC module ground circuit is open or has high resistance ● IC module reference low circuit is open or has high resistance ● IC module reference low circuit shorted to VREF or power (B+) ● This code may set if the PCM is not installed properly (e.g., it is not grounded correctly) ● IC module is damaged or has failed ● TSB 71-65-69 contains a repair procedure for this code
DTC P1377 **1T CCM, MIL: Yes** 1996, 1997, 1998, 1999 Aurora, DeVille, Eldorado & Seville 4.0L VIN C, 4.6L VIN 9 & 4.6L VIN Y engines Transmissions: A/T	**ICM CMP Pulse To 4X Reference Pulse Comparison Conditions:** DTC P1376 not set, engine started, engine speed over 496 rpm, at least 7 CMP pulses received since startup, and the PCM detected that the number of 4X REF pulses it received during 2 CMP pulse periods did not equal 16 at least twice in a row during the CCM test. **Possible Causes** ● IC fuel control 4X signal circuit is open, shorted to ground or B+ ● CMP sensor signal circuit is open or has high resistance ● CMP high or low reference circuit is open or high resistance ● This code may set if all (4) camshafts are off by one tooth, 10 degrees before top dead center (driveability complaints exist) ● IC module is damaged or has failed
DTC P1380 **1T CCM, MIL: No** 1997, 1998, 1999, 2000, 2001, 2002, 2003, 2004, 2005 Bonneville, Century, Cutlass Supreme, Grand Prix, LeSabre, LSS, Lumina, Monte Carlo, 88', Regal, Park Avenue, 98' & Rendezvous 3.1L VIN J, 3.1L VIN M, 3.4L VIN E, 3.4L VIN X, 3.5L VIN H, 3.8L VIN 1, 3.8L VIN K engines Transmissions: All	**Misfire Detected, Rough Road Data Unusable Conditions:** Engine started, vehicle driven to a speed over 10 mph at an engine speed under 5000 rpm at an engine load less than 100%, Misfire code (DTC P0300) with MIL requested "on", and the PCM detected that the rough road data transmitted from the Electronic Brake Controller Module was lost or unusable for over 10 seconds. **Possible Causes** ● The presence of this code (P1380) indicates the ABS/TCS system detected a malfunction that would not allow the EBTCM to transmit the correct rough road data to the PCM ● Refer to Diagnostic System Check for Antilock Brake System ● Refer to Diagnostic System Check for the Engine Controls

OBD II Trouble Code List (P1xxx Codes)

DTC	Trouble Code Title, Conditions & Possible Causes
DTC P1380 **1T CCM, MIL: No** 1996, 1997, 1998, 1999 Achieva, Alero, Beretta, Cavalier, Ciera, Century, Corsica, Grand Am, Sunfire, Skyhawk, S/T Blazer & S/T Pickup 2.2L VIN 4, 2.4L VIN T Transmissions: All	**EBTCM DTC Detected, Rough Road Data Unusable Conditions:** Engine started, vehicle driven to over 1 mph at an engine speed of less than 3406 rpm, MAP sensor less than 99.7 kPa, Misfire code (DTC P0300) set with the MIL requested "on", and the PCM received a message from the EBCM indicating a rough road sensing error. **Possible Causes** ● Service the ABS before diagnosing a misfire because an actual engine misfire may or may not exist. Also, an actual engine misfire may have occurred during an ABS malfunction. ● Determine if the vehicle was driven on a rough road, and the ABS could not detect this due to a malfunction. The PCM may interpret variations in crankshaft speed caused by the rough road as a misfire without an actual engine misfire present. ● If U1016 is set, check for a fault in the IPC serial data line, its connections or in the IPC assembly or controller.
DTC P1380 **1T CCM, MIL: No** 1996, 1997, 1998, 1999, 2000, 2001, 2002, 2003, 2004, 2005 Cab & Chassis, C/K Truck, G Van, L/M Van, S/T Blazer & S/T Pickup 4.3L VIN W, 4.3L VIN X, 5.0L VIN M, 5.7L VIN K, 5.7L VIN R, 7.4L VIN J Transmissions: All	**Misfire Detected, Rough Road Data Not Available Conditions:** DTC P0101, P0102, P0103, P0335, P0336, P0341, P0342, P0343, P0500, P0502, P0503, P1120, P1220 and P1221 not set, engine started, vehicle driven to a speed over 10 mph at an engine load over 60%, engine speed less than 3200 rpm, Misfire code (P0300) set with MIL requested "on", and the PCM/VCM detected a malfunction occurred that prevented it from receiving rough road detection data from the EBCM. The PCM detects engine misfire events by monitoring variations in the crankshaft rotation speed. Wheel speed changes caused by rough road conditions can cause changes in crankshaft speed. The ABS monitors the wheel speed sensors to determine if the vehicle is operating on a rough road. **Possible Causes** ● Use the Freeze Frame/Failure Records data to help find the cause on an intermittent fault. If the code cannot be duplicated, the data in the Freeze Frame/Failure Records can determine how many miles since the code set. The Fail Counter and Pass Counter can also help determine how many ignition cycles the diagnostic reported a pass or a fail. Operate the vehicle within the Freeze Frame conditions (i.e., load, engine and vehicle speed, temperature etc.). This will isolate when the code set. ● Service the ABS before diagnosing a misfire because an actual engine misfire may or may not exist. An actual engine misfire may have occurred during an ABS malfunction. ● Determine if the vehicle was driven on a rough road, and the ABS could not detect this due to a malfunction. The PCM may interpret variations in crankshaft speed caused by the rough road as a misfire without an actual engine misfire present. ● Refer to Diagnostic System Check for Antilock Brake System ● Refer to Diagnostic System Check for the Engine Controls
DTC P1380 **1T CCM, MIL: No** 1999, 2000, 2001, 2002, 2003, 2004, 2005 C/K Truck, Envoy, G Van, Escalade & TrailBlazer 4.8L VIN V, 5.3L VIN P, 5.3L VIN T, 5.3L VIN Z, 6.0L VIN N, 6.0L VIN U, 8.1L VIN G engines Transmissions: All	**Misfire Detected, Rough Road Data Not Available Conditions:** DTC P0101, P0102, P0103, P0120, P0335, P0336 and P0742 not set, engine started, vehicle driven to over 10 mph at an engine load over 60%, engine speed less than 3200 rpm, Misfire code (P0300) set with MIL requested "on", and the PCM detected a malfunction occurred that prevented it from receiving rough road detection data from the EBCM. The PCM detects engine misfire events by monitoring variations in the crankshaft rotation speed. Wheel speed changes caused by rough road conditions can cause changes in crankshaft speed. The ABS (system) monitors the wheel speed sensors to determine when the vehicle is operating on a rough road. **Possible Causes** ● Use the Freeze Frame/Failure Records data to help find the cause on an intermittent fault. If the code cannot be duplicated, the data in the Freeze Frame/Failure Records can determine how many miles since the code set. The Fail Counter and Pass Counter can also help determine how many ignition cycles the diagnostic reported a pass or a fail. Operate the vehicle within the Freeze Frame conditions (i.e., load, engine and vehicle speed, temperature etc.). This will isolate when the code set. ● Service the ABS before diagnosing a misfire because an actual engine misfire may or may not exist. Also, an actual engine misfire may have occurred during an ABS malfunction. ● Determine if the vehicle was driven on a rough road, and the ABS could not detect this due to a malfunction. The PCM may interpret variations in crankshaft speed caused by the rough road as a misfire without an actual engine misfire present. ● Refer to Diagnostic System Check for Antilock Brake System ● Refer to Diagnostic System Check for the Engine Controls

OBD II Trouble Code List (P1xxx Codes)

DTC	Trouble Code Title, Conditions & Possible Causes
DTC P1380 **1T CCM, MIL: No** 1996, 1997, 1998, 1999, 2000, 2001, 2002, 2003, 2004, 2005 Aurora, DeVille, Eldorado & Seville 4.0L VIN C, 4.6L VIN 9 & 4.6L VIN Y engines Transmissions: A/T	**Misfire Detected, Rough Road Data Not Available Conditions:** Engine started engine speed less than 6500-7969 rpm, vehicle driven over 1 mph at an engine load of less than 100%, Misfire code (DTC P0300) set and the MIL requested "on", and the PCM detected a Rough Road Indicator sensing error message from the EBTCM. **Possible Causes** ● The presence of this code (P1380) indicates the ABS/TCS system detected a fault that would not allow the EBTCM to transmit the correct rough road data to the PCM ● Refer to Diagnostic System Check for Antilock Brake System
DTC P1380 **1T CCM, MIL: No** 1996, 1997 Camaro, Caprice, Corvette, Firebird, Fleetwood & Roadmaster 4.3L VIN W, 5.7L VIN 5, 5.7L VIN P engines Transmissions: All	**EBTCM DTC Detected, Rough Road Data Unusable Conditions:** DTC P0101, P0102, P0103, P0122, P0123, P0335, P0336 and P0742 not set, engine started, vehicle driven to a speed over 5 mph, engine at light to moderate load, Misfire code (P0300) set with the MIL requested "on", and the PCM did not detect any Rough Road Data from the EBTCM (module) during the CCM test. **Possible Causes** ● The presence of DTC P1380 indicates the ABS/TCS system detected a malfunction that would not allow the EBTCM to transmit the correct rough road data to the PCM ● Refer to Diagnostic System Check for Antilock Brake System ● Refer to Diagnostic System Check for the Engine Controls
DTC P1380 **1T CCM, MIL: No** 1997, 1998, 1999, 2000, 2001, 2002, 2003, 2004, 2005 Camaro, Corvette & Firebird 3.8L VIN K engine, 5.7L VIN G, 5.7L VIN S Transmissions: All	**EBTCM DTC Detected, Rough Road Data Unusable Conditions:** DTC P0101-P0103, P0121-P0123, P0335, P0336, P0742, P1120, P1121, P1220 and P1221 not set, vehicle speed over 10 mph at an engine speed under 3200 rpm (under 5000 rpm on 3.8L VIN K), engine load less than 60%, DTC P0300 set with the MIL requested "on", and the PCM received a message from the EBCM indicating a rough road sensing error. **Possible Causes** ● Service the ABS before diagnosing a misfire because an actual engine misfire may or may not exist. Also, an actual engine misfire may have occurred during an ABS fault. ● Determine if the vehicle was driven on a rough road, and the ABS could not detect this due to a malfunction. The PCM may interpret variations in crankshaft speed caused by the rough road as a misfire without an actual engine misfire present. ● Refer to Diagnostic System Check for Antilock Brake System
DTC P1380 **1T CCM, MIL: No** 2000 S/T Pickup 2.2L VIN 4 Transmissions: All	**EBTCM DTC Detected, Rough Road Data Unusable Conditions:** Engine started, vehicle driven to over 5 mph at an engine speed of less than 3406 rpm, MAP sensor less than 99.7 kPa, Misfire code (DTC P0300) set with the MIL requested "on", and the PCM received a message from the EBCM indicating a rough road sensing error. **Possible Causes** ● Service the ABS before diagnosing a misfire because an actual engine misfire may or may not exist. Also, an actual engine misfire may have occurred during an ABS fault. ● Determine if the vehicle was driven on a rough road, and the ABS could not detect this due to a malfunction. The PCM may interpret variations in crankshaft speed caused by the rough road as a misfire without an actual engine misfire present. ● If U1016 is set, check for a fault in the IPC serial data line, its connections or in the IPC assembly or controller.
DTC P1380 **1T CCM, MIL: No** 2002, 2003, 2004, 2005 Envoy & TrailBlazer 4.2L VIN S, 5.3L VIN P Transmissions: All	**EBTCM DTC Detected, Rough Road Data Unusable Conditions:** Engine started, engine speed less than 6,500 rpm, engine load less than 100%, vehicle speed more than 0.6 mph, Misfire code (DTC P0300) set with the MIL requested "on", and the PCM received a message from the EBCM indicating a rough road sensing error. **Possible Causes** ● Service the ABS before diagnosing a misfire because an actual engine misfire may or may not exist. Also, an actual engine misfire may have occurred during an ABS fault. ● Determine if the vehicle was driven on a rough road, and the ABS could not detect this due to a malfunction. The PCM may interpret variations in crankshaft speed caused by the rough road as a misfire without an actual engine misfire present. ● If U1016 is set, check for a fault in the IPC serial data line, its connections or in the IPC assembly or controller.

OBD II Trouble Code List (P1xxx Codes)

DTC	Trouble Code Title, Conditions & Possible Causes
DTC P1381 **1T CCM, MIL: No** 1996, 1997, 1998, 1999, 2000, 2001, 2002, 2003, 2004, 2005 Bonneville, Century, Cutlass Supreme, Grand Prix, LeSabre, LSS, Lumina, Monte Carlo, 88', Regal, Park Avenue & 98' 3.1L VIN J, 3.1L VIN M, 3.4L VIN E, 3.4L VIN X, 3.5L VIN H, 3.8L VIN 1, 3.8L VIN K engines Transmissions: All	**Misfire Detected - No Communication With The EBCM Conditions:** Engine started, vehicle driven to over 10 mph with the engine speed under 5000 rpm at an engine load of less than 87%, DTC P0300 (Misfire) set with the MIL requested "on", and the PCM detected that it could not communicate with the EBCM for 10 seconds (possible serial data circuit fault present). The PCM detects engine misfire events by monitoring variations in the crankshaft rotation speed. Wheel speed changes due to rough roads can cause changes in crankshaft speed. By monitoring the wheel speed sensors, the ABS detects if the vehicle is operating on a rough road. If it detects a severe rough road condition enough to affect misfire detection, the ABS sends a rough road signal to the PCM on the serial data circuit. **Possible Causes** ● Service the ABS before diagnosing a misfire because an actual engine misfire may or may not exist. Also, an actual engine misfire may have occurred during a period of ABS malfunction. ● Determine if the vehicle was driven on a rough road, and the ABS could not detect this due to a malfunction. The PCM may interpret variations in crankshaft speed caused by the rough road as a misfire without an actual engine misfire present. ● Refer to the Diagnostic System Check for ABS and the PCM
DTC P1381 **2T OOM, MIL: Yes** 1996, 1997 Camaro, Caprice, Corvette, Firebird, Fleetwood & Roadmaster 4.3L VIN W, 5.7L VIN 5, 5.7L VIN P engines Transmissions: All	**Misfire Detected - No Communication With the EBCM Conditions:** Engine started, vehicle driven to a speed over 10 mph at an engine speed under 3500 rpm at light to moderate load, Misfire code (DTC P0300) set with MIL requested "on", and the PCM detected a loss of communication with the EBCM (no Rough Road Detection data from the ABS controller for 20 seconds). The ABS (system) can detect if the vehicle is on a rough road based on wheel acceleration or deceleration data supplied to it by the wheel speed sensors. If the ABS detects rough road above a predetermined threshold, it sends a message to the PCM via serial data. This allows the PCM to take the rough road data into account when calculating a misfire condition. Even if the ABS is malfunctioning and cannot detect rough roads, Misfire diagnostics will continue to run. However, if a misfire trouble code is set, DTC P1381 is also set to indicate that rough road data was not available during misfire calculation due to a serial data fault. **Possible Causes** ● Service the UART data before diagnosing a misfire because an actual engine misfire may not exist. An engine misfire may have occurred in a period of ABS malfunction. ● Determine if the vehicle was driven on a rough road, and the ABS could not detect this due to a malfunction. The PCM may interpret variations in crankshaft speed caused by the rough road as a misfire without an actual engine misfire present.
DTC P1381 **2T CCM, MIL: Yes** 1996, 1997, 1998, 1999, 2000, 2001, 2002, 2003, 2004, 2005 Cab & Chassis, C/K Truck, G Van, L/M Van, S/T Blazer & S/T Pickup 4.3L VIN W, 4.3L VIN X, 5.0L VIN M, 5.7L VIN K, 5.7L VIN R, 7.4L VIN J Transmissions: All	**Misfire Detected - No Communication With The EBCM Conditions:** Engine speed less than 3200 rpm (5800 rpm on some models) with the engine load less than 60%, VSS more than 10 mph, engine misfire detected (P0300 set), and the PCM/VCM detected a serial data problem that prevented it from receiving rough road detection data, condition met for 20 seconds. The PCM detects engine misfire by detecting variations in crankshaft deceleration between firing strokes. For accurate detection of engine misfire, the PCM must distinguish between crankshaft deceleration caused by actual misfire and deceleration caused by rough road conditions. The ABS can detect if the vehicle is on a rough road based on wheel acceleration or deceleration data supplied by the wheel speed sensors. If the ABS detects rough road above a certain threshold, it sends a message to the PCM via serial data. The PCM can take the rough road into account when calculating misfire. **Possible Causes** ● Service the ABS before diagnosing a misfire because an actual engine misfire may not exist. Also, an actual engine misfire may have occurred during a period of ABS fault. ● Determine if the vehicle was driven on a rough road, and the ABS could not detect this due to a malfunction. The PCM may interpret variations in crankshaft speed caused by the rough road as a misfire without an actual engine misfire present. ● Refer to Diagnostic System Check for Antilock Brake System ● Refer to Diagnostic System Check for the Engine Controls
DTC P1381 **1T CCM, MIL: Yes** 1996, 1997, 1998, 1999 Achieva, Alero, Beretta, Cavalier, Ciera, Century, Corsica, Grand Am, Sunfire, Skyhawk, S/T Blazer & S/T Pickup 2.2L VIN 4, 2.4L VIN T Transmissions: All	**Misfire Detected - No Communication With The EBCM Conditions:** Engine started, engine speed of less than 3406 rpm, MAP sensor less than 99.7 kPa, vehicle speed more than 5 mph, Misfire code (DTC P0300-P304) set with the MIL requested "on", and the PCM detected it did not receive any ABS to IPC data for 2.5 seconds. **Possible Causes** ● When this code is set, this indicates that a misfire was detected and that the PCM could not determine if the detected misfire was true or due to operating the vehicle on a rough surface. A misfire can be a true misfire with or without setting this code. ● Check the IPC and EBCM for poor electrical connections at the UART serial data terminals. Verify that no true misfire exists after repairing the cause of this trouble code. If DTC U1016 is also set, this indicates that the malfunction may lie within the IPC serial data line, its electrical connections, or within the IPC.

OBD II Trouble Code List (P1xxx Codes)

DTC P1381	Misfire Detected - No Communication With The EBCM Conditions:
1T CCM, MIL: No 1999, 2000, 2001, 2002, 2003, 2004, 2005 C/K Truck, Envoy, G Van, Escalade & TrailBlazer 4.8L VIN V, 5.3L VIN P, 5.3L VIN T, 5.3L VIN Z, 6.0L VIN N, 6.0L VIN U, 8.1L VIN G engines Transmissions: All	Engine started, engine speed less than 3200 rpm with the engine load less than 60%, VSS more than 10 mph, engine misfire detected (P0300 set), and the PCM/VCM detected a serial data problem that prevented it from receiving rough road detection data, condition met for 20 seconds. The PCM detects engine misfire by detecting variations in crankshaft deceleration between firing strokes. For accurate detection of engine misfire, the PCM must distinguish between crankshaft deceleration caused by actual misfire and deceleration caused by rough road conditions. The ABS (system) can detect if the vehicle is on a rough road based on wheel acceleration or deceleration data supplied by the wheel speed sensors. If the ABS detects rough road above a certain threshold, it sends a message to the PCM via serial data. The PCM can then take the rough road into account when calculating misfire. Even if the ABS is malfunctioning and cannot detect rough roads, Misfire diagnostics will continue to run. However, if a misfire trouble code is set, DTC P1381 is also set to indicate that rough road data was not available during misfire calculation due to a serial data fault. **Possible Causes** ● Service the ABS before diagnosing a misfire because an actual engine misfire may not exist. Also, an actual engine misfire may have occurred during a period of ABS fault. ● Determine if the vehicle was driven on a rough road, and the ABS could not detect this due to a malfunction. The PCM may interpret variations in crankshaft speed caused by the rough road as a misfire without an actual engine misfire present. ● Refer to Diagnostic System Check for Antilock Brake System ● Refer to Diagnostic System Check for the Engine Controls
DTC P1381 **1T CCM, MIL: No** 1996, 1997, 1998, 1999, 2000, 2001, 2002, 2003, 2004, 2005 Aurora, DeVille, Eldorado & Seville 4.0L VIN C, 4.6L VIN 9 & 4.6L VIN Y engines Transmissions: A/T	**Misfire Detected, No Communication With The EBCM Conditions:** Engine started, vehicle driven to over 10 mph at an engine speed under 5000 rpm (under 7969 rpm on 4.0L VIN C) at an engine load less than 81%, Misfire code (DTC P0300) set and MIL requested "on", and the PCM detected it lost communication with the EBCM after receiving data for 5 seconds. Rough roads can cause torque on the tires to slow down the crankshaft and give the appearance of a misfire. The EBCM detects rough road surfaces using inputs from the wheel speed sensors and sends this information to the PCM via Class 2 serial data. The PCM uses the data to enhance Misfire diagnostics so that it can distinguish crankshaft speed variations caused by rough road surfaces from those variations caused by a true misfire. If the PCM loses communication with the EBCM while the MIL is requested "on" due to a Misfire, it will set DTC P1381. **Possible Causes** ● Verify with a Scan Tool that the EBCM can transmit serial data on the Class 2 serial data circuit. Service the EBCM as needed before diagnosing a misfire because an actual engine misfire may or may not exist. An engine misfire may have occurred during a period when the EBCM could not communicate. ● Refer to Diagnostic System Check for Antilock Brake System
DTC P1381 **1T CCM, MIL: No** 1996, 1997, 1998, 1999, 2000, 2001, 2002, 2003, 2004, 2005 Camaro, Corvette & Firebird 3.8L VIN K engine, 5.7L VIN G, 5.7L VIN S Transmissions: All	**Misfire Detected - No Communication With The EBCM Conditions:** Engine started, vehicle driven to over 10 mph at an engine speed under 3200 (5000 rpm on VIN K) at an engine load less than 60%, Misfire code (DTC P0300) set with MIL requested "on", and the PCM detected a loss of communication with the EBCM for 20 seconds. The PCM receives rough road data from the EBCM via the serial data circuit. The PCM uses the rough road data to enhance Misfire diagnostics by distinguishing crankshaft speed variations caused by driving on rough road surfaces from those caused by true misfires. The EBCM transmits rough road data based on inputs from the wheel speed sensors. If the PCM loses communication with the EBCM while the MIL is requested "on", it will set DTC P1381. **Possible Causes** ● Verify with a Scan Tool that the EBCM can transmit serial data on the Class 2 serial data circuit. Service the EBCM as needed before diagnosing a misfire because an actual engine misfire may or may not exist. An engine misfire may have occurred during a period when the EBCM could not communicate. ● Refer to Diagnostic System Check for Antilock Brake System
DTC P1381 **1T CCM, MIL: No** 2000 S/T Blazer & S/T Pickup 2.2L VIN 4 Transmissions: All	**Misfire Detected - No Communication With The EBCM Conditions:** Engine started, vehicle driven to over 5 mph at an engine speed of less than 3406 rpm, MAP sensor less than 99.7 kPa, Misfire code (DTC P0300-P304) set with the MIL requested "on", and the PCM detected it did not receive any ABS to IPC data for 2.5 seconds. Note: Refer to the information in the Failure Records as needed. **Possible Causes** ● When this code is set, this indicates that a misfire was detected and that the PCM could not determine if the detected misfire was true or due to operating the vehicle on a rough surface. A misfire can be a true misfire with or without setting this code. ● Check the IPC and EBCM for poor electrical connections at the UART serial data terminals. Verify that no true misfire exists after repairing the cause of this trouble code. If DTC U1016 is also set, this indicates that the malfunction may lie within the IPC serial data line, its electrical connections, or within the IPC.

OBD II Trouble Code List (P1xxx Codes)

DTC	Trouble Code Title, Conditions & Possible Causes
DTC P1381 **1T CCM, MIL: No** 2002, 2003, 2004, 2005 Envoy & TrailBlazer 4.2L VIN S, 5.3L VIN P Transmissions: All	**Misfire Detected - No Communication With The EBCM Conditions:** Engine started, engine speed less than 3200 rpm, engine load less than 60%, vehicle speed more than 10 mph, Misfire code (DTC P0300) set with the MIL requested "on", and the PCM detected it lost communication with the EBCM for less than 5 seconds. **Possible Causes** ● When this code is set, this indicates that a misfire was detected and that the PCM could not determine if the detected misfire was true or due to operating the vehicle on a rough surface. A misfire can be a true misfire with or without setting this code. ● Check the IPC and EBCM for poor electrical connections at the UART serial data terminals. Verify that no true misfire exists after repairing the cause of this trouble code. If DTC U1016 is also set, this indicates that the malfunction may lie within the IPC serial data line, its electrical connections, or within the IPC.
DTC P1404 **2T EGR, MIL: Yes** 1998, 1999, 2000, 2001, 2002 Aurora, Bonneville, Century, Cutlass Supreme, Grand Prix, Intrigue, LeSabre, LSS, Lumina, Monte Carlo, 88', Regal, Park Avenue, 98', Montana, Silhouette, Rendezvous, Venture 3.1L VIN J, 3.1L VIN M, 3.4L VIN E, 3.8L VIN 1, 3.8L VIN K engines Transmissions: All	**EGR Valve Closed Position Signal Range/Performance Conditions:** DTC P0101-P0103, P0106-P0108, P0112, P0113, P0117, P0118, P0121, P0122, P0123, P0200, P0300, P0301-P0306, P0335, P0502, P0506, P0507 and P1441 not set, engine started, system voltage from 11-18v, and the PCM detected the EGR Position sensor parameter was 0.2v more than the EGR Learned Minimum Position parameter with the Desired EGR Position parameter is commanded to 0% for 20 seconds, or the EGR Position sensor signal was more than 40% and steady for 500 ms after a test failure and before the next test will be run. These conditions must be met four (4) times. The PCM monitors the EGR valve pintle position signal to ensure the valve responds properly to its commands. When the key on, the PCM learns the EGR learned minimum position, and then compares the EGR learned minimum position to the EGR Position sensor signal when the EGR valve is commanded closed. If the EGR Position sensor indicates the EGR valve is still open after the PCM commands the EGR valve closed, it will set DTC P1404. **Possible Causes** ● EGR sensor signal circuit is shorted to VREF ● EGR sensor ground circuit is open or has high resistance ● EGR sensor is damaged or has failed (sensor/solenoid unit) ● EGR valve pintle or valve seat contains carbon deposits ● PCM has failed
DTC P1403 **1T CCM, MIL: Yes** 1995 Grand Prix, Lumina, Monte Carlo & Regal Engines: 3.4L VIN X Transmissions: All	**Exhaust Gas Recirculation Error (EGR Solenoid 1) Conditions:** Engine started; ECT sensor more than 183°F, vehicle driven to over 20 mph in 1st or 2nd gear to over 20 mph at a speed of 1000-1488 rpm, A/C Clutch "off", then while in coast-down mode, HO2S signal more than 566 mv, the PCM detected that the engine speed did not decrease 4 out of 5 times during the test. **Possible Causes** ● EGR tube is clogged or restricted, or the EGR valve is sticking ● EGR Solenoid 1 control circuit is open or shorted to ground
DTC P1404 **2T EGR, MIL: Yes** 2003, 2004, 2005 Alero, Aurora, Bonneville, Century, Grand Prix, LeSabre, Monte Carlo, Regal, Park Avenue, Montana, Silhouette, Rendezvous, Venture 3.1L VIN J, 3.4L VIN E, 3.8L VIN 1, 3.8L VIN K, 4.0L VIN C Transmissions: All	**EGR Valve Closed Position Signal Range/Performance Conditions:** Engine started; system voltage from 11-18v, then after the EGR valve was commanded open at more than a 40% duty cycle for 500 ms, and then commanded to 0% (closed) for over 20 seconds, the PCM detected the EGR Position Sensor signal was 0.2 volt more than the EGR Learned Minimum Position value after the Desired EGR Position was commanded to 0% for 20 seconds, or the EGR Position Sensor signal was more than 40% and steady for 0.5 seconds after a test failure and before the next test will be run. The PCM must detect this condition 4 times to set this trouble code. The PCM monitors the EGR valve position sensor. The 5v VREF, low reference and EGR valve position signal circuits are used by the PCM to determine the EGR valve position. With the ignition switch on, the PCM records the EGR Learned Minimum Position and compares this value to the EGR Position Sensor parameter. **Possible Causes** ● EGR sensor signal circuit is open, shorted to ground or VREF ● EGR sensor ground circuit is open or it has high resistance ● EGR sensor is damaged or it has failed (sensor/solenoid unit) ● EGR valve pintle or valve seat contains carbon deposits ● PCM has failed
DTC P1404 **2T EGR, MIL: Yes** 1996, 1997, 1998, 1999, 2000 C/K Truck, S, T 4.3L VIN W, 4.3L VIN X Transmissions: All	**EGR Valve Closed Position Performance Conditions:** Engine started; ECT sensor from 176-248°F, IAT sensor below 212°F, VSS over 10 mph with the Desired EGR position over 15%, then the PCM detected the EGR Position was 0.29v over the EGR learned minimum position with the desired EGR position at 0% (3 failures). **Possible Causes** ● EGR sensor signal circuit is shorted to VREF or system voltage ● EGR sensor VREF circuit is shorted to system power ● EGR sensor ground circuit is open or has high resistance ● EGR sensor is damaged or has failed (sensor/solenoid unit) ● EGR valve pintle or valve seat contains carbon deposits ● TSB 86-65-03A contains a repair procedure for this code

OBD II Trouble Code List (P1xxx Codes)

DTC	Trouble Code Title, Conditions & Possible Causes
DTC P1404 **2T EGR, MIL: Yes** 1996, 1997, 1998, 1999, 2000, 2001, 2002 C/K Truck, G Van, L, M, P 5.0L VIN M, 5.7L VIN K, 5.7L VIN R, 7.4L VIN J Transmissions: All	**EGR Valve Closed Position Performance Conditions:** Engine started; ECT sensor from 176-248°F, IAT sensor less than 212°F, vehicle speed over 10 mph with the Desired EGR position more than 15%, and the PCM detected the EGR Position sensor signal was 0.29v more than the EGR learned minimum position with the desired EGR position commanded to 0% (fault occurred 3 times on one trip to set this code). **Possible Causes** ● EGR sensor signal circuit is shorted to VREF or system voltage ● EGR sensor VREF circuit is shorted to system power ● EGR sensor ground circuit is open or has high resistance ● EGR sensor is damaged or has failed (sensor/solenoid unit) ● EGR valve pintle or valve seat contains carbon deposits ● PCM has failed ● TSB 01-06-04-043 contains a repair procedure for this code
DTC P1404 **2T EGR, MIL: Yes** 2001, 2002, 2003, 2004, 2005 C/K Truck, G/H Van, Envoy, Escalade & TrailBlazer 4.3L VIN W, 4.3L VIN X, 4.8L VIN V, 5.3L VIN T, 6.0L VIN U, 8.1L VIN G Transmissions: All	**EGR Valve Closed Position Performance Conditions:** Engine started; system voltage from 11-18v, EGR valve enabled at least (6) times, and the PCM detected the EGR position sensor was 0.29v more than the EGR learned minimum position when the desired EGR position was commanded to 0% for over 2 seconds, or the EGR position sensor command is over 30% and steady for 2 seconds after a test failure and before the next test. **Possible Causes** ● EGR sensor signal circuit is shorted to VREF (5v) ● EGR sensor ground circuit is open or has high resistance ● EGR sensor is damaged or has failed (sensor/solenoid unit) ● EGR valve pintle or valve seat contains carbon deposits ● PCM has failed ● TSB 01-06-04-043 contains a repair procedure for this code
DTC P1404 **2T EGR, MIL: Yes** 1998, 1999, 2000, 2001, 2002, 2003, 2004, 2005 Aurora, DeVille, Eldorado & Seville 4.0L VIN C, 4.6L VIN 9 & 4.6L VIN Y engines Transmissions: A/T	**EGR Valve Closed Position Range/Performance Conditions:** No CKP, ECT, IAT, Idle Speed, Injector, MAF, MAP, TP or VSS codes set, engine running, system voltage over 10.0v, EGR commanded "on" (after an initial test failure a 30% valve opening must be achieved and maintained for 5 seconds), and the PCM detected the EGR position sensor signal was 0.40v more than the EGR learned position with the EGR valve commanded to 0% (fault must be detected 4 times in a row before DTC P1474 will set). **Possible Causes** ● EGR sensor signal circuit is shorted to VREF or system voltage ● EGR sensor VREF circuit is shorted to system power ● EGR sensor ground circuit is open or has high resistance ● EGR sensor is damaged or has failed (sensor/solenoid unit) ● EGR valve pintle or valve seat contains carbon deposits ● PCM has failed
DTC P1404 **2T EGR, MIL: Yes** 1998, 1999, 2000 Camaro & Firebird Engines: 5.7L VIN G Transmissions: All	**EGR Valve Closed Position Performance Conditions:** Engine started; system voltage over 10v, and the PCM detected the Actual EGR Position was 290 mv above the Desired EGR Position with the EGR command 0% for 20 seconds. **Possible Causes** ● EGR sensor signal circuit is shorted to VREF or system voltage ● EGR sensor VREF circuit is shorted to system power ● EGR sensor ground circuit is open or has high resistance ● EGR sensor is damaged or has failed (sensor/solenoid unit) ● EGR valve pintle or valve seat contains carbon deposits ● PCM has failed
DTC P1404 **1T EGR, MIL: Yes** 1997, 1998 Achieva, Cavalier, Grand Am, Skyhawk, Sunfire, S/T Blazer & S/T Pickup 2.2L VIN 4, 2.4L VIN T Transmissions: All	**EGR Valve Pintle Stuck Open Conditions:** DTC P0106-P0108, P0112, P0113, P0117, P0118, P0121, P0122, P0123, P0200, P0300, P0301-P0304, P0335, P0502, P0506, P0507 and P1441 not set, engine started, engine running in closed loop, system voltage over 10.0v, EGR solenoid commanded "off" (closed), and the PCM detected the Actual EGR Position indicated open (over 15%) at idle speed with the throttle closed for over 20 seconds. **Possible Causes** ● EGR sensor signal circuit is shorted to VREF or system voltage ● EGR sensor VREF circuit is shorted to system power ● EGR sensor ground circuit is open or has high resistance ● EGR sensor is damaged or has failed (sensor/solenoid unit) ● EGR valve pintle or valve seat contains carbon deposits ● TSB 61-65-58 (12/96) contains a repair procedure for this code

OBD II Trouble Code List (P1xxx Codes)

DTC	Trouble Code Title, Conditions & Possible Causes
DTC P1404 **2T EGR, MIL: Yes** 2001, 2002, 2003, 2004, 2005 C/K Truck 6.6L VIN 1 Diesel engine Transmissions: All	**EGR Valve Vacuum Sensor Range/Performance Conditions:** DTC P0101, P0102, P0103, P0106, P0107, P0108, P0112, P0113, P0116, P0117, P0118, P0405, P0406, P0489, P0490, P0500, P0651, P2142, P2144, P2144 and P2145 not set, engine runtime over 5 seconds, engine speed from 610-820 rpm (± 50 rpm) for 3 seconds, EGR valve vacuum control solenoid commanded to over 70%, ECM not in Reduced Engine Power mode, BARO sensor more than 72 kPa, ECT sensor is from 140-212ºF. Calculated Fuel Rate value between 3-20 mm³, Power Takeoff (PTO) is disabled, IAT sensor more than 32ºF, APP Indicated Angle sensor less than 1%, system voltage from 11-18v, VSS less than 0.25 mph, and the ECM detected a higher than desired EGR vacuum sensor signal along with a lower than expected MAF sensor value for over 3 seconds. **Possible Causes** ● Check for leaking or restricted vacuum line connections ● EGR vacuum sensor is damaged or it has failed ● EGR vacuum vent solenoid is stuck open ● EGR valve is damaged or it has failed ● EGR valve vacuum control solenoid is stuck open ● EGR vent solenoid is damaged or it has failed ● Exhaust system may be restricted causing high back pressure
DTC P1404 **1T CCM, MIL: Yes** 1995 Grand Prix, Lumina, Monte Carlo & Regal Engines: 3.4L VIN X Transmissions: All	**Exhaust Gas Recirculation Error (EGR Solenoid 2) Conditions:** Engine started; ECT sensor more than 183ºF, vehicle driven while in 1st or 2nd gear to a speed over 20 mph at an engine speed of 1000-1488 rpm, then vehicle in coast-down mode, HO2S signal over 566 mv, A/C Clutch "off", and the PCM detected that the engine speed did not decrease 4 out of 5 times in the test. **Possible Causes** ● EGR tube is clogged or restricted, or the EGR valve is sticking ● EGR Solenoid 2 control circuit is open or shorted to ground
DTC P1405 **1T CCM, MIL: Yes** 1995 Grand Prix, Lumina, Monte Carlo & Regal Engines: 3.4L VIN X Transmissions: All	**Exhaust Gas Recirculation Error (EGR Solenoid 3) Conditions:** Engine started; ECT sensor over 183ºF, vehicle driven in 1st or 2nd gear to a speed over 20 mph at a speed of 1000-1488 rpm, then vehicle in coast-down mode, HO2S signal over 566 mv, A/C Clutch "off", and the PCM detected the engine speed did not decrease 4 of 5 times. **Possible Causes** ● EGR tube is clogged or restricted, or the EGR valve is sticking ● EGR Solenoid 3 control circuit is open or shorted to ground
DTC P1406 **2T CCM, MIL: Yes** 1995 Century, Cutlass Ciera, Cutlass Supreme, Grand Prix, Lumina, Monte Carlo Park Avenue Regal, Riviera, 88' & 98' 3.1L VIN M, 3.4L VIN X, 3.8L VIN 1, 3.8L VIN K engines, 3.8L VIN L Transmissions: All	**EGR Valve Pintle Position Circuit Malfunction Conditions:** Engine started; and the PCM detected the EGR Pintle signal indicated 10% more than the Desired EGR position with the EGR commanded position at 0% for 20 seconds. **Possible Causes** ● EGR solenoid control circuit is open, shorted to ground or to B+ ● EGR solenoid power circuit is open (check the C/C 15A fuse) ● EGR solenoid is damaged or has failed ● PCM has failed
DTC P1406 **2T CCM, MIL: Yes** 1996, 1997 Aurora, Bonneville, Century, Cutlass Supreme, Grand Prix, Intrigue, LeSabre, LSS, Lumina, Monte Carlo, 88', Regal, Park Avenue & 98' 3.1L VIN M, 3.4L VIN X, 3.8L VIN 1, 3.8L VIN K Transmissions: All	**EGR Valve Pintle Position Circuit Malfunction Conditions:** Engine started; and PCM detected the EGR Feedback signal was below 0.14v, or was 0.40v more or less than the EGR closed valve position with Desired EGR command at 0%. **Possible Causes** ● EGR solenoid control circuit is open, shorted to ground or to B+ ● EGR solenoid power circuit is open (check the IGNFD 5A fuse) ● EGR solenoid is damaged or has failed ● PCM has failed
DTC P1406 **2T CCM, MIL: Yes** 1996, 1997 Cab & Chassis, C/K Truck, G Van, L/M Van, S/T Blazer & S/T Pickup 4.3L VIN W, 4.3L VIN X, 4.8L VIN V, 5.0L VIN M, 5.7L VIN G, 5.7L VIN R, 7.4L VIN J engines Transmissions: All	**EGR Valve Pintle Position Circuit Malfunction Conditions:** Engine started; and the PCM/CM detected the EGR Position sensor signal was out-of-range EGR Position, or that the EGR Position sensor signal indicated more than 10% above the Desired EGR Position with the EGR solenoid commanded to 0%. **Possible Causes** ● EGR solenoid control circuit is open, shorted to ground or to B+ ● EGR solenoid power circuit is open (check the ENG1 fuse) ● EGR solenoid is damaged or has failed ● PCM has failed ● TSB 76-65-04 contains a repair procedure for this code

OBD II Trouble Code List (P1xxx Codes)

DTC	Trouble Code Title, Conditions & Possible Causes
DTC P1406 **2T CCM, MIL: Yes** 1996 Achieva, Beretta, Cavalier, Ciera, Century, Corsica, Grand Am, Sunfire, Skyhawk, S/T Blazer & S/T Pickup 2.2L VIN 4, 2.4L VIN T Transmissions: All	**EGR Valve Pintle Position Circuit Malfunction Conditions:** DTC P0106-P0108, P0112, P0113, P0117, P0118, P0121, P0122, P0123, P0200, P0300, P0301-P0304, P0335, P0502, P0506, P0507 and P1441 not set, engine started, EGR commanded "off", and the PCM detected the EGR command was over 3%, or the EGR position disagreed with Desired EGR by over 6%, or with EGR commanded "on", the EGR position disagreed with Desired EGR by over 9% with the Desired EGR less than 99%, or the EGR position differed from Desired EGR by over 20% with Desired EGR over 99%. **Possible Causes** ● EGR solenoid control circuit is open, shorted to ground or to B+ ● EGR solenoid power circuit is open (check the ERLS fuse) ● EGR solenoid is damaged or has failed ● PCM has failed
DTC P1406 **2T CCM, MIL: Yes** 1996 DeVille, Eldorado, Seville 4.6L VIN 9, 4.6L VIN Y Transmissions: A/T	**EGR Valve Pintle Position Circuit Malfunction Conditions:** Engine running at idle speed, and the PCM detected the EGR Position sensor was 0.18v during Test 1; or the EGR Position sensor was 2.06v or higher during Test 2; or the maximum change in Desired EGR was less than 0.10v with the signal error over 0.20v for 1 second, or the Desired EGR was 0v with the EGR signal over 0.40v (fault occurred twice). **Possible Causes** ● EGR solenoid (valve) is binding or sticking ● EGR solenoid control circuit is open, shorted to ground or to B+ ● EGR solenoid power circuit is open (check the ECS 10A fuse) ● EGR solenoid is damaged or has failed ● PCM has failed
DTC P1406 **2T CCM, MIL: Yes** 1995, 1996, 1997 Camaro & Firebird 3.8L VIN K engine Transmissions: All	**EGR Valve Pintle Position Circuit Malfunction Conditions:** Engine started, and the PCM detected the EGR Feedback signal was less than 0.14v, or with the EGR Position commanded to 0%, the EGR Feedback signal was 0.40v more than the EGR Closed Valve Pintle Position, or that the Actual EGR Position was 10% more or less than Desired EGR Position for 20 seconds. **Possible Causes** ● EGR solenoid control circuit is open, shorted to ground or to B+ ● EGR solenoid power circuit is open (test ACTUATORS fuse) ● EGR solenoid is damaged or has failed ● PCM has failed
DTC P1406 **2T CCM, MIL: Yes** 1995 S/T Pickup 4.3L VIN W engine Transmissions: All	**EGR Valve Pintle Position Circuit Malfunction Conditions:** Engine running, then the VCM detected that the EGR Position signal was out-of-range, or it detected an EGR signal 10% more or less than the VCM commanded position. **Possible Causes** ● EGR solenoid control circuit is open, shorted to ground or to B+ ● EGR solenoid power circuit is open (test power from the PCM) ● EGR solenoid is damaged or has failed ● PCM has failed
DTC P1415 **2T AIR, MIL: Yes** 1996, 1997, 1998, 1999, 2000, 2001, 2002, 2003, 2004, 2005 Cab & Chassis, C/K Truck, G Van, L/M Van, S/T Blazer & S/T Pickup 4.3L VIN W, 4.3L VIN X, 4.8L VIN V, 5.0L VIN M, 5.7L VIN G, 5.7L VIN K, 5.7L VIN R, 7.4L VIN J Transmissions: All	**Secondary Air Injection System (Bank 1) Malfunction Conditions:** DTC P0137, 0138 P0140-P0147, P0151-P0158, P0160, P0161, P0171, P0172, P0174, P0175, P0300, P0500, P1106, P1107, P1111-P1115, P1121, P1122, P1133, P1134, P1153, P1154, P1351 and P1361 not set, engine started, vehicle driven to an engine speed over 900 rpm at an engine load less than 33.25%, airflow less than 22 g/sec, A/F ratio at 13.125:1, ECT sensor from 158-230ºF, system voltage over 10.0v, and the PCM detected the Bank 1 HO2S signal was less than 222 mv for 1 second with the AIR pump on while in closed loop. A secondary air injection (AIR) pump is used to reduce the tailpipe emissions during startup. The PCM supplies a ground to the AIR pump relay control circuit, and this action energizes the AIR pump. The PCM monitors the front HO2S signal in order to diagnose the AIR system. During the AIR test, the PCM activates the AIR pump during closed loop operation. Once the AIR pump is "on", the PCM monitors the HO2S signal and the Short Term fuel trim values of both banks of the engine. If the AIR system is operating properly, the HO2S signal should go low, and the Short Term fuel trim value should go high. If the PCM detects the HO2S signals for both banks did not respond as expected during the tests, it will set DTC P0410. If only one sensor responds, the PCM sets either a DTC P1415 or P1416 to indicate the bank where the AIR system failed. **Possible Causes** ● Air hoses disconnected, loose, kinked or failed (a burnt hose) ● AIR pump is damaged or has failed (inspect air pump for water) ● AIR system check valves and/or pipes are damaged or leaking ● PCM has failed

OBD II Trouble Code List (P1xxx Codes)

DTC	Trouble Code Title, Conditions & Possible Causes
DTC P1415 **2T AIR, MIL: Yes** 2000, 2001, 2002, 2003, 2004, 2005 Aurora, DeVille, Eldorado & Seville 4.0L VIN C, 4.6L VIN 9 & 4.6L VIN Y engines Transmissions: All	**Secondary Air Injection System (Bank 1) Malfunction Conditions:** DTC P0101, P0102, P0103, P0106, P0107, P0108, P0112, P0113, P0116, P0117, P0118, P0121, P0122, P0123, P0130, P0131, P0132, P0133, P0134, P0135, P0137, P0138, P0140, P0141, P0150, P0151, P0152, P0153, P0154, P0155, P0201, P0202, P0203, P0204, P0205, P0206, P0207, P0208, P0300, P0335, P0336, P0351, P0352, P0353, P0354, P0355, P0356, P0357, P0358, P0412, P0418, P0506, P0507, P1133 and P1134 not set. Passive Test: Engine started, engine speed over 600 rpm, TP angle steady, engine load less than 80, system voltage over 10.0v, MAF sensor less than 35 g/sec, A/F more than 12.5:1, ECT sensor from 32ºF-226ºF, IAT sensor signal from 32ºF-140ºF, Power Enrichment and Decel Fuel Cutoff inactive, then with the AIR pump(s) operating, the PCM detected the front HO2S signal remained above 300 mv for 12 seconds (or 350 mv for 9 seconds on a hot restart); or with the AIR pumps "off", the PCM detected the front Bank 1 HO2S signal remained below 700 mv for 25 seconds. Active Test: Engine started, engine speed over 600 rpm in closed loop, TP angle steady, engine load less than 80, system voltage over 10.0v, EVAP Purge active, ECT sensor more than 154ºF, then after a failed or an indeterminate passive test, the PCM detected the front HO2S signal remained above 35 mv, or it detected a failed Passive Test or two failed Active Tests, or the PCM recorded an indeterminate Passive test and three Active Tests. The secondary air injection (AIR) pump is used to lower the tailpipe emissions during startup. The AIR system includes the AIR pump, AIR shutoff valves, vacuum control solenoid valve, AIR hoses and pipes, AIR relay, fuses and the related wiring. The PCM uses the AIR relay to control the AIR pump. The PCM also controls the AIR vacuum control solenoid valve that supplies vacuum to the AIR shutoff valves. With the AIR system inactive, the AIR shutoff valves prevent airflow in either direction. With the AIR system active, the PCM applies ground to the AIR relay and to the vacuum control solenoid valve. Fresh air flows from the pump, through the system hoses, past the shutoff valves and into the exhaust stream. The extra air helps the catalyst to quickly get to a working temperature, thus lowering the tail pipe emissions right after startup. The PCM runs a Passive test and an Active test to diagnose the AIR system. Both tests involve a response from the front HO2S-11 and HO2S-12. The Passive test consists of 2 parts. If both Passive tests pass, the PCM takes no further action. If either part of the Passive test fails or is inconclusive, the PCM initiates the Active tests. If the Active test does not pass, the PCM may set one or more related trouble codes. **Possible Causes** • Air hoses disconnected, loose, kinked or failed (a burnt hose) • AIR pump is damaged or has failed (inspect air pump for water) • AIR system check valves and/or pipes are damaged or leaking • AIR check valve (in hose to vacuum bleed valve) stuck closed • PCM has failed
DTC P1415 **2T AIR, MIL: Yes** 1996, 1997 Camaro, Caprice, Corvette, Firebird, Fleetwood & Roadmaster 4.3L VIN W, 5.7L VIN 5, 5.7L VIN P engines Transmissions: All	**Secondary Air Injection System (Bank 1) Malfunction Conditions:** Engine runtime over 2 seconds in closed loop, ECT sensor over 239ºF, TP angle stable, engine load under 6.6%, airflow below 26 g/sec, and the PCM detected the HO2S signal was less than 400 mv for a time period based on the ECT sensor signal at startup (Test 1), or ECT sensor from 140-239ºF, engine speed over 600 rpm, engine runtime over 200 seconds, IAT sensor over 50ºF, engine load less than 5.5%, maximum airflow at 17 g/sec, and the PCM detected the HO2S-11 signal was more than 222 mv a 1.5 second period, and that the SHRTFT value was more than a predetermined amount (Test 2) during the AIR Monitor test. **Possible Causes** • Air hoses disconnected, loose, kinked or failed (a burnt hose) • AIR pump is damaged or has failed (inspect air pump for water) • AIR system check valves and/or pipes are damaged or leaking • AIR check valve (in hose to vacuum bleed valve) stuck closed • PCM has failed
DTC P1415 **2T AIR, MIL: Yes** 1997, 1998, 1999, 2000, 2001, 2002, 2003, 2004, 2005 Camaro, Corvette & Firebird 5.7L VIN G, 5.7L VIN S Transmissions: All	**Secondary Air Injection System (Bank 1) Malfunction Conditions:** DTC P0101-P0103, P0107, P0108, P0112-P0118, P0121-P0123, P0131-P0138, P0140, P0141, P0151-P0155, P0171-P0175, P0200, P0300, P0325, P0327, P0332, P0335-P0336, P0351-P0358, P0430, P0500, P0502, P0503, P1133, P1134, P1153 and P1154 not set, engine started, engine speed over 850 rpm in closed loop, maximum airflow less than 22 g/sec, ECT sensor from 14-255ºF, IAT sensor from 14-212ºF, system voltage over 10.0v, engine load under 40%, A/F ratio at 13.125:1, P/E and DFCO mode inactive, Short Term fuel trim operating in Fuel Trim cells 1, 2, 5 or 20, and the PCM detected the front HO2S signal remained above 222 mv for 0.9 seconds, or the SHRTFT value did not change enough with the AIR pump "on". An AIR pump is used to reduce the tailpipe emissions during startup. **Possible Causes** • Air hoses disconnected, loose, kinked or failed (a burnt hose) • AIR pump is damaged or has failed (inspect air pump for water) • AIR system check valves and/or pipes are damaged or leaking • PCM has failed

OBD II Trouble Code List (P1xxx Codes)

DTC	Trouble Code Title, Conditions & Possible Causes
DTC P1416 **2T AIR, MIL: Yes** 1996, 1997 Camaro, Caprice, Corvette, Firebird, Fleetwood & Roadmaster 4.3L VIN W, 5.7L VIN 5, 5.7L VIN P engines Transmissions: All	**Secondary Air Injection System (Bank 2) Malfunction Conditions:** AIR Test 1: Engine started, ECT sensor less than 239°F, engine runtime over 2 seconds, TP angle stable, A/F ratio more than 13:1, engine load under 6.6%, airflow less than 26 g/sec, HO2S bias voltage from 365-513 mv, and the PCM detected the front HO2S signal was less than 400 mv for a time period based on the ECT sensor signal at the start of the test. AIR Test 2: Engine started, ECT sensor from 140-239°F, engine speed over 600 rpm, engine runtime over 200 seconds, IAT sensor more than 50°F, engine load less than 5.5%, maximum airflow at 17 g/sec, and the PCM detected the HO2S signal was more than 222 mv a 1.5 second period, and that the SHRTFT value was more than a predetermined amount during the AIR Monitor test. A secondary air injection (AIR) pump is used to reduce the tailpipe emissions during startup. The PCM supplies a ground to the AIR pump relay control circuit, and this action energizes the AIR pump. The PCM monitors the front HO2S signal in order to diagnose the AIR system. During the AIR test, the PCM activates the AIR pump during closed loop operation. Once the AIR pump is "on", the PCM monitors the HO2S signal and the Short Term fuel trim values of both banks of the engine. If the AIR system is operating properly, the HO2S signal should go low, and the Short Term fuel trim value should go high. If the PCM detects the HO2S signals for both banks did not respond as expected during the tests, it will set DTC P0410. If only one sensor responds, the PCM sets either a DTC P1415 or P1416 to indicate the bank where the AIR system failed. **Possible Causes** ● Air hoses disconnected, loose, kinked or failed (a burnt hose) ● AIR pump is damaged or has failed (inspect air pump for water) ● AIR system check valves and/or pipes are damaged or leaking ● AIR check valve (in hose to vacuum bleed valve) stuck closed ● PCM has failed
DTC P1416 **2T AIR, MIL: Yes** 1996, 1997, 1998, 1999, 2000, 2001, 2002, 2003, 2004, 2005 Cab & Chassis, C/K Truck, G Van, L/M Van, S/T Blazer & S/T Pickup 4.3L VIN W, 4.3L VIN X, 4.8L VIN V, 5.0L VIN M, 5.7L VIN G, 5.7L VIN K, 5.7L VIN R, 7.4L VIN J Transmissions: All	**Secondary Air Injection System (Bank 2) Malfunction Conditions:** DTC P0137, 0138 P0140-P0147, P0151-P0158, P0160, P0161, P0171, P0172, P0174, P0175, P0300, P0500, P1106, P1107, P1111-P1115, P1121, P1122, P1133, P1134, P1153, P1154, P1351 and P1361 not set, engine started, vehicle driven to an engine speed over 900 rpm at an engine load less than 33.25%, airflow less than 22 g/sec, A/F ratio at 13.125:1, ECT sensor from 158-230°F, system voltage over 10.0v, and the PCM detected the Bank 2 HO2S signal was less than 222 mv for 1 second with the AIR pump on while in closed loop. The PCM supplies a ground to the AIR pump relay control circuit, and this action energizes the AIR pump. The PCM monitors the front HO2S signal in order to diagnose the AIR system. During the AIR test, the PCM activates the AIR pump during closed loop operation. Once the AIR pump is "on", the PCM monitors the HO2S signal and the Short Term fuel trim values of both banks of the engine. If the AIR system is operating properly, the HO2S signal should go low, and the Short Term fuel trim value should go high. If the PCM detects the HO2S signals for both banks did not respond as expected during the tests, it will set DTC P0410. If only one sensor responds, the PCM sets either a DTC P1415 or P1416 to indicate the bank where the AIR system failed. **Possible Causes** ● Air hoses disconnected, loose, kinked or failed (a burnt hose) ● AIR pump is damaged or has failed (inspect air pump for water) ● AIR system check valves and/or pipes are damaged or leaking ● PCM has failed
DTC P1416 **2T AIR, MIL: Yes** 2000, 2001, 2002, 2003, 2004, 2005 DeVille, Eldorado, Seville 4.6L VIN 9, 4.6L VIN Y Transmissions: All	**Secondary Air Injection System (Bank 2) Malfunction Conditions:** DTC P0101, P0102, P0103, P0106, P0107, P0108, P0112, P0113, P0116, P0117, P0118, P0121, P0122, P0123, P0130, P0131, P0132, P0133, P0134, P0135, P0137, P0138, P0140, P0141, P0150, P0151, P0152, P0153, P0154, P0155, P0201, P0202, P0203, P0204, P0205, P0206, P0207, P0208, P0300, P0335, P0336, P0351, P0352, P0353, P0354, P0355, P0356, P0357, P0358, P0412, P0418, P0506, P0507, P1133 and P1134 not set. Passive Test: Engine speed over 600 rpm, TP angle steady, engine load less than 80, system voltage over 10.0v, MAF sensor less than 35 g/sec, A/F more than 12.5:1, ECT sensor from 32°F-226°F, IAT sensor signal from 32°F-140°F, P/E and DFCO "off", then with the AIR pump operating, the PCM detected the HO2S signal remained above 300 mv for 12 seconds (or 350 mv for 9 seconds on a hot restart); or with the AIR pumps "off", the PCM detected the HO2S signal remained below 700 mv for 25 seconds. Active Test: engine speed over 600 rpm in closed loop, TP angle steady, engine load less than 80, system voltage over 10.0v, EVAP Purge active, ECT sensor more than 154°F, then after a failed or indeterminate passive test, the PCM detected the HO2S signal remained above 35 mv, or it detected a failed Passive Test or two failed Active Tests, or the PCM recorded a faulty Passive test and three Active Tests. **Possible Causes** ● Air hoses disconnected, loose, kinked or failed (a burnt hose) ● AIR pump is damaged or has failed (inspect air pump for water) ● AIR system check valves and/or pipes are damaged or leaking ● PCM has failed

OBD II Trouble Code List (P1xxx Codes)

DTC	Trouble Code Title, Conditions & Possible Causes
DTC P1416 **2T AIR, MIL: Yes** 1997, 1998, 1999, 2000, 2001, 2002, 2003, 2004, 2005 Camaro, Corvette & Firebird 5.7L VIN G, 5.7L VIN S Transmissions: All	**Secondary Air Injection System (Bank 2) Malfunction Conditions:** DTC P0101-P0103, P0107, P0108, P0112-P0118, P0121-P0123, P0131-P0138, P0140, P0141, P0151-P0155, P0171-P0175, P0200, P0300, P0325, P0327, P0332, P0335-P0336, P0351-P0358, P0430, P0500, P0502, P0503, P1133, P1134, P1153 and P1154 not set, engine started, engine speed over 850 rpm in closed loop, maximum airflow less than 22 g/sec, ECT sensor from 14-255°F, IAT sensor from 14-212°F, system voltage over 10.0v, engine load under 40%, A/F ratio at 13.125:1, P/E and DFCO mode inactive, Short Term fuel trim operating in Fuel Trim cells 1, 2, 5 or 20, and the PCM detected the front HO2S signal remained above 222 mv for 0.9 seconds, or the SHRTFT value did not change enough with the AIR pump "on". The PCM monitors the front HO2S signal in order to diagnose the AIR system. During the AIR test, the PCM activates the AIR pump during closed loop operation. Once the AIR pump is "on", the PCM monitors the HO2S signal and the Short Term fuel trim values of both banks of the engine. If the AIR system is operating properly, the HO2S signal should go low, and the Short Term fuel trim value should go high. If the PCM detects the HO2S signals for both banks did not respond as expected during the tests, it will set DTC P0410. If only one sensor responds, the PCM sets either a DTC P1415 or P1416 to indicate the bank where the AIR system failed. **Possible Causes** ● Air hoses disconnected, loose, kinked or failed (a burnt hose) ● AIR pump is damaged or has failed (inspect air pump for water) ● AIR system check valves and/or pipes are damaged or leaking ● PCM has failed
DTC P1431 **1T CCM, MIL: No** 1997, 1998, 1999, 2000, 2001, 2002, 2003, 2004, 2005 Corvette Engines: 5.7L VIN G Transmissions: All	**Fuel Level Sensor 2 Signal Range/Performance Conditions:** Engine running, and the PCM detected one of these 3 conditions: 1. With the secondary fuel tank not empty, more than 150 miles have been accumulated and the PCM did not detect that the fuel level in the right fuel tank moved by at least 0.80 gallons. 2. With the secondary fuel tank not empty and the primary fuel tank not full, engine runtime over 60 minutes, that the primary tank did not achieve the top of its range. Observe, after operating the engine for greater than 60 minutes that the fuel in the right fuel tank will transfer to the left fuel tank. 3. With the secondary fuel tank empty and the primary fuel tank full, that the fuel level in both fuel tanks did not change after traveling more than 200 miles. With the secondary fuel tank empty, observe if the primary fuel level decreased after 200 miles. The in-tank fuel pump supplies fuel via an in pipe fuel filter/pressure regulator assembly to the fuel rail. The fuel pressure supplied by the fuel pump exceeds the fuel injectors required pressure. The fuel pressure regulator regulates the fuel pressure supplied to the fuel injectors. Excess fuel returns from the fuel filter/regulator, through a separate fuel return pipe, to the left fuel tank. The fuel pump delivers a constant flow of fuel to the engine even during low fuel conditions and aggressive vehicle maneuvers. The Left Tank fuel pump also supplies a small amount of pressurized fuel to the right fuel tank siphon jet pump through the auxiliary fuel feed rear pipe. The pressurized fuel creates a venturi action inside the siphon jet pump. The venturi action causes the fuel to be drawn out of the right fuel tank. Fuel is then transferred from the right fuel tank to the left fuel tank through the auxiliary fuel return rear pipe. The fuel system maintains a higher level in the left fuel tank then the right fuel tank when the pump is "on". The fuel transfer rate from the left fuel tank to the right fuel tank is less than the transfer rate from the right fuel tank to the left fuel tank. Therefore, with the electric fuel pump "on", the left fuel tank level should be higher than the right fuel tank level. **Possible Causes** ● The fuel tanks over filled. ● Fuel level almost empty, then re-fueled with only a few gallons ● The Fuel Level sender became un-stuck during refueling ● The Fuel Level sender un-stuck on rough road or while turning ● Fuel level sender circuit is open or shorted to ground
DTC P1432 **1T CCM, MIL: No** 1999, 2000, 2001, 2002, 2003, 2004, 2005 Cavalier & C/K Truck 2.2L VIN 4, 2.2L VIN F, 6.0L VIN U CNG engines Transmissions: All	**Fuel Pressure Sensor Circuit Low Input Conditions:** DTC P1207 not set, engine started, engine operating on alternative fuel, and the AF ECU detected the Fuel Pressure Sensor (FPS) signal was less than 0.10v for more than 1 second during the test. **Possible Causes** ● Fuel tank pressure sensor is sticking or it is damaged ● Fuel tank pressure sensor signal circuit is shorted to ground ● Fuel tank pressure sensor is damaged or it has failed ● PCM has failed

OBD II Trouble Code List (P1xxx Codes)

DTC	Trouble Code Title, Conditions & Possible Causes
DTC P1432 **2T CCM, MIL: Yes** 2003, 2004, 2005 C/K Series Truck, G Van 6.0L VIN U CNG engine Transmissions: All	**Fuel Pressure Sensor Circuit Low Input Conditions:** DTC P0182 and P0183 not set; engine started, Fuel Tank temperature below 149ºF, and the PCM detected the Fuel Tank Pressure (FTP) sensor was less than 0.45v for over 1 second. **Possible Causes** ● Fuel tank pressure sensor is sticking or it is damaged ● Fuel tank pressure sensor signal circuit is shorted to ground ● Fuel tank pressure sensor is damaged or it has failed ● PCM has failed
DTC P1432 **1T CCM, MIL: No** 1997, 1998, 1999, 2000, 2001, 2002, 2003, 2004, 2005 Corvette Engines: 5.7L VIN G Transmissions: All	**Fuel Level Sensor 2 Circuit Low Input Conditions:** Engine started; electric fuel pump not operating, fuel level more than 50%, and the PCM detected the Fuel Pressure sensor signal was less than 0.39v for 2 minutes during the test. **Possible Causes** ● Fuel level sensor connector is damaged or shorted ● Fuel level sensor signal circuit is shorted to ground ● Fuel level sensor is damaged or has failed ● PCM has failed
DTC P1433 **1T CCM, MIL: No** 1997, 1998, 1999, 2000, 2001, 2002, 2003, 2004, 2005 Corvette Engines: 5.7L VIN G Transmissions: All	**Fuel Level Sensor (Right Side) Circuit High Input Conditions:** Engine started; electric fuel pump not operating, fuel level more than 50%, and the PCM detected the Fuel Pressure sensor signal was more than 2.90v for 2 minutes during the test. **Possible Causes** ● Fuel level sensor connector is damaged, loose or open ● Fuel level sensor signal circuit is open or shorted to VREF ● Fuel level sensor is damaged or has failed ● PCM has failed
DTC P1433 **1T CCM, MIL: No** 1999, 2000, 2001, 2002, 2003, 2004, 2005 C/K Truck, Cavalier 2.2L VIN 4, 2.2L VIN F, 6.0L VIN U CNG engine Transmissions: All	**Fuel Pressure Sensor Circuit High Input Conditions:** DTC P0182 and P0183 not set, Fuel Tank temperature below 149ºF, engine started, and the PCM detected the Fuel Tank Pressure (FTP) sensor was over 4.95v for more than 1 second. **Possible Causes** ● Fuel pressure sensor is sticking or it is damaged ● Fuel pressure sensor signal circuit is open or shorted to VREF ● Fuel pressure sensor is damaged or it has failed ● PCM has failed
DTC P1433 **2T CCM, MIL: Yes** 2003, 2004, 2005 C/K Series Truck, G Van 6.0L VIN U CNG engine Transmissions: All	**Fuel Pressure Sensor Circuit Low Input Conditions:** DTC P0182 and P0183 not set, Fuel Tank temperature below 149ºF, engine started, and the PCM detected the Fuel Tank Pressure (FTP) sensor was less than 0.45v for over 1 second. **Possible Causes** ● Fuel tank pressure sensor is sticking or it is damaged ● Fuel tank pressure sensor signal circuit is shorted to ground ● Fuel tank pressure sensor is damaged or it has failed ● PCM has failed
DTC P1441 **2T EVAP, MIL: Yes** 1996, 1997, 1998, 1999 Aurora, Bonneville, Century, Cutlass Supreme, Grand Prix, Intrigue, LeSabre, LSS, Lumina, Monte Carlo, 88', Regal, Park Avenue & 98' 3.1L VIN J, 3.1L VIN M, 3.4L VIN E, 3.5L VIN H, 3.8L VIN 1, 3.8L VIN K Transmissions: All	**EVAP System Flow During Non-Purge Conditions:** DTC P0107, P0108, P0112, P0113, P0117, P0118, P0122, P0123, P0125, P0440, P0442, P0446, P0452 and P0453 not set, P0440, P0442 and P0446 all passed, and the PCM detected a continuous open purge flow condition (fuel tank pressure at less than -11" H2O). **Possible Causes** ● EVAP purge valve, purge pipe, or the charcoal canister has a temporary blockage (this fault could cause an intermittent fault) ● FTP sensor is out-of-calibration, damaged or it is skewed ● EVAP purge solenoid valve is damaged or has failed ● PCM has failed ● TSB 00-06-04-022 contains a repair procedure for this code
DTC P1441 **2T EVAP, MIL: Yes** 2000, 2001, 2002, 2003, 2004, 2005 Aurora, Bonneville, Century, Cutlass Supreme, Grand Prix, Intrigue, LeSabre, LSS, Lumina, Monte Carlo, 88', Regal, Park Avenue & 98' 3.1L VIN J, 3.4L VIN E, 3.5L VIN H, 3.8L VIN 1, 3.8L VIN K, 4.0L VIN G Transmissions: All	**EVAP System Flow During Non-Purge Conditions:** DTC P0107, P0108, P0112, P0113, P0116- P0118, P0121-P0123, P0125, P0440-P0446, P0449, P0452, P0453, P1106-P1107, P1112, P1114, P1115, P1121 and P1122 not set, engine started, ECT and IAT sensors from 39-86ºF and within 16ºF at startup, BARO sensor over 75 kPa, Fuel Level from 15-85%, VSS under 75 mph, and the PCM detected continuous vacuum during a non-purge condition. **Possible Causes** ● EVAP purge valve, purge pipe, or the charcoal canister has a temporary blockage (this fault could cause an intermittent fault) ● FTP sensor is out-of-calibration, damaged or it is skewed ● EVAP purge solenoid valve is damaged or has failed ● PCM has failed

OBD II Trouble Code List (P1xxx Codes)

DTC	Trouble Code Title, Conditions & Possible Causes
DTC P1441 **2T EVAP, MIL: Yes** 1996, 1997, 1998, 1999, 2000, 2001, 2002, 2003, 2004, 2005 Achieva, Alero, Beretta, Cavalier, Ciera, Century, Grand Am, Sunfire Skyhawk, S/T Blazer & S/T Pickup 2.2L VIN 4, 2.2L VIN 5, 2.2L VIN F, 2.2L VIN H, 2.4L VIN T engines Transmissions: All	**EVAP System Flow During Non-Purge Conditions:** DTC P0107, P0108, P0112, P0113, P0117, P0118, P0121-P0123, P0125, P0440, P0442, P0443, P0446, P0449, P0452, P0453, P1106, P1107, P1111, P1112, P1114, P1115, P1121 and P1122 not set, engine started, vehicle speed less than 75 mph, system voltage over 10.0v, ECT and IAT sensors from 39-86ºF and within 16ºF at startup, BARO sensor over 75 kPa, fuel level from 15-85%, PCM detected a continuous open purge flow condition with the fuel tank pressure decreasing to less than -11" H2O during the test. The PCM seals the EVAP system by commanding the EVAP Purge valve "off" and the EVAP Canister Vent valve "on". The PCM monitors the fuel tank pressure (FTP) sensor to detect if a vacuum is being drawn on the EVAP system. If vacuum in the EVAP system is more than a set value for a certain time, this code is set. **Possible Causes** ● An improperly installed or damaged EVAP canister purge valve ● A temporary blockage in the EVAP canister purge valve, purge pipe, or EVAP canister could cause an intermittent condition ● FTP sensor is out-of-calibration, damaged or "skewed" ● Purge solenoid valve is damaged or has failed ● PCM has failed ● TSB 02 06 04 040 contains a repair procedure for this code
DTC P1441 **2T EVAP, MIL: Yes** 1996, 1997 Camaro, Caprice, Corvette, Firebird, Fleetwood & Roadmaster 4.3L VIN W, 5.7L VIN 5, 5.7L VIN P engines Transmissions: All	**EVAP System Flow During Non-Purge Conditions:** DTC P0100, P0101, P0102, P0103, P0106, P0323, P0325, P0327, P0332, P0335, P0336, P0372, P0401, P0403, P0410, P0412, P0420, P0430, P0441, P0443, P0500, P0502, P0503, P0506, P0507, P0530, P0531, P0719, P0758, P1415, P1642 and P1652 not set, engine started, vehicle driven at a speed of 500-3000 rpm, ECT sensor less than 239ºF, IAT sensor more than 37ºF, BARO sensor more than 77 kPa, TP angle less than 50%, MAP sensor less than 70 kPa, Purge valve commanded "off", and the PCM detected the Vacuum Switch signal indicated open (high vacuum) for 5 seconds. **Possible Causes** ● A temporary blockage in the EVAP canister purge valve, purge pipe, or EVAP canister could cause an intermittent condition ● EVAP canister is full of moisture or clogged (leaking charcoal) ● Incorrect vacuum hose routing to EVAP switch or Purge valve ● Purge solenoid valve is damaged or has failed ● EVAP vacuum switch is damaged (the switch could be sticking)
DTC P1441 **2T EVAP, MIL: Yes** 1996, 1997, 1998, 1999, 2000, 2001, 2002, 2003, 2004, 2005 Cab & Chassis, C/K Truck, G Van, L/M Van, S/T Blazer & S/T Pickup 4.3L VIN W, 4.3L VIN X, 5.0L VIN M, 5.7L VIN K, 5.7L VIN R, 7.4L VIN J, 8.1L VIN G engines Transmissions: All	**EVAP System Flow During Non-Purge Conditions:** DTC P0107, P0108, P0112, P0113, P0116, P0117, P0118, P0125, P0440, P0442, P0443, P0446, P0449, P0452, P0453, P1111, P1112, P1114, P1115, P1120, P1220 and P1221 not set, engine started, vehicle driven to a speed of less than 75 mph, ECT and IAT sensors from 39-86ºF and within 16ºF at startup, system voltage over 10.0v, BARO sensor more than 75 kPa, fuel level from 15-85%, and the PCM detected a vacuum condition present during a non-purge operating condition. The PCM seals the EVAP system by commanding the EVAP Purge valve "off" and commands the EVAP Canister Vent valve "on" (closed). The PCM monitors the FTP sensor to determine if a vacuum is being drawn on the EVAP system. If the vacuum in the EVAP system is more than a predetermined value within a calculated time, DTC P1441 will set. **Possible Causes** ● An improperly installed or damaged EVAP canister purge valve ● A temporary blockage in the EVAP canister purge valve, purge pipe, or EVAP canister could cause an intermittent condition ● FTP sensor is out-of-calibration, damaged or "skewed" ● Purge solenoid valve is damaged or has failed ● PCM has failed
DTC P1441 **2T EVAP, MIL: Yes** 1995 S/T Blazer & S/T Pickup 4.3L VIN W engine Transmissions: All	**EVAP System Flow During Non-Purge Conditions:** DTC P0106-P0108, P0112, P0113, P0117, P0118, P0121-P0123, P0401, P0506, P0507 and P1442 not set, ECT sensor under 230ºF, IAT sensor from -4ºF to 194ºF, ECT and IAT sensor within 194ºF of each other, engine started, BARO sensor over 75 kPa, TP angle at 5-60%, Purge command at 0%, MAP sensor under 20 kPa, and the PCM detected the EVAP Vacuum switch was open for 4 seconds. **Possible Causes** ● Base engine mechanical problem causing erratic/low vacuum ● Vacuum switch vacuum line leaking or vacuum source clogged ● Vacuum switch is damaged, disconnected or it failed ● Purge solenoid control circuit is open or shorted to power (B+) ● Purge solenoid is damaged or sticking (it may be stuck closed) ● PCM has failed

OBD II Trouble Code List (P1xxx Codes)

DTC	Trouble Code Title, Conditions & Possible Causes
DTC P1441 **2T EVAP, MIL: Yes** 1996, 1997, 1998, 1999, 2000, 2001, 2002, 2003, 2004, 2005 Aurora, DeVille, Eldorado & Seville 4.0L VIN C, 4.6L VIN 9 & 4.6L VIN Y engines Transmissions: A/T	**EVAP System Flow During Non-Purge Conditions:** DTC P0107, P0108, P0112, P0113, P0117, P0118, P0122, P0123, P0125, P0131, P0132, P0133, P0134, P0135, P0141, P0151, P0152, P0153, P0154, P0452, P0453, P0503, P1111, P1112, P1114, P1115, P1133 and P1153 not set, Engine started; system voltage over 10.0v, BARO sensor more than 72 kPa, fuel level from 10-85% for 3.2 seconds (time limit is required because fuel sloshing within the tank may cause the fuel level indication to vary outside the fuel level limits), ECT sensor from 35.6°F-86°F and within 57.56°F at startup, and the PCM detected the vacuum level in the EVAP system increased during the test. *Note: Although this diagnostic is considered type A (1T), it acts like a type B (2T) diagnostic under certain conditions. Whenever the EVAP diagnostics reports that the system passed, or the code has cleared, the test must fail two consecutive trips to set a code.* **Possible Causes** ● An improperly installed or damaged EVAP canister purge valve ● A temporary blockage in the EVAP canister purge valve, purge pipe, or EVAP canister ● FTP sensor is out-of-calibration, damaged or "skewed" ● Purge solenoid valve is damaged or has failed ● PCM has failed ● TSB 71-65-26 contains a repair procedure for this code
DTC P1441 **2T EVAP, MIL: Yes** 1997, 1998, 1999, 2000, 2001, 2002, 2003, 2004, 2005 Camaro, Corvette & Firebird 5.7L VIN G, 5.7L VIN S Transmissions: All	**EVAP System Flow During Non-Purge Conditions:** DTC P0107, P0108, P0112, P0113, P0116, P0117, P0118, P0125, P0443, P0449, P0452, P0453, P1106, P1107, P1112, P1114, P1115, P1120, P1220 and P1221, engine started, ECT and IAT sensors from 39-86°F and within 16°F at startup, BARO sensor more than 75 kPa, fuel level from 15-85%, system voltage over 10.0v, vehicle driven to a speed less than 75 mph, and the PCM detected vacuum in the system during a non-purge condition. The PCM tests for invalid intake manifold vacuum flow to the EVAP system. The PCM seals the EVAP system by commanding the EVAP Purge valve "off" and Canister Vent valve "on". **Possible Causes** ● An improperly installed or damaged EVAP canister purge valve ● A temporary blockage in the EVAP canister purge valve, purge pipe, or EVAP canister ● FTP sensor is out-of-calibration, damaged or "skewed" ● Purge solenoid valve is damaged or has failed ● PCM has failed
DTC P1441 **2T EVAP, MIL: Yes** 1995, 1996, 1997, 1998, 1999, 2000, 2001, 2002 Camaro & Firebird 3.8L VIN K engine Transmissions: All	**EVAP System Flow During Non-Purge Conditions:** DTC P0107, P0108, P0112, P0113, P0116-P0118, P0125, P0440-P0449, P0452, P0453, P1106, P1107, P1112-P1115, P1120, P1220 and P1221 not set, engine started, ECT and IAT sensors from 39-86°F and within 16°F at startup, BARO more than 75 kPa, system voltage over 10.0v, fuel level from 15-85%, vehicle driven to a speed of less than 75 mph, and the PCM detected vacuum in the system during a non-purge operating condition. The PCM tests for undesired intake manifold vacuum flow to the EVAP system. The PCM seals the EVAP system by commanding the EVAP Purge valve "off" and the EVAP Canister Vent valve "on". The PCM monitors the fuel tank pressure (FTP) sensor to determine if a vacuum is being drawn on the EVAP system. If vacuum in the EVAP system is more than a predetermined value within a predetermined time, this code is set. **Possible Causes** ● An improperly installed or damaged EVAP canister purge valve ● A temporary blockage in the EVAP canister purge valve, purge pipe, or EVAP canister could cause an intermittent condition ● FTP sensor is out-of-calibration, damaged or "skewed" ● Purge solenoid valve is damaged or has failed ● PCM has failed
DTC P1441 **2T EVAP, MIL: Yes** 1996, 1997, 1998, 1999, 2000, 2001, 2002, 2003, 2004, 2005 Aztek, Montana, Venture, Silhouette & Rendezvous 3.4L VIN E engine Transmissions: All	**EVAP System Flow During Non-Purge Conditions:** DTC P0107, P0108, P0110, P0112-P0118, P0121-P0123, P0125, P0440, P0442-P0449, P0452, P0453, P1106, P1107, P1111, P1112, P1114, P1115, P1121 and P1122 not set, engine started, ECT and IAT signals from 38-86°F and with 16°F at startup, vehicle driven to a speed of less than 75 mph, BARO sensor more than 75 kPa, system voltage over 10.0v, fuel level from 15-85%, and the PCM detected vacuum in the EVAP system during a not purge condition. The PCM tests for undesired intake manifold vacuum flow to the EVAP system. The PCM seals the EVAP system by commanding the EVAP Purge valve "off" and the EVAP Canister Vent valve "on". The PCM monitors the fuel tank pressure (FTP) sensor to determine if a vacuum is being drawn on the EVAP system. If vacuum in the EVAP system is more than a preset value within a predetermined time, DTC P1441 is set. **Possible Causes** ● An improperly installed or damaged EVAP canister purge valve ● A temporary blockage in the EVAP canister purge valve, purge pipe, or EVAP canister ● FTP sensor is out-of-calibration, damaged or "skewed" ● Purge solenoid valve is damaged or has failed ● PCM has failed

OBD II Trouble Code List (P1xxx Codes)

DTC	Trouble Code Title, Conditions & Possible Causes
DTC P1441 **2T EVAP, MIL: Yes** 1996, 1997 Grand Prix, Lumina & Monte Carlo Engines: 3.4L VIN X Transmissions: All	**EVAP System Flow During Non-Purge Conditions:** DTC P0101-P0103, P0107, P0108, P0112, P0113, P0116-P0118, P0121-P0123, P0125, P0201-P0206, P0441 and P1441 not set, engine started, engine started, ECT and IAT sensors from 32-156°F and within 18°F at startup, engine speed from 650-5000 rpm, BARO sensor over 70 kPa, ECT sensor less than 237°F, TP angle from 2.5-40%, Purge command less than 3%, and the PCM detected the EVAP switch signal indicated "off" (high vacuum). **Possible Causes** ● A temporary blockage in the EVAP canister purge valve, purge pipe, or EVAP canister ● EVAP canister is full of moisture or clogged (leaking charcoal) ● Incorrect vacuum hose routing to EVAP switch or Purge valve ● Purge solenoid valve is damaged or has failed ● EVAP vacuum switch is damaged (the switch could be sticking)
DTC P1441 **2T EVAP, MIL: Yes** 2002, 2003, 2004, 2005 Envoy & TrailBlazer 4.2L VIN S engine Transmissions: All	**EVAP System Flow During Non-Purge Conditions:** DTC P0107, P0108, P0112, P0113, P0116- P0118, P0122, P0123, P0125, P0440, P0442, P0446, P0452 and P0453 not set, vehicle speed less than 75 mph, system voltage from 10-18v, ECT and IAT sensors from 39-86°F and within 16°F at startup, BARO sensor over 75 kPa, fuel level at 15-85%, and the PCM detected vacuum present in a non-purge event. **Possible Causes** ● An improperly installed or damaged EVAP canister purge valve ● A temporary blockage in the EVAP canister purge valve, purge pipe, or EVAP canister ● FTP sensor is out-of-calibration, damaged or "skewed" ● Purge solenoid valve is damaged or has failed ● PCM has failed ● TSB 02-06-04-046 contains a repair procedure for this code
DTC P1442 **2T CCM, MIL: Yes** 1995 S/T Blazer & S/T Pickup 4.3L VIN W engine Transmissions: All	**EVAP Purge Solenoid Switch Static Test Malfunction Conditions:** Key on, engine off, Purge solenoid inhibit criteria met, and the PCM detected the EVAP Purge Vacuum Switch indicated a high state (i.e., switch stuck open indicating high vacuum). **Possible Causes** ● EVAP vacuum switch signal circuit is shorted to ground ● EVAP switch ground circuit is open (test ground at the switch) ● EVAP vacuum switch is damaged, or the PCM has failed
DTC P1442 **2T CCM, MIL: Yes** 1996 Bonneville, Century, Cutlass Supreme, Grand Prix, LeSabre, LSS, Lumina, Monte Carlo, 88', Regal, Park Avenue & 98' 3.1L VIN M, 3.4L VIN X, 3.8L VIN 1, 3.8L VIN K Transmissions: All	**EVAP Vacuum Switch Circuit Malfunction Conditions:** DTC P0106, P0107, P0108, P0112, P0113, P0121, P0122, P0123 and P1655 not set, key on, engine off (test is performed prior to engine startup), IAT sensor from 32°F-158°F, and the PCM detected the EVAP Vacuum switch indicated "open" during the test. **Possible Causes** ● EVAP vacuum switch signal circuit is shorted to ground ● EVAP switch power circuit is open (test the MISC ENG fuse) ● EVAP vacuum switch is damaged or has failed in open position ● PCM has failed
DTC P1460 **1T MISFIRE, MIL: No** 1997, 1998, 1999, 2000, 2001 Catera 3.0L VIN R engine Transmissions: All	**Misfire Detected With Low Fuel Conditions:** DTC P0171, P0174, P0300, P0301-P0306 not set, engine started, and the PCM detected the fuel level signal was less than 2.9 gallons (11 liters) for 30 seconds. This code is used to indicate that the fuel level was low during a misfire or a lean air fuel ratio condition. **Possible Causes** ● If this DTC is present, and no misfire condition or lean condition currently exists, the vehicle may have simply run low on fuel. ● If the fuel level is low, add some fuel to the tank prior a retest ● Drive the vehicle under the conditions present in Freeze Frame to verify the original trouble code does not reset ● Perform the Diagnostic Circuit Check for Engine Controls
DTC P1483 **1T CCM, MIL: Yes** 1998, 1999 Century, Lumina, Monte Carlo, Montana, Silhouette & Venture 3.1L VIN M, 3.4L VIN E Transmissions: All	**Engine Cooling System Range/Performance Conditions:** DTC P0101, P0102, P0103, P0105, P0107, P0108, P0117, P0118, P0125, P0502 and P1108 not set, ECT sensor under 230°F at startup, or with the ECT sensor under 219°F, engine load, IAT sensor, engine speed and VSS within a specific range, the PCM detected the ECT sensor was more than the calculated engine coolant temperature. The PCM richens the A/F ratio when high coolant temperature is detected. This function, referred to as Hot Fuel Enrichment (HFE), can cause high exhaust emissions on these vehicles. This code is required in order to prevent HFE from being active when the ECT sensor is "skewed". **Possible Causes** ● Cooling system problem (check for restricted radiator airflow) ● Coolant level is too low or the coolant mixture is not correct ● ECT sensor or the IAT sensor is out-of-calibration or "skewed" ● Thermostat is damaged or has failed

OBD II Trouble Code List (P1xxx Codes)

DTC	Trouble Code Title, Conditions & Possible Causes
DTC P1501 **1T CCM, MIL: No** 1997, 1998, 1999, 2000 Catera 3.0L VIN R engine Transmissions: A/T	**Theft Deterrent System Malfunction Conditions:** Engine cranking; and the PCM determined the Theft Deterrent password had not been programmed properly into the PCM. The ignition key contains a transponder that transmits a signal to the theft deterrent module when the ignition switch is turned to the "on" position. The TDM (module) compares the signal from the ignition key and then determines if an authorized key was actually used. **Possible Causes** ● Attempt to program the ECM with the correct frequency code. If the original trouble code appears after the reprogramming step, the PCM may need to be replaced.
DTC P1502 **1T CCM, MIL: No** 1997, 1998, 1999, 2000 Catera 3.0L VIN R engine Transmissions: A/T	**Theft Deterrent System Signal Not Received Conditions:** Engine cranking; and the PCM determined that it did not receive the correct frequency code from the Theft Deterrent Module. The theft deterrent module will detect an unexpected voltage condition on the VSS/theft deterrent signal circuit and set DTC P1502 in memory. **Possible Causes** ● VSS/Theft Deterrent signal circuit is open, shorted to ground or shorted to power (B+) ● An inoperative theft deterrent module will also set this code
DTC P1503 **1T CCM, MIL: No** 1997, 1998, 1999, 2000 Catera 3.0L VIN R engine Transmissions: A/T	**Theft Deterrent System Password Incorrect Conditions:** Engine cranking; and the PCM detected a frequency code that did not match the code it received from the theft deterrent module. **Possible Causes** ● The wrong ignition key, the use of a key without a transponder, or an incorrectly programmed theft deterrent module will cause this code to set in the theft deterrent module. Perform the Diagnostic System Check for the theft deterrent system.
DTC P1508 **2T CCM, MIL: Yes** 1996, 1997 Camaro, Caprice, Corvette, Firebird, Fleetwood & Roadmaster 4.3L VIN W, 5.7L VIN 5, 5.7L VIN P engines Transmissions: All	**Idle Speed Low RPM Conditions:** DTC P0107, P0108, P0122, P0123, P0174, P0175, P0300 and P0325 not set, engine started, engine runtime over 25 seconds, vehicle not moving, ECT sensor more than 140ºF, IAT sensor more than 14ºF, BARO sensor over 77 kPa, TP angle at 0%, idle speed less than Desired Idle Speed, then vehicle driven to 20-77 mph with the MAF sensor from 13-35 g/sec, TP angle and RPM stable for 5 seconds, IAC motor opened a specified amount, and the PCM did not detect a 6 g/sec change in the MAF signal in the Active Test. **Possible Causes** ● Base engine problem causing a slow or unstable idle speed ● Fuel system is too rich or too lean ● IAC motor is damaged or has failed (it may be sticking) ● Throttle body plate or bore contaminated with foreign material ● Throttle body adjustment screw tampered with or damaged ● Perform the IAC Reset procedure with the Scan Tool
DTC P1508 **2T CCM, MIL: Yes** 1996, 1997, 1998, 1999, 2000 Cab & Chassis, C/K Truck, G Van, L/M Van, S/T Blazer & S/T Pickup 4.3L VIN W, 4.3L VIN X, 5.0L VIN M, 5.7L VIN K, 5.7L VIN R, 7.4L VIN J Transmissions: All	**Idle Speed Low - IAC System Not Responding Conditions:** DTC P0107, P0108, P0117, P0118, P0121, P0122, P0123 and P0500 not set, engine started, vehicle not moving, ECT sensor more than 122ºF, IAT sensor more than -13ºF, system voltage over 10.0v, engine runtime over 30 seconds, BARO sensor over 70 kPa, MAF sensor from 17.5-50 g/sec, TP angle steady (± 1%), engine speed stable (50 rpm), then vehicle driven to a speed of 25-85 mph, IAC valve commanded to move a specified number of steps for over 2 seconds, and the PCM detected the change in the MAF sensor signal indicated more than 3 g/sec during the Active Test period. **Possible Causes** ● Base engine problem causing a slow or unstable idle speed ● Fuel system is too rich or too lean ● IAC motor is damaged or has failed (it may be sticking) ● Throttle body plate or bore contaminated with foreign material ● Throttle body adjustment screw tampered with or damaged ● Perform the IAC Reset procedure with the Scan Tool
DTC P1508 **2T CCM, MIL: Yes** 1995 Camaro & Firebird 3.8L VIN K engine Transmissions: All	**Idle Speed Low RPM Conditions:** Engine started, engine runtime over 2 minutes, ECT sensor more than 132ºF, IAT sensor more than 68ºF, vehicle speed less than 3 mph with the throttle closed, and the PCM detected the Actual idle speed was 175 rpm lower than the Desired idle speed. **Possible Causes** ● Base engine problem causing a slow or unstable idle speed ● Fuel system is too rich or too lean ● IAC motor is damaged or has failed (it may be sticking) ● Throttle body plate or bore contaminated with foreign material ● Throttle body adjustment screw tampered with or damaged ● Perform the IAC Reset procedure with the Scan Tool

OBD II Trouble Code List (P1xxx Codes)

DTC	Trouble Code Title, Conditions & Possible Causes
DTC P1508 **1T CCM, MIL: No** 1996, 1997, 1998, 1999 Aurora Engines: 4.0L VIN C Transmissions: A/T	**Idle Speed Low - IAC System Not Responding Conditions:** DTC P0101-P0103, P0106-P0108, P0112-P0113, P0117-P0118, P0121-P0123, P0201-P0208 and P502 not set, DTC P1370, P1371, P1406 and P1441 not testing, PCM not checking EGR operation, system voltage from 10-16v, BARO input at least 65 kPa, ECT input from -4°F to 230°F, engine runtime over 10 seconds, in closed loop, MAF input from 2-35 g/sec, Actual idle speed less than desired idle speed by at least 80 RPM, VSS input from 30-45 mph, IAC position commanded to 100-205 counts, TP angle stable (± 1%), engine speed stable (± 75 rpm), a P/N to D/R or a D/R to P/N transition did not occur within the last 60 seconds, and the PCM detected the MAF sensor changed less than 1.5 g/sec during the 2 second period right after the IAC motor position was increased (opened). **Possible Causes** ● Base engine problem causing a slow or unstable idle speed ● Fuel system is too rich or too lean ● IAC motor is damaged or has failed (it may be sticking) ● IAC motor circuit open, shorted to ground or shorted to power ● Throttle body plate or bore contaminated with foreign material ● Throttle body adjustment screw tampered with or damaged ● Perform the IAC Reset procedure with the Scan Tool
DTC P1508 **1T CCM, MIL: Yes** 1996, 1997, 1998, 1999 DeVille, Eldorado, Seville 4.6L VIN 9, 4.6L VIN Y Transmissions: All	**Idle Speed Low - IAC Control System Not Responding Conditions:** DTC P1370, P1371, P1406, P1441 and P1442 not set, engine started, vehicle not moving, ECT sensor from -4°F to 230°F, BARO sensor over 65 kPa, engine runtime over 10 seconds in closed loop, MAF sensor from 2-35 g/sec, Actual idle speed 80 rpm less than the Desired idle speed, then vehicle driven to a speed of 30-45 mph, TP angle and engine speed stable, IAC counts from 100-205, EGR test inactive, and the PCM detected the MAF sensor changed less than 1.5 g/sec during the 2 second period that the IAC position increased. **Possible Causes** ● Base engine problem causing a slow or unstable idle speed ● Fuel system is too rich or too lean ● IAC motor is damaged or has failed (it may be sticking) ● Throttle body plate or bore contaminated with foreign material ● Throttle body adjustment screw tampered with or damaged ● Perform the IAC Reset procedure with the Scan Tool
DTC P1508 **2T CCM, MIL: Yes** 1995 S/T Blazer & S/T Pickup 4.3L VIN W engine Transmissions: All	**Idle Speed Lower Than Expected Conditions:** DTC P0107, P0108, P0121, P0122, P0123 and P0500 not set, engine started, engine runtime over 30 seconds, IAC valve position more than 93 counts, VSS under 2 mph, ECT sensor more than 161°F, IAT sensor more than -13°F, system voltage over 10.0v, BARO sensor over 70 kPa, MAF sensor from 17.5-50 g/sec, Actual Idle Speed more than 75 rpm below the Desired Idle Speed for 3 seconds (Passive Test failed), then vehicle driven to a speed of 35-85 mph, IAC valve commanded to 36-116 93 counts, calculated airflow from 17.5-37.5 g/sec, TP angle stable (± 1%) with any change in engine speed less than 50 rpm, IAC motor commanded to zero (0) counts, and the PCM detected the change in the Calculated airflow was less than 2 g/sec during the Active Test period. **Possible Causes** ● Base engine problem causing a slow or unstable idle speed ● Fuel system is too rich or too lean ● IAC motor is damaged or has failed (it may be sticking) ● IAC motor circuit open, shorted to ground or shorted to power ● Throttle body plate or bore contaminated with foreign material ● Throttle body adjustment screw tampered with or damaged ● Perform the IAC Reset procedure with the Scan Tool
DTC P1508 **2T CCM, MIL: Yes** 1996, 1997 Camaro, Caprice, Corvette, Firebird, Fleetwood & Roadmaster 4.3L VIN W, 5.7L VIN 5, 5.7L VIN P engines Transmissions: All	**Idle Speed High RPM Conditions:** DTC P0107, P0108, P0122, P0123, P0174, P0175, P0300 and P0325 not set, engine runtime over 25 seconds, vehicle not moving, ECT sensor more than 140°F, IAT sensor more than 14°F, BARO sensor more than 77 kPa, TP angle at 0%, idle speed less than the Desired Idle Speed, all conditions met for 1 second, then the vehicle was driven to a speed of 20-77 mph, MAF sensor from 13-35 g/sec, engine speed stable and TP angle stable, conditions met for 5 seconds, and the PCM did not detect a 6 g/sec change in the MAF signal in the IAC Active test (valve not moving). **Possible Causes** ● Base engine problem (engine vacuum leak or PCV valve leak) ● IAC motor is damaged or has failed (it may be sticking) ● Throttle body adjustment screw tampered with or damaged ● Perform the IAC Reset procedure with the Scan Tool

OBD II Trouble Code List (P1xxx Codes)

DTC	Trouble Code Title, Conditions & Possible Causes
DTC P1509 **1T CCM, MIL: No** 1996, 1997, 1998, 1999 Aurora Engines: 4.0L VIN C Transmissions: A/T	**Idle Speed High - IAC System Not Responding Conditions:** DTC P0101-P0103, P0106-P0108, P0112-P0113, P0117-P0118, P0121-P0123, P0201-P0208 and P502 not set, DTC P1370, P1371, P1406 and P1441 not testing, PCM not checking EGR operation, system voltage from 10-16v, BARO input at least 65 kPa, ECT input from -4°F to 230°F, engine runtime over 10 seconds, in closed loop, MAF input from 2-35 g/sec, Actual idle speed less than desired idle speed by at least 80 RPM, VSS input from 30-45 mph, IAC position commanded to 100-205 counts, TP angle stable (± 1%), engine speed stable (± 75 rpm), a P/N to D/R or a D/R to P/N transition did not occur within the last 60 seconds, and the PCM detected the MAF sensor changed less than 1.5 g/sec during the 2 second period right after the IAC motor position was increased (opened). **Possible Causes** • Base engine problem (engine vacuum leak or PCV valve leak) • IAC motor circuit open, shorted to ground or shorted to power • Fuel system is too rich or too lean • Throttle body adjustment screw tampered with or damaged • Perform the IAC Reset procedure with the Scan Tool
DTC P1509 **2T CCM, MIL: Yes** 1996, 1997, 1998, 1999, 2000 Cab & Chassis, C/K Truck, G Van, L/M Van, S/T Blazer & S/T Pickup 4.3L VIN W, 4.3L VIN X, 5.0L VIN M, 5.7L VIN K, 5.7L VIN R, 7.4L VIN J Transmissions: All	**Idle Speed High - IAC System Not Responding Conditions:** DTC P0107, P0108, P0117, P0118, P0121, P0122, P0123 and P0500 not set, engine started, vehicle not moving, ECT sensor more than 122°F, IAT sensor more than -13°F, system voltage over 10.0v, engine runtime over 30 seconds, BARO sensor over 70 kPa, MAF sensor from 17.5-50 g/sec, TP angle steady (± 1%), engine speed stable (50 rpm), then vehicle driven to a speed of 25-85 mph, IAC valve commanded to move a specified number of steps for over 2 seconds, and the PCM detected the change in the MAF sensor signal indicated more than 3 g/sec during the Active Test period. **Possible Causes** • Base engine problem (engine vacuum leak or PCV valve leak) • Fuel system is too rich or too lean • IAC motor circuit open, shorted to ground or shorted to power • Throttle body adjustment screw tampered with or damaged • Perform the IAC Reset procedure with the Scan Tool
DTC P1509 **2T CCM, MIL: Yes** 1995 Camaro & Firebird 3.8L VIN K engine Transmissions: All	**Idle Speed High RPM Conditions:** Engine started, engine runtime over 2 minutes, ECT sensor more than 132°F, IAT sensor more than 68°F, vehicle speed less than 3 mph with the throttle closed, and the PCM detected the Actual idle speed was 175 rpm higher than the Desired idle speed. **Possible Causes** • Base engine problem (engine vacuum leak or PCV valve leak) • IAC motor is damaged or has failed (it may be sticking) • IAC motor circuit open, shorted to ground or shorted to power • Throttle body adjustment screw tampered with or damaged
DTC P1509 **1T CCM, MIL: Yes** 1996, 1997, 1998, 1999 DeVille, Eldorado, Seville 4.6L VIN 9, 4.6L VIN Y Transmissions: A/T	**Idle Speed High - IAC System Not Responding Conditions:** DTC P1370, P1371, P1406, P1441 and P1442 not set, engine started, vehicle not moving, ECT sensor from -4°F to 230°F, BARO sensor over 65 kPa, engine runtime over 10 seconds in closed loop, MAF sensor from 2-35 g/sec, Actual idle speed 80 rpm less than the Desired idle speed, then vehicle driven to a speed of 30-45 mph, TP angle and engine speed stable, IAC counts from 100-205, EGR test inactive, and the PCM detected the MAF sensor changed less than 1.5 g/sec during the 2 second period that the IAC position increased. **Possible Causes** • Base engine problem (engine vacuum leak or PCV valve leak) • Fuel system is too rich or too lean • IAC motor circuit open, shorted to ground or shorted to power • Throttle body adjustment screw tampered with or damaged • Perform the IAC Reset procedure with the Scan Tool
DTC P1509 **2T CCM, MIL: Yes** 1995 S/T Blazer & S/T Pickup 4.3L VIN W engine Transmissions: All	**Idle Speed Higher Than Expected Conditions:** DTC P0107, P0108, P0121, P0122, P0123 and P0500 not set, engine runtime over 30 seconds, IAC valve counts over 93, VSS under 2 mph, system voltage over 10v, ECT sensor over 161°F, IAT sensor over -13°F, BARO sensor over 70 kPa, MAF sensor at 17.5-50 g/sec, Actual Idle Speed at least 75 rpm below the Desired Idle Speed for 3 seconds, then vehicle driven to a speed of 35-85 mph, IAC valve counts from 36-116, calculated airflow at 17.5-37.5 g/sec, TP angle and engine speed stable, IAC motor at zero counts, and the PCM detected the Calculated airflow changed less than 2 g/sec. **Possible Causes** • Base engine problem (engine vacuum leak or PCV valve leak) • IAC motor circuit open, shorted to ground or shorted to power • Fuel system is too rich or too lean • Throttle body adjustment screw tampered with or damaged

OBD II Trouble Code List (P1xxx Codes)

DTC	Trouble Code Title, Conditions & Possible Causes
DTC P1551 **2T CCM, MIL: Yes** 2003, 2004, 2005 CTS (Cadillac) 2.6L VIN M, 3.2L VIN N Transmissions: All	**Secondary Or Primary Fuel Level Sensor Malfunction Conditions:** Key on, engine speed less than 40 rpm, ECT sensor from 41-212ºF, IAT sensor over 41ºF, system voltage over 10v, APP sensor signal under 15%, and the PCM detected the throttle blade was at "rest" position with the TP angle under 2.1% or over 9.8%. The ECM controls the throttle blades by applying a varying voltage to the throttle valve motor. The ECM monitors the actual throttle blade position using throttle position (TP) sensor 1 and 2. **Possible Causes** • TAC motor circuit (one or more) shorted to ground • Throttle valves not in a "rest" position due to stuck open/closed • Throttle valves moving open or closed without spring pressure • Throttle body assembly is damaged or it has failed • PCM has failed
DTC P1510 **1T CCM, MIL: Yes** 1999, 2000, 2001 Catera 3.0L VIN R engine Transmissions: A/T	**Throttle Control System - Throttle Limitation Active Conditions:** Engine started, system voltage over 10.0v, Idle Learn procedure inactive, and the PCM detected the Throttle Valve Motor pulsewidth modulation (PWM) signal was over 80 percent for more than 6 seconds during the CCM test. The PCM controls the throttle blade by applying varying voltage to the Throttle Valve Motor. The PCM monitors the pulsewidth modulation (PWM) signal that is required to move the throttle blade. If an incorrect signal is required to operate the throttle valve motor, the PCM will set DTC P1510. **Possible Causes** • Throttle valve motor circuit(s) open, shorted to ground or to B+ • Throttle valves that at the "rest" position (due to binding or dirt) • Throttle valves that are free to move open or closed • Throttle body assembly is damaged or has failed • PCM has failed
DTC P1511 **1T CCM, MIL: Yes** 1999, 2000, 2001 Catera 3.0L VIN R engine Transmissions: A/T	**Throttle Control System - Backup System Performance Conditions:** Engine started, engine speed less than 250 rpm, ECT sensor from 41-212ºF, IAT sensor more than 41ºF, vehicle not moving, APP sensor less than 15%, and the PCM detected the TP angle indicated less than 2.1%, or it indicated more than 9.8% with the throttle closed. The PCM controls the throttle valves by applying a varying voltage to the Throttle Valve Motor. The PCM monitors the actual throttle blade position using Throttle Position (TP) Sensor 1 and Sensor 2. The PCM tests the ability of the throttle blade to operate in the limp home throttle angle opening. If the PCM commands the throttle blade to a limp home throttle angle opening and the throttle blade do not operate in the desired range, it sets the code. **Possible Causes** • Throttle valve motor circuit(s) open, shorted to ground or to B+ • Throttle valves that at the "rest" position (due to binding or dirt) • Throttle valves that are free to move open or closed • Throttle body assembly is damaged or has failed • PCM has failed
DTC P1512 **1T CCM, MIL: Yes** 2002, 2003, 2004, 2005 Envoy & TrailBlazer Engines: 4.2L VIN S Transmissions: All	**Control Module Throttle Actuator Position Performance Conditions:** Key on or engine running; and the PCM detected the difference between the Actual and Predicted throttle position was more than a calibrated amount. The PCM compares the commanded throttle position to the actual throttle position based on the accelerator pedal position and other possible limiting factors. Both values should be within a calibrated range of each other. The PCM continuously monitors the Actual and Commanded throttle positions. The PCM will set this code if the values are greater than the calibrated range. **Possible Causes** • Throttle valve motor circuit(s) open, shorted to ground or to B+ • Throttle valves that at the "rest" position (due to binding or dirt) • Throttle valves that are free to move open or closed • Throttle body assembly is damaged or has failed • PCM has failed
DTC P1514 **1T CCM, MIL: Yes** 1997, 1998, 1999, 2000, 2001, 2002, 2003, 2004, 2005 Corvette 5.7L VIN G, 5.7L VIN S Transmissions: All	**Throttle Body Performance Conditions:** DTC P0601, P0602, P0606, P1515, P1516, P1517 and P1518 not set, P1120, P1220 and P1221 not active at this time, or P1120 and P1220 not set at the same time, engine speed over 500 rpm, and the PCM detected the difference between Actual (MAF) airflow and Speed Density Calculated airflow was more than expected for less than 1 second. **Possible Causes** • Inspect the throttle blade for damage and/or proper installation • Inspect the TAC module connectors for signs of water intrusion. When water intrusion occurs, multiple codes can set with no circuit or component faults apparent at the time. • Move throttle blade from closed to wide open position without applying much force. It should move smoothly through the full range and return to a slightly open position. • If the TAC module detects a fault in the system, it may set more than one related code because of the many redundant tests that run continuously on this system.

OBD II Trouble Code List (P1xxx Codes)

DTC	Trouble Code Title, Conditions & Possible Causes
DTC P1514 **1T CCM, MIL: Yes** 1999, 2000, 2001, 2002, 2003, 2004, 2005 Camaro, C/K Series Truck, G Van, Envoy, Escalade & TrailBlazer 3.8L VIN K, 4.8L VIN V, 5.3L VIN P, 5.3L VIN T, 5.3L VIN Z, 6.0L VIN N, 6.0L VIN U, 7.4L VIN J, 8.1L VIN G engines Transmissions: All	**Throttle Body Performance Conditions:** DTC P0601, P0602, P0606, P1515, P1516, P1517 and P1518 not set, P1120, P1220 and P1221 not active at the time this code set, or P1120 and P1220 not set at the same time, engine speed over 500 rpm, and the PCM detected the difference between Actual (MAF) airflow and Speed Density Calculated airflow was more than expected for 1 second. The Reduced Engine Power message displays on the Driver Information Center if this code sets. **Possible Causes** ● Inspect the throttle blade for damage and/or proper installation ● Inspect the TAC module connectors for signs of water intrusion. When water intrusion occurs, multiple codes can set with no circuit or component faults apparent during diagnostic testing. ● Physically and visually inspect the throttle body assembly, and throttle position sensor for damage and/or a loose mounting. Move the throttle blade from closed to wide open position without applying too much force. The throttle blade should move smoothly through the full range and should return to a slightly open position on its own. ● If the TAC module detects a fault in the system, it may set more than one related code because of the many redundant tests that run continuously on this system. Locating and repairing one individual condition may fix more than one code.
DTC P1514 **1T CCM, MIL: Yes** 2002, 2003, 2004, 2005 Envoy & TrailBlazer Engines: 4.2L VIN S Transmissions: All	**Throttle Body Performance Conditions:** DTC P1512, P1515, P1516, P1523, P1635 and P1639 not set, engine started, engine speed over 600 rpm, and the PCM detected the Calculated airflow was more than expected. The Reduced Engine Power text appears on Driver Info Center if this code sets. **Possible Causes** ● Inspect the throttle blade for damage and/or proper installation ● Inspect the TAC module connectors for signs of water intrusion. When water intrusion occurs, multiple codes can set with no circuit or component faults found during testing. ● Physically and visually inspect the throttle body assembly, and throttle position sensor for damage and/or a loose mounting. Manually move the throttle blade from closed to wide open position without applying too much force. The throttle blade should move smoothly through the full range and should return to a slightly open position on its own. ● If the TAC module detects a fault in the system, it may set more than one related code because of the many redundant tests that run continuously on this system. Locating and repairing one individual condition may fix more than one code.
DTC P1515 **1T CCM, MIL: Yes** 1997, 1998, 1999, 2000, 2001, 2002, 2003, 2004, 2005 Corvette 5.7L VIN G, 5.7L VIN S Transmissions: All	**Control Module Throttle Actuator Position Performance Conditions:** DTC P0601, P0602, P0606, P1515, P1516, P1517 and P1518 not set, P1120, P1220 and P1221 not active at the time this code set, or P1120 and P1220 not set at the same time, key in crank or run mode, ETC or TAC system not in Battery Saver Mode, and the PCM detected the Actual and Commanded throttle positions were out-of-range for under 1 second. **Possible Causes** ● Throttle actuator motor CKT 1 or CKT 2 is open, shorted to ground or B+ ● Throttle actuator motor is damaged or has failed ● Throttle actuator motor control module has failed
DTC P1515 **1T CCM, MIL: Yes** 1999, 2000, 2001, 2002, 2003, 2004, 2005 Camaro, C/K Series Truck, G Van, Envoy, Escalade & TrailBlazer 3.8L VIN K engine, 4.8L VIN V, 5.7L VIN P, 5.3L VIN T, 5.3L VIN Z, 6.0L VIN N, 6.0L VIN U, 7.4L VIN J, 8.1L VIN G Transmissions: All	**Control Module Throttle Actuator Position Performance Conditions:** DTC P0601, P0602, P0606, P1515, P1516, P1517 and P1518 not set, P1120, P1220 and P1221 not active at the time this code set, or P1120 and P1220 not set at the same time, key in crank or run mode, ETC or TAC system not in Battery Saver Mode, and the PCM detected the Actual and Commanded throttle positions were out-of-range for under 1 second. **Possible Causes** ● Throttle actuator motor CKT 1 is open, shorted to ground or B+ ● Throttle actuator motor CKT 2 is open, shorted to ground or B+ ● Throttle actuator motor is damaged or has failed ● Throttle actuator motor control module has failed ● TSB 00-06-04-035 contains a repair procedure for this code
DTC P1515 **1T CCM, MIL: Yes** 2002, 2003, 2004, 2005 Envoy & TrailBlazer Engines: 4.2L VIN S Transmissions: All	**Control Module Throttle Actuator Position Performance Conditions:** DTC P0601, P0602, P0606, P1515, P1516, P1517 and P1518 not set, P1120, P1220 and P1221 not active at the time this code set, or P1120 and P1220 not set at the same time, key in crank or run mode, ETC or TAC system not in Battery Saver Mode, and the PCM detected the difference between the Predicted and Actual Throttle Position was more than a calibrated amount for one second. **Possible Causes** ● Throttle actuator motor CKT 1 is open, shorted to ground or B+ ● Throttle actuator motor CKT 2 is open, shorted to ground or B+ ● Throttle actuator motor is damaged or has failed ● Throttle actuator motor control module has failed ● TSB 00-06-04-035 contains a repair procedure for this code

OBD II Trouble Code List (P1xxx Codes)

DTC	Trouble Code Title, Conditions & Possible Causes
DTC P1516 **1T CCM, MIL: Yes** 1999, 2000, 2001, 2002, 2003, 2004, 2005 Corvette 5.7L VIN G, 5.7L VIN S Transmissions: All	**TAC Module Throttle Actuator Position Performance Conditions:** DTC P1518 not set, key in crank or run mode, ETC or TAC system not in Battery Saver Mode, then the ETC/TAC module detected the predicted and actual throttle positions were not within a calibrated range of each other or the PCM and ETC/TAC could not determine the throttle position or that both TP sensors signals were invalid. **Possible Causes** • Throttle actuator motor CKT 1 is open, shorted to ground or B+ • Throttle actuator motor CKT 2 is open, shorted to ground or B+ • Throttle actuator motor CKT 1 is shorted to CKT 2 • Throttle actuator motor is damaged or has failed • Throttle actuator motor control module has failed
DTC P1516 **1T CCM, MIL: Yes** 1999, 2000, 2001, 2002 Camaro, C/K Series Truck, G Van, Envoy, Escalade & TrailBlazer 3.8L VIN K, 4.8L VIN V, 5.3L VIN P, 5.3L VIN T, 5.3L VIN Z, 6.0L VIN N, 6.0L VIN U, 7.4L VIN J, 8.1L VIN G engines Transmissions: All	**TAC Module Throttle Actuator Position Performance Conditions:** DTC P1518 not set, key in crank or run mode, ETC or TAC system not in Battery Saver Mode, then the ETC/TAC module detected the predicted and actual throttle positions were not within a calibrated range of each other or the PCM and ETC/TAC could not determine the throttle position or that both TP sensors signals were invalid. **Possible Causes** • TAC motor CKT 1 or CKT 2 is open, shorted to ground or B+ • Throttle actuator motor CKT 1 is shorted to CKT 2 • Throttle actuator motor control module has failed • TSB 03-04-06-034 contains a repair procedure for this code
DTC P1516 **1T CCM, MIL: Yes** 2003, 2004, 2005 C/K Series Truck, G Van, Envoy, Escalade & TrailBlazer 3.8L VIN K, 4.8L VIN V, 5.3L VIN P, 5.3L VIN T, 5.3L VIN Z, 6.0L VIN N, 6.0L VIN U, 8.1L VIN G Transmissions: All	**TAC Module Throttle Actuator Position Performance Conditions:** DTC P1518 not set, key in crank or run mode, ETC or TAC system not in Battery Saver Mode, then the ETC/TAC module detected the predicted and actual throttle positions were not within a calibrated range of each other or the PCM and ETC/TAC could not determine the throttle position or that both TP sensors signals were invalid. **Possible Causes** • TAC motor CKT 1 or CKT 2 is open, shorted to ground or B+ • Throttle actuator motor CKT 1 is shorted to CKT 2 • Throttle actuator motor control module has failed • TSB 03-04-06-032 contains a repair procedure for this code
DTC P1516 **1T CCM, MIL: Yes** 2002, 2003, 2004, 2005 Envoy & TrailBlazer Engines: 4.2L VIN S Transmissions: All	**TAC Module Throttle Actuator Position Performance Conditions:** DTC P1514 not set, engine running, TAC system not in Battery Saver Mode, and the TAC module detected the Predicted and Actual throttle positions were not in the calibrated range. **Possible Causes** • TAC motor CKT 1 or CKT 2 is open, shorted to ground or B+ • Throttle actuator motor CKT 1 is shorted to CKT 2 • Throttle actuator control (TAC) module is damaged or failed • Verify the throttle closes completely, and that it is not binding
DTC P1516 **1T CCM, MIL: Yes** 1999, 2000, 2001 Catera Engines: 3.0L VIN R Transmissions: A/T	**TAC System - Throttle Actuator Position Performance Conditions:** Engine started, and the PCM detected the Actual Throttle Position was 0% with a Calculated Throttle Position of 4%, or the Actual Throttle Position was 0-3% with a Calculated Throttle Position of 6%, or the Actual Throttle Position was 0-3% with a Calculated Throttle Position of 6%, or the Actual Throttle Position was 1% with a Calculated Throttle Position of 11%, or the Actual Throttle Position was 5% with a Calculated Throttle Position of 20%, or the Actual Throttle Position was 15% with a Calculated Throttle Position of 30% during the test period. **Possible Causes** • Throttle actuator motor CKT 1 or CKT 2 is open, shorted to ground or B+ • Throttle actuator motor control module has failed
DTC P1517 **1T CCM, MIL: Yes** 1999, 2000, 2001, 2002, 2003, 2004, 2005 Camaro & Firebird, C/K Truck, G Van, Envoy, Escalade & TrailBlazer 3.8L VIN K, 4.2L VIN S, 4.8L VIN V, 5.3L VIN T, 5.3L VIN Z, 6.0L VIN N, 6.0L VIN U, 7.4L VIN J, 8.1L VIN G engines Transmissions: All	**Throttle Actuator Control Module Performance Conditions:** DTC P1518 not set, key in the crank or run mode, system voltage over 5.23v, and the ETC or TAC module detected that an internal data test failed (did not pass) for a time period of less than 1 second. **Possible Causes** • Test the charging system output (low voltage can set this code) • Inspect the TAC module connectors for signs of water intrusion. If water intrusion occurs, multiple codes may set without any circuit or component conditions found during diagnostic testing. • When the TAC module detects a fault condition, several TAC related codes set because there are redundant tests running. • TAC module has failed

OBD II Trouble Code List (P1xxx Codes)

DTC	Trouble Code Title, Conditions & Possible Causes
DTC P1517 **1T CCM, MIL: Yes** 1999, 2000, 2001, 2002, 2003, 2004, 2005 Corvette 5.7L VIN G, 5.7L VIN S Transmissions: All	**Throttle Actuator Control Module Performance Conditions:** DTC P1518 not set, key in crank or run mode, system voltage over 5.23v, and the ETC or TAC module detected that an internal data test failed (did not pass) for less than 1 second. **Possible Causes** ● Test the charging system output (low voltage can set this code) ● Inspect the TAC module connectors for signs of water intrusion. If water intrusion occurs, multiple codes may set without any circuit or component conditions present. ● When the TAC module detects a fault condition, several TAC related codes set because there are redundant tests running. ● TAC module has failed
DTC P1518 **1T CCM, MIL: Yes** 1999, 2000, 2001, 2002 Camaro, Corvette, Envoy, Firebird, C/K Truck, G Van, Escalade & TrailBlazer 3.8L VIN K engine, 4.2L VIN S, 4.8L VIN V, 5.3L VIN P, 5.3L VIN T, 5.3L VIN Z, 5.7L VIN G, 6.0L VIN N, 6.0L VIN U, 7.4L VIN J, 8.1L VIN G Transmissions: All	**Throttle Actuator Control Module Serial Data Malfunction Conditions:** Key in the crank or run mode, system voltage over 5.23v, and the ETC or TAC module detected invalid or missing serial data present for a specified amount of time, condition met for less than 1 second. **Possible Causes** ● DTC P1518 sets if the battery voltage is low. If the customer's concern is slow cranking or no crank due to low battery voltage, ignore the DTC P1518. Clear codes and retest. ● DTC P1518 also sets when there is a short to B+ on the TAC module ground circuit. Inspect the fuses for the circuits that are in the TAC module harness (e.g., Brake & Cruise) for the fault. ● DTC P1518 sets if the TAC module power circuit is shorted to a B+ supply circuit. The TAC module stays powered-up when the ignition switch is turned off. When the key on, the TAC module is powered-up before the PCM gets power. DTC 1518 is set because the TAC module did not detect signals from the PCM. ● TAC module has failed
DTC P1518 **1T CCM, MIL: Yes** 2003, 2004, 2005 Corvette, Envoy, C/K Series Truck, G Van, Escalade & TrailBlazer 3.8L VIN K, 4.2L VIN S, 4.8L VIN V, 5.3L VIN P, 5.3L VIN T, 5.3L VIN Z, 5.7L VIN G, 6.0L VIN N, 6.0L VIN U, 8.1L VIN G Transmissions: All	**Throttle Actuator Control Module Serial Data Malfunction Conditions:** Key in the crank or run mode, system voltage over 5.23v, and the ETC or TAC module detected invalid or missing serial data present for a specified amount of time, condition met for less than 1 second. **Possible Causes** ● DTC P1518 sets if the battery voltage is low. If the customer's concern is slow cranking or no crank due to low battery voltage, ignore the DTC P1518. Clear codes and retest. ● DTC P1518 also sets when there is a short to B+ on the TAC module ground circuit. Inspect the Brake & Cruise fuses first. ● TSB 03-04-06-032 contains a repair procedure for this code
DTC P1519 **1T CCM, MIL: Yes** 1999, 2000, 2001, 2002 Camaro & Firebird 3.8L VIN K engine Transmissions: All	**Throttle Actuator Control Module Range/Performance Conditions:** DTC P0606, P1517 and P1518 not set, key in crank or run position, and the PCM received a message that the AC module had detected an invalid signal from the high-resolution TP sensor. The TP Sensor 1 signal connects to a 5X multiplier to produce a high-resolution calculated signal that is used in actuator control situations that occur at low throttle angle. **Possible Causes** ● APP sensor assembly is damaged or has failed. It must be replaced to properly repair this trouble code.
DTC P1519 **1T CCM, MIL: Yes** 1999, 2000, 2001 Catera Engines: 3.0L VIN R Transmissions: A/T	**Throttle Actuator Control Module Internal Circuit Conditions:** Engine started; Idle Learn procedure inactive, TAC duty cycle range switch indicating over 80%, and the PCM detected an unexpected voltage condition on the Throttle Valve Motor circuit, or it detected a problem in one of the TAC module internal circuits. **Possible Causes** ● Throttle valve motor circuit is open, shorted to ground or shorted to system power ● Throttle valve motor is damaged or failed ● TAC module has failed (an internal circuit may have failed)
DTC P1520 **1T CCM, MIL: No** 1995 Century, Grand Prix, 88', Park Avenue, 98' Regal, Riviera, Lumina APV, Trans Sport, Silhouette 3.8L VIN 1, 3.8L VIN K, 3.8L VIN L engines Transmissions: A/T	**Park Neutral Position Switch Circuit Malfunction Conditions:** DTC P0703 not set, Engine started; and the PCM detected an unexpected high voltage condition on the PNP Switch circuit for three consecutive startups, or with Transaxle in 3rd or 4th gear, TCC in lockup, it detected a low voltage signal for 1 second. **Possible Causes** ● PNP switch signal circuit is open or shorted to ground ● PNP switch ground circuit is open ● PNP switch is out-of-adjustment, loose or damaged ● PCM has failed

OBD II Trouble Code List (P1xxx Codes)

DTC	Trouble Code Title, Conditions & Possible Causes
DTC P1520 **1T CCM, MIL: No** 1996, 1997, 1998 Body Codes: A, L, S, T 2.2L VIN 4 engine Achieva, Beretta, Cavalier, Ciera, Corsica, Grand Am, Sunfire, Skylark & S/T Pickup Transmissions: A/T	**Park Neutral Position Switch Circuit Malfunction Conditions:** DTC P0121, P0122, P0123, P0335, P0502 and P0503 not set, engine started, vehicle driven to a speed over 40 mph at an engine speed from 1700-3000 rpm, TP angle from 7-20%, and the PCM detected the PNP switch signal indicated Park/Neutral position. **Possible Causes** ● PNP switch signal circuit is open or shorted to ground ● PNP switch ground circuit is open ● PNP switch is out-of-adjustment, loose or damaged ● PCM has failed
DTC P1520 **1T CCM, MIL: No** 1996, 1997, 1998, 1999 DeVille, Eldorado, Seville 4.6L VIN 9, 4.6L VIN Y Transmissions: A/T	**Transmission Range Switch Circuit Malfunction Conditions:** DTC P0502 or P0503 not set, Engine started; and the PCM detected the Transaxle Range (TR) switch indicated an illegal combination, or with DTP P1810 diagnostic passed and the engine running, that the TR switch signal did not match the correct signal combination for 3.3 seconds during the CCM test. **Possible Causes** ● One or more TR switch circuits is open or shorted to ground ● One or more TR switch circuits is shorted to VREF or B+ ● One or more TR switch circuits is shorted together ● TR switch is out of adjustment, damaged or it has failed ● PCM has failed
DTC P1520 **1T CCM, MIL: No** 1996, 1997, 1998, 1999 Aurora Engines: 4.0L VIN C Transmissions: A/T	**Closed Throttle Position Performance Conditions:** DTC P0121, P0122, P0123, P0502, P0503 and P0730 not set, engine started, vehicle driven to a speed over 40 mph, TP angle at 10 degrees or more, Delivered Torque at 84% or more, Gear Box ratio at 3.11 or less, and the PCM detected the switch indicated P/N. **Possible Causes** ● PNP switch signal circuit is open or shorted to ground ● PNP switch ground circuit is open ● PNP switch is out-of-adjustment, loose or damaged ● PCM has failed
DTC P1523 **1T CCM, MIL: No** 1999, 2000, 2001, 2002 Camaro & Firebird 3.8L VIN K Transmissions: All	**TAC Module Return Range/Performance Conditions:** DTC P1518 not set, then with the key on and engine off, system voltage over 8.5v, Battery Saver Mode is active, the TAC module detected the Actual and Commanded throttle positions were not within a calibrated range of each other during the CCM test. During Battery Saver Mode, the TAC module determines if the throttle plate is returning to the correct de-energized position. The TAC module sends a message to the PCM across serial data when this fault is detected, and the PCM sets DTC P1523. **Possible Causes** ● Check for mechanical problems or binding that are temperature related. Components do not move freely in extreme heat or cold due to the presence of contaminants or ice. ● Throttle body assembly is damaged or has failed
DTC P1523 **1T CCM, MIL: No** 2002, 2003, 2004, 2005 Envoy & TrailBlazer Engines: 4.2L VIN S Transmissions: All	**TAC Module Return Range/Performance Conditions:** Key on and engine off, system voltage over 8.5v, Battery Saver Mode active, and the TAC module detected the Commanded and Actual throttle positions were not within a calibrated range of each other. During Battery Saver Mode, the TAC module determines if the throttle plate is returning to the correct de-energized position. The TAC module sends a message to the PCM across serial data when this fault is detected, and the PCM sets DTC P1523. **Possible Causes** ● Check for mechanical (binding) conditions in the throttle body ● If no problems are found, inspect for mechanical conditions or binding that may be temperature related. ● Components may not move freely in extreme heat or cold due to the presence of contaminants or ice formation. ● Throttle body assembly is damaged or has failed
DTC P1523 **1T CCM, MIL: Yes** 1999, 2000, 2001 Catera 3.0L VIN R engine Transmissions: All	**Throttle Closed Position Range/Performance Conditions:** Engine speed under 250 rpm, vehicle not moving, ECT sensor from 41ºF-212ºF, IAT sensor over 41ºF, and the PCM detected the throttle valves were not in their closed position. The ECM controls the throttle valves by applying a varying voltage to the throttle valve motor. The ECM monitors the actual throttle blade position using TP1 Sensor and TP2 Sensor signals. **Possible Causes** ● This code is set when the throttle is unable to return to a calibrated "stop" position. Inspect for a condition where the throttle valves have been held open (binding, sticking). ● Check for signs that ice may have formed in the throttle bore. ● Throttle motor circuit(s) is open, shorted to ground or to B+ ● Throttle body assembly is damaged ● PCM has failed

OBD II Trouble Code List (P1xxx Codes)

DTC	Trouble Code Title, Conditions & Possible Causes
DTC P1524 **1T CCM, MIL: No** 1996, 1997, 1998 DeVille, Eldorado, Seville 4.6L VIN 9, 4.6L VIN Y Transmissions: A/T	**Throttle Closed Position Range/Performance Conditions:** Engine started; TP Sensor already "learned" this key cycle, and the PCM detected the closed TP angle correction was 6.8 degrees. The PCM "learns" the closed throttle position when the ignition is turned to Lock position. This position is always corrected to 0 degrees. The correction is performed to allow for deposit buildup on the throttle valve. When the ignition is turned to Lock, the PCM enables the TP Sensor "leaning" routine. If the same correction factor occurs on two consecutive key Lock cycles, the TP sensor is corrected to 0 degrees using the correction factor just learned. If the value needs correction by more than 6.8 degrees, DTC P1524 is set on the very next Key "on" cycle. **Possible Causes** ● TP sensor signal is out-of-range "low", or out-of-range "high" ● Throttle linkage is damaged, binding or out of adjustment ● Throttle body bore contains excessive debris or deposits ● Throttle body is damaged
DTC P1526 **1T CCM, MIL: Yes** 1999, 2000, 2001 Catera 3.0L VIN R engine Transmissions: A/T	**Minimum Throttle Position Not Learned Conditions:** Engine started, engine speed under 250 rpm, vehicle not moving, system voltage over 10.0v, ECT sensor input 41°F-212°F, IAT sensor input over 41°F, APP sensor less than 15%, Minimum Closed Throttle position "not learned", and the PCM detected the TP Sensor 1 signal was not from 0.24v-0.82v, and the TP Sensor 2 signal was not from 4.2v-4.8v after the throttle valves were commanded to closed position. When the key is turned to the "on" position, the PCM operates the throttle valves to determine the integrity of the Throttle Control system prior to engine startup. The PCM commands the throttle valves open momentarily and then commands the throttle valves to the closed position. Using data from the TP sensors, the PCM detects if the throttle valves respond correctly (as commanded). The PCM controls the throttle valves by applying a varying voltage to the throttle valve motor. **Possible Causes** ● Inspect for throttle valves that are "not" in the rest position ● Verify the throttle valves are free to move open or closed ● Throttle motor circuit(s) is open, shorted to ground or to B+ ● Throttle body assembly is damaged or has failed ● PCM is damaged or has failed
DTC P1526 **1T CCM, MIL: No** 1996 DeVille, Eldorado, Seville 4.6L VIN 9, 4.6L VIN Y Transmissions: All	**TP Sensor Learn Not Completed Conditions:** Engine running and the PCM detected the TP Sensor Learn Value was not stored. *Note: Do not move the throttle during the TP Sensor Learn procedure or this code will reset.* **Possible Causes** ● Throttle body valves or linkage is binding or sticking ● TP sensor is damaged or has failed (erratic TP sensor signals)
DTC P1527 **1T CCM, MIL: No** 1996, 1997, 1998, 1999 DeVille, Eldorado, Seville 4.6L VIN 9, 4.6L VIN Y Transmissions: A/T	**Transmission Range To Pressure Switch Correlation Conditions:** Engine started, engine runtime over 5 seconds, DTC P1810 test has run and passed, system voltage over 10.0v, Transmission Range (TR) switch does not indicate an illegal position, TFP Manual Valve Position switch does not indicate an illegal position, and the PCM detected the TR and TFP switches indicated different gear positions for 7.8 seconds. This vehicle uses an electronic gear indicator display. Thee four inputs are switches to ground (each has a unique switch pattern). All four switches are housed in the PNP switch and operate together as the Transaxle Range (TR) Switch. The PCM uses the combination of the switch patterns to determine the gear selected by the driver. This information is sent by serial data to the IPC for gear indicator display. Once these conditions are met the PCM compares the gear indicated by the TR switch to the gear selected from the TFP Manual Valve Position Switch (TFP switch). If the gear indicated is not the same for both, the PCM sets this code. **Possible Causes** ● TR switch signal circuit is open or shorted to ground ● TR switch ground circuit is open ● One or more transmission pressure switches have failed ● PCM has failed
DTC P1530 **1T CCM, MIL: Yes** 1995 Century, Cutlass Ciera, Cutlass Supreme, Grand Prix, Lumina, Monte Carlo Park Avenue, Regal, Riviera, 88' & 98' Engines: 3.1L VIN M, 3.4L VIN X, 3.8L VIN 1, 3.8L VIN K, 3.8L VIN L Transmissions: All	**Air Conditioning Head Pressure Switch Circuit Malfunction Conditions:** No ECT, IAT or MAF codes set; engine speed under 1800 rpm, ECT sensor less than 88°F, IAT sensor less than 82°F, and the PCM detected the A/C Pressure switch was grounded for 5 seconds after engine startup. **Possible Causes** ● A/C pressure sensor circuit is open or shorted to ground ● A/C pressure sensor circuit shorted to VREF or system power ● A/C pressure sensor ground circuit is open ● A/C pressure sensor VREF circuit is open or shorted to ground ● A/C pressure sensor is damaged or has failed ● PCM has failed

OBD II Trouble Code List (P1xxx Codes)

DTC	Trouble Code Title, Conditions & Possible Causes
DTC P1530 **1T CCM, MIL: Yes** 1999, 2000, 2001 Catera 3.0L VIN R engine Transmissions: A/T	**Throttle Actuator Control Module Internal Circuit Conditions:** Engine started, engine speed under 250 rpm, vehicle not moving, ECT sensor from 41ºF-212ºF, IAT sensor more than 41ºF, Pedal Position angle less than 15%, and the PCM determined he TAC system was outside of its calibrated operating range. When the key is first turned "on", the PCM operates the throttle valves to determine the integrity of the Throttle Control system prior to engine startup. The PCM commands the throttle valves open momentarily and then commands the throttle valves to closed position. Using data from the TP sensors, the PCM determines if the throttle valves responded as directed. The PCM controls the throttle valves by applying a varying voltage to the throttle valve motor. The throttle control system AMP Adjustment diagnostic trouble code (DTC) is internal to the PCM **Possible Causes** ● Inspect for throttle valves that are "not" in the rest position ● Verify the throttle valves are free to move open or closed ● Throttle motor circuit(s) is open, shorted to ground or to B+ ● Throttle body assembly is damaged or has failed ● PCM is damaged or has failed
DTC P1530 **1T CCM, MIL: Yes** 1995 Regal, Lumina APV, Silhouette, Trans Sport 3.8L VIN L engine Transmissions: All	**Low Air Conditioning Refrigerant Charge Low Conditions:** Engine started; and the PCM detected the A/C compressor cycled "on" 10 or more times within 15 minutes, and that the amount of time the compressor was "on" for each cycle was less than 1.5 seconds (i.e., the A/C clutch turned "off" each time). **Possible Causes** ● A/C high pressure cutout switch is damaged or has failed ● A/C pressure cycling switch is damaged or has failed ● A/C orifice tube is clogged or plugged ● Check for loose connections at A/C cutout and pressure switch ● Service the A/C system (may be low on refrigerant)
DTC P1532 **1T CCM, MIL: No** 1996, 1997 Camaro, Caprice, Corvette, Firebird, Fleetwood & Roadmaster 4.3L VIN W, 5.7L VIN 5, 5.7L VIN P engines Transmissions: All	**A/C Evaporative Temp Sensor Circuit Low Input Conditions:** Engine started; A/C switch and A/C Clutch "on", and the PCM detected the A/C Evaporator Temperature sensor indicated less than 2ºF for 5 seconds during the test. **Possible Causes** ● A/C pressure switch circuit is shorted to ground ● A/C pressure switch VREF circuit is open or shorted to ground ● A/C pressure switch is damaged or has failed ● PCM has failed
DTC P1533 **1T CCM, MIL: No** 1996, 1997 Camaro, Caprice, Corvette, Firebird, Fleetwood & Roadmaster 4.3L VIN W, 5.7L VIN 5, 5.7L VIN P engines Transmissions: All	**A/C Evaporative Temp Sensor Circuit High Input Conditions:** Engine started; A/C switch and A/C Clutch "on", and the PCM detected the A/C Evaporator Temperature sensor was more than 126ºF for 5 seconds during the test. **Possible Causes** ● A/C pressure switch circuit is open or shorted to VREF ● A/C pressure switch ground circuit is open ● A/C pressure switch is damaged or has failed ● PCM has failed
DTC P1533 **1T CCM, MIL: No** 1996, 1997, 1998, 1999 Aurora Engines: 4.0L VIN C Transmissions: A/T	**A/C Low Side Temperature Sensor Circuit Malfunction Conditions:** No Ambient Air Temperature sensor codes set in the HVAC module, engine running, ambient air temperature from 39ºF-210ºF, and the PCM detected the A/C Low Side signal was less than -35ºF, or it was more than 210ºF. The A/C low side temperature sensor uses a thermistor to control the signal to the PCM. The PCM applies 5v on circuit 731 to the sensor. **Possible Causes** ● A/C low side sensor circuit is open or shorted to ground ● A/C low side sensor ground circuit is open ● A/C low side temperature sensor is damaged or has failed ● PCM has failed
DTC P1535 **1T CCM, MIL: No** 1996, 1997, 1998, 1999 Aurora Engines: 4.0L VIN C Transmissions: A/T	**A/C High Side Temperature Sensor Circuit Malfunction Conditions:** No Ambient Air Temperature Sensor codes set in HVAC module, Engine started; ambient air temperature from 40ºF-210ºF, and the PCM detected the A/C High Side Temperature sensor indicated less than 10ºF, or that it indicated more than 410ºF. The A/C high side temperature sensor uses a thermistor to control the signal voltage to the PCM. The PCM applies 5 volts on circuit 732 to the sensor. When the sensor is cold, its resistance is high; and the PCM will read a high signal voltage. As the sensor warms, its resistance becomes less and the signal voltage decreases. **Possible Causes** ● A/C high side sensor circuit is open or shorted to ground ● A/C high side temperature sensor ground circuit is open ● A/C high side temperature sensor is damaged or has failed ● PCM has failed

OBD II Trouble Code List (P1xxx Codes)

DTC	Trouble Code Title, Conditions & Possible Causes
DTC P1536 **1T CCM, MIL: No** 1996, 1997, 1998, 1999 Aurora Engines: 4.0L VIN C Transmissions: All	**Engine Over-Temperature - A/C System Disabled Conditions:** DTC P0117 and P0118 not set, Engine started; A/C compressor requested by HVAC module, and the PCM detected the ECT sensor indicated more than 257°F; or with A/C compressor operation requested or not requested by the HVAC controller, that the ECT signal indicated more than 248°F during the CCM test. **Possible Causes** ● Check the coolant level and the coolant mixture ● Check the operation of the cooling system (i.e., the cooling system fan controls and the engine thermostat operation)
DTC P1537 **1T CCM, MIL: No** 1996, 1997, 1998, 1999 Aurora Engines: 4.0L VIN C Transmissions: All	**Air Conditioning Request Circuit Low Input Conditions:** DTC P1605 not set, Engine started; A/C compressor requested "on" by the HVAC module, and the PCM detected the A/C request line was in a low state for 20 continuous seconds. The PCM tests for very low refrigerant pressure or an open low-pressure switch/circuit by comparing the A/C request discrete line, which contains the low pressure switch, to the HVAC serial data A/C request status. If the A/C request line is low, with no voltage present, while the HVAC module requests A/C operation, this code is set. **Possible Causes** ● A/C low pressure switch (A/C request circuit) is open or shorted to ground, or the A/C low pressure switch ground circuit is open ● A/C low pressure switch is damaged or has failed ● A/C refrigerant pressure is too low (service the A/C system) ● PCM has failed
DTC P1538 **1T CCM, MIL: No** 1996, 1997, 1998, 1999 Aurora Engines: 4.0L VIN C Transmissions: All	**Air Conditioning Request Circuit Voltage High Conditions:** DTC P1605 not set, Engine started; A/C compressor requested "on" by the HVAC module, and the PCM detected the A/C request line was in a high state for 20 continuous seconds. The PCM tests the A/C request circuit for a short to voltage. This test is accomplished by comparing the A/C Request line (circuit 603) to the serial data received from the HVAC module. If the module has not requested the A/C "on", and the A/C Request line is high, the PCM will set this code during the CCM test. **Possible Causes** ● A/C low pressure switch (A/C request circuit) shorted to VREF ● PCM has failed
DTC P1539 **1T CCM, MIL: No** 1996, 1997 Camaro, Caprice, Corvette, Firebird, Fleetwood & Roadmaster 4.3L VIN W, 5.7L VIN 5, 5.7L VIN P engines Transmissions: All	**Air Conditioning Clutch Status High Input Conditions:** Engine started; A/C Clutch relay disengaged, and the PCM detected the A/C status (voltage) was high for 5 seconds. **Possible Causes** ● A/C clutch status circuit is shorted to system power (B+) ● A/C clutch relay is damaged or has failed (contacts are stuck) ● PCM has failed
DTC P1539 **1T CCM, MIL: No** 1998, 1999, 2000 Camaro & Firebird Engines: 5.7L VIN G Transmissions: All	**Air Conditioning Clutch Status High Input Conditions:** Engine started; A/C operation "not" requested, and the PCM detected a voltage signal on the A/C Status circuit for more than 15 seconds (i.e., with the A/C Clutch relay turned off). **Possible Causes** ● A/C clutch status circuit is shorted to system power (B+) ● A/C clutch relay is damaged or has failed (contacts are stuck) ● PCM has failed
DTC P1539 **1T CCM, MIL: No** 1997, 1998, 1999, 2000 Corvette Engines: 5.7L VIN G Transmissions: All	**A/C Clutch Feedback Circuit High Input Conditions:** Key on or engine running; A/C Clutch relay disabled, and the PCM detected a voltage on the A/C status circuit for more than 15 seconds after it had disengaged the A/C Clutch relay. **Possible Causes** ● A/C clutch status circuit is shorted to system power (B+) ● A/C clutch relay is damaged or has failed (contacts are stuck)
DTC P1540 **1T CCM, MIL: No** 1996, 1997, 1998, 1999 Aurora Engines: 4.0L VIN C Transmissions: A/T	**A/C Refrigerant Over-Pressure, A/C Disabled Conditions:** Engine started, engine running with A/C requested "on", and the PCM detected the A/C Clutch was disabled 10 times by the A/C High Pressure switch with an A/C High Side Temperature sensor indicating less than 109°F. This diagnostic detects an over pressure condition in the A/C system. The PCM sets this code if the A/C high-pressure switch opens at 430 psi, and the A/C clutch is disabled at least 10 times with the high side temperature below its threshold. This test is designed to detect blockages in the A/C system. The A/C clutch is disabled during the key cycle in which the code sets. **Possible Causes** ● A/C high pressure switch is damaged or has failed (failed open) ● A/C high pressure switch circuit is open ● Locate and repair the cause of high pressure in the A/C system

OBD II Trouble Code List (P1xxx Codes)

DTC	Trouble Code Title, Conditions & Possible Causes
DTC P1542 **1T CCM, MIL: No** 1996, 1997, 1998, 1999 Aurora Engines: 4.0L VIN C Transmissions: A/T	**A/C System Over-Pressure/Over-Temperature, A/C Disabled Conditions:** DTC P1535 not set and P1542 not testing, Engine started; and the PCM detected the A/C system was disabled due to a High Pressure switch signal. The actual fault depends on one or more of the System States (i.e., conditions 0, 1 or 2). The PCM tests for an over temperature/over pressure condition in the A/C system. This code is set if the high-pressure switch disables the A/C clutch and the high side temperature is above a threshold. Under extremely hot ambient conditions, a rapid cycling situation could occur as this code is set and cleared. If this code continues to set and then clear, a lower high side temperature is used to clear the code. If after this occurs and rapid cycling is still detected, a third, lower temperature is used to clear the problem and code, and to enable the A/C clutch operation. **Possible Causes** ● It is normal for this condition to occur if sufficient cooling of the A/C system does not happen due to high ambient temperature. ● Determine if the code set under high outside temperature s
DTC P1543 **1T CCM, MIL: No** 1996, 1997 Camaro, Caprice, Corvette, Firebird, Fleetwood & Roadmaster 4.3L VIN W, 5.7L VIN 5, 5.7L VIN P engines Transmissions: All	**Air Conditioning System Range/Performance Conditions:** Engine started; and the PCM detected the A/C refrigerant charge was less than a specified value (due to Evaporator Temperature and Refrigerant Pressure signals). **Possible Causes** ● A/C refrigerant pressure sensor is out-of-calibration or skewed ● A/C evaporator temperature sensor out-of-calibration or skewed ● A problem in the A/C compressor or A/C thermal expansion valve can also cause this code to set - check the A/C system.
DTC P1545 **1T CCM, MIL: No** 1996, 1997 Camaro, Caprice, Corvette, Firebird, Fleetwood & Roadmaster 4.3L VIN W, 5.7L VIN 5, 5.7L VIN P engines Transmissions: All	**Air Conditioning Clutch Relay Control Circuit Malfunction Conditions:** DTC P0106, P0113, P0500, P0502, P0560, P0562, P1133, P1135, P1136, P1153, P1154, P1532, P1539 and P1543 not set, engine started, engine speed over 600 rpm, system voltage over 10.0v, and the PCM detected the Actual and Commanded state of the A/C Clutch control circuit did not match for 10 seconds. This relay is used to control the high current flow to the A/C clutch circuit. **Possible Causes** ● A/C compressor clutch coil is damaged (it may be shorted) ● A/C clutch relay is damaged or has failed
DTC P1545 **1T CCM, MIL: No** 2000, 2001, 2002, 2003, 2004, 2005 DeVille, Eldorado, Seville 4.6L VIN 9, 4.6L VIN Y Transmissions: A/T	**Air Conditioning Clutch Relay Control Circuit Malfunction Conditions:** Engine started; system voltage over 10.0v, A/C Clutch requested "on", and the PCM detected an unexpected voltage on the A/C Clutch control circuit for 10 seconds. **Possible Causes** ● A/C clutch relay control circuit is open, shorted to ground or shorted to system power ● A/C clutch relay power circuit is open (test the AC CLU fuse) ● A/C clutch relay is damaged or has failed ● PCM has failed
DTC P1545 **1T CCM, MIL: No** 1999, 2000, 2001 Catera 3.0L VIN R engine Transmissions: A/T	**Air Conditioning Clutch Relay Control Circuit Malfunction Conditions:** Engine started, vehicle driven to a speed of over 40 mph, system voltage over 10.0v, A/C clutch requested "on", and the PCM detected an unexpected voltage condition on the A/C clutch relay control circuit during the CCM continuous test. **Possible Causes** ● A/C clutch relay control circuit is open, shorted to ground or shorted to system power ● A/C clutch relay power circuit is open (test HTR BLOWER fuse) ● A/C clutch relay is damaged or has failed ● A/C compressor clutch coil is damaged (it may be shorted) ● PCM has failed
DTC P1546 **1T CCM, MIL: No** 1996, 1997 Camaro, Caprice, Corvette, Firebird, Fleetwood & Roadmaster 4.3L VIN W, 5.7L VIN 5, 5.7L VIN P engines Transmissions: All	**A/C Clutch Status Circuit Low Input Conditions:** Engine started, A/C Clutch requested "on", and the PCM detected an unexpected low voltage condition on the A/C Clutch status circuit for over 5 seconds during the CCM test. **Possible Causes** ● A/C clutch status line is open or shorted to ground ● A/C clutch relay is damaged or has failed (contains not closing) ● A/C clutch relay power circuit is open (test AC CRUISE fuse) ● PCM has failed
DTC P1546 **1T CCM, MIL: No** 1999, 2000 Camaro & Firebird 3.8L VIN K engine Transmissions: All	**A/C Clutch Relay Control Circuit Malfunction Conditions:** Engine running; A/C requested "on", and the PCM detected an unexpected voltage on the output circuit that controls the A/C compressor clutch relay for 30 seconds during the test. **Possible Causes** ● A/C clutch status line is open or shorted to ground ● A/C clutch relay is damaged or has failed (contains not closing) ● A/C clutch relay power circuit is open (test AC CRUISE fuse) ● PCM has failed

OBD II Trouble Code List (P1xxx Codes)

DTC	Trouble Code Title, Conditions & Possible Causes
DTC P1546 **1T CCM, MIL: No** 1999, 2000 Bonneville, Century, Grand Prix, Intrigue, LeSabre, LSS, Lumina, Monte Carlo, Regal, Park Avenue & Eighty Eight 3.8L VIN 1, 3.8L VIN K Transmissions: All	**A/C Clutch Feedback Circuit Low Input Conditions:** Engine started, system voltage over 10.0v, A/C requested "on", and the PCM detected an unexpected "low" voltage condition on the ODM 'A' circuit (Output 1) used to control the A/C Compressor Clutch Relay circuit for 30 seconds in the CCM test. The PCM uses an output driver module (ODM) to control the operation of the A/C clutch relay coil by applying ground to this device. The PCM will set this cod if it detects an improper voltage level on the circuit used to control the A/C compressor control relay **Possible Causes** ● A/C clutch status line is open or shorted to ground ● A/C clutch relay is damaged or has failed (contains not closing) ● A/C clutch relay power circuit is open (test the HVAC fuse) ● PCM has failed
DTC P1546 **1T CCM, MIL: No** 1998, 1999, 2000 Camaro & Firebird Engines: 5.7L VIN G Transmissions: All	**A/C Clutch Feedback Circuit Low Input Conditions:** Engine started; A/C requested "on", and the PCM detected an unexpected low voltage condition on the A/C Clutch Status circuit for 5 seconds with the A/C clutch enabled. **Possible Causes** ● A/C clutch status line is open or shorted to ground ● A/C clutch relay is damaged or has failed (contains not closing) ● A/C clutch relay power circuit is open (test the HVAC fuse) ● PCM has failed
DTC P1546 **1T CCM, MIL: No** 1997, 1998, 1999, 2000 Corvette Engines: 5.7L VIN G Transmissions: All	**A/C Clutch Feedback Circuit Low Input Conditions:** Engine started; A/C requested "on", and the PCM detected an unexpected low voltage condition on the A/C Clutch status circuit for over 5 seconds during the CCM test period. **Possible Causes** ● A/C clutch status line is open or shorted to ground ● A/C clutch relay is damaged or has failed (contains not closing) ● A/C clutch relay power circuit is open (test AC CRUISE fuse) ● PCM has failed
DTC P1546 **1T CCM, MIL: Yes** 1999, 2000, 2001, 2002, 2003, 2004, 2005 Body Codes: B, N, U, W, X 3.1L VIN J, 3.1L VIN M, 3.4L VIN E engine Transmissions: All N & W Cars, U & X Minivans (Montana, etc)	**A/C Clutch Relay Control Circuit Conditions:** Key on or engine running; and the PCM detected an improper voltage level on the output circuit that controls the A/C compressor clutch relay, condition met for at least 30 seconds. **Possible Causes** ● A/C clutch status line is open or shorted to ground ● A/C clutch relay is damaged or has failed (contains not closing) ● A/C clutch relay power circuit is open (test the AC CLU fuse) ● PCM has failed
DTC P1550 **1T CCM, MIL: No** 1995 Regal, Lumina APV, Silhouette, Trans Sport 3.8L VIN L engine Transmissions: All	**Stepper Motor Cruise Control Circuit Malfunction Conditions:** Engine started; SMCC Cruise commanded off, and the PCM detected the Cruise Status signal indicated that the Cruise Control was engaged under these conditions. **Possible Causes** ● Cruise control inhibit circuit is open or shorted to ground ● SMCC II cruise control module is damaged or has failed ● PCM has failed
DTC P1550 **1T CCM, MIL: No** 2001, 2002, 2003, 2004, 2005 Body Codes: C, K 6.6L VIN 1 Diesel engine Transmissions: All	**Fuel Injector Control Module Power Circuit Malfunction Conditions:** U1800 and U2104 not set, key on, and the PCM detected a low supply voltage present at the Fuel Injector Control Module (FICM). The FICM activates the fuel injector. The ECM commands the FICM to turn "on" the injectors through each fuel injector control circuit. **Possible Causes** ● FICM ground circuit has a high resistance condition ● FICM power circuit has a high resistance condition ● FICM (module) is damaged or it has failed
DTC P1554 **1T CCM, MIL: No** 1999, 2000 Bonneville, Century, Grand Prix, Intrigue, LeSabre, LSS, Lumina, Monte Carlo, Regal, Park Avenue & Eighty Eight 3.8L VIN 1, 3.8L VIN K Transmissions: All	**Cruise Control Status Circuit Malfunction Conditions:** Engine started; then after a SMCC command was sent to inhibit Cruise control, the PCM detected the Cruise Status signal indicated the Cruise Control remained engaged for more than 1 second. **Possible Causes** ● Cruise control inhibit circuit is open or shorted to ground ● SMCC II cruise control module is damaged or has failed ● PCM has failed

OBD II Trouble Code List (P1xxx Codes)

DTC	Trouble Code Title, Conditions & Possible Causes
DTC P1554 **1T CCM, MIL: No** 1996, 1997, 1998, 1999, 2000, 2001, 2002, 2003, 2004, 2005 Aurora, DeVille, Eldorado & Seville 4.0L VIN C, 4.6L VIN 9, 4.6L VIN Y engines Transmissions: All	**Cruise Control Feedback Circuit Malfunction Conditions:** Engine started; Cruise Control module signal sent to inhibit cruise control operation, and the PCM detected a cruise "engage" signal that indicated that cruise control was still engaged. The PCM will not allow the cruise control to operate under certain conditions (e.g., if the vehicle speed is too low). The PCM controls this function by grounding the 12v cruise inhibit signal (CKT 83) sent by the cruise control module. When the customer requests cruise control, the cruise control module grounds the cruise engage signal. **Possible Causes** ● Cruise control inhibit circuit is open or shorted to ground ● SMCC II cruise control module is damaged or has failed ● PCM has failed
DTC P1554 **1T CCM, MIL: No** 1996, 1997, 1998 Camaro & Firebird 3.8L VIN K engine Transmissions: All	**Cruise Control Status Circuit Malfunction Conditions:** Engine started; SMCC command sent to inhibit Cruise control, and the PCM detected the Cruise Status signal indicated that Cruise Control was still engaged over 1 second. **Possible Causes** ● Cruise control inhibit circuit is open or shorted to ground ● SMCC II cruise control module is damaged or has failed ● PCM has failed
DTC P1554 **1T CCM, MIL: Yes** 1999, 2000, 2001, 2002, 2003, 2004, 2005 Alero, Aurora, Bonneville, Century, Grand Prix, LeSabre, Monte Carlo, Regal, Park Avenue, Montana, Silhouette 3.1L VIN J, 3.1L VIN M, 3.4L VIN E engines Transmissions: All	**Cruise Engaged Circuit High Input Conditions:** Engine started; Cruise Control commanded "off" (the Cruise Inhibit circuit is grounded), and the PCM detected the Cruise Status signal indicated that Cruise Control was enabled for one second under these operating conditions. **Possible Causes** ● Cruise control inhibit circuit is open or shorted to ground ● SMCC II cruise control module is damaged or has failed ● PCM has failed
DTC P1558 **1T CCM, MIL: Yes** 1995 Century, Grand Prix, 88', Park Avenue, 98' Regal, Riviera, Lumina APV, Trans Sport, Silhouette 3.8L VIN 1, 3.8L VIN K, 3.8L VIN L engines Transmissions: All	**Cruise Control System, SPS Circuit Indicated Low Conditions:** Engine started; TP angle less than 50%, and the PCM detected the Desired Servo Position Sensor position was 90%, and the Actual Servo Position was less than 2% for 3 seconds. **Possible Causes** ● Brake switch is damaged or failed (switch resistance too high) ● Cruise control cable binding or sticking (may be intermittent) ● EMI/RFI interference affecting Cruise Control module (antenna) ● Vacuum portion of the Cruise Control system is malfunctioning
DTC P1560 **1T CCM, MIL: No** 1996, 1997, 1998, 1999, 2000, 2001, 2002, 2003, 2004, 2005 Aurora, DeVille, Eldorado & Seville 4.0L VIN C, 4.6L VIN 9, 4.6L VIN Y engines Transmissions: All	**Transaxle Not In Drive - Cruise Control Disabled Conditions:** Engine started; Cruise Control requested and engaged, and the PCM detected the shift lever indicated Neutral, or that the Transaxle Range (TR) Switch indicated "Neutral" position. **Possible Causes** ● Review proper procedures to use to verify the shift lever is not placed in Neutral with Cruise Control "on" with the customer.
DTC P1560 **1T CCM, MIL: No** 1997, 1998 Catera 3.0L VIN R engine Transmissions: A/T	**TCM Supply Voltage Loss Conditions:** Key on or off, and the TCM detected an unexpected loss of direct battery power under these operating conditions. **Possible Causes** ● Check the battery positive (+) voltage circuit between the battery, starter relay and the TCM for high resistance or loose connections ● This code will set if the battery is dead, or if it is removed
DTC P1561 **1T CCM, MIL: No** 1995 Century, Grand Prix, 88', Park Avenue, 98' Regal, Riviera, Lumina APV, Trans Sport, Silhouette 3.8L VIN 1, 3.8L VIN K, 3.8L VIN L engines Transmissions: All	**Cruise Vent Solenoid Circuit Malfunction Conditions:** Engine started; and the PCM detected an unexpected voltage on the Vent Solenoid control circuit with the Cruise Control engaged, or with it disengaged for 50 ms. **Possible Causes** ● Vent solenoid circuit is open or shorted to ground (solenoid on) ● Vent solenoid circuit is shorted to system power (solenoid off) ● C/C vent solenoid is damaged or has failed ● PCM has failed

OBD II Trouble Code List (P1xxx Codes)

DTC	Trouble Code Title, Conditions & Possible Causes
DTC P1562 **1T CCM, MIL: No** 1995 Century, Grand Prix, 88', Park Avenue, 98' Regal, Riviera, Lumina APV, Trans Sport, Silhouette 3.8L VIN 1, 3.8L VIN K, 3.8L VIN L engines Transmissions: All	**Cruise Vacuum Solenoid Circuit Malfunction Conditions:** Engine started; and the PCM detected an unexpected voltage on the Vacuum Solenoid control circuit with the Cruise Control engaged, or with it disengaged for 50 ms. **Possible Causes** ● Vacuum solenoid circuit open, shorted to ground (solenoid on) ● Vacuum solenoid circuit shorted to system power (solenoid off) ● C/C vent solenoid is damaged or has failed ● PCM has failed
DTC P1564 **1T CCM, MIL: No** 1996, 1997, 1998, 1999, 2000, 2001, 2002, 2003, 2004, 2005 Aurora, DeVille, Eldorado & Seville 4.0L VIN C, 4.6L VIN 9, 4.6L VIN Y engines Transmissions: A/T	**Vehicle Acceleration Too High - Cruise Control Disabled Conditions:** Engine started, engine running with Cruise Control requested and engaged, and the PCM detected an extremely rapid rate of vehicle acceleration during certain driving conditions. **Possible Causes** ● This code sets after a sudden rate of increase in vehicle speed occurs (i.e., vehicle speed sensing problem or traveling on icy roads) while the cruise is engaged, so it is necessary to check with the customer to verify the conditions present when it set.
DTC P1564 **1T CCM, MIL: No** 1997, 1998 Catera 3.0L VIN R engine Transmissions: All	**ECM Battery Voltage Loss Conditions:** Engine started; and the PCM detected an unexpected loss of direct battery power. It should be noted that a No Start (with the MIL Off) condition exists under these conditions. **Possible Causes** ● Check the battery positive (+) voltage circuit between the Keep Alive Memory (KAM) for high resistance or loose connections ● Check the fuse (Fuse 50) in power distribution box near PCM
DTC P1565 **1T CCM, MIL: No** 1995 Century, Grand Prix, 88', Park Avenue, 98' Regal, Riviera, Lumina APV, Trans Sport, Silhouette 3.8L VIN 1, 3.8L VIN K, 3.8L VIN L engines Transmissions: All	**Cruise Servo Position Sensor Circuit Malfunction Conditions:** Engine started; and the PCM detected an unexpected voltage condition on the Servo Position Sensor High or Low circuit during the CCM test. **Possible Causes** ● C/C sensor "high" circuit is open, shorted to ground or to power ● C/C sensor "low" circuit is open ● C/C servo is damaged or has failed ● PCM has failed
DTC P1566 **1T CCM, MIL: No** 1996, 1997, 1998, 1999, 2000, 2001, 2002, 2003, 2004, 2005 Aurora, DeVille, Eldorado & Seville 4.0L VIN C, 4.6L VIN 9, 4.6L VIN Y engines Transmissions: A/T	**Engine Speed Too High - Cruise Control Disabled Conditions:** Engine started; Cruise Control system requested and engaged, and the PCM detected the engine speed was more than 6300 rpm under these operating conditions. **Possible Causes** ● Verify that there are not ABS/TCM trouble codes set ● Verify that there are no TCM related trouble codes set ● Discuss with the customer what conditions were present when this trouble code set so that it will not reoccur.
DTC P1567 **1T CCM, MIL: No** 1995 Century, Grand Prix, 88', Park Avenue, 98' Regal, Riviera, Lumina APV, Trans Sport, Silhouette 3.8L VIN 1, 3.8L VIN K, 3.8L VIN L engines Transmissions: All	**Cruise Switches Circuit Malfunction Conditions:** Engine running, and the PCM detected the Cruise On/Off Switch circuit was open between the switch and PCM, or the Set/Coast Switch or Resume/Accelerator Switch was closed for 7.5 minutes, or that no change occurred in the Cruise Brake switch status after vehicle made 5 stops from 35 to 0 mph, condition met for 1 second. **Possible Causes** ● C/C brake switch is damaged or has failed ● C/C set switch circuit is open or shorted to ground ● C/C resume switch circuit is open or shorted to ground ● C/C switch assembly is damaged or has failed ● PCM has failed
DTC P1567 **1T CCM, MIL: No** 1998, 1999, 2000, 2001, 2002, 2003, 2004, 2005 Aurora, DeVille, Eldorado & Seville 4.0L VIN C, 4.6L VIN 9, 4.6L VIN Y engines Transmissions: All	**Active Braking Control Active - Cruise Control Disabled Conditions:** Engine started; Cruise Control requested and engaged, and the PCM detected the Active Braking Control system was applying the brakes under these conditions. **Possible Causes** ● C/C brake switch is damaged or has failed ● C/C set switch circuit is open or shorted to ground ● C/C resume switch circuit is open or shorted to ground ● C/C switch assembly is damaged or has failed ● PCM has failed

OBD II Trouble Code List (P1xxx Codes)

DTC	Trouble Code Title, Conditions & Possible Causes
DTC P1568 **1T CCM, MIL: No** 1995 Century, Grand Prix, 88', Park Avenue, 98' Regal, Riviera, Lumina APV, Trans Sport, Silhouette 3.8L VIN 1, 3.8L VIN K, 3.8L VIN L engines Transmissions: All	**Cruise Control System - SPS Circuit Indicated High Conditions:** Engine started; and the PCM detected the Actual Servo Position Sensor position was more than 15% over the Desired Servo Position of 0% for 1 second under these conditions. **Possible Causes** ● Brake vacuum release valve may be out of adjustment ● C/C throttle cable binding or sticking (fault may be intermittent) ● C/C vacuum solenoid is damaged (it may be stuck closed) ● C/C servo is damaged or has failed ● EMI/RFI interference affecting Cruise Control module (antenna) ● PCM has failed
DTC P1570 **1T CCM, MIL: No** 1996, 1997, 1998, 1999, 2000, 2001, 2002, 2003, 2004, 2005 Aurora, DeVille, Eldorado & Seville 4.0L VIN C, 4.6L VIN 9, 4.6L VIN Y engines Transmissions: A/T	**Traction Control Active - Cruise Control Disabled Conditions:** Engine started; Cruise Control requested and engaged, and the PCM detected that the Traction Control System was enabled. The C/C module uses the Cruise Engage circuit to request Cruise Control. If improper conditions exist, the PCM will disable Cruise Control via the cruise control inhibit circuit. During cruise control, if the vehicle enters active Traction Control, the PCM will inhibit the cruise control system, and set DTC P1570. **Possible Causes** ● Verify that there are no ABS/TCS trouble codes in memory ● This trouble code is usually caused by conditions that cause traction control to activate during cruise control. Check with the customer to verify the conditions present when the code set.
DTC P1571 **1T CCM, MIL: No** 1995 Century, Park Avenue, Regal, Riviera, 98' 3.8L VIN 1, 3.8L VIN K, 3.8L VIN L engines Transmissions: All	**Traction Control Desired Torque Out-Of-Range Conditions:** Engine started; and the PCM detected an EBTCM Desired Torque PWM signal that was less than 5%, or a signal that was more than 95% under these operating conditions. It should be noted the TRACTION OFF Lamp is "on" when this code is set. **Possible Causes** ● T/C desired torque "out" circuit is open or shorted to ground ● T/C desired torque "in" circuit is open ● EBCM has failed ● PCM has failed
DTC P1571 **1T CCM, MIL: No** 1996, 1997, 1998, 1999, 2000 Bonneville, Century, Cutlass Supreme, Grand Prix, LeSabre, LSS, Lumina, Monte Carlo, 88', Regal, Park Avenue & 98' 3.8L VIN 1, 3.8L VIN K Transmissions: All	**Traction Control Torque Request Circuit Malfunction Conditions:** Engine started, engine speed over 500 rpm for 20 seconds, Traction Control system enabled, and the PCM detected the Desired Torque signal was less than 5%, or the Desired Torque Signal was more than 95% for up to 10 seconds. To determine the amount of torque reduction required, the PCM monitors the Traction Control System (TCS) Desired Torque PWM signal from the EBTCM. The PCM also provides a TCS Delivered Torque PWM informing the EBCM how much torque is actually produced by the engine. The Desired Torque display on the Scan Tool indicates the amount of engine torque requested by the EBCM. With Traction Control inactive, the Desired Torque should be approximately 90%. *Note: The Desired Torque display on the Scan Tool is not the PWM duty cycle.* **Possible Causes** ● T/C desired torque circuit is open or shorted to ground ● T/C desired torque circuit is shorted to system power (B+) ● EBTCM has failed ● PCM has failed
DTC P1571 **1T CCM, MIL: No** 1996, 1997, 1998, 1999, 2000 Aurora, DeVille, Eldorado & Seville 4.0L VIN C, 4.6L VIN 9, 4.6L VIN Y engines Transmissions: All	**Traction Control System - PWM Circuit No Frequency Conditions:** DTC P1602 not set, engine started, engine speed over 500 rpm for 20 seconds, Traction Control active (not failed), and the PCM did not detect a PWM signal on the Desired torque circuit for 3 seconds. The PCM tests the Traction Control System (TCS) Desired Torque circuit. The PCM sends a 5v signal on the Desired Torque on Circuit 463 to the EBCM. The EBCM rapidly switches this circuit to ground creating a PWM frequency signal to request a specific amount of torque reduction during a Traction Control event. **Possible Causes** ● TCS desired torque circuit is open or shorted to ground ● TCS desired torque circuit is shorted to system power (B+) ● EBCM has failed ● PCM has failed
DTC P1571 **1T CCM, MIL: No** 1999, 2000 Camaro & Firebird 3.8L VIN K engine Transmissions: All	**Traction Control Torque Request Circuit Malfunction Conditions:** Traction Control system not failed, engine speed over 500 rpm for 20 seconds, and the PCM did not receive a Desired Torque signal from the ETC or TAC Module for 3 seconds. **Possible Causes** ● TCS desired torque signal circuit is open or shorted to ground ● TCS desired torque signal circuit is shorted to system power ● PCM has failed

OBD II Trouble Code List (P1xxx Codes)

DTC	Trouble Code Title, Conditions & Possible Causes
DTC P1571 **1T CCM, MIL: No** 1996, 1997, 1998, 1999, 2000 Montana, Silhouette, Venture 3.4L VIN E engine Transmissions: All	**Traction Control System Desired Torque Out-Of-Range Conditions:** Engine started, engine speed over 500 rpm for 20 seconds, Traction Control system enabled, and the PCM detected the Desired Torque signal was less than 5%, or the Desired Torque Signal was more than 95% for up to 10 seconds. To determine the amount of torque reduction required, the PCM monitors the Traction Control System (TCS) Desired Torque PWM signal from the EBTCM. The PCM also provides a TCS Delivered Torque PWM informing the EBCM how much torque is actually produced by the engine. The Desired Torque display on the Scan Tool indicates the amount of engine torque requested by the EBCM. With Traction Control inactive, the Desired Torque should be approximately 90%. *Note: The Desired Torque display on the Scan Tool is not the PWM duty cycle.* **Possible Causes** ● TCS desired torque circuit is open or shorted to ground ● TCS desired torque circuit is shorted to system power ● EBTCM has failed ● PCM has failed
DTC P1572 **1T CCM, MIL: No** 1996 Camaro, Corvette & Firebird 5.7L VIN 5, 5.7L VIN P Transmissions: All	**Traction Control System - Circuit Low Too Long Conditions:** DTC P1810 not set, Engine started; Traction Control and ABS active with no faults detected, vehicle driven to over 20 mph (non-drive wheel speed), and then did to less than 4 mph, and the PCM detected the Traction Control Active circuit remained in a low state for 6.4 seconds. **Possible Causes** ● TCS active circuit is open or shorted to ground ● EBTCM has failed ● PCM has failed
DTC P1573 **1T CCM, MIL: No** 1995 Century, Grand Prix, 88', Park Avenue, 98' Regal, Riviera, Lumina APV, Trans Sport, Silhouette 3.8L VIN 1, 3.8L VIN K, 3.8L VIN L engines Transmissions: All	**Traction Control System, Loss Of ABS/TCS Serial Data Conditions:** P1630 not set, key on or engine running and the PCM detected an unexpected voltage condition on the serial data line to the EBTCM for over 500 ms. **Possible Causes** ● Serial data circuit is open between the EBTCM and the PCM ● EBTCM has failed
DTC P1573 **2T CCM, MIL: Yes** 1996, 1997, 1998 Aurora 3.8L VIN 1, 3.8L VIN K Transmissions: A/T	**PCM/EBTCM Serial Data Circuit Malfunction Conditions:** Key on or engine running; and the PCM detected an unexpected voltage condition on the serial data circuit between the PCM to EBTCM circuit for 500 ms. **Possible Causes** ● Serial data circuit is open between the EBTCM and the PCM ● EBTCM has failed
DTC P1573 **1T CCM, MIL: Yes** 1997, 1998, 1999, 2000, 2001, 2002, 2003, 2004, 2005 Montana, Silhouette, Venture 3.4L VIN E engine Transmissions: All	**PCM & EBTCM Serial Data Circuit Malfunction Conditions:** P1630 not set, Key on or engine running; and the PCM detected an unexpected voltage condition on the serial data circuit between the PCM and EBTCM (modules) for 500 ms. **Possible Causes** ● Serial data circuit is open between the EBTCM and the PCM ● EBTCM has failed
DTC P1574 **1T CCM, MIL: No** 1996, 1997, 1998, 1999, 2000, 2001, 2002, 2003, 2004, 2005 Aurora, DeVille, Eldorado & Seville 4.0L VIN C, 4.6L VIN 9, 4.6L VIN Y engines Transmissions: All	**Stoplamp Switch Circuit Malfunction Conditions:** DTC P0703, P1575 and P1602 not set, engine started, ABS and Traction Control inactive or failed, VSS over 20 mph, then remained over 3 mph during the test, difference in speed between the non-driven wheels and driven wheels less than 6 mph, and the PCM detected a signal transition on the TCC or Electronic Brake (E/T) switches without a transition occurring on the Stoplamp switch. The Stoplamp switch is a N.O. switch that receives fused battery voltage, and grounds via a pull-up resistor in the Turn/Hazard flasher module. The flasher module will detect a low voltage on the Stoplamp switch circuit with the brake not applied. **Possible Causes** ● Refer to Diagnostic Circuit Check for Antilock Brake Systems
DTC P1574 **1T CCM, MIL: No** 1999, 2000 Camaro & Firebird 3.8L VIN K engine Transmissions: All	**Stoplamp Switch Circuit Malfunction Conditions:** Engine started, engine speed over 700 rpm and the wheel speed indicates more than 20 mph to enable the test (test will disable if the wheel speed is less than 5 mph), then the vehicle speed decreased at a rate greater than 0.5 mph in 500 ms, and the EBTCM did not detect any Brake switch transitions after 20 Accel/Decel cycles. **Possible Causes** ● Brake switch circuit is open, shorted to ground or shorted to B+ ● Brake switch power circuit is open (test STOP/HAZARD fuse) ● Brake switch is out-of-adjustment, damaged or has failed ● EBTCM has failed

OBD II Trouble Code List (P1xxx Codes)

DTC	Trouble Code Title, Conditions & Possible Causes
DTC P1574 **1T CCM, MIL: No** 1997, 1998, 1999, 2000 Corvette Engines: 5.7L VIN G Transmissions: All	**Stoplamp Switch Circuit Malfunction Conditions:** Engine speed over 700 rpm (wheel speed indicates more than 30 mph in order to enable the test, and the test will disable when wheel speed is less than 10 mph), then vehicle speed decreased at a rate greater than 10.4 mph in 1 second, then the Throttle Actuator Control (TAC) module did not detect voltage on the Brake Lamp Switch circuit with the Brake Lamp switch closed for 1.5 seconds. **Possible Causes** ● Brake switch circuit is open, shorted to ground or shorted to B+ ● Brake switch power circuit is open (test STOP/HAZARD fuse) ● Brake switch is out-of-adjustment, damaged or has failed ● EBTCM has failed
DTC P1575 **1T CCM, MIL: No** 1996, 1997, 1998, 1999, 2000, 2001 Aurora, DeVille, Eldorado & Seville 4.0L VIN C, 4.6L VIN 9, 4.6L VIN Y engines Transmissions: A/T	**Extended Travel Brake Switch Circuit Malfunction Conditions:** DTC P0703, P1574 and P1604 not set, engine started, Traction Control and ABS inactive, vehicle driven with the non-drive wheel speed over 20 mph, and then to below 4 mph, and the PCM detected at least a 6 mph decrease in the non-drive wheel speed in 400 ms and the Stop/BTSI/TCC brake Switch indicated the brakes were applied and did not detect a change in extended travel contacts of the TCC Brake Switch to indicate the brakes were applied). **Possible Causes** ● Brake switch circuit is open, shorted to ground or shorted to B+ ● Brake switch power circuit is open (test the IGN1 fuse) ● Brake switch is out-of-adjustment, damaged or has failed ● EBTCM has failed ● TSB 01-06-03-009 contains a repair procedure for this code
DTC P1575 **1T CCM, MIL: No** 1997, 1998, 1999, 2000, 2001, 2002 Corvette Engines: 5.7L VIN G Transmissions: All	**Extended Travel Brake Switch Circuit Malfunction Conditions:** Engine started, engine speed over 700 rpm, vehicle driven with the wheel speed indicating over 30 mph (test will stop if the wheel speed falls below 10 mph), then with the wheel speed decreasing at a rate greater than 2.6-10.5 mph in 250 ms, the PCM detected voltage on the Extended Travel Brake Switch circuit for 2 seconds when the switch circuit should have been open. **Possible Causes** ● Brake switch circuit is open, shorted to ground or shorted to B+ ● Brake switch power circuit is open (test the ENGIGN1 fuse) ● Brake switch is out-of-adjustment, damaged or has failed ● EBTCM has failed
DTC P1576 **1T CCM, MIL: No** 1996, 1997 DeVille, Eldorado, Seville 4.6L VIN 9, 4.6L VIN Y Transmissions: A/T	**Brake Booster Vacuum Sensor Circuit High Input Conditions:** DTC P0107, P0108, P0703 and P1106 not set, Engine started; Extended Travel Switch open for one second, MAP value at 30 kPa or more during the period Extended Travel Switch was open, and the PCM detected the Brake Booster Vacuum was 82 kPa or more for 5 straight occurrences under these conditions. **Possible Causes** ● BBV sensor signal circuit is shorted to VREF or system power ● BBV sensor ground circuit is open ● BBV sensor is damaged or has failed ● PCM has failed ● TSB 71-65-25 contains a repair procedure for this code
DTC P1577 **1T CCM, MIL: No** 1996, 1997 DeVille, Eldorado, Seville 4.6L VIN 9, 4.6L VIN Y Transmissions: A/T	**Brake Booster Vacuum Sensor Circuit Low Input Conditions:** DTC P0107, P0108, P0703 and P1106 not set, Engine started; and the PCM detected the Brake Booster Vacuum indicated less than -8 kPa or less for 2 seconds. **Possible Causes** ● BBV sensor signal circuit is open or shorted to ground ● BBV sensor VREF is open or shorted to ground ● BBV sensor is damaged or has failed ● PCM has failed ● TSB 71-65-25 contains a repair procedure for this code
DTC P1578 **1T CCM, MIL: No** 1996, 1997 DeVille, Eldorado, Seville 4.6L VIN 9, 4.6L VIN Y Transmissions: A/T	**Brake Booster Vacuum Sensor Circuit Low Vacuum Conditions:** DTC P0107, P0108, P1107, P1577 and P1602 not set, engine started, engine speed over 500 rpm for 20 seconds, BARO sensor over 72 kPa, MAP sensor more than 60 kPa, brake pedal not applied, and the PCM detected the Brake Booster Vacuum signal indicated less than 5 kPa for 16 seconds. **Possible Causes** ● BBV sensor signal circuit is open or shorted to ground ● BBV sensor VREF is open or shorted to ground ● BBV sensor is damaged or has failed ● Base engine problem (i.e., engine vacuum leaks, PCV leaks) ● Leaking vacuum components or vacuum lines ● PCM has failed

OBD II Trouble Code List (P1xxx Codes)

DTC P1579 **1T CCM, MIL: No** 1996, 1997, 1998, 1999, 2000 Aurora, DeVille, Eldorado & Seville 4.0L VIN C, 4.6L VIN 9, 4.6L VIN Y engines Transmissions: A/T	**P/N To D/R At High TP Angle - Power Reduction Mode On Conditions:** DTC P0122 and P0123 code not set, gear selection in Park or Neutral to Drive or Reverse, VSS input less than 6 mph, engine speed more than 2000 rpm, TP angle at 20 degrees or more, and the PCM determined that it should retard spark timing and disable fuel to individual cylinders to reduce the engine power. **Possible Causes** ● Vehicle shifted into Drive or Reverse under improper conditions ● Transaxle range switch is out-of-adjustment or it is damaged
DTC P1585 **1T CCM, MIL: No** 1999, 2000 Bonneville, Century, Grand Prix, Intrigue, LeSabre, LSS, Lumina, Monte Carlo, Regal, Park Avenue & Eighty Eight 3.1L VIN M, 3.8L VIN 1, 3.8L VIN K engines Transmissions: All	**Cruise Control Inhibit Output Circuit Malfunction Conditions:** Engine started; system voltage over 10.0v and the PCM detected an unexpected voltage condition on the Cruise Control Inhibit driver circuit for at least 30 seconds. **Possible Causes** ● Cruise control inhibit circuit is shorted to system voltage ● Cruise control module power circuit is open (test CR CNT fuse) ● Cruise control stepper motor power circuit is shorted to ground ● Cruise control module is damaged or has failed ● PCM has failed
DTC P1585 **1T CCM, MIL: No** 2000, 2001, 2002, 2003, 2004, 2005 DeVille, Eldorado, Seville 4.6L VIN 9, 4.6L VIN Y Transmissions: A/T	**Cruise Control Inhibit Output Circuit Malfunction Conditions:** Engine started; system voltage over 10.0v and the PCM detected an unexpected voltage condition on the Cruise Control Inhibit driver circuit for at least 10 seconds. **Possible Causes** ● Cruise control inhibit circuit is shorted to system voltage ● Cruise control inhibit circuit is open or shorted to ground ● Cruise control module power circuit is open (test CR CNT fuse) ● Cruise control module is damaged or has failed ● PCM has failed
DTC P1585 **1T CCM, MIL: No** 1999, 2000, 2001, 2002, 2003, 2004, 2005 Century, Lumina, Monte Carlo, Montana, Silhouette & Venture 3.1L VIN M, 3.4L VIN E Transmissions: All	**Cruise Control Inhibit Output Circuit Malfunction Conditions:** Engine started; system voltage over 10.0v and the PCM detected an unexpected voltage condition on the Cruise Control Inhibit driver circuit for at least 30 seconds. **Possible Causes** ● Cruise control inhibit circuit is shorted to system voltage ● Cruise control inhibit circuit is open or shorted to ground ● Cruise control module power circuit is open (test CR CNT fuse) ● Cruise control module is damaged or has failed ● PCM has failed
DTC P1585 **1T CCM, MIL: No** 1999, 2000 Camaro & Firebird 3.8L VIN K engine Transmissions: All	**Cruise Control Brake Switch 2 Circuit Malfunction Conditions:** Engine started, engine speed over 1000 rpm with the wheel speed indicating more than 20 mph to enable the test (test will disable if the wheel speed drops below 5 mph), and the PCM detected the vehicle speed decreased by 1 mph in 500 ms without the PCM detecting a Brake switch transition for 10 Accel/Decel cycles. The brake switch indicates brake pedal status to the PCM. The brake switch is a N.O. switch that supplies battery voltage to the PCM. Applying the brake pedal opens the switch, interrupting voltage to the PCM. **Possible Causes** ● C/C brake switch circuit is open, shorted to ground or shorted to system power ● C/C brake switch is out-of-adjustment or damaged ● C/C brake switch power circuit is open (test ENG SEN fuse) ● PCM has failed
DTC P1599 **1T CCM, MIL: No** 1995 Century, Grand Prix, 88', Park Avenue, 98' Regal, Riviera, Lumina APV, Trans Sport, Silhouette 3.8L VIN 1, 3.8L VIN K, 3.8L VIN L engines Transmissions: All	**Cruise Power Management Mode Enabled Conditions:** Engine started; and the PCM detected the Cruise Control System was operating in Power Management Mode (where the PCM shuts off fuel to three (3) cylinders to prevent the engine from over revving in case the throttle is held open for too long a time. **Possible Causes** ● Brake vacuum release valve may be out of adjustment ● C/C throttle cable binding or sticking (fault may be intermittent) ● C/C vacuum solenoid is damaged (it may be stuck closed) ● C/C servo is damaged or has failed ● EMI/RFI interference affecting Cruise Control module (antenna) ● PCM has failed
DTC P1599 **1T CCM, MIL: No** 1996, 1997, 1998, 1999, 2000 Aurora, DeVille, Eldorado & Seville 4.0L VIN C, 4.6L VIN 9, 4.6L VIN Y engines Transmissions: All	**Engine Stall Or Near Stall Detected Conditions:** Engine started; and the PCM detected the engine speed dropped below 152 rpm after it has exceeded 400 rpm on the current cycle, and that no ignition reference pulses had been received for 75 ms. This code sets if the engine speed drops below 400 rpm and recovers **Possible Causes** ● Many conditions can cause stalling - use snapshot taken from DTC P1599. Compare the values when the code set to Typical Scan Data values. Look for conditions that could cause a stall.

OBD II Trouble Code List (P1xxx Codes)

DTC	Trouble Code Title, Conditions & Possible Causes
DTC P1600 **1T CCM, MIL: Yes** 1997, 1998, 1999, 2000, 2001, 2002, 2003, 2004, 2005 Aurora & Catera 3.0L VIN R, 3.5L VIN H, 4.0L VIN C engines Transmissions: A/T	**TCM Internal Watchdog Operation (4L60-E) Conditions:** Key on or engine running; TCM High side driver open, and the PCM received a signal that indicated the TCM detected an unexpected voltage condition on the internal TCM Power Control relay. **Possible Causes** ● Power control relay circuit is open, shorted to ground or to B+ ● Power control relay is damaged or has failed ● TCM has failed
DTC P1601 **1T CCM, MIL: No** 1996, 1997 Achieva, Beretta, Cavalier, Ciera, Century, Corsica, Grand Am, Sunfire & Skyhawk 2.2L VIN 4, 2.4L VIN T Transmissions: All	**PCM Serial Communications Malfunction Conditions:** Key on, and the PCM did not receive 25 valid responses from the IPC system, condition met for 5-10 seconds during the CCM test. **Possible Causes** ● Inspect the serial data line for an open or shorted condition ● Determine if the BCM (module) is pulling the data line down ● Determine if the EBCM (module) is pulling the data line down ● TSB 61 65 67 contains a repair procedure for this code
DTC P1601 **1T CCM, MIL: No** 1997, 1998 Catera 3.0L VIN R engine Transmissions: A/T	**ECM (PCM) Over-Temperature Malfunction Conditions:** Engine started; and the PCM detected an internal PCM temperature of more than 221°F, or an internal PCM temperature less than -58°F or over 284°F. **Possible Causes** ● Verify that the relay center cover is properly installed ● Check for a damaged relay center ● Check for air blockage/obstructions preventing proper airflow
DTC P1602 **1T CCM, MIL: No** 1996, 1997, 1998, 1999 Aurora, DeVille, Eldorado Regal & Seville 3.8L VIN 1, 3.8L VIN K, 4.0L VIN C, 4.6L VIN 9, 4.6L VIN Y Transmissions: All	**Loss Of EBTCM Serial Data Conditions:** DTC P0322 not set, started, engine speed 500 rpm or more for 10 seconds, and the PCM did not detect any serial data or it detected invalid serial data from EBTCM for 11 seconds during the CCM test. **Possible Causes** ● Check the serial data circuit for a poor connection at the PCM ● Perform the EBTCM Diagnostic Circuit Check ● Perform the Powertrain Onboard Diagnostic System Check
DTC P1602 **1T CCM, MIL: No** 1999, 2000, 2001 Catera 3.0L VIN R engine Transmissions: A/T	**Knock Sensor Module Performance Conditions:** Engine started; ECT sensor more than 104°F and the PCM detected an unexpected voltage condition on the Knock Sensor module (internal) circuit during the CCM test. **Possible Causes** ● The PCM has failed and must be replaced and reprogrammed. ● This vehicle has a Theft Deterrent system that interfaces with the PCM. The new PCM must be programmed with the frequency code from the current Theft Deterrent Module.
DTC P1603 **1T CCM, MIL: No** 1996, 1997, 1998, 1999 Aurora, DeVille, Eldorado & Seville 4.0L VIN C, 4.6L VIN 9, 4.6L VIN Y engines Transmissions: All	**Loss Of Sensing Diagnostic Mode Data Conditions:** DTC P0322 not set, engine started with the engine speed over 500 rpm for 10 seconds, and the PCM did not detect any Sensing Diagnostic Mode (SDM) data or it received invalid SDM data for a one second period during the CCM test. **Possible Causes** ● Check the serial data circuit for a poor connection at the SIR ● Perform the SIR System Check ● Perform the Powertrain Onboard Diagnostic System Check
DTC P1604 **1T CCM, MIL: No** 1996, 1997, 1998, 1999 Aurora, DeVille, Eldorado & Seville 4.0L VIN C, 4.6L VIN 9, 4.6L VIN Y engines Transmissions: A/T	**Loss Of IPC Serial Data Conditions:** DTC P0322 not set, engine started with the engine speed more than 500 rpm for 10 seconds, and the PCM did not detect any Instrument Panel Cluster (IPC) serial data or it detected invalid serial data from the IPC for 11 seconds during the CCM test. **Possible Causes** ● Check the serial data circuit for a poor connection at the IPC ● Perform the IPC System Check ● Perform the Powertrain Onboard Diagnostic System Check
DTC P1604 **1T CCM, MIL: No** 1996, 1997, 1998, 1999 DeVille, Eldorado, Seville & Regal 3.8L VIN 1, 3.8L VIN K, 4.6L VIN 9, 4.6L VIN Y Transmissions: All	**Loss Of HVAC Serial Data Conditions:** DTC P0322 not set, engine started with the engine speed more than 500 rpm for 10 seconds, and the PCM did not detect any HVAC Programmer serial data or it detected invalid Heater and A/C (HVAC) Programmer serial data for 11 seconds in the CCM test. **Possible Causes** ● Check the serial data circuit for a poor connection at the HVAC ● Perform the ECC System Check ● Perform the Powertrain Onboard Diagnostic System Check

OBD II Trouble Code List (P1xxx Codes)

DTC	Trouble Code Title, Conditions & Possible Causes
DTC P1610 **1T CCM, MIL: No** 1996, 1997, 1998, 1999 DeVille, Eldorado, Seville 4.6L VIN 9, 4.6L VIN Y Transmissions: All	**Loss of Platform Zone Module Serial Data Conditions:** DTC P0322 not set; key on for 5 seconds and the PCM did not detect serial data from the Platform Zone Module (PZM) or it was unable to communicate with the PZM for 11 seconds. **Possible Causes** ● Check the serial data circuit for a poor connection at the PZM ● Perform the PZM System Check ● Perform the Powertrain Onboard Diagnostic System Check
DTC P1610 **1T CCM, MIL: No** 1997, 1998, 1999 Regal 3.8L VIN 1, 3.8L VIN K Transmissions: All	**Loss of Standard Body Module Serial Data Conditions:** DTC P0322 not set; key on for 5 seconds, system voltage over 8.23v, and the PCM did not detect any serial data from the Body Control Module (BCM) or for 5 seconds during the test. **Possible Causes** ● Check the serial data circuit for a poor connection at the BCM ● Perform the BCM System Check ● Perform the Powertrain Onboard Diagnostic System Check
DTC P1611 **1T CCM, MIL: No** 1996, 1997, 1998, 1999 DeVille, Eldorado, Seville 4.6L VIN 9, 4.6L VIN Y Transmissions: All	**Loss of CVRTD Serial Data Conditions:** DTC P0322 not set; key on for 10 seconds and the PCM did not detect any Continuous Variable Road Sensing Suspension serial data or it detected invalid data for 11 seconds. **Possible Causes** ● Check the serial data circuit for poor connections at the CVRSS ● Perform the CVRSS System Diagnosis ● Perform the Powertrain Onboard Diagnostic System Check
DTC P1617 **1T CCM, MIL: No** 1998, 1999, 2000 Aurora, DeVille, Eldorado & Seville 4.0L VIN C, 4.6L VIN 9, 4.6L VIN Y engines Transmissions: A/T	**Engine Oil Level Switch Circuit Malfunction Code Conditions:** Engine started, engine speed over 2000 rpm, throttle position over 5%, VSS over 3 mph, vehicle in steady state cruise (no lateral acceleration), and the PCM detected a low EOL switch signal voltage, and/or no transition of the EOL switch for at 8 sec's. **Possible Causes** ● Engine oil level too low or too high ● Engine oil level switch circuit shorted to ground ● Engine oil level switch is damaged or has failed ● PCM has failed
DTC P1619 **1T CCM, MIL: No** 1995 Century, Grand Prix, 88', Park Avenue, 98' Regal, Riviera, Lumina APV, Trans Sport, Silhouette 3.8L VIN 1, 3.8L VIN K, 3.8L VIN L engines Transmissions: All	**Engine Oil Reset Monitor Circuit Malfunction Conditions:** Key on or engine running; and the PCM detected an unexpected "low voltage condition on the Oil Life Monitor circuit for 60 seconds. **Possible Causes** ● Oil life monitor reset switch signal circuit is shorted to ground ● Oil life monitor reset switch signal circuit is shorted to ground ● PCM has failed
DTC P1621 **1T CCM, MIL: No** 1998, 1999, 2000, 2001, 2002, 2003, 2004, 2005 C/K Cab & Chassis, C/K Truck, G Van, M/L Van, S/T Blazer & S/T Pickup 4.3L VIN W, 4.3L VIN X, 5.0L VIN M, 5.7L VIN K, 5.7L VIN R, 7.4L VIN J Transmissions: All	**Control Module Long Term Memory Performance Conditions:** Engine cranking or running; and the PCM/VCM determined that it was unable to read data correctly from the EEPROM memory. **Possible Causes** ● Serial data circuit is open between VCM and VTD (Passlock) control module, or serial data circuit is shorted to ground or B+ ● Perform the Theft Deterrent System Check ● Perform the Powertrain Onboard Diagnostic System Check
DTC P1621 **1T CCM, MIL: No** 2002, 2003, 2004, 2005 C/K Series Truck, Envoy & TrailBlazer 4.2L VIN S, 5.3L VIN P, 6.0L VIN N, 6.0L VIN U Transmissions: All	**Control Module Long Term Memory Performance Conditions:** Key on, and the PCM detected a problem with its internal microprocessor integrity during the initial power-up phase. **Possible Causes** ● Perform the Diagnostic System Check for Engine Controls ● The PCM needs to be replaced and then reprogrammed
DTC P1621 **1T CCM, MIL: No** 1996, 1997, 1998, 1999, 2000, 2001, 2002, 2003, 2004, 2005 C/K Series Truck, G Van 6.5L VIN F, 6.5L VIN S, 6.5L VIN Y Diesel engine Transmissions: All	**Control Module Long Term Memory Performance Conditions:** Key on, and the PCM determined that it was unable to read data correctly from the EEPROM memory in the initial power-up phase. **Possible Causes** ● Perform the Powertrain Onboard Diagnostic System Check ● The PCM needs to be replaced

OBD II Trouble Code List (P1xxx Codes)

DTC P1621 **1T PCM, MIL: Yes** 2003, 2004, 2005 CTS (Cadillac) 2.6L VIN M, 3.2L VIN N Transmissions: All	**Control Module Long Term Memory Performance Conditions:** Key on, system voltage at 8-18v, and the TCM detects a checksum error when the RAM and EEPROM test data were compared. The engine and transmission data is accumulated and stored in RAM, which cannot store data reliably after the TCM powers down. Before the TCM powers down, the data is copied from RAM to an EEPROM device. One function of TCM programming is to test the EEPROM at TCM power-down by copying test data from RAM to EEPROM and then comparing the two copies. This code is set if they do not match. **Possible Causes** ● Perform the Diagnostic System Check for Engine Controls ● TCM needs to be replaced
DTC P1621 **1T PCM, MIL: Yes** 1996, 1997, 1998, 1999, 2000, 2001, 2002, 2003, 2004, 2005 Aurora, DeVille, Eldorado & Seville 4.0L VIN C, 4.6L VIN 9, 4.6L VIN Y engines Transmissions: A/T	**Control Module Long Term Memory Performance Conditions:** Key on, and the PCM detected a malfunction occurred while reading the non-volatile memory during the initial power-up. **Possible Causes** ● Perform the Diagnostic System Check for the Engine Controls ● PCM may have failed. Clear codes and then retest for the same code. If this codes resets, replace the PCM and reprogram it.
DTC P1621 **1T PCM, MIL: No** 1997, 1998, 1999, 2000, 2001, 2002, 2003, 2004, 2005 Achieva, Alero, Cavalier, Ciera, Century, Grand Am, Sunfire, Skyhawk, S/T Blazer & S/T Pickup, Envoy & TrailBlazer 2.2L VIN 4, 2.4L VIN T, 4.2L VIN S engines Transmissions: All	**Control Module Long Term Memory Performance Conditions:** Key on, and the PCM determined that it could not communicate internally during the initial power-up phase. **Possible Causes** ● Perform the OBD System Check for Engine Controls ● The PCM needs to be replaced
DTC P1621 **1T PCM, MIL: Yes** 1999, 2000, 2001 Catera 3.0L VIN R engine Transmissions: A/T	**TCM Long Term Memory Performance (4L30-E) Conditions:** Key on and the TCM was not able to read EEPROM data during the initial power-up phase. **Possible Causes** ● Perform the Onboard Diagnostic Circuit Check ● TCM has failed and needs to be replaced
DTC P1623 **1T PCM, MIL: Yes** 1995 Century, Grand Prix, 88', Park Avenue, 98' Regal, Riviera, Lumina APV, Trans Sport, Silhouette 3.8L VIN 1, 3.8L VIN K, 3.8L VIN L engines Transmissions: All	**PCM PROM Error Conditions:** Key on or engine running; and the PCM detected a PROM problem. **Possible Causes** ● Check PROM installation and for correct PROM part number ● If the PROM is installed properly, the PCM needs replacing
DTC P1624 **1T PCM, MIL: Yes** 2000, 2001, 2002, 2003, 2004, 2005 Aurora, DeVille, Eldorado & Seville 4.0L VIN C, 4.6L VIN 9, 4.6L VIN Y engines Transmissions: A/T	**Customer Snapshot - Data Available Conditions:** Key on or engine running; PCM receives a Customer Snapshot Request, and the PCM will illuminate the MIL for 40 seconds and store the conditions present when the customer requested the Snapshot and Fail Records Data. The feature permits during each key cycle. **Possible Causes** ● Read and record any diagnostic trouble codes that are stored ● Perform the Diagnostic System Check for Engine Controls
DTC P1625 **1T PCM, MIL: Yes** 1999, 2000, 2001 Catera 3.0L VIN R engine Transmissions: A/T	**TCM System Reset (4L30-E) Conditions:** Engine started; and the TCM detected an unexpected voltage condition on the TCM Power Control relay circuit (the relay is internal to the ECM) during the CCM test. **Possible Causes** ● TCM power control relay control circuit is open or shorted to B+ ● TCM power control relay has failed, or the TCM has failed
DTC P1626 **1T CCM, MIL: No** 1996, 1997 Camaro, Caprice, Corvette, Firebird, Fleetwood & Roadmaster 4.3L VIN W, 5.7L VIN 5, 5.7L VIN P engines Transmissions: All	**Theft Deterrent Signal Not Present, Engine Running Conditions:** Engine cranking or running; and the PCM did not receive the correct signal from the Theft Deterrent module for 1.7 seconds. **Possible Causes** ● If the engine will not crank with this code set, there is a problem in the Theft Deterrent System or an invalid key or starting procedure was used. ● If the engine starts and then stalls, and the Theft Deterrent Module that generates a signal to the PCM is inoperative or Circuit 229 is open or shorted, this code can be set. ● If the Theft Deterrent System is okay, the PCM may have failed ● TSB 61-65-48 contains a repair procedure for this code

OBD II Trouble Code List (P1xxx Codes)

DTC	Trouble Code Title, Conditions & Possible Causes
DTC P1626 **1T CCM, MIL: No** 1996, 1997, 1998 C/K Series Truck, G Van 6.5L VIN F, 6.5L VIN S, 6.5L VIN Y Diesel engine Transmissions: All	**Theft Deterrent Fuel Enable Signal Lost Conditions:** Key on, Vehicle Theft Deterrent (VTD) system enabled, "fuel enable" decision point reached (during cranking), and the VCM could not communicate with the VTD (Pass Lock) control module and did not received a valid password before reaching the fuel enable decision point. **Possible Causes** ● Check the serial data circuit for poor connections at the PCM ● Perform the Powertrain Onboard Diagnostic System Check ● Perform the Theft Deterrent Diagnostic System Check
DTC P1626 **1T CCM, MIL: No** 1998, 1999, 2000 C/K Cab & Chassis, C/K Truck, G Van, M/L Van, S/T Blazer & S/T Pickup Engines: 4.3L VIN W, 4.3L VIN X, 4.8L VIN V, 5.0L VIN M, 5.3L VIN T, 5.7L VIN K, 5.7L VIN R, 6.0L VIN U, 6.5L VIN F, 6.5L VIN S, 7.4L VIN J Transmissions: All	**Theft Deterrent Fuel Enable Signal Not Received Conditions:** Key on, Vehicle Theft Deterrent (VTD) system enabled, "fuel enable" decision point reached during cranking, and the VCM could not communicate with the VTD (Pass Lock) module and did not received a valid password before reaching the fuel enable decision point. Once the BCM confirms the proper voltage from the ignition switch, it sends a password to the PCM via the Class 2 circuit. When this password matches the password stored in the PCM, the system enables the fuel. If the BCM does not send a password or if the PCM does not receive a password, the vehicle will not start unless the PCM is in VTD Fail-Enabled mode. If the BCM and PCM lose communication with each other after the system has received the correct password, the PCM goes into VTD Fail-Enable mode. This allows the driver to restart the vehicle on future ignition cycles until communication is restored. If the BCM and PCM were to lose communication before the PCM receives the BCM password, the PCM disables the fuel until it is restored to prevent vehicle theft. The PCM will not disable fuel delivery once it enables the fuel injectors within a given key cycle to prevent engine stall. **Possible Causes** ● Check the serial data circuit for poor connections at the PCM ● Perform the Powertrain Onboard Diagnostic System Check ● Perform the Theft Deterrent Diagnostic System Check ● TSB 77-65-31 contains a repair procedure for this code
DTC P1626 **1T CCM, MIL: No** 1995 Century, Grand Prix, 88', Park Avenue, 98' Regal, Riviera, Lumina APV, Trans Sport, Silhouette 3.8L VIN 1, 3.8L VIN K, 3.8L VIN L engines Transmissions: All	**Pass Key® II Fuel Enable Circuit Malfunction Conditions:** Key on, and the PCM did not detect a Fuel Enable signal from the Theft Deterrent Module immediately after engine startup. When the ignition is turned "on", the Theft Deterrent Module reads the key resistor pellet. If the module recognizes the proper resistance, it sends a digital "fuel enable" signal to the PCM on circuit 229. The PCM looks for this signal during cranking and allows fuel delivery by enabling the fuel injectors when the signal is recognized. **Possible Causes** ● If the engine will not crank with DTC P1626 or P1629 stored, the problem affects the entire Theft Deterrent System and is not isolated to the "fuel enable" circuit. ● The PCM supplies 5v on circuit 229, and the Theft Deterrent Module pulses this signal to ground if the correct key resistance is detected. Test for an open or grounded circuit. ● The PCM ignores the presence of a "fuel enable" signal only if DTC P1626 or P1629 is stored. The vehicle should not start if the fault exists with either of the codes stored.
DTC P1626 **1T PCM, MIL: No** 1996, 1997, 1998, 1999, 2000, 2001, 2002, 2003, 2004, 2005 Body Codes: C, G, H, L, Achieva, Alero, Beretta, Bonneville, Century, LSS, Corsica, Cutlass Supreme, Grand Am, Grand Prix, LeSabre, Lumina, Monte Carlo, 88', Regal, Park Avenue & 98' 3.1L VIN J, 3.1L VIN M, 3.4L VIN E engine, 3.4L VIN X, 3.5L VIN H, 3.8L VIN 1, 3.8L VIN K Transmissions: All	**Theft Deterrent System Fuel Enable Circuit Malfunction Conditions:** DTC P01631 not set, engine cranking (during an attempt to start the engine and the Antitheft System allowed fuel deliver to occur, and the PCM detected a loss of the "state of health" serial data message from the Theft Deterrent System. The VTD system is part of the BCM. The PCM monitors the "state of health" message from the VTD module to ensure that PCM to BCM communications are present. If the PCM detects a loss of the "state of health" message with the engine running, DTC P1626 will be set. DTC P1626 can cause a non-start condition or normal engine operation (depending on when the loss of VTD System communication was detected). The engine will continue to start and run if the fault that set DTC P1626 occurred "after" the PCM received a valid theft deterrent password from the BCM and already allowed fuel during the ignition cycle. The engine will start and immediately stall if the fault that set DTC P1626 occurred "before" the PCM received a valid theft deterrent password. With this condition present, the PCM will inhibit fuel delivery and disable the starter until a valid theft deterrent password is detected. **Possible Causes** ● Perform the Onboard Diagnostic System Check ● If the PCM needs replacing, it must be reprogrammed ● TSB 02-08-56-002 contains a repair procedure for this code
DTC P1626 **1T PCM, MIL: No** 1996, 1997, 1998, 1999 Aurora, DeVille, Eldorado & Seville 4.0L VIN C, 4.6L VIN 9, 4.6L VIN Y engines Transmissions: All	**Theft Deterrent System Fuel Enable Circuit Malfunction Conditions:** Key on, Vehicle Theft Deterrent (VTD) System enabled and Fuel Decision Point reached, and the PCM detected a loss of "fuel enabled" due to a loss of communications with the engine running, or the PCM did not detect a password message form the Instrument Panel Cluster (IPC) prior to reaching the VTD fuel decision point. **Possible Causes** ● Perform the Powertrain Onboard Diagnostic System Check ● If DTC P1604 is also set, refer to the repair chart for the Loss of IPC Serial Data (the most likely cause is lost communications)

OBD II Trouble Code List (P1xxx Codes)

DTC	Trouble Code Title, Conditions & Possible Causes
DTC P1626 **1T PCM, MIL: No** 1997, 1998, 1999, 2000 Camaro, Corvette & Firebird Engines: 5.7L VIN G Transmissions: All	**Theft Deterrent Fuel Enable Signal Lost Conditions:** Key on, Vehicle Theft Deterrent (VTD) system enabled, "fuel enable" decision point reached (engine cranking), and the VCM could not establish communications with the VTD (Pass Lock) control module and did not received a valid password before reaching the fuel decision point. The VTD system is designed to disable engine operation if the incorrect key or starting procedure is used. The BCM enables the crank circuit to the starter and sends a signal to the PCM if the correct key is being used. If the proper signal (40-60 Hz) does not reach the PCM on the fuel enable circuit, the PCM will not pulse the injectors "on" and thus not allow the vehicle to continue to operate, even if the crank circuit is bypassed. If a circuit problem occurs after the engine is running, the PCM will not disable the injectors. It continues to pulse the injectors because a correct signal from the BCM was received by the PCM at startup. **Possible Causes** ● Check the "fuel enable " circuit for an open, short to ground or short to system power (B+) condition ● Perform the Powertrain Onboard Diagnostic System Check ● BCM or PCM may have failed
DTC P1626 **1T PCM, MIL: No** 1995, 1996, 1997, 1998, 1999, 2000 Camaro & Firebird 3.8L VIN K engine Transmissions: All	**Theft Deterrent System Fuel Enable Circuit Malfunction Conditions:** Engine started; and the PCM detected a loss of the "fuel enable" signal or an incorrect signal from the VTD module. The BCM produces the Theft Deterrent "fuel enable" signal when the key is "on" and the proper key resistor pellet is sensed. The PCM monitors the Fuel Enable signal during cranking. If the proper signal is present on the Theft Deterrent "fuel enable" circuit, the PCM enables fuel delivery to allow the engine to start. If the PCM detects the fuel "enable signal" is not present or incorrect while the engine is running, DTC P1626 will be set. DTC P1626 can cause a no-start condition or normal operation depending on when the loss of the fuel enable signal was detected. The engine will continue to start and run if the condition that set DTC P1626 occurred after the BCM sensed the proper key resistor pellet and signaled the PCM to continue fuel delivery. The engine will start and immediately stall if the condition that set DTC P1626 occurred before the BCM sensed the proper key resistor pellet. With this condition present, the PCM will inhibit fuel delivery and the BCM will disable the starter. If the problem affects inputs to the BCM, the starter motor may be disabled. **Possible Causes** ● The most likely cause of this code is lost communications ● If the engine starts and stalls, refer to VTD system diagnosis ● Perform the Powertrain Onboard Diagnostic System Check
DTC P1626 **1T PCM, MIL: No** 1998, 1999, 2000 Achieva, Alero, Cavalier, Ciera, Grand Am, Sunfire, Skyhawk, S/T Blazer & S/T Pickup 2.2L VIN 4, 2.4L VIN T, 3.4L VIN E engines Transmissions: All	**Theft Deterrent Fuel Enable Signal Lost Conditions:** Engine started; system voltage over 10.0v and the PCM determined that it lost communication with the Body Control Module (BCM) after it received a "Theft Passed" message from the BCM. If the PCM does not receive an identifier message from a module, the PCM assumes that there is a loss of communication and a code is stored. DTC P1626 will set if, after the engine has been started, the PCM loses communication with the BCM. Since DTC P1626 set after the engine was started, the theft protection is bypassed at subsequent key "on" periods. The vehicle will start, but as long as DTC P1626 is set there is no theft protection available. **Possible Causes** ● The most likely cause of this code is lost communications ● If the engine starts and stalls, refer to VTD system diagnosis ● Perform the Powertrain Onboard Diagnostic System Check
DTC P1626 **1T PCM, MIL: No** 1997, 1998, 1999, 2000, 2001, 2002, 2003, 2004, 2005 Aztek, Montana, Rendezvous, Silhouette, Venture 3.4L VIN E engine Transmissions: All	**Theft Deterrent System Fuel Enable Signal Lost Conditions:** DTC P1631 not set, Engine cranking or running; VTD system has allowed fuel delivery, and the PCM lost the health status message from the Vehicle Theft Deterrent (VTD) system. The VTD module produces the theft deterrent "fuel enable" signal with the ignition "on" and after the proper key resistor pellet is detected. The PCM monitors the fuel enable signal during cranking. If the proper signal is present on the serial data circuit, the PCM enables fuel delivery to allow the engine to start. If the PCM detects the fuel enable signal is not present or incorrect with the engine running, DTC P1626 is set. The engine will continue to start and run as long as DTC P1626 is set. If the problem affects inputs to the VTD signal, the starter motor may be disabled. **Possible Causes** ● The most likely cause of this code is lost communications ● If the engine starts and stalls, refer to VTD system diagnosis ● Perform the Powertrain Onboard Diagnostic System Check

OBD II Trouble Code List (P1xxx Codes)

DTC	Trouble Code Title, Conditions & Possible Causes
DTC P1626 **1T PCM, MIL: No** 1996, 1997, 1998, 1999, 2000, 2001, 2002, 2003, 2004, 2005 C/K Series Truck, G Van Engines: 3.5L VIN H, 4.0L VIN C, 6.0L VIN N, 6.0L VIN U, 6.5L VIN F, 6.5L VIN S, 6.5L VIN Y Transmissions: All	**Analog To Digital (A/D) Performance Conditions:** Key on, and the PCM detected the ECM internal Analog to Digital converter failed. The PCM monitors internal circuits for problems. If it detects a problem in the one or more circuits, DTC P1627 will set. **Possible Causes** ● Perform the Diagnostic System Check for Engine Controls ● The PCM needs to be replaced
DTC P1629 **1T CCM, MIL: No** 1995 Century, Grand Prix, 88', Park Avenue, 98' Regal, Riviera, Lumina APV, Trans Sport, Silhouette 3.8L VIN 1, 3.8L VIN K, 3.8L VIN L engines Transmissions: All	**Pass Key® II Fuel Enable Circuit Malfunction Conditions:** Key on, and the PCM did not detect a Fuel Enable signal from the Theft Deterrent Module immediately after engine startup. When the ignition is turned "on", the Theft Deterrent Module reads the key resistor pellet. If the module recognizes the proper resistance, it sends a digital "fuel enable" signal to the PCM on circuit 229. The PCM looks for this signal during cranking and then allows fuel delivery by enabling the injectors when the signal is detected. **Possible Causes** ● If the engine will not crank with DTC P1626 or P1629 stored, the problem affects the entire Theft Deterrent System and is not isolated to the "fuel enable" circuit. ● The PCM supplies 5v on circuit 229, and the Theft Deterrent Module pulses this signal to ground if the correct key resistance is detected. Test for an open or grounded circuit. ● The PCM ignores the presence of a "fuel enable" signal only if DTC P1626 or P1629 are stored. The vehicle should not start if the problem is present and either code is set.
DTC P1629 **1T CCM, MIL: No** 1996, 1997, 1998, 1999, 2000 Achieva, Alero, Beretta, Bonneville, Century, LSS, Corsica, Cutlass Supreme, Grand Am, Grand Prix, LeSabre, Lumina, Monte Carlo, 88', Regal, Park Avenue & 98' 3.1L VIN M, 3.4L VIN E, 3.4L VIN X, 3.8L VIN 1, 3.8L VIN K engines Transmissions: All	**Theft Deterrent Signal Not Present, Engine Cranking Conditions::** P1626 not set, after an attempt to start the engine, the PCM detected an incorrect signal on the Theft Deterrent Fuel Enable circuit during cranking for 2 seconds. The VTD module produces the Theft Deterrent "fuel enable" signal when the ignition is "on" and the proper key resistor pellet is sensed by the Theft Deterrent system. The PCM monitors the "fuel enable" signal during cranking. If the proper signal is present on the Theft Deterrent "fuel enable" circuit, the PCM enables fuel deliver to allow the engine to start. If the PCM detects that the "fuel enable" signal is lost or it detects an invalid "fuel enable" signal while an attempt is made to start the engine, DTC P1629 is set. The engine will not start as long as this condition is present. If the problem also affects inputs to the Theft Deterrent system, the starter motor will be disabled. **Possible Causes** ● Perform the Onboard Diagnostic System Check ● Refer to the diagnostics for the Theft Deterrent System ● If the PCM needs replacing, it must be reprogrammed ● TSB 02-08-56-002 contains a repair procedure for this code
DTC P1629 **1T CCM, MIL: No** 1995, 1996, 1997, 1998, 1999, 2000 Camaro & Firebird 3.8L VIN K engine Transmissions: All	**Theft Deterrent Fuel Enable Signal Not Received Conditions:** DTC P1626 not set, an attempt was made to start the engine, and the PCM detected an incorrect signal on the Theft Deterrent Fuel Enable circuit, conditions met for more than 2 seconds. The BCM produces the Theft Deterrent "fuel enable" signal with the key "on" and after the proper key resistor pellet is sensed. The PCM monitors the "fuel enable" signal during cranking. If it detects the proper signal on the Theft Deterrent "fuel enable" circuit, the PCM enables fuel delivery and starter operation to allow the engine to start. If the PCM detects the fuel enable signal is not present or invalid while an attempt is made to start the engine, DTC P1629 will be set. The engine will not start and the starter motor will be disabled. This condition will occur if the problem also affects inputs to the BCM. **Possible Causes** ● Perform the Onboard Diagnostic System Check ● Refer to the diagnostics for the Theft Deterrent System ● If the PCM needs replacing, it must be reprogrammed
DTC P1629 **1T CCM, MIL: No** 1998, 1999 Achieva, Alero, Cavalier, Ciera, Grand Am, Sunfire, Skyhawk, S/T Blazer & S/T Pickup 2.2L VIN 4, 2.4L VIN T, 3.4L VIN E Transmissions: All	**Theft Deterrent Fuel Enable Signal Not Received Conditions:** Key on or engine cranking; and the PCM did not detect a password or it detected an unrecognized password from the Instrument Panel Cluster (IPC). When starting the engine, the PCM looks for a password from the IPC on the UART serial data circuit. If the password is not recognized or is not present, the PCM will disable the engine. The two modes of tamper detection are Short Tamper Mode (the engine may start and stall quickly) and the Theft System telltale will flash on the IPC for 4 seconds, or the Long Tamper Mode (more than three starting attempts or an invalid password is received). In this mode, the engine is disabled for 10 minutes and the Theft System telltale will flash on the IPC for the full 10 minutes. After the vehicle has passed theft detection the PCM will continue normal operation. **Possible Causes** ● Perform the Onboard Diagnostic System Check ● Refer to the diagnostics for the Theft Deterrent System ● If the PCM needs replacing, it must be reprogrammed

OBD II Trouble Code List (P1xxx Codes)

DTC	Trouble Code Title, Conditions & Possible Causes
DTC P1630 **1T CCM, MIL: No** 1997, 1998, 1999, 2000, 2001, 2002, 2003, 2004, 2005 Achieva, Alero, Beretta, Bonneville, Century, LSS, Corsica, Cutlass Supreme, Grand Am, Grand Prix, LeSabre, Lumina, Monte Carlo, 88', Regal, Park Avenue & 98' Engines: 3.1L VIN M, 3.1L VIN J, 3.4L VIN E, 3.8L VIN 1, 3.8L VIN K Transmissions: All	**Theft Deterrent Learn Mode Active Conditions:** Key on or engine cranking; VTD system has allowed fuel delivery, PCM ready to learn a new password from the BCM, and the PCM detected it remained in "password learn mode" for over 2 seconds. The VTD system is designed to prevent vehicle theft by disabling the engine until a mechanical key correct engages the lock cylinder. The TFD system utilizes a lock cylinder, ignition switch, the body function controller (BFC) and the PCM. During startup, the PCM looks for a password from the BFC on the serial data circuit. If the password is not recognized or is not present, the PCM will disable the engine. If no password is received, the engine will start and stall quickly. The Theft System telltale will flash on the IPC. If an incorrect or disable password is received (i.e., over 3 invalid passwords are received), the engine is disabled for 10 minutes and the Theft System telltale will illuminate solid on the IPC for 10 minutes. After the vehicle has passed theft detection, the PCM will operate normally. **Possible Causes** ● Perform the Onboard Diagnostic System Check ● Refer to the diagnostics for the Theft Deterrent System ● If the PCM needs replacing, it must be reprogrammed
DTC P1630 **1T CCM, MIL: No** 1996, 1997, 1998, 1999, 2000, 2001, 2002, 2003, 2004, 2005 C/K Series Truck, G Van 6.5L VIN F, 6.5L VIN S, 6.5L VIN Y Diesel engine Transmissions: All	**Theft Deterrent System - Learn Mode Enable Conditions:** Key on or engine running; and the PCM determined the Theft Leaning Flag was enabled. The PCM checks for the Enable Password Learning Flag indicating it has entered learn password mode. This function during assembly when the PCM needs to learn the password from the Passlock module if either module is replaced. **Possible Causes** ● This DTC (PCM in learn mode) is used at the assembly plant, dealership or by other service personnel to indicate learn mode is enabled (i.e., the PCM is ready to lean a new password). ● Perform Diagnostic System Check for Theft Deterrent System ● Perform the Powertrain Onboard Diagnostic System Check ● TSB 77-65-31 contains a repair procedure for this code
DTC P1630 **1T CCM, MIL: No** 1996, 1997, 1998, 1999, 2000, 2001, 2002, 2003, 2004, 2005 DeVille, Eldorado, Seville 4.6L VIN 9, 4.6L VIN Y Transmissions: All	**Theft Deterrent System - Learn Mode Active Conditions:** Engine started; Vehicle Theft Deterrent system has allowed fuel delivery to occur, PCM in theft deterrent password learn mode, and it remained in theft deterrent password learn mode for more than 2 seconds. When the VTD system has been replaced, the password must be relearned using approved diagnostic equipment. If the PCM is replaced, the replacement PCM will learn the password within a few seconds after the ignition is turned ON. DTC P1630 is a code indicating the PCM is ready to learn the VTD password. The engine will start and continue to run with DTC P1630 set. **Possible Causes** ● Refer to Programming the Theft Deterrent System ● Perform the Powertrain Onboard Diagnostic System Check
DTC P1630 **1T CCM, MIL: Yes** 1995 Century, Grand Prix, 88', Park Avenue, 98' Regal, Riviera, Lumina APV, Trans Sport, Silhouette 3.8L VIN 1, 3.8L VIN K, 3.8L VIN L engines Transmissions: All	**System Voltage Error Conditions:** Engine started; and the PCM detected the system voltage was less than 9v, or it was more than 17v for 10 seconds. **Possible Causes** ● Check for high resistance at the battery connections or at the Underhood Fuse Block power circuit connection to the PCM ● Check the drive belt for excessive wear and the proper tension ● Check the condition of the battery and the Generator output ● Vehicle may have been used to jump-start another vehicle
DTC P1630 **1T CCM, MIL: Yes** 1998, 1999, 2000, 2001, 2002, 2003, 2004, 2005 Achieva, Alero, Cavalier, Ciera, Grand Am, Sunfire, Skyhawk, S/T Blazer & S/T Pickup, Montana, Aztek, Rendezvous, Silhouette, Venture 2.2L VIN 4, 2.4L VIN T, 3.4L VIN E engines Transmissions: All	**Theft Deterrent Learn Mode Active Conditions:** Key on, PCM ready to learn new Pass Lock II password after it finished the 10-minute learn pending timer with the key left on for 10 minutes, and the PCM received an incorrect password from BCM or the Vehicle Theft Detection (VDC) module. If the Theft Deterrent system is replaced, the password must be relearned using approved diagnostic equipment. If the PCM is replaced, the replacement PCM should learn the password within a few seconds after the ignition is turned "on". DTC P1630 indicates the PCM is ready to learn the VTD password. The engine will start and run with DTC P1630 set. **Possible Causes** ● Refer to Programming the Theft Deterrent System ● Perform the Powertrain Onboard Diagnostic System Check

OBD II Trouble Code List (P1xxx Codes)

DTC	Trouble Code Title, Conditions & Possible Causes
DTC P1630 **1T CCM, MIL: Yes** 1997, 1998, 1999, 2000, 2001, 2002, 2003, 2004, 2005 Corvette Engines: 5.7L VIN G Transmissions: All	**Theft Deterrent Learn Mode Active Conditions:** Key on or engine cranking; the PCM is ready to learn a new password from the BCM, and the PCM determined that the BCM did not send a valid password or that it did not send any password. The PCM tests for the Enable Password Learning Flag indicating that it is in learn password mode. This mode allows the PCM to learn the password from the BCM at assembly or whenever it is serviced. The password needs to be learned whenever the PCM or the BCM is replaced or if a replacement key with a different resistance is used **Possible Causes** ● Refer to Diagnostic System Check for the Body Control Module ● Perform the Powertrain Onboard Diagnostic System Check
DTC P1631 **1T CCM, MIL: No** 1997, 1998, 1999, 2000, 2001, 2002, 2003, 2004, 2005 Achieva, Alero, Beretta, Bonneville, Century, LSS, Corsica, Cutlass Supreme, Grand Am, Grand Prix, LeSabre, Lumina, Monte Carlo, 88', Regal, Park Avenue & 98' Engines: 3.1L VIN M, 3.1L VIN J, 3.4L VIN E, 3.8L VIN 1, 3.8L VIN K Transmissions: All	**Theft Deterrent Enable Signal Not Correct Conditions:** Key on or engine cranking; and the PCM detected an incorrect Vehicle Theft Deterrent password from the VTD system for less than one second. The VTD module produces a Theft Deterrent "fuel enable" signal when ignition is "on" and after the proper key resistor pellet is detected. The PCM monitors the fuel enable signal during cranking. If the proper signal is present on the serial data circuit, the PCM enables fuel delivery to allow the engine to start. If the PCM detects an incorrect password (theft deterrent system failure or attempted vehicle theft), DTC 1631 is set. The engine may not start or crank as long as this failure is present. **Possible Causes** ● This code indicates that the VTD password that the PCM has learned does not agree with the password being received from the theft deterrent system. This condition can occur if an incorrect key is being used when attempting to start the vehicle or if the Theft Deterrent Module has been replaced and the PCM password learn function has not been enabled. ● Refer to Programming the Theft Deterrent System ● Perform the Powertrain Onboard Diagnostic System Check
DTC P1631 **1T CCM, MIL: No** 1998, 1999, 2000, 2001, 2002, 2003, 2004, 2005 C/K Cab & Chassis, C/K Truck, G Van, M/L Van, S/T Blazer & S/T Pickup Engines: 4.3L VIN W, 4.3L VIN X, 4.8L VIN V, 5.0L VIN M, 5.3L VIN T, 5.3L VIN Z, 5.7L VIN K, 5.7L VIN R, 6.0L VIN U, 6.5L VIN F, 6.5L VIN S, 7.4L VIN J, 8.1L VIN G Transmissions: All	**Theft Deterrent - Start Enable Signal Not Correct Conditions:** DTC P1626 not active, engine cranking with the PCM not in "password learn mode", VTD (Pass Lock) system enabled, and the PCM did not receive a valid password before the fuel disable decision point was reached. When the Passlock portion of the VTD system has sensed the proper operation of the ignition switch and lock, or determined that the switch and lock have not been tampered with, the VTD (Passlock) module transmits a password to the PCM. Fuel delivery is enabled if this password matches the password stored in the PCM memory. If a component in the Theft Deterrent system has been replaced, the two modules need to relearn the password of the new components. If the relearn procedure has not been performed, DTC P1631 will set. If a VTD failure occurs during an ignition cycle on which the PCM has enabled fuel, then the PCM will enter Fail Safe mode (VTD System Failure with Fuel Enabled). The PCM remains in Fail Enable Mode for the current and future ignition cycles, until the fault is corrected, a valid password is received, or until the battery is disconnected. If the codes are cleared, the vehicle will lose its Fail Enable status and will not start until the fault is corrected or the ten minute timer expires. At this point, the PCM receives the correct fuel delivery password. **Possible Causes** ● Refer to Diagnostic System Check for Theft Deterrent Module ● Perform the Powertrain Onboard Diagnostic System Check ● TSB 77-65-31 contains a repair procedure for this code
DTC P1631 **1T CCM, MIL: No** 1996, 1997, 1998, 1999, 2000, 2001, 2002, 2003, 2004, 2005 C/K Series Truck, G Van 6.5L VIN F, 6.5L VIN S, 6.5L VIN Y Diesel engine Transmissions: All	**Theft Deterrent System - Learn Mode Enable Conditions:** Engine cranking, Vehicle Antitheft System enabled, and the PCM detected the Fuel Disable Flag was set because an incorrect "Fuel Continue Password has been received. The PCM checks for mismatched passwords between the VTD control module and the PCM. When the VTD control module or PCM is replaced the theft "learn" procedure must be followed so a new password is learned. **Possible Causes** ● Refer to Diagnostic System Check for Theft Deterrent Module ● Perform the Powertrain Onboard Diagnostic System Check
DTC P1631 **1T CCM, MIL: No** 1996, 1997, 1998, 1999, 2000, 2001, 2002, 2003, 2004, 2005 DeVille, Eldorado, Seville 4.6L VIN 9, 4.6L VIN Y Transmissions: All	**Theft Deterrent System - Start Enable Signal Not Current Conditions:** Engine cranking and the PCM detected an incorrect VTD password from the VTD system for one second. The PCM controls fuel injector and starter operation based on a vehicle theft deterrent (VTD) password from the Theft Deterrent module. When the ignition is first turned "on", the module sends a programmed password to the PCM. The PCM acknowledges the password and responds to the module that normal fuel injector and starter operation should continue. If the PCM detects an incorrect password due to a VTD system failure or attempted vehicle theft situation, DTC 1631 will set in memory. **Possible Causes** ● Refer to Programming the Theft Deterrent System ● Perform the Powertrain Onboard Diagnostic System Check

OBD II Trouble Code List (P1xxx Codes)

DTC	Trouble Code Title, Conditions & Possible Causes
DTC P1631 **1T CCM, MIL: No** 2000, 2001, 2002, 2003, 2004, 2005 Achieva, Alero, Cavalier, Ciera, Grand Am, Sunfire, Skyhawk, S/T Pickup, Montana, Aztek, Rendezvous, Silhouette, Venture 2.2L VIN 4, 2.4L VIN T, 3.4L VIN E engine Transmissions: All	**Theft Deterrent Fuel Enable Signal Not Correct Conditions:** Engine cranking and the PCM received a "Fuel Disabled" or "Undecided: password that does not match from the BCM. When the ignition switch is first turned ON, the BCM sends a password to the PCM through the serial data circuit. If the BCM password does not match the current password stored in the PCM, the PCM will disable the engine. The engine will start and stall, or it will not start. The Theft System telltale will flash on the IPC and the engine will be disabled until a matching password is received. The password is checked every 4 seconds. The engine is disabled for at least 10 minutes, and during that period, the Theft System telltale will flash on the IPC for approximately 4 seconds then will illuminate solid for 10 minutes or until a correct password is received. After the vehicle has passed theft detection, the PCM will continue a normal engine operation. If the PCM loses the BCM communication within the same ignition cycle, the vehicle will continue to run. This mode is called the fail enable mode **Possible Causes** ● Refer to Diagnostic System Check for Body Control Systems ● Perform the Powertrain Onboard Diagnostic System Check
DTC P1631 **1T CCM, MIL: No** 1997, 1998, 1999, 2000, 2001, 2002, 2003, 2004, 2005 Corvette Engines: 5.7L VIN G Transmissions: All	**Theft Deterrent Enable Signal Not Correct Conditions:** Key on or engine cranking; and the PCM detected mismatched passwords between itself and the BCM. The PCM tests for mismatched passwords between the BCM and itself. Whenever replacing the BCM or PCM, follow the theft learn procedure so that the VTD system can learn the new password. **Possible Causes** ● Refer to Programming the Theft Deterrent System ● Perform the Powertrain Onboard Diagnostic System Check
DTC P1631 **1T CCM, MIL: No** 1996, 1997, 1998, 1999 DeVille, Eldorado, Seville 4.6L VIN 9, 4.6L VIN Y Transmissions: All	**Theft Deterrent System, Fuel Disabled Conditions:** Engine cranking and the PCM detected a Fuel Disable password. This code sets when the PCM receives a Fuel Disable or a Fuel Undecided Passwords from the Vehicle Theft Deterrent controller. It is set as a service tool to indicate the VTD is not allowing fuel due to a possible theft attempt or a worn out or incorrect ignition key. **Possible Causes** ● Refer to Programming the Theft Deterrent System ● Perform the Powertrain Onboard Diagnostic System Check
DTC P1632 **1T CCM, MIL: No** 1998, 1999, 2000, 2001, 2002, 2003, 2004, 2005 Achieva, Alero, Cavalier, Ciera, Envoy, Grand Am, Malibu, Sunfire, Skyhawk, S/T Pickup & TrailBlazer 2.2L VIN 4, 2.4L VIN T, 4.2L VIN S engines Transmissions: All	**Theft Deterrent Fuel Disable Signal Received Conditions:** Engine cranking, and the PCM received a Fuel Disabled password from the Body Control Module (BCM), or an "undecided" password was sent from the PCM to the BCM. The VTD system (the Passlock II System) is designed to prevent vehicle theft by disabling the engine unless a mechanical key is used to correctly engage the Passlock lock cylinder. This system utilizes a lock cylinder, ignition switch, the BCM and the PCM. When starting the engine, the PCM looks for a password from the BCM through the Class 2 serial data circuit. If the password is not recognized or not present, the PCM will disable the engine. If an incorrect or no Password is received, this indicates that the engine will start and stall and the Theft System telltale will flash on the IPC for 4 seconds. If an incorrect or disable password is received (more than three invalid passwords received) the engine will be disabled for at least 10 minutes and the Theft System telltale will turn to "solid" on the IPC for 3 seconds, and then flash on the IPC for 10 minutes. After the vehicle passes theft detection, the PCM will continue normal engine operation. If the PCM loses the BCM communication within the same ignition cycle, the vehicle will continue to run on the following ignition cycles. This mode is called the fail enable mode. **Possible Causes** ● Perform the Powertrain Onboard Diagnostic System Check ● Password Learn Procedure - Attempt to start the vehicle (leave the ignition "on"). The Theft System telltale will flash for 4 seconds and then remain "on" for 10 minutes. Theft System Learn Mode will display Disabled on the Scan Tool ● Turn the ignition "off" after the Theft System telltale goes "out". ● Repeat steps 1 and 2 two more times. After the Theft System telltale turns "off" on the 3rd key cycle, and Theft System Learn Mode will display Enabled on the Scan Tool, attempt to start the engine. Once the engine starts the password is learned.
DTC P1633 **1T CCM, MIL: No** 1996, 1997, 1998, 1999, 2000, 2001, 2002, 2003, 2004, 2005 Aurora, DeVille, Eldorado & Seville 4.0L VIN C, 4.6L VIN 9, 4.6L VIN Y engines Transmissions: All	**Ignition Supplement Power (0) Switch Circuit Malfunction Conditions:** Key on, Class 2 serial data received normally, and the PCM detected an improper Supplement Power Circuit (Circuit 0) voltage. The PCM receives two ignition switch IGN 0 and IGN 1 inputs. It also receives a Class II message from the power mode master (PMM) module. The PCM receives the IGN 0 input with the ignition switch in OFF and RUN. The PCM uses a Class 2 message and voltage from the ignition switch to test the IGN 0 circuit. **Possible Causes** ● Ignition OFF/RUN/CRANK circuit is open or shorted to power ● Ignition switch is damaged or has failed ● PCM has failed

OBD II Trouble Code List (P1xxx Codes)

DTC	Trouble Code Title, Conditions & Possible Causes
DTC P1633 **1T CCM, MIL: No** 2002, 2003, 2004, 2005 Envoy & TrailBlazer Engines: 4.2L VIN S Transmissions: All	**Ignition Supplement Power (0) Switch Circuit Malfunction Conditions:** Key on, Class 2 serial data transmission normal and the PCM detected an improper Supplement Power Circuit (Circuit 0) voltage. The PCM receives two ignition switch IGN 0 and IGN 1 inputs. It also receives a Class II message from the power mode master (PMM) module. The PCM receives the IGN 0 input with the ignition switch in OFF and RUN. The PCM uses a Class 2 message and voltage from the ignition switch to test the IGN 0 circuit. **Possible Causes** ● Ignition OFF/RUN/CRANK circuit is open ● Ignition OFF/RUN/CRANK circuit is shorted to power ● Ignition switch is damaged or has failed ● PCM has failed
DTC P1633 **1T CCM, MIL: No** 1996, 1997, 1998, 1999, 2000, 2001, 2002, 2003, 2004, 2005 Aurora, DeVille, Eldorado & Seville 4.0L VIN C, 4.6L VIN 9, 4.6L VIN Y engines Transmissions: All	**Ignition One (1) Power (0) Circuit Low Input Conditions:** Key on, Class 2 serial data received normally, and the PCM detected an improper Ignition One (1) Power circuit voltage. The PCM receives two ignition switch IGN 0 and IGN 1 inputs. It also receives a Class II message from the power mode master (PMM) module. The PCM receives the IGN 0 input with the ignition switch in OFF and RUN. The PCM uses a PMM Class 2 message, and the voltage input from the ignition switch to test the IGN 1 circuit. If the PCM detects an improper IGN 1 input, DTC P1634 will set. **Possible Causes** ● Ignition OFF/RUN/CRANK power circuit is open (test IGN fuse) ● Ignition switch is damaged or has failed ● PCM has failed
DTC P1635 **2T CCM, MIL: Yes** 1996, 1997, 1998, 1999, 2000, 2001, 2002, 2003, 2004, 2005 Achieva, Alero, Beretta, Bonneville, Century, LSS, Ciera, Corsica, Cutlass Supreme, Grand Am, Grand Prix, LeSabre, Lumina, Monte Carlo, 88', Regal, Park Avenue, 98', Aztek, Rendezvous, Montana, Silhouette 3.1L VIN M, 3.1L VIN J, 3.4L VIN E, 3.5L VIN H, 3.8L VIN 1, 3.8L VIN K Transmissions: All	**5-Volt Reference 'A' Circuit Malfunction Conditions:** Key on or engine running; and the PCM detected the 5v REF 'A' circuit was less than 3.5v or more than 5.5v for 10 seconds. The PCM uses the 5v Reference 1 circuit as a sensor feed to the TP sensor, MAP sensor, EGR valve pintle position and FTP sensor. **Possible Causes** ● 5v VREF circuit is shorted to sensor ground or chassis ground ● 5v VREF circuit, MAP or FTP sensor circuit is shorted to (B+) ● 5v VREF circuit shorted to EGR sensor High signal circuit ● EGR solenoid valve (and sensor) is damaged or has failed ● PCM has failed
DTC P1635 **2T CCM, MIL: Yes** 1999, 2000, 2001, 2002, 2003, 2004, 2005 C/K Cab & Chassis, C/K Series Truck, G Van, S/T Blazer & S/T Pickup 2.2L VIN 4, 4.3L VIN W, 4.3L VIN X, 5.0L VIN M, 5.7L VIN R, 7.4L VIN J Transmissions: All	**5-Volt Reference 1 Circuit Malfunction Conditions:** Key on or engine running; and the PCM detected the 5v Reference 1 circuit was out of tolerance for 2 seconds. This circuit provides power to the EGR Pintle and MAP sensor. **Possible Causes** ● 5v VREF circuit is shorted to sensor ground or chassis ground ● 5v VREF circuit, MAP or EOP sensor circuit is shorted to (B+) ● 5v VREF circuit shorted to EGR sensor High signal circuit ● EGR solenoid valve (and sensor) is damaged or has failed ● PCM has failed
DTC P1635 **2T CCM, MIL: Yes** 1996, 1997, 1998, 1999, 2000, 2001, 2002, 2003, 2004, 2005 C/K Series Truck, G Van 6.5L VIN F, 6.5L VIN S, 6.5L VIN Y, 6.6L VIN 1 Transmissions: All	**5-Volt Reference Circuit Low Input Conditions:** Key on, and the PCM detected the 5v Reference circuit indicated less than 4.0v during the CCM (continuous monitor) test. The PCM provides a 5v supply for provide power to various engine sensors. The PCM monitors the voltage present at terminals BRD13, shared by the Boost and Crankshaft Position (CKP) sensors, and BRD14 that connects to the optical/fuel temperature sensor (Cam/HI. RES). **Possible Causes** ● 5v VREF circuit is shorted to sensor ground or chassis ground ● 5v VREF circuit, Boost or CKP sensor circuit is shorted to (B+) ● 5v VREF circuit shorted to the Boost sensor signal circuit ● EGR solenoid valve (and sensor) is damaged or has failed ● PCM has failed

OBD II Trouble Code List (P1xxx Codes)

DTC	Trouble Code Title, Conditions & Possible Causes
DTC P1635 **2T CCM, MIL: Yes** 1999, 2000, 2001, 2002, 2003, 2004, 2005 C/K Series Truck, G Van, Cargo Van, M/L Van 4.8L VIN V, 5.3L VIN T, 5.3L VIN Z, 6.0L VIN N, 6.0L VIN U, 8.1L VIN G Transmissions: All	**5-Volt Reference 1 Circuit Malfunction Conditions:** Key on, and the PCM detected the 5v VREF 1 circuit was out of tolerance for 2 seconds. The EGR Pintle, Fuel Level, Fuel Tank Pressure, MAP and TP sensors all connect to this circuit. **Possible Causes** ● 5v VREF circuit is shorted to sensor ground or chassis ground ● 5v VREF circuit, MAP or EGR sensor circuit is shorted to (B+) ● 5v VREF circuit shorted to EGR sensor High signal circuit ● EGR solenoid valve (and sensor) is damaged or has failed ● PCM has failed
DTC P1635 **2T CCM, MIL: Yes** 2000, 2001, 2002, 2003, 2004, 2005 Aurora, DeVille, Eldorado & Seville 4.0L VIN C, 4.6L VIN 9, 4.6L VIN Y engines Transmissions: All	**5-Volt Reference 1 Circuit Malfunction Conditions:** Engine started; and the PCM detected an out-of-range voltage condition on the 5v Reference 1 circuit for 1 second. The PCM uses the 5v VREF 1 circuit connects to the MAP sensor, EGR Valve Pintle position sensor and Throttle Position sensor. **Possible Causes** ● 5v VREF circuit is shorted to sensor ground or chassis ground ● 5v VREF circuit or MAP sensor circuit is shorted to power (B+) ● 5v VREF circuit shorted to EGR sensor High signal circuit ● EGR solenoid valve (and sensor) is damaged or has failed ● PCM has failed
DTC P1635 **2T CCM, MIL: Yes** 1995 Century, Grand Prix, 88', Park Avenue, 98' Regal, Riviera, Lumina APV, Trans Sport, Silhouette 3.8L VIN 1, 3.8L VIN K, 3.8L VIN L engines Transmissions: All	**5-Volt Reference 1 Circuit Malfunction Conditions:** Key on or engine running; and the PCM detected the 5v VREF 1 circuit was out of range for 10 seconds. The 5v VREF circuit is used to provide power to the EGR Pintle Position sensor, MAP sensor, Engine Oil Pressure sensor and the TP sensor on this vehicle. **Possible Causes** ● 5v VREF circuit is shorted to sensor ground or chassis ground ● 5v VREF circuit, MAP or EOP sensor circuit is shorted to (B+) ● 5v VREF circuit shorted to EGR sensor High signal circuit ● EGR solenoid valve (and sensor) is damaged or has failed ● PCM has failed
DTC P1635 **2T CCM, MIL: Yes** 1996, 1997, 1998, 1999, 2000, 2001, 2002, 2003, 2004, 2005 Camaro, Corvette & Firebird 3.8L VIN K engine, 5.7L VIN G, 5.7L VIN S Transmissions: All	**5-Volt Reference 1 Circuit Malfunction Conditions:** Key on or engine running; and the PCM detected the 5v Reference 1 circuit was out of range for 2 seconds. The 5v VREF 1 circuit is used to provide power to the EGR Pintle Position sensor, MAP sensor, Engine Oil Pressure sensor and the TP sensor on this vehicle. **Possible Causes** ● 5v VREF circuit is shorted to sensor ground or chassis ground ● 5v VREF circuit, TP or MAP sensor circuit is shorted to (B+) ● 5v VREF circuit shorted to EGR sensor High signal circuit ● EGR solenoid valve (and sensor) is damaged or has failed ● PCM has failed
DTC P1635 **2T CCM, MIL: Yes** 2002, 2003, 2004, 2005 Envoy & TrailBlazer 4.2L VIN S, 5.3L VIN P Transmissions: All	**5-Volt Reference 1 Circuit Malfunction Conditions:** Key on or engine running; and the PCM detected the 5v Reference 1 circuit was out of range for 10 seconds. The 5v VREF circuit is used to provide power to the A/C Pressure, Engine Oil Pressure sensor, Fuel Tank Pressure, Cooling Fan, TP Sensor 1 and APP Sensor 2. **Possible Causes** ● 5v VREF circuit is shorted to sensor ground or chassis ground ● 5v VREF circuit, FTP or EOP sensor circuit is shorted to (B+) ● 5v VREF circuit is shorted to FTP sensor signal circuit ● 5v VREF circuit is shorted to the Cooling Fan circuit ● 5v VREF circuit is shorted to TP Sensor 1 and APP Sensor 2 ● PCM has failed
DTC P1637 **1T CCM, MIL: No** 1999, 2000, 2001, 2002, 2003, 2004, 2005 C/K Series Truck, G Series Van, Cargo Van, Envoy & TrailBlazer 4.8L VIN V, 5.3L VIN P, 5.3L VIN T, 5.3L VIN Z, 6.0L VIN N, 6.0L VIN U, 8.1L VIN G Transmissions: All	**Generator 'L' Terminal Circuit Malfunction Conditions:** Engine started; and the PCM detected an incorrect voltage on the Generator 'L' terminal during the CCM test. The PCM supplies the ignition voltage to the generator lamp feed. This voltage is pulled low by the generator once the circuit is supplied voltage. Once the generator begins to turn, the PCM detects ignition voltage. If there are no Charging system faults, the lamp terminal circuit will be low (0 volts) with the ignition switch "on" and then change to the system voltage after engine startup. If the Charging system detects this circuit is shorted to ground) the IPC will display a fault message. **Possible Causes** ● A Scan Tool should display Inactive for the 'L' Terminal and 10-40% for the 'F' Terminal with the ignition "on". With the engine running, the display should indicate the 'L' Terminal is Active and the 'F' Terminal is higher than 5% on the tool display. ● Generator 'L terminal circuit shorted to ground or to power (B+) ● Generator 'F' terminal circuit is open or shorted to ground ● Generator is damaged or has failed or the PCM has failed

OBD II Trouble Code List (P1xxx Codes)

DTC	Trouble Code Title, Conditions & Possible Causes
DTC P1637 **1T CCM, MIL: No** 1996, 1997, 1998, 1999 Aurora Engines: 4.0L VIN C Transmissions: All	**Generator 'L' Terminal Circuit Malfunction Conditions:** Engine started; and the PCM detected the current state of 'L' terminal feedback circuit indicated "low" Generator voltage for 15 seconds. DTC P1637 and a low battery charge may be caused by a generator problem (i.e., a shorted output diode, faulty regulator, open or shorted condition in the rotor, or open sense lead. This code will not set if Circuit 225 is shorted to voltage but may cause a parasitic load on the battery with the key turned "off". **Possible Causes** ● Generator 'L terminal circuit is open or shorted to ground ● Generator 'S' terminal circuit is open ● Battery temperature sensor circuit is open or high resistance ● Generator is damaged or has failed or the PCM has failed
DTC P1637 **1T CCM, MIL: No** 1997, 1998 Camaro, Corvette & Firebird Engines: 5.7L VIN G Transmissions: All	**Generator 'L' Terminal Circuit Malfunction Conditions:** The L-terminal circuit from the generator is a discrete circuit into the PCM. The PCM applies ignition voltage to the generator L-terminal circuit. A small amount of current flows from this circuit through the generator windings to ground to create a magnetic field that starts the generator process. When the generator is at operating speed and producing voltage, a solid state switch for the L-terminal circuit in the generator opens and the PCM detects that the initial startup current flow has stopped. The PCM expects to detect low voltage on the L-terminal circuit prior to the generator rotating at operating speed and conversely expects the circuit to be at ignition voltage potential when the generator is operational. **Possible Causes** ● A Scan Tool should display Inactive for the 'L' Terminal and 10-40% for the 'F' Terminal with the ignition "on". With the engine running, the display should indicate the 'L' Terminal is Active and the 'F' Terminal is higher than 5% on the tool display. ● Generator 'L terminal circuit shorted to ground or to power (B+) ● Generator 'F' terminal circuit is open or shorted to ground ● Generator is damaged or has failed or the PCM has failed
DTC P1637 **1T CCM, MIL: No** 1999, 2000 Camaro, Corvette & Firebird Engines: 5.7L VIN G Transmissions: All	**Generator 'L' Terminal Circuit Malfunction Conditions:** Engine started; and the PCM detected that the 'L' terminal was Active with the key on and engine off, or it detected the 'L' terminal was Inactive with the engine running for 6 seconds. The L-terminal circuit from the generator is a discrete circuit (no splices) into the PCM. The PCM applies voltage to the generator L-terminal circuit. A small amount of current flows from this circuit through the generator windings to ground to create a magnetic field that starts the generating process. When the generator is at operating speed and producing voltage, a solid state switch for the L-terminal circuit in the generator opens and the PCM detects that the initial startup current flow has stopped. The PCM expects to detect low voltage on the L-terminal circuit prior to the generator rotating at operating speed and conversely expects the circuit to be at ignition voltage potential when the generator is operational. **Possible Causes** ● DTC P1637 and P1638 may set together. The L-terminal circuit can cause DTC P1638. ● A Scan Tool should display Inactive for the 'L' Terminal and 10-40% for the 'F' Terminal when the ignition is "on". When the engine is running, the display should indicate the 'L' Terminal is Active and the 'F' Terminal is higher than 5% on the tool display. ● Generator 'L terminal circuit shorted to ground or to power (B+) ● Generator 'F' terminal circuit is open or shorted to ground ● Generator is damaged or has failed or the PCM has failed
DTC P1638 **1T CCM, MIL: No** 1997, 1998, 1999, 2000, 2001, 2002, 2003, 2004, 2005 Body Codes: C, F, K, Y Camaro, Corvette, Firebird, C/K Series Truck 5.7L VIN G, 6.0L VIN N, 6.0L VIN U engines Transmissions: All	**Generator 'F' Terminal Circuit Malfunction Conditions:** No CKP, CMP or Generator codes set, key on and the PCM detected the PWM signal was from 10-40% for over 6 seconds; or with the engine speed under 3000 rpm, the PCM detected the PWM signal was less than 5% for 6 seconds. The PCM uses the generator field duty cycle signal circuit to monitor the duty cycle of the generator. The generator field duty cycle signal circuit connects to the high side of the field winding in the generator. A pulse width modulated (PWM) high side driver in the voltage regulator turns the field winding on/off. When the key is in "run" position and the engine is off, the PCM should detect a duty cycle near 0%. However, when the engine is running, the duty cycle should be from 5-100%. The PCM monitors the PWM signal using a key "on" test and a "run" test. During the tests, if the PCM detects an out of range PWM signal, DTC P1638 will set. When the DTC sets, the PCM will send a class 2 serial data message to the IPC to illuminate the charge indicator. **Possible Causes** ● Generator connector is damaged or has high resistance ● Generator field duty cycle signal circuit is open or shorted ● Generator is damaged or has failed ● PCM has failed

OBD II Trouble Code List (P1xxx Codes)

DTC	Trouble Code Title, Conditions & Possible Causes
DTC P1638 **1T CCM, MIL: No** 1996, 1997, 1998, 1999 Aurora Engines: 4.0L VIN C Transmissions: All	**Generator 'F' Terminal Circuit Malfunction Conditions:** Engine started, engine speed at 500-3000 rpm for over 60 seconds, and the PCM detected the Generator 'F' terminal PWM signal was 5% or less for 20 seconds during the CCM test. The PCM monitors the field voltage of the generator in order to produce the desired voltage by varying the duty cycle of the field current. At low engine speeds, the field may have a duty cycle of 90%. At higher engine speeds and lower electrical loads, the duty cycle will be less. The generator produces a PWM signal that is monitored by the PCM. **Possible Causes** ● Generator 'F' terminal is open, shorted to ground or to B+ ● Generator is damaged or has failed ● PCM has failed
DTC P1639 **2T CCM, MIL: Yes** 1996, 1997, 1998, 1999, 2000, 2001, 2002, 2003, 2004, 2005 Achieva, Alero, Beretta, Bonneville, Century, LSS, Ciera, Corsica, Cutlass Supreme, Grand Am, Grand Prix, LeSabre, Lumina, Monte Carlo, 88', Regal, Park Avenue, 98', Aztek, Rendezvous, Montana, Silhouette 3.1L VIN M, 3.1L VIN J, 3.4L VIN E, 3.5L VIN H, 3.8L VIN 1, 3.8L VIN K Transmissions: All	**5-Volt Reference 'B' Circuit Malfunction Conditions:** Key on or engine running; and the PCM detected the 5v Reference 'B' circuit that connects to the A/C Refrigerant pressure sensor was less than 3.5v or more than 5.5v for 30 seconds. The PCM provides a 5v VREF circuit to the Air Conditioning (A/C) pressure sensor and the Fuel Tank Pressure (FTP) sensor. These 5-volt reference circuits are independent of each other outside the PCM, but are connected together inside the PCM. Therefore, a circuit condition on one sensor 5v VREF circuit may affect the other sensor 5v reference circuit. The PCM monitors the voltage on the 5v VREF circuit. If the PCM detects that the voltage is out of tolerance, it sets DTC P1639. **Possible Causes** ● A/C refrigerant pressure sensor VREF circuit is open, shorted to ground or shorted to system power ● A/C refrigerant pressure sensor is damaged or PCM has failed ● TSB 61-65-61 contains a repair procedure for this code
DTC P1639 **2T CCM, MIL: Yes** 1999, 2000, 2001, 2002, 2003, 2004, 2005 C/K Cab & Chassis, C/K Truck, G Van, M/L Van, S/T Blazer & S/T Pickup Engines: 4.3L VIN W, 4.3L VIN X, 5.0L VIN M, 5.7L VIN R, 7.4L VIN J Transmissions: All	**5-Volt Reference 2 Circuit Malfunction Conditions:** Key on or engine running; and the PCM detected the 5v Reference No. 2 circuit was out of tolerance for 2 seconds. This circuit connects to the Fuel Tank Pressure (FTP) sensor. **Possible Causes** ● 5v VREF circuit is shorted to sensor ground or chassis ground ● 5v VREF circuit or FTP sensor circuit is shorted to (B+) ● 5v VREF circuit shorted to the FTP sensor signal circuit ● FTP sensor is damaged or has failed ● PCM has failed
DTC P1639 **1T CCM, MIL: Yes** 1999, 2000, 2001, 2002, 2003, 2004, 2005 C/K Truck, G Van, Cargo Van, Envoy & TrailBlazer 4.8L VIN V, 5.3L VIN P, 5.3L VIN T, 5.3L VIN Z, 6.0L VIN N, 6.0L VIN U, 8.1L VIN G Transmissions: All	**5-Volt Reference 2 Circuit Malfunction Conditions:** Key on or engine running; and the PCM detected the 5v Reference No. 2 circuit was out of tolerance for 2 seconds during the CCM test. This circuit is connected to the Fuel Tank Pressure (FTP) and TP sensor. **Possible Causes** ● 5v VREF circuit is shorted to sensor ground or chassis ground ● 5v VREF circuit, FTP or TP sensor circuit is shorted to (B+) ● 5v VREF circuit shorted to FTP or TP sensor signal circuit ● FTP sensor or TP sensor is damaged or PCM has failed
DTC P1639 **1T CCM, MIL: Yes** 1996, 1997, 1998, 1999, 2000, 2001, 2002 C/K Series Truck, G Van 6.5L VIN F, 6.5L VIN S, 6.5L VIN Y, 6.6L VIN 1 Transmissions: All	**5-Volt Reference Circuit Low Input Conditions:** Key on, and the PCM detected the 5v Reference circuit was out of range for 10 seconds during the test. This circuit connects to the APP2, BARO, Boost, and EGR Vacuum sensor. **Possible Causes** ● 5v VREF circuit is shorted to sensor ground or chassis ground ● 5v VREF circuit, Boost or CKP sensor circuit is shorted to (B+) ● 5v VREF circuit shorted to the Boost sensor signal circuit ● EGR solenoid valve (and sensor) is damaged or has failed ● PCM has failed
DTC P1639 **2T CCM, MIL: Yes** 1996, 1997, 1998, 1999, 2000, 2001, 2002, 2003, 2004, 2005 Camaro, Corvette & Firebird 3.8L VIN K, 5.7L VIN G, 5.7L VIN S engines Transmissions: All	**5-Volt Reference 2 Circuit Malfunction Conditions:** Engine started and the PCM detected the 5v VREF 2 circuit was out of range for 10 seconds. The 5v Reference 2 circuit connects to the A/C Pressure and Fuel Tank Pressure sensors. **Possible Causes** ● A/C refrigerant pressure sensor VREF circuit is open or shorted to ground ● A/C refrigerant pressure sensor VREF circuit is shorted to system power ● A/C pressure sensor or FTP sensor is damaged or has failed ● PCM has failed

OBD II Trouble Code List (P1xxx Codes)

DTC	Trouble Code Title, Conditions & Possible Causes
DTC P1639 **2T CCM, MIL: Yes** 1995 Camaro & Firebird 3.8L VIN K engine Transmissions: All	**5-Volt Reference 2 Circuit Malfunction Conditions:** Key on or engine running; and the PCM detected the 5v Reference No. 2 circuit was out of range for 2 seconds. The 5v VREF 2 circuit connects to the A/C Refrigerant pressure sensor. **Possible Causes** ● 5v VREF circuit is shorted to sensor ground or chassis ground ● 5v VREF circuit, A/C pressure sensor circuit is shorted to (B+) ● 5v VREF circuit shorted to A/C pressure sensor signal circuit ● A/C refrigerant pressure sensor is damaged or PCM has failed
DTC P1639 **2T CCM, MIL: Yes** 2000, 2001, 2002, 2003, 2004, 2005 Aurora, DeVille, Eldorado & Seville 4.0L VIN C, 4.6L VIN 9, 4.6L VIN Y engines Transmissions: All	**5-Volt Reference 2 Circuit Malfunction Conditions:** Key on or engine running; and the PCM detected that the voltage range of the 5v VREF 2 circuit was out of tolerance for 10 seconds during the CCM continuous test. The 5v VREF 2 circuit connects to the A/C Pressure and Fuel Tank Pressure (FTP) sensor circuits. **Possible Causes** ● ACP sensor VREF or FTP sensor VREF circuit is open, shorted to ground or to power ● A/C refrigerant pressure or the Fuel Tank pressure sensor is damaged or has failed ● PCM has failed
DTC P1639 **2T CCM, MIL: Yes** 2002, 2003, 2004, 2005 Envoy & TrailBlazer 4.2L VIN S, 5.3L VIN P Transmissions: All	**5-Volt Reference 2 Circuit Malfunction Conditions:** Engine started; and the PCM detected the 5v Reference No. 2 circuit was out of range for 10 seconds. The 5v VREF circuit is connects to the APP1, MAP sensor and the TP sensor 2. **Possible Causes** ● 5v VREF circuit is shorted to sensor ground or chassis ground ● 5v VREF circuit, APP1, MAP or TP sensor circuit shorted to B+ ● 5v VREF circuit shorted to APP1, MAP or FTP signal circuit ● APP1, MAP or TP sensor is damaged or PCM has failed
DTC P1640 **2T CCM, MIL: Yes** 2001, 2002, 2003, 2004, 2005 Alero, Bonneville, Century, Grand Am, Grand Prix, Impala, Intrigue, LeSabre, Lumina, Monte Carlo, Regal, Aztek, Park Avenue, Rendezvous, Montana, Silhouette 3.1L VIN J, 3.4L VIN E, 3.8L VIN 1, 3.8L VIN K	**Output Driver 1 Input Voltage High Input Conditions:** Key on or engine running; and the PCM detected an unexpected high voltage condition (over 33 volts) on the Output Driver 1 circuit. The ODM (modules) are, located inside the PCM, provides grounds for output circuits that control various devices. Each output has an internal feedback circuit that connects to the PCM. The ODM 1 monitors the voltage and current condition on circuits that could cause damage to the PCM. The PCM monitors voltage through the ignition 1 input. It sets this code if it detects a fault on the ODM 1 circuit. **Possible Causes** ● EVAP Vent control circuit open, shorted to ground or to B+ ● Fan relay 1 control circuit is open, shorted to ground or to B+ ● Fan relay 2 or 3 control circuit is open, shorted to ground or B+ ● PCM has failed
DTC P1640 **1T CCM, MIL: No** 1995 Century, Grand Prix, 88', Park Avenue, 98' Regal, Riviera, Lumina APV, Trans Sport, Silhouette 3.8L VIN 1, 3.8L VIN K, 3.8L VIN L engines Transmissions: All	**Output Driver Module 'A' Circuit Malfunction Conditions:** Engine started; brakes pedal not applied, and the PCM detected an unexpected voltage condition on the ODM 'A' fault line for 5 seconds during the CCM test. The controls to the EVAP, Fan Control and TCC Solenoid may not work if this code is set. **Possible Causes** ● EVAP solenoid control circuit open, shorted to ground or to B+ ● Fan relay 2 control circuit is open, shorted to ground or to B+ ● TCC solenoid control circuit is open, shorted to ground or to B+ ● PCM has failed
DTC P1640 **1T CCM, MIL: Yes** 1996, 1997, 1998, 1999 Aurora, DeVille, Eldorado & Seville 4.0L VIN C, 4.6L VIN 9, 4.6L VIN Y engines	**Output Driver 2 Input Voltage High Input Conditions:** Engine started; and the PCM detected an unexpected "high" voltage condition on the ignition feed circuit (Circuit 639) for Output Driver Module 1 circuit during the CCM test. **Possible Causes** ● Generator is over charging - check the generator output ● Generator 'L' terminal is loose (an intermittent fault) ● PCM has failed
DTC P1640 **1T CCM, MIL: No** 1996, 1997 Achieva, Alero, Beretta, Bonneville, Century, LSS, Ciera, Corsica, Cutlass Supreme, Grand Am, Grand Prix, LeSabre, Lumina, Monte Carlo, 88', Regal, Park Avenue, 98' 3.1L VIN M, 3.4L VIN X, 3.4L VIN E, 3.8L VIN 1, 3.8L VIN K engines	**MIL Control Circuit Malfunction Conditions:** Key on, and the PCM detected an unexpected voltage condition on the MIL driver control circuit for 30 seconds during the CCM test. **Possible Causes** ● MIL control circuit is open, shorted to ground or to power (B+) ● MIL (lamp) is damaged or failed (open circuit or missing) ● MIL power circuit is open (check Fuse 39 in U/H fuse block) ● PCM has failed

OBD II Trouble Code List (P1xxx Codes)

DTC	Trouble Code Title, Conditions & Possible Causes
DTC P1641 **1T CCM, MIL: No** 1996, 1997 Camaro, Caprice, Corvette, Firebird, Fleetwood & Roadmaster 4.3L VIN W, 5.7L VIN 5, 5.7L VIN P engines Transmissions: All	**Fan Control Relay 1 Control Circuit Malfunction Conditions:** DTC P0117, P0118 and P1539 not set, engine started, engine speed over 600 rpm, and the PCM detected the commanded state of the FC Relay 1 driver and Actual state did not match for 5 seconds. **Possible Causes** ● Fan relay control circuit is open, shorted to ground or to B+ ● Fan relay power circuit is open (check the FAN fuse) ● Fan relay is damaged or has failed ● PCM has failed
DTC P1641 **1T CCM, MIL: No** 1998, 1999 Achieva, Alero, Bonneville, Century, LSS, Ciera, Cutlass Supreme, Grand Am, Grand Prix, LeSabre, Lumina, Monte Carlo, Regal, 88', 98', Park Avenue, Montana, Silhouette & Venture 3.1L VIN M, 3.4L VIN E, 3.8L VIN 1, 3.8L VIN K Transmissions: All	**A/C Fan Relay Control Circuit Malfunction Conditions:** Key on, and the PCM detected an unexpected voltage condition on the A/C Fan Relay control circuit for 30 seconds. **Possible Causes** ● A/C relay circuit is open, shorted to ground or to power (B+) ● A/C relay is damaged or failed (open circuit or missing) ● A/C relay power circuit is open (test the A/C BFC fuse) ● PCM has failed
DTC P1641 **1T CCM, MIL: No** 1996, 1997, 1998, 1999, 2000, 2001, 2002 C/K Series Truck, G Van 6.5L VIN F, 6.5L VIN S, 6.5L VIN Y Diesel engine Transmissions: All	**MIL Control Circuit Malfunction Conditions:** Engine started; MIL requested on, the PCM detected the voltage on the MIL control circuit was high, or with the MIL requested off, the voltage on the MIL control circuit was 0 volts. **Possible Causes** ● MIL control circuit is open, shorted to ground or to power (B+) ● MIL (lamp) is damaged or failed (open circuit or missing) ● MIL power circuit is open (check GAUGES fuse in fuse block) ● PCM has failed
DTC P1641 **1T CCM, MIL: No** 1996, 1997, 1998, 1999 Aurora, DeVille, Eldorado & Seville 4.0L VIN C, 4.6L VIN 9, 4.6L VIN Y engines Transmissions: All	**MIL Control Circuit Malfunction Conditions:** DTC P1640 not set, engine started, engine runtime 10 seconds, and the PCM detected the MIL fault line was low for 2.5 seconds. **Possible Causes** ● MIL control circuit is open, shorted to ground or to power (B+) ● MIL (lamp) is damaged or failed (located in Instrument Cluster) ● MIL power circuit is in the Instrument cluster ● PCM has failed ● TSB 71-65-26 contains a repair procedure for this code
DTC P1641 **1T CCM, MIL: No** 1995, 1996, 1997 Camaro & Firebird 3.8L VIN K engine Transmissions: All	**MIL Control Circuit Malfunction Conditions:** Key on or engine running; and the PCM received an unexpected voltage condition on the MIL control circuit for 30 seconds. **Possible Causes** ● MIL control circuit is open, shorted to ground or to power (B+) ● MIL (lamp) is damaged or failed (open circuit or missing) ● MIL power circuit is open (check GAUGES fuse in fuse block) ● PCM has failed
DTC P1642 **1T CCM, MIL: No** 1996, 1997 Camaro, Caprice, Corvette, Firebird, Fleetwood & Roadmaster 4.3L VIN W, 5.7L VIN 5, 5.7L VIN P engines Transmissions: All	**Fan Control Relay 2/3 Control Circuit Malfunction Conditions:** DTC P1640 not set, engine started, engine runtime 10 seconds, and the PCM detected the commanded state of the FC Relay 2/3 driver and Actual state did not match for 5 seconds. **Possible Causes** ● Fan relay control circuit is open, shorted to ground or to B+ ● Fan relay power circuit is open (check the FAN fuse) ● Fan relay is damaged or has failed ● PCM has failed
DTC P1642 **1T CCM, MIL: No** 1996, 1997, 1998, 1999 Aurora, DeVille, Eldorado & Seville 4.0L VIN C, 4.6L VIN 9, 4.6L VIN Y engines Transmissions: A/T	**Vehicle Speed Output Signal Circuit Malfunction Conditions:** DTC P1640 not set; engine runtime over 10 seconds and the PCM detected an unexpected "low" voltage on the VSS Output Line for 2.5 seconds during the CCM test. DTC P1642 is used to monitor the fault line of the VSS output signal circuit. The IPC sends a 12v signal on the VSS output circuit to the PCM. The PCM sends a PWM input to the IPC, Cruise Control module and the RSS or CVRSS modules. This signal is used to determine vehicle speed. **Possible Causes** ● VSS output signal circuit is open, shorted to ground or to B+ ● PCM has failed

OBD II Trouble Code List (P1xxx Codes)

DTC	Trouble Code Title, Conditions & Possible Causes
DTC P1642 **2T CCM, MIL: Yes** 1996 Lumina, Monte Carlo, Cutlass Supreme 3.4L VIN X engine Transmissions: All	**Secondary AIR System Control Circuit Malfunction Conditions:** Key on or engine running; and the PCM detected an unexpected voltage condition on the Air Injection Relay driver control circuit for 30 seconds during the CCM test. **Possible Causes** ● AIR relay control circuit is open, shorted to ground or to B+ ● AIR relay power circuit is open (check the ECM IGN fuse) ● Air injection relay is damaged or has failed ● PCM has failed
DTC P1643 **1T CCM, MIL: No** 1996, 1997 Camaro, Caprice, Corvette, Firebird, Fleetwood & Roadmaster 4.3L VIN W, 5.7L VIN 5, 5.7L VIN P engines Transmissions: All	**Engine Speed (RPM) Output Circuit Malfunction Conditions:** Engine started, engine speed over 600 rpm, and the PCM detected the commanded state of Engine Speed signal driver and Actual state of the driver did not match for 25.5 seconds. **Possible Causes** ● Engine speed signal circuit is open, shorted to ground or to B+ ● Instrument Cluster is damaged or has failed ● PCM has failed
DTC P1643 **1T CCM, MIL: Yes** 1996, 1997 Bonneville, Century, LSS, Cutlass Supreme, Grand Prix, LeSabre, Regal, Eighty Eight, Ninety Eight 3.8L VIN 1, 3.8L VIN K Transmissions: All	**Fuel Pump PWM Control Circuit Malfunction Conditions:** Key on, and the PCM detected an unexpected voltage condition on the Fuel Pump PWM control circuit between the Fuel Pump Control module and the PCM for 30 seconds. **Possible Causes** ● F/P PWM signal circuit is open, shorted to ground or to B+ ● Fuel pump control module is damaged or has failed ● PCM has failed
DTC P1643 **1T CCM, MIL: Yes** 1996, 1997, 1998, 1999, 2000, 2001, 2002 C/K Series Truck, G Van 6.5L VIN F, 6.5L VIN S, 6.5L VIN Y Diesel engine Transmissions: All	**Wait To Start Lamp Control Circuit Malfunction Conditions:** Key on or engine running; Wait To Start lamp requested "on" in Test 1, and the PCM detected the voltage on the Wait To Start lamp control circuit was high, or with the lamp requested "off", the voltage on the lamp control circuit was near zero (0) volts during the test. **Possible Causes** ● Wait to start lamp control circuit is open, shorted to ground or shorted to power (B+) ● Wait to start lamp power circuit is open (check GAUGES fuse) ● Wait to start "bulb" is open or is missing ● PCM has failed
DTC P1644 **1T CCM, MIL: No** 1996, 1997, 1998, 1999 Aurora, DeVille, Eldorado & Seville 4.0L VIN C, 4.6L VIN 9, 4.6L VIN Y engines Transmissions: A/T	**Traction Control Delivered Torque Output Circuit Malfunction Conditions:** Key on or engine running in closed loop, brake pedal not applied, and the PCM detected an unexpected voltage condition on the Delivered Torque Output circuit for 10 seconds during the CCM test. The EBCM controls the PWM signal on the Desired Torque circuit while monitoring the wheel speed sensors to detect slippage. The PCM monitors the PWM signal and reduces engine torque as needed. The PCM sends a Delivered Torque PWM signal in response to the Desired Torque PWM signal from EBCM. If the PCM detects a problem on the Delivered Torque circuit, DTC P1644 will set. The EBCM disables traction control when this code sets. An ABS/TCS DTC may also be set at the time that this code sets. **Possible Causes** ● Delivered torque circuit is open, shorted to ground or to B+ ● EBCM is damaged or has failed ● PCM has failed

OBD II Trouble Code List (P0xxx Codes)

DTC	Trouble Code Title, Conditions & Possible Causes
DTC P1645 **1T CCM, MIL: Yes** 1996, 1997, 1998, 1999 Aurora, DeVille, Eldorado & Seville 4.0L VIN C, 4.6L VIN 9, 4.6L VIN Y engines Transmissions: A/T	**EVAP Solenoid Control Circuit Malfunction Conditions:** DTC P1640 not set, engine runtime over 10 seconds, and the PCM detected an unexpected "low" voltage condition on the EVAP Solenoid control circuit for 2 seconds during the test. **Possible Causes** ● EVAP solenoid control circuit is open or shorted to ground ● EVAP solenoid power circuit is open (check the ECS fuse) ● EVAP solenoid is damaged or has failed ● PCM has failed
DTC P1646 **1T CCM, MIL: Yes** 1996, 1997, 1998 Bonneville, LSS, Cutlass Supreme, Grand Prix, LeSabre, LSS, Regal, Park Avenue, 88' & 98' Engines: 3.8L VIN 1 Transmissions: All	**Boost Control Solenoid Circuit Malfunctions Conditions:** Engine started, and the PCM detected an unexpected voltage condition on the Boost Solenoid control circuit for 30 seconds. **Possible Causes** ● Boost control solenoid circuit is open or shorted to ground ● Boost control solenoid circuit is shorted to power ● Boost control solenoid power circuit is open (test AC CLU fuse) ● Boost control solenoid is damaged or has failed ● PCM has failed
DTC P1646 **1T CCM, MIL: Yes** 1998, 1999 Aurora, DeVille, Eldorado & Seville 4.0L VIN C, 4.6L VIN 9, 4.6L VIN Y engines Transmissions: A/T	**EVAP Vent Solenoid Control Circuit Malfunction Conditions:** DTC P1640 not set, engine runtime over 10 seconds, and the PCM detected an unexpected "low" voltage condition on the EVAP Vent Solenoid control circuit for 2 seconds. **Possible Causes** ● EVAP vent solenoid circuit is open or shorted to ground ● EVAP vent solenoid power circuit is open (test SHIFTSOL fuse) ● EVAP vent solenoid is damaged or has failed ● PCM has failed
DTC P1650 **1T CCM, MIL: No** 1995 Century, Grand Prix, 88', Park Avenue, 98' Regal, Riviera, Lumina APV, Trans Sport, Silhouette 3.8L VIN 1, 3.8L VIN K, 3.8L VIN L engines Transmissions: All	**Output Driver Module 'B' Circuit Malfunction Conditions:** Engine started, brake pedal not applied, and the PCM detected an unexpected voltage condition on the ODM 'B' fault line for 5 seconds during the CCM test. The controls to Shift Solenoid 'A' and Shift Solenoid 'B' may not work if this code is set. **Possible Causes** ● EGR solenoid control circuit open, shorted to ground or to B+ ● TCC solenoid control circuit is open, shorted to ground or to B+ ● EGR or TCC solenoid is damaged or has failed ● PCM has failed
DTC P1650 **1T CCM, MIL: Yes** 2001, 2002, 2003, 2004, 2005 Aztek, Century, LeSabre, Impala, Malibu, Montana, Rendezvous, Silhouette & S/T Pickup 2.2L VIN 4, 3.1L VIN J, 3.4L VIN E, 3.8L VIN 1, 3.8L VIN K engines Transmissions: All	**Output Driver 2 Input Voltage High Input Conditions:** Key on or engine running; and the PCM detected an unexpected high voltage condition (over 33 volts) on the Output Driver 2 circuit. The ODM (modules) are, located inside the PCM, provides grounds for output circuits that control various devices. Each output has an internal feedback circuit that connects to the PCM. The ODM 2 monitors the voltage and current condition on circuits that could cause damage to the PCM. The PCM monitors the voltage through the ignition 1 circuit for an incorrect current value to the ODM 2. **Possible Causes** ● EGR solenoid control circuit open, shorted to ground or to B+ ● TCC solenoid control circuit is open, shorted to ground or to B+ ● EGR or TCC solenoid is damaged or has failed ● PCM has failed
DTC P1650 **1T CCM, MIL: Yes** 1996, 1997, 1998, 1999 Aurora, DeVille, Eldorado & Seville 4.0L VIN C, 4.6L VIN 9, 4.6L VIN Y engines Transmissions: A/T	**Output Driver 2 Input Voltage High Input Conditions:** Engine started, and the PCM detected an unexpected high voltage condition on the ignition feed circuit (Circuit 639) for Output Driver Module 2 circuit during the CCM test. **Possible Causes** ● Generator is over charging - check the generator output ● Generator 'L' terminal is loose (an intermittent fault) ● PCM has failed
DTC P1650 **1T CCM, MIL: Yes** 1999, 2000, 2001, 2002 Cavalier & Sunbird Engines: 2.2L VIN 4 Transmissions: All	**Output Driver 2 Input Voltage High Input Conditions:** Key on, system voltage from 6-18v, and the Alternative Fuels ECM detected the Actual and Commanded state of the Output Driver did not match for two seconds during the CCM continuous test. **Possible Causes** ● AF fuel pump relay control circuit is open or shorted to ground ● AF fuel pump relay power circuit is open (test the IGN E fuse) ● AF fuel pump relay is damaged or it has failed ● AF ECM has failed

OBD II Trouble Code List (P0xxx Codes)

DTC	Trouble Code Title, Conditions & Possible Causes
DTC P1651 **2T CCM, MIL: Yes** 1996, 1997, 1998 Achieva, Beretta, Bonneville, Century, Corsica, Cutlass Supreme, Grand Am, Grand Prix, Lumina, Monte Carlo & Regal 3.1L VIN M, 3.4L VIN E, 3.4L VIN X engines Transmissions: All	**Cooling Fan 1 Relay Control Circuit Malfunction Conditions:** Key on or engine running; and the PCM detected an unexpected voltage on the Cooling Fan 1 Relay driver circuit for 30 seconds. **Possible Causes** ● Cooling fan relay 1 circuit is open, shorted to ground or to B+ ● Cooling fan relay 1 power circuit is open (test COOL FAN1 fuse or the related Fusible Link 'G' or MAXI fuse) ● Cooling fan relay 1 is damaged or has failed ● PCM has failed
DTC P1651 **2T CCM, MIL: Yes** 1996, 1997, 1998 Bonneville, LSS, Cutlass Supreme, Grand Prix, LeSabre, LSS, Regal, Park Avenue, 88' & 98' 3.8L VIN 1, 3.8L VIN K Transmissions: All	**Fan 1 Control Relay Control Circuit Malfunction Conditions:** Key on or engine running; and the PCM detected an unexpected voltage condition on the Cooling Fan 1 control circuit for 30 seconds. **Possible Causes** ● Cooling fan relay 1 control circuit is open or shorted to ground ● Cooling fan relay 1 control circuit is shorted to system power ● Cooling fan relay 1 power circuit is open (test COOL FAN1 fuse or the MAXI fuse) ● Cooling fan relay 1 is damaged or has failed ● PCM has failed
DTC P1651 **2T CCM, MIL: Yes** 1995, 1996, 1997, 1998 Camaro & Firebird 3.8L VIN K engine Transmissions: All	**Cooling Fan 1 Relay Control Circuit Malfunctions Conditions:** Key on or engine running; and the PCM detected an unexpected voltage condition on the Cooling Fan Relay 2 control circuit for 30 seconds during the CCM test. **Possible Causes** ● Cooling fan relay 1 control circuit is open, shorted to ground or to system power ● Cooling fan relay 1 power circuit is open (test COOL FAN1 fuse or the MAXI fuse) ● Cooling fan relay 1 is damaged or has failed ● PCM has failed
DTC P1652 **2T CCM, MIL: Yes** 1996, 1997, 1998 Achieva, Beretta, Bonneville, Century, Corsica, Cutlass Supreme, Grand Am, Grand Prix, Lumina, Monte Carlo & Regal 3.1L VIN M, 3.4L VIN E, 3.4L VIN X engines Transmissions: All	**Cooling Fan 2 Relay Control Circuit Malfunction Conditions:** Key on or engine running; and the PCM detected an unexpected voltage condition on the Cooling Fan 2 Relay driver control circuit for 30 seconds during the CCM test. **Possible Causes** ● Cooling fan relay 2 control circuit is open or shorted to ground ● Cooling fan relay 2 control circuit is shorted to system power ● Cooling fan relay 2 power circuit is open (test COOL FAN2 fuse or the related Fusible Link 'G' or MAXI fuse) ● Cooling fan relay 2 is damaged or has failed ● PCM has failed
DTC P1652 **1T CCM, MIL: Yes** 1996, 1997 Caprice, Camaro, Corvette, Firebird, Fleetwood & Roadmaster 4.3L VIN W, 5.7L VIN 5, 5.7L VIN P engines Transmissions: All	**Vehicle Speed Sensor Output Circuit Malfunction Conditions:** Engine speed over 600 rpm, and the PCM detected the Actual state of the VSS Output driver and Commanded state of the control circuit did not match for 25.5 seconds during the test. **Possible Causes** ● VSS output signal circuit is open, shorted to ground or to B+ ● Body Control, Cruise Control, Instrument Cluster, Radio module is damaged (disconnect them one at a time to find the fault) ● PCM has failed
DTC P1652 **1T CCM, MIL: Yes** 1996, 1997, 1998, 1999 DeVille, Eldorado & Seville 4.6L VIN 9, 4.6L VIN Y Transmissions: A/T	**Powertrain Induced Chassis Pitch Output Circuit Malfunction Conditions:** P1650 not set, engine started, engine running, system voltage over 10.0v, and the PCM detected the Actual state and the Commanded state of the output driver module (ODM) control of the Suspension Lift/Drive output circuit did not match for over 10 seconds. **Possible Causes** ● Chassis pitch output circuit is open, shorted to ground or to B+ ● Electronic Suspension Control Module has failed ● PCM has failed ● TSB 71-65-25 contains a repair procedure for this code
DTC P1652 **2T CCM, MIL: Yes** 1997, 1998, 1999, 2000 Corvette 5.7L VIN G engine Transmissions: All	**Powertrain Induced Chassis Pitch Output Circuit Malfunction Conditions:** Engine started, system voltage over 10.0v, and the PCM detected the Commanded and Actual state of the Induced Pitch Output driver control circuit did not match for 5 seconds. **Possible Causes** ● Chassis pitch output circuit is open, shorted to ground or to B+ ● Electronic Suspension Control Module has failed ● PCM has failed

OBD II Trouble Code List (P0xxx Codes)

DTC	Trouble Code Title, Conditions & Possible Causes
DTC P1652 **2T CCM, MIL: Yes** 1995, 1996, 1997, 1998 Camaro & Firebird 3.8L VIN K engine Transmissions: All	**Cooling Fan 2 Relay Control Circuit Malfunctions Conditions:** Key on or engine running; and the PCM detected an unexpected voltage on the Cooling Fan Relay 2 control circuit for 30 seconds. **Possible Causes** ● Cooling fan relay 2 control circuit open, shorted to ground or B+ ● Cooling fan relay 2 power circuit is open (test COOL FAN2 fuse or the MAXI fuse) ● Cooling fan relay 2 is damaged or has failed ● PCM has failed
DTC P1653 **1T CCM, MIL: No** 1996 Fleetwood, Roadmaster 5.7L VIN P engine Transmissions: All	**Oil Level Lamp Control Circuit Malfunction Conditions:** Engine speed over 600 rpm, and the PCM detected the Actual and Commanded state of the Oil Level Lamp driver control circuit did not match for 5 seconds during the CCM test. **Possible Causes** ● Oil level lamp control circuit is open, shorted to ground or to B+ ● Oil level lamp power circuit is open (test the IP INDC fuse) ● Oil level lamp (bulb) is open or missing ● PCM has failed
DTC P1653 **1T CCM, MIL: No** 1996, 1997 Bonneville, LSS, Cutlass Supreme, Grand Prix, LeSabre, LSS, Regal, Park Avenue, 88' & 98' 3.8L VIN 1, 3.8L VIN H Transmissions: All	**TCS Torque Control Circuit Malfunction Conditions:** Key on or engine running; and the PCM detected an unexpected voltage condition on the Traction Control System control circuit for 30 seconds during the CCM test. **Possible Causes** ● Torque control circuit is open, shorted to ground or to B+ ● EBTCM has failed ● PCM has failed
DTC P1653 **2T CCM, MIL: Yes** 1996, 1997, 1998, 1999 C/K Series Truck, G Van 6.5L VIN F, 6.5L VIN S, 6.5L VIN Y Diesel engine Transmissions: All	**EGR Vent Solenoid Control Circuit Malfunction Conditions:** Engine started; Vent solenoid requested "on" during Test 1, and the PCM detected a high voltage on the solenoid control circuit, or with the solenoid "off" it detected a low voltage (0v). **Possible Causes** ● Vent solenoid control circuit is open, shorted to ground or to B+ ● Vent solenoid power circuit is open (check the ENG1 fuse) ● Vent solenoid is damaged, or the PCM has failed
DTC P1653 **2T CCM, MIL: Yes** 1998, 1999 Camaro & Firebird 3.8L VIN K engine Transmissions: All	**Fuel Level Output Control Circuit Malfunction Conditions:** Engine started; and the PCM detected an unexpected voltage on the Fuel Level output control circuit for 30 seconds. **Possible Causes** ● Fuel level control circuit is open, shorted to ground or to system power (B+) ● Instrument panel cluster (IPC) has failed ● PCM has failed
DTC P1653 **2T CCM, MIL: Yes** 1998, 1999 Montana, Silhouette, Trans Sport & Venture 3.8L VIN E engine Transmissions: All	**Fuel Level Output Control Circuit Malfunction Conditions:** Key on or engine running; and the PCM detected an unexpected voltage condition on the Fuel Level output control circuit 30 seconds. **Possible Causes** ● Fuel level control circuit is open, shorted to ground or shorted to system power (B+) ● Instrument panel cluster (IPC) has failed ● PCM has failed
DTC P1653 **1T CCM, MIL: No** 1997 Aurora 4.0L VIN C engine Transmissions: A/T	**A/C Relay Control Circuit Malfunction Conditions:** Key on or engine running; and the PCM detected an unexpected low voltage condition on the A/C Relay control circuit for 2.5 seconds. **Possible Causes** ● A/C relay control circuit is open, shorted to ground or to system power (B+) ● A/C relay power circuit is open (check HVAC RELAY fuse) ● A/C relay is damaged or has failed ● PCM has failed ● TSB 71-65-60 contains a repair procedure for this code
DTC P1654 **1T CCM, MIL: No** 1996, 1997, 1998 Achieva, Beretta, Bonneville, Century, Corsica, Cutlass Supreme, Grand Am, Grand Prix, Lumina, Monte Carlo & Regal 3.1L VIN M, 3.4L VIN X Transmissions: All	**A/C Relay Control Circuit Malfunction Conditions:** Key on or engine running; and the PCM detected an unexpected voltage condition on the A/C Relay control circuit for 20 seconds. **Possible Causes** ● A/C relay control circuit is open, shorted to ground or to B+ ● A/C relay power circuit is open (check the A/C CONT fuse) ● A/C relay is damaged or has failed ● PCM has failed

OBD II Trouble Code List (P0xxx Codes)

DTC	Trouble Code Title, Conditions & Possible Causes
DTC P1654 **1T CCM, MIL: No** 1996, 1997 Bonneville, LSS, Cutlass Supreme, Grand Prix, LeSabre, LSS, Regal, Park Avenue, 88' & 98' 3.8L VIN 1, 3.8L VIN K Transmissions: All	**A/C Relay Control Circuit Malfunction Conditions:** Key on or engine running; and the PCM detected an unexpected voltage condition on the A/C Relay control circuit for 30 seconds during the CCM test. **Possible Causes** ● A/C relay control circuit is open, shorted to ground or to B+ ● A/C relay power circuit is open (check the HVAC RLY fuse) ● A/C relay is damaged or has failed ● PCM has failed
DTC P1654 **1T CCM, MIL: No** 1996, 1997, 1998, 1999, 2000, 2001, 2002 C/K Series Truck, G Van 6.5L VIN F, 6.5L VIN S, 6.5L VIN Y Diesel engine Transmissions: All	**Service Throttle Soon Lamp Control Circuit Malfunction Conditions:** Key on or engine running; and the PCM detected an unexpected voltage condition on the Service Throttle Soon (STS) lamp control circuit during the CCM test. The PCM turns "on" the STS dash lamp when it detects certain problems related to the accelerator pedal position (APP) sensor. When the PCM is commands the STS lamp "on", the voltage potential of the circuit will be low (near 0 volts). When the PCM commands the STS lamp "off", the voltage potential of the circuit will be high (near 12 volts). The primary function of the PCM in this circuit is to supply the ground for the STS lamp. **Possible Causes** ● STS lamp control circuit is open, shorted to ground or to system power (B+) ● STS lamp (bulb) is open or missing ● STS lamp power circuit is open (power is through the IPC) ● PCM has failed
DTC P1654 **1T CCM, MIL: No** 1996, 1997, 1998, 1999 Aurora, DeVille, Eldorado & Seville 4.0L VIN C, 4.6L VIN 9, 4.6L VIN Y engines Transmissions: A/T	**Cruise Inhibit Output Circuit Malfunction Conditions:** DTC P1650 not set, engine runtime 10 seconds, and the PCM detected an unexpected voltage condition on the Cruise Inhibit Output control circuit for 2 seconds during the test. **Possible Causes** ● Cruise inhibit output control circuit is open, shorted to ground or shorted to power (B+) ● Cruise control module has failed ● PCM has failed ● TSB 71-65-25 contains a repair procedure for this code
DTC P1654 **1T CCM, MIL: No** 1995, 1996, 1997 Camaro & Firebird 3.8L VIN K engine Transmissions: All	**A/C Relay Control Circuit Malfunction Conditions:** Key on or engine running; and the PCM detected an unexpected voltage condition on the A/C Relay control circuit for 30 seconds during the CCM test. **Possible Causes** ● A/C relay control circuit is open, shorted to ground or to system power (B+) ● A/C relay power circuit is open (check the AC CRUISE fuse) ● A/C relay is damaged or has failed ● PCM has failed
DTC P1654 **1T CCM, MIL: No** 1996, 1997 Montana, Silhouette, Trans Sport & Venture 3.8L VIN E engine Transmissions: All	**A/C Relay Control Circuit Malfunction Conditions:** Key on or engine running; and the PCM detected an unexpected voltage condition on the A/C Relay control circuit for 30 seconds during the CCM test. **Possible Causes** ● A/C relay control circuit is open, shorted to ground or to system power (B+) ● A/C relay power circuit is open (check the AC CLU fuse) ● A/C relay is damaged or has failed ● PCM has failed
DTC P1655 **2T CCM, MIL: Yes** 1996, 1997, 1998 Achieva, Beretta, Bonneville, Century, Corsica, Cutlass Supreme, Grand Am, Grand Prix, Lumina, Monte Carlo & Regal 3.1L VIN M, 3.4L VIN X Transmissions: All	**EVAP Purge Solenoid Control Circuit Malfunction Conditions:** Key on or engine running; and the PCM received an unexpected voltage condition on the EVAP Purge Solenoid control circuit for 20 seconds during the CCM test. **Possible Causes** ● EVAP solenoid control circuit is open or shorted to ground ● EVAP solenoid control circuit is shorted to system power (B+) ● EVAP solenoid power circuit is open (check the ECM or ENG EMISS fuse in the U/H relay or accessory junction block) ● EVAP solenoid is damaged or has failed ● PCM has failed
DTC P1655 **2T CCM, MIL: Yes** 1996, 1997, 1998 Bonneville, LSS, Cutlass Supreme, Grand Prix, LeSabre, LSS, Regal, Park Avenue, 88' & 98' 3.8L VIN 1, 3.8L VIN K Transmissions: All	**EVAP Purge Solenoid Control Circuit Malfunction Conditions:** Key on or engine running; and the PCM detected an unexpected voltage condition on the EVAP Purge Solenoid control circuit for 30 seconds during the CCM test. **Possible Causes** ● EVAP solenoid control circuit is open, shorted to ground or shorted to power (B+) ● EVAP solenoid power circuit is open (check MISC ENG fuse) ● EVAP solenoid is damaged or has failed ● PCM has failed ● TSB 61-65-41 contains a repair procedure for this code

OBD II Trouble Code List (P0xxx Codes)

DTC	Trouble Code Title, Conditions & Possible Causes
DTC P1655 **2T CCM, MIL: Yes** 1996, 1997, 1998 C/K Series Truck, G Van 6.5L VIN F, 6.5L VIN S, 6.5L VIN Y Diesel engine Transmissions: All	**EGR Control Solenoid Control Circuit Malfunction Conditions:** Engine started, engine running, EGR Control solenoid commanded "on" (Test 1), and the PCM detected the voltage on the EGR solenoid control circuit was near 12v, or with solenoid commanded "off" (Test 2), the voltage on the solenoid control circuit remained near 0v for 2 seconds (fault must occur twice in a row to set a code). **Possible Causes** ● EGR solenoid control circuit is open, shorted to ground or to system power (B+) ● EGR solenoid power circuit is open (check the ENG1 fuse) ● EGR solenoid is damaged or has failed ● PCM has failed
DTC P1655 **2T CCM, MIL: Yes** 1995, 1996, 1997 Camaro & Firebird 3.8L VIN K engine Transmissions: All	**EVAP Purge Solenoid Control Circuit Malfunction Conditions:** Key on or engine running; and the PCM detected an unexpected voltage condition on the EVAP Purge solenoid control circuit for 30 seconds during the CCM test. **Possible Causes** ● EVAP solenoid control circuit is open, shorted to ground or shorted to power (B+) ● EVAP solenoid power circuit is open (check PCM IGN1 fuse) ● EVAP solenoid is damaged or has failed ● PCM has failed
DTC P1655 **2T CCM, MIL: Yes** 1996, 1997 Montana, Silhouette, Trans Sport & Venture 3.8L VIN E engine Transmissions: All	**EVAP Purge Solenoid Control Circuit Malfunction Conditions:** Key on or engine running; and the PCM detected an unexpected voltage condition on EVAP Purge solenoid control circuit for 30 seconds during the CCM test. **Possible Causes** ● EVAP solenoid control circuit is open, shorted to ground or shorted to power (B+) ● EVAP solenoid power circuit is open (check PCM IGN1 fuse) ● EVAP solenoid is damaged or has failed ● PCM has failed
DTC P1656 **2T CCM, MIL: Yes** 1996, 1997, 1998, 1999, 2000, 2001, 2002 C/K Series Truck, G Van 6.5L VIN F, 6.5L VIN S, 6.5L VIN Y Diesel engine Transmissions: All	**Wastegate Solenoid Control Circuit Malfunction Conditions:** Engine started, engine running, Wastegate Solenoid commanded "on" (Test 1), and the PCM detected the voltage on the solenoid control circuit remained near 12v, or with solenoid commanded "off" (Test 2), the voltage on the solenoid control circuit remained near 0v for 2 seconds (fault must occur twice in a row to set the code). **Possible Causes** ● Waste solenoid circuit is open, shorted to ground or to power (B+) ● Wastegate solenoid power circuit is open (check ENG1 fuse) ● Wastegate solenoid is damaged or has failed ● PCM has failed
DTC P1657 **1T CCM, MIL: Yes** 1996, 1997 Caprice, Camaro, Corvette, Firebird, Fleetwood & Roadmaster 4.3L VIN W, 5.7L VIN 5, 5.7L VIN P engines Transmissions: All	**Skip Shift 1-4 Upshift Solenoid Control Circuit Malfunction Conditions:** Engine started, engine speed over 600 rpm, and the PCM detected the Actual state and the Commanded state of the Skip Shift control circuit did not match during the CCM test. **Possible Causes** ● Skip shift solenoid control circuit is open, shorted to ground or shorted to power (B+) ● Skip shift solenoid power circuit is open (check ENG SEN fuse) ● Skip shift solenoid is damaged or has failed ● PCM has failed
DTC P1658 **1T CCM, MIL: Yes** 2002, 2002, 2003, 2004, 2005 C/K Series Pickup 6.6L VIN 1 Diesel engine Transmissions: All	**Fuel Injection Control Module Internal Malfunction Conditions:** U1800, U2104 and U2106 not set; key on or engine running, and the FICM detected current through one of the Fuel Injector Control Module (FICM) internal drivers with that particular driver turned "off". **Possible Causes** ● FICM has failed. There is no external failure that can cause DTC P1658 to set. This code is due to an internal circuit failure (i.e., this code indicates a FICM replacement).
DTC P1660 **1T CCM, MIL: Yes** 2001, 2002, 2003, 2004, 2005 Aztek, Century, Impala, LeSabre, Malibu, Montana, Rendezvous, Silhouette & TrailBlazer 3.1L VIN J, 3.4L VIN E, 3.8L VIN 1, 3.8L VIN K Transmissions: All	**Output Driver 3 Input Voltage High Input Conditions:** Key on or engine running; and the PCM detected an unexpected high voltage condition (over 33 volts) on the Output Driver 3 circuit. The ODM (modules) are, located inside the PCM, provides grounds for output circuits that control various devices. Each output has an internal feedback circuit that connects to the PCM. The ODM 3 monitors the voltage and current condition on circuits that could cause damage to the PCM. The PCM monitors voltage through the ignition 1 input. Any incorrect current detected on a circuit to the ODM 3 will cause the ODM to report this trouble code. **Possible Causes** ● Cooling fan relay control circuit is open or shorted to ground ● Cooling fan relay control circuit is shorted to system power ● Cooling fan relay power circuit is open (check the power fuse) ● Cooling fan relay is damaged or has failed ● PCM has failed

OBD II Trouble Code List (P0xxx Codes)

DTC	Trouble Code Title, Conditions & Possible Causes
DTC P1660 **1T CCM, MIL: Yes** 1996, 1997, 1998, 1999 Aurora, DeVille, Eldorado & Seville 4.0L VIN C, 4.6L VIN 9, 4.6L VIN Y engines Transmissions: All	**Cooling Fan Relay Control Circuit Malfunction Conditions:** Engine started, engine runtime over 10 seconds in closed loop, system voltage over 10.0v, and the PCM detected an unexpected "low" voltage condition on the Cooling Fan Relay control circuit for 2 seconds during the CCM test. **Possible Causes** ● Cooling fan relay control circuit is open or shorted to ground ● Cooling fan relay control circuit is shorted to system power ● Cooling fan relay power circuit is open (check CLNC FAN fuse) ● Cooling fan relay is damaged or has failed ● PCM has failed
DTC P1661 **1T CCM, MIL: Yes** 1996, 1997 Caprice, Camaro, Corvette, Firebird, Fleetwood & Roadmaster 4.3L VIN W, 5.7L VIN 5, 5.7L VIN P engines Transmissions: All	**MIL Control Circuit Malfunction Conditions:** Key on or engine running; and the PCM detected the Actual and Commanded state of the MIL control circuit did not match for over 5 seconds during the CCM test. **Possible Causes** ● MIL control circuit is open, shorted to ground or to power (B+) ● MIL power circuit is open (check the GAUGES fuse) ● MIL (bulb) is open or missing ● PCM has failed
DTC P1662 **1T CCM, MIL: No** 1995, 1996, 1997, 1998 Camaro & Firebird Engines: 3.8L VIN F Transmissions: All	**Cruise Control Inhibit Control Circuit Malfunction Conditions:** Engine started, engine running, and the PCM detected an unexpected voltage condition on the Cruise Control Inhibit circuit for 30 seconds during the CCM test. **Possible Causes** ● Cruise control inhibit circuit is open, shorted to ground or to B+ ● Cruise control module has failed ● PCM has failed
DTC P1662 **1T CCM, MIL: No** 1996, 1997, 1998 Achieva, Beretta, Bonneville, Century, Corsica, Cutlass Supreme, Grand Am, Grand Prix, Lumina, Monte Carlo & Regal 3.1L VIN M, 3.4L VIN X Transmissions: All	**Cruise Control Inhibit Control Circuit Malfunction Conditions:** Engine started, engine running, and the PCM detected an unexpected voltage condition on the Cruise Control Inhibit circuit for 30 seconds during the CCM test. **Possible Causes** ● Cruise control inhibit circuit is open, shorted to ground or to B+ ● Cruise control module has failed ● PCM has failed
DTC P1662 **1T CCM, MIL: No** 1996, 1997, 1998 Bonneville, LSS, Cutlass Supreme, Grand Prix, LeSabre, LSS, Regal, Park Avenue, 88' & 98' 3.8L VIN 1, 3.8L VIN K Transmissions: All	**Cruise Control Inhibit Control Circuit Malfunction Conditions:** Engine started, engine running, and the PCM detected an unexpected voltage condition on the Cruise Control Inhibit circuit for 30 seconds during the CCM test. **Possible Causes** ● Cruise control inhibit circuit is open, shorted to ground or to B+ ● Cruise control module has failed ● PCM has failed
DTC P1662 **1T CCM, MIL: No** 1999, 2000, 2001 Catera 3.0L VIN R engine Transmissions: A/T	**Cruise Lamp Control Circuit Malfunction Conditions:** Engine started, vehicle driven to a speed of over 40 rpm, system voltage over 10.0v, Cruise Control engaged, and the PCM detected an unexpected voltage on the Cruise Lamp driver control circuit. The PCM turns the CRUISE lamp "on" whenever cruise operation is present. The PCM controls the lamp by grounding the control circuit with an internal solid state driver. **Possible Causes** ● Cruise lamp control circuit is open, shorted to ground or to B+ ● Cruise lamp power circuit is open or the bulb is open (in IPC) ● IPC had failed or the PCM has failed
DTC P1663 **1T CCM, MIL: No** 1997, 1998 Camaro & Firebird 3.8L VIN K engine Transmissions: All	**Generator Lamp Control Circuit Malfunction Conditions:** Engine started, engine running, and the PCM detected an unexpected voltage condition on the Generator Lamp control circuit for 30 seconds during the CCM test. **Possible Causes** ● Generator lamp control circuit is open or shorted to ground ● Generator lamp control circuit is shorted to system power (B+) ● Generator lamp power circuit is open or bulb is open (in IPC) ● IPC had failed or the PCM has failed

OBD II Trouble Code List (P0xxx Codes)

DTC	Trouble Code Title, Conditions & Possible Causes
DTC P1663 **1T CCM, MIL: No** 1996 Fleetwood & Roadmaster 5.7L VIN P engine Transmissions: All	**Change Oil Lamp Control Circuit Malfunction Conditions:** Engine speed over 600 rpm, and the PCM detected the Actual and Commanded state of the Change Oil Lamp driver control circuit did not match for over 5 seconds during the CCM test. **Possible Causes** ● Change oil lamp control circuit is open or shorted to ground ● Change oil lamp control circuit is shorted to system power (B+) ● Change oil lamp power circuit is open (test the IP INDC fuse) ● Change oil lamp (bulb) is open or missing (check the IPC) ● PCM has failed
DTC P1664 **1T CCM, MIL: No** 1996, 1997 Caprice, Camaro, Corvette, Firebird 4.3L VIN W, 5.7L VIN 5, 5.7L VIN P engines Transmissions: M/T	**Skip Shift 1-4 Upshift Lamp Control Circuit Malfunction Conditions:** Engine started, engine speed over 600 rpm, and the PCM detected the Actual state and the Commanded state of the Skip Shift Lamp control circuit did not match during the CCM test. **Possible Causes** ● Skip shift lamp control circuit is open, shorted to ground or to system power (B+) ● Skip shift lamp power circuit is open (check the GAGES fuse) ● Skip shift lamp (bulb) is open or missing (check in the IPC) ● PCM has failed
DTC P1665 **1T CCM, MIL: No** 1998, 1999 Camaro & Firebird 3.8L VIN K engine Transmissions: All	**EVAP Vent Solenoid Control Circuit Malfunction Conditions:** Key on or engine running; and the PCM detected an unexpected voltage on the EVAP Vent solenoid control circuit for 30 seconds. **Possible Causes** ● Vent solenoid control circuit is open, shorted to ground or to system power ● Vent solenoid power circuit is open (check the ENG CNTL fuse) ● EVAP vent solenoid is damaged or has failed ● PCM has failed
DTC P1665 **1T CCM, MIL: No** 1998, 1999 Bonneville, LSS, Cutlass Supreme, Grand Prix, LeSabre, LSS, Regal, Park Avenue, 88' & 98' 3.8L VIN 1, 3.8L VIN K Transmissions: All	**EVAP Vent Solenoid Control Circuit Malfunction Conditions:** Key on or engine running; and the PCM detected an unexpected voltage on the EVAP Vent solenoid control circuit for 30 seconds. **Possible Causes** ● Vent solenoid control circuit is open, shorted to ground or to system power ● Vent solenoid power circuit is open (test the ENG EMISS fuse) ● EVAP vent solenoid is damaged or has failed ● PCM has failed ● TSB 71-65-66 contains a repair procedure for this code
DTC P1667 **1T CCM, MIL: No** 1996, 1997 Camaro, Corvette & Firebird 4.3L VIN W, 5.7L VIN 5, 5.7L VIN P engines Transmissions: M/T	**Reverse Inhibit Solenoid Control Circuit Malfunction Conditions:** Engine speed over 600 rpm, and the PCM detected the Actual and Commanded state of the Reverse Inhibit Solenoid driver control circuit did not match for over 5 seconds. **Possible Causes** ● Reverse inhibit solenoid control circuit is open, shorted to ground or to power (B+) ● Reverse solenoid power circuit is open (test ENG SENS fuse) ● Reverse solenoid is damaged or has failed ● PCM has failed
DTC P1670 **1T CCM, MIL: No** 1995 Century, Grand Prix, 88', Park Avenue, 98' Regal, Riviera, Lumina APV, Trans Sport, Silhouette 3.8L VIN 1, 3.8L VIN K, 3.8L VIN L engines Transmissions: All	**Output Driver Module 4 Circuit Malfunction Conditions:** Engine started, engine running, brakes pedal not applied, and the PCM detected an unexpected voltage condition on the ODM 'D' fault line for 5 seconds during the CCM test. The controls to A/C Relay, Cooling Fan 1 and 2, and the MIL will not work with this code set. **Possible Causes** ● A/C relay control circuit open, shorted to ground or to B+ ● Fan 1, Fan 2 control circuit is open, shorted to ground or to B+ ● MIL (lamp) control circuit is open, shorted to ground or to B+ ● A/C Relay, Fan 1, Fan 2 and MIL power circuit(s) are open ● PCM has failed
DTC P1670 **1T CCM, MIL: No** 2001, 2002, 2003, 2004, 2005 Aztek, Century, Impala, LeSabre, Malibu, Montana, Regal, Rendezvous, Silhouette 3.1L VIN J, 3.4L VIN E, 3.8L VIN 1, 3.8L VIN K Transmissions: All	**Output Driver 4 Input Voltage High Input Conditions:** Key on or engine running; and the PCM detected an unexpected high voltage condition (over 33 volts) on the Output Driver 4 circuit. The ODM (modules) are, located inside the PCM, provides grounds for output circuits that control various devices. Each output has an internal feedback circuit that connects to the PCM. The ODM 4 monitors the voltage and current condition on circuits that could cause damage to the PCM. The PCM monitors voltage through the ignition 1 input. Any incorrect current detected on a circuit to the ODM 4 will cause the ODM to report this trouble code. **Possible Causes** ● A/C relay control, Fan 1 or Fan 2 control circuit is open, shorted to ground or to power ● MIL (lamp) control circuit is open, shorted to ground or to B+ ● A/C Relay, Fan 1, Fan 2 and MIL power circuit(s) are open ● Check for a possible battery over-charge condition ● PCM has failed

OBD II Trouble Code List (P0xxx Codes)

DTC	Trouble Code Title, Conditions & Possible Causes
DTC P1671 **1T CCM, MIL: No** 1998, 1999 Camaro & Firebird 3.8L VIN K engine Transmissions: All	**MIL Control Circuit Malfunction Conditions:** Key on or engine running; and the PCM detected the Actual and Commanded state of the MIL control circuit did not match for 30 seconds during the CCM test. **Possible Causes** ● MIL control circuit is open, shorted to ground or to power (B+) ● MIL power circuit is open (check the GAUGES fuse) ● MIL (bulb) is open or missing ● PCM has failed
DTC P1671 **1T CCM, MIL: No** 1998, 1999 Regal & Lumina 3.8L VIN K engine Transmissions: All	**MIL Control Circuit Malfunction Conditions:** Key on or engine running; and the PCM detected the Actual and Commanded state of the MIL control circuit did not match for 30 seconds during the CCM test. **Possible Causes** ● MIL control circuit is open, shorted to ground or to power (B+) ● MIL power circuit is open (check the CLUSTER fuse) ● MIL (bulb) is open or missing ● PCM has failed
DTC P1672 **1T CCM, MIL: No** 1996, 1997, 1998 Achieva, Beretta, Bonneville, Century, Corsica, Cutlass Supreme, Grand Am, Grand Prix, Lumina, Monte Carlo & Regal 3.1L VIN M, 3.4L VIN X, 3.4L VIN E engines Transmissions: All	**Low Oil Level Lamp Control Circuit Malfunction Conditions:** Engine started, engine speed over 600 rpm, and the PCM detected the Actual and Commanded state of the Low Oil Level Lamp driver control circuit did not match for 5 seconds during the CCM test. **Possible Causes** ● Oil level lamp control circuit is open, shorted to ground or to B+ ● Oil level lamp power circuit is open (test Fuse 39 in fuse block) ● Oil level lamp (bulb) is open or missing ● PCM has failed
DTC P1672 **1T CCM, MIL: No** 1998, 1999 Camaro & Firebird 3.8L VIN K engine Transmissions: All	**Low Engine Oil Lamp Control Circuit Malfunction Conditions:** Engine started, engine running, and the PCM detected an unexpected voltage condition on the Low Oil Lamp driver control circuit for 30 seconds during the CCM test. **Possible Causes** ● Oil level lamp control circuit is open, shorted to ground or to B+ ● Oil level lamp power circuit is open (check the GAUGES fuse) ● Oil level lamp (bulb) is open or missing ● PCM has failed
DTC P1673 **1T CCM, MIL: No** 1996 Achieva, Beretta, Century, Ciera, Corsica, Grand Am, Grand Prix, Lumina & Skylark Engines: 3.1L VIN M Transmissions: All	**Engine Hot Lamp Control Circuit Malfunction Conditions:** Key on or engine running; and the PCM detected an unexpected voltage condition on the Engine Hot Lamp control circuit for 30 seconds during the CCM test. **Possible Causes** ● Engine hot lamp control circuit is open or shorted to ground ● Engine hot lamp is shorted to system power (B+) ● Engine hot lamp power circuit is open (test the GAUGES fuse) ● Engine hot lamp (bulb) is open or missing ● PCM has failed
DTC P1673 **1T CCM, MIL: No** 1997, 1998, 1999 Silhouette, Trans Sport & Venture Engines: 3.4L VIN E Transmissions: All	**Engine Hot Lamp Control Circuit Malfunction Conditions:** Key on or engine running; and the PCM detected an unexpected voltage condition on the Engine Hot Lamp control circuit for 30 seconds during the CCM test. **Possible Causes** ● Engine hot lamp control circuit is open, shorted to ground or shorted to power (B+) ● Engine hot lamp power circuit is open (test the GAUGES fuse) ● Engine hot lamp (bulb) is open or missing ● PCM has failed
DTC P1675 **2T CCM, MIL: Yes** 1998, 1999 Bonneville, LSS, Cutlass Supreme, Grand Prix, LeSabre, LSS, Regal, Park Avenue, 88' & 98' 3.8L VIN 1, 3.8L VIN K Transmissions: All	**EVAP Vent Solenoid Control Circuit Malfunction Conditions:** Key on or engine running; and the PCM detected an unexpected voltage condition on the Vent solenoid control circuit for 30 seconds. **Possible Causes** ● Vent solenoid control circuit is open, shorted to ground or shorted to power (B+) ● Vent solenoid power circuit is open (check the IGN1 fuse) ● Vent solenoid is damaged or has failed ● PCM has failed

OBD II Trouble Code List (P0xxx Codes)

DTC	Trouble Code Title, Conditions & Possible Causes
DTC P167 **2T CCM, MIL: Yes** 1998, 1999 Bonneville, LSS, Cutlass Supreme, Grand Prix, LeSabre, LSS, Regal, Park Avenue, 88' & 98' 3.8L VIN 1, 3.8L VIN K Transmissions: All	**EVAP Purge Solenoid Control Circuit Malfunction Conditions:** Key on or engine running; and the PCM detected an unexpected voltage condition on the Purge solenoid control circuit for 30 seconds. **Possible Causes** ● Purge solenoid control circuit is open, shorted to ground or shorted to power (B+) ● Purge solenoid power circuit is open (check ENG EMISS fuse) ● Purge solenoid is damaged, or the PCM has failed ● TSB 71-65-66 contains a repair procedure for this code
DTC P1680 **1T PCM, MIL: Yes** 2001, 2002, 2003, 2004, 2005 C/K Truck, Aurora, DeVille, Eldorado, Seville 4.0L VIN C, 4.6L VIN 9, 4.6L VIN Y, 6.0L VIN N, 6.0L VIN U engines Transmissions: All	**Powertrain Control Module Internal Malfunction Conditions:** Key on, and the PCM detected an internal malfunction had occurred. **Possible Causes** ● Clear the codes and then recheck for this trouble code. If the same code resets, the PCM is damaged and must be replaced (and the new PCM must be programmed) to repair this code.
DTC P1681 **1T PCM, MIL: Yes** 2001, 2002, 2003, 2004, 2005 C/K Truck, Aurora, DeVille, Eldorado, Seville 4.0L VIN C, 4.6L VIN 9, 4.6L VIN Y, 6.0L VIN N, 6.0L VIN U engines Transmissions: All	**Powertrain Control Module Internal Malfunction Conditions:** Key on, and the PCM detected an internal malfunction had occurred. **Possible Causes** ● Clear the codes and then recheck for this trouble code. If the same code resets, the PCM is damaged and must be replaced (and the new PCM must be programmed) to repair this code.
DTC P1682 **2T CCM, MIL: Yes** 2002, 2003, 2004, 2005 Envoy & TrailBlazer 4.2L VIN S engine Transmissions: All	**Ignition 1 Voltage Low Input Conditions:** Engine started; and the PCM detected the Ignition 1 voltage was below 10.0v. The electronic throttle control (ETC) system uses an ignition voltage separate from other power circuits. **Possible Causes** ● Ignition 1 voltage circuit is open, high resistance or it is shorted to ground ● Ignition 1 power supply is open (test ETC1 fuse in fuse block) ● PCM has failed
DTC P1683 **1T PCM, MIL: Yes** 2001, 2002, 2003, 2004, 2005 C/K Truck, Aurora, DeVille, Eldorado, Seville 4.0L VIN C, 4.6L VIN 9, 4.6L VIN Y, 6.0L VIN N, 6.0L VIN U engines Transmissions: All	**Powertrain Control Module Internal Malfunction Conditions:** Key on, and the PCM detected an internal malfunction had occurred. **Possible Causes** ● Clear the codes and then recheck for this trouble code. If the same code resets, the PCM is damaged and must be replaced (and the new PCM must be programmed) to repair this code.
DTC P1689 **1T CCM, MIL: No** 2000 Aztek, Century, Impala, LeSabre, Malibu, Montana, Regal, Rendezvous, Silhouette 3.4L VIN E, 3.8L VIN 1, 3.8L VIN K engines Transmissions: All	**Traction Control Delivered Torque Output Circuit Malfunction Conditions:** Engine started; and the PCM detected an unexpected voltage on the Traction Control delivered torque output circuit for 30 seconds. The EBTCM controls the PWM signal on the Desired Torque circuit while monitoring the wheel speed sensors to detect slippage. The PCM monitors the PWM signal and reduces engine torque as needed by retarding ignition timing, decreasing boost duty cycle, increasing the air/fuel ratio, or, in severe cases, shutting "off" up to three fuel injectors. The PCM sends a signal to the EBTCM on the delivered torque circuit informing the EBTCM of response made to the desired torque signal. **Possible Causes** ● Delivered torque circuit is open, shorted to ground or to B+ ● EBTCM as failed, or the PCM has failed ● TSB 00-06-04-023 contains a repair procedure for this code
DTC P1689 **1T CCM, MIL: No** 1999, 2000 Corvette 5.7L VIN G engine Transmissions: All	**Traction Control Delivered Torque Output Circuit Malfunction Conditions:** Engine speed over 500 rpm for 20 seconds, and the PCM detected an unexpected voltage on the Traction Control Delivered Torque Output circuit for 3 seconds. The EBTCM supplies 12v on the Delivered Torque circuit to the PCM. The PCM toggles this 12v signal to a ground to produce a duty cycle signal proportional to the amount of engine output torque reduction. In order to reduce engine output torque, the PCM reduces the amount of spark advance, and in some cases, the PCM disables up to three fuel injectors. **Possible Causes** ● Delivered torque circuit is open, shorted to ground or to B+ ● EBTCM is damaged or has failed ● PCM has failed

OBD II Trouble Code List (P0xxx Codes)

DTC	Trouble Code Title, Conditions & Possible Causes
DTC P1700 **1T CCM, MIL: Yes** 1997, 1998, 1999, 2000, 2001 Catera 3.0L VIN R engine Transmissions: A/T	**TCM Requested Malfunction Indicator Lamp (4L30-E) Conditions:** Key on or engine running; and the PCM detected a signal from the TCM that indicated an emission related fault had been detected. The TCM flashes the Sport Mode Lamp when a fault occurs on certain non-emissions related TCM systems. The TCM has no direct control of the MIL. However if a transmission fault occurs that is emissions related, the MIL must illuminate. The TCM transmits a MIL request signal on the CAN network to the PCM and the PCM turns "on" the MIL. In effect, the MIL is turned "on" even with a TCM related fault. **Possible Causes** ● Perform the Powertrain Onboard Diagnostic System Check ● Perform the Automatic Transmission System Check
DTC P1701 **1T CCM, MIL: Yes** 1997, 1998 Catera 3.0L VIN R engine Transmissions: A/T	**MIL Request Circuit Malfunction (4L30-E) Conditions:** Engine started; and the PCM detected an unexpected voltage on the Mil Request circuit for 2.5 seconds. This code indicates that a transmission related OBD II failure has occurred. The TCM controls a dedicated Service Transmission Soon (STS) lamp that illuminates when a failure occurs with certain non-emissions related TCM diagnostics. The TCM has no direct control of the engine MIL, but if a transmission fault occurs that is emissions related, the engine MIL must illuminate. A MIL request circuit between the PCM and the TCM provides a means for illuminating the MIL, even though the fault was detected by the TCM. This circuit is pulled up to system voltage (B+) within the PCM. In order to illuminate the MIL, the TCM pulls the circuit "low". The PCM detects this fact and sets DTC P1700, which turns "on" the MIL. DTC P1701 indicates the PCM detected a problem with the MIL request circuit. **Possible Causes** ● MIL request circuit is open, shorted to ground or shorted to B+ ● TCM has failed ● PCM has failed
DTC P1705 **1T CCM, MIL: No** 1997, 1998 Catera 3.0L VIN R engine Transmissions: All	**Park Neutral Output Circuit Malfunction (4L30-E) Conditions:** Key on or engine running; and the PCM detected the Actual and Commanded state of the Park Neutral Output signal did not match. **Possible Causes** ● P/N signal circuit is open, shorted to ground or shorted to B+ ● TCM has failed ● PCM has failed
DTC P1740 **1T CCM, MIL: No** 1997, 1998, 1999, 2000, 2001 Catera 3.0L VIN R engine Transmissions: All**1740.bmp**	**Torque Control Circuit Malfunction (4L30-E) Conditions:** Key on or engine running; and the PCM detected that a request for engine torque reduction had failed. During shift events, the TCM sends requests to the PCM o reduce the engine torque to improve shift feel. The PCM reduces engine torque by retarding the spark timing in response to the TCM request. The TCM request is sent to the PCM through the controller area network (CAN). Two circuits are used to communicate CAN data between the PCM and TCM. A fault in the CAN will not cause DTC P1740 to set by itself. If a CAN fault occurs, other codes will set before DTC P1740. When the PCM detects the request for torque reduction failed, it sends a message to the TCM. When the TCM receives a torque reduction failure message from the PCM, then DTC P1740 will set. **Possible Causes** ● Torque control circuit is open, shorted to ground or to power ● EBTCM has failed ● TCM flashes the Sports Lamp when this code is set ● TCM has failed or the PCM has failed
DTC P1743 **1T CCM, MIL: No** 1997, 1998, 1999 Catera 3.0L VIN R engine Transmissions: All	**Throttle Position Signal Invalid Data Received Conditions:** Key on or engine running; and the TCM detected a throttle position pulsewidth signal that was more than 96% during the CCM test. The TCM receives throttle position data from the PCM. This data is part of the PWM PCM data signal to the PCM. This signal is processed in the PCM. Several parameter values, contained within the PWM signal, are separated out during processing. If the entire PWM signal is not received due to a circuit fault, DTC P1890 is set. If the TCM receives the signal and the TP value is invalid, it will set this code. **Possible Causes** ● Perform the Powertrain Onboard Diagnostics System Check ● Refer to the repair information for DTC P0120
DTC P1760 **1T CCM, MIL: No** 1999, 2000, 2001 Catera 3.0L VIN R engine Transmissions: All	**TCM Supply Voltage Interrupted (4L30-E) Conditions:** Key on or engine running; and the TCM detected a Keep Alive Memory circuit was interrupted. Circuits A828 and X837 are the power supply feeds for the TCM Keep Alive Memory. These fused circuits supply constant voltage on terminals 26 and 54 of the TCM. The TCM sets this code if detects that this circuit(s) is interrupted. **Possible Causes** ● Perform the Powertrain Onboard Diagnostics System Check ● Refer to the repair information for DTC P0120 ● TCM flashes the Sports Lamp when this code is set

OBD II Trouble Code List (P0xxx Codes)

DTC	Trouble Code Title, Conditions & Possible Causes
DTC P1780 **2T CCM, MIL: Yes** 2003, 2004, 2005 CTS (Cadillac) 2.6L VIN M, 3.2L VIN N Transmissions: All	**Engine Torque Signal Circuit Malfunction Conditions:** Engine started, system voltage at 8-18v, no other CAN codes set, and the PCM notified the TCM that a torque reduction request had failed for 2 seconds. To improve shift feel, the TCM may request that the PCM reduce engine torque during shift events. When such a request is received, the PCM responds by retarding the base ignition timing and notifying the TCM that the request has succeeded. If the PCM is unable to comply with the request, the PCM sends the TCM a message that the request has failed. The torque reduction request is sent to the PCM through the CAN network. Two circuits are used to communicate CAN data between the PCM and TCM. A fault in the CAN will not cause DTC P1780 to set by itself. If a CAN fault occurs, other codes will set before DTC P1780. When the TCM receives a torque reduction failure message from the PCM, DTC P1780 is set. **Possible Causes** ● Perform the Diagnostic System Check for the Engine Controls ● TCM needs to be replaced
DTC P1781 **1T CCM, MIL: Yes** 1999, 2000, 2001 Catera 3.0L VIN R engine Transmissions: All	**Engine Torque Signal Circuit (4L30-E) Conditions:** Key on or engine running; and the PCM detected the engine torque signal was below the lower or over the upper limit, or that the torque signal was unreliable for one second. The Engine Torque signal is sent to the TCM by the PCM via the controller area network (CAN). Circuits XR888 and XY888 are used to communicate CAN data between the ECM and TCM. **Possible Causes** ● Engine torque signal circuit is open, shorted to ground or to B+ ● EBTCM has failed ● TCM flashes the Sports Lamp when this code is set ● TCM has failed
DTC P1791 **2T CCM, MIL: Yes** 2003, 2004, 2005 CTS (Cadillac) 2.6L VIN M, 3.2L VIN N Transmissions: All	**Accelerator Pedal Position Information Not Received Conditions:** Engine started, system voltage at 8-18v, no other CAN codes set, and the TCM receives no valid accelerator pedal data from the PCM for 2 seconds. The PCM sends accelerator pedal position data to the TCM. The TCM uses this data to modify the shift speeds. The data is sent to the TCM through the CAN (network). Two circuits are used to communicate CAN data between the PCM and TCM. A fault in the CAN will not cause DTC P1791 to set by itself. If a CAN fault occurs, other codes will set before DTC P1791. If the TCM receives invalid accelerator pedal data from the ECM, DTC P1791 is set. **Possible Causes** ● Perform the Diagnostic System Check for the Engine Controls ● TCM may need to be replaced to repair this trouble code
DTC P1792 **1T CCM, MIL: No** 2003, 2004, 2005 CTS (Cadillac) 2.6L VIN M, 3.2L VIN N Transmissions: All	**Engine Coolant Temperature Invalid Data Received Conditions:** Engine started; system voltage at 8-18v, no other CAN codes set, and the TCM received invalid engine coolant temperature data from the PCM for 2 seconds. The PCM sends engine coolant temperature data to the TCM. The TCM uses this data to initiate warm-up shift patterns and to establish default TFT sensor values. The data is sent to the TCM via the CAN (network). Two circuits are used to communicate the data between the PCM and TCM. A fault in the CAN will not cause DTC P1792 to set. If a CAN fault occurs, other codes will set before DTC P1792. The TCM sets P1792 if it receives invalid engine temperature data. **Possible Causes** ● Perform the Diagnostic System Check for the Engine Controls ● TCM may need to be replaced to repair this trouble code
DTC P1792 **1T CCM, MIL: No** 1999, 2000, 2001 Catera 3.0L VIN R engine Transmissions: All	**ECM to TCM Engine Coolant Signal (4L30-E) Conditions:** Key on or engine running; and the PCM detected the ECT sensor signal was below the lower or over the upper limit, or that the ECT signal was unreliable for one second. The engine temperature signal is sent to the TCM by the PCM via a controller area network (CAN). Circuits XR888 and XY888 are used to communicate CAN data between the PCM and TCM. **Possible Causes** ● Perform the Onboard Diagnostic System Check ● Read and record any the codes. For multiple code situations - repair this code (P1792) before repairing any other codes.
DTC P1793 **1T CCM, MIL: No** 2003, 2004, 2005 CTS (Cadillac) 2.6L VIN M, 3.2L VIN N Transmissions: All	**Wheel Speed Sensor Invalid Data Received Conditions:** Engine started; system voltage at 8-18v, no other CAN codes set, and the TCM received invalid wheel speed sensor data from the PCM for 2 seconds. The PCM sends engine coolant temperature data to the TCM. The TCM uses this data to initiate warm-up shift patterns and to establish default transmission fluid temperature (TFT) values. The data is sent to the TCM through the CAN (network). Two circuits are used to communicate CAN data between the PCM and TCM. A fault in the CAN will not cause DTC P1792 to set by itself. If a CAN fault occurs, other codes will set before DTC P1792. The TCM sets this code if it receives invalid engine coolant temperature data. **Possible Causes** ● Perform the Diagnostic System Check for the Engine Controls ● TCM may need to be replaced to repair this trouble code

OBD II Trouble Code List (P0xxx Codes)

DTC	Trouble Code Title, Conditions & Possible Causes
DTC P1795 **2T CCM, MIL: Yes** 2003, 2004, 2005 CTS (Cadillac) 2.6L VIN M, 3.2L VIN N Transmissions: All	**Throttle Plate Position Invalid Data Received Conditions:** Engine started, system voltage at 8-18v, no other CAN codes set, and the TCM received invalid throttle plate position data from the PCM for 2 seconds. The PCM sends throttle plate position data to the TCM. The data is sent to the TCM via the CAN. Two circuits are used to communicate CAN data between the PCM and TCM. A fault in the CAN will not cause DTC P1795 to set by itself. If a CAN fault occurs, other codes will set before DTC P1795. The TCM sets this code if it receives invalid throttle plate position data from the PCM. **Possible Causes** ● Perform the Diagnostic System Check for the Engine Controls ● TCM may need to be replaced to repair this trouble code
DTC P1800 **21T CCM, MIL: Yes** 1999, 2000, 2001 Catera 3.0L VIN R engine Transmissions: A/T	**Transmission Power Supply Relay Malfunction (4L30-E) Conditions:** DTC P0743, P0753, P0758 and P1850 not set, key on or engine running and the TCM detected the power supply to TCM terminals 52 and 53 was interrupted. The Shift solenoids, the Band-apply solenoid, and TCC solenoid receive their power through the internal transmission relay. The TCM has fault detection circuitry for the power supply circuit that powers all of the solenoids in the transmission. If the TCM detects a problem in this circuit, DTC P1800 will set, and shut off the transmission power relay. Any time a TCM DTC is set that requires the transmission to operate in default mode, the TCM turns off the power relay, effectively disabling all of the transmission components. **Possible Causes** ● TCM flashes the Sports Lamp when this code is set ● Transmission power supply relay is open, shorted to ground or shorted to power (B+)
DTC P1810 **2T CCM, MIL: Yes** 1996, 1997, 1998, 1999, 2000, 2001, 2002, 2003, 2004, 2005 Aurora, Bonneville, Century, Cutlass Supreme, Grand Prix, Intrigue, LeSabre, LSS, Lumina, Monte Carlo, 88', Regal, Park Avenue & 98' 3.1L VIN M, 3.4L VIN X, 3.8L VIN 1, 3.8L VIN K Transmissions: A/T	**Transmission Pressure Switch Manifold Circuit Malfunction Conditions:** DTC P0500, P0502, P0503 and P0560 not set, engine running, and the PCM detected an illegal TR switch combination for 4 seconds; or after the engine speed and vehicle speed met certain parameters, it detected the TR switch input of D2, D4 or Reverse for 2 seconds; or with the detected gear range of P/N, the selected gear range at D4 with the TCC engaged for 2 seconds, the PCM detected the transmission speed ratio value was 29-33. **Possible Causes** ● TFP valve position switch signal circuit is open or grounded ● TFP valve position switch circuit shorted to another signal ● TFP valve position switch is damaged or has failed ● PCM has failed
DTC P1810 **2T CCM, MIL: Yes** 1995, 1996, 1997, 1998, 1999, 2000, 2001, 2002 Camaro & Firebird 3.8L VIN K engine Transmissions: A/T	**Transmission Pressure Switch Malfunction (4L60-E) Conditions:** DTC P0502 and P0503 not set, engine started, and the PCM detected an illegal TR switch combination for 4 seconds; or with the VSS within specific values, it detected the TR switch indicated D2, D4 or Reverse for 2 seconds; or with the selected gear range of D4 and TCC enabled for 2 seconds, the PCM detected the speed ratio value was from 29-33. **Possible Causes** ● Inspect the transmission linkage (gear select lever to manual valve) adjustment ● TFP valve position switch signal circuit is open or grounded ● TFP valve position switch circuit shorted to another signal ● TFP valve position switch is damaged or has failed ● PCM has failed
DTC P1810 **2T CCM, MIL: Yes** 1996, 1997, 1998, 1999, 2000, 2001, 2002, 2003, 2004, 2005 C/K Series Truck, G Van, Cargo Van, L/M Van, S/T Blazer & S/T Pickup 4.3L VIN W, 4.3L VIN X, 4.8L VIN V, 5.0L VIN M, 5.3L VIN T, 5.3L VIN Z, 5.7L VIN K, 5.7L VIN R, 6.0L VIN N, 6.0L VIN U, 6.6L VIN 1, 7.4L VIN J, 8.1L VIN G engines Transmissions: A/T	**TFP Valve Position Switch Assembly (4L60-E, 4L65-E, 4L80-E) Conditions:** DTC P0502 and P0503 not set, system voltage over 10.0v, engine running for 5 seconds, Fuel Cutoff inactive, engine torque from 40-400 ft-lbs, engine vacuum from 0-105 kPa, then during Condition 1 the PCM detected an illegal TFP manual valve position switch state for 60 seconds; or during Condition 2 with the engine speed less than 80 rpm for 0.1 second, then the engine speed from 80-550 rpm for 100 ms, then the engine speed was greater than 550 rpm; then the vehicle speed was less than 2 mph, and the PCM detected the gear range was D2, D4 or Reverse during startup for 5 seconds; or during Condition 3 with the TP angle from 10-50%, fourth gear commanded "on", TCC engaged, speed ratio from 0.6-0.75, and the PCM detected the gear range indicated Park or Neutral with the vehicle is operating in D4 for 10 seconds. The TFP manual valve position switch assembly cannot distinguish between P/N because the monitored valve body pressures are identical in both cases. **Possible Causes** ● TFP valve position switch signal circuit is open, grounded or shorted to another signal ● TFP valve position switch is damaged or has failed ● This code can set during fluid refilling. After refilling the fluid, cycle the key "off", then idle the engine for 20 seconds. Turn the key "off" and allow the PCM to power down. ● This code can set due to low pump pressure or due to a stuck pressure regulator. ● This code can set due to a rolled forward clutch piston seal. It may allow the PCM to see a 2.08:1 ratio (reverse) when the manual valve position is actually indicated in D4. ● PCM has failed

OBD II Trouble Code List (P0xxx Codes)

DTC	Trouble Code Title, Conditions & Possible Causes
DTC P1810 **2T CCM, MIL: Yes** 1996, 1997, 1998, 1999, 2000, 2001, 2002 Aurora, DeVille, Eldorado & Seville 4.0L VIN C, 4.6L VIN 9, 4.6L VIN Y engines Transmissions: A/T	**TFP Valve Position Switch Malfunction (4T80-E) Conditions:** DTC P0101, P0102, P0103, P0121, P0122, P0123, P0502, P0503, P0716 and P0717 not set, engine started, engine runtime over 5 seconds, system voltage over 10.0v, engine runtime over 5 seconds, Fuel Cutoff inactive, then during Condition 1 the PCM detected an invalid TFP manual valve position switch combination of ON/ON/ON or ON/OFF/ON existed for 4 minutes and 10 seconds; or during Condition 2 with the vehicle speed over 5 mph, TP angle over 11% and the delivered torque at 80-200 lb ft, it detected the TFP switch indicated P/N and the gear ratio indicated Reverse or Drive for 5 seconds; or during Condition 3 with the vehicle speed over 5 mph, TP angle over 11%, delivered torque from 80-200 lb ft, then the TFP switch indicated REVERSE and the gear ratio indicated DRIVE for 7 seconds; then with the vehicle speed over 5 mph, TP angle over 7%, delivered torque from 30-150 lb ft, it detected the TFP switch indicated D4, D3, D2 or D1 and the ratio indicated Reverse; or with the engine "off", it detected the TFP manual switch indicated D2 initially and then did not indicate P/N within 5 seconds after engine startup (test stops if the engine is running for 5 seconds, vehicle speed is below 5 mph and the TFT sensor indicates more than 0ºF). **Possible Causes** ● Inspect the transmission linkage from the gear select lever to the manual valve for proper adjustment ● TFP valve position switch signal circuit is open or grounded ● TFP valve position switch circuit shorted to another signal ● TFP valve position switch is damaged or has failed ● PCM has failed
DTC P1810 **2T CCM, MIL: Yes** 1997, 1998, 1999, 2000, 2001, 2002, 2003, 2004, 2005 Cavalier, Sunbird & S/T Pickup 2.2L VIN 4, 2.4L VIN T Transmissions: A/T	**TFP Valve Position Switch Malfunction (4T40-E/4T45-E) Conditions:** DTC P0502 and P0503 not set, engine started, engine running, system voltage over 10.0v, then during Condition 1 the PCM detected an invalid gear range for 60 seconds; or with the engine speed from 0-600 rpm with the vehicle speed less than 2 mph, it detected a gear position of D2, D4 or Reverse gear after startup for 2 seconds; or with P0121, P0122, P0123, P0502, P0503, P0716, P0717, P0751, P0753, P0756 and P0758 not set, TP angle over 10%, vehicle speed over 5 mph, engine torque more than 10 lb ft., it detected the TFP manual valve position switch indicated P/N with the gear ratio less than 0.72:1 (4th gear), or it indicated Reverse with the gear ratio more than 2.23:1, or less than 2.02:1; or the switch indicated D4, D3, D2 or D1 when the gear ratio indicated Reverse. **Possible Causes** ● Inspect the transmission linkage from the gear select lever to the manual valve for proper adjustment ● TFP valve position switch signal circuit is open or grounded ● TFP valve position switch circuit shorted to another signal ● TFP valve position switch is damaged or has failed ● PCM has failed
DTC P1810 **2T CCM, MIL: Yes** 2002, 2003, 2004, 2005 Envoy & TrailBlazer 4.2L VIN S engine Transmissions: A/T	**TFP Valve Position Switch Assembly (4L60-E) Conditions:** DTC P0502 and P0503 not set, engine runtime over 5 seconds, system voltage over 10.0v, engine not in Fuel Cutoff mode, then during Condition 1: The PCM detected an invalid TFP manual valve position switch state for 60 seconds; or during Condition 2: With the engine speed less than 80 rpm for 0.1 second; then with the engine speed from 80-550 rpm for 0.07 second; then with the engine speed over 550 rpm, vehicle speed less than 2 mph, and the PCM detected the gear range indicated D2, D4 or Reverse during engine startup, all conditions met for 5 seconds; or during Condition 3: With the TP angle from 10-50%, the PCM commanded 4th gear on with the TCC locked "on", speed ratio from 0.6-0.75 (speed ratio is engine speed divided by transmission output speed), and the PCM detected the gear range was Park or Neutral with the vehicle operating in D4 range, all conditions met for 10 seconds. **Possible Causes** ● Check for sediment in the valve body as it can cause improper operation of the TFP manual valve position switch. If sediment intrusion is suspected, clean the valve body and replace the TFP manual valve position switch. ● TFP valve position switch signal circuit is open or grounded ● TFP valve position switch circuit shorted to another signal ● TFP valve position switch is damaged or has failed ● PCM has failed

OBD II Trouble Code List (P0xxx Codes)

DTC	Trouble Code Title, Conditions & Possible Causes
DTC P1810 **2T CCM, MIL: Yes** 2003, 2004, 2005 Envoy & TrailBlazer 5.3L VIN P engine Transmissions: A/T	**TFP Valve Position Switch Assembly (4L65-E) Conditions:** DTC P0502 and P0503 not set, engine runtime over 5 seconds, system voltage over 10.0v, engine not in Fuel Cutoff mode, engine torque from 40-400 lb ft., engine vacuum from 0-105 kPa, then during Condition 1: The PCM detected an invalid TFP manual valve position switch state for 60 seconds; or during Condition 2: With the engine speed less than 80 rpm for 0.1 second; then with the engine speed from 80-550 rpm for 0.07 second; then with the engine speed over 550 rpm, vehicle speed was less than 2 mph, and the PCM detected the gear range indicated D2, D4 or Reverse during engine startup, all conditions met for 5 seconds; or during Condition 3: With the TP angle from 10-50%, the PCM commanded 4th gear on with the TCC locked "on", speed ratio from 0.6-0.75 (speed ratio is engine speed divided by transmission output speed), and the PCM detected the gear range indicated Park or Neutral with the vehicle operating in D4 range, all conditions met for 10 seconds. **Possible Causes** ● Check for sediment in the valve body as it can cause improper operation of the TFP manual valve position switch. If sediment intrusion is suspected, clean the valve body and replace the TFP manual valve position switch. ● TFP valve position switch signal circuit is open or grounded ● TFP valve position switch circuit shorted to another signal ● TFP valve position switch is damaged or has failed ● PCM has failed
DTC P1810 **2T CCM, MIL: Yes** 1997, 1998, 1999, 2000, 2001, 2002, 2003, 2004, 2005 Corvette 5.7L VIN G engine, 5.7L VIN S Transmissions: A/T	**TFP Valve Position Switch Malfunction (4L60-E) Conditions:** DTC P0502 and P0503 not set, system voltage over 10.0v, engine runtime over 5 seconds, Fuel Cutoff inactive, engine torque from 40-400 ft-lbs, engine vacuum from 0-105 kPa, then during Condition 1 the PCM detected an illegal TFP manual valve position switch state for 60 seconds; or during Condition 2 with the engine speed less than 80 rpm for 0.1 second, then the engine speed from 80-550 rpm for 100 ms, then the engine speed was greater than 550 rpm; then the vehicle speed was less than 2 mph, and the PCM detected the gear range was D2, D4 or Reverse during startup for 5 seconds; or during Condition 3 with the TP angle from 10-50%, fourth gear commanded "on", TCC engaged, speed ratio from 0.6-0.75, and the PCM detected the gear range indicated Park or Neutral with the vehicle is operating in D4 for 10 seconds. **Possible Causes** ● TFP valve position switch signal circuit is open or grounded ● TFP valve position switch circuit shorted to another signal ● TFP valve position switch is damaged or has failed ● PCM has failed
DTC P1810 **2T CCM, MIL: Yes** 1999, 2000, 2001, 2002, 2003, 2004, 2005 Aurora, Alero, Bonneville, Century, LSS, Cutlass Supreme, Grand Prix, LeSabre, LSS, Regal, Park Avenue, 88' & 98' 3.1L VIN J, 3.1L VIN M, 3.4L VIN E, 3.5L VIN H, 3.8L VIN 1, 3.8L VIN K Transmissions: All	**TFP Valve Position Switch Assembly (4T65-E) Conditions:** DTC P0502 and P0503 not set, engine started, engine speed over 500 rpm for 5 seconds, system voltage over 10.0v, Fuel Cutoff inactive, then during Condition 1 the PCM detected an invalid TFP manual valve position switch state for 60 seconds; or during Condition 2 with the engine speed under 10 mph, it detected the TFP switch indicated D2, D4 or Reverse for 7 seconds at startup; or during Condition 3 with DTC P0121, P0122, P0123, P0502, P0503, P0716, P0717, P0751, P0753, P0756 and P0758 not set, TP angle over 9%, vehicle speed over 5 mph, engine torque over 50 lb ft, it detected the gear ratio indicated Reverse, D4, D3, D2 or D1 and the TFP manual position switch indicated Park or Neutral for 5 seconds, or it detected the gear ratio indicated D4, D3, D2 or D1 and the TFP manual valve position switch indicated Reverse for 7 seconds, or it detected the gear ratio indicated Reverse and the TFP manual valve position switch indicated D4, D3, D2 or D1 for 5 seconds. **Possible Causes** ● Inspect the transmission linkage from the gear select lever to the manual valve for proper adjustment ● TFP valve position switch signal circuit is open or grounded ● TFP valve position switch circuit shorted to another signal ● TFP valve position switch is damaged or has failed ● PCM has failed

OBD II Trouble Code List (P0xxx Codes)

DTC	Trouble Code Title, Conditions & Possible Causes
DTC P1811 **1T CCM, MIL: No** 1996, 1997, 1998, 1999, 2000, 2001, 2002, 2003, 2004, 2005 Aurora, DeVille, Eldorado & Seville 4.0L VIN C, 4.6L VIN 9, 4.6L VIN Y engines Transmissions: All	**Maximum Adaptive and Long Term Shift (4T80-E) Conditions:** Engine started, engine running and after an adaptable shift occurred, the PCM detected during Condition 1 the 1-2 Shift required more than 0.65 seconds and 1-2 Shift adaptive reached its limit; or during Condition 2, it detected the 2-3 Shift required more than 0.65 seconds and the 2-3 Shift adaptive reached its limit; or during Condition 3, it detected the 3-4 Shift required more than 0.65 seconds and the 3-4 Shift adaptive reached its limit (fault must occur twice per vehicle trip to set the trouble code). **Possible Causes** ● Ask the customer about overloading the vehicle, exceeding the trailer-towing limit, or towing while in Overdrive position. ● ATF level is low, or the fluid is contaminated or burnt ● The line pressure is too low ● The pressure control solenoid valve is stuck "on" or pressure control solenoid valve has a broken clamp causing leakage. ● The clutch plates (431 through 435) are burned or damaged. ● The 2nd clutch piston assembly is cracked or damaged. ● The 2nd clutch return spring and retainer assembly is damaged or out of position, or the 2nd clutch piston assembly seals are rolled, damaged or leaking. ● The 2nd sprag clutch assembly is damaged or not holding. ● The 1-2 shift solenoid valve stuck "on", or 1-2 shift valves (A & B) are stuck in 1st gear.
DTC P1811 **1T CCM, MIL: No** 1999, 2000, 2001, 2002, 2003, 2004, 2005 Aurora, Alero, Bonneville, Century, LSS, Cutlass Supreme, Grand Prix, Impala, Lumina, LeSabre, LSS, Regal, Park Avenue, Eighty Eight, Ninety Eight 3.1L VIN J, 3.1L VIN M, 3.4L VIN E, 3.5L VIN H, 3.8L VIN 1, 3.8L VIN K Transmissions: A/T	**Maximum Adaptive & Long Term Shift (4T65-E) Conditions:** Engine started, engine running and with shift adaptable enabled, the 1-2, 2-3 and 3-4 adaptive cells have reached their limits, and the PCM detected the 1-2 shift, 2-3 shift and 3-4 shifts were longer than 65 ms. The fault must occur twice per vehicle trip to set this code. **Possible Causes** ● Ask customer about overloading vehicle, exceeding trailer-towing limit, or towing in O/D ● ATF level is low, or the fluid is contaminated or burnt ● 1-2 accumulator piston seals rolled or damaged ● 1-2 accumulator piston and pin missing, binding or damaged ● Forward servo assembly damaged or misassembled ● Oil pump assembly damaged or missing components ● Spacer plate and gaskets damaged or misassembled ● Driven sprocket support seals damaged or missing ● 2nd clutch piston and seal assembly binding or damaged ● 2nd clutch fiber and steel plates misassembled or damaged ● Second clutch spring assembly damaged or misassembled ● Forward band burned, damaged or misassembled ● 1-2 support roller clutch assembly damaged or misassembled
DTC P1811 **1T CCM, MIL: No** 1997, 1998, 1999, 2000, 2001, 2002, 2003, 2004, 2005 Cavalier, Envoy, Sunbird, S/T Pickup & TrailBlazer 2.2L VIN 4, 2.4L VIN T, 4.2L VIN S engines Transmissions: A/T	**Maximum Adaptive and Long Term Shift (4T40/4T45-E) Conditions:** Engine running with the shift adaptable and shift adaptive is at their limit, and the PCM detected the shift time was more than 65 ms (fault must occur twice on one trip to set code). **Possible Causes** ● Ask customer about overloading vehicle, exceeding trailer-towing limit, or towing in O/D ● ATF level is low, or the fluid is contaminated or burnt ● Low fluid level caused by external leaks ● Out-of-position fluid filter or clogged fluid filter, or Internal fluid passage leaks ● Casting porosity or damage, or damaged gasket or spacer plate ● Out-of-position gasket or spacer plate ● Pressure control solenoid is contaminated, stuck or damaged ● Leaking or stuck pressure regulator valve train ● Stuck torque signal valve train, or leaking torque signal valve train ● Damaged or leaking oil pump or inadequate oil pump suction ● Oil pump cavitation
DTC P1812 **1T CCM, MIL: No** 1996 Achieva, Beretta, Bonneville, Century, Corsica, Cutlass Supreme, Grand Am, Grand Prix, Lumina, Monte Carlo & Regal Engines: 3.1L VIN M, 3.4L VIN E, 3.8L VIN 1, 3.8L VIN K Transmissions: A/T	**Transmission Fluid Over-Temperature (4T60-E) Conditions:** DTC P0712 and P0713 not set, engine started, engine running and the PCM detected the TFT sensor was more than 266°F for over 6 minutes and 48 seconds during the CCM test. **Possible Causes** ● ATF level is low or the fluid is burnt or contaminated ● ATF filter is clogged or not properly installed/seated ● A/T line pressure is low (check for clogged cooler lines) ● ECT sensor is contaminated or out-of-calibration (skewed) ● Engine is overheated ● Radiator is restricted ● Vehicle used for towing or driven while overloaded

OBD II Trouble Code List (P0xxx Codes)

DTC	Trouble Code Title, Conditions & Possible Causes
DTC P1812 **1T CCM, MIL: No** 1996 C/K Series Truck, G Van, Cargo Van, L/M Van, S/T Blazer & S/T Pickup 4.3L VIN W, 4.3L VIN X, 5.0L VIN M, 5.7L VIN R, 7.4L VIN J Transmissions: A/T	**Transmission Fluid Over-Temperature Conditions:** DTC P0712 and P0713 not set, engine started, engine running and the PCM detected the TFT sensor was more than 272ºF for over 6 minutes and 50 seconds during the CCM test. **Possible Causes** ● ATF level is low or the fluid is burnt or contaminated ● ATF filter is clogged or not properly installed/seated ● A/T line pressure is low (check for clogged cooler lines) ● ECT sensor is contaminated or out-of-calibration (skewed) ● Engine is overheated ● Radiator is restricted ● Vehicle used for towing or driven while overloaded
DTC P1812 **1T CCM, MIL: No** 1995, 1996 Camaro & Firebird 3.8L VIN K engine Transmissions: A/T	**Transmission Fluid Over-Temperature Conditions:** DTC P0712 and P0713 not set, engine started, and the PCM detected the Transmission Fluid Temperature (TFT) sensor indicated more than 266ºF for 7 minutes during the test. **Possible Causes** ● ATF level is low or the fluid is burnt or contaminated ● ATF filter is clogged or not properly installed/seated ● A/T line pressure is low (check for clogged cooler lines) ● ECT sensor is contaminated or out-of-calibration (skewed) ● Engine is overheated ● Radiator is restricted ● Vehicle used for towing or driven while overloaded
DTC P1812 **1T CCM, MIL: No** 1999, 2000, 2001, 2002, 2003, 2004, 2005 Aurora, Alero, Bonneville, Century, LSS, Cutlass Supreme, Grand Prix, Impala, Lumina, LeSabre, LSS, Regal, Park Avenue, Eighty Eight, Ninety Eight 3.1L VIN E, 3.4L VIN J, 3.1L VIN M, 3.8L VIN 1, 3.8L VIN K engines Transmissions: All	**Torque Converter Overstressed (4T65-E) Conditions:** DTC P0121, P0122, P0123, P0502, P0503 and P1810 not set, engine started, gear selector in Drive or Reverse, vehicle speed less than 7 mph, TP angle more than 70%, and the PCM detected the TCC slip speed was more than 2100 rpm for 12 seconds. The PCM checks for unusually high throttle angle and low vehicle speed when the transmission is in Drive or Reverse. The purpose of this code is to record the Failure Records as this condition could damage the Powertrain or create an unsafe condition. The code is set if the PCM detects an unusually high TP angle at low speed in Drive or reverse. **Possible Causes** ● ATF level is low or the fluid is burnt or contaminated ● ATF filter is clogged or not properly installed/seated ● Vehicle used for towing or driven while overloaded
DTC P1814 **1T CCM, MIL: No** 1996, 1997, 1998, 1999, 2000, 2001, 2002, 2003, 2004, 2005 Aurora, Deville, Eldorado, Seville, Impala, Intrigue, Silhouette, Montana, Trans Sport & Ventura 3.4L VIN E, 3.5L VIN H, 4.0L VIN C, 4.6L VIN 9, 4.6L VIN Y engines Transmissions: A/T	**Torque Converter Overstressed (4T65-E, 4T80-E) Conditions:** DTC P0121, P0122, P0123, P0502, P0503, P1820, P1822, P1823 and P1825 not set, engine started, gearshift selector indicating Drive or Reverse, TP angle over 70%, vehicle speed less than 7 mph, and the PCM detected the TCC slip speed was over 2200 rpm for 12 seconds. A recommended step is to replace the ATF fluid and clean or replace the scavenger screens, then recheck for the same code. **Possible Causes** ● ATF level is low or the fluid is burnt or contaminated ● ATF filter is clogged or not properly installed/seated ● Vehicle used for towing or driven while overloaded
DTC P1815 **2T CCM, MIL: Yes** 2003, 2004, 2005 CTS (Cadillac) 2.6L VIN M, 3.2L VIN N Transmissions: A/T	**A/T Manual Shaft Shift Switch Circuit Malfunction (5L40-E) Conditions:** DTC P0722 and P0723 not set, engine started, system voltage from 8-18v, transmission output speed less than 100 rpm, and the PCM detected the Manual Shift Position Switch indicated a transitional state during this sequence: Engine speed less than 60 rpm for over 250 ms, then engine speed at 81-625 rpm for over 0.01875 seconds, then engine speed over 651 rpm for 4 seconds, then the OSS speed was more than 200 rpm for 4 seconds. The Transmission Manual Shift shaft switch assembly is a sliding contact switch attached to the control valve body in the transmission. The four inputs to the TCM from the switch indicate which position has been selected with the transmission selector lever. The switch voltage is high with the switch open and low with it closed. The Scan Tool displays each of the switch states as IMS, and are represented as transmission range Signal A, Signal B, Signal C and Signal P (Parity). **Possible Causes** ● Manual shaft shift switch connector is open or shorted ● Manual shaft shift switch circuit is open or signals are shorted ● Manual shaft shift switch is damaged or out of adjustment ● TCM has failed

OBD II Trouble Code List (P1xxx Codes)

DTC	Trouble Code Title, Conditions & Possible Causes
DTC P1815 **1T CCM, MIL: No** 2003, 2004, 2005 Alero, Grand Am, Malibu 3.1L VIN J engine Transmissions: All	**A/T Range Sensor Circuit Malfunction (4T40-E/4T45-E) Conditions:** DTC P0502, P0503, P0716, P0717 and P1810 not set, engine started, engine speed in transition from 0 rpm to over 600 rpm, vehicle speed less than 5 mph, and the PCM detected that gear shift position D2, D4 or Reverse was indicated for 250 ms after startup. **Possible Causes** ● TFP manual valve position switch circuits are open or shorted ● TFP manual valve position switch is damaged or it has failed ● Transmission gear selector linkage is out of adjustment ● PCM has failed
DTC P1816 **1T CCM, MIL: No** 1997, 1998, 1999, 2000, 2001, 2002, 2003, 2004, 2005 Century, Lumina 3.1L VIN J engine Transmissions: A/T	**A/T Input Shaft Speed Sensor Malfunction (4T40-E/4T45-E) Conditions:** DTC P0121, P0122, P0123, P0502, P0503, P0717, P0751, P0753 and P0758 not set, engine runtime over 5 seconds, TP angle more than 15%, vehicle speed over 5 mph, and the PCM detected the Input Shaft Speed sensor changed by over 1,300 rpm within 800 ms. **Possible Causes** ● A/T ISS connector is damaged or it has failed ● A/T ISS circuit is open or shorted to system power ● A/T ISS assembly is damaged or it has failed ● PCM has failed
DTC P1817 **1T CCM, MIL: No** 2000, 2001, 2002, 2003, 2004, 2005 Century, Lumina 3.1L VIN J engine Transmissions: A/T	**A/T Input Shaft Speed Sensor Malfunction (4T40-E/4T45-E) Conditions:** DTC P0502, P0503 and P1810 not set, engine runtime over 5 seconds, TFP manual valve position switch signal not in Park or Neutral, VSS more than 5 mph, and the PCM detected the Input Shaft Speed sensor changed by over 100 rpm for 5 seconds. **Possible Causes** ● A/T ISS connector is damaged or it has failed ● A/T ISS circuit is shorted to ground ● A/T ISS assembly is damaged or it has failed ● PCM has failed
DTC P1818 **2T CCM, MIL: Yes** 2003, 2004, 2005 CTS (Cadillac) 2.6L VIN M, 3.2L VIN N Transmissions: All	**A/T Manual Shaft Shift Switch Circuit Malfunction (5L40-E) Conditions:** DTC P0716, P0717, P0722 or P0723, P0751, P0752, P0756, P0757, P0761, P0762, P1779, P1815, P1818, P1820, P1822, P1823, P1825 and P1826 not set, engine runtime over 5 seconds, system voltage from 8-18v, APP angle over 5%, vehicle speed over 5 mph, engine torque from 15-258 lb ft., gear ratio from 2.95:1 to 3.10:1, and the PCM detected the IMS indicates the gear selector was in a forward range while the gear ratio was 2.95-3.10 for 3 seconds. The (4) inputs to the TCM from the switch indicate the position selected with the transmission selector lever. The switch voltage is high with it open and low with it closed. **Possible Causes** ● Manual shaft shift switch connector is open or shorted ● Manual shaft shift switch circuit is open or signals are shorted ● Manual shaft shift switch is damaged or out of adjustment ● TCM has failed
DTC P1819 **1T CCM, MIL: No** 1997, 1998, 1999, 2000, 2001, 2002, 2003, 2004, 2005 Aurora, Alero, Bonneville, Century, LSS, Cutlass Supreme, Grand Prix, Impala, Lumina, LeSabre, LSS, Regal, Park Avenue, Eighty Eight, Ninety Eight, Montana, Silhouette, Rendezvous & Venture 3.1L VIN E, 3.4L VIN J, 3.1L VIN M, 3.8L VIN 1, 3.8L VIN K engines Transmissions: All	**Internal Mode Switch - No Start (4T65-E) Conditions:** Engine cranking, system voltage over 10.0v, and the PCM detected an invalid combination from the IMS switch, or a set of signals that indicated a transitional state between gear positions for 500 ms. **Possible Causes** ● Inspect the transmission linkage from the range selector to the manual shift shaft for proper adjustment ● TFP valve position switch signal circuit is open or grounded ● TFP valve position switch circuit shorted to another signal ● TFP valve position switch is damaged or has failed ● PCM has failed
DTC P1819 **1T CCM, MIL: No** 2001, 2002 Aurora 3.5L VIN H engine Transmissions: All	**Internal Mode Switch - No Start (4T65-E) Conditions:** Engine cranking, system voltage over 10.0v, and the PCM detected an invalid combination of Internal Mode Switch signals, or signals that indicated a transitional state between gear positions for 500 ms. **Possible Causes** ● Inspect transmission linkage (range selector to manual shift shaft) for proper adjustment ● TFP valve position switch signal circuit is open or grounded ● TFP valve position switch circuit shorted to another signal ● TFP valve position switch is damaged or has failed ● PCM has failed

OBD II Trouble Code List (P1xxx Codes)

DTC P1820 **1T CCM, MIL: No** 1997, 1998, 1999, 2000, 2001, 2002, 2003, 2004, 2005 Aurora, Alero, Bonneville, Century, LSS, Cutlass Supreme, Grand Prix, Impala, Lumina, LeSabre, LSS, Regal, Park Avenue, Eighty Eight, Ninety Eight, Montana, Silhouette, Rendezvous & Venture 3.1L VIN J, 3.4L VIN E, 3.1L VIN M, 3.5L VIN H, 3.8L VIN 1, 3.8L VIN K engines Transmissions: All	**Internal Mode Switch Circuit 'A' Low (4T65-E) Conditions:** DTC P0107 and P0108 not set, engine speed over 500 rpm for 5 seconds, system voltage over 10.0v, Fuel Shutoff not active, and the PCM detected the IMS 'A' signal was in a continuously low state, and the IMS indicated Park for 2 seconds, then the IMS indicated transitional position D4-D3 with the engine torque from 70-300 lb ft., with no engine torque defaults detected for 6 seconds. **Possible Causes** ● IMS Signal 'A' circuit is shorted to ground ● The lever assembly-manual shaft detent with internal mode switch may be damaged or have failed ● Transmission has internal problems (it may need an overhaul) ● PCM has failed
DTC P1820 **1T CCM, MIL: No** 2003, 2004, 2005 CTS (Cadillac) 2.6L VIN M, 3.2L VIN N Transmissions: All	**A/T Manual Shaft Shift Switch Circuit Malfunction (5L40-E) Conditions:** DTC P1779 not set, engine runtime over 5 seconds, system voltage from 8-18v, engine torque from 18-258 lb ft., and the PCM detected the switch indicated Park for 2 seconds, and the switch indicated a transitional state between D4 and D3 for more than 5 seconds. The four (4) inputs to the TCM from the switch indicate the position selected with the transmission selector lever. The switch voltage is high with the switch open and low with it closed. The Scan Tool displays each of the switch states as IMS, and are represented as TR Signal A, Signal B, Signal C and Signal P (Parity). **Possible Causes** ● Manual shaft shift switch connector is open or shorted ● Manual shaft shift switch circuit is open or signals are shorted ● Manual shaft shift switch is damaged or out of adjustment ● TCM has failed
DTC P1820 **1T CCM, MIL: No** 2000, 2001, 2002, 2003, 2004, 2005 Aurora, DeVille, Eldorado & Seville 4.0L VIN C, 4.6L VIN 9, 4.6L VIN Y engines Transmissions: All	**Internal Mode Switch 'A' Circuit Low Input (4T80-E) Conditions:** Engine started, engine runtime over 5 seconds, system voltage over 10.0v, Fuel Cutoff inactive, engine torque from 30-148 lb ft, and the PCM detected the IMS signal indicated PARK for 2 seconds, and then the IMS indicated a transitional state between position D4 and D3 for 5 seconds. The transmission IMS is a sliding contact switch attached to the lower control valve body, within the transmission. The manual valve link assembly then connects the IMS to the manual valve. The four inputs to the PCM from the IMS indicate which position is has been selected by the transmission selector lever. Each input can be read on the Scan Tool. The parameters are Signal 'A', Signal 'B', Signal 'C' and Signal 'P' (Parity). **Possible Causes** ● IMS Signal 'A' circuit is shorted to ground ● Transmission has internal problems / may need an overhaul
DTC P1822 **1T CCM, MIL: No** 1997, 1998, 1999, 2000, 2001, 2002, 2003, 2004, 2005 Aurora, Alero, Bonneville, Century, LSS, Cutlass Supreme, Grand Prix, Impala, Lumina, LeSabre, LSS, Regal, Park Avenue, Eighty Eight, Ninety Eight, Montana, Silhouette, Rendezvous & Venture 3.1L VIN J, 3.4L VIN E, 3.1L VIN M, 3.5L VIN H, 3.8L VIN 1, 3.8L VIN K Transmissions: All	**Internal Mode Switch 'B' Circuit Low Input (4T65-E) Conditions:** DTC P0107 and P0108 not set, system voltage from 9-18v, engine speed at least 500 rpm for 5 seconds, Fuel Shutoff inactive, and the PCM detected that the Internal Mode Switch 'B' signal was in a continuously high state for the current key cycle, and that the IMS indicated Park for 2 seconds, then the IMS indicated transitional position D2-D1 and the engine torque was 70-300 ft-lbs with no engine torque defaults for 6 seconds. **Possible Causes** ● IMS Signal 'B' circuit is open or has a high resistance condition ● Transmission has internal problems / may need an overhaul
DTC P1822 **2T CCM, MIL: Yes** 2003, 2004, 2005 CTS (Cadillac) 2.6L VIN M, 3.2L VIN N Transmissions: All	**A/T Manual Shaft Shift Switch Circuit Malfunction (5L40-E) Conditions:** DTC P1779 not set, engine runtime over 5 seconds, system voltage from 8-18v, engine torque from 18-258 lb ft., and the PCM detected the switch indicated Park for 1 second, and the switch indicated a transitional state between D2 and D3 for more than 4 seconds. The four (4) inputs to the TCM from the switch indicate the position selected with the transmission selector lever. The switch voltage is high with the switch open and low with it closed. The Scan Tool displays each of the switch states as IMS, and are represented as TR Signal A, Signal B, Signal C and Signal P (Parity). **Possible Causes** ● Manual shaft shift switch connector is open or shorted ● Manual shaft shift switch circuit is open or signals are shorted ● Manual shaft shift switch is damaged or out of adjustment ● TCM has failed

OBD II Trouble Code List (P0xxx Codes)

DTC	Trouble Code Title, Conditions & Possible Causes
DTC P1822 **2T CCM, MIL: Yes** 2000, 2001, 2002, 2003, 2004, 2005 Aurora, DeVille, Eldorado & Seville 4.0L VIN C, 4.6L VIN 9, 4.6L VIN Y engines Transmissions: All	**Internal Mode Switch 'B' Circuit Low Input (4T80-E) Conditions:** Engine started, engine runtime over 5 seconds, system voltage over 10.0v, Fuel Cutoff inactive, engine torque from 30-148 lb ft, and the PCM detected the IMS signal indicated PARK for 2 seconds, and then the IMS indicated a transitional state between position D2 and D1 for 5 seconds during the CCM test. The transmission internal mode switch (IMS) is a sliding contact switch attached to the lower control valve body, within the transmission. The manual valve link assembly then connects the IMS to the manual valve. **Possible Causes** ● IMS Signal 'B' circuit is open or has a high resistance condition ● Transmission has internal problems / may need an overhaul
DTC P1823 **1T CCM, MIL: No** 1997, 1998, 1999, 2000, 2001, 2002, 2003, 2004, 2005 Aurora, Alero, Bonneville, Aztek, Century, Cutlass Supreme, Grand Prix, Impala, Lumina, LeSabre, LSS, Regal, Park Avenue, Eighty Eight, Ninety Eight, Montana, Silhouette, Rendezvous & Venture 3.1L VIN J, 3.4L VIN E, 3.1L VIN M, 3.5L VIN H, 3.8L VIN 1, 3.8L VIN K Transmissions: A/T	**Internal Mode Switch 'P' Circuit Low Input (4T65-E) Conditions:** DTC P0107 and P0108 not set, engine started, engine speed over 500 rpm for 5 seconds, system voltage over 10.0v, Fuel Shutoff inactive, and the PCM detected the Internal Mode Switch 'P' signal was in a continuously low state for the current key cycle, and that the IMS indicated Park for 2 seconds, then the IMS indicated transitional position 'N' to D4 and the engine torque was 70-300 lb ft, with no engine torque defaults for 6 seconds during the CCM test. **Possible Causes** ● IMS Signal 'P' circuit is shorted to sensor or chassis ground ● Transmission has internal problems / may need an overhaul
DTC P1823 **2T CCM, MIL: Yes** 2003, 2004, 2005 CTS (Cadillac) 2.6L VIN M, 3.2L VIN N Transmissions: A/T	**A/T Manual Shaft Shift Switch Circuit Malfunction (5L40-E) Conditions:** DTC P1779 not set, engine runtime over 5 seconds, system voltage from 8-18v, engine torque from 18-258 lb ft., and the PCM detected the switch indicated Park for 1 second, and then the switch indicated a transitional state between Neutral and D5 for over 4 seconds. The four (4) inputs to the TCM from the switch indicate the position selected with the transmission selector lever. The switch voltage is high with the switch open and low with it closed. The Scan Tool displays each of the switch states as IMS, and are represented as TR Signal A, Signal B, Signal C and Signal P (Parity). **Possible Causes** ● Manual shaft shift switch connector is open or shorted ● Manual shaft shift switch circuit is open or signals are shorted ● Manual shaft shift switch is damaged or out of adjustment ● TCM has failed
DTC P1825 **2T CCM, MIL: Yes** 2000, 2001, 2002, 2003, 2004, 2005 Aurora, DeVille, Eldorado & Seville 4.0L VIN C, 4.6L VIN 9, 4.6L VIN Y engines Transmissions: A/T	**Internal Mode Switch 'P' Circuit Low Input (4T80-E) Conditions:** Engine started, engine runtime over 5 seconds, system voltage over 10.0v, Fuel Cutoff inactive, engine torque from 30-148 ft-lb (40-200 Nm), and the PCM received an IMS signal indicating PARK for 2 seconds, and then the IMS indicated a transitional state between Neutral and D4, conditions met for 5 seconds. The transmission IMS is a sliding contact switch attached to the lower control valve body. The manual valve link assembly then connects the IMS to the manual valve. The four inputs to the PCM from the IMS indicate the position selected by the transmission selector lever. This data is used for engine controls as well as determining the transmission shift patterns. **Possible Causes** ● IMS Signal 'P' circuit is shorted to sensor or chassis ground ● Transmission has internal problems / may need an overhaul
DTC P1825 **1T CCM, MIL: No** 1997, 1998, 1999, 2000, 2001, 2002, 2003, 2004, 2005 Aurora, Alero, Bonneville, Century, LSS, Cutlass Supreme, Grand Prix, Impala, Lumina, LeSabre, LSS, Regal, Park Avenue, Eighty Eight, Ninety Eight, Montana, Silhouette, Rendezvous & Venture 3.1L VIN J, 3.4L VIN E, 3.1L VIN M, 3.5L VIN H, 3.8L VIN 1, 3.8L VIN K Transmissions: A/T	**Internal Mode Switch - Invalid Range (4T65-E) Conditions:** DTC P0107 and P0108 not set, engine started, engine speed over 500 rpm for 5 seconds, system voltage over 10.0v, Fuel Shutoff inactive, and the PCM detected an invalid combination of Internal Mode Switch signals for 500 ms during the CCM test. **Possible Causes** ● IMS Signal 'A', 'B', 'C' and 'P' possible short to ground condition ● IMS Signal 'A', 'B', 'C' and 'P' possible open or high resistance ● Transmission has internal problems / may need an overhaul ● PCM has failed

OBD II Trouble Code List (P0xxx Codes)

DTC	Trouble Code Title, Conditions & Possible Causes
DTC P1825 **2T CCM, MIL: Yes** 2003, 2004, 2005 CTS (Cadillac) 2.6L VIN M, 3.2L VIN N Transmissions: All	**A/T Manual Shaft Shift Switch Invalid Signal (5L40-E) Conditions:** Engine runtime over 5 seconds, system voltage from 8-18v, and the PCM detected an invalid switch range for 5 seconds. The four (4) inputs to the TCM from the switch indicate the position selected with the transmission selector lever. The switch voltage is high with the switch open and low with it closed. The Scan Tool displays each of the switch states as IMS, and are represented as TR Signal A, Signal B, Signal C and Signal P (Parity). **Possible Causes** ● Manual shaft shift switch circuit is open or signals are shorted ● Manual shaft shift switch is damaged or out of adjustment ● TCM has failed
DTC P1825 **1T CCM, MIL: No** 2000, 2001, 2002, 2003, 2004, 2005 Aurora, DeVille, Eldorado & Seville 4.0L VIN C, 4.6L VIN 9, 4.6L VIN Y engines Transmissions: All	**Internal Mode Switch - Invalid Range (4T80-E) Conditions:** Engine running, system voltage over 10.0v, and the PCM detected an invalid IMS range signal for 5 seconds. The transmission internal mode switch is a sliding contact switch that is attached to the selector detent inside the transmission side cover. The four inputs to the PCM indicate the position of the transmission range selector. The input voltage level at the PCM is high (B+) when the IMS is open and low when the switch is closed to ground. **Possible Causes** ● IMS Signal 'A', 'B', 'C' and 'P' possible short to ground condition ● IMS Signal 'A', 'B', 'C' and 'P' possible open or high resistance ● Transmission has internal problems / may need an overhaul ● PCM has failed
DTC P1826 **1T CCM, MIL: No** 1999, 2000, 2001, 2002, 2003, 2004, 2005 Aurora, Alero, Bonneville, Century, LSS, Cutlass Supreme, Grand Prix, Impala, Lumina, LeSabre, LSS, Regal, Park Avenue, Eighty Eight, Ninety Eight, Montana, Silhouette, Rendezvous & Venture 3.1L VIN J, 3.4L VIN E, 3.1L VIN M, 3.5L VIN H, 3.8L VIN 1, 3.8L VIN K Transmissions: A/T	**Internal Mode Switch - Invalid Range (4T65-E) Conditions:** DTC P0502 and P0503 not set, DTC P1826 has passed this key cycle, system voltage from 9-18v, engine speed at least 500 rpm for 5 seconds, Fuel Shutoff inactive, engine torque more than 20 ft-lbs, and the PCM detected that the Internal Mode Switch 'C' signal was in a high state while the gear ratio indicated 1st, 2nd or 3rd gear, condition met for 0.5 seconds. **Possible Causes** ● IMS Signal 'C' circuit is open or has a high resistance condition ● Transmission has internal problems / may need an overhaul
DTC P1826 **2T CCM, MIL: Yes** 2003, 2004, 2005 CTS (Cadillac) 2.6L VIN M, 3.2L VIN N Transmissions: A/T	**A/T Manual Shaft Shift Switch Invalid Signal (5L40-E) Conditions:** DTC P0722, P0723 and P1779 not set, engine started, system voltage from 8-18v, engine torque over 15 lb ft., VSS over 5 mph, then with the gear ratio within one of these ranges: 3.33:1 to 3.50:1 for 1st gear, 2.16:1 to 2.27:1 for 2nd gear, 1.56:1 to 1.64:1 for 3rd gear, 0.98:1 to 1.03:1 for 4th gear or 0.73:1 to 0.77:1 for 5th gear; the TCM detected Signal C was high, while the gear ratio indicated 1st, 2nd, 3rd, 4th or 5th gear for 3 seconds. The switch voltage is high with the switch open and low with it closed. **Possible Causes** ● Manual shaft shift switch circuit is open or signals are shorted ● Manual shaft shift switch is damaged or out of adjustment ● TCM has failed
DTC P1826 **1T CCM, MIL: No** 2001, 2002 Aurora 3.5L VIN H engine Transmissions: All	**Internal Mode Switch - Invalid Range (4T80-E) Conditions:** DTC P0502 and P0503 not set, DTC P1826 not passed this key cycle, engine speed over 500 rpm for 5 seconds, system voltage over 10.0v, Fuel Shutoff "off", engine torque over 20 lb ft, and the PCM detected the IMS Signal C was high, while the gear ratio indicated 1st, 2nd or 3rd gear for 500 ms. The input voltage level at the PCM is high (B+) when the IMS is open and low when the switch is closed to ground. **Possible Causes** ● IMS Signal 'C' circuit is open or has a high resistance condition ● Transmission has internal problems / may need an overhaul
DTC P1831 **2T CCM, MIL: Yes** 2003, 2004, 2005 CTS (Cadillac) 2.6L VIN M, 3.2L VIN N Transmissions: A/T	**Pressure Control Solenoid Circuit Low Input (5L40-E) Conditions:** Engine started, system voltage from 8-18v, and the TCM detected a short to ground on the HSD1 circuit with the HSD1 commanded "on". The TCM provides power to the pressure control solenoid, the shift solenoids, and the TCC PWM solenoid through separate high side drivers, called HSD1 and HSD2. HSD1 provides power to pressure control solenoid. HSD2 provides power to the shift solenoids. The TCM tests each driver to ensure it is functioning. **Possible Causes** ● Pressure control solenoid control circuit is shorted to ground ● Pressure control solenoid is damaged or it has failed ● Transmission internal harness assembly may be shorted, or the TCM has failed

OBD II Trouble Code List (P0xxx Codes)

DTC	Trouble Code Title, Conditions & Possible Causes
DTC P1832 **2T CCM, MIL: Yes** 2003, 2004, 2005 CTS (Cadillac) 2.6L VIN M, 3.2L VIN N Transmissions: A/T	**Pressure Control Solenoid Circuit High Input (5L40-E) Conditions:** Engine started, system voltage from 8-18v, and the TCM detected a voltage over 4.2v on the HSD1 circuit prior to commanding the HSD1 on. The TCM provides power to the pressure control solenoid, the shift solenoids, and the TCC PWM solenoid through separate high side drivers, called HSD1 and HSD2. HSD1 provides power to pressure control solenoid. HSD2 provides power to the shift solenoids. The TCM tests each driver to ensure that it is functioning. If the TCM detects a short to power on HSD1, DTC P1832 is set. **Possible Causes** ● Pressure control solenoid control circuit is shorted to power ● Pressure control solenoid is damaged or it has failed ● Transmission internal harness assembly may be shorted ● TCM has failed
DTC P1833 **2T CCM, MIL: Yes** 2003, 2004, 2005 CTS (Cadillac) 2.6L VIN M, 3.2L VIN N Transmissions: A/T	**Pressure Control Solenoid Circuit Low Input (5L40-E) Conditions:** Engine started, system voltage from 8-18v, and the TCM detected a short to ground on the HSD2 circuit with the HSD2 commanded on. The TCM provides power to the pressure control solenoid, the shift solenoids, and the TCC PWM solenoid through separate high side drivers, called HSD1 and HSD2. HSD1 provides power to pressure control solenoid. HSD2 provides power to the shift solenoids. The TCM tests each driver to ensure that it is functioning. If the TCM detects a short to ground on HSD2, DTC P1833 is set. **Possible Causes** ● Pressure control solenoid control circuit is shorted to ground ● Pressure control solenoid is damaged or it has failed ● Transmission internal harness assembly may be shorted ● TCM has failed
DTC P1834 **2T CCM, MIL: Yes** 2003, 2004, 2005 CTS (Cadillac) 2.6L VIN M, 3.2L VIN N Transmissions: A/T	**Pressure Control Solenoid Circuit High Input (5L40-E) Conditions:** Engine started, system voltage from 8-18v, and the TCM detected a voltage over 4.2v on the HSD2 circuit prior to commanding the HSD2 on. The TCM provides power to the pressure control solenoid, the shift solenoids, and the TCC PWM solenoid through separate high side drivers, called HSD1 and HSD2. HSD1 provides power to pressure control solenoid. HSD2 provides power to the shift solenoids. The TCM tests each driver to ensure that it is functioning. If the TCM detects a short to power on HSD2, DTC P1834 is set. **Possible Causes** ● Pressure control solenoid control circuit is shorted to power ● Pressure control solenoid is damaged or it has failed ● Transmission internal harness assembly may be shorted ● TCM has failed
DTC P1835 **1T CCM, MIL: Yes** 1997, 1998, 1999, 2000, 2001 Catera 3.0L VIN R engine Transmissions: A/T	**Kickdown Switch Circuit Malfunction (4L30-E) Conditions:** DTC P1890 not set; engine speed from 500-8160, and the PCM detected a Kickdown request sent to TCM with TP angle under 89%. The PCM transmits a downshift request to the TCM if the accelerator pedal position exceeds 89%. The PCM sends the downshift request to the TCM on the CAN line. A fault in the CAN cannot cause P1835 to set by itself. **Possible Causes** ● Perform the Powertrain Diagnostic System Check for the Automatic Transmission ● Check for DTC U2100, U2104, U2105 and U2108 - if these CAN codes are not set, the ● PCM has failed
DTC P1842 **2T CCM, MIL: Yes** 2003, 2004, 2005 CTS (Cadillac) 2.6L VIN M, 3.2L VIN N Transmissions: A/T	**1-2 Shift Solenoid Circuit Low Input (5L40-E) Conditions:** Engine runtime over 5 seconds, system voltage from 8-18v, and the TCM detected an open in the 1-2 shift solenoid valve circuit with the HSD2 commanded "on", or a short to ground in the 1-2 shift solenoid valve circuit with the HSD2 commanded "off". The TCM provides power to the solenoid through the high side driver 2 (HSD2). A second driver is used to control the solenoid ground circuit. The controlled ground driver reports feedback voltage to the TCM. **Possible Causes** ● A/T 1-2 shift solenoid control circuit open or shorted to ground ● A/T 1-2 shift solenoid is damaged or has failed ● Transmission internal harness assembly is open or shorted ● TCM has failed
DTC P1842 **2T CCM, MIL: Yes** 2000, 2001, 2002, 2003, 2004, 2005 Aurora, DeVille, Eldorado & Seville 4.0L VIN C, 4.6L VIN 9, 4.6L VIN Y engines Transmissions: A/T	**1-2 Shift Solenoid Circuit Low Input (4T80-E) Conditions:** Engine runtime over 5 seconds, system voltage over 10.0v, and the PCM detected an unexpected voltage on the 1-2 Shift Solenoid feedback circuit for 4.3 seconds. The 1-2 shift solenoid valve controls transmission fluid pressure acting on the 1-2 and 3-4 shift valves to position the valves for the correct gear. The 1-2 SS valve is a normally open (N.O.) exhaust valve located on the lower control valve body. When the 1-2 SS valve is "on", the PCM should sense low voltage. When the 1-2 SS valve is "off", the PCM should sense high signal. **Possible Causes** ● A/T 1-2 SS control circuit is open or shorted to ground ● A/T 1-2 SS power circuit is open (check the TRANS fuse) ● A/T 1-2 SS (solenoid) has failed, or the PCM has failed

OBD II Trouble Code List (P0xxx Codes)

DTC	Trouble Code Title, Conditions & Possible Causes
DTC P1843 **2T CCM, MIL: Yes** 2003, 2004, 2005 CTS (Cadillac) 2.6L VIN M, 3.2L VIN N Transmissions: A/T	**1-2 Shift Solenoid Circuit High Input (5L40-E) Conditions:** Engine runtime over 5 seconds, system voltage from 8-18v, and the TCM detected a short to power on the 1-2 shift solenoid valve circuit. The TCM provides power to the solenoid through the high side driver 2 (HSD2). The TCM uses a second driver to control the solenoid ground circuit. The controlled ground driver reports feedback voltage to the TCM. When the TCM commands the 1-2 shift solenoid valve "on", the control circuit voltage should be 0v. When the 1-2 shift solenoid valve "off", the control circuit voltage should be 12-14v. **Possible Causes** • A/T 1-2 shift solenoid control circuit is shorted to power • A/T 1-2 shift solenoid is damaged or has failed • DTC U2100, U2104, U2105, U2106 can cause this code to set • Transmission internal harness assembly is shorted to power • TCM has failed
DTC P1843 **2T CCM, MIL: Yes** 2000, 2001, 2002, 2003, 2004, 2005 Aurora, DeVille, Eldorado & Seville 4.0L VIN C, 4.6L VIN 9, 4.6L VIN Y engines Transmissions: All	**1-2 Shift Solenoid Circuit High Input (4T80-E) Conditions:** Engine runtime over 5 seconds, system voltage over 10.0v, vehicle driven with the 1-2 Shift solenoid commanded "on" by the PCM, and the PCM detected an unexpected "high" voltage condition on the 1-2 Shift Solenoid feedback circuit for 4.3 seconds. **Possible Causes** • A/T 1-2 SS control circuit is shorted to system power (B+) • A/T 1-2 SS (solenoid) is damaged or has failed • PCM has failed
DTC P1845 **2T CCM, MIL: Yes** 2003, 2004, 2005 CTS (Cadillac) 2.6L VIN M, 3.2L VIN N Transmissions: A/T	**A/T 2-3 Shift Solenoid Circuit Malfunction (5L40-E) Conditions:** Engine runtime over 5 seconds, system voltage from 8-18v, and the TCM detected an open in the 2-3 shift solenoid valve circuit with the HSD2 commanded "on", a short to ground in the 2-3 shift solenoid valve circuit with the HSD2 commanded "off". The TCM provides voltage to the solenoid through High Side Driver 2 (HSD2). The TCM uses a second driver to control the solenoid ground circuit. The controlled ground driver reports feedback signal to the TCM. When the TCM commands the 2-3 Shift Solenoid valve "on", the control circuit voltage should be 0v. When the TCM commands the 2-3 Shift Solenoid valve "off", the control circuit voltage should be near 12v. **Possible Causes** • A/T 2-3 shift solenoid control circuit open or shorted to ground • A/T 2-3 shift solenoid is damaged or has failed • Transmission internal harness assembly open or shorted • TCM has failed
DTC P1845 **2T CCM, MIL: Yes** 2000, 2001, 2002, 2003, 2004, 2005 Aurora, DeVille, Eldorado & Seville 4.0L VIN C, 4.6L VIN 9, 4.6L VIN Y engines Transmissions: A/T	**2-3 Shift Solenoid Circuit Low Input (4T80-E) Conditions:** Engine runtime over 5 seconds, system voltage over 10.0v, and the PCM detected an unexpected voltage condition on the 2-3 Shift Solenoid feedback circuit for 4.3 seconds. The 2-3 Shift Solenoid valve controls the transmission fluid pressure acting on the 1-2 and 2-3 shift valves to position the valves for the correct gear. The 2-3 SS valve is a normally open exhaust valve located on the lower control valve body. The PCM commands the solenoid "on" or "off" by providing a ground path. When the 2-3 SS valve is "on", the PCM should sense low voltage. When the 2-3 SS valve is "off", the PCM should sense high voltage. If the PCM detects a problem on the 2-3 SS circuit during continuous testing, it sets DTC P1845. **Possible Causes** • A/T 2-3 SS control circuit is open or shorted to ground • A/T 2-3 SS power circuit is open (check the TRANS fuse) • A/T 2-3 SS (solenoid) is damaged or has failed • PCM has failed
DTC P1845 **1T CCM, MIL: Yes** 1999, 2000, 2001 Catera 3.0L VIN R engine Transmissions: A/T	**Torque Limit Management Malfunction (4L30-E) Conditions:** Engine started; and the PCM detected the engine output torque was more than a calculated value for 600 seconds. The PCM will set this code if there is an engine torque limitation caused by the traction control system, the transmission or an engine speed limitation **Possible Causes** • Perform the Powertrain Onboard Diagnostic System Check • PCM may need to be replaced
DTC P1847 **2T CCM, MIL: Yes** 2003, 2004, 2005 CTS (Cadillac) 2.6L VIN M, 3.2L VIN N Transmissions: A/T	**A/T 2-3 Shift Solenoid Circuit Malfunction (5L40-E) Conditions:** Engine runtime over 5 seconds, system voltage from 8-18v, and the TCM detected a short to power in the 2-3 Shift Solenoid valve circuit. The TCM provides voltage to the solenoid through High Side Driver 2 (HSD2). The TCM uses a second driver to control the solenoid ground circuit. The controlled ground driver reports feedback signal to the TCM. When the TCM commands the 2-3 Shift Solenoid valve "on", the control circuit voltage should be 0v. When the 2-3 Shift Solenoid valve is "off", the control circuit voltage should be near 12v. **Possible Causes** • A/T 2-3 shift solenoid control circuit is shorted to power • A/T 2-3 shift solenoid is damaged or has failed • Transmission internal harness assembly is shorted to power, or the TCM has failed

OBD II Trouble Code List (P0xxx Codes)

DTC	Trouble Code Title, Conditions & Possible Causes
DTC P1847 **1T CCM, MIL: Yes** 2000, 2001, 2002, 2003, 2004, 2005 Aurora, DeVille, Eldorado & Seville 4.0L VIN C, 4.6L VIN 9, 4.6L VIN Y engines Transmissions: A/T	**2-3 Shift Solenoid Circuit High Input (4T80-E) Conditions:** Engine started, engine runtime over 5 seconds, system voltage over 10.0v, vehicle driven with the 2-3 Shift Solenoid commanded "on" by the PCM, and the PCM detected an unexpected "high" voltage on the 2-3 Shift Solenoid feedback circuit for 4.3 seconds. **Possible Causes** ● A/T 2-3 SS control circuit is shorted to system power (B+) ● A/T 2-3 SS (solenoid) is damaged or has failed ● PCM has failed
DTC P1850 **2T CCM, MIL: Yes** 1997, 1998, 1999, 2000, 2001 Catera 3.0L VIN R engine Transmissions: A/T	**Brake Band Apply Solenoid Circuit Malfunction (4L30-E) Conditions:** Engine running with the TCM High Side driver closed, then with the A/T Band Control solenoid commanded "off", the TCM detected the solenoid feedback voltage below 2.45v; or with the A/T Band Control solenoid commanded "off", the feedback voltage was from 2.45 -3.53v; or with the solenoid commanded "on", the feedback voltage was more than 4.02v. **Possible Causes** ● A/T band solenoid control circuit is open or shorted to ground ● A/T band control solenoid circuit is shorted to system power ● A/T band control solenoid power circuit (from PCM) is open ● A/T band control solenoid is damaged or has failed ● PCM has failed
DTC P1860 **1T CCM, MIL: Yes** 1996, 1997, 1998, 1999, 2000, 2001, 2002, 2003, 2004, 2005 Achieva, Alero, Century, Bonneville, LSS, Cutlass Supreme, Grand Am, Grand Prix, Impala, Lumina, LeSabre, LSS, Regal, Park Avenue, 88', 98', Montana, Silhouette, Rendezvous & Venture 3.1L VIN M, 3.4L VIN X, 3.4L VIN E, 3.8L VIN 1, 3.8L VIN K engines Transmissions: A/T	**TCC PWM Solenoid Circuit Malfunction (4T60-E) Conditions:** DTC P0560 not set, engine started, engine runtime over 5 seconds, Fuel Shutoff inactive, and the PCM detected an unexpected "low" voltage with a 10% command or an unexpected "high" voltage with a 95% command. The Torque Converter Clutch (TCC) PWM solenoid controls fluid acting on the converter clutch valve. The clutch valve controls the application and release of the torque converter clutch. **Possible Causes** ● TCC solenoid control circuit is open or shorted to ground ● TCC solenoid control circuit is shorted to system power (B+) ● TCC solenoid power circuit is open (test ENG or TRANS fuse) ● TCC solenoid is damaged or has failed ● PCM has failed
DTC P1860 **2T CCM, MIL: Yes** 1995, 1996, 1997, 1998, 1999, 2000, 2001, 2002 Camaro & Firebird 3.8L VIN K engine Transmissions: A/T	**TCC PWM Solenoid Circuit Malfunction (4L60-E) Conditions:** Engine started, engine runtime over 5 seconds, system voltage over 10.0v, Fuel Cutoff inactive, 1st gear commanded "on", and the PCM detected a high voltage with the TCC solenoid commanded to 90%, or a low voltage with the TCC commanded to 0% during the test. **Possible Causes** ● TCC solenoid control circuit is open or shorted to ground ● TCC solenoid control circuit is shorted to system power (B+) ● TCC solenoid power circuit is open (test the ENG CNTL fuse) ● TCC solenoid is damaged or has failed ● PCM has failed
DTC P1860 **1T CCM, MIL: Yes** 1999, 2000, 2001, 2002, 2003, 2004, 2005 Aurora, Alero, Bonneville, Century, LSS, Cutlass Supreme, Grand Prix, Impala, Lumina, LeSabre, LSS, Regal, Park Avenue, Eighty Eight, Ninety Eight, Montana, Silhouette, Rendezvous & Venture 3.1L VIN J, 3.4L VIN E, 3.1L VIN M, 3.5L VIN H, 3.8L VIN 1, 3.8L VIN K Transmissions: A/T	**TCC PWM Solenoid Circuit Malfunction (4T65-E) Conditions:** Engine started, engine runtime over 5 seconds, Fuel Shutoff inactive, and the PCM detected an unexpected "low" voltage with a 10% command or unexpected "high" voltage with a 95% command. The Torque Converter Clutch (TCC) PWM solenoid controls fluid acting on the converter clutch valve. The clutch valve controls the application and release of the torque converter clutch. Ignition voltage is provided to the torque converter clutch (TCC) solenoid valve. The PCM controls the solenoid with a negative duty cycle in order to control application and release of the TCC. When the solenoid is commanded "off", the PCM senses high voltage. When it is commanded "on", the PCM senses low voltage **Possible Causes** ● TCC solenoid control circuit is open, shorted to ground or shorted to system power (B+) ● TCC solenoid power circuit is open (check the TRANS fuse) ● TCC solenoid is damaged or has failed ● PCM has failed ● TSB 02-07-30-022A contains a repair procedure for this code

OBD II Trouble Code List (P0xxx Codes)

DTC	Trouble Code Title, Conditions & Possible Causes
DTC P1860 **1T CCM, MIL: Yes** 1996, 1997, 1998, 1999, 2000, 2001, 2002, 2003, 2004, 2005 Aurora, DeVille, Eldorado & Seville 4.0L VIN C, 4.6L VIN 9, 4.6L VIN Y engines Transmissions: A/T	**TCC PWM Solenoid Circuit Malfunction (4T80-E) Conditions:** Engine started, engine runtime over 5 seconds, system voltage over 10.0v, Fuel Cutoff inactive, and the PCM detected a high voltage with the TCC solenoid commanded to 90%, or a low voltage with the TCC commanded to 0%. Ignition voltage is provided to the TCC PWM solenoid valve. The PCM controls the "on" and "off" time of the solenoid by providing a ground path through the output driver module (ODM) to control application and release of the TCC. With the solenoid commanded "off", the PCM should sense high voltage. With the solenoid commanded "on", it should sense low voltage. **Possible Causes** ● TCC solenoid control circuit is open or shorted to ground ● TCC solenoid control circuit is shorted to system power (B+) ● TCC solenoid power circuit is open (check the TRANS fuse) ● TCC solenoid is damaged or has failed ● PCM has failed
DTC P1860 **1T CCM, MIL: Yes** 1996, 1997, 1998, 1999, 2000, 2001, 2002, 2003, 2004, 2005 C/K Series Truck, G Van, Cargo Van, L/M Van, S/T Blazer & S/T Pickup 4.3L VIN W, 4.3L VIN X Transmissions: A/T	**TCM PWM Solenoid Circuit Malfunction (4L60-E) Conditions:** Engine started, engine runtime over 5 seconds, system voltage over 10.0v, Fuel Cutoff inactive, 1st gear commanded "on", and the PCM detected a high voltage with the TCC solenoid commanded to 90%, or a low voltage with the TCC commanded to 0%. The TCC PWM solenoid controls fluid acting on the converter clutch valve that controls the application and release of the torque converter clutch. The solenoid attaches to the control valve body in the transmission. **Possible Causes** ● TCC solenoid control circuit is open or shorted to ground ● TCC solenoid control circuit is shorted to system power (B+) ● TCC solenoid power circuit is open (test TRANS or IGN fuse) ● TCC solenoid is damaged or has failed ● PCM has failed
DTC P1860 **1T CCM, MIL: Yes** 1996, 1997, 1998, 1999, 2000, 2001, 2002, 2003, 2004, 2005 C/K Truck, G Series Van, M/L Series Van, S/T Blazer & S/T Pickup 4.3L VIN W, 4.3L VIN X, 4.8L VIN V, 5.0L VIN M, 5.3L VIN T, 5.3L VIN Z, 5.7L VIN K, 5.7L VIN R, 6.0L VIN N, 6.0L VIN U, 7.4L VIN J, 8.1L VIN G Transmissions: A/T	**TCM PWM Solenoid Circuit Malfunction (4L60-E, 4L65-E, 4L80-E) Conditions:** Engine started, engine runtime over 5 seconds, system voltage over 10.0v, Fuel Cutoff inactive, 1st gear commanded "on", and the PCM detected a high voltage with the TCC solenoid commanded to 90%, or a low voltage with the TCC commanded to 0%. The TCC PWM solenoid controls fluid acting on the converter clutch valve that controls the application and release of the torque converter clutch. The solenoid attaches to the control valve body in the transmission. **Possible Causes** ● TCC solenoid control circuit is open or shorted to ground ● TCC solenoid control circuit is shorted to system power (B+) ● TCC solenoid power circuit is open (test TRANS or IGN fuse) ● TCC solenoid is damaged or has failed ● PCM has failed
DTC P1860 **2T CCM, MIL: Yes** 1997, 1998, 1999, 2000, 2001, 2002, 2003, 2004, 2005 Camaro, Corvette & Firebird 5.7L VIN G engine, 5.7L VIN S Transmissions: A/T	**TCM PWM Solenoid Circuit Malfunction (4L80-E) Conditions:** Engine runtime over 5 seconds, system voltage over 10.0v, Fuel Cutoff mode "off", 1st gear commanded "on", and the PCM detected a high voltage with the TCC solenoid command at 90%, or a low voltage with the TCC command at 0%. The TCC PWM solenoid controls fluid acting on the converter clutch valve to control the On/Off application of the converter clutch. **Possible Causes** ● TCC solenoid control circuit is open, shorted to ground and to power (B+) ● TCC solenoid power circuit is open (test the ENG IGN fuse) ● TCC solenoid is damaged or has failed ● PCM has failed
DTC P1860 **2T CCM, MIL: Yes** 1997, 1998, 1999, 2000, 2001, 2002, 2003, 2004, 2005 Achieva, Alero, Cavalier, Malibu, Skylark & Sunbird 2.2L VIN 4, 2.4L VIN T Transmissions: A/T	**TCC PWM Solenoid Electrical (4T40/4T45-E) Conditions:** Engine runtime over 5 seconds, system voltage over 10.0v, Fuel Cutoff inactive, 1st gear commanded "on", and the PCM detected a high voltage with the TCC solenoid commanded to 90%, or a low voltage with the TCC commanded to 0%. The TCC PWM solenoid controls fluid acting on the converter clutch valve that controls the application and release of the torque converter clutch. The solenoid attaches to the control valve body in the transmission. **Possible Causes** ● TCC solenoid control circuit is open, shorted to ground or shorted to system power (B+) ● TCC solenoid power circuit is open (check the TRANS fuse) ● TCC solenoid is damaged or has failed ● PCM has failed ● TSB 02-07-30-022A contains a repair procedure for this code

OBD II Trouble Code List (P0xxx Codes)

DTC	Trouble Code Title, Conditions & Possible Causes
DTC P1860 **1T CCM, MIL: Yes** 1996, 1997, 1998, 1999, 2000, 2001, 2002, 2003, 2004, 2005 S/T Blazer 2.2L VIN 4, 2.2L VIN 5 Transmissions: A/T	**TCM PWM Solenoid Circuit Malfunction (4L60-E) Conditions:** Engine runtime over 5 seconds, system voltage over 10.0v, Fuel Cutoff inactive, 1st gear commanded "on", and the PCM detected a high voltage with the TCC solenoid command at 90%, or a low voltage with the TCC command at 0%. The TCC PWM solenoid controls fluid acting on the converter clutch valve to control the On/Off application of the converter clutch. **Possible Causes** • TCC solenoid control circuit is open, shorted to ground or shorted to system power (B+) • TCC solenoid power circuit is open (check the TRANS fuse) • TCC solenoid is damaged or has failed • PCM has failed
DTC P1860 **1T CCM, MIL: Yes** 2002, 2003, 2004, 2005 Envoy & TrailBlazer 4.2L VIN S, 5.3L VIN P Transmissions: A/T	**TCM PWM Solenoid Circuit Malfunction (4L60-E, 4L65-E) Conditions:** Engine started, engine runtime over 5 seconds, system voltage from 10-18vv, engine not in Fuel Cutoff mode, 1st gear commanded "on", and the PCM detected a high voltage with the solenoid commanded to 90%, or a low voltage with the solenoid commanded to 0%. The TCC PWM solenoid controls fluid that acts on converter clutch valve to controls the application and release of the torque converter clutch. **Possible Causes** • TCC solenoid control circuit is open or shorted to ground • TCC solenoid control circuit is shorted to system power (B+) • TCC solenoid power circuit is open (check the IGN 0 fuse) • TCC solenoid is damaged or has failed • PCM has failed
DTC P1864 **1T CCM, MIL: Yes** 1996 Aurora, Alero, Bonneville, Camaro, Century, LSS, Corvette, Firebird, Cutlass Supreme, Grand Prix, Impala, Lumina, LeSabre, LSS, Regal, Park Avenue, Eighty Eight, Ninety Eight, Montana, Silhouette, Rendezvous & Venture 3.1L VIN M, 3.4L VIN X, 3.4L VIN E, 3.8L VIN 1, 3.8L VIN K, 4.3L VIN W, 5.7L VIN 5, 5.7L VIN P Transmissions: All	**TCC Enable Solenoid Circuit Malfunction (4T60-E) Conditions:** DTC P0560 not set, engine started, engine speed over 500 rpm for 5 seconds, Fuel Cutoff inactive, and the PCM detected an unexpected voltage condition on the TCC Solenoid control circuit during the test. In conjunction with the TCC PWM solenoid valve, the TCC solenoid valve controls the fluid that acts on the TCC converter clutch valve in order to control the TCC apply and release. The PCM monitors the several sensor input to determine when to energize this solenoid. **Possible Causes** • TCC enable solenoid circuit is open or shorted to ground • TCC enable solenoid circuit is shorted to system power (B+) • TCC enable solenoid B+ circuit is open (test fuse in fuse block) • TCC enable solenoid is damaged or has failed • PCM has failed
DTC P1864 **1T CCM, MIL: Yes** 1996 Cab & Chassis, C/K Truck, G Van, L/M Van, S/T Blazer & S/T Pickup 4.3L VIN W, 4.3L VIN X, 5.0L VIN M, 5.7L VIN R, 7.4L VIN J engines Transmissions: A/T	**TCC Enable Solenoid Circuit Malfunction (4T60-E) Conditions:** DTC P0560 not set, engine started, engine speed over 500 rpm for 5 seconds, Fuel Cutoff inactive, and the PCM detected an unexpected voltage condition on the TCC Solenoid control circuit during the test. In conjunction with the TCC PWM solenoid valve, the TCC solenoid valve controls the fluid that acts on the TCC converter clutch valve in order to control the TCC apply and release. The PCM monitors the several sensor input to determine when to energize this solenoid. **Possible Causes** • TCC enable solenoid circuit is open or shorted to ground • TCC enable solenoid circuit is shorted to system power (B+) • TCC enable solenoid B+ circuit is open (test fuse in fuse block) • TCC enable solenoid is damaged or has failed • PCM has failed
DTC P1864 **2T CCM, MIL: Yes** 2003, 2004, 2005 CTS (Cadillac) 2.6L VIN M, 3.2L VIN N Transmissions: A/T	**A/T 4-5 Shift Solenoid Circuit Malfunction (5L40-E) Conditions:** Engine runtime over 5 seconds, system voltage from 8-18v, and the TCM detected an open in the 4-5 shift solenoid valve circuit with the HSD2 commanded "on", a short to ground in the 4-5 shift solenoid valve circuit with the HSD2 commanded "off". The TCM provides voltage to the solenoid through High Side Driver 2 (HSD2). The TCM uses a second driver to control the solenoid ground circuit. The controlled ground driver reports feedback signal to the TCM. When the TCM commands the 4-5 Shift Solenoid valve "on", the control circuit voltage should be 0v. When the TCM commands the 4-5 Shift Solenoid valve "off", the control circuit voltage should be near 12v. **Possible Causes** • A/T 4-5 shift solenoid control circuit open or shorted to ground • A/T 4-5 shift solenoid is damaged or has failed • Transmission internal harness assembly open or shorted • TCM has failed

OBD II Trouble Code List (P0xxx Codes)

DTC	Trouble Code Title, Conditions & Possible Causes
DTC P1864 **1T CCM, MIL: Yes** 1995 Camaro & Firebird 3.8L VIN K engine Transmissions: A/T	**TCC Enable Solenoid Circuit Malfunction (4T60-E) Conditions:** DTC P0560 not set, engine started, engine speed over 500 rpm for 5 seconds, Fuel Cutoff inactive, and the PCM detected an unexpected voltage condition on the TCC Solenoid control circuit during the test. The TCC and TCC PWM solenoids control the fluid that acts on the TCC converter clutch valve to control the TCC apply and release. **Possible Causes** ● TCC enable solenoid circuit is open or shorted to ground ● TCC enable solenoid B+ circuit is open (test fuse in fuse block) ● TCC enable solenoid is damaged or has failed ● PCM has failed
DTC P1865 **2T CCM, MIL: Yes** 2003, 2004, 2005 CTS (Cadillac) 2.6L VIN M, 3.2L VIN N Transmissions: A/T	**A/T 4-5 Shift Solenoid Circuit Malfunction (5L40-E) Code Conditions:** Engine runtime over 5 seconds, system voltage from 8-18v, and the TCM detected a short to power in the 4-5 Shift Solenoid valve circuit. The TCM provides voltage to the solenoid through High Side Driver 2 (HSD2). The TCM uses a second driver to control the solenoid ground circuit. The controlled ground driver reports feedback signal to the TCM. When the TCM commands the 4-5 Shift Solenoid valve "on", the control circuit voltage should be 0v. When the TCM commands the 4-5 Shift Solenoid valve "off", the control circuit voltage should be near 12v. **Possible Causes** ● A/T 4-5 shift solenoid control circuit is shorted to power ● A/T 4-5 shift solenoid is damaged or has failed ● Transmission internal harness assembly is shorted to power ● TCM has failed
DTC P1865 **2T CCM, MIL: Yes** 2003, 2004, 2005 CTS (Cadillac) 2.6L VIN M, 3.2L VIN N Transmissions: A/T	**A/T TCC Solenoid PWM Circuit Low Input (5L40-E) Conditions:** Engine runtime over 5 seconds, system voltage from 8-18v, HSD2 commanded "on", and the TCM detected an open in the TCC PWM solenoid valve circuit when the PWM duty cycle is over 20%, or it detected a short to ground in the solenoid valve circuit with the PWM duty cycle from 20-50%. The TCM provides power to the solenoid through High Side Driver 2 (HSD2). The TCM provides power to the torque converter clutch pulse width modulated (TCC PWM) solenoid valve through High Side Driver 2 (HSD2). **Possible Causes** ● TCC PWM solenoid control circuit is open or shorted to ground ● TCC PWM solenoid is damaged or has failed ● Transmission internal harness assembly is open or shorted ● TCM has failed
DTC P1867 **2T CCM, MIL: Yes** 2003, 2004, 2005 CTS (Cadillac) 2.6L VIN M, 3.2L VIN N Transmissions: A/T	**A/T TCC Solenoid PWM Circuit High Input (5L40-E) Conditions:** Engine runtime over 5 seconds, system voltage from 8-18v, HSD2 commanded "on", and the TCM detected a short to ground on the TCC PWM solenoid valve circuit with the PWM duty cycle over 45%. The TCM provides power to the solenoid through High Side Driver 2 (HSD2). The TCM provides power to the torque converter clutch pulse width modulated (TCC PWM) solenoid valve through High Side Driver 2 (HSD2). **Possible Causes** ● TCC PWM solenoid control circuit is open or shorted to ground ● TCC PWM solenoid is damaged, or the TCM has failed ● Transmission internal harness assembly is open or shorted
DTC P1868 **1T CCM, MIL: No** 2003, 2004, 2005 CTS (Cadillac) 2.6L VIN M, 3.2L VIN N Transmissions: A/T	**A/T TCC Solenoid PWM Circuit High Input (5L40-E) Conditions:** DTC P0218, P0711, P0712 and P0713 not set, engine started, system voltage from 8-18v, and the TCM detected the calculated transmission oil life remaining was less than 10%. Transmission fluid life is determined by monitoring transmission fluid temperature. The TCM calculates transmission fluid life based on a set of temperature ranges stored in the TCM memory. As the vehicle is driven, a counter increments. The rate at which the counter increases depends on the temperature range under current conditions. The higher the temperature range, the faster the counter increments. As the counts increment, the TCM will begin to decrease the fluid life percentage from 100, high fluid life, to 0, low fluid life. If the TCM detects a calculated transmission oil life of 10% or less, DTC P1868 is set. **Possible Causes** ● Replace the transmission fluid and filter ● Change Transmission Fluid appear Driver Information Center ● Reset the Oil Life Monitor after the fluid and filter are changed
DTC P1868 **1T CCM, MIL: No** 2000, 2001, 2002, 2003, 2004, 2005 Aurora, DeVille, Eldorado & Seville 4.0L VIN C, 4.6L VIN 9, 4.6L VIN Y engines Transmissions: A/T	**Transmission Fluid Life (4T80-E) Conditions:** DTC P0711, P0712 and P0713 not set, engine running, system voltage over 10.0v, and the PCM detected the calculated transmission oil life remaining was 10% or less. Also, the Driver Information Center (DIC) will display TRANSMISSION FLUID QUALITY POOR. **Possible Causes** ● Change the ATF (fluid) and transmission filter and refill the fluid to the correct fluid level ● Reset Transmission Oil Life Monitor to 100% with a Scan Tool ● Check for transmission cooling system blockage and restriction

OBD II Trouble Code List (P0xxx Codes)

DTC	Trouble Code Title, Conditions & Possible Causes
DTC P1870 **2T CCM, MIL: Yes** 1995, 1996, 1997, 1998, 1999, 2000, 2001, 2002, 2003, 2004, 2005 Camaro, Corvette & Firebird 3.8L VIN K engine, 5.7L VIN 5, 5.7L VIN P, 5.7L VIN G, 5.7L VIN S Transmissions: A/T	**Transaxle Component Slipping (4T60-E) Conditions:** DTC P0502, P0503, P0740, P0753, P0758, P0785 and P1860 not set, engine started, vehicle driven to a speed of 30-82 at an engine speed of 1000-3000 rpm, Fuel Shutoff inactive, TP angle from 8-35%, commanded gear not in 1st gear and the gear range is D4, TFT sensor from 68ºF-266ºF, engine torque from 40-450 lb ft, The speed ratio is 0.65-0.98 (speed ratio is engine speed divided by transmission output speed), shift solenoid performance diagnostic counters are zero, TCC commanded "on" with maximum apply for 5 seconds, and the PCM detected the TCC slip speed was from 130-800 rpm, fault detected 3 times with TCC commanded "off" in between cycles. The PCM monitors the difference between engine and transmission output speed. In D3 drive range with the TCC engaged, the engine speed should closely match the transmission output speed. In D4 drive range with the TCC engaged, the TCC slip speed should be from -20 to +40 rpm. If the PCM detects excessive TCC slip when the TCC should be engaged, it will set DTC P1870. **Possible Causes** ● 1-2 shift solenoid valve has sediment, damage or leaking seals ● 2-3 shift solenoid valve has sediment, damage or leaking seals ● 3-2 shift solenoid valve has sediment, damage or leaking seals ● Valve body regulator apply valve stuck or regulator is scored ● Torque converter front stator shaft bushing is worn, the stator roller clutch is not holding ● Converter clutch valve is stuck or it is installed backwards ● Converter clutch valve retaining ring is not positioned properly ● Converter clutch outer valve spring is cocked ● Pump to case gasket is not positioned properly ● Orifice cup plugs are restricted or damaged ● Over-tightened, or unevenly tightened pump body to cover bolts
DTC P1870 **1T CCM, MIL: Yes** 1996, 1997, 1998, 1999, 2000, 2001, 2002, 2003, 2004, 2005 Achieva, Alero, Century, Bonneville, Corvette, Corsica, LSS, Cutlass Supreme, Grand Prix, Impala, Lumina, LeSabre, LSS, Regal, Park Avenue, Eighty Eight, Ninety Eight, Montana, Silhouette, Rendezvous & Venture Engines: 3.1L VIN M, 3.1L VIN J, 3.4L VIN X, 3.4L VIN E, 3.8L VIN 1, 3.8L VIN K, 4.3L VIN W, 4.3L VIN X engines Transmissions: A/T	**Transaxle Component Slipping (4T60-E) Conditions:** DTC P0502, P0503, P0740, P0753, P0758, P0785 and P1860 not set, engine started, vehicle driven to a speed of 30-82 mph at an engine speed of 1000-3000 rpm, Fuel Shutoff inactive, TP angle from 8-35%, Transaxle not in First gear, Transaxle gear range is D4, TFT sensor 68ºF-266ºF, engine torque from 50-200 lb ft, engine vacuum 0-105 kPa, TCC commanded "on" with maximum apply for 5 seconds, and the PCM detected the TCC slip speed was from 200-1500 rpm for over 5 seconds. The fault must be detected three times with the TCC commanded "off" each time in between cycles. **Possible Causes** ● Check the ATF level and condition (look for burnt fluid) ● Transmission may have internal damage (a mechanical fault)
DTC P1870 **1T CCM, MIL: Yes** 1996, 1997, 1998, 1999, 2000, 2001, 2002, 2003, 2004, 2005 Cab & Chassis, C/K Truck, G Van, L/M Van, S/T Blazer & S/T Pickup 4.3L VIN W, 4.3L VIN X, 4.8L VIN V, 5.0L VIN M, 5.3L VIN T, 5.3L VIN Z, 5.7L VIN K, 5.7L VIN R, 6.0L VIN U, 6.6L VIN 1, 7.4L VIN J, 8.1L VIN G Transmissions: A/T	**Transmission Component Slipping (4L60-E, 4L65-E, 4L80-E) Conditions:** DTC P0122, P0123, P0502, P0503, P0711-P0713, P0740, P0753, P0758, P1810 and P1860 not set, vehicle driven at a speed of 30-70 mph at an engine speed of 1500-3000 rpm, Fuel Cutoff inactive, TP angle from 9-35%, engine vacuum 0-150 kPa, speed ratio is 0.69-0.88, Transmission not 1st gear, gear range is D4, TFT sensor from 68ºF-266ºF, shift solenoid diagnostic counter at zero, then with the TCC solenoid commanded "on" at a 95% duty cycle for 5 seconds, the PCM detected the TCC slip speed was 130-180 rpm for 7 seconds. The fault must be detected three times with the TCC commanded "off" each time between cycles. **Possible Causes** ● 1-2 shift solenoid valve has sediment, damage or leaking seals ● 2-3 shift solenoid valve has sediment, damage or leaking seals ● 3-2 shift solenoid valve has sediment, damage or leaking seals ● Valve body regulator apply valve stuck or regulator is scored ● Torque converter front stator shaft bushing is worn, the stator roller clutch is not holding or it has external damage/leaks ● Converter clutch valve is stuck or it is installed backwards ● Converter clutch valve retaining ring is not positioned properly ● Converter clutch outer valve spring is cocked ● Pump to case gasket is not positioned properly ● Orifice cup plugs are restricted or damaged ● Over-tightened, or unevenly tightened pump body to cover bolts ● TSB 02-07-30-001 contains a repair procedure for this code

OBD II Trouble Code List (P0xxx Codes)

DTC	Trouble Code Title, Conditions & Possible Causes
DTC P1870 **2T CCM, MIL: Yes** 2002, 2003, 2004, 2005 Envoy & TrailBlazer 4.2L VIN S engine Transmissions: A/T	**Transmission Component Slipping (4L60-E) Conditions:** DTC P0502, P0503, P0740, P0742, P0743, P0748, P1120, P1220, P1810 and P1860 not set, engine runtime over 5 seconds, engine not in Fuel Cutoff mode, engine speed from 1,500-3,000 rpm, vehicle speed from 30-82 mph, TP angle from 20-99%, speed ratio from 0.64-1.35, TFT sensor from 68-302ºF, engine vacuum from 0-105 kPa, engine torque from 50-400 lb ft., commanded gear not in 1st gear, gear range is D4, Shift Solenoid diagnostic counters at zero, TCC commanded "on" at a 40% duty cycle with the TCC slip speed from 130-800 rpm for 7 seconds, then during Condition 1: If the TCC slip speed is 130-800 RPM for 7 seconds, then the PCM commanded maximum line pressure and prevent the freeze shift adapts from being updated, or during Condition 2: then with the TCC slip speed 130-800 rpm for 7 seconds, the PCM commanded the TCC off for 1.5 seconds, then during Condition 3: with the TCC slip speed at 130-800 rpm for 7 seconds, then the fail counter on the current key cycle incremented. Note that the above slip conditions and actions may be disregarded if the TCC is commanded off due to certain unusual driving actions. The PCM monitors the difference between the engine speed and transmission output speed. In D3 drive range with the TCC engaged, the engine speed should closely match the transmission output speed. The TCC slip speed should be -20 to +50 rpm In D4 range with torque converter clutch engaged. **Possible Causes** ● ATF is burnt or contaminated ● Transmission has an internal malfunction ● TCC PWM solenoid seals are damaged or the TCC PWM solenoid is damaged or failed ● Transmission has failed
DTC P1870 **2T CCM, MIL: Yes** 1996, 1997, 1998, 1999, 2000, 2001, 2002 Achieva, Alero, Cavalier, Malibu, Skylark & Sunbird 2.2L VIN 4, 2.2L VIN 5, 2.2L VIN F, 2.4L VIN T Transmissions: A/T	**Transaxle Component Slipping (3T40) Conditions:** DTC P0121, P0122, P0123, P0502 and P0503 not set, vehicle driven to a speed of 35-75 mph at an engine speed of 1500-3600 rpm, TP angle from 10-40%, engine torque more than 20 lb ft, TCC commanded "on" at a 95% duty cycle for 5 seconds, TCC slip speed more than 140 rpm for 8 seconds, speed ratio from 0.90-1.375, and the PCM detected the Transmission slip counter was equal to or higher than one count (fault must occur 3 times). **Possible Causes** ● Check the ATF level and condition (look for burnt fluid) ● Transmission may have internal damage (a mechanical fault)
DTC P1870 **2T CCM, MIL: Yes** 1997, 1998, 1999, 2000, 2001 Catera 3.0L VIN R engine Transmissions: A/T	**Transmission Component Slipping (4L30-E) Conditions:** DTC P0730 not set, P0751, P0752 and P0757 tests enabled and Fail Cases 1 and 2 Conditions not met yet, engine running with the transmission not in Park, Reverse or Neutral position and not in Default mode, TCC commanded "on", and the TCM detected the transmission slip rate was more than 288 rpm for 3.285 seconds. The TCM monitors the transmission output speed, engine speed and the current gear. The TCM calculates turbine shaft speed and TCC slip speed by using these inputs. The engine speed should be close to the transmission output speed in 3rd Gear with TCC engaged, and the TCC slip speed should be -20 to +64 rpm in 4th gear with TCC engaged. If not, the TCM sets this code. *Note: The TCM flashes the Sports Lamp when this code is set.* **Possible Causes** ● Check the ATF level and condition (look for burnt fluid) ● Transmission may have internal damage (a mechanical fault)
DTC P1875 **2T CCM, MIL: Yes** 1996, 1997, 1998, 1999, 2000, 2001, 2002, 2003, 2004, 2005 Cab & Chassis, C/K Truck, G Van, L/M Van, S/T Blazer & S/T Pickup 4.3L VIN W, 4.3L VIN X, 4.8L VIN V, 5.0L VIN M, 5.3L VIN T, 5.3L VIN Z, 5.7L VIN K, 5.7L VIN R, 6.0L VIN N, 6.0L VIN U, 6.6L VIN 1, 7.4L VIN J, 8.1L VIN G engines Transmissions: A/T	**4WD Low Switch Circuit Fault (4L60-E, 4L65-E, 4L80-E) Conditions:** DTC P0122, P0123, P0502, P0503, P0740, P0742, P0751, P0752, P0756, P0758, P1810, P1860 and P1870 not set, engine started, vehicle driven to a speed over 7 mph for 5 seconds, gear range is D4, Fuel Cutoff not active, TP angle from 17-50%, engine torque from 50-400 lb ft, engine vacuum from 0-105 kPa, shift solenoid performance counters at zero, TFT sensor from 68-266ºF, then during Condition 1 with the 4WD Low switch in 4WD low, transfer case not in 4WD low, TCC slip speed from -3000 to -50 RPM, the PCM detected the speed ratio was 0.8-1.2; or during Condition 2 with the 4WD Low switch not in 4WD low, transfer case in 4WD low, TCC commanded "on", TCC slip speed was 100 to 3000 rpm, the PCM detected the speed ratio was 2.5-2.9 for 10 seconds. **Possible Causes** ● 4WD low switch signal circuit is open, shorted to ground or B+ ● 4WD low switch is damaged or has failed ● PCM has failed

OBD II Trouble Code List (P0xxx Codes)

DTC	Trouble Code Title, Conditions & Possible Causes
DTC P1875 **2T CCM, MIL: Yes** 2002, 2003, 2004, 2005 Envoy & TrailBlazer 5.3L VIN P engine Transmissions: A/T	**4WD Low Switch Circuit Malfunction (4L60-E, 4L80-E) Conditions:** DTC P0120, P0220, P0502, P0503, P0740, P0742, P0751, P0753, P0756, P0758, P1810, P1860 and P1870 not set, engine runtime over 5 seconds, Fuel Cutoff mode "off", vehicle speed over 7 mph, TP angle from 17-50%, engine torque from 40-400 lb ft., engine vacuum from 0-105 kPa, gear range is D4, Shift Solenoid performance counters at zero, TFT sensor from 68-266ºF, then during Condition 1 with the 4WD Low switch in 4WD low, Transfer Case not in 4WD low, TCC slip speed from -3000 to -50 rpm, and the PCM detected the Transfer Case ratio was 0.8-1.2; or during Condition 2 with the 4WD Low switch not in 4WD low, Transfer Case in 4WD low, TCC commanded "on", TCC slip speed from 100 to 3000 rpm, the PCM detected the Transfer Case ratio was 2.5-2.9, all conditions met for 10 seconds. **Possible Causes** ● 4WD switch low signal circuit is open, or shorted to ground ● 4WD low switch is damaged or it has failed ● PCM has failed
DTC P1886 **1T CCM, MIL: Yes** 1995, 1996 Camaro, Caprice, Firebird & Rivera 3.8L VIN K, 4.3L VIN W Transmissions: A/T	**A/T3-2 Shift Solenoid Circuit Malfunction (4T60-E) Conditions:** DTC P0560 not set, engine runtime over 5 seconds, Fuel Shutoff off, and the PCM detected an unexpected voltage condition on the A/T 3-2 Shift Solenoid control circuit for 5 seconds. **Possible Causes** ● 3-2 shift solenoid control circuit is open or shorted to ground ● 3-2 shift solenoid control circuit is shorted to system power (B+) ● 3-2 shift solenoid power circuit is open (test the PCM/IGN fuse) ● 3-2 shift solenoid is damaged or has failed ● PCM has failed
DTC P1887 **2T CCM, MIL: Yes** 1997, 1998, 1999, 2000, 2001, 2002 Achieva, Alero, Cavalier, Malibu, Skylark & Sunbird 2.2L VIN 4, 2.4L VIN T Transmissions: A/T	**TCC Release Switch Circuit Malfunction (4T40/4T45-E) Conditions:** DTC P0716, P0717, P0741 and P0742 not set, vehicle driven to a speed of 30-70 mph, TFT manual valve position switch indicating D4 range, engine torque over 33 lb ft, TCC commanded "on", TCC slip speed from -20 to +40 rpm, and the PCM detected the TCC release pressure was present for 6 seconds (fault must occur twice during the active trip). **Possible Causes** ● TCC release switch signal circuit is open or the PCM has failed ● Turbine shaft O-ring seal leaks, oil seal rings missing/damaged. ● TCC control valve damaged or No. 1 check ball is damaged ● Spacer plate release exhaust blocked with debris ● Case cover or spacer plate gaskets improperly installed ● TSB 02-07-30-022A contains a repair procedure for this code
DTC P1887 **2T CCM, MIL: Yes** 1997, 1998, 1999, 2000, 2001, 2002, 2003, 2004, 2005 Aurora, Alero, Bonneville, Aztek, Century, Cutlass Supreme, Grand Prix, Impala, Lumina, LeSabre, LSS, Regal, Park Avenue, Eighty Eight, Ninety Eight, Montana, Silhouette, Rendezvous & Venture 3.1L VIN M, 3.1L VIN J, 3.4L VIN X, 3.4L VIN E, 3.5L VIN H, 3.8L VIN 1, 3.8L VIN K engines Transmissions: A/T	**TCC Release Switch Circuit Malfunction (4T65-E) Conditions:** DTC P0716, P0717, P0741, P0742 and P1810 not set, engine started, Fuel Cutoff inactive, engine driven to a speed of 30-70 mph, engine torque from 30-300 lb ft, Transmission gear is D4 with the TCC commanded "on", TCC pressure from 15-120 psi, TCC slip speed from -20 to +60 rpm, and the PCM detected the pressure switch was open for 6 seconds. The fault must occur twice in 1 trip to set this code. The TCC release switch is normally closed (N.C.) switch that signals the PCM that the TCC is released. This is accomplished by torque converter release fluid pressure acting on the switch contacts that open the circuit. When the circuit voltage is high, the PCM detects the TCC is no longer engaged. If the PCM determines the TCC release switch is open (indicating the TCC is not applied) and the TCC slip speed indicates the TCC is applied, then DTC P1887 sets **Possible Causes** ● TCC release switch signal circuit is open ● Turbine shaft O-ring seal leaks, oil seal rings missing/damaged. ● TCC control valve damaged or No. 1 check ball is damaged ● Spacer plate release exhaust blocked or case cover or spacer plate gaskets damaged ● TSB 02-07-30-022A contains a repair procedure for this code
DTC P1890 **2T CCM, MIL: Yes** 1997, 1998, 1999, 2000, 2001 Catera 3.0L VIN R engine Transmissions: A/T	**ECM Data Input Circuit Malfunction (4L30-E) Conditions:** Engine started, and the PCM detected an unexpected voltage condition on the ECM Data Input circuit for one second. The PCM sends throttle position information to the TCM through the CAN. Two circuits are used to communicate CAN data between the ECM and TCM. A fault in the CAN will not cause DTC P1890 to set by itself. Check for other CAN codes. **Possible Causes** ● ECM data input circuit is open, shorted to ground or B+ ● TCM has failed or the PCM has failed
DTC P1895 **1T CCM, MIL: No** 1997, 1998 Catera 3.0L VIN R engine Transmissions: A/T	**Engine Torque Delivered Circuit Malfunction (4L30-E) Conditions:** Engine started, engine running, and the PCM detected an unexpected voltage condition on the Engine Torque Delivered control circuit during the CCM test. **Possible Causes** ● Engine torque delivered circuit is open, shorted to ground or B+ ● EBTCM has failed ● TCM has failed

OBD II Trouble Code List (P2xxx Codes)

DTC	Trouble Code Title, Conditions & Possible Causes
DTC P2008 **1T CCM, MIL: No** 2003, 2004, 2005 CTS (Cadillac) 2.6L VIN M, 3.2L VIN N Transmissions: A/T	**IMRC Solenoid Control Circuit Malfunction Conditions:** Engine started, system voltage from 8-18v, IMRC solenoid control state changed from open to closed position, and the PCM detected the voltage on the IMRC solenoid control circuit was from 2.4-4.6v with the solenoid "off". The intake manifold upper plenum is divided into two separate runners, one for each bank, by an intake manifold runner control (IMRC) valve. During low speed, high load conditions the IMRC valve is closed creating a longer runner path inside the plenum that increases torque. During period when the engine speed is over 3,200 rpm, the IMRC valve opens creating a shorter runner path inside the plenum to increase the horsepower. The IMRC valve is controlled by an IMRC vacuum solenoid. Manifold vacuum is supplied to the IMRC solenoid. A 12v supply from the Manifold fuse to the IMRC solenoid is applied when the ignition key is turned on. This is a normally closed (N.C.) solenoid that will not allow vacuum to pass through it. When the engine speed and load are increased above a calibrated threshold, the ECM provides a ground to the solenoid, energizing the solenoid and allowing a vacuum to be applied to the IMRC valve. With the IMRC solenoid "off", the control circuit voltage is 12v, and it is near 0v with it commanded "on". The ECM applies a pullup voltage (2.6-4.6v) on the control circuit to differentiate between an open, short to ground, or short to voltage. **Possible Causes** • IMRC solenoid control connector is damaged, open or shorted • IMRC solenoid control circuit is open or shorted to ground • IMRC solenoid power circuit is open (check the Manifold fuse) • IMRC solenoid is damaged or it has failed • PCM has failed
DTC P2009 **1T CCM, MIL: No** 2003, 2004, 2005 CTS (Cadillac) 2.6L VIN M, 3.2L VIN N Transmissions: All	**IMRC Solenoid Control Circuit Low Input Conditions:** Engine started, system voltage from 8-18v, IMRC solenoid control state changed open to closed position, and the PCM detected the voltage on the IMRC solenoid control circuit was near 0v with the solenoid "off". A 12v supply from the Manifold fuse to the IMRC solenoid is applied when the ignition key is turned on. This is a normally closed (N.C.) solenoid that will not allow vacuum to pass through it. As the engine speed and load reach a certain threshold, the PCM grounds the solenoid to allow vacuum to be applied to the IMRC valve. When the IMRC solenoid is commanded off, the voltage of the control circuit is near 12v. When the IMRC solenoid is commanded on, the voltage of the control circuit is near 0v. The ECM applies a pullup voltage (2.6-4.6v) on the control circuit to differentiate between an open, short to ground, or short to voltage. **Possible Causes** • IMRC solenoid control circuit is open or shorted to ground • IMRC solenoid power circuit is open (check the Manifold fuse) • IMRC solenoid is damaged or it has failed • PCM has failed
DTC P2066 **1T CCM, MIL: No** 2003, 2004, 2005 C/K Series Pickup 6.0L VIN N, 6.0L VIN U Transmissions: All	**Secondary Fuel Sensor Signal Range/Performance Conditions:** Engine started; and the PCM did not detect a change in the Secondary fuel level of at least 0.79 gallon after the vehicle traveled a distance of 200 miles. Low fuel indicator will be "on". **Possible Causes** • Perform the I/P Cluster Diagnostic System Check • Check the fuel tank for signs of foreign material (i.e., ice) • Fuel level sender is malfunctioning (check the fuel strainer to determine if it is interfering with the sender float arm - sticking) • Fuel level sensor is damaged or it has failed
DTC P2067 **1T CCM, MIL: No** 2003, 2004, 2005 C/K Series Pickup 6.0L VIN N, 6.0L VIN U Transmissions: All	**Secondary Fuel Level Sensor Circuit Low Input Conditions:** Key on or engine running; and the PCM detected that the Secondary fuel level sensor indicated less than 3.5% for 20 seconds. **Possible Causes** • Perform the I/P Cluster Diagnostic System Check • Fuel level sensor connector is damaged or shorted • Fuel level sensor signal circuit is shorted to ground • Fuel level sensor is damaged or it has failed
DTC P2068 **1T CCM, MIL: No** 2003, 2004, 2005 C/K Series Pickup 6.0L VIN N, 6.0L VIN U Transmissions: All	**Secondary Fuel Level Sensor Circuit High Input Conditions:** Key on or engine running; and the PCM detected that the Secondary fuel level sensor indicated more than 98% for 20 seconds. **Possible Causes** • Perform the I/P Cluster Diagnostic System Check • Fuel level sensor connector is damaged, open or shorted • Fuel level sensor signal circuit is open or shorted to power • Fuel level sensor is damaged or it has failed

OBD II Trouble Code List (P2xxx Codes)

DTC	Trouble Code Title, Conditions & Possible Causes
DTC P2096 **1T CCM, MIL: No** 2003, 2004, 2005 CTS (Cadillac) 2.6L VIN M, 3.2L VIN N Transmissions: All	**HO2S Fuel Trim Bias Commanded Lean (Bank 1) Conditions:** DTC P0030, P0031, P0032, P0036, P0037, P0038, P0050, P0051, P0052, P0056, P0057, P0058, P0101, P0102, P0103, P0116, P0117, P0118, P0125, P0128, P0135, P0141, P0155, P0161, P0171, P0172, P0174, P0175, P0300, P0301-P0306, P0313, P0341, P0342, P0343, P0420, P0430, P0443, P0444 and P0445 not set, engine running in closed loop mode, and the PCM detected the HO2S-11 Fuel Trim bias was commanded lean for over 1.3 seconds of a 200-second monitoring period. Heated oxygen sensors are used for fuel control and post catalyst monitoring. Each HO2S compares the oxygen content of the surrounding air with the oxygen content of the exhaust stream. When the vehicle is first started, the PCM operates in open loop mode, ignoring the HO2S signal voltage when calculating the air fuel ratio. The PCM supplies the HO2S with a reference (bias) voltage of about 450 mv. The HO2S generates a voltage from 0-1000 mv that fluctuates above and below bias voltage once in closed loop. Heating elements in the HO2S minimize the time required for the sensor to reach normal operating temperature, and to provide an accurate voltage signal. **Possible Causes** ● HO2S is contaminated or deteriorated due to poor fuel usage ● Compare the activity of the Bank 1 HO2S-11 and HO2S-12 to a known good bank using the live plot feature of the Scan Tool. Drive the vehicle under various engine load and vehicle speed conditions. Look for any indication of improper operation (due to contamination) of the Bank 1 HO2S-11 or HO2S-12 sensors.
DTC P2097 **1T CCM, MIL: No** 2003, 2004, 2005 CTS (Cadillac) 2.6L VIN M, 3.2L VIN N Transmissions: All	**HO2S Fuel Trim Bias Commanded Rich (Bank 1) Conditions:** DTC P0030, P0031, P0032, P0036, P0037, P0038, P0050, P0051, P0052, P0056, P0057, P0058, P0101, P0102, P0103, P0116, P0117, P0118, P0125, P0128, P0135, P0141, P0155, P0161, P0171, P0172, P0174, P0175, P0300, P0301-P0306, P0313, P0341, P0342, P0343, P0420, P0430, P0443, P0444 and P0445 not set, engine running in closed loop mode, and the PCM detected the Bank 1 HO2S-11 Fuel Trim bias was commanded rich for over 1.3 seconds of a 200-second period. One HO2S is used for fuel control and one for post catalyst monitoring. The PCM uses Fuel Trim Biasing to keep the post catalyst HO2S voltage close to 600 mv to achieve for optimal catalyst efficiency. **Possible Causes** ● HO2S is contaminated or deteriorated due to poor fuel usage ● Compare the activity of the Bank 1 HO2S-11 and HO2S-12 to a known good bank using the live plot feature of the Scan Tool. Drive the vehicle under various engine load and vehicle speed conditions. Look for an indication of improper operation (due to contamination) of the Bank 1 HO2S-11 or HO2S-12 (sensors).
DTC P2098 **1T CCM, MIL: No** 2003, 2004, 2005 CTS (Cadillac) 2.6L VIN M, 3.2L VIN N Transmissions: All	**HO2S Fuel Trim Bias Commanded Lean (Bank 2) Conditions:** DTC P0030, P0031, P0032, P0036, P0037, P0038, P0050, P0051, P0052, P0056, P0057, P0058, P0101, P0102, P0103, P0116, P0117, P0118, P0125, P0128, P0135, P0141, P0155, P0161, P0171, P0172, P0174, P0175, P0300, P0301-P0306, P0313, P0341, P0342, P0343, P0420, P0430, P0443, P0444 and P0445 not set, engine running in closed loop mode, and the PCM detected the Bank 2 HO2S-21 Fuel Trim bias was commanded lean for over 1.3 seconds of a 200 second period. The PCM supplies the HO2S with a bias voltage of 450 mv. The HO2S generates a voltage from 0-1000 mv that fluctuates above and below bias voltage once in closed loop. Heating elements inside the HO2S minimize the time required for the sensor to reach normal operating temperature, and to provide an accurate voltage signal. **Possible Causes** ● HO2S is contaminated or deteriorated due to poor fuel usage ● Compare the activity of the Bank 2 HO2S-21 and HO2S-22 to a known good bank using the live plot feature of the Scan Tool. Drive the vehicle under various engine load and vehicle speed conditions. Look for an indication of improper operation (due to contamination) of the Bank 2 HO2S-21 or HO2S-22 (sensors).
DTC P2099 **2T CCM, MIL: No** 2003, 2004, 2005 CTS (Cadillac) 2.6L VIN M, 3.2L VIN N Transmissions: All	**HO2S Fuel Trim Bias Commanded Rich (Bank 2) Conditions:** DTC P0030, P0031, P0032, P0036, P0037, P0038, P0050, P0051, P0052, P0056, P0057, P0058, P0101, P0102, P0103, P0116, P0117, P0118, P0125, P0128, P0135, P0141, P0155, P0161, P0171, P0172, P0174, P0175, P0300, P0301-P0306, P0313, P0341, P0342, P0343, P0420, P0430, P0443, P0444 and P0445 not set, engine running in closed loop mode, and the PCM detected the Bank 2 HO2S-21 Fuel Trim bias was commanded rich for longer than 1.3 seconds of a 200-second monitoring period. The PCM operates in open loop at initial startup, ignoring the HO2S signal voltage when calculating the air fuel ratio. The PCM supplies the HO2S with a reference (bias) voltage of about 450 mv. The PCM uses Fuel Trim Biasing to keep the post catalyst HO2S voltage as close to 600 mv as possible. This allows for optimal catalyst efficiency under light load conditions (e.g., at idle or steady cruise). **Possible Causes** ● HO2S is contaminated or deteriorated due to poor fuel usage ● Compare the activity of the Bank 2 HO2S-21 and HO2S-22 to a known good bank using the live plot feature of the Scan Tool. Drive the vehicle under various engine load and vehicle speed conditions. Look for an indication of improper operation (due to contamination) of the Bank 2 HO2S-21 or HO2S-22 (sensors).

OBD II Trouble Code List (P2xxx Codes)

DTC	Trouble Code Title, Conditions & Possible Causes
DTC P2100 **1T CCM, MIL: Yes** 2003, 2004, 2005 CTS (Cadillac) 2.6L VIN M, 3.2L VIN N Transmissions: All	**Throttle Air Control Motor Position Too High Conditions:** DTC P0121, P0122, P0123, P0221, P0222 and P0223 not set, key on, system voltage over 10v, ECT sensor signal from 41-212ºF, IAT sensor more than 41ºF, and the PCM determined that the PWM command to move the throttle blades exceeded a predetermined value. The PCM opens the throttle blades by applying a pulsewidth modulated (PWM) signal to the throttle valve motor. The PCM reverses the polarity on the Throttle Valve motor control circuits in order to close the throttle blades. The PCM increases the pulse width as necessary to open the throttle blades. The PCM monitors TP Sensors 1 and 2 to determine the actual blade position. If the PCM detects the required pulsewidth modulation to move the throttle valves exceeded a predetermined value, DTC 2100 is set. **Possible Causes** ● TAC motor control circuit (one or more) is shorted to ground, or shorted to power (B+) ● Throttle body connector is damaged, open or shorted ● Throttle valves are not in the "rest" position ● Throttle valves are binding open or binding closed ● Throttle valves moving open or closed without spring pressure ● Throttle body assembly is damaged or it has failed ● PCM has failed
DTC P2101 **1T CCM, MIL: Yes** 2003, 2004, 2005 CTS (Cadillac) 2.6L VIN M, 3.2L VIN N Transmissions: All	**Throttle Air Control Motor Position Too Low Conditions:** DTC P0121, P0122, P0123, P0221, P0222 and P0223 not set, key on, and the PCM detected the difference between the calculated and the actual throttle position was more than a predetermined value. The PCM controls the throttle blades by applying a varying voltage to the throttle valve motor. The PCM monitors the actual throttle blade position using TP Sensor 1 and 2. If the PCM cannot detect the minimum throttle position, DTC P2101 is set. **Possible Causes** ● TAC motor control circuit (one or more) is open, shorted to ground or to power (B+) ● Throttle body connector is damaged, open or shorted ● Throttle body assembly is damaged or it has failed ● PCM has failed
DTC P2108 **1T CCM, MIL: Yes** 2003, 2004, 2005 C/K Truck, Escalade, Envoy, G Van, TrailBlazer 4.8L VIN V, 5.3L VIN P, 5.3L VIN T, 5.3L VIN Z, 6.0L VIN N, 6.0L VIN U, 8.1L VIN G engines Transmissions: All	**Throttle Actuator Control Module Internal Data Test Failed Conditions:** DTCP1518 not set, engine cranking or running, system voltage over 6.0v, and the TAC determined that its internal data test did not pass, condition met for 1 second. The TAC module contains data that is essential for proper TAC system operation. The TAC module continuously tests the integrity of this data. When the TAC module is unable to write or read data to and from random access memory, or the TAC module was unable to correctly read data from the flash memory or internal TAC processor fault is detected, it sets P2108. **Possible Causes** ● TAC module is damaged or it has failed
DTC P2105 **1T CCM, MIL: Yes** 2003, 2004, 2005 CTS (Cadillac) 2.6L VIN M, 3.2L VIN N Transmissions: All	**Throttle Air Control System Malfunction Conditions:** DTC P0121, P0122, P0123, P0221, P0222, P0223, P0638, P1551, P2100, P2101, P2107, P2119, P2122, P2123, P2127, P2128, P2135, P2138 and P2176 not set, engine started, and the PCM detected a fault in the TAC system with the engine speed over 1,120 rpm for 2 seconds. If the PCM detects a fault condition in the TAC system, the PCM disables the TAC motor. The APP sensor will have no affect on the throttle position and the throttle will remain in the "rest" position. The engine speed will vary due to limited fuel and spark changes based on the APP sensor signal. If the engine speed is over 1300 rpm, the PCM will turn the fuel off for short intervals until the engine speed is below 1,300 rpm. If the PCM detects TAC system engine speed is beyond its normal limits, DTC P2105 is set. **Possible Causes** ● Clear the trouble codes, and if this same code resets, the PCM has failed and must be replaced to repair this trouble code.
DTC P2107 **1T CCM, MIL: Yes** 2003, 2004, 2005 CTS (Cadillac) 2.6L VIN M, 3.2L VIN N Transmissions: All	**Throttle Air Control System Malfunction Conditions:** DTC P0121, P0122, P0123, P0221, P0222 and P0223 not set, key on, system voltage over 10v, ECT sensor signal from 41-212ºF, IAT sensor more than 41ºF, and the PCM determined the system was operating outside of the calibrated range. The PCM controls the throttle blades by applying a varying voltage to the throttle valve motor. The PCM monitors the actual throttle blade position using the TP Sensor 1 and 2. If the PCM detects the TAC system does not operate in the desired throttle (operating) range, DTC P2107 is set. **Possible Causes** ● TAC motor control circuit (one or more) is open, shorted to ground or to power (B+) ● Throttle valves binding open or closed, or moving too freely ● Throttle body connector is damaged, open or shorted ● Throttle body assembly has failed, or the PCM has failed

OBD II Trouble Code List (P2xxx Codes)

DTC	Trouble Code Title, Conditions & Possible Causes
DTC P2110 **2T CCM, MIL: Yes** 2003, 2004, 2005 CTS (Cadillac) 2.6L VIN M, 3.2L VIN N Transmissions: All	**IMRC Solenoid Control Circuit High Input Conditions:** Engine started, system voltage from 8-18v, IMRC solenoid control state changed open to closed position, and the PCM detected the voltage on the IMRC solenoid control circuit was near 12v with the solenoid "on". A 12v supply from the Manifold fuse to the IMRC solenoid is applied when the ignition key is turned on. This is a normally closed (N.C.) solenoid that will not allow vacuum to pass through it. When the engine speed and load are increased above a calibrated threshold, the ECM provides a ground to the solenoid, energizing the solenoid and allowing a vacuum to be applied to the IMRC valve. When the IMRC solenoid is commanded off, the voltage of the control circuit is near 12v. When the IMRC solenoid is commanded on, the voltage of the control circuit is near 0v. The ECM applies a pullup voltage (2.6-4.6v) on the control circuit to differentiate between an open, short to ground, or short to voltage. **Possible Causes** ● IMRC solenoid control connector is damaged or shorted to B+ ● IMRC solenoid control circuit is open or shorted to power ● IMRC solenoid is damaged or it has failed ● PCM has failed
DTC P2119 **1T CCM, MIL: Yes** 2003, 2004, 2005 CTS (Cadillac) 2.6L VIN M, 3.2L VIN N Transmissions: All	**Throttle Air Control System Malfunction Conditions:** DTC P0121, P0122, P0123, P0221, P0222 and P0223 not set, key on or engine running, system voltage over 10v, ECT sensor from 41-212°F, IAT sensor more than 41°F, APP sensor less than 10%, and the PCM determined the system was operating outside of the calibrated range. The PCM controls the throttle blades by applying a varying voltage to the throttle valve motor. The PCM monitors the actual throttle blade position using signals from the TP Sensor 1 and 2. If the PCM detects the TAC system does not return to the "rest" position within a calibrated time, DTC P2119 is set. **Possible Causes** ● TAC motor control circuit (one or more) is open or shorted (B+) ● TAC motor control circuit (one or more) is shorted to ground ● Throttle valves binding open or closed, or moving too freely ● Throttle body connector is damaged, open or shorted ● Throttle body assembly is damaged or it has failed ● PCM has failed
DTC P2120 **1T CCM, MIL: Yes** 2003, 2004, 2005 C/K Truck, Escalade, G Van, Envoy & TrailBlazer 4.8L VIN V, 5.3L VIN P, 5.3L VIN T, 5.3L VIN Z, 6.0L VIN N, 6.0L VIN U, 8.1L VIN G engines Transmissions: All	**Accelerator Pedal Position Sensor 1 Signal Performance Conditions:** DTC P0601, P0602, P0606, P1518 and P2108 not set; engine cranking or running, system voltage more than 5.23v, and the PCM detected the APP Sensor 1 signal circuit voltage was less than 0.24v or more than 4.49v, or that the APP VREF (5v) circuit was less than 4.54v or more than 5.21v. The PCM provides the APP sensor with a 5v reference circuit and a low reference circuit. The APP sensor provides the control module a signal voltage proportional to pedal movement. The APP sensor 1 signal voltage is low at rest and increases as the pedal is depressed. When the control module detects that the APP sensor 1 signal or APP sensor 5-volt reference voltage is outside the predetermined range, it sets DTC P2120. **Possible Causes** ● APP sensor connector is damaged, open or shorted ● APP1 sensor signal circuit is open or shorted to ground ● APP1 sensor signal circuit is shorted to APP sensor 2 circuit ● APP1 sensor signal circuit is open or shorted to VREF (5v) ● APP sensor is damaged or it has failed ● TAC module is damaged or it has failed
DTC P2121 **1T CCM, MIL: Yes** 2003, 2004, 2005 C/K Truck, Escalade, G Van, Envoy & TrailBlazer 4.8L VIN V, 5.3L VIN P, 5.3L VIN T, 5.3L VIN Z, 6.0L VIN N, 6.0L VIN U, 8.1L VIN G engines Transmissions: All	**Accelerator Pedal Position Sensor 1-2 Correlation Malfunction Conditions:** DTC P0606, P1518 and P2108 not set, engine cranking or running, system voltage over 5.23v, and the PCM detected the APP Sensor 1 signal disagreed with the APP Sensor 2 signal by over 10.5% for 1 second. The PCM provides the APP sensor with a 5v reference circuit and a low reference circuit. The APP sensor provides the PCM a signal proportional to pedal movement. The APP sensor 1 signal voltage is low at rest and increases as the pedal is depressed. When the control module detects that the APP sensor 1 signal or APP sensor 5-volt reference voltage is outside the predetermined range, it sets DTC P2120. **Possible Causes** ● APP sensor connector is damaged, open or shorted ● APP1 sensor low reference circuit is open or high resistance ● APP1 sensor VREF circuit is open or shorted to ground ● APP sensor is damaged or it has failed ● TAC module is damaged or it has failed

OBD II Trouble Code List (P2xxx Codes)

DTC P2121 **1T CCM, MIL: No** 2002, 2002, 2003, 2004, 2005 C/K Series Pickup 6.6L VIN 1 Diesel engine Transmissions: All	**Accelerator Pedal Position Sensor 1 Low Input Conditions:** DTC P0641 and P0651 not set, engine cranking or running, and the ECM detected the APP Sensor 1 was less than 0.25v. The accelerator pedal position (APP) sensor is made up of 3 individual sensors in a single housing. Three separate signal, low reference, and 5-volt reference circuits are used in order to interface the APP sensor with the ECM. Each sensor has a unique functionality to determine the pedal position. The ECM uses the APP sensor to determine the desired amount of acceleration or deceleration. **Possible Causes** ● APP sensor connector is damaged or shorted ● APP sensor 1 signal circuit is open or shorted to ground ● APP sensor is damaged or it has failed ● ECM has failed
DTC P2122 **1T CCM, MIL: Yes** 2003, 2004, 2005 CTS (Cadillac) 2.6L VIN M, 3.2L VIN N Transmissions: All	**Accelerator Pedal Position Sensor 1 Circuit Low Input Conditions:** Engine started, system voltage over 10v, and the PCM detected the APP1 sensor signal is less than 0.84v for more than 140 ms. The accelerator pedal position (APP) sensor is made up of two sensors that are housed inside one assembly. The APP Sensor 1 sends a signal from the sensor to the PCM indicating the accelerator pedal position. The PCM actuates the throttle plates based on this information. The APP Sensor 1 signal increases as the pedal is depressed, from about 0.8v at rest to above 3.5v fully depressed. If the PCM detects that the signal voltage was excessively low, DTC P2122 is set. **Possible Causes** ● APP1 sensor low reference circuit open or has high resistance ● APP1 sensor signal circuit is open, shorted to ground ● APP1 sensor signal circuit is shorted to low reference circuit ● APP1 5v VREF circuit is open or shorted to ground ● APP1 sensor is damaged or it has failed ● MAF sensor or FTP sensor 5v VREF circuit shorted to ground ● PCM has failed
DTC P2123 **1T CCM, MIL: Yes** 2003, 2004, 2005 CTS (Cadillac) 2.6L VIN M, 3.2L VIN N Transmissions: All	**Accelerator Pedal Position Sensor 1 Circuit High Input Conditions:** Engine started, system voltage over 10v, and the PCM detected the APP1 sensor signal is more than 4.80v for more than 140 ms. The accelerator pedal position (APP) sensor is made up of two sensors that are housed inside one assembly. The PCM supplies a separate 5v reference and the low reference circuit for each of the sensors. The 5v VREF for the APP Sensor 1 (APP1) is supplied from the same source in the PCM as the 5v VREF for the MAF sensor and the FTP sensor. The 5v VREF for all of the sensors is supplied on separate PCM terminals, but the terminals are connected internally to a voltage supply. The APP Sensor 1 sends a signal from the sensor to the PCM indicating the accelerator pedal position. The PCM actuates the throttle plates based on this information. The APP Sensor 1 signal increases as the pedal is depressed, from about 0.8v at rest to above 3.5v fully depressed. If the PCM detects that the signal voltage was excessively high, DTC P2123 is set. **Possible Causes** ● APP1 sensor low reference circuit open or has high resistance ● APP1 sensor signal circuit is open, shorted to low reference or shorted to power ● APP1 5v VREF circuit is open or shorted to power ● APP1 sensor is damaged or it has failed ● MAF sensor or FTP sensor 5v VREF circuit shorted to power ● PCM has failed
DTC P2123 **1T CCM, MIL: Yes** 2002, 2002, 2003, 2004, 2005 C/K Series Pickup 6.6L VIN 1 Diesel engine Transmissions: All	**Accelerator Pedal Position Sensor 1 High Input Conditions:** DTC P0641 and P0651 not set; engine cranking or running and the ECM detected the APP Sensor 1 was more than 4.75v. The accelerator pedal position (APP) sensor is made up of 3 individual sensors in a single housing. Three separate signal, low reference, and 5-volt reference circuits are used in order to interface the APP sensor with the ECM. Each sensor has a unique functionality to determine the pedal position. The ECM uses the APP sensor to determine the desired amount of acceleration or deceleration. **Possible Causes** ● APP sensor 1 signal circuit is shorted to VREF ● APP sensor low reference circuit is open or shorted to VREF ● APP sensor is damaged or it has failed ● ECM has failed
DTC P2125 **1T CCM, MIL: Yes** 2003, 2004, 2005 C/K Truck, Escalade, G Van, Envoy & TrailBlazer 4.8L VIN V, 5.3L VIN P, 5.3L VIN T, 5.3L VIN Z, 6.0L VIN N, 6.0L VIN U, 8.1L VIN G engines Transmissions: All	**Accelerator Pedal Position Sensor 2 Circuit Malfunction Conditions:** DTC P0601, P0602, P0606, P1518 and P2108 not set, engine cranking or running, system voltage over 5.23v, and the PCM detected the APP Sensor 1 signal was less than 0.24v or more than 4.49v, or the APP VREF (5v) circuit was less than 4.54v or more than 5.21v. The APP sensor provides the ECM with a signal voltage proportional to pedal movement. The APP sensor 1 is low at rest and increases as the pedal is depressed. **Possible Causes** ● APP1 sensor signal circuit is open or shorted to ground ● APP1 sensor signal circuit is shorted to APP sensor 2 circuit ● APP1 sensor signal circuit is open or shorted to VREF (5v) ● APP sensor is damaged, or the TAC module is damaged or it has failed

OBD II Trouble Code List (P2xxx Codes)

DTC	Trouble Code Title, Conditions & Possible Causes
DTC P2127 **1T CCM, MIL: Yes** 2002, 2002, 2003, 2004, 2005 C/K Series Pickup 6.6L VIN 1 Diesel engine Transmissions: All	**Accelerator Pedal Position Sensor 2 Low Input Conditions:** DTC P0641 and P0651 not set; engine cranking or running, and the ECM detected the APP Sensor 2 was less than 0.25v. The accelerator pedal position (APP) sensor is made up of 3 individual sensors in a single housing. Three separate signal, low reference, and 5-volt reference circuits are used in order to interface the APP sensor with the ECM. Each sensor has a unique functionality to determine the pedal position. The ECM uses the APP sensor to determine the amount of desired acceleration or deceleration. **Possible Causes** ● APP sensor connector is damaged or shorted ● APP sensor 2 signal circuit is open or shorted to ground ● APP sensor is damaged or it has failed ● ECM has failed
DTC P2127 **1T CCM, MIL: Yes** 2003, 2004, 2005 CTS (Cadillac) 2.6L VIN M, 3.2L VIN N Transmissions: All	**Accelerator Pedal Position Sensor 2 Circuit Low Input Conditions:** Engine started, system voltage over 10v, accelerator pedal not at "rest" position, and the PCM detected the APP Sensor No. 2 signal was less than 0.66v for more than 140 ms. The accelerator pedal position (APP) sensor is made up of 2 sensors that are housed inside one assembly. The PCM supplies a separate 5v VREF and the low reference circuit for each of the sensors. The 5v VREF for the APP Sensor 2 (APP2) is supplied from the same source in the PCM as the 5v VREF of the TP sensors. The 5v VREF voltage for all of the sensors is supplied on separate PCM terminals, but the terminals are connected internally to a voltage supply. The APP2 sends a signal indicating the accelerator pedal position and the PCM actuates the throttle plates based on this information The APP Sensor 2 signal increases as the pedal is depressed, from about 0.2v at rest to above 1.8v fully depressed. If the PCM detects that the signal voltage was excessively low, DTC P2127 is set. **Possible Causes** ● APP2 sensor low reference circuit open or has high resistance ● APP2 sensor signal circuit is open, shorted to ground ● APP2 sensor signal circuit is shorted to low reference circuit ● APP2 5v VREF circuit is open or shorted to ground ● APP2 sensor is damaged or it has failed ● TP Sensor 5v VREF circuit shorted to ground ● PCM has failed
DTC P2128 **1T CCM, MIL: Yes** 2002, 2002, 2003, 2004, 2005 C/K Series Pickup 6.6L VIN 1 Diesel engine Transmissions: All	**Accelerator Pedal Position Sensor 2 High Input Conditions:** DTC P0641 and P0651 not se, engine cranking or running, and the ECM detected the APP Sensor 2 was more than 4.75v. The accelerator pedal position (APP) sensor is made up of 3 individual sensors in a single housing. Three separate signal, low reference, and 5-volt reference circuits are used to interface the APP sensor with the ECM. Each sensor has a unique functionality to determine the pedal position so that the ECM can determine the amount of acceleration or deceleration desired by the driver of the vehicle. **Possible Causes** ● APP sensor connector is damaged or shorted ● APP sensor 2 signal circuit is open or shorted to ground ● APP sensor is damaged or it has failed ● ECM has failed
DTC P2128 **1T CCM, MIL: Yes** 2003, 2004, 2005 CTS (Cadillac) 2.6L VIN M, 3.2L VIN N Transmissions: All	**Accelerator Pedal Position Sensor 2 Circuit High Input Conditions:** Engine started, system voltage over 10v, and the PCM detected the APP2 sensor signal is more than 4.60v for more than 140 ms. The accelerator pedal position (APP) sensor is made up of two sensors that are housed inside one assembly. The PCM supplies a separate 5v reference and the low reference circuit for each of the sensors. The 5v VREF for the APP Sensor 2 (APP2) is supplied from the same source in the PCM as the 5v VREF for the TP sensors. The 5v VREF for all of the sensors is supplied on separate PCM terminals, but the terminals are connected internally to a voltage supply. The APP Sensor 2 sends a signal from the sensor to the PCM indicating the accelerator pedal position. The PCM actuates the throttle plates based on this information. The APP Sensor 2 signal increases as the pedal is depressed, from about 0.2v at rest to above 1.8v fully depressed. If the PCM detects that the signal voltage was excessively high, DTC P2128 is set. **Possible Causes** ● APP2 sensor low reference circuit open or has high resistance ● APP2 sensor signal circuit is open, shorted to power ● APP2 sensor signal circuit is shorted to low reference circuit ● APP2 5v VREF circuit is open or shorted to power ● APP2 sensor is damaged or it has failed ● TP Sensor 5v VREF circuit shorted to power ● PCM has failed

OBD II Trouble Code List (P2xxx Codes)

DTC	Trouble Code Title, Conditions & Possible Causes
DTC P2132 **1T CCM, MIL: Yes** 2002, 2002, 2003, 2004, 2005 C/K Series Pickup 6.6L VIN 1 Diesel engine Transmissions: All	**Accelerator Pedal Position Sensor 3 Low Input Conditions:** DTC P0641 and P0651 not set; engine cranking or running; and the ECM detected the APP Sensor 3 was less than 1.49v. The accelerator pedal position (APP) sensor is made up of 3 individual sensors in a single housing. Three separate signal, low reference, and 5-volt reference circuits are used in order to interface the APP sensor with the ECM. Each sensor has a unique functionality to determine the pedal position. The ECM uses the APP sensor to determine the amount of desired acceleration or deceleration. **Possible Causes** ● APP sensor connector is damaged or shorted ● APP sensor 3 signal circuit is open or shorted to ground ● APP sensor is damaged or it has failed ● ECM has failed
DTC P2133 **1T CCM, MIL: Yes** 2002, 2002, 2003, 2004, 2005 C/K Series Pickup 6.6L VIN 1 Diesel engine Transmissions: All	**Accelerator Pedal Position Sensor 3 High Input Conditions:** DTC P0641 and P0651 not se, engine cranking or running, and the ECM detected the APP Sensor 3 was more than 4.75v. The accelerator pedal position (APP) sensor is made up of 3 individual sensors in a single housing. Three separate signal, low reference, and 5-volt reference circuits are used to interface the APP sensor with the ECM. Each sensor has a unique functionality to determine the pedal position so that the ECM can determine the amount of acceleration or deceleration desired by the driver of the vehicle. **Possible Causes** ● APP sensor connector is damaged or shorted ● APP sensor 3 signal circuit is open or shorted to ground ● APP sensor is damaged or it has failed ● ECM has failed
DTC P2135 **1T CCM, MIL: Yes** 2003, 2004, 2005 C/K Truck, Escalade, G Van, Envoy & TrailBlazer 4.8L VIN V, 5.3L VIN P, 5.3L VIN T, 5.3L VIN Z, 6.0L VIN N, 6.0L VIN U, 8.1L VIN G engines Transmissions: All	**Throttle Position Sensor 1-2 Correlation Error Conditions:** DTC P1518 and P2108 not set, key in crank or run mode, system voltage more than 5.23v, and the PCM detected the TP Sensor 2 signal disagreed with the TP Sensor 1 signal by more than 7.5% for one second. The TP sensors are used to determine the throttle plate angle for various engine management systems. The TP sensor signals are both low at closed throttle and increase as the throttle opens. When the PCM detects that TP sensor 1 and TP sensor 2 signals disagree or signal voltages are too far apart, this code is set. **Possible Causes** ● TP1 sensor signal circuit shorted to TP sensor signal 2 circuit ● TP1 sensor signal circuit is shorted to the low reference circuit ● TP2 sensor signal circuit is shorted to the low reference circuit ● Throttle body assembly is damaged or it has failed
DTC P2135 **1T CCM, MIL: Yes** 2003, 2004, 2005 CTS (Cadillac) 2.6L VIN M, 3.2L VIN N Transmissions: All	**Throttle Position Sensor 1-2 Correlation Error Conditions:** DTC P0101, P0121, P0122, P0123, P0221, P0222, P0223, P2122, P2123, P2127 and P2128 not set, Key on or engine running; system voltage over 10v, and the PCM detected the difference between TP Sensor 1 and TP Sensor 2 was more than the predicted value. The accelerator pedal position (APP) sensor is made up of two sensors that are housed inside one assembly. The TP Sensors 1 and 2 are located within the throttle body assembly. The TP sensors share a common 5v VREF circuit and a common low reference circuit. The 5v VREF circuit is also shared with accelerator pedal position (APP) Sensor 2. When the throttle plate is in the closed position, the TP Sensor 1 signal voltage is low and it increases as the throttle plate is opened. The TP Sensor 2 signal voltage at closed throttle is near the 5v VREF and decreases as the throttle plate is opened. The PCM compares the signal of the TP sensors to the predicted TP sensor signal, based on calculations from the MAF sensor. The PCM also compares the signal of the TP1 and TP2 sensor through the entire range. If the PCM detects a difference from the predicted range of more than 6% between TP1 and TP2, DTC P2135 is set. **Possible Causes** ● TP Sensor 1 signal, low reference or 5v VREF circuit is open ● TP Sensor 2 signal, low reference or 5v VREF circuit is open ● TP Sensor 1 signal circuit is shorted to the TP2 signal circuit ● Throttle body assembly is damaged or it has failed
DTC P2138 **1T CCM, MIL: No** 2002, 2002, 2003, 2004, 2005 C/K Series Pickup 6.6L VIN 1 Diesel engine Transmissions: All	**Accelerator Pedal Position Sensor 1-2 Correlation Error Conditions:** DTC P0641, P0651, P2122, P2123, P2132 and P2133 not set, engine cranking or running, system voltage from 7-16v, and the ECM detected the APP Sensor 1 and APP sensor 2 was more than 10% out-of-range with each other. The APP sensor is made up of 3 sensors in one housing that connect a signal, low reference, and 5v VREF circuit to the ECM. Each sensor has a unique functionality to detect the pedal position so that the ECM can determine the amount of acceleration or deceleration desired by the driver of the vehicle. **Possible Causes** ● APP sensor 3 signal circuit is open or shorted to ground ● APP sensor is damaged or it has failed ● ECM has failed

OBD II Trouble Code List (P2xxx Codes)

DTC	Trouble Code Title, Conditions & Possible Causes
DTC P2138 **1T CCM, MIL: Yes** 2003, 2004, 2005 CTS (Cadillac) 2.6L VIN M, 3.2L VIN N Transmissions: All	**Accelerator Pedal Position Sensor 1-2 Correlation Error Conditions:** DTC P0101, P0121, P0122, P0123, P0221, P0222, P0223, P2122, P2123, P2127 and P2128 not set, key on, system voltage over 10v, accelerator pedal not at "rest" position, and the PCM detected the difference between APP Sensor 1 and APP Sensor 2 was more than the predicted value. The accelerator pedal position (APP) sensor is made up of two sensors that are housed inside one assembly. The PCM supplies a separate 5v VREF and low reference circuit for each of the sensors. The 5v VREF for APP Sensor 1 is supplied from the same source in the PCM as the 5v VREF for the MAF sensor and the FTP sensor. The 5v VREF for APP Sensor 2 is supplied from the same source in the PCM as the 5v VREF for the TP sensors. The 5v VREF voltage for all of the sensors is supplied on separate PCM terminals, but the terminals are connected internally to the shared voltage supply. The APP sensor sends a signal from each of the sensors to the PCM indicating the accelerator pedal position. The PCM actuates the throttle plates based on this information. The APP Sensor 1 signal voltage increases as the throttle opens, from about 0.4v at closed throttle to above 3.5v at wide-open throttle (WOT). The APP Sensor 2 signal voltage increases as the throttle opens, but at a slightly different rate. The APP Sensor 2 signal increases from about 0.2v at closed throttle to about 1.8v at WOT. The PCM compares the signal of each of the APP sensors, when the pedal is not in the rest position. If the PCM detects the APP sensors are not within a predicted value from each other, DTC P2138 is set. **Possible Causes** • APP Sensor 1 signal circuit is shorted to the APP2 signal circuit • APP sensor is damaged or it has failed
DTC P2139 **1T CCM, MIL: No** 2002, 2002, 2003, 2004, 2005 C/K Series Pickup 6.6L VIN 1 Diesel engine Transmissions: All	**Accelerator Pedal Position Sensor 2-3 Correlation Conditions:** DTC P0641, P0651, P2122, P2123, P2132 and P2133 not set, engine cranking or running, system voltage from 7-16v, and the ECM detected the APP Sensor 2 and APP Sensor 3 were more than 10% out-of-range with each other. The APP sensor is made up of 3 sensors in one housing that connect a signal, low reference, and 5v VREF circuit to the ECM. Each sensor has a unique functionality to detect the pedal position so that the ECM can determine the amount of acceleration or deceleration desired by the driver of the vehicle. **Possible Causes** • APP2 or APP3 signal circuit has a high resistance condition • APP2 VREF circuit has a high resistance condition • APP sensor is damaged or it has failed • ECM has failed
DTC P2140 **1T CCM, MIL: No** 2002, 2002, 2003, 2004, 2005 C/K Series Pickup 6.6L VIN 1 Diesel engine Transmissions: All	**Accelerator Pedal Position Sensor 1-3 Correlation Conditions:** DTC P0641, P0651, P2122, P2123, P2132 and P2133 not set, engine cranking or running, system voltage from 7-16v, and the PCM detected the APP Sensor 1 and APP Sensor 3 were over 10% out-of-range. The APP sensor includes 3 sensors in a single housing. Each sensor has a unique functionality to detect the pedal position so that the ECM can determine the amount of acceleration or deceleration desired by the driver of the vehicle. **Possible Causes** • If DTC P2138 is also set, refer to that code repair information • If DTC P2139 is also set, refer to that code repair information
DTC P2141 **2T CCM, MIL: Yes** 2002, 2002, 2003, 2004, 2005 C/K Series Pickup 6.6L VIN 1 Diesel engine Transmissions: All	**EGR Throttle Valve Vacuum Control Solenoid Low Input Conditions:** Key on for 500 ms, EGR Valve Throttle Vacuum Control solenoid commanded "off", system voltage from 11-18v, and the PCM detected a low voltage on the control circuit with the EGR throttle valve vacuum control solenoid commanded "off". Diesel engines do not create sufficient engine vacuum to allow EGR gases into the combustion process. Once the ECM commands the EGR valve open, the EGR throttle valve is closed. **Possible Causes** • EGR vacuum solenoid circuit is open or has high resistance • EGR vacuum solenoid power circuit is open (test Fuel HT fuse) • EGR valve throttle solenoid is damaged or it has failed • ECM has failed
DTC P2142 **2T CCM, MIL: Yes** 2002, 2002, 2003, 2004, 2005 C/K Series Pickup 6.6L VIN 1 Diesel engine Transmissions: All	**EGR Throttle Valve Vacuum Control Solenoid High Input Conditions:** Key on for 500 ms, EGR Valve Throttle Vacuum Control solenoid commanded "on", system voltage from 11-18v, and the PCM detected a high voltage on the control circuit with the EGR throttle valve vacuum control solenoid "on". Diesel engines do not create sufficient engine vacuum to allow EGR gases into the combustion process. Once the PCM commands the EGR valve open, the EGR throttle valve is commanded closed. The EGR throttle valve creates a restriction in the incoming fresh air to create engine vacuum. With the EGR throttle valve closed, engine vacuum develops and allows the exhaust gases to enter the engine. **Possible Causes** • EGR vacuum solenoid control circuit is shorted to power (B+) • EGR vacuum solenoid power circuit is open (test Fuel HT fuse) • EGR valve throttle solenoid is damaged or it has failed • PCM has failed

OBD II Trouble Code List (P2xxx Codes)

DTC	Trouble Code Title, Conditions & Possible Causes
DTC P2144 **2T CCM, MIL: Yes** 2002, 2002, 2003, 2004, 2005 C/K Series Pickup 6.6L VIN 1 Diesel engine Transmissions: All	**EGR Vacuum Vent Solenoid Low Input Conditions:** Key on for 500 ms, EGR Vacuum Vent solenoid commanded "off", system voltage from 11-18v, and the PCM detected a low voltage on the control circuit with the EGR vacuum vent solenoid commanded "off". Diesel engines do not create sufficient engine vacuum to allow EGR gases into the combustion process. Once the PCM commands the EGR valve open, the EGR throttle valve is commanded closed. The EGR throttle valve creates a restriction in the incoming fresh air to create engine vacuum. With the EGR throttle valve closed, engine vacuum develops and allows exhaust gases to enter the engine. **Possible Causes** ● EGR vacuum vent solenoid circuit open or has high resistance ● EGR vacuum solenoid power circuit is open (test Fuel HT fuse) ● EGR valve throttle solenoid is damaged or it has failed ● ECM has failed
DTC P2145 **2T CCM, MIL: Yes** 2002, 2002, 2003, 2004, 2005 C/K Series Pickup 6.6L VIN 1 Diesel engine Transmissions: All	**EGR Vacuum Vent Solenoid Low Input Conditions:** Key on for 500 ms, EGR Vacuum Vent solenoid commanded "off", system voltage from 11-18v, and the ECM detected a low voltage condition on the control circuit of the EGR vacuum vent solenoid with the ECM with the EGR vacuum vent solenoid commanded "off". **Possible Causes** ● EGR vacuum vent solenoid connector is damaged or shorted ● EGR vacuum vent solenoid circuit is shorted to system power ● EGR vacuum vent solenoid is damaged or it has failed ● ECM has failed
DTC P2145 **2T CCM, MIL: Yes** 2002, 2002, 2003, 2004, 2005 C/K Series Pickup 6.6L VIN 1 Diesel engine Transmissions: All	**EGR Vacuum Vent Solenoid High Input Conditions:** Key on for 500 ms, EGR Vacuum Vent solenoid commanded "on", system voltage from 11-18v, and the ECM detected a high voltage condition on the control circuit of the EGR vacuum vent solenoid with the ECM with the EGR vacuum vent solenoid commanded "on". **Possible Causes** ● EGR vacuum solenoid circuit has a high resistance condition ● EGR vacuum solenoid power circuit is open (test Fuel HT fuse) ● EGR valve throttle solenoid is damaged or it has failed ● ECM has failed
DTC P2146 **2T CCM, MIL: Yes** 2003, 2004, 2005 C/K Series Pickup 6.0L VIN U, 6.6L VIN 1 Transmissions: All	**Fuel Pump Relay Circuit Malfunction (Injectors 1, 4, 6 or 7) Conditions:** DTC U1800 and U2104 not set; key on or engine running, and the PCM detected the Actual state of the Fuel Injector control circuit did not match the Commanded state for 5 seconds. The fuel injection control module (FICM) supplies high voltage to each fuel injector on the ignition voltage circuits. The FICM energizes each fuel injector by grounding the command circuit between the FICM and the fuel injector. The FICM monitors the status of the ignition voltage circuits and the fuel injector command circuits. When a fuel injector circuit fault condition is detected by the FICM, all of the fuel injectors on the affected ignition voltage circuit will be disabled. If a circuit condition is detected on a fuel injector circuit for cylinders 1, 4, 6, or 7, DTC P0201, P0204, P0206, P0207 will set, along with DTC P2146. **Possible Causes** ● Refer to repair information for P0201, P0204, P0206 or P0207 ● FICM (module) has failed (when P2146 is set all by itself)
DTC P2149 **1T CCM, MIL: Yes** 2003, 2004, 2005 C/K Series Pickup 6.0L VIN U, 6.6L VIN 1 Transmissions: All	**Fuel Pump Relay Circuit Malfunction (Injectors 2, 3, 5 or 8) Conditions:** U1800 and U2104 not set; engine started, and the PCM detected an unexpected voltage condition on a Fuel Injector control circuit for 5 seconds. The fuel injection control module (FICM) supplies high voltage to each fuel injector on the ignition voltage circuits. The FICM energizes each fuel injector by grounding the command circuit between the FICM and the fuel injector. The FICM monitors the status of the ignition voltage circuits and the fuel injector command circuits. When a fuel injector circuit fault condition is detected by the FICM, all of the fuel injectors on the affected ignition voltage circuit will be disabled. If a circuit condition is detected on a fuel injector circuit for cylinders 2, 3, 5, or 8, DTC P0202, P0203, P0205, P0208 will set, along with this trouble code (DTC P2149). **Possible Causes** ● Refer to repair information for P0202, P0203, P0205 or P0208 ● FICM (module) has failed (when P2149 is set all by itself)

OBD II Trouble Code List (P2xxx Codes)

DTC	Trouble Code Title, Conditions & Possible Causes
DTC P2176 **2T CCM, MIL: Yes** 2003, 2004, 2005 CTS (Cadillac) 2.6L VIN M, 3.2L VIN N Transmissions: All	**Throttle Position Sensor Minimum Position Not Detected Conditions:** DTC P0121, P0122, P0123, P0221, P0222 and P0223 not set, key on, engine speed less than 40 rpm, system voltage over 10v, ECT sensor from 41-185°F, IAT sensor less than 122°F, accelerator pedal position under 15%, VSS less than 0 mph, and the PCM could not determine the minimum throttle position. The PCM controls the throttle blades by applying a varying voltage to the throttle valve motor. The PCM monitors the actual throttle blade position using throttle position (TP) Sensor 1 and 2. If the PCM cannot detect the minimum throttle position, DTC P2176 is set. **Possible Causes** ● TAC motor control circuit (one or more) open or shorted to B+ ● TAC motor control circuit (one or more) shorted to ground ● Throttle valves not at the "rest" position ● Throttle valve are binding open, or are binding closed ● Throttle valves moving open or closed without spring pressure ● TP connector is damaged, open or shorted ● Throttle body assembly is damaged or it has failed ● PCM has failed
DTC P2227 **1T CCM, MIL: No** 2002, 2002, 2003, 2004, 2005 C/K Series Pickup 6.6L VIN 1 Diesel engine Transmissions: All	**BARO Sensor Signal Range/Performance Conditions:** DTC P0101, P0102, P0103, P0116, P0117, P0118, P0236, P0237, P0238, P0335, P0336, P0500, P2228 and P2229 not set, engine speed at 500-900 rpm for 20 seconds, Accelerator pedal angle under 20%, ECT sensor over 176°F, VSS under 25 mph, MAF sensor under 50 g/sec, Power Takeoff off, conditions met for 5 seconds, and the ECM detected the difference between the BARO and Boost Pressure sensor was 10-20 kPa for 2 seconds. **Possible Causes** ● BARO sensor signal circuit is shorted to VREF ● BARO sensor is damaged or it has failed ● ECM has failed
DTC P2228 **2T CCM, MIL: Yes** 2002, 2002, 2003, 2004, 2005 C/K Series Pickup 6.6L VIN 1 Diesel engine Transmissions: All	**BARO Sensor Circuit Low Input Conditions:** DTC P0461not set, key on, and the ECM detected the BARO sensor was less than 44 kPa for two seconds during the CCM test period. **Possible Causes** ● BARO sensor signal circuit is open or shorted to ground ● BARO sensor is damaged or it has failed ● ECM has failed
DTC P2228 **2T CCM, MIL: Yes** 2002, 2002, 2003, 2004, 2005 C/K Series Pickup 6.6L VIN 1 Diesel engine Transmissions: All	**BARO Sensor Circuit High Input Conditions:** DTC P0461not set, key on, and the ECM detected the BARO sensor was more than 110 kPa for two seconds during the CCM test period. **Possible Causes** ● BARO sensor signal circuit is shorted to VREF ● BARO sensor is damaged or it has failed ● ECM has failed
DTC P2279 **2T CCM, MIL: Yes** 2002, 2002, 2003, 2004, 2005 C/K Series Pickup 6.6L VIN 1 Diesel engine Transmissions: All	**Air Intake Leak Detected Conditions:** DTC P0101, P0102, P0103, P0112, P0113, P0116, P0117, P0118, P0405, P0406, P0489, P0490, P0500, P0651, P2141, P2142, P2144, P2145, P2227, P2228 and P2229 not set, engine runtime over 60 seconds, engine speed from 600-700 rpm, BARO sensor more than 72 kPa (10 psi), system voltage from 9-16v, vehicle speed less than 1 mph, ECT sensor from 32-176°F, and the PCM detected an engine intake leak that was over the calibrated value stored in the PCM. The MAF sensor measures the amount of air that enters the engine at any given time. The PCM uses the MAF sensor to monitor the EGR flow rate. The PCM can detect an intake leak using the MAF sensor and EGR Control Pressure sensor signals. **Possible Causes** ● Air Intake Duct is leaking somewhere after the MAF sensor ● Check for an air leak in Air Induction system after the MAF sensor, in the engine intake gaskets, or the EGR tower gasket
DTC P2500 **1T CCM, MIL: No** 2003, 2004, 2005 CTS (Cadillac) 2.6L VIN M, 3.2L VIN N Transmissions: All	**Generator Turn "On" Circuit Low Input Conditions:** DTC P0335, P0336, P0337, P0341, P0342, P0343 and P0626 not set, engine started, Generator not commanded "off" by a Scan Tool, and the PCM detected the Generator turn "on" signal circuit was low for over 15 seconds. The PCM uses the Generator turn "on" signal to control the load placed on the engine by the Generator. A high side driver in the ECM applies a voltage to the voltage regulator to signal the voltage regulator to turn "on" and "off" the field circuit. When the PCM turns the high side driver "on", the voltage regulator turns "on" the field circuit. When the PCM turns the high side driver "off", the voltage regulator turns off the Generator control circuit. **Possible Causes** ● Generator connector is damaged or shorted to ground ● Generator turn "on" signal circuit is open or shorted to ground ● PCM has failed

OBD II Trouble Code List (P2xxx Codes)

DTC	Trouble Code Title, Conditions & Possible Causes
DTC P2501 **1T CCM, MIL: No** 2003, 2004, 2005 CTS (Cadillac) 2.6L VIN M, 3.2L VIN N Transmissions: All	**Generator Turn "On" Circuit High Input Conditions:** DTC P0335, P0336, P0337, P0341, P0342, P0343 and P0626 not set, key on, engine off, Generator not commanded "off" by a Scan Tool, and the PCM detected the Generator turn "on" signal circuit was high for over 5 seconds. The PCM uses the Generator turn "on" signal to control the load placed on the engine by the Generator. A high side driver in the ECM applies a voltage to the voltage regulator to signal the voltage regulator to turn "on" and "off" the field circuit. When the PCM turns the high side driver "on", the voltage regulator turns "on" the field circuit. When the PCM turns the high side driver "off", the voltage regulator turns off the Generator control circuit. **Possible Causes** ● Generator connector is damaged or shorted to power ● Generator turn "on" signal circuit is shorted to system power ● PCM has failed
DTC P2610 **2T PCM, MIL: Yes** 2001, 2002, 2003, 2004, 2005 C/K Pickup, Escalade 6.0L VIN N, 6.6L VIN 1 Transmissions: All	**Powertrain Control Module Internal Malfunction Conditions:** Key on, and the PCM detected an internal malfunction had occurred. **Possible Causes** ● Clear the codes and then recheck for this trouble code. If the same code resets, the PCM is damaged and must be replaced (and the new PCM must be programmed) to repair this code.
DTC P2610 **2T PCM, MIL: Yes** 2001, 2002, 2003, 2004, 2005 CTS (Cadillac), E, G, K 4.0L VIN C, 4.6L VIN 9, 4.6L VIN Y engines Transmissions: All	**Powertrain Control Module Internal Malfunction Conditions:** Key on, and the PCM detected an internal malfunction had occurred. **Possible Causes** ● Clear the codes and then recheck for this trouble code. If the same code resets, the PCM is damaged and must be replaced (and the new PCM must be programmed) to repair this code.
DTC P2665 **2T CCM, MIL: Yes** 2003, 2004, 2005 C/K Series Pickup Engines: 6.0L VIN U Transmissions: All	**AF Low Pressure Lock-Off Solenoid Circuit Malfunction Conditions:** Engine started, engine running, system voltage from 6-18v, and the PCM detected the Actual and Commanded state of the LPL solenoid control circuit did not match for two seconds. Ignition voltage is supplied to the Low Pressure Lock-Off (LPL) solenoid through the AF Lock-Off relay. The PCM controls the LPL solenoid by grounding the control circuit via an internal switch called a driver. The primary function of the driver is to supply ground for a controlled component. The PCM monitors the status of the driver. If it detects an incorrect voltage for the commanded state of the driver, it sets DTC P2665. **Possible Causes** ● LPL solenoid control circuit is open or shorted to ground ● LPL solenoid relay power circuit is open (test the AFS fuse) ● LPL solenoid relay is damaged or it has failed ● AF control module has failed
DTC P2665 **2T CCM, MIL: Yes** 2003, 2004, 2005 G Van, Van Cargo Engines: 6.0L VIN U Transmissions: All	**Low Pressure Lock-Off Relay Circuit Malfunction Conditions:** Key on or engine running, and the PCM detected the Actual state of the LPL solenoid and the Commanded stated did not match for one seconds. Ignition voltage is supplied to the Low Pressure Lock-Off (LPL) solenoid through the LPL Relay. The PCM controls the LPL Relay by grounding the control circuit via an internal switch called a driver. The primary function of the driver is to supply the ground for the controlled component. The PCM monitors the status of the driver. If the PCM detects an incorrect voltage for the commanded state of the solenoid driver, DTC P2665 is set. **Possible Causes** ● LPL relay control circuit is open or shorted to power ● LPL relay control circuit is shorted to ground ● LPL relay is damaged or it has failed ● PCM has failed
DTC P2668 **1T CCM, MIL: Yes** 2003, 2004, 2005 G Van, Van Cargo Engines: 6.0L VIN U Transmissions: All	**Fuel Level Indicator Lamp Circuit Malfunction Conditions:** Key on or engine running, and the PCM detected the Actual state and the Commanded state of the FIL Lamp did not match for five seconds. The fuel indicator lamp (FIL) is located within the headlamp switch assembly. The FIL indicates which fuel system is utilized to operate the vehicle. The FIL illuminates for a few seconds at startup for a bulb check. The FIL remains "on" while the vehicle operates on gasoline. The PCM controls the FIL by grounding the control circuit via an internal switch called a driver. The primary function of the driver is to supply the ground for the controlled component. The PCM monitors the status of the driver. If the PCM detects an invalid voltage for the commanded state of the driver, DTC P2668 is set. **Possible Causes** ● FIL lamp control circuit is open or shorted to power ● FIL lamp control circuit is shorted to ground ● Headlight switch is damaged or it has failed ● PCM has failed

OBD II Trouble Code List (U1xxx Codes)

DTC	Trouble Code Title, Conditions & Possible Causes
DTC U1000 **1T PCM, MIL: Yes** 1997, 1998, 1999, 2000, 2001, 2002, 2003, 2004, 2005 All Engines: All Transmissions: All	**Class 2 Communication Malfunction Conditions:** Modules connected to the Class 2 circuit monitor for serial data communications during normal vehicle operation. Operating information and commands are exchanged among the modules. When a module receives a message for a critical operating parameter, the module records the identification number of the module that sent the message. These Node Alive messages are used for State of Health monitoring. A critical operating parameter is one which, when not received, requires that the module use a default value for that parameter. When a module does not associate an identification number with at least one critical parameter within 5 seconds of starting data communication, DTC U1000 or U1255 is set. When more than one critical parameter does not have an identification number associated with it, the code will only set once. **Possible Causes** ● Class 2 circuit is open, shorted to ground or shorted to power ● PCM ignition power circuit(s) has a high resistance condition ● PCM main ground circuit(s) has a high resistance condition ● SDM (module) could be shorted pulling the voltage low
DTC U1016 **1T PCM, MIL: Yes** 1998, 1999, 2000, 2001, 2002, 2003, 2004, 2005 All Engines: All Transmissions: All	**No Communication With Powertrain Control Module Conditions:** Key on, and a message from a learned ID number was not detected for the five seconds. Modules on the Class 2 circuit monitor for data communications during vehicle operation. When a module receives a message for critical data, the module records the identification number of the module sending the message for State of Health monitoring (Node Alive messages). Once a module learns an ID number, it checks for that module's Node Alive message. *Note: Look for this code in all modules. The one without the code is the module that has a problem, and it may have failed.* **Possible Causes** ● PCM Class 2 circuit is open, shorted to ground or to B+ ● PCM ignition power circuit(s) has a high resistance condition ● PCM main ground circuit(s) has a high resistance condition ● PCM (module) may have failed and is pulling the circuit low
DTC U1026 **1T PCM, MIL: Yes** 1998, 1999, 2000, 2001, 2002, 2003, 2004, 2005 C/K Series Truck, G Van, Escalade, M/L Series Van S/T Blazer & S/T Pickup 4.3L VIN W, 4.3L VIN X, 5.7L VIN R, 7.4L VIN J, 8.1L VIN G, 4.8L VIN V, 5.3L VIN T, 6.0L VIN U, 6.5L VIN F, 6.5L VIN S Transmissions: All	**Loss of ATC Class 2 Communication Conditions:** Key on or engine running; and a module detected that it could not communicate with the ATC controller for 1 second. Modules connected to the Class 2 circuit monitor for data communications during normal vehicle operation. Operating information and commands are exchanged among the modules. When a module receives a message for a critical operating parameter, the module records the identification number of the module that sent the message for State of Health monitoring (Node Alive messages). Once a module learns an identification number, it will monitor for that module's Node Alive message. Each module on the Class 2 circuit that is powered and performing functions that require detection of a communications malfunction is required to send a Node Alive message every two seconds. When no message is detected from a learned identification number for five seconds, a DTC U1xxx (XXX is equal to the 3-digit identification number) is set. **Possible Causes** ● Check for a loose connection at the ATC module ● Test the main power and ground circuits to the ATC module ● Check the Class 2 serial data circuit to the ATC module ● ATC module may have failed
DTC U1026 **2T PCM, MIL: Yes** 1998, 1999, 2000, 2001, 2002, 2003, 2004, 2005 All Other Models Engines: All Others Transmissions: All Others	**No Communication With Transfer Case Shift Control Module Conditions:** Key on, and a message from a learned ID number was not detected for the five seconds. Modules on the Class 2 circuit monitor for data communications during vehicle operation. When a module receives a message for critical data, the module records the identification number of the module sending the message for State of Health monitoring (Node Alive messages). Once a module learns an ID number, it checks for that module's Node Alive message. Note: Look for this code in all modules. The one without the code is the module that has a problem, and it may have failed. **Possible Causes** ● TCSCM Class 2 circuit is open, shorted to ground or to B+ ● TCSCM ignition power circuit(s) has a high resistance condition ● TCSCM main ground circuit(s) has a high resistance condition ● TCSCM (module) may have failed and is pulling the circuit low

OBD II Trouble Code List (U1xxx Codes)

DTC	Trouble Code Title, Conditions & Possible Causes
DTC U1041 **2T PCM, MIL: Yes** 1998, 1999, 2000, 2001, 2002, 2003, 2004, 2005 C/K Series Truck, G Van, Escalade, M/L Series Van S/T Blazer & S/T Pickup 4.3L VIN W, 4.3L VIN X, 5.7L VIN R, 7.4L VIN J, 8.1L VIN G, 4.8L VIN V, 5.3L VIN T, 6.0L VIN U, 6.5L VIN F, 6.5L VIN S Transmissions: All	**Loss of Electronic Brake Controller Communication Conditions:** Key on or engine running; and a module detected that it could not communicate with the EBCM controller for 1 second. Modules connected to the Class 2 circuit monitor for data communications during normal vehicle operation. Operating information and commands are exchanged among the modules. When a module receives a message for a critical operating parameter, the module records the identification number of the module that sent the message for State of Health monitoring (Node Alive messages). Once a module learns an identification number, it will monitor for that module's Node Alive message. Each module on the Class 2 circuit that is powered and performing functions that require detection of a communications malfunction is required to send a Node Alive message every two seconds. When no message is detected from a learned identification number for five seconds, a DTC U1xxx (XXX is equal to the 3-digit identification number) is set. **Possible Causes** ● Check for a loose connection at the EBCM (module) ● Test the main power and ground circuits to the EBCM (module) ● Check the Class 2 serial data circuit to the EBCM (module) ● EBCM (module) may have failed
DTC U1041 **2T PCM, MIL: Yes** 1998, 1999, 2000, 2001, 2002, 2003, 2004, 2005 All Other Models All Other engines Transmissions: All Others	**No Communication With Electronic Brake Control Module Conditions:** Key on, and a message from a learned ID number was not detected for the five seconds. Modules on the Class 2 circuit monitor for data communications during vehicle operation. When a module receives a message for critical data, the module records the identification number of the module sending the message for State of Health monitoring (Node Alive messages). Once a module learns an ID number, it checks for that module's Node Alive message. *Note: Look for this code in all modules. The one without the code is the module that has a problem, and it may have failed.* **Possible Causes** ● EBCM Class 2 circuit is open, shorted to ground or to B+ ● EBCM ignition power circuit(s) has a high resistance condition ● EBCM main ground circuit(s) has a high resistance condition ● EBCM (module) may have failed and is pulling the circuit low
DTC U1048 **2T PCM, MIL: Yes** 1998, 1999, 2000, 2001, 2002, 2003, 2004, 2005 All Models All engines Transmissions: All	**No Communication With Rear Wheel Steering Control Module Conditions:** Key on, and a message from a learned ID number was not detected for the five seconds. Modules on the Class 2 circuit monitor for data communications during vehicle operation. When a module receives a message for critical data, the module records the identification number of the module sending the message for State of Health monitoring (Node Alive messages). Once a module learns an ID number, it checks for that module's Node Alive message. *Note: Look for this code in all modules. The one without the code is the module that has a problem, and it may have failed.* **Possible Causes** ● RWSCM Class 2 circuit is open, shorted to ground or to B+ ● RWSCM ignition power circuit has a high resistance condition ● RESCM main ground circuit(s) has a high resistance condition ● RESCM (module) may have failed and is pulling the circuit low
DTC U1064 **2T PCM, MIL: Yes** 1998, 1999, 2000, 2001, 2002, 2003, 2004, 2005 All Engines: All Transmissions: All	**No Communication With Body Control Module Conditions:** Key on, and a message from a learned ID number was not detected for the five seconds. Modules on the Class 2 circuit monitor for data communications during vehicle operation. When a module receives a message for critical data, the module records the identification number of the module sending the message for State of Health monitoring (Node Alive messages). Look for this code in all modules. The one without this code may have failed **Possible Causes** ● BCM Class 2 circuit is open, shorted to ground or to B+ ● BCM ignition power circuit has a high resistance condition ● BCM main ground circuit(s) has a high resistance condition ● BCM (module) may have failed and is pulling the circuit low
DTC U1088 **2T PCM, MIL: Yes** 1998, 1999, 2000, 2001, 2002, 2003, 2004, 2005 All Engines: All Transmissions: All	**No Communication With SDM (Restraint Module) Conditions:** Key on, and a message from a learned ID number was not detected for the five seconds. Modules on the Class 2 circuit monitor for data communications during vehicle operation. When a module receives a message for critical data, the module records the identification number of the module sending the message for State of Health monitoring (Node Alive messages). Look for this code in all modules. The one without this code may have failed **Possible Causes** ● SDM Class 2 circuit is open, shorted to ground or to B+ ● SDM ignition power circuit has a high resistance condition ● SDM main ground circuit(s) has a high resistance condition ● SDM (module) may have failed and is pulling the circuit low

OBD II Trouble Code List (U1xxx Codes)

DTC	Trouble Code Title, Conditions & Possible Causes
DTC U1092 **2T PCM, MIL: Yes** 1998, 1999, 2000, 2001, 2002, 2003, 2004, 2005 C/K Series Truck, G Van, M, L, S, T 4.3L VIN W, 4.3L VIN X, 5.7L VIN R, 7.4L VIN J, 8.1L VIN G, 4.8L VIN V, 5.3L VIN T, 6.0L VIN U, 6.5L VIN F, 6.5L VIN S Transmissions: All	**Loss of VTD (Pass Lock) Communication Conditions:** Key on or engine running; and a module detected that it could not communicate with the VTD controller for 1 second. Modules connected to the Class 2 circuit monitor for data communications during normal vehicle operation. Operating information and commands are exchanged among the modules. When a module receives a message for a critical operating parameter, the module records the identification number of the module that sent the message for State of Health monitoring (Node Alive messages). Once a module learns an identification number, it will monitor for that module's Node Alive message. Each module on the Class 2 circuit that is powered and performing functions that require detection of a communications malfunction is required to send a Node Alive message every two seconds. When no message is detected from a learned identification number for five seconds, a DTC U1xxx (the X's identify the 3-digit identification number) is set. **Possible Causes** ● Check for a loose connection at the VTD module ● Test the main power and ground circuits to the VTD module ● Check the Class 2 serial data circuit to the VTD module ● VTD module may have failed
DTC U1096 **2T PCM, MIL: Yes** 1998, 1999, 2000, 2001, 2002, 2003, 2004, 2005 All Engines: All Transmissions: All	**No Communication With Instrument Panel Cluster Conditions:** Key on, and a message from a learned ID number was not detected for the five seconds. Modules on the Class 2 circuit monitor for data communications during vehicle operation. When a module receives a message for critical data, the module records the identification number of the module sending the message for State of Health monitoring (Node Alive messages). Once a module learns an ID number, it checks for that module's Node Alive message. Note: Look for this code in all modules. The one without the code is the module that has a problem, and it may have failed. **Possible Causes** ● IPC Class 2 circuit is open, shorted to ground or to B+ ● IPC ignition power circuit has a high resistance condition ● IPC main ground circuit(s) has a high resistance condition ● IPC (module) may have failed and is pulling the circuit low
DTC U1097 **2T PCM, MIL: Yes** 1998, 1999, 2000, 2001, 2002, 2003, 2004, 2005 All Engines: All Transmissions: All	**No Communication With Driver Information Center Conditions:** Key on, and a message from a learned ID number was not detected for the five seconds. Modules on the Class 2 circuit monitor for data communications during vehicle operation. When a module receives a message for critical data, the module records the identification number of the module sending the message for State of Health monitoring (Node Alive messages). Once a module learns an ID number, it checks for that module's Node Alive message. *Note: Look for this code in all modules. The one without the code is the module that has a problem, and it may have failed.* **Possible Causes** ● DIC Class 2 circuit is open, shorted to ground or to B+ ● DIC ignition power circuit has a high resistance condition ● DIC main ground circuit(s) has a high resistance condition ● DIC (module) may have failed and is pulling the circuit low
DTC U1128 **2T PCM, MIL: Yes** 1999, 2000, 2001 Catera 3.0L VIN R engine Transmissions: All	**Loss Of Serial Communications With IRC Conditions:** U1300 and U1301 not set, key on or engine started, system voltage over 9.0v, and a module detected it did not receive a message from a "learned" Source ID for 5 seconds. Modules connected to the Class 2 line monitor the bus for data communications during normal vehicle operation. Both report messages and load messages (commands) are exchanged among various modules. When a module receives a "report" or "load" message for a critical operating parameter, the module records the Source ID of the module that sent the message for State of Health monitoring (Node Alive messages). Once a module learns a Source ID, it will monitor for that Source ID's Node Alive message. Each module on the Class 2 line which is powered and performing functions that require detection a communication fault is required to send a Node Alive message every two seconds. When no message is detected from a "learned" Source ID for 5 seconds, DTC U1128 is set. **Possible Causes** ● Inspect the connector on Remote Control Door Lock Receiver ● Check the power and ground circuits to the IRC module

OBD II Trouble Code List (U1xxx Codes)

DTC	Trouble Code Title, Conditions & Possible Causes
DTC U1151 **2T PCM, MIL: Yes** 1998, 1999, 2000, 2001, 2002, 2003, 2004, 2005 All Engines: All Transmissions: All	**No Communication With Vehicle Interface Unit Conditions:** Key on, and a message from a learned ID number was not detected for the five seconds. Modules on the Class 2 circuit monitor for data communications during vehicle operation. When a module receives a message for critical data, the module records the identification number of the module sending the message for State of Health monitoring (Node Alive messages). Once a module learns an ID number, it checks for that module's Node Alive message. The module without this code is the module with a problem (it has failed). **Possible Causes** ● VIU Class 2 circuit is open, shorted to ground or to B+ ● VIU ignition power circuit has a high resistance condition ● VIU main ground circuit(s) has a high resistance condition ● VIU (module) may have failed and is pulling the circuit low
DTC U1152 **2T PCM, MIL: Yes** 1998, 1999, 2000, 2001, 2002, 2003, 2004, 2005 All Engines: All Transmissions: All	**No Communication With HVAC Control Module Conditions:** Key on, and a message from a learned ID number was not detected for the five seconds. Modules on the Class 2 circuit monitor for data communications during vehicle operation. When a module receives a message for critical data, the module records the identification number of the module sending the message for State of Health monitoring (Node Alive messages). Once a module learns an ID number, it checks for that module's Node Alive message. The module without this code is the module with a problem (it has failed). **Possible Causes** ● HVAC Class 2 circuit is open, shorted to ground or to B+ ● HVAC ignition power circuit has a high resistance condition ● HVAC main ground circuit(s) has a high resistance condition ● HVAC (module) may have failed and is pulling the circuit low
DTC U1193 **2T PCM, MIL: Yes** 1998, 1999, 2000, 2001, 2002, 2003, 2004, 2005 C/K Series Truck, G Van, Escalade, M/L Series Van S/T Blazer & S/T Pickup 4.3L VIN W, 4.3L VIN X, 5.7L VIN R, 7.4L VIN J, 8.1L VIN G, 4.8L VIN V, 5.3L VIN T, 6.0L VIN U, 6.5L VIN F, 6.5L VIN S Transmissions: All	**Loss of Vehicle Immobilizer Module Communications Conditions:** Key on or engine running; and a module detected that it could not communicate with the VIM controller for 1 second. Modules connected to the Class 2 circuit monitor for data communications during normal vehicle operation. Operating information and commands are exchanged among the modules. When a module receives a message for a critical operating parameter, the module records the identification number of the module that sent the message for State of Health monitoring (Node Alive messages). Once a module learns an identification number, it will monitor for that module's Node Alive message. Each module on the Class 2 circuit that is powered and performing functions that require detection of a communications malfunction is required to send a Node Alive message every two seconds. When no message is detected from a learned identification number for five seconds, a DTC U1xxx (XXX is equal to the 3-digit identification number) is set. **Possible Causes** ● Test the main power and ground circuits to the VIM module for a loose connection ● Check the Class 2 serial data circuit to the VIM module ● VTD module may have failed
DTC U1255 **2T PCM, MIL: Yes** 1997, 1998, 1999, 2000, 2001, 2002, 2003, 2004, 2005 All Engines: All Transmissions: Alls	**Class 2 Communications Malfunction Conditions:** Modules connected to the Class 2 circuit monitor for serial data communications during normal vehicle operation. Operating data and commands are exchanged among modules. When a module receives a message for a critical operating parameter, the module records the identification number of the module that sent the message. These Node Alive messages are used for State of Health monitoring. A critical operating parameter is one which, when not received, requires the module use a default value for that parameter. If a module does not associate an ID number with at least one critical parameter in 5 seconds after starting communication, U1000 or U1255 is set. If two or more are missing, the code sets at once. **Possible Causes** ● Class 2 circuit is open, shorted to ground or shorted to power ● PCM ignition power circuit(s) has a high resistance condition ● PCM main ground circuit(s) has a high resistance condition
DTC U1255 **2T PCM, MIL: Yes** 1999, 2000, 2001 Catera 3.0L VIN R engine Transmissions: All	**Generic Loss of Class 2 Data Conditions:** Key on or engine started, system voltage over 9.0v, then with the Scan Tool connected and powered up, a Could Not Communicate With The Class 2 Data Line" message is displayed. Modules connected to the Class 2 line monitor the data bus for serial data communications during normal vehicle operation. Operating information and commands are exchanged between the modules. In addition, each module transmits Node Alive or State of Health messages on the data bus about once every two seconds. When the Body Control Module (BCM) and Radio Receiver do not detect any messages on the Class 2 line, U1255 is set. **Possible Causes** ● Check the Scan Tool connection to the data link connector ● Verify that the Scan Tool works on another vehicle ● Check the Class 2 circuit for an intermittent short to ground ● Check for battery power at the data link connector (Pin 16) ● Check for main ground at the data link connector (Pins 4 and 5)

OBD II Trouble Code List (U1xxx Codes)

DTC	Trouble Code Title, Conditions & Possible Causes
DTC U1300 **1T PCM, MIL: Yes** 1999, 2000, 2001 Catera 3.0L VIN R engine Transmissions: All	**Class 2 Circuit Short To Ground Conditions:** Key on or engine started, system voltage over 9.0v, then with the Scan Tool connected and powered up, a Could Not Communicate With The Class 2 Data Line" message. Modules connected to the Class 2 line monitor the data bus for serial data communications during normal vehicle operation. Operating information and commands are exchanged between the modules. In addition, each module transmits Node Alive or State of Health messages on the data bus about once every two seconds. When the Body Control Module (BCM) and Radio Receiver do not detect messages and the Class 2 line is "low" for 3 seconds, U1300 is set. **Possible Causes** ● Check the Class 2 circuit for a short to ground condition ● One or more modules on the Class 2 line is shorted to ground
DTC U1300 **1T PCM, MIL: Yes** 1999, 2000, 2001, 2002, 2003, 2004, 2005 All Other Models Engines: All engines Transmissions: All	**Class 2 Circuit Short to Ground Conditions:** Key on or engine running; system voltage supplied to the module is in the normal operating voltage range, vehicle power mode requires serial data communication to occur, and the PCM did no detect any valid messages on the Class 2 circuit, or the voltage condition detected on the Class 2 circuit was low for 3 seconds. Modules connected to the Class 2 circuit check for data communications during normal vehicle operation. Operating information and commands are exchanged among the modules. Each module transmits Node Alive messages on the Class 2 data circuit once every 2 seconds. When the module detects a low voltage condition on the Class 2 serial data circuit for approximately 3 seconds, it sets U1300 or U1305 if it cannot identify the problem. Note: This code is set by loss of communication. Look in all of the modules for this trouble code - the one without it has a problem **Possible Causes** ● Class 2 serial data line was in a low state for 3 seconds due to a short to sensor ground or chassis ground ● One or more modules on the Class 2 line has a short to ground
DTC U1301 **2T PCM, MIL: Yes** 1999, 2000, 2001 Catera 3.0L VIN R engine Transmissions: All	**Class 2 Circuit Short To Power Conditions:** Key on or engine started, system voltage over 9.0v, then with the Scan Tool connected and powered up, a Could Not Communicate With The Class 2 Data Line" message is displayed. Modules connected to the Class 2 line monitor the data bus for serial data communications during normal operation. Operating information and commands are exchanged between the modules. In addition, each module transmits Node Alive or State of Health messages on the data bus about once every two seconds. When the Body Control Module and Radio Receiver do not detect any messages and the Class 2 line is "high" for 3 seconds, U1300 is set. **Possible Causes** ● Check the Class 2 circuit for a short to power condition ● One or more modules on the Class 2 line is shorted to power
DTC U1301 **1T PCM, MIL: Yes** 1999, 2000, 2001, 2002, 2003, 2004, 2005 All Other Models Engines: All engines Transmissions: All	**Class 2 Circuit Short to Battery Conditions:** Key on or engine running; system voltage supplied to the module is in the normal operating voltage range, vehicle power mode requires serial data communication to occur, and the PCM did no detect any valid messages on the Class 2 circuit, or the voltage condition detected on the Class 2 circuit was low for 3 seconds. Modules connected to the Class 2 circuit check for data communications during normal vehicle operation. Operating information and commands are exchanged among the modules. In addition, each module transmits Node Alive messages on the Class 2 data circuit once every 2 seconds. If the module detects a high voltage condition on the Class 2 serial data circuit for 3 seconds, it sets U1300. Note: This code is set by loss of communication. Look in all of the modules for this trouble code - the one without it has a problem. **Possible Causes** ● Class 2 serial data line was in a high state for 3 seconds due to a short to VREF or system power ● One or more modules on Class 2 line has an short to power
DTC U1305 **1T PCM, MIL: Yes** 1999, 2000, 2001, 2002, 2003, 2004, 2005 All Engines: All Transmissions: All	**Class 2 Data Link High or Low Conditions:** Key on or engine running; system voltage supplied to the module is in the normal operating voltage range, vehicle power mode requires serial data communication to occur, and the PCM did no detect any valid messages on the Class 2 circuit, or the voltage condition detected on the Class 2 circuit was low for 3 seconds. Modules connected to the Class 2 circuit check for data communications during normal vehicle operation. Operating information and commands are exchanged among the modules. In addition, each module transmits Node Alive messages on the Class 2 data circuit about once every 2 seconds. When the module detects a high voltage condition on the Class 2 serial data circuit for approximately 3 seconds, it sets U1300 or U1305 if it cannot identify the problem. **Possible Causes** ● Class 2 serial data line has either a high or low voltage condition on the circuit, and the module cannot identify the fault ● One or more modules on Class 2 line has an short to power ● One or more modules on the Class 2 line has a short to ground

OBD II Trouble Code List (U1xxx – U2xxx)

DTC	Trouble Code Title, Conditions & Possible Causes
DTC U1800 **2T PCM, MIL: Yes** 2001, 2002, 2003, 2004, 2005 C/K Series Pickup 6.6L VIN 1 Diesel engine Transmissions: All	**CAN Bus Messages Missing From FICD Conditions:** Key on, and the ECM did not receive any CAN messages from the FICD for 45 milliseconds. The ECM, FICD and the TCM communicate control and diagnostic data via a SAE J1939 CAN bus circuit. The ECM monitors CAN operational status by expecting a constant flow of messages from the FICM and the TCM. If the ECM fails to receive an expected message from one of these modules, U1800, U2104, or U2106 will set depending on the lost data. **Possible Causes** ● Modules that communicate on the SAE J1939 CAN system are wired parallel to each other until the respective circuits are spliced together at the ECM. An open in the CAN circuit of one module will not affect other modules. A short to ground or short to power affects all modules no matter where the failure occurs. ● FICD power and/or ground circuit are open or shorted ● FICD (module) has failed ● ECM or TCM has failed
DTC U2100 **2T PCM, MIL: Yes** 1999, 2000, 2001 Catera 3.0L VIN R engine Transmissions: All	**CAN Bus Communication Malfunction Conditions:** Key on or engine running; and the module that set the code has attempted to establish communications on the CAN circuits more than 7 times without connecting. The controller area network (CAN) is a high-speed network for sharing data. The circuit consists of a two (2) twisted wire pairs - one pair connects the PCM (ECM) to the TCM and a second pair connects the PCM (ECM) to the EBCM. **Possible Causes** ● Check for a loose connection at the EBCM, PCM or the TCM ● Class 2 circuit problem (one or more wires open or shorted) ● EBCM has an internal short to ground or short to power problem ● PCM has an internal short to ground or short to power problem ● TCM has an internal short to ground or short to power problem
DTC U2103 **2T PCM, MIL: Yes** 1999, 2000, 2001 Catera 3.0L VIN R engine Transmissions: All	**Fewer Controllers On The Bus Than Programmed Conditions:** Key on or engine running; and after each module on the CAN circuit monitored activity on the bus, one module set this code because it did not receive CAN formatted messages from all of the modules that it should have after (7) attempts (there are less modules communicating on the bus than programmed). The CAN is a high-speed network for sharing data. **Possible Causes** ● Check for a loose connection at the EBCM, PCM or the TCM ● Read the data lists from the EBCM, PCM and TCM with a Scan Tool. If no data is received from a module, check its connector! ● The module that did not communicate may have failed
DTC U2104 **2T PCM, MIL: Yes** 1999, 2000, 2001 Catera 3.0L VIN R engine Transmissions: All	**CAN Bus Reset Counter Overrun Conditions:** Key on or engine running; and a module set this code because it was unable to establish communications on the CAN circuit over 7 times. After the seventh attempt, the module sets this code (i.e., there fewer modules communicating on the CAN bus than there should be. The CAN bus is a high-speed network for sharing data. **Possible Causes** ● Check for a loose connection at the EBCM, PCM or the TCM ● Read the data lists from the EBCM, PCM and TCM with a Scan Tool. If no data is received from a module, check that module ● The module that did not communicate all its data has failed
DTC U2104 **2T PCM, MIL: Yes** 2001, 2002, 2003, 2004, 2005 C/K Series Pickup 6.6L VIN 1 Diesel engine Transmissions: All	**CAN Bus Messages Missing From FICD & TCM Conditions:** Key on, and the ECM did not receive any CAN messages from the FICD and TCM for 45 ms. The ECM, FICD and the TCM send control and diagnostic information via a SAE J1939 CAN bus circuit. The ECM monitors CAN operational status by expecting a constant flow of messages from the FICM and the TCM. If the ECM fails to receive an expected message from one of these modules, U1800, U2104, or U2106 will set; depending on the lost data. **Possible Causes** ● Modules that communicate on the SAE J1939 CAN system are wired parallel to each other until the respective circuits are spliced together at the ECM. An open in the CAN circuit of one module will not affect other modules. A short to ground or short to power affects all modules no matter where the failure occurs. ● FICD power and/or ground circuit are open or shorted ● FICD (module) has failed ● ECM or TCM has failed

OBD II Trouble Code List (U2xxx Codes)

DTC	Trouble Code Title, Conditions & Possible Causes
DTC U2105 **2T PCM, MIL: Yes** 2003, 2004, 2005 CTS (Cadillac) 2.6L VIN M, 3.2L VIN N Transmissions: All Trouble Code ID: **U2105** Number of Trips to Set Code: **2T** OBD II Monitor Type: **CCM** MIL: YES Schematic: **GMU2105.bmp** ; Cadillac CTS models	**TCM - No Communication With Engine Control Module Conditions:** Key on or engine running; and the TCM did not receive messages from the PCM for 40 ms. The CAN (network) consists of 2 separate conductors between the ECM and the transmission control module (TCM). Each module sends information by toggling the circuits high or low to produce pulse width signals, which the other modules decode. The CAN data stream contains information such as status messages for relevant operating parameters, device requests and acknowledgments, and state-of-health messages. If the TCM does not receive PCM data for a certain amount of time, U2105 is set. **Possible Causes** ● Perform the Diagnostic System Check for Powertrain Module ● Test the main power and ground circuits to the PCM ● Check the CAN serial data circuit to the TCM ● TCM may have failed
DTC U2105 **2T PCM, MIL: Yes** 1999, 2000, 2001 Catera 3.0L VIN R engine Transmissions: All	**Lost Communication With The Engine Control System Conditions:** Key on or engine running; then after all module had initiated for the ignition cycle, one module (PCM or ECM) stopped communicating and other modules set a code specific to the module not communicating. The CAN bus is a high-speed network for sharing data. The circuit consists of a two (2) twisted wire pairs - one pair connects the PCM to the TCM and the other the PCM to the EBCM. **Possible Causes** ● Check for a loose connection at the PCM (ECM) ● Test the main power and ground circuits to the PCM ● Check the CAN serial data circuit to the PCM ● PCM may have failed
DTC U2106 **2T PCM, MIL: Yes** 2001, 2002, 2003, 2004, 2005 C/K Series Pickup 6.6L VIN 1 Diesel engine Transmissions: All	**CAN Bus Messages Missing From TCM Conditions:** Key on, and the ECM did not receive any CAN messages from the TCM for 45 ms. The ECM, FICD and the TCM send control and diagnostic information via a SAE J1939 CAN bus circuit. The ECM monitors CAN operational status by expecting a constant flow of messages from the FICM and the TCM. If the ECM fails to receive an expected message from one of these modules, DTC U1800, U2104, or U2106 will set; depending on what data is lost. **Possible Causes** ● Modules that communicate on the SAE J1939 CAN system are wired parallel to each other until the respective circuits are spliced together at the ECM. An open in the CAN circuit of one module will not affect other modules. A short to ground or short to power affects all modules no matter where the failure occurs. ● FICD power and/or ground circuit are open or shorted ● FICD (module) has failed ● ECM or TCM has failed
DTC U2106 **2T PCM, MIL: Yes** 1999, 2000, 2001 Catera 3.0L VIN R engine Transmissions: All	**Lost Communication With Transmission Control System Conditions:** Key on or engine running; then after all module had initiated for the ignition cycle, one module (TCM) stopped communicating and other modules set a code specific to the module not communicating. The controller area network (CAN) is a high-speed network for sharing data. The circuit consists of a two (2) twisted wire pairs - one pair connects the PCM (ECM) to the TCM and a second pair connects the PCM (ECM) to the EBCM. **Possible Causes** ● Check for a loose connection at the TCM ● Test the main power and ground circuits to the TCM ● Check the CAN serial data circuit to the TCM ● TCM may have failed
DTC U2108 **2T PCM, MIL: Yes** 1999, 2000, 2001 Catera 3.0L VIN R engine Transmissions: All	**Lost Communication With Electronic Brake Control Module Conditions:** Key on or engine running; then after all module had initiated for the ignition cycle, one module (EBCM) stopped communicating and other modules set a code specific to the module not communicating. The controller area network (CAN) is a high-speed network for sharing data. The circuit consists of a two (2) twisted wire pairs - one pair connects the PCM (ECM) to the TCM and a second pair connects the PCM (ECM) to the EBCM. **Possible Causes** ● Check for a loose connection at the EBCM ● Test the main power and ground circuits to the EBCM ● Check the CAN serial data circuit to the EBCM ● PCM may have failed

OBD II Trouble Code List (B1xxx – B2xxx Codes)

DTC	Trouble Code Title, Conditions & Possible Causes
DTC B1327 **1T CCM, MIL: No** 2003, 2004, 2005 CTS (Cadillac) 2.6L VIN M, 3.2L VIN N Transmissions: All	**Dash Integration Module Power Circuit Low Input Conditions:** Key on or engine running; DIM has power, ground and ignition not in Start mode. This DTC executes regardless of the battery voltage, except when the DTC B1390 is current, and the DIM (module) detected the voltage fell below 9.0v for 1200 ms. When the vehicle exits Start, the DIM will delay checking the voltage for two seconds. The Dash Integration Module (DIM) has an internal voltage sensor with a dedicated circuit that checks the battery positive circuit and battery negative circuit voltage to determine if it is more than 9.0v. **Possible Causes** ● A Message is sent over the Class 2 circuit to all other modules ● Battery power circuit to the DIM is open or has a high resistance condition ● DIM has failed
DTC B1328 **1T CCM, MIL: No** 2003, 2004, 2005 CTS (Cadillac) 2.6L VIN M, 3.2L VIN N Transmissions: All	**Dash Integration Module Power Circuit High Input Conditions:** Key on or engine running; DIM has power, ground and ignition not in Start mode. This DTC executes regardless of the battery voltage, except when the DTC B1390 is current, and the DIM (module) detected the voltage was more than 16.5v for 1200 ms. When the vehicle exits Start, the DIM will delay checking the voltage for two seconds. The Dash Integration Module (DIM) has an internal voltage sensor with a dedicated circuit that checks the battery positive circuit and battery negative circuit voltage to detect if it is less than 15.5v. **Possible Causes** ● A Message is sent over the Class 2 circuit to all other modules ● Battery power circuit to the DIM is open or has a high resistance condition ● DIM has failed
DTC B1513 **1T CCM, MIL: No** 2003, 2004, 2005 CTS (Cadillac) 2.6L VIN M, 3.2L VIN N Transmissions: All	**Instrument Panel Cluster Voltage Low Input Conditions:** DTC P1327 not set, engine started, engine speed over 1,500 rpm, and the PCM received a "low voltage" message on the Class 2 data line from the Instrument panel cluster (IPC). The IPC detected the system voltage was less than 10.5v for 30 seconds at the IPC. **Possible Causes** ● Check the operation of the Charging system (Generator) ● The charge indicator and driver warning message are set in the driver information center (DIC) ● IPC is damaged or has failed
DTC B1514 **1T CCM, MIL: No** 2003, 2004, 2005 CTS (Cadillac) 2.6L VIN M, 3.2L VIN N Transmissions: All	**Instrument Panel Cluster Voltage High Input Conditions:** DTC P1327 not set, engine started, engine speed over 1,500 rpm, and the PCM received a "high voltage" message on the Class 2 data line from the Instrument panel cluster (IPC). The IPC detected the system voltage was more than 16.0v for 30 seconds at the IPC. **Possible Causes** ● Check the operation of the Charging system (Generator) ● The charge indicator and driver warning message are set in the driver information center (DIC) ● IPC is damaged or has failed
DTC B2722 **1T CCM, MIL: No** 2001, 2002, 2003, 2004, 2005 C/K Series Pickup 6.6L VIN 1 Diesel engine Transmissions: All	**Powertrain Control Module Internal Malfunction Conditions:** Key on or engine running; system voltage from 9-16v, Tow/Haul switch has been activated, and the ECM detected the Tow/Haul switch signal circuit was low for around 3 minutes. Tow/Haul mode enables the operator to achieve enhanced shift performance when towing or hauling a load. When tow/haul is selected, the tow/haul switch input signal to the BCM is momentarily toggled to zero volts. The BCM sends a serial data message to the PCM and instrument panel controller (IPC). The PCM extends the length of time between an upshift, increases transmission line pressure and the IPC illuminates the tow/haul indicator lamp. Cycling the tow/haul switch a second time disables the Tow/Haul mode and returns the transmission to a normal shift pattern. **Possible Causes** ● Tow/Haul switch is shorted to ground ● Tow/Haul switch is damaged or it has failed ● PCM has failed

Geo OBD II Trouble Codes

OBD II Trouble Code List (P0xxx Codes)

DTC	Trouble Code Title, Conditions & Possible Causes
DTC P0101 **2T CCM, MIL: Yes** 1996, 1997 Tracker 1.6L VIN 6 engine Transmissions: All	**MAF Sensor Range/Performance Conditions:** DTC P0102, P0103, P0335, P1408 and P1451 not set, BARO sensor more than 75 kPa, IAT signal from 7-158ºF, engine speed less than 2000 rpm, TP angle less than 20 degrees, and the PCM detected the maximum/minimum flow rate was less than 0.1 gm/s. **Possible Causes** ● Base engine vacuum leak, PCV valve leaking or stuck open ● Engine oil dipstick missing or not fully seated ● MAF sensor element (wire) is contaminated or dirty ● MAF sensor signal or ground circuit fault or sensor has failed ● PCM has failed
DTC P0102 **1T CCM, MIL: Yes** 1996, 1997 Tracker 1.6L VIN 6 engine Transmissions: All	**MAF Sensor Circuit Low Frequency Conditions:** Key on or engine running; and the PCM detected the MAF sensor current was less than 0.64 milliamps for 5 seconds during the test. **Possible Causes** ● Base engine vacuum leak, PCV valve leaking or stuck open ● Engine oil dipstick missing or not fully seated ● MAF sensor element hot wire contaminated or the sensor failed ● MAF sensor signal shorted to ground or ground circuit problem ● MAF sensor wiring routed close to ignition wires or generator ● PCM has failed
DTC P0103 **1T CCM, MIL: Yes** 1996, 1997 Tracker 1.6L VIN 6 engine Transmissions: All	**Title: MAF Sensor Circuit High Frequency** **Trouble Code Conditions** Key on or engine running; and the PCM detected the MAF sensor output current was more than 4.90 mA for 5 seconds. *Note: The PCM will enter the Fail-Safe Function and determine injector drive time according to TP sensor and engine speed as well as IAC valve position and injector frequency when this code sets.* **Possible Causes** ● MAF sensor element hot wire is contaminated or dirty ● MAF sensor is damaged or has failed ● MAF sensor signal circuit is open between the sensor and PCM ● MAF sensor wiring routed too close to the ignition wires ● MAF sensor wiring routed to close to the back of the Generator ● PCM has failed
DTC P0105 **1T CCM, MIL: Yes** 1996, 1997 Prizm All engines Transmissions: All	**MAP Sensor Circuit Malfunction Conditions:** Engine started, engine speed under 1000 rpm, throttle closed, ECT sensor more than 158ºF, and the PCM detected the MAP sensor differed from the MAP sensor signal at initial startup by more than 3.3v; or with the engine speed under 2500 rpm and TP angle over 1.0v, the PCM detected the MAP sensor signal was less than 1.0v **Possible Causes** ● MAP sensor signal circuit is open or shorted to ground ● MAP sensor VREF circuit is open or shorted to ground ● MAP sensor ground circuit is open (other codes will be set) ● MAP sensor is damaged or has failed ● PCM has failed
DTC P0105 **2T CCM, MIL: Yes** 1996, 1997 Metro All engines	**MAP Sensor Performance Conditions:** DTC P0105, P0107 and P0108 not set, engine speed and TP angle both steady, and the PCM detected the MAP sensor (obtained during engine cranking) differed from the MAP input (obtained at engine startup) by less than 1.3 kPa, or the BARO pressure differed from the MAP sensor by less than 33 kPa for 2 minutes. **Possible Causes** ● MAP sensor source vacuum line is leaking or disconnected ● MAP sensor is damaged, out-of-calibration or has failed ● PCM has failed
DTC P0105 **1T CCM, MIL: Yes** 1996, 1997 Models: Prism All engines Transmissions: All	**MAP Sensor Range/Performance Conditions:** DTC P0105, P0107 and P0108 not set, engine started, engine speed less than 1000 rpm with TP sensor near 0v, ECT sensor more than 158ºF, and the PCM detected the MAP sensor indicated more than 3.3v, or with the engine speed near 2500 rpm, that the MAP input was less than 1.0v with the TP sensor more than 1.0v. **Possible Causes** ● MAP sensor source vacuum line is leaking or disconnected ● MAP sensor is damaged, out-of-calibration or has failed ● PCM has failed

OBD II Trouble Code List (P0xxx Codes)

DTC	Trouble Code Title, Conditions & Possible Causes
DTC P0110 **1T CCM, MIL: Yes** 1996, 1997 Prizm All engines Transmissions: All	**Intake Air Temperature Sensor Circuit Malfunction Conditions:** Key on or engine running; and the PCM detected the IAT sensor signal was from -40ºF to -67ºF, or it indicated a value from 240-262ºF for 500 ms during the CCM test period. **Possible Causes** ● IAT sensor signal circuit is open or shorted to ground ● IAT sensor signal circuit is shorted to VREF or system power ● IAT sensor is damaged or has failed (out of calibration) ● PCM has failed
DTC P0112 **1T CCM, MIL: Yes** 1996, 1997 Metro & Tracker All engines Transmissions: All	**Intake Air Temperature Sensor Circuit Low Input Conditions:** Engine started; and the PCM detected the IAT sensor was over 246ºF to 282ºF for 500 ms. **Possible Causes** ● IAT sensor signal circuit is shorted to sensor ground ● IAT sensor signal circuit is shorted to chassis ground ● IAT sensor is damaged or has failed ● PCM has failed
DTC P0113 **1T CCM, MIL: Yes** 1996, 1997 Metro & Tracker All engines Transmissions: All	**Intake Air Temperature Sensor Circuit High Input Conditions:** Engine started; and the PCM detected the IAT sensor indicated less than -40ºF for 500 ms. **Possible Causes** ● IAT sensor signal circuit is open between sensor and the PCM ● IAT sensor signal circuit is shorted to system power ● IAT sensor is damaged or has failed ● PCM has failed
DTC P0115 **1T CCM, MIL: Yes** 1996, 1997 Prizm All engines Transmissions: All	**Engine Coolant Temperature Sensor Circuit Malfunction Conditions:** Key on or engine running; and the PCM detected the ECT sensor that indicated less than -40 to -67ºF, or the ECT sensor indicated a value from 240ºF to 262ºF, condition met for 500 ms. **Possible Causes** ● ECT sensor circuit is open, shorted to ground or to VREF ● ECT sensor is damaged or has failed (out of calibration) ● PCM has failed
DTC P0116 **1T CCM, MIL: Yes** 1996, 1997 Prizm All engines Transmissions: All	**Engine Coolant Temperature Sensor Range/Performance Conditions:** Engine started, ECT sensor at startup less than 20ºF, engine runtime over 20 minutes, and the PCM detected the ECT sensor indicated less than 95ºF, or with the ECT and IAT sensors from 20ºF to 50ºF at startup with the engine runtime over 5 minutes, it detected the ECT sensor was less than 95ºF, or with the ECT and IAT sensors more than 50ºF at startup with the engine runtime over 2 minutes, it detected the ECT sensor signal was below 95ºF. **Possible Causes** ● ECT sensor signal circuit is open or shorted to ground ● ECT sensor is contaminated, skewed or it has failed ● PCM has failed
DTC P0117 **1T CCM, MIL: Yes** 1996, 1997 Metro & Tracker All engines Transmissions: All	**Engine Coolant Temperature Sensor Circuit Low Input Conditions:** Key on or engine running; and the PCM detected the ECT sensor indicated more than 279ºF for 500 ms during the CCM test. **Possible Causes** ● ECT sensor signal circuit shorted to sensor or chassis ground ● ECT sensor is damaged or has failed (it may be shorted) ● PCM has failed
DTC P0118 **1T CCM, MIL: Yes** 1996, 1997 Metro & Tracker All engines	**Engine Coolant Temperature Sensor Circuit High Input Conditions:** Key on or engine running; and the PCM detected the IAT sensor indicated more than -40ºF for 500 ms during the CCM test. **Possible Causes** ● ECT sensor signal circuit is open, or shorted to VREF or system power ● ECT sensor ground circuit is open between sensor and PCM ● ECT sensor is damaged or has failed (may be open internally) ● PCM has failed
DTC P0120 **1T CCM, MIL: Yes** 1996, 1997 Prizm All engines Transmissions: All	**Title: Throttle Position Sensor Circuit Malfunction** **Trouble Code Conditions** Key on or engine running; and the PCM detected the TP sensor was more than 1.0v with the throttle closed, or that the TP sensor was less than 4.90v during a wide open throttle event. **Possible Causes** ● TP sensor signal circuit open or shorted to ground ● TP sensor ground circuit is open ● TP sensor power circuit is open (check VREF circuit at PCM) ● TP sensor is damaged or has failed ● PCM has failed

OBD II Trouble Code List (P0xxx Codes)

DTC	Trouble Code Title, Conditions & Possible Causes
DTC P0121 **2T CCM, MIL: Yes** 1996, 1997 Metro All engines	**Throttle Position Sensor Range/Performance Conditions:** DTC P0122 and P0123 not set, engine started, engine speed from 1500-3500 rpm, ECT sensor more than 158ºF, IAT sensor from 14ºF to 122ºF, fuel level over 25%, BARO sensor more than 75 kPa, MAP sensor change less than 13 kPa during 16 firing events, and the PCM detected the Expected and Actual TP sensor values differed by more than 11 degrees for 3 seconds in the CCM Rationality test. **Possible Causes** ● TP sensor signal circuit is open to the PCM (intermittent fault) ● TP sensor ground circuit is open (an intermittent fault) ● MAP sensor is out of calibration ● Throttle body is damaged or throttle linkage is bent or binding ● TP sensor is damaged or has failed
DTC P0121 **2T CCM, MIL: Yes** 1996, 1997 Prizm All engines	**Throttle Position Sensor Range/Performance Conditions:** Engine started, vehicle speed over 30 mph while in closed loop, and the PCM detected the TP sensor was out of the acceptable range while operating under these conditions. **Possible Causes** ● TP sensor signal circuit is open to the PCM (intermittent fault) ● TP sensor ground circuit is open (an intermittent fault) ● MAP sensor is out of calibration ● Throttle body is damaged or throttle linkage is bent or binding ● TP sensor is damaged or has failed
DTC P0121 **2T CCM, MIL: Yes** 1996, 1997 Tracker All engines Transmissions: All	**Throttle Position Sensor Range/Performance Conditions:** DTC P0101, P0102, P0103, P0116, P0117, P0118, P0122, P0123, P0335 and P1451 not set, engine speed under 3500 rpm, MAF sensor indicating a maximum to minimum airflow rate variation of over 25.5 gm/s, MAP sensor under 26 kPa with the TP sensor signal less than 35 degrees, or with the MAP sensor over 67 kPa and the TP angle over 35 degrees, the PCM detected too much difference in the Expected and Actual TP values for over 3 seconds. **Possible Causes** ● TP sensor signal circuit is open to the PCM (intermittent fault) ● TP sensor ground circuit is open (an intermittent fault) ● MAP sensor is out of calibration ● Throttle body is damaged or throttle linkage is bent or binding ● TP sensor is damaged or has failed
DTC P0122 **1T CCM, MIL: Yes** 1996, 1997 Metro & Tracker All engines Transmissions: All	**Throttle Position Sensor Circuit Low Input Conditions:** DTC P0123 not set, engine started, engine running at hot idle speed, and the PCM detected the TP sensor indicated less than a 2% throttle opening for 5 seconds during the CCM test. **Possible Causes** ● TP sensor signal circuit is shorted to ground ● TP sensor power circuit is open (check VREF from the PCM) ● TP sensor is damaged or failed (it may be shorted internally) ● PCM has failed
DTC P0123 **1T CCM, MIL: Yes** 1996, 1997 Metro & Tracker All engines	**Throttle Position Sensor Circuit High Input Conditions:** DTC P0122 not set, engine started, engine running at hot idle speed, and the PCM detected the TP sensor indicated more than a 96% throttle opening for 5 seconds during the test. **Possible Causes** ● TP sensor signal circuit is open between sensor and the PCM ● TP sensor ground circuit is open between sensor and the PCM ● TP sensor is damaged or has failed (it may be open internally) ● PCM has failed
DTC P0125 **2T CCM, MIL: Yes** 1996, 1997 Metro All engines Transmissions: All	**ECT Excessive Time to Enter Closed Loop Conditions:** Engine started, fuel level over 25%, BARO sensor over 75 KPa, HO2S-11 signal varying, IAT sensor from 14-122ºF, and the PCM detected the ECT sensor did not indicate closed loop. **Possible Causes** ● Check the operation of the thermostat (it may be stuck open) ● ECT sensor signal circuit has high resistance, or the ECT sensor has failed ● Inspect for low coolant level or an incorrect coolant mixture
DTC P0125 **2T CCM, MIL: Yes** 1996, 1997 Prizm, Tracker All engines Transmissions: All	**ECT Excessive Time To Enter Closed Loop Conditions:** DTC P0117, P0118 and P0335 not set, engine runtime over 10 minutes, ECT sensor below 86ºF at startup, BARO sensor over 75 kPa, IAT sensor from 7-158ºF, and the PCM detected the ECT sensor did not indicate a closed loop temperature value after another 2-5 minutes. **Possible Causes** ● Check the operation of the thermostat (it may be stuck open) ● ECT sensor signal circuit has high resistance ● ECT sensor has failed ● Inspect for low coolant level or an incorrect coolant mixture

OBD II Trouble Code List (P0xxx Codes)

DTC	Trouble Code Title, Conditions & Possible Causes
DTC P0130 **2T CCM, MIL: Yes** 1996, 1997 Prizm All engines Transmissions: All	**O2S-11 or HO2S-11 (Bank 1 Sensor 1) Circuit Malfunction Conditions:** Engine started, engine running at idle speed, ECT sensor more than 158ºF, and the PCM detected the O2S or HO2S signal was fixed between 400 and 550 mv, or the O2S or HO2S signal was fixed at less than 550 mv, or the O2S signal was fixed at more than 440 mv. **Possible Causes** ● HO2S signal circuit is open between the sensor and the PCM ● HO2S signal circuit is shorted to sensor or chassis ground ● HO2S signal circuit is shorted to VREF or system power (B+) ● HO2S is damaged, contaminated or it has failed ● PCM has failed
DTC P0131 **2T CCM, MIL: Yes** 1996, 1997 Metro All engines Transmissions: All	**HO2S-11 (Bank 1 Sensor 1) Circuit Low Input Conditions:** Engine started, IAT sensor from 14-158ºF, ECT sensor more than 176ºF, BARO sensor more than 75 kPa, vehicle driven to a speed of over 35 mph for 2 minutes, followed by an idle period with the engine speed steady for 2 minutes, and the PCM detected the maximum HO2S signal was less than 300 mv during the CCM test. **Possible Causes** ● HO2S signal circuit is shorted to ground ● HO2S is damaged or it has failed ● PCM has failed
DTC P0131 **2T CCM, MIL: Yes** 1996, 1997 Tracker All engines Transmissions: All	**HO2S-11 (Bank 1 Sensor 1) Circuit Low Input Conditions:** Engine started, ECT and IAT sensors more than 19ºF at startup, BARO sensor over 75 kPa, vehicle driven to a speed of over 35 mph for 2 minutes, followed by an idle period with the engine speed steady for 1 minute, and the PCM detected the maximum HO2S signal was less than 500 mv with the minimum HO2S signal less than 70 mv, or that an internal check of the HO2S circuit indicated too much voltage drop was present during the CCM test. **Possible Causes** ● HO2S signal circuit is shorted to ground ● HO2S is damaged or it has failed ● PCM has failed
DTC P0132 **2T CCM, MIL: Yes** 1996, 1997 Metro All engines Transmissions: All	**HO2S-11 (Bank 1 Sensor 1) Circuit High Input Conditions:** Engine started, IAT sensor from 14-158ºF, ECT sensor more than 176ºF, BARO sensor over 75 kPa, vehicle driven to a speed of over 35 mph for 2 minutes, followed by an idle period with the engine speed steady for 2 minutes, and the PCM detected the maximum HO2S signal was more than 600 mv during the CCM test. **Possible Causes** ● HO2S is contaminated, damaged or it has failed ● HO2S signal circuit shorted to power (check the heater circuit) ● PCM has failed
DTC P0132 **2T CCM, MIL: Yes** 1998, 1999, 2000, 2001, 2002 Tracker All engines Transmissions: All	**HO2S-11 (Bank 1 Sensor 1) Circuit High Input Conditions:** ECT sensor and IAT sensors more than 19ºF at startup, BARO sensor over 75 kPa, engine started, vehicle driven to a speed over 35 mph for 2 minutes, followed by an idle period with the engine speed steady for 1 minute, and the PCM detected the minimum HO2S signal was more than 300 mv and the maximum average HO2S signal was more than 950 mv, or the minimum average HO2S signal was 400 mv during the HO2S Diagnostic Test function. **Possible Causes** ● HO2S is contaminated, damaged or it has failed ● HO2S signal circuit shorted to power (check the heater circuit) ● PCM has failed
DTC P0133 **2T O2S, MIL: Yes** 1996, 1997 Metro All engines Transmissions: All	**HO2S-11 (Bank 1 Sensor 1) Slow Response Conditions:** Engine started, IAT sensor from 14-158ºF, ECT sensor over 176ºF, BARO sensor over 75 kPa, vehicle driven to a speed of over 35 mph for 2 minutes, followed by an idle period with the engine speed steady for 2 minutes, and the PCM detected the response time of the HO2S-11 was more than 1 second, or that the switch cycle average was over 5 seconds. **Possible Causes** ● Exhaust leak present in the exhaust manifold or exhaust pipes ● HO2S element is fuel contaminated or has deteriorated ● PCM has failed
DTC P0133 **2T O2S, MIL: Yes** 1996, 1997 Prizm All engines Transmissions: All	**HO2S-11 (Bank 1 Sensor 1) Slow Response Conditions:** Engine started, engine running a idle speed in closed loop, ECT sensor more than 158ºF, and the PCM detected the average response rate of the HO2S to change from rich-to-lean or lean-to-rich was more than 1.1 seconds during the HO2S Diagnostic Test. **Possible Causes** ● Exhaust leak present in the exhaust manifold or exhaust pipes ● HO2S element is fuel contaminated or has deteriorated ● PCM has failed

OBD II Trouble Code List (P0xxx Codes)

DTC	Trouble Code Title, Conditions & Possible Causes
DTC P0133 **2T O2S, MIL: Yes** 1996, 1997 Tracker All engines Transmissions: All	**HO2S-11 (Bank 1 Sensor 1) Slow Response Conditions:** Engine started, ECT and IAT sensors more than 19ºF at startup, BARO sensor over 75 kPa, vehicle driven to a speed of over 35 mph for 2 minutes, then back to a stable idle speed for 1 minute, and the PCM detected the average HO2S response rate was over 2.5 seconds, or the average HO2S switch cycle was over 6 seconds. **Possible Causes** ● Exhaust leak present in the exhaust manifold or exhaust pipes ● HO2S element is fuel contaminated or has deteriorated ● PCM has failed
DTC P0134 **2T O2S, MIL: Yes** 1996, 1997 Metro All engines Transmissions: All	**HO2S-11 (Bank 1 Sensor 1) Insufficient Activity Conditions:** Engine started, IAT sensor from 14ºF to 122ºF at startup, BARO sensor more than 75 kPa, vehicle driven at a speed of over 35 mph, and then returned to idle speed and allowed to idle for over 1 minute, and the PCM detected the average response rate of the HO2S signal was over 2.5 seconds, or the switch cycle average of the HO2S signal was over 6 seconds. **Possible Causes** ● Exhaust leak present in exhaust manifold or exhaust pipes ● HO2S element fuel contamination or has deteriorated ● HO2S signal circuit or the ground circuit has high resistance ● HO2S heater element has failed, or the heater circuit is open ● PCM has failed
DTC P0134 **2T O2S, MIL: Yes** 1996, 1997 Tracker All engines Transmissions: All	**HO2S-11 (Bank 1 Sensor 1) Insufficient Activity Conditions:** Engine started with the ECT and IAT sensors more than 19ºF at startup, BARO sensor more than 75 kPa, engine running in closed loop, and the PCM detected the minimum and maximum HO2S signals were less than or more than 450 mv for 30 seconds. **Possible Causes** ● Exhaust leak present in exhaust manifold or exhaust pipes ● HO2S element fuel contamination or has deteriorated ● HO2S signal circuit or the ground circuit has high resistance ● HO2S heater element has failed, or the heater circuit is open ● PCM has failed
DTC P0135 **2T CCM, MIL: Yes** 1996, 1997 Metro All engines	**HO2S-11 (Bank 1 Sensor 1) Heater Circuit Malfunction Conditions:** Engine started, engine runtime 1 minute, then after an acceleration period of over 5 seconds, the PCM detected the heater circuit was less than 2.5v with heater commanded "off", or that it was less than 0.31v with heater turned commanded "on" during the CCM test. **Possible Causes** ● HO2S heater control circuit is open or shorted to ground ● HO2S heater control circuit is shorted to power ● HO2S heater power circuit is open (test power to IG Coil fuse) ● HO2S heater is damaged or has failed ● PCM has failed
DTC P0135 **2T CCM, MIL: Yes** 1996, 1997 Tracker All engines Transmissions: All	**HO2S-11 (Bank 1 Sensor 1) Heater Circuit Malfunction Conditions:** Engine started, engine running, and the PCM detected the HO2S heater current level exceeded 5.3 amps or indicated less than 0.15 amps with the heater "on", or the HO2S heater signal was more than 13.8v or was less than 8.7v with the heater "on", or the HO2S heater signal was 6v with the heater commanded "off" for 3 seconds. **Possible Causes** ● HO2S heater control circuit is open or shorted to ground ● HO2S heater control circuit is shorted to power ● HO2S heater power circuit is open (check power from IG fuse) ● HO2S heater is damaged or has failed ● PCM has failed
DTC P0136 **2T CCM, MIL: Yes** 1996, 1997 Metro All engines Transmissions: All	**HO2S-12 (Bank 1 Sensor 2) Circuit Malfunction Conditions:** Engine started, vehicle driven to 20-50 mph at an engine speed over 1500 rpm in closed loop, ECT sensor more than 176ºF, IAT sensor from 14-158ºF, BARO sensor more than 75 kPa, fuel level over 25%, and the PCM detected the HO2S signal was over 600 mv for 8 minutes, or the minimum HO2S signal was greater than or equal to 300 mv after a Maximum Voltage Check while operating under Decel Fuel Shutoff conditions for 5 seconds during the CCM test. **Possible Causes** ● HO2S signal circuit is open between the sensor and the PCM ● HO2S signal circuit is shorted to sensor or chassis ground ● HO2S signal circuit is shorted to VREF or system power (B+) ● HO2S is damaged, contaminated or it has failed ● PCM has failed

OBD II Trouble Code List (P0xxx Codes)

DTC	Trouble Code Title, Conditions & Possible Causes
DTC P0136 **2T CCM, MIL: Yes** 1996, 1997 Prizm All engines Transmissions: All	**HO2S-12 (Bank 1 Sensor 2) Circuit Malfunction Conditions:** Engine started, vehicle driven at a steady speed of over 1600 rpm in closed loop, ECT sensor more than 158°F, and the PCM detected the HO2S signal indicated more than or equal to 400 mv, or the HO2S signal indicated less than or equal to 500 mv for 5 seconds. **Possible Causes** ● HO2S signal circuit is open between the sensor and the PCM ● HO2S signal circuit is shorted to sensor or chassis ground ● HO2S signal circuit is shorted to VREF or system power (B+) ● HO2S is damaged, contaminated or it has failed ● PCM has failed
DTC P0136 **2T CCM, MIL: Yes** 1996, 1997 Tracker All engines Transmissions: All	**HO2S-12 (Bank 1 Sensor 2) Circuit Malfunction Conditions:** Engine started, vehicle driven at a steady speed of over 1600 rpm in closed loop, ECT sensor more than 158°F, BARO sensor over 75 kPa, fuel level over 15%, and the PCM detected the average HO2S signal was less than 120 mv or more than 900 mv, or that the HO2S Pullup signal indicated more than 4.5v for 2 seconds. **Possible Causes** ● HO2S signal circuit is open between the sensor and the PCM ● HO2S signal circuit is shorted to sensor or chassis ground ● HO2S signal circuit is shorted to VREF or system power (B+) ● HO2S is damaged, contaminated or it has failed ● PCM has failed
DTC P0141 **2T CCM, MIL: Yes** 1996, 1997 Metro All engines Transmissions: All	**HO2S-12 (Bank 1 Sensor 2) Heater Circuit Malfunction Conditions:** Engine speed over 2000 rpm for 2 minutes, ECT sensor more than 176°F, and the PCM detected the HO2S heater circuit indicated less than 2.5v with HO2S heater turned "off", or that the heater circuit was more than 0.31v with HO2S heater turned "on" for 5 seconds. **Possible Causes** ● HO2S heater control circuit is open or shorted to ground ● HO2S heater control circuit is shorted to power ● HO2S heater power circuit is open (check power from the fuse) ● HO2S heater is damaged or has failed ● PCM has failed
DTC P0141 **2T CCM, MIL: Yes** 1996, 1997 Prizm All engines	**HO2S-12 (Bank 1 Sensor 2) Heater Circuit Malfunction Conditions:** Engine started, engine running, and the PCM detected the HO2S heater current level was more than 2.0 amps or was less than 0.2 amps with the heater "on" during the CCM test. **Possible Causes** ● HO2S heater control circuit is open or shorted to ground ● HO2S heater control circuit is shorted to power ● HO2S heater power circuit is open (check power from the relay) ● HO2S heater is damaged or has failed ● PCM has failed
DTC P0141 **2T CCM, MIL: Yes** 1996, 1997 Tracker All engines	**HO2S-12 (Bank 1 Sensor 2) Heater Circuit Malfunction Conditions:** Engine started, engine running, and the PCM detected the HO2S heater current was more than 5.3 amps or was less than 0.15 amps with the heater "on"; or the HO2S heater signal was more than 13.8v or less than 8.7v with the heater "on"; or the HO2S heater signal indicated about 6.0v with the heater turned "off" for 3 seconds. **Possible Causes** ● HO2S heater control circuit is open or shorted to ground ● HO2S heater control circuit is shorted to power ● HO2S heater power circuit is open (check power from IG fuse) ● HO2S heater is damaged or has failed ● PCM has failed
DTC P0153 **2T CCM, MIL: Yes** 2001, 2002 Tracker 2.5L VIN 4 engine Transmissions: All	**HO2S-21 (Bank 2 Sensor 1) Circuit Slow Response Conditions:** ECT sensor and IAT sensors more than 19°F at startup, BARO sensor over 75 kPa, vehicle driven to a speed of over 35 mph for 2 minutes, then returned to a stable idle speed for 1 minute, and the PCM detected the average HO2S response rate was over 2.5 seconds, or the average HO2S switch cycle was over 6 seconds. **Possible Causes** ● Exhaust leak present in the exhaust manifold or exhaust pipes ● HO2S element fuel contamination ● HO2S element has deteriorated ● PCM has failed

OBD II Trouble Code List (P0xxx Codes)

DTC	Trouble Code Title, Conditions & Possible Causes
DTC P0171 **2T FUEL, MIL: Yes** 1996, 1997 Metro All engines Transmissions: All	**Fuel Trim System Lean (Bank 1) Conditions:** Engine started, engine running in closed loop, IAT sensor from 14-122°F, BARO sensor over 75 kPa, fuel level over 25%, and the PCM detected the Long Term fuel trim value was +30% for 5 seconds, or the Short Term fuel trim value was +20% for 45 seconds. **Possible Causes** ● Air leaks after the MAP sensor, or in the EGR or PCV system ● Base engine "mechanical" fault affecting one or more cylinders ● Exhaust leaks located in front of the HO2S location ● Fuel control sensor is out of calibration (i.e., ECT, IAT or MAP) ● Fuel delivery system supplying too little fuel during cruise or idle periods (e.g., faulty fuel pump or dirty, restricted fuel filter) ● Fuel injector (one or more) dirty or pressure regulator has failed ● HO2S is contaminated, deteriorated or it has failed ● Vehicle driven low on fuel or until it ran out of fuel
DTC P0171 **2T FUEL, MIL: Yes** 1996, 1997 Prizm, Tracker All engines Transmissions: All	**Fuel Trim System Lean (Bank 1) Conditions:** Engine started, ECT sensor from 18-203°F and less than 230°F during testing, IAT sensor from 18-140°F at startup, BARO sensor more than 75 kPa, and the PCM detected the Short Term fuel trim was more than 15% for 128 firing events, or the Long Term fuel trim was more than or equal to 20% for 128 firing events, or the total Fuel Trim was more than or equal to 33% for 128 firing events. **Possible Causes** ● Air leaks after the MAF sensor, or in the EGR or PCV system ● Base engine "mechanical" fault affecting one or more cylinders ● Exhaust leaks located in front of the HO2S location ● Fuel control sensor is out of calibration (i.e., ECT, IAT or MAP) ● Fuel delivery system supplying too little fuel during cruise or idle periods (e.g., faulty fuel pump or dirty, restricted fuel filter) ● Fuel injector (one or more) dirty or pressure regulator has failed ● HO2S is contaminated, deteriorated or it has failed ● Vehicle driven low on fuel or until it ran out of fuel
DTC P0172 **2T FUEL, MIL: Yes** 1996, 1997 Metro All engines Transmissions: All	**Fuel Trim System Rich (Bank 1) Conditions:** Engine started, engine running in closed loop, IAT sensor from 14-122°F, BARO sensor more than 75 kPa, fuel level over 25%, and the PCM detected the Long Term fuel trim value was less than -30% for 5 seconds, or the Short Term fuel trim value was less than -20% for 45 seconds during the Fuel System Monitor test. **Possible Causes** ● Base engine "mechanical" fault affecting one or more cylinders ● EVAP system component has failed or canister fuel saturated ● Exhaust leaks located in front of the A/FS or HO2S location ● Fuel control sensor is out of calibration (i.e., ECT, IAT or MAF) ● Fuel delivery system supplying too much fuel during cruise or idle periods (e.g., faulty fuel pump, or faulty pressure regulator) ● Fuel injector(s) is leaking or stuck partially open (one or more) ● HO2S is contaminated, deteriorated or it has failed
DTC P0172 **2T FUEL, MIL: Yes** 1996, 1997 Prizm, Tracker All engines Transmissions: All	**Fuel Trim System Rich (Bank 1) Conditions:** Engine started, ECT sensor between 18-203°F and less than 230°F during testing, IAT sensor from 18-140°F at startup, BARO sensor more than 75 kPa, engine running in closed loop, and the PCM detected the Short Term fuel trim value was less than -11% for 128 firing events, or the Long Term fuel trim value was less than -11% for 128 firing events, or the total Fuel Trim was less than -30% in 128 firing events during the Fuel System Monitor test. **Possible Causes** ● Base engine "mechanical" fault affecting one or more cylinders ● EVAP system component has failed or canister fuel saturated ● Exhaust leaks located in front of the HO2S location ● Fuel control sensor is out of calibration (i.e., ECT, IAT or MAF) ● Fuel delivery system supplying too much fuel during cruise or idle periods (e.g., faulty fuel pump, or faulty pressure regulator) ● Fuel injector(s) is leaking or stuck partially open (one or more) ● HO2S is contaminated, deteriorated or it has failed

OBD II Trouble Code List (P0xxx Codes)

DTC	Trouble Code Title, Conditions & Possible Causes
DTC P0300 **2T MISFIRE, MIL: Yes** 1996, 1997 Metro All engines Transmissions: All	**Multiple Engine Misfire Detected Conditions:** Engine speed less than 4500 rpm, ECT sensor more than 14°F, IAT sensor from 14-122°F, BARO sensor more than 75 kPa, fuel level over 25%, and the PCM detected the engine speed changed over 200 rpm in 50 m/sec while the MAP sensor changed less than 1.3 kPa in 16 firing events in more than one cylinder for 5 seconds after startup, or too high for 1 second of fuel shutoff. *Note: If the misfire is severe, the MIL will flash on/off on the 1st trip!* **Possible Causes** ● Air leak in the intake manifold, or in the EGR or PCM system ● Base engine mechanical fault affecting 2 or more cylinders ● Erratic or interrupted CKP or CMP sensor signals ● Fuel delivery component fault affecting 2 or more cylinders (i.e., a contaminated, dirty or sticking fuel injector) ● Ignition system problem (coil or plug) affecting over 2 cylinders ● Vehicle driven with low fuel pressure or while very low on fuel
DTC P0301 **2T MISFIRE, MIL: Yes** 1996, 1997 Metro All engines Transmissions: All	**Cylinder 1 Misfire Detected Conditions:** Engine speed under 4500 rpm, ECT sensor more than 14°F, IAT sensor from 14-122°F, BARO sensor over 75 kPa, fuel level over 25%, and the PCM detected the Misfire rate at 200 or 1000 engine revolutions was too high in one cylinder for 5 seconds, or it was too high for 1 second of fuel shutoff. *Note: If the misfire is severe, the MIL will flash on/off on the 1st trip!* **Possible Causes** ● Air leak in the intake manifold, or in the EGR or PCM system ● Base engine mechanical fault that affects only one cylinder ● Fuel delivery component fault that affects only one cylinder (i.e., a contaminated, dirty or sticking fuel injector) ● Ignition system problem (coil or plug) that affects one cylinder
DTC P0302 **2T MISFIRE, MIL: Yes** 1996, 1997 Metro All engines Transmissions: All	**Cylinder 2 Misfire Detected Conditions:** Engine speed under 4500 rpm, ECT sensor more than 14°F, IAT sensor from 14-122°F, BARO sensor over 75 kPa, fuel level over 25%, and the PCM detected the Misfire rate at 200 or 1000 engine revolutions was too high in one cylinder for 5 seconds, or it was too high for 1 second of fuel shutoff. *Note: If the misfire is severe, the MIL will flash on/off on the 1st trip!* **Possible Causes** ● Air leak in the intake manifold, or in the EGR or PCM system ● Base engine mechanical fault that affects only one cylinder ● Fuel delivery component fault that affects only one cylinder (i.e., a contaminated, dirty or sticking fuel injector) ● Ignition system problem (coil or plug) that affects one cylinder
DTC P0303 **2T MISFIRE, MIL: Yes** 1996, 1997 Metro All engines Transmissions: All	**Cylinder 3 Misfire Detected Conditions:** Engine speed under 4500 rpm, ECT sensor more than 14°F, IAT sensor from 14-122°F, BARO sensor over 75 kPa, fuel level over 25%, and the PCM detected the Misfire rate at 200 or 1000 engine revolutions was too high in one cylinder for 5 seconds, or it was too high for 1 second of fuel shutoff. *Note: If the misfire is severe, the MIL will flash on/off on the 1st trip!* **Possible Causes** ● Air leak in the intake manifold, or in the EGR or PCM system ● Base engine mechanical fault that affects only one cylinder ● Fuel delivery component fault that affects only one cylinder (i.e., a contaminated, dirty or sticking fuel injector) ● Ignition system problem (coil or plug) that affects one cylinder
DTC P0304 **2T MISFIRE, MIL: Yes** 1996, 1997 Metro 1.3L VIN 9 engine Transmissions: All	**Cylinder 4 Misfire Detected Conditions:** Engine started, engine speed under 4500 rpm, ECT sensor more than 14°F, IAT sensor from 14-122°F, BARO sensor more than 75 kPa, fuel level over 25%, and the PCM detected the Misfire rate at 200 or 1000 engine revolutions was too high in one cylinder for 5 seconds, or it was too high for 1 second of fuel shutoff. *Note: If the misfire is severe, the MIL will flash on/off on the 1st trip!* **Possible Causes** ● Air leak in the intake manifold, or in the EGR or PCM system ● Base engine mechanical fault that affects only one cylinder ● Fuel delivery component fault that affects only one cylinder (i.e., a contaminated, dirty or sticking fuel injector) ● Ignition system problem (coil or plug) that affects one cylinder

OBD II Trouble Code List (P0xxx Codes)

DTC	Trouble Code Title, Conditions & Possible Causes
DTC P0300 **2T MISFIRE, MIL: Yes** 1996, 1997 Prizm All engines	**Multiple Cylinder Misfire Detected Conditions:** Engine speed from 200-4000 rpm, ECT sensor from 20-254°F, BARO sensor over 75 kPa, TP angle change less than 1.9° in 16 firing events, and the PCM detected the misfire rate during 200 or 1000 engine revolutions was over a specified value in more than one cylinder during the Misfire test. *Note: If the misfire is severe, the MIL will flash on/off on the 1st trip!* **Possible Causes** ● Air leak in the intake manifold, or in the EGR or PCM system ● Base engine mechanical fault that affects only one cylinder ● Fuel delivery component fault that affects only one cylinder (i.e., a contaminated, dirty or sticking fuel injector) ● Ignition system problem (coil or plug) that affects one cylinder ● Vehicle driven with low fuel pressure or while very low on fuel
DTC P0301 **2T MISFIRE, MIL: Yes** 1996, 1997 Prizm All engines Transmissions: All	**Cylinder 1 Misfire Detected Conditions:** Engine speed from 200-4000 rpm, ECT sensor from 20-254°F, BARO sensor over 75 kPa, TP angle change less than 1.9° in 16 firing events, and the PCM detected the misfire rate during 200 or 1000 engine revolutions was more than a specified value in one cylinder during the Misfire test. *Note: If the misfire is severe, the MIL will flash on/off on the 1st trip!* **Possible Causes** ● Air leak in the intake manifold, or in the EGR or PCM system ● Base engine mechanical fault that affects only one cylinder ● Fuel delivery component fault that affects only one cylinder (i.e., a contaminated, dirty or sticking fuel injector) ● Ignition system problem (coil or plug) that affects one cylinder
DTC P0301 **2T MISFIRE, MIL: Yes** 1996, 1997 Prizm All engines Transmissions: All	**Cylinder 2 Misfire Detected Conditions:** Engine speed from 200-4000 rpm, ECT sensor from 20-254°F, BARO sensor over 75 kPa, TP angle change less than 1.9° in 16 firing events, and the PCM detected the misfire rate during 200 or 1000 engine revolutions was more than a specified value in one cylinder during the Misfire test. *Note: If the misfire is severe, the MIL will flash on/off on the 1st trip!* **Possible Causes** ● Air leak in the intake manifold, or in the EGR or PCM system ● Base engine mechanical fault that affects only one cylinder ● Fuel delivery component fault that affects only one cylinder (i.e., a contaminated, dirty or sticking fuel injector) ● Ignition system problem (coil or plug) that affects one cylinder
DTC P0303 **2T MISFIRE, MIL: Yes** 1996, 1997 Prizm All engines Transmissions: All	**Cylinder 3 Misfire Detected Conditions:** Engine speed from 200-4000 rpm, ECT sensor from 20-254°F, BARO sensor over 75 kPa, TP angle change less than 1.9° in 16 firing events, and the PCM detected the misfire rate during 200 or 1000 engine revolutions was more than a specified value in one cylinder during the Misfire test. *Note: If the misfire is severe, the MIL will flash on/off on the 1st trip!* **Possible Causes** ● Air leak in the intake manifold, or in the EGR or PCM system ● Base engine mechanical fault that affects only one cylinder ● Fuel delivery component fault that affects only one cylinder (i.e., a contaminated, dirty or sticking fuel injector) ● Ignition system problem (coil or plug) that affects one cylinder
DTC P0304 **2T MISFIRE, MIL: Yes** 1996, 1997 Prizm All engines Transmissions: All	**Cylinder 4 Misfire Detected Conditions:** Engine speed from 200-4000 rpm, ECT sensor from 20-254°F, BARO sensor over 75 kPa, TP angle change less than 1.9° in 16 firing events, and the PCM detected the misfire rate during 200 or 1000 engine revolutions was more than a specified value in one cylinder during the Misfire test. *Note: If the misfire is severe, the MIL will flash on/off on the 1st trip!* **Possible Causes** ● Air leak in the intake manifold, or in the EGR or PCM system ● Base engine mechanical fault that affects only one cylinder ● Fuel delivery component fault that affects only one cylinder (i.e., a contaminated, dirty or sticking fuel injector) ● Ignition system problem (coil or plug) that affects one cylinder

OBD II Trouble Code List (P0xxx Codes)

DTC	Trouble Code Title, Conditions & Possible Causes
DTC P0300 **2T MISFIRE, MIL: Yes** 1996, 1997 Tracker All engines	**Multiple Engine Misfire Detected Conditions:** Engine speed less than 4000 rpm, ECT and IAT sensors more than 18ºF, BARO sensor over 75 kPa, fuel level over 15%, TP angle change less than 1º in 10 ms, and the PCM detected the misfire rate during 200 or 1000 engine revolutions was more too high in more than one cylinder for 5 seconds after engine startup, or it was too high for one second of fuel shutoff. *Note: If the misfire is severe, the MIL will flash on/off on the 1st trip!* **Possible Causes** ● Air leak in the intake manifold, or in the EGR or PCM system ● Base engine mechanical fault affecting 2 or more cylinders ● Erratic or interrupted CKP or CMP sensor signals ● Fuel delivery component fault affecting 2 or more cylinders (i.e., a contaminated, dirty or sticking fuel injector) ● Ignition system problem (coil or plug) affecting over 2 cylinders ● Vehicle driven with low fuel pressure or while very low on fuel
DTC P0301 **2T MISFIRE, MIL: Yes** 1996, 1997 Tracker All engines Transmissions: All	**Cylinder 1 Misfire Detected Conditions:** Engine speed under 4000 rpm, ECT and IAT sensors over 18ºF, BARO sensor over 75 kPa, fuel level over 15%, TP angle change under 1º in 10 ms, and the PCM detected the misfire rate during 200 or 1000 engine revolutions was too high in Cylinder 1 for 5 seconds for 1 second from fuel shutoff. *Note: If the misfire is severe, the MIL will flash on/off on the 1st trip!* **Possible Causes** ● Air leak in the intake manifold, or in the EGR or PCM system ● Base engine mechanical fault that affects only one cylinder ● Fuel delivery component fault that affects only one cylinder (i.e., a contaminated, dirty or sticking fuel injector) ● Ignition system problem (coil or plug) that affects one cylinder
DTC P0302 **2T MISFIRE, MIL: Yes** 1996, 1997 Tracker All engines Transmissions: All	**Cylinder 2 Misfire Detected Conditions:** Engine speed under 4000 rpm, ECT and IAT sensors over 18ºF, BARO sensor over 75 kPa, fuel level over 15%, TP angle change under 1º in 10 ms, and the PCM detected the misfire rate during 200 or 1000 engine revolutions was too high in Cylinder 2 for 5 seconds for 1 second from fuel shutoff. *Note: If the misfire is severe, the MIL will flash on/off on the 1st trip!* **Possible Causes** ● Air leak in the intake manifold, or in the EGR or PCM system ● Base engine mechanical fault that affects only one cylinder ● Fuel delivery component fault that affects only one cylinder (i.e., a contaminated, dirty or sticking fuel injector) ● Ignition system problem (coil or plug) that affects one cylinder
DTC P0303 **2T MISFIRE, MIL: Yes** 1996, 1997 Tracker All engines Transmissions: All	**Cylinder 3 Misfire Detected Conditions:** Engine speed under 4000 rpm, ECT and IAT sensors over 18ºF, BARO sensor over 75 kPa, fuel level over 15%, TP angle change under 1º in 10 ms, and the PCM detected the misfire rate during 200 or 1000 engine revolutions was too high in Cylinder 3 for 5 seconds for 1 second from fuel shutoff. *Note: If the misfire is severe, the MIL will flash on/off on the 1st trip!* **Possible Causes** ● Air leak in the intake manifold, or in the EGR or PCM system ● Base engine mechanical fault that affects only one cylinder ● Fuel delivery component fault that affects only one cylinder (i.e., a contaminated, dirty or sticking fuel injector) ● Ignition system problem (coil or plug) that affects one cylinder
DTC P0304 **2T MISFIRE, MIL: Yes** 1996, 1997 Tracker All engines Transmissions: All	**Cylinder 4 Misfire Detected Conditions:** Engine speed under 4000 rpm, ECT and IAT sensors over 18ºF, BARO sensor over 75 kPa, fuel level over 15%, TP angle change under 1º in 10 ms, and the PCM detected the misfire rate during 200 or 1000 engine revolutions was too high in Cylinder 4 for 5 seconds for 1 second from fuel shutoff. *Note: If the misfire is severe, the MIL will flash on/off on the 1st trip!* **Possible Causes** ● Air leak in the intake manifold, or in the EGR or PCM system ● Base engine mechanical fault that affects only one cylinder ● Fuel delivery component fault that affects only one cylinder (i.e., a contaminated, dirty or sticking fuel injector) ● Ignition system problem (coil or plug) that affects one cylinder

OBD II Trouble Code List (P0xxx Codes)

DTC	Trouble Code Title, Conditions & Possible Causes
DTC P0325 **1T CCM, MIL: Yes** 1996, 1997 Prizm 1.8L VIN 8 engine Transmissions: All	**Knock Sensor 1 Circuit Malfunction Conditions:** Engine started, engine speed over 1200 rpm, TP sensor and VSS signals indicating the vehicle is accelerating, and the PCM detected the Knock sensor 1 signal amplitude was less than specified for over 20 seconds during the CCM test. **Possible Causes** ● Knock sensor signal circuit is open or shorted to ground ● Knock sensor signal circuit is shorted to VREF or system power ● Knock sensor is damaged or has failed ● PCM has failed
DTC P0335 **1T CCM, MIL: Yes** 1996, 1997 Metro All engines Transmissions: All	**Crankshaft Position Sensor Circuit Malfunction Conditions:** Engine started, engine speed over 700 rpm, and the PCM did not detect any CKP sensor pulses during 2 complete engine revolutions. **Possible Causes** ● CKP (Magnetic) sensor signal (+) or (-) circuit is open, shorted to ground or to power ● CKP sensor is damaged or has failed ● PCM has failed
DTC P0335 **2T CCM, MIL: Yes** 1996, 1997 Prizm All engines Transmissions: All	**Crankshaft Position Sensor Circuit Malfunction Conditions:** Engine cranking for 5 seconds, or engine speed over 600 rpm, and the PCM did not detect any CKP sensor pulses during 2 complete engine revolutions. **Possible Causes** ● CKP (Magnetic) sensor signal (+) or (-) circuit is open, shorted to ground or to power ● CKP sensor is damaged or has failed ● PCM has failed
DTC P0335 **1T CCM, MIL: Yes** 1996, 1997 Tracker All engines Transmissions: All	**Crankshaft Position Sensor Circuit Malfunction Conditions:** Engine started, engine running, and the PCM did not detect any CKP sensor signals after receiving 20 CMP sensor pulses. **Possible Causes** ● CKP (Magnetic) sensor signal (+) circuit is open, shorted to ground or shorted to power ● CKP (Magnetic) sensor signal (-) circuit is open, shorted to ground or shorted to power ● CKP sensor is damaged or has failed ● PCM has failed
DTC P0336 **2T CCM, MIL: Yes** 1996, 1997 Prizm All engines Transmissions: All	**Crankshaft Position Sensor Range/Performance Conditions:** Engine started, engine speed over 600 rpm, and the PCM detected an erratic CKP sensor signal during 2 complete engine revolutions. **Possible Causes** ● CKP (Magnetic) sensor signal (+) circuit has high resistance ● CKP (Magnetic) sensor signal (-) circuit has high resistance ● CKP (Magnetic) sensor is damaged or has failed ● PCM has failed
DTC P0340 **1T CCM, MIL: Yes** 1996, 1997 Metro All engines Transmissions: All	**Camshaft Position Sensor Circuit Malfunction Conditions:** Engine cranking for 2 seconds, and the PCM did not receive any CMP sensor signals for 2 seconds during the CCM test. **Possible Causes** ● CMP (Magnetic) sensor signal (+) or (-) circuit is open, shorted to ground or shorted to system power ● CMP sensor is damaged or has failed ● PCM has failed
DTC P0340 **1T CCM, MIL: Yes** 1996, 1997 Prizm All engines Transmissions: All	**Camshaft Position Sensor Circuit Malfunction Conditions:** Engine cranking or engine running, and the PCM did not receive any CMP sensor signals for 3 seconds during the CCM test. **Possible Causes** ● CMP (Magnetic) sensor signal (+) or (-) circuit is open, shorted to ground or shorted to system power ● CMP sensor is damaged or has failed ● PCM has failed
DTC P0340 **1T CCM, MIL: Yes** 1996, 1997 Tracker All engines Transmissions: All	**Camshaft Position Sensor Circuit Malfunction Conditions:** Engine cranking or running; and the PCM did not receive any CMP signals after it detected a Start signal for a period of 3 seconds. **Possible Causes** ● CMP (Hall) sensor signal circuit is open or shorted to ground ● CMP (Hall) sensor ground circuit is open ● CMP (Hall) sensor power circuit open (test power at MFI relay) ● CMP sensor is damaged or has failed ● PCM has failed (PCM provides a VREF signal to the sensor)

OBD II Trouble Code List (P0xxx Codes)

DTC	Trouble Code Title, Conditions & Possible Causes
DTC P0400 **2T EGR, MIL: Yes** 1996, 1997 Metro Engines: 1.0L VIN 6 Transmissions: All	**EGR System Performance Conditions:** Engine started, engine runtime over 290 seconds, ECT sensor from 176-230°F, IAT sensor over 18°F, BARO sensor over 75 kPa, fuel level over 25%, vehicle driven to over 60 mph for 5 minutes with any throttle change less than 25 degrees, then during a deceleration period with the throttle closed, the PCM detected the difference in intake pressure indicated was less a specified value. *Note: The test of the intake pressure change lasts for 1 second.* **Possible Causes** ● EGR back pressure transducer is damaged or has failed ● EGR bypass valve is damaged or has failed ● EGR solenoid vacuum valve is damaged or has failed ● EGR boost sensor is damaged or has failed ● EGR valve assembly is leaking, stuck open or stuck closed ● EGR valve vacuum hose(s) lose, damaged or disconnected ● Exhaust system is clogged or restricted
DTC P0400 **2T EGR, MIL: Yes** 1996, 1997 Tracker 2.0L VIN C engine Transmissions: All	**EGR System Performance Conditions:** Engine started, ECT sensor from 158-230°F, IAT sensor over 18°F, BARO sensor over 75 kPa, vehicle driven at over 60 mph for 7 minutes, followed by a deceleration period (throttle closed an brakes "off") with Fuel Cutoff "on", and the PCM detected the difference in intake manifold pressure between when the EGR valve was open and when it was closed was less than a specified value in the test. **Possible Causes** ● EGR stepper motor is binding, stuck partially open or closed ● EGR stepper motor is damaged or has failed ● PCM has failed
DTC P0401 **2T EGR, MIL: Yes** 1996, 1997 Models: Prism All engines Transmissions: All	**Insufficient EGR Flow Detected Conditions:** Engine started, ECT sensor from 158-230°F, IAT sensor over 18°F, BARO sensor over 75 kPa, vehicle driven to over 60 mph an engine speed over 4400 rpm, followed by a deceleration period with the throttle closed, and the PCM detected the difference in intake manifold pressure between when the EGR valve was open and when it was closed was too low a value during the flow test. **Possible Causes** ● EGR stepper motor is binding, or stuck in closed position ● EGR stepper motor is damaged or has failed ● PCM has failed
DTC P0402 **2T EGR, MIL: Yes** 1998, 1999, 2000, 2001, 2002 Models: Prism All engines Transmissions: All	**Excessive EGR Flow Detected Conditions:** Engine started, ECT sensor from 158-230°F, IAT sensor over 18°F, BARO sensor over 75 kPa, vehicle driven to over 60 mph an engine speed over 4400 rpm, followed by a deceleration period with the throttle closed, and the PCM detected the difference in intake manifold pressure between when the EGR valve was open and when it was closed was too high a value during the flow test. **Possible Causes** ● EGR stepper motor is binding, or stuck in open position ● EGR stepper motor is damaged or has failed ● PCM has failed
DTC P0420 **2T CAT, MIL: Yes** 1996, 1997 Metro All engines Transmissions: All	**Catalyst System Low Efficiency (Bank 1) Conditions:** Engine started, ECT sensor from 158-230°F, IAT sensor from 14-122°F, BARO sensor over 75 kPa, fuel level over 25%, vehicle driven at a constant engine speed of 2300-4500 rpm for 3 minutes in closed loop, calculated load value from 26-80% and the PCM detected the switch rates of the rear HO2S and front HO2S were similar for 96 seconds during the test. **Possible Causes** ● Air leaks in the exhaust manifold or exhaust pipes ● Catalytic converter is contaminated, damaged or has failed ● Front HO2S is older (aged) than the rear HO2S (HO2S is lazy) ● PCM has failed
DTC P0420 **2T CAT, MIL: Yes** 1996, 1997 Prizm, Tracker All engines Transmissions: All	**Catalyst System Low Efficiency (Bank 1) Conditions:** Engine started, ECT sensor from 18-230°F, IAT sensor more than 18°F, BARO sensor over 75 kPa, vehicle driven with the engine in closed loop, and the PCM detected the voltage swings (and switch rates) of the rear HO2S and front HO2S were similar for 96 seconds. **Possible Causes** ● Air leaks in the exhaust manifold or exhaust pipes ● Catalytic converter is contaminated, damaged or has failed ● Front HO2S is older (aged) than the rear HO2S (HO2S is lazy) ● PCM has failed

OBD II Trouble Code List (P0xxx Codes)

DTC	Trouble Code Title, Conditions & Possible Causes
DTC P0440 **2T EVAP, MIL: Yes** 1996, 1997 Metro All engines Transmissions: All	**EVAP System No Flow During Purge Detected Conditions:** Engine started, the DTC P0455 Diagnostic has run and passed, ECT sensor from 158-230°F, IAT sensor from 14-158°F, fuel level less than 75%, vehicle driven for several minutes, and the PCM detected a change in fuel tank pressure indicating the EVAP purge accumulation time was over 200 seconds (test can take 20 minutes). **Possible Causes** ● Charcoal canister is loaded with fuel or moisture ● Fuel filler cap loose, cross-threaded, incorrect part or damaged ● Fuel tank pressure sensor is damaged or has failed ● Fuel tank vapor line(s) blocked, damaged or disconnected ● Fuel Tank pressure control solenoid is damaged or has failed ● PCM has failed
DTC P0440 **2T EVAP, MIL: Yes** 1996, 1997 Tracker All engines Transmissions: All	**EVAP System Incorrect Purge Flow Detected Conditions:** Engine started, ECT sensor from 158-230°F, IAT sensor from 14-122°F, BARO sensor over 75 kPa, fuel level from 25-75%, vehicle driven with the engine speed from 1000-3000 rpm, and the PCM detected a change in fuel tank pressure indicating the EVAP purge accumulation time took over 6 minutes (test can take 20 minutes). **Possible Causes** ● Charcoal canister is loaded with fuel or moisture ● Fuel filler cap loose, cross-threaded, incorrect part or damaged ● Fuel tank pressure sensor is damaged or has failed ● Fuel tank vapor line(s) blocked, damaged or disconnected ● Purge solenoid valve is damaged or has failed ● PCM has failed
DTC P0443 **2T CCM, MIL: Yes** 1996, 1997 Metro & Tracker All engines Transmissions: All	**EVAP System Incorrect Purge Flow Detected Conditions:** ECT sensor from 158-230°F, IAT sensor from 14-158°F, BARO sensor over 75 kPa, fuel level from 25-75%, engine started and running at idle in closed loop, and the PCM detected an incorrect engine speed change after the Purge solenoid was cycled off and on at a duty cycle of over 50% during a 20 second test period. **Possible Causes** ● EVAP purge solenoid control circuit open or shorted to ground ● EVAP purge solenoid power (B+) circuit open ● EVAP purge solenoid is damaged or has failed ● PCM has failed
DTC P0450 **2T CCM, MIL: Yes** 1996, 1997 Metro & Tracker 1.3L VIN 9 engine, 1.6L VIN 6 engines Transmissions: All	**Fuel Tank Pressure Sensor Circuit Malfunction Conditions:** Engine started, and the PCM detected the FTP sensor signal indicated 16.3" H2O (-30.48 mm Hg), or the FTP signal indicated 8.16" H2O (15.24 mm Hg) for 7 seconds during the CCM test. **Possible Causes** ● FTP sensor circuit is open or shorted to ground ● FTP sensor ground or power circuit is open ● FTP sensor is damaged or has failed ● PCM has failed
DTC P0451 **2T CCM, MIL: Yes** 1996, 1997 Metro 1.3L VIN 9 engine Transmissions: All	**Fuel Tank Pressure Sensor Performance Conditions:** ECT sensor from 158-230°F, IAT sensor from 14-122°F, BARO sensor over 75 kPa, fuel level from 25-75%, engine runtime over 2 minutes, and the PCM detected the fuel tank pressure was lower than a certain value after completing the EVAP Leak Check. **Possible Causes** ● Fuel tank pressure sensor vent hole is clogged or restricted ● Fuel tank pressure sensor is damaged or out-of-calibration ● PCM has failed
DTC P0455 **2T EVAP, MIL: Yes** 1996, 1997 Metro & Tracker 1.3L VIN 9 engine, 1.6L VIN 6 engines Transmissions: All	**EVAP System Gross Leak (0.080") Detected Conditions:** DTC P0451 and P0452 not set, engine started, ECT sensor from 158-230°F, IAT sensor from 14-144°F, BARO sensor over 75 kPa, fuel level less than 75%, vehicle driven at a steady throttle, and the PCM detected the difference between the maximum and minimum internal fuel tank pressure was less than the specified value. *Note: This EVAP Diagnostic test can take up to 20 minutes to run.* **Possible Causes** ● Canister Purge valve is damaged, leaking or it has failed ● Charcoal canister is loaded with fuel or moisture ● Fuel filler cap loose, cross-threaded, incorrect part or damaged ● Fuel tank is cracked (leaking), or a leak exists in the 'O' ring ● Fuel tank pressure sensor is damaged or has failed ● Fuel vapor line(s), fuel pipes or hoses damaged or leaking ● PCM has failed

OBD II Trouble Code List (P0xxx Codes)

DTC	Trouble Code Title, Conditions & Possible Causes
DTC P0461 **1T CCM, MIL: Yes** 1996, 1997 Metro & Tracker All engines Transmissions: All	**Fuel Level Sensor Range/Performance Conditions:** DTC P0463 not set, engine started, vehicle driven for several miles, and the PCM detected during the Fuel Level Range check, a maximum to minimum reading of less than 2 gallons (once per trip). **Possible Causes** • Fuel level sensor is stuck in one position • Fuel level sensor is damaged or has failed • High resistance in the fuel level sensor ground circuit (G400) • Loose fuel tank baffle interfering with the fuel level sensor • PCM has failed
DTC P0463 **2T CCM, MIL: Yes** 1996, 1997 Metro & Tracker All engines Transmissions: All	**Fuel Level Sensor Circuit High Input Conditions:** Engine started, engine running, and PCM detected an unexpected "high" voltage condition on the Fuel Level sensor circuit for 500 ms. **Possible Causes** • Fuel level sensor circuit is shorted to VREF or system power • Fuel level sensor ground circuit is open • Fuel level sensor is damaged or has failed • PCM has failed
DTC P0480 **2T CCM, MIL: Yes** 1996, 1997 Metro All engines Transmissions: All	**Cooling Fan Control Circuit Malfunction Conditions:** Key on or engine running; and the PCM detected an unexpected voltage condition on the Cooling Fan control circuit for 5 seconds. **Possible Causes** • Cooling fan control circuit is open or shorted to ground • Cooling fan control circuit is shorted to system power • Cooling fan relay is damaged or has failed • PCM has failed
DTC P0500 **2T CCM, MIL: Yes** 1996, 1997 Metro, Prism, Tracker All engines Transmissions: All	**Vehicle Speed Sensor Circuit Malfunction Conditions:** Engine started, vehicle driven to cruise speed, followed by a deceleration period while in Fuel Cutoff mode, and the PCM did not detect any VSS signals for 4 seconds during the test. **Possible Causes** • VSS signal circuit is open or shorted to ground • VSS is damaged or has failed (behind the Instrument Cluster) • PCM has failed (the PCM sends a 5v signal to the reed switch)
DTC P0505 **2T CCM, MIL: Yes** 1996, 1997 Metro All engines Transmissions: All	**Idle Speed Control System Conditions:** DTC P0510 not set, engine started, engine at idle speed with the closed throttle switch indicating "on", ECT sensor more than 84ºF, system voltage over 10.0v, throttle valve feedback check active, and the PCM detected the ISC motor movement was less than 0.25 degrees, or with the ISC Motor enabled, the ISC motor did not move. **Possible Causes** • ISC motor control circuit is open or shorted to ground • ISC motor control circuit is shorted to system power (B+) • ISC motor power circuit is open • ISC motor is damaged or has failed • PCM has failed
DTC P0505 **2T CCM, MIL: Yes** 1996, 1997 Prizm All engines Transmissions: All	**Idle Control System Performance Conditions:** Engine running at a steady idle speed in closed loop and the PCM detected the Actual speed was too low or too high by a calibrated amount when compared to the Target idle speed. **Possible Causes** • IAC motor control circuit is open or shorted to ground • IAC motor control circuit is shorted to system power (B+) • IAC motor power circuit is open (check power from EFI relay) • IAC motor is damaged or has failed • PCM has failed
DTC P0505 **2T CCM, MIL: Yes** 1996, 1997 Tracker All engines Transmissions: All	**Idle Control System Performance Conditions:** Engine started, ECT sensor more than 158ºF, IAT sensor more than 18ºF, BARO sensor over 75 kPa, engine running in closed loop with the throttle closed, A/C and Power Steering switches both indicating "off", and the PCM detected the Actual idle speed was over 100 rpm lower than the Target idle speed, or the Actual idle speed was over 200 rpm higher than the Target idle speed for 20 seconds. **Possible Causes** • IAC motor control circuit is open, shorted to ground or shorted to power (B+) • IAC motor power circuit is open (check power from EFI relay) • IAC motor is damaged or has failed • PCM has failed

OBD II Trouble Code List (P0xxx Codes)

DTC	Trouble Code Title, Conditions & Possible Causes
DTC P0506 **2T CCM, MIL: Yes** 1996, 1997 Metro All engines Transmissions: All	**Idle Speed Too Low Conditions:** ECT sensor from 158-230°F, engine running with the closed throttle position switch indicating the throttle is closed, and the PCM detected the Actual idle speed at least 100 rpm less than the Desired idle speed during the IAC Diagnostic test. **Possible Causes** ● Throttle valve is dirty or sticking (it may need to be cleaned) ● ISC motor is damaged or has failed ● PCM has failed
DTC P0507 **2T CCM, MIL: Yes** 1996, 1997 Metro All engines Transmissions: All	**Idle Speed Too High Conditions:** ECT sensor from 158-230°F, engine running with the closed throttle position switch indicating the throttle is closed, and the PCM detected the Actual idle speed at least 200 rpm more than the Desired idle speed during the IAC Diagnostic test. **Possible Causes** ● Throttle valve is dirty or sticking (it may need to be cleaned) ● ISC motor is damaged or has failed ● PCM has failed
DTO P0510 **1T CCM, MIL: Yes** 1996, 1997 Metro, Prism, Tracker All engines Transmissions: All	**Closed Throttle Position Switch Circuit Malfunction Conditions:** Engine running, and then the PCM detected a VSS signal change of 0-20 mph during a TP switch off-to-on time, or a VSS signal change of 0-20 mph during a switch on-to-off time, condition met 16 times. **Possible Causes** ● Closed throttle position switch signal circuit is open or grounded ● Closed throttle position switch signal circuit is shorted to power ● Closed throttle position switch or TP sensor damaged or failed ● PCM has failed
DTC P0601 **2T PCM, MIL: Yes** 1996, 1997 Metro & Tracker All engines Transmissions: All	**PCM Read Only Memory Error Conditions:** Key on, and the PCM detected an internal check sum data error. *Note: Refer to a code repair chart for instructions on how to replace a PCM and how to reprogram the replacement unit.* **Possible Causes** ● Check the PCM for possible water contamination ● Clear the trouble codes and retest for this trouble code. If the same trouble code resets, the PCM has failed and must be replaced to repair this problem.
DTC P0603 **2T PCM, MIL: Yes** 1996, 1997 Metro & Tracker All engines Transmissions: All	**PCM Long Term Memory Reset Conditions:** Key on, and the PCM detected an EEPROM data readout error. *Note: Refer to a code repair chart for instructions on how to replace a PCM and how to reprogram the replacement unit.* **Possible Causes** ● Check the PCM for possible water contamination ● Clear the trouble codes and retest for this trouble code. If the same trouble code resets, the PCM has failed and must be replaced to repair this problem.
DTC P0705 **2T CCM, MIL: Yes** 1996, 1997 Metro & Tracker 1.3L VIN 9 engine, 1.6L VIN 6 Transmissions: A/T	**Transmission Range Switch Circuit Malfunction Conditions:** Engine started, vehicle driven in Drive to 38 mph or higher, and the PCM did not detect any Transmission Range (TR) switch signals (Reverse, Neutral, Drive, 2nd or Low) for a period of 25 seconds, or the PCM detected multiple signals from the TR switch simultaneously. **Possible Causes** ● TR switch signal circuit is open or shorted to ground ● TR switch signal circuit is shorted to VREF or system power ● TR switch is damaged or has failed ● PCM has failed
DTC P0720 **2T CCM, MIL: Yes** 1996, 1997 Metro & Tracker 1.3L VIN 9 engine, 1.6L VIN 6 Transmissions: A/T	**Output Speed Sensor Circuit Malfunction Conditions:** Engine started, then with the 4WD switch indicating open, the PCM did not detect and VSS signals from the transmission, or it detected that the 4WD switch open with the engine running in Drive at higher than the torque converter stall speed (2200-2600 rpm), the PCM did not detect any VSS signals during the CCM test. **Possible Causes** ● VSS signal circuit is open or shorted to ground ● VSS signal circuit is shorted to VREF or system power ● VSS is damaged or has failed
DTC P0741 **2T CCM, MIL: Yes** 1996, 1997 Tracker All engines Transmissions: A/T	**Torque Converter Clutch Mechanical Performance Conditions:** Engine started, vehicle driven in 3rd or 4th gear or in Drive, and the PCM detected the TCC control command did not agree with Actual driving conditions. *Note: Torque Converter Clutch does not have an electrical problem.* **Possible Causes** ● TCC solenoid valve is stuck or the TCC control valve is stuck ● Valve body fluid passages are clogged or restricted ● TCC solenoid is damaged or has failed

OBD II Trouble Code List (P0xxx Codes)

DTC	Trouble Code Title, Conditions & Possible Causes
DTC P0743 **2T CCM, MIL: Yes** 1996, 1997 Tracker All engines Transmissions: All	**Torque Converter Clutch Circuit Malfunction Conditions:** Engine started, vehicle driven in 3rd or 4th gear, and the PCM detected the TCC control circuit was less than 2.0v with the solenoid "off", or more than 5.5v with the solenoid "on". **Possible Causes** ● TCC solenoid control circuit is open, shorted to ground or shorted to system power ● TCC solenoid is damaged or has failed ● PCM has failed
DTC P0751 **2T CCM, MIL: Yes** 1996, 1997 Metro & Tracker 1.3L VIN 9 engine, 1.6L VIN 6 engines Transmissions: A/T	**Title: A/T Shift Solenoid 1 Mechanical Performance** **Trouble Code Conditions** Engine started, vehicle driven in Drive for over 1 minute, and the PCM detected the Shift Solenoid 1 control signal does not agree with the Actual gear position. *Note: Shift Solenoid #1 does not have an electrical problem.* **Possible Causes** ● Shift solenoid #1 is stuck, or the 2-3 Shift valve is stuck ● Valve body fluid passages are clogged or restricted ● Direct clutch is damaged or has failed
DTC P0753 **2T CCM, MIL: Yes** 1996, 1997 Metro & Tracker 1.3L VIN 9 engine, 1.6L VIN 6 engines Transmissions: A/T	**A/T Shift Solenoid #1 Circuit Malfunction Conditions:** Engine started; vehicle driven in 1st or 2nd gear, and the PCM detected the Shift Solenoid control circuit was below 2.0v with the solenoid commanded "on", or over 5.5v or it "off". **Possible Causes** ● Shift solenoid #1 control circuit is open, shorted to ground or shorted to system power ● Shift solenoid #1 is damaged or has failed ● PCM has failed
DTC P0756 **2T CCM, MIL: Yes** 1996, 1997 Metro & Tracker 1.3L VIN 9 engine, 1.6L VIN 6 engines Transmissions: A/T	**A/T Shift Solenoid 2 Mechanical Performance Conditions:** Engine started, vehicle driven in Drive for over 1 minute, and the PCM detected the Shift Solenoid 2 control signal does not agree with the Actual gear position. *Note: Shift Solenoid #2 does not have an electrical problem.* **Possible Causes** ● Shift solenoid 2 is stuck, or the 1-2 or 2-3 Shift valve is stuck ● Valve body fluid passages are clogged or restricted ● 2nd Brake is damaged or has failed
DTC P0758 **2T CCM, MIL: Yes** 1996, 1997 Metro & Tracker 1.3L VIN 9 engine, 1.6L VIN 6 engines Transmissions: A/T	**A/T Shift Solenoid 2 Circuit Malfunction Conditions:** Engine started, vehicle driven in 1st or 2nd gear for over 1 minute, and the PCM detected the Shift Solenoid 2 control circuit was 2.0v or lower with the solenoid "off", or the Shift Solenoid 2 control circuit was 5.5v or higher with the solenoid "on" during the CCM test. **Possible Causes** ● Shift solenoid 2 control circuit is open, shorted to ground or shorted to system power ● Shift solenoid 2 has failed, or the PCM has failed
DTC P0770 **1T CCM, MIL: Yes** 1996, 1997 Prizm 1.8L VIN 8 engine Transmissions: A/T	**Torque Converter Clutch Lockup Performance Conditions:** Engine started, vehicle driven to over 30 mph at a steady speed for over one minute, and the PCM detected the TCC system did not engage when commanded "on", or it did not disengage when it was commanded "off" during the CCM test. **Possible Causes** ● ATF level is too low, or fluid is burnt or contaminated ● TCC solenoid valve is stuck "off" due to sediment or binding ● TCC solenoid valve is stuck "on due to sediment or binding ● TCC solenoid is damaged or has failed ● Transmission has internal component damage or failure
DTC P0773 **1T CCM, MIL: Yes** 1996, 1997 Prizm 1.8L VIN 8 engine Transmissions: A/T	**Torque Converter Clutch Circuit Malfunction Conditions:** Engine started, vehicle driven to over 30 mph at a steady speed for over one minute, and the PCM detected an unexpected voltage condition on the TCC control circuit with the TCC commanded "on", or with it commanded "off". **Possible Causes** ● TCC control circuit is open, shorted to ground or shorted to system power ● TCC ground circuit is open (at the transmission location) ● TCC solenoid has failed, or the PCM has failed

OBD II Trouble Code List (P1xxx Codes)

DTC	Trouble Code Title, Conditions & Possible Causes
DTC P1250 1T CCM, MIL: Yes 1996, 1997 Metro All engines Transmissions: All	**Early Fuel Evaporation Heater Circuit Malfunction Conditions:** Engine started; IAT sensor from 14-122ºF, BARO sensor over 75 kPa, fuel level over 25%, and the PCM detected the EFE heater resistor "on" signal was less than 2.5v with the heater off, or the heater "on" signal was over 0.3v with the heater on for 5 seconds. **Possible Causes** ● EFE heater control circuit is open or shorted to ground ● EFE heater power circuit is open (check the power source) ● EFE heater is damaged or has failed ● PCM has failed
DTC P1300 1T CCM, MIL: Yes 1996, 1997 Prizm All engines Transmissions: All	**Ignition Coil 1 Primary Feedback Circuit Malfunction Conditions:** Engine cranking or running; and the PCM did not receive any failsafe signal from Ignition Coil 1primary after two consecutive ignition trigger signal cycles were detected. **Possible Causes** ● Ignition coil primary circuit is open or shorted to ground ● Ignition coil is damaged or has failed ● Ignition system "noise" filter may be shorted to ground ● PCM has failed
DTC P1335 2T CCM, MIL: Yes 1996, 1997 Prizm All engines Transmissions: All	**Crankshaft Position Sensor Circuit Malfunction Conditions:** Engine speed more than 1000 rpm, and the PCM did not receive any Crankshaft Position (CKP) sensor pulses for more than 50 ms. **Possible Causes** ● CKP sensor signal (+) or (-) circuit is open, shorted to ground or power (intermittent) ● CKP sensor is damaged or has failed (an intermittent fault) ● Crankshaft reluctor wheel contains debris or is damaged ● PCM has failed
DTC P1408 2T CCM, MIL: Yes 1996, 1997 Tracker All engines Transmissions: All	**MAP Sensor Circuit Malfunction Conditions:** Engine started, engine running with the TP sensor indicating above 0.20v, ECT sensor over 113ºF, IAT sensor over 41ºF, and the PCM detected the MAP sensor was more than 4.80v. **Possible Causes** ● MAP sensor signal circuit is open or shorted to VREF ● MAP sensor ground circuit is open ● MAP sensor is damaged or has failed ● PCM has failed
DTC P1410 2T CCM, MIL: Yes 1996, 1997 Metro & Tracker 1.3L VIN 9 engine, 1.6L VIN 6 engines Transmissions: All	**Fuel Tank Pressure Sensor Range/Performance Conditions:** Engine running, fuel level from 25-75%, and the PCM detected an unexpected voltage on the Fuel Tank Pressure Control signal circuit with the fuel level higher than a value in memory. **Possible Causes** ● Fuel tank pressure control solenoid control circuit is open, shorted to ground or to power ● Fuel tank pressure control solenoid power circuit is open ● Fuel tank pressure control solenoid is damaged or has failed
DTC P1450 1T CCM, MIL: Yes 1996, 1997 Metro All engines Transmissions: All	**Barometric Pressure Sensor Circuit Malfunction Conditions:** Engine cranking or engine is running in closed loop at a steady speed, and the PCM detected the BARO sensor signal indicated less than 33 kPa, or it detected it indicated over 135 kPa. **Possible Causes** ● BARO pressure sensor circuit is open, shorted to ground or shorted to VREF ● BARO pressure sensor is damaged or has failed ● PCM has failed
DTC P1450 1T CCM, MIL: Yes 1996, 1997 Tracker All engines Transmissions: All	**Barometric Pressure Sensor Circuit Malfunction Conditions:** Engine started, engine running in closed loop, and the PCM detected the BARO sensor was below 38 kPa, or it detected the BARO sensor was more than 113 kPa for over 5 seconds. **Possible Causes** ● BARO pressure sensor circuit is open, shorted to ground or shorted to VREF ● BARO pressure sensor has failed, or the PCM has failed
DTC P1451 2T CCM, MIL: Yes 1996, 1997 Metro All engines Transmissions: All	**Barometric Pressure Sensor Range/Performance Conditions:** Engine cranking for 2 seconds, or with the engine running in closed loop at a stable speed, the PCM detected the difference between the BARO and MAP sensors was 26 kPa or more. **Possible Causes** ● BARO pressure sensor is damaged, "skewed" or it has failed ● MAP sensor is out-of-calibration or the sensor is "skewed" ● PCM has failed
DTC P1451 2T CCM, MIL: Yes 1996, 1997 Tracker All engines Transmissions: All	**Barometric Pressure Sensor Range/Performance Conditions:** Engine running in closed loop, then while operating in Fuel Cutoff mode, the PCM detected the difference between the detected BARO sensor and the MAP sensor was out-of-range. **Possible Causes** ● BARO pressure sensor is damaged, "skewed" or it has failed ● MAP sensor is out-of-calibration or the sensor is "skewed" ● PCM has failed

OBD II Trouble Code List (P1xxx Codes)

DTC	Trouble Code Title, Conditions & Possible Causes
DTC P1500 **2T CCM, MIL: Yes** 1996, 1997 Metro All engines Transmissions: All	**Starter Signal Circuit Malfunction Conditions:** Engine cranking, and the PCM did not detect any starter signals, or it detected a low starter signal while the engine was cranking, or a high signal after the engine started. *Note: This test occurs once per drive cycle for 400 milliseconds.* **Possible Causes** ● Starter signal circuit is open, shorted to ground or to power ● Starter is damaged or has failed ● PCM has failed
DTC P1500 **1T CCM, MIL: Yes** 1996, 1997 Tracker All engines Transmissions: All	**Starter Signal Circuit Malfunction Conditions:** Engine cranking, and the PCM did not detect any Starter signal, or it detected a low voltage on the Starter signal during cranking, or it detected a high voltage condition on the Starter signal after the engine started, condition met for 3 seconds. **Possible Causes** ● Starter signal circuit is open, shorted to ground or to power ● Starter is damaged or has failed ● PCM has failed
DTC P1510 **1T CCM, MIL: Yes** 1996, 1997 Metro & Tracker All engines Transmissions: All	**Backup Power Supply Malfunction Conditions:** Key off, and the PCM did not detect any voltage on the Backup Power circuit for 5 seconds during the CCM test. *Note: This code can only be read at key on, engine off.* **Possible Causes** ● Backup Power Supply circuit is open (check the Dome fuse) ● Battery connections are corroded, dirty or loose ● Battery has been disconnected or removed ● PCM has failed
DTC P1520 **2T CCM, MIL: Yes** 1998, 1999, 2000 Tracker All engines Transmissions: All	**Stop Lamp Switch Circuit Malfunction Conditions:** Engine started, vehicle driven to a speed of over 19 mph, and the PCM detected a Stop Lamp switch indicated the switch was in the "on" position continuously during the CCM test. **Possible Causes** ● Stop lamp switch signal circuit is shorted to system power (B+) ● Stop lamp switch is damaged or has failed (in closed position) ● Stop lamp switch is out-of-adjustment ● PCM has failed
DTC P1530 **2T CCM, MIL: Yes** 1996, 1997 Metro & Tracker All engines Transmissions: All	**Ignition Timing Adjustment Switch Circuit Malfunction Conditions:** Engine started, vehicle driven to a speed of over 1 mph, and the PCM detected the Test Switch signal to the DLC circuit indicated "on" for 5 seconds during the CCM test. **Possible Causes** ● Ignition timing adjustment switch circuit is shorted to ground ● PCM has failed
DTC P1600 **1T CCM, MIL: Yes** 1996, 1997 Prizm All engines Transmissions: All	**PCM Battery Circuit Malfunction Conditions:** Key on, and the PCM detected that there was no power to the battery direct feed connection at the PCM (C1-1). **Possible Causes** ● Backup Power Supply circuit is open (check the Dome fuse) ● Battery connections are corroded, dirty or loose ● Battery has been disconnected or removed ● PCM has failed
DTC P1780 **1T CCM, MIL: Yes** 1996, 1997 Prizm 1.8L VIN 8 engine Transmissions: A/T	**Park Neutral Position Switch Circuit Malfunction Conditions:** Engine started, vehicle driven to over 50 mph at an engine speed from 2000-4000 rpm, MAP sensor indicating more than 300 mmHg, and the PCM detected the P/N switch indicated Neutral for 30 seconds or two different P/N switch signals at the same time. **Possible Causes** ● P/N switch signal circuit is open, shorted to ground or to power ● P/N switch power or ground circuit is open ● P/N switch is out of adjustment, damaged or has failed ● PCM has failed
DTC P1875 **2T CCM, MIL: Yes** 1996, 1997 Tracker All engines Transmissions: All	**4WD Low Switch Circuit Malfunction Conditions:** Engine started; vehicle driven and then the PCM detected a VSS signal that indicated the vehicle speed was more than 10 km/h lower than the speed determined from the transmission output shaft speed sensor during the CCM test period. **Possible Causes** ● 4WD signal circuit open or grounded between the switch and the PCM connector ● 4WD signal circuit shorted to VREF or system power ● 4WD switch is damaged or has failed

OBD II Trouble Code List (P0xxx Codes)

DTC	Trouble Code Title, Conditions & Possible Causes
DTC P0031 **2T CCM, MIL: Yes** 2002, 2003, 2004, 2005 Tracker 2.0L VIN C engine Transmissions: All	**HO2S-11 (Bank 1 Sensor 1) Heater Circuit Low Input Conditions:** Engine started, IAT sensor over 18ºF, ECT sensor from 18-230ºF, BARO sensor over 75 kPa, and the PCM detected less than 0.12 amps on the heater circuit with the heater commanded "on", or the voltage potential between the HO2S heater power and ground circuits was more than 13.9v with the heater "on". **Possible Causes** • HO2S heater control circuit is open or is shorted to ground • HO2S is damaged (heater resistance is 4.5-5.7 ohms at 68ºF) • PCM has failed
DTC P0032 **2T CCM, MIL: Yes** 2002, 2003, 2004, 2005 Tracker 2.0L VIN C engine Transmissions: All	**HO2S-11 (Bank 1 Sensor 1) Heater Circuit High Input Conditions:** Engine started, IAT sensor over 18ºF, ECT sensor from 18-230ºF, BARO sensor over 75 kPa, and the PCM detected more than 6.92 amps on the heater circuit with the heater commanded on, or the voltage potential between the HO2S heater power and ground circuits was more than 11.7v with the heater on. **Possible Causes** • HO2S heater control circuit is shorted to vehicle power • HO2S is damaged (heater resistance is 4.5-5.7 ohms at 68ºF) • PCM has failed
DTC P0037 **2T CCM, MIL: Yes** 2002, 2003, 2004, 2005 Tracker 2.0L VIN C engine Transmissions: All	**HO2S-12 (Bank 1 Sensor 2) Heater Circuit Low Input Conditions:** Engine started, IAT sensor over 18ºF, ECT sensor from 18-230ºF, BARO sensor over 75 kPa, and the PCM detected less than 0.12 amps on the heater circuit with the heater commanded "on", or the voltage potential between the HO2S heater power and ground circuits was more than 13.9v with the heater "on". **Possible Causes** • HO2S heater control circuit is open or shorted to ground • HO2S is damaged (heater resistance is 4.5-5.7 ohms at 68ºF) • PCM has failed
DTC P0038 **2T CCM, MIL: Yes** 2002, 2003, 2004, 2005 Tracker 2.0L VIN C engine Transmissions: All	**HO2S-12 (Bank 1 Sensor 2) Heater Circuit High Input Conditions:** Engine started, IAT sensor over 18ºF, ECT sensor from 18-230ºF, BARO sensor over 75 kPa, and the PCM detected the HO2S heater current was more than 4.90 amps with the heater commanded on, or the voltage potential between the HO2S heater power and ground circuits was less than 8.72v with the heater on. **Possible Causes** • HO2S heater control circuit is open or shorted to ground • HO2S is damaged (heater resistance is 4.5-5.7 ohms at 68ºF) • PCM has failed
DTC P0100 **2T CCM, MIL: Yes** 1998, 1999, 2000, 2001, 2002 Prizm All engines Transmissions: All	**MAF Sensor Circuit Malfunction Conditions:** Engine started, vehicle driven at an engine speed under 3000 rpm, and the PCM detected an unexpected voltage condition on the MAF sensor circuit for 3 seconds during the CCM test. **Possible Causes** • MAF sensor signal circuit is open or shorted to ground • MAF sensor signal circuit is shorted to VREF or system power • MAF sensor power circuit is open (check power from MFI relay) • MAF sensor is damaged or has failed • PCM has failed
DTC P0100 **2T CCM, MIL: Yes** 2000, 2001, 2002 Prizm All engines Transmissions: All	**MAF Sensor Performance Conditions:** Engine started, ECT sensor more than 176ºF, engine speed below 900 rpm with the throttle valve closed, and the PCM detected the MAF sensor was more than 2.20v, or with the engine speed above 1500 rpm and the TP sensor signal at least 0.63v, it detected the MAF sensor indicated less than 1.06v for 10 seconds. **Possible Causes** • Intake air leak between the MAF sensor and the throttle body • MAF sensor signal circuit is open or shorted to ground • MAF sensor is damaged or has failed • PCM has failed
DTC P0101 **2T CCM, MIL: Yes** 1998, 1999, 2000, 2001, 2002, 2003, 2004, 2005 Tracker 1.6L VIN 6, 2.0L VIN C, 2.5L VIN 4 engines Transmissions: All	**MAF Sensor Range/Performance Conditions:** DTC P0102, P0103, P0335, P1408 and P1451 not set, engine speed below 2000 rpm, BARO sensor more than 75 kPa, IAT signal from 7-158ºF, TP angle less than 20 degrees, and the PCM detected the maximum - minimum airflow rate was less than 0.1 gm/s. **Possible Causes** • Base engine vacuum leak, PCV valve leaking or stuck open • Engine oil dipstick missing or not fully seated • MAF sensor element (wire) is contaminated or dirty • MAF sensor signal or ground circuit fault or sensor has failed • PCM has failed

OBD II Trouble Code List (P0xxx Codes)

DTC	Trouble Code Title, Conditions & Possible Causes
DTC P0102 **1T CCM, MIL: Yes** 1998, 1999, 2000, 2001, 2002, 2003, 2004, 2005 Tracker 1.6L VIN 6, 2.0L VIN C, 2.5L VIN 4 engines Transmissions: All	**MAF Sensor Circuit Low Frequency Conditions:** Key on or engine running; and the PCM detected the MAF sensor current was less than 0.64 milliamps for 5 seconds during the test. **Possible Causes** • Base engine vacuum leak, PCV valve leaking or stuck open • Engine oil dipstick missing or not fully seated • MAF sensor element hot wire contaminated or the sensor failed • MAF sensor signal shorted to ground or ground circuit problem • MAF sensor wiring routed close to ignition wires or generator • PCM has failed
DTC P0103 **1T CCM, MIL: Yes** 1998, 1999, 2000, 2001, 2002, 2003, 2004, 2005 Tracker 1.6L VIN 6, 2.0L VIN C, 2.5L VIN 4 engines Transmissions: All	**MAF Sensor Circuit High Frequency Conditions:** Key on or engine running; and the PCM detected the MAF sensor output current was more than 4.90 mA for 5 seconds. *Note: The PCM enters Fail-Safe Function and determines injector drive time using the TP sensor, engine speed, IAC position and injector frequency.* **Possible Causes** • MAF sensor element hot wire is contaminated or dirty • MAF sensor is damaged or has failed • MAF sensor signal circuit is open between the sensor and PCM • MAF sensor wiring routed too close to the ignition wires • MAF sensor wiring routed to close to the back of the Generator • PCM has failed
DTC P0106 **1T CCM, MIL: Yes** 1998, 1999, 2000, 2001, 2002 Prizm All engines Transmissions: All	**MAP Sensor Range/Performance Conditions:** DTC P0105, P0107 and P0108 not set; engine speed below 1000 rpm with TP sensor at 0v, ECT sensor over 158°F, and the PCM detected the MAP sensor was over 3.3v, or with the engine speed near 2500 rpm, the MAP sensor was less than 1.0v with TP sensor over 1.0v. **Possible Causes** • MAP sensor source vacuum line is leaking or disconnected • MAP sensor is damaged, out-of-calibration or has failed • PCM has failed
DTC P0107 **1T CCM, MIL: Yes** 1998, 1999, 2000, 2001 Metro All engines Transmissions: All	**MAP Sensor Circuit Low Input Conditions:** Engine started, in closed loop with engine speed and TP angle both steady, and the PCM detected a low voltage (Scan Tool reads 5 kPa) on the MAP sensor circuit for 500 ms. **Possible Causes** • MAP sensor signal circuit is shorted to ground • MAP sensor power (VREF) circuit from the PCM is open • MAP sensor is damaged or has failed • PCM has failed
DTC P0107 **2T CCM, MIL: Yes** 2002, 2003, 2004, 2005 Tracker 2.0L VIN C engine, 2.5L VIN 4 engines Transmissions: All	**MAP Sensor Circuit Low Input Conditions:** DTC P0121, P0122 and P0123 not set, engine started, IAT sensor over 41°F, ECT sensor from 18-230°F, BARO sensor over 75 kPa, throttle valve open with a VSS signal detected for 5 seconds, and the PCM detected the MAP sensor was under 0.20v for 40 seconds. **Possible Causes** • MAP sensor signal circuit shorted to sensor or chassis ground • MAP sensor power circuit open between the sensor and PCM • MAP sensor is damaged or has failed • PCM has failed
DTC P0108 **1T CCM, MIL: Yes** 1998, 1999, 2000, 2001 Metro All engines Transmissions: All	**MAP Sensor Circuit High Input Conditions:** Engine started, in closed loop with engine speed and TP angle steady, and the PCM detected a high voltage (Scan Tool reads 130 kPa) on the MAP sensor circuit for 500 ms. **Possible Causes** • MAP sensor signal circuit is open between the sensor and PCM • MAP sensor signal circuit is shorted to VREF or system power • MAP sensor ground circuit is open between sensor and PCM • MAP sensor is damaged or has failed • PCM has failed
DTC P0108 **2T CCM, MIL: Yes** 2002, 2003, 2004, 2005 Tracker 2.0L VIN C engine, 2.5L VIN 4 engines Transmissions: All	**MAP Sensor Circuit High Input Conditions:** Engine started, IAT sensor over 41°F, ECT sensor from 18-230°F, BARO sensor over 75 kPa, throttle valve open, VSS signal present for 5 seconds, and the PCM detected the MAP sensor was above 2.70v for 60 seconds. **Possible Causes** • MAP sensor signal circuit is open between the sensor and PCM • MAP sensor signal circuit is shorted to VREF or system power • MAP sensor ground circuit is open between sensor and PCM • MAP sensor is damaged or has failed • PCM has failed

OBD II Trouble Code List (P0xxx Codes)

DTC	Trouble Code Title, Conditions & Possible Causes
DTC P0105 **1T CCM, MIL: Yes** 1998, 1999, 2000, 2001, 2002 Prizm All engines Transmissions: All	**MAP Sensor Circuit Malfunction Conditions:** Engine speed under 1000 rpm with the throttle closed, ECT sensor more than 158°F, and the PCM detected the MAP sensor was more than 3.3v, or with engine speed less than 2500 rpm and TP angle over 1.0v, that the MAP input was less than 1.0v during the test. **Possible Causes** ● MAP sensor signal circuit to the PCM interrupted (intermittent) ● MAP sensor is damaged or has failed ● PCM has failed
DTC P0106 **1T CCM, MIL: Yes** 1998, 1999, 2000, 2001 Metro All engines Transmissions: All	**MAP Sensor Range/Performance Conditions:** DTC P0105, P0107 and P0108 not set, engine speed and TP angle both steady, and the PCM detected the MAP sensor (obtained during engine cranking) differed from the MAP input (obtained at engine startup) by less than 1.3 kPa, or the BARO pressure differed from the MAP sensor by less than 33 kPa for 2 minutes. **Possible Causes** ● MAP sensor signal circuit to the PCM interrupted (intermittent) ● MAP sensor is damaged or has failed ● PCM has failed
DTC P0110 **1T CCM, MIL: Yes** 1998, 1999, 2000, 2001, 2002 Prizm All engines Transmissions: All	**Intake Air Temperature Sensor Circuit Malfunction Conditions:** Engine started; and the PCM detected the IAT sensor indicated a value from -40°F to -67°F, or it indicated a value from 240°F to 262°F, either condition met for 500 ms during the test. **Possible Causes** ● IAT sensor signal circuit is open or shorted to ground ● IAT sensor signal circuit is shorted to VREF or system power ● IAT sensor is damaged or has failed (out of calibration) ● PCM has failed
DTC P0111 **1T CCM, MIL: Yes** 1998, 1999, 2000, 2001, 2002, 2003, 2004, 2005 Metro, Tracker All engines Transmissions: All	**Intake Air Temperature Sensor Range/Performance Conditions:** DTC P0101, P0102, P0103, P0112, P0113, P0116, P0117, P0118, and P0125 not set, ECT sensor at startup from 14°F to 86°F, vehicle driven at normal operating temperature for over 10 minutes, and the PCM detected a smaller than expected amount of change in the IAT sensor signal during the CCM Rationality test. **Possible Causes** ● IAT sensor signal circuit is open or shorted to ground ● IAT sensor is contaminated, out-of-calibration or has failed ● PCM has failed
DTC P0112 **1T CCM, MIL: Yes** 1998, 1999, 2000, 2001, 2002, 2003, 2004, 2005 Metro, Tracker All engines Transmissions: All	**Intake Air Temperature Sensor Circuit Low Input Conditions:** Key on or engine running; and the PCM detected that the IAT sensor was less than 0.10v (Scan Tool reads over 282°F) for 500 ms. **Possible Causes** ● IAT sensor signal circuit is shorted to sensor ground ● IAT sensor signal circuit is shorted to chassis ground ● IAT sensor is damaged or has failed ● PCM has failed
DTC P0113 **1T CCM, MIL: Yes** 1998, 1999, 2000, 2001, 2002, 2003, 2004, 2005 Metro, Tracker All engines Transmissions: All	**Intake Air Temperature Sensor Circuit High Input Conditions:** Key on or engine running; and the PCM detected that the IAT sensor was more than 4.90v (Scan Tool reads less than -40°F for 500 ms. **Possible Causes** ● IAT sensor ground circuit is open ● IAT sensor signal circuit is open between sensor and the PCM ● IAT sensor signal circuit is shorted to system power ● IAT sensor is damaged or has failed ● PCM has failed
DTC P0115 **1T CCM, MIL: Yes** 1998, 1999, 2000, 2001, 2002 Prizm All engines Transmissions: All	**Engine Coolant Temperature Sensor Circuit Malfunction Conditions:** Key on or engine running; and the PCM detected the ECT sensor that indicated less than -40 to -67°F, or the ECT sensor indicated a value from 240°F to 262°F, condition met for 500 ms during the test. **Possible Causes** ● ECT sensor circuit is open, shorted to ground or to VREF ● ECT sensor is damaged or has failed (out of calibration) ● PCM has failed

OBD II Trouble Code List (P0xxx Codes)

DTC	Trouble Code Title, Conditions & Possible Causes
DTC P0116 **1T CCM, MIL: Yes** 1998, 1999, 2000, 2001, 2002 Prizm All engines Transmissions: All	**Engine Coolant Temperature Sensor Range/Performance Conditions:** Engine started, ECT sensor at startup less than 20ºF, engine runtime over 20 minutes, and the PCM detected the ECT sensor indicated less than 95ºF, or with the ECT and IAT sensors from 20ºF to 50ºF at startup with the engine runtime over 5 minutes, it detected the ECT sensor was less than 95ºF, or with the ECT and IAT sensors more than 50ºF at startup with the engine runtime over 2 minutes, it detected the ECT sensor was less than 95ºF. **Possible Causes** ● ECT sensor signal circuit is open or shorted to ground ● ECT sensor is contaminated, out-of-calibration or has failed ● PCM has failed
DTC P0116 **2T CCM, MIL: Yes** 1998, 1999, 2000, 2001, 2002, 2003, 2004, 2005 Metro, Tracker All engines Transmissions: All	**Engine Coolant Temperature Sensor Range/Performance Conditions:** DTC P0117, P0118 and P0335 not set, engine runtime over 20 minutes, vehicle driven to a speed over 30 mph at least once since engine startup, and the PCM detected too small an amount of change in the ECT sensor signal after this period of time expired. **Possible Causes** ● ECT sensor signal circuit is open or shorted to ground ● ECT sensor is contaminated, out-of-calibration or has failed ● PCM has failed
DTC P0117 **1T CCM, MIL: Yes** 1998, 1999, 2000, 2001, 2002, 2003, 2004, 2005 Metro, Tracker All engines Transmissions: All	**Engine Coolant Temperature Sensor Circuit Low Input Conditions:** Key on or engine running; and the PCM detected the ECT sensor was less than 0.10v (Scan Tool reads over 279ºF) for 500 ms. **Possible Causes** ● ECT sensor connector is damaged or shorted ● ECT sensor signal circuit is shorted to ground ● ECT sensor is damaged or has failed (it may be shorted) ● PCM has failed
DTC P0118 **1T CCM, MIL: Yes** 1998, 1999, 2000, 2001, 2002, 2003, 2004, 2005 Metro, Tracker All engines Transmissions: All	**Engine Coolant Temperature Sensor Circuit High Input Conditions:** Key on or engine running; and the PCM detected the IAT sensor was more than 4.90v (Scan Tool reads less than -40ºF for 500 ms. **Possible Causes** ● ECT sensor connector is damaged, loose or open ● ECT sensor signal circuit is open between sensor and the PCM ● ECT sensor ground circuit is open between sensor and PCM ● ECT sensor is damaged or has failed (an internal short) ● PCM has failed
DTC P0120 **1T CCM, MIL: Yes** 1998, 1999, 2000, 2001, 2002 Prizm All engines Transmissions: All	**Throttle Position Sensor Circuit Malfunction Conditions:** Engine started; and the PCM detected the TP sensor was more than 1.0v with the throttle closed, or that the TP sensor indicated less than 4.90v during a wide open throttle event. **Possible Causes** ● TP sensor signal circuit open or shorted to ground ● TP sensor ground circuit is open ● TP sensor power circuit is open (check VREF circuit at PCM) ● TP sensor is damaged or has failed ● PCM has failed
DTC P0121 **2T CCM, MIL: Yes** 1998, 1999, 2000, 2001 Metro All engines Transmissions: All	**Throttle Position Sensor Range/Performance Conditions:** DTC P0122 and P0123 not set, engine speed from 1500-3500 rpm, ECT sensor over 158ºF, IAT sensor from 14ºF to 122ºF, fuel level over 25%, BARO sensor over 75 kPa, MAP sensor change less than 13 kPa in 16 firing events, and the PCM detected the Expected and Actual TP sensor values were more than 11 degrees apart for 3 seconds during the CCM test. **Possible Causes** ● MAP sensor may be out-of-calibration ● TP sensor signal circuit is open to the PCM (intermittent fault) ● TP sensor ground circuit is open (an intermittent fault) ● Throttle body is damaged or throttle linkage is bent or binding ● TP sensor is damaged or has failed
DTC P0121 **1T CCM, MIL: Yes** 1998, 1999, 2000, 2001, 2002 Prizm All engines Transmissions: All	**Throttle Position Sensor Range/Performance Conditions:** Engine started, vehicle speed over 30 mph while in closed loop, and the PCM detected the TP sensor signal was out of the acceptable range for these operating conditions. **Possible Causes** ● MAP sensor may be out-of-calibration ● TP sensor signal circuit is open to the PCM (intermittent fault) ● TP sensor ground circuit is open (an intermittent fault) ● Throttle body is damaged or throttle linkage is bent or binding ● TP sensor is damaged or has failed

OBD II Trouble Code List (P0xxx Codes)

DTC	Trouble Code Title, Conditions & Possible Causes
DTC P0121 **2T CCM, MIL: Yes** 1998, 1999, 2000, 2001, 2002, 2003, 2004, 2005 Tracker All engines Transmissions: All	**Throttle Position Sensor Range/Performance Conditions:** DTC P0101, P0102, P0103, P0116, P0117, P0118, P0122, P0123, P0335 and P1451 not set, engine speed less than 3500 rpm, MAF sensor maximum to minimum airflow rate variation more than 25.5 gm/s, MAP sensor under 26 kPa with the TP sensor signal less than 35 degrees, then with the MAP sensor over 67 kPa and the TP angle over 35 degrees, the PCM detected too much difference between the Expected and the Actual TP values, condition met for 3 seconds. **Possible Causes** ● MAP sensor may be out-of-calibration ● TP sensor signal circuit is open to the PCM (intermittent fault) ● TP sensor ground circuit is open (an intermittent fault) ● Throttle body is damaged or throttle linkage is bent or binding ● TP sensor is damaged or has failed
DTC P0122 **1T CCM, MIL: Yes** 1998, 1999, 2000, 2001, 2002, 2003, 2004, 2005 Metro, Tracker All engines Transmissions: All	**Throttle Position Sensor Circuit Low Input Conditions:** DTC P0123 not set, engine running at hot idle speed, and the PCM detected the TP sensor indicated less than a 2% throttle opening for a period of over 5 seconds during the CCM test. **Possible Causes** ● TP sensor signal circuit is shorted to ground ● TP sensor power circuit is open (check VREF from the PCM) ● TP sensor is damaged or failed (it may be shorted internally) ● PCM has failed
DTC P0123 **1T CCM, MIL: Yes** 1998, 1999, 2000, 2001, 2002, 2003, 2004, 2005 Metro, Tracker All engines Transmissions: All	**Throttle Position Sensor Circuit High Input Conditions:** DTC P0122 not set, engine running at hot idle speed, and the PCM detected the TP sensor indicated more than a 96% throttle opening for a period of over 5 seconds during the CCM continuous test. **Possible Causes** ● TP sensor signal circuit is open between sensor and the PCM ● TP sensor ground circuit is open between sensor and the PCM ● TP sensor is damaged or has failed (it may be open internally) ● PCM has failed
DTC P0125 **2T CCM, MIL: Yes** 1998, 1999, 2000, 2001 Metro All engines Transmissions: All	**ECT Excessive Time to Enter Closed Loop Conditions:** Engine started, fuel level over 25%, BARO sensor over 75 KPa, front HO2S-11 signal varying, IAT sensor from 14-122ºF, and the PCM detected the ECT sensor indicated the engine did not reach closed loop during a specified amount of time. **Possible Causes** ● Check the operation of the thermostat (it may be stuck open) ● ECT sensor signal circuit has high resistance ● ECT sensor has failed ● Inspect for low coolant level or an incorrect coolant mixture
DTC P0125 **2T CCM, MIL: Yes** 1998, 1999, 2000, 2001, 2002, 2003, 2004, 2005 Prizm, Tracker All engines Transmissions: All	**ECT Excessive Time To Enter Closed Loop Conditions:** DTC P0117, P0118 and P0335 not set, ECT sensor less than 86ºF at startup, engine runtime over 10 minutes, BARO sensor over 75 kPa, IAT sensor from 7-158ºF, and the PCM detected the ECT sensor did not indicate at least 68ºF (i.e., a closed loop temperature value) within a predetermined amount of time during the CCM Rationality test period. **Possible Causes** ● Check the operation of the thermostat (it may be stuck open) ● ECT sensor signal circuit has high resistance ● ECT sensor has failed ● Inspect for low coolant level or an incorrect coolant mixture
DTC P0128 **2T CCM, MIL: Yes** 2000, 2001, 2002 Prizm All engines Transmissions: All	**ECT Below Thermostat Regulating Temperature Conditions:** Engine started, ECT and IAT sensors from 14-97ºF at startup, engine running for a calibrated period of time, and the PCM detected the ECT sensor failed to reach 167ºF within the amount of time determined by the Water Temperature Counter diagnostic. **Possible Causes** ● Check the operation of the thermostat (it may be stuck open) ● ECT sensor is damaged or out-of-calibration (it is "skewed") ● PCM has failed
DTC P0128 **2T CCM, MIL: Yes** 2002, 2003, 2004, 2005 Tracker 2.0L VIN C engine Transmissions: All	**ECT Below Thermostat Regulating Temperature Conditions:** Engine started, ECT sensor from 18ºF to 113ºF at startup, IAT sensor more than 18ºF, and the PCM detected the ECT sensor failed to reach 167ºF within a period of time determined by the Water Temperature Counter diagnostic test function. **Possible Causes** ● Check the operation of the thermostat (it may be stuck open) ● ECT sensor is damaged or out-of-calibration (it is "skewed") ● PCM has failed

OBD II Trouble Code List (P0xxx Codes)

DTC	Trouble Code Title, Conditions & Possible Causes
DTC P0128 **2T CCM, MIL: Yes** 2002, 2003, 2004, 2005 Tracker 2.5L VIN 4 engine Transmissions: All	**ECT Below Thermostat Regulating Temperature Conditions:** Engine started, ECT sensor from 14°F to 113°F at startup, IAT sensor more than 7°F, and the PCM detected the ECT sensor failed to reach 167°F within the amount of time determined by the Water Temperature Counter diagnostic test function. **Possible Causes** ● ECT sensor is out of calibration ● Check for low coolant level or incorrect coolant mixture ● Cooling system component failure (thermostat stuck open)
DTC P0130 **2T CCM, MIL: Yes** 1998, 1999, 2000, 2001, 2002 Prizm All engines Transmissions: All	**O2S-11 or HO2S-11 (Bank 1 Sensor 1) Circuit Malfunction Conditions:** Engine started, at idle speed, ECT sensor more than 158°F, and the PCM detected the O2S or HO2S signal was fixed between 400 and 550 mv, or the O2S or HO2S signal was fixed at less than 550 mv, or the O2S signal was fixed at more than 440 mv. **Possible Causes** ● HO2S signal circuit is open between the sensor and the PCM ● HO2S signal circuit is shorted to sensor or chassis ground ● HO2S signal circuit is shorted to VREF or system power (B+) ● HO2S is damaged, contaminated or it has failed ● PCM has failed
DTC P0131 **2T CCM, MIL: Yes** 1998, 1999, 2000, 2001 Metro All engines Transmissions: All	**HO2S-11 (Bank 1 Sensor 1) Circuit Low Input Conditions:** Engine started, IAT sensor from 14-158°F, ECT sensor more than 176°F, BARO sensor over 75 kPa, vehicle driven to a speed of over 35 mph for 2 minutes, followed by an idle period with the engine speed steady for 2 minutes, and the PCM detected the maximum HO2S signal was less than 300 mv during the CCM test. **Possible Causes** ● HO2S signal circuit is open or shorted to ground ● HO2S ground circuit is open ● HO2S is damaged or it has failed ● PCM has failed
DTC P0131 **2T CCM, MIL: Yes** 1998, 1999, 2000, 2001, 2002, 2003, 2004, 2005 Tracker 2.0L VIN C engine Transmissions: All	**Title: HO2S-11 (Bank 1 Sensor 1) Circuit Low Input** **Trouble Code Conditions** Engine started, IAT sensor more than 18°F at startup, ECT sensor from 18-230°F, BARO sensor over 75 kPa, vehicle driven to a speed over 35 mph for 2 minutes, followed by a period at steady idle speed for 1 minute, and the PCM detected the maximum average voltage was less than 600 mv and the minimum average voltage was less than 300 m, or PCM internal voltage check of the HO2S circuit while idling indicated a higher than expected voltage drop during the test. **Possible Causes** ● HO2S signal circuit is open or shorted to ground ● HO2S ground circuit is open ● HO2S is damaged or it has failed ● PCM has failed
DTC P0131 **2T CCM, MIL: Yes** 1998, 1999, 2000, 2001, 2002, 2003, 2004, 2005 Tracker 2.5L VIN 4 engine Transmissions: All	**HO2S-11 (Bank 1 Sensor 1) Circuit Low Input Conditions:** IAT sensor more than 18°F at startup, engine started, ECT sensor from 18-230°F, BARO sensor over 75 kPa, vehicle driven to a speed over 35 mph for 2 minutes, followed by an idle period with the engine speed steady for 1 minute, and the PCM detected the maximum average HO2S signal was under 600 mv and the minimum average voltage was under 300 mv, or that an internal check of the HO2S circuit indicated too large of voltage drop across the signal circuit. **Possible Causes** ● HO2S signal circuit is open or shorted to ground ● HO2S ground circuit is open ● HO2S is damaged or it has failed ● PCM has failed
DTC P0132 **2T CCM, MIL: Yes** 1998, 1999, 2000, 2001 Metro All engines Transmissions: All	**HO2S-11 (Bank 1 Sensor 1) Circuit High Input Conditions:** IAT sensor more than 18°F at startup, engine started, ECT sensor from 18-230°F, BARO sensor over 75 kPa, vehicle driven to a speed over 35 mph for 2 minutes, followed by a period at idle speed for 1 minute, and the PCM detected the maximum HO2S signal was more than 1200 mv and the minimum HO2S signal was less than 600 mv. **Possible Causes** ● HO2S is contaminated, damaged or it has failed ● HO2S signal circuit shorted to power (check the heater circuit) ● PCM has failed

OBD II Trouble Code List (P0xxx Codes)

DTC	Trouble Code Title, Conditions & Possible Causes
DTC P0132 **2T CCM, MIL: Yes** 1998, 1999, 2000, 2001, 2002, 2003, 2004, 2005 Tracker All engines Transmissions: All	**HO2S-11 (Bank 1 Sensor 1) Circuit High Input Conditions:** IAT sensor more than 18ºF at startup, engine started, ECT sensor from 18-230ºF, BARO sensor over 75 kPa, vehicle driven to a speed over 35 mph for 2 minutes, followed by a period at steady idle speed for at least 1 minute, and the PCM detected the minimum voltage was over 300 mv, or the maximum average voltage was at least 740 mv and the minimum average voltage was at least 330 mv. **Possible Causes** ● HO2S is contaminated, damaged or it has failed ● HO2S signal circuit shorted to power (check the heater circuit) ● PCM has failed
DTC P0132 **2T CCM, MIL: Yes** 2001, 2002, 2003, 2004, 2005 Tracker 2.5L VIN 4 engine Transmissions: All	**HO2S-11 (Bank 1 Sensor 1) Circuit High Input Conditions:** Engine started, IAT sensor more than 18ºF at startup, ECT sensor from 18-230ºF, BARO sensor over 75 kPa, vehicle driven to a speed over 35 mph for 2 minutes, followed by a period at idle speed for 1 minute, and the PCM detected the maximum HO2S signal was more than 1200 mv and the minimum HO2S signal was less than 600 mv. **Possible Causes** ● Exhaust leak present in the exhaust manifold or exhaust pipes ● HO2S signal circuit shorted to power (check the heater circuit) ● HO2S element is fuel contaminated or it has deteriorated ● PCM has failed
DTC P0133 **2T O2S, MIL: Yes** 1998, 1999, 2000, 2001 Metro All engines Transmissions: All	**HO2S-11 (Bank 1 Sensor 1) Slow Response Conditions:** Engine started, IAT sensor from 14-158ºF, ECT sensor over 176ºF, BARO sensor over 75 kPa, vehicle speed over 35 mph for 2 minutes, followed by an idle period with the engine speed steady for 2 minutes, and the PCM detected the response time of the HO2S-11 was more than 1 second, or that the switch cycle average was over 5 seconds during the test. **Possible Causes** ● Exhaust leak present in the exhaust manifold or exhaust pipes ● HO2S element is fuel contaminated or has deteriorated ● PCM has failed
DTC P0133 **2T O2S, MIL: Yes** 1998, 1999, 2000, 2001, 2002 Prizm All engines Transmissions: All	**HO2S-11 (Bank 1 Sensor 1) Slow Response Conditions:** Engine started, engine running a idle speed in closed loop, ECT sensor more than 158ºF, and the PCM detected the average response rate of the HO2S to change from rich-to-lean or lean-to-rich was more than 1.1 seconds during the HO2S Diagnostic Test. **Possible Causes** ● Exhaust leak present in the exhaust manifold or exhaust pipes ● HO2S element is fuel contaminated or has deteriorated ● PCM has failed
DTC P0133 **2T O2S, MIL: Yes** 1998, 1999, 2000, 2001, 2002, 2003, 2004, 2005 Tracker 2.0L VIN C engine Transmissions: All	**HO2S-11 (Bank 1 Sensor 1) Slow Response Conditions:** Engine started; IAT sensor over 18ºF at startup, ECT sensor from 18-230ºF, BARO sensor over 75 kPa, vehicle speed over 35 mph for 2 minutes, followed by an idle period at a steady speed for at least 1 minute, and the PCM detected the average HO2S response rate was over 2.5 seconds, or the average HO2S switch cycle was over 6 seconds during the test. **Possible Causes** ● Exhaust leak present in the exhaust manifold or exhaust pipes ● HO2S element is fuel contaminated or has deteriorated ● PCM has failed
DTC P0133 **2T O2S, MIL: Yes** 2001, 2002, 2003, 2004, 2005 Tracker 2.5L VIN 4 engine Transmissions: All	**HO2S-11 (Bank 1 Sensor 1) Circuit Slow Response Conditions:** Engine started, IAT sensor more than 18ºF at startup, ECT sensor from 18-230ºF, BARO sensor over 75 kPa, vehicle driven to a speed of over 35 mph for 2 minutes, then back to a idle speed for 1 minute, and the PCM detected the average HO2S response rate was over 2.5 seconds, or the HO2S switch cycle was more than 6 seconds. **Possible Causes** ● Exhaust leak present in the exhaust manifold or exhaust pipes ● HO2S element has fuel contamination or it has deteriorated ● PCM has failed
DTC P0134 **2T O2S, MIL: Yes** 1998, 1999, 2000, 2001 Metro All engines Transmissions: All	**HO2S-11 (Bank 1 Sensor 1) Insufficient Activity Conditions:** Engine started, IAT sensor from 14ºF to 158ºF, ECT sensor over 176ºF, BARO sensor over 75 kPa, vehicle speed over 35 mph, and then returned to idle speed and allowed to idle for over 1 minute, and the PCM detected the average response rate of the HO2S was over 2.5 seconds, or that the switch cycle average of the HO2S was over 6 seconds during the test. **Possible Causes** ● Exhaust leak present in exhaust manifold or exhaust pipes ● HO2S element fuel contamination or has deteriorated ● HO2S signal circuit or the ground circuit has high resistance ● HO2S heater element has failed, or the heater circuit is open ● PCM has failed

OBD II Trouble Code List (P0xxx Codes)

DTC	Trouble Code Title, Conditions & Possible Causes
DTC P0134 **2T O2S, MIL: Yes** 1998, 1999, 2000, 2001, 2002, 2003, 2004, 2005 Tracker 2.0L VIN C engine Transmissions: All	**HO2S-11 (Bank 1 Sensor 1) Insufficient Activity Conditions:** Engine started, IAT sensor over 18°F at startup, ECT sensor from 18-230°F, BARO sensor over 75 kPa, vehicle speed over 35 mph for 2 minutes, followed by an idle period at a steady speed for at least 1 minute, and the PCM detected the HO2S signal remained less than or equal to 450 mv for 30 seconds, or the HO2S signal voltage remained less than 450 mv. **Possible Causes** • Exhaust leak present in exhaust manifold or exhaust pipes • HO2S element fuel contamination or has deteriorated • HO2S signal circuit or the ground circuit has high resistance • HO2S heater element has failed, or the heater circuit is open • PCM has failed
DTC P0134 **2T O2S, MIL: Yes** 2001, 2002, 2003, 2004, 2005 Tracker 2.5L VIN 4 engine Transmissions: All	**HO2S-11 (Bank 1 Sensor 1) Insufficient Activity Detected Conditions:** Engine started, IAT sensor more than 18°F at startup, ECT sensor from 18-230°F, BARO sensor over 75 kPa, and the PCM detected the minimum HO2S signal was less than 450 mv for 30 seconds, or that the maximum HO2S signal was less than 450 mv. **Possible Causes** • Exhaust leak present in exhaust manifold or exhaust pipes • HO2S element is fuel contaminated or it has deteriorated • HO2S signal circuit or the ground circuit has high resistance • HO2S heater element has failed, or the heater circuit is open • PCM has failed
DTC P0135 **2T CCM, MIL: Yes** 1998, 1999, 2000, 2001 Metro All engines Transmissions: All	**HO2S-11 (Bank 1 Sensor 1) Heater Circuit Malfunction Conditions:** Engine started, engine runtime 1 minute, then after an acceleration period of over 5 seconds, the PCM detected the heater resistor was less than 2.5v with heater commanded "off", or that it was less than 0.31v with heater turned commanded "on" during the CCM test. **Possible Causes** • HO2S heater control circuit is open or shorted to ground • HO2S heater control circuit is shorted to power • HO2S heater power circuit is open (test power to IG Coil fuse) • HO2S heater is damaged or has failed • PCM has failed
DTC P0135 **2T CCM, MIL: Yes** 1998, 1999, 2000, 2001, 2002 Prizm All engines Transmissions: All	**HO2S-11 (Bank 1 Sensor 1) Heater Circuit Malfunction Conditions:** Engine started, and the PCM detected the HO2S heater current level exceeded 2.0 amps, or was under 0.2 amps for 3 seconds with the heater commanded "on" during the Heater test. **Possible Causes** • HO2S heater control circuit is open or shorted to ground • HO2S heater control circuit is shorted to power • HO2S heater power circuit is open (check power from the relay) • HO2S heater is damaged or has failed • PCM has failed
DTC P0135 **2T CCM, MIL: Yes** 1998, 1999, 2000, 2001 Tracker 2.0L VIN C engine Transmissions: All	**HO2S-11 (Bank 1 Sensor 1) Heater Circuit Malfunction Conditions:** Engine started, and the PCM detected the HO2S heater current level was more than 5.3 amps or less than 0.15 amps with the heater "on", or the HO2S heater signal was more than 13.8v or was less than 8.7v with the heater "on", or the HO2S heater signal was 6v with the heater commanded "off" for 3 seconds during the CCM test. **Possible Causes** • HO2S heater control circuit is open, shorted to ground or power • HO2S heater power circuit is open (check power from IG fuse) • HO2S heater is damaged or has failed • PCM has failed
DTC P0135 **2T CCM, MIL: Yes** 1998, 1999, 2000, 2001, 2002, 2003, 2004, 2005 Tracker 2.5L VIN 4 engine Transmissions: All	**HO2S-11 (Bank 1 Sensor 1) Heater Circuit Malfunction Conditions:** Engine started, and the PCM detected the HO2S heater current level was more than 5.3 amps or less than 0.15 amps with the heater "on", or the HO2S heater signal was more than 13.8v or was less than 8.7v with the heater "on", or the HO2S heater signal was 6v with the heater commanded "off" for 3 seconds during the CCM test. **Possible Causes** • HO2S heater control circuit is open, shorted to ground or power • HO2S heater power circuit is open (check power from IG fuse) • HO2S heater is damaged or has failed • PCM has failed

OBD II Trouble Code List (P0xxx Codes)

DTC	Trouble Code Title, Conditions & Possible Causes
DTC P0136 **2T CCM, MIL: Yes** 1998, 1999, 2000, 2001 Metro All engines Transmissions: All	**HO2S-12 (Bank 1 Sensor 2) Circuit Malfunction Conditions:** ECT sensor more than 176°F, IAT sensor from 14-158°F, BARO sensor over 75 kPa, engine speed steady at over 1500 rpm in closed loop at 20-50 mph, fuel level over 25%, and the PCM detected the HO2S signal was over 600 mv for 8 minutes, or the minimum HO2S signal was greater than or equal to 300 mv after a Maximum Voltage Check during Decel Fuel Shutoff for 5 seconds. **Possible Causes** ● HO2S signal circuit is open between the sensor and the PCM ● HO2S signal circuit is shorted to sensor or chassis ground ● HO2S signal circuit is shorted to VREF or system power (B+) ● HO2S is damaged, contaminated or it has failed ● PCM has failed
DTC P0136 **2T CCM, MIL: Yes** 1998, 1999, 2000, 2001, 2002 Prizm All engines Transmissions: All	**HO2S-12 (Bank 1 Sensor 2) Circuit Malfunction Conditions:** Engine speed steady at over 1600 rpm in closed loop, ECT sensor more than 158°F, engine running at a steady speed, and the PCM detected the HO2S signal was greater than or equal to 400 mv, or the HO2S signal was less than or equal to 500 mv for 5 seconds. **Possible Causes** ● HO2S signal circuit is open between the sensor and the PCM ● HO2S signal circuit is shorted to sensor or chassis ground ● HO2S signal circuit is shorted to VREF or system power (B+) ● HO2S is damaged, contaminated or it has failed ● PCM has failed
DTC P0136 **2T CCM, MIL: Yes** 1998, 1999, 2000, 2001, 2002 Tracker All engines Transmissions: All	**HO2S-12 (Bank 1 Sensor 2) Circuit Malfunction Conditions:** ECT sensor more than 158°F, BARO sensor over 75 kPa, engine speed steady at over 1600 rpm in closed loop, fuel level over 15%, conditions stable, and the PCM detected the average HO2S signal was less than 120 mv or more than 900 mv, or that the HO2S Pullup signal indicated more than 4.5v for 2 seconds. **Possible Causes** ● HO2S signal circuit is open between the sensor and the PCM ● HO2S signal circuit is shorted to sensor or chassis ground ● HO2S signal circuit is shorted to VREF or system power (B+) ● HO2S is damaged, contaminated or it has failed ● PCM has failed
DTC P0137 **2T CCM, MIL: Yes** 2002, 2003, 2004, 2005 Tracker 2.0L VIN C engine Transmissions: All	**HO2S-12 (Bank 1 Sensor 2) Circuit Low Input Conditions:** DTC P0141 not set, IAT sensor more than 7°F at startup, engine started, ECT sensor from 18-230°F, BARO sensor over 75 kPa, vehicle driven at a speed of over 35 mph for 2 minutes with the throttle steady, and the PCM detected the HO2S signal remained at a very low voltage level during the test period. **Possible Causes** ● HO2S signal circuit is shorted to ground ● HO2S is damaged or it has failed ● PCM has failed
DTC P0137 **2T CCM, MIL: Yes** 2002, 2003, 2004, 2005 Tracker 2.0L VIN C engine Transmissions: All	**HO2S-12 (Bank 1 Sensor 2) Circuit Low Input Conditions:** DTC P0141 not set, IAT sensor more than 18°F at startup, engine started, ECT sensor from 18-230°F, BARO sensor over 75 kPa, vehicle driven at a speed from 20-40 mph for 3 minutes with the throttle steady, and the PCM detected the HO2S signal remained at an excessive low voltage during the CCM continuous test period. **Possible Causes** ● HO2S signal circuit is shorted to ground ● HO2S is damaged or it has failed ● PCM has failed
DTC P0137 **2T CCM, MIL: Yes** 2001, 2002, 2003, 2004, 2005 Tracker 2.5L VIN 4 engine Transmissions: All	**HO2S-12 (Bank 1 Sensor 2) Circuit Low Input Conditions:** DTC P0161 not set, engine started, IAT sensor more than 7°F at startup, ECT sensor from 18-230°F, BARO sensor over 75 kPa, vehicle driven at a speed of over 35 mph for 2 minutes with the throttle steady, and the PCM detected the maximum HO2S signal was less than 350 mv for 4 seconds while in the Fuel Cutoff mode. **Possible Causes** ● HO2S signal circuit is open or shorted to ground ● HO2S ground circuit is open ● HO2S is damaged or it has failed ● PCM has failed

OBD II Trouble Code List (P0xxx Codes)

DTC	Trouble Code Title, Conditions & Possible Causes
DTC P0138 **2T CCM, MIL: Yes** 2002, 2003, 2004, 2005 Tracker 2.0L VIN C engine Transmissions: All	**HO2S-12 (Bank 1 Sensor 2) Circuit High Input Conditions:** DTC P0141 not set, IAT sensor more than 7°F at startup, engine started, ECT sensor from 18-230°F, BARO sensor over 75 kPa, engine operating in closed loop, Fuel Level over 25%, and the PCM detected the HO2S signal PCM detected an average signal voltage that exceeded a specified value, or the lowest voltage produced by the HO2S was more than a specified value during the test period. **Possible Causes** • HO2S signal circuit is shorted to ground • HO2S is damaged or it has failed • PCM has failed
DTC P0138 **2T CCM, MIL: Yes** 2001, 2002, 2003, 2004, 2005 Tracker 2.5L VIN 4 engine Transmissions: All	**HO2S-12 (Bank 1 Sensor 2) Circuit High Input Conditions:** DTC P0161 not set, engine started, IAT sensor more than 7°F at startup, ECT sensor from 18-230°F, BARO sensor over 75 kPa, vehicle driven at a speed of over 35 mph for 2 minutes with the throttle steady, and the PCM detected the minimum HO2S signal over 500 mv for 4 seconds while operating in the Fuel Cutoff mode. **Possible Causes** • HO2S signal circuit shorted to power (check the heater circuit) • HO2S is damaged or it has failed • PCM has failed
DTC P0140 **2T CCM, MIL: Yes** 2002, 2003, 2004, 2005 Tracker 2.0L VIN C engine Transmissions: All	**HO2S-12 (Bank 1 Sensor 2) Circuit Malfunction Conditions:** IAT sensor more than 18°F at startup, engine started, ECT sensor from 18-230°F, BARO sensor over 75 kPa, engine running in closed loop, Fuel Level over 25%, and the PCM detected that the average HO2S signal was less than a specified value, or the highest voltage produced by the HO2S was less than a specified value. **Possible Causes** • HO2S is contaminated, damaged or it has failed • HO2S signal circuit shorted to power (check the heater circuit) • PCM has failed
DTC P0141 **2T CCM, MIL: Yes** 1998, 1999, 2000, 2001 Metro All engines Transmissions: All	**HO2S-12 (Bank 1 Sensor 2) Heater Circuit Malfunction Conditions:** Engine started, ECT sensor more than 176°F, engine speed over 2000 rpm for 2 minutes, and the PCM detected the HO2S heater resistor signal was less than 2.5v with heater turned "off", or that the signal was more than 0.31v with heater turned "on" for 5 seconds. **Possible Causes** • HO2S heater control circuit is open, shorted to ground or power • HO2S heater power circuit is open (check power from the fuse) • HO2S heater is damaged or has failed • PCM has failed
DTC P0141 **2T CCM, MIL: Yes** 1998, 1999, 2000, 2001, 2002 Prizm All engines Transmissions: All	**HO2S-12 (Bank 1 Sensor 2) Heater Circuit Malfunction Conditions:** Engine started, and the PCM detected the HO2S heater current level was more than 2.0 amps or was less than 0.2 amps with the heater on" during the HO2S Heater diagnostic. **Possible Causes** • HO2S heater control circuit is open, shorted to ground or power • HO2S heater power circuit is open (check power from the relay) • HO2S heater is damaged or has failed • PCM has failed
DTC P0141 **2T CCM, MIL: Yes** 1998, 1999, 2000, 2001, 2002, 2003, 2004, 2005 Tracker All engines Transmissions: All	**HO2S-12 (Bank 1 Sensor 2) Heater Circuit Malfunction Conditions:** Engine started, and the PCM detected the HO2S heater current was more than 5.3 amps or was less than 0.15 amps with the heater "on", or the HO2S heater circuit was over 13.8v or was less than 8.7v with the heater "on", or the HO2S heater circuit was 6v with the heater turned "off", condition met for 3 seconds. **Possible Causes** • HO2S heater control circuit is open, shorted to ground or power • HO2S heater power circuit is open (check power from IG fuse) • HO2S heater is damaged or has failed • PCM has failed
DTC P0151 **2T CCM, MIL: Yes** 2001, 2002, 2003, 2004, 2005 Tracker 2.5L VIN 4 engine Transmissions: All	**Title: HO2S-21 (Bank 2 Sensor 1) Circuit Low Input** **Trouble Code Conditions** Engine started, IAT sensor over 18°F at startup, ECT sensor from 18-230°F, BARO sensor over 75 kPa, vehicle speed over 35 mph for 2 minutes, followed by a period at steady idle speed for 1 minute and the PCM detected the maximum HO2S signal was less than 400 mv. **Possible Causes** • HO2S signal circuit is open or shorted to ground • HO2S ground circuit is open • HO2S is damaged or it has failed • PCM has failed

OBD II Trouble Code List (P0xxx Codes)

DTC	Trouble Code Title, Conditions & Possible Causes
DTC P0152 **2T CCM, MIL: Yes** 2001, 2002, 2003, 2004, 2005 Tracker 2.5L VIN 4 engine Transmissions: All	**HO2S-21 (Bank 2 Sensor 1) Circuit High Input Conditions:** Engine started, IAT sensor more than 18ºF at startup, ECT sensor from 18-230ºF, BARO sensor over 75 kPa, vehicle driven to a speed over 35 mph for 2 minutes, followed by a period at idle speed for 1 minute, and the PCM detected the maximum HO2S signal was more than 1200 mv and the minimum HO2S signal was less than 600 mv. **Possible Causes** ● Exhaust leak present in the exhaust manifold or exhaust pipes ● HO2S signal circuit shorted to power (check the heater circuit) ● HO2S element is fuel contaminated or it has deteriorated ● PCM has failed
DTC P0153 **2T O2S, MIL: Yes** 2001, 2002, 2003, 2004, 2005 Tracker 2.5L VIN 4 engine Transmissions: All	**HO2S-21 (Bank 2 Sensor 1) Circuit Slow Response Conditions:** Engine started, IAT sensor more than 18ºF at startup, ECT sensor from 18-230ºF, BARO sensor over 75 kPa, vehicle driven to a speed of over 35 mph for 2 minutes, then back to a idle speed for 1 minute, and the PCM detected the average HO2S response rate was over 2.5 seconds, or the HO2S switch cycle was more than 6 seconds. **Possible Causes** ● Exhaust leak present in the exhaust manifold or exhaust pipes ● HO2S element has fuel contamination or it has deteriorated ● PCM has failed
DTC P0154 **2T O2S, MIL: Yes** 2001, 2002, 2003, 2004, 2005 Tracker 2.5L VIN 4 engine Transmissions: All	**HO2S-21 (Bank 2 Sensor 1) Insufficient Activity Detected Conditions:** Engine started, IAT sensor more than 18ºF at startup, ECT sensor from 18-230ºF, BARO sensor over 75 kPa, and the PCM detected the minimum HO2S signal was less than 450 mv for 30 seconds, or that the maximum HO2S signal was less than 450 mv. **Possible Causes** ● Exhaust leak present in exhaust manifold or exhaust pipes ● HO2S element is fuel contaminated or it has deteriorated ● HO2S signal circuit or the ground circuit has high resistance ● HO2S heater element has failed, or the heater circuit is open ● PCM has failed
DTC P0155 **2T CCM, MIL: Yes** 2001, 2002, 2003, 2004, 2005 Tracker 2.5L VIN 4 engine Transmissions: All	**HO2S-21 (Bank 2 Sensor 1) Heater Circuit Malfunction Conditions:** Engine started, and the PCM detected the HO2S heater current level was more than 5.3 amps or was less than 0.15 amps with the heater "on", or the HO2S heater signal was more than 13.8v or was less than 8.7v with the heater "on", or the HO2S heater signal was 6v with the heater commanded "off" for 3 seconds during the CCM test. **Possible Causes** ● HO2S heater control circuit is open, shorted to ground or power ● HO2S heater power circuit is open (check power from IG fuse) ● HO2S heater is damaged or has failed ● PCM has failed
DTC P0156 **2T O2S, MIL: Yes** 2001, 2002, 2003, 2004, 2005 Tracker 2.5L VIN 4 engine Transmissions: All	**HO2S-22 (Bank 2 Sensor 2) Circuit Malfunction Conditions:** Engine speed over 1600 rpm in closed loop and stable, ECT sensor over 158ºF, BARO sensor over 75 kPa, fuel level over 15%, and the PCM detected the average HO2S signal was below 120 mv or over 900 mv, or the HO2S Pullup circuit was over 4.5v for 2 seconds. **Possible Causes** ● HO2S signal circuit is open between the sensor and the PCM ● HO2S signal circuit is shorted to sensor or chassis ground ● HO2S signal circuit is shorted to VREF or system power (B+) ● PCM has failed
DTC P0157 **2T CCM, MIL: Yes** 2001, 2002, 2003, 2004, 2005 Tracker 2.5L VIN 4 engine Transmissions: All	**HO2S-22 (Bank 2 Sensor 2) Circuit Low Input Conditions:** DTC P0161 not set; vehicle speed over 35 mph for 2 minutes at steady throttle, IAT sensor over 7ºF at startup, BARO sensor over 75 kPa, ECT sensor from 18-230ºF, and the PCM detected the maximum HO2S signal was under 350 mv for 4 seconds while in Fuel Cutoff. **Possible Causes** ● HO2S signal circuit is open or shorted to ground ● HO2S ground circuit is open ● HO2S is damaged or it has failed ● PCM has failed
DTC P0158 **2T CCM, MIL: Yes** 2001, 2002, 2003, 2004, 2005 Tracker 2.5L VIN 4 engine Transmissions: All	**HO2S-22 (Bank 2 Sensor 2) Circuit High Input Conditions:** DTC P0161 not set, vehicle speed over 35 mph for 2 minutes at steady throttle, IAT sensor over 7ºF at startup, BARO sensor over 75 kPa, ECT sensor from 18-230ºF, and the PCM detected the minimum HO2S signal over 500 mv for 4 seconds while in Fuel Cutoff mode. **Possible Causes** ● HO2S signal circuit shorted to power (check the heater circuit) ● HO2S is damaged or it has failed ● PCM has failed

OBD II Trouble Code List (P0xxx Codes)

DTC	Trouble Code Title, Conditions & Possible Causes
DTC P0161 **2T CCM, MIL: Yes** 2001, 2002, 2003, 2004, 2005 Tracker 2.5L VIN 4 engine Transmissions: All	**HO2S-22 (Bank 2 Sensor 2) Heater Circuit Malfunction Conditions:** Engine started, and the PCM detected the HO2S heater current level was more than 5.3 amps or less than 0.15 amps with the heater "on", or the HO2S heater signal was more than 13.8v or was less than 8.7v with the heater "on", or the HO2S heater signal was 6v with the heater commanded "off" for 3 seconds. **Possible Causes** ● HO2S heater control circuit is open, shorted to ground or shorted to power ● HO2S heater power circuit is open (check power from IG fuse) ● HO2S heater is damaged or has failed ● PCM has failed
DTC P0171 **2T FUEL, MIL: Yes** 1998, 1999, 2000, 2001 Metro All engines Transmissions: All	**Fuel Trim System Lean (Bank 1) Conditions:** Engine started, in closed loop, IAT sensor from 14-122°F, BARO sensor over 75 kPa, fuel level more than 25%, and the PCM detected the LONGFT reading was +30% for 5 seconds, or the SHRTFT reading was +20% for 45 seconds in the Fuel System test. **Possible Causes** ● Air leaks after the MAF sensor, or in the EGR or PCV system ● Base engine "mechanical" fault affecting one or more cylinders ● Exhaust leaks located in front of the HO2S location ● Fuel delivery system supplying too little fuel during cruise or idle periods (e.g., faulty fuel pump or dirty, restricted fuel filter) ● Fuel injector (one or more) dirty or pressure regulator has failed ● HO2S is contaminated, deteriorated or it has failed ● Vehicle driven low on fuel or until it ran out of fuel
DTC P0171 **2T FUEL, MIL: Yes** 1998, 1999, 2000, 2001, 2002 Prizm All engines Transmissions: All	**Fuel Trim System Lean (Bank 1) Conditions:** Engine started, ECT sensor from 18-203°F and IAT sensor from 18-140°F at startup, BARO sensor over 75 kPa, ECT sensor under 230°F during testing, and the PCM detected the SHRTFT was over or equal to 15% in 128 firing events, or the LONGFT was over or equal to 20% in 128 firing events, or the total Fuel Trim was over or equal to 33% in 128 firing events. **Possible Causes** ● Air leaks after the MAF sensor, or in the EGR or PCV system ● Base engine "mechanical" fault affecting one or more cylinders ● Exhaust leaks located in front of the HO2S location ● Fuel delivery system supplying too little fuel during cruise or idle periods (e.g., faulty fuel pump or dirty, restricted fuel filter) ● Fuel injector (one or more) dirty or pressure regulator has failed ● HO2S is contaminated, deteriorated or it has failed ● Vehicle driven low on fuel or until it ran out of fuel
DTC P0171 **2T FUEL, MIL: Yes** 1998, 1999, 2000, 2001, 2002, 2003, 2004, 2005 Tracker All engines Transmissions: All	**Fuel Trim System Lean (Bank 1) Conditions:** Engine started, ECT sensor from 18-230°F, IAT sensor from 18-140°F at startup, BARO sensor over 75 kPa, ECT sensor under 230°F during testing, and the PCM detected the SHRTFT was over or equal to 15% in 128 firing events, or the LONGFT was over or equal to 20% in 128 firing events, or the total Fuel Trim was over or equal to 33% in 128 firing events. **Possible Causes** ● Air leaks after the MAF sensor, or in the EGR or PCV system ● Base engine "mechanical" fault affecting one or more cylinders ● Exhaust leaks located in front of the HO2S location ● Fuel control sensor is out of calibration (i.e., ECT, IAT or MAP) ● Fuel delivery system supplying too little fuel during cruise or idle periods (e.g., faulty fuel pump or dirty, restricted fuel filter) ● Fuel injector (one or more) dirty or pressure regulator has failed ● HO2S is contaminated, deteriorated or it has failed ● Vehicle driven low on fuel or until it ran out of fuel
DTC P0172 **2T FUEL, MIL: Yes** 1998, 1999, 2000, 2001 Metro All engines Transmissions: All	**Fuel Trim System Rich (Bank 1) Conditions:** Engine started, in closed loop, IAT sensor from 14-122°F, BARO sensor over 75 kPa, fuel level over 25%, and the PCM detected the LONGFT reading was less than -30% for 5 seconds, or the SHRTFT reading indicated less than -20% for 45 seconds. **Possible Causes** ● Base engine "mechanical" fault affecting one or more cylinders ● EVAP system component has failed or canister fuel saturated ● Exhaust leaks located in front of the A/FS or HO2S location ● Fuel control sensor is out of calibration (i.e., ECT, IAT or MAF) ● Fuel delivery system supplying too much fuel during cruise or idle periods (e.g., faulty fuel pump, or faulty pressure regulator) ● Fuel injector(s) is leaking or stuck partially open (one or more) ● HO2S is contaminated, deteriorated or it has failed

OBD II Trouble Code List (P0xxx Codes)

DTC	Trouble Code Title, Conditions & Possible Causes
DTC P0172 **2T FUEL, MIL: Yes** 1998, 1999, 2000, 2001, 2002 Prizm All engines Transmissions: All	**Fuel Trim System Rich (Bank 1) Conditions:** ECT sensor between 18-203°F and IAT sensor from 18-140°F at startup, BARO sensor over 75 kPa, engine running, ECT sensor less than 230°F during testing, engine running in closed loop, and the PCM detected the SHRTFT indicated more than -11% in 128 firing events, or the LONGFT was more than -11% in 128 firing events, or the total Fuel Trim was more than -30% in 128 firing events. **Possible Causes** ● Base engine "mechanical" fault affecting one or more cylinders ● EVAP system component has failed or canister fuel saturated ● Exhaust leaks located in front of the HO2S location ● Fuel control sensor is out of calibration (i.e., ECT, IAT or MAF) ● Fuel delivery system supplying too much fuel during cruise or idle periods (e.g., faulty fuel pump, or faulty pressure regulator) ● Fuel injector(s) is leaking or stuck partially open (one or more) ● HO2S is contaminated, deteriorated or it has failed
DTC P0172 **2T FUEL, MIL: Yes** 1998, 1999, 2000, 2001, 2002, 2003, 2004, 2005 Tracker All engines Transmissions: All	**Fuel Trim System Rich (Bank 1) Conditions:** Engine started, ECT sensor from 18-230°F at startup, IAT sensor from 18-140°F at startup, BARO sensor over 75 kPa, and the PCM detected (ECT input less than 230°F during testing) the Bank 1 the SHRTFT indicated more than -11% in 128 firing events, or the LONGFT was more than -11% during 128 firing events, or the total Fuel Trim was more than -30% during 128 firing events. **Possible Causes** ● Base engine "mechanical" fault affecting one or more cylinders ● EVAP system component has failed or canister fuel saturated ● Exhaust leaks located in front of the HO2S location ● Fuel control sensor is out of calibration (i.e., ECT, IAT or MAF) ● Fuel delivery system supplying too much fuel during cruise or idle periods (e.g., faulty fuel pump, or faulty pressure regulator) ● Fuel injector(s) is leaking or stuck partially open (one or more) ● HO2S is contaminated, deteriorated or it has failed
DTC P0174 **2T FUEL, MIL: Yes** 2001, 2002, 2003, 2004, 2005 Tracker 2.5L VIN 4 engine Transmissions: All	**Fuel Trim System Lean (Bank 2) Conditions:** Engine started, ECT sensor from 18-230°F at startup, IAT sensor from 18-140°F at startup, BARO sensor over 75 kPa, and the PCM detected (ECT input less than 230°F during testing) the Bank 2 SHRTFT was more than +15% in 128 firing events, or the LONGFT was more than +20% in 128 firing events, or the total Fuel Trim was more than or equal to 33% in 128 firing events during the test. **Possible Causes** ● Base engine "mechanical" fault affecting one or more cylinders ● EVAP system component has failed or canister fuel saturated ● Exhaust leaks located in front of the HO2S location ● Fuel control sensor is out of calibration (i.e., ECT, IAT or MAF) ● Fuel delivery system supplying too much fuel during cruise or idle periods (e.g., faulty fuel pump, or faulty pressure regulator) ● Fuel injector(s) is leaking or stuck partially open (one or more) ● HO2S is contaminated, deteriorated or it has failed
DTC P0175 **2T FUEL, MIL: Yes** 2001, 2002, 2003, 2004, 2005 Tracker 2.5L VIN 4 engine Transmissions: All	**Fuel Trim System Rich (Bank 2) Conditions:** Engine started, ECT sensor from 18-230°F at startup, IAT sensor from 18-140°F at startup, BARO sensor over 75 kPa, and the PCM detected (ECT input less than 230°F during testing) the Bank 2 SHRTFT was less than -11% in 128 firing events, or the LONGFT was less than -11% in 128 firing events, or the total Fuel Trim was more than or equal to -30% in 128 firing events during the test. **Possible Causes** ● Base engine "mechanical" fault affecting one or more cylinders ● EVAP system component has failed or canister fuel saturated ● Exhaust leaks located in front of the HO2S location ● Fuel control sensor is out of calibration (i.e., ECT, IAT or MAF) ● Fuel delivery system supplying too much fuel during cruise or idle periods (e.g., faulty fuel pump, or faulty pressure regulator) ● Fuel injector(s) is leaking or stuck partially open (one or more) ● HO2S is contaminated, deteriorated or it has failed

OBD II Trouble Code List (P0xxx Codes)

DTC	Trouble Code Title, Conditions & Possible Causes
DTC P0300 **2T MISFIRE, MIL: Yes** 1998, 1999, 2000, 2001, 2002 Metro All engines Transmissions: All	**Multiple Engine Misfire Detected Conditions:** Engine speed less than 6000 rpm (± 200 rpm), ECT sensor from 14-230ºF, IAT sensor from 14-158ºF, BARO sensor over 75 kPa, MAP sensor stable (± 1.3 kPa), change in TP sensor less than 1.9º in 16 firing events, and the PCM detected the misfire rate in two or more cylinders was too high during the 200 engine revolution test, or the misfire rate was too high during the 1000 engine revolution test. *Note: If the misfire is severe, the MIL will flash on/off on the 1st trip!* **Possible Causes** ● Air leak in the intake manifold, or in the EGR or PCM system ● Base engine mechanical fault affecting 2 or more cylinders ● Erratic or interrupted CKP or CMP sensor signals ● Fuel delivery component fault affecting 2 or more cylinders (i.e., a contaminated, dirty or sticking fuel injector) ● Ignition system problem (coil or plug) affecting over 2 cylinders
DTC P0301 **2T MISFIRE, MIL: Yes** 1998, 1999, 2000, 2001, 2002 Metro All engines Transmissions: All	**Cylinder 1 Misfire Detected Conditions:** Engine speed less than 6000 rpm (± 200 rpm), ECT sensor from 14-230ºF, IAT sensor from 14-158ºF, BARO sensor over 75 kPa, MAP sensor stable (± 1.3 kPa) in 60 firing events, change in TP sensor less than 1.9º in 16 firing events, and the PCM detected the misfire rate in Cylinder 1 was too high during the 200 engine revolution test, or the misfire rate was too high in the 1000 engine revolution test. *Note: If the misfire is severe, the MIL will flash on/off on the 1st trip!* **Possible Causes** ● Air leak in the intake manifold, or in the EGR or PCM system ● Base engine mechanical fault that affects only Cylinder 1 ● Fuel delivery component fault that affects only Cylinder 1 (i.e., a contaminated, dirty or sticking fuel injector) ● Ignition system fault (coil or plug) that affects only Cylinder 1
DTC P0302 **2T MISFIRE, MIL: Yes** 1998, 1999, 2000, 2001, 2002 Metro All engines Transmissions: All	**Cylinder 2 Misfire Detected Conditions:** Engine speed less than 6000 rpm (± 200 rpm), ECT sensor from 14-230ºF, IAT sensor from 14-158ºF, BARO sensor over 75 kPa, MAP sensor stable (± 1.3 kPa) in 60 firing events, change in TP sensor less than 1.9º in 16 firing events, and the PCM detected the misfire rate in Cylinder 2 was too high during the 200 engine revolution test, or the misfire rate was too high in the 1000 engine revolution test. *Note: If the misfire is severe, the MIL will flash on/off on the 1st trip!* **Possible Causes** ● Air leak in the intake manifold, or in the EGR or PCM system ● Base engine mechanical fault that affects only Cylinder 2 ● Fuel delivery component fault that affects only Cylinder 2 (i.e., a contaminated, dirty or sticking fuel injector) ● Ignition system fault (coil or plug) that affects only Cylinder 2
DTC P0303 **2T MISFIRE, MIL: Yes** 1998, 1999, 2000, 2001, 2002 Metro All engines Transmissions: All	**Cylinder 3 Misfire Detected Conditions:** Engine speed less than 6000 rpm (± 200 rpm), ECT sensor from 14-230ºF, IAT sensor from 14-158ºF, BARO sensor over 75 kPa, MAP sensor stable (± 1.3 kPa) in 60 firing events, change in TP sensor less than 1.9º in 16 firing events, and the PCM detected the misfire rate in Cylinder 3 was too high during the 200 engine revolution test, or the misfire rate was too high in the 1000 engine revolution test. *Note: If the misfire is severe, the MIL will flash on/off on the 1st trip!* **Possible Causes** ● Air leak in the intake manifold, or in the EGR or PCM system ● Base engine mechanical fault that affects only Cylinder 3 ● Fuel delivery component fault that affects only Cylinder 3 (i.e., a contaminated, dirty or sticking fuel injector) ● Ignition system fault (coil or plug) that affects only Cylinder 3
DTC P0304 **2T MISFIRE, MIL: Yes** 1998, 1999, 2000, 2001, 2002 Metro Engines: 1.3L VIN 2 Transmissions: All	**Cylinder 4 Misfire Detected Conditions:** Engine speed less than 6000 rpm (± 200 rpm), ECT sensor from 14-230ºF, IAT sensor from 14-158ºF, BARO sensor over 75 kPa, MAP sensor stable (± 1.3 kPa) in 60 firing events, change in TP sensor less than 1.9º in 16 firing events, and the PCM detected the misfire rate in Cylinder 1 was too high during the 200 engine revolution test, or the misfire rate was too high in the 1000 engine revolution test. *Note: If the misfire is severe, the MIL will flash on/off on the 1st trip!* *Possible Causes* ● Air leak in the intake manifold, or in the EGR or PCM system ● Base engine mechanical fault that affects only Cylinder 4 ● Fuel delivery component fault that affects only Cylinder 4 (i.e., a contaminated, dirty or sticking fuel injector) ● Ignition system fault (coil or plug) that affects only Cylinder 4

OBD II Trouble Code List (P0xxx Codes)

DTC	Trouble Code Title, Conditions & Possible Causes
DTC P0300 **2T MISFIRE, MIL: Yes** 1998, 1999, 2000, 2001, 2002 Prizm All engines Transmissions: All	**Multiple Cylinder Misfire Detected Conditions:** Engine started, engine speed between 200-4000 rpm, ECT sensor from 20-254ºF, BARO sensor more than 75 kPa, TP angle change less than 1.9º in 16 firing events, and the PCM detected the misfire rate during 200 or 1000 engine revolutions was more than a specified value in two or more cylinders during the Misfire test. *Note: If the misfire is severe, the MIL will flash on/off on the 1st trip!* **Possible Causes** ● Air leak in the intake manifold, or in the EGR or PCM system ● Base engine mechanical fault that affects only one cylinder ● Fuel delivery component fault that affects only one cylinder (i.e., a contaminated, dirty or sticking fuel injector) ● Ignition system problem (coil or plug) that affects one cylinder
DTC P0301 **2T MISFIRE, MIL: Yes** 1998, 1999, 2000, 2001, 2002 Prizm All engines Transmissions: All	**Cylinder 1 Misfire Detected Conditions:** Engine started, engine speed between 200-4000 rpm, ECT sensor from 20-254ºF, BARO sensor more than 75 kPa, system voltage from 9-17v, and the PCM detected the misfire rate during the 200 revolution test or during the 1000 engine revolution test was more than a specified value in Cylinder 1 during the Misfire Monitor test. *Note: If the misfire is severe, the MIL will flash on/off on the 1st trip!* **Possible Causes** ● Air leak in the intake manifold, or in the EGR or PCM system ● Base engine mechanical fault that affects only Cylinder 1 ● Fuel delivery component fault that affects only Cylinder 1 (i.e., a contaminated, dirty or sticking fuel injector) ● Ignition system fault (coil or plug) that affects only Cylinder 1
DTC P0302 **2T MISFIRE, MIL: Yes** 1998, 1999, 2000, 2001, 2002 Prizm All engines Transmissions: All	**Cylinder 2 Misfire Detected Conditions:** Engine started, engine speed between 200-4000 rpm, ECT sensor from 20-254ºF, BARO sensor more than 75 kPa, system voltage from 9-17v, and the PCM detected the misfire rate during the 200 revolution test or during the 1000 engine revolution test was more than a specified value in Cylinder 2 during the Misfire Monitor test. *Note: If the misfire is severe, the MIL will flash on/off on the 1st trip!* **Possible Causes** ● Air leak in the intake manifold, or in the EGR or PCM system ● Base engine mechanical fault that affects only Cylinder 2 ● Fuel delivery component fault that affects only Cylinder 2 (i.e., a contaminated, dirty or sticking fuel injector) ● Ignition system fault (coil or plug) that affects only Cylinder 2
DTC P0303 **2T MISFIRE, MIL: Yes** 1998, 1999, 2000, 2001, 2002 Prizm All engines Transmissions: All	**Cylinder 3 Misfire Detected Conditions:** Engine started, engine speed between 200-4000 rpm, ECT sensor from 20-254ºF, BARO sensor more than 75 kPa, system voltage from 9-17v, and the PCM detected the misfire rate during the 200 revolution test or during the 1000 engine revolution test was more than a specified value in Cylinder 3 during the Misfire Monitor test. *Note: If the misfire is severe, the MIL will flash on/off on the 1st trip!* **Possible Causes** ● Air leak in the intake manifold, or in the EGR or PCM system ● Base engine mechanical fault that affects only Cylinder 3 ● Fuel delivery component fault that affects only Cylinder 3 (i.e., a contaminated, dirty or sticking fuel injector) ● Ignition system fault (coil or plug) that affects only Cylinder 3
DTC P0304 **2T MISFIRE, MIL: Yes** 1998, 1999, 2000, 2001, 2002 Prizm All engines Transmissions: All	**Cylinder 4 Misfire Detected Conditions:** Engine started, engine speed between 200-4000 rpm, ECT sensor from 20-254ºF, BARO sensor more than 75 kPa, system voltage from 9-17v, and the PCM detected the misfire rate during the 200 revolution test or during the 1000 engine revolution test was more than a specified value in Cylinder 4 during the Misfire Monitor test. *Note: If the misfire is severe, the MIL will flash on/off on the 1st trip!* **Possible Causes** ● Air leak in the intake manifold, or in the EGR or PCM system ● Base engine mechanical fault that affects only Cylinder 4 ● Fuel delivery component fault that affects only Cylinder 4 (i.e., a contaminated, dirty or sticking fuel injector) ● Ignition system fault (coil or plug) that affects only Cylinder 4

OBD II Trouble Code List (P0xxx Codes)

DTC	Trouble Code Title, Conditions & Possible Causes
DTC P0300 **2T MISFIRE, MIL: Yes** 1998, 1999, 2000, 2001, 2002 Tracker All engines Transmissions: All	**Multiple Engine Misfire Detected Conditions:** Engine started, engine speed less than 4000 rpm, ECT and IAT sensors more than 18ºF, BARO sensor more than 75 kPa, Fuel Level over 15%, TP angle less than 1º in 10 ms, and the PCM detected the misfire rate during the 200 or 1000 engine revolution test was too high in two or more cylinders for 5 seconds after engine startup, or it was too high for one second of fuel shutoff. *Note: If the misfire is severe, the MIL will flash on/off on the 1st trip!* **Possible Causes** • Air leak in the intake manifold, or in the EGR or PCM system • Base engine mechanical fault affecting 2 or more cylinders • Erratic or interrupted CKP or CMP sensor signals • Fuel delivery component fault affecting 2 or more cylinders (i.e., a contaminated, dirty or sticking fuel injector) • Ignition system problem (coil or plug) affecting over 2 cylinders
DTC P0301 **2T MISFIRE, MIL: Yes** 1998, 1999, 2000, 2001, 2002, 2003, 2004, 2005 Tracker All engines Transmissions: All	**Cylinder 1 Misfire Detected Conditions:** Engine started, ECT and IAT sensors over 18ºF, engine speed less than 4000 rpm, BARO sensor over 75 kPa, Fuel Level over 15%, TP angle change less than 1º in 10 ms, and the PCM detected the misfire rate in the 200 or 1000 engine revolution test was too high in Cylinder 1, conditions met for 1-5 seconds from fuel shutoff point. *Note: If the misfire is severe, the MIL will flash on/off on the 1st trip!* **Possible Causes** • Air leak in the intake manifold, or in the EGR or PCM system • Base engine mechanical fault that affects only Cylinder 1 • Fuel delivery component fault that affects only Cylinder 1 (i.e., a contaminated, dirty or sticking fuel injector) • Ignition system fault (coil or plug) that affects only Cylinder 1
DTC P0302 **2T MISFIRE, MIL: Yes** 1998, 1999, 2000, 2001, 2002, 2003, 2004, 2005 Tracker All engines Transmissions: All02.bmp ; Tracker	**Cylinder 2 Misfire Detected Conditions:** Engine started, ECT and IAT sensors over 18ºF, engine speed less than 4000 rpm, BARO sensor over 75 kPa, Fuel Level over 15%, TP angle change less than 1º in 10 ms, and the PCM detected the misfire rate in the 200 or 1000 engine revolution test was too high in Cylinder 2, conditions met for 1-5 seconds from fuel shutoff point. *Note: If the misfire is severe, the MIL will flash on/off on the 1st trip!* **Possible Causes** • Air leak in the intake manifold, or in the EGR or PCM system • Base engine mechanical fault that affects only Cylinder 2 • Fuel delivery component fault that affects only Cylinder 2 (i.e., a contaminated, dirty or sticking fuel injector) • Ignition system fault (coil or plug) that affects only Cylinder 2
DTC P0303 **2T MISFIRE, MIL: Yes** 1998, 1999, 2000, 2001, 2002, 2003, 2004, 2005 Tracker All engines Transmissions: All	**Cylinder 3 Misfire Detected Conditions:** Engine started, ECT and IAT sensors over 18ºF, engine speed less than 4000 rpm, BARO sensor over 75 kPa, Fuel Level over 15%, TP angle change less than 1º in 10 ms, and the PCM detected the misfire rate in the 200 or 1000 engine revolution test was too high in Cylinder 3, conditions met for 1-5 seconds from fuel shutoff point. *Note: If the misfire is severe, the MIL will flash on/off on the 1st trip!* **Possible Causes** • Air leak in the intake manifold, or in the EGR or PCM system • Base engine mechanical fault that affects only Cylinder 3 • Fuel delivery component fault that affects only Cylinder 3 (i.e., a contaminated, dirty or sticking fuel injector) • Ignition system fault (coil or plug) that affects only Cylinder 3
DTC P0304 **2T MISFIRE, MIL: Yes** 1998, 1999, 2000, 2001, 2002, 2003, 2004, 2005 Tracker All engines Transmissions: All	**Cylinder 4 Misfire Detected Conditions:** Engine started, ECT and IAT sensors over 18ºF, engine speed less than 4000 rpm, BARO sensor over 75 kPa, Fuel Level over 15%, TP angle change less than 1º in 10 ms, and the PCM detected the misfire rate in the 200 or 1000 engine revolution test was too high in Cylinder 4, conditions met for 1-5 seconds from fuel shutoff point. *Note: If the misfire is severe, the MIL will flash on/off on the 1st trip!* **Possible Causes** • Air leak in the intake manifold, or in the EGR or PCM system • Base engine mechanical fault that affects only Cylinder 4 • Fuel delivery component fault that affects only Cylinder 4 (i.e., a contaminated, dirty or sticking fuel injector) • Ignition system fault (coil or plug) that affects only Cylinder 4

OBD II Trouble Code List (P0xxx Codes)

DTC	Trouble Code Title, Conditions & Possible Causes
DTC P0305 **2T MISFIRE, MIL: Yes** 2001, 2002, 2003, 2004, 2005 Tracker 2.5L VIN 4 engine Transmissions: All	**Cylinder 5 Misfire Detected Conditions:** Engine started, ECT and IAT sensors over 18°F, engine speed less than 4000 rpm, BARO sensor over 75 kPa, Fuel Level over 15%, TP angle change less than 1° in 10 ms, and the PCM detected the misfire rate in the 200 or 1000 engine revolution test was too high in Cylinder 5, conditions met for 1-5 seconds from fuel shutoff point. Note: If the misfire is severe, the MIL will flash on/off on the 1st trip! **Possible Causes** ● Air leak in the intake manifold, or in the EGR or PCM system ● Base engine mechanical fault that affects only Cylinder 5 ● Fuel delivery component fault that affects only Cylinder 5 (i.e., a contaminated, dirty or sticking fuel injector) ● Ignition system fault (coil or plug) that affects only Cylinder 5
DTC P0306 **2T MISFIRE, MIL: Yes** 2001, 2002, 2003, 2004, 2005 Tracker 2.5L VIN 4 engine Transmissions: All	**Cylinder 6 Misfire Detected Conditions:** Engine started, ECT and IAT sensors over 18°F, engine speed less than 4000 rpm, BARO sensor over 75 kPa, Fuel Level over 15%, TP angle change less than 1° in 10 ms, and the PCM detected the misfire rate in the 200 or 1000 engine revolution test was too high in Cylinder 6, conditions met for 1-5 seconds from fuel shutoff point. Note: If the misfire is severe, the MIL will flash on/off on the 1st trip! **Possible Causes** ● Air leak in the intake manifold, or in the EGR or PCM system ● Base engine mechanical fault that affects only Cylinder 6 ● Fuel delivery component fault that affects only Cylinder 6 (i.e., a contaminated, dirty or sticking fuel injector) ● Ignition system fault (coil or plug) that affects only Cylinder 6
DTC P0325 **1T CCM, MIL: Yes** 1998, 1999, 2000, 2001, 2002 Prizm All engines Transmissions: All	**Knock Sensor 1 Circuit Malfunction Conditions:** Engine started, engine speed over 1200 rpm, TP sensor and VSS signals indicate the vehicle is accelerating, and the PCM detected the Knock sensor 1 signal amplitude was less than specifications for over 20 seconds during the CCM test. **Possible Causes** ● Knock sensor signal circuit is open or shorted to ground ● Knock sensor signal circuit is shorted to VREF or system power ● Knock sensor is damaged or has failed ● PCM has failed
DTC P0327 **1T CCM, MIL: Yes** 2001, 2002, 2003, 2004, 2005 Tracker 2.0L VIN C engine Transmissions: All	**Knock Sensor Circuit Low Input Conditions:** Engine started, engine running, and the PCM detected the Knock sensor signal was less than 0.90v for 10 seconds during the test. **Possible Causes** ● Knock sensor connector is damaged or shorted to ground ● Knock sensor signal circuit is shorted to ground ● Knock sensor is damaged or has failed ● PCM has failed
DTC P0328 **1T CCM, MIL: Yes** 2001, 2002, 2003, 2004, 2005 Tracker 2.0L VIN C engine Transmissions: All	**Knock Sensor Circuit High Input Conditions:** Engine started, engine running, and the PCM detected the Knock sensor signal was more than 3.98v for 10 seconds during the test. **Possible Causes** ● Knock sensor connector is damaged, open or shorted to power ● Knock sensor signal circuit is open or shorted to power ● Knock sensor is damaged or has failed ● PCM has failed
DTC P0335 **1T CCM, MIL: Yes** 1998, 1999, 2000, 2001 Metro All engines Transmissions: All	**Crankshaft Position Sensor Circuit Malfunction Conditions:** Engine started, engine speed over 700 rpm, and the PCM did not receive any CKP sensor signals during 2 engine revolutions. **Possible Causes** ● CKP sensor connector is damaged, open or shorted ● CKP (Magnetic) sensor signal (+) or (-) circuit is open, shorted to ground or to power ● CKP sensor is damaged or has failed ● PCM has failed
DTC P0335 **1T CCM, MIL: Yes** 1998, 1999, 2000, 2001, 2002 Prizm All engines Transmissions: All	**Crankshaft Position Sensor Circuit Malfunction Conditions:** Engine cranking for 5 seconds, or engine speed over 600 rpm, and the PCM did not detect any CKP signals during 2 engine revolutions. **Possible Causes** ● CKP sensor connector is damaged, open or shorted ● CKP (Magnetic) sensor signal (+) or (-) circuit is open, shorted to ground or to power ● CKP sensor is damaged or has failed ● PCM has failed

OBD II Trouble Code List (P0xxx Codes)

DTC	Trouble Code Title, Conditions & Possible Causes
DTC P0335 **1T CCM, MIL: Yes** 1998, 1999, 2000, 2001, 2002, 2003, 2004, 2005 Tracker All engines Transmissions: All	**Crankshaft Position Sensor Circuit Malfunction Conditions:** Engine started, and the PCM did not detect any CKP signals after it had received 20 signal pulses from the CMP sensor. **Possible Causes** ● CKP sensor connector is damaged, open or shorted ● CKP (Magnetic) sensor signal (+) or (-) circuit is open, shorted to ground or to power ● CKP sensor is damaged or has failed ● PCM has failed
DTC P0340 **1T CCM, MIL: Yes** 1998, 1999, 2000, 2001 Metro All engines Transmissions: All	**Camshaft Position Sensor Circuit Malfunction Conditions:** Engine cranking for 2 seconds, and the PCM did not receive any CMP sensor signals, condition met for more than 2 seconds. **Possible Causes** ● CMP sensor connector is damaged, open or shorted ● CMP (Magnetic) sensor signal (+) or (-) circuit is open, shorted to ground or to power ● CMP sensor is damaged or has failed ● PCM has failed
DTC P0340 **1T CCM, MIL: Yes** 1998, 1999, 2000, 2001, 2002 Prizm All engines Transmissions: All	**Camshaft Position Sensor Circuit Malfunction Conditions:** Engine cranking or engine running, and the PCM did not receive any CMP sensor signals for 3 seconds during the CCM test. **Possible Causes** ● CMP sensor connector is damaged, open or shorted ● CMP sensor signal (+) circuit is open or shorted to ground ● CMP sensor signal (-) circuit is open or shorted to ground ● CMP sensor is damaged or has failed ● PCM has failed
DTC P0340 **1T CCM, MIL: Yes** 1998, 1999, 2000, 2001, 2002, 2003, 2004, 2005 Tracker All engines Transmissions: All	**Camshaft Position Sensor Circuit Malfunction Conditions:** Engine cranking or running; and the PCM did not detect any CMP signals after it received a Start signal for a period of 3 seconds. **Possible Causes** ● CMP (Hall) sensor signal circuit is open or shorted to ground ● CMP (Hall) sensor ground circuit is open ● CMP sensor power circuit is open (test power from MFI relay) ● CMP sensor is damaged or has failed ● PCM has failed (PCM provides a VREF signal to the sensor)
DTC P0400 **2T EGR, MIL: Yes** 1998, 1999, 2000, 2001 Metro 1.0L VIN 6 engine Transmissions: All	**EGR System Performance Conditions:** Engine runtime over 290 seconds, ECT sensor from 176-230ºF, IAT sensor over 18ºF, BARO sensor over 75 kPa, fuel level over 25%, vehicle speed over 60 mph for 5 minutes, any throttle change under 25 degrees, followed by a decel period with the throttle closed, and the PCM detected the difference in indicated intake pressure was less a specified value. **Possible Causes** ● EGR back pressure transducer is damaged or has failed ● EGR bypass valve is damaged or has failed ● EGR solenoid vacuum valve is damaged or has failed ● EGR boost sensor is damaged or has failed ● EGR valve assembly is leaking, stuck open or stuck closed ● EGR valve vacuum hose(s) lose, damaged or disconnected ● Exhaust system is clogged or restricted
DTC P0400 **2T EGR, MIL: Yes** 1998, 1999, 2000, 2001, 2002 Tracker 2.0L VIN C engine Transmissions: All	**EGR System Performance Conditions:** Engine started, ECT sensor from 158-230ºF, IAT sensor over 18ºF, BARO sensor over 75 kPa, vehicle speed over 60 mph for 7 minutes, followed by a deceleration period (brakes "off" and throttle closed") with Fuel Cutoff "on", and the PCM detected the difference in intake manifold pressure between when the EGR valve was open and closed was not within range. **Possible Causes** ● EGR stepper motor is binding, stuck partially open or closed ● EGR stepper motor is damaged or has failed ● PCM has failed
DTC P0401 **2T EGR, MIL: Yes** 2002, 2003, 2004, 2005 Tracker 2.5L VIN 4 engine Transmissions: All	**EGR System Insufficient EGR Flow Detected Conditions:** DTC P0107 and P0108 not set, ECT sensor from 158-230ºF, IAT sensor over 18ºF, BARO sensor over 75 kPa, vehicle driven at 30-40 mph at an engine speed of 2000-3000 rpm for 3 minutes, throttle valve opening constant, and the PCM detected the difference in intake manifold pressure between when the EGR valve was open and closed was not within range. **Possible Causes** ● EGR stepper motor is binding, or stuck in closed position ● EGR stepper motor is damaged or has failed ● PCM has failed

Chevrolet GEO OBD II Trouble Codes

OBD II Trouble Code List (P0xxx Codes)

DTC	Trouble Code Title, Conditions & Possible Causes
DTC P0402 **2T EGR, MIL: Yes** 2002, 2003, 2004, 2005 Tracker All engines Transmissions: All	**EGR System Excessive Flow Detected Conditions:** Engine started, ECT sensor from 158-230°F, IAT sensor over 18°F, BARO sensor over 75 kPa, vehicle speed over 60 mph for 5 minutes with the throttle open and steady, followed by a decel period with no change in the accelerator or brake switch, and the PCM detected the difference in intake manifold pressure between when the EGR valve was open and when the valve was closed was too high a value. **Possible Causes** ● Check for an obstruction or blockage in EGR valve passages ● Check the EGR valve for physical damage or carbon deposits ● EGR stepper motor circuit (one or more) is open or shorted ● EGR stepper motor is binding, stuck open or closed, or failed ● MAP sensor is out-of-calibration or "skewed" ● PCM has failed
DTC P0403 **2T CCM, MIL: Yes** 1998, 1999, 2000, 2001, 2002, 2003, 2004, 2005 Tracker 2.0L VIN C engine Transmissions: All	**EGR Solenoid Control Circuit Malfunction Conditions:** Engine started; system voltage over 10v, and the PCM detected an unexpected voltage on the EGR solenoid control circuit with the solenoid commanded "on" or "off" during the test. **Possible Causes** ● EGR solenoid control circuit is open or shorted to ground ● EGR solenoid power circuit is open (test power from MFI relay) ● EGR solenoid is damaged or has failed ● PCM has failed
DTC P0403 **2T CCM, MIL: Yes** 1998, 1999, 2000, 2001, 2002, 2003, 2004, 2005 Tracker 2.5L VIN 4 engine Transmissions: All	**EGR Solenoid Control Circuit Malfunction Conditions:** Engine started, IAT sensor over 7°F, and the PCM detected an unexpected low voltage on one of the EGR solenoid control circuits. **Possible Causes** ● EGR solenoid connector is damaged, open or shorted ● EGR solenoid control circuit is open or shorted to ground ● EGR solenoid power circuit is open (test power from MFI relay) ● EGR solenoid is damaged or has failed ● PCM has failed
DTC P0420 **2T CAT, MIL: Yes** 1998, 1999, 2000, 2001 Metro All engines Transmissions: All	**Catalyst System Low Efficiency (Bank 1) Conditions:** Engine started, ECT sensor from 158-230°F, IAT sensor from 14-122°F, BARO sensor over 75 kPa, fuel level over 25%, vehicle driven at a constant engine speed of 2300-4500 rpm for 3 minutes in closed loop with a calculated load value of 26-80%, and the PCM detected the switch rates of the rear HO2S and front HO2S were similar for 96 seconds during the test. **Possible Causes** ● Air leaks in the exhaust manifold or exhaust pipes ● Catalytic converter is contaminated, damaged or has failed ● Front HO2S is older (aged) than the rear HO2S (HO2S is lazy) ● PCM has failed
DTC P0420 **2T CAT, MIL: Yes** 1998, 1999, 2000, 2001, 2002 Prizm All engines Transmissions: All	**Catalyst System Low Efficiency (Bank 1) Conditions:** Engine started, ECT sensor from 18-230°F, IAT sensor more than 18°F, BARO sensor over 75 kPa, vehicle driven with the engine in closed loop, and the PCM detected the voltage swings (and switch rates) of the rear HO2S and front HO2S were similar for 96 seconds. **Possible Causes** ● Air leaks in the exhaust manifold or exhaust pipes ● Catalytic converter is contaminated, damaged or has failed ● Front HO2S is older (aged) than the rear HO2S (HO2S is lazy) ● PCM has failed
DTC P0420 **2T CAT, MIL: Yes** 1998, 1999, 2000, 2001, 2002, 2003, 2004, 2005 Tracker All engines Transmissions: All	**Catalyst System Low Efficiency (Bank 1) Conditions:** Engine started, ECT sensor from 18-230°F, IAT sensor over 18°F, BARO sensor over 75 kPa, Fuel Level over 15%, engine running in closed loop, and the PCM detected the high and low voltage swings of the front HO2S-11 and rear HO2S-12 were similar for 96 seconds. **Possible Causes** ● Air leaks in the exhaust manifold or exhaust pipes ● Catalytic converter is contaminated, damaged or has failed ● Front HO2S and/or rear HO2S is loose in the mounting hole ● Front HO2S is older (aged) than the rear HO2S (HO2S is lazy) ● Rear HO2S is contaminated due to poor fuel or water in the fuel ● PCM has failed

OBD II Trouble Code List (P0xxx Codes)

DTC	Trouble Code Title, Conditions & Possible Causes
DTC P0430 **2T CAT, MIL: Yes** 2001, 2002, 2003, 2004, 2005 Tracker 2.5L VIN 4 engine Transmissions: All	**Catalyst System Low Efficiency (Bank 2) Conditions:** Engine started, ECT sensor from 18-230ºF, IAT sensor over 18ºF, BARO sensor over 75 kPa, Fuel Level over 15%, engine running in closed loop, and the PCM detected the high and low voltage swings of the front HO2S-21 and rear HO2S-22 were similar for 96 seconds. **Possible Causes** ● Air leaks in the exhaust manifold or exhaust pipes ● Catalytic converter is contaminated, damaged or has failed ● Front HO2S and/or rear HO2S is loose in the mounting hole ● Front HO2S is older (aged) than the rear HO2S (HO2S is lazy) ● Rear HO2S is contaminated due to poor fuel or water in the fuel ● PCM has failed
DTC P0440 **2T EVAP, MIL: Yes** 1998, 1999, 2000, 2001 Metro All engines Transmissions: All	**EVAP System No Flow During Purge Detected Conditions:** Engine running, the DTC P0455 Diagnostic has run and passed, ECT sensor from 158-230ºF, IAT sensor from 14-158ºF, fuel level less than 75%, vehicle driven for several minutes, and the PCM detected a change in fuel tank pressure indicating the EVAP purge accumulation time was over 200 seconds (test can take 20 minutes). **Possible Causes** ● Charcoal canister is loaded with fuel or moisture ● Fuel filler cap loose, cross-threaded, incorrect part or damaged ● Fuel tank pressure sensor is damaged or has failed ● Fuel tank vapor line(s) blocked, damaged or disconnected ● Fuel Tank pressure control solenoid is damaged or has failed ● PCM has failed
DTC P0440 **2T EVAP, MIL: Yes** 1998, 1999, 2000, 2001, 2002 Prizm All engines Transmissions: A/T	**EVAP System Incorrect Purge Flow Detected Conditions:** ECT sensor from 50-95ºF and the ECT and IAT sensors within a few degrees of each other at startup (cold engine), engine runtime over 20 minutes, BARO sensor over 75 kPa, fuel level from 25-75%, ECT sensor over 158ºF during testing, and the PCM detected a change in fuel tank pressure with the EVAP purge and Vent solenoid both closed during the test. **Possible Causes** ● Charcoal canister is loaded with fuel or moisture ● Fuel filler cap loose, cross-threaded, incorrect part or damaged ● Fuel tank pressure sensor is damaged or has failed ● Fuel tank vapor line(s) blocked, damaged or disconnected ● Purge or Vent solenoid control circuit open or shorted to ground ● PCM has failed
DTC P0440 **2T EVAP, MIL: Yes** 1998, 1999, 2000, 2001 Tracker All engines Transmissions: All	**EVAP System Incorrect Purge Flow Detected Conditions:** Engine started, ECT sensor from 158-230ºF, IAT sensor from 19-158ºF, BARO sensor over 75 kPa, Fuel Level less than 75%, engine running in closed loop at a speed from 1000-3000 rpm, and the PCM detected a change in fuel tank pressure indicating the EVAP purge accumulation time took over 6 minutes (test can take 20 minutes). **Possible Causes** ● Charcoal canister is loaded with fuel or moisture ● Fuel filler cap loose, cross-threaded, incorrect part or damaged ● Fuel tank pressure sensor is damaged or has failed ● Fuel tank vapor line(s) blocked, damaged or disconnected ● Purge solenoid valve is damaged or has failed ● PCM has failed
DTC P0440 **2T EVAP, MIL: Yes** 1998, 1999, 2000, 2001 Tracker All engines Transmissions: M/T	**EVAP System Incorrect Purge Flow Detected Conditions:** Engine started, ECT sensor from 158-230ºF, IAT sensor from 19-158ºF, BARO sensor over 75 kPa, Fuel Level less than 75%, engine running in closed loop at a speed from 1000-3500 rpm, and the PCM detected a change in fuel tank pressure indicating the EVAP purge accumulation time took over 6 minutes (test can take 20 minutes). **Possible Causes** ● Charcoal canister is loaded with fuel or moisture ● Fuel filler cap loose, cross-threaded, incorrect part or damaged ● Fuel tank pressure sensor is damaged or has failed ● Fuel tank vapor line(s) blocked, damaged or disconnected ● Purge solenoid valve is damaged or has failed ● PCM has failed

OBD II Trouble Code List (P0xxx Codes)

DTC	Trouble Code Title, Conditions & Possible Causes
DTC P0441 **2T EVAP, MIL: Yes** 1998, 1999, 2000, 2001, 2002 Prizm All engines Transmissions: All	**EVAP System Performance Conditions:** Cold engine startup (ECT and IAT sensors from 50-95ºF and within a few degrees of each other at startup), ECT sensor more than 158ºF during testing, fuel level at 25-75%, then while the vehicle was driven under city traffic conditions, the PCM detected the pressure indicated in the canister did not drop when the Purge solenoid was opened during the test. **Possible Causes** ● Charcoal canister is loaded with fuel or moisture ● Fuel filler cap loose, cross-threaded, incorrect part or damaged ● Fuel tank pressure sensor is damaged or has failed ● Fuel tank vapor line(s) blocked, damaged or disconnected ● Purge or Vent solenoid control circuit open or shorted to ground ● PCM has failed
DTC P0441 **2T EVAP, MIL: Yes** 2002, 2003, 2004, 2005 Tracker 2.0L VIN C engine Transmissions: All	**EVAP System Performance Conditions:** Engine started, IAT sensor over 18ºF at startup, ECT sensor from 18-230ºF, BARO sensor over 75 kPa, Fuel Level less than 75%, vehicle speed over 35 mph for 20 minutes, followed by an idle period of over 1 minute, the PCM detected the Idle air control (IAC) and engine load values did not change as expected during purge routine. **Possible Causes** ● Charcoal canister is loaded with fuel or moisture ● Fuel filler cap loose, cross-threaded, incorrect part or damaged ● Fuel tank pressure sensor is damaged or has failed ● Fuel tank vapor line(s) blocked, damaged or disconnected ● Purge solenoid valve is damaged or has failed ● PCM has failed
DTC P0442 **2T EVAP, MIL: Yes** 2001, 2002 Prizm All engines Transmissions: All	**EVAP System Small Leak (0.040") Detected Conditions:** Cold engine startup (ECT and IAT sensors from 50-95ºF and within a few degrees of each other at startup), ECT sensor more than 158ºF during testing, fuel level at 25-75%, vehicle driven for over 20 minutes, then with the EVAP purge and vent solenoids both commanded "on", the PCM detected a change in fuel tank pressure during the EVAP System Monitor test. *Note: This test can take up to 20 minutes to run and complete.* **Possible Causes** ● Canister Purge valve is damaged, leaking or it has failed ● Charcoal canister is loaded with fuel or moisture ● Fuel filler cap loose, cross-threaded, incorrect part or damaged ● Fuel tank is cracked (leaking), or a leak exists in the 'O' ring ● Fuel tank pressure sensor is damaged or has failed ● Fuel vapor line(s), fuel pipes or hoses damaged or leaking ● PCM has failed
DTC P0442 **2T EVAP, MIL: Yes** 2001, 2002, 2003, 2004, 2005 Tracker 2.0L VIN C engine Transmissions: All	**EVAP System Small Leak (0.040") Detected Conditions:** Engine started, ECT sensor from 18-230ºF, IAT sensor over 18ºF, BARO sensor over 75 kPa, fuel level less than 75%, engine running in closed loop with a VSS signal detected, EVAP Purge and Vent solenoids closed, and the PCM detected the EVAP system was not able to maintain the specified vacuum pressure with the system sealed. *Note: This test can take up to 20 minutes to run and complete.* **Possible Causes** ● Canister Purge valve is damaged, leaking or it has failed ● Charcoal canister is loaded with fuel or moisture ● Fuel filler cap loose, cross-threaded, incorrect part or damaged ● Fuel tank is cracked (leaking), or a leak exists in the 'O' ring ● Fuel tank pressure sensor is damaged or has failed ● Fuel vapor line(s), fuel pipes or hoses damaged or leaking ● PCM has failed
DTC P0442 **2T EVAP, MIL: Yes** 2001, 2002, 2003, 2004, 2005 Tracker 2.5L VIN 4 engine Transmissions: All	**EVAP System Small Leak (0.040") Detected Conditions:** DTC P0444, P0445, P0447, P0448, P0451, P0452, P0453, P0455 and P0496 not set, ECT sensor from 18-230ºF, IAT sensor from 14-144ºF, BARO sensor over 75 kPa, engine running in closed loop, Fuel Level less than 75%, TP sensor signal steady, EVAP Purge and Vent solenoids closed, and the PCM detected the increase in the internal pressure of the fuel tank was more than a specified value. *This test can take up to 20 minutes to run and complete.* **Possible Causes** ● Canister Purge valve is damaged, leaking or it has failed ● Charcoal canister is loaded with fuel or moisture ● Fuel filler cap loose, cross-threaded, incorrect part or damaged ● Fuel tank is cracked (leaking), or a leak exists in the 'O' ring ● Fuel tank pressure sensor is damaged or has failed ● Fuel vapor line(s), fuel pipes or hoses damaged or leaking ● PCM has failed

OBD II Trouble Code List (P0xxx Codes)

DTC	Trouble Code Title, Conditions & Possible Causes
DTC P0443 **2T CCM, MIL: Yes** 1998 Metro, Tracker All engines Transmissions: All	**EVAP System Incorrect Purge Flow Detected Conditions:** ECT sensor from 158-230ºF, IAT sensor from 14-122ºF, BARO sensor over 75 kPa, fuel level from 25-75%, engine started and running at idle in closed loop, and the PCM detected the engine speed changed too much after the Purge solenoid was cycled off and on at a duty cycle of over 50% during a 20 second test period. **Possible Causes** ● EVAP purge solenoid control circuit open or shorted to ground ● EVAP purge solenoid power (B+) circuit open ● EVAP purge solenoid is damaged or has failed ● PCM has failed
DTC P0443 **2T CCM, MIL: Yes** 1999, 2000, 2001, 2002, 2003, 2004, 2005 Tracker 1.6L VIN 6, 2.0L VIN C engines Transmissions: All	**EVAP Purge Solenoid Circuit Malfunction Conditions:** Engine started; and the PCM detected an unexpected low or high voltage condition on the EVAP purge control circuit for 2 seconds. **Possible Causes** ● EVAP purge solenoid connector is damaged, open or shorted ● EVAP purge solenoid control circuit open or shorted to ground ● EVAP purge solenoid power circuit is open (test from the relay) ● EVAP purge solenoid is damaged or has failed ● PCM has failed
DTC P0444 **2T CCM, MIL: Yes** 2001, 2002, 2003, 2004, 2005 Tracker 2.5L VIN 4 engine Transmissions: All	**EVAP Purge Solenoid Circuit Low Input Conditions:** Engine started, EVAP solenoid commanded "off", and the PCM detected the EVAP purge solenoid circuit was near 0v 2 seconds. **Possible Causes** ● EVAP purge solenoid connector is damaged or shorted ● EVAP purge solenoid control circuit shorted to ground ● EVAP purge solenoid power (B+) circuit open ● EVAP purge solenoid is damaged or has failed ● PCM has failed
DTC P0445 **2T CCM, MIL: Yes** 2001, 2002, 2003, 2004, 2005 Tracker 2.5L VIN 4 engine Transmissions: All	**EVAP Purge Solenoid Circuit High Input Conditions:** Engine started, ECT sensor more than 176ºF, IAT sensor over 14ºF, BARO sensor over 75 kPa, Purge solenoid commanded "on", and the PCM detected the EVAP purge solenoid control signal voltage was near system voltage (12-14v) for 2 seconds. **Possible Causes** ● EVAP purge solenoid control circuit shorted to power (B+) ● EVAP purge solenoid is damaged or has failed ● PCM has failed
DTC P0446 **2T EVAP, MIL: Yes** 1998, 1999, 2000, 2001, 2002 Prizm All engines Transmissions: All	**EVAP System Vent System Performance Conditions:** Engine running in closed loop, ECT sensor from 18-230ºF, BARO sensor over 75 kPa, IAT sensor over 18ºF at startup, and the PCM detected the canister pressure was out of range with the Pressure Switching solenoid commanded "on" and "off", or the pressure in the canister was equal to atmospheric pressure after a purge cutoff event (purge solenoid "off"). **Possible Causes** ● EVAP vent solenoid control circuit is open or shorted to ground ● EVAP vent solenoid is stuck partially open ● EVAP vent solenoid is damaged or has failed ● PCM has failed
DTC P0446 **2T EVAP, MIL: Yes** 1998, 1999, 2000, 2001, 2002, 2003, 2004, 2005 Tracker All engines Transmissions: All	**EVAP System Vent System Performance Conditions:** Engine started, ECT sensor from 18-230ºF, BARO sensor over 75 kPa, IAT sensor over 18ºF at startup, and the PCM detected the FTP sensor indicated a lower than expected pressure with the EVAP vent valve commanded "off". **Possible Causes** ● EVAP vent solenoid connector is damaged, open or shorted ● EVAP vent solenoid control circuit is open or shorted to ground ● EVAP vent solenoid is stuck partially open ● EVAP vent solenoid is damaged or has failed ● PCM has failed
DTC P0446 **2T CCM, MIL: Yes** 2001, 2002, 2003, 2004, 2005 Tracker 2.5L VIN 4 engine Transmissions: All	**EVAP Vent Solenoid Circuit Low Input Conditions:** Engine started, EVAP vent solenoid commanded "off", and the PCM detected the voltage at the EVAP vent solenoid control circuit was near 0.0v for 2 seconds. **Possible Causes** ● EVAP vent solenoid connector is damaged or shorted ● EVAP vent solenoid control circuit shorted to ground ● EVAP vent solenoid power (B+) circuit open ● EVAP vent solenoid is damaged or has failed ● PCM has failed

OBD II Trouble Code List (P0xxx Codes)

DTC	Trouble Code Title, Conditions & Possible Causes
DTC P0448 **2T CCM, MIL: Yes** 2001, 2002, 2003, 2004, 2005 Tracker 2.5L VIN 4 engine Transmissions: All	**EVAP Vent Solenoid Circuit High Input Conditions:** Engine started, ECT sensor over 176ºF, IAT sensor over 14ºF, BARO sensor over 75 kPa, EVAP Purge solenoid commanded "on", and the PCM detected a high voltage on the solenoid control circuit. **Possible Causes** ● EVAP vent solenoid connector is open or shorted to power ● EVAP vent solenoid control circuit shorted to power (B+) ● EVAP vent solenoid is damaged or has failed ● PCM has failed
DTC P0449 **2T CCM, MIL: Yes** 2002, 2003, 2004, 2005 Tracker 2.0L VIN C engine Transmissions: All	**EVAP Vent System Performance Conditions:** Engine started, ECT sensor from 18-230ºF, IAT sensor more than 18ºF, and the PCM detected an unexpected high voltage condition on the EVAP vent valve control circuit. **Possible Causes** ● EVAP vent solenoid control circuit is open or shorted to power ● EVAP vent solenoid control circuit shorted to power (B+) ● EVAP vent solenoid is damaged or has failed ● PCM has failed
DTO P0430 **2T CCM, MIL: Yes** 1999, 2000, 2001 Metro All engines Transmissions: All	**Fuel Tank Pressure Sensor Circuit Malfunction Conditions:** Engine started, VSS over 10 mph for 1 minute, Fuel Level from 25-75%, and the PCM detected the FTP sensor was below 0.5v, or the FTP signal was above 4.50v for 500 ms. **Possible Causes** ● FTP sensor circuit is open or shorted to ground ● FTP sensor ground or power circuit is open ● FTP sensor is damaged or has failed ● PCM has failed
DTC P0450 **2T CCM, MIL: Yes** 1999, 2000, 2001, 2002 Prizm All engines Transmissions: All	**Fuel Tank Pressure Sensor Circuit Malfunction Conditions:** Engine started, engine runtime over 10 seconds, and the PCM detected the FTP sensor was less than 16.3" H2O (-30.48 mm Hg), or the FTP signal was over 8.16" H2O (15.24 mm Hg) for 7 seconds. **Possible Causes** ● FTP sensor circuit is open, shorted to ground or to power ● FTP sensor ground circuit is open ● FTP sensor power circuit is open ● FTP sensor is damaged or it has failed ● PCM has failed
DTC P0450 **2T CCM, MIL: Yes** 1999, 2000, 2001 Tracker 1.6L VIN 6, 2.0L VIN C engines Transmissions: All	**Fuel Tank Pressure Sensor Circuit Malfunction Conditions:** Engine started, ECT sensor from 18-230ºF, IAT sensor from 18-104ºF, BARO sensor over 75 kPa, Fuel Level from 25-75%, and the PCM detected the FTP sensor signal was less than 0.5v or more than 4.50v for a period of 500 ms. **Possible Causes** ● FTP sensor signal circuit is open or shorted to ground ● FTP sensor ground circuit is open ● FTP sensor VREF circuit is open ● FTP sensor is damaged or it has failed ● PCM has failed
DTC P0450 **2T CCM, MIL: Yes** 2001 Tracker 2.5L VIN 4 engine Transmissions: All	**Fuel Tank Pressure Sensor Circuit Malfunction Conditions:** Engine started, ECT sensor from 18-230ºF, IAT sensor from 18-104ºF, BARO sensor over 75 kPa, Fuel Level from 25-75%, and the PCM detected the FTP sensor signal was less than 0.5v, or that it was more than 4.50v for several seconds. **Possible Causes** ● FTP sensor signal circuit is open or shorted to ground ● FTP sensor ground circuit is open ● FTP sensor VREF circuit is open ● FTP sensor is damaged or it has failed ● PCM has failed
DTC P0451 **2T CCM, MIL: Yes** 1998, 1999, 2000, 2001 Metro All engines Transmissions: All	**Fuel Tank Pressure Sensor Range/Performance Conditions:** Engine runtime over 2 minutes, ECT sensor from 158-230ºF, IAT sensor from 14-122ºF, BARO sensor over 75 kPa, Fuel Level from 25-75%, and the PCM detected the fuel tank pressure was lower than a certain value after completing the EVAP Leak Check test. **Possible Causes** ● Fuel tank pressure sensor vent hole is clogged or restricted ● Fuel tank pressure sensor is damaged or out-of-calibration ● PCM has failed

OBD II Trouble Code List (P0xxx Codes)

DTC	Trouble Code Title, Conditions & Possible Causes
DTC P0451 **2T CCM, MIL: Yes** 1999, 2000, 2001, 2002 Prizm All engines Transmissions: All	**Fuel Tank Pressure Sensor Range/Performance Conditions:** Engine started, vehicle not moving, EVAP Vent Pressure solenoid commanded "off", and with the PCM detected the FTP sensor voltage shifted rapidly from inside to outside of a specified range during the CCM test period. **Possible Causes** ● Fuel tank pressure sensor connector is damaged or shorted ● Fuel tank pressure sensor vacuum hoses loose or damaged ● Fuel tank pressure sensor is damaged or out-of-calibration ● PCM has failed
DTC P0451 **2T CCM, MIL: Yes** 1999, 2000, 2001, 2002, 2003, 2004, 2005 Tracker 2.0L VIN C engine Transmissions: All	**Fuel Tank Pressure Sensor Range/Performance Conditions:** DTC P1450 and P1451 not set, Engine runtime at idle 2 minutes, ECT sensor from 18-230ºF, IAT sensor from 7-158ºF, BARO sensor over 75 kPa, Fuel Level from 25-75%, then with the FTP sensor indicating a value more than or less than atmospheric pressure, the PCM detected the fuel tank pressure was less than -14.05 in H2O (-26.10 mm Hg), or over 26.40 in H2O (51.75 mm Hg) for 5 seconds. **Possible Causes** ● Fuel tank pressure sensor vacuum hoses loose or damaged ● Fuel tank pressure sensor is damaged or out-of-calibration ● PCM has failed
DTC P0451 **2T CCM, MIL: Yes** 2001, 2002, 2003, 2004, 2005 Tracker 2.5L VIN 4 engine Transmissions: All	**Fuel Tank Pressure Sensor Range/Performance Conditions:** DTC P0452, DTC P0453, DTC P2227, DTC P2228, and DTC P2229 not set, engine runtime at idle 2 minutes, ECT sensor at 18-230ºF, IAT sensor from 7-158ºF, BARO sensor over 75 kPa, Fuel Level from 25-75%, then with the FTP sensor indicating a value more than or less than atmospheric pressure, the PCM detected the fuel tank pressure was below -14.05 in H2O (-26.10 mm Hg), or that it was more than 26.40 in H2O (51.75 mm Hg) for 5 seconds. **Possible Causes** ● Fuel tank pressure sensor vacuum hoses loose or damaged ● Fuel tank pressure sensor is damaged or out-of-calibration ● PCM has failed
DTC P0452 **2T CCM, MIL: Yes** 1999, 2000, 2001, 2002, 2003, 2004, 2005 Tracker 2.0L VIN C engine Transmissions: All	**Fuel Tank Pressure Sensor Low Input Conditions:** Engine started; engine runtime at idle speed 2 minutes, IAT sensor from 7-158ºF, ECT sensor from 18-230ºF, and the PCM detected the FTP sensor signal was less than 0.5v (i.e., less than -26.5 in H2O)for 5 seconds. **Possible Causes** ● FTP sensor connector is damaged or shorted to ground ● FTP sensor signal circuit is shorted to ground ● FTP sensor power (VREF) circuit is open ● FTP sensor is damaged or has failed ● PCM has failed
DTC P0452 **2T CCM, MIL: Yes** 2001, 2002, 2003, 2004, 2005 Tracker 2.5L VIN 4 engine Transmissions: All	**Fuel Tank Pressure Sensor Low Input Conditions:** Engine started, IAT sensor over 7ºF, BARO sensor over 75 kPa, and the PCM detected the FTP sensor signal was less than 0.10v for over 5 seconds. **Possible Causes** ● FTP sensor signal circuit is shorted to ground ● FTP sensor power (VREF) circuit is open ● FTP sensor is damaged or has failed ● PCM has failed
DTC P0453 **2T CCM, MIL: Yes** 1999, 2000, 2001, 2002, 2003, 2004, 2005 Tracker 2.5L VIN 4 engine Transmissions: All	**Fuel Tank Pressure Sensor High Input Conditions:** Engine started; engine runtime at idle speed 2 minutes, IAT sensor from 7-158ºF, ECT sensor from 18-230ºF, and the PCM detected the FTP sensor signal was less than 0.5v (less than -26.5 in H2O) for 5 seconds. **Possible Causes** ● FTP sensor signal circuit is shorted to VREF or system power ● FTP sensor ground circuit is open ● FTP sensor is damaged or has failed ● PCM has failed
DTC P0453 **2T CCM, MIL: Yes** 2001, 2002, 2003, 2004, 2005 Tracker 2.5L VIN 4 engine Transmissions: All	**Fuel Tank Pressure Sensor High Input Conditions:** Engine started, IAT sensor over 7ºF, BARO sensor over 75 kPa, and the PCM detected the FTP sensor signal was more than 4.90v for over 5 seconds. **Possible Causes** ● FTP sensor signal circuit is shorted to VREF or system power ● FTP sensor ground circuit is open ● FTP sensor is damaged or has failed ● PCM has failed

OBD II Trouble Code List (P0xxx Codes)

DTC	Trouble Code Title, Conditions & Possible Causes
DTC P0455 **2T EVAP, MIL: Yes** 1998, 1999, 2000, 2001, 2002 Metro, Prizm All engines Transmissions: All	**EVAP System Gross Leak (0.080") Detected Conditions:** DTC P0444, P0445, P0447, P0448, P0451, P0452, P0453 and P0496 not set, ECT sensor from 158-230ºF, IAT sensor from 14-144ºF, engine started, BARO sensor over 75 kPa, Fuel Level less than 75%, vehicle driven at a steady throttle, and the PCM detected the difference between the maximum internal pressure of the fuel tank and the minimum pressure was less than a specified value. *Note: This EVAP Leak test can take up to 20 minutes to run.* **Possible Causes** ● Canister Purge valve is damaged, leaking or it has failed ● Charcoal canister is loaded with fuel or moisture ● Fuel filler cap loose, cross-threaded, incorrect part or damaged ● Fuel tank is cracked (leaking), or a leak exists in the 'O' ring ● Fuel tank pressure sensor is damaged or has failed ● Fuel vapor line(s), fuel pipes or hoses damaged or leaking ● PCM has failed
DTC P0455 **2T EVAP, MIL: Yes** 1998, 1999, 2000, 2001, 2002, 2003, 2004, 2005 Tracker All engines Transmissions: All	**EVAP System Gross Leak (0.080") Detected Conditions:** DTC P0441, P0443, P0446, P0449, P0452, P0453, P0496, P2026 and P2027 not set, engine running in closed loop, ECT sensor from 18-230ºF, IAT sensor over 18ºF, BARO sensor over 75 kPa, TP sensor steady, Fuel Level below 75%, VSS signal present, and the PCM detected the difference between the maximum and minimum FTP sensor values was less than a stored value. *Note: The EVAP Leak test can take up to 20 minutes to run.* **Possible Causes** ● Charcoal canister is loaded with fuel or moisture ● EVAP Purge valve is damaged, leaking or it has failed ● Fuel filler cap loose, cross-threaded, incorrect part or damaged ● Fuel tank is cracked (leaking), or a leak exists in the 'O' ring ● Fuel tank pressure control valve is damaged or it has failed ● Fuel tank pressure sensor is damaged or it has failed ● Fuel vapor line(s), fuel pipes or hoses damaged or leaking ● PCM has failed
DTC P0455 **2T EVAP, MIL: Yes** 2003, 2004, 2005 Tracker All engines Transmissions: All	**EVAP System Very Small Leak (0.020") Detected Conditions:** DTC P0441, P0442, P0443, P0446, P0449, P0452, P0453, P0455, P0463, P0464, P0496, P2026 and P2027 not set, engine started, ECT sensor from 18-230ºF, IAT sensor over 18ºF, BARO sensor over 75 kPa, TP sensor steady, Fuel Level below 75%, engine running in closed loop, VSS signal present, and the PCM detected the expected increase in the internal pressure of the fuel tank was more than a specified value. Note: This test can take up to 20 minutes to run. **Possible Causes** ● Charcoal canister is loaded with fuel or moisture ● EVAP Purge valve is damaged, leaking or it has failed ● Fuel filler cap loose, cross-threaded, incorrect part or damaged ● Fuel tank is cracked (leaking), or a leak exists in the 'O' ring ● Fuel tank pressure control valve is damaged or it has failed ● Fuel tank pressure sensor is damaged or it has failed ● Fuel vapor line(s), fuel pipes or hoses damaged or leaking ● PCM has failed
DTC P0461 **2T CCM, MIL: Yes** 1998, 1999, 2000, 2001, 2002, 2003, 2004, 2005 Tracker All engines Transmissions: All	**Fuel Level Sensor Range/Performance Conditions:** Engine started, engine speed over 1000 rpm, VSS signal present, PCM calculating the fuel volume usage based on the fuel injector activity, and the PCM detected the change in the Fuel Level reading was less than expected when compared to the calculated fuel volume usage. The Fuel Level Sensor Performance test monitors the fuel level sensor for accuracy. The PCM uses the fuel level input from the Fuel Level sensor in order to calculate the expected vapor pressures in the fuel system. Vapor pressures vary as the fuel level changes. Vapor pressure is critical to determining if the EVAP system is operating properly. Fuel level information is used in determining whether to run an EVAP control system test. A fuel level that is too high or too low may prevent the EVAP control system diagnostics from accurately detecting a fault. **Possible Causes** ● Fuel level sensor is stuck in one position ● Fuel level sensor is damaged or has failed ● High resistance in the fuel level sensor ground circuit (G400) ● Loose fuel tank baffle interfering with the fuel level sensor ● PCM has failed

OBD II Trouble Code List (P0xxx Codes)

DTC	Trouble Code Title, Conditions & Possible Causes
DTC P0461 **2T CCM, MIL: Yes** 1998, 1999, 2000, 2001 Metro All engines Transmissions: All	**Fuel Level Sensor Performance Conditions:** DTC P0463 not set, vehicle driven several miles, then during the Fuel Level Range check, the PCM detected a maximum to minimum reading of less than 2 gallons (test run once). **Possible Causes** ● Fuel level sensor is stuck in one position ● Fuel level sensor is damaged or has failed ● High resistance in the fuel level sensor ground circuit (G400) ● Loose fuel tank baffle interfering with the fuel level sensor ● PCM has failed
DTC P0463 **2T CCM, MIL: Yes** 1998, 1999, 2000, 2001 Metro All engines Transmissions: All	**Fuel Level Sensor Circuit High Input Conditions:** Engine started and PCM detected an unexpected voltage condition on the Fuel Level sensor circuit for 500 milliseconds during the test. **Possible Causes** ● Fuel level sensor circuit is shorted to VREF or system power ● Fuel level sensor ground circuit is open ● Fuel level sensor is damaged or has failed ● PCM has failed
DTC P0463 **2T CCM, MIL: Yes** 1998, 1999, 2000, 2001, 2002, 2003, 2004, 2005 Tracker All engines Transmissions: All	**Fuel Level Sensor Circuit High Input Conditions:** Key on or engine running, and the PCM detected an unexpected high voltage condition on the Fuel Level sensor circuit for 500 ms. **Possible Causes** ● Fuel level sensor circuit is shorted to VREF or system power ● Fuel level sensor ground circuit is open ● Fuel level sensor is damaged or has failed ● PCM has failed
DTC P0464 **2T CCM, MIL: Yes** 2002, 2003, 2004, 2005 Tracker All engines Transmissions: All	**Fuel Level Sensor Circuit Low Input Conditions:** Engine speed over 1000 rpm, VSS signals present, PCM calculating the fuel volume usage based on fuel injector activity, and the PCM detected the change in the Fuel Level reading did not exceed the specified volume. The Fuel Level Sensor Performance diagnostic monitors the fuel level sensor for accuracy. The PCM uses the fuel level input from the Fuel Level sensor to calculate expected vapor pressures within the Fuel system. Vapor pressures vary as the fuel level changes. Vapor pressure is critical in determining if the EVAP control system is operating properly. Fuel level information is used in determining whether to run an EVAP control system diagnostic test. **Possible Causes** ● Fuel level sensor connector is damaged or it is shorted ● Fuel level sensor circuit is shorted to ground ● Fuel level sensor ground circuit is open ● Fuel sensor assembly is damaged or it has failed ● PCM has failed
DTC P0494 **2T EVAP, MIL: Yes** 2002, 2003, 2004, 2005 Tracker All engines Transmissions: All Trouble Code ID: **P0496** Number of Trips to Set Code: **2T** OBD II Monitor Type: **CCM** MIL: YES Schematic: **GeoP0496.bmp** ; Tracker	**EVAP System Flow During Non-Purge Detected Conditions:** DTC P0441, P0442, P0443, P0446, P0449, P0452, P0453, P0455, P0463, P0464, P0496, P2026 and P2027 not set, engine started, ECT sensor from 18-230ºF, IAT sensor over 18ºF, BARO sensor over 75 kPa, TP sensor steady, Fuel Level below 75%, engine running in closed loop, vehicle speed over 35 mph for 20 minutes, then vehicle speed held from 30-40 mph with a constant throttle position for 3 minutes, and the PCM detected the FTP sensor was lower than the expected pressure with the EVAP system sealed. The EVAP System Flow During Non-Purge test checks the performance of the EVAP canister purge routine. System voltage is supplied directly to the EVAP canister purge valve solenoid. The PCM opens the purge valve by grounding the solenoid control circuit using a driver circuit. When the valve is open, the Fuel Tank Pressure (FTP) sensor will indicate a drop in pressure in the EVAP system. If FTP sensor values do not correspond with purge activity, P0496 will set. **Possible Causes** ● Charcoal canister is loaded with fuel or moisture, or it has failed ● EVAP purge valve is damaged, leaking or it has failed ● EVAP purge valve circuit is open or shorted to ground ● EVAP purge valve supply circuit is open (check the FL fuse) ● EVAP purge line from intake manifold to the canister is blocked ● Fuel tank pressure sensor is damaged or it has failed ● Fuel vapor line(s), fuel pipes or hoses damaged or leaking ● PCM has failed

OBD II Trouble Code List (P0xxx Codes)

DTC	Trouble Code Title, Conditions & Possible Causes
DTC P0500 **2T CCM, MIL: Yes** 1998, 1999, 2000, 2001 Metro All engines Transmissions: All	**Vehicle Speed Sensor Circuit Malfunction Conditions:** Engine started, vehicle driven and while decelerating in Fuel Cutoff mode, the PCM did not detect any VSS signals for 4 seconds. The VSS is located behind the Instrument Panel cluster assembly. The PCM supplies 5 volts to the VSS. The VSS incorporates a reed switch that opens and closes four times per revolution of the speedometer cable. The VSS signal at the PCM toggles high and low as the reed switch opens and closes. The PCM converts the toggled signal into a vehicle speed reading that is displayed in mph. **Possible Causes** ● VSS connector is damaged, open or shorted ● VSS signal circuit is open or shorted to ground ● VSS is damaged or has failed (behind the Instrument Cluster) ● PCM has failed (the PCM sends a 5v signal to the reed switch)
DTC P0500 **1T CCM, MIL: Yes** 1998, 1999, 2000, 2001, 2002, 2003, 2004, 2005 Tracker All engines Transmissions: All	**Vehicle Speed Sensor Circuit Malfunction Conditions:** Engine started, vehicle driven and while decelerating in Fuel Cutoff mode, the PCM did not detect any VSS signals for 5 seconds. The VSS is located in the transmission assembly. The VSS incorporates a hall switch that generates a digital signal. The VSS sends this signal to the instrument cluster or the speedometer, the PCM, and the Cruise Control servo. The PCM converts the VSS signal into vehicle speed for use during engine and transmission operation. **Possible Causes** ● VSS connector is damaged, open or shorted ● VSS signal circuit is open or shorted to ground ● VSS is damaged or has failed (behind the Instrument Cluster) ● PCM has failed
DTC P0500 **2T CCM, MIL: Yes** 1998, 1999, 2000, 2001, 2002 Models: Prism All engines Transmissions: All	**Vehicle Speed Sensor Signal Range/Performance Conditions:** Engine started, vehicle driven in a forward gear, engine speed from 2000-5000 rpm, and the PCM did not detect any VSS signals. The VSS is mounted on the transaxle. As the transaxle turns, the VSS provides the speedometer with voltage pulses (4 per revolution). **Possible Causes** ● VSS signal circuit is open, shorted to ground or to power ● VSS power circuit is open (check Gauge fuse in Junction block) ● VSS is damaged or is has failed ● PCM has failed
DTC P0505 **2T CCM, MIL: Yes** 1998, 1999, 2000, 2001 Metro All engines Transmissions: All	**Idle Speed Control System Conditions:** DTC P0510 not set, engine started, engine at idle speed with the closed throttle switch "on", ECT sensor more than 84°F, system voltage over 12v, throttle valve feedback check active, and the PCM detected the ISC motor movement was less than 0.25 degrees, or with the ISC Motor enabled, the ISC motor did not move at all. **Possible Causes** ● ISC motor control circuit is open or shorted to ground or to B+ ● ISC motor power circuit is open ● ISC motor is damaged or has failed ● PCM has failed
DTC P0505 **1T CCM, MIL: Yes** 1998, 1999, 2000, 2001, 2002 Models: Prism All engines Transmissions: All	**Idle Control System Performance Conditions:** Engine started, engine running at hot idle speed, and the PCM detected the Actual idle speed varied greatly from the Target idle speed. To increase the idle speed, the PCM sends a PWM signal to "open" or "close" the solenoid inside the IAC valve. This action turns a cylinder in the IAC valve, allowing air to bypass the throttle plate. **Possible Causes** ● Throttle valve is dirty or sticking (it may need to be cleaned) ● ISC motor is damaged or has failed ● PCM has failed
DTC P0505 **2T CCM, MIL: Yes** 1998, 1999, 2000 Tracker Engines: 1.6L VIN 6 Transmissions: All	**Idle Control System Performance Conditions:** Engine started, ECT sensor more than 158°F, IAT sensor more than 18°F, BARO sensor over 75 kPa, engine running in closed loop with the throttle closed, A/C and Power Steering switches both indicating "off", and the PCM detected the Actual idle speed was over 100 rpm lower than the Target idle speed, or the Actual idle speed was over 200 rpm higher than the Target idle speed for 20 seconds. **Possible Causes** ● IAC motor control circuit is open, shorted to ground or to power ● IAC motor power circuit is open (check power from EFI relay) ● IAC motor is damaged or has failed ● PCM has failed

OBD II Trouble Code List (P0xxx Codes)

DTC	Trouble Code Title, Conditions & Possible Causes
DTC P0505 **2T CCM, MIL: Yes** 1998, 1999, 2000, 2001 Tracker 2.0L VIN C engine Transmissions: All	**Idle Control System Performance Conditions:** Engine started, ECT between from 158-230ºF, IAT sensor between 19-158ºF, BARO sensor over 75 kPa, and the PCM detected the actual engine speed was over 100 rpm below the Target idle speed, or it was more than 200 rpm above than the Target idle speed. **Possible Causes** ● IAC valve control is open or shorted to ground or to power (B+) ● IAC valve power circuit is open (test for power from main relay) ● IAC valve is damaged or it has failed ● PCM has failed
DTC P0505 **2T CCM, MIL: Yes** 2001, 2002 Tracker 2.5L VIN 4 engine Transmissions: All	**Idle Control System Performance Conditions:** Key on and the PCM detected an unexpected voltage on the IAC motor coil control circuit for 5 seconds; or with the engine running, ECT sensor over 176ºF, TP angle at a 0% opening, and the PCM detected the MAF sensor indicated a larger than expected air flow reading during deceleration. The PCM controls the IAC valve to regulate the air flow through the idle air bypass to control idle speed. **Possible Causes** ● Stepper motor Coil A1 or A2 circuit is open or shorted to ground ● Stepper motor Coil B1 or B2 circuit is open or shorted to ground ● Stepper motor coil circuit(s) shorted to system power (B+) ● Stepper motor power circuit is open (check power at MFI relay) ● Stepper motor is damaged or has failed ● PCM has failed
DTC P0506 **2T CCM, MIL: Yes** 1998, 1999, 2000, 2001 Metro All engines Transmissions: All	**Idle Speed Too Low Conditions:** ECT sensor from 158-230ºF, engine running with the closed throttle position switch indicating the throttle is closed, and the PCM detected the Actual idle speed at least 100 rpm less than the Desired idle speed during the IAC Diagnostic test. **Possible Causes** ● Throttle valve is dirty or sticking (it may need to be cleaned) ● ISC motor is damaged or has failed ● PCM has failed
DTC P0506 **2T CCM, MIL: Yes** 1998, 1999, 2000, 2001, 2002, 2003, 2004, 2005 Tracker 2.0L VIN C, 2.5L VIN 4 engines Transmissions: All	**Idle Speed Too Low Conditions:** DTC P0500 not set, engine started, ECT sensor more than 158ºF, IAT sensor from 7-158ºF, BARO sensor over 75 kPa, A/C and Power Steering switches both indicating "off", engine running in closed loop at idle speed with throttle valve closed, VSS indicating 0 mph, and the PCM detected the Actual idle speed was 100 rpm less than the Desired idle speed for 20 seconds. **Possible Causes** ● Throttle valve is dirty or sticking (it may need to be cleaned) ● Stepper motor is damaged or has failed ● PCM has failed
DTC P0507 **2T CCM, MIL: Yes** 1998, 1999, 2000, 2001 Metro All engines Transmissions: All	**Idle Speed Too High Conditions:** ECT sensor from 158-230ºF, Engine started, at hot idle with the closed throttle position switch indicating closed throttle, and the PCM detected the Actual idle speed at least 200 rpm more than the Desired idle speed during the IAC Diagnostic test. **Possible Causes** ● Throttle valve is dirty or sticking (it may need to be cleaned) ● IAC motor is damaged or has failed ● PCM has failed
DTC P0507 **2T CCM, MIL: Yes** 1998, 1999, 2000, 2001, 2002, 2003, 2004, 2005 Tracker 2.0L VIN C, 2.5L VIN 4 engines Transmissions: All	**Idle Speed Too High Conditions:** DTC P0500 not set, engine running in closed loop at idle speed with throttle valve closed, ECT sensor more than 158ºF, IAT sensor from 7-158ºF, BARO sensor over 75 kPa, A/C and Power Steering switches both indicating "off", VSS indicating 0 mph, and the PCM detected the Actual idle speed was 200 rpm more than the Desired idle speed for 20 seconds. **Possible Causes** ● Throttle valve is dirty or sticking (it may need to be cleaned) ● IAC motor is damaged or has failed ● PCM has failed
DTC P0510 **2T CCM, MIL: Yes** 1998, 1999, 2000, 2001 Metro All engines Transmissions: All	**Closed Throttle Position Switch Circuit Malfunction Conditions:** Engine started; and the PCM detected a VSS signal change of 0-20 mph during a TP switch off-to-on time, or a VSS signal change of 0-20 mph during a switch on-to-off time (16 times). **Possible Causes** ● Closed throttle position switch signal circuit is open or grounded ● Closed throttle position switch signal circuit is shorted to power ● Closed throttle position switch or TP sensor damaged or failed ● PCM has failed

OBD II Trouble Code List (P0xxx Codes)

DTC	Trouble Code Title, Conditions & Possible Causes
DTC P0601 **1T PCM, MIL: Yes** 1998, 1999, 2000, 2001, 2002, 2003, 2004, 2005 Metro, Prizm, Tracker All engines Transmissions: All	**PCM Read Only Memory Error Conditions:** Key on, and the PCM detected an internal check sum data error. Note: Refer to a code repair chart for instructions on how to replace a PCM and how to reprogram the replacement unit. **Possible Causes** ● Check the PCM for possible water contamination ● Clear the trouble codes and retest for this trouble code. If the same trouble code resets, the PCM has failed and must be replaced to repair this problem.
DTC P0603 **1T PCM, MIL: Yes** 1998, 1999, 2000, 2001 Metro All engines Transmissions: All	**PCM Long Term Memory Reset Conditions:** Key on, and the PCM detected an EEPROM data readout error. Note: Refer to a code repair chart for instructions on how to replace a PCM and how to reprogram the replacement unit. **Possible Causes** ● Check the PCM for possible water contamination ● Clear the trouble codes and retest for this trouble code. If the same trouble code resets, the PCM has failed and must be replaced to repair this problem.
DTC P0616 **2T CCM, MIL: Yes** 2003, 2004, 2005 Tracker All engines Transmissions: All	**Starter Relay Control Circuit Low Input Conditions:** Engine started, engine running, and the PCM detected the engine was running but that no starter signal was detected during cranking. The Starter Relay Control Circuit High Voltage test monitors the voltage signal that is sent from the ignition switch when starting the engine. When the engine is being cranked, battery voltage is applied to the PCM. On a cold engine, the PCM can increase fuel injection volume and modify ignition timing when receiving the start signal. This signal passes through the transmission range (TR) switch or the clutch pedal position (CPP) switch. **Possible Causes** ● PCM connector is damaged or open ● Start signal to the PCM is open ● PCM has failed
DTC P0617 **2T CCM, MIL: Yes** 2003, 2004, 2005 Tracker All engines Transmissions: All	**Starter Relay Control Circuit High Input Conditions:** Engine started, engine running, and the PCM detected a high voltage present at the PCM for 3 minutes after the engine started. The Starter Relay Control Circuit High Voltage test monitors the voltage signal that is sent from the ignition switch when starting the engine. When the engine is being cranked, battery voltage is applied to the PCM. On a cold engine, the PCM can increase fuel injection volume and modify ignition timing when receiving the start signal. This signal passes through the transmission range (TR) switch or the clutch pedal position (CPP) switch. **Possible Causes** ● Start signal to the PCM is shorted to system power ● PCM connector is damaged or shorted ● PCM has failed
DTC P0705 **2T CCM, MIL: Yes** 1998, 1999, 2000, 2001, 2002, 2003, 2004, 2005 Tracker All engines Transmissions: A/T	**Transmission Range Switch Circuit Malfunction Conditions:** Engine started, engine running in Drive at a vehicle speed over 38 mph, and the PCM did not detect any Transmission Range (TR) switch signals (Reverse, Neutral, Drive, 2nd or Low) for a period of 25 seconds, or the PCM detected multiple signals from the switch simultaneously. The TR switch sends gear position signals to the PCM according to the manual selector range position. The PCM uses this information in order to manage transmission operation. **Possible Causes** ● TR switch signal circuit is open or shorted to ground ● TR switch signal circuit is shorted to VREF (5v) ● TR switch is damaged, failed or out-of-adjustment ● PCM has failed
DTC P0707 **2T CCM, MIL: Yes** 2002, 2003, 2004, 2005 Tracker All engines Transmissions: A/T	**Transmission Range Switch Circuit Malfunction Conditions:** Engine started, engine running in Drive at a vehicle speed over 38 mph, and the PCM did not detect any Transmission Range (TR) switch signals (Reverse, Neutral, Drive, 2nd or Low) for a period of 25 seconds, or the PCM detected multiple signals from the switch simultaneously. The TR switch sends gear position signals to the PCM according to the manual selector range position. The PCM uses this information in order to manage transmission operation. **Possible Causes** ● TR switch signal circuit is open or shorted to ground ● TR switch signal circuit is shorted to VREF (5v) ● TR switch is damaged, failed or out-of-adjustment ● PCM has failed

OBD II Trouble Code List (P0xxx Codes)

DTC	Trouble Code Title, Conditions & Possible Causes
DTC P0715 **1T CCM, MIL: Yes** 1999, 2000 Tracker All engines Transmissions: A/T	**Input Shaft Speed Sensor Circuit Malfunction (4S-M41) Conditions:** Engine started; vehicle driven in a Drive with the 4L switch indicating "open", and the PCM did not detect any input shaft speed (ISS) signals with the VSS (speedometer) signal received; or with the 4L switch "open", it did not detect any ISS signals with the engine speed higher than the torque converter stall speed (2300-2600 rpm). The input shaft speed sensor provides the PCM with an AC voltage signal whenever the drive wheels rotate. The AC voltage signal frequency increases with wheel speed. The PCM calculates vehicle speed from this signal, and uses the data to determine shift points. **Possible Causes** ● ISS speed sensor connector is damaged or has failed ● ISS speed sensor (+) circuit is open or shorted to ground ● ISS speed sensor (-) circuit is open or shorted to ground ● ISS is damaged or it has failed ● PCM has failed
DTC P0717 **1T CCM, MIL: Yes** 2002, 2003, 2004, 2005 Tracker All engines Transmissions: A/T	**Transmission Range Switch Circuit Malfunction Conditions:** Engine started, vehicle driven in a forward gear, then after the PCM compared the vehicle speed with the Input/Turbine speed sensor, it detected the Input/Turbine speed sensor speed was below 63 rpm with the vehicle speed over 7 mph with 1st gear held in Drive range for 1 second, or the vehicle speed was more than 13 mph with 2nd gear held in Drive range for 2 seconds, or the vehicle speed was more than 20 mph with 3rd gear held in Drive range for 2 seconds. **Possible Causes** ● TR switch signal circuit is open or shorted to ground ● TR switch signal circuit is shorted to VREF (5v) ● TR switch is damaged, failed or out-of-adjustment ● PCM has failed
DTC P0720 **1T CCM, MIL: Yes** 1998, 1999, 2000, 2001 Metro All engines Transmissions: A/T	**Output Speed Sensor Circuit Malfunction Conditions:** Engine started; 4WD switch indicating open, the PCM did not detect any VSS signals from the transmission, or with the 4WD switch open and the gear selector in Drive at more than the torque converter stall speed (2100-2400 rpm), the PCM did not detect any VSS signals. **Possible Causes** ● VSS signal circuit is open or shorted to ground ● VSS signal circuit is shorted to VREF or system power ● VSS is damaged or has failed
DTC P0720 **1T CCM, MIL: Yes** 1998, 1999, 2000, 2001, 2002, 2003, 2004, 2005 Tracker All engines Transmissions: A/T	**Output Speed Sensor Circuit Malfunction Conditions:** Engine started; 4WD switch indicating open, the PCM did not detect any VSS signals from the transmission or with the 4WD switch open and the gear selector in Drive at more than the torque converter stall speed (2300-2600 rpm), the PCM did not detect any VSS signals. **Possible Causes** ● OSS signal circuit is open or shorted to ground ● OSS signal circuit is shorted to VREF or system power ● OSS is damaged or has failed ● PCM has failed
DTC P0722 **1T CCM, MIL: Yes** 2002, 2003, 2004, 2005 Tracker All engines Transmissions: A/T	**Output Shaft Speed Sensor Circuit Malfunction Conditions:** Engine started; 4WD switch indicating "off", the PCM did not detect any Output Shaft Speed (OSS) sensor with the engine speed more than the specified engine speed with the transmission in Drive. **Possible Causes** ● OSS signal circuit is open or shorted to ground ● OSS signal circuit is shorted to VREF or system power ● OSS is damaged or has failed ● PCM has failed
DTC P0741 **2T CCM, MIL: Yes** 1998, 1999, 2000, 2001, 2002, 2003, 2004, 2005 Tracker All engines Transmissions: A/T	**Torque Converter Clutch System Performance Conditions:** Engine started, ECT sensor more than 80ºF, vehicle driven in a forward gear (2nd, 3rd, 4th Gear, or in Drive), and the PCM detected that the TCC control command from the PCM did not agree with the actual operation of the Torque Converter Clutch system. **Possible Causes** ● Control valve assembly is damaged (check for any restrictions) ● Torque converter assembly is damaged or it has failed ● Torque converter solenoid may be stuck in closed position ● Torque converter clutch solenoid may be stuck closed ● PCM has failed

OBD II Trouble Code List (P0xxx Codes)

DTC	Trouble Code Title, Conditions & Possible Causes
DTC P0742 **2T CCM, MIL: Yes** 2002, 2003, 2004, 2005 Tracker All engines Transmissions: A/T	**Torque Converter Clutch System Performance Conditions:** Engine started, ECT sensor more than 80ºF, vehicle driven in a forward gear (2nd, 3rd, 4th Gear, or in Drive), and the PCM detected that the TCC control command from the PCM did not agree with the actual operation of the Torque Converter Clutch system. **Possible Causes** ● Control valve assembly is damaged (check for any restrictions) ● Torque converter assembly is damaged or it has failed ● Torque converter solenoid may be stuck in closed position ● Torque converter clutch solenoid may be stuck closed ● PCM has failed
DTC P0743 **1T CCM, MIL: Yes** 1998, 1999, 2000, 2001, 2002, 2003, 2004, 2005 Tracker All engines Transmissions: A/T	**Torque Converter Clutch Circuit Malfunction Conditions:** Engine started, ECT sensor more than 80ºF, vehicle driven in a forward gear (2nd, 3rd, 4th Gear, or in Drive), and the PCM detected the TCC control circuit voltage was less than 2.0v with the TCC solenoid commanded "on", or it was more than 6.0v with the TCC solenoid commanded "off". The PCM supplies the power to engage the TCC lockup clutch. The TCC solenoid controls the hydraulic pressure to the lockup valve in the transmission. The TCC solenoid remains "off" while in 1st and 2nd gear. The TCC remains "on" in the 3rd and 4th gears when the PCM deems it appropriate. The TCC solenoid is never "on" if the ECT signal is less than 80ºF. **Possible Causes** ● TCC solenoid connector is damaged, open or shorted ● TCC solenoid control circuit is open, shorted to ground or to B+ ● TCC solenoid is damaged or it has failed ● PCM has failed
DTC P0750 **2T CCM, MIL: Yes** 1998, 1999, 2000, 2001, 2002 Models: Prism All engines Transmissions: A/T	**A/T Shift Solenoid 1 Circuit Malfunction Conditions:** Engine started, vehicle driven in a forward gear, and the PCM detected the commanded gear was different from the Actual gear position. The PCM compares the actual gear with the shift schedule in its memory to detect a mechanical problem in the shift solenoid 1. **Possible Causes** ● Control valve body clogged or damaged (check for a restriction) ● Shift solenoid valve 1 is clogged or leaking ● Shift select cable out of adjustment ● Transaxle mechanical malfunction may be present
DTC P0751 **2T CCM, MIL: Yes** 1998, 1999, 2000, 2001 Metro All engines Transmissions: A/T	**A/T Shift Solenoid 1 Performance Conditions:** Engine started, gear selector in Drive, and the PCM detected the Shift Solenoid 1 command did not agree with the Actual gear position (with no electrical faults in Shift Solenoid 1). **Possible Causes** ● Control valve assembly is damaged (check for any restrictions) ● Direct clutch is damaged or it has failed ● Shift solenoid valve 1 is clogged or leaking ● Shift select cable out of adjustment ● Throttle Valve (TV) cable out of adjustment ● Torque converter clutch is damaged or it has failed ● Transaxle may have a mechanical problem and need repairing ● PCM has failed
DTC P0751 **2T CCM, MIL: Yes** 1998, 1999, 2000, 2001, 2002, 2003, 2004, 2005 Tracker All engines Transmissions: A/T	**A/T Shift Solenoid 2-3 Performance Conditions:** Engine started, gear selector in Drive, and the PCM detected that the gear change control command sent to the transmission did not agree with the actual gear position (with no electrical faults detected in Shift Solenoid 2-3). **Possible Causes** ● Control valve assembly is damaged (check for any restrictions) ● Direct clutch is damaged or it has failed ● Shift solenoid valve 1 is stuck off ● 2-3 shift valve may be stuck ● PCM has failed
DTC P0752 **2T CCM, MIL: Yes** 1998, 1999, 2000, 2001, 2002, 2003, 2004, 2005 Tracker All engines Transmissions: A/T	**A/T Shift Solenoid 3-4 Performance Conditions:** Engine started, vehicle driven in Drive, and the PCM detected that the gear change control command sent to the transmission did not agree with the actual gear position. No electrical faults detected. **Possible Causes** ● Control valve assembly is damaged (check for any restrictions) ● Overdrive brake is malfunctioning ● Shift solenoid valve 2 is stuck on or it may be leaking ● 3-4 shift valve may be stuck ● PCM has failed

OBD II Trouble Code List (P0xxx Codes)

DTC	Trouble Code Title, Conditions & Possible Causes
DTC P0753 **2T CCM, MIL: Yes** 1998, 1999, 2000, 2001 Metro All engines Transmissions: A/T	**A/T Shift Solenoid 1 Circuit Malfunction Conditions:** Engine started, vehicle driven in a forward gear, and the PCM detected the Shift Solenoid 1 control circuit was less than 2.0v with the solenoid "on", or it was more than 5.5v with the solenoid "off". **Possible Causes** ● Shift solenoid 1 control circuit is open, shorted to ground or shorted to system power ● Shift solenoid 1 is damaged (resistance is 11-16 ohms at 75ºF) ● PCM has failed
DTC P0753 **1T CCM, MIL: Yes** 1998, 1999, 2000, 2001, 2002, 2003, 2004, 2005 Models: Prism, Tracker All engines Transmissions: A/T	**A/T Shift Solenoid 1 Circuit Malfunction Conditions:** Engine started, vehicle driven in Drive, and the PCM detected the Shift Solenoid 1 voltage was less than 2.0v with the solenoids commanded "on", or the Shift Solenoid voltage was more than 6.0v with the solenoids commanded "off". The Shift Solenoid 1 and Shift Solenoid 2 operate together in order to electronically control the shifting of the transmission from 1st gear to 4th gear (Overdrive). **Possible Causes** ● Shift solenoid 1-2 connector is damaged, open or shorted ● Shift solenoid 1-2 control circuit is open or shorted to ground ● Shift solenoid is damaged (resistance is 11-16 ohms at 75ºF) ● PCM has failed
DTC P0755 **2T CCM, MIL: Yes** 1998, 1999, 2000, 2001, 2002 Models: Prism All engines Transmissions: A/T	**A/T Shift Solenoid 2 Performance Conditions:** Engine started, vehicle driven in a forward gear, and the PCM detected the commanded gear was different from the Actual gear position. The PCM compares the actual gear with the shift schedule in its memory to detect a mechanical problem in the shift solenoid 2. **Possible Causes** ● Control valve body clogged or damaged (check for a restriction) ● Shift solenoid valve 2 is clogged or leaking ● Shift select cable out of adjustment ● Transaxle mechanical malfunction may be present
DTC P0756 **2T CCM, MIL: Yes** 1998, 1999, 2000, 2001, 2002, 2003, 2004, 2005 Metro, Tracker All engines Transmissions: A/T	**A/T Shift Solenoid 1-2 Performance Conditions:** Engine started, vehicle driven in Drive, and the PCM detected that the gear change control command sent to the transmission did not agree with the actual gear position (with no electrical faults detected in Shift Solenoid 1-2). **Possible Causes** ● Control valve assembly is damaged (check for any restrictions) ● 1-2 shift valve or the 3-4 shift valve may be stuck ● 2nd Brake malfunction present ● PCM has failed
DTC P0756 **2T CCM, MIL: Yes** 1998, 1999, 2000, 2001, 2002 Models: Prism All engines Transmissions: A/T	**A/T Shift Solenoid 1-2 Performance (MB3, MS7) Conditions:** Engine started, vehicle driven in Drive, and the PCM detected that the gear change control command sent to the transmission did not agree with the actual gear position (with no electrical faults detected in Shift Solenoid 1-2). **Possible Causes** ● Control valve assembly is damaged (check for any restrictions) ● 1-2 shift valve or the 3-4 shift valve may be stuck ● 2nd Brake malfunction present ● PCM has failed ● TSB 01-07-30-036B contains a repair procedure for this code
DTC P0757 **2T CCM, MIL: Yes** 2002, 2003, 2004, 2005 Tracker All engines Transmissions: A/T	**A/T Shift Solenoid 3-4 Performance Conditions:** Engine started, vehicle driven in Drive, and the PCM detected that the gear change control command sent to the transmission did not agree with the actual gear position (with no electrical faults detected in Shift Solenoid 3-4). **Possible Causes** ● Control valve assembly is damaged (check for any restrictions) ● Overdrive brake is malfunctioning ● Shift solenoid valve 2 is stuck, or the 3-4 shift valve may be stuck ● PCM has failed
DTC P0758 **2T CCM, MIL: Yes** 1998, 1999, 2000, 2001 Metro All engines Transmissions: A/T	**A/T Shift Solenoid 2 Circuit Malfunction Conditions:** Engine started, vehicle driven in a forward gears, and the PCM detected the Shift Solenoid 2 control circuit was less than 2.0v with the solenoid "on", or it was more than 5.5v with the solenoid "off". **Possible Causes** ● Shift solenoid 2 control circuit is open or shorted to ground ● Shift solenoid 2 control circuit is shorted to system power ● Shift solenoid 2 is damaged (resistance is 11-16 ohms at 75ºF) ● PCM has failed

OBD II Trouble Code List (P0xxx Codes)

DTC	Trouble Code Title, Conditions & Possible Causes
DTC P0758 **1T CCM, MIL: Yes** 1998, 1999, 2000, 2001, 2002, 2003, 2004, 2005 Models: Prism, Tracker All engines Transmissions: A/T	**A/T Shift Solenoid 2 Circuit Malfunction Conditions:** Engine started, vehicle driven in Drive, and the PCM detected the Shift Solenoid 2 voltage was less than 2.0v with the solenoids commanded "on", or the Shift Solenoid voltage was more than 6.0v with the solenoids commanded "off". The Shift Solenoid 1 and Shift Solenoid 2 operate together in order to electronically control the shifting of the transmission from 1st gear to 4th gear (Overdrive). **Possible Causes** ● Shift solenoid 2 connector is damaged, open or shorted ● Shift solenoid 2 control circuit is open or shorted to ground ● Shift solenoid is damaged (resistance is 11-16 ohms at 75ºF) ● PCM has failed
DTC P0770 **1T CCM, MIL: Yes** 1998, 1999, 2000, 2001, 2002 Prizm All engines Transmissions: A/T	**Torque Converter Clutch Lockup Performance Conditions:** Engine started, vehicle driven to over 30 mph at a steady speed for over one minute, and the PCM detected the TCC system did not engage when commanded "on", or it did nod disengage when it was commanded "off" during the CCM test. **Possible Causes** ● ATF level is too low, or fluid is burnt or contaminated ● TCC solenoid valve is stuck "off" due to sediment or binding ● TCC solenoid valve is stuck "on due to sediment or binding ● TCC solenoid is damaged or has failed ● Transmission has internal component damage or failure
DTC P0773 **1T CCM, MIL: Yes** 1998, 1999, 2000, 2001, 2002 Models: Prism Engines: 1.8L VIN 8 Transmissions: A/T	**Torque Converter Clutch Circuit Malfunction Conditions:** Engine started, vehicle driven to over 30 mph at a steady speed for over one minute, and the PCM detected an unexpected voltage condition on the TCC control circuit with the TCC "on" or with it "off". **Possible Causes** ● P/N switch signal circuit is open, shorted to ground or to power ● P/N switch power or ground circuit is open ● P/N switch is out of adjustment, damaged or has failed ● PCM has failed
DTC P0973 **2T CCM, MIL: Yes** 2002, 2003, 2004, 2005 Tracker All engines Transmissions: A/T	**A/T Shift Solenoid 1 Circuit Low Input Conditions:** Engine started, gear selector in Drive, and the PCM detected the Shift Solenoid 1 signal was less than 2.0v with the solenoids commanded "on". The Shift Solenoid 1 and Shift Solenoid 2 operate together to control transmission shifts from 1st gear to 4th gear (Overdrive). **Possible Causes** ● Shift solenoid 1 connector is damaged or shorted ● Shift solenoid 1 control circuit is shorted to ground ● Shift solenoid is damaged (resistance is 11-16 ohms at 75ºF) ● PCM has failed
DTC P0974 **2T CCM, MIL: Yes** 2002, 2003, 2004, 2005 Tracker All engines Transmissions: A/T	**A/T Shift Solenoid 1 Circuit High Input Conditions:** Engine started, gear selector in Drive, and the PCM detected the Shift Solenoid 1 signal was over 5.5v with the solenoids commanded "off". The Shift Solenoid 1 and Shift Solenoid 2 operate together to control transmission shifts from 1st gear to 4th gear (Overdrive). **Possible Causes** ● Shift solenoid 1 connector is shorted to power ● Shift solenoid 1 control circuit is shorted to power ● Shift solenoid is damaged (resistance is 11-16 ohms at 75ºF) ● PCM has failed
DTC P0976 **2T CCM, MIL: Yes** 2002, 2003, 2004, 2005 Tracker All engines Transmissions: A/T	**A/T Shift Solenoid 2 Circuit Low Input Conditions:** Engine started, vehicle driven in Drive, Shift Solenoid 2 commanded "on", and the PCM detected the SS2 control signal was under 2.0v. Shift Solenoids 1 and operate together to electronically control the transmission shifts from 1st gear to 4th gear (Overdrive). **Possible Causes** ● Shift solenoid 1 connector is damaged or shorted ● Shift solenoid 1 control circuit is shorted to ground ● Shift solenoid is damaged (resistance is 11-16 ohms at 75ºF) ● PCM has failed
DTC P0977 **2T CCM, MIL: Yes** 2002, 2003, 2004, 2005 Tracker All engines Transmissions: All	**A/T Shift Solenoid 2 Circuit High Input Conditions:** Engine started, vehicle driven in Drive, Shift Solenoid 2 commanded "off", and the PCM detected the SS2 control signal was over 5.5v. Shift Solenoids 1 and operate together to electronically control the shifting of the transmission from 1st gear to 4th gear (Overdrive). **Possible Causes** ● Shift solenoid 2 connector is shorted to power ● Shift solenoid 2 control circuit is shorted to power ● Shift solenoid is damaged (resistance is 11-16 ohms at 75ºF) ● PCM has failed

OBD II Trouble Code List (P1xxx Codes)

DTC	Trouble Code Title, Conditions & Possible Causes
DTC P1132 **2T CCM, MIL: Yes** 2002 Tracker 2.5L VIN 4 engine Transmissions: All	**HO2S-11 (Bank 1 Sensor 1) Heater Circuit Malfunction Conditions:** P0171, P0172, P0174, P0175, P0300, P0301, P0302, P0303, P0304, P0305, P0306 and P0400 not set, ECT sensor from 18-230ºF, IAT sensor over 18ºF at engine startup, BARO sensor over 75 kPa, vehicle driven to a speed of over 35 mph for 2 minutes followed by an idle period of 1 minute, and the PCM detected the HO2S signal was over 2.0v for too long. **Possible Causes** ● HO2S signal circuit is shorted to power (check heater circuit) ● HO2S is damaged or has failed) ● PCM has failed
DTC P1152 **2T CCM, MIL: Yes** 2002 Tracker 2.5L VIN 4 engine Transmissions: All	**HO2S-21 (Bank 2 Sensor 1) Heater Circuit Malfunction Conditions:** IAT sensor over 18ºF at startup, VSS over 35 mph, then at idle, ECT sensor from 18-230ºF, BARO sensor over 75 kPa and the PCM detected the HO2S input was over 2.0v for too long. **Possible Causes** ● HO2S signal circuit is shorted to power (check heater circuit) ● HO2S is damaged or has failed ● PCM has failed
DTC P1250 **1T CCM, MIL: Yes** 1998, 1999, 2000, 2001 Metro All engines Transmissions: All	**Early Fuel Evaporation Heater Circuit Malfunction Conditions:** Engine started, IAT sensor from 14-122ºF, BARO sensor over 75 kPa, fuel level over 25%, and the PCM detected the EFE heater resistor "on" signal was less than 2.5v with the heater off, or the heater "on" signal was over 0.3v with the heater on for 5 seconds. **Possible Causes** ● EFE heater control circuit is open or shorted to ground ● EFE heater power circuit is open (check the power source) ● EFE heater is damaged or has failed ● PCM has failed
DTC P1300 **1T CCM, MIL: Yes** 1998, 1999, 2000, 2001, 2002 Prizm All engines Transmissions: All	**Ignition Coil 1 Primary Feedback Circuit Malfunction Conditions:** Engine cranking or running; and the PCM did not detect any failsafe signals from Coil 1 after two ignition trigger signal cycles occurred. **Possible Causes** ● Ignition coil primary circuit is open or shorted to ground ● Ignition coil is damaged or has failed ● Ignition system "noise" filter may be shorted to ground ● PCM has failed
DTC P1305 **1T CCM, MIL: Yes** 1998, 1999, 2000, 2001, 2002 Prizm All engines Transmissions: All	**Ignition Coil 2 Primary Feedback Circuit Malfunction Conditions:** Engine cranking or running; and the PCM did not detect any failsafe signals from Coil 2 after two ignition trigger signal cycles occurred. **Possible Causes** ● Ignition coil primary circuit is open or shorted to ground ● Ignition coil is damaged or has failed ● Ignition system "noise" filter may be shorted to ground ● PCM has failed
DTC P1310 **1T CCM, MIL: Yes** 1998, 1999, 2000, 2001, 2002 Prizm All engines Transmissions: All	**Ignition Coil 3 Primary Feedback Circuit Malfunction Conditions:** Engine cranking or running; and the PCM did not detect any failsafe signals from Coil 3 after two ignition trigger signal cycles occurred. **Possible Causes** ● Ignition coil primary circuit is open or shorted to ground ● Ignition coil is damaged or has failed ● Ignition system "noise" filter may be shorted to ground ● PCM has failed
DTC P1315 **1T CCM, MIL: Yes** 1998, 1999, 2000, 2001, 2002 Prizm All engines Transmissions: All	**Ignition Coil 4 Primary Feedback Circuit Malfunction Conditions:** Engine cranking or running; and the PCM did not detect any failsafe signals from Coil 4 after two ignition trigger signal cycles occurred. **Possible Causes** ● Ignition coil primary circuit is open or shorted to ground ● Ignition coil is damaged or has failed ● Ignition system "noise" filter may be shorted to ground ● PCM has failed
DTC P1335 **2T CCM, MIL: Yes** 1998, 1999, 2000, 2001, 2002 Prizm All engines Transmissions: All	**Crankshaft Position Sensor Circuit Malfunction Conditions:** Engine speed more than 1000 rpm, and the PCM did not receive any Crankshaft Position (CKP) sensor pulses for more than 50 ms. **Possible Causes** ● CKP sensor signal (+) or (-) circuit is open, shorted to ground or shorted to power ● CKP sensor is damaged or has failed (an intermittent fault) ● Crankshaft reluctor wheel contains debris or is damaged ● PCM has failed

OBD II Trouble Code List (P1xxx Codes)

DTC	Trouble Code Title, Conditions & Possible Causes
DTC P1346 **2T CCM, MIL: Yes** 1999, 2000, 2001, 2002 Prizm All engines Transmissions: All	**Intake Camshaft Position System Performance Conditions:** Vehicle speed over 5 minutes, ECT sensor over 158°F, and the PCM detected unexpected deviation between the relationship of the CMP and CKP sensor signals for 20 seconds. **Possible Causes** ● Camshaft is incorrectly installed or incorrectly timed ● CMP System actuator is damaged or has failed ● CMP sensor is damaged, out of alignment or has failed ● Timing chain has jumper or is worn out
DTC P1349 **2T CCM, MIL: Yes** 1999, 2000, 2001, 2002 Prizm All engines Transmissions: All	**Intake Camshaft Position System Performance Conditions:** Engine started, ECT sensor more than 158°F, VSS more than 5 mph, and PCM did not detect any change in camshaft timing after the activating the CMP actuator, or it detected that the camshaft timing was fixed, either condition met for 20 seconds. Note: If DTC P1346 and P1349 are both set, the solenoid has failed. **Possible Causes** ● Camshaft is incorrectly installed or incorrectly timed ● CMP System actuator is damaged or has failed ● Check the oil passages to the actuator for a restriction ● Timing chain has jumper or is worn out
DTC P1408 **2T CCM, MIL: Yes** 1998, 1999, 2000, 2001 Metro All engines Transmissions: All	**MAP Sensor Circuit Malfunction Conditions:** Key on or engine running; ECT sensor more than 113°F, IAT sensor more than 41°F, TP sensor less than 0.20v, and the PCM detected the MAP sensor was more than 4.80v. **Possible Causes** ● MAP sensor signal circuit is open ● MAP sensor signal circuit is shorted to VREF ● MAP sensor ground circuit is open ● MAP sensor is damaged or has failed ● PCM has failed
DTC P1408 **2T CCM, MIL: Yes** 1998, 1999, 2000, 2001, 2002 Tracker Engines: 1.6L VIN 4 Transmissions: All	**MAP Sensor Circuit Malfunction Conditions:** Engine started; ECT sensor more than 113°F, IAT sensor more than 41°F, throttle angle less than 10 degrees, and the PCM detected the MAP sensor was more than 4.80v, or it detected the MAP sensor was less than 0.2v with the throttle angle more than 0.5% during the test. **Possible Causes** ● MAP sensor signal circuit is open or shorted to ground ● MAP sensor signal circuit is shorted to VREF or system power ● MAP sensor power (VREF) circuit is open ● MAP sensor ground circuit is open ● MAP sensor is damaged or has failed ● PCM has failed
DTC P1410 **2T CCM, MIL: Yes** 1998, 1999, 2000, 2001 Metro All engines Transmissions: All	**Fuel Tank Pressure System Range/Performance Conditions:** Engine started, fuel level from 25-75%, and the PCM detected an unexpected voltage on the Fuel Tank Pressure Control signal circuit with the fuel level over a value stored in memory. **Possible Causes** ● Fuel tank pressure control solenoid control circuit is open, shorted to ground or to B+ ● Fuel tank pressure control solenoid power circuit is open ● Fuel tank pressure control solenoid is damaged or has failed ● PCM has failed
DTC P1450 **1T CCM, MIL: Yes** 1998, 1999, 2000, 2001 Metro All engines Transmissions: All	**Barometric Pressure Sensor Circuit Malfunction Conditions:** Engine cranking or engine is running in closed loop at a steady speed, and the PCM detected the BARO sensor signal indicated less than 33 kPa, or it detected that it indicated more than 135 kPa. **Possible Causes** ● BARO pressure sensor circuit is open or shorted to ground ● BARO pressure sensor signal circuit is shorted to VREF ● BARO pressure sensor power (VREF) circuit is open ● BARO pressure sensor is damaged or has failed ● PCM has failed
DTC P1450 **1T CCM, MIL: Yes** 1998, 1999, 2000, 2001, 2002 Tracker Engines: 1.6L VIN 6 Transmissions: All	**Barometric Pressure Sensor Circuit Malfunction Conditions:** Engine running in closed loop, and the PCM detected the BARO sensor was less than 38 kPa, or it detected the BARO sensor was more than 113 kPa for over 5 seconds. **Possible Causes** ● BARO pressure sensor circuit is open or shorted to ground ● BARO pressure sensor signal circuit is shorted to VREF ● BARO pressure sensor power (VREF) circuit is open ● BARO pressure sensor is damaged or has failed ● PCM has failed

OBD II Trouble Code List (P1xxx Codes)

DTC	Trouble Code Title, Conditions & Possible Causes
DTC P1451 **2T CCM, MIL: Yes** 1998, 1999, 2000, 2001 Metro All engines Transmissions: All	**Barometric Pressure Sensor Range/Performance Conditions:** Engine cranking for 2 seconds, or engine running in closed loop at a stable speed, the PCM detected the difference between the BARO and MAP sensors was 26 kPa or higher. **Possible Causes** ● BARO pressure sensor is damaged, "skewed" or it has failed ● MAP sensor is out-of-calibration or the sensor is "skewed" ● PCM has failed
DTC P1451 **2T CCM, MIL: Yes** 1998, 1999, 2000, 2001, 2002 Tracker All engines Transmissions: All	**Barometric Pressure Sensor Range/Performance Conditions:** Engine started, engine running in closed loop, then while operating in Fuel Cutoff mode, the PCM detected the difference between the detected BARO input and the MAP sensor was not within a specified value for 5 seconds during the CCM test. **Possible Causes** ● BARO pressure sensor is damaged, "skewed" or it has failed ● MAP sensor is out-of-calibration or the sensor is "skewed" ● PCM has failed
DTC P1460 **2T CCM, MIL: Yes** 1998, 1999, 2000, 2001 Metro All engines Transmissions: All	**Cooling Fan Control System Malfunction Conditions:** Key on, and the PCM detected the fan control circuit failed the output voltage check with the fan "on", conditions met for 5 seconds. **Possible Causes** ● Cooling fan relay control circuit is open ● Cooling fan relay control circuit is shorted to ground ● Cooling fan relay power circuit is open ● Cooling fan relay is damaged or has failed ● PCM has failed
DTC P1500 **2T CCM, MIL: Yes** 1998, 1999, 2000, 2001 Metro All engines Transmissions: All	**Starter Signal Circuit Malfunction Conditions:** Engine cranking, and the PCM did not detect any starter signals, or it detected a low starter signal while the engine was cranking, or a high signal after the engine started. *Note: This test occurs once per drive cycle for 400 milliseconds.* **Possible Causes** ● Starter signal circuit is open, shorted to ground or to power ● Starter is damaged or has failed ● PCM has failed
DTC P1500 **1T CCM, MIL: Yes** 1998, 1999, 2000, 2001, 2002 Tracker All engines Transmissions: All	**Starter Signal Circuit Malfunction Conditions:** Engine cranking, and the PCM did not detect any Starter signal, or it detected a low voltage on the Starter signal during cranking, or it detected a high voltage condition on the Starter signal after the engine started, condition met for 3 seconds. **Possible Causes** ● Starter signal circuit is open, shorted to ground or to power ● Starter is damaged or has failed ● PCM has failed
DTC P1510 **1T CCM, MIL: Yes** 1998, 1999, 2000, 2001 Metro All engines Transmissions: All	**Backup Power Supply Malfunction Conditions:** Key off, and the PCM did not detect any voltage on the Backup Power circuit, condition met for 5 seconds. *Note: This code can only be read at key on, engine off.* **Possible Causes** ● Backup Power Supply circuit is open (check the Dome fuse) ● Battery connections are corroded, dirty or loose ● Battery has been disconnected or removed ● PCM has failed
DTC P1510 **12T CCM, MIL: Yes** 1998, 1999, 2000, 2001, 2002, 2003, 2004, 2005 Tracker All engines Transmissions: All	**Backup Power Supply Conditions:** Engine started and the PCM detected there was no Backup Power being supplied to it after engine startup for 5 seconds. The battery positive voltage circuit supplies power to the PCM for the retention of long term memory data. Trouble codes and learned parameters are kept in the PCM memory even with the ignition off. The PCM sets this code and loses long term memory if this circuit is interrupted. **Possible Causes** ● Backup Power Supply circuit is open (check the Dome fuse) ● Battery connections are corroded, dirty or loose ● Battery has been disconnected or removed ● PCM has failed

OBD II Trouble Code List (P1xxx Codes)

DTC	Trouble Code Title, Conditions & Possible Causes
DTC P1520 **2T CCM, MIL: Yes** 1998, 1999, 2000, 2001, 2002 Models: Prism All engines Transmissions: A/T	**Stop Lamp Switch Circuit Malfunction Conditions:** Engine started, VSS over 19 mph, and the PCM detected the Stop Lamp switch indicated a continuous "on" condition. If the PCM detects that the Stop Lamp switch remains closed (i.e., that it never opens), this code is set. The PCM disables the TCC when it closes. **Possible Causes** ● Stop lamp switch signal circuit is shorted to system power (B+) ● Stop lamp switch is damaged or has failed (in closed position) ● Stop lamp switch is out-of-adjustment ● PCM has failed
DTC P1520 **2T CCM, MIL: Yes** 1998, 1999, 2000 Tracker All engines Transmissions: A/T	**Stop Lamp Switch Circuit Malfunction Conditions:** Engine started, vehicle speed over 19 mph, and the PCM detected the Stop Lamp switch indicated a continuously "on" condition. **Possible Causes** ● Stop lamp switch signal circuit is shorted to system power (B+) ● Stop lamp switch is damaged or has failed (in closed position) ● Stop lamp switch is out-of-adjustment ● PCM has failed
DTC P1530 **2T CCM, MIL: Yes** 1998, 1999, 2000, 2001 Metro All engines Transmissions: All	**Ignition Timing Adjustment Switch Circuit Malfunction Conditions:** Engine started, vehicle driven to a speed of over 1 mph, and the PCM detected the Test Switch signal to the DLC circuit indicated "on" for 5 seconds during the CCM continuous test. **Possible Causes** ● Ignition timing adjustment switch circuit is shorted to ground ● PCM has failed
DTC P1600 **1T CCM, MIL: Yes** 1998, 1999, 2000, 2001, 2002 Prizm All engines Transmissions: All	**PCM Battery Circuit Conditions:** Key on, and the PCM detected that there was no power to the battery direct feed connection (connector C1-1) of the PCM. **Possible Causes** ● Backup Power Supply circuit is open (check the Dome fuse) ● Battery connections are corroded, dirty or loose ● Battery has been disconnected or removed ● PCM has failed
DTC P1621 **1T CCM, MIL: Yes** 1998, 1999, 2000, 2001, 2002 Prizm All engines Transmissions: All	**PCM Long Term Memory Error Conditions:** Key on or engine running; and the PCM detected an internal memory reset error had occurred. **Possible Causes** ● Record any related trouble codes in memory ● Reprogram the PCM and recheck for the same trouble code ● If the same trouble code resets, replace & reprogram the PCM ● Perform the Security System (Passlock) Relearn Procedure
DTC P1656 **1T CCM, MIL: Yes** 1998, 1999, 2000, 2001, 2002 Prizm All engines Transmissions: All	**Intake CMP Actuator Solenoid Control Circuit Malfunction Conditions:** Engine started and PCM detected an unexpected voltage condition on the Intake CMP Actuator Solenoid Control circuit during the test. **Possible Causes** ● Intake Actuator solenoid control circuit is open, shorted to ground or shorted to system power (B+) ● Intake Actuator solenoid power circuit is open ● Intake Actuator solenoid is damaged or has failed ● PCM has failed
DTC P1740 **1T CCM, MIL: Yes** 2002, 2003, 2004, 2005 Tracker All engines Transmissions: All	**Cruise Control Overdrive Cut Control Circuit Malfunction Conditions:** Engine started, vehicle speed over 10 mph, and the PCM detected a low voltage (grounded circuit) condition on the Cruise Control O/D Cut Control system. During normal operation of the cruise control system, when the Cruise Control system detects long grades or hilly conditions, it will send a ground signal to the PCM to disengage the overdrive. If the Cruise Control Cutoff signal circuit experienced a short to ground condition when the vehicle is traveling less than 7 mph, the PCM will disengage overdrive and generate the set P1740. **Possible Causes** ● Cruise control servo connector is damaged or shorted ● Cruise control cutoff signal circuit is shorted to ground ● Cruise control servo is damaged or it has failed ● PCM has failed

OBD II Trouble Code List (P1xxx - P2xxx Codes)

DTC	Trouble Code Title, Conditions & Possible Causes
DTC P1780 **1T CCM, MIL: Yes** 1998, 1999, 2000, 2001, 2002 Prizm All engines Transmissions: A/T	**Park Neutral Position Switch Circuit Malfunction Conditions:** Engine started, vehicle driven to over 50 mph at an engine speed from 2000-4000 rpm, MAP sensor indicating more than 300 mmHg, and the PCM detected the P/N switch indicated Neutral for 30 seconds or two different P/N switch signals at the same time. **Possible Causes** • P/N switch signal circuit is open, shorted to ground or to power • P/N switch power or ground circuit is open • P/N switch is out of adjustment, damaged or has failed • PCM has failed
DTC P1875 **2T CCM, MIL: Yes** 1998, 1999, 2000, 2001, 2002, 2003, 2004, 2005 Tracker All engines Transmissions: All	**4WD Low Switch Circuit Malfunction Conditions:** Engine started, vehicle driven in a forward gear, and the PCM detected the VSS signal indicated 6 mph less than the output shaft speed sensor signal. If the VSS transmission does not function as designed, the transmission may set P1875. The above action may occur even when the 4L signal circuit is mechanically and electrically in good condition. With the transfer case shifted into the 4L position, the 4L switch closes, and the 4L switch grounds the circuit to the PCM. The PCM uses the above information in order to prevent torque converter clutch (TCC) lock-up and 4th gear engagement. **Possible Causes** • 4WD signal circuit open or grounded between the switch and the PCM connector • 4WD signal circuit shorted to VREF or system power • 4WD switch is damaged or has failed
DTC P2025 **2T CCM, MIL: Yes** 2003, 2004, 2005 Tracker All engines Transmissions: All	**Fuel Vapor Temperature Sensor Range/Performance Conditions:** Engine started, ECT sensor less than 86ºF at startup, ECT and IAT sensors within 12ºF of each at startup, and the PCM detected the signal from the fuel vapor temperature sensor indicated that there was a smaller change in the voltage than expected during warmup. The Fuel Vapor Temperature Sensor is thermistor whose value varies with the temperature. The sensor is wired in series with a fixed resistor inside the PCM. The PCM applies 5 volts to the vapor temperature sensor. The PCM monitors the voltage change across the sensor and converts the voltage into a temperature reading. **Possible Causes** • Fuel vapor temperature sensor ground circuit high resistance • Fuel vapor temperature sensor signal circuit high resistance • Fuel vapor temperature sensor is damaged or it has failed • PCM has failed
DTC P2026 **1T CCM, MIL: Yes** 2003, 2004, 2005 Tracker All engines Transmissions: All	**Fuel Vapor Temperature Sensor Circuit Low Input Conditions:** Engine started, and the PCM detected the Fuel Vapor Temperature sensor voltage indicated a temperature over 271ºF for 10 seconds. The sensor is a thermistor whose value varies with temperature. The sensor is wired in series to a fixed resistor in the PCM. The PCM applies 5v to the vapor temperature sensor. The PCM monitors the change in voltage across the sensor and converts it to temperature. **Possible Causes** • Fuel vapor temperature sensor connector is shorted • Fuel vapor temperature sensor signal circuit shorted to ground • Fuel vapor temperature sensor is damaged or it has failed • PCM has failed
DTC P2027 **2T CCM, MIL: Yes** 2003, 2004, 2005 Tracker All engines Transmissions: All	**Fuel Vapor Temperature Sensor Circuit High Input Conditions:** Engine started, and the PCM detected the Fuel Vapor Temperature sensor voltage indicated a temperature of less than -40ºF. This sensor is a thermistor whose value varies with the temperature. The sensor is wired in series with a fixed resistor in the PCM. The PCM applies 5v to the vapor temperature sensor. The PCM monitors the change in voltage across the sensor and converts it to temperature. **Possible Causes** • Fuel vapor temperature sensor signal circuit open • Fuel vapor temperature ground circuit is open • Fuel vapor temperature sensor is damaged or it has failed • PCM has failed

OBD II Trouble Code List (P2xxx Codes)

DTC	Trouble Code Title, Conditions & Possible Causes
DTC P2027 **2T CCM, MIL: Yes** 2003, 2004, 2005 Tracker All engines Transmissions: All	**BARO Sensor Signal Range/Performance Conditions:** DTC P0107 and P0108 not set, engine running in closed loop, but not in Fuel Cutoff mode, and the PCM detected the difference between the observed barometric pressure reading and the MAP sensor reading was not within a specified value. The BARO sensor and circuitry are located inside the PCM. Barometric pressure varies with weather conditions and changes in the altitude from sea level. The PCM modifies fuel and spark delivery in response to barometric pressure changes. The PCM compares the BARO sensor reading with the MAP sensor reading when running the test for this code. **Possible Causes** ● BARO sensor is damaged, out-of-calibration or "skewed" ● MAP sensor is out-of-calibration or "skewed" ● MAP sensor vacuum supply line is blocked or restricted ● MAP sensor signal circuit has a high resistance condition ● PCM has failed
DTC P2028 **2T CCM, MIL: Yes** 2003, 2004, 2005 Tracker All engines Transmissions: All	**BARO Sensor Signal Circuit Low Input Conditions:** Engine started, and the PCM detected the BARO sensor signal was less than 38 kPa for 5 seconds. The BARO sensor and circuitry are located inside the PCM. Barometric pressure varies with weather conditions and changes in the altitude from sea level. The PCM modifies fuel and spark delivery in response to barometric pressure changes. The PCM compares the BARO sensor reading with the MAP sensor reading when running the test for this code. **Possible Causes** ● BARO sensor VREF supply circuit is shorted to ground ● BARO sensor shares the VREF (5v) supply from the PCM with the MAP sensor, the TP sensor, and the FTP sensor ● PCM has failed
DTC P2029 **1T CCM, MIL: Yes** 2003, 2004, 2005 Tracker All engines Transmissions: All	**BARO Sensor Signal Circuit High Input Conditions:** Engine started, and the PCM detected the BARO sensor signal was more than 113 kPa for 5 seconds. The BARO sensor and circuitry are located inside the PCM. Barometric pressure varies with weather conditions and changes in the altitude from sea level. The PCM modifies fuel and spark delivery in response to barometric pressure changes. The PCM compares the BARO sensor reading with the MAP sensor reading when running the test for this code. **Possible Causes** ● BARO sensor is damaged, out-of-calibration or "skewed" ● PCM has failed
DTC P2769 **1T CCM, MIL: Yes** 2003, 2004, 2005 Tracker All engines Transmissions: A/T	**Torque Converter Clutch Solenoid Circuit Low Input Conditions:** Engine started, vehicle driven in a forward gear at a speed of over 35 mph, and the PCM detected the TCC solenoid voltage was less than 2v after the PCM commanded the TCC control solenoid "on". The TCC solenoid engages the lockup clutch. The PCM supplies system voltage to turn "on" (energize) or remove voltage to turn "off" (de-energize the solenoid). The TCC solenoid controls the hydraulic pressure to the lockup valve in the transmission. The TCC solenoid remains "off" in the 1st and 2nd gears. The TCC remains "on" in the 3rd and 4th gears when the PCM deems appropriate. The TCC solenoid is never "on" during 4L operation or when the engine coolant temperature is less than 80ºF. **Possible Causes** ● TCC solenoid connector is damaged or shorted to ground ● TCC solenoid control circuit is shorted to ground ● TCC solenoid is damaged or it has failed ● PCM has failed
DTC P2770 **1T CCM, MIL: Yes** 2003, 2004, 2005 Tracker All engines Transmissions: A/T	**Torque Converter Clutch Solenoid Circuit High Input Conditions:** Engine started, vehicle driven in a forward gear at a speed of over 35 mph, and the PCM detected the TCC solenoid voltage was more than 5.5v after the PCM commanded the TCC control solenoid "off". The TCC solenoid engages the lockup clutch. The PCM supplies system voltage to turn "on" (energize) or remove voltage to turn "off" (de-energize the solenoid). The TCC solenoid controls the hydraulic pressure to the lockup valve in the transmission. The TCC solenoid remains "off" in the 1st and 2nd gears. The TCC remains "on" in the 3rd and 4th gears when the PCM deems appropriate. The TCC solenoid is never "on" during 4L operation or when the engine coolant temperature is less than 80ºF. **Possible Causes** ● TCC solenoid connector is damaged or shorted to ground ● TCC solenoid control circuit is shorted to ground ● TCC solenoid is damaged or it has failed ● PCM has failed

Pontiac Vibe OBD II Trouble Codes

OBD II Trouble Code List (P0xxx Codes)

DTC	Trouble Code Title, Conditions & Possible Causes
DTC P0100 **2T CCM, MIL: Yes** 2003, 2004, 2005 Vibe All engines Transmissions: All	**MAF Sensor Circuit Malfunction Conditions:** Engine started, vehicle driven at an engine speed under 4000 rpm, and the PCM detected an unexpected voltage condition on the MAF sensor circuit for 3 seconds. The MAF Sensor Circuit diagnostic detects an out of range sensor signal. The MAF sensor measures changes in the intake air volume that result from the changes in throttle opening and air density. Air flow measurements are used by the PCM in order to determine engine fueling requirements. **Possible Causes** ● MAF sensor signal circuit is open or shorted to ground ● MAF sensor signal circuit is shorted to VREF or system power ● MAF sensor power circuit is open (check power from MFI relay) ● MAF sensor is damaged or has failed ● PCM has failed
DTC P0101 **2T CCM, MIL: Yes** 2003, 2004, 2005 Vibe All engines Transmissions: All	**MAF Sensor Signal Range/Performance Conditions:** Engine started, ECT sensor more than 158°F, engine speed below 900 rpm with the throttle valve closed, and the PCM detected the MAF sensor was more than 2.20v, or with the engine speed above 1500 rpm and the TP sensor signal at least 0.63v, it detected the MAF sensor indicated less than 1.06v for 10 seconds. **Possible Causes** ● Intake air leak between the MAF sensor and the throttle body ● MAF sensor signal circuit is open or shorted to ground ● MAF sensor is damaged or has failed ● PCM has failed
DTC P0110 **2T CCM, MIL: Yes** 2003, 2004, 2005 Vibe All engines Transmissions: All	**Intake Air Temperature Sensor Circuit Malfunction Conditions:** Key on or engine running; and the PCM detected the IAT sensor indicated a value less than -40°F for 2 seconds, or indicated a value of more than 284°F for 2 seconds during the test. **Possible Causes** ● IAT sensor signal circuit is open or shorted to ground ● IAT sensor ground circuit is open ● IAT sensor is damaged or has failed (out of calibration) ● PCM has failed
DTC P0115 **2T CCM, MIL: Yes** 2003, 2004, 2005 Vibe All engines Transmissions: All	**Engine Coolant Temperature Sensor Circuit Malfunction Conditions:** Engine started, and the PCM detected the ECT sensor indicated a value less than -40°F for 2 seconds, or indicated a value more than 284°F for 2 seconds. The ECT sensor is a thermistor whose value varies with temperature. The sensor resistance is high if the coolant temperature is cold; and low if the coolant temperature is warm. The ECT sensor is wired in series with a fixed resistor inside the PCM. **Possible Causes** ● ECT sensor signal circuit is open or shorted to ground ● ECT sensor ground circuit is open ● ECT sensor is damaged or has failed (out of calibration) ● PCM has failed
DTC P0116 **1T CCM, MIL: Yes** 2003, 2004, 2005 Vibe All engines Transmissions: All	**Engine Coolant Temperature Sensor Range/Performance Conditions:** ECT sensor less than 20°F at started, engine runtime more than 20 minutes, and the PCM detected the ECT sensor was less than 68°F, or with the ECT and IAT sensors from 20-50°F at startup and the engine runtime over 5 minutes, and the PCM detected the ECT sensor indicated less than 68°F at the end of the test period. **Possible Causes** ● ECT sensor signal circuit is open or shorted to ground ● ECT sensor is contaminated, out-of-calibration or has failed ● PCM has failed
DTC P0120 **2T CCM, MIL: Yes** 2003, 2004, 2005 Vibe All engines Transmissions: All	**Throttle Position Sensor Circuit Malfunction Conditions:** Key on or engine running; and the PCM detected the TP sensor signal was more than 1.0v with the throttle closed, or the TP sensor signal was more than 4.90v at any time in the CCM continuous test. **Possible Causes** ● TP sensor signal circuit open or shorted to ground ● TP sensor ground circuit is open ● TP sensor power circuit is open (test VREF circuit to the PCM) ● TP sensor is damaged or has failed ● PCM has failed

OBD II Trouble Code List (P0xxx Codes)

DTC	Trouble Code Title, Conditions & Possible Causes
DTC P0121 **1T CCM, MIL: Yes** 2003, 2004, 2005 Vibe All engines Transmissions: All	**Throttle Position Sensor Signal Range/Performance Conditions:** Engine started, vehicle driven to a speed of over 30 mph while in closed loop, and the PCM detected the TP sensor signal was out of the acceptable range for these operating conditions. **Possible Causes** ● MAP sensor may be out-of-calibration ● \P sensor signal circuit is open to the PCM (intermittent fault) ● TP sensor ground circuit is open (an intermittent fault) ● Throttle body is damaged or throttle linkage is bent or binding ● TP sensor is damaged or has failed
DTC P0125 **2T CCM, MIL: Yes** 2003, 2004, 2005 Vibe All engines Transmissions: All	**ECT Excessive Time To Enter Closed Loop Conditions:** Engine started, ECT sensor more than 140°F at startup, engine runtime over 180 seconds, vehicle speed from 25-62 mph with the throttle valve indicating open, and the PCM detected the HO2S-11 signal did not indicate a rich condition for more than 1.5 minutes. **Possible Causes** ● Check the operation of the thermostat (it may be stuck open) ● ECT sensor signal circuit has high resistance ● ECT sensor has failed ● Inspect for low coolant level or an incorrect coolant mixture
DTC P0128 **1T CCM, MIL: Yes** 2003, 2004, 2005 Vibe All engines Transmissions: All	**ECT Below Thermostat Regulating Temperature Conditions:** ECT and IAT sensors from 14-97°F at startup, engine running for a calibrated period of time, and the PCM detected the ECT sensor failed to reach 167°F within the amount of time determined by the Water Temperature Counter diagnostic. **Possible Causes** ● Check the operation of the thermostat (it may be stuck open) ● ECT sensor is damaged or out-of-calibration (it is "skewed") ● PCM has failed
DTC P0130 **2T CCM, MIL: Yes** 2003, 2004, 2005 Vibe All engines Transmissions: All	**HO2S-11 (Bank 1 Sensor 1) Circuit Malfunction Conditions:** Engine started, engine running at idle speed, ECT sensor more than 158°F, and the PCM detected the HO2S signal was fixed between 400 and 550 mv, or the HO2S signal was fixed at less than 550 mv, or the HO2S signal was fixed at more than 440 mv during the test. **Possible Causes** ● HO2S signal circuit is open between the sensor and the PCM ● HO2S signal circuit is shorted to sensor or chassis ground ● HO2S signal circuit is shorted to VREF or system power (B+) ● HO2S is damaged, contaminated or it has failed ● PCM has failed
DTC P0133 **2T O2S, MIL: Yes** 2003, 2004, 2005 Vibe All engines Transmissions: All	**HO2S-11 (Bank 1 Sensor 1) Slow Response Conditions:** Engine started, engine running a idle speed in closed loop, ECT sensor more than 158°F, and the PCM detected the response time necessary for the HO2S voltage to change from rich-to-lean, or from lean-to-rich, was 1.1 seconds or more during the HO2S test period. **Possible Causes** ● Exhaust leak present in the exhaust manifold or exhaust pipes ● HO2S element is fuel contaminated or has deteriorated ● PCM has failed
DTC P0135 **2T CCM, MIL: Yes** 2003, 2004, 2005 Vibe All engines Transmissions: All	**HO2S-11 (Bank 1 Sensor 1) Heater Circuit Malfunction Conditions:** Engine started, and the PCM detected the HO2S heater current level exceeded 2.0 amps, or was under 0.2 amps for 3 seconds with the heater commanded "on". The HO2S Heater Performance Sensor diagnostic monitors the operation of the HO2S-11 heater circuit. The heaters in the oxygen sensors increase the amount of time the sensor spends in Closed Loop fuel control operation or catalyst monitoring operation. The HO2S heater greatly reduces the amount of time required for the fuel control HO2S-11 to become active. **Possible Causes** ● HO2S heater control circuit is open or shorted to ground ● HO2S heater control circuit is shorted to power ● HO2S heater power circuit is open (check power from the relay) ● HO2S heater is damaged or has failed ● PCM has failed

OBD II Trouble Code List (P0xxx Codes)

DTC	Trouble Code Title, Conditions & Possible Causes
DTC P0136 **2T CCM, MIL: Yes** 2003, 2004, 2005 Vibe All engines Transmissions: All	**Title: HO2S-12 (Bank 1 Sensor 2) Circuit Malfunction** **Trouble Code Conditions** Engine started, ECT sensor more than 158°F, engine speed more over 1600 rpm in closed loop, and the PCM detected the HO2S signal was more than or equal to 400 mv, or that the HO2S signal was less than or equal to 500 mv for 5 seconds during the test. **Possible Causes** • HO2S signal circuit is open between the sensor and the PCM • HO2S signal circuit is shorted to sensor or chassis ground • HO2S signal circuit is shorted to VREF or system power (B+) • HO2S is damaged, contaminated or it has failed • PCM has failed
DTC P0141 **2T CCM, MIL: Yes** 2003, 2004, 2005 Vibe All engines Transmissions: All	**HO2S-12 (Bank 1 Sensor 2) Heater Circuit Malfunction Conditions:** Engine started, and the PCM detected the HO2S heater current level was more than 2.0 amps or was less than 0.2 amps with the heater commanded "on". The HO2S Heater Performance Sensor diagnostic monitors the operation of the HO2S heater circuit. The heaters in the oxygen sensors increase the amount of time the sensor spends in Closed Loop fuel control operation or catalyst monitoring operation. **Possible Causes** • HO2S heater control circuit is open, shorted to ground or power • HO2S heater power circuit is open (check power from the relay) • HO2S heater is damaged or has failed • PCM has failed
DTC P0171 **2T FUEL, MIL: Yes** 2003, 2004, 2005 Vibe All engines Transmissions: All	**Fuel Trim System Lean (Bank 1) Conditions:** Engine started, ECT sensor more than 158°F, engine running at a steady speed in closed loop, and the PCM detected the Long Term fuel trim value was constantly high during a stable engine operation. The Fuel Trim test monitors HO2S indications of the air/fuel mixture. The PCM uses a Closed Loop air/fuel metering system to achieve the optimum fuel economy and emission control. The PCM monitors the HO2S signal voltage and when in Closed Loop adjusts fuel delivery based on the HO2S signal voltage. Changes in fuel delivery are indicated by the Long Term and Short Term fuel trim values that are displayed on the Scan Tool. The ideal fuel trim value is around 0%. The PCM adds fuel when the HO2S signal indicates a lean condition. Additional fuel is indicated by fuel trim values that are above 0%. The PCM reduces the amount of fuel delivered if a rich condition is indicated by the HO2S. Fuel trim values below 0% indicate a less fuel. P0171 is set if exhaust emissions reach excessive levels due to a lean condition. **Possible Causes** • Air leaks after the MAF sensor, or in the EGR or PCV system • Base engine "mechanical" fault affecting one or more cylinders • Exhaust leaks located in front of the HO2S location • Fuel control sensor is out of calibration (i.e., ECT, IAT or MAF) • Fuel delivery system supplying too little fuel during cruise or idle periods (e.g., faulty fuel pump or dirty, restricted fuel filter) • Fuel injector (one or more) dirty or pressure regulator has failed • HO2S is contaminated, deteriorated or it has failed • Vehicle driven low on fuel or until it ran out of fuel
DTC P0172 **2T FUEL, MIL: Yes** 2003, 2004, 2005 Vibe All engines Transmissions: All	**Fuel Trim System Rich (Bank 1) Conditions:** Engine started, ECT sensor more than 158°F, engine running at a steady speed in closed loop, and the PCM detected the Long Term fuel trim value was constantly low during stable engine operation. The PCM will add fuel when the HO2S signal is indicating a lean condition. Additional fuel is indicated by fuel trim values above 0%. The PCM reduces the amount of fuel delivered if a rich condition is indicated by the HO2S. Fuel trim values below 0% indicate a less fuel. P0172 is set if exhaust emissions reach excessive levels due to a rich condition. **Possible Causes** • Base engine "mechanical" fault affecting one or more cylinders • EVAP system component has failed or canister fuel saturated • Exhaust leaks located in front of the HO2S location • Fuel control sensor is out of calibration (i.e., ECT, IAT or MAF) • Fuel delivery system supplying too much fuel during cruise or idle periods (e.g., faulty fuel pump, or faulty pressure regulator) • Fuel injector(s) is leaking or stuck partially open (one or more) • HO2S is contaminated, deteriorated or it has failed

OBD II Trouble Code List (P0xxx Codes)

DTC	Trouble Code Title, Conditions & Possible Causes
DTC P0300 **2T MISFIRE, MIL: Yes** 2003, 2004, 2005 Vibe All engines Transmissions: All	**Multiple Cylinder Misfire Detected Conditions:** Engine started, engine speed between 200-4000 rpm, ECT sensor from 20-254°F, system voltage from 9-17v, and the PCM detected the misfire rate during 200 or 1000 engine revolutions was more than a specified value in 2 or more cylinders during the Misfire test. *Note: If the misfire is severe, the MIL will flash on/off on the 1st trip!* **Possible Causes** ● Air leak in the intake manifold, or in the EGR or PCM system ● Base engine mechanical fault that affects only one cylinder ● Fuel delivery component fault that affects only one cylinder (i.e., a contaminated, dirty or sticking fuel injector) ● Ignition system problem (coil or plug) that affects one cylinder
DTC P0301 **2T MISFIRE, MIL: Yes** 2003, 2004, 2005 Vibe All engines Transmissions: All	**Cylinder 1 Misfire Detected Conditions:** Engine started, engine speed between 200-4000 rpm, ECT sensor from 20-254°F, system voltage from 9-17v, and the PCM detected the misfire rate during 200 or 1000 engine revolutions was more than a specified value in Cylinder 1 during the Misfire test period. *Note: If the misfire is severe, the MIL will flash on/off on the 1st trip!* **Possible Causes** ● Air leak in the intake manifold, or in the EGR or PCM system ● Base engine mechanical fault that affects only Cylinder 1 ● Fuel delivery component fault that affects only Cylinder 1 (i.e., a contaminated, dirty or sticking fuel injector) ● Ignition system fault (coil or plug) that affects only Cylinder 1
DTC P0302 **2T MISFIRE, MIL: Yes** 2003, 2004, 2005 Vibe All engines Transmissions: All2.bmp ; 4D Sedan	**Cylinder 2 Misfire Detected Conditions:** Engine started, engine speed between 200-4000 rpm, ECT sensor from 20-254°F, system voltage from 9-17v, and the PCM detected the misfire rate during 200 or 1000 engine revolutions was more than a specified value in Cylinder 2 during the Misfire test period. *Note: If the misfire is severe, the MIL will flash on/off on the 1st trip!* **Possible Causes** ● Air leak in the intake manifold, or in the EGR or PCM system ● Base engine mechanical fault that affects only Cylinder 2 ● Fuel delivery component fault that affects only Cylinder 2 (i.e., a contaminated, dirty or sticking fuel injector) ● Ignition system fault (coil or plug) that affects only Cylinder 2
DTC P0303 **2T MISFIRE, MIL: Yes** 2003, 2004, 2005 Vibe All engines Transmissions: All	**Cylinder 3 Misfire Detected Conditions:** Engine started, engine speed between 200-4000 rpm, ECT sensor from 20-254°F, system voltage from 9-17v, and the PCM detected the misfire rate during 200 or 1000 engine revolutions was more than a specified value in Cylinder 3 during the Misfire test period. *Note: If the misfire is severe, the MIL will flash on/off on the 1st trip!* **Possible Causes** ● Air leak in the intake manifold, or in the EGR or PCM system ● Base engine mechanical fault that affects only Cylinder 3 ● Fuel delivery component fault that affects only Cylinder 3 (i.e., a contaminated, dirty or sticking fuel injector) ● Ignition system fault (coil or plug) that affects only Cylinder 3
DTC P0304 **2T MISFIRE, MIL: Yes** 2003, 2004, 2005 Vibe All engines Transmissions: All	**Cylinder 4 Misfire Detected Conditions:** Engine started, engine speed between 200-4000 rpm, ECT sensor from 20-254°F, system voltage from 9-17v, and the PCM detected the misfire rate during 200 or 1000 engine revolutions was more than a specified value in Cylinder 4 during the Misfire test period. *Note: If the misfire is severe, the MIL will flash on/off on the 1st trip!* **Possible Causes** ● Air leak in the intake manifold, or in the EGR or PCM system ● Base engine mechanical fault that affects only Cylinder 4 ● Fuel delivery component fault that affects only Cylinder 4 (i.e., a contaminated, dirty or sticking fuel injector) ● Ignition system fault (coil or plug) that affects only Cylinder 4
DTC P0325 **1T CCM, MIL: Yes** 2003, 2004, 2005 Vibe All engines Transmissions: All	**Knock Sensor 1 Circuit Malfunction Conditions:** Engine started, and the PCM detected a bias voltage on the KS circuit, or with the engine speed over 2000 rpm, TP sensor and VSS inputs indicating the vehicle is accelerating, the PCM did not detect change in the Knock Sensor 1 signal for 20 seconds. **Possible Causes** ● Knock sensor signal circuit is open or shorted to ground ● Knock sensor signal circuit is shorted to VREF or system power ● Knock sensor is damaged or has failed ● PCM has failed

OBD II Trouble Code List (P0xxx Codes)

DTC	Trouble Code Title, Conditions & Possible Causes
DTC P0335 **2T CCM, MIL: Yes** 2003, 2004, 2005 Vibe All engines Transmissions: All	**Crankshaft Position Sensor Circuit Malfunction Conditions:** Engine cranking for 5 seconds, or engine speed over 600 rpm, and the PCM did not detect any CKP signals during 2 engine revolutions. **Possible Causes** • CKP sensor connector is damaged, open or shorted • CKP (Magnetic) sensor signal (+) or (-) circuit is open, shorted to ground or to system power • CKP sensor is damaged or has failed • PCM has failed
DTC P0340 **1T CCM, MIL: Yes** 2003, 2004, 2005 Vibe All engines Transmissions: All	**Camshaft Position Sensor Circuit Malfunction Conditions:** Engine cranking for 5 seconds, or engine speed over 600 rpm, and the PCM did not detect any CMP sensor signals. **Possible Causes** • CMP sensor connector is damaged, open or shorted • CMP sensor signal (+) circuit is open or shorted to ground • CMP sensor signal (-) circuit is open or shorted to ground • CMP sensor is damaged or has failed • PCM has failed
DTC P0420 **2T CAT, MIL: Yes** 2003, 2004, 2005 Vibe All engines Transmissions: All	**Catalyst System Low Efficiency (Bank 1) Conditions:** Engine started, ECT sensor more than 158°F, engine speed from 2,500-3,000 rpm for at least 3 minutes with the engine running in closed loop, and the PCM detected the voltage swings (and switch rates) of the rear HO2S and front HO2S were similar for 96 seconds. **Possible Causes** • Air leaks in the exhaust manifold or exhaust pipes • Catalytic converter is contaminated, damaged or has failed • Front HO2S is older (aged) than the rear HO2S (HO2S is lazy) • PCM has failed
DTC P0440 **2T EVAP, MIL: Yes** 2003, 2004, 2005 Vibe All engines Transmissions: A/T	**EVAP System Incorrect Purge Flow Detected Conditions:** ECT and IAT sensors from 50-95°F and within a few degrees at startup, engine started, ECT sensor at least 165°F, engine runtime over 20 minutes, BARO sensor over 75 kPa, Fuel Level from 25-75%, and the PCM detected the EVAP system was not able to achieve or maintain the expected vacuum pressure during the test. **Possible Causes** • Charcoal canister is loaded with fuel or moisture • Fuel filler cap loose, cross-threaded, incorrect part or damaged • Fuel tank pressure sensor is damaged or has failed • Fuel tank vapor line(s) blocked, damaged or disconnected • Purge or Vent solenoid control circuit open or shorted to ground • PCM has failed
DTC P0441 **2T EVAP, MIL: Yes** 2003, 2004, 2005 Vibe All engines Transmissions: All	**EVAP System No Flow During Purge Conditions:** ECT and IAT sensors from 50-95°F and within a few degrees at startup, engine started, ECT sensor more than 165°F, Fuel Level at 25-75%, then after the vehicle was driven in city traffic conditions, the PCM detected the pressure indicated in the EVAP canister and the fuel tank did not drop with the EVAP purge solenoid valve "on"; or the pressure indicated in the EVAP canister remains too low when compared to atmospheric pressure with the purge valve "off". **Possible Causes** • Charcoal canister is loaded with fuel or moisture • Fuel filler cap loose, cross-threaded, incorrect part or damaged • Fuel tank pressure sensor is damaged or has failed • Fuel tank vapor line(s) blocked, damaged or disconnected • Purge or Vent solenoid control circuit open or shorted to ground • PCM has failed
DTC P0442 **2T EVAP, MIL: Yes** 2003, 2004, 2005 Vibe All engines Transmissions: All	**EVAP System Small Leak (0.040") Detected Conditions:** ECT and IAT sensors from 50-95°F and within a few degrees of each other at startup, engine started, ECT sensor at least 165°F, engine runtime over 20 minutes, BARO sensor over 75 kPa, Fuel Level at 25-75%, and the PCM detected the EVAP system achieved a vacuum, but that a vacuum decay was detected during the diagnostic test. **Possible Causes** • Canister purge solenoid is damaged or it has failed • Charcoal canister is loaded with fuel or moisture • Fuel filler cap loose, cross-threaded, incorrect part or damaged • Fuel tank is cracked (leaking), or a leak exists in the 'O' ring • Fuel tank pressure sensor is damaged or has failed • Fuel vapor line(s), fuel pipes or hoses damaged or leaking • PCM has failed

OBD II Trouble Code List (P0xxx Codes)

DTC	Trouble Code Title, Conditions & Possible Causes
DTC P0446 **2T CCM, MIL: Yes** 2003, 2004, 2005 Vibe All engines Transmissions: All	**EVAP System Vent System Performance Conditions:** ECT and IAT sensors from 50-95ºF and within a few degrees at startup, engine started, ECT sensor at least 165ºF, engine runtime over 20 minutes, BARO sensor over 75 kPa, Fuel Level at 25-75%, and the PCM detected the pressure indicated in the EVAP canister and the fuel tank did not drop with the EVAP purge solenoid "on"; or the pressure indicated in the EVAP canister remained very low when compared to atmospheric pressure with the purge solenoid valve off. **Possible Causes** ● EVAP vent solenoid control circuit is open or shorted to ground ● EVAP vent solenoid is stuck partially open ● EVAP vent solenoid is damaged or has failed ● PCM has failed
DTC P0450 **2T CCM, MIL: Yes** 2003, 2004, 2005 Vibe All engines Transmissions: All	**Fuel Tank Pressure Sensor Circuit Malfunction Conditions:** Engine started, engine runtime over 10 seconds, and the PCM detected the FTP sensor was less than 16.3" H2O (-30.48 mm Hg), or the FTP signal was over 8.16" H2O (15.24 mm Hg) for 7 seconds. The FTP Sensor circuit test monitors the FTP sensor voltage. The EVAP control system prevents the emission of the fuel vapors. The fuel vapor generated in the fuel tank while driving or idling enters the EVAP canister where a charcoal element absorbs and stores fuel vapors. When certain operating conditions are met, the PCM purges the canister of stored vapors by turning "on" the EVAP canister purge valve in order to purge the system. **Possible Causes** ● FTP sensor circuit is open, shorted to ground or to power ● FTP sensor ground circuit is open ● FTP sensor power circuit is open ● FTP sensor is damaged or it has failed ● PCM has failed
DTC P0451 **2T CCM, MIL: Yes** 2003, 2004, 2005 Vibe All engines Transmissions: All	**Fuel Tank Pressure Sensor Range/Performance Conditions:** Engine started, vehicle not moving, EVAP Vent Pressure solenoid commanded "off", and the PCM detected the FTP sensor signal shifted rapidly from inside to outside a specified range. **Possible Causes** ● Fuel tank pressure sensor connector is damaged or shorted ● Fuel tank pressure sensor vacuum hoses loose or damaged ● Fuel tank pressure sensor is damaged or out-of-calibration ● PCM has failed
DTC P0500 **1T CCM, MIL: Yes** 2003, 2004, 2005 Vibe All engines Transmissions: All	**Vehicle Speed Sensor Circuit Malfunction Conditions:** Engine started, Park Neutral switch indicating off, TP angle over 13 degrees, engine speed over 2350 rpm, and the PCM did not detect any VSS inputs for 5 seconds. The VSS is mounted to the transaxle. As the transaxle turns the VSS, the VSS provides the speedometer with a VSS input through voltage pulses. Each revolution of the axle shaft equals 4 pulses. This input is used to drive the speedometer. **Possible Causes** ● VSS connector is damaged, open or shorted ● VSS signal circuit is open or shorted to ground ● VSS is damaged or has failed (behind the Instrument Cluster) ● PCM has failed
DTC P0505 **1T CCM, MIL: Yes** 2003, 2004, 2005 Vibe All engines Transmissions: All	**Idle Control System Performance Conditions:** Engine started, and the PCM detected that the Actual idle speed varied greatly from the Desired idle speed. The IAC System test observes the engine speed to check the operation of the IAC valve. The PCM controls engine idle speed with the idle air control valve. **Possible Causes** ● Throttle valve is dirty or sticking (it may need to be cleaned) ● ISC motor is damaged or has failed ● PCM has failed
DTC P0601 **2T PCM, MIL: Yes** 2003, 2004, 2005 Vibe All engines Transmissions: All	**PCM Read Only Memory Error Conditions:** Key on, and the PCM detected an internal check sum data error. Note: Refer to a code repair chart for instructions on how to replace a PCM and how to reprogram the replacement unit. **Possible Causes** ● Check the PCM for possible water contamination ● Clear the trouble codes and retest for this trouble code. If the same trouble code resets, the PCM has failed and must be replaced to repair this problem.

OBD II Trouble Code List (P1xxx Codes)

DTC	Trouble Code Title, Conditions & Possible Causes
DTC P0710 **1T CCM, MIL: Yes** 2003, 2004, 2005 Vibe All engines Transmissions: A/T	**Transmission Range Switch Circuit Malfunction Conditions:** Engine started, engine running in Drive at a vehicle speed over 38 mph, and the PCM detected the Transmission Fluid Temperature sensor resistance was less than 79 ohms; or after the engine had been operating for 15 minutes or more, it detected the resistance of the transmission temperature sensor was more than 156,000 ohms. **Possible Causes** ● TFT sensor connector is damaged, open or shorted to ground ● TFT sensor signal circuit is open, shorted to ground or to VREF ● TFT sensor is damaged, failed or out-of-calibration (skewed) ● PCM has failed
DTC P0711 **1T CCM, MIL: Yes** 2003, 2004, 2005 Vibe All engines Transmissions: A/T	**Transmission Range Switch Circuit Malfunction Conditions:** Engine started, engine runtime over 12 seconds, and the PCM detected the difference between the ambient air temperature and the TFT sensor temperature was more than 14ºF; or the PCM detected the change in transmission fluid temperature was less than 50ºF after 20 minutes and 6.2 miles of normal driving conditions. When the transmission fluid is cold, the sensor resistance is high. As the transmission fluid warms up, the sensor resistance decreases. This test monitors the TFT sensor circuit to determine if it is not in the normal range. This test can detect if the signal is erratic, intermittent, or skewed. The normal TFT sensor range is -40ºF to +437ºF. **Possible Causes** ● TFT sensor connector is damaged, open or shorted to ground ● TFT sensor signal circuit is open, shorted to ground or to VREF ● TFT sensor is damaged, failed or out-of-calibration (skewed) ● PCM has failed
DTC P0750 **1T CCM, MIL: Yes** 2003, 2004, 2005 Vibe All engines Transmissions: A/T	**A/T Shift Solenoid 1 Range/Performance Conditions:** Engine started, vehicle driven in Drive, and the PCM detected the control of Shift Solenoid 1 did not agree with the Actual gear position. A perquisite for this test is that Shift Solenoid 1 does not have an electrical problem. **Possible Causes** ● Control valve assembly is damaged (check for any restrictions) ● Shift select cable out of adjustment. ● Shift solenoid 1 is clogged or leaking ● Transaxle mechanical malfunction. ● PCM has failed
DTC P0753 **1T CCM, MIL: Yes** 2003, 2004, 2005 Vibe All engines Transmissions: A/T	**A/T Shift Solenoid 1 Circuit Malfunction Conditions:** Engine started, vehicle driven in a forward gear, and the PCM detected an unexpected voltage condition on the Shift Solenoid 1 control circuit. These two solenoids are used to electronically control the shifting of the transaxle from 1st gear to 4th gear. The PCM outputs system voltage to turn "on" (energize) or remove voltage to turn "off" (de-energize) either solenoid when it deems necessary. The transaxle shifts with the combination of both solenoids being turned on and off. Shift solenoid 1 is normally "on" in 1st and 2nd gear and never "on" in 3rd or 4th gear. Shift solenoid 2 is normally "on" in 2nd and 3rd gear, and never "on" in 1st or 4th gear. 4th gear/overdrive is obtained when both shift solenoids are turned "on" at the same time. **Possible Causes** ● Shift solenoid 1 control circuit is open or shorted to ground ● Shift solenoid 1 control circuit is shorted to system power ● Shift solenoid 1 is damaged (resistance is 11-16 ohms at 75ºF) ● PCM has failed
DTC P0755 **2T CCM, MIL: Yes** 2003, 2004, 2005 Vibe All engines Transmissions: A/T	**A/T Shift Solenoid 2 Range/Performance Conditions:** Engine started, vehicle driven in Drive, and the PCM detected the control of Shift Solenoid 2 did not agree with the Actual gear position. A prerequisite for this test is that Shift Solenoid 2 does not have an electrical problem. **Possible Causes** ● Control valve assembly is damaged (check for any restrictions) ● Shift select cable out of adjustment. ● Shift solenoid 2 is clogged or leaking ● Transaxle mechanical malfunction. ● PCM has failed

OBD II Trouble Code List (P0xxx – P1xxx Codes)

DTC	Trouble Code Title, Conditions & Possible Causes
DTC P0758 **1T CCM, MIL: Yes** 2003, 2004, 2005 Models: Vibe All engines Transmissions: A/T	**A/T Shift Solenoid 2 Circuit Malfunction Conditions:** Engine started, vehicle driven in a forward gear, and the PCM detected an unexpected voltage condition on the Shift Solenoid 2 control circuit. Shift Solenoid 1 and 2 are used to electronically control the shifting of the transaxle from 1st gear to 4th gear. The PCM outputs system voltage to turn 'on' (energize) or remove voltage to turn "off" (de-energize) either solenoid when necessary. The transaxle shifts with the combination of both solenoids being turned on and off. Shift solenoid 1 is normally "on" in 1st and 2nd gear and never "on" in 3rd or 4th gear. Shift solenoid 2 is normally "on" in 2nd and 3rd gear, and never "on" in 1st or 4th gear. 4th gear/overdrive is obtained with both shift solenoids energized at the same time. **Possible Causes** ● Shift solenoid 2 control circuit is open or shorted to ground ● Shift solenoid 2 control circuit is shorted to system power ● Shift solenoid 2 is damaged (resistance is 11-16 ohms at 75ºF) ● PCM has failed
DTC P0770 **1T CCM, MIL: Yes** 2003, 2004, 2005 Vibe All engines Transmissions: A/T	**Torque Converter Clutch Lockup Performance Conditions:** Engine started, vehicle driven to over 30 mph at a steady speed for over one minute, and the PCM detected the TCC system did not engage when commanded "on", or it did nod disengage when it was commanded "off" during the CCM test. **Possible Causes** ● ATF level is too low, or fluid is burnt or contaminated ● TCC solenoid valve is stuck "off" due to sediment or binding ● TCC solenoid valve is stuck "on due to sediment or binding ● TCC solenoid is damaged or has failed ● Transmission has internal component damage or failure
DTC P0773 **1T CCM, MIL: Yes** 2003, 2004, 2005 Vibe All engines Transmissions: A/T	**Torque Converter Clutch Circuit Malfunction Conditions:** Engine started, vehicle driven to over 30 mph at a steady speed for over one minute, and the PCM detected an unexpected voltage condition on the TCC control circuit with the TCC commanded "on" or with it commanded "off". **Possible Causes** ● P/N switch signal circuit is open, shorted to ground or to power ● P/N switch power or ground circuit is open ● P/N switch is out of adjustment, damaged or has failed ● PCM has failed
DTC P1300 **1T CCM, MIL: Yes** 2003, 2004, 2005 Vibe All engines Transmissions: All	**Ignition Coil 1 Primary Feedback Circuit Malfunction Conditions:** Engine cranking or running; and the PCM did not receive any failsafe signal from Ignition Coil 1primary after two consecutive ignition trigger signal cycles were detected. **Possible Causes** ● Ignition coil primary circuit is open or shorted to ground ● Ignition coil is damaged or has failed ● Ignition system "noise" filter may be shorted to ground ● PCM has failed
DTC P1305 **1T CCM, MIL: Yes** 2003, 2004, 2005 Vibe All engines Transmissions: All	**Ignition Coil 2 Primary Feedback Circuit Malfunction Conditions:** Engine cranking or running; and the PCM did not receive any failsafe signal from Ignition Coil 2 primary after two consecutive ignition trigger signal cycles were detected. **Possible Causes** ● Ignition coil primary circuit is open or shorted to ground ● Ignition coil is damaged or has failed ● Ignition system "noise" filter may be shorted to ground ● PCM has failed
DTC P1310 **1T CCM, MIL: Yes** 2003, 2004, 2005 Vibe All engines Transmissions: All	**Ignition Coil 3 Primary Feedback Circuit Malfunction Conditions:** Engine cranking or running; and the PCM did not receive any failsafe signal from Ignition Coil 3 primary after two consecutive ignition trigger signal cycles were detected. **Possible Causes** ● Ignition coil primary circuit is open or shorted to ground ● Ignition coil is damaged or has failed ● Ignition system "noise" filter may be shorted to ground ● PCM has failed
DTC P1315 **1T CCM, MIL: Yes** 2003, 2004, 2005 Vibe All engines Transmissions: All	**Ignition Coil 4 Primary Feedback Circuit Malfunction Conditions:** Engine cranking or running; and the PCM did not receive any failsafe signal from Ignition Coil 4 primary after two consecutive ignition trigger signal cycles were detected. **Possible Causes** ● Ignition coil primary circuit is open or shorted to ground ● Ignition coil is damaged or has failed ● Ignition system "noise" filter may be shorted to ground ● PCM has failed

OBD II Trouble Code List (P1xxx Codes)

DTC	Trouble Code Title, Conditions & Possible Causes
DTC P1335 **2T CCM, MIL: Yes** 2003, 2004, 2005 Vibe All engines Transmissions: All	**Crankshaft Position Sensor Circuit Malfunction Conditions:** Engine speed more than 1000 rpm, and the PCM did not receive any Crankshaft Position (CKP) sensor pulses for more than 50 ms. **Possible Causes** • CKP sensor signal (+) or (-) circuit is open, shorted to ground or shorted to system power (an intermittent fault) • CKP sensor is damaged or has failed (an intermittent fault) • Crankshaft reluctor wheel contains debris or is damaged • PCM has failed
DTC P1346 **2T CCM, MIL: Yes** 1999, 2000, 2001, 2002 Vibe All engines Transmissions: All	**Intake Camshaft Position System Performance Conditions:** Engine started, ECT sensor more than 158°F, VSS over 5 mph, and the PCM detected unexpected deviation between the relationship of the CMP and CKP sensor signals, condition for 20 seconds. **Possible Causes** • Camshaft is incorrectly installed or incorrectly timed • CMP System actuator is damaged or has failed • CMP sensor is damaged, out of alignment or has failed • Timing chain has jumper or is worn out
DTC P1349 **2T CCM, MIL: Yes** 2003, 2004, 2005 Vibe All engines Transmissions: All	**Intake Camshaft Position System Performance Conditions:** Engine started, ECT sensor more than 158°F, VSS more than 5 mph, and PCM did not detect any change in camshaft timing after the activating the CMP actuator, or it detected that the camshaft timing was fixed, either condition met for 20 seconds. Note: If DTC P1346 and P1349 are both set, the solenoid has failed. **Possible Causes** • Camshaft is incorrectly installed or incorrectly timed • CMP System actuator is damaged or has failed • Check the oil passages to the actuator for a restriction • Timing chain has jumper or is worn out
DTC P1520 **2T CCM, MIL: Yes** 2003, 2004, 2005 Vibe All engines Transmissions: All	**Stop Lamp Switch Circuit Malfunction Conditions:** Engine started, VSS over 19 mph, and the PCM detected the Stop Lamp switch indicated a continuous "on" condition. If the PCM detects that the Stop Lamp switch remains closed (i.e., that it never opens), this code is set. The PCM disables the TCC when the switch closes. **Possible Causes** • Stop lamp switch signal circuit is shorted to system power (B+) • Stop lamp switch is damaged or has failed (in closed position) • Stop lamp switch is out-of-adjustment • PCM has failed
DTC P1600 **1T CCM, MIL: Yes** 2003, 2004, 2005 Vibe All engines Transmissions: All	**PCM Battery Circuit Conditions:** Key on, and the PCM did not detect any power on its battery power circuit (connector C1-1). **Possible Causes** • Backup Power Supply circuit is open (check the Dome fuse) • Battery connections are corroded, dirty or loose • Battery has been disconnected or removed • PCM has failed
DTC P1621 **1T PCM, MIL: Yes** 2003, 2004, 2005 Vibe All engines Transmissions: All	**PCM Long Term Memory Error Conditions:** Engine started and the PCM detected an internal memory reset error had occurred. **Possible Causes** • Record any related trouble codes in memory • Reprogram the PCM and recheck for the same trouble code • If the same trouble code resets, replace & reprogram the PCM • Perform the Security System (Passlock) Relearn Procedure
DTC P1656 **1T CCM, MIL: Yes** 2003, 2004, 2005 Vibe All engines Transmissions: All	**Intake CMP Actuator Solenoid Control Circuit Malfunction Conditions:** Engine started and PCM detected an unexpected voltage condition on the Intake CMP Actuator Solenoid Control circuit during the test. **Possible Causes** • Intake Actuator solenoid control circuit is open, shorted to ground or shorted to system power (B+) • Intake Actuator solenoid power circuit is open • Intake Actuator solenoid is damaged or has failed • PCM has failed

OBD II Trouble Code List (P1xxx Codes)

DTC	Trouble Code Title, Conditions & Possible Causes
DTC P1725 **1T CCM, MIL: Yes** 2003, 2004, 2005 Vibe All engines Transmissions: All	**A/T Input Shaft Speed Sensor Circuit Malfunction Conditions:** Engine started, transmission in 2nd, 3rd or O/D position, PNP and Shift Solenoids operating normally, vehicle driven to over 20 mph, and PCM detected an unexpected voltage condition on the Input Shaft Speed (ISS) sensor circuit. The ISS sensor is mounted on the top of transaxle. As the transaxle turns, the ISS detects the rotation speed of the input turbine and outputs a signal to the PCM. The PCM uses the ISS signal to control the engine torque, shift timing and torque converter clutch operation. **Possible Causes** ● ISS connector is damaged, open or shorted ● ISS low signal circuit is open or shorted to ground ● ISS high signal circuit is open or shorted to ground ● ISS low and high signal circuit are shorted together ● ISS assembly is damaged or it has failed ● PCM has failed
DTC P1760 **1T CCM, MIL: Yes** 2003, 2004, 2005 Vibe All engines Transmissions: All	**A/T Line Pressure Solenoid Control Circuit Malfunction Conditions:** Engine started, vehicle driven in a forward gear, and the PCM detected the SLT circuit terminal was 0v for 1 second or more; or the SLT circuit terminal was 12v for 1 second or more. **Possible Causes** ● SLT control connector is damaged, open or shorted ● SLT control circuit is open or shorted to ground ● SLT control circuit is shorted to system power ● SLT assembly is damaged or it has failed ● PCM has failed
DTC P1780 **1T CCM, MIL: Yes** 2003, 2004, 2005 Vibe All engines Transmissions: A/T	**Park Neutral Position Switch Circuit Malfunction Conditions:** Engine started, vehicle driven, and the PCM detected engine speed changed from below 300 rpm to over 500 rpm when the vehicle speed was 0 mph and the PNP switch is open (i.e., in R, D, 2 or L gear position); or with the PNP switch closed, the PCM detected a P/N gear position for 10 seconds while the engine was running at over 2500 RPM and with more than 38.5% of calculated load value. The PCM determines when the vehicle is in Park or Neutral gear position by detecting the Park/Neutral Position (PNP) switch input. The RED/BLK wire to the PCM is grounded by the PNP switch when the transmission manual selector is in Park or Neutral gear position. **Possible Causes** ● P/N switch signal circuit is open, shorted to ground or to power ● P/N switch power or ground circuit is open ● P/N switch is out of adjustment, damaged or has failed ● PCM has failed
DTC P1790 **1T CCM, MIL: Yes** 2003, 2004, 2005 Vibe All engines Transmissions: A/T	**A/T Second Brake Control Solenoid Performance Conditions:** Engine started, vehicle driven in a forward gear, and the PCM detected the Second Brake Control solenoid resistance was 30 ohms or less (indicating a short circuit condition) with the solenoid energized; or the solenoid resistance was 100, 000 ohms or more (indicating an open circuit condition) with the solenoid de-energized. The 2nd brake control solenoid is energized and de-energized by the PCM. The solenoid controls the orifice control valve that engages and disengages the O/D and 2nd brake. The 2nd brake solenoid is activated when the reverse clutch is applied or released. The combined action of the solenoid is to control shift timing and feel. **Possible Causes** ● 2nd brake solenoid connector is damaged, open or shorted ● 2nd brake solenoid control circuit is open or shorted to ground ● 2nd brake solenoid control circuit is shorted to system power ● 2nd brake control solenoid is damaged or it has failed ● PCM has failed

Contents

GM CARS (EXC. CADILLAC)

2.0L, 2.2L, 2.4L 4-Cylinder Engines ....Page 4-2
3.1L, 3.4L, 3.5L, 3.6L, 3.8L V6 Engines...Page 4-39
4.0L, 4.2L, 4.3L, 4.6L, 4.8L, 5.3L, 5.7L, 6.0L V8 Engines ....Page 4-94

CADILLAC

Cadillac Catera 3.0L V6 Engines..Page 4-173
Cadillac DeVille 4.6L V8 Engines ....Page 4-203
Cadillac Eldorado 4.6L V8 Engines ....Page 4-232
Cadillac Seville 4.6L V8 Engines..Page 4-250

GM TRUCKS & VANS

4.3L, 5.7L, 6.0L V8 Engines...Page 4-297

GENERAL MOTORS CARS: 2000-05 BUICK, CHEVROLET, PONTIAC, OLDSMOBILE
4-CYLINDER ENGINES: 2.0L, 2.2L, 2.4L

CAMSHAFT POSITION (CMP) SENSOR

Description & Operation

2.2L (LN2) & 2.4L ENGINES

The camshaft position (CMP) sensor signal is a digital ON/OFF pulse, output once per revolution of the camshaft. The CMP sensor does not directly affect the operation of the ignition system. The CMP sensor information is used by the powertrain control module (PCM) to determine the position of the valve train relative to the crankshaft. By monitoring the CMP and CKP signals the PCM can accurately time the operation of the fuel injectors. The CMP sensor is connected to the PCM by a 12-volt, low reference, and signal circuit.

Component Locations

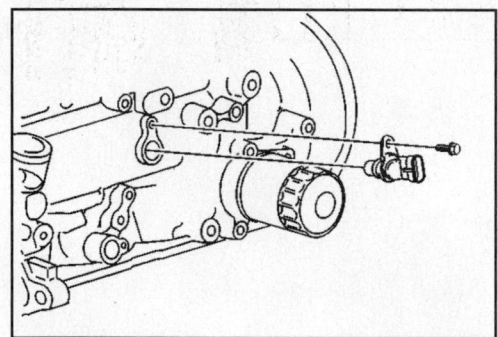

Location of the CMP sensor on 2.2L (LN2) engine

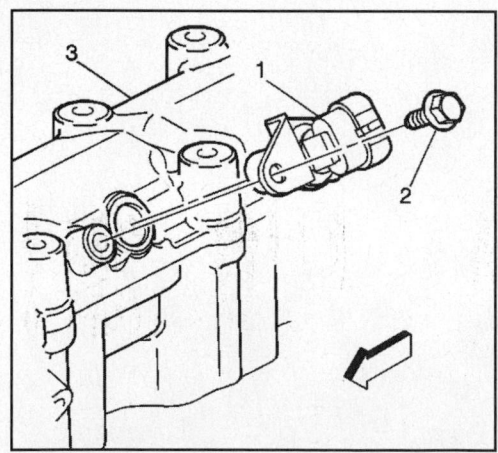

Showing the CMP sensor (1) location on 2.4L engine

Removal & Installation

2.2L (LN2)

1. Remove the air cleaner outlet resonator.
2. Remove the camshaft position (CMP) sensor electrical connector.
3. Remove the CMP sensor retaining bolt.
4. Remove the CMP sensor.
5. To install, reverse the removal procedure. Tighten the CMP sensor bolt to 89 inch lbs. (10 Nm).

2.4L ENGINE

1. Disconnect the camshaft position (CMP) sensor electrical connector.

2. Remove the CMP sensor fastener (2).

3. Remove the CMP sensor (1) from the camshaft housing (3).

4. Installation is the reverse of the removal procedure. Tighten the CMP sensor bolt to 11 ft. lbs. (15 Nm).

Connector Pinouts

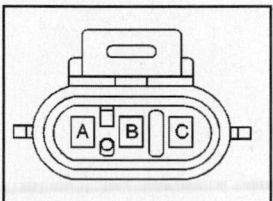

PIN	WIRE COLOR	CIRCUIT NO.	FUNCTION
A	RED	631	12 VOLT REFERENCE
B	PNK/BLK	632	LOW REFERENCE
C	BRN/WHT	633	CMP SENSOR SIGNAL

2.2L (LN2) Engine CMP Sensor Connector Pinout

PIN	WIRE COLOR	CIRCUIT NO.	FUNCTION
A	BRN/WHT	633	CMP SENSOR SIGNAL
B	PNK/BLK	632	LOW REFERENCE
C	RED	631	12 VOLT REFERENCE

2.4L Engine CMP Sensor Connector Pinout

CRANKSHAFT POSITION (CKP) SENSOR

Description & Operation

2.2L & 2.4L ENGINES

The crankshaft position (CKP) sensor is a permanent magnet generator, known as a variable reluctance sensor. The CKP sensor produces an AC voltage of different amplitude and frequency. The frequency depends on the velocity of the crankshaft. The AC voltage output depends on the crankshaft position and the battery voltage. The CKP sensor works in conjunction with a 7X reluctor wheel attached to the crankshaft. The CKP sensor produces 7 pulses for each revolution of the crankshaft. The pulse from the 10-degree notch is known as the sync pulse. The sync pulse is used to synchronize the coil firing sequence with the crankshaft position. The CKP sensor is used for ignition timing, the fuel injector timing, misfire diagnostics and tachometer display. The CKP sensor is connected to the ECM by a signal circuit and a low reference circuit.

Component Locations

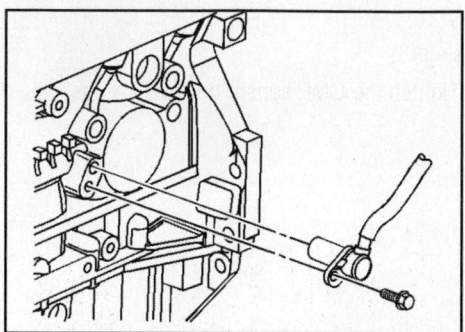

Showing location of CKP sensor on 2.2L (L61) engines

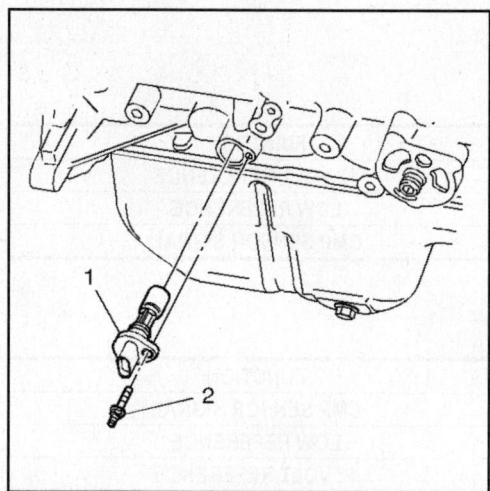

Showing location of CKP sensor (1) on 2.2L (LN2) engines

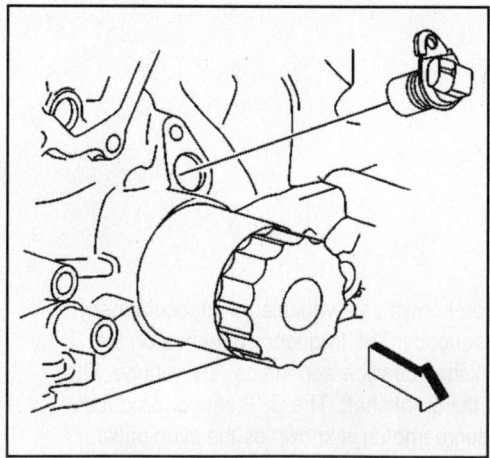

Showing the CKP Sensor Location on 2.4L Engines

Removal & Installation

2.2L (L61) ENGINES

1. Disconnect the battery.
2. Raise the vehicle.
3. Remove the front air deflector, if applicable.
4. Remove the starter.
5. Disconnect the Crankshaft Position (CKP) sensor electrical connector.
6. Remove the CKP sensor bolt.
7. Remove the CKP sensor.

To install:

8. Inspect the CKP sensor O-ring and lubricate with a mineral-based grease.
9. Gently insert the CKP sensor into the block.
10. Install the CKP sensor bolt. Tighten the CKP sensor bolt to 71 inch lbs. (8 Nm).
11. Reconnect the CKP sensor electrical connector.
12. Reinstall the starter.
13. Reinstall the front air deflector, if removed in previous step.
14. Lower the vehicle.
15. Reconnect the battery.
16. Perform the Crankshaft Variation Learn Procedure.

2.2L (LN2) ENGINES

1. Raise the vehicle
2. Remove the crankshaft position (CKP) sensor harness connector.
3. Remove the CKP sensor mounting bolt (2).
4. Remove the CKP sensor from the engine (1).
5. Inspect the CKP sensor O-ring for wear, cracks or leakage. Replace the O-ring if necessary
6. Installation is the reverse of the removal procedure. Tighten the sensor mounting bolt to 71 inch lbs. (8 Nm).

2.4L ENGINES

1. Raise and support the vehicle.
2. Disconnect the crankshaft position (CKP) sensor harness connector.
3. Remove the CKP sensor retaining bolt.
4. Remove the CKP sensor from the engine block.
5. Installation is the reverse of the removal procedure. Lubricate the O-ring. Tighten the sensor mounting bolt to 89 inch lbs. (10 Nm).

Crankshaft Variation Learn Procedure

2.2L ENGINES

1. Install a scan tool.
2. With a scan tool, monitor the powertrain control module (PCM) for DTCs. If other DTCs are set, except DTC P1336, refer to Diagnostic Trouble Code (DTC) List for the applicable DTC that set.

3. With a scan tool, select the "Crankshaft Position Variation Learn" procedure.

4. Observe the fuel cut-off for the engine that you are performing the learn procedure on.

5. The scan tool instructs you to perform the following:

- Block the drive wheels.

- Apply the vehicles parking brake.

- Cycle the ignition from OFF to ON.

- Apply and hold the brake pedal.

- Start and idle the engine.

- Turn OFF the A/C.

- Place the vehicle's transmission in Park (A/T) or Neutral (M/T).

- The scan tool monitors certain component signals to determine if all the conditions are met to continue with the procedure. The scan tool only displays the condition that inhibits the procedure. The scan tool monitors the following components:

  - Crankshaft position (CKP) sensors activity--If there is a CKP sensor condition, refer to the applicable DTC that set.

  - Camshaft position (CMP) sensor activity--If there is a CMP sensor condition, refer to the applicable DTC that set.

  - Engine coolant temperature (ECT)--If the engine coolant temperature is not warm enough, idle the engine until the engine coolant temperature reaches the correct temperature.

6. With the scan tool, enable the crankshaft position system variation learn procedure.

NOTE: *While the learn procedure is in progress, release the throttle immediately when the engine starts to decelerate. The engine control is returned to the operator and the engine responds to throttle position after the learn procedure is complete.*

7. Slowly increase the engine speed to the RPM that you observed.

8. Immediately release the throttle when fuel cut-off is reached.

9. The scan tool displays Learn Status: Learned this ignition. If the scan tool does NOT display this message and no additional DTCs set, refer to Symptom Diagnosis – No Codes section. If a DTC set, refer to the Diagnostic Trouble Code (DTC) List for the applicable DTC that set.

10. Turn OFF the ignition for 30 seconds after the learn procedure is completed successfully.

2.4L ENGINES

A crankshaft position (CKP) system variation learn procedure must be performed any time a change is made to the crankshaft sensor to crankshaft relationship. Changing the crank sensor to crankshaft relationship will not allow the powertrain control module (PCM) to detect misfire at all speeds and loads accurately, resulting in a possible false misfire diagnostic trouble code (DTC) being set.

Removing a part for inspection and then reinstalling the same part is considered a disturbance. A false DTC P0300 could be set if this procedure is not performed.

The learn procedure is required after the following service procedures have been performed, regardless of whether or not DTC P1336 is set:

- A PCM replacement
- An engine replacement
- A crankshaft replacement
- A CKP sensor replacement

- Any engine repair which disturbs the crankshaft/harmonic balancer relationship to the crankshaft position sensor

A fully warmed up engine is critical to learning the variation correctly. If a valid learn occurs, no other learns can be completed on that ignition cycle.

If the engine cuts out before the specified learn procedure engine speed or at normal fuel cut-off RPM, the PCM is not in the learn procedure mode. Review the crankshaft position system variation learn procedure and re-enable the learn procedure. Verify that the scan tool displays Test in Progress.

WARNING: Before performing the Crankshaft Position System Variation Learning Procedure always set the vehicle parking brake and block the drive wheels in order to prevent personal injury. Release the throttle immediately when the engine starts to decelerate in order to eliminate over revving the engine. Once the learn procedure is completed, the control module will return engine control to the operator and the engine will respond to the throttle position.

CAUTION: The battery must be fully charged and in good condition. Verify that the scan tool connection at the data link connector (DLC) is clean and tight before starting the crankshaft position (CKP) system variation learn procedure.

1. Close the hood.
2. Block the drive wheels and set the vehicle parking brake.
3. Put the vehicle in Park or Neutral.
4. Turn all the accessories OFF.
5. Connect a scan tool.
6. Start and run the engine until the engine is at normal operating temperature 85°C (185°F).
7. With the engine still running, use the scan tool in order to enable the crankshaft position (CKP) system variation learn procedure.
8. Press and hold the brake pedal firmly and raise the engine speed to the specified value of 3,920 RPM, RELEASING the throttle as soon as the engine cuts out.
9. Verify with the scan tool that the crankshaft variation has been learned.
10. A fully warmed up engine is critical to learning the variation correctly. If a valid learn occurs, no other learns can be completed on that ignition cycle.
11. If the engine cuts out before the specified learn procedure engine speed or at normal fuel cut-off RPM, the PCM is not in the learn procedure mode. Review the crankshaft position system variation learn procedure and re-enable the learn procedure. Verify that the scan tool displays Test in Progress.
12. If the variation will not learn, perform this procedure up to 10 times.

SCAN TOOL DISPLAY	POSSIBLE CAUSES
FACTORS OUT OF RANGE	RELUCTOR WHEEL-MACHINING QUALITY, A RUN OUT, OR AN INCORRECT AIR GAP.
OPPOSING FACTORS OUT OF RANGE	DISTURBANCE-NOISE ON THE CKP SENSOR CIRCUIT. REATTEMPT THE LEARN PROCEDURE.
SUM OUT OF RANGE	ENGINE IS TOO COLD. REATTEMPT THE LEARN PROCEDURE.
CRANK PULSE COUNT ERROR.	CRANK OR CAM SENSOR DTCS IS SET. REPAIR THE AFFECTED DTC FIRST

Excessive Crankshaft Variation Symptoms on 2.4L Engines

Connector Pinouts

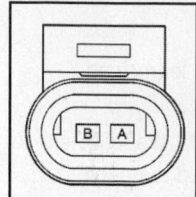

PIN	WIRE COLOR	CIRCUIT NO.	FUNCTION
A	YEL	573	CKP SENSOR SIGNAL
B	PPL	574	LOW REFERENCE

2.2L (LN2) Engine CKP Sensor Connector Pinout

PIN	WIRE COLOR	CIRCUIT NO.	FUNCTION
A	PNK	574	LOW REFERENCE
B	YEL	573	CKP SENSOR SIGNAL

2.2L (LN2) Engine CKP Sensor Connector Pinout

PIN	WIRE COLOR	CIRCUIT NO.	FUNCTION
A	PPL	574	LOW REFERENCE
B	YEL	573	CKP SENSOR SIGNAL

2.4L Engine CKP Sensor Connector Pinout

ENGINE COOLANT TEMPERATURE (ECT) SENSOR

Component Locations

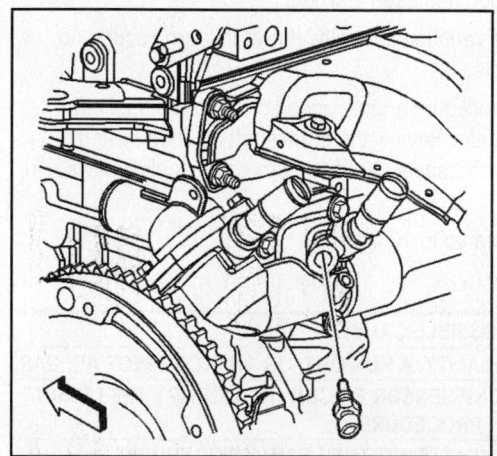

Location of the ECT sensor on 2.2L (L61) engines

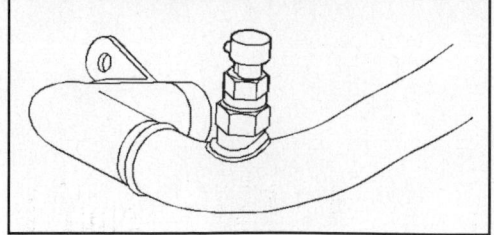

Showing ECT sensor location on 2.2L (LN2) engines

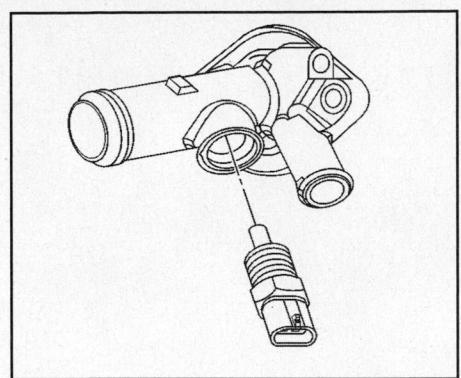

ECT sensor location on 2.4L engines

Removal & Installation

2.2L & 2.4L ENGINES

1. Turn OFF the ignition.
2. Drain the coolant system to below the ECT sensor.
3. Disconnect the engine coolant temperature (ECT) sensor electrical connector.
4. Carefully remove the ECT sensor.
5. Installation is the reverse of the removal procedure. Coat sensor threads with thread lock sealant and tighten sensor to 89 inch lbs. (10 Nm).

Connector Pinouts

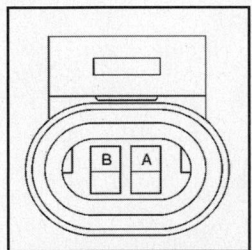

PIN	WIRE COLOR	CIRCUIT NO.	FUNCTION
A	BLK	2761	LOW REFERENCE
B	YEL	410	ECT SENSOR SIGNAL

2.2L (L61 & LN2) Engine ECT Sensor Connector Pinout

PIN	WIRE COLOR	CIRCUIT NO.	FUNCTION
A	BRN	718	LOW REFERENCE
B	YEL	410	ECT SENSOR SIGNAL

2.4L Engine ECT Sensor Connector Pinout

ENGINE COOLING FAN CONNECTORS

Component Locations

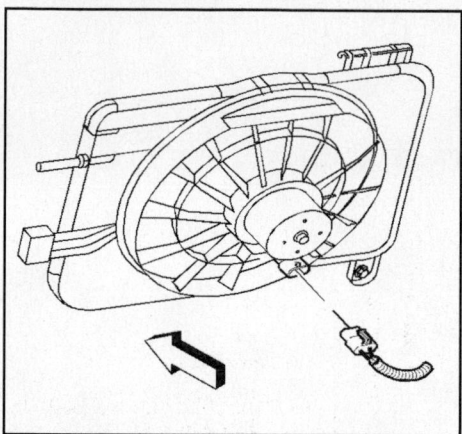

2.2L & 2.4L engine electric cooling fan and connector

Removal & Installation

2.2L & 2.4L ENGINES

1. Disconnect the negative battery cable.
2. Remove the hood close out filler retainers.
3. Remove the hood close out filler panel.
4. Remove the upper radiator mount bolts.
5. Remove the upper two hood latch support bolts.
6. Remove the two lower hood latch support bolts.
7. Raise the vehicle.
8. Remove the cooling fan mounting bolt.
9. Disconnect the electrical connector from the cooling fan.
10. Remove the cooling fan assembly.

To install:

11. Install the cooling fan assembly.
12. Connect the electrical connector to the cooling fan.
13. Install the cooling fan mounting bolt. Tighten the bolt to 53 inch lbs. (6 Nm).
14. Lower the vehicle.
15. Install the lower two hood latch support bolts. Tighten the bolts to 20 ft. lbs. (27 Nm).
16. Install the two upper hood latch support bolts. Tighten the two upper hood latch support bolts to 80 inch lbs. (9 Nm).
17. Install the upper radiator mount bolts. Tighten bolts to 89 inch lbs. (10 Nm).
18. Install the hood close out filler panel.
19. Install the hood close out filler panel retainers.

Connector Pinouts

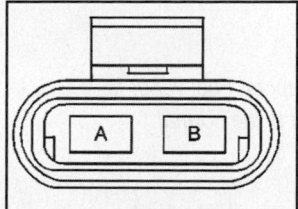

PIN	WIRE COLOR	CIRCUIT NO.	FUNCTION
A	BLK/WHT	151	GROUND
B	LT BLU	409	COOLANT FAN MOTOR SUPPLY VOLTAGE

2.2L & 2.4L Engine Electric Cooling Fan Connector Pinout

EVAP CANISTER VENT & PURGE SOLENOIDS

Description & Operation

The evaporative emission (EVAP) control system limits fuel vapors from escaping into the atmosphere. Fuel tank vapors are allowed to move from the fuel tank, due to pressure in the tank, through the vapor pipe, into the EVAP canister. Carbon in the canister absorbs and stores the fuel vapors. Excess pressure is vented through the vent line and EVAP vent valve to atmosphere. The EVAP canister stores the fuel vapors until the engine is able to use them. At an appropriate time, the control module will command the EVAP purge valve ON, open, allowing engine vacuum to be applied to the EVAP canister. With the EVAP vent valve OFF, open, fresh air will be drawn through the valve and vent line to the EVAP canister. Fresh air is drawn through the canister, pulling fuel vapors from the carbon. The air/fuel vapor mixture continues through the EVAP purge pipe and EVAP purge valve into the intake manifold to be consumed during normal combustion. The control module uses several tests to determine if the EVAP system is leaking.

LARGE LEAK TEST

This tests for large leaks and blockages in the EVAP system. The control module will command the EVAP vent valve ON, closed, and command the EVAP purge valve ON, open, with the engine running, allowing engine vacuum into the EVAP system. The control module monitors the fuel tank pressure (FTP) sensor to verify that the system is able to reach a predetermined level of vacuum within a set amount of time. The control module then commands the EVAP purge valve OFF, closed, sealing the system and monitors the vacuum level for decay. If the control module does not detect that the predetermined vacuum level was achieved, or the vacuum decay is more than a calibrated level on 2 consecutive tests, a DTC P0440 will set.

SMALL LEAK TEST

If the large leak test passes, the control module will test for small leaks by continuing to monitor the FTP sensor for a change in voltage over a period of time. If the decay rate is more than a calibrated value, the control module will rerun the test. If the test fails again, a DTC P0442 will set.

CANISTER VENT RESTRICTION TEST

If the EVAP vent system is restricted, fuel vapors will not be properly purged from the EVAP canister. The control module tests this by commanding the EVAP purge valve ON, open, and commanding the EVAP vent valve OFF, open, and monitoring the FTP sensor for an increase in vacuum. If vacuum increases more than a calibrated value, DTC P0446 will set.

PURGE VALVE LEAK TEST

If the EVAP purge valve does not seal properly, fuel vapors could enter the engine at an undesired time causing driveability concerns. The control module tests for this by commanding the EVAP purge valve OFF, closed, and vent valve OFF, open, sealing the system, and monitoring the FTP for an increase in vacuum. If the control module detects that EVAP system vacuum increases above a calibrated value, DTC P1441 will set.

EVAP SYSTEM COMPONENTS

The EVAP system consists of the following components:

EVAP Canister - The canister is filled with carbon pellets used to absorb and store fuel vapors. Fuel vapor is stored in the canister until the control module determines that the vapor can be consumed in the normal combustion process.

EVAP Purge Valve - The EVAP purge valve controls the flow of vapors from the EVAP system to the intake manifold. This normally closed valve is pulse width modulated (PWM) by the control module to precisely control the flow of fuel vapor to the engine. The valve will also be opened during some portions of the EVAP testing, allowing engine vacuum to enter the EVAP system.

EVAP Vent Valve - The EVAP vent valve controls fresh airflow into the EVAP canister. The valve is normally open. The control module will command the valve closed during some EVAP tests, allowing the system to be tested for leaks.

Fuel Tank Pressure Sensor - The FTP sensor measures the difference between the pressure or vacuum in the fuel tank and outside air pressure. The control module provides a 5-volt reference and a ground to the FTP sensor. The FTP sensor provides a signal voltage back to the control module that can vary between 0.1-4.9 volts. As FTP increases, FTP sensor voltage decreases, high pressure = low voltage. As FTP decreases, FTP voltage increases, low pressure or vacuum = high voltage.

EVAP Service Port - The EVAP service port is located in the EVAP purge pipe between the EVAP purge valve and the EVAP canister. The service port is identified by a green colored cap.

Component Locations

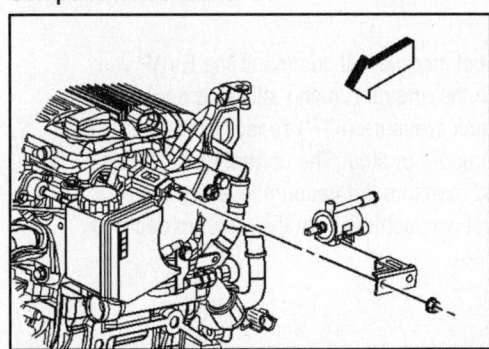

EVAP canister purge solenoid location on 2.2L (L61) engines

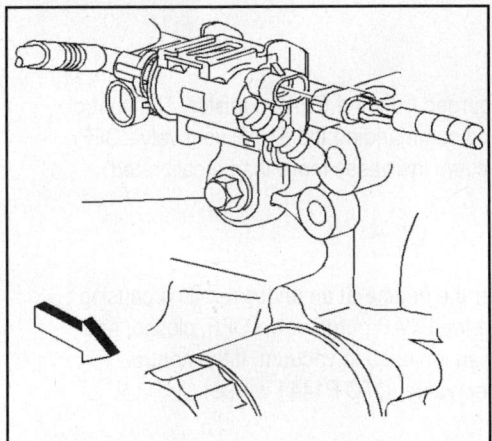

Location of EVAP purge solenoid valve on 2.2L (LN2) engines

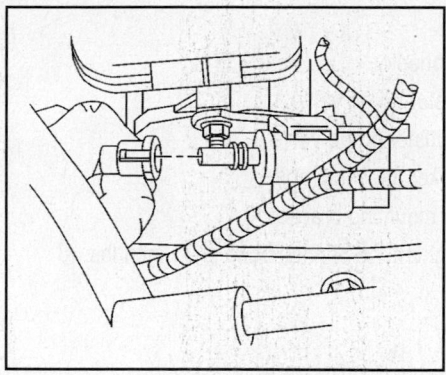

EVAP purge solenoid location on 2.4L engines

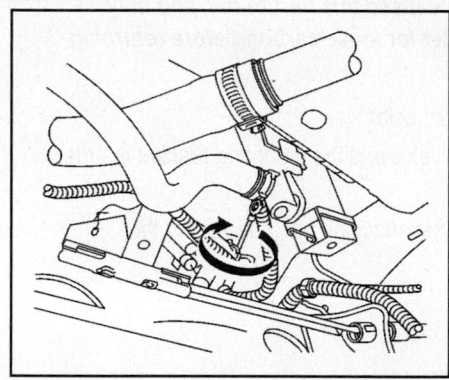

Showing the location of the EVAP vent solenoid valve on 2.2L & 2.4L engines

Removal & Installation

PURGE SOLENOID VALVE

2.2L (L61) Engine

1. Disconnect the evaporative emission (EVAP) canister purge valve harness connector.
2. Disconnect the vacuum pipe from the EVAP canister purge valve.
3. Disconnect the purge pipe from the EVAP canister purge valve.
4. Remove the EVAP canister purge valve and bracket.
5. Remove the EVAP canister purge valve from the purge bracket.
6. Inspect for carbon release in the EVAP canister purge valve ports.

To install:

7. Install the EVAP canister purge valve on to the purge bracket.
8. Install the EVAP canister purge valve and bracket. Tighten the purge bracket nut to 71 inch lbs. (8 Nm).
9. Connect the purge pipe to the EVAP canister purge valve.
10. Connect the vacuum pipe to the EVAP canister purge valve.
11. Connect the EVAP canister purge valve harness connector.

2.2L (LN2) Engine

1. Remove the air cleaner outlet resonator.
2. Disconnect the EVAP canister purge valve harness connector.
3. Disconnect the engine purge pipe from the EVAP canister purge valve.
4. Disconnect the engine vacuum pipe from the EVAP canister purge valve.
5. Remove the EVAP canister purge valve mounting bracket attaching bolt.
6. Remove the EVAP canister purge valve along with the mounting bracket.
7. Installation is the reverse of the removal procedure. Tighten the bracket bolts to 71 inch lbs. (8 Nm).

2.4L Engines

1. Remove the front purge pipe from the evaporative emission (EVAP) canister purge valve.
2. Remove the rear purge pipe from the EVAP canister purge valve.

NOTE: The EVAP canister may have released carbon particles, which caused this part to fail, and may cause damage to other components. Check the EVAP canister for loose carbon before returning the vehicle to service.

3. Disconnect the EVAP canister purge valve electrical connector.
4. Insert a screwdriver between the EVAP canister purge valve and the mounting bracket in order to release the lock tab.
5. Remove the EVAP canister purge valve mounting bracket attaching nut, if replacing the bracket.
6. Remove the bracket.

To install:

7. Install the mounting bracket and the attaching nut for the EVAP canister purge valve. Tighten the nut to 71 inch lbs. (8 Nm).
8. Slide the EVAP canister purge valve onto the mounting bracket. Make sure the lock tab locks the valve in place.
9. Connect the EVAP canister purge valve electrical connector.
10. Connect the rear purge pipe to the EVAP canister purge valve.
11. Connect the front purge pipe to the EVAP canister purge valve.

VENT SOLENOID VALVE

2.2L & 2.4L Engines

1. Raise the vehicle
2. Loosen the EVAP canister vent valve hose clamp and slide the clamp down the hose away from the EVAP vent valve hose connection.
3. Remove the hose from the EVAP canister vent valve.
4. Remove the EVAP canister vent valve by placing a pry bar between the valve and mounting surface, and moving the valve forward.
5. Remove the clip from the EVAP canister vent valve harness connector
6. Disconnect the EVAP canister vent valve harness connector.

To install:

7. Connect the EVAP canister vent valve harness connector
8. Install the clip into the EVAP canister vent valve harness connector.
9. Install the EVAP canister vent valve by pushing the mounts into the original holes.

10. Lubricate the hose end with lubricant.

11. Install the hose to the EVAP canister vent valve.

12. Slide the EVAP canister vent valve hose clamp down on to the hose connection at the EVAP canister vent valve.

13. Lower the vehicle.

Connector Pinouts

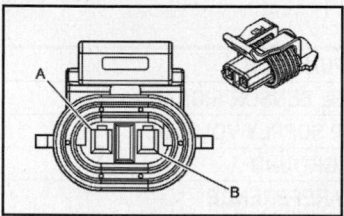

PIN	WIRE COLOR	CIRCUIT NO.	FUNCTION
A	PNK	439	IGNITION 1 VOLTAGE
B	DK GRN/WHT	428	EVAP CANISTER PURGE SOLENOID CONTROL

2.2L & 2.4L Engines EVAP Purge Solenoid Connector Pinout

PIN	WIRE COLOR	CIRCUIT NO.	FUNCTION
A	PNK	439	IGNITION 1 VOLTAGE
B	WHT	1310	EVAP CANISTER VENT SOLENOID CONTROL

2.2L & 2.4L Engines EVAP Vent Solenoid Connector Pinout

FUEL PUMP & SENDER

Component Locations

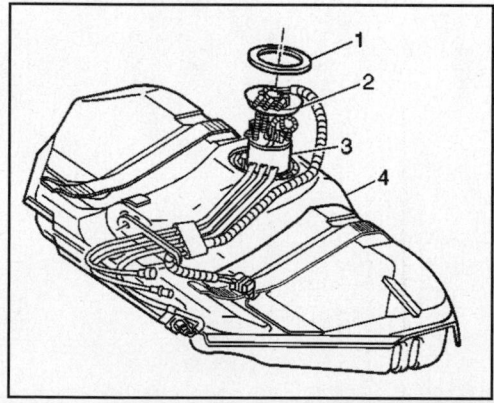

Fuel pump (1) and sender assembly (2) on 2.2L & 2.4L engines

Connector Pinouts

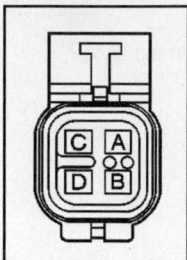

PIN	WIRE COLOR	CIRCUIT NO.	FUNCTION
A	PPL	1589	FUEL LEVEL SENSOR SIGNAL
B	GRY	120	FUEL PUMP SUPPLY VOLTAGE
C	BLK	150	GROUND
D	BLK	2759	LOW REFERENCE

2.2L & 2.4L Engines Fuel Pump and Sender Connector Pinout

FUEL TANK PRESSURE (FTP) SENSOR

Component Location

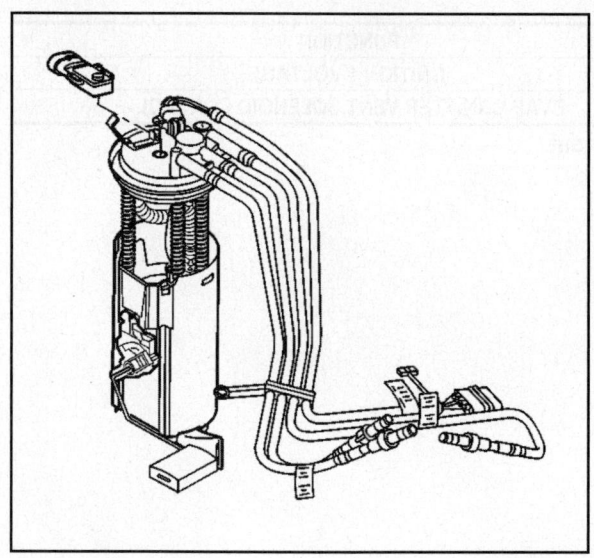

Fuel tank pressure sensor on 2.2L & 2.4L engines

Connector Pinouts

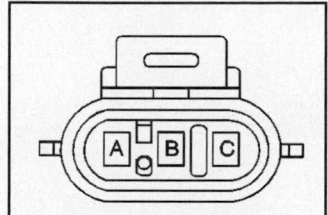

PIN	WIRE COLOR	CIRCUIT NO.	FUNCTION
A	BLK	2759	LOW REFERENCE
B	DK GRN	890	FUEL TANK PRESSURE SENSOR SIGNAL
C	GRY	2709	5 V REFERENCE A

2.2L & 2.4L Engines FTP Sensor Connector Pinout

GENERATOR CONNECTOR

Component Locations

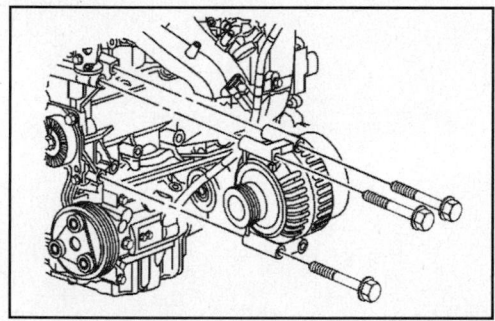

Generator mounting on 2.2L (L61 & LN2) engine (2.4L engine similar)

Connector Pinouts

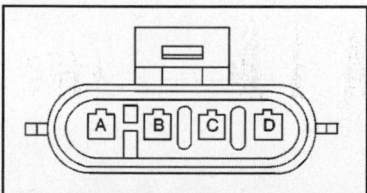

PIN	WIRE COLOR	CIRCUIT NO.	FUNCTION
A	--	--	NOT USED
B	RED	225	GENERATOR TURN ON SIGNAL
C	GRY	23	GENERATOR FIELD DUTY CYCLE SIGNAL
D	--	--	NOT USED

2.2L & 2.4L Engines Generator Connector Pinout

HEATED OXYGEN SENSORS (HO2S)

Component Locations

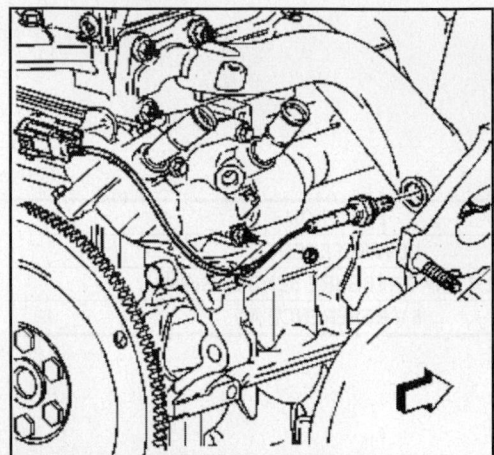

HO2S1 (Bank 1) location on 2.2L (L61) engines (exhaust manifold heat shield removed)

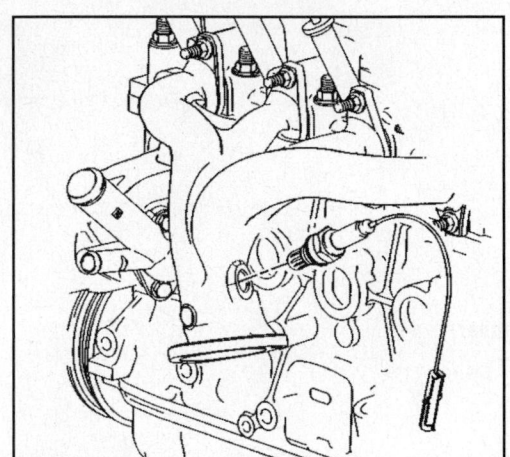

Showing HO2S1 (Bank 1) location on 2.2L (LN2) engines

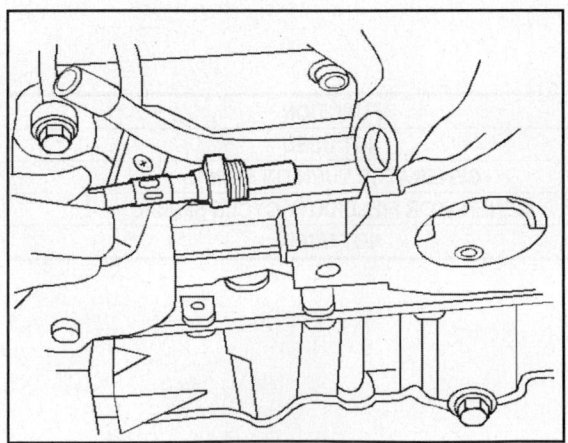

Showing the Oxygen Sensor on 2.4L engines

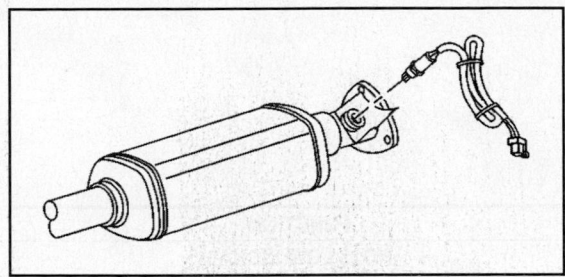

HO2S2 (Bank 2) location on 2.2L & 2.4L engines

Connector Pinouts

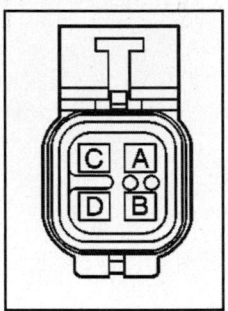

PIN	WIRE COLOR	CIRCUIT NO.	FUNCTION
A	TAN	1667	HO2S LOW SIGNAL
B	PPL	1666	HO2S HIGH SIGNAL
C	DK GRN	676	HO2S HEATER LOW CONTROL
D	PNK	439	IGNITION 1 VOLTAGE

2.2L (L61 & LN2) Engine HO2S1 Connector Pinout

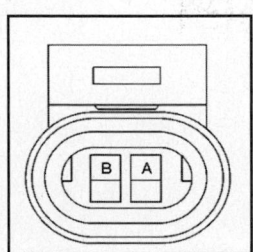

PIN	WIRE COLOR	CIRCUIT NO.	FUNCTION
A	TAN	413	O2S LOW REFERENCE
B	PPL	412	O2S HIGH SIGNAL

2.4L Engine Oxygen Sensor 1 Connector Pinout

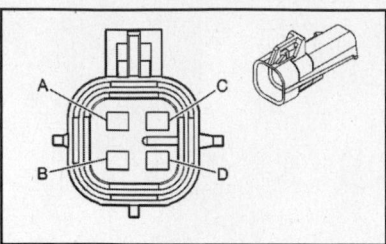

PIN	WIRE COLOR	CIRCUIT NO.	FUNCTION
A	TAN/WHT	1669	HO2S LOW SIGNAL
B	PPL/WHT	1668	HO2S HIGH SIGNAL
C	BLK/WHT	1423	HO2S HEATER LOW CONTROL
D	BRN	241	IGNITION 3 VOLTAGE

2.2L & 2.4L Engines HO2S2 Connector Pinout

IDLE AIR CONTROL (IAC) VALVE

Component Locations

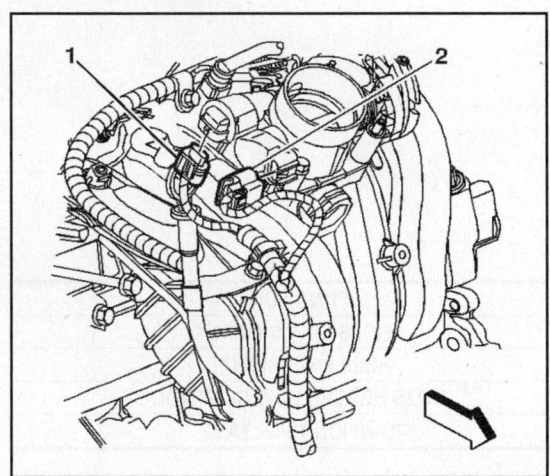

Showing the IAC Sensor and connector (1) and TP sensor (2) on 2.2L (L61) engines (air cleaner and duct removed)

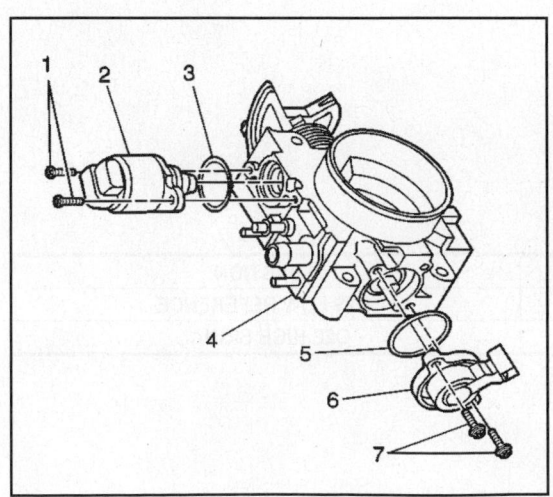

Exploded view of the IAC motor (2) mounting on the throttle body on 2.2L (LN2) & 2.4L engines

Removal & Installation

2.4L ENGINES

1. Remove the idle air control (IAC) valve connector.
2. Remove the IAC valve attaching screws (1).

NOTE: If the IAC valve has been in service: DO NOT push or pull on the IAC valve pintle. The force required to move the pintle may damage the threads on the worm drive. Also, DO NOT soak the IAC valve in any liquid cleaner or solvent, as damage may result.

3. Remove the IAC valve assembly (2).
4. Clean the IAC valve O-ring sealing surface, the pintle valve seat, and the air passage.

Measurement Procedure

NOTE: This procedure is only for installing a new IAC valve. If installing a new IAC valve, replace the IAC valve with an identical part. The IAC valve pintle shape and diameter are designed for the specific application.

5. Measure the distance between tip of the IAC valve pintle and the mounting flange.
6. If the distance is more than 28 mm (1.1 in), use finger pressure to slowly retract the pintle. The force required to retract the pintle of a new valve will not cause damage to the valve.

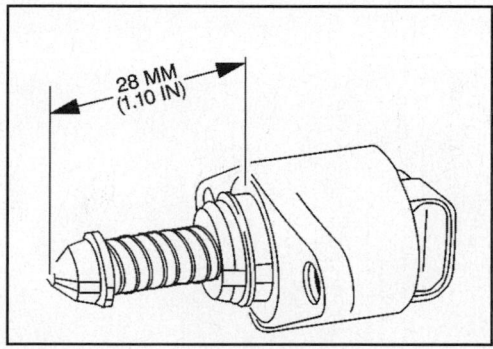

Measuring the IAC valve pintle

To install:

7. Lubricate the IAC valve O-ring (3) with clean engine oil.
8. Install the IAC valve assembly (2).
9. Install the IAC valve attaching screws (1). Tighten the screws to 27 inch lbs. (3.0 Nm).
10. Reconnect the IAC connector.
11. Use the following procedure to reset the IAC valve pintle position.
12. Turn ON the ignition, with the engine OFF.
13. Turn OFF the ignition for 10 seconds.
14. Start the engine and verify the correct idle operation.

Connector Pinouts

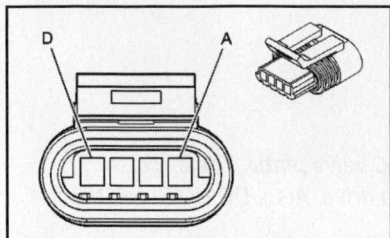

PIN	WIRE COLOR	CIRCUIT NO.	FUNCTION
A	LT GRN/BLK	444	IAC COIL B LOW CONTROL
B	LT GRN/WHT	1749	IAC COIL B HIGH CONTROL
C	LT BLU/BLK	1748	IAC COIL A LOW CONTROL
D	LT BLU/WHT	1747	IAC COIL A HIGH CONTROL

2.2L & 2.4L Engines IAC Valve Connector Pinout

IGNITION CONTROL MODULE (ICM)

Component Locations

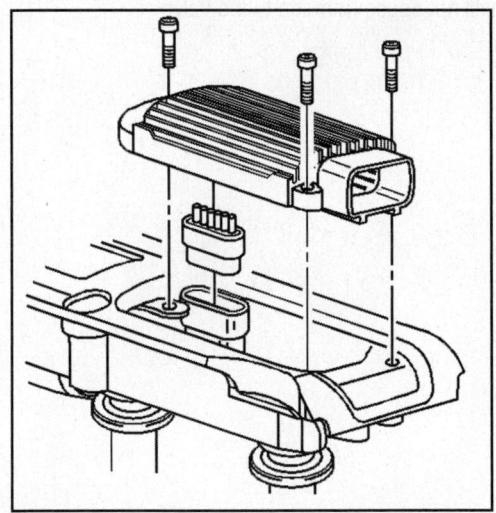

Ignition control module mounting on 2.2L (L61) engines

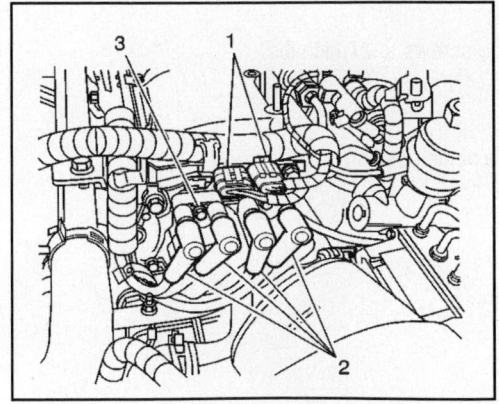

Ignition Control Module (ICM) location on 2.2L (LN2) engines

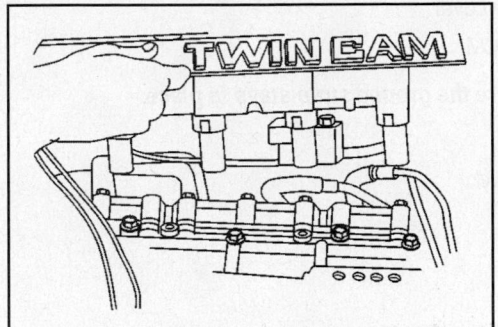

The ICM is located under the camshaft cover on 2.4L engines

Removal & Installation

2.2L (L61) ENGINE

1. Turn OFF the ignition.
2. Remove the accelerator cable from the bracket.
3. Remove the accelerator cable bracket bolt.
4. Remove the accelerator cable bracket.
5. Disconnect the ignition control module (ICM) harness connector.
6. Remove the ICM retaining screws.
7. Remove the ICM from the ignition coil housing.

To install:

8. Install the ignition control module in the ignition coil housing.
9. Install the ICM retaining screws. Tighten the retaining screws.
10. Connect the ICM harness connector.
11. Install the accelerator cable bracket.
12. Install the accelerator cable bracket bolt. Tighten the retaining screws.
13. Install the accelerator cable to the bracket.

2.2L (LN2) ENGINE

1. Remove the air cleaner outlet from the air cleaner.
2. Remove the ICM electrical connectors (1) and spark plug wires (2).
3. Remove the ignition coils bolts (3).
4. Remove the ignition coils and ICM assembly.
5. Installation is the reverse of the removal procedure.

2.4L ENGINE

1. Disconnect the negative battery cable.
2. Remove the accelerator cable from the hold down clip.
3. Remove the cruise control cable, if applicable.
4. Remove the bolt from the fuel line retaining clip.
5. Disconnect the 11-pin harness connector for the ignition control module (ICM).
6. Remove the bolts from the ignition coil and the ICM assembly-to-camshaft housing.

7.   Remove the ignition coil and ICM assembly from the engine.

8.   Remove the screws that retain the housing to the cover.

9.   Disconnect the coil harness connector from the ICM.

***CAUTION**: When removing the housing from the cover, make sure the ground strap stays in place.*

10.   Remove the housing from the cover.

11.   Remove the screws that retain the ICM to the cover.

12.   Remove the ICM from the cover.

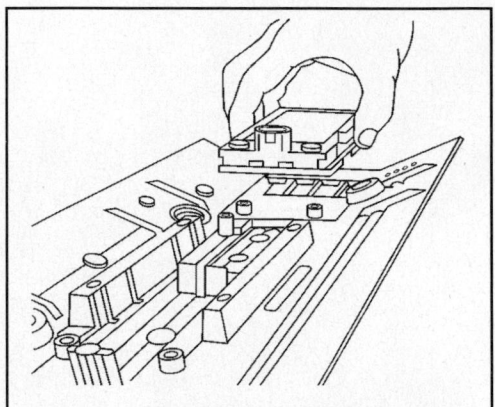

Removing the ICM from the camshaft cover

To install:

***CAUTION**: DO NOT wipe grease from the module or coil if the same module is to be replaced. If a new module is to be installed, a package of silicone grease will be included with the module. Spread the grease on the metal face of the module and on the cover where the module seats. This grease is necessary for module cooling.*

13.   Install the ICM to the cover.

14.   Install the screws that retain the ICM to the cover.

15.   Install the ground strap, if necessary.

16.   Connect the ignition coils connector to the ICM.

***CAUTION**: When installing the housing to the cover, make sure the ground strap stays in place.*

17.   Install the housing to the cover.

18.   Install the screws that retain the housing to the cover.

19.   Install the spark plug boots and the retainers to the housing, if necessary.

20.   Install the ICM assembly to the engine while carefully aligning the spark boots to the spark plug terminals.

***CAUTION**: The ICM cover bolts must be installed using isolator washers with the rubber side facing down.*

21.   Install the bolts that retain the ICM assembly to the camshaft housing after coating the bolt threads with LOCTITE®, or equivalent. Tighten the bolts to 16 ft. lbs. (22 Nm).

22.   Connect the ICM 11 pin harness connector.

23.   Install the bolt to the fuel line retainer clip.

24.   Install the accelerator cable into the hold down clip.

25.   Install the cruise control cable, if applicable.

26.   Connect the negative battery cable.

Connector Pinouts

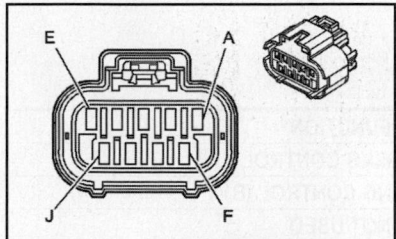

PIN	WIRE COLOR	CIRCUIT NO.	FUNCTION
A	PNK	239	IGNITION 1 VOLTAGE
B	WHT	423	IC TIMING CONTROL (A)
C-D	--	--	NOT USED
E	BLK/WHT	451	GROUND
F	BRN/WHT	633	CMP SENSOR SIGNAL
G	ORN	406	IC TIMING CONTROL B
H-J	--	--	NOT USED

2.2L (L61) Engine ICM Connector Pinout

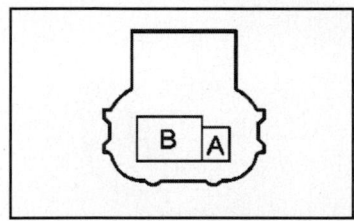

PIN	WIRE COLOR	CIRCUIT NO.	FUNCTION
A	BLK	750	GROUND
B	PNK	239	IGNITION 1 VOLTAGE

2.2L (LN2) Engine ICM1 Connector Pinout

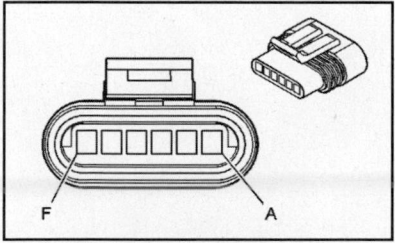

PIN	WIRE COLOR	CIRCUIT NO.	FUNCTION
A	--	--	NOT USED
B	WHT	423	IC TIMING CONTROL
C	ORN	406	IC TIMING CONTROL (B)
D-E	--	--	NOT USED
F	RED/BLK	453	LOW REFERENCE

2.2L (LN2) Engine ICM2 Connector Pinout

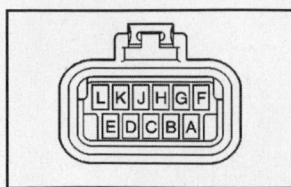

PIN	WIRE COLOR	CIRCUIT NO.	FUNCTION
A	WHT	423	IC TIMING CONTROL
B	ORN	406	IC TIMING CONTROL (B)
C-E	--	--	NOT USED
F	YEL	573	CKP SENSOR SIGNAL
G	PPL/WHT	430	LOW RESOLUTION ENGINE SPEED SIGNAL
H	RED/BLK	453	LOW REFERENCE
J	PPL	574	LOW REFERENCE
K	BLK	750	GROUND
L	PNK	239	IGNITION 1 VOLTAGE

2.4L Engines ICM Connector Pinout

INTAKE AIR TEMPERATURE (IAT) SENSOR

Component Locations

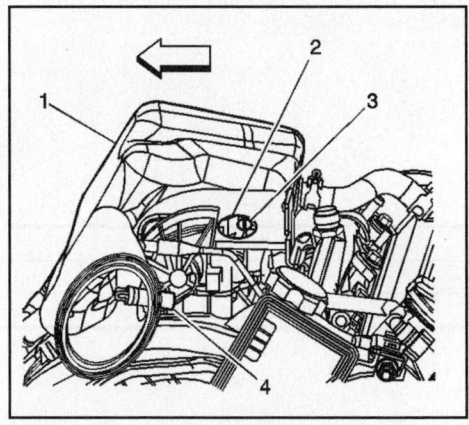

Showing IAT sensor (4) location on 2.2L (L61) engines

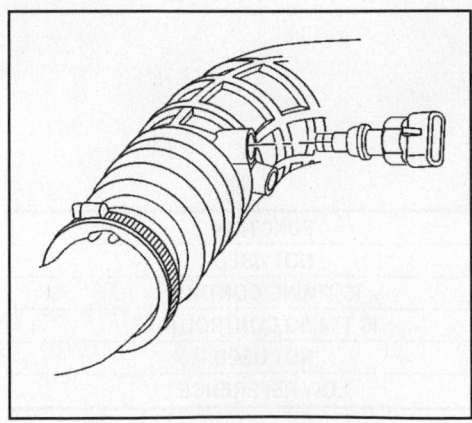

Showing IAT sensor location on 2.2L (LN2) engines (2.4L engines similar)

Connector Pinouts

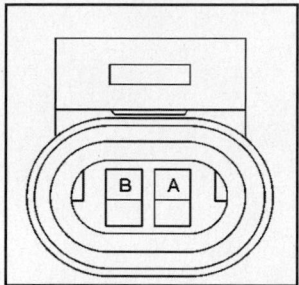

PIN	WIRE COLOR	CIRCUIT NO.	FUNCTION
A	BLK	2760	LOW REFERENCE
B	TAN	472	IAT SENSOR SIGNAL

2.2L (L61 & LN2) Engine IAT Sensor Connector Pinout

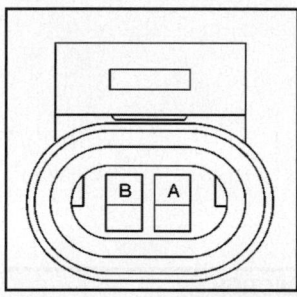

PIN	WIRE COLOR	CIRCUIT NO.	FUNCTION
A	GRY	720	LOW REFERENCE
B	TAN	472	IAT SENSOR SIGNAL

2.4L Engine IAT Sensor Connector Pinout

KNOCK SENSOR

Component Locations

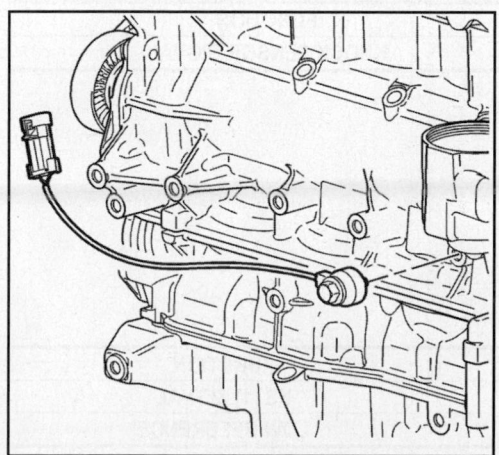

Knock sensor location on 2.2L (L61) engines

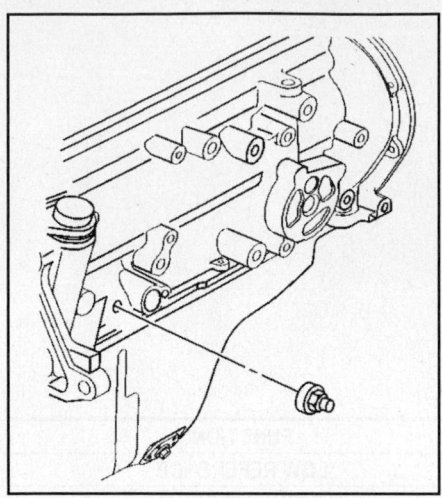

Knock sensor location on 2.2L (LN2) engines (similar on 2.4L engines)

Connector Pinouts

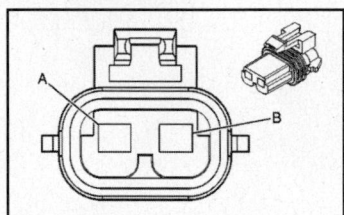

PIN	WIRE COLOR	CIRCUIT NO.	FUNCTION
A	DK BLU	496	KNOCK SENSOR SIGNAL
B	GRY	1716	LOW REFERENCE

2.2L (L61) Engine Knock Sensor Connector Pinout

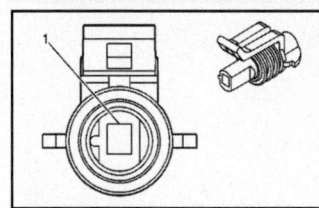

PIN	WIRE COLOR	CIRCUIT NO.	FUNCTION
1	DK BLU	496	KNOCK SENSOR SIGNAL

2.2L (LN2) Engine Knock Sensor Connector Pinout

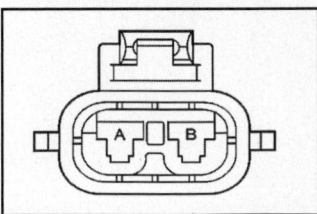

PIN	WIRE COLOR	CIRCUIT NO.	FUNCTION
A	DK BLU	496	KS [1] SIGNAL
B	GRY	1716	LOW REFERENCE

2.4L Engine Knock Sensor Connector Pinout

MANIFOLD ABSOLUTE PRESSURE (MAP) SENSOR

Component Locations

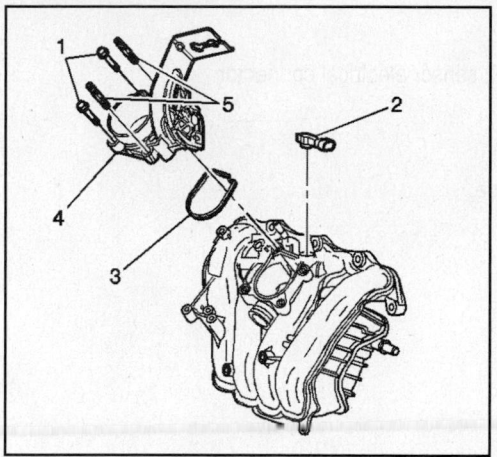

MAP sensor (2) location on 2.2L (L61) engines

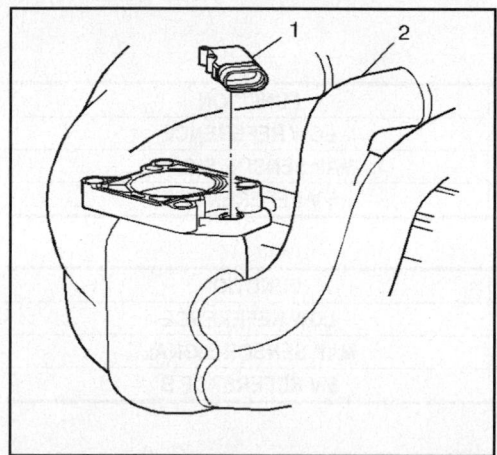

Showing the MAP sensor location on the intake manifold on 2.2L (LN2) engines

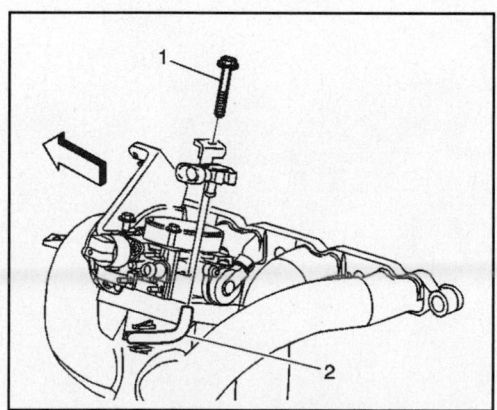

MAP sensor location on 2.4L engines

Removal & Installation

2.2L & 2.4L ENGINES

1. Remove the air cleaner outlet resonator.
2. Disconnect the manifold absolute pressure (MAP) sensor electrical connector.
3. Remove the MAP sensor attaching screw (1).
4. Remove the MAP sensor vacuum hose (2).
5. Installation is the reverse of the removal procedure.

Connector Pinouts

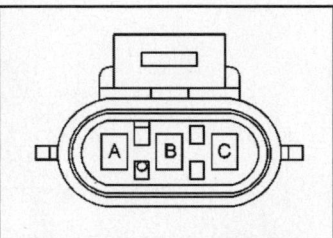

PIN	WIRE COLOR	CIRCUIT NO.	FUNCTION
A	ORN/BLK	469	LOW REFERENCE
B	LT GRN	432	MAP SENSOR SIGNAL
C	GRY	2704	5 V REFERENCE B

2.2L (L61 & LN2) Engine MAP Sensor Connector Pinout

PIN	WIRE COLOR	CIRCUIT NO.	FUNCTION
A	BLK/WHT	1704	LOW REFERENCE
B	LT GRN	432	MAP SENSOR SIGNAL
C	GRY	474	5 V REFERENCE B

2.4L Engine MAP Sensor Connector Pinout

POWERTRAIN CONTROL MODULE (PCM) CONNECTORS

Component Locations

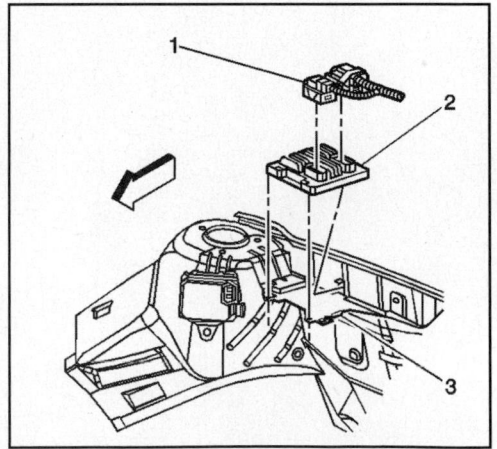

Showing PCM (2) location on 2.2L & 2.4L engines

Connector Pinouts

C1 CONNECTOR ON 2.2L ENGINES

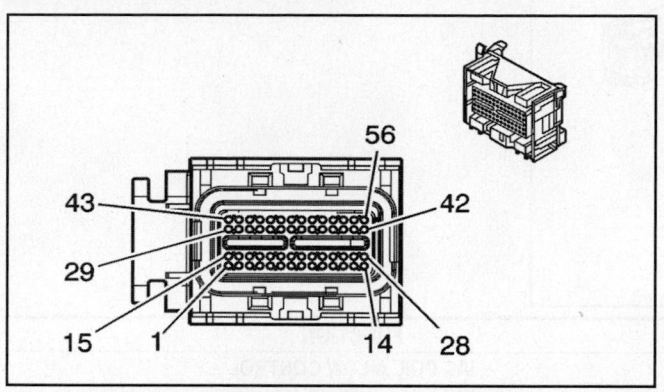

PIN	WIRE COLOR	CIRCUIT NO.	FUNCTION
1-3	--	--	NOT USED
4	PPL	420	TCC RELEASE/ A/T SHIFT LOCK RELEASE SIGNAL (MXO W/O K34) OR TCC RELEASE/ A/T SHIFT LOCK RELEASE/ CRUISE CONTROL RELEASE SIGNAL (MXO W/ K34) OR CLUTCH ANTICIPATE SIGNAL (MM5 W/O K34) OR CLUTCH ANTICIPATE/ CRUISE CONTROL RELEASE SIGNAL (MM5 W/ K34)
5-6	--	--	NOT USED
7	BLK	2760	LOW REFERENCE
8-10	--	--	NOT USED
11	BLK	2759	LOW REFERENCE
12-14	--	--	NOT USED
15	PPL	1807	CLASS 2 SERIAL DATA
16	PPL	1807	CLASS 2 SERIAL DATA
17-18	--	--	NOT USED
19	PNK	639	IGNITION 1 VOLTAGE
20	ORN	480	BATTERY POSITIVE VOLTAGE
21	TAN	472	IAT SENSOR SIGNAL
22	PPL	1589	FUEL LEVEL SENSOR SIGNAL
23	GRY	2700	5 V REFERENCE A (C60)
24	--	--	NOT USED
25	GRY	2709	5 V REFERENCE A
26-33	--	--	NOT USED
33	BRN/WHT	419	MIL CONTROL
34	DK GRN/WHT	817	VEHICLE SPEED SIGNAL (K34)
35-36	--	--	NOT USED
37	RED/BLK	380	A/C REFRIGERANT PRESSURE SENSOR SIGNAL (C60)
38	--	--	NOT USED
39	DK GRN	890	FUEL TANK PRESSURE SENSOR SIGNAL
40-42	--	--	NOT USED
43	DK GRN	335	COOLING FAN RELAY CONTROL
44	--	--	NOT USED
45	WHT	1310	EVAP CANISTER VENT SOLENOID CONTROL
46	DK GRN/WHT	465	FUEL PUMP RELAY CONTROL
47	DK GRN/WHT	459	A/C COMPRESSOR CLUTCH RELAY CONTROL (C60)
48	--	--	NOT USED
49	DK GRN	83	CRUISE CONTROL INHIBIT SIGNAL (K34)
50	--	--	NOT USED
51	BLK	2751	LOW REFERENCE (C60)
52-56	--	--	NOT USED

PCM C1 Connector Pinout on 2.2L (L61 & LN2) Engines

C2 CONNECTOR ON 2.2L ENGINES

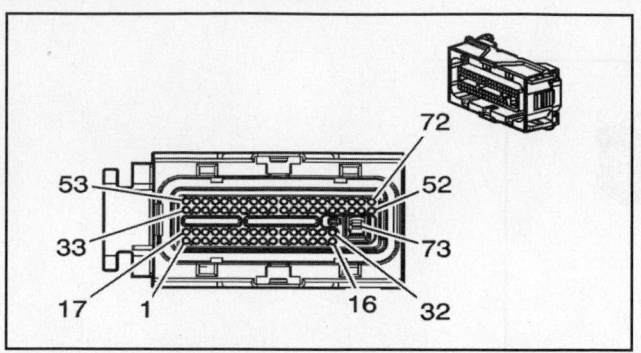

PIN	WIRE COLOR	CIRCUIT NO.	FUNCTION
1	LT BLU/BLK	1748	IAC COIL A LOW CONTROL
2	LT BLU/WHT	1747	IAC COIL A HIGH CONTROL
3-4	--	--	NOT USED
5	DK GRN/WHT	428	EVAP CANISTER PURGE SOLENOID CONTROL
6	RED/BLK	1228	PC SOLENOID VALVE HIGH CONTROL (MX0)
7	LT BLU/WHT	1229	PC SOLENOID VALVE LOW CONTROL (MX0)
8-9	--	--	NOT USED
10	LT GRN	1222	1-2 SHIFT SOLENOID VALVE CONTROL (MX0)
11	YEL/BLK	1223	2-3 SHIFT SOLENOID VALVE CONTROL (MX0)
12	BRN	418	TCC PWM SOLENOID VALVE CONTROL (MX0)
13	PPL/WHT	1668	HO2S HIGH SIGNAL
14	TAN/WHT	1669	HO2S LOW SIGNAL
15	TAN	1667	HO2S LOW SIGNAL
16	PPL	1666	HO2S HIGH SIGNAL
17	LT GRN/WHT	1749	IAC COIL B HIGH CONTROL
18	LT GRN/BLK	444	IAC COIL B LOW CONTROL
19	WHT	423	IC TIMING CONTROL
20	ORN	406	IC TIMING CONTROL (B)
21-25	--	--	NOT USED
26	BRN/WHT	633	CMP SENSOR SIGNAL
27	YEL	573	CKP SENSOR SIGNAL
28	PPL	574	LOW REFERENCE
29	LT GRN	432	MAP SENSOR SIGNAL
30	--	--	NOT USED
31	DK BLU	417	TP SENSOR SIGNAL
32-33	--	--	NOT USED
34	WHT	776	TRANSMISSION RANGE SWITCH SIGNAL P (MX0)
35	GRY	773	TRANSMISSION RANGE SWITCH SIGNAL C (MX0)
36	YEL	772	TRANSMISSION RANGE SWITCH SIGNAL B (MX0)
37	BLK/WHT	771	TRANSMISSION RANGE SWITCH SIGNAL A (MX0)
38	--	--	NOT USED
39	BLK	2761	LOW REFERENCE
40	BLK	2762	LOW REFERENCE (MX0)
41	DK BLU/WHT	1231	A/T ISS SENSOR LOW (MX0)
42	PPL	401	SIGNAL LOW-FRONT
43	--	--	NOT USED
44	GRY	1716	LOW REFERENCE
45	ORN/BLK	469	LOW REFERENCE
46	--	--	NOT USED
47	BLK	2752	LOW REFERENCE
48	--	--	NOT USED
49	GRY	23	GENERATOR FIELD DUTY CYCLE SIGNAL

PCM C2 Connector Pinout on 2.2L (L61 & LN2) Engines (1 of 2)

C2 CONNECTORON 2.2L ENGINES (CONTINUED)

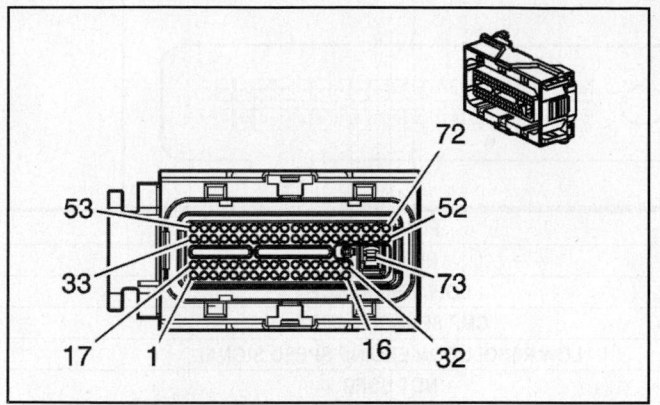

PIN	WIRE COLOR	CIRCUIT NO.	FUNCTION
50	RED	225	GENERATOR TURN ON SIGNAL
51	BLK	1744	FUEL INJECTOR 1 CONTROL
52	LT GRN/BLK	1745	FUEL INJECTOR 2 CONTROL
53	LT BLU/BLK	396	CRUISE CONTROL ENGAGED SIGNAL (K34)
54	TAN/BLK	231	OIL PRESSURE SWITCH SIGNAL
55	PNK	1224	TRANSMISSION FLUID PRESSURE SWITCH SIGNAL A (MX0)
56	DK BLU	1225	TRANSMISSION FLUID PRESSURE SWITCH SIGNAL B (MX0)
57	RED	1226	TRANSMISSION FLUID PRESSURE SWITCH SIGNAL C (MX0)
58	YEL	657	TCC RELEASE SWITCH SIGNAL (MX0)
59	YEL	410	ECT SENSOR SIGNAL
60	YEL/BLK	1227	TFT SENSOR SIGNAL (MX0)
61	RED/BLK	1230	A/T ISS SENSOR HIGH (MX0)
62	YEL	400	SIGNAL-HIGH FRONT
63	--	--	NOT USED
64	DK BLU	496	KNOCK SENSOR SIGNAL
65	GRY	2704	5 V REFERENCE B
66	--	--	NOT USED
67	GRY	2701	5 V REFERENCE A
68	--	--	NOT USED
69	DK GRN	676	HO2S HEATER LOW CONTROL
70	BLK/WHT	1423	HO2S HEATER LOW CONTROL
71	LT BLU/BLK	844	FUEL INJECTOR 4 CONTROL
72	PNK/BLK	1746	FUEL INJECTOR 3 CONTROL
73	BLK/WHT	451	GROUND

PCM C2 Connector Pinout on 2.2L (L61 & LN2) Engines (2 of 2)

C1 CONNECTOR ON 2.4L ENGINES

PIN	WIRE COLOR	CIRCUIT NO.	FUNCTION
1-5	--	--	NOT USED
6	BLK	2759	LOW REFERENCE
7	BRN/WHT	633	CMP SENSOR SIGNAL
8	PPL/WHT	430	LOW RESOLUTION ENGINE SPEED SIGNAL
9	--	--	NOT USED
10	GRY	2709	5 V REFERENCE A
11-17	--	--	NOT USED
18	BLK/WHT	451	GROUND
19	PNK	639	IGNITION 1 VOLTAGE
20	ORN	480	BATTERY POSITIVE VOLTAGE
21	DK BLU	417	TP SENSOR SIGNAL
22	BLK	150	GROUND
23-24	--	--	NOT USED
25	ORN	406	IC TIMING CONTROL (B)
26	WHT	423	IC TIMING CONTROL
27	TAN/BLK	231	OIL PRESSURE SWITCH SIGNAL
28	RED	1226	TRANSMISSION FLUID PRESSURE SWITCH SIGNAL C (MX0)
29	BLK/WHT	771	TRANSMISSION RANGE SWITCH SIGNAL A (MX0)
30-31	--	--	NOT USED
32	GRY	1716	LOW REFERENCE
33	--	--	NOT USED
34	DK BLU	496	KS [1] SIGNAL
35	PPL	420	TCC RELEASE/ A/T SHIFT LOCK RELEASE SIGNAL (MX0 W/O K34) OR TCC RELEASE/ A/T SHIFT LOCK RELEASE/ CRUISE CONTROL RELEASE SIGNAL (MX0 W/ K34) OR CLUTCH ANTICIPATE SIGNAL (MM5 W/O K34) OR CLUTCH ANTICIPATE/ CRUISE CONTROL RELEASE SIGNAL (MM5 W/K34)
36-44	--	--	NOT USED
45	DK GRN	890	FUEL TANK PRESSURE SENSOR SIGNAL
46	YEL/BLK	1227	TFT SENSOR SIGNAL (MX0)
47-48	--	--	NOT USED
49	RED/BLK	380	A/C REFRIGERANT PRESSURE SENSOR SIGNAL (C60)
50-51	--	--	NOT USED
52	BLK	552	LOW REFERENCE (C60)
53	GRY	23	GENERATOR FIELD DUTY CYCLE SIGNAL
54	YEL	410	ECT SENSOR SIGNAL
55	PPL	1589	FUEL LEVEL SENSOR SIGNAL
56	--	--	NOT USED
57	BLK	452	LOW REFERENCE
58	PPL	1807	CLASS 2 SERIAL DATA
59	PPL	1807	CLASS 2 SERIAL DATA
60	TAN/WHT	551	GROUND
61	--	--	NOT USED
62	LT GRN	432	MAP SENSOR SIGNAL
63	RED/BLK	1230	A/T ISS SENSOR HIGH (MX0)
64	YEL	400	SIGNAL HIGH - FRONT
65-66	--	--	NOT USED

PCM C1 Connector Pinout on 2.4L Engines (1 of 2)

C1 CONNECTOR ON 2.4L ENGINES (CONTINUED)

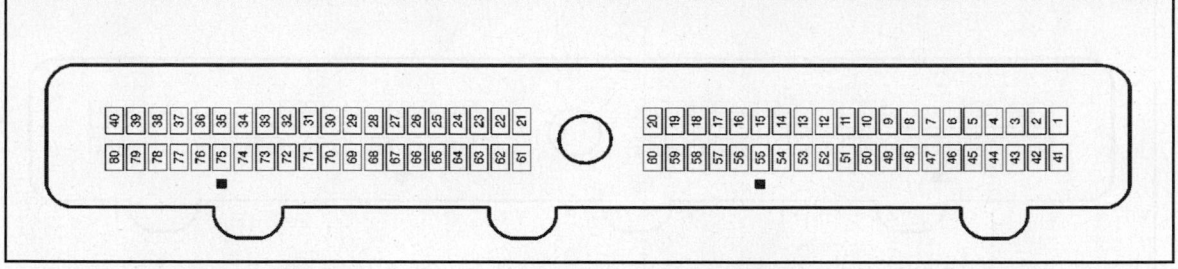

PIN	WIRE COLOR	CIRCUIT NO.	FUNCTION
67	DK BLU	1225	TRANSMISSION FLUID PRESSURE SWITCH SIGNAL B (MX0)
68	PNK	1224	TRANSMISSION FLUID PRESSURE SWITCH SIGNAL A (MX0)
69	YEL	772	TRANSMISSION RANGE SWITCH SIGNAL B (MX0)
70-71	--	--	NOT USED
72	GRY	773	TRANSMISSION RANGE SWITCH SIGNAL C (MX0)
73	WHT	776	TRANSMISSION RANGE SWITCH SIGNAL P (MX0)
74	RED	225	GENERATOR TURN ON SIGNAL
75	BLK	1744	FUEL INJECTOR 1 CONTROL
76	LT GRN/BLK	1745	FUEL INJECTOR 2 CONTROL
77-80	--	--	NOT USED

PCM C1 Connector Pinout on 2.4L Engines (2 of 2)

C2 CONNECTOR ON 2.4L ENGINES

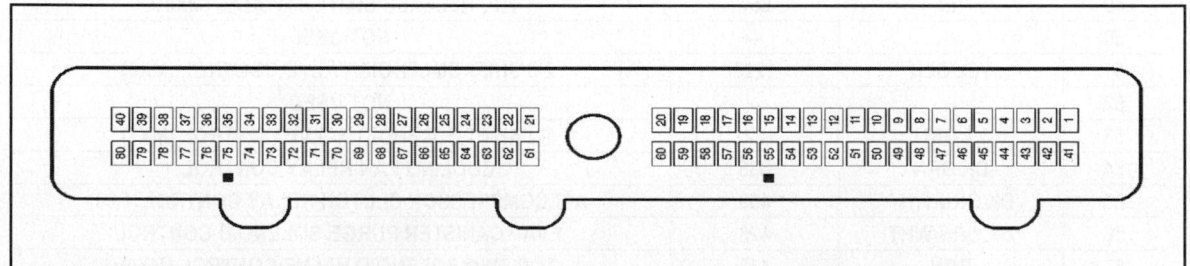

PIN	WIRE COLOR	CIRCUIT NO.	FUNCTION
1-5	--	--	NOT USED
6	RED/BLK	1228	PC SOLENOID VALVE HIGH CONTROL (MX0)
7	--	--	NOT USED
8	DK GRN/WHT	465	FUEL PUMP RELAY CONTROL
9	LT BLU/WHT	1229	PC SOLENOID VALVE LOW CONTROL (MX0)
10-13	--	--	NOT USED
14	BLK/WHT	1704	LOW REFERENCE
15	--	--	NOT USED
16	DK GRN/WHT	817	VEHICLE SPEED SIGNAL (K34)
17-19	--	--	NOT USED
20	TAN	472	IAT SENSOR SIGNAL
21	TAN	413	O2S LOW SIGNAL
22	PPL	412	O2S HIGH SIGNAL
23	TAN/WHT	1669	HO2S LOW SIGNAL
24	PPL/WHT	1668	HO2S HIGH SIGNAL
25-26	--	--	NOT USED
27	BRN	718	LOW REFERENCE

PCM C2 Connector Pinout on 2.4L Engines (1 of 2)

C2 CONNECTOR ON 2.4L ENGINES (CONTINUED)

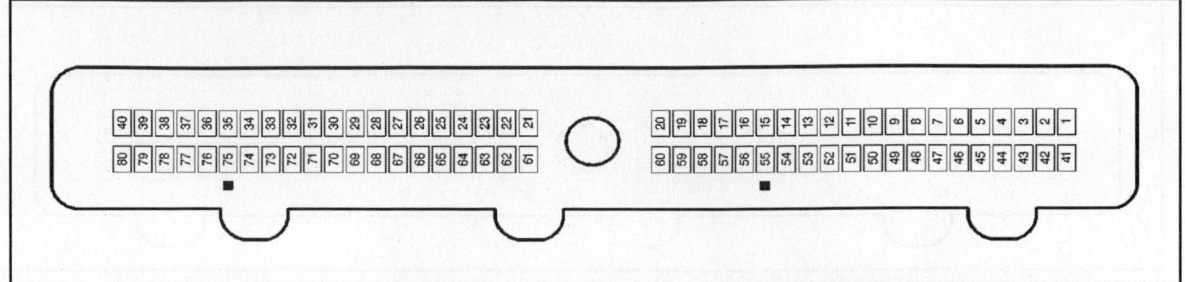

PIN	WIRE COLOR	CIRCUIT NO.	FUNCTION
28	GRY	720	LOW REFERENCE
29	BLK	407	LOW REFERENCE (MX0)
30	--	--	NOT USED
31	RED/BLK	453	LOW REFERENCE
32	--	--	NOT USED
33	GRY	416	5 V REFERENCE A
34	GRY	596	5 V REFERENCE A (C60)
35	GRY	474	5 V REFERENCE B
36	RED	631	12 VOLT REFERENCE
37-44	--	--	NOT USED
45	DK GRN	83	CRUISE CONTROL INHIBIT SIGNAL (K34)
46	BRN/WHT	419	MIL CONTROL
47	--	--	NOT USED
48	LT BLU/BLK	396	CRUISE CONTROL ENGAGED SIGNAL (K34)
49	YEL	657	TCC RELEASE SWITCH SIGNAL (MX0)
50	--	--	NOT USED
51	YEL/BLK	1223	2-3 SHIFT SOLENOID VALVE CONTROL (MX0)
52	--	--	NOT USED
53	LT GRN	1222	1-2 SHIFT SOLENOID VALVE CONTROL (MX0)
54	DK GRN	335	COOLING FAN RELAY CONTROL
55	DK GRN/WHT	459	A/C COMPRESSOR CLUTCH RELAY CONTROL (C60)
56	DK GRN/WHT	428	EVAP CANISTER PURGE SOLENOID CONTROL
57	BRN	418	TCC PWM SOLENOID VALVE CONTROL (MX0)
58	WHT	1310	EVAP CANISTER VENT SOLENOID CONTROL
59-60	--	--	NOT USED
61	PPL	401	SIGNAL LOW - FRONT
62	DK BLU/WHT	1231	A/T ISS SENSOR LOW (MX0)
63	PNK/BLK	632	LOW REFERENCE
64	BLK/WHT	451	GROUND
65-70	--	--	NOT USED
71	LT BLU/WHT	1747	IAC COIL A HIGH CONTROL
72	LT GRN/WHT	1749	IAC COIL B HIGH CONTROL
73	LT GRN/BLK	444	IAC COIL B LOW CONTROL
74	LT BLU/BLK	1748	IAC COIL A LOW CONTROL
75	LT BLU/BLK	844	FUEL INJECTOR 4 CONTROL
76	PNK/BLK	1746	FUEL INJECTOR 3 CONTROL
77-80	--	--	NOT USED

PCM C2 Connector Pinout on 2.4L Engines (2 of 2)

THROTTLE POSITION (TP) SENSOR

Component Locations

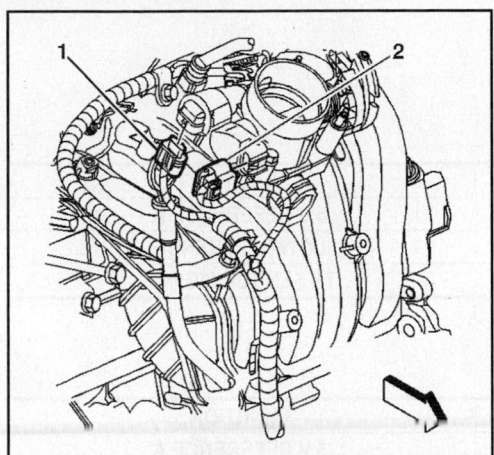

Showing the TP sensor location (2) and the IAC motor location (1) on 2.2L (L61) engines

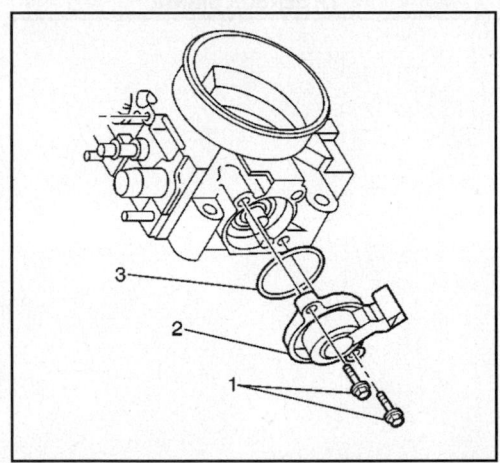

Exploded view of the TP sensor (2) on 2.2L (LN2) & 2.4L engines

Connector Pinouts

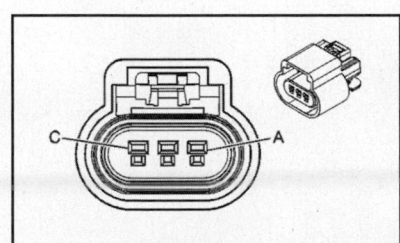

PIN	WIRE COLOR	CIRCUIT NO.	FUNCTION
A	GRY	2701	5 V REFERENCE A
B	BLK	2752	LOW REFERENCE
C	DK BLU	417	TP SENSOR SIGNAL

2.2L (L61) Engine TP Sensor Connector Pinout

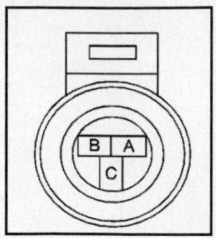

PIN	WIRE COLOR	CIRCUIT NO.	FUNCTION
A	GRY	2701	5 V REFERENCE A
B	BLK	2752	LOW REFERENCE
C	DK BLU	417	TP SENSOR SIGNAL

2.2L (LN2) Engine TP Sensor Connector Pinout

PIN	WIRE COLOR	CIRCUIT NO.	FUNCTION
A	GRY	416	5 V REFERENCE A
B	BLK	452	LOW REFERENCE
C	DK BLU	417	TP SENSOR SIGNAL

2.4L Engine TP Sensor Connector Pinout

VEHICLE SPEED SENSOR (VSS)

Component Locations

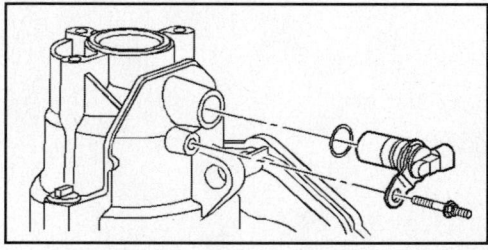

Showing VSS location on rear housing of transmission on 2.2L (L61 & LN2) engines

GENERAL MOTORS CARS: 2000-05 BUICK, CHEVROLET, PONTIAC, OLDSMOBILE V6 ENGINES: 3.1L, 3.4L, 3.5L, 3.8L

A/T (4T65-E) INLINE CONNECTORS

Connector Pinouts

ENGINE SIDE: 2000-02

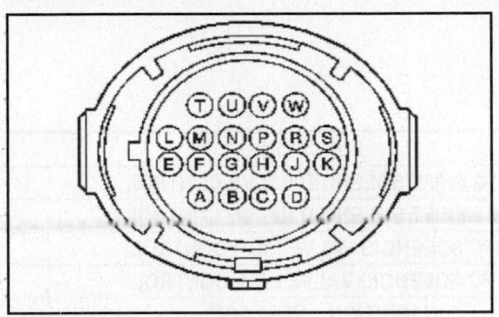

PIN	WIRE COLOR	CIRCUIT NO.	FUNCTION
A	LT GRN	1222	1-2 SHIFT SOLENOID VALVE CONTROL
B	YEL/BLK	1223	2-3 SHIFT SOLENOID VALVE CONTROL
C	RED/BLK	1228	PC SOLENOID VALVE HIGH CONTROL
D	LT BLU/WHT	1229	PC SOLENOID VALVE LOW CONTROL
E	PNK	1039	IGNITION 1 VOLTAGE
F-K	--	--	NOT USED
L	YEL/BLK	1227	TFT SENSOR SIGNAL
M	BLK	2762	LOW REFERENCE
N	PNK	1224	TFP SWITCH SIGNAL A
P	RED	1226	TFP SWITCH SIGNAL C
R	DK BLU	1225	TFP SWITCH SIGNAL B
S	RED/BLK	1230	AT ISS HIGH SIGNAL
T	BRN	418	TCC PWM SOLENOID VALVE CONTROL
U	YEL	657	TCC RELEASE SWITCH SIGNAL
V	DK BLU/WHT	1231	AT ISS LOW SIGNAL
W	--	--	NOT USED

2000-02 w/4T65-E A/T Engine Side Connector Pinout

ENGINE SIDE: 2003-05

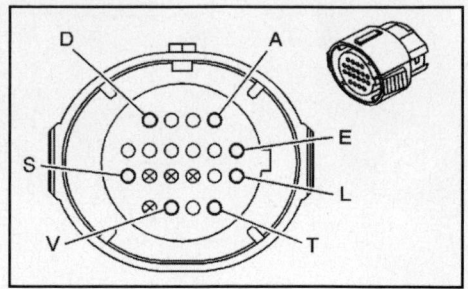

PIN	WIRE COLOR	CIRCUIT NO.	FUNCTION
A	LT GRN	1222	1-2 SHIFT SOLENOID VALVE CONTROL
B	YEL/BLK	1223	2-3 SHIFT SOLENOID VALVE CONTROL
C	RED/BLK	1228	PC SOLENOID VALVE HIGH CONTROL
D	LT BLU/WHT	1229	PC SOLENOID VALVE LOW CONTROL
E	PNK	1039	IGNITION 1 VOLTAGE
F	BLK/WHT	771	TRANSMISSION RANGE SIGNAL A
G	YEL	772	TRANSMISSION RANGE SIGNAL B
H	GRY	773	TRANSMISSION RANGE SIGNAL C
J	WHT	776	TRANSMISSION RANGE SIGNAL P
K	BLK/WHT	451	GROUND
L	YEL/BLK	1227	TFT SENSOR SIGNAL
M	BLK	2762	LOW REFERENCE
N-R	--	--	NOT USED
S	RED/BLK	1230	AT ISS HIGH SIGNAL
T	BRN	418	TCC PWM SOLENOID VALVE CONTROL
U	YEL	657	TCC RELEASE SWITCH SIGNAL
V	DK BLU/WHT	1231	AT ISS LOW SIGNAL
W	--	--	NOT USED

2003-05 w/4T65-E A/T Engine Side Connector Pinout

Transmission Side: 2000-02

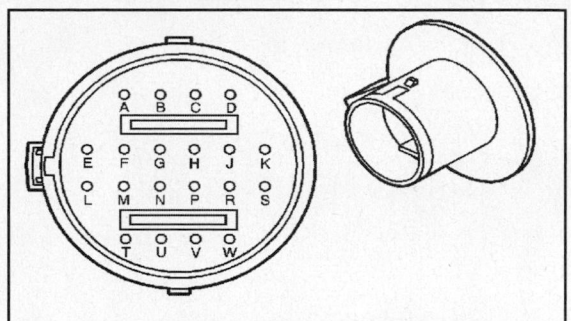

PIN	WIRE COLOR	CIRCUIT NO.	FUNCTION
A	LT GRN	1222	1-2 SHIFT SOLENOID VALVE CONTROL
B	YEL	1223	2-3 SHIFT SOLENOID VALVE CONTROL
C	PPL	1228	PC SOLENOID VALVE HIGH CONTROL
D	LT BLU	1229	PC SOLENOID VALVE LOW CONTROL
E	RED	839	IGNITION 1 VOLTAGE
F-K	--	--	NOT USED
L	BRN	1227	TFT SENSOR SIGNAL
M	GRY	452	LOW REFERENCE
N	PNK	1224	TFP SWITCH SIGNAL A
P	ORN	1226	TFP SWITCH SIGNAL C
R	DK BLU	1225	TFP SWITCH SIGNAL B
S	BLK	1230	AT ISS HIGH SIGNAL
T	TAN	418	TCC PWM SOLENOID VALVE CONTROL
U	WHT	1804	TCC RELEASE SWITCH SIGNAL
V	DK GRN	1231	AT ISS LOW SIGNAL
W	--	--	NOT USED

2000-02 w/4T65-E A/T Transmission Side Connector Pinout

Transmission Side: 2003-05

PIN	WIRE COLOR	CIRCUIT NO.	FUNCTION
A	LT GRN	1222	1-2 SHIFT SOLENOID VALVE CONTROL
B	YEL	1223	2-3 SHIFT SOLENOID VALVE CONTROL
C	PPL	1228	PC SOLENOID VALVE HIGH CONTROL
D	LT BLU	1229	PC SOLENOID VALVE LOW CONTROL
E	RED	839	IGNITION 1 VOLTAGE
F	RED/BLK	771	TRANSMISSION RANGE SIGNAL A
G	DK GRN/WHT	772	TRANSMISSION RANGE SIGNAL B
H	YEL/BLK	773	TRANSMISSION RANGE SIGNAL C
J	GRY/WHT	776	TRANSMISSION RANGE SIGNAL P
K	BLK/WHT	1050	GROUND
L	BRN	1227	TFT SENSOR SIGNAL
M	GRY	452	LOW REFERENCE
N-R	--	--	NOT USED
S	BLK	1230	AT ISS HIGH SIGNAL
T	TAN	418	TCC PWM SOLENOID VALVE CONTROL
U	WHT	1804	TCC RELEASE SWITCH SIGNAL
V	DK GRN	1231	AT ISS LOW SIGNAL
W	--	--	NOT USED

2003-05 w/4T65-E A/T Transmission Side Connector Pinout

A/T RANGE SWITCH

Connector Pinouts

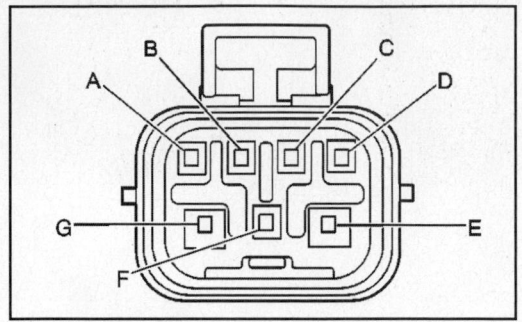

PIN	WIRE COLOR	CIRCUIT NO.	FUNCTION
A	ORN/BLK	434	NEUTRAL SAFETY SWITCH SIGNAL
B	ORN	840	BATTERY POSITIVE VOLTAGE (3.1 L)
B	--	--	NOT USED (3.8 L)
C	--	--	NOT USED
D	BLK	1050	GROUND
E	PPL	6	TRANSMISSION RANGE OUTPUT TO STARTER
F	LT GRN	24	BACKUP LAMP SUPPLY VOLTAGE (3.1 L)
F	--	--	NOT USED (3.8 L)
G	YEL	1737	CRANK RELAY OUTPUT TO TRANSMISSION RANGE SWITCH

2000-02 w/4T65-E A/T Range Switch C1 Connector Pinout

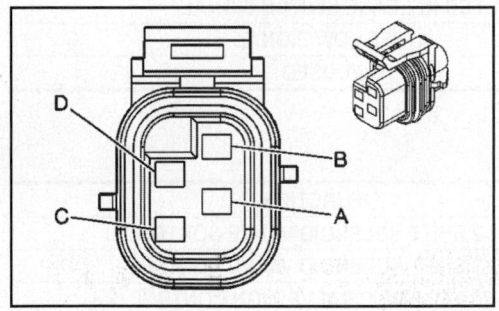

PIN	WIRE COLOR	CIRCUIT NO.	FUNCTION
A	WHT	776	TRANSMISSION RANGE SIGNAL P
B	YEL	772	TRANSMISSION RANGE SIGNAL B
C	GRY	773	TRANSMISSION RANGE SIGNAL C
D	BLK/WHT	771	TRANSMISSION RANGE SIGNAL A

2000-02 w/4T65-E A/T Range Switch C2 Connector Pinout

A/T SHIFT LOCK CONTROL SWITCH

Component Locations

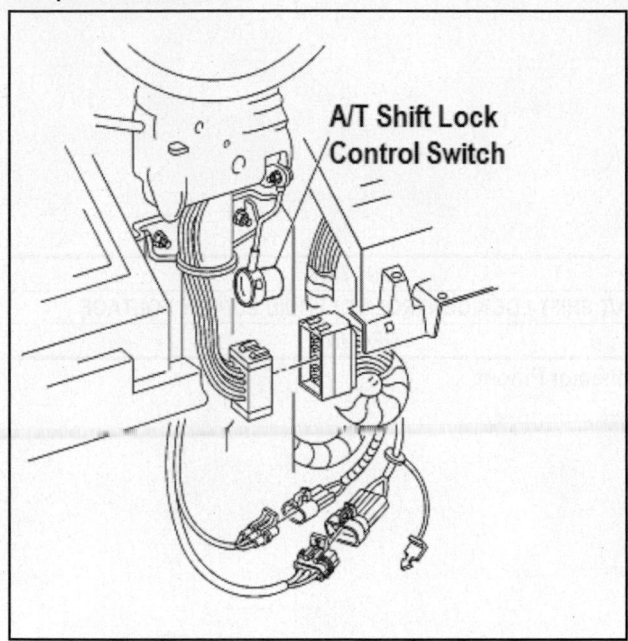

Automatic Transmission Shift Lock Control Switch (1) Steering Column Mounted

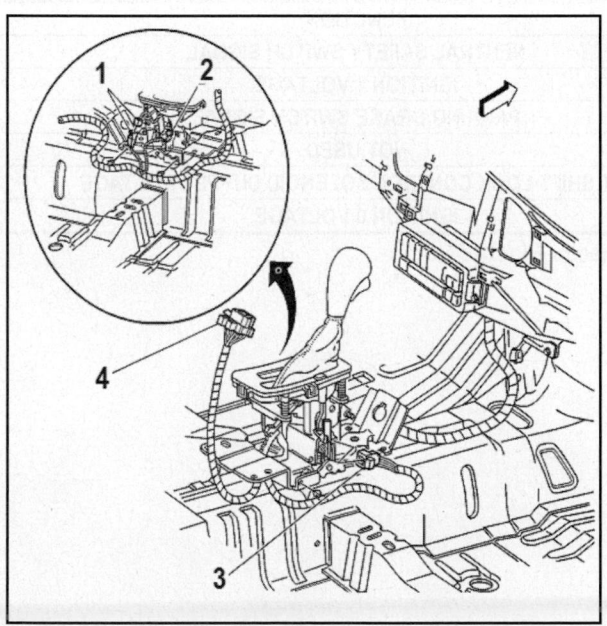

Showing Shift lock control switch (3); with floor shifter on Regal model. (1); Performance shift switch (2); transmission control position indicator bulb

Connector Pinouts

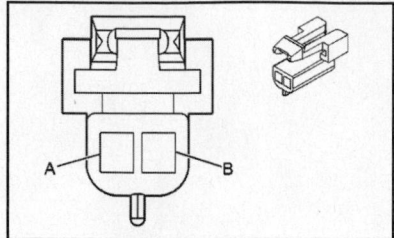

PIN	WIRE COLOR	CIRCUIT NO.	FUNCTION
A	DK GRN/ WHT	1135	A/T SHIFT LOCK CONTROL SOLENOID SUPPLY VOLTAGE
B	BLK	150	GROUND

2000-02 w/4T65-E A/T Shift Lock Control Switch Connector Pinout

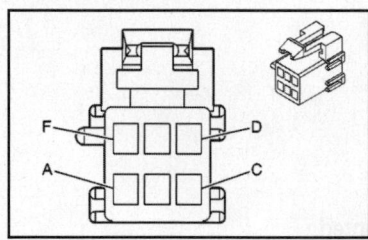

PIN	WIRE COLOR	CIRCUIT NO.	FUNCTION
A	ORN/BLK	434	NEUTRAL SAFETY SWITCH SIGNAL
B	PNK	1439	IGNITION 1 VOLTAGE
C	LT BLU	1134	PARKING BRAKE SWITCH SIGNAL
D	--	--	NOT USED
E	DK GRN/WHT	1135	A/T SHIFT LOCK CONTROL SOLENOID SUPPLY VOLTAGE
F	PPL	1600	IGNITION 0 VOLTAGE

2000-02 w/4T65-E A/T Shift Lock Control Switch Connector Pinout

A/T STOP LAMP (SHIFT INTERLOCK/TCC) SWITCH

Connector Pinouts

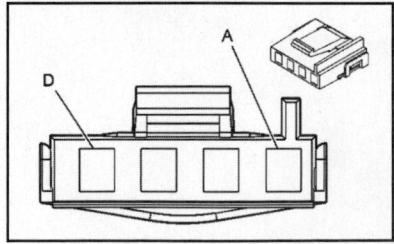

PIN	WIRE COLOR	CIRCUIT NO.	FUNCTION
A	PNK	1439	IGNITION 1 VOLTAGE
B	DK GRN/WHT	1135	BRAKE SWITCH SHIFT INTERLOCK INPUT
C	PNK	339	IGNITION 1 VOLTAGE
D	PPL	420	TCC BRAKE SWITCH SIGNAL

2000-05 w/4T65-E A/T Stop Lamp (Shift Interlock/TCC) Switch Connector Pinout

CAMSHAFT POSITION (CMP) SENSOR

Description & Component Location

3.1L AND 3.4L

During cranking, the ignition control (IC) module monitors the 7X crankshaft position sensor signal. Once the IC module determines spark synchronization, 3X reference signals are sent to the PCM. The PCM will command all 6 injectors on for one priming shot of fuel in all cylinders. After the priming, the injectors are left off for the next 6 fuel control reference signals, 2 crankshaft revolutions. This allows each cylinder a chance to use the fuel from the priming shot. During this waiting period, a cam pulse will have been received by the PCM. The PCM uses the Cam signal pulses to initiate sequential fuel injection. The PCM constantly monitors the number of pulses on the Cam signal circuit and compares the number of Cam pulses to the number of 24X reference pulses and the number of 3X reference pulses being received. If the PCM receives an incorrect number of pulses on the Cam reference circuit, DTC P0341 will set and the PCM will initiate injector sequence without the Cam signal with a one in 6 chance that injector sequence is correct. The engine will continue to start and run normally, although the misfire diagnostic will be affected if a misfiring condition occurs.

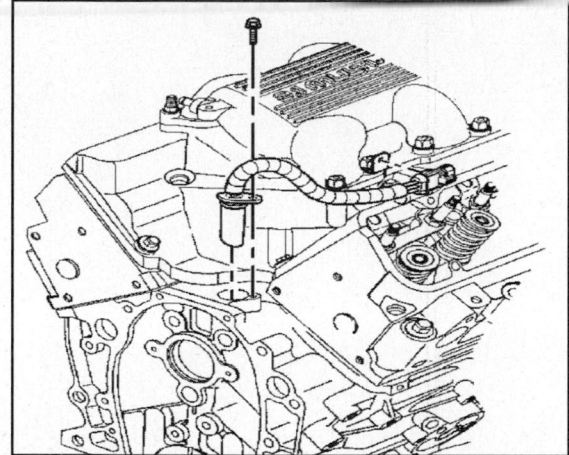

CMP sensor location on 3.1L and 3.4L

3.5L

The CMP sensor signal, when combined with the CKP sensor signal, enables the PCM to determine exactly which cylinder is on a firing stroke. The PCM can then properly synchronize the ignition system, the fuel injectors, and the knock control. The CMP sensor has a power, ground, and signal circuit. The PCM supplies 12 volts to the sensor. The PCM provides the ground path, or sensor return circuit, from the sensor. The power and ground circuits are also connected to the CKP sensor. If a problem is detected with the CMP circuit, a DTC P0340 will set.

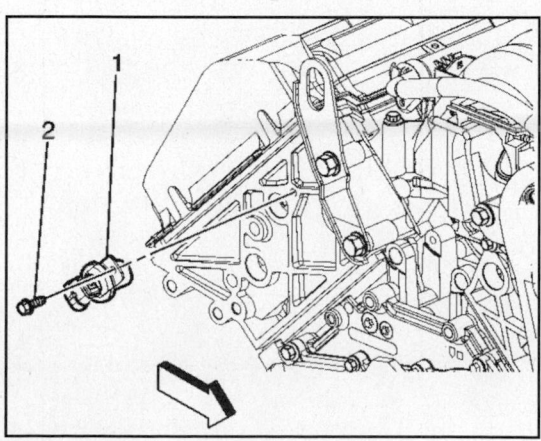

CMP sensor location on 2000-02 3.5L

3.8L

The camshaft position PCM input is produced by the ignition control module (ICM). The ICM produces the camshaft position PCM input by filtering the camshaft position (CMP) sensor pulses when the engine is running and CKP sync pulses are also being received. The powertrain control module (PCM) uses the camshaft position PCM input pulses to initiate sequential fuel injection and to determine crankshaft position for the misfire diagnostic. The PCM constantly monitors the number of pulses on the camshaft position PCM input circuit and compares the number of camshaft position PCM input pulses to the number of 18X reference pulses and the number of 3X reference pulses being received. If the PCM receives an incorrect number of pulses on the camshaft position PCM input circuit, the PCM will initiate injector sequence without the camshaft position PCM input with a one in six chance that injector sequence is correct. The engine will continue to start and run normally, although the misfire diagnostic will be disabled. DTC P0341 will set.

CMP sensor location on 3.8L (2000-02)

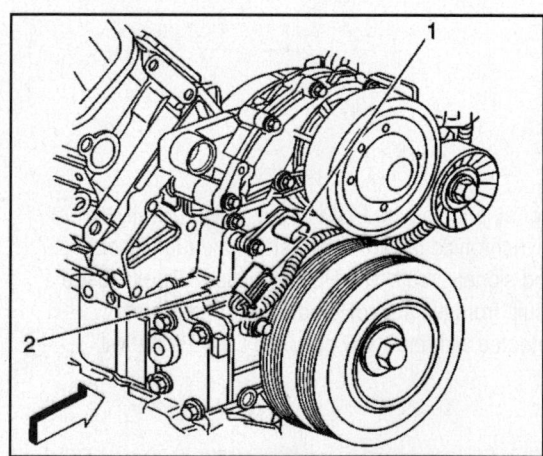

CMP sensor location on 3.8L (2003-05)

Removal & Installation

3.1L AND 3.4L

1. Turn OFF the ignition.
2. Remove the serpentine drive belt.
3. Disconnect the sensor electrical connector.
4. Remove the power steering pump, if required.
5. Remove the attaching bolt.
6. Remove the sensor.
7. Inspect the sensor for wear, cracks or leakage if the sensor is not being replaced.

To install:

8. Lubricate the O-ring with clean engine oil and replace if damaged.
9. Install the camshaft Position Sensor.
10. Tighten the retaining bolt to 8 ft. lbs. (10 Nm).
11. Connect the sensor electrical connector.
12. Install the power steering pump, if removed.
13. Reinstall the serpentine drive belt.

3.5L

1. Remove the mounting nuts retaining the coolant recovery tank.
2. Lift the coolant recovery tank to gain access to the camshaft position (CMP) sensor.
3. Disconnect the electrical connector from the CMP sensor.
4. Remove the CMP sensor mounting bolt.
5. Remove the CMP sensor from the cylinder head.
6. Remove the sensor by pulling on the sensor connector. Do not pry on the bracket.
7. Do not rotate or twist the sensor or damage the mounting bracket if the sensor is being reused.

To install:

8. Install and fully seat the CMP sensor into the cylinder head.
9. Install the CMP sensor mounting bolt. Tighten the bolt to 80 inch lbs. (9 Nm).
10. Connect the electrical connector to the CMP sensor.
11. Lower the coolant recovery tank onto the mounting studs.
12. Install the 2 mounting nuts to retain the coolant recovery tank. Tighten the nuts.

3.8L

2000-02

1. Turn OFF the ignition.
2. Disconnect the electrical connector from the camshaft position sensor.
3. Remove the camshaft position sensor retaining screw.
4. Remove the camshaft position (CMP) sensor from the engine front cover.

To install:

5. Install the camshaft position sensor to the engine front cover.
6. Reinstall the camshaft position sensor retaining bolt.
7. Tighten the bolt to 44 inch lbs. (5 Nm).
8. Connect the camshaft position sensor electrical connector.

2003-05

1. Remove the coolant recovery reservoir
2. Remove the drive belt.
3. Disconnect the electrical connector from the camshaft position sensor.
4. Remove the camshaft position sensor bolt.
5. Remove the camshaft position sensor from the engine front cover.

To install:

6. Install the camshaft position sensor to the engine front cover.

7. Install the camshaft position sensor bolt.

8. Tighten the bolt to 89 inch lbs. (10 Nm).

9. Connect the electrical connector to the camshaft position sensor.

10. Install the drive belt.

11. Install the coolant recovery reservoir.

Connector Pinouts

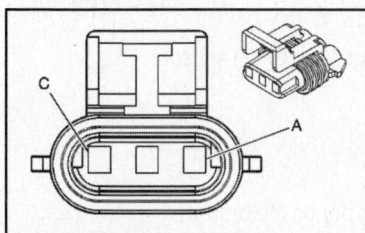

PIN	WIRE COLOR	CIRCUIT NO.	FUNCTION
A	RED/WHT	812	12 VOLT REFERENCE
B	BRN/WHT	633	CMP SENSOR SIGNAL
C	BLK	407	LOW REFERENCE

2000-02 3.1L. 3.4L & 3.8L CMP Sensor Connector Pinout

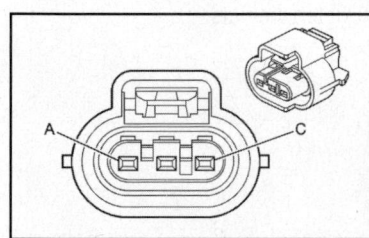

PIN	WIRE COLOR	CIRCUIT NO.	FUNCTION
A	BRN/WHT	633	CMP SENSOR SIGNAL
B	BLK	407	LOW REFERENCE
C	RED/WHT	812	12-VOLT REFERENCE

2003-05 3.1L, 3.4L & 3.8L CMP Sensor Connector Pinout

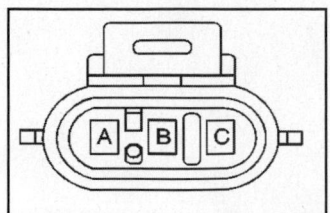

PIN	WIRE COLOR	CIRCUIT NO.	FUNCTION
A	ORN	1799	CAMSHAFT POSITION (CMP) SENSOR SIGNAL - INPUT
B	PNK/BLK	632	REFERENCE LOW
C	RED	631	IGNITION FEED

2000-02 3.5L CMP Sensor Connector Pinout

CRANKSHAFT POSITION (CKP) SENSOR

Description & Component Location

3.1L

The 7X crankshaft position sensor provides a signal used by the ignition control module. The 24X Crankshaft Position (CKP) Sensor is used to improve idle spark control at engine speeds up to approximately 1600 RPM. The camshaft position (CMP) sensor is also a magneto resistive sensor, with the same type of circuits as the crankshaft position (CKP) sensor. The CMP sensor signal is a digital ON/OFF pulse, output once per revolution of the camshaft. The CMP sensor information is used by the PCM to determine the position of the valve train relative to the crankshaft position.

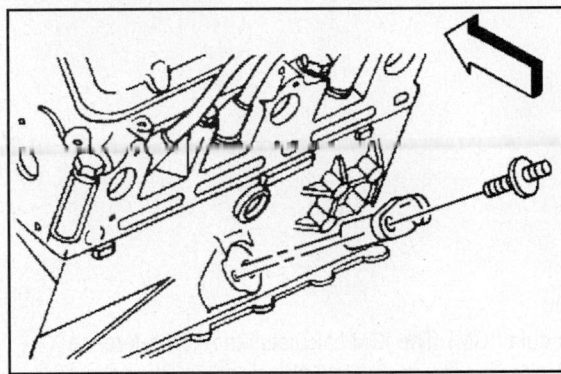

7x type CKP sensor location on 3.1L & 3.4L

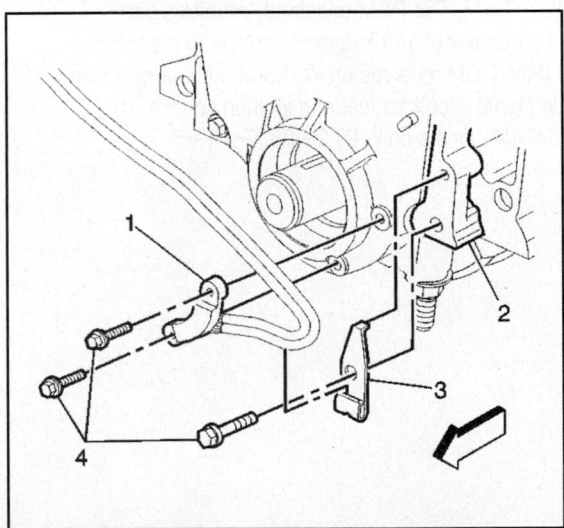

24x type CKP sensor (1) location on 3.1L & 3.4L, showing mounting location (2), holding bracket (3), and mounting bolts (4)

3.5L ENGINES

The crankshaft position (CKP) sensor is a 3-wire sensor based on the magneto resistive principle. A magneto resistive sensor uses two magnetic pickups between a permanent magnet. As an element such as a reluctor wheel passes the magnets the resulting change in the magnetic field is used by the sensor electronics to produce a digital output pulse. This system uses two sensors within the same housing for the V6 engine, and two separate sensors for the V8 engine. The PCM supplies each sensor a 12-volt reference, low reference, and a signal circuit. The signal circuit returns a digital ON/OFF pulse 24 times per crankshaft revolution.

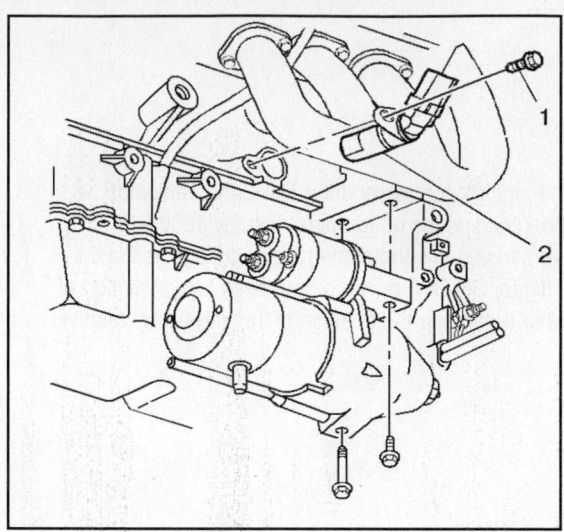

CKP sensor (2) location on 3.5L

3.8L

The 18X reference signal is produced by the ignition control module (ICM). The ICM calculates the 18X reference signal by filtering the crankshaft position (CKP) sensor 18X pulses when the engine is running and CKP sync pulses are also being received. The powertrain control module (PCM) uses the 18X reference signal to calculate engine RPM and crankshaft position at engine speeds below 1200 RPM. The PCM constantly monitors the number of pulses on the 18X Reference circuit and compares the number of 18X reference pulses to the number of 3X reference pulses and CAM signal pulses being received. If the PCM receives an incorrect number of pulses on the 18X Reference circuit, the PCM will use the 3X reference signal circuit for fuel and ignition control. The engine will continue to start and run using the 3X reference and CAM signals only. DTC P0336 will set.

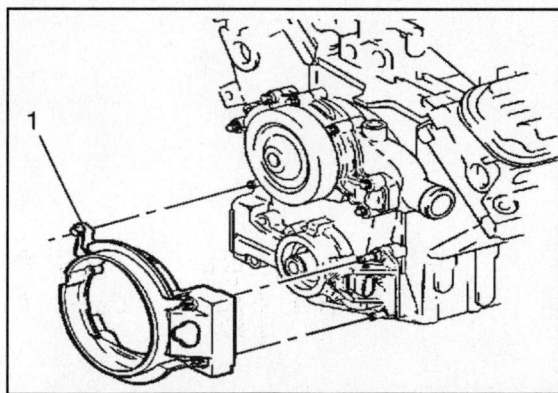

CKP sensor location on 3.8L

Removal & Installation

3.1L & 3.4L

7x TYPE

1. Remove the negative battery cable.
2. Turn the steering wheel fully to the left.
3. Raise vehicle on hoist.
4. Disconnect the electrical connector.
5. Remove the retaining bolt.

6. Remove the sensor from the engine.

7. If the sensor is not being replaced inspect for wear, cracks, or leakage. Replace the O-ring if necessary.

8. Lubricate the new O-ring with clean engine oil before installing.

To install:

9. Install the sensor into the block.

10. Reinstall the retainer bolt to hold sensor to block face. Tighten the bolt to 71 inch lbs. (8 Nm).

11. Connect the electrical connector.

12. Lower the vehicle. Reconnect the negative battery cable.

24X TYPE

1. Turn OFF the ignition.

2. Remove the serpentine drive belt from crankshaft pulley.

3. Raise the vehicle on hoist.

4. Remove the crankshaft harmonic balancer retaining bolt.

5. Remove the crankshaft harmonic balancer.

6. Note the routing of sensor harness before removal.

7. Remove the harness retaining clip with bolt.

8. Disconnect the sensor electrical connector.

9. Remove the sensor bolts.

10. Remove the sensor.

To install:

11. Install the 24X Crankshaft Position Sensor with bolts and route harness as noted during removal.

12. Install the harness retaining clip with bolt. Tighten the bolts to 8 ft. lbs. (10 Nm).

13. Connect the sensor electrical connector.

14. Install the balancer on the crankshaft.

15. Apply thread sealer to threads of the crankshaft harmonic balancer bolt. Tighten the bolt to 110 ft. lbs. (150 Nm).

16. Lower vehicle.

17. Reinstall the serpentine drive belt.

18. Perform the CKP System Variation Learn Procedure.

3.5L

1. Disconnect the negative battery cable.

2. Raise and support the vehicle.

3. Remove the starter motor.

4. Disconnect the crankshaft position (CKP) sensor electrical connector.

5. Remove the CKP sensor bolt (1).

6. Remove the CKP sensor (2) from the engine block.

7. Remove the sensor by pulling on the sensor connector. DO NOT pry on the bracket.

9. DO NOT rotate or twist the sensor or damage the mounting bracket if the sensor is being reused.

To install:

10. Install and fully seat the CKP sensor (2) into the engine block.
11. Install the CKP sensor bolt. Tighten the bolt to 80 inch lbs. (9 Nm).
12. Ensure that the CKP sensor mounting flange is contacting the engine block.
13. Connect the CKP sensor electrical connector.
14. Install the starter motor.
15. Connect the negative battery cable.
16. Lower the vehicle.
17. Perform CKP System Variation Learn Procedure.

3.8L

1. Turn OFF the ignition.
2. Remove the serpentine belts from the crankshaft pulley.
3. Raise the vehicle.
4. Remove the right front wheel and tire assembly.
5. Remove the right inner fender access cover.
6. Hold the flywheel with a (J 37096) flywheel holding tool.
7. Using 28 mm socket, remove the crankshaft harmonic balancer retaining bolt.
8. Remove the crankshaft harmonic balancer, using a (J 38197-A) balancer remover.
9. Remove the crankshaft position sensor shield (1). Do not use a pry bar.
10. Disconnect the sensor electrical connector.
11. Remove the attaching bolts from the crankshaft position sensor.
12. Remove the crankshaft position sensor from the block face.

To install:

13. Position the crankshaft position (CKP) sensor to the block.
14. Install the bolts to hold the CKP sensor to the block face. Tighten the bolts to 14-28 ft. lbs. (20-40 Nm).
15. Reinstall the crankshaft position sensor shield.
16. Connect the CKP sensor electrical connector.
17. Position the crankshaft harmonic balancer on the crankshaft.
18. Hold the flywheel with a (J 37096) flywheel holding tool.
19. Using a (J 36660-A) torque/angle meter, tighten the harmonic balancer bolt to 110 ft. lbs. (150 Nm), plus an additional 76 degrees.
20. Install the right inner fender access cover.
21. Install the right front wheel and tire assembly.
22. Tighten the wheel nuts to 104 ft. lbs. (140 Nm.).
23. Lower the vehicle.
24. Reinstall the serpentine belts.
25. Perform the Crankshaft Variation Learn Procedure.

Crankshaft Variation Learn Procedure

The crankshaft position system variation compensating values are stored in the PCM non-volatile memory after a learn procedure has been performed. If the actual crankshaft position system variation is not within the crankshaft position system variation compensating values stored in the PCM, DTC P0300 may set. Refer to Diagnostic Aids for DTC P0300. The Crankshaft Position System Variation Learn Procedure should be performed if any of the following conditions are true:

- DTC P1336 is set.
- The PCM has been replaced.
- The engine has been replaced.
- The crankshaft has been replaced.
- The crankshaft harmonic balancer has been replaced.
- The crankshaft position sensor has been replaced.

CAUTION: The scan tool crankshaft position system variation learn function will be inhibited if engine coolant temperature is less than 70°C (158°F). Allow the engine to warm to at least 70°C (158°F) before attempting the crankshaft position system variation learn procedure. The scan tool crankshaft position system variation learn function will be inhibited if any Powertrain DTCs other than DTC P1336 are set before or during the crankshaft position system variation learn procedure. Diagnose and repair any DTCs if set. Refer to applicable DTCs. The crankshaft position system variation learn function will be inhibited if the PCM detects a malfunction involving the camshaft position signal circuit, the 3X reference circuit, or the 24X reference circuit. If a malfunction has been indicated, refer to the following list to diagnoses the system or sensor.

- DTC P0336 Crankshaft Position (CKP) Sensor Circuit.
- DTC P0341 Camshaft Position (CMP) Sensor Performance.
- DTC P1374 Crankshaft Position (CKP) High to Low Resolution Frequency Correlation.

The scan tool crankshaft position system variation learn function will not be enabled until engine coolant temperature reaches 158°F (70°C). Selecting the crankshaft position system variation learn procedure on the scan tool will command the PCM to enable CKP system variation learn fuel cutoff and allow the crankshaft position system variation compensating values to be stored in the PCM. The PCM must detect an engine speed of 5150 RPM (CKP system variation learn fuel cutoff) during the crankshaft position system variation learn procedure to store the crankshaft position system variation compensating values and complete the procedure.

CAUTION: Set the vehicle parking brake and block the drive wheels when performing the Crankshaft Position System Variation Learning Procedure in order to prevent personal injury. Release the throttle when the engine reaches the second fuel cut off. Leaving the throttle open during the fuel cut off will allow the engine to decel at an even rate. Once the learn procedure is completed, the PCM will return the engine control to the operator and the engine will respond to the throttle position.

1. Set the parking brake.
2. Block the drive wheels.
3. Ensure the hood is closed.
4. Start the engine and allow engine coolant temperature to reach at least 70°C (158°F).
5. Turn off the ignition switch.
6. Select and enable the crankshaft position variation learn procedure with the scan tool.
7. Start the vehicle.
8. Apply and hold the service brake pedal firmly.
9. Ensure the transaxle is in park.

10. Increase accelerator pedal position and hold until the fuel cutoff is reached at 5150 RPM. Release the accelerator pedal after the second fuel cutoff has been reached.

11. The crankshaft position system variation compensating values are learned when the RPM decreases back to idle. If the procedure terminates.

12. Observe DTC status for DTC P1336.

13. If the scan tool indicates that DTC P1336 ran and passed, the crankshaft position system variation learn procedure is complete. If the scan tool indicates DTC P1336 failed or not run, determine if other DTCs have set. If DTCs other than P1336 are not set, repeat the crankshaft position system variation learn procedure as necessary.

Connector Pinouts

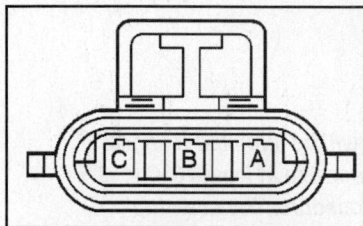

PIN	WIRE COLOR	CIRCUIT NO.	FUNCTION
A	LT GRN	1867	CKP SENSOR FEED-12 VOLT REFERENCE
B	LT BLU/BLK	647	CKP SIGNAL/MED. RESOLUTION ENGINE SPEED SIGNAL
C	YEL/BLK	1868	CKP SENSOR RETURN/LOW REFERENCE

2000-05 3.1L, 3.4L & 3.8L CKP Sensor 3-Pin Connector Pinout

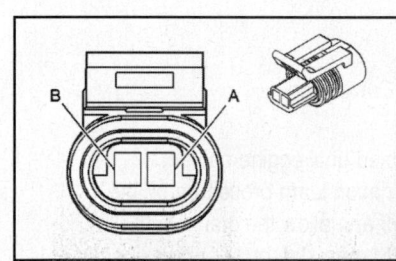

PIN	WIRE COLOR	CIRCUIT NO.	FUNCTION
A	YEL	573	CKP SENSOR SIGNAL
B	PPL	574	CKP SENSOR RETURN (LOW REFERENCE)

2000-05 3.1L, 3.4L & 3.8L CKP Sensor 2-Pin Connector Pinout

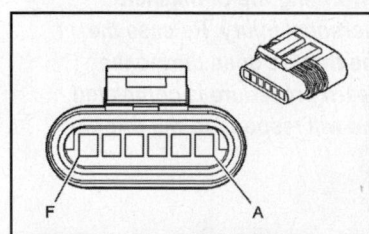

PIN	WIRE COLOR	CIRCUIT NO.	FUNCTION
A	LT GRN/WHT	2867	12 VOLT REFERENCE
B	YEL/WHT	2868	LOW REFERENCE
C	YEL	573	CKP SENSOR 1 SIGNAL
D	LT GRN	1867	12-VOLT REFERENCE
E	YEL/BLK	1868	LOW REFERENCE
F	LT BLU/WHT	1800	CKP SENSOR 2 SIGNAL

2000-02 3.5L CKP Sensor Connector Pinout

EGR SOLENOID VALVE

Component Locations

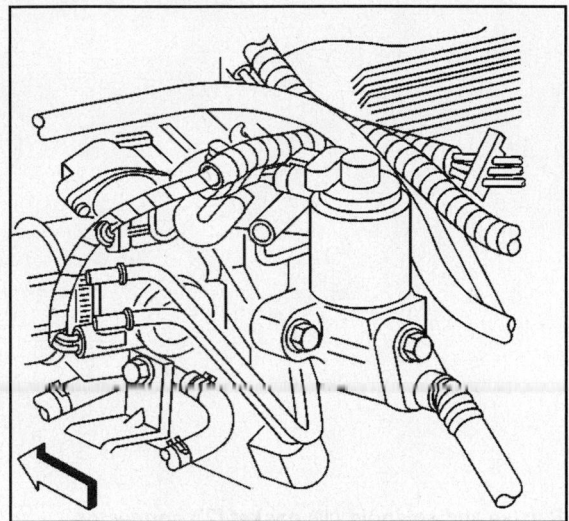

Showing the EGR Solenoid location on 3.1L & 3.4L

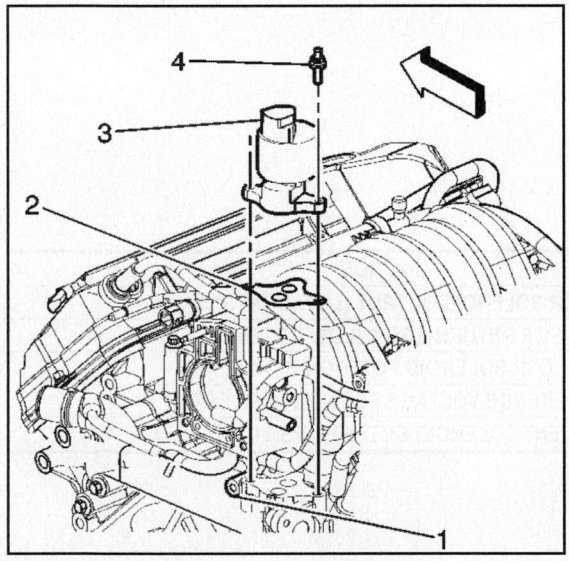

EGR valve and solenoid valve (3) location on 3.5L

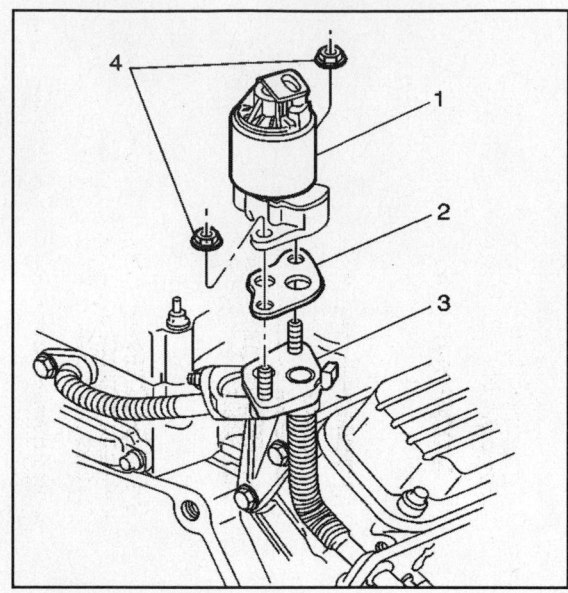

Exploded view of the EGR valve and solenoid on 3.8L: EGR valve and solenoid (1); gasket (2); connector (3); nuts (4)

Connector Pinouts

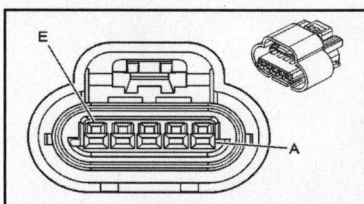

PIN	WIRE COLOR	CIRCUIT NO.	FUNCTION
A	GRY	435	EGR SOLENOID OUTPUT (LOW CONTROL)
B	BLK	2753	SENSOR RETURN-EGR 1 (LOW REFERENCE)
C	BRN	1456	EGR SOLENOID POSITION SIGNAL
D	GRY	2702	REFERENCE VOLTAGE FEED-5 VOLTS-EGR 1
E	WHT	257	EGR VENT SOLENOID OUTPUT (HIGH CONTROL)

2000-05 3.1L, 3.4L & 3.8L EGR Solenoid Connector Pinout

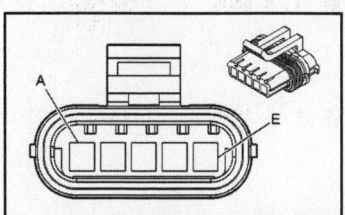

PIN	WIRE COLOR	CIRCUIT NO.	FUNCTION
A	GRY	435	EGR SOLENOID CONTROL
B	BLK	2753	SENSOR RETURN EXHAUST GAS RECIRCULATION (EGR) 1
C	BRN	1456	EGR VALVE POSITION SIGNAL
D	GRY	2702	5-VOLT REFERENCE
E	RED	1676	EGR VALVE SUPPLY VOLTAGE

2001-02 3.5L EGR Solenoid Connector Pinout

ENGINE COOLANT LEVEL SENSOR

Component Locations

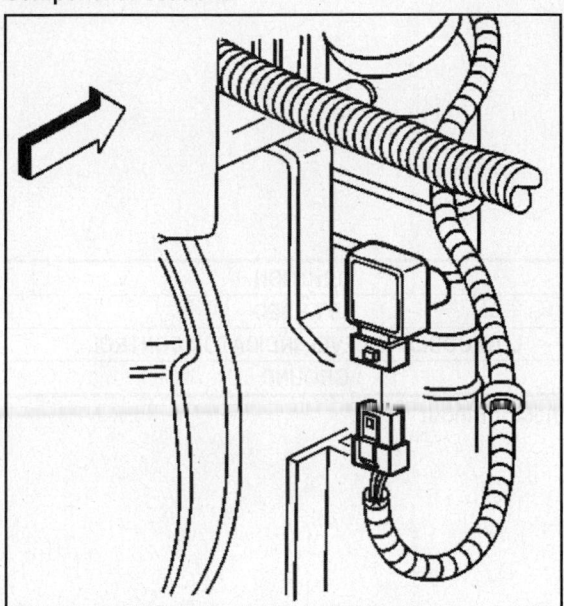

Engine coolant level sensor location on 3.1L

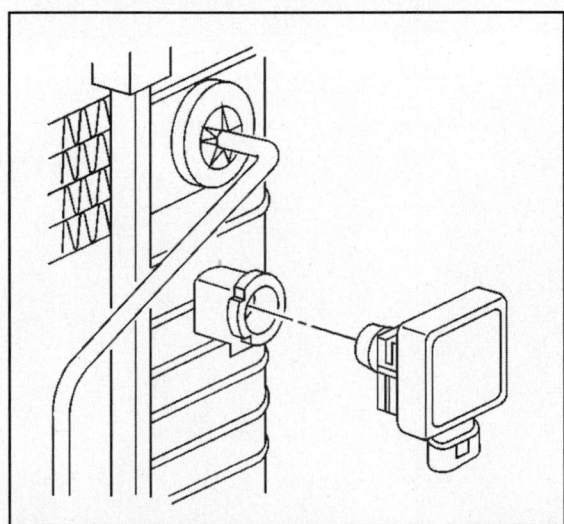

Engine coolant level sensor on 3.4L & 3.8L

Connector Pinouts

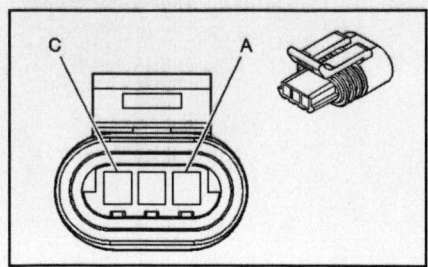

PIN	WIRE COLOR	CIRCUIT NO.	FUNCTION
A	--	--	NOT USED
B	YEL/BLK	68	LOW COOLANT LEVEL INDICATOR CONTROL
C	BLK OR BLK/WHT	1350	GROUND

2000-05 3.1L, 3.4L, 3.8L Engine Coolant Level Sensor Connector Pinout

ENGINE COOLANT TEMPERATURE (ECT) SENSOR

Component Locations

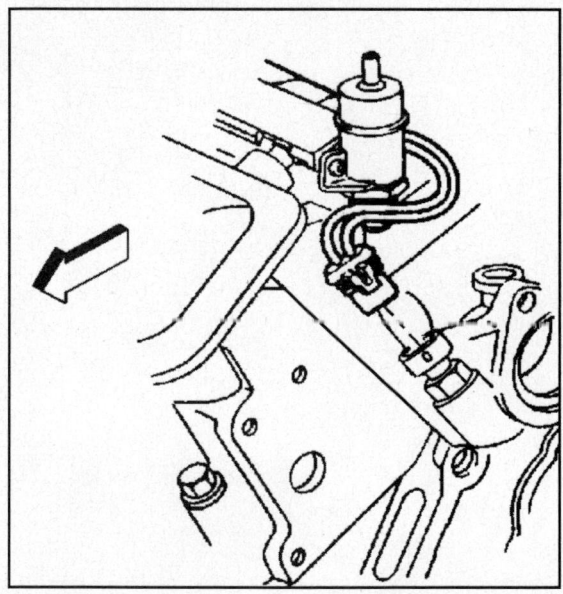

Indicating the ECT sensor on 3.1L & 3.4L

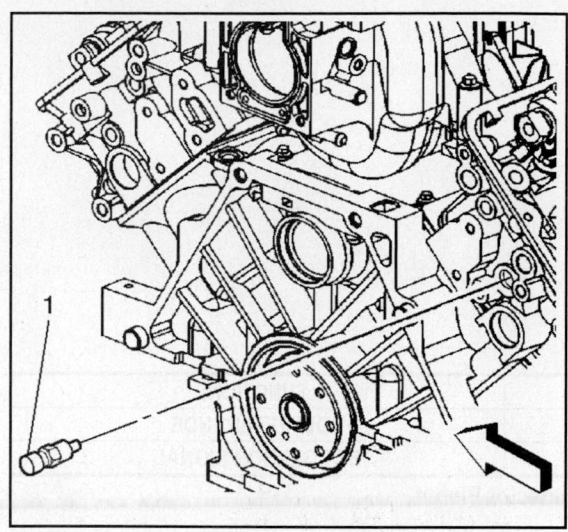

Identifying the ECT sensor (1) location on 3.5L

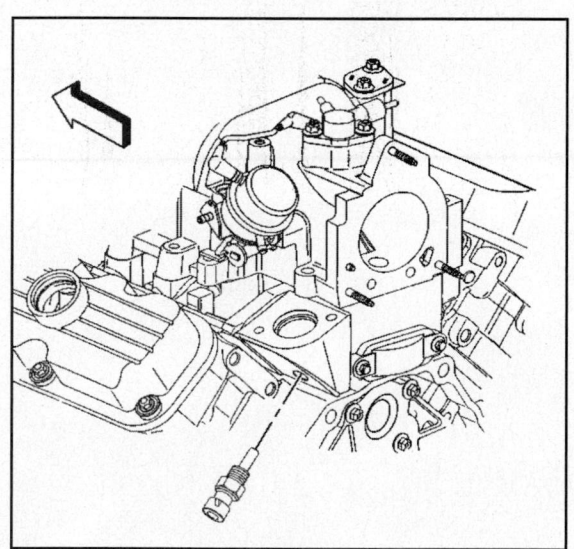

Showing the location of the ECT sensor on the 3.8L

Connector Pinouts

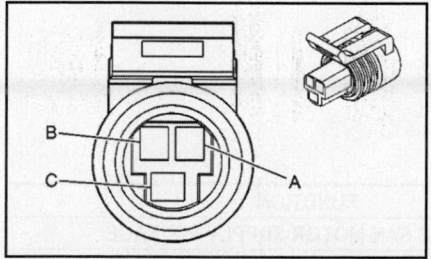

PIN	WIRE COLOR	CIRCUIT NO.	FUNCTION
A	ORN/BLK	469	LOW REFERENCE
B	YEL	410	ECT SENSOR SIGNAL
C	DK GRN	135	ECT SENSOR GAGE SIGNAL

2000-02 3.1L & 3.8L ECT Sensor Connector Pinout

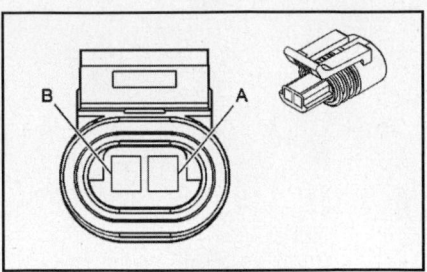

PIN	WIRE COLOR	CIRCUIT NO.	FUNCTION
A	ORN/BLK	469	LOW REFERENCE
B	YEL	410	ECT SENSOR SIGNAL

2003-05 3.1L, 2000-05 3.4L & 2000-05 3.8L ECT Sensor Connector Pinout

ENGINE COOLING FAN CONNECTORS

Component Locations

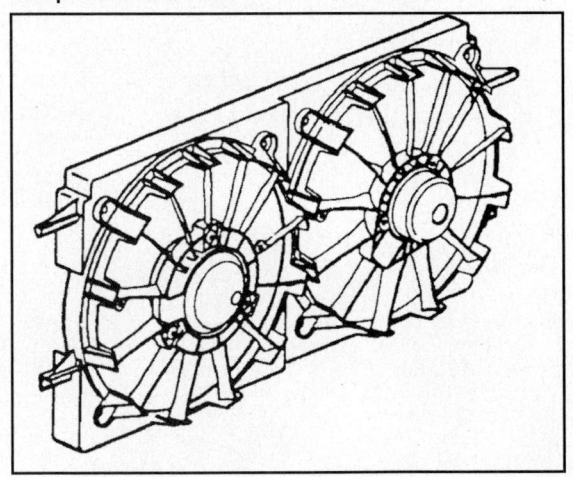

Left and right cooling fans on 3.1L (similar on all others)

Connector Pinouts

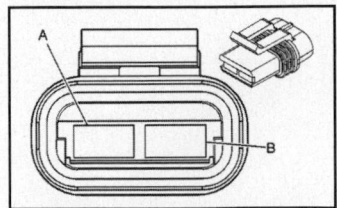

PIN	WIRE COLOR	CIRCUIT NO.	FUNCTION
A	BLK	250, 532 OR 1050	COOLING FAN MOTOR SUPPLY VOLTAGE
B	LT BLU	409	COOLING FAN MOTOR SUPPLY VOLTAGE

2000-05 3.1L, 2000-02 3.4L, 2001-02 3.5L & 2000-05 3.8L Left Cooling Fan Connector Pinout

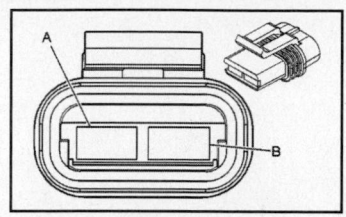

PIN	WIRE COLOR	CIRCUIT NO.	FUNCTION
A	WHT	504	COOLING FAN LOW REFERENCE
B	GRY	532	COOLING FAN MOTOR SUPPLY VOLTAGE

2003-05 3.4L Left Cooling Fan Connector Pinout

PIN	WIRE COLOR	CIRCUIT NO.	FUNCTION
A	GRY	1050	GROUND
B	WHT	504	COOLING FAN LOW REFERENCE

2000-05 3.1L & 2000-05 3.8L Right Cooling Fan Connector Pinout

PIN	WIRE COLOR	CIRCUIT NO.	FUNCTION
A	WHT	504	COOLING FAN MOTOR GROUND
B	GRY	532	COOLING FAN MOTOR SUPPLY VOLTAGE

2003-05 3.4L, 2001-02 3.5L Right Cooling Fan Connector Pinout

ENGINE OIL LEVEL SENSOR

Component Locations

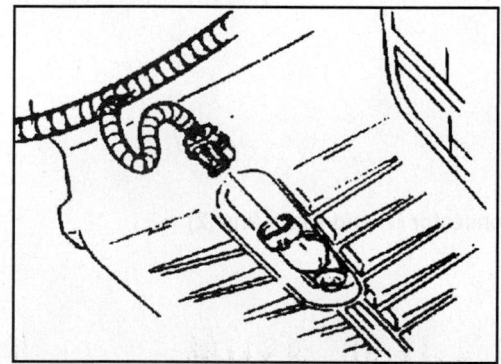

Oil level sensor location 3.1L & 3.4L

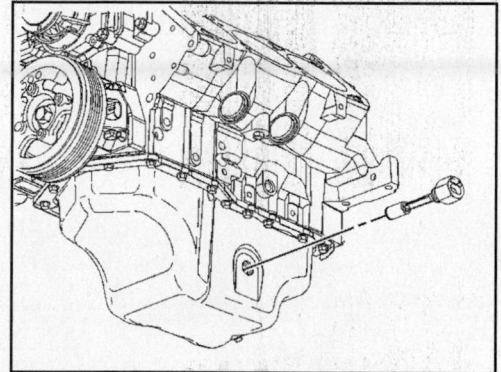

Oil level sensor location 3.8L

Connector Pinouts

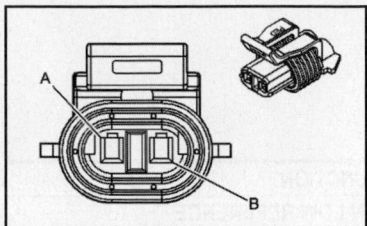

PIN	WIRE COLOR	CIRCUIT NO.	FUNCTION
A	BRN	1174	OIL LEVEL SENSOR SIGNAL
B	BLK/WHT	451	GROUND-CLEAN

2000-05 3.1L, 3.4L & 3.8L Engine Oil Level Sensor Connector Pinout

EVAP CANISTER VENT & PURGE SOLENOIDS

Component Locations

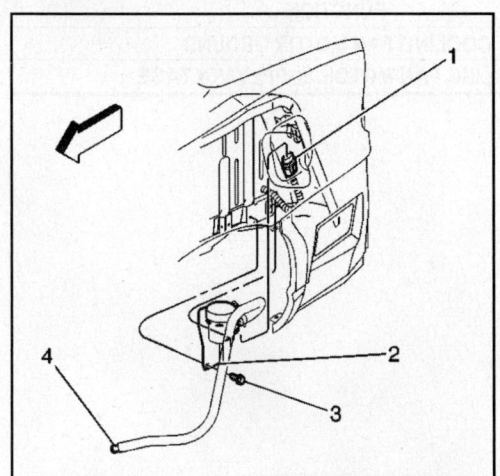

EVAP vent solenoid (2) on 3.1L, 3.4L, 3.5L & 3.8L, showing the connector (1) and vapor line (2)

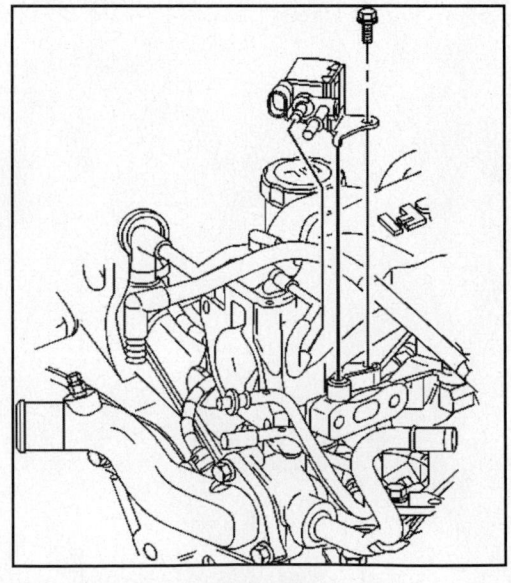

EVAP canister purge solenoid location on 3.1L & 3.4L

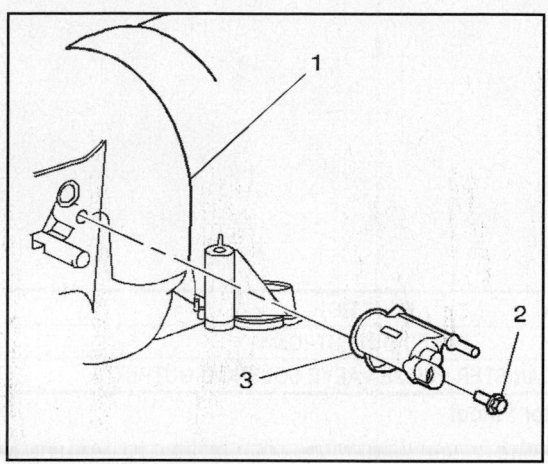

EVAP purge solenoid valve (3) location on 3.5L

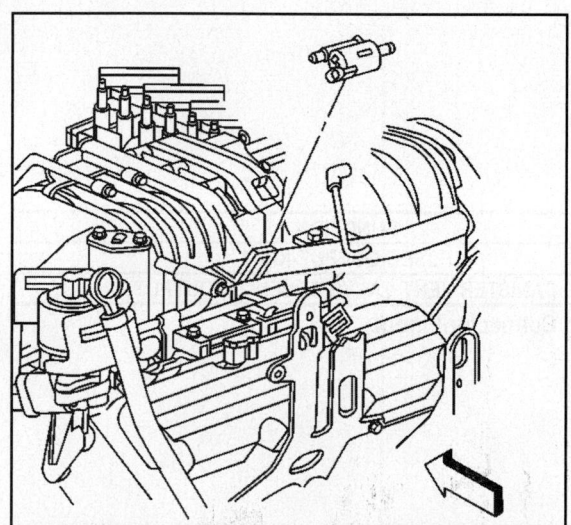

EVAP purge solenoid location on 3.8L (L67)

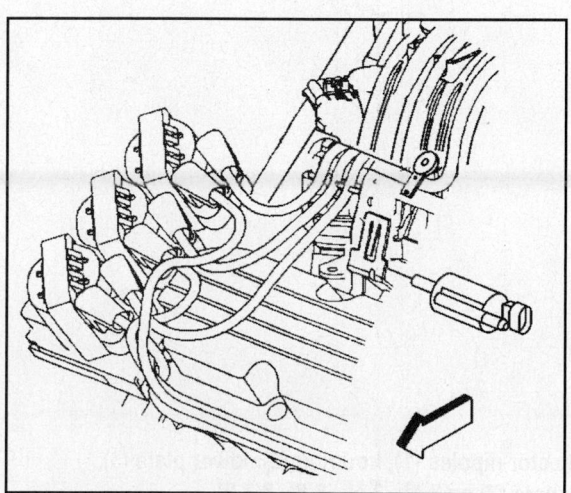

EVAP purge solenoid location on 3.8L (L36)

Connector Pinouts

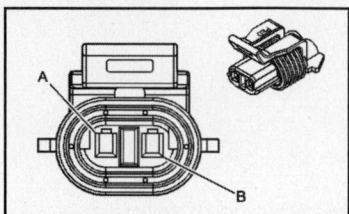

PIN	WIRE COLOR	CIRCUIT NO.	FUNCTION
A	PNK	339 OR 739	FUSED OUTPUT-IGN 1
B	DK GRN/WHT	428	EVAP CANISTER PURGE VALVE SOLENOID OUTPUT

3.1L, 3.4L, 3.5L & 3.8L EVAP Purge Valve Solenoid Connector Pinout

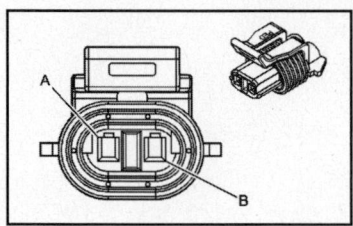

PIN	WIRE COLOR	CIRCUIT NO.	FUNCTION
A	PNK	239 OR 339	FUSED OUTPUT-IGN 1
B	WHT	1310	CANISTER VENT VALVE SOLENOID OUTPUT

3.1L, 3.4L, 3.5L & 3.8L EVAP Canister Vent Valve Solenoid Connector Pinout

FUEL PUMP & SENDER

Component Locations

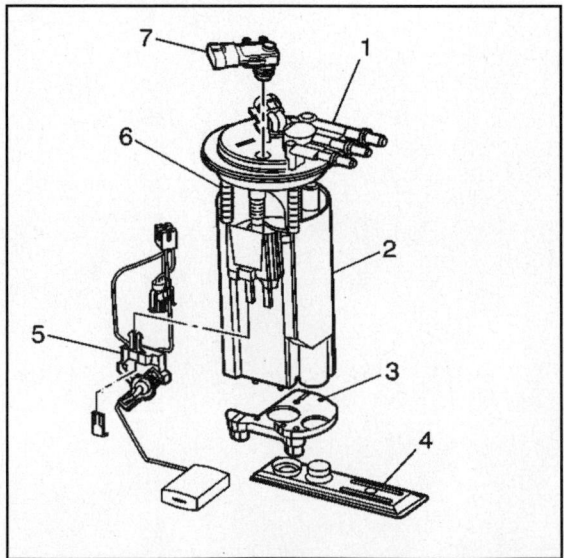

Exploded view of the fuel pump (5) and showing the connector nipples (1), housing (2), lower plate (3), sensor plate (4), mounting studs (6), fuel tank pressure sensor (7) on 3.1L, 3.4L, 3.5L & 3.8L

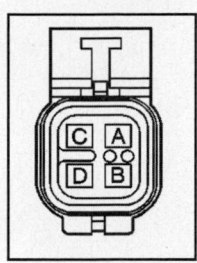

PIN	WIRE COLOR	CIRCUIT NO.	FUNCTION
A	PPL	30	FUEL GAUGE SENSOR SIGNAL
B	GRY	120	FUEL PUMP MOTOR FEED
C	BLK	150/650	GROUND
D	BLK/WHT	651	PCM SENSOR GROUND

2000-05 3.1L, 3.4L, 3.5L & 3.8L Fuel Tank Connector Pinout

FUEL TANK PRESSURE (FTP) SENSOR

Component Location

NOTE: See illustration above.

Connector Pinouts

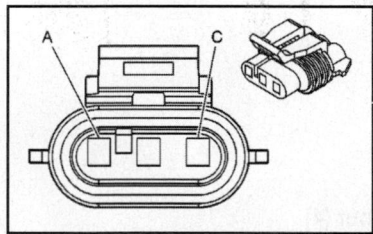

PIN	WIRE COLOR	CIRCUIT NO.	FUNCTION
A	ORN/BLK	469	LOW REFERENCE
B	DK GRN	890	FUEL TANK PRESSURE SENSOR SIGNAL
C	GRY/BLK	416	5 VOLT REFERENCE

2001-05 3.1L, 3.4L, 3.5L & 3.8L FTP Sensor Connector Pinout

GENERATOR CONNECTOR

Component Locations

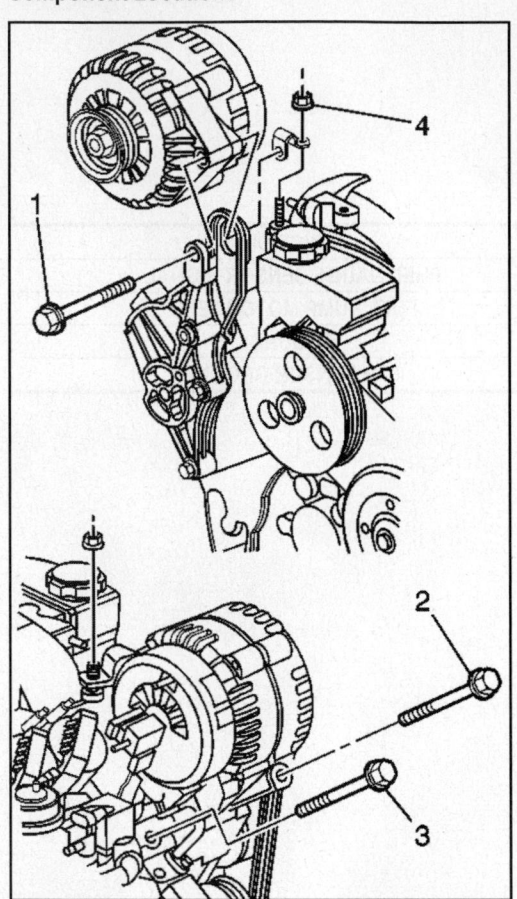

Generator mounting on 3.1L & 3.4L with mounting bolts (2, 3) and attaching nut (4)

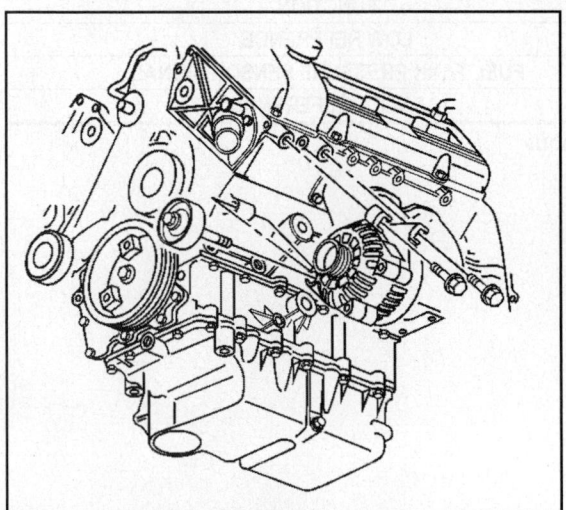

Generator mounting on 3.1L

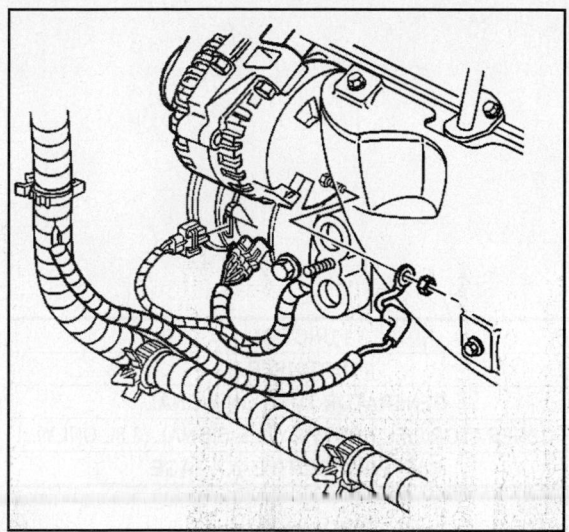

Generator mounting and electrical connectors on 3.5L

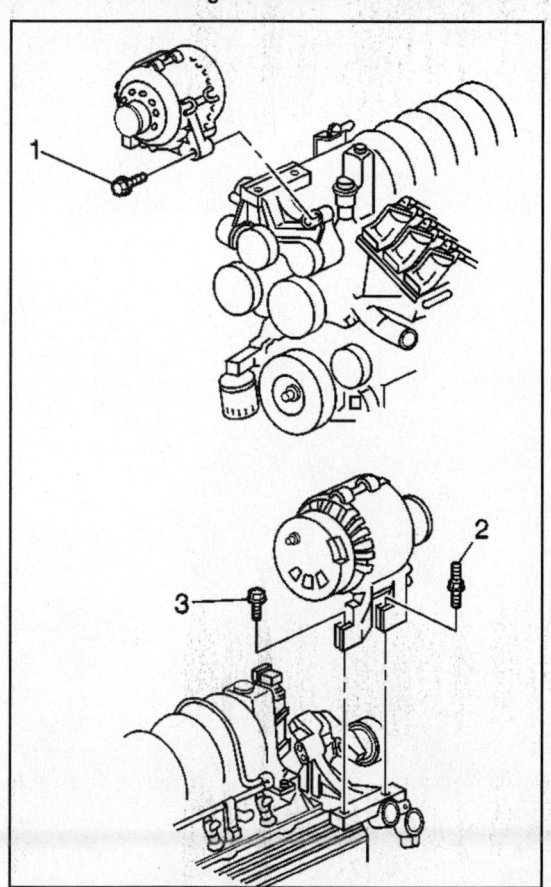

Showing the Generator mounting on the 3.8L, indicating mounting bolts (1, 2, 3)

Connector Pinouts

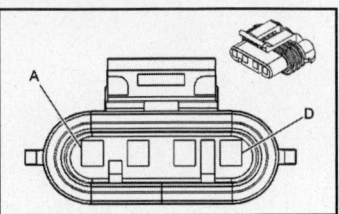

PIN	WIRE COLOR	CIRCUIT NO.	FUNCTION
A	--	--	NOT USED
B	RED	225	GENERATOR TURN ON SIGNAL
C	GRY	23	GENERATOR FIELD DUTY CYCLE SIGNAL (3.8L ONLY)
D	ORN	140 OR 2740	BATTERY POSITIVE VOLTAGE

3.1L, 3.4L, 3.5L & 3.8L Generator Connector Pinout

HEATED OXYGEN SENSORS (HO2S)

Component Locations

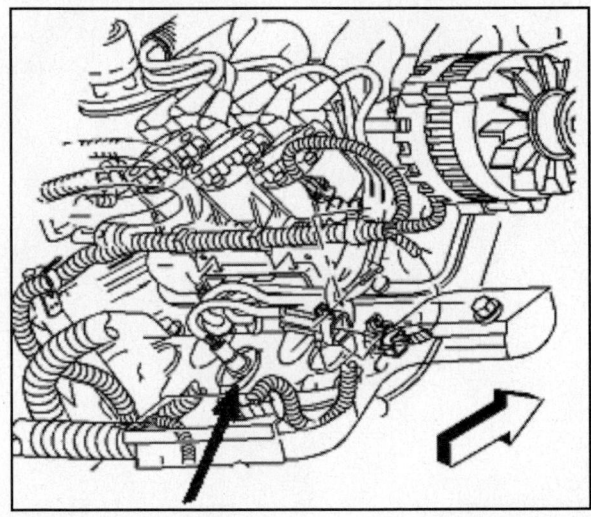

HO2S1 location on 3.1L & 3.4L

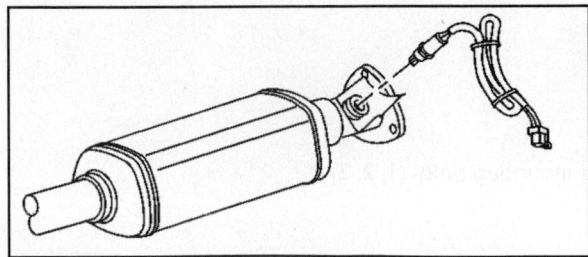

HO2S2 location on 3.1L & 3.4L

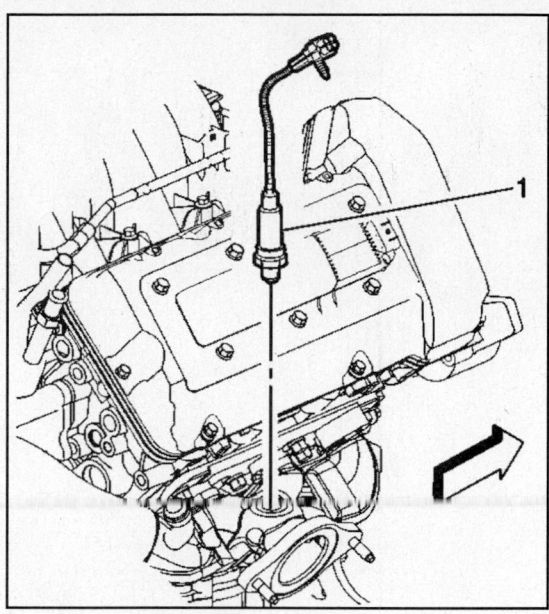

HO2S1 location (1) on 3.5L

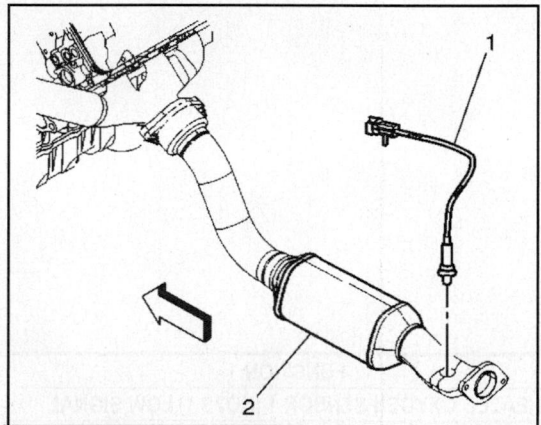

HO2S2 location (1) on 3.5L

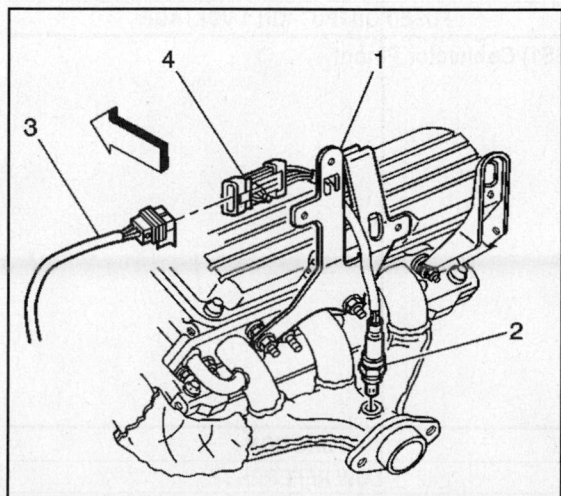

HO2S1 location (2) on 3.8L

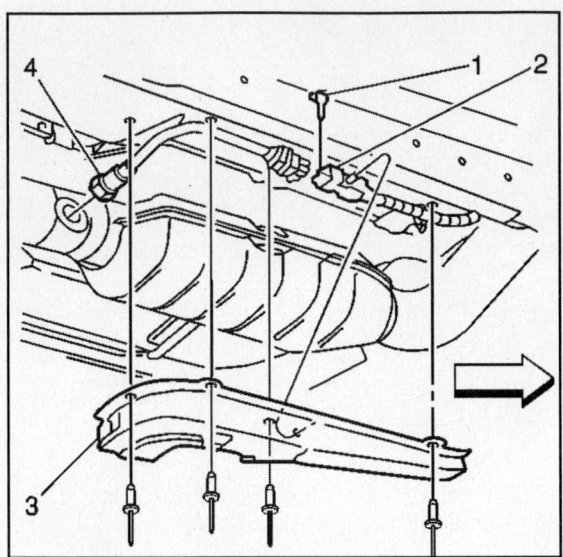

HO2S2 (4) location on 3.8L

Connector Pinouts

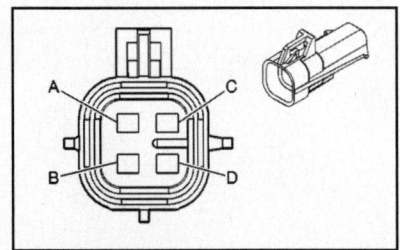

PIN	WIRE COLOR	CIRCUIT NO.	FUNCTION
A	TAN	413	HEATED OXYGEN SENSOR 1 (HO2S 1) LOW SIGNAL
B	PPL	412	HEATED OXYGEN SENSOR 1 (HO2S 1) HIGH SIGNAL
C	BLK OR BLK/WHT	350, 1050 OR 3113	GROUND
D	BRN OR PNK	839 OR 1039	FUSED OUTPUT-IGN 1 VOLTAGE

2000-01 3.1L, 2002-05 3.4L & 2000-05 3.8L O2 Sensor (HO2S1) Connector Pinout

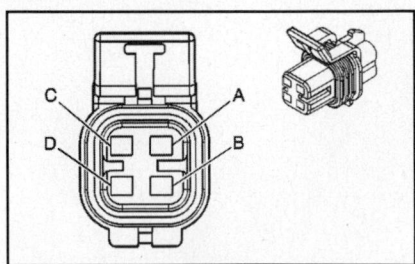

PIN	WIRE COLOR	CIRCUIT NO.	FUNCTION
A	DK GRN	676	LOW REFERENCE
B	PPL	412	HO2S HIGH SIGNAL
C	BLK/WHT	3113	HO2S HEATER LOW CONTROL
D	PNK	839	IGNITION 1 VOLTAGE

2002-05 3.1L O2 Sensor (HO2S1) Connector Pinout

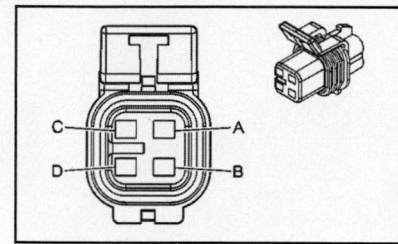

3.8L

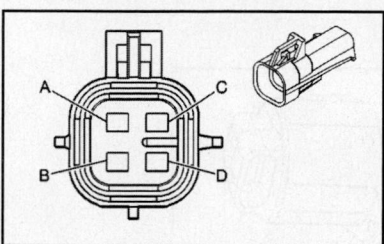

3.1L

PIN	WIRE COLOR	CIRCUIT NO.	FUNCTION
A	TAN/WHT	1669	HO2S LOW SIGNAL
B	PPL/WHT	1668	HO2S HIGH SIGNAL
C	BLK	350 OR 1050	GROUND
D	PNK	339 OR 1039	IGNITION I VOLTAGE

2000-01 3.1L, 2000-05 3.4L & 2000-05 3.8L O2 Sensor (HO2S2) Connector Pinout

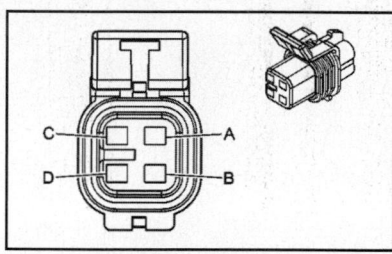

PIN	WIRE COLOR	CIRCUIT NO.	FUNCTION
A	BLK	808	LOW REFERENCE
B	PPL/WHT	1665	HO2S HIGH SIGNAL - BANK 1 SENSOR 1
C	BLK/WHT	3113	HO2S HEATER LOW CONTROL-BANK1 SENSOR2
D	PNK	839	IGNITION 1 VOLTAGE

3.5L O2 Sensor 1 (HO2S1) Connector Pinout

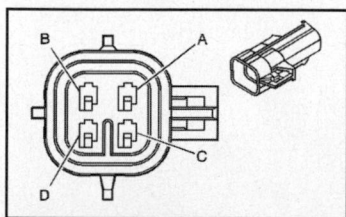

PIN	WIRE COLOR	CIRCUIT NO.	FUNCTION
A	BLK	808	LOW REFERENCE
B	PPL/WHT	1668	HO2S HIGH SIGNAL BANK 1 SENSOR 2
C	BRN	2391	HO2S HEATER LOW CONTROL-BANK 1 SENSOR 2
D	PNK	839	IGNITION 1 VOLTAGE

3.5L O2 Sensor 2 (HO2Ss) Connector Pinout

IDLE AIR CONTROL (IAC) VALVE

Component Locations

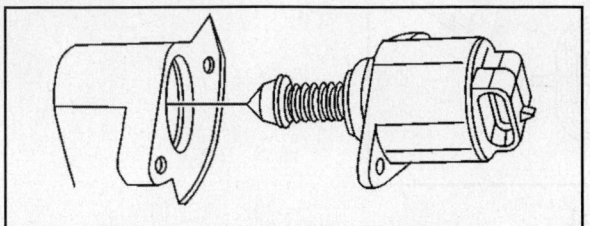

Indicating the IAC valve on the 3.1L & 3.4L

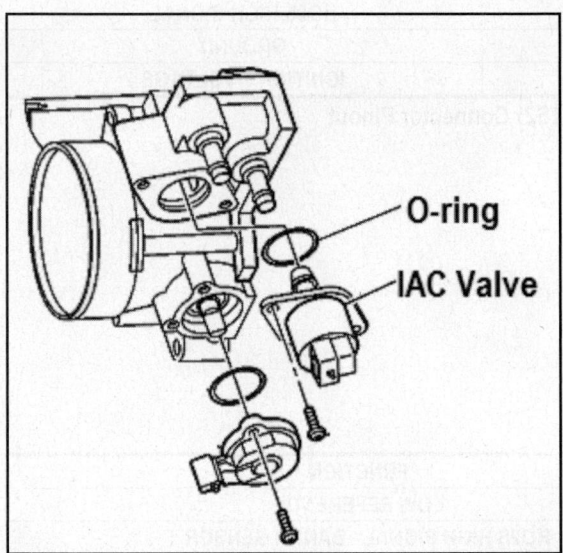

O-ring

IAC Valve

Exploded view of the mounting of the IAC valve on 3.5L

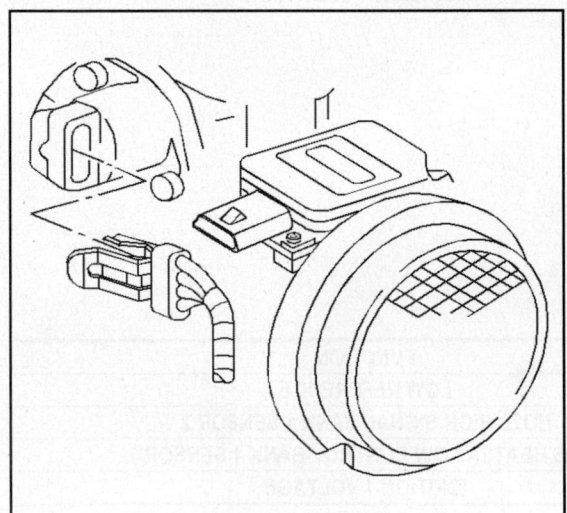

IAC valve location on the 3.8L

Connector Pinouts

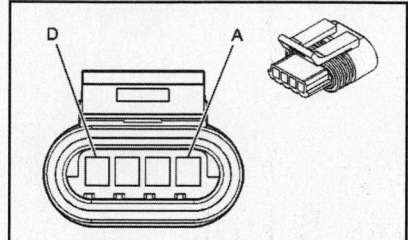

PIN	WIRE COLOR	CIRCUIT NO.	FUNCTION
A	LT GRN/BLK	444	IAC COIL B LOW
B	LT GRN/WHT	1749	IAC COIL B HIGH
C	LT BLU/BLK	1748	IAC COIL A LOW
D	LT BLU/WHT	1747	IAC COIL A HIGH

3.1L, 3.4L, 3.5L & 3.8L IAC Valve Connector Pinout

IGNITION CONTROL MODULE (ICM)

Component Locations

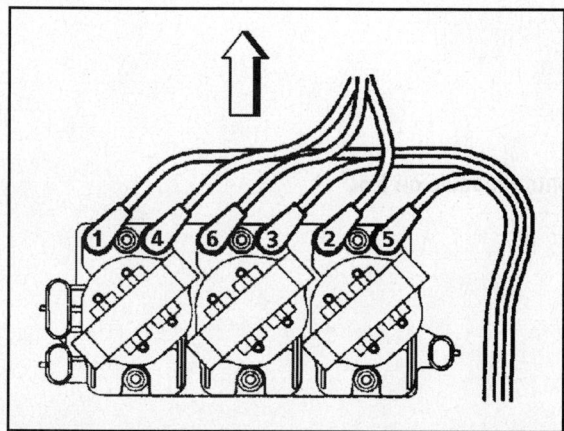

Ignition control module locations on 3.1L & 3.4L

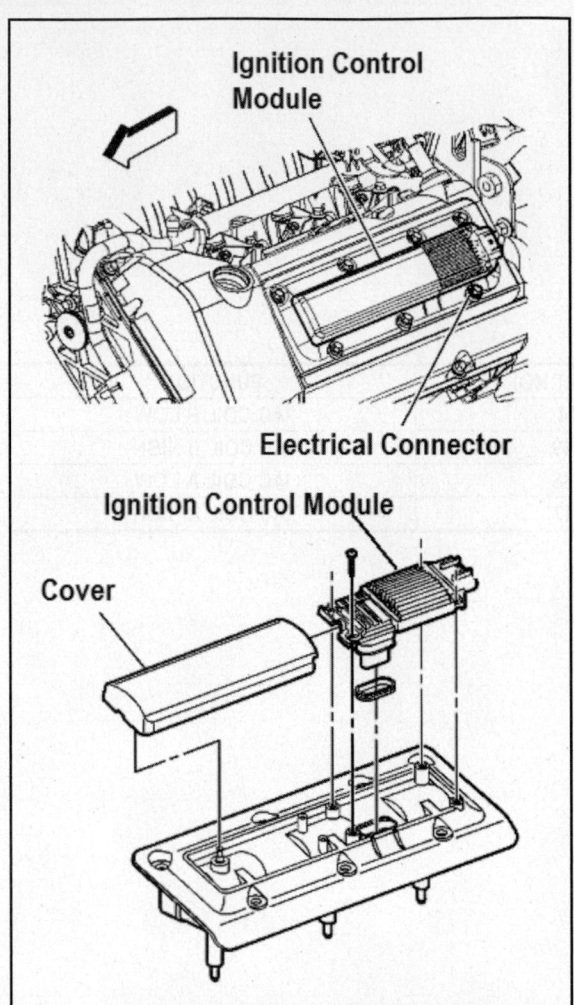

Showing the location and exploded view of the ignition control module on 3.5L

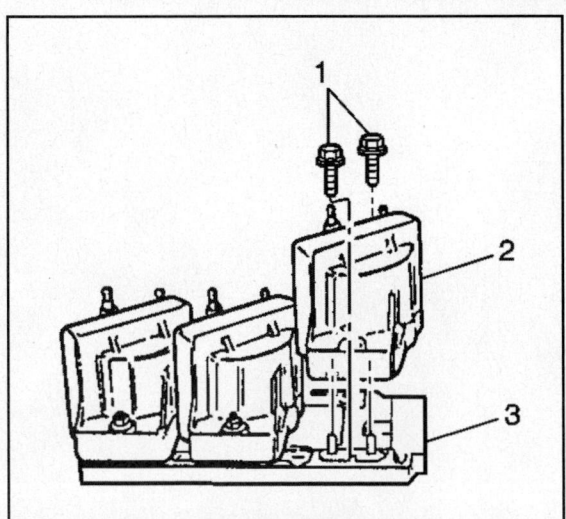

Ignition control module (3); ignition coils (2); 3.8L

Connector Pinouts

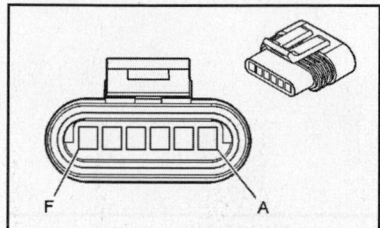

PIN	WIRE COLOR	CIRCUIT NO.	FUNCTION
A	TAN/BLK	424	BYPASS CONTROL
B	TAN/BLK OR WHT	423	IGNITION CONTROL
C-D	--	--	NOT USED
E	PPL/WHT	430	3 X REFERENCE SIGNAL
F	RED/BLK	453	REFERENCE LOW

2000-05 3.1L, 3.4L & 2000 3.8L ICM C1 Connector Pinout

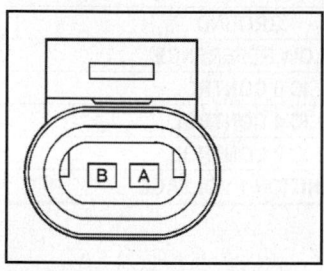

PIN	WIRE COLOR	CIRCUIT NO.	FUNCTION
A	BLK OR BLK/WHT	51 OR 350	GROUND
B	PNK	239 OR 539	FUSE IGNITION 1 FEED

2000-05 3.1L, 3.4L & 2000 3.8L ICM C2 Connector Pinout

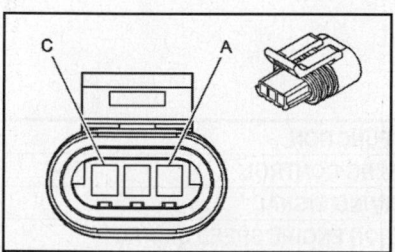

PIN	WIRE COLOR	CIRCUIT NO.	FUNCTION
A	YEL	573	CKP SENSOR 1 SIGNAL
B	--	--	NOT USED
C	PPL	574	LOW REFERENCE

2003-05 3.1L & 2001-04 3.4L ICM C3 Connector Pinout

PIN	WIRE COLOR	CIRCUIT NO.	FUNCTION
A	BLK	350	GROUND
B	PNK	539	IGNITION 1 VOLTAGE

2005 3.4L ICM C3 Connector Pinout

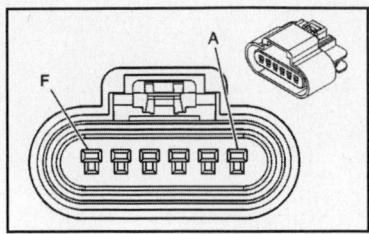

PIN	WIRE COLOR	CIRCUIT NO.	FUNCTION
A	BLK/WHT	51	GROUND
B	YEL/BLK	2174	LOW REFERENCE
C	PPL	2121	IC 1 CONTROL
D	LT BLU	2123	IC 3 CONTROL
E	DK GRN	2125	IC 5 CONTROL
F	PNK	639	IGNITION 1 VOLTAGE

3.5L Bank 1 ICM Connector Pinout

PIN	WIRE COLOR	CIRCUIT NO.	FUNCTION
A	BLK/WHT	51	GROUND
B	YEL/BLK	2174	LOW REFERENCE
C	LT BLU/WHT	2126	IC 6 CONTROL
D	DK GRN/WHT	2124	IC 4 CONTROL
E	RED/WHT	2122	IC 2 CONTROL
F	PNK	239	IGNITION 1 VOLTAGE

3.5L Bank 2 ICM Connector Pinout

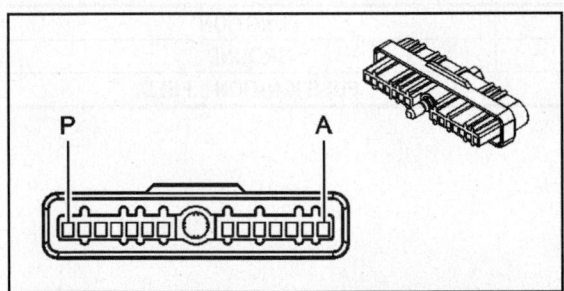

PIN	WIRE COLOR	CIRCUIT NO.	FUNCTION
A	WHT	423	IC TIMING CONTROL
B	TAN/BLK	424	IC TIMING SIGNAL
C	LT BLU/BLK	647	MEDIUM RESOLUTION ENGINE SPEED SIGNAL
D	PPL/WHT	430	LOW RESOLUTION ENGINE SPEED SIGNAL
E	WHT	121	ENGINE SPEED SIGNAL
F	BLK	630	CAMSHAFT POSITION SIGNAL
G	YEL	573	CKP SENSOR 1 SIGNAL
H	LT BLU/WHT	1800	CKP SENSOR 2 SIGNAL
J	BRN/WHT	633	CMP SENSOR SIGNAL
K	BLK/WHT	451	GROUND
L	RED/BLK	453	LOW REFERENCE
M	BLK/WHT	836	LOW REFERENCE
N	WHT/BLK	644	12-VOLT REFERENCE
P	PNK	239	IGNITION 1 VOLTAGE

2001-05 3.8L ICM Connector Pinout

INTAKE AIR TEMPERATURE (IAT) SENSOR

Component Locations

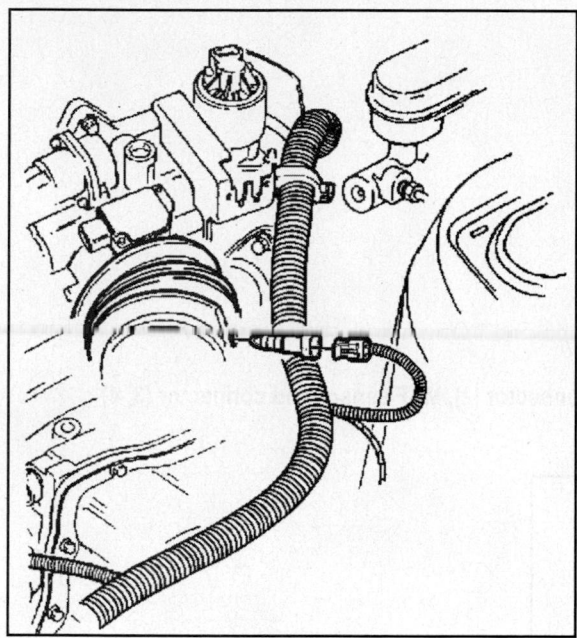

IAT sensor location on 2000-05 3.1L & 2000-02 3.4L

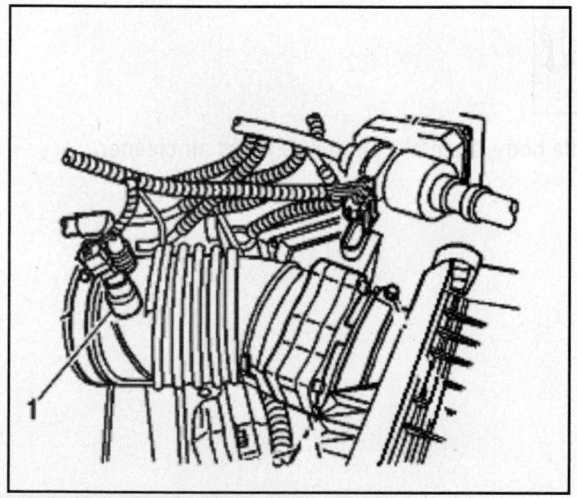

IAT sensor (1) location on 2003-05 3.4L

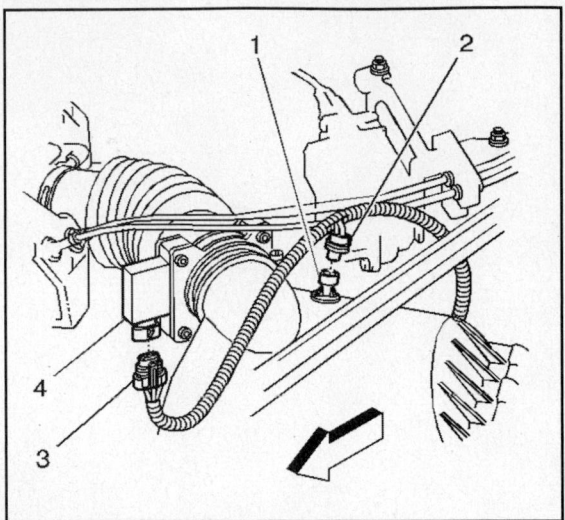

IAT sensor (1) location on 3.5L, showing IAT electrical connector (2), MAF sensor and connector (3, 4)

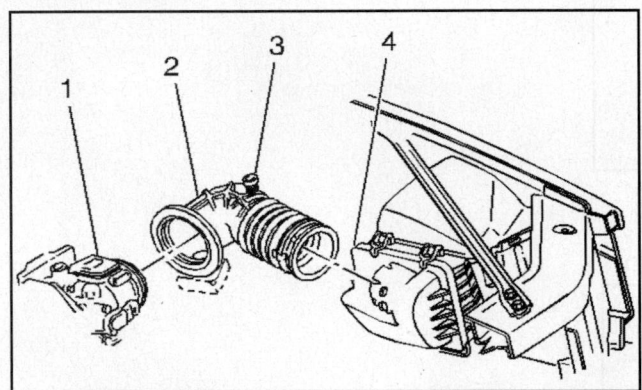

Intake air temperature sensor 3.8L (3), showing the throttle body (1), intake air duct (2), and air cleaner housing (4)

Connector Pinouts

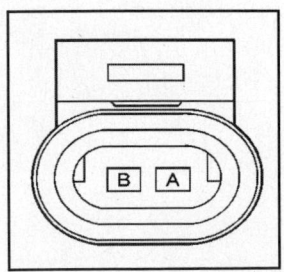

PIN	WIRE COLOR	CIRCUIT NO.	FUNCTION
A	BLK	2760	FUSED OUTPUT-BATTERY FEED
B	TAN	472	INTAKE AIR TEMPERATURE SENSOR SIGNAL

2000-05 3.1L, 3.4L, 3.5L & 3.8L IAT Sensor Connector Pinout

KNOCK SENSOR
Component Locations

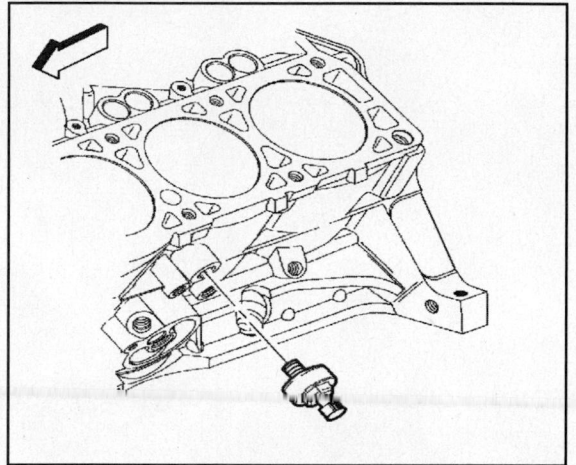

Indicating the Knock Sensor location on the 3.1L & 3.4L

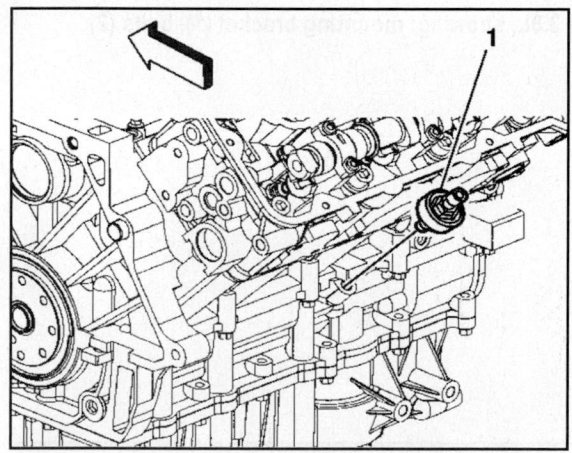

Knock sensor (1) location on 3.5L

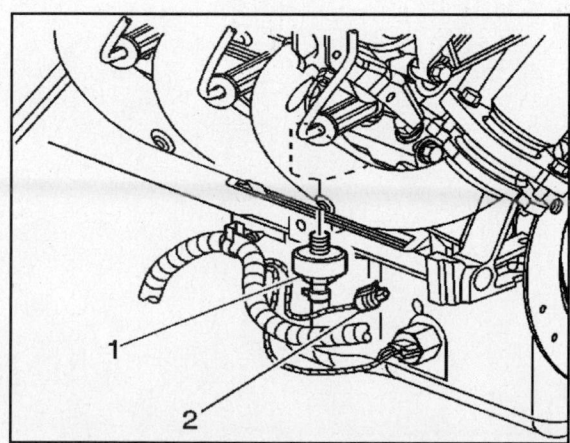

Indicating the Knock Sensor (1) left side location on the 3.8L, showing the electrical connector (2)

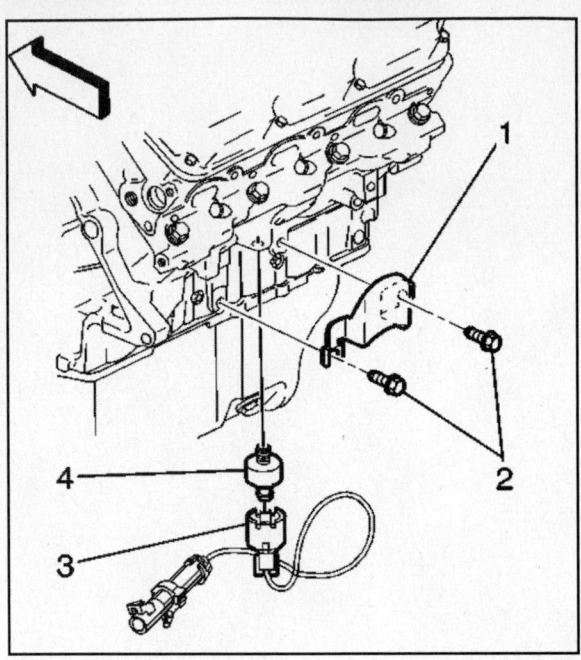

Indicating the Knock Sensor (4) right side location on the 3.8L, showing: mounting bracket (1), bolts (2), electrical connector (3)

Connector Pinouts

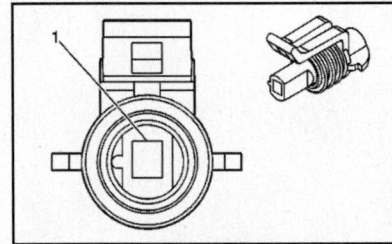

PIN	WIRE COLOR	CIRCUIT NO.	FUNCTION
A	DK BLU	496	KNOCK SENSOR SIGNAL 1

2000-05 3.1L, 3.4L, 3.5L & 3.8L Knock Sensor Connector Pinout

MASS AIR FLOW (MAF) SENSOR

Component Locations

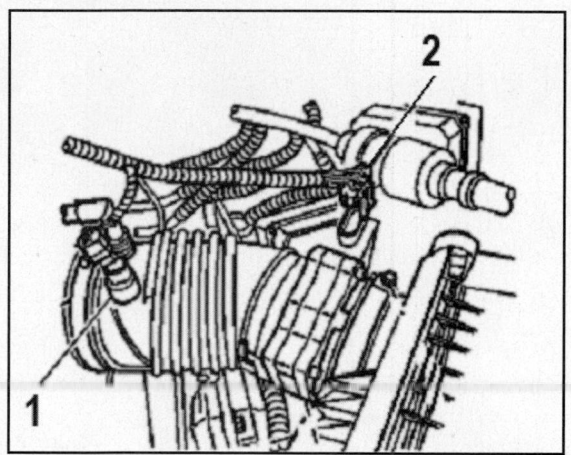

MAF sensor (2) location on 2000-05 3.1L & 2003-05 3.4L

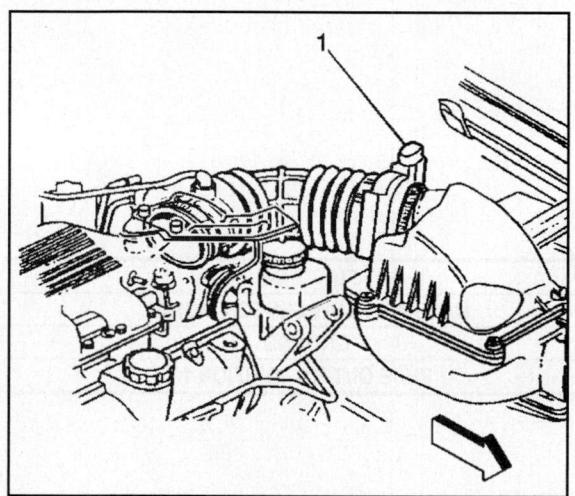

MAF sensor (1) location on 2000-02 3.4L

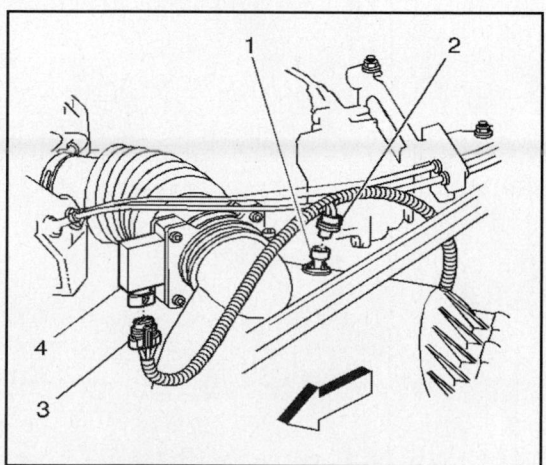

Showing the MAF sensor (4) location, electrical connector (3) and IAT sensor and connector (1, 2) on 3.5L

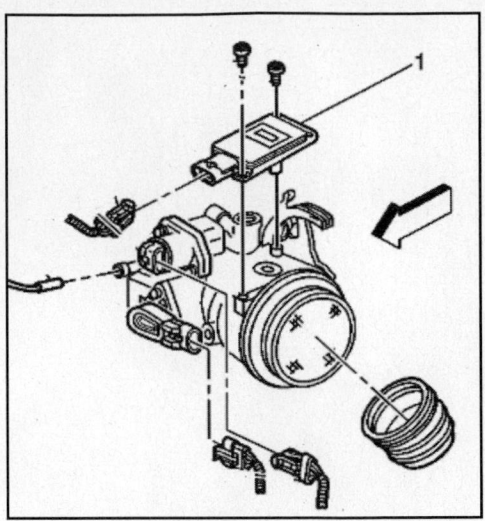

MAF sensor (1) location 3.8L

Connector Pinouts

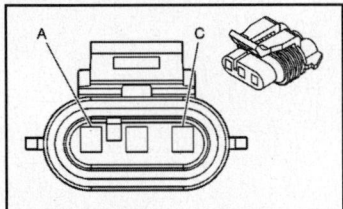

PIN	WIRE COLOR	CIRCUIT NO.	FUNCTION
A	YEL	492	MASS AIR FLOW SENSOR SIGNAL
B	BLK OR BLK/WHT	451	GROUND
C	PNK	339 OR 739	FUSE OUTPUT-IGNITION 1

3.1L, 3.4L, 3.5L & 3.8L MAF Sensor Connector Pinout

MANIFOLD ABSOLUTE PRESSURE (MAP) SENSOR

Component Locations

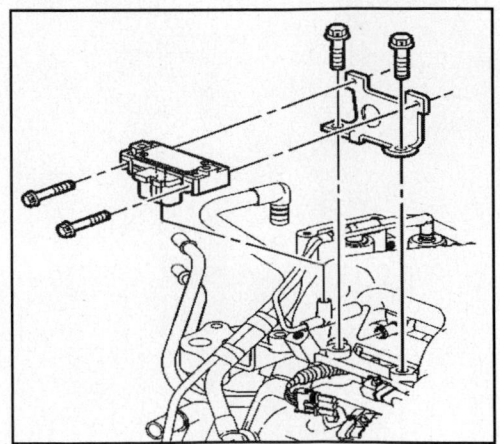

MAP sensor location on 3.1L & 3.4L

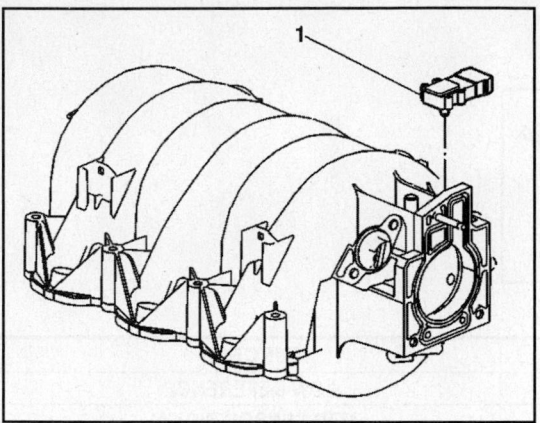

Showing MAP sensor (1) location on 3.5L

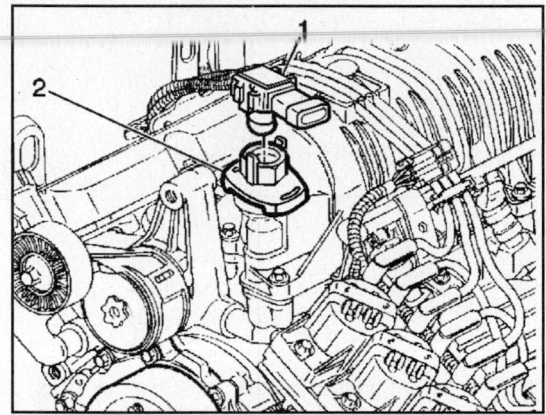

MAP sensor (1) location on 3.8L (L36)

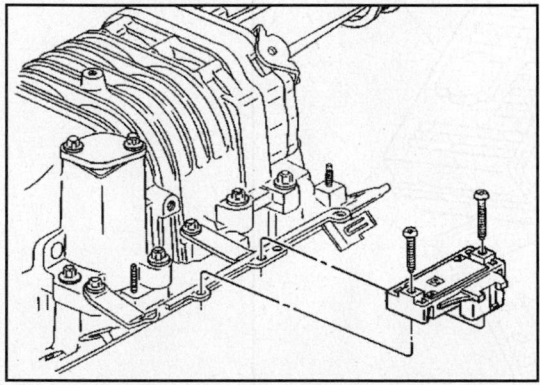

MAP sensor location on 3.8L (L67)

Connector Pinouts

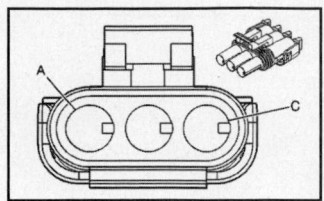 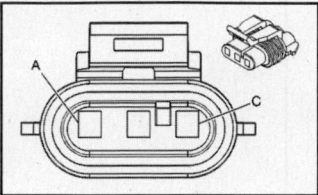

PIN	WIRE COLOR	CIRCUIT NO.	FUNCTION
A	ORN/BLK	469	LOW REFERENCE
B	LT GRN	432	MAP SENSOR SIGNAL
C	GRY	2704	5 VOLT REFERENCE

3.1L, 3.4L, 3.5L & 3.8L MAP Sensor Connector Pinout

POWERTRAIN CONTROL MODULE (PCM) CONNECTORS

Component Locations

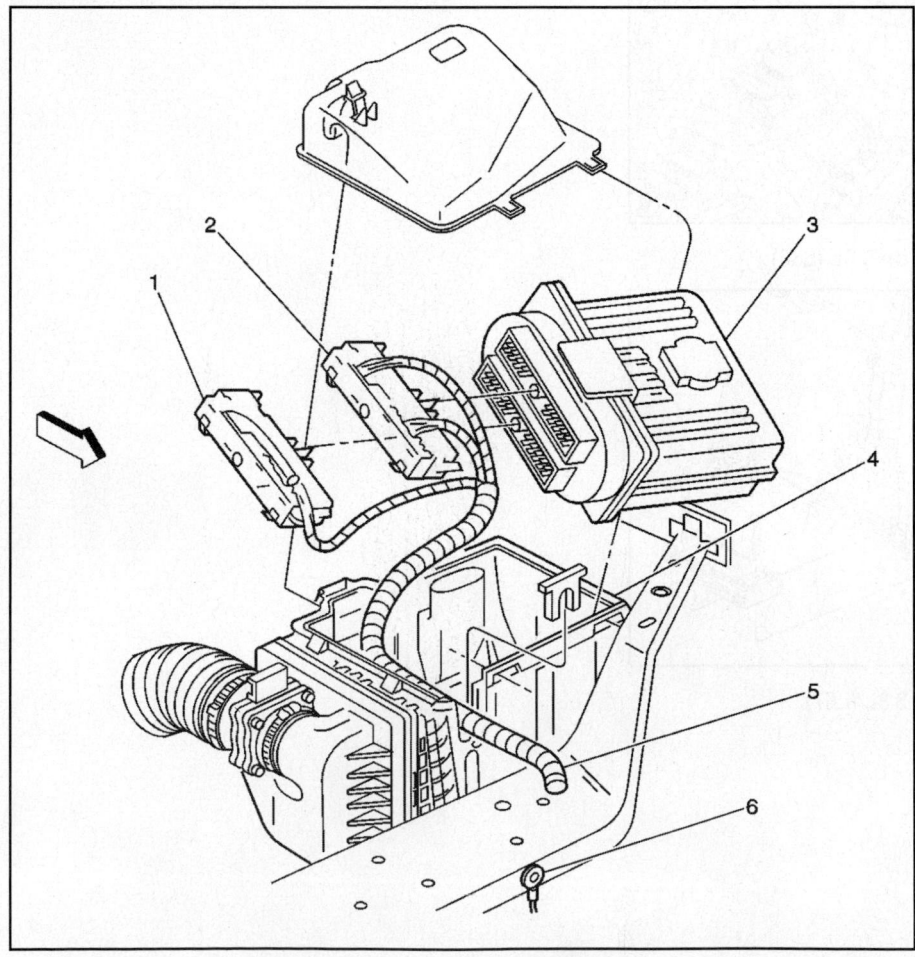

Powertrain control module (3) inside air cleaner housing in left side of engine compartment, showing the C2 connector (1), C1 connector (2), air cleaner housing (4), wiring harness (5), and forward lamp ground (6)

Connector Pinouts

C1 CONNECTOR

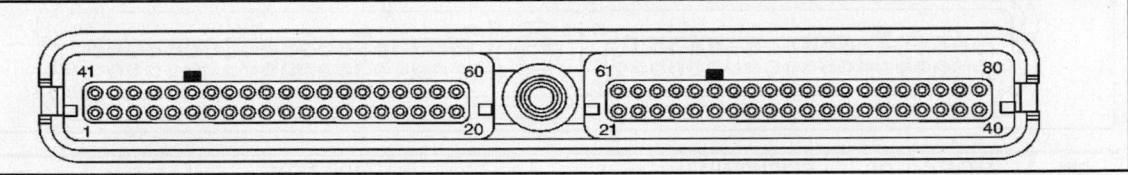

PIN	WIRE COLOR	CIRCUIT NO.	FUNCTION
1	BLK	2762	ATF TEMPERATURE (TFT) SENSOR GROUND
2-3	--	--	NOT USED
4	LT GRN	1222	1-2 SHIFT SOLENOID VALVE
5	DK BLU	473	HIGH SPEED FANS CONTROL
6	DK GRN	335	LOW SPEED FANS CONTROL
7	BRN/WHT	633	CAMSHAFT POSITION PCM INPUT SIGNAL
8	PPL/WHT	430	3X REFERENCE HIGH
9	LT BLU/BLK	647	24X CKP SENSOR SIGNAL (REFERENCE)
10	GRY	2704	MAP 5 VOLT REFERENCE A
11	--	--	NOT USED
12	DK GRN	676	HO2S HEATER HIGH CONTROL (BANK 1 SENSOR 1)
13	ORN BLK	469	MAP/ECT SENSOR GROUND
14-15	--	--	NOT USED
16	BLK/WHT	451	PCM GROUND
17	BLK	2760	IAT SENSOR GROUND
18	--	--	NOT USED
19	PPL	1500	IGNITION WAKE UP POWER (HOT IN RUN / BULB TEST / START)
20	ORN	540	BATTERY POSITIVE VOLTAGE
21	--	--	NOT USED
22	PNK	1224	TRANSMISSION FLUID PRESSURE SWITCH SIGNAL A (2000-02 ONLY)
23-27	-	-	NOT USED
28	TAN/WHT	1669	HO2 SENSOR 2 LOW
29	TAN	413	HO2 SENSOR 1 LOW
30	PPL	420	TCC BRAKE SWITCH INPUT
31	BLK	2753	EGR PINTLE POSITION SENSOR GROUND
32	GRY	435	EGR VALVE CONTROL
33	DK BLU	496	KNOCK SENSOR (KS) SIGNAL
34-37	--	--	NOT USED
38	LT GRN/WHT	1749	IAC VALVE COIL B HIGH
39-42	--	--	NOT USED
43	BLK	1744	FUEL INJECTOR #1 CONTROL
44	YEL/BLK	1223	2-3 SHIFT SOLENOID VALVE CONTROL
45	-	-	NOT USED
46	PNK/BLK	1746	FUEL INJECTOR #3 CONTROL
47	LT GRN/BLK	1745	FUEL INJECTOR #2 CONTROL
48	RED/BLK	453	3X REFERENCE LOW
49-52	--	--	NOT USED
53	TAN/BLK	424	BYPASS CONTROL
54	WHT	423	IGNITION CONTROL (IC)
55	DK GRN/WHT	817	VEHICLE SPEED SIGNAL OUTPUT
56	BLK/WHT	451	PCM GROUND
57	BLK/WHT	451	PCM GROUND
58	DK GRN	1049	CLASS 2 SERIAL DATA
59	--	--	NOT USED

3.1L, 3.4L, 3.5L & 3.8L PCM C1 Connector Pinout (1 of 2)

C1 Connector - Continued

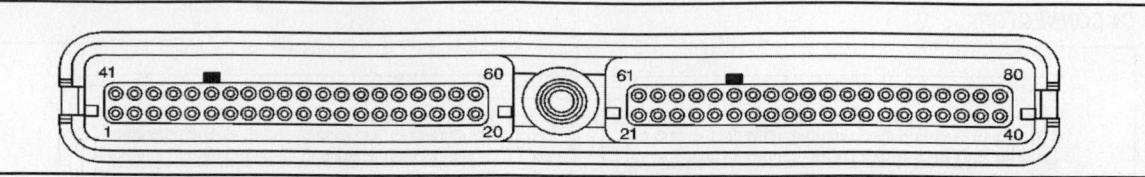

PIN	WIRE COLOR	CIRCUIT NO.	FUNCTION
60	BLK/WHT	451	PCM GROUND
61	BLK	2752	THROTTLE POSITION (TP) SENSOR GROUND
62	DK BLU/WHT	1231	AT ISS LOW SIGNAL
63	RED/BLK	1230	AT ISS HIGH SIGNAL
64	YEL	400	VSS HIGH
65	PPL	401	VSS LOW
66-67	--	--	NOT USED
68	YEL	772	TRANSAXLE RANGE SWITCH B
69	YEL	492	MAF SENSOR INPUT (SIGNAL)
70-71	--	--	NOT USED
72	DK GRN	83	CRUISE DISABLE/INHIBIT CONTROL
73	BLK/WHT	845	FUEL INJECTOR #5 CONTROL
74-75	--	--	-
76	DK GRN/WHT	428	EVAP CANISTER PURGE VALVE CONTROL
77-78	--	--	NOT USED
79	YEL/BLK	846	FUEL INJECTOR #6 CONTROL
80	BLK	2751	AC PRESSURE SENSOR GROUND

3.1L, 3.4L, 3.5L & 3.8L PCM C1 Connector Pinout (2 of 2)

C2 CONNECTOR

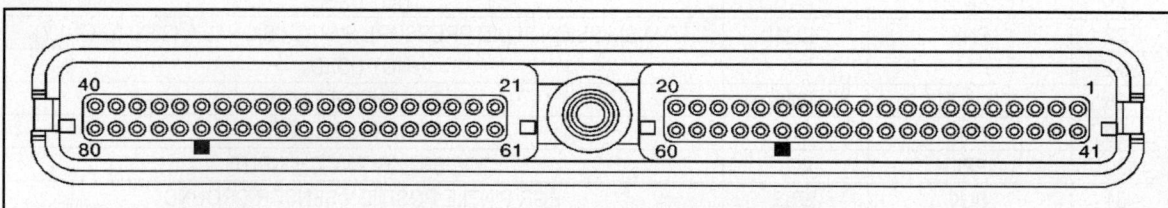

PIN	WIRE COLOR	CIRCUIT NO.	FUNCTION
1	--	--	NOT USED
2	WHT	1310	EVAP CANISTER VACUUM SWITCH INPUT (VENT VALVE SOLENOID CONTROL)
3	DK GRN/WHT	465	FUEL PUMP RELAY CONTROL
4	WHT	257	EGR VALVE IGNITION POSITIVE VOLTAGE
5	BRN/WHT	419	MIL CONTROL
6	--	--	NOT USED
7	LT BLU/WHT	1747	IAC VALVE COIL A HIGH
8-9	--	--	NOT USED
10	PPL	412	HO2 SENSOR 1 SIGNAL
11	PPL/WHT	1668	HO2 SENSOR 2 SIGNAL
12-15	--	--	NOT USED
16	WHT	776	TRANSAXLE RANGE SWITCH P
17	RED	1226	TRANSMISSION FLUID PRESSURE SWITCH SIGNAL C
18	BLK/WHT	771	TRANSAXLE RANGE SWITCH A
19	TAN/BLK	231	ENGINE OIL PRESSURE SWITCH INPUT
20-21	--	--	NOT USED
22	DK GRN/WHT	762	A/C REQUEST SIGNAL (C60/CJ3)

3.1L, 3.4L, 3.5L & 3.8L PCM C2 Connector (1 of 2)

C2 CONNECTOR - CONTINUED

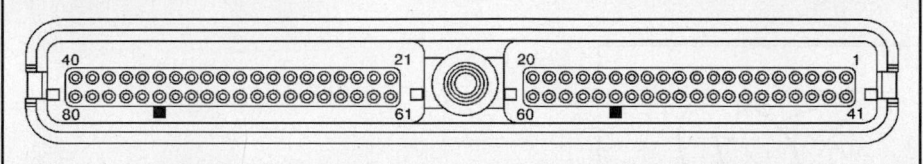

PIN	WIRE COLOR	CIRCUIT NO.	FUNCTION
23	PPL	806	CRANK INPUT (HOT IN START ONLY)
24	PNK	439	IGNITION 1 POSITIVE VOLTAGE (HOT IN RUN / BULB TEST / START)
25	LT GRN	432	MAP SENSOR SIGNAL
26	YEL	410	ECT SENSOR SIGNAL
27	RED/BLK	380	A/C REFRIGERANT PRESSURE SENSOR SIGNAL
28	BRN	1456	EGR PINTLE POSITION SIGNAL
29	-	--	NOT USED
30	GRY/BLK	2702	EGR 5 VOLT REFERENCE A
31-32	--	--	NOT USED
33	GRY	2701	TP SENSOR 5 VOLT REFERENCE A
34	GRY	474	FUEL TANK PRESSURE (FTP)/AC PRESSURE SENSOR 5 VOLT REFERENCE B
35	BLK	452	SENSOR GROUND
36	--	--	NOT USED
37	BLK/WHT	3113	O2
38	--	--	NOT USED
39	DK GRN/WHT	459	A/C COMPRESSOR CLUTCH RELAY CONTROL
40-41	--	--	NOT USED
42	LT BLU/BLK	844	FUEL INJECTOR #4 CONTROL
43	--	--	NOT USED
44	LT BLU/BLK	1748	IAC VALVE COIL A LOW
45	RED/BLK	1228	PC SOLENOID VALVE HIGH CONTROL (SOL, A)
46	LT BIU/WHT	1229	PC SOLENOID VALVE LOW CONTROL (SOL, A)
47-48	--	--	NOT USED
49	LT GRN/BLK	444	IAC VALVE COIL B LOW
50	TAN	472	INTAKE AIR TEMPERATURE (IAT) SENSOR
51-54	--	--	NOT USED
55	DK GRN	890	FUEL TANK PRESSURE SENSOR INPUT (SIGNAL)
56	GRY	773	TRANSAXLE RANGE SWITCH C
57	DK BLU	1225	TRANSMISSION FLUID PRESSURE SWITCH SIGNAL B
58	BRN	1174	ENGINE OIL LEVEL SWITCH INPUT
59	WHT	85	CRUISE ENGAGED INPUT (STATUS)
60	--	--	NOT USED
61	RED	225	GENERATOR L TERMINAL CONTROL
62	--	--	NOT USED
63	YEL	657	TCC RELEASE SWITCH SIGNAL
64-65	--	--	NOT USED
66	DK BLU	417	THROTTLE POSITION (TP) SENSOR SIGNAL
67	--	--	NOT USED
68	YEL/BLK	1227	TRANSAXLE FLUID TEMPERATURE (TFT) SENSOR SIGNAL
69	PPL	1589	FUEL LEVEL SENSOR INPUT
70	LT GRN	1867	CKP BATTERY POSITIVE VOLTAGE
71	--	--	NOT USED
72	RED/WHT	812	CMP SENSOR BATTERY POSITIVE VOLTAGE
73	BLK	407	CMP SENSOR GROUND
74	YEL/BLK	1868	CKP SENSOR GROUND
75	BRN	436	AIR PUMP RELAY CONTROL
76	YEL/BLK	625	STARTER ENABLE CONTROL
77	--	--	NOT USED
78	BRN	418	TCC PWM SOLENOID CONTROL
79	--	--	NOT USED
80	TAN	413	HO2S LOW SIGNAL

3.1L, 3.4L, 3.5L & 3.8L PCM C2 Connector (2 of 2)

SECONDARY AIR INJECTION REACTION (AIR) PUMP & SOLENOID
Component Locations

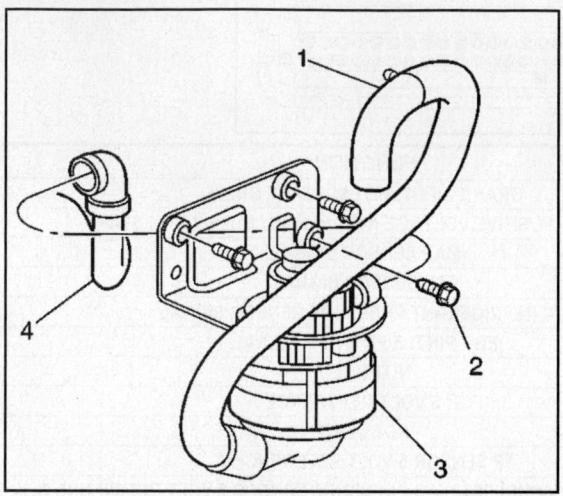

AIR pump (3) on 3.1L and 3.8L engines, showing the air tube (1), bracket bolts (2), and air hose (4)

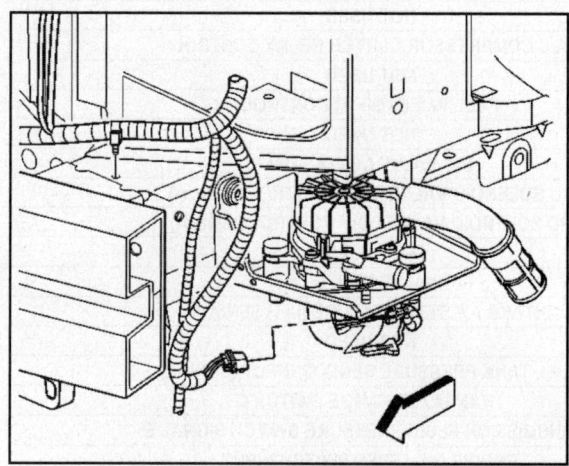

AIR pump location on 3.4L engines (3.5L similar)

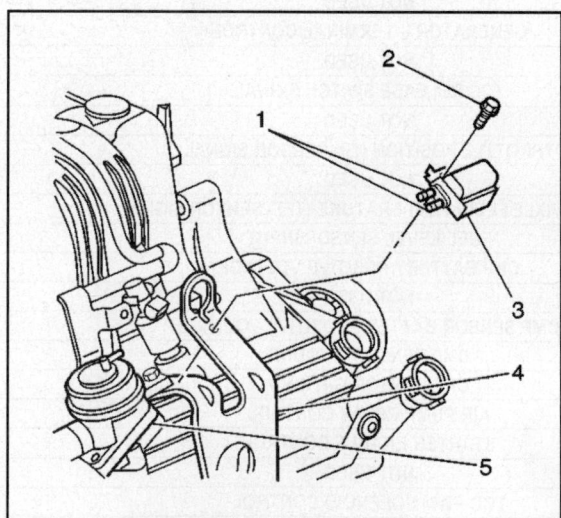

AIR pump vacuum solenoid location on 3.8L engines, showing the connectors (1), retaining bolt (2), mounting bracket (4) and EGR valve

Connector Pinout

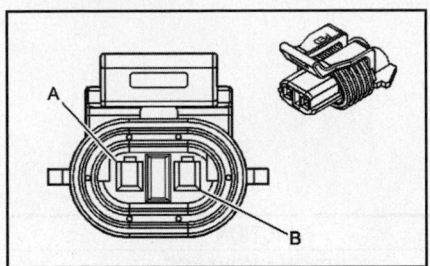

PIN	WIRE COLOR	CIRCUIT NO.	FUNCTION
A	PNK	339 OR 739	FUSE OUTPUT-IGNITION 1
B	PNK/BLK	429	AIR INJECTION REACTION SOLENOID OUTPUT

2000 3.1L, 2000 3.4L & 2000-01 3.8L AIR Solenoid Valve Connector Pinout

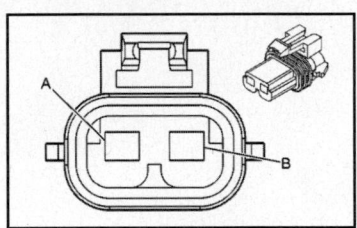

PIN	WIRE COLOR	CIRCUIT NO.	FUNCTION
A	RED	78	AIR PUMP SUPPLY VOLTAGE
B	BLK	250 OR 1250	GROUND

2000 3.1L, 2000 3.4L & 2000-01 3.8L Secondary Air Pump Connector Pinout

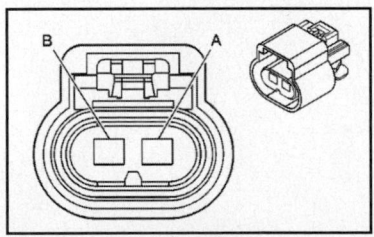

PIN	WIRE COLOR	CIRCUIT NO.	FUNCTION
A	BLK	1350	GROUND
B	RED	78	AIR PUMP SUPPLY VOLTAGE

3.5L Secondary Air Pump Connector Pinout

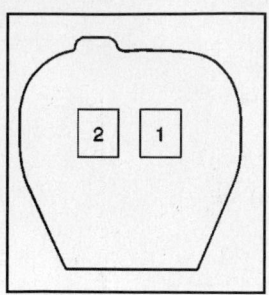

PIN	WIRE COLOR	CIRCUIT NO.	FUNCTION
1	PNK	339	IGNITION 1 VOLTAGE
2	PNK/BLK	429	AIR SOLENOID CONTROL

3.5L Secondary Air Pump Solenoid Connector Pinout

THROTTLE POSITION (TP) SENSOR CONNECTOR

Component Locations

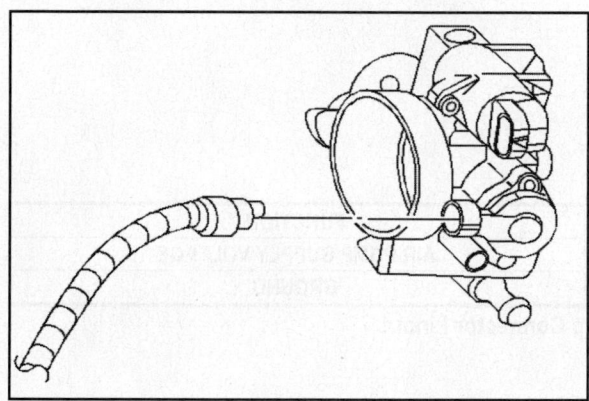

Throttle position sensor location 3.1L & 3.4L

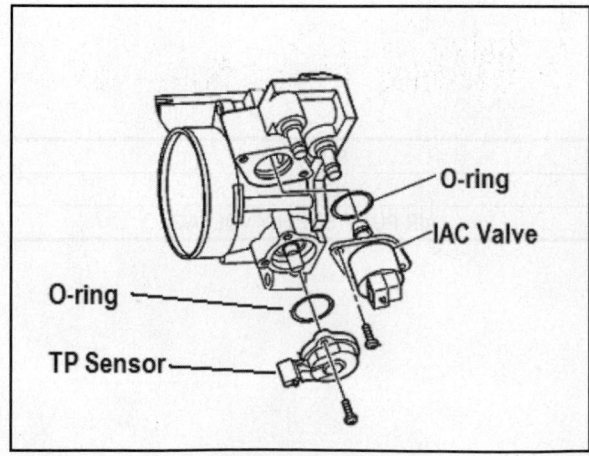

Throttle Position (TP) sensor location on 3.5L

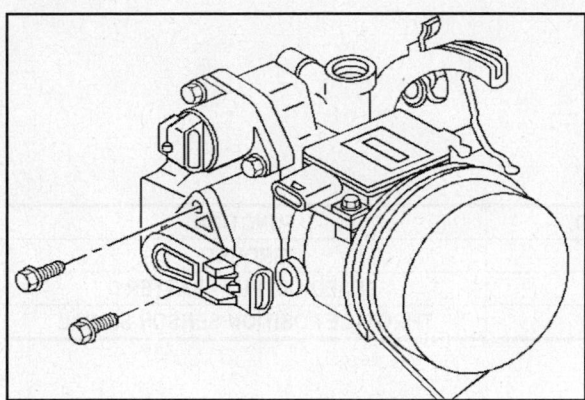

Throttle position sensor location 3.8L

Connector Pinouts

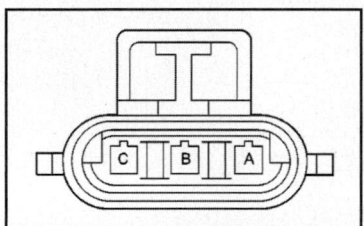

PIN	WIRE COLOR	CIRCUIT NO.	FUNCTION
A	GRY	2701	REFERENCE VOLTAGE FEED-12 VOLTS
B	BLK	2752	SENSOR RETURN (LOW)
C	DK BLU	417	THROTTLE POSITION SENSOR SIGNAL

2000-01 3.1L & 2000-05 3.4L TP Sensor Connector Pinout

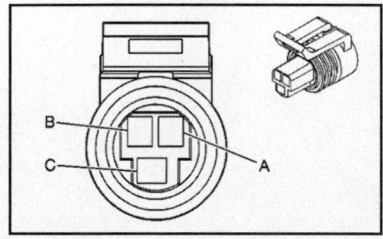

PIN	WIRE COLOR	CIRCUIT NO.	FUNCTION
A	GRY	2701	5-VOLT REFERENCE
B	BLK	2752	LOW REFERENCE
C	DK BLU	417	TP SENSOR SIGNAL

2002-05 3.1L & 2001-02 3.5L TP Sensor Connector Pinout

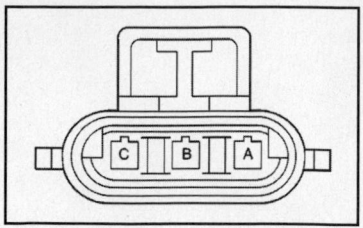

PIN	WIRE COLOR	CIRCUIT NO.	FUNCTION
A	DK BLU	2752	GROUND
B	BLK	2701	FUSED OUTPUT - BATTERY
C	GRY	417	THROTTLE POSITION SENSOR SIGNAL

2000 3.8L TP Sensor Connector Pinout

PIN	WIRE COLOR	CIRCUIT NO.	FUNCTION
A	BLK	2752	SENSOR RETURN (LOW)
B	DK BLU	417	5-VOLT REFERENCE
C	GRY	2701	THROTTLE POSITION SENSOR SIGNAL

2001 3.8L TP Sensor Connector Pinout

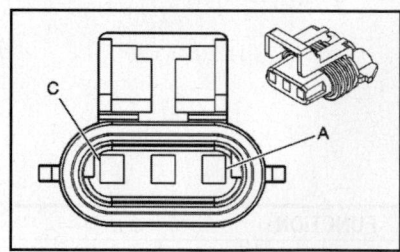

PIN	WIRE COLOR	CIRCUIT NO.	FUNCTION
A	BLK	2752	LOW REFERENCE
B	DK BLU	417	TP SENSOR SIGNAL
C	GRY	2701	5-VOLT REFERENCE

2002-05 3.8L TP Sensor Connector Pinout

VEHICLE SPEED SENSOR (VSS) CONNECTOR

Component Locations

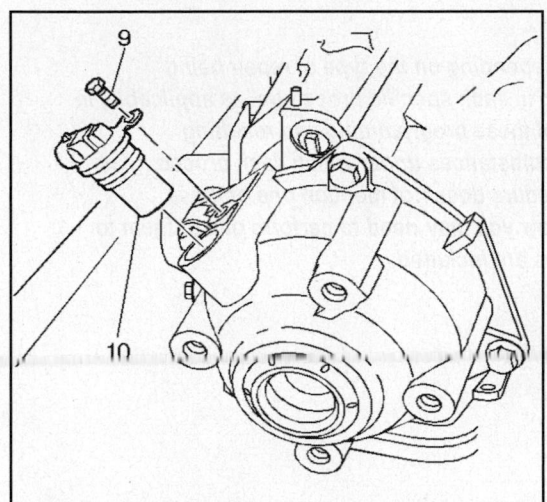

Vehicle speed sensor (10) location 4T-65-E

Connector Pinouts

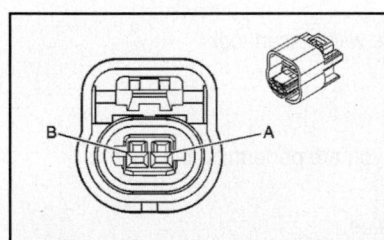

PIN	WIRE COLOR	CIRCUIT NO.	FUNCTION
A	PPL	401	VSS LOW SIGNAL
B	YEL	400	VSS HIGH SIGNAL

2000-05 w/4T65-E A/T Vehicle Speed Sensor Connector Pinout

GENERAL MOTORS CARS: 2000-05 CAMARO, CORVETTE & FIREBIRD
V8 ENGINES: 5.7L & 6.0L

LEARNING, PROGRAMMING & RESETTING PROCEDURES

NOTE: These procedures will apply in different instances, depending on the type of repair being performed. The specific procedures are mentioned in each specific procedure, as applicable, in this section. It is suggested you read through all of these programming and resetting procedures to ensure you are familiar with the circumstances under which each procedure is required. Then, even if a specific component procedure does not mention one of these programming or resetting procedures, you will know you may need to perform one of them to restore engine operation. The following procedures are included:

- Crankshaft Variation Learn Procedure
- Engine Oil Life Monitor Reset Procedure
- Idle Learn Procedure
- PCM Programming Procedure

Crankshaft Variation Learn Procedure

CORVETTE 5.7L & 6.0L V8 ENGINES

1. Install a scan tool.
2. Monitor the powertrain control module for DTCs with a scan tool. If other DTCs are set, except DTC P1336 (5.7L) or P0315 (6.0L), refer to Diagnostic Trouble Code (DTC) List for the applicable DTC that set.
3. Select the crankshaft position (CKP) variation learn procedure with a scan tool.
4. The scan tool instructs you to perform the following:
   a. Accelerate to wide-open throttle (WOT).
   b. Release and observe the fuel cut-off for the engine that you are performing the learn procedure on.
   c. Engine should not accelerate beyond calibrated RPM value.
   d. Release throttle immediately if value is exceeded.
   e. Block drive wheels.
   f. Set parking brake.
   g. DO NOT apply brake pedal.
   h. Cycle the ignition from OFF to ON.
   i. Apply and hold the brake pedal.
   j. Start and idle engine.
   k. Turn A/C OFF.
   l. Vehicle must remain in Park or Neutral.
   m. The scan tool monitors certain component signals to determine if all the conditions are met to continue with the procedure. The scan tool only displays the condition that inhibits the procedure. The scan tool monitors the following components:
   n. Crankshaft position (CKP) sensors activity--If there is a CKP sensor condition, refer to the applicable DTC that set.
   o. Camshaft position (CMP) sensor activity--If there is a CMP sensor condition, refer to the applicable DTC that set.
   p. Engine coolant temperature (ECT)--If the ECT is not warm enough, idle the engine until the engine coolant temperature reaches the correct temperature.
5. Enable the CKP system variation learn procedure with a scan tool.

6. While the learn procedure is in progress, release the throttle immediately when the engine starts to decelerate. The engine control is returned to the operator and the engine responds to throttle position after the learn procedure is complete.

7. Accelerate to WOT.

8. Release when the fuel cut-off occurs.

9. The scan tool displays Learn Status: Learned this ignition. If the scan tool indicates that DTC P1336 (5.7L) or P0315 (6.0L) ran and passed, the CKP variation learn procedure is complete. If the scan tool indicates DTC P1336 (5.7L) or P0315 (6.0L) failed or did not run, refer to DTC P1336 (5.7L) or P0315 (6.0L). If any other DTCs set, refer to Diagnostic Trouble Code (DTC) List for the applicable DTC that set.

10. Turn OFF the ignition for 30 seconds after the learn procedure is completed successfully.

CAMARO & FIREBIRD 5.7L ENGINES

CAUTION: While the learn procedure is in progress, release the throttle immediately when the engine starts to decelerate. The engine control is returned to the operator and the engine will respond to throttle position after the learn procedure is complete.

1. Install the scan tool.

2. Apply the vehicle's parking brake.

3. Block the drive wheels.

4. Close the vehicle's hood.

5. Place the vehicle's transmission in Park (A/T) or Neutral (M/T).

6. Idle the engine until the engine coolant temperature reaches 65°C (150°F).

7. Turn OFF all the accessories.

8. Apply the brakes for the duration of the procedure.

9. If the CKP System Variation Learn Procedure cannot be completed successfully, refer to DTC P1336 Crankshaft Position (CKP) System Variation Not Learned, in "Diagnostic Trouble Codes" section for additional diagnostic information.

10. Enable the Crankshaft Position System Variation Learn Procedure with the scan tool.

11. Slowly raise the engine speed to 4000 RPM.

12. Immediately release the throttle when the engine speed decreases.

13. Turn OFF the ignition for 15 seconds after the learn procedure is completed successfully.

Engine Oil Life Monitor Reset Procedure

The PCM monitors various engine parameters to determine when an oil change should be done. The PCM sends a message to the Instrument Panel Cluster (IPC) when an oil change is needed. The PCM monitors the following to determine when an oil change should occur:

- Engine Speed
- Engine coolant temperature
- Engine load
- Intake Air Temperature
- Vehicle mileage

1. Repair any TP sensor DTCs before proceeding with the reset procedure.

2. If the Engine Oil Life Monitor is reset due to a PCM replacement, advise the customer . The customer has two options:

3. Change the engine oil and filter.

4. Determine when the last engine oil change occurred. Monitor the vehicle mileage and change the engine oil and filter at the next recommended service interval.

5. Turn ON the ignition leaving the engine OFF.

6. Cycle the accelerator pedal from a closed throttle to a wide open throttle (WOT) 3 times within 5 seconds. The Change Oil message flashes for 5 seconds when reset.

7. Change the engine oil and filter.

Idle Learn Procedure

Anytime the powertrain control module (PCM) or the battery is disconnected, the PCM loses power, or the PCM is reprogrammed, the PCM's learned idle position is lost. The engine idle is unstable when the learned idle position is lost.

Perform the following procedure in order to return the learned idle to the correct position:

AUTOMATIC TRANSMISSION

1. Turn OFF the ignition.
2. Restore the PCM battery feed.
3. Turn OFF the A/C controls.
4. Set the parking brake and block the drive wheels.
5. Start the engine.
6. Ensure that the engine coolant temperature is more than 80°C (176°F).
7. Shift the transmission into Drive.
8. Allow the engine to idle for 5 minutes.
9. Shift the transmission into Park.
10. Allow the engine to idle for 5 minutes.
11. Turn OFF the engine for 30 seconds.

MANUAL TRANSMISSION

1. Turn OFF the ignition.
2. Restore the PCM battery feed.
3. Turn OFF the A/C controls.
4. Set the parking brake and block the drive wheels.
5. Place the transmission in Neutral.
6. Start the engine.
7. Ensure that the engine coolant temperature is more than 80°C (176°F).
8. Allow the engine to idle for 5 minutes.
9. Turn OFF the engine for 30 seconds.

PCM Programming Procedure

NOTE: Programming procedures may vary, depending on type of scan tool being used and on vehicle-specific conditions. If in doubt, contact an authorized service agent for information before proceeding with an attempt at programming the PCM.

1. Ensure the battery is fully charged, the ignition is ON, the cable connection at the Data Link Connector (DLC) is secure.

2. Turn off or disable any feature that may put a load on the battery.

3. Program the PCM using the latest software matching the vehicle.

4. If the PCM fails to program, proceed as follows:

5. Ensure that all PCM connections are OK.

6. Check with the manufacturer for the latest software version.

7. Attempt to program the PCM. If the PCM still cannot be programmed properly, replace the PCM. The replacement PCM must be programmed.

8. Program the replacement PCM.

9. Perform the CKP System Variation Learn Procedure if a new PCM is installed. Refer to "CKP System Variation Learn Procedure."

10. Perform the Idle Learn Procedure. Refer to "Idle Learn Procedure."

11. Perform the Engine Oil Life Reset procedure if a new PCM is installed. Refer to "Engine Oil Life Monitor Reset Procedure."

FUNCTIONAL CHECK AFTER PCM PROGRAMMING

1. Clear the Diagnostic Trouble Codes (DTCs).

2. Perform the On-Board Diagnostic System Check.

3. Start the engine and idle for one minute.

4. Scan for DTCs using the scan tool.

ACCELERATOR PEDAL POSITION (APP) SENSOR

Component Locations

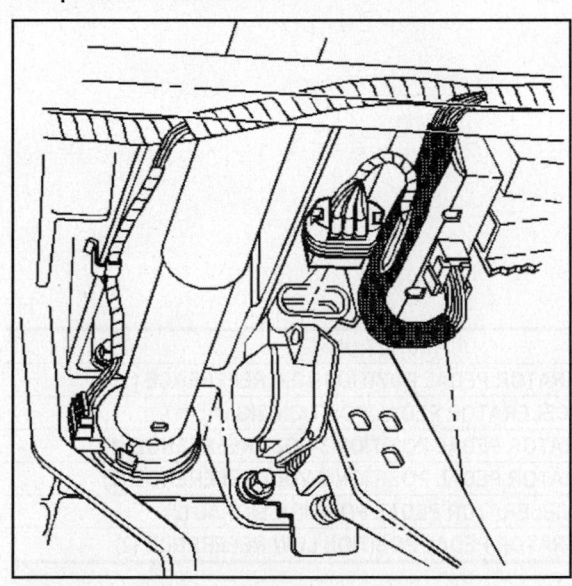

APP sensor is located on accelerator pedal assembly on 5.7L & 6.0L V8 engines

Connector Pinouts

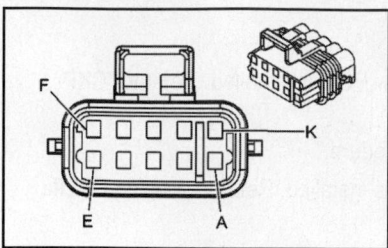

PIN	WIRE COLOR	CIRCUIT NO.	FUNCTION
A	GRY	1273	LOW REFERENCE
B	PPL	1272	LOW REFERENCE
C	LT BLU	1162	APP SENSOR 2 SIGNAL
D	TAN	1274	5 VOLT REFERENCE
E	YEL/BLK	1275	5 VOLT REFERENCE
F	DK BLU	1161	APP SENSOR 1 SIGNAL
G	LT BLU	1276	5 VOLT REFERENCE
H	--	--	NOT USED
J	BRN	1271	LOW REFERENCE
K	DK GRN	1163	APP SENSOR 3 SIGNAL

5.7L V8 Engine APP Sensor Connector Pinout

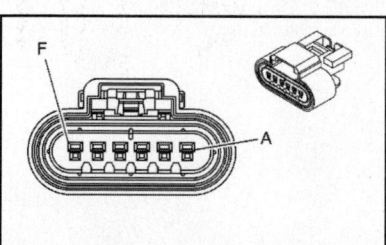

PIN	WIRE COLOR	CIRCUIT NO.	FUNCTION
A	BN	1271	ACCELERATOR PEDAL POSITION LOW REFERENCE (1)
B	D-BU	1161	ACCELERATOR PEDAL POSITION SIGNAL (1)
C	WH/BK	1164	ACCELERATOR PEDAL POSITION 5-VOLT REFERENCE (1)
D	TN	1274	ACCELERATOR PEDAL POSITION 5-VOLT REFERENCE (2)
E	L-BU	1162	ACCELERATOR PEDAL POSITION SIGNAL (2)
F	PU	1272	ACCELERATOR PEDAL POSITION LOW REFERENCE (2)

2005 6.0L V8 Engine APP Sensor Connector Pinout

CAMSHAFT POSITION (CMP) SENSOR

Description & Operation

5.7L & 6.0L V8 ENGINES

The CMP sensor is also a magneto resistive sensor, with the same type of circuits as the CKP sensor. The CMP sensor signal is a digital ON/OFF pulse, output once per revolution of the camshaft. The CMP sensor information is used by the PCM to determine the position of the valve train relative to the CKP.

The camshaft reluctor wheel is either pressed onto the camshaft or part of the camshaft gear depending on the application. The feature-or target- is read in a radial or axial fashion respectively. The wheel is a smooth track, half of which is of a lower profile than the other half. This feature allows the camshaft position (CMP) sensor to supply a signal as soon as the key is turned ON, since the CMP sensor reads the track profile, instead of a notch.

Component Locations

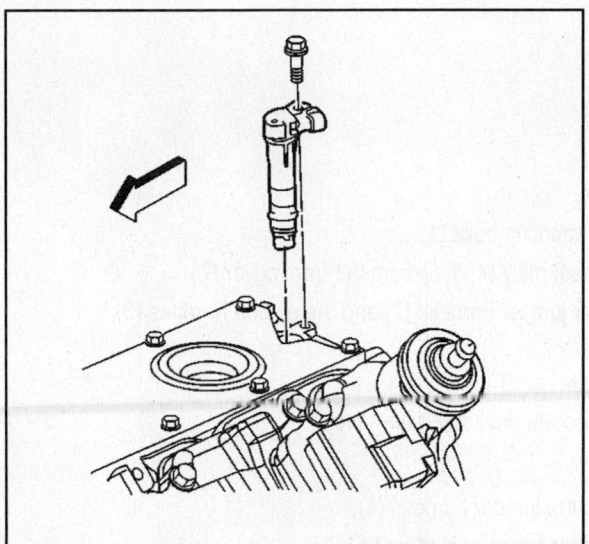

Location of the CMP sensor on 5.7L V8 engine

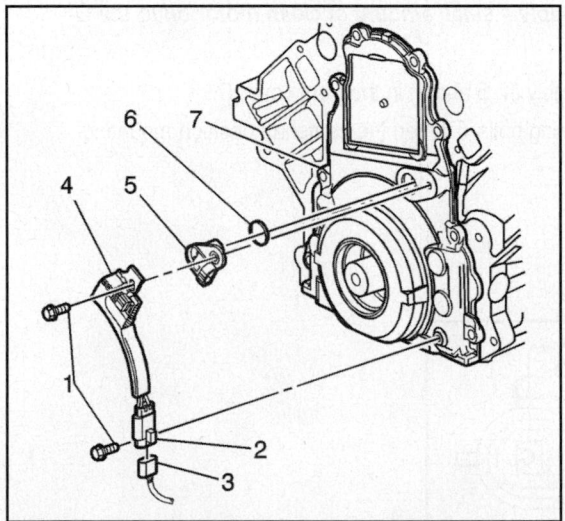

Location of CMP sensor on 6.0L V8 engines

Removal & Installation

5.7L V8

1. Disconnect the negative battery cable.
2. Remove the intake manifold.
3. Remove the camshaft position (CMP) sensor electrical connector
4. Remove the CMP sensor retaining bolt.
5. Remove the CMP sensor.

To install:

6. Install the sensor and the CMP sensor retaining bolt. Tighten the CMP to 18 ft. lbs. (25 Nm).

7. Connect the CMP sensor electrical connector.

8. Install the intake manifold.

9. Connect the negative battery cable.

10. Program the transmitters.

6.0L V8

1. Remove the generator bracket assembly.

2. Remove the camshaft position sensor mounting bolts (1).

3. Remove the camshaft position sensor assembly (4, 5, 6) from the front cover (7).

4. Disconnect the camshaft position sensor jumper harness (2) and the engine harness (3) electrical connectors.

5. Remove the camshaft sensor assembly (4, 5, 6).

6. Disconnect camshaft position sensor (5) from the jumper harness (4).

To install:

7. Reconnect the camshaft sensor (5) and the jumper harness (4).

8. Install the O-ring (6) on the camshaft sensor assembly (4 and 5).

9. Reconnect the camshaft position sensor assembly (4, 5 and 6) and the engine harness connector (3).

NOTE: Before installing the camshaft sensor assembly, apply a small amount of clean motor oil to the O-ring (6).

10. Install the camshaft position sensor assembly (4, 5 and 6) in the front cover (7).

11. Install the camshaft position sensor mounting bolts. Tighten the camshaft position mounting bolts 25 Nm (18 ft. lbs.).

12. Install the generator assembly.

Connector Pinouts

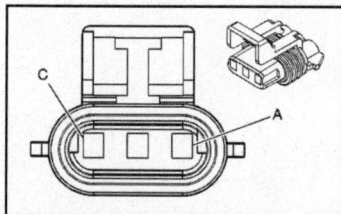

Corvette Applications

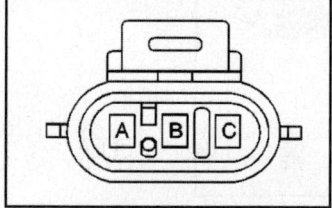

Camaro & Firebird Applications

PIN	WIRE COLOR	CIRCUIT NO.	FUNCTION
A	BRN/WHT	633	CMP SENSOR SIGNAL
B	PNK/BLK	632	LOW REFERENCE
C	RED	631	12 VOLT REFERENCE

5.7L V8 Engine CMP Sensor Connector Pinout

PIN	WIRE COLOR	CIRCUIT NO.	FUNCTION
A	OG	631	12-VOLT REFERENCE
B	PK/BK	632	LOW REFERENCE
C	BN/WH	633	CMP SENSOR SIGNAL

6.0L V8 Engine CMP Sensor Connector Pinout

CRANKSHAFT POSITION (CKP) SENSOR

Description & Operation

5.7L & 6.0L V8 ENGINES

CKP Sensor

The crankshaft position (CKP) sensor is a three-wire sensor based on the magneto resistive principle. A magneto resistive sensor uses two magnetic pickups between a permanent magnet. As an element such as a reluctor wheel passes the magnets the resulting change in the magnetic field is used by the sensor electronics to produce a digital output pulse. The PCM supplies a 12-volt, low reference, and signal circuit to the CKP sensor. The sensor returns a digital ON/OFF pulse 24 times per crankshaft revolution.

Crankshaft Reluctor Wheel

The crankshaft reluctor wheel is mounted on the rear of the crankshaft. The wheel is comprised of four 90-degree segments. Each segment represents a pair of cylinders at TDC, and is further divided into six 15-degree segments. Within each 15 degree segment is a notch of 1 of 2 different sizes. Each 90-degree segment has a unique pattern of notches. This is known as pulse width encoding. This pulse width encoded pattern allows the PCM to quickly recognize which pair of cylinders is at top dead center (TDC). The reluctor wheel is also a dual track-or mirror image-design. This means there is an additional wheel pressed against the first, with a gap of equal size to each notch of the mating wheel. When one sensing element of the CKP sensor is reading a notch, the other is reading a set of teeth. The resulting signals are then converted into a digital square wave output by the circuitry within the CKP sensor.

Component Locations

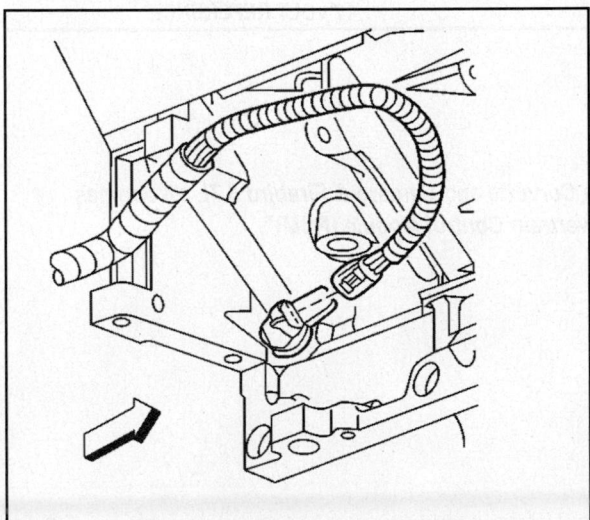

Showing location of CKP sensor on 5.7L & 6.0L V8 engines

Removal & Installation

5.7L & 6.0L V8 ENGINES

1. Disconnect the negative battery cable.
2. Raise the vehicle.
3. Remove the starter.
4. Disconnect the crankshaft position (CKP) sensor electrical connector.

5. Remove the CKP sensor retaining bolt.

6. Remove the CKP sensor.

To install:

7. Install the CKP sensor.

8. Install the CKP sensor retaining bolt. Tighten the CKP sensor to 25 Nm (18 ft. lbs.).

9. Connect the CKP sensor electrical connector.

10. Install the starter.

11. Lower the vehicle.

12. Connect the negative battery cable.

13. Program the transmitters.

14. Perform the CKP System Variation Learn Procedure.

Connector Pinouts

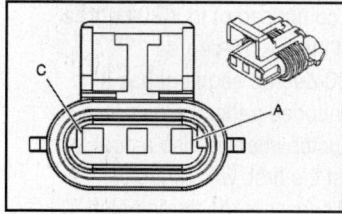

Corvette Applications

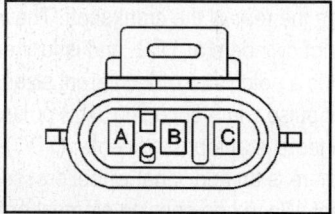

Camaro & Firebird Applications

PIN	WIRE COLOR	CIRCUIT NO.	FUNCTION
A	DK BLU/WHT	1869	CKP SENSOR SIGNAL
B	YEL/BLK	1868	LOW REFERENCE
C	LT GRN	1867	12 VOLT REFERENCE

5.7L & 6.0L V8 Engine CKP Sensor Connector Pinout

ENGINE CONTROL MODULE (ECM)

NOTE: The Corvette 6.0L V8 engine uses an ECM, while Corvette and Camaro & Firebird 5.7L V8 engines use a PCM. For 5.7L V8 applications, see "Powertrain Control Module (PCM)".

Component Locations

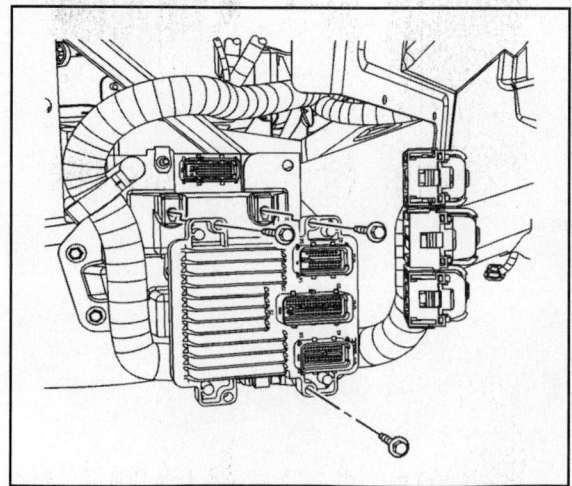

ECM location on 6.0L V8 engines

Removal & Installation

Engine control module (ECM) service should normally consist of either ECM replacement or electrically erasable programmable read only memory (EEPROM) programming. If the diagnostic procedures require ECM replacement, check the ECM first to see if the correct part is being used.

In order to prevent internal ECM damage, the ignition must be OFF when you disconnect or reconnect the power to the ECM. For example, disconnect the power when you work with the following components:

- A battery cable
- The ECM pigtail
- The ECM fuse
- The jumper cables

WARNING: When you diagnose or replace the ECM, remove any debris from the ECM connector surfaces before servicing the ECM module connector gaskets. Ensure that the gaskets are installed correctly. The gaskets prevent intrusion into the ECM. The replacement ECM MUST be programmed.

CAUTION: It is necessary to record the remaining engine oil life. If the replacement module is not programmed with the remaining engine oil life, the engine oil life will default to 100%. If the replacement module is not programmed with the remaining engine oil life, the engine oil will need to be changed at 5000 km (3,000 mi) from the last engine oil change.

1. Using a scan tool, retrieve the percentage of remaining engine oil. Record the remaining engine oil life.
2. Remove the wheelhouse filler panel.
3. Disconnect the ECM electrical harness connectors.
4. Loosen but do not remove the ECM rear retaining fastener. Use the rear retaining fastener as an anchor for the outer bracket.
5. Remove the front retaining fastener from the ECM.
6. Reposition the ECM outer bracket.
7. Remove the ECM from the bracket and the vehicle.

To install:

8. Install the ECM to the ECM rear bracket.
9. Position the ECM front bracket.
10. Install the ECM front retaining fasteners. Tighten the ECM retaining fasteners to 2.0 Nm (17 inch lbs.).
11. Connect the electrical connectors to the ECM. Tighten the ECM electrical connectors to 8 Nm (70 inch lbs.).
12. Install the wheelhouse filler panel.
13. If a replacement ECM is being installed, it must be reprogrammed.

Connector Pinouts

ECM C1 CONNECTOR (MM6)

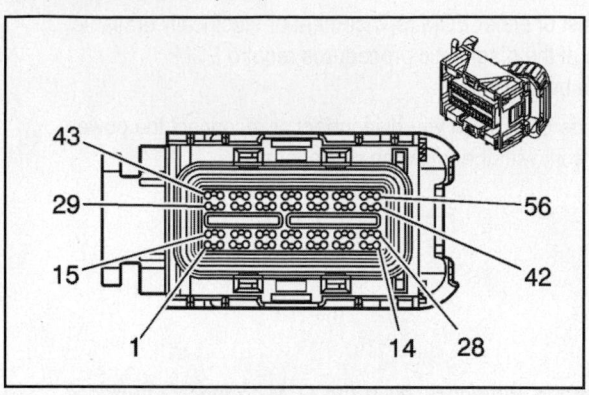

PIN	WIRE COLOR	CIRCUIT NO.	FUNCTION
1	TN/BK	2500	HIGH SPEED GMLAN SERIAL DATA BUS+
2	TN	2501	HIGH SPEED GMLAN SERIAL DATA BUS-
3	--	--	NOT USED
4	BN/WH	419	MIL CONTROL
5-6	--	--	NOT USED
7	BN	1271	LOW REFERENCE
8	PU	1272	LOW REFERENCE
9	--	--	NOT USED
10	WH/BK	448	IGNITION 1 RELAY CONTROL
11-12	--	--	NOT USED
13	YE/BK	625	STARTER ENABLE RELAY CONTROL
14	PK	439	IGNITION 1 VOLTAGE
15-17	--	--	NOT USED
18	YE	43	ACCESSORY VOLTAGE
19	PK	1439	IGNITION 1 VOLTAGE
20	RD/WH	2440	BATTERY POSITIVE VOLTAGE
21	D-GN/WH	817	VEHICLE SPEED SIGNAL
22	D-BU	1161	APP SENSOR 1 SIGNAL
23	--	--	NOT USED
24	D-GN	890	FUEL TANK PRESSURE SENSOR SIGNAL
25	--	--	NOT USED
26	OG/BK	380	A/C REFRIGERANT PRESSURE SENSOR SIGNAL
27-28	--	--	NOT USED
29	TN	1274	5-VOLT REFERENCE
30	GY	2700	5-VOLT REFERENCE
31	GY	2709	5-VOLT REFERENCE
32	WH	17	STOP LAMP SWITCH SIGNAL
33-34	--	--	NOT USED
35	WH/BK	1164	5-VOLT REFERENCE
36	D-GN	335	COOLING FAN SPEED CONTROL
37	D-BU	1936	FUEL LEVEL SENSOR SIGNAL - PRIMARY
38	--	--	NOT USED
39	L-BU	1937	FUEL LEVEL SENSOR SIGNAL - SECONDARY

6.0L V8 ECM C1 (MM6) Connector Pinout (1 of 2)

ECM C1 CONNECTOR (MM6)

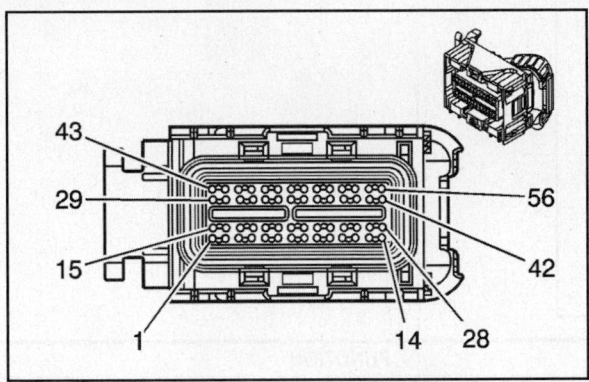

PIN	WIRE COLOR	CIRCUIT NO.	FUNCTION
40	--	--	NOT USED
41	L-BU	1162	APP SENSOR 2 SIGNAL
42	L-GN	1652	REVERSE LOCKOUT SOLENOID CONTROL
43	GY	48	CPP SWITCH SIGNAL
44	--	--	NOT USED
45	D-GN/WH	465	FUEL PUMP RELAY CONTROL - PRIMARY
46-47	--	--	NOT USED
48	WH	121	ENGINE SPEED SIGNAL
49-50	--	--	NOT USED
51	PU/WH	1035	STARTER RELAY COIL SUPPLY VOLTAGE
52-53	--	--	NOT USED
54	WH	1310	EVAP CANISTER VENT SOLENOID CONTROL
55	BK/WH	2751	LOW REFERENCE
56	--	--	NOT USED

6.0L V8 ECM C1 (MM6) Connector Pinout (2 of 2)

ECM C1 CONNECTOR (M32)

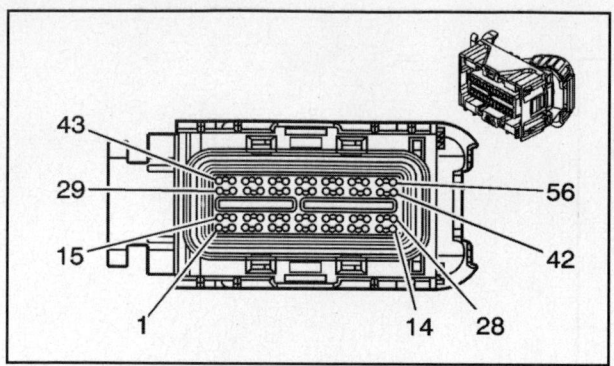

PIN	WIRE COLOR	CIRCUIT NO.	FUNCTION
1	TN/BK	2500	HIGH SPEED GMLAN SERIAL DATA BUS+
2	TN	2501	HIGH SPEED GMLAN SERIAL DATA BUS-
3	--	--	NOT USED
4	BN/WH	419	MIL CONTROL
5-6	--	--	NOT USED
7	BN	1271	LOW REFERENCE
8	PU	1272	LOW REFERENCE
9	--	--	NOT USED
10	WH/BK	448	IGNITION 1 RELAY CONTROL
11-12	--	--	NOT USED
13	YE/BK	625	STARTER ENABLE RELAY CONTROL
14	PK	439	IGNITION 1 VOLTAGE
15-17	--	--	NOT USED
18	YE	43	ACCESSORY VOLTAGE
19	PK	1439	IGNITION 1 VOLTAGE
20	RD/WH	2440	BATTERY POSITIVE VOLTAGE
21	D-GN/WH	817	VEHICLE SPEED SIGNAL
22	D-BU	1161	APP SENSOR 1 SIGNAL
23	--	--	NOT USED
24	D-GN	890	FUEL TANK PRESSURE SENSOR SIGNAL
25	--	--	NOT USED
26	OG/BK	380	A/C REFRIGERANT PRESSURE SENSOR SIGNAL
27-28	--	--	NOT USED
29	TN	1274	5-VOLT REFERENCE
30	GY	2700	5-VOLT REFERENCE
31	GY	2709	5-VOLT REFERENCE
32	WH	17	STOP LAMP SWITCH SIGNAL
33-34	--	--	NOT USED
35	WH/BK	1164	5-VOLT REFERENCE
36	D-GN	335	COOLING FAN SPEED CONTROL
37	D-BU	1936	FUEL LEVEL SENSOR SIGNAL – PRIMARY
38	--	--	NOT USED
39	L-BU	1937	FUEL LEVEL SENSOR SIGNAL – SECONDARY
40	--	--	NOT USED
41	L-BU	1162	APP SENSOR 2 SIGNAL
42-44	--	--	NOT USED
45	D-GN/WH	465	FUEL PUMP RELAY CONTROL – PRIMARY

6.0L V8 ECM C1 (M32) Connector Pinout (1 of 2)

ECM C1 CONNECTOR (M32)

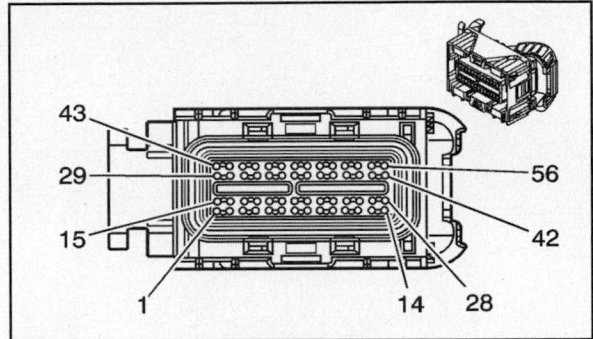

PIN	WIRE COLOR	CIRCUIT NO.	FUNCTION
46-47	--	--	NOT USED
48	WH	121	ENGINE SPEED SIGNAL
49-53	--	--	NOT USED
54	WH	1310	EVAP CANISTER VENT SOLENOID CONTROL
55	BK/WH	2751	LOW REFERENCE
56	--	--	NOT USED

6.0L V8 ECM C1 (M32) Connector Pinout (2 of 2)

ECM C2 CONNECTOR

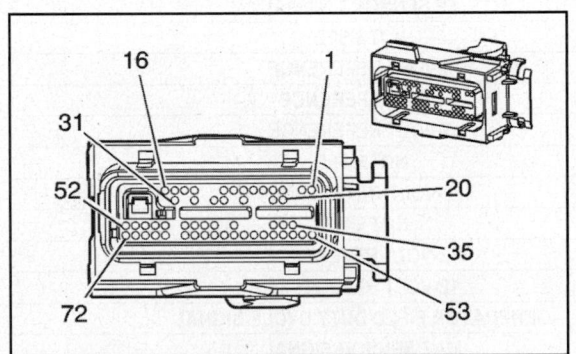

PIN	WIRE COLOR	CIRCUIT NO.	FUNCTION
1	L-GN	3212	HO2S HEATER LOW CONTROL – BANK 2 SENSOR 1
2	GY/WH	3113	HO2S HEATER LOW CONTROL – BANK 1 SENSOR 1
3	--	--	NOT USED
4	D-BU/WH	878	FUEL INJECTOR 8 CONTROL
5	TN	1744	FUEL INJECTOR 1 CONTROL
6	L-GN/BK	1745	FUEL INJECTOR 2 CONTROL
7	OG/BK	877	FUEL INJECTOR 7 CONTROL
8	L-BU/BK	844	FUEL INJECTOR 4 CONTROL
9	PK/BK	1746	FUEL INJECTOR 3 CONTROL
10	TN/WH	845	FUEL INJECTOR 5 CONTROL
11-12	--	--	NOT USED
13	YE/BK	846	FUEL INJECTOR 6 CONTROL
14	YE/BK	1868	LOW REFERENCE
15	PK/BK	632	LOW REFERENCE
16	BN	2129	LOW REFERENCE
17-19	--	--	NOT USED
20	OG/BK	469	LOW REFERENCE

6.0L V8 ECM C2 Connector Pinout (1 of 3)

ECM C2 CONNECTOR

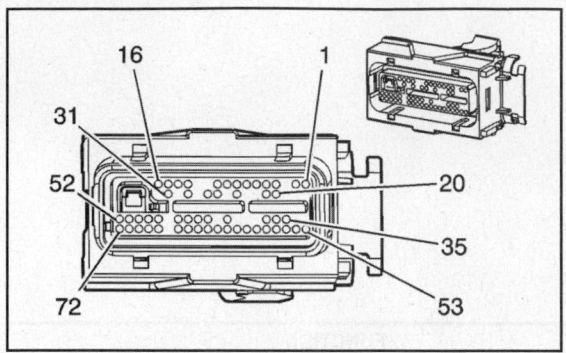

PIN	WIRE COLOR	CIRCUIT NO.	FUNCTION
21	BK	2755	LOW REFERENCE
22-23	--	--	NOT USED
24	TN/WH	331	OIL PRESSURE SENSOR SIGNAL
25	--	--	NOT USED
26	BN/WH	633	CMP SENSOR SIGNAL
27	BN	1174	OIL LEVEL SWITCH SIGNAL
28	--	--	NOT USED
29	PU	486	TP SENSOR 2 SIGNAL
30	--	--	NOT USED
31	D-GN	485	TP SENSOR 1 SIGNAL
32-34	--	--	NOT USED
35	GY	2704	5-VOLT REFERENCE
36	TN/WH	1704	LOW REFERENCE
37	GY	2705	5-VOLT REFERENCE
38-40	--	--	NOT USED
41	L-BU/BK	1688	5-VOLT REFERENCE
42	--	--	NOT USED
43	L-GN	1867	12-VOLT REFERENCE
44	OG	631	12-VOLT REFERENCE
45	GY	23	GENERATOR FIELD DUTY CYCLE SIGNAL
46	L-GN	432	MAP SENSOR SIGNAL
47	--	--	NOT USED
48	PU/WH	1665	HO2S HIGH SIGNAL - BANK 1 SENSOR 1
49	TN	1664	HO2S LOW SIGNAL - BANK 1 SENSOR 1
50	D-GN/WH	428	EVAP CANISTER PURGE SOLENOID CONTROL
51	PU	1666	HO2S HIGH SIGNAL - BANK 2 SENSOR 1
52	TN	2761	LOW REFERENCE
53	OG	2127	IC 7 CONTROL
54	PU	2121	IC 1 CONTROL
55	YE	410	ECT SENSOR SIGNAL

6.0L V8 ECM C2 Connector Pinout (2 of 3)

ECM C2 CONNECTOR (CONTINUED)

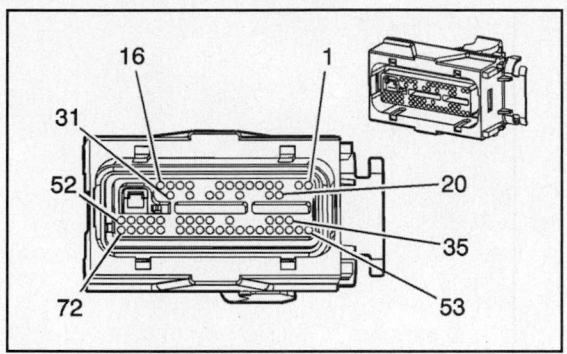

PIN	WIRE COLOR	CIRCUIT NO.	FUNCTION
56	OG	225	GENERATOR TURN ON SIGNAL
57	L-BU/WH	2126	IC 0 CONTROL
58	D-GN/WH	2124	IC 4 CONTROL
59	D-GN	2125	IC 5 CONTROL
60	BN	582	TAC MOTOR CONTROL - 2
61	L-BU	2123	IC 3 CONTROL
62	D-BU/WH	1869	CKP SENSOR SIGNAL
63	YE	581	TAC MOTOR CONTROL - 1
64	BN/WH	2130	LOW REFERENCE
65	PU/WH	2128	IC 8 CONTROL
66	OG/WH	2122	IC 2 CONTROL
67	--	--	NOT USED
68	D-BU	496	KNOCK SENSOR 1 SIGNAL
69	GY	1716	LOW REFERENCE
70	TN	1667	HO2S LOW SIGNAL - BANK 2 SENSOR 1
71	GY	2303	LOW REFERENCE
72	L-BU	1876	KNOCK SENSOR 2 SIGNAL
73	BK/WH	451	GROUND

6.0L V8 ECM C2 Connector Pinout (3 of 3)

ECM C3 CONNECTOR (MM6)

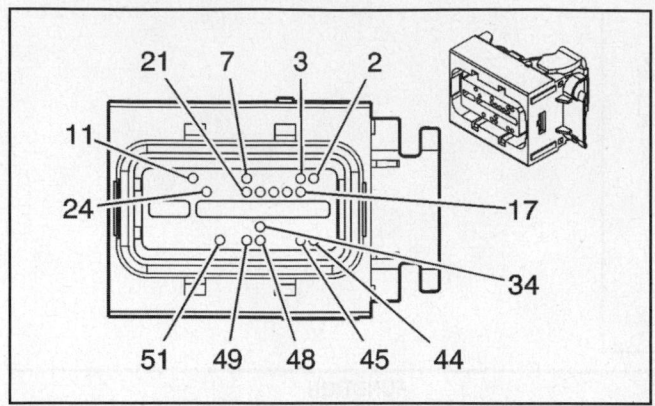

PIN	WIRE COLOR	CIRCUIT NO.	FUNCTION
1	--	--	NOT USED
2	BN	2391	HO2S HEATER LOW CONTROL - BANK 1 SENSOR 2
3	OG/WH	3223	HO2S HEATER LOW CONTROL - BANK 2 SENSOR 2
4-6	--	--	NOT USED
7	YE/BK	1227	LOW REFERENCE
8-10	--	--	NOT USED
11	YE	492	MAF SENSOR SIGNAL
12-16	--	--	NOT USED
17	TN	452	LOW REFERENCE
18	TN/WH	1669	HO2S LOW SIGNAL BANK 1 SENSOR 2
19	PU/WH	1668	HO2S HIGH SIGNAL BANK 1 SENSOR 2
20	PU	1670	HO2S HIGH SIGNAL BANK 2 SENSOR 2
21	TN	1671	HO2S LOW SIGNAL BANK 2 SENSOR 2
22-23	--	--	NOT USED
24	TN	2760	LOW REFERENCE
25-33	--	--	NOT USED
34	TN	472	IAT SENSOR SIGNAL
35-43	--	--	NOT USED
44	PU	401	VSS LOW SIGNAL
45	YE	400	VSS HIGH SIGNAL
46-47	--	--	NOT USED
48	GY	587	SKIP/SHIFT SOLENOID CONTROL
49	TN	2759	LOW REFERENCE
50	--	--	NOT USED
51	D-GN/WH	459	A/C COMPRESSOR CLUTCH RELAY CONTROL
52-56	--	--	NOT USED

6.0L V8 ECM C3 (MM6) Connector Pinout

ECM C3 CONNECTOR (M32)

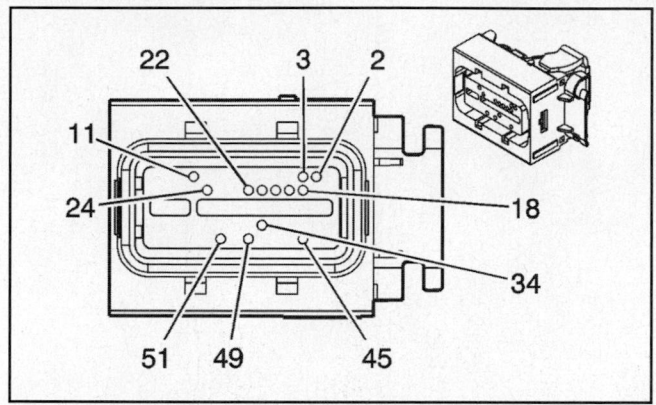

PIN	WIRE COLOR	CIRCUIT NO.	FUNCTION
1	--	--	NOT USED
2	BN	2391	HO2S HEATER LOW CONTROL - BANK 1 SENSOR 2
3	OG/WH	3223	HO2S HEATER LOW CONTROL - BANK 2 SENSOR 2
4-10	--	--	NOT USED
11	YE	492	MAF SENSOR SIGNAL
12-17	--	--	NOT USED
18	TN/WH	1669	HO2S LOW SIGNAL BANK 1 SENSOR 2
19	PU/WH	1668	HO2S HIGH SIGNAL BANK 1 SENSOR 2
20	PU	1670	HO2S HIGH SIGNAL BANK 2 SENSOR 2
21	TN	1671	HO2S LOW SIGNAL BANK 2 SENSOR 2
22	OG/BK	1786	PARK/NEUTRAL SIGNAL
23	--	--	NOT USED
24	TN	2760	LOW REFERENCE
25-33	--	--	NOT USED
34	TN	472	IAT SENSOR SIGNAL
35-44	--	--	NOT USED
45	OG	381	VEHICLE SPEED SIGNAL
46-48	--	--	NOT USED
49	TN	2759	LOW REFERENCE
50	--	--	NOT USED
51	D-GN/WH	459	A/C COMPRESSOR CLUTCH RELAY CONTROL
52-56	--	--	NOT USED

6.0L V8 ECM C3 (M32) Connector Pinout

ENGINE COOLANT TEMPERATURE (ECT) SENSOR

Component Locations

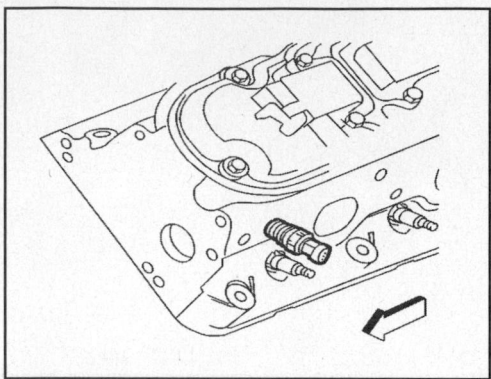

Location of the ECT sensor on 5.7L V8 engines

Removal & Installation

V8 ENGINES

1. Raise the vehicle.
2. Drain the engine coolant below the level of the engine coolant temperature (ECT) sensor.
3. Lower the vehicle.
4. Disconnect the harness connector from the ECT sensor.
5. Remove the ECT sensor.

To install:

CAUTION: Use care when handling the coolant sensor. Damage to the coolant sensor will affect the operation of the fuel control system.

6. Coat the ECT sensor threads with sealer P/N 12346004 or the equivalent
7. Install the ECT sensor. Tighten the ECT sensor to 20 Nm (15 ft. lbs.).
8. Connect the ECT sensor electrical connector.
9. Refill the engine coolant.

Connector Pinouts

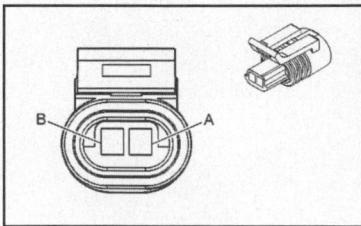

PIN	WIRE COLOR	CIRCUIT NO.	FUNCTION
A	BRN	718	LOW REFERENCE
B	YEL	410	ECT SENSOR SIGNAL

Corvette 5.7L V8 Engine ECT Sensor Connector Pinout

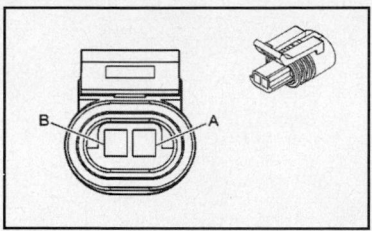

PIN	WIRE COLOR	CIRCUIT NO.	FUNCTION
A	BLK OR TAN	407	SENSOR GROUND
B	YEL	410	ENGINE COOLANT TEMPERATURE SENSOR (ECT) SIGNAL

Camaro & Firebird 5.7L V8 Engine ECT Sensor Connector Pinout

PIN	WIRE COLOR	CIRCUIT NO.	FUNCTION
A	TN	2761	COOLANT TEMPERATURE SENSOR LOW REFERENCE
B	YE	410	ENGINE COOLANT TEMPERATURE SENSOR SIGNAL

6.0L V8 Engine ECT Sensor Connector Pinout

ENGINE COOLING FAN CONNECTORS

Component Locations

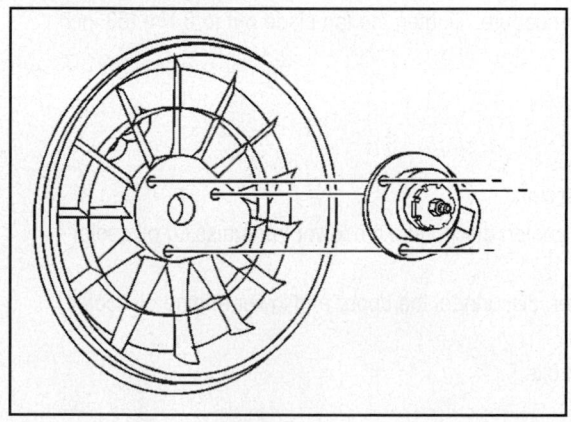

5.7L V8 engine electric cooling fan motor location

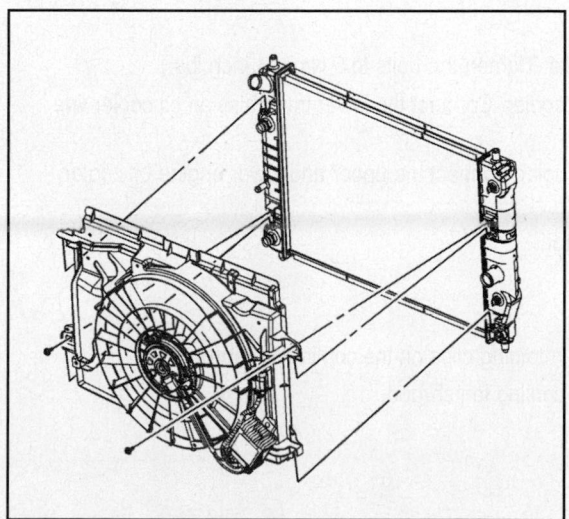

6.0L V8 engine electric cooling fan assembly location

Removal & Installation

CORVETTE 5.7L V8 ENGINES

1. Remove the cooling fan after removing the fan shroud.
2. Remove the fan motor bolts.
3. Remove the fan motor from the shroud.
4. Installation is the reverse of the removal procedure. Tighten the fan motor bolts to 6 Nm (53 inch lbs.).

CAMARO & FIREBIRD 5.7L V8 ENGINES

1. Disconnect the intake air temperature (IAT) and mass airflow (MAF) sensors electrical connectors, if equipped with the 5.7 L engine.
2. Remove the air intake duct resonator, if equipped with the 5.7 L engine.
3. Disconnect the rosebud clips from the fans.
4. Remove the air cleaner bolts.
5. Remove the air cleaner.
6. Remove the electric engine coolant fan. The fan will slide off of the radiator.
7. Remove the fan blade nut.
8. Installation is the reverse of the removal procedure. Tighten the fan blade nut to 6 Nm (53 inch lbs.).

6.0L V8 ENGINES

1. Disconnect the cooling fan electrical connector.
2. Vehicles equipped with transmission fluid cooler, disconnect the lower transmission oil cooler line from the radiator.
3. Vehicles equipped with an engine oil cooler, disconnect the upper and lower engine oil cooler pipes.
4. Remove the cooling fan shroud retaining bolts.
5. Remove the cooling fan and shroud.

To install:

6. Install the cooling fan and shroud.
7. Install the cooling fan shroud retaining bolts. Tighten the bolts to 5 Nm (44 inch lbs.).
8. Vehicles equipped with transmission fluid cooler, Connect the lower transmission oil cooler line to the radiator.
9. On vehicles equipped with an engine oil cooler, connect the upper and lower engine oil cooler pipes.
10. Connect the cooling fan electrical connector.
11. Install the stabilizer shaft.
12. Lower the vehicle.
13. Connect the surge tank outlet hose to the retaining clips on the cooling fan shroud.
14. Connect the engine wiring harness to the cooling fan shroud.
15. Install the radiator support.

Connector Pinouts

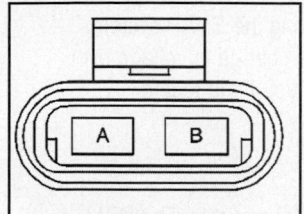

PIN	WIRE COLOR	CIRCUIT NO.	FUNCTION
A	GRY	532	COOLING FAN MOTOR LOW REFERENCE
B	LT BLU	409	COOLING FAN MOTOR SUPPLY VOLTAGE

5.7L V8 Engine Left Electric Cooling Fan Connector Pinout

PIN	WIRE COLOR	CIRCUIT NO.	FUNCTION
A	BLK	150 OR 250	COOLING FAN MOTOR LOW REFERENCE
B	WHT	504	COOLING FAN MOTOR SUPPLY VOLTAGE

5.7L V8 Engine Right Electric Cooling Fan Connector Pinout

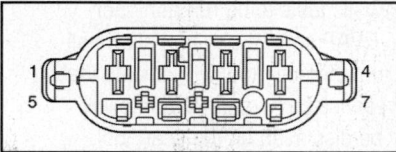

PIN	WIRE COLOR	CIRCUIT NO.	FUNCTION
1	BK	--	COOLING FAN LOW REFERENCE
2	BK	1250	GROUND
3	RD	342	BATTERY POSITIVE VOLTAGE
4	RD	--	COOLING FAN MOTOR SUPPLY VOLTAGE
5	--	--	NOT USED
6	RD/OG	335	COOLING FAN SPEED CONTROL
7	--	--	NOT USED

6.0L V8 Engine Fan Control Module Connector Pinout

EVAP CANISTER VENT & PURGE SOLENOIDS

Description & Operation

The EVAP control system limits fuel vapors from escaping into the atmosphere. Fuel tank vapors are allowed to move from the fuel tank, due to pressure in the tank, through the vapor pipe, into the EVAP canister. Carbon in the canister absorbs and stores the fuel vapors. Excess pressure is vented through the vent line and the EVAP vent solenoid to the atmosphere. The EVAP canister stores the fuel vapors until the engine is able to use the vapors. At an appropriate time, the PCM will command the EVAP purge valve open, allowing engine vacuum to be applied to the EVAP canister. With the EVAP vent valve open, fresh air will be drawn through the valve and vent line to the EVAP canister. Fresh air is drawn through the canister, pulling fuel vapors from the carbon. The air/fuel vapor mixture continues through the EVAP purge pipe and EVAP purge valve into the intake manifold to be consumed during normal combustion. The EVAP system requires the PCM be able to detect a leak as small as 1 mm (0.040 inch) in the EVAP system. The PCM uses several tests to determine if the EVAP system is sealed.

WEAK VACUUM TEST

This tests for large leaks and blockages in the EVAP system. The PCM will command ON and close the EVAP vent valve and will command ON and open the EVAP purge valve with the engine running, allowing engine vacuum into the EVAP system. The PCM monitors the FTP sensor in order to verify that the system is able to reach a predetermined level of vacuum within a set amount of time. If the PCM does not detect the predetermined vacuum level on 2 consecutive tests, a DTC P0440 will set.

VACUUM DECAY TEST

If the weak vacuum test passes, the PCM will command OFF the EVAP purge valve, sealing the EVAP system. The PCM tests for vacuum decay in the EVAP system by monitoring the FTP sensor for a change in voltage over a period of time. If the decay rate is more than a calibrated value, the PCM will rerun the test. If the test fails again, a DTC P0442 will set.

CANISTER VENT RESTRICTION TEST

If the EVAP vent system is restricted, fuel vapors will not be properly purged from the EVAP canister. The PCM tests this condition by commanding the EVAP purge valve CLOSED and commanding the EVAP vent valve OPEN and monitoring the FTP sensor for an increase in vacuum. If an increase in vacuum is detected, DTC P0446 will set.

PURGE VALVE LEAK TEST

If the EVAP purge valve does not seal properly, fuel vapors could enter the engine at an undesired time, causing driveability concerns. The PCM tests for this condition by commanding the EVAP purge and vent valves closed, sealing the system, and monitoring the FTP for an increase in vacuum. If the PCM detects an increase in vacuum, DTC P1441 will set.

EVAP SERVICE BAY TEST

The EVAP service bay test is accessed with a scan tool, and allows EVAP diagnostic tests to be run at higher engine coolant temperatures (ECT) than are allowed during normal testing. The EVAP service bay test allows all of the above tests to be run on demand. When the EVAP service bay tests are run, the scan tool will indicate that the tests have passed, or will indicate which specific DTC has failed. If a EVAP service bay test fails, DTCs will not be recorded in the PCM Freeze Frame/Failure Records. The DTCs will only be displayed on the scan tool. The EVAP service bay test is useful in determining if a fault is present, and for verifying a repair.

The EVAP purge valve controls the flow of vapors from the EVAP system to the intake manifold. This normally closed valve is pulse width modulated by the PCM in order to precisely control the flow of fuel vapor to the engine. The valve will also be opened during some portions of the EVAP testing, allowing engine vacuum to enter the EVAP system.

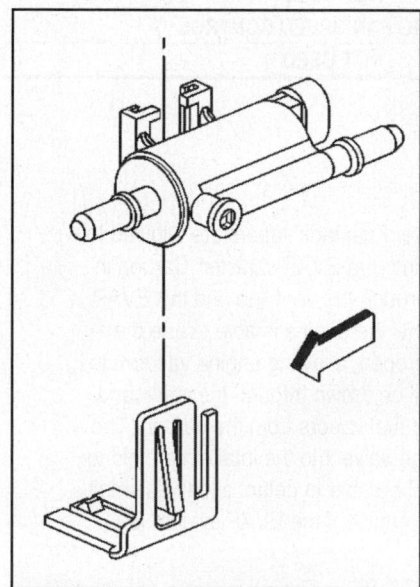

EVAP canister purge solenoid on 5.7L & 6.0L V8 engines

The EVAP vent valve (1) controls the fresh airflow into the EVAP canister. The EVAP is a normally open valve. The PCM will command the valve closed during some EVAP tests, allowing the system to be tested for leaks

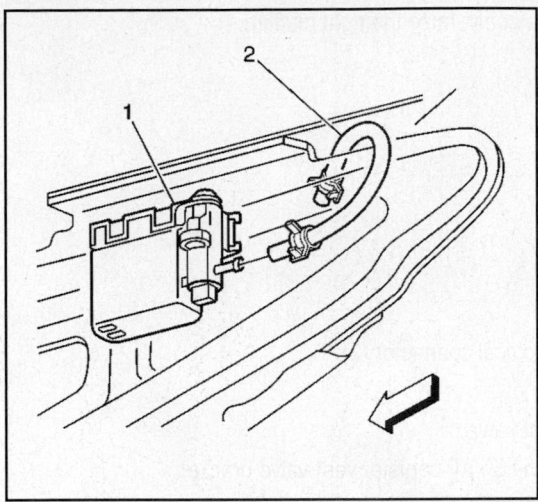

Location of EVAP vent solenoid valve (1) on Corvette 5.7L & 6.0L V8 engines

Removal & Installation

PURGE SOLENOID VALVE

5.7L & 6.0L V8 Engine

1. Remove the left fuel rail cover, if equipped.
2. Disconnect the engine vacuum pipe from the EVAP canister purge solenoid valve to intake manifold EVAP pipe.
3. Disconnect the engine purge pipe from the EVAP canister purge valve.
4. Disconnect the EVAP canister purge valve harness connector.
5. Remove the EVAP canister purge valve from the purge bracket.

To install:

6. Install the EVAP canister purge valve to the purge bracket.
7. Connect the harness connector to the EVAP canister purge valve.
8. Connect the engine purge pipe to the chassis purge pipe.
9. Connect the engine vacuum pipe to the EVAP canister purge valve.
10. Install the left fuel rail cover.

VENT SOLENOID VALVE

Corvette 5.7L & 6.0L V8 Engines

1. Raise the vehicle.
2. Lower the right muffler for automatic transmission equipped vehicles only.
3. Disconnect the EVAP canister valve harness connector.
4. Disconnect the vent hose (2) from the EVAP canister vent valve (1). See illustration above.
5. Remove the EVAP canister vent valve from the vent bracket.

To install:

6. Install the EVAP canister vent valve to the vent bracket.
7. Connect the vent hose (2) to the EVAP canister vent valve (1).
8. Connect the EVAP vent valve electrical connector.

9. For automatic transmission- equipped vehicles only, raise the right muffler.

10. Lower the vehicle.

CAMARO & FIREBIRD 5.7L V8 ENGINES

1. Disconnect the negative battery cable.

2. Relieve the fuel system pressure.

3. Drain the fuel tank.

4. Remove the fuel tank.

5. Disconnect the EVAP canister vent valve electrical connector (2).

6. Slide the vent hose clamp (1) back.

7. Disconnect the vent hose from the EVAP vent valve.

8. Remove the EVAP canister vent valve from the EVAP canister vent valve bracket.

To install:

9. Install the EVAP canister vent valve to the EVAP canister vent valve bracket.

10. Connect the vent hose to the EVAP canister vent valve.

11. Slide the vent hose clamp (1) back into position.

12. Connect the EVAP canister vent valve electrical connector (2).

13. Install the fuel tank.

14. Refill the fuel tank.

15. Install the fuel filler cap.

16. Connect the negative battery cable.

17. Inspect for leaks.

18. Turn the ignition switch ON for 2 seconds.

19. Turn the ignition switch OFF for 10 seconds.

20. Turn the ignition switch ON.

21. Inspect for fuel leaks.

22. Perform the Idle Learn Procedure.

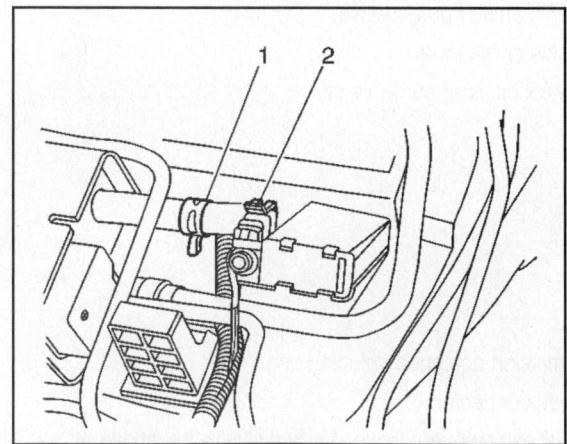

Location of EVAP vent solenoid valve on Camaro & Firebird 5.7L V8 engines

Connector Pinouts

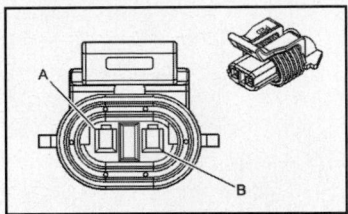

PIN	WIRE COLOR	CIRCUIT NO.	FUNCTION
A	PNK	339 OR 539	IGNITION (HOT W/IGN RELAY ENERGIZED)
B	DK GRN/WHT	428	EVAP CANISTER PURGE SOLENOID CONTROL

Corvette 5.7L & 6.0L V8 Engines EVAP Purge Solenoid Connector Pinout

PIN	WIRE COLOR	CIRCUIT NO.	FUNCTION
A	PNK	239	FUSED IGNITION FEED
B	DK GRN/WHT	428	EVAP CANISTER PURGE VALVE CONTROL

Camaro & Firebird 5.7L V8 Engines EVAP Purge Solenoid Connector Pinout

PIN	WIRE COLOR	CIRCUIT NO.	FUNCTION
A	PNK	239 OR 339	BATTERY POSITIVE FEED
B	WHT	1310	EVAP CANISTER VENT SOLENOID CONTROL

5.7L V8 Engines EVAP Vent Solenoid Connector Pinout

PIN	WIRE COLOR	CIRCUIT NO.	FUNCTION
A	RD/WH	2840	BATTERY POSITIVE VOLTAGE
B	WH	1310	EVAP CANISTER VENT SOLENOID CONTROL

6.0L V8 Engines EVAP Vent Solenoid Connector Pinout

EXHAUST GAS RECIRCULATION (EGR) VALVE

Component Locations

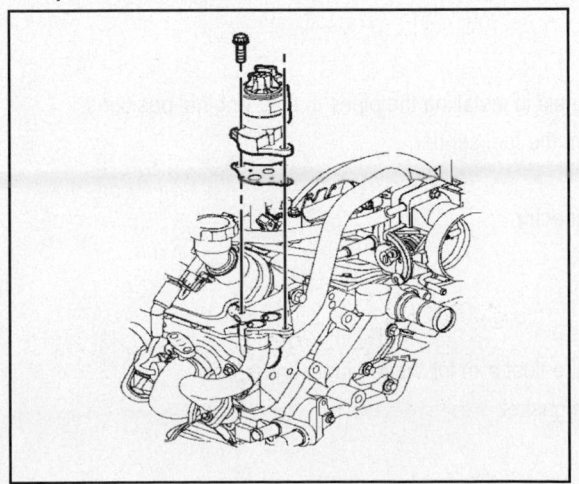

EGR valve location on 2000 Firebird with 5.7L V8 engine

Connector Pinouts

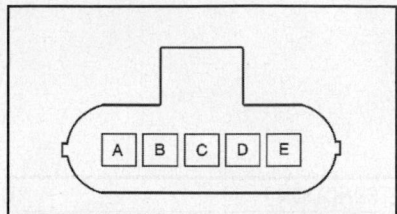

PIN	WIRE COLOR	CIRCUIT NO.	FUNCTION
A	GRY	435	EXHAUST GAS RECIRCULATION (EGR) VALVE GROUND
B	BLK	407	EXHAUST GAS RECIRCULATION (EGR) SENSOR GROUND
C	BRN	1456	EXHAUST GAS RECIRCULATION (EGR) PINTLE POSITION
D	GRY	416	EXHAUST GAS RECIRCULATION (EGR) 5V REFERENCE
E	RED	1676	EXHAUST GAS RECIRCULATION (EGR) VALVE CONTROL

2000 Camaro & Firebird 5.7L V8 Engine EGR Valve Connector Pinout

FUEL PUMP & SENDER

Removal & Installation

2000-03 CORVETTE W/O FFS

1. Disconnect the negative battery cable.
2. Relieve the fuel system pressure.

CAUTION: *To avoid any vehicle damage, serious personal injury or death when major components are removed from the vehicle and when a hoist supports the vehicle, support the vehicle with jack stands at the opposite end from which the components are being removed.*

3. Raise the vehicle.
4. Remove the rear wheel and tire assembly.
5. Clean all of the fuel connections and the surrounding areas before disconnecting the fuel pipes in order to avoid possible contamination of the fuel system.
6. Drain the fuel tanks.
7. Remove the fuel tank shield.
8. Mark or identify each fuel pipe in order to aid in installing the pipes in their original positions.
9. Disconnect the quick-connect fittings from the fuel sender.
10. Cap all of the fuel pipes.
11. Disconnect the fuel sender electrical connector.
12. Remove the fuel tank strap.
13. Support the fuel tank.
14. Remove the fuel sender attaching bolts.
15. Remove the float arm retaining clip and the float arm for the left fuel sender only.
16. Carefully remove the fuel sender with the gasket.
17. Clean the gasket sealing surfaces.

To install:

RIGHT

CAUTION: Do not bend or twist the float arm.

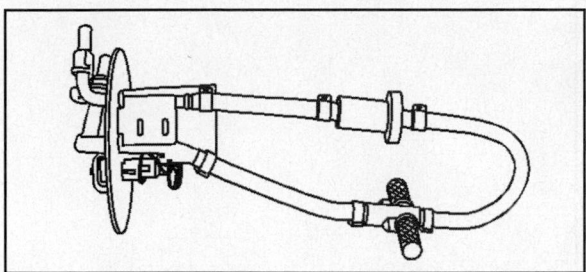

Showing the fuel sending unit assembly on 5.7L V8 engines

18. Inspect the fuel sender gasket for damage and replace if necessary.
19. Insert the float arm through the fuel tank opening.
20. The fuel sender may need to be rotated in order to facilitate the installation.
21. Align the fuel sender gasket tab (2) with the fuel sender cover mark (1).
22. Align the fuel sender cover mark (1) with the fuel tank mark.

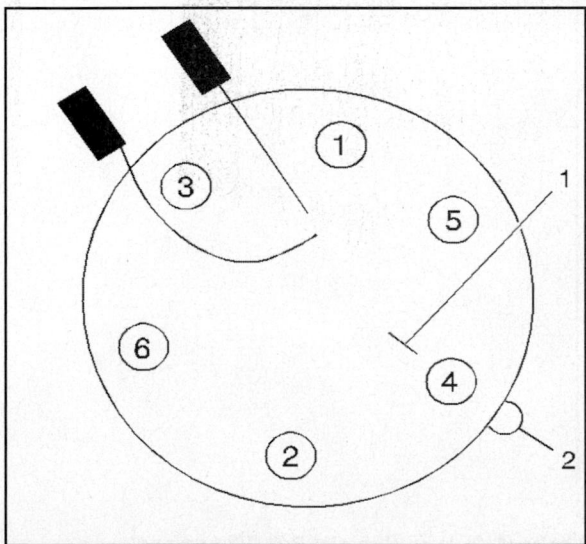

Showing fuel sender assembly bolt tightening sequence

23. Hand tighten the fuel sender attaching bolts until finger tight.
24. Tighten the fuel sender assembly attaching bolts in sequence. Tighten the bolt to 7 Nm (62 inch lbs.).
25. Connect the fuel sender fuel feed pipe (1) (from the jet pump to the left tank, and the fuel feed rear crossover pipe (2) from the left tank to the jet pump).
26. Connect the fuel sender electrical connector.

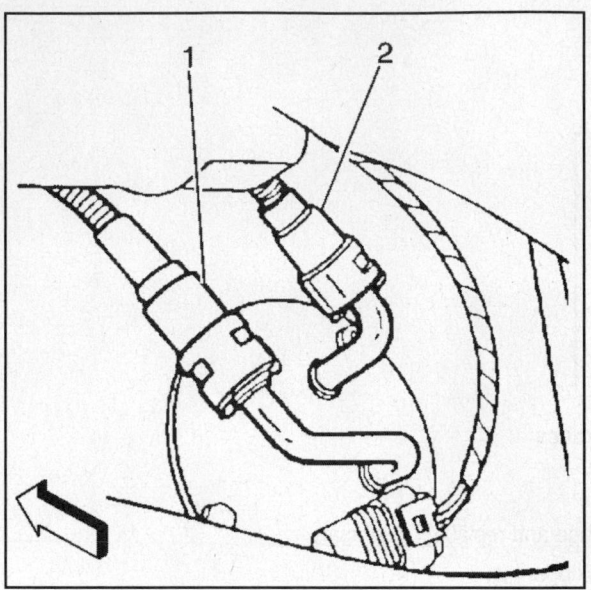

Showing the fuel sender fuel feed pipe (1) and fuel feed crossover pipe (2) on 5.7L V8 engines

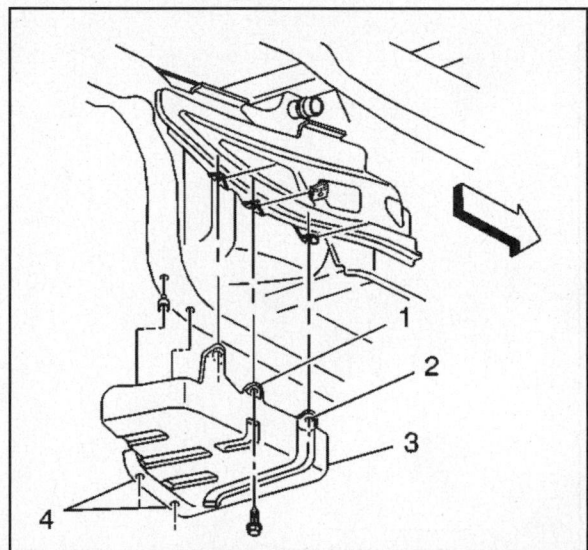

Showing the location of fuel tank strap (3) and retaining bolts (1, 2, 4) on Corvette 5.7L V8 engines

27. Install the fuel tank strap (3).

28. Install the fuel tank strap bolts (1, 2, 4). Tighten as follows:

- Tighten the bolt (2) to 25 Nm (18 ft. lbs.).
- Tighten the bolt (1) to 25 Nm (18 ft. lbs.).
- Tighten the remaining bolts (4) to 25 Nm (18 ft. lbs.).

29. Install the fuel tank shield (1).

30. Install the fuel tank shield mount bolt (2).

31. Install the fuel tank shield mount nut (3). Tighten retainers as follows:

- Tighten the fuel tank shield mount bolt to 25 Nm (18 ft. lbs.).

- Tighten the fuel tank shield mount nut to 12 Nm (106 inch lbs.).

32. Install the rear wheel and tire assembly.
33. Lower the vehicle.
34. Refill the fuel system.
35. Install the fuel filler cap.
36. Connect the negative battery cable.
37. Perform the following procedure in order to inspect for leaks:
38. Turn the ignition switch ON for 2 seconds.
39. Turn the ignition switch OFF for 10 seconds.
40. Turn the ignition switch ON.
41. Inspect for fuel leaks.
42. Program the transmitters.

LEFT

CAUTION: Do not bend or twist the float arm during installation.

1. Install fuel pump strainer in the same position as noted during disassembly. Push on the outer edge of the strainer ferrule until the strainer is fully seated.
2. Install a sender gasket on the fuel sender.
3. Insert the fuel sender through the fuel tank opening.
4. The fuel sender may need to be rotated in order to facilitate the installation.
5. Insert the float and the float arm into the fuel tank opening.
6. Install the float arm retaining clip.
7. Align the fuel sender gasket tab with the fuel sender cover mark.
8. Align the fuel sender cover mark with the fuel tank mark.
9. Tighten the fuel sender attaching bolts until the bolts are finger tight.

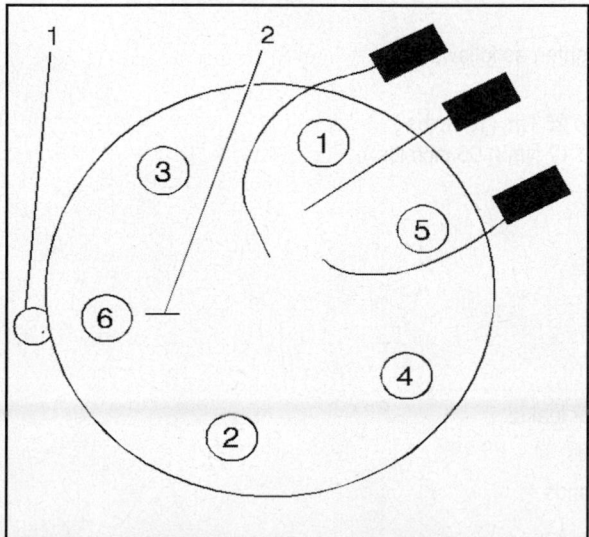

Showing fuel sender assembly bolt tightening sequence

10. Tighten the fuel sender assembly attaching bolts in proper sequence, as shown. Tighten bolts to 7 Nm (62 inch lbs.).

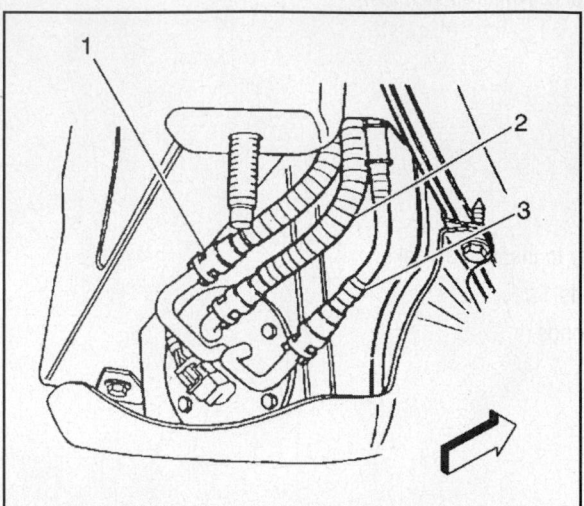

Showing fuel line arrangement

11. Connect the fuel sender fuel feed pipe (1) from the jet pump to the left tank, to the fuel return rear pipe (2), and to the fuel feed rear pipe (3).

12. Connect the fuel sender electrical connector.

13. Install the fuel tank strap (3).

14. Install the fuel tank strap bolts (1, 2, 4). Tighten the bolts as follows:

- Tighten the bolt (2) to 25 Nm (18 ft. lbs.).
- Tighten the bolt (1) to 25 Nm (18 ft. lbs.).
- Tighten the remaining bolts (4) to 25 Nm (18 ft. lbs.).

15. Install the fuel tank shield.

16. Install the fuel tank shield mount bolt.

17. Install the fuel tank shield mount nut. Tighten as follows:

- Tighten the fuel tank shield mount bolt to 25 Nm (18 ft. lbs.).
- Tighten the fuel tank shield mount nut to 12 Nm (106 inch lbs.).

18. Install the rear wheel and tire assembly.

19. Lower the vehicle.

20. Refill the fuel system.

21. Install the fuel filler cap.

22. Connect the negative battery cable.

23. Perform the following steps to inspect for leaks:

24. Turn the ignition switch ON for 2 seconds.

25. Turn the ignition switch OFF for 10 seconds.

26. Turn the ignition switch ON.

27. Inspect for fuel leaks.

28. Program the transmitters.

RH SIDE FUEL TANK – 2000-05 CORVETTE (ALL OTHER CONFIGURATIONS)

1. Disconnect the negative battery cable.
2. Remove the right or left fuel tank.
3. Place the fuel tank on a suitable work surface.
4. Disconnect the EVAP purge line from the fuel pump module.
5. Disconnect the fuel pump module harness connector.
6. Disconnect the FTP sensor harness connector.
7. If replacing the fuel pump module, remove the FTP sensor.
8. Disconnect the jet line insert connector from the crossover tube to fuel tank opening.

CAUTION: The fuel pump module is spring-loaded and will spring upward when the locking ring is removed.

9. Using the J39765-A, remove the fuel pump module locking ring.
10. Carefully remove the fuel pump module from the fuel tank, with the jet lines connected. Take care not to damage the fuel sender float arm.
11. Disconnect the jet line quick-connect connectors from the fuel pump module, noting the location of the lines for installation.
12. Remove the fuel pump module O-ring from the fuel tank opening.
13. Remove the jet line insert through the crossover tube to fuel tank opening.

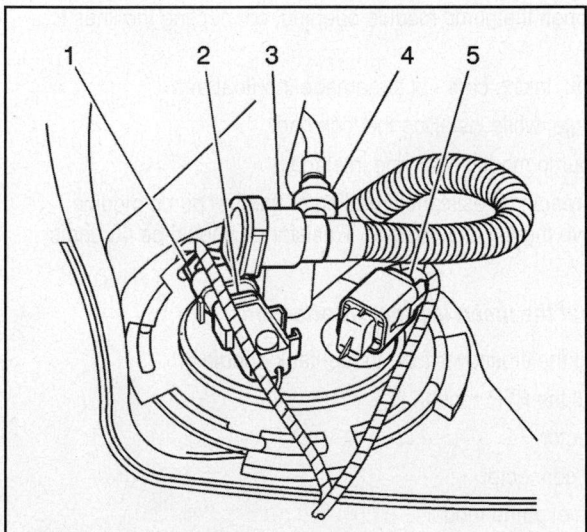

Showing fuel line arrangement on 5.7L with FFS, showing the FTP harness (1) and sensor (2), purge line connector (3), EVAP purge line (4), and fuel pump module harness connector (5)

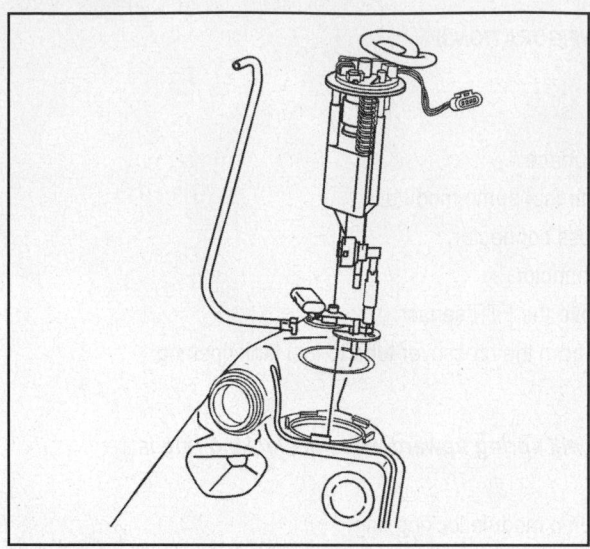

Showing fuel sending assembly removed from tank (right side shown; left side similar)

To install:

14. Inspect the jet line insert for damage and replace if necessary.

15. Install the jet line insert through the crossover tube to fuel tank opening.

16. Install a new fuel pump module O-ring to the fuel tank opening.

CAUTION: Pull on each connector to ensure that the connectors are properly latched.

17. Pull the jet line quick-connectors up through the pump module opening, connecting the lines to the pump module as previously noted.

18. Install the pump module into the fuel tank, taking care not to damage the float arm.

19. Compress and align the fuel pump module, while installing the lock ring.

20. Using the J39765-A, fully lock the fuel pump module lock ring in place.

21. Using a DMM, verify the full and empty readings resistance reading of the fuel pump module. Turn the fuel tank upside down to achieve the full tank reading. Resistance should be 40 ohms when empty to 250 ohms when full.

CAUTION: Pull the jet line insert connector to ensure that the insert is properly attached.

22. Connect the jet line insert connector into the crossover tube to fuel tank opening.

23. If replacing the fuel pump module, install the FTP sensor.

24. Connect the FTP sensor harness connector.

25. Connect the fuel pump module harness connector.

26. Connect the EVAP purge line from the fuel pump module.

27. Install the fuel tank.

LH SIDE FUEL TANK – 2000-05 CORVETTE (ALL OTHER CONFIGURATIONS)

1. Disconnect the negative battery cable.

2. Remove the left fuel tank.

3. Place the fuel tank on a suitable work surface.

4. Disconnect the fuel pump jumper harness from the fuel pump module.

5. Disconnect the jet line insert connector from the crossover tube to fuel tank opening.

6. Disconnect the fuel feed line from the welded clip on the side of the fuel tank.

CAUTION: The fuel pump module is spring-loaded and will spring upward when the locking ring is removed.

    7.   Using the J39765-A, remove the fuel pump module locking ring.

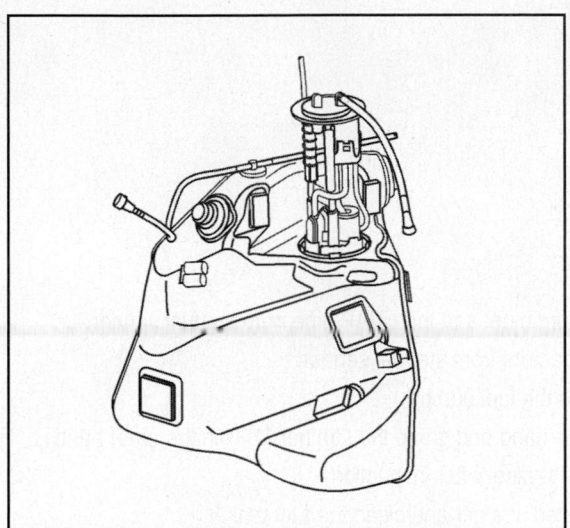

Removing fuel sending unit from LH fuel tank

    8.   Carefully remove the fuel pump module from the fuel tank, with the jet lines connected. Take care not to damage the fuel sender float arm.

    9.   Disconnect the jet line quick-connect connectors from the fuel pump module inner port.

    10.   Remove the jet line from the module retainer cup. This line has no attached connector.

    11.   Remove the fuel pump module O-ring from the fuel tank opening.

    12.   Remove the jet line insert through the crossover tube to fuel tank opening.

To install:

    13.   Inspect the jet line insert for damage and replace if necessary.

    14.   Install the jet line insert through the crossover tube to fuel tank opening.

    15.   Install a new fuel pump module O-ring to the fuel tank opening.

    16.   Place tape around the jet line with the connector. This will permit line access once the pump module is inserted into the fuel tank.

    17.   Install the pump module into the fuel tank half way , taking care not to damage the float arm.

    18.   Using the tape as a guide, gently pull the jet line up through the fuel pump module opening.

    19.   Place the jet line with no connector in the module retainer cup.

    20.   Secure the line into the module retaining clip.

    21.   Remove the tape from the jet line with a connector.

NOTE: Pull on each connector to ensure that the connectors are properly latched.

    22.   Connect the jet line quick-connect connectors to the fuel pump module inner port.

    23.   Compress and align the fuel pump module into the fuel tank, while taking care not to damage the float arm.

    24.   Install the fuel pump module lock ring.

    25.   Using the J39765-A, fully lock the fuel pump module lock ring in place.

    26.   Connect the fuel supply line into the weld clip on the side of the fuel tank.

    27.   Using a DMM, verify the full and empty readings resistance reading of the fuel pump module. Turn the fuel tank upside down to achieve the full tank reading.

NOTE: Correct resistance readings are 40 ohms at empty and 250 at full tank.

28. Connect the jet line insert connector into the crossover tube to fuel tank opening.
29. Connect the fuel pump jumper harness to the fuel pump module.
30. Install the left fuel tank.

CAMARO & FIREBIRD

1. Disconnect the negative battery cable.
2. Relieve the fuel system pressure.
3. Drain the fuel tank.
4. Remove the fuel tank.
5. Remove the fuel sender assembly:
6. Disconnect the fuel feed pipe, the fuel return pipe, and the EVAP pipe from the fuel sender.
7. Disconnect the fuel sender electrical connectors from the fuel sender.
8. Note the position of the fuel strainer (4) on the fuel pump.
9. Support the fuel sender assembly with one hand and grasp the strainer (4) with the other hand.
10. Pull the strainer off the pump. Discard the strainer after inspection.
11. Inspect the strainer. Replace a contaminated strainer and clean the fuel tank.
12. Remove the fuel pressure regulator retaining clip (2).
13. Remove the fuel pressure regulator (3) from the housing on the fuel return pipe.
14. Disconnect the fuel pump electrical connector (7).
15. Remove the fuel level sensor electrical connector retaining clip (9).
16. Disconnect the fuel level sensor electrical connector (8) from under the fuel sender cover.
17. Remove the fuel level sensor retaining clip (6).
18. Squeeze the locking tangs and remove the fuel level sensor (5).
19. Remove the fuel pressure sensor (1).

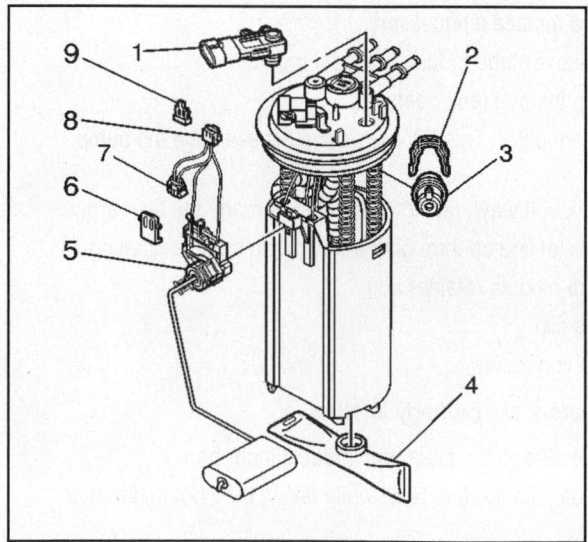

Exploded view of fuel tank pump & sending unit on Camaro & Firebird 5.7L V8 engines

To install:

20. Install the fuel pressure sensor (1).
21. Install the fuel level sensor (5).
22. Install the fuel level sensor retaining clip (6).
23. Connect the fuel level sensor electrical connector (8).
24. Connect the fuel level sensor electrical connector retaining clip (9).
25. Connect the fuel pump electrical connector (7).
26. If needed, clean the fuel pressure regulator filter screen with gasoline.
27. Lubricate the fuel pressure regulator large and small O-rings with clean engine oil.
28. Install the fuel pressure regulator (3) into the housing on the fuel return pipe.
29. Install the fuel pressure regulator retaining clip (2).

NOTE: Always install a new fuel strainer when replacing the fuel tank fuel pump module.

30. Install a new fuel strainer (4) in the same position as noted during disassembly. Push the strainer on the bottom of the fuel sender until the strainer is fully seated.
31. Install the fuel sender assembly.
32. Perform the fuel tank leak check.
33. Install the fuel tank.
34. Refill the fuel tank.
35. Install the fuel filler cap.
36. Connect the negative battery cable.
37. Inspect for leaks.
38. Turn the ignition switch ON for 2 seconds.
39. Turn the ignition switch OFF for 10 seconds.
40. Turn the ignition switch ON.
41. Check for fuel leaks.
42. Perform the idle learn procedure.

Connector Pinouts

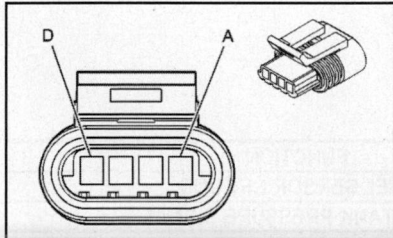

PIN	WIRE COLOR	CIRCUIT NO.	FUNCTION
A	GRY	120	FUEL PUMP SUPPLY VOLTAGE
B	GRY	720	LOW REFERENCE
C	DK BLU	1936	FUEL LEVEL SENSOR SIGNAL [- SECONDARY]
D	BLK	150	GROUND

2000-03 Corvette 5.7L V8 Engines Fuel Pump and Sender (w/o FFS) Connector Pinout

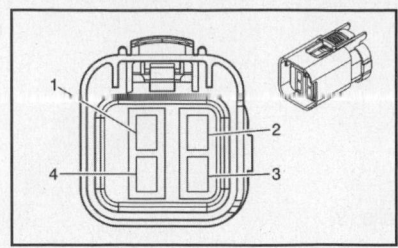

PIN	WIRE COLOR	CIRCUIT NO.	FUNCTION
1	DK BLU	1936	FUEL LEVEL SENSOR SIGNAL [- SECONDARY]
2	GRY	120	FUEL PUMP SUPPLY VOLTAGE
3	BLK	9531	GROUND
4	BLK	808	LOW REFERENCE

Corvette 5.7L & 6.0L V8 Engines Fuel Pump and Sender (2000-03 w/FFS & 2004-05) Connector Pinout

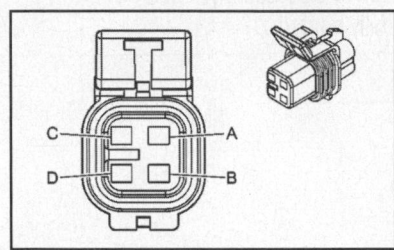

PIN	WIRE COLOR	CIRCUIT NO.	FUNCTION
A	PPL	30	FUEL LEVEL INPUT
B	GRY	120	FUEL PUMP CONTROL
C	BLK	150	FUEL PUMP GROUND
D	BLK/WHT	651	FUEL LEVEL SENSOR GROUND

Camaro & Firebird 5.7L V8 Engine Fuel Pump & Sender Connector Pinout

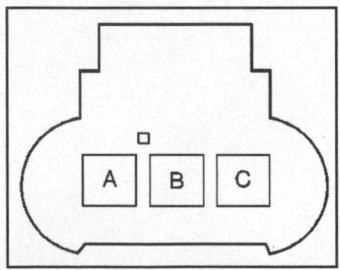

PIN	WIRE COLOR	CIRCUIT NO.	FUNCTION
A	ORN/BLK	469	FUEL SENSOR GROUND
B	DK GRN	890	FUEL TANK PRESSURE SIGNAL
C	GRY/BLK	416	5V SENSOR FEED

Camaro & Firebird 5.7L V8 Engines FTP Sensor Connector Pinout

FUEL TANK PRESSURE (FTP) SENSOR

Component Location

CORVETTE

NOTE: For Firebird applications, see ""Fuel Pump & Sender" above.

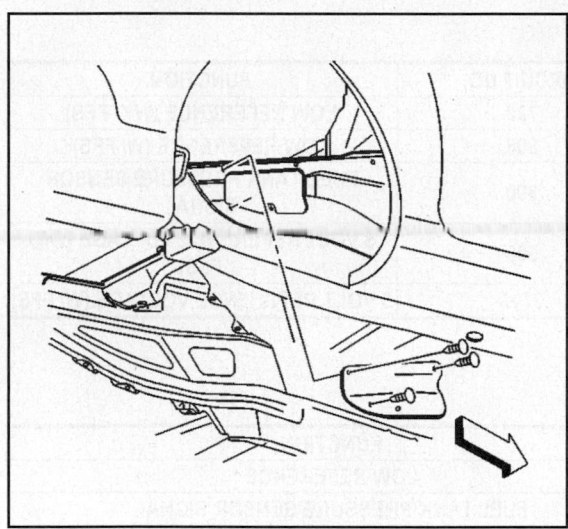

Fuel tank pressure sensor on Corvette 5.7L & 6.0L V8 engines

Removal & Installation

1. Remove the right rear wheelhouse panel.
2. Remove the EVAP canister access cover.
3. Disconnect the fuel tank pressure sensor electrical connector.
4. Remove the fuel tank pressure sensor by carefully prying the sensor out of the fuel tank with a screwdriver.

To install:

5. Lubricate the fuel tank pressure sensor rubber grommet with clean engine oil in order to aid in installation.
6. Install the fuel tank pressure sensor into the top of the fuel tank.
7. Connect the fuel tank pressure sensor electrical connector.
8. Install the EVAP canister access cover.
9. Install the right rear wheelhouse panel.

Connector Pinouts

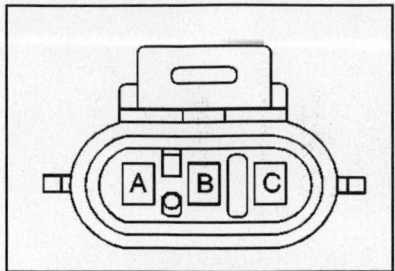

PIN	WIRE COLOR	CIRCUIT NO.	FUNCTION
A	GRY	720	LOW REFERENCE (W/O FFS)
	BLK	808	LOW REFERENCE (W/ FFS)
B	DK GRN	890	FUEL TANK PRESSURE SENSOR SIGNAL
C	GRY/BLK	598	5 VOLT REFERENCE VOLTAGE (W/O FFS)
	BRN/ LT GRN	2709	5 VOLT REFERENCE VOLTAGE (W/ FFS)

Corvette 5.7L V8 Engines FTP Sensor Connector Pinout

PIN	WIRE COLOR	CIRCUIT NO.	FUNCTION
A	TN	2759	LOW REFERENCE
B	D-GN	890	FUEL TANK PRESSURE SENSOR SIGNAL
C	GY	2709	5-VOLT REFERENCE

6.0L V8 Engines FTP Sensor Connector Pinout

GENERATOR

Component Locations

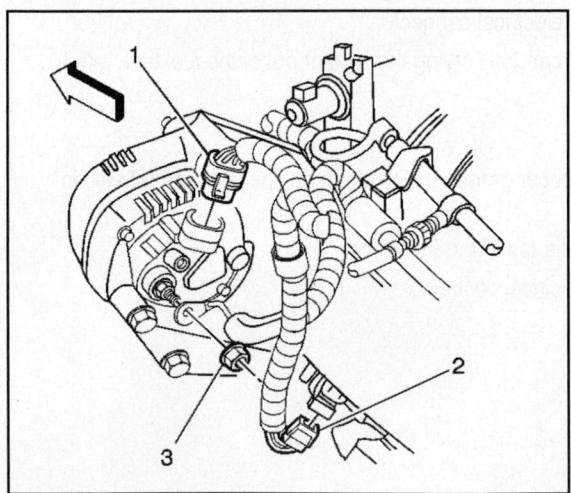

Generator mounting on 5.7L & 6.0L V8 engines

Removal & Installation

CORVETTE

1. Remove the accessory drive belt.
2. Disconnect the negative battery cable.
3. Disconnect the generator electrical connector (1).
4. Remove the engine harness terminal from the generator by sliding the boot back along the cable, removing the engine harness cable nut (3), and the engine harness terminal from the stud.
5. Remove the generator bolts and generator.

To install:

6. Install the generator.
7. Install the generator bolts. Tighten the generator bolts to 50 Nm (37 ft. lbs.).
8. Install the engine harness terminal to the generator:
9. Install the engine harness terminal to the stud.
10.
11. Install the engine harness cable nut (3). Tighten the engine harness cable nut to 13 Nm (10 ft. lbs.).
12. Slide the boot over the generator stud.
13. Connect the generator electrical connector.

CAMARO & FIREBIRD

1. Disconnect the negative battery cable.
2. Remove the accessory drive belt.
3. Raise and suitably support the vehicle.
4. Remove the positive cable from the generator:
5. Slide the boot down revealing the positive terminal stud.
6. Remove the positive cable nut from the generator output stud.
7. Remove the positive cable.
8. Remove the generator rear bracket mounting bolt.
9. Remove the transmission oil cooler lines from the oil cooler clip.
10. Remove the front generator mounting bolts and oil cooler clip.
11. Disconnect the generator electrical connector.
12. Remove the generator from the vehicle.

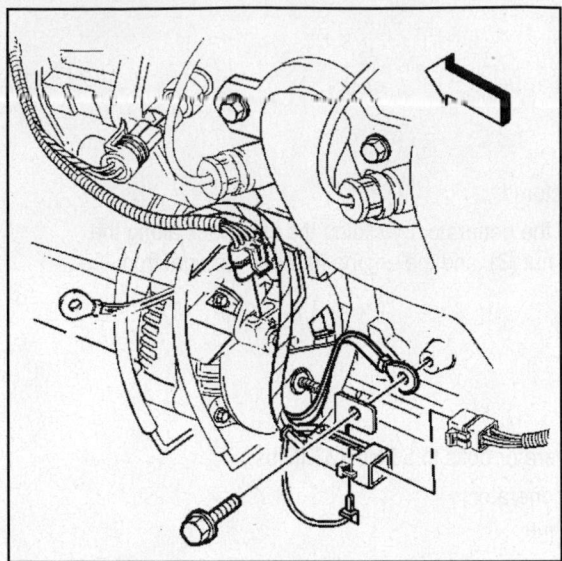

Showing location of generator on Camaro & Firebird with 5.7L V8 engines

To install:

13. Install the generator to the vehicle.
14. Connect the generator electrical connector.
15. Install the front generator mounting bolts and oil cooler clip. Tighten the generator mounting bolts to 50 Nm (37 ft. lbs.).
16. Install the transmission oil cooler lines to the oil cooler clip.
17. Install the generator rear bracket mounting bolt. Tighten the rear bracket mounting bolt to 25 Nm (18 ft. lbs.).
18. Install the positive cable to the generator:
19. Install the positive cable.
20. Install the positive cable nut to the generator output stud. Tighten the positive cable nut to the generator output stud to 22 Nm (16 ft. lbs.).
21. Slide the boot over the positive terminal stud.
22. Lower the vehicle.
23. Install the accessory drive belt.
24. Connect the negative battery cable.

Connector Pinouts

PIN	WIRE COLOR	CIRCUIT NO.	FUNCTION
A	--	--	NOT USED
B	RED	225	GENERATOR TURN ON SIGNAL
C	GRY	23	GENERATOR FIELD DUTY CYCLE SIGNAL
D	RED	2	BATTERY POSITIVE VOLTAGE

Corvette 5.7L V8 Engines Generator Connector Pinout

PIN	WIRE COLOR	CIRCUIT NO.	FUNCTION
A	--	--	NOT USED
B	RED	225	GENERATOR REGULATOR REFERENCE VOLTAGE
C-D	--	--	NOT USED

Camaro & Firebird 5.7L V8 Engines Generator Connector Pinout

PIN	WIRE COLOR	CIRCUIT NO.	FUNCTION
A	--	--	NOT USED
B	OG	225	GENERATOR TURN ON SIGNAL
C	GY	23	GENERATOR FIELD DUTY CYCLE SIGNAL
D	RD/WH	2540	BATTERY POSITIVE VOLTAGE

6.0L V8 Engines Generator Connector Pinout

HEATED OXYGEN SENSORS (HO2S)

Component Locations

CORVETTE 5.7L V8 ENGINES

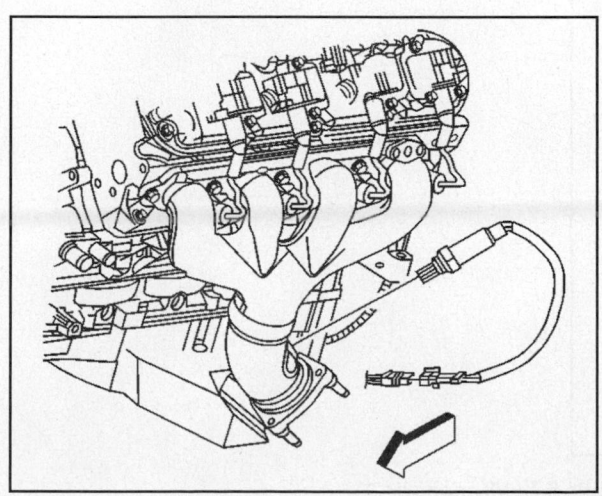

HO2S1/1 (Bank 1 Sensor 1) location on Corvette 5.7L V8 engines

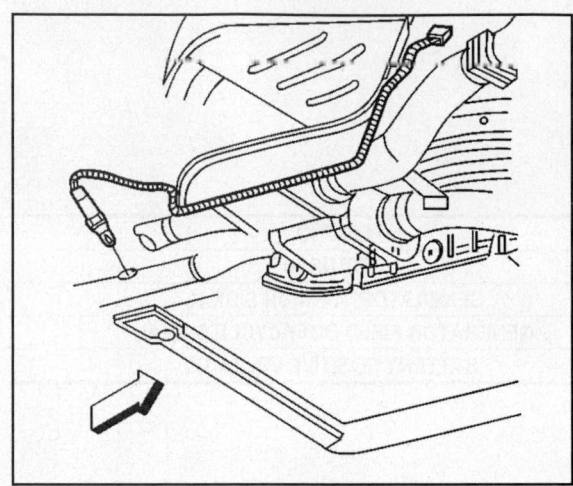

Showing HO2S1/2 (Bank 1 Sensor 2) location on Corvette 5.7L V8 engines

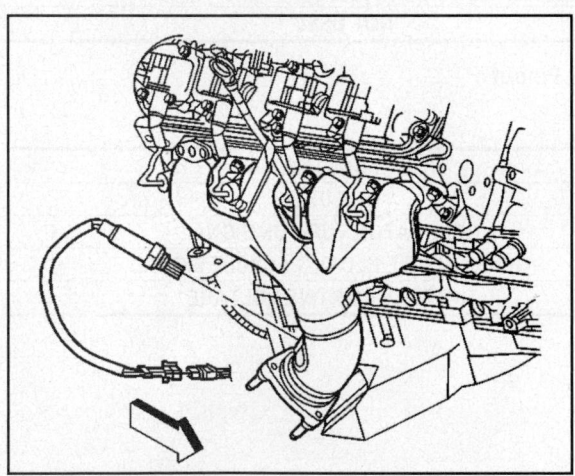

Showing HO2S2/1 (Bank 2 Sensor 1) on Corvette 5.7L V8 engines

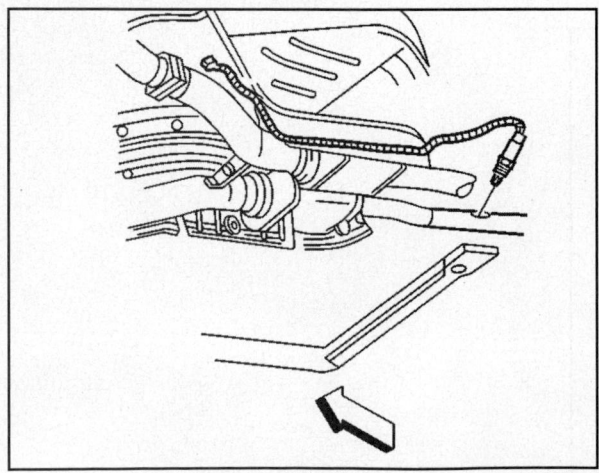

Showing HO2S2/1 (Bank 2 Sensor 2) location on Corvette 5.7L V8 engines

CAMARO & FIREBIRD 5.7L V8 ENGINES

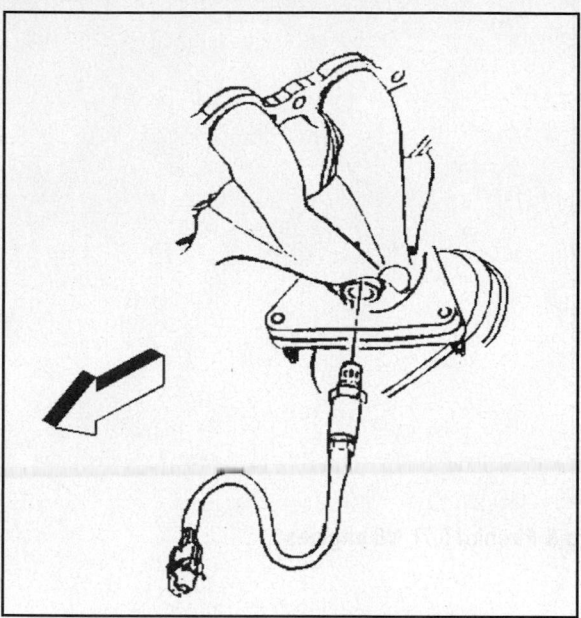

Showing HO2S1/1 (Bank 1 Sensor 1) location on Camaro & Firebird 5.7L V8 engines

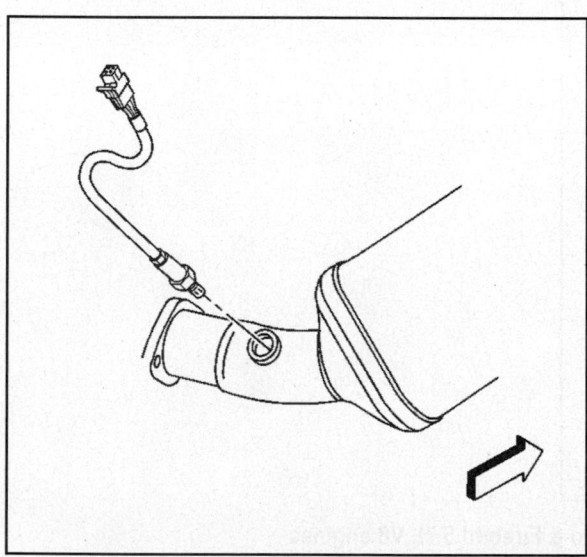

Showing HO2S1/2 (Bank 1 Sensor 2) location on Camaro & Firebird 5.7L V8 engines

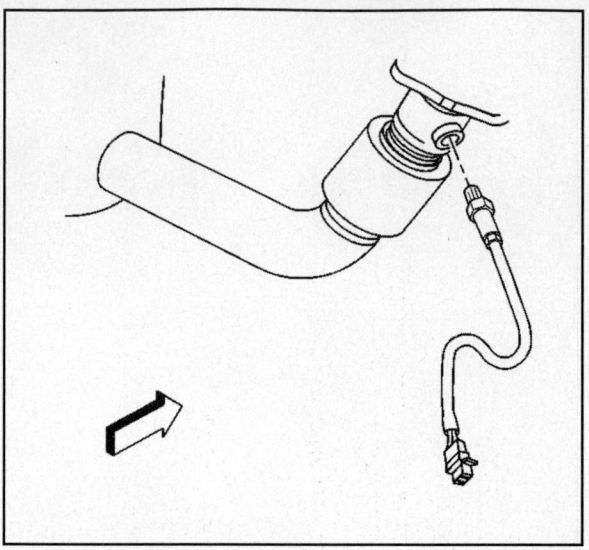

Showing HO2S2/1 (Bank 2 Sensor 1) location on Camaro & Firebird 5.7L V8 engines

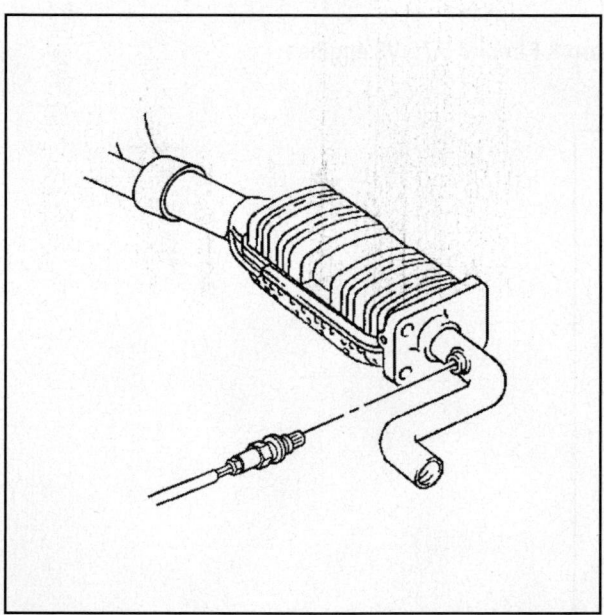

Showing HO2S2/2 (Bank 2 Sensor 2) location on Camaro & Firebird 5.7L V8 engines

CORVETTE 6.0L V8 ENGINES

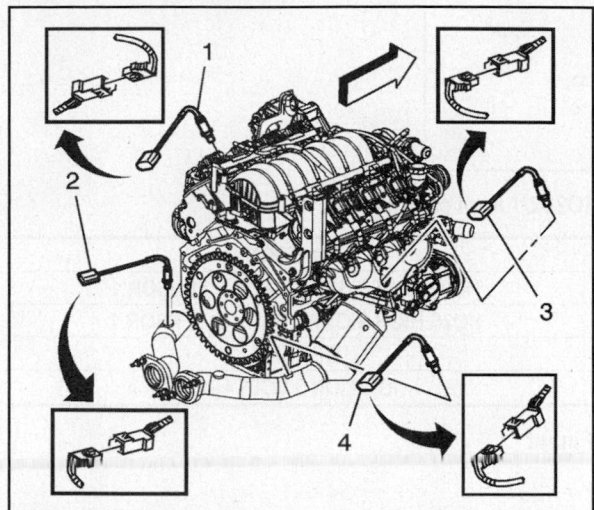

Showing all HO2S locations on 6.0L V8 engines: HO2S1/1 (1), HO2S1/2 (2), HO2S2/1 (3), HO2S2/2 (4)

Connector Pinouts

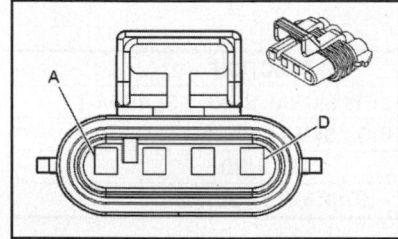

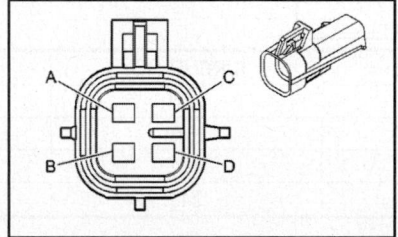

Corvette 5.7L HO2S2/1 Applications   Camaro & Firebird 5.7L HO2S2/1 Applications

IN	WIRE COLOR	CIRCUIT NO.	FUNCTION
A	TAN/WHT	1653	HO2S LOW SIGNAL BANK 1 SENSOR 1
B	PPL/WHT	1665	HO2S HIGH SIGNAL BANK 1 SENSOR 1
C	BLK	150	GROUND
D	BRN	241	IGNITION 3 VOLTAGE

5.7L V8 Engine HO2S1/1 (Bank 1 Sensor 1) Connector Pinout

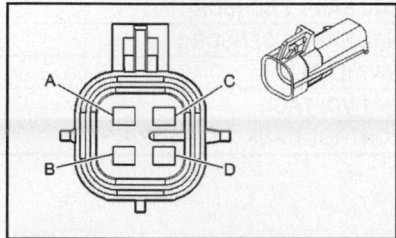

PIN	WIRE COLOR	CIRCUIT NO.	FUNCTION
A	TAN/WHT	1669	HO2S LOW SIGNAL BANK 1 SENSOR 2
B	PPL/WHT	1668	HO2S HIGH SIGNAL BANK 1 SENSOR 2
C	BLK	150	GROUND
D	BRN	241	IGNITION 3 VOLTAGE

5.7L V8 Engine HO2S1/2 (Bank 1 Sensor 2) Connector Pinout

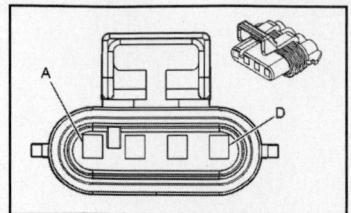 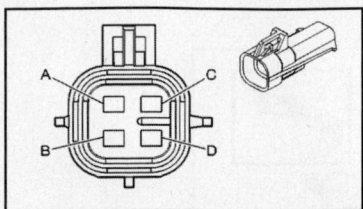

Corvette 5.7L HO2S2/1 Applications   **Firebird 5.7L HO2S2/1 Applications**

PIN	WIRE COLOR	CIRCUIT NO.	FUNCTION
A	TAN	1667	HO2S LOW SIGNAL BANK 2 SENSOR 1
B	PPL	1666	HO2S HIGH SIGNAL BANK 2 SENSOR 1
C	BLK	150	GROUND
D	BRN	241	IGNITION 3 VOLTAGE

5.7L V8 Engine HO2S2/1 (Bank 2 Sensor 1) Connector Pinout

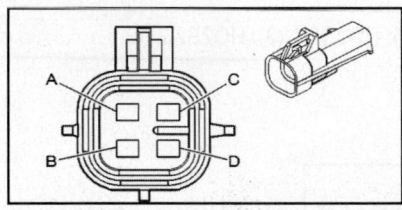

PIN	WIRE COLOR	CIRCUIT NO.	FUNCTION
A	TAN	1671	HO2S LOW SIGNAL BANK 2 SENSOR 2
B	PPL	1670	HO2S HIGH SIGNAL BANK 2 SENSOR 2
C	BLK	150	GROUND
D	BRN	241	IGNITION 3 VOLTAGE

5.7L V8 Engine HO2S2/2 (Bank 2 Sensor 2) Connector Pinout

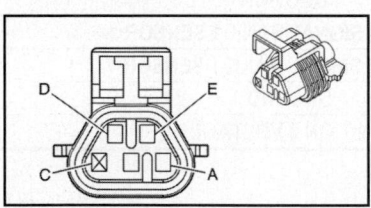

PIN	WIRE COLOR	CIRCUIT NO.	FUNCTION
A	TN	1664	HO2S LOW SIGNAL BANK 1 SENSOR 1
B	PU/WH	1665	HO2S HIGH SIGNAL BANK 1 SENSOR 1
C	--	--	NOT AVAILABLE
D	PK	339	IGNITION 1 VOLTAGE
E	GY/WH	3113	HO2S HEATER LOW CONTROL BANK 1 SENSOR 1

6.0L V8 HO2S1/1 (Bank 1 Sensor 1) Connector Pinout

PIN	WIRE COLOR	CIRCUIT NO.	FUNCTION
A	TN	1667	HO2S LOW SIGNAL BANK 2 SENSOR 1
B	PU	1666	HO2S HIGH SIGNAL BANK 2 SENSOR 1
C	--	--	NOT AVAILABLE
D	PK	339	IGNITION 1 VOLTAGE
E	L-GN	3212	HO2S HEATER LOW CONTROL BANK 2 SENSOR 1

6.0L V8 HO2S2/1 (Bank 2 Sensor 1) Connector Pinout

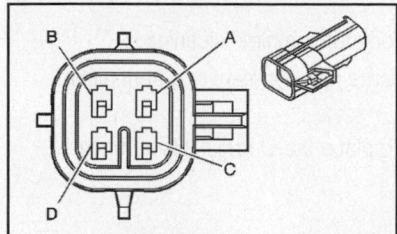

PIN	WIRE COLOR	CIRCUIT NO.	FUNCTION
A	TN/WH	1669	HO2S LOW SIGNAL BANK 1 SENSOR 2
B	PU/WH	1668	HO2S HIGH SIGNAL BANK 1 SENSOR 2
C	BN	2391	HO2S HEATER LOW CONTROL BANK 1 SENSOR 2
D	PK	339	IGNITION 1 VOLTAGE

6.0L V8 HO2S1/2 (Bank 1 Sensor 2) Connector Pinout

PIN	WIRE COLOR	CIRCUIT NO.	FUNCTION
A	TN	1671	HO2S LOW SIGNAL BANK 2 SENSOR 2
B	PU	1670	HO2S HIGH SIGNAL BANK 2 SENSOR 2
C	OG/WH	3223	HO2S HEATER LOW CONTROL BANK 2 SENSOR 2
D	PK	339	IGNITION 1 VOLTAGE

6.0L V8 HO2S2/2 (Bank 2 Sensor 2) Connector Pinout

IDLE AIR CONTROL (IAC) VALVE

Component Locations

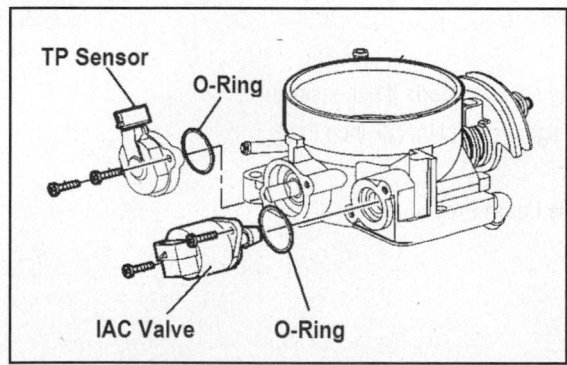

Showing the location of the IAC Valve on Camaro & Firebird 5.7L V8 engine

Removal & Installation

1. Disconnect the IAC valve electrical connector.
2. Remove the IAC valve attaching screws.
3. Remove the IAC valve.
4. Remove the IAC valve O-ring seal.

NOTE: If the IAC valve has been in service: DO NOT push or pull on the IAC valve pintle. The force required to move the pintle may damage the threads on the worm drive. Also, DO NOT soak the IAC valve in any liquid cleaner or solvent, as damage may result.

5. Clean the IAC valve O-ring sealing surface, the pintle valve seat, and the air passage.

6. Clean the IAC valve using GM cleaner 1052626 or GM X-66A. Use a shop towel or parts brush to remove heavy deposits.

7. If the air passage has heavy deposits, remove the throttle body for complete cleaning.

8. Shiny spots on the pintle or seat are normal, and do not indicate misalignment or a bent pintle shaft.

9. Inspect the IAC valve O-ring for cuts, cracks, or distortion. Replace the O-ring if it is damaged.

To install:

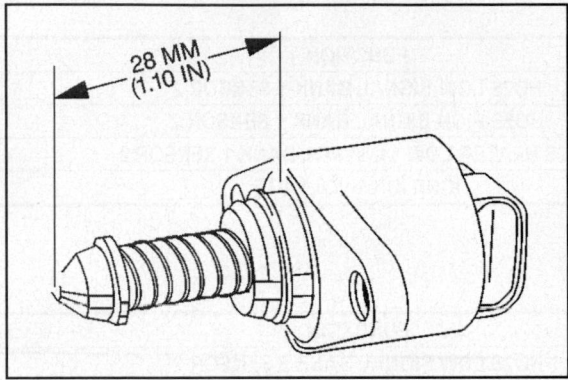

Measuring IAC valve pintle extension

NOTE: If installing a new IAC valve, be sure to replace it with an identical part. The pintle shape of the IAC valve and the diameter of the IAC valve are designed for the specific application.

10. Measure the distance between the tip of the IAC valve pintle and the mounting surface. If the distance is more than 28 mm, use finger pressure to slowly retract the pintle. The force required to retract the pintle of a new valve will not cause damage to the valve.

11. Lubricate the IAC valve O-ring with clean engine oil.

12. Install the IAC valve O-ring on the IAC valve.

13. Install the IAC valve.

14. Apply Loctite 262 to the IAC valve attaching screw threads if necessary.

15. Install the IAC valve attaching screws and tighten to 3 Nm (27 inch lbs.).

16. Connect the IAC valve electrical connector.

17. Perform the Idle Learn Procedure. See "Idle Learn Procedure."

Connector Pinouts

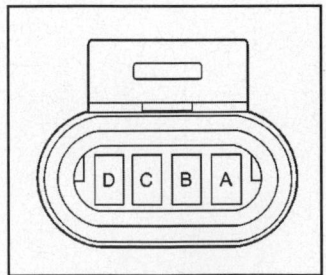

PIN	WIRE COLOR	CIRCUIT NO.	FUNCTION
A	LT GRN/BLK	444	IDLE AIR CONTROL (IAC) VALVE COIL B SIGNAL - LOW
B	LT GRN/WHT	1749	IDLE AIR CONTROL (IAC) VALVE COIL B SIGNAL - HIGH
C	LT BLU/BLK	1748	IDLE AIR CONTROL (IAC) VALVE COIL A SIGNAL - LOW
D	LT BLU/WHT	1747	IDLE AIR CONTROL (IAC) VALVE COIL A SIGNAL - HIGH

Camaro & Firebird 5.7L V8 Engine IAC Valve Connector Pinout

IGNITION CONTROL MODULE (ICM)

Removal & Installation

CORVETTE

1. Disconnect the negative battery cable.
2. Remove the fuel rail cover.
3. Disconnect the ignition coil harness connector.
4. Disconnect the spark plug wire at the ignition coil.
5. Remove the ignition coil mounting bolts.
6. Remove the ignition coil.

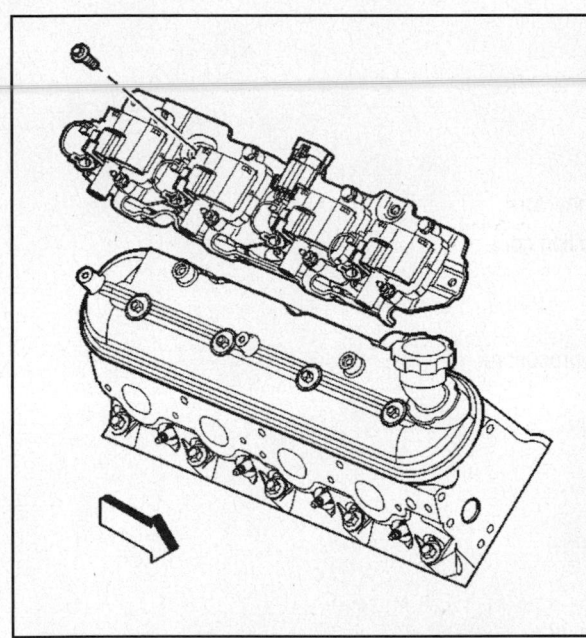

Removing the ignition coil assembly on Corvette 5.7L & 6.0L V8 engines

To install:

7. Install the ignition coil.
8. Install the ignition coil mounting bolts. Tighten the ignition coil mounting bolts to 12 Nm (106 inch lbs.).
9. Connect the spark plug wire at the ignition coil.
10. Connect the ignition coil harness connector. Tighten the harness bolt to 12 Nm (106 inch lbs.).
11. Install the fuel rail cover.
12. Connect the negative battery cable.
13. Program the transmitters.

CAMARO & FIREBIRD

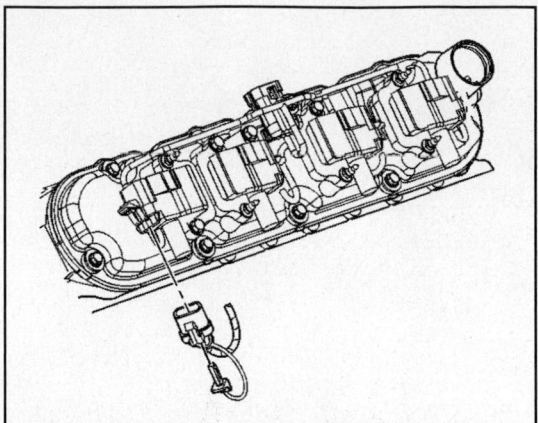

Disconnect the ignition coil connectors

1. Disconnect the ignition coils harness connectors.
2. Disconnect the spark plug wire at the ignition coils.
3. Remove the ignition coil mounting bolts.
4. Remove the ignition coil.
5. Installation is the reverse of the removal procedure.

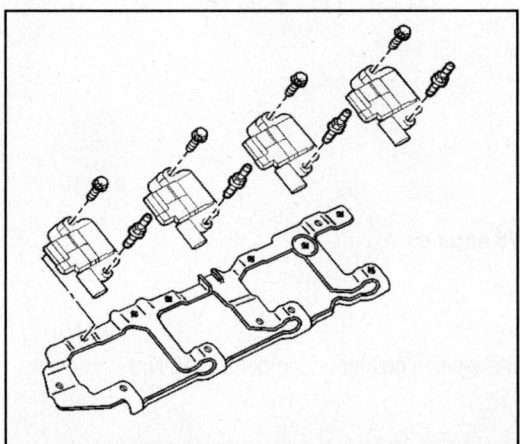

Remove the ignition coils

Connector Pinouts

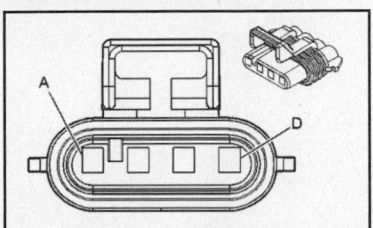

PIN	WIRE COLOR	CIRCUIT NO.	FUNCTION
A	BLK	151	GROUND
B	BRN	2129	LOW REFERENCE
C	PPL	2121	IC 1 CONTROL
C	RED	2127	IC 2 CONTROL
C	LT BLU	2123	IC 3 CONTROL
C	DK GRN	2125	IC 4 CONTROL
C	DK GRN	2125	IC 5 CONTROL
C	LT BLU	2123	IC 6 CONTROL
C	RED	2127	IC 7 CONTROL
C	PPL	2121	IC 8 CONTROL
D	PNK	39	IGNITION 1 VOLTAGE

5.7L & 6.0L V8 Engine Ignition Coil Connector Pinouts

INTAKE AIR TEMPERATURE (IAT) SENSOR

Note: Also see "Mass Air Flow Sensor/Intake Air Temperature (MAF/IAT) Sensor" for 2001 and later models.

Component Locations

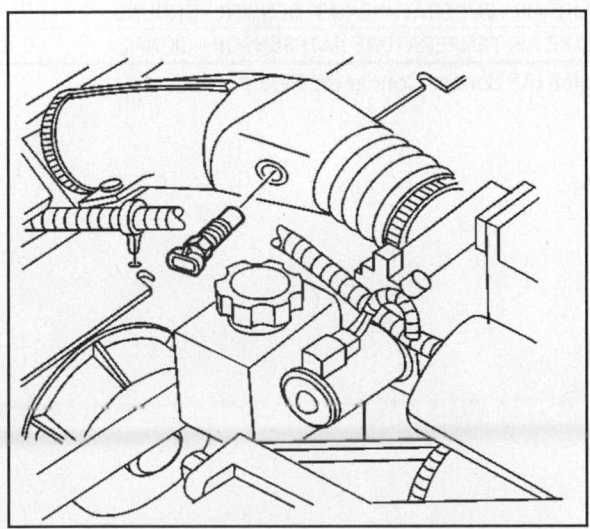

2000 Corvette 5.7L V8 Engine IAT sensor location

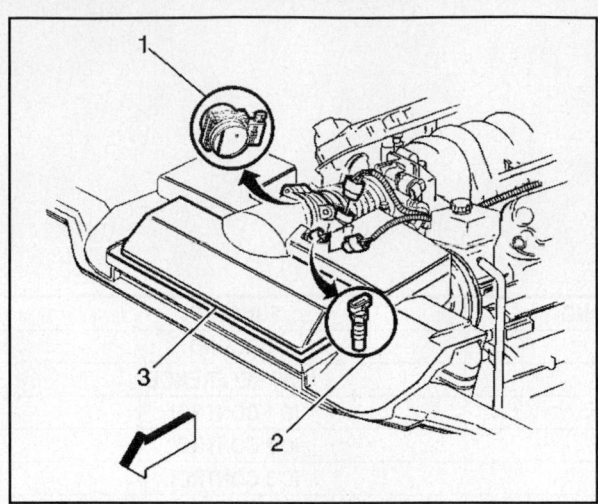

2000-03 Camaro & Firebird 5.7L V8 Engine IAT sensor location

Connector Pinouts

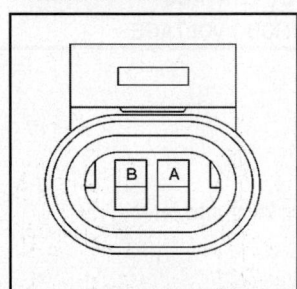

PIN	WIRE COLOR	CIRCUIT NO.	FUNCTION
A	PPL	719	INTAKE AIR TEMPERATURE (IAT) SENSOR - GROUND
B	TAN	472	INTAKE AIR TEMPERATURE (IAT) SENSOR - SIGNAL

2000 Corvette & 2000-03 Camaro & Firebird 5.7L V8 Engine IAT Sensor Connector Pinout

KNOCK SENSOR

Component Locations

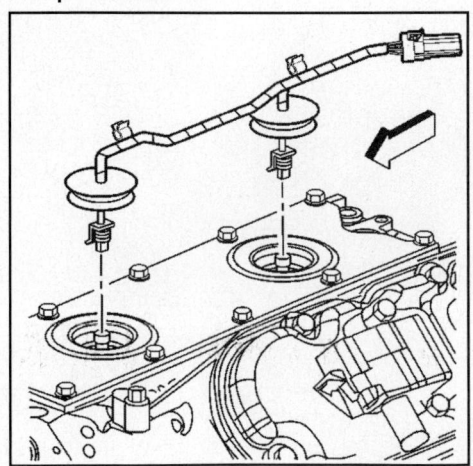

Knock sensor location on 5.7L V8 engines

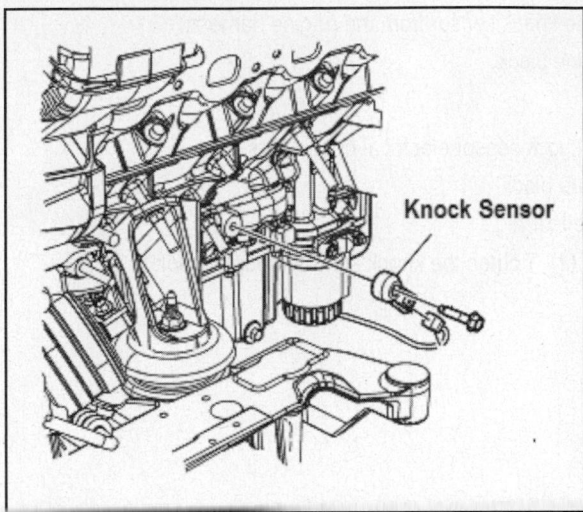

Left knock sensor location of 6.0L V8 engines

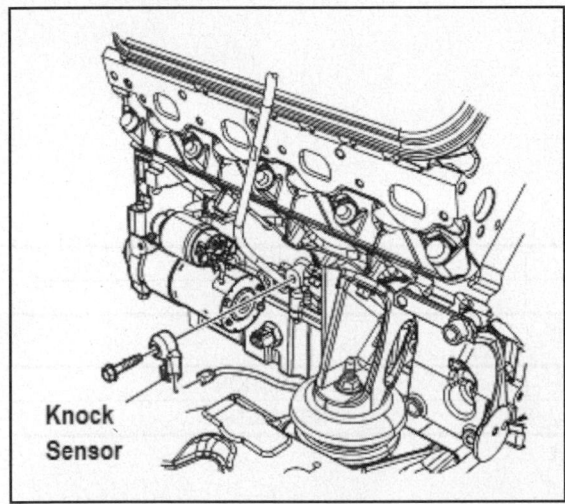

Right knock sensor location of 6.0L V8 engines

Removal & Installation

5.7L V8

1. Remove the intake manifold.
2. Remove the knock sensor wiring harness assembly.
3. Remove the knock sensor.
4. Installation is the reverse of the removal procedure. Tighten the knock sensor to 20 Nm (15 ft. lbs.).

6.0L V8

1. Disconnect the negative battery cable.
2. Remove the left exhaust manifold.
3. Remove the left catalytic converter.
4. On right side, remove starter.
5. Remove the mounting bolt for the knock sensor 1.

6. Disconnect the electrical connector of the knock sensor from the engine harness.

7. Remove the knock sensor from the engine block.

To install:

8. Reconnect the engine harness and the knock sensor electrical connectors.

9. Position the knock sensor 2 on the engine block.

10. Install the mounting bolt for the knock sensor 2.

11. Tighten the knock sensor mounting bolt (1). Tighten the knock sensor mounting bolt to 20 Nm (15 ft. lbs.).

12. On right side, install starter.

13. Install the left catalytic converter.

14. Install the left exhaust manifold.

15. Connect the negative battery cable.

16. Program the transmitters.

Connector Pinouts

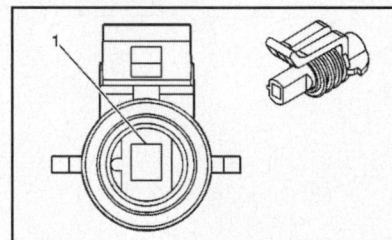

PIN	WIRE COLOR	CIRCUIT NO.	FUNCTION
1	DK BLU	496	KS 1 SIGNAL

Corvette 5.7L V8 Engine Knock Sensor 1 Connector Pinout

PIN	WIRE COLOR	CIRCUIT NO.	FUNCTION
1	LT BLU	1876	KS 2 SIGNAL

Corvette 5.7L V8 Engine Knock Sensor 2 Connector Pinout

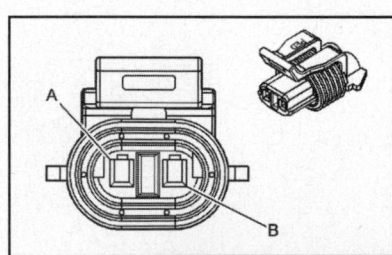

PIN	WIRE COLOR	CIRCUIT NO.	FUNCTION
A	DK BLU	496	KNOCK SENSOR (KS) SIGNAL - FRONT
B	LT BLU	1876	KNOCK SENSOR (KS) SIGNAL - REAR

Camaro & Firebird 5.7L V8 Engine Knock Sensor Connector Pinout

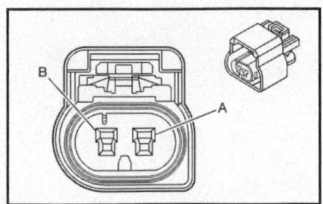

PIN	WIRE COLOR	CIRCUIT NO.	FUNCTION
A	D-BU	496	KS SIGNAL BANK 1
B	GY	1716	KS LOW REFERENCE BANK 1

6.0L V8 Engine Left Knock Sensor Connector Pinout

PIN	WIRE COLOR	CIRCUIT NO.	FUNCTION
A	L-BU	1876	KS SIGNAL BANK 2
B	GY	2303	KS LOW REFERENCE BANK 2

6.0L V8 Engine Right Knock Sensor Connector Pinout

MANIFOLD ABSOLUTE PRESSURE (MAP) SENSOR

Component Locations

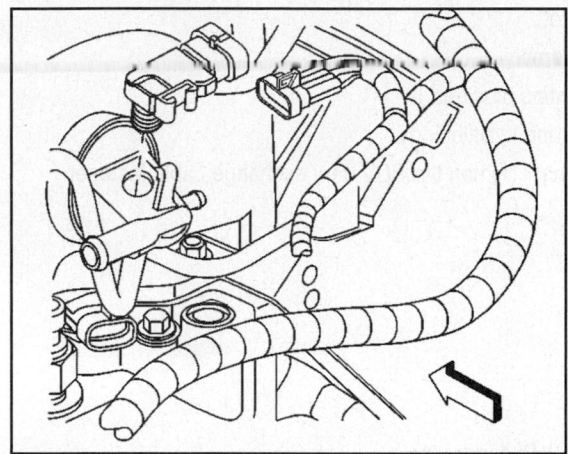

MAP sensor location on 5.7L V8 engines

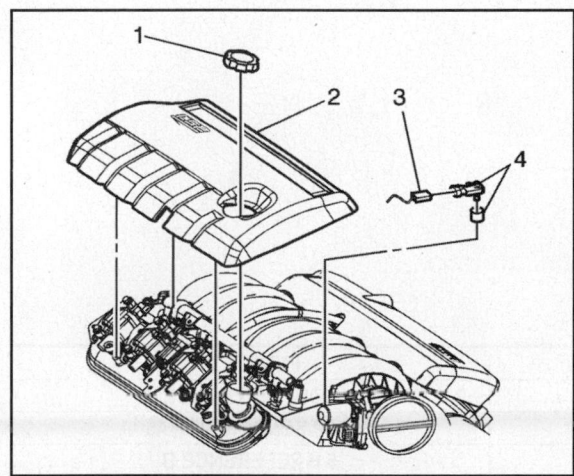

MAP sensor location on 6.0L V8 engines

Removal & Installation

5.7L V8

1. Remove the engine sight shields, if equipped.
2. Remove the fuel rail assembly.

NOTE: Access to the MAP sensor is limited, but does not require the removal of the intake manifold.

3. Remove the PCV heat exchange cable fastener.

4. Remove the PCV hose from the throttle body and right bank port.

5. Release the PCV hose assembly from the mounting brackets.

6. Move the PCV hose assembly aside.

7. Disconnect the MAP sensor harness connector.

8. Pull the MAP sensor forward in order to release the sensor from the retainer.

9. Lift the MAP sensor upward.

To install:

10. Inspect the MAP sensor seal for damage. Replace the seal if necessary.

11. Lightly coat the MAP sensor seal with clean engine oil before installing the sensor.

12. Install the MAP sensor. Push down the sensor in order to engage the sensor into the retainer.

13. Connect the MAP sensor harness connector.

14. Position the PCV hose assembly for reassembly.

15. Insert the PCV hose assembly to the mounting brackets.

16. Install the PCV hose to the right bank port and throttle body.

17. Install the PCV heat exchange cable fastener. Tighten the PCV heat exchange cable fastener to 12 Nm (106 inch lbs.).

18. Install the fuel rail.

19. Install the engine sight shields.

6.0L V8

1. Remove the oil filler cap (1).

2. Remove the right engine sight shield (2).

3. Remove the manifold absolute pressure (MAP) sensor (4).

4. Disconnect the electrical connector (3) for the MAP sensor.

5. Installation is the reverse of the removal procedure.

Connector Pinouts

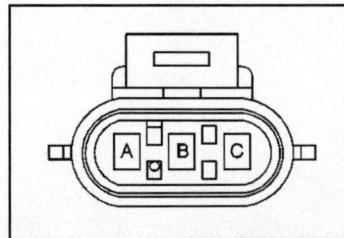

PIN	WIRE COLOR	CIRCUIT NO.	FUNCTION
A	TAN OR ORN/BLK	407 OR 469	LOW REFERENCE
B	LT GRN	432	MAP SENSOR SIGNAL
C	GRY	416 OR 2704	5 V REFERENCE B

5.7L & 6.0L V8 Engine MAP Sensor Connector Pinout

MASS AIR FLOW (MAF) SENSOR

Removal & Installation

2000-03 CAMARO & FIREBIRD 5.7L V8 ENGINES

CAUTION: Take care when handling the MAF sensor. Do not dent, puncture, or otherwise damage the Honeycell located at the air inlet end of the MAF sensor. Do not touch the sensing elements or

allow anything including cleaning solvents and lubricants to come in contact with them. Use a small amount of a non-silicone based lubricant on the air duct only, to aid in installation. Do not drop or roughly handle the MAF sensor.

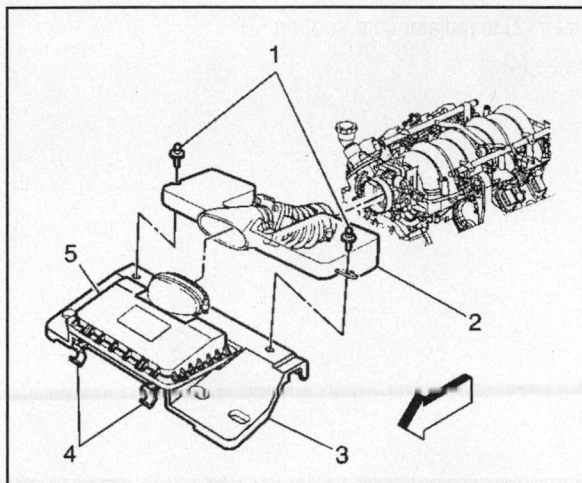

Removing air cleaner housing from radiator support and engine

1. Release Air Cleaner Housing front latches (4).
2. Remove the retainers (1) holding the resonator to radiator core support (3).
3. Disconnect the MAF sensor (1) electrical connector.
4. Remove the clamp from the MAF sensor (1) at the resonator.
5. Remove the clamp from the MAF sensor (1) at the air intake duct.
6. Remove the MAF sensor (1).

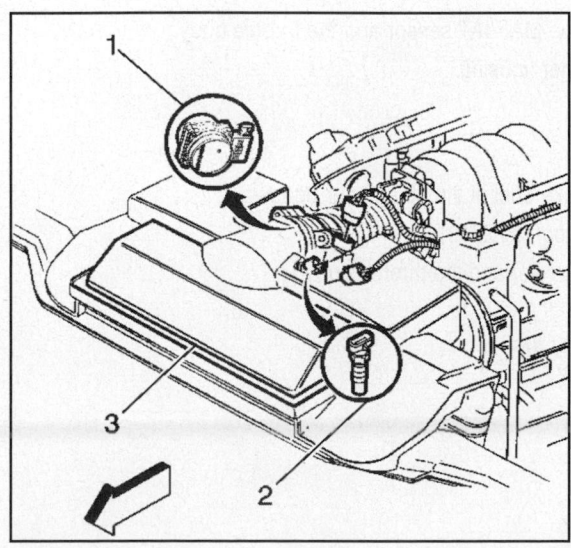

Removing the MAF sensor (1)

To install:

NOTE: The embossed arrows on the MAF sensor indicate the proper air flow direction. The arrows must point toward the engine.

7. Install the MAF sensor (1) into the air intake duct and resonator.

8. Install the clamp at the MAF sensor (1) to the air intake duct.

9. Install the clamp at the MAF sensor (1) to the resonator.

10. Install the retainers (1) holding the resonator (2) to radiator core support (3).

11. Secure the Air Cleaner Housing front latches (4).

12. Connect the electrical connector.

Connector Pinouts

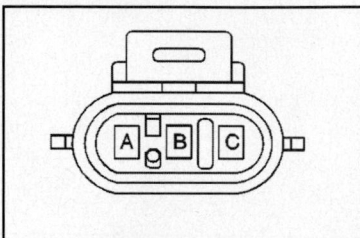

PIN	WIRE COLOR	CIRCUIT NO.	FUNCTION
A	YEL	492	MASS AIR FLOW (MAF) SENSOR SIGNAL
B	BLK/WHT	451	GROUND
C	PNK	539	FUSED IGNITION FEED

Camaro & Firebird 5.7L V8 Engine MAF Sensor Connector Pinout

MASS AIR FLOW/INTAKE AIR TEMPERATURE (MAF/IAT) SENSOR

Removal & Installation

2001-05 CORVETTE 5.7L V8

1. Disconnect the electrical connector of the MAF/IAT sensor.

2. Remove the air intake duct clamps from the MAF/IAT sensor and the throttle body.

3. Remove the air duct clamp at the air cleaner housing.

4. Remove the MAF/IAT sensor.

To install:

5. Install the MAF/IAT sensor into the air intake duct at the air cleaner housing.

6. Install the new clamp using the clamping tool (J 22610).

7. Install the air intake duct to the MAF/IAT sensor and the throttle body.

8. Install the clamps.

9. Connect the electrical connector of the MAF/IAT.

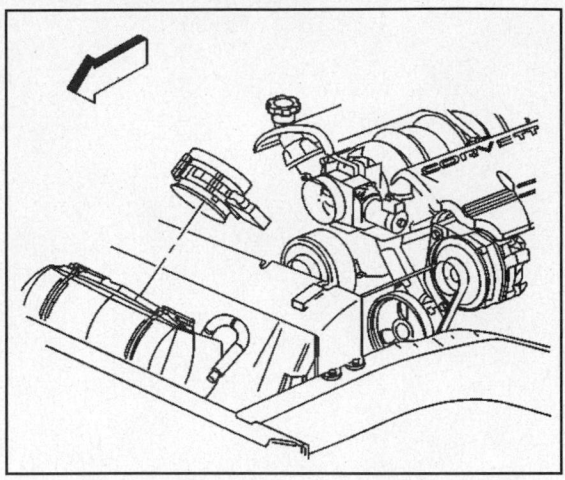

Showing the location of the MAF/IAT sensor (in intake air duct) on 2001-03 5.7L V8 engines

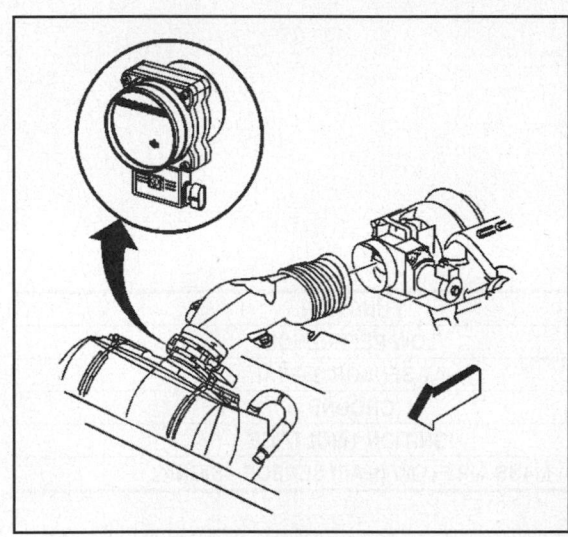

Installing the MAF/IAT sensor, with clamp in proper position

6.0L V8

1. Disconnect the electrical connector (1) for the MAF/IAT sensor (5).
2. Remove the lines (2) from the air duct (4).
3. Loosen the air intake duct clamps (3).
4. Remove the air intake duct (4) from the throttle body.
5. Remove the MAF/IAT sensor (5) from the air cleaner housing (6).
6. Installation is the reverse of the removal procedure.

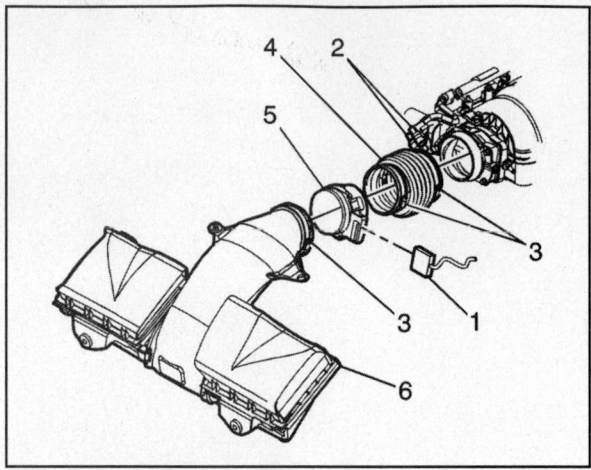

Showing the location of the MAF/IAT sensor (1) on 6.0L V8 engines

Connector Pinouts

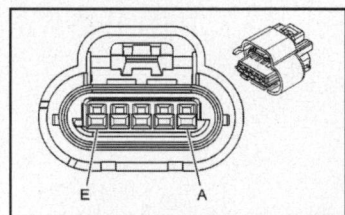

PIN	WIRE COLOR	CIRCUIT NO.	FUNCTION
A	PPL	719	LOW REFERENCE
B	TAN	472	IAT SENSOR SIGNAL
C	BLK/WHT	451	GROUND
D	PNK	339	IGNITION 1 VOLTAGE
E	YEL	492	MASS AIR FLOW (MAF) SENSOR - SIGNAL

5.7L V8 Engine MAF/IAT Sensor Connector Pinout

PIN	WIRE COLOR	CIRCUIT NO.	FUNCTION
A	TN	2760	INTAKE AIR TEMPERATURE SENSOR LOW REFERENCE
B	TN	472	INTAKE AIR TEMPERATURE SENSOR SIGNAL
C	BK/WH	451	GROUND
D	PK	539	12-VOLT REFERENCE
E	YE	492	MASS AIR FLOW SENSOR SIGNAL

6.0L V8 Engine MAF/IAT Sensor Connector Pinout

POWERTRAIN CONTROL MODULE (PCM)

***NOTE: the Corvette 5.7L V8 engine uses a PCM, while 6.0L V8 engine uses an ECM. For 6.0L V8
applications, see "Engine Control Module (ECM)".***

Component Locations

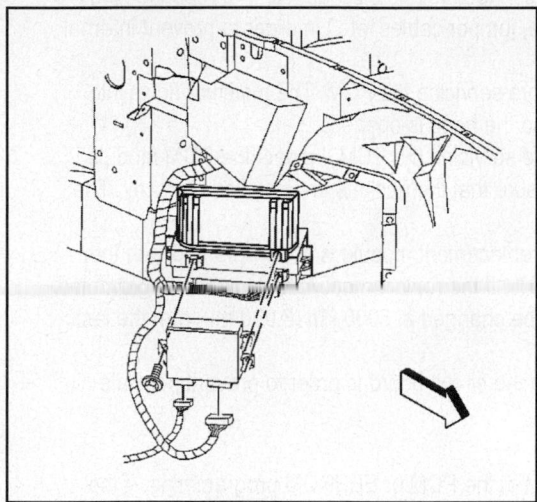

Showing PCM location on 5.7L V8 engines

Removal & Installation

CORVETTE 5.7L V8 ENGINES

1. Using a scan tool, retrieve the percentage of remaining engine oil. Record the remaining engine oil life.
2. Remove the wheelhouse filler panel.
3. Remove the throttle actuator control (TAC) module. See "Throttle Actuator Control (TAC) Module".
4. Disconnect the powertrain control module (PCM) electrical harness connectors.
5. Loosen but do not remove the PCM rear retaining fastener. Use the rear retaining fastener as an anchor for the outer bracket.
6. Remove the front retaining fastener from the PCM.
7. Reposition the PCM outer bracket.
8. Remove the PCM from the bracket and the vehicle.

To install:

9. Install the PCM to the PCM rear bracket.
10. Position the PCM front bracket.
11. Install the PCM front retaining fasteners. Tighten the PCM retaining fasteners to 2.0 Nm (17 inch lbs.).
12. Connect the electrical connectors to the PCM. Tighten the PCM electrical connectors to 8 Nm (70 inch lbs.).
13. Install the TAC module. See "Throttle Actuator Control (TAC) Module".
14. Install the wheelhouse filler panel.
15. If a new PCM is being installed, program the PCM.

CAMARO & FIREBIRD 5.7L V8 ENGINES

NOTES AND CAUTIONS:

- THE SERVICE PCM EEPROM WILL NOT BE PROGRAMMED. DTC P0601 and P0602 indicate the EEPROM is not programmed or has malfunctioned.
- Turn the ignition OFF when installing or removing the PCM connectors and disconnecting or reconnecting the power to the PCM (battery cable, PCM pigtail, PCM fuse, jumper cables, etc.) in order to prevent internal PCM damage.
- Ensure that the hood is free of contaminates (moisture) before servicing the PCM. The moisture flows into the PCM connector body when the PCM is disconnected and the hood is opened.
- Remove any debris from the PCM connector surfaces before servicing the PCM. Inspect the PCM module connector gaskets when diagnosing/replacing the PCM. Ensure that the gaskets are installed correctly. The gaskets prevent contaminate intrusion into the PCM.
- It is necessary to record the remaining engine oil life. If the replacement module is not programed with the remaining engine oil life, the engine oil life will default to 100%. If the replacement module is not programmed with the remaining engine oil life, the engine oil will need to be changed at 5000 km (3,000 mi) from the last engine oil change.
- Do not touch the connector pins or soldered components on the circuit board in order to prevent possible electrostatic discharge (ESD) damage to the PCM.

Service of the PCM should normally consist of either replacement of the PCM or EEPROM programming. If the diagnostic procedures call for the PCM to be replaced, the PCM should be checked first to see if it is the correct part. If it is, remove the faulty PCM and install the new service PCM.

The following must be performed anytime the PCM is replaced:

- Program the PCM
- The Idle Learn Procedure
- The CKP System Variation Learn Procedure
- The Engine Oil Life Reset Procedure
- The Functional Check

The following must be performed anytime the PCM is disconnected, reprogrammed or loses power:

- The Idle Learn Procedure
- The Functional Check

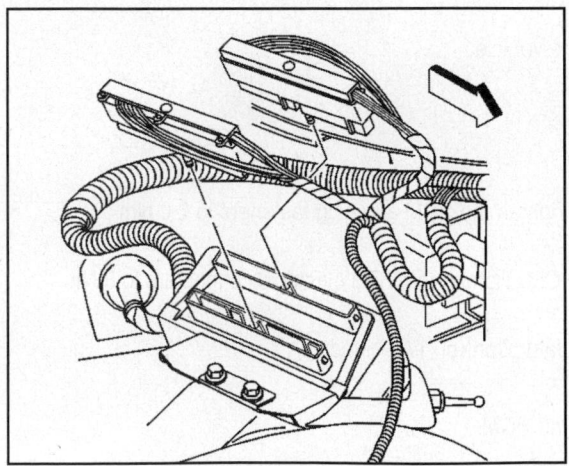

Removing PCM connectors on Firebird with 5.7L V8 engine

Removal & Installation

1. Using a scan tool, retrieve the percentage of remaining engine oil. Record the remaining engine oil life.
2. Disconnect the negative battery cable.
3. Disconnect the PCM connectors.
4. Remove the mounting fasteners from the PCM bracket.
5. Remove the PCM and mounting bracket assembly from the engine compartment.
6. Remove the PCM from the mounting bracket.

To install:

WARNING: Do not touch the connector pins or soldered components on the circuit board in order to prevent possible electrostatic discharge (ESD) damage to the PCM.

7. Install the PCM to the mounting bracket.
8. Install the PCM and mounting bracket assembly into the vehicle. Tighten the mounting bracket fasteners to 7 Nm (62 inch lbs.).
9. Reconnect the PCM connectors. Tighten the PCM connector end fasteners to 8 Nm (70 inch lbs.).
10. Connect the negative battery cable.
11. If a new PCM is being installed, reprogram the PCM. See "PCM Reprogramming."

Connector Pinouts

2000-2003 CORVETTE 5.7L V8 C1 CONNECTOR

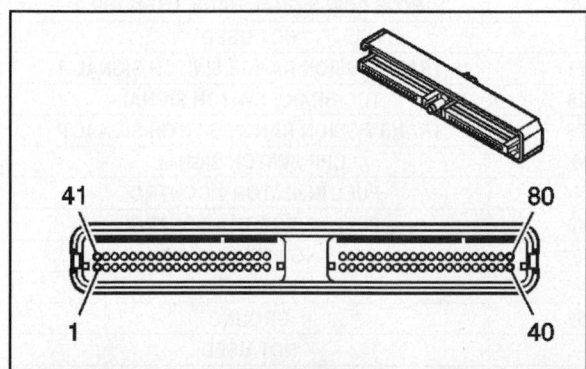

PIN	WIRE COLOR	CIRCUIT NO.	FUNCTION
1	BLK/WHT	451	GROUND
2	LT GRN	1867	12 VOLT REFERENCE
3	PNK/BLK	1746	FUEL INJECTOR 3 CONTROL
4	LT GRN/BLK	1745	FUEL INJECTOR 2 CONTROL
5-6	--	--	NOT USED
7	GRY	596	5 VOLT REFERENCE
8-10	--	--	NOT USED
11	LT BLU	1876	KNOCK SENSOR 2 SIGNAL
12	DK BLU/WHT	1869	CRANKSHAFT POSITION (CKP) SENSOR SIGNAL
13	ORN/BLK	463	REQUESTED TORQUE SIGNAL
14	TAN	800	THROTTLE ACTUATOR CONTROL SERIAL DATA
15	ORN/BLK	1061	THROTTLE ACTUATOR CONTROL SERIAL DATA

2000-03 5.7L V8 PCM C1 Connector Pinout (1 of 3)

2000-2003 CORVETTE 5.7L V8 C1 CONNECTOR

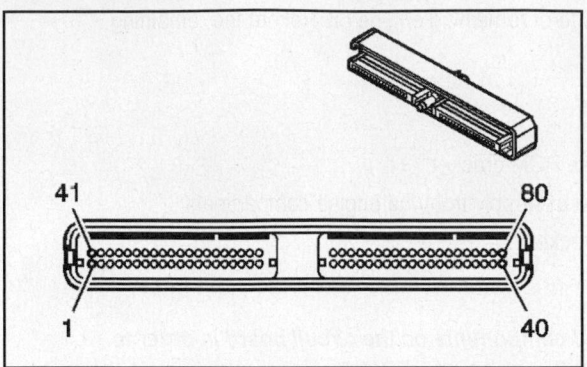

PIN	WIRE COLOR	CIRCUIT NO.	FUNCTION
16	--	--	NOT USED
17	DK BLU	1225	TFP SWITCH SIGNAL B
18	RED	1226	TFP SWITCH SIGNAL C
19	PNK	239	IGNITION 1 VOLTAGE
20	ORN	340	BATTERY POSITIVE VOLTAGE
21	YEL/BLK	1868	LOW REFERENCE
22	--	--	NOT USED
23	GRY	720	LOW REFERENCE
24	--	--	NOT USED
25	TAN	1671	HO2S LOW SIGNAL BANK 2 SENSOR 2
26	TAN	1667	HO2S LOW SIGNAL BANK 2 SENSOR 1
27	--	--	NOT USED
28	TAN/WHT	1669	HO2S LOW SIGNAL BANK 1 SENSOR 2
29	TAN/WHT	1653	HO2S LOW SIGNAL BANK 1 SENSOR 1
30-31	--	--	NOT USED
32	BLK/WHT	771	TRANSMISSION RANGE SWITCH SIGNAL A
33	PPL	420	TCC BRAKE SWITCH SIGNAL
34	WHT	776	TRANSMISSION RANGE SWITCH SIGNAL P
35	GRY	48	CPP SWITCH SIGNAL
36	BLK	1744	FUEL INJECTOR 1 CONTROL
37	YEL/BLK	846	FUEL INJECTOR 6 CONTROL
38	PNK/WHT	1101	DAMPING LIFT/DIVE SIGNAL
39	--	--	NOT USED
40	BLK/WHT	451	GROUND
41	--	--	NOT USED
42	DK GRN	335	LOW SPEED COOLING FAN RELAY CONTROL
43	RED/BLK	877	FUEL INJECTOR 7 CONTROL
44	LT BLU/BLK	844	FUEL INJECTOR 4 CONTROL
45	GRY	474	5 VOLT REFERENCE
46	GRY	598	5 VOLT REFERENCE
47	--	--	NOT USED
48	GRY	416	5 VOLT REFERENCE
49-50	--	--	NOT USED
51	DK BLU	496	KNOCK SENSOR 1 SIGNAL
52	--	--	NOT USED
53	ORN/BLK	1057	LOW REFERENCE
54	ORN/BLK	469	LOW REFERENCE

2000-03 5.7L V8 PCM C1 Connector Pinout (2 of 3)

2000-2003 CORVETTE 5.7l v8 C1 Connector (Continued)

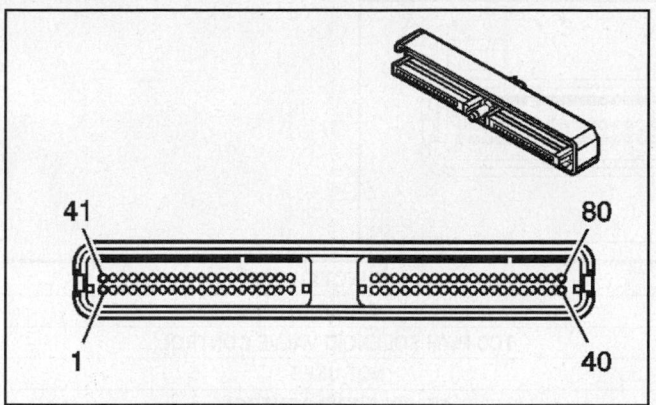

PIN	WIRE COLOR	CIRCUIT NO.	FUNCTION
55-56	--	--	NOT USED
57	ORN	340	BATTERY POSITIVE VOLTAGE
58	DK GRN	1049	PCM CLASS 2 SERIAL DATA
59	--	--	NOT USED
60	BLK	407	LOW REFERENCE
61	PNK/BLK	632	LOW REFERENCE
62	GRY	847	EXTENDED TRAVEL BRAKE SWITCH SIGNAL
63	BLK	452	LOW REFERENCE
64	--	--	NOT USED
65	PPL	1670	HO2S HIGH SIGNAL BANK 2 SENSOR 2
66	PPL	1666	HO2S HIGH SIGNAL BANK 2 SENSOR 1
67	--	--	NOT USED
68	PPL/WHT	1668	HO2S HIGH SIGNAL BANK 1 SENSOR 2
69	PPL/WHT	1665	HO2S HIGH SIGNAL BANK 1 SENSOR 1
70	BRN	1174	OIL LEVEL SWITCH SIGNAL
71	--	--	NOT USED
72	YEL	772	TRANSMISSION RANGE SWITCH SIGNAL B
73	BRN/WHT	633	CAMSHAFT POSITION (CMP) SENSOR SIGNAL
74	YEL	410	ECT SENSOR SIGNAL
75	--	--	NOT USED
76	BLK/WHT	845	FUEL INJECTOR 5 CONTROL
77	DK BLU/WHT	878	FUEL INJECTOR 8 CONTROL
78	--	--	NOT USED
79	GRY	587	SKIP SHIFT SOLENOID CONTROL
79	WHT	687	3-2 SHIFT SOLENOID VALVE CONTROL
80	BRN	718	LOW REFERENCE

2000-03 5.7L V8 PCM C1 Connector Pinout (3 of 3)

2000-2003 CORVETTE 5.7L V8 C2 CONNECTOR

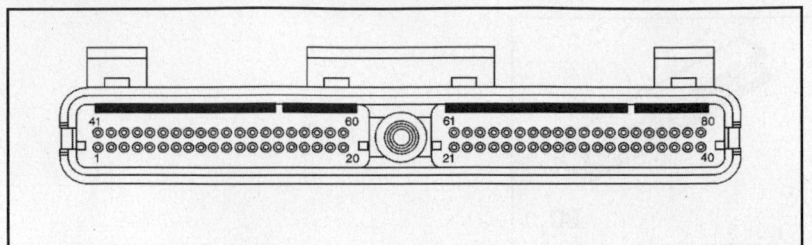

PIN	WIRE COLOR	CIRCUIT NO.	FUNCTION
1	BLK/WHT	451	GROUND
2	BRN	418	TCC PWM SOLENOID VALVE CONTROL
3	--	--	NOT USED
4	PPL	421	AIR SOLENOID CONTROL
5	TAN/BLK	464	DELIVERED TORQUE SIGNAL
6	RED/BLK	1228	PC SOLENOID VALVE HIGH CONTROL
7	--	--	NOT USED
8	LT BLU/WHT	1229	PC SOLENOID VALVE LOW CONTROL
9	DK GRN/WHT	465	FUEL PUMP RELAY CONTROL
10	WHT	121	ENGINE SPEED SIGNAL
11-13	--	--	NOT USED
14	RED/BLK	380	A/C REFRIGERANT PRESSURE SENSOR SIGNAL
15	RED	225	GENERATOR TURN ON SIGNAL
16	--	--	NOT USED
17	DK GRN/WHT	762	A/C REQUEST SIGNAL (C60 ONLY)
18	DK GRN	59	A/C COMPRESSOR CLUTCH SUPPLY VOLTAGE
19	--	--	NOT USED
20	PPL	401	SIGNAL LOW - FRONT
21	YEL	400	SIGNAL HIGH - FRONT
22-24	--	--	NOT USED
25	TAN	472	IAT SENSOR SIGNAL
26	PPL	2121	IC 1 CONTROL
27	RED	2127	IC 7 CONTROL
28	LT BLU/WHT	2126	IC 6 CONTROL
29	DK GRN/WHT	2124	IC 4 CONTROL
30	--	--	NOT USED
31	YEL	492	MAF SENSOR SIGNAL
32	LT GRN	432	MAP SENSOR SIGNAL
33	DK BLU	473	HIGH SPEED COOLING FAN RELAY CONTROL
34	DK GRN/WHT	428	EVAP CANISTER PURGE SOLENOID CONTROL
35	--	--	NOT USED
36	BRN	436	AIR PUMP RELAY CONTROL
37-38	--	--	NOT USED
39	RED	631	12 VOLT REFERENCE
40	BLK/WHT	451	GROUND
41	--	--	NOT USED
42	TAN/BLK	422	TCC SOLENOID VALVE CONTROL
43	DK GRN/WHT	459	A/C COMPRESSOR CLUTCH RELAY CONTROL
44	LT GRN	1652	REVERSE LOCKOUT SOLENOID CONTROL
45	WHT	1310	EVAP CANISTER VENT SOLENOID CONTROL
46	BRN/WHT	419	MALFUNCTION INDICATOR LAMP (MIL) CONTROL
47	WHT	375	SKIP SHIFT INDICATOR CONTROL (MM6)
47	YEL/BLK	1223	2-3 SHIFT SOLENOID VALVE CONTROL (M30)
48	LT GRN	1222	1-2 SHIFT SOLENOID VALVE CONTROL
49	--	--	NOT USED
50	DK GRN/WHT	817	VEHICLE SPEED SIGNAL

2000-03 5.7L V8 PCM C2 Connector Pinout (1 of 2)

2000-2003 CORVETTE 5.7L V8 C2 CONNECTOR (CONTINUED)

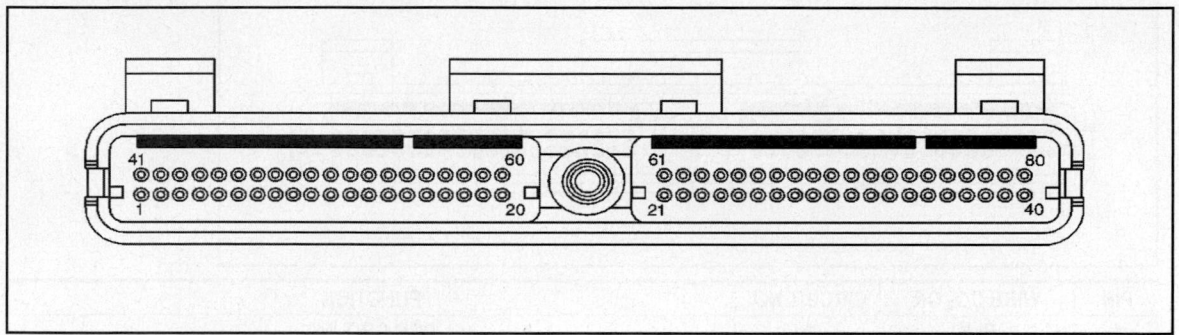

PIN	WIRE COLOR	CIRCUIT NO.	FUNCTION
51	YEL/BLK	1227	TFT SENSOR SIGNAL
52	GRY	23	GENERATOR FIELD DUTY CYCLE SIGNAL
53	--	--	NOT USED
54	DK BLU	1936	FUEL LEVEL SENSOR SIGNAL
55-56	--	--	NOT USED
57	PPL	719	LOW REFERENCE
58	TAN/WHT	331	OIL PRESSURE SENSOR SIGNAL
59	--	--	NOT USED
60	BRN	2129	LOW REFERENCE
61	BRN/WHT	2130	LOW REFERENCE
62	GRY	773	TRANSMISSION RANGE SWITCH SIGNAL C
63	PNK	1224	TRANSMISSION FLUID PRESSURE SWITCH SIGNAL A
64	DK GRN	890	FUEL TANK PRESSURE SENSOR SIGNAL
65	--	--	NOT USED
66	PPL/WHT	2128	IC 8 CONTROL
67	RED/WHT	2122	IC 2 CONTROL
68	DK GRN	2125	IC 5 CONTROL
69	LT BLU	2123	IC 3 CONTROL
70-72	--	--	NOT USED
73	LT BLU	1937	FUEL LEVEL SENSOR SIGNAL (SECONDARY)
74-80	--	--	NOT USED

2000-03 5.7L V8 PCM C2 Connector Pinout (2 of 2)

2000-2003 CAMARO & FIREBIRD 5.7L V8 C1 CONNECTOR

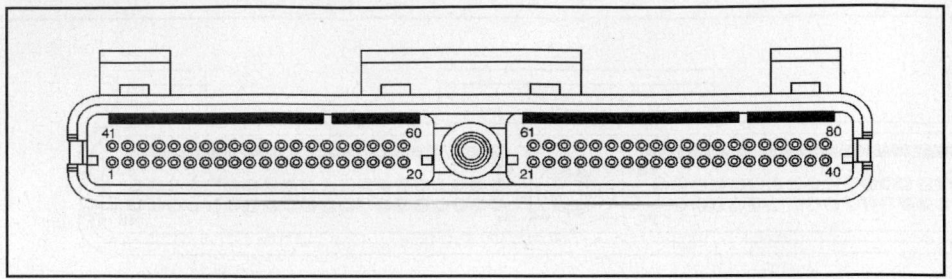

PIN	WIRE COLOR	CIRCUIT NO.	FUNCTION
1	BLK	451	PCM GROUND
2	LT GRN	1867	CRANKSHAFT POSITION SENSOR B+ SUPPLY
3	PNK/BLK	1746	INJECTOR 3 CONTROL
4	LT GRN/BLK	1745	INJECTOR 2 CONTROL
5-7	--	--	NOT USED
8	GRA	596	TP SENSOR 5V REFERENCE
9-10	--	--	NOT USED
11	LT BLU	1876	KNOCK SENSOR SIGNAL REAR
12	DK BLU/WHT	1869	CRANKSHAFT POSITION SENSOR SIGNAL
13-16	--	--	NOT USED
17	DK BLU	1225	TRANSMISSION RANGE SIGNAL B
18	RED	1226	TRANSMISSION RANGE SIGNAL C
19	PNK	439	IGNITION POSITIVE VOLTAGE
20	ORN	340	BATTERY POSITIVE VOLTAGE
21	YEL/BLK	1868	CRANKSHAFT POSITION SENSOR REFERENCE LOW
22	--	--	NOT USED
23	GRA	720	FUEL TANK PRESSURE SENSOR/FUEL TANK SENDER GROUND
24	--	--	NOT USED
25	TAN	1671	HO2S SIGNAL LOW BANK 2 SENSOR 2
26	TAN	1667	HO2S SIGNAL LOW BANK 2 SENSOR 1
27	--	--	NOT USED
28	TAN/WHT	1669	HO2S SIGNAL LOW BANK 1 SENSOR 2
29	TAN/WHT	1653	HO2S SIGNAL LOW BANK 1 SENSOR 1
30-31	--	--	NOT USED
32	GRY	48	CLUTCH PEDAL POSITION SWITCH SIGNAL
33	PPL	420	TCC BRAKE SWITCH
34	ORN/BLK	434	PNP SWITCH SIGNAL
35	--	--	NOT USED
36	BLK	1744	INJECTOR 1 CONTROL
37	YEL/BLK	846	INJECTOR 6 CONTROL
38-39	--	--	NOT USED
40	BLK	451	PCM GROUND
41	BLK	407	EGR PINTLE POSITION SENSOR GROUND
42	DK GRN	335	ENGINE COOLING FAN RELAY 1 CONTROL
43	RED/BLK	877	INJECTOR 7 CONTROL
44	LT BLU/BLK	844	INJECTOR 4 CONTROL
45	GRA	474	A/C REFRIGERANT PRESSURE SENSOR 5V REFERENCE
46	GRA	474	FUEL TANK PRESSURE SENSOR 5V REFERENCE
47	GRA	416	EGR PINTLE POSITION SENSOR 5V REFERENCE
48	GRA	416	MAP SENSOR 5V REFERENCE
49-50	--	--	NOT USED
51	DK BLU	496	KNOCK SENSOR SIGNAL FRONT
52	--	--	NOT USED
53	BLK	407	TRANSMISSION TEMPERATURE SENSOR GROUND
54	ORN/BLK	407	MAP SENSOR GROUND

Camaro & Firebird 5.7L V8 Engine PCM C1 Connector Pinout (1 of 2)

2000-2003 CAMARO & FIREBIRD 5.7L V8 C1 CONNECTOR (CONTINUED)

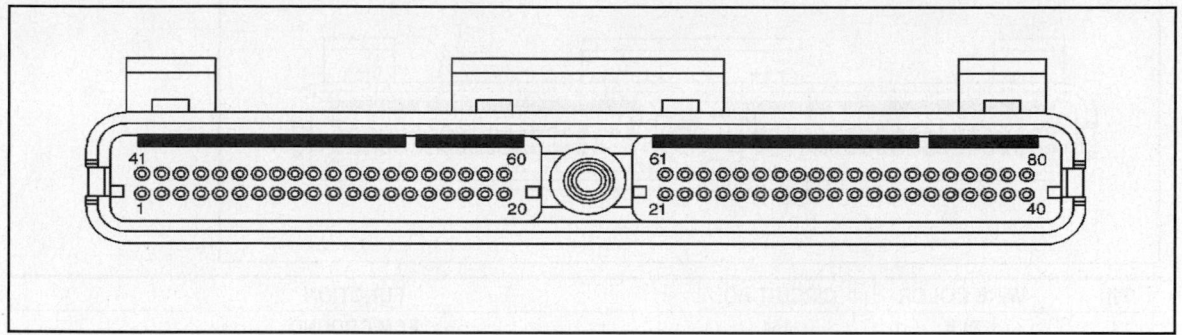

PIN	WIRE COLOR	CIRCUIT NO.	FUNCTION
55	BRN	1456	EGR PINTLE POSITION SENSOR SIGNAL
56	--	--	NOT USED
57	ORN	340	BATTERY POSITIVE VOLTAGE
58	DK GRN	1049	SERIAL DATA
59	--	--	NOT USED
60	BLK	452	TP SENSOR GROUND
61	PNK/BLK	632	CAMSHAFT POSITION SENSOR REFERENCE LOW
62-64	--	--	NOT USED
65	PPL	1670	HO2S SIGNAL HIGH BANK 2 SENSOR 2
66	PPL	1666	HO2S SIGNAL HIGH BANK 2 SENSOR 1
67	--	--	NOT USED
68	PPL/WHT	1668	HO2S SIGNAL HIGH BANK 1 SENSOR 2
69	PPL/WHT	1665	HO2S SIGNAL HIGH BANK 1 SENSOR 1
70	BRN	1174	LOW OIL LEVEL SWITCH
71-72	--	--	NOT USED
73	BRN/WHT	633	CAMSHAFT POSITION (CMP) SENSOR SIGNAL
74	YEL	410	ENGINE COOLANT TEMPERATURE (ECT) SENSOR SIGNAL
75	--	--	NOT USED
76	BLK/WHT	845	INJECTOR 5 CONTROL
77	DK BLU/WHT	878	INJECTOR 8 CONTROL
78	--	--	NOT USED
79	GRA OR WHT	587 OR 687	SKIP SHIFT SOLENOID CONTROL (M/T) OR 3-2 SHIFT SOLENOID CONTROL (A/T)
80	BLK	407	ENGINE COOLANT TEMPERATURE (ECT) SENSOR GROUND

Camaro & Firebird 5.7L V8 Engine PCM C1 Connector Pinout (2 of 2)

2000-2003 CAMARO & FIREBIRD 5.7L V8 C2 CONNECTOR

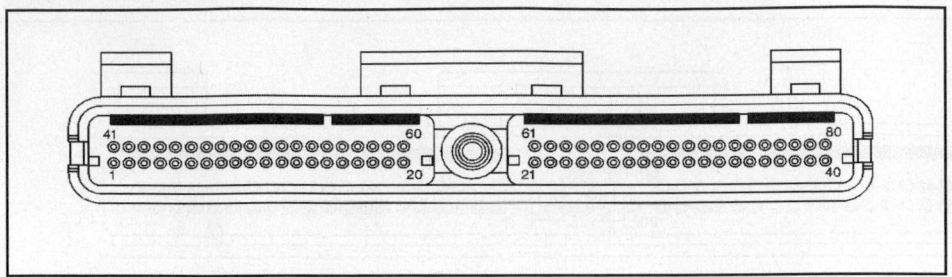

PIN	WIRE COLOR	CIRCUIT NO.	FUNCTION
1	BLK	451	PCM GROUND
2	BRN	418	TCC CONTROL SOLENOID
3	--	--	NOT USED
4	PPL	421	AIR SOLENOID RELAY CONTROL
5	--	--	NOT USED
6	RED/BLK	1228	TRANSMISSION FLUID PRESSURE CONTROL SOLENOID HIGH
7	RED	1676	EGR CONTROL
8	LT BLU/WHT	1229	TRANSMISSION FLUID PRESSURE CONTROL SOLENOID LOW
9	DK GRN/WHT	465	FUEL PUMP RELAY CONTROL
10	WHT	121	ENGINE SPEED (TACH) OUTPUT SIGNAL
11-12	--	--	NOT USED
13	WHT	85	CRUISE CONTROL ENABLE SIGNAL
14	RED/BLK	380	A/C REFRIGERANT PRESSURE SENSOR SIGNAL
15	RED	225	ALTERNATOR L TERMINAL
16	--	--	NOT USED
17	DK GRN/WHT	762	A/C REQUEST SIGNAL
18	DK GRN	59	A/C STATUS SIGNAL
19	--	--	NOT USED
20	LT GRN/BLK	822	VEHICLE SPEED SENSOR (VSS) REFERENCE LOW
21	PPL/WHT	821	VEHICLE SPEED SENSOR (VSS) SIGNAL
22-23	--	--	NOT USED
24	DK BLU	417	TP SENSOR SIGNAL
25	TAN	472	IAT SENSOR SIGNAL
26	PPL	2121	IGNITION CONTROL 1
27	RED	2127	IGNITION CONTROL 7
28	LT BLU/WHT	2126	IGNITION CONTROL 6
29	DK GRN/WHT	2124	IGNITION CONTROL 4
30	DK BLU	229	VTD FUEL ENABLE SIGNAL
31	YEL	492	MAF SENSOR SIGNAL
32	LT GRN	432	MAP SENSOR SIGNAL
33	DK BLU	473	ENGINE COOLING FAN RELAY 2 AND 3 CONTROL
34	DK GRN/WHT	428	EVAP CANISTER PURGE VALVE CONTROL
35	--	--	NOT USED
36	BRN	436	AIR PUMP RELAY CONTROL
37	DK GRN	83	CRUISE CONTROL INHIBIT
38	--	--	NOT USED
39	RED	631	CAMSHAFT POSITION SENSOR B+ SUPPLY
40	BLK	451	PCM GROUND
41	GRA	435	EGR POSITION SENSOR GROUND
42	TAN/BLK	422	TCC ENABLE CIRCUIT
43	DK GRN/WHT	459	A/C CLUTCH RELAY CONTROL
44	LT GRN	1652	REVERSE INHIBIT SOLENOID CONTROL
45	WHT	1310	EVAP CANISTER VENT VALVE CONTROL
46	BRN/WHT	419	MALFUNCTION INDICATOR LAMP (MIL) CONTROL
47	YEL/BLK	1223	TRANSMISSION SHIFT SOLENOID B

Camaro & Firebird 5.7L V8 Engine PCM C2 Connector Pinout (1 of 2)

2000-2003 CAMARO & FIREBIRD 5.7L V8 C2 Connector (CONTINUED)

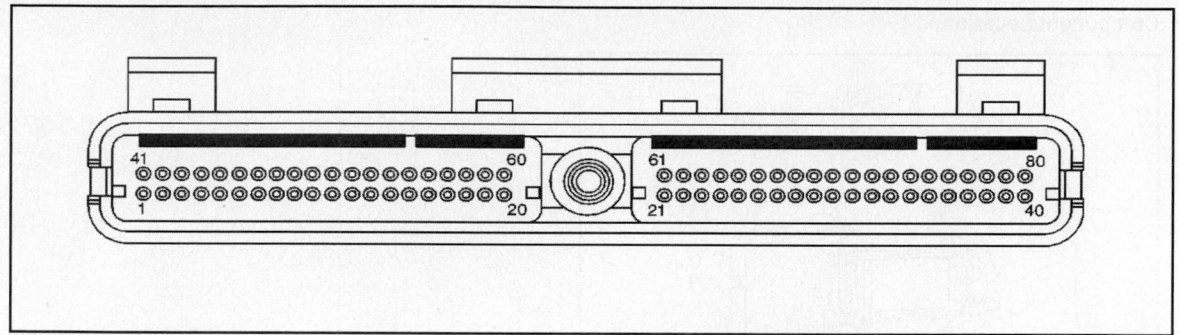

PIN	WIRE COLOR	CIRCUIT NO.	FUNCTION
48	LT GRN	1222	TRANSMISSION SHIFT SOLENOID A
49	--	--	NOT USED
50	DK GRN/WHT	817	VEHICLE SPEED OUTPUT CIRCUIT
51	YEL/BLK	1227	TRANSMISSION TEMPERATURE SENSOR SIGNAL
52	--	--	NOT USED
53	GRA/BLK	1687	SPARK RETARD SIGNAL
54	PPL	1589	FUEL LEVEL SENSOR SIGNAL
55-56	--	--	NOT USED
57	PPL	719	IAT SENSOR GROUND
58-59	--	--	NOT USED
60	BRN	2129	IGNITION CONTROL REFERENCE LOW BANK 1
61	BRN/WHT	2130	IGNITION CONTROL REFERENCE LOW BANK 2
62	--	--	NOT USED
63	PNK	1224	TRANSMISSION RANGE SIGNAL A
64	DK GRN	890	FUEL TANK PRESSURE SENSOR SIGNAL
65	--	--	NOT USED
66	PPL/WHT	2128	IGNITION CONTROL 8
67	RED/WHT	2122	IGNITION CONTROL 2
68	DK GRN	2125	IGNITION CONTROL 5
69	LT BLU	2123	IGNITION CONTROL 3
70-75	--	--	NOT USED
76	LT GRN/WHT	1749	IAC COIL B HIGH
77	LT GRN/BLK	444	IAC COIL B LOW
78	LT BLU/BLK	1748	IAC COIL A LOW
79	LT BLU/WHT	1747	IAC COIL A HIGH
80	--	--	NOT USED

Camaro & Firebird 5.7L V8 Engine PCM C2 Connector Pinout (2 of 2)

THROTTLE ACTUATOR CONTROL (TAC) MODULE

Component Locations

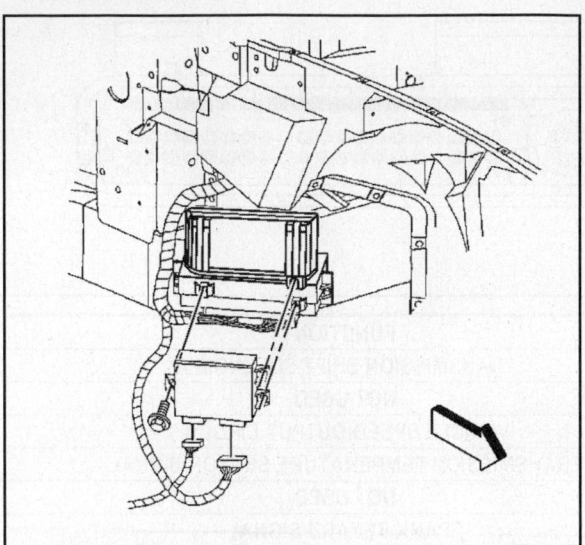

TAC module is located in same mounting with PCM on Corvette 5.7L V8 engines

Removal & Installation

1. Remove the wheelhouse filler panel.

2. Remove the fasteners retaining the throttle actuator control (TAC) module to the powertrain control module (PCM) mounting bracket.

3. Remove any debris from the TAC module connector surfaces before servicing the TAC module. Inspect the TAC module connector gaskets if you diagnose or replace the TAC module. Verify that the gaskets are installed correctly. The gaskets prevent contaminate intrusion into the TAC module.

CAUTION: Do not touch the connector pins in order to prevent possible electrostatic discharge (ESD) damage to the TAC module.

4. The ignition should always be OFF if you install or remove the TAC module connectors.

5. Remove the electrical connectors from the TAC module.

6. Remove the TAC module from the vehicle.

To install:

7. Connect the electrical connectors to the TAC module.

8. Align the hole in the rear mounting tab of the TAC module to the corresponding hole in the PCM mounting bracket. Position the TAC module below the rear mounting tab for greater clearance in order to install the rear retaining fastener.

9. Install the TAC module rear retaining fastener.

10. Align the front mounting holes of the TAC module.

11. Install the TAC module front retaining fasteners. Tighten the TAC module retaining fasteners to 2.0 Nm (17 inch lbs.).

12. Install the wheelhouse filler panel.

Connector Pinouts

C1 CONNECTOR

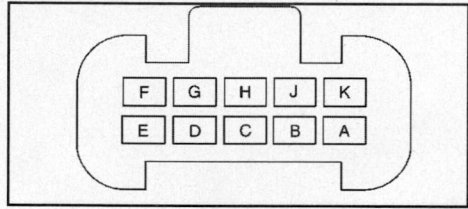

PIN	WIRE COLOR	CIRCUIT NO.	FUNCTION
A	GRY	1273	LOW REFERENCE
B	PPL	1272	LOW REFERENCE
C	LT BLU	1162	APP SENSOR 2 SIGNAL
D	TAN	1274	5 VOLT REFERENCE
E	YEL/BLK	1275	5 VOLT REFERENCE
F	DK BLU	1161	APP SENSOR 1 SIGNAL
G	LT BLU	1276	5 VOLT REFERENCE
H	--	--	NOT USED
J	BRN	1271	LOW REFERENCE
K	DK GRN	1163	APP SENSOR 3 SIGNAL

Corvette 5.7L V8 Engine TAC Module C1 Connector Pinout

C2 CONNECTOR

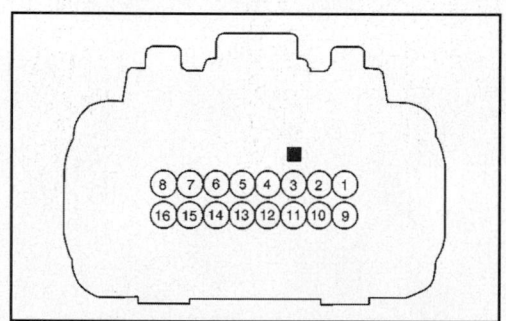

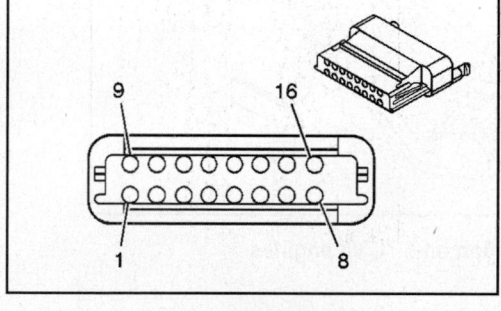

2000-02 Corvette TAC C2 Connector **2003 Corvette TAC C2 Connector**

PIN	WIRE COLOR	CIRCUIT NO.	FUNCTION
1	DK BLU	417	TP SENSOR 1 SIGNAL
2	DK GRN	485	5 VOLT REFERENCE
3	PPL	486	LOW REFERENCE
4	DK BLU	84	CRUISE CONTROL SET/COAST SWITCH SIGNAL
5	GRY/BLK	87	CRUISE CONTROL RESUME/ACCEL SWITCH SIGNAL
6	LT BLU	20	STOPLAMP SUPPLY VOLTAGE
7	PNK	539	IGNITION 1 VOLTAGE
8	BRN	582	TAC MOTOR CONTROL - 2
9	YEL/BLK	487	5 VOLT REFERENCE
10	WHT	484	LOW REFERENCE
11	PNK	427	TP SENSOR 2 SIGNAL
12	TAN	800	UART SERIAL DATA PRIMARY
13	ORN/BLK	1061	UART SERIAL DATA SECONDARY
14	GRY	397	CRUISE CONTROL ON SIGNAL
15	BLK/WHT	451	GROUND
16	YEL	581	TAC MOTOR CONTROL - 1

Corvette 5.7L V8 TAC Module C2 Connector Pinout

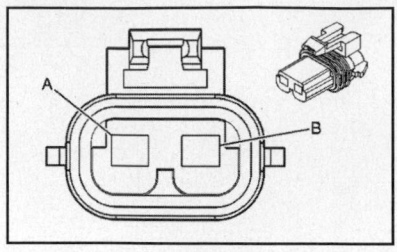

PIN	WIRE COLOR	CIRCUIT NO.	FUNCTION
A	YEL	581	TAC MOTOR CONTROL - 1
B	BRN	582	TAC MOTOR CONTROL - 2

Corvette 5.7L V8 TAC Module Motor Connector Pinout

SECONDARY AIR INJECTOR (AIR) PUMP

Component Locations

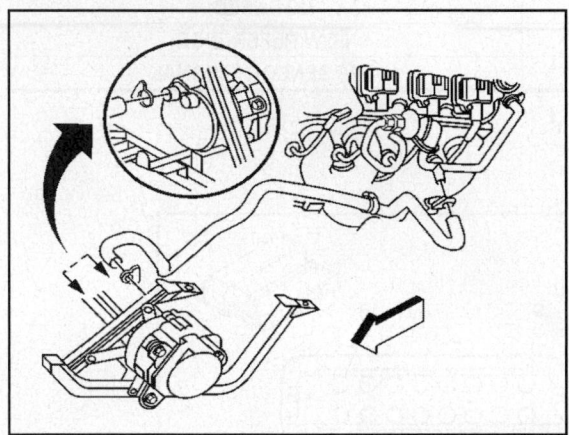

AIR Pump location on 5.7L V8 engines

Removal & Installation

1. Raise the vehicle.
2. Remove the left hand side lower close-out panel.
3. Remove the hoses from the AIR pump.
4. Disconnect the electrical connector from the AIR pump.
5. Remove the AIR pump mounting bolts from the bracket.
6. Remove the AIR pump.

To install:

7. Install the AIR pump.
8. Install the mounting bolts. Tighten the AIR Pump bolts to 9 Nm (80 inch lbs.).
9. Connect the electric connector to AIR pump.
10. Connect the hoses to the AIR pump.
11. Install the left hand side close-out panel.
12. Lower the vehicle.
13. Verify the AIR system for proper operation.

Connector Pinouts

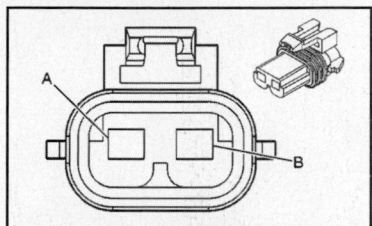

PIN	WIRE COLOR	CIRCUIT NO.	FUNCTION
A	RED	78	AIR PUMP SUPPLY VOLTAGE
B	BLK	150	GROUND

Corvette 5.7L V8 AIR Pump Connector Pinout

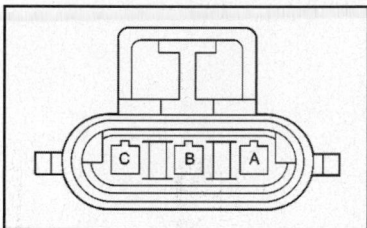

PIN	WIRE COLOR	CIRCUIT NO.	FUNCTION
A	RED OR ORN	78	SECONDARY AIR INJECTION PUMP MOTOR - FEED
B	PNK/BLK	429	SECONDARY AIR INJECTION SOLENOID - FEED
C	BLK	150	GROUND

Camaro & Firebird 5.7L V8 AIR Pump Connector Pinout

SECONDARY AIR INJECTION (AIR) SOLENOID

Component Locations

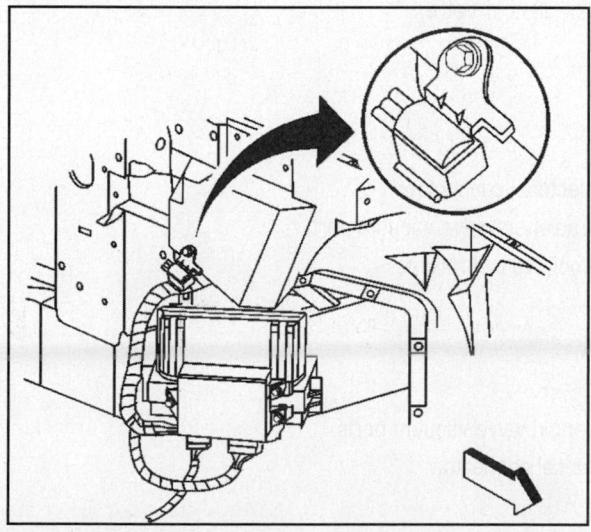

AIR system solenoid location on Corvette 5.7L V8 engines

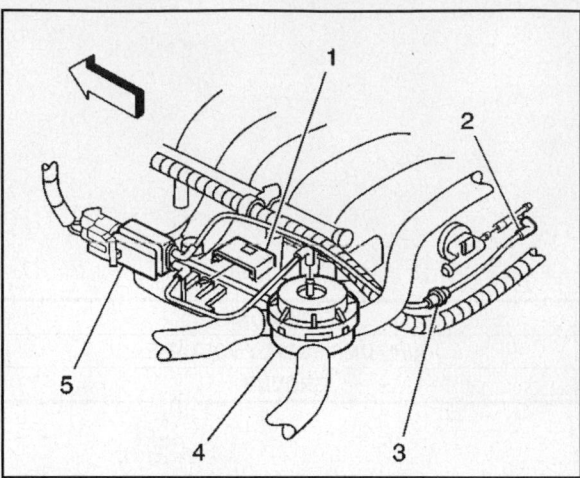

AIR system solenoid location on Camaro & Firebird 5.7L V8 engines

Removal & Installation

CORVETTE

1. Remove the right wheelhouse filler panel.
2. Remove the vacuum hoses from the Secondary Air Injection (AIR) Solenoid Valve.
3. Remove the electrical connector from the AIR Solenoid Valve.
4. Remove the AIR Solenoid Valve retaining nut.
5. Remove the AIR Solenoid Valve.

To install:

6. Install the AIR Solenoid Valve.
7. Install the AIR Solenoid Valve retaining nut. Tighten the AIR Solenoid Valve retaining nut to 7 Nm (62 inch lbs.).
8. Install the vacuum hoses to the AIR Solenoid Valve.
9. Install the electrical connector to the AIR Solenoid Valve.
10. Install the right wheelhouse filler panel.

CAMARO & FIREBIRD

1. Disconnect the AIR solenoid valve (5) electrical connector.
2. Remove the vacuum hose from the AIR solenoid valve vacuum port.
3. Depress the locking tab for the AIR solenoid valve bracket.
4. Remove the AIR solenoid valve.

To install:

5. Install the AIR solenoid valve.
6. Install the vacuum hoses to the AIR Solenoid valve vacuum ports.
7. Connect the AIR solenoid valve (5) electrical connector.

Connector Pinouts

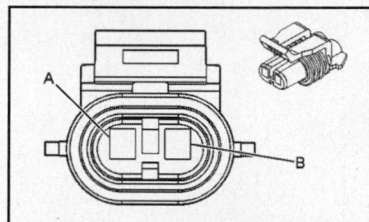

PIN	WIRE COLOR	CIRCUIT NO.	FUNCTION
A	PNK	339	IGNITION 1 VOLTAGE
B	PPL	421	AIR SOLENOID CONTROL

Corvette 5.7L V8 AIR Solenoid Connector Pinout

PIN	WIRE COLOR	CIRCUIT NO.	FUNCTION
A	PNK/BLK	429	SECONDARY AIR INJECTION BLEED VALVE SOLENOID FEED
B	BLK	150	GROUND

Camaro & Firebird 5.7L V8 AIR Solenoid Connector Pinout

THROTTLE POSITION (TP) SENSOR

Component Locations

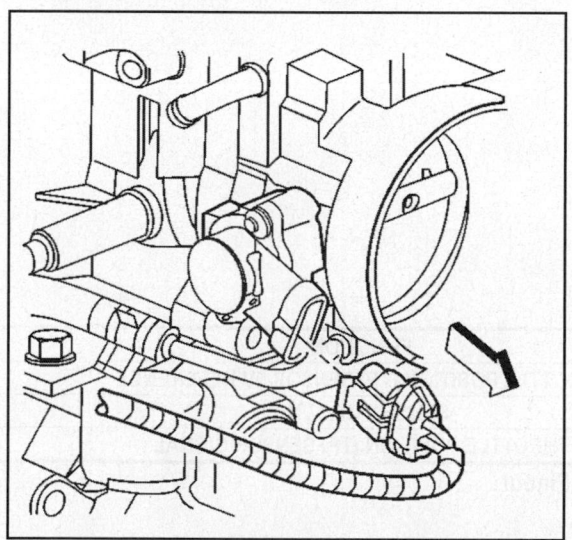

TP Sensor location on Corvette 5.7L V8 engines

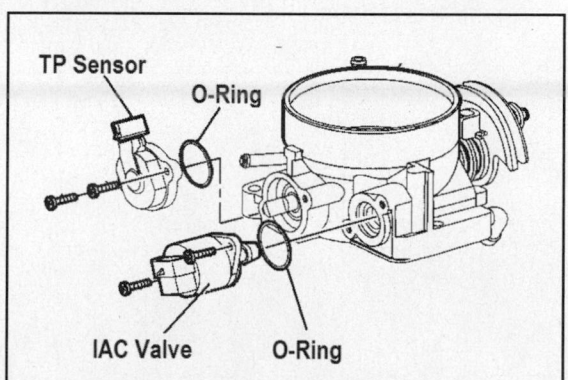

TP Sensor location on Camaro & Firebird 5.7L V8 engines

Connector Pinouts

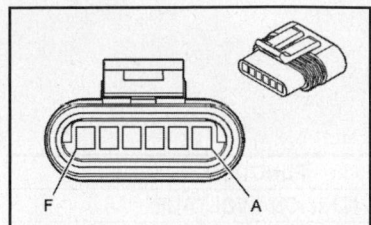

PIN	WIRE COLOR	CIRCUIT NO.	FUNCTION
A	DK GRN	485	5 VOLT REFERENCE
B	PPL	486	LOW REFERENCE
C	DK BLU	417	TP SENSOR 1 SIGNAL
D	YEL/BLK	487	5 VOLT REFERENCE
E	WHT	484	LOW REFERENCE
F	PNK	427	TP SENSOR 2 SIGNAL

Corvette 5.7L V8 TP Sensor Connector Pinout

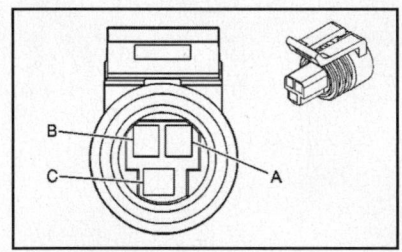

PIN	WIRE COLOR	CIRCUIT NO.	FUNCTION
A	GRY	596	THROTTLE POSITION (TP) SENSOR 5V REFERENCE
B	BLK OR TAN	452	SENSOR GROUND
C	DK BLU	417	THROTTLE POSITION (TP) SENSOR SIGNAL

Camaro & Firebird 5.7L V8 Engine TP Sensor Connector Pinout

GENERAL MOTORS CARS: 2000-01 CADILLAC CATERA

A/C REFRIGERANT PRESSURE SWITCH

Component Locations

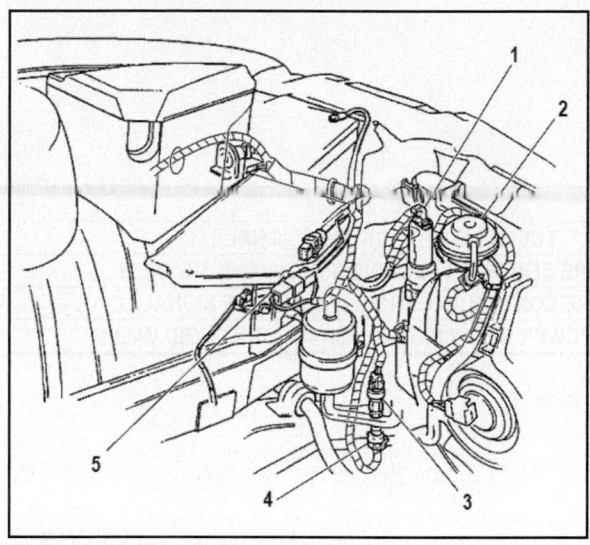

(1) Water Auxiliary Pump

(2) Secondary Air Injection Switch-Over Valve

(3) A/C Load Switch

(4) A/C Compressor Refrigerant Pressure Switch

(5) Engine Coolant Fan Resistor

Showing the location of the A/C compressor pressure switch and related components

Connector Pinouts

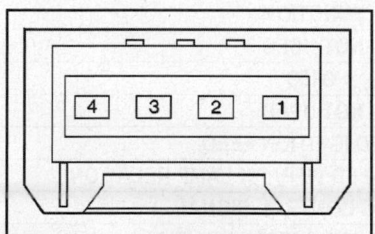

PIN	WIRE COLOR	CIRCUIT NO.	FUNCTION
1	BLK/WHT	FM99	A/C REQUEST OUTPUT SIGNAL
2	BLK/WHT	FM7	A/C REQUEST SIGNAL
3	BRN/WHT	XM120	B+ FROM RELAY K28
4	BLK	F57	GROUND TO G103

2000-01 3.0L A/C Refrigerant Pressure Switch Connector Pinout

A/T ADAPTER CASE

Connector Pinouts

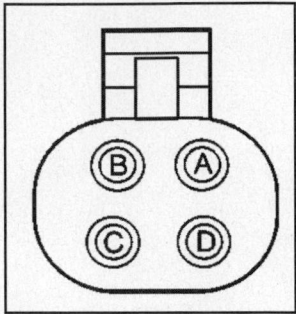

PIN	WIRE COLOR	CIRCUIT NO.	FUNCTION
A	BLK/YEL	FB738	TCC SOLENOID CONTROL SIGNAL
B	BLK/PPL	FY740	PRESSURE CONTROL SOLENOID (PCS) VALVE SIGNAL HI
C	GRY/RED	RA741	PRESSURE CONTROL SOLENOID (PCS) VALVE SIGNAL LO
D	BLU/BLK	PF854	12 V TCM POWER RELAY SIGNAL FOR TCC SOLENOID VALVE

2000-01 A/T Adapter Case Connector Pinout

A/T CONTROL INDICATOR

Connector Pinouts

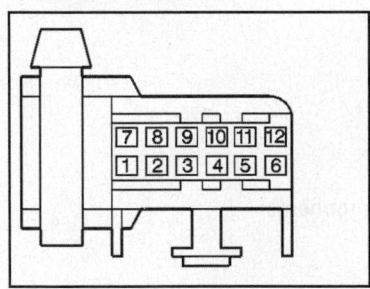

PIN	WIRE COLOR	CIRCUIT NO.	FUNCTION
1	--	--	NOT USED
2	BLK	F970	G103
3-4	--	--	NOT USED
5	BRN	X90	FUSED IGNITION FEED
6	GRY/GRN	RU500	FUSED BATTERY FEED W/ PARK LAMP RELAY ON
7	GRY/YEL	RB70	DIMMING CONTROL SIGNAL
8	BLK/BLU	FP400	TRANS RANGE SWITCH INPUT A
9	BLK/GRN	FU400	TRANS RANGE SWITCH INPUT B
10	BLK/YEL	FB400	TRANS RANGE SWITCH INPUT C
11	GRY	R400	TRANS RANGE SWITCH INPUT P
12	--	--	NOT USED

2000-01 A/T Control Indicator Connector Pinout

A/T MAIN CASE

Connector Pinouts

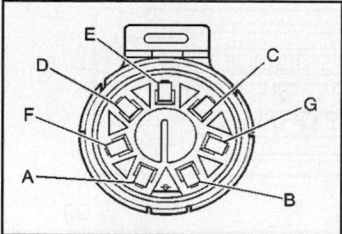

PIN	WIRE COLOR	CIRCUIT NO.	FUNCTION
A	BLU/YEL	PR743	2-3 SHIFT SOLENOID CONTROL
B	GRY/BLK	RF745	BAND APPLY SOLENOID CONTROL
C	BLU/WHT	PM748	1-2 AND 3-4 SHIFT SOLENOID CONTROL
D	BLU/BLK	PF954	12 V TCM POWER RELAY SIGNAL FOR SOLENOIDS
E	BRN/GRN	XU716	TFT SENSOR RETURN
F	BLK/BLU	FP722	5 V REFERENCE SIGNAL FOR TFT SENSOR
G	--	--	NOT USED

2000-01 A/T Main Case Connector Pinout

A/T RANGE SWITCH

Connector Pinouts

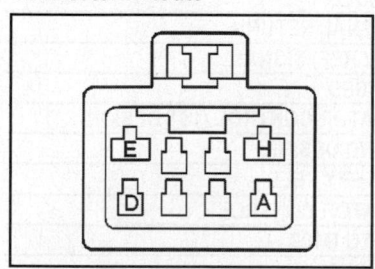

PIN	WIRE COLOR	CIRCUIT NO.	FUNCTION
A	BLK/BLU	FP733	TRANS RANGE SWITCH OUTPUT A
B	BLK/GRN	FU726	TRANS RANGE SWITCH OUTPUT B
C	BLK/YEL	FB708	TRANS RANGE SWITCH OUTPUT C
D	BRN	X756	FUSED IGNITION FEED
E	BLK/RED	FA501	P/N SWITCH OUTPUT TO STARTER SOLENOID
F	WHT/BLK	MF757	R SIGNAL TO REAR VIEW MIRROR
G	GRY	R733	TRANS RANGE SWITCH OUTPUT P
H	BLK/RED	FA500	IGNITION SWITCH START (POSITION III) SIGNAL

2000-01 A/T Range Switch Connector Pinout

A/T TRANSMISSION CONTROL MODULE (TCM)

Connector Pinouts

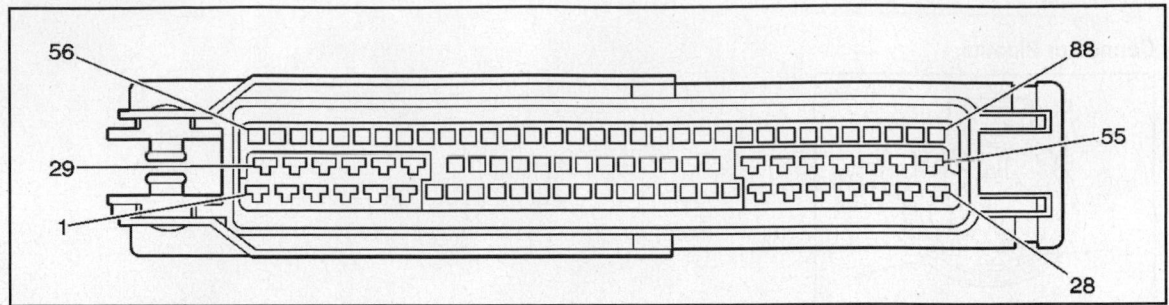

PIN	WIRE COLOR	CIRCUIT NO.	FUNCTION
1-3	--	--	NOT USED
4	GRY/BLK	RF745	BAND APPLY SOLENOID CONTROL
5	GRY/RED	RA741	PRESSURE CONTROL SOLENOID (PCS) LO SIGNAL
6	BLK	F819	GROUND TO G103
7	--	--	NOT USED
8	BLK/GRN	FU400	TRANS RANGE SWITCH INPUT B
9	GRY	R400	TRANS RANGE SWITCH INPUT P
10-11	--	--	NOT USED
12	BLU	P100	SPORT MODE SWITCH INPUT
13	GRN	U102	WINTER MODE SWITCH INPUT
14	BLU/GRN	PU714	VSS LO SIGNAL
15	BLK	S40	VSS SHIELDED GROUND
16	--	--	NOT USED
17	BLK/YEL	FB800	SPORT MODE & TRANS MALFUNCTION INDICATOR CONTROL (2ND DESIGN)
18-20	--	--	NOT USED
21	BRN/GRN	XU716	TFT SENSOR SIGNAL RETURN
22	BLK/BLU	FP722	TFT SENSOR (5V REF) SIGNAL
23-24	--	--	NOT USED
25	GRY	R133	TRANSMISSION SERVICE INDICATOR CONTROL (1ST DESIGN)
25	--	--	NOT USED (2ND DESIGN)
26	RED	A828	FUSED BATTERY FEED
27	BRN/GRN	XU100	3RD GEAR INDICATOR CONTROL
28	BLK	F836	GROUND TO G103
29	--	--	NOT USED
30	BLU/WHT	PM748	1-2 AND 3-4 SHIFT SOLENOID CONTROL
31	--	--	NOT USED
32	BLK/YEL	FB738	TCC SOLENOID CONTROL
33	BLU/YEL	PB743	2-3 SHIFT SOLENOID CONTROL
34-35	--	--	NOT USED
36	BLK/BLU	FP400	TRANS RANGE SWITCH INPUT A
37	BLK/YEL	FB709	TRANS RANGE SWITCH INPUT C
38-41	--	--	NOT USED
42	BLK/YEL	FB720	VSS HI SIGNAL
43-48	--	--	NOT USED
49	BRN/GRY	XR888	CAN INTERFACE LO
50	BRN/PPL	XY888	CAN INTERFACE HI
51	BRN/WHT	XM12	SERIAL DATA (KW2000)
52	BLK/PPL	FY740	PRESSURE CONTROL SOLENOID (PCS) VALVE HI SIGNAL
53	BLU/BLK	PF754	POWER RELAY (12V REF) OUTPUT
54	BRN	X837	FUSED IGNITION FEED
55-88	--	--	NOT USED

2000-01 3.0L w/4L30-E A/T TCM Connector Pinout

A/T WINTER MODE SWITCH

Connector Pinouts

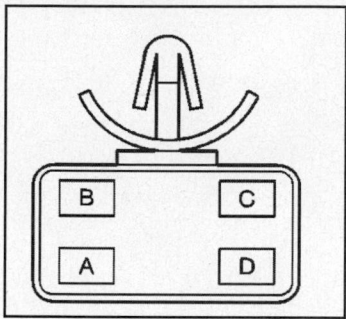

PIN	WIRE COLOR	CIRCUIT NO.	FUNCTION
A	GRN	U102	WINTER MODE SWITCH OUTPUT SIGNAL
R	BLK	F060	G103
C	BRN/GRN	XU100	3RD GEAR INDICATOR CONTROL
D	BRN	X14	FUSED IGNITION FEED

2000-01 A/T Winter Mode Switch Connector Pinout

ACCELERATOR PEDAL POSITION (APP) SENSOR

Connector Pinouts

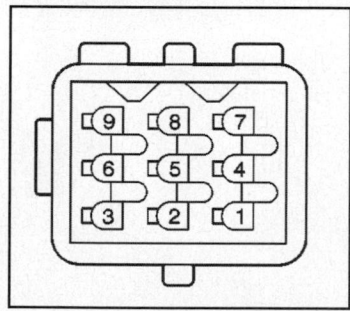

PIN	WIRE COLOR	CIRCUIT NO.	FUNCTION
1	BRN/RED	XA40	PEDAL POSITION 2 SIGNAL RETURN
2	BRN/GRY	XR24	PEDAL POSITION 1 REFERENCE GROUND
3	--	--	NOT USED
4	BLK/GRY	FR56	PEDAL POSITION 1 5V REFERENCE
5	BRN/BLK	XF8	PEDAL POSITION 1 SIGNAL RETURN
6	BRN/GRN	XU800	GROUND
7	--	--	NOT USED
8	BRN/RED	FA7	PEDAL POSITION 2 5V REFERENCE
9	--	--	NOT USED

2000-01 APP Sensor Connector Pinout

AUXILIARY HEATER WATER PUMP RELAY

Connector Pinouts

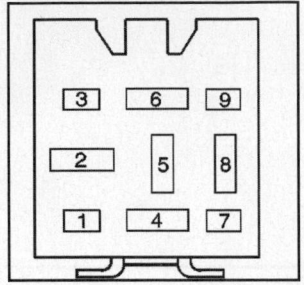

PIN	WIRE COLOR	CIRCUIT NO.	FUNCTION
1	--	--	NOT USED
2	RED/BLU	AP3	B+ FROM RELAY K26
3	--	--	NOT USED
4	BRN	X6	B+ FROM FUSE #10
5	BRN/GRN	XU2	AUXILIARY WATER PUMP ON SIGNAL
6	BLK	F5	GROUND TO G103
7-9	--	--	NOT USED

2000-01 3.0L Auxiliary Heater Water Pump Relay Connector Pinout

CAMSHAFT POSITION (CMP) SENSOR

Component Locations

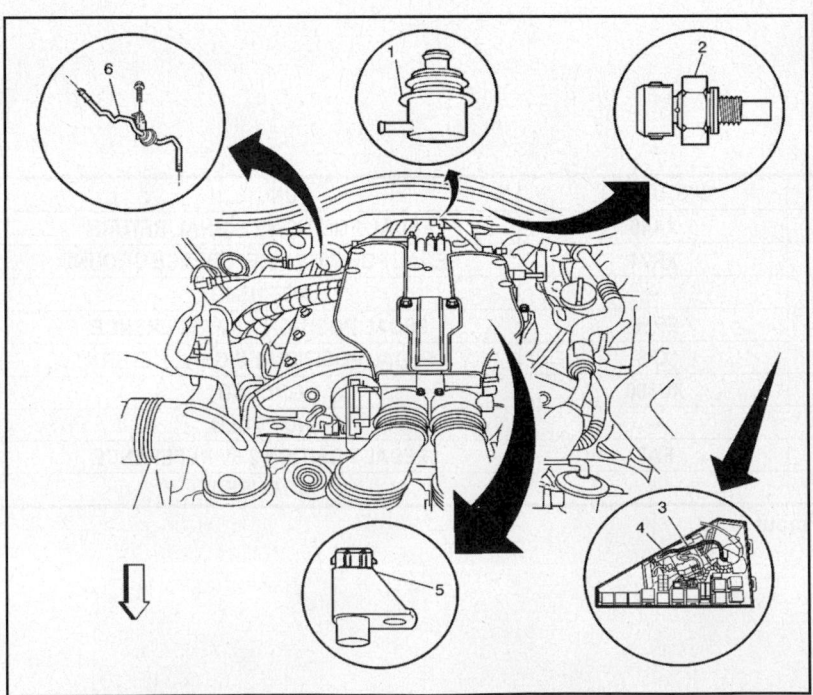

Top view of the engine, showing: 1. Fuel pressure regulator; 2. Engine coolant temperature (ECT) sensor; 3. Engine control module (ECM); 4. Underhood fuse & relay center; 5. Camshaft position (CMP) sensor; 6. EVAP canister purge valve

Connector Pinouts

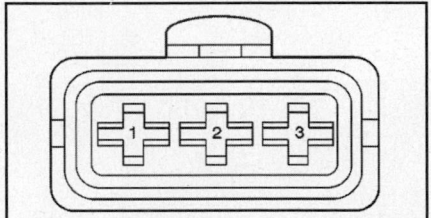

PIN	WIRE COLOR	CIRCUIT NO.	FUNCTION
1	GRY/BLK	RF78	CKP SIGNAL HIGH OUTPUT
2	GRY/RED	RA20	CKP SIGNAL LOW OUTPUT
3	BLK	S03	SHIELDED GROUND

2000-01 CMP Sensor Connector Pinout

CRANKSHAFT POSITION (CKP) SENSOR

Connector Pinouts

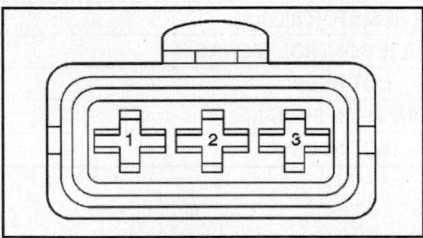

PIN	WIRE COLOR	CIRCUIT NO.	FUNCTION
1	GRY/BLK	RF78	CKP SIGNAL HIGH OUTPUT
2	GRY/RED	RA20	CKP SIGNAL LOW OUTPUT
3	BLK	S03	SHIELDED GROUND

2000-01 3.0L CKP Sensor Connector Pinout

ELECTRIC COOLING FAN

Connector Pinouts

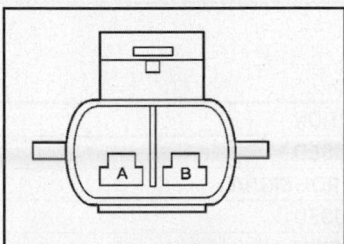

PIN	WIRE COLOR	CIRCUIT NO.	FUNCTION
A	RED/WHT	AM17	FAN ON SIGNAL
B	BLK	F100	GROUND TO G103

2000-01 3.0L Electric Cooling Fan Connector Pinout

ELECTRIC COOLING FAN CONTROL RELAY

Connector Pinouts

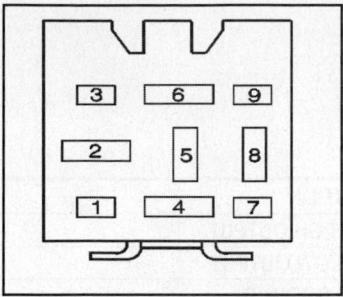

PIN	WIRE COLOR	CIRCUIT NO.	FUNCTION
1	--	--	NOT USED
2	RED	A222	B+ FROM FUSE #52
3	RED	A6	B+ FROM FUSE #50
4	RED	A223	B+ FROM FUSE #52
5	RED/BLU	AP1	FAN RESISTOR SIGNAL
6	BRN/WHT	XP3	RELAY K26 CONTROL SIGNAL
7	--	--	NOT USED
8	RED/WHT	AM1	FAN #1 ON SIGNAL
9	--	--	NOT USED

2000-01 3.0L Electric Cooling Fan Control Relay K26 Connector Pinout

PIN	WIRE COLOR	CIRCUIT NO.	FUNCTION
1	--	--	NOT USED
2	BRN	X400	B+ FROM FUSE #10
3	--	--	NOT USED
4	RED	A14	B+ FROM FUSE #40
5	BRN/WHT	XM10	RELAY K28 CONTROL SIGNAL
6	BRN/WHT	XP124	RELAY K28 CONTROL SIGNAL
7	--	--	NOT USED
8	RED/WHT	AM4	FAN #2 ON SIGNAL
9	--	--	NOT USED

2000-01 3.0L Electric Cooling Fan Control Relay K28 Connector Pinout

PIN	WIRE COLOR	CIRCUIT NO.	FUNCTION
1	--	--	NOT USED
2	BRN/GRN	XU3	FAN #1 CONTROL SIGNAL
3	--	--	NOT USED
4	BRN	X30	B+ FORM FUSE #10
5	RED/WHT	AM3	FAN #2 CONTROL SIGNAL
6	BRN/WHT	XM123	RELAY K52 CONTROL SIGNAL
7	--	--	NOT USED
8	BLK	F7	GROUND TO G103
9	--	--	NOT USED

2000-01 3.0L Electric Cooling Fan Control Relay K52 Connector Pinout

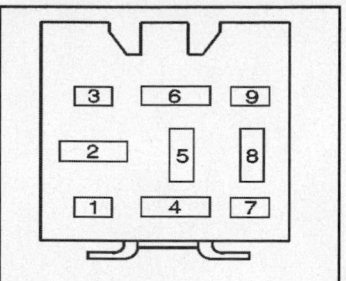

PIN	WIRE COLOR	CIRCUIT NO.	FUNCTION
1	--	--	NOT USED
2	BRN	X2	B+ FROM FUSE #10
3	--	--	NOT USED
4	RED	A710	B+ FROM FUSE #42
5	--	--	NOT USED
6	BRN/WHT	XM5/XM998	RELAY K67 CONTROL SIGNAL
7	--	--	NOT USED
8	RED/WHT	AM16	COOLANT FAN ON SIGNAL
9	--	--	NOT USED

2000-01 3.0L Electric Cooling Fan Control Relay K67 Connector Pinout

RELAY K87

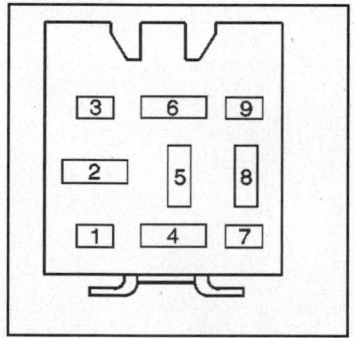

PIN	WIRE COLOR	CIRCUIT NO.	FUNCTION
1	--	--	NOT USED
2	GRN	U4	COMPRESSOR CONTROL SIGNAL
3	--	--	NOT USED
4	RED	A221	B+ FROM FUSE #52
5	--	--	NOT USED
6	BLK	F2	GROUND TO G103
7	--	--	NOT USED
8	RED/WHT	AM12	FAN #1 CONTROL SIGNAL
9	--	--	NOT USED

2000-01 3.0L Electric Cooling Fan Control Relay K87 Connector Pinout

ELECTRIC COOLING FAN TEMPERATURE SWITCH

Connector Pinouts

PRIMARY

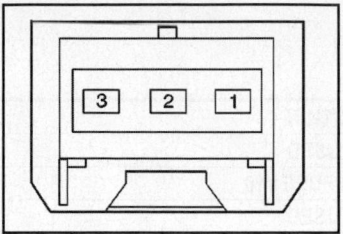

PIN	WIRE COLOR	CIRCUIT NO.	FUNCTION
1	BLK	F600	GROUND TO G103
2	BRN/BLU	XP30	RELAY K26 CONTROL
3	BRN/WHT	XM5	RELAY K67 CONTROL

2000-01 3.0L Electric Cooling Fan (Primary) Temperature Switch Connector Pinout

SECONDARY

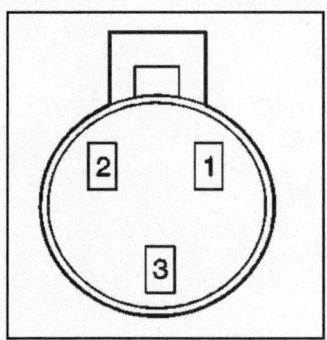

PIN	WIRE COLOR	CIRCUIT NO.	FUNCTION
1	BRN/BLK	XF101	COMPRESSOR CONTROL SIGNAL
2	BRN/WHT	XM122	RELAY K28 & K52 CONTROL
3	BLK	F60	GROUND TO G103

2000-01 3.0L Electric Cooling Fan (Secondary) Temperature Switch Connector Pinout

ENGINE CONTROL MODULE (ECM)

Component Locations

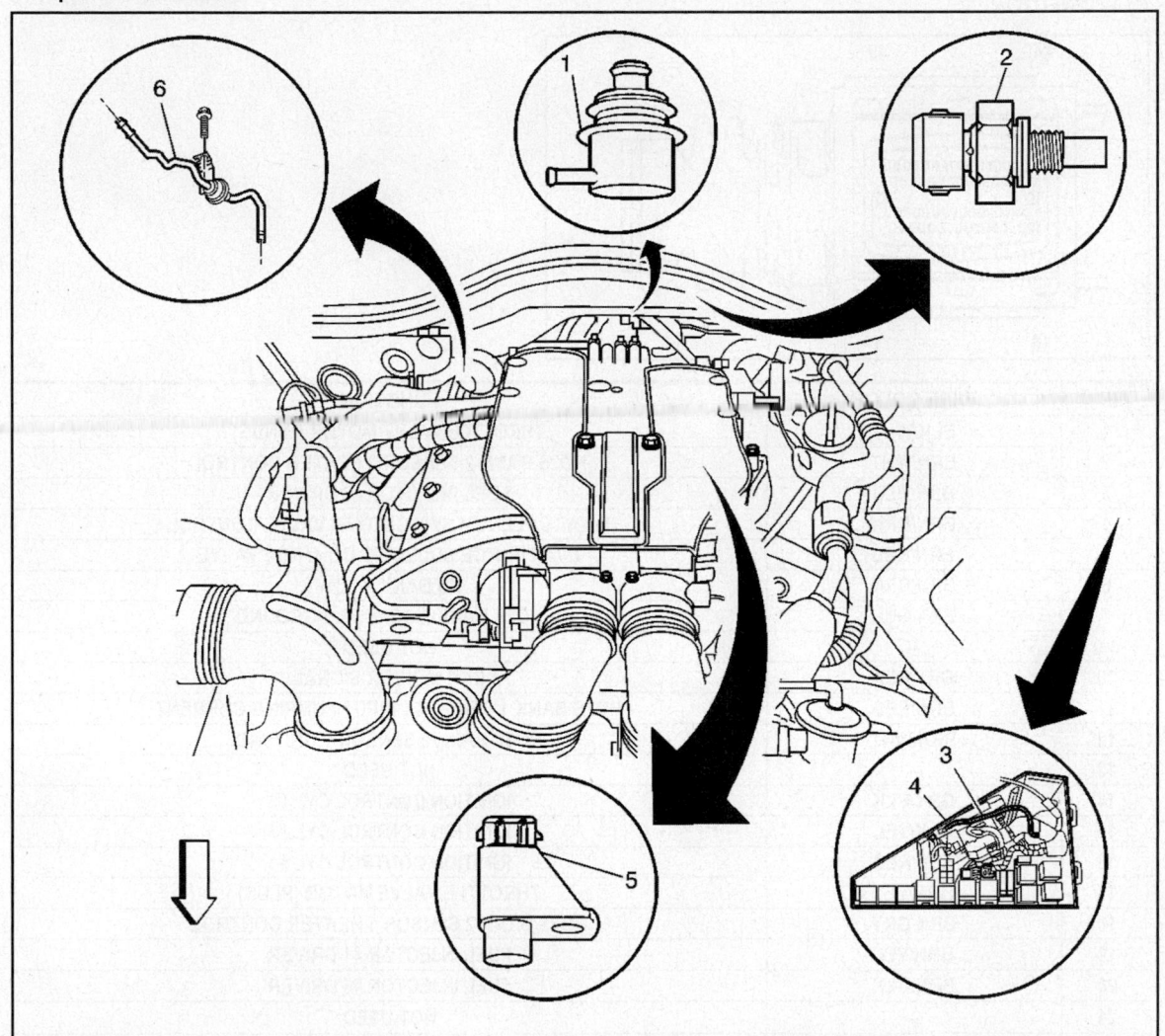

(1) Fuel Pressure Regulator

(2) Engine Coolant Temperature (ECT) Sensor

(3) Engine Control Module (ECM)

(4) Underhood Fuse and Relay Center

(5) Camshaft Position (CMP) Sensor

(6) EVAP Canister Purge Valve

Showing top engine view, with location of ECM and other electronic components

Connector Pinouts

C1 CONNECTOR

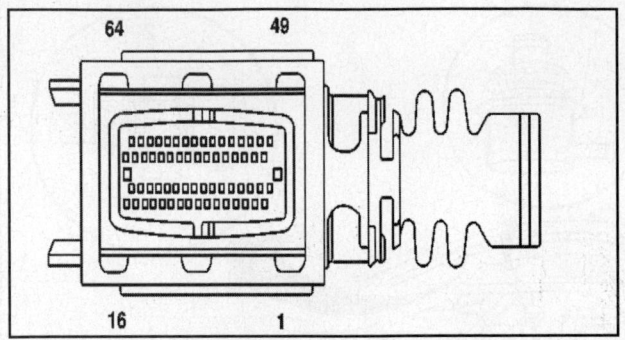

PIN #	WIRE COLOR	FUNCTION
1	BLK/WHT	THROTTLE VALVE MOTOR (MINUS)
2	BRN/WHT	HO2S BANK 2 SENSOR 2 HEATER CONTROL
3	BRN/BLU	FUEL INJECTOR #3 DRIVER
4	BRN/GRN	INTAKE PLENUM SWITCHOVER VALVE CONTROL
5	BRN/RED	EVAP PURGE SOLENOID CONTROL VALVE
6	BRN/GRN	KS BANK 2 LOW
7	BRN/GRN	HO2S BANK 2 SENSOR 2 GROUND
8-9	--	NOT USED
10	GRY/BLK	CKP SENSOR SIGNAL
11	BRN/RED	HO2S BANK 1 SENSOR 1 INPUT PUMPING CURRENT
12	BRN/WHT	HO2S BANK 2 SENSOR 1 GROUND
13	--	NOT USED
14	GRY/BLK	IGNITION CONTROL CYL #5
15	BLK/YEL	IGNITION CONTROL CYL #3
16	BLK/RED	IGNITION CONTROL CYL #1
17	BRN/YEL	THROTTLE VALVE MOTOR (PLUS)
18	BRN/GRY	HO2S BANK 2 SENSOR 1 HEATER CONTROL
19	BRN/YEL	FUEL INJECTOR #1 DRIVER
20	BRN/BLK	FUEL INJECTOR #5 DRIVER
21	--	NOT USED
22	BRN/WHT	KS BANK 1 SIGNAL
23	BRN/BLU	HO2S BANK 1 SENSOR 2 SIGNAL
24	WHT	TP SENSOR 1 SIGNAL
25	BRN/BLU	ECT SENSOR SIGNAL
26	BRN/GRN	SENSOR GROUND (MAF/IAT, ECT, TP)
27	--	NOT USED
28	BRN/BLU	HO2S BANK 1 SENSOR 1 SIGNAL
29	BRN/GRN	HO2S BANK 2 SENSOR 1 OUTPUT PUMPING CIRCUIT
30	GRY/BLK	IGNITION CONTROL CYL #6
31	BLK/YEL	IGNITION CONTROL CYL #4
32	BLK/RED	IGNITION CONTROL CYL #2
33	BLK/WHT	THROTTLE VALVE MOTOR (MINUS)
34	BRN/WHT	HO2S BANK 1 SENSOR 2 HEATER CONTROL
35	BRN/GRN	FUEL INJECTOR #2 DRIVER
36	BRN/PPL	FUEL INJECTOR #6 DRIVER
37	--	NOT USED
38	BRN/WHT	KS BANK 2 SIGNAL
39	BRN/BLU	HO2S BANK 2 SENSOR 2 SIGNAL

2000-01 ECM C1 Connector Pinout (1 of 2)

C1 CONNECTOR - CONTINUED

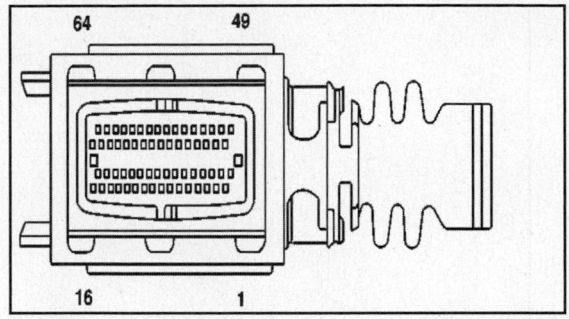

PIN #	WIRE COLOR	FUNCTION
40	BRN/BLU	TP SENSOR 2 SIGNAL
41	--	NOT USED
42	WHT	CMP SENSOR SIGNAL
43	BRN/GRN	HO2S BANK 1 SENSOR 1 OUTPUT PUMPING CIRCUIT
44	BRN/BLU	HO2S BANK 2 SENSOR 1 SIGNAL
45-48	--	NOT USED
49	BRN/YEL	THROTTLE VALVE MOTOR (PLUS)
50	BRN/GRY	HO2S BANK 1 SENSOR 1 HEATER CONTROL
51	BLU/BLK	MAF 5 VOLT REFERENCE
52	BRN/RED	FUEL INJECTOR #4 DRIVER
53	BRN/GRN	INTAKE RESONANCE SWITCHOVER VALVE CONTROL
54	BRN/GRN	KS BANK 1 LOW
55	BRN/GRN	HO2S BANK 1 SENSOR 2 GROUND
56	GRY/RED	MAF SIGNAL
57	BRN/BLK	IAT SENSOR SIGNAL
58	BLU/BLK	TP SENSOR 5 VOLT REFERENCE
59	GRY/RED	CKP REFERENCE LOW
60	BRN/WHT	HO2S BANK 1 SENSOR 1 GROUND
61	BRN/RED	HO2S BANK 2 SENSOR 1 INPUT PUMPING CURRENT
62	BLK	KS 1, KS 2, AND CKP SENSOR SHIELDED GROUND
63-64	--	NOT USED

2000-01 ECM C1 Connector Pinout (2 of 2)

C2 CONNECTOR

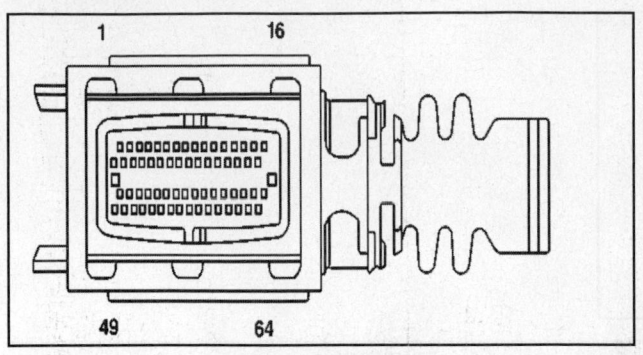

PIN	WIRE COLOR	FUNCTION
1	BRN/WHT	CRUISE INDICATOR
2-5	--	NOT USED
6	BLK/BLU	A/C ON/OFF SIGNAL
7	BLK/RED	APP 2 5 VOLT REFERENCE
8	BRN/BLK	APP 1 SIGNAL
9	--	NOT USED
10	BLK/BRN	CRUISE ON/OFF SIGNAL
11	--	NOT USED
12	BRN/GRY	EBTCM CAN INTERFACE LOW
13	BRN/PPL	A/C RELAY CONTROL
14	--	NOT USED
15	RED	B+
16	RED/BLU	SWITCHED B+
17	--	NOT USED
18	BRN/BLK	EVAP VENT SOLENOID CONTROL
19-21	--	NOT USED
22	BLU/BLK	FUEL LEVEL INPUT
23	--	NOT USED
24	BRN/GRY	APP 1 REFERENCE GROUND
25	BRN/GRN	FUEL TANK PRESSURE SENSOR GROUND
26	BLK/BRN	CRUISE RELEASE SIGNAL
27	BLK/YEL	STOP LAMP SIGNAL
28	--	NOT USED
29	BRN/PPL	EBTCM CAN INTERFACE HIGH
30	BRN/BLU	FUEL PUMP RELAY CONTROL
31	BRN/GRY	ENGINE CONTROLS POWER RELAY CONTROL
32	RED/BLU	SWITCHED B+
33	GRY	FILLER CAP LAMP
34	BRN/YEL	SECONDARY AIR PUMP SOLENOID CONTROL
35	GRN	RPM SIGNAL OUTPUT
36	--	NOT USED
37	BRN/YEL	FUEL TANK PRESSURE SENSOR SIGNAL
38-39	--	NOT USED
40	BRN/RED	APP 2 SIGNAL
41	--	NOT USED
42	BLK/RED	CRUISE RESUME/DECEL SIGNAL
43	BLK/YEL	CRUISE SET/ACCEL SIGNAL
44	BLU/RED	VSS/THEFT DETERRENT SIGNAL
45	BRN/PPL	TCM CAN INTERFACE HIGH

2000-01 ECM C2 Connector Pinout (1 of 2)

C2 CONNECTOR (CONTINUED)

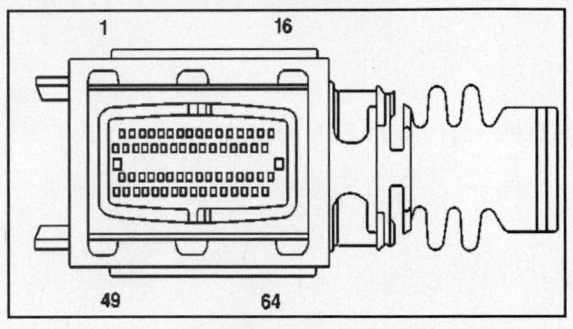

PIN	WIRE COLOR	FUNCTION
46	BRN/BLU	MIL CONTROL
47	BRN	IGNITION SWITCH
48	RED/BLU	SWITCHED B+
49-52	--	NOT USED
53	BRN/WHT	SERIAL DATA/KW 2000/DLC
54-55	--	NOT USED
56	BLK/GRY	APP 1 5 VOLT REFERENCE
57	--	NOT USED
58	BLK/WHT	A/C REQUEST
59-60	--	NOT USED
61	BRN/GRY	TCM CAN INTERFACE LOW
62	BRN/GRY	SECONDARY AIR PUMP RELAY CONTROL
63	BLU/BLK	FUEL TANK PRESSURE SENSOR 5 VOLT
64	--	NOT USED

2000-01 ECM C2 Connector Pinout (2 of 2)

ENGINE CONTROL POWER RELAY

Connector Pinout

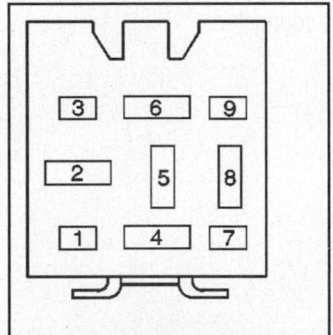

PIN	WIRE COLOR	CIRCUIT NO.	FUNCTION
1	--	--	NOT USED
2	RED	A71	FUSED BATTERY FEED
3	--	--	NOT USED
4	RED	A73	FUSED BATTERY FEED
5	--	--	NOT USED
6	BRN/GRY	XR31	RELAY K43 CONTROL
7	--	--	NOT USED
8	RED/BLU	AP75	SWITCHED B+ FROM RELAY K43
9	--	--	NOT USED

2000-01 3.0L Engine Controls Power Relay Connector Pinout

ENGINE COOLANT FAN TEST CONNECTOR

Connector Locations

The engine coolant fan test connector is located in the left side of the engine in the ECM housing.

Connector Pinout

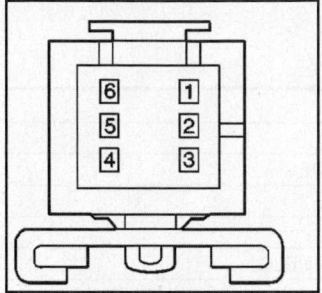

PIN	WIRE COLOR	CIRCUIT NO.	FUNCTION
1	BRN/BLU	XP999	TEST PIN FOR RELAY K26 AND RELAY K48
2	BRN/WHT	XM999	TEST PIN FOR RELAY K28
3	BRN	X999	B+ FROM FUSE #13
4	BRN/WHT	XM998	TEST PIN FOR RELAY K67
5	BRN/WHT	XM997	TEST PIN FOR RELAY K28 AND RELAY K52
6	BLK	F999	GROUND TO G103

2000-01 3.0L Engine Coolant Fan Test Connector Pinout

ENGINE COOLANT TEMPERATURE (ECT) SENSOR

Connector Locations

The ECT sensor is mounted on coolant bridge rear of engine near fuel injector harness connector.

Connector Pinouts

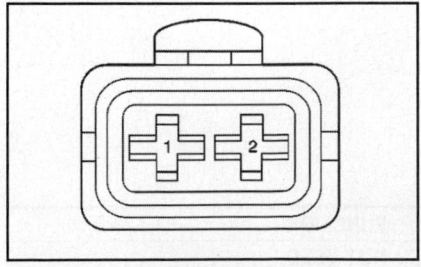

PIN	WIRE COLOR	CIRCUIT NO.	FUNCTION
1	BRN/BLU	XP74	ECT SENSOR SIGNAL
2	BRN/GRN	XU715	SENSOR GROUND

2000-01 3.0L ECT Sensor Connector Pinout

ENGINE COOLING FAN

Connector Pinouts

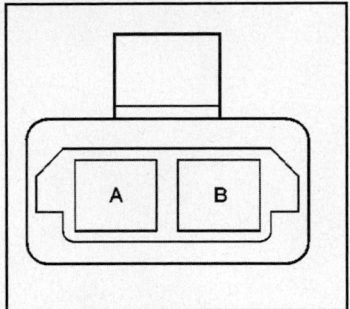

PIN	WIRE COLOR	CIRCUIT NO.	FUNCTION
A	RED/WHT	AM13	B+ FROM RELAY K26
B	BRN/GRN	XU3	GROUND FROM RELAY K52

2000-01 3.0L Engine Cooling Fan #1 Connector Pinout

PIN	WIRE COLOR	CIRCUIT NO.	FUNCTION
A	RED/WHT	AM5	FAN ON SIGNAL
B	BLK	F4	GROUND TO G103

2000-01 3.0L Engine Cooling Fan #2 Connector Pinout

EVAP CANISTER PURGE SOLENOID VALVE

Description & Operation

The evaporative emission (EVAP) control system used on all vehicles is the charcoal canister storage method. This method transfers fuel vapor from the fuel tank to an activated carbon, charcoal, storage device, canister, which stores the vapors when the vehicle is not operating. When the engine is running, the fuel vapor is purged from the carbon element by the engine vacuum and is consumed in the normal combustion process. The enhanced EVAP system uses software within the engine control module (ECM) and several additional components that allow the ECM to monitor the system performance and to perform the comprehensive on-board diagnostics.

The enhanced EVAP system consists of the following components:

- The EVAP canister vent valve
- The EVAP canister purge valve
- The fuel tank pressure sensor
- The EVAP canister
- The fuel pipes and hoses
- The fuel vapor line
- The purge line
- The vent line
- The fuel tank
- The fuel filler cap

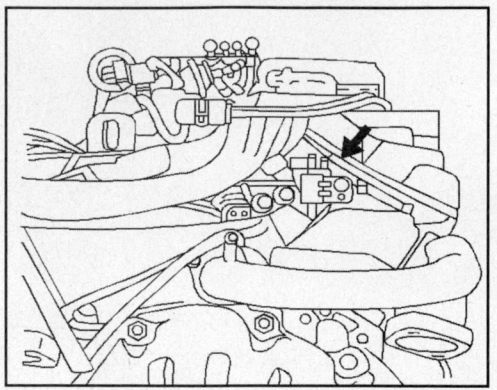

Showing the EVAP canister purge valve location

The EVAP canister purge valve controls the engine vacuum to the canister. Under the appropriate conditions, the ECM commands the purge valve open. This allows engine vacuum to draw fresh air into the canister through the EVAP vent valve. The fuel vapors exit the canister and are consumed during the normal combustion process.

The EVAP canister vent valve is used for certain EVAP system performance tests that are performed by the ECM. The ECM can close the vent which effectively seals the system. The ECM can then evaluate pressure changes within the system by monitoring the fuel tank pressure sensor signal.

The fuel tank pressure sensor is used for certain EVAP system performance tests performed by the ECM. The fuel tank pressure sensor contains a diaphragm which changes the resistance based on pressure. When EVAP system pressure is low, during purge, sensor output voltage is low. When the system pressure is high, the sensor output voltage is high. The ECM monitors the pressure changes within the system by using the fuel tank pressure sensor signal. This information can be used in order to detect any leaks within the system or to verify the operation of the system components.

Connector Pinouts

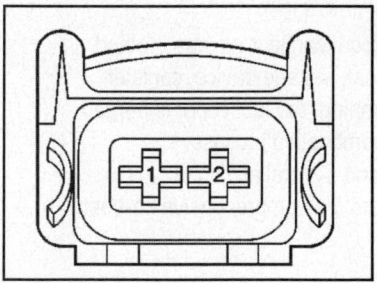

PIN	WIRE COLOR	CIRCUIT NO.	FUNCTION
1	RED/BLU	AP108	B+ FROM RELAY K43
2	BRN/RED	XA61	EVAP PURGE SOLENOID VALVE CONTROL

2000-01 3.0L EVAP Purge Solenoid Valve Connector Pinout

EVAP TANK SOLENOID VALVE CONNECTOR

Connector Pinouts

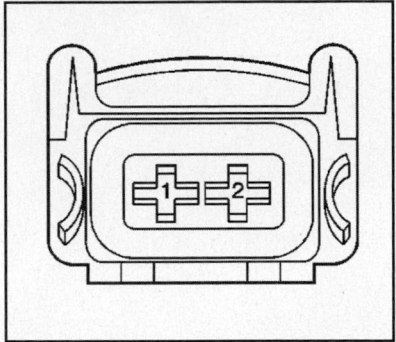

PIN	WIRE COLOR	CIRCUIT NO.	FUNCTION
1	RED/BLU	AP88	B+ FROM RELAY K43
2	BRN/BLK	XF888	EVAP VENT SOLENOID VALVE

2000-01 3.0L EVAP Tank Solenoid Valve Connector Pinout

FUEL INJECTOR CONNECTORS

Connector Pinouts

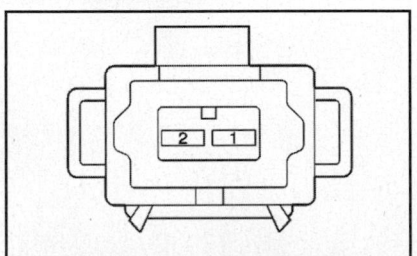

PIN	WIRE COLOR	CIRCUIT NO.	FUNCTION
1	RED/BLU	AP61	FUEL INJECTOR #1 B+ FROM RELAY K43
2	BRN/YEL	XB17	FUEL INJECTOR #1 CONTROL
1	RED/BLU	AP63	FUEL INJECTOR #2 B+ FROM RELAY K43
2	BRN/GRN	XU16	FUEL INJECTOR #2 CONTROL
1	RED/BLU	AP64	FUEL INJECTOR # 3 B+ FROM RELAY K43
2	BRN/BLU	XP35	FUEL INJECTOR #3 CONTROL
1	RED/BLU	AP62	FUEL INJECTOR # 4 B+ FROM RELAY K43
2	BRN/RED	XA34	FUEL INJECTOR #4 CONTROL
1	RED/BLU	AP65	FUEL INJECTOR #5 B+ FROM RELAY K43
2	BRN/BLK	XF15	FUEL INJECTOR #5 CONTROL
1	RED/BLU	AP66	FUEL INJECTOR # 6 B+ FROM RELAY K43
2	BRN/PPL	XY33	FUEL INJECTOR #6 CONTROL

2000-01 3.0L Fuel Injector Connector Pinouts

FUEL INJECTOR IN-LINE CONNECTOR

Connector Pinouts

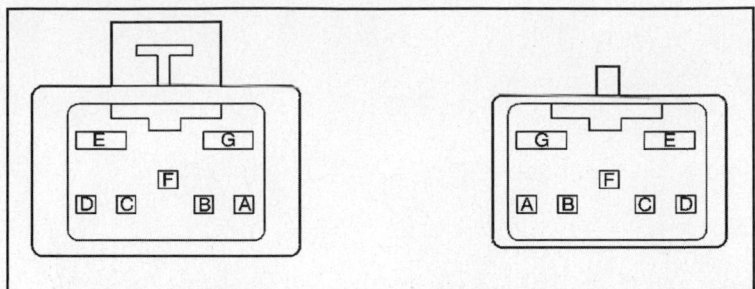

PIN	WIRE COLOR	CIRCUIT NO.	FUNCTION
A	BRN/YEL	XB17	FUEL INJECTOR # 1 CONTROL
B	BRN/GRN	XU16	FUEL INJECTOR # 2 CONTROL
C	BRN/BLU	XP35	FUEL INJECTOR # 3 CONTROL
D	BRN/RED	XA34	FUEL INJECTOR # 4 CONTROL
E	BRN/BLK	XF17	FUEL INJECTOR # 5 CONTROL
F	BRN/PPL	XY33	FUEL INJECTOR # 6 CONTROL
G	RED/BLU	ZP751	B+ FROM RELAY K43

2000-01 3.0L Fuel Injector In-Line Connector Pinout

FUEL PUMP CONNECTOR

Connector Pinouts

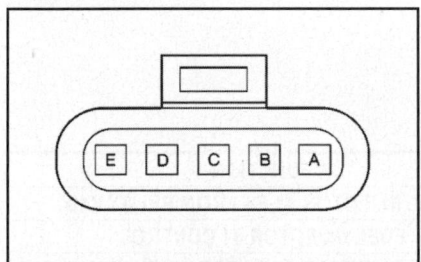

PIN	WIRE COLOR	CIRCUIT NO.	FUNCTION
A	RED/BLU	AP10	B+ FROM FUEL PUMP FUSE
B	BLK	F224	GROUND TO G103
C	BLK	F30	GROUND TO G400
D	BLU/BLK	PF11	FUEL LEVEL SIGNAL

2000-01 3.0L Fuel Pump Connector Pinout

FUEL PUMP RELAY CONNECTOR

Connector Pinouts

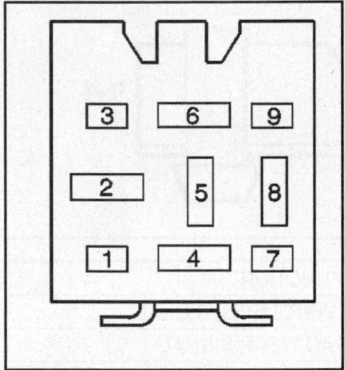

PIN	WIRE COLOR	CIRCUIT NO.	FUNCTION
1	--	--	NOT USED
2	RED	A72	FUSED BATTERY FEED
3	--	--	NOT USED
4	RED/BLU	AP76	B+ FROM RELAY K43
5	--	--	NOT USED
6	BRN/BLU	XR3	RELAY CONTROL SIGNAL
7	--	--	NOT USED
8	RED/BLU	AP77	B+ FEED TO FUEL PUMP FUSE
9	--	--	NOT USED

2000-01 3.0L Fuel Pump Relay Connector Pinout

FUEL TANK PRESSURE (FTP) SENSOR CONNECTOR

Connector Pinouts

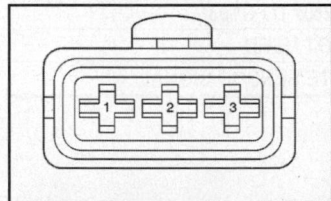

PIN	WIRE COLOR	CIRCUIT NO.	FUNCTION
1	BLU/BLK	PF1	5V REFERENCE
2	BRN/GRN	XU1	SENSOR GROUND
3	BRN/YEL	XB1	FUEL PRESSURE SENSOR SIGNAL

2000-01 3.0L FTP Sensor Connector Pinout

GAUGE CLUSTER CONNECTOR

Connector Pinouts

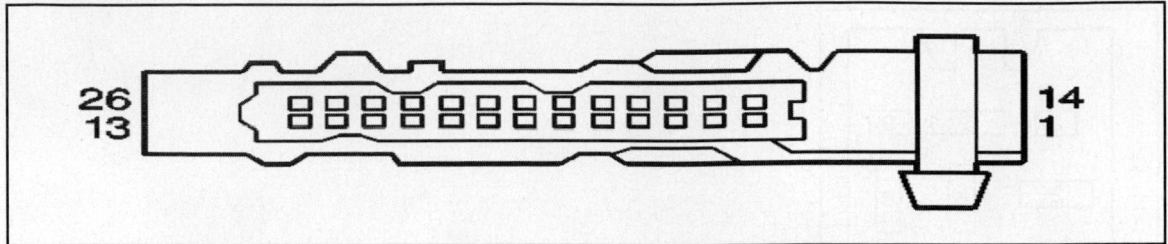

PIN	WIRE COLOR	CIRCUIT NO.	FUNCTION
1	BRN/YEL	XB3	WSWA SOLVENT LEVEL SIGNAL
2	YEL/RED	BA2	BRAKE PAD WEAR SIGNAL
3	BRN/BLU	XP2	COOLANT LEVEL SIGNAL
4	BRN/GRN	XU600	OIL LEVEL SIGNAL
5	RED	A900	B+ FROM FUSE #24
6	BLK	F8	GROUND TO G103
7-8	--	--	NOT USED
9	BLU/RED	PA600	VEHICLE SPEED SIGNAL
10-12	--	--	NOT USED
13	GRN	U5	TACHOMETER SIGNAL
14	BRN/BLK	XF20	SEAT BELT SIGNAL
15	BLU/GRN	PU1	OIL PRESSURE SIGNAL
16	GRY/WHT	RM900	HEADLAMP SWITCH SIGNAL
17	GRY/YEL	RB103	INTERIOR LAMPS DIMMING SIGNAL
18	BLU/WHT	PM1	GENERATOR SIGNAL
19	--	--	NOT USED
20	BRN	X70	B+ FROM FUSE #15
21	BRN/PPL	XY2	ALC SENSOR SIGNAL
22	RED	A16	GROUND SIGNAL
23	BRN/BLU	XP840	SECURITY LAMP INDICATOR
24	BLK	F11	GROUND TO G103
25	--	--	NOT USED
26	BLU/YEL	PB90	OIL PRESSURE SENSOR SIGNAL

2000-01 3.0L Gauge Cluster Connector Pinout

HEATER WATER AUXILIARY PUMP CONNECTOR

Connector Pinouts

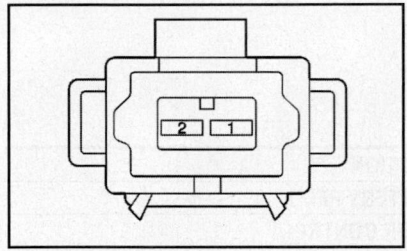

PIN	WIRE COLOR	CIRCUIT NO.	FUNCTION
1	BLK	F75	GROUND TO G103
2	BRN	X70	B+ FROM FUSE #10

2000-01 3.0L Heater Water Auxiliary Pump Connector Pinout

HEATED OXYGEN SENSOR (HO2S) CONNECTORS

Connector Pinouts

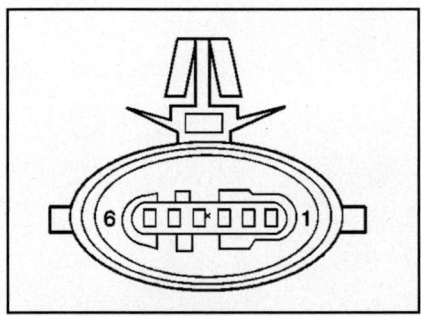

PIN	WIRE COLOR	CIRCUIT NO.	FUNCTION
1	BRN/RED	XA43	LAMBDA SENSOR PUMP CURRENT INPUT
2	BRN/BLU	XP19	HO2S SIGNAL
3	BRN/WHT	XM30	HO2S REFERENCE GROUND
4	BRN/GRY	XR11	HO2S HEATER CONTROL
5	RED/BLU	AP122	FUSED BATTERY FEED
6	BRN/GRN	XU462	LAMBDA SENSOR PUMP CURRENT OUTPUT
PIN	WIRE COLOR	CIRCUIT NO.	FUNCTION
1	BRN/RED	XA43	LAMBDA SENSOR PUMP CURRENT INPUT

2000-01 3.0L O2 Sensor (1/1) Connector Pinout

PIN	WIRE COLOR	CIRCUIT NO.	FUNCTION
1	BRN/RED	XA610	LAMBDA SENSOR PUMP CURRENT INPUT
2	BRN/BLU	XP18	HO2S SIGNAL
3	BRN/WHT	XM309	HO2S REFERENCE GROUND
4	BRN/GRY	XR61	HO2S HEATER CONTROL
5	RED/BLU	AP124	FUSED BATTERY FEED
6	BRN/GRN	XU464	LAMBDA SENSOR PUMP CURRENT OUTPUT
PIN	WIRE COLOR	CIRCUIT NO.	FUNCTION
1	BRN/RED	XA610	LAMBDA SENSOR PUMP CURRENT INPUT

2000-01 3.0L O2 Sensor (2/1) Connector Pinout

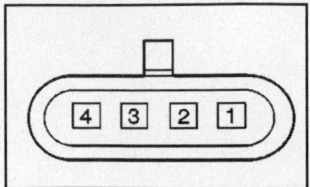

PIN	WIRE COLOR	CIRCUIT NO.	FUNCTION
1	RED/BLU	AP951	FUSED BATTERY FEED
2	BRN/WHT	XM951	HO2S HEATER CONTROL
3	BRN/GRN	XU950	HO2S REFERENCE GROUND
4	BRN/BLU	XP961	HO2S SIGNAL

2000-01 3.0L O2 Sensor (1/2) Connector Pinout

PIN	WIRE COLOR	CIRCUIT NO.	FUNCTION
1	RED/BLU	AP950	FUSED BATTERY FEED
2	BRN/WHT	XM950	HO2S HEATER CONTROL
3	BRN/GRN	XU960	HO2S REFERENCE GROUND
4	BRN/BLU	XP960	HO2S SIGNAL

2000-01 3.0L O2 Sensor (2/2) Connector Pinout

IGNITION COIL CONNECTORS

Connector Pinouts

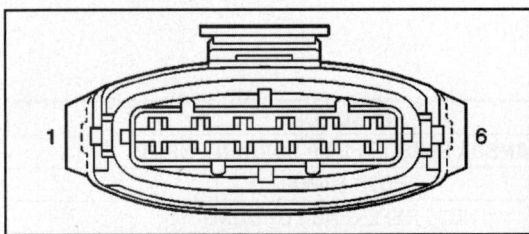

PIN	WIRE COLOR	CIRCUIT NO.	FUNCTION
1	BLK	F60	GROUND
2	BRN	X69	SWITCHED IGNITION FEED
3	GRY/BLK	RF51	CYL. 6 IGNITION CONTROL
4	BLK/YEL	FB61	CYL. 4 IGNITION CONTROL
5	BLK/RED	FA71	CYL. 2 IGNITION CONTROL
6	--	--	NOT USED

2000-01 3.0L Ignition Coil (Left Side) Connector Pinout

PIN	WIRE COLOR	CIRCUIT NO.	FUNCTION
1	BLK	F50	GROUND
2	BRN	X68	SWITCHED IGNITION FEED
3	GRY/BLK	RF50	CYL. 5 IGNITION CONTROL
4	BLK/YEL	FB60	CYL. 3 IGNITION CONTROL
5	BLK/RED	FA70	CYL. 1 IGNITION CONTROL
6	--	--	NOT USED

2000-01 3.0L Ignition Coil (Right Side) Connector Pinout

IGNITION SWITCH & LIFTING MAGNET CONNECTOR

Connector Pinouts

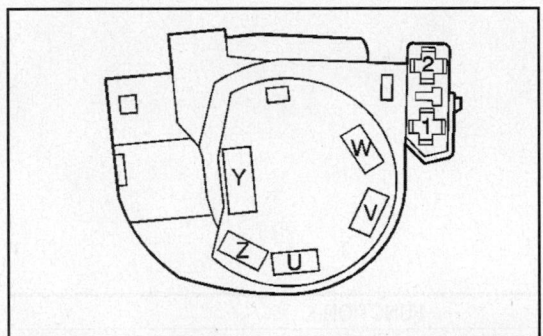

PIN	WIRE COLOR	CIRCUIT NO.	FUNCTION
1	RED	A905	B+ FROM FUSE #24
2	GRN	U900	BRAKE TO SHIFT SIGNAL
U	BRN	X826	B+ IN THE ON POSITION
V	BRN	X1	B+ IN THE ON OR START POSITION
W	RED	A19	B+ WHEN THE KEY IS INSERTED
Y	RED	A503	B+ FUSED
Z	BLK/RED	FA700	STARTER SOLENOID SIGNAL

2000-01 3.0L Ignition Switch & Lifting Magnet Connector Pinout

INTAKE PLENUM SWITCH-OVER SOLENOID VALVE CONNECTOR

Connector Pinouts

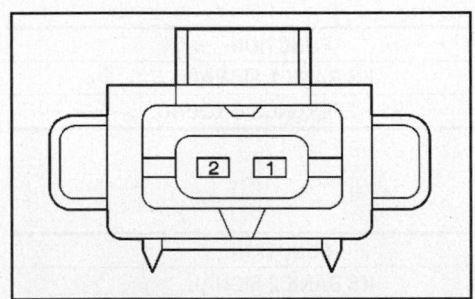

PIN	WIRE COLOR	CIRCUIT NO.	FUNCTION
1	BRN/GRN	XU35	INTAKE PLENUM SWITCH-OVER SOLENOID VALVE CONTROL
2	RED/BLU	AP105	B+ FROM RELAY K43

2000-01 3.0L Intake Plenum Switch-Over Solenoid Valve Connector Pinout

INTAKE RESONANCE SWITCH-OVER SOLENOID VALVE CONNECTOR

Connector Pinout

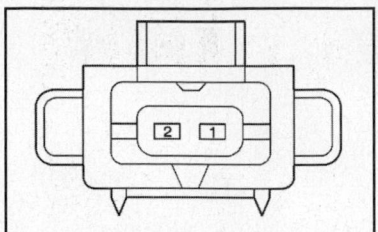

PIN	WIRE COLOR	CIRCUIT NO.	FUNCTION
1	BRN/GRN	XU7	INTAKE RESONANCE SWITCH-OVER SOLENOID VALVE CONTROL
2	RED/BLU	AP104	B+ FROM RELAY K43

2000-01 3.0L Intake Resonance Switch-Over Solenoid Valve Connector Pinout

KNOCK SENSOR CONNECTOR

Connector Pinouts

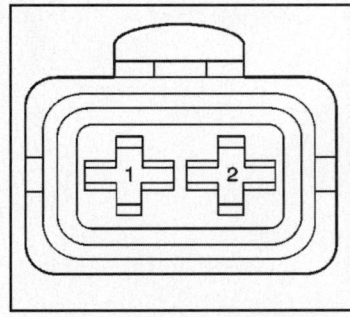

PIN	WIRE COLOR	CIRCUIT NO.	FUNCTION
1	BRN/WHT	XM70	KS BANK 1 SIGNAL
2	BRN/GRN	XU420	KS REFERENCE GROUND

2000-01 3.0L Bank 1 Knock Sensor Connector Pinout

PIN	WIRE COLOR	CIRCUIT NO.	FUNCTION
1	BRN/WHT	XM40	KS BANK 2 SIGNAL
2	BRN/GRN	XU410	KS REFERENCE GROUND

2000-01 3.0L Bank 2 Knock Sensor Connector Pinout

MASS AIR FLOW (MAF) SENSOR CONNECTOR

Connector Pinouts

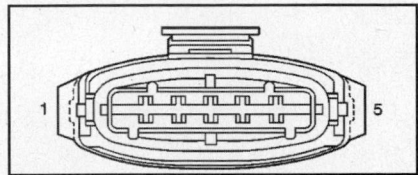

PIN	WIRE COLOR	CIRCUIT NO.	FUNCTION
1	BRN/BLK	XF30	IAT SENSOR SIGNAL
2	RED/BLU	AP84	B+ FROM RELAY K43
3	BRN/GRN	XU304	SENSOR GROUND
4	BLU/BLK	PF107	5 V REFERENCE
5	GRY/RED	RA40	MAF SENSOR SIGNAL

2000-01 3.0L MAF Sensor Connector Pinout

SECONDARY AIR INJECTION (AIR) PUMP CONNECTOR

Connector Pinouts

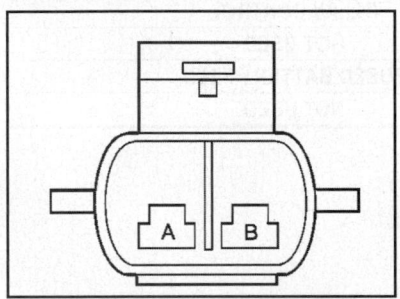

PIN	WIRE COLOR	CIRCUIT NO.	FUNCTION
A	PPL	Y110	PUMP ON SIGNAL
B	BLK	F110	GROUND TO G104

2000-01 3.0L Secondary Air Injection Pump Connector Pinout

SECONDARY AIR INJECTION (AIR) PUMP RELAY CONNECTOR

Connector Pinouts

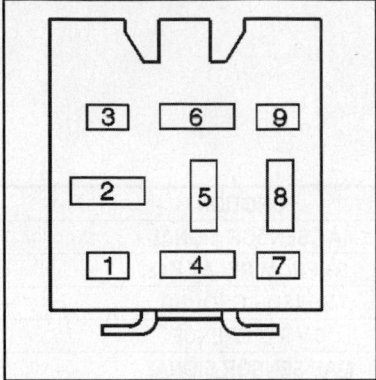

PIN	WIRE COLOR	CIRCUIT NO.	FUNCTION
1	--	--	NOT USED
2	PPL	Y110	PUMP ON SIGNAL
3	--	--	NOT USED
4	RED/BLU	AP200	B+ FROM RELAY K43
5	BLK	F200	GROUND TO G103
6	BRN/GRY	XR200	RELAY CONTROL
7	--	--	NOT USED
8	RED	A730	FUSED BATTERY FEED
9	--	--	NOT USED

2000-01 3.0L Secondary Air Injection Pump Relay Connector Pinout

SECONDARY AIR INJECTION (AIR) SOLENOID VALVE CONNECTOR

Connector Pinouts

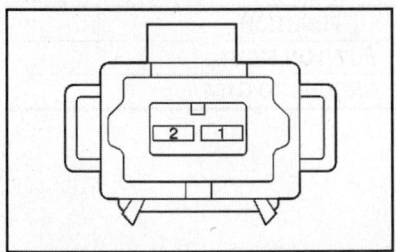

PIN	WIRE COLOR	CIRCUIT NO.	FUNCTION
1	RED/BLU	AP952	B+ FROM RELAY K43
2	BRN/YEL	XB952	SOLENOID VALVE CONTROL

2000-01 3.0L Secondary Air Injection Pump Solenoid Valve Connector Pinout

THROTTLE POSITION (TP) SENSOR CONNECTOR

Connector Pinouts

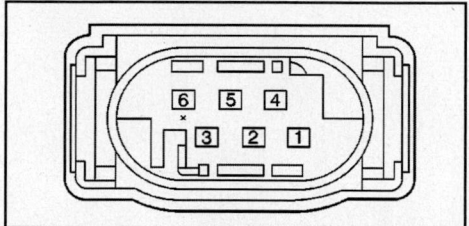

PIN	WIRE COLOR	CIRCUIT NO.	FUNCTION
1	BLK/WHT	FM10	THROTTLE VALVE MOTOR GROUND
2	BRN/GRN	XU30	SENSOR GROUND
3	BLU/BLK	PF102	5 V REFERENCE
4	BRN/YEL	XB10	THROTTLE VALVE MOTOR POWER
5	BRN/BLU	XP53	TP SENSOR 2 SIGNAL
6	WHT	M24	TP SENSOR 1 SIGNAL

2000-01 3.0L TP Sensor Connector Pinout

VEHICLE SPEED SENSOR (VSS) CONNECTOR

Connector Locations

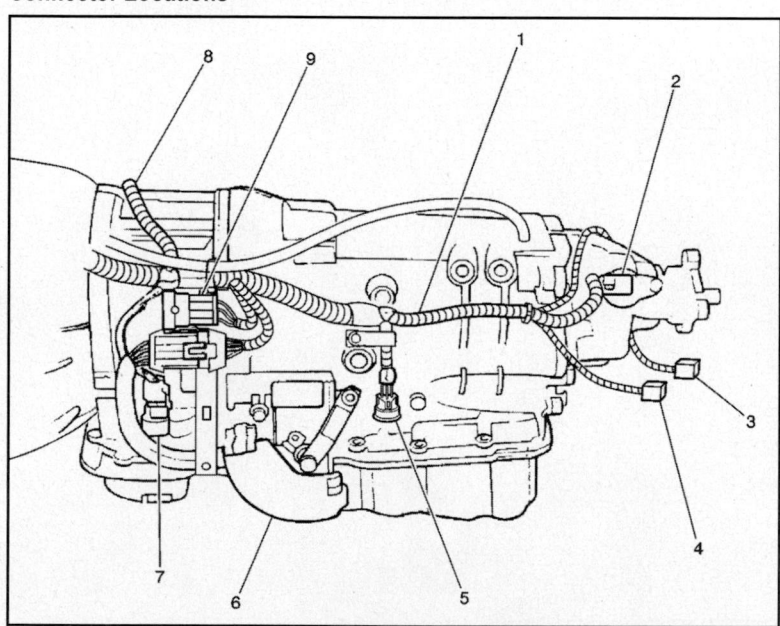

(1) Engine Control Module (ECM) Harness

(2) Vehicle Speed Sensor (VSS)

(3) Connector to Rear HO2S (Bank 1, Position 2)

(4) Connector to Rear HO2S (Bank 2, Position 2)

(5) Main Case Connector

(6) Automatic Transmission Manual Shaft Shift Position Switch

(7) Case Adapter

(8) Harness to Front HO2S (Bank 1, Position 1)

(9) Connector to Front HO2s (Bank 1, Position 1)

Indicating the locations of A/T electronic components on left side of transmission

Connector Pinouts

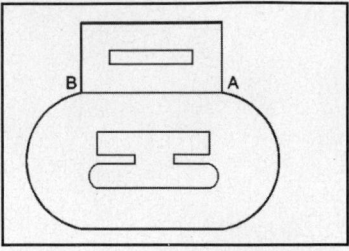

PIN	WIRE COLOR	CIRCUIT NO.	FUNCTION
A	BLK/YEL	FB703	VSS SIGNAL HI
B	BLU/GRN	PU705	VSS SIGNAL LO

2000-01 VSS Connector Pinout

GENERAL MOTORS CARS: 2000-05 CADILLAC DEVILLE

<u>ABS/TCC SWITCH CONNECTOR</u>

Connector Pinouts

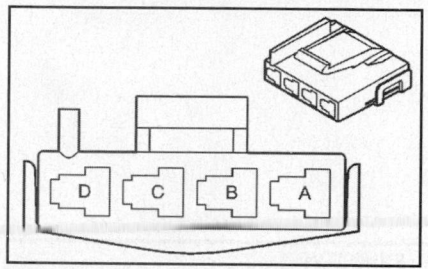

PIN	WIRE COLOR	CIRCUIT NO.	FUNCTION
A	PNK	1039	TCC BRAKE SWITCH POWER FEED
B	PPL	420	TCC BRAKE SWITCH INPUT
C	PNK	1039	ABS BRAKE SWITCH POWER FEED
D	GRY	847	ABS BRAKE SWITCH INPUT

2000-05 4.6L 4T80-E ABS/TCC Switch Connector Pinout

<u>A/C PRESSURE SENSOR CONNECTOR</u>

Connector Pinouts

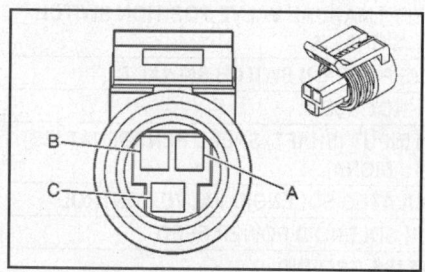

PIN	WIRE COLOR	CIRCUIT NO.	FUNCTION
A	BLK	2751	A/C PRESSURE SENSOR GROUND
B	GRY	2700	A/C PRESSURE 5 VOLT REFERENCE
C	RED/BLK	380	A/C PRESSURE SENSOR SIGNAL

2000 4.6L 4T80-E A/C Pressure Sensor Connector Pinout

A/T INLINE CONNECTORS

Connector Pinouts

ENGINE SIDE: 2000 MODELS

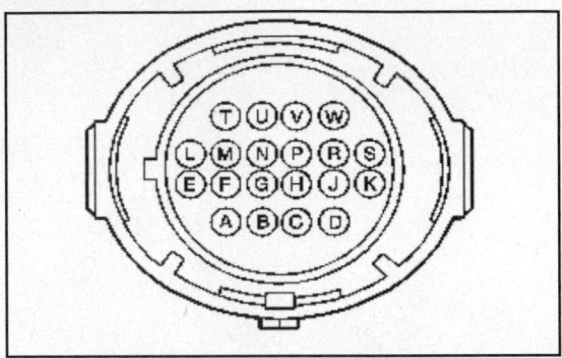

PIN	WIRE COLOR	CIRCUIT NO.	FUNCTION
A	LT GRN	1222	1-2 SHIFT SOLENOID VALVE CONTROL
B	YEL/BLK	1223	2-3 SHIFT SOLENOID VALVE CONTROL
C	RED/BLK	1228	PRESSURE CONTROL (PC) SOLENOID VALVE CONTROLLED POWER FEED
D	LT BLU/WHT	1229	PC SOLENOID VALVE CONTROLLED GROUND
E	PNK	739	TRANSMISSION SOLENOID POWER FEED
F	BLK/WHT	771	INTERNAL MODE SWITCH (IMS) SIGNAL A
G	YEL	772	IMS SIGNAL B
H	GRY	773	IMS SIGNAL C
J	WHT	776	IMS SIGNAL P
K	BLK/WHT	1551	IMS GROUND
L	YEL/BLK	1227	TRANSMISSION FLUID TEMPERATURE (TFT) SENSOR SIGNAL
M	BLK	2762	TFT SENSOR GROUND
N	PNK	1224	TRANS FLUID PRESSURE (TFP) MANUAL VALVE POSITION SWITCH SIGNAL A
P	RED	1226	TFP MANUAL VALVE POSITION SWITCH SIGNAL C
R	--	--	NOT USED
S	RED/BLK	1230	AUTOMATIC TRANSMISSION INPUT (SHAFT) SPEED SENSOR (AT ISS) SIGNAL
T	TAN/BLK	422	TCC PULSE WIDTH MODULATED SOLENOID VALVE CONTROL
U	PNK	739	TRANSMISSION SOLENOID POWER FEED
V	DK BLU/WHT	1231	AT ISS GROUND
W	--	--	NOT USED

2000 4.6L 4T80-E A/T Engine Side Inline Connector Pinout

TRANSMISSION SIDE: 2000 MODELS

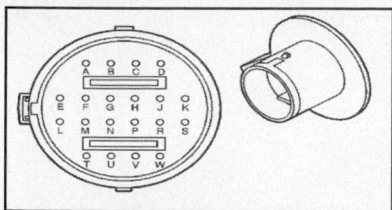

PIN	WIRE COLOR	CIRCUIT NO.	FUNCTION
A	LT GRN	1222	1-2 SHIFT SOLENOID VALVE CONTROL
B	YEL	1223	2-3 SHIFT SOLENOID VALVE CONTROL
C	PPL	1228	PRESSURE CONTROL (PC) SOLENOID VALVE CONTROLLED POWER FEED
D	LT BLU	1229	PC SOLENOID VALVE CONTROLLED GROUND
E	RED	839	TRANSMISSION SOLENOID POWER FEED
F	RED/BLK	771	INTERNAL MODE SWITCH (IMS) SIGNAL A
G	DK GRN/WHT	772	IMS SIGNAL B
H	YEL/BLK	773	IMS SIGNAL C
J	GRY/WHT	776	IMS SIGNAL P
K	BLK/WHT	1050	IMS GROUND
L	BRN	1227	TRANSMISSION FLUID TEMPERATURE (TFT) SENSOR SIGNAL
M	GRY	452	TFT SENSOR GROUND
N	PNK	1224	TRANS FLUID PRESSURE (TFP) MANUAL VALVE POSITION SWITCH SIGNAL A
P	ORN	1226	TFP MANUAL VALVE POSITION SWITCH SIGNAL C
R	--	--	NOT USED
S	BLK	1230	AUTOMATIC TRANSMISSION INPUT (SHAFT) SPEED SENSOR (AT ISS) SIGNAL
T	TAN	418	TCC PULSE WIDTH MODULATED SOLENOID VALVE CONTROL
U	WHT	1135	TRANSMISSION SOLENOID POWER FEED
V	DK GRN	1231	AT ISS GROUND
W	--	--	NOT USED

2000 4.6L 4T80-E A/T Transmission Side Inline Connector Pinout

ENGINE SIDE: 2001-05 MODELS

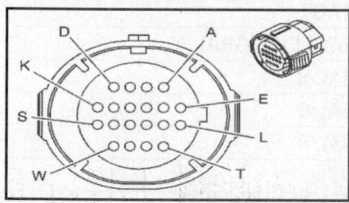

PIN	WIRE COLOR	CIRCUIT NO.	FUNCTION
A	LT GRN	1222	1-2 SHIFT SOLENOID VALVE CONTROL
B	YEL/BLK	1223	2-3 SHIFT SOLENOID VALVE CONTROL
C	RED/BLK	1228	PRESSURE CONTROL (PC) SOLENOID VALVE HIGH CONTROL
D	LT BLU/WHT	1229	PC SOLENOID VALVE LOW CONTROL
E	PNK	739	IGNITION 1 VOLTAGE
F	BLK/WHT	771	TRANSMISSION RANGE (TR) SWITCH SIGNAL A
G	YEL	772	TR SWITCH SIGNAL B
H	GRY	773	TR SWITCH SIGNAL C
J	WHT	776	TR SWITCH SIGNAL P
K	BLK	1550	GROUND
L	YEL/BLK	1227	TRANSMISSION FLUID TEMPERATURE SENSOR SIGNAL
M	BLK	2762	LOW REFERENCE
N	PNK	1224	TRANSMISSION FLUID PRESSURE (TFP) MANUAL VALVE POSITION SWITCH SIGNAL A
P	RED	1226	TFP MANUAL VALVE POSITION SWITCH SIGNAL C
R	--	--	NOT USED

2001-05 4T80-E A/T Engine Side In-Line Connector Pinout (1 of 2)

ENGINE SIDE: 2001-05 MODELS (CONTINUED)

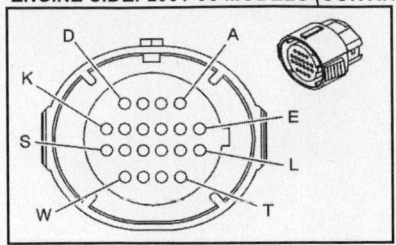

S	RED/BLK	1230	A/T INPUT (SHAFT) SPEED SENSOR (AT ISS) HIGH SIGNAL
T	TAN/BLK	422	TCC PULSE WIDTH MODULATED SOLENOID VALVE CONTROL
U	PNK	739	IGNITION 1 VOLTAGE
V	DK BLU/WHT	1231	AT ISS LOW SIGNAL
W	--	--	NOT USED

2001-05 4.6L 4T80-E A/T Engine Side In-Line Connector Pinout (2 of 2)

TRANSMISSION SIDE: 2001-05 MODELS

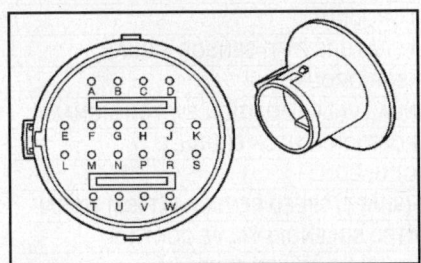

PIN	WIRE COLOR	CIRCUIT NO.	FUNCTION
A	LT GRN	1222	1-2 SHIFT SOLENOID VALVE CONTROL
B	YEL	1223	2-3 SHIFT SOLENOID VALVE CONTROL
C	PPL	1228	PRESSURE CONTROL (PC) SOLENOID VALVE HIGH CONTROL
D	LT BLU	1229	PC SOLENOID VALVE LOW CONTROL
E	RED	839	IGNITION 1 VOLTAGE
F	RED/BLK	771	TRANSMISSION RANGE (TR) SWITCH SIGNAL A
G	DK GRN/WHT	772	TR SWITCH SIGNAL B
H	YEL/BLK	773	TR SWITCH SIGNAL C
J	GRY/WHT	776	TR SWITCH SIGNAL P
K	BLK/WHT	1050	GROUND
L	BRN	1227	TRANSMISSION FLUID TEMPERATURE SENSOR SIGNAL
M	GRY	452	LOW REFERENCE
N	PNK	1224	TFP MANUAL VALVE POSITION SWITCH SIGNAL A
P	ORN	1226	TFP MANUAL VALVE POSITION SWITCH SIGNAL C
R	--	--	NOT USED
S	BLK	1230	A/T INPUT, SHAFT, SPEED SENSOR (AT ISS) HIGH SIGNAL
T	TAN	418	TCC PULSE WIDTH MODULATED SOLENOID VALVE CONTROL
U	WHT	1135	IGNITION 1 VOLTAGE
V	DK GRN	1231	AT ISS LOW SIGNAL
W	--	--	NOT USED

2001-05 4.6L 4T80-E A/T Transmission Side In-Line Connector Pinout

CAMSHAFT POSITION (CMP) SENSOR CONNECTOR

Connector Pinouts

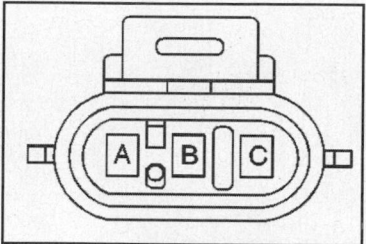

PIN	WIRE COLOR	CIRCUIT NO.	FUNCTION
A	ORN	1799	CMP SENSOR SIGNAL
B	PNK/BLK	632	CMP SENSOR GROUND
C	RED	631	CMP SENSOR IGNITION FEED

2000-05 4.6L CMP Sensor Connector Pinout

CRANKSHAFT POSITION (CKP) SENSOR CONNECTORS

Connector Pinouts

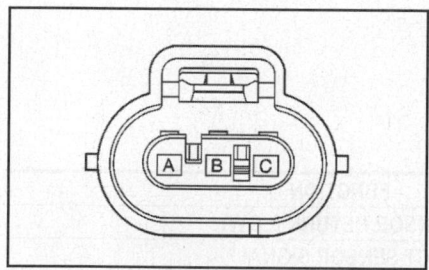

PIN	WIRE COLOR	CIRCUIT NO.	FUNCTION
A	LT GRN	1867	CKP SENSOR A IGNITION FEED
B	YEL/BLK	1868	CKP SENSOR A GROUND
C	LT BLU/WHT	1800	CKP SENSOR A SIGNAL

2000-05 4.6L CKP Sensor A Connector Pinout

PIN	WIRE COLOR	CIRCUIT NO.	FUNCTION
A	LT GRN	2867	CKP SENSOR B IGNITION FEED
B	YEL	2868	CKP SENSOR B GROUND
C	YEL/BLK	573	CKP SENSOR B SIGNAL

2000-05 4.6L CKP Sensor B Connector Pinout

ENGINE COOLANT LEVEL SWITCH CONNECTOR

Connector Pinouts

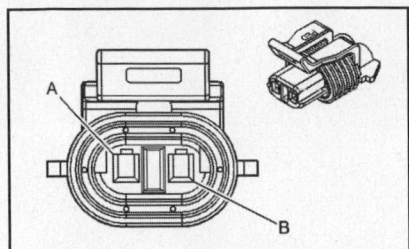

PIN	WIRE COLOR	CIRCUIT NO.	FUNCTION
A	LT GRN	1478	ENGINE COOLANT LEVEL SWITCH
B	BLK	450	GROUND

2000-05 4.6L Engine Coolant Level Switch Connector Pinout

ENGINE COOLANT TEMPERATURE (ECT) SENSOR CONNECTOR

Connector Pinouts

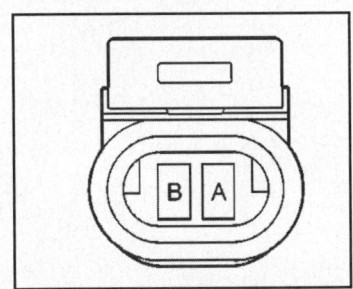

PIN	WIRE COLOR	CIRCUIT NO.	FUNCTION
A	BLK	470 OR 2761	SENSOR RETURN (LOW)
B	YEL	410	ECT SENSOR SIGNAL

2000-05 4.6L ECT Sensor Connector Pinout

ENGINE COOLING FAN CONNECTORS

Connector Pinouts

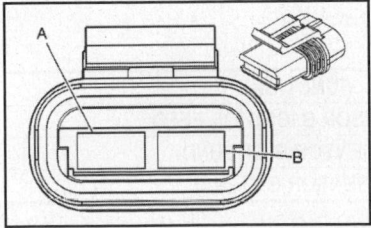

PIN	WIRE COLOR	CIRCUIT NO.	FUNCTION
A	WHT	504	COOLING FAN MOTOR OUTPUT
B	GRY	532	COOLING FAN MOTOR FEED - SECONDARY

2000-05 4.6L Right Engine Cooling Fan Connector Pinout

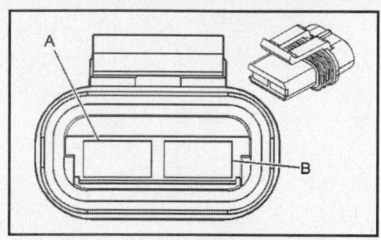

PIN	WIRE COLOR	CIRCUIT NO.	FUNCTION
A	BLK	250	GROUND
B	LT BLU	409	COOLING FAN MOTOR FEED

2000-05 4.6L Left Engine Cooling Fan Connector Pinout

EVAP CANISTER PURGE SOLENOID CONNECTOR

Connector Pinouts

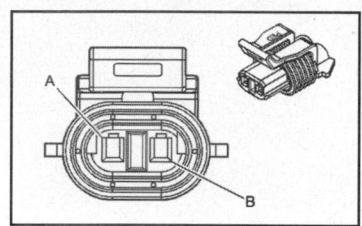

PIN	WIRE COLOR	CIRCUIT NO.	FUNCTION
A	PNK	139	IGN ITION VOLTAGE
B	DK GRN/WHT	428	EVAP CANISTER PURGE VALVE SOLENOID OUTPUT

2000-05 4.6L EVAP Purge Valve Solenoid Connector Pinout

EVAP CANISTER VENT VALVE CONNECTOR

Connector Pinouts

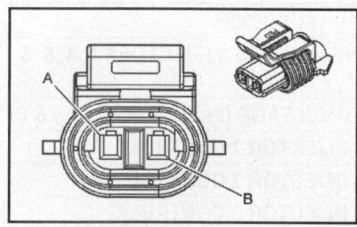

PIN	WIRE COLOR	CIRCUIT NO.	FUNCTION
A	PNK	1239	IGNITION VOLTAGE
B	WHT	1310	EVAP CANISTER VENT VALVE OUTPUT

2000-05 4.6L EVAP Vent Valve Connector Pinout

EXHAUST GAS RECIRCULATION (EGR) VALVE CONNECTOR

Connector Pinouts

PIN	WIRE COLOR	CIRCUIT NO.	FUNCTION
A	GRY	435	EGR VALVE GROUND
B	BLK	2753	EGR PINTLE POSITION GROUND
C	BRN	1456	EGR POSITION SIGNAL
D	GRY	2702	5 VOLT REFERENCE
E	RED	1676	EGR VALVE CONTROL

2000-05 4.6L EGR Valve Connector Pinout

FUEL INJECTOR CONNECTORS

Connector Pinouts

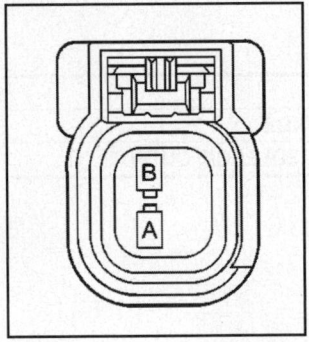

PIN	WIRE COLOR	CIRCUIT NO.	FUNCTION
A	PNK	239	IGNITION POSITIVE VOLTAGE (INJECTORS 1, 4, 6 & 7)
A	PNK	339	IGNITION POSITIVE VOLTAGE (INJECTORS 2, 3, 5 & 8)
B	BLK	1744	FUEL INJECTOR 1 CONTROL
B	LT GRN	1745	FUEL INJECTOR 2 CONTROL
B	PNK/BLK	1746	FUEL INJECTOR 3 CONTROL
B	LT BLU/BLK	844	FUEL INJECTOR 4 CONTROL
B	BLK/WHT	845	FUEL INJECTOR 5 CONTROL
B	YEL/BLK	846	FUEL INJECTOR 6 CONTROL
B	RED/BLK	877	FUEL INJECTOR 7 CONTROL
B	DK BLU/WHT	878	FUEL INJECTOR 8 CONTROL

2000-05 4.6L Fuel Injector Connector Pinouts

FUEL PUMP SENDER CONNECTOR

Connector Pinouts

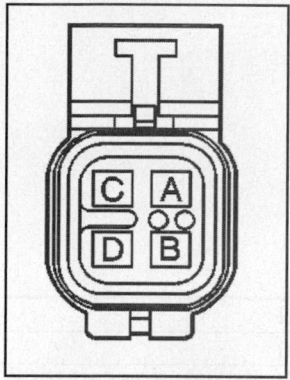

PIN	WIRE COLOR	CIRCUIT NO.	FUNCTION
A	PPL	1589	FUEL LEVEL SIGNAL
B	GRY	120	FUEL PUMP MOTOR FEED
C	BLK	850	FUEL PUMP MOTOR GROUND
D	BLK/WHT	2759	SENDER GROUND

2000-05 4.6L Fuel Pump Sender Connector Pinout

FUEL TANK PRESSURE (FTP) SENSOR CONNECTOR

Connector Pinouts

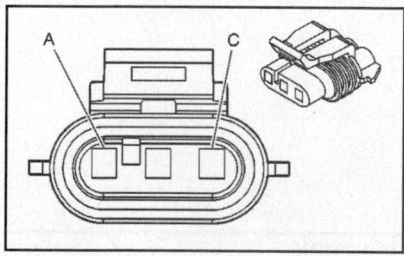

PIN	WIRE COLOR	CIRCUIT NO.	FUNCTION
A	BLK	2759	FUEL TANK PRESSURE SENSOR GROUND
B	DK GRN	890	FUEL TANK PRESSURE SENSOR SIGNAL
C	GRY	2709	5 VOLT REFERENCE

2000-05 4.6L FTP Sensor Connector Pinout

GENERATOR CONNECTOR

Connector Pinouts

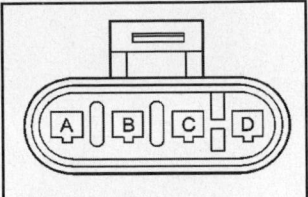

PIN	WIRE COLOR	CIRCUIT NO.	FUNCTION
A	--	--	NOT USED
B	RED	225	GENERATOR L TERMINAL TURN-ON SIGNAL/MONITOR FROM PCM
C	GRY	23	GENERATOR F TERMINAL FIELD DUTY SIGNAL/MONITOR FOR PCM
D	ORN	3340	SYSTEM VOLTAGE SENSE

2000-05 4.6L Generator Connector Pinout

HEATED OXYGEN SENSOR (HO2S) CONNECTORS

Connector Pinouts

HO2S 1/1

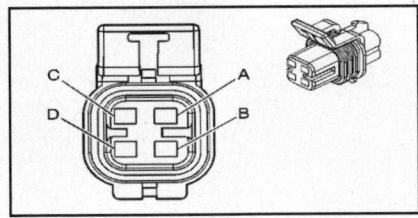

PIN	WIRE COLOR	CIRCUIT NO.	FUNCTION
A	TAN	1664	HEATED OXYGEN SENSOR (HO2S) LOW
B	PPL/WHT	1665	HEATED OXYGEN SENSOR (HO2S) HIGH
C	BLK	3112	HEATER LOW
D	DK GRN	676	HEATER HIGH

2000-03 4.6L O2 Sensor (Bank 1 Sensor 1) Connector Pinout

PIN	WIRE COLOR	CIRCUIT NO.	FUNCTION
A	TAN/WHT	3111	HO2S LOW SIGNAL (BANK 1 SENSOR 1)
B	PPL/WHT	3110	HO2S HIGH SIGNAL (BANK 1 SENSOR 1)
C	BLK/WHT	3113	HO2S HEATER LOW CONTROL (BANK 1 SENSOR 1)
D	PNK	539	IGNITION 1 VOLTAGE

2004-05 4.6L O2 Sensor (Bank 1 Sensor 1) Connector Pinout

HO2S 1/2

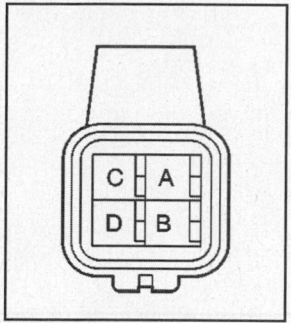

PIN	WIRE COLOR	CIRCUIT NO.	FUNCTION
A	DK BLU/WHT	1677	HEATED OXYGEN SENSOR (HO2S) LOW
B	DK GRN/WHT	1678	HEATED OXYGEN SENSOR (HO2S) HIGH
C	BLK/WHT	1423	HEATER LOW
D	PNK	539	IGNITION POSITIVE VOLTAGE

2000-03 4.6L O2 Sensor (Bank 1 Sensor 2) Connector Pinout

PIN	WIRE COLOR	CIRCUIT NO.	FUNCTION
A	TAN/WHT	3211	HO2S LOW SIGNAL (BANK 1 SENSOR 2)
B	PPL OR PPL/WHT	3210	HO2S HIGH SIGNAL (BANK 1 SENSOR 2)
C	BRN OR LT GRN	3212	HO2S HEATER LOW CONTROL (BANK 1 SENSOR 2)
D	PNK	539	IGNITION 1 VOLTAGE

2004-05 4.6L O2 Sensor (Bank 1 Sensor 2) Connector Pinout

HO2S 2/1

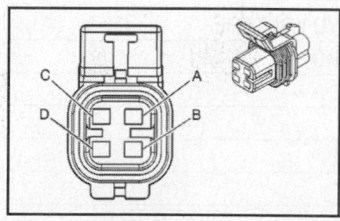

PIN	WIRE COLOR	CIRCUIT NO.	FUNCTION
A	TAN/WHT	1669	HEATED OXYGEN SENSOR (HO2S) LOW
B	PPL/WHT	1668	HEATED OXYGEN SENSOR (HO2S) HIGH
C	BRN	3122	HEATER LOW
D	PNK	539	IGNITION POSITIVE VOLTAGE

2000-03 4.6L O2 Sensor (Bank 2 Sensor 1) Connector Pinout

PIN	WIRE COLOR	CIRCUIT NO.	FUNCTION
A	TAN/WHT	3211	HO2S LOW SIGNAL (BANK 2 SENSOR 1)
B	PPL/WHT	3210	HO2S HIGH SIGNAL (BANK 2 SENSOR 1)
C	LT GRN	3212	HO2S HEATER LOW CONTROL (BANK 2 SENSOR 1)
D	PNK	539	IGNITION 1 VOLTAGE

2004 4.6L O2 Sensor (Bank 2 Sensor 1) Connector Pinout

IDLE AIR CONTROL (IAC) VALVE CONNECTOR

Connector Pinouts

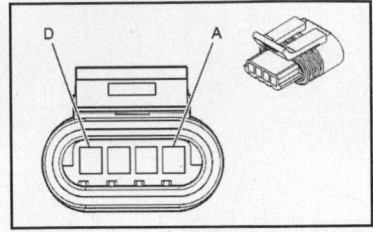

PIN	WIRE COLOR	CIRCUIT NO.	FUNCTION
A	LT GRN/BLK	444	IAC COIL B LOW
B	LT GRN/WHT	1749	IAC COIL B HIGH
C	LT BLU/BLK	1748	IAC COIL A LOW
D	LT BLU/WHT	1747	IAC COIL A HIGH

2000-05 6.4L IAC Valve Connector Pinout

IGNITION COIL/CONTROL MODULE (ICM) CONNECTORS

Connector Pinouts

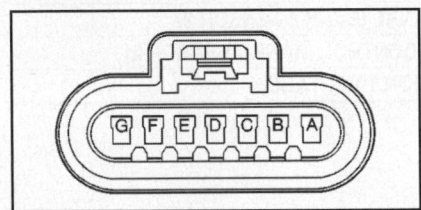

PIN	WIRE COLOR	CIRCUIT NO.	FUNCTION
A	PNK	839	IGNITION POSITIVE VOLTAGE
B	BLK/WHT	1551	GROUND
C	GRY	2175	REF LOW
D	DK GRN	2125	IC 5
E	LT BLU	2123	IC 3
F	PPL	2121	IC 1
G	RED	2127	IC 7

2000-03 4.6L ICM Bank 1 Connector Pinout

PIN	WIRE COLOR	CIRCUIT NO.	FUNCTION
A	PNK	839	IGNITION POSITIVE VOLTAGE
B	BLK/WHT	1551	GROUND
C	YEL	2174	REF LOW
D	DK GRN/WHT	2124	IC 4
E	LT BLU/WHT	2126	IC 6
F	PPL/WHT	2128	IC 8
G	RED/WHT	2122	IC 2

2000-03 4.6L ICM Bank 2 Connector Pinout

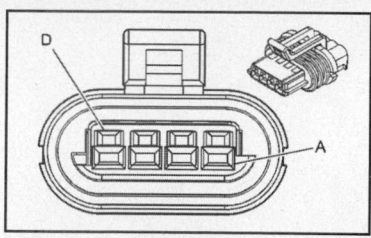

PIN	WIRE COLOR	CIRCUIT NO.	FUNCTION
A	PNK	839	IGNITION 1 VOLTAGE
B	RED	2121	IC 1 CONTROL (IGNITION COIL/MODULE 1)
	PPL	2122	IC 2 CONTROL (IGNITION COIL/MODULE 2)
	DK GRN	2123	IC 3 CONTROL (IGNITION COIL/MODULE 3)
	LT BLU	2124	IC 4 CONTROL (IGNITION COIL/MODULE 4)
	LT BLU	2125	IC 5 CONTROL (IGNITION COIL/MODULE 5)
	DK GRN	2126	IC 6 CONTROL (IGNITION COIL/MODULE 6)
	PPL	2127	IC 7 CONTROL (IGNITION COIL/MODULE 7)
	RED	2128	IC 8 CONTROL (IGNITION COIL/MODULE 8)
C	BRN	2129	LOW REFERENCE (BANK 1)
	BRN	2130	LOW REFERENCE (BANK 2)
D	BLK	1551	GROUND

2004-05 Ignition Coil/Module Connector Pinout

KNOCK SENSOR CONNECTOR

Connector Pinouts

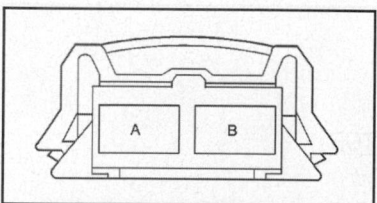

PIN	WIRE COLOR	CIRCUIT NO.	FUNCTION
A	DK BLU	496	KNOCK SENSOR SIGNAL
B	GRY	1716	KNOCK SENSOR RETURN

2000-03 4.6L Knock Sensor Connector Pinout

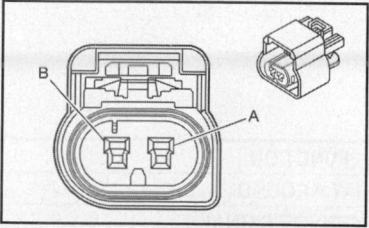

PIN	WIRE COLOR	CIRCUIT NO.	FUNCTION
A	DK BLU	496	KNOCK SENSOR 1 SIGNAL
B	GRY	1716	LOW REFERENCE

2004-05 4.6L Knock Sensor 1 Connector Pinout

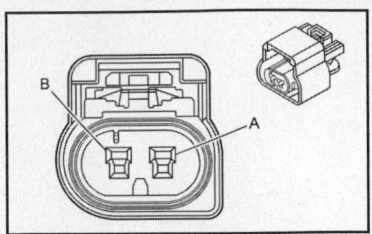

PIN	WIRE COLOR	CIRCUIT NO.	FUNCTION
A	LT BLU	1876	KNOCK SENSOR 2 SIGNAL
B	GRY	2303	LOW REFERENCE

2004-05 4.6L Knock Sensor 2 Connector Pinout

MANIFOLD ABSOLUTE PRESSURE (MAP) SENSOR CONNECTOR

Connector Pinouts

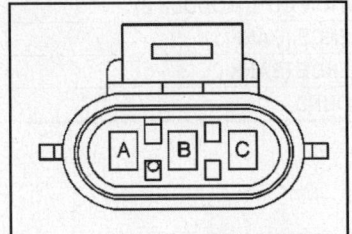

PIN	WIRE COLOR	CIRCUIT NO.	FUNCTION
A	ORN/BLK	469	MAP SENSOR GROUND
B	LT GRN	432	MAP SENSOR SIGNAL
C	GRY	2704	5 VOLT REFERENCE

2000-05 4.6L MAP Sensor Connector Pinout

MASS AIR FLOW/INTAKE AIR TEMPERATURE (MAF/IAT) SENSOR CONNECTOR

Connector Pinouts

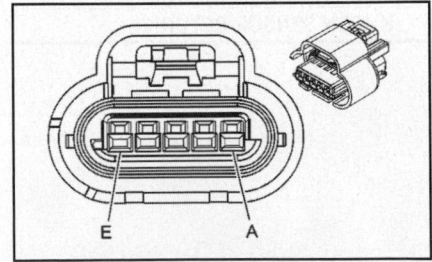

PIN	WIRE COLOR	CIRCUIT NO.	FUNCTION
A	BLK	2760	IAT GROUND
B	TAN	472	IAT SENSOR SIGNAL
C	BLK OR BLK/WHT	451 OR 1550	GROUND
D	PNK	139	IGNITION POSITIVE VOLTAGE
E	YEL	492	MAF SENSOR SIGNAL

2000-05 4.6L MAF/IAT Sensor Connector Pinout

POWERTRAIN CONTROL MODULE (PCM) CONNECTORS

Connector Pinouts

C1 CONNECTOR: 2000 MODELS

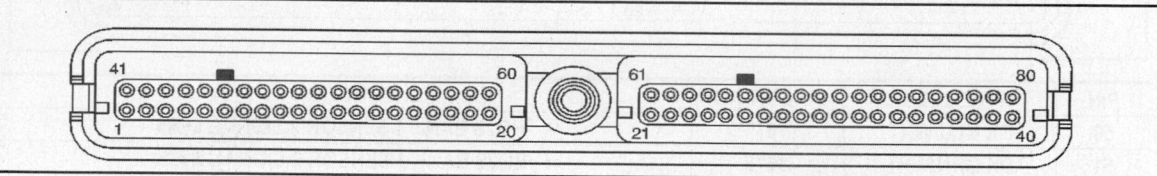

PIN	WIRE COLOR	CIRCUIT NO.	FUNCTION
1	ORN	1799	CAMSHAFT POSITION PCM INPUT
2	LT BLU/WHT	1800	CRANKSHAFT POSITION SENSOR A SIGNAL
3-4	--	--	NOT USED
5	DK GRN	890	FUEL TANK PRESSURE SENSOR SIGNAL
6	PPL	806	CRANK REQUEST
7-9	--	--	NOT USED
10	TAN	472	INTAKE AIR TEMPERATURE (IAT) SENSOR SIGNAL
11	PPL	773	TRANSAXLE RANGE SWITCH C INPUT
12	YEL	772	TRANSAXLE RANGE SWITCH B INPUT
13	BLK/WHT	771	TRANSAXLE RANGE SWITCH A INPUT
14	YEL	573	CRANKSHAFT POSITION SENSOR B SIGNAL
15	--	--	NOT USED
16	DK BLU/WHT	1231	TRANSAXLE INPUT SPEED SENSOR LOW
17	RED/BLK	1230	TRANSAXLE INPUT SPEED SENSOR HIGH
18	PPL	1807	CLASS 2 SERIAL DATA (BETWEEN EBTCM AND PCM)
19	PNK	1390	PCM IGNITION 0 FEED
20	ORN	2240	PCM BATTERY FEED
21	ORN	2240	PCM BATTERY FEED
22	RED	2127	CYLINDER 7 IC CONTROL
23	LT GRN	432	MANIFOLD ABSOLUTE PRESSURE (MAP) SENSOR SIGNAL
24	--	--	NOT USED
25	DK GRN	335	LOW SPEED COOLING FANS CONTROL CIRCUIT
26	DK GRN/WHT	817	4K/MI SPEED
27	LT GRN/BLK	444	IAC B LOW
28	LT GRN/WHT	1749	IAC B HIGH
29	LT BLU/WHT	1747	IAC A HIGH
30	LT BLU/BLK	1748	IAC A LOW
31	DK GRN	676	HO2S BANK 1 SENSOR 1 HEATER CONTROL CIRCUIT HIGH
32	BRN/WHT	419	MIL CONTROL
33	YEL/BLK	3122	HO2S BANK 2 SENSOR 1 HEATER CONTROL CIRCUIT LOW
34	TAN/BLK	464	DELIVERED TORQUE OUTPUT
35	DK GRN	83	CRUISE INHIBIT OUTPUT
36	--	--	NOT USED
37	BLK	1744	FUEL INJECTOR #1 CONTROL
38	PNK/BLK	1746	FUEL INJECTOR #3 CONTROL
39	LT GRN/BLK	1745	FUEL INJECTOR #2 CONTROL
40	BLK/WHT	845	FUEL INJECTOR #5 CONTROL
41	WHT	776	TRANSAXLE RANGE SWITCH PARITY INPUT
42	YEL	492	MASS AIR FLOW (MAF) SENSOR SIGNAL
43	ORN/BLK	463	DESIRED TORQUE INPUT
44	GRY	847	EXTENDED TRAVEL BRAKE SWITCH INPUT
45-49	--	--	NOT USED

2000 4.6L PCM C1 Connector Pinout (1 of 2)

C1 CONNECTOR: 2000 MODELS – CONTINUED

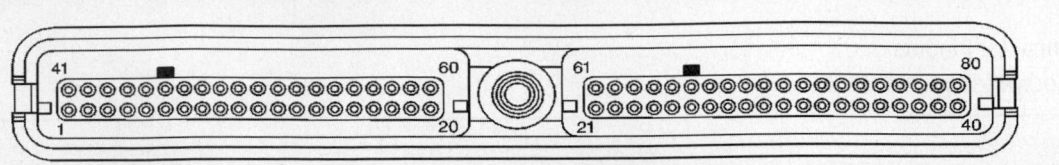

PIN	WIRE COLOR	CIRCUIT NO.	FUNCTION
50	DK BLU/WHT	1677	HO2S BANK 1 SENSOR 2 SIGNAL LOW
51	DK GRN/WHT	1678	HO2S BANK 1 SENSOR 2 SIGNAL HIGH
52	TAN/WHT	1664	HO2S BANK 1 SENSOR 1 SIGNAL LOW
53	PPL/WHT	1665	HO2S BANK 1 SENSOR 1 SIGNAL HIGH
54	TAN/WHT	1669	HO2S BANK 2 SENSOR 1 SIGNAL LOW
55	PPL/WHT	1668	HO2S BANK 2 SENSOR 1 SIGNAL HIGH
56	YEL	400	VSS SENSOR INPUT HIGH
57	PPL	401	VSS SENSOR INPUT LOW
58	RED	631	CAMSHAFT POSITION SENSOR IGNITION FEED
59	PPL	1807	CLASS 2 SERIAL DATA (BETWEEN IPC, DLC, AND PCM)
60	BLK/WHT	451	PCM POWER GROUND
61	PPL	2121	CYLINDER 1 IC CONTROL
62	DK BLU	417	THROTTLE POSITION (TP) SENSOR SIGNAL
63	PNK	39	PCM IGNITION 1 FEED
64-65	--	--	NOT USED
66	PNK/BLK	429	AIR PUMP SOLENOID CONTROL
67	LT BLU/WHT	1229	PRESSURE CONTROL SOLENOID LOW
68	YEL/BLK	3112	HO2S BANK 1 SENSOR 1 HEATER CONTROL CIRCUIT LOW
69	RED/BLK	1228	PRESSURE CONTROL SOLENOID HIGH
70	DK GRN/WHT	465	FUEL PUMP RELAY CONTROL
71	RED	1676	EGR VALVE CONTROL
72	PPL	1490	POWERTRAIN INDUCED CHASSIS PITCH STATUS OUTPUT (RTD LIFT/DIVE)
73	WHT	1310	EVAP VENT SOLENOID CONTROL
74	--	--	NOT USED
75	BRN	436	AIR PUMP A RELAY CONTROL
76	LT BLU/BLK	844	FUEL INJECTOR #4 CONTROL
77	YEL/BLK	846	FUEL INJECTOR #6 CONTROL
78	--	--	NOT USED
79	DK BLU/WHT	878	FUEL INJECTOR #8 CONTROL
80	RED/BLK	877	FUEL INJECTOR #7 CONTROL

2000 4.6L PCM C1 Connector Pinout (2 of 2)

C2 CONNECTOR: 2000 MODELS

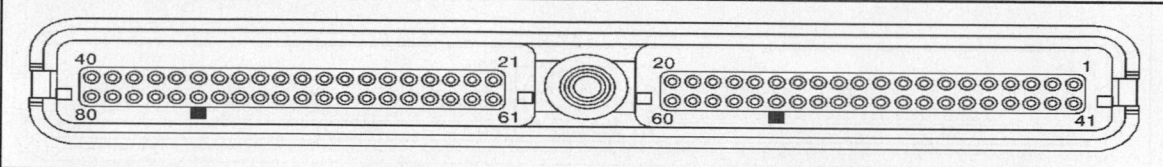

PIN	WIRE COLOR	CIRCUIT NO.	FUNCTION
1-2	--	--	NOT USED
3	BLK/WHT	451	PCM POWER GROUND
4	BLK	2760	INTAKE AIR TEMPERATURE (IAT) GROUND
5	--	--	NOT USED
6	ORN/BLK	469	MANIFOLD ABSOLUTE PRESSURE (MAP) SENSOR GROUND
7-8	--	--	NOT USED
9	YEL	2174	REFERENCE LOW (FRONT/EVEN BANK IGNITION ASSEMBLY)
10	--	--	NOT USED
11	BLK	2762	TRANSAXLE FLUID TEMPERATURE (TFT) SENSOR GROUND
12	BLK	2753	EGR PINTLE POSITION SENSOR GROUND
13	BLK	2759	FUEL LEVEL SENSOR AND FUEL TANK PRESSURE SENSOR GROUND
14	BLK	2751	AC PRESSURE SENSOR GROUND
15	BLK/WHT	451	PCM POWER GROUND
16	--	--	NOT USED
17	GRY	2709	5 VOLT REFERENCE B – FUEL TANK PRESSURE SENSOR FEED
18	--	--	NOT USED
19	DK BLU	496	KNOCK SENSOR SIGNAL
20	TAN/BLK	231	ENGINE OIL PRESSURE SWITCH INPUT
21	LT BLU	2123	CYLINDER #3 IC CONTROL
22	RED/WHT	2122	CYLINDER #2 IC CONTROL
23	BRN	1174	LOW OIL LEVEL SWITCH INPUT
24	LT BLU/BLK	396	CRUISE ENGAGED INPUT
25	LT GRN	2867	CRANKSHAFT POSITION SENSOR B IGNITION FEED
26	BLK/WHT	451	PCM POWER GROUND
27	--	--	NOT USED
28	YEL	2868	CRANKSHAFT POSITION SENSOR B GROUND
29-30	--	--	NOT USED
31	YEL	410	ENGINE COOLANT TEMPERATURE (ECT) SENSOR SIGNAL
32	YEL/BLK	1227	TRANSAXLE FLUID TEMPERATURE (TFT) SENSOR SIGNAL
33-34	--	--	NOT USED
35	GRY	1716	KNOCK SENSOR GROUND
36	--	--	NOT USED
37	RED	1226	TRANSAXLE PRESSURE MANIFOLD SWITCH C INPUT
38-39	--	--	NOT USED
40	RED	225	GENERATOR L TERMINAL CONTROL
41	--	--	NOT USED
42	DK GRN/WHT	428	EVAP CANISTER PURGE VALVE CONTROL
43	LT GRN	1222	SHIFT SOLENOID A CONTROL
44	TBD	447	STARTER RELAY CONTROL
45	YEL/BLK	1223	SHIFT SOLENOID B CONTROL
46	TAN/BLK	422	TORQUE CONVERTER CLUTCH (TCC) CONTROL
47	DK GRN/WHT	459	A/C COMPRESSOR CLUTCH CONTROL
48	BLK/WHT	1423	HO2S BANK 1 SENSOR 2 HEATER CONTROL CIRCUIT LOW
49	GRY	2704	5 VOLT REFERENCE A – MAP SENSOR FEED

2000 4.6L PCM C2 Connector (1 of 2)

C2 CONNECTOR: 2000 MODELS – CONTINUED

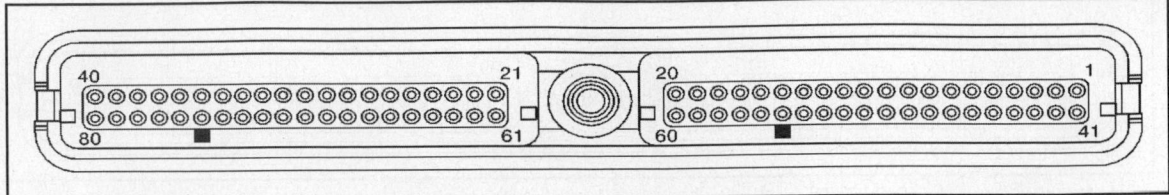

PIN	WIRE COLOR	CIRCUIT NO.	FUNCTION
50	GRY	2702	5 VOLT REFERENCE A - EGR PINTLE POSITION SENSOR FEED
51	GRY	2701	5 VOLT REFERENCE A - TP SENSOR FEED
52	BRN	1456	EGR PINTLE POSITION SENSOR SIGNAL
53	--	--	NOT USED
54	BRN	436	AIR PUMP 2 RELAY CONTROL
55	DK BLU	473	HIGH SPEED COOLING FANS CONTROL
56	WHT	121	TACHOMETER INPUT
57	GRY	2700	5 VOLT REFERENCE B - A/C PRESSURE SENSOR FEED
58	--	--	NOT USED
59	DK GRN/WHT	2124	CYLINDER #4 IC CONTROL
60	LT BLU/WHT	2126	CYLINDER #6 IC CONTROL
61	PPL/WHT	2128	CYLINDER #8 IC CONTROL
62	DK GRN	2125	CYLINDER #5 IC CONTROL
63-64	--	--	NOT USED
65	LT GRN	1867	CRANKSHAFT POSITION SENSOR A IGNITION FEED
66	YEL/BLK	1868	CRANKSHAFT POSITION SENSOR A GROUND
67	PNK/BLK	632	CAMSHAFT POSITION SENSOR GROUND
68	BLK	2761	ENGINE COOLANT TEMPERATURE (ECT) SENSOR GROUND
69	BLK	2752	THROTTLE POSITION (TP) SENSOR GROUND
70	GRY	435	EGR VALVE CONTROL GROUND
71	GRY	2175	REFERENCE LOW (REAR/ODD BANK IGNITION ASSEMBLY)
72	PPL	1589	FUEL LEVEL SENSOR SIGNAL
73	PNK	1224	TRANSAXLE PRESSURE MANIFOLD SWITCH A INPUT
74	RED/BLK	380	A/C PRESSURE SENSOR SIGNAL
75-78	--	--	NOT USED
79	GRY	23	GENERATOR F TERMINAL FEEDBACK
80	PPL	420	TORQUE CONVERTER CLUTCH (TCC) BRAKE SWITCH INPUT

2000 4.6L PCM C2 Connector (2 of 2)

C1 CONNECTOR: 2001-03 MODELS

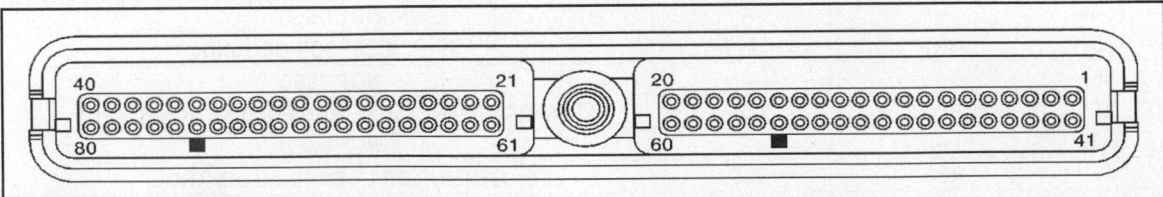

PIN	WIRE COLOR	CIRCUIT NO.	FUNCTION
1	ORN	1799	HIGH RESOLUTION SIGNAL
2	LT BLU/WHT	1800	CKP SENSOR A SIGNAL
3-4	--	--	NOT USED
5	DK GRN	890	FUEL TANK PRESSURE SENSOR SIGNAL
6	YEL	5	CRANK VOLTAGE
7-9	--	--	NOT USED

2001-03 4.6L PCM C1 Connector (1 of 3)

C1 CONNECTOR: 2001-03 MODELS (CONTINUED)

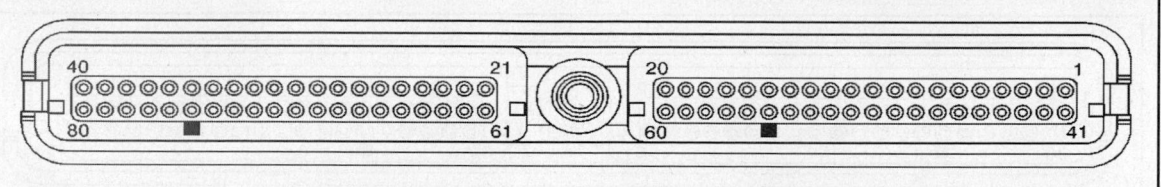

PIN	WIRE COLOR	CIRCUIT NO.	FUNCTION
10	TAN	472	IAT SENSOR SIGNAL
11	GRY	773	TRANSMISSION RANGE SWITCH SIGNAL C
12	YEL	772	TRANSMISSION RANGE SWITCH SIGNAL B
13	BLK/WHT	771	TRANSMISSION RANGE SWITCH SIGNAL A
14	YEL/BLK	573	CKP SENSOR B SIGNAL
15	--	--	NOT USED
16	DK BLU/WHT	1231	AT ISS LOW SIGNAL
17	RED/BLK	1230	AT ISS HIGH SIGNAL
18	PPL	1807	CLASS 2 SERIAL DATA
19	WHT	1390	OFF/RUN/CRANK VOLTAGE
20	ORN	240	BATTERY POSITIVE VOLTAGE
21	ORN	240	BATTERY POSITIVE VOLTAGE
22	RED	2127	IC 7 CONTROL
23	LT GRN	432	MAP SENSOR SIGNAL
24	--	--	NOT USED
25	DK GRN	335	LOW SPEED COOLING FAN RELAY CONTROL
26	DK GRN/WHT	817	VEHICLE SPEED SIGNAL
27	LT GRN/BLK	444	IAC COIL B LOW CONTROL
28	LT GRN/WHT	1749	IAC COIL B HIGH CONTROL
29	LT BLU/WHT	1747	IAC COIL A HIGH CONTROL
30	LT BLU/BLK	1748	IAC COIL A LOW CONTROL
31	DK GRN	676	HO2S HEATER HIGH CONTROL (BANK 1 SENSOR 1)
32	BRN/WHT	419	MIL CONTROL
33	BRN	3122	HO2S HEATER LOW CONTROL (BANK 2 SENSOR 1)
34	TAN/BLK	464	DELIVERED TORQUE SIGNAL
35	DK GRN	83	CRUISE CONTROL INHIBIT SIGNAL
36	--	--	NOT USED
37	BLK	1744	FUEL INJECTOR 1 CONTROL
38	PNK/BLK	1746	FUEL INJECTOR 3 CONTROL
39	LT GRN	1745	FUEL INJECTOR 2 CONTROL
40	BLK/WHT	845	FUEL INJECTOR 5 CONTROL
41	WHT	776	TRANSMISSION RANGE SWITCH SIGNAL P
42	YEL	492	MAF SENSOR SIGNAL
43	ORN/BLK	463	REQUESTED TORQUE SIGNAL
44	GRY	847	EXTENDED TRAVEL BRAKE SWITCH SIGNAL
45-49	--	--	NOT USED
50	DK BLU/WHT	1677	HO2S LOW SIGNAL (BANK 1 SENSOR 2)
51	DK GRN/WHT	1678	HO2S HIGH SIGNAL (BANK 1 SENSOR 2)
52	TAN	1664	HO2S LOW SIGNAL (BANK 1 SENSOR 1)
53	PPL/WHT	1665	HO2S HIGH SIGNAL (BANK 1 SENSOR 1)
54	TAN/WHT	1669	HO2S LOW SIGNAL (BANK 2 SENSOR 1)
55	PPL/WHT	1668	HO2S HIGH SIGNAL (BANK 2 SENSOR 1)
56	YEL	400	SIGNAL HIGH - FRONT
57	PPL	401	SIGNAL LOW - FRONT
58	RED	631	12 VOLT REFERENCE

2001-03 4.6L PCM C1 Connector (1 of 3)

C1 CONNECTOR: 2001-03 MODELS (CONTINUED)

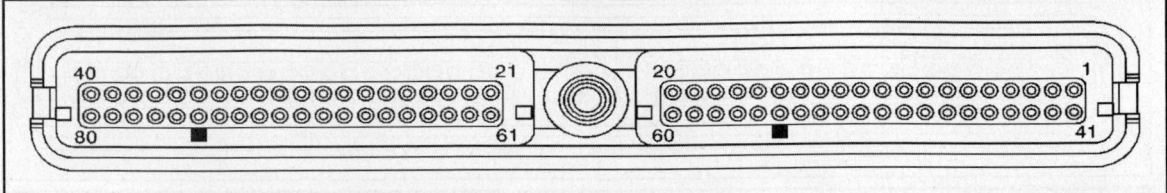

PIN	WIRE COLOR	CIRCUIT NO.	FUNCTION
59	PPL	1807	CLASS 2 SERIAL DATA
60	BLK/WHT	451	GROUND
61	PPL	2121	IC 1 CONTROL
62	DK BLU	417	TP SENSOR SIGNAL
63	PNK	39	IGNITION 1 VOLTAGE
64-65	--	--	NOT USED
66	PNK/BLK	429	AIR SOLENOID CONTROL
67	LT BLU/WHT	1229	PC SOLENOID VALVE LOW CONTROL
68	BLK	3112	H2OS HEATER LOW CONTROL (BANK 1 SENSOR 1)
69	RED/BLK	1228	PC SOLENOID VALVE HIGH CONTROL
70	DK GRN/WHT	465	FUEL PUMP RELAY CONTROL
71	RED	1676	EGR SOLENOID HIGH CONTROL
72	PPL	1490	CHASSIS PITCH SIGNAL
73	WHT	1310	EVAP CANISTER VENT SOLENOID CONTROL
74	--	--	NOT USED
75	BRN	436	AIR PUMP RELAY CONTROL
76	LT BLU/BLK	844	FUEL INJECTOR 4 CONTROL
77	YEL/BLK	846	FUEL INJECTOR 6 CONTROL
78	--	--	NOT USED
79	DK BLU/WHT	878	FUEL INJECTOR 8 CONTROL
80	RED/BLK	877	FUEL INJECTOR 7 CONTROL

2001-03 4.6L PCM C1 Connector (3 of 3)

C2 CONNECTOR: 2001-03 MODELS

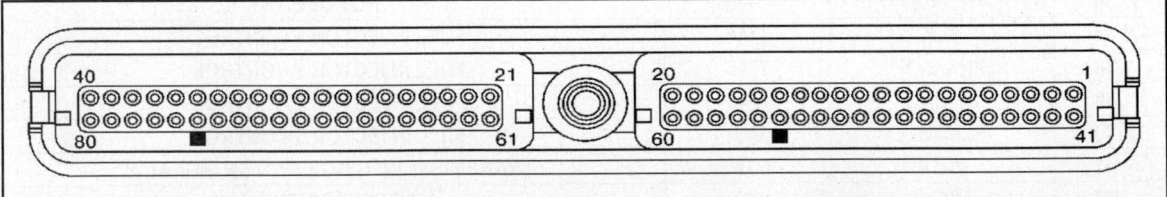

PIN	WIRE COLOR	CIRCUIT NO.	FUNCTION
1-2	--	--	NOT USED
3	BLK/WHT	451	GROUND
4	BLK	2760	LOW REFERENCE
5	--	--	NOT USED
6	ORN/BLK	469	LOW REFERENCE
7-8	--	--	NOT USED
9	YEL	2174	LOW REFERENCE
10	--	--	NOT USED

2001-03 4.6L PCM C2 Connector (1 of 3)

C2 CONNECTOR: 2001-03 MODELS – CONTINUED

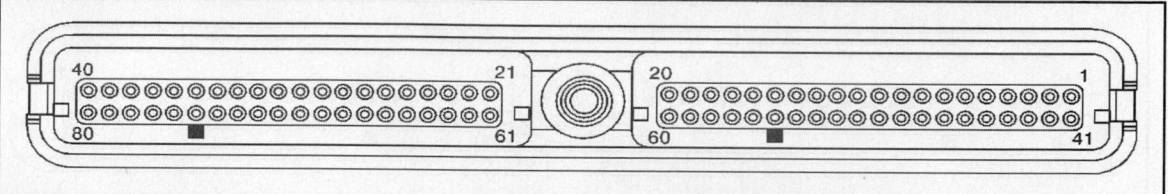

PIN	WIRE COLOR	CIRCUIT NO.	FUNCTION
11	BLK	2762	LOW REFERENCE
12	BLK	2753	LOW REFERENCE
13	BLK	2759	LOW REFERENCE
14	BLK	2751	LOW REFERENCE
15	BLK/WHT	451	GROUND
16	--	--	NOT USED
17	GRY	2709	5 VOLT REFERENCE
18	--	--	NOT USED
19	DK BLU	496	KNOCK SENSOR [1] SIGNAL
20	TAN/BLK	231	OIL PRESSURE SWITCH SIGNAL
21	LT BLU	2123	IC 3 CONTROL
22	RED/WHT	2122	IC 2 CONTROL
23	BRN	1174	OIL LEVEL SWITCH SIGNAL
24	LT BLU/BLK	396	CRUISE CONTROL ENGAGED SIGNAL
25	LT GRN	2867	12 VOLT REFERENCE
26	BLK/WHT	451	GROUND
27	--	--	NOT USED
28	YEL	2868	LOW REFERENCE
29-30	--	--	NOT USED
31	YEL	410	ECT SENSOR SIGNAL
32	YEL/BLK	1227	TFT SENSOR SIGNAL
33-34	--	--	NOT USED
35	GRY	1716	LOW REFERENCE
36	--	--	NOT USED
37	RED	1226	TRANSMISSION FLUID PRESSURE SWITCH SIGNAL C
38-39	--	--	NOT USED
40	RED	225	GENERATOR TURN ON SIGNAL
41	--	--	NOT USED
42	DK GRN/WHT	428	EVAP CANISTER PURGE SOLENOID CONTROL
43	LT GRN	1222	1-2 SHIFT SOLENOID VALVE CONTROL
44	YEL	447	STARTER RELAY COIL CONTROL
45	YEL/BLK	1223	2-3 SHIFT SOLENOID VALVE CONTROL
46	TAN/BLK	422	TCC SOLENOID VALVE CONTROL
47	DK GRN/WHT	459	A/C COMPRESSOR CLUTCH RELAY CONTROL
48	BLK/WHT	1423	HO2S HEATER LOW CONTROL (BANK 1 SENSOR 2)
49	GRY	2704	5 VOLT REFERENCE
50	GRY	2702	5 VOLT REFERENCE
51	GRY	2701	5 VOLT REFERENCE
52	BRN	1456	EGR VALVE POSITION SIGNAL
53	--	--	NOT USED
54	GRY	904	IGNITION COIL SUPPLY VOLTAGE
55	DK BLU	473	HIGH SPEED COOLING FAN RELAY CONTROL
56	WHT	121	ENGINE SPEED SIGNAL
57	GRY	2700	5 VOLT REFERENCE
58	--	--	NOT USED
59	DK GRN/WHT	2124	IC 4 CONTROL

2001-03 4.6L PCM C2 Connector (2 of 3)

C2 CONNECTOR: 2001-03 MODELS – CONTINUED

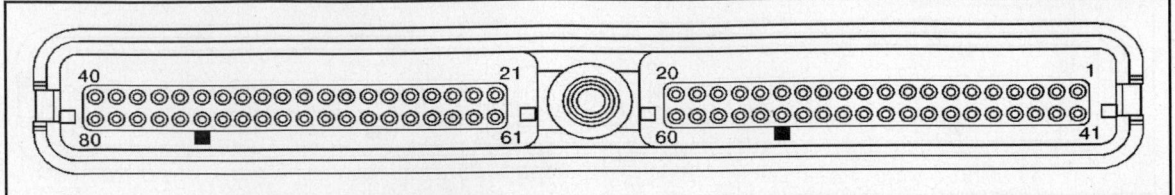

PIN	WIRE COLOR	CIRCUIT NO.	FUNCTION
60	LT BLU/WHT	2126	IC 6 CONTROL
61	PPL/WHT	2128	IC 8 CONTROL
62	DK GRN	2125	IC 5 CONTROL
63-64	--	--	NOT USED
65	LT GRN	1867	12 VOLT REFERENCE
66	YEL/BLK	1868	LOW REFERENCE
67	PNK/BLK	632	LOW REFERENCE
68	BLK	2761	LOW REFERENCE
69	BLK	2752	LOW REFERENCE
70	GRY	435	EGR SOLENOID CONTROL
71	GRY	2175	LOW REFERENCE
72	PPL	1589	FUEL LEVEL SENSOR SIGNAL
73	PNK	1224	TRANSMISSION FLUID PRESSURE SWITCH SIGNAL A
74	RED/BLK	380	A/C REFRIGERANT PRESSURE SENSOR SIGNAL
75-78	--	--	NOT USED
79	GRY	23	GENERATOR FIELD DUTY CYCLE SIGNAL
80	PPL	420	TCC BRAKE SWITCH/CRUISE CONTROL RELEASE SIGNAL

2001-03 4.6L PCM C2 Connector (3 of 3)

C1 CONNECTOR: 2004-05 MODELS

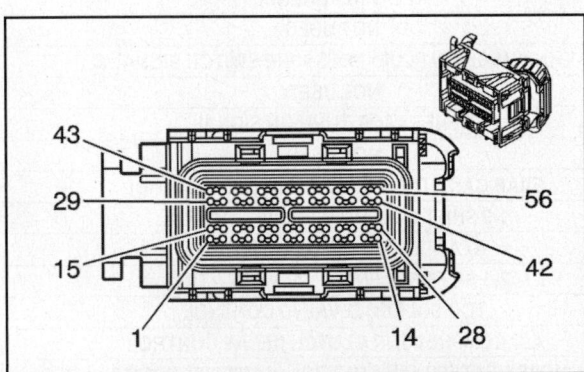

PIN	WIRE COLOR	CIRCUIT NO.	FUNCTION
1-2	--	--	NOT USED
3	BRN	436	AIR PUMP RELAY CONTROL
4	DK BLU	473	HIGH SPEED COOLING FAN RELAY CONTROL
5-7	--	--	NOT USED
8	DK GRN/ WHT	817	VEHICLE SPEED SIGNAL
9-10	--	--	NOT USED
11	RED/BLK	380	A/C REFRIGERANT PRESSURE SENSOR SIGNAL
12	PPL	1589	FUEL LEVEL SENSOR SIGNAL - PRIMARY
13-14	--	--	NOT USED
15-16	PPL	1807	CLASS 2 SERIAL DATA (PRIMARY)

2004-05 4.6L PCM C1 Connector (1 of 2)

C1 CONNECTOR: 2004-05 MODELS (CONTINUED)

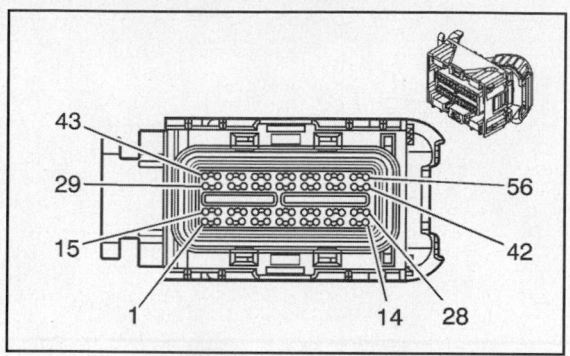

PIN	WIRE COLOR	CIRCUIT NO.	FUNCTION
17	--	--	NOT USED
18	WHT	1310	EVAP CANISTER VENT SOLENOID CONTROL
19	PNK	39	IGNITION 1 VOLTAGE
20	ORN	240	BATTERY POSITIVE VOLTAGE
21	--	--	NOT USED
22	DK GRN	83	CRUISE CONTROL INHIBIT SIGNAL
23	WHT	121	ENGINE SPEED SIGNAL
24-30	--	--	NOT USED
31	GRY	2709	5-VOLT REFERENCE
32	BLK	2751	LOW REFERENCE
33-34	--	--	NOT USED
35	BLK	2759	LOW REFERENCE
36-37	--	--	NOT USED
38	WHT	1390	OFF/RUN/CRANK VOLTAGE
39	GRY	847	EXTENDED TRAVEL BRAKE SWITCH SIGNAL
40	--	--	NOT USED
41	DK GRN	890	FUEL TANK PRESSURE SENSOR SIGNAL
42	--	--	NOT USED
43	GRY	2700	5-VOLT REFERENCE
44	DK GRN/ WHT	465	FUEL PUMP RELAY CONTROL - PRIMARY
45	YEL	447	STARTER RELAY COIL CONTROL
46	DK GRN	335	LOW SPEED COOLING FAN RELAY CONTROL
47	--	--	NOT USED
48	LT BLU/BLK	396	CRUISE CONTROL ENGAGED SIGNAL
49-50	--	--	NOT USED
51	PPL	420	TCC BRAKE SWITCH/CRUISE CONTROL RELEASE SIGNAL
52	YEL	5	CRANK VOLTAGE
53	PNK/ WHT	1101	DAMPING LIFT/DIVE SIGNAL
54	--	--	NOT USED
55	BRN/ WHT	419	MIL CONTROL
56	DK GRN/ WHT	459	A/C COMPRESSOR CLUTCH RELAY CONTROL

2004-05 4.6L PCM C1 Connector Pinout (1 of 2)

C2 CONNECTOR: 2004-05 MODELS

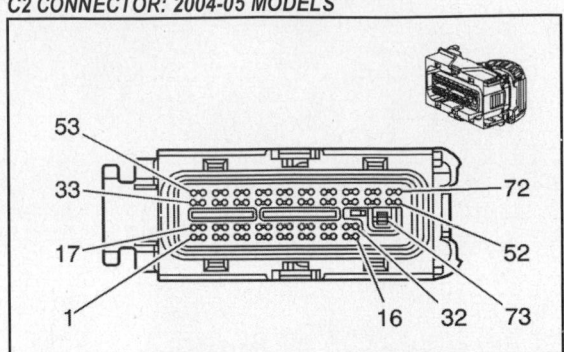

PIN	WIRE COLOR	CIRCUIT NO.	FUNCTION
1	BRN	1174	OIL LEVEL SWITCH SIGNAL
2	RED	1676	EGR SOLENOID HIGH CONTROL
3	PPL/WHT	421	REAR AIR PUMP RELAY CONTROL
4	--	--	NOT USED
5	TAN/ WHT	3111	HO2S LOW SIGNAL (BANK 1 SENSOR 1)
6	GRY	2702	5-VOLT REFERENCE
7	--	--	NOT USED
8	GRY	2701	5-VOLT REFERENCE
9	GRY	435	EGR SOLENOID LOW CONTROL OR EGR SOLENOID CONTROL
10	RED	225	GENERATOR TURN ON SIGNAL
11	LT BLU	1876	KNOCK SENSOR 2 SIGNAL
12	GRY	23	GENERATOR FIELD DUTY CYCLE SIGNAL
13	GRY	1716	KNOCK SENSOR 1 (KS1) LOW REFERENCE
14	--	--	NOT USED
15	PPL	3120	HO2S HIGH SIGNAL (BANK 1 SENSOR 2)
16	BRN	1456	EGR VALVE POSITION SIGNAL
17	LT GRN/ WHT	1749	IAC COIL B HIGH CONTROL
18	YEL	410	ECT SENSOR SIGNAL
19	DK BLU	496	KNOCK SENSOR 1 SIGNAL
20	TAN/ WHT	3121	HO2S LOW SIGNAL (BANK 1 SENSOR 2)
21	PNK/BLK	632	LOW REFERENCE
22	DK BLU	417	TP SENSOR SIGNAL
23	PPL	2121	IC 1 CONTROL
24	LT BLU/WHT	2126	IC 6 CONTROL
25	ORN	1799	HIGH RESOLUTION SIGNAL
26	--	--	NOT USED
27	GRY	2303	LOW REFERENCE KNOCK SENSOR 2 (KS2)
28	YEL/BLK	573	CRANK SENSOR B SIGNAL
29	PPL/WHT	3110	HO2S HIGH SIGNAL (BANK 1 SENSOR 1)
30	GRY	2704	5-VOLT REFERENCE (MAP)
31	LT BLU/WHT	1800	CRANK SENSOR A SIGNAL
32	--	--	NOT USED
33	BLK	2761	LOW REFERENCE (COOLANT SENSOR)
34	RED	2127	IC 7 CONTROL
35	BLK	2753	LOW REFERENCE (EGR)
36	ORN/BLK	469	LOW REFERENCE (MAP)
37	BLK	2752	LOW REFERENCE (TPS)
38	--	--	NOT USED
39	BRN	2129	LOW REFERENCE (EST BANK 1)
40	--	--	NOT USED
41	YEL/BLK	1868	LOW REFERENCE
42	BRN/ WHT	2130	LOW REFERENCE (EST BANK 2)
43	LT BLU	2123	IC 3 CONTROL

C2 CONNECTOR: 2004-05 MODELS (continued)

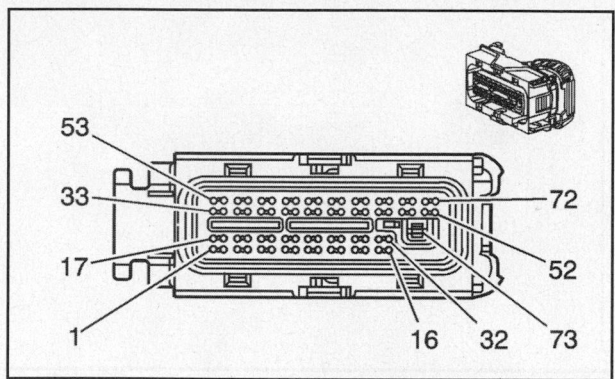

PIN	WIRE COLOR	CIRCUIT NO.	FUNCTION
44-45	--	--	NOT USED
46	YEL	2868	LOW REFERENCE (CRANK RETURN)
47	TAN/BLK	231	OIL PRESSURE SWITCH SIGNAL
48	--	--	NOT USED
49	RED	631	12-VOLT REFERENCE
50	LT GRN	1867	12-VOLT REFERENCE
51	LT GRN	2867	12-VOLT REFERENCE
52	--	--	NOT USED
53	LT BLU/BLK	1748	IAC COIL A LOW CONTROL
54	BLK/WHT	3113	HO2S HEATER LOW CONTROL (BANK 1 SENSOR 1)
55	BLK/WHT	845	FUEL INJECTOR 5 CONTROL
56	BRN	3122	HO2S HEATER LOW CONTROL (BANK 1 SENSOR 2)
57	DK BLU/WHT	878	FUEL INJECTOR 8 CONTROL
58	LT GRN/BLK	444	IAC COIL B LOW CONTROL
59	LT BLU/BLK	844	FUEL INJECTOR 4 CONTROL
60	YEL/BLK	846	FUEL INJECTOR 6 CONTROL
61	LT BLU/WHT	1747	IAC COIL A HIGH CONTROL
62	DK GRN/WHT	2124	IC 4 CONTROL
63	PPL/WHT	2128	IC 8 CONTROL
64	DK GRN	2125	IC 5 CONTROL
65	RED/WHT	2122	IC 2 CONTROL
66	PNK/BLK	1746	FUEL INJECTOR 3 CONTROL
67	LT GRN	432	MAP SENSOR SIGNAL
68	--	--	NOT USED
69	RED/BLK	877	FUEL INJECTOR 7 CONTROL
70	DK GRN/WHT	428	EVAP CANISTER PURGE SOLENOID CONTROL
71	BLK	1744	FUEL INJECTOR 1 CONTROL
72	LT GRN	1745	FUEL INJECTOR 2 CONTROL
73	BLK	451	GROUND

2004-05 4.6L PCM C2 Connector Pinout

C3 CONNECTOR: 2004-05 MODELS

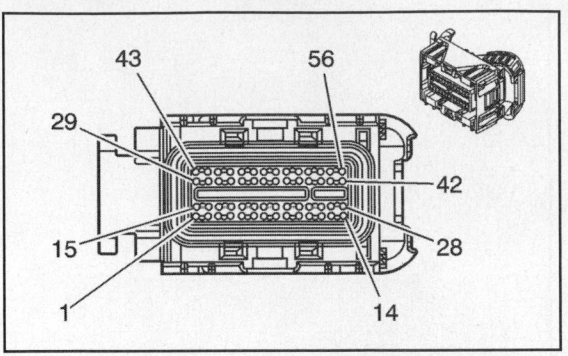

PIN	WIRE COLOR	CIRCUIT NO.	FUNCTION
1	PNK	1224	TRANSMISSION FLUID PRESSURE SWITCH SIGNAL A
2	YEL/BLK	1223	2-3 SHIFT SOLENOID OR SHIFT SOLENOID B VALVE CONTROL
3	LT BLU/WHT	1229	PC SOLENOID VALVE LOW CONTROL (SOL. A)
4	TAN/BLK	464	DELIVERED TORQUE SIGNAL
5	YEL	772	TRANSMISSION RANGE SWITCH SIGNAL B
6	GRY	773	TRANSMISSION RANGE SWITCH SIGNAL C
7	WHT	776	TRANSMISSION RANGE SWITCH SIGNAL P
8	--	--	NOT USED
9	--	--	NOT USED
10	YEL/BLK	1227	TFT SENSOR SIGNAL
11	YEL	492	MAF SENSOR SIGNAL
12-15	--	--	NOT USED
16	LT GRN	3212	HO2S HEATER LOW CONTROL (BANK 2 SENSOR 1)
17-18	--	--	NOT USED
19	TAN/ WHT	3211	HO2S LOW SIGNAL (BANK 2 SENSOR 1)
20-21	--	--	NOT USED
22	PPL	401	SIGNAL LOW - FRONT/VSS LOW SIGNAL/OSS LOW SIGNAL
23-26	--	--	NOT USED
27	ORN/BLK	463	REQUESTED TORQUE SIGNAL
28-31	--	--	NOT USED
32	BLK/WHT	771	TRANSMISSION RANGE SWITCH SIGNAL A
33	DK BLU/WHT	1231	AT ISS LOW SIGNAL
34	BLK	2760	LOW REFERENCE (IAT)
35	BLK	2762	LOW REFERENCE (TRANS FLUID TEMP SENSOR)
36-42	--	--	NOT USED
43	PPL/WHT	3210	HO2S HIGH SIGNAL (BANK 2 SENSOR 1)
44	LT GRN	1222	VALVE CONTROL (1-2 SHIFT SOLENOID OR SHIFT SOLENOID A)
45	RED/BLK	1228	PC SOLENOID VALVE HIGH CONTROL (SOL. A)
46	--	--	NOT USED
47	YEL	400	SIGNAL HIGH - FRONT/VSS HIGH SIGNAL/OSS HIGH SIGNAL
48	RED	1226	TRANSMISSION FLUID PRESSURE SWITCH SIGNAL C
49-50	--	--	NOT USED
51	TAN	472	IAT SENSOR SIGNAL
52-53	--	--	NOT USED
54	RED/BLK	1230	AT ISS HIGH SIGNAL
55	TAN/BLK	422	TCC SOLENOID VALVE CONTROL
56	--	--	NOT USED

2004-05 PCM C3 Connector Pinout

SECONDARY AIR INJECTION (AIR) PUMP CONNECTORS

Connector Pinouts

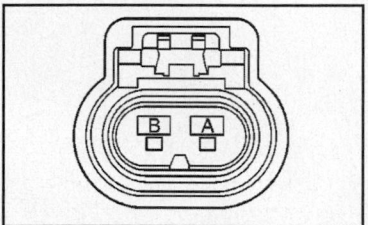

PIN	WIRE COLOR	CIRCUIT NO.	FUNCTION
A	BLK	1650	SECONDARY AIR INJECTION (AIR) PUMP MOTOR 1 GROUND
B	RED	78	SECONDARY AIR INJECTION (AIR) PUMP MOTOR 1 FEED

2000-05 4.6L Secondary Air Pump Motor Connector Pinout (Motor 1 on 2000 Only)

PIN	WIRE COLOR	CIRCUIT NO.	FUNCTION
A	BLK	1650	SECONDARY AIR INJECTION (AIR) PUMP MOTOR 2 GROUND
B	DK BLU	2202	SECONDARY AIR INJECTION (AIR) PUMP MOTOR 2 FEED

2000 4.6L Secondary Air Pump Motor 2 Connector Pinout (2000 Only)

SECONDARY AIR INJECTION (AIR) PUMP RELAYS

Connector Pinouts

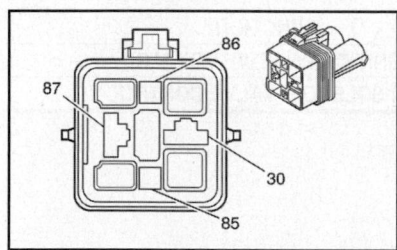

PIN	WIRE COLOR	CIRCUIT NO.	FUNCTION
30	RED	742	BATTERY POSITIVE VOLTAGE
85	PNK	139	IGNITION POSITIVE VOLTAGE - COIL FEED
86	BRN	436	SECONDARY AIR INJECTION (AIR) PUMP RELAY 1 CONTROL
87	RED	78	SECONDARY AIR INJECTION (AIR) PUMP MOTOR 1 FEED

2000-05 4.6L Secondary Air Injection Pump Relay 1 Connector Pinout

PIN	WIRE COLOR	CIRCUIT NO.	FUNCTION
30	RED	842	BATTERY POSITIVE VOLTAGE
85	PNK	139	IGNITION POSITIVE VOLTAGE - COIL FEED
86	BRN	1408	SECONDARY AIR INJECTION (AIR) PUMP RELAY 2 CONTROL
87	DK BLU	2202	SECONDARY AIR INJECTION (AIR) PUMP MOTOR 2 FEED

2000 4.6L Secondary Air Injection Pump Relay 2 Connector Pinout (2000 Only)

SECONDARY AIR INJECTION (AIR) SHUTOFF VALVE ASSEMBLY CONNECTOR

Connector Pinouts

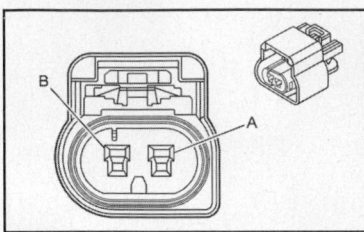

PIN	WIRE COLOR	CIRCUIT NO.	FUNCTION
A	PNK/BLK	415	SECONDARY AIR INJECTION (AIR) CHECK VALVE SUPPLY VOLTAGE
B	BLK	1650	GROUND

2004-05 4.6L AIR Shutoff Valve Assembly Connector Pinout

SECONDARY AIR INJECTION (AIR) SOLENOID VALVE

Connector Pinouts

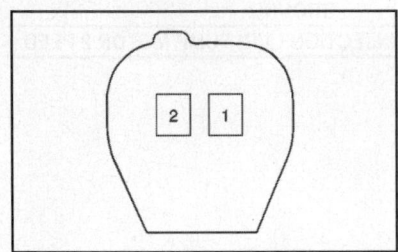

PIN	WIRE COLOR	CIRCUIT NO.	FUNCTION
1	PNK	139	IGNITION POSITIVE VOLTAGE
2	PNK/BLK	429	AIR SOLENOID VALVE CONTROL

2000-03 4.6L Secondary Air Injection Solenoid Valve Connector Pinout

THROTTLE POSITION (TP) SENSOR CONNECTOR

Connector Pinouts

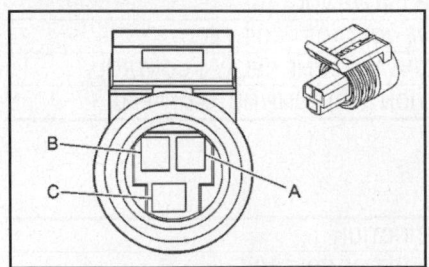

PIN	WIRE COLOR	CIRCUIT NO.	FUNCTION
A	GRY	2701	5 VOLT REFERENCE
B	BLK	2752	THROTTLE POSITION SENSOR GROUND
C	DK BLU	417	THROTTLE POSITION SENSOR SIGNAL

2000-05 4.6L TP Sensor Connector Pinout

TRACTION CONTROL SWITCH CONNECTOR

Connector Pinouts

W/ CHASSIS CONTINUOUS VALVE REAL TIME DAMPING (F45)

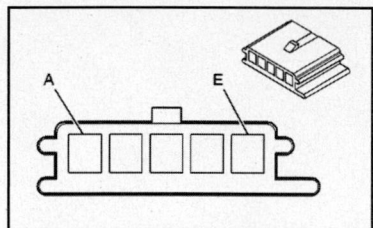

PIN	WIRE COLOR	CIRCUIT NO.	FUNCTION
A	--	--	NOT USED
B	BLK	450	GROUND
C	LT BLU	1788	TRACTION CONTROL SWITCH SIGNAL
D	BLK/WHT	451	GROUND
E	YEL	1491	BACKLIGHT LAMPS CONTROL

2001-02 4.6L (w/F45) Traction Control Switch Connector Pinout

W/O CHASSIS CONTINUOUS VALVE REAL TIME DAMPING (F45)

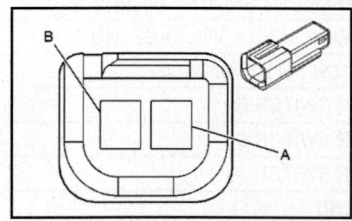

PIN	WIRE COLOR	CIRCUIT NO.	FUNCTION
A	LT BLU	1788	TRACTION CONTROL SWITCH SIGNAL
B	BLK	450	GROUND

2001-02 4.6L (w/o F45) Traction Control Switch Connector Pinout

VEHICLE SPEED SENSOR (VSS) CONNECTOR

Connector Pinouts

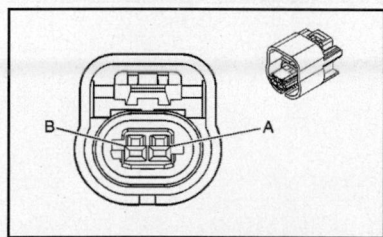

PIN	WIRE COLOR	CIRCUIT NO.	FUNCTION
A	PPL	401	VSS LOW SIGNAL (GROUND)
B	YEL	400	VSS HIGH SIGNAL

2000-05 4.6L 4T80-E A/T Vehicle Speed Sensor Connector Pinout

GENERAL MOTORS CARS: 2000-02 CADILLAC ELDORADO

A/T INLINE CONNECTORS

Connector Pinouts

ENGINE SIDE

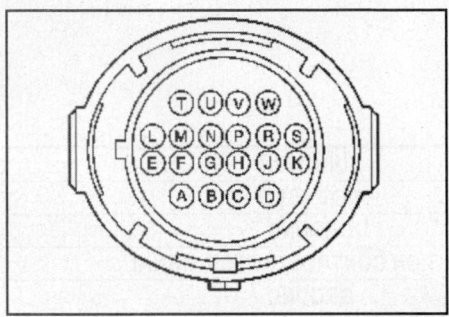

PIN	WIRE COLOR	CIRCUIT NO.	FUNCTION
A	LT GRN	1222	1-2 SHIFT SOLENOID VALVE CONTROL
B	YEL/BLK	1223	2-3 SHIFT SOLENOID VALVE CONTROL
C	RED/BLK	1228	PRESS. CONTROL (PC) SOLENOID VALVE CONTROLLED POWER FEED
D	LT BLU/WHT	1229	PC SOLENOID VALVE CONTROLLED GROUND (LOW)
E	PNK	739 OR 839	TRANSMISSION SOLENOID POWER FEED – VOLTAGE IGN 1
F	BLK/WHT	771	INTERNAL MODE SWITCH (IMS) SIGNAL A
G	YEL	772	IMS SIGNAL/TR SWITCH B
H	GRY	773	IMS SIGNAL/TR SWITCH C
J	WHT	776	IMS SIGNAL/TR SWITCH P
K	BLK/WHT	1551 OR 451	GROUND
L	YEL/BLK OR YEL	1227	TRANSMISSION FLUID TEMPERATURE (TFT) SENSOR SIGNAL
M	BLK	2762	TFT SENSOR GROUND (LOW REFERENCE)
N	PNK	1224	TFP MANUAL VALVE POSITION SWITCH SIGNAL A
P	RED	1226	TFP MANUAL VALVE POSITION SWITCH SIGNAL C
R	--	--	NOT USED
S	RED/BLK	1230	A/T INPUT (SHAFT) SPEED SENSOR (AT ISS) SIGNAL
T	TAN/BLK	422	TCC PULSE WIDTH MODULATED SOLENOID VALVE CONTROL
U	PNK	739 OR 839	TRANSMISSION SOLENOID POWER FEED – VOLTAGE IGN 1
V	DK BLU/WHT	1231	AT ISS GROUND (LOW SIGNAL)
W	--	--	NOT USED

2000-02 4.6L 4T80-E A/T Engine Side Inline Connector Pinout

TRANSMISSION SIDE

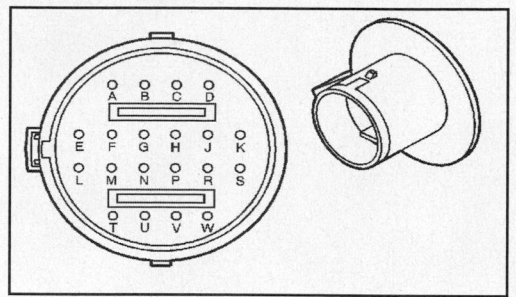

PIN	WIRE COLOR	CIRCUIT NO.	FUNCTION
A	LT GRN	1222	1-2 SHIFT SOLENOID VALVE CONTROL
B	YEL	1223	2-3 SHIFT SOLENOID VALVE CONTROL
C	PPL	1228	PRESSURE CONTROL (PC) SOLENOID VALVE CONTROLLED POWER FEED (HIGH)
D	LT BLU	1229	PC SOLENOID VALVE CONTROLLED GROUND (LOW)
E	RED	839	TRANSMISSION SOLENOID POWER FEED (IGN 1 VOLTAGE)
F	RED/BLK	771	INTERNAL MODE SWITCH (IMS)/TRANS SWITCH (TS) SIGNAL A
G	DK GRN/WHT	772	IMS/TS SIGNAL B
H	YEL/BLK	773	IMS/TS SIGNAL C
J	GRY/WHT	776	IMS/TS SIGNAL P
K	BLK/WHT	1050	GROUND
L	BRN	1227	TFT SENSOR SIGNAL
M	GRY	452	TFT SENSOR GROUND (LOW REFERENCE)
N	PNK	1224	TFP MANUAL VALVE POSITION SWITCH SIGNAL A
P	ORN	1226	TFP MANUAL VALVE POSITION SWITCH SIGNAL C
R	--	--	NOT USED
S	BLK	1230	AUTOMATIC TRANSMISSION INPUT (SHAFT) SPEED SENSOR (AT ISS) SIGNAL
T	TAN	418	TCC PULSE WIDTH MODULATED SOLENOID VALVE CONTROL
U	WHT	1135	TRANSMISSION SOLENOID POWER FEED (IGN 1 VOLTAGE)
V	DK GRN	1231	AT ISS GROUND
W	--	--	NOT USED

2000-02 4.6L 4T80-E A/T Transmission Side Inline Connector Pinout

CAMSHAFT POSITION (CMP) SENSOR CONNECTOR

Connector Pinouts

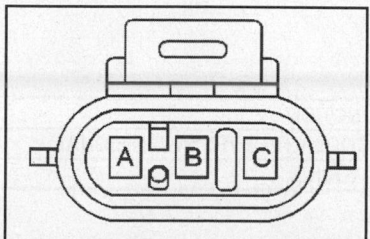

PIN	WIRE COLOR	CIRCUIT NO.	FUNCTION
A	ORN	1799	CMP SENSOR SIGNAL
B	PNK/BLK	632	CMP SENSOR GROUND
C	RED	631	CMP SENSOR IGNITION FEED

2000-02 4.6L CMP Sensor Connector Pinout

CRANKSHAFT POSITION (CKP) SENSOR CONNECTORS

Connector Pinouts

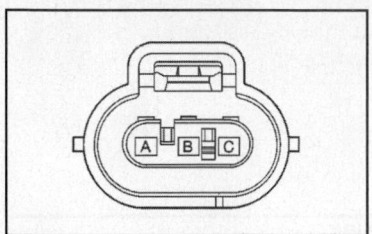

PIN	WIRE COLOR	CIRCUIT NO.	FUNCTION
A	LT GRN	1867	CKP SENSOR A IGNITION FEED
B	YEL/BLK	1868	CKP SENSOR A GROUND (LOW REFERENCE)
C	LT BLU/WHT	1800	CKP SENSOR A SIGNAL

2000-02 4.6L CKP Sensor A Connector Pinout

PIN	WIRE COLOR	CIRCUIT NO.	FUNCTION
A	LT GRN	2867	CKP SENSOR B IGNITION FEED
B	YEL	2868	CKP SENSOR B GROUND (LOW REFERENCE)
C	YEL/BLK OR YEL	573	CKP SENSOR B SIGNAL

2000-02 4.6L CKP Sensor B Connector Pinout

ENGINE COOLANT LEVEL SWITCH CONNECTOR

Connector Pinouts

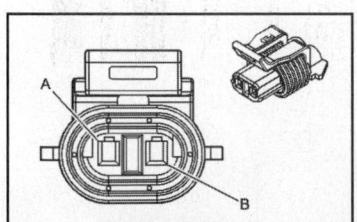

PIN	WIRE COLOR	CIRCUIT NO.	FUNCTION
A	LT GRN	147 OR 1478	FUSE OUTPUT - OFF/RUN/COOLANT LEVEL SWITH SIGNAL
B	BLK	650	GROUND

2000-02 4.6L Engine Coolant Level Switch Connector Pinout

ENGINE COOLANT TEMPERATURE (ECT) SENSOR CONNECTOR

Connector Pinouts

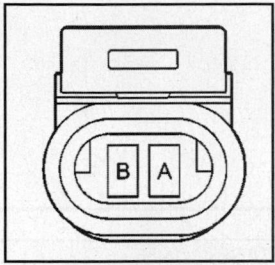

PIN	WIRE COLOR	CIRCUIT NO.	FUNCTION
A	BLK OR BRN	718	SENSOR GROUND
B	YEL	410	ECT SENSOR SIGNAL

2000-02 4.6L ECT Sensor Connector Pinout

ENGINE COOLING FAN CONNECTORS

Connector Pinouts

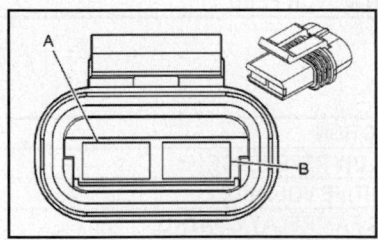

PIN	WIRE COLOR	CIRCUIT NO.	FUNCTION
A	WHT	504	COOLING FAN MOTOR OUTPUT
B	GRY	532	COOLING FAN MOTOR FEED

2000-02 4.6L Left Engine Cooling Fan Connector Pinout

PIN	WIRE COLOR	CIRCUIT NO.	FUNCTION
A	BLK	350	GROUND
B	LT BLU	409	COOLING FAN MOTOR FEED

2000-02 4.6L Right Engine Cooling Fan Connector Pinout

ENGINE COOLING FAN RELAY CONNECTOR

Connector Pinouts

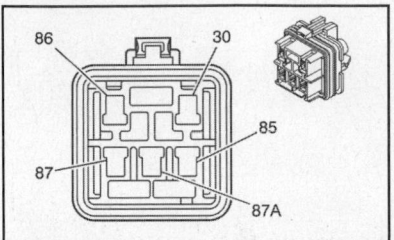

PIN	WIRE COLOR	CIRCUIT NO.	FUNCTION
30	RED	1142	BATTERY POSITIVE VOLTAGE
85	ORN	2340	BATTERY POSITIVE VOLTAGE
86	DK GRN	335	LOW SPEED COOLING FAN RELAY CONTROL
87	GRY	532	COOLING FAN MOTOR SUPPLY VOLTAGE
87A	YEL	998	12 VOLT REFERENCE

2000-02 4.6L Engine Cooling Fan Relay 1 Connector Pinout

PIN	WIRE COLOR	CIRCUIT NO.	FUNCTION
30	WHT	504	COOLING FAN MOTOR OUTPUT
85	ORN	2340	FUSED OUTPUT - BATTERY
86	DK BLU	473	COOLING FAN RELAY OUTPUT - COIL - SECONDARY
87	BLK	350	GROUND
87A	LT BLU	409	COOLING FAN MOTOR FEED

2000 4.6L Engine Cooling Fan Relay 2 Connector Pinout

PIN	WIRE COLOR	CIRCUIT NO.	FUNCTION
30	WHT	504	COOLING FAN LOW REFERENCE
85	ORN	2340	BATTERY POSITIVE VOLTAGE
86	DK BLU	473	HIGH SPEED COOLING FAN RELAY CONTROL
87	BLK	350	GROUND
87A	LT BLU	409	COOLING FAN MOTOR SUPPLY VOLTAGE

2001-02 4.6L Engine Cooling Fan S/P Relay Connector Pinout

PIN	WIRE COLOR	CIRCUIT NO.	FUNCTION
30	RED	1142	FUSED OUTPUT - BATTERY
85	ORN	2340	FUSED OUTPUT - BATTERY
86	DK BLU	473	COOLING FAN RELAY OUTPUT - COIL - SECONDARY
87	LT BLU	409	COOLING FAN MOTOR FEED
87A	YEL	998	COOLING FAN MOTOR FEED DISABLE

2000 4.6L Engine Cooling Fan Relay 3 Connector Pinout

PIN	WIRE COLOR	CIRCUIT NO.	FUNCTION
30	RED	1142	BATTERY POSITIVE VOLTAGE
85	ORN	2340	BATTERY POSITIVE VOLTAGE
86	DK BLU	473	HIGH SPEED COOLING FAN RELAY CONTROL
87	LT BLU	409	COOLING FAN MOTOR SUPPLY VOLTAGE
87A	YEL	998	12 VOLT REFERENCE

2001-02 4.6L Engine Cooling Fan Relay 2 Connector Pinout

EVAP CANISTER PURGE VALVE CONNECTOR

Connector Pinouts

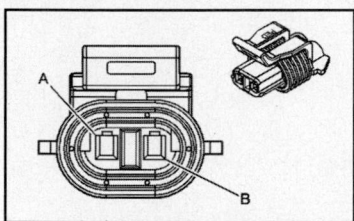

PIN	WIRE COLOR	CIRCUIT NO.	FUNCTION
A	PNK	839	IGN ITION VOLTAGE
B	DK GRN/WHT	428	EVAP CANISTER PURGE VALVE CONTROL

2000-02 4.6L EVAP Purge Valve Connector Pinout

EVAP CANISTER VENT VALVE CONNECTOR

Connector Pinouts

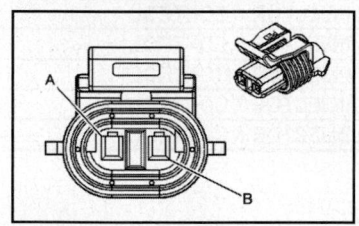

PIN	WIRE COLOR	CIRCUIT NO.	FUNCTION
A	PNK	439	IGNITION VOLTAGE
B	WHT	1310	EVAP CANISTER VENT VALVE CONTROL

2000-02 4.6L EVAP Vent Valve Connector Pinout

EXHAUST GAS RECIRCULATION (EGR) VALVE CONNECTOR

Connector Pinouts

PIN	WIRE COLOR	CIRCUIT NO.	FUNCTION
A	GRY	435	EGR VALVE GROUND
B	BLK	2753	EGR PINTLE POSITION GROUND
C	BRN	1456	EGR POSITION SIGNAL
D	GRY	2702	5 VOLT REFERENCE
E	RED	1676	EGR VALVE CONTROL

2000-02 4.6L EGR Valve Connector Pinout

FUEL INJECTOR CONNECTORS

Connector Pinouts

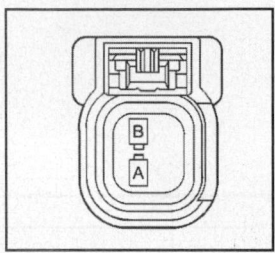

PIN	WIRE COLOR	CIRCUIT NO.	FUNCTION
A	PNK	739	IGNITION POSITIVE VOLTAGE (INJECTORS 1, 4, 6 & 7)
A	PNK	1239	IGNITION POSITIVE VOLTAGE (INJECTORS 2, 3, 5 & 8)
B	BLK	1744	FUEL INJECTOR 1 CONTROL
B	LT GRN OR LT GRN/BLK	1745	FUEL INJECTOR 2 CONTROL
B	PNK/BLK	1746	FUEL INJECTOR 3 CONTROL
B	LT BLU/BLK	844	FUEL INJECTOR 4 CONTROL
B	BLK/WHT	845	FUEL INJECTOR 5 CONTROL
B	YEL/BLK	846	FUEL INJECTOR 6 CONTROL
B	RED/BLK	877	FUEL INJECTOR 7 CONTROL
B	DK BLU/WHT	878	FUEL INJECTOR 8 CONTROL

2000-02 4.6L Fuel Injector Connector Pinouts

FUEL PUMP & SENDER CONNECTOR

Connector Pinouts

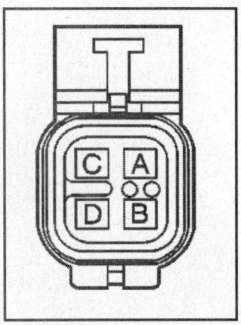

PIN	WIRE COLOR	CIRCUIT NO.	FUNCTION
A	PPL	1589 OR 30	FUEL LEVEL SIGNAL
B	GRY	120	FUEL PUMP MOTOR FEED (SUPPLY VOLTAGE)
C	BLK	1250	FUEL PUMP MOTOR GROUND
D	BLK/WHT	2759 OR 552	SENDER GROUND (LOW REFERENCE)

2000-02 4.6L Fuel Pump & Sender Connector Pinout

FUEL TANK PRESSURE (FTP) SENSOR CONNECTOR

Connector Pinouts

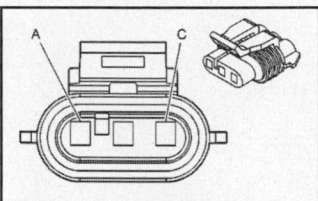

PIN	WIRE COLOR	CIRCUIT NO.	FUNCTION
A	BLK	2759	FUEL TANK PRESSURE SENSOR GROUND (LOW REFERENCE)
B	DK GRN	890	FUEL TANK PRESSURE SENSOR SIGNAL
C	GRY	2709	5 VOLT REFERENCE

2000-02 4.6L FTP Sensor Connector Pinout

GENERATOR CONNECTOR

Connector Pinouts

2000 MODELS

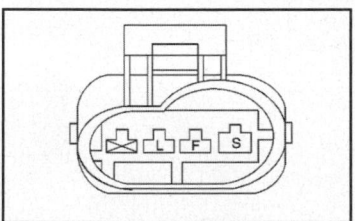

PIN	WIRE COLOR	CIRCUIT NO.	FUNCTION
F	GRY	23	GENERATOR FIELD DUTY CYCLE SIGNAL
L	RED	225	GENERATOR OUTPUT - REGULATOR REFERENCE VOLTAGE
S	--	--	NOT USED

2000 4.6L Generator Connector Pinout

2001-02 MODELS

PIN	WIRE COLOR	CIRCUIT NO.	FUNCTION
S	--	--	NOT USED
F	GRY	23	GENERATOR FIELD DUTY CYCLE SIGNAL
L	RED	225	GENERATOR TURN ON SIGNAL
P	--	--	NOT USED

2001-02 4.6L Generator Connector Pinout

HEATED OXYGEN SENSOR (HO2S) CONNECTORS

Connector Pinouts

HO2S 1/1

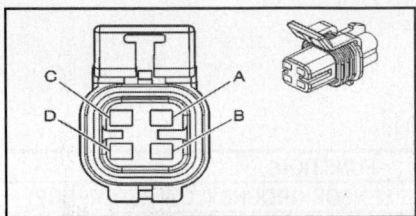

PIN	WIRE COLOR	CIRCUIT NO.	FUNCTION
A	TAN	1664	HEATED OXYGEN SENSOR (HO2S) LOW
B	PPL/WHT	1665	HEATED OXYGEN SENSOR (HO2S) HIGH
C	TAN/WHT	1312	HEATER LOW
D	DK GRN	676	HEATER HIGH

2000-02 4.6L O2 Sensor (Bank 1 Sensor 1) Connector Pinout

HO2S 1/2

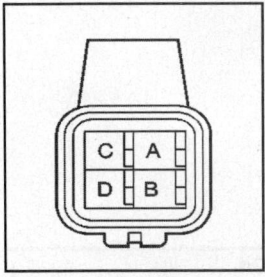

PIN	WIRE COLOR	CIRCUIT NO.	FUNCTION
A	DK BLU/WHT	1677	HEATED OXYGEN SENSOR (HO2S) LOW
B	DK GRN/WHT	1678	HEATED OXYGEN SENSOR (HO2S) HIGH
C	BLK/WHT	1423	HEATER LOW
D	PNK	139	IGNITION POSITIVE VOLTAGE

2000-02 4.6L O2 Sensor (Bank 1 Sensor 2) Connector Pinout

HO2S 2/1

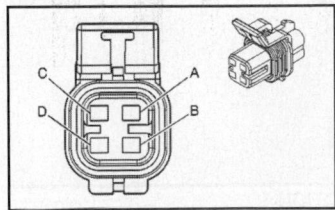

PIN	WIRE COLOR	CIRCUIT NO.	FUNCTION
A	TAN/WHT	1669	HEATED OXYGEN SENSOR (HO2S) LOW
B	PPL/WHT	1668	HEATED OXYGEN SENSOR (HO2S) HIGH
C	BRN	3122	HEATER LOW
D	PNK	139 OR 39	IGNITION POSITIVE VOLTAGE (IGN 1 VOLTAGE)

2000-02 4.6L O2 Sensor (Bank 2 Sensor 1) Connector Pinout

IDLE AIR CONTROL (IAC) VALVE CONNECTOR

Connector Pinouts

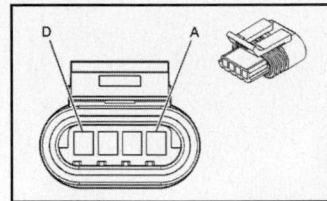

PIN	WIRE COLOR	CIRCUIT NO.	FUNCTION
A	LT GRN/BLK	444	IAC COIL B LOW
B	LT GRN/WHT	1749	IAC COIL B HIGH
C	LT BLU/BLK	1/48	IAO OOIL A LOW
D	LT BLU/WHT	1747	IAC COIL A HIGH

2000-02 6.4L IAC Valve Connector Pinout

IGNITION COIL/CONTROL MODULE (ICM) CONNECTORS

Connector Pinouts

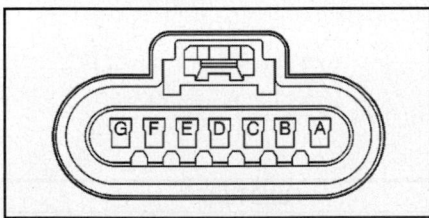

PIN	WIRE COLOR	CIRCUIT NO.	FUNCTION
A	PNK	539	IGNITION POSITIVE VOLTAGE
B	BLK/WHT	1551	GROUND
C	YEL	2174	REF LOW
D	DK GRN/WHT	2124	IC 4
E	LT BLU/WHT	2126	IC 6
F	PPL/WHT	2128	IC 8
G	RED/WHT	2122	IC 2

2000-02 4.6L ICM Front/Bank 2 Connector Pinout

PIN	WIRE COLOR	CIRCUIT NO.	FUNCTION
A	PNK	539	IGNITION POSITIVE VOLTAGE
B	BLK/WHT	1551	GROUND
C	GRY	2175	REF LOW
D	DK GRN	2125	IC 5
E	LT BLU	2123	IC 3
F	PPL	2121	IC 1
G	RED	2127	IC 7

2000-02 4.6L ICM Rear/Bank 1 Connector Pinout

<u>KNOCK SENSORS</u>

Connector Pinouts

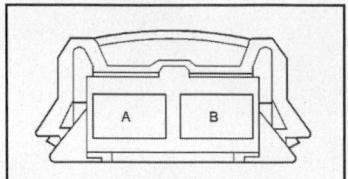

PIN	WIRE COLOR	CIRCUIT NO.	FUNCTION
A	DK BLU	496	KNOCK SENSOR SIGNAL
B	GRY	1716	KNOCK SENSOR RETURN

2000-02 4.6L Knock Sensor Connector Pinout

<u>MANIFOLD ABSOLUTE PRESSURE (MAP) SENSOR</u>

Connector Pinouts

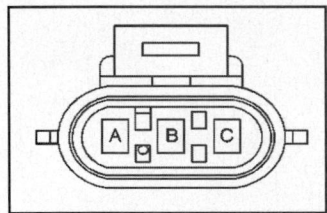

PIN	WIRE COLOR	CIRCUIT NO.	FUNCTION
A	ORN/BLK	469	MAP SENSOR GROUND
B	LT GRN	432	MAP SENSOR SIGNAL
C	GRY	474	5 VOLT REFERENCE

2000-02 4.6L MAP Sensor Connector Pinout

<u>MASS AIR FLOW/INTAKE AIR TEMPERATURE (MAF/IAT) SENSOR</u>

Connector Pinouts

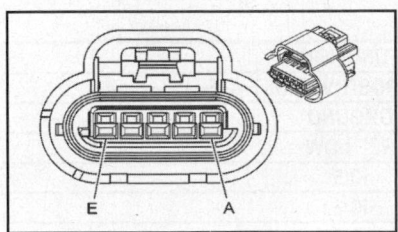

PIN	WIRE COLOR	CIRCUIT NO.	FUNCTION
A	BLK	2760	IAT GROUND (LOW REFERENCE)
B	TAN	472	IAT SENSOR SIGNAL
C	BLK OR BLK/WHT	1550 OR 451	GROUND
D	PNK	139 OR 839	IGNITION POSITIVE VOLTAGE
E	YEL	492	MAF SENSOR SIGNAL

2000-02 4.6L MAF/IAT Sensor Connector Pinout

POWERTRAIN CONTROL MODULE (PCM) CONNECTORS

Connector Pinouts

C1 CONNECTOR

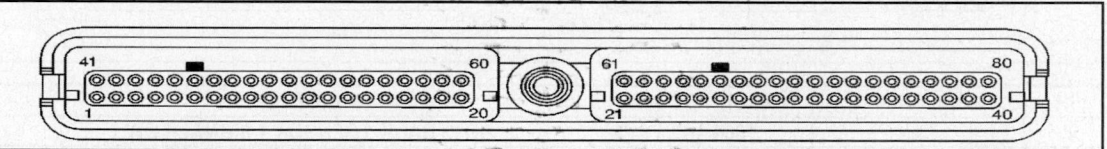

PIN	WIRE COLOR	CIRCUIT NO.	FUNCTION
1	ORN	1799	CMP SENSOR SIGNAL
2	LT BLU/WHT	1800	CKP SENSOR A SIGNAL
3-4	--	--	NOT USED
5	DK GRN	890	FTP SENSOR SIGNAL
6	YEL	5	CRANK REQUEST
7-9	--	--	NOT USED
10	TAN	472	IAT SENSOR SIGNAL
11	GRY	773	TRANSAXLE RANGE SWITCH C INPUT
12	YEL	772	TRANSAXLE RANGE SWITCH B INPUT
13	BLK/WHT	771	TRANSAXLE RANGE SWITCH A INPUT
14	BLK	573	CKP SENSOR B SIGNAL
15	--	--	NOT USED
16	DK BLU/WHT	1231	TRANSAXLE INPUT SPEED SENSOR LOW
17	RED/WHT	1230	TRANSAXLE INPUT SPEED SENSOR HIGH
18	PPL	1807	CLASS 2 SERIAL DATA (BETWEEN EBTCM AND PCM)
19	LT GRN	147	PCM IGNITION 0 POSITIVE VOLTAGE
20	ORN	240 OR 2240	PCM BATTERY FEED
21	ORN	240 OR 2240	PCM BATTERY FEED
22	RED	2127	CYLINDER 7 IC CONTROL
23	LT GRN	432	MAP SENSOR SIGNAL
24	--	--	NOT USED
25	DK GRN	335	LOW SPEED COOLING FANS CONTROL CIRCUIT
26	DK GRN/WHT	817	4K/MI SPEED
27	LT GRN/BLK	444	IAC A LOW
28	LT GRN/WHT	1749	IAC A HIGH
29	LT BLU/WHT	1747	IAC B HIGH
30	LT BLU/BLK	1748	IAC B LOW
31	DK GRN	676	HO2S BANK 1 SENSOR 1 HEATER CONTROL CIRCUIT HIGH
32	BRN/WHT	419	MIL CONTROL
33	BRN	3122	HO2S BANK 2 SENSOR 1 HEATER CONTROL CIRCUIT LOW
34	TAN/BLK	464	DELIVERED TORQUE OUTPUT
35	DK GRN	83	CRUISE INHIBIT OUTPUT
36	--	--	NOT USED
37	BLK	1744	FUEL INJECTOR 1 CONTROL
38	PNK/BLK	1746	FUEL INJECTOR 3 CONTROL
39	LT GRN/BLK	1745	FUEL INJECTOR 2 CONTROL
40	BLK/WHT	845	FUEL INJECTOR 5 CONTROL
41	WHT	776	TRANSAXLE RANGE SWITCH PARITY INPUT
42	YEL	492	MAF SENSOR SIGNAL
43	ORN/BLK	463	DESIRED TORQUE INPUT
44	GRY	847	EXTENDED TRAVEL BRAKE SWITCH INPUT
45-49	--	--	NOT USED
50	DK BLU/WHT	1677	HO2S BANK 1 SENSOR 2 SIGNAL LOW
51	DK GRN/WHT	1678	HO2S BANK 1 SENSOR 2 SIGNAL HIGH

2000-02 4.6L PCM C1 Connector Pinout (1 of 2)

C1 CONNECTOR (CONTINUED)

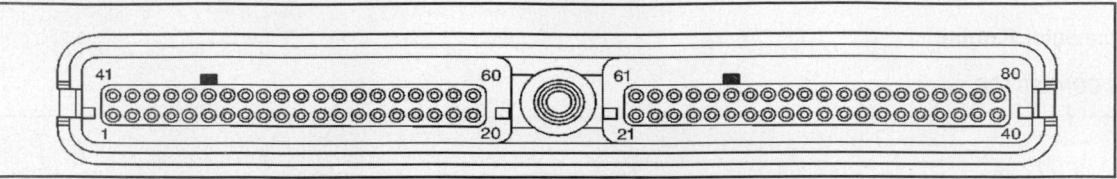

PIN	WIRE COLOR	CIRCUIT NO.	FUNCTION
52	TAN	1664	HO2S BANK 1 SENSOR 1 SIGNAL LOW
53	PPL/WHT	1665	HO2S BANK 1 SENSOR 1 SIGNAL HIGH
54	TAN/WHT	1669	HO2S BANK 2 SENSOR 1 SIGNAL LOW
55	PPL/WHT	1668	HO2S BANK 2 SENSOR 1 SIGNAL HIGH
56	YEL	400	VSS SENSOR INPUT HIGH
57	PPL	401	VSS SENSOR INPUT LOW
58	RED	631	CMP SENSOR IGNITION FEED
59	PPL	1807	CLASS 2 SERIAL DATA
60	BLK/WHT	451	PCM GROUND
61	PPL	2121	CYLINDER 1 IC CONTROL
62	DK BLU	417	TP SENSOR SIGNAL
63	PNK	39 OR 639	PCM IGNITION 1 POSITIVE VOLTAGE
64-65	--	--	NOT USED
66	PNK/BLK	429	AIR SOLENOID CONTROL (NOT USED ON 2000)
67	LT BLU/WHT	1229	PRESSURE CONTROL SOLENOID LOW
68	TAN/WHT	1312	HO2S BANK 1 SENSOR 1 HEATER CONTROL CIRCUIT LOW
69	RED/BLK	1228	PRESSURE CONTROL SOLENOID HIGH
70	DK GRN/WHT	465	FUEL PUMP RELAY CONTROL
71	RED	1676	EGR VALVE CONTROL
72	PPL	1490	POWERTRAIN INDUCED CHASSIS PITCH STATUS CONTROL CIRCUIT
73	WHT	1310	EVAP VENT SOLENOID CONTROL
74	--	--	NOT USED
75	BRN	436	AIR PUMP RELAY CONTROL (NOT USED ON 2000)
76	LT BLU/BLK	844	FUEL INJECTOR 4 CONTROL
77	YEL/BLK	846	FUEL INJECTOR 6 CONTROL
78	--	--	NOT USED
79	DK BLU/WHT	878	FUEL INJECTOR 8 CONTROL
80	RED/BLK	877	FUEL INJECTOR 7 CONTROL

2000-02 4.6L PCM C1 Connector Pinout (2 of 2)

C2 CONNECTOR

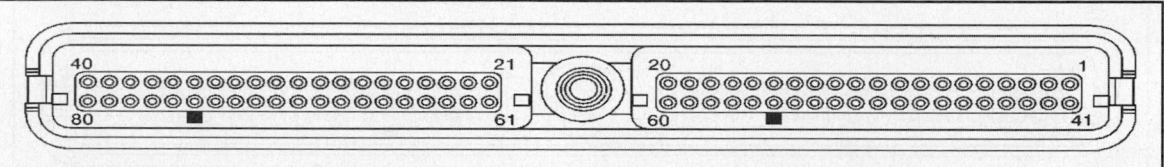

PIN	WIRE COLOR	CIRCUIT NO.	FUNCTION
1-2	--	--	NOT USED
3	BLK/WHT	451	PCM GROUND
4	BLK	2760	IAT GROUND
5	--	--	NOT USED
6	ORN/BLK	469	MAP SENSOR GROUND
7-8	--	--	NOT USED
9	YEL	2174	REFERENCE LOW (FRONT BANK IGNITION COIL CASSETTE)
10	--	--	NOT USED
11	BLK	2762	TFT SENSOR GROUND
12	BLK	2753	EGR PINTLE POSITION SENSOR GROUND
13	BLK	2759	FUEL LEVEL SENSOR AND FUEL TANK PRESSURE SENSOR GROUND
14	--	--	NOT USED
15	BLK/WHT	451	PCM GROUND
16	--	--	NOT USED
17	GRY	2709	5 VOLT REFERENCE B - FUEL TANK PRESSURE SENSOR FEED
18	--	--	NOT USED
19	DK BLU	496	KNOCK SENSOR SIGNAL
20	TAN/BLK	231	ENGINE OIL PRESSURE SWITCH INPUT
21	LT BLU	2123	CYLINDER 3 IC CONTROL
22	RED/WHT	2122	CYLINDER 2 IC CONTROL
23	BRN	1174	ENGINE OIL LEVEL SWITCH INPUT
24	LT BLU/BLK	396	CRUISE ENGAGED INPUT
25	LT GRN	2867	CKP SENSOR B IGNITION FEED
26	BLK/WHT	451	PCM GROUND
27	--	--	NOT USED
28	YEL	2868	CKP SENSOR B GROUND (LOW REFERENCE)
29-30	--	--	NOT USED
31	YEL	410	ECT SENSOR SIGNAL
32	YEL	1227	TFT SENSOR SIGNAL
33-34	--	--	NOT USED
35	GRY	1716	KNOCK SENSOR GROUND
36	--	--	NOT USED
37	RED	1226	TRANSAXLE PRESSURE MANIFOLD SWITCH C INPUT
38-39	--	--	NOT USED
40	RED	225	GENERATOR L TERMINAL CONTROL
41	--	--	NOT USED
42	DK GRN/WHT	428	EVAP CANISTER PURGE VALVE CONTROL
43	LT GRN	1222	SHIFT SOLENOID A CONTROL
44	YEL	447	STARTER RELAY CONTROL
45	YEL/BLK	1223	SHIFT SOLENOID B CONTROL
46	TAN/BLK	422	TCC CONTROL
47	DK GRN/WHT	459	A/C COMPRESSOR CLUTCH CONTROL
48	BLK/WHT	1423	HO2S BANK 1 SENSOR 2 HEATER CONTROL CIRCUIT LOW
49	GRY	474	5 VOLT REFERENCE A - MAP SENSOR FEED

2000-02 4.6L PCM C2 Connector (1 of 2)

C2 CONNECTOR – CONTINUED

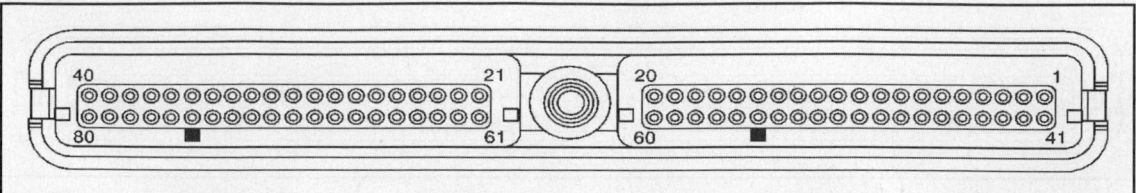

PIN	WIRE COLOR	CIRCUIT NO.	FUNCTION
50	GRY	2702	5 VOLT REFERENCE A - EGR PINTLE POSITION SENSOR FEED
51	GRY	2701	5 VOLT REFERENCE A - TP SENSOR FEED
52	BRN	1456	EGR PINTLE POSITION SENSOR SIGNAL
53	--	--	NOT USED
54	--	--	NOT USED
55	DK BLU	473	HIGH SPEED COOLING FANS CONTROL
56	WHT	121	TACHOMETER CONTROL
57-58	--	--	NOT USED
59	DK GRN/WHT	2124	CYLINDER 4 IC CONTROL
60	LT BLU/WHT	2126	CYLINDER 6 IC CONTROL
61	PPL/WHT	2128	CYLINDER 8 IC CONTROL
62	DK GRN	2125	CYLINDER 5 IC CONTROL
63-64	--	--	NOT USED
65	LT GRN OR LT GRN/BLK	1867	CKP SENSOR A IGNITION FEED
66	YEL/BLK	1868	CKP SENSOR A GROUND
67	PNK/BLK	632	CMP SENSOR GROUND
68	BRN	718	ECT SENSOR GROUND
69	BLK	2752	TP SENSOR GROUND
70	GRY	435	EGR VALVE CONTROL GROUND
71	GRY	2175	REFERENCE LOW (REAR BANK IGNITION COIL CASSETTE)
72	-	-	NOT USED
73	PNK	1224	TRANSAXLE PRESSURE MANIFOLD SWITCH A INPUT
74-78	--	--	NOT USED
79	GRY	23	GENERATOR F TERMINAL FEEDBACK
80	PPL	420	TCC BRAKE SWITCH INPUT

2000-02 4.6L PCM C2 Connector (2 of 2)

SECONDARY AIR INJECTION (AIR) PUMP CONNECTORS

Connector Pinouts

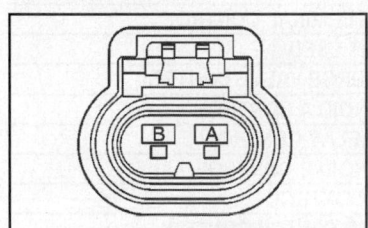

PIN	WIRE COLOR	CIRCUIT NO.	FUNCTION
A	BLK	1650	GROUND
B	RED	78	AIR PUMP SUPPLY VOLTAGE

2001-02 4.6L Secondary Air Pump Motor Connector Pinout

SECONDARY AIR INJECTION (AIR) PUMP RELAY

Connector Pinouts

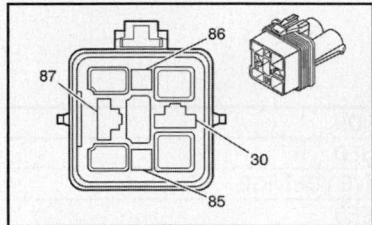

PIN	WIRE COLOR	CIRCUIT NO.	FUNCTION
30	RED	1642	BATTERY POSITIVE VOLTAGE
85	PNK	839	IGNITION 1 VOLTAGE
00	BRN	436	AIR PUMP RELAY CONTROL
87	RED	78	AIR PUMP SUPPLY VOLTAGE

2001-02 4.6L Secondary Air Injection Pump Relay Connector Pinout

SECONDARY AIR INJECTION (AIR) SOLENOID VALVE

Connector Pinouts

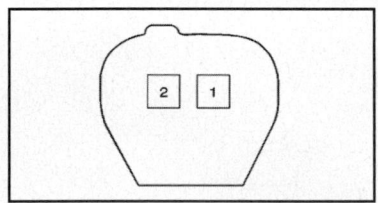

PIN	WIRE COLOR	CIRCUIT NO.	FUNCTION
1	PNK	839	IGNITION POSITIVE VOLTAGE
2	PNK/BLK	429	AIR SOLENOID VALVE CONTROL

2001-02 4.6L Secondary Air Injection Solenoid Valve Connector Pinout

STARTER RELAY

Connector Pinouts

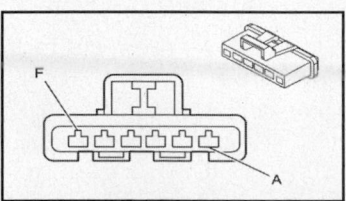

PIN	WIRE COLOR	CIRCUIT NO.	FUNCTION
A	--	--	NOT USED
B	RED	142	BATTERY POSITIVE VOLTAGE
C	--	--	NOT USED
D	YEL	625	THEFT DETERRENT STARTER ENABLE RELAY OUTPUT
E	PPL	6	STARTER SOLENOID FEED (VOLTAGE)
F	PNK	1737	PARK/NEUTRAL POSITION SWITCH OUTPUT

2000 4.6L Starter Enable Relay Connector Pinout

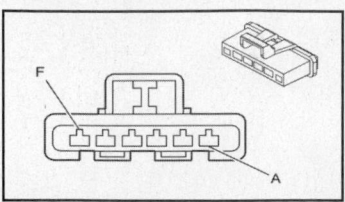

PIN	WIRE COLOR	CIRCUIT NO.	FUNCTION
A	--	--	NOT USED
B	RED	142	BATTERY POSITIVE VOLTAGE
C	--	--	NOT USED
D	YEL	447	STARTER RELAY CONTROL
E	PPL	6	STARTER SOLENOID CRANK VOLTAGE
F	PNK	139	IGN 1 VOLTAGE

2001-02 4.6L Starter Relay Connector Pinout

THROTTLE POSITION (TP) SENSOR

Connector Pinouts

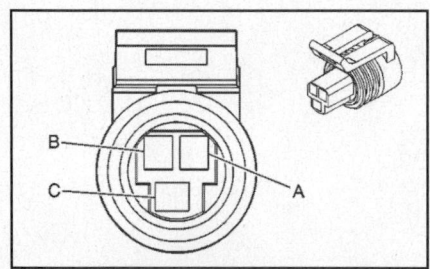

PIN	WIRE COLOR	CIRCUIT NO.	FUNCTION
A	GRY	2701	5 VOLT REFERENCE
B	BLK	2752	THROTTLE POSITION SENSOR GROUND
C	DK BLU	417	THROTTLE POSITION SENSOR SIGNAL

2000-02 4.6L TP Sensor Connector Pinout

TRACTION CONTROL SWITCH

Connector Pinouts

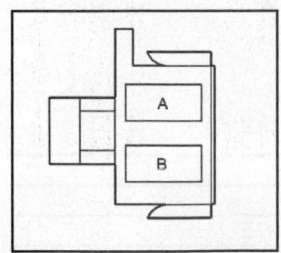

PIN	WIRE COLOR	CIRCUIT NO.	FUNCTION
A	BRN/WHT	1571	TRACTION CONTROL SWITCH SIGNAL
B	BLK/WHT	151	TRACTION CONTROL SWITCH GROUND

2000-01 4.6L 4T80-E A/T Traction Control Switch Connector Pinout

VEHICLE SPEED SENSOR (VSS) CONNECTOR

Connector Pinouts

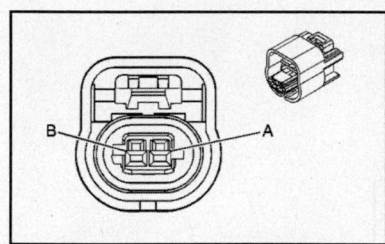

PIN	WIRE COLOR	CIRCUIT NO.	FUNCTION
A	PPL	401	VSS LOW SIGNAL (GROUND)
B	YEL	400	VSS HIGH SIGNAL

2000-02 4.6L 4T80-E A/T Vehicle Speed Sensor Connector Pinout

GENERAL MOTORS CARS: 2000-04 CADILLAC SEVILLE

<u>A/T IN-LINE CONNECTORS</u>

Connector Pinouts

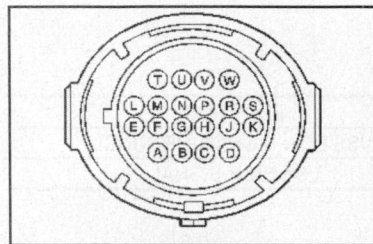

2000-2001

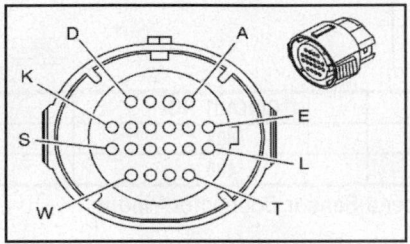

2002-04

PIN	WIRE COLOR	CIRCUIT NO.	FUNCTION
A	LT GRN	1222	1-2 SHIFT SOLENOID VALVE CONTROL
B	YEL/BLK	1223	2-3 SHIFT SOLENOID VALVE CONTROL
C	RED/BLK	1228	PRESSURE CONTROL (PC) SOLENOID VALVE CONTROLLED POWER FEED
D	LT BLU/WHT	1229	PC SOLENOID VALVE CONTROLLED GROUND (LOW)
E	PNK	739	TRANSMISSION SOLENOID POWER FEED – VOLTAGE IGN 1
F	BLK/WHT	771	INTERNAL MODE SWITCH (IMS) SIGNAL A
G	YEL	772	IMS SIGNAL/TR SWITCH B
H	GRY	773	IMS SIGNAL/TR SWITCH C
J	WHT	776	IMS SIGNAL/TR SWITCH P
K	BLK/WHT	1551	GROUND (2000-01)
K	BLK OR BLK/WHT	1550	GROUND (FROM 2002)
L	YEL/BLK	1227	TRANSMISSION FLUID TEMPERATURE (TFT) SENSOR SIGNAL
M	BLK	2762	TFT SENSOR GROUND (LOW REFERENCE)
N	PNK	1224	TFP MANUAL VALVE POSITION SWITCH SIGNAL A
P	RED	1226	TFP MANUAL VALVE POSITION SWITCH SIGNAL C
R	--	--	NOT USED
S	RED/BLK	1230	A/T INPUT (SHAFT) SPEED SENSOR (AT ISS) SIGNAL
T	TAN/BLK	422	TCC PULSE WIDTH MODULATED SOLENOID VALVE CONTROL
U	PNK OR PNK/BLK	739	TRANSMISSION SOLENOID POWER FEED – VOLTAGE IGN 1
V	DK BLU/WHT	1231	AT ISS GROUND (LOW SIGNAL)
W	--	--	NOT USED

2000-04 4.6L 4T80-E A/T Engine Side In-Line Connector Pinout

A/T IN-LINE CONNECTORS – CONTINUED

Connector Pinouts

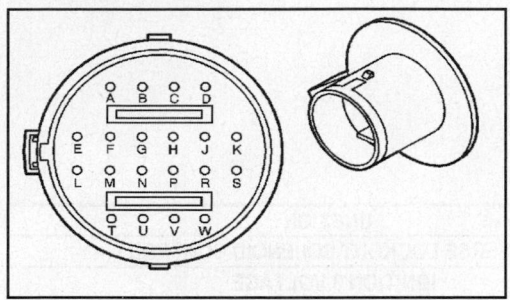

PIN	WIRE COLOR	CIRCUIT NO.	FUNCTION
A	LT GRN	1222	1-2 SHIFT SOLENOID VALVE CONTROL
B	YEL	1223	2-3 SHIFT SOLENOID VALVE CONTROL
C	PPL	1228	PRESSURE CONTROL (PC) SOLENOID VALVE CONTROLLED POWER FEED (HIGH)
D	LT BLU	1229	PC SOLENOID VALVE CONTROLLED GROUND (LOW)
E	RED	839	TRANSMISSION SOLENOID POWER FEED (IGN 1 VOLTAGE)
F	RED/BLK	771	INTERNAL MODE SWITCH (IMS)/TRANS SWITCH (TS) SIGNAL A
G	DK GRN/WHT	772	IMS/TS SIGNAL B
H	YEL/BLK	773	IMS/TS SIGNAL C
J	GRY/WHT	776	IMS/TS SIGNAL P
K	BLK/WHT	1050	GROUND
L	BRN	1227	TFT SENSOR SIGNAL
M	GRY	452	TFT SENSOR GROUND (LOW REFERENCE)
N	PNK	1224	TFP MANUAL VALVE POSITION SWITCH SIGNAL A (2000-01 ONLY)
P	ORN	1226	TFP MANUAL VALVE POSITION SWITCH SIGNAL C (2000-01 ONLY)
R	--	--	NOT USED
S	BLK	1230	AUTOMATIC TRANSMISSION INPUT (SHAFT) SPEED SENSOR (AT ISS) SIGNAL
T	TAN	418	TCC PULSE WIDTH MODULATED SOLENOID VALVE CONTROL
U	WHT	1135	TRANSMISSION SOLENOID POWER FEED (IGN 1 VOLTAGE)
V	DK GRN	1231	AT ISS GROUND (LOW)
W	--	--	NOT USED

2000-04 4.6L 4T80-E A/T Transmission Side In-Line Connector Pinout

A/T REVERSE LOCKOUT SOLENOID **Description & Operation**

The reverse lockout system is a safety device that prevents an inadvertent shift from a forward or neutral gear into reverse. The system consists of the following components.

- The reverse lockout solenoid.
- The dash integration module (DIM).

When the ignition switch is turned ON, battery voltage is supplied to the feed circuit of the reverse lockout solenoid. The DIM supplies the ground for the control side circuit. When the DIM determines that the ignition is ON and the vehicle has been traveling at over 5 mph for 15 seconds or longer, it grounds the control circuit of the reverse lockout solenoid. This energizes the reverse lockout solenoid and mechanically prevents the shift lever from being moved into the REVERSE position.

Connector Pinouts

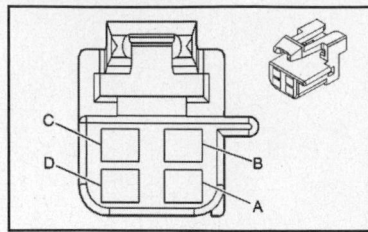

PIN	WIRE COLOR	CIRCUIT NO.	FUNCTION
A	LT GRN	1652	REVERSE LOCKOUT SOLENOID CONTROL
B	BRN	141	IGNITION 3 VOLTAGE
C	YEL	1491	BACKLIGHT LAMPS
D	BLK	350	GROUND

2001-04 4.6L 4T80-E A/T Reverse Lockout Solenoid Connector Pinout

A/T SHIFT LOCK CONTROL

Description & Operation

The automatic transmission shift lock control system is a safety device that prevents an inadvertent shift out of PARK when the engine is running. The driver must press the brake pedal before moving the shift lever out of the PARK position. The system consists of the following components.

- The park relay.
- The automatic transmission shift lock control solenoid.
- The automatic transmission shift lock control switch.
- The rear integration module (RIM).
- The powertrain control module (PCM).

With the ignition ON, positive voltage is supplied from the ignition switch to the coil side feed circuit of the park relay. The RIM receives a gear position input signal from the PCM through a class 2 serial data message. When the RIM receives a PARK position message from the PCM, the RIM applies a ground to the control circuit of the park relay. This energizes the relay, closing the switch contacts, battery voltage flows through the relay to the automatic transmission shift lock control switch. With the brake pedal released voltage flows through the normally closed contacts of the automatic transmission shift lock control switch to the automatic transmission shift lock control solenoid. The automatic transmission shift lock control solenoid is permanently grounded. This energizes the automatic transmission shift lock control solenoid and mechanically locks the shift lever in the PARK position. When the driver presses the brake pedal the contacts of the automatic transmission shift lock control switch open, de-energizing the automatic transmission shift lock control solenoid. This allows the shift lever to be moved from the PARK position.

When the shift lever is moved out of the PARK position, the PCM sends a class 2 serial data message to the RIM. The RIM opens the ground path of the park relay control circuit, de-energizing the park relay and disabling the automatic transmission shift lock control solenoid. Once the vehicle is out of PARK, the automatic transmission shift lock control solenoid remains de-energized until the RIM receives a PARK signal from the PCM.

Removal & Installation

1. Remove the console trim plate.
2. If equipped, remove the anti-theft plate and tamper proof fastener from the shifter using removal tool J 43146.

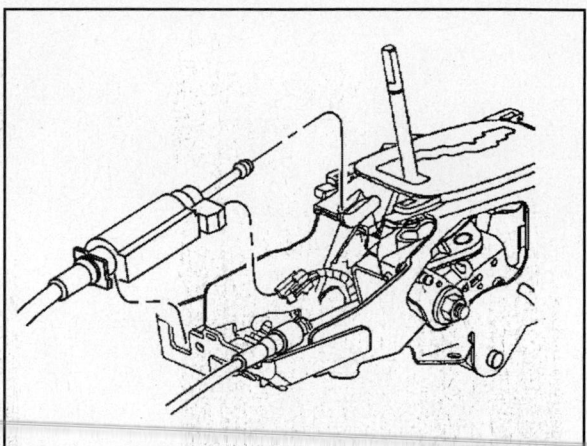

Removing shift lock actuator wiring from shift lock actuator solenoid connector

3. Disconnect the automatic transmission shift lock actuator wire harness connector, from the automatic transmission shift lock actuator solenoid.

4. Disengage the locking tab on the park lock cable and lift the cable out of the slot in the shifter.

5. Unhook the park lock cable from the park lock cam on the shifter.

6. Turn the ignition switch to the run position.

7. Depress the locking tab (1) on the cable and pull the cable out of the back of the ignition switch. The locking tab is located on the bottom of the ignition switch facing in the 6 o'clock position.

8. Remove the park lock cable from the vehicle.

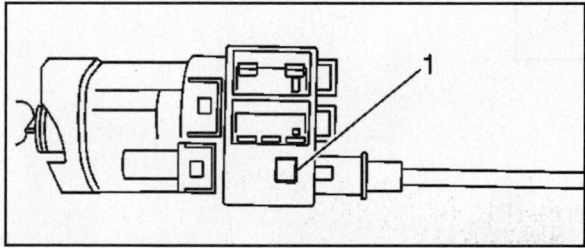

Identifying locking tab (1) on shift lock actuator cable

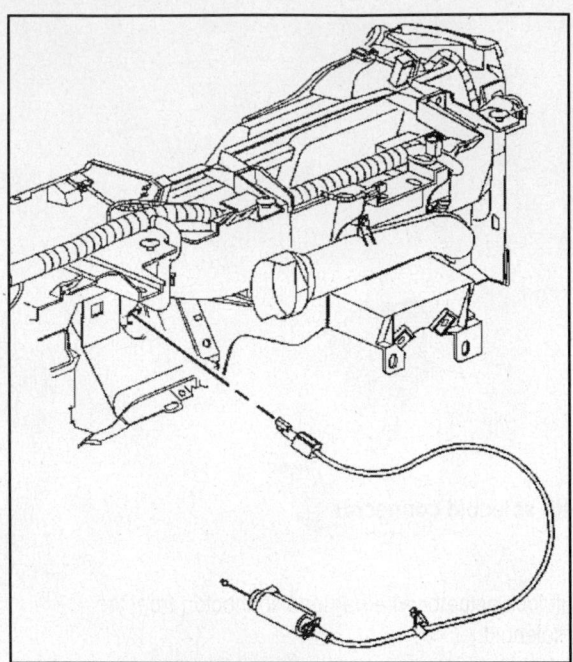

Removing the park lock cable

To install:

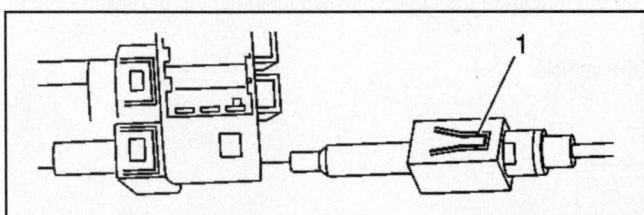

Aligning cable locking tab (1) for installation

9. Turn the ignition switch to the run position.

10. Install the park lock cable into the back of the I/P.

11. Align the park lock cable locking tab (1) with its slot in the ignition switch and snap the cable into the switch. The locking tab should be facing downward at the 6 o'clock position.

12. Route the park lock cable back to its original position.

13. Connect the park lock cable to the park lock cam on the shifter and slide the cable housing into the shifter.

14. If equipped, install the anti-theft plate and tamper proof fastener to the shifter using the special tool.

15. Install the console trim plate.

Connector Pinouts

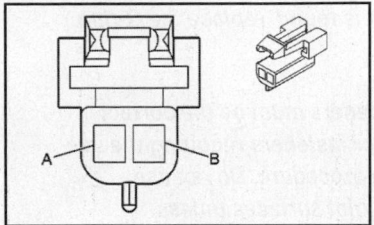

PIN	WIRE COLOR	CIRCUIT NO.	FUNCTION
A	DK GRN/WHT	1135	BRAKE TRANSMISSION SHIFT INTERLOCK SOLENOID FEED
B	BLK	350	GROUND

2000-04 4.6L 4T80-E A/T Shift Lock Control Connector Pinout

CAMSHAFT POSITION (CMP) SENSOR

Description

The camshaft position (CMP) sensor is also a magneto resistive sensor, with the same type of circuits as the crankshaft position (CKP) sensor. The CMP sensor signal is a digital ON/OFF pulse, output once per revolution of the camshaft. The CMP sensor information is used by the PCM to determine the position of the valve train relative to the crankshaft position.

The camshaft reluctor wheel is part of the camshaft sprocket. The reluctor wheel profile is a smooth track, half of which is of a lower profile than the other half. This allows the CMP sensor to supply a signal as soon as the key is turned ON, since the CMP sensor reads the track profile, instead of a notch.

Removal & Installation

1. Remove the camshaft sensor electrical connector.
2. Remove the camshaft retaining bolt.
3. Remove the camshaft sensor.

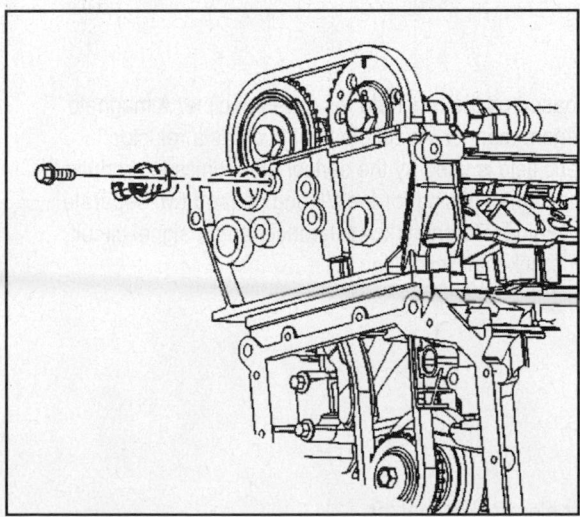

Removing CMP Sensor

To install:

NOTE: Inspect the camshaft sensor O-ring for wear or damage. If a problem is found, replace the O-ring.

4. Lubricate the camshaft O-ring with engine oil.

CAUTION: Use the correct fastener in the correct location. Replacement fasteners must be the correct part number for that application. Fasteners requiring replacement or fasteners requiring the use of thread locking compound or sealant are identified in the service procedure. Do not use paints, lubricants, or corrosion inhibitors on fasteners or fastener joint surfaces unless specified. These coatings affect fastener torque and joint clamping force and may damage the fastener. Use the correct tightening sequence and specifications when installing fasteners in order to avoid damage to parts and systems.

5. Install the camshaft sensor and retaining bolt. Tighten the retaining bolt to 10 Nm (89 inch lbs.).
6. Reconnect the camshaft sensor electrical connector.
7. Operate the engine and inspect the camshaft sensor for engine oil leaks.

Connector Pinouts

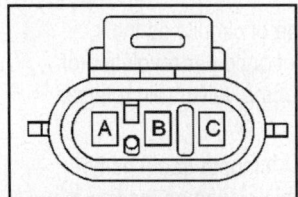

PIN	WIRE COLOR	CIRCUIT NO.	FUNCTION
A	ORN	1799	CMP SENSOR SIGNAL
B	PNK/BLK	632	CMP SENSOR GROUND
C	RED	631	CMP SENSOR IGNITION FEED

2000-04 4.6L CMP Sensor Connector Pinout

CRANKSHAFT POSITION (CKP) SENSOR

Description

The crankshaft position (CKP) sensor is a three wire sensor based on the magneto resistive principle. A magneto resistive sensor uses two magnetic pickups between a permanent magnet. As an element such as a reluctor wheel passes the magnets the resulting change in the magnetic field is used by the sensor electronics to produce a digital output pulse. This system uses two sensors within the same housing for the V6 engine, and two separate sensors for the V8 engine. The PCM supplies each sensor a 12-volt reference, low reference, and a signal circuit. The signal circuit returns a digital ON/OFF pulse 24 times per crankshaft revolution.

Removal & Installation

1. Raise and support the vehicle.
2. Remove the air deflector.
3. Remove the oil filter adaptor.
4. Disconnect the crankshaft position sensor electrical connector.
5. Remove the crankshaft sensor retaining bolt.
6. Remove the crankshaft sensor.

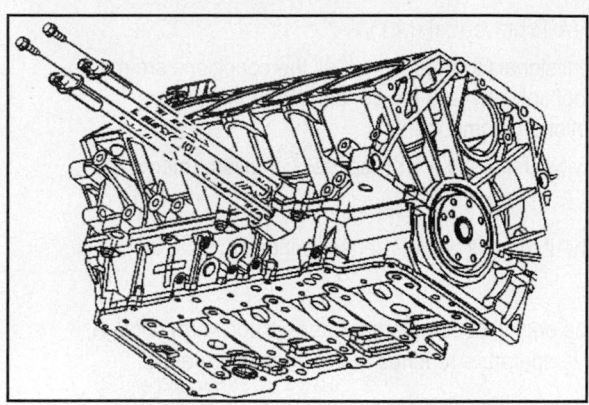

Removing CKP Sensors

To install:

NOTE: Inspect the crankshaft sensor O-ring for wear or damage. If a problem is found, replace the O-ring.

7. Lubricate the crankshaft sensor O-ring with clean engine oil.

Notice: Use the correct fastener in the correct location. Replacement fasteners must be the correct part number for that application. Fasteners requiring replacement or fasteners requiring the use of thread locking compound or sealant are identified in the service procedure. Do not use paints, lubricants, or corrosion inhibitors on fasteners or fastener joint surfaces unless specified. These coatings affect fastener torque and joint clamping force and may damage the fastener. Use the correct tightening sequence and specifications when installing fasteners in order to avoid damage to parts and systems.

8. Install the crankshaft sensor and the retaining bolt. Tighten the retaining bolt between 10 Nm (89 inch lbs.).
9. Connect the crankshaft position sensor electrical connector.
10. Install the oil filter adaptor.
11. Install the air deflector.
12. Lower the vehicle.
13. Operate the engine and inspect crankshaft sensor for engine oil leaks.
14. Perform the CKP System Variation Learn Procedure.

CKP SYSTEM VARIATION LEARN PROCEDURE

NOTE: For additional diagnostic information, refer to DTC P1336.

1. Install a scan tool.
2. With a scan tool, monitor the powertrain control module for DTCs. If other DTCs are set, except DTC P1336, refer to Diagnostic Trouble Code (DTC) List for the applicable DTC that set.
3. With a scan tool, select the crankshaft position variation learn procedure.
4. Observe the fuel cut-off for the engine that you are performing the learn procedure on.
5. The scan tool instructs you to perform the following:
6. Block the drive wheels.
7. Apply the vehicles parking brake.
8. Cycle the ignition from OFF to ON.
9. Apply and hold the brake pedal.
10. Start and idle the engine (with A/C OFF).

11. Place the vehicle's transmission in Park (A/T) or Neutral (M/T).

12. The scan tool monitors certain component signals to determine if all the conditions are met to continue with the procedure. The scan tool only displays the condition that inhibits the procedure. The scan tool monitors the following components:

- Crankshaft position (CKP) sensors activity--If there is a CKP sensor condition, refer to the applicable DTC that set.

- Camshaft position (CMP) sensor activity--If there is a CMP sensor condition, refer to the applicable DTC that set.

- Engine coolant temperature (ECT)--If the engine coolant temperature is not warm enough, idle the engine until the engine coolant temperature reaches the correct temperature.

13. With the scan tool, enable the crankshaft position system variation learn procedure.

NOTE: While the learn procedure is in progress, release the throttle immediately when the engine starts to decelerate. The engine control is returned to the operator and the engine responds to throttle position after the learn procedure is complete.

14. Slowly increase the engine speed to the RPM that you observed.

15. Immediately release the throttle when fuel cut-off is reached.

16. The scan tool displays Learn Status: Learned this ignition. If the scan tool does NOT display this message and no additional DTCs set, refer to Symptom Diagnosis (No Codes) section . If a DTC set, refer to Diagnostic Trouble Code (DTC) List for the applicable DTC that set.

17. Turn OFF the ignition for 30 seconds after the learn procedure is completed successfully.

Component Locations

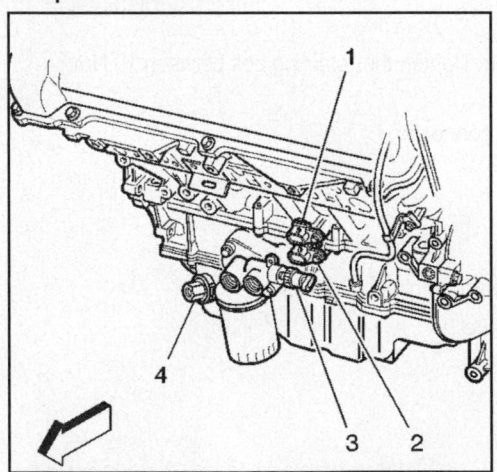

1. Crankshaft Position Sensor (CKP) Sensor B
2. Crankshaft Position Sensor (CKP) Sensor A
3. Engine Oil Pressure (EOP) Switch
4. Engine Oil Level Switch

Showing engine side locations of the CKP sensors, EOP switch, and oil level switch

Connector Pinouts

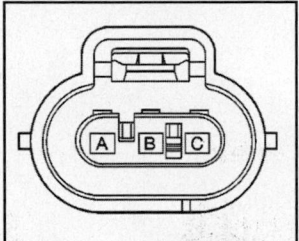

PIN	WIRE COLOR	CIRCUIT NO.	FUNCTION
A	LT GRN	1867	CKP SENSOR A IGNITION FEED
B	YEL/BLK	1868	CKP SENSOR A GROUND (LOW REFERENCE)
C	LT BLU/WHT	1800	CKP SENSOR A SIGNAL

2000 04 4.6L CKP Sensor A Connector Pinout

PIN	WIRE COLOR	CIRCUIT NO.	FUNCTION
A	LT GRN	2867	CKP SENSOR B IGNITION FEED
B	LT BLU/WHT OR YEL	1800 OR 2868	CKP SENSOR B GROUND (LOW REFERENCE)
C	YEL OR YEL/BLK	573	CKP SENSOR B SIGNAL

2000-04 4.6L CKP Sensor B Connector Pinout

ENGINE COOLANT LEVEL SWITCH

Description

The engine cooling system contains an engine coolant level switch, located on the bottom of the surge tank, that alerts the driver in the event of a coolant loss. When the engine coolant level switch reads a low coolant level in the surge tank, the switch opens. This sends a coolant loss signal to the IPC by the coolant level switch signal circuit.

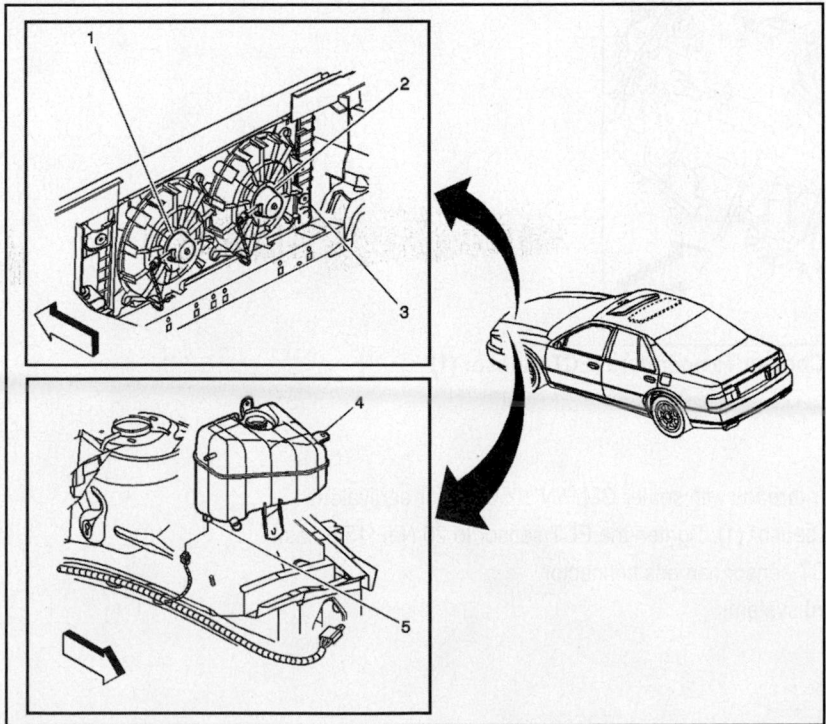

Identifying location of cooling system components: (1) Cooling Fan – Left; (2) Cooling Fan – Right; (3) Radiator; (4) Coolant Reservoir; (5) Coolant Level Switch

Connector Pinouts

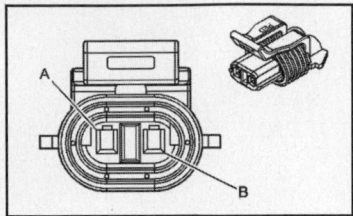

PIN	WIRE COLOR	CIRCUIT NO.	FUNCTION
A	LT GRN	1478	FUSE OUTPUT - OFF/RUN/COOLANT LEVEL SWITH SIGNAL
B	BLK	150	GROUND

2000-04 4.6L Engine Coolant Level Switch Connector Pinout

ENGINE COOLANT TEMPERATURE (ECT) SENSOR

Removal & Installation

1. Turn OFF the ignition
2. Drain the coolant to below the level of the engine coolant temperature (ECT) sensor.
3. Disconnect the ECT sensor harness connector.
4. Remove the ECT sensor.

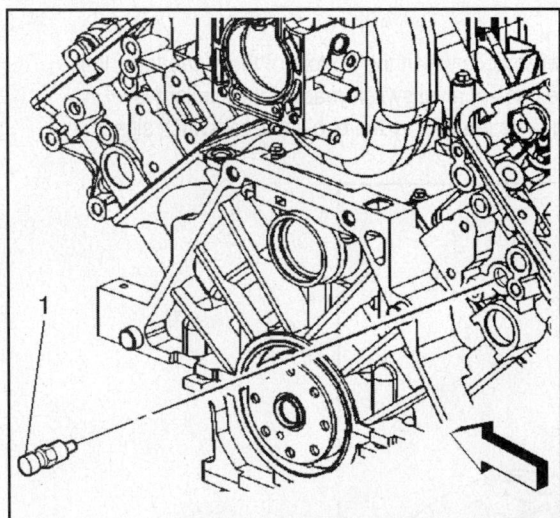

Showing position of the Engine Coolant Temperature (ECT) Sensor (1)

To install:

1. Coat the sensor threads with sealer GM P/N 1050805, or equivalent.
2. Install the ECT sensor (1). Tighten the ECT sensor to 20 Nm (15 ft. lbs.).
3. Connect the ECT sensor harness connector.
4. Refill the coolant system.

Connector Pinouts

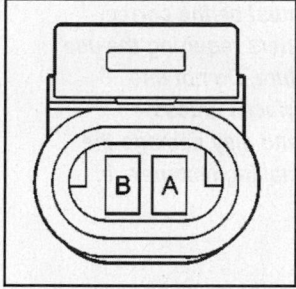

PIN	WIRE COLOR	CIRCUIT NO.	FUNCTION
A	BLK	2761	SENSOR GROUND
B	YEL	410	ECT SENSOR SIGNAL

2000-03 4.6L ECT Sensor Connector Pinout

ENGINE COOLING FAN CONNECTORS

Removal & Installation

1. Remove the cooling fan assembly.

CAUTION: The fan blade retaining nut is left hand thread, damage may occur to motor threads if attempting to remove as a right hand thread nut.

2. Remove the fan blade retaining nut (5) from the motor (2).
3. Remove the fan blade (4) from the motor shaft.
4. Remove the motor retaining screws (3).
5. Remove the motor (2) from the shroud (1).

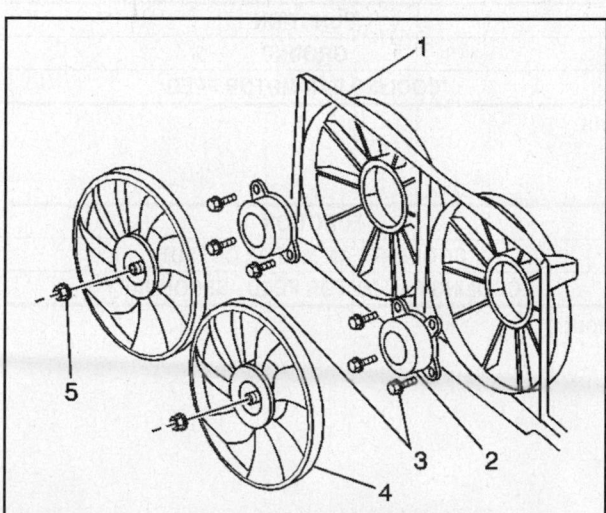

Exploded view of cooling fan assemblies

To install:

CAUTION: Use the correct fastener in the correct location. Replacement fasteners must be the correct part number for that application. Fasteners requiring replacement or fasteners requiring the use of thread locking compound or sealant are identified in the service procedure. Do not use paints, lubricants, or corrosion inhibitors on fasteners or fastener joint surfaces unless specified. These coatings affect fastener torque and joint clamping force and may damage the fastener. Use the correct tightening sequence and specifications when installing fasteners in order to avoid damage to parts and systems.

6. Install the motor (2) to the shroud (1).

7. Install the motor retaining screws (3). Tighten the electric cooling fan motor mounting screws to 6 Nm (53 inch lbs.).

8. Install the fan blade (4) to the motor shaft.

9. Install the fan blade retaining nut (5) to the motor (2). Tighten the electric cooling fan blade retaining nut to 6 Nm (53 inch lbs.).

10. Install the cooling fan assembly.

Connector Pinouts

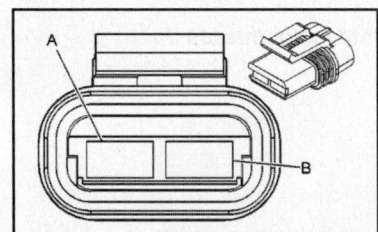

PIN	WIRE COLOR	CIRCUIT NO.	FUNCTION
A	BLK	250	GROUND
B	LT BLU	409	COOLING FAN MOTOR FEED

2000-04 4.6L Left Engine Cooling Fan Connector Pinout

PIN	WIRE COLOR	CIRCUIT NO.	FUNCTION
A	WHT	504	COOLANT FAN MOTOR OUTPUT
B	GRY	532	COOLING FAN MOTOR FEED - SECONDARY

2000-04 4.6L Right Engine Cooling Fan Connector Pinout

ENGINE COOLING FAN RELAY

Connector Pinouts

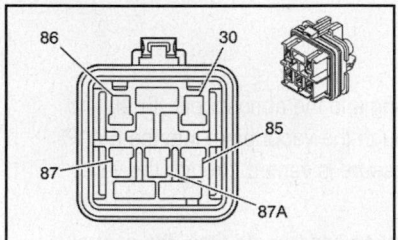

PIN	WIRE COLOR	CIRCUIT NO.	FUNCTION
30	RED	1142	BATTERY POSITIVE VOLTAGE
85	ORN	2340	BATTERY POSITIVE VOLTAGE
86	DK GRN	335	LOW SPEED COOLING FAN RELAY CONTROL
87	GRY	532	COOLING FAN MOTOR SUPPLY VOLTAGE
87A	YEL	998	12 VOLT REFERENCE

2000-04 4.6L Engine Cooling Fan Relay 1 Connector Pinout

PIN	WIRE COLOR	CIRCUIT NO.	FUNCTION
30	WHT	504	COOLING FAN MOTOR OUTPUT
85	ORN	2340	FUSED OUTPUT - BATTERY
86	DK BLU	473	COOLING FAN RELAY OUTPUT - COIL - SECONDARY
87	BLK	350	GROUND
87A	LT BLU	409	COOLING FAN MOTOR FEED

2000 4.6L Engine Cooling Fan Relay 2 Connector Pinout

PIN	WIRE COLOR	CIRCUIT NO.	FUNCTION
30	RED	1142	BATTERY POSITIVE VOLTAGE
85	ORN	2340	BATTERY POSITIVE VOLTAGE
86	DK BLU	473	HIGH SPEED COOLING FAN RELAY CONTROL
87	LT BLU	409	COOLING FAN MOTOR SUPPLY VOLTAGE
87A	YEL	998	12 VOLT REFERENCE

2001-04 4.6L Engine Cooling Fan Relay 2 Connector Pinout

PIN	WIRE COLOR	CIRCUIT NO.	FUNCTION
30	WHT	504	COOLING FAN LOW REFERENCE
85	ORN	2340	BATTERY POSITIVE VOLTAGE
86	DK BLU	473	HIGH SPEED COOLING FAN RELAY CONTROL
87	BLK	350	GROUND
87A	LT BLU	409	COOLING FAN MOTOR SUPPLY VOLTAGE

2001-04 4.6L Engine Cooling Fan S/P Relay Connector Pinout

PIN	WIRE COLOR	CIRCUIT NO.	FUNCTION
30	RED	1142	FUSED OUTPUT - BATTERY
85	ORN	2340	FUSED OUTPUT - BATTERY
86	DK BLU	473	COOLING FAN RELAY OUTPUT - COIL - SECONDARY
87	LT BLU	409	COOLING FAN MOTOR FEED
87A	YEL	998	COOLING FAN MOTOR FEED DISABLE

2000 4.6L Engine Cooling Fan Relay 3 Connector Pinout

EVAP CANISTER PURGE VALVE & VENT VALVE

Description & Operation

EVAP CONTROL SYSTEM

The evaporative emission (EVAP) control system limits fuel vapors from escaping into the atmosphere. Fuel tank vapors are allowed to move from the fuel tank, due to pressure in the tank, through the vapor pipe, into the EVAP canister. Carbon in the canister absorbs and stores the fuel vapors. Excess pressure is vented through the vent line and EVAP vent solenoid to atmosphere.

The EVAP canister stores the fuel vapors until the engine is able to use them. At an appropriate time, the control module will command the EVAP purge solenoid ON, open, allowing engine vacuum to be applied to the EVAP canister. With the EVAP vent solenoid OFF, open, fresh air will be drawn through the solenoid and vent line to the EVAP canister. Fresh air is drawn through the canister, pulling fuel vapors from the carbon.

The air/fuel vapor mixture continues through the EVAP purge pipe and EVAP purge solenoid into the intake manifold to be consumed during normal combustion. The control module uses several tests to determine if the EVAP system is leaking.

The EVAP system consists of the following components:

EVAP Canister

The canister is filled with carbon pellets used to absorb and store fuel vapors. Fuel vapor is stored in the canister until the control module determines that the vapor can be consumed in the normal combustion process.

EVAP Purge Solenoid

The EVAP purge solenoid controls the flow of vapors from the EVAP system to the intake manifold. This normally closed solenoid is pulse width modulated (PWM) by the control module to precisely control the flow of fuel vapor to the engine. The solenoid will also be opened during some portions of the EVAP testing, allowing engine vacuum to enter the EVAP system.

EVAP Vent Solenoid

The EVAP vent solenoid controls fresh airflow into the EVAP canister. The solenoid is normally open. The control module will command the solenoid closed during some EVAP tests, allowing the system to be tested for leaks.

Fuel Tank Pressure Sensor

The FTP sensor measures the difference between the pressure or vacuum in the fuel tank and outside air pressure. The control module provides a 5-volt reference and a ground to the FTP sensor. The FTP sensor provides a signal voltage back to the control module that can vary between 0.1-4.9 volts. As FTP increases, FTP sensor voltage decreases, high pressure = low voltage. As FTP decreases, FTP voltage increases, low pressure or vacuum = high voltage.

EVAP Service Port

The EVAP service port is located in the EVAP purge pipe between the EVAP purge solenoid and the EVAP canister. The service port is identified by a green colored cap.

TESTING

Large Leak Test

This tests for large leaks and blockages in the EVAP system. The control module will command the EVAP vent solenoid ON, closed, and command the EVAP purge solenoid ON, open, with the engine running, allowing engine vacuum into the EVAP system. The control module monitors the fuel tank pressure (FTP) sensor voltage to verify that the system is able to reach a predetermined level of vacuum within a set amount of time. The control module then commands the EVAP purge solenoid OFF, closed, sealing the system and monitors the vacuum level for decay. If the control module does not detect that the predetermined vacuum level was achieved, or the vacuum decay rate is more than a calibrated level on 2 consecutive tests, a DTC P0440 will set.

Small Leak Test

If the large leak test passes, the control module will test for small leaks by continuing to monitor the FTP sensor for a change in voltage over a period of time. If the decay rate is more than a calibrated value, the control module will rerun the test. If the test fails again, a DTC P0442 will set.

Canister Vent Restriction Test

If the EVAP vent system is restricted, fuel vapors will not be properly purged from the EVAP canister. The control module tests this by commanding the EVAP purge solenoid ON, open; and commanding the EVAP vent solenoid OFF, open; and monitoring the FTP sensor for an increase in vacuum. If vacuum increases more than a calibrated value, DTC P0446 will set.

Purge Solenoid Leak Test

If the EVAP purge solenoid does not seal properly, fuel vapors could enter the engine at an undesired time, causing driveability concerns. The control module tests for this by commanding the EVAP purge solenoid OFF, closed; and vent solenoid ON, closed; sealing the system, and monitoring the FTP for an increase in vacuum. If the control module detects that EVAP system vacuum increases above a calibrated value, DTC P1441 will set.

"Check Gas Cap" Message

The PCM sends a class 2 message to the Driver Information Center (DIC) illuminating the Check Gas Cap message when any of the following occur:

- A malfunction in the EVAP system and a large leak test fails
- A malfunction in the EVAP system and a small leak test fails

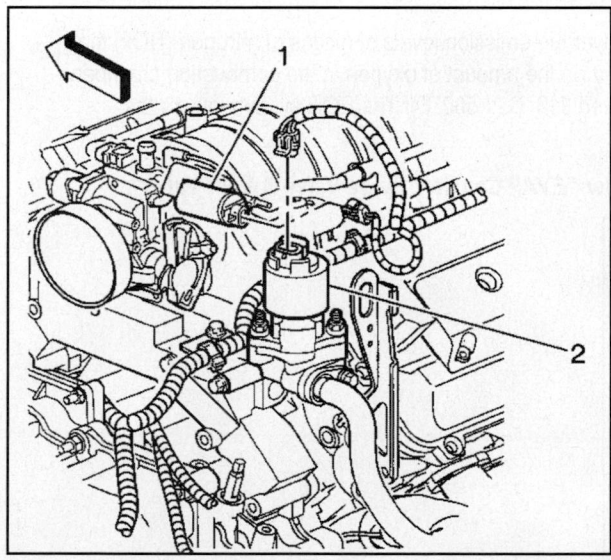

Showing location of (1) Evaporative Emissions (EVAP) Canister Purge Solenoid and (2) Exhaust Gas Recirculation (EGR) Valve

Connector Pinouts

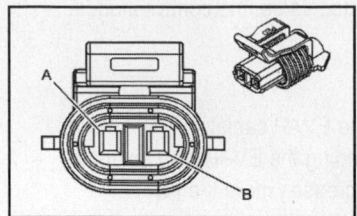

PIN	WIRE COLOR	CIRCUIT NO.	FUNCTION
A	PNK	139	IGN ITION VOLTAGE
B	DK GRN/WHT	428	EVAP CANISTER PURGE VALVE CONTROL

2000-04 4.6L EVAP Purge Valve Connector Pinout

PIN	WIRE COLOR	CIRCUIT NO.	FUNCTION
A	PNK	1239	IGNITION VOLTAGE
B	WHT	1310	EVAP CANISTER VENT VALVE CONTROL

2000-04 4.6L EVAP Vent Valve Connector Pinout

EXHAUST GAS RECIRCULATION (EGR) VALVE

Description & Operation

The exhaust gas recirculation (EGR) system is used to lower the emission levels of oxides of nitrogen (NOx) that form during the combustion process. NOx levels are based on the amount of oxygen in the combustion chamber, and the length of time that combustion temperatures exceed 816°C (1,500°F). The EGR system lowers the combustion temperatures, thus lowering the level of NOx.

NOTE: For location of EGR Valve, see illustration under "EVAP Canister Purge Valve & Vent Valve".

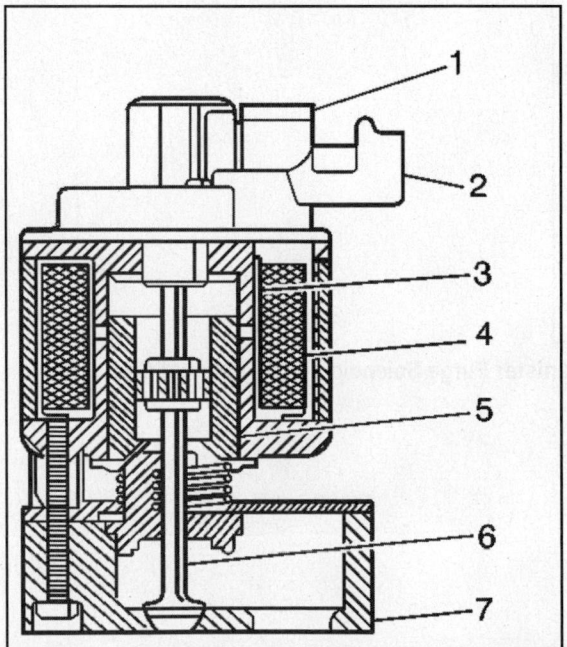

Showing the components of the EGR Valve: (1) Cap-Sensor, (2) Sensor-EGR Pintle Position, (3) Pole Piece-Primary, (4) Bobbin and Coil Assembly, (5) Sleeve-Armature, (6) Valve-Pintle, and (7) Armature and Base Assembly

The EGR valve is designed to supply EGR to the engine. The EGR valve consists of a sealed bobbin and coil assembly, or solenoid. Inside the solenoid is a sleeve-armature assembly that contains a pintle and valve, two seals, retaining washers, a seal spring, and armature spring, and a bearing. The bearing seals the pintle valve shaft from the exhaust chamber. Also, a shield, held in place by a compression spring, deflects exhaust gas from the shaft and armature.

The powertrain control module (PCM) uses the pintle valve to control the EGR flow into the engine. Exhaust gas is routed from the intermediate exhaust pipe to the valve through a feed pipe. When the PCM commands the EGR valve open, the exhaust flows through the EGR valve, past the pintle and into the engine via the crossover water pump housing. When the throttle valve opens, the exhaust gas mixes with the incoming air. When the combination of air/fuel/exhaust gas is burned in the chamber, a portion of the heat energy is absorbed by the exhaust gas. This helps to lower the level of NOx emissions.

The PCM monitors the pintle position using the signal from the EGR pintle position sensor. The sensor is an integral part of the EGR valve. The PCM supplies the pintle position sensor with a 5-volt reference and a ground. The pintle position sensor provides a signal voltage to the PCM. By monitoring the voltage on the signal line, the PCM is able to determine if the EGR valve responds properly to commands from the PCM. As the EGR valve position changes, the pintle position signal voltage will change. With the EGR valve closed, the signal voltage is near 0 volts. However, the pintle position signal voltage increases as the EGR valve opens.

Testing

Too much EGR flow tends to weaken combustion, causing the engine to run rough and/or stall. With too much EGR flow at idle, cruise, or cold operation, any of the following conditions may occur:

- Engine stalls after cold start.
- Vehicle surge during cruise operation
- Engine stalls during closed throttle conditions
- Rough idle

DTC P0300 MISFIRE DETECTED

If the EGR valve is stuck open, the engine may not run. Too little or no EGR flow allows combustion temperatures to get too high during acceleration and load conditions. This could cause:

- Spark knock/detonation, especially on light acceleration
- Engine overheating
- Emission test failure

DTC P0401

Poor fuel economy. EGR flow diagnosis is included in the DTC P0401 diagnostic table. Pintle position error and control circuit diagnosis is covered in DTCs P0403, P0404, P0405, and P1404. If EGR DTCs are encountered, refer Diagnostic Trouble Code (DTC) List in the appropriate section.

Component Locations

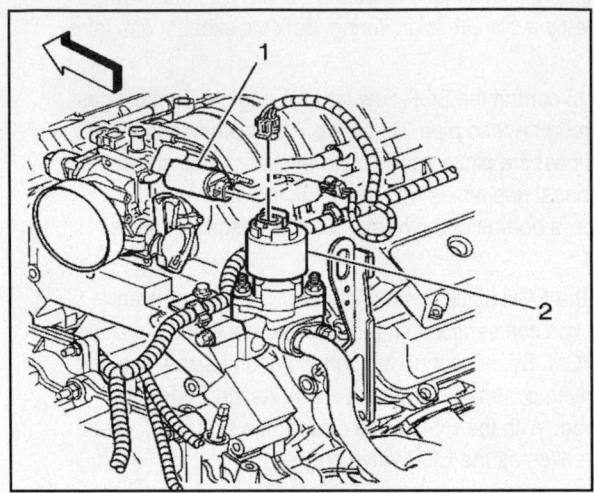

1. Evaporative Emissions (EVAP) Canister Purge Solenoid
2. Exhaust Gas Recirculation (EGR) Valve

Indicating the location of the EVAP purge solenoid and EGR valves

Connector Pinouts

PIN	WIRE COLOR	CIRCUIT NO.	FUNCTION
A	GRY	435	EGR VALVE GROUND
B	BLK	2753	EGR PINTLE POSITION GROUND
C	BRN	1456	EGR POSITION SIGNAL
D	GRY OR GRY/WHT	2702	5 VOLT REFERENCE
E	RED	1676	EGR VALVE CONTROL

2000-03 4.6L EGR Valve Connector Pinout

FUEL INJECTOR CONNECTORS

Description & Operation

The top-feed fuel injector assembly is a solenoid operated device, controlled by the PCM, that meters pressurized fuel to a single engine cylinder. The PCM energizes the injector solenoid, which opens a ball valve, allowing fuel to flow past the ball valve, and through a recessed flow director plate. The director plate has multiple machined holes that control the fuel flow, generating a conical spray pattern of finely atomized fuel at the injector tip. Fuel is directed at the intake valve, causing it to become further atomized and vaporized before entering the combustion chamber. An injector stuck partly open can cause a loss of pressure after engine shutdown. Consequently, long cranking times would be noticed on some engines.

Connector Pinouts

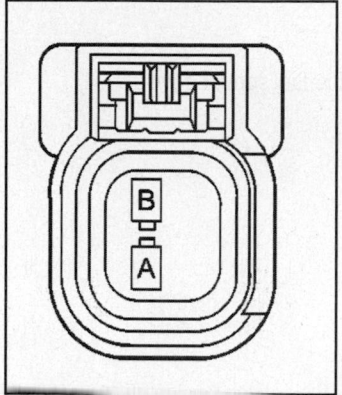

PIN	WIRE COLOR	CIRCUIT NO.	FUNCTION
A	PNK	239	IGNITION POSITIVE VOLTAGE (INJECTORS 1, 4, 6 & 7)
A	PNK	339	IGNITION POSITIVE VOLTAGE (INJECTORS 2, 3, 5 & 8)
B	BLK	1744	FUEL INJECTOR 1 CONTROL
B	LT GRN/BLK	1745	FUEL INJECTOR 2 CONTROL
B	PNK/BLK	1746	FUEL INJECTOR 3 CONTROL
B	LT BLU/BLK	844	FUEL INJECTOR 4 CONTROL
B	BLK/WHT	845	FUEL INJECTOR 5 CONTROL
B	YEL/BLK	846	FUEL INJECTOR 6 CONTROL
B	RED/BLK	877	FUEL INJECTOR 7 CONTROL
B	DK BLU/WHT	878	FUEL INJECTOR 8 CONTROL

2000-04 4.6L Fuel Injector Connector Pinouts

FUEL PUMP & SENDER CONNECTOR

Description & Operation

The fuel level sensor consists of a float, a wire float arm, and a ceramic resistor card. The position of the float arm indicates the fuel level. The fuel level sensor contains a variable resistor that changes resistance in correspondence with the amount of fuel in the fuel tank. The PCM sends the fuel level information via the Class II circuit to the Instrument Panel (I/P) cluster. This information is used for the I/P fuel gage and the low fuel warning indicator, if applicable. The PCM also monitors the fuel level input for various diagnostics.

Caution: Fuel Vapors can collect while servicing fuel system parts in enclosed areas such as a trunk. To reduce the risk of fire and increased exposure to vapors:

- Use forced air ventilation such as a fan set outside of the trunk.
- Plug or cap any fuel system openings in order to reduce fuel vapor formation.
- Clean up any spilled fuel immediately.
- Avoid sparks and any source of ignition.
- Use signs to alert others in the work area that fuel system work is in process.

Removal & Installation

1. Remove the fuel sender assembly.
2. Disconnect the fuel pump electrical connector.
3. Remove the fuel level sensor electrical connector retaining clip.
4. Disconnect the fuel level sensor electrical connector from under the fuel sender cover.
5. Remove the fuel level sensor retaining clip.
6. Squeeze the locking tangs and remove the fuel level sensor.

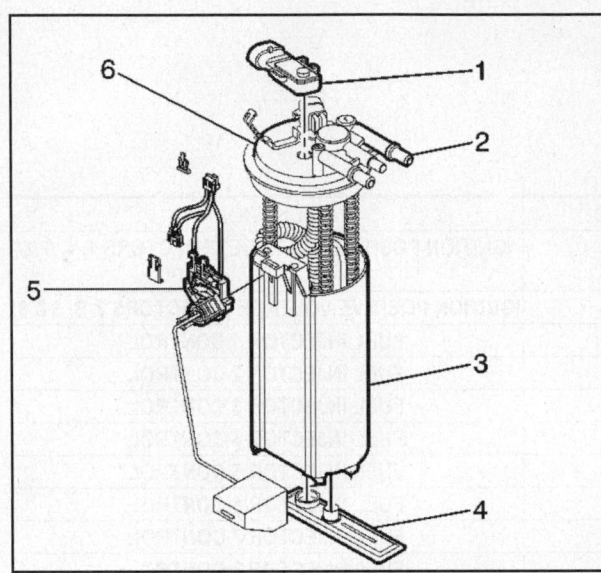

The fuel sender assembly consists of the following major components: fuel tank pressure sensor (1); fuel line fittings (2); fuel tank fuel pump module (3); fuel strainer (4); fuel level sensor (5)

To install:

7. Install the fuel level sensor.
8. Install the fuel level sensor retaining clip.
9. Connect the fuel level sensor electrical connector.
10. Install the fuel level sensor electrical connector retaining clip.
11. Connect the fuel pump electrical connector.
12. Install the fuel sender assembly.

Connector Pinouts

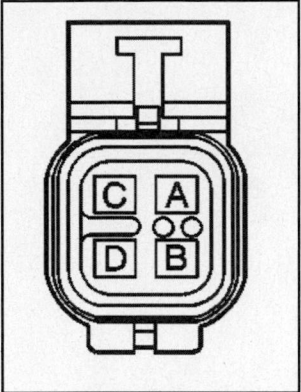

PIN	WIRE COLOR	CIRCUIT NO.	FUNCTION
A	PPL	1589	FUEL LEVEL SENSOR SIGNAL
B	GRY	120	FUEL PUMP MOTOR FEED (SUPPLY VOLTAGE)
C	BLK	1250	FUEL PUMP MOTOR GROUND
D	BLK/WHT	2759	FUEL LEVEL SIGNAL RETURN

2000-03 4.6L Fuel Pump & Sender Connector Pinout

FUEL TANK PRESSURE (FTP) SENSOR CONNECTOR

Removal & Installation

1. Remove the rear compartment trim panel.
2. Remove the fuel sender access panel bolts.
3. Remove the fuel sender access panel.
4. Disconnect the fuel tank pressure sensor electrical connector.
5. Remove the fuel tank pressure sensor.

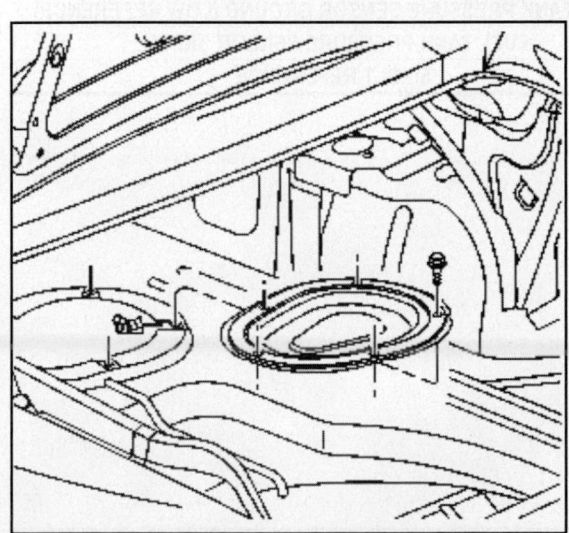

Removing access panel for fuel tank pressure sensor

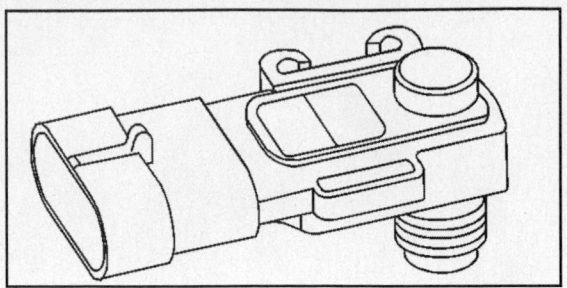

Identifying Fuel Tank Pressure Sensor

To install:

6.   Install the fuel tank pressure sensor.

7.   Connect the fuel tank pressure sensor electrical connector.

8.   Install the fuel sender access panel.

9.   Install the fuel sender access panel bolts. Tighten the fuel sender access panel bolts to 2 Nm (18 inch lbs.).

10. Install the rear compartment trim panel.

Connector Pinouts

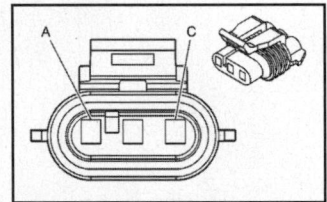

PIN	WIRE COLOR	CIRCUIT NO.	FUNCTION
A	BLK	2759	FUEL TANK PRESSURE SENSOR GROUND (LOW REFERENCE)
B	DK GRN	890	FUEL TANK PRESSURE SENSOR SIGNAL
C	GRY	2709	5 VOLT REFERENCE

2000-04 4.6L FTP Sensor Connector Pinout

GENERATOR CONNECTOR

Connector Pinouts

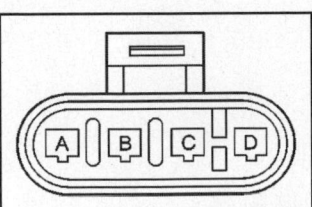

PIN	WIRE COLOR	CIRCUIT NO.	FUNCTION
A	--	--	NOT USED
B	RED	225	GENERATOR L TERMINAL TURN-ON SIGNAL/MONITOR FROM PCM
C	GRY	23	GENERATOR F TERMINAL FIELD DUTY SIGNAL/MONITOR FOR PCM
D	ORN	3340	SYSTEM VOLTAGE SENSE

2000-04 4.6L Generator Connector Pinout

HEATED OXYGEN SENSOR (HO2S)

Removal & Installation

HEATED OXYGEN SENSOR BANK 1 SENSOR 1 (HO2S 1/1)

WARNING: Handle the oxygen sensors carefully in order to prevent damage to the component. Keep the electrical connector and the exhaust inlet end free of contaminants. Do not use cleaning solvents on the sensor. Do not drop or mishandle the sensor.

CAUTION: Remove oxygen sensors with the engine temperature above 48°C (120°F). Otherwise the oxygen sensors may be difficult to remove. A special anti-seize compound is used on the oxygen sensor threads. New service sensors should already have the compound applied to the threads. Coat the threads of a reused sensor with anti-seize compound P/N 5613695 or equivalent.

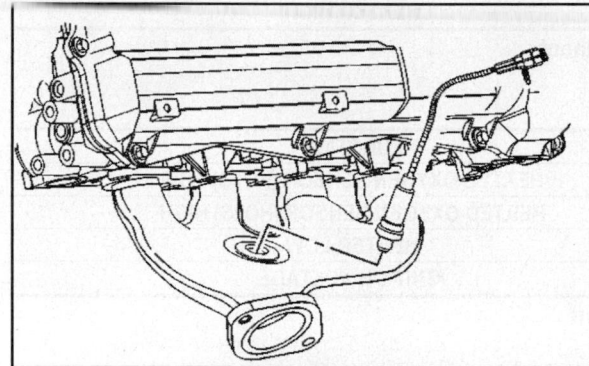

Showing the location of the HO2S Bank 1 Sensor 1

Removal & Installation

1. Raise and support the vehicle.
2. Support rear of the powertrain frame.
3. Remove four rear frame bolts.
4. Lower the frame no more than 73.2 mm (3 inches).
5. Disconnect the rear oxygen sensor electrical connector.
6. Remove the rear oxygen sensor from the rear exhaust manifold.

To install:

7. Install the rear oxygen sensor in the rear exhaust manifold. Tighten the rear heated oxygen sensor (HO2S) sensor to 41 Nm (30 ft. lbs.).
8. Connect the rear oxygen sensor connector.
9. Raise the frame.
10. Install the four rear frame bolts. Tighten the frame bolts to 100 Nm (74 ft. lbs.).
11. Lower the vehicle.

Connector Pinouts

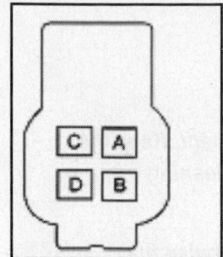

PIN	WIRE COLOR	CIRCUIT NO.	FUNCTION
A	TAN	1664	HEATED OXYGEN SENSOR (HO2S) LOW
B	PPL/WHT	1665	HEATED OXYGEN SENSOR (HO2S) HIGH
C	BLK	3112	HEATER LOW
D	LT GRN	9102	HEATER HIGH

2000-03 4.6L O2 Sensor (Bank 1 Sensor 1) Connector Pinout

PIN	WIRE COLOR	CIRCUIT NO.	FUNCTION
A	TAN/WHT	3111	HEATED OXYGEN SENSOR (HO2S) LOW
B	PPL/WHT	3110	HEATED OXYGEN SENSOR (HO2S) HIGH
C	BLK/WHT	3113	HEATER LOW
D	PNK	539	IGNITION VOLTAGE

2004 4.6L O2 Sensor (Bank 1 Sensor 1) Connector Pinout

Removal & Installation

HEATED OXYGEN SENSOR BANK 1 SENSOR 2 (HO2S 1/2)

CAUTION: Handle the oxygen sensors carefully in order to prevent damage to the component. Keep the electrical connector and the exhaust inlet end free of contaminants. Do not use cleaning solvents on the sensor. Do not drop or mishandle the sensor.

CAUTION: Remove oxygen sensors with the engine temperature above 48°C (120°F). Otherwise the oxygen sensors may be difficult to remove.

WARNING: A special anti-seize compound is used on the oxygen sensor threads. New service sensors should already have the compound applied to the threads. Coat the threads of a reused sensor with anti-seize compound P/N 5613695 or equivalent.

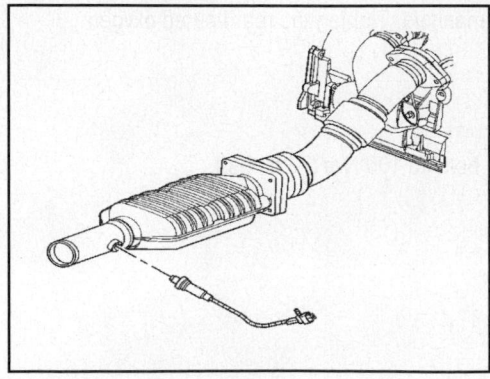

Showing the location of the HO2S Bank 1 Sensor 2

1. Raise and support the vehicle.
2. Remove the heat shield.
3. Disconnect the heated oxygen sensor electrical connector.
4. Remove the heated oxygen sensor.

To install:

5. Install the heated oxygen sensor. Tighten the heated oxygen sensor (HO2S) to 41 Nm (30 ft. lbs.).
6. Connect the heated oxygen sensor electrical connector.

CAUTION: Ensure connectors are securely installed beneath harness heat shield to prevent damage.

7. Install the heat shield.
8. Lower the vehicle.

Connector Pinouts

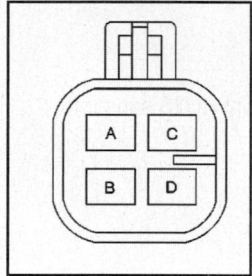

PIN	WIRE COLOR	CIRCUIT NO.	FUNCTION
A	TAN	1664	HEATED OXYGEN SENSOR (HO2S) LOW
B	PPL/WHT	1665	HEATED OXYGEN SENSOR (HO2S) HIGH
C	BLK	3112	HEATER LOW
D	LT GRN	9102	HEATER HIGH

2000-03 4.6L O2 Sensor (Bank 1 Sensor 2) Connector Pinout

PIN	WIRE COLOR	CIRCUIT NO.	FUNCTION
A	TAN/WHT	3121	HEATED OXYGEN SENSOR (HO2S) LOW
B	PPL	1320	HEATED OXYGEN SENSOR (HO2S) HIGH
C	BRN	3121	HEATER LOW
D	PNK	539	IGNITION VOLTAGE

2004 4.6L O2 Sensor (Bank 1 Sensor 2) Connector Pinout

Removal & Installation

HEATED OXYGEN SENSOR BANK 2 SENSOR 1 (HO2S 2/1)

CAUTION: Handle the oxygen sensors carefully in order to prevent damage to the component. Keep the electrical connector and the exhaust inlet end free of contaminants. Do not use cleaning solvents on the sensor. Do not drop or mishandle the sensor.

CAUTION: Remove oxygen sensors with the engine temperature above 48°C (120°F). Otherwise the oxygen sensors may be difficult to remove.

WARNING: A special anti-seize compound is used on the oxygen sensor threads. New service sensors should already have the compound applied to the threads. Coat the threads of a reused sensor with anti-seize compound P/N 5613695 or equivalent.

1. Raise and support the vehicle.
2. Remove the air deflector.
3. Disconnect the front oxygen sensor electrical connector.
4. Remove the front oxygen sensor from the front exhaust manifold.

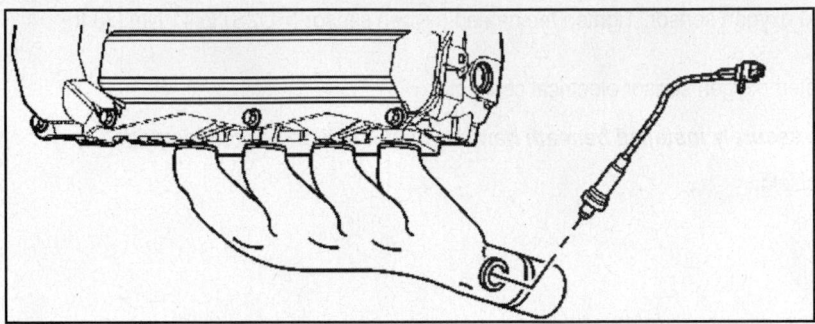

Showing the location of the HO2S Bank 2 Sensor 1

To install:

5. Install the front oxygen sensor in the front exhaust manifold. Tighten the front HO2S sensor to 41 Nm (30 ft. lbs.).
6. Connect the front oxygen sensor electrical connector.
7. Install the air deflector.
8. Lower the vehicle.

Connector Pinouts

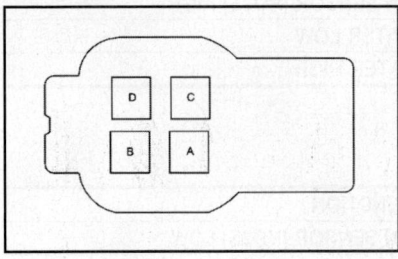

PIN	WIRE COLOR	CIRCUIT NO.	FUNCTION
A	TAN/WHT	1669	HEATED OXYGEN SENSOR (HO2S) LOW
B	PPL/WHT	1668	HEATED OXYGEN SENSOR (HO2S) HIGH
C	BLK	3122	HEATER LOW
D	PNK	539	IGNITION POSITIVE VOLTAGE

2000-03 4.6L O2 Sensor (Bank 2 Sensor 1) Connector Pinout

PIN	WIRE COLOR	CIRCUIT NO.	FUNCTION
A	TAN/WHT	3211	HEATED OXYGEN SENSOR (HO2S) LOW
B	PPL/WHT	3210	HEATED OXYGEN SENSOR (HO2S) HIGH
C	LT GRN	3212	HEATER LOW
D	PNK	539	IGNITION POSITIVE VOLTAGE

2004 4.6L O2 Sensor (Bank 2 Sensor 1) Connector Pinout

IDLE AIR CONTROL (IAC) VALVE

Removal & Installation

1. Remove the electrical connector (1) from the IAC valve (4).
2. Remove the IAC valve attaching screws (2) from the idle air control valve (4).
3. Remove the IAC valve from the throttle body.
4. Remove the IAC valve O-ring.

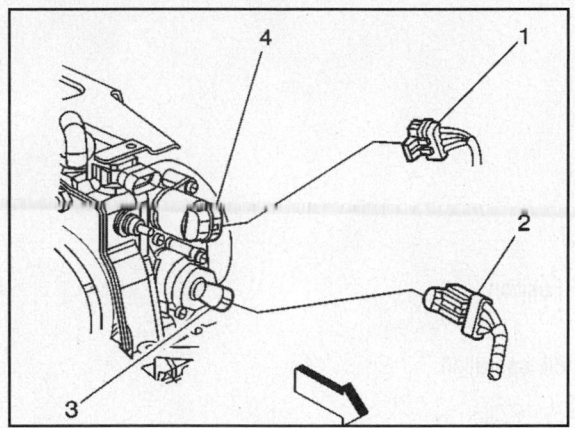

Identifying the components for removing the IAC Valve

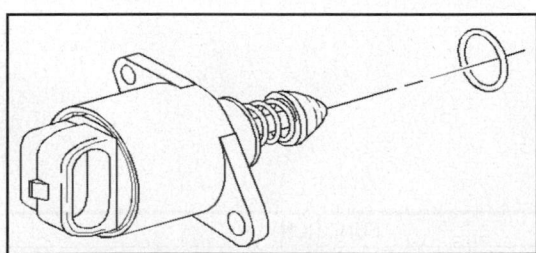

Showing the IAC Valve and O-ring

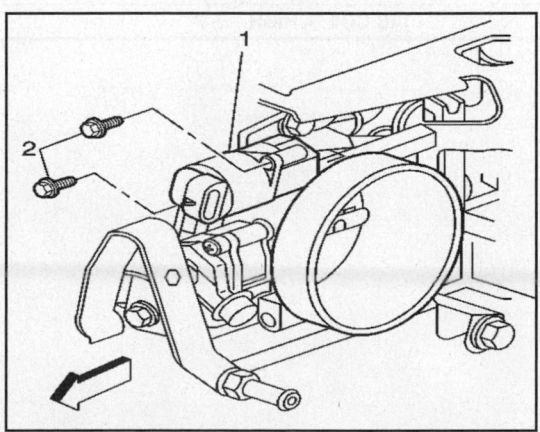

Remove the IAC Valve attaching screws

To install:

5. Install the new IAC valve O-ring.

6. Install the IAC valve (1) into the throttle body.

7. Install the IAC valve attaching screws (2) into the IAC valve. Tighten the IAC valve attaching screws to 3 Nm (27 inch lbs.).

8. Install the electrical connector into the idle air control valve.

9. The powertrain control module (PCM) will reset the idle air control valve whenever the ignition switch is turned ON, then OFF.

10. Turn the ignition switch ON.

11. Turn the ignition switch OFF.

12. Perform the IAC Valve learn procedure.

IAC Valve Learn Procedure

1. Start and idle the engine for 15 seconds.

2. Turn the ignition switch to the LOCK/OFF position.

3. Wait 15 seconds.

4. Restart the engine, and check for proper idle operation.

Connector Pinouts

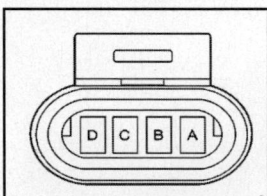

PIN	WIRE COLOR	CIRCUIT NO.	FUNCTION
A	LT GRN/BLK	444	IAC COIL B LOW
B	LT GRN/WHT	1749	IAC COIL B HIGH
C	LT BLU/BLK	1748	IAC COIL A LOW
D	LT BLU/WHT	1747	IAC COIL A HIGH

2000-04 4.6L IAC Valve Connector Pinout

IGNITION COIL/CONTROL MODULE (ICM)

Description & Operation

Each ignition control module (ICM) has the following circuits:

- An ignition 1 voltage circuit
- A chassis ground
- An ignition control circuit for each cylinder
- A low reference circuit

The PCM controls spark by pulsing the ignition control circuits to the ICM to trigger the coils and fire the spark plugs. The PCM and ICM are internally protected against shorts to power and ground on the ignition control circuits.

The spark plugs are connected to each coil by a short boot. The boot contains a spring that conducts the spark energy from the coil to the spark plug. The spark plugs are tipped with platinum for long wear and higher efficiency.

Connector Pinouts

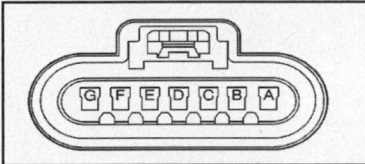

PIN	WIRE COLOR	CIRCUIT NO.	FUNCTION
A	PNK	539	IGNITION POSITIVE VOLTAGE
B	BLK/WHT	1551	GROUND
C	YEL	2174	REF LOW
D	DK GRN/WHT	2124	IC 4
E	LT BLU/WHT	2126	IC 6
F	PPL/WHT	2128	IC 8
G	RED/WHT	2122	IC 2

2000-03 4.6L ICM Front/Bank 2 Connector Pinout

PIN	WIRE COLOR	CIRCUIT NO.	FUNCTION
A	PNK	839	IGNITION POSITIVE VOLTAGE
B	BLK/WHT	1551	GROUND
C	GRY	2175	REF LOW
D	DK GRN	2125	IC 5
E	LT BLU	2123	IC 3
F	PPL	2121	IC 1
G	RED	2127	IC 7

2000-03 4.6L ICM Rear/Bank 1 Connector Pinout

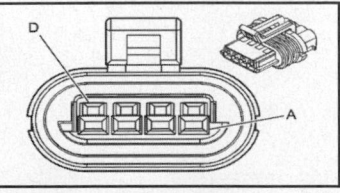

PIN	WIRE COLOR	CIRCUIT NUMBER	FUNCTION
A	PNK	839	IGNITION 1 VOLTAGE
B	PPL	2121	IC 1 CONTROL
B	ORN	2122	IC 2 CONTROL
B	DK GRN	2123	IC 3 CONTROL
B	DK GRN/WHT	2124	IC 4 CONTROL
B	LT BLU	2125	IC 5 CONTROL
B	DK GRN	2126	IC 6 CONTROL
B	PPL	2127	IC 7 CONTROL
B	RED	2128	IC 8 CONTROL
C	BRN	2129	LOW REFERENCE (IC 1, IC 3, IC 5, IC 7)
C	BRN	2130	LOW REFERENCE (IC 2, IC 4, IC 6, IC 8)
D	BLK	1551	GROUND

2004 4.6L Ignition Coil/Modules 1-8 Connector Pinouts

KNOCK SENSOR

DESCRIPTION & OPERATION

The knock sensor (KS) system enables the powertrain control module (PCM) to control the ignition timing advance for the best possible performance while protecting the engine from potentially damaging levels of detonation. The sensors in the KS system are used by the PCM as microphones to listen for abnormal engine noise that may indicate pre-ignition/detonation.

There are 2 types of KS currently being used:

- The broadband single wire sensor
- The flat response 2-wire sensor

Both sensors use piezo-electric crystal technology to produce and send signals to the PCM. The amplitude and frequency of this signal will vary constantly depending on the vibration level within the engine. Flat response and broadband KS signals are processed differently by the PCM. The major differences are outlined below:

All broadband sensors use a single wire circuit. Some types of controllers will output a bias voltage on the KS signal wire. The bias voltage creates a voltage drop the PCM monitors and uses to help diagnose KS faults. The KS noise signal rides along this bias voltage, and due to the constantly fluctuating frequency and amplitude of the signal, will always be outside the bias voltage parameters. Another way to use the KS signals is for the PCM to learn the average normal noise output from the KS. The PCM uses this noise channel, and KS signal that rides along the noise channel, in much the same way as the bias voltage type does. Both systems will constantly monitor the KS system for a signal that is not present or falls within the noise channel.

The flat response KS uses a 2-wire circuit. The KS signal rides within a noise channel which is learned and output by the PCM. This noise channel is based upon the normal noise input from the KS and is known as background noise. As engine speed and load change, the noise channel upper and lower parameters will change to accommodate the KS signal, keeping the signal within the channel. If there is knock, the signal will range outside the noise channel and the PCM will reduce spark advance until the knock is reduced. These sensors are monitored in much the same way as the broadband sensors, except that an abnormal signal will stay outside of the noise channel or will not be present.

KS diagnostics can be calibrated to detect faults with the KS diagnostic inside the PCM, the KS wiring, the sensor output, or constant knocking from an outside influence such as a loose or damaged component. In order to determine which cylinders are knocking, the PCM uses KS signal information when the cylinders are near top dead center (TDC) of the firing stroke.

Connector Pinouts

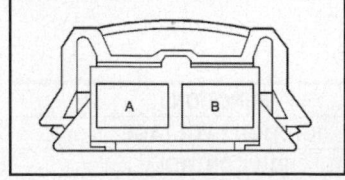

PIN	WIRE COLOR	CIRCUIT NO.	FUNCTION
A	DK BLU	496	KNOCK SENSOR SIGNAL
B	GRY	1716	KNOCK SENSOR RETURN

2000-03 4.6L Knock Sensor Connector Pinout

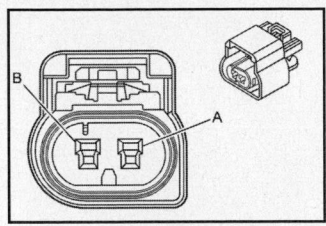

PIN	WIRE COLOR	CIRCUIT NO.	FUNCTION
A	DK BLU	496	KNOCK SENSOR [1] SIGNAL
B	GRY	1716	LOW REFERENCE

2004 4.6L Knock Sensor 1 Connector Pinout

PIN	WIRE COLOR	CIRCUIT NO.	FUNCTION
A	LT BLU	1876	KNOCK SENSOR [2] SIGNAL
B	GRY	2303	LOW REFERENCE

2004 4.6L Knock Sensor 2 Connector Pinout

MANIFOLD ABSOLUTE PRESSURE (MAP) SENSOR

Component Locator

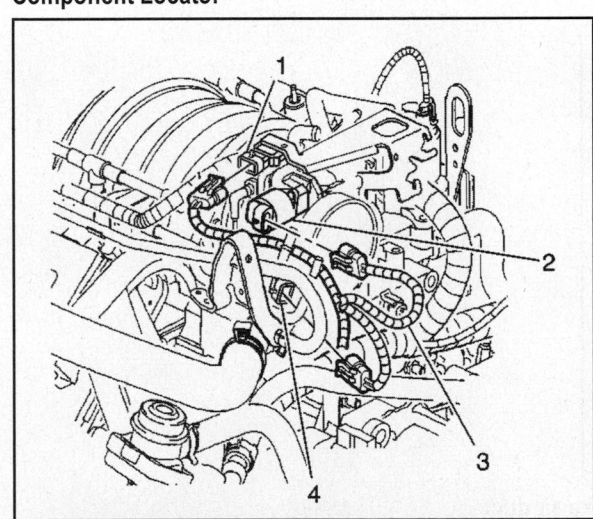

1. Manifold Absolute Pressure (MAP) Sensor

2. Idle Air Control (IAC) Valve

3. Engine Coolant Temperature (ECT) Sensor

4. Throttle Position (TP) Sensor

Showing the location of MAP Sensor and other components on the front of the engine

Connector Pinouts

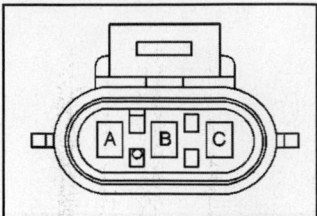

PIN	WIRE COLOR	CIRCUIT NO.	FUNCTION
A	ORN/BLK	469	MAP SENSOR GROUND
B	LT GRN	432	MAP SENSOR SIGNAL
C	GRY	2704	5 VOLT REFERENCE

2000-04 4.6L MAP Sensor Connector Pinout

MASS AIR FLOW/INTAKE AIR TEMPERATURE (MAF/IAT) SENSOR

Removal & Installation

1. Disconnect the mass air flow/intake air temperature (MAF/IAT) sensor electrical connector (1).
2. Remove the clamps securing the MAF/IAT sensor assembly to the air cleaner and intake air duct.
3. Carefully remove the MAF/IAT sensor from the air cleaner and intake air duct.

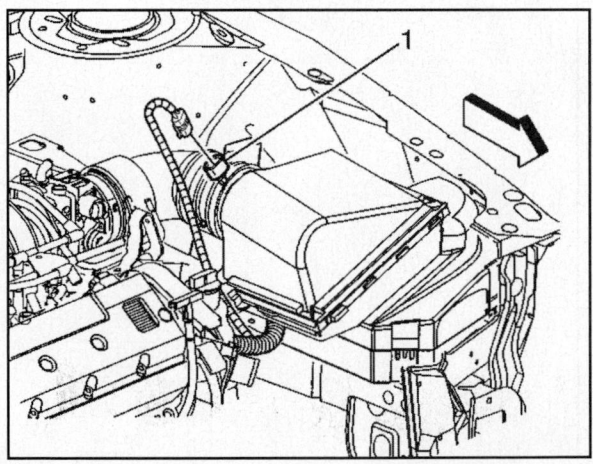

Showing the location of the MAF/IAT sensor on the intake air duct

WARNING: Handle the MAF sensor carefully. Do not drop the MAF sensor in order to prevent damage to the MAF sensor. Do not damage the screen located on the air inlet end of the MAF. Do not touch the sensing elements. Do not allow solvents and lubricants to come in contact with the sensing elements. Use a small amount of a soap based solution in order to aid in the installation.

To install:

4. Carefully install the MAF/IAT sensor (1) into the air cleaner and intake air duct.
5. Install the clamps securing the MAF/IAT sensor assembly to the air cleaner and intake air duct. Tighten the clamps to 3 Nm (27 inch lbs.).
6. Reconnect the MAF/IAT sensor electrical connector (1).

Connector Pinouts

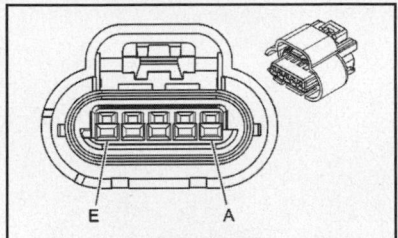

PIN	WIRE COLOR	CIRCUIT NO.	FUNCTION
A	BLK	2760	IAT GROUND (LOW REFERENCE)
B	TAN	472	IAT SENSOR SIGNAL
C	BLK OR BLK/WHT	1550 OR 451	GROUND
D	PNK	139	IGNITION POSITIVE VOLTAGE
E	YEL	492	MAF SENSOR SIGNAL

2000-04 4.6L MAF/IAT Sensor Connector Pinout

POWERTRAIN CONTROL MODULE (PCM)

Description & Operation

The top-feed fuel injector assembly is a solenoid operated device, controlled by the PCM, that meters pressurized fuel to a single engine cylinder. The PCM energizes the injector solenoid, which opens a ball valve, allowing fuel to flow past the ball valve, and through a recessed flow director plate. The director plate has multiple machined holes that control the fuel flow, generating a conical spray pattern of finely atomized fuel at the injector tip. Fuel is directed at the intake valve, causing it to become further atomized and vaporized before entering the combustion chamber. An injector stuck partly open can cause a loss of pressure after engine shutdown. Consequently, long cranking times would be noticed on some engines.

Removal & Installation

NOTE: It is necessary to record the remaining engine oil life and/or ATF life. If the replacement module is not programmed with the remaining oil life, the oil life reading will default to 100%. If the replacement module is not programmed with the remaining oil life, the fluids will need to be changed at 3,000 miles (engine oil) or 50,000 miles (ATF) from the last fluid change.

CAUTION: Before servicing any electrical component, the ignition key must be in the OFF or LOCK position and all electrical loads must be OFF, unless instructed otherwise in these procedures. If a tool or equipment could easily come in contact with a live exposed electrical terminal, also disconnect the negative battery cable. Failure to follow these precautions may cause personal injury and/or damage to the vehicle or its components.

1. Using a scan tool, retrieve the percentage of remaining engine oil and the remaining automatic transmission fluid life. Record the remaining engine oil and the remaining automatic transmission fluid life.
2. Turn OFF the ignition.
3. Disconnect the negative battery cable.
4. Remove the screw retaining the intake air duct/resonator (3) to the PCM housing.
5. Unsnap the air filter housing clamps on top of the air cleaner housing assembly.
6. Pull the intake air duct/resonator assembly (2) up enough to remove the air cleaner cover (3) and the PCM.
7. Slide the air cleaner element assembly toward the fender to separate the air cleaner element assembly from the air inlet assembly.

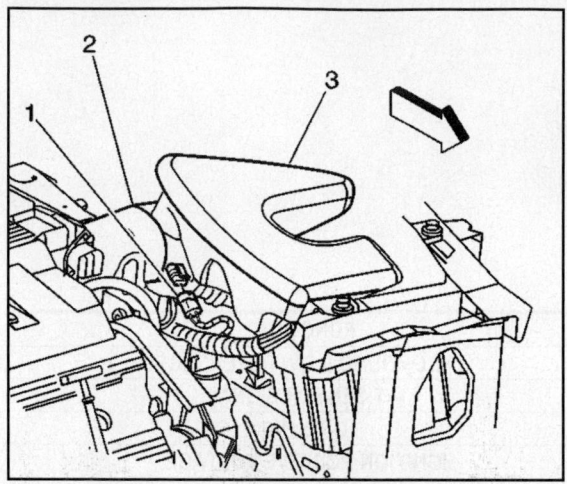

Removing the intake air duct/resonator

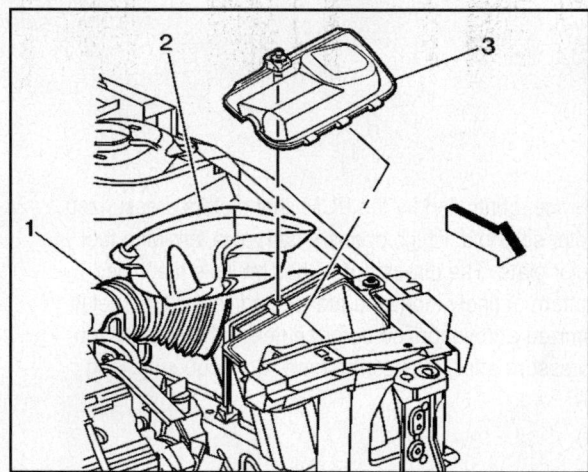

Removing the air cleaner cover (3)

8. Loosen the PCM connector attaching bolts.

9. Disconnect both PCM connectors.

10. Remove the PCM (1).

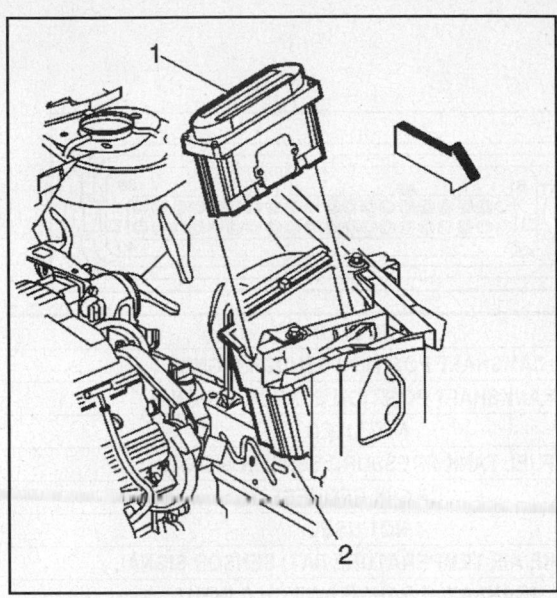

Removing the PCM (1)

Important: If the PCM is being programmed using the Off-Board method, perform the programming before continuing with the steps listed below. Refer to the Off-Board procedure in Powertrain Control Module (PCM) Programming.

To install:

11. Install the PCM (1).

12. Install the PCM connectors. Tighten the PCM connectors screws to 8 Nm (71 inch lbs.).

13. Install the air cleaner cover (3) onto the air inlet housing.

14. Snap the assembly clamps into place.

15. Align the intake air duct/resonator assembly (3) into place.

16. Install the screw onto the intake air duct/resonator assembly.

17. Tighten the intake air duct clamp to 3 Nm (27 inch lbs.).

18. Connect the negative battery cable.

Connector Pinouts

C1 CONNECTOR

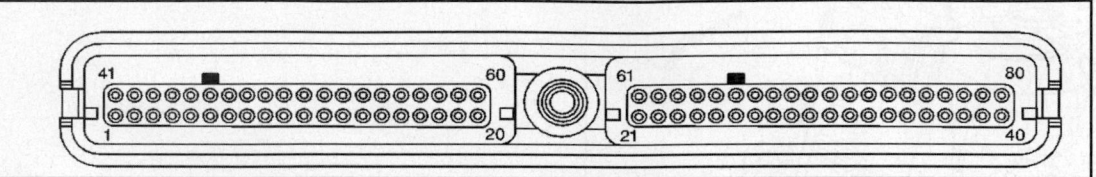

PIN	WIRE COLOR	CIRCUIT NO.	FUNCTION
1	ORN	1799	CAMSHAFT POSITION SENSOR SIGNAL
2	LT BLU/WHT	1800	CRANKSHAFT POSITION SENSOR A SIGNAL
3-4	--	--	NOT USED
5	DK GRN	890	FUEL TANK PRESSURE SENSOR SIGNAL
6	YEL	5	CRANK REQUEST
7-9	--	--	NOT USED
10	TAN	472	INTAKE AIR TEMPERATURE (IAT) SENSOR SIGNAL
11	GRY	773	TRANSAXLE RANGE SWITCH C INPUT
12	YEL	772	TRANSAXLE RANGE SWITCH B INPUT
13	BLK/WHT	771	TRANSAXLE RANGE SWITCH A INPUT
14	YEL/BLK	573	CRANKSHAFT POSITION SENSOR B SIGNAL
15	--	--	NOT USED
16	DK BLU/WHT	1231	TRANSAXLE INPUT SPEED SENSOR LOW
17	RED/BLK	1230	TRANSAXLE INPUT SPEED SENSOR HIGH
18	PPL	1807	CLASS 2 SERIAL DATA
19	WHT	1390	PCM IGNITION 0 POSITIVE VOLTAGE
20	ORN	240	PCM BATTERY FEED (B+)
21	ORN	240	PCM BATTERY FEED (B+)
22	RED	2127	CYLINDER 7 IC CONTROL
23	LT GRN	432	MANIFOLD ABSOLUTE PRESSURE (MAP) SENSOR SIGNAL
24	--	--	NOT USED
25	DK GRN	335	LOW SPEED COOLING FANS CONTROL CIRCUIT
26	DK GRN/WHT	817	VEHICLE SPEED OUTPUT
27	LT GRN/BLK	444	IAC B LOW
28	LT GRN/WHT	1749	IAC B HIGH
29	LT BLU/WHT	1747	IAC A HIGH
30	LT BLU/BLK	1748	IAC A LOW
31	DK GRN	676	HO2S BANK 1 SENSOR 1 HEATER CONTROL CIRCUIT HIGH
32	BRN/WHT	419	MIL CONTROL
33	BRN	3122	HO2S BANK 2 SENSOR 1 HEATER CONTROL CIRCUIT LOW
34	TAN/BLK	464	DELIVERED TORQUE OUTPUT
35	DK GRN	83	CRUISE INHIBIT OUTPUT
36	--	--	NOT USED
37	BLK	1744	FUEL INJECTOR 1 CONTROL
38	PNK/BLK	1746	FUEL INJECTOR 3 CONTROL
39	LT GRN	1745	FUEL INJECTOR 2 CONTROL
40	BLK/WHT	845	FUEL INJECTOR 5 CONTROL
41	WHT	776	TRANSAXLE RANGE SWITCH PARITY INPUT
42	YEL	492	MASS AIR FLOW (MAF) SENSOR SIGNAL
43	ORN/BLK	463	REQUESTED TORQUE INPUT
44	GRY	847	EXTENDED TRAVEL BRAKE SWITCH INPUT
45-49	--	--	NOT USED

2000-03 4.6L PCM C1 Connector Pinout (1 of 2)

C1 CONNECTOR – CONTINUED

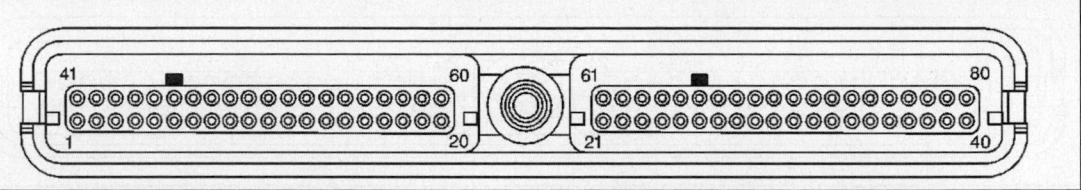

PIN	WIRE COLOR	CIRCUIT NO.	FUNCTION
50	DK BLU/WHT	1677	HO2S BANK 1 SENSOR 2 SIGNAL LOW
51	DK GRN/WHT	1678	HO2S BANK 1 SENSOR 2 SIGNAL HIGH
52	TAN	1664	HO2S BANK 1 SENSOR 1 SIGNAL LOW
53	PPL/WHT	1665	HO2S BANK 1 SENSOR 1 SIGNAL HIGH
54	TAN/WHT	1669	HO2S BANK 2 SENSOR 1 SIGNAL LOW
55	PPL/WHT	1668	HO2S BANK 2 SENSOR 1 SIGNAL HIGH
56	YEL	400	VSS SENSOR INPUT HIGH
57	PPL	401	VSS SENSOR INPUT LOW
58	RED	631	CAMSHAFT POSITION SENSOR IGNITION FEED
59	PPL	1807	CLASS 2 SERIAL DATA
60	BLK/WHT	451	PCM GROUND
61	PPL	2121	CYLINDER 1 IC CONTROL
62	DK BLU	417	THROTTLE POSITION (TP) SENSOR SIGNAL
63	PNK	39	PCM IGNITION 1 FEED
64-65	--	--	NOT USED
66	GRY	429	AIR PUMP SOLENOID CONTROL (IF EQUIPPED
67	LT BLU/WHT	1229	PRESSURE CONTROL SOLENOID LOW
68	BLK	3112	HO2S BANK 1 SENSOR 1 HEATER CONTROL CIRCUIT LOW
69	RED/BLK	1228	PRESSURE CONTROL SOLENOID HIGH
70	DK GRN/WHT	465	FUEL PUMP RELAY CONTROL
71	RED	1676	EGR VALVE CONTROL
72	PPL	1490	LIFT/DIVE SIGNAL OUTPUT
73	WHT	1310	EVAP VENT SOLENOID CONTROL
74	--	--	NOT USED
75	GRY	436	AIR PUMP A RELAY CONTROL (IF EQUPPED)
76	LT BLU/BLK	844	FUEL INJECTOR 4 CONTROL
77	YEL/BLK	846	FUEL INJECTOR 6 CONTROL
78	--	--	NOT USED
79	DK BLU/WHT	878	FUEL INJECTOR 8 CONTROL
80	RED/BLK	877	FUEL INJECTOR 7 CONTROL

2000-03 4.6L PCM C1 Connector Pinout (2 of 2)

C2 CONNECTOR

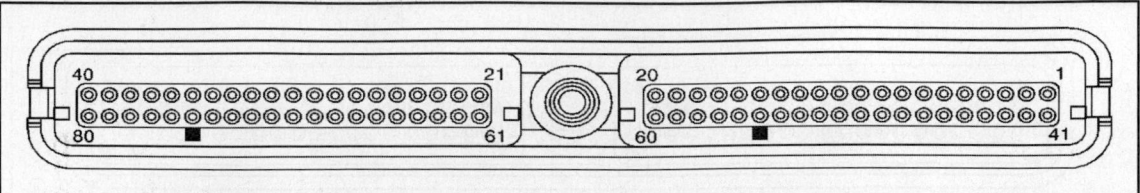

PIN	WIRE COLOR	CIRCUIT NO.	FUNCTION
1-2	--	--	NOT USED
3	BLK/WHT	451	PCM GROUND
4	BLK	2760	INTAKE AIR TEMPERATURE (IAT) GROUND
5	--	--	NOT USED
6	ORN/BLK	469	MANIFOLD ABSOLUTE PRESSURE (MAP) SENSOR GROUND
7-8	--	--	NOT USED
9	YEL	2174	REFERENCE LOW (FRONT BANK IGNITION COIL CASSETTE)
10	--	--	NOT USED
11	BLK	2762	TRANSAXLE FLUID TEMPERATURE (TFT) SENSOR GROUND
12	BLK	2753	EGR PINTLE POSITION SENSOR GROUND
13	BLK	2759	FUEL LEVEL SENSOR AND FUEL TANK PRESSURE SENSOR GROUND
14	BLK	2751	AC PRESSURE SENSOR GROUND
15	BLK/WHT	451	PCM GROUND
16	--	--	NOT USED
17	GRY	2709	5 VOLT REFERENCE B - FUEL TANK PRESSURE SENSOR FEED
18	--	--	NOT USED
19	DK BLU	496	KNOCK SENSOR SIGNAL
20	TAN/BLK	231	ENGINE OIL PRESSURE SWITCH INPUT
21	LT BLU	2123	CYLINDER 3 IC CONTROL
22	RED/WHT	2122	CYLINDER 2 IC CONTROL
23	BRN	1174	ENGINE OIL LEVEL SWITCH INPUT
24	LT BLU/BLK	396	CRUISE ENGAGED INPUT
25	LT GRN	2867	CRANKSHAFT POSITION SENSOR B IGNITION FEED
26	BLK/WHT	451	PCM GROUND
27	--	--	NOT USED
28	YEL	2868	CRANKSHAFT POSITION SENSOR B GROUND
29-30	--	--	NOT USED
31	YEL	410	ENGINE COOLANT TEMPERATURE (ECT) SENSOR SIGNAL
32	YEL/BLK	1227	TRANSAXLE FLUID TEMPERATURE (TFT) SENSOR SIGNAL
33-34	--	--	NOT USED
35	GRY	1716	KNOCK SENSOR GROUND
36	--	--	NOT USED
37	RED	1226	TRANSAXLE PRESSURE MANIFOLD SWITCH C INPUT
38-39	--	--	NOT USED
40	RED	225	GENERATOR L TERMINAL CONTROL
41	--	--	NOT USED
42	DK GRN/WHT	428	EVAP CANISTER PURGE VALVE CONTROL
43	LT GRN	1222	SHIFT 1-2 SOLENOID VALVE CONTROL
44	YEL	447	STARTER RELAY CONTROL
45	YEL/BLK	1223	SHIFT 2-3 SOLENOID VALVE CONTROL
46	TAN/BLK	422	TORQUE CONVERTER CLUTCH (TCC) CONTROL
47	DK GRN/WHT	459	A/C COMPRESSOR CLUTCH RELAY CONTROL
48	BLK/WHT	1423	HO2S BANK 1 SENSOR 2 HEATER CONTROL CIRCUIT LOW
49	GRY	2704	5 VOLT REFERENCE A - MAP SENSOR FEED

2000-03 4.6L PCM C2 Connector (1 of 2)

C2 CONNECTOR – CONTINUED

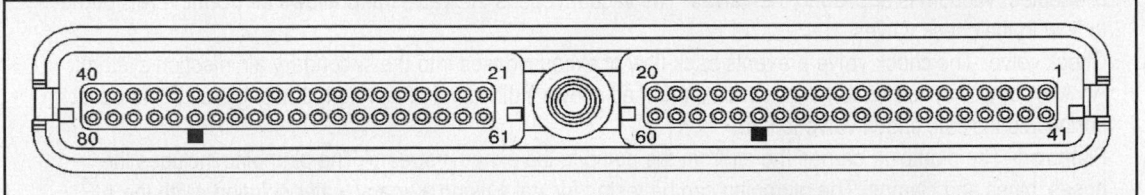

PIN	WIRE COLOR	CIRCUIT NO.	FUNCTION
50	GRY	2702	5 VOLT REFERENCE A - EGR PINTLE POSITION SENSOR FEED
51	GRY	2701	5 VOLT REFERENCE A - TP SENSOR FEED
52	BRN	1456	EGR PINTLE POSITION SENSOR SIGNAL
53	--	--	NOT USED
54	GRN	1408	AIR PUMP 2 RELAY CONTROL (IF EQUIPPED)
55	DK BLU	473	HIGH SPEED COOLING FANS CONTROL
56	WHT	121	TACHOMETER OUTPUT
57	GRY	2700	5 VOLT REFERENCE B - A/C PRESSURE SENSOR FEED
58	--	--	NOT USED
59	DK GRN/WHT	2124	CYLINDER 4 IC CONTROL
60	LT BLU/WHT	2126	CYLINDER 6 IC CONTROL
61	PPL/WHT	2128	CYLINDER 8 IC CONTROL
62	DK GRN	2125	CYLINDER 5 IC CONTROL
63-64	--	--	NOT USED
65	LT GRN	1867	CKP SENSOR A IGNITION FEED
66	YEL/BLK	1868	CKP SENSOR A GROUND
67	PNK/BLK	632	CMP SENSOR GROUND
68	BLK	2761	ECT SENSOR GROUND
69	BLK	2752	TP SENSOR GROUND
70	GRY	435	EGR VALVE CONTROL GROUND
71	GRY	2175	REFERENCE LOW (REAR BANK IGNITION COIL CASSETTE)
72	PPL	1589	FUEL LEVEL SENSOR SIGNAL
73	PNK	1224	TRANSAXLE PRESSURE MANIFOLD SWITCH A INPUT
74	RED/BLK	380	A/C PRESSURE SENSOR SIGNAL
75-78	--	--	NOT USED
79	GRY	23	GENERATOR F TERMINAL FEEDBACK
80	PPL	420	TCC BRAKE SWITCH INPUT

2000-03 4.6L PCM C2 Connector (2 of 2)

SECONDARY AIR INJECTION (AIR) PUMP

Description & Operation

The secondary air injection (AIR) system helps reduce exhaust emissions. The system forces fresh filtered air into the exhaust stream in order to accelerate catalyst operation.

The system includes the following:

- AIR pump--The AIR pump supplies filtered air through the secondary air injection system into the exhaust stream. The control module provides ground for the pump relay. Battery voltage is then applied to the pump. The filter is the only serviceable part of the pump.
- AIR vacuum control solenoid--The AIR vacuum control solenoid controls the AIR shut-off valves. When the secondary air injection system is enabled, the control module provides a ground to the solenoid. Enabling the solenoid, allows engine vacuum to be applied to the AIR shut-off valves.

- AIR shut-off valves--The AIR shut-off valves are vacuum operated. When the secondary air injection system is enabled, vacuum is applied to the valves. The vacuum opens the valves and allows air from the AIR pump to flow to the check valves.
- Check valve--The check valve prevents back flow of exhaust gases into the secondary air injection system. An AIR pump that had become inoperative and had shown indications of having exhaust gases in the outlet port would indicate check valve failure.
- Plumbing--The plumbing carries the air from the pump to the exhaust stream. The plumbing includes the hoses, pipes and clamps. The plumbing can be tested for leaks using a soapy water solution. With the AIR pump running, bubbles will form if a leak exists.

RESULTS OF INCORRECT OPERATION

If no air (oxygen) flow enters the exhaust stream, the start up emission levels will rise. The control module can detect a system flow problem using the pre-catalyst heated oxygen sensor (HO2S) and short term FT. If a system flow problem is present the HO2S voltage will not indicate an expected lean condition, short term FT will not increase and a diagnostic trouble code (DTC) will set.

If incorrect voltage is present on the vacuum control solenoid or the pump relay control circuits the device will not operate. This will be detected by the control module and a DTC will set.

The following DTCs can set if a secondary air injection fault is detected:

- DTC P0410-- A system flow problem has been detected.
- DTC P0412-- A vacuum control solenoid control circuit problem has been detected.
- DTC P0418-- A pump relay control circuit problem has been detected.
- DTC P1415-- A system flow problem has been detected relating to cylinder bank 1.
- DTC P1416-- A system flow problem has been detected relating to cylinder bank 2.

Removal & Installation

1. Raise the vehicle.
2. Remove the retaining screws from the front portion of the left front wheel well/splash shield.
3. Lower the splash shield far enough to gain access to the secondary air injection (AIR) pump (2) assembly.
4. Loosen and lower the AIR outlet hose pinch clamp.
5. Remove the AIR inlet hose from the AIR pump (2).
6. Disconnect the AIR pump electrical connector.
7. Remove the upper AIR pump to cradle bolt.
8. Lift up and pull out to remove the AIR pump (2) from the cradle bracket.

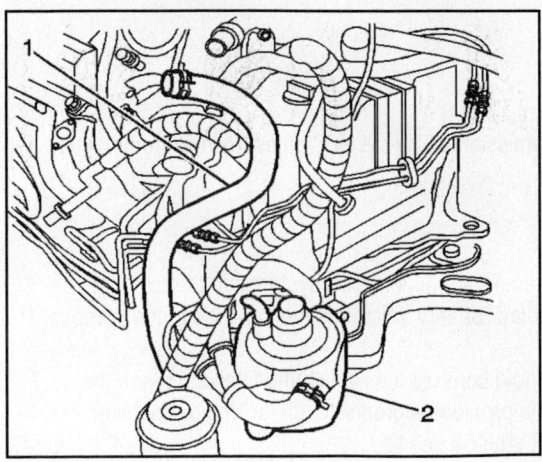

Indicating components for removal of the AIR Pump

To install:

9. Install the lower AIR pump bolts/nuts into the cradle bracket and push down.

10. Install the upper AIR pump to cradle bolt. Tighten the upper and lower AIR pump to cradle bolts to 9 Nm (80 inch lbs.).

11. Reconnect the AIR pump electrical connector.

12. Install the AIR outlet hose and pinch clamp to the AIR pump (2).

13. Install the AIR inlet hose to the AIR pump (2).

14. Lower the vehicle.

Connector Pinouts

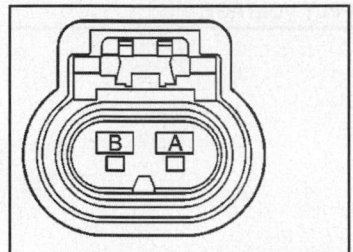

PIN	WIRE COLOR	CIRCUIT NO.	FUNCTION
A	BLK	1650	AIR PUMP MOTOR 1 GROUND
B	RED	78	AIR PUMP MOTOR 1 SUPPLY VOLTAGE

2000-03 4.6L AIR Pump Motor Connector Pinout

SECONDARY AIR INJECTION (AIR) PUMP RELAY

Component Locations

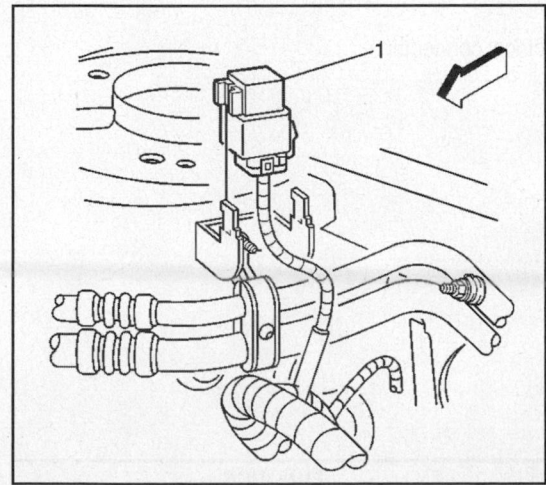

Showing the location of the AIR Pump Relay (near the shock tower)

Connector Pinouts

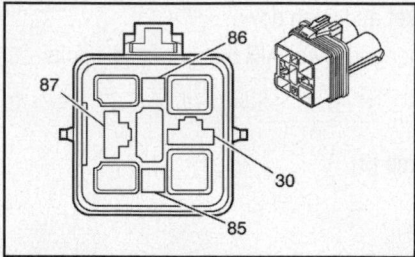

PIN	WIRE COLOR	CIRCUIT NO.	FUNCTION
30	RED	742	BATTERY POSITIVE VOLTAGE
85	PNK	139	IGNITION 1 VOLTAGE
86	BRN	436	AIR PUMP RELAY 1 CONTROL
87	RED	78	AIR PUMP MOTOR 1 SUPPLY VOLTAGE

2000-03 4.6L AIR Pump Relay Connector Pinout

SECONDARY AIR INJECTION (AIR) SOLENOID VALVE

Removal & Installation

1. Remove the fuel injector sight shield.
2. Disconnect the AIR vacuum control valve electrical connector.
3. Disconnect the vacuum lines.
4. Remove the nut securing the AIR vacuum control valve to the bank 1 AIR shut-off valve bracket.
5. Remove the AIR vacuum control valve.

To install:

6. Install the AIR vacuum control valve to the bank 1 AIR shut-off valve bracket. Tighten the nut to 9 Nm (80 inch lbs.).
7. Reconnect the vacuum lines.
8. Reconnect the AIR vacuum control valve electrical connector.
9. Install the fuel injector sight shield.

Connector Pinouts

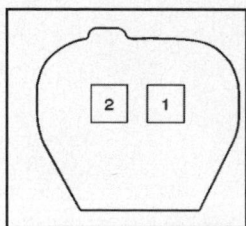

PIN	WIRE COLOR	CIRCUIT NO.	FUNCTION
1	PNK	139	IGNITION POSITIVE VOLTAGE
2	PNK/BLK	429	AIR SOLENOID VALVE CONTROL

2000-03 4.6L Secondary Air Injection Solenoid Valve Connector Pinout

THROTTLE POSITION (TP) SENSOR

Removal & Installation

1. Remove the idle air control (IAC) electrical connector (1) from the IAC sensor (4).
2. Remove the manifold absolute pressure (MAP) sensor electrical connector from the MAP sensor.
3. Remove the water pump belt cover.
4. Remove the TP sensor retaining screws electrical connector (2) from the TP sensor (3).
5. Remove the TP sensor electrical connector (2) from the TP sensor (3).

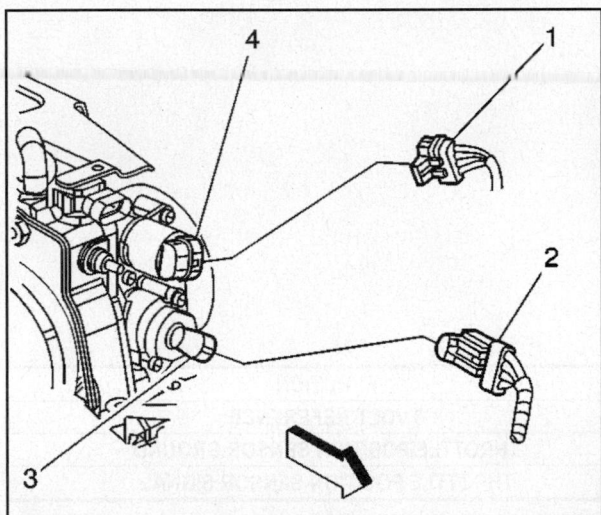

Detaching electrical connectors prior to TP sensor removal

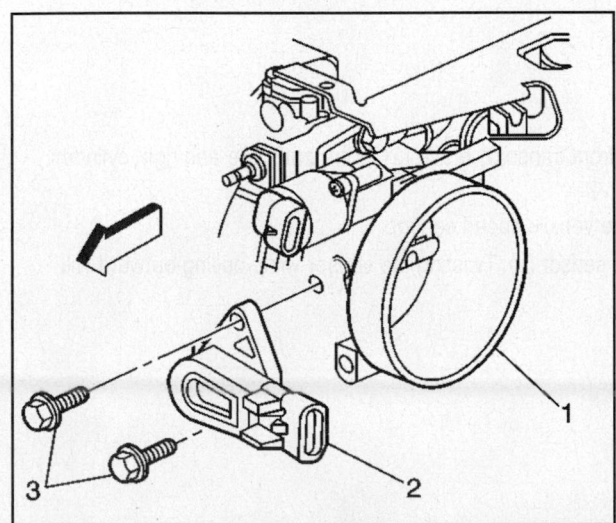

Identifying components for removal of the TP Sensor

To install:

6. Install the new O-ring on to the TP sensor.
7. Connect the TP Sensor electrical connector (2) to the TP sensor (3).

8. Install the TP sensor (2) on to the throttle body (1). Tighten the TP sensor screws to 2.3 Nm (20 inch lbs.).

9. Connect the IAC electrical connector (1) to the IAC sensor (4).

10. Connect the MAP sensor electrical connector to the MAP sensor.

11. Install the water pump belt and tensioner cover.

12. Perform the TP Sensor learn procedure:

13. Turn the ignition to the RUN/ON position.

14. Wait 1 minute..

15. Turn the ignition to the LOCK/OFF position.

Connector Pinouts

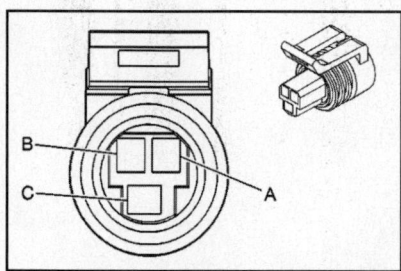

PIN	WIRE COLOR	CIRCUIT NO.	FUNCTION
A	GRY	2701	5 VOLT REFERENCE
B	BLK	2752	THROTTLE POSITION SENSOR GROUND
C	DK BLU	417	THROTTLE POSITION SENSOR SIGNAL

2000-03 4.6L TP Sensor Connector Pinout

VEHICLE SPEED SENSOR (VSS)

Removal & Installation

1. Raise the vehicle.

2. Remove the bolts (1, 3) securing the front transaxle brace (2) to the transaxle and right cylinder head.

3. Remove the electrical connector at the vehicle speed sensor.

4. Remove the retaining bolt (1) and the sensor (2). Twisting the sensor while pulling outward will aid in removal.

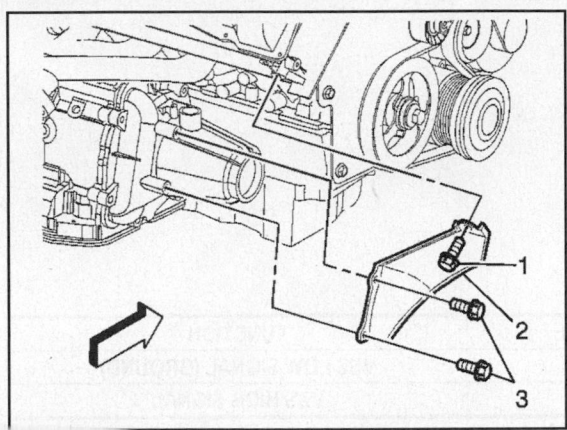

Removing front transaxle brace (2) after removing bolts (1 and 3)

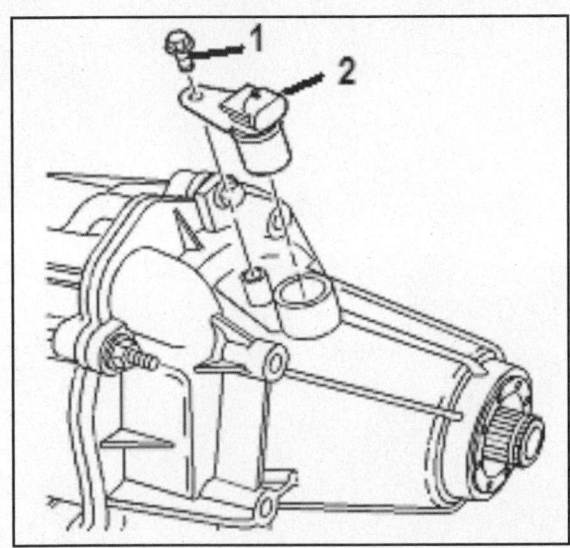

Removing vehicle speed sensor (2)

To install:

CAUTION: *Use the correct fastener in the correct location. Replacement fasteners must be the correct part number for that application. Fasteners requiring replacement or fasteners requiring the use of thread locking compound or sealant are identified in the service procedure. Do not use paints, lubricants, or corrosion inhibitors on fasteners or fastener joint surfaces unless specified. These coatings affect fastener torque and joint clamping force and may damage the fastener. Use the correct tightening sequence and specifications when installing fasteners in order to avoid damage to parts and systems.*

5. Install the vehicle speed sensor and the retaining bolt. Install a new O-ring seal on sensor if necessary. Tighten the bolt to 10 Nm (89 inch lbs.).

6. Install the electrical connector at the sensor.
   Install the bolts securing the front transaxle brace to the transaxle and right cylinder head.

7. Lower the vehicle.

Connector Pinouts

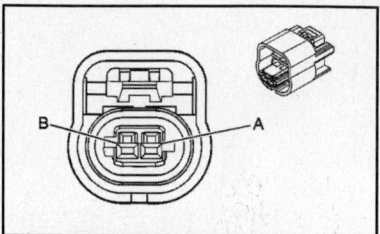

PIN	WIRE COLOR	CIRCUIT NO.	FUNCTION
A	PPL	401	VSS LOW SIGNAL (GROUND)
B	YEL	400	VSS HIGH SIGNAL

2000-04 4.6L 4T80-E A/T Vehicle Speed Sensor Connector Pinout

GENERAL MOTORS TRUCKS & VANS: 2000-05 C & K SERIES
4.8L, 5.3L, 6.0L ENGINES

CAMSHAFT POSITION (CMP) SENSOR CONNECTOR

Component Locations

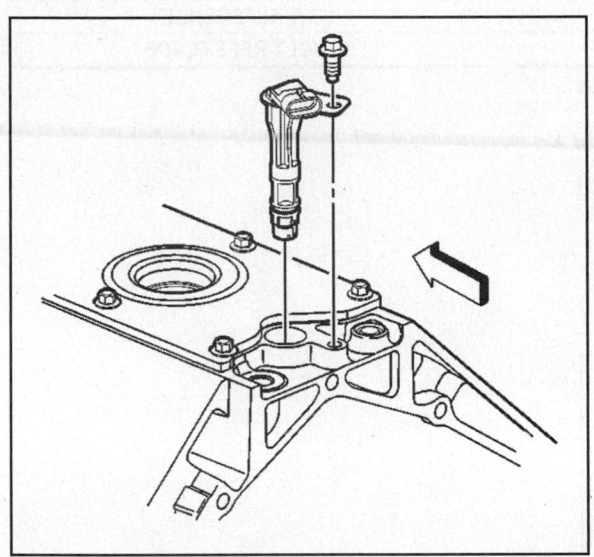

CMP Sensor location of 2000-05 4.8L, 5.3L, 6.0L Engines

Connector Pinouts

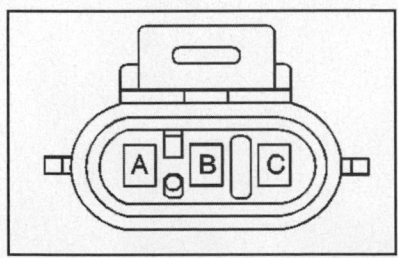

PIN	WIRE COLOR	CIRCUIT NO.	FUNCTION
A	BRN/WHT	633	CAMSHAFT POSITION (CMP) SENSOR SIGNAL
B	PNK/BLK	632	CAMSHAFT POSITION (CMP) SENSOR RETURN
C	RED	631	CAMSHAFT SENSOR FEED - 12 VOLT

2000 4.8L, 5.3L, 6.0L CMP sensor connector pinout

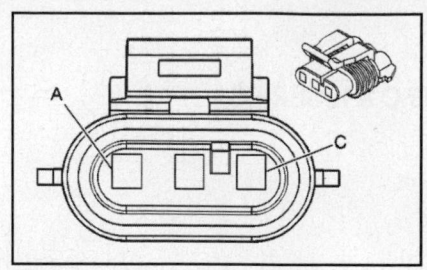

PIN	WIRE COLOR	CIRCUIT NO.	FUNCTION
A	BRN/WHT	633	CMP SENSOR SIGNAL
B	PNK/BLK	632	LOW REFERENCE
C	RED	631	12 VOLT REFERENCE

2001-2005 4.8L, 5.3L, 6.0L CMP sensor connector pinout

CRANKSHAFT POSITION (CKP) SENSOR CONNECTORS

Component Locations

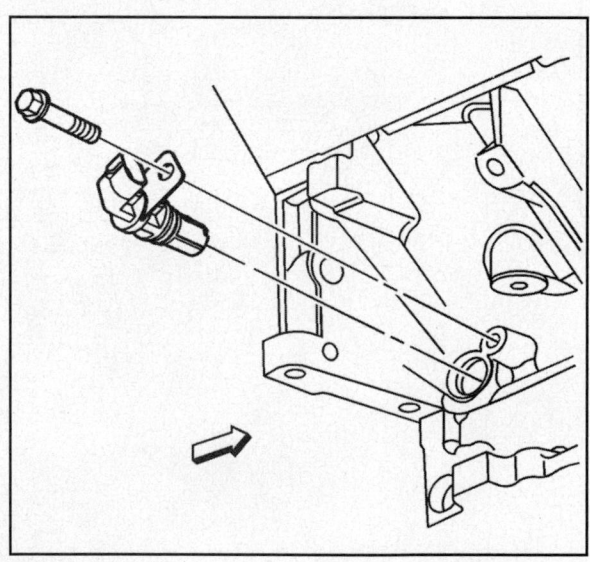

CKP sensor on 2000-05 4.8L, 5.3L, 6.0L engines

Connector Pinouts

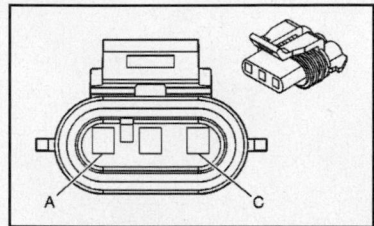

PIN	WIRE COLOR	CIRCUIT NO.	FUNCTION
A	DK BLU/WHT	1869	CRANKSHAFT POSITION SENSOR SIGNAL
B	YEL/BLK	1868	CRANKSHAFT POSITION SENSOR RETURN
C	LT GRN	1867	CRANKSHAFT POSITION SENSOR FEED - 12 VOLT

2000-2005 4.8L, 5.3L, 6.0L CKP sensor connector pinout

EGR SOLENOID VALVE CONNECTOR

Component Locations

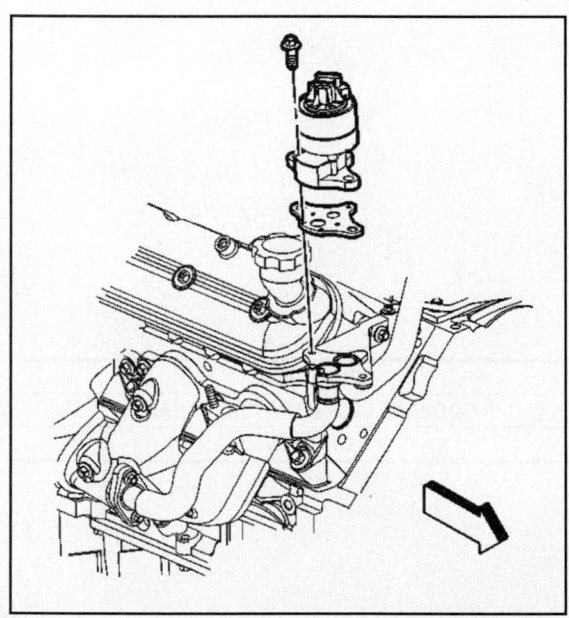

EGR valve and solenoid location on 2000-2002 4.8L, 5.3L, 6.0L Engines

Connector Pinouts

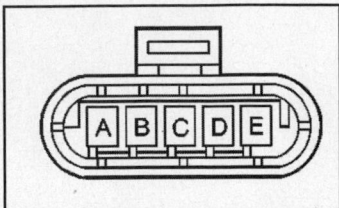

PIN	WIRE COLOR	CIRCUIT NO.	FUNCTION
A	WHT	257	EXHAUST GAS RECIRCULATION (EGR) SOLENOID CONTROL
B	PPL	719	SENSOR RETURN
C	BRN	1456	EXHAUST GAS RECIRCULATION SOLENOID POSITION SIGNAL
D	GRY	596	REFERENCE VOLTAGE FEED (5 VOLT)
E	RED	1676	EXHAUST GAS RECIRCULATION SOLENOID FEED

EGR solenoid connector pinout on 2000-2002 4.8L, 5.3L, 6.0L

ENGINE COOLANT LEVEL SENSOR CONNECTOR

Connector Pinouts

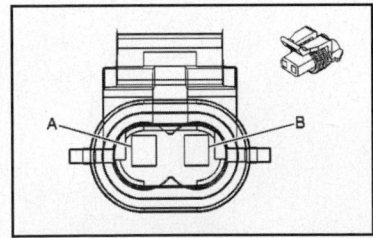

PIN	WIRE COLOR	CIRCUIT NO.	FUNCTION
A	LT GRN	1478	COOLANT LEVEL SWITCH SIGNAL
B	BLK	550	GROUND

Coolant level connector pinout on 2000-2005

ENGINE COOLANT TEMPERATURE (ECT) SENSOR CONNECTOR

Component Locations

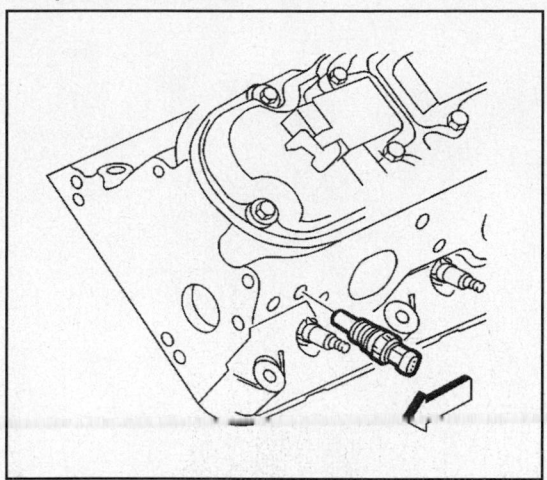

Coolant temperature sensor location on 2000-2005 4.8L, 5.3L, 6.0L engines

Connector Pinouts

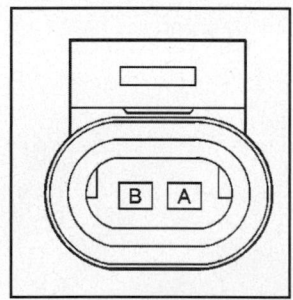

PIN	WIRE COLOR	CIRCUIT NO.	FUNCTION
A	GRY	720	SENSOR RETURN
B	YEL	410	COOLANT TEMPERATURE SENSOR SIGNAL

Coolant temperature sensor connector on 2000-2002 4.8L, 5.3L, 6.0L

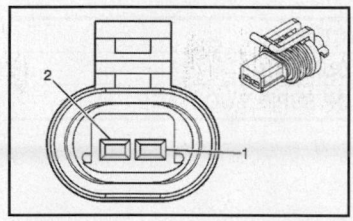

PIN	WIRE COLOR	CIRCUIT NO.	FUNCTION
1	BLK	407	LOW REFERENCE
2	YEL	410	ECT SENSOR SIGNAL

Coolant temperature sensor connector on 2003-2005 4.8L, 5.3L, 6.0L

ENGINE COOLING FAN CONNECTORS

Component Locations

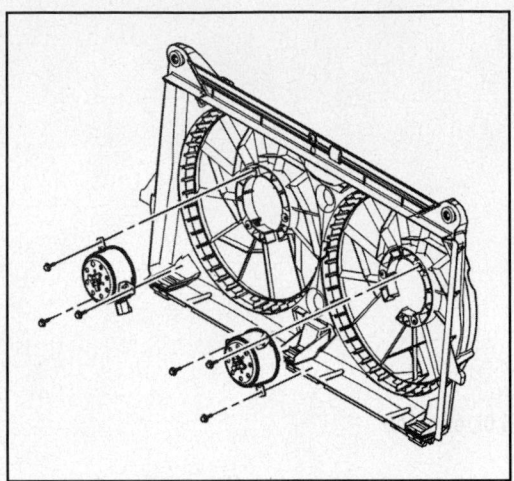

Electric cooling fan motors for 2005 10 Series only

Connector Pinouts

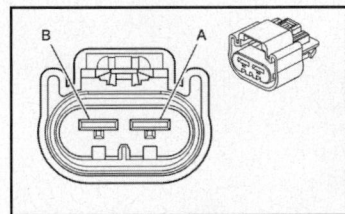

PIN	WIRE COLOR	CIRCUIT NO.	FUNCTION
A	YE	5358	COOLING FAN MOTOR CONTROL
B	GY	532	COOLING FAN MOTOR SUPPLY VOLTAGE

Left cooling fan connector pinout on 2005 10 series

PIN	WIRE COLOR	CIRCUIT NO.	FUNCTION
A	BK	250	GROUND
B	L-BU	409	COOLING FAN MOTOR SUPPLY VOLTAGE

Right cooling fan connector pinout on 2005 10 series

ENGINE OIL LEVEL SENSOR CONNECTOR

Component Locations

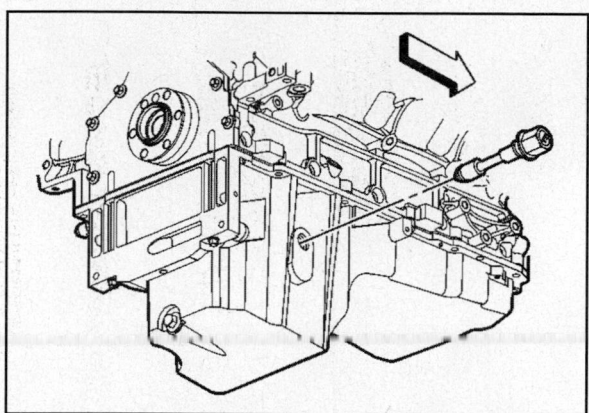

Oil level sensor location on 2002-2005 4.8L, 5.3L, 6.0L

Connector Pinouts

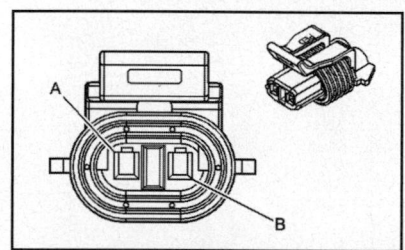

PIN	WIRE COLOR	CIRCUIT NO.	FUNCTION
A	BRN	1174	OIL LEVEL SWITCH SIGNAL
B	BLK	550	GROUND

Engine oil level sensor connector pinout

EVAP CANISTER PURGE CONNECTORS

Component Locations

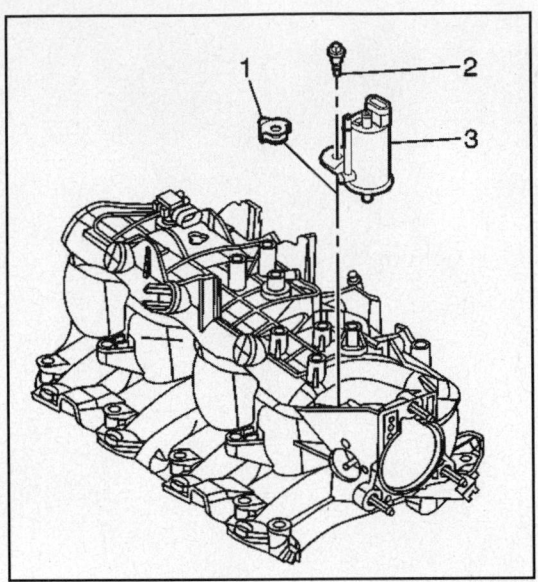

EVAP Purge valve solenoid (3) mounting bolt (2) and insulator (1) on 2000-2005 4.8L, 5.3L, 6.0L Engines

Connector Pinouts

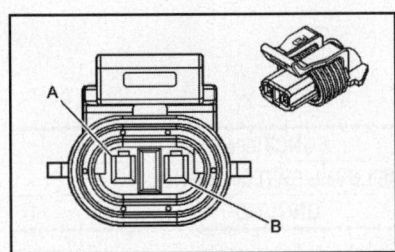

PIN	WIRE COLOR	CIRCUIT NO.	FUNCTION
A	PNK	239	FUSED IGNITION POSITIVE VOLTAGE
B	DK GRN/WHT	428	CANISTER PURGE SOLENOID OUTPUT

Evaporative emission canister purge valve connector pinout on 2000 4.8L, 5.3L, 6.0L Engines

PIN	WIRE COLOR	CIRCUIT NO.	FUNCTION
A	WHT	1310	EVAP CANISTER VENT SOLENOID CONTROL
B	PNK	239	IGNITION 1 VOLTAGE -NM2

Evaporative emission canister purge valve connector pinout on 2001-2002 4.8L, 5.3L, 6.0L Engines

PIN	WIRE COLOR	CIRCUIT NO.	FUNCTION
A	PNK	439	IGNITION 1 VOLTAGE
B	DK GRN/WHT	428	EVAP CANISTER PURGE SOLENOID CONTROL

Evaporative emission canister purge valve connector pinout on 2003-2005 4.8L, 5.3L, 6.0L Engines

FUEL PUMP & GAUGE SENDER CONNECTOR

Component Locations

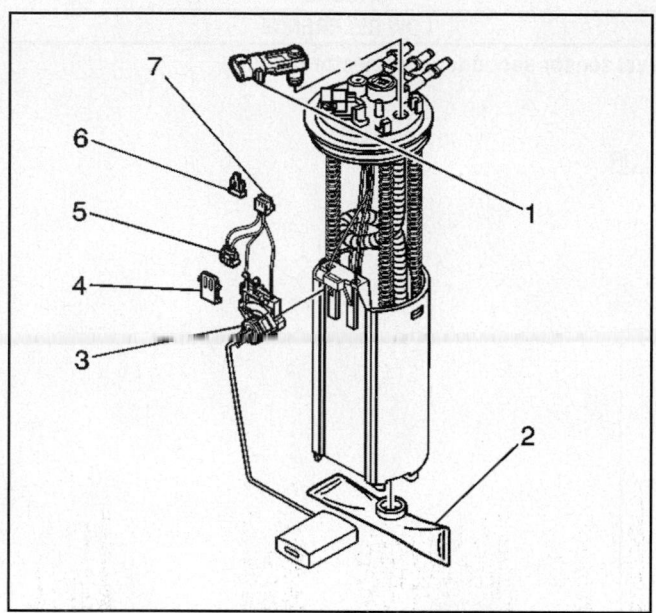

Illustrating fuel pump and sending unit sensor (3) on 2000-2004 4.8L, 5.3L, 6.0L Engines

Connector Pinouts

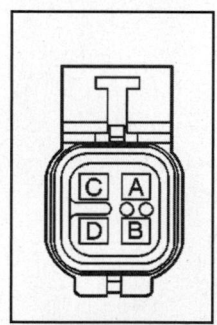

PIN	WIRE COLOR	CIRCUIT NO.	FUNCTION
A	PPL	1589	FUEL LEVEL SENSOR SIGNAL - PRIMARY
B	GRY	120	FUEL PUMP SUPPLY VOLTAGE
C	BLK	1650 OR 2150	GROUND
D	BLK OR ORN/BLK	470 OR 510	LOW REFERENCE

2000-2002 4.8L, 5.3L, 6.0L Engines fuel pump & level sensor primary connector pinout

PIN	WIRE COLOR	CIRCUIT NO.	FUNCTION
A	DK BLU	1936	FUEL LEVEL SENSOR SIGNAL-SECONDARY
B	LT GRN	1058	FUEL BALANCE PUMP SUPPLY VOLTAGE
C	BLK	1650	GROUND (10/20 SERIES)
C	BLK	2150	GROUND (30 SERIES)
D	ORN/BLK	510	LOW REFERENCE

2000-2002 4.8L, 5.3L, 6.0L Engines fuel pump & level sensor secondary connector pinout

PIN	WIRE COLOR	CIRCUIT NO.	FUNCTION
A	DK BLU	1936	FUEL LEVEL SENSOR SIGNAL - SECONDARY
B	LT GRN	1058	FUEL BALANCE PUMP SUPPLY VOLTAGE
C	BLK	2150	GROUND
D	BLK	470	LOW REFERENCE

2003-2004 4.8L, 5.3L, 6.0L Engines fuel pump & level sensor secondary connector pinout

FUEL TANK PRESSURE (FTP) SENSOR CONNECTOR

Component Locations

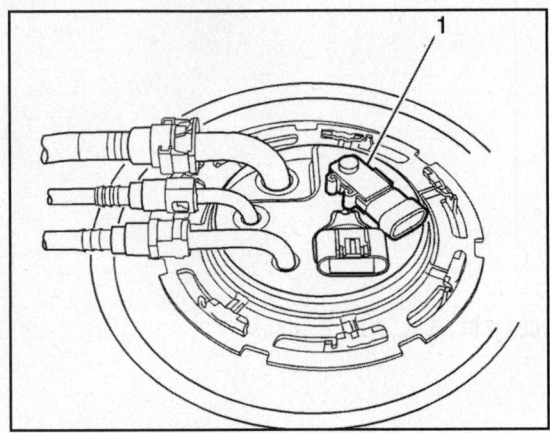

Illustrating the fuel tank pressure sensor (1) for 2000-04 4.8L, 5.3L, 6.0L engines

Connector Pinouts

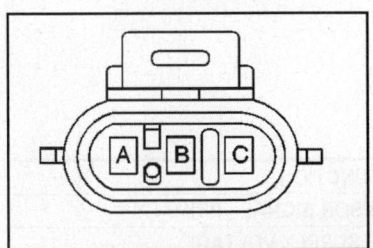

PIN	WIRE COLOR	CIRCUIT NO.	FUNCTION
A	BLK	470	SENSOR RETURN
B	DK GRN	890	FUEL TANK PRESSURE SENSOR SIGNAL
C	GRY	474	REFERENCE VOLTAGE FEED - 5 VOLT

Fuel tank pressure sensor connector pinout on 2000-04 4.8L, 5.3L, 6.0L Engines

GENERATOR CONNECTOR

Component Locations

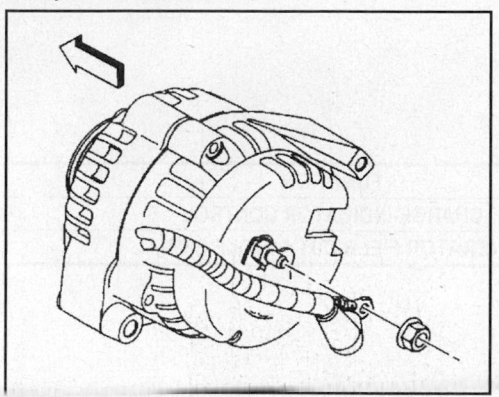

Generator wiring location on 2000-2005 4.8L, 5.3L, 6.0L Engines

Connector Pinouts

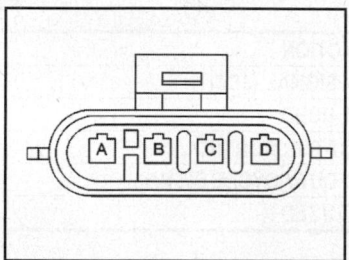

PIN	WIRE COLOR	CIRCUIT NO.	FUNCTION
A	--	--	NOT USED
B	BRN	25	CHARGE INDICATOR LAMP OUTPUT
C	GRY	23	GENERATOR FIELD DUTY CYCLE SIGNAL
D	--	--	NOT USED

Generator Connector Pinout On 2000-2004 4.8L, 5.3L, 6.0L Engines Without SBA

PIN	WIRE COLOR	CIRCUIT NO.	FUNCTION
A	DK BLU	5668	ENGINE ON SIGNAL
B	BRN	25	CHARGE INDICATOR CONTROL
C	GRY	23	GENERATOR FIELD DUTY CYCLE SIGNAL
D	--	--	NOT USED

Generator Connector Pinout On 2003-2004 4.8L, 5.3L, 6.0L Engines with SBA

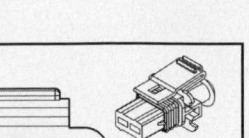

PIN	WIRE COLOR	CIRCUIT NO.	FUNCTION
1	BN	25	CHARGE INDICATOR CONTROL
2	GY	23	GENERATOR FIELD DUTY CYCLE SIGNAL

2005 Gas 10 Series Generator Connector Pinout

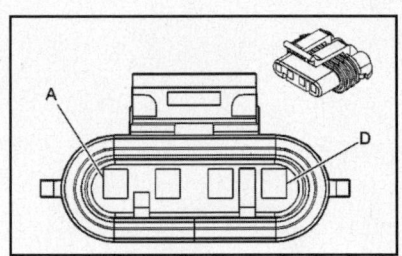

PIN	WIRE COLOR	CIRCUIT NO.	FUNCTION
A	D-BU	5668	ENGINE ON SIGNAL (JC4)
	--	--	NOT USED
B	BN	25	CHARGE INDICATOR CONTROL
C	GY	23	GENERATOR FIELD DUTY CYCLE SIGNAL
D	--	--	NOT USED

2005 Gas 20/30 Series Generator Connector Pinout

KNOCK SENSOR CONNECTOR

Component Locations

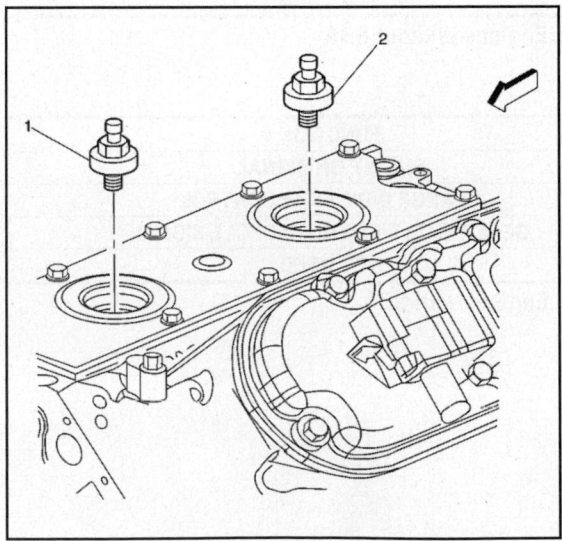

Knock sensor locations (1, 2) for 2000-05 4.8L, 5.3L, 6.0L Engines

Connector Pinouts

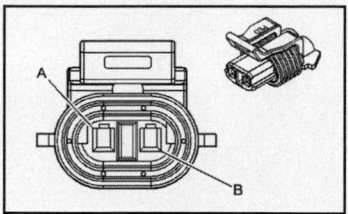

PIN	WIRE COLOR	CIRCUIT NO.	FUNCTION
A	DK BLU	496	KNOCK SENSOR SIGNAL # 1
B	LT BLU	1876	KNOCK SENSOR SIGNAL # 2

Knock sensor connector pinout on 2000-2004 4.8L, 5.3L, 6.0L Engines

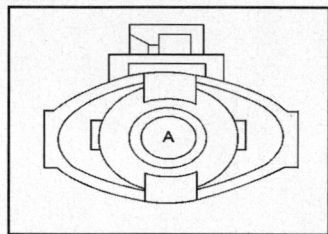

PIN	WIRE COLOR	CIRCUIT NO.	FUNCTION
A	D-BU	496	KNOCK SENSOR 1 SIGNAL

2005 4.8L, 5.3L, 6.0L Engines Knock sensor (KS1) Connector Pinout

PIN	WIRE COLOR	CIRCUIT NO.	FUNCTION
A	L-BU	1876	KNOCK SENSOR 2 SIGNAL

2005 4.8L, 5.3L, 6.0L Engines Knock sensor (KS2) Connector Pinout

IDLE AIR CONTROL (IAC) VALVE CONNECTOR

Component Locations

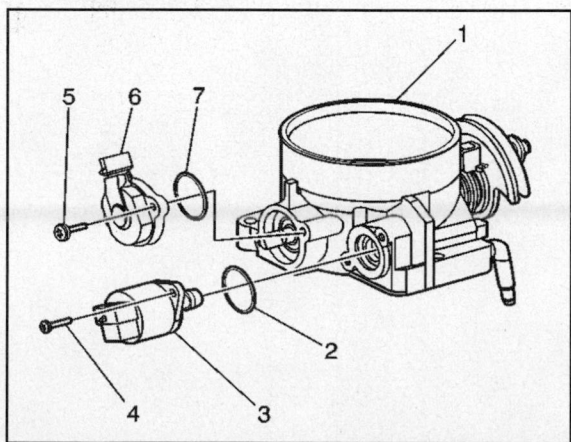

Air passage breakdown, mounting screws (4) IAC valve (3) O-ring seal (2) on 2000-02 4.8L, 5.3L, 6.0L engines

Connector Pinouts

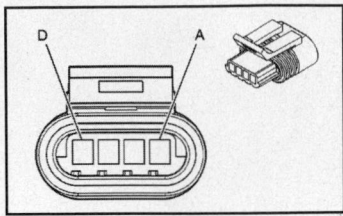

PIN	WIRE COLOR	CIRCUIT NO.	FUNCTION
A	LT GRN/BLK	444	IDLE AIR CONTROL (IAC) VALVE FEED (COIL B LOW)
B	LT GRN/WHT	1749	IDLE AIR CONTROL (IAC) VALVE FEED (COIL B HIGH)
C	LT BLU/BLK	1748	IDLE AIR CONTROL (IAC) VALVE FEED (COIL A LOW)
D	LT BLU/WHT	1747	IDLE AIR CONTROL (IAC) VALVE FEED (COIL A HIGH)

Idle air control valve connector pinout on 2000-2002 4.8L, 5.3L, 6.0L Engines

IGNITION CONTROL MODULE (ICM) CONNECTORS

Component Locations

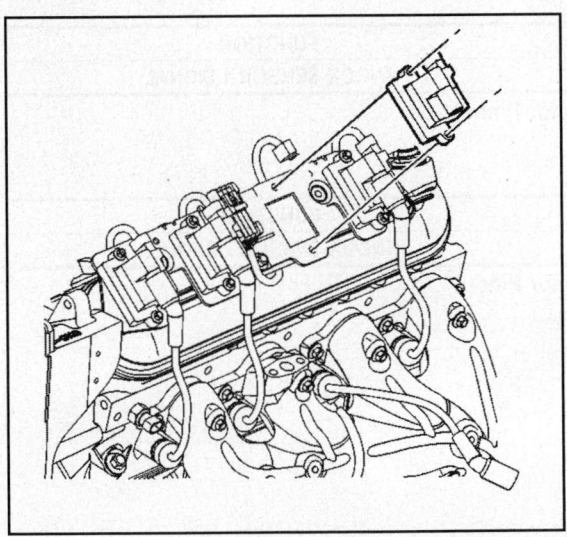

Ignition coil locations 2001-2004 4.8L, 5.3L, 6.0L Engines

Connector Pinouts

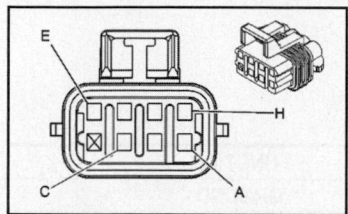

PIN	WIRE COLOR	CIRCUIT NO.	FUNCTION
A	BLK	550	GROUND
B	RED	2127	IC 7 CONTROL
C	GRN DK	2125	IC 5 CONTROL
E	BRN	2129	LOW REFERENCE
F	BLU LT	2123	IC 3 CONTROL
G	PPL	2121	IC 1 CONTROL
H	PNK	1039	IGNITION 1 VOLTAGE

Ignition coil connectors 1,3,5,7 2001-2002 4.8L, 5.3L, 6.0L Engines

PIN	WIRE COLOR	CIRCUIT NO.	FUNCTION
A	BLK	550	GROUND
B	RED/WHT	2122	IC 2 CONTROL
C	GRN DK/WHT	2124	IC 4 CONTROL
E	BRN/WHT	2130	LOW REFERENCE
F	BLU LT/WHT	2126	IC 6 CONTROL
G	PPL/WHT	2128	IC 8 CONTROL
H	PNK	1239	IGNITION 1 VOLTAGE

Ignition coil connectors 2,4,6,8 2001-2002 4.8L, 5.3L, 6.0L Engines

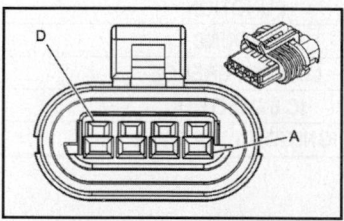

PIN	WIRE COLOR	CIRCUIT NO.	FUNCTION
A	BLK	550	GROUND
B	BRN	2129	LOW REFERENCE
C	PPL	2121	IC 1 CONTROL
D	PNK	1039	IGNITION 1 VOLTAGE

Ignition coil connector coil #1 on 2003-2005 4.8L, 5.3L, 6.0L Engines

PIN	WIRE COLOR	CIRCUIT NO.	FUNCTION
A	BLK	550	GROUND
B	BRN	2130	LOW REFERENCE
C	RED	2122	IC 2 CONTROL
D	PNK	1239	IGNITION 1 VOLTAGE

Ignition coil connector coil #2 on 2003-2005 4.8L, 5.3L, 6.0L Engines

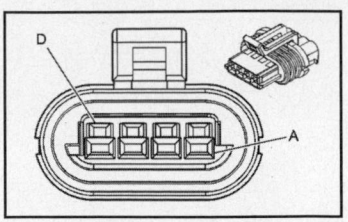

PIN	WIRE COLOR	CIRCUIT NO.	FUNCTION
A	BLK	550	GROUND
B	BRN	2129	LOW REFERENCE
C	LT BLU	2123	IC 3 CONTROL
D	PNK	1039	IGNITION 1 VOLTAGE

Ignition coil connector coil #3 on 2003-2005 4.8L, 5.3L, 6.0L Engines

PIN	WIRE COLOR	CIRCUIT NO.	FUNCTION
A	BLK	550	GROUND
B	BRN	2130	LOW REFERENCE
C	DK GRN	2124	IC 4 CONTROL
D	PNK	1239	IGNITION 1 VOLTAGE

Ignition coil connector coil #4 on 2003-2005 4.8L, 5.3L, 6.0L Engines

PIN	WIRE COLOR	CIRCUIT NO.	FUNCTION
A	BLK	550	GROUND
B	BRN	2129	LOW REFERENCE
C	DK GRN	2125	IC 5 CONTROL
D	PNK	1039	IGNITION 1 VOLTAGE

Ignition coil connector coil #5 on 2003-2005 4.8L, 5.3L, 6.0L Engines

PIN	WIRE COLOR	CIRCUIT NO.	FUNCTION
A	BLK	550	GROUND
B	BRN	2130	LOW REFERENCE
C	LT BLU	2126	IC 6 CONTROL
D	PNK	1239	IGNITION 1 VOLTAGE

Ignition coil connector coil #6 on 2003-2005 4.8L, 5.3L, 6.0L Engines

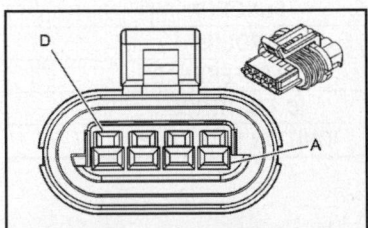

PIN	WIRE COLOR	CIRCUIT NO.	FUNCTION
A	BLK	550	GROUND
B	BRN	2129	LOW REFERENCE
C	RED	2127	IC 7 CONTROL
D	PNK	1039	IGNITION 1 VOLTAGE

Ignition coil connector coil #7 on 2003-2005 4.8L, 5.3L, 6.0L Engines

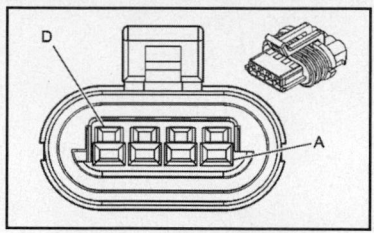

PIN	WIRE COLOR	CIRCUIT NO.	FUNCTION
A	BLK	550	GROUND
B	BRN	2130	LOW REFERENCE
C	PPL	2128	IC 8 CONTROL
D	PNK	1239	IGNITION 1 VOLTAGE

Ignition coil connector coil #8 on 2003-2005 4.8L, 5.3L, 6.0L Engines

MASS AIR FLOW (MAF) SENSOR CONNECTOR

Component Locations

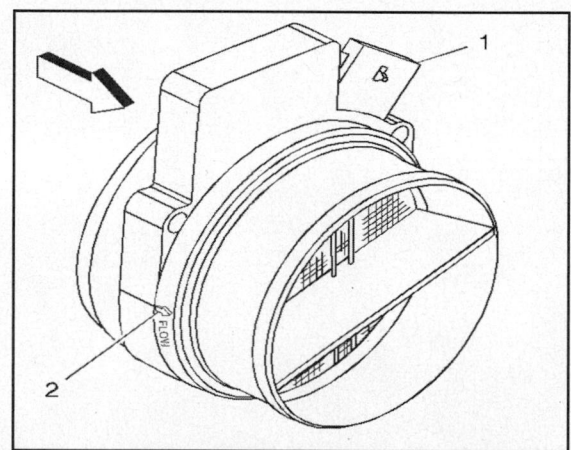

MAF sensor connector location (1) and flow direction (2) for 2000-2005 4.8L, 5.3L, 6.0L engines

Connector Pinouts

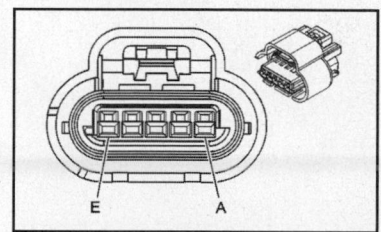

PIN	WIRE COLOR	CIRCUIT NO.	FUNCTION
A	BLK	552	SENSOR RETURN
B	TAN	472	INTAKE AIR TEMPERATURE (IAT) SENSOR SIGNAL
C	BLK/WHT	451	ENGINE CONTROL MODULE GROUND
D	PNK	239	FUSED IGNITION POSITIVE VOLTAGE
E	YEL	492	MASS AIR FLOW (MAF) SENSOR SIGNAL

MAF Sensor Connector Pinout on 2000-2002 4.8L, 5.3L, 6.0L Engines

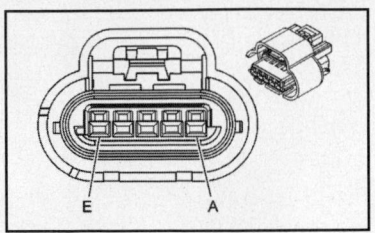

PIN	WIRE COLOR	CIRCUIT NO.	FUNCTION
A	BLK	552	LOW REFERENCE
B	TAN	472	IAT SENSOR SIGNAL
C	BLK/WHT	451	GROUND
D	PNK	439	IGNITION 1 VOLTAGE
E	YEL	492	MAF SENSOR SIGNAL

MAF Sensor Connector Pinout on 2003-2005 4.8L, 5.3L, 6.0L Engines

MANIFOLD ABSOLUTE PRESSURE (MAP) SENSOR CONNECTOR

Component Locations

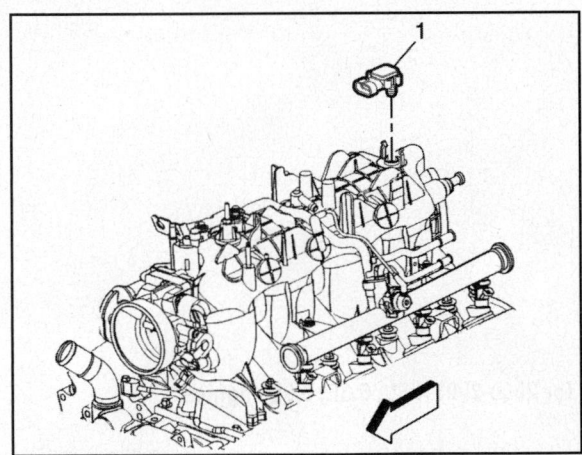

MAP sensor location on 2000-2005 4.8L, 5.3L, 6.0L Engines

Connector Pinouts

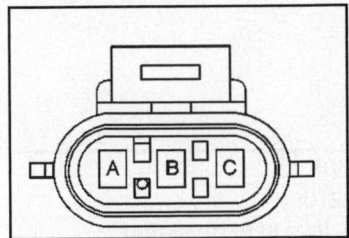

PIN	WIRE COLOR	CIRCUIT NO.	FUNCTION
A	ORN/BLK	469	MANIFOLD ABSOLUTE PRESSURE (MAP) SENSOR RETURN
B	LT GRN	432	MANIFOLD ABSOLUTE PRESSURE (MAP) SENSOR SIGNAL
C	GRY	597	REFERENCE VOLTAGE FEED (5 VOLT)

Manifold absolute pressure sensor connector pinout on 2000-2005 4.8L, 5.3L, 6.0L Engines

OXYGEN SENSOR CONNECTORS

Connector Pinouts

2000-01 ALL ENGINES

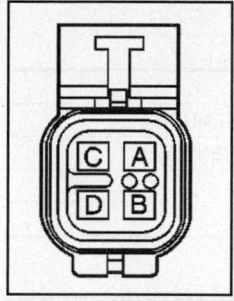

PIN	WIRE COLOR	CIRCUIT NO.	FUNCTION
A	TAN/WHT	1653	OXYGEN SENSOR RETURN - LEFT - FORWARD
B	PPL/WHT	1665	OXYGEN SENSOR SIGNAL - LEFT - FORWARD
C	BLK	550	GROUND
D	PNK	539	FUSED IGNITION POSITIVE VOLTAGE

2000 4.8L, 5.3L, 6.0L Engines HO2S1/1 (Bank 1 Sensor 1) Connector Pinout

PIN	WIRE COLOR	CIRCUIT NO.	FUNCTION
A	TAN/WHT	1669	OXYGEN SENSOR RETURN
B	PPL/WHT	1668	OXYGEN SENSOR SIGNAL
C	BLK	550	GROUND
D	PNK	1539	FUSED IGNITION POSITIVE VOLTAGE

2000 4.8L, 5.3L, 6.0L Engines HO2S1/2 (Bank 1 Sensor 2) Connector Pinout

PIN	WIRE COLOR	CIRCUIT NO.	FUNCTION
A	TAN	1667	OXYGEN SENSOR RETURN - RIGHT - FORWARD
B	PPL	1666	OXYGEN SENSOR SIGNAL - RIGHT - FORWARD
C	BLK	550	GROUND
D	PNK	539	FUSED IGNITION POSITIVE VOLTAGE

2000 4.8L, 5.3L, 6.0L Engines HO2S2/1 (Bank 2 Sensor 1) Connector Pinout

PIN	WIRE COLOR	CIRCUIT NO.	FUNCTION
A	TAN	413	GROUND
B	PPL/WHT	1665	OXYGEN SENSOR SIGNAL - LEFT - FORWARD
C	PNK	539	FUSED IGNITION POSITIVE VOLTAGE
D	BLK	550	GROUND

2000 4.8L, 5.3L, 6.0L Engines HO2S2/2 (Bank 2 Sensor 2) Connector Pinout

4.8L & 5.3L DELPHI TYPE

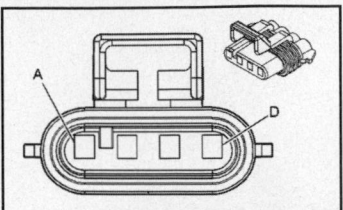

PIN	WIRE COLOR	CIRCUIT NO.	FUNCTION
A	TAN/WHT	1653	HO2S LOW SIGNAL - BANK 1 SENSOR 1
B	PPL/WHT	1665	HO2S HIGH SIGNAL - BANK 1 SENSOR 1
C	BLK	550	GROUND
D	PNK	539	IGNITION 1 VOLTAGE

2001-2002 4.8L, 5.3L Engines HO2S1/1 (Bank 1 Sensor 1) Connector Pinouts – Delphi Type

PIN	WIRE COLOR	CIRCUIT NO.	FUNCTION
A	TAN	1667	HO2S LOW SIGNAL - BANK 2 SENSOR 1
B	PPL	1666	HO2S HIGH SIGNAL - BANK 2 SENSOR 1
C	BLK	550	GROUND
D	PNK	539	HO2S HIGH SIGNAL - BANK 2 SENSOR 1

2001-2002 4.8L, 5.3L Engines HO2S2/1 (Bank 2 Sensor 1) Connector Pinouts – Delphi Type

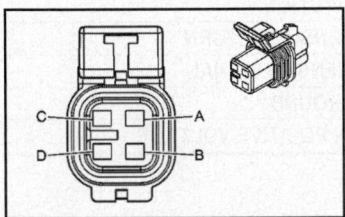

PIN	WIRE COLOR	CIRCUIT NO.	FUNCTION
A	TAN/WHT	1669	HO2S LOW SIGNAL - BANK 1 SENSOR 2
B	PPL/WHT	1668	HO2S HIGH SIGNAL - BANK 1 SENSOR 2
C	BLK	550	GROUND
D	PNK	1539	IGNITION 1 VOLTAGE

2001-2002 4.8L, 5.3L Engines HO2S1/2 (Bank 1 Sensor 2) Connector Pinouts – Delphi Type

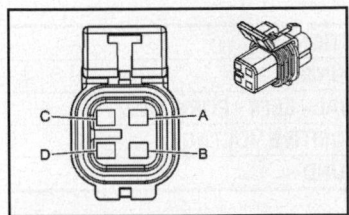

PIN	WIRE COLOR	CIRCUIT NO.	FUNCTION
A	TAN	1671	HO2S LOW SIGNAL - BANK 2 SENSOR 2
B	PPL	1670	HO2S HIGH SIGNAL - BANK 2 SENSOR 2
C	BLK	550	GROUND
D	PNK	1539	IGNITION 1 VOLTAGE

2001-2002 4.8L, 5.3L Engines HO2S2/2 (Bank 2 Sensor 2) Connector Pinouts – Delphi Type

4.8L & 5.3L DENSO TYPE

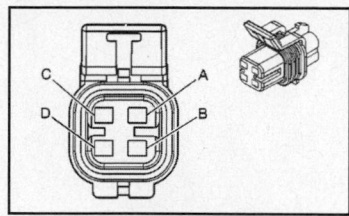

PIN	WIRE COLOR	CIRCUIT NO.	FUNCTION
A	TAN	413	HO2S LOW SIGNAL
B	PPL/WHT	1665	HO2S HIGH SIGNAL - BANK 1 SENSOR 1
C	PNK	539	HO2S HIGH SIGNAL - BANK 1 SENSOR 1
D	BLK	550	GROUND

2001-2002 4.8L, 5.3L Engines HO2S1/1 (Bank 1 Sensor 1) Connector Pinouts – Denso Type

PIN	WIRE COLOR	CIRCUIT NO.	FUNCTION
A	TAN	413	HO2S LOW SIGNAL
B	PPL/WHT	1668	HO2S HIGH SIGNAL [- BANK 1 SENSOR 2]
C	PNK	1539	IGNITION 1 VOLTAGE
D	BLK	550	GROUND

2001-2002 4.8L, 5.3L Engines HO2S1/2 (Bank 1 Sensor 2) Connector Pinouts – Denso Type

PIN	WIRE COLOR	CIRCUIT NO.	FUNCTION
A	TAN	413	HO2S LOW SIGNAL
B	PPL	1666	HO2S HIGH SIGNAL - BANK 2 SENSOR 1
C	PNK	539	HO2S HIGH SIGNAL - BANK 2 SENSOR 1
D	BLK	550	GROUND

2001-2002 4.8L, 5.3L Engines HO2S2/1 (Bank 2 Sensor 1) Connector Pinouts – Denso Type

PIN	WIRE COLOR	CIRCUIT NO.	FUNCTION
A	TAN	413	HO2S LOW SIGNAL
B	PPL	1670	HO2S HIGH SIGNAL - BANK 2 SENSOR 2
C	PNK	1539	IGNITION 1 VOLTAGE
D	BLK	550	GROUND

2001-2002 4.8L, 5.3L Engines HO2S2/2 (Bank 2 Sensor 2) Connector Pinouts – Denso Type

2001-02 6.0L ENGINES

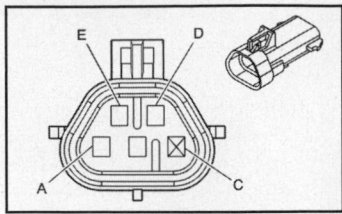

PIN	WIRE COLOR	CIRCUIT NO.	FUNCTION
A	TAN	413	HO2S LOW REFERENCE
B	PPL/WHT	1665	HO2S HIGH SIGNAL [- BANK 1 SENSOR 1]
D	LT GRN/WHT	3213	HO2S HEATER HIGH CONTROL [- BANK 1 SENSOR 1]
E	LT GRN	3212	HO2S HEATER LOW CONTROL [- BANK 1 SENSOR 1]

2001-02 6.0L Engine HO2S1/1 (Bank 1 Sensor 1) Connector Pinout

PIN	WIRE COLOR	CIRCUIT NO.	FUNCTION
A	TAN	413	HO2S LOW REFERENCE
B	PPL	1666	HO2S HIGH SIGNAL [- BANK 2 SENSOR 1]
D	LT GRN/WHT	3213	HO2S HEATER HIGH CONTROL [- BANK 2 SENSOR 1]
E	LT GRN	3212	HO2S HEATER LOW CONTROL [- BANK 2 SENSOR 1]

2001-02 6.0L Engine HO2S2/1 (Bank 2 Sensor 1) Connector Pinout

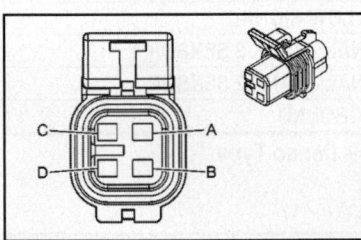

PIN	WIRE COLOR	CIRCUIT NO.	FUNCTION
A	TAN/WHT	1669	HO2S LOW SIGNAL [- BANK 1 SENSOR 2]
B	PPL/WHT	1668	HO2S HIGH SIGNAL [- BANK 1 SENSOR 2]
C	BLK	550	GROUND
D	PNK	1539	IGNITION 1 VOLTAGE

2001-02 6.0L Engine HO2S1/2 (Bank 1 Sensor 2) Connector Pinout

PIN	WIRE COLOR	CIRCUIT NO.	FUNCTION
A	TAN	1671	HO2S LOW SIGNAL [- BANK 2 SENSOR 2]
B	PPL	1670	HO2S HIGH SIGNAL [- BANK 2 SENSOR 2]
C	BLK	550	GROUND
D	PNK	1539	IGNITION 1 VOLTAGE

2001-02 6.0L Engine HO2S2/2 (Bank 2 Sensor 2) Connector Pinout

2003-05 6.0L (L59/LM7) ENGINES

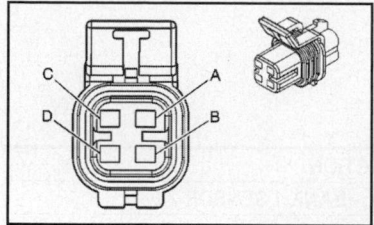

PIN	WIRE COLOR	CIRCUIT NO.	FUNCTION
A	TAN	1664	HO2S LOW SIGNAL - BANK 1 SENSOR 1
B	PPL/WHT	1665	HO2S HIGH SIGNAL - BANK 1 SENSOR 1
C	BLK/WHT	3113	HO2S HEATER LOW CONTROL - BANK 1 SENSOR 1
D	PNK	539	IGNITION 1 VOLTAGE

2003-05 6.0L (L59/LM7) Engine HO2S1/1 (Bank 1 Sensor 1) Connector Pinout

PIN	WIRE COLOR	CIRCUIT NO.	FUNCTION
A	TAN/WHT	1669	HO2S LOW SIGNAL - BANK 1 SENSOR 2
B	PPL/WHT	1668	HO2S HIGH SIGNAL - BANK 1 SENSOR 2
C	BRN	2391	HO2S HEATER LOW CONTROL - BANK 1 SENSOR 2
D	PNK	1539	IGNITION 1 VOLTAGE

2003-05 6.0L (L59/LM7) Engine HO2S1/2 (Bank 1 Sensor 2) Connector Pinout

A	TAN	1667	HO2S LOW SIGNAL - BANK 2 SENSOR 1
B	PPL	1666	HO2S HIGH SIGNAL - BANK 2 SENSOR 1
C	LT GRN	3212	HO2S HEATER LOW CONTROL - BANK 2 SENSOR 1
D	PNK	539	HO2S HIGH SIGNAL - BANK 2 SENSOR 1

2003-05 6.0L (L59/LM7) Engine HO2S2/1 (Bank 2 Sensor 1) Connector Pinout

PIN	WIRE COLOR	CIRCUIT NO.	FUNCTION
A	TAN	1671	HO2S LOW SIGNAL - BANK 2 SENSOR 2
B	PPL	1670	HO2S HIGH SIGNAL - BANK 2 SENSOR 2
C	RED/WHT	3223	HO2S HEATER LOW CONTROL - BANK 2 SENSOR 2
D	PNK	1539	IGNITION 1 VOLTAGE

2003-05 6.0L (L59/LM7) Engine HO2S2/2 (Bank 2 Sensor 2) Connector Pinout

2003-05 6.0L (LR4) ENGINES

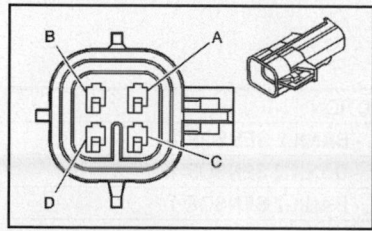

PIN	WIRE COLOR	CIRCUIT NO.	FUNCTION
A	TAN	1664	HO2S LOW SIGNAL - BANK 1 SENSOR 1
B	PPL/WHT	1665	HO2S HIGH SIGNAL - BANK 1 SENSOR 1
C	BLK/WHT	3113	HO2S HEATER LOW CONTROL - BANK 1 SENSOR 1
D	PNK	539	IGNITION 1 VOLTAGE

2003-05 6.0L (LR4) Engine HO2S1/1 (Bank 1 Sensor 1) Connector Pinout

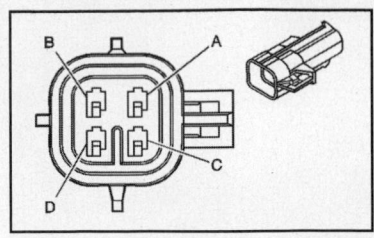

PIN	WIRE COLOR	CIRCUIT NO.	FUNCTION
A	TAN/WHT	1669	HO2S LOW SIGNAL - BANK 1 SENSOR 2
B	PPL/WHT	1668	HO2S HIGH SIGNAL - BANK 1 SENSOR 2
C	BRN	2391	HO2S HEATER LOW CONTROL - BANK 1 SENSOR 2
D	PNK	1539	IGNITION 1 VOLTAGE

2003-05 6.0L (LR4) Engine HO2S1/2 (Bank 1 Sensor 2) Connector Pinout

PIN	WIRE COLOR	CIRCUIT NO.	FUNCTION
A	TAN	1667	HO2S LOW SIGNAL - BANK 2 SENSOR 1
B	PPL	1666	HO2S HIGH SIGNAL - BANK 2 SENSOR 1
C	LT GRN	3212	HO2S HEATER LOW CONTROL - BANK 2 SENSOR 1
D	PNK	539	HO2S HIGH SIGNAL - BANK 2 SENSOR 1

2003-05 6.0L (LR4) Engine HO2S2/1 (Bank 2 Sensor 1) Connector Pinout

2003-05 6.0L (LQ4) ENGINES

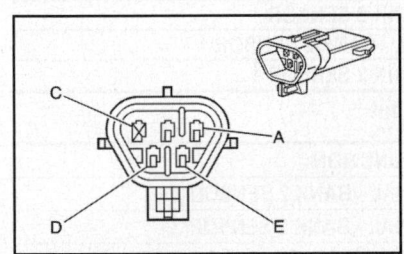

PIN	WIRE COLOR	CIRCUIT NO.	FUNCTION
A	TAN	1664	HO2S LOW SIGNAL - BANK 1 SENSOR 1
B	PPL/WHT	1665	HO2S HIGH SIGNAL - BANK 1 SENSOR 1
D	PNK	539	IGNITION 1 VOLTAGE
E	BLK/WHT	3113	HO2S HEATER LOW CONTROL - BANK 1 SENSOR 1

2003-2005 6.0L (LQ4) Engine HO2S1/1 (Bank 1 Sensor 1) Connector Pinout

PIN	WIRE COLOR	CIRCUIT NO.	FUNCTION
A	TAN	1667	HO2S LOW SIGNAL - BANK 2 SENSOR 1
B	PPL	1666	HO2S HIGH SIGNAL - BANK 2 SENSOR 1
D	PNK	539	HO2S HIGH SIGNAL - BANK 2 SENSOR 1
E	LT GRN	3212	HO2S HEATER LOW CONTROL - BANK 2 SENSOR 1

2003-2005 6.0L (LQ4) Engine HO2S2/1 (Bank 2 Sensor 1) Connector Pinout

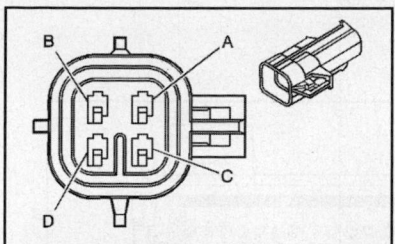

PIN	WIRE COLOR	CIRCUIT NO.	FUNCTION
A	TAN	1671	HO2S LOW SIGNAL - BANK 2 SENSOR 2
B	PPL	1670	HO2S HIGH SIGNAL - BANK 2 SENSOR 2
C	RED/WHT	3223	HO2S HEATER LOW CONTROL - BANK 2 SENSOR 2
D	PNK	1539	IGNITION 1 VOLTAGE

2003-2005 6.0L (LR4 or LQ4) Engine HO2S2/2 (Bank 2 Sensor 2) Connector Pinout

POWERTRAIN CONTROL MODULE (PCM)

Component Locations

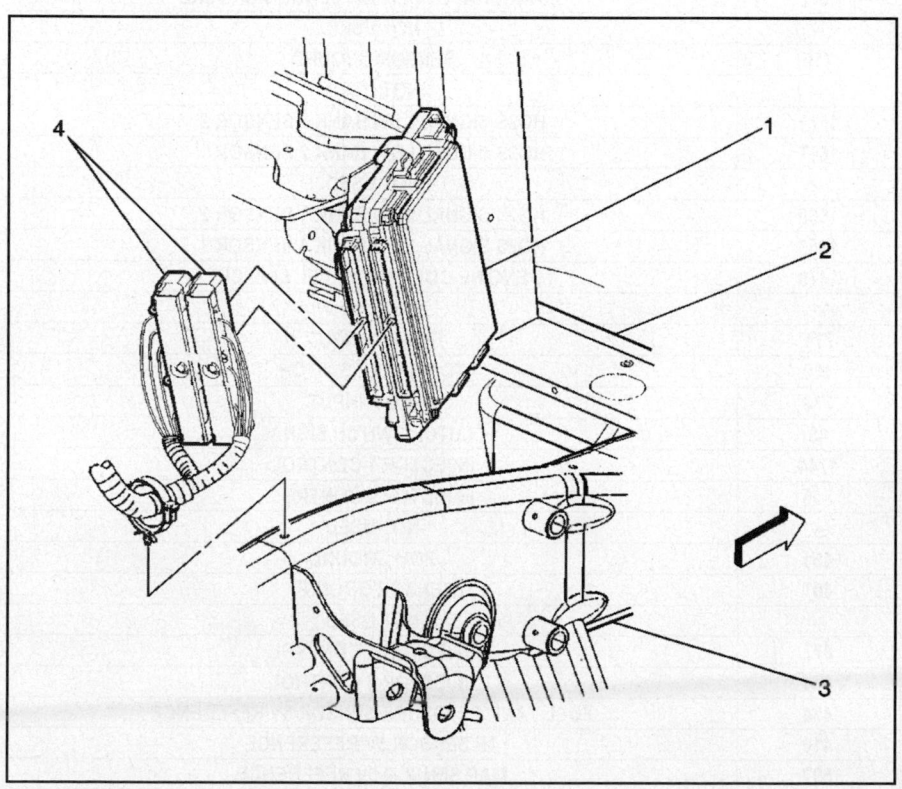

Powertrain control module (1) radiator support (2) left frame rail (3) wiring harness connectors (4)

Connector Pinouts

2000 C1
CONNECTOR

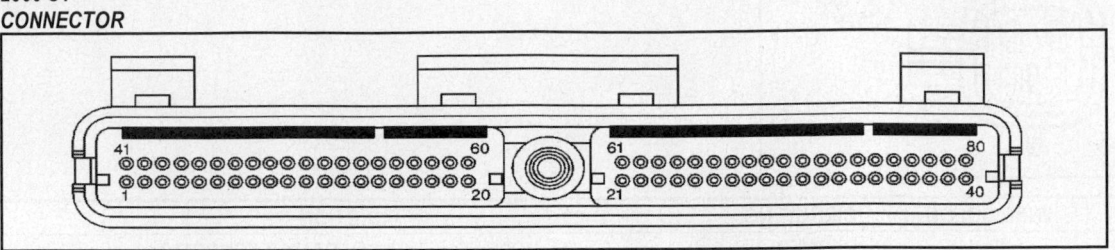

PIN	WIRE COLOR	CIRCUIT NO.	FUNCTION
1	BLK/WHT	451	PCM GROUND
2	LT GRN	1867	CRANKSHAFT POSITION SENSOR B+ SUPPLY
3	PNK/BLK	1746	INJECTOR 3 CONTROL
4	LT GRN/BLK	1745	INJECTOR 2 CONTROL
5-10	--	--	NOT USED
11	LT BLU	1876	KNOCK SENSOR SIGNAL REAR
12	DK BLU/WHT	1869	CRANKSHAFT POSITION SENSOR SIGNAL
13-16	--	--	NOT USED
17	DK BLU	1225	TRANSMISSION RANGE SIGNAL B
18	RED	1226	TRANSMISSION RANGE SIGNAL C
19	PNK	439	PCM IGNITION SUPPLY
20	ORN	440	PCM BATTERY SUPPLY
21	YEL/BLK	1868	CRANKSHAFT POSITION SENSOR GROUND
22	--	--	NOT USED
23	PPL	719	SENSOR GROUND
24	--	--	NOT USED
25	TAN	1671	HO2S SIGNAL LOW BANK 2 SENSOR 2
26	TAN	1667	HO2S SIGNAL LOW BANK 2 SENSOR 1
27	--	--	NOT USED
28	TAN/WHT	1669	HO2S SIGNAL LOW BANK 1 SENSOR 2
29	TAN/WHT	1653	HO2S SIGNAL LOW BANK 1 SENSOR 1
30	LT GRN	1478	ENGINE COOLANT LEVEL SWITCH
31	--	--	NOT USED
32	BLK/WHT	771	PRND A INPUT
33	PPL	420	TCC BRAKE SWITCH
34	WHT	776	PRND P INPUT
35	GRA	48	CLUTCH SWITCH SIGNAL
36	BLK	1744	INJECTOR 1 CONTROL
37	YEL/BLK	846	INJECTOR 6 CONTROL
38-39	--	--	NOT USED
40	BLK/WHT	451	PCM GROUND
41	BLK	407	SENSOR GROUND
42	--	--	NOT USED
43	RED/BLK	877	INJECTOR 7 CONTROL
44	LT BLU/BLK	844	INJECTOR 4 CONTROL
45	GRA	474	FUEL TANK PRESSURE SENSOR 5V REFERENCE
46	GRA	416	TP SENSOR 5V REFERENCE
47	GRA	597	MAP SENSOR 5V REFERENCE
48	GRA	596	EGR PINTLE POSITION SENSOR 5V REFERENCE
49-50	--	--	NOT USED
51	DK BLU	496	KNOCK SENSOR SIGNAL FRONT
52	--	--	NOT USED
53	BLK	470	FUEL TANK PRESSURE SENSOR GROUND
54	BLK	452	TP SENSOR GROUND
55	BRN	1456	EGR PINTLE POSITION SENSOR SIGNAL

2000 4.8L, 5.3L, 6.0L Engine PCM C1 Connector Pinout (1 of 2)

2000 C1 CONNECTOR
(CONTINUED)

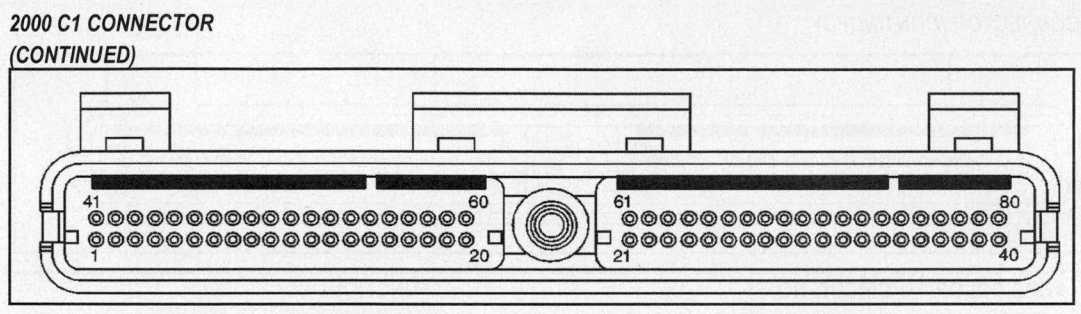

PIN	WIRE COLOR	CIRCUIT NO.	FUNCTION
56	--	--	NOT USED
57	ORN	440	PCM BATTERY SUPPLY
58	DK GRN	1049	SERIAL DATA
59	YEL	710	SERIAL DATA
60	ORN/BLK	469	MAP SENSOR GROUND
61	PNK/BLK	632	CMP SENSOR GROUND
62	--	--	NOT USED
63	GRA	720	ENGINE COOLANT TEMPERATURE (ECT) SENSOR GROUND
64	--	--	NOT USED
65	PPL	1670	HO2S SIGNAL HIGH BANK 2 SENSOR 2
66	PPL	1666	HO2S SIGNAL HIGH BANK 2 SENSOR 1
67	--	--	NOT USED
68	PPL/WHT	1668	HO2S SIGNAL HIGH BANK 1 SENSOR 2
69	PPL/WHT	1665	HO2S SIGNAL HIGH BANK 1 SENSOR 1
70	BRN	1174	LOW OIL LEVEL SWITCH
71	--	--	NOT USED
72	YEL	772	PRND B INPUT
73	BRN/WHT	633	CAMSHAFT POSITION (CMP) SENSOR SIGNAL
74	YEL	410	ENGINE COOLANT TEMPERATURE (ECT) SENSOR SIGNAL
75	PNK	1020	PCM IGNITION SUPPLY
76	BLK/WHT	845	INJECTOR 5 CONTROL
77	DK BLU/WHT	878	INJECTOR 8 CONTROL
78	--	--	NOT USED
79	WHT	687	3-2 SHIFT SOLENOID CONTROL
80	ORN/BLK	510	FUEL LEVEL SENSOR GROUND

2000 4.8L, 5.3L, 6.0L Engine PCM C2 Connector Pinout (2 of 2)

2000 C2 CONNECTOR

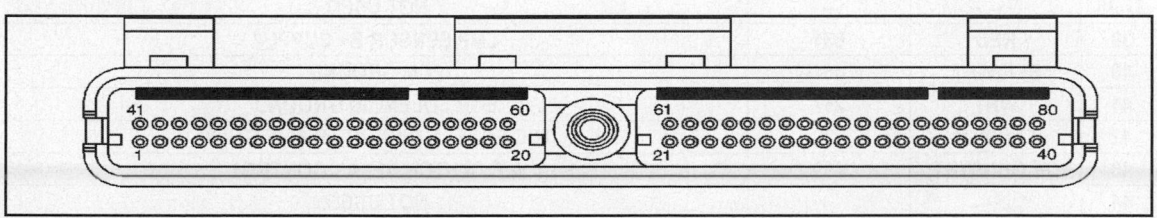

PIN	WIRE COLOR	CIRCUIT NO.	FUNCTION
1	BLK/WHT	451	PCM GROUND
2	BRN	418	TCC CONTROL SOLENOID
3-5	--	--	NOT USED
6	RED/BLK	1228	TRANSMISSION FLUID PRESSURE CONTROL SOLENOID HIGH
7	RED	1676	EGR SOLENOID SUPPLY

2000 4.8L, 5.3L, 6.0L Engine PCM C2 Connector Pinout (1 of 3)

2000 C2 CONNECTOR (CONTINUED)

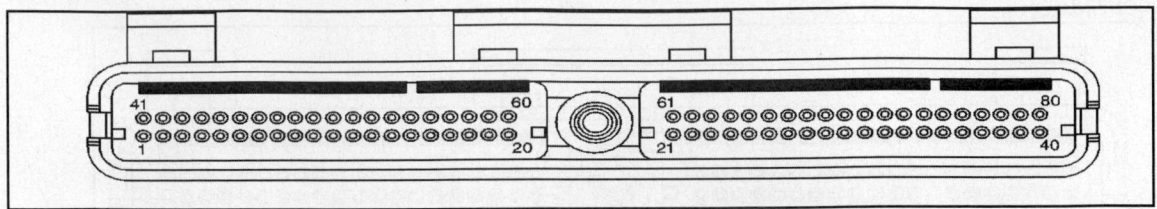

PIN	WIRE COLOR	CIRCUIT NO.	FUNCTION
8	LT BLU/WHT	1229	TRANSMISSION FLUID PRESSURE CONTROL SOLENOID LOW
9	DK GRN/WHT	465	FUEL PUMP RELAY CONTROL
10	WHT	121	ENGINE SPEED (TACH) OUTPUT SIGNAL
11	DK BLU	604	SECONDARY A/C HIGH PRESSURE SWITCH
12	--	--	NOT USED
13	LT BLU/BLK	396	CRUISE CONTROL ENGAGE SIGNAL
14	--	--	NOT USED
15	BRN	25	GENERATOR L TERMINAL
16	GRA/BLK	1694	4 WHEEL DRIVE SWITCH SIGNAL LOW
17	DK GRN/WHT	762	A/C REQUEST SIGNAL
18	--	--	NOT USED
19	BLK/WHT	1695	4 WHEEL DRIVE FRONT AXLE SWITCH
20	LT GRN/BLK	822	VEHICLE SPEED SENSOR (VSS) REFERENCE LOW
21	PPL/WHT	821	VEHICLE SPEED SENSOR (VSS) SIGNAL
22	RED/BLK	1230	VEHICLE SPEED SENSOR (VSS) SIGNAL (6.0L ONLY)
23	DK BLU/WHT	1231	VEHICLE SPEED SENSOR (VSS) REFERENCE LOW (6.0L ONLY)
24	DK BLU	417	TP SENSOR SIGNAL
25	TAN	472	IAT SENSOR SIGNAL
26	PPL	2121	IGNITION CONTROL 1
27	RED	2127	IGNITION CONTROL 7
28	LT BLU/WHT	2126	IGNITION CONTROL 6
29	DK GRN/WHT	2124	IGNITION CONTROL 4
30	--	--	NOT USED
31	YEL	492	MAF SENSOR SIGNAL
32	LT GRN	432	MAP SENSOR SIGNAL
33	DK GRN	1614	HVAC RECIRCULATION DOOR CONTROL
34	DK GRN/WHT	428	EVAP CANISTER PURGE SOLENOID CONTROL
35	--	--	NOT USED
36	BRN	436	AIR PUMP RELAY CONTROL (W/NC1)
37-38	--	--	NOT USED
39	RED	631	CMP SENSOR B+ SUPPLY
40	BLK/WHT	451	PCM GROUND
41	WHT	257	EGR SOLENOID GROUND
42	TAN/BLK	422	TCC ENABLE CIRCUIT (4L60-E ONLY)
43	DK GRN/WHT	459	A/C CLUTCH RELAY CONTROL
44	--	--	NOT USED
45	WHT	1310	EVAP CANISTER VENT VALVE CONTROL
46	BRN/WHT	419	MALFUNCTION INDICATOR LAMP (MIL) CONTROL
47	YEL/BLK	1223	TRANSMISSION SHIFT SOLENOID B
48	LT GRN	1222	TRANSMISSION SHIFT SOLENOID A
49	YEL/BLK	1827	VEHICLE SPEED OUTPUT CIRCUIT 128K
50	DK GRN/WHT	817	VEHICLE SPEED OUTPUT CIRCUIT 4K

2000 4.8L, 5.3L, 6.0L Engine PCM C2 Connector Pinout (2 of 3)

2000 C2 CONNECTOR (CONTINUED)

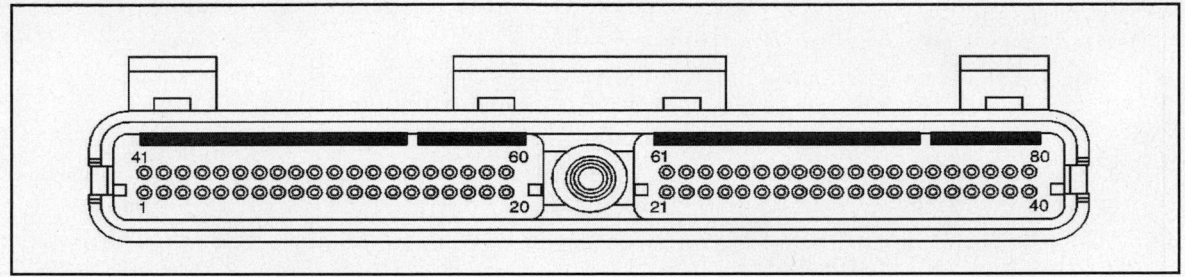

PIN	WIRE COLOR	CIRCUIT NO.	FUNCTION
51	YEL/BLK	1227	TRANSMISSION TEMPERATURE SENSOR SIGNAL
52-53	--	--	NOT USED
54	PPL	1589	FUEL LEVEL SENSOR SIGNAL
55	DK GRN	603	A/C COMPRESSOR CYCLING SWITCH SIGNAL
56	--	--	NOT USED
57	BLK	552	IAT SENSOR GROUND
58-59	--	--	NOT USED
60	BRN	2129	IGNITION CONTROL REFERENCE LOW BANK 1
61	BRN/WHT	2130	IGNITION CONTROL REFERENCE LOW BANK 2
62	GRA	773	PRND C
63	PNK	1224	TRANSMISSION RANGE SIGNAL A
64	DK GRN	890	FUEL TANK PRESSURE SENSOR SIGNAL
65	--	--	NOT USED
66	PPL/WHT	2128	IGNITION CONTROL 8
67	RED/WHT	2122	IGNITION CONTROL 2
68	DK GRN	2125	IGNITION CONTROL 5
69	LT BLU	2123	IGNITION CONTROL 3
70-75	--	--	NOT USED
76	LT GRN/WHT	1749	IAC COIL B HIGH
77	LT GRN/BLK	444	IAC COIL B LOW
78	LT BLU/BLK	1748	IAC COIL A LOW
79	LT BLU/WHT	1747	IAC COIL A HIGH
80	--	--	NOT USED

2000 4.8L, 5.3L, 6.0L Engine PCM C2 Connector Pinout (3 of 3)

2001-02 C1 CONNECTOR

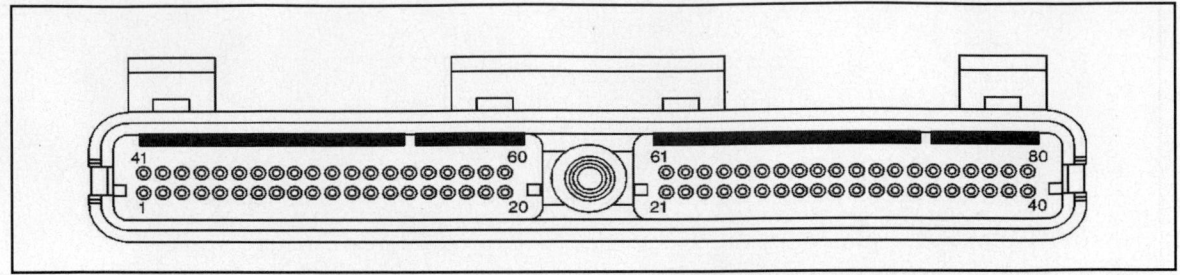

PIN	WIRE COLOR	CIRCUIT NO.	FUNCTION
1	BLK/WHT	451	GROUND
2	LT GRN	1867	12 VOLT REFERENCE
3	PNK/BLK	1746	FUEL INJECTOR 3 CONTROL
4	LT GRN/BLK	1745	FUEL INJECTOR 2 CONTROL
5-10	--	--	NOT USED
11	LT BLU	1876	KNOCK SENSOR 2 SIGNAL
12	DK BLU/WHT	1869	CKP SENSOR SIGNAL
13	ORN/BLK	463	REQUESTED TORQUE SIGNAL
14	ORN/BLK	1061	UART SERIAL DATA [SECONDARY]
15	DRK BLU/WHT	774	UART SERIAL DATA [TERTIARY]
16	--	--	NOT USED
17	DK BLU	1225	TRANSMISSION FLUID PRESSURE SWITCH SIGNAL B
18	RED	1226	TRANSMISSION FLUID PRESSURE SWITCH SIGNAL C
19	PNK	439	IGNITION 1 VOLTAGE
20	ORN	440	BATTERY POSITIVE VOLTAGE
21	YEL	1868	LOW REFERENCE
22	--	--	NOT USED
23	PPL	719	LOW REFERENCE
24	--	--	NOT USED
25	TAN	1671	HO2S LOW SIGNAL [- BANK 2 SENSOR 1]
25	TAN	413	HO2S LOW REFERENCE (DENSO SENSORS) HO2S LOW SIGNAL (DELPHI SENSORS)
26	TAN	1667	HO2S LOW SIGNAL [- BANK 2 SENSOR 2]
26	TAN	413	HO2S LOW REFERENCE (DENSO SENSORS) HO2S LOW SIGNAL (DELPHI SENSORS)
27	--	--	NOT USED
28	TAN/WHT	1669	HO2S LOW SIGNAL [- BANK 1 SENSOR 2]
28	TAN	413	HO2S LOW REFERENCE (DENSO SENSORS) HO2S LOW SIGNAL (DELPHI SENSORS)
29	TAN/WHT	1653	HO2S LOW SIGNAL [- BANK 1 SENSOR 1]
29	TAN	413	HO2S LOW REFERENCE (DENSO SENSORS) HO2S LOW SIGNAL (DELPHI SENSORS)
30	LT GRN	1478	COOLANT LEVEL SWITCH SIGNAL
31	--	--	NOT USED
32	BLK/WHT	771	TRANSMISSION RANGE SWITCH SIGNAL A
33	PPL	420	TCC BRAKE SWITCH SIGNAL
34	WHT	776	TRANSMISSION RANGE SWITCH SIGNAL P
35	GRA	48	CPP SWITCH SIGNAL
36	BLK	1744	INJECTOR 1 CONTROL
37	YEL/BLK	846	INJECTOR 6 CONTROL
38	PNK/WHT	1101	DAMPING LIFT/DIVE SIGNAL
39	--	--	NOT USED
40	BLK/WHT	451	GROUND

2001-02 4.8L, 5.3L, 6.0L Engine PCM C1 Connector Pinout (1 of 2)

2001-02 C1 CONNECTOR (CONTINUED)

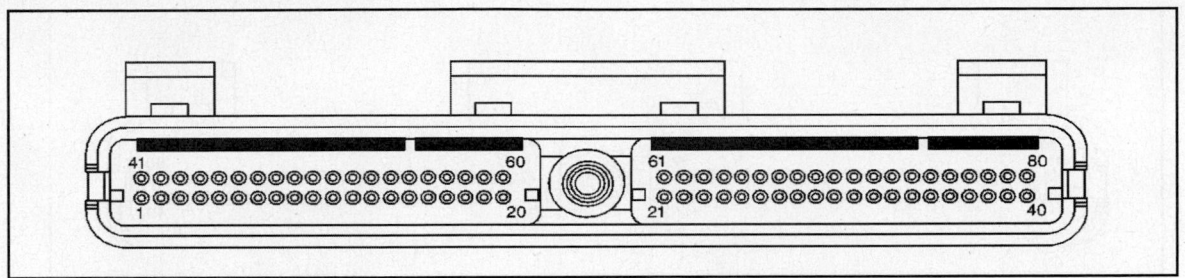

PIN	WIRE COLOR	CIRCUIT NO.	FUNCTION
41	BLK	407	LOW REFERENCE
41	GRA	720	LOW REFERENCE (W/O DELPHI HO2S)
42	--	--	NOT USED
43	RED/BLK	877	FUEL INJECTOR 7 CONTROL
44	LT BLU/BLK	844	FUEL INJECTOR 4 CONTROL
45	GRA	474	5 VOLT REFERENCE (FTP SENSOR)
46	GRY	474	5 VOLT REFERENCE (TPS SENSOR)
47	GRY	597	5 VOLT REFERENCE (MAP SENSOR)
48	GRA	596	5 VOLT REFERENCE (EGR SENSOR)
49 50	--	--	NOT USED
51	DK BLU	496	KS [1] SIGNAL
52	--	--	NOT USED
53	BLK	470	LOW REFERENCE
54	BLK	452	LOW REFERENCE
55	BRN	1456	EGR VALVE POSITION SIGNAL
56	--	--	NOT USED
57	ORN	440	BATTERY POSITIVE VOLTAGE
58	DK GRN	1049	ECM/PCM/VCM CLASS 2 SERIAL DATA
59	YEL	710	CLASS 2 SERIAL DATA
60	ORN/BLK	469	LOW REFERENCE
61	PNK/BLK	632	LOW REFERENCE
62	--	--	NOT USED
63	GRA	720	LOW REFERENCE
63	TAN	413	HO2S LOW REFERENCE (DENSO SENSORS) HO2S LOW SIGNAL (DELPHI SENSORS)
64	--	--	NOT USED
65	PPL	1670	HO2S HIGH SIGNAL [- BANK 2 SENSOR 2]
66	PPL	1666	HO2S HIGH SIGNAL [- BANK 2 SENSOR 1]
67	--	--	NOT USED
68	PPL/WHT	1668	HO2S HIGH SIGNAL [- BANK 1 SENSOR 2]
69	PPL/WHT	1665	HO2S HIGH SIGNAL [- BANK 1 SENSOR 1]
70	BRN	1174	OIL LEVEL SWITCH SIGNAL
71	--	--	NOT USED
72	YEL	772	TRANSMISSION RANGE SWITCH SIGNAL B
73	BRN/WHT	633	CMP SENSOR SIGNAL
74	YEL	410	ECT SENSOR SIGNAL
75	PNK	1020	OFF/RUN/CRANK VOLTAGE
76	BLK/WHT	845	FUEL INJECTOR 5 CONTROL
77	DK BLU/WHT	878	FUEL INJECTOR 8 CONTROL
78	--	--	NOT USED
79	WHT	687	3-2 SHIFT SOLENOID VALVE CONTROL
80	ORN/BLK	510	LOW REFERENCE

2001-02 4.8L, 5.3L, 6.0L Engine PCM C1 Connector Pinout (2 of 2)

2001-02 C2 CONNECTOR

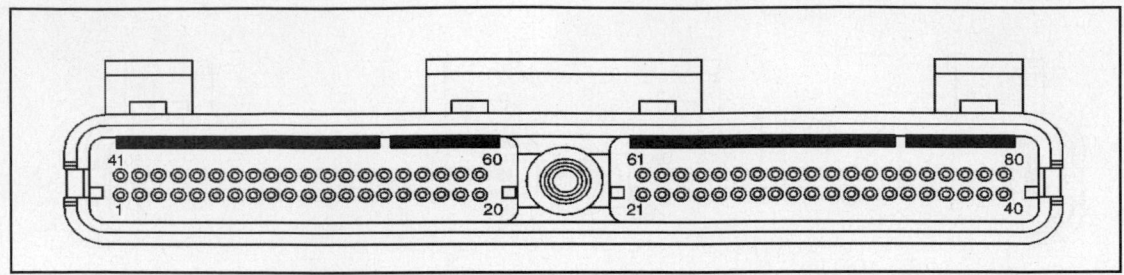

PIN	WIRE COLOR	CIRCUIT NO.	FUNCTION
1	BLK/WHT	451	GROUND
2	BRN	418	TCC PWM SOLENOID VALVE CONTROL
3	TAN	1465	FUEL PUMP RELAY CONTROL [- SECONDARY]
4	--	--	NOT USED
5	TAN/BLK	464	DELIVERED TORQUE SIGNAL
6	RED/BLK	1228	PC SOLENOID VALVE HIGH CONTROL (SOL. A)
7	RED	1676	EGR VALVE SUPPLY VOLTAGE (5 VOLT)
8	LT BLU/WHT	1229	PC SOLENOID VALVE LOW CONTROL (SOL. A)
9	DK GRN/WHT	465	FUEL PUMP RELAY CONTROL [- PRIMARY]
10	WHT	121	ENGINE SPEED SIGNAL
11	DK BLU	604	A/C HIGH PRESSURE RECIRCULATION SWITCH SIGNAL
12	--	--	NOT USED
13	LT BLU/BLK	396	CRUISE CONTROL ENGAGED SIGNAL
14	--	--	NOT USED
15	BRN	25	CHARGE INDICATOR CONTROL
16	GRY/BLK	1694	4WD LOW SIGNAL
17	DK GRN/WHT	2523	A/C REQUEST
18	--	--	NOT USED
19	BLK/WHT	1695	AXLE SWITCH SIGNAL
20	LT GRN/BLK	822	VSS LOW SIGNAL
21	PPL/WHT	821	VSS HIGH SIGNAL
22	RED/BLK	1230	AT ISS HIGH SIGNAL
23	DK BLU/WHT	1231	AT ISS LOW SIGNAL
24	--	--	NOT USED
25	TAN	472	IAT SENSOR SIGNAL
26	PPL	2121	IC 1 CONTROL
27	RED	2127	IC 7 CONTROL
28	LT BLU/WHT	2126	IC 6 CONTROL
29	DK GRN/WHT	2124	IC 4 CONTROL
30	--	--	NOT USED
31	YEL	492	MAF SENSOR SIGNAL
32	LT GRN	432	MAP SENSOR SIGNAL
33	DK GRN	1614	RECIRCULATION ACTUATOR CONTROL
34	DK GRN/WHT	428	EVAP CANISTER PURGE SOLENOID CONTROL
35	--	--	NOT USED
36	BRN	436	AIR PUMP RELAY CONTROL
37-38	--	--	NOT USED
39	RED	631	12 VOLT REFERENCE
40	BLK/WHT	451	GROUND
41	WHT	257	EGR SOLENOID CONTROL
42	TAN/BLK	422	TCC SOLENOID VALVE CONTROL
43	DK GRN/WHT	459	A/C COMPRESSOR CLUTCH RELAY CONTROL

2001-02 4.8L, 5.3L, 6.0L Engine PCM C2 Connector Pinout (1 of 2)

2001-02 C2 CONNECTOR (CONTINUED)

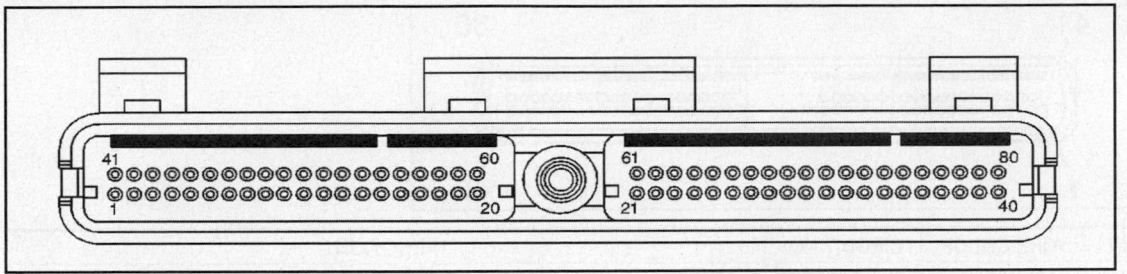

PIN	WIRE COLOR	CIRCUIT NO.	FUNCTION
44	--	--	NOT USED
45	WHT	1310	EVAP CANISTER VENT SOLENOID CONTROL
46	BRN/WHT	419	MIL CONTROL
47	YEL/BLK	1223	(2-3 SHIFT SOLENOID OR SHIFT SOLENIOD B) VALVE CONTROL
48	LT GRN	1222	(1-2 SHIFT SOLENOID OR SHIFT SOLENIOD A) VALVE CONTROL
49	YEL/BLK	1827	VEHICLE SPEED SIGNAL
50	DK GRN/WHT	817	VSS SIGNAL
51	YEL/BLK	1227	TFT SENSOR SIGNAL
52	GRA	23	GENERATOR FIELD DUTY CYCLE SIGNAL
53	--	--	NOT USED
54	PPL	1589	FUEL LEVEL SENSOR SIGNAL [- PRIMARY]
55	DK GRN	603	A/C LOW PRESSURE SWITCH SIGNAL
56	--	--	NOT USED
57	BLK	552	LOW REFERENCE
58-59	--	--	NOT USED
60	BRN	2129	LOW REFERENCE
61	BRN/WHT	2130	LOW REFERENCE
62	GRA	773	TRANSMISSION RANGE SWITCH SIGNAL C
63	PNK	1224	TRANSMISSION FLUID PRESSURE SWITCH SIGNAL A
64	DK GRN	890	FUEL TANK PRESSURE SENSOR SIGNAL
65	--	--	NOT USED
66	PPL/WHT	2128	IC 8 CONTROL
67	RED/WHT	2122	IC 2 CONTROL
68	DK GRN	2125	IC 5 CONTROL
69	LT BLU	2123	IC 3 CONTROL
70-71	--	--	NOT USED
72	BLK/WHT	3122	HO2S HEATER LOW CONTROL [- BANK 2 SENSOR 1]
73	DK BLU	1936	FUEL LEVEL SENSOR SIGNAL [- SECONDARY]
74	LT GRN/WHT	3213	HO2S HEATER HIGH CONTROL [- BANK 2 SENSOR 1]
75	--	--	NOT USED
76	LT GRN/WHT	1749	IAC COIL B HIGH CONTROL
77	LT GRN/BLK	444	IAC COIL B LOW CONTROL
78	LT BLU/BLK	1748	IAC COIL A LOW CONTROL
79	LT BLU/WHT	1747	IAC COIL A HIGH CONTROL
80	LT GRN	3212	HO2S HEATER LOW CONTROL [- BANK 2 SENSOR 1]

2001-02 4.8L, 5.3L, 6.0L Engine PCM C2 Connector Pinout (2 of 2)

2003-05 C1 CONNECTOR

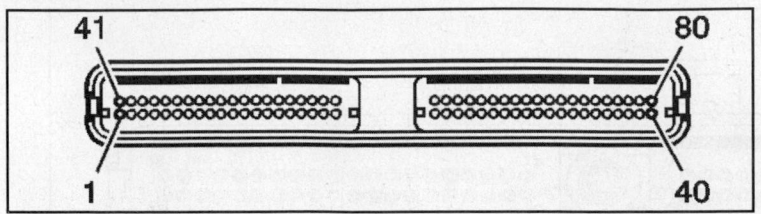

PIN	WIRE COLOR	CIRCUIT NO.	FUNCTION
1	BLK/WHT	451	GROUND
2	LT GRN	1867	12 VOLT REFERENCE
3	PNK/BLK	1746	FUEL INJECTOR 3 CONTROL
4	LT GRN/BLK	1745	FUEL INJECTOR 2 CONTROL
5-6	--	--	NOT USED
7	GRY	2705	5 VOLT REFERENCE
8-10	--	--	NOT USED
11	LT BLU	1876	KNOCK SENSOR 2 SIGNAL
12	DK BLU/WHT	1869	CKP SENSOR SIGNAL
13	ORN/BLK	463	REQUESTED TORQUE SIGNAL (NW7)
14	ORN/BLK	1061	UART SERIAL DATA
15	DK BLU/WHT	774	UART SERIAL DATA
16	--	--	NOT USED
17	DK BLU	1225	TRANSMISSION FLUID PRESSURE SWITCH SIGNAL B (A/T)
18	RED	1226	TRANSMISSION FLUID PRESSURE SWITCH SIGNAL C (A/T)
19	PNK	439	IGNITION 1 VOLTAGE
20	ORN	440	BATTERY POSITIVE VOLTAGE
21	YEL/BLK	1868	LOW REFERENCE
22	--	--	NOT USED
23	BLK	470	LOW REFERENCE
24	BLK/WHT	451	GROUND
25	TAN	1671	HO2S LOW SIGNAL - BANK 2 SENSOR 2
26	TAN	1667	HO2S LOW SIGNAL - BANK 2 SENSOR 1
27	BLK/WHT	451	GROUND
28	TAN/WHT	1669	HO2S LOW SIGNAL - BANK 1 SENSOR 2
29	TAN	1664	HO2S LOW SIGNAL - BANK 1 SENSOR 1
30	LT GRN	1478	COOLANT LEVEL SWITCH SIGNAL
31	--	--	NOT USED
32	BLK/WHT	771	TRANSMISSION RANGE SWITCH SIGNAL A (A/T)
33	PPL	420	TCC BRAKE SWITCH/CRUISE CONTROL RELEASE SWITCH SIGNAL
34	WHT	776	TRANSMISSION RANGE SWITCH SIGNAL P (A/T)
35	GRY	48	CPP SWITCH SIGNAL (M/T)
36	BLK	1744	FUEL INJECTOR 1 CONTROL
37	YEL/BLK	846	FUEL INJECTOR 6 CONTROL
38	PNK/WHT	1101	DAMPING LIFT/DIVE SIGNAL
39	YEL/BLK	625	STARTER ENABLE RELAY CONTROL
40	BLK/WHT	451	GROUND
41-42	--	--	NOT USED
43	RED/BLK	877	FUEL INJECTOR 7 CONTROL
44	LT BLU/BLK	844	FUEL INJECTOR 4 CONTROL
45	GRY	2700	5 VOLT REFERENCE (A/C)
46	GRY	474	5 VOLT REFERENCE (EVA)
47	--	--	NOT USED

2003-05 4.8L, 5.3L, 6.0L Engine PCM C1 Connector Pinout (1 of 2)

2003-05 C1 CONNECTOR (CONTINUED)

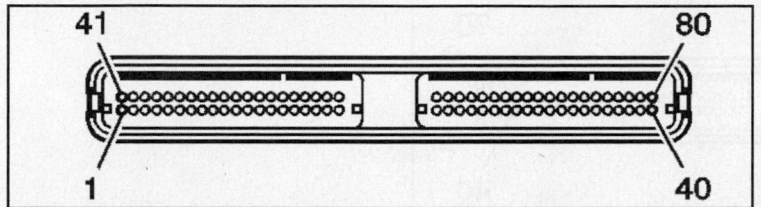

PIN	WIRE COLOR	CIRCUIT NO.	FUNCTION
48	GRY	597	5 VOLT REFERENCE
49-50	--	--	NOT USED
51	DK BLU	496	KNOCK SENSOR 1 SIGNAL
52	--	--	NOT USED
53	GRY	720	LOW REFERENCE (A/T)
54	ORN/BLK	469	LOW REFERENCE
55	--	--	NOT USED
56	WHT	1579	FUEL TEMPERATURE/COMPOSITION SIGNAL (L59)
57	ORN	440	BATTERY POSITIVE VOLTAGE
58	DK GRN	1049	CLASS 2 SERIAL DATA
59	YEL	710	CLASS 2 SERIAL DATA
60	--	--	NOT USED
61	PNK/BLK	632	LOW REFERENCE
62	--	--	NOT USED
63	BLK	2755	LOW REFERENCE
64	BLK/WHT	451	GROUND
65	PPL	1670	HO2S HIGH SIGNAL - BANK 2 SENSOR 2
66	PPL	1666	HO2S HIGH SIGNAL - BANK 2 SENSOR 1
67	BLK/WHT	451	GROUND
68	PPL/WHT	1668	HO2S HIGH SIGNAL - BANK 1 SENSOR 2
69	PPL/WHT	1665	HO2S HIGH SIGNAL - BANK 1 SENSOR 1
70	BRN	1174	OIL LEVEL SWITCH SIGNAL
71	--	--	NOT USED
72	YEL	772	TRANSMISSION RANGE SWITCH SIGNAL B (A/T)
73	BRN/WHT	633	CMP SENSOR SIGNAL
74	YEL	410	ECT SENSOR SIGNAL
75	PNK	1020	OFF/RUN/CRANK VOLTAGE
76	BLK/WHT	845	FUEL INJECTOR 5 CONTROL
77	DK BLU/WHT	878	FUEL INJECTOR 8 CONTROL
78	--	--	NOT USED
79	WHT	687	3-2 SHIFT SOLENOID VALVE CONTROL (M30/M32)
80	BLK	407	LOW REFERENCE

2003-05 4.8L, 5.3L, 6.0L Engine PCM C1 Connector Pinout (2 of 2)

2003-05 C2 CONNECTOR

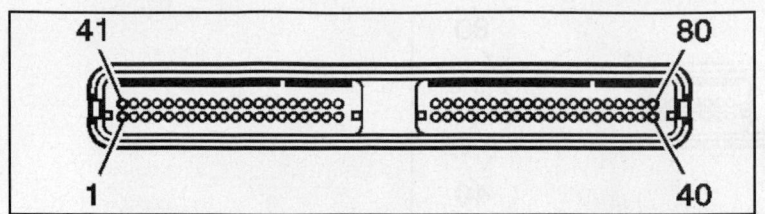

PIN	WIRE COLOR	CIRCUIT NO.	FUNCTION
1	BLK/WHT	451	GROUND
2	BRN	418	TCC PWM SOLENOID VALVE CONTROL (A/T)
3	TAN	1465	FUEL PUMP RELAY CONTROL - SECONDARY (DUAL TANKS)
4	--	--	NOT USED
5	TAN/BLK	464	DELIVERED TORQUE SIGNAL (NW7)
6	RED/BLK	1228	PC SOLENOID VALVE HIGH CONTROL (A/T)
7	--	--	NOT USED
8	LT BLU/WHT	1229	PC SOLENOID VALVE LOW CONTROL (A/T)
9	DK GRN/WHT	465	FUEL PUMP RELAY CONTROL - PRIMARY
10	WHT	121	ENGINE SPEED SIGNAL
11-13	--	--	NOT USED
14	RED/BLK	380	A/C REFRIGERANT PRESSURE SENSOR SIGNAL (A/C)
15	BRN	25	CHARGE INDICATOR CONTROL
16	GRY/BLK	1694	4WD LOW SIGNAL
17	--	--	NOT USED
18	DK GRN	1433	CLUTCH START SWITCH SIGNAL (M/T)
19	BLK/WHT	1695	AXLE SWITCH SIGNAL (NP2)
20	LT GRN/BLK	822	VSS LOW SIGNAL
21	PPL/WHT	821	VSS HIGH SIGNAL
22	RED/BLK	1230	AT ISS HIGH SIGNAL (4WD W/MT1)
23	DK BLU/WHT	1231	AT ISS LOW SIGNAL (4WD W/MT1)
24	--	--	NOT USED
25	TAN	472	IAT SENSOR SIGNAL
26	PPL	2121	IC 1 CONTROL
27	RED	2127	IC 7 CONTROL
28	LT BLU/WHT	2126	IC 6 CONTROL
29	DK GRN/WHT	2124	IC 4 CONTROL
30	--	--	NOT USED
31	YEL	492	MAF SENSOR SIGNAL
32	LT GRN	432	MAP SENSOR SIGNAL
33	--	--	NOT USED
34	DK GRN/WHT	428	EVAP CANISTER PURGE SOLENOID CONTROL
35-38	--	--	NOT USED
39	RED	631	12 VOLT REFERENCE
40	BLK/WHT	451	GROUND
41	--	--	NOT USED
42	TAN/BLK	422	TCC SOLENOID VALVE CONTROL (M30/M32)
43	DK GRN/WHT	459	A/C COMPRESSOR CLUTCH RELAY CONTROL (A/C)
44	--	--	NOT USED
45	WHT	1310	EVAP CANISTER VENT SOLENOID CONTROL (EVA)
46	BRN/WHT	419	MIL CONTROL
47	YEL/BLK	1223	2-3 SHIFT SOLENOID VALVE CONTROL (A/T)

2003-05 4.8L, 5.3L, 6.0L Engine PCM C2 Connector Pinout (1 of 2)

2003-05 C2 CONNECTOR (CONTINUED)

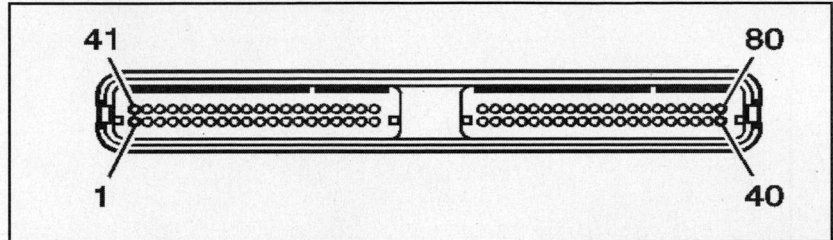

PIN	WIRE COLOR	CIRCUIT NO.	FUNCTION
48	LT GRN	1222	1-2 SHIFT SOLENOID VALVE CONTROL (A/T)
49	YEL/BLK	1827	VEHICLE SPEED SIGNAL
50	DK GRN/WHT	817	VEHICLE SPEED SIGNAL
51	YEL/BLK	1227	TFT SENSOR SIGNAL (A/T)
52	BRN	2391	HO2S HEATER LOW CONTROL BANK 1 SENSOR 2
53	RED/WHT	3223	HO2S HEATER LOW CONTROL BANK 2 SENSOR 2
54	PPL	1589	FUEL LEVEL SENSOR SIGNAL - PRIMARY
55-56	--	--	NOT USED
57	BLK	552	LOW REFERENCE
58	TAN/WHT	332	OIL PRESSURE SENSOR SIGNAL
59	PPL	806	CRANK VOLTAGE
60	BRN	2129	LOW REFERENCE
61	BRN/WHT	2130	LOW REFERENCE
62	GRY	773	TRANSMISSION RANGE SWITCH SIGNAL C (A/T)
63	PNK	1224	TRANSMISSION FLUID PRESSURE SWITCH SIGNAL A (A/T)
64	DK GRN	890	FUEL TANK PRESSURE SENSOR SIGNAL (EVA)
65	--	--	NOT USED
66	PPL/WHT	2128	IC 8 CONTROL
67	RED/WHT	2122	IC 2 CONTROL
68	DK GRN	2125	IC 5 CONTROL
69	LT BLU	2123	IC 3 CONTROL
70-71	--	--	NOT USED
72	BLK/WHT	3113	HO2S HEATER LOW CONTROL BANK 1 SENSOR 1
73	DK BLU	1936	FUEL LEVEL SENSOR SIGNAL - SECONDARY
74	LT GRN	3212	HO2S HEATER LOW CONTROL BANK 2 SENSOR 1
75	GRY	23	GENERATOR FIELD DUTY CYCLE SIGNAL
76-79	--	--	NOT USED
80	BLK	2751	LOW REFERENCE (A/C)

2003-05 4.8L, 5.3L, 6.0L Engine PCM C2 Connector Pinout (2 of 2)

SECONDARY AIR INJECTION REACTION (AIR) BYPASS VALVE

Connector Pinouts

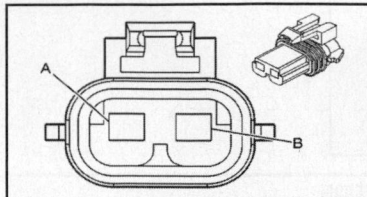

PIN	WIRE COLOR	CIRCUIT NO.	FUNCTION
A	RED	78	AIR INJECTION REACTION PUMP MOTOR FEED
B	BLK	550	GROUND

2000-2002 4.8L, 5.3L, 6.0L Engines AIR Bypass Valve Connector Pinout

THROTTLE POSITION (TP) SENSOR CONNECTOR

Component Locations

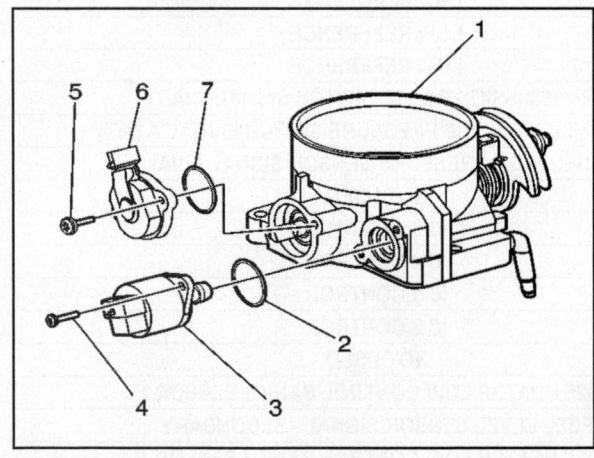

2000-2003 4.8L, 5.3L, 6.0L TP sensor (6) and IAC motor (3)

Connector Pinouts

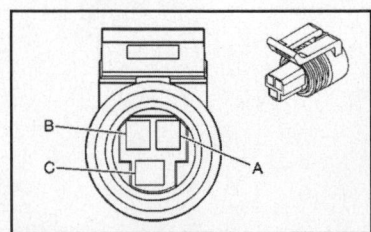

PIN	WIRE COLOR	CIRCUIT NO.	FUNCTION
A	GRY	416	REFERENCE VOLTAGE FEED (5 VOLTS)
B	BLK	452	THROTTLE POSITION (TP) SENSOR RETURN
C	DK BLU	417	THROTTLE POSITION (TP) SENSOR SIGNAL

TP Sensor Connector Pinout On 2000-2002 4.8L, 5.3L, 6.0L Engines

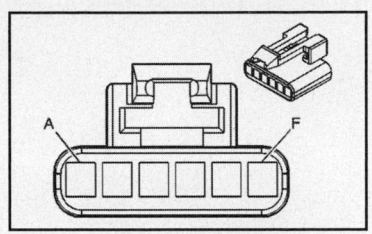

PIN	WIRE COLOR	CIRCUIT NO.	FUNCTION
A	GRY	416	REFERENCE VOLTAGE FEED (5 VOLTS)
B	BLK	452	THROTTLE POSITION (TP) SENSOR # 1 RETURN
C	DK GRN	485	THROTTLE POSITION (TP) SENSOR SIGNAL # 1
D	LT BLK	1688	DUAL TRACK POSITION SENSOR REFERENCE VOLTAGE FEED - 5 VOLT
E	BLK/WHT	1704	THROTTLE POSITION (TP) SENSOR # 2 RETURN
F	PPL	486	THROTTLE POSITION (TP) SENSOR SIGNAL # 2

TP Sensor Connector (dual track) Pinout on 2000-2002 4.8L, 5.3L, 6.0L Engines

VEHICLE SPEED SENSOR (VSS) CONNECTOR

Component Locations

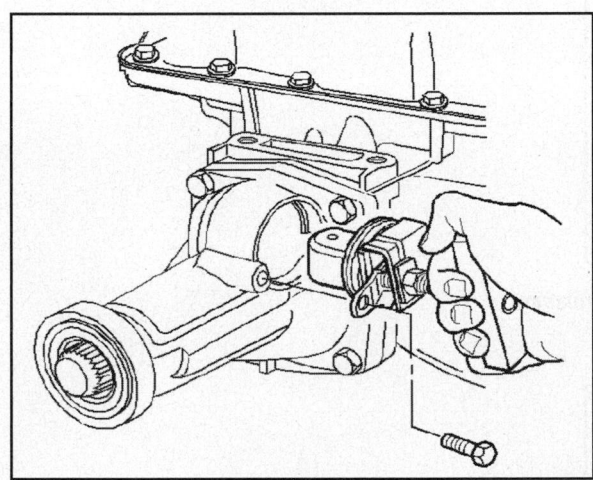

Speed sensor location on 4L60-E/4L65-E automatic transmission

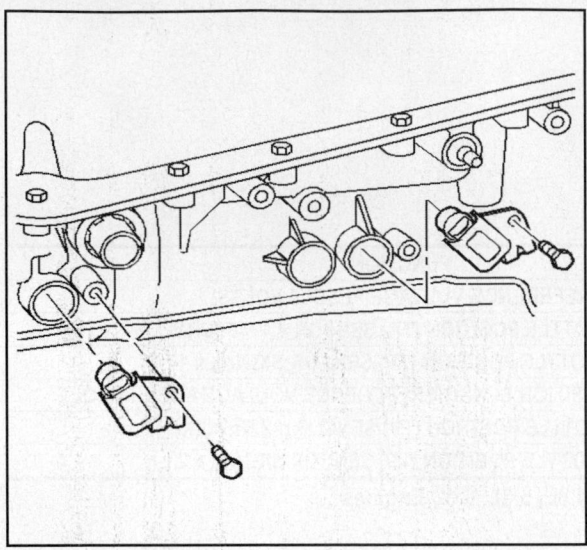

Speed sensors location on 4L80-E/4L85-E automatic transmission

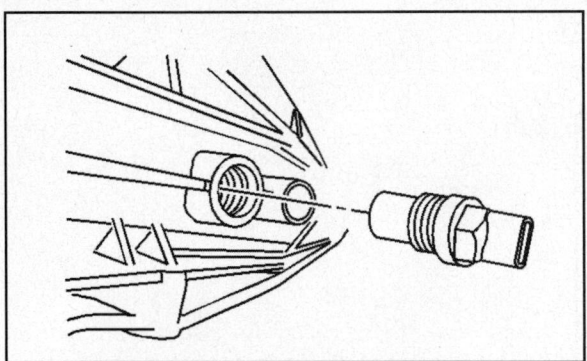

Vehicle speed sensor location on NV3500 manual transmission

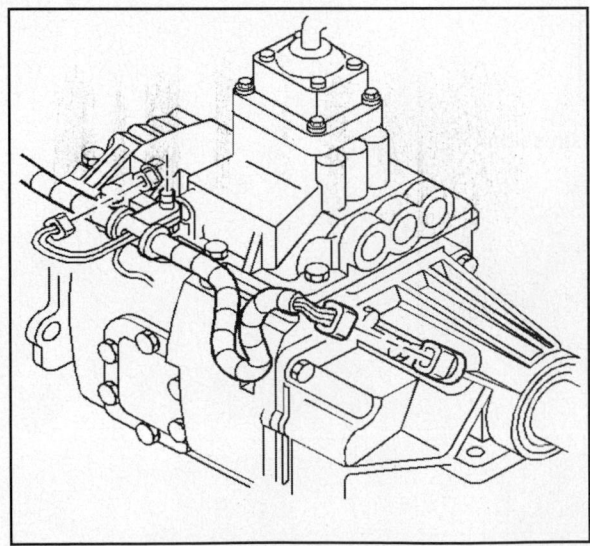

Vehicle speed sensor location on NV4500 manual transmission

Connector Pinouts

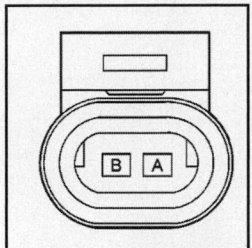

PIN	WIRE COLOR	CIRCUIT NO.	FUNCTION
A	PPL/WHT	821	VEHICLE SPEED SENSOR (VSS) SIGNAL
B	LT GRN/BLK	822	VEHICLE SPEED SENSOR (VSS) RETURN

VSS Connector Pinout on 2000-2005 (All Transmissions)

Contents

What To Do When There Are No DTCs ....Page 5-2
    Driveability Symptom Index Table ....Page 5-2
Symptom Diagnosis Tests ....Page 5-3
    Test 1: No Start, Hard Start Condition ....Page 5-3
    Test 2: Rough, Low or High Idle Speed Condition ....Page 5-4
    Test 3: Runs Rough Condition ....Page 5-6
    Test 4: Cuts-out or Misses Condition ....Page 5-8
    Test 5: Surge Condition ....Page 5-10
Intermittent Fault Tests ....Page 5-11
    Test for Loose Connectors ....Page 5-11
    The Wiggle Test ....Page 5-12
Other Diagnosis & Testing ....Page 5-12
    Vehicle Does Not Fill ....Page 5-12

WHAT TO DO WHEN THERE ARE NO DTCS

Do not attempt to diagnose a Drivability Symptoms without having a logical plan to use to determine which Engine Control system is the cause of the symptom - this plan should include a way to determine which systems do not have a problem! *Drivability symptom diagnosis is a part of an organized approach to problem solving and repair.*

Drivability Symptom Index Table

To use this list, locate the symptom that matches a particular problem and refer to the areas to test. The items listed under each symptom may not apply to all models, engines or vehicle systems. The repair steps indicate what vehicle component or system to test.

Note: *The Drivability Symptoms in this list are intended to be generic. While they apply to most vehicles, some vehicles may not have all of the components listed. Refer to other Chilton repair manuals and electronic media for specific tests.*

Symptom Test Table

Symptom Description	Suggested Areas to Test
Test 1 - No Start, Hard Start Condition ● No Crank ● Hard Start, Long Crank, Erratic Crank ● Stall After Start ● No Start, Normal Crank ● No Start, MIL is off (if the VREF shorts to ground)	- Check battery, battery circuits to starter - Check for a damaged flywheel, engine compression, base timing and minimum air rate - Check for a failed fuel pump relay - Check for distributor rotor "punch-through" - Check for a faulty ignition control module (ICM) - Check for a VREF circuit shorted to ground - Check SKIM (security system) with a Scan Tool
Test 2 - Rough Idle or Stalls Condition ● Low or slow idle speed ● Fast idle speed ● Hunting or rolling idle speed ● Slow return to idle speed ● Stalls or almost stalls	- Check for engine vacuum leaks - Check the condition of the PCV valve and lines - Check for excessive carbon buildup - Check for a restricted exhaust (in Section 2) - Check base idle speed, check for low fuel pressure - Check the throttle linkage for sticking or binding
Test 3 - Runs Rough Condition ● At idle speed ● During acceleration ● At cruise speed ● During deceleration	- Check for engine vacuum leaks at intake manifold - Check condition of ignition secondary components - Check base timing and idle speed settings - Check for low or high fuel pressure - Check for dirty, leaking or shorted fuel injectors - Check for excessive carbon buildup on valves
Test 4 - Cuts-out, Misses Condition ● At idle speed ● During acceleration ● At cruise speed ● During deceleration	- Check for engine vacuum leaks at intake manifold - Check condition of ignition secondary components - Check that spark timing advance is available - Check for low or high fuel pressure - Check for dirty, leaking or shorted fuel injectors - Check for excessive carbon buildup on valves
Test 5 - Bucks, Jerks Condition ● During acceleration ● At cruise speed ● During deceleration	- Check for engine vacuum leaks at intake manifold - Check condition of ignition secondary components - Check that spark timing advance is available - Check for low or high fuel pressure - Check for dirty, leaking or shorted fuel injectors - Check operation of the TCC solenoid, brake switch

SYMPTOM DIAGNOSIS TESTS

Test 1: No Start, Hard Start Condition

Note: *If there is no spark output or fuel pressure available, check for a failed fuel pump relay, no power to the PCM, or loss of the ignition reference signal to the PCM.*

PRELIMINARY CHECKS

Prior to starting this symptom test routine, inspect these underhood items:

1. Check battery charge and condition, starter current draw.
2. Verify the starter relay operation and that the engine cranks (turns over).
3. Verify the check engine light (MIL) operation - if it does not activate, check the PCM power and ground circuits, and check for 5v supply at the MAP or TP sensor.
4. Check Air Intake system for restrictions (inspect air inlet tubes, air filter for dirt, etc.).
5. Check the status of the Smart Key Immobilizer System (SKIM) with the Scan Tool.

Test 1 Chart

Step	Action	Yes	No
1	Step Description: No Start Condition Only » Check battery cables, state of charge. » If the engine does not rotate, inspect for a locked engine (hydrostatic lockup condition). » Does the engine crank normally?	Go to Step 2.	Repair the fault in the battery, starter, or Base Engine. Retest for the symptom when all repairs are done.
2	Step Description: Check the Fuel System » Verify that the pump operates at key on. » Check the fuel pump relay operation. If the relay does not operate, check for blown fuse. » Inspect pump for a leak-down condition » Test fuel pressure, volume and quality. » Test the operation of the fuel regulator. » Are there any faults in the Fuel system?	Make needed repairs. Fuel Pressure Gauge / Fuel Rail Test Port	Go to Step 3.
3	Step Description: Check the Ignition System » Inspect ignition secondary components for damage (look for rotor "punch-through"). » Inspect the coils for signs of spark leakage at coil towers or primary connections. » Check the spark output with a spark tester. » Test Ignition system with an engine analyzer. » Are there any faults in the Ignition system?	Make repairs to the Ignition system. Then retest the symptom. CABLE / SPARK TESTER	Go to Step 4.
4	Step Description: Check the Exhaust System » Check Exhaust system for leaks or damage. » Check the Exhaust system for a restriction using the Vacuum or Pressure Gauge Test (e.g., exhaust backpressure reading should not exceed 1.5 psi at cruise speeds). » Are there any faults in the Exhaust system?	Make repairs to the Exhaust system. Then retest the symptom. **Inspect for Damage** 	Go to Step 5.
5	Step Description: Check the MAP Sensor » Disconnect the MAP sensor and attempt to start the engine. » Does the engine start and run normally?	Replace the MAP sensor. Retest for the symptom when repairs are completed.	Go to Step 6.

Test 1: No Start, Hard Start Condition (Continued)

Test 1 Chart (Continued)

Step	Action (Hard Start Only)	Yes	No
6	Step Description: Check for a Hot Engine » Check for signs of an engine overheating condition related to a Hard Start Symptom. » Does the engine appear to be overheated?	Make the repairs to correct the hot engine and then retest for the symptom when done.	Go to Step 7.
7	Step Description: Check ECT Sensor PID » Connect a Scan Tool and turn the key to on. » Read the ECT sensor (compare to chart). » Has the ECT sensor shifted out of range?	Replace the ECT sensor. Then retest for the symptom when all repairs are completed.	Go to Step 8.
8	Step Description: Check the PCV System » Inspect the PCV system components for broken parts or loose connections. » Test the operation of the PCV valve. » Are there any faults in the PCV system?	Repair the PCV system. Refer to the PCV system tests in this manual. Retest the symptom when all repairs are done.	Go to Step 9.
9	Step Description: Check the EVAP System » Inspect for damaged or disconnected EVAP system components. » Inspect for a fuel saturated charcoal canister. » Are there any faults in the EVAP system?	Refer to the EVAP system tests in this manual. Retest for the symptom when all repairs are completed.	Go to Step 10.
10	Step Description: Test the Base Engine » Check the engine compression. » Test valve timing and timing chain condition. » Check for a worn camshaft or valve train. » Check for any large intake manifold leaks. » Are there any faults in the Base Engine?	Repair the Base Engine. Refer to the Base Engine Tests in this manual. Retest symptom when done.	Return to Step 2 to repeat the test steps in this series to locate and repair the "No Start, Hard Start" condition.

Test 2: Rough, Low or High Idle Speed Condition

Note: *If the vehicle has a rough idle and the base timing, idle speed and the IAC (or AIS) motor operates properly, check the engine for excessive carbon buildup.*

PRELIMINARY CHECKS
Prior to starting this symptom test routine, inspect these underhood items:

1. All related vacuum lines for proper routing and integrity.
2. All related electrical connectors and wiring harnesses for faults (Wiggle Test).
3. Check the throttle linkage for a sticking or binding condition.
4. Air Intake system for restrictions (air inlet tubes, dirty air filter, etc.).
5. Search for any technical service bulletins related to this symptom.
6. Turn the key to off. Unplug the MAP sensor connection and restart the engine to recheck for the idle concern. If the condition is gone, replace the MAP sensor.

Test 2: Rough, Low or High Idle Speed Condition (Continued)

Test 2 Chart

Step	Action	Yes	No
1	Step Description: Verify the rough idle or stall » Does the engine have a warm engine rough idle, low idle or high idle condition in P or N?	Go to Step 2.	Fault is intermittent. Return to the Symptom List and select another fault.
2	Step Description: Verify idle speed & timing » Verify the base timing is within specifications » Verify that the base idle speed is set properly » Are the timing and idle speed set properly?	Go to Step 3.	Set the base idle speed and timing to the specifications and then retest for the symptom.
3	Step Description: Check AIS / IAC Operation » Check the AIS or IAC motor operation » Inspect the AIS/IAC housing in throttle body for restricted passages. Clean as needed. » Set the parking brake, block the drive wheels and turn the A/C off. Install the Scan Tool. » IAC Motor Tester - Turn the key off and then connect the IAC tester to the IAC valve. » Start the engine and use the IAC tester to extend and retract the IAC valve. » ATM Test - Start the engine. Use the tool to change the speed from min-idle to 1500 rpm. » Did the idle speed change as commanded?	Install an Aftermarket Noid light and check the operation of the PCM and AIS or IAC motor circuits. Check the motor for signs of open or shorted circuits. Replace the IAC motor or PCM as needed or make repairs to the IAC motor wiring. If all are okay, go to Step 4.	If the AIS/IAC motor passages are clean and engine speed did not change as described when the AIS/IAC motor was extended and retracted, replace the AIS/IAC motor. Then retest for the condition.
4	Step Description: Check/compare PID values » Connect Scan Tool & turn off all accessories. » Start the engine and allow it to fully warmup. » Monitor all related PIDs on the Scan Tool. » Verify the P/N switch input in gear and Park. » Check the O2S operation with a Lab Scope. » Are all PIDs within normal range?	Go to Step 5. Note: An IAC motor count of over 80 indicates the pintle is extended and an IAC count of (0) indicates the pintle is retracted.	One or more of the PIDs are out of range when compared to "known good" values. Make repairs to the system that is out of range, then retest for the symptom.
5	Step Description: Check the Ignition System » Inspect the coils for signs of spark leakage at coil towers or primary connections. » Check the spark output with a spark tester. » Test Ignition system with an engine analyzer. » Were any faults found in the Ignition system?	Make repairs as needed 	Go to Step 6.
6	Step Description: Check the Fuel System » Inspect the Fuel delivery system for leaks. » Test the fuel pressure, quality and volume. » Test the operation of the pressure regulator. » Were any faults found in the Fuel system?	Make repairs as needed Fuel Pressure Gauge Fuel Rail Test Port	Go to Step 7.
7	Step Description: Check the Exhaust System » Check Exhaust system for leaks or damage. » Check the Exhaust system for a restriction using the Vacuum or Pressure Gauge Test (e.g., exhaust backpressure reading should not exceed 1.5 psi at cruise speeds). » Were any faults found in Exhaust System?	Make repairs to the Exhaust system. Then retest the symptom. Inspect for Damage 	Go to Step 8.

Rough, Low or High Idle Speed Condition (Continued)

Test 2 Chart (Continued)

Step	Action	Yes	No
8	Step Description: Check the PCV System » Inspect the PCV system components for broken parts or loose connections. » Test the operation of the PCV valve. » Were any faults found in the PCV system?	Make repairs to the PCV system. Refer to the PCV system tests in this manual. Then retest for the condition.	Go to Step 9.
9	Step Description: Check the EVAP System » Inspect for damaged or disconnected EVAP system components or a saturated canister. » Were any faults found in the EVAP system?	Make repairs to EVAP system (use the EVAP tests in this manual). Retest for the condition.	Go to Step 10.
10	Step Description: Check the Base Engine » Test the engine compression. » Test valve timing and timing chain condition. » Check for a worn camshaft or valve train. » Check for any large intake manifold leaks. » Were any faults found in the Base Engine?	Make repairs as needed to the Base Engine. Refer to the Base Engine tests in this manual. Then retest for the condition when repairs are completed.	Go to Step 2 and repeat the tests from the beginning to locate and repair the cause of the "Rough, Low or High Idle Speed" condition.

Test 3: Runs Rough Condition

PRELIMINARY CHECKS

Prior to starting this symptom test routine, inspect these underhood items:

1. All related vacuum lines for proper routing and integrity
2. Air Intake system for restrictions (air inlet tubes, dirty air filter, etc.)
3. Search for any technical service bulletins related to this symptom.

Test 3 Chart

Step	Action	Yes	No
1	Step Description: Verify engine runs rough » Start the engine and allow it to idle in P or N. » Does the engine run rough when warm in Park or Neutral position?	Check for any stored codes. If codes are set, repair codes and retest. If no codes are set, go to Step 3.	Go to Step 2.
2	Step Description: Condition does not exist! » Inspect various underhood items that could cause an intermittent Runs Rough condition (i.e., dirt in the throttle body, vacuum leaks, IAC motor connections, etc.). » Were any problems located in this step?	Correct the problems. Do a PCM reset and engine "idle relearn" procedure. Then verify the "runs rough" condition is repaired.	The problem is not present at this time. It may be an intermittent problem.
3	Step Description: Check/compare PID values » Connect a Scan Tool to the test connector. » Turn off all accessories. » Start the engine and allow it to fully warmup. » Monitor all related PIDs on the Scan Tool. » Were all PIDs within their normal range?	Go to Step 4. Note: The IAC motor should read from 5-50 counts. Check the LONGFT reading for a large shift into the negative range (due to a rich condition).	One or more of the PIDs are out of range when compared to "known good" values. Make repairs to the system that is out of range, then retest for the symptom.

Test 3: Runs Rough Condition (Continued)

Test 3 Chart (Continued)

Step	Action	Yes	No
4	Step Description: Check the Ignition System » Inspect the coils for signs of spark leakage at coil towers or primary connections. » Check the spark output with a spark tester. » Test Ignition system with an engine analyzer. » Were any faults found in the Ignition system?	Make repairs as needed	Go to Step 5.
5	Step Description: Check the Fuel System » Inspect the Fuel delivery system for leaks. » Test the fuel pressure, quality and volume. » Test the operation of the pressure regulator. » Were any faults found in the Fuel system?	Make repairs as needed	Go to Step 6.
6	Step Description: Check the Exhaust System » Check Exhaust system for leaks or damage. » Check the Exhaust system for a restriction using the Vacuum or Pressure Gauge Test (e.g., exhaust backpressure reading should not exceed 1.5 psi at cruise speeds). » Were any faults found in Exhaust System?	Make repairs to the Exhaust system. Then retest the symptom.	Go to Step 7.
7	Step Description: Check the PCV System » Inspect the PCV system components for broken parts or loose connections. » Test the operation of the PCV valve. » Were any faults found in the PCV system?	Make repairs to the PCV system. Refer to the PCV system tests in this manual. Then retest for the condition.	Go to Step 9.
8	Step Description: Check the EVAP System » Inspect for damaged or disconnected EVAP system components or a saturated canister. » Were any faults found in the EVAP system?	Make repairs to EVAP system (use the EVAP tests in this manual). Retest for the condition.	Go to Step 10.
9	Step Description: Check Engine Condition » Test the engine compression. » Test valve timing and timing chain condition. » Check for a worn camshaft or valve train. » Check for any large intake manifold leaks. » Were any faults found in the Base Engine?	Make repairs as needed to the Base Engine. Refer to the Base Engine tests in this manual. Then retest for the condition when repairs are completed.	Return to Step 2 and repeat the tests from the beginning to locate and repair the cause of the "Runs Rough" condition.

Example EVAP System Graphic

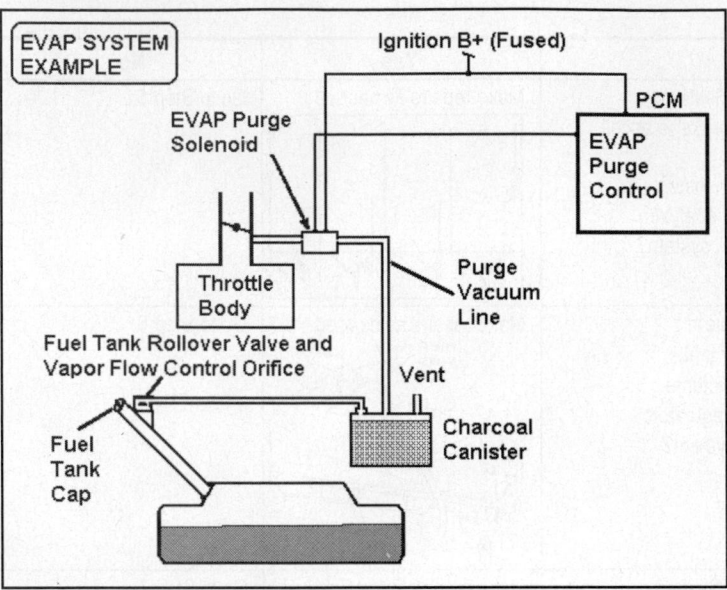

Test 4: Cuts-out or Misses Condition

PRELIMINARY CHECKS

Prior to starting this symptom test routine, inspect these underhood items:

1. All related vacuum lines for proper routing and integrity
2. Search for any technical service bulletins related to this symptom.

Test 4 Chart

Step	Action	Yes	No
1	Step Description: Verify Cuts-out condition » Start the engine and attempt to verify the Cuts-out or misses condition. » Does the engine have a cuts-out condition?	Check for any stored codes. If codes are set, repair codes and retest. If no codes are set, go to Step 3.	Go to Step 2.
2	Step Description: Condition does not exist! » Inspect various underhood items that could cause an intermittent Cuts-out condition (i.e., EVAP, Fuel or Ignition system components). » Were any problems located in this step?	Correct the problems. Do a PCM reset and "Fuel Trim Relearn" procedure. Then verify condition is repaired.	The problem is not present at this time. It may be an intermittent problem.
3	Step Description: Check/compare PID values » Connect a Scan Tool to the test connector. » Turn off all accessories. » Start the engine and allow it to fully warmup. » Monitor all related PIDs on the Scan Tool (i.e., ECT IAC Counts and LONGFT at idle). » Were all PIDs within their normal range?	Go to Step 4. Note: The IAC motor should be from 5-50 counts. Watch fuel trim (%) for a large shift into the negative (-) range (due to a rich condition).	One or more of the PIDs are out of range when compared to "known good" values. Make repairs to the system that is out of range, then retest for the symptom.

Test 4: Cuts Out or Misses Condition (Continued)

Test 4 Chart (Continued)

Step	Action	Yes	No
4	Step Description: Check the Ignition System » Inspect the coils for signs of spark leakage at coil towers or primary connections. » Check the spark output with a spark tester. » Test Ignition system with an engine analyzer. » Were any faults found in the Ignition system?	Make repairs as needed	Go to Step 5.
5	Step Description: Check the Fuel System » Inspect the Fuel delivery system for leaks. » Test the fuel pressure, quality and volume. » Test the operation of the pressure regulator. » Were any faults found in the Fuel system?	Make repairs as needed Fuel Pressure Gauge Fuel Rail Test Port	Go to Step 6.
6	Step Description: Check the Exhaust System » Check Exhaust system for leaks or damage. » Check the Exhaust system for a restriction using the Vacuum or Pressure Gauge Test (e.g., exhaust backpressure reading should not exceed 1.5 psi at cruise speeds). » Were any faults found in Exhaust System?	Make repairs to the Exhaust system. Then retest the symptom. Inspect for Damage	Go to Step 7.
7	Step Description: Check the PCV System » Inspect the PCV system components for broken parts or loose connections. » Test the operation of the PCV valve. » Were any faults found in the PCV system?	Make repairs to the PCV system. Refer to the PCV system tests in this manual. Then retest for the condition.	Go to Step 8.
8	Step Description: Check the EVAP System » Inspect for damaged or disconnected EVAP system components » Check for a saturated EVAP canister. » Were any faults found in the EVAP system?	Make repairs to EVAP system (use the EVAP tests in this manual). Retest for the condition.	Go to Step 9.
9	Step Description: Check the AIR system » Inspect AIR system for broken parts, leaking valves or disconnected hoses (see graphic). » Test the operation of Secondary AIR system. » Were any faults found in the AIR system?	Make repairs as needed. Refer to the Secondary AIR system tests in this manual. Retest for the condition.	Go to Step 10.
10	Step Description: Check Engine Condition » Test the engine compression. » Test valve timing and timing chain condition. » Check for a worn camshaft or valve train. » Check for any large intake manifold leaks. » Were any faults found in the Base Engine?	Make repairs as needed to the Base Engine. Refer to the Base Engine tests in this manual. Then retest for the condition when repairs are completed.	Go to Step 2 and repeat the tests from the beginning to locate and repair the cause of the "Cuts Out or Misses" condition.

Typical Secondary Air System Graphic

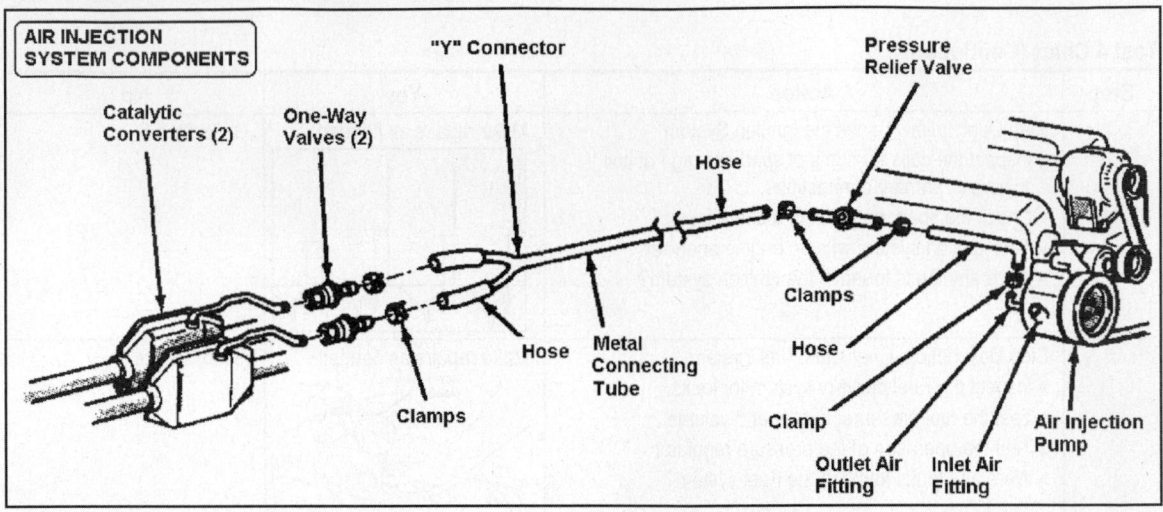

Test 5: Surge Condition

PRELIMINARY CHECKS
1. Discuss how the operation of the torque converter clutch (TCC) or air conditioning compressor can affect the "feel" of the vehicle during normal operation. Refer to the information in the Owner's Manual to explain how these devices normally operate.
2. Search for any technical service bulletins related to this symptom.

Test 5 Chart

Step	Action	Yes	No
1	Step Description: Verify the surge condition » Drive the vehicle and attempt to verify that the vehicle surges at cruise speeds. » Does the engine have a surge condition?	Check for any stored codes. If codes are set, repair codes and retest. If no codes are set, go to Step 3.	Go to Step 2.
2	Step Description: Condition does not exist! » Inspect various underhood items that could cause an intermittent surge condition (check for leaks in the MAP sensor vacuum lines). » Were any problems located in this step?	Correct the problems. Do a PCM reset and "Fuel Trim Relearn" procedure. Then verify condition is repaired.	The problem is not present at this time. It may be an intermittent problem.
3	Step Description: Check/compare PID values » Connect a Scan Tool to the test connector. » Start the engine and allow it to fully warmup. » Monitor all related PIDs on Scan Tool (HO2S switching, LONGFT, and the TCC operation) » Compare VSS PID reading to speedometer. » Were all PIDs within their normal range?	Go to Step 4. Note: Verify that the front HO2S responds quickly to throttle changes. Check for silicon contamination on the front HO2S (this can cause a rich A/F signal).	One or more of the PIDs are out of range when compared to "known good" values. Make repairs to the system that is out of range, then retest for the symptom.

Test 5: Surge Condition (Continued)

Test 5 Chart (Continued)

Step	Action	Yes	No
4	Step Description: Check the Ignition System » Inspect the coils for signs of spark leakage at coil towers or primary connections. » Check the spark output with a spark tester. » Test Ignition system with an engine analyzer. » Were any faults found in the Ignition system?	Make repairs as needed 	Go to Step 5.
5	Step Description: Check the Fuel System » Inspect the Fuel delivery system for leaks. » Test the fuel pressure, quality and volume. » Test the operation of the pressure regulator. » Were any faults found in the Fuel system?	Make repairs as needed Fuel Pressure Gauge Fuel Rail Test Port	Go to Step 6.
6	Step Description: Check the Exhaust System » Check Exhaust system for leaks or damage. » Check the Exhaust system for a restriction using the Vacuum or Pressure Gauge Test (e.g., exhaust backpressure reading should not exceed 1.5 psi at cruise speeds). » Were any faults found in Exhaust System?	Make repairs to the Exhaust system. Then retest the symptom. **Inspect for Damage** 	Return to Step 2 and repeat the tests from the beginning to locate and repair the cause of the "Surge" condition.

INTERMITTENT FAULT TESTS

Many trouble code repair charts end with a result that reads "Fault Not Present at this Time." What this expression means is that the conditions that were present when a code set or drivability symptom occurred are no longer there or were not met. In effect, the problem was present at least once, but is not present at this time. However, it is likely to return in the future, so it should be diagnosed and repaired if at all possible.

One way to find an intermittent problem is to gather the information that was present when the problem occurred. In the case of a Code Fault, this can be done in two ways: by capturing the data in Snapshot or Movie mode or by driver observations.

The PCM has to detect the fault for a specific period of time before a trouble code will set. While intermittent problems may appear to be occasional in nature, they usually occur under specific conditions. Therefore, you should identify and duplicate these conditions. Since intermittent faults are difficult to duplicate, a logical routine (checklist) must be followed when attempting to find the faulty component, system or circuit. The tests on the next page can be used to help find the cause of an intermittent fault.

Some intermittent faults occur due to a loose connection, wiring problem or warped circuit board. An intermittent fault can also be caused by poor test techniques that cause damage to the male or female ends of a connector.

Test for Loose Connectors
To test for a loose or damaged connection, take the male end of a connector from another wiring harness and carefully push it into the "suspect" female terminal to verify that the opening is tight. There should be some resistance felt as the male connector is inserted in the terminal connection.

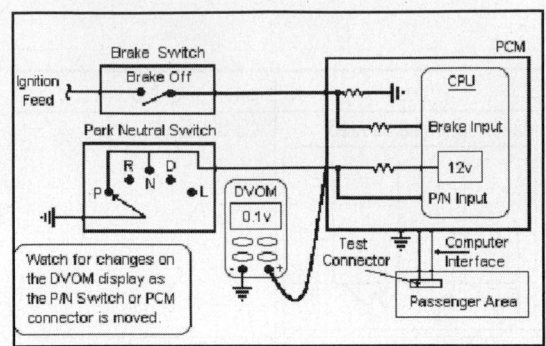
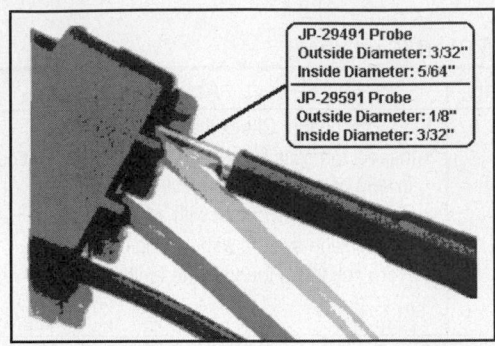

The Wiggle Test

A wiggle test can be used to locate the cause of some intermittent faults. The sensor, switch or the PCM wiring can be back-probed, as shown, while the test is done.

During testing, move or wiggle the suspect device, connector or wiring while watching for a change.
If the DVOM has a Min/Max record mode, use this mode during the test.

OTHER DIAGNOSIS & TESTING

Vehicle Does Not Fill

Test Chart

CONDITION	POSSIBLE CAUSES	CORRECTION
Pre-Mature Nozzle Shut-Off	Defective fuel tank assembly components.	Fill tube improperly installed (sump)
		Fill tube hose pinched.
		Check valve stuck shut.
		Control valve stuck shut.
	Defective vapor/vent components.	Vent line from control valve to canister pinched.
		Vent line from canister to vent filter pinched.
		Canister vent valve failure (requires double failure, plugged to NVLD and atmosphere).
		Leak detection pump failed closed.
		Leak detection pump filter plugged.
	On-Board diagnostics evaporative system leak test just conducted.	Canister vent valve vent plugged to atmosphere.
		Engine still running when attempting to fill (System designed not to fill).
	Defective fill nozzle.	Try another nozzle.
Fuel Spits Out Of Filler Tube.	During fill.	See Pre-Mature Shut-Off.
	At conclusion of fill.	Defective fuel handling component. (Check valve stuck open).
		Defective vapor/vent handling component.
		Defective fill nozzle.

SECTION 6 CONTENTS

Buick

CENTURY
2.2L I4 MFI VIN 4 **(1996-97)** .... Page 6-6
3.1L V6 MFI VIN M **(1995)** .... Page 6-10
3.1L V6 MFI VIN M **(1996-99)** .... Page 6-11
3.1L V6 MFI VIN J **(2000-05)** .... Page 6-13

LACROSSE
3.6L V6 MFI VIN 7 **(2005)** .... Page 6-17
3.8L V6 MFI VIN 1 **(2005)** .... Page 6-20

LESABRE
3.8L V6 MFI VIN L, VIN 1**(1995)** .... Page 6-23
3.8L V6 MFI VIN K, VIN 1 **(1996-99)** .... Page 6-24
3.8L V6 MFI VIN K, VIN 1 **(2000-05)** .... Page 6-28

PARK AVENUE
3.8L V6 MFI VIN L, VIN 1**(1995)** .... Page 6-23
3.8L V6 MFI VIN K, VIN 1 **(1996-99)** .... Page 6-24
3.8L V6 MFI VIN K, VIN 1 **(2000-05)** .... Page 6-28

REGAL
3.8L V6 MFI VIN L, VIN 1**(1995)** .... Page 6-23
3.8L V6 MFI VIN K, VIN 1 **(1996-99)** .... Page 6-33
3.8L V6 MFI VIN K, VIN 1 **(2000-03)** .... Page 6-37

RENDEZVOUS
3.4L V6 MFI VIN E **(2002-05)** .... Page 6-38

RIVIERA
3.8L V6 MFI VIN L, VIN 1**(1995)** .... Page 6-23
3.8L V6 MFI VIN K, VIN 1 **(1996-99)** .... Page 6-33

ROADMASTER
5.7L V8 MFI VIN P **(1995)** .... Page 6-42
5.7L V8 MFI VIN P **(1996)** .... Page 6-43

SKYLARK
2.4L I4 MFI VIN T **(1996-97)** .... Page 6-47
3.1L V6 MFI VIN M **(1996-98)** .... Page 6-51

TERRAZA
3.5L V6 MFI VIN L **(2005)** .... Page 6-55

Cadillac

CATERA
3.0L V6 MFI VIN R **(1996-99)** .... Page 6-59
3.0L V6 MFI VIN R **(2000-01)** .... Page 6-61

CTS
2.6L V6 MFI VIN N, M/T or A/T **(2003-04)** .... Page 6-63
3.2L V6 MFI VIN M, A/T **(2003-04)** .... Page 6-68

DEVILLE
4.6L V8 MFI VIN Y, VIN 9 **(1996-99)** .... Page 6-73
4.6L V8 MFI VIN Y, VIN 9 **(2000-05)** .... Page 6-80

ELDORADO
4.6L V8 MFI VIN Y, VIN 9 **(1996-99)** .... Page 6-73
4.6L V8 MFI VIN Y, VIN 9 **(2000-02)** .... Page 6-80

ESCALADE, ESCALADE EXT, ESCALADE ESV
5.7L V8 MFI VIN R (1999-2000) ....Page 6-86
5.3L V8 MFI VIN T (2002-05) ....Page 6-90
6.0L V8 MFI VIN N (2002-05) ....Page 6-90

FLEETWOOD, FLEETWOOD BROUGHAM
5.7L V8 MFI VIN P (1996) ....Page 6-95

SEVILLE
4.6L V8 MFI VIN Y, VIN 9 (1996-99) ....Page 6-73
4.6L V8 MFI VIN Y, VIN 9 (2000-04) ....Page 6-80

STS
3.6L V6 MFI VIN 7 (2005) ....Page 6-99
4.6L V8 MFI VIN A (2005) ....Page 6-102

Chevrolet

BERETTA, CORSICA
2.2L I4 MFI VIN 4 (1996) ....Page 6-105
3.1L V6 MFI VIN M (1996) ....Page 6-108

CAMARO
3.8L V6 MFI VIN K (1995-99) ....Page 6-110
3.8L V6 MFI VIN K (2000-02) ....Page 6-114
5.7L V8 MFI VIN P (1995) ....Page 6-119
5.7L V8 MFI VIN P (1996-97) ....Page 6-121
5.7L V8 MFI VIN G (1998-99) ....Page 6-126
5.7L V8 MFI VIN G (2000-02) ....Page 6-132

CAPRICE
4.3L V6 MFI VIN W (1996) ....Page 6-136
5.7L V8 MFI VIN P (1996) ....Page 6-140

CAVALIER
2.2L I4 MFI VIN 4 (1996-2005) ....Page 6-144
2.2L I4 MFI VIN F (2002-05) ....Page 6-144
2.4L I4 MFI VIN T (1996-2005) ....Page 6-144

COBALT
2.2L I4 MFI VIN F (2005) ....Page 6-148
2.0L I4 MFI VIN P (2005) ....Page 6-151

CORVETTE
5.7L V8 MFI VIN P (1995) ....Page 6-154
5.7L V8 MFI VIN P, VIN 5 (1996) ....Page 6-156
5.7L V8 MFI VIN G (1997-99) ....Page 6-160
5.7L V8 MFI VIN G, VIN S (2000-05) ....Page 6-168

IMPALA
3.8L V6 MFI VIN K, VIN 1 (2000-2005) ....Page 6-173

LUMINA & MONTE CARLO
3.1L V6 MFI VIN M (1995) ....Page 6-178
3.4L V6 MFI VIN X (1995) ....Page 6-179
3.1L V6 MFI VIN M (1996-99) ....Page 6-180
3.4L V6 MFI VIN X (1996-97) ....Page 6-186
3.8L V6 MFI VIN K (1998-99) ....Page 6-189

LUMINA
3.1L V6 MFI VIN J (2000-01) ....Page 6-195

MALIBU
2.4L I4 MFI VIN T (1997-99) ....Page 6-200
3.1L V6 MFI VIN M (1997-99) ....Page 6-202
3.1L V6 MFI VIN J (1999-2005) ....Page 6-207

MONTE CARLO
   3.4L V6 MFI VIN E **(2000-05)** .... Page 6-212
   3.8L V6 MFI VIN K **(2000-05)** .... Page 6-215
UPLANDER
   3.5L V6 MFI VIN L **(2005)** .... Page 6-220
C/K, M/L & S/T SERIES VEHICLES
Chevrolet Models:

Avalanche, Cab & Chassis, Chevy Express, Cutaway, Pickup, RV Cutaway, Silverado, Suburban & Tahoe
GMC Models:

Cab & Chassis, Envoy, Forward Control, Pickup, Rally, Safari, Savana, Sierra, Sonoma, Suburban, Vandura,
Yukon
   4.3L V6 MFI VIN W **(1996-2005)** .... Page 6-224
   4.8L V8 MFI VIN V **(1999-2005)** .... Page 6-228
   5.0L V8 MFI VIN M **(1996-2005)** .... Page 6-232
   5.3L V8 MFI VIN T **(1999-2005)** .... Page 6-228
   5.7L V8 MFI VIN R **(1996-2005)** .... Page 6-232
   6.0L V8 MFI VIN U **(1999-2005)** .... Page 6-228
   6.5L V8 Diesel VIN F, VIN S **(1996-2002)** .... Page 6-236
   6.6L V8 Diesel VIN 1 **(1999-2005)** .... Page 6-237
   7.4L V8 MFI VIN J **(1996-2000)** .... Page 6-240
   8.1L V8 MFI VIN G **(2001-05)** .... Page 6-244
ASTRO & SAFARI
   4.3L V6 MFI VIN W **(1996-2001)** .... Page 6-224
   4.3L V6 MFI VIN X **(2002-05)** .... Page 6-224
BLAZER, JIMMY, SONOMA, S10 PICKUP
   2.2L I4 MFI VIN 4 **(1996-2000)** .... Page 6-249
   2.2L I4 MFI VIN 5 **(2000-04)** .... Page 6-252
   4.3L V6 MFI VIN W **(1996-2004)** .... Page 6-8
   4.3L V6 MFI VIN X **(1996-99)** .... Page 6-8
ENVOY & TRAILBLAZER UTILITY VEHICLE
   4.2L I6 MFI VIN X **(2002-05)** .... Page 6-254
   5.3L V8 MFI VIN T **(1999-2005)** .... Page 6-228
EQUINOX UTILITY VEHICLE
   3.4L V6 MFI VIN F **(2005)** .... Page 6-257
SSR
   6.0L V8 MFI VIN H **(2004-05)** .... Page 6-261
VENTURE
   3.4L V6 MFI VIN E **(1997-2005)** .... Page 6-265
Oldsmobile
ACHIEVA
   2.4L I4 MFI VIN T **(1996-98)** .... Page 6-268
   3.1L V6 MFI VIN M **(1996-98)** .... Page 6-272
ALERO
   2.4L I4 MFI VIN T **(1999-2004)** .... Page 6-276
   3.4L V6 MFI VIN E **(1999-2004)** .... Page 6-281
AURORA
   3.5L V6 MFI VIN H **(2001-03)** .... Page 6-285
   4.0L V8 MFI VIN C **(1996-99)** .... Page 6-289
   4.0L V8 MFI VIN C **(2001-03)** .... Page 6-296
CUTLASS
   3.1L V6 MFI VIN M **(1997-99)** .... Page 6-301
   3.1L V6 MFI VIN J **(1999)** .... Page 6-303

CUTLASS CIERA
  2.2L I4 MFI VIN 4 **(1996)** ....Page 6-307
  3.1L V6 MFI VIN M **(1996)** ....Page 6-310

CUTLASS CIERA, CRUISER
  3.1L V6 MFI VIN M **(1995)** ....Page 6-312

CUTLASS SUPREME
  3.1L V6 MFI VIN M **(1995)** ....Page 6-313
  3.1L V6 MFI VIN M **(1996-97)** ....Page 6-314
  3.4L V6 MFI VIN X **(1995)** ....Page 6-316
  3.4L V6 MFI VIN X **(1996)** ....Page 6-317

EIGHTY-EIGHT, NINETY EIGHT
  3.8L V6 MFI VIN K, 1 **(1995)** ....Page 6-320
  3.8L V6 MFI VIN K **(1996-99)** ....Page 6-321

INTRIGUE
  3.8L V6 MFI VIN K **(1998-99)** ....Page 6-326
  3.5L V6 MFI VIN H **(2000-02)** ....Page 6-330

LSS, REGENCY
  3.8L V6 MFI VIN K, 1 **(1996-98)** ....Page 6-334

SILHOUETTE
  3.4L V6 MFI VIN E **(1996-2004)** ....Page 6-338

Pontiac
AZTEK
  3.4L V6 MFI VIN E **(2001-05)** ....Page 6-341

BONNEVILLE
  3.8L V6 MFI VIN K, VIN 1 **(1995)** ....Page 6-344
  3.8L V6 MFI VIN K, VIN 1 **(1996-99)** ....Page 6-345
  3.8L V6 MFI VIN K, VIN 1 **(2000-05)** ....Page 6-347

FIREBIRD
  3.8L V6 MFI VIN K **(1995)** ....Page 6-352
  3.8L V6 MFI VIN K **(1996-99)** ....Page 6-357
  3.8L V6 MFI VIN K **(2000-02)** ....Page 6-362
  5.7L V8 MFI VIN P **(1996-97)** ....Page 6-367
  5.7L V8 MFI VIN G **(1998-2002)** ....Page 6-372

G6
  3.5L V6 MFI VIN 8 **(2005)** ....Page 6-378

GRAND AM
  2.4L I4 MFI VIN T **(1996-99)** ....Page 6-382
  2.2L I4 MFI VIN F **(2002-05)** ....Page 6-386
  2.4L I4 MFI VIN T **(2000-05)** ....Page 6-386
  3.1L V6 MFI VIN M **(1996-98)** ....Page 6-389
  3.4L V6 MFI VIN E **(1999-05)** ....Page 6-394

GRAND PRIX
  3.1L V6 MFI VIN M **(1995)** ....Page 6-401
  3.1L V6 MFI VIN M **(1996-99)** ....Page 6-402
  3.1L V6 MFI VIN J **(2000-05)** ....Page 6-406
  3.4L V6 MFI VIN X **(1995)** ....Page 6-411
  3.4L V6 MFI VIN X **(1996)** ....Page 6-412
  3.8L V6 MFI VIN K, VIN 1 **(1997-2000)** ....Page 6-415
  3.8L V6 MFI VIN K, VIN 1 **(2001-05)** ....Page 6-421

GTO
  5.7L V8 MFI VIN G **(2004)** ....Page 6-426
  6.0L V8 MFI VIN U **(2005)** ....Page 6-428

MONTANA SV6
  3.5L V6 MFI VIN L **(2005)** .... Page 6-432
MONTANA
  3.4L V6 MFI VIN E **(1999-2005)** .... Page 6-436
SUNFIRE
  2.2L I4 MFI VIN 4, VIN F, VIN T **(1996-2005)** .... Page 6-439
TRANS SPORT
  3.4L V6 MFI VIN E **(1996-98)** .... Page 6-436

Saturn

L100 4-DOOR SEDAN
  2.2L I4 MFI VIN F **(2001-02)** .... Page 6-443
L200 4-DOOR SEDAN
  2.2L I4 MFI VIN F **(2001-03)** .... Page 6-443
L300 4-DOOR SEDAN
  3.0L V6 MFI VIN R **(2001 05)** .... Page 0-440
LS 4-DOOR SEDAN
  2.2L I4 MFI VIN F **(2000)** .... Page 6-443
LS1 4-DOOR SEDAN
  2.2L I4 MFI VIN F **(2000)** .... Page 6-443
LS2 4-DOOR SEDAN
  3.0L V6 MFI VIN R **(2000)** .... Page 6-443
LW1 4-DOOR WAGON
  2.2L I4 MFI VIN F **(2000)** .... Page 6-443
LW2 4-DOOR WAGON
  2.2L I4 MFI VIN F **(2000)** .... Page 6-443
LW200 4-DOOR WAGON
  2.2L I4 MFI VIN F **(2001-03)** .... Page 6-443
LW300 4-DOOR WAGON
  3.0L V6 MFI VIN R **(2001-05)** .... Page 6-446
RELAY
  3.5L V6 MFI VIN L **(2005)** .... Page 6-449
SC1 2-DOOR COUPE
  1.9L I4 MFI VIN 8 **(1996-2002)** .... Page 6-453
SC2 2-DOOR COUPE
  1.9L I4 MFI VIN 7 **(1996-2002)** .... Page 6-453
SL 4-DOOR SEDAN
  1.9L I4 MFI VIN 8 **(1996-2002)** .... Page 6-453
SL1 4-DOOR SEDAN
  1.9L I4 MFI VIN 8 **(1996-2002)** .... Page 6-453
SL2 4-DOOR SEDAN
  1.9L I4 MFI VIN 7 **(1996-2002)** .... Page 6-453
SW1 4-DOOR WAGON
  1.9L I4 MFI VIN 8 **(1996-99)** .... Page 6-453
SW2 4-DOOR WAGON
  1.9L I4 MFI VIN 7 **(1996-2001)** .... Page 6-453
VUE 4-DOOR UTILITY
  2.2L I4 MFI VIN F **(2002-05)** .... Page 6-443
  3.0L V6 MFI VIN R **(2002-05)** .... Page 6-446

CENTURY & REGAL PID DATA - OBD II

1996-97 Century 2.2L I4 MFI VIN 4 (A/T) - Engine Data List 1

Parameter Identification (PID)	PID Value Range	PID Value at Hot Idle	PID Value at 30 mph	Data List Type
BARO Sensor (V)	0-5.1 volts	3.5-4.5	3.5-4.5	Engine Data 1
Calculated Airflow	0-512 g/sec	3-7	Varies	Engine Data 1
CKP Active Counter	0-255 counts	Increments	Increments	Engine Data 1
CMP Active Counter	0-255 counts	Increments	Increments	Engine Data 1
CMP Resync Counter	0-255 counts	0	0	Engine Data 1
Desired Idle Speed	0-3187	PCM control	---	Engine Data 1
ECT Sensor (°F)	-40 to 419°F	185-239	185-239	Engine Data 1
Engine Load	0-100%	22-26	Varies	Engine Data 1
Engine Run Time	Hr: Min: Sec	00:00:00	00:00:00	Engine Data 1
Engine Speed	0-9999 rpm	±50 Desired	---	Engine Data 1
EVAP Purge Solenoid	0-100%	0	0-100	Engine Data 1
EVAP Vent Solenoid	On/Off	Off	Off	Engine Data 1
Fuel Pump Relay	On/Off	On	On	Engine Data 1
Fuel Trim Cell	Cell #0-21	18-21	Varies	Engine Data 1
Fuel Trim Index	0-255 counts	58-198	58-198	Engine Data 1
Fuel Trim Index	-10% to 10%	Varies	Varies	Engine Data 1
Generator F-Terminal	0-100%	0-100	0-100	Engine Data 1
Generator L-Terminal	Active/Inactive	Inactive	Inactive	Engine Data 1
HO2S-12 (B1 S2)	0-1132 mv	10-1000	10-1000	Engine Data 1
IAC Motor	0-255 counts	5-60	---	Engine Data 1
IAT Sensor (°F)	-40 to 304°F	50-176	50-176	Engine Data 1
Ignition 1 Signal	0.0-25.5v	14.2	14.2	Engine Data 1
Knock Retard (°)	0-128°	0	0	Engine Data 1
Knock Signal	Yes/No	No	No	Engine Data 1
KS Noise Channel	Yes/No	No	No	Engine Data 1
KS Active Counts	0-255 counts	35-50	35-50	Engine Data 1
Low Octane Spark Modifier	0-90	Varies	Varies	Engine Data 1
Long Term Fuel Trim	-10% to 10%	0 (± 5%)	0 (± 5%)	Engine Data 1
Long Term F/T Average	-10% to 10%	0 (± 5%)	0 (± 5%)	Engine Data 1
Loop Status	Closed Loop / Open Loop	Closed Loop	Closed Loop	Engine Data 1
MAP Sensor (kPa)	10-110 kPa	25-48	Varies	Engine Data 1
Medium Resolution Sync Counter	0-255 counts	0	0	Engine Data 1
Medium Resolution Resync Counter	0-255 counts	0	0	Engine Data 1
Number of DTC(s)	0-255 counts	0	0	Engine Data 1
O2S-11 (B1 S1)	0-1132 mv	10-1000	10-1000	Engine Data 1
Power Enrichment	Active/Inactive	Inactive	Inactive	Engine Data 1
Purge Learn Memory	0-100%	0	0-100	Engine Data 1
Short Term Fuel Trim	-10% to 10%	0 (± 5%)	0 (± 5%)	Engine Data 1
Short Term F/T Average	-10% to 10%	0 (± 5%)	0 (± 5%)	Engine Data 1
Spark Advance (°)	-64° to 64°	Varies	Varies	Engine Data 1
TCC Brake Switch	On/Off	Off	Off	Engine Data 1
TP Angle (%)	0-100%	0%	0-100%	Engine Data 1
TP Sensor (V)	0-5.1 volts	0.20-0.90	Varies	Engine Data 1
Vehicle Speed	0-155 mph	0	30	Engine Data 1

CENTURY PID DATA - OBD II

1996-97 Century 2.2L I4 MFI VIN 4 (A/T) - Engine Data 2 List

Parameter Identification (PID)	PID Value Range	PID Value at Hot Idle	PID Value at 30 mph	Data List Type
A/C High Side (psi)	0-459 psi	139-399	139-399	Engine Data 2
A/C High Side (V)	0-5.1 volts	1-4	1-4	Engine Data 2
A/C Relay	On/Off	Off	Off	Engine Data 2
A/C Request	Yes/No	No	No	Engine Data 2
Air Fuel Ratio	0-25:5	14.7:1	14.7:1	Engine Data 2
BARO Sensor (V)	0-5.1 volts	3.5-4.5	3.5-4.5	Engine Data 2
Cruise Engaged	Yes/No	No	No	Engine Data 2
Desired Idle Speed	0-3187	PCM control	---	Engine Data 2
ECT Sensor (°F)	-40 to 419°F	185-239	185-239	Engine Data 2
Engine Oil Level	LOW / Okay	Okay	Okay	Engine Data 2
Engine Oil Pressure	LOW / Okay	Okay	Okay	Engine Data 2
Engine Speed	0-9999 rpm	±50 Desired	Varies	Engine Data 2
IAT Sensor (°F)	-40 to 304°F	50-176	50-176	Engine Data 2
INJ pulsewidth Cylinder 1-4	0-985 ms	1-4	Varies	Engine Data 2
MAP Sensor (V)	0-5.1 volts	1.5	1-3	Engine Data 2
MPH / KPH	0-155 mph	0	30	Engine Data 2
Output Driver 1, 2 Open	00000000	00000000	00000000	Engine Data 2
Output Driver 1, 2 Short	00000000	00000000	00000000	Engine Data 2
Total Misfire Counts	0-255 counts	0	0	Engine Data 2
Stepper Cruise	Disabled/Enabled	Disabled	Disabled	Engine Data 2
TP Angle (%)	0-100%	0	Varies	Engine Data 2
TP Sensor (V)	0-5.1 volts	0.20-0.90	Varies	Engine Data 2
TR Range Switch	PN-R-O-3-2	PN	3	Engine Data 2
TR Switch (P)	HI / LOW	LOW	HI	Engine Data 2
TR Switch (A)	HI / LOW	LOW	HI	Engine Data 2
TR Switch (B)	HI / LOW	HI	LOW	Engine Data 2
TR Switch (C)	HI / LOW	HI	LOW	Engine Data 2

1996-97 Century 2.2L I4 MFI VIN 4 (A/T) - Misfire Data List

Parameter Identification (PID)	PID Value Range	PID Value at Hot Idle	PID Value at 30 mph	Data List Type
BARO Sensor (kPa)	10-110 kPa	11-105	11-105	Misfire
Calculated Airflow	0-512 g/sec	Varies	Varies	Misfire
ECT Sensor (°F)	-40 to 419°F	185-239	185-239	Misfire
Desired Idle Speed	0-3187	PCM control	---	Misfire
Engine Run Time	Hr: Min: Sec	00:00:00	00:00:00	Misfire
Engine Speed	0-9999 rpm	±50 Desired	Varies	Misfire
IAT Sensor (°F)	-40 to 304°F	50-176	50-176	Misfire
INJ pulsewidth Cyl 1-4	0-985 ms	1-4	Varies	Misfire
Engine Load	0-100%	22-26	Varies	Misfire
MAP Sensor (kPa)	10-110 kPa	25-48	Varies	Misfire
Misfire current Cyl 1-4	0-255 counts	0	0	Misfire
Misfire History Cyl 1-4	0-255 counts	0	0	Misfire
TP Angle (%)	0-100%	0	Varies	Misfire
TP Sensor (V)	0-5.1 volts	0.20-0.90	Varies	Misfire
Total Misfire	0-255 counts	0	0	Misfire

1996-97 Century 2.2L I4 MFI VIN 4 (A/T) - Specific EGR & EVAP

Parameter Identification (PID)	PID Value Range	PID Value at Hot Idle	PID Value at 30 mph	Data List Type
Actual EGR Position	0-100%	0	0-100	EGR, EVAP
BARO Sensor (V)	0-5.1 volts	3.5-4.5	3.5-4.5	EGR, EVAP
Decel EWMA	-10 - 10 kPa	-3 to - 4	Varies	EGR, EVAP
Desired Idle Speed	0-3187	PCM control	---	EGR, EVAP
Delta MAP Change	-20 - 20 kPa	0 / - number	0	EGR, EVAP
Desired EGR Position	0-100%	0	0-100	EGR, EVAP
ECT Sensor (°F)	-40 to 419°F	185-239	185-239	EGR, EVAP
EGR Trip Samples	0-255 counts	Varies	Varies	EGR, EVAP
Engine Run Time	Hr: Min: Sec	00:00:00	00:00:00	EGR, EVAP
Engine Speed	0-9999 rpm	±50 Desired	Varies	EGR, EVAP
EVAP Solenoid	0-100%	0	0-100	EGR, EVAP
EVAP Vent Solenoid	0=off / 1=on	0	0	EGR, EVAP
Fuel Level Sensor (V)	0-5.1 volts	Varies	Varies	EGR, EVAP
Fuel Level Sensor (%)	0-100%	Varies	Varies	EGR, EVAP
FTP Sensor (V)	0-5.1 volts	2.5: cap off	Varies	EGR, EVAP
Fuel Trim Cell	Cell #0-21	18-21	---	EGR, EVAP
HO2S-12 (B1 S2)	0-1132 mv	10-1000	10-1000	EGR, EVAP
IAC Motor	0-255 counts	5-60	---	EGR, EVAP
IAT Sensor (°F)	-40 to 304°F	50-176	50-176	EGR, EVAP
Long Term Fuel Trim	0-255 counts	58-198	58-198	EGR, EVAP
Long Term Fuel Trim	-10% to 10%	0 (± 5%)	0 (± 5%)	EGR, EVAP
MAP Sensor (kPa)	10-110 kPa	25-48	Varies	EGR, EVAP
Number of DTC(s)	Number	0	0	EGR, EVAP
O2S-11 (B1 S1)	0-1132 mv	10-1000	10-1000	EGR, EVAP
Short Term Fuel Trim	-10% to 10%	Varies	Varies	EGR, EVAP
TP Angle (%)	0-100%	0%	0-100%	EGR, EVAP
TP Sensor (V)	0-5.1 volts	0.20-0.90	---	EGR, EVAP

1996-97 Century 2.2L I4 MFI VIN 4 - F/F Data & Failure Records

Parameter Identification (PID)	PID Value Range	PID Value at Hot Idle	PID Value at 30 mph	Data List Type
BARO Sensor (V)	0-5.1 volts	3.5-4.5	3.5-4.5	F/F & F/R
Calculated Airflow	0-512 g/sec	Varies	Varies	F/F & F/R
Desired Idle Speed	0-3187 rpm	PCM control	---	F/F & F/R
ECT Sensor (°F)	-40 to 419°F	185-239	185-239	F/F & F/R
Engine Speed	0-9999 rpm	±50 Desired	Varies	F/F & F/R
INJ pulsewidth Cyl 1-4	0-985 ms	1-4	Varies	F/F & F/R
Long Term Fuel Trim	0-255 counts	58-198	58-198	F/F & F/R
Long Term Fuel Trim	-10% to 10%	0 (± 5%)	0 (± 5%)	F/F & F/R
Engine Load	0-100%	22-26	Varies	F/F & F/R
Loop Status	Closed Loop / Open Loop	Closed Loop	Closed Loop	F/F & F/R
MAP Sensor (kPa)	10-110 kPa	25-48	Varies	F/F & F/R
Short Term Fuel Trim	0-255 counts	58-198	58-198	F/F & F/R
Short Term Fuel Trim	-10% to 10%	0 (± 5%)	0 (± 5%)	F/F & F/R
TP Angle (%)	0-100%	0	Varies	F/F & F/R
Vehicle Speed	0-155 mph	0	30	F/F & F/R

1996-97 Century 2.2L I4 MFI VIN 4 (A/T) - Fuel Trim List

Parameter Identification (PID)	PID Value Range	PID Value at Hot Idle	PID Value at 30 mph	Data List Type
BARO Sensor (V)	0-5.1 volts	3.5-4.5	3.5-4.5	Fuel Trim
Calculated Airflow	0-512 g/sec	Varies	Varies	Fuel Trim
Desired Idle Speed	0-3187	PCM control	---	Fuel Trim
ECT Sensor (°F)	-40 to 419°F	185-239	185-239	Fuel Trim
Engine Run Time	Hr: Min: Sec	00:00:00	00:00:00	Fuel Trim
Engine Speed	0-9999 rpm	±50 Desired	---	Fuel Trim
Fuel Trim Cell	Cell #0-21	18-21	---	Fuel Trim
Fuel Trim Index	0-255 counts	58-198	Varies	Fuel Trim
HO2S-11 L/R to R/L Ratio	0:1-15.93:1	Varies	Varies	Fuel Trim
HO2S-12 (B1 S2)	0-1132 mv	10-1000	10-1000	Fuel Trim
IAT Sensor (°F)	-40 to 304°F	50-176	50-176	Fuel Trim
Lean / Rich Average	0-249 ms	0	0	Fuel Trim
Lean / Rich Transition	0-255 counts	0	0	Fuel Trim
Engine Load	0-100%	22-26	Varies	Fuel Trim
Long Term Fuel Trim	0-255 counts	58-198	Varies	Fuel Trim
Long Term Fuel Trim	-10% to 10%	Varies	Varies	Fuel Trim
Long Term F/T Average	0-255 counts	58-198	Varies	Fuel Trim
Loop Status	Closed Loop / Open Loop	Closed Loop	Closed Loop	Fuel Trim
MAP Sensor (kPa)	10-110 kPa	25-48	Varies	Fuel Trim
O2S-11 (B1 S1)	0-1132 mv	10-1000	10-1000	Fuel Trim
Rich / Lean Average	0-249 ms	0	0	Fuel Trim
Rich / Lean Transition	0-255 counts	0	0	Fuel Trim
Short Term Fuel Trim	0-255 counts	58-198	Varies	Fuel Trim
Short Term Fuel Trim	-10% to 10%	Varies	Varies	Fuel Trim
Short Term F/T Average	0-255 counts	58-198	Varies	Fuel Trim
TP Angle (%)	0-100%	0	Varies	Fuel Trim
TP Sensor (V)	0-5.1 volts	0.20-0.90	---	Fuel Trim
Vehicle Speed	0-155 mph	0	30	Fuel Trim

1995 Century 3.1L V6 MFI VIN M (A/T) - Engine Data List

Parameter Identification (PID)	PID Value Range	PID Value at Hot Idle	PID Value at 30 mph	Data List Type
2nd Gear Start (3T40)	On/Off	On	Off	Engine Data 1
2nd, 3rd, 4th Gear Switch	Yes/No	No	Yes/No	Engine Data 1
A/C Clutch	On/Off	Off	Off	Engine Data 1
A/C Pressure	0-459 psi	139-399	139-399	Engine Data 1
A/C Request	Yes/No	No	Off	Engine Data 1
Camshaft Signal	0 or 1	Varies	Varies	Engine Data 1
Commanded Gear	1-2-3-4	1	3	Engine Data 1
Cranking RPM	0-999	0	0	Engine Data 1
Cruise Control	On/Off	Off	Off	Engine Data 1
Current Weak Cylinder Number	CYL #/None	Not Enabled	Not Enabled	Engine Data 1
Desired Idle Speed	0-3187	800	---	Engine Data 1
ECT Sensor (°F)	-40 to 419°F	185-228	185-228	Engine Data 1
EGR Solenoid 1, 2, 3	On/Off	Off	On/Off	Engine Data 1
Engine Speed	0-9999 rpm	±50 Desired	Varies	Engine Data 1
Fan Relays	On/Off	Off	Off	Engine Data 1
Fuel Trim Cell	Cell #0-4	0	0-4	Engine Data 1
HO2S Cross Counts	0-255 counts	0-20	0-20	Engine Data 1
HO2S-11 (B1 S1)	0-1132 mv	10-1000	10-1000	Engine Data 1
HO2S-21 (B2 S1)	0-1132 mv	10-1000	10-1000	Engine Data 1
IAC Motor	0-255 counts	35	---	Engine Data 1
IAT Sensor (°F)	-40 to 304°F	50-176	50-176	Engine Data 1
Injector Pulsewidth	0-985 ms	2.9	3.1	Engine Data 1
Knock Retard (°)	0-128°	0	Varies	Engine Data 1
Long Term Fuel Trim	0-255 counts	128	1276	Engine Data 1
Loop Status	Closed Loop / Open Loop	Closed Loop	Closed Loop	Engine Data 1
Low Oil Light	On/Off	Off	Off	Engine Data 1
LV8 Engine Load	0-100%	50-60	Varies	Engine Data 1
MAF Sensor (g/sec)	0-512 g/sec	5-9	Varies	Engine Data 1
MAP Sensor (kPa)	10-110 kPa	35	44	Engine Data 1
Quad Driver 1, 2	HI or LOW	LOW	---	Engine Data 1
PROM ID Number	Number	0000	0000	Engine Data 1
PSP Switch	HI/Normal	Normal	Normal	Engine Data 1
PRNDL 'P', 'A', 'B', 'C'	HI or LOW	HI / LOW	HI / LOW	Engine Data 1
Purge Duty Cycle	0-100%	0	20	Engine Data 1
Rich/Lean Flag	L or R	L-R-L-R	L-R-L-R	Engine Data 1
Short Term Fuel Trim	0-255	126	127	Engine Data 1
Spark Advance (°)	-90° to 90°	18-24	30-33	Engine Data 1
TCC Solenoid	On/Off	Off	Off	Engine Data 1
TCC B Switch	APPY/REL	Released	Released	Engine Data 1
TCC Slip Speed	-255/+255	720-770	3	Engine Data 1
TCC Duty Cycle	0-100%	0	0-100	Engine Data 1
Throttle Angle (%)	0-100%	0	4	Engine Data 1
Time From Startup	Hr: Min: Sec	00:00:00	00:00:00	Engine Data 1
TP Sensor (V)	0-5.1 volts	0.54	0.74	Engine Data 1
TFT Sensor (°F)	-40 to 304°F	177	177	Engine Data 1
Transmission Range Switch	P, A, B, C	Park	Drive 2/3	Engine Data 1

1996-99 Century 3.1L V6 MFI VIN M (A/T) - Engine Data List

Parameter Identification (PID)	PID Value Range	PID Value at Hot Idle	PID Value at 30 mph	Data List Type
3X Crank Sensor	Number	Varies	Varies	Engine Data 1
24X Crank Sensor	0-1280	Varies	Varies	Engine Data 1
Abuse Management	Active/Inactive	Inactive	Inactive	Engine Data 1
A/C HI Side Pressure	0-5.1 volts	Varies	Varies	Engine Data 1
A/C Off for WOT	On/Off	Off	Off	Engine Data 1
A/C Pressure Disable	Yes/No	No	No	Engine Data 1
A/C Request	Yes/No	No	No	Engine Data 1
Actual EGR Position	0-100%	0	0-100	Engine Data 1
Air Fuel Ratio	0.0-25:1	14.7:1	14.7:1	Engine Data 1
BARO Sensor (kPa)	10-110 kPa	65-105	65-105	Engine Data 1
Brake Switch	Apply/Released	Released	Released	Engine Data 1
Cam Signal Present	Yes/No	Yes	Yes	Engine Data 1
Commanded A/C	On/Off	Off	Off	Engine Data 1
Fuel Pump Command	On/Off	On	On	Engine Data 1
Commanded GEN	On/Off	On	On	Engine Data 1
Cruise Mode	On/Off	Off	On	Engine Data 1
Cruise Inhibited	Yes/No	Yes	Yes	Engine Data 1
Current Gear	1-2-3-4	1	3	Engine Data 1
Decel Fuel Mode	Active/Inactive	Inactive	Inactive	Engine Data 1
Desired EGR Position	0-100%	0	0-100	Engine Data 1
Desired Idle Speed	Number	PCM control	---	Engine Data 1
ECT Sensor (°F)	-40 to 304°F	185-220	185-220	Engine Data 1
EGR Closed Valve Pintle Position	0-5.1 volts	0.14-1.00	Varies	Engine Data 1
EGR Duty Cycle	0-100%	0	0-100	Engine Data 1
EGR Feedback	0-5.1 volts	0.14-1.00	Varies	Engine Data 1
EGR Position Error	0-100%	0-9	Varies	Engine Data 1
EGR Flow Test	0-255 counts	0-10	0-10	Engine Data 1
Engine Load	0-100%	2-5	Varies	Engine Data 1
Engine Run Time	Hr: Min: Sec	00:00:00	00:00:00	Engine Data 1
Engine Speed	0-9999 rpm	±50 Desired	---	Engine Data 1
EVAP Canister Purge	0-100%	0-25	0-75	Engine Data 1
EVAP Vent Solenoid	Open/Closed	Open	Open	Engine Data 1
Fan - Low Speed	On/Off	Off	Off	Engine Data 1
Fan - High Speed	On/Off	Off	Off	Engine Data 1
Fuel Trim Cell	Cell #0-4	0	0-4	Engine Data 1
Fuel Trim Learn	Disabled/Enabled	Enabled	Enabled	Engine Data 1
Generator Lamp	On/Off	Off	Off	Engine Data 1
HO2S (B1 S1, B2 S1)	0-1132 mv	10-1000	10-1000	Engine Data 1
HO2S-11 Status	Not Ready/Ready	Ready	Ready	Engine Data 1
HO2S-11 XCount	0-255 counts	Varies	Varies	Engine Data 1
IAC Position	0-255 counts	10-40	10-40	Engine Data 1
IAT Sensor (°F)	-40 to 304°F	50-194	50-194	Engine Data 1
Ignition 1 Signal	0.0-25.5v	14.2	14.2	Engine Data 1
Ignition Mode	IC/Bypass	IC	IC	Engine Data 1
Injector Pulsewidth	0-985 ms	1.5-3.5	Varies	Engine Data 1
Injector Status	Okay/Stuck	Okay	Okay	Engine Data 1

1996-99 Century 3.1L V6 MFI VIN M (A/T) - Engine Data List

Parameter Identification (PID)	PID Value Range	PID Value at Hot Idle	PID Value at 30 mph	Data List Type
Knock Retard (°)	0-128°	0	0	Engine Data 1
Long Term Fuel Trim	-10% to 10%	0 (± 5%)	0 (± 5%)	Engine Data 1
Loop Status	Closed Loop / Open Loop	Closed Loop	Closed Loop	Engine Data 1
Low Oil Lamp	On/Off	Off	Off	Engine Data 1
MAF Sensor (g/sec)	0-512 g/sec	4-6	Varies	Engine Data 1
MAF Sensor (Hz)	0-31,999hz	1200-3000	Varies	Engine Data 1
MAP Sensor (kPa)	10-110 kPa	29-48	Varies	Engine Data 1
MAP Sensor (V)	0-5.1 volts	1-2	Varies	Engine Data 1
MIL (lamp) Command	On/Off	Off	Off	Engine Data 1
Misfire Current Cylinder 1-6	0-198	0 - 4	Varies	Engine Data 1
Misfire History Cylinder 1-6	0-65535	0	0	Engine Data 1
Non-Volatile Memory	Pass/Fail	Pass	Pass	Engine Data 1
Power Enrichment	Active/Inactive	Active	Active	Engine Data 1
Rich / Lean	L or R	L-R-L-R	L-R-L-R	Engine Data 1
Short Term Fuel Trim	-10% to 10%	Varies	Varies	Engine Data 1
Spark Advance (°)	-64° to +64°	20°	Varies	Engine Data 1
Startup ECT Sensor	-40 to 304°F	Varies	Varies	Engine Data 1
Startup IAT Sensor	-40 to 304°F	Varies	Varies	Engine Data 1
TCC Engaged	Disengage/Engage	Disengaged	Engaged	Engine Data 1
Total Misfire Failures	0-65,535 counts	0	0	Engine Data 1
TP Angle (%)	0-100%	0	Varies	Engine Data 1
TP Sensor	0-5.1 volts	0.20-0.74	Varies	Engine Data 1
Transmission Hot Mode	On/Off	Off	Off	Engine Data 1
TWC Protection	Active/Inactive	Inactive	Inactive	Engine Data 1
Total Misfire Current Count	0-99 counts	0-5	0-5	Engine Data 1
Total Misfire Passes Since First Failure	0-65,535 counts	0	0	Engine Data 1
Transmission Range	P-R-N-4-3-2	P	3	Engine Data 1
TR Switch P-A-B-C	HI/LO	LO/LO/HI/HI	---	Engine Data 1
TWC Protection	Active/Inactive	Inactive	Inactive	Engine Data 1
Vehicle Speed	0-155 mph	0	30	Engine Data 1
VTD Fuel Disable	Active/Inactive	Inactive	Inactive	Engine Data 1

2000-05 Century 3.1L V6 MFI VIN J (A/T) - PID Data Lists

Parameter Identification (PID)	PID Value Range	PID Value at Hot Idle	PID Value at 30 mph	Data List Type
3X Crank Sensor	0-9999 rpm	Varies	Varies	Engine Data 2
24X Crank Sensor	0-1600 rpm	Varies	Varies	Engine Data 2
A/C HI Side Pressure	0-5.1 volts	Varies	Varies	Engine Data 2
A/C Off for WOT	Yes/No	No	No	Engine Data 2
A/C Pressure Disable	Yes/No	No	No	Engine Data 2
A/C Relay Control Circuit Status	Okay/Fault	Okay	Okay	Output Driver
A/C Relay Command	On/Off	Off	Off	Engine Data 1, Data 2, EGR
A/C Request Signal	Yes/No	No	No	Engine Data 2
Air Fuel Ratio	0.0-25.5:1	14.7:1	14.7:1	Engine Data 2, Fuel Trim
BARO Sensor (kPa)	10-110 kPa	98	98	Engine Data 1, EGR, EVAP, Fuel Trim
CMP Sensor Signal	Yes/No	Yes	Yes	Engine Data 2
Crank Request Signal	Yes/No	No	No	Engine Data 2
Cruise Control Active	Yes/No	No	No	Engine Data 1
Cruise Control Inhibit	On/Off	On	On	Engine Data 1
Cruise Inhibit Reason	Yes/No	Yes	Yes	Engine Data 1
Current Gear	1-2-3-4	1	4	Engine Data 1-2, EGR, Fuel Trim
Cycles of Misfire Data	0-99	Varies	Varies	Misfire
Cylinder 1-6 Injector Circuit Status	Okay/Fault	Okay	Okay	Output Driver
Cylinder 1-6 Injector Circuit History	Okay/Fault	Okay	Okay	Output Driver
Desired Idle Speed	0-3187 rpm	PCM control	---	Engine Data 2
Decel Fuel Cutoff	Active/Inactive	Inactive	Inactive	EGR, Fuel Trim
Desired EGR Position	0-100%	0	Varies	Engine Data 1, EGR, Misfire
Desired Idle Speed	0-3187 rpm	± 50 rpm of Actual Speed	Varies	Engine Data 1, Data 2, EVAP
Driver Module 1-4 Status	Enabled/DIS	Enabled	Enabled	Output Driver
ECT Sensor (°F)	-40 to 304°F	185-220	185-220	Engine Data 1-2, EGR, Fuel Trim, EVAP, Misfire
EGR Flow Test Count	0-255 counts	0	Varies	EGR
EGR Learned Minimum Position	0-5.1 volts	0	Varies	EGR
EGR Position Sensor	0-100%, or 0-5.1volts	0.0	Varies	Engine Data 1, EGR, Misfire
EGR Position Variance	0-5.1 volts	0.0	Varies	EGR
EGR Solenoid Control	0-100%	0	Varies	EGR
Engine Load (%)	0-100%	17	Varies	Engine Data 1-2, EGR, Fuel Trim, EVAP, Misfire

2000-05 Century 3.1L V6 MFI VIN J (A/T) - PID Data Lists

Parameter Identification (PID)	PID Value Range	PID Value at Hot Idle	PID Value at 30 mph	Data List Type
Engine Oil Level Switch	Okay/Not Okay	Okay	Okay	Engine Data 2
Engine Oil Life Left	0-100%	Varies	Varies	Engine Data 2
EOP Switch Signal	Okay/Not Okay	Okay	Okay	Engine Data 2
Engine Run Time	Hr: Min: Sec	00:00:00	00:00:00	Engine Data 1-2, EGR, Fuel Trim, EVAP, Misfire
Engine Speed	0-9999 rpm	± 50 rpm of Desired Speed	Varies	Engine Data 1-2, EGR, Fuel Trim, EVAP, Misfire
EVAP Fault History	No Fault / Excess VAC / Purge Valve Leak / Small Leak / Weak Vacuum	No Fault	No Fault	EVAP
EVAP Purge Solenoid Control Circuit Status	Okay/Fault	Okay	Okay	Output Driver
EVAP Purge Control	0-100%	19	Varies	Engine Data 1, EVAP, Fuel Trim
EVAP Test Abort Reason	Not Aborted / Lost Enable / Small Leak / Vehicle Not at Rest	Not Aborted	Not Aborted	EVAP
EVAP Test Result	No Result / Passed / Aborted / Fail DTC PXXXX	Passed	Passed	EVAP
EVAP Test State	Wait for Purge / Test Running/Test Completed	Test Completed	Test Completed	EVAP
EVAP Vent Solenoid Control Circuit Status	Okay/Fault	Okay	Okay	Output Driver
EVAP Vent Solenoid	Venting/Not Venting	Not Venting	Not Venting	Engine Data 1, EVAP, Fuel Trim
Fan Control Relay 1 Circuit Status	Okay/Fault	Okay	Okay	Output Driver
Fan Control Relay 1 Command	On/Off	Off	Off	Engine Data 2
Fan Control Relay 2-3 Circuit Status	Okay/Fault	Okay	Okay	Output Driver
Fan Control Relay 2-3 Command	On/Off	Okay	Okay	Engine Data 2
Fuel Level Sensor	0-5.1 volts	0.8-2.5	0.8-2.5	EVAP
Fuel Pump Relay Control Circuit Status	Okay/Fault	Okay	Okay	Output Driver
Fuel Pump Relay Control Circuit History	Okay/Fault	Okay	Okay	Output Driver

2000-05 Century 3.1L V6 MFI VIN J (A/T) - PID Data Lists

Parameter Identification (PID)	PID Value Range	PID Value at Hot Idle	PID Value at 30 mph	Data List Type
F/P Relay Command	On/Off	On	On	Engine Data 1
Fuel Tank Level Remaining	0-100%	Varies	Varies	EVAP
Fuel Tank Pressure (FTP) Sensor	Inches H2O, mm Hg, volts	-17 to +7.5 Inches H2O	-17 to +7.5 Inches H2O	Engine Data 1, EVAP
Fuel Trim Cell	Cell 0-10	Varies	Varies	Engine Data 1, EVAP, Fuel Trim
Fuel Trim Learn	Enabled or Disabled	Enabled	May Toggle	Engine Data 1, EVAP, Fuel Trim
Generator F-Terminal	0-100%	Varies	Varies	Engine Data 2
Generator L-Terminal	On/Off	On	On	Engine Data 2
HO2S-11 (B1 S1)	0-1132 mv	10-1000	10-1000	Engine Data 1, EVAP, Fuel Trim
HO2S-12 (B1 S2)	0-1132 mv	10-1000	10-1000	Engine Data 1, EVAP, Fuel Trim
HO2S-11 Heater	On/Off	On	On	Fuel Trim
IAC Position	0-255 counts	15	15	Engine Data 1, EGR, Fuel Trim
IAT Sensor (^{0}F)	-40 to 304^{0}F	50-194	50-194	Engine Data 1-2, EGR, EVAP, Fuel Trim
Ignition 1 Signal	0.0-25.5v	14.2	14.2	Engine Data 1-2, EGR, EVAP, Fuel Trim
Ignition Mode	Bypass/IC	IC	IC	Engine Data 2
Injector Pulsewidth	0-1000 ms	2.9	Varies	Engine Data 2, Fuel Trim, Misfire
Knock Retard (^{0})	0-25.5^0	0^0	0^0	Engine Data 1, EGR
Long Term Fuel Trim	-27% to 27%	0 (± 5%)	0 (± 5%)	Engine Data 1-2 EVAP, Fuel Trim
Loop Status	Closed Loop / Open Loop	Closed Loop	Closed Loop	Engine Data 1-2, EGR, EVAP, Fuel Trim
MAF Sensor (g/sec)	0-512 g/sec	3.41	Varies	Engine Data 1-2, EGR, Fuel Trim, EVAP, Misfire
MAF Sensor (Hz)	0-32,000 Hz	2000	Varies	Engine Data 1-2, EGR, Fuel Trim, EVAP, Misfire
MAP Sensor (kPa)	10-105 kPa	35	Varies	Engine Data 1-2, EGR, EVAP, Fuel Trim, Misfire
MAP Sensor (V)	0-5.1 volts	1.35	Varies	Engine Data 1-2, EGR, EVAP, Fuel Trim, Misfire
MIL Status	Okay/Fault	Okay	Okay	Output Driver
MIL Command	On/Off	Off	Off	Engine Data 2

2000-05 Century 3.1L V6 MFI VIN J (A/T) - PID Data Lists

Parameter Identification (PID)	PID Value Range	PID Value at Hot Idle	PID Value at 30 mph	Data List Type
Misfire Current Cyl 1-6	0-198 counts	0	0	Misfire
M/Fire History Cyl 1-6	0-65,535	0	0	Misfire
Number of DTC(s)	Number	0	0	Engine Data 1, EVAP, Fuel Trim
O2 Heater Current	0-5.0 amps	0.54	0.54	Fuel Trim
PCM Reset	Yes/No	No	No	Engine Data 1-2, EGR, EVAP, Fuel Trim, Output Driver
PCM/VCM in VTD Fail Enable	Yes/No	Yes	Yes	Engine Data 1-2, EGR, EVAP, Fuel Trim, Output Driver
PCM Reset	Yes/No	No	No	Engine Data 1
Power Enrichment	Active/Inactive	Inactive	Inactive	Engine Data 2, Misfire
Short Term Fuel Trim	-27% to 27%	0 (± 5%)	0 (± 5%)	Engine Data 1-2, EVAP, Fuel Trim
Spark Advance (°)	-64° to 64°	18	Varies	Engine Data 1-2 Fuel Trim, Misfire
Startup ECT	-40 to 419°F	Varies	Varies	Engine Data 2, EVAP, Fuel Trim
Startup IAT	-40 to 419°F	80°F	80°F	Engine Data 2, EVAP, Fuel Trim
Starter Enable Relay Control Circuit Status	Okay/Fault	Okay	Okay	Output Driver
Starter Enable Command	On/Off	Off	Off	Engine Data 2
TCC Brake Pedal Switch	Applied/REL	Released	Released	Engine Data 1-2
TFP Switch A/B/C	P, R, N, D, 4, 3, 2, 1	Park	4	Engine Data 2, EGR, Fuel Trim
Torque Delivered	0-100%	86	Varies	Engine Data 2
Torque Request	0-100%	100%	Varies	Engine Data 2
TP Angle (%)	0-100%	0	0-100	Engine Data 1-2, EGR, Fuel Trim, EVAP, Misfire
TP Sensor (V)	0-5.1 volts	0.59	Varies	Engine Data 1-2, EGR, Fuel Trim, EVAP, Misfire
Traction Control	Active/Inactive	Inactive	Inactive	Engine Data 2
Vehicle Speed	0-155 mph	0	30	Engine Data 1-2, EGR, Fuel Trim, EVAP, Misfire
VTD Auto Learn timer	Active/Inactive	Inactive	Inactive	Engine Data 1
VTD Fuel Disable	Active/Inactive	Inactive	Inactive	Engine Data 1
VTD Fuel Disable Until Key Off	Yes/No	No	No	Engine Data 1

LACROSSE PID DATA

2005 LaCrosse 3.6L V6 VIN 7

Scan Tool Parameter	Data List	Parameter Range/Units	Typical Data Values
A/C Relay Circuit Status	ODM	OK/Fault/Indeterminate	OK
A/C Relay Command	Eng 1, 2, MF	On/Off	Off
APP Indicated Angle	Eng 1, 2, CMP, EVAP, HO2S, FT, TAC	0-100%	0%
APP Sensor 1	TAC	0.9-4.5 Volts	1 Volt
APP Sensor 2	TAC	0.40-2.25 Volts	0.4 Volt
APP Sensor 1 and 2	TAC	Agree/Disagree	Agree
BARO	Eng 1, 2, CMP, EVAP, HO2S, FT, MF, TAC	65-104 kPa (8-16 psi)	Varies with altitude
BARO	Eng 2, TAC	Volts	Varies
Calculated ECT - Closed Loop Fuel Control	Eng 2	-39°C to +40°C (-38°F to +104°F)	40°C (104°F)
Calculated ECT - Thermostat Diagnosis	Eng 2	-39°C to +140°C (-38°F to +284°F)	89°C (192°F)
Catalytic Converter Protection Active	FT, MF	Yes/No	No
Catalyst Monitor Complete This Ignition	IM	Yes/No	Varies
Catalyst Monitor Enabled this Ignition	IM	Yes/No	Varies
CKP Resync Counter	Eng 2, MF	Counts	0 Counts
Component Monitor Complete This Ignition	IM	Yes/No	Varies
Component Monitor Enable This Ignition	IM	Yes/No	Varies
Current Gear	Eng 1, 2, MF	P/N/Reverse/1st-5th	P/N
Cycles of Misfire Data	MF	0-200 counts	Varies
Cylinder 1-6 Injector Circuit Status	ODM	OK/Fault/Indeterminate	OK
Cylinder 1-6 Knock Retard	Eng 1	Degree	0 Degrees
Decel. Fuel Cutoff	Eng 2, FT, HO2S	Active/Inactive	Inactive
Desired Exh. CMP Bank 1 or 2	CMP	Degree	Varies
Desired HO2S 1	HO2S	Lambda	Varies
Desired Idle Speed	Eng 1, 2, EVAP, CMP, TAC	0-7,000 RPM	650 RPM
Desired Int. CMP Bank 1 or 2	CMP	Degrees	Varies
DTC Set This Ignition	Eng 1, CMP, EVAP, FT, HO2S, TAC	Yes/No	No
EC Ignition Relay Command	Eng 1, TAC	On/Off	On
EC Ignition Relay Feedback Signal	Eng 1, TAC	0-25.5 Volts	12.0-14.5 Volts
ECM Reset	Eng 1, EVAP, FT	Yes/No	No
ECT Sensor	Eng 1, 2, CMP, EVAP, HO2S, FF, FR, FT, MF, TAC	-39°C to +140°C (-38°F to +284°F)	Varies
Engine Load	Eng 1, 2, CMP, EVAP, HO2S, FF, FR, FT, MF, TAC	0-99%	0-5% - Idle 3-7% - 2500 RPM
Engine OFF EVAP Test Conditions Met	EVAP	Yes/No	Varies
Engine Off Time	EVAP	0:00:00 Seconds	Varies
Engine Oil Level Switch	CMP	OK/Low	OK
Engine Oil Life Remaining	CMP	%	Varies
Engine Oil Pressure Sensor	MF, CMP	kPa/psi	Varies
Engine Run Time	Eng 1, 2, EVAP, FT, HO2S, MF, TAC, CMP, ODM	0:00:00	Varies
Engine Speed	ALL	RPM	575-700 RPM
EVAP Monitor Complete This Ignition	IM	Yes/No	Varies
EVAP Monitor Enabled This Ignition	IM	Yes/No	Varies
EVAP Purge Solenoid Circuit Status	ODM	OK/Fault/Indeterminate	OK

2005 LaCrosse 3.6L V6 VIN 7 (CONT)

Scan Tool Parameter	Data List	Parameter Range/Units	Typical Data Values
EVAP Purge Solenoid Command	Eng 1, EVAP, FT	0-100%	Varies
EVAP Vent Solenoid Circuit Status	ODM	OK/Fault/Indeterminate	OK
EVAP Vent Solenoid Command	Eng 1, EVAP, FT	Venting/Not Venting	Venting
Exh. CMP Angle Bank 1 or 2	CMP	Degrees	Varies
Exh. CMP Bank 1 or 2 Command	CMP	0-99%	Varies
Exh. CMP Solenoid Bn. 1 or 2 Circuit Status	ODM	OK/Fault/Indeterminate	OK
FC Relay 1 Circuit Status	ODM	OK/Fault/Indeterminate	OK
FC Relay 1 Command	Eng 2	ON/OFF	Varies
FC Relay 2 and 3 Command	Eng 2	ON/OFF	Varies
FC Relay 2 and 3 Circuit Status	ODM	OK/Fault/Indeterminate	OK
Fuel Level Sensor	EVAP	0-5 Volts	Varies
Fuel Pump Relay Circuit Status	ODM	OK/Fault/Indeterminate	OK
Fuel Pump Relay Command	Eng 1, 2	On/Off	On
Fuel Tank Level Remaining	Eng 1, EVAP, MF	Liter/Gallon	Varies
Fuel Tank Pressure Sensor	EVAP	in H2O/mm Hg	Varies
Fuel Tank Pressure Sensor	EVAP	0-5 Volts	Varies
Fuel Tank Rated Capacity	EVAP	Liter/Gallon	17.8 Gallons (67.3 Liters)
Fuel Trim Learn	Eng 1, EVAP, FT, HO2S	Enabled/Disabled	Enabled
HO2S 1	Eng 1, EVAP, FT, HO2S	Lambda	1.0 Lambda
HO2S 2	Eng 1, FT, HO2S	0-1,006 mV	Varies 400-700 mV
HO2S Heater Monitor Complete This Ignition	IM	Yes/No	Varies
HO2S Heater Monitor Enabled This Ignition	IM	Yes/No	Varies
HO2S 1 or 2 Heater Circuit Status	ODM	OK/Fault/Indeterminate	OK
HO2S 1 or 2 Heater Command	HO2S	0-99%	Varies
HO2S/O2S Monitor Complete This Ignition	IM	Yes/No	Varies
HO2S/O2S Monitor Enabled This Ignition	IM	Yes/No	Varies
HO2S 1 or 2 Sensing Element	HO2S	Ohms	70-90 Ohms Depends on Temperature
IAT Sensor	Eng 1, 2, EVAP, FT, FF, FR, HO2S, TAC, CMP	-39° to +140°C (-38° to +284°F)	Varies
Ignition 0 Signal	Eng 1, 2, EVAP, TAC	On/Off	On
Ignition 1 Signal	Eng 1, 2, EVAP, FT, MF, TAC, CMP	On/Off	On
IMRC Solenoid Circuit Status	ODM	OK/Fault/Indeterminate	OK
IMRC Solenoid Command	Eng 1	On/Off	Off
Initial Brake Apply Signal	TAC	Applied/Released	Released
Injector 1-6 Command	Eng 2, FT, MF	ms	Varies
Injector PWM Bank 1 or 2 Average	FT, MF	0-9 ms	4.6-5.4 ms
Int. CMP Angle Bank 1 or 2	CMP	Degrees	Varies
Int. CMP Bank 1 or 2 Command	CMP	0-99%	Varies
Int. CMP Solenoid Bn. 1 or 2 Circuit Status	ODM	OK/Fault/Indeterminate	OK
Knock Retard	Eng 1	Degrees	Varies
KS 1 or 2 Signal	Eng 1	Volts	Varies
Loop Status Sensor 1	Eng 1, 2, EVAP, FT, HO2S, FF, FR	Open Loop/Closed Loop	Closed Loop
Loop Status Sensor 2	Eng 1, 2, EVAP, FT, HO2S, FF, FR	Open Loop/Closed Loop	Open Loop
LT FT Cruise/Accel.	Eng 1, 2, EVAP, FT, HO2S	-100% to +100%	0%
LT FT Idle/Decel.	Eng 1, 2, EVAP, FT, HO2S	-100% to +100%	0%

2005 LaCrosse 3.6L V6 VIN 7 (CONT)

Scan Tool Parameter	Data List	Parameter Range/Units	Typical Data Values
MAF Sensor	Eng 1, 2, FT, HO2S, MF, TAC, CMP, FF, FR	0-655 g/s	3.4-6.3 g/s
MAF Sensor	Eng 2, EVAP, FT, HO2S, MF, TAC, CMP	0-5 Volts	0.9-1.7 Volts
MIL Circuit Status	ODM	OK/Fault/Indeterminate	OK
MIL Command	Eng 1	On/Off	Off
Mileage Since DTC Cleared	Eng 2, FF, FR	km - miles	Varies
Mileage Since First Failure	FF, FR	km - miles	Varies
Mileage Since Last Failure	FF, FR	km - miles	Varies
Mileage Since MIL Requested	FF, FR	km - miles	Varies
Misfire Current Cyl. 1-6	MF	0-255 Counts	0
Misfire History Cyl. 1-6	MF	0-65,535 Counts	0
Misfire Monitor Complete This Ignition	IM	Yes/No	Varies
PNP Switch	Eng 2	Park/Neutral/In Gear	P/N
Power Enrichment	Eng 2, FT, HO2S, MF	Active/Inactive	Inactive
Reduced Engine Power	Eng 1, 2, TAC	Active/Inactive	Inactive
Short Term FT	Eng 1, 2, EVAP, FT, HO2S, FF, FR	-25% to +25%	-3% to +3%
Spark	Eng 1, 2, FT, HO2S, MF, FF, FR	-20 to +40 Degrees	2-10 Degrees
Starter Relay Circuit Status	ODM	OK/Fault/Indeterminate	OK
Start Up ECT	Eng 2, EVAP, FT, HO2S	-39° to +140°C (-38° to +284°F)	Varies
Start Up IAT	Eng 2, EVAP, FT, HO2S	-39° to +140°C (-38° to +284°F)	Varies
Stoplamp Pedal Switch	Eng 1, TAC	Applied/Released	Released
System Power Mode	Eng 2, EVAP, TAC	Off, Accessory, RAP, Run, or Crank Request	Run
TAC System Learned Counts	TAC	0-11 Counts	0
Torque Delivered Signal	FT, TAC	0-100%	Varies
Torque Request Signal	FT, TAC	0-100%	90-98%
Total Fuel Trim Average	EVAP, FT, HO2S	-100 to +100%	Varies
Total Misfire Count	MF	Counts	0
TP Desired Angle	Eng 1, 2, EVAP, TAC	0-100%	Varies
TP Indicated Angle	Eng 1, 2, EVAP, FT, HO2S, MF, TAC, CMP, FF, FR	0-100%	Varies
TP Sensor 1	TAC	0-5 Volts	0.5-0.8 Volts
TP Sensor 1 Learned Minimum	TAC	0-5 Volts	0-0.6 Volts
TP Sensor 2	TAC	0-5 Volts	4.7-4.1 Volts
TP Sensor 2 Learned Minimum	TAC	0-5 Volts	5-3.7 Volts
TP Sensors 1 and 2	TAC	Agree/Disagree	Agree
Traction Control Status	TAC	Active/Inactive	Inactive
Transmission Gear Selector Signal	Eng 2	Valid/Invalid	Valid
TWC Temperature Calculated	FT, MF	Temperature	Varies
Vehicle Security Status	Eng 2	Password Learning, VTD Fail-Enable, Auto Learn Timer, Fuel Disable Until Ign Off, Fuel Disable Timeout, and Fuel Continue	Fuel Continued
Vehicle Speed Sensor	Eng 1, 2, EVAP, FT, HO2S, MF, TAC, CMP, FF, FR	Km/h mph	Varies
Volumetric Efficiency	Eng 1, 2, EVAP, FT, HO2S, MF, TAC, CMP	0-100%	5-25% - Idle and at 2,500 RPM

2005 LaCrosse 3.8L V6 VIN 1

Scan Tool Parameter	Data List	Units Displayed	Typical Data Value
3X Crank Sensor	Eng 2	RPM	Varies
18X Crank Sensor	Eng 2	RPM	Varies
A/C High Side Pressure Sensor	Eng 2	KPa/PSI/Volts	Varies
A/C Off For WOT	Eng 2	Yes/No	No
A/C Pressure Disable	Eng 2	Yes/No	No
A/C Relay Command	Eng 1, 2, Misfire, EGR	On/Off	Off
A/C Relay Circuit Status	ODD	OK/Fault/Invalid State	OK
A/C Request Signal	Eng 2	Yes/No	No
Air Fuel Ratio	Eng 2, Fuel Trim	Ratio	14.2:1-14.7:1
APP Average	TAC	Number	0
APP Indicated Angle	Eng 1, 2, EGR, EVAP, TAC, Fuel Trim, CC	Percent	0
APP Sensor 1	TAC	Volts	Varies
APP Sensor 2	TAC	Volts	Varies
APP Sensor 1	TAC	Percent	Varies
APP Sensor 2	TAC	Percent	Varies
APP Sensor 1 and 2	TAC	Agree/Disagree	Agree
BARO	Eng 1, EGR, EVAP, Fuel Trim	kPa	65-110 kPa
Boost Solenoid Circuit Status	ODD	OK/Fault/Invalid State	OK
Boost Solenoid Command	Engine 2	Percent	100%
CMP Sensor Signal Present	Eng 2	Yes/No	Yes
Cruise Control Active	Eng 1, CC, TAC	Yes/No	No
Cruise Disengage History 1-6	CC	None	None
Cruise Mode Signal	CC	Volts	0
Cruise ON/OFF Switch	CC, TAC	On/Off	Off
Cruise Resume/Accel Switch	CC, TAC	On/Off	Off
Cruise Set/Coast Switch	CC, TAC	On/Off	Off
Current Gear	Eng 1, Eng 2, EGR, Fuel Trim	1, 2, 3, 4	1
Cycles of Misfire Data	Misfire	Counts	0-99
Cyl. 1-6 Injector Circuit Status	ODD	OK/Stuck Low (open)/Stuck High	OK
Decel Fuel Cut-off	EGR, Fuel Trim	Active/Inactive	Inactive
Desired EGR Position	Eng 1, EGR, Misfire	Percent	0
Desired EGR Position	EGR	Volts	0
Desired Idle Speed	Eng 1, 2, EVAP, TAC	RPM	Varies
Driver Module 1 Status	ODD	Enabled/Off-High Volts/Off High Temp/Invalid State	Enabled
Driver Module 2 Status	ODD	Enabled/Off-High Volts/Off High Temp/Invalid State	Enabled
Driver Module 3 Status	ODD	Enabled/Off-High Volts/Off High Temp/Invalid State	Enabled
EC Ignition Relay Circuit	ODD	OK/ Fault/ Invalid State	OK
ECT Sensor	Eng 1, 2, EGR, EVAP, Misfire, Fuel Trim	°C/°F	Varies
EGR Flow Test Count	EGR	Counts	0-12
EGR Learned Minimum Position	EGR	Volts	0.14-1.0
EGR Position Sensor	Eng 1, EGR, Misfire	Percent	Varies
EGR Position Sensor	EGR	Volts	0
EGR Position Variance	EGR	Percent	0
EGR Solenoid Circuit Status	ODD	OK/Fault/ Invalid State	OK
EGR Solenoid Command	EGR	Percent	0
Engine Load	Eng 1, 2, Misfire, Fuel Trim, EVAP, EGR, CC, TAC	Percent	0
Engine Oil Level Switch	Eng 2	OK/Low	OK
Engine Oil Life Remaining	Eng 2	Percent	0-100% (Varies)
Engine Oil Pressure Switch	Eng 2	Low/OK	OK
Engine Run Time	Eng 1, 2, EVAP, Misfire, Fuel Trim, EGR, CC, TAC	Hr: Min: Sec	Varies

2005 LaCrosse 3.8L V6 VIN 1 (CONT)

Scan Tool Parameter	Data List	Units Displayed	Typical Data Value
Engine Speed	Eng 1, 2, EGR, EVAP, Misfire, Fuel Trim, CC, TAC	RPM	Varies
EVAP Fault History	EVAP	No Fault/Excess Vacuum/Purge Valve Leak/Small Leak/Weak Vacuum	No Fault
EVAP Purge Solenoid Circuit Status	ODD	OK/Fault/Invalid State	OK
EVAP Purge Solenoid Command	Eng 1, EVAP, Fuel Trim	Percent	Varies
EVAP Vent Solenoid Command	Eng 1, EVAP, Fuel Trim	Venting/Not Venting	Varies
EVAP Vent Solenoid Circuit Status	ODD	OK/Fault/Invalid State	OK
FC Relay 1 Circuit Status	ODD	OK/Fault/Invalid State	OK
FC Relay 1 Command	Eng 2	On/Off	Off
FC Relay 2 and 3 Circuit Status	ODD	OK/Fault/Invalid State	OK
FC Relay 2 and 3 Command	Eng 2	On/Off	Off
Fuel Pump Relay Circuit Status	ODD	OK/Fault/Invalid State	OK
Fuel Pump Relay Command	Eng 1	On/Off	On
Fuel Tank Level Remaining	EVAP	Percent	0-104
Fuel Tank Pressure Sensor	Eng 1, EVAP	mm Hg/in H2O	Varies
Fuel Tank Pressure Sensor	EVAP	Volts	Varies
Fuel Trim Cell	EVAP, Eng 1, Fuel Trim	0-4	0
Fuel Trim Learn	Eng 1, EVAP, Fuel Trim	Enabled/Disabled	Varies
GEN L-Terminal Signal Command	Eng 2	On/Off	Off
HO2S 1	EVAP, Eng 1, Fuel Trim	millivolts	0-1000 and Varying
HO2S 2	Eng 1, Fuel Trim	millivolts	0-1000 and Varying
HO2S 1 Heater Circuit Status	ODD	OK/ Fault	OK
HO2S 2 Heater Circuit Status	ODD	OK/ Fault	OK
HO2S 1 Heater Command	Eng 2	On/Off	On
HO2S 2 Heater Command	Eng 2	On/Off	On
HO2S 1 Heater Current	Eng 2	Amps	Varies
HO2S 2 Heater Current	Eng 2	Amps	Varies
IAT Sensor	Eng 1, 2, EVAP, EGR, Fuel Trim	°C/°F	Varies
Ignition 1 Signal	EVAP, Eng 1, 2, EGR, Fuel Trim, CC, TAC	Volts	Varies
Ignition Mode	Eng 2	IC/Bypass	IC
IMS	Eng 1, 2, EGR, Fuel Trim, CC	P, R, N, D4, D, 3, D2, D1	P
Initial Brake Apply Signal	CC, TAC	Released/ Applied	Released
Injector PWM	Eng 2, Fuel Trim, Misfire	milliseconds	Varies
Knock Retard	Eng 1, EGR	Degrees	0
Long Term FT	Eng 1, 2, EVAP, Fuel Trim	Percent	-10% to +10%
Loop Status	Eng 1, 2, Fuel Trim, EVAP, EGR	Open/Closed	Closed
MAF Sensor	Eng 1, 2, EGR, Misfire, EVAP, Fuel Trim, TAC	g/s	3-6 (Varies with altitude)
MAF Sensor	Eng 2	Hz	1,200-3,000 (depends on altitude)
MAP Sensor	Eng 1, 2	Volts	0.4-2.0 Volts
MAP Sensor	Eng 1, 2, EVAP, Fuel Trim, Misfire, EGR, TAC	kPa	20-48 kPa (Varies with altitude)
MIL Command	Eng 2	On/Off	Off
MIL Circuit Status	ODD	OK/Fault/Invalid State	OK
Misfire Current Cyl. 1-6	Misfire	Counts	0-4
Misfire History Cyl. 1-6	Misfire	Counts	0
Number of DTCs	Eng 1, EVAP, Fuel Trim, TAC	Number	0
PCM/VCM in VTD Fail/Enable	Eng 1	Yes/No	No
Power Enrichment	Eng 2, Misfire	Active/Inactive	Inactive
Reduced Engine Power	Eng 1, TAC, Misfire, CC	Active/Inactive	Inactive
Short Term FT	Eng 1, 2, EVAP, Fuel Trim	Percent	-10 to +10
Spark	Eng 1, 2, Misfire, Fuel Trim	Degrees	22 (varies)
Starter Enable Relay Circuit Status	ODD	OK/ Fault	OK
Starter Relay Command	Eng 2	On/Off	Off
Start Up ECT	EVAP, Fuel Trim	°C/°F	Varies
Start Up IAT	EVAP, Fuel Trim	°C/°F	Varies

2005 LaCrosse 3.8L V6 VIN 1 (CONT)

Scan Tool Parameter	Data List	Units Displayed	Typical Data Value
Stop Lamp Signal	CC, TAC	On/Off	Off
TAC/PCM Communications Signal	Eng 1, CC, TAC	OK/ Fault	OK
TCC Brake Pedal Switch	Eng 2	Applied/Released	Released
TCC PWM Solenoid Circuit Status	ODD	OK/ Fault/ Invalid State	OK
TCC PWM Solenoid Command	Eng 2, Misfire, CC	On/Off	Off
TCS Delivered Torque Circuit Status	ODD	OK/ Fault	OK
Torque Request Signal (If Equipped)	Eng 2, TAC	Percent	100%
Torque Delivered Signal (If Equipped)	Eng 2, TAC	Percent	75%
TP Desired Angle	Eng 1, 2, EGR, TAC, Fuel Trim, EVAP, CC	Percent	Varies
TP Indicated Angle	Eng 1, 2, EGR, Fuel Trim, EVAP, Misfire, CC, TAC	Percent	%
TP Sensor 1	TAC	Volts	Varies
TP Sensor 2	TAC	Volts	Varies
TP Sensor 1	TAC	Percent	Varies
TP Sensor 2	TAC	Percent	Varies
TP Sensor 1 and 2	TAC	Agree/ Disagree	Agree
Traction Control Status (If Equipped)	Eng 2, CC, TAC	Active/Inactive	Inactive
Vehicle Speed Sensor	Eng 1, 2, EGR, EVAP, Fuel Trim, Misfire, CC, TAC	mph/ km/h	0
VTD Auto Learn Timer	Eng 1	Active/ Inactive	Inactive
VTD Fuel Disable	Eng 1	Active/ Inactive	Inactive
VTD Fuel Disable until Ignition Off	Eng 1	Yes/No	No

LESABRE, PARK AVENUE, REGAL & RIVIERA - PID DATA

1995 LeSabre, Park Avenue, Regal, Riviera 3.8L V6 MFI VIN L, 1

Parameter Identification (PID)	PID Value Range	PID Value at Hot Idle	PID Value at 30 mph	Data List Type
A/C Clutch Control	On/Off	Off	Off	Engine Data 1
A/C High Pressure	Not Okay/Okay	Okay	Okay	Engine Data 1
A/C Request	Yes/No	No	No	Engine Data 1
Actual EGR Position	0-100%	0	0	Engine Data 1
Boost PWM (VIN 1)	0-100%	0%	100%	Engine Data 1
Cam/Crank Error	0-255 counts	0-2	1	Engine Data 1
C/C Brake Switch	Apply/Released	Released	Released	Engine Data 1
Commanded Gear	1-2-3-4	1	3	Engine Data 1
Cooling Fan 1, 2	On/Off	Off	Off	Engine Data 1
Cruise Mode	Off, 1-7	Off	Off	Engine Data 1
Current Weak Cyl #	CYL #/None	Not Enabled	Not Enabled	Engine Data 1
Desired Idle Speed	0-3187 rpm	620	620	Engine Data 1
Desired EGR Position	0-100%	0	0	Engine Data 1
ECT Sensor (°F)	-40 to 419°F	185-220	185-220	Engine Data 1
EGR Pintle Position	0-5.1 volts	0.55	0.76	Engine Data 1
Engine Load	0-100%	70-80	45	Engine Data 1
Engine Speed	0-9999 rpm	±50 Desired	Varies	Engine Data 1
Fuel Trim Cell	Cell #0-4	0	0-4	Engine Data 1
HO2S-11 (B1 S1)	0-1132 mv	10-1000	480mv	Engine Data 1
HO2S Cross Counts	0-255 counts	Varies	Varies	Engine Data 1
IAC Motor	0-255 counts	16-20	---	Engine Data 1
IAT Sensor (°F)	-38-284°F	50-176	50-176	Engine Data 1
Ignition 1 Signal	0.0-25.5v	14.2	14.2	Engine Data 1
Knock Retard (°)	0-128°	0	0	Engine Data 1
Knock Signal	Yes/No	No	No	Engine Data 1
Long Term Fuel Trim	0-255 counts	95-138	95-138	Engine Data 1
Loop Status	Closed Loop / Open Loop	Closed Loop	Closed Loop	Engine Data 1
MAF Sensor (g/sec)	0-512 g/sec	4-9	Varies	Engine Data 1
Pass-Key Fuel Enable	Disabled/Enabled	Enabled	Enabled	Engine Data 1
Park Neutral Position	PN / R-DL	PN	R-DL	Engine Data 1
PRNDL 'P', 'A', 'B', 'C'	HI / LOW	HIGH	LOW	Engine Data 1
Purge Duty Cycle	0-100%	0%	0-100%	Engine Data 1
QDM 1, 2 & 4	HIGH/LOW	LOW	LOW	Engine Data 1
Rich/Lean Status	L or R	L-R-L-R	L-R-L-R	Engine Data 1
Raw Servo Position	0-255 counts	15-85	47	Engine Data 1
Shift Solenoid 'A', 'B'	On/Off	On	Off	Engine Data 1
Short Term Fuel Trim	0-255 counts	95-138	95-138	Engine Data 1
Spark Advance (°)	-90° to 90°	16	26	Engine Data 1
TCC Brake Switch	Apply/Released	Released	Released	Engine Data 1
TCC Duty Cycle	0-100%	0	0-100	Engine Data 1
TCC Mode	On/Applied/Off	Off	Applied	Engine Data 1
TCC Slip Speed	-4080 - 4079	+254	Varies	Engine Data 1
Throttle Angle (%)	0-100%	0	2	Engine Data 1
Time From Startup	Hr: Min: Sec	00:00:00	00:00:00	Engine Data 1
TFT Sensor (°F)	40-304°F	198	198	Engine Data 1
Transaxle N/V Ratio	0-255 counts	0	42	Engine Data 1
Transmission Range Switch	P-R-N-4-3-2	Park	4	Engine Data 1
2nd, 3rd, 4th Gear Switch	No	No	Yes: 2nd	Engine Data 1

1996-99 LeSabre, Park Avenue 3.8L V6 VIN K, VIN 1 - Engine Data 1

Parameter Identification (PID)	PID Value Range	PID Value at Hot Idle	PID Value at 30 mph	Data List Type
Actual EGR Position	0-100%	0	0-100	Engine Data 1
Air Fuel Ratio	0.1-25.5:1	14.7:1	14.7:1	Engine Data 1
BARO Sensor (kPa)	10-110 kPa	65-105	65-105	Engine Data 1
Generator Command	On/Off	On	On	Engine Data 1
Cruise Control	Disabled/Enabled	Disabled	Disabled	Engine Data 1
Cruise Inhibited	Yes/No	Yes	Yes	Engine Data 1
Current Gear	1-2-3-4	1	3	Engine Data 1
Decel Fuel Mode	Active/Inactive	Inactive	Inactive	Engine Data 1
Desired EGR Position	0-100%	0	0-100	Engine Data 1
Desired Idle Speed	0-3187 rpm	± 50 rpm	---	Engine Data 1
ECT Sensor (°F)	-40 to 304°F	185-220	185-220	Engine Data 1
Engine Load	0-100%	2-5	Varies	Engine Data 1
Engine Run Time	Hr: Min: Sec	00:00:00	00:00:00	Engine Data 1
Engine Speed	0-9999 rpm	±50 Desired	Varies	Engine Data 1
Fuel Trim Cell	Cell #0-4	0	0-4	Engine Data 1
Fuel Trim Learn	Disabled/Enabled	Enabled	Enabled	Engine Data 1
HO2S-11 (B1 S1)	0-1132 mv	10-1000	10-1000	Engine Data 1
HO2S-12 (B1 S2)	0-1132 mv	10-1000	10-1000	Engine Data 1
HO2S-11 Status	Not Ready	Ready	Ready	Engine Data 1
IAC Position	0-255 counts	10-40	---	Engine Data 1
IAT Sensor (°F)	-40 to 304°F	50-194	50-194	Engine Data 1
Ignition 1 Signal	0.0-25.5v	14.2	14.2	Engine Data 1
Injector Pulsewidth	0-985 ms	1.5-3.5	Varies	Engine Data 1
Long Term Fuel Trim	-10% to 10%	0 (± 5%)	0 (± 5%)	Engine Data 1
Loop Status	Closed Loop / Open Loop	Closed Loop	Closed Loop	Engine Data 1
MAF Sensor (g/sec)	0-512 g/sec	3-6	Varies	Engine Data 1
MAF Sensor (Hz)	0-32,000 Hz	1200-3000	Varies	Engine Data 1
MAP Sensor (kPa)	10-110 kPa	29-48	Varies	Engine Data 1
MAP Sensor (V)	0-5.1 volts	1-2	Varies	Engine Data 1
Power Enrichment	Active/Inactive	Inactive	Inactive	Engine Data 1
Rich / Lean	L or R	L-R-L-R	L-R-L-R	Engine Data 1
Short Term Fuel Trim	-10% to 10%	0 (± 5%)	0 (± 5%)	Engine Data 1
Throttle at Idle	Yes/No	Yes	No	Engine Data 1
TP Angle (%)	0-100%	0	Varies	Engine Data 1
TP Sensor (V)	0-5.1 volts	0.20-0.74	Varies	Engine Data 1
TWC Protection	Active/Inactive	Inactive	Inactive	Engine Data 1

1996-99 LeSabre, Park Avenue 3.8L V6 VIN K, VIN 1 - Engine Data 2

Parameter Identification (PID)	PID Value Range	PID Value at Hot Idle	PID Value at 30 mph	Data List Type
A/C HI Side Pressure	0-5.1 volts	Varies	Varies	Engine Data 2
A/C Pressure Out of Range	Yes/No	No	No	Engine Data 2
A/C Request	Yes/No	No	No	Engine Data 2
BARO Sensor (kPa)	10-110 kPa	65-105	65-105	Engine Data 2
BARO Sensor (V)	0-5.1 volts	3.5-4.5	3.5-4.5	Engine Data 2
Boost Solenoid VIN 1	0-100%	100	0-100	Engine Data 2

1996-99 LeSabre, Park Avenue 3.8L V6 VIN K, VIN 1 - Engine Data 2

Parameter Identification (PID)	PID Value Range	PID Value at Hot Idle	PID Value at 30 mph	Data List Type
Commanded A/C	On/Off	Off	Off	Engine Data 2
Commanded Fan 1, 2	On/Off	Off	Off	Engine Data 2
Commanded Starter	Disabled/Enabled	Disabled	Disabled	Engine Data 2
Decel Fuel Mode	Active/Inactive	Inactive	Inactive	Engine Data 2
Desired Idle Speed	0-3187 rpm	PCM control	---	Engine Data 2
ECT Sensor (°F)	-40 to 304°F	185-220	185-220	Engine Data 2
Engine Load	0-100%	2-5	Varies	Engine Data 2
Engine Oil Level	LOW / Okay	Okay	Okay	Engine Data 2
Engine Oil Life	0-100%	Varies	Varies	Engine Data 2
Engine Run Time	Hr: Min: Sec	00:00:00	00:00:00	Engine Data 2
Engine Speed	0-9999 rpm	±50 Desired	Varies	Engine Data 2
Fuel Pump	On/Off	On	On	Engine Data 2
Fuel Pump Speed	HI/Normal	Normal	Normal	Engine Data 2
Generator Signal	0-100%	Varies	Varies	Engine Data 2
IAC Position	0-255 counts	10-40	10-40	Engine Data 2
IAT Sensor (°F)	-40 to 304°F	50-194	50-194	Engine Data 2
Knock Retard (°)	0-25.5°	0	0	Engine Data 2
KS Minimum Learned Noise	0-5.1 volts	0.1-3.0	Varies	Engine Data 2
Low Oil Lamp	Off or On	Off	Off	Engine Data 2
MAF Sensor (g/sec)	0-512 g/sec	3-6	Varies	Engine Data 2
MAP Sensor (kPa)	10-110 kPa	29-48	Varies	Engine Data 2
MIL Command	On/Off	Off	Off	Engine Data 2
Power Enrichment	Active/Inactive	Active	Active	Engine Data 2
Spark Advance (°)	-64° - 64°	20	Varies	Engine Data 2
TP Angle (%)	0-100%	0	Varies	Engine Data 2
Traction Control	Active/Inactive	Inactive	Inactive	Engine Data 2
TCM Desired Torque	0-100%	80-95	0-100	Engine Data 2
T/C System Torque	0-100%	5-20	0-100	Engine Data 2
Transmission Range	P-R-N-4-3-2	P	4	Engine Data 2
TR Switch P-A-B-C	HI/LO/HI/LO	LO/LO/HI/HI	HI/HI/LO/LO	Engine Data 2
TWC Protection	Active/Inactive	Inactive	Inactive	Engine Data 2
VTD Fuel Disable	Active/Inactive	Inactive	Inactive	Engine Data 2

1996-99 LeSabre, Park Avenue 3.8L V6 VIN K, VIN 1 - EVAP Data List

Parameter Identification (PID)	PID Value Range	PID Value at Hot Idle	PID Value at 30 mph	Data List Type
BARO Sensor (kPa)	10-110 kPa	65-105	65-105	EVAP
ECT Sensor (°F)	-40 to 304°F	185-220	185-220	EVAP
Engine Run Time	Hr: Min: Sec	00:00:00	00:00:00	EVAP
Engine Speed	0-9999 rpm	±50 Desired	---	EVAP
EVAP Canister Purge	0-100%	0-25	0-50	EVAP
EVAP Vacuum Switch	Purge/No	No Purge	Purge	EVAP
IAT Sensor (°F)	-40 to 304°F	50-194	50-194	EVAP
MAF Sensor (g/sec)	0-512 g/sec	3-6	Varies	EVAP
MAP Sensor (kPa)	10-110 kPa	29-48	Varies	EVAP
Startup ECT Sensor	-40 to 304°F	Varies	Varies	EVAP
Startup IAT Sensor	-40 to 304°F	Varies	Varies	EVAP

1996-99 LeSabre, Park Avenue 3.8L V6 VIN K, VIN 1 - EGR Data List

Parameter Identification (PID)	PID Value Range	PID Value at Hot Idle	PID Value at 30 mph	Data List Type
Actual EGR Position	0-100%	0	0-100	EGR
BARO Sensor (kPa)	10-110 kPa	65-105	65-105	EGR
BARO Sensor (V)	0-5.1 volts	3.5-4.5	3.5-4.5	EGR
Desired EGR Position	0-100%	0	0-100	EGR
ECT Sensor (°F)	-38-304°F	185-220	185-220	EGR
EGR Sensor Closed Pintle Position	0-5.1 volts	0.14-1.0	Varies	EGR
EGR Duty Cycle	0-100%	0	0-100	EGR
EGR Feedback	0-5.1 volts	0.14-1.0	---	EGR
EGR Flow Test	0-255 counts	0-10	0-10	EGR
EGR Position Error	0-100%	0	0-100	EGR
Engine Speed	0-9999 rpm	±50 Desired	Varies	EGR
IAC Position	0-255 counts	10-40	10-40	EGR
Ignition 1 Signal	0.0-25.5v	14.2	14.2	EGR
MAP Sensor (kPa)	10-110 kPa	29-48	Varies	EGR
MAP Sensor (V)	0-5.1 volts	1-2	Varies	EGR
TP Angle (%)	0-100%	0	Varies	EGR
Vehicle Speed	0-155 mph	0	30	EGR

1996-99 LeSabre, Park Avenue 3.8L V6 VIN K, VIN 1 - Misfire Data List

Parameter Identification (PID)	PID Value Range	PID Value at Hot Idle	PID Value at 30 mph	Data List Type
Actual EGR Position	0-100%	0	0-100	Misfire
Decel Fuel Mode	Active/Inactive	Inactive	Inactive	Misfire
Desired EGR Position	0-100%	0	0-100	Misfire
ECT Sensor (°F)	-40 to 304°F	185-220	185-220	Misfire
Engine Load	0-100%	2-5	Varies	Misfire
Engine Speed	0-9999 rpm	±50 Desired	Varies	Misfire
IAC Position	0-255 counts	10-40	---	Misfire
Long Term Fuel Trim	-10% to 10%	Varies	Varies	Misfire
Loop Status	Closed Loop / Open Loop	Closed Loop	Closed Loop	Misfire
Misfire Current CYL #	0-198	0	0	Misfire
Misfire History CYL #	0-65535	0	0	Misfire
Misfiring Cylinder	CYL 1-6	0	0	Misfire
Power Enrichment	Active/Inactive	Active	Active	Misfire
Short Term Fuel Trim	-10% to 10%	Varies	Varies	Misfire
Total Misfire Current Count	0-99	0-5	0-5	Misfire
Total Misfire Fails Since First Failure	0-65,535	0	0	Misfire
Total Misfire Passes Since First Fail	0-65,535	0	0	Misfire
TP Angle (%)	0-100%	0	Varies	Misfire
TWC Protection	Active/Inactive	Inactive	Inactive	Misfire
Vehicle Speed	0-155 mph	0	30	Misfire

1996-99 LeSabre 3.8L V6 MFI VIN K, VIN 1 (A/T) - HO2S Data List

Parameter Identification (PID)	PID Value Range	PID Value at Hot Idle	PID Value at 30 mph	Data List Type
Air Fuel Ratio	0.1-25.5:1	14.7:1	14.7:1	14.7:1
Decel Fuel Mode	Active/Inactive	Inactive	Inactive	Inactive
ECT Sensor (°F)	-40 to 304°F	185-220	185-220	185-220
EVAP Canister Purge	0-100%	0-25	0-100	0-100
Engine Speed	0-9999 rpm	±50 Desired	Varies	Varies
HO2S-11 (B1 S1)	0-1132 mv	10-1000	10-1000	10-1000
HO2S-21 (B2 S1)	0-1132 mv	10-1000	10-1000	10-1000
HO2S-11 Status	Not Ready	Ready	Ready	Ready
HO2S1 Warmup	Min: Sec	Varies	Varies	Varies
HO2S2 Warmup	Min: Sec	Varies	Varies	Varies
HO2S2 XCounts	0-255 counts	Varies	Varies	Varies
IAT Sensor (°F)	-40 to 304°F	50-194	50-194	50-194
Ignition 1 Signal	0.0-25.5v	14.2	14.2	14.2
Loop Status	Closed Loop / Open Loop	Closed Loop	Closed Loop	Closed Loop
MAF Sensor (g/sec)	0-512 g/sec	3-6	Varies	Varies
Power Enrichment	Active/Inactive	Active	Active	Active
Startup ECT Sensor	-40 to 304°F	Varies	Varies	Varies
Startup IAT Sensor	-40 to 304°F	Varies	Varies	Varies
TP Angle (%)	0-100%	0	Varies	Varies
TWC Protection	Active/Inactive	Inactive	Inactive	Inactive

1996-99 LeSabre 3.8L V6 MFI VIN K, VIN 1 (A/T) - Catalyst Data List

Parameter Identification (PID)	PID Value Range	PID Value at Hot Idle	PID Value at 30 mph	Data List Type
Decel Fuel Mode	Active/Inactive	Inactive	Inactive	Catalyst
ECT Sensor (°F)	-40 to 304°F	185-220	185-220	Catalyst
Engine Load	0-100%	2-5	Varies	Catalyst
Engine Run Time	Hr: Min: Sec	00:00:00	00:00:00	Catalyst
Engine Speed	0-9999 rpm	±50 Desired	---	Catalyst
HO2S-11 (B1 S1)	0-1132 mv	10-1000	10-1000	Catalyst
HO2S-21 (B2 S1)	0-1132 mv	10-1000	10-1000	Catalyst
IAC Position	0-255 counts	10-40	---	Catalyst
IAT Sensor (°F)	-40 to 304°F	50-194	50-194	Catalyst
Loop Status	Closed Loop / Open Loop	Closed Loop	Closed Loop	Catalyst
MAF Sensor (g/sec)	0-512 g/sec	3-6	Varies	Catalyst
Power Enrichment	Active/Inactive	Active	Active	Catalyst
TP Angle (%)	0-100%	0	Varies	Catalyst
TWC Diagnostic	Disabled/Enabled	Enabled	Enabled	Catalyst
TWC Monitor Test Counter	0-255 counts	0	0	Catalyst
TWC Protection	Active/Inactive	Inactive	Inactive	Catalyst
Vehicle Speed	0-155 mph	0	30	Catalyst

2000-05 LeSabre, Park Avenue, Regal 3.8L V6 VIN K, VIN 1 - Data List

Parameter Identification (PID)	PID Value Range	PID Value at Hot Idle	PID Value at 30 mph	Data List Type
3X Crank Sensor	0-9999 rpm	Varies	Varies	Engine Data 2
18X Crank Sensor	0-1600 rpm	Varies	Varies	Engine Data 2
A/T 1-2 Solenoid Circuit Status	Okay / Fault / INVALID	Okay	Okay	Output Driver
A/T 2-3 Solenoid Circuit Status	Okay / Fault / INVALID	Okay	Okay	Output Driver
A/C HI Side Pressure	0-5.1 volts	Varies	Varies	Engine Data 2
A/C Off for WOT	Yes/No	No	No	Engine Data 2
A/C Pressure Disable	Yes/No	No	No	Engine Data 2
A/C Relay Control Circuit Status	Okay / Fault / INVALID	Okay	Okay	Output Driver
A/C Relay Command	On/Off	Off	Off	Engine Data 1-2, EGR, Misfire
A/C Request Signal	Yes/No	No	No	Engine Data 2
Air Fuel Ratio	0.0-25.5:1	14.2-14.7:1	14.2-14.7:1	Engine Data 2, Fuel Trim
AIR Active Test Air Injection	Yes/No	No	No	AIR
AIR Active Test Inhibit	Yes/No	No	No	AIR
AIR Active Test Passed	Yes/No	Yes	Yes	AIR
AIR Passive Test 1 Pass	Yes/No	No	No	AIR
AIR Passive Test 2 Failed	Yes/No	No	No	AIR
AIR Passive Test In Progress	Yes/No	No	No	AIR
AIR Passive Test Inhibit	Yes/No	No	No	AIR
AIR Passive Test Passed	Yes/No	No	No	AIR
Air Pump Relay Circuit Status	Okay / Fault / Invalid State	Okay	Okay	Output Driver
Air Pump Relay Command	On/Off	Off	Off	Engine Data 1, Fuel Trim
Air Pump Solenoid Circuit Status	Okay / Fault / Invalid State	Okay	Okay	Output Driver
Air Pump Solenoid Command	On/Off	Off	Off	Engine Data 1, Fuel Trim
BARO Sensor (kPa)	10-110 kPa	98	98	Engine Data 1, EGR, EVAP, Fuel Trim
Boost Solenoid Circuit Status (VIN 1 only)	Okay / Fault / INVALID	Okay	Okay	Output Driver
Boost Solenoid Control (VIN 1 only)	0-100%	0	0	Engine Data 2

2000-05 LeSabre, Park Avenue, Regal 3.8L V6 VIN K, VIN 1 - Data List

Parameter Identification (PID)	PID Value Range	PID Value at Hot Idle	PID Value at 30 mph	Data List Type
CMP Sensor Signal	Yes/No	Yes	Yes	Engine Data 2
Cruise Control Active	Yes/No	No	No	Engine Data 1
Cruise Inhibit Reason	VSS (speed)	0	30	Engine Data 1
Cruise Inhibit Signal Circuit Status	Okay / Stuck Low (open) / Stuck High	Okay	Okay	Engine Data 1
Current Gear	1-2-3-4	1	4	Engine Data 1-2, EGR, Fuel Trim
Cycles of Misfire Data	0-99	Varies	Varies	Misfire
Cylinder 1-6 Injector Circuit History	Okay / Stuck Low / Stuck High / Fault	Okay	Okay	Output Driver
Cylinder 1-6 Injector Circuit Status	Okay / Stuck Low (open) / Stuck High	Okay	Okay	Output Driver
Desired EGR Position	0-100%	0	Varies	Engine Data 1, EGR, Misfire
Decel Fuel Cutoff	Active/Inactive	Inactive	Inactive	EGR, Fuel Trim
Desired Idle Speed	0-3187 rpm	± 50 rpm of Actual Speed	Varies	Engine Data 1, Data 2, EVAP
Driver Module 1-4 Status	Enabled/Off/ High Volts / High TEMP / Invalid State	Enabled	Enabled	Output Driver
ECT Sensor (°F)	-40 to 304°F	185-220	185-220	Engine Data 1-2, EGR, Fuel Trim, EVAP, Misfire
EGR Flow Test Count	0-255 counts	0-10	Varies	EGR
EGR Learned Minimum Position	0-5.1 volts	0.14-1.0	Varies	EGR
EGR Position Sensor	0-100%	0	Varies	Engine Data 1, EGR, Misfire
EGR Position Sensor	0-5.1volts	0.0	Varies	Engine Data 1, EGR, Misfire
EGR Position Variance	0-100%	0-9	Varies	EGR
EGR Solenoid Circuit History	Okay/Fault	Okay	Okay	Output Driver
EGR Solenoid Circuit Status	Enabled/Off/ High Volts / High TEMP / Invalid State	0	Varies	EGR
Engine Load (%)	0-100%	17	Varies	Engine Data 1-2, EGR, Fuel Trim, EVAP, Misfire
Engine Oil Level Switch	Okay/Not Okay	Okay	Okay	Engine Data 2
Engine Oil Life Left	0-100%	Varies	Varies	Engine Data 2
EOP Sensor	0-5.1 volts	2-3	2-4	Engine Data 2

2000-05 LeSabre, Park Avenue, Regal 3.8L V6 VIN K, VIN 1 - Data List

Parameter Identification (PID)	PID Value Range	PID Value at Hot Idle	PID Value at 30 mph	Data List Type
Engine Run Time	Hr: Min: Sec	00:00:00	00:00:00	Engine Data 1-2, EGR, Fuel Trim, EVAP, Misfire
Engine Speed	0-9999 rpm	± 50 rpm of Desired Speed	Varies	Engine Data 1-2, EGR, Fuel Trim, EVAP, Misfire
EVAP Fault History	No Fault / Excess VAC / Purge Valve Leak / Small Leak / Weak Vacuum	No Fault	No Fault	EVAP
EVAP Purge Solenoid Control Circuit Status	Okay / Fault / Invalid State	Okay	Okay	Output Driver
EVAP Purge Solenoid Command	0-100%	18	Varies	Engine Data 1, EVAP, Fuel Trim
EVAP Vent Solenoid Circuit Status	Okay / Fault / Invalid State	Okay	Okay	Output Driver
EVAP Vent Solenoid	Venting/Not Venting	Not Venting	Not Venting	Engine Data 1, EVAP, Fuel Trim
Extended Travel Brake Pedal Switch	Engaged / Released	Released	Released	Engine Data 2
Fan Control Relay 1 Circuit Status	Okay / Fault / Invalid State	Okay	Okay	Output Driver
Fan Control Relay 1 Command	On/Off	Off	Off	Engine Data 2
Fan Control Relay 2-3 Command	On/Off	Off	Off	Engine Data 2
Fan Control Relay 2-3 Circuit Status	Okay / Fault / Invalid State	Okay	Okay	Output Driver
Fuel Pump Relay Circuit Status	Okay / Fault / Invalid State	Okay	Okay	Output Driver
Fuel Pump Relay History Status	Okay/Fault	Okay	Okay	Output Driver
Fuel Pump Relay Command	On/Off	On	On	Engine Data 1
Fuel Tank Level Remaining	0-100%	Varies	Varies	EVAP
Fuel Tank Pressure (FTP) Sensor	Inches H2O, mm Hg	-17 to +7.5 Inches H2O	-17 to +7.5 Inches H2O	Engine Data 1, EVAP
Fuel Tank Pressure (FTP) Sensor	0-5.1 volts	Varies	Varies	Engine Data 1, EVAP
Fuel Trim Cell	Cell 0-9	0	1	Engine Data 1, EVAP, Fuel Trim
Fuel Trim Learn	Enabled or Disabled	Enabled	May Toggle	Engine Data 1, EVAP, Fuel Trim
Generator F-Terminal	0-100%	Varies	Varies	Engine Data 2
Generator L-Terminal	On/Off	On	On	Engine Data 2

2000-05 LeSabre, Park Avenue, Regal 3.8L V6 VIN K, VIN 1 - Data List

Parameter Identification (PID)	PID Value Range	PID Value at Hot Idle	PID Value at 30 mph	Data List Type
HO2S-11 (B1 S1)	0-1132 mv	10-1000	10-1000	Engine Data 1, EVAP, Fuel Trim
HO2S-12 (B1 S2)	0-1132 mv	10-1000	10-1000	Engine Data 1, EVAP, Fuel Trim
IAC Position	0-255 counts	10-40	10-40	Engine Data 1, EGR, Fuel Trim
IAT Sensor (°F)	-40 to 304°F	50-194	50-194	Engine Data 1-2, EGR, EVAP, Fuel Trim
Ignition 1 Signal	0.0-25.5v	14.2	14.2	Engine Data 1-2, EGR, EVAP, Fuel Trim
Ignition Mode	Bypass/IC	IC	IC	Engine Data 2
Injector Pulsewidth	0-1000 ms	1.5-3.5	Varies	Engine Data 2, Fuel Trim, Misfire
Knock Retard (°)	0-25.5°	0°	0°	Engine Data 1, EGR
Long Term Fuel Trim	-27% to 27%	0 (± 5%)	0 (± 5%)	Engine Data 1-2 EVAP, Fuel Trim
Loop Status	Closed Loop / Open Loop	Closed Loop	Closed Loop	Engine Data 1-2, EGR, EVAP, Fuel Trim
MAF Sensor (g/sec)	0-512 g/sec	3-6	Varies	Engine Data 1-2, EGR, Fuel Trim, EVAP, Misfire
MAF Sensor (Hz)	0-32,000 Hz	1200-3000	Varies	Engine Data 2
MAP Sensor (kPa)	10-105 kPa	20-48	Varies	EGR, EVAP, Fuel Trim, Misfire
MAP Sensor (V)	0-5.1 volts	0.75-2.0	Varies	Engine Data 1-2
MIL Status	Okay / Fault / Invalid State	Okay	Okay	Output Driver
MIL Command	On/Off	Off	Off	Engine Data 2
Misfire Current Cyl 1-6	0-198 counts	0-4	0-4	Misfire
M/Fire History Cyl 1-6	0-65,535	0	0	Misfire
Number of DTC(s)	Number	0	0	Engine Data 1, EVAP, Fuel Trim
PCM/VCM in VTD Fail Enable	Yes/No	Yes	Yes	Engine Data 1
PCM Reset	Yes/No	No	No	Engine Data 1
Power Enrichment	Active/Inactive	Inactive	Inactive	Engine Data 2, Misfire
Short Term Fuel Trim	-27% to 27%	0 (± 5%)	0 (± 5%)	Engine Data 1-2 EVAP, Fuel Trim
Spark Advance (°)	-64° to 64°	20	Varies	Engine Data 1-2 Fuel Trim, Misfire

2000-05 LeSabre, Park Avenue, Regal 3.8L V6 VIN K, VIN 1 - Data List

Parameter Identification (PID)	PID Value Range	PID Value at Hot Idle	PID Value at 30 mph	Data List Type
Startup ECT	-40 to 419°F	Varies	Varies	EVAP, Fuel Trim
Startup IAT	-40 to 419°F	80°F	80°F	EVAP, Fuel Trim
Starter Enable Relay Control Circuit Status	Okay / Fault / Invalid State	Okay	Okay	Output Driver
Starter Enable Command	On/Off	Off	Off	Engine Data 2
TCC Brake Pedal Switch	Applied/REL	Released	Released	Engine Data 1-2
TFP Switch A/B/C	P, R, N, D, 4, 3, 2, 1	Park	4	Engine Data 2, EGR, Fuel Trim
Torque Delivered	0-100%	86	Varies	Engine Data 2
Torque Request	0-100%	100%	Varies	Engine Data 2
TP Angle (%)	0-100%	0	Varies	Engine Data 1-2, EGR, Fuel Trim, EVAP, Misfire
TP Sensor (V)	0-5.1 volts	0.5-1.2	Varies	Engine Data 1-2
Traction Control Circuit History	Okay/Fault	Okay	Okay	Output Driver
Traction Control Circuit Status	Okay/Fault	Okay	Okay	Output Driver
Traction Control Status	Active/Inactive	Inactive	Inactive	Engine Data 2
TWC Protection	Active/Inactive	Inactive	Inactive	Engine Data 1-2 HO2S, Misfire
Vehicle Speed	0-155 mph	0	30	Engine Data 1-2, EGR, Fuel Trim, EVAP, Misfire
VTD Auto Learn timer	Active/Inactive	Inactive	Inactive	Engine Data 1
VTD Fuel Disable	Active/Inactive	Inactive	Inactive	Engine Data 1
VTD Fuel Disable Until Key Off	Yes/No	No	No	Engine Data 1

REGAL & RIVIERA PID DATA - OBD II

1996-99 Regal & Riviera 3.8L V6 MFI VIN K, VIN 1 - Engine Data 1 List

Parameter Identification (PID)	PID Value Range	PID Value at Hot Idle	PID Value at 30 mph	Data List Type
Actual EGR Position	0-100%	0	0-100	Engine Data 1
Air Fuel Ratio	0.1-25.5:1	14.7:1	14.7:1	Engine Data 1
BARO Sensor (kPa)	10-110 kPa	65-105	65-105	Engine Data 1
BARO Sensor (V)	0-5.1 volts	3.5-4.5	3.5-4.5	Engine Data 1
Generator Command	On/Off	On	On	Engine Data 1
Commanded TCC	Disengage/Engage	Disengaged	Disengaged	Engine Data 1
Cruise Engaged	No / Yes	No	No	Engine Data 1
Cruise Inhibited	Yes/No	Yes	Yes	Engine Data 1
Current Gear	1-2-3-4	1	3	Engine Data 1
Decel Fuel Mode	Active/Inactive	Inactive	Inactive	Engine Data 1
Desired EGR Position	0-100%	0	0-100	Engine Data 1
ECT Sensor (°F)	-40 to 304°F	185-220	185-220	Engine Data 1
Engine Load	0-100%	2-5	Varies	Engine Data 1
Engine Run Time	Hr: Min: Sec	00:00:00	00:00:00	Engine Data 1
Engine Speed	0-9999 rpm	±50 Desired	Varies	Engine Data 1
Fuel Trim Cell	Cell #0-4	0	0-4	Engine Data 1
Fuel Trim Learn	Disabled/Enabled	Enabled	Enabled	Engine Data 1
Hot Open Loop	Active/Inactive	Inactive	Inactive	Engine Data 1
HO2S (B1 S1)	10-1000	10-1000	10-1000	Engine Data 1
HO2S (B1 S2)	10-1000	10-1000	10-1000	Engine Data 1
HO2S-11 Status	Not Ready	Ready	Ready	Engine Data 1
IAC Position	0-255 counts	10-40	---	Engine Data 1
IAT Sensor (°F)	-40 to 304°F	50-194	50-194	Engine Data 1
Injector Pulsewidth	0-985 ms	1.5-3.5	Varies	Engine Data 1
Long Term Fuel Trim	-10% to 10%	Varies	Varies	Engine Data 1
Loop Status	Closed Loop / Open Loop	Closed Loop	Closed Loop	Engine Data 1
MAF Sensor (g/sec)	0-512 g/sec	3-6	Varies	Engine Data 1
MAF Sensor (Hz)	0-32000 Hz	1200-3000	Varies	Engine Data 1
MAP Sensor (kPa)	10-110 kPa	29-48	Varies	Engine Data 1
Power Enrichment	Active/Inactive	Active	Active	Engine Data 1
Rich / Lean	L or R	L-R-L-R	L-R-L-R	Engine Data 1
Short Term Fuel Trim	-10% to 10%	Varies	Varies	Engine Data 1
Throttle at Idle	Yes/No	Yes	No	Engine Data 1
TP Angle (%)	0-100%	0	0-100	Engine Data 1
TWC Protection	Active/Inactive	Inactive	Inactive	Engine Data 1
Vehicle Speed	0-155 mph	0	30	Engine Data 1

1996-99 Regal & Riviera 3.8L V6 MFI VIN K, VIN 1 - Engine Data 2 List

Parameter Identification (PID)	PID Value Range	PID Value at Hot Idle	PID Value at 30 mph	Data List Type
Actual EGR Position	0-100%	0	0-100	Engine Data 2
A/C Off for WOT	No / Yes	No	No	Engine Data 2
A/C HI Side Pressure	0-5.1 volts	Varies	Varies	Engine Data 2
A/C Press. Out of Range	Yes/No	No	No	Engine Data 2
A/C Request	Yes/No	No	No	Engine Data 2
A/C Slugging	0-100%	0	0	Engine Data 2
BARO Sensor (kPa)	10-110 kPa	65-105	65-105	Engine Data 2
Boost Solenoid (VIN 1)	0-100%	100	0-100	Engine Data 2

1996-99 Regal & Riviera 3.8L V6 MFI VIN K, VIN 1 - Engine Data 2 List

Parameter Identification (PID)	PID Value Range	PID Value at Hot Idle	PID Value at 30 mph	Data List Type
Commanded A/C	On/Off	Off	Off	Engine Data 2
Commanded Fan 1, 2	On/Off	Off	Off	Engine Data 2
Commanded TCC	Disengage/Engage	Disengaged	Disengaged	Engine Data 2
Desired EGR Position	0-100%	0	0-100	Engine Data 2
Desired Idle Speed	0-3187 rpm	PCM control	---	Engine Data 2
ECT Sensor (°F)	-40 to 304°F	185-220	185-220	Engine Data 2
Engine Load	0-100%	2-5	Varies	Engine Data 2
Engine Speed	0-9999 rpm	±50 Desired	Varies	Engine Data 2
EVAP Purge PWM	0-100%	0-25	0-100	Engine Data 2
EVAP Vacuum Switch	Purge/No	No Purge	Purge	Engine Data 2
Fuel Pump	On/Off	On	On	Engine Data 2
Generator Lamp	Off / On	Off	Off	Engine Data 2
Hot Open Loop	Active/Inactive	Inactive	Inactive	Engine Data 2
IAC Position	0-255 counts	10-40	---	Engine Data 2
IAT Sensor (°F)	-40 to 304°F	50-194	50-194	Engine Data 2
Idle Speed Error	0-3187	Under 100	---	Engine Data 2
Ignition Mode	IC/Bypass	IC	IC	Engine Data 2
Knock Retard (°)	0-25.5°	0	Varies	Engine Data 2
KS Min. Learn Noise	0-5.1 volts	0.1-3.0	Varies	Engine Data 2
Low Oil Lamp	Off or On	Off	Off	Engine Data 2
MAF Sensor (g/sec)	0-512 g/sec	3-6	Varies	Engine Data 2
MAP Sensor (kPa)	450 kPa	29-48	Varies	Engine Data 2
MIL (lamp) Command	On/Off	Off	Off	Engine Data 2
Power Enrichment	Active/Inactive	Active	Active	Engine Data 2
Spark Advance (°F)	-64° to 64°	20	Varies	Engine Data 2
TP Angle (%)	0-100%	0	Varies	Engine Data 2
Transaxle Hot Mode	Active/Inactive	Inactive	Inactive	Engine Data 2
Transmission Range	P-N-R-4-3-2	P	4	Engine Data 2
TR Switch P-A-B-C	HI/LO	LO/LO/HI/HI	---	Engine Data 2
TWC Protection	Active/Inactive	Inactive	Inactive	Engine Data 2
VTD Fuel Disable	Active/Inactive	Inactive	Inactive	Engine Data 2

1996-99 Regal & Riviera 3.8L V6 MFI VIN K, VIN 1 EVAP Data List

Parameter Identification (PID)	PID Value Range	PID Value at Hot Idle	PID Value at 30 mph	Data List Type
ECT Sensor (°F)	-40 to 304°F	185-220	185-220	EVAP
Engine Load	0-100%	2-5	Varies	EVAP
Engine Speed	0-9999 rpm	±50 Desired	Varies	EVAP
EVAP Canister Purge	0-100%	0-25	0-100	EVAP
EVAP Vacuum Switch	Fault/No	No Purge	Purge	EVAP
Hot Open Loop	Active/Inactive	Inactive	Inactive	EVAP
Power Enrichment	Active/Inactive	Active	Active	EVAP
Startup ECT Sensor	-40 to 304°F	Varies	Varies	EVAP
Startup IAT Sensor	-40 to 304°F	Varies	Varies	EVAP
TP Angle (%)	0-100%	0	Varies	EVAP
TWC Protection	Active/Inactive	Inactive	Inactive	EVAP
Vehicle Speed	0-155 mph	0	30	EVAP

1996-99 Regal & Riviera 3.8L V6 MFI VIN K, VIN 1 - Catalyst Data List

Parameter Identification (PID)	PID Value Range	PID Value at Hot Idle	PID Value at 30 mph	Data List Type
Decel Fuel Mode	Active/Inactive	Inactive	Inactive	Catalyst
ECT Sensor (°F)	-40 to 304°F	185-220	185-220	Catalyst
Engine Load	0-100%	2-5	Varies	Catalyst
Engine Speed	0-9999 rpm	±50 Desired	Varies	Catalyst
Hot Open Loop	Active/Inactive	Inactive	Inactive	Catalyst
HO2S-11 (B1 S1)	0-1132 mv	10-1000	10-1000	Catalyst
HO2S-21 (B2 S1)	0-1132 mv	10-1000	10-1000	Catalyst
IAC Position	0-255 counts	10-40	10-40	Catalyst
IAT Sensor (°F)	-40 to 304°F	50-194	50-194	Catalyst
Loop Status	Closed Loop / Open Loop	Closed Loop	Closed Loop	Catalyst
MAF Sensor (g/sec)	0-512 g/sec	3-6	Varies	Catalyst
Power Enrichment	Active/Inactive	Inactive	Inactive	Catalyst
Throttle at Idle	Yes/No	Yes	No	Catalyst
TP Angle (%)	0-100%	0	Varies	Catalyst
TP Sensor (V)	0-5.1 volts	0.20-0.74	Varies	Catalyst
TWC Diagnostic	Disabled/Enabled	Enabled	Enabled	Catalyst
TWC Monitor Counter	0-49	0	0	Catalyst
TWC Protection	Active/Inactive	Inactive	Inactive	Catalyst
Vehicle Speed	0-155 mph	0	30	Catalyst

1996-99 Regal & Riviera 3.8L V6 MFI VIN K, VIN 1 - EGR Data List

Parameter Identification (PID)	PID Value Range	PID Value at Hot Idle	PID Value at 30 mph	Data List Type
Actual EGR Position	0-100%	0	0-100	EGR
BARO Sensor (kPa)	10-110 kPa	65-105	65-105	EGR
Commanded TCC	Disengage/Engage	Disengaged	Disengaged	EGR
Current Gear	1-2-3-4	1	3	EGR
Decel Fuel Mode	Active/Inactive	Inactive	Inactive	EGR
Desired EGR Position	0-100%	0	0-100	EGR
ECT Sensor (°F)	-38-304°F	185-220	185-220	EGR
EGR Closed Pintle Position	0-5.1 volts	0.14-1.0	Varies	EGR
EGR Duty Cycle	0-100%	0	0-100	EGR
EGR Feedback	0-5.1 volts	0.14-1.0	---	EGR
EGR Flow Test	0-255 counts	0-10	0-10	EGR
EGR Position Error	0-100%	0	0-100	EGR
Engine Load	0-100%	2-5	Varies	EGR
Engine Speed	0-9999 rpm	±50 Desired	---	EGR
Hot Open Loop	Active/Inactive	Inactive	Inactive	EGR
IAC Position	0-255 counts	10-40	---	EGR
IAT Sensor (°F)	-40 to 304°F	50-194	50-194	EGR
Loop Status	Closed Loop / Open Loop	Closed Loop	Closed Loop	EGR
MAP Sensor (kPa)	10-110 kPa	29-48	Varies	EGR
Rich / Lean Flag	L or R	L-R-L-R	L-R-L-R	EGR
TP Angle (%)	0-100%	0%	Varies	EGR
Transaxle Hot Mode	Active/Inactive	Inactive	Inactive	EGR
TWC Protection	Active/Inactive	Inactive	Inactive	EGR
Vehicle Speed	0-155 mph	0	30	EGR

1996-99 Regal & Riviera 3.8L V6 MFI VIN K, VIN 1 - HO2S Data List

Parameter Identification (PID)	PID Value Range	PID Value at Hot Idle	PID Value at 30 mph	Data List Type
Decel Fuel Mode	Active/Inactive	Inactive	Inactive	HO2S
ECT Sensor (°F)	-40 to 304°F	185-220	185-220	HO2S
Engine Load	0-100%	2-5	Varies	HO2S
Engine Speed	0-9999 rpm	±50 Desired	Varies	HO2S
Hot Open Loop	Active/Inactive	Inactive	Inactive	HO2S
HO2S (B1 S1)	0-1132 mv	10-1000	10-1000	HO2S
HO2S (B1 S2)	0-1132 mv	10-1000	10-1000	HO2S
HO2S-11 Status	Not Ready	Ready	Ready	HO2S
HO2S-11 Warmup	Min: Sec	00:00	00:00	HO2S
HO2S-12 Warmup	Min: Sec	00:00	00:00	HO2S
HO2S-12 XCount	0-255 counts	Varies	Varies	HO2S
IAT Sensor (°F)	-40 to 304°F	50-194	50-194	HO2S
Loop Status	Closed Loop / Open Loop	Closed Loop	Closed Loop	HO2S
MAF Sensor (g/sec)	0-512 g/sec	3-6	Varies	HO2S
Power Enrichment	Active/Inactive	Active	Active	HO2S
Startup ECT Sensor	-40 to 304°F	Varies	Varies	HO2S
Startup IAT Sensor	-40 to 304°F	Varies	Varies	HO2S
TP Angle (%)	0-100%	0	Varies	HO2S
TWC Protection	Active/Inactive	Inactive	Inactive	HO2S
Vehicle Speed	0-155 mph	0	30	HO2S

1996-99 Regal & Riviera 3.8L V6 MFI VIN K, VIN 1 - Misfire Data List

Parameter Identification (PID)	PID Value Range	PID Value at Hot Idle	PID Value at 30 mph	Data List Type
Actual EGR Position	0-100%	0%	0-100%	Misfire
Current Gear	1-2-3-4	1	3	Misfire
Decel Fuel Mode	Active/Inactive	Inactive	Inactive	Misfire
Desired EGR Position	0-100%	0%	0-100%	Misfire
ECT Sensor (°F)	-40 to 304°F	185-220	185-220	Misfire
Engine Load	0-100%	2-5	Varies	Misfire
Engine Speed	0-9999 rpm	±50 Desired	Varies	Misfire
Hot Open Loop	Active/Inactive	Inactive	Inactive	Misfire
IAC Position	0-255 counts	10-40	10-40	Misfire
Long Term Fuel Trim	-10% to 10%	0 (± 5%)	0 (± 5%)	Misfire
Loop Status	Closed Loop / Open Loop	Closed Loop	Closed Loop	Misfire
Misfire current Cyl 1-6	0-198	0-4	0-4	Misfire
Misfire History Cyl 1-6	0-65535	0	0	Misfire
Misfires Since 1st Fail	0-65535	0	0	Misfire
Misfire Since 1st Pass	0-65535	0	0	Misfire
Power Enrichment	Active/Inactive	Inactive	Inactive	Misfire
Short Term Fuel Trim	-10% to 10%	0 (± 5%)	0 (± 5%)	Misfire
Spark	-64° to 64°	20	Varies	Misfire
TCC Enable	No / Yes	No	No	Misfire
Total Misfire Current	0-99	0	0	Misfire
TP Angle (%)	0-100%	0	Varies	Misfire
TWC Protection	Active/Inactive	Inactive	Inactive	Misfire
Vehicle Speed	0-155 mph	0	30	Misfire

REGAL PID DATA - OBD II

2000-03 Regal 3.8L V6 MFI VIN K, VIN 1 (A/T) - Engine Data 1 List

Parameter Identification (PID)	PID Value Range	PID Value at Hot Idle	PID Value at 30 mph	Data List Type
A/C Relay Command	On/Off	Off	Off	Off
Actual EGR Position	0-100%	0	0-100	0-100
AIR Relay Command	On/Off	Off	Off	Off
AIR Solenoid Command	On/Off	Off	Off	Off
Air Fuel Ratio	0.0-25:1	14.7:1	14.7:1	14.7:1
BARO Sensor (kPa)	10-110 kPa	65-105	65-105	65-105
Brake Switch	Engage/REL	Released	Released	Released
Commanded Generator	On/Off	On	On	On
Cruise Control Active	Yes/No	No	No	No
Cruise Inhibit Command	Yes/No	Yes	Yes	Yes
Cruise Inhibit Reason	Several	Not Enabled	Not Enabled	Not Enabled
Current Gear	1-2-3-4	1	3	4
Decel Fuel Mode	Active/Inactive	Inactive	Inactive	Inactive
Desired EGR Position	0-100%	0	0-100	0-100
Desired Idle Speed	0-3187 rpm	PCM control	----	----
ECT Sensor (°F)	-40 to 304°F	185-220	185-220	185-220
EGR Position Sensor	0-100%	0-100	0-100	0-100
Engine Load	0-100%	2-5	Varies	Varies
Engine Run Time	Hr: Min: Sec	00:00:00	00:00:00	00:00:00
Engine Speed	0-9999 rpm	±50 Desired	Varies	Varies
EVAP Purge Command	0-100%	0-100	0-100	0-100
EVAP Vent Command	Open/closed	Open	Open	Open
F/P Relay Command	On/Off	On	On	On
FTP Sensor (In. H2O)	-17.5 to 7.5"	Varies	Varies	Varies
Fuel Trim Cell	Cell #0-4	0	0-4	0-4
Fuel Trim Learn	Enabled/DIS	Varies	Varies	Varies
HO2S-11 (B1 S1)	0-1132 mv	10-1000	10-1000	10-1000
HO2S-12 (B1 S2)	0-1132 mv	10-1000	10-1000	10-1000
HO2S-11 (B1 S1) Status	Not Ready/Ready	Ready	Ready	Ready
IAC Position (Counts)	0-255 counts	10-40	Varies	Varies
IAT Sensor (°F)	-40 to 304°F	50-194	50-194	50-194
Injector Pulsewidth	0-985 ms	1.5-3.5	Varies	Varies
Knock Retard (°)	0-25.5°	0	0	0
Long Term Fuel Trim	-10% to 10%	Varies	Varies	Varies
Loop Status	Closed Loop / Open Loop	Closed Loop	Closed Loop	Closed Loop
MAF Sensor (g/sec)	0-512 g/sec	3-6	Varies	Varies
MAF Sensor (Hz)	0-31,999Hz	1200-3000	Varies	Varies
MAP Sensor (kPa)	10-110 kPa	29-48	Varies	Varies
MAP Sensor (V)	0-5.1 volts	0.4-2	Varies	Varies
Number of Trouble Codes	Number	0	0	0
PCM/VCM VTD Fail	Yes/No	No	No	No
Power Enrichment	Active/Inactive	Inactive	Inactive	Inactive
Rich/Lean Status	Rich/Lean	R/L/R/L/R/L	R/L/R/L/R/L	R/L/R/L/R/L
Short Term Fuel Trim	-10% to 10%	Varies	Varies	Varies
Spark Advance (°)	-64 to 64°	20	Varies	Varies
TCC Brake Pedal Switch	Apply/Released	Released	Released	Released
Throttle At Idle	Yes/No	Yes	No	No

RENDEZVOUS PID DATA

2002-05 Rendezvous 3.4L VIN E V6 MFI - PID Data Lists (All)

Parameter Identification (PID)	PID Value Range	PID Value at Hot Idle	PID Value at 30 mph	Data List Type
3X Crank Sensor	0-9,999 rpm	Varies	Varies	Engine Data 2
24X Crank Sensor	0-1,600 rpm	Varies	Varies	Engine Data 2
A/C HI Side Pressure	0-459 kPa	90	90	Engine Data 2
A/C HI Side Pressure	0-5.1 volts	0.98	0.98	Engine Data 2
A/C Off for WOT	Yes/No	No	No	Engine Data 2
A/C Pressure Disable	Yes/No	No	No	Engine Data 2
A/C Relay CKT Status	Okay/Fault	Okay	Okay	Output Driver
A/C Relay Command	On/Off	Off	Off	Engine Data 1-2, EGR, Misfire
A/C Request	Yes/No	No	No	Engine Data 2
Air Fuel Ratio	0.0-25.5:1	14.7:1	14.7:1	Engine Data 2, Fuel Trim
BARO Sensor (kPa)	10-105 kPa	Varies w/ALT	Varies	Engine Data 1, EGR, EVAP, Fuel Trim
BARO Sensor (V)	0-5.1 volts	Varies w/ALT	Varies	Engine Data 1, EGR, EVAP, Fuel Trim
CMP Sensor Signal	Yes/No	Yes	Yes	Engine Data 2
Crank Request Signal	Yes/No	No	No	Engine Data 2
Cruise Control Active	Yes/No	No	No	Engine Data 1
Cruise Inhibit Reason	Several	Reason	Reason	Engine Data 1
Cruise Inhibit Signal Circuit Status	Okay/Fault	Okay	Okay	Output Driver
Cruise Inhibit Signal Command	On/Off	Off	On	Engine Data 1
Current Gear	1, 2, 3, 4	1	4	Engine Data 1-2, EGR, Fuel Trim
Cycles of Misfire Data	0-99 counts	Varies	Varies	Misfire
Cylinder 1-6 Injector Circuit Status	Okay/Fault	Okay	Okay	Output Driver
Cylinder 1-6 Injector Circuit History	Okay/Fault	Okay	Okay	Output Driver
Decel Fuel Cutoff	Active/Inactive	Inactive	Inactive	EGR, Fuel Trim
Desired EGR Position	0-100%	Varies	Varies	Engine Data 1, EGR, Misfire
Desired EGR Position	0-5.1v	Varies	Varies	Engine Data 1, EGR, Misfire
Desired Idle Speed	0-3187 rpm	Commanded by the PCM	Varies	Engine Data 1, Data 2, EVAP
Driver Module 1 Status	Enabled/DIS	Enabled	Enabled	Output Driver
Driver Module 2 Status	Enabled/DIS	Enabled	Enabled	Output Driver
Driver Module 3 Status	Enabled/DIS	Enabled	Enabled	Output Driver
Driver Module 4 Status	Enabled/DIS	Enabled	Enabled	Output Driver

2002-05 Rendezvous 3.4L VIN E V6 MFI - PID Data Lists (All)

Parameter Identification (PID)	PID Value Range	PID Value at Hot Idle	PID Value at 30 mph	Data List Type
ECT Sensor (°F)	-40 to 304°F	190-221°F	190-221°F	Engine Data 1-2, EGR, Fuel Trim, EVAP, Misfire
Engine Load	0-100%	1-4	Varies	Engine Data 1-2 EVAP, Fuel Trim HO2S, Misfire
EGR Flow Test Count	0-255 counts	0	Varies	EGR
EGR Learned Minimum Position	0-5.1 volts	0.70	Varies	EGR
EGR Position Sensor	0-100%	0	Varies	Engine Data 1, EGR, Misfire
EGR Position Sensor	0-5.1 volts	0.71	Varies	Engine Data 1, EGR, Misfire
EGR Position Variance	0-100%	0	0	EGR
EGR Solenoid Command	0-100%	0	Varies	EGR
Engine Oil Level Switch	Okay/Fault	Okay	Okay	Engine Data 2
Engine Oil Life Remaining	0-100%	Varies	Varies	Engine Data 2
Engine Oil Pressure Switch	Okay/Fault	Okay	Okay	Engine Data 2
Engine Run Time	Hr: Min: Sec 00:00:00 to 99:99:99	00:00:00	00:00:00	Engine Data 1-2, EGR, Fuel Trim EVAP, Misfire
Engine Speed	0-9999 rpm	± 50 rpm of actual speed	Varies	Engine Data 1-2, EGR, Fuel Trim EVAP, Misfire
EVAP Fault History	No Fault / Excess VAC / Purge Valve Leak / Small Leak / Weak VAC	No Fault	No Fault	EVAP
EVAP Purge Solenoid Circuit Status	Okay/Fault	Okay	Okay	Output Driver
EVAP Purge Solenoid Command	0-100%	19%	Varies	Engine Data 1, EVAP, Fuel Trim
EVAP Vent Solenoid Circuit Status	Okay/Fault	Okay	Okay	Output Driver
EVAP Vent Solenoid Command	Venting/Not Venting	Not Venting	Not Venting	Engine Data 1, EVAP, Fuel Trim
Fan Control Relay 1 Circuit Status	Okay/Fault	Okay	Okay	Output Driver
Fan Control Relay 1 Command	On/Off	Off	Off	Engine Data 2
Fan Control Relay 2-3 Circuit Status	Okay/Fault	Okay	Okay	Output Driver

2002-05 Rendezvous 3.4L VIN E V6 MFI - PID Data Lists (All)

Parameter Identification (PID)	PID Value Range	PID Value at Hot Idle	PID Value at 30 mph	Data List Type
Fan Control Relay 2-3 Command	On/Off	Off	Off	Engine Data 2
Fuel Tank Sensor	0-5.1 volts	0.5-2.8	0.5-2.8	EVAP
Fuel Pump Relay Circuit Status	Okay/Fault	Okay	Okay	Output Driver
Fuel Pump Relay Circuit History Status	Okay/Fault	Okay	Okay	Output Driver
Fuel Pump Relay Command	On/Off	On	On	Engine Data 1
Fuel Tank Level Remaining	0-100%	Varies	Varies	EVAP
Fuel Tank Pressure Sensor	Inches H2O, mm Hg, volts	-17 to +7.5 Inches H2O	-17 to +7.5 Inches H2O	EVAP
Fuel Trim Cell	Cell 0-10	Varies	Varies	Engine Data 1, EVAP, Fuel Trim
Fuel Trim Learn	Enabled or Disabled	Enabled	May Toggle	Engine Data 1, EVAP, Fuel Trim
Generator L-Terminal	On/Off	On	On	Engine Data 2
HO2S (B1 S1) Signal	0-1132 mv	10-1000	10-1000	Engine Data 1, EVAP, Fuel Trim
HO2S (B1 S2) Signal	0-1132 mv	10-1000	10-1000	Engine Data 1, EVAP, Fuel Trim
HO2S (B1 S1) Heater	On/Off	On	On	Fuel Trim
IAC Position	0-255 counts	15	15	Engine Data 1, EGR, Fuel Trim
IAT Sensor (°F)	-40 to 304°F	91	92	Engine Data 1-2, EGR, EVAP, Fuel Trim
Ignition 1	0.0-25.5v	13.5	13.6	Engine Data 1-2, EGR, EVAP, Fuel Trim
Ignition Mode	Bypass/IC	IC	IC	Engine Data 2
Injector Pulsewidth	0-999 ms	2.90	Varies	Engine Data 2, Fuel Trim, Misfire
Knock Retard (°)	0-25.5°	0°	0°	Engine Data 1, EGR
Long Term Fuel Trim	-23 to +16	0 (± 5%)	0 (± 5%)	Engine Data 1-2 EVAP, Fuel Trim
Loop Status	Closed Loop / Open Loop	Closed Loop	Closed Loop	Engine Data 1-2, EGR, EVAP, Fuel Trim
MAF Sensor (g/sec)	0-512 g/sec	3.41	Varies	Engine Data 1-2, EGR, EVAP, Fuel Trim, Misfire
MAF Sensor (Hz)	0-32,000 Hz	2000	Varies	Engine Data 1-2

2002-05 Rendezvous 3.4L VIN E V6 MFI - PID Data Lists (All)

Parameter Identification (PID)	PID Value Range	PID Value at Hot Idle	PID Value at 30 mph	Data List Type
MAP Sensor (kPa)	10-105 kPa	35	Varies	Engine Data 1-2, EGR, EVAP, Fuel Trim, Misfire
MAP Sensor (V)	0-5.1 volts	1.35	Varies	Engine Data 1-2
MIL Status	Okay/Fault	Okay	Okay	Output Driver
MIL Command	On/Off	Off	Off	Engine Data 2
Misfire Current Cyl 1-6	0-198 counts	0	0	Misfire
M/Fire History Cyl 1-6	0-65,535	0	0	Misfire
Number of DTC(s)	Number	0	No	Engine Data 1, EVAP, Fuel Trim
O2 Heater Current	0-5.0 amps	0.54	0.54	Fuel Trim
PCM Reset	Yes/No	No	No	Engine Data 1-2, EGR, EVAP, Fuel Trim, Output Driver
PCM/VCM in VTD Fail	Yes/No	No	No	Engine Data 1
Power Enrichment	Active/Inactive	Inactive	Inactive	Engine Data 2, Misfire
Short Term Fuel Trim	-10 to +10%	0 (± 5%)	0 (± 5%)	Fuel Trim
Spark Advance	-64° to +64°	18°	Varies	Engine Data 1-2 Fuel Trim, M/Fire
Startup ECT	-40 to 419°F	Varies	Varies	Engine Data 2, EVAP, Fuel Trim
Startup IAT	-40 to 419°F	80°F	80°F	HO2S
Starter Enable Relay Circuit Status	Okay/Fault	Okay	Okay	Output Driver
Starter Enable Command	On/Off	Off	Off	Engine Data 2
TCC Brake Pedal Switch	Apply/Released	Released	Released	Engine Data 1-2
TCC PWM Solenoid Command	On/Off	Off	On	Engine Data 2, Misfire
TFP Switch A/B/C	P, R, N, D, 4, 3, 2, 1	Park	4	Engine Data 2, EGR, Fuel Trim
Torque Delivered Signal	0-100%	86	Varies	Engine Data 2
Torque Request Signal	0-100%	100%	Varies	Engine Data 2
TP Desired Angle	0-100%	Varies	Varies	Engine Data 1-2 EGR, Fuel Trim, EVAP, Misfire
Traction Control	Active/Inactive	Inactive	Inactive	Engine Data 2
Vehicle Speed	Km/h, MPH	0	30	Engine Data 1-2 EGR, Fuel Trim, EVAP, Misfire
VTD Auto Learn timer	Active/Inactive	Inactive	Inactive	Engine Data 1
VTD Fuel Disable	Active/Inactive	Inactive	Inactive	Engine Data 1
VTD Fuel Disable Until Ignition Off	Yes/No	No	No	Engine Data 1

ROADMASTER PID DATA

1995 Roadmaster 5.7L V8 MFI VIN P (A/T) - Engine Data 1 List

Parameter Identification (PID)	PID Value Range	PID Value at Hot Idle	PID Value at 30 mph	Data List Type
A/C Clutch	Off / On	Off	Off	Engine Data 1
A/C Head Pressure	0-459 psi	139-399	139-399	Engine Data 1
A/C Request	No / Yes	No	No	Engine Data 1
A/C Status	Off / On	Off	Off	Engine Data 1
AIR System Control	Off / On	On: Cold	Off	Engine Data 1
BARO Sensor (kPa)	10-110 kPa	70-110	70-100	Engine Data 1
Cooling Fan 1, 2	Off / On	Off	Off	Engine Data 1
Desired Idle Speed	0-3187 rpm	800	---	Engine Data 1
ECT Sensor (°F)	-40 to 419°F	185-221	185-221	Engine Data 1
EGR Duty Cycle	0-100%	0	0-50	Engine Data 1
Engine Speed	0-9999 rpm	±50 Desired	---	Engine Data 1
Fuel Injector Fault	No / Yes	No	No	Engine Data 1
Fuel Trim Cell	Cell #0-21	16-18	0-21	Engine Data 1
Fuel Trim Enable	No / Yes	Yes	Yes	Engine Data 1
High Resolution signal	No / Yes	Yes	Yes	Engine Data 1
HO2S-11 (B1 S1)	0-1132 mv	10-1000	10-1000	Engine Data 1
HO2S-21 (B2 S1)	0-1132 mv	10-1000	10-1000	Engine Data 1
IAC Motor	0-255 counts	5-50	---	Engine Data 1
IAC Motor Learned	0-255 counts	5-50	---	Engine Data 1
IAT Sensor (°F)	-40 to 304°F	50-194	50-194	Engine Data 1
Ignition 1 Signal	0.0-25.5v	14.2	14.2	Engine Data 1
INJ Pulsewidth B1, B2	0-985 ms	1-4	Varies	Engine Data 1
Knock Signal	Yes/No	No	No	Engine Data 1
Knock Retard (°)	0-128°	0	0	Engine Data 1
Knock Sensors	Okay-Open	Okay	Okay	Engine Data 1
Long Term F/T B1, B2	0-255 counts	118-138	118-138	Engine Data 1
Low Resolution Signal	0-985 ms	Varies	Varies	Engine Data 1
MAF Sensor (g/sec)	0-512 g/sec	Varies	Varies	Engine Data 1
MAP Sensor (kPa)	10-110 kPa	20-48	Varies	Engine Data 1
MAP Sensor (V)	0-5.1 volts	1-2	Varies	Engine Data 1
Loop Status Bank 1, 2	Closed Loop / Open Loop	Closed Loop	Closed Loop	Engine Data 1
PCM or TCM Codes	No / Yes	No	No	Engine Data 1
PSP Switch	HI/Normal	Normal	Normal	Engine Data 1
PROM ID Number	Number	0000	0000	Engine Data 1
Purge Duty Cycle	0-100%	0-15	0-50	Engine Data 1
Short Term F/T B1-B2	0-255 counts	Varies	Varies	Engine Data 1
1-2 Shift Solenoid	Off / On	On	---	Engine Data 1
2-3 Shift Solenoid	Off / On	On	---	Engine Data 1
Spark Advance (°)	-90° to 90°	Varies	Varies	Engine Data 1
TCC Brake Switch	Apply/Released	Released	Released	Engine Data 1
TCC Solenoid	Off / On	Off	On or Off	Engine Data 1
TCC Temperature Switch	Closed/open	Closed Loop	Closed Loop	Engine Data 1
Throttle Angle (%)	0-100%	0%	0-100%	Engine Data 1
TP Sensor (V)	0-5.1 volts	0.3-0.9v	---	Engine Data 1
Time From Start	Hr: Min: Sec	Varies	Varies	Engine Data 1
Transmission Range Switch	PN-R-D-L	PN	D	Engine Data 1
Vehicle Speed	0-155 mph	0	30	Engine Data 1

1996 Roadmaster 5.7L V8 MFI VIN P (A/T) - Engine Data 1 List

Parameter Identification (PID)	PID Value Range	PID Value at Hot Idle	PID Value at 30 mph	Data List Type
A/C Clutch	Off or On	Off	Off	Engine Data 1
A/C HI Side Pressure	0-450 psi	139-399	139-399	Engine Data 1
A/C High Side Volts	0-5.1 volts	1-3	1-3	Engine Data 1
A/C Request	No or Yes	No	No	Engine Data 1
A/C Status	Off or On	Off	Off	Engine Data 1
AIR Control System	Disabled/Enabled	Disabled	Disabled	Engine Data 1
BARO Sensor (kPa)	10-110 kPa	10-105	10-105	Engine Data 1
Cold Engine Startup	No or Yes	Yes: If cold	No	Engine Data 1
Current Gear	1-2-3-4	1	3	Engine Data 1
Desired Idle Speed	0-3187 rpm	PCM control	---	Engine Data 1
ECT Sensor (°F)	-40 to 304°F	185-226	185-226	Engine Data 1
EGR Duty Cycle	0-100%	0	0-100	Engine Data 1
Engine Run Time	Hr: Min: Sec	00:00:00	00:00:00	Engine Data 1
Engine Speed	0-9999 rpm	±50 Desired	---	Engine Data 1
EVAP Canister Purge	0-100%	0-25	0-50	Engine Data 1
EVAP Vacuum Switch	No/Purge	No Purge	Purge	Engine Data 1
Fan Control Relay 1-2	Off or On	Off	Off	Engine Data 1
Fuel Trim Cell	Cell #0-17	16	0-17	Engine Data 1
Fuel Trim Learn	Disabled/Enabled	Enabled	Enabled	Engine Data 1
High Resolution signal	Active/Inactive	Active	Active	Engine Data 1
HO2S (B1 S1, B2 S1)	0-1132 mv	10-1000	10-1000	Engine Data 1
IAC Learned	0-255 counts	10-50	---	Engine Data 1
IAC Position	0-255 counts	10-50	---	Engine Data 1
IAT Sensor (°F)	-40 to 304°F	50-194	50-194	Engine Data 1
INJ Pulsewidth B1, B2	0-985 ms	1-4	Varies	Engine Data 1
Knock Retard (°)	0.0 - 25.5°	0	0	Engine Data 1
Knock Activity	Yes/No	No	No	Engine Data 1
Long Term F/T B1, B2	-10% to 10%	Varies	Varies	Engine Data 1
Loop Status	Closed Loop / Open Loop	Closed Loop	Closed Loop	Engine Data 1
LO Resolution Signal	0-332.7 ms	Varies	Varies	Engine Data 1
MAF Sensor (g/sec)	0-512 g/sec	5-9	Varies	Engine Data 1
MAP Sensor (kPa)	10-110 kPa	20-48	Varies	Engine Data 1
OIL Level	Not Okay/Okay	Okay	Okay	Engine Data 1
Pass Key® Fuel	Disabled/Okay	Okay	Okay	Engine Data 1
PCM Reset	No or Yes	No	No	Engine Data 1
PNP Switch	P-R-0-3-2-1	P	3	Engine Data 1
PSP Switch	HI/Normal	Normal	Normal	Engine Data 1
Short Term F/T B1-B2	-10% to 10%	0%	Varies	Engine Data 1
Spark Advance (°)	-64° to 64°	20	Varies	Engine Data 1
Startup ECT Sensor	-40 to 304°F	Varies	Varies	Engine Data 1
TCC Brake Switch	Apply/Released	Released	Released	Engine Data 1
TCC Duty Cycle	Disabled/Enabled	Enabled	Enabled	Engine Data 1
TCC Enable	Disabled/Enabled	Disabled	Enabled	Engine Data 1
TCC Stator Temperature	HI/Normal	Normal	Normal	Engine Data 1
TP Angle (%)	0-100%	0	Varies	Engine Data 1
TP Sensor	0-5.1 volts	0.55-0.90	Varies	Engine Data 1
TR Switch	P/N/Drive	Park	Drive	Engine Data 1

1996 Roadmaster 5.7L V8 MFI VIN P (A/T) - Engine Data 2 List

Parameter Identification (PID)	PID Value Range	PID Value at Hot Idle	PID Value at 30 mph	Data List Type
A/C Clutch	Off or On	Off	Off	Engine Data 2
AIR Control Pump	Disabled/Enabled	Disabled	Disabled	Engine Data 2
BARO Sensor (kPa)	10-110 kPa	10-105	10-105	Engine Data 2
CKP Engine Speed	0-9999 rpm	±50 Desired	Varies	Engine Data 2
CKP LO Resolution Angle	-30 to +60	0 Deviation	0 Deviation	Engine Data 2
Cold Engine Startup	No or Yes	Yes: If cold	No	Engine Data 2
Desired Idle Speed	0-3187 rpm	PCM control	Varies	Engine Data 2
DTC Set this Ignition	No or Yes	No	No	Engine Data 2
ECT Sensor (°F)	-40 to 304°F	185-226	185-226	Engine Data 2
EGR Duty Cycle	0-100%	0%	0-100%	Engine Data 2
Engine Run Time	Hr: Min: Sec	00:00:00	00:00:00	Engine Data 2
Engine Speed	0-9999 rpm	±50 Desired	Varies	Engine Data 2
EVAP Purge	0-100%	0-25	0-50	Engine Data 2
EVAP Vacuum Switch	No/Purge	No Purge	Purge	Engine Data 2
Fan Control Relay 1	Off or On	Off	Off	Engine Data 2
Fan Control Relay 2	Off or On	Off	Off	Engine Data 2
Fuel Trim Cell	Cell #	16	Varies	Engine Data 2
Fuel Trim Learn	Disabled/Enabled	Enabled	Enabled	Engine Data 2
High Resolution Signal	Active/Inactive	Active	Active	Engine Data 2
HO2S-11 (B1 S1)	0-1132 mv	10-1000	10-1000	Engine Data 2
HO2S-21 (B2 S1)	0-1132 mv	10-1000	10-1000	Engine Data 2
IAC Learned	0-255 counts	10-50	10-50	Engine Data 2
IAC Position	0-255 counts	10-50	10-50	Engine Data 2
IAT Sensor (°F)	-40 to 304°F	50-194	50-194	Engine Data 2
Ignition 1 Signal	0.0-25.5v	14.2	14.2	Engine Data 2
Knock Retard (°)	0.0 - 25.5°	0	Varies	Engine Data 2
Knock Activity	Yes/No	No	No	Engine Data 2
Long Term F/T B1, B2	-10% to 10%	0%	Varies	Engine Data 2
Loop Status	Closed Loop / Open Loop	Closed Loop	Closed Loop	Engine Data 2
LO Resolution Signal	0-332.7 ms	Varies	Varies	Engine Data 2
MAF Sensor (g/sec)	0-512 g/sec	5-9	Varies	Engine Data 2
MAP Sensor (kPa)	10-110 kPa	20-48	Varies	Engine Data 2
MAP Sensor (V)	0-5.1 volts	1-2	Varies	Engine Data 2
MPH or KPH	0-155 mph	0	30	Engine Data 2
Non-Emission Cycle Count	0-255 counts	0	0	Engine Data 2
Pass Key® Fuel	Disabled/Okay	Okay	Okay	Engine Data 2
PCM Reset	No or Yes	No	No	Engine Data 2
Short Term F/T B1	-10% to 10%	0%	Varies	Engine Data 2
Short Term F/T B2	-10% to 10%	0%	Varies	Engine Data 2
Spark Advance (°)	-64° to 64°	20	Varies	Engine Data 2
Startup ECT Sensor	-40 to 304°F	Varies	Varies	Engine Data 2
TP Angle (%)	0-100%	0	Varies	Engine Data 2
TP Sensor (V)	0-5.1 volts	0.55-0.90	Varies	Engine Data 2

1996 Roadmaster 5.7L V8 MFI VIN P - F/F Data, Failure Records

Parameter Identification (PID)	PID Value Range	PID Value at Hot Idle	PID Value at 30 mph	Data List Type
Desired Idle Speed	0-3187 rpm	PCM control	Varies	F/F, F/R
DTC # P0001-P1999	PXXXX	PXXXX	PXXXX	F/F, F/R
ECT Sensor (°F)	-40 to 304°F	185-226	185-226	F/F, F/R
Emission Fault Counter	0-255 counts	0	0	F/F, F/R
Engine Load	0-100%	22-26	Varies	F/F, F/R
Engine Speed	0-9999 rpm	±50 Desired	---	F/F, F/R
Fail Counter	0-255 counts	Varies	Varies	F/F, F/R
Fuel Trim Cell	Cell #0-17	16	0-17	F/F, F/R
Fuel Trim Learn	Disabled/Enabled	Enabled	Enabled	F/F, F/R
HO2S (B1 S1, B2 S1)	0-1132 mv	10-1000	10-1000	F/F, F/R
IAT Sensor (°F)	-40 to 304°F	50-194	50-194	F/F, F/R
Long Term F/T B1, B2	-10% to 10%	Varies	Varies	F/F, F/R
MAF Sensor (g/sec)	0-512 g/sec	5-9	Varies	F/F, F/R
Miles Since DTC Cleared	0-65655	Varies	Varies	F/F, F/R
Miles Since First Failure	0-65535	Varies	Varies	F/F, F/R
Miles Since MIL Request	0-65535	Varies	Varies	F/F, F/R
No Run Counter	0-255 counts	0	0	F/F, F/R
Pass Counter	0-255 counts	Varies	Varies	F/F, F/R
Short Term F/T B1-B2	-10% to 10%	Varies	Varies	F/F, F/R
TP Angle (%)	0-100%	0	Varies	F/F, F/R

1996 Roadmaster 5.7L V8 MFI VIN P (A/T) - Fuel Trim Data List

Parameter Identification (PID)	PID Value Range	PID Value at Hot Idle	PID Value at 30 mph	Data List Type
AIR Control Pump	Disabled/Enabled	Disabled	Disabled	Fuel Trim
ECT Sensor (°F)	-40 to 304°F	185-226	185-226	Fuel Trim
EGR Duty Cycle	0-100%	0	0-100	Fuel Trim
EVAP Purge	0-100%	0	0-50	Fuel Trim
EVAP Vacuum Switch	No/Purge	No Purge	Purge	Fuel Trim
Fuel Trim Cell	Cell #	16	Varies	Fuel Trim
Fuel Trim Learn	Disabled/Enabled	Enabled	Enabled	Fuel Trim
HO2S (B1 S1, B2 S1)	0-1132 mv	10-1000	10-1000	Fuel Trim
IAC Learned	0-255 counts	10-50	---	Fuel Trim
IAC Position	0-255 counts	10-50	---	Fuel Trim
IAT Sensor (°F)	-40 to 304°F	50-194	50-194	Fuel Trim
INJ Pulsewidth B1, B2	0-985 ms	1-4m	Varies	Fuel Trim
Long Term F/T B1, B2	-10% to 10%	Varies	Varies	Fuel Trim
Long Term F/T Average	-13 to 21%	Varies	Varies	Fuel Trim
MAF Sensor (g/sec)	0-512 g/sec	5-9	Varies	Fuel Trim
MAP Sensor (kPa)	10-110 kPa	20-48	Varies	Fuel Trim
Short Term F/T B1-B2	-10% to 10%	Varies	Varies	Fuel Trim
Short Term F/T Average	-9% to +7%	Varies	Varies	Fuel Trim
Spark Advance (°)	-64° to 64°	20	Varies	Fuel Trim
TP Angle (%)	0-100%	0	Varies	Fuel Trim

1996 Roadmaster 5.7L V8 MFI VIN P (A/T) - Misfire Data

Parameter Identification (PID)	PID Value Range	PID Value at Hot Idle	PID Value at 30 mph	Data List Type
A/C Request	No or Yes	No	No	Misfire
A/C Status	Off or On	Off	Off	Misfire
CKP Engine Speed	0-9999 rpm	±50 Desired	Varies	Misfire
CKP LO Resolution Angle	-30 to +60	0 Deviation	0 Deviation	Misfire
Cycles of Misfire Data	0-100	0	0	Misfire
CYL Mode Misfire Index	0-65535	0	0	Misfire
Desired Idle Speed	0-3187 rpm	PCM control	---	Misfire
ECT Sensor (°F)	-40 to 304°F	185-226	185-226	Misfire
Engine Load	0-100%	22-26	Varies	Misfire
Engine Speed	0-9999 rpm	±50 Desired	Varies	Misfire
IAT Sensor (°F)	-40 to 304°F	50-194	50-194	Misfire
MAF Sensor (g/sec)	0-512 g/sec	5-9	Varies	Misfire
Misfire Current Cylinders 1-8	0-255 counts	0	0	Misfire
Misfire History Cylinders 1-8	0-65535	0	0	Misfire
Misfire Per Cycle Status	PRI# 1-8	Varies	Varies	Misfire
Misfire Per Cycle Status	SEC# 1-8	Varies	Varies	Misfire
Misfire Rough Road	No or Yes	No	No	Misfire
MPH / KPH	0-155 mph	0	30	Misfire
Total Misfire Counts	0-255 counts	0	0	Misfire
Total Misfires Per Test	0-255 counts	0	0	Misfire
TP Angle (%)	0-100%	0	Varies	Misfire

SKYLARK PID DATA - OBD II

1996-97 Skylark 2.4L I4 MFI VIN T (A/T) - Engine Data List 1

Parameter Identification (PID)	PID Value Range	PID Value at Hot Idle	PID Value at 30 mph	Data List Type
BARO Sensor (V)	0-5.1 volts	3.5-4.5	3.5-4.5	Engine Data 1
Calculated Airflow	0-512	3-7	Varies	Engine Data 1
CKP Active Counter	0-255 counts	Counts Up	Counts Up	Engine Data 1
CMP Active Counter	0-255 counts	Counts Up	Counts Up	Engine Data 1
CMP Resync Counter	0-255 counts	0	0	Engine Data 1
Desired Idle Speed	0-3187	PCM control	---	Engine Data 1
ECT Sensor (°F)	-40 to 419°F	185-239	185-239	Engine Data 1
Engine Load	0-100%	22-26	Varies	Engine Data 1
Engine Run Time	Hr: Min: Sec	00:00:00	00:00:00	Engine Data 1
Engine Speed	0-9999 rpm	±50 Desired	Varies	Engine Data 1
EVAP Purge Solenoid	0-100%	0	Varies	Engine Data 1
EVAP Vent Solenoid	On/Off	Off	Off	Engine Data 1
Fan Control Relay	On/Off	Off	Off	Engine Data 1
Fuel Trim Cell	Cell #	18-21	18-21	Engine Data 1
Fuel Trim Index	0-255 counts	58-198	Varies	Engine Data 1
Fuel Trim Index	-10% to 10%	Varies	Varies	Engine Data 1
Generator F-Terminal	0-100%	0-100%	0-100%	Engine Data 1
Generator L-Terminal	Active/Inactive	Inactive	Inactive	Engine Data 1
HO2S-12 (B1 S2)	0-1132 mv	10-1000	10-1000	Engine Data 1
IAC Position	0-255 counts	5-60	5-60	Engine Data 1
IAT Sensor (°F)	-40 to 304°F	50-176	50-176	Engine Data 1
Ignition 1 Signal	0.0-25.5v	14.2	14.2	Engine Data 1
Knock Retard (°)	0-128°	0	0	Engine Data 1
KS Noise Channel	Yes/No	No	No	Engine Data 1
KS Active Counts	0-255 counts	35-50	35-50	Engine Data 1
Low Octane Spark Modifier	0-90	Varies	Varies	Engine Data 1
Long Term Fuel Trim	0-255 counts	58-198	Varies	Engine Data 1
Long Term F/T (%)	-10% to 10%	0 (± 5%)	0 (± 5%)	Engine Data 1
Long Term F/T Average	-10% to 10%	0 (± 5%)	0 (± 5%)	Engine Data 1
Loop Status	Closed Loop / Open Loop	Closed Loop	Closed Loop	Engine Data 1
MAP Sensor (V)	0-5.1 volts	1.5	1-3	Engine Data 1
Medium Resolution Resync Counter	0-255 counts	0	0	Engine Data 1
Medium Resolution Engine Sync Counter	Yes/No	Yes	Yes	Engine Data 1
Number of DTC(s)	0-255 counts	0	0	Engine Data 1
O2S-11 (B1 S1)	0-1132 mv	10-1000	10-1000	Engine Data 1
Power Enrichment	Active/Inactive	Inactive	Inactive	Engine Data 1
Purge Learn Memory	0-100%	0%	0-100%	Engine Data 1
Short Term F/T	0-255 counts	58-198	Varies	Engine Data 1
Short Term Fuel Trim	-10% to 10%	0 (± 5%)	0 (± 5%)	Engine Data 1
Short Term F/T Average	-10% to 10%	0 (± 5%)	0 (± 5%)	Engine Data 1
Spark Advance (°)	-64° - 64°	Varies	Varies	Engine Data 1
TCC Brake Switch	On/Off	Off	Off	Engine Data 1
TP Angle (%)	0-100%	0	Varies	Engine Data 1
TP Sensor (V)	0-5.1 volts	0.20-0.90	Varies	Engine Data 1

1996-97 Skylark 2.4L I4 MFI VIN T (A.T) - Engine Data 2 List

Parameter Identification (PID)	PID Value Range	PID Value at Hot Idle	PID Value at 30 mph	Data List Type
A/C High Side (psi)	0-459 psi	139-399	139-399	Engine Data 2
A/C High Side (V)	0-5.1 volts	1-4	1-4	Engine Data 2
A/C Relay	On/Off	Off	Off	Engine Data 2
A/C Request	Yes/No	No	Off	Engine Data 2
Adaptive Knock retard	0-128°	0	Varies	Engine Data 2
Air Fuel Ratio	0.0-25.5:1	14.7:1	14.7:1	Engine Data 2
BARO Sensor (kPa)	10-110 kPa	65-105	65-105	Engine Data 2
Cruise Engaged	Yes/No	No	No	Engine Data 2
Desired Idle Speed	0-3187	PCM control	----	Engine Data 2
Engine Coolant Level	Okay / LOW	Okay	Okay	Engine Data 2
ECT Sensor (°F)	-40 to 419°F	185-239	185-239	Engine Data 2
Engine Oil Level	LOW / Okay	Okay	Okay	Engine Data 2
Engine Oil Pressure	LOW / Okay	Okay	Okay	Engine Data 2
Engine Speed	0-9999 rpm	±50 Desired	----	Engine Data 2
IAT Sensor (°F)	-40 to 304°F	50-176	50-176	Engine Data 2
INJ pulsewidth Cyl 1-4	0-985 ms	1-4	Varies	Engine Data 2
MAP Sensor (V)	0-5.1 volts	1.5	Varies	Engine Data 2
MPH / KPH	0-155 mph	0	30	Engine Data 2
Output DRV 1-2 Open	00000000	00000000	00000000	Output Driver
Output DRV 1-2 Short	00000000	00000000	00000000	Output Driver
Total Misfire Counts	0-255 counts	0	0	Engine Data 2
Stepper Cruise	Disabled/Enabled	Disabled	Disabled	Engine Data 2
TP Angle (%)	0-100%	0	Varies	Engine Data 2
TP Sensor (V)	0-5.1 volts	0.20-0.90	Varies	Engine Data 2
TR Range	PN-R-O-3-2	PN	3	Engine Data 2
TR Switch P, A	HI / LOW	HIGH	LOW	Engine Data 2
TR Switch B, C	HI / LOW	LOW	HIGH	Engine Data 2

1996-97 Skylark 2.4L I4 MFI VIN T (A.T) - Misfire Data List

Parameter Identification (PID)	PID Value Range	PID Value at Hot Idle	PID Value at 30 mph	Data List Type
BARO Sensor (V)	0-5.1 volts	3.5-4.5	3.5-4.5	Misfire
Calculated Airflow	0-512 g/sec	Varies	Varies	Misfire
ECT Sensor (°F)	-40 to 419°F	185-239	185-239	Misfire
Engine Load	0-100%	22-26	Varies	Misfire
Engine Run Time	Hr: Min: Sec	Varies	Varies	Misfire
IAT Sensor (°F)	-40 to 304°F	50-176	50-176	Misfire
INJ pulsewidth Cyl 1-4	0-985 ms	1-4	Varies	Misfire
MAP Sensor (V)	0-5.1 volts	1.5	Varies	Misfire
Mileage Since First Failure	Number	Varies	Varies	Misfire
Mileage Since Last Failure	Number	Varies	Varies	Misfire
Misfire current Cyl 1-4	0-255 counts	0	0	Misfire
Misfire History Cyl 1-4	0-255 counts	0	0	Misfire
TP Angle (%)	0-100%	0	Varies	Misfire
TP Sensor (V)	0-5.1 volts	0.20-0.90	Varies	Misfire
Total Misfire	0-255 counts	0	0	Misfire

1996-97 Skylark 2.4L I4 MFI VIN T (A/T) - EGR & EVAP Data List

Parameter Identification (PID)	PID Value Range	PID Value at Hot Idle	PID Value at 30 mph	Data List Type
Actual EGR Position	0-100%	0%	0-100%	EGR, EVAP
BARO Sensor (V)	0-5.1 volts	3.5-4.5	3.5-4.5	EGR, EVAP
Catalyst Converter TT	Not Okay/Okay	Okay	Okay	EGR, EVAP
Decel EWMA	-10 - 10 kPa	-3 to - 4	Varies	EGR, EVAP
Delta MAP Change	-20 - 20 kPa	0 / - number	0	EGR, EVAP
Desired Idle Speed	0-3187	PCM control	---	EGR, EVAP
Desired EGR Position	0-100%	0	0-100	EGR, EVAP
ECT Sensor (°F)	-40 to 419°F	185-239	185-239	EGR, EVAP
EGR Decel Filter	-10 - 10 kPa	-3 to - 4	Varies	EGR, EVAP
Engine Speed	0-9999 rpm	±50 Desired	Varies	EGR, EVAP
EGR Trip Samples	0-255 counts	Varies	Varies	EGR, EVAP
Engine Run Time	Hr: Min: Sec	00:00:00	00:00:00	EGR, EVAP
EVAP Solenoid	0-100%	0	0-100	EGR, EVAP
EVAP Vent Solenoid	0=off / 1=on	0	0	EGR, EVAP
Fuel Level	0-5.1 volts	Varies	Varies	EGR, EVAP
Fuel Level Sensor (%)	0-100%	Varies	Varies	EGR, EVAP
FTP Sensor	0-5.1 volts	2.5: cap off	Varies	EGR, EVAP
Fuel Trim Cell	Cell #0-21	18-21	0-21	EGR, EVAP
HO2S-12 (B1 S2)	0-1132 mv	10-1000	10-1000	EGR, EVAP
IAC Position	0-255 counts	5-60	5-60	EGR, EVAP
IAT Sensor (°F)	-40 to 304°F	50-176	50-176	EGR, EVAP
Long Term Fuel Trim	0-255 counts	58-198	Varies	EGR, EVAP
Long Term Fuel Trim	-10% to 10%	Varies	Varies	EGR, EVAP
MAP Sensor (V)	0-5.1 volts	1.5	1-3	EGR, EVAP
Number of Catalyst Tests	0-255 counts	Varies	Varies	EGR, EVAP
Number of DTC(s)	0-255 counts	0	0	EGR, EVAP
O2S-11 (B1 S1)	0-1132 mv	10-1000	10-1000	EGR, EVAP
Output Driver 1 Open	00000000	00000000	00000000	EGR, EVAP
Output Driver 2 Open	00000000	00000000	00000000	EGR, EVAP
Output Driver 1 Short	00000000	00000000	00000000	EGR, EVAP
Output Driver 2 Short	00000000	00000000	00000000	EGR, EVAP
Short Term Fuel Trim	0-255 counts	58-198	Varies	EGR, EVAP
Short Term Fuel Trim	-10% to 10%	Varies	Varies	EGR, EVAP
TP Angle (%)	0-100%	0	Varies	EGR, EVAP
TP Sensor (V)	0-5.1 volts	0.20-0.90	Varies	EGR, EVAP

1996-97 Skylark 2.4L I4 MFI VIN T (A/T) - Fuel Trim Data List

Parameter Identification (PID)	PID Value Range	PID Value at Hot Idle	PID Value at 30 mph	Data List Type
BARO Sensor (V)	0-5.1 volts	3.5-4.5	3.5-4.5	Fuel Trim
Calculated Airflow	0-512 g/sec	Varies	Varies	Fuel Trim
Desired Idle Speed	0-3187	PCM control	---	Fuel Trim
ECT Sensor (°F)	-40 to 419°F	185-239	185-239	Fuel Trim
Engine Load	0-100%	22-26	Varies	Fuel Trim
Engine Run Time	Hr: Min: Sec	Varies	Varies	Fuel Trim
Engine Speed	0-9999 rpm	±50 Desired	---	Fuel Trim
Fuel Trim Cell	Cell #0-21	18-21	0-21	Fuel Trim
Fuel Trim Index	0-255 counts	58-198	Varies	Fuel Trim
HO2S-12 (B1 S2)	0-1132 mv	10-1000	10-1000	Fuel Trim
IAT Sensor (°F)	-40 to 304°F	50-176	50-176	Fuel Trim
Lean / Rich Average	0-249 ms	0	0	Fuel Trim
Lean / Rich Transition	0-255 counts	0	0	Fuel Trim
HO2S1 Lean-Rich or to Rich-Lean Ratio	0:1-15.93:1	Varies	Varies	Fuel Trim
Long Term Fuel Trim	0-255 counts	58-198	Varies	Fuel Trim
Long Term Fuel Trim	-10% to 10%	Varies	Varies	Fuel Trim
Long Term Fuel Trim Average Percentage	0-100%	0 (± 5%)	0 (± 5%)	Fuel Trim
Loop Status	Closed Loop / Open Loop	Closed Loop	Closed Loop	Fuel Trim
MAP Sensor (V)	0-5.1 volts	1.5	Varies	Fuel Trim
O2S-11 (B1 S1)	0-1132 mv	10-1000	10-1000	Fuel Trim
Rich / Lean Average	0-249 ms	0	0	Fuel Trim
Rich / Lean Transition	0-255 counts	0	0	Fuel Trim
Short Term Fuel Trim	0-255 counts	58-198	Varies	Fuel Trim
Short Term Fuel Trim	-10% to 10%	Varies	Varies	Fuel Trim
Short Term Fuel Trim Average Percentage	0-100%	0 (± 5%)	0 (± 5%)	Fuel Trim
TP Angle (%)	0-100%	0	Varies	Fuel Trim
TP Sensor (V)	0-5.1 volts	0.20-0.90	Varies	Fuel Trim

1996-97 Skylark 2.4L I4 MFI VIN T (A/T) - Freeze Frame & Failure Records

Parameter Identification (PID)	PID Value Range	PID Value at Hot Idle	PID Value at 30 mph	Data List Type
Air Fuel Ratio	0-25.5:1	14.7:1	14.7:1	F/F, F/R
BARO Sensor (V)	0-5.1 volts	3.5-4.5	3.5-4.5	F/F, F/R
Calculated Airflow	0-512 g/sec	Varies	Varies	F/F, F/R
Desired Idle Speed	0-3187	PCM control	---	F/F, F/R
DTC # P0001- P1999	PXXXX	PXXXX	PXXXX	F/F, F/R
ECT Sensor (°F)	-40 to 419°F	185-239	185-239	F/F, F/R
INJ Pulsewidth	0-985 ms	1-4	Varies	F/F, F/R
Long Term Fuel Trim	-10% to 10%	Varies	Varies	F/F, F/R
Engine Load	0-100%	22-26	Varies	F/F, F/R
Loop Status	Closed Loop / Open Loop	Closed Loop	Closed Loop	F/F, F/R
MAP Sensor (V)	0-5.1 volts	1.5	Varies	F/F, F/R
MPH / KPH	0-155 mph	0	30	F/F, F/R
Short Term Fuel Trim	-10% to 10%	Varies	Varies	F/F, F/R
TP Angle (%)	0-100%	0	Varies	F/F, F/R

1996-98 Skylark 3.1L V6 MFI VIN M (A/T) - Engine Data List

Parameter Identification (PID)	PID Value Range	PID Value at Hot Idle	PID Value at 30 mph	Data List Type
Abuse Management	Active/Inactive	Inactive	Inactive	Engine Data 1
A/C HI Side Pressure	0-459 psi	Varies	Varies	Engine Data 1
A/C Pressure Sensor Out of Range	Yes/No	No	No	Engine Data 1
A/C Pressure Disable	Yes/No	No	No	Engine Data 1
A/C Clutch	Yes/No	No	No	Engine Data 1
A/C Request	Yes/No	No	No	Engine Data 1
A/C Off for WOT	Yes/No	No	No	Engine Data 1
A/C Slugging	Active/Inactive	Inactive	Inactive	Engine Data 1
Actual EGR Position	0-100%	0	0-100	Engine Data 1
Air Fuel Ratio	0.1-25.5:1	14.7:1	14.7:1	Engine Data 1
BARO Sensor (kPa)	10-110 kPa	65-105	65-105	Engine Data 1
Cam Signal Present	Yes/No	Yes	Yes	Engine Data 1
Commanded A/C	On/Off	Off	Off	Engine Data 1
Commanded Fan 1, 2	On/Off	Off	Off	Engine Data 1
Cruise Engaged	Disengage/Engage	Disabled	Disabled	Engine Data 1
Cruise Inhibit	Yes/No	Yes	Yes	Engine Data 1
Cruise Mode	On/Off	Off	On	Engine Data 1
Current Gear	1-2-3-4	1	3	Engine Data 1
Decel Fuel Mode	Active/Inactive	Inactive	Inactive	Engine Data 1
Desired EGR Position	0-100%	0%	0-100%	Engine Data 1
Desired Idle Speed	0-3187 rpm	PCM control	Varies	Engine Data 1
ECT Sensor (°F)	-40 to 304°F	185-220	185-220	Engine Data 1
Engine Load	0-100%	2-5	Varies	Engine Data 1
Engine Run Time	Hr: Min: Sec	Varies	Varies	Engine Data 1
Engine Speed	0-9999 rpm	±50 Desired	Varies	Engine Data 1
EGR Closed Pintle Position	0-5.1 volts	0.14-1.0	Varies	Engine Data 1
EGR Duty Cycle	0-100%	0	0-50	Engine Data 1
EGR Feedback	0-5.1 volts	0.14-1.0	Varies	Engine Data 1
EVAP Vacuum Switch	Purge/No	No Purge	No Purge	Engine Data 1
EVAP Canister Purge	0-100%	0-25	0-100	Engine Data 1
EVAP Vent Solenoid	Open/closed	Open	Open	Engine Data 1
Fan - Low Speed	On/Off	Off	Off	Engine Data 1
Fan - High Speed	On/Off	Off	Off	Engine Data 1
Fuel Pump Command	On/Off	On	On	Engine Data 1
Fuel Trim Cell	Cell #0-4	0	0-4	Engine Data 1
Fuel Trim Learn	Disabled/Enabled	Enabled	Enabled	Engine Data 1
Fuel Pump	On/Off	On	On	Engine Data 1
Generator Command	On/Off	On	On	Engine Data 1
Generator Lamp	Off / On	Off	Off	Engine Data 1
HO2S (B1 S1)	0-1132 mv	10-1000	10-1000	Engine Data 1
HO2S (B1 S2)	0-1132 mv	10-1000	10-1000	Engine Data 1
HO2S-11 Status	Not Ready	Ready	Ready	Engine Data 1
IAC Position	0-255 counts	5-50	---	Engine Data 1
IAT Sensor (°F)	-40 to 304°F	50-194	50-194	Engine Data 1
IC Mode	Active/Inactive	Active	Active	Engine Data 1
Idle Speed Error	0-12800 rpm	<100 (DI)	Varies	Engine Data 1

1996-98 Skylark 3.1L V6 MFI VIN M (A/T) - Engine Data List

Parameter Identification (PID)	PID Value Range	PID Value at Hot Idle	PID Value at 30 mph	Data List Type
Ignition Bypass	Active/Inactive	Inactive	Inactive	Engine Data 1
Ignition Mode	IC/Bypass	IC	IC	Engine Data 1
Injector Fault	Yes/No	No	No	Engine Data 1
Injector Pulsewidth	0-985 ms	1.5-3.5	Varies	Engine Data 1
Injector Status	Okay / Stuck	Okay	Okay	Engine Data 1
Knock Retard (°)	0-25.5°	0	0	Engine Data 1
KS Active Counter	0-255 counts	Varies	Varies	Engine Data 1
KS Noise Channel	0-5.1 volts	0.1-3.0	Varies	Engine Data 1
Long Term Fuel Trim	Varies	0%	1%	Engine Data 1
Loop Status	Closed Loop / Open Loop	Closed Loop	Closed Loop	Engine Data 1
Low Oil Lamp	On/Off	Off	Off	Engine Data 1
MAF Sensor (g/sec)	0-512 g/sec	3-6	Varies	Engine Data 1
MAF Sensor (Hz)	0-31,999	1200-3000	Varies	Engine Data 1
MAP Sensor (kPa)	10-110 kPa	29-48	Varies	Engine Data 1
MAP Sensor (V)	0-5.1 volts	1-2	Varies	Engine Data 1
MIL Status	Off / On	Off	Off	Engine Data 1
Non-Volatile Memory	Pass/Fail	Pass	Pass	Engine Data 1
Power Enrichment	Active/Inactive	Active	Active	Engine Data 1
Spark Advance (°)	-64° to 64°	20	Varies	Engine Data 1
TCC Brake Switch	Apply/Released	Released	Released	Engine Data 1
TCC Engaged	Yes/No	No	No	Engine Data 1
TP Angle (%)	0-100%	0	Varies	Engine Data 1
TP Sensor (V)	0-5.1 volts	0.20-0.74	Varies	Engine Data 1
Trans. Hot Mode	Active/Inactive	Inactive	Inactive	Engine Data 1
Transmission Range	X / O	O X X O	Varies	Engine Data 1
Transmission Range	PN-R-4-3-2	Park	3	Engine Data 1
TR Switch P-A-B-C	HI/LO	LO/LO/HI/HI	---	Engine Data 1
TWC Protection	Active/Inactive	Inactive	Inactive	Engine Data 1
Vehicle Speed	0-155 mph	0	30	Engine Data 1
VTD Fuel Disable	Disabled/Enabled	Disabled	Disabled	Engine Data 1
3X Crank Sensor	Number	Rpm>1280	Varies	Engine Data 1
24X Crank Sensor	0-1280	Varies	Varies	Engine Data 1

1996-98 Skylark 3.1L V6 MFI VIN M (A/T) - EVAP System Data

Parameter Identification (PID)	PID Value Range	PID Value at Hot Idle	PID Value at 30 mph	Data List Type
Engine Run Time	Hr: Min: Sec	Varies	Varies	EVAP
Engine Speed	0-9999 rpm	±50 Desired	Varies	EVAP
EVAP Canister Purge	0-100%	0-25	0-100	EVAP
EVAP Vent Solenoid	Open/closed	Open	Open	EVAP
FTP Sensor (V)	0-5.1 volts	2.5: cap off	Varies	EVAP
IAT Sensor (°F)	-40 to 304°F	50-194	50-194	EVAP
MAF Sensor (g/sec)	0-512 g/sec	3-6	Varies	EVAP
MAP Sensor (kPa)	10-110 kPa	29-48	Varies	EVAP
MAP Sensor (V)	0-5.1 volts	1-2	Varies	EVAP
Startup ECT Sensor	-40 to 304°F	Varies	Varies	EVAP
Startup IAT Sensor	-40 to 304°F	Varies	Varies	EVAP

1996-98 Skylark 3.1L V6 MFI VIN M (A/T) - EGR System Data

Parameter Identification (PID)	PID Value Range	PID Value at Hot Idle	PID Value at 30 mph	Data List Type
Actual EGR Position	0-100%	0	0-100	EGR
BARO Sensor (kPa)	10-110 kPa	65-105	65-105	EGR
Current Gear	1-2-3-4	1	3	EGR
Decel Fuel Mode	Active/Inactive	Inactive	Inactive	EGR
Desired EGR Position	0-100%	0	0-100	EGR
Desired Idle Speed	0-3187 rpm	PCM control	---	EGR
ECT Sensor (°F)	-38-304°F	185-220	185-220	EGR
EGR Closed Pintle Position	0-5.1 volts	0.14-1.0	---	EGR
EGR Duty Cycle	0-100%	0	0-100	EGR
EGR Feedback	0-5.1 volts	0.14-1.0	---	EGR
EGR Flow Test	0-255 counts	0-10	0-10	EGR
EGR Position Error	0-100%	0-9	---	EGR
Engine Load	0-100%	2-5	Varies	EGR
Engine Run Time	Hr: Min: Sec	00:00:00	00:00:00	EGR
Engine Speed	0-9999 rpm	±50 Desired	---	EGR
Hot Open Loop	Active/Inactive	Inactive	Inactive	EGR
IAC Position	0-255 counts	10-40	---	EGR
IAT Sensor (°F)	-40 to 304°F	50-194	50-194	EGR
Knock Retard (°)	0.0-25.5°	0	0	EGR
KS Active Counter	0-255 counts	Varies	Varies	EGR
Loop Status	Closed Loop / Open Loop	Closed Loop	Closed Loop	EGR
MAF Sensor (g/sec)	0-512 g/sec	3-6	Varies	EGR
MAP Sensor (kPa)	10-110 kPa	29-48	Varies	EGR
Power Enrichment	Active/Inactive	Inactive	Inactive	EGR
TCC Engaged	Yes/No	No	No	EGR
TP Angle (%)	0-100%	0	Varies	EGR
Transmission Range	PN-R-4-3-2	P	3	EGR
TWC Protection	Active/Inactive	Inactive	Inactive	EGR
Vehicle Speed	0-155 mph	0	30	EGR

1996-98 Skylark 3.1L V6 MFI VIN M (A/T) - HO2S Data List

Parameter Identification (PID)	PID Value Range	PID Value at Hot Idle	PID Value at 30 mph	Data List Type
Air Fuel Ratio	0.1-25.5:1	14.7:1	14.7:1	HO2S
Air Pump	On/Off	Off	Off	HO2S
Air Pump Relay	On/Off	Off	Off	HO2S
Decel Fuel Mode	Active/Inactive	Inactive	Inactive	HO2S
ECT Sensor (°F)	-40 to 304°F	185-220	185-220	HO2S
Engine Load	0-100%	2-5	Varies	HO2S
Engine Run Time	Hr: Min: Sec	00:00:00	00:00:00	HO2S
Engine Speed	0-9999 rpm	±50 Desired	---	HO2S
EVAP Canister Purge	0-100%	0-25%	0-100%	HO2S
Hot Open Loop	Active/Inactive	Inactive	Inactive	HO2S
HO2S (B1 S1, B2 S1)	0-1132 mv	10-1000	10-1000	HO2S
HO2S-13 (B1 S3)	0-1132 mv	10-1000	10-1000	HO2S
HO2S (B1 S1) Status	Not Ready	Ready	Ready	HO2S
HO2S-11 XCounts	0-255 counts	Varies	Varies	HO2S

1996-98 Skylark 3.1L V6 MFI VIN M (A/T) - HO2S Data List

Parameter Identification (PID)	PID Value Range	PID Value at Hot Idle	PID Value at 30 mph	Data List Type
HO2S-11 Warmup B1	Min / Sec	Varies	Varies	HO2S
HO2S-12 Warmup B1	Min / Sec	Varies	Varies	HO2S
IAT Sensor (°F)	-40 to 304°F	50-194	50-194	HO2S
Injector Pulsewidth	0-985 ms	1.5-3.5	Varies	HO2S
Loop Status	Closed Loop / Open Loop	Closed Loop	Closed Loop	HO2S
Long Term Fuel Trim	0-100%	Varies	Varies	HO2S
MAF Sensor (g/sec)	0-512 g/sec	3-6	Varies	HO2S
Power Enrichment	Active/Inactive	Active	Active	HO2S
Short Term Fuel Trim	0-100%	Varies	Varies	HO2S
Startup ECT Sensor	-40 to 304°F	Varies	Varies	HO2S
Startup IAT Sensor	-40 to 304°F	Varies	Varies	HO2S
TP Angle (%)	0-100%	0%	Varies	HO2S
TWC Protection	Active/Inactive	Inactive	Inactive	HO2S

1996-98 Skylark 3.1L V6 MFI VIN M (A/T) - Misfire Data List

Parameter Identification (PID)	PID Value Range	PID Value at Hot Idle	PID Value at 30 mph	Data List Type
3X Crank Sensor	Number	> 1280 rpm	Varies	Misfire
24X Crank Sensor	0-1280	Varies	Varies	Misfire
Abuse Management	Active/Inactive	Inactive	Inactive	Misfire
Actual EGR Position	0-100%	0%	0-100%	Misfire
Air Active Test	Active/Inactive	Inactive	Inactive	Misfire
CMP Since Last 3X	Yes/No	Yes	Yes	Misfire
Decel Fuel Mode	Active/Inactive	Inactive	Inactive	Misfire
Desired EGR Position	0-100%	0%	0-100%	Misfire
ECT Sensor (°F)	-40 to 304°F	185-220	185-220	Misfire
Engine Load	0-100%	2-5	Varies	Misfire
Engine Speed	0-9999 rpm	±50 Desired	---	Misfire
HO2S-11 XCount	0-255 counts	Varies	Varies	Misfire
IAC Position	0-255 counts	10-40	---	Misfire
Injector Pulsewidth	0-985 ms	1.5-3.5	Varies	Misfire
MAF Sensor (g/sec)	0-512 g/sec	5-9	Varies	Misfire
Misfire Current Cyl 1-6	0-198	0-4	0-4	Misfire
Misfire History Cyl 1-6	0-65535	0	0	Misfire
Misfiring Cylinder	CYL 1-6	0	0	Misfire
Non-Volatile Memory	Pass/Fail	Pass	Pass	Misfire
Power Enrichment	Active/Inactive	Active	Active	Misfire
Total Misfire Current Count	0-99	0-5	0-5	Misfire
Misfire Since 1st Fail	0-65535	0	0	Misfire
Misfire Since 1st Pass	0-65535	0	0	Misfire
Spark Advance (°)	-64° to 64°	-20	Varies	Misfire
TCC Engaged	Yes/No	No	No	Misfire
TP Angle (%)	0-100%	0	Varies	Misfire
Trans. Hot Mode	Active/Inactive	Inactive	Inactive	Misfire
TWC Protection	Active/Inactive	Inactive	Inactive	Misfire
Vehicle Speed	0-155 mph	0	30	Misfire

TERRAZA PID DATA

2005 Terraza 3.5L VIN L (ALL) – PID Data List

Scan Tool Parameter	Data List	Units Displayed	Typical Data Value
24X Crank Sensor	Ignition	RPM	0-6,400
A/C Relay Circuit Status	ODD	OK/Fault/Invalid State	OK
A/C Relay Circuit History	ODD	OK/Fault/Invalid State	OK
A/C Relay Command	Eng, Misfire, EGR, TAC	On/Off	Off
Air Fuel Ratio	Eng, Fuel Trim, HO2S	Ratio	14.2:1-14.7:1
APP Indicated Angle	Eng, EGR, EVAP, HO2S, Ignition, Misfire, TAC, Fuel Trim	Percent	0
APP Sensor 1	TAC	Volts	Varies
APP Sensor 1 Circuit	TAC	OK/Not OK	OK
APP Sensor 2	TAC	Volts	Varies
APP Sensor 2 Circuit	TAC	OK/Not OK	OK
APP Sensor 1	TAC	Percent	Varies
APP Sensor 2	TAC	Percent	Varies
APP Sensor 1 and 2	TAC	Agree/Disagree	Agree
BARO	Eng, EGR, EVAP, Fuel Trim, HO2S, Ignition	kPa	65-110 kPa
Catalytic Converter Protection Active	Fuel Trim, HO2S	Yes/No	No
CMP Sensor	Ignition	RPM	Varies
CMP Sensor Signal Present	Ignition, Misfire	Yes/No	Yes
Cruise Control Active	TAC	Yes/No	No
Cycles of Misfire Data	Misfire	Counts	0-99
Cyl. 1-6 Injector Circuit Status	ODD	OK/Stuck Low (open)/Stuck High	OK
Cylinder 1 and 4 IC Circuit Status	ODD	OK/Not OK	OK
Cylinder 2 and 5 IC Circuit Status	ODD	OK/Not OK	OK
Cylinder 3 and 6 IC Circuit Status	ODD	OK/Not OK	OK
Cylinder 1-6 Injector Circuit History	ODD	OK/Stuck Low (open)/Stuck High	OK
Decel Fuel Cut-off	Eng, EGR, Fuel Trim, HO2S	Active/Inactive	Inactive
Desired EGR Position	EGR, Misfire	Percent	0
Desired Idle Speed	Eng, EVAP, TAC	RPM	Varies
Driver Module 1 Status	ODD	Enabled/Off-High Volts/Off High Temp/Invalid State	Enabled
Driver Module 2 Status	ODD	Enabled/Off-High Volts/Off High Temp/Invalid State	Enabled
Driver Module 3 Status	ODD	Enabled/Off-High Volts/Off High Temp/Invalid State	Enabled
EC Ignition Relay Circuit Status	ODD	OK/Fault/Invalid State	OK
EC Relay Circuit History	ODD	OK/Fault/Invalid State	OK
ECT Sensor	Eng, EGR, EVAP, HO2S, Ignition, Misfire, Fuel Trim, TAC	°C/°F	Varies
EGR Flow Test Count	EGR	Counts	0-12
EGR Learned Minimum Position	EGR	Volts	0.14-1.0
EGR Position Sensor	Eng, EGR, Misfire	Percent	Varies
EGR Position Sensor	EGR	Volts	0
EGR Position Variance	EGR	Percent	0
EGR Solenoid Circuit History	ODD	OK/Open/Fault/Invalid State	OK
EGR Solenoid Circuit Status	ODD	OK/Fault/Invalid State	OK
EGR Solenoid Command	EGR	Percent	0

2005 Terraza 3.5L VIN L (ALL) – PID Data List (CONT)

Scan Tool Parameter	Data List	Units Displayed	Typical Data Value
Engine Load	Eng, Misfire, Fuel Trim, EVAP, EGR, HO2S, Ignition, TAC	Percent	0
Engine Oil Pressure Switch	Eng	Low/OK	OK
Engine Run Time	Eng, EVAP, Misfire, Fuel Trim, EGR, HO2S, Ignition, TAC, ODD	Hr: Min: Sec	Varies
Engine Speed	Eng, EGR, EVAP, Ignition, Misfire, Fuel Trim, TAC	RPM	Varies
EVAP Fault History	EVAP	No Fault/Excess Vacuum/Purge Valve Leak/Small Leak/Weak Vacuum/No Test Result	No Fault
EVAP Purge Solenoid Circuit History	ODD	OK/Open/Short	OK
EVAP Purge Solenoid Circuit Status	ODD	OK/Open/Short	OK
EVAP Purge Solenoid Command	Eng, EVAP, Fuel Trim, HO2S	Percent	Varies
EVAP Test Abort Reason	EVAP	No Aborted/Lost Enable/Small Leak/Veh. Not At Rest	Varies
EVAP Test Result	EVAP	• No Result/Passed/Aborted OR • Fail-DTC P0455 OR • Fail-DTC P0442 OR • Fail-DTC P0446 OR • Fail-DTC P0496	Varies
EVAP Test State	EVAP	Waiting for Purge/Test Running/Test Completed	Varies
EVAP Vent Solenoid Circuit History	ODD	OK/Open/Short	OK
EVAP Vent Solenoid Circuit Status	ODD	OK/Open/Short	OK
EVAP Vent Solenoid Command	Eng, EVAP	Venting/Not Venting	Varies
FC Relay 1 Circuit History	ODD	OK/Fault/Invalid State	OK
FC Relay 1 Circuit Status	ODD	OK/Fault/Invalid State	OK
FC Relay 2 and 3 Circuit History	ODD	OK/Fault/Invalid State	OK
FC Relay 2 and 3 Circuit Status	ODD	OK/Fault/Invalid State	OK
Fuel Level Sensor	EVAP	Volts	Varies
Fuel Pump Relay Circuit History	ODD	OK/Fault/Invalid State	OK
Fuel Pump Relay Circuit Status	ODD	OK/Fault/Invalid State	OK
Fuel Pump Relay Command	Eng	On/Off	On
Fuel Tank Level Remaining	EVAP	Percent	0-104
Fuel Tank Pressure Sensor	Eng, EVAP	mm Hg/in H2O	Varies
Fuel Tank Pressure Sensor	EVAP	Volts	Varies
Fuel Trim Cell	EVAP, Eng, Fuel Trim	0-4	0
Fuel Trim Learn	Eng, EVAP, Fuel Trim	Enabled/Disabled	Varies
HO2S Bank 1 Sensor 1	EVAP, Eng, Fuel Trim, HO2S	millivolts	0-1000 and Varying
HO2S Bank 1 Sensor 1 Ready	Fuel Trim, HO2S	Yes/No	Yes
HO2S Bank 1 Sensor 2	Eng, Fuel Trim, HO2S	millivolts	0-1000 and Varying
HO2S Bank 2 Sensor 1 Ready	Fuel Trim, HO2S	Yes/No	Yes
HO2S Bank 2 Sensor 1	Eng, EVAP, Fuel Trim, HO2S	millivolts	0-1000 and Varying
HO2S Bank 2 Sensor 2	Eng, Fuel Trim, HO2S	millivolts	0-1000 and Varying
HO2S Htr. Bn. 1 Sen. 1	HO2S	On/Off	On

2005 Terraza 3.5L VIN L (ALL) – PID Data List (CONT)

Scan Tool Parameter	Data List	Units Displayed	Typical Data Value
HO2S Htr. Bank 1 Sensor 1 Command	HO2S	On/Off	On
HO2S Htr. Bn. 1 Sen. 2	HO2S	On/Off	On
HO2S Htr. Bank 1 Sensor 2 Command	HO2S	On/Off	On
HO2S Heater Bn. 1 Sen. 1	HO2S	Amps	Varies
HO2S Heater Bn. 1 Sen. 2	HO2S	Amps	Varies
HO2S Heater Bn. 2 Sen. 1	HO2S	Amps	Varies
HO2S Heater Bn. 2 Sen. 2	HO2S	Amps	Varies
HO2S Htr. Bn. 1 Sen. 1 Circuit History	ODD	OK/Fault/Invalid State	OK
HO2S Htr. Bn. 1 Sen. 2 Circuit History	ODD	OK/Fault/Invalid State	OK
HO2S Htr. Bn. 2 Sen. 1	HO2S	On/Off	On
HO2S Htr. Bn. 2 Sen. 1 Circuit History	ODD	OK/Fault/Invalid State	OK
HO2S Htr. Bn. 2 Sen. 2	HO2S	On/Off	On
HO2S Htr. Bn. 2 Sen. 2 Circuit History	ODD	OK/Fault/Invalid State	OK
HO2S Htr. Bn. 1 Sen. 1 Circuit Status	ODD	OK/Fault/Invalid State	OK
HO2S Htr. Bn. 1 Sen. 2 Circuit Status	ODD	OK/Fault/Invalid State	OK
HO2S Htr. Bn. 2 Sen. 1 Circuit Status	ODD	OK/Fault/Invalid State	OK
HO2S Htr. Bn. 2 Sen. 2 Circuit Status	ODD	OK/Fault/Invalid State	OK
Hot Open Loop	Fuel Trim, HO2S	Active/Inactive	Inactive
IAT Sensor	Eng, EVAP, EGR, Fuel Trim, HO2S, Ignition, Misfire, TAC	°C/°F	Varies
Ignition 1 Signal	EVAP, Eng, EGR, Fuel Trim, HO2S, Ignition, Misfire, ODD, TAC	Volts	Varies
Ignition Off Time	EVAP, Eng	Minutes/Varies	Varies
Initial Brake Apply Signal	TAC	Released/Applied	Released
Injector PWM	Fuel Trim, Misfire	milliseconds	Varies
Knock Retard	EGR, Ignition	Degrees	0
Long Term FT Bank 1	Eng, EVAP, Fuel Trim, HO2S	Percent	-10% to +10%
Long Term FT Bank 2	Eng, EVAP, Fuel Trim, HO2S	Percent	-10% to +10%
Loop Status	Eng, Fuel Trim, EVAP, HO2S	Open/Closed	Closed
MAF Sensor	Eng, EGR, Misfire, EVAP, Fuel Trim, HO2S, Ignition, TAC	g/s	3-6 (Varies with altitude)
MAF Sensor	EVAP, Fuel Trim, HO2S, Misfire, TAC	Hz	1,200-3,000 (depends on altitude)
MAP Sensor	Eng, EGR, Fuel Trim, HO2S, Misfire, TAC	Volts	0.4-2.0 Volts
MAP Sensor	Eng, EVAP, Fuel Trim, HO2S, Ignition, Misfire, EGR, TAC	kPa	20-48 kPa (Varies with altitude)
MIL Circuit History	ODD	OK/Fault/Invalid State	OK
MIL Command	EVAP, Eng, EGR, Ignition	On/Off	Off
MIL Circuit Status	ODD	OK/Fault/Invalid State	OK
Mileage Since DTC Cleared	Eng	km/h/mph	Varies
Misfire Current Cyl. 1-6	Misfire	Counts	0
Misfire History Cyl. 1-6	Misfire	Counts	0
Moderate Brake Apply Signal	TAC	Applied/Released	Released
Number of DTCs	Eng EVAP, Fuel Trim, TAC	Number	0
PCM Reset	Eng	Yes/No	No

2005 Terraza 3.5L VIN L (ALL) – PID Data List (CONT)

Scan Tool Parameter	Data List	Units Displayed	Typical Data Value
Power Enrichment	Eng, Fuel Trim, HO2S, Misfire	Active/Inactive	Inactive
Reduced Engine Power	TAC	Active/Inactive	Inactive
Short Term FT Bank 1	Eng, EVAP, Fuel Trim, HO2S,	Percent	-10 to +10
Short Term FT Bank 2	Eng, EVAP, Fuel Trim, HO2S,	Percent	-10 to +10
Spark	Eng, Misfire, Fuel Trim, HO2S, Ignition	Degrees	22 (varies)
Start Up ECT	EVAP, Fuel Trim, HO2S	°C/°F	Varies
Start Up IAT	EVAP, Fuel Trim, HO2S	°C/°F	Varies
Starter Relay Circuit Status	ODD	OK/Fault	OK
Starter Relay Circuit History	ODD	OK/Fault/Invalid State	OK
Stop Lamp Pedal Switch	Eng, TAC	Released/Applied	Released
TAC/PCM Communications Signal	TAC	OK/Fault	OK
TAC Stop Lamp Pedal Switch	Eng	Released/Applied	Released
TAC Stop Lamp Pedal Switch	TAC	Applied/Released	Released
TAC Vehicle Speed Signal	TAC	OK/Fault	OK
TCS Delivered Torque Circuit Status	ODD	OK/Fault	OK
Torque Request Signal	TAC	Percent	100%
Torque Delivered Signal	TAC	Percent	75%
Total Misfire Count	Misfire	1-255	0
TP Desired Angle	TAC	Percent	Varies
TP Indicated Angle	Eng, EGR, Fuel Trim, HO2S, Ignition, EVAP, Misfire, TAC	Percent	%
TP Sensor 1	TAC	Percent	Varies
TP Sensor 1	TAC	Volts	Varies
TP Sensors 1 and 2	TAC	Agree/Disagree	Agree

CATERA PID DATA

1996-99 Catera 3.0L V6 MFI VIN R (A/T) - Engine Data List 1

Parameter Identification (PID)	PID Value Range	PID Value at Hot Idle	PID Value at 30 mph	Data List Type
A/C Compressor	On / Off	Yes	Yes	Engine Data 1
A/C Request	Yes / No	No	No	Engine Data 1
A/C Load Signal	Heavy/Norm	Varies	Varies	Engine Data 1
Actual MAF g/sec	0.0-35.4	Varies	Varies	Engine Data 1
Air Pump Relay	On / Off	Off	Off	Engine Data 1
Air Pump Ready	Yes / No	Yes	Yes	Engine Data 1
Average INJ Pulsewidth	0.0-6.1 ms	2.0-2.5	Varies	Engine Data 1
Cam H/L Signal	L or H	L-H-L-H	L-H-L-H	Engine Data 1
Calc. Full Position	Active / Inactive	Inactive	Inactive	Engine Data 1
Calc. IAC Position	Active / Inactive	Inactive	Inactive	Engine Data 1
Calc. torque reduction	0-255 Count	0	Varies	Engine Data 1
Desired MAF (Idle)	0.0-35.4 g/s	Varies	Varies	Engine Data 1
Desired Idle Speed	0-255 Count	15-35	Varies	Engine Data 1
ECT Sensor (°F)	-40 to 304°F	194-280	194-280	Engine Data 1
ECT Sensor (V)	0-5.1v	0.89	0.80	Engine Data 1
Engine Load Signal	0-100%	2-5	5-8	Engine Data 1
Engine Run Time	Hr: Min: Sec	00:00:00	00:00:00	Engine Data 1
Engine Speed	0-9999 rpm	±50 Desired	1310	Engine Data 1
EVAP Canister Vent	Active / Inactive	Inactive	Inactive	Engine Data 1
EVAP Leakage Factor	0-1992	Varies	Varies	Engine Data 1
EVAP Purge Valve	0-100%	0-15	0-100	Engine Data 1
EVAP System Ready	No / Yes	Yes	Yes	Engine Data 1
Fuel Level Output	0-10.1v	0v: near full	1.86v	Engine Data 1
Fuel Pump Relay	On / Off	On	On	Engine Data 1
FTP Sensor (In. H2O)	-17.5 to 7.5"	Varies	Varies	Engine Data 1
FTP Sensor (V)	0-5v	2.5v: cap off	Varies	Engine Data 1
HO2S Heaters Ready	No / Yes	Yes	Yes	Engine Data 1
HO2S-11 (B1 S1)	0-1132 mv	10-1000	10-1000	Engine Data 1
HO2S-12 (B1 S2)	0-1132 mv	10-1000	10-1000	Engine Data 1
HO2S-21 (B2 S1)	0-1132 mv	10-1000	10-1000	Engine Data 1
HO2S-22 (B2 S2)	0-1132 mv	10-1000	10-1000	Engine Data 1
HO2S-11 (B1 S1)	No / Yes	Yes	Yes	Engine Data 1
HO2S-11 (B1 S1) L/R	L or R	L-R-L-R	L-R-L-R	Engine Data 1
HO2S-21 (B2 S1) L/R	L or R	L-R-L-R	L-R-L-R	Engine Data 1
HO2S-11 Avg. B1	0-655 sec	Varies	Varies	Engine Data 1
HO2S-21 Avg. B2	0-655 sec	Varies	Varies	Engine Data 1
IAC Long Term Air Trim	-100 to 100%	2	Varies	Engine Data 1
IAC Short Term Air Trim	-100-100%	2	Varies	Varies
IAC Pulse Ratio	0-65%	4	---	Engine Data 1
IAT Sensor (°F)	-40 to 304°F	32-194	32-194	Engine Data 1
IAT Sensor (V)	0.3-4.9v	1.4	1.4	Engine Data 1
Ignition 1 Signal	0.0-25.5v	14.1	14.1	Engine Data 1
Ignition Coil 1 & 4	Active / Inactive	Toggles A-I	Toggles A-I	Engine Data 1
Ignition Coil 2 & 5	Active / Inactive	Toggles A-I	Toggles A-I	Engine Data 1
Ignition Coil 3 & 6	Active / Inactive	Toggles A-I	Toggles A-I	Engine Data 1
Intake Plenum Switch	Open / Closed	Open	Open	Engine Data 1
Intake Resonance Switch	Open / Closed	Closed	Closed	Engine Data 1

1996-99 Catera 3.0L V6 MFI VIN R (A/T) - Engine Data List 1

Parameter Identification (PID)	PID Value Range	PID Value at Hot Idle	PID Value at 30 mph	Data List Type
Knock Retard (°)	0-90°	0	0	Engine Data 1
KS Activity	Yes / No	Yes: Knock	No	Engine Data 1
Long FT Adaptive B1, B2	-2.05-2.05	15 ms	Varies	Engine Data 1
Long FT Multiplier B1, B2	-100-100%	15%	Varies	Engine Data 1
Loop Status Bank 1	Closed Loop / Open Loop	Closed Loop	Closed Loop	Engine Data 1
Loop Status Bank 2	Closed Loop / Open Loop	Closed Loop	Closed Loop	Engine Data 1
MAF Sensor (V)	0-5.1v	1-4	Varies	Engine Data 1
MIL (lamp)	On / Off	Off	Off	Engine Data 1
MIL Request	Active / Inactive	Inactive	Inactive	Engine Data 1
Number of Current DTC(s)	0-255 Count	0	0	Engine Data 1
Park Neutral Switch	P-N / R-D-L	P-N	R-D-L	Engine Data 1
Short Term F/T B1	-10% to 10%	Varies	Varies	Engine Data 1
Short Term F/T B2	-10% to 10%	Varies	Varies	Engine Data 1
Spark Advance (°)	-133° to 78°	Varies	Varies	Engine Data 1
TC Torque Request	0-100%	Varies	Varies	Engine Data 1
TCM Torque Reduction	0-100%	0%	0%	Engine Data 1
TCM Torque Reduction	Yes / No	No	No	Engine Data 1
Theft Deterrent Status	Several	Correct	Correct	Engine Data 1
TP Normalized	0-100%	0	1	Engine Data 1
TP Angle (%)	0-100%	0	12	Engine Data 1
TP Sensor (V)	0-5v	0.63	0.68	Engine Data 1
Traction Control	Active / Inactive	Inactive	Inactive	Engine Data 1
Traction Control Status	Active / Inactive	Inactive	Inactive	Engine Data 1
Transmission Torque Reduction	No / Yes	No	No	Engine Data 1
TWC Ready Bank 1	No / Yes	Yes	Yes	Engine Data 1
TWC Ready Bank 2	No / Yes	Yes	Yes	Engine Data 1
Vehicle Speed	0-198 mph	0	30	Engine Data 1
Vehicle Speed Pulse	Active / Inactive	Inactive	Active	Engine Data 1

1996-99 Catera 3.0L V6 MFI VIN R (A/T) - Misfire Data List

Parameter Identification (PID)	PID Value Range	PID Value at Hot Idle	PID Value at 30 mph	Data List Type
Calculated Converter Temperature	-40 - 1742°F	---	---	Misfire
Compensation Gradient	0-9.18 H2O"	---	---	Misfire
Misfire Current Cyl. No. 1	0-255 Count	0	0	Misfire
Misfire Current Cyl. No. 2	0-255 Count	0	0	Misfire
Misfire Current Cyl. No. 3	0-255 Count	0	0	Misfire
Misfire Current Cyl. No. 4	0-255 Count	0	0	Misfire
Misfire Current Cyl. No. 5	0-255 Count	0	0	Misfire
Misfire Current Cyl. No. 6	0-255 Count	0	0	Misfire
Vacuum Decay Gradient	0-9.18" H2O	Varies	Varies	Misfire

2000-01 Catera 3.0L V6 MFI VIN R (A/T) - Engine Data List 1

Parameter Identification (PID)	PID Value Range	PID Value at Hot Idle	PID Value at 30 mph	Data List Type
A/C Clutch Relay	On / Off	Off	Off	Engine Data 1
AIR Pump Relay	On / Off	On	Off	Engine Data 1
AIR Solenoid	On / Off	On	Off	Engine Data 1
A/C Load Signal	Heavy/Norm	Varies	Varies	Engine Data 1
APP Sensor 1	0.0-5.1v	0.4v or less	Below 2.5v	Engine Data 1
APP Sensor 2	0.0-5.1v	0.2v or less	Below 1.25v	Engine Data 1
Battery Voltage	0-25.5v	14.1	14.1	Engine Data 1
Calc. Full Position Switch	Active / Inactive	Inactive	Inactive	Engine Data 1
Calc Idle Position Switch	Active / Inactive	Active	Active	Engine Data 1
CMP Resync Counter	0-255 Count	0	Varies	Engine Data 1
Catalyst Monitor BK 1	Yes / No	Yes or No	Yes	Engine Data 1
Catalyst Monitor BK 2	Yes / No	Yes or No	Yes	Engine Data 1
Cruise Brake Pedal	On / Off	Off	Off	Engine Data 1
Cruise Switch	On / Off	Off	Off	Engine Data 1
Desired Idle Speed	0-255 Count	720-1200	---	Engine Data 1
Drive Cycle Finished	Yes / No	No	Yes	Engine Data 1
DTC Number 1	Number	00000	00000	Engine Data 1
DTC Number 2	Number	00000	00000	Engine Data 1
ECT Sensor (°F)	-40 to 284°F	194-280	194-280	Engine Data 1
Engine Load	0-100%	2-3	5-7	Engine Data 1
Engine Power Relay	On / Off	On	On	Engine Data 1
Engine Run Time	Hr: Min: Sec	00:00:00	00:00:00	Engine Data 1
Engine Speed	0-9999 rpm	720-1200	1310	Engine Data 1
EVAP Canister Purge	0-100%	0-5	Varies	Engine Data 1
EVAP Canister Vent	On / Off	Off	Off	Engine Data 1
Fuel Level	Liters / Gallons	Varies	Varies	Engine Data 1
Fuel Level (V)	0-5.1v	Varies	Varies	Engine Data 1
FTP Sensor (In. H2O)	-10.4 to 10.9"	Varies	Varies	Engine Data 1
FTP Sensor (V)	0.0 to 5.1v	Varies	Varies	Engine Data 1
HO2S-11 (B1 S1)	0-1275 mv	Varies	Varies	Engine Data 1
HO2S-12 (B1 S2)	0-1275 mv	Varies	Varies	Engine Data 1
HO2S-21 (B2 S1)	0-1275 mv	Varies	Varies	Engine Data 1
HO2S-22 (B2 S2))	0-1275 mv	Varies	Varies	Engine Data 1
HO2S-11 Heater	High / Low	Low	High	Engine Data 1
HO2S-12 Heater	High / Low	Low	High	Engine Data 1
HO2S-21 Heater	High / Low	Low	High	Engine Data 1
HO2S-22 Heater	High / Low	Low	High	Engine Data 1
IAT Sensor (°F)	-40 to 284°F	Varies	Varies	Engine Data 1
Ignition 1 Signal On	0-24v	14.1	14.1	Engine Data 1
Injector Pulsewidth	0-985 ms	3.75	4.12	Engine Data 1
Intake Switchover	Open / Closed	Open	Open	Engine Data 1
Intake Resonance Valve	Open / Closed	Closed	Closed	Engine Data 1
Intake Plenum Switch	Open / Closed	Open	Open	Engine Data 1
Knock Retard CYL 1-6	0-190°	0.0	0.0	Engine Data 1
Knock Sensor #1, #2	Active / Inactive	Inactive	Inactive	Engine Data 1
Long Term F/T BK 1	-10% to 10%	Varies	Varies	Engine Data 1
Long Term F/T BK 2	-10% to 10%	Varies	Varies	Engine Data 1

2000-01 Catera 3.0L V6 MFI VIN R (A/T) - Engine Data List 1

Parameter Identification (PID)	PID Value Range	PID Value at Hot Idle	PID Value at 30 mph	Data List Type
Loop Status (B1 S1)	Open / Closed	Open	Closed	Engine Data 1
Loop Status (B1 S2)	Open / Closed	Open	Closed	Engine Data 1
Loop Status (B2 S1)	Open / Closed	Open	Closed	Engine Data 1
Loop Status (B2 S2)	Open / Closed	Open	Closed	Engine Data 1
MAF Sensor (g/sec)	0-72 g/sec	4-6	12-16	Engine Data 1
MIL (lamp) Control	On / Off	Off	Off	Engine Data 1
MIL Flashing	Yes / No	No	No	Engine Data 1
MIL Request	Yes / No	No	No	Engine Data 1
Mileage Since DTC Cleared	0-65535	Varies	Varies	Engine Data 1
Park Neutral Switch	P-N / R-D-L	P-N	R-D-L	Engine Data 1
Set Switch (Cruise)	On / Off	Off	Off	Engine Data 1
Short Term Fuel Trim Bank 1	-10% to 10%	Varies	Varies	Engine Data 1
Short Term Fuel Trim Bank 2	-10% to 10%	Varies	Varies	Engine Data 1
Spark Advance (°)	-133 to 78°	Varies	Varies	Engine Data 1
Stoplamp Pedal Switch Signal	Pressed or Released	Released	Released	Engine Data 1
TAC Learn Counter	0-255 Count	0	0	Engine Data 1
Theft Deterrent System	Several	Correct	Correct	Engine Data 1
TP Angle (%)	0-100%	0-3	Varies	Engine Data 1
TP Sensor 1 (V)	0-5.1v	> 0.8	1.6	Engine Data 1
TP Sensor 2 (V)	0-5.1v	< 4.0	2.8	Engine Data 1
Vehicle Speed	0-155 mph	0	30	Engine Data 1
Warmup Cycle Done	Yes / No	Yes	Yes	Engine Data 1

2000-01 Catera 3.0L V6 MFI VIN R (A/T) - Misfire Data List

Parameter Identification (PID)	PID Value Range	PID Value at Hot Idle	PID Value at 30 mph	Data List Type
Engine Run Time	Hr: Min: Sec	00:00:00	00:00:00	Misfire
Misfire Current CYL #1	0-255 Count	0	0	Misfire
Misfire Current CYL #2	0-255 Count	0	0	Misfire
Misfire Current CYL #3	0-255 Count	0	0	Misfire
Misfire Current CYL #4	0-255 Count	0	0	Misfire
Misfire Current CYL #5	0-255 Count	0	0	Misfire
Misfire Current CYL #6	0-255 Count	0	0	Misfire
Misfire History CYL #1	0-255 Count	0	0	Misfire
Misfire History CYL #2	0-255 Count	0	0	Misfire
Misfire History CYL #3	0-255 Count	0	0	Misfire
Misfire History CYL #4	0-255 Count	0	0	Misfire
Misfire History CYL #5	0-255 Count	0	0	Misfire
Misfire History CYL #6	0-255 Count	0	0	Misfire

CTS PID DATA

2003-04 CTS 2.6L V6 MFI VIN N (M/T, A/T) - PID Data (All Lists)

Parameter Identification (PID)	PID Value Range	PID Value at Hot Idle	PID Value at 30 mph	Data List Type
A/C HI Side Pressure	0-450 kPa (*)	Varies	Varies	Engine Data 1-2
A/C HI Side Pressure	0-500 psi (*)	Varies	Varies	Engine Data 1-2
A/C Pressure Sensor	0-5.1 volts	Varies	Varies	Engine Data 1-2
A/C Off or WOT	Yes / No	No	No	Engine Data 1
A/C Off or WOT	Yes / No	No	No	TAC Data
A/C Pressure Disable	Yes / No	No	No	Engine Data 1
A/C Relay Command	On / Off	Off	Off	Engine Data 1-3
A/C Request	Yes / No	No	No	Engine Data 1-2
A/F Ratio Bank 1	0-25.5:1	14.7:1	14.7:1	Fuel Trim
APP Angle	0-99% (*)	0	11	Engine Data 1-2, EVAP, Fuel Trim, Misfire, TAC
APP at Idle	0-99% (*)	Varies	Varies	TAC
APP Sensor 1	0.90-4.50v	1v	Varies	TAC
APP Sensor 2	0.45-2.25v	0.5v	Varies	TAC
Battery Voltage	0-25.5v	12.8-14.2v	12.8-14.2v	Engine Data 1-3, EVAP, TAC
Calculated BARO	65-104 kPa	Varies w/Alt.	Varies w/Alt.	Engine Data 1, EVAP, Fuel Trim
Calculated BARO	0-16 psi	Varies w/Alt.	Varies w/Alt.	Engine Data 1, EVAP Fuel Trim
Calculated Converter Temperature (Cruise)	-40°F to +1742°F	Varies	Varies	Engine Data 1, Fuel Trim
Catalyst Over-Temperature Bank 1	Yes / No	No	No	Fuel Trim
Catalyst Over-Temperature Bank 2	Yes / No	No	No	Fuel Trim
Clutch Start Switch	On / Off	Off	Off	Engine Data 1-2, TAC
CMP Active Counter	0-255 counts	Varies	Varies	Engine Data 2
Cooling Fan After Run (*)	Request/Not Requested	Not Requested	Not Requested	Engine Data 3
Commanded Cooling Fan (*)	Off / Low / High Auxiliary if used	Off	Off	Engine Data 2-3
Crank Request	Yes / No	No	No	Engine Data 1, 3
Cruise Clutch Switch	Applied or Released	Released	Released	Engine Data 1, Data 3, TAC
Cruise Engaged (*)	Yes / No	No	No	Engine Data 1-3
Cruise Engaged (*)	Yes / No	No	No	TAC
Cruise Release Brake Pedal Switch	Applied or Released	Released	Released	Engine Data 1, Data 3, TAC
Cruise Resume and Accel Switch (*)	On / Off	Off	Off	TAC
Cruise Set and Coast Switch (*)	On / Off	Off	Off	TAC
Cruise VSS Out of Range (*)	Yes / No	No	No	TAC

(*) Indicates this PID can be included in the related Freeze Frame Data.

2003-04 CTS 2.6L V6 MFI VIN N (M/T, A/T) - PID Data (All Lists)

Parameter Identification (PID)	PID Value Range	PID Value at Hot Idle	PID Value at 30 mph	Data List Type
Current Gear (*)	P/N/R/1-5	P/N	1-5	Engine Data 1, 2 & 3, Fuel Trim
Decel Fuel Cutoff (*)	Active / Inactive	Inactive	Inactive	Engine Data 2, Fuel Trim
Delivered Torque PWM Duty Cycle	0-100%	10%	90%	Engine Data 2
Desired Idle Speed	0-7000 rpm	650	Varies	Engine Data 1-3, EVAP, TAC
Distance Since 1st Failure	Km/h, mph	Varies	Varies	Failure Record
DTC No. 1	PXXXX	Varies	Varies	Engine Data 1, EVAP, TAC
DTC No. 2	PXXXX	Varies	Varies	Engine Data 1, EVAP, TAC
ECM in VTD Fail Enable VTD	Yes / No	No	No	Engine Data 3
ECT Sensor (V)	-38 to 284°F	185-220	185-220	Engine Data 1, 2 & 3, EVAP, Misfire
Engine at Operating Temperature	Yes / No	Yes	Yes	Failure Record
Engine Drag Control	Yes / No	Yes	Yes	TAC
Engine Idling	Yes / No	Yes	No	Failure record
Engine Load (*)	0-100%	2	5-8	Engine Data 1
Engine Oil Life Left	0-100%	Varies	Varies	Engine Data 3
Engine Oil Level Switch	Okay/Low	Okay	Okay	Engine Data 2-3
Engine Oil Press. Switch	Okay/Low	Okay	Okay	Engine Data 3
Engine Run Time	Hr: Min: Sec	00:00:00	00:00:00	00:00:00
Engine Speed	0-9999 rpm	650	1310	Engine Data 1, 2 & 3, EVAP, Fuel Trim HO2S, TAC
EVAP Canister Purge (*)	0-100%	Varies	Varies	Engine Data 1, Fuel Trim, HO2S
EVAP Canister Vent (*)	On / Off	Off	Off	Engine Data 1, Fuel Trim, EVAP
Extended Travel Brake Pedal Switch	Apply/Release	Released	Released	Engine Data 1-2 Data 3, TAC
Fast Idle for A/C Performance	Active, Inactive	Inactive	Inactive	Engine Data 1, TAC
Fast Idle for Cold start	Active / Inactive	Inactive	Inactive	TAC
Fast Idle for Generator	Active, Inactive	Inactive	Inactive	Engine Data 2, TAC
Fast Idle for Heater Performance	Active, Inactive	Inactive	Inactive	TAC
Fast Idle for Hot Start	Active / Inactive	Inactive	Inactive	TAC
Fuel Level (*)	Gallons, Liters	Varies	Varies	Engine Data 1, Fuel Trim, Misfire

(*) Indicates this PID can be included in the related Freeze Frame Data.

2003-04 CTS 2.6L V6 MFI VIN N (M/T, A/T) - PID Data (All Lists)

Parameter Identification (PID)	PID Value Range	PID Value at Hot Idle	PID Value at 30 mph	Data List Type
Fuel Level Sensor Left Tank	0-5.1 volts	Varies	Varies	Engine Data 1, EVAP
Fuel Level Sensor Right Tank	0-5.1 volts	Varies	Varies	Engine Data 1, EVAP
Fuel Pump Relay (*)	On / Off	On	On	Engine Data 1
Fuel Tank Pressure	In. H20 or mm Hg	Varies	Varies	Engine Data 1, EVAP
FTP Sensor (V)	0-5.1v	Cap off: 2.5v	Varies	Engine Data 1, EVAP
Generator L-Terminal	On / Off	On	On	Engine Data 2
Generator F-Terminal	0-100%	20	35	Engine Data 2
HO2S-11 (B1 S1)	0-4998 mv	30-800 mv	30-800 mv	Engine Data 1-2, EVAP, Fuel Trim
HO2S-11 (B1 S2)	0-4998 mv	400-700 mv	400-700 mv	Engine Data 1-2, EVAP, Fuel Trim
HO2S-11 (B2 S1)	0-4998 mv	30-800 mv	30-800 mv	Engine Data 1-2, EVAP, Fuel Trim
HO2S-11 (B2 S2)	0-4998 mv	400-700 mv	400-700 mv	Engine Data 1-2, EVAP, Fuel Trim
HO2S-11 (B1 S1) Heater	High/Low	Varies	Varies	Fuel Trim
HO2S-12 (B1 S2) Heater	High/Low	Varies	Varies	Fuel Trim
HO2S-21 (B2 S1) Heater	High/Low	Varies	Varies	Fuel Trim
HO2S-22 (B2 S2) Heater	High/Low	Varies	Varies	Fuel Trim
IAT Sensor (V)	-38 to 284°F	32-194	32-194	Engine Data 1, 2 & 3, EVAP, Fuel Trim, Misfire
Ignition 0	0-25.5v	12.8-14.2	12.8-14.2	Engine Data 1, 3, EVAP, TAC
Ignition 1	0-25.5v	12.8-14.2	12.8-14.2	Engine Data 1, 2 & 3, EVAP Fuel Trim, Misfire, TAC
Ignition Dwell	1.0-4.0 ms	1.9	1.9	Engine Data 1
Ignition State	On / Cranking / Off	On	On	Engine Data 1
Injector Pulsewidth Bank 1 or Bank 2	0-9.0 ms	3.0-5.4	3.0-5.4	Engine Data 2-3 EVAP, Fuel Trim
Intake Manifold Tuning Valve Sol. (*)	Open or Closed	Closed	Closed	ENG 2
Intake Manifold Runner Control Sol (*)	Open or Closed	Open	Open	ENG 2
Knock Control	Active / Inactive	Inactive	Inactive	Engine Data 1
Knock Sensor 1	0.25-5.40v	4.6-5.4	4.6-5.4	Engine Data 1
Knock Sensor 2	0.25-5.40v	4.6-5.4	4.6-5.4	Engine Data 1

(*) Indicates this PID can be included in the related Freeze Frame Data.

2003-04 CTS 2.6L V6 MFI VIN N (M/T, A/T) - PID Data (All Lists)

Parameter Identification (PID)	PID Value Range	PID Value at Hot Idle	PID Value at 30 mph	Data List Type
Knock Retard Cylinder 1-6	0-191 (°) degrees	0°	0°	Engine Data 1
Long Term Fuel Trim Bank 1 (*)	-100% to +100%	0 (± 5%)	0 (± 5%)	Engine Data 1, 2 & 3, EVAP, Fuel Trim
Long Term Fuel Trim Bank 2 (*)	-100% to +100%	0 (± 5%)	0 (± 5%)	Engine Data 1, 2 & 3, EVAP, Fuel Trim
Long Term Fuel Trim Idle/Decel Bank 1 (*)	-100% to +100%	0 (± 5%)	0 (± 5%)	Engine Data 1, 2 & 3, EVAP, Fuel Trim
Long Term Fuel Trim Idle/Decel Bank 2 (*)	-100% to +100%	0 (± 5%)	0 (± 5%)	Engine Data 1, 2 & 3, EVAP, Fuel Trim
Loop Status Bank 1 (*)	O/L or C/L	Closed	Closed	Engine Data 1, 2 & 3, EVAP, Fuel Trim
Loop Status Bank 2 (*)	O/L or C/L	Closed	Closed	Engine Data 1, 2 & 3, EVAP, Fuel Trim
MAF Sensor Grams Per Second (*)	0-655 g/sec	3.4-6.3	8.2-9.8	Engine Data 1, 2 & 3, EVAP, Fuel Trim Misfire, TAC
MAF Sensor Volts (*)	0-5.1 volts	1.2-1.5	1.6-1.8v	Engine Data 1, 2 & 3, EVAP, Fuel Trim Misfire, TAC
Main Relay Command	On / Off	On	On	Engine Data 1
MIL Request	On / Off	Off	Off	Engine Data 1
MIL Flashing	Yes / No	No	No	Engine Data 1
Mileage Since DTC Cleared	Kilometers or Miles	Varies	Varies	Engine Data 1, Data 3
Mileage Since MIL Request	Kilometers or Miles	Varies	Varies	Engine Data 1, Data 3
Misfire Current Cylinder 1-6	0-255 counts	0	0	Misfire
Misfire History Cylinder 1-6	0-255 counts	0	0	Misfire
Number of DTC(s)	Number	0	0	Engine Data 1, EVAP, Fuel Trim, TAC
Park Neutral Position Switch	Park/Neutral or In Gear	Park / Neutral	In Gear	Engine Data 2, Data 3, TAC
Password Leaning Ability	On / Off	Off	Off	Engine Data 3
Power Enrichment (*)	Active / Inactive	Inactive	Inactive	Engine Data 2, Fuel Trim
Power Management (TAC Error)	Active / Inactive	Inactive	Inactive	Engine Data 3, TAC
Reduced Power (TAC Error)	Active / Inactive	Inactive	Inactive	Engine Data 3, TAC
Requested Torque PWM Duty Cycle	0-99%	90%	90%	Engine Data 2, TAC
Rough Road Detection	Present/Not Present	Not Present	Not Present	Engine Data 2, Misfire

(*) Indicates this PID can be included in the related Freeze Frame Data.

2003-04 CTS 2.6L V6 MFI VIN N (M/T, A/T) - PID Data (All Lists)

Parameter Identification (PID)	PID Value Range	PID Value at Hot Idle	PID Value at 30 mph	Data List Type
Short Term Fuel Trim Bank 1 (*)	-100 to +100%	0 (± 5%)	0 (± 5%)	Engine Data 1, 2 & 3, EVAP, Fuel Trim
Short Term Fuel Trim Bank 1 (*)	-100 to +100%	0 (± 5%)	0 (± 5%)	Engine Data 1, 2 & 3, EVAP, Fuel Trim
Spark Advance	-133° to +78°	2-5°	2-5°	Engine Data 1, 2 & 3, Fuel Trim
Startup ECT Sensor (*)	-38°F to +284°F	Varies	Varies	Engine Data 2, EVAP
Startup IAT Sensor (*)	-38°F to +284°F	Varies	Varies	Engine Data 2, EVAP
Stoplamp Pedal Switch (*)	Applied or Released	Released	Released	Engine Data 1-3 TAC
TAC Forced Engine Idle Mode	Active or Inactive	Inactive	Inactive	Engine Data 3, TAC
TAC Learn Counter	0-255 counts	Varies	Varies	TAC
TAC Limited Power	Yes / No	No	No	Engine Data 1, TAC
Throttle Actuator Duty Cycle	0-99%	Varies	Varies	Engine Data 1, TAC
Throttle Plate at Idle (*)	Yes / No	Yes	No	TAC
Throttle Plate at WOT (*)	Yes / No	No	No	TAC
TP Sensor 1 Learned Minimum	0-5.0 volts	0-0.5	0.0.5	TAC
TP Sensor 2 Learned Minimum	5.0-0 volts	5.0-3.7	5.0-3.7	TAC
TP Angle (*)	0-99%	0	8-9	Engine Data 1, 2 & 3, EVAP, Fuel Trim Misfire, TAC
TP Sensor 1	0-5.0 volts	0.5-0.8	8.1-9.1	Engine Data 1-3 TAC
TP Sensor 2	0-5.0 volts	5.0-0	4.1-4.5	Engine Data 1-3 TAC
Vehicle Speed (*)	Kilometers or Miles	0	30	Engine Data 1-3, EVAP, TAC
Volumetric Efficiency	0-100%	13	22	Engine Data 1-3, EVAP, TAC

(*) Indicates this PID can be included in the related Freeze Frame Data.

2003-04 CTS 3.2L V6 MFI VIN M (A/T) - PID Data (All Lists)

Parameter Identification (PID)	PID Value Range	PID Value at Hot Idle	PID Value at 30 mph	Data List Type
A/C HI Side Pressure	0-450 kPa (*)	Varies	Varies	Engine Data 1-2
A/C HI Side Pressure	0-500 psi (*)	Varies	Varies	Engine Data 1-2
A/C Pressure Sensor	0-5.1 volts	Varies	Varies	Engine Data 1-2
A/C Off or WOT	Yes / No	No	No	Engine Data 1
A/C Off or WOT	Yes / No	No	No	TAC Data
A/C Pressure Disable	Yes / No	No	No	Engine Data 1
A/C Relay Command	On / Off	Off	Off	Engine Data 1-3
A/C Request	Yes / No	No	No	Engine Data 1-2
A/F Ratio Bank 1	0-25.5:1	14.7:1	14.7:1	Fuel Trim
APP Angle	0-99% (*)	0	11	Engine Data 1-2, EVAP, Fuel Trim, Misfire, TAC
APP at Idle	0-99% (*)	Varies	Varies	TAC
APP Sensor 1	0.90-4.50v	1v	Varies	TAC
APP Sensor 2	0.45-2.25v	0.5v	Varies	TAC
Battery Voltage	0-25.5v	12.8-14.2v	12.8-14.2v	Engine Data 1-3, EVAP, TAC
Calculated BARO	65-104 kPa	Varies w/Alt.	Varies w/Alt.	Engine Data 1, EVAP, Fuel Trim
Calculated BARO	65-104 kPa	Varies w/Alt.	Varies w/Alt.	EVAP
Calculated BARO	65-104 kPa	Varies w/Alt.	Varies w/Alt.	Fuel Trim
Calculated BARO	0-16 psi	Varies w/Alt.	Varies w/Alt.	Engine Data 1
Calculated BARO	0-16 psi	Varies w/Alt.	Varies w/Alt.	EVAP
Calculated BARO	0-16 psi	Varies w/Alt.	Varies w/Alt.	Fuel Trim
Calculated Converter Temperature (Cruise)	-40°F to +1742°F	Varies	Varies	Engine Data 1, Fuel Trim
Catalyst Over-Temperature Bank 1	Yes / No	No	No	Fuel Trim
Catalyst Over-Temperature Bank 2	Yes / No	No	No	Fuel Trim
CMP Active Counter	0-255 counts	Varies	Varies	Engine Data 2
Cooling Fan After Run (*)	Request/Not Requested	Not Requested	Not Requested	Engine Data 3
Commanded Cooling Fan (*)	Off / Low / High Auxiliary if used	Off	Off	Engine Data 2-3
Crank Request	Yes / No	No	No	Engine Data 1, 3
Cruise Engaged (*)	Yes / No	No	No	Engine Data 1-3
Cruise Engaged (*)	Yes / No	No	No	TAC
Cruise Release Brake Pedal Switch	Applied or Released	Released	Released	Engine Data 1, Data 3, TAC
Cruise Resume and Accel Switch (*)	On / Off	Off	Off	TAC
Cruise Set and Coast Switch (*)	On / Off	Off	Off	TAC
Cruise VSS Out of Range (*)	Yes / No	No	No	TAC

(*) Indicates this PID can be included in the related Freeze Frame Data.

2003-04 CTS 3.2L V6 MFI VIN M (A/T) - PID Data (All Lists)

Parameter Identification (PID)	PID Value Range	PID Value at Hot Idle	PID Value at 30 mph	Data List Type
Current Gear (*)	P/N/R/1-5	P/N	1-5	Engine Data 1, 2 & 3, Fuel Trim
Decel Fuel Cutoff (*)	Active / Inactive	Inactive	Inactive	Engine Data 2, Fuel Trim
Delivered Torque PWM Duty Cycle	0-100%	10%	90%	Engine Data 2
Desired Idle Speed	0-7000 rpm	650	Varies	Engine Data 1-3, EVAP, TAC
Distance Since 1st Failure	Km/h, mph	Varies	Varies	Failure Record
DTC No. 1	PXXXX	Varies	Varies	Engine Data 1, EVAP, TAC
DTC No. 2	PXXXX	Varies	Varies	Engine Data 1, EVAP, TAC
ECM in VTD Fail Enable VTD	Yes / No	No	No	Engine Data 3
ECT Sensor (V)	-38 to 284°F	185-220	185-220	Engine Data 1, 2 & 3, EVAP, Misfire
Engine at Operating Temperature	Yes / No	Yes	Yes	Failure Record
Engine Drag Control	Yes / No	Yes	Yes	TAC
Engine Idling	Yes / No	Yes	No	Failure record
Engine Load (*)	0-100%	2	5-8	Engine Data 1
Engine Oil Life Left	0-100%	Varies	Varies	Engine Data 3
Engine Oil Level Switch	Okay/Low	Okay	Okay	Engine Data 2-3
Engine Oil Press. Switch	Okay/Low	Okay	Okay	Engine Data 3
Engine Run Time	Hr: Min: Sec	00:00:00	00:00:00	00:00:00
Engine Speed	0-9999 rpm	650	1310	Engine Data 1, 2 & 3, EVAP, Fuel Trim HO2S, TAC
EVAP Canister Purge (*)	0-100%	Varies	Varies	Engine Data 1, Fuel Trim, HO2S
EVAP Canister Vent (*)	On / Off	Off	Off	Engine Data 1, Fuel Trim, EVAP
Extended Travel Brake Pedal Switch	Apply/Release	Released	Released	Engine Data 1-2 Data 3, TAC
Fast Idle for A/C Performance	Active, Inactive	Inactive	Inactive	Engine Data 1, TAC
Fast Idle for Cold start	Active / Inactive	Inactive	Inactive	TAC
Fast Idle for Generator	Active, Inactive	Inactive	Inactive	Engine Data 2, TAC
Fast Idle for Heater Performance	Active, Inactive	Inactive	Inactive	TAC
Fast Idle for Hot Start	Active / Inactive	Inactive	Inactive	TAC
Fuel Level (*)	Gallons, Liters	Varies	Varies	Engine Data 1, Fuel Trim, Misfire

(*) Indicates this PID can be included in the related Freeze Frame Data.

2003-04 CTS 3.2L V6 MFI VIN M (A/T) - PID Data (All Lists)

Parameter Identification (PID)	PID Value Range	PID Value at Hot Idle	PID Value at 30 mph	Data List Type
Fuel Level Sensor Left Tank	0-5.1 volts	Varies	Varies	Engine Data 1, EVAP
Fuel Level Sensor Right Tank	0-5.1 volts	Varies	Varies	Engine Data 1, EVAP
Fuel Pump Relay (*)	On / Off	On	On	Engine Data 1
Fuel Tank Pressure	In. H20 or mm Hg	Varies	Varies	Engine Data 1, EVAP
FTP Sensor (V)	0-5.1v	Cap off: 2.5v	Varies	Engine Data 1, EVAP
Generator L-Terminal	On / Off	On	On	Engine Data 2
Generator F-Terminal	0-100%	20	35	Engine Data 2
HO2S-11 (B1 S1)	0-4998 mv	30-800 mv	30-800 mv	Engine Data 1-2, EVAP, Fuel Trim
HO2S-11 (B1 S2)	0-4998 mv	400-700 mv	400-700 mv	Engine Data 1-2, EVAP, Fuel Trim
HO2S-11 (B2 S1)	0-4998 mv	30-800 mv	30-800 mv	Engine Data 1-2, EVAP, Fuel Trim
HO2S-11 (B2 S2)	0-4998 mv	400-700 mv	400-700 mv	Engine Data 1-2, EVAP, Fuel Trim
HO2S-11 (B1 S1) Heater	High/Low	Varies	Varies	Fuel Trim
HO2S-12 (B1 S2) Heater	High/Low	Varies	Varies	Fuel Trim
HO2S-21 (B2 S1) Heater	High/Low	Varies	Varies	Fuel Trim
HO2S-22 (B2 S2) Heater	High/Low	Varies	Varies	Fuel Trim
IAT Sensor (V)	-38 to 284°F	32-194	32-194	Engine Data 1, 2 & 3, EVAP, Fuel Trim, Misfire
Ignition 0	0-25.5v	12.8-14.2	12.8-14.2	Engine Data 1, 3, EVAP, TAC
Ignition 1	0-25.5v	12.8-14.2	12.8-14.2	Engine Data 1, 2 & 3, EVAP Fuel Trim, Misfire, TAC
Ignition Dwell	1.0-4.0 ms	1.9	1.9	Engine Data 1
Ignition State	On / Cranking / Off	On	On	Engine Data 1
Injector Pulsewidth Bank 1 or Bank 2	0-9.0 ms	3.0-5.4	3.0-5.4	Engine Data 2-3 EVAP, Fuel Trim
Intake Manifold Tuning Valve Sol. (*)	Open or Closed	Closed	Closed	ENG 2
Intake Manifold Runner Control Sol (*)	Open or Closed	Open	Open	ENG 2
Knock Control	Active / Inactive	Inactive	Inactive	Engine Data 1
Knock Sensor 1	0.25-5.40v	4.6-5.4	4.6-5.4	Engine Data 1
Knock Sensor 2	0.25-5.40v	4.6-5.4	4.6-5.4	Engine Data 1

(*) Indicates this PID can be included in the related Freeze Frame Data.

2003-04 CTS 3.2L V6 MFI VIN M (A/T) - PID Data (All Lists)

Parameter Identification (PID)	PID Value Range	PID Value at Hot Idle	PID Value at 30 mph	Data List Type
Knock Retard Cylinder 1-6	0-191 (°) degrees	0°	0°	Engine Data 1
Long Term Fuel Trim Bank 1 (*)	-100% to +100%	0 (± 5%)	0 (± 5%)	Engine Data 1, 2 & 3, EVAP, Fuel Trim
Long Term Fuel Trim Bank 2 (*)	-100% to +100%	0 (± 5%)	0 (± 5%)	Engine Data 1, 2 & 3, EVAP, Fuel Trim
Long Term Fuel Trim Idle/Decel Bank 1 (*)	-100% to +100%	0 (± 5%)	0 (± 5%)	Engine Data 1, 2 & 3, EVAP, Fuel Trim
Long Term Fuel Trim Idle/Decel Bank 2 (*)	-100% to +100%	0 (± 5%)	0 (± 5%)	Engine Data 1, 2 & 3, EVAP, Fuel Trim
Loop Status Bank 1 (*)	O/L or C/L	Closed	Closed	Engine Data 1, 2 & 3, EVAP, Fuel Trim
Loop Status Bank 2 (*)	O/L or C/L	Closed	Closed	Engine Data 1, 2 & 3, EVAP, Fuel Trim
MAF Sensor Grams Per Second (*)	0-655 g/sec	3.4-6.3	8.2-9.8	Engine Data 1, 2 & 3, EVAP, Fuel Trim Misfire, TAC
MAF Sensor Volts (*)	0-5.1 volts	1.2-1.5	1.6-1.8v	Engine Data 1, 2 & 3, EVAP, Fuel Trim Misfire, TAC
Main Relay Command	On / Off	On	On	Engine Data 1
MIL Request	On / Off	Off	Off	Engine Data 1
MIL Flashing	Yes / No	No	No	Engine Data 1
Mileage Since DTC Cleared	Kilometers or Miles	Varies	Varies	Engine Data 1, Data 3
Mileage Since MIL Request	Kilometers or Miles	Varies	Varies	Engine Data 1, Data 3
Misfire Current Cylinder 1-6	0-255 counts	0	0	Misfire
Misfire History Cylinder 1-6	0-255 counts	0	0	Misfire
Number of DTC(s)	Number	0	0	Engine Data 1, EVAP, Fuel Trim, TAC
Password Leaning Ability	On / Off	Off	Off	Engine Data 3
Power Enrichment (*)	Active / Inactive	Inactive	Inactive	Engine Data 2, Fuel Trim
Power Management (TAC Error)	Active / Inactive	Inactive	Inactive	Engine Data 3, TAC
Reduced Power (TAC Error)	Active / Inactive	Inactive	Inactive	Engine Data 3, TAC
Requested Torque PWM Duty Cycle	0-99%	90%	90%	Engine Data 2, TAC
Rough Road Detection	Present/Not Present	Not Present	Not Present	Engine Data 2, Misfire

(*) Indicates this PID can be included in the related Freeze Frame Data.

2003-04 CTS 3.2L V6 MFI VIN M (A/T) - PID Data (All Lists)

Parameter Identification (PID)	PID Value Range	PID Value at Hot Idle	PID Value at 30 mph	Data List Type
Short Term Fuel Trim Bank 1 (*)	-100 to +100%	0 (± 5%)	0 (± 5%)	Engine Data 1, 2 & 3, EVAP, Fuel Trim
Short Term Fuel Trim Bank 1 (*)	-100 to +100%	0 (± 5%)	0 (± 5%)	Engine Data 1, 2 & 3, EVAP, Fuel Trim
Spark Advance	-133° to +78°	2-5°	2-5°	Engine Data 1, 2 & 3, Fuel Trim
Startup ECT Sensor (*)	-38°F to +284°F	Varies	Varies	Engine Data 2, EVAP
Startup IAT Sensor (*)	-38°F to +284°F	Varies	Varies	Engine Data 2, EVAP
Stoplamp Pedal Switch (*)	Applied or Released	Released	Released	Engine Data 1-3 TAC
TAC Forced Engine Idle Mode	Active or Inactive	Inactive	Inactive	Engine Data 3, TAC
TAC Learn Counter	0-255 counts	Varies	Varies	TAC
TAC Limited Power	Yes / No	No	No	Engine Data 1, TAC
Throttle Actuator Duty Cycle	0-99%	Varies	Varies	Engine Data 1, TAC
Throttle Plate at Idle (*)	Yes / No	Yes	No	TAC
Throttle Plate at WOT (*)	Yes / No	No	No	TAC
TP Sensor 1 Learned Minimum	0-5.0 volts	0-0.5	0.0.5	TAC
TP Sensor 2 Learned Minimum	5.0-0 volts	5.0-3.7	5.0-3.7	TAC
TP Angle (*)	0-99%	0	8-9	Engine Data 1, 2 & 3, EVAP, Fuel Trim Misfire, TAC
TP Sensor 1	0-5.0 volts	0.5-0.8	8.1-9.1	Engine Data 1-3 TAC
TP Sensor 2	0-5.0 volts	5.0-0	4.1-4.5	Engine Data 1-3 TAC
Vehicle Speed (*)	Kilometers or Miles	0	30	Engine Data 1-3, EVAP, TAC
Volumetric Efficiency	0-100%	13	22	Engine Data 1-3, EVAP, TAC

(*) Indicates this PID can be included in the related Freeze Frame Data.

DEVILLE, ELDORADO & SEVILLE PID DATA

1996-99 DeVille-Eldorado-Seville 4.6L V8 MFI VIN Y, VIN 9 Data List 1

Parameter Identification (PID)	PID Value Range	PID Value at Hot Idle	PID Value at 30 mph	Data List Type
Actual EGR Position	0-100%	0	35	Engine Data 1
BARO Sensor (kPa)	10-110 kPa	97	97	Engine Data 1
Commanded EGR	0-100%	0	35	Engine Data 1
Desired Idle Speed	0-2040 rpm	600	---	Engine Data 1
ECT Sensor (°F)	-40 to 419°F	185-221	185-221	Engine Data 1
EGR Closed Pintle Pos.	0.0-5.0v	0.75	Varies	Engine Data 1
EGR Pintle Position	0.0-5.0v	0.75	Varies	Engine Data 1
Engine Load	0-100%	3	10	Engine Data 1
Engine Run Time	Hr: Min: Sec	00:00:00	00:00:00	Engine Data 1
Engine Speed	0-9999 rpm	±50 Desired	Varies	Engine Data 1
Extended Travel Brake	Apply/Release	Released	Released	Engine Data 1
Fan Control Relay 1	Off or On	Off	Off	Engine Data 1
Fan Control Relay 2, 3	Off or On	Off	Off	Engine Data 1
Fuel Pump Voltage	0-25.5v	14.1	14.1	Engine Data 1
Fuel Pump Relay	On / Off	On	On	Engine Data 1
Fuel Trim Cell	1-16	14	4	Engine Data 1
Fuel Trim Learn	Disabled/Enabled	Enabled	Enabled	Engine Data 1
Generator PWM	0-100%	0-100	0-100	Engine Data 1
High Electrical Load	Yes / No	No	No	Engine Data 1
HO2S (B1 S1, B1 S2)	0-1132 mv	10-1000	10-1000	Engine Data 1
HO2S-13 (B1 S3)	0-1132 mv	10-1000	10-1000	Engine Data 1
IAC Position	0-255 Count	43	56	Engine Data 1
IAT Sensor (°F)	-40 to 304°F	105	105	Engine Data 1
I/C Mode	ICM / PCM	PCM	PCM	Engine Data 1
Ignition-0, Ignition 1signal	0.0-25.5v	14.1, 12.6	14.1, 12.6	Engine Data 1
Injector Pulsewidth B1	0-985 ms	3.4	4.6	Engine Data 1
Injector Pulsewidth B2	0-985 ms	3.4	4.6	Engine Data 1
Knock Retard (°)	0-45°	0	0	Engine Data 1
Knock Too Long	Yes / No	No	No	Engine Data 1
KS Active Counter	0-255 Count	0	0	Engine Data 1
Loop Status	Closed Loop / Open Loop	Closed Loop	Closed Loop	Engine Data 1
Long Term Fuel Trim B1, B2	-10% to 10%	Varies	Varies	Engine Data 1
MAF Sensor (g/sec)	0-512 g/sec	5-7	Varies	Engine Data 1
MAP Sensor (kPa)	10-110 kPa	39	53	Engine Data 1
MPH or KPH	0-155 mph	0	30	Engine Data 1
P/NP Switch	P/N-R-OD-21	P/N	R-OD-21	Engine Data 1
PSP Switch	HI/Normal	Normal	Normal	Engine Data 1
HO2S-11 (B1 S1) R / L	Rich/Lean	L-R-L-R	L-R-L-R	Engine Data 1
HO2S-12 (B1 S2) R / L	Rich/Lean	L-R-L-R	L-R-L-R	Engine Data 1
HO2S-21 (B2 S1) R / L	Rich/Lean	L-R-L-R	L-R-L-R	Engine Data 1
Short Term Fuel Trim B1, B2	-10% to 10%	1.2%	1.5%	Engine Data 1
Spark Advance (°)	-90° to 90°	18	24	Engine Data 1
TCC Brake Switch	Apply/Release	Released	Released	Engine Data 1
TCC Slip Speed rpm	-4096-4096	620	+25	Engine Data 1
Throttle At Idle	Yes / No	Yes	No	Engine Data 1
TP Normalized	0-100%	0	8	Engine Data 1
TP Normalized	0.0-81.6°	0	7	Engine Data 1
TP Sensor (V)	0-5v	0.51	1.0	Engine Data 1

1996-99 DeVille-Eldorado-Seville 4.6L V8 MFI VIN Y, VIN 9 Data List 2

Parameter Identification (PID)	PID Value Range	PID Value at Hot Idle	PID Value at 30 mph	Data List Type
A/C Clutch	Yes / No	No	No	Engine Data 2
Air Fuel Ratio	0:1-25.5:1	14.6	14.5	Engine Data 2
Current Gear	1-2-3-4	1	3	Engine Data 2
Desired Idle Speed	0-2040 rpm	600	---	Engine Data 2
Desired Torque ABS/TCS	0-100%	100	Varies	Engine Data 2
ECT Sensor (°F)	-40 to 419°F	185-221	185-221	Engine Data 2
ECT At Startup °F	-40 to 419°F	Varies	Varies	Engine Data 2
Engine Load	0-100%	3	10	Engine Data 2
Engine Run Time	Hr: Min: Sec	00:00:00	00:00:00	Engine Data 2
Engine Speed	0-9999 rpm	±50 Desired	Varies	Engine Data 2
Engine Oil Level	Okay/Low	Okay	Okay	Engine Data 2
Engine Oil Life	0-100%	Varies	Varies	Engine Data 2
EVAP Canister Purge	0-100%	0	30	Engine Data 2
HO2S-11 (B1 S1)	0-1132 mv	10-1000	10-1000	Engine Data 2
HO2S-12 (B1 S2)	0-1132 mv	10-1000	10-1000	Engine Data 2
IAC Position	0-255 Count	43	65	Engine Data 2
I/C Mode	ICM / PCM	PCM	PCM	Engine Data 2
Ignition Cycle Counter	0-255 Count	5	5	Engine Data 2
Ignition 1 Signal	0.0-25.5v	14.1	14.1	Engine Data 2
Loop Status	Closed Loop / Open Loop	Closed Loop	Closed Loop	Engine Data 2
MAF Sensor (g/sec)	0-512 g/sec	5.5	16.8	Engine Data 2
MAP Sensor (kPa)	10-110 kPa	42	53	Engine Data 2
MAP Sensor (V)	0-5v	1.70	2.11	Engine Data 2
MPH or KPH	0-155 mph	0	30	Engine Data 2
Non-Driven Wheel Speed	0-155 mph	0	30	Engine Data 2
Outside Air Temperature	-40-215°F	90	90	Engine Data 2
P/NP Switch	P/N-R-OD-21	P/N	R-OD-21	Engine Data 2
Power Enrichment	Active / Inactive	Inactive	Inactive	Engine Data 2
Reference Volts Low	0-5v	0.05	0.05	Engine Data 2
Spark Advance (°)	0-90°	Varies	Varies	Engine Data 2
Shift Solenoid 1	Off / On	On	Off	Engine Data 2
Shift Solenoid 2	Off / On	On	Off	Engine Data 2
TCC Brake Switch	Apply/Release	Released	Released	Engine Data 2
TCC Duty Cycle	0-100%	0	50-100	Engine Data 2
TCC Enable	Yes / No	No	Yes	Engine Data 2
TCC Slip Speed (rpm)	-4096-4096	80	+25	Engine Data 2
TCM Calibration Number	Number	0000	0000	Engine Data 2
TCM Software Number	Number	0000	0000	Engine Data 2
TFT Sensor (°)	-40 to 419°F	168	170	Engine Data 2
TFP Switch A/B/C	Several	Several	Several	Engine Data 2
TP Angle (°)	0-100%	0	6	Engine Data 2
TP Sensor (V)	0-5v	0.51	1.0	Engine Data 2
Traction Control	Active / Inactive	Inactive	Inactive	Engine Data 2
Traction Control Torque	0-100%	8-18	Varies	Engine Data 2
Transmission Oil Life	0-100%	Varies	Varies	Engine Data 2
4X Ref. Pulse Counter	0-255 Count	8	8	Engine Data 2

1996-99 DeVille-Eldorado-Seville 4.6L V8 MFI VIN Y, VIN 9 Data List 3

Parameter Identification (PID)	PID Value Range	PID Value at Hot Idle	PID Value at 30 mph	Data List Type
A/C Clutch	Yes / No	No	No	Engine Data 3
A/C Relay Driver	Okay/Fault 1-3	Okay	Okay	Engine Data 3
Auto Learn Timer	Active / Inactive	Inactive	Inactive	Engine Data 3
Cruise	Enabled/Disabled	Disabled	Disabled	Engine Data 3
Cruise Enable Driver	Okay/Fault 1-3	Okay	Okay	Engine Data 3
Cruise Requested	Yes / No	No	No	Engine Data 3
Current Octane Level	87-93	93	93	Engine Data 3
Delivered Torque Driver	Okay/Fault 1-3	Okay	Okay	Engine Data 3
Driver No. 1	Okay/Fault	Okay	Okay	Engine Data 3
Driver 1, 2 Ground	Okay/Fault	Okay	Okay	Engine Data 3
ECT Sensor (ºF)	-40 to 304ºF	185-221	185-221	Engine Data 3
Engine Load	0-100%	3	10	Engine Data 3
Engine Run Time	Hr: Min: Sec	00:00:00	00:00:00	Engine Data 3
Engine Speed	0-9999 rpm	±50 Desired	Varies	Engine Data 3
Engine Oil Pressure	Okay/Low	Okay	Okay	Engine Data 3
EVAP Canister Purge	0-100%	0	30	Engine Data 3
EVAP Purge Solenoid	Okay/Fault 1-3	Okay	Okay	Engine Data 3
EVAP Vent Solenoid	Open / Closed	Open	Open	Engine Data 3
EVAP Vent Valve Driver	Okay/Fault 1-3	Okay	Okay	Engine Data 3
Fan Control Relay 1	Off or On	Off	Off	Engine Data 3
Fan Control Relay 2, 3	Off or On	Off	Off	Engine Data 3
Generator PWM	0-100%	28	27	Engine Data 3
High Electrical Load	Yes / No	No	No	Engine Data 3
HO2S (B1 S1, B1 S2)	0-1132 mv	10-1000	10-1000	Engine Data 3
IAC Position	0-255 Count	43	56	Engine Data 3
IAT Sensor (ºF)	-40 to 304ºF	95	95	Engine Data 3
Ignition-0, Ignition -1	0.0-25.5v	14.1, 12.6	14.1, 12.7	Engine Data 3
Ignition 1 Supplement	Okay/Low	Okay	Okay	Engine Data 3
Injector 1-8 Fault	Yes / No	No	No	Engine Data 3
IPC Fuel Disable	Active / Inactive	Inactive	Inactive	Engine Data 3
Loop Status	Closed Loop / Open Loop	Closed Loop	Closed Loop	Engine Data 3
MAF Sensor (g/sec)	0-512 g/sec	5.5	16.8	Engine Data 3
MAF Sensor (Hz)	0-31,999 Hz	2800	3400	Engine Data 3
MAP Sensor (kPa)	10-110 kPa	39	53	Engine Data 3
MIL (lamp) Command	On / Off	Off	Off	Engine Data 3
MIL Driver Status	Okay/Fault 1-3	Okay	Okay	Engine Data 3
MPH or KPH	0-155 mph	0	30	Engine Data 3
Odometer Miles	0-1677722	Varies	Varies	Engine Data 3
Outside Air Temperature	-40-215ºF	90	90	Engine Data 3
P/NP Switch	P/N-R-OD-21	P/N	R-OD-21	Engine Data 3
Power Enrichment	Active / Inactive	Inactive	Inactive	Engine Data 3
Shift Solenoid 1 Driver	Okay/Fault 1-3	Okay	Okay	Engine Data 3
Shift Solenoid 2 Driver	Okay/Fault 1-3	Okay	Okay	Engine Data 3
Shift Solenoid 1, 2	Off / On	Off	Off	Engine Data 3
Tachometer Driver	Okay/Fault 1-3	Okay	Okay	Engine Data 3
TCC Brake Switch	Apply/Release	Apply: On	Released	Engine Data 3
TCC Duty Cycle	0-100%	0	0-50	Engine Data 3

1996-99 DeVille-Eldorado-Seville 4.6L V8 MFI VIN Y, VIN 9 Data List 3

Parameter Identification (PID)	PID Value Range	PID Value at Hot Idle	PID Value at 30 mph	Data List Type
TCC Enable	Yes / No	No	No	Engine Data 3
TCC Solenoid Driver	Okay/Fault 1-3	Okay	Okay	Engine Data 3
TFP Switch	Several	Park	Several	Engine Data 3
TP Sensor (V)	0-5v	0.51	1.0	Engine Data 3
Traction Control Torque	0-100%	8-18	10-12	Engine Data 3
Transmission Range	P/N-R-D432L	P/N	D3	Engine Data 3
TR Switch A, B, C, P	High / Low	L, H, H, L	---	Engine Data 3
Vehicle Speed Driver	Okay/Fault 1-3	Okay	Okay	Engine Data 3
VTD Fuel Disable	Active / Inactive	Inactive	Inactive	Engine Data 3
VTD Password	Okay/Incorrect	Okay	Okay	Engine Data 3
VTD P/W Learn Mode	Active / Inactive	Inactive	Inactive	Engine Data 3

1996-99 DeVille-Eldorado-Seville 4.6L V8 MFI VIN Y, VIN 9 Catalyst

Parameter Identification (PID)	PID Value Range	PID Value at Hot Idle	PID Value at 30 mph	Data List Type
Air Fuel Ratio	0:1-25:1	14.6:1	14.6:1	Catalyst
BARO Sensor (kPa)	10-110 kPa	85-103	85-103	Catalyst
Calculated Air Flow g/sec	0-511.99	Varies	Varies	Catalyst
Calculated Converter Temperature	32-1409°F	680	750	Catalyst
Catalyst EWMA Samples	0-255 Count	2	2	Catalyst
Catalyst Test Fail Limit	-25-25.5 sec	0.05	0.05	Catalyst
Cat Test Time Diff EWMA	-25-25.5 sec	---	---	Catalyst
Current TWC Test Comp.	Yes / No	No	No	Catalyst
Desired Idle Speed	0-2040 rpm	600	---	Catalyst
ECT Sensor (°F)	-40 to 304°F	195	195	Catalyst
Engine Load	0-100%	3	10	Catalyst
Engine Run Time	Hr: Min: Sec	Varies	Varies	Catalyst
Engine Speed	0-9999 rpm	±50 Desired	Varies	Catalyst
HO2S-11 (B1 S1)	0-1132 mv	10-1000	10-1000	Catalyst
HO2S-12 (B1 S2)	0-1132 mv	10-1000	10-1000	Catalyst
HO2S-21 (B2 S1)	0-1132 mv	10-1000	10-1000	Catalyst
HO2S-11 (B1 S1) XCount	0-255 Count	Varies	Varies	Catalyst
HO2S-21 (B2 S1) XCount	0-255 Count	Varies	Varies	Catalyst
IAC Position	0-255 Count	43	56	Catalyst
IAT Sensor (°F)	-40 to 304°F	88	88	Catalyst
Injector Pulsewidth B1, B2	0-955 ms	3.4	4.6	Catalyst
Loop Status	Closed Loop / Open Loop	Closed Loop	Closed Loop	Catalyst
Long Term Fuel Trim B1, B2	-10% to 10%	Varies	Varies	Catalyst
MAF Sensor (g/sec)	0-512 g/sec	5.5	16.8	Catalyst
MAP Sensor (kPa)	10-110 kPa	43	53	Catalyst
No. Catalyst Tests Done	0-255 Count	5	Varies	Catalyst
OSC Test Complete	Yes / No	Varies	No	Catalyst
Short Term Fuel Trim B1, B2	-10% to 10%	Varies	Varies	Catalyst
TP Angle (%)	0-100%	0	Varies	Catalyst
TP Sensor (V)	0-5v	0.55	1.0	Catalyst
TWC Stage 1, 2 Started	Yes / No	Varies	No	Catalyst
TWC Test Enabled	Yes / No	Varies	No	Catalyst
TWC Test Initialization	Yes / No	Varies	No	Catalyst

1996-99 DeVille-Eldorado-Seville 4.6L V8 MFI VIN Y, VIN 9 EVAP List

Parameter Identification (PID)	PID Value Range	PID Value at Hot Idle	PID Value at 30 mph	Data List Type
BARO Sensor (kPa)	10-110 kPa	85-103	85-103	EVAP
Calculated ECT°F	-40 to 304°F	Varies	Varies	EVAP
Desired Idle Speed	0-2040 rpm	720	---	EVAP
ECT Sensor (°F)	-40 to 304°F	185-221	185-221	EVAP
Engine Load	0-100%	1-5	6-10	EVAP
Engine Run Time	Hr: Min: Sec	00:00:00	00:00:00	EVAP
Engine Speed	0-9999 rpm	±50 Desired	Varies	EVAP
EVAP Canister Purge	0-100%	10-20	Varies	EVAP
EVAP Vacuum Decay	0-2" H2O	> 0.10	> 0.10	EVAP
EVAP Test Result	Pass/Fail	Pass	Pass	EVAP
EVAP Test State	Wait/Done	Done	Done	EVAP
EVAP Test Abort Reason	Several	Several	Several	EVAP
EVAP Vent Solenoid	Open / Closed	Open	Open	EVAP
Fuel Control State Bank 1	Normal/Fault	Normal	Normal	EVAP
Fuel Control State Bank 2	Normal/Fault	Normal	Normal	EVAP
FTP Sensor (In. H2O)	-17.5 to 7.5"	Varies	Varies	EVAP
FTP Sensor (V)	0-5.1v	2.5: cap off	Varies	EVAP
HO2S (B1 S1, B1 S2)	0-1132 mv	10-1000	10-1000	EVAP
IAT Sensor (°F)	-40 to 304°F	50-194	50-194	EVAP
Injector Pulsewidth B1	0-985 ms	3.7	4.6	EVAP
Injector Pulsewidth B2	0-985 ms	3.7	4.6	EVAP
Loop Status	Closed Loop / Open Loop	Closed Loop	Closed Loop	EVAP
Long Term Fuel Trim B1, B2	-10% to 10%	Varies	Varies	EVAP
MAF Sensor (g/sec)	0-512 g/sec	5-7	16-20	EVAP
MAP Sensor (kPa)	10-110 kPa	30-50	53	EVAP
MAP Sensor (V)	0-5v	1-2	1.5-2.1	EVAP
Short Term Fuel Trim B1, B2	-10% to 10%	Varies	Varies	EVAP
TP Angle (%)	0-100%	0	Varies	EVAP
TP Sensor (V)	0-5v	0.55	Varies	EVAP
Vehicle Speed	0-155 mph	0	30	EVAP

1996-99 DeVille-Eldorado-Seville 4.6L V8 MFI VIN Y, VIN 9 Fuel Trim 1

Parameter Identification (PID)	PID Value Range	PID Value at Hot Idle	PID Value at 30 mph	Data List Type
Air Fuel Ratio	0:1-25:1	14.6:1	14.6:1	Fuel Trim 1
Desired Idle Speed	0-2040 rpm	720	---	Fuel Trim 1
ECT Sensor (°F)	-40 to 304°F	185-226	185-226	Fuel Trim 1
Engine Load	0-100%	1-5	Varies	Fuel Trim 1
Engine Run Time	Hr: Min: Sec	00:00:00	00:00:00	Fuel Trim 1
Engine Speed	0-9999 rpm	±50 Desired	Varies	Fuel Trim 1
HO2S (B1 S1) Bank 1	0-1132 mv	10-1000	10-1000	Fuel Trim 1
HO2S (B2 S1) Bank 2	0-1132 mv	10-1000	10-1000	Fuel Trim 1
HO2S-13 (B1 S3)	0-1132 mv	10-1000	10-1000	Fuel Trim 1
HO2S1 L/R Trans. Bank 1	0-3187.5 ms	Varies	Varies	Fuel Trim 1
HO2S2 L/R Trans. Bank 1	0-3187.5 ms	Varies	Varies	Fuel Trim 1
HO2S1 R/L Status Bank 1	Rich/Lean	L-R-L-R	L-R-L-R	Fuel Trim 1
HO2S3 R/L Status Bank 1	Rich/Lean	L-R-L-R	L-R-L-R	Fuel Trim 1
HO2S1 R/L Status Bank 2	Rich/Lean	L-R-L-R	L-R-L-R	Fuel Trim 1

1996-99 DeVille-Eldorado-Seville 4.6L V8 MFI VIN Y, VIN 9 Fuel Trim 1

Parameter Identification (PID)	PID Value Range	PID Value at Hot Idle	PID Value at 30 mph	Data List Type
HO2S-11 (B1 S1) warmup	0:00-4:15	30-50	30-50	Fuel Trim 1
HO2S-21 (B2 S1) warmup	0:00-4:15	30-50	30-50	Fuel Trim 1
HO2S-13 (B1 S3) warmup	0:00-4:15	30-50	30-50	Fuel Trim 1
HO2S XCounts B1, B2	0-255 Count	Varies	Varies	Fuel Trim 1
IAC Position	0-255 Count	30-80	---	Fuel Trim 1
IAT Sensor (°F)	-40 to 304°F	50-194	50-194	Fuel Trim 1
Ignition 1 Signal	0.0-25.5v	14.1	14.1	Fuel Trim 1
Injector Pulsewidth B1, B2	0-985 ms	3.7	4.6	Fuel Trim 1
Loop Status	Closed Loop / Open Loop	Closed Loop	Closed Loop	Fuel Trim 1
Long Term Fuel Trim B1, B2	-10% to 10%	Varies	Varies	Fuel Trim 1
MAF Sensor (g/sec)	0-512 g/sec	5-7	Varies	Fuel Trim 1
MAP Sensor (kPa)	10-110 kPa	30-50	53	Fuel Trim 1
Power Enrichment	Active / Inactive	Inactive	Inactive	Fuel Trim 1
PSP Switch	HI/Normal	Normal	Normal	Fuel Trim 1
Short Term Fuel Trim B1, B2	-10% to 10%	Varies	Varies	Fuel Trim 1
Spark Advance (°)	-90° to 90°	20	0-60	Fuel Trim 1
TFP Switch	Several	Park	Several	Fuel Trim 1
Throttle At Idle	Yes / No	Yes	No	Fuel Trim 1
TP Sensor (V)	0-5v	0.55	1.0	Fuel Trim 1
Vehicle Speed	0-155 mph	0	30	Fuel Trim 1

1996-99 DeVille-Eldorado-Seville 4.6L V8 MFI VIN Y, VIN 9 Fuel Trim 2

Parameter Identification (PID)	PID Value Range	PID Value at Hot Idle	PID Value at 30 mph	Data List Type
Air Fuel Ratio	0:1-25:1	14.6:1	Varies	Fuel Trim 2
Calculated Converter Temperature	32-1409°F	Varies	Varies	Fuel Trim 2
Desired Idle Speed	0-2040 rpm	720	---	Fuel Trim 2
ECT Sensor (°F)	-40 to 304°F	185-226	185-226	Fuel Trim 2
Engine Load	0-100%	1-5	Varies	Fuel Trim 2
Engine Speed	0-9999 rpm	±50 Desired	---	Fuel Trim 2
HO2S (B1 S1, B2 S1)	0-1132 mv	10-1000	10-1000	Fuel Trim 2
HO2S-13 (B1 S3)	0-1132 mv	10-1000	10-1000	Fuel Trim 2
HO2S L/R Switch B1, B2	0-65535	< 12	< 12	Fuel Trim 2
HO2S R/L Status B1, B2	Rich/Lean	R or L	R or L	Fuel Trim 2
HO2S-13 R/L Status B1	Rich/Lean	R or L	R or L	Fuel Trim 2
HO2S R/L Switch B1, B2	0-65535	< 12	< 12	Fuel Trim 2
HO2S-11 (B1 S1) warmup	0:00-4:15	30-50	30-50	Fuel Trim 2
HO2S XCounts Bank 1, 2	0-255 Count	Varies	Varies	Fuel Trim 2
IAC Position	0-255 Count	30-80	---	Fuel Trim 2
Ignition 1 Signal	0.0-25.5v	14.1	14.1	Fuel Trim 2
Long Term Fuel Trim B1, B2	0-255 Count	0	Varies	Fuel Trim 2
Loop Status	Closed Loop / Open Loop	Closed Loop	Closed Loop	Fuel Trim 2
MAF Sensor (g/sec)	0-512 g/sec	5-7	---	Fuel Trim 2
Short Term Fuel Trim B1, B2	-10% to 10%	-8-8%	Varies	Fuel Trim 2
TFP Switch	Several	Park	Several	Fuel Trim 2
Throttle At Idle	Yes / No	Yes	No	Fuel Trim 2
TP Sensor (V)	0-5v	0.55	1.0	Fuel Trim 2
Vehicle Speed	0-155 mph	0	30	Fuel Trim 2

1996-99 DeVille-Eldorado-Seville 4.6L V8 MFI VIN Y, VIN 9 Misfire List

Parameter Identification (PID)	PID Value Range	PID Value at Hot Idle	PID Value at 30 mph	Data List Type
Cycles Of Misfire Data	0-100	Varies	Varies	Misfire
Desired Idle Speed	0-2040 rpm	720	---	Misfire
ECT Sensor (°F)	-40 to 304°F	185-226	185-226	Misfire
Engine Load	0-100%	1-5	Varies	Misfire
Engine Speed	0-9999 rpm	±50 Desired	---	Misfire
Engine Speed At Misfire	0-9999 rpm	0	0	Misfire
Knock Retard (°)	0° to 45°	0	0	Misfire
Load At Misfire	0-100%	0	0	Misfire
Loop Status	Closed Loop / Open Loop	Closed Loop	Closed Loop	Misfire
MAF Sensor (g/sec)	0-512 g/sec	5-7	Varies	Misfire
MAP Sensor (kPa)	10-110 kPa	30-50	53	Misfire
MAP Sensor (V)	0-5v	1-2	Varies	Misfire
Misfire Current Cyl 1 - 8	0-255 Count	0	0	Misfire
Misfire History Cyl 1 - 8	0-65535	0	0	Misfire
Misfire Delay Counter	0-255 Count	0	0	Misfire
Number of Failed Misfire Tests out of the last 16	0-16	0	0	Misfire
HO2S-11 (B1 S1)	0-1132 mv	10-1000	10-1000	Misfire
HO2S-12 (B1 S2)	0-1132 mv	10-1000	10-1000	Misfire
Spark Advance (°)	-90° to 90°	Varies	Varies	Misfire
Misfire Fail since 1st Failure	0-65535	0	0	Misfire
Misfire Pass since 1st Failure	0-65535	0	0	Misfire
TP Sensor	0-5v	0.55	1.0	Misfire
Vehicle Speed	0-155 mph	0	30	Misfire

2000-05 DeVille-Eldorado-Seville 4.6L V8 MFI VIN Y, VIN 9 Data List 1

Parameter Identification (PID)	PID Value Range	PID Value at Hot Idle	PID Value at 30 mph	Data List Type
A/C Relay Command	On / Off	Off	Off	Engine Data 1
Air Pump Relay command	On / Off	On	Off	Engine Data 1
Air Solenoid Command	On / Off	On	Off	Engine Data 1
BARO Sensor (kPa)	10-110 kPa	85-105	85-105	Engine Data 1
Current Gear	1-2-3-4	1	3	Engine Data 1
Desired Idle Speed	0-3187 rpm	650	---	Engine Data 1
ECT Sensor (°F)	-40 to 304°F	185-221	185-221	Engine Data 1
Engine Load	0-100%	2-5	6-8	Engine Data 1
Engine Run Time	Hr: Min: Sec	00:00:00	00:00:00	Engine Data 1
Engine Speed	0-9999 rpm	650	Varies	Engine Data 1
EVAP Purge Solenoid	On / Off	Off	Off	Engine Data 1
EVAP Vent Solenoid	Open / Closed	Open	Open	Engine Data 1
Extended Travel Brake Pressure Switch	Apply/Release	Released	Released	Engine Data 1
Fuel Pump Relay Control	On / Off	On	On	Engine Data 1
FTP Sensor (In. H2O)	-17 to +7.5	Varies	Varies	Engine Data 1
Fuel Trim Cell	Cell #0-4	0	0-4	Engine Data 1
HO2S-11 (B1 S1)	0-1275 mv	10-1000	10-1000	Engine Data 1
HO2S-12 (B1 S2)	0-1275 mv	10-1000	10-1000	Engine Data 1
HO2S-21 (B2 S1)	0-1275 mv	10-1000	10-1000	Engine Data 1
HO2S-11 (B1 S1) warmup	Seconds	Varies	Varies	Engine Data 1
IAT Sensor (°F)	-40 to 304°F	91	91	Engine Data 1
Ignition 1 Signal	0.0-25.5v	14.1	14.1	Engine Data 1
Knock Retard (°)	0° to 45°	0°	Varies	Engine Data 1
Long Term Fuel Trim B1	-99 to +99%	-25 to 25%	-25 to 25%	Engine Data 1
Long Term Fuel Trim B2	-99 to +99%	-25 to 25%	-25 to 25%	Engine Data 1
Loop Status	Closed Loop / Open Loop	Closed Loop	Closed Loop	Engine Data 1
MAF Sensor (g/sec)	0-512 g/sec	5-7	Varies	Engine Data 1
MAP Sensor (kPa)	10-110 kPa	38	Varies	Engine Data 1
MAP Sensor (V)	0-5.1v	1.5	Varies	Engine Data 1
MIL (lamp) Command	On / Off	Off	Off	Engine Data 1
Short Term Fuel Trim B1	-99 to 99%	-8% to 8%	-8% to 8%	Engine Data 1
Short Term Fuel Trim B2	-99 to 99%	-8% to 8%	-8% to 8%	Engine Data 1
Spark Advance (°)	0° to 64°	20°	24°	Engine Data 1
TCC Enable Solenoid	On / Off	Off	Off	Engine Data 1
Throttle Angle (%)	0-100%	0	Varies	Engine Data 1
TP Sensor (V)	0-5.1v	0.55	Varies	Engine Data 1
Vehicle Speed Signal	0-155 mph	0	30	Engine Data 1
5 Volt Reference 1	Okay/Fault	Okay	Okay	Engine Data 1
5 Volt Reference 2	Okay/Fault	Okay	Okay	Engine Data 1
12 Volt Reference	Okay/Fault	Okay	Okay	Engine Data 1

2000-05 DeVille-Eldorado-Seville 4.6L V8 MFI VIN Y, VIN 9 Data List 2

Parameter Identification (PID)	PID Value Range	PID Value at Hot Idle	PID Value at 30 mph	Data List Type
A/C High Side Press. (V)	0-5.1v	Varies	Varies	Engine Data 2
A/C Relay Command	On / Off	Off	Off	Engine Data 2
CMP Sensor (rpm)	0-8192	Varies	Varies	Engine Data 2
Cruise Control Active	Yes / No	No	No	Engine Data 2
Current Gear	1-2-3-4	1	3	Engine Data 2
Desired EGR Position	0-100%	0%	Varies	Engine Data 2
Desired Idle Speed	0-3187 rpm	650	---	Engine Data 2
ECT Sensor (°F)	-40 to 304°F	185-221	185-221	Engine Data 2
EGR Position Sensor (%)	0-100%	0%	Varies	Engine Data 2
EGR Position Sensor (V)	0-5.1v	0.60v	Varies	Engine Data 2
Engine Load	0-100%	2-5	12	Engine Data 2
Engine Oil Level Switch	Okay/Low	Okay	Okay	Engine Data 2
Engine Run Time	Hr: Min: Sec	00:00:00	00:00:00	Engine Data 2
Engine Speed	0-9999 rpm	650	Varies	Engine Data 2
Extended Travel Brake Pressure Switch	Apply/Release	Released	Released	Engine Data 2
Fan Control Relay 1	On / Off	Off	Off	Engine Data 2
Fan Control Relay 2 & 3	On / Off	Off	Off	Engine Data 2
Generator 'F' Signal	0-100%	Varies	Varies	Engine Data 2
Generator 'L' Signal	Okay / Not Okay	Okay	Okay	Engine Data 2
IAT Sensor (°F)	-40 to 304°F	50-194	50-194	Engine Data 2
Ignition 1 Signal	0.0-25.5v	14.1	14.1	Engine Data 2
Injector Pulsewidth B1	0-985 ms	3.6	Varies	Engine Data 2
Injector Pulsewidth B2	0-985 ms	3.5	Varies	Engine Data 2
Long Term Fuel Trim B1	-99 to +99%	-25 to 25%	-25 to 25%	Engine Data 2
Long Term Fuel Trim B2	-99 to +99%	-25 to 25%	-25 to 25%	Engine Data 2
Loop Status	Closed Loop / Open Loop	Closed Loop	Closed Loop	Engine Data 2
MAF Sensor (g/sec)	0-512 g/sec	5-7	Varies	Engine Data 2
MAP Sensor (kPa)	10-110 kPa	38	Varies	Engine Data 2
MAP Sensor (V)	0-5.1v	1.5	Varies	Engine Data 2
Non-Driven Wheel Speed	0-155 mph	0	30	Engine Data 2
Short Term Fuel Trim B1	-99 to 99%	-8% to 8%	-8% to 8%	Engine Data 2
Short Term Fuel Trim B2	-99 to 99%	-8% to 8%	-8% to 8%	Engine Data 2
Spark Advance (°)	0° to 64°	20	24	Engine Data 2
Startup ECT Sensor (°F)	-40 to 304°F	Varies	Varies	Engine Data 2
TCC Duty Cycle	0-100%	0	0-50	Engine Data 2
TFP Switch Position	Several	P-N	Drive	Engine Data 2
Throttle Angle (%)	0-100%	0	Varies	Engine Data 3
TP Sensor (V)	0-5.1v	0.55	Varies	Engine Data 2
Torque Delivered Signal	0-100%	8 - 18	---	Engine Data 2
Traction Control Torque	0-100%	8 - 18	---	Engine Data 2
Vehicle Speed Signal	0-155 mph	0	30	Engine Data 2
VTD Password	Okay / Not Okay	Okay	Okay	Engine Data 2
24X Crank Sensor	0-9999 rpm	Varies	Varies	Engine Data 2

2000-05 DeVille-Eldorado-Seville 4.6L V8 MFI VIN Y, VIN 9 Data List 3

Parameter Identification (PID)	PID Value Range	PID Value at Hot Idle	PID Value at 30 mph	Data List Type
A/C Relay Command	On / Off	Off	Off	Engine Data 3
Air Fuel Ratio	0:1-25:1	14.6:1	14.6:1	Engine Data 3
Cruise Control Active	Yes / No	No	No	Engine Data 3
Cruise Inhibited Signal	Yes / No	No	No	Engine Data 3
Cruise Request Signal	Yes / No	No	No	Engine Data 3
Desired Idle Speed	0-3187 rpm	650	---	Engine Data 3
ECT Sensor (°F)	-40 to 304°F	185-221	185-221	Engine Data 3
Engine Oil Level Switch	Okay/Low	Okay	Okay	Engine Data 3
Engine Oil Press. Switch	Okay/Low	Okay	Okay	Engine Data 3
Engine Run Time	Hr: Min: Sec	00:00:00	00:00:00	Engine Data 3
Engine Speed	0-9999 rpm	650	Varies	Engine Data 3
Extended Travel Brake Pressure Switch	Apply/Release	Released	Released	Engine Data 3
Fan Control Relay 1	On / Off	Off	Off	Engine Data 3
Fan Control Relay 2, 3	On / Off	Off	Off	Engine Data 3
IAT Sensor (°F)	-40 to 304°F	50-194	50-194	Engine Data 3
Ignition 1 Signal	0.0-25.5v	14.1	14.1	Engine Data 3
IMS	Several	Park	Drive 3	Engine Data 3
Long Term Fuel Trim B1	-99 to +99%	-25 to 25%	-25 to 25%	Engine Data 3
Long Term Fuel Trim B2	-99 to +99%	-25 to 25%	-25 to 25%	Engine Data 3
Loop Status	Closed Loop / Open Loop	Closed Loop	Closed Loop	Engine Data 3
MAF Sensor (g/sec)	0-512 g/sec	5-7	Varies	Engine Data 3
MAF Sensor (Hz)	0-65535 Hz	2800	Varies	Engine Data 3
MAP Sensor (kPa)	10-110 kPa	38	Varies	Engine Data 3
Miles Since DTC Cleared	0-1677722	Varies	Varies	Engine Data 3
Odometer Miles	0-1677722	Varies	Varies	Engine Data 3
PCM Induced Chassis	Normal/Firm	Normal	Normal	Engine Data 3
PCM/VCM in VTD Fail	Yes / No	No	No	Engine Data 3
Short Term Fuel Trim B1	-99 to 99%	-8% to 8%	-8% to 8%	Engine Data 3
Short Term Fuel Trim B2	-99 to 99%	-8% to 8%	-8% to 8%	Engine Data 3
Spark Advance (°)	0° to 64°	20°	24°	Engine Data 3
Starter Relay Command	On / Off	Off	Off	Engine Data 3
TCC Brake Pedal Switch	Apply/Release	Released	Released	Engine Data 3
TFP Switch Position	Several	P-N	Drive	Engine Data 3
Throttle Angle (%)	0-100%	0%	Varies	Engine Data 3
TP Sensor (V)	0-5.1v	0.55	Varies	Engine Data 3
TFT Sensor (°F)	-40 to 304°F	195	195	Engine Data 3
Vehicle Speed (mph)	0-155 mph	0	30	Engine Data 3
VTD Auto Learn Timer	Active / Inactive	Inactive	Inactive	Engine Data 3
VTD Fuel Disable	Enabled/Disabled	Disabled	Disabled	Engine Data 3
Warmups w/o E/faults	0-255 Count	0	0	Engine Data 3
Warmups w/o non/E/faults	0-255 Count	0	0	Engine Data 3

2000-05 DeVille-Eldorado-Seville 4.6L V8 MFI VIN Y, VIN 9 - Fuel Trim

Parameter Identification (PID)	PID Value Range	PID Value at Hot Idle	PID Value at 30 mph	Data List Type
Air Fuel Ratio	0:1-25:1	14.6:1	14.6:1	Fuel Trim
Air Pump Relay command	On / Off	On	Off	Fuel Trim
Air Solenoid Command	On / Off	On	Off	Fuel Trim
BARO Sensor (kPa)	10-110 kPa	85-105	85-105	Fuel Trim
Current Gear	1-2-3-4	1	3	Fuel Trim
ECT Sensor (°F)	-40 to 304°F	185-221	185-221	Fuel Trim
Engine Load	0-100%	3	10	Fuel Trim
Engine Run Time	Hr: Min: Sec	00:00:00	00:00:00	Fuel Trim
Engine Speed	0-9999 rpm	±50 Desired	----	Fuel Trim
EVAP Purge Solenoid	On / Off	Off	Off	Fuel Trim
EVAP Vent Solenoid	On / Off	Off	Off	Fuel Trim
Fuel Trim Cell	Cell #0-4	0	0-4	Fuel Trim
HO2S-11 (B1 S1)	0-1275 mv	10-1000	10-1000	Fuel Trim
HO2S-12 (B1 S2)	0-1275 mv	10-1000	10-1000	Fuel Trim
HO2S-21 (B2 S1)	0-1275 mv	10-1000	10-1000	Fuel Trim
HO2S-11 (B1 S1) warmup	Seconds	Varies	Varies	Fuel Trim
IAC Position	0-255 Count	10-30	Varies	Fuel Trim
IAT Sensor (°F)	-40 to 304°F	50-194	50-194	Fuel Trim
Ignition 1 Signal	0.0-25.5v	14.1	14.1	Fuel Trim
Injector Pulsewidth B1	0-985 ms	3.6	Varies	Fuel Trim
Injector Pulsewidth B2	0-985 ms	3.5	Varies	Fuel Trim
Long Term Fuel Trim B1	-99 to +99%	-25 to 25%	-25 to 25%	Fuel Trim
Long Term Fuel Trim B2	-99 to +99%	-25 to 25%	-25 to 25%	Fuel Trim
Long Term F/T Avg. B1	-99 to +99%	-25 to 25%	-25 to 25%	Fuel Trim
Long Term F/T Avg. B2	-99 to +99%	-25 to 25%	-25 to 25%	Fuel Trim
Loop Status	Closed Loop / Open Loop	Closed Loop	Closed Loop	Fuel Trim
MAF Sensor (g/sec)	0-512 g/sec	5-7	Varies	Fuel Trim
Short Term Fuel Trim B1	-99 to 99%	-8% to 8%	-8% to 8%	Fuel Trim
Short Term Fuel Trim B2	-99 to 99%	-8% to 8%	-8% to 8%	Fuel Trim
Short Term F/T Avg. B1	-99 to 99%	-8% to 8%	-8% to 8%	Fuel Trim
Short Term F/T Avg. B2	-99 to 99%	-8% to 8%	-8% to 8%	Fuel Trim
Spark Advance (°)	0° to 64°	20°	24°	Fuel Trim
Startup ECT Sensor (°F)	-40 to 304°F	Varies	Varies	Fuel Trim
TFP Switch Position	Several	P-N	Drive	Fuel Trim
Throttle Angle (%)	0-100%	0%	Varies	Fuel Trim
Vehicle Speed (mph)	0-155 mph	0	30	Fuel Trim

2000-05 DeVille-Eldorado-Seville 4.6L V8 MFI VIN Y, VIN 9 EVAP List

Parameter Identification (PID)	PID Value Range	PID Value at Hot Idle	PID Value at 30 mph	Data List Type
Air Fuel Ratio	0:1-25:1	14.6:1	14.6:1	EVAP
BARO Sensor (kPa)	10-110 kPa	85-105	85-105	EVAP
Desired Idle Speed	0-3187 rpm	650	---	EVAP
ECT Sensor (°F)	-40 to 304°F	185-221	185-221	EVAP
Engine Load	0-100%	2-5	12	EVAP
Engine Run Time	Hr: Min: Sec	00:00:00	00:00:00	EVAP
Engine Speed	0-9999 rpm	650	Varies	EVAP
EVAP Test Abort	Several	---	---	EVAP
EVAP Test Result	Several	---	---	EVAP
EVAP Test State	Wait/Done	Done	Done	EVAP
EVAP Vent Solenoid	On / Off	Off	Off	EVAP
Fuel Level mmHg/"H2O	-17.5 to 7.5"	Varies	Varies	EVAP
Fuel Level Sensor	0-5.1v	Varies	Varies	EVAP
FTP Sensor (V)	0-5.1v	Varies	Varies	EVAP
FTP Sensor (In. H2O)	-17.5 to 7.5"	Varies	Varies	EVAP
Fuel Trim Cell	Cell #0-4	0	0-4	EVAP
HO2S-11 (B1 S1)	0-1275 mv	10-1000	10-1000	EVAP
HO2S-12 (B1 S2)	0-1275 mv	10-1000	10-1000	EVAP
HO2S-21 (B2 S1)	0-1275 mv	10-1000	10-1000	EVAP
HO2S-11 (B1 S1) warmup	Seconds	Varies	Varies	EVAP
IAT Sensor (°F)	-40 to 304°F	50-194	50-194	EVAP
Ignition 1 Signal	0-25.5v	14.1	14.1	EVAP
Long Term Fuel Trim B1, B2	-99 to +99%	-25 to 25%	-25 to 25%	EVAP
MAF Sensor (g/sec)	0-512 g/sec	5-7	Varies	EVAP
Short Term Fuel Trim B1, B2	-99 to 99%	-8% to 8%	-8% to 8%	EVAP
Startup ECT Sensor (°F)	-40 to 304°F	195°F	194°F	EVAP
Startup IAT Sensor (°F)	-40 to 304°F	50-194	50-194	EVAP
Vehicle Speed	0-155 mph	0	30	EVAP

2000-05 DeVille-Eldorado-Seville 4.6L V8 MFI VIN Y, VIN 9 Misfire List

Parameter Identification (PID)	PID Value Range	PID Value at Hot Idle	PID Value at 30 mph	Data List Type
A/C Relay Command	On / Off	Off	Off	Misfire
CKP Sensor Status	Angle, A, B	Angle	Angle	Misfire
CMP Sensor (rpm)	0-8192	Varies	Varies	Misfire
CMP Sensor Signal	Yes / No	Yes	Yes	Misfire
Cycles of Misfire Data	0-100	Varies	Varies	Misfire
ECT Sensor (°F)	-40 to 304°F	185-221	185-221	Misfire
Engine Load	0-100%	2-5	Varies	Misfire
Engine Run Time	Hr: Min: Sec	00:00:00	00:00:00	Misfire
Engine Speed	0-9999	650	Varies	Misfire
Injector Pulsewidth B1, B2	0-985 ms	3.6	Varies	Misfire
MAF Sensor (g/sec)	0-512 g/sec	5-7	Varies	Misfire
Misfire Current Cyl 1-8	0-255 Count	0	0	Misfire
Misfire History Cyl 1-8	0-65535	0	0	Misfire
Spark Advance (°)	0° to 64°	20	24	Misfire
TP Angle (%)	0-100%	0	Varies	Misfire
24X Crank Sensor	0-9999 rpm	Varies	Varies	Misfire

ESCALADE

1999-2000 Escalade 5.7L V8 MFI VIN R (A/T) - PID Data (All Lists)

Parameter Identification (PID)	PID Value Range	PID Value at Hot Idle	PID Value at 30 mph	Data List Type
A/T 1-2 Solenoid	On / Off	On	On	Engine Data 2
A/T 2-3 Solenoid	On / Off	On	On	Engine Data 2
A/T 3-2 Downshift Sol.	On / Off	On	On	Engine Data 2
4WD Switch	Enable/Disabled	Disabled	Disabled	Engine Data 2
4WD Low Switch	Enable/Disabled	Disabled	Disabled	Engine Data 2
A/C Evaporator Switch	Open / Closed	Closed	Closed	Engine Data 2
A/C Sec HI Press Switch	Open / Closed	Open	Open	Engine Data 2
A/C Relay Driver	On / Off	Off	Off	Engine Data 2
A/C Relay Driver	On / Off	Off	Off	Misfire
A/C Request	Yes / No	No	No	Engine Data 2, Misfire
Actual EGR Position	0-100%	0	35	Engine Data 1-2
Actual EGR Position	0-100%	0	35	HO2S
Air Fuel Ratio	0:1-25.5:1	14.6	14.5	Engine Data 1
Air Fuel Ratio	0:1-25.5:1	14.6	14.5	HO2S
BARO Sensor (kPa)	10-110 kPa	99	97	Engine Data 1-2
BARO Sensor (kPa)	10-110 kPa	99	97	EVAP
BARO Sensor (V)	0-5.1 volts	4.75v	4.75v	Engine Data 1-2
BARO Sensor (V)	0-5.1 volts	4.75v	4.75v	EVAP
Brake Switch	Applied/Released	Released	Released	Engine Data 2, HO2S
Closed Loop / Startup	Yes / No	Yes	Yes	Engine Data 2
Closed Loop / Startup	Yes / No	Yes	Yes	HO2S
CMP Retard	0-99.9°F	0° - 2°	0° - 2°	Engine Data 1
CRUISE	Enable/Disabled	Disabled	Disabled	Engine Data 2
Deceleration Fuel Mode	Active / Inactive	Inactive	Inactive	Engine Data 2
Desired EGR Position	0-100%	0%	25%	Engine Data 1, HO2S
Desired EGR Position	0-100%	0%	25%	Engine Data 1
Desired IAC Position	0-250 counts	15-35	---	Engine Data 1-2 HO2S
Desired Idle Speed	0-1100 rpm	625	2150	Engine Data 1-2, EVAP, HO2S
DTC Set This Ignition	Yes / No	No	No	Engine Data 2, HO2S
ECT Sensor (°F)	-40 to 419°F	200	185-221	Engine Data 1-2, EVAP, HO2S, Misfire
ECT Sensor (V)	0-5.1 volts	1.90	1.89	Engine Data 1-2, EVAP, HO2S, Misfire
EGR Duty Cycle	0-100%	0	25	Engine Data 1-2 HO2S
EGR Sensor	0.0-5.1v	0.73	Varies	Engine Data 1

1999-2000 Escalade 5.7L V8 MFI VIN R (A/T) - PID Data (All Lists)

Parameter Identification (PID)	PID Value Range	PID Value at Hot Idle	PID Value at 30 mph	Data List Type
Engine Load	0-100%	2	11	Engine Data 1-2, EVAP, HO2S, Misfire
Engine Run Time	Hr: Min: Sec	00:00:00	00:00:00	Engine Data 1-2, EVAP, HO2S, Misfire
Engine Speed	0-9999 rpm	625	Varies	Engine Data 1-2, EVAP, HO2S, Misfire
EVAP Canister Purge	On / Off	On	On	Engine Data 1-2, EVAP, HO2S
Excess Vacuum Test	Pass/Run/fail	Not Run/Fail	Not Run/Fail	EVAP
EVAP Duty Cycle	0-100%	17	Varies	Engine Data 1-2
EVAP Canister Purge	On / Off	On	On	EVAP, HO2S
EVAP Duty Cycle	0-100%	17	Varies	EVAP, HO2S
EVAP Vent Solenoid	Open / Closed	Open	Open	Engine Data 2
EVAP Vent Solenoid	Open / Closed	Open	Open	EVAP
Fuel Level	0-100%	Varies	Varies	Engine Data 2
Fuel Level	0-100%	Varies	Varies	EVAP
Fuel Level Sensor	0-5.1 volts	Varies	Varies	EVAP
Fuel Level	0-100%	Varies	Varies	EVAP
Fuel Tank Pressure	Volts	Varies	Varies	EVAP
Fuel Tank Pressure	Inches H2O	Varies	Varies	EVAP
Fuel Trim Cell	Cell Number	Cell #	Cell #	Engine Data 1
Fuel Trim Cell	Cell Number	Cell #	Cell #	EVAP, HO2S
Fuel Trim Enable	Yes / No	Yes	Yes	Engine Data 1-2 HO2S
HO2S (B1 S1)	0-1132 mv	10-1000	10-1000	HO2S
HO2S (B1 S2)	0-1132 mv	10-1000	10-1000	HO2S
HO2S (B2 S1)	0-1132 mv	10-1000	10-1000	HO2S
HO2S (B2 S2)	0-1132 mv	10-1000	10-1000	HO2S
HO2S (B1 S3)	0-1132 mv	10-1000	10-1000	HO2S
HO2S XCounts B1	0-255 counts	Varies	Varies	HO2S
HO2S XCounts B2	0-255 counts	Varies	Varies	HO2S
HO2S XCounts B1	0-255 counts	Varies	Varies	HO2S
HVAC RECIRL mode	Yes / No	No	No	Engine Data 1
IAC Position	0-255 counts	15-35	---	Engine Data 1-2
IAC Position	0-255 counts	15-35	---	HO2S
IAT Sensor (°F)	-40 to 304°F	105	105	Engine Data 1-2 EVAP
IAT Sensor (V)	0-5.1 volts	1.2v	1.1v	Engine Data 1-2 EVAP
Ignition 1	0.0-25.5v	13.5v	13.6v	Engine Data 1-2
Ignition 1 Low	0.0-15.5v	Varies	Varies	Engine Data 2
Ignition 1High	0.0-15.5v	Varies	Varies	Engine Data 2
Ignition 1 On	Yes / No	Yes	Yes	Engine Data 2

1999-2000 Escalade 5.7L V8 MFI VIN R (A/T) - PID Data (All Lists)

Parameter Identification (PID)	PID Value Range	PID Value at Hot Idle	PID Value at 30 mph	Data List Type
Injector Pulsewidth B1	0-985 ms	3.3	4.7	Engine Data 2, HO2S
Injector Pulsewidth B2	0-985 ms	3.3	4.7	Engine Data 2, HO2S
Knock Adjust Factor	0-99.9°F	0°	0°	EVAP
Knock Retard (°)	0-45°	0°	0°	Engine Data 1
Knock Sensor	0-5.1v	4.9	4.9	Engine Data 1
Knock Sensor Activity	0-255 counts	0	0	Engine Data 1
Long Term FT Bank 1	Counts	127	126	Engine Data 1
Long Term FT Bank 1	Counts	127	126	HO2S
Long Term FT Bank 1	-10% to 10%	0 (± 5%)	0 (± 5%)	Engine Data 1
Long Term FT Bank 1	-10% to 10%	0 (± 5%)	0 (± 5%)	HO2S
Long Term FT Bank 2	Counts	127	126	Engine Data 1
Long Term FT Bank 2	Counts	127	126	HO2S
Long Term FT Bank 2	-10% to 10%	0 (± 5%)	0 (± 5%)	Engine Data 1
Long Term FT Bank 2	-10% to 10%	0 (± 5%)	0 (± 5%)	HO2S
Loop Status	Closed Loop / Open Loop	Closed Loop	Closed Loop	Engine Data 1-2
Loop Status	Closed Loop / Open Loop	Closed Loop	Closed Loop	EVAP, HO2S
Loop Status	Closed Loop / Open Loop	Closed Loop	Closed Loop	Misfire
MAF Sensor (g/sec)	0-512 g/sec	5-7	Varies	Engine Data 1-2
MAF Sensor (g/sec)	0-512 g/sec	5-7	Varies	EVAP, HO2S
MAF Sensor (g/sec)	0-512 g/sec	5-7	Varies	Misfire
MAF Sensor (Hz)	0-31,999 Hz	2255	Varies	Engine Data 1-2
MAF Sensor (Hz)	0-31,999 Hz	2255	Varies	EVAP, HO2S
MAF Sensor (Hz)	0-31,999 Hz	2255	Varies	Misfire
MAP Sensor (kPa)	10-110 kPa	33	Varies	Engine Data 1-2
MAP Sensor (kPa)	10-110 kPa	33	Varies	EVAP, HO2S
MAP Sensor (kPa)	10-110 kPa	33	Varies	Misfire
MAP Sensor (V)	0-5.1 volts	1.2	Varies	Engine Data 1-2
MAP Sensor (V)	0-5.1 volts	1.2	Varies	EVAP, HO2S
MAP Sensor (V)	0-5.1 volts	1.2	Varies	Misfire
Mileage Since DTC Cleared	0-255 Counts	Varies	Varies	Engine Data 2, HO2S
Misfire Current Cyl 1-8	Counts	0	0	Misfire
Misfire History Cyl 1-8	Counts	0	0	Misfire
PCM/VCM in VTD fail	Yes / No	No	No	Engine Data 2
Pre-HO2S	Ready/Not	Ready	Ready	Engine Data 1
Pre-HO2S	Ready/Not	Ready	Ready	HO2S
Post-HO2S	Ready/Not	Ready	Ready	Engine Data 1
Post-HO2S	Ready/Not	Ready	Ready	HO2S
Power Enrichment	Active / Inactive	Inactive	Inactive	Engine Data 2
Purge Leak Test	Pass/Run/fail	Pass	Pass	EVAP
REF Pulse Occurred	Yes / No	Yes	Yes	Engine Data 1

1999-2000 Escalade 5.7L V8 MFI VIN R (A/T) - PID Data (All Lists)

Parameter Identification (PID)	PID Value Range	PID Value at Hot Idle	PID Value at 30 mph	Data List Type
Rich/Lean Bank 1	Rich/Lean	Toggle R/L	Toggle R/L	Engine Data 1-2
Rich/Lean Bank 1	Rich/Lean	Toggle R/L	Toggle R/L	EVAP, HO2S
Rich/Lean Bank 1	Rich/Lean	Toggle R/L	Toggle R/L	Misfire
Rich/Lean Bank 2	Rich/Lean	Toggle R/L	Toggle R/L	Engine Data 1-2
Rich/Lean Bank 2	Rich/Lean	Toggle R/L	Toggle R/L	EVAP, HO2S
Rich/Lean Bank 2	Rich/Lean	Toggle R/L	Toggle R/L	Misfire
Short Term FT Bank 1	Counts	127	125	Engine Data 1
Short Term FT Bank 1	Counts	127	125	HO2S
Short Term FT Bank 1	-10% to 10%	0 (± 5%)	0 (± 5%)	Engine Data 1
Short Term FT Bank 1	-10% to 10%	0 (± 5%)	0 (± 5%)	HO2S
Short Term FT Bank 2	Counts	127	125	Engine Data 1
Short Term FT Bank 2	Counts	127	125	HO2S
Short Term FT Bank 2	-10% to 10%	0 (± 5%)	0 (± 5%)	Engine Data 1
Short Term FT Bank 2	-10% to 10%	0 (± 5%)	0 (± 5%)	HO2S
Small Leak Test	Pass/Run/fail	Not Run/Fail	Not Run/Fail	EVAP
Spark Advance (°)	-90° to 90°	21-24°	26-30°	Engine Data 1
Spark Advance (°)	-90° to 90°	21-24°	26-30°	Misfire
Spark Control	ADV/Retard	Advance	Advance	Engine Data 1
Startup ECT	-40 to 419°F	80°F	80°F	Engine Data 1-2 HO2S, Misfire
Startup IAT	-40 to 419°F	81°F	81°F	Engine Data 1-2 HO2S, Misfire
TCC Enable	Yes / No	No	Yes	Engine Data 2
TCC Duty Cycle	0-100%	0	95	Engine Data 2, Misfire
TFP Switch A/B/C	On / Off A/B/C	Off/On / Off	Off/Off/On	Engine Data 2
Total Misfire Current Count	0-255 counts	Varies	Varies	Misfire
Total Misfire Failures Since First Fail	0-255 counts	0	0	Misfire
Total Misfire Failures Since First Pass	0-255 counts	0	0	Misfire
TP Angle	0-100%	0	8	Engine Data 1-2
TP Angle	0-100%	0	8	EVAP, HO2S
TP Angle	0-100%	0	8	Misfire
TP Sensor (V)	0-5.1 volts	0.51	1.1	Engine Data 1-2
TP Sensor (V)	0-5.1 volts	0.51	1.1	EVAP, HO2S
TP Sensor (V)	0-5.1 volts	0.51	1.1	Misfire
TR Switch	Park/Neutral	Park	In Gear	Engine Data 2
Vehicle Speed	Km/h, MPH	0	30	Engine Data 1-2 HO2S, Misfire
VTD Auto Learn Timer	Active or Inactive	Inactive	Inactive	Engine Data 2
VTD Fuel Disable	Active or Inactive	Inactive	Inactive	Engine Data 2
VTD Fuel Disable Until Ignition Off	Yes / No	No	No	Engine Data 2
Weak Vacuum Test	Pass/Run/fail	Not Run/Fail	Not Run/Fail	EVAP

2002-05 Escalade 5.3L VIN T, 6.0L VIN N (All Lists)

Parameter Identification (PID)	PID Value Range	PID Value at Hot Idle	PID Value at 30 mph	Data List Type
4WD Signal	Enabled/Disabled	Disabled	Disabled	Engine Data 2
4WD Low Signal	Enabled/Disabled	Disabled	Disabled	Engine Data 2
A/C Pressure Sensor	0-5.1 volts	0.98	0.98	Engine Data 2
A/C Pressure Sensor	0-500 psi	90	90	Engine Data 2
A/C Relay Command	On / Off	Off	Off	Engine Data 1-2 Misfire
A/C Request	Yes / No	No	No	Engine Data 2
APP Average	0-100%	0%	Varies	TAC
APP Indicated Angle	0-100%	0%	Varies	Engine Data 1-2 C/C, EVAP, Fuel Trim, HO2S, TAC
APP Sensor 1	0.90-4.50v	0.4-0.9	Varies	TAC
APP Sensor 1	0-100%	0	Varies	TAC
APP Sensor 2	5.0-0.0 volts	4.5-4.1	Varies	TAC
APP Sensor 2	0-100%	0	Varies	TAC
APP Sensor 1 and 2	Agree/Disabled	Agree	Agree	TAC
BARO Sensor (kPa)	50-104 kPa	Varies with Altitude	Varies	Engine Data 1, EVAP, Fuel Trim
CMP Sensor HI / Low	0-255 counts	Varies	Varies	Engine Data 2
CMP Sensor Low/HI	0-255 counts	Varies	Varies	Engine Data 2
Clutch Pedal Switch	Applied/Released	Released	Released	Engine Data 2
Coolant Level Switch	Okay/Fault	Okay	Okay	Engine Data 2
Cruise Control Active	Yes / No	No	No	Engine Data 1, C/C, TAC
Cruise Disengage History 1-8	DTC SET / None	None	None	Cruise Control (C/C)
Cruise Inhibit Signal	On / Off	Off	Off	Engine Data 1
Cruise On / Off Switch	On / Off	Off	On	C/C, TAC
Cruise Release Brake Pedal Switch	Applied / Released	Released	Released	Cruise Control (C/C)
Cruise Resume / Accel Switch	On / Off	Off	Off	Cruise Control, TAC
Cruise Set/Coast Switch	On / Off	Off	Off	C/C, TAC
Current Gear	0, 1, 2, 3, 4	0	4	Engine Data 1-2 Fuel Trim
Cycles of Misfire Data	0-100 counts	Varies	Varies	Misfire
Decel Fuel Cutoff	Active / Inactive	Inactive	Inactive	HO2S
Desired IAC Airflow	0-64 g/sec	Varies	Varies	Engine Data 1
Desired Idle Speed	0-9999 rpm	Varies	Varies	Engine Data 1-2, EVAP, TAC
Desired IAC Position	0-250 counts	15-35	---	HO2S
DTC Set This Ignition	Yes / No	No	No	Engine Data 1-2 C/C, EVAP, HO2S, Fuel Trim
ECT Sensor (ºF)	-38 to 284ºF	190-221ºF	190-221ºF	Engine Data 1-2, EVAP, Fuel Trim HO2S, Misfire

2002-05 Escalade 5.3L VIN T, 6.0L VIN N (All Data Lists)

Parameter Identification (PID)	PID Value Range	PID Value at Hot Idle	PID Value at 30 mph	Data List Type
Engine Load	0-100%	1-4	Varies	Engine Data 1-2, EVAP, Fuel Trim HO2S, Misfire
Engine Oil Level Switch	Okay/Low	Okay	Okay	Engine Data 2
EOP Sensor	0-5.1 volts	1.5	Varies	Engine Data 2
Engine Run Time	Hr: Min: Sec	00:00:00	00:00:00	Engine Data 1-2, EVAP, Fuel Trim HO2S, Misfire
Engine Speed	0-10,000 rpm	600	Varies	Engine Data 1-2
Engine Speed	0-9999 rpm	625	Varies	EVAP, HO2S
Engine Speed	0-9999 rpm	625	Varies	Misfire
EVAP Purge Solenoid	0-100%	10-25%	Varies	Engine Data 1, EVAP, Fuel Trim
EVAP Test Result	No Result, Passed, Aborted, Fail	No Result	Passed	EVAP
EVAP Test Result	Waiting for Purge, Test Running, Test Done	Waiting for Purge	Test Done	EVAP
EVAP Vent Solenoid	Venting/Not	Venting	Not Venting	Engine Data 1, EVAP, Fuel Trim
Fuel Level Sensor	0-5.1 volts	0.7-2.5	0.7-2.5	Engine Data 1, EVAP
Fuel Level Sensor, Rear Tank (if used)	0-5.1 volts	0.7-2.5	0.7-2.5	Engine Data 1, EVAP
Fuel Tank Level Remaining	Gallons, Liters	Varies	Varies	EVAP
Fuel Tank Level Remaining	0-100%	Varies	Varies	EVAP
Fuel Tank Pressure	Inches H2O	-17.5 to +7.5	-17.5 to +7.5	Engine Data 1, EVAP
Fuel Tank Pressure	0-5.1 volts	Varies	Varies	EVAP
Fuel Tank Rated Capacity	25.9-34 Gal., 98-129 Liters	Varies	Varies	EVAP
Fuel Trim Cell	Cell 0-23	19	19	Engine Data 1, EVAP, Fuel Trim
Fuel Trim Learn	Enabled/Disabled	Enabled	May Toggle	Engine Data 1, EVAP, Fuel Trim
Generator F-Terminal	0-100%	Varies	Varies	Engine Data 2
Generator L-Terminal	On / Off	On	On	Engine Data 2
HO2S (B1 S1)	0-1106 mv	10-1000	10-1000	Engine Data 1, EVAP, Fuel Trim, HO2S
HO2S (B1 S2)	0-1106 mv	10-1000	10-1000	Engine Data 1, Fuel Trim, HO2S

2002-05 Escalade 5.3L VIN T, 6.0L VIN N (All Data Lists)

Parameter Identification (PID)	PID Value Range	PID Value at Hot Idle	PID Value at 30 mph	Data List Type
HO2S (B2 S1)	0-1106 mv	10-1000	10-1000	Engine Data 1, EVAP, Fuel Trim, HO2S
HO2S (B2 S2)	0-1106 mv	10-1000	10-1000	Engine Data 1, Fuel Trim, HO2S
HO2S-11 Heater B1	0-1.38 amps	0.25-1.375	0.25-1.375	HO2S
HO2S-12 Heater B1	0-1.38 amps	0.25-1.375	0.25-1.375	HO2S
HO2S-21 Heater B2	0-1.38 amps	0.25-1.375	0.25-1.375	HO2S
HO2S-22 Heater B2	0-1.38 amps	0.25-1.375	0.25-1.375	HO2S
IAT Sensor (°F)	-38 to 284°F	91	92	Engine Data 1-2, EVAP, Fuel Trim, HO2S
Ignition 1	0.0-25.5v	11.5-14.5	11.5-14.5	Engine Data 1-2 C/C, EVAP, Fuel Trim, TAC
Ignition 1 On	Yes / No	Yes	Yes	Engine Data 2
Injector Pulsewidth Bank 1 Average	0-985 ms	2-6	3-9	Engine Data 2, Fuel Trim, Misfire
Injector Pulsewidth Bank 2 Average	0-985 ms	2-6	3-9	Engine Data 2, Fuel Trim, Misfire
Knock Adjust Factor	0-99.9°F	0°	0°	EVAP
Knock Retard (°)	0-16°	0°	0°	Engine Data 1
Knock Sensor	0-5.1v	4.9	4.9	Engine Data 1
Long Term Fuel Trim Average Bank 1	0-100%	0 (± 5%)	0 (± 5%)	Fuel Trim
Long Term Fuel Trim Average Bank 2	0-100%	0 (± 5%)	0 (± 5%)	Fuel Trim
Long Term Fuel Trim Bank 1	-10% to 10%	0 (± 5%)	0 (± 5%)	Engine Data 1-2, EVAP, Fuel Trim, HO2S
Long Term Fuel Trim Bank 2	-10% to 10%	0 (± 5%)	0 (± 5%)	Engine Data 1-2, EVAP, Fuel Trim, HO2S
Loop Status	Closed Loop / Open Loop	Closed Loop	Closed Loop	Engine Data 1-2, EVAP, HO2S, Misfire
Low Oil Lamp Control	On / Off	Off	Off	Engine Data 2
MAF Sensor (g/sec)	0-512 g/sec	1-9	15-26s	Engine Data 1-2, EVAP, Fuel Trim, HO2S, Misfire, TAC
MAF Sensor (Hz)	0-31,999 Hz	2000-3000	Varies	Engine Data 2
MAP Sensor (kPa)	10-110 kPa	20-48	Varies	Engine Data 1-2, EVAP, Fuel Trim, HO2S, Misfire, TAC
MAP Sensor (V)	0-5.1 volts	1.2	Varies	Engine Data 1-2

2002-05 Escalade 5.3L VIN T, 6.0L VIN N (All Data Lists)

Parameter Identification (PID)	PID Value Range	PID Value at Hot Idle	PID Value at 30 mph	Data List Type
Mileage Since DTC Cleared	Kilometers, Miles	Varies	Varies	Engine Data 2
Misfire Counter Status	0-65,535	Counts	Varies	Misfire
Misfire Current Cyl 1-8	0-200 counts	0	0	Misfire
Misfire History Cyl 1-8	0-65,535	Counts	Varies	Misfire
PCM Reset	Yes / No	No	No	Engine Data 1-2 EVAP
PCM/VCM in VTD Fail	Yes / No	No	No	Engine Data 1
Power Enrichment	Active / Inactive	Inactive	Inactive	Engine Data 1, HO2S
Reduced Engine Power	Active / Inactive	Inactive	Inactive	Engine Data 1, C/C, TAC
Short Term Fuel Trim Average Bank 1	0-100%	0 (± 5%)	0 (± 5%)	Fuel Trim
Short Term Fuel Trim Average Bank 2	0-100%	0 (± 5%)	0 (± 5%)	Fuel Trim
Short Term Fuel Trim Bank 1	-10% to 10%	0 (± 5%)	0 (± 5%)	Engine Data 1-2, EVAP, Fuel Trim, HO2S
Short Term Fuel Trim Bank 2	-10% to 10%	0 (± 5%)	0 (± 5%)	Engine Data 1-2, EVAP, Fuel Trim, HO2S
Spark Advance	-90° to +90°	15°-20°	Varies	Engine Data 1-2 Fuel Trim, HO2S, Misfire
Startup ECT	-40 to 419°F	Varies	Varies	Engine Data 2, EVAP, Fuel Trim
Startup ECT	-40 to 419°F	80°F	80°F	HO2S
Stoplamp Pedal Switch	Applied/Released	Released	Released	Engine Data 2, C/C, TAC
TAC/PCM Communication signal	Okay/Fault	Okay	Okay	C/C, TAC
TCC Enable Solenoid	On / Off	Off	On	Engine Data 1
TCC PWM Solenoid	0-100%	Varies	Varies	Engine Data 2
TFP Switch A/B/C	On / Off A/B/C	Off/On / Off	Off/Off/On	Engine Data 2, C/C, Fuel Trim
Torque Delivered Signal	N.M., ft lbs	Varies	Varies	Engine Data 2, TAC
Torque Request Signal	N.M., ft lbs	Varies	Varies	Engine Data 2, TAC
TP Desired Angle	0-100%	Varies	Varies	Engine Data 1-2 C/C, EVAP, Fuel Trim, HO2S, Misfire, TAC

2002-05 Escalade 5.3L VIN T, 6.0L VIN N (All Data Lists)

Parameter Identification (PID)	PID Value Range	PID Value at Hot Idle	PID Value at 30 mph	Data List Type
TP Sensor 1	0-5.0 volts	0.4-0.9v	Varies	TAC
TP Sensor 1	0-100%	0%	Varies	TAC
TP Sensor 2	5.0-0 volts	4.5-0.0v	Varies	TAC
TP Sensor 2	100-0%	100%	Varies	TAC
TP Sensors 1 and 2	Agree/Disabled	Agree	Agree	TAC
TR Switch	Park/Neutral	Park	In Gear	Engine Data 2, C/C, Fuel Trim
Traction Control Signal	Active or Inactive	Inactive	Inactive	Engine Data 2, C/C, TAC
Vehicle Speed	Km/h, MPH	0	30	Engine Data 1-2 C/C, EVAP, Fuel Trim, HO2S, Misfire, TAC
VTD Auto Learn Timer	Active or Inactive	Inactive	Inactive	Engine Data 1
VTD Fuel Disable	Active or Inactive	Inactive	Inactive	Engine Data 1
VTD Fuel Disable Until Ignition Off	Yes / No	No	No	Engine Data 1
Warmups Without Emission Faults	0-255 counts	Varies	Varies	Engine Data 2
Warmups Without Non-Emission Faults	0-255 counts	Varies	Varies	Engine Data 2

2002-05 Escalade 5.3L VIN T, 6.0L VIN N (Freeze Frame)

Parameter Identification (PID)	PID Value Range	PID Value at Hot Idle	PID Value at 30 mph	Data List Type
DTC Number	Number	Number	Number	Freeze Frame
ECT Sensor (°F)	-38 to 284°F	190-221°F	190-221°F	Freeze Frame
Engine Load	0-100%	1-4	Varies	Freeze Frame
Engine Speed	0-10,000 rpm	600	Varies	Freeze Frame
IAT Sensor (°F)	-38 to 284°F	91°F	92°F	Freeze Frame
HO2S (B1 S1)	0-1106 mv	10-1000	10-1000	Freeze Frame
HO2S (B1 S2)	0-1106 mv	10-1000	10-1000	Freeze Frame
HO2S (B2 S1)	0-1106 mv	10-1000	10-1000	Freeze Frame
HO2S (B2 S2)	0-1106 mv	10-1000	10-1000	Freeze Frame
Ignition 1	0.0-25.5v	11.5-14.5	11.5-14.5	Freeze Frame
Long FT Bank 1	-10% to 10%	0 (± 5%)	0 (± 5%)	Freeze Frame
Long FT Bank 2	-10% to 10%	0 (± 5%)	0 (± 5%)	Freeze Frame
MAF Sensor (g/sec)	0-512 g/sec	1-9	15-26s	Freeze Frame
MAP Sensor (kPa)	10-110 kPa	20-48	Varies	Freeze Frame
SHRTFT Bank 1	-10% to 10%	0 (± 5%)	0 (± 5%)	Freeze Frame
SHRTFT Bank 2	-10% to 10%	0 (± 5%)	0 (± 5%)	Freeze Frame
Spark Advance	-90° to 90°	15°-20°	Varies	Freeze Frame
TP Sensor 1	0-5.0 volts	0.4-0.9v	Varies	Freeze Frame
Vehicle Speed	Km/h, MPH	0	30	Freeze Frame

FLEETWOOD PID DATA

1996 Fleetwood 5.7L V8 MFI VIN P (A/T) - Engine Data 1 List

Parameter Identification (PID)	PID Value Range	PID Value at Hot Idle	PID Value at 30 mph	Data List Type
A/C Clutch	On / Off	Off	Off	Engine Data 1
A/C High Side Press. (psi)	-15 - 425 psi	139-399	139-399	Engine Data 1
A/C High Side (V)	0.0-5.0v	1-2	1-2	Engine Data 1
A/C Request	Yes / No	No	No	Engine Data 1
A/C Status	On / Off	Off	Off	Engine Data 1
AIR Pump	Disabled/Enabled	Disabled	Disabled	Engine Data 1
BARO Sensor (kPa)	10-110 kPa	65-104	65-104	Engine Data 1
BARO Sensor (V)	0.0-5.0v	3.5-4.5	3.5-4.5	Engine Data 1
Cold Engine Startup	No or Yes	Yes: If cold	No	Engine Data 1
Current Gear	1-2-3-4	1	3	Engine Data 1
Desired Idle Speed	0-3187 rpm	PCM control	---	Engine Data 1
DTC Set This Ignition	Yes / No	No	No	Engine Data 1
ECT Sensor (°F)	-40 to 304°F	185-226	185-226	Engine Data 1
EGR Duty Cycle	0-100%	0	20	Engine Data 1
Engine Run Time	Hr: Min: Sec	00:00:00	00:00:00	Engine Data 1
Engine Speed	0-9999 rpm	±50 Desired	Varies	Engine Data 1
EVAP Purge (%)	0-100%	0	0-50	Engine Data 1
EVAP Vacuum Switch	No/Purge	No Purge	Purge	Engine Data 1
Fan Control Relay 1	Off or On	Off	Off	Engine Data 1
Fan Control Relay 2	Off or On	Off	Off	Engine Data 1
Fuel Trim Cell	0-255 Count	16	Varies	Engine Data 1
Fuel Trim Learn	Disabled/Enabled	Enabled	Enabled	Engine Data 1
High Resolution Signal	Active / Inactive	Active	Active	Engine Data 1
HO2S-11 (B1 S1)	0-1132 mv	10-1000	10-1000	Engine Data 1
HO2S-12 (B1 S2)	0-1132 mv	10-1000	10-1000	Engine Data 1
HO2S-21 (B2 S1)	0-1132 mv	10-1000	10-1000	Engine Data 1
HO2S-22 (B2 S2)	0-1132 mv	10-1000	10-1000	Engine Data 1
IAC Learned	0-255 Count	10-50	---	Engine Data 1
IAC Position	0-255 Count	10-50	---	Engine Data 1
IAT Sensor (°F)	-40 to 304°F	50-194	50-194	Engine Data 1
Ignition 1 Signal	0.0-25.5v	14.1	14.1	Engine Data 1
Injector Pulsewidth B1, B2	0-985 ms	3.7	4.5	Engine Data 1
Knock Retard (°)	0° to 25.5°	0	0	Engine Data 1
Knock Activity	Yes / No	No	No	Engine Data 1
Long Term Fuel Trim B1, B2	-10% to 10%	0%	Varies	Engine Data 1
Loop Status	Closed Loop / Open Loop	Closed Loop	Closed Loop	Engine Data 1
LO Resolution Signal	0-332.7 ms	Varies	Varies	Engine Data 1
MAF Sensor (g/sec)	0-512 g/sec	5-9	Varies	Engine Data 1
MAP Sensor (kPa)	10-110 kPa	30-70	Varies	Engine Data 1
MAP Sensor (V)	0-5v	1-2	Varies	Engine Data 1
MPH or KPH	0-155 mph	0	30	Engine Data 1
OIL Level	Low/Normal	Okay	Okay	Engine Data 1
Pass Key® Fuel	Disabled/Okay	Okay	Okay	Engine Data 1
PCM Reset	No or Yes	No	No	Engine Data 1
P/NP Switch	P-R-0-3-2-1	P-R	3	Engine Data 1
PSP Switch	HI/Normal	Normal	Normal	Engine Data 1
Short Term Fuel Trim B1, B2	-10% to 10%	Varies	Varies	Engine Data 1
Spark Advance (°)	-64° to 64°	20	Varies	Engine Data 1

1996 Fleetwood 5.7L V8 MFI VIN P (A/T) - Engine Data 1 List

Parameter Identification (PID)	PID Value Range	PID Value at Hot Idle	PID Value at 30 mph	Data List Type
Startup ECT Sensor (°F)	-40 to 304°F	Varies	Varies	Engine Data 1
TCC Brake Switch	Apply/Release	Apply: On	Released	Engine Data 1
TCC Duty Cycle	Disabled/Enabled	Enabled	Enabled	Engine Data 1
TCC Enable	Disabled/Enabled	Disabled	Enabled	Engine Data 1
TCC Stator Temperature	HI/Normal	Normal	Normal	Engine Data 1
TP Angle (%)	0-100%	0	Varies	Engine Data 1
TP Sensor (V)	0-5v	0.55-0.90	Varies	Engine Data 1
TR Switch	P/N / DRIVE	PARK	DRIVE	Engine Data 1
Traction Control	Active / Inactive	Inactive	Inactive	Engine Data 1
Vehicle Speed	0-155 mph	0	30	Engine Data 1

1996 Fleetwood 5.7L V8 MFI VIN P (A/T) - Engine Data 2 List

Parameter Identification (PID)	PID Value Range	PID Value at Hot Idle	PID Value at 30 mph	Data List Type
A/C Clutch	Off or On	On: AC on	Off	Engine Data 2
AIR Pump	Disabled/Enabled	Disabled	Disabled	Engine Data 2
ASR/TCS Retard (°)	Yes / No	No	No	Engine Data 2
BARO Sensor (kPa)	10-110 kPa	65-104	65-104	Engine Data 2
BARO Sensor (V)	0.0-5.1v	3.5-4.5	3.5-4.5	Engine Data 2
CKP Engine Speed	0-9999 rpm	±50 Desired	---	Engine Data 2
CKP Low Resolution Angle	-30°to +60°	0° deviation	0° deviation	Engine Data 2
Cold Engine Startup	No or Yes	Yes: If cold	No	Engine Data 2
Desired Idle Speed	0-3187 rpm	PCM control	---	Engine Data 2
DTC Set This Ignition	Yes / No	No	No	Engine Data 2
ECT Sensor (°F)	-40 to 304°F	185-226	185-226	Engine Data 2
EGR Solenoid	0-100%	0	20	Engine Data 2
Engine Run Time	Hr: Min: Sec	00:00:00	00:00:00	Engine Data 2
Engine Speed	0-9999 rpm	±50 Desired	Varies	Engine Data 2
EVAP Purge	0-100%	0	0-50	Engine Data 2
EVAP Vacuum Switch	No/Purge	No Purge	Purge	Engine Data 2
Fan Control Relay 1	Off or On	Off	Off	Engine Data 2
Fan Control Relay 2	Off or On	Off	Off	Engine Data 2
Fuel Trim Cell	Cell #0-17	16	Varies	Engine Data 2
Fuel Trim Learn	Disabled/Enabled	Enabled	Enabled	Engine Data 2
High Resolution Signal	Active / Inactive	Active	Active	Engine Data 2
HO2S-11 (B1 S1)	0-1132 mv	10-1000	10-1000	Engine Data 2
HO2S-12 (B1 S2)	0-1132 mv	10-1000	10-1000	Engine Data 2
HO2S-21 (B2 S1)	0-1132 mv	10-1000	10-1000	Engine Data 2
HO2S-22 (B2 S2)	0-1132 mv	10-1000	10-1000	Engine Data 2
IAC Learned	0-255 Count	10-50	---	Engine Data 2
IAC Position	0-255 Count	10-50	---	Engine Data 2
IAT Sensor (°F)	-40 to 304°F	50-194	50-194	Engine Data 2
Ignition 1 Signal	0.0-25.5v	14.1	14.1	Engine Data 2
Knock Retard	0° to 25.5°	0	0	Engine Data 2
Knock Sensor Activity	Yes / No	No	No	Engine Data 2
Long Term Fuel Trim B1, B2	-10% to 10%	Varies	Varies	Engine Data 2
Loop Status	Closed Loop / Open Loop	Closed Loop	Closed Loop	Engine Data 2
LO Resolution Signal	0-332.7 ms	Varies	Varies	Engine Data 2

1996 Fleetwood 5.7L V8 MFI VIN P (A/T) - Engine Data 2 List

Parameter Identification (PID)	PID Value Range	PID Value at Hot Idle	PID Value at 30 mph	Data List Type
MAF Sensor (g/sec)	0-512 g/sec	5-9	Varies	Engine Data 2
MAP Sensor (kPa)	10-110 kPa	30-70	Varies	Engine Data 2
MAP Sensor (V)	0-5v	1-2	Varies	Engine Data 2
Pass Key® Fuel	Okay/Disabled	Okay	Okay	Engine Data 2
PCM Reset	Yes / No	No	No	Engine Data 2
Short Term Fuel Trim B1, B2	-10% to 10%	Varies	Varies	Engine Data 2
Spark Advance (°)	-64°to 64°	20	Varies	Engine Data 2
Startup ECT Sensor (°F)	-40 to 304°F	Varies	Varies	Engine Data 2
TP Angle (%)	0-100%	0	Varies	Engine Data 2
Traction Control	Active / Inactive	Inactive	Inactive	Engine Data 2
Warmup w/o Emission Fault	0-40	Varies	Varies	Engine Data 2
Warmup w/o Non Emission Fault	0-40	Varies	Varies	Engine Data 2

1996 Fleetwood 5.7L V8 VIN P - Freeze Frame, Failure Records

Parameter Identification (PID)	PID Value Range	PID Value at Hot Idle	PID Value at 30 mph	Data List Type
DTC Number	Number	00000	00000	F/F, F/R
ECT Sensor (°F)	-40 to 304°F	185-226	185-226	F/F, F/R
Engine Load	0-100%	22-26	Varies	F/F, F/R
Engine Speed	0-9999 rpm	±50 Desired	Varies	F/F, F/R
Fail Counter	0-255 Count	Varies	Varies	F/F, F/R
Long Term Fuel Trim B1, B2	-10% to 10%	Varies	Varies	F/F, F/R
MAF Sensor (g/sec)	0-512 g/sec	5-9	Varies	F/F, F/R
Miles Since First Failure	0-65535	Varies	Varies	F/F, F/R
Miles Since Last Failure	0-65655	Varies	Varies	F/F, F/R
Miles Since MIL Request	0-65535	Varies	Varies	F/F, F/R
No Run Counter	0-255 Count	0	0	F/F, F/R
Pass Counter	0-255 Count	Varies	Varies	F/F, F/R
Short Term Fuel Trim B1, B2	-10% to 10%	Varies	Varies	F/F, F/R
TP Angle (%)	0-100%	0%	Varies	F/F, F/R

1996 Fleetwood 5.7L V8 MFI VIN P (A/T) - Fuel Trim Data List

Parameter Identification (PID)	PID Value Range	PID Value at Hot Idle	PID Value at 30 mph	Data List Type
AIR Pump	Disabled/Enabled	Disabled	Disabled	Fuel Trim
ECT Sensor (°F)	-40 to 304°F	185-226	185-226	Fuel Trim
EGR Solenoid	0-100%	0	20	Fuel Trim
Engine Speed	0-9999 rpm	±50 Desired	Varies	Fuel Trim
EVAP Purge	0-100%	0	0-50	Fuel Trim
EVAP Vacuum Switch	No/Purge	No Purge	Purge	Fuel Trim
Fuel Trim Cell	Cell #1-17	16	Varies	Fuel Trim
Fuel Trim Learn	Disabled/Enabled	Enabled	Enabled	Fuel Trim
HO2S-11 (B1 S1)	0-1132 mv	10-1000	10-1000	Fuel Trim
HO2S-12 (B1 S2)	0-1132 mv	10-1000	10-1000	Fuel Trim
HO2S-21 (B2 S1)	0-1132 mv	10-1000	10-1000	Fuel Trim
HO2S-22 (B2 S2)	0-1132 mv	10-1000	10-1000	Fuel Trim
IAC Learned	0-255 Count	10-50	---	Fuel Trim
IAC Position	0-255 Count	10-50	---	Fuel Trim

1996 Fleetwood 5.7L V8 MFI VIN P (A/T) - Fuel Trim Data List

Parameter Identification (PID)	PID Value Range	PID Value at Hot Idle	PID Value at 30 mph	Data List Type
IAT Sensor (°F)	-40 to 304°F	50-194	50-194	Fuel Trim
Ignition 1 Signal	0.0-25.5v	14.1	14.1	Fuel Trim
Injector Pulsewidth B1, B2	0-985 ms	3.7	4.5	Fuel Trim
Long Term Fuel Trim B1	-10% to 10%	0 (± 5%)	0 (± 5%)	Fuel Trim
Long Term Fuel Trim B2	-10% to 10%	0 (± 5%)	0 (± 5%)	Fuel Trim
Loop Status	Closed Loop / Open Loop	Closed Loop	Closed Loop	Fuel Trim
Long FT Average B1	-10% to 10%	0 (± 5%)	0 (± 5%)	Fuel Trim
Long FT Average B2	-10% to 10%	0 (± 5%)	0 (± 5%)	Fuel Trim
MAF Sensor (g/sec)	0-512 g/sec	5-9	Varies	Fuel Trim
MAP Sensor (kPa)	10-110 kPa	30-70	Varies	Fuel Trim
MAP Sensor (V)	0-5v	1-2	Varies	Fuel Trim
Short Term Fuel Trim B1	-10% to 10%	0 (± 5%)	0 (± 5%)	Fuel Trim
Short Term Fuel Trim B2	-10% to 10%	0 (± 5%)	0 (± 5%)	Fuel Trim
SHRTFT Average Bank 1	-10% to 10%	0 (± 5%)	0 (± 5%)	Fuel Trim
SHRTFT Average Bank 2	-10% to 10%	0 (± 5%)	0 (± 5%)	Fuel Trim
Spark Advance (°)	-64° to 64°	20	Varies	Fuel Trim
TP Angle (%)	0-100%	0	Varies	Fuel Trim

1996 Fleetwood 5.7L V8 MFI VIN P (A/T) - Misfire Data

Parameter Identification (PID)	PID Value Range	PID Value at Hot Idle	PID Value at 30 mph	Data List Type
A/C Request	Yes / No	No	No	Misfire
A/C Status	On / Off	Off	Off	Misfire
CKP Engine Speed	0-9999 rpm	±50 Desired	---	Misfire
CKP Low Resolution Angle	-30º to +60º	0º deviation	0º deviation	Misfire
Cycles of Misfire Data	0-100	0	0	Misfire
CYL Mode Misfire Index	0-65535	0	0	Misfire
ECT Sensor (ºF)	-40 to 304ºF	185-226	185-226	Misfire
Engine Load	0-100%	22-26	Varies	Misfire
Engine Speed	0-9999 rpm	±50 Desired	Varies	Misfire
MAF Sensor (g/sec)	0-512 g/sec	5-9	Varies	Misfire
Misfire Current CYL 1-8	0-255 Count	0	0	Misfire
Misfire Failures	0-65535	0	0	Misfire
Misfire History CYL 1-8	0-65535	0	0	Misfire
Misfire Passes	0-65535	0	0	Misfire
Misfire Per Cycle Status	PRI# 1-8	Varies	Varies	Misfire
Misfire Per Cycle Status	SEC# 1-8	Varies	Varies	Misfire
Misfire Revolution Status	Accept/Reject	Accept	Accept	Misfire
Revolutions Within Misfire	0-200	0	0	Misfire
Total Misfire Counts	0-255 Count	0	0	Misfire
TP Angle (%)	0-100%	0	Varies	Misfire

STS PID DATA

2005 STS 3.6L V6 VIN 7

Scan Tool Parameter	Data List	Parameter Range/Units	Typical Data Values
A/C Relay Circuit Status	ODM	OK/Fault/Indeterminate	OK
A/C Relay Command	Eng 1, 2, MF	On/Off	Off
APP Indicated Angle	Eng 1, 2, CMP, EVAP, HO2S, FT, TAC	0-100%	0%
APP Sensor 1	TAC	0.9-4.5 Volts	1 Volt
APP Sensor 2	TAC	0.40-2.25 Volts	0.4 Volt
APP Sensor 1 and 2	TAC	Agree/Disagree	Agree
BARO	Eng 1, 2, CMP, EVAP, HO2S, FT, MF, TAC	65-104 kPa (8-16 psi)	Varies with altitude
BARO	Eng 2, TAC	Volts	Varies
Calculated ECT - Closed Loop Fuel Control	Eng 2	-39°C to +40°C (-38°F to +104°F)	40°C (104°F)
Calculated ECT - Thermostat Diagnosis	Eng 2	-39°C to +140°C (-38°F to +284°F)	89°C (192°F)
Catalytic Converter Protection Active	FT, MF	Yes/No	No
Catalyst Monitor Complete This Ignition	IM	Yes/No	Varies
Catalyst Monitor Enabled this Ignition	IM	Yes/No	Varies
CKP Resync Counter	Eng 2, MF	Counts	0 Counts
Component Monitor Complete This Ignition	IM	Yes/No	Varies
Component Monitor Enable This Ignition	IM	Yes/No	Varies
Current Gear	Eng 1, 2, MF	P/N/Reverse/1st-5th	P/N
Cycles of Misfire Data	MF	0-200 counts	Varies
Cylinder 1-6 Injector Circuit Status	ODM	OK/Fault/Indeterminate	OK
Cylinder 1-6 Knock Retard	Eng 1	Degree	0 Degrees
Decel. Fuel Cutoff	Eng 2, FT, HO2S	Active/Inactive	Inactive
Desired Exh. CMP Bank 1 or 2	CMP	Degree	Varies
Desired HO2S 1	HO2S	Lambda	Varies
Desired Idle Speed	Eng 1, 2, EVAP, CMP, TAC	0-7,000 RPM	650 RPM
Desired Int. CMP Bank 1 or 2	CMP	Degrees	Varies
DTC Set This Ignition	Eng 1, CMP, EVAP, FT, HO2S, TAC	Yes/No	No
EC Ignition Relay Command	Eng 1, TAC	On/Off	On
EC Ignition Relay Feedback Signal	Eng 1, TAC	0-25.5 Volts	12.0-14.5 Volts
ECM Reset	Eng 1, EVAP, FT	Yes/No	No
ECT Sensor	Eng 1, 2, CMP, EVAP, HO2S, FF, FR, FT, MF, TAC	-39°C to +140°C (-38°F to +284°F)	Varies
Engine Load	Eng 1, 2, CMP, EVAP, HO2S, FF, FR, FT, MF, TAC	0-99%	0-5% - Idle 3-7% - 2500 RPM
Engine OFF EVAP Test Conditions Met	EVAP	Yes/No	Varies
Engine Off Time	EVAP	0:00:00 Seconds	Varies
Engine Oil Level Switch	CMP	OK/Low	OK
Engine Oil Life Remaining	CMP	%	Varies
Engine Oil Pressure Sensor	MF, CMP	kPa/psi	Varies
Engine Run Time	Eng 1, 2, EVAP, FT, HO2S, MF, TAC, CMP, ODM	0:00:00	Varies
Engine Speed	ALL	RPM	575-700 RPM
EVAP Monitor Complete This Ignition	IM	Yes/No	Varies
EVAP Monitor Enabled This Ignition	IM	Yes/No	Varies
EVAP Purge Solenoid Circuit Status	ODM	OK/Fault/Indeterminate	OK

2005 STS 3.6L V6 VIN 7 (CONT)

Scan Tool Parameter	Data List	Parameter Range/Units	Typical Data Values
EVAP Purge Solenoid Command	Eng 1, EVAP, FT	0-100%	Varies
EVAP Vent Solenoid Circuit Status	ODM	OK/Fault/Indeterminate	OK
EVAP Vent Solenoid Command	Eng 1, EVAP, FT	Venting/Not Venting	Venting
Exh. CMP Angle Bank 1 or 2	CMP	Degrees	Varies
Exh. CMP Bank 1 or 2 Command	CMP	0-99%	Varies
Exh. CMP Solenoid Bn. 1 or 2 Circuit Status	ODM	OK/Fault/Indeterminate	OK
FC Relay 1 Circuit Status	ODM	OK/Fault/Indeterminate	OK
FC Relay 1 Command	Eng 2	ON/OFF	Varies
FC Relay 2 and 3 Command	Eng 2	ON/OFF	Varies
FC Relay 2 and 3 Circuit Status	ODM	OK/Fault/Indeterminate	OK
Fuel Level Sensor	EVAP	0-5 Volts	Varies
Fuel Pump Relay Circuit Status	ODM	OK/Fault/Indeterminate	OK
Fuel Pump Relay Command	Eng 1, 2	On/Off	On
Fuel Tank Level Remaining	Eng 1, EVAP, MF	Liter/Gallon	Varies
Fuel Tank Pressure Sensor	EVAP	in H2O/mm Hg	Varies
Fuel Tank Pressure Sensor	EVAP	0-5 Volts	Varies
Fuel Tank Rated Capacity	EVAP	Liter/Gallon	17.8 Gallons (67.3 Liters)
Fuel Trim Learn	Eng 1, EVAP, FT, HO2S	Enabled/Disabled	Enabled
HO2S 1	Eng 1, EVAP, FT, HO2S	Lambda	1.0 Lambda
HO2S 2	Eng 1, FT, HO2S	0-1,006 mV	Varies 400-700 mV
HO2S Heater Monitor Complete This Ignition	IM	Yes/No	Varies
HO2S Heater Monitor Enabled This Ignition	IM	Yes/No	Varies
HO2S 1 or 2 Heater Circuit Status	ODM	OK/Fault/Indeterminate	OK
HO2S 1 or 2 Heater Command	HO2S	0-99%	Varies
HO2S/O2S Monitor Complete This Ignition	IM	Yes/No	Varies
HO2S/O2S Monitor Enabled This Ignition	IM	Yes/No	Varies
HO2S 1 or 2 Sensing Element	HO2S	Ohms	70-90 Ohms Depends on Temperature
IAT Sensor	Eng 1, 2, EVAP, FT, FF, FR, HO2S, TAC, CMP	-39° to +140°C (-38° to +284°F)	Varies
Ignition 0 Signal	Eng 1, 2, EVAP, TAC	On/Off	On
Ignition 1 Signal	Eng 1, 2, EVAP, FT, MF, TAC, CMP	On/Off	On
IMRC Solenoid Circuit Status	ODM	OK/Fault/Indeterminate	OK
IMRC Solenoid Command	Eng 1	On/Off	Off
Initial Brake Apply Signal	TAC	Applied/Released	Released
Injector 1-6 Command	Eng 2, FT, MF	ms	Varies
Injector PWM Bank 1 or 2 Average	FT, MF	0-9 ms	4.6-5.4 ms
Int. CMP Angle Bank 1 or 2	CMP	Degrees	Varies
Int. CMP Bank 1 or 2 Command	CMP	0-99%	Varies
Int. CMP Solenoid Bn. 1 or 2 Circuit Status	ODM	OK/Fault/Indeterminate	OK
Knock Retard	Eng 1	Degrees	Varies
KS 1 or 2 Signal	Eng 1	Volts	Varies
Loop Status Sensor 1	Eng 1, 2, EVAP, FT, HO2S, FF, FR	Open Loop/Closed Loop	Closed Loop
Loop Status Sensor 2	Eng 1, 2, EVAP, FT, HO2S, FF, FR	Open Loop/Closed Loop	Open Loop
LT FT Cruise/Accel.	Eng 1, 2, EVAP, FT, HO2S	-100% to +100%	0%
LT FT Idle/Decel.	Eng 1, 2, EVAP, FT, HO2S	-100% to +100%	0%

2005 STS 3.6L V6 VIN 7 (CONT)

Scan Tool Parameter	Data List	Parameter Range/Units	Typical Data Values
MAF Sensor	Eng 1, 2, FT, HO2S, MF, TAC, CMP, FF, FR	0-655 g/s	3.4-6.3 g/s
MAF Sensor	Eng 2, EVAP, FT, HO2S, MF, TAC, CMP	0-5 Volts	0.9-1.7 Volts
MIL Circuit Status	ODM	OK/Fault/Indeterminate	OK
MIL Command	Eng 1	On/Off	Off
Mileage Since DTC Cleared	Eng 2, FF, FR	km - miles	Varies
Mileage Since First Failure	FF, FR	km - miles	Varies
Mileage Since Last Failure	FF, FR	km - miles	Varies
Mileage Since MIL Requested	FF, FR	km - miles	Varies
Misfire Current Cyl. 1-6	MF	0-255 Counts	0
Misfire History Cyl. 1-6	MF	0-65,535 Counts	0
Misfire Monitor Complete This Ignition	IM	Yes/No	Varies
PNP Switch	Eng 2	Park/Neutral/In Gear	P/N
Power Enrichment	Eng 2, FT, HO2S, MF	Active/Inactive	Inactive
Reduced Engine Power	Eng 1, 2, TAC	Active/Inactive	Inactive
Short Term FT	Eng 1, 2, EVAP, FT, HO2S, FF, FR	-25% to +25%	-3% to +3%
Spark	Eng 1, 2, FT, HO2S, MF, FF, FR	-20 to +40 Degrees	2-10 Degrees
Starter Relay Circuit Status	ODM	OK/Fault/Indeterminate	OK
Start Up ECT	Eng 2, EVAP, FT, HO2S	-39° to +140°C (-38° to +284°F)	Varies
Start Up IAT	Eng 2, EVAP, FT, HO2S	-39° to +140°C (-38° to +284°F)	Varies
Stoplamp Pedal Switch	Eng 1, TAC	Applied/Released	Released
System Power Mode	Eng 2, EVAP, TAC	Off, Accessory, RAP, Run, or Crank Request	Run
TAC System Learned Counts	TAC	0-11 Counts	0
Torque Delivered Signal	FT, TAC	0-100%	Varies
Torque Request Signal	FT, TAC	0-100%	90-98%
Total Fuel Trim Average	EVAP, FT, HO2S	-100 to +100%	Varies
Total Misfire Count	MF	Counts	0
TP Desired Angle	Eng 1, 2, EVAP, TAC	0-100%	Varies
TP Indicated Angle	Eng 1, 2, EVAP, FT, HO2S, MF, TAC, CMP, FF, FR	0-100%	Varies
TP Sensor 1	TAC	0-5 Volts	0.5-0.8 Volts
TP Sensor 1 Learned Minimum	TAC	0-5 Volts	0-0.6 Volts
TP Sensor 2	TAC	0-5 Volts	4.7-4.1 Volts
TP Sensor 2 Learned Minimum	TAC	0-5 Volts	5-3.7 Volts
TP Sensors 1 and 2	TAC	Agree/Disagree	Agree
Traction Control Status	TAC	Active/Inactive	Inactive
Transmission Gear Selector Signal	Eng 2	Valid/Invalid	Valid
TWC Temperature Calculated	FT, MF	Temperature	Varies
Vehicle Security Status	Eng 2	Password Learning, VTD Fail-Enable, Auto Learn Timer, Fuel Disable Until Ign Off, Fuel Disable Timeout, and Fuel Continue	Fuel Continued
Vehicle Speed Sensor	Eng 1, 2, EVAP, FT, HO2S, MF, TAC, CMP, FF, FR	Km/h mph	Varies
Volumetric Efficiency	Eng 1, 2, EVAP, FT, HO2S, MF, TAC, CMP	0-100%	5-25% - Idle and at 2,500 RPM

2005 STS 4.6L V6 VIN A

Scan Tool Parameter	Data List	Parameter Range/Units	Typical Data Values
A/C Relay Circuit Status	ODM	OK/Fault	OK
A/C Relay Command	Eng 1, MF	On/Off	Off
APP Indicated Angle	CMP, EVAP, HO2S, Ign, TAC, FT	0-99%	0 percent at idle
APP Sensor 1	TAC	0.9-4.5 Volts	1 Volt
APP Sensor 1	TAC	%	2%
APP Sensor 2	TAC	0.45-2.25 Volts	0.5 Volt
APP Sensor 2	TAC	%	2%
APP Sensor 1 and 2	TAC	Agree/Disagree	Agree
Baro	Eng 1, EVAP, FT, HO2S, Ign	65-104 kPa -16 psi	Varies with altitude
Catalytic Converter Protection Active	FT, HO2S	Yes/No	No
Commanded Gear	Eng 1, FT, MF	P/N/R/1-5	P/N
Cruise Control Active	Eng 1, TAC	Active/Inactive	Inactive
Current Gear Signal	Eng 1, FT	Valid/Invalid	Valid
Cycles of Misfire Data	MF	0-9999	0
Cylinder 1-8 IC Circuit Status	ODM	Fault/OK	OK
Cylinder 1-8 Injector Circuit Status	ODM	Fault/OK	OK
Cylinder 1-8 Knock Retard	Ign	Degrees	0
Decel. Fuel Cutoff	Eng 1, FT, HO2S	Active/Inactive	Inactive
Desired Exhaust CMP Bn 1 or Bn 2	CMP	Degrees	0
Desired HO2S Bank 1 or Bank 2 Sensor 1	FT, HO2S	Lambda	Varies
Desired Idle Speed	CMP, Eng 1, EVAP, TAC	0-7,000 RPM	650 RPM
Desired Intake CMP Bn 1 or Bn 2	CMP	Degrees	0
Desired Ignition Spark	Ign	0-100%	68%
Desired Torque	Ign, FT	N m	45-55
EC Ignition Relay Command	Eng 1, TAC	On/Off	On
EC Ignition Relay Feedback Signal	Eng 1, TAC	volts	13.5 V
ECM Reset	Eng 1, EVAP, FT, Ign, ODM,	Yes/NO	No
ECT Sensor	CMP, Eng 1, EVAP, FT, Ign 1, HO2S, MF, TAC	-39 to +140°C/-38 to +284°F	85-105°C/185-220°F
Engine Load	CMP, Eng 1, EVAP, FT, HO2S, Ign, MF, TAC	0-99%	2%
Engine Off EVAP Test Conditions Met	EVAP	Yes/No	No
Engine Off EVAP Test Ran Last Ignition Cycle	EVAP	Yes/No	No
Engine Off Time	EVAP	Seconds	0
Engine Oil Level Switch	Eng 1, CMP	OK/Low	OK
Engine Oil Life Remaining	CMP	%	Varies
Engine Oil Pressure Sensor	CMP, Eng1, MF	kPa	180 kPa
Engine Oil Temperature	CMP	°C/°F	95°C
Engine Run Time	CMP, Eng 1, EVAP, FT, HO2S, Ign, MF, TAC	0:00:00	Varies
Engine Speed	CMP, Eng 1, EVAP, FT, HO2S, Ign, MF, TAC	RPM	650 RPM
Engine Torque	Ign, FT	N m	45-55
EVAP Fault History	EVAP	Fault/No Fault	No Fault
EVAP Purge Solenoid Command	Eng 1, EVAP, FT, HO2S	0-100%	Varies
EVAP Purge Solenoid Circuit Status	ODM	Fault/OK	OK
EVAP Vent Solenoid Command	Eng 1, EVAP, FT	On/Off	Off
EVAP Vent Solenoid Circuit Status	ODM	Fault/OK	OK
EVAP Test Abort Reason	EVAP	Aborted/Not Aborted	Not Aborted
EVAP Test Result	EVAP	No Result	No Result
EVAP Test State	EVAP	Purge/wait for purge	Wait for purge
Exh. CMP Angle Bank 1 or Bank 2	CMP, Eng 1, Ign	Degrees	0
FC Relay 1 Circuit Status	ODM	Fault/OK	OK

2005 STS 4.6L V6 VIN A (CONT)

Scan Tool Parameter	Data List	Parameter Range/Units	Typical Data Values
FC Relay 2 Circuit Status	ODM	Fault/OK	OK
FC Relay 1 Command	Eng 1	ON/OFF	ON
FC Relay 2 Command	Eng 1	ON/OFF	ON
Fuel Level Sensor Left Tank	EVAP	0-5 Volts	Varies
Fuel Level Sensor Right Tank	EVAP	0-5 Volts	Varies
Fuel Pump Command	Eng 1	On/Off	On
Fuel Pump Relay Circuit Status	ODM	Fault/OK	OK
Fuel Tank Level Remaining	Eng 1, EVAP , MF	Gal/L	Varies
Fuel Tank Pressure	EVAP	Volts	0-5 Volts
Fuel Tank Pressure Sensor	Eng 1, EVAP	in H2O/mm Hg	Varies
Fuel Tank Rated Capacity	EVAP	Liters/Gal	Varies
Fuel Trim Cell	Eng 1, EVAP, FT	Varies	2
Fuel Trim Learn	Eng 1, EVAP, FT	Enabled/Disabled	Enabled
Fuel Volatility	EVAP, FT	Low/High	Low
Gen L Terminal Signal Command	Eng 1	On/Off	On
HO2S Bank 1 or Bank 2 Sensor 1	Eng 1, EVAP, FT, HO2S	Lambda	Varies
HO2S Bank 1 or Bank 2 Sensor 2	Eng 1, FT, HO2S	0-4,998 mV	Varies 400-800 mV
HO2S Heater Bn. 1 or Bn. 2 Sen. 1	FT, HO2S	0-99%	Varies
HO2S Heater Bn. 1 or Bn. 2 Sen. 2	FT, HO2S	0-99%	Varies
Hot Open Loop	HO2S, FT	Active/Inactive	Inactive
IAT Sensor	CMP, Eng 1, EVAP, FT, HO2S, Ign, TAC	-39 to +140°C/-38 to +284°F	Varies
Ignition 0 Signal	Eng 1, TAC	ON/OFF	ON
Ignition 1 Signal	CMP, Eng 1, EVAP, FT, HO2S, Ign, MF, TAC	0-25.5 Volts	12.8-14.2 Volts
Ignition Spark Efficiency	Ign	0-100%	Varies
Initial Brake Apply Signal	Eng 1, TAC	Applied/Released	Released
Injector Bn. 1 or Bn. 2 PWM Average	Eng 1, FT, HO2S, Ign, MF	0-9 ms	3.0-5.4 ms
Injector 1-8 Command	FT, MF	mS	Varies
Instant FT Bank 1 or Bank 2	FT	%	Varies
Int CMP Angle Bank 1 or Bank 2 Signal	Eng 1, Ign	Degrees	0
Int CMP angle Bank 1 or Bank 2 Command	CMP	0-100%	7%
Integral FT Bank 1 or Bank 2	FT	%	Varies
Knock Retard	Eng1, Ign	0-191 Degrees	0
Long Term FT Bank 1 or Bank. 2	Eng 1, EVAP, FT, HO2S	-100 to +100%	9%
LT FT High RPM/Load Bn. 1 or Bn. 2	FT	-100 to +100%	-50%
LT FT Low RPM/Load Bn. 1 or Bn. 2	FT	mg	255
Loop Status	Eng 1, EVAP, FT, HO2S	Open/Closed	Closed
MAF Sensor	CMP, Eng 1, EVAP, HO2S, Ign, MF, TAC, FT	0-655 g/s	3-7 g/s
MAF Sensor	CMP, Eng 1, EVAP, HO2S, MF, TAC	0-9999 Hz	28 Hz
MAP Sensor	CMP, Eng 1, EVAP, FT, HO2S, Ign, MF, TAC, FT	kPa/psi	33
MIL Circuit Status	ODM	Fault/OK	OK
MIL Command	Eng 1, Ign	On/Off	Off
MIL Requested by DTC	Eng 1, Ign	Yes/No	No
Mileage Since DTC Cleared	Eng 1	km/mi	Varies
Misfire Detects Rough Road	MF	Yes/No	No
Misfire History Cyl. 1-8	MF	0-255 Counts	0
Moderate Brake Apply Signal	Eng 1, TAC,	Applied/Released	Released
PNP Switch	Eng 1	In Gear/Park/Neutral	Park/Neutral
Power Enrichment	Eng 1, FT, HO2S, MF	Active/Inactive	Inactive
Reduced Engine Power	Eng 1, TAC	Active/Inactive	Inactive

2005 STS 4.6L V6 VIN A (CONT)

Scan Tool Parameter	Data List	Parameter Range/Units	Typical Data Values
Reference Voltage 1 Signal	Eng 1, Ign, TAC	0-5 Volts	5 V
Reference Voltage 2 Signal	Eng 1, Ign, TAC	0-5 Volts	5 V
Short Term FT Bank 1 or Bank 2	Eng 1, EVAP, FT, HO2S	-100 to +100%	-3% to +3%
Spark	Eng 1, FT, HO2S, Ign, MF	-133 to +78 Degrees	2-5 Degrees
Start Up ECT	EVAP, FT, HO2S	-39 to +140°C/-38 to +284°F	Varies
Start Up IAT	EVAP, FT, HO2S	-39 to +140°C/-38 to +284°F	Varies
Starter Relay Circuit Status	ODM	OK/Fault	OK
Starter Relay Command	Eng 1	On/Off	Off
Stoplamp Pedal Switch	Eng 1, TAC	Applied/Released	Released
TAC Motor Command	TAC	%	12-26%
TAC Direction	TAC	Closed/Open	Closed
TCS Control Signal	TAC	None/Reduced	None
Torque Request Signal	TAC	Valid/Invalid	Valid
Total Misfire Count	MF	0-9999	0
TP Sensor 1 Learned Min	TAC	0-5 Volts	0-0.5 Volts
TP Sensor 2 Learned Min	TAC	5 0 Volts	5 3.7 Volts
TP Desired Angle	Eng 1, EVAP, TAC	0-99%	Varies
TP Indicated Angle	CMP, Eng 1, EVAP, FT, HO2S, Ign, MF, TAC	0-99%	2 percent
TP Learned	TAC	Yes/No	No
TP Learn Enables	TAC	Yes/No	No
TP Learn Required	TAC	Yes/No	No
TP Sensor 1	TAC	0-5 Volts	0.5-0.8 Volts
TP Sensor 1	TAC	%	1
TP Sensors 1 and 2	TAC	Agree/Disagree	Agree
TP Sensor 2	TAC	0-5 Volts	4.1-4.5 Volts
TP Sensor 2	TAC	%	1
Transmission Gear Selector Signal	Eng 1	Valid/Invalid	Valid
TWC Temperature Calculated	FT, MF	C/F	Varies
Vehicle Security Status	Eng 1	Fuel Continue/Fuel Disable	Fuel Continue
Vehicle Speed Sensor	CMP, Eng 1, EVAP, FT, HO2S, Ign, MF, TAC	km/mi	0
VTD Password Learn Enabled	Eng 1	Yes/No	No

BERETTA & CORSICA PID DATA - OBD II

1996 Beretta, Corsica 2.2L I4 MFI VIN 4 (A/T) - PID Data List

Parameter Identification (PID)	PID Value Range	PID Value at Hot Idle	PID Value at 30 mph	Data List Type
BARO Sensor (V)	0-5.0v	3.5-4.5	3.5-4.5	Engine Data 1
Calculated Airflow	0-512 g/sec	3-7	Varies	Engine Data 1
CKP Active Counter	0-255 counts	Counts up	Counts up	Engine Data 1
CMP Active Counter	0-255 counts	Counts up	Counts up	Engine Data 1
CMP Resync Counter	0-255 counts	0	0	Engine Data 1
Desired Idle Speed	0-3187	PCM control	---	Engine Data 1
ECT Sensor (°F)	-40-419°F	185-239	185-239	Engine Data 1
Engine Load	0-100%	22-26	Varies	Engine Data 1
Engine Run Time	Hr: Min: Sec	00:00:00	00:00:00	Engine Data 1
Engine Speed	0-9999 rpm	Varies	---	Engine Data 1
EVAP Purge Solenoid	0-100%	0%	0-100%	Engine Data 1
EVAP Vacuum Switch	On / Off	On	On or Off	Engine Data 1
Fan Control Relay	On / Off	Off	Off	Engine Data 1
Fuel Trim Cell	Cell # 1-21	18-21	18-21	Engine Data 1
Fuel Trim Index	0-255 counts	58-198	58-198	Engine Data 1
HO2S-12 (B1 S2)	0-1132 mv	10-1000	10-1000	Engine Data 1
IAC Position	0-255 counts	5-60	5-60	Engine Data 1
IAT Sensor (°F)	-40 to 304°F	50-176	50-176	Engine Data 1
Ignition 1 Signal	0.0-25.5v	14.2	14.2	Engine Data 1
Knock Retard (°)	0-128°	0	0	Engine Data 1
KS Noise Channel	Yes / No	No	No	Engine Data 1
KS Active Counter	0-255 counts	35-50	35-50	Engine Data 1
Loop Status	Closed Loop / Open Loop	Closed Loop	Closed Loop	Engine Data 1
Long Term Fuel Trim	-10% to 10%	Varies	Varies	Engine Data 1
Low Octane Modifier	0-90	Varies	Varies	Engine Data 1
MAP Sensor (V)	0-5.0v	1.5	1-3	Engine Data 1
Medium Resolution Resync Counter	0-255 counts	0	0	Engine Data 1
Medium Resolution Engine Sync	Yes/No	Yes	Yes	Engine Data 1
Number of Current DTC(s)	0-255 counts	0	0	Engine Data 1
O2S-11 (B1 S1)	0-1132 mv	10-1000	10-1000	Engine Data 1
Power Enrichment	Active/Inactive	Inactive	Inactive	Engine Data 1
Purge Learn Memory	0-100%	0	0-50	Engine Data 1
Short Term Fuel Trim	-10% to 10%	0 (± 5%)	0 (± 5%)	Engine Data 1
Short Term Fuel Trim Average	-10% to 10%	0 (± 5%)	0 (± 5%)	Engine Data 1
Spark Advance (°)	-64° to 64°	Varies	Varies	Engine Data 1
TCC Brake Switch	On / Off	On: If on	Off	Engine Data 1
TP Angle (%)	0-100%	0	Varies	Engine Data 1
TP Sensor (V)	0-5.0v	0.20-0.90	Varies	Engine Data 1

1996 Beretta, Corsica 2.2L I4 MFI VIN 4 (A/T) - PID Data List

Parameter Identification (PID)	PID Value Range	PID Value at Hot Idle	PID Value at 30 mph	Data List Type
A/C High Side PSI	0-459 psi	139-399	139-399	Engine Data 2
A/C Relay	On / Off	Off	Off	Engine Data 2
A/C Request	Yes / No	No	Off	Engine Data 2
Air Fuel Ratio	0-25.5:1	14.6:1	14.6:1	Engine Data 2
BARO Sensor (V)	0-5.0v	3.5-4.5	3.5-4.5	Engine Data 2
Cruise Engaged	Yes / No	No	No	Engine Data 2
Desired Idle Speed	0-3187	PCM control	---	Engine Data 2
ECT Sensor (°F)	-40-419°F	185-239	185-239	Engine Data 2
Engine Speed	0-9999 rpm	Varies	---	Engine Data 2
Fan Control Relay	On / Off	Off	Off	Engine Data 2
IAT Sensor (°F)	-40 to 304°F	50-176	50-176	Engine Data 2
INJ Pulsewidth	0-985 ms	1-4	Varies	Engine Data 2
MAP Sensor (V)	0-5.0v	1.5	1-3	Engine Data 2
Output Driver 1, 2 Open	00000000	00000000	00000000	Engine Data 2
Output Driver 1, 2 Short	00000000	00000000	00000000	Engine Data 2
Output Speed Sensor	0-10,000	Varies	Varies	Engine Data 2
PNP Switch	PN / R-DL	PN (1)	R-DL (0)	Engine Data 2
Stepper Cruise	Disable/Enable	Disabled	Disabled	Engine Data 2
TCC Enable	Active/Inactive	Inactive	Inactive	Engine Data 2
TP Angle (%)	0-100%	0%	0-100%	Engine Data 2
TP Sensor (V)	0-5.0v	0.20-0.90	---	Engine Data 2
Total Misfire Counts	0-255 counts	0	0	Engine Data 2
Vehicle Speed	0-155 mph	0	30	Engine Data 2

1996 Beretta, Corsica 2.2L I4 MFI VIN 4 (A/T) - PID Data List

Parameter Identification (PID)	PID Value Range	PID Value at Hot Idle	PID Value at 30 mph	Data List Type
Actual EGR Position	0-100%	0	0-100	EGR, EVAP
BARO Sensor (V)	0-5.0v	3.5-4.5	3.5-4.5	EGR, EVAP
Decel EWMA	-10 - 10 kPa	-3 to - 4	Varies	EGR, EVAP
Delta MAP Change	-20 - 20 kPa	0 / - number	0	EGR, EVAP
Desired EGR Position	0-100%	0%	0-100%	EGR, EVAP
Desired Idle Speed	0-3187	PCM control	---	EGR, EVAP
ECT Sensor (°F)	-40-419°F	185-239	185-239	EGR, EVAP
Engine Speed	0-9999 rpm	Varies	---	EGR, EVAP
Fuel Trim Cell	Cell #1-21	18-21	18-21	EGR, EVAP
HO2S-12 (B1 S2)	0-1132 mv	10-1000	10-1000	EGR, EVAP
IAC Position	0-255 counts	5-60	---	EGR, EVAP
IAT Sensor (°F)	-40 to 304°F	50-176	50-176	EGR, EVAP
Long Term Fuel Trim	-10% to 10%	0 (± 5%)	0 (± 5%)	EGR, EVAP
MAP Sensor (V)	0-5.0v	1.5	1-3	EGR, EVAP
O2S-11 (B1 S1)	0-1132 mv	10-1000	10-1000	EGR, EVAP
Short Term Fuel Trim	0-255 counts	58-198	58-198	EGR, EVAP
Short Term Fuel Trim	-10% to 10%	0 (± 5%)	0 (± 5%)	EGR, EVAP
TP Angle (%)	0-100%	0%	0-100%	EGR, EVAP
TP Sensor (V)	0-5.0v	0.20-0.90	---	EGR, EVAP

1996 Beretta, Corsica 2.2L I4 MFI VIN 4 (A/T) - PID Data List

Parameter Identification (PID)	PID Value Range	PID Value at Hot Idle	PID Value at 30 mph	Data List Type
BARO Sensor (V)	0-5.0v	3.5-4.5	3.5-4.5	Fuel Trim
Desired Idle Speed	0-3187	PCM control	Varies	Fuel Trim
ECT Sensor (°F)	-40-419°F	185-239	185-239	Fuel Trim
Engine Run Time	Hr: Min: Sec	00:00:00	00:00:00	Fuel Trim
Engine Speed	0-9999 rpm	Varies	Varies	Fuel Trim
Fuel Trim Index	0-255 counts	58-198	Varies	Fuel Trim
IAT Sensor (°F)	-40 to 304°F	50-176	50-176	Fuel Trim
HO2S L/R to R/L ratio	0:1-15.93:1	Varies	Varies	Fuel Trim
Long Term Fuel Trim	-10% to 10%	Varies	Varies	Fuel Trim
Long Term F/T Average.	0-255 counts	58-198	Varies	Fuel Trim
Loop Status	Closed Loop / Open Loop	Closed Loop	Closed Loop	Fuel Trim
MAP Sensor (V)	0-5.0v	1.5	1-3	Fuel Trim
O2S-11 (B1 S1)	0-1132 mv	10-1000	10-1000	Fuel Trim
Rich / Lean Average	0-249 ms	0	0	Fuel Trim
Short Term Fuel Trim	-10% to 10%	Varies	Varies	Fuel Trim
Short Term F/T Average.	0-255 counts	58-198	Varies	Fuel Trim
TP Angle (%)	0-100%	0	0-100	Fuel Trim

1996 Beretta & Corsica 2.2L I4 MFI (A/T) - PID Data List

Parameter Identification (PID)	PID Value Range	PID Value at Hot Idle	PID Value at 30 mph	Data List Type
Air Fuel Ratio	0-25.5:1	14.6:1	14.6:1	F/F, F/R
BARO Sensor (V)	0-5.0v	3.5-4.5	3.5-4.5	F/F, F/R
DTC Freeze Frame	DTC #	DTC #	DTC #	F/F, F/R
Calculated Airflow	0-512 g/sec	Varies	Varies	F/F, F/R
ECT Sensor (°F)	-40-419°F	185-239	185-239	F/F, F/R
Engine Speed	0-9999 rpm	Varies	Varies	F/F, F/R
INJ Pulsewidth	0-985 ms	1-4	Varies	F/F, F/R
Engine Load	0-100%	22-26	Varies	F/F, F/R
Long Term Fuel Trim	-10% to 10%	Varies	Varies	F/F, F/R
MAP Sensor (V)	0-5.0v	1.5	1-3	F/F, F/R
Short Term Fuel Trim	-10% to 10%	Varies	Varies	F/F, F/R
TP Angle (%)	0-100%	0%	0-100%	F/F, F/R

1996 Beretta & Corsica 2.2L I4 MFI VIN 4 (A/T) - PID Data List

Parameter Identification (PID)	PID Value Range	PID Value at Hot Idle	PID Value at 30 mph	Data List Type
BARO Sensor (V)	0-5.0v	3.5-4.5	3.5-4.5	Misfire
Desired Idle Speed	0-3187	PCM control	Varies	Misfire
ECT Sensor (°F)	-40-419°F	185-239	185-239	Misfire
Engine Speed	0-9999 rpm	Varies	Varies	Misfire
IAT Sensor (°F)	-40 to 304°F	50-176	50-176	Misfire
INJ Pulsewidth	0-985 ms	1-4	Varies	Misfire
MAP Sensor (V)	0-5.0v	1.5	1-3	Misfire
Misfire current Cyl 1-4	0-255 counts	0	0	Misfire
Misfire History Cyl 1-4	0-255 counts	0	0	Misfire
Total Misfire	0-255 counts	0	0	Misfire
TP Angle (%)	0-100%	0	Varies	Misfire

1996 Beretta, Corsica 3.1L V6 MFI VIN M (A/T) - PID Data List

Parameter Identification (PID)	PID Value Range	PID Value at Hot Idle	PID Value at 30 mph	Data List Type
Abuse Management	Active/Inactive	Inactive	Inactive	Engine Data 1
A/C Clutch	Yes / No	No	No	Engine Data 1
A/C High Side PSI	0-459 psi	139-399	139-399	Engine Data 1
A/C Off for WOT	On / Off	Off	Off	Engine Data 1
A/C Request	Yes / No	No	No	Engine Data 1
Actual EGR Position	0-100%	0	0-100	Engine Data 1
Air Fuel Ratio	0.0-25.5:1	14.6:1	14.6:1	Engine Data 1
AIR Active Test	Active/Inactive	Inactive	Inactive	Engine Data 1
AIR System Pump	Off / On	On: Cold	Off	Engine Data 1
AIR Pump Relay	Off / On	On: Cold	Off	Engine Data 1
BARO Sensor (kPa)	10-110 kPa	65-110	65-110	Engine Data 1
CKP 3X Signal	1280-9999	Above 1280	Varies	Engine Data 1
CKP 24X Signal	0-1280	Below 1280	Varies	Engine Data 1
Cam Signal Present	Yes / No	Yes	Yes	Engine Data 1
CMP 4ince Last 3X	Yes / No	Yes	Yes	Engine Data 1
Commanded Fan 1	On / Off	Off	Off	Engine Data 1
Commanded Fan 2	On / Off	Off	Off	Engine Data 1
Cruise Engaged Mode	On / Off	Off	On	Engine Data 1
Cruise Inhibited	Yes / No	Yes	Yes	Engine Data 1
Current Gear	1-2-3-4	1	3	Engine Data 1
Decel Fuel Mode	Active/Inactive	Inactive	Inactive	Engine Data 1
Desired EGR Position	0-100%	0	0-100	Engine Data 1
Desired Idle Speed	0-3187	PCM control	Varies	Engine Data 1
ECT Sensor (°F)	-40 to 304°F	185-220	185-220	Engine Data 1
EGR Closed Valve Position	0-5.0v	0.14-1.00	Varies	Engine Data 1
EGR Duty Cycle	0-100%	0	0-100	Engine Data 1
EGR Feedback	0-5.0v	0.14-1.00	Varies	Engine Data 1
EGR Position Error	0-100%	0-9%	Varies	Engine Data 1
EGR Flow Test Count	0-255 counts	0-10	0-10	Engine Data 1
Engine Load	0-100%	2-5	Varies	Engine Data 1
Engine Oil Level	Not Okay/Okay	Okay	Okay	Engine Data 1
Engine Run Time	Hr: Min: Sec	00:00:00	00:00:00	Engine Data 1
Engine Speed	0-9999 rpm	Varies	Varies	Engine Data 1
EVAP Purge PWM	0-100%	0-25	0-75	Engine Data 1
EVAP Vacuum Switch	Open/Close	Open	Closed	Engine Data 1
Fuel Pump	On / Off	On	On	Engine Data 1
Fuel Trim Cell	Cell #	10	10	Engine Data 1
Fuel Trim Learn	Disable/Enable	Enabled	Enabled	Engine Data 1
Generator Command	No / Yes	No	No	Engine Data 1
Generator L-Terminal	Fault/No	No Fault	No Fault	Engine Data 1
Generator Lamp	On / Off	Off	Off	Engine Data 1
HO2S-11 (B1 S1)	0-1132 mv	10-1000	10-1000	Engine Data 1
HO2S-12 (B1 S2)	0-1132 mv	10-1000	10-1000	Engine Data 1
HO2S-11 Status	Not / Ready	Ready	Ready	Engine Data 1
HO2S-11 Warmup	Min: Sec	00:00	00:00	Engine Data 1
HO2S-12 Warmup	Min: Sec	00:00	00:00	Engine Data 1
HO2S-11 XCounts	0-255 counts	Varies	Varies	Engine Data 1

1996 Beretta, Corsica 3.1L V6 MFI VIN M (A/T) - PID Data List

Parameter Identification (PID)	PID Value Range	PID Value at Hot Idle	PID Value at 30 mph	Data List Type
IAC Position	0-255 counts	5-50	---	Engine Data 1
IAT Sensor (°F)	-40 to 304°F	50-194	50-194	Engine Data 1
Ignition 1 Signal	0.0-25.5v	14.2	14.2	Engine Data 1
Ignition Bypass	Active/Inactive	Inactive	Inactive	Engine Data 1
IC Mode	Active/Inactive	Active	Active	Engine Data 1
Injector Fault	No / Yes	No	No	Engine Data 1
Injector Pulsewidth	0-985 ms	1.5-3.5	Varies	Engine Data 1
Knock Retard (°)	0-22°	0	0	Engine Data 1
KS Activity	Yes / No	No	No	Engine Data 1
KS Noise Channel	0-5.0v	1-2	Varies	Engine Data 1
Long Term Fuel Trim	-23 to 16%	Varies	Varies	Engine Data 1
Long Term F/T Average.	-23 to 16%	Varies	Varies	Engine Data 1
Loop Status	Closed Loop / Open Loop	Closed Loop	Closed Loop	Engine Data 1
Low Oil Lamp	On / Off	Off	Off	Engine Data 1
MAF Sensor (g/sec)	0-512 g/sec	3-5	Varies	Engine Data 1
MAF Sensor (Hz)	0-31,999 Hz	2000-2500	Varies	Engine Data 1
MAP Sensor (kPa)	10-110 kPa	29-48	Varies	Engine Data 1
MIL Status	On / Off	Off	Off	Engine Data 1
Misfire Current Cylinder 1-6	0-255 counts	0	0	Engine Data 1
Misfire History Cylinder 1-6	0-65535	0	0	Engine Data 1
Misfire Fail Since 1st Fail	0-65535	0	0	Engine Data 1
Misfire Pass since 1st Fail	0-65535	0	0	Engine Data 1
Non-Volatile Memory	Pass/Fail	Pass	Pass	Engine Data 1
Power Enrichment	Active/Inactive	Active	Active	Engine Data 1
Short Term Fuel Trim	-11-20%	Varies	Varies	Engine Data 1
Short Term F/T Average.	-11-20%	Varies	Varies	Engine Data 1
Spark Advance (°)	-64° to 64°	16	Varies	Engine Data 1
Startup ECT Sensor	-40 to 304°F	Varies	Varies	Engine Data 1
Startup IAT Sensor	-40 to 304°F	Varies	Varies	Engine Data 1
TCC Brake Switch	Apply/Release	Released	Released	Engine Data 1
TCC Engaged	Yes / No	No	Yes or No	Engine Data 1
Total Misfire Current Count	0-255 counts	0	0	Engine Data 1
TP Angle (%)	0-100%	0	Varies	Engine Data 1
TP Sensor (V)	0-5.0v	0.20-0.74	Varies	Engine Data 1
Transmission Hot Mode	Active/Inactive	Inactive	Inactive	Engine Data 1
Transmission Range	X-0	0-X-X-0	Varies	Engine Data 1
TWC Protection	Active/Inactive	Inactive	Inactive	Engine Data 1
Vehicle Speed	0-155 mph	0	30	Engine Data 1
VTD Fuel Enable	Disable/Enable	Disabled	Disabled	Engine Data 1

CAMARO PID DATA

1995-99 Camaro 3.8L V6 MFI VIN K (All) - PID Data List

Parameter Identification (PID)	PID Value Range	PID Value at Hot Idle	PID Value at 30 mph	Data List Type
Actual EGR Position	0-100%	0%	0-100%	Engine Data 1
Air Fuel Ratio	0:1-25.5:1	14.6:1	14.6:1	Engine Data 1
BARO Sensor (kPa)	10-110 kPa	65-110	65-110	Engine Data 1
BARO Sensor (V)	0-5.0v	3.5-4.5	3.5-4.5	Engine Data 1
Cruise Engaged	Disable/Enable	Disabled	Disabled	Engine Data 1
Cruise Inhibit	Yes / No	Yes	Yes	Engine Data 1
Current Gear	1-2-3-4	1	3	Engine Data 1
Decel Fuel Mode	Active/Inactive	Inactive	Inactive	Engine Data 1
Desired EGR Position	0-100%	0%	0-100%	Engine Data 1
Desired Idle Speed	0-3187 rpm	PCM control	---	Engine Data 1
ECT Sensor (°F)	-40 to 304°F	185-220	185-220	Engine Data 1
Engine Load	0-100%	2-5	Varies	Engine Data 1
Engine Run Time	Hr: Min: Sec	00:00:00	00:00:00	Engine Data 1
Engine Speed	0-9999 rpm	Varies	---	Engine Data 1
Fuel Trim Cell	Cell #	0	1	Engine Data 1
Fuel Trim Learn	Enabled/DIS	Enabled	Enabled	Engine Data 1
Generator Command	On / Off	On	On	Engine Data 1
Hot Open Loop	Active/Inactive	Inactive	Inactive	Engine Data 1
HO2S-11 (B1 S1)	0-1132 mv	10-1000	10-1000	Engine Data 1
HO2S-12 (B1 S2)	0-1132 mv	10-1000	10-1000	Engine Data 1
HO2S-21 (B2 S1)	0-1132 mv	10-1000	10-1000	Engine Data 1
HO2S-22 (B2 S2)	0-1132 mv	10-1000	10-1000	Engine Data 1
HO2S-13 (B1 S3)	0-1132 mv	10-1000	10-1000	Engine Data 1
HO2S-11 Status	Not Ready/Ready	Ready	Ready	Engine Data 1
HO2S-12 Status	Not Ready/Ready	Ready	Ready	Engine Data 1
HO2S-21 Status	Not Ready/Ready	Ready	Ready	Engine Data 1
HO2S-22 Status	Not Ready/Ready	Ready	Ready	Engine Data 1
IAC Position	0-255 counts	10-40	---	Engine Data 1
IAT Sensor (°F)	-40 to 304°F	50-194	50-194	Engine Data 1
Ignition 1 Signal	0.0-25.5v	14.2	14.2	Engine Data 1
Injector Pulsewidth	0-985 ms	1.5-3.5 ms	Varies	Engine Data 1
Long Term F/T B1, B2	Varies	0%	1%	Engine Data 1
Loop Status	Closed Loop / Open Loop	Closed Loop	Closed Loop	Engine Data 1
MAF Sensor (g/sec)	0-512 g/sec	3 - 6	Varies	Engine Data 1
MAF Sensor (Hz)	0-31,999	1200-3000	Varies	Engine Data 1
MAP Sensor (kPa)	10-110 kPa	29-48	Varies	Engine Data 1
MAP Sensor (V)	0-5.0v	1-2	Varies	Engine Data 1
Power Enrichment	Active/Inactive	Active	Active	Engine Data 1
Rich / Lean Flag	L or R	L-R-L-R	L-R-L-R	Engine Data 1
Short Term F/T B1	Varies	1%	0%	Engine Data 1
Short Term F/T B2	Varies	1%	0%	Engine Data 1
TCC Command	Engage/DIS	Disengaged	Disengaged	Engine Data 1
Throttle at Idle	Yes / No	Yes	---	Engine Data 1
TP Angle (%)	0-100%	0%	0-100%	Engine Data 1
TP Sensor (V)	0-5.0v	0.20-0.74	---	Engine Data 1
TWC Protection	Active/Inactive	Inactive	Inactive	Engine Data 1
Vehicle Speed	0-155 mph	0	30	Engine Data 1
VTD Fuel Disable	Disable/Enable	Disabled	Disabled	Engine Data 1

1995-99 Camaro 3.8L V6 MFI VIN K (All) - PID Data List

Parameter Identification (PID)	PID Value Range	PID Value at Hot Idle	PID Value at 30 mph	Data List Type
A/C HI Side Pressure	0-5.0v	Varies	Varies	Engine Data 2
A/C Pressure Out of Range	Yes / No	No	No	Engine Data 2
A/C Request	Yes / No	No	No	Engine Data 2
A/C Off for WOT	Yes / No	No	No	Engine Data 2
A/C Slugging	Active/Inactive	Inactive	Inactive	Engine Data 2
Actual EGR Position	0-100%	0%	0-100%	Engine Data 2
BARO Sensor (kPa)	10-110 kPa	65-110	65-110	Engine Data 2
Clutch Pedal Switch	Apply/Release	APPLY: If in	Released	Engine Data 2
Commanded A/C	On / Off	Off	Off	Engine Data 2
Commanded Fan 1, 2	On / Off	Off	Off	Engine Data 2
Cruise Engaged	DIS/Engage	Disabled	Disabled	Engine Data 2
Cruise Inhibit	Yes / No	Yes	Yes	Engine Data 2
Current Gear	1-2-3-4	1	3	Engine Data 2
Decel Fuel Mode	Active/Inactive	Inactive	Inactive	Engine Data 2
Desired EGR Position	0-100%	0%	0-100%	Engine Data 2
Desired Idle Speed	0-3187 rpm	PCM control	---	Engine Data 2
ECT Sensor (°F)	-40 to 304°F	185-220	185-220	Engine Data 2
Engine Load	0-100%	2-5	Varies	Engine Data 2
Engine Run Time	Hr: Min: Sec	00:00:00	00:00:00	Engine Data 2
Engine Speed	0-9999 rpm	Varies	---	Engine Data 2
EVAP Purge PWM	0-100%	0-25	0-100%	Engine Data 2
EVAP Vacuum Switch	Purge/No	No Purge	No Purge	Engine Data 2
Fuel Pump	On / Off	On	On	Engine Data 2
Generator Lamp	Off / On	On: If fault	Off	Engine Data 2
Hot Open Loop	Active/Inactive	Inactive	Inactive	Engine Data 2
IAC Position	0-255 counts	10-40	---	Engine Data 2
IAT Sensor (°F)	-40 to 304°F	50-194	50-194	Engine Data 2
Idle Speed Error	0-12800 rpm	<100 (DI)	---	Engine Data 2
Ignition 1 Signal	0.0-25.5v	14.2	14.2	Engine Data 2
Ignition Mode	IC / Bypass	IC	IC	Engine Data 2
Knock Retard (°)	0.0-25.5%	0%	Varies	Engine Data 2
KS Active Counter	0-255 counts	Varies	Varies	Engine Data 2
KS Min. Learn Noise	0-5.0v	0.1-3.0v	Varies	Engine Data 2
KS Noise Channel	0-5.0v	0.1-3.0v	Varies	Engine Data 2
MAF Sensor (g/sec)	0-512 g/sec	3 – 6	Varies	Engine Data 2
MAP Sensor (kPa)	10-110 kPa	29-48	Varies	Engine Data 2
MIL Status	Off / On	On: If fault	Off	Engine Data 2
Power Enrichment	Active/Inactive	Active	Active	Engine Data 2
Spark Advance (°)	-64° to 64°	-20	Varies	Engine Data 2
TCC Command	DIS/Engage	Disengaged	Disengaged	Engine Data 2
TP Angle (%)	0-100%	0%	Varies	Engine Data 2
Trans. Hot Mode	Active/Inactive	Inactive	Inactive	Engine Data 2
Transmission Range	PN-R-4-3-2	P	3	Engine Data 2
TR Switch P/A/B/C	HI/LO	LO/LO/HI/HI	---	Engine Data 2
TWC Protection	Active/Inactive	ACT: if on	Inactive	Engine Data 2
Vehicle Speed	0-155 mph	0	30	Engine Data 2
VTD Fuel Disable	Disable/Enable	Disabled	Disabled	Engine Data 2

1995-99 Camaro 3.8L V6 MFI VIN K (All) - PID Data List

Parameter Identification (PID)	PID Value Range	PID Value at Hot Idle	PID Value at 30 mph	Data List Type
Air Fuel Ratio	0:1-25.5:1	14.6:1	14.6:1	Catalyst
Decel Fuel Mode	Active/Inactive	Inactive	Inactive	Catalyst
ECT Sensor (°F)	-40 to 304°F	185-220	185-220	Catalyst
Engine Load	0-100%	2-5	Varies	Catalyst
Engine Run Time	Hr: Min: Sec	00:00:00	00:00:00	Catalyst
Engine Speed	0-9999 rpm	±50 rpm	---	Catalyst
Hot Open Loop	Active/Inactive	Inactive	Inactive	Catalyst
HO2S3 Bank 1	0-1132 mv	10-1000	10-1000	Catalyst
IAC Position	0-255 counts	10-40	---	Catalyst
IAT Sensor (°F)	-40 to 304°F	50-194	50-194	Catalyst
Loop Status	Closed Loop / Open Loop	Closed Loop	Closed Loop	Catalyst
MAF Sensor (g/sec)	0-512 g/sec	5-9	Varies	Catalyst
Power Enrichment	Active/Inactive	Active	Active	Catalyst
TP Angle (%)	0-100%	0%	Varies	Catalyst
TWC Diagnostic	Disable/Enable	Enabled	Enabled	Catalyst
TWC Monitor Test Counter	0-49	0	0	Catalyst
TWC Protection	Active/Inactive	ACT: if on	Inactive	Catalyst
Vehicle Speed	0-155 mph	0	30	Catalyst

1995-99 Camaro 3.8L V6 MFI VIN K (All) - PID Data List

Parameter Identification (PID)	PID Value Range	PID Value at Hot Idle	PID Value at 30 mph	Data List Type
Actual EGR Position	0-100%	0%	0-100%	EGR
BARO Sensor (kPa)	10-110 kPa	65-110	65-110	EGR
BARO Sensor (V)	0-5.0v	3.5-4.5	3.5-4.5	EGR
Current Gear	1-2-3-4	1	3	EGR
Decel Fuel Mode	Active/Inactive	Inactive	Inactive	EGR
Desired EGR Position	0-100%	0%	0-100%	EGR
Desired Idle Speed	0-3187 rpm	PCM control	Varies	EGR
ECT Sensor (°F)	-38-304°F	185-220	185-220	EGR
EGR Closed Pintle Position	0-5.0v	0.14-1.0v	Varies	EGR
EGR Duty Cycle	0-100%	0%	0-100%	EGR
EGR Feedback	0-5.0v	0.14-1.0v	Varies	EGR
Engine Speed	0-9999 rpm	Varies	Varies	EGR
Hot Open Loop	Active/Inactive	Inactive	Inactive	EGR
IAC Position	0-255 counts	10-40	10-40	EGR
IAT Sensor (°F)	-40 to 304°F	50-194	50-194	EGR
Ignition 1 Signal	0.0-25.5v	14.2	14.2	EGR
MAP Sensor (kPa)	10-110 kPa	29-48	Varies	EGR
MAP Sensor (V)	0-5.0v	1-2	Varies	EGR
TP Angle (%)	0-100%	0%	Varies	EGR
Vehicle Speed	0-155 mph	0	30	EGR

1995-99 Camaro 3.8L V6 MFI VIN K (All) - PID Data List

Parameter Identification (PID)	PID Value Range	PID Value at Hot Idle	PID Value at 30 mph	Data List Type
Engine Run Time	Hr: Min: Sec	00:00:00	00:00:00	EVAP
Engine Speed	0-9999 rpm	Varies	Varies	EVAP
EVAP Canister Purge	0-100%	0-25	0-100%	EVAP
IAT Sensor (°F)	-40 to 304°F	50-194	50-194	EVAP
MAF Sensor (g/sec)	0-512 g/sec	3 - 6	Varies	EVAP
MAP Sensor (kPa)	10-110 kPa	29-48	Varies	EVAP
MAP Sensor (V)	0-5.0v	1-2	Varies	EVAP
Startup ECT Degrees	-40 to 304°F	Varies	Varies	EVAP
Startup IAT Degrees	-40 to 304°F	Varies	Varies	EVAP

1995-99 Camaro 3.8L V6 MFI VIN K (All) - PID Data List

Parameter Identification (PID)	PID Value Range	PID Value at Hot Idle	PID Value at 30 mph	Data List Type
Decel Fuel Mode	Active/Inactive	Inactive	Inactive	HO2S
ECT Sensor (°F)	-40 to 304°F	185-220	185-220	HO2S
Engine Speed	0-9999 rpm	Varies	---	HO2S
Engine Load	0-100%	2-5	Varies	HO2S
Hot Open Loop	Active/Inactive	Inactive	Inactive	HO2S
HO2S-11 (B1 S1)	0-1132 mv	10-1000	10-1000	HO2S
HO2S-12 (B1 S2)	0-1132 mv	10-1000	10-1000	HO2S
HO2S-21 (B2 S1)	0-1132 mv	10-1000	10-1000	HO2S
HO2S-22 (B2 S2)	0-1132 mv	10-1000	10-1000	HO2S
HO2S-13 (B1 S3)	0-1132 mv	10-1000	10-1000	HO2S
HO2S-11 Status	Not Ready/Ready	Ready	Ready	HO2S
HO2S-12 Status	Not Ready/Ready	Ready	Ready	HO2S
HO2S-21 Status	Not Ready/Ready	Ready	Ready	HO2S
HO2S-22 Status	Not Ready/Ready	Ready	Ready	HO2S
HO2S-11 Warmup	Min: Sec	00:00	00:00	HO2S
HO2S-12 Warmup	Min: Sec	00:00	00:00	HO2S
HO2S-21 Warmup	Min: Sec	00:00	00:00	HO2S
HO2S-22 Warmup	Min: Sec	00:00	00:00	HO2S
HO2S-13 Warmup	Min / Sec	Varies	Varies	HO2S
HO2S-11 XCounts	0-255 counts	Varies	Varies	HO2S
IAT Sensor (°F)	-40 to 304°F	50-194	50-194	HO2S
Ignition 1 Signal	0.0-25.5v	14.2	14.2	HO2S
Loop Status	Closed Loop / Open Loop	Closed Loop	Closed Loop	HO2S
MAF Sensor (g/sec)	0-512 g/sec	5-9	Varies	HO2S
Power Enrichment	Active/Inactive	Inactive	Inactive	HO2S
Startup ECT Degrees	-40 to 304°F	Varies	Varies	HO2S
Startup IAT Degrees	-40 to 304°F	Varies	Varies	HO2S
TP Angle (%)	0-100%	0%	Varies	HO2S
TWC Protection	Active/Inactive	Inactive	Inactive	HO2S

1995-99 Camaro 3.8L V6 MFI VIN K (All) - PID Data List

Parameter Identification (PID)	PID Value Range	PID Value at Hot Idle	PID Value at 30 mph	Data List Type
Actual EGR Position	0-100%	0%	0-100%	Misfire
Commanded TCC	Engage/DIS	Disengaged	Disengaged	Misfire
Current Gear	1-2-3-4	1	3	Misfire
Decel Fuel Mode	Active/Inactive	Inactive	Inactive	Misfire
Desired EGR Position	0-100%	0%	0-100%	Misfire
ECT Sensor (°F)	-40 to 304°F	185-220	185-220	Misfire
Engine Load	0-100%	2-5	Varies	Misfire
Engine Speed	0-9999 rpm	Varies	---	Misfire
Fuel Trim Cell	Cell #	0	1	Misfire
Hot Open Loop	Active/Inactive	Inactive	Inactive	Misfire
IAC Position	0-255 counts	10-40	---	Misfire
Long Term F/T B1, B2	Varies	0%	1%	Misfire
Loop Status	Closed Loop / Open Loop	Closed Loop	Closed Loop	Misfire
Misfire Current Cylinder 1-6	0-198	0-4	0-4	Misfire
Misfire History Cylinder 1-6	0-65535	0	0	Misfire
Misfiring Cylinder	CYL 1-6	0	0	Misfire
Power Enrichment	Active/Inactive	Active	Active	Misfire
Short Term Fuel Trim Bank 2	-10% to 10%	0 (± 5%)	0 (± 5%)	Misfire
Short Term Fuel Trim Bank2	-10% to 10%	0 (± 5%)	0 (± 5%)	Misfire
Total Misfire current count	0-99	0-5	0-5	Misfire
Total Misfire Failures Since First Failure	0-65535	0	0	Misfire
Total Misfire Passes Since First Failure	0-65535	0	0	Misfire
TP Angle (%)	0-100%	0	Varies	Misfire
Vehicle Speed	0-155 mph	0	30	Misfire
TWC Protection	Active/Inactive	Inactive	Inactive	Misfire

1995-99 Camaro 3.8L V6 MFI VIN K (All) - PID Data List

Parameter Identification (PID)	PID Value Range	PID Value at Hot Idle	PID Value at 30 mph	Data List Type
IAC Position	0-255 counts	10-40	10-40	Catalyst
IAT Sensor (ºF)	-40 to 304ºF	50-194	50-194	Catalyst
Loop Status	Closed Loop / Open Loop	Closed Loop	Closed Loop	Catalyst
MAF Sensor (g/sec)	0-512 g/sec	5-9	Varies	Catalyst
TP Angle (%)	0-100%	0%	0-100%	Catalyst
TWC Diagnostic	Disable/Enable	Enabled	Enabled	Catalyst
TWC Monitor Test Counter	0-49	0	0	Catalyst
TWC Protection	Active/Inactive	Inactive	Inactive	Catalyst

2000-02 Camaro 3.8L V6 MFI VIN K (A/T) - PID Data List

Parameter Identification (PID)	PID Value Range	PID Value at Hot Idle	PID Value at 30 mph	Data List Type
3X Crank Sensor	0-9,999 rpm	Varies	Varies	Engine Data 2
18X Crank Sensor	0-1,600 rpm	Varies	Varies	Engine Data 2
1-2 Shift Solenoid Circuit Status	Okay / Fault/ Invalid State	Okay	Okay	Output Driver
2-3 Shift Solenoid Circuit Status	Okay / Fault/ Invalid State	Okay	Okay	Output Driver
A/C HI Side Pressure	0-459 kPa	90	90	Engine Data 2
A/C HI Side Pressure	0-5.1 volts	0.98	0.98	Engine Data 2
A/C Off for WOT	Yes/No	No	No	Engine Data 2
A/C Pressure Disable	Yes/No	No	No	Engine Data 2
A/C Relay Circuit Status	Okay / Fault/ Invalid State	Okay	Okay	Output Driver
A/C Relay Command	On/Off	Off	Off	Engine Data 1-2 EGR, Misfire
A/C Request Signal	Yes/No	No	No	Engine Data 2
Air Fuel Ratio	0.0-25.5:1	14.7:1	14.7:1	Engine Data 2, Fuel Trim
APP Average	0-150	5	WOT: 122	TAC
APP Indicated Angle	0-100%	0	WOT: 100%	Engine Data 1-2 EGR, EVAP, Fuel Trim, Cruise, TAC
APP Sensor 1 (V)	0-5.0 volts	0.8-1.0	Varies	TAC
APP Sensor 2 (V)	0-5.0 volts	3.9-4.8	Varies	TAC
APP Sensor 3 (V)	0-5.0 volts	3.2-4.5	Varies	TAC
APP Sensor 1 (%)	0-100%	0	Varies	TAC
APP Sensor 2 (%)	0-100%	0	Varies	TAC
APP Sensor 3 (%)	0-100%	0	Varies	TAC
BARO Sensor (kPa)	10-105 kPa	Varies w/ALT	Varies	Engine Data 1, EGR, EVAP, Fuel Trim
BARO Sensor (V)	0-5.1 volts	Varies w/ALT	Varies	Engine Data 1, EGR, EVAP, Fuel Trim
Clutch Pedal Switch	Applied/Released	Released	Released	EGR, Cruise
CMP Sensor Signal	Yes/No	Yes	Yes	Engine Data 2
Cruise Control Active	Yes/No	No	No	Engine Data 1, Cruise, TAC
Cruise Disengage History 1-6	30 possible causes	Reason	Reason	Cruise
Cruise On/Off Switch	On/Off	Off	Off	Cruise, TAC
Cruise Release Brake Pedal Switch	Applied/Released	Released	Released	Cruise, TAC
Cruise Resume/Accel Switch	On/Off	Off	Off	Cruise, TAC
Cruise Set/Coast Switch	On/Off	Off	Off	Cruise, TAC
Current Gear	0, 1, 2, 3, 4	3	4	Engine Data 1-2 EGR, F/Trim

2000-02 Camaro 3.8L V6 MFI VIN K (A/T) PID Data List

Parameter Identification (PID)	PID Value Range	PID Value at Hot Idle	PID Value at 30 mph	Data List Type
Cycles of Misfire Data	0-99 counts	0	0	Misfire
Cylinder 1-6 Injector Circuit Status	Okay / Stuck Low/ Stuck High / Fault	Okay	Okay	Output Driver
Cylinder 1-6 Injector Circuit History	Okay / Stuck Low/ Stuck High / Fault	Okay	Okay	Output Driver
Decel Fuel Cutoff	Active/Inactive	Inactive	Inactive	EGR, F/Trim
Desired EGR Position	0-100%	0	Varies	EGR
Desired EGR Position	0-5.1v	0.0	Varies	EGR
Desired Idle Speed	0-3187 rpm	Commanded by the PCM	Varies	Engine Data 1-2 EVAP, TAC
Driver Module 1 Status	Enabled / Off- HI Volts/ Off-HI TEMP/ Invalid State	Enabled	Enabled	Output Driver
Driver Module 2 Status	Enabled / Off- HI Volts/ Off-HI TEMP/ Invalid State	Enabled	Enabled	Output Driver
Driver Module 3 Status	Enabled / Off- HI Volts/ Off-HI TEMP/ Invalid State	Enabled	Enabled	Output Driver
Driver Module 4 Status	Enabled / Off- HI Volts/ Off-HI TEMP/ Invalid State	Enabled	Enabled	Output Driver
ECT Sensor (°F)	-40 to 304°F	190-221°F	190-221°F	Engine Data 1-2 EGR, F/Trim, EVAP, Misfire
EGR Flow Test Count	0-255 counts	0-12	Varies	EGR
EGR Learned Minimum Position	0-5.1 volts	0.16-1.0	Varies	EGR
EGR Position Sensor	0-100%	0	Varies	Engine Data 1, Misfire
EGR Position Sensor	0-5.1 volts	0.58-0.85	Varies	EGR
EGR Position Variance	0-100%	0	0	EGR
EGR Solenoid Circuit Status	Okay / Fault/ Invalid State	Okay	Okay	Output Driver
EGR Solenoid Command	0-100%	0	Varies	EGR
Engine Load	0-100%	1-4	Varies	Engine Data 1-2 EGR, F/Trim, TAC, Misfire, Cruise, EVAP
Engine Oil Level Switch	Okay / LOW	Okay	Okay	Engine Data 2

2000-02 Camaro 3.8L V6 MFI VIN K (A/T) - PID Data List

Parameter Identification (PID)	PID Value Range	PID Value at Hot Idle	PID Value at 30 mph	Data List Type
Engine Run Time	Hr: Min: Sec 00:00:00 to 99:99:99	00:00:00	00:00:00	Engine Data 1-2 EGR, F/Trim, TAC, Misfire, Cruise, EVAP
Engine Speed	0-9999 rpm	± 50 rpm of actual speed	Varies	Engine Data 1-2 EGR, F/Trim, TAC, Misfire, Cruise, EVAP
EVAP Fault History	No Fault / Excess VAC /Purge Valve Leak / Small Leak / Weak Vacuum	No Fault	No Fault	EVAP
EVAP Purge Solenoid Circuit Status	Okay / Fault/ Invalid State	Okay	Okay	Output Driver
EVAP Purge Solenoid Command	0-100%	15	Varies	Engine Data 1, EVAP, F/Trim
EVAP Test Abort Reason	Not Aborted/ Lost Enable/ Small Leak/ Not at Rest	Not Aborted	Not Aborted	EVAP
EVAP Test Result	No Result / Passed/ Fail/ P0440/442/ P0446/1441	Okay	Okay	EVAP
EVAP Test State	Waiting for Purge/ Test Running/Test Completed	Test Completed	Test Completed	EVAP
EVAP Vent Solenoid Circuit Status	Okay / Fault/ Invalid State	Okay	Okay	Output Driver
EVAP Vent Solenoid Command	Venting / Not Venting	Not Venting	Not Venting	Engine Data 1, EVAP, F/Trim
Fan Control Relay 1 Circuit Status	Okay / Fault/ Invalid State	Okay	Okay	Output Driver
Fan Control Relay 1 Command	On/Off	Off	Off	Engine Data 2
Fan Control Relay 2-3 Circuit Status	Okay/Fault	Okay	Okay	Output Driver
Fan Control Relay 2-3 Command	On/Off	Off	Off	Engine Data 2
Fuel Pump Relay Circuit Status	Okay / Fault/ Invalid State	Okay	Okay	Output Driver
Fuel Pump Relay Circuit History Status	Okay / Fault	Okay	Okay	Output Driver
Fuel Pump Relay Command	On/Off	On	On	Engine Data 2
Fuel Tank level	0-100%	Varies	Varies	EVAP

2000-02 Camaro 3.8L V6 MFI VIN K (A/T) - PID Data List

Parameter Identification (PID)	PID Value Range	PID Value at Hot Idle	PID Value at 30 mph	Data List Type
Fuel Tank Pressure Sensor	Inches H2O, mm Hg	-17 to +7.5 Inches H2O	-17 to +7.5 Inches H2O	Engine Data 1
Fuel Tank Pressure Sensor	0-5.0 volts	Varies	Varies	EVAP
Fuel Trim Cell	Cell 0-10	Varies	Varies	Engine Data 1, EVAP, F/Trim
Fuel Trim Learn	Enabled or Disabled	Enabled	May Toggle	Engine Data 1, EVAP, F/Trim
Generator L-Terminal	On/Off	On	On	Engine Data 2
HO2S (B1 S1) Signal	0-1132 mv	10-1000	10-1000	Engine Data 1, EVAP, F/Trim
HO2S (B2 S1) Signal	0-1132 mv	10-1000	10-1000	Engine Data 1, EVAP, F/Trim
HO2S (B1 S2) Signal	0-1132 mv	10-1000	10-1000	Engine Data 1, EVAP, F/Trim
IAT Sensor (°F)	-40 to 304°F	91	92	Engine Data 1-2 EGR, EVAP, Fuel Trim
Ignition 1	0.0-25.5v	13.5	13.6	Engine Data 1-2 EGR, EVAP, Fuel Trim
Ignition Mode	Bypass/IC	IC	IC	Engine Data 2
Injector Pulsewidth	0-1000 ms	1.5-3.5	Varies	Engine Data 2, F/Trim, M/fire
Knock Retard (°)	0-25.5°	0°	0°	Engine Data 1, EGR
Long Term Fuel Trim Bank 1	-23 to +16	0 (± 5%)	0 (± 5%)	Engine Data 1-2 EVAP, F/Trim
Long Term Fuel Trim Bank 2	-23 to +16	0 (± 5%)	0 (± 5%)	Engine Data 1-2 EVAP, F/Trim
Loop Status	Open / Closed	Closed	Closed	Engine Data 1-2 EGR, EVAP, Fuel Trim
MAF Sensor (g/sec)	0-512 g/sec	3-6	Varies	Engine Data 1-2 EGR, EVAP, Fuel Trim, Misfire, TAC
MAF Sensor (Hz)	0-32,000 Hz	2000-3000	Varies	Engine Data 2
MAP Sensor (kPa)	10-105 kPa	20-48	Varies	Engine Data 1-2 EGR, EVAP, Fuel Trim, Misfire, TAC
MAP Sensor (V)	0-5.1 volts	0.75-2.0	Varies	Engine Data 1-2
MIL Status	Okay / Fault/ Invalid State	Okay	Okay	Output Driver
MIL Command	On/Off	Off	Off	Engine Data 2
Misfire Current Cylinder 1-6	0-198 counts	0-4	0	Misfire
Misfire History Cylinder 1-6	0-65,535	0	0	Misfire

2000-02 Camaro 3.8L V6 MFI VIN K (A/T) PID Data List

Parameter Identification (PID)	PID Value Range	PID Value at Hot Idle	PID Value at 30 mph	Data List Type
Number of DTC(s)	Number	0	No	Engine Data 1, EVAP, Fuel Trim, TAC
Power Enrichment	Active/Inactive	Inactive	Inactive	Engine Data 2, Misfire
Reduced Engine Power	Active/Inactive	Inactive	Inactive	Engine Data 1, EGR, TAC, Cruise
Short Term Fuel Trim Bank 1	-10 to +10%	0 (± 5%)	0 (± 5%)	Engine Data 1-2 EVAP, F/Trim
Short Term Fuel Trim Bank 2	-10 to +10%	0 (± 5%)	0 (± 5%)	Engine Data 1-2 EVAP, F/Trim
Spark Advance	-64° to +64°	-20°	Varies	Engine Data 1-2 F/Trim, M/fire
Starter Enable Relay Circuit Status	Okay / Fault/ Invalid State	Okay	Okay	Output Driver
Starter Relay Control	On/Off	Off	Off	Engine Data 2
Startup ECT	-40 to 419°F	Varies	Varies	EVAP, F/Trim
Startup IAT	-40 to 419°F	80°F	80°F	EVAP, F/Trim
Stop Lamp Pedal Switch	Apply/REL	Released	Released	Engine Data 2, Cruise, TAC
TAC/PCM Communication signal	Okay/Fault	Okay	Okay	Engine Data 1, Cruise, TAC
TCC/Cruise Brake Pedal Switch	Apply/REL	Released	Released	Engine Data 2, Cruise, TAC
TCC Enable Solenoid Circuit Status	Okay / Fault/ Invalid State	Okay	Okay	Output Driver
TCC Enable Solenoid Command	On/Off	Off	Off	Engine Data 2, Cruise, M/Fire
TCS Circuit History	Okay / Fault	Okay	Okay	Output Driver
TCS Circuit Status	Okay / Fault	Okay	Okay	Output Driver
Torque Delivered	0-100%	70	Varies	ENG 2, TAC
Torque Request	0-100%	100	Varies	ENG 2, TAC
TP Desired Angle	0-100%	2-6	Varies	Engine Data 1-2 TAC EGR, F/Trim, EVAP, Misfire
TP Indicated Angle	0-100%	2-6	Varies	Engine Data 1-2 TAC EGR, F/Trim, EVAP, Misfire
TP Sensor 1	0-100%	0-6	Varies	TAC
TP Sensor 2	0-100%	0-6	Varies	TAC
TP Sensor 1-2	Agree/DIS	Agree	Agree	TAC
Traction Control Status	Active/Inactive	Inactive	Inactive	Engine Data 2, Cruise, TAC
Transaxle Range Switch	Park/Neutral/ Reverse	Park	---	ENG 2, EGR, Cruise, F/Trim
Vehicle Speed	Km/h, MPH	0	30	Engine Data 1-2 EGR F/Trim, EVAP Cruise, M/fire
VTD Fuel Enable	Active/Inactive	Inactive	Inactive	Engine Data 1

1995 Camaro 5.7L V8 MFI VIN P (All) - PID Data List

Parameter Identification (PID)	PID Value Range	PID Value at Hot Idle	PID Value at 30 mph	Data List Type
A/C Clutch	On / Off	Off	Off	Engine Data 1
A/C EVAP Temperature	19° to 90	Varies	Varies	Engine Data 1
A/C Refrigerant Pressure	0-499 psi	139-399	139-399	Engine Data 1
A/C Request	Yes / No	No	No	Engine Data 1
AC Status	Off / On	Off	Off	Engine Data 1
AIR Control System	On / Off	On: cold	Off	Engine Data 1
BARO Sensor (kPa)	10-110 kPa	70-100	70-100	Engine Data 1
Cooling Fan 1, 2	On / Off	Off	Off	Engine Data 1
Desired Idle Speed	0-3187	800	---	Engine Data 1
ECT Sensor (°F)	-40-419°F	185-221	185-221	Engine Data 1
EGR Duty Cycle	0-100%	0	0-100	Engine Data 1
Engine Speed	0-9999 rpm	Varies	Varies	Engine Data 1
Fuel Trim Cell	Cell #0-16	0	0-16	Engine Data 1
Fuel Trim Enable	No / Yes	Yes	Yes	Engine Data 1
IAT Sensor (°F)	-40 to 304°F	50-176	50-176	Engine Data 1
HI Resolution Signal	Active/Inactive	Active	Active	Engine Data 1
HO2S (B1 S1, B1 S2)	0-1132 mv	10-1000	10-1000	Engine Data 1
HO2S (B2 S1, B2 S2)	0-1132 mv	10-1000	10-1000	Engine Data 1
Long Term F/T B1, B2	0-100%	0 (± 5%)	0 (± 5%)	Engine Data 1
Knock Sensor	Fault/Okay	Okay	Okay	Engine Data 1
Knock Signal	No / Yes	No	No	Engine Data 1
Knock Retard (°)	0-90°	0	Varies	Engine Data 1
IAC Motor	0-255 counts	5-50	---	Engine Data 1
Injector Fault	No / Yes	No	No	Engine Data 1
Injector Pulsewidth	0-985 ms	1-4	Varies	Engine Data 1
Learned IAC	0-255 counts	5-50	---	Engine Data 1
LO Resolution Signal	0-332.7ms	Varies	Varies	Engine Data 1
Loop Status	Closed Loop / Open Loop	Closed Loop	Closed Loop	Engine Data 1
MAF Sensor (g/sec)	0-512 g/sec	5-7	Varies	Engine Data 1
MAP Sensor (kPa)	10-110 kPa	20-48	Varies	Engine Data 1
MPH / KPH	0-155 mph	0	30	Engine Data 1
PCM or TCM DTC	No / Yes	No	No	Engine Data 1
Performance Mode	Closed/open	Open	Open	Engine Data 1
PNP Switch	PN / R-DL	PN	R-DL	Engine Data 1
Purge Duty Cycle	0-100%	0	0-100	Engine Data 1
Short Term F/T B1-B2	0-100%	0 (± 5%)	0 (± 5%)	Engine Data 1
Skip Shift Active	No / Yes	No	No	Engine Data 1
Spark Advance (°)	-90° to 90°	Varies	Varies	Engine Data 1
TCC Brake Switch	On / Off	Off	Off	Engine Data 1
TCC Solenoid	On / Off	Off	Off	Engine Data 1
TCC Solenoid PWM	0-100%	0	0-50	Engine Data 1
TCS / ASR Active	Yes / No	No	No	Engine Data 1
Throttle Angle (%)	0-100%	0	Varies	Engine Data 1
TP Sensor (V)	0-5.0v	0.3-0.9	Varies	Engine Data 1
Trans. Range Switch	PN-R-4-3-2	PN	3	Engine Data 1
1-2, 2-3 Shift Solenoid	On / Off	On	Off	Engine Data 1

1995 Camaro 5.7L V8 MFI VIN P (A/T) - PID Data List

Parameter Identification (PID)	PID Value Range	PID Value at Hot Idle	PID Value at 30 mph	Data List Type
1/2 2/3 CTR Feedback	Okay/Fault	Okay	Okay	Transmission
1-2 Shift Time	0-100 sec	0	Varies	Transmission
1-2 Shift Time Error	0-100 sec	0	Varies	Transmission
2-3 Shift Time	0-100 sec	0	Varies	Transmission
1-2, 2-3 Solenoid	On / Off	On / On	Varies	Transmission
3-2 Control Solenoid	0-100%	0	0-100	Transmission
3-2 Control Feedback	Okay/Fault	Okay	Okay	Transmission
ABC RNG	Off / On	Off/On/Off	Off/On/Off	Transmission
Actual PCS (amps)	-10 to 10	1.01	Varies	Transmission
Adaptable Shift	No / Yes	No	No	Transmission
Commanded Gear	1-2-3-4	1	3	Transmission
Cruise Engaged	No / Yes	No	No	Transmission
Current Adaptive Cell	% TP	<25%	Varies	Transmission
Desired PCS (amps)	-10 to 10	1.01	Varies	Transmission
ECT Sensor (°F)	-40-419°F	185-223	185-223	Transmission
Engine Speed	0-9999 rpm	Varies	---	Transmission
Kickdown Enabled	No / Yes	No	No	Transmission
MPH / KPH	0-155 mph	0	30	Transmission
PCM or TCM DTC	No / Yes	No	No	Transmission
Performance Lamp Feedback	Okay/Fault	Okay	Okay	Transmission
Performance Mode Switch	Close/Open	Open	Open	Transmission
Performance Mode Status	Normal / Performance	Normal	Normal	Transmission
PNP Switch	PN / R-DL	PN	R-DL	Transmission
Purge Duty Cycle	0-100%	0	0-100	Transmission
Ignition 1 Signal	0-25.5v	14.2	14.2	Transmission
TCS / ASR Active	No / Yes	No	No	Transmission
Throttle Angle (%)	0-100%	0	0-100	Transmission
TP Sensor (V)	0-5.0v	0.3-0.9	---	Transmission
TFT Sensor (°)	-40 to 304°F	160-200	160-200	Transmission
Trans. Hot Mode	No / Yes	No	No	Transmission
Trans. Output Speed	0-8192	0	Varies	Transmission
Trans. Range Switch	PN-R-4-3-2	PN	3	Transmission
TCC Brake Switch	On / Off	Off	Off	Transmission
TCC Slip Speed rpm	-4080 - 4079	730-768	2-4	Transmission
TCC Solenoid	Off / On	Off	Off or On	Transmission
TCC Solenoid CTR Feedback	Okay/Fault	Okay	Okay	Transmission
TCC Solenoid PWM	Off/ On	Off	Off	Transmission
TCC Solenoid PWM	0-100%	0	0-50	Transmission
TCC PWM CTR Feedback	Okay/Fault	Okay	Okay	Transmission

1996-97 Camaro 5.7L V8 MFI VIN P (All) - PID Data List

Parameter Identification (PID)	PID Value Range	PID Value at Hot Idle	PID Value at 30 mph	Data List Type
A/C Clutch	On / Off	Off	Off	Engine Data 1
A/C EVAP Temp.	19° - 90F	Varies	Varies	Engine Data 1
A/C High Side PSI	-15-452 psi	139-399	139-399	Engine Data 1
A/C Request	Yes / No	No	No	Engine Data 1
A/C Status	On / Off	Off	Off	Engine Data 1
Actual EGR Position	0-100%	0	0-100	Engine Data 1
AIR Pump	Disable/Enable	Disabled	Disabled	Engine Data 1
ASR/TSR Spark Advance or Retard	No / Yes	No	No	Engine Data 1
BARO Sensor (kPa)	10-110 kPa	65-104	65-104	Engine Data 1
BARO Sensor (V)	0-5.0v	3.5-4.5	3.5-4.5	Engine Data 1
CKP Engine Speed	0-9999 rpm	Varies	---	Engine Data 1
CKP LO Resolution Angle	-30 to +60	0 Deviation	0 Deviation	Engine Data 1
Cold Startup	No / Yes	Yes: Cold	No	Engine Data 1
Current Gear	1-2-3-4	1	3	Engine Data 1
Desired Idle Speed	0-3187	PCM control	---	Engine Data 1
DTC Number	Number	00000	00000	Engine Data 1
DTC Set this Ignition	No / Yes	No	No	Engine Data 1
EGR Duty Cycle	0-100%	0	0-100	Engine Data 1
ECT Sensor (°F)	-40 to 304°F	185-226	185-226	Engine Data 1
Engine Run Time	Hr: Min: Sec	00:00:00	00:00:00	Engine Data 1
Engine Speed	0-9999 rpm	Varies	---	Engine Data 1
EVAP Purge %	0-100%	0-25	0-100	Engine Data 1
EVAP Vacuum Switch	No/Purge	No Purge	No Purge	Engine Data 1
FC Relay 1	On / Off	On: If hot	Off	Engine Data 1
FC Relay 2, 3	On / Off	Off	Off	Engine Data 1
Fuel Trim Cell	Cell #0-16	0	0-16	Engine Data 1
Fuel Trim Learn	Disable/Enable	Enabled	Enabled	Engine Data 1
HI Resolution Signal	Active/Inactive	Active	Active	Engine Data 1
HO2S (B1 S1, B1 S2)	0-1132 mv	10-1000	10-1000	Engine Data 1
HO2S (B2 S1, B2 S2)	0-1132 mv	10-1000	10-1000	Engine Data 1
IAC Learned	0-255 counts	10-50	---	Engine Data 1
IAC Position	0-255 counts	10-50	---	Engine Data 1
IAT Sensor (°F)	-40 to 304°F	50-194	50-194	Engine Data 1
Ignition 1 Signal	0.0-25.5v	14.2	14.2	Engine Data 1
Injector Pulsewidth	0-985 ms	1-4	Varies	Engine Data 1
Knock Retard (°)	0-25.5°	0	Varies	Engine Data 1
Long Term F/T B1, B2	-8% to 5%	0 (± 5%)	0 (± 5%)	Engine Data 1
Long Term F/T Average	-13 to 22%	0 (± 5%)	0 (± 5%)	Engine Data 1
Loop Status	Closed Loop / Open Loop	Closed Loop	Closed Loop	Engine Data 1
LO Resolution Signal	0-332.7ms	Varies	Varies	Engine Data 1
MAF Sensor (g/sec)	0-512 g/sec	5-9	Varies	Engine Data 1
MAP Sensor (V)	0-5.0v	1-2	Varies	Engine Data 1
MAP Sensor (kPa)	10-110 kPa	20-48 kPa	Varies	Engine Data 1
MIL Status	On / Off	Off	Off	Engine Data 1
Pass Key® Fuel	Disabled/Okay	Okay	Okay	Engine Data 1
PCM Reset	Yes / No	Yes: if reset	No	Engine Data 1

1996-97 Camaro 5.7L V8 MFI VIN P (All) - PID Data List

Parameter Identification (PID)	PID Value Range	PID Value at Hot Idle	PID Value at 30 mph	Data List Type
Performance Mode	Normal / Performance	Normal	Normal	Engine Data 1
Performance Switch	Active/Inactive	Inactive	Inactive	Engine Data 1
PNP Switch	P-R-0-3-2-1	Park	3	Engine Data 1
PSP Switch	HI/Normal	Normal	Normal	Engine Data 1
Reverse Inhibit Switch	Yes / No	Yes: in R	No	Engine Data 1
SHRTFT Average. B1, B2	-10% to 10%	0 (± 5%)	0 (± 5%)	Engine Data 1
Short Term F/T B1-B2	-10% to 10%	0 (± 5%)	0 (± 5%)	Engine Data 1
Skip Shift Solenoid	Disable/Enable	Disabled	Disabled	Engine Data 1
Spark Advance (°)	-64° to 64°	Varies	Varies	Engine Data 1
Startup ECT Sensor	-40 to 304°F	Varies	Varies	Engine Data 1
TCC Brake Switch	Apply/Release	Released	Released	Engine Data 1
TCC Duty Cycle	0-100%	0%	0%	Engine Data 1
TCC Enable	Disable/Enable	Disabled	Disabled	Engine Data 1
TP Angle (%)	0-100%	0%	Varies	Engine Data 1
TP Sensor (V)	0-5.0v	0.55-0.90	---	Engine Data 1
Trans. Range Switch	PN-D-L-R	PN	D	Engine Data 1
Traction Control	Active/Inactive	Inactive	Inactive	Engine Data 1
Vehicle Speed	0-155 mph	0	30	Engine Data 1

1996-97 Camaro 5.7L V8 MFI VIN P (All) - PID Data List

Parameter Identification (PID)	PID Value Range	PID Value at Hot Idle	PID Value at 30 mph	Data List Type
A/C Clutch	On / Off	Off	Off	Engine Data 2
AIR Pump	Disable/Enable	Disabled	Disabled	Engine Data 2
ASR/TSR Spark Advance or Retard	No / Yes	No	No	Engine Data 2
BARO Sensor (kPa)	10-110 kPa	65-104	65-104	Engine Data 2
BARO Sensor (V)	0-5.0v	3.5-4.5	3.5-4.5	Engine Data 2
CKP Engine Speed	0-9999 rpm	Varies	Varies	Engine Data 2
CKP LO Resolution	-30 to +60	0 Deviation	Varies	Engine Data 2
Cold Startup	No / Yes	Yes: Cold	No	Engine Data 2
Desired Idle Speed	0-3187	PCM control	---	Engine Data 2
DTC Set This Ignition	No / Yes	No	No	Engine Data 2
ECT Sensor (°F)	-40 to 304°F	185-226	185-226	Engine Data 2
EGR Duty Cycle	0-100%	0%	0-100%	Engine Data 2
Engine Run Time	Hr: Min: Sec	00:00:00	00:00:00	Engine Data 2
Engine Speed	0-9999 rpm	Varies	Varies	Engine Data 2
EVAP Purge %	0-100%	0%	0-100%	Engine Data 2
EVAP Vacuum Switch	No/Purge	No Purge	No Purge	Engine Data 2
FC Relay 1	On / Off	Off	Off	Engine Data 2
FC Relay 2, 3	On / Off	Off	Off	Engine Data 2
Fuel Trim Cell	Cell #	16	Varies	Engine Data 2
Fuel Trim Learn	Disable/Enable	Enabled	Enabled	Engine Data 2
HI Resolution Signal	Active/Inactive	Active	Active	Engine Data 2
HO2S-11 (B1 S1)	0-1132 mv	10-1000	10-1000	Engine Data 2
HO2S-12 (B1 S2)	0-1132 mv	10-1000	10-1000	Engine Data 2
HO2S-21 (B2 S1)	0-1132 mv	10-1000	10-1000	Engine Data 2
HO2S-22 (B2 S2)	0-1132 mv	10-1000	10-1000	Engine Data 2

1996-97 Camaro 5.7L V8 MFI VIN P (All) - PID Data List

Parameter Identification (PID)	PID Value Range	PID Value at Hot Idle	PID Value at 30 mph	Data List Type
IAC Learned	0-255 counts	10-50	---	Engine Data 2
IAC Position	0-255 counts	10-50	---	Engine Data 2
IAT Sensor (°F)	-40 to 304°F	50-194	50-194	Engine Data 2
Ignition 1 Signal	0.0-25.5v	14.2	14.2	Engine Data 2
Knock Activity	No / Yes	No	No	Engine Data 2
Knock Retard (°)	0-128°	0-16	Varies	Engine Data 2
LO Resolution Signal	0-332.7ms	Varies	Varies	Engine Data 2
Long Term F/T B1	-8% to 5%	Varies	Varies	Engine Data 2
Long Term F/T B2	-8% to 5%	Varies	Varies	Engine Data 2
Long Term F/T Average	-13 to 22%	0 (± 5%)	0 (± 5%)	Engine Data 2
Loop Status	Closed Loop / Open Loop	Closed Loop	Closed Loop	Engine Data 2
MAF Sensor (g/sec)	0-512 g/sec	5-9	Varies	Engine Data 2
MAP Sensor (kPa)	10-110 kPa	20-48	Varies	Engine Data 2
MAP Sensor (V)	0-5.0v	1-2	Varies	Engine Data 2
Miles Since DTC Cleared	MPH / KPH	Varies	Varies	Engine Data 2
MPH / KPH	0-155	0	30	Engine Data 2
Pass Key® Fuel	Disabled/Okay	Okay	Okay	Engine Data 2
PCM Reset	Yes / No	No	No	Engine Data 2
Short Term F/T B1-B2	-10% to 10%	0 (± 5%)	0 (± 5%)	Engine Data 2
Spark Advance (°)	-64° to 64°	Varies	Varies	Engine Data 2
Startup ECT Sensor	-40 to 304°F	Varies	Varies	Engine Data 2
TP Angle (%)	0-100%	0%	0-100%	Engine Data 2
TP Sensor (V)	0-5.0v	0.55-0.90	---	Engine Data 2
Traction Control	Active/Inactive	Inactive	Inactive	Engine Data 2
Warmup Cycles w/o an Emission Fault	0-40	Varies	Varies	Engine Data 2
Warmup Cycles w/o a Non-Emission Fault	0-40	Varies	Varies	Engine Data 2

1996-97 Camaro 5.7L V8 MFI VIN P (All) - PID Data List

Parameter Identification (PID)	PID Value Range	PID Value at Hot Idle	PID Value at 30 mph	Data List Type
Air Pump	On / Off	On: cold	Off	Fuel Trim
BARO Sensor (kPa)	10-110 kPa	65-104	65-104	Fuel Trim
BARO Sensor (V)	0-5.0v	3.5-4.5	3.5-4.5	Fuel Trim
ECT Sensor (°F)	-40 to 304°F	185-226	185-226	Fuel Trim
EGR Duty Cycle	0-100%	0	0-100	Fuel Trim
Engine Run Time	Hr: Min: Sec	00:00:00	00:00:00	Fuel Trim
Engine Speed	0-9999 rpm	Varies	Varies	Fuel Trim
EVAP Purge %	0-100%	0-100%	0-100%	Fuel Trim
EVAP Vacuum Switch	No/Purge	No Purge	No Purge	Fuel Trim
Fuel Trim Cell	Cell # 1-17	16-17	16-17	Fuel Trim
Fuel Trim Learn	Disable/Enable	Enabled	Enabled	Fuel Trim
HO2S-11 (B1 S1)	0-1132 mv	10-1000	10-1000	Fuel Trim
HO2S-12 (B1 S2)	0-1132 mv	10-1000	10-1000	Fuel Trim
HO2S-21 (B2 S1)	0-1132 mv	10-1000	10-1000	Fuel Trim
HO2S-22 (B2 S2)	0-1132 mv	10-1000	10-1000	Fuel Trim

1996-97 Camaro 5.7L V8 MFI VIN P (All) - PID Data List

Parameter Identification (PID)	PID Value Range	PID Value at Hot Idle	PID Value at 30 mph	Data List Type
IAC Learned	0-255 counts	10-50	10-50	Fuel Trim
IAC Position	0-255 counts	10-50	10-50	Fuel Trim
IAT Sensor (°F)	-40 to 304°F	50-194	50-194	Fuel Trim
Ignition 1 Signal	0.0-25.5v	14.2	14.2	Fuel Trim
Injector Pulsewidth B1, B2	0-985 ms	1-4	Varies	Fuel Trim
Long Term F/T B1, B2	-8% to 5%	0 (± 5%)	0 (± 5%)	Fuel Trim
Long Term F/T Average	-13 to 22%	Varies	Varies	Fuel Trim
Loop Status	Closed Loop / Open Loop	Closed Loop	Closed Loop	Fuel Trim
MAF Sensor (g/sec)	0-512 g/sec	5-9	Varies	Fuel Trim
MAP Sensor (kPa)	10-110 kPa	20-48	Varies	Fuel Trim
MAP Sensor (V)	0-5.0v	1-2	Varies	Fuel Trim
MPH / KPH	0-155	0	30	Fuel Trim
Short Term F/T B1-B2	-10% to 10%	0 (± 5%)	0 (± 5%)	Fuel Trim
Short Term F/T Average	-10% to 10%	0 (± 5%)	0 (± 5%)	Fuel Trim
Spark Advance (°)	-64° to 64°	Varies	Varies	Fuel Trim
TP Angle (%)	0-100%	0%	0-100%	Fuel Trim

1996-97 Camaro 5.7L V8 VIN P (All) - PID Data List

Parameter Identification (PID)	PID Value Range	PID Value at Hot Idle	PID Value at 30 mph	Data List Type
Desired Idle Speed	0-3187	PCM control	Varies	F/F, F/R
DTC # (P0001-P1)	Number	00000	00000	F/F, F/R
ECT Sensor (°F)	-40 to 304°F	185-226	185-226	F/F, F/R
Emission Failure	0-255 counts	0	0	F/F, F/R
Engine Load	0-100%	3%	Varies	F/F, F/R
Engine Speed	0-9999 rpm	Varies	Varies	F/F, F/R
Failure Counter	0-255 counts	0	0	F/F, F/R
FC Relay 1	On / Off	On: If hot	Off	F/F, F/R
FC Relay 2, 3	On / Off	Off	Off	F/F, F/R
Fuel Trim Cell	Cell #	16	Varies	F/F, F/R
Fuel Trim Learn	Disable/Enable	Enabled	Enabled	F/F, F/R
HO2S-11 (B1 S1)	0-1132 mv	10-1000	10-1000	F/F, F/R
HO2S-12 (B1 S2)	0-1132 mv	10-1000	10-1000	F/F, F/R
HO2S-21 (B2 S1)	0-1132 mv	10-1000	10-1000	F/F, F/R
HO2S-22 (B2 S2)	0-1132 mv	10-1000	10-1000	F/F, F/R
IAT Sensor (°F)	-40 to 304°F	50-194	50-194	F/F, F/R
Long Term F/T B1, B2	-8% to 5%	0 (± 5%)	0 (± 5%)	F/F, F/R
MAF Sensor (g/sec)	0-512 g/sec	5-9	Varies	F/F, F/R
Miles Since DTC Cleared	0-10,000	Varies	Varies	F/F, F/R
Miles Since First Fault	0-10,000	Varies	Varies	F/F, F/R
Miles Since MIL Req.	0-10,000	Varies	Varies	F/F, F/R
MPH / KPH	0-155	0	30	F/F, F/R
Not Run Counter	0-255 counts	0	0	F/F, F/R
Pass Counter	0-255 counts	0	0	F/F, F/R
Short Term F/T B1-B2	-10% to 10%	0 (± 5%)	0 (± 5%)	F/F, F/R
TP Angle (%)	0-100%	0	0-100	F/F, F/R

1996-97 Camaro 5.7L V8 MFI VIN P (All) - PID Data List

Parameter Identification (PID)	PID Value Range	PID Value at Hot Idle	PID Value at 30 mph	Data List Type
A/C Request	Yes / No	No	No	Misfire
A/C Status	On / Off	Off	Off	Misfire
CKP Engine Speed	0-9999 rpm	Varies	Varies	Misfire
CKP LO Resolution Angle	-30 to +60	0 Deviation	Varies	Misfire
Cycles of Misfire Data	0-100 count	Varies	Varies	Misfire
Cylinder Mode Misfire Index	0-65535	Varies	Varies	Misfire
Desired Idle Speed	0-3187	PCM control	Varies	Misfire
ECT Sensor (°F)	-40 to 304°F	185-226	185-226	Misfire
EGR Duty Cycle	0-100%	0	0-100	Misfire
Engine Load	0-100%	2-5	Varies	Misfire
Engine Speed	0-9999 rpm	Varies	Varies	Misfire
IAT Sensor (°F)	-40 to 304°F	50-194	50-194	Misfire
MAF Sensor (g/sec)	0-512 g/sec	5-9	Varies	Misfire
Misfire Current Cylinder 1-8	0-65535	Varies	Varies	Misfire
Misfire Failures	0-65535	0	0	Misfire
Misfiring History Cylinder 1-8	0-65535	0	0	Misfire
Misfire Passes	0-65535	0	0	Misfire
Misfire Per Cycle Status	0-255 counts	Varies	Varies	Misfire
Misfire Revolution Status	Accept/REJ	ACCEPT	ACCEPT	Misfire
Revolutions within Misfire	0-65535	Varies	Varies	Misfire
TP Angle (%)	0-100%	0	Varies	Misfire
Total Misfire	0-255 counts	0	0	Misfire
Vehicle Speed	0-155	0	30	Misfire

1998-1999 Camaro 5.7L V8 MFI VIN G (All) - PID Data List

Parameter Identification (PID)	PID Value Range	PID Value at Hot Idle	PID Value at 30 mph	Data List Type
A/C Relay Command	Off or On	Off	Off	Engine Data 1
A/C Request	No or Yes	No	No	Engine Data 1
A/C Status	Off or On	No	Off	Engine Data 1
Actual EGR Position	0-100%	0	Varies	Engine Data 1
AIR Pump Relay	On / Off	Off	Off	Engine Data 1
AIR Solenoid	On / Off	Off	Off	Engine Data 1
BARO Sensor (kPa)	10-110 kPa	65-104	65-104	Engine Data 1
Cam Signal HI to Low	0-65535	Varies	Varies	Engine Data 1
Clutch Pedal Switch MT	DEP/REL	Released	Released	Engine Data 1
Cold Startup	Yes/No	No	No	Engine Data 1
Commanded EGR	0.0-5.0v	< 0.13	Varies	Engine Data 1
Cruise Control Active	Yes / No	No	No	Engine Data 1
Cruise Control Inhibit	Yes / No	Yes	Yes	Engine Data 1
Cruise Requested	Yes / No	No	No	Engine Data 1
Desired IAC Airflow	0-64 g/sec	Varies	Varies	Engine Data 1
Desired EGR Position	0-100%	0	0-100	Engine Data 1
Desired Idle Speed	0-3187 rpm	PCM control	Varies	Engine Data 1
DTC Set This Ignition	Yes / No	No	No	Engine Data 1
ECT Sensor (°F)	-38 to 284°F	185-221	185-221	Engine Data 1
EGR Closed Pintle Position	0.0-5.0v	Varies	Varies	Engine Data 1
EGR Pintle Position	0.0-5.0v	0.13	Varies	Engine Data 1
Engine Run Time	Hr: Sec: Min	00:00:00	00:00:00	Engine Data 1
Engine Speed	0-9999 rpm	Varies	Varies	Engine Data 1
EVAP Canister Purge	0-100%	0-25	0-50	Engine Data 1
EVAP Canister Vent	Not Venting / Venting	VENTING	VENTING	Engine Data 1
Fan Relay 1, 2-3	Off or On	Off	Off	Engine Data 1
FTP Sensor (In. H2O)	-17.5 to 7.5"	Varies	Varies	Engine Data 1
Fuel Trim Cell	Cell #1-23	16, 17 & 20	Varies	Engine Data 1
Fuel Trim Learn	Disable/Enable	Enabled	Enabled	Engine Data 1
HO2S-11 (B1 S1)	0-1132 mv	10-1000	10-1000	Engine Data 1
HO2S-12 (B1 S2)	0-1132 mv	10-1000	10-1000	Engine Data 1
HO2S-21 (B2 S1)	0-1132 mv	10-1000	10-1000	Engine Data 1
HO2S-22 (B2 S2)	0-1132 mv	10-1000	10-1000	Engine Data 1
IAC Position	0-255 counts	Varies	Varies	Engine Data 1
IAT Sensor (°F)	-38 to 284°F	50-194	50-194	Engine Data 1
Knock Retard (°)	0° to 16°	0	Varies	Engine Data 1
Long Term F/T B1, B2	-10% to 10%	0 (± 5%)	0 (± 5%)	Engine Data 1
Loop Status	Closed Loop / Open Loop	Closed Loop	Closed Loop	Engine Data 1
MAF Sensor (g/sec)	0-512 g/sec	5-9	Varies	Engine Data 1
MAP Sensor (kPa)	10-110 kPa	20-48	Varies	Engine Data 1
MIL Status	Off / On	Off	Off	Engine Data 1
PCM Reset	Yes / No	No	No	Engine Data 1
Reverse Inhibit (M/T)	Yes / No	Yes: In Reverse	No	Engine Data 1
Short Term F/T B1-B2	-10% to 10%	0 (± 5%)	0 (± 5%)	Engine Data 1
Skip Shift Lamp (M/T)	On / Off	Off	Off	Engine Data 1
Skip Shift Solenoid	DIS/Enable	Disabled	Disabled	Engine Data 1
Spark Advance (°)	-64° to 64°	Varies	Varies	Engine Data 1

1998-1999 Camaro 5.7L V8 MFI VIN G (All) - PID Data List

Parameter Identification (PID)	PID Value Range	PID Value at Hot Idle	PID Value at 30 mph	Data List Type
Startup ECT Sensor	-40 to 304°F	Varies	Varies	Engine Data 1
TCC Brake Switch	Apply/Release	Released	Released	Engine Data 1
TCC PWM Solenoid	Disable/Enable	Disabled	Enabled	Engine Data 1
TCC Enable Solenoid	Disable/Enable	Disabled	Enabled	Engine Data 1
TP Angle (%)	0-100%	0	Varies	Engine Data 1
Traction Control	Active/Inactive	Inactive	Inactive	Engine Data 1
Transmission Range	0-4	0	4	Engine Data 1
VTD Fuel Disable	Active/Inactive	Inactive	Inactive	Engine Data 1

1998-1999 Camaro 5.7L V8 MFI VIN G (All) - PID Data List

Parameter Identification (PID)	PID Value Range	PID Value at Hot Idle	PID Value at 30 mph	Data List Type
A/C Clutch Feedback	On / Off	Off	Off	Engine Data 2
A/C HI Side Pressure	0-450 psi	139-399	139-399	Engine Data 2
A/C High Side (V)	0.0-5.0v	1-4	1-4	Engine Data 2
A/C Request Signal	No or Yes	No	No	Engine Data 2
A/C Status	Off or On	Off	Off	Engine Data 2
AIR Relay Control	Off / On	Off	Off	Engine Data 2
AIR Solenoid Control	Off / On	Off	Off	Engine Data 2
Clutch Pedal Switch MT	DEP/REL	Released	Released	Engine Data 2
Cold Startup	No / Yes	No	No	Engine Data 2
Cruise Control Inhibit	Yes / No	Yes	Yes	Engine Data 2
Cruise Requested	Yes / No	No	No	Engine Data 2
Current Gear	1-2-3-4	1	3	Engine Data 2
Desired Idle Speed	0-3187 rpm	PCM control	---	Engine Data 2
DTC Set this Ignition	No or Yes	No	No	Engine Data 2
ECT Sensor (°F)	-38 to 284°F	185-221	185-221	Engine Data 2
Engine Run Time	Hr: Sec: Min	00:00:00	00:00:00	Engine Data 2
Engine Speed	0-9999 rpm	Varies	Varies	Engine Data 2
EVAP Canister Purge	0-100%	0%	0-50%	Engine Data 2
EVAP Canister Vent	No/VENT	VENT	VENT	Engine Data 2
Fan Relay 1, 2-3	Off or On	Off	Off	Engine Data 2
Fuel Gauge Control	0-100%	Varies	Varies	Engine Data 2
Fuel Level Sensor (V)	0.0-5.0v	Varies	Varies	Engine Data 2
Fuel Tank level	0-16.6 gal.	Varies	Varies	Engine Data 2
Fuel Tank level	0-100%	Varies	Varies	Engine Data 2
Fuel Tank Capacity	0-16.6 gal.	Varies	Varies	Engine Data 2
FTP Sensor (In. H2O)	-17.5 to 7.5"	Varies	Varies	Engine Data 2
FTP Sensor (V)	0.0-5.0v	Varies	Varies	Engine Data 2
Fuel Trim Learn	Disable/Enable	Enabled	Enabled	Engine Data 2
Generator L-Terminal	Active/Inactive	Active	Active	Engine Data 2
IAT Sensor (°F)	-38 to 284°F	50-194	50-194	Engine Data 2
Injector Pulsewidth	0-985 ms	1-4	Varies	Engine Data 2
Long Term F/T B1, B2	-10% to 10%	0 (± 5%)	0 (± 5%)	Engine Data 2
Loop Status	Closed Loop / Open Loop	Closed Loop	Closed Loop	Engine Data 2
Low Oil Lamp	On / Off	On: If low	Off	Engine Data 2
Low Oil Level	Yes / No	Yes: If low	No	Engine Data 2
MAF Sensor (g/sec)	0-512 g/sec	5-9	Varies	Engine Data 2

1998-1999 Camaro 5.7L V8 MFI VIN G (All) - PID Data List

Parameter Identification (PID)	PID Value Range	PID Value at Hot Idle	PID Value at 30 mph	Data List Type
MAF Sensor (Hz)	0-31,999 Hz	2100-3000	Varies	Engine Data 2
MAP Sensor (kPa)	10-110 kPa	20-48	Varies	Engine Data 2
MAP Sensor (V)	0.0-5.0v	1-2	Varies	Engine Data 2
MIL Status	On/Off	Off	Off	Engine Data 2
Mile Since DTC Clear	Km/Miles	Varies	Varies	Engine Data 2
PCM Reset	Yes / No	No	No	Engine Data 2
Park Neutral Position	PN-R-D321	PN	3	Engine Data 2
PCM/VCM in VTD Fail	DIS/Enable	Disabled	Disabled	Engine Data 2
Reverse Inhibit (M/T)	Yes / No	Yes: In Reverse	No	Engine Data 2
Short Term F/T B1-B2	-10% to 10%	Varies	Varies	Engine Data 2
Skip Shift Lamp (M/T)	On / Off	Off	Off	Engine Data 2
Skip Shift Solenoid	Disable/Enable	Disabled	Disabled	Engine Data 2
Spark Advance (°)	-64° to 64°	Varies	Varies	Engine Data 2
TCC Brake Switch	Apply/Release	Released	Released	Engine Data 2
TCC PWM Solenoid	Disable/Enable	Disabled	Enabled	Engine Data 2
TCC Enable Solenoid	Disable/Enable	Disabled	Enabled	Engine Data 2
TP Angle (%)	0-100%	0	Varies	Engine Data 2
TP Sensor (V)	0.0-5.0v	0.4-0.9	Varies	Engine Data 2
Traction Control	Active/Inactive	Inactive	Inactive	Engine Data 2
Transmission Range	0-4	1	3	Engine Data 2
VTD Auto Learn timer	Min: Sec	00:00	00:00	Engine Data 2
Warmups w/o E/Fault	0-255	0	0	Engine Data 2
Warmup w/o N/E fault	0-255	0	0	Engine Data 2

1998-1999 Camaro 5.7L V8 MFI VIN G (All) - PID Data List

Parameter Identification (PID)	PID Value Range	PID Value at Hot Idle	PID Value at 30 mph	Data List Type
A/C Relay Command	On / Off	Off	Off	EVAP
A/C Request Signal	Yes / No	No	No	EVAP
A/C Status	Off or On	Off	Off	EVAP
BARO Sensor (kPa)	10-110 kPa	65-104	65-104	EVAP
Cold Startup	Yes / No	No	No	EVAP
DTC Set This Ignition	No or Yes	No	No	EVAP
ECT Sensor (°F)	-38 to 284°F	185-221	185-221	EVAP
Engine Run Time	Hr: Sec: Min	00:00:00	00:00:00	EVAP
Engine Speed	0-9999 rpm	Varies	Varies	EVAP
EVAP Canister Purge	0-100%	0	0-50	EVAP
EVAP Canister Vent	No/VENT	VENTING	VENTING	EVAP
Fan Relay 1, 2-3	Off or On	On: Fan on	Off	EVAP
Fuel Gauge Control	0-100%	Varies	Varies	EVAP
Fuel Level Sensor (V)	0.0-5.0v	Varies	Varies	EVAP
Fuel Tank level	0-100%	Varies	Varies	EVAP
FTP Sensor (In. H2O)	-17.5 to 7.5"	Varies	Varies	EVAP
FTP Sensor (V)	0.0-5.0v	2.5v; cap off	Varies	EVAP
Fuel Tank Capacity	0-16.5 gal.	Varies	Varies	EVAP
Fuel Trim Learn	Disable/Enable	Enabled	Enabled	EVAP
IAT Sensor (°F)	-38 to 284°F	50-194	50-194	EVAP
Injector Pulsewidth	0-512 ms	1-4	Varies	EVAP

1998-1999 Camaro 5.7L V8 MFI VIN G (All) - PID Data List

Parameter Identification (PID)	PID Value Range	PID Value at Hot Idle	PID Value at 30 mph	Data List Type
Knock Retard (°)	0° to 16°	0	0	EVAP
Long Term F/T B1, B2	-10% to 10%	Varies	Varies	EVAP
MAF Sensor (g/sec)	0-512 g/sec	5-9	Varies	EVAP
MAP Sensor (kPa)	10-110 kPa	20-48	Varies	EVAP
MIL Status	On / Off	Off	Off	EVAP
PCM Reset	Yes / No	No	No	EVAP
Short Term F/T B1-B2	-10% to 10%	Varies	Varies	EVAP
Startup ECT Sensor	-40 to 304°F	Varies	Varies	EVAP
TP Angle (%)	0-100%	0	Varies	EVAP
Vehicle Speed	0-155 mph	0	30	EVAP

1998-1999 Camaro 5.7L V8 MFI VIN G (All) - PID Data List

Parameter Identification (PID)	PID Value Range	PID Value at Hot Idle	PID Value at 30 mph	Data List Type
Air Fuel Ratio	0.0-25.5:1	14.6:1	14.6:1	F/F, F/R
BARO Sensor (kPa)	10-110 kPa	65-104	65-104	F/F, F/R
Current Gear	1-2-3-4	1	3	F/F, F/R
Desired Idle Speed	0-3187 rpm	PCM control	---	F/F, F/R
ECT Sensor (°F)	-38 to 284°F	185-221	185-221	F/F, F/R
Engine Load	0-100%	2-5	Varies	F/F, F/R
Engine Run Time	Hr: Sec: Min	00:00:00	00:00:00	F/F, F/R
Engine Speed	0-9999 rpm	Varies	Varies	F/F, F/R
EVAP Canister Purge	0-100%	0	0-50	F/F, F/R
Failure Counter	0-65535	0	0	F/F, F/R
Faults Since 1st Fault	0-65535	0	0	F/F, F/R
Injector Pulsewidth	0-985 ms	1-4	Varies	F/F, F/R
Long Term F/T B1, B2	-10% to 10%	Varies	Varies	F/F, F/R
Loop Status	Closed Loop / Open Loop	Closed Loop	Closed Loop	F/F, F/R
MAF Sensor (g/sec)	0-512 g/sec	5-9	Varies	F/F, F/R
MAP Sensor (kPa)	10-110 kPa	20-48	Varies	F/F, F/R
MAP Sensor (V)	0-5.0v	1-2	Varies	F/F, F/R
Miles Since 1st Fault	0-9999 rpm9	Varies	Varies	F/F, F/R
Miles Since Last Fault	0-9999 rpm9	Varies	Varies	F/F, F/R
Miles Since MIL Req.	0-9999 rpm9	Varies	Varies	F/F, F/R
Not Run Counter	0-65535	Varies	Varies	F/F, F/R
Pass Counter	0-65535	Varies	Varies	F/F, F/R
Passes Since 1st Fail	0-65535	Varies	Varies	F/F, F/R
Short Term F/T B1-B2	-10% to 10%	Varies	Varies	F/F, F/R
Startup ECT Sensor	-40 to 304°F	Varies	Varies	F/F, F/R
TCC Brake Switch	Apply/Release	Released	Released	F/F, F/R
TCC PWM Solenoid	DIS/Enable	Disabled	Disabled	F/F, F/R
TCC Enable Solenoid	Disable/Enable	Disabled	Disabled	F/F, F/R
TP Angle (%)	0-100%	0	Varies	F/F, F/R
Traction Control	Active/Inactive	Inactive	Inactive	F/F, F/R
T/C Output Speed	0-10,000	Varies	Varies	F/F, F/R
Transmission Range	0-4	1	3	F/F, F/R
Vehicle Speed	0-155 mph	0	30	F/F, F/R
VTD Fuel Disable	Active/Inactive	Inactive	Inactive	F/F, F/R

1998-1999 Camaro 5.7L V8 MFI VIN G (All) - PID Data List

Parameter Identification (PID)	PID Value Range	PID Value at Hot Idle	PID Value at 30 mph	Data List Type
AIR Relay Control	On / Off	Off	Off	Fuel Trim
AIR Solenoid Control	On / Off	Off	Off	Fuel Trim
BARO Sensor (kPa)	10-110 kPa	65-104	65-104	Fuel Trim
Current Gear	1-2-3-4	1	3	Fuel Trim
DTC Set This Ignition	No or Yes	No	No	Fuel Trim
ECT Sensor (°F)	-38 to 284°F	185-221	185-221	Fuel Trim
Engine Load	0-100%	0-9%	Varies	Fuel Trim
Engine Speed	0-9999 rpm	Varies	---	Fuel Trim
EVAP Purge Solenoid PWM Command	0-100%	0-25	0-50	Fuel Trim
EVAP Canister Vent Command	Venting/Not	Not Venting	Not Venting	Fuel Trim
Fuel Tank level	0-100%	Varies	Varies	Fuel Trim
Fuel Tank Capacity	0-16.6 gal.	Varies	Varies	Fuel Trim
Fuel Trim Cell	Cell #	16-20	16-20	Fuel Trim
Fuel Trim Learn	Disable/Enable	Enabled	Enabled	Fuel Trim
F/T Diagnostic Inhibit	Disable/Enable	Disabled	Disabled	Fuel Trim
HO2S-11 (B1 S1)	0-1132 mv	10-1000	10 1000	Fuel Trim
HO2S-12 (B1 S2)	0-1132 mv	10-1000	10-1000	Fuel Trim
HO2S-21 (B2 S1)	0-1132 mv	10-1000	10-1000	Fuel Trim
HO2S-22 (B2 S2)	0-1132 mv	10-1000	10-1000	Fuel Trim
IAC Position	0-255 counts	Varies	Varies	Fuel Trim
IAT Sensor (°F)	-40 to 284°F	50-194	50-194	Fuel Trim
Ignition 1 Signal	0.0-25.5v	14.2	14.2	Fuel Trim
Injector Pulsewidth B1	0-985 ms	1-4	Varies	Fuel Trim
Injector Pulsewidth B2	0-985 ms	1-4	Varies	Fuel Trim
Knock Retard (°)	0° to 16°°	0	0	Fuel Trim
Long Term F/T B1	-10% to 10%	0 (± 5)	0 (± 5)	Fuel Trim
Long Term F/T B2	-10% to 10%	0 (± 5)	0 (± 5)	Fuel Trim
Long Term F/T Average Bank 1 & 2	-10% to 10%	0 (± 5)	0 (± 5)	Fuel Trim
Loop Status	Closed Loop / Open Loop	Closed Loop	Closed Loop	Fuel Trim
MAF Sensor (g/sec)	0-512 g/sec	5-9	Varies	Fuel Trim
MAP Sensor (kPa)	10-110 kPa	20-48	Varies	Fuel Trim
MAP Sensor (V)	0-5.0v	1-2	Varies	Fuel Trim
PCM Reset	No / Yes	No	No	Fuel Trim
Short Term Fuel Trim Bank 1	-10% to 10%	0 (± 5)	0 (± 5)	Fuel Trim
Short Term Fuel Trim Bank 2	-10% to 10%	0 (± 5)	0 (± 5)	Fuel Trim
Short Term F/T Average Bank 1 & 2	-10% to 10%	0 (± 5)	0 (± 5)	Fuel Trim
Spark Advance (°)	-64° to 64°	Varies	Varies	Fuel Trim
Startup ECT Sensor	-40 to 304°F	Varies	Varies	Fuel Trim
TFT Sensor (°F)	-40 to 304°F	Varies	Varies	Fuel Trim
TP Angle (%)	0-100%	0	Varies	Fuel Trim
Transmission Range	0-4	1	3	Fuel Trim
Vehicle Speed	0-155 mph	0	30	Fuel Trim

1998-1999 Camaro 5.7L V8 MFI VIN G (All) - PID Data List

Parameter Identification (PID)	PID Value Range	PID Value at Hot Idle	PID Value at 30 mph	Data List Type
A/C Relay Command	On / Off	Off	Off	Misfire
A/C Request Signal	Yes / No	No	No	Misfire
A/C Status	On / Off	Off	Off	Misfire
Clutch Pedal Switch (MT)	DEP/REL	Released	Released	Misfire
Cycles of Misfire Data	0-100	Varies	Varies	Misfire
Cylinder Mode M/F Index	0-65535	Varies	Varies	Misfire
ECT Sensor (°F)	-38 to 284°F	185-221	185-221	Misfire
Engine Load	0-100%	0-9	Varies	Misfire
Engine Speed	0-9999 rpm	Varies	Varies	Misfire
IC Circuit Cylinder 1-8	Fault/Okay	Okay	Okay	Misfire
Loop Status	Closed Loop / Open Loop	Closed Loop	Closed Loop	Misfire
MAF Sensor (g/sec)	0-512 g/sec	5-9	Varies	Misfire
MAP Sensor (kPa)	10-110 kPa	20-48	Varies	Misfire
MAP Sensor (V)	0-5.0v	1-2	Varies	Misfire
Misfire Current Cylinder 1-8	0-200	0	0	Misfire
Misfire History Cylinder 1-8	0-65535	0	0	Misfire
Misfiring Cylinder 1-8	CYL 1-8	0	0	Misfire
Misfire Revolution Status	Accept/REJ (not in range)	Accept	Accept	Misfire
PCM Reset	Yes/No	No	No	Misfire
Revolutions With Misfire Present	0-65535	Varies	Varies	Misfire
Spark Advance (°)	-64° to 64°	Varies	Varies	Misfire
TCC Enable Solenoid	Disable/Enable	Disabled	Enabled	Misfire
TCC PWM Solenoid	0-100%	DIS/Enable	Enabled	Misfire
Total Misfire Count	0-65535	Varies	Varies	Misfire
Total Misfire Failures Since First Fail	0-65535	Varies	Varies	Misfire
Total Misfire Passes Since First Fail	0-65535	Varies	Varies	Misfire
TP Angle (%)	0-100%	0	Varies	Misfire
TP Sensor (V)	0-5.0v	0.4-0.9	Varies	Misfire
Vehicle Speed	0-155 mph	0	30	Misfire

2000-02 Camaro 5.7L V8 MFI VIN G (All) - PID Data List

Parameter Identification (PID)	PID Value Range	PID Value at Hot Idle	PID Value at 30 mph	Data List Type
A/C Clutch Feedback	On/Off	Off	Off	Engine Data 2
A/C HI Side Pressure	0-459 kPa	90	90	Engine Data 2
A/C HI Side Pressure	0-5.1 volts	0.98	0.98	Engine Data 2
A/C Relay Command	On/Off	Off	Off	Engine Data 1, 2 & 3, Misfire
A/C Request Signal	Yes/No	No	No	Engine Data 2
Air Fuel Ratio	0.0-25.5:1	14.7:1	14.7:1	Engine Data 2, Data 3
Air Pump Relay	On/Off	Off	Off	Engine Data 1, Fuel Trim
AIR Solenoid Control	On/Off	Off	Off	Engine Data 1, Fuel Trim
BARO Sensor (kPa)	10-105 kPa	65-104	Varies with changes in the Altitude	Engine Data 1, EVAP, Fuel Trim
BARO Sensor (V)	0-5.1 volts	3.5-4.9	Varies with changes in the Altitude	Engine Data 1, EVAP, Fuel Trim
Clutch Pedal Switch	Applied/Released	Released	Released	Engine Data 1-2
CMP Sensor High to Low Signal	0-65,535 counts	Varies	Varies	Engine Data 2
CMP Sensor Low to High Signal	0-65,535 counts	Varies	Varies	Engine Data 2
Cold Engine Startup	Yes/No	Yes	Yes	Engine Data 2, EVAP
Cruise Control Active	Yes/No	No	No	Engine Data 1
Cruise Control Inhibit	Yes/No	No	No	Engine Data 1
Current Gear	0, 1, 2, 3, 4	1	4	Engine Data 1, Fuel Trim
Cycles of Misfire Data	0-100 counts	0	0	Misfire
DTC Set This Ignition	Yes/No	No	No	Engine Data 1-2 EVAP, F/Trim
Desired IAC Airflow	0-64 g/sec	Varies	Varies	Engine Data 1
Desired Idle Speed	0-3187 rpm	Commanded by the PCM	Varies	Engine Data 1-2 EVAP
ECT Sensor (°F)	-40 to 304°F	190-221°F	190-221°F	Engine Data 1-3 EGR, F/Trim, EVAP, Misfire
Engine Load	0-100%	1-4	5-9	Engine Data 1-3 EGR, F/Trim, EVAP, Misfire
Engine Oil Level Switch	Okay / LOW	Okay	Okay	Engine Data 2
Engine Oil Life Remaining	0-100%	Varies	Varies	Engine Data 2

2000-02 Camaro 5.7L V8 MFI VIN G (All) - PID Data List

Parameter Identification (PID)	PID Value Range	PID Value at Hot Idle	PID Value at 30 mph	Data List Type
Engine Run Time	Hr: Min: Sec 00:00:00 to 99:99:99	00:00:00	00:00:00	Engine Data 1, Data 2-3, EGR, F/Trim, EVAP, Misfire
Engine Speed	0-10,000 rpm	± 50 rpm of actual speed	Varies	Engine Data 1, Data 2-3, EGR, F/Trim, EVAP, Misfire
EVAP Purge Solenoid Command	0-100%	Okay	Okay	Engine Data 1, EVAP, F/Trim
EVAP Vent Solenoid Command	Venting/Not	Not Venting	Not Venting	Engine Data 1, EVAP, F/Trim
Fan Control Relay 1 Command	On/Off	Off	Off	Engine Data 2
Fan Control Relay 2-3 Command	On/Off	Off	Off	Engine Data 2
Fuel Level Sensor	0-5.0 volts	Varies	Varies	EVAP
Fuel Tank Level Remaining	Gallons or Liters	Varies	Varies	EVAP
Fuel Tank Level Remaining	0-100%	Varies	Varies	EVAP
Fuel Tank Pressure Sensor	Inches H2O, mm Hg	-17 to +7.5 Inches H2O	-17 to +7.5 Inches H2O	Engine Data 1, EVAP
Fuel Tank Pressure Sensor	0-5.0 volts	Varies	Varies	EVAP
Fuel Tank Capacity	16.6 Gallons, 62.8 Liters	16.6 Gallons, 62.8 Liters	16.6 Gallons, 62.8 Liters	EVAP
Fuel Trim Cell	Cell 0-23	16, 17 or 20	16, 17 or 20	Engine Data 1, EVAP, F/Trim
Fuel Trim Learn	Enabled or Disabled	Enabled	May Toggle	Engine Data 1, EVAP, F/Trim
Generator L-Terminal	Voltage/No	Voltage	Voltage	Engine Data 2
HO2S (B1 S1) Signal	0-1000 mv	10-1000	10-1000	Engine Data 1, EVAP, F/Trim
HO2S (B2 S1) Signal	0-1000 mv	10-1000	10-1000	Engine Data 1, EVAP, F/Trim
HO2S (B1 S2) Signal	0-1000 mv	10-1000	10-1000	Engine Data 1, EVAP, F/Trim
HO2S (B2 S2) Signal	0-1000 mv	10-1000	10-1000	Engine Data 1, EVAP, F/Trim
IAC Position	0-1024 counts	10-40	Varies	Engine Data 1, Fuel Trim
IAT Sensor (°F)	-40 to 304°F	91	92	Engine Data 1-2 EGR, EVAP, Fuel Trim
Ignition 1 Signal	0.0-25.5v	13.5	13.6	Engine Data 1, Data 2, Fuel Trim, EVAP

2000-02 Camaro 3.8L V6 MFI VIN K (All) - PID Data List

Parameter Identification (PID)	PID Value Range	PID Value at Hot Idle	PID Value at 30 mph	Data List Type
Ignition Mode	Bypass/IC	IC	IC	Engine Data 1, Data 2, Fuel Trim, EVAP
Injector Pulsewidth Average Bank 1	0-1000 ms	1-4	Varies	Engine Data 2, F/Trim, M/fire
Injector Pulsewidth Average Bank 2	0-1000 ms	1-4	Varies	Engine Data 2, F/Trim, M/fire
Knock Retard (°)	0-25.5°	0.0° to 16°	Varies	Engine Data 1
Long Term Fuel Trim Bank 1	-23 to +16	0 (± 5%)	0 (± 5%)	Engine Data 1-2 EVAP, F/Trim
Long Term Fuel Trim Bank 2	-23 to +16	0 (± 5%)	0 (± 5%)	Engine Data 1-2 EVAP, F/Trim
Long Term Fuel Trim Average Bank 1	-23 to +16	0 (± 5%)	0 (± 5%)	Engine Data 1-2 EVAP, F/Trim
Long Term Fuel Trim Average Bank 2	-23 to +16	0 (± 5%)	0 (± 5%)	Engine Data 1-2 EVAP, F/Trim
Loop Status	Open / Closed	Closed	Closed	Engine Data 1-2 EVAP, F/Trim
MAF Sensor (g/sec)	0-512 g/sec	5-9	20-26	Engine Data 1-3 EGR, F/Trim, EVAP, Misfire
MAF Sensor (Hz)	0-31,999 Hz	2000-3000	Varies	Engine Data 2
MAP Sensor (kPa)	10-105 kPa	20-48	Varies	Engine Data 1-3 EGR, F/Trim, EVAP, Misfire
MAP Sensor (V)	0-5.1 volts	1.0-2.0	Varies	Engine Data 1-3 EGR, F/Trim, EVAP, Misfire
MIL Command	On/Off	Off	Off	Engine Data 2
Mileage Since DTC Cleared	0-65,535 miles	Varies	Varies	Misfire
Misfire Current Cylinder 1-8	0-200 counts	0-4	0	Misfire
Misfire History Cylinder 1-8	0-65,535	0	0	Misfire
PCM Reset	Yes/No	No	No	Engine Data 1-2 EVAP, F/Trim
PCM/VCM in VTD Fail Enable Mode	Enable / Disable	Enable	Enable	Engine Data 1
Pass Counter	0-256 counts	Varies	Varies	Misfire
Park Neutral Position	Park/Neutral or R-D321	Park/Neutral	R-D321	Engine Data 2
Power Enrichment	Active/Inactive	Inactive	Inactive	Engine Data 2, Fuel Trim
Reverse Inhibit Solenoid (M/T only)	Yes/No	No	No	Engine Data 1

2000-02 Camaro 3.8L V6 MFI VIN K (All) - PID Data List

Parameter Identification (PID)	PID Value Range	PID Value at Hot Idle	PID Value at 30 mph	Data List Type
Short Term Fuel Trim Bank 1	-10 to +10%	0 (± 5%)	0 (± 5%)	Fuel Trim
Short Term Fuel Trim Bank 2	-10 to +10%	0 (± 5%)	0 (± 5%)	Fuel Trim
Short Term Fuel Trim Average Bank 1	-10 to +10%	0 (± 5%)	0 (± 5%)	Engine Data 1-2 EVAP, F/Trim
Short Term Fuel Trim Average Bank 2	-10 to +10%	0 (± 5%)	0 (± 5%)	Engine Data 1-2 EVAP, F/Trim
Skip Shift Solenoid (M/T only)	Enabled or Disabled	Disabled	Disabled	Engine Data 1
Skip Shift Lamp (M/T)	On/Off	Off	Off	Engine Data 1
Spark Advance	-64° to +64°	-20°	Varies	Engine Data 1-2 F/Trim, M/fire
Startup ECT	-40 to 419°F	Varies	Varies	Engine Data 2, EVAP, F/Trim
Startup IAT	-40 to 419°F	80°F	80°F	Engine Data 2, EVAP, F/Trim
TCC/Cruise Brake Pedal Switch	Applied or Released	Released	Released	Engine Data 1, Data 2
TCC Enable Solenoid Command	Enabled or Disabled	Disabled	Disabled	Engine Data 2
TCC PWM Solenoid Command	Enabled or Disabled	Disabled	Disabled	Engine Data 1-2 Misfire
TP Sensor 1	0-100%	0-6	Varies	Engine Data 1-3 EGR, F/Trim, EVAP, Misfire
TP Sensor 1	0-5.0 volts	0.4-0.9	Varies	Engine Data 1-3 EGR, F/Trim, EVAP, Misfire
Traction Control Status	Active/Inactive	Inactive	Inactive	Engine Data 2
Transaxle Range Switch	0, 1, 2, 3, 4	1	4	Fuel Trim
Vehicle Speed	Km/h, MPH	0	30	Engine Data 1-3 EGR, F/Trim, EVAP, Misfire
VTD Auto Learn Timer	Hr: Min: Sec 00:00:00 to 99:99:99	00:00:00	00:00:00	Engine Data 1
VTD Fuel Disable	Active/Inactive	Inactive	Inactive	Engine Data 1
VTD Fuel Disable Until Ignition Off	Yes/No	No	No	Engine Data 1
Warmups Without an Emission Fault	0-255 counts	Varies	Varies	Engine Data 2
Warmups Without a Non-Emission Fault	0-255 counts	Varies	Varies	Engine Data 2

CAPRICE PID DATA

1996 Caprice 4.3L V8 MFI VIN W (A/T) - PID Data List

Parameter Identification (PID)	PID Value Range	PID Value at Hot Idle	PID Value at 30 mph	Data List Type
A/C Clutch	Off or On	Off	Off	Engine Data 1
A/C HI Side Pressure	0 to 425 psi	139-399	139-399	Engine Data 1
A/C Request	No or Yes	No	No	Engine Data 1
A/C Status	Off or On	Off	Off	Engine Data 1
AIR Control Pump	Disable/Enable	Disabled	Disabled	Engine Data 1
BARO Sensor (kPa)	10-110 kPa	10-105	10-105	Engine Data 1
Cold Engine Startup	No or Yes	No	No	Engine Data 1
Current Gear	1-2-3-4	1	3	Engine Data 1
Desired Idle Speed	0-3187 rpm	PCM control	---	Engine Data 1
DTC Set This Ignition	Yes/No	No	No	Engine Data 1
ECT Sensor (°F)	-40 to 304°F	185-226	185-226	Engine Data 1
EGR Duty Cycle	0-100%	0%	0-100%	Engine Data 1
Engine Run Time	Hr: Min: Sec	00:00:00	00:00:00	Engine Data 1
Engine Speed	0-9999 rpm	Varies	---	Engine Data 1
EVAP Purge Control	0-100%	0-15	0-50	Engine Data 1
EVAP Vacuum Switch	No/Purge	No Purge	Purge	Engine Data 1
FC Relay 1, Relay 2	On/Off	Off	Off	Engine Data 1
Fuel Trim Cell	Cell #	16	Varies	Engine Data 1
Fuel Trim Learn	Disable/Enable	Enabled	Enabled	Engine Data 1
HI Resolution Signal	Active/Inactive	Active	Active	Engine Data 1
HO2S (B1 S1, B1 S2)	0-1132 mv	10-1000	10-1000	Engine Data 1
HO2S (B2 S1, B2 S2)	0-1132 mv	10-1000	10-1000	Engine Data 1
IAC Learned Position	0-255 counts	10-50	10-50	Engine Data 1
IAC Position	0-255 counts	10-50	10-50	Engine Data 1
IAT Sensor (°F)	-40 to 304°F	50-194	50-194	Engine Data 1
Ignition 1 Signal	0.0-25.5v	14.2	14.2	Engine Data 1
Injector Pulsewidth	0-1000 ms	1-4	Varies	Engine Data 1
Knock Retard (°)	0.0 - 25.5°	0	Varies	Engine Data 1
Long Term F/T B1, B2	-10% to 10%	0 (± 5%)	0 (± 5%)	Engine Data 1
Loop Status	Closed Loop / Open Loop	Closed Loop	Closed Loop	Engine Data 1
LO Resolution Signal	0-332.7 ms	Varies	Varies	Engine Data 1
MAF Sensor (g/sec)	0-512 g/sec	5-9	Varies	Engine Data 1
MAP Sensor (kPa)	10-110 kPa	20-48	Varies	Engine Data 1
OIL Level	Not Okay/Okay	Okay	Okay	Engine Data 1
Pass Key® Fuel	Disabled/Okay	Okay	Okay	Engine Data 1
PCM Reset	No or Yes	No	No	Engine Data 1
PNP Switch	P-R-0-3-2-1	P-R	3	Engine Data 1
PSP Switch	HI/Normal	Normal	Normal	Engine Data 1
Short Term F/T B1-B2	-10% to 10%	0 (± 5%)	0 (± 5%)	Engine Data 1
Spark Advance (°)	-64° to 64°	-20	Varies	Engine Data 1
Startup ECT Sensor	-40 to 304°F	Varies	Varies	Engine Data 1
TCC Brake Switch	Applied/Released	Released	Released	Engine Data 1
TCC Duty Cycle	Disable/Enable	Enabled	Enabled	Engine Data 1
TCC Enable	Disable/Enable	Disabled	Disabled	Engine Data 1
TCC Stator Temp.	HI/Normal	Normal	Normal	Normal
TP Angle (%)	0-100%	0	Varies	Varies
TP Sensor (V)	0-5.0v	0.55-0.90	Varies	Varies
TR Switch	P/N/Drive	Park	Drive	Drive

1996 Caprice 4.3L V8 MFI VIN W (A/T) - PID Data List

Parameter Identification (PID)	PID Value Range	PID Value at Hot Idle	PID Value at 30 mph	Data List Type
A/C Clutch	Off or On	Off	Off	Engine Data 2
AIR Control Pump	Enabled/DIS	Disabled	Disabled	Engine Data 2
BARO Sensor (kPa)	10-110 kPa	Varies	Varies	Engine Data 2
BARO Sensor (V)	0-5.0v	3.5-4.5	3.5-4.5	Engine Data 2
CKP Engine Speed	0-9999 rpm	Varies	----	Engine Data 2
CKP Low Resolution Signal Angle	-30 to +60	0 Deviation	0 Deviation	Engine Data 2
Cold Engine Startup	No or Yes	No	No	Engine Data 2
Desired Idle Speed	0-3187 rpm	PCM control	----	Engine Data 2
DTC Set this Ignition	No or Yes	No	No	Engine Data 2
ECT Sensor (°F)	-40 to 304°F	185-226	185-226	Engine Data 2
EGR Duty Cycle	0-100%	0%	0-100%	Engine Data 2
Engine Run Time	Hr: Min: Sec	00:00:00	00:00:00	Engine Data 2
Engine Speed	0-9999 rpm	Varies	----	Engine Data 2
EVAP Purge	0-100%	0%	0-50	Engine Data 2
EVAP Vacuum Switch	No/Purge	No Purge	Purge	Engine Data 2
Fan Control Relay 1	Off or On	Off	Off	Engine Data 2
Fan Control Relay 2	Off or On	Off	Off	Engine Data 2
Fuel Trim Cell	Cell #	16	Varies	Engine Data 2
Fuel Trim Learn	Disable/Enable	Enabled	Enabled	Engine Data 2
HI Resolution Signal	Active/Inactive	Active	Active	Engine Data 2
HO2S (B1 S1, B1 S2)	0-1132 mv	10-1000	10-1000	Engine Data 2
HO2S (B2 S1, B2 S2)	0-1132 mv	10-1000	10-1000	Engine Data 2
IAC Learned	0-255 counts	10-50	----	Engine Data 2
IAC Position	0-255 counts	10-50	----	Engine Data 2
IAT Sensor (°F)	-40 to 304°F	50-194	50-194	Engine Data 2
Ignition 1 Signal	0.0-25.5v	14.2	14.2	Engine Data 2
Knock Retard (°)	0.0 - 25.5°	0	0	Engine Data 2
Knock Activity	Yes / No	No	No	Engine Data 2
Long Term F/T B1, B2	-10% to 10%	0 (± 5%)	0 (± 5%)	Engine Data 2
Loop Status	Closed Loop / Open Loop	Closed Loop	Closed Loop	Engine Data 2
LO Resolution Signal	0-332.7 ms	Varies	Varies	Engine Data 2
MAF Sensor (g/sec)	0-512 g/sec	5-9	Varies	Engine Data 2
MAP Sensor (kPa)	10-110 kPa	20-48	Varies	Engine Data 2
MAP Sensor (V)	0-5.0v	1-2	Varies	Engine Data 2
MPH or KPH	0-155 mph	0	30	Engine Data 2
Non-Emission Cycle Count	0-255 counts	0	0	Engine Data 2
Pass Key® Fuel	Disabled/Okay	Okay	Okay	Engine Data 2
PCM Reset	No or Yes	No	No	Engine Data 2
Short Term F/T B1	-10% to 10%	0 (± 5%)	0 (± 5%)	Engine Data 2
Short Term F/T B2	-10% to 10%	0 (± 5%)	0 (± 5%)	Engine Data 2
Spark Advance (°)	-64° to 64°	-20	Varies	Engine Data 2
Startup ECT Sensor	-40 to 304°F	Varies	Varies	Engine Data 2
TP Angle (%)	0-100%	0	Varies	Engine Data 2
TP Sensor (V)	0-5.0v	0.55-0.90	Varies	Engine Data 2
Vehicle Speed	0-155 mph	0	30	Engine Data 2

1996 Caprice 4.3L V8 MFI VIN W (A/T) - PID Data List

Parameter Identification (PID)	PID Value Range	PID Value at Hot Idle	PID Value at 30 mph	Data List Type
AIR Control Pump	Enabled/DIS	Disabled	Disabled	Fuel Trim
Desired Idle Speed	0-3187 rpm	PCM control	---	Fuel Trim
ECT Sensor (°F)	-40 to 304°F	185-226	185-226	Fuel Trim
EGR Duty Cycle	0-100%	0%	0-100%	Fuel Trim
Engine Speed	0-9999 rpm	Varies	---	Fuel Trim
EVAP Purge	0-100%	0%	0-50	Fuel Trim
EVAP Vacuum Switch	No/Purge	No Purge	Purge	Fuel Trim
Fuel Trim Cell	Cell #	16	Varies	Fuel Trim
Fuel Trim Learn	Disable/Enable	Enabled	Enabled	Fuel Trim
HO2S (B1 S1)	0-1132 mv	10-1000	10-1000	Fuel Trim
HO2S (B1 S2)	0-1132 mv	10-1000	10-1000	Fuel Trim
HO2S (B2 S1)	0-1132 mv	10-1000	10-1000	Fuel Trim
HO2S (B2 S2)	0-1132 mv	10-1000	10-1000	Fuel Trim
IAC Learned	0-255 counts	10-50	10-50	Fuel Trim
IAC Position	0-255 counts	10-50	10-50	Fuel Trim
IAT Sensor (°F)	-40 to 304°F	50-194	50-194	Fuel Trim
Ignition 1 Signal	0.0-25.5v	14.2	14.2	Fuel Trim
Injector Pulsewidth Bank 1	0-1000 ms	1-4	Varies	Fuel Trim
Injector Pulsewidth Bank 2	0-1000 ms	1-4	Varies	Fuel Trim
Long Term Fuel Trim Bank 1	-10% to 10%	0 (± 5%)	0 (± 5%)	Fuel Trim
Long Term Fuel Trim Bank 2	-10% to 10%	0 (± 5%)	0 (± 5%)	Fuel Trim
Loop Status	Closed Loop / Open Loop	Closed Loop	Closed Loop	Fuel Trim
Long Term Fuel Trim Average Bank 1	-13 to 21%	0 (± 5%)	0 (± 5%)	Fuel Trim
Long Term Fuel Trim Average Bank 2	-13 to 21%	0 (± 5%)	0 (± 5%)	Fuel Trim
MAF Sensor (g/sec)	0-512 g/sec	5-9	Varies	Fuel Trim
MAP Sensor (kPa)	10-110 kPa	20-48	Varies	Fuel Trim
MAP Sensor (V)	0-5.0v	1-2	Varies	Fuel Trim
MPH or KPH	0-155 mph	0	30	Fuel Trim
Short Term Fuel Trim Bank 1	-10% to 10%	Varies	Varies	Fuel Trim
Short Term Fuel Trim Bank 2	-10% to 10%	Varies	Varies	Fuel Trim
Short Term Fuel Trim Average Bank 1	-9% to 7%	0 (± 5%)	0 (± 5%)	Fuel Trim
Short Term Fuel Trim Average Bank 2	-9% to 7%	0 (± 5%)	0 (± 5%)	Fuel Trim
Spark Advance (°)	-64° to 64°	-20	Varies	Fuel Trim
TP Angle (%)	0-100%	0%	0-100%	Fuel Trim
Vehicle Speed	0-155 mph	0	30	Fuel Trim

1996 Caprice 4.3L V8 MFI VIN W (A/T) - PID Data List

Parameter Identification (PID)	PID Value Range	PID Value at Hot Idle	PID Value at 30 mph	Data List Type
Desired Idle Speed	0-3187 rpm	PCM control	Varies	F/F, F/R
DTC NUMBER	Pxxxx	Pxxxx	Pxxxx	F/F, F/R
ECT Sensor (°F)	-40 to 304°F	185-226	185-226	F/F, F/R
Emission Fault Count	0-255 counts	Varies	Varies	F/F, F/R
Engine Load	0-100%	22-26	Varies	F/F, F/R
Engine Speed	0-9999 rpm	Varies	---	F/F, F/R
Fail Counter	0-255 counts	Varies	Varies	F/F, F/R
Fuel Trim Cell	Cell #	16	Varies	F/F, F/R
Fuel Trim Learn	Disable/Enable	Enabled	Enabled	F/F, F/R
HO2S (B1 S1, B1 S2)	0-1132 mv	10-1000	10-1000	F/F, F/R
HO2S (B2 S1, B2 S2)	0-1132 mv	10-1000	10-1000	F/F, F/R
IAT Sensor (°F)	-40 to 304°F	50-194	50-194	F/F, F/R
Long Term F/T B1, B2	-10% to 10%	Varies	Varies	F/F, F/R
MAF Sensor (g/sec)	0-512 g/sec	5-9	Varies	F/F, F/R
Miles Since DTC Cleared	0-65655	Varies	Varies	F/F, F/R
Miles Since 1st Fail	0-65535	Varies	Varies	F/F, F/R
Miles Since MIL Req.	0-65535	Varies	Varies	F/F, F/R
MPH / KPH	0-155 mph	0	30	F/F, F/R
No Run Counter	0-255 counts	0	0	F/F, F/R
Pass Counter	0-255 counts	Varies	Varies	F/F, F/R
Short Term F/T B1-B2	-10% to 10%	0	Varies	F/F, F/R
TP Angle (%)	0-100%	0	Varies	F/F, F/R

1996 Caprice 4.3L V8 MFI VIN W (A/T) - PID Data List

Parameter Identification (PID)	PID Value Range	PID Value at Hot Idle	PID Value at 30 mph	Data List Type
A/C Request	No or Yes	No	No	Misfire
A/C Status	Off or On	Off	Off	Misfire
CKP Engine Speed	0-9999 rpm	Varies	Varies	Misfire
CKP LO Resolution Angle	-30 to +60	0 Deviation	0 Deviation	Misfire
Cycles of Misfire Data	0-100	0	0	Misfire
CYL Mode Misfire	0-65535	0	0	Misfire
Desired Idle Speed	0-3187 rpm	PCM control	---	Misfire
ECT Sensor (°F)	-40 to 304°F	185-226	185-226	Misfire
Engine Load	0-100%	22-26	Varies	Misfire
Engine Speed	0-9999 rpm	Varies	---	Misfire
IAT Sensor (°F)	-40 to 304°F	50-194	50-194	Misfire
MAF Sensor (g/sec)	0-512 g/sec	5-9	Varies	Misfire
Misfire current Cyl 1-8	0-255 counts	0	0	Misfire
Misfire History Cyl 1-8	0-65535	0	0	Misfire
Misfire Per Cycle	PRI # 1-8	Varies	Varies	Misfire
Misfire Per Cycle	SEC # 1-8	Varies	Varies	Misfire
Misfire Rough Road	No or Yes	No	No	Misfire
Total Misfire Counts	0-255 counts	0	0	Misfire
Total Misfire Per Test	0-255 counts	0	0	Misfire
TP Angle (%)	0-100%	0%	0-100%	Misfire

1996 Caprice 5.7L V8 MFI VIN P (A/T) - PID Data List

Parameter Identification (PID)	PID Value Range	PID Value at Hot Idle	PID Value at 30 mph	Data List Type
A/C Clutch	Off or On	Off	Off	Engine Data 1
A/C HI Side Pressure	0 to 450 psi	199-399	199-399	Engine Data 1
A/C Request	No or Yes	No	No	Engine Data 1
A/C Status	Off or On	Off	Off	Engine Data 1
AIR Control Pump	Enabled/DIS	Disabled	Disabled	Engine Data 1
BARO Sensor (kPa)	10-110 kPa	10-105	10-105	Engine Data 1
Cold Engine Startup	No or Yes	No	No	Engine Data 1
Current Gear	1-2-3-4	1	3	Engine Data 1
Desired Idle Speed	0-3187 rpm	PCM control	---	Engine Data 1
DTC Set This Ignition	No or Yes	No	No	Engine Data 1
ECT Sensor (°F)	-40 to 304°F	185-226	185-226	Engine Data 1
EGR Duty Cycle	0-100%	0%	0-100%	Engine Data 1
Engine Run Time	Hr: Min: Sec	00:00:00	00:00:00	Engine Data 1
Engine Speed	0-9999 rpm	Varies	---	Engine Data 1
EVAP Purge	0-100%	0%	0-50	Engine Data 1
EVAP Vacuum Switch	No/Purge	No Purge	Purge	Engine Data 1
FC Relay 1, 2	Off or On	Off	Off	Engine Data 1
Fuel Trim Cell	Cell #	16	Varies	Engine Data 1
Fuel Trim Learn	Disable/Enable	Enabled	Enabled	Engine Data 1
HI Resolution Signal	Active/Inactive	Active	Active	Engine Data 1
HO2S (B1 S1, B1 S2)	0-1132 mv	10-1000	10-1000	Engine Data 1
HO2S (B2 S1, B2 S2)	0-1132 mv	10-1000	10-1000	Engine Data 1
IAC Learned	0-255 counts	10-50	---	Engine Data 1
IAC Position	0-255 counts	10-50	---	Engine Data 1
IAT Sensor (°F)	-40 to 304°F	50-194	50-194	Engine Data 1
Ignition 1 Signal	0.0-25.5v	14.2	14.2	Engine Data 1
Injector Pulsewidth	0-1000 ms	1-4	Varies	Engine Data 1
Knock Retard (°)	0.0 - 25.5°	0	0	Engine Data 1
Knock Activity	Yes / No	No	No	Engine Data 1
Long Term F/T B1, B2	-10% to 10%	0	Varies	Engine Data 1
LO Resolution Signal	0-332.7 ms	Varies	Varies	Engine Data 1
MAF Sensor (g/sec)	0-512 g/sec	5-9	Varies	Engine Data 1
MAP Sensor (kPa)	10-110 kPa	20-48	Varies	Engine Data 1
OIL Level	Not Okay/Okay	Okay	Okay	Engine Data 1
Pass Key® Fuel	Disabled/Okay	Okay	Okay	Engine Data 1
PCM Reset	No or Yes	No	No	Engine Data 1
PNP Switch	P-R-0-3-2-1	P-R	3	Engine Data 1
PSP Switch	HI/Normal	Normal	Normal	Engine Data 1
Short Term F/T B1-B2	-10% to 10%	Varies	Varies	Engine Data 1
Spark Advance (°)	-64° to 64°	20	Varies	Engine Data 1
Startup ECT Sensor	-40 to 304°F	Varies	Varies	Engine Data 1
TCC Brake Switch	Apply/Release	Released	Released	Engine Data 1
TCC Duty Cycle	0-100%	0%	0%	Engine Data 1
TCC Enable	Disable/Enable	Disabled	Disabled	Engine Data 1
TCC Stator Temp.	HI/Normal	Normal	Normal	Engine Data 1
TP Angle (%)	0-100%	0%	0-100%	Engine Data 1
TP Sensor	0-5.0v	0.55-0.90	Varies	Engine Data 1
TR Switch	P/N/Drive	Park	Drive	Engine Data 1

1996 Caprice 5.7L V8 MFI VIN P (A/T) - PID Data List

Parameter Identification (PID)	PID Value Range	PID Value at Hot Idle	PID Value at 30 mph	Data List Type
A/C Clutch	Off or On	Off	Off	Engine Data 2
AIR Control Pump	Enabled/DIS	Disabled	Disabled	Engine Data 2
BARO Sensor (kPa)	10-110 kPa	Varies	Varies	Engine Data 2
BARO Sensor (V)	0-5.0v	3.5-4.5	3.5-4.5	Engine Data 2
CKP Engine Speed	0-9999 rpm	Varies	---	Engine Data 2
CKP LO Resolution Angle	-30 to +60	0 Deviation	0 Deviation	Engine Data 2
Cold Engine Startup	No or Yes	No	No	Engine Data 2
Desired Idle Speed	0-3187 rpm	PCM control	---	Engine Data 2
DTC Set this Ignition	No or Yes	No	No	Engine Data 2
ECT Sensor (°F)	-40 to 304°F	185-226	185-226	Engine Data 2
EGR Duty Cycle	0-100%	0	0-50	Engine Data 2
Engine Run Time	Hr: Min: Sec	00:00:00	00:00:00	Engine Data 2
Engine Speed	0-9999 rpm	Varies	---	Engine Data 2
EVAP Purge PWM	0-100%	0	0-50	Engine Data 2
EVAP Vacuum Switch	No/Purge	No Purge	Purge	Engine Data 2
Fan Control Relay 1	Off or On	Off	Off	Engine Data 2
Fan Control Relay 2	Off or On	Off	Off	Engine Data 2
Fuel Trim Cell	Cell #0-22	16	0-22	Engine Data 2
Fuel Trim Learn	Disable/Enable	Enabled	Enabled	Engine Data 2
HI Resolution Signal	Active/Inactive	Active	Active	Engine Data 2
HO2S (B1 S1, B1 S2)	0-1132 mv	10-1000	10-1000	Engine Data 2
HO2S (B2 S1, B2 S2)	0-1132 mv	10-1000	10-1000	Engine Data 2
IAC Learned	0-255 counts	10-50	---	Engine Data 2
IAC Position	0-255 counts	10-50	---	Engine Data 2
IAT Sensor (°F)	-40 to 304°F	50-194	50-194	Engine Data 2
Ignition 1 Signal	0.0-25.5v	14.2	14.2	Engine Data 2
Knock Retard (°)	0.0 - 25.5°	0	0	Engine Data 2
Knock Activity	Yes / No	No	No	Engine Data 2
Long Term F/T B1	-10% to 10%	0 (± 5%)	0 (± 5%)	Engine Data 2
Long Term F/T B2	-10% to 10%	0 (± 5%)	0 (± 5%)	Engine Data 2
Loop Status	Closed Loop / Open Loop	Closed Loop	Closed Loop	Engine Data 2
LO Resolution Signal	0-332.7 ms	Varies	Varies	Engine Data 2
MAF Sensor (g/sec)	0-512 g/sec	5-9	Varies	Engine Data 2
MAP Sensor (kPa)	10-110 kPa	20-48	Varies	Engine Data 2
MAP Sensor (V)	0-5.0v	1-2	Varies	Engine Data 2
MPH or KPH	0-155 mph	0	30	Engine Data 2
Non-Emission Cycle Count	0-255 counts	0	0	Engine Data 2
Pass Key® Fuel	Disabled/Okay	Okay	Okay	Engine Data 2
PCM Reset	No or Yes	No	No	Engine Data 2
Short Term F/T B1	-10% to 10%	0 (± 5%)	0 (± 5%)	Engine Data 2
Short Term F/T B2	-10% to 10%	0 (± 5%)	0 (± 5%)	Engine Data 2
Spark Advance (°)	-64° to 64°	20	Varies	Engine Data 2
Startup ECT Sensor	-40 to 304°F	Varies	Varies	Engine Data 2
TP Angle (%)	0-100%	0	Varies	Engine Data 2
TP Sensor	0-5.0v	0.55-0.90	Varies	Engine Data 2
Vehicle Speed	0-155 mph	0	30	Engine Data 2

1996 Caprice 5.7L V8 MFI VIN P (A/T) - PID Data List

Parameter Identification (PID)	PID Value Range	PID Value at Hot Idle	PID Value at 30 mph	Data List Type
AIR Control Pump	Enabled/DIS	Disabled	Disabled	Fuel Trim
Desired Idle Speed	0-3187 rpm	PCM control	---	Fuel Trim
ECT Sensor (°F)	-40 to 304°F	185-226	185-226	Fuel Trim
EGR Duty Cycle	0-100%	0	0-100	Fuel Trim
Engine Speed	0-9999 rpm	Varies	---	Fuel Trim
EVAP Purge PWM	0-100%	0	0-50	Fuel Trim
EVAP Vacuum Switch	No/Purge	No Purge	Purge	Fuel Trim
Fuel Trim Cell	Cell #	16	Varies	Fuel Trim
Fuel Trim Learn	Disable/Enable	Enabled	Enabled	Fuel Trim
HO2S (B1 S1, B1 S2)	0-1132 mv	10-1000	10-1000	Fuel Trim
HO2S (B2 S1, B2 S2)	0-1132 mv	10-1000	10-1000	Fuel Trim
IAC Learned	0-255 counts	10-50	---	Fuel Trim
IAC Position	0-255 counts	10-50	---	Fuel Trim
IAT Sensor (°F)	-40 to 304°F	50-194	50-194	Fuel Trim
Ignition 1 Signal	0.0-25.5v	14.2	14.2	Fuel Trim
Injector Pulsewidth Bank 1	0-1000 ms	1-4	Varies	Fuel Trim
Injector Pulsewidth Bank 2	0-1000 ms	1-4	Varies	Fuel Trim
Long Term Fuel Trim Bank 1	-10% to 10%	0 (± 5%)	0 (± 5%)	Fuel Trim
Long Term Fuel Trim Bank 2	-10% to 10%	0 (± 5%)	0 (± 5%)	Fuel Trim
Loop Status	Closed Loop / Open Loop	Closed Loop	Closed Loop	Fuel Trim
Long Term Fuel Trim Average Bank 1	-13 to 21%	0 (± 5%)	0 (± 5%)	Fuel Trim
Long Term Fuel Trim Bank 2	-13 to 21%	0 (± 5%)	0 (± 5%)	Fuel Trim
MAF Sensor (g/sec)	0-512 g/sec	5-9	Varies	Fuel Trim
MAP Sensor (kPa)	10-110 kPa	20-48	Varies	Fuel Trim
MAP Sensor (V)	0-5.0v	1-2	Varies	Fuel Trim
MPH or KPH	0-155 mph	0	30	Fuel Trim
Short Term Fuel Trim Bank 1	-10% to 10%	0 (± 5%)	0 (± 5%)	Fuel Trim
Short Term Fuel Trim Bank 2	-10% to 10%	0 (± 5%)	0 (± 5%)	Fuel Trim
Short Term Fuel Trim Average Bank 1	-9% to 7%	0 (± 5%)	0 (± 5%)	Fuel Trim
Short Term F/T Average Bank 2	-9% to 7%	0 (± 5%)	0 (± 5%)	Fuel Trim
Spark Advance (°)	-64° to 64°	-20	Varies	Fuel Trim
TP Angle (%)	0-100%	0	0-100	Fuel Trim

1996 Caprice 5.7L V8 MFI VIN P (A/T) - PID Data List

Parameter Identification (PID)	PID Value Range	PID Value at Hot Idle	PID Value at 30 mph	Data List Type
Desired Idle Speed	0-3187 rpm	PCM control	Varies	F/F, F/R
DTC NUMBER	Pxxxx	Pxxxx	Pxxxx	F/F, F/R
ECT Sensor (°F)	-40 to 304°F	185-226	185-226	F/F, F/R
Emission Fault Count	0-255 counts	Varies	Varies	F/F, F/R
Engine Load	0-100%	22-26	Varies	F/F, F/R
Engine Speed	0-9999 rpm	Varies	---	F/F, F/R
Fail Counter	0-255 counts	Varies	Varies	F/F, F/R
Fuel Trim Cell	Cell #0-22	16	0-22	F/F, F/R
Fuel Trim Learn	Disable/Enable	Enabled	Enabled	F/F, F/R
HO2S (B1 S1, B1 S2)	0-1132 mv	10-1000	10-1000	F/F, F/R
HO2S (B2 S1, B2 S2)	0-1132 mv	10-1000	10-1000	F/F, F/R
IAT Sensor (°F)	-40 to 304°F	50-194	50-194	F/F, F/R
Long Term F/T B1, B2	-10% to 10%	0 (± 5%)	0 (± 5%)	F/F, F/R
MAF Sensor (g/sec)	0-512 g/sec	5-9	Varies	F/F, F/R
Miles Since DTC Cleared	0-65655	Varies	Varies	F/F, F/R
Miles Since 1st Fail	0-65535	Varies	Varies	F/F, F/R
Miles Since MIL Req.	0-65535	Varies	Varies	F/F, F/R
MPH / KPH	0-155 mph	0	30	F/F, F/R
No Run Counter	0-255 counts	0	0	F/F, F/R
Pass Counter	0-255 counts	Varies	Varies	F/F, F/R
Short Term F/T B1-B2	-10% to 10%	0 (± 5%)	0 (± 5%)	F/F, F/R
TP Angle (%)	0-100%	0	Varies	F/F, F/R

1996 Caprice 5.7L V8 MFI VIN P (A/T) - PID Data List

Parameter Identification (PID)	PID Value Range	PID Value at Hot Idle	PID Value at 30 mph	Data List Type
A/C Request	No or Yes	No	No	Misfire
A/C Status	Off or On	Off	Off	Misfire
CKP Engine Speed	0-9999 rpm	Varies	---	Misfire
CKP LO Resolution Angle	-30 to +60	0 Deviation	0 Deviation	Misfire
Cycles of Misfire Data	0-100	0	0	Misfire
CYL Mode Misfire	0-65535	0	0	Misfire
Desired Idle Speed	0-3187 rpm	PCM control	---	Misfire
ECT Sensor (°F)	-40 to 304°F	185-226	185-226	Misfire
Engine Load	0-100%	22-26	Varies	Misfire
Engine Speed	0-9999 rpm	Varies	---	Misfire
IAT Sensor (°F)	-40 to 304°F	50-194	50-194	Misfire
MAF Sensor (g/sec)	0-512 g/sec	5-9	Varies	Misfire
Misfire current Cyl 1-8	0-255 counts	0	0	Misfire
Misfire History Cyl 1-8	0-65535	0	0	Misfire
Misfire Per Cycle	PRI # 1-8	Varies	Varies	Misfire
Misfire Per Cycle	SEC # 1-8	Varies	Varies	Misfire
Misfire Rough Road	No or Yes	No	No	Misfire
Total Misfire Counts	0-255 counts	0	0	Misfire
Total Misfire Per Test	0-255 counts	0	0	Misfire
TP Angle (%)	0-100%	0	Varies	Misfire

<u>CAVALIER PID DATA</u>

1996-2005 Cavalier 2.2L VIN 4, 2.2L VIN F, 2.4L VIN T - PID List

Parameter Identification (PID)	PID Value Range	PID Value at Hot Idle	PID Value at 30 mph	Data List Type
A/C Relay Command	On / Off	Off	Off	Engine Data 1
Airflow Calculated	0-512 g/sec	3-7	Varies	Engine Data 1
BARO Sensor (V)	10-110 kPa	65-104	65-104	Engine Data 1
Catalyst Converter TT	32-1409°F	Varies	Varies	Engine Data 1
Cruise Control Active	Yes / No	No	No	Engine Data 1
C/C & TCC Brake Switch	Applied/Not	Not Applied	Not Applied	Engine Data 1
Current Gear	1-6	6	3	Engine Data 1
Desired Idle Speed	0-3187	900	---	Engine Data 1
ECT Sensor (°F)	-40 to 304°F	190	190	Engine Data 1
Engine Load	0-100%	11-25	Varies	Engine Data 1
Engine Run Time	Hr: Sec: Min	00:00:00	00:00:00	Engine Data 1
Engine Speed	0-9999 rpm	Varies	---	Engine Data 1
EVAP Purge Solenoid	0-100%	0	0-100	Engine Data 1
EVAP Vent Solenoid	Venting / Not Venting	Not Venting	Not Venting	Engine Data 1
Fuel Level Sensor	0-100%	Varies	Varies	Engine Data 1
Fuel Pump Relay	On / Off	On	On	Engine Data 1
FTP Sensor (In. H2O)	-17.5 to 7.5"	Varies	Varies	Engine Data 1
Fuel Trim Cell	Cell #0-22	18-21	18-21	Engine Data 1
Generator F-Terminal	0-100%	Varies	Varies	Engine Data 1
Generator L-Terminal	Voltage/No	Voltage	Voltage	Engine Data 1
HO2S-12 (B1 S2)	0-1132 mv	10-1000	10-1000	Engine Data 1
IAC Position	0-255 counts	42	---	Engine Data 1
IAT Sensor (°F)	-40 to 304°F	50-176	50-176	Engine Data 1
Ignition 1 Signal	0.0-25.5v	13.9	13.9	Engine Data 1
Knock Retard (°)	0-90°	0	0	Engine Data 1
Long Term Fuel Trim	-10% to 10%	0 (± 5%)	0 (± 5%)	Engine Data 1
Loop Status	Closed Loop / Open Loop	Closed Loop	Closed Loop	Engine Data 1
MAP Sensor (kPa)	10-110 kPa	33	Varies	Engine Data 1
MAP Sensor (V)	0.0-5.0v	1-2	Varies	Engine Data 1
MIL (lamp) Command	On / Off	Off	Off	Engine Data 1
Number Current DTC	0-255	0	0	Engine Data 1
O2S-11 (B1 S1)	0-1132 mv	10-1000	10-1000	Engine Data 1
Short Term Fuel Trim	-10% to 10%	0 (± 5%)	0 (± 5%)	Engine Data 1
Spark Advance (°)	-64° to 64°	18	Varies	Engine Data 1
TCC Enable Solenoid	On / Off	On	Off	Engine Data 1
Throttle Angle (%)	0-100%	0	Varies	Engine Data 1
TP Sensor (V)	0.0-5.0v	0.20-0.90	Varies	Engine Data 1
Vehicle Speed	0-155 mph	0	30	Engine Data 1

1996-2005 Cavalier 2.2L VIN 4, 2.2L VIN F, 2.4L VIN T - PID List

Parameter Identification (PID)	PID Value Range	PID Value at Hot Idle	PID Value at 30 mph	Data List Type
A/C HI Side Pressure	0-450 psi	139-399	139-399	Engine Data 2
A/C Relay Command	On / Off	Off	Off	Engine Data 2
A/C Request Signal	Yes / No	Off	Off	Engine Data 2
Airflow Calculated	0-512 g/sec	3-7	Varies	Engine Data 2
Air Fuel Ratio	0.0-25.5:1	14.6:1	14.6:1	Engine Data 2
BARO Sensor (kPa)	10-110 kPa	65-104	65-104	Engine Data 2

1996-2005 Cavalier 2.2L VIN 4, 2.2L VIN F, 2.4L VIN T - PID List

Parameter Identification (PID)	PID Value Range	PID Value at Hot Idle	PID Value at 30 mph	Data List Type
CKP Active Counter	0-255 counts	Counts up	Counts up	Engine Data 2
CMP Active Counter	0-255 counts	Counts up	Counts up	Engine Data 2
CMP Resync Counter	0-255	0	0	Engine Data 2
Cruise Control Active	Yes / No	No	No	Engine Data 2
C/C & TCC Brake Switch	Applied/Not	Not Applied	Not Applied	Engine Data 2
Current Gear	1-4	1	3	Engine Data 2
Desired Idle Speed	0-3187	900	---	Engine Data 2
ECT Sensor (°F)	-40 to 419°F	185-239	185-239	Engine Data 2
Engine Load	0-100%	11-25	Varies	Engine Data 2
Engine Run Time	Hr: Sec: Min	00:00:00	00:00:00	Engine Data 2
Engine Speed	0-9999 rpm	Varies	---	Engine Data 2
FC Relay 1 Command	On / Off	Off	Off	Engine Data 2
Generator F-Terminal	0-100%	37	Varies	Engine Data 2
Generator L-Terminal	Voltage/No Voltage	Voltage	Voltage	Engine Data 2
IAT Sensor (°F)	-40 to 304°F	50-176	50-176	Engine Data 2
Ignition 1 Signal	0-25.5v	14.2	14.2	Engine Data 2
Injector Pulsewidth	0-985 ms	3-4	Varies	Engine Data 2
Long Term Fuel Trim	-10% to 10%	0 (± 5%)	0 (± 5%)	Engine Data 2
Loop Status	Closed Loop / Open Loop	Closed Loop	Closed Loop	Engine Data 2
MAP Sensor (kPa)	10-110 kPa	33	Varies	Engine Data 2
Medium Resolution Resync Counter	0-255 counts	0	0	Engine Data 2
Power Enrichment	Active/Inactive	Inactive	Inactive	Engine Data 2
Short Term Fuel Trim	-10% to 10%	0 (± 5%)	0 (± 5%)	Engine Data 2
Spark Advance (°)	-64° to 64°	18	Varies	Engine Data 2
Startup ECT Sensor	-40 to 419°F	Varies	Varies	Engine Data 2
TCC Enable Solenoid	On / Off	On	Off	Engine Data 2
Throttle Angle (%)	0-100%	0	Varies	Engine Data 2
TR Range Switch	Several	Several	Several	Engine Data 2
Vacuum Calculated	10-110 kPa	Varies	Varies	Engine Data 2

1996-2005 Cavalier 2.2L VIN 4, 2.2L VIN F, 2.4L VIN T - PID List

Parameter Identification (PID)	PID Value Range	PID Value at Hot Idle	PID Value at 30 mph	Data List Type
A/C Relay Command	On / Off	Off	Off	Misfire
Airflow Calculated	0-512 g/sec	Varies	Varies	Misfire
Cycles of Misfire Data	0-99	Varies	Varies	Misfire
ECT Sensor (°F)	-40 to 419°F	185-239	185-239	Misfire
Engine Load	0-100%	11-25	Varies	Misfire
Engine Run Time	Hr: Min: Sec	00:00:00	00:00:00	Misfire
Engine Speed	0-16384	Varies	---	Misfire
Injector Pulsewidth	0-985 ms	3-4	Varies	Misfire
MAP Sensor (kPa)	11-104	33	Varies	Misfire
Misfire current Cyl 1-4	0-255	0	0	Misfire
Misfire History Cyl 1-4	0-255	0	0	Misfire
Power Enrichment	Active/Inactive	Inactive	Inactive	Misfire
Spark Advance (°)	-64° to 64°	18	Varies	Misfire
Throttle Angle (%)	0-100%	0	Varies	Misfire

1996-2005 Cavalier 2.2L VIN 4, 2.2L VIN F, 2.4L VIN T - PID List

Parameter Identification (PID)	PID Value Range	PID Value at Hot Idle	PID Value at 30 mph	Data List Type
A/C Relay Command	On / Off	Off	Off	Engine Data 3
Air Fuel Ratio	0.0-25.5:1	14.6:1	14.6:1	Engine Data 3
Cruise Control Active	Yes / No	No	No	Engine Data 3
C/C & TCC Brake Switch	Applied/Not	Not Applied	Not Applied	Engine Data 3
Desired Idle Speed	0-3187	900	---	Engine Data 3
ECT Sensor (°F)	-40 to 419°F	185-239	185-239	Engine Data 3
Engine Load	0-100%	11-25	Varies	Engine Data 3
Engine Oil Life	0-100%	Varies	Varies	Engine Data 3
Engine Oil Pressure	LOW / Okay	Okay	Okay	Engine Data 3
Engine Run Time	Hr: Min: Sec	00:00:00	00:00:00	Engine Data 3
Engine Speed	0-9999 rpm	Varies	Varies	Engine Data 3
FC Relay 1 Command	On / Off	Off	Off	Engine Data 3
IAT Sensor (°F)	-40 to 304°F	50-176	50-176	Engine Data 3
Ignition 1 Signal	0-25.5v	14.2	14.2	Engine Data 3
Injector 1-4 Command	0-985 ms	3-4	Varies	Engine Data 3
Long Term Fuel Trim	-10% to 10%	0 (± 5%)	0 (± 5%)	Engine Data 3
Loop Status	Closed Loop / Open Loop	Closed Loop	Closed Loop	Engine Data 3
MAP Sensor (kPa)	10-110 kPa	33	Varies	Engine Data 3
MAP Sensor (V)	0.0-5.0v	1-2	Varies	Engine Data 3
Short Term Fuel Trim	-10% to 10%	0 (± 5%)	0 (± 5%)	Engine Data 3
Spark Advance (°)	-64° to 64°	18	Varies	Engine Data 3
TP Sensor (V)	0.0-5.0v	0.20-0.90	Varies	Engine Data 3
TR Range Switch	Several	Several	Several	Engine Data 3
Vehicle Speed	0-155 mph	0	30	Engine Data 3
VTD Fuel Disable	Active/Inactive	Inactive	Inactive	Engine Data 3
VTD Auto Learn Timer	Active/Inactive	Inactive	Inactive	Engine Data 3

1996-2005 Cavalier 2.2L VIN 4, 2.2L VIN F, 2.4L VIN T - PID List

Parameter Identification (PID)	PID Value Range	PID Value at Hot Idle	PID Value at 30 mph	Data List Type
BARO Sensor (kPa)	11-105	65-104	65-104	EVAP
Desired Idle Speed	0-3187	900	---	EVAP
ECT Sensor (°F)	-40 to 419°F	185-239	185-239	EVAP
Engine Load	0-100%	11-25	Varies	EVAP
Engine Run Time	Hr: Min: Sec	00:00:00	00:00:00	EVAP
Engine Speed	0-9999 rpm	Varies	Varies	EVAP
EVAP Purge Solenoid	0-100%	Varies	Varies	EVAP
EVAP Test Aborted	Abort/Not	Not Aborted	Not Aborted	EVAP
EVAP Test Results	Passed/No	Passed	Passed	EVAP
EVAP Test State	Several	Several	Several	EVAP
EVAP Vent Solenoid	Venting / Not Venting	Not Venting	Not Venting	EVAP
Fuel Level Sensor	0-100%	Varies	Varies	EVAP
FTP Sensor (In. H2O)	-17.5 to 7.5"	Varies	Varies	EVAP
FTP Sensor (V)	0-5.0v	Varies	Varies	EVAP
Fuel Trim Cell	Cell #0-22	18-21	---	EVAP
Fuel Trim Index	-10% to 10%	0 (± 5%)	0 (± 5%)	EVAP
IAT Sensor (°F)	-40 to 304°F	50-176	50-176	EVAP

1996-2005 Cavalier 2.2L VIN 4, 2.2L VIN F, 2.4L VIN T - PID List

Parameter Identification (PID)	PID Value Range	PID Value at Hot Idle	PID Value at 30 mph	Data List Type
Ignition 1 Signal	0-25.5v	14.2	14.2	EVAP
Long Term Fuel Trim	-10% to 10%	0 (± 5%)	0 (± 5%)	EVAP
Loop Status	Closed Loop / Open Loop	Closed Loop	Closed Loop	EVAP
MAP Sensor (kPa)	11-105	33	Varies	EVAP
Number Current DTC	0-255	0	0	EVAP
O2S-11 (B1 S1)	0-1132 mv	10-1000	10-1000	EVAP
Short Term Fuel Trim	-10% to 10%	0 (± 5%)	0 (± 5%)	EVAP
Startup ECT Sensor	-40 to 419°F	Varies	Varies	EVAP
Throttle Angle (%)	0-100%	0	Varies	EVAP
Vehicle Speed	0-159	0	30	EVAP

1996-2005 Cavalier 2.2L VIN 4, 2.2L VIN F, 2.4L VIN T - PID List

Parameter Identification (PID)	PID Value Range	PID Value at Hot Idle	PID Value at 30 mph	Data List Type
Airflow Calculated	0-512 g/sec	3-7	Varies	Fuel Trim
Air Fuel Ratio	0.0-25.5:1	14.6:1	14.6:1	Fuel Trim
BARO Sensor (kPa)	11-104 kPa	65-104	65-104	Fuel Trim
Current Gear	1-4	1	3	Fuel Trim
ECT Sensor (°F)	-40 to 419°F	185-239	185-239	Fuel Trim
Engine Load	0-100%	11-25	Varies	Fuel Trim
Engine Run Time	Hr: Min: Sec	00:00:00	00:00:00	Fuel Trim
Engine Speed	0-9999 rpm	Varies	Varies	Fuel Trim
EVAP Purge Solenoid	0-100%	Varies	Varies	Fuel Trim
EVAP Vent Solenoid	Venting/Not	Not Venting	Not Venting	Fuel Trim
Fuel Trim Cell	Cell #0-22	18-21	0-22	Fuel Trim
HO2S-12 (B1 S2)	0-1132 mv	10-1000	10-1000	Fuel Trim
IAC Position	0-255 counts	42	---	Fuel Trim
IAT Sensor (°F)	-40 to 304°F	50-176	50-176	Fuel Trim
Injector Pulsewidth	0-985 ms	3-4	Varies	Fuel Trim
Long Term Fuel Trim	-10% to 10%	0 (± 5%)	0 (± 5%)	Fuel Trim
Loop Status	Closed Loop / Open Loop	Closed Loop	Closed Loop	Fuel Trim
MAP Sensor (kPa)	11-105	25-48	Varies	Fuel Trim
Number of Current DTC(s)	0-255	0	0	Fuel Trim
O2S-11 (B1 S1)	0-1132 mv	10-1000	10-1000	Fuel Trim
Short Term Fuel Trim	-10% to 10%	0 (± 5%)	0 (± 5%)	Fuel Trim
Spark Advance (°)	-64° to 64°	18	Varies	Fuel Trim
Throttle Angle (%)	0-100%	0	Varies	Fuel Trim
TR Range Switch	Several	Several	Several	Fuel Trim
TWC Calculated Temperature	32-1409°F	Varies	Varies	Fuel Trim
Vehicle Speed	0-155 mph	0	30	Fuel Trim

COBALT PID DATA

2005 Cobalt 2.2L VIN F (All) PID Data List

Parameter Identification (PID)	Data List	Parameter Range/Units	Typical Data Values
AC OFF for WOT	HVAC	Yes/No	No
AC Pressure Disable	HVAC	Yes/No	No
A/C Relay Command	IND, ED, MF, HVAC	On/Off	Off
A/C Request	MF, HVAC	Yes/No	No
Air Fuel Ratio	ED, FT, HO2S, IND	Ratio	14.7:1
APP Indicated Angle	ED, EVAP, FT, HO2S, CC, ID, MF, TAC, IND	0-97%	0%
APP Sensor 1	TAC	Volts	1.02 Volts
APP Sensor 1	TAC	0-99%	0%
APP Sensor 2	TAC	Volts	4.04 Volts
APP Sensor 2	TAC	0-99%	0%
APP Sensor 1 and 2	TAC	Agree/Disagree	Agree
APP Sensor 1 Circuit Status	TAC	OK/Fault	OK
APP Sensor 2 Circuit Status	TAC	OK/Fault	OK
BARO	IND	Volts	Varies with Altitude
BARO	ED, EVAP, FT, HO2S, ID, IND	kPa	65-104 kPa/ Varies w/Altitude
SC Inlet Pressure	IND	kPa	72 kPa
SC Inlet Pressure Sensor	IND	Volts	0.45 Volts
Boost Duty Cycle	IND	0-99%	99%
Catalytic Converter Protection Active	FT, HO2S	Yes/No	No
CKP Active Counter	ID	1-250	9-250
CMP Active Counter	ID	Counts	Varies
CMP Sensor	ED, ID, MF	RPM	500-700
Cold Start Up	ED, EVAP	Yes/No	Varies
Crank Request Signal	ET	Yes/No	No
Cruise Control Active	TAC, IPC, CC	Yes/No	No
Cruise Disengage 1 History	CC	Various messages	None
Cruise Disengage 2 History	CC	Various messages	None
Cruise Disengage 3 History	CC	Various messages	None
Cruise Disengage 4 History	CC	Various messages	None
Cruise Disengage 5 History	CC	Various messages	None
Cruise Disengage 6 History	CC	Various messages	None
Cruise Disengage 7 History	CC	Various messages	None
Cruise Disengage 8 History	CC	Various messages	None
Cruise ON/OFF Switch	CC	On/Off	Off
Cruise Resume/Accel. Switch	CC	On/Off	Off
Cruise Set/Coast Switch	CC	On/Off	Off
Cycles of Misfire Data	MF	0-100 Counts	Varies
Decel Fuel Cutoff	ED, FT, HO2S	Active/Inactive	Inactive
Desired Idle Speed	CC, ED, EVAP, TAC, IND	RPM	PCM Controlled
EC Ignition Relay Feedback Signal	ED, TAC	Volts	14 Volts
Engine Oil Life Remaining	IPC	0-100%	Varies
ECT Sensor	ED, EVAP, FT, HO2S, ID, MF, TAC, IND, HVAC, IPC	-40 to +150°C (-40 to +302°F)	88-105°C (190-221°F) Depends on ambient temperature
Engine OFF Time	EVAP	Hours, Minutes, Seconds	Varies
Engine Load	ED, EVAP, FT, HO2S, ID, MF, TAC, IND, HVAC	0-100%	23-24% @ Idle 42% @ 2500 RPM
Engine Oil Pressure Switch	ED, MF, IPC	OK/Low	OK
Engine Run Time	ED, EVAP, HO2S, ID, CC, MF, TAC, IND, HVAC	Hrs, Min, Sec	Varies
Engine Speed	ED, EVAP, FT, HO2S, ID, MF, TAC, IPC, IND, CC, ET, HVAC	0-10,000 RPM	500-700 RPM
EVAP Fault History	EVAP	Various Messages	No Fault
EVAP Purge Solenoid Command	ED, EVAP, FT, HO2S	0-100%	10-30%
EVAP Vent Solenoid Command	ED, EVAP	Not Venting/Venting	Venting
FC Relay 1 Command	HVAC	On/Off	Off
FC Relay 2 Command	HVAC	On/Off	Off
Fuel Tank Level Remaining	ED, EVAP, MF, IPC	0-100%	Varies
Fuel Tank Pressure Sensor	ED, EVAP	mm/Hg / In H2O	Varies

2005 Cobalt 2.2L VIN F (All) PID Data List (CONT.)

Parameter Identification (PID)	Data List	Parameter Range/Units	Typical Data Values
Fuel Tank Pressure Sensor	EVAP	Volts	Varies
Fuel Trim Cell	ED, EVAP, FT	0-23	16, 17, 21
Fuel Trim Learn	ED, EVAP, FT	Able/Disable	Disabled
Fuel Level Sensor	EVAP, IPC	0-2.5 Volts	Varies
Fuel Pump Relay Command	ED	On/Off	On
Gen F - Terminal Signal	ET	%	35%
Gen L - Terminal Signal	ET	Voltage/No Voltage	No Voltage
HO2S 1	ED, EVAP, FT, HO2S, IND	mV	10-1,000 mV and Varying
HO2S 2	ED, FT, HO2S	mV	10-1,000 mV and Varying
HO2S 1 Heater	HO2S	Amps	Varies
HO2S 2 Heater	HO2S	Amps	Varies
Hot Open Loop	FT, HO2S, HVAC	Active/Inactive	Inactive
IAT Sensor	ED, EVAP, FT, ID, MF, HO2S, IND, TAC, HVAC	-40 to +152°C (-40 to +305°F)	Depends on ambient temperature
Ignition Accessory Signal	ED, ET, IPC, HVAC	On/Off	Off
Ignition 1 Signal	ED, EVAP, FT, HO2S, ID, MF, TAC, HVAC, IND, CC, ET, IPC	0-25 Volts	11.5-14.5 Volts
Intercooler Pump	ID	On/Off	On
Initial Brake Apply Signal	TAC, CC, ED	Applied/Released	Released
Injector PWM	FT, MF	Milliseconds	2-6 ms
Knock Retard	ID, HVAC	Degrees	0°
Long Term FT	ED, EVAP, FT, HO2S	%	Near 0%
Long Term FT Avg	FT	%	%
Loop Status	ED, EVAP, FT, HO2S	Open/Closed	Closed
MAF Sensor	EVAP, FT, HO2S, MF, IND	Hz	Varies
MAF Sensor	ED, EVAP, FT, HO2S, ID, MF, TAC, IND	g/s	Varies
MAP Sensor	ED, EVAP, FT, HO2S, ID, MF, TAC, IND	kPa	20-48 kPa
Medium Resolution Resync Counter	ID	Counts	0
MIL Command	ED, EVAP, ID, IPC	Off/On	Off
MIL Requested By DTC	ED, EVAP, ID	No/Yes	No
Mileage Since DTC Clear	ED	M/Km	Varies
Misfire Current Cyl 1	MF	Varies	0
Misfire Current Cyl 2	MF	Varies	0
Misfire Current Cyl 3	MF	Varies	0
Misfire Current Cyl 4	MF	Varies	0
Misfire History Cyl 1	MF	Varies	0
Misfire History Cyl 2	MF	Varies	0
Misfire History Cyl 3	MF	Varies	0
Misfire History Cyl 4	MF	Varies	0
Non-Driven Wheel Speed	CC, IPC	km/h	0 km/h
PCM Reset	ED, EVAP, ID	Yes/No	No
Power Enrichment	ED, FT, HO2S, MF	Active/Inactive	Inactive
Reduced Engine Power	TAC, IPC	Active/Inactive	Inactive
Reference Voltage 1 Signal	ED, EVAP, ID, TAC, HVAC	Volts	4.98 Volts
Reference Voltage 2 Signal	ED, EVAP, ID, TAC, HVAC	Volts	4.98 Volts
Short Term FT	ED, EVAP, FT, HO2S	%	Near 0%
Short Term FT Avg	FT	%	0%
Spark	ED, FT, HO2S, ID, MF, CC, IND, HVAC	Degrees	11-20°
Start Up ECT	EVAP, FT, HO2S, HVAC	°C/°F	Varies
Start Up IAT	ET, EVAP, FT, HO2S	°C/°F	Varies
Starter Relay Command	ET	ON/OFF	OFF
Stop Lamp Pedal Switch	TAC, CC, ET	Released/Applied	Released
TFT Sensor	HVAC	°C/°F	-40°C to +140°C (-40 to +284°F)
Total Misfire Count	MF	Count	0

2005 Cobalt 2.2L VIN F (All) PID Data List (CONT.)

Parameter Identification (PID)	Data List	Parameter Range/Units	Typical Data Values
TP Desired Angle	TAC, CC	%	4.2%
TP Sensor 1 Learned Minimum	TAC	Volts	0.55 Volts
TP Sensor 2 Learned Minimum	TAC	Volts	0.55 Volts
Torque Request Signal	TAC, CC	%	Varies
Torque Delivered Signal	TAC	%	Varies
TP Indicated Angle	ED, EVAP, FT, HO2S, CC, ID, MF, TAC, IND	%	4.2%
TP Sensors 1 and 2	TAC	Agree/Disagree	Agree
TP Sensor 1	TAC	0-100%	4%
TP Sensor 2	TAC	0-100%	4%
TP Sensor 1	TAC	0-5.0 Volts	4.24 Volts
TP Sensor 2	TAC	0-5.0 Volts	0.78 Volts
Traction Control Status	TAC, CC	Active/Inactive	Inactive
TWC Temperature Calculated	MF	C/F	350°C
Vehicle Security Status	FT	Fuel Enabled/Disabled	Fuel Enabled
Vehicle Speed Sensor	ED, EVAP, HO2S, ID, IPC, CC, MF, TAC, IND, HVAC	km/h mph	0
Warm-ups w/o Non-Emissions Faults	ED	Counts	0
Warm-ups w/o Emissions Faults	ED	Counts	0

2005 Cobalt 2.0L VIN P (All) PID Data List

Parameter Identification (PID)	Data List	Parameter Range/Units	Typical Data Values
AC OFF for WOT	HVAC	Yes/No	No
AC Pressure Disable	HVAC	Yes/No	No
A/C Relay Command	IND, ED, MF, HVAC	On/Off	Off
A/C Request	MF, HVAC	Yes/No	No
Air Fuel Ratio	ED, FT, HO2S, IND	Ratio	14.7:1
APP Indicated Angle	ED, EVAP, FT, HO2S, CC, ID, MF, TAC, IND	0-97%	0%
APP Sensor 1	TAC	Volts	1.02 Volts
APP Sensor 1	TAC	0-99%	0%
APP Sensor 2	TAC	Volts	4.04 Volts
APP Sensor 2	TAC	0-99%	0%
APP Sensor 1 and 2	TAC	Agree/Disagree	Agree
APP Sensor 1 Circuit Status	TAC	OK/Fault	OK
APP Sensor 2 Circuit Status	TAC	OK/Fault	OK
BARO	IND	Volts	Varies with Altitude
BARO	ED, EVAP, FT, HO2S, ID, IND	kPa	65-104 kPa/ Varies w/Altitude
SC Inlet Pressure	IND	kPa	72 kPa
SC Inlet Pressure Sensor	IND	Volts	0.45 Volts
Boost Duty Cycle	IND	0-99%	99%
Catalytic Converter Protection Active	FT, HO2S	Yes/No	No
CKP Active Counter	ID	1-250	9-250
CMP Active Counter	ID	Counts	Varies
CMP Sensor	ED, ID, MF	RPM	500-700
Cold Start Up	ED, EVAP	Yes/No	Varies
Crank Request Signal	ET	Yes/No	No
Cruise Control Active	TAC, IPC, CC	Yes/No	No
Cruise Disengage 1 History	CC	Various messages	None
Cruise Disengage 2 History	CC	Various messages	None
Cruise Disengage 3 History	CC	Various messages	None
Cruise Disengage 4 History	CC	Various messages	None
Cruise Disengage 5 History	CC	Various messages	None
Cruise Disengage 6 History	CC	Various messages	None
Cruise Disengage 7 History	CC	Various messages	None
Cruise Disengage 8 History	CC	Various messages	None
Cruise ON/OFF Switch	CC	On/Off	Off
Cruise Resume/Accel. Switch	CC	On/Off	Off
Cruise Set/Coast Switch	CC	On/Off	Off
Cycles of Misfire Data	MF	0-100 Counts	Varies
Decel Fuel Cutoff	ED, FT, HO2S	Active/Inactive	Inactive
Desired Idle Speed	CC, ED, EVAP, TAC, IND	RPM	PCM Controlled
EC Ignition Relay Feedback Signal	ED, TAC	Volts	14 Volts
Engine Oil Life Remaining	IPC	0-100%	Varies
ECT Sensor	ED, EVAP, FT, HO2S, ID, MF, TAC, IND, HVAC, IPC	-40 to +150°C (-40 to +302°F)	88-105°C (190-221°F) Depends on ambient temperature
Engine OFF Time	EVAP	Hours, Minutes, Seconds	Varies
Engine Load	ED, EVAP, FT, HO2S, ID, MF, TAC, IND, HVAC	0-100%	23-24% @ Idle 42% @ 2500 RPM
Engine Oil Pressure Switch	ED, MF, IPC	OK/Low	OK
Engine Run Time	ED, EVAP, HO2S, ID, CC, MF, TAC, IND, HVAC	Hrs, Min, Sec	Varies
Engine Speed	ED, EVAP, FT, HO2S, ID, MF, TAC, IPC, IND, CC, ET, HVAC	0-10,000 RPM	500-700 RPM
EVAP Fault History	EVAP	Various Messages	No Fault
EVAP Purge Solenoid Command	ED, EVAP, FT, HO2S	0-100%	10-30%
EVAP Vent Solenoid Command	ED, EVAP	Not Venting/Venting	Venting
FC Relay 1 Command	HVAC	On/Off	Off
FC Relay 2 Command	HVAC	On/Off	Off
Fuel Tank Level Remaining	ED, EVAP, MF, IPC	0-100%	Varies
Fuel Tank Pressure Sensor	ED, EVAP	mm/Hg / In H2O	Varies
Fuel Tank Pressure Sensor	EVAP	Volts	Varies
Fuel Trim Cell	ED, EVAP, FT	0-23	16, 17, 21
Fuel Trim Learn	ED, EVAP, FT	Able/Disable	Disabled

2005 Cobalt 2.0L VIN P (All) PID Data List (CONT.)

Parameter Identification (PID)	Data List	Parameter Range/Units	Typical Data Values
Fuel Level Sensor	EVAP, IPC	0-2.5 Volts	Varies
Fuel Pump Relay Command	ED	On/Off	On
Gen F - Terminal Signal	ET	%	35%
Gen L - Terminal Signal	ET	Voltage/No Voltage	No Voltage
HO2S 1	ED, EVAP, FT, HO2S, IND	mV	10-1,000 mV and Varying
HO2S 2	ED, FT, HO2S	mV	10-1,000 mV and Varying
HO2S 1 Heater	HO2S	Amps	Varies
HO2S 2 Heater	HO2S	Amps	Varies
Hot Open Loop	FT, HO2S, HVAC	Active/Inactive	Inactive
IAT Sensor	ED, EVAP, FT, ID, MF, HO2S, IND, TAC, HVAC	-40 to +152°C (-40 to +305°F)	Depends on ambient temperature
Ignition Accessory Signal	ED, ET, IPC, HVAC	On/Off	Off
Ignition 1 Signal	ED, EVAP, FT, HO2S, ID, MF, TAC, HVAC, IND, CC, ET, IPC	0-25 Volts	11.5-14.5 Volts
Intercooler Pump	ID	On/Off	On
Initial Brake Apply Signal	TAC, CC, ED	Applied/Released	Released
Injector PWM	FT, MF	Milliseconds	2-6 ms
Knock Retard	ID, HVAC	Degrees	0°
Long Term FT	ED, EVAP, FT, HO2S	%	Near 0%
Long Term FT Avg	FT	%	%
Loop Status	ED, EVAP, FT, HO2S	Open/Closed	Closed
MAF Sensor	EVAP, FT, HO2S, MF, IND	Hz	Varies
MAF Sensor	ED, EVAP, FT, HO2S, ID, MF, TAC, IND	g/s	Varies
MAP Sensor	ED, EVAP, FT, HO2S, ID, MF, TAC, IND	kPa	20-48 kPa
Medium Resolution Resync Counter	ID	Counts	0
MIL Command	ED, EVAP, ID, IPC	Off/On	Off
MIL Requested By DTC	ED, EVAP, ID	No/Yes	No
Mileage Since DTC Clear	ED	M/Km	Varies
Misfire Current Cyl 1	MF	Varies	0
Misfire Current Cyl 2	MF	Varies	0
Misfire Current Cyl 3	MF	Varies	0
Misfire Current Cyl 4	MF	Varies	0
Misfire History Cyl 1	MF	Varies	0
Misfire History Cyl 2	MF	Varies	0
Misfire History Cyl 3	MF	Varies	0
Misfire History Cyl 4	MF	Varies	0
Non-Driven Wheel Speed	CC, IPC	km/h	0 km/h
PCM Reset	ED, EVAP, ID	Yes/No	No
Power Enrichment	ED, FT, HO2S, MF	Active/Inactive	Inactive
Reduced Engine Power	TAC, IPC	Active/Inactive	Inactive
Reference Voltage 1 Signal	ED, EVAP, ID, TAC, HVAC	Volts	4.98 Volts
Reference Voltage 2 Signal	ED, EVAP, ID, TAC, HVAC	Volts	4.98 Volts
Short Term FT	ED, EVAP, FT, HO2S	%	Near 0%
Short Term FT Avg	FT	%	0%
Spark	ED, FT, HO2S, ID, MF, CC, IND, HVAC	Degrees	11-20°
Start Up ECT	EVAP, FT, HO2S, HVAC	°C/°F	Varies
Start Up IAT	ET, EVAP, FT, HO2S	°C/°F	Varies
Starter Relay Command	ET	ON/OFF	OFF
Stop Lamp Pedal Switch	TAC, CC, ET	Released/Applied	Released
TFT Sensor	HVAC	°C/°F	-40°C to +140°C (-40 to +284°F)
Total Misfire Count	MF	Count	0

2005 Cobalt 2.0L VIN P (All) PID Data List (CONT.)

Parameter Identification (PID)	Data List	Parameter Range/Units	Typical Data Values
TP Desired Angle	TAC, CC	%	4.2%
TP Sensor 1 Learned Minimum	TAC	Volts	0.55 Volts
TP Sensor 2 Learned Minimum	TAC	Volts	0.55 Volts
Torque Request Signal	TAC, CC	%	Varies
Torque Delivered Signal	TAC	%	Varies
TP Indicated Angle	ED, EVAP, FT, HO2S, CC, ID, MF, TAC, IND	%	4.2%
TP Sensors 1 and 2	TAC	Agree/Disagree	Agree
TP Sensor 1	TAC	0-100%	4%
TP Sensor 2	TAC	0-100%	4%
TP Sensor 1	TAC	0-5.0 Volts	4.24 Volts
TP Sensor 2	TAC	0-5.0 Volts	0.78 Volts
Traction Control Status	TAC, CC	Active/Inactive	Inactive
TWC Temperature Calculated	MF	C/F	350°C
Vehicle Security Status	ET	Fuel Enabled/Disabled	Fuel Enabled
Vehicle Speed Sensor	ED, EVAP, HO2S, ID, IPC, CC, MF, TAC, IND, HVAC	km/h mph	0
Warm-ups w/o Non-Emissions Faults	ED	Counts	0
Warm-ups w/o Emissions Faults	ED	Counts	0

CORVETTE PID DATA

1995 Corvette 5.7L V8 MFI VIN P (All) - PID Data List

Parameter Identification (PID)	PID Value Range	PID Value at Hot Idle	PID Value at 30 mph	Data List Type
A/C Clutch	On / Off	Off	Off	Engine Data 1
A/C Head Pressure	0-459 psi	139-399	139-399	Engine Data 1
Air Pump Relay	On / Off	Off	Off	Engine Data 1
A/C Request	Yes / No	No	No	Engine Data 1
A/C Status	On / Off	Off	Off	Engine Data 1
ASR Control	Yes / No	No	No	Engine Data 1
Air Switch Control	Divert/Port	Divert	Divert	Engine Data 1
BARO Sensor (V)	0-5.0v	3.5-4.5	3.5-4.5	Engine Data 1
Block Learn Count 1-2	0-255 counts	118-138	118-138	Engine Data 1
Block Learn Cell	0-10	4	4	Engine Data 1
Block Learn Enable	No / Yes	Yes	Yes	Engine Data 1
Converter Protection	Yes / No	No	No	Engine Data 1
Cooling Fan 1, 2	On / Off	On: If on	Off	Engine Data 1
Current EST	Okay/Open	Okay	Okay	Engine Data 1
Desired Idle Speed	0-3187	800	---	Engine Data 1
ECT Sensor (°F)	-40-419°F	185-221	185-221	Engine Data 1
EGR Duty Cycle	0-100%	0	0-100	Engine Data 1
Engine Speed	0-9999 rpm	800 ±50	---	Engine Data 1
ESC Sensors	Okay / Not Okay	Okay	Okay	Engine Data 1
Fan Relays 1, 2	On / Off	On: If on	Off	Engine Data 1
IAC Motor	0-255 counts	5-50	---	Engine Data 1
IAC Learned	0-255 counts	5-50	---	Engine Data 1
IAT Sensor (°F)	-40 to 304°F	50-194	50-194	Engine Data 1
Injector Fault 1, 2	Yes / No	No	No	Engine Data 1
INJ Pulsewidth 1, 2	0-985 ms	1-4	Varies	Engine Data 1
Fuel Integrator 1, 2	0-255 counts	110-145	110-145	Engine Data 1
Knock Retard (°)	0-128°	0	0	Engine Data 1
Knock Signal	Yes / No	No	No	Engine Data 1
Loop Status	Closed Loop / Open Loop	Closed Loop	Closed Loop	Engine Data 1
MAP Sensor (kPa)	10-110 kPa	29-48	Varies	Engine Data 1
Oil Temp. Sensor	-40 to 304°F	160-200	160-200	Engine Data 1
HO2S (B1 S1, B2 S1)	0-1132 mv	10-1000	10-1000	Engine Data 1
PNP Switch	PN / R-DL	PN	R-DL	Engine Data 1
Park Throttle Solenoid	On / Off	Off	Off	Engine Data 1
Pass-Key® / FEDS	Disable/Enable	Enabled	Enabled	Engine Data 1
Purge Duty Cycle	0-100%	0	0-50	Engine Data 1
QDM 1, 2 & 3	On/Fault	Okay	Okay	Engine Data 1
Res. Signal - Low	0-1000 ms	Varies	Varies	Engine Data 1
Res. Signal - High	0-255 counts	Varies	Varies	Engine Data 1
Serial Data	Okay / Failed	Okay	Okay	Engine Data 1
Skip Shift Active	Yes / No	No	No	Engine Data 1
Spark Advance (°)	-90° to 90°	Varies	Varies	Engine Data 1
TCC 2nd Solenoid	On / Off	Off	Off	Engine Data 1
Throttle Angle	0-100%	0	Varies	Engine Data 1
Time From Start	Hr: Min: Sec	00:00:00	00:00:00	Engine Data 1
TP Sensor (V)	0-5.0v	0.36-0.62	---	Engine Data 1
Vehicle Speed	0-155 mph	0	30	55
3rd, 4th Gear Switch	Yes / No	No	Yes: 3rd	Yes: 4th

1995 Corvette 5.7L V8 MFI VIN P (All) - PID Data List

Parameter Identification (PID)	PID Value Range	PID Value at Hot Idle	PID Value at 30 mph	Data List Type
A/C Clutch	On / Off	Off	Off	Engine Data 2
AC Head Pressure	0-459 psi	139-399	139-399	Engine Data 2
A/C Request	Yes / No	No	No	Engine Data 2
AC Status	On / Off	Off	Off	Engine Data 2
AIR Pump Control	Divert/Port	Divert	Divert	Engine Data 2
BARO Sensor (V)	0-5.0v	3.5-4.5	3.5-4.5	Engine Data 2
Cooling Fan 1, 2	On / Off	On: If on	Off	Engine Data 2
ECT Sensor (°F)	-40-419°F	185-221	185-221	Engine Data 2
EGR Duty Cycle	0-100%	0%	0-50	Engine Data 2
Engine Oil Temperature	-40 to 304°F	180-200	180-200	Engine Data 2
Engine Speed	0-9999 rpm	700 ±50	----	Engine Data 2
Fuel Trim Cell	Cell #	16-18	Varies	Engine Data 2
Fuel Trim Enable	No / Yes	Yes	Yes	Engine Data 2
HO2S-11 (B1 S1)	0-1132 mv	10-1000	10-1000	Engine Data 2
HO2S-21 (B2 S1)	0-1132 mv	10-1000	10-1000	Engine Data 2
IAT Sensor (°F)	-40 to 304°F	50-194	50-194	Engine Data 2
Knock Retard (°)	0-128°	0	0	Engine Data 2
Knock Signal 1, 2	Yes / No	No	No	Engine Data 2
HI Resolution Signal	No / Yes	Yes	Yes	Engine Data 2
IAC Learned	0-255 counts	5-50	----	Engine Data 2
IAC Motor	0-255 counts	5-50	----	Engine Data 2
Ignition 1 Signal	0.0-25.5v	14.2	14.2	Engine Data 2
Injector Fault 1, 2	Yes / No	No	No	Engine Data 2
INJ Pulsewidth 1	0-985 ms	1-4	Varies	Engine Data 2
INJ Pulsewidth 2	0-985 ms	1-4	Varies	Engine Data 2
Knock Sensors	Okay / Not Okay	Okay	Okay	Engine Data 2
Long Term F/T B1, B2	0-255 counts	118-138	118-138	Engine Data 2
LO Resolution Signal	0-1000	Varies	Varies	Engine Data 2
Loop Status	Closed Loop / Open Loop	Closed Loop	Closed Loop	Engine Data 2
MAF Sensor (g/sec)	0-512 g/sec	5-7	Varies	Engine Data 2
MAP Sensor (kPa)	10-110 kPa	29-48	Varies	Engine Data 2
MPH / KPH	0-155 mph	0	30	Engine Data 2
Pass-Key® / FEDS	Disable/Enable	Enabled	Enabled	Engine Data 2
PNP Switch	PN / R-DL	PN	R-DL	Engine Data 2
PROM ID Number	Number	0000	0000	Engine Data 2
Purge Duty Cycle	0-100%	0-25	0-50	Engine Data 2
Shift Solenoid 1-2, 2-3	Off / On	On	Off or On	Engine Data 2
Short Term F/T B1	0-255 counts	118-138	118-138	Engine Data 2
Short Term F/T B2	0-255 counts	118-138	118-138	Engine Data 2
Spark Advance (°)	0-128°	Varies	Varies	Engine Data 2
TCC Brake Switch	On / Off	On: If on	Off	Engine Data 2
TCC Solenoid	On / Off	Off	On or Off	Engine Data 2
Throttle Angle	0-100%	0	0-100	Engine Data 2
Time From Start	Hr: Min: Sec	00:00:00	00:00:00	Engine Data 2
TP Sensor (V)	0-5.0v	0.36-0.62	----	Engine Data 2
Trans. Range Switch	PN-R-D4-D3	PN	D3	Engine Data 2
TSR / ASR Control	Yes / No	No	No	Engine Data 2

1996 Corvette 5.7L V8 MFI VIN P (All) - PID Data List

Parameter Identification (PID)	PID Value Range	PID Value at Hot Idle	PID Value at 30 mph	Data List Type
A/C Clutch	Off/On	Off	Off	Engine Data 1
A/C HI Side Pressure	0-459 psi	139-399	139-399	Engine Data 1
A/C Request	Yes/No	No	No	Engine Data 1
A/C Status	Off or On	Off	Off	Engine Data 1
AIR Control Pump	Disable/Enable	Disabled	Disabled	Engine Data 1
BARO Sensor (kPa)	10-110 kPa	10-105	10-105	Engine Data 1
Cold Engine Startup	Yes/No	No	No	Engine Data 1
Current Gear	1-2-3-4	1	3	Engine Data 1
Desired Idle Speed	0-3187 rpm	PCM control	Varies	Engine Data 1
DTC Set This Ignition	Yes/No	No	No	Engine Data 1
ECT Sensor (°F)	-40 to 304°F	185-226	185-226	Engine Data 1
EGR Solenoid	0-100%	0	0-100	Engine Data 1
Engine Run Time	Hr: Min: Sec	00:00:00	00:00:00	Engine Data 1
Engine Speed	0-9999 rpm	Varies	Varies	Engine Data 1
EVAP Purge Solenoid	0-100%	0	0-50	Engine Data 1
EVAP Vacuum Switch	No/Purge	No Purge	Purge	Engine Data 1
FC Relay 1 ?	Off or On	Off	Off	Engine Data 1
Fuel Trim Cell	Cell #	16-18	16-18	Engine Data 1
Fuel Trim Learn	Disable/Enable	Enabled	Enabled	Engine Data 1
HI Resolution Signal	Active/Inactive	Active	Active	Engine Data 1
HO2S (B1 S1, B1 S2)	0-1132 mv	10-1000	10-1000	Engine Data 1
HO2S (B2 S1, B2 S2)	0-1132 mv	10-1000	10-1000	Engine Data 1
IAC Learned	0-255 counts	10-50	10-50	Engine Data 1
IAC Position	0-255 counts	10-50	10-50	Engine Data 1
IAT Sensor (°F)	-40 to 304°F	50-194	50-194	Engine Data 1
Injector Pulsewidth	0-1000ms	1-4	Varies	Engine Data 1
Knock Retard (°)	0.0 - 25.5°	0	0	Engine Data 1
Long Term F/T B1, B2	-10% to 10%	0%	Varies	Engine Data 1
Loop Status	Closed Loop / Open Loop	Closed Loop	Closed Loop	Engine Data 1
LO Resolution Signal	0-332.7 ms	Varies	Varies	Engine Data 1
MAF Sensor (g/sec)	0-512 g/sec	5-9	Varies	Engine Data 1
MAP Sensor (kPa)	10-110 kPa	20-48	Varies	Engine Data 1
MAP Sensor (V)	0-5.0v	1-2	Varies	Engine Data 1
OIL Level	Okay/Not Okay	Okay	Okay	Engine Data 1
Pass Key® Fuel	Disabled/Okay	Okay	Okay	Engine Data 1
PCM Reset	Yes/No	No	No	Engine Data 1
PNP Switch	P-R-0-3-2-1	P	3	Engine Data 1
PSP Switch	HI/Normal	Normal	Normal	Engine Data 1
Short Term F/T B1-B2	-10% to 10%	0%	Varies	Engine Data 1
Spark Advance (°)	-64° to 64°	20	Varies	Engine Data 1
Startup ECT Sensor	-40 to 304°F	Varies	Varies	Engine Data 1
TCC Brake Switch	Applied/Released	Released	Released	Engine Data 1
TCC Duty Cycle	Disable/Enable	Enabled	Enabled	Engine Data 1
TCC Enable	Disable/Enable	Disabled	Disabled	Engine Data 1
TCC Stator Temp.	HI/Normal	Normal	Normal	Engine Data 1
TP Angle (%)	0-100%	0%	0-100%	Engine Data 1
TP Sensor	0-5.0v	0.55-0.90	Varies	Engine Data 1
TR Switch	P/N/Drive	Park	Drive	Engine Data 1

1996 Corvette 5.7L V8 MFI VIN P (All) - PID Data List

Parameter Identification (PID)	PID Value Range	PID Value at Hot Idle	PID Value at 30 mph	Data List Type
A/C Clutch	Off or On	Off	Off	Engine Data 2
AIR Control Pump	Enabled/DIS	Disabled	Disabled	Engine Data 2
BARO Sensor (kPa)	10-110 kPa	Varies	Varies	Engine Data 2
BARO Sensor (V)	0-5.0v	3.5-4.5	3.5-4.5	Engine Data 2
CKP Engine Speed	0-9999 rpm	Varies	---	Engine Data 2
CKP LO Resolution Angle	-30 to +60	0 Deviation	0 Deviation	Engine Data 2
Cold Engine Startup	No or Yes	No	No	Engine Data 2
Desired Idle Speed	0-3187 rpm	PCM control	---	Engine Data 2
DTC Set this Ignition	No or Yes	No	No	Engine Data 2
ECT Sensor (°F)	-40 to 304°F	185-226	185-226	Engine Data 2
EGR Duty Cycle	0-100%	0%	0-100%	Engine Data 2
Engine Run Time	Hr: Min: Sec	00:00:00	00:00:00	Engine Data 2
Engine Speed	0-9999 rpm	Varies	Varies	Engine Data 2
EVAP Purge Solenoid	0-100%	0	0-50	Engine Data 2
EVAP Vacuum Switch	No/Purge	No Purge	Purge	Engine Data 2
Fan Control Relay 1	Off or On	Off	Off	Engine Data 2
Fan Control Relay 2	Off or On	Off	Off	Engine Data 2
Fuel Trim Cell	Cell #1-17	16	Varies	Engine Data 2
Fuel Trim Learn	Disable/Enable	Enabled	Enabled	Engine Data 2
HI Resolution Signal	Active/Inactive	Active	Active	Engine Data 2
HO2S (B1 S1, B1 S2)	0-1132 mv	10-1000	10-1000	Engine Data 2
HO2S (B2 S1, B2 S2)	0-1132 mv	10-1000	10-1000	Engine Data 2
IAC Learned	0-255 counts	10-50	---	Engine Data 2
IAC Position	0-255 counts	10-50	---	Engine Data 2
IAT Sensor (°F)	-40 to 304°F	50-194	50-194	Engine Data 2
Ignition 1 Signal	0.0-25.5v	14.2	14.2	Engine Data 2
Knock Retard (°)	0.0 - 25.5°	0	Varies	Engine Data 2
Knock Activity	Yes / No	No	No	Engine Data 2
Long Term F/T B1	-10% to 10%	0%	Varies	Engine Data 2
Long Term F/T B2	-10% to 10%	0%	Varies	Engine Data 2
Loop Status	Closed Loop / Open Loop	Closed Loop	Closed Loop	Engine Data 2
LO Resolution Signal	0-332.7 ms	Varies	Varies	Engine Data 2
MAF Sensor (g/sec)	0-512 g/sec	5-9	Varies	Engine Data 2
MAP Sensor (kPa)	10-110 kPa	20-48	Varies	Engine Data 2
MAP Sensor (V)	0-5.0v	1-2	Varies	Engine Data 2
MPH or KPH	0-155 mph	0	30	Engine Data 2
Non-Emission Cycle Count	0-255 counts	0	0	Engine Data 2
Pass Key® Fuel	Disabled/Okay	Okay	Okay	Engine Data 2
PCM Reset	No or Yes	No	No	Engine Data 2
Short Term F/TB1	-10% to 10%	0%	Varies	Engine Data 2
Short Term F/T B2	-10% to 10%	0%	Varies	Engine Data 2
Spark Advance (°)	-64° to 64°	-20	Varies	Engine Data 2
Startup ECT Sensor	-40 to 304°F	Varies	Varies	Varies
TP Angle (%)	0-100%	0%	0-100%	0-100%
TP Sensor (V)	0-5.0v	0.55-0.90	Varies	Varies
Vehicle Speed	0-155 mph	0	30	55

1996 Corvette 5.7L V8 MFI VIN P (All) - PID Data List

Parameter Identification (PID)	PID Value Range	PID Value at Hot Idle	PID Value at 30 mph	Data List Type
AIR Control Pump	Enabled/DIS	Disabled	Disabled	Fuel Trim
Desired Idle Speed	0-3187 rpm	PCM control	Varies	Fuel Trim
ECT Sensor (°F)	-40 to 304°F	185-226	185-226	Fuel Trim
EGR Duty Cycle	0-100%	0	0-100	Fuel Trim
Engine Speed	0-9999 rpm	Varies	Varies	Fuel Trim
EVAP Purge Solenoid	0-100%	0	0-50	Fuel Trim
EVAP Vacuum Switch	No/Purge	No Purge	Purge	Fuel Trim
Fuel Trim Cell	Cell #	16	Varies	Fuel Trim
Fuel Trim Learn	Disable/Enable	Enabled	Enabled	Fuel Trim
HO2S-11 (B1 S1)	0-1132 mv	10-1000	10-1000	Fuel Trim
HO2S-12 (B1 S2)	0-1132 mv	10-1000	10-1000	Fuel Trim
HO2S-21 (B2 S1)	0-1132 mv	10-1000	10-1000	Fuel Trim
HO2S-22 (B2 S2)	0-1132 mv	10-1000	10-1000	Fuel Trim
IAC Learned	0-255 counts	10-50	---	Fuel Trim
IAC Position	0-255 counts	10-50	---	Fuel Trim
IAT Sensor (°F)	-40 to 304°F	50-194	50-194	Fuel Trim
Ignition 1 Signal	0.0-25.5v	14 ?	14 ?	Fuel Trim
Injector Pulsewidth Bank 1	0-1000 ms	1-4	Varies	Fuel Trim
Injector Pulsewidth Bank 2	0-1000 ms	1-4	Varies	Fuel Trim
Long Term Fuel Trim Bank 1	-10% to 10%	0 (± 5%)	0 (± 5%)	Fuel Trim
Long Term Fuel Trim Bank 2	-10% to 10%	0 (± 5%)	0 (± 5%)	Fuel Trim
Long Term Fuel Trim Average Bank 1	-10% to 10%	0 (± 5%)	0 (± 5%)	Fuel Trim
Long Term Fuel Trim Average Bank 2	-10% to 10%	0 (± 5%)	0 (± 5%)	Fuel Trim
Loop Status	Closed Loop, Open Loop	Closed Loop	Closed Loop	Fuel Trim
MAF Sensor (g/sec)	0-512 g/sec	5-9	Varies	Fuel Trim
MAP Sensor (kPa)	10-110 kPa	20-48	Varies	Fuel Trim
MAP Sensor (V)	0-5.0v	1-2	Varies	Fuel Trim
MPH or KPH	0-155 mph	0	30	Fuel Trim
Short Term Fuel Trim Bank 1	-10% to 10%	0 (± 5%)	0 (± 5%)	Fuel Trim
Short Term Fuel Trim Bank 2	-10% to 10%	0 (± 5%)	0 (± 5%)	Fuel Trim
Short Term F/T Average Bank 1	-10% to 10%	0 (± 5%)	0 (± 5%)	Fuel Trim
Short Term F/T Average Bank 2	-10% to 10%	0 (± 5%)	0 (± 5%)	Fuel Trim
Spark Advance (°)	-64° to 64°	20	Varies	Fuel Trim
TP Angle (%)	0-100%	0%	0-100%	Fuel Trim
Vehicle Speed	0-155 mph	0	30	Fuel Trim

1996 Corvette 5.7L V8 VIN P (All) - PID Data List

Parameter Identification (PID)	PID Value Range	PID Value at Hot Idle	PID Value at 30 mph	Data List Type
Desired Idle Speed	0-3187 rpm	PCM control	---	F/F, F/R
DTC Number	Number	00000	00000	F/F, F/R
ECT Sensor (°F)	-40 to 304°F	185-226	185-226	F/F, F/R
Emission Fault Count	0-255 counts	Varies	Varies	F/F, F/R
Engine Load	0-100%	22-26	Varies	F/F, F/R
Engine Speed	0-9999 rpm	Varies	Varies	F/F, F/R
Fail Counter	0-255 counts	Varies	Varies	F/F, F/R
Fuel Trim Cell	Cell #1-17	16	Varies	F/F, F/R
Fuel Trim Learn	Disable/Enable	Enabled	Enabled	F/F, F/R
HO2S (B1 S1, B1 S2)	0-1132 mv	10-1000	10-1000	F/F, F/R
HO2S (B2 S1, B2 S2)	0-1132 mv	10-1000	10-1000	F/F, F/R
IAT Sensor (°F)	-40 to 304°F	50-194	50-194	F/F, F/R
Long Term F/T B1-B2	-10% to 10%	0 (± 5%)	0 (± 5%)	F/F, F/R
MAF Sensor (g/sec)	0-512 g/sec	5-9	Varies	F/F, F/R
Miles Since DTC Cleared	0-65655	Varies	Varies	F/F, F/R
Miles Since 1st Fault	0-65535	Varies	Varies	F/F, F/R
Miles Since MIL Req.	0-65535	Varies	Varies	F/F, F/R
MPH / KPH	0-155 mph	0	30	F/F, F/R
No Run Counter	0-255 counts	0	0	F/F, F/R
Pass Counter	0-255 counts	Varies	Varies	F/F, F/R
Short Term F/T B1-B2	-10% to 10%	0 (± 5%)	0 (± 5%)	F/F, F/R
TP Angle (%)	0-100%	0	Varies	F/F, F/R

1996 Corvette 5.7L V8 MFI VIN P (All) - PID Data List

Parameter Identification (PID)	PID Value Range	PID Value at Hot Idle	PID Value at 30 mph	Data List Type
A/C Request	No or Yes	No	No	Misfire
A/C Status	Off or On	Off	Off	Misfire
CKP Engine Speed	0-9999 rpm	Varies	Varies	Misfire
CKP LO Resolution Angle	-30 to +60	0 Deviation	0 Deviation	Misfire
Cycles of Misfire Data	0-100	0	0	Misfire
CYL Mode Misfire	0-65535	0	0	Misfire
Desired Idle Speed	0-3187 rpm	PCM control	----	Misfire
ECT Sensor (°F)	-40 to 304°F	185-226	185-226	Misfire
Engine Load	0-100%	22-26	Varies	Misfire
Engine Speed	0-9999 rpm	Varies	Varies	Misfire
IAT Sensor (°F)	-40 to 304°F	50-194	50-194	Misfire
MAF Sensor (g/sec)	0-512 g/sec	5-9	Varies	Misfire
Misfire current Cyl 1-8	0-255 counts	0	0	Misfire
Misfire History Cyl 1-8	0-65535	0	0	Misfire
Misfire Per Cycle	PRI # 1-8	Varies	Varies	Misfire
Misfire Per Cycle	SEC # 1-8	Varies	Varies	Misfire
Misfire Rough Road	No or Yes	No	No	Misfire
Total Misfire Counts	0-255 counts	0	0	Misfire
Total Misfire Per Test	0-255 counts	0	0	Misfire
TP Angle (%)	0-100%	0	Varies	Misfire

1997-99 Corvette 5.7L V8 MFI VIN G (All) - PID Data List

Parameter Identification (PID)	PID Value Range	PID Value at Hot Idle	PID Value at 30 mph	Data List Type
1-4 Shift Lamp	On / Off	Off	On or Off	Engine Data 1
1-4 Shift Solenoid	Disable/Enable	Disabled	Enabled	Engine Data 1
A/C Clutch	On/Off	Off	Off	Engine Data 1
A/C HI Side Pressure	0-459 psi	139-399	139-399	Engine Data 1
A/C High Side Volts	0-5.0v	Varies	Varies	Engine Data 1
A/C Request	Yes/No	No	No	Engine Data 1
A/C Status	On/Off	Off	Off	Engine Data 1
AIR Pump Relay	On/Off	Off	Off	Engine Data 1
AIR Solenoid Relay	On/Off	Off	Off	Engine Data 1
APP Average Counts	0-125	Varies	Varies	Engine Data 1
APP Indicated Angle	0-100%	Varies	Varies	Engine Data 1
APP Sensor 1	0-5.0v	0.25-1.1v	Varies	Engine Data 1
APP Sensor 2	0-5.0v	3.9-4.8v	Varies	Engine Data 1
APP Sensor 3	0-5.0v	3.2-4.5v	Varies	Engine Data 1
BARO Sensor (kPa)	10-110 kPa	10-105	10-105	Engine Data 1
Cam Signal HI to LO	0-65535	Varies	Varies	Engine Data 1
Cam Signal LO to HI	0-65535	Varies	Varies	Engine Data 1
Clutch Switch	DEP/REL	Released	Released	Engine Data 1
Cruise Requested	No / Yes	No	No	Engine Data 1
Cruise Resume/Accel	Off / On	Off	Off	Engine Data 1
Cruise Set/Coast	Off / On	Off	Off	Engine Data 1
Cruise Switch	Off / On	Off	Off	Engine Data 1
Desired Idle Speed	0-3187 rpm	PCM control	---	Engine Data 1
DTC Set This Ignition	No or Yes	No	No	Engine Data 1
Drive 1	Drive 1	Off/off/on/on	Off/off/on/on	Engine Data 1
Drive 2	Drive 2	On/off/on/Off	On/off/on/Off	Engine Data 1
Drive 3/D	Drive 3/D	On/on/on/on	On/on/on/on	Engine Data 1
Drive 4/OD	Drive 4/OD	Off/on/on/off	Off/on/on/off	Engine Data 1
ECT Sensor (°F)	-38-284°F	185-226	185-226	Engine Data 1
Engine Load	0-100%	2%	Varies	Engine Data 1
Engine Oil Pressure	0-992 kPa	Varies	Varies	Engine Data 1
Engine Oil Pressure	0-144 psi	Varies	Varies	Engine Data 1
Engine Oil Pressure	0-5.0v	Varies	Varies	Engine Data 1
Engine Run Time	Hr: Min: Sec	00:00:00	00:00:00	Engine Data 1
Engine Speed	0-10000	Varies	Varies	Engine Data 1
EVAP Purge Solenoid	0-100%	0-25	0-50	Engine Data 1
EVAP Vacuum Switch	No/Purge	No Purge	Purge	Engine Data 1
Extended Brake Switch	Apply/Release	Released	Released	Engine Data 1
Fan Control Relay 1	Off or On	Off	Off	Engine Data 1
FC Relay 2, 3	Off or On	Off	Off	Engine Data 1
Fuel Trim Cell	Cell #	16	Varies	Engine Data 1
Fuel Trim Learn	Disable/Enable	Enabled	Enabled	Engine Data 1
HI Resolution Signal	Active/Inactive	Active	Active	Engine Data 1
HO2S (B1 S1)	0-1132 mv	10-1000	10-1000	Engine Data 1
HO2S (B1 S2)	0-1132 mv	10-1000	10-1000	Engine Data 1
HO2S (B2 S1, B2 S2)	0-1132 mv	10-1000	10-1000	Engine Data 1
IAT Sensor (°F)	-38-284°F	50-194	50-194	Engine Data 1

1997-99 Corvette 5.7L V8 MFI VIN G (All) - PID Data List

Parameter Identification (PID)	PID Value Range	PID Value at Hot Idle	PID Value at 30 mph	Data List Type
Ignition 1 Signal	0.0-25.5v	14.2	14.2	Engine Data 1
Knock Retard (°)	0.0 - 16°	0	0	Engine Data 1
Knock Signal Present	Yes / No	No	No	Engine Data 1
Long Term F/T B1, B2	-10% to 10%	0 (± 5%)	0 (± 5%)	Engine Data 1
Loop Status	Closed Loop / Open Loop	Closed Loop	Closed Loop	Engine Data 1
MAF Sensor (g/sec)	0-512 g/sec	5-9	Varies	Engine Data 1
MAP Sensor (kPa)	10-110 kPa	20-48	Varies	Engine Data 1
MIL Status	Off / On	Off	Off	Engine Data 1
PCM Reset	No or Yes	Yes: if reset	No	Engine Data 1
PT Induced Chassis Pitch	Active/Inactive	Inactive	Inactive	Engine Data 1
Reduce Engine Power	Active/Inactive	Inactive	Inactive	Engine Data 1
Short Term F/T B1-B2	-10% to 10%	0 (± 5%)	0 (± 5%)	Engine Data 1
Spark Advance (°)	-64° to 64°	-20	Varies	Engine Data 1
Startup ECT Sensor	-40 to 304°F	Varies	Varies	Engine Data 1
Stop Lamp Switch	Applied/RED	Released	Released	Engine Data 1
TAC/PCM Comm.	Okay/Fault	Okay	Okay	Engine Data 1
TCC/CC Brake Switch	Applied/Released	Released	Released	Engine Data 1
TCC Duty Cycle	0-100%	0%	0%	Engine Data 1
TCC Enable Solenoid	Disable/Enable	Disabled	Disabled	Engine Data 1
TCC Stator Temp.	HI/Normal	Normal	Normal	Engine Data 1
TP Desired Angle	0-100%	0 - 10%	0-100%	Engine Data 1
TP Indicated Angle	0-100%	0 - 9%	0-100%	Engine Data 1
TP Sensor 1, 2 Angle	0-100%	Varies	Varies	Engine Data 1
TP Sensor 1 Volts	0-5.0v	0.25-1.5	Varies	Engine Data 1
TP Sensor 2 Volts	0-5.0v	3.0-1.5	Varies	Engine Data 1
Traction Control	Active/Inactive	Inactive	Inactive	Engine Data 1
Transmission Range	PN-R-D4-D3	PN	D3	Engine Data 1
Vehicle Speed	0-155 mph	0	30	Engine Data 1

1997-99 Corvette 5.7L V8 MFI VIN G (All) - PID Data List

Parameter Identification (PID)	PID Value Range	PID Value at Hot Idle	PID Value at 30 mph	Data List Type
1-4 Shift Lamp	On / Off	Off	On or Off	Engine Data 2
1-4 Shift Solenoid	Disable/Enable	Disabled	Enabled	Engine Data 2
A/C Clutch	Off or On	Off	Off	Engine Data 2
A/C HI Side Pressure	0-459 psi	139-399	139-399	Engine Data 2
A/C High Side Volts	0-5.0v	Varies	Varies	Engine Data 2
A/C Request	No or Yes	No	No	Engine Data 2
A/C Status	Off or On	Off	Off	Engine Data 2
APP Average Counts	0-125	Varies	Varies	Engine Data 2
APP Indicated Angle	0-100%	Varies	Varies	Engine Data 2
Clutch Switch	DEP/REL	Released	Released	Engine Data 2
Column Lock BCM/PCM	Okay/Fault	Okay	Okay	Engine Data 2
Column Lock Disable	Yes/No	No	No	Engine Data 2
Column lock Disable	Yes/No	No	No	Engine Data 2
Current Gear	1-2-3-4	1	3	Engine Data 2

1997-99 Corvette 5.7L V8 MFI VIN G (All) - PID Data List

Parameter Identification (PID)	PID Value Range	PID Value at Hot Idle	PID Value at 30 mph	Data List Type
Delivered Torque lbs	-349 to 349	Varies	Varies	Engine Data 2
Desired Torque lbs	-349 to 349	Varies	Varies	Engine Data 2
Drive 1	Drive 1	Off/off/on/on	Off/off/on/on	Engine Data 2
Drive 2	Drive 2	On/off/on/Off	On/off/on/Off	Engine Data 2
Drive 3/D	Drive 3/D	On/on/on/on	On/on/on/on	Engine Data 2
Drive 4/OD	Drive 4/OD	Off/on/on/off	Off/on/on/off	Engine Data 2
DTC Set This Ignition	Yes/No	No	No	Engine Data 2
ECT Sensor (°F)	-38 - 284°F	185-226	185-226	Engine Data 2
Engine Run Time	Hr: Min: Sec	00:00:00	00:00:00	Engine Data 2
Engine Speed	0-10000	Varies	Varies	Engine Data 2
Extended Brake Switch Signal	Applied/Released	Released	Released	Engine Data 2
Fan Control Relay 1	Off or On	Off	Off	Engine Data 2
Fan Control Relay 2-3	Off or On	Off	Off	Engine Data 2
Fuel Trim Learn	Disable/Enable	Enabled	Enabled	Engine Data 2
IAT Sensor (°F)	-38-284°F	50-194	50-194	Engine Data 2
Ignition 1 Signal	0.0-25.5v	14.2	14.2	Engine Data 2
INJ Pulsewidth B1	0-512 ms	1-4	Varies	Engine Data 2
INJ Pulsewidth B2	0-512 ms	1-4	Varies	Engine Data 2
Knock Signal Present	No / Yes	No	No	Engine Data 2
Long Term F/T B1, B2	-10% to 10%	0 (± 5%)	0 (± 5%)	Engine Data 2
Loop Status	Closed Loop / Open Loop	Closed Loop	Closed Loop	Engine Data 2
Low Oil Level	Yes/No	Yes: If low	No	Engine Data 2
MAF Sensor (g/sec)	0-512 g/sec	5-9	Varies	Engine Data 2
MAP Sensor (kPa)	10-110 kPa	20-48	Varies	Engine Data 2
MAP Sensor (V)	0-5.0v	1-2	Varies	Engine Data 2
MIL Status	Off / On	Off	Off	Engine Data 2
Misfire Since DTC Cleared	Miles/km	Varies	Varies	Engine Data 2
PCM Reset	No / Yes	Yes: if reset	No	Engine Data 2
PRND Position	A-B-C-P	P	C	Engine Data 2
PT Induced Chassis Pitch	Active/Inactive	Inactive	Inactive	Engine Data 2
Reduce Engine Power	Active/Inactive	Inactive	Inactive	Engine Data 2
Short Term F/T B1-B2	-10% to 10%	0 (± 5%)	0 (± 5%)	Engine Data 2
Spark Advance (°)	-64° to 64°	-20	Varies	Engine Data 2
TCC/CC Brake Switch	Apply/Release	Released	Released	Engine Data 2
TCC Duty Cycle	0-100%	0%	0%	Engine Data 2
TCC Enable Solenoid	Disable/Enable	Disabled	Disabled	Engine Data 2
TP Desired Angle	0-100%	0 - 10%	0-100%	Engine Data 2
TP Indicated Angle	0-100%	0 - 9%	0-100%	Engine Data 2
Transmission Range	PN-R-D4-D3	PN	D3	Engine Data 2
Vehicle Speed	0-155 mph	0	30	Engine Data 2
Warmup Cycles w/o an Emission Fault	0-255 counts	0	0	Engine Data 2
Warmup Cycles w/o a Non-Emission Fault	0-255 counts	0	0	Engine Data 2

1997-99 Corvette 5.7L V8 MFI VIN G (All) - PID Data List

Parameter Identification (PID)	PID Value Range	PID Value at Hot Idle	PID Value at 30 mph	Data List Type
1-4 Shift Lamp	On/Off	Off	On or Off	Engine Data 3
1-4 Shift Solenoid	Disable/Enable	Disabled	Enabled	Engine Data 3
APP Indicated Angle	0-100%	Varies	Varies	Engine Data 3
BARO Sensor (kPa)	10-110 kPa	10-105	10-105	Engine Data 3
Cam Signal HI to LO	0-65535	Varies	Varies	Engine Data 3
Cam Signal LO to HI	0-65535	Varies	Varies	Engine Data 3
Column Lock BCM/PCM	Fault/Okay	Okay	Okay	Engine Data 3
Column Lock Disable ABS	Yes/No	No	No	Engine Data 3
Column Lock Disable BCM	Yes/No	No	No	Engine Data 3
Cruise Requested	No / Yes	No	No	Engine Data 3
Current Gear	1-2-3-4	1	3	Engine Data 3
Cruise Resume/Accel	Off / On	Off	Off	Engine Data 3
Cruise Set/Coast	Off / On	Off	Off	Engine Data 3
Cruise Switch	Off / On	Off	Off	Engine Data 3
Delivered Torque lbs	-349 to 349	Varies	Varies	Engine Data 3
Drive 1	Drive 1	Off/off/on/on	Off/off/on/on	Engine Data 3
Drive 2	Drive 2	On/off/on/Off	On/off/on/Off	Engine Data 3
Drive 3/D	Drive 3/D	On/on/on/on	On/on/on/on	Engine Data 3
Drive 4/OD	Drive 4/OD	Off/on/on/off	Off/on/on/off	Engine Data 3
DTC Set This Ignition	No or Yes	No	No	Engine Data 3
ECT Sensor (°F)	-38 - 284°F	185-226	185-226	Engine Data 3
Engine Load	0-100%	2%	Varies	Engine Data 3
Engine Oil Life	0-100%	Varies	Varies	Engine Data 3
Engine Oil Pressure	0-992 kPa	Varies	Varies	Engine Data 3
Engine Oil Press. (V)	0-5.0v	Varies	Varies	Engine Data 3
Engine Run Time	Hr: Min: Sec	00:00:00	00:00:00	Engine Data 3
Engine Speed	0-10000	Varies	Varies	Engine Data 3
Extended Brake Switch	Apply/Release	Released	Released	Engine Data 3
Fuel Level Sensor (L)	0-5.0v	Varies	Varies	Engine Data 3
Fuel Level Sensor ®	0-5.0v	Varies	Varies	Engine Data 3
Fuel Tank level	0-19 gallons	Varies	Varies	Engine Data 3
Fuel Tank level	0-100%	Varies	Varies	Engine Data 3
Generator F Terminal	0-255 counts	Varies	Varies	Engine Data 3
Generator L-Terminal	Active/Inactive	Active	Active	Engine Data 3
Low Oil Level	No / Yes	Yes: If low	No	Engine Data 3
MAF Sensor (g/sec)	0-512 g/sec	5-9	Varies	Engine Data 3
MAP Sensor (kPa)	10-110 kPa	20 - 48	Varies	Engine Data 3
Mileage Since Last DTC Cleared	0-1677722	Varies	Varies	Engine Data 3
PCM VTD Fail Enable	No / Yes	Varies	Varies	Engine Data 3
Reduce Engine Power	Active/Inactive	Inactive	Inactive	Engine Data 3
PRND A, B, C, P	P-R-N-D	P	A	Engine Data 3
PRND Position	P-R-N-D	P	D	Engine Data 3
Reverse Inhibit (M/T)	Disable/Enable	Disabled	Disabled	Engine Data 3
Spark Advance (°)	-64° to 64°	-20	Varies	Engine Data 3

1997-99 Corvette 5.7L V8 MFI VIN G (All) - PID Data List

Parameter Identification (PID)	PID Value Range	PID Value at Hot Idle	PID Value at 30 mph	Data List Type
Stop Lamp Switch	Applied/Released	Released	Released	Engine Data 3
TAC/PCM Comm.	Okay/Fault	Okay	Okay	Engine Data 3
TCC/CC Brake Switch	Applied/RED	Released	Released	Engine Data 3
TP Desired Angle	0-100%	0 - 10%	0-100%	Engine Data 3
TP Indicated Angle	0-100%	0 - 9%	0-100%	Engine Data 3
Traction Control	Active/Inactive	Inactive	Inactive	Engine Data 3
Transmission Range	PN-R-D4-D3	PN	D3	Engine Data 3
Vehicle Speed	0-155 mph	0	30	Engine Data 3
VTD Auto Learn Time	Active/Inactive	Inactive	Inactive	Engine Data 3
VTD Fuel Disable	Active/Inactive	Inactive	Inactive	Engine Data 3
VTD Fuel Disable to Next Key Off	Yes/No	No	No	Engine Data 3

1997-99 Corvette 5.7L V8 MFI VIN G (A/T) - PID Data List

Parameter Identification (PID)	PID Value Range	PID Value at Hot Idle	PID Value at 30 mph	Data List Type
AIR Pump Relay	Off / On	Off	Off	Fuel Trim
AIR Solenoid Relay	Off / On	Off	Off	Fuel Trim
APP Average Counts	0-125	Varies	Varies	Varies
APP Indicated Angle	0-100%	Varies	Varies	Varies
BARO Sensor (kPa)	10-110 kPa	10-105	10-105	10-105
DTC Set This Ignition	No or Yes	No	No	No
ECT Sensor (°F)	-38 - 284°F	185-226	185-226	185-226
Engine Speed	0-10000	Varies	---	---
EVAP Purge Solenoid	0-100%	0%	0-50	0-100%
EVAP Vacuum Switch	No/Purge	No Purge	Purge	Purge
Fuel Tank level	0-19 gallons	Varies	Varies	Varies
Fuel Tank level	0-100%	Varies	Varies	Varies
Fuel Trim Cell	Cell #	16	Varies	Varies
Fuel Trim Test Inhibit	Disable/Enable	Disabled	Disabled	Disabled
Fuel Trim Learn	Disable/Enable	Enabled	Enabled	Enabled
HO2S (B1 S1, B1 S2)	0-1132 mv	10-1000	10-1000	10-1000
HO2S (B2 S1, B2 S2)	0-1132 mv	10-1000	10-1000	10-1000
IAT Sensor (°F)	-40 to 304°F	50-194	50-194	50-194
Ignition 1 Signal	0.0-25.5v	14.2	14.2	14.2
Injector Pulsewidth	0-1000 ms	1-4	Varies	Varies
Knock Signal Present	Yes / No	No	No	No
Long Term F/T B1, B2	-10% to 10%	0 (± 5%)	0 (± 5%)	Varies
Loop Status	Closed Loop / Open Loop	Closed Loop	Closed Loop	Closed Loop
MAF Sensor (g/sec)	0-512 g/sec	5-9	Varies	Varies
MAP Sensor (kPa)	10-110 kPa	20 - 48	Varies	Varies
Reduce Engine Power	Active/Inactive	Inactive	Inactive	Inactive
Short Term F/T B1-B2	-10% to 10%	0 (± 5%)	0 (± 5%)	Varies
SHRTFT Average	-9% to 7%	0 (± 5%)	0 (± 5%)	Varies
Spark Advance (°)	-64° to 64°	20	Varies	Varies
TP Desired Angle	0-100%	0	0-100	0-100
TP Indicated Angle	0-100%	0	0-100	0-100
Vehicle Speed	0-155 mph	0	30	55

1997-99 Corvette 5.7L V8 MFI VIN G (All) - PID Data List

Parameter Identification (PID)	PID Value Range	PID Value at Hot Idle	PID Value at 30 mph	Data List Type
APP Average Counts	0-125	Varies	Varies	Cruise
APP Indicated Angle	0-100%	Varies	Varies	Cruise
Clutch Switch	DEP/REL	Released	Released	Cruise
Cruise Disengage	History 1-8	Varies	Varies	Cruise
Cruise Requested	No / Yes	No	No	Cruise
Cruise Resume/Accel	Off / On	Off	Off	Cruise
Cruise Set/Coast	Off / On	Off	Off	Cruise
Cruise Switch	Off / On	Off	Off	Cruise
Drive 1, 2	Drive 1, 2	Off/off/on/on	On/off/on/off	Cruise
Drive 3/D, 4/OD	Drive 3/D	---	Off/on/on/off	Cruise
Extended Brake Switch	Applied/REF	Released	Released	Cruise
PRND Position	P-R-N-D	P	D	Cruise
Reduce Engine Power	Active/Inactive	Inactive	Inactive	Cruise
Stop Lamp Switch	Apply/Release	Released	Released	Cruise
TAC/PCM Comm.	Okay / Fault	Okay	Okay	Cruise
TCC/CC Brake Switch	Apply/Release	Released	Released	Cruise
TCC Enable Solenoid	Disable/Enable	Disabled	Disabled	Cruise
TP Desired Angle	0-100%	0%	0-100%	Cruise
TP Indicated Angle	0-100%	0%	0-100%	Cruise
Traction Control	Active/Inactive	Inactive	Inactive	Cruise
Transmission Range	PN-R-D4-D3	PN	D3	Cruise
Vehicle Speed	0-155 mph	0	30	Cruise

1997-99 Corvette 5.7L V8 VIN G (All) - PID Data List

Parameter Identification (PID)	PID Value Range	PID Value at Hot Idle	PID Value at 30 mph	Data List Type
APP Indicated Angle	0-100%	Varies	Varies	F/F, F/R
Air Fuel Ratio	0.0-25.5:1	14.6:1	14.6:1	F/F, F/R
BARO Sensor (kPa)	10-110 kPa	10-105	10-105	F/F, F/R
Clutch Switch	DEP/REL	Released	Released	F/F, F/R
Desired Idle Speed	0-3187 rpm	PCM control	Varies	F/F, F/R
Desired Torque lbs	-349 to 349	Varies	Varies	F/F, F/R
Drive 1, Drive 2	Drive 1	Drive 1	Off/off/on/on	F/F, F/R
Drive 3/D, D4/OD	Drive 2	Drive 2	On/on/on/on	F/F, F/R
ECT Sensor (°F)	-38 - 284°F	185-226	185-226	F/F, F/R
Engine Load	0-100%	2%	Varies	F/F, F/R
Engine Speed	0-10000	Varies	Varies	F/F, F/R
Extended Brake Switch	Applied/Released	Released	Released	F/F, F/R
Failure Counter	0-65535	0	0	F/F, F/R
Injector Pulsewidth	0-1000 ms	1-4	Varies	F/F, F/R
Long Term F/T B1, B2	-10% to 10%	0 (± 5%)	0 (± 5%)	F/F, F/R
Loop Status	Closed Loop / Open Loop	Closed Loop	Closed Loop	F/F, F/R
MAF Sensor (g/sec)	0-512 g/sec	5-9	Varies	F/F, F/R
MAP Sensor (kPa)	10-110 kPa	20-48	Varies	F/F, F/R
Misfire Since DTC Clear	Miles/km	Varies	Varies	F/F, F/R
Miles Since 1st Fail	0-9999 rpm9	Varies	Varies	F/F, F/R
Miles Since Last Fail	0-9999 rpm9	Varies	Varies	F/F, F/R

1997-99 Corvette 5.7L V8 VIN G (All) - PID Data List

Parameter Identification (PID)	PID Value Range	PID Value at Hot Idle	PID Value at 30 mph	Data List Type
PT Induced Chassis Pitch	Active/Inactive	Inactive	Inactive	F/F, F/R
Reverse Inhibit	Disable/Enable	Disabled	Disabled	F/F, F/R
Short Term F/T B1-B2	-10% to 10%	0 (± 5%)	0 (± 5%)	F/F, F/R
Startup ECT Sensor	-40 to 304°F	Varies	Varies	F/F, F/R
TCC/CC Brake Switch	Applied/Released	Released	Released	F/F, F/R
TCC Duty Cycle	0-100%	0%	0%	F/F, F/R
TCC Enable Solenoid	Disable/Enable	Disabled	Disabled	F/F, F/R
TP Desired Angle	0-100%	0%	0-100%	F/F, F/R
TP Indicated Angle	0-100%	0%	0-100%	F/F, F/R
Transmission OSS	0-9999 rpm	Varies	Varies	F/F, F/R
Vehicle Speed	0-155 mph	0	30	F/F, F/R
VTD Fuel Disable	Active/Inactive	Inactive	Inactive	F/F, F/R
VTD Fuel Disable To Key Off	Yes/No	No	No	F/F, F/R

1997-99 Corvette 5.7L V8 MFI VIN G (All) - PID Data List

Parameter Identification (PID)	PID Value Range	PID Value at Hot Idle	PID Value at 30 mph	Data List Type
A/C Clutch	On/Off	Off	Off	Misfire
A/C Request	Yes/No	No	No	Misfire
A/C Status	Off or On	Off	Off	Misfire
Clutch Switch (M/T)	DEP/REL	Released	Released	Misfire
Cycles of Misfire Data	0-100	Varies	Varies	Misfire
CYL Mode Misfire	0-65535	Varies	Varies	Misfire
ECT Sensor (°F)	-38-284°F	185-226	185-226	Misfire
Engine Load	0-100%	2%	Varies	Misfire
Engine Speed	0-10000	Varies	---	Misfire
IC CKT Cylinder 1-8	Okay/Fault	Okay	Okay	Misfire
Loop Status	Closed Loop / Open Loop	Closed Loop	Closed Loop	Misfire
MAF Sensor (g/sec)	0-512 g/sec	5-9	Varies	Misfire
Misfire current Cyl 1-8	0-255 counts	Varies	Varies	Misfire
Misfire Failures	0-65535	Varies	Varies	Misfire
Misfiring Cylinder 1-8	0-65535	Varies	Varies	Misfire
Misfire History Cyl 1-8	0-65535	Varies	Varies	Misfire
Misfire Passes	0-65535	Varies	Varies	Misfire
Misfire Revolution	Status	Except	Except	Misfire
PCM Reset 1, 2	No / Yes	No	No	Misfire
Revolutions w/Misfire	0-65535	Varies	Varies	Misfire
Spark Advance (°)	-64° to 64°	-20	Varies	Misfire
TCC Duty Cycle	0-100%	0%	0%	Misfire
TCC Enable Solenoid	Disable/Enable	Disabled	Disabled	Misfire
Total Misfire Current Count	0-255 counts	Varies	Varies	Misfire
Misfires Since 1st Fail	0-65535	Varies	Varies	Misfire
Passes Since 1st fault	0-65535	Varies	Varies	Misfire
TP Indicated Angle	0-100%	0%	0-100%	Misfire
Vehicle Speed	0-155 mph	0	30	Misfire

1997-99 Corvette 5.7L V8 MFI VIN G (All) - PID Data List

Parameter Identification (PID)	PID Value Range	PID Value at Hot Idle	PID Value at 30 mph	Data List Type
APP Average Counts	0-125	Varies	Varies	TAC
APP Indicated Angle	0-100%	Varies	Varies	TAC
APP Sensor 1	0-5.0v	0.25-1.1v	Varies	TAC
APP Sensor 2	0-5.0v	3.9-4.8v	Varies	TAC
APP Sensor 3	0-5.0v	3.2-4.5v	Varies	TAC
APP Sensor 1, 2, 3 Angle	0-100%	Varies	Varies	TAC
APP 1, 2, 3 Out of Range	Yes/No	No	No	TAC
APP Sensor 1, 2 Disagree	Yes/No	No	No	TAC
APP Sensor 1, 3 Disagree	Yes/No	No	No	TAC
APP Sensor 2, 3 Disagree	Yes/No	No	No	TAC
Clutch Switch (M/T)	DEP/REL	Released	Released	TAC
Cruise Requested	No / Yes	No	No	TAC
Cruise Resume/Accel	Off / On	Off	Off	TAC
Cruise Set/Coast	Off / On	Off	Off	TAC
Cruise Switch	Off / On	Off	Off	TAC
DTC Set This Ignition	No or Yes	No	No	TAC
Delivered Torque lbs	-349 to 349	Varies	Varies	TAC
Desired Idle Speed	0-3187 rpm	PCM control	---	TAC
Desired Torque lbs	-349 to 349	Varies	Varies	TAC
Engine Speed	0-10000	Varies	---	TAC
Extended Brake Switch	Applied/Released	Released	Released	TAC
MAF Sensor (g/sec)	0-512 g/sec	5-9	Varies	TAC
MAP Sensor (kPa)	10-110 kPa	20-48	Varies	TAC
Reduced Engine Power	Active/Inactive	Inactive	Inactive	TAC
Stop Lamp Switch	Applied/Released	Released	Released	TAC
TAC/PCM Communication	Okay/Fault	Okay	Okay	TAC
TCC/CC Brake Switch	Applied/Released	Released	Released	TAC
TCC Enable Solenoid	Disable/Enable	Disabled	Enabled	TAC
TP Desired Angle	0-100%	0%	0-100%	TAC
TP Indicated Angle	0-100%	0%	0-100%	TAC
TP Sensor 1, 2 Angle	0-100%	Varies	Varies	TAC
TP Sensor 1 Volts	0-5.0v	0.25-1.5	Varies	TAC
TP Sensor 2 Volts	0-5.0v	3.0-1.5	Varies	TAC
TP Sensors Disagree	No / Yes	No	No	TAC
TP Sensor 1 or 2 Out of Limit	No / Yes	No	No	TAC
Traction Control	Active/Inactive	Inactive	Inactive	TAC
Vehicle Speed	0-155 mph	0	30	TAC

2000-05 Corvette 5.7L V8 VIN G, VIN S (All) - PID Data List

Parameter Identification (PID)	PID Value Range	PID Value at Hot Idle	PID Value at 30 mph	Data List Type
A/C Clutch Feedback	On/Off	Off	Off	Engine Data 2
A/C HI Side Pressure	0-459 kPa	90	90	Engine Data 2
A/C HI Side Pressure	0-5.1 volts	0.98	0.98	Engine Data 2
A/C Relay Command	On/Off	Off	Off	Engine Data 1, 2 & 3, Misfire
A/C Request Signal	Yes/No	No	No	Engine Data 2
Air Fuel Ratio	0.0-25.5:1	14.7:1	14.7:1	Engine Data 2, Data 3
Air Pump Relay	On/Off	Off	Off	Engine Data 1, Fuel Trim
AIR Solenoid Control	On/Off	Off	Off	Engine Data 1, Fuel Trim
APP Average	0-125	Varies	Varies	TAC
APP Indicated Angle	0-100%	0	WOT: 100%	Engine Data 1-2 EVAP, Cruise F/Trim, TAC
APP Sensor 1 (V)	0-5.0 volts	0.25-1.10	Varies	TAC
APP Sensor 2 (V)	0-5.0 volts	3.9-4.8	Varies	TAC
APP Sensor 3 (V)	0-5.0 volts	3.2-4.5	Varies	TAC
APP Sensor 1 (%)	0-100%	0	Varies	TAC
APP Sensor 2 (%)	0-100%	0	Varies	TAC
APP Sensor 3 (%)	0-100%	0	Varies	TAC
APP Sensor 1 and 2	Agee/DIS	Agree	Agree	TAC
APP Sensor 1 and 3	Agee/DIS	Agree	Agree	TAC
APP Sensor 2 and 3	Agee/DIS	Agree	Agree	TAC
BARO Sensor (kPa)	10-105 kPa	65-104	Varies with changes in the Altitude	Engine Data 1, EVAP, Fuel Trim
BARO Sensor (V)	0-5.1 volts	3.5-4.9	Varies with changes in the Altitude	Engine Data 1, EVAP, Fuel Trim
Clutch Pedal Switch (M/T only)	Applied/Released	Released	Released	Engine Data 1, Cruise
CMP Sensor High to Low Signal	0-65,535 counts	Varies	Varies	Engine Data 2
CMP Sensor Low to High Signal	0-65,535 counts	Varies	Varies	Engine Data 2
Cold Engine Startup	Yes/No	Yes	Yes	EVAP
Column Lock Fuel Disable - ABS DTC	Yes/No (starts/stalls)	No	No	Engine Data 3
Column Lock Fuel Disable - BCM DTC	Yes/No (starts/stalls)	No	No	Engine Data 3
Column Lock - BCM to PCM Comm.	Okay/Fault (starts/stalls)	Okay	Okay	Engine Data 3
Cruise Control Active	Yes/No	No	No	Engine Data 1-3 Cruise, TAC
Cruise Disengage History 1-8	30 possible causes	Varies	Varies	Cruise

2000-05 Corvette 5.7L V8 VIN G, VIN S (All) - PID Data List

Parameter Identification (PID)	PID Value Range	PID Value at Hot Idle	PID Value at 30 mph	Data List Type
C/C On/Off Switch	On/Off	Off	Off	Cruise, TAC
C/C Release Brake Pedal Switch	Applied/Released	Released	Released	Engine Data 2-3 Cruise, TAC
C/C Resume/Accel Switch	On/Off	Off	Off	Cruise, TAC
C/C Set/Coast Switch	On/Off	Off	Off	Cruise, TAC
Current Gear	1, 2, 3, 4	1	4	Engine Data 2
Cycles of Misfire Data	0-100 counts	0	0	Misfire
Desired Idle Speed	0-3187 rpm	Commanded by the PCM	Varies	Engine Data 1-2 EVAP
DTC Set This Ignition	Yes/No	No	No	Engine Data 1-2 EVAP, F/Trim
ECT Sensor (°F)	-38 to 284°F	194-230°F	194-230°F	Engine Data 1-3 TAC, F/Trim, EVAP, Misfire
Engine Load	0-100%	1-4	5-9	Engine Data 1-3 Fuel Trim, EVAP, Misfire Cruise, TAC
Engine Oil Level Switch	Okay / LOW	Okay	Okay	Engine Data 2, Data 3
Engine Oil Life Remaining	0-100%	Varies	Varies	Engine Data 2, Data 3
EOP Sensor (kPa)	0-992 kPa	Varies	Varies	Engine Data 3
EOP Sensor (V)	0-5.0 volts	Varies	Varies	Engine Data 3
Engine Run Time	Hr: Min: Sec 00:00:00 to 99:99:99	00:00:00	00:00:00	Engine Data 1-3 Fuel Trim, EVAP, Misfire Cruise, TAC
Engine Speed	0-10,000 rpm	± 50 rpm of actual speed	Varies	Engine Data 1-3 Fuel Trim, EVAP, Misfire Cruise, TAC
EVAP Purge Solenoid Command	0-100%	Okay	Okay	Engine Data 1, EVAP, F/Trim
EVAP Vent Solenoid Command	Venting/Not	Not Venting	Not Venting	Engine Data 1, EVAP, F/Trim
Extended Travel Brake Pedal Switch	Applied/Released	Released	Released	Engine Data 1, Data 2-3
Fan Control Relay 1 Command	On/Off	Off	Off	Engine Data 2, Data 3
Fan Control Relay 2-3 Command	On/Off	Off	Off	Engine Data 2, Data 3
Fuel Level Sensor - Left Tank	0-5.0 volts	0.7-2.5v	0.7-2.5v	EVAP
Fuel Level Sensor - Right Tank	0-5.0 volts	0.7-2.5v	0.7-2.5v	EVAP

2000-05 Corvette 5.7L V8 VIN G, VIN S (All) - PID Data List

Parameter Identification (PID)	PID Value Range	PID Value at Hot Idle	PID Value at 30 mph	Data List Type
Fuel Tank Level Remaining	Gallons: 0-19 Liters: 0-73	Varies	Varies	EVAP
Fuel Tank Level Remaining	0-100%	Varies	Varies	EVAP
Fuel Tank Pressure Sensor	Inches H2O, or mm Hg	-17 to +7.5 Inches H2O	-17 to +7.5 Inches H2O	Engine Data 1, EVAP
Fuel Tank Pressure Sensor	0-5.0 volts	Varies	Varies	EVAP
Fuel Tank Capacity	19.0 Gallons, 73.0 Liters	19.0 Gallons, 73.0 Liters	19.0 Gallons, 73.0 Liters	EVAP
Fuel Trim Cell	Cell 0-23	Varies	Varies	Engine Data 1, EVAP, F/Trim
Fuel Trim Learn	Enabled or Disabled	Enabled	May Toggle	Engine Data 1, EVAP, F/Trim
Generator L-Terminal	Voltage/No	Voltage	Voltage	Engine Data 2
Generator T-Terminal	0-100%	Varies	Varies	Engine Data 2
HO2S (B1 S1) Signal	0-1000 mv	10-1000	10-1000	Engine Data 1, EVAP, F/Trim
HO2S (B2 S1) Signal	0-1000 mv	10-1000	10-1000	Engine Data 1, EVAP, F/Trim
HO2S (B1 S2) Signal	0-1000 mv	10-1000	10-1000	Engine Data 1, EVAP, F/Trim
HO2S (B2 S2) Signal	0-1000 mv	10-1000	10-1000	Engine Data 1, EVAP, F/Trim
IAC Position	0-1024 counts	10-40	Varies	Engine Data 1, Fuel Trim
IAT Sensor (°F)	-38 to 284°F	91	92	Engine Data 1, Data 2-3, Fuel Trim, EVAP
Ignition 1 Signal	0.0-25.5v	13.5	13.6	Engine Data 1, Data 2-3, Fuel Trim, EVAP, Cruise, TAC
Ignition Mode	Bypass/IC	IC	IC	Engine Data 1, Data 2, Fuel Trim, EVAP
Injector Pulsewidth Average Bank 1	0-1000 ms	1-4	Varies	Engine Data 2, F/Trim, M/fire
Injector Pulsewidth Average Bank 2	0-1000 ms	1-4	Varies	Engine Data 2, F/Trim, M/fire
Knock Retard (°)	0-25.5°	0.0° to 16°	Varies	Engine Data 1
Long Term Fuel Trim Bank 1	-23 to +16	0 (± 5%)	0 (± 5%)	Engine Data 1-3 EVAP, F/Trim
Long Term Fuel Trim Bank 2	-23 to +16	0 (± 5%)	0 (± 5%)	Engine Data 1-3 EVAP, F/Trim
Long Term Fuel Trim Average Bank 1	-23 to +16	0 (± 5%)	0 (± 5%)	Engine Data 1-3 EVAP, F/Trim
Long Term Fuel Trim Average Bank 2	-23 to +16	0 (± 5%)	0 (± 5%)	Engine Data 1-3 EVAP, F/Trim

2000-05 Corvette 5.7L V8 VIN G, VIN S (All) - PID Data List

Parameter Identification (PID)	PID Value Range	PID Value at Hot Idle	PID Value at 30 mph	Data List Type
Loop Status	Open / Closed	Closed	Closed	Engine Data 1-3 Fuel Trim, EVAP, Misfire
MAF Sensor (g/sec)	0-512 g/sec	5-9	20-26	Engine Data 1-3 Fuel Trim, EVAP, Misfire
MAF Sensor (Hz)	0-31,999 Hz	2000-3000	Varies	Engine Data 3
MAP Sensor (kPa)	10-105 kPa	20-48	Varies	Engine Data 1-3 Fuel Trim, EVAP, Misfire
MAP Sensor (V)	0-5.1 volts	1.0-2.0	Varies	Engine Data 1-3 F/Trim, TAC EVAP, Misfire
MIL Command	On/Off	Off	Off	Engine Data 1
Mileage Since DTC Cleared	0-65,535 miles	Varies	Varies	Engine Data 3
Misfire Current Cylinder 1-8	0-200 counts	0-4	0	Misfire
Misfire History Cylinder 1-8	0-65,535	0	0	Misfire
PCM Reset	Yes/No	No	No	Engine Data 1, EVAP, F/Trim
PCM/VCM in VTD Fail Enable Mode	Enable / Disable	Enable	Enable	Engine Data 3
Power Enrichment	Active/Inactive	Inactive	Inactive	Engine Data 2, Fuel Trim
Powertrain Induced Chassis Pitch	Active/Inactive	Inactive	Inactive	Engine Data 3
Reduced Engine Power	Active/Inactive	Inactive	Inactive	Engine Data 1-3 Cruise, TAC
Reverse Inhibit Solenoid (M/T only)	Yes/No	No	No	Engine Data 1
Short Term Fuel Trim Bank 1	-10 to +10%	0 (± 5%)	0 (± 5%)	Engine Data 1-3 EVAP, F/Trim
Short Term Fuel Trim Bank 2	-10 to +10%	0 (± 5%)	0 (± 5%)	Engine Data 1-3 EVAP, F/Trim
Short Term Fuel Trim Average Bank 1	-10 to +10%	0 (± 5%)	0 (± 5%)	Fuel Trim
Short Term Fuel Trim Average Bank 2	-10 to +10%	0 (± 5%)	0 (± 5%)	Fuel Trim
Skip Shift Solenoid (M/T only)	Skip/No Skip	No Skip	No Skip	Engine Data 1
Skip Shift Lamp (M/T)	On/Off	Off	Off	Engine Data 1
Spark Advance	-64° to +64°	-20°	Varies	Engine Data 1-2 F/Trim, M/fire
Startup ECT	-38 to 284°F	Varies	Varies	Engine Data 2, EVAP, F/Trim
Startup IAT	-38 to 284°F	Varies	Varies	Engine Data 2, EVAP, F/Trim

2000-05 Corvette 5.7L V8 VIN G, VIN S (All) - PID Data List

Parameter Identification (PID)	PID Value Range	PID Value at Hot Idle	PID Value at 30 mph	Data List Type
Stop Lamp Pedal Switch	Applied or Released	Released	Released	Engine Data 1-3 Cruise, TAC
TAC/PCM Comm.	Okay/Fault	Okay	Okay	Cruise, TAC
TCC/Cruise Brake Switch	Applied or Released	Released	Released	Engine Data 2
TCC Enable Solenoid Command	Enabled or Disabled	Disabled	Disabled	Engine Data 2, Misfire
TCC PWM Solenoid Command	Enabled or Disabled	Disabled	Disabled	Engine Data 1-2 Misfire
TFP Switch	P/N, Reverse Drive 4, 3, 2, 1, Invalid	Park/Neutral	Drive 4	Engine Data 2-3
TFT Sensor	-38 to 284°F	Varies	Varies	Engine Data 3
Torque Delivered	NM, ft lbs	Varies	Varies	Engine Data 2, TAC
Torque Requested	NM, ft lbs	Varies	Varies	Engine Data 2, TAC
TP Desired Angle	0-100%	Varies	Varies	Engine Data 1-2 EVAP, TAC, Cruise
TP Desired Angle	0-100%	Varies	Varies	Engine Data 1-3 Cruise, TAC Fuel Trim, EVAP, Misfire
TP Sensor 1	0-100%	Varies	Varies	TAC
TP Sensor 1	0-5.0 volts	0.25-1.50	Varies	TAC
TP Sensor 2	0-100%	Varies	Varies	TAC
TP Sensor 2	5.0-0 volts	4.0-1.5	Varies	TAC
TP Sensor 1 and 2	Agree/DIS	Agree	Agree	TAC
Traction Control	Active/Inactive	Inactive	Inactive	Cruise
Traction Control Status	Active/Inactive	Inactive	Inactive	Cruise
Transaxle Range Switch	1, 2, 3, 4	1	4	Engine Data 2-3
Vehicle Speed	Km/h, MPH	0	30	Engine Data 1-3 Cruise, TAC Fuel Trim, EVAP, Misfire
VTD Auto Learn Timer	Hr: Min: Sec 00:00:00 to 99:99:99	00:00:00	00:00:00	Engine Data 3
VTD Fuel Disable	Active/Inactive	Inactive	Inactive	Engine Data 3
VTD Fuel Disable Until Ignition Off	Yes/No	No	No	Engine Data 3
Warmups Without an Emission Fault	0-255 counts	Varies	Varies	Engine Data 3
Warmups Without a Non-Emission Fault	0-255 counts	Varies	Varies	Engine Data 3

IMPALA PID DATA

2000-05 Impala 3.8L V6 MFI VIN K (A/T) - PID Data List

Parameter Identification (PID)	PID Value Range	PID Value at Hot Idle	PID Value at 30 mph	Data List Type
3X Crank Sensor	0-9,999 rpm	Varies	Varies	Engine Data 2
18X Crank Sensor	0-1,600 rpm	Varies	Varies	Engine Data 2
1-2 Shift Solenoid Circuit Status	Okay / Fault/ Invalid State	Okay	Okay	Output Driver
2-3 Shift Solenoid Circuit Status	Okay / Fault/ Invalid State	Okay	Okay	Output Driver
A/C HI Side Pressure	0-459 kPa	90	90	Engine Data 2
A/C HI Side Pressure	0-5.1 volts	0.98	0.98	Engine Data 2
A/C Off for WOT	Yes/No	No	No	Engine Data 2
A/C Pressure Disable	Yes/No	No	No	Engine Data 2
A/C Relay Circuit Status	Okay / Fault/ Invalid State	Okay	Okay	Output Driver
A/C Relay Command	On/Off	Off	Off	Engine Data 1-2 EGR, Misfire
A/C Request Signal	Yes/No	No	No	Engine Data 2
Air Fuel Ratio	0.0-25.5:1	14.2-14.7	14.2-14.7	Engine Data 2, Fuel Trim
Air Pump Circuit Status	Okay/Stuck Low/Stuck High/Fault	Okay	Okay	Output Driver
Air Pump Relay Command	On/Off	Off	Off	Engine Data 1, Fuel Trim
Air Solenoid Circuit Status	Okay/Stuck Low/Stuck High/Fault	Okay	Okay	Output Driver
AIR Solenoid Command	On/Off	Off	Off	Engine Data 1, Fuel Trim
BARO Sensor (kPa)	10-105 kPa	Varies w/ALT	Varies	Engine Data 1, EGR, EVAP, Fuel Trim
BARO Sensor (V)	0-5.1 volts	Varies w/ALT	Varies	Engine Data 1, EGR, EVAP, Fuel Trim
CMP Sensor Signal	Yes/No	Yes	Yes	Engine Data 2
Cruise Control Active	Yes/No	No	No	Engine Data 1
Cruise Inhibit Reason	Brake/Clutch VSS	Vehicle Speed	Vehicle Speed	ENG Data
Cruise Inhibit Signal Circuit Status	Okay/Stuck Low/Stuck High/Fault	Okay	Okay	Output Driver
Cruise Inhibit Signal Command	On/Off	On	On	Engine Data 1
Current Gear	1, 2, 3, 4	1	4	Engine Data 1-2 EGR, F/Trim
Cycles of Misfire Data	0-99 counts	0	0	Misfire
Cylinder 1-6 Injector Circuit History	Okay/Fault	Okay	Okay	Output Driver

2000-05 Impala 3.8L V6 MFI VIN K (A/T) - PID Data List

Parameter Identification (PID)	PID Value Range	PID Value at Hot Idle	PID Value at 30 mph	Data List Type
Cylinder 1-6 Injector Circuit Status	Okay / Stuck Low/ Stuck High / Fault	Okay	Okay	Output Driver
Decel Fuel Cutoff	Active/Inactive	Inactive	Inactive	EGR, F/Trim
Desired EGR Position	0-100%	0	Varies	Engine Data 1, EGR, Misfire
Desired EGR Position	0-5.1v	0.0	Varies	EGR
Desired Idle Speed	0-3187 rpm	Commanded by the PCM	Varies	Engine Data 1, Data 2, EVAP
Driver Module 1 Status	Enabled / Off- HI Volts/ Off-HI TEMP/ Invalid State	Enabled	Enabled	Output Driver
Driver Module 2 Status	Enabled / Off- HI Volts/ Off-HI TEMP/ Invalid State	Enabled	Enabled	Output Driver
Driver Module 3 Status	Enabled / Off- HI Volts/ Off-HI TEMP/ Invalid State	Enabled	Enabled	Output Driver
Driver Module 4 Status	Enabled / Off- HI Volts/ Off-HI TEMP/ Invalid State	Enabled	Enabled	Output Driver
ECT Sensor (°F)	-40 to 304°F	190-221°F	190-221°F	Engine Data 1-2 EGR, F/Trim, EVAP, Misfire
EGR Flow Test Count	0-255 counts	0-12	Varies	EGR
EGR Learned Minimum Position	0-5.1 volts	0.16-1.0	Varies	EGR
EGR Position Sensor	0-100%	0	Varies	Engine Data 1, EGR, Misfire
EGR Position Sensor	0-5.1 volts	0.14-1.0	Varies	EGR
EGR Position Variance	0-100%	0	0	EGR
EGR Solenoid Circuit History	Okay / Fault/ Invalid State	Okay	Okay	Output Driver
EGR Solenoid Circuit Status	Okay / Fault/ Invalid State	Okay	Okay	Output Driver
EGR Solenoid Command	0-100%	0	Varies	EGR
Engine Load	0-100%	1-4	Varies	Engine Data 1-2 EGR, F/Trim, Misfire, EVAP
Engine Oil Level Switch	Okay / LOW	Okay	Okay	Engine Data 2
Engine Oil Life Remaining	0-100%	Varies	Varies	Engine Data 2

2000-05 Impala 3.8L V6 MFI VIN K (A/T) - PID Data List

Parameter Identification (PID)	PID Value Range	PID Value at Hot Idle	PID Value at 30 mph	Data List Type
Engine Oil Pressure Switch	Okay/LOW	Okay	Okay	Engine Data 2
Engine Run Time	Hr: Min: Sec 00:00:00 to 99:99:99	00:00:00	00:00:00	Engine Data 1-2 EGR, F/Trim, TAC, Misfire, Cruise, EVAP
Engine Speed	0-9999 rpm	± 50 rpm of actual speed	Varies	Engine Data 1-2 EGR, F/Trim, Misfire, EVAP
EVAP Fault History	No Fault / Excess VAC /Purge Valve Leak / Small Leak / Weak Vacuum	No Fault	No Fault	EVAP
EVAP Purge Solenoid Circuit Status	Okay / Fault/ Invalid State	Okay	Okay	Output Driver
EVAP Purge Solenoid Command	0-100%	15	Varies	Engine Data 1, EVAP, F/Trim
EVAP Test Abort Reason	Not Aborted/ Lost Enable/ Small Leak/ Not at Rest	Not Aborted	Not Aborted	EVAP
EVAP Test Result	No Result / Passed/ Fail/ P0440/442/ P0446/1441	Okay	Okay	EVAP
EVAP Test State	Waiting for Purge/ Test Running/Test Completed	Test Completed	Test Completed	EVAP
EVAP Vent Solenoid Circuit Status	Okay / Fault/ Invalid State	Okay	Okay	Output Driver
EVAP Vent Solenoid Command	Venting / Not Venting	Not Venting	Not Venting	Engine Data 1, EVAP, F/Trim
Fan Control Relay 1 Circuit Status	Okay / Fault/ Invalid State	Okay	Okay	Output Driver
Fan Control Relay 1 Command	On/Off	Off	Off	Engine Data 2
Fan Control Relay 2-3 Circuit Status	Okay/Fault	Okay	Okay	Output Driver
Fan Control Relay 2-3 Command	On/Off	Off	Off	Engine Data 2
Fuel Pump Relay Circuit Status	Okay / Fault/ Invalid State	Okay	Okay	Output Driver
Fuel Pump Relay Circuit History Status	Okay / Fault	Okay	Okay	Output Driver
Fuel Pump Relay Command	On/Off	On	On	Engine Data 2
Fuel Tank Level	0-100%	Varies	Varies	EVAP

2000-05 Impala 3.8L V6 MFI VIN K (A/T) - PID Data List

Parameter Identification (PID)	PID Value Range	PID Value at Hot Idle	PID Value at 30 mph	Data List Type
Fuel Tank Pressure Sensor	Inches H2O, mm Hg	-17 to +7.5 Inches H2O	-17 to +7.5 Inches H2O	Engine Data 1, EVAP
Fuel Tank Pressure Sensor	0-5.0 volts	Varies	Varies	EVAP
Fuel Trim Cell	Cell 0-10	Varies	Varies	Engine Data 1, EVAP, F/Trim
Fuel Trim Learn	Enabled or Disabled	Enabled	May Toggle	Engine Data 1, EVAP, F/Trim
Generator L-Terminal	On/Off	On	On	Engine Data 2
HO2S (B1 S1) Signal	0-1132 mv	10-1000	10-1000	Engine Data 1, EVAP, F/Trim
HO2S (B1 S2) Signal	0-1132 mv	10-1000	10-1000	Engine Data 1, EVAP, F/Trim
IAC Position	0-255 counts	10-60	Varies	Engine Data 1, EGR, F/Trim
IAT Sensor (°F)	-40 to 304°F	91	92	Engine Data 1-2 EGR, EVAP, Fuel Trim
Ignition 1	0.0-25.5v	13.5	13.6	Engine Data 1-2 EGR, EVAP, Fuel Trim
Ignition Mode	Bypass/IC	IC	IC	Engine Data 2
Injector Pulsewidth	0-1000 ms	1.5-3.5	Varies	Engine Data 2, F/Trim, M/fire
Knock Retard (°)	0-25.5°	0°	0°	Engine Data 1, EGR
Long Term Fuel Trim Bank 1	-23 to +16	0 (± 5%)	0 (± 5%)	Engine Data 1-2 EVAP, F/Trim
Long Term Fuel Trim Bank 2	-23 to +16	0 (± 5%)	0 (± 5%)	Engine Data 1-2 EVAP, F/Trim
Loop Status	Open / Closed	Closed	Closed	Engine Data 1-2 EGR, EVAP, Fuel Trim
MAF Sensor (g/sec)	0-512 g/sec	3-6	Varies	Engine Data 1-2 EGR, EVAP, Fuel Trim, Misfire
MAF Sensor (Hz)	0-32,000 Hz	2000-3000	Varies	Engine Data 2
MAP Sensor (kPa)	10-105 kPa	20-48	Varies	Engine Data 1-2 EGR, F/Trim, EVAP, Misfire
MAP Sensor (V)	0-5.1 volts	0.40-2.0	Varies	Engine Data 1-2
MIL Status	Okay / Fault/ Invalid State	Okay	Okay	Output Driver
MIL Command	On/Off	Off	Off	Engine Data 2
Misfire Current Cylinder 1-6	0-198 counts	0-4	0	Misfire
Misfire History Cylinder 1-6	0-65,535	0	0	Misfire

2000-02 Impala 3.8L V6 MFI VIN K (All) - PID Data List

Parameter Identification (PID)	PID Value Range	PID Value at Hot Idle	PID Value at 30 mph	Data List Type
Number of DTC(s)	Number	0	No	Engine Data 1, EVAP, F/Trim
PCM/VCM in VTD Fail Enable	Yes/No	Yes	Yes	Engine Data 1
Power Enrichment	Active/Inactive	Inactive	Inactive	Engine Data 2, Misfire
Short Term Fuel Trim Bank 1	-10 to +10%	0 (± 5%)	0 (± 5%)	Engine Data 1-2 EVAP, F/Trim
Short Term Fuel Trim Bank 2	-10 to +10%	0 (± 5%)	0 (± 5%)	Engine Data 1-2 EVAP, F/Trim
Spark Advance	-64° to +64°	22	Varies	Engine Data 1-2 F/Trim, M/fire
Starter Enable Relay Circuit Status	Okay / Fault/ Invalid State	Okay	Okay	Output Driver
Starter Relay Control	On/Off	Off	Off	Engine Data 2
Startup ECT	-40 to 419°F	Varies	Varies	EVAP, F/Trim
Startup IAT	-40 to 419°F	80°F	80°F	EVAP, F/Trim
TCC Brake Pedal Switch	Apply/REL	Released	Released	Engine Data 1, Data 2
TCC PWM Solenoid Command	On/Off	Off	Off	Engine Data 2
TCS Circuit Status	Okay/Fault/ Invalid State	Okay	Okay	Output Driver
TCC Circuit History	Okay/Fault/ Invalid State	Okay	Okay	Output Driver
TFP Switch	Park/Neutral/ Drive 4-3-2-1	1	4	Engine Data 2, EGR, F/Trim
Torque Delivered Signal	0-100%	85	85	Engine Data 2
Torque Request Signal	0-100%	100	100	Engine Data 2
TP Sensor (%)	0-100%	0	Varies	Engine Data 1-2 EGR, F/Trim, EVAP, M/Fire
TP Sensor (V)	0-5.0 volts	0.2-0.8	Varies	Engine Data 1-2
Traction Control Status	Active/Inactive	Inactive	Inactive	Engine Data 2
Vehicle Speed	Km/h, MPH	0	30	Engine Data 1-2 EGR, F/Trim, EVAP M/fire
VTD Auto Learn Timer	Active/Inactive	Inactive	Inactive	Engine Data 1
VTD Fuel Enable	Active/Inactive	Inactive	Inactive	Engine Data 1
VTD Fuel Disable Until Ignition Off	Active/Inactive	Inactive	Inactive	Engine Data 1

LUMINA & MONTE CARLO PID DATA

1995 Lumina, Monte Carlo 3.1L V6 VIN M (A/T) - PID Data List

Parameter Identification (PID)	PID Value Range	PID Value at Hot Idle	PID Value at 30 mph	Data List Type
A/C Clutch	On / Off	Off	Off	Engine Data 1
A/C Pressure psi	0-459 psi	139-399	139-399	Engine Data 1
A/C Request	Yes / No	No	Off	Off
Engine Load (Calc.)	0-100%	50-60	Varies	Varies
Camshaft Signal	0 or 1	0-1	0-1	0-1
Commanded Gear	1-2-3-4	1	3	4
Cooling Fan 1, 2	On / Off	Off	Off	Off
Cranking RPM	0-999	N/A	N/A	N/A
Cruise Control	On / Off	Off	Off	On or Off
Current Weak CYL	CYL# / None	None	None	None
Desired Idle Speed	0-3187	800	---	---
ECT Sensor (°F)	-40-419°F	185-228	185-228	185-228
EGR Solenoid 1, 2, 3	On / Off	Off	On	On
Engine Speed	0-9999 rpm	Varies	---	---
Fuel Trim Cell	Cell #0-4	0	4	4
HO2S-11 (B1 S1)	0-1132 mv	10-1000	10-1000	10-1000
HO2S-21 (B2 S1)	0-1132 mv	10-1000	10-1000	10-1000
HO2S Cross Counts	0-255 counts	0-20	0-20	0-20
IAC Motor	0-255 counts	5-50	---	---
IAT Sensor (°F)	-40 to 304°F	50-176	50-176	50-176
Injector Pulsewidth	0-985 ms	2.9	3.0	4.8
Knock Signal	Yes / No	No	No	No
Long Term Fuel Trim	0-255 counts	110-155	110-155	110-155
Loop Status	Closed Loop / Open Loop	Closed Loop	Closed Loop	Closed Loop
Low Oil Light	On / Off	Off	Off	Off
MAF Sensor (g/sec)	0-512 g/sec	5-7	Varies	Varies
MAP Sensor (kPa)	10-110 kPa	26-48	Varies	Varies
Quad Driver 1, 2	HI or LOW	LOW	---	---
PROM ID Number	Number	0000	0000	0000
PSP Switch	HI/Normal	Normal	Normal	Normal
PRNDL (P/A/B/C)	HIGH/LOW	HIGH	LOW	LOW
Purge Duty Cycle	0-100%	0%	20%	35%
Rich/Lean Flag	L or R	L-R-L-R	L-R-L-R	L-R-L-R
Short Term Fuel Trim	0-255	110-155	110-155	110-155
Spark Advance (°)	-90° to 90°	12-22	30-34	29-33
Spark Retard (°)	0-128°	0	0	0
TCC Command	On/Off	Off	Off	On
TCC Solenoid	On/Off	Off	Off	On
TCC Slip Speed	-255/+255	700-750	5	4
TCC Duty Cycle	0-100%	0	0	75
Throttle Angle (%)	0-100%	0	5	15
Time From Startup	Hr: Min: Sec	00:00:00	00:00:00	00:00:00
TP Sensor (V)	0.0--5.0v	0.56v	0.74	0.87v
TFT Sensor (°)	-40 to 304°F	194	194	194
Trans. Mode Switch	Off / On	On	Off: 2nd	On
Trans. Range Switch	P, A, B, C	P	C	C
2nd Gear Start	On / Off	On	Off	On
2nd, 3rd, 4th Gear Switch	Yes / No	No	Yes/No	Yes/No

1995 Lumina, Monte Carlo 3.4L V6 VIN X (A/T) - PID Data List

Parameter Identification (PID)	PID Value Range	PID Value at Hot Idle	PID Value at 30 mph	Data List Type
1st Gear Switch	Yes / No	No	No	Engine Data 1
4th Gear Switch	Yes / No	No	Yes	Engine Data 1
A/C Clutch	On / Off	Off	Off	Engine Data 1
A/C Pressure psi	0-459 psi	139-399	139-399	Engine Data 1
A/C Request	Yes / No	No	No	Engine Data 1
AIR Pump Relay	Off / On	On: Cold	Off	Engine Data 1
BARO Sensor (V)	0-5.0v	3.5-4.5	3.5-4.5	Engine Data 1
Block Learn Counts	0-255 counts	110-155	110-155	Engine Data 1
Block Learn Cell	0-21	0 or 1	0 or 2	Engine Data 1
Clutch Switch	Off / On	Off	Off	Engine Data 1
Cooling Fan 1, 2	On / Off	Off	Off	Engine Data 1
Crank RPM	0-999 rpm	N/A	N/A	Engine Data 1
Desired Idle Speed	0-3187 rpm	800	Varies	Engine Data 1
Engine Speed	0-9999 rpm	800	Varies	Engine Data 1
ECT Sensor (°F)	-40-419°F	185-221	185-221	Engine Data 1
EGR Solenoid 1, 2	On / Off	Off	On or Off	Engine Data 1
EGR Solenoid 3	On / Off	Off	On or Off	Engine Data 1
Fuel Integrator Counts	0-255 counts	118-138	118-138	Engine Data 1
Fuel Pump Signal	0.0-25.5v	14.2	14.2	Engine Data 1
Fuel Trim Cell	Cell #	0 or 1	0 or 2	Engine Data 1
IAC Motor	0-255 counts	10-50	10-50	Engine Data 1
IAT Sensor (°F)	-40 to 304°F	50-176	50-176	Engine Data 1
Ignition 1 Signal	0.0-25.5v	14.2	14.2	Engine Data 1
Injector Pulsewidth	0-985 ms	1-4	Varies	Engine Data 1
Knock Retard (°)	0-128°	0	0	Engine Data 1
Knock Signal	Yes / No	No	No	Engine Data 1
Loop Status	Closed Loop / Open Loop	Closed Loop	Closed Loop	Engine Data 1
MAP Sensor (kPa)	10-110 kPa	29-48	Varies	Engine Data 1
MAP Sensor (V)	0-5.0v	1.5	Varies	Engine Data 1
O2S-11 (B1 S1)	0-1132 mv	10-1000	10-1000	Engine Data 1
PNP Switch	PN / R-DL	PN	R-DL	Engine Data 1
PSP Switch	HI/Normal	Normal	Normal	Engine Data 1
PROM ID Number	Number	0000	0000	Engine Data 1
Purge Duty Cycle	0-100%	0%	0-50	Engine Data 1
Shift Solenoid 'A'	On / Off	On	Off	Engine Data 1
Shift Solenoid 'B'	On / Off	On	Off	Engine Data 1
Shift Light	On / Off	Off	On: Shift	Engine Data 1
Spark Advance (°)	-90° to 90°	Varies	Varies	Engine Data 1
TCC Solenoid	On / Off	Off	Off	Engine Data 1
TP Sensor (V)	0-5.0v	0.29-0.98	---	Engine Data 1
Throttle Angle	0-100%	0	0-100	Engine Data 1
Time From Start	Hr: Min: Sec	00:00:00	00:00:00	Engine Data 1
Trans. Mode Switch	On / Off	On	Off: 2nd	Engine Data 1
Vehicle Speed	0-155 mph	0	30	Engine Data 1

1996-99 Lumina, Monte Carlo 3.1L VIN M (A/T) - PID Data List

Parameter Identification (PID)	PID Value Range	PID Value at Hot Idle	PID Value at 30 mph	Data List Type
A/C HI Side Pressure	0-5v	1-4	1-4	Engine Data 1
A/C Off for WOT	Yes / No	No	No	Engine Data 1
A/C Pressure Disable	Yes / No	No	No	Engine Data 1
A/C Request Signal	Yes / No	No	No	Engine Data 1
Actual EGR Position	0-100%	0	0-100	Engine Data 1
AIR Relay Command	On/Off	Off	Off	Engine Data 1
AIR Solenoid Control	On/Off	Off	Off	Engine Data 1
BARO Sensor (kPa)	10-110 kPa	65-110	65-110	Engine Data 1
Brake Switch	Applied/RE	Released	Released	Engine Data 1
Cam Signal Present	Yes / No	Yes	Yes	Engine Data 1
Commanded A/C	On / Off	Off	Off	Engine Data 1
Commanded Starter	Enabled/DIS	Disabled	Disabled	Engine Data 1
Cruise Active Mode	On/Off	Off	Off	Engine Data 1
Cruise Inhibit Signal	On/Off	On	On	Engine Data 1
Current Gear	1-2-3-4	1	3	Engine Data 1
Desired EGR Position	0-100%	0	0-100	Engine Data 1
Desired Idle Speed	0-3107 rpm	ROM control		Engine Data 1
ECT Sensor (°F)	-40 to 304°F	185-220	185-220	Engine Data 1
EGR Duty Cycle	0-100%	0	0-100	Engine Data 1
EGR Position Sensor	0-100%	0	0-100	Engine Data 1
Engine Load	0-100%	2-10	Varies	Engine Data 1
Engine Run Time	Hr: Min: Sec	00:00:00	00:00:00	Engine Data 1
Engine Speed	0-9999 rpm	Varies	---	Engine Data 1
EVAP Purge Solenoid	0-100%	0-25	0-100	Engine Data 1
EVAP Vent Solenoid	Open/Close	Open	Open	Engine Data 1
FTP Sensor (In. H2O)	-17.5 to 7.5"	Varies	Varies	Engine Data 1
FC Relay Low & High	On / Off	Off	Off	Engine Data 1
Fuel Pump Command	On / Off	On	On	Engine Data 1
Fuel Trim Cell	Cell #0-4	0	0-4	Engine Data 1
Fuel Trim Learn	Enabled/DIS	Enabled	Enabled	Engine Data 1
HO2S-11 (B1 S1)	0-1132 mv	10-1000	10-1000	Engine Data 1
HO2S-12 (B1 S2)	0-1132 mv	10-1000	10-1000	Engine Data 1
HO2S-11 Status	Not / Ready	Ready	Ready	Engine Data 1
HO2S-11 XCounts	0-255 counts	Varies	Varies	Engine Data 1
IAC Position	0-255 counts	10-60	---	Engine Data 1
IAT Sensor (°F)	-40 to 304°F	50-194	50-194	Engine Data 1
Ignition 1 Signal	0.0-25.5v	14.2	14.2	Engine Data 1
Ignition Mode	IC / Bypass	IC	IC	Engine Data 1
Injector Pulsewidth	0-985 ms	1.5-3.5	Varies	Engine Data 1
Knock Retard (°)	0-128°	0	0	Engine Data 1
Long Term Fuel Trim	-10% to 10%	0 (± 5%)	0 (± 5%)	Engine Data 1
Loop Status	Closed Loop / Open Loop	Closed Loop	Closed Loop	Engine Data 1
MAF Sensor (g/sec)	0-512 g/sec	3-6	Varies	Engine Data 1
MAF Sensor (Hz)	0-31,999hz	1200-3000	Varies	Engine Data 1
MAP Sensor (kPa)	10-110 kPa	20-48	Varies	Engine Data 1
Number of Trouble Codes	Number	00000	00000	Engine Data 1
Power Enrichment	Active/Inactive	Inactive	Inactive	Engine Data 1
Short Term Fuel Trim	-10% to 10%	0 (± 5%)	0 (± 5%)	Engine Data 1

1996-99 Lumina, Monte Carlo 3.1L VIN M (A/T) - PID Data List

Parameter Identification (PID)	PID Value Range	PID Value at Hot Idle	PID Value at 30 mph	Data List Type
Spark Advance (º)	-64º to 64º	20	Varies	Engine Data 1
TCC Engaged	Engage/DIS	Disengaged	Disengaged	Engine Data 1
TP Angle (%)	0-100%	0	Varies	Engine Data 1
Traction Control Mode	Active/Inactive	Inactive	Inactive	Engine Data 1
Traction Control Desired	0-100%	100	0-100	Engine Data 1
Traction Control Torque	0-100%	80-90	0-100	Engine Data 1
Transmission Range	P-R-N-4-3-2	P	3	Engine Data 1
TR Switch P-A-B-C	Several	Several	Several	Engine Data 1
Vehicle Speed	0-155 mph	0	30	Engine Data 1
VTD Fuel Disable	Active/Inactive	Inactive	Inactive	Engine Data 1
24X Crank Sensor	0-1600	Varies	Varies	Engine Data 1

1996-99 Lumina, Monte Carlo 3.1L VIN M (A/T) - PID Data List

Parameter Identification (PID)	PID Value Range	PID Value at Hot Idle	PID Value at 30 mph	Data List Type
Air Fuel Ratio	0:1-25.5:1	14.6:1	14.6:1	Engine Data 2
A/C HI Side Pressure	0-5v	1-4	1-4	Engine Data 2
A/C Off for WOT	Yes / No	No	No	Engine Data 2
A/C Pressure Disable	Yes / No	No	No	Engine Data 2
A/C Request Signal	Yes / No	No	No	Engine Data 2
Actual EGR Position	0-100%	0	0-100	Engine Data 2
AIR Relay Command	On / Off	Off	Off	Engine Data 2
AIR Solenoid Control	On / Off	Off	Off	Engine Data 2
CMP Signal Present	Yes / No	Yes	Yes	Engine Data 2
Current Gear	1-2-3-4	1	3	Engine Data 2
Desired EGR Position	0-100%	0	0-100	Engine Data 2
Desired Idle Speed	0-3187 rpm	PCM control	---	Engine Data 2
Engine Hot Lamp	On / Off	Off	Off	Engine Data 2
Engine Load	0-100%	2-10	Varies	Engine Data 2
Engine Oil Level Switch	Okay/LOW	Okay	Okay	Engine Data 2
Engine Oil Life	0-100%	Varies	Varies	Engine Data 2
Engine Run Time	Hr: Min: Sec	00:00:00	00:00:00	Engine Data 2
Engine Speed	0-9999 rpm	Varies	---	Engine Data 2
Fan Control Relay 1	On / Off	Off	Off	Engine Data 2
Fan Control Relay 2-3	On / Off	Off	Off	Engine Data 2
Generator Lamp	On / Off	Off	Off	Engine Data 2
Generator 'L' Signal	On / Off	On	On	Engine Data 2
IAC Position	0-255 counts	10-60	---	Engine Data 2
IAT Sensor (ºF)	-40 to 304ºF	50-194	50-194	Engine Data 2
Ignition 1 Signal	0.0-25.5v	14.2	14.2	Engine Data 2
Ignition Mode	IC / Bypass	IC	IC	Engine Data 2
Injector Pulsewidth	0-985 ms	1.5-3.5	Varies	Engine Data 2
Knock Retard (º)	0.0-25.5º	0	0	Engine Data 2
Long Term Fuel Trim	-10% to 10%	Varies	Varies	Engine Data 2
Loop Status	Open Loop or Closed Loop	Closed Loop	Closed Loop	Engine Data 2
Low Oil Lamp Control	On / Off	Off	Off	Engine Data 2

1996-99 Lumina, Monte Carlo 3.1L VIN M (A/T) - PID Data List

Parameter Identification (PID)	PID Value Range	PID Value at Hot Idle	PID Value at 30 mph	Data List Type
MAF Sensor (g/sec)	0-512 g/sec	3-6	Varies	Engine Data 2
MAF Sensor (Hz)	0-31,999hz	1200-3000	Varies	Engine Data 2
MAP Sensor (kPa)	10-110 kPa	20-48	Varies	Engine Data 2
MIL (lamp) Command	On / Off	Off	Off	Engine Data 2
Power Enrichment	Active/Inactive	Inactive	Inactive	Engine Data 2
Short Term Fuel Trim	-10% to 10%	Varies	Varies	Engine Data 2
Spark Advance (°)	-64° to 64°	20	Varies	Engine Data 2
TCC Brake Pedal Switch	Apply/REL	Released	Released	Engine Data 2
TCC PWM Solenoid	On / Off	Off	On	Engine Data 2
TFP Switch	Several	P/N	Several	Engine Data 2
TP Angle (%)	0-100%	0	Varies	Engine Data 2
Traction Control State	Active/Inactive	Inactive	Inactive	Engine Data 2
Traction Requested	0-100%	100%	Varies	Engine Data 2
Traction Delivered	75%	100%	Varies	Engine Data 2
3X Crank Sensor	1600-10,000	Varies	Varies	Engine Data 2
24X Crank Sensor	0-1600	Varies	Varies	Engine Data 2

1996-99 Lumina, Monte Carlo 3.1L VIN M (A/T) - PID Data List

Parameter Identification (PID)	PID Value Range	PID Value at Hot Idle	PID Value at 30 mph	Data List Type
A/C Request	Yes / No	No	No	EGR
Actual EGR Position	0-100%	0	0-100	EGR
BARO Sensor (kPa)	10-110 kPa	65-110	65-110	EGR
Commanded A/C	On / Off	Off	Off	EGR
Current Gear	1-2-3-4	1	3	EGR
Decel Fuel Mode	Active/Inactive	Inactive	Inactive	EGR
Desired EGR Position	0-100%	0	0-100	EGR
ECT Sensor (°F)	-40 to 304°F	185-220	185-220	EGR
EGR Closed Valve Pintle	0-5v	0.14-1.00	Varies	EGR
EGR Duty Cycle	0-100%	0	0-100	EGR
EGR Feedback	0-5v	0.14-1.00	Varies	EGR
EGR Position Error	0-100%	0	0	EGR
EGR Flow Test Count	0-255	0-10	0-10	EGR
Engine Load	0-100%	2-10	Varies	EGR
Engine Run Time	Hr: Min: Sec	00:00:00	00:00:00	EGR
Engine Speed	0-9999 rpm	Varies	---	EGR
Hot Open Loop	Active/Inactive	Inactive	Inactive	EGR
IAT Sensor (°F)	-40 to 304°F	50-194	50-194	EGR
Ignition 1 Signal	0.0-25.5v	14.2	14.2	EGR
Knock Retard (°)	0.0-25.5°	0	0	EGR
Loop Status	Open Loop or Closed Loop	Closed Loop	Closed Loop	EGR
MAF Sensor (g/sec)	0-512 g/sec	3-6	Varies	EGR
MAP Sensor (kPa)	10-110 kPa	20-48	Varies	EGR
Non-Volatile Memory	Pass/Fail	Pass	Pass	EGR
Power Enrichment	Active/Inactive	Inactive	Inactive	EGR
TCC Engaged	Engage/DIS	Disengaged	Disengaged	EGR
TP Angle (%)	0-100%	0	Varies	EGR
Transmission Range	P-R-N-4-3-2	P	3	EGR

1996-99 Lumina, Monte Carlo 3.1L VIN M (A/T) - PID Data List

Parameter Identification (PID)	PID Value Range	PID Value at Hot Idle	PID Value at 30 mph	Data List Type
Air Fuel Ratio	0.0-25:1	14.6:1	14.6:1	Fuel Trim
AIR Active Test Inhibit	Yes / No	Yes	Yes	Fuel Trim
AIR Active Test	Yes / No	No	No	Fuel Trim
AIR Test Complete	Yes / No	Yes	Yes	Fuel Trim
AIR Passive test 2 (F)	Yes / No	No	No	Fuel Trim
AIR Passive test 1 (P)	Yes / No	Yes	Yes	Fuel Trim
AIR Passive Test (P)	Yes / No	Yes	Yes	Fuel Trim
AIR Passive Test Inhibit	Yes / No	Yes	Yes	Fuel Trim
AIR Passive Test Act.	Yes / No	No	No	Fuel Trim
AIR Pump Relay	On / Off	Off	Off	Fuel Trim
AIR Pump Solenoid	On / Off	Off	Off	Fuel Trim
BARO Sensor (kPa)	10-110 kPa	65-110	65-110	Fuel Trim
ECT Sensor (°F)	-40 to 304°F	185-220	185-220	Fuel Trim
Engine Load	0-100%	2-10	Varies	Fuel Trim
Engine Run Time	Hr: Min: Sec	00:00:00	00:00:00	Fuel Trim
Engine Speed	0-9999 rpm	Varies	---	Fuel Trim
EVAP Purge Solenoid	0-100%	0-25	0-100	Fuel Trim
Fuel Trim Cell	Cell # 0-4	0	0-4	Fuel Trim
Fuel Trim Learn	Disable/Enable	Enabled	Enabled	Fuel Trim
HO2S (B1 S1, B1 S2)	0-1132 mv	10-1000	10-1000	Fuel Trim
IAT Sensor (°F)	-40 to 304°F	50-194	50-194	Fuel Trim
Long Term Fuel Trim	-10% to 10%	0 (± 5%)	0 (± 5%)	Fuel Trim
Loop Status	Closed Loop / Open Loop	Closed Loop	Closed Loop	Fuel Trim
MAF Sensor (g/sec)	0-512 g/sec	3-6	Varies	Fuel Trim
Number of Trouble Codes	Number	00000	00000	Fuel Trim
Power Enrichment	Active/Inactive	Inactive	Inactive	Fuel Trim
Startup ECT Sensor	-40 to 304°F	Varies	Varies	Fuel Trim
Startup IAT Sensor	-40 to 304°F	Varies	Varies	Fuel Trim
TP Angle (%)	0-100%	0	Varies	Fuel Trim

1996-99 Lumina, Monte Carlo 3.1L VIN M (A/T) - PID Data List

Parameter Identification (PID)	PID Value Range	PID Value at Hot Idle	PID Value at 30 mph	Data List Type
1-2, 2-3 solenoid state	Several	Okay	Okay	Output Device
A/C Relay Status	Several	Okay	Okay	Output Device
AIR Pump Relay state	Several	Okay	Okay	Output Device
AIR Solenoid Status	Several	Okay	Okay	Output Device
Cruise Inhibit Status	Several	Okay	Okay	Output Device
Cyl 1-6 INJ Circuit	Okay/Fault	Okay	Okay	Output Device
Cyl 1-6 INJ History	Several	Okay	Okay	Output Device
Driver Module 1-4	Okay/Fault	Okay	Okay	Output Device
EGR Solenoid Status	Several	Okay	Okay	Output Device
EVAP Purge Solenoid	Several	Okay	Okay	Output Device
EVAP Vent Solenoid	Several	Okay	Okay	Output Device
FC Relay 1, 2-3	Several	Okay	Okay	Output Device
MIL (lamp) Status	Several	Okay	Okay	Output Device
TCC Solenoid Status	Several	Okay	Okay	Output Device

1996-99 Lumina, Monte Carlo 3.1L VIN M (A/T) - PID Data List

Parameter Identification (PID)	PID Value Range	PID Value at Hot Idle	PID Value at 30 mph	Data List Type
Air Fuel Ratio	0.0-25:1	14.6:1	14.6:1	Catalyst
BARO Sensor (kPa)	10-110 kPa	65-110	65-110	Catalyst
BARO Sensor (V)	0.0-5.0v	3.5-4.5	---	Catalyst
Commanded A/C	On / Off	Off	Off	Catalyst
Current Gear	1-2-3-4	1	3	Catalyst
ECT Sensor (°F)	-40 to 304°F	185-220	185-220	Catalyst
Engine Load	0-100%	2-10	Varies	Catalyst
Engine Run Time	Hr: Min: Sec	00:00:00	00:00:00	Catalyst
Engine Speed	0-9999 rpm	Varies	---	Catalyst
Fan High Speed	On / Off	Off	Off	Catalyst
Fan Low Speed	On / Off	Off	Off	Catalyst
Fuel Trim Cell	Cell #0-4	0	0-4	Catalyst
Fuel Trim Learn	Enabled/DIS	Enabled	Enabled	Catalyst
HO2S-11 (B1 S1)	0-1132 mv	10-1000	10-1000	Catalyst
HO2S-12 (B1 S2)	0-1132 mv	10-1000	10-1000	Catalyst
HO2S-11 Status	Ready/Not	Ready	Ready	Catalyst
HO2S-11 X Counts	0-255 counts	Varies	Varies	Catalyst
Hot Open Loop	Active/Inactive	Inactive	Inactive	Catalyst
IAC Position	0-255 counts	10-40	---	Catalyst
IAT Sensor (°F)	-40 to 302°F	50-194	50-194	Catalyst
Ignition 1 Signal	0.0-25.5v	14.2	14.2	Catalyst
Injector Pulsewidth	0-985 ms	1.5-3.5	Varies	Catalyst
Loop Status	Open Loop or Closed Loop	Closed Loop	Closed Loop	Catalyst
MAF Sensor (g/sec)	0-512 g/sec	4-6	Varies	Catalyst
MAP Sensor (kPa)	10-110 kPa	20-48	Varies	Catalyst
MAP Sensor (V)	0-5v	0.75-2.0	Varies	Catalyst
Non-Volatile Memory	Pass/Fail	Pass	Pass	Catalyst
Power Enrichment	Active/Inactive	Inactive	Inactive	Catalyst
Startup ECT Sensor	-40 to 304°F	Varies	Varies	Catalyst
Startup IAT Sensor	-40 to 304°F	Varies	Varies	Catalyst
TP Angle (%)	0-100%	0	Varies	Catalyst
TP Sensor (V)	0-5v	0.20-0.74	Varies	Catalyst
Transmission Range	P-R-N-4-3-2	P	3	Catalyst
TWC Protection	Active/Inactive	Inactive	Inactive	Catalyst
Vehicle Speed	0-155 mph	0	30	Catalyst

1996-99 Lumina, Monte Carlo 3.1L VIN M (A/T) - PID Data List

Parameter Identification (PID)	PID Value Range	PID Value at Hot Idle	PID Value at 30 mph	Data List Type
Engine Oil Level	Okay/LOW	Okay	Okay	IPC
Engine Oil Life	0-100%	0-100%	0-100%	IPC
Engine Oil Pressure	Okay/LOW/HI	Okay	Okay	IPC
Fuel Tank Level	0-100%	Varies	Varies	IPC
Generator Command	On / Off	On	On	IPC
Generator PWM	0-100%	0-100	0-100	IPC
Ignition 1 Signal	0.0-25.5v	14.2	14.2	IPC
MIL (lamp)	On / Off	Off	Off	IPC
Non-Volatile Memory	Pass/Fail	Pass	Pass	IPC

1996-99 Lumina, Monte Carlo 3.1L VIN M (A/T) - PID Data List

Parameter Identification (PID)	PID Value Range	PID Value at Hot Idle	PID Value at 30 mph	Data List Type
BARO Sensor (kPa)	10-110 kPa	65-110	65-110	EVAP
ECT Sensor (°F)	-40 to 304°F	185-220	185-220	EVAP
Engine Load	0-100%	2-10	Varies	EVAP
Engine Run Time	Hr: Min: Sec	00:00:00	00:00:00	EVAP
Engine Speed	0-9999 rpm	Varies	---	EVAP
EVAP Canister Purge	0-100%	0-100	0-100	EVAP
EVAP Fault History	Several	No Fault	No Fault	EVAP
EVAP Purge Solenoid	Several	Okay	Okay	EVAP
EVAP Vent Solenoid	Open/Close	Open	Open	EVAP
Fuel Tank Level	0-100%	Varies	Varies	EVAP
FTP Sensor (In. H2O)	-17.5 to 7.5"	Varies	Varies	EVAP
Fuel Trim Learn	Enabled/DIS	Enabled	Enabled	EVAP
IAT Sensor (°F)	-40 to 304°F	50-194	50-194	EVAP
Long Term Fuel Trim	-10% to 10%	Varies	Varies	EVAP
Loop Status	Closed Loop / Open Loop	Closed Loop	Closed Loop	EVAP
MAF Sensor (g/sec)	0-512 g/sec	3-6	Varies	EVAP
MAP Sensor (kPa)	0-450 psi	20-48 kPa	Varies	EVAP
Non-Volatile Memory	Pass/Fail	Pass	Pass	EVAP
Startup ECT Sensor	-40 to 304°F	Varies	Varies	EVAP
Startup IAT Sensor	-40 to 304°F	Varies	Varies	EVAP
TP Angle (%)	0-100%	0	Varies	EVAP

1996-99 Lumina, Monte Carlo 3.1L VIN M (A/T) - PID Data List

Parameter Identification (PID)	PID Value Range	PID Value at Hot Idle	PID Value at 30 mph	Data List Type
Abuse Management	Active/Inactive	Inactive	Inactive	Misfire
A/C Request Signal	Yes / No	No	No	Misfire
Actual EGR Position	0-100%	0	0-100	Misfire
Cam Signal Present	Yes / No	Yes	Yes	Misfire
Commanded A/C	On / Off	On: AC on	Off	Misfire
Cycles of Misfire Data	0-255	0-99	Varies	Misfire
Decel Fuel Mode	Active/Inactive	Inactive	Inactive	Misfire
Desired EGR Position	0-100%	0	0-100	Misfire
ECT Sensor (°F)	-40 to 302°F	185-220	185-220	Misfire
Engine Load	0-100%	2-10	Varies	Misfire
Engine Speed	0-9999 rpm	Varies	---	Misfire
Hot Mode	On / Off	Off	Off	Misfire
HO2S-11 (B1 S1)	0-1132 mv	10-1000	10-1000	Misfire
Injector Pulsewidth	0-985 ms	1.5-3.5	Varies	Misfire
MAF Sensor (g/sec)	0-512 g/sec	4-6	Varies	Misfire
Misfire current Cyl 1-6	0-198	Varies	Varies	Misfire
Misfire History Cyl 1-6	0-65535	0	0	Misfire
Non-Volatile Memory	Pass/Fail	Pass	Pass	Misfire
Total Misfire Current	0-99 counts	Varies	Varies	Misfire
Total Misfire Failures	0-65535	Varies	Varies	Misfire
Total Misfire Passes	0-65535	Varies	Varies	Misfire
3X Crank Sensor	1600-10,000	Varies	Varies	Varies
24X Crank Sensor	0-1600	Varies	Varies	Varies

1996-97 Lumina, Monte Carlo 3.4L V6 VIN X (A/T) - PID Data List

Parameter Identification (PID)	PID Value Range	PID Value at Hot Idle	PID Value at 30 mph	Data List Type
3X Crank Sensor	0-9999 rpm	Varies	Varies	Engine Data 1
24X Crank Sensor	0-1600	Varies	Varies	Engine Data 1
A/C Clutch	On/Off	Off	Off	Engine Data 1
A/C HI Side Pressure	0-459 psi	139-399	139-399	Engine Data 1
A/C Off for WOT	On / Off	Off	Off	Engine Data 1
A/C Pressure Disable	Yes / No	No	No	Engine Data 1
A/C Request	Yes / No	No	No	Engine Data 1
Actual EGR Position	0-100%	0%	0-100%	Engine Data 1
AIR Pump	On / Off	Off	Off	Engine Data 1
BARO Sensor (kPa)	10-110 kPa	65-110	65-110	Engine Data 1
Cam Signal Present	Yes / No	Yes	Yes	Engine Data 1
Commanded A/C	On / Off	Off	Off	Engine Data 1
Commanded Fan 1, 2	On / Off	Off	On	Engine Data 1
Cruise Engaged	Yes / No	No	No	Engine Data 1
Cruise Inhibited	Yes / No	Yes	Yes	Engine Data 1
Desired EGR Position	0-100%	0	0-100	Engine Data 1
Desired Idle Speed	0-9999 rpm	PCM control	---	Engine Data 1
ECT Sensor (°F)	-40 to 304°F	185-220	185-220	Engine Data 1
EGR Duty Cycle	0-100%	0	0-100	Engine Data 1
Engine Load	0-100%	2-5	Varies	Engine Data 1
Engine Run Time	Hr: Min: Sec	00:00:00	00:00:00	Engine Data 1
Engine Speed	0-9999 rpm	Varies	---	Engine Data 1
EVAP Purge PWM	0-100%	0-25	0-75%	Engine Data 1
EVAP Vacuum Switch	Open/closed	Open	Open	Engine Data 1
Fuel Pump	On / Off	On	On	Engine Data 1
Fuel Trim Cell	Cell # 0-4	0	0-4	Engine Data 1
Fuel Trim Learn	Disable/Enable	Enabled	Enabled	Engine Data 1
HO2S-11 Status	Ready/Not	Ready	Ready	Engine Data 1
HO2S (B1 S1)	0-1132 mv	10-1000	10-1000	Engine Data 1
HO2S (B1 S2)	0-1132 mv	10-1000	10-1000	Engine Data 1
HO2S-11 XCounts	0-255 counts	0-20	0-20	Engine Data 1
IAC Position	0-255 counts	10-40	---	Engine Data 1
IAT Sensor (°F)	-40 to 304°F	50-194	50-194	Engine Data 1
Ignition Bypass	Active/Inactive	Inactive	Inactive	Engine Data 1
Ignition IC Mode	Active/Inactive	Active	Active	Engine Data 1
Injector Fault	Yes / No	No	No	Engine Data 1
Injector Pulsewidth	0-985 ms	1.5-3.5	Varies	Engine Data 1
Knock Retard (°)	0-128°	0	0	Engine Data 1
KS Noise Channel	0-5.0v	1-2	Varies	Engine Data 1
Long Term Fuel Trim	-10 to 10%	0 (± 5%)	0 (± 5%)	Engine Data 1
MAF Sensor (g/sec)	0-512 g/sec	3 - 6	Varies	Engine Data 1
MAF Sensor (Hz)	0-31,999hz	2000-2500	Varies	Engine Data 1
MAP Sensor (kPa)	10-110 kPa	25-34	Varies	Engine Data 1
Short Term Fuel Trim	-10 to 10%	0 (± 5%)	0 (± 5%)	Engine Data 1
Spark Advance (°)	-64° to 64°	16	Varies	Engine Data 1
Startup ECT/IAT	-40 to 304°F	Varies	Varies	Engine Data 1
TCC Engaged	Yes / No	No	No	Engine Data 1

1996-97 Lumina, Monte Carlo 3.4L V6 VIN X (A/T) - PID Data List

Parameter Identification (PID)	PID Value Range	PID Value at Hot Idle	PID Value at 30 mph	Data List Type
Transmission Range	PN/R/Drive	Park/Neutral	Drive 4	Engine Data 1
TWC Diagnostic	Enabled/DIS	Enabled	Enabled	Engine Data 1
TWC Protection	Active/Inactive	Inactive	Inactive	Engine Data 1
TWC Monitor Test	0-255	0	0	Engine Data 1
Vehicle Speed	0-155 mph	0	30	Engine Data 1
VTD Fuel Enable	Disable/Enable	Disabled	Disabled	Engine Data 1

1996-97 Lumina, Monte Carlo 3.4L V6 VIN X (A/T) - PID Data List

Parameter Identification (PID)	PID Value Range	PID Value at Hot Idle	PID Value at 30 mph	Data List Type
A/C Clutch	On / Off	Off	Off	EGR
A/C Request	Yes / No	No	No	EGR
Actual EGR Position	0-100%	0%	0-100%	EGR
BARO Sensor (kPa)	10-110 kPa	65-110	65-110	EGR
Current Gear	1-2-3-4	1	3	EGR
Decel Fuel Mode	Active/Inactive	Inactive	Inactive	EGR
Desired EGR Position	0-100%	0%	0-100%	EGR
ECT Sensor (°F)	-40 to 304°F	185-220	185-220	EGR
EGR Closed Valve Pintle Position	0-5.0v	0.14-1.00	---	EGR
EGR Duty Cycle	0-100%	0%	0-100%	EGR
EGR Position Error	0-100%	0 - 9%	---	EGR
EGR Flow Test Count	0-255 counts	0-10	0-10	EGR
Engine Load	0-100%	2-5	Varies	EGR
Engine Run Time	Hr: Min: Sec	00:00:00	00:00:00	EGR
Engine Speed	0-9999 rpm	Varies	---	EGR
Hot Open Loop	Active/Inactive	Inactive	Inactive	EGR
Knock Activity	Yes / No	No	No	EGR
Knock Retard (°)	0-128°	0	0	EGR
Loop Status	Closed Loop / Open Loop	Closed Loop	Closed Loop	EGR
MAF Sensor (g/sec)	0-512 g/sec	3 – 6	Varies	EGR
MAP Sensor (kPa)	10-110 kPa	25-34	Varies	EGR
MAP Sensor (V)	0-5.0v	0.78-1.30	Varies	EGR
Power Enrichment	Active/Inactive	Active	Active	EGR
TCC Engaged	DIS/Engage	Disengaged	Disengaged	EGR
TP Angle (%)	0-100%	0%	0-100%	EGR
Transmission Range	X or O	OOXX	Varies	EGR
Vehicle Speed	0-155 mph	0	30	EGR

1996-97 Lumina, Monte Carlo 3.4L V6 VIN X (A/T) - PID Data List

Parameter Identification (PID)	PID Value Range	PID Value at Hot Idle	PID Value at 30 mph	Data List Type
Engine Oil Level	Okay / LOW	Okay	Okay	IPC
Generator Control	On / Off	On	On	IPC
Generator Lamp	On / Off	Off	Off	IPC
Ignition 1 Signal	0.0-25.5v	14.2	14.2	IPC
Low Oil Lamp	On / Off	Off	Off	IPC
MIL Status	On / Off	Off	Off	IPC

1996-97 Lumina, Monte Carlo 3.4L V6 VIN X (A/T) - PID Data List

Parameter Identification (PID)	PID Value Range	PID Value at Hot Idle	PID Value at 30 mph	Data List Type
Air Fuel Ratio	0.0-25.5:1	14.6:1	14.6:1	HO2S
AIR Pump	On / Off	Off	Off	HO2S
Decel Fuel Mode	Active/Inactive	Inactive	Inactive	HO2S
ECT Sensor (°F)	-40 to 304°F	185-220	185-220	HO2S
Engine Run Time	Hr: Min: Sec	00:00:00	00:00:00	HO2S
Engine Speed	0-9999 rpm	Varies	---	HO2S
EVAP Purge PWM	0-100%	0-25	0-75%	HO2S
HO2S (B1 S1, B1 S2)	0-1132 mv	10-1000	10-1000	HO2S
HO2S-11 Status	Not / Ready	Ready	Ready	HO2S
HO2S-11 XCounts	0-255 counts	Varies	Varies	HO2S
IAT Sensor (°F)	-40 to 304°F	50-194	50-194	HO2S
Injector Pulsewidth	0-985 ms	1.5-3.5	Varies	HO2S
Long Term Fuel Trim	-10 to 10%	0 (± 5%)	0 (± 5%)	HO2S
Loop Status	Closed Loop / Open Loop	Closed Loop	Closed Loop	HO2S
MAF Sensor (g/sec)	0-512 g/sec	3 – 6	Varies	HO2S
Power Enrichment	Active/Inactive	Inactive	Inactive	HO2S
Short Term Fuel Trim	-10 to 10%	0 (± 5%)	0 (± 5%)	HO2S
Startup ECT Sensor	-40 to 304°F	Varies	Varies	HO2S
Startup IAT Sensor	-40 to 304°F	Varies	Varies	HO2S

1996-97 Lumina, Monte Carlo 3.4L V6 VIN X (A/T) - PID Data List

Parameter Identification (PID)	PID Value Range	PID Value at Hot Idle	PID Value at 30 mph	Data List Type
3X, 24X Crank Sensor	0-9999 rpm	Varies	Varies	Misfire
Abuse Management	Active/Inactive	Inactive	Inactive	Misfire
A/C Clutch	On / Off	Off	Off	Misfire
A/C Request	Yes / No	No	No	Misfire
Actual EGR Position	0-100%	0%	0-100%	Misfire
AIR Active Test	Active/Inactive	Inactive	Inactive	Misfire
Cam Signal Present	Yes / No	Yes	Yes	Misfire
Current Gear	1-2-3-4	1	3	Misfire
Decel Fuel Mode	Active/Inactive	Inactive	Inactive	Misfire
Desired EGR Position	0-100%	0%	0-100%	Misfire
ECT Sensor (°F)	-40 to 304°F	185-220	185-220	Misfire
HO2S-11 (B1 S1)	0-1132 mv	10-1000	10-1000	Misfire
MAF Sensor (g/sec)	0-512 g/sec	3 - 6	Varies	Misfire
Misfire current Cyl 1-6	0-198	0 - 4	Varies	Misfire
Misfire History Cyl 1-6	0-65535	0	0	Misfire
Misfire Since 1st Fail	0-65535	0	0	Misfire
Misfire Passes Since First Failure	0-65535	0	0	Misfire
Non-Volatile Memory	Pass/Fail	Pass	Pass	Misfire
Power Enrichment	Active/Inactive	Inactive	Inactive	Misfire
Spark Advance (°)	-64° to 64°	16	Varies	Misfire
TCC Engaged	Yes / No	No	No	Misfire
Total Misfire Count	0-255 counts	0	0	Misfire
TP Angle (%)	0-100%	0	0-100	Misfire
Trans. Hot Mode	Active/Inactive	Inactive	Inactive	Misfire

1998-99 Lumina 3.8L V6 MFI VIN K (A/T) - PID Data List

Parameter Identification (PID)	PID Value Range	PID Value at Hot Idle	PID Value at 30 mph	Data List Type
A/C Relay Command	On/Off	Off	Off	Engine Data 1
Actual EGR Position	0-100%	0	Varies	Engine Data 1
Air Fuel Ratio	0.0-25:1	14.6:1	Varies	Engine Data 1
AIR Relay Command	On / Off	On	Off	Engine Data 1
BARO Sensor (kPa)	10-110 kPa	65-110	65-110	Engine Data 1
Cruise Control Active	Yes / No	No	No	Engine Data 1
Cruise Inhibit Control	Yes / No	Yes	Yes	Engine Data 1
Cruise Inhibit Reason	Several	None	None	Engine Data 1
Cruise Inhibit Signal	On / Off	Off	Off	Engine Data 1
Current Gear	1-2-3-4	1	3	Engine Data 1
Decel Fuel Mode	Active/Inactive	Inactive	Inactive	Engine Data 1
Desired EGR Position	0-100%	0	0-100	Engine Data 1
Desired Idle Speed	0-3187 rpm	PCM control	Varies	Engine Data 1
ECT Sensor (°F)	-40 to 304°F	195	195	Engine Data 1
EGR Position Sensor	0-100%	0	Varies	Engine Data 1
Engine Load	0-100%	2-5	Varies	Engine Data 1
Engine Run Time	Hr: Min: Sec	00:00:00	00:00:00	Engine Data 1
Engine Speed	0-9999 rpm	Varies	Varies	Engine Data 1
EVAP Purge Control	0-100%	0-10	0-100	Engine Data 1
EVAP Vent Command	Open/Closed	Open	Open	Engine Data 1
F/P Relay Command	On / Off	On	On	Engine Data 1
FTP Sensor (In. H2O)	-17.5 to 7.5"	Varies	Varies	Engine Data 1
Fuel Trim Cell	Cell# 0-9	0	0-9	Engine Data 1
Fuel Trim Learn	Yes / No	Yes	Yes	Engine Data 1
Generator Control	On / Off	On	On	Engine Data 1
HO2S-11 (B1 S1)	0-1132 mv	10-1000	10-1000	Engine Data 1
HO2S-12 (B1 S2)	0-1132 mv	10-1000	10-1000	Engine Data 1
HO2S-11 Status	Ready/Not	Ready	Ready	Engine Data 1
IAC Position (Counts)	0-255	10-40	Varies	Engine Data 1
IAT Sensor (°F)	-40 to 304°F	50-194	50-194	Engine Data 1
Ignition 1 Signal	0.0-25.5v	14.2	14.2	Engine Data 1
Injector Pulsewidth	0-985 ms	1.5-3.5	Varies	Engine Data 1
Knock Retard (°)	0.0-25.5°	0	Varies	Engine Data 1
Long Term Fuel Trim	-10% to 10%	Varies	Varies	Engine Data 1
MAF Sensor (g/sec)	0-512 g/sec	3-6	Varies	Engine Data 1
MAF Sensor (Hz)	0-32000 Hz	1200-3000	Varies	Engine Data 1
MAP Sensor (kPa)	10-110 kPa	20-48	Varies	Engine Data 1
Number of Trouble Codes	Number	0	0	Engine Data 1
PCM/VCM in VTD Fail	Yes / No	No	No	Engine Data 1
Short Term Fuel Trim	-10 to 26%	Varies	Varies	Engine Data 1
Spark Advance (°)	-64° to 64°	19	Varies	Engine Data 1
TCC Brake Pedal Switch	Applied/Released	Released	Released	Engine Data 1
TP Angle (%)	0-100%	1-4	0-100%	Engine Data 1
TWC Protection	Active/Inactive	Inactive	Inactive	Engine Data 1
VTD Auto Learn Time	Active/Inactive	Inactive	Inactive	Engine Data 1
VTD Fuel Disable	Active/Inactive	Inactive	Inactive	Engine Data 1
VTD Fuel Disable Until Ignition Off	Yes/No	No	No	Engine Data 1

1998-99 Lumina 3.8L V6 MFI VIN K (A/T) - PID Data List

Parameter Identification (PID)	PID Value Range	PID Value at Hot Idle	PID Value at 30 mph	Data List Type
3X Crank Sensor	0-9999 rpm	Varies	Varies	Engine Data 2
18X Crank Sensor	0-1600 rpm	Varies	Varies	Engine Data 2
A/C HI Side Pressure	0-5.0v	Varies	Varies	Engine Data 2
A/C Off for WOT	Yes / No	No	No	Engine Data 2
A/C Pressure Disable	Yes / No	No	No	Engine Data 2
A/C Pressure (Range)	Yes / No	No	No	Engine Data 2
A/C Relay Command	On / Off	Off	Off	Engine Data 2
A/C Request Signal	Yes / No	No	No	Engine Data 2
Actual EGR Position	0-100%	0	Varies	Engine Data 2
Air Fuel Ratio	0.0-25:1	14.6:1	14.6:1	Engine Data 2
BARO Sensor (kPa)	10-110 kPa	65-110	65-110	Engine Data 2
CMP Signal Present	Yes / No	Yes	Yes	Engine Data 2
Current Gear	1-2-3-4	1	3	Engine Data 2
Desired EGR Position	0-100%	0	Varies	Engine Data 2
Desired Idle Speed	0-3187 rpm	PCM control	---	Engine Data 2
ECT Sensor (°F)	-40 to 304°F	195	195	Engine Data 2
Engine Load	0-100%	2-5	Varies	Engine Data 2
Engine Oil Level Switch	Okay / LOW	Okay	Okay	Engine Data 2
Engine Oil Life	0-100%	Varies	Varies	Engine Data 2
Engine Oil Pressure	0-5v	2-3	2-3	Engine Data 2
Engine Run Time	Hr: Min: Sec	00:00:00	00:00:00	Engine Data 2
Engine Speed	0-9999 rpm	Varies	---	Engine Data 2
Extended Travel Brake Pedal Switch	Engage/REL	Released	Released	Engine Data 2
FC Relay 1, 2-3	On / Off	On	Off	Engine Data 2
Generator F-Terminal	0-100%	5-95	5-95	Engine Data 2
Generator L-Terminal	On / Off	Off	Off	Engine Data 2
IAT Sensor (°F)	-40 to 304°F	85	85	Engine Data 2
Ignition Mode	Bypass / IC	IC	IC	Engine Data 2
Injector Pulsewidth	0-985 ms	1.5-3.5	Varies	Engine Data 2
Knock Retard (°)	0.0-25.5°	0	Varies	Engine Data 2
Long Term Fuel Trim	-10% to 10%	Varies	Varies	Engine Data 2
MAF Sensor (g/sec)	0-512 g/sec	3-6	Varies	Engine Data 2
MAF Sensor (Hz)	0-32000 Hz	1200-3000	Varies	Engine Data 2
MAP Sensor (kPa)	10-110 kPa	20-48	Varies	Engine Data 2
MIL (lamp) Command	On / Off	Off	Off	Engine Data 2
Power Enrichment	Active/Inactive	Inactive	Inactive	Engine Data 2
Short Term Fuel Trim	-10% to 10%	Varies	Varies	Engine Data 2
Spark Advance (°)	-64° to 64°	19	Varies	Engine Data 2
TCC Brake Pedal Switch	Apply/Release	Released	Released	Engine Data 2
TCC PWM Solenoid	On / Off	Off	Off	Engine Data 2
TFP Switch	Several	P-N	Drive 3	Engine Data 2
TP Angle (%)	0-100%	1-4	Varies	Engine Data 2
TP Sensor (V)	0-5v	0.5-1.2	Varies	Engine Data 2
Torque Delivered (%)	0-100%	70	Varies	Engine Data 2
Torque Request (%)	0-100%	100	Varies	Engine Data 2
TR Switch Position	P-A-B-C	P	B	Engine Data 2
Traction Control State	Active/Inactive	Inactive	Inactive	Engine Data 2

1998-99 Lumina 3.8L V6 MFI VIN K (A/T) - PID Data List

Parameter Identification (PID)	PID Value Range	PID Value at Hot Idle	PID Value at 30 mph	Data List Type
AIR Active Test Inhibit	Yes / No	Yes	Yes	AIR
AIR Active Test Pass	Yes / No	No	No	AIR
AIR Passive Enabled	Yes / No	No	No	No
AIR Passive Test Inhibit	Yes / No	No	No	No
AIR Passive test Pass	Yes / No	No	No	No
AIR Passive 1 Fail	Yes / No	No	No	No
AIR Passive 2 Fail	Yes / No	No	No	No
AIR Relay, Solenoid	On / Off	On	Off	Off
Air Pump Relay State	Several	Okay	Okay	Okay
Air Solenoid Status	Several	Okay	Okay	Okay
ECT Sensor (°F)	-40 to 304°F	195	195	196
Engine Load	0-100%	2-10	Varies	Varies
Engine Run Time	Hr: Min: Sec	00:00:00	00:00:00	00:00:00
EVAP Canister Purge	0-100%	Varies	Varies	Varies
Fuel Trim Cell	Cell# 0-9	0-9	0-9	0-9
HO2S-11 (B1 S1)	0-1132 mv	10-1000	10-1000	10-1000
IAT Sensor (°F)	-40 to 304°F	85	85	86
Long, Short Term F/T	-10% to 10%	Varies	Varies	Varies
MAF Sensor (g/sec)	0-512 g/sec	3-6	Varies	Varies
Startup ECT Sensor	-40 to 304°F	Varies	Varies	Varies
Startup IAT Sensor	-40 to 304°F	Varies	Varies	Varies
TP Angle (%)	0-100%	0	Varies	Varies

1998-99 Lumina 3.8L V6 MFI VIN K (A/T) - PID Data List

Parameter Identification (PID)	PID Value Range	PID Value at Hot Idle	PID Value at 30 mph	Data List Type
A/C Request Signal	Yes / No	No	No	Catalyst
Air Fuel Ratio	0.0-25:1	14.6:1	14.6:1	Catalyst
BARO Sensor (kPa)	10-110 kPa	65-110	65-110	Catalyst
Commanded A/C	On / Off	Off	Off	Catalyst
Decel Fuel Mode	Active/Inactive	Inactive	Inactive	Catalyst
ECT Sensor (°F)	-40 to 304°F	195	195	Catalyst
Engine Load	0-100%	2-10	Varies	Catalyst
Engine Run Time	Hr: Min: Sec	00:00:00	00:00:00	Catalyst
Engine Speed	0-9999 rpm	Varies	---	Catalyst
FC Relay 1 & 2	On / Off	Off	Off	Catalyst
Fuel Trim Learn	Enabled/DIS	Enabled	Enabled	Catalyst
HO2S-11, HO2S-12	0-1132 mv	10-1000	10-1000	Catalyst
IAC Position	0-255	10-40	---	Catalyst
IAT Sensor (°F)	-40 to 304°F	50-194	50-194	Catalyst
Injector Pulsewidth	0-999 ms	1.5-3.5	Varies	Catalyst
Loop Status	Open Loop or Closed Loop	Closed Loop	Closed Loop	Catalyst
MAF Sensor (g/sec)	0-512 g/sec	3-6	Varies	Catalyst
Power Enrichment	Active/Inactive	Inactive	Inactive	Catalyst
Short Term Fuel Trim	-10% to 10%	0 (± 5%)	0 (± 5%)	Catalyst
TP Angle (%)	0-100%	0	Varies	Catalyst
TWC Protection	Active/Inactive	Inactive	Inactive	Catalyst

1998-99 Lumina 3.8L V6 MFI VIN K (A/T) - PID Data List

Parameter Identification (PID)	PID Value Range	PID Value at Hot Idle	PID Value at 30 mph	Data List Type
Air Fuel Ratio	0.0-25:1	14.6:1	14.6:1	Fuel Trim
AIR Relay Command	On / Off	Off	Off	Fuel Trim
AIR Solenoid Control	On / Off	Off	Off	Fuel Trim
BARO Sensor (kPa)	10-110 kPa	65-110	65-110	Fuel Trim
Current Gear	1-2-3-4	1	3	Fuel Trim
Decel Fuel Cutoff	Active/Inactive	Inactive	Inactive	Fuel Trim
ECT Sensor (°F)	-40 to 304°F	195	195	Fuel Trim
Engine Load	0-100%	2-5	Varies	Fuel Trim
Engine Run Time	Hr: Min: Sec	00:00:00	00:00:00	Fuel Trim
Engine Speed	0-9999 rpm	Varies	---	Fuel Trim
EVAP Purge Control	0-100%	0-50	50-100	Fuel Trim
EVAP Vent Command	Open/Close	Open	Open	Fuel Trim
Fuel Trim Cell	Cell 0-9	0-9	0-9	Fuel Trim
Fuel Trim Learn	Enabled/DIS	Enabled	Enabled	Fuel Trim
HO2S-11 (B1 S1)	0-1132 mv	10-1000	10-1000	Fuel Trim
HO2S-12 (B1 S2)	0-1132 mv	10-1000	10-1000	Fuel Trim
HO2S-11 Status	Ready/Not	Ready	Ready	Fuel Trim
IAC Position (Counts)	0-255	10-40	Varies	Fuel Trim
IAT Sensor (°F)	-40 to 304°F	85	85	Fuel Trim
Ignition 1 Signal	0.0-25.5v	14.2	14.2	Fuel Trim
Injector Pulsewidth	0-985 ms	1.5-3.5	Varies	Fuel Trim
Long Term Fuel Trim	-10% to 10%	0 (± 5%)	0 (± 5%)	Fuel Trim
MAF Sensor (g/sec)	0-512 g/sec	3-6	Varies	Fuel Trim
MAP Sensor (kPa)	10-110 kPa	20-48	Varies	Fuel Trim
Number of Trouble Codes	Number	0	0	Fuel Trim
Power Enrichment	Active/Inactive	Inactive	Inactive	Fuel Trim
Short Term Fuel Trim	-10% to 10%	0 (± 5%)	0 (± 5%)	Fuel Trim
Spark Advance (°)	-64° to 64°	19	Varies	Fuel Trim
Startup ECT Sensor	-40 to 304°F	Varies	Varies	Fuel Trim
Startup IAT Sensor	-40 to 304°F	Varies	Varies	Fuel Trim
TFP Switch	Several	P-N	Varies	Fuel Trim
TP Angle (%)	0-100%	1-4	Varies	Fuel Trim
TWC Protection	Active/Inactive	Inactive	Inactive	Fuel Trim

1998-99 Lumina 3.8L V6 MFI VIN K (A/T) - PID Data List

Parameter Identification (PID)	PID Value Range	PID Value at Hot Idle	PID Value at 30 mph	Data List Type
Actual EGR Position	0-100%	0	Varies	EGR
AIR Relay Command	On / Off	Off	Off	EGR
BARO Sensor (kPa)	10-110 kPa	65-110	65-110	EGR
Current Gear	1-2-3-4	1	3	EGR
Decel Fuel Cutoff	Active/Inactive	Inactive	Inactive	EGR
Desired EGR Position	0-100%	0	Varies	EGR
Desired EGR Position	0-5.0v	0	Varies	EGR
Desired Idle Speed	0-3187 rpm	PCM control	---	EGR
ECT Sensor (°F)	-40 to 304°F	195	195	EGR
EGR Closed Valve Position	0-5.0v	0.14-1.0	---	EGR

1998-99 Lumina 3.8L V6 MFI VIN K (A/T) - PID Data List

Parameter Identification (PID)	PID Value Range	PID Value at Hot Idle	PID Value at 30 mph	Data List Type
EGR Flow Test Count	0-255	0-10	0-10	EGR
EGR Learn Minimum	0-5.0v	0.14-1.0	---	EGR
EGR Feedback Signal	0-5.0v	0.14-1.0	---	EGR
EGR Position Error	0-100%	0-9	Varies	EGR
EGR Solenoid Control	0-100%	0	Varies	EGR
Engine Load	0-100%	2-5	Varies	EGR
Engine Run Time	Hr: Min: Sec	00:00:00	00:00:00	EGR
Engine Speed	0-9999 rpm	Varies	---	EGR
IAC Position	0-255 counts	10-40	10-40	EGR
IAT Sensor (ºF)	-40 to 304ºF	85	85	EGR
Loop Status	Open Loop or Closed Loop	Closed Loop	Closed Loop	EGR
MAF Sensor (g/sec)	0-512 g/sec	3-6	Varies	EGR
MAP Sensor (kPa)	10-110 kPa	20-48	Varies	EGR
TFP Switch	Several	P-N	Varies	EGR
TP Angle (%)	0-100%	1-4	Varies	EGR

1998-99 Lumina 3.8L V6 MFI VIN K (A/T) - PID Data List

Parameter Identification (PID)	PID Value Range	PID Value at Hot Idle	PID Value at 30 mph	Data List Type
BARO Sensor (kPa)	10-110 kPa	65-110	65-110	EVAP
Desired Idle Speed	0-3187 rpm	PCM control	---	EVAP
ECT Sensor (ºF)	-40 to 304ºF	195	195	EVAP
Engine Load	0-100%	2-5	Varies	EVAP
Engine Run Time	Hr: Min: Sec	00:00:00	00:00:00	EVAP
Engine Speed	0-9999 rpm	Varies	---	EVAP
EVAP Fault History	Several	No Fault	No Fault	EVAP
EVAP Purge Control	0-100%	Varies	Varies	EVAP
EVAP Test Abort	Several	Not Aborted	Not Aborted	EVAP
EVAP Test Result	Several	Passed	Passed	EVAP
EVAP Test State	Several	Completed	Completed	EVAP
EVAP Vent Command	Open/closed	Open	Open	EVAP
Fuel Tank Level	0-100%	Varies	Varies	EVAP
FTP Sensor (In. H2O)	-17.5 to 7.5"	Varies	Varies	EVAP
FTP Sensor (V)	0-5.0v	2.5v cap off	Varies	EVAP
Fuel Trim Cell	Cell #0-4	0	0-9	EVAP
Fuel Trim Learn	Enabled/DIS	Enabled	Enabled	EVAP
HO2S-11 (B1 S1)	0-1132 mv	10-1000	10-1000	EVAP
IAT Sensor (ºF)	-40 to 304ºF	85	85	EVAP
Long Term Fuel Trim	-10 to 10	Varies	Varies	EVAP
Loop Status	Open Loop or Closed Loop	Closed Loop	Closed Loop	EVAP
MAF Sensor (g/sec)	0-512 g/sec	3-6	Varies	EVAP
MAP Sensor (kPa)	10-110 kPa	20-48	Varies	EVAP
Number of Trouble Codes	Number	0	0	EVAP
Short Term Fuel Trim	-10 to 10	Varies	Varies	EVAP
Startup ECT Sensor	-40 to 304ºF	Varies	Varies	EVAP
Startup IAT Sensor	-40 to 304ºF	Varies	Varies	EVAP
TP Angle (%)	0-100%	0%	Varies	EVAP
Vehicle Speed	0-155 mph	0	30	EVAP

1998-99 Lumina 3.8L V6 MFI VIN K (A/T) - PID Data List

Parameter Identification (PID)	PID Value Range	PID Value at Hot Idle	PID Value at 30 mph	Data List Type
A/C Relay circuit state	Several	Okay	Okay	Output Driver
AIR Relay Status	Several	Okay	Okay	Output Driver
AIR Solenoid Status	Several	Okay	Okay	Output Driver
Cruise Inhibit Status	Several	Okay	Okay	Output Driver
Cyl 1-6 Injector Circuit	Several	Okay	Okay	Output Driver
Cyl 1-6 Injector history	Several	Okay	Okay	Output Driver
Driver Module 1-4 Circuit Status	Several	Enabled	Enabled	Output Driver
EGR Solenoid Circuit	Several	Enabled	Enabled	Output Driver
EGR Solenoid History	Several	Enabled	Enabled	Output Driver
EVAP Purge Solenoid	Several	Okay	Okay	Output Driver
EVAP Vent Solenoid	Several	Okay	Okay	Output Driver
F/C Relay 1, 2-3 State	Several	Okay	Okay	Output Driver
F/P Relay Circuit state	Several	Okay	Okay	Output Driver
F/P Relay History	Okay/Fault	Okay	Okay	Output Driver
MIL Control Circuit	Several	Okay	Okay	Output Driver
Starter Enable Circuit	Several	Okay	Okay	Output Driver
TCC Solenoid Circuit	Several	Okay	Okay	Output Driver
TCS Circuit Status	Fault / Okay	Okay	Okay	Output Driver
TCS Circuit History	Fault / Okay	Okay	Okay	Output Driver
1-2, 2-3 Solenoid	Several	Okay	Okay	Output Driver

1998-99 Lumina 3.8L V6 MFI VIN K (A/T) - PID Data List

Parameter Identification (PID)	PID Value Range	PID Value at Hot Idle	PID Value at 30 mph	Data List Type
3X Crank Sensor	0-9999 rpm	Varies	Varies	Misfire
18X Crank Sensor	0-1600 rpm	Varies	Varies	Misfire
A/C Request	Yes / No	No	No	Misfire
Air Relay Command	On / Off	Off	Off	Misfire
CMP Signal Present	Yes / No	Yes	Yes	Misfire
Commanded A/C	On / Off	Off	Off	Misfire
Cycles of Misfire Data	0-99	Varies	Varies	Misfire
Decel Fuel Mode	Active/Inactive	Inactive	Inactive	Misfire
Desired EGR Position	0-100%	0	0-100	Misfire
ECT Sensor (°F)	-40 to 304°F	195	195	Misfire
EGR Position Sensor	0-100%	0	Varies	Misfire
Engine Load	0-100%	2-5	Varies	Misfire
Engine Speed	0-9999 rpm	Varies	---	Misfire
Injector Pulsewidth	0-985 ms	1.5-3.5	Varies	Misfire
MAF Sensor (g/sec)	0-512 g/sec	3-6	Varies	Misfire
MAP Sensor (kPa)	10-110 kPa	20-48	Varies	Misfire
Misfire current Cyl 1-6	0-198	Varies	Varies	Misfire
Misfire History Cyl 1-6	0-65535	Varies	Varies	Misfire
Spark Advance (°)	-64° to 64°	19°	Varies	Misfire
Total Misfire Failures	0-65535	0	0	Misfire
Total Misfire Passes	0-65535	0	0	Misfire
Total Misfire Current	0-99 counts	0	0	Misfire
TP Angle (%)	0-100%	1-4	Varies	Misfire

LUMINA PID DATA

2000-01 Lumina 3.1L V6 MFI VIN J (A/T) - PID Data List

Parameter Identification (PID)	PID Value Range	PID Value at Hot Idle	PID Value at 30 mph	Data List Type
3X Crank Sensor	0-9999 rpm	Varies	Varies	Engine Data 2
24X Crank Sensor	0-1600 rpm	Varies	Varies	Engine Data 2
1-2 Solenoid Control	Okay/Fault/ Invalid State	Varies	Varies	Output Driver
2-3 Solenoid Control	Okay/Fault/ Invalid State	Varies	Varies	Output Driver
1-2 Solenoid Control	Okay/Fault/ Invalid State	Varies	Varies	Output Driver
Abuse Management	Active/Inactive	Inactive	Inactive	Misfire
A/C HI Side Pressure	0-5.0v	Varies	Varies	Engine Data 1
A/C Relay Circuit Status	Okay/Fault/ Invalid State	Varies	Varies	Output Driver
A/C Request Signal	Yes/No	No	No	Engine Data 2
Actual EGR Position	0-100%	0	Varies	Engine Data 1, EGR, Misfire
Air Active Test Inhibit	Yes/No	No	No	AIR
Air Active Test Injection	Yes/No	No	No	AIR
Air Diagnostic Complete	Yes/No	No	No	AIR
Air Passive Test 1 Pass	Yes/No	No	No	AIR
Air Passive Test 2 Fail	Yes/No	No	No	AIR
Air Passive Test Inhibit	Yes/No	No	No	AIR
Air Pump Relay Circuit Status	Okay/Fault/ Invalid State	Okay	Okay	AIR, Output Driver
Air Pump Relay Command	On/Off	Off	Off	Engine Data 1 Data, AIR, HO2S
Air Solenoid Circuit Status	Okay/Fault/ Invalid State	Okay	Okay	AIR, Output Driver
Air Solenoid Command	On/Off	Off	Off	Engine Data 1 Data, AIR, HO2S
Air Fuel Ratio	0.0-25.5:1	14.7:1	14.7:1	Engine Data 2, Fuel Trim
BARO Sensor (kPa)	10-110 kPa	98	98	Engine Data 1, EGR, EVAP, Catalyst
BARO Sensor (V)	0-5.0 volts	3.5-4.5	3.5-4.5	Engine Data 1, EGR, EVAP, Catalyst
Brake Switch Signal	Applied/Released	Released	Released	Engine Data 1
CMP Sensor Signal	Yes/No	Yes	Yes	Engine Data 1, Misfire
Commanded A/C	On/Off	Off	Off	Engine Data 1, EGR, Misfire, Catalyst
Commanded Fuel Pump	On/Off	On	On	Engine Data 1

2000-01 Lumina 3.1L V6 MFI VIN J (A/T) - PID Data List

Parameter Identification (PID)	PID Value Range	PID Value at Hot Idle	PID Value at 30 mph	Data List Type
Commanded Generator	On/Off	On	On	IPC
Cruise Mode	Yes/No	No	No	Engine Data 1
Cruised Inhibited	Okay/Fault/ Invalid State	Okay	Okay	Output Driver
Cruised Inhibited	On/Off	On	On	Engine Data 1
Current Gear	1-2-3-4	1	4	EGR, Catalyst
Cycles of Misfire Data	0-99 counts	Varies	Varies	Misfire
Cylinder 1-6 Injector Circuit Status	Okay/Stuck Low/Stuck High/Fault	Okay	Okay	Output Driver
Cylinder 1-6 Injector Circuit History	Okay/Stuck Low/Stuck High/Fault	Okay	Okay	Output Driver
Decel Fuel Mode	Active/Inactive	Inactive	Inactive	EGR, HO2S, Misfire
Desired Idle Speed	0-3187 rpm	PCM control	---	Engine Data 2
Desired EGR Position	0-100%	0	Varies	Engine Data 1, EGR, Misfire
Desired Idle Speed	0-3187 rpm	± 50 rpm of Actual Speed	Varies	Engine Data 1
Driver Module 1 Status	Enabled/Off/ High Temp/ Invalid state	Enabled	Enabled	Output Driver
Driver Module 2 Status	Enabled/Off/ High Temp/ Invalid state	Enabled	Enabled	Output Driver
Driver Module 3 Status	Enabled/Off/ High Temp/ Invalid state	Enabled	Enabled	Output Driver
Driver Module 4 Status	Enabled/Off/ High Temp/ Invalid state	Enabled	Enabled	Output Driver
ECT Sensor (°F)	-40 to 304°F	185-220	185-220	Engine Data 1, AIR, Catalyst, EGR EVAP, Misfire
EGR Closed Valve Pintle Position	0-5.1volts	0.14-1.0	Varies	EGR
EGR Duty Cycle	0-100	0	Varies	Engine Data 1, EGR
EGR Feedback	0-5.1 volts	0.14-1.0	Varies	EGR
EGR Flow Test Count	0-255 counts	0	Varies	EGR
EGR Position Error	0-100%	0	Varies	EGR
Engine Load (%)	0-100%	17	Varies	Engine Data 1, AIR, Catalyst, EGR EVAP, Misfire
Engine Oil Level Switch	Okay/LOW	Okay	Okay	IPC
Engine Oil Life Left	0-100%	Varies	Varies	IPC
EOP Switch Signal	Okay/LOW/HI	Okay	Okay	IPC

2000-01 Lumina 3.1L V6 MFI VIN J (A/T) - PID Data List

Parameter Identification (PID)	PID Value Range	PID Value at Hot Idle	PID Value at 30 mph	Data List Type
Engine Run Time	Hr: Min: Sec	00:00:00	00:00:00	Engine Data 1, AIR, Catalyst, EGR EVAP, Misfire
Engine Speed	0-9999 rpm	± 50 rpm of Desired Speed	Varies	Engine Data 1, AIR, Catalyst, EGR EVAP, Misfire
EVAP Canister Purge	0-100%	19	Varies	Engine Data 1, AIR, EVAP, HO2S
EVAP Fault History	No Fault / Excess VAC / Purge Valve Leak / Small Leak / Weak Vacuum	No Fault	No Fault	EVAP
EVAP Purge Solenoid Control Circuit Status	Okay/Fault/ Invalid state	Okay	Okay	Output Driver
EVAP Vent Solenoid Control Circuit Status	Okay/Fault/ Invalid state	Okay	Okay	Output Driver, EVAP
EVAP Vent Solenoid	Open/Closed	Open	Open	Engine Data 1, EVAP, HO2S
Fans High Speed Control	On/Off	Off	Off	Engine Data 1, Catalyst
Fans Low Speed Control	On/Off	Off	Off	Engine Data 1, Catalyst
Fans High Speed Circuit Status	Okay/Fault/ Invalid State	Okay	Okay	Output Driver
Fans Low Speed Circuit Status	Okay/Fault/ Invalid State	Okay	Okay	Output Driver
Fuel Pump Relay Control Circuit Status	Okay/Fault	Okay	Okay	Output Driver
Fuel Pump Relay Control Circuit History	Okay/Fault	Okay	Okay	Output Driver
Fuel Tank Level	0-100%	Varies	Varies	EVAP, IPC
Fuel Tank Pressure (FTP) Sensor	0-5.0 volts	Varies	Varies	EVAP
Fuel Tank Pressure (FTP) Sensor	Inches H2O, mm Hg, volts	-17 to +7.5 Inches H2O	-17 to +7.5 Inches H2O	EVAP
Fuel Trim Cell	Cell 0-10	Varies	Varies	Engine Data 1, AIR, Catalyst
Fuel Trim Learn	Enabled or Disabled	Enabled	May Toggle	Engine Data 1, AIR, Catalyst
Generator PWM	0-100%	Varies	Varies	IPC

2000-01 Lumina 3.1L V6 MFI VIN J (A/T) - PID Data List

Parameter Identification (PID)	PID Value Range	PID Value at Hot Idle	PID Value at 30 mph	Data List Type
HO2S-11 (B1 S1) Ready	Ready/Not Ready	Ready	Ready	Engine Data 1, HO2S, Catalyst
HO2S-11 (B1 S1)	0-1132 mv	10-1000	10-1000	Engine Data 1, EVAP, F/Trim
HO2S-12 (B1 S2)	0-1132 mv	10-1000	10-1000	Engine Data 1, HO2S, Catalyst
HO2S-11 XCounts	0-233	Varies	Varies	Engine Data 1, HO2S
Hot Mode	On/Off	Off	Off	Misfire
Hot Open Loop	Active/Inactive	Inactive	Inactive	Catalyst, EGR
IAC Position	0-255 counts	15	15	Engine Data 1, Catalyst
IAT Sensor (°F)	-40 to 304°F	50-194	50-194	Engine Data 1, AIR, Catalyst, EGR EVAP, Misfire
Ignition 1 Signal	0.0-25.5v	14.2	14.2	Engine Data 1, AIR, Catalyst, EGR EVAP, Misfire
Ignition Mode	Bypass/IC	IC	IC	Engine Data 1
Injector Pulsewidth	0-1000 ms	1.5-3.5	Varies	Engine Data 1, CAT, HO2S, M/Fire
Knock Retard (°)	0-25.5°	0°	0°	Engine Data 1, EGR
Long Term Fuel Trim	-27% to 27%	0 (± 5%)	0 (± 5%)	Engine Data 1, AIR, EVAP, HO2S
Loop Status	Closed Loop / Open Loop	Closed Loop	Closed Loop	Engine Data 1, AIR, Catalyst, EGR EVAP, Misfire
MAF Sensor (g/sec)	0-512 g/sec	3.41	Varies	Engine Data 1, AIR, Catalyst, EGR EVAP, Misfire
MAF Sensor (Hz)	0-32,000 Hz	2000	Varies	Engine Data 1
MAP Sensor (kPa)	10-105 kPa	35	Varies	Engine Data 1, EGR, EVAP, Catalyst
MAP Sensor (V)	0-5.1 volts	1.35	Varies	Engine Data 1, EGR, EVAP, Catalyst
MIL Status	Okay/Fault/ Invalid State	Okay	Okay	Output Driver
MIL Command	On/Off	Off	Off	IPC
Misfire Current Cylinder 1-6	0-198 counts	0	0	Misfire
Misfire History Cylinder 1-6	0-65,535	0	0	Misfire

2000-01 Lumina 3.1L V6 MFI VIN J (A/T) - PID Data List

Parameter Identification (PID)	PID Value Range	PID Value at Hot Idle	PID Value at 30 mph	Data List Type
Non-Volatile Memory	Pass/Fail	Pass	Pass	Engine Data 1, AIR, Catalyst, EGR EVAP, IPC, Misfire, ODD
Power Enrichment	Active/Inactive	Inactive	Inactive	CAT, EGR, HO2S, M/Fire
Short Term Fuel Trim	-27% to 27%	0 (± 5%)	0 (± 5%)	Engine Data 1, AIR, HO2S
Spark Advance (°)	-64° to 64°	20	Varies	Engine Data 1, Misfire
Startup ECT	-40 to 419°F	Varies	Varies	AIR, Catalyst, EVAP, HO2S
Startup IAT	-40 to 419°F	80°F	80°F	AIR, Catalyst, EVAP, HO2S
TCC Engaged	Engaged, Disengaged	Disengaged	Disengaged	Engine Data 1, EGR, Misfire
TCC Solenoid Circuit Status	Okay/Fault/ Invalid State	Okay	Okay	Output Driver
Total Misfire Current Count	0-99	Varies	Varies	Misfire
Total Misfire Failures	0-65,535 counts	Varies	Varies	Misfire
Total Misfire Passed	0-65,535 counts	Varies	Varies	Misfire
TP Angle (%)	0-100%	0	0-100	Engine Data 1, AIR, Catalyst, EGR EVAP, M/Fire
TP Sensor (V)	0-5.1 volts	0.20-0.74	Varies	Engine Data 1
Transmission Range	Park/Neutral, Drive 4-3-2-1	Park/Neutral	Drive 4	Engine Data 1, AIR, EGR, Catalyst
TR Switch 'A' Signal	HIGH/LOW	LOW	HIGH	Engine Data 1
TR Switch 'B' Signal	HIGH/LOW	HIGH	LOW	Engine Data 1
TR Switch 'C' Signal	HIGH/LOW	HIGH	LOW	Engine Data 1
TR Switch 'P' Signal	HIGH/LOW	LOW	HIGH	Engine Data 1
TWC Protection	Active/Inactive	Inactive	Inactive	Catalyst, EGR
Vehicle Speed	0-155 mph	0	30	Engine Data 1, AIR, EGR, Catalyst
VTD Fuel Disable	Active/Inactive	Inactive	Inactive	Engine Data 1

MALIBU PID DATA

1997-99 Malibu 2.4L I4 MFI VIN T (A/T) - PID Data List

Parameter Identification (PID)	PID Value Range	PID Value at Hot Idle	PID Value at 30 mph	Data List Type
A/C HI Side Pressure	0-459 psi	139-399	139-399	Engine Data 1
A/C HI Side Pressure	0-5.0 volts	Varies	Varies	Engine Data 1
A/C Relay Command	On/Off	Off	Off	Engine Data 1
A/C Request Signal	Yes/No	No	No	Engine Data 1
Actual EGR Position	0-100%	0%	0-100%	Engine Data 1
Adaptive Knock retard	0-90°	0	Varies	Engine Data 1
Air Fuel Ratio	0.1-25.5:1	14.6:1	14.6:1	Engine Data 1
Another TWC Test	Yes/No	No	No	Engine Data 1
BARO Sensor (kPa)	10-110 kPa	65-104	65-105	Engine Data 1
Base Pulsewidth	0-985 ms	1-4	Varies	Engine Data 1
Calculated Airflow	0-512 g/sec	Varies	Varies	Engine Data 1
Calculated Vacuum	10-110 kPa	65-104	65-105	Engine Data 1
Calc. Convert Temp.	30-1409°F	Varies	Varies	Engine Data 1
CKP 7X Active Count	0-255 counts	Counts up	Counts up	Engine Data 1
CMP Active Counter	0-255 counts	Counts up	Counts up	Engine Data 1
CMP Resync Counter	0-255 counts	0	0	Engine Data 1
Cruise Control Enable	Disable/Enable	Disabled	Disabled	Engine Data 1
Cruise Control Engaged	Engaged / Disengaged	Disengaged	Disengaged	Engine Data 1
Current TWC Test Completed	Yes/No	Yes	Yes	Engine Data 1
Desired Idle Speed	0-3187	PCM control	Varies	Engine Data 1
Closed Loop Sensor (coolant)	Okay/LOW	Okay	Okay	Engine Data 1
ECT Sensor (°F)	-40 to 304°F	185-239	185-239	Engine Data 1
EGR Delta MAP Chg.	-20 to 20 kPa	0 or - No.	Varies	Engine Data 1
EGR Decel Filter	-10 to 10 kPa	-3 to -4	Varies	Engine Data 1
EGR Desired Position	0-100%	0	Varies	Engine Data 1
EGR Flow Test Count	0-255 counts	0	0	Engine Data 1
Engine Load	0-100%	11-21	16-29	Engine Data 1
Engine Oil Pressure	Okay/LOW	Okay	Okay	Engine Data 1
Engine Run Time	Hr: Min: Sec	00:00:00	00:00:00	Engine Data 1
Engine Speed	0-9999 rpm	Varies	Varies	Engine Data 1
EVAP Canister Purge	0-100%	0	0-50	Engine Data 1
EVAP Vent Solenoid	Open/Closed	Open	Open	Engine Data 1
FC Relay 1	On/Off	Off	Off	Engine Data 1
FC Relay 2	On/Off	Off	Off	Engine Data 1
FTP Sensor (In. H2O)	-17.5 to 7.5"	Varies	Varies	Engine Data 1
FTP Sensor (V)	0-5.0 volts	Varies	Varies	Engine Data 1
Fuel Level Sensor	0-100%	Varies	Varies	Engine Data 1
Fuel Trim Cell	Cell # 0-21	18-21	Varies	Engine Data 1
Fuel Trim Index	-10% to 10%	0 (± 5%)	0 (± 5%)	Engine Data 1
Generator L-Terminal	Active/Inactive	Active	Active	Engine Data 1
Generator PWM	0-100%	Varies	Varies	Engine Data 1
HO2S-12 (B1 S2)	0-1132 mv	10-1000	10-1000	Engine Data 1
IAC Position	0-255 counts	5-60	5-60	Engine Data 1
IAT Sensor (°F)	-40 to 304°F	50-176	50-176	Engine Data 1
Ignition 1 Signal	0.0-25.5v	12-15	12-15	Engine Data 1
KS Active Counter	0-255 counts	Counts up	Counts up	Engine Data 1

1997-99 Malibu 2.4L I4 MFI VIN T (A/T) - PID Data List

Parameter Identification (PID)	PID Value Range	PID Value at Hot Idle	PID Value at 30 mph	Data List Type
KS Noise Channel	Yes/No	No	No	Engine Data 1
Knock Retard (°)	0-90	0	0	Engine Data 1
Lean-Rich Average. Time	0-999 ms	0	0	Varies
Lean-Rich Transition	0-255 counts	0	0	Varies
Long Term Fuel Trim	-10% to 10%	0 (± 5%)	0 (± 5%)	Varies
Loop Status	Closed Loop / Open Loop	Closed Loop	Closed Loop	Closed Loop
MAP Sensor (kPa)	10-110 kPa	25-48	Varies	Varies
MAP Sensor (V)	0-5.0 volts	1.5	Varies	Varies
Medium Resolution Resync Counter	0-255 counts	0	0	0
Med Resolution Engine Sync	Yes/No	Yes	Yes	Yes
Misfire Current Cylinder 1-4	0-255 counts	0	0	Varies
Misfire History Cylinder 1-4	0-255 counts	0	0	Varies
Number of DTC(s)	0-255	0	0	Engine Data 1
Number of Catalyst Monitor Test complete	0-255	1	1	Engine Data 1
O2S-11 (B1 S1)	0-1132 mv	10-1000	10-1000	Engine Data 1
Output Driver 1 Open	00000000	00000000	00000000	Engine Data 1
Output Driver 1 Short	00000000	00000000	00000000	Engine Data 1
Output Driver 2 Open	00000000	00000000	00000000	Engine Data 1
Output Driver 2 Short	00000000	00000000	00000000	Engine Data 1
Power Enrichment	Active/Inactive	Inactive	Inactive	Engine Data 1
Purge Learn Memory	0.00-1.00	1.00	1.00	Engine Data 1
Rich-Lean Average. Time	0-999 ms	0	0	Engine Data 1
Rich-Lean Transition	0-255 counts	0	0	Engine Data 1
Rich-Lean to Lean-Rich Average Ratio	0:1 to 15.93:1	Varies	Varies	Engine Data 1
Short Term Fuel Trim	-10% to 10%	Varies	Varies	Engine Data 1
Spark Advance (°)	-64° to 64°	20	Varies	Engine Data 1
TCC Brake Switch	Open/Closed	Closed	Closed	Engine Data 1
TCC Engaged	Engage/DIS	Disengaged	Disengaged	Engine Data 1
Total Misfire Current	0-255 counts	0	Varies	Engine Data 1
TP Angle (%)	0-100%	0	Varies	Engine Data 1
TP Sensor (V)	0-5.0 volts	0.2-0.9	Varies	Engine Data 1
Transmission Range	P-R-N-4-3-2	Park/Neutral	Drive 4	Engine Data 1
TR Switch P-A-B-C	L/L/H/H (several)	L/L/H/H	L/H/H/L	Engine Data 1
TWC Stage 1 Started	Yes/No	Yes	Yes	Engine Data 1
TWC Stage 2 Started	Yes/No	Yes	Yes	Engine Data 1
TWC System Related DTC is Set	Yes/No	No	No	Engine Data 1
TWC Test Conditions	Yes/No	Yes	Yes	Engine Data 1
TWC Test Enabled	Yes/No	No	No	Engine Data 1
TWC Test Warmup	Yes/No	Yes	Yes	Engine Data 1
Vehicle Speed	0-155 mph	0	30	Engine Data 1

1997-99 Malibu 3.1L V6 MFI VIN M (A/T) - PID Data List

Parameter Identification (PID)	PID Value Range	PID Value at Hot Idle	PID Value at 30 mph	Data List Type
A/C HI Side Pressure	0-5v	1-4	1-4	Engine Data 1
A/C Off for WOT	Yes / No	No	No	Engine Data 1
A/C Pressure Disable	Yes / No	No	No	Engine Data 1
A/C Request Signal	Yes / No	No	No	Engine Data 1
Actual EGR Position	0-100%	0	0-100	Engine Data 1
AIR Relay Command	On / Off	Off	Off	Engine Data 1
AIR Solenoid Control	On / Off	Off	Off	Engine Data 1
BARO Sensor (kPa)	10-110 kPa	65-110	65-110	Engine Data 1
Brake Switch	Applied/Released	Released	Released	Engine Data 1
Cam Signal Present	Yes / No	Yes	Yes	Engine Data 1
Commanded A/C	On / Off	Off	Off	Engine Data 1
Commanded Starter	Enabled/DIS	Disabled	Disabled	Engine Data 1
Cruise Active Mode	On / Off	Off	Off	Engine Data 1
Cruise Inhibit Signal	On / Off	On	On	Engine Data 1
Current Gear	1-2-3-4	1	3	Engine Data 1
Desired EGR Position	0-100%	0	0-100	Engine Data 1
Desired Idle Speed	0-3187 rpm	PCM control	Varies	Engine Data 1
ECT Sensor (°F)	-40 to 304°F	185-220	185-220	Engine Data 1
EGR Duty Cycle	0-100%	0	0-100	Engine Data 1
EGR Position Sensor	0-100%	0	0-100	Engine Data 1
Engine Load	0-100%	2-10	Varies	Engine Data 1
Engine Run Time	Hr: Min: Sec	00:00:00	00:00:00	Engine Data 1
Engine Speed	0-9999 rpm	Varies	Varies	Engine Data 1
EVAP Purge Solenoid	0-100%	0-25	0-100	Engine Data 1
EVAP Vent Solenoid	Open/Close	Open	Open	Engine Data 1
FTP Sensor (In. H2O)	-17.5 to 7.5"	Varies	Varies	Engine Data 1
FC Relay (low/high)	On / Off	Off	Off	Engine Data 1
Fuel Pump Control	On/Off	On	On	Engine Data 1
Fuel Pump Relay	On/Off	On	On	Engine Data 1
Fuel Trim Cell	Cell # 0-4	0	0-4	Engine Data 1
Fuel Trim Learn	Enabled/DIS	Enabled	Enabled	Engine Data 1
HO2S-11 (B1 S1)	0-1132 mv	10-1000	10-1000	Engine Data 1
HO2S-12 (B1 S2)	0-1132 mv	10-1000	10-1000	Engine Data 1
HO2S-11 Status	Not / Ready	Ready	Ready	Engine Data 1
HO2S-11 XCounts	0-255 counts	Varies	Varies	Engine Data 1
IAC Position	0-255 counts	10-60	---	Engine Data 1
IAT Sensor (°F)	-40 to 304°F	50-194	50-194	Engine Data 1
Ignition 1 Signal	0.0-25.5v	14.2	14.2	Engine Data 1
Ignition Mode	IC / Bypass	IC	IC	Engine Data 1
Injector Pulsewidth	0-985 ms	1.5-3.5	Varies	Engine Data 1
Knock Retard (°)	0-128°	0	0	Engine Data 1
Long Term Fuel Trim	-10% to 10%	0 (± 5%)	0 (± 5%)	Engine Data 1
Loop Status	Closed Loop / Open Loop	Closed Loop	Closed Loop	Engine Data 1
MAF Sensor (g/sec)	0-512 g/sec	3-6	Varies	Engine Data 1
MAF Sensor (Hz)	0-31,999hz	1200-3000	Varies	Engine Data 1
MAP Sensor (kPa)	10-110 kPa	20-48	Varies	Engine Data 1
Number of Trouble Codes	Number	00000	00000	Engine Data 1
Power Enrichment	Active/Inactive	Inactive	Inactive	Engine Data 1

1997-99 Malibu 3.1L V6 MFI VIN M (A/T) - PID Data List

Parameter Identification (PID)	PID Value Range	PID Value at Hot Idle	PID Value at 30 mph	Data List Type
Short Term Fuel Trim	-10% to 10%	0 (± 5%)	0 (± 5%)	Engine Data 1
Spark Advance (°)	-64° to 64°	20	Varies	Engine Data 1
TCC Engaged	Engage/DIS	Disengaged	Disengaged	Engine Data 1
TP Angle (%)	0-100%	0	Varies	Engine Data 1
Traction Control Mode	Active/Inactive	Inactive	Inactive	Engine Data 1
T/C Desired	0-100%	100	0-100	Engine Data 1
T/C Torque	0-100%	80-90	0-100	Engine Data 1
Transmission Range	P-R-N-4-3-2	P	4	Engine Data 1
TR Switch P-A-B-C	Several	Several	Several	Engine Data 1
Vehicle Speed	0-155 mph	0	30	Engine Data 1
VTD Fuel Disable	Active/Inactive	Inactive	Inactive	Engine Data 1
24X Crank Sensor	0-1600	Varies	Varies	Engine Data 1

1997-99 Malibu 3.1L V6 MFI VIN M (A/T) - PID Data List

Parameter Identification (PID)	PID Value Range	PID Value at Hot Idle	PID Value at 30 mph	Data List Type
3X Crank Sensor	1600-10,000	Varies	Varies	Engine Data 2
24X Crank Sensor	0-1600	Varies	Varies	Engine Data 2
Air Fuel Ratio	0:1-25.5:1	14.6:1	14.6:1	Engine Data 2
A/C HI Side Pressure	0-5v	1-4	1-4	Engine Data 2
A/C Off for WOT	Yes / No	No	No	Engine Data 2
A/C Pressure Disable	Yes / No	No	No	Engine Data 2
A/C Request Signal	Yes / No	No	No	Engine Data 2
Actual EGR Position	0-100%	0	0-100	Engine Data 2
AIR Relay Command	On / Off	Off	Off	Engine Data 2
AIR Solenoid Control	On / Off	Off	Off	Engine Data 2
CMP Signal Present	Yes / No	Yes	Yes	Engine Data 2
Current Gear	1-2-3-4	1	4	Engine Data 2
Desired EGR Position	0-100%	0	0-100	Engine Data 2
Desired Idle Speed	0-3187 rpm	PCM control	Varies	Engine Data 2
Engine Hot Lamp	On/Off	Off	Off	Engine Data 2
Engine Load	0-100%	2-10	Varies	Engine Data 2
Engine Oil Level Switch	Okay/LOW	Okay	Okay	Engine Data 2
Engine Oil Life	0-100%	Varies	Varies	Engine Data 2
Engine Run Time	Hr: Min: Sec	00:00:00	00:00:00	Engine Data 2
Engine Speed	0-9999 rpm	Varies	Varies	Engine Data 2
Fan Control Relay 1	On / Off	Off	Off	Engine Data 2
Fan Control Relay 2-3	On / Off	Off	Off	Engine Data 2
Generator Lamp	On/Off	Off	Off	Engine Data 2
Generator 'L' Signal	On/Off	On	On	Engine Data 2
IAC Position	0-255 counts	10-60	10-60	Engine Data 2
IAT Sensor (°F)	-40 to 304°F	50-194	50-194	Engine Data 2
Ignition 1 Signal	0.0-25.5v	14.2	14.2	Engine Data 2
Ignition Mode	IC / Bypass	IC	IC	Engine Data 2
Injector Pulsewidth	0-985 ms	1.5-3.5	Varies	Engine Data 2
Knock Retard (°)	0.0-25.5°	0	0	Engine Data 2
Long Term Fuel Trim	-10% to 10%	0 (± 5%)	0 (± 5%)	Engine Data 2
Loop Status	Open Loop or Closed Loop	Closed Loop	Closed Loop	Engine Data 2

1997-99 Malibu 3.1L V6 MFI VIN M (A/T) - PID Data List

Parameter Identification (PID)	PID Value Range	PID Value at Hot Idle	PID Value at 30 mph	Data List Type
Low Oil Lamp Control	On / Off	Off	Off	Engine Data 2
MAF Sensor (g/sec)	0-512 g/sec	3-6	Varies	Engine Data 2
MAF Sensor (Hz)	0-31,999hz	1200-3000	Varies	Engine Data 2
MAP Sensor (kPa)	10-110 kPa	20-48	Varies	Engine Data 2
MIL (lamp) Command	On / Off	Off	Off	Engine Data 2
Power Enrichment	Active/Inactive	Inactive	Inactive	Engine Data 2
Short Term Fuel Trim	-10% to 10%	0 (± 5%)	0 (± 5%)	Engine Data 2
Spark Advance (°)	-64° to 64°	20	Varies	Engine Data 2
TCC Brake Pedal Switch	Applied/Released	Released	Released	Engine Data 2
TCC PWM Solenoid	On / Off	Off	On	Engine Data 2
TFP Switch Signal	Several	P/N	Several	Engine Data 2
TP Angle (%)	0-100%	0	Varies	Engine Data 2
Traction Control State	Active/Inactive	Inactive	Inactive	Engine Data 2
Traction Request (%)	0-100%	100%	Varies	Engine Data 2
Traction Delivered (%)	75%	100%	Varies	Engine Data 2

1997-99 Malibu 3.1L V6 MFI VIN M (A/T) - PID Data List

Parameter Identification (PID)	PID Value Range	PID Value at Hot Idle	PID Value at 30 mph	Data List Type
A/C Request	Yes / No	No	No	EGR
Actual EGR Position	0-100%	0	0-100	EGR
BARO Sensor (kPa)	10-110 kPa	65-110	65-110	EGR
Commanded A/C	On / Off	Off	Off	EGR
Current Gear	1-2-3-4	1	3	EGR
Decel Fuel Mode	Active/Inactive	Inactive	Inactive	EGR
Desired EGR Position	0-100%	0	0-100	EGR
ECT Sensor (°F)	-40 to 304°F	185-220	185-220	EGR
EGR Closed Valve Pintle	0-5v	0.14-1.00	Varies	EGR
EGR Duty Cycle	0-100%	0	0-100	EGR
EGR Feedback	0-5v	0.14-1.00	Varies	EGR
EGR Position Error	0-100%	0	0	EGR
EGR Flow Test Count	0-255	0-10	0-10	EGR
Engine Load	0-100%	2-10	Varies	EGR
Engine Run Time	Hr: Min: Sec	00:00:00	00:00:00	EGR
Engine Speed	0-9999 rpm	Varies	---	EGR
Hot Open Loop	Active/Inactive	Inactive	Inactive	EGR
IAT Sensor (°F)	-40 to 304°F	50-194	50-194	EGR
Ignition 1 Signal	0.0-25.5v	14.2	14.2	EGR
Knock Retard (°)	0.0-25.5°	0	0	EGR
Loop Status	Open Loop or Closed Loop	Closed Loop	Closed Loop	EGR
MAF Sensor (g/sec)	0-512 g/sec	3-6	Varies	EGR
MAP Sensor (kPa)	10-110 kPa	20-48	Varies	EGR
Non-Volatile Memory	Pass/Fail	Pass	Pass	EGR
Power Enrichment	Active/Inactive	Inactive	Inactive	EGR
TCC Engaged	Engage/DIS	Disengaged	Disengaged	EGR
TP Angle (%)	0-100%	0	Varies	EGR
Transmission Range	P-R-N-4-3-2	P	3	EGR

1997-99 Malibu 3.1L V6 MFI VIN M (A/T) - PID Data List

Parameter Identification (PID)	PID Value Range	PID Value at Hot Idle	PID Value at 30 mph	Data List Type
Air Fuel Ratio	0.0-25:1	14.6:1	14.6:1	Fuel Trim
AIR Active Test Inhibit	Yes / No	Yes	Yes	Fuel Trim
AIR Active Test	Yes / No	No	No	Fuel Trim
AIR Diagnostic Done	Yes / No	Yes	Yes	Fuel Trim
AIR Passive Test (P)	Yes / No	Yes	Yes	Fuel Trim
AIR Passive test 1 (P)	Yes / No	Yes	Yes	Fuel Trim
AIR Passive test 2 (F)	Yes / No	No	No	Fuel Trim
AIR Pass. Test Inhibit	Yes / No	Yes	Yes	Fuel Trim
AIR Passive Test (A)	Yes / No	No	No	Fuel Trim
AIR Pump Relay	On / Off	Off	Off	Fuel Trim
AIR Pump Solenoid	On / Off	Off	Off	Fuel Trim
BARO Sensor (kPa)	10-110 kPa	65-110	65-110	Fuel Trim
ECT Sensor (°F)	-40 to 304°F	185-220	185-220	Fuel Trim
Engine Load	0-100%	2-10	Varies	Fuel Trim
Engine Run Time	Hr: Min: Sec	00:00:00	00:00:00	Fuel Trim
Engine Speed	0-9999 rpm	Varies	---	Fuel Trim
EVAP Purge Solenoid	0-100%	0-25	0-100	Fuel Trim
Fuel Trim Cell	Cell # 0-4	0	0-4	Fuel Trim
Fuel Trim Learn	Disable/Enable	Enabled	Enabled	Fuel Trim
HO2S-11 (B1 S1)	0-1132 mv	10-1000	10-1000	Fuel Trim
HO2S-12 (B1 S2)	0-1132 mv	10-1000	10-1000	Fuel Trim
IAT Sensor (°F)	-40 to 304°F	50-194	50-194	Fuel Trim
Long Term Fuel Trim	-10% to 10%	Varies	Varies	Fuel Trim
Loop Status	Closed Loop / Open Loop	Closed Loop	Closed Loop	Fuel Trim
MAF Sensor (g/sec)	0-512 g/sec	3-6	Varies	Fuel Trim
Number of Trouble Codes	Number	00000	00000	Fuel Trim
Power Enrichment	Active/Inactive	Inactive	Inactive	Fuel Trim
Startup ECT Sensor	-40 to 304°F	Varies	Varies	Fuel Trim
Startup IAT Sensor	-40 to 304°F	Varies	Varies	Fuel Trim
TP Angle (%)	0-100%	0	Varies	Fuel Trim

1997-99 Malibu 3.1L V6 MFI VIN M (A/T) - PID Data List

Parameter Identification (PID)	PID Value Range	PID Value at Hot Idle	PID Value at 30 mph	Data List Type
1-2, 2-3 Solenoid	Okay/Fault	Okay	Okay	Output Driver
A/C Relay Status	Okay/Fault	Okay	Okay	Output Driver
AIR Pump Relay	Okay/Fault	Okay	Okay	Output Driver
AIR Solenoid Status	Okay/Fault	Okay	Okay	Output Driver
Cruise Inhibit Status	Okay/Fault	Okay	Okay	Output Driver
CYL 1-6 INJ Circuit	Okay/Fault	Okay	Okay	Output Driver
CYL 1-6 INJ History	Okay/Fault	Okay	Okay	Output Driver
Driver Module 1-4	Okay/Fault	Okay	Okay	Output Driver
EGR Solenoid Status	Okay/Fault	Okay	Okay	Output Driver
EVAP Purge Solenoid	Okay/Fault	Okay	Okay	Output Driver
EVAP Vent Solenoid	Okay/Fault	Okay	Okay	Output Driver
FC Relay 1, 2-3	Okay/Fault	Okay	Okay	Output Driver
MIL (lamp) Status	Okay/Fault	Okay	Okay	Output Driver
TCC Solenoid Status	Okay/Fault	Okay	Okay	Output Driver

1997-99 Malibu 3.1L V6 MFI VIN M (A/T) - PID Data List

Parameter Identification (PID)	PID Value Range	PID Value at Hot Idle	PID Value at 30 mph	Data List Type
Air Fuel Ratio	0.0-25:1	14.6:1	14.6:1	Catalyst
BARO Sensor (kPa)	10-110 kPa	65-110	65-110	Catalyst
BARO Sensor (V)	0.0-5.0v	3.5-4.5	---	Catalyst
Commanded A/C	On / Off	Off	Off	Catalyst
Current Gear	1-2-3-4	1	3	Catalyst
ECT Sensor (ºF)	-40 to 304ºF	185-220	185-220	Catalyst
Engine Load	0-100%	2-10	Varies	Catalyst
Engine Run Time	Hr: Min: Sec	00:00:00	00:00:00	Catalyst
Engine Speed	0-9999 rpm	Varies	---	Catalyst
Fan High Speed	On / Off	Off	Off	Catalyst
Fan Low Speed	On / Off	Off	Off	Catalyst
Fuel Trim Cell	Cell #0-4	0	0-4	Catalyst
Fuel Trim Learn	Enabled/DIS	Enabled	Enabled	Catalyst
HO2S-11 (B1 S1)	0-1132 mv	10-1000	10-1000	Catalyst
HO2S-12 (B1 S2)	0-1132 mv	10-1000	10-1000	Catalyst
HO2S-11 Status	Ready/Not	Ready	Ready	Catalyst
HO2S-11 XCounts	0-255 counts	Varies	Varies	Catalyst
Hot Open Loop	Active/Inactive	Inactive	Inactive	Catalyst
IAC Position	0-255 counts	10-40	---	Catalyst
IAT Sensor (ºF)	-40 to 302ºF	50-194	50-194	Catalyst
Ignition 1 Signal	0.0-25.5v	14.2	14.2	Catalyst
Injector Pulsewidth	0-985 ms	1.5-3.5	Varies	Catalyst
Loop Status	Open Loop or Closed Loop	Closed Loop	Closed Loop	Catalyst
MAF Sensor (g/sec)	0-512 g/sec	4-6	Varies	Catalyst
MAP Sensor (kPa)	10-110 kPa	20-48	Varies	Catalyst
MAP Sensor (V)	0-5v	0.75-2.0	Varies	Catalyst
Non-Volatile Memory	Pass/Fail	Pass	Pass	Catalyst
Power Enrichment	Active/Inactive	Inactive	Inactive	Catalyst
Startup ECT Sensor	-40 to 304ºF	Varies	Varies	Catalyst
Startup IAT Sensor	-40 to 304ºF	Varies	Varies	Catalyst
TP Angle (%)	0-100%	0	Varies	Catalyst
TP Sensor (V)	0-5v	0.20-0.74	Varies	Catalyst
Transmission Range	P-R-N-4-3-2	P	3	Catalyst
TWC Protection	Active/Inactive	Inactive	Inactive	Catalyst
Vehicle Speed	0-155 mph	0	30	Catalyst

1997-99 Malibu 3.1L V6 MFI VIN M (A/T) - PID Data List

Parameter Identification (PID)	PID Value Range	PID Value at Hot Idle	PID Value at 30 mph	Data List Type
Engine Oil Level	Okay/LOW	Okay	Okay	IPC
Engine Oil Life	0-100%	0-100%	0-100%	IPC
Engine Oil Pressure	Okay/LOW/HI	Okay	Okay	IPC
Fuel Tank Level	0-100%	Varies	Varies	IPC
Generator Control	On / Off	On	On	IPC
Generator PWM	0-100%	0-100	0-100	IPC
Ignition 1 Signal	0.0-25.5v	14.2	14.2	IPC
MIL (lamp)	On / Off	Off	Off	IPC
Non-Volatile Memory	Pass/Fail	Pass	Pass	IPC

1997-99 Malibu 3.1L V6 MFI VIN M (A/T) - PID Data List

Parameter Identification (PID)	PID Value Range	PID Value at Hot Idle	PID Value at 30 mph	Data List Type
BARO Sensor (kPa)	10-110 kPa	65-110	65-110	EVAP
ECT Sensor (°F)	-40 to 304°F	185-220	185-220	EVAP
Engine Load	0-100%	2-10	Varies	EVAP
Engine Run Time	Hr: Min; Sec	00:00:00	00:00:00	EVAP
Engine Speed	0-9999 rpm	Varies	---	EVAP
EVAP Canister Purge	0-100%	0-100	0-100	EVAP
EVAP Fault History	Fault/No	No Fault	No Fault	EVAP
EVAP Purge Solenoid	Several	Okay	Okay	EVAP
EVAP Vent Solenoid	Open/Close	Open	Open	EVAP
Fuel Tank Level	0-100%	Varies	Varies	EVAP
FTP Sensor (In. H2O)	-17.5 to 7.5"	Varies	Varies	EVAP
Fuel Trim Learn	Enabled/DIS	Enabled	Enabled	EVAP
IAT Sensor (°F)	-40 to 304°F	50-194	50-194	EVAP
Long Term Fuel Trim	-10% to 10%	Varies	Varies	EVAP
Loop Status	Closed Loop / Open Loop	Closed Loop	Closed Loop	EVAP
MAF Sensor (g/sec)	0-512 g/sec	3-6	Varies	EVAP
MAP Sensor (kPa)	0-450 psi	20-48 kPa	Varies	EVAP
Non-Volatile Memory	Pass/Fail	Pass	Pass	EVAP
Startup ECT Sensor	-40 to 304°F	Varies	Varies	EVAP
Startup IAT Sensor	-40 to 304°F	Varies	Varies	EVAP
TP Angle (%)	0-100%	0	Varies	EVAP

1997-99 Malibu 3.1L V6 MFI VIN M (A/T) - PID Data List

Parameter Identification (PID)	PID Value Range	PID Value at Hot Idle	PID Value at 30 mph	Data List Type
3X Crank Sensor	1600-10,000	Varies	Varies	Misfire
24X Crank Sensor	0-1600	Varies	Varies	Misfire
Abuse Management	Active/Inactive	Inactive	Inactive	Misfire
A/C Request Signal	Yes / No	No	No	Misfire
Actual EGR Position	0-100%	0	0-100	Misfire
Cam Signal Present	Yes / No	Yes	Yes	Misfire
Commanded A/C	On / Off	On: AC on	Off	Misfire
Cycles of Misfire Data	0-255	0-99	Varies	Misfire
Decel Fuel Mode	Active/Inactive	Inactive	Inactive	Misfire
Desired EGR Position	0-100%	0	0-100	Misfire
ECT Sensor (°F)	-40 to 302°F	185-220	185-220	Misfire
Engine Load	0-100%	2-10	Varies	Misfire
Engine Speed	0-9999 rpm	Varies	Varies	Misfire
Hot Mode	On / Off	Off	Off	Misfire
HO2S-11 (B1 S1)	0-1132 mv	10-1000	10-1000	Misfire
Injector Pulsewidth	0-985 ms	1.5-3.5	Varies	Misfire
MAF Sensor (g/sec)	0-512 g/sec	4-6	Varies	Misfire
Misfire current Cyl 1-6	0-198	Varies	Varies	Misfire
Misfire History Cyl 1-6	0-65535	0	0	Misfire
Non-Volatile Memory	Pass/Fail	Pass	Pass	Misfire
Total Misfire Current	0-99 counts	Varies	Varies	Misfire
Total Misfire Failures	0-65535	Varies	Varies	Misfire
Total Misfire Passes	0-65535	Varies	Varies	Misfire

1999-2005 Malibu 3.1L V6 MFI VIN J (A/T) - PID Data List

Parameter Identification (PID)	PID Value Range	PID Value at Hot Idle	PID Value at 30 mph	Data List Type
3X Crank Sensor	0-9999 rpm	Varies	Varies	Engine Data 2
24X Crank Sensor	0-1600 rpm	Varies	Varies	Engine Data 2
1-2 Solenoid Control	Okay/Fault/ Invalid State	Varies	Varies	Output Driver
2-3 Solenoid Control	Okay/Fault/ Invalid State	Varies	Varies	Output Driver
1-2 Solenoid Control	Okay/Fault/ Invalid State	Varies	Varies	Output Driver
Abuse Management	Active/Inactive	Inactive	Inactive	Misfire
A/C HI Side Pressure	0-5.0v	Varies	Varies	Engine Data 1
A/C Pressure Disable	Yes/No	No	No	Engine Data 2
A/C Relay Circuit Status	Okay/Fault/ Invalid State	Varies	Varies	Output Driver
A/C Relay Command	On/Off	Off	Off	Engine Data 1-2 EGR, Misfire
A/C Request Signal	Yes/No	No	No	Engine Data 2
Actual EGR Position	0-100%	0	Varies	Engine Data 1, EGR, Misfire
Air Active Test Inhibit	Yes/No	No	No	AIR
Air Active Test Injection	Yes/No	No	No	AIR
Air Diagnostic Complete	Yes/No	No	No	AIR
Air Passive Test 1 Pass	Yes/No	No	No	AIR
Air Passive Test 2 Fail	Yes/No	No	No	AIR
Air Passive Test Inhibit	Yes/No	No	No	AIR
Air Pump Relay Circuit Status	Okay/Fault/ Invalid State	Okay	Okay	AIR, Output Driver
Air Pump Relay Command	On/Off	Off	Off	Engine Data 1 Data, AIR, HO2S
Air Solenoid Circuit Status	Okay/Fault/ Invalid State	Okay	Okay	AIR, Output Driver
Air Solenoid Command	On/Off	Off	Off	Engine Data 1 Data, AIR, HO2S
Air Fuel Ratio	0.0-25.5:1	14.7:1	14.7:1	Engine Data 2, Fuel Trim
BARO Sensor (kPa)	10-110 kPa	98	98	Engine Data 1, EGR, EVAP, Fuel Trim
BARO Sensor (V)	0-5.0 volts	3.5-4.5	3.5-4.5	Engine Data 1, EGR, EVAP, Fuel Trim
Brake Switch Signal	Applied/Released	Released	Released	Engine Data 1
CMP Sensor Signal	Yes/No	Yes	Yes	Engine Data 1, Misfire

1999-2005 Malibu 3.1L V6 MFI VIN J (A/T) - PID Data List

Parameter Identification (PID)	PID Value Range	PID Value at Hot Idle	PID Value at 30 mph	Data List Type
Commanded A/C	On/Off	Off	Off	Engine Data 1, EGR, Misfire, Fuel Trim
Command Fuel Pump	On/Off	On	On	Engine Data 1
Commanded Generator	On/Off	On	On	IPC
Crank Request Signal	Yes/No	Yes	Yes	Engine Data 2
Cruise Control Active	Yes/No	No	No	Engine Data 1
Cruise Mode	Yes/No	No	No	Engine Data 1
Cruised Inhibit Circuit Status	Okay/Fault/ Invalid State	Okay	Okay	Output Driver
Cruise Inhibit Signal	On/Off	On	On	Engine Data 1
Cruise Inhibit Reason	Several Descriptions	Varies	Varies	Engine Data 1
Current Gear	1-2-3-4	1	4	EGR, Catalyst
Cycles of Misfire Data	0-99 counts	Varies	Varies	Misfire
Cylinder 1-6 Injector Circuit Status	Okay/Stuck Low/Stuck High/Fault	Okay	Okay	Output Driver
Cylinder 1-6 Injector Circuit History	Okay/Stuck Low/Stuck High/Fault	Okay	Okay	Output Driver
Decel Fuel Cutoff	Active/Inactive	Inactive	Inactive	EGR, Fuel Trim, Misfire
Desired Idle Speed	0-3187 rpm	PCM control	---	Engine Data 2
Desired EGR Position	0-100%	0	Varies	Engine Data 1-2 EGR, EVAP
Driver Module 1 Status	Enabled/ Disabled	Enabled	Enabled	Output Driver
Driver Module 2 Status	Enabled/ Disabled	Enabled	Enabled	Output Driver
Driver Module 3 Status	Enabled/ Disabled	Enabled	Enabled	Output Driver
Driver Module 4 Status	Enabled/ Disabled	Enabled	Enabled	Output Driver
ECT Sensor (°F)	-40 to 304°F	185-220	185-220	Engine Data 1-2 EGR F/Trim, EVAP, Misfire
EGR Closed Valve Pintle Position	0-5.1volts	0.14-1.0	Varies	EGR
EGR Solenoid Duty Cycle Command	0-100	0	Varies	Engine Data 1, EGR
EGR Feedback	0-5.1 volts	0.14-1.0	Varies	EGR
EGR Flow Test Count	0-255 counts	0	Varies	EGR
EGR Learned Position	0-5.1 volts	0.71	Varies	EGR
EGR Position Error	0-100%	0	Varies	EGR
EGR Pos. Variance	0-100%	0	Varies	EGR

1999-2005 Malibu 3.1L V6 MFI VIN J (A/T) - PID Data List

Parameter Identification (PID)	PID Value Range	PID Value at Hot Idle	PID Value at 30 mph	Data List Type
Engine Load (%)	0-100%	17	Varies	Engine Data 1-2 EGR, F/Trim EVAP, Misfire
Engine Oil Level Switch	Okay/LOW	Okay	Okay	Engine Data 2
EOP Switch Signal	Okay/LOW/HI	Okay	Okay	Engine Data 2
Engine Run Time	Hr: Min: Sec	00:00:00	00:00:00	Engine Data 1-2 EGR, F/Trim EVAP, Misfire
Engine Speed	0-9999 rpm	± 50 rpm of Desired Speed	Varies	Engine Data 1-2 EGR, F/Trim EVAP, Misfire
EVAP Purge Solenoid Circuit Status	Okay/Fault	Okay	Okay	Output Driver
EVAP Purge Solenoid Control	0-100%	19	Varies	Engine Data 1, EVAP, HO2S
EVAP Fault History	No Fault / Excess VAC / Purge Valve Leak / Small Leak / Weak Vacuum	No Fault	No Fault	EVAP
EVAP Purge Solenoid Control Circuit Status	Okay/Fault/ Invalid state	Okay	Okay	Output Driver
EVAP Test Abort Reason	Not Aborted, Lost Enable, Small Leak, Vehicle Not at Rest	Not Aborted	Not Aborted	EVAP
EVAP Test Results	Aborted, No Test Results	No Test Results	No Test Results	EVAP
EVAP Test State	Completed/ Test running, Wait to purge	Test Completed	Test Completed	EVAP
EVAP Vent Solenoid Control Circuit Status	Okay/Fault/ Invalid state	Okay	Okay	Output Driver, EVAP
EVAP Vent Solenoid Command	Venting/Not Venting	Venting	Venting	Engine Data 1, EVAP, F/Trim
FC Relay 1 Circuit Status	Okay/Fault	Okay	Okay	Output Driver
FC Relay 1 Command	On/Off	Off	Off	Engine Data 2
FC Relay 2 and 3 Circuit Status	Okay/Fault	Okay	Okay	Output Driver
FC Relay 2 and 3 Command	On/Off	Off	Off	Engine Data 2
Fuel Level Sensor	0-5.0 volts	0.8-2.5	Varies	EVAP
Fuel Pump Command	On/Off	On	On	Engine Data 1
Fuel Pump Relay Circuit Status	Okay/Fault	Okay	Okay	Output Driver
Fuel Pump Relay Circuit History	Okay/Fault	Okay	Okay	Output Driver

1999-2005 Malibu 3.1L V6 MFI VIN J (A/T) - PID Data List

Parameter Identification (PID)	PID Value Range	PID Value at Hot Idle	PID Value at 30 mph	Data List Type
Fuel Tank Level Remaining	0-100%	Varies	Varies	EVAP, IPC
Fuel Tank Pressure Sensor	Inches H2O, mm Hg, volts	-17 to +7.5 Inches H2O	-17 to +7.5 Inches H2O	Engine Data 1, EVAP
Fuel Tank Pressure Sensor	0-5.0 volts	Varies	Varies	Engine Data 1, EVAP
Fuel Trim Cell	Cell 0-10	Varies	Varies	Engine Data 1, EVAP, F/Trim
Fuel Trim Learn	Enabled or Disabled	Enabled	May Toggle	Engine Data 1, EVAP, F/Trim
Generator Lamp	On/Off	Off	Off	Engine Data 2
Generator F-Terminal	0-100%	27	Varies	Engine Data 2
Generator L-Terminal	On/Off	On	On	Engine Data 2
HO2S-11 (B1 S1)	0-1132 mv	10-1000	10-1000	Engine Data 1, EVAP, F/Trim
HO2S-12 (B1 S2)	0-1132 mv	10-1000	10-1000	Engine Data 1, EVAP, F/Trim
HO2S-11 XCounts	0-255	Varies	Varies	Engine Data 1, HO2S
HO2S-11 Heater Command	On/Off	On	On	Fuel Trim
Hot Mode	On/Off	Off	Off	Misfire
IAC Position	0-255 counts	15	Varies	Engine Data 1, EGR, F/Trim
IAT Sensor (°F)	-40 to 304°F	50-194	50-194	Engine Data 1-2 EGR, EVAP, Fuel Trim
Ignition 1 Signal	0.0-25.5v	12.8-14.2	12.8-14.2	Engine Data 1-2 EGR, EVAP, Fuel Trim
Ignition Mode	Bypass/IC	IC	IC	Engine Data 2
Injector Pulsewidth	0-1000 ms	2.90	Varies	Engine Data 2, F/Trim, M/Fire
Knock Retard (°)	0-25.5°	0	0	Engine Data 1, EGR
Long Term Fuel Trim	-27% to 27%	0 (± 5%)	0 (± 5%)	Engine Data 1-2 EVAP, F/Trim
Loop Status	Closed Loop or Open Loop	Closed Loop	Closed Loop	Engine Data 1-2 EGR, EVAP, Fuel Trim
MAF Sensor (g/sec)	0-512 g/sec	3.41	Varies	Engine Data 1-2 EGR, EVAP, F/Trim, M/Fire
MAF Sensor (Hz)	0-32,000 Hz	2000	Varies	Engine Data 1-2 EGR, EVAP, F/Trim, M/Fire
MAP Sensor (kPa)	10-105 kPa	35	Varies	Engine Data 1-2 EGR, EVAP, F/Trim, M/Fire

1999-2005 Malibu 3.1L V6 MFI VIN J (A/T) - PID Data List

Parameter Identification (PID)	PID Value Range	PID Value at Hot Idle	PID Value at 30 mph	Data List Type
MAP Sensor (V)	0-5.1 volts	1.35	Varies	Engine Data 1-2 EGR, EVAP, F/Trim, M/Fire
MIL Circuit Status	Okay/Fault	Okay	Okay	Output Driver
MIL Command	On/Off	Off	Off	Engine Data 2
M/Fire Current Cyl 1-6	0-198 counts	0	0	Misfire
M/Fire History Cyl 1-6	0-65,535	0	0	Misfire
Non-Volatile Memory	Pass/Fail	Pass	Pass	Engine Data 1, EGR EVAP, Misfire, ODD
Number of DTC(s)	0-255 counts	0	0	Engine Data 1, EVAP, F/Trim
O2S Heater Current	0-10.0 amps	0.54	0.54	Fuel Trim
PCM/VCM in VTD Fuel Enable	Yes/No	No	No	Engine Data 1
Power Enrichment	Active/Inactive	Inactive	Inactive	CAT, EGR, HO2S, M/Fire
Short Term Fuel Trim	-27% to 27%	0 (± 5%)	0 (± 5%)	Engine Data 1-2 EGR, EVAP, F/Trim, M/Fire
Spark Advance (°)	-64° to 64°	18	Varies	Engine Data 1-2 F/Trim, M/Fire
Startup ECT	-40 to 419°F	Varies	Varies	Engine Data 2, EVAP, F/Trim
Startup IAT	-40 to 419°F	80°F	80°F	Engine Data 2, EVAP, F/Trim
TCC Brake Pedal Switch	Applied/Released	Released	Released	Engine Data 1-2
TCC Solenoid Circuit Status	Okay/Fault/ Invalid State	Okay	Okay	Output Driver
TCC Solenoid Control	On/Off	Off	On	ENG 2 M/Fire
TFP Switch Signal	P/R/N/D4-1	Park	Drive 4	Engine Data 2, EGR, F/Trim
Total Misfire Current	0-99 counts	Varies	Varies	Misfire
Total Misfire Failures	0-65,535	Varies	Varies	Misfire
Total Misfire Passed	0-65,535	Varies	Varies	Misfire
TP Angle (%)	0-100%	0	0-100	Engine Data 1-2 EGR, EVAP
TP Sensor (V)	0-5.1 volts	0.20-0.74	Varies	Engine Data 1
Transmission Range	Park/Neutral, Drive 4-3-2-1	Park/Neutral	Drive 4	Engine Data 1, AIR, EGR, Catalyst
TR Switch P/A/B/C	HIGH/LOW	HIGH	LOW	Engine Data 1
Vehicle Speed	0-155 mph	0	30	Engine Data 1-2 EGR, EVAP, F/Trim, M/Fire
VTD Auto Learn	Active/Inactive	Inactive	Inactive	Engine Data 1
VTD Fuel Disable	Active/Inactive	Inactive	Inactive	Engine Data 1
VTD Fuel Disable Until Ignition Off	Yes/No	No	No	Engine Data 1

MONTE CARLO PID DATA

2000-2005 Monte Carlo 3.4L V6 MFI VIN E (A/T) - PID Data List

Parameter Identification (PID)	PID Value Range	PID Value at Hot Idle	PID Value at 30 mph	Data List Type
3X Crank Sensor	0-9999 rpm	Varies	Varies	Engine Data 2
24X Crank Sensor	0-1600 rpm	Varies	Varies	Engine Data 2
Abuse Management	Active/Inactive	Inactive	Inactive	Misfire
A/C HI Side Pressure	0-5.0v	Varies	Varies	Engine Data 1
A/C Pressure Disable	Yes/No	No	No	Engine Data 2
A/C Pressure Disable	Yes/No	No	No	Engine Data 2
A/C Off for WOT	Yes/No	No	No	Engine Data 2
A/C Pressure Disable	Yes/No	No	No	Engine Data 2
A/C Relay Circuit Status	Okay/Fault	Okay	Okay	Output Driver
A/C Relay Command	On/Off	Off	Off	Engine Data 1-2 EGR, Misfire
A/C Request Signal	Yes/No	No	No	Engine Data 2
Actual EGR Position	0-100%	0	Varies	Engine Data 1, EGR, Misfire
Air Fuel Ratio	0.0-25.5:1	14.7:1	14.7:1	Engine Data 2, Fuel Trim
BARO Sensor (kPa)	10-110 kPa	98	98	Engine Data 1, EGR, EVAP, Fuel Trim
BARO Sensor (V)	0-5.0 volts	3.5-4.5	3.5-4.5	Engine Data 1, EGR, EVAP, Fuel Trim
Brake Switch Signal	Applied/Released	Released	Released	Engine Data 1
CMP Sensor Signal	Yes/No	Yes	Yes	Engine Data 1, Misfire
Crank Request Signal	Yes/No	No	No	Engine Data 2
Cruise Control Active	Yes/No	No	No	Engine Data 1
Cruise Inhibit Reason	Several	Varies	Varies	Engine Data 1
Cruised Inhibit Circuit Status	Okay/Fault/ Invalid State	Okay	Okay	Output Driver
Cruise Inhibit Signal Command	On/Off	On	On	Engine Data 1
Current Gear	1-2-3-4	1	4	Engine Data 1-2 EGR, F/Trim
Cycles of Misfire Data	0-99 counts	Varies	Varies	Misfire
Cylinder 1-6 Injector Circuit Status	Okay/Stuck Low/Stuck High/Fault	Okay	Okay	Output Driver
Cylinder 1-6 Injector Circuit History	Okay/Stuck Low/Stuck High/Fault	Okay	Okay	Output Driver
Decel Fuel Cutoff	Active/Inactive	Inactive	Inactive	EGR, Fuel Trim, Misfire
Desired Idle Speed	0-3187 rpm	PCM control	---	Engine Data 2
Desired EGR Position	0-100%	0	Varies	Engine Data 1-2 EGR, EVAP

2000-2005 Monte Carlo 3.4L V6 MFI VIN E (A/T) - PID Data List

Parameter Identification (PID)	PID Value Range	PID Value at Hot Idle	PID Value at 30 mph	Data List Type
Driver Module 1 Status	Enabled/ Disabled	Enabled	Enabled	Output Driver
Driver Module 2 Status	Enabled/ Disabled	Enabled	Enabled	Output Driver
Driver Module 3 Status	Enabled/ Disabled	Enabled	Enabled	Output Driver
Driver Module 4 Status	Enabled/ Disabled	Enabled	Enabled	Output Driver
ECT Sensor (ºF)	-40 to 304ºF	185-220	185-220	Engine Data 1-2 EGR F/Trim, EVAP, Misfire
EGR Closed Valve Pintle Position	0-5.1volts	0.14-1.0	Varies	EGR
EGR Solenoid Duty Cycle Command	0-100	0	Varies	Engine Data 1, EGR
EGR Feedback	0-5.1 volts	0.14-1.0	Varies	EGR
EGR Flow Test Count	0-255 counts	0	Varies	EGR
EGR Learned Position	0-5.1 volts	0.68	Varies	EGR
EGR Position Error	0-100%	0	Varies	EGR
EGR Pos. Variance	0-100%	0	Varies	EGR
Engine Load (%)	0-100%	17	Varies	Engine Data 1-2 EGR, F/Trim EVAP, Misfire
Engine Oil Level Switch	Okay/LOW	Okay	Okay	Engine Data 2
Engine Oil Life Remaining	0-100%	Varies	Varies	Engine Data 2
EOP Switch Signal	Okay/LOW/HI	Okay	Okay	Engine Data 2
Engine Run Time	Hr: Min: Sec	00:00:00	00:00:00	Engine Data 1-2 EGR, F/Trim EVAP, Misfire
Engine Speed	0-9999 rpm	± 50 rpm of Desired Speed	Varies	Engine Data 1-2 EGR, F/Trim EVAP, Misfire
EVAP Fault History	No Fault / Excess VAC / Purge Valve Leak / Small Leak / Weak Vacuum	No Fault	No Fault	EVAP
EVAP Purge Solenoid Circuit Status	Okay/Fault	Okay	Okay	Output Driver
EVAP Purge Solenoid Command	0-100%	19	Varies	Engine Data 1, EVAP, HO2S
EVAP Vent Solenoid Control Circuit Status	Okay/Fault/ Invalid state	Okay	Okay	Output Driver
EVAP Vent Solenoid Command	Venting/Not Venting	Venting	Venting	Engine Data 1, EVAP, F/Trim

2000-2005 Monte Carlo 3.4L V6 MFI VIN E (A/T) - PID Data List

Parameter Identification (PID)	PID Value Range	PID Value at Hot Idle	PID Value at 30 mph	Data List Type
FC Relay 1 Circuit Status	Okay/Fault	Okay	Okay	Output Driver
FC Relay 1 Command	On/Off	Off	Off	Engine Data 2
FC Relay 2 and 3 Circuit Status	Okay/Fault	Okay	Okay	Output Driver
FC Relay 2 and 3 Command	On/Off	Off	Off	Engine Data 2
Fuel Level Sensor	0-5.0 volts	0.8-2.5	Varies	EVAP
Fuel Pump Command	On/Off	On	On	Engine Data 1
Fuel Pump Relay Circuit Status	Okay/Fault	Okay	Okay	Output Driver
Fuel Pump Relay Circuit History	Okay/Fault	Okay	Okay	Output Driver
Fuel Pump Relay Command	On/Off	On	On	Engine Data 1
Fuel Tank Level Remaining	0-100%	Varies	Varies	EVAP
Fuel Tank Pressure Sensor	Inches H2O, mm Hg, volts	-17 to +7.5 Inches H2O	-17 to +7.5 Inches H2O	Engine Data 1, EVAP
Fuel Tank Pressure Sensor	0-5.0 volts	Varies	Varies	Engine Data 1, EVAP
Fuel Trim Cell	Cell 0-10	Varies	Varies	Engine Data 1, EVAP, F/Trim
Fuel Trim Learn	Enabled or Disabled	Enabled	May Toggle	Engine Data 1, EVAP, F/Trim
Generator Lamp	On/Off	Off	Off	Engine Data 2
Generator L-Terminal	On/Off	On	On	Engine Data 2
HO2S-11 (B1 S1)	0-1132 mv	10-1000	10-1000	Engine Data 1, EVAP, F/Trim
HO2S-12 (B1 S2)	0-1132 mv	10-1000	10-1000	Engine Data 1, EVAP, F/Trim
HO2S-11 XCounts	0-255	Varies	Varies	Engine Data 1, HO2S
HO2S-11 Heater Command	On/Off	On	On	Fuel Trim
Hot Mode	On/Off	Off	Off	Misfire
IAC Position	0-255 counts	15	Varies	Engine Data 1, EGR, F/Trim
IAT Sensor (°F)	-40 to 304°F	50-194	50-194	Engine Data 1-2 EGR, EVAP, Fuel Trim
Ignition 1 Signal	0.0-25.5v	12.8-14.2	12.8-14.2	Engine Data 1-2 EGR, EVAP, Fuel Trim
Ignition Mode	Bypass/IC	IC	IC	Engine Data 2
IMS	P/R/N/D4-1	Park	Drive 4	Engine Data 1
Injector Pulsewidth	0-1000 ms	2.90	Varies	Engine Data 2, F/Trim, M/Fire
Knock Retard (°)	0-25.5°	0	0	Engine Data 1, EGR

2000-2005 Monte Carlo 3.4L V6 MFI VIN E (A/T) - PID Data List

Parameter Identification (PID)	PID Value Range	PID Value at Hot Idle	PID Value at 30 mph	Data List Type
Long Term Fuel Trim	-27% to 27%	0 (± 5%)	0 (± 5%)	Engine Data 1-2 EVAP, F/Trim
Loop Status	Closed Loop or Open Loop	Closed Loop	Closed Loop	Engine Data 1-2 EGR, EVAP, Fuel Trim
MAF Sensor (g/sec)	0-512 g/sec	4.25	Varies	Engine Data 1-2 EGR, EVAP, F/Trim, M/Fire
MAF Sensor (Hz)	0-32,000 Hz	2210	Varies	Engine Data 1-2 EGR, EVAP, F/Trim, M/Fire
MAP Sensor (kPa)	10-105 kPa	38	Varies	Engine Data 1-2 EGR, EVAP, F/Trim, M/Fire
MAP Sensor (V)	0-5.1 volts	1.45	Varies	Engine Data 2
MIL Circuit Status	Okay/Fault	Okay	Okay	Output Driver
MIL Command	On/Off	Off	Off	Engine Data 2
M/Fire Current Cyl 1-6	0-198 counts	0	0	Misfire
M/Fire History Cyl 1-6	0-65,535	0	0	Misfire
Non-Volatile Memory	Pass/Fail	Pass	Pass	Engine Data 1
Number of DTC(s)	0-255 counts	0	0	Engine Data 1, EVAP, F/Trim
HO2S Heater Current	0-10.0 amps	0.54	0.54	Fuel Trim
PCM Reset	Yes/No	No	No	Engine Data 1-2 EGR, EVAP, F/Trim, ODD
PCM/VCM in VTD Fuel Enable	Yes/No	No	No	Engine Data 1
Power Enrichment	Active/Inactive	Inactive	Inactive	CAT, EGR, HO2S, M/Fire
Short Term Fuel Trim	-27% to 27%	0 (± 5%)	0 (± 5%)	Engine Data 1-2 EVAP, F/Trim
Spark Advance (°)	-64° to 64°	20	Varies	Engine Data 1-2 F/Trim, M/Fire
Startup ECT	-40 to 419°F	Varies	Varies	Engine Data 2, EVAP, F/Trim
Startup IAT	-40 to 419°F	80°F	80°F	Engine Data 2, EVAP, F/Trim
Starter Enable Relay Circuit Status	Okay / Fault/ Invalid State	Okay	Okay	Output Driver
Starter Relay Control	On/Off	Off	Off	Engine Data 2
TCC Brake Pedal Switch	Applied/Released	Released	Released	Engine Data 1-2
TCC Solenoid Circuit Status	Okay/Fault/ Invalid State	Okay	Okay	Output Driver
TCC Solenoid Control	On/Off	Off	On	Engine Data 2, Misfire
TFP Switch Signal	P/R/N/D4-1	Park	Drive 4	Engine Data 2, EGR, F/Trim

2000-2005 Monte Carlo 3.4L V6 MFI VIN E (A/T) - PID Data List

Parameter Identification (PID)	PID Value Range	PID Value at Hot Idle	PID Value at 30 mph	Data List Type
Torque Delivered Signal	0-100%	86	86	Engine Data 2
Torque Requested Signal	0-100%	100	100	Engine Data 2
Total Misfire Current Count	0-65,535	Varies	Varies	Misfire
TP Angle (%)	0-100%	0	0-100	Engine Data 1-2 EGR, EVAP, F/Trim, M/Fire
TP Sensor (V)	0-5.1 volts	0.20-0.74	Varies	Engine Data 1-2 EGR, EVAP, F/Trim, M/Fire
Traction Control Status	Active or Inactive	Inactive	Inactive	Engine Data 2
Transmission Range	Park/Neutral, Drive 4-3-2-1	Park/Neutral	Drive 4	Engine Data 1, AIR, EGR, Catalyst
TR Switch P/A/B/C	HIGH/LOW	HIGH	LOW	Engine Data 1
TWC Protection	Active or Inactive	Inactive	Inactive	Engine Data 2, Misfire
Vehicle Speed	0-155 mph	0	30	Engine Data 1-2 EGR, EVAP, F/Trim, M/Fire
VTD Auto Learn Timer	Active/Inactive	Inactive	Inactive	Engine Data 1
VTD Fuel Disable	Active/Inactive	Inactive	Inactive	Engine Data 1
VTD Fuel Disable Until Ignition Off	Yes/No	No	No	Engine Data 1

2000-2005 Monte Carlo 3.8L V6 MFI VIN K (A/T) - PID Data List

Parameter Identification (PID)	PID Value Range	PID Value at Hot Idle	PID Value at 30 mph	Data List Type
3X Crank Sensor	0-9,999 rpm	Varies	Varies	Engine Data 2
14X Crank Sensor	0-1,600 rpm	Varies	Varies	Engine Data 2
1-2 Shift Solenoid Circuit Status	Okay / Fault/ Invalid State	Okay	Okay	Output Driver
2-3 Shift Solenoid Circuit Status	Okay / Fault/ Invalid State	Okay	Okay	Output Driver
A/C HI Side Pressure	0-459 kPa	90	90	Engine Data 2
A/C HI Side Pressure	0-5.1 volts	0.98	0.98	Engine Data 2
A/C Off for WOT	Yes/No	No	No	Engine Data 2
A/C Pressure Disable	Yes/No	No	No	Engine Data 2
A/C Relay Circuit Status	Okay / Fault/ Invalid State	Okay	Okay	Output Driver
A/C Relay Command	On/Off	Off	Off	Engine Data 1-2 EGR, Misfire
A/C Request Signal	Yes/No	No	No	Engine Data 2
Air Fuel Ratio	0.0-25.5:1	14.2-14.7:1	14.2-14.7:1	Engine Data 2, Fuel Trim
Air Pump Relay Circuit Status	Okay/Fault/ Invalid State	Okay	Okay	AIR, Output Driver
Air Pump Relay Command	On/Off	Off	Off	Engine Data 1 Data, AIR, HO2S
Air Solenoid Circuit Status	Okay/Fault/ Invalid State	Okay	Okay	Output Driver
Air Solenoid Command	On/Off	Off	Off	Engine Data 1 Data, AIR, HO2S
Air Fuel Ratio	0.0-25.5:1	14.7:1	14.7:1	Engine Data 2, Fuel Trim
BARO Sensor (kPa)	10-105 kPa	Varies w/ALT	Varies	Engine Data 1, EGR, EVAP, Fuel Trim
BARO Sensor (V)	0-5.1 volts	Varies w/ALT	Varies	Engine Data 1, EGR, EVAP, Fuel Trim
CMP Sensor Signal	Yes/No	Yes	Yes	Engine Data 2
Cruise Control Active	Yes/No	No	No	Engine Data 1
Cruise Inhibit Reason	Vehicle Speed/Brake/ Clutch	Vehicle Speed	Vehicle Speed	Engine Data 1
Cruise Inhibit Signal Circuit Status	Okay/Stuck Low or High	Okay	Okay	Output Driver
Cruise Inhibit Signal Command	On/Off	Off	Off	Engine Data 1
Current Gear	1, 2, 3, 4	1	4	Engine Data 1-2 EGR, F/Trim
Cycles of Misfire Data	0-99 counts	0	0	Misfire
Cylinder 1-6 Injector Circuit Status	Okay / Stuck Low/ Stuck High / Fault	Okay	Okay	Output Driver

2000-2005 Monte Carlo 3.8L V6 MFI VIN K (A/T) - PID Data List

Parameter Identification (PID)	PID Value Range	PID Value at Hot Idle	PID Value at 30 mph	Data List Type
Cylinder 1-6 Injector Circuit History	Okay / Stuck Low/ Stuck High / Fault	Okay	Okay	Output Driver
Decel Fuel Cutoff	Active/Inactive	Inactive	Inactive	EGR, F/Trim
Desired EGR Position	0-100%	0	Varies	Engine Data 1, EGR, F/Trim
Desired EGR Position	0-5.1v	0.0	Varies	EGR
Desired Idle Speed	0-3187 rpm	Commanded by the PCM	Varies	Engine Data 1-2 EVAP
Driver Module 1 Status	Enabled / Off- HI Volts/ Off-HI TEMP/ Invalid State	Enabled	Enabled	Output Driver
Driver Module 2 Status	Enabled / Off- HI Volts/ Off-HI TEMP/ Invalid State	Enabled	Enabled	Output Driver
Driver Module 3 Status	Enabled / Off- HI Volts/ Off-HI TEMP/ Invalid State	Enabled	Enabled	Output Driver
Driver Module 4 Status	Enabled / Off- HI Volts/ Off-HI TEMP/ Invalid State	Enabled	Enabled	Output Driver
ECT Sensor (°F)	-40 to 304°F	190-221°F	190-221°F	Engine Data 1-2 EGR, F/Trim, EVAP, Misfire
EGR Flow Test Count	0-255 counts	0-12	Varies	EGR
EGR Learned Minimum Position	0-5.1 volts	0.14-1.0	Varies	EGR
EGR Position Sensor	0-100%	Varies	Varies	Engine Data 1, Misfire
EGR Position Sensor	0-5.1 volts	Varies	Varies	EGR
EGR Position Variance	0-100%	0	0	EGR
EGR Solenoid Circuit History	Okay / Fault/ Invalid State	Okay	Okay	Output Driver
EGR Solenoid Circuit Status	Okay / Fault/ Invalid State	Okay	Okay	Output Driver
EGR Solenoid Command	0-100%	Varies	Varies	EGR
Engine Load	0-100%	1-4	Varies	Engine Data 1-2 EGR, F/Trim, EVAP, Misfire
Engine Oil Level Switch	Okay / LOW	Okay	Okay	Engine Data 2
Engine Oil Life	0-100%	Varies	Varies	Engine Data 2
EOP Switch Signal	Okay/LOW	Okay	Okay	Engine Data 2

2000-2005 Monte Carlo 3.8L V6 MFI VIN K (A/T) - PID Data List

Parameter Identification (PID)	PID Value Range	PID Value at Hot Idle	PID Value at 30 mph	Data List Type
Engine Run Time	Hr: Min: Sec 00:00:00 to 99:99:99	00:00:00	00:00:00	Engine Data 1-2 EGR, F/Trim, EVAP, Misfire
Engine Speed	0-9999 rpm	± 50 rpm of actual speed	Varies	Engine Data 1-2 EGR, F/Trim, EVAP, Misfire
EVAP Fault History	No Fault / Excess VAC /Purge Valve Leak / Small Leak / Weak Vacuum	No Fault	No Fault	EVAP
EVAP Purge Solenoid Circuit Status	Okay / Fault/ Invalid State	Okay	Okay	Output Driver
EVAP Purge Solenoid Command	0-100%	15	Varies	Engine Data 1, EVAP, F/Trim
EVAP Test Abort Reason	Not Aborted/ Lost Enable/ Small Leak/ Not at Rest	Not Aborted	Not Aborted	EVAP
EVAP Test Result	No Result / Passed/ Fail/ P0440/442/ P0446/1441	No Test Results	No Test Results	EVAP
EVAP Test State	Waiting for Purge/ Test Running/Test Completed	Test Completed	Test Completed	EVAP
EVAP Vent Solenoid Circuit Status	Okay / Fault/ Invalid State	Okay	Okay	Output Driver
EVAP Vent Solenoid Command	Venting / Not Venting	Not Venting	Not Venting	Engine Data 1, EVAP, F/Trim
Fan Control Relay 1 Circuit Status	Okay / Fault/ Invalid State	Okay	Okay	Output Driver
Fan Control Relay 1 Command	On/Off	Off	Off	Engine Data 2
Fan Control Relay 2-3 Circuit Status	Okay/Fault	Okay	Okay	Output Driver
Fan Control Relay 2-3 Command	On/Off	Off	Off	Engine Data 2
Fuel Pump Relay Circuit Status	Okay / Fault/ Invalid State	Okay	Okay	Output Driver
Fuel Pump Relay Circuit History Status	Okay / Fault	Okay	Okay	Output Driver
Fuel Pump Relay Command	On/Off	On	On	Engine Data 1
Fuel Tank level	0-100%	Varies	Varies	EVAP
Fuel Tank Pressure Sensor	Inches H2O, mm Hg	-17 to +7.5 Inches H2O	-17 to +7.5 Inches H2O	Engine Data 1, EVAP

2000-2005 Monte Carlo 3.8L V6 MFI VIN K (A/T) - PID Data List

Parameter Identification (PID)	PID Value Range	PID Value at Hot Idle	PID Value at 30 mph	Data List Type
Fuel Tank Pressure Sensor	0-5.0 volts	Varies	Varies	EVAP
Fuel Trim Cell	Cell 0-9	0-4	Varies	Engine Data 1, EVAP, F/Trim
Fuel Trim Learn	Enabled or Disabled	Enabled	May Toggle	Engine Data 1, EVAP, F/Trim
Generator L-Terminal	On/Off	On	On	Engine Data 2
HO2S (B1 S1) Signal	0-1132 mv	10-1000	10-1000	Engine Data 1, EVAP, F/Trim
HO2S (B2 S1) Signal	0-1132 mv	10-1000	10-1000	Engine Data 1, EVAP, F/Trim
HO2S (B1 S2) Signal	0-1132 mv	10-1000	10-1000	Engine Data 1, EVAP, F/Trim
IAC Position	0-255 counts	Varies	Varies	Engine Data 1, EGR, EVAP, Fuel Trim
IAT Sensor (°F)	-40 to 304°F	91	92	Engine Data 1-2 EGR, EVAP, Fuel Trim
Ignition 1	0.0-25.5v	13.5	13.6	Engine Data 1-2 EGR, EVAP, Fuel Trim
Ignition Mode	Bypass/IC	IC	IC	Engine Data 2
Injector Pulsewidth	0-1000 ms	1.5-3.5	Varies	Engine Data 2, F/Trim, M/fire
Knock Retard (°)	0-25.5°	0°	0°	Engine Data 1, EGR
Long Term Fuel Trim Bank 1	-23 to +16	0 (± 5%)	0 (± 5%)	Engine Data 1-2 EVAP, F/Trim
Long Term Fuel Trim Bank 2	-23 to +16	0 (± 5%)	0 (± 5%)	Engine Data 1-2 EVAP, F/Trim
Loop Status	Open / Closed	Closed	Closed	Engine Data 1-2 EGR, EVAP, Fuel Trim
MAF Sensor (g/sec)	0-512 g/sec	3-6	Varies	Engine Data 1-2 EGR, F/Trim, EVAP, Misfire
MAF Sensor (Hz)	0-32,000 Hz	2000-3000	Varies	Engine Data 2
MAP Sensor (kPa)	10-105 kPa	20-48	Varies	Engine Data 1-2 EGR, F/Trim, EVAP, Misfire
MAP Sensor (V)	0-5.1 volts	0.75-2.0	Varies	Engine Data 1-2
MIL Circuit Status	Okay / Fault/ Invalid State	Okay	Okay	Output Driver
MIL Command	On/Off	Off	Off	Engine Data 2
Misfire Current Cylinder 1-6	0-198 counts	0-4	0	Misfire
Misfire History Cylinder 1-6	0-65,535	0	0	Misfire

2000-2005 Monte Carlo 3.8L V6 MFI VIN K (A/T) - PID Data List

Parameter Identification (PID)	PID Value Range	PID Value at Hot Idle	PID Value at 30 mph	Data List Type
Number of DTC(s)	Number	0	No	Engine Data 1, EVAP, F/Trim
PCM/VCM in VTD Fail Enable	Yes/No	Yes	Yes	Engine Data 1
Power Enrichment	Active/Inactive	Inactive	Inactive	Engine Data 2, Misfire
Short Term Fuel Trim Bank 1	-10 to +10%	0 (± 5%)	0 (± 5%)	Engine Data 1-2 EVAP, F/Trim
Short Term Fuel Trim Bank 2	-10 to +10%	0 (± 5%)	0 (± 5%)	Engine Data 1-2 EVAP, F/Trim
Spark Advance	-64° to +64°	-20°	Varies	Engine Data 1-2 F/Trim, M/fire
Starter Enable Relay Circuit Status	Okay / Fault/ Invalid State	Okay	Okay	Output Driver
Starter Relay Control	On/Off	Off	Off	Engine Data 2
Startup ECT	-40 to 419°F	Varies	Varies	EVAP, F/Trim
Startup IAT	-40 to 419°F	80°F	80°F	EVAP, F/Trim
TCC/Cruise Brake Pedal Switch	Apply/REL	Released	Released	Engine Data 1, Data 2
TCC PWM Solenoid Command	0-100%	0	Varies	Engine Data 2
TCS Circuit Status	Okay/Fault/ Invalid State	Okay	Okay	Output Driver
TCS Circuit History	Okay/Fault/ Invalid State	Okay	Okay	Output Driver
TFP Switch	P/N, Reverse Drive 4, 3, 2, 1, Invalid	Park/Neutral	Drive 4	Engine Data 2, EGR, Fuel Trim
TP Desired Angle	0-100%	2-6	Varies	Engine Data 1-2 TAC EGR, F/Trim, EVAP, Misfire
TP Sensor (%)	0-100%	2-6	Varies	Engine Data 1-2 EGR, F/Trim, EVAP, Misfire
TP Sensor (V)	0-5.0 volts	0.2-0.8	Varies	Engine Data 1-2
Traction Control Status	Active/Inactive	Inactive	Inactive	Engine Data 2
Torque Delivered Signal	0-100%	85	85	Engine Data 2
Torque Requested Signal	0-100%	100	100	Engine Data 2
Vehicle Speed	Km/h, MPH	0	30	Engine Data 1-2 EGR, F/Trim, EVAP, M/Fire
VTD Auto Learn Timer	Active/Inactive	Inactive	Inactive	Engine Data 1
VTD Fuel Enable	Active/Inactive	Inactive	Inactive	Engine Data 1
VTD Fuel Disable Until Ignition Off	Yes/No	No	No	Engine Data 1

UPLANDER PID DATA

2005 Uplander 3.5L VIN L (ALL) – PID Data List

Scan Tool Parameter	Data List	Units Displayed	Typical Data Value
24X Crank Sensor	Ignition	RPM	0-6,400
A/C Relay Circuit Status	ODD	OK/Fault/Invalid State	OK
A/C Relay Circuit History	ODD	OK/Fault/Invalid State	OK
A/C Relay Command	Eng, Misfire, EGR, TAC	On/Off	Off
Air Fuel Ratio	Eng, Fuel Trim, HO2S	Ratio	14.2:1-14.7:1
APP Indicated Angle	Eng, EGR, EVAP, HO2S, Ignition, Misfire, TAC, Fuel Trim	Percent	0
APP Sensor 1	TAC	Volts	Varies
APP Sensor 1 Circuit	TAC	OK/Not OK	OK
APP Sensor 2	TAC	Volts	Varies
APP Sensor 2 Circuit	TAC	OK/Not OK	OK
APP Sensor 1	TAC	Percent	Varies
APP Sensor 2	TAC	Percent	Varies
APP Sensor 1 and 2	TAC	Agree/Disagree	Agree
BARO	Eng, EGR, EVAP, Fuel Trim, HO2S, Ignition	kPa	65-110 kPa
Catalytic Converter Protection Active	Fuel Trim, HO2S	Yes/No	No
CMP Sensor	Ignition	RPM	Varies
CMP Sensor Signal Present	Ignition, Misfire	Yes/No	Yes
Cruise Control Active	TAC	Yes/No	No
Cycles of Misfire Data	Misfire	Counts	0-99
Cyl. 1-6 Injector Circuit Status	ODD	OK/Stuck Low (open)/Stuck High	OK
Cylinder 1 and 4 IC Circuit Status	ODD	OK/Not OK	OK
Cylinder 2 and 5 IC Circuit Status	ODD	OK/Not OK	OK
Cylinder 3 and 6 IC Circuit Status	ODD	OK/Not OK	OK
Cylinder 1-6 Injector Circuit History	ODD	OK/Stuck Low (open)/Stuck High	OK
Decel Fuel Cut-off	Eng, EGR, Fuel Trim, HO2S	Active/Inactive	Inactive
Desired EGR Position	EGR, Misfire	Percent	0
Desired Idle Speed	Eng, EVAP, TAC	RPM	Varies
Driver Module 1 Status	ODD	Enabled/Off-High Volts/Off High Temp/Invalid State	Enabled
Driver Module 2 Status	ODD	Enabled/Off-High Volts/Off High Temp/Invalid State	Enabled
Driver Module 3 Status	ODD	Enabled/Off-High Volts/Off High Temp/Invalid State	Enabled
EC Ignition Relay Circuit Status	ODD	OK/Fault/Invalid State	OK
EC Relay Circuit History	ODD	OK/Fault/Invalid State	OK
ECT Sensor	Eng, EGR, EVAP, HO2S, Ignition, Misfire, Fuel Trim, TAC	°C/°F	Varies
EGR Flow Test Count	EGR	Counts	0-12
EGR Learned Minimum Position	EGR	Volts	0.14-1.0
EGR Position Sensor	Eng, EGR, Misfire	Percent	Varies
EGR Position Sensor	EGR	Volts	0
EGR Position Variance	EGR	Percent	0
EGR Solenoid Circuit History	ODD	OK/Open/Fault/Invalid State	OK
EGR Solenoid Circuit Status	ODD	OK/Fault/Invalid State	OK
EGR Solenoid Command	EGR	Percent	0

2005 Uplander 3.5L VIN L (ALL) – PID Data List (CONT)

Scan Tool Parameter	Data List	Units Displayed	Typical Data Value
Engine Load	Eng, Misfire, Fuel Trim, EVAP, EGR, HO2S, Ignition, TAC	Percent	0
Engine Oil Pressure Switch	Eng	Low/OK	OK
Engine Run Time	Eng, EVAP, Misfire, Fuel Trim, EGR, HO2S, Ignition, TAC, ODD	Hr: Min: Sec	Varies
Engine Speed	Eng, EGR, EVAP, Ignition, Misfire, Fuel Trim, TAC	RPM	Varies
EVAP Fault History	EVAP	No Fault/Excess Vacuum/Purge Valve Leak/Small Leak/Weak Vacuum/No Test Result	No Fault
EVAP Purge Solenoid Circuit History	ODD	OK/Open/Short	OK
EVAP Purge Solenoid Circuit Status	ODD	OK/Open/Short	OK
EVAP Purge Solenoid Command	Eng, EVAP, Fuel Trim, HO2S	Percent	Varies
EVAP Test Abort Reason	EVAP	No Aborted/Lost Enable/Small Leak/Veh. Not At Rest	Varies
EVAP Test Result	EVAP	• No Result/Passed/Aborted OR • Fail-DTC P0455 OR • Fail-DTC P0442 OR • Fail-DTC P0446 OR • Fail-DTC P0496	Varies
EVAP Test State	EVAP	Waiting for Purge/Test Running/Test Completed	Varies
EVAP Vent Solenoid Circuit History	ODD	OK/Open/Short	OK
EVAP Vent Solenoid Circuit Status	ODD	OK/Open/Short	OK
EVAP Vent Solenoid Command	Eng, EVAP	Venting/Not Venting	Varies
FC Relay 1 Circuit History	ODD	OK/Fault/Invalid State	OK
FC Relay 1 Circuit Status	ODD	OK/Fault/Invalid State	OK
FC Relay 2 and 3 Circuit History	ODD	OK/Fault/Invalid State	OK
FC Relay 2 and 3 Circuit Status	ODD	OK/Fault/Invalid State	OK
Fuel Level Sensor	EVAP	Volts	Varies
Fuel Pump Relay Circuit History	ODD	OK/Fault/Invalid State	OK
Fuel Pump Relay Circuit Status	ODD	OK/Fault/Invalid State	OK
Fuel Pump Relay Command	Eng	On/Off	On
Fuel Tank Level Remaining	EVAP	Percent	0-104
Fuel Tank Pressure Sensor	Eng, EVAP	mm Hg/in H2O	Varies
Fuel Tank Pressure Sensor	EVAP	Volts	Varies
Fuel Trim Cell	EVAP, Eng, Fuel Trim	0-4	0
Fuel Trim Learn	Eng, EVAP, Fuel Trim	Enabled/Disabled	Varies
HO2S Bank 1 Sensor 1	EVAP, Eng, Fuel Trim, HO2S	millivolts	0-1000 and Varying
HO2S Bank 1 Sensor 1 Ready	Fuel Trim, HO2S	Yes/No	Yes
HO2S Bank 1 Sensor 2	Eng, Fuel Trim, HO2S	millivolts	0-1000 and Varying
HO2S Bank 2 Sensor 1 Ready	Fuel Trim, HO2S	Yes/No	Yes
HO2S Bank 2 Sensor 1	Eng, EVAP, Fuel Trim, HO2S	millivolts	0-1000 and Varying
HO2S Bank 2 Sensor 2	Eng, Fuel Trim, HO2S	millivolts	0-1000 and Varying
HO2S Htr. Bn. 1 Sen. 1	HO2S	On/Off	On

2005 Uplander 3.5L VIN L (ALL) – PID Data List (CONT)

Scan Tool Parameter	Data List	Units Displayed	Typical Data Value
HO2S Htr. Bank 1 Sensor 1 Command	HO2S	On/Off	On
HO2S Htr. Bn. 1 Sen. 2	HO2S	On/Off	On
HO2S Htr. Bank 1 Sensor 2 Command	HO2S	On/Off	On
HO2S Heater Bn. 1 Sen. 1	HO2S	Amps	Varies
HO2S Heater Bn. 1 Sen. 2	HO2S	Amps	Varies
HO2S Heater Bn. 2 Sen. 1	HO2S	Amps	Varies
HO2S Heater Bn. 2 Sen. 2	HO2S	Amps	Varies
HO2S Htr. Bn. 1 Sen. 1 Circuit History	ODD	OK/Fault/Invalid State	OK
HO2S Htr. Bn. 1 Sen. 2 Circuit History	ODD	OK/Fault/Invalid State	OK
HO2S Htr. Bn. 2 Sen. 1	HO2S	On/Off	On
HO2S Htr. Bn. 2 Sen. 1 Circuit History	ODD	OK/Fault/Invalid State	OK
HO2S Htr. Bn. 2 Sen. 2	HO2S	On/Off	On
HO2S Htr. Bn. 2 Sen. 2 Circuit History	ODD	OK/Fault/Invalid State	OK
HO2S Htr. Bn. 1 Sen. 1 Circuit Status	ODD	OK/Fault/Invalid State	OK
HO2S Htr. Bn. 1 Sen. 2 Circuit Status	ODD	OK/Fault/Invalid State	OK
HO2S Htr. Bn. 2 Sen. 1 Circuit Status	ODD	OK/Fault/Invalid State	OK
HO2S Htr. Bn. 2 Sen. 2 Circuit Status	ODD	OK/Fault/Invalid State	OK
Hot Open Loop	Fuel Trim, HO2S	Active/Inactive	Inactive
IAT Sensor	Eng, EVAP, EGR, Fuel Trim, HO2S, Ignition, Misfire, TAC	°C/°F	Varies
Ignition 1 Signal	EVAP, Eng, EGR, Fuel Trim, HO2S, Ignition, Misfire, ODD, TAC	Volts	Varies
Ignition Off Time	EVAP, Eng	Minutes/Varies	Varies
Initial Brake Apply Signal	TAC	Released/Applied	Released
Injector PWM	Fuel Trim, Misfire	milliseconds	Varies
Knock Retard	EGR, Ignition	Degrees	0
Long Term FT Bank 1	Eng, EVAP, Fuel Trim, HO2S	Percent	-10% to +10%
Long Term FT Bank 2	Eng, EVAP, Fuel Trim, HO2S	Percent	-10% to +10%
Loop Status	Eng, Fuel Trim, EVAP, HO2S	Open/Closed	Closed
MAF Sensor	Eng, EGR, Misfire, EVAP, Fuel Trim, HO2S, Ignition, TAC	g/s	3-6 (Varies with altitude)
MAF Sensor	EVAP, Fuel Trim, HO2S, Misfire, TAC	Hz	1,200-3,000 (depends on altitude)
MAP Sensor	Eng, EGR, Fuel Trim, HO2S, Misfire, TAC	Volts	0.4-2.0 Volts
MAP Sensor	Eng, EVAP, Fuel Trim, HO2S, Ignition, Misfire, EGR, TAC	kPa	20-48 kPa (Varies with altitude)
MIL Circuit History	ODD	OK/Fault/Invalid State	OK
MIL Command	EVAP, Eng, EGR, Ignition	On/Off	Off
MIL Circuit Status	ODD	OK/Fault/Invalid State	OK
Mileage Since DTC Cleared	Eng	km/h/mph	Varies
Misfire Current Cyl. 1-6	Misfire	Counts	0
Misfire History Cyl. 1-6	Misfire	Counts	0
Moderate Brake Apply Signal	TAC	Applied/Released	Released
Number of DTCs	Eng EVAP, Fuel Trim, TAC	Number	0
PCM Reset	Eng	Yes/No	No

2005 Uplander 3.5L VIN L (ALL) – PID Data List (CONT)

Scan Tool Parameter	Data List	Units Displayed	Typical Data Value
Power Enrichment	Eng, Fuel Trim, HO2S, Misfire	Active/Inactive	Inactive
Reduced Engine Power	TAC	Active/Inactive	Inactive
Short Term FT Bank 1	Eng, EVAP, Fuel Trim, HO2S,	Percent	-10 to +10
Short Term FT Bank 2	Eng, EVAP, Fuel Trim, HO2S,	Percent	-10 to +10
Spark	Eng, Misfire, Fuel Trim, HO2S, Ignition	Degrees	22 (varies)
Start Up ECT	EVAP, Fuel Trim, HO2S	°C/°F	Varies
Start Up IAT	EVAP, Fuel Trim, HO2S	°C/°F	Varies
Starter Relay Circuit Status	ODD	OK/Fault	OK
Starter Relay Circuit History	ODD	OK/Fault/Invalid State	OK
Stop Lamp Pedal Switch	Eng, TAC	Released/Applied	Released
TAC/PCM Communications Signal	TAC	OK/Fault	OK
TAC Stop Lamp Pedal Switch	Eng	Released/Applied	Released
TAC Stop Lamp Pedal Switch	TAC	Applied/Released	Released
TAC Vehicle Speed Signal	TAC	OK/Fault	OK
TCS Delivered Torque Circuit Status	ODD	OK/Fault	OK
Torque Request Signal	TAC	Percent	100%
Torque Delivered Signal	TAC	Percent	75%
Total Misfire Count	Misfire	1-255	0
TP Desired Angle	TAC	Percent	Varies
TP Indicated Angle	Eng, EGR, Fuel Trim, HO2S, Ignition, EVAP, Misfire, TAC	Percent	%
TP Sensor 1	TAC	Percent	Varies
TP Sensor 1	TAC	Volts	Varies
TP Sensors 1 and 2	TAC	Agree/Disagree	Agree

C/K, M/L & S/T SERIES VEHICLES

1996-2005 Trucks & Vans 4.3L V6 VIN W, X (All) - PID Data List

Parameter Identification (PID)	PID Value Range	PID Value at Hot Idle	PID Value at 30 mph	Data List Type
4WD Switch Signal	Enable/Disable	Disabled	Disabled	Engine Data 2
4WD Low Signal	Enable/Disable	Disabled	Disabled	Engine Data 2
A/C Clutch Feedback Circuit	Yes/No	No	No	Engine Data 2
A/C Compressor Cycling Switch	Open/Closed	OPEN	OPEN	Engine Data 2
A/C Relay Command	On/Off	Off	Off	Engine Data 1-2, EGR, Misfire
A/C Request Signal	Yes/No	No	No	Engine Data 2
A/C Secondary High Pressure Switch	Open/Closed	OPEN	OPEN	Engine Data 2
Air Pump Relay Command	On/Off	Off	Off	Engine Data 1, Fuel Trim
Air Fuel Ratio	0.0-25.5:1	14.7:1	14.7:1	Engine Data 2, Fuel Trim
BARO Sensor (kPa)	10-110 kPa	65-104	65-104	Engine Data 1, EGR, EVAP, Fuel Trim
BARO Sensor (V)	0-5.0 volts	3.5-4.9	3.5-4.9	Engine Data 1, EGR, EVAP, Fuel Trim
Brake Switch Signal	Applied/REL	Released	Released	Engine Data 1-2
CMP Retard	-90° to 90°	-3	Varies	Engine Data 1
CMP Sensor LO to HI	0-65,535 counts	Varies	Varies	Engine Data 2
CMP Sensor HI to LO	0-65,535 counts	Varies	Varies	Engine Data 2
Clutch Pedal Switch	Applied / Released	Released	Released	Engine Data 1, Data 2, EGR
Cruise Control Active	Yes/No	No	No	Engine Data 1
Cruise Inhibit Reason	Several	Varies	Varies	Engine Data 1
Cruise Inhibit Signal	On/Off	Off	Off	Engine Data 1
Current Gear	0-1-2-3-4	1	4	Engine Data 1-2, EGR, Fuel Trim
Cycles of Misfire Data	0-100 counts	Varies	Varies	Misfire
Desired EGR Position	0-100%	0	Varies	Engine Data 1, EGR, Misfire
Desired EGR Position	0-5.0 volts	< 1.3	Varies	EGR
Desired IAC Airflow	0-64 g/sec	Varies	Varies	Engine Data 1
Desired IAC Position	0-255 counts	10-40	Varies	Engine Data 1
Desired Idle Speed	0-3187 rpm	Varies	Varies	Engine Data 1-2, EGR, EVAP
DTC Set This Ignition	Yes/No	No	No	Engine Data 1-2, EVAP, Fuel Trim
ECT Sensor (°F)	-40° to 304°F	185-220	185-220	Engine Data 1-2, EGR, Fuel Trim, EVAP, Misfire

1996-2005 Trucks & Vans 4.3L V6 MFI VIN W, X (All) - PID Data List

Parameter Identification (PID)	PID Value Range	PID Value at Hot Idle	PID Value at 30 mph	Data List Type
EGR Learned Minimum Position	0-5.1 volts	< 1.3	Varies	EGR
EGR Position Sensor	0-5.1 volts	< 1.3	Varies	Engine Data 1, EGR, Fuel Trim
Engine Load (%)	0-100%	17	Varies	Engine Data 1-2, EGR, Fuel Trim EVAP, Misfire
Engine Oil Life Remaining	0-100%	Varies	Varies	Engine Data 2
Engine Run Time	Hr: Min: Sec	00:00:00	00:00:00	Engine Data 1-2, EGR, Fuel Trim EVAP, Misfire
Engine Speed	0-10,000 rpm	500-700 rpm	Varies	Engine Data 1-2, EGR, Fuel Trim EVAP, Misfire
EVAP Purge Solenoid Command	0-100%	10-25	Varies	Engine Data 1, EVAP, Fuel Trim
EVAP Vent Solenoid Command	Venting/Not Venting	Venting	Venting	Engine Data 1, EVAP, Fuel Trim
EVAP Test Result	No Result / Passed/ Fail/ P0440/442/ P0446/1441	No Test Results	No Test Results	EVAP
EVAP Test State	Waiting for Purge/ Test Running/Test Completed	Test Completed	Test Completed	EVAP
Fail Counter	0-255 counts	0	0	F/F & F/R
Fuel Level Sensor	0-5.0 volts	0.8-2.5	Varies	Engine Data 1, Data 2, EVAP
Fuel Tank Level Remaining	Gallons or Liters	Varies	Varies	Engine Data 1-2 EVAP/Fuel Trim
Fuel Tank Level Remaining	0-100%	Varies	Varies	Engine Data 1-2, EVAP, Fuel Trim
Fuel Tank Pressure Sensor	Inches H2O, mm Hg	-17.5 to +7.5 Inches H2O	-17.5 to +7.5 Inches H2O	Engine Data 1, EVAP
Fuel Tank Pressure Sensor	0-5.0 volts	Varies	Varies	EVAP
Fuel Tank Rated Capacity	Gallons or Liters	25.9 / 34.0 or 98/128 Liters	25.9 / 34.0 or 98/128 Liters	EVAP
Fuel Trim Cell	Cell 0-23	16-17	18-21	Engine Data 1, EVAP, Fuel Trim
Fuel Trim Learn	Enabled or Disabled	Enabled	May Toggle	Engine Data 1, EVAP, Fuel Trim
Generator F-Terminal	0-100%	Varies	Varies	Engine Data 2
Generator L-Terminal	On/Off	On	On	Engine Data 2
HO2S-11 (B1 S1)	0-1106 mv	10-1000	10-1000	Engine Data 1, EVAP, Fuel Trim
HO2S-12 (B1 S2)	0-1106 mv	10-1000	10-1000	Engine Data 1, Fuel Trim

1996-2005 Trucks & Vans 4.3L V6 MFI VIN W, X (All) - PID Data List

Parameter Identification (PID)	PID Value Range	PID Value at Hot Idle	PID Value at 30 mph	Data List Type
HO2S-21 (B2 S1)	0-1106 mv	10-1000	10-1000	Engine Data 1, EVAP, Fuel Trim
HO2S-22 (B2 S2)	0-1106 mv	10-1000	10-1000	Engine Data 1, Fuel Trim
IAC Position	0-1024 counts	Varies	Varies	Engine Data 1, EGR, Fuel Trim
IAT Sensor (°F)	-38 to 284°F	50-194	50-194	Engine Data 1-2, EGR, EVAP, Fuel Trim
Ignition 1 Signal	0.0-25.5v	12.8-14.2	12.8-14.2	Engine Data 1-2, EGR, EVAP, Fuel Trim
Injector Pulsewidth Bank 1	0-1000 ms	2-6	2-6	Engine Data 2, Fuel Trim, Misfire
Injector Pulsewidth Bank 2	0-1000 ms	2-6	2-6	Engine Data 2, Fuel Trim, Misfire
Knock Retard (°)	0-25.5°	0.0-16.0	0	Engine Data 1, EGR
Long Term Fuel Trim Bank 1	-10% to 10%	0 (± 5%)	0 (± 5%)	Engine Data 1-2, EVAP, Fuel Trim
Long Term Fuel Trim Bank 2	-10% to 10%	0 (± 5%)	0 (± 5%)	Engine Data 1-2, EVAP, Fuel Trim
Long Term Fuel Trim Average Bank 1	-10% to 10%	0 (± 5%)	0 (± 5%)	Fuel Trim
Long Term Fuel Trim Average Bank 2	-10% to 10%	0 (± 5%)	0 (± 5%)	Fuel Trim
Loop Status	Closed Loop / Open Loop	Closed Loop	Closed Loop	Engine Data 1-2, EGR, EVAP, Fuel Trim
MAF Sensor (g/sec)	0-512 g/sec	1.9	14-26	Engine Data 1-2, EGR, EVAP, Fuel Trim, Misfire
MAF Sensor (Hz)	0-32,000 Hz	2000-3000	Varies	Engine Data 2
MAP Sensor (kPa)	10-105 kPa	20-48	Varies	Engine Data 1-2, EGR, EVAP, Fuel Trim, Misfire
MAP Sensor (V)	0-5.1 volts	1.0-2.0	Varies	Engine Data 1-2
MIL Command	On/Off	Off	Off	Engine Data 2
Mileage Since DTC Cleared	Kilometers or Miles	Varies	Varies	Engine Data 2
Misfire Counter Status	Valid / Invalid	Valid	Valid	Misfire
Misfire Current Cylinder 1-6	0-200 counts	0	0	Misfire
Misfire History Cylinder 1-6	0-65,535	0	0	Misfire
PCM Reset	Yes/No	No	No	Engine Data 1-2, EGR, EVAP, Fuel Trim

1996-2005 Trucks & Vans 4.3L V6 MFI VIN W, X (All) - PID Data List

Parameter Identification (PID)	PID Value Range	PID Value at Hot Idle	PID Value at 30 mph	Data List Type
PCM/VCM in VTD Fuel Enable	Yes/No	No	No	Engine Data 1
Reduced Engine Power	Active or Inactive	Inactive	Inactive	Engine Data 1
Short Term Fuel Trim Bank 1	-10% to 10%	0 (± 5%)	0 (± 5%)	Engine Data 1-2, EVAP, Fuel Trim
Short Term Fuel Trim Bank 2	-10% to 10%	0 (± 5%)	0 (± 5%)	Engine Data 1-2, EVAP, Fuel Trim
Short Term Fuel Trim Average Bank 1	-10% to 10%	0 (± 5%)	0 (± 5%)	Fuel Trim
Short Term Fuel Trim Average Bank 2	-10% to 10%	0 (± 5%)	0 (± 5%)	Fuel Trim
Spark Advance (º)	-64º to 64º	15-20	Varies	Engine Data 1-2 Fuel Trim, Misfire
Startup ECT	-38 to 284ºF	Varies	Varies	Engine Data 2, EVAP, Fuel Trim
Startup IAT	-38 to 284ºF	80ºF	80ºF	Engine Data 2, EVAP, Fuel Trim
Stoplamp Brake Pedal Switch	Applied or Released	Released	Released	Engine Data 1, Data 2
TCC Enable Solenoid Command	On/Off	Off	On	Engine Data 1-2, Misfire
TCC PWM Solenoid	0-100%	0	Varies	Engine Data 2
TFP Switch Signal	P/R/N/D4-1	Park	Drive 4	Engine Data 2, EGR, Fuel Trim
Torque Delivered Signal	Nm / ft lbs	14 / 15-25	Varies	Engine Data 2
Torque Request Signal	Nm., ft lbs	300 / 400-410	Varies	Engine Data 2
TP Angle (%)	0-100%	0	Varies	Engine Data 1-2, EGR, EVAP, Fuel Trim, Misfire
TP Sensor (V)	0-5.1 volts	0.40-0.90	Varies	Engine Data 1-2
Transmission Range Switch	P, N, D4, D3, D2, D1 / Invalid	Park	D4	Engine Data 1-2 Catalyst, Fuel Trim, EGR
TR Switch A/B/C	HI/Low	High/Low/High	High/Low/Low	Engine Data 1-2
Vehicle Speed	0-155 km/h or miles per hour (mph)	0	30	Engine Data 1-2, EGR, EVAP, Fuel Trim, Misfire
VTD Auto Learn Timer	Active/Inactive	Inactive	Inactive	Engine Data 1
VTD Fuel Disable	Active/Inactive	Inactive	Inactive	Engine Data 1
VTD Fuel Disable Until Ignition Off	Yes/No	No	No	Engine Data 1
Warmups without an Emission Fault	0-255 counts	Varies	Varies	Engine Data 2
Warmups without a Non-Emission Fault	0-255 counts	Varies	Varies	Engine Data 2

1999-05 C/K & S/T 4.8L VIN V, 5.3L VIN T, 6.0L VIN U V8 - PID List

Parameter Identification (PID)	PID Value Range	PID Value at Hot Idle	PID Value at 30 mph	Data List Type
4WD Switch Signal	Enabled or Disabled	Disabled	Disabled	Engine Data 2
4WD Low Signal	Enabled or Disabled	Disabled	Disabled	Engine Data 2
A/C Pressure Sensor	0-450 psi	90	90	Engine Data 2
A/C Pressure Sensor	0-5.0 volts	0.98	0.98	Engine Data 2
A/C Relay Command	On/Off	Off	Off	Engine Data 1-2, Misfire
A/C Request Signal	Yes/No	No	No	Engine Data 2
APP Average	0-125 counts	0-100%	Varies	TAC
APP Sensor 1 (%)	0-100%	0	Varies	TAC
APP Sensor 1 (V)	0-5.0 volts	0.4-0.9	Varies	TAC
APP Sensor 2 (%)	0-100%	0	Varies	TAC
APP Sensor 2 (V)	5.0-0 volts	4.5-4.1	Varies	TAC
APP Sensor 1 and 2	Agree or Disagree	Agree	Agree	TAC
BARO Sensor (kPa)	10-110 kPa	50-104	50-104	Engine Data 1, EVAP, Fuel Trim
BARO Sensor (V)	0-5.0 volts	3.5-4.9	3.5-4.9	Engine Data 1, EVAP, Fuel Trim
CMP Actual Angle	0-25 degrees	0	Varies	Engine Data 2
CMP Desired Angle	0-25 degrees	0	Varies	Engine Data 2
CMP Angle Variance	0-25 degrees	0	Varies	Engine Data 2
Camshaft Actuator Solenoid Command	0-100%	0	Varies	Engine Data 2
Clutch Pedal Switch	Applied / Released	Released	Released	Engine Data 2
Coolant Level Switch	Okay / Fault	Okay	Okay	Engine Data 2
Cruise Control Active	Yes/No	No	No	Engine Data 1, Cruise, TAC
Cruise Disengage 1-8 History	20 Possible Causes	None	None	Cruise Control
Cruise Inhibit Signal	On/Off	Off	Off	Engine Data 1
Cruise On/Off Switch	On/Off	Off	Off	Cruise, TAC
Cruise Release Brake Pedal Switch	Applied / Released	Released	Released	Cruise Control
Cruise Resume / Accel Switch Signal	On/Off	Off	Off	Cruise, TAC
Cruise Set / Coast Switch Signal	On/Off	Off	Off	Cruise, TAC
Current Gear	0-1-2-3-4	1	4	Engine Data 1-2 Fuel Trim
Cycles of Misfire Data	0-100 counts	Varies	Varies	Misfire
Decel*Fuel Shutoff	Active or Inactive	Inactive	Inactive	Fuel Trim
Desired IAC Airflow	0-64 g/sec	Varies	Varies	Engine Data 1-3 EVAP, TAC
DTC Set This Ignition	0-255	Yes/No	Yes/No	Engine Data 1-2, EVAP, TAC

1999-05 C/K & S/T 4.8L VIN V, 5.3L VIN T, 6.0L VIN U V8 - PID List

Parameter Identification (PID)	PID Value Range	PID Value at Hot Idle	PID Value at 30 mph	Data List Type
ECT Sensor (°F)	-38 to 284°F	190-221	190-221	Engine Data 1-2, EVAP, Fuel Trim, Misfire
Engine Load (%)	0-100%	1-4	7-9	Engine Data 1-2, Cruise, EVAP Fuel Trim, Misfire, TAC
Engine Oil Level Switch	Okay / Low	Okay	Okay	Engine Data 2
Engine Oil Life Remaining	0-100%	Varies	Varies	Engine Data 2
EOP Sensor (V)	0-5.0 volts	Varies	Varies	Engine Data 2
Engine Run Time	Hr: Min: Sec	00:00:00	00:00:00	Engine Data 1-2, Cruise, EVAP Fuel Trim, Misfire, TAC
Engine Speed	0-9999 rpm	500-700 rpm	Varies	Engine Data 1-2, Cruise, EVAP Fuel Trim, Misfire, TAC
EVAP Purge Solenoid Command	0-100%	10-25	Varies	Engine Data 1, EVAP, Fuel Trim
EVAP Test Result	No Result / Passed/ Fail/ P0440/442/ P0446/1441	No Test Results	No Test Results	EVAP
EVAP Test State	Waiting for Purge/ Test Running/Test Completed	Test Completed	Test Completed	EVAP
EVAP Vent Solenoid Command	Venting/Not Venting	Venting	Venting	Engine Data 1, EVAP, Fuel Trim
Fuel Level Sensor	0-5.0 volts	0.7-2.5	Varies	Engine Data 1, EVAP
Fuel Level Sensor - Rear Tank	0-5.0 volts	0.7-2.5	Varies	Engine Data 1, EVAP
Fuel Tank Level Remaining	Gallons or Liters	Varies	Varies	EVAP
Fuel Tank Level Remaining	0-100%	0	Varies	EVAP
Fuel Tank Pressure Sensor	Inches H2O, mm Hg	-17.5 to +7.5 Inches H2O	-17.5 to +7.5 Inches H2O	Engine Data 1, EVAP
Fuel Tank Pressure Sensor	0-5.0 volts	Varies	Varies	EVAP
Fuel Tank Rated Capacity	Gallons or Liters	25.9 / 34.0 or 98/128 Liters	25.9 / 34.0 or 98/128 Liters	EVAP
Fuel Trim Cell	Cell 0-23	19	Varies	Engine Data 1, EVAP, Fuel Trim
Fuel Trim Learn	Enabled or Disabled	Enabled	May Toggle	Engine Data 1, EVAP, Fuel Trim

1999-05 C/K & S/T 4.8L VIN V, 5.3L VIN T, 6.0L VIN U V8 - PID List

Parameter Identification (PID)	PID Value Range	PID Value at Hot Idle	PID Value at 30 mph	Data List Type
Generator F-Terminal	0-100%	Varies	Varies	Engine Data 2
Generator L-Terminal	Active/Inactive	Active	Active	Engine Data 2
HO2S-11 (B1 S1)	0-1106 mv	10-1000	10-1000	Engine Data 1, EVAP, Fuel Trim, HO2S
HO2S-12 (B1 S2)	0-1106 mv	10-1000	10-1000	Engine Data 1, Fuel Trim, HO2S
HO2S-21 (B2 S1)	0-1106 mv	10-1000	10-1000	Engine Data 1, EVAP, Fuel Trim, HO2S
HO2S-22 (B2 S2)	0-1106 mv	10-1000	10-1000	Engine Data 1, Fuel Trim, HO2S
HO2S-11 Heater Current	0-5.0 amps	0.6	0.6	HO2S
HO2S-12 Heater Current	0-5.0 amps	0.6	0.6	HO2S
HO2S-21 Heater Current	0-5.0 amps	0.6	0.6	HO2S
HO2S-22 Heater Current	0-5.0 amps	0.6	0.6	HO2S
IAT Sensor (°F)	-38 to 284°F	Varies	Varies	Engine Data 1-2, Cruise, EVAP Fuel Trim, HO2S
Ignition 1 Signal	0.0-25.5v	12.8-14.2	12.8-14.2	Engine Data 1-2, Cruise, EVAP Fuel Trim, TAC
Injector 1-6 Command	0-1000 ms	2.49-2.59	Varies	Engine Data 3
Injector Pulsewidth Bank 1	0-1000 ms	2-6	2-6	Engine Data 2, Fuel Trim, Misfire
Injector Pulsewidth Bank 2	0-1000 ms	2-6	2-6	Engine Data 2, Fuel Trim, Misfire
Knock Retard (°)	0-25.5°	0.0-16.0	0	Engine Data 1
Long Term Fuel Trim Bank 1 & Bank 2	-10% to 10%	0 (± 5%)	0 (± 5%)	Engine Data 1-2, EVAP, Fuel Trim, HO2S
Loop Status	Closed Loop / Open Loop	Closed Loop	Closed Loop	Engine Data 1-2, EVAP, Fuel Trim, HO2S
MAF Sensor (g/sec)	0-512 g/sec	1-9	15-26	Engine Data 1-2, EVAP, Fuel Trim Misfire, TAC
MAF Sensor (Hz)	0-31,999 Hz	2000-3000	Varies	Engine Data 1-2
MAP Sensor (kPa)	10-105 kPa	20-48	Varies	Engine Data 1-2, EVAP, Fuel Trim Misfire, TAC
MAP Sensor (V)	0-5.1 volts	1.0-2.0	Varies	Engine Data 1-2
MIL Command	On/Off	Off	Off	Engine Data 1
Mileage Since DTC Cleared	Kilometers or Miles	Varies	Varies	Engine Data 2
MIL Counter Status	0-65,535	Varies	Varies	Misfire

1999-05 C/K & S/T 4.8L VIN V, 5.3L VIN T, 6.0L VIN U V8 - PID List

Parameter Identification (PID)	PID Value Range	PID Value at Hot Idle	PID Value at 30 mph	Data List Type
Misfire Current Cylinder 1-8	0-200 counts	0	0	Misfire
Misfire History Cylinder 1-8	0-65,535	0	0	Misfire
PCM Reset	Yes/No	No	No	Engine Data 1-2, EVAP, Fuel Trim
PCM/VCM in VTD Fuel Enable	Yes/No	No	No	Engine Data 1
Power Enrichment	Yes/No	No	No	Engine Data 1, HO2S
Reduced Engine Power	Active or Inactive	Inactive	Inactive	Engine Data 1, Cruise, TAC
Short Term Fuel Trim Bank 1 and Bank 2	-10% to 10%	0 (± 5%)	0 (± 5%)	Engine Data 1-2, EVAP, Fuel Trim
Spark Advance (°)	-64° to 64°	15-20	Varies	Engine Data 1-3 Fuel Trim, Misfire
Startup ECT	-40 to 419°F	Varies	Varies	Engine Data 2, EVAP, Fuel Trim
Stoplamp Pedal Switch	On/Off	Off	Off	Engine Data 2, Cruise, TAC
TCC Enable Solenoid	Enable/Disable	Disabled	Enabled	Engine Data 1
TCC PWM Solenoid	0-100%	0	Varies	Engine Data 2
Torque Delivered	N/m., ft lbs	Varies	Varies	ENG 2, TAC
Torque Requested	0-100%	0	Varies	ENG 2, TAC
TP Desired Angle	0-100%	5.5	Varies	Engine Data 1, TAC, Cruise, EVAP
TP Indicated Angle	4.2	0	Varies	Engine Data 1-2, Cruise, EVAP Misfire, TAC
TP Sensor 1 (%)	0-100	0	Varies	TAC
TP Sensor 1 (V)	5.0-0.0 volts	0.4-0.9	Varies	TAC
TP Sensor 2	0-100	0	Varies	TAC
TP Sensor 2 (V)	5.0-0 volts	4.78-4.10	Varies	TAC
TP Sensor 1 and 2	Agree/DIS	Agree	Agree	TAC
Transmission Range Switch	Park/Neutral, Drive 4-3-2-1	Park/Neutral	Drive 4	Engine Data 2, Cruise, Fuel Trim
Traction Control Signal	Active/Inactive	Inactive	Inactive	Engine Data 2, Cruise, TAC
Vehicle Speed	0-155 km/h or miles per hour (mph)	0	30	Engine Data 1-2, Cruise, EVAP Misfire, TAC
VTD Auto Learn Timer	Active/Inactive	Inactive	Inactive	Engine Data 1
VTD Fuel Disable	Active/Inactive	Inactive	Inactive	Engine Data 1
VTD Fuel Disable Until Ignition Off	Yes/No	No	No	Engine Data 1
Warmups w/o an Emission Fault	0-255 counts	Varies	Varies	Engine Data 2
Warmups w/o a Non-Emission Fault	0-255 counts	Varies	Varies	Engine Data 2

1996-2002 C/K Series 5.0L VIN M, 5.7L VIN R V8 - PID Data List

Parameter Identification (PID)	PID Value Range	PID Value at Hot Idle	PID Value at 30 mph	Data List Type
A/C Clutch Feedback Circuit	Yes/No	No	No	Engine Data 2
A/C Compressor Cycling Switch	Open/Closed	OPEN	OPEN	Engine Data 2
A/C Relay Command	On/Off	Off	Off	Engine Data 1-2, EGR, Misfire
A/C Request Signal	Yes/No	No	No	Engine Data 2
A/C Relay Command	On/Off	Off	Off	Engine Data 1-2, EGR, Misfire
A/C Request Signal	Yes/No	No	No	Engine Data 2
A/C Secondary High Pressure Switch	Open/Closed	OPEN	OPEN	Engine Data 2
Air Pump Relay Command	On/Off	Off	Off	Engine Data 1, Fuel Trim
Air Fuel Ratio	0.0-25.5:1	14.7:1	14.7:1	Engine Data 2, Fuel Trim
BARO Sensor (kPa)	10-110 kPa	65-104	65-104	Engine Data 1, EGR, EVAP, Fuel Trim
BARO Sensor (V)	0-5.0 volts	3.5-4.9	3.5-4.9	Engine Data 1, EGR, EVAP, Fuel Trim
Brake Switch Signal	Applied/REL	Released	Released	Engine Data 1-2
CMP Sensor LO to HI	0-65,535 counts	Varies	Varies	Engine Data 2
CMP Sensor HI to LO	0-65, 535 counts	Varies	Varies	Engine Data 2
Coolant Level Switch	Yes/No	No	No	Engine Data 2
Cruise Control Active	Yes/No	No	No	Engine Data 1
Cruise Disengage History 1-8	Several	Varies	Varies	Cruise Control
Cruise Inhibit Signal	On/Off	Off	Off	Engine Data 1
Cruise Control On/Off Switch	On/Off	Off	Off	Cruise Control
Cruise Control release Brake Pedal Switch	On/Off	Off	Off	Cruise Control
Cruise Resume / Accel Switch	Resume / Accel	Resume	Resume	Cruise Control
Cruise Set / Coast Switch	Set / Coast	Coast	Coast	Cruise Control
Current Gear	0-1-2-3-4	1	4	Engine Data 1-2, EGR, Misfire
Cycles of Misfire Data	0-100 counts	Varies	Varies	Misfire
Desired EGR Position	0-100%	0	Varies	Engine Data 1, EGR, Fuel Trim
Desired EGR Position	0-5.0 volts	< 1.3	Varies	EGR
Desired IAC Airflow	0-64 g/sec	Varies	Varies	Engine Data 1
Desired IAC Position	0-255 counts	10-40	Varies	Engine Data 1
DTC Set This IGN	Yes/No	No	No	Engine Data 1-2 C/C, Fuel Trim

1996-2005 C/K Series 5.0L VIN M, 5.7L VIN R V8 - PID Data List

Parameter Identification (PID)	PID Value Range	PID Value at Hot Idle	PID Value at 30 mph	Data List Type
ECT Sensor (°F)	-40° to 304°F	185-220	185-220	Engine Data 1-2, EGR, Fuel Trim, EVAP, Misfire
EGR Learned Minimum Position	0-5.0 volts	< 1.3	Varies	EGR
EGR Position Sensor	0-5.0 volts	< 1.3	Varies	Engine Data 1, EGR, Fuel Trim
Engine Load (%)	0-100%	1-4	7-9	Engine Data 1-2, EGR, Fuel Trim EVAP, Misfire
Engine Oil Level Switch	Yes/No	No	No	Engine Data 2
Engine Oil Life Remaining	0-100%	Varies	Varies	Engine Data 2
Engine Run Time	Hr: Min: Sec	00:00:00	00:00:00	Engine Data 1-2, EGR, Fuel Trim EVAP, Misfire
Engine Speed	0-10,000 rpm	500-700 rpm	Varies	Engine Data 1-2, EGR, Fuel Trim EVAP, Misfire
EVAP Purge Solenoid Command	0-100%	10-25	Varies	Engine Data 1, EVAP, Fuel Trim
EVAP Vent Solenoid Command	Venting/Not Venting	Venting	Venting	Engine Data 1, Fuel Trim
EVAP Test Abort Reason	No Result / Passed/ Fail/ P0440/442/ P0446/1441	No Test Results	No Test Results	EVAP
EVAP Test State	Waiting for Purge/ Test Running/Test Completed	Test Completed	Test Completed	EVAP
Fail Counter	0-255 counts	0	0	Freeze Frame / Failure Record
Fuel Level Sensor	0-5.0 volts	0.7-2.5	0.7-2.5	Engine Data 1, Data 2, EVAP
Fuel Level Sensor - Rear Tank	0-5.0 volts	0.7-2.5	0.7-2.5	Engine Data 1, Data 2, EVAP
Fuel Tank Level Remaining	Gallons or Liters	Varies	Varies	Engine Data 2, EVAP/Fuel Trim
Fuel Tank Level Remaining	0-100%	Varies	Varies	Engine Data 2, EVAP, Fuel Trim
Fuel Tank Pressure Sensor	Inches H2O, mm Hg	-17.5 to +7.5 Inches H2O	-17.5 to +7.5 Inches H2O	Engine Data 1, EVAP
Fuel Tank Pressure Sensor	0-5.0 volts	Varies	Varies	EVAP
Fuel Tank Rated Capacity	Gallons or Liters	25.9 / 34.0 or 98/128 Liters	25.9 / 34.0 or 98/128 Liters	EVAP
Fuel Trim Cell	Cell 0-23	16-17	18-21	Engine Data 1, EVAP, Fuel Trim

1996-2005 C/K Series 5.0L VIN M, 5.7L VIN R V8 - PID Data List

Parameter Identification (PID)	PID Value Range	PID Value at Hot Idle	PID Value at 30 mph	Data List Type
Fuel Trim Learn	Enabled or Disabled	Enabled	May Toggle	Engine Data 1, EVAP, Fuel Trim
Generator F-Terminal	0-100%	Varies	Varies	Engine Data 2
Generator L-Terminal	Active/Inactive	Active	Active	Engine Data 2
HO2S-11 (B1 S1)	0-1132 mv	10-1000	10-1000	Engine Data 1, EVAP, Fuel Trim
HO2S-12 (B1 S2)	0-1132 mv	10-1000	10-1000	Engine Data 1, EVAP, Fuel Trim
HO2S-21 (B2 S1)	0-1132 mv	10-1000	10-1000	Engine Data 1, EVAP, Fuel Trim
HO2S-22 (B2 S2)	0-1132 mv	10-1000	10-1000	Engine Data 1, EVAP, Fuel Trim
IAC Position	0-1024 counts	Varies	Varies	Engine Data 1, EGR, Fuel Trim
IAT Sensor (°F)	-38 to 284°F	50-194	50-194	Engine Data 1-2, EGR, EVAP, Fuel Trim
Ignition 1 Signal	0.0-25.5v	12.8-14.2	12.8-14.2	Engine Data 1-2, EGR, EVAP, Fuel Trim
Injector Pulsewidth Bank 1	0-1000 ms	2-6	2-6	Engine Data 2, Fuel Trim, Misfire
Injector Pulsewidth Bank 2	0-1000 ms	2-6	2-6	Engine Data 2, Fuel Trim, Misfire
Knock Retard (°)	0-25.5°	0.0-16.0	0	Engine Data 1, EGR
Long Term Fuel Trim Bank 1	-10% to 10%	0 (± 5%)	0 (± 5%)	Engine Data 1-2, EVAP, Fuel Trim
Long Term Fuel Trim Bank 2	-10% to 10%	0 (± 5%)	0 (± 5%)	Engine Data 1-2, EVAP, Fuel Trim
Long Term Fuel Trim Average Bank 1	-10% to 10%	0 (± 5%)	0 (± 5%)	Engine Data 1-2, EVAP, Fuel Trim
Long Term Fuel Trim Average Bank 2	-10% to 10%	0 (± 5%)	0 (± 5%)	Engine Data 1-2, EVAP, Fuel Trim
Loop Status	Closed Loop / Open Loop	Closed Loop	Closed Loop	Engine Data 1-2, EGR, EVAP, Fuel Trim
MAF Sensor (g/sec)	0-512 g/sec	1-9	15-26	Engine Data 1-2, EGR, EVAP, Fuel Trim, Misfire
MAF Sensor (Hz)	0-31,999 Hz	2000-3000	Varies	Engine Data 2
MAP Sensor (kPa)	10-105 kPa	20-48	Varies	Engine Data 1-2, EGR, EVAP, Fuel Trim, Misfire
MAP Sensor (V)	0-5.1 volts	1.0-2.0	Varies	Engine Data 1-2
MIL Command	On/Off	Off	Off	Engine Data 2
Mileage Since DTC Cleared	Kilometers or Miles	Varies	Varies	Engine Data 2
Misfire Current Cylinder 1-6	0-200 counts	0	0	Misfire

1996-2005 C/K Series 5.0L VIN M, 5.7L VIN R V8 - PID Data List

Parameter Identification (PID)	PID Value Range	PID Value at Hot Idle	PID Value at 30 mph	Data List Type
Misfire History Cylinder 1-6	0-65,535	0	0	Misfire
Misfire Not Run Counter	0-65, 535 counts	Varies	Varies	Freeze Frame / Failure Record
Misfire Pass Counter	0-65, 535 counts	Varies	Varies	Freeze Frame / Failure Record
PCM Reset	Yes/No	No	No	Freeze Frame / Failure Record
PCM/VCM in VTD Fuel Enable	Yes/No	No	No	Engine Data 1
Reduced Engine Power	Yes/No	No	No	Engine Data 1, EGR
Short Term Fuel Trim Bank 1	-10% to 10%	0 (± 5%)	0 (± 5%)	Engine Data 1-2, EVAP, Fuel Trim
Short Term Fuel Trim Bank 2	-10% to 10%	0 (± 5%)	0 (± 5%)	Engine Data 1-2, EVAP, Fuel Trim
Short Term Fuel Trim Average Bank 1	-10% to 10%	0 (± 5%)	0 (± 5%)	Engine Data 1-2, EVAP, Fuel Trim
Short Term Fuel Trim Average Bank 2	-10% to 10%	0 (± 5%)	0 (± 5%)	Engine Data 1-2, EVAP, Fuel Trim
Spark Advance (°)	-64° to 64°	15-20	Varies	Engine Data 1-2 Fuel Trim, Misfire
Startup ECT	-40 to 419°F	Varies	Varies	Engine Data 2, EVAP, Fuel Trim
Stoplamp Pedal Switch	Applied/REL	Released	Released	Engine Data 2
TCC Enable Solenoid Command	Enabled or Disabled	Disabled	Enabled	Engine Data 1-2, Misfire
TCC PWM Solenoid	0-100%	0	Varies	Engine Data 2
TFP Switch Signal	P/R/N/D4-1	Park	Drive 4	Engine Data 2, EGR, Fuel Trim
TP Angle (%)	0-100%	0	Varies	Engine Data 1-2, EGR, EVAP, Fuel Trim, Misfire
TP Sensor (V)	0-5.1 volts	0.40-0.90	Varies	Engine Data 1-2
Transmission Range Switch	Park/Neutral, Drive 4-3-2-1	Park/Neutral	Drive 4	Engine Data 2, EGR, Fuel Trim
Vehicle Speed	0-155 km/h or miles per hour (mph)	0	30	Engine Data 1-2, EGR, EVAP, Fuel Trim, Misfire
VTD Auto Learn Timer	Active/Inactive	Inactive	Inactive	Engine Data 1
VTD Fuel Disable	Active/Inactive	Inactive	Inactive	Engine Data 1
VTD Fuel Disable Until Ignition Off	Yes/No	No	No	Engine Data 1
Warmups without an Emission Fault	0-255 counts	Varies	Varies	Engine Data 2
Warmups without a Non-Emission Fault	0-255 counts	Varies	Varies	Engine Data 2

1996-2002 C/K Series 6.5L V8 Diesel VIN F, VIN S - PID Data List

Parameter Identification (PID)	PID Value Range	PID Value at Hot Idle	PID Value at 30 mph	Data List Type
A/T 1-2 Solenoid Command	On/Off	On	Off	Engine Data 1
A/T 2-3 Solenoid Command	On/Off	Off	On	Engine Data 1
A-B-C Range Switch	On/Off/On	On-On/Off-Off/Off-Off	Off-Off/On-On/Off-Off	Engine Data 1
A/C Compressor	Engaged or Disengaged	Disengaged	Disengaged	Engine Data 1
A/C Relay Command	On/Off	Off	Off	Engine Data 1
A/C Request Signal	Yes/No	No	No	Engine Data 1
APP Angle	0-100%	0	Varies	Engine Data 1
Actual Injector Pump Timing	0-25.0 degrees	4-13.1	Varies	Engine Data 1
APP Position 1 (V)	0-5.0 volts	0.44-0.95	Varies	Engine Data 1
APP Position 2 (V)	0-5.0 volts	3.90-4.50	Varies	Engine Data 1
APP Position 3 (V)	0-5.0v	3.6-4.1	Varies	Engine Data 1
BARO Sensor (kPa)	65-104 kPa	Varies with Altitude	Varies with Altitude	Engine Data 1
Boost Pressure Sensor	10-200 kPa	60-170	Varies	Engine Data 1
Boost Pressure Sensor	0-5.0 volts	Varies	Varies	Engine Data 1
Brake Switch	Open/Closed	Open	Open	Engine Data 1
Calculated A/C Load	0-255 counts	0	Varies	Engine Data 1
Crank Reference Missed	0-8 counts	0	0	Engine Data 1
Cruise Control Active	On/Off	Off	Off	Engine Data 1
Cruise Brake Switch	Open/Closed	Closed	Closed	Engine Data 1
Cruise Switch	On/Off	Off	Off	Engine Data 1
Desired Idle Speed	0-3187 rpm	630-650	630-650	Engine Data 1
DTC Set This Ignition	0-255	Yes/No	Yes/No	Engine Data 1
ECT Sensor (°F)	-40° to 304°F	185-220	185-220	Engine Data 1
ECT Sensor (V)	0-5.0 volts	Varies	Varies	Engine Data 1
Engine Load (%)	0-100%	4-8	Varies	Engine Data 1
Engine Run Time	Hr: Min: Sec	00:00:00	00:00:00	Engine Data 1
Engine Speed	0-9999 rpm	± 100 rpm of Desired rpm	Varies	Engine Data 1
ESO Solenoid	On/Off	On	On	Engine Data 1
Fail Counter	0-255 counts	0	0	F/F & F/R
First Odometer	km/h or mph	Varies	Varies	F/F & F/R
Glow Plug (V)	0-25.5v	0.0	0.0	Engine Data 1
Glow Plug System (request to turn "on")	Disabled / Enabled	Enabled	Enabled	Engine Data 1
Glow Plug System Type	California or Federal	Varies	Varies	Engine Data 1
IAT Sensor (°F)	-40° to 304°F	Varies	Varies	Engine Data 1
Ignition 1 Signal	0.0-25.5v	8-16	8-16	Engine Data 1

1996-2002 C/K Series 6.5L V8 Diesel VIN F, VIN S - PID Data List

Parameter Identification (PID)	PID Value Range	PID Value at Hot Idle	PID Value at 30 mph	Data List Type
Injector Pump CAM Reference Missed	0-8	0	0	Engine Data 1
Injector Pump Solenoid Closure time	0.0-4.0 ms	1.70-1.90	Varies	Engine Data 1
Last Odometer	km/h or mph	Varies	Varies	F/F & F/R
Lift Pump (V)	0-25.5 volts	0	0	Engine Data 1
Lift Pump System	Disabled or Enabled	Enabled	Enabled	Engine Data 1
MIL Command	On/Off	Off	Off	Engine Data 1
Not Run Counter	0-255 counts	Varies	Varies	F/F & F/R
Number of Current DTC(s)	0-255	0	0	Engine Data 1
Pass Counter	0-255 counts	Varies	Varies	F/F & F/R
PCM/VCM in VTD Fuel Enable	Yes/No	No	No	Engine Data 1
Resume Switch (Cruise Control)	On/Off	Off	Off	Engine Data 1
Service Throttle Soon Lamp Command	On/Off	Off	Off	Engine Data 1
Set Switch Signal (Cruise Control)	On/Off	Off	Off	Engine Data 1
Startup ECT	-40° to 304°F	Varies	Varies	Engine Data 1
TDC Offset Angle	-2.50° to +2.50°	-1.75 to +1.75	Varies	Engine Data 1
Transmission Fluid Temperature	-40° to 304°F	122-158	122-158	Engine Data 1
Transmission Range	Park/Neutral/ Reverse/ Overdrive/ Drive 3, 2, 1	Park/Neutral	Drive 3	Engine Data 12
Vehicle Speed	0-155 km/h or miles per hour (mph)	0	30	Engine Data 1
VTD Auto Learn Timer	Active/Inactive	Inactive	Inactive	Engine Data 1
VTD Fuel Disable	Active/Inactive	Inactive	Inactive	Engine Data 1
VTD Fuel Disable Until Ignition Off	Yes/No	No	No	Engine Data 1
Wastegate Solenoid Duty Cycle Command	0-100%	0	0	Engine Data 1

1999-2005 C/K Series 6.6L V8 Diesel VIN 1 - PID Data List

Parameter Identification (PID)	PID Value Range	PID Value at Hot Idle	PID Value at 30 mph	Data List Type
4WD Switch Signal	Disabled or Enabled	Disabled	Disabled	Engine Data 1
4WD Low Signal	Disabled or Enabled	Disabled	Disabled	Engine Data 1
A/C Clutch Relay Feedback Signal	On/Off	Off	Off	Engine Data 1
A/C Relay Command	On/Off	Off	Off	Engine Data 1
A/C Request Signal	Yes/No	No	No	Engine Data 1
A/C Secondary High Pressure Switch	Open/Closed	Closed	Closed	Engine Data 1
Actual Fuel Rail Pressure	0-10,000 psi	5200-5800	5200-5800	Engine Data 1, EGR, FS
APP Indicated Angle	0-100%	0	Varies	Engine Data 1, F/F, F/R, FS
APP Position 1 (V)	0-5.0 volts	0.44-0.95	0.75-1.25	Engine Data 1
APP Position 2 (V)	0-5.0 volts	3.90-4.50	3.7-4.3	Engine Data 1
APP Position 3 (V)	0-5.0v	3.6-4.1	3.5-4.0	Engine Data 1
Balancing Rate for Cylinder 1-8	Millimeters (mm)	-2.5 to +2.5	Varies	Engine Data 2, F/F & F/R
BARO Sensor (kPa)	65-104 kPa	Varies with Altitude	Varies with Altitude	Engine Data 1
Battery Voltage	0.0-25.5v	12-14	12-14	Engine Data 1
Boost Pressure Sensor	10-200 kPa	60-170	Varies	Engine Data 1
Calculated Fuel Rate	Millimeters	7-9	Varies	Engine Data 1
CAM Reference Signal Missed	0-8 counts	0	0	Engine Data 1
CHECK TRANS Lamp	On/Off	Off	Off	Engine Data 1
CKP Reference Pulses Missed	0-8 counts	0	0	Engine Data 1
Clutch Pedal Switch	Applied or Released	Released	Released	Engine Data 1
Cruise Control Active	On/Off	Off	Off	Engine Data 1
Cruise On/Off Switch	On/Off	Off	Off	Engine Data 1
Cruise Resume/Accel Switch	On/Off	Off	Off	Engine Data 1
Cruise Set/Coast Switch	On/Off	Off	Off	Engine Data 1
Desired Fuel Rail Pressure Regulator	Milliamps	1200-1400	1200-1400	Fuel System (FS)
Desired Fuel Rail Pressure	37-40 MPa	37-40	37-40	Engine Data 1, EGR, FS
Desired Idle Speed	0-3187 rpm	630-700	630-700	Engine Data 1
DTC Set This Ignition	0-255	Yes/No	Yes/No	Engine Data 1
ECT Sensor (°F)	-40° to 304°F	185-220	185-220	Engine Data 1, EGR, FS
ECT Sensor (V)	0-5.0 volts	Varies	Varies	Engine Data 1, EGR, FS

1999-2005 C/K Series 6.6L V8 Diesel VIN 1 - PID Data List

Parameter Identification (PID)	PID Value Range	PID Value at Hot Idle	PID Value at 30 mph	Data List Type
EGR Vacuum Sensor	0-450 psi	18-158	Varies	EGR
EGR Solenoid Command	0-100%	35-45	Varies	EGR
EGR Throttle Valve Solenoid Command	Open/Closed	Closed	Closed	EGR
EGR Vent Solenoid Command	Venting / Not Venting	Not Venting	Not Venting	EGR
Engine Oil Level Switch	Okay / Fault	Okay	Okay	Engine Data 1
Engine Oil Life Remaining	0-100%	Varies	Varies	Engine Data 1
Engine Run Time	Hr: Min: Sec	00:00:00	00:00:00	Engine Data 1, EGR, FS
Engine Speed	0-9999 rpm	± 100 rpm of Desired rpm	Varies	Engine Data 1, EGR, FS
Fail Counter	0-255 counts	0	0	F/R
Fuel Pressure Regulator Command	0-100%	40-50	40-50	Engine Data 1, EGR, FS
Fuel Pressure Regulator Command	Milliamps	Varies	Varies	Engine Data 1, FS, F/F, F/R
Fuel Press. Regulator Fuel Flow Command	Millimeter	Varies	Varies	Engine Data 1
Front Axle Switch	Lock/Unlock	Unlock	Unlock	Engine Data 1
Fuel Level Sensor	0-100%	Varies	Varies	Fuel System
Fuel Temp. Sensor	-18° to 285°	50-194	50-194	Engine Data 1, Data 2, FS
Generator L-Terminal	Voltage or No Voltage	No Voltage	No Voltage	Engine Data 1
Glow Plug Feedback	0-25.5v	0.0-0.2	0.0-0.2	Engine Data 1
Glow Plug Relay	On/Off	Off	Off	Engine Data 1
Glow Plug System Type	California or Federal	Varies	Varies	Engine Data 1
High Idle Switch	On/Off	Off	Off	Engine Data 1
IAT Sensor (°F)	-40° to 304°F	Varies	Varies	Engine Data 1, EGR, F/F
Ignition 1 Signal	0.0-25.5v	8-16	8-16	Engine Data 1, EGR, FS
Injector 1-8 Command	0-99.9 ms	0.3-0.5	Varies	Engine Data 2
Intake Air Heater Relay Feedback	On/Off	Off	Off	Engine Data 1
Intake Air Heater	0-25.5 volts	0	0	Engine Data 1
Intake Air Heater Relay Command	On/Off	Off	Off	Engine Data 1
Low Coolant Level	Yes/No	No	No	Engine Data 1
Main Injection Command	Microsecond	Varies	Varies	Fuel System
Main Injection Fuel Rate	Millimeters	Varies	Varies	Fuel System
Main Injection Timing	-90 to 90°	Varies	Varies	Engine Data 1, Data 2, FS

1999-2005 C/K Series 6.6L V8 Diesel VIN 1 - PID Data List

Parameter Identification (PID)	PID Value Range	PID Value at Hot Idle	PID Value at 30 mph	Data List Type
MIL Command	On/Off	Off	Off	Engine Data 1
Mileage Since First Failure	Kilometers or Miles	Varies	Varies	F/F or F/R
Mileage Since Last Failure	Kilometers or Miles	Varies	Varies	F/F or F/R
Not Run Counter	0-255	0	0	F/R
Pass Counter	0-255 counts	Varies	Varies	F/R
PCM Reset	Yes/No	No	No	Engine Data 1, EGR
PCM/VCM in VTD Fuel Enable	Yes/No	No	No	Engine Data 1
Pilot Injection Control	Milliseconds	Varies	Varies	Fuel System
Pilot Inject. Fuel Rate	Millimeters	0.8-1.0	Varies	Fuel System
Pilot Injection Timing	Degrees	Varies	Varies	Fuel System
PNP Switch Signal	Park/Neutral/ In Gear	Park/Neutral	In Gear	Engine Data 1, EGR
PCM Reset	Yes/No	No	No	Engine Data 1, EGR
PTO Enable	Yes/No	No	No	Engine Data 1
PTO Engage Relay	On/Off	Off	Off	Engine Data 1
PTO Engine Shutdown Signal	Yes/No	No	No	Engine Data 1
PTO Relay Feedback Signal	On/Off	Off	Off	Engine Data 1
Reverse Enable	Yes/No	No	No	Engine Data 1
Startup ECT Signal	-40º to 304ºF	Varies	Varies	Engine Data 1, F/F or F/R
TCC/Cruise Brake Switch Signal	Applied or Released	Released	Released	Engine Data 1
Vehicle Speed	0-155 km/h or miles per hour (mph)	0	30	Engine Data 1, EGR, Fuel System, F/F
VTD Auto Learn Timer	Active/Inactive	Inactive	Inactive	Engine Data 1
VTD Fuel Disable	Active/Inactive	Inactive	Inactive	Engine Data 1
VTD Fuel Disable Until Ignition Off	Yes/No	No	No	Engine Data 1
Wait To Start	On/Off	Off	Off	Engine Data 1
Water In Fuel	On/Off	Off	Off	Engine Data 1

1996-2000 C/K Series 7.4L V8 MFI VIN J (All) - PID Data List

Parameter Identification (PID)	PID Value Range	PID Value at Hot Idle	PID Value at 30 mph	Data List Type
A/T 1-2 Shift Solenoid	On/Off	On	Off	Engine Data 2
A/T 2-3 Shift Solenoid	On/Off	Off	On	Engine Data 2
4WD Switch Signal	Enable/Disable	Disabled	Disabled	Engine Data 2
4WD Low Signal	Enable/Disable	Disabled	Disabled	Engine Data 2
A/C Evaporator Switch	Open/Closed	Closed	Closed	Engine Data 2
A/C Secondary High Pressure Switch	Open/Closed	OPEN	OPEN	Engine Data 2
A/C Relay Command	On/Off	Off	Off	Engine Data 2
A/C Request Signal	Yes/No	No	No	Engine Data 2
A/C Relay Command	On/Off	Off	Off	Engine Data 1-2, EGR, Misfire
Actual EGR Position	0-100%	0	Varies	Engine Data 1, Data 2, HO2S
Air Fuel Ratio	0.0-25.5:1	14.7:1	14.7:1	Engine Data 1, HO2S
Air Pump Relay Command	On/Off	Off	Off	HO2S
A/C Auto Recirculation Mode	On/Off	Off	Off	Engine Data 2
BARO Sensor (kPa)	10-110 kPa	100	100	Engine Data 1, Data 2, EVAP
BARO Sensor (V)	0-5.0 volts	1.94	1.94	Engine Data 1, Data 2, EVA
Brake Switch Signal	Applied/REL	Released	Released	Engine Data 2, HO2S, Misfire
Closed Loop Since Restart	Yes/No	Yes	Yes	Engine Data 2, HO2S
Clutch Pedal Switch	On/Off	Off	Off	Engine Data 2, Misfire
CMP Sensor Retard (CMP / CKP variance)	0-90 degrees	0	Varies	Engine Data 1
Cruise On/Off Switch	Enable/Disable	Disabled	Disabled	Engine Data 2
Cycles of Misfire Data	0-100 counts	Varies	Varies	Misfire
Decel Fuel Mode	Active or Inactive	Inactive	Inactive	Engine Data 1, Data 2, HO2S
Desired EGR Position	0-100%	0	Varies	Engine Data 1, EGR, Fuel Trim
Desired Engine Speed	0-3187 rpm	650	Varies	EVAP
Desired IAC Position	0-255 counts	60	Varies	Engine Data 1, Data 2, HO2S
DTC Set This IGN	Yes/No	No	No	Engine Data 1-2 C/C, Fuel Trim
Desired Idle Speed	0-3187 rpm	650	Varies	Engine Data 1, Data 2, HO2S
DTC Set This Ignition	Yes/No	No	No	HO2S
ECT Sensor (°F)	-40 to 303°F	200	200	Engine Data 1-2, EVAP, HO2S, Misfire

1996-2000 C/K Series 7.4L V8 MFI VIN J (All) - PID Data List

Parameter Identification (PID)	PID Value Range	PID Value at Hot Idle	PID Value at 30 mph	Data List Type
ECT Sensor (V)	0-5.0 volts	0.64	0.64	Engine Data 1
EGR Solenoid Duty Cycle Command	0-100%	0	Varies	Engine Data 1, Data 2, HO2S
Engine Load (%)	0-100%	2	11	Engine Data 1-2, EVAP, HO2S, Misfire
Engine Run Time	Hr: Min: Sec	00:00:00	00:00:00	Engine Data 1-2, EVAP, HO2S, Misfire
Engine Speed	0-10,000 rpm	650	Varies	Engine Data 1-2, EVAP, HO2S, Misfire
EVAP Canister Purge	On/Off	On	On	Engine Data 1-2, EVAP, HO2S
EVAP Purge Duty Cycle Command	0-100%	17	Varies	Engine Data 1-2, EVAP, HO2S
EVAP Vent Solenoid Command	Open or Closed	Open	Open	Engine Data 2, EVAP
EVAP Excess Vacuum Test	Pass / Not Run / Fail	Pass	Pass	EVAP
Fuel Level	0-100%	Varies	Varies	Engine Data 2, EVAP
Fuel Level Sensor	0-5.0 volts	Varies	Varies	EVAP
Fuel Tank Pressure Sensor	Inches H2O, mm Hg	-17.5 to +7.5 Inches H2O	-17.5 to +7.5 Inches H2O	EVAP
Fuel Trim Cell	Cell 0-22	17	18-21	Engine Data 1, EVAP, HO2S
Fuel Trim Enable	Yes/No	Yes	Yes	Engine Data 1, Data 2, HO2S
HO2S-11 (B1 S1)	0-1000 mv	10-1000	10-1000	HO2S
HO2S-12 (B1 S2)	0-1000 mv	10-1000	10-1000	HO2S
HO2S-21 (B2 S1)	0-1000 mv	10-1000	10-1000	HO2S
HO2S-22 (B2 S2)	0-1000 mv	10-1000	10-1000	HO2S
IAC Position	0-1024 counts	60	60	Engine Data 1, Data 2, HO2S
IAT Sensor (°F)	-40 to 303°F	113	113	Engine Data 1, Data 2, EVAP
IAT Sensor (V)	0-5.0 volts	1.10	1.10	Engine Data 1
Ignition 1 Signal	0.0-25.5v	12.8-14.2	12.8-14.2	Engine Data 1-2
Ignition 1 Low Signal	Yes/No	No	No	Engine Data 2
Ignition 1 High Signal	Yes/No	No	No	Engine Data 2
Ignition 1 On	Yes/No	Yes	Yes	Engine Data 2
Injector Pulsewidth Average Bank 1	0-1000 ms	4.5	2-6	Engine Data 2, HO2S, Misfire
Injector Pulsewidth Average Bank 2	0-1000 ms	4.5	2-6	Engine Data 2, HO2S, Misfire
Knock Adjust Factor	0-25.5°	0	0	EVAP
Knock Retard (°)	0-25.5°	0	0	Engine Data 1
Knock Sensor (V)	0-5.0 volts	2.6	2.6	Engine Data 1

1996-2000 C/K Series 7.4L V8 MFI VIN J (All) - PID Data List

Parameter Identification (PID)	PID Value Range	PID Value at Hot Idle	PID Value at 30 mph	Data List Type
Knock Sensor Activity	0-255 counts	0	0	Engine Data 1
Long Term Fuel Trim Bank 1	-10% to 10%	0 (± 5%)	0 (± 5%)	Engine Data 1, HO2S
Long Term Fuel Trim Bank 1	0-255 counts	127	127	Engine Data 1, HO2S
Long Term Fuel Trim Bank 2	-10% to 10%	0 (± 5%)	0 (± 5%)	Engine Data 1, HO2S
Long Term Fuel Trim Bank 2	0-255 counts	127	127	Engine Data 1, HO2S
Loop Status	Closed Loop / Open Loop	Closed Loop	Closed Loop	Engine Data 1, EVAP, HO2S, Misfire
MAF Sensor (g/sec)	0-512 g/sec	7.05	Varies	Engine Data 1, EVAP, HO2S, Misfire
MAF Sensor (Hz)	0-31,999 Hz	2540	Varies	Engine Data 1, Data 2, HO2S
MAP Sensor (kPa)	10-105 kPa	36	Varies	Engine Data 1-2, EVAP, HO2S, Misfire
MAP Sensor (V)	0-5.1 volts	1.2	Varies	Engine Data 1-2, EVAP, HO2S, Misfire
Mileage Since DTC Cleared	Kilometers or Miles	Varies	Varies	HO2S
Misfire Current Cylinder 1-8	0-200 counts	0	0	Misfire
Misfire History Cylinder 1-8	0-65,535	0	0	Misfire
Misfire Not Run Counter	0-65, 535 counts	Varies	Varies	Freeze Frame / Failure Record
Misfire Pass Counter	0-65, 535 counts	Varies	Varies	Freeze Frame / Failure Record
PCM/VCM in VTD Fuel Enable	Yes/No	No	No	Engine Data 2
Pre-HO2S	Ready / Not Ready	Ready	Ready	Engine Data 1, HO2S
Post-HO2S	Ready / Not Ready	Ready	Ready	Engine Data 1, HO2S
Reference Pulse Occurred	Yes/No	Yes	Yes	Engine Data 1
Power Enrichment	Active or Inactive	Inactive	Inactive	Engine Data 2
Purge Leak Test	Pass / Not Run / Failed	Pass	Pass	EVAP
Rich-Lean Bank 1	Rich / Lean	Varying from Rich to Lean	Varying from Rich to Lean	Engine Data 1-2, EVAP, HO2S
Rich-Lean Bank 2	Rich / Lean	Varying from Rich to Lean	Varying from Rich to Lean	Engine Data 1-2, EVAP, HO2S

1996-2000 C/K Series 7.4L V8 MFI VIN J (All) - PID Data List

Parameter Identification (PID)	PID Value Range	PID Value at Hot Idle	PID Value at 30 mph	Data List Type
Short Term Fuel Trim Bank 1	-10% to 10%	0 (± 5%)	0 (± 5%)	Engine Data 1, HO2S
Short Term Fuel Trim Bank 1	0-255 counts	127	127	Engine Data 1, HO2S
Short Term Fuel Trim Bank 2	-10% to 10%	0 (± 5%)	0 (± 5%)	Engine Data 1, HO2S
Short Term Fuel Trim Bank 2	0-255 counts	127	127	Engine Data 1, HO2S
Small Leak Test	Pass / Not Run / Failed	Pass	Pass	EVAP
Spark Advance (°)	-64° to 64°	21	Varies	Engine Data 1, Misfire
Spark Control	Advance / Retard	Advance	Advance	Engine Data 1
Startup ECT	-40 to 303°F	Varies	Varies	Engine Data 1, Data 2,
Startup IAT	-40 to 303°F	Varies	Varies	Engine Data 1, Data 2, HO2S
TCC Enable Solenoid Command	Yes/No	No	No	Engine Data 2, Misfire
TCC PWM Solenoid	0-100%	0	Varies	Engine Data 2, Misfire
TFP Switch A-B-C	On / Off / On / Off	Off/On/Off	Off/On/On	Engine Data 2
Total Misfire Current	0-255 counts	0	0	Misfire
Total Misfire Current	0-255 counts	0	0	Misfire
Total Misfire Passes	0-255 counts	0	0	Misfire
TP Angle (%)	0-100%	0	Varies	Engine Data 1-2, EVAP, HO2S, Misfire
TP Sensor (V)	0-5.1 volts	0.47	Varies	Engine Data 1-2, EVAP, HO2S
Transmission Range Switch	P-R-N, D4, D3, D2, D1	Park	D4	Engine Data 2
Valid Reference Pulse Occurred	Yes/No	Yes	Yes	Engine Data 1
Vehicle Speed	Kilometers or MPH	0	30	Engine Data 1-2 HO2S, Misfire
VTD Auto Learn Timer	Active/Inactive	Inactive	Inactive	Engine Data 2
VTD Fuel Disable	Active/Inactive	Inactive	Inactive	Engine Data 2
VTD Fuel Disable Until Ignition Off	Yes/No	No	No	Engine Data 2
Warmups without an Emission Fault	0-255 counts	Varies	Varies	Engine Data 2
Warmups without a Non-Emission Fault	0-255 counts	Varies	Varies	Engine Data 2
Weak Vacuum Leak Test	Pass / Not Run / Failed	Pass	Pass	EVAP

2001-05 C/K Series 8.1L V8 VIN G (A/T) - PID Data List

Parameter Identification (PID)	PID Value Range	PID Value at Hot Idle	PID Value at 30 mph	Data List Type
A/C Clutch Feedback Circuit	On/Off	Off	Off	Engine Data 2
A/C Compressor Cycling Switch	Normal / Low Pressure	Normal	Normal	Engine Data 2
A/C Relay Command	On/Off	Off	Off	Engine Data 1-2, EGR, Misfire
A/C Request Signal	Yes/No	No	No	Engine Data 2
A/C Relay Command	On/Off	Off	Off	Engine Data 1-2, EGR, Misfire
A/C Request Signal	Yes/No	No	No	Engine Data 2
Air Fuel Ratio	0.0-25.5:1	14.7:1	14.7:1	Engine Data 2, Fuel Trim
APP Average	0-125 counts	Varies	Varies	TAC
APP Indicated Angle	0-100%	0	WOT: 100%	Engine Data 1-2, EGR, EVAP, Fuel Trim, Cruise, TAC
APP Sensor 1 (%)	0-100%	0	Varies	TAC
APP Sensor 1 (V)	0-5.0 volts	0.4-0.9	Varies	TAC
APP Sensor 2 (%)	0-100%	0	Varies	TAC
APP Sensor 2 (V)	0-5.0 volts	4.5-4.1	Varies	TAC
APP Sensor 3 (%)	0-100%	0	Varies	TAC
APP Sensor 3 (V)	0-5.0 volts	4.2-3.7	Varies	TAC
APP Sensor 1 and 2	Agree or Disagree	Agree	Agree	TAC
APP Sensor 1 and 3 Disagree	Agree or Disagree	Agree	Agree	TAC
APP Sensor 2 and 3 Disagree	Agree or Disagree	Agree	Agree	TAC
Auxiliary Fan Request	On/Off	Off	Off	Engine Data 2
BARO Sensor (kPa)	10-110 kPa	65-104	65-104	Engine Data 1, EGR, EVAP, Fuel Trim
BARO Sensor (V)	0-5.0 volts	3.5-4.9	3.5-4.9	Engine Data 1, EGR, EVAP, Fuel Trim
CMP Sensor LO to HI	0-65,535 counts	Varies	Varies	Engine Data 2
CMP Sensor HI to LO	0-65, 535 counts	Varies	Varies	Engine Data 2
Cold Startup	Yes/No	No	No	Engine Data 2, EVAP
Coolant Level Switch	Yes/No	No	No	Engine Data 2
Cruise Control Active	Yes/No	No	No	Engine Data 1, Cruise, TAC
Cruise Disengage History 1-8	Several	Varies	Varies	Cruise Control
Cruise On/Off Switch	On/Off	Off	Off	Cruise Control

2001-05 C/K Series 8.1L V8 VIN G (A/T) - PID Data List

Parameter Identification (PID)	PID Value Range	PID Value at Hot Idle	PID Value at 30 mph	Data List Type
Cruise Release Brake Pedal Switch	Applied or Released	Released	Released	Cruise Control
Cruise Resume / Accel Switch	On/Off	Off	Off	Cruise Control, TAC
Cruise Set / Coast Switch	On/Off	Off	Off	Cruise Control, TAC
Current Gear (4-Speed Transmission)	0-1-2-3-4	1	4	Engine Data 1-2, EGR, Misfire
Cycles of Misfire Data	0-100 counts	Varies	Varies	Misfire
Desired EGR Position	0-100%	0	Varies	Engine Data 1-2, EGR, Fuel Trim
Desired EGR Position	0-5.0 volts	< 1.3	Varies	EGR
Desired IAC Airflow	0-64 g/sec	Varies	Varies	Engine Data 1
Desired Idle Speed	0-3187 rpm	Varies	Varies	Engine Data 1-2, EVAP, TAC
DTC Set This Ignition	Yes/No	No	No	Engine Data 1-2 C/C, Fuel Trim
ECT Sensor (°F)	-38 to 284°F	Varies	Varies	Engine Data 1-2, EGR, Fuel Trim, EVAP, Misfire
EGR Closed Valve Position Sensor	0-5.0 volts	< 1.3	Varies	EGR
EGR Position Sensor	0-100%	0	Varies	Engine Data 1, EGR, Misfire
EGR Position Sensor	0-5.0 volts	< 1.3	Varies	EGR
Engine Load (%)	0-100%	1-4	7-9	Engine Data 1-2, EGR, Fuel Trim EVAP, Misfire
Engine Oil Level Switch	Okay / Low	Okay	Okay	Engine Data 2
Engine Oil Life Remaining	0-100%	Varies	Varies	Engine Data 2
Engine Run Time	Hr: Min: Sec	00:00:00	00:00:00	Engine Data 1-2, EGR, Fuel Trim EVAP, Misfire
Engine Speed	0-10,000 rpm	500-700 rpm	Varies	Engine Data 1-2, EGR, Fuel Trim EVAP, Misfire
EVAP Purge Solenoid Command	0-100%	0-25	Varies	Engine Data 1, EVAP, Fuel Trim
EVAP Test Result	No Result / Passed/ Fail/ P0440/442/ P0446/1441	No Test Results	No Test Results	EVAP
EVAP Vent Solenoid Command	Venting/Not Venting	Venting	Venting	Engine Data 1, EVAP, Fuel Trim

2001-05 C/K Series 8.1L V8 VIN G (A/T) - PID Data List

Parameter Identification (PID)	PID Value Range	PID Value at Hot Idle	PID Value at 30 mph	Data List Type
Fuel Level Sensor	0-5.0 volts	0.7-2.5	0.7-2.5	Engine Data 1, Data 2, EVAP
Fuel Level Sensor - Rear Tank	0-5.0 volts	0.7-2.5	0.7-2.5	Engine Data 1, Data 2, EVAP
Fuel Tank Level Remaining	Gallons or Liters	Varies	Varies	EVAP
Fuel Tank Level Remaining	0-100%	Varies	Varies	EVAP
Fuel Tank Pressure Sensor	Inches H2O, mm Hg	-17.5 to +7.5 Inches H2O	-17.5 to +7.5 Inches H2O	Engine Data 1, EVAP
Fuel Tank Pressure Sensor	0-5.0 volts	Varies	Varies	EVAP
Fuel Tank Rated Capacity	Gallons or Liters	25.9 / 34.0 or 98/128 Liters	25.9 / 34.0 or 98/128 Liters	EVAP
Fuel Trim Cell	Cell 0-23	16, 20	16, 20	Engine Data 1, EVAP, Fuel Trim
Fuel Trim Learn	Enabled or Disabled	Enabled	May Toggle	Engine Data 1, EVAP, Fuel Trim
Generator F-Terminal	0-100%	Varies	Varies	Engine Data 2
Generator L-Terminal	On/Off	On	On	Engine Data 2
HO2S-11 (B1 S1)	0-1132 mv	10-1000	10-1000	Engine Data 1, EVAP, Fuel Trim
HO2S-12 (B1 S2)	0-1132 mv	10-1000	10-1000	Engine Data 1, EVAP, Fuel Trim
HO2S-21 (B2 S1)	0-1132 mv	10-1000	10-1000	Engine Data 1, EVAP, Fuel Trim
HO2S-22 (B2 S2)	0-1132 mv	10-1000	10-1000	Engine Data 1, EVAP, Fuel Trim
IAT Sensor (°F)	-38 to 284°F	50-194	50-194	Engine Data 1-2, EGR, EVAP, Fuel Trim
Ignition 1 Signal	0.0-25.5v	12.8-14.2	12.8-14.2	Engine Data 1-2, EGR, EVAP, Fuel Trim, TAC
Injector Pulsewidth Bank 1	0-1000 ms	2-6	2-6	Engine Data 2, Fuel Trim, Misfire
Injector Pulsewidth Bank 2	0-1000 ms	2-6	2-6	Engine Data 2, Fuel Trim, Misfire
Knock Retard (°)	0-25.5°	0.0-16.0	0	Engine Data 1, EGR
Long Term Fuel Trim Bank 1	-10% to 10%	0 (± 5%)	0 (± 5%)	Engine Data 1-2, EVAP, Fuel Trim
Long Term Fuel Trim Bank 2	-10% to 10%	0 (± 5%)	0 (± 5%)	Engine Data 1-2, EVAP, Fuel Trim
Long Term Fuel Trim Average Bank 1	-10% to 10%	0 (± 5%)	0 (± 5%)	Engine Data 1-2, EVAP, Fuel Trim
Long Term Fuel Trim Average Bank 2	-10% to 10%	0 (± 5%)	0 (± 5%)	Engine Data 1-2, EVAP, Fuel Trim

2001-05 C/K Series 8.1L V8 VIN G (A/T) - PID Data List

Parameter Identification (PID)	PID Value Range	PID Value at Hot Idle	PID Value at 30 mph	Data List Type
Loop Status	Closed Loop / Open Loop	Closed Loop	Closed Loop	Engine Data 1-2, EGR, EVAP, Fuel Trim
Low Oil Lamp Control	On/Off	Off	Off	Engine Data 2
MAF Sensor (g/sec)	0-512 g/sec	1-9	15-26	Engine Data 1-2, EGR, EVAP, Fuel Trim, Misfire, TAC
MAF Sensor (Hz)	0-31,999 Hz	2000-3000	Varies	Engine Data 2
MAP Sensor (kPa)	10-105 kPa	20-48	Varies	Engine Data 1-2, EGR, EVAP, Fuel Trim, Misfire, TAC
MAP Sensor (V)	0-5.1 volts	1.0-2.0	Varies	Engine Data 1-2
MIL Command	On/Off	Off	Off	Engine Data 2
Mileage Since DTC Cleared	Kilometers or Miles	Varies	Varies	Engine Data 2
Misfire Current Cylinder 1-6	0-200 counts	0	0	Misfire
Misfire History Cylinder 1-6	0-65,535	0	0	Misfire
Park/Neutral Switch (5-Speed A/T only)	In-Gear, Park, Neutral	Park	In-Gear	Engine Data 2, EGR, Fuel Trim
PCM Reset	Yes/No	No	No	Engine Data 1-2, EGR, EVAP, Fuel Trim
PCM/VCM in VTD Fuel Enable	Yes/No	No	No	Engine Data 1
Power Enrichment	Yes/No	No	No	Engine Data 2
PTO Enable Signal	Yes/No	No	No	TAC
Reduced Engine Power	Yes/No	No	No	Engine Data 1, EGR
Short Term Fuel Trim Bank 1	-10% to 10%	0 (± 5%)	0 (± 5%)	Engine Data 1-2, EVAP, Fuel Trim
Short Term Fuel Trim Bank 2	-10% to 10%	0 (± 5%)	0 (± 5%)	Engine Data 1-2, EVAP, Fuel Trim
Short Term Fuel Trim Average Bank 1	-10% to 10%	0 (± 5%)	0 (± 5%)	Engine Data 1-2, EVAP, Fuel Trim
Short Term Fuel Trim Average Bank 2	-10% to 10%	0 (± 5%)	0 (± 5%)	Engine Data 1-2, EVAP, Fuel Trim
Spark Advance (°)	-64° to 64°	15-20	Varies	Engine Data 1-2 Fuel Trim, Misfire
Startup ECT	-40 to 419°F	Varies	Varies	Engine Data 2, EVAP, Fuel Trim
Stoplamp Pedal Switch	Applied/REL	Released	Released	Engine Data 2, Cruise, TAC
TAC/PCM Communication signal	Okay/Fault	Okay	Okay	Engine Data 1, Cruise, TAC
TCC Brake Pedal Switch	Applied/REL	Released	Released	Engine Data 1-2

2001-05 C/K Series 8.1L V8 VIN G (A/T) - PID Data List

Parameter Identification (PID)	PID Value Range	PID Value at Hot Idle	PID Value at 30 mph	Data List Type
TCC Enable Solenoid Command (4L80-E)	Enabled or Disabled	Disabled	Enabled	Engine Data 1-2, Misfire
TCC PWM Solenoid Command (4L80-E)	0-100%	0	Varies	Engine Data 2
TFP Switch Signal (4L80-E only)	Park, Neutral Drive 4, 3, 2 and Low	Park	Drive 4	Engine Data 2, EGR, Fuel Trim
TP Indicated Angle	0-100%	3-4.5	Varies	Engine Data 1-2, EGR, EVAP, Cruise, TAC
TP Desired Angle	0-100%	3-7	Varies	Engine Data 1-2, EGR, EVAP, Cruise, TAC, Fuel Trim, Misfire
TP Sensor 1	0-100%	3	Varies	TAC
TP Sensor 1	0-5.0v	0.4-0.9	Varies	TAC
TP Sensor 2	0-100%	3	Varies	TAC
TP Sensor 1	5.0-0.0	4.8-4.3	Varies	TAC
TP Sensors 1 and 2	Agree or Disagree	Agree	Agree	TAC
Transmission Range Switch (4L80-E only)	Park/Neutral, Drive 4-3-2-1	Park/Neutral	Drive 4	Engine Data 2, EGR, Fuel Trim
Vehicle Speed	0-155 km/h or miles per hour (mph)	0	30	Engine Data 1-2, EGR, EVAP, Cruise, TAC, Fuel Trim, Misfire
VTD Auto Learn Timer	Active/Inactive	Inactive	Inactive	Engine Data 1
VTD Fuel Disable	Active/Inactive	Inactive	Inactive	Engine Data 1
VTD Fuel Disable Until Ignition Off	Yes/No	No	No	Engine Data 1
Warmups without an Emission Fault	0-255 counts	Varies	Varies	Engine Data 2
Warmups without a Non-Emission Fault	0-255 counts	Varies	Varies	Engine Data 2

CHEVROLET & GMC S/T TRUCK APPLICATIONS

1996-2000 S/T Series 2.2L I4 VIN 4 (All) - PID Data List

Parameter Identification (PID)	PID Value Range	PID Value at Hot Idle	PID Value at 30 mph	Data List Type
A/C HI Side Pressure	0-450 psi	Varies	Varies	Engine Data 2
A/C HI Side Pressure	0-5.0 volts	Varies	Varies	Engine Data 2
A/C Relay Command	On/Off	Off	Off	Engine Data 1-2, Misfire
A/C Request Signal	Yes/No	No	No	Engine Data 2
Adaptive Spark Retard	0° to 90°	Varies	Varies	Engine Data 1-2
Airflow (calculated)	0-64 g/sec	Varies	Varies	Fuel Trim, Freeze Frame & Failure Record
Air Fuel Ratio	0.0-25.5:1	14.7:1	14.7:1	Engine Data 2-3 Fuel Trim
BARO Sensor (kPa)	10-110 kPa	65-104	65-104	Engine Data 1, EVAP, Fuel Trim
BARO Sensor (V)	0-5.0 volts	3.5-4.9	3.5-4.9	Engine Data 1, EVAP, Fuel Trim
Base Injector Pulsewidth	0-1000 ms	1-4	1-4	Engine Data 2, Fuel Trim, Misfire
Catalytic Converter Temperature (calc.)	32-1409°F	Varies	Varies	Fuel Trim
CKP (7X) Active Counter	0-255 counts	Varies	Varies	Engine Data 2
CMP Active Counter	0-255 counts	Varies	Varies	Engine Data 2
CMP Resync Counter	0-255 counts	Varies	Varies	Engine Data 2
Cruise Control Command	Engaged or Released	Released	Released	Cruise
Cruise On/Off Switch	On/Off	Off	Off	Cruise
Desired Idle Speed	0-3187 rpm	Varies	Varies	Engine Data 1-2 EVAP
ECT Sensor (°F)	-38 to 284°F	185-239	185-239	Engine Data 1-2, EVAP, Fuel Trim Misfire
Engine Load (%)	0-100%	1-4	7-9	Engine Data 1-2 EVAP Fuel Trim, Misfire
Engine Run Time	Hr: Min: Sec	00:00:00	00:00:00	Engine Data 1-2 EVAP Fuel Trim, Misfire
Engine Speed	0-9999 rpm	500-700 rpm	Varies	Engine Data 1-3, Cruise, EVAP Fuel Trim, Misfire, TAC
EVAP Purge Solenoid Command	0-100%	10-25	Varies	Engine Data 1, EVAP, Fuel Trim
EVAP Vent Solenoid Command	Open / Closed	Open	Open	Engine Data 1, EVAP, Fuel Trim
Fuel Level	0-100%	Varies	Varies	EVAP

1996-2000 S/T Series 2.2L I4 VIN 4 (All) - PID Data List

Parameter Identification (PID)	PID Value Range	PID Value at Hot Idle	PID Value at 30 mph	Data List Type
Fuel Tank Pressure Sensor	Inches H2O, mm Hg	-17.5 to +7.5 Inches H2O	-17.5 to +7.5 Inches H2O	Engine Data 1, EVAP
Fuel Tank Pressure Sensor	0-5.0 volts	Varies	Varies	Engine Data 1, EVAP
Fuel Trim Cell	Cell 0-23	18-21	Varies	Engine Data 1, EVAP, Fuel Trim
Fuel Trim Index	-100 to 100%	0% (± 5%)	0% (± 5%)	Fuel Trim
Generator L-Terminal	Active/Inactive	Active	Active	Engine Data 2
HO2S-12 (B1 S2)	0-1106 mv	10-1000	10-1000	Engine Data 1, EVAP, Fuel Trim
IAT Sensor (°F)	-38 to 284°F	50-176	50-176	Engine Data 1-2, EVAP, Fuel Trim
Ignition 1 Signal	0.0-25.5v	12.8-14.2	12.8-14.2	Engine Data 1-2, EVAP, Fuel Trim, Misfire
Knock Sensor Active Counter	0-255	0-2	0-2	ENG 1
Knock Sensor Noise Channel	Yes/No	No	Yes	ENG 1
Knock Retard	0-90 degrees	0	9	ENG 1
Lean-Rich Transition	0-255 counts	0	-	Engine Data 1-2, EVAP, Fuel Trim
Lean-Rich Average	0-249 ms	0	0	Engine Data 1-2, EVAP, Fuel Trim
Long Term Fuel Trim	-10% to 10%	0 (± 5%)	0 (± 5%)	Engine Data 1-2, EVAP, Fuel Trim
Loop Status	Closed Loop / Open Loop	Closed Loop	Closed Loop	Engine Data 1-2, EVAP, Fuel Trim
MAP Sensor (kPa)	10-105 kPa	25-48	Varies	Engine Data 1-2, EVAP, Fuel Trim, Misfire
MAP Sensor (V)	0-5.0 volts	1.0-3.0	Varies	Engine Data 1-2
Medium Resolution Engine	Yes/No	Yes	Yes	Engine Data 1-2
Medium Resolution Resync Counter	0-255 counts	1.0-3.0	Varies	Engine Data 1-2
MIL Command	On/Off	Off	Off	Engine Data 1
Misfire (total) Current Counter	0-255	0	0	Misfire
Misfire Current Cylinder 1-4	0-200 counts	0	0	Misfire
Misfire History Cylinder 1-4	0-65,535	0	0	Misfire
Mileage Since First Failure	Kilometers or Miles	Varies	Varies	Misfire
Mileage Since Last Failure	Kilometers or Miles	Varies	Varies	Misfire

1996-2000 S/T Series 2.2L I4 VIN 4 (All) - PID Data List

Parameter Identification (PID)	PID Value Range	PID Value at Hot Idle	PID Value at 30 mph	Data List Type
Number of Catalyst Tests Completed	0-255 counts	0	0	Catalyst, Fuel Trim
Number of Current DTC(s)	0-255	0	0	Engine Data 1, EVAP, TAC, Fuel Trim
O2S-11 (B1 S1)	0-1106 mv	10-1000	10-1000	Engine Data 1, EVAP, Fuel Trim
Output Driver 1 Open	00000000 (01 with fault)	00000000	00000000	Output Driver
Output Driver 1 Short	00000000	00000000	00000000	Output Driver
Output Driver 2 Open	(01 with fault)	00000000	00000000	Output Driver
Output Driver 2 Short	00000000 (01 with fault)	00000000	00000000	Output Driver
Power Enrichment	Yes/No	No	No	Engine Data 2
Purge Learned Memory	0-255 counts	0	0	Engine Data 2, EVAP, Fuel Trim
Rich-Lean Transition	0-255 counts	0	0	Engine Data 1
Rich-Lean to Lean-Rich Average Time	0-249 ms	Varies	Varies	Engine Data 1
Rich-Lean / Lean-Rich Ratio	0:1 to 15.93:1	Varies	Varies	Engine Data 1
Short Term Fuel Trim	-10% to 10%	0 (± 5%)	0 (± 5%)	Engine Data 1-2, EVAP, Fuel Trim
Spark Advance (°)	-64° to 64°	Varies	Varies	Engine Data 1-2 Fuel Trim, Misfire
Startup ECT	-40 to 419°F	Varies	Varies	Engine Data 2, EVAP, Fuel Trim
Theft System Learn Mode	Enabled or Disabled	Enabled	Enabled	Engine Data 1-2
TP Angle (%)	0-100%	0	Varies	Engine Data 1-2, EVAP, Misfire
TP Sensor (V)	0-5.0 volts	0.20-0.90	Varies	Engine Data 2
Transmission Range Switch	P, N, D4, D3, D2/D1/invalid	Park/Neutral	D4	ENG Data1-2 Cruise, Fuel Trim
TR Switch P/A/B/C	L/H/L/H	L/L/H/H	L/H/H/L	Engine Data 1-2
Vehicle Speed	0-155 km/h / 0-155 mph	0	30	Engine Data 1-2, EVAP, Misfire

2000-04 S/T Series 2.2L I4 Gas/Ethanol VIN 5 - PID Data List

Parameter Identification (PID)	PID Value Range	PID Value at Hot Idle	PID Value at 30 mph	Data List Type
A/C HI Side Pressure	0-450 psi	Varies	Varies	Engine Data 2
A/C HI Side Pressure	0-5.0 volts	Varies	Varies	Engine Data 2
A/C Relay Command	On/Off	Off	Off	Engine Data 1-3 Misfire
A/C Request Signal	Yes/No	No	No	Engine Data 2
Adaptive Spark Retard	0° to 90°	Varies	Varies	Engine Data 1-2
Airflow (Calc.)	0-64 g/sec	Varies	Varies	Engine Data 1-3 Fuel Trim, Misfire
Air Fuel Ratio	0.0-25.5:1	14.6:1	14.6:1	Engine Data 2-3 Fuel Trim
Air Pump Relay	On/Off	Off	Off	Engine Data 1
BARO Sensor (kPa)	10-110 kPa	65-104	65-104	Engine Data 1, EVAP, Fuel Trim
BARO Sensor (V)	0-5.0 volts	3.5-4.9	3.5-4.9	Engine Data 1, EVAP, Fuel Trim
Base Injector Pulsewidth	0-1000 ms	1-4	1-4	Engine Data 2, Fuel Trim, Misfire
CKP (7X) Active Counter	0-255 counts	Varies	Varies	Engine Data 2
CMP Active Counter	0-255 counts	Varies	Varies	Engine Data 2
CMP Resync Counter	0-255 counts	Varies	Varies	Engine Data 2
Cruise Control Active	Yes/No	No	No	Engine Data 1-3
Cruise Release Clutch / TCC Pedal Switch	Applied or Released	Released	Released	Engine Data 1-3
Current Gear	1, 2, 3, 4	1	4	Engine Data 1-2 Fuel Trim
Desired Idle Speed	0-3187 rpm	Varies	Varies	Engine Data 1-3 EVAP
ECT Sensor (°F)	-38 to 284°F	185-239	185-239	Engine Data 1-3 EVAP, Fuel Trim, Misfire
Engine Load (%)	0-100%	11-20	30-36	Engine Data 1-3 EVAP Fuel Trim, Misfire
Engine Oil Live (%)	0-100%	Varies	Varies	Engine Data 3
Engine Oil Press. Switch	Okay / Low	Okay	Okay	Engine Data 3
Engine Run Time	Hr: Min: Sec	00:00:00	00:00:00	Engine Data 1-3 EVAP Fuel Trim, Misfire
Engine Speed	0-9999 rpm	500-700 rpm	Varies	Engine Data 1-3, Cruise, EVAP Fuel Trim, Misfire, TAC
EVAP Purge Solenoid Command	0-100%	10-25	Varies	Engine Data 1, EVAP, Fuel Trim
EVAP Vent Solenoid Command	Open / Closed	Open	Open	Engine Data 1, EVAP, Fuel Trim

2000-04 S10 Pickup 2.2L I4 Gas/Ethanol VIN 5 - PID Data List

Parameter Identification (PID)	PID Value Range	PID Value at Hot Idle	PID Value at 30 mph	Data List Type
EVAP Test Abort Reason	No Result / Passed/ Fail/ P0440/442/ P0446/1441	No Test Results	No Test Results	EVAP
EVAP Test State	Waiting for Purge/ Test Running/Test Completed	Test Completed	Test Completed	EVAP
EVAP Vent Solenoid	Venting or Not Venting	Venting	Venting	Engine Data 1, EVAP, Fuel Trim
Fuel Level Sensor	0-100%	Varies	Varies	Engine Data 1, EVAP
Fuel Pump Relay	On/Off	On	On	Engine Data 1
Fuel Tank Pressure Sensor	Inches H2O, mm Hg	-17.5 to +7.5 Inches H2O	-17.5 to +7.5 Inches H2O	Engine Data 1, EVAP
Fuel Tank Pressure	0-5.0 volts	Varies	Varies	EVAP
Fuel Trim Cell	Cell 0-23	18-21	Varies	Engine Data 1, EVAP, Fuel Trim
Fuel Temp. Sensor	-38 to 310ºF	Varies	Varies	Engine Data 3
Generator F-Terminal	0-100%	Varies	Varies	Engine Data 2
Generator L-Terminal	Active/Inactive	Active	Active	Engine Data 2
HO2S-12 (B1 S2)	0-1106 mv	10-1000	10-1000	Engine Data 1, Fuel Trim
IAC Position	0-255 counts	5-60	5-60	Engine Data 1, Fuel Trim
IAT Sensor (ºF)	-38 to 284ºF	50-176	50-176	Engine Data 1-3 EVAP, Fuel Trim
Ignition 1 Signal	0.0-25.5v	12.0-15.0	12.0-15.0	Engine Data 1-3 EVAP, Fuel Trim, Misfire
Injector Pulsewidth	0-999.9 ms	2-4	2-4	Engine Data 2, Fuel Trim, Misfire
Injector 1-4 Command	0-999.9 ms	2-4	2-4	Engine Data 3
Knock Retard	0-90 degrees	0	0	Engine Data 1
Long Term Fuel Trim	-10% to 10%	0 (± 5%)	0 (± 5%)	Engine Data 1-2, EVAP, Fuel Trim
Loop Status	Closed Loop / Open Loop	Closed Loop	Closed Loop	Engine Data 1-3 EVAP, Fuel Trim
MAP Sensor (kPa)	10-105 kPa	25-48	Varies	Engine Data 1-3 EVAP, Fuel Trim, Misfire
MAP Sensor (V)	0-5.0 volts	1.0-3.0	Varies	Engine Data 1-3
Medium Resolution Resync Counter	Yes/No	Yes	Yes	Engine Data 1-2
MIL Command	On/Off	Off	Off	Engine Data 1
Misfire Current Cylinder 1-4	0-255 counts	0	0	Misfire
Misfire History Cylinder 1-4	0-255 counts	0	0	Misfire

2000-04 S10 Pickup 2.2L I4 Gas/Ethanol VIN 5 - PID Data List

Parameter Identification (PID)	PID Value Range	PID Value at Hot Idle	PID Value at 30 mph	Data List Type
Number of Current DTC(s)	0-255	0	0	Engine Data 1, EVAP, TAC, Fuel Trim
O2S-11 (B1 S1)	0-1106 mv	10-1000	10-1000	Engine Data 1, EVAP, Fuel Trim
Power Enrichment	Yes/No	No	No	Engine Data 2, Misfire
Short Term Fuel Trim	-10% to 10%	0 (± 5%)	0 (± 5%)	Engine Data 1-3 EVAP, Misfire
Spark Advance (º)	-64º to 64º	Varies	Varies	Engine Data 1-3 Fuel Trim, Misfire
Startup ECT	-40 to 419ºF	Varies	Varies	Engine Data 2, EVAP, Fuel Trim
TCC Enable Solenoid Command	On/Off	Off	On	Engine Data 1-2, Misfire
TP Angle (%)	0-100%	0	Varies	Engine Data 1-2, EVAP, Fuel Trim
TP Sensor (V)	0-5.0 volts	0.20-0.90	Varies	Engine Data 1, Data 2 and 3
TR Switch A/B/C	H/H/L/L	HI/Low/HI	HI/Low/Low	Engine Data 2-3 Fuel Trim
Vehicle Speed	0-155 km/h / 0-155 mph	0	30	Engine Data 1-3 EVAP, Fuel Trim, Misfire
VTD Auto Learn Timer	Active or Inactive	Inactive	Inactive	Engine Data 3
VTD Fuel Disable	Active or Inactive	Inactive	Inactive	Engine Data 3

2002-05 Envoy, Trailblazer 4.2L I6 MFI VIN X (A/T) - PID Data List

Parameter Identification (PID)	PID Value Range	PID Value at Hot Idle	PID Value at 30 mph	Data List Type
A/C HI Side Pressure	0-450 psi	Varies	Varies	Engine Data 2
A/C HI Side Pressure	0-5.0 volts	Varies	Varies	Engine Data 2
A/C Relay Command	On/Off	Off	Off	Engine Data 1, 2-3, Misfire
A/C Request Signal	Yes/No	No	No	Engine Data 2
Airflow (calculated)	0-64 g/sec	Varies	Varies	Engine Data 1-3 EVAP, Fuel Trim Misfire, TAC
Air Fuel Ratio	0.0-25.5:1	14.7:1	14.7:1	Engine Data 2-3 Fuel Trim
APP Average	0-125 counts	0-100%	Varies	TAC
APP Indicated Angle	0-100%	0	WOT: 100%	Engine Data 1-2, Cruise, EVAP Fuel Trim, TAC
APP Sensor 1 (%)	0-100%	0	Varies	TAC
APP Sensor 1 (V)	0-5.0 volts	1.02	Varies	TAC
APP Sensor 2 (%)	0-100%	0	Varies	TAC
APP Sensor 2 (V)	0-5.0 volts	4.04	Varies	TAC
APP Sensor 1 and 2	Agree or Disagree	Agree	Agree	TAC
BARO Sensor (kPa)	10-110 kPa	65-104	65-104	Engine Data 1, EVAP, Fuel Trim
BARO Sensor (V)	0-5.0 volts	3.5-4.9	3.5-4.9	Engine Data 1, EVAP, Fuel Trim
Camshaft Actual Angle	0-25 degrees	0	Varies	Engine Data 2
Camshaft Desired Angle	0-25 degrees	0	Varies	Engine Data 2
Camshaft Angle Variance	0-25 degrees	0	Varies	Engine Data 2
Camshaft Actuator Solenoid Command	0-100%	0	Varies	Engine Data 2
CKP Active Counter	1-250 counts	9-250	9-250	Engine Data 2
CMP Active Counter	0-255 counts	Varies	Varies	Engine Data 2
CMP Resync Counter	0-255 counts	Varies	Varies	Engine Data 2
Crank Request Signal	Yes/No	No	No	Engine Data 3
Cruise Control Active	Yes/No	No	No	Engine Data 1-3, Cruise, TAC
Cruise Disengage History 1-8	20 Possible Causes	Varies	Varies	Cruise Control
Cruise On/Off Switch	On/Off	Off	Off	Cruise, TAC
Cruise Release Brake Pedal Switch	Engaged or Released	Released	Released	TAC
Cruise Release Clutch / TCC Pedal Switches	Engaged or Released	Released	Released	Engine Data 1-3, Cruise, TAC
Cruise Resume / Accel Switch	On/Off	Off	Off	Cruise, TAC
Cruise Set / Coast Switch	On/Off	Off	Off	Cruise, TAC

2002-05 Envoy, Trailblazer 4.2L I6 MFI VIN X (A/T) - PID Data List

Parameter Identification (PID)	PID Value Range	PID Value at Hot Idle	PID Value at 30 mph	Data List Type
Current Gear	0-1-2-3-4	1	4	Engine Data 1-2, EGR, Fuel Trim
Cycles of Misfire Data	0-100 counts	Varies	Varies	Misfire
Desired Fan Speed	0-9999 rpm	Varies	Varies	Engine Data 2-3
Desired Idle Speed	0-3187 rpm	Varies	Varies	Engine Data 1-3 EVAP, TAC
ECT Sensor (°F)	-38 to 284°F	190-221	190-221	Engine Data 1-3 EVAP, Fuel Trim Misfire
Engine Load (%)	0-100%	1-4	7-9	Engine Data 1-3, Cruise, EVAP Fuel Trim, Misfire, TAC
Engine Oil Level Switch	Okay / Low	Okay	Okay	Engine Data 3
Engine Oil Life Remaining	0-100%	Varies	Varies	Engine Data 3
EOP Sensor (kPa)	0-450 psi	Varies	Varies	Engine Data 3
EOP Sensor (V)	0-5.0 volts	Varies	Varies	Engine Data 3
EOT Sensor	-38 to 284°F	156	156	Engine Data 3
Engine Run Time	Hr: Min: Sec	00:00:00	00:00:00	Engine Data 1-3, Cruise, EVAP Fuel Trim, Misfire, TAC
Engine Speed	0-9999 rpm	500-700 rpm	Varies	Engine Data 1-3, Cruise, EVAP Fuel Trim, Misfire, TAC
Engine Torque	0-100%	Varies	Varies	TAC
EVAP Purge Solenoid Command	0-100%	10-25	Varies	Engine Data 1, EVAP, Fuel Trim
EVAP Test Result	No Result / Passed/ Fail/ P0440/442/ P0446/1441	No Test Results	No Test Results	EVAP
EVAP Test State	Waiting for Purge/ Test Running/Test Completed	Test Completed	Test Completed	EVAP
EVAP Vent Solenoid Command	Venting/Not Venting	Venting	Venting	Engine Data 1, EVAP, Fuel Trim
Fail Counter	0-255 counts	0	0	F/F & F/R
Fan Speed	0-9999 rpm	Varies	Varies	Engine Data 2-3
Fuel Level Sensor	0-5.0 volts	0.7-2.5	Varies	Engine Data 1
Fuel Pump Relay	On/Off	On	On	Engine Data 1
Fuel Tank Pressure Sensor	Inches H2O, mm Hg	-17.5 to +7.5 Inches H2O	-17.5 to +7.5 Inches H2O	Engine Data 1, EVAP
Fuel Tank Pressure Sensor	0-5.0 volts	Varies	Varies	Engine Data 1, EVAP
Fuel Trim Cell	Cell 0-23	16, 17	Varies	Engine Data 1, EVAP, Fuel Trim
Fuel Trim Learn	Enabled or Disabled	Enabled	May Toggle	Engine Data 1, EVAP, Fuel Trim
Generator F-Terminal	0-100%	Varies	Varies	Engine Data 2
Generator L-Terminal	Active/Inactive	Active	Active	Engine Data 2
HO2S-11 (B1 S1)	0-1132 mv	10-1000	10-1000	Engine Data 1, EVAP, Fuel Trim
HO2S-12 (B1 S2)	0-1132 mv	10-1000	10-1000	Engine Data 1, Fuel Trim
HO2S-11 Heater Command	On/Off	On	On	Fuel Trim
HO2S-12 Heater Command	On/Off	On	On	Fuel Trim
HO2S-11 Heater Current	0-5.0 amps	0.6	0.6	Fuel Trim
HO2S-12 Heater Current	0-5.0 amps	0.6	0.6	Fuel Trim
IAT Sensor (°F)	-38 to 284°F	Varies	Varies	Engine Data 1-3 EVAP, Fuel Trim
Ignition 1 Signal	0.0-25.5v	12.8-14.2	12.8-14.2	Engine Data 1-2, Cruise, EVAP Fuel Trim, TAC
Injector 1-6 Command	0-1000 ms	2.49-2.59	Varies	Engine Data 3
Injector Pulsewidth	0-1000 ms	2-6	2-6	Engine Data 2, Fuel Trim, Misfire

2002-05 Envoy, Trailblazer 4.2L I6 MFI VIN X (A/T) - PID Data List

Parameter Identification (PID)	PID Value Range	PID Value at Hot Idle	PID Value at 30 mph	Data List Type
Knock Retard (º)	0-25.5º	0.0-16.0	0	ENG 1
Knock Sensor 1	0-5.0v	0.8-0.9	0.8-0.9	ENG 1
Knock Sensor 2	0-5.0v	0.8-0.9	0.8-0.9	ENG 1
Long Term Fuel Trim	-10% to 10%	0 (± 5%)	0 (± 5%)	Engine Data 1-3 EVAP, Fuel Trim
Loop Status	Closed Loop / Open Loop	Closed Loop	Closed Loop	Engine Data 1-3 EVAP, Fuel Trim
MAP Sensor (kPa)	10-105 kPa	20-48	Varies	Engine Data 1-2, EVAP, Fuel Trim Misfire, TAC
MAP Sensor (V)	0-5.1 volts	1.0-2.0	Varies	Engine Data 1-3
Medium Resolution Resync Counter	0-255	0	0	Engine Data 2
MIL Command	On/Off	Off	Off	Engine Data 1
Misfire Current Cylinder 1-6	0-200 counts	0	0	Misfire
Misfire History Cylinder 1-6	0-65,535	0	0	Misfire
Number of DTC(s)	0-255	0	0	Engine Data 1, EVAP, TAC, Fuel Trim
PCM/VCM in VTD Fuel Enable	Yes/No	No	No	Engine Data 3
Power Enrichment	Yes/No	No	No	Engine Data 2
Reduced Engine Power	Yes/No	No	No	Engine Data 1-3, Cruise, TAC
Reference Voltage 1	4.9-5.1 volts	4.57	4.57	TAC
Reference Voltage 2	4.9-5.1 volts	4.57	4.57	TAC
Short Term Fuel Trim	-10% to 10%	0 (± 5%)	0 (± 5%)	Engine Data 1-3 EVAP, Fuel Trim
Spark Advance (º)	-64º to 64º	15-20	Varies	Engine Data 1-3 Fuel Trim, Misfire
Startup ECT	-40 to 419ºF	Varies	Varies	Engine Data 2, EVAP, Fuel Trim
Starter Relay Command	On/Off	Off	Off	Engine Data 3
TCC Enable Solenoid Command	Enabled or Disabled	Disabled	Enabled	Engine Data 1-2, Cruise, Misfire
TCC PWM Solenoid	0-100%	0	Varies	Engine Data 1-2
Torque Delivered Signal	N/m., ft lbs	Varies	Varies	TAC
Torque Requested Signal	0-100%	0	Varies	TAC
TP Desired Angle	0-100%	4.2	Varies	Engine Data 1-2, EVAP, TAC
TP Indicated Angle	0-100%	4.2	Varies	Engine Data 1-3, Cruise, EVAP Misfire, TAC
TP Sensor 1 (%)	0-100	0	Varies	TAC, F/F, F/R
TP Sensor 1 (V)	5.0-0.0 volts	4.24	Varies	TAC, F/F, F/R
TP Sensor 2	0-100	0	Varies	TAC, F/F, F/R
TP Sensor 2 (V)	0-5.0 volts	0.78	Varies	TAC, F/F, F/R
TP Sensor 1 and 2	Agree or Disagree	Agree	Agree	TAC
Transmission Range Switch	Park/Neutral, Drive 4-3-2-1	Park/Neutral	Drive 4	Engine Data 2-3 Cruise, Fuel Trim
Traction Control System	Active or Inactive	Inactive	Inactive	Cruise, TAC
TWC Temperature	32-1409ºF	843	850	Fuel Trim
Vehicle Speed	0-155 km/h or miles per hour (mph)	0	30	Engine Data 1-3, Cruise, EVAP Misfire, TAC
VTD Auto Learn Timer	Active/Inactive	Inactive	Inactive	Engine Data 3
VTD Fuel Disable	Active/Inactive	Inactive	Inactive	Engine Data 3
VTD Fuel Disable Until Ignition Off	Yes/No	No	No	Engine Data 3
VTD Password	Okay / Fault	Okay	Okay	Engine Data 3

2005 Equinox 3.4L I6 MFI VIN F (A/T) - PID Data List

Scan Tool Parameter	Data List	Units Displayed	Typical Data Value
A/C Disengage Reason	Cooling/HVAC	Swithed off/Pressure too high/pressure too low	Switched Off
A/C High side Pressure	Cooling/HVAC	PSI	varies
A/C High side Pressure	Cooling/HVAC	volts	varies
A/C Off for WOT	Cooling/HVAC	Yes/No	No
A/C Pressure Disable	Cooling/HVAC	Yes/No	No
A/C Relay Circuit Status	ODD	OK/Fault/Invalid State	OK
A/C Relay Circuit History	ODD, Cooling/HVAC	OK/Fault/Invalid State	OK
A/C Relay Command	Eng, Misfire, EGR, Cooling/HVAC, TAC	On/Off	Off
A/C Request Signal	Cooling/HVAC	Yes/No	No
Air Fuel Ratio	Eng, Fuel Trim, HO2S	Ratio	14.2:1-14.7:1
APP Indicated Angle	Cruise/Traction, Eng, EGR, EVAP, HO2S, Ignition, Misfire, TAC, Fuel Trim	Percent	0
APP Sensor 1	TAC	Volts	Varies
APP Sensor 1 Circuit Status	TAC	OK/Not OK	OK
APP Sensor 2	TAC	Volts	Varies
APP Sensor 2 Circuit Status	TAC	OK/Not OK	OK
APP Sensor 1	TAC	Percent	Varies
APP Sensor 2	TAC	Percent	Varies
APP Sensor 1 and 2	TAC	Agree/Disagree	Agree
BARO	Eng, EGR, EVAP, Fuel Trim, HO2S, Ignition	kPa	65-110 kPa
Catalytic Converter Protection Active	Fuel Trim, HO2S	Yes/No	No
CKP Sensor	Ignition Data	RPM	varies
CMP Sensor	Ignition	RPM	Varies
CMP Sensor Signal Present	Ignition, Misfire	Yes/No	Yes
Crank Request Signal	Electrical/Theft	Yes/No	No
Cruise On/Off switch	Cruise/Traction	On/Off	Off
Cruise Control Active	Cruise/Traction, IPC, TAC	Yes/No	No
Cruise Disengage 1-6 History	Cruise/Traction	engage/disengage	--
Cruise Mode Signal	Cruise/Traction	Volts	varies
Cruise Resume/Accel. Switch	Cruise/Traction	On/Off	Off
Cruise Set/Coast Switch	Cruise/Traction	On/Off	Off
Cycles of Misfire Data	Misfire	Counts	0-99
Cyl. 1-6 Injector Circuit Status	ODD	OK/Stuck Low (Open)/Stuck High	OK
Cylinder 1 and 4 IC Circuit Status	ODD	OK/Not OK	OK
Cylinder 2 and 5 IC Circuit Status	ODD	OK/Not OK	OK
Cylinder 3 and 6 IC Circuit Status	ODD	OK/Not OK	OK
Cylinder 1-6 Injector Circuit History	ODD	OK/Stuck Low (Open)/Stuck High	OK
Decel Fuel Cut-off	Eng, EGR, Fuel Trim, HO2S	Active/Inactive	Inactive
Desired EGR Position	EGR, Misfire	Percent	0
Desired Idle Speed	Cruise/Traction, Eng, EVAP, TAC	RPM	Varies
Driver Module 1 Status	ODD	Enabled/Off-High Volts/Off High Temp/Invalid State	Enabled
Driver Module 2 Status	ODD	Enabled/Off-High Volts/Off High Temp/Invalid State	Enabled
Driver Module 3 Status	ODD	Enabled/Off-High Volts/Off High Temp/Invalid State	Enabled
EC Ignition Relay Circuit Status	ODD	OK/ Fault/ Invalid State	OK
EC Ignition Relay Circuit History	Electrical Theft, ODD	OK/ Fault/ Invalid State	OK
ECT Sensor	Cooling/HVAC, Eng, EGR, EVAP, HO2S, IPC, Ignition, Misfire, Fuel Trim, TAC	°C/°F	Varies
EGR Flow Test Count	EGR	Counts	0
EGR Learned Minimum Position	EGR	Volts	0.14-1.0
EGR Position Sensor	Eng, EGR, Misfire	Percent	0%
EGR Position Sensor	EGR	Volts	0.6--0.7 volts
EGR Position Variance	EGR	Percent	0

2005 Equinox 3.4L I6 MFI VIN F (A/T) - PID Data List (CONT)

Scan Tool Parameter	Data List	Units Displayed	Typical Data Value
EGR Solenoid Circuit History	ODD	OK/Open/Fault/ Invalid State	OK
EGR Solenoid Circuit Status	ODD	OK/Fault/ Invalid State	OK
EGR Solenoid Command	EGR	Percent	0
Engine Hot Lamp Command	Cooling/HVAC	On/Off	Off
Engine Load	Cooling/HVAC, Eng, Misfire, Fuel Trim, EVAP, EGR, HO2S, Ignition, TAC	Percent	Varies
Engine Oil Life Remaining	IPC	Percent	Varies
Engine Oil Pressure Switch	Eng, IPC	Low/OK	OK
Engine Run Time	Cruise/Traction, Cooling/HVAC, Eng, EVAP, Misfire, Fuel Trim, EGR, HO2S, Ignition, TAC, ODD	Hr: Min: Sec	Varies
Engine Speed	Cooling/HVAC, Cruise/TractionEng, EGR, Electrical/Theft, EVAP, Fuel Trim, HO2S, Ignition, IPC, Misfire, TAC	RPM	Varies
EVAP Fault History	EVAP	Fault/No Fault	No Fault
EVAP Purge Solenoid Circuit History	ODD	OK/Open/Short	OK
EVAP Purge Solenoid Circuit Status	ODD	OK/Open/Short	OK
EVAP Purge Solenoid Command	Eng, EVAP, Fuel Trim, HO2S	Percent	Varies
EVAP Vent Solenoid Circuit History	ODD	OK/Open/Short	OK
EVAP Vent Solenoid Circuit Status	ODD	OK/Open/Short	OK
EVAP Vent Solenoid Command	Eng, EVAP	Venting/Not Venting	Venting
FC Relay 1 Circuit History	Cooling/HVAC, ODD	OK/Fault/Invalid State	OK
FC Relay 1 Command	Cooling/HVAC	On/Off	Off
FC Relay 1 Circuit Status	ODD	OK/Fault/Invalid State	OK
FC Relay 2 and 3 Circuit History	Cooling/HVAC, ODD	OK/Fault/Invalid State	OK
FC Relay 2 and 3 Circuit Status	ODD	OK/Fault/Invalid State	OK
FC Relay 2 and 3 Command	Cooling/HVAC	On/Off	Off
Fuel Level Sensor	EVAP, IPC	Volts	Varies
Fuel Pump Relay Circuit History	ODD	OK/Fault/Invalid State	OK
Fuel Pump Relay Circuit Status	ODD	OK/Fault/Invalid State	OK
Fuel Pump Relay Command	Eng	On/Off	On
Fuel Tank Level Remaining	EVAP, IPC	Percent	0-104
Fuel Tank Pressure Sensor	Eng, EVAP	mm Hg/in H2O	Varies
Fuel Tank Pressure Sensor	EVAP	Volts	Varies
Fuel Trim Cell	EVAP, Eng, Fuel Trim	0-4	0
Fuel Trim Learn	Eng, EVAP, Fuel Trim	Enabled/Disabled	Enabled
Generator L-Terminal Signal Command	Electrical/Theft	On/Off	On
Generator Lamp Command	IPC Data, Electrical/Theft	On/Off	Off
HO2S 1	EVAP, Eng, Fuel Trim, HO2S	millivolts	0-1000 and Varying
HO2S 1 Ready	HO2S	Yes/No	Yes
HO2S 2	Eng, Fuel Trim, HO2S	millivolts	0-1000 and Varying
HO2S 1 Heater Command	HO2S	On/Off	On
HO2S 2 Heater Command	HO2S	On/Off	On
HO2S 1 Heater	HO2S	Amps	Varies
HO2S 2 Heater	HO2S	Amps	Varies
HO2S 1 Heater Circuit History	ODD	OK/Fault/ Invalid State	OK
HO2S 2 Heaster Circuit History	ODD	OK/Fault/ Invalid State	OK
HO2S 1 Heater Circuit Status	ODD	OK/Fault/ Invalid State	OK
HO2S 2 Heater Circuit Status	ODD	OK/Fault/ Invalid State	OK
Hot Open Loop	Cooling/HVAC, Fuel Trim, HO2S	Active/Inactive	Inactive
IAT Sensor	Cooling/HVAC, Eng, EVAP, EGR, Fuel Trim, HO2S, Ignition, Misfire, TAC	°C/°F	Varies

2005 Equinox 3.4L I6 MFI VIN F (A/T) - PID Data List (CONT)

Scan Tool Parameter	Data List	Units Displayed	Typical Data Value
Ignition 1 Signal	Cooling/HVAC, Cruise/Traction, Electical/Theft, EVAP, Eng, EGR, Fuel Trim, HO2S, Ignition, IPC, Misfire, TAC	Volts	12.6 volts or more (varies)
Ignition Off Time	EVAP, Eng	Minutes/ Varies	Varies
Initial Brake Apply Signal	TAC	Released/ Applied	Released
Injector PWM	Fuel Trim, HO2S, Misfire	milliseconds	Varies
Knock Retard	EGR, Ignition	Degrees	0
KS Activity	Ignition	Yes/No	No
Long Term FT	Eng, EVAP, Fuel Trim, HO2S	Percent	-10 to +10%
Long Term FT Average	Fuel Trim	Percent	varies
Loop Status	Eng, Fuel Trim, EVAP, HO2S	Open/Closed	Closed
MAF Sensor	Eng, EGR, Misfire, EVAP, Fuel Trim, HO2S, Ignition, TAC	g/s	3-6 (Varies with altitude)
MAF Sensor	EVAP, Fuel Trim, HO2S, Misfire, TAC	Hz	1,200-3,000 (Depends on altitude)
MAP Sensor	Eng, EGR, Fuel Trim, HO2S, Misfire, TAC	Volts	0.4-2.0 Volts
MAP Sensor	Eng, EVAP, Fuel Trim, HO2S, Ignition, Misfire, EGR, TAC	kPa	20-48 kPa (Varies with altitude)
MIL Circuit History	IPC, ODD	OK/Fault/Invalid State	OK
MIL Command	EVAP, Eng, EGR, Ignition, IPC	On/Off	Off
MIL Circuit Status	ODD	OK/Fault/Invalid State	OK
Mileage Since DTC Cleared	Eng	km/h /mph	Varies
Misfire Current Cyl. 1-6	Misfire	Counts	0
Misfire History Cyl. 1-6	Misfire	Counts	0
Moderate Brake Apply Signal	TAC	Applied/Released	Released
Number of DTCs	Eng	Number	0
PCM Reset	EGR, Eng, EVAP, Ign, ODD	Yes/No	No
Power Enrichment	Eng, Fuel Trim, HO2S, Misfire	Active/Inactive	Inactive
Reduced Engine Power	HO2S, TAC	Active/Inactive	Inactive
Short Term FT	Eng, EVAP, Fuel Trim, HO2S,	Percent	-10 to +10%
Spark	Cooling/HVAC, Cruise/Traction, Eng, Misfire, Fuel Trim, HO2S, Ignition	Degrees	16° (varies)
Start Up ECT	Cooling/HVAC, EVAP, Fuel Trim, HO2S	°C/°F	Varies
Start Up IAT	EVAP, Fuel Trim, HO2S	°C/°F	Varies
Starter Relay Circuit Status	ODD	OK/ Fault	OK
Starter Relay Circuit History	Electrical/Theft, ODD	OK/ Fault/ Invalid State	OK
Starter Relay Command	Electrical/Theft	On/Off	Off
Stop Lamp Pedal Switch	Cruise/Traction, Eng, Electrical/Theft, TAC	Released/Applied	Released
Stop Lamp Switch Signal	Electrical/Theft, Cruise/Traction, TAC	volts	Varies
TAC/PCM Communications Signal	Cruise/Traction, TAC	OK/ Fault	OK
TAC Stop Lamp Pedal Switch	Cruise/Traction, Electrical/Theft, TAC	Applied/Released	Released
TAC Vehicle Speed Signal	TAC	OK/Fault	OK
TCS Delivered Torque Circuit Status	ODD	OK/ Fault	OK
Torque Request Signal	Cruise/Traction, TAC	Percent	100%
Torque Delivered Signal	TAC	Percent	75%
Total Misfire Count	Misfire	1-255	0
TP Desired Angle	Cruise/Traction, TAC	Percent	Varies

2005 Equinox 3.4L I6 MFI VIN F (A/T) - PID Data List (CONT)

Scan Tool Parameter	Data List	Units Displayed	Typical Data Value
TP Indicated Angle	Cruise/Traction, EGR, Eng , Fuel Trim, HO2S, Ignition, EVAP, Misfire, TAC	Percent	%
TP Sensor 1	TAC	Percent	Varies
TP Sensor 1	TAC	Volts	Varies
TP Sensors 1 and 2	TAC	Agree/ Disagree	Agree
TP Sensor 1 Circuit Status	TAC	OK/Not OK	OK
TP Sensor 2	TAC	Percent	Varies
TP Sensor 2	TAC	Volts	Varies
TP Sensor 2 Circuit Status	TAC	OK/Not OK	OK
Traction Control Status	Cruise/Traction, TAC	Active/Inactive	Inactive
TWC Temperature Calculated	Fuel Trim, Misfire	°C/°F	varies
Vehicle Security Status	Electrical/Theft	OK/Not OK	OK
Vehicle Speed Circuit History	ODD	OK/ Fault/ Invalid State	OK
Vehicle Speed Circuit Status	ODD	OK/ Fault/ Invalid State	OK
Vehicle Speed Sensor	Cooling HVAC, Cruise/Traction, Eng, EGR, EVAP, HO2S, Ignition, IPC, Fuel Trim, Misfire, TAC	mph/ km/h	0
Warm-ups w/o Emission Faults	Eng	1-10	varies
Warm-ups w/o Non-Emission Faults	Eng	1-10	varies

2004-05 SSR 6.0L VIN H PID Data List

Scan Tool Parameter	Data List	Parameter Range/Units	Typical Data Values
5-Volt Reference 1 Circuit Status	EE, Eng, Ign, TAC	OK/Fault	OK
5-Volt Reference 2 Circuit Status	EE, Eng, Ign, TAC	OK/Fault	OK
5-Volt Reference 1	EE, Eng, Ign, TAC	Volts	4.5 V
5-Volt Reference 2	EE, Eng, Ign, TAC	Volts	4.5 V
A/C Relay Circuit Status	OD	OK, Incomplete, Short B+, Short Gnd/Open	OK/Incomplete
A/C Relay Command	Eng , MF, TAC	On/Off	Varies
Air Flow Calculated	EE, Eng, FT, HO2S, MF, TAC	Grams Per Seconds (g/s)	1-9 g/s @ idle
Air Fuel Ratio	Eng, FT, HO2S	X:1	14.7:1
Ambient Air Temperature	Eng	°C/°F	Varies
APP Indicated Angle	EE, Eng, FT, HO2S, Ign, MF, TAC	0-100%	0
APP Sensor 1	TAC	0-5.0 Volts	0.4-1.0 Volt
APP Sensor 2	TAC	5.0-0 Volts	4.5-4.1 Volts
APP Sensor 1	TAC	0-100%	0%
APP Sensor 2	TAC	0-100%	0%
APP Sensor 1 and 2	TAC	Agree/Disagree	Agree
APP Sensor 1 Indicated Position	TAC	%	0%
APP Sensor 2 Indicated Position	TAC	%	--
APP Sensors	TAC	%	0%
BARO	EE, Eng, FT, HO2S, Ign	kPa	50-104 kPa/ Varies w/Altitude
BPP Circuit Signal	TAC	Applied/Released	--
Calculated TWC Temp Bank 1	FT, MF	Degrees C/F	315-427°C (600-800°F)
Calculated TWC Temp Bank 2	FT, MF	Degrees C/F	315-427°C (600-800°F)
Catalytic Converter Protection Active	FT, HO2S	Yes/No	No
CKP Active Counter	Ign	0-250 Counts	Varies
CKP Resync Counter	Ign	Counts	0
CKP Sensor	Eng, Ign	RPM	500-700 RPM
CMP Active Counter	Ign	0-250 Counts	Varies
CMP Sensor	Eng, Ign, MF	RPM	1,000-1,400 RPM
Cold Start-Up	Eng, EE	Yes/No	Varies
Cruise Control Active	Eng, TAC	Active/Inactive	Inactive
Cycles of Misfire Data	MF	0-100 Counts	Varies
Cylinder 1-8 IC Circuit Status	OD	OK, Incomplete, Short B+, Short Gnd/Open	OK
Cylinder 1-8 Injector Circuit Status	OD	OK, Incomplete, Short B+, Short Gnd/Open	OK
Decel Fuel Cutoff	Eng, FT, HO2S	Active/Inactive	Inactive
Desired Idle Speed	EE, Eng, TAC	RPM	ECM Controlled
Distance Since DTC Cleared	Eng	km/miles	Varies
EC Ignition Relay Circuit Status	OD	OK, Incomplete, Short B+, Short Gnd/Open	OK/Incomplete
EC Ignition Relay Command	Eng, TAC	On/Off	On
EC Ignition Relay Feedback	Eng, TAC	Volts	2.25-2.95 Volts
ECM Reset	EE, Eng, Ign	Yes/No	No
ECT Sensor	All	-39 to +140°C (-38 to +284°F)	88-105°C (190-221°F)

2004-05 SSR 6.0L VIN H PID Data List (CONT)

Scan Tool Parameter	Data List	Parameter Range/Units	Typical Data Values
Engine Load	All	0-100%	18% @ Idle 21% @ 2500 RPM
Engine Off EVAP Test Connections Met	EE	Yes/No	Yes
Engine Oil Level Switch	Eng	OK/Low	OK
Engine Oil Pressure Sensor	Eng, MF	PSI	30
Engine Oil Pressure Sensor	EVAP	Volts	1.5
Engine Run Time	All	Hrs, Min, Sec	Varies
Engine Speed	All	0-10,000 RPM	500-700 RPM
Engine Speed Circuit Status	OD	OK, Incomplete, Short B+, Short Gnd/Open	OK
EVAP Purge Solenoid Circuit Status	OD	OK, Incomplete, Short B+, Short Gnd/Open	OK/Incomplete
EVAP Purge Solenoid Command	EE, Eng, FT, HO2S	0-100%	10-25%
EVAP Vent Solenoid Circuit Status	OD	OK, Incomplete, Short B+, Short Gnd/Open	OK/Incomplete
EVAP Vent Solenoid Command	EE, Eng, FT	Not Venting/Venting	Venting
FC Circuit Status	OD	OK, Incomplete, Short B+, Short Gnd/Open	OK/Incomplete
Fuel Level Sensor	EE	0-5 Volts	0.7-2.5 Volts
Fuel Pump Relay Circuit Status	OD	OK, Incomplete, Short B+, Short Gnd/Open	OK/Incomplete
Fuel Pump Relay Command	Eng, FT	On/Off	ON
Fuel Tank Level Remaining	Eng, MF	0-100%	Varies
Fuel Tank Pressure Sensor	EE, Eng	-32.7 to +14.0 mm/Hg (-17.5 to +7.5 in/H2O)	Varies
Fuel Tank Pressure Sensor	EE, Eng	0-5.0 Volts	Varies
Fuel Tank Rated Capacity	EE	98 Liters (25.9 Gallons) or 96.2 Liters (25 Gallons)	Varies with Fuel Tank Option
Fuel Trim Cell	EE, Eng, FT	0-23	19
Fuel Trim Learn	EE, Eng, FT	Enabled/Disabled	Enabled, May Toggle
Fuel Volatility	EE, FT, HO2S	Low/Moderate/High	Varies
HO2S Bank 1 Sensor 1	EE, Eng, FT, HO2S	Millivolts	10-1,000 mV and Varying
HO2S Bank 1 Sensor 2	Eng, FT, HO2S	Millivolts	10-1,000 mV and Varying
HO2S Bank 2 Sensor 1	EE, Eng, FT, HO2S	Millivolts	10-1,000 mV and Varying
HO2S Bank 2 Sensor 2	Eng, FT, HO2S	Millivolts	10-1,000 mV and Varying
HO2S BN 1 Sensor 1 Heater	HO2S	Amps	0.7-0.9 Amps
HO2S BN 1 Sensor 2 Heater	HO2S	Amps	0.4-0.6 Amps
HO2S BN 2 Sensor 1 Heater	HO2S	Amps	0.7-0.9 Amps
HO2S BN 2 Sensor 2 Heater	HO2S	Amps	0.4-0.6 Amps
HO2S BNK 1 Sen 1 Heater Circuit Status	OD	OK, Incomplete, Short B+, Short Gnd/Open	OK/Incomplete
HO2S BNK 1 Sen 2 Heater Circuit Status	OD	OK, Incomplete, Short B+, Short Gnd/Open	OK/Incomplete
HO2S BNK 2 Sen 1 Heater Circuit Status	OD	OK, Incomplete, Short B+, Short Gnd/Open	OK/Incomplete
HO2S BNK 2 Sen 2 Heater Circuit Status	OD	OK, Incomplete, Short B+, Short Gnd/Open	OK/Incomplete
HO2S Bnk 1 Sensor 1 Heater Command	HO2S	On/Off	ON
HO2S Bnk 1 Sensor 1 Heater Command	HO2S	%	100%
HO2S Bnk 1 Sensor 2 Heater Command	HO2S	On/Off	ON
HO2S Bnk 1 Sensor 2 Heater Command	HO2S	%	100%
HO2S Bnk 2 Sensor 1 Heater Command	HO2S	On/Off	ON
HO2S Bnk 2 Sensor 1 Heater Command	HO2S	%	100%
HO2S Bnk 2 Sensor 2 Heater Command	HO2S	On/Off	ON
HO2S Bnk 2 Sensor 2 Heater Command	HO2S	%	100%

2004-05 SSR 6.0L VIN H PID Data List (CONT)

Scan Tool Parameter	Data List	Parameter Range/Units	Typical Data Values
Hot Open Loop	FT, HO2S	Active/Inactive	Inactive
IAT Sensor	All	-39 to +140°C (-38 to +284°F)	35°C (91°F) Depends on Ambient Temperature
Ignition Accessory Signal	Eng, TAC	On/Off	ON
Ignition 1 Signal	All	0-25 Volts	11.5-14.5 Volts
Ignition Off Timer	Eng	Seconds, Minutes, Hours	Varies
Inj. PWM Bank 1	Eng, FT, HO2S	Milliseconds	6-Feb
Inj. PWM Bank 2	Eng, FT, HO2S	Milliseconds	6-Feb
Knock Retard	Ign	0.0-16°	0°
KS Active Counter	Ign	Counts	Varies
KS Bank 1 Circuit Status	Ign	OK/Fault/Incomplete	OK/Incomplete
KS Bank 2 Circuit Status	Ign	OK/Fault/Incomplete	OK/Incomplete
KS Module Status	Ign	OK/Fault/Incomplete	OK/Incomplete
Long Term FT Bank 1	EE, Eng, FT, HO2S	Percentage	Near 0%
Long Term FT Bank 2	EE, Eng, FT, HO2S	Percentage	Near 0%
Loop Status	EE, Eng, FT, HO2S	Open/Closed	Closed
MAF Sensor	EE, Eng, FT, HO2S, Ign, MF, TAC,	Grams Per Seconds (g/s)	1-9 g/s @ Idle, Depends on Altitude
			15-26 g/s @ 2,500 RPM, Depends on Altitude
MAF Sensor	EE, Eng, FT, HO2S, MF	0-31,999 Hz	1,150-1,350 Hz
MAP Sensor	EE, Eng, FT, HO2S, Ign, MF, TAC,	kPa	20-48 kPa
MAP Sensor	Eng, FT, HO2S, MF, TAC	Volts	1.0-2.0 Volts
			Varies with Altitude
MIL Circuit Status	OD	OK, Incomplete, Short B+, Short Gnd/Open	OK/Incomplete
MIL Command	EE, Eng, Ign	Off/On	Off
MIL Requested by DTC	EE, Eng, Ign	Yes/No	No
Misfire Current Cyl. 1-8	MF	0-200 Counts	0
Misfire History Cyl. 1-8	MF	0-65,535 Counts	0
PNP Switch	Eng	P/N	P/N
		Yes/No in gear	
Power Enrichment	Eng, FT, HO2S, MF	Active/Inactive	Inactive
Reduced Engine Power	TAC	Active/Inactive	Inactive
Short Term FT Bank 1	EE, Eng, FT, HO2S	Percentage	Near 0%
Short Term FT Bank 2	EE, Eng, FT, HO2S	Percentage	Near 0%
Spark	Eng, FT, HO2S, Ign, MF	Degrees	10-17°
Starter Relay Circuit Status	OD	OK, Incomplete, Short B+, Short Gnd/Open	OK/Incomplete
Start-Up ECT	EE, FT	°C/°F	Varies
Start-Up IAT	EE, HO2S	°C/°F	Varies
TAC Forced Engine Shutdown	TAC	Yes/No	No
TAC Motor	TAC	Enabled/Disabled	Enabled
TAC Motor Command	TAC	0-100%	15-35%
TCC/Cruise Brake Pedal	Eng	Applied/Released/Invalid	Released
TCC/Cruise Brake Pedal Switch	TAC	Applied/Released/Invalid	Released
TCS Circuit Status	OD	OK, Incomplete, Short B+, Short Gnd/Open	OK/Incomplete
TCS Torque Request Signal	TAC	0-100%	100%
Torque Delivered Signal	TAC	0-100%	5-15%
Torque Management Spark Retard	Ign	Degrees	0°
Total Knock Retard	Ign	Degrees	0°
Total Misfire	MF	Counts	Varies
TP Desired Angle	TAC	0-100%	5.50%

2004-05 SSR 6.0L VIN H PID Data List (CONT)

Scan Tool Parameter	Data List	Parameter Range/Units	Typical Data Values
TP Indicated Angle	EE, Eng, FT, HO2S, Ign, MF, TAC	0-100%	5.50%
TP Sensor 1	TAC	0-5.0 Volts	4.1-4.95 Volts
TP Sensor 1	TAC	0-100%	Varies near 5%
TP Sensor 1 Learned Minimum	TAC	Volts	0.55
TP Sensor 2	TAC	5.0-0 Volts	0.4-0.85 V
TP Sensor 2	TAC	100-0%	Varies near 5%
TP Sensor 2 Learned minimum	TAC	Volts	0.55
TP Sensors 1 and 2	TAC	Agree/Disagree	Agree
TP Sensor 1 Indicated Position	TAC	%	5%
TP Sensor 2 Indicated Position	TAC	%	5%
TWC Temperature Calculated	FT, MF	Degrees C/F	--570°C (1,075°F)
Vacuum Calculated	EE, Eng, FT, HO2S	kPa/in Hg	59 kPa/16 in Hg
Vehicle Speed Circuit Status	OD	OK, Incomplete, Short B+, Short Gnd/Open	OK/Incomplete
Vehicle Speed Circuit 2 Status	OD	OK, Incomplete, Short B+, Short Gnd/Open	OK/Incomplete
Vehicle Speed Sensor	EE, Eng, FT, HO2S, Ign, MF, TAC	km/h mph	0
Warm-Ups w/o Emission Faults	Eng	0-255 Counts	Varies
Warm-Ups w/o Non-Emission Faults	Eng	0-255 Counts	Varies
Warm-Ups Since DTC Cleared	Eng	Counts	Varies
Wide Open Throttle	TAC	Yes/No	No

VENTURE

1997-2005 Venture Light Van 3.4L VIN E - PID Data List

Parameter Identification (PID)	PID Value Range	PID Value at Hot Idle	PID Value at 30 mph	Data List Type
3X Crank Sensor	0-9999 rpm	Varies	Varies	Engine Data 2
24X Crank Sensor	0-1600 rpm	Varies	Varies	Engine Data 2
A/C HI Side Pressure	0-5.0v	Varies	Varies	Engine Data 1
A/C Pressure Disable	Yes/No	No	No	Engine Data 2
A/C Pressure Disable	Yes/No	No	No	Engine Data 2
A/C Off for WOT	Yes/No	No	No	Engine Data 2
A/C Pressure Disable	Yes/No	No	No	Engine Data 2
A/C Relay Circuit Status	Okay/Fault	Okay	Okay	Output Driver
A/C Relay Command	On/Off	Off	Off	Engine Data 1-2, EGR, Misfire
A/C Request Signal	Yes/No	No	No	Engine Data 2
Actual EGR Position	0-100%	0	Varies	Engine Data 1, EGR, Misfire
Air Fuel Ratio	0.0-25.5:1	14.7:1	14.7:1	Engine Data 2, Fuel Trim
BARO Sensor (kPa)	10 110 kPa	08	08	Engine Data 1, EGR, EVAP, Fuel Trim
BARO Sensor (V)	0-5.0 volts	3.5-4.5	3.5-4.5	Engine Data 1, EGR, EVAP, Fuel Trim
Brake Switch Signal	Applied/REL	Released	Released	Engine Data 1
CMP Sensor Signal	Yes/No	Yes	Yes	Engine Data 1, Misfire
Crank Request Signal	Yes/No	No	No	Engine Data 2
Cruise Control Active	Yes/No	No	No	Engine Data 1
Cruise Inhibit Reason	Several	Varies	Varies	Engine Data 1
Cruised Inhibit Signal Circuit Status	Okay/Fault/ Invalid State	Okay	Okay	Output Driver
Cruise Inhibit Signal Command	On/Off	On	On	Engine Data 1
Current Gear	1-2-3-4	1	4	Engine Data 1-2, EGR, Fuel Trim
Cycles of Misfire Data	0-99 counts	Varies	Varies	Misfire
Cylinder 1-6 Injector Circuit Status	Okay/Stuck Low/Stuck High/Fault	Okay	Okay	Output Driver
Cylinder 1-6 Injector Circuit History	Okay/Stuck Low/Stuck High/Fault	Okay	Okay	Output Driver
Deceleration Fuel Cutoff	Active/Inactive	Inactive	Inactive	EGR, Fuel Trim, Misfire
Desired EGR Position	0-100%	0	Varies	Engine Data 1-2, EGR, Misfire
Desired Idle Speed	0-3187 rpm	PCM control	---	Engine Data 1, Data 2, EVAP
Driver Module 1 Status	Enabled/ Disabled	Enabled	Enabled	Output Driver
Driver Module 2 Status	Enabled/ Disabled	Enabled	Enabled	Output Driver
Driver Module 3 Status	Enabled/ Disabled	Enabled	Enabled	Output Driver
Driver Module 4 Status	Enabled/ Disabled	Enabled	Enabled	Output Driver
ECT Sensor (°F)	-40° to 304°F	185-220	185-220	Engine Data 1-2, EGR, Fuel Trim, EVAP, Misfire
EGR Flow Test Count	0-255 counts	0	Varies	EGR
EGR Position Sensor	0-100%	0	Varies	Engine Data 1, EGR, Misfire
EGR Position Sensor	0-5.1 volts	0.14-1.0	Varies	Engine Data 1, EGR, Misfire
EGR Learned Minimum Position	0-5.1 volts	0.71	Varies	EGR
EGR Position Error	0-100%	0	Varies	EGR
EGR Pos. Variance	0-100%	0	Varies	EGR
EGR Solenoid Command	0-100%	0	Varies	Engine Data 1, EGR, Misfire
Engine Load (%)	0-100%	17	Varies	Engine Data 1-2, EGR, Fuel Trim EVAP, Misfire
Engine Oil Level Switch	Okay/Low	Okay	Okay	Engine Data 2
Engine Oil Life Remaining	0-100%	Varies	Varies	Engine Data 2

1997-2005 Venture Light Van 3.4L VIN E - PID Data List

Parameter Identification (PID)	PID Value Range	PID Value at Hot Idle	PID Value at 30 mph	Data List Type
EOP Switch Signal	Okay/Low/HI	Okay	Okay	Engine Data 2
Engine Run Time	Hr: Min: Sec	00:00:00	00:00:00	Engine Data 1-2, EGR, Fuel Trim EVAP, Misfire
Engine Speed	0-9999 rpm	± 50 rpm of Desired Speed	Varies	Engine Data 1-2, EGR, Fuel Trim EVAP, Misfire
EVAP Fault History	No Fault / Excess VAC / Purge Valve Leak / Small Leak / Weak Vacuum	No Fault	No Fault	EVAP
EVAP Purge Solenoid Circuit Status	Okay / Fault	Okay	Okay	Output Driver
EVAP Purge Solenoid Command	0-100%	19	Varies	Engine Data 1, EVAP, HO2S
EVAP Vent Solenoid Control Circuit Status	Okay / Fault	Okay	Okay	Output Driver
EVAP Vent Solenoid Command	Venting/Not Venting	Venting	Venting	Engine Data 1, EVAP, Fuel Trim
FC Relay 1 Command	On/Off	Off	Off	Engine Data 2
FC Relay 2 and 3 Circuit Status	Okay/Fault	Okay	Okay	Output Driver
FC Relay 2 and 3 Command	On/Off	Off	Off	Engine Data 2
Fuel Level Sensor	0-5.0 volts	0.8-2.5	Varies	EVAP
Fuel Pump Command	On/Off	On	On	Engine Data 1
Fuel Pump Relay Circuit Status	Okay/Fault	Okay	Okay	Output Driver
Fuel Pump Relay Circuit History	Okay/Fault	Okay	Okay	Output Driver
Fuel Pump Relay Command	On/Off	On	On	Engine Data 1
Fuel Tank Level Remaining	0-100%	Varies	Varies	EVAP
Fuel Tank Pressure Sensor	Inches H2O, mm Hg, volts	-17 to +7.5 Inches H2O	-17 to +7.5 Inches H2O	Engine Data 1, EVAP
Fuel Tank Pressure Sensor	0-5.0 volts	Varies	Varies	Engine Data 1, EVAP
Fuel Trim Cell	Cell 0-10	Varies	Varies	Engine Data 1, EVAP, Fuel Trim
Fuel Trim Learn	Enabled or Disabled	Enabled	May Toggle	Engine Data 1, EVAP, Fuel Trim
FC Relay 1 Circuit Status	Okay / Fault	Okay	Okay	Output Driver
Generator Lamp	On/Off	Off	Off	Engine Data 2
Generator L-Terminal	On/Off	On	On	Engine Data 2
HO2S-11 (B1 S1)	0-1132 mv	10-1000	10-1000	Engine Data 1, EVAP, Fuel Trim
HO2S-12 (B1 S2)	0-1132 mv	10-1000	10-1000	Engine Data 1, EVAP, Fuel Trim
HO2S-11 XCounts	0-255	Varies	Varies	Engine Data 1, HO2S
HO2S-11 Heater Command	On/Off	On	On	Fuel Trim
Hot Mode	On/Off	Off	Off	Misfire
IAC Position	0-255 counts	15	Varies	Engine Data 1, EGR, Fuel Trim
IAT Sensor (ºF)	-40º to 304ºF	50-194	50-194	Engine Data 1-2, EGR, EVAP, Fuel Trim
Ignition 1 Signal	0.0-25.5v	12.8-14.2	12.8-14.2	Engine Data 1-2, EGR, EVAP, Fuel Trim
Ignition Mode	Bypass/IC	IC	IC	Engine Data 2
IMS	P/R/N/D4-1	Park	Drive 4	Engine Data 1
Injector Pulsewidth	0-1000 ms	2.90	Varies	Engine Data 2, Fuel Trim, Misfire
Knock Retard (º)	0-25.5º	0	0	Engine Data 1, EGR
Long Term Fuel Trim	-27% to 27%	0 (± 5%)	0 (± 5%)	Engine Data 1-2, EVAP, Fuel Trim
Loop Status	Closed Loop / Open Loop	Closed Loop	Closed Loop	Engine Data 1-2, EGR, EVAP, Fuel Trim

1997-2005 Venture Light Van 3.4L VIN E - PID Data List

Parameter Identification (PID)	PID Value Range	PID Value at Hot Idle	PID Value at 30 mph	Data List Type
MAF Sensor (g/sec)	0-512 g/sec	3.41	Varies	Engine Data 1-2, EGR, EVAP, Fuel Trim, Misfire
MAF Sensor (Hz)	0-32,000 Hz	2210	Varies	Engine Data 1-2, EGR, EVAP, Fuel Trim, Misfire
MAP Sensor (kPa)	10-105 kPa	38	Varies	Engine Data 1-2, EGR, EVAP, Fuel Trim, Misfire
MAP Sensor (V)	0-5.1 volts	1.45	Varies	Engine Data 2
MIL Circuit Status	Okay/Fault	Okay	Okay	Output Driver
MIL Command	On/Off	Off	Off	Engine Data 2
Misfire Current Cylinder 1-6	0-198 counts	0	0	Misfire
Misfire History Cylinder 1-6	0-65,535	0	0	Misfire
Non-Volatile Memory	PASS/FAIL	PASS	PASS	Engine Data 1
Number of DTC(s)	0-255 counts	0	0	Engine Data 1, EVAP, Fuel Trim
HO2S Heater Current	0-10.0 amps	0.54	0.54	Fuel Trim
PCM Reset	Yes/No	No	No	Engine Data 1-2, EGR, EVAP, Fuel Trim, ODD
PCM/VCM in VTD Fuel Enable	Yes/No	No	No	Engine Data 1
Power Enrichment	Active/Inactive	Inactive	Inactive	Engine Data 2, Misfire
Reduced Engine Power	Active/Inactive	Inactive	Inactive	EGR
Short Term Fuel Trim	-27% to 27%	0 (± 5%)	0 (± 5%)	Engine Data 1-2, EVAP, Fuel Trim
Spark Advance (º)	-64º to 64º	20	Varies	Engine Data 1-2 Fuel Trim, Misfire
Startup ECT	-40 to 419ºF	Varies	Varies	Engine Data 2, EVAP, Fuel Trim
Startup IAT	-40 to 419ºF	80ºF	80ºF	Engine Data 2, EVAP, Fuel Trim
Starter Enable Relay Circuit Status	Okay / Fault/ Invalid State	Okay	Okay	Output Driver
Starter Relay Control	On/Off	Off	Off	Engine Data 2
TCC Brake Pedal Switch	Applied / Released	Released	Released	Engine Data 1-2
TCC PWM Solenoid Command	On/Off	Off	On	Engine Data 2, Misfire
TFP Switch Signal	P/R/N/D4-1	Park/Neutral	Drive 4	Engine Data 2, EGR, Fuel Trim
Torque Delivered Signal	0-100%	86	86	Engine Data 2
Torque Request Signal	0-100%	100	100	Engine Data 2
Total Misfire Current Count	0-65,535	Varies	Varies	Misfire
TP Angle (%)	0-100%	0	0-100	Engine Data 1-2, EGR, EVAP, Fuel Trim, Misfire
TP Sensor (V)	0-5.1 volts	0.20-0.74	Varies	Engine Data 1-2, EGR, EVAP, Fuel Trim, Misfire
Traction Control Status	Active or Inactive	Inactive	Inactive	Engine Data 2
Transmission Range	Park/Neutral, Drive 4-3-2-1	Park/Neutral	Drive 4	Engine Data 1, AIR, EGR, Catalyst
TR Switch P/A/B/C	High/Low	High	Low	Engine Data 1
TWC Protection	Active or Inactive	Inactive	Inactive	Engine Data 2, Misfire
Vehicle Speed	0-155 mph	0	30	Engine Data 1-2, EGR, EVAP, Fuel Trim, Misfire
VTD Auto Learn Timer	Active/Inactive	Inactive	Inactive	Engine Data 1
VTD Fuel Disable	Active/Inactive	Inactive	Inactive	Engine Data 1
VTD Fuel Disable Until Ignition Off	Yes/No	No	No	Engine Data 1

ACHIEVA PID DATA - OBD II

1996-98 Achieva 2.4L I4 MFI VIN T (A/T) - PID Data List

Parameter Identification (PID)	PID Value Range	PID Value at Hot Idle	PID Value at 30 mph	Data List Type
BARO Sensor (V)	0-5.0 volts	3.5-4.5	3.5-4.5	Engine Data 1
Calc. Airflow g/sec	0-512	3-7	Varies	Engine Data 1
CKP Active Counter	0-255 counts	Increments	Increments	Engine Data 1
CMP Active Counter	0-255 counts	Increments	Increments	Engine Data 1
CMP Resync Counter	0-255 counts	0	0	Engine Data 1
Desired Idle Speed	0-3187	PCM control	---	Engine Data 1
ECT Sensor (ºF)	-40-419ºF	185-239	185-239	Engine Data 1
Engine Load	0-100%	22-26	Varies	Engine Data 1
Engine Run Time	Hr: Min: Sec	00:00:00	00:00:00	Engine Data 1
Engine Speed	0-9999 rpm	±50 Desired	Varies	Engine Data 1
EVAP Purge Solenoid	0-100%	0	0-100	Engine Data 1
EVAP Vent Solenoid	On / Off	Off	Off	Engine Data 1
Fan Control Relay	On / Off	Off	Off	Engine Data 1
Fuel Trim Cell	Cell #0-22	18-21	0-21	Engine Data 1
Fuel Trim Index	0-255 counts	58-198	Varies	Engine Data 1
Fuel Trim Index	-10% to 10%	Varies	Varies	Engine Data 1
Generator F-Terminal	0-100%	0-100%	0-100%	Engine Data 1
Generator L-Terminal	Active / Inactive	Inactive	Inactive	Engine Data 1
HO2S-12 (B1 S2)	0-1132 mv	10-1000	10-1000	Engine Data 1
IAC Position	0-255 counts	5-60	---	Engine Data 1
IAT Sensor (ºF)	-40 to 304ºF	50-176	50-176	Engine Data 1
Ignition Voltage	0.0-25.5v	14.2	14.2	Engine Data 1
Knock Retard	0-128º	0	0	Engine Data 1
KS Noise Channel	Yes / No	No	No	Engine Data 1
KS Active Counts	0-255 counts	35-50	35-50	Engine Data 1
Low Octane Spark Modifier	0-90º	0	0	Engine Data 1
Long Term Fuel Trim	0-255 counts	58-198	58-198	Engine Data 1
Long Term Fuel Trim	-10% to 10%	Varies	Varies	Engine Data 1
Long Term Fuel Trim Average	-10% to 10%	Varies	Varies	Engine Data 1
Loop Status	Closed Loop / Open Loop	Closed Loop	Closed Loop	Engine Data 1
MAP Sensor (V)	0.0-5.0 volts	1.5	1-3	Engine Data 1
Medium Resolution Resync Counter	0-255 counts	0	0	Engine Data 1
Medium Resolution Engine Sync	Yes / No	Yes	Yes	Engine Data 1
Number of Current Trouble Codes	0-255 counts	0	0	Engine Data 1
O2S-11 (B1 S1)	0-1132 mv	10-1000	10-1000	Engine Data 1
Power Enrichment	Active / Inactive	Inactive	Inactive	Engine Data 1
Purge Learn Memory	0-100%	0%	0-100%	Engine Data 1
Short Term Fuel Trim	0-255 counts	58-198	Varies	Engine Data 1
Short Term Fuel Trim	-10% to 10%	Varies	Varies	Engine Data 1
Short Term F/T Avg.	-10% to 10%	Varies	Varies	Engine Data 1
Spark Advance	-64º-64º	Varies	Varies	Engine Data 1
TCC Brake Switch	On / Off	Off	Off	Engine Data 1
TP Angle (%)	0-100%	0%	0-100%	Engine Data 1
TP Sensor (V)	0.0-5.0 volts	0.20-0.90	---	Engine Data 1
Vehicle Speed	0-155 mph	0	30	Engine Data 1

1996-98 Achieva 2.4L I4 MFI VIN T (A/T) - PID Data List

Parameter Identification (PID)	PID Value Range	PID Value at Hot Idle	PID Value at 30 mph	Data List Type
A/C High Side (psi)	0-459 psi	139-399	139-399	Engine Data 2
A/C High Side (V)	0-5.0 volts	1-4	1-4	Engine Data 2
A/C Relay	On / Off	Off	Off	Engine Data 2
A/C Request	Yes / No	No	Off	Engine Data 2
Adaptive Knock retard	0-128º	0	0	Engine Data 2
Air Fuel Ratio	0.0-25.5:1	14.7:1	14.7:1	Engine Data 2
BARO Sensor (kPa)	10-110 kPa	65-105	65-105	Engine Data 2
Cruise Engaged	Yes / No	No	No	Engine Data 2
Desired Idle Speed	0-3187	PCM control	---	Engine Data 2
Engine Coolant Level	Okay / Low	Okay	Okay	Engine Data 2
ECT Sensor (ºF)	-40-419ºF	185-239	185-239	Engine Data 2
Engine Oil Level	Low / Okay	Okay	Okay	Engine Data 2
Engine Oil Pressure	Low / Okay	Okay	Okay	Engine Data 2
Engine Speed	0-9999 rpm	±50 Desired	---	Engine Data 2
IAT Sensor (ºF)	-40 to 304ºF	50-176	50-176	Engine Data 2
INJ Pulsewidth 1-4	0-985 ms	1-4	Varies	Engine Data 2
MAP Sensor (V)	0.0-5.0 volts	1.5	1-3	Engine Data 2
MPH / KPH	0-155 mph	0	30	Engine Data 2
Output Driver 1, 2 Open	00000000	00000000	00000000	Engine Data 2
Output Driver 1, 2 Short	00000000	00000000	00000000	Engine Data 2
Total Misfire Counts	0-255 counts	0	0	Engine Data 2
Stepper Cruise	Disabled / Enabled	Disabled	Disabled	Engine Data 2
TP Angle (%)	0-100%	0	Varies	Engine Data 2
TP Sensor (V)	0.0-5.0 volts	0.20-0.90	Varies	Engine Data 2
TR Range	P/N-R-O-3-2	P/N	3	Engine Data 2
TR Switch P/A/B/C	High / Low	Low	High	Engine Data 2

1996-98 Achieva 2.4L I4 MFI VIN T (A/T) - Misfire Data List

Parameter Identification (PID)	PID Value Range	PID Value at Hot Idle	PID Value at 30 mph	Data List Type
BARO Sensor (V)	0-5.0 volts	3.5-4.5	3.5-4.5	Misfire
Calculated Airflow	0-512 g/sec	Varies	Varies	Misfire
ECT Sensor (ºF)	-40-419ºF	185-239	185-239	Misfire
Engine Load	0-100%	22-26	Varies	Misfire
Engine Run Time	Hr: Min: Sec	00:00:00	00:00:00	Misfire
IAT Sensor (ºF)	-40 to 304ºF	50-176	50-176	Misfire
INJ Pulsewidth 1-4	0-985 ms	1-4	Varies	Misfire
MAP Sensor (V)	0.0-5.0 volts	1.5	1-3	Misfire
Miles Since First Fail	0-9999 rpm	Varies	Varies	Misfire
Miles Since Last Fail	0-9999 rpm	Varies	Varies	Misfire
Misfire current Cyl 1-4	0-255 counts	0	0	Misfire
Misfire History Cyl 1-4	0-255 counts	0	0	Misfire
TP Angle (%)	0-100%	0	Varies	Misfire
TP Sensor (V)	0.0-5.0 volts	0.20-0.90	Varies	Misfire
Total Misfire	0-255 counts	0	0	Misfire
Vehicle Speed	0-155 mph	0	30	Misfire

1996-98 Achieva 2.4L I4 MFI VIN T (A/T) - Specific EGR & EVAP

Parameter Identification (PID)	PID Value Range	PID Value at Hot Idle	PID Value at 30 mph	Data List Type
Actual EGR Position	0-100%	0	0-100	EGR, EVAP
BARO Sensor (V)	0-5.0 volts	3.5-4.5	3.5-4.5	EGR, EVAP
Catalyst Converter TT	32-1409°F	Varies	Varies	EGR, EVAP
Decel EWMA	-10 - 10 kPa	-3 to - 4	Varies	EGR, EVAP
Delta MAP Change	-20 - 20 kPa	0 / - number	0	EGR, EVAP
Desired EGR Position	0-100%	0	0-100	EGR, EVAP
Desired Idle Speed	0-3187	PCM control	---	EGR, EVAP
ECT Sensor (°F)	-40-419°F	185-239	185-239	EGR, EVAP
EGR Decel Filter	-10 - 10 kPa	-3 to - 4	Varies	EGR, EVAP
EGR Trip Samples	0-255 counts	Varies	Varies	EGR, EVAP
Engine Run Time	Hr: Min: Sec	00:00:00	00:00:00	EGR, EVAP
Engine Speed	0-9999 rpm	±50 Desired	Varies	EGR, EVAP
EVAP Solenoid	0-100%	0	0-100	EGR, EVAP
EVAP Vent Solenoid	0=off / 1=on	0	0	EGR, EVAP
Fuel Tank Level (V)	0-5.0 volts	Varies	Varies	EGR, EVAP
Fuel Tank Level (%)	0-100%	Varies	Varies	EGR, EVAP
FTP Sensor (V)	0-5.0 volts	2.5v: cap off	Varies	EGR, EVAP
Fuel Trim Cell	Cell #0-22	18-21	0-22	EGR, EVAP
HO2S-12 (B1 S2)	0-1132 mv	10-1000	10-1000	EGR, EVAP
IAC Position	0-255 counts	5-60	---	EGR, EVAP
IAT Sensor (°F)	-40 to 304°F	50-176	50-176	EGR, EVAP
Long Term Fuel Trim	-10% to 10%	Varies	Varies	EGR, EVAP
MAP Sensor (V)	0.0-5.0 volts	1.5	1-3	EGR, EVAP
No. Of Catalyst Tests	0-255 counts	Varies	Varies	EGR, EVAP
Number of Current Trouble Codes	0-255 counts	0	0	EGR, EVAP
O2S-11 (B1 S1)	0-1132 mv	10-1000	10-1000	EGR, EVAP
ODD 1-2 Open Circuit	00000000	00000000	00000000	EGR, EVAP
ODD 1-2 Short Circuit	00000000	00000000	00000000	EGR, EVAP
Short Term Fuel Trim	-10% to 10%	Varies	Varies	EGR, EVAP
TP Angle (%)	0-100%	0	Varies	EGR, EVAP
TP Sensor (V)	0.0-5.0 volts	0.20-0.90	Varies	EGR, EVAP

1996-98 Achieva 2.4L I4 MFI VIN T - Freeze Frame Data & Failure Records

Parameter Identification (PID)	PID Value Range	PID Value at Hot Idle	PID Value at 30 mph	Data List Type
Air Fuel Ratio	0.0-25.5:1	14.7:1	14.7:1	14.7:1
BARO Sensor (V)	0-5.0 volts	3.5-4.5	3.5-4.5	3.5-4.5
Calculated Airflow	0-512 g/sec	Varies	Varies	Varies
Desired Idle Speed	0-3187	PCM control	---	---
DTC # P0001- P1999	PXXXX	PXXXX	PXXXX	PXXXX
ECT Sensor (°F)	-40-419°F	185-239	185-239	185-239
INJ Pulsewidth 1-4	0-985 ms	1-4	Varies	Varies
Long, Short Term F/T	-10% to 10%	Varies	Varies	Varies
Engine Load	0-100%	22-26	Varies	Varies
Loop Status	Closed Loop / Open Loop	Closed Loop	Closed Loop	Closed Loop
MAP Sensor (V)	0.0-5.0 volts	1.5	1-3	1-3
MPH / KPH	0-155 mph	0	30	55
TP Angle (%)	0-100%	0%	0-100%	0-100%

1996-98 Achieva 2.4L I4 MFI VIN T (A/T) - Fuel Trim List

Parameter Identification (PID)	PID Value Range	PID Value at Hot Idle	PID Value at 30 mph	Data List Type
BARO Sensor (V)	0-5.0 volts	3.5-4.5	3.5-4.5	Fuel Trim
Calculated Airflow	0-512 g/sec	Varies	Varies	Fuel Trim
Desired Idle Speed	0-3187	PCM control	---	Fuel Trim
ECT Sensor (°F)	-40-419°F	185-239	185-239	Fuel Trim
Engine Load	0-100%	22-26	Varies	Fuel Trim
Engine Run Time	Hr: Min: Sec	00:00:00	00:00:00	Fuel Trim
Engine Speed	0-9999 rpm	±50 Desired	Varies	Fuel Trim
Fuel Trim Cell	Cell #0-21	18-21	0-22	Fuel Trim
Fuel Trim Index	0-255 counts	58-198	Varies	Fuel Trim
HO2S-12 (B1 S2)	0-1132 mv	10-1000	10-1000	Fuel Trim
IAT Sensor (°F)	-40 to 304°F	50-176	50-176	Fuel Trim
Lean / Rich Average	0-249ms	0	0	Fuel Trim
Lean / Rich Transition	0-255 counts	0	0	Fuel Trim
Lean-Rich to Rich-Lean Ratio HO2S-11	0:1 - 15.93:1	Varies	Varies	Fuel Trim
Long Term Fuel Trim	0-255 counts	58-198	Varies	Fuel Trim
Long Term Fuel Trim	-10% to 10%	Varies	Varies	Fuel Trim
Long Term Fuel Trim Average	0-255 counts	58-198	Varies	Fuel Trim
Loop Status	Closed Loop / Open Loop	Closed Loop	Closed Loop	Fuel Trim
MAP Sensor (V)	0.0-5.0 volts	1.5	1-3	Fuel Trim
O2S-11 (B1 S1)	0-1132 mv	10-1000	10-1000	Fuel Trim
Rich / Lean Average	0-249ms	0	0	Fuel Trim
Rich / Lean Transition	0-255 counts	0	0	Fuel Trim
Short Term Fuel Trim	0-255 counts	58-198	Varies	Fuel Trim
Short Term Fuel Trim	-10% to 10%	Varies	Varies	Fuel Trim
Short Term Fuel Trim Average	0-255 counts	58-198	Varies	Fuel Trim
TP Angle (%)	0-100%	0%	0-100%	Fuel Trim
TP Sensor (V)	0.0-5.0 volts	0.20-0.90	---	Fuel Trim

1996-98 Achieva 3.1L V6 MFI VIN M (A/T) - Engine Data List

Parameter Identification (PID)	PID Value Range	PID Value at Hot Idle	PID Value at 30 mph	Data List Type
3X Crank Sensor	0-9999 rpm	Varies	Varies	Engine Data 1
24X Crank Sensor	0-1280 rpm	Varies	Varies	Engine Data 1
Abuse Management	Active / Inactive	Inactive	Inactive	Engine Data 1
A/C High Side Pressure	0-459 psi	199-399	199-399	Engine Data 1
A/C High Side Pressure	0-5.0 volts	Varies	Varies	Engine Data 1
A/C Press. To High / LO	Yes / No	No	No	Engine Data 1
A/C Pressure Disable	Yes / No	No	No	Engine Data 1
A/C Clutch	Yes / No	No	No	Engine Data 1
A/C Request	Yes / No	No	No	Engine Data 1
A/C Off for WOT	Yes / No	No	No	Engine Data 1
A/C Slugging	Active / Inactive	Inactive	Inactive	Engine Data 1
Actual EGR Position	0-100%	0	0-100	Engine Data 1
Air Fuel Ratio	0.1-25.5:1	14.7:1	14.7:1	Engine Data 1
BARO Sensor (kPa)	10-110 kPa	65-110	65-110	Engine Data 1
CMP Signal Present	Yes / No	Yes	Yes	Engine Data 1
Commanded A/C	On / Off	Off	Off	Engine Data 1
Commanded Fan 1, 2	On / Off	Off	Off	Engine Data 1
Cruise Engaged	Engaged / Disengaged	Disabled	Disabled	Engine Data 1
Cruise Inhibit	Yes / No	Yes	Yes	Engine Data 1
Cruise Mode	On / Off	Off	On	Engine Data 1
Current Gear	1-2-3-4	1	3	Engine Data 1
Decel Fuel Mode	Active / Inactive	Inactive	Inactive	Engine Data 1
Desired EGR Position	0-100%	0	0-100	Engine Data 1
Desired Idle Speed	0-3187 rpm	PCM control	----	Engine Data 1
ECT Sensor (°F)	-40 to 304°F	185-220	185-220	Engine Data 1
Engine Load	0-100%	2-5	Varies	Engine Data 1
Engine Run Time	Hr: Min: Sec	00:00:00	00:00:00	Engine Data 1
Engine Speed	0-9999 rpm	±50 Desired	----	Engine Data 1
EGR Closed Pintle Position	0-5.0 volts	0.14-1.0v	----	Engine Data 1
EGR Duty Cycle	0-100%	0	0-100	Engine Data 1
EGR Feedback	0-5.0 volts	0.14-1.0v	----	Engine Data 1
EVAP Purge PWM	0-100%	0-25	0-100	Engine Data 1
EVAP Vacuum Switch	Purge/No	No Purge	No Purge	Engine Data 1
EVAP Canister Purge	0-100%	0-25%	0-50%	Engine Data 1
EVAP Vent Solenoid	Open/Close	Open	Open	Engine Data 1
Fan - Low Speed	On / Off	Off	Off	Engine Data 1
Fan - High Speed	On / Off	Off	Off	Engine Data 1
Fuel Pump Command	On / Off	On	On	Engine Data 1
Fuel Trim Cell	Cell #0-4	0	0-4	Engine Data 1
Fuel Trim Learn	Disabled / Enabled	Enabled	Enabled	Engine Data 1
Fuel Pump	On / Off	On	On	Engine Data 1
Generator Command	On / Off	On	On	Engine Data 1
Generator Lamp	On / Off	Off	Off	Engine Data 1
HO2S-11 Status	Not Ready/Ready	Ready	Ready	Engine Data 1
HO2S-11 (B1 S1)	0-1132 mv	10-1000	10-1000	Engine Data 1
HO2S-12 (B1 S2)	0-1132 mv	10-1000	10-1000	Engine Data 1

1996-98 Achieva 3.1L V6 MFI VIN M (A/T) - Engine Data List

Parameter Identification (PID)	PID Value Range	PID Value at Hot Idle	PID Value at 30 mph	Data List Type
IAC Position	0-255 counts	5-50	---	Engine Data 1
IAT Sensor (°F)	-40 to 304°F	50-194	50-194	Engine Data 1
IC Mode	Active / Inactive	Active	Active	Engine Data 1
Idle Speed Error	0-12800 rpm	< 100 rpm	---	Engine Data 1
Ignition Bypass	Active / Inactive	Inactive	Inactive	Engine Data 1
Ignition Mode	IC/Bypass	IC	IC	Engine Data 1
Injector Fault	Yes / No	No	No	Engine Data 1
Injector Pulsewidth	0-985 ms	1.5- 3.5	Varies	Engine Data 1
Injector Status	Okay / Stuck	Okay	Okay	Engine Data 1
Knock Retard	0.0-25.5%	0	0	Engine Data 1
KS Active Counter	0-255 counts	Varies	Varies	Engine Data 1
KS Noise Channel	0-5.0 volts	0.1-3.0	Varies	Engine Data 1
Long Term F/T Bank 1 (%)	Varies	0%	1%	Engine Data 1
Loop Status	Closed Loop / Open Loop	Closed Loop	Closed Loop	Engine Data 1
Low Oil Lamp	On / Off	Off	Off	Engine Data 1
MAF Sensor (g/sec)	0-512 g/sec	3-6	Varies	Engine Data 1
MAF Sensor (Hz)	0-31,999	1200-3000	Varies	Engine Data 1
MAP Sensor (kPa)	10-110 kPa	29-48	Varies	Engine Data 1
MIL Command	On / Off	Off	Off	Engine Data 1
Non-Volatile Memory	Pass / Fail	Pass	Pass	Engine Data 1
Power Enrichment	Active / Inactive	Active	Active	Engine Data 1
Spark Advance	-64° to 64°	20	Varies	Engine Data 1
TCC Brake Switch	Applied / Released	Released	Released	Engine Data 1
TCC Engaged	Yes / No	No	No	Engine Data 1
TP Angle (%)	0-100%	0	Varies	Engine Data 1
TP Sensor (V)	0-5.0 volts	0.20-0.74	Varies	Engine Data 1
Trans. Hot Mode	Active / Inactive	Inactive	Inactive	Engine Data 1
Transmission Range	X / O	O / X / X /O	Varies	Engine Data 1
Transmission Range	P/N-R-4-3-2	Park	3	Engine Data 1
TR Switch P-A-B-C	High / Low	LO/LO/High/High	---	Engine Data 1
TWC Protection	Active / Inactive	Inactive	Inactive	Engine Data 1
Vehicle Speed	0-155 mph	0	30	Engine Data 1
VTD Fuel Disable	Disabled / Enabled	Disabled	Disabled	Engine Data 1

1996-98 Achieva 3.1L V6 MFI VIN M (A/T) - EVAP System Data

Parameter Identification (PID)	PID Value Range	PID Value at Hot Idle	PID Value at 30 mph	Data List Type
Engine Run Time	Hr: Min: Sec	00:00:00	00:00:00	EVAP
Engine Speed	0-9999 rpm	±50 Desired	---	EVAP
EVAP Canister Purge	0-100%	0-25%	0-100%	EVAP
EVAP Vent Solenoid	Open/Close	Open	Open	EVAP
Fuel Tank VAC/Press	0-5.0 volts	2.5v: cap off	Varies	EVAP
IAT Sensor (°F)	-40 to 304°F	50-194	50-194	EVAP
MAF Sensor (g/sec)	0-512 g/sec	3-6	Varies	EVAP
MAP Sensor (kPa)	10-110 kPa	29-48	Varies	EVAP
MAP Sensor (V)	0-5.0 volts	1-2	Varies	EVAP
Startup ECT Sensor	-40 to 304°F	Varies	Varies	EVAP
Startup IAT Sensor	-40 to 304°F	Varies	Varies	EVAP

1996-98 Achieva 3.1L V6 MFI VIN M (A/T) - EGR System Data

Parameter Identification (PID)	PID Value Range	PID Value at Hot Idle	PID Value at 30 mph	Data List Type
Actual EGR Position	0-100%	0%	0-100%	EGR
BARO Sensor (kPa)	10-110 kPa	65-110	65-110	EGR
Current Gear	1-2-3-4	1	3	EGR
Decel Fuel Mode	Active / Inactive	Inactive	Inactive	EGR
Desired EGR Position	0-100%	0%	0-100%	EGR
Desired Idle Speed	0-3187 rpm	PCM control	---	EGR
ECT Sensor (°F)	-40 to 304°F	185-220	185-220	EGR
EGR Closed Pintle Position	0-5.0 volts	0.14-1.0v	---	EGR
EGR Duty Cycle	0-100%	0%	0-100%	EGR
EGR Feedback	0-5.0 volts	0.14-1.0v	---	EGR
EGR Flow Test Count	0-255 counts	0-10	0-10	EGR
EGR Position Error	0-100%	0-9	---	EGR
Engine Load	0-100%	2-5	Varies	EGR
Engine Run Time	Hr: Min: Sec	00:00:00	00:00:00	EGR
Engine Speed	0-9999 rpm	±50 Desired	---	EGR
Hot Open Loop	Active / Inactive	Inactive	Inactive	EGR
IAC Position	0-255 counts	10-40	---	EGR
IAT Sensor (°F)	-40 to 304°F	50-194	50-194	EGR
Ignition 1 Signal	0.0-25.5v	14.2	14.2	EGR
Knock Retard	0.0-25.5%	0	0	EGR
KS Active Counter	0-255 counts	Varies	Varies	EGR
Loop Status	Closed Loop / Open Loop	Closed Loop	Closed Loop	EGR
MAF Sensor (g/sec)	0-512 g/sec	3-6	Varies	EGR
MAP Sensor (kPa)	10-110 kPa	29-48	Varies	EGR
Power Enrichment	Active / Inactive	Active	Active	EGR
TCC Engaged	Yes / No	No	No	EGR
TP Angle (%)	0-100%	0%	Varies	EGR
Transmission Range	P/N-R-4-3-2	Park	3	EGR
TWC Protection	Active / Inactive	Inactive	Inactive	EGR
Vehicle Speed	0-155 mph	0	30	EGR

1996-98 Achieva 3.1L V6 MFI VIN M (A/T) - HO2S Data List

Parameter Identification (PID)	PID Value Range	PID Value at Hot Idle	PID Value at 30 mph	Data List Type
Air Fuel Ratio	0.1-25.5:1	14.7:1	14.7:1	HO2S
Decel Fuel Mode	Active / Inactive	Inactive	Inactive	HO2S
ECT Sensor (°F)	-40 to 304°F	185-220	185-220	HO2S
Engine Load	0-100%	2-5	Varies	HO2S
Engine Run Time	Hr: Min: Sec	00:00:00	00:00:00	HO2S
Engine Speed	0-9999 rpm	±50 Desired	---	HO2S
EVAP Purge PWM	0-100%	0-25%	0-100%	HO2S
Hot Open Loop	Active / Inactive	Inactive	Inactive	HO2S
HO2S-11 (B1 S1)	0-1132 mv	10-1000	10-1000	HO2S
HO2S-12 (B2 S1)	0-1132 mv	10-1000	10-1000	HO2S
HO2S (B1 S1) Status	Not Ready/Ready	Ready	Ready	HO2S
HO2S-11 Warmup	Min: Sec	00:00	00:00	HO2S
HO2S-12 Warmup	Min: Sec	00:00	00:00	HO2S

1996-98 Achieva 3.1L V6 MFI VIN M (A/T) - HO2S Data List

Parameter Identification (PID)	PID Value Range	PID Value at Hot Idle	PID Value at 30 mph	Data List Type
HO2S-11 XCounts	0-255 counts	Varies	Varies	HO2S
IAT Sensor (ºF)	-40 to 304ºF	50-194	50-194	HO2S
Ignition 1 Signal	0.0-25.5v	14.2	14.2	HO2S
Injector Pulsewidth	0-985 ms	1.5-3.5	Varies	HO2S
Loop Status	Closed Loop / Open Loop	Closed Loop	Closed Loop	HO2S
Long Term Fuel Trim Average	0-100%	Varies	Varies	HO2S
MAF Sensor (g/sec)	0-512 g/sec	3-6	Varies	HO2S
Power Enrichment	Active / Inactive	Inactive	Inactive	HO2S
Short Term F/T Avg.	0-100%	Varies	Varies	HO2S
Startup ECT Sensor	-40 to 304ºF	Varies	Varies	HO2S
Startup IAT Sensor	-40 to 304ºF	Varies	Varies	HO2S
TP Angle (%)	0-100%	0%	Varies	HO2S
TWC Protection	Active / Inactive	Inactive	Inactive	HO2S

1996-98 Achieva 3.1L V6 MFI VIN M (A/T) - Misfire Data List

Parameter Identification (PID)	PID Value Range	PID Value at Hot Idle	PID Value at 30 mph	Data List Type
3X Crank Sensor	0-9999 rpm	Varies	Varies	Varies
24X Crank Sensor	0-1280 rpm	Varies	Varies	Varies
Abuse Management	Active / Inactive	Inactive	Inactive	Misfire
Actual EGR Position	0-100%	0%	0-100%	Misfire
Air Active Test	Active / Inactive	Inactive	Inactive	Inactive
Cam Signal Since 3X	Yes / No	Yes	Yes	Misfire
Decel Fuel Mode	Active / Inactive	Inactive	Inactive	Misfire
Desired EGR Position	0-100%	0%	0-100%	Misfire
ECT Sensor (ºF)	-40 to 304ºF	185-220	185-220	Misfire
Engine Load	0-100%	2-5	Varies	Misfire
Engine Speed	0-9999 rpm	±50 Desired	---	Misfire
HO2S-11 XCounts	0-255 counts	Varies	Varies	Misfire
IAC Position	0-255 counts	10-40	---	Misfire
Injector Pulsewidth	0-985 ms	1.5-3.5	Varies	Misfire
Loop Status	Closed Loop / Open Loop	Closed Loop	Closed Loop	Misfire
MAF Sensor (g/sec)	0-512 g/sec	3-6	Varies	Misfire
Misfire current Cyl 1-6	0-198	0	0	Misfire
Misfire History Cyl 1-6	0-65535	0	0	Misfire
Misfiring Cylinder	CYL 1-6	0	0	Misfire
Non-Volatile Memory	Pass / Fail	Pass	Pass	Misfire
Power Enrichment	Active / Inactive	Inactive	Inactive	Misfire
Total Misfire Current	0-99 counts	0-5	0-5	Misfire
Misfire Fail Since 1st Fail	0-65535	0	0	Misfire
Misfire Pass Since 1st Fail	0-65535	0	0	Misfire
Spark Advance	-64º to 64º	20	Varies	Misfire
TCC Engaged	Yes / No	No	No	Misfire
TP Angle (%)	0-100%	0	Varies	Misfire
Trans. Hot Mode	Active / Inactive	Inactive	Inactive	Misfire
TWC Protection	Active / Inactive	Inactive	Inactive	Misfire

ALERO PID DATA

1999-2004 Alero 2.4L I4 MFI VIN T (A/T) - Engine Data List

Parameter Identification (PID)	PID Value Range	PID Value at Hot Idle	PID Value at 30 mph	Data List Type
A/C Relay Command	On / Off	Off	Off	Engine Data 1
Adaptive Knock retard	0-90°	0	0	Engine Data 1
BARO Sensor (kPa)	10-110 kPa	65-104	65-104	Engine Data 1
Calc. Airflow (g/sec)	0-512 g/sec	Increments	Increments	Engine Data 1
Calc. Vacuum (kPa)	0-80 kPa	Varies	Varies	Engine Data 1
CKP Active Counter	0-255 counts	Increments	Increments	Engine Data 1
CMP Active Counter	0-255 counts	Increments	Increments	Engine Data 1
CMP Resync Counter	0-255 counts	0	0	Engine Data 1
Cruise Control Active	Yes / No	No	No	Engine Data 1
Cruise Release Clutch	Applied/UN	Unapplied	Unapplied	Engine Data 1
Current Gear	1-6	6	3	Engine Data 1
Desired Idle Speed	0-3187 rpm	PCM control	---	Engine Data 1
ECT Sensor (°F)	-40 to 304°F	185-239	185-239	Engine Data 1
Engine Load	0-100%	11-21	Varies	Engine Data 1
Engine Run Time	Hr: Min: Sec	00:00:00	00:00:00	Engine Data 1
Engine Speed	0-9999 rpm	±99 Desired	---	Engine Data 1
EVAP Canister Purge	0-100%	0-100	0-100	Engine Data 1
EVAP Vent Solenoid	Venting/Non	Varies	Varies	Engine Data 1
F/P Relay Command	On / Off	On	On	Engine Data 1
Fuel Level Sensor	0-100%	0-100%	0-100%	Engine Data 1
FTP Sensor (In. H2O)	-17.5 to 7.5"	Varies	Varies	Engine Data 1
Fuel Trim Cell	Cell #0-22	18-21	0-22	Engine Data 1
Generator L-Terminal	Active / Inactive	Inactive	Inactive	Engine Data 1
Generator PWM	0-100%	0-100	0-100	Engine Data 1
HO2S-11 (B1 S1)	0-1132 mv	10-1000	10-1000	Engine Data 1
HO2S-12 (B1 S2)	0-1132 mv	10-1000	10-1000	Engine Data 1
IAC Position (Counts)	0-255 counts	5-60	---	Engine Data 1
IAT Sensor (°F)	-40 to 304°F	50-176	50-176	Engine Data 1
Ignition 1 Signal	0.0-25.5v	14.2	14.2	Engine Data 1
Knock Retard (°)	0-90°	0	0	Engine Data 1
KS Noise Channel (V)	0.0-5.0 volts	Varies	Varies	Engine Data 1
Long Term Fuel Trim	-10% to 10%	Varies	Varies	Engine Data 1
Loop Status	Closed Loop / Open Loop	Closed Loop	Closed Loop	Engine Data 1
MAP Sensor (kPa)	10-110 kPa	25-48	Varies	Engine Data 1
MAP Sensor (V)	0-5.0 volts	1.0-3.0	Varies	Engine Data 1
Medium Resolution Resync Counter	0-255 counts	0	0	Engine Data 1
Medium Resolution Sync	Yes / No	Yes	Yes	Engine Data 1
MIL Command	On / Off	Off	Off	Engine Data 1
Number of Trouble Codes	NUMBER	0	0	Engine Data 1
Power Enrichment	Active / Inactive	Inactive	Inactive	Engine Data 1
Short Term Fuel Trim	-10% to 10%	Varies	Varies	Engine Data 1
Spark Advance (°)	-64° to 64°	20	Varies	Engine Data 1
TCC Enable Solenoid	On / Off	On	On	Engine Data 1
TP Angle (%)	0-100%	0	Varies	Engine Data 1
TP Sensor (V)	0-5.0 volts	0.20-0.90	Varies	Engine Data 1
Vehicle Speed	0-155 mph	0	30	Engine Data 1

1999-2004 Alero 2.4L I4 MFI VIN T (A/T) - Engine Data List

Parameter Identification (PID)	PID Value Range	PID Value at Hot Idle	PID Value at 30 mph	Data List Type
A/C High Side Pressure	0-5.0 volts	Varies	Varies	Engine Data 2
A/C High Side Pressure	0-3100 kPa	Varies	Varies	Engine Data 2
A/C High Side Pressure	0-459 psi	Varies	Varies	Engine Data 2
A/C Relay Command	On / Off	Off	Off	Engine Data 2
A/C Request	Yes / No	No	No	Engine Data 2
Calculated Airflow	0-512 g/sec	Varies	Varies	Engine Data 2
Air Fuel Ratio	0.0-25:1	14.7:1	14.7:1	Engine Data 2
BARO Sensor (kPa)	10-110 kPa	65-104	65-104	Engine Data 2
CKP Active Counter	0-255 counts	Increments	Increments	Engine Data 2
CMP Active Counter	0-255 counts	Increments	Increments	Engine Data 2
CMP Resync Counter	0-255 counts	0	0	Engine Data 2
Cruise Control Active	Yes / No	No	No	Engine Data 2
Cruise Release Clutch	Applied/UN	Unapplied	Unapplied	Engine Data 2
Current Gear	1-6	6	3	Engine Data 2
Cruise Control Active	Engage/DIS	Disengaged	Disengaged	Engine Data 2
Desired Idle Speed	0-3187 rpm	PCM control	---	Engine Data 2
ECT Sensor (°F)	-40 to 304°F	185-239	185-239	Engine Data 2
Engine Coolant Low	Okay / Low	Okay	Okay	Engine Data 2
Engine Load	0-100%	11-21	Varies	Engine Data 2
Engine Oil Pressure	Low / Okay	Okay	Okay	Engine Data 2
Engine Run Time	Hr: Min: Sec	00:00:00	00:00:00	Engine Data 2
Engine Speed	0-9999 rpm	±99 Desired	---	Engine Data 2
FC Relay 1 & 2	On / Off	Off	Off	Engine Data 2
Generator F-Terminal	0-100%	0-100	0-100	Engine Data 2
Generator L-Terminal	Voltage/No	Voltage	Voltage	Engine Data 2
IAT Sensor (°F)	-40 to 304°F	50-176	50-176	Engine Data 2
Ignition 1 Signal	0.0-25.5v	14.2	14.2	Engine Data 2
Injector Pulsewidth	0-985 ms	3-4	Varies	Engine Data 2
Long Term Fuel Trim	-10% to 10%	Varies	Varies	Engine Data 2
Loop Status	Closed Loop / Open Loop	Closed Loop	Closed Loop	Engine Data 2
Medium Resolution Resync	0-255 counts	0	Varies	Engine Data 2
MAP Sensor (kPa)	10-110 kPa	25-48	Varies	Engine Data 2
MAP Sensor (V)	0-5.0 volts	1.0-3.0	Varies	Engine Data 2
Power Enrichment	Active / Inactive	Inactive	Inactive	Engine Data 2
Short Term Fuel Trim	-10% to 10%	Varies	Varies	Engine Data 2
Spark Advance (°)	-64° to 64°	20	Varies	Engine Data 2
Startup ECT Sensor	-40 to 304°F	Varies	Varies	Engine Data 2
TCC Brake Pedal Switch	Open/closed	Closed	Closed	Engine Data 2
TCC Enable Solenoid	On / Off	On	On	Engine Data 2
TP Angle (%)	0-100%	0	0-100	Engine Data 2
TP Sensor (V)	0-5.0 volts	0.20-0.90	---	Engine Data 2
Transmission Range	Several	P/N	D2	Engine Data 2
TR Switch P-A-B-C	Low / High	LO/LO/High/High	---	Engine Data 2
Vacuum Calculated	10-110 kPa	Varies	Varies	Engine Data 2
Vehicle Speed	0-155 mph	0	30	Engine Data 2

1999-2004 Alero 2.4L I4 MFI VIN T (A/T) - Engine Data List

Parameter Identification (PID)	PID Value Range	PID Value at Hot Idle	PID Value at 30 mph	Data List Type
A/C Relay Command	On / Off	Off	Off	Engine Data 3
Air Fuel Ratio	0.0-25:1	14.7:1	Varies	Engine Data 3
Calculated Airflow	0-512 g/sec	Varies	Varies	Engine Data 3
Cruise Control Active	Yes / No	No	No	Engine Data 3
Cruise Release Clutch	On / Off	Off	Off	Engine Data 3
Desired Idle Speed	0-3187 rpm	PCM control	---	Engine Data 3
ECT Sensor (°F)	-40 to 304°F	185-239	185-239	Engine Data 3
Engine Load	0-100%	11-21	Varies	Engine Data 3
Engine Oil Life	0-100%	Varies	Varies	Engine Data 3
Engine Oil Pressure	Low / Okay	Okay	Okay	Engine Data 3
Engine Run Time	Hr: Min: Sec	00:00:00	00:00:00	Engine Data 3
Engine Speed	0-9999 rpm	±99 Desired	---	Engine Data 3
Fan Control 1 Relay	On / Off	Varies	Varies	Engine Data 3
IAT Sensor (°F)	-40 to 304°F	50-176	50-176	Engine Data 3
Ignition 1 Signal	0.0-25.5v	14.2	14.2	Engine Data 3
Injector 1-4 Command	0-985 ms	3-4	Varies	Engine Data 3
Long Term Fuel Trim	-10% to 10%	Varies	Varies	Engine Data 3
Loop Status	Closed Loop / Open Loop	Closed Loop	Closed Loop	Engine Data 3
MAP Sensor (kPa)	10-110 kPa	25-48	Varies	Engine Data 3
MAP Sensor (V)	0-5.0 volts	1.0-3.0	Varies	Engine Data 3
Short Term Fuel Trim	-10% to 10%	Varies	Varies	Engine Data 3
Spark Advance (°)	-64° to 64°	20	Varies	Engine Data 3
TP Sensor (V)	0-5.0 volts	0.20-0.90	Varies	Engine Data 3
TR Switch P-A-B-C	Several	LO/LO/High/High	Several	Engine Data 3
VTD Fuel Disable	Active / Inactive	Inactive	Inactive	Engine Data 3
VTD Auto Learn Timer	Active / Inactive	Inactive	Inactive	Engine Data 3

1999-2004 Alero 2.4L I4 MFI VIN T (A/T) - EVAP Data List

Parameter Identification (PID)	PID Value Range	PID Value at Hot Idle	PID Value at 30 mph	Data List Type
BARO Sensor (kPa)	10-110 kPa	65-104	65-104	EVAP
Desired Idle Speed	0-3187 rpm	PCM control	---	EVAP
ECT Sensor (°F)	-40 to 304°F	185-239	185-239	EVAP
Engine Load	0-100%	11-21	Varies	EVAP
Engine Run Time	Hr: Min: Sec	00:00:00	00:00:00	EVAP
Engine Speed	0-9999 rpm	±99 Desired	---	EVAP
EVAP Canister Purge	0-100%	0-100	0-100	EVAP
EVAP Test Abort	Aborted/Not	Not Aborted	Not Aborted	EVAP
EVAP Test Result	Passed/No	No Result	No Result	EVAP
EVAP Test State	Completed / Running	Running	Running	EVAP
EVAP Vent Solenoid	Open/closed	Open	Open	EVAP
Fuel Level Sensor	0-100%	0-100	0-100	EVAP
FTP Sensor (In. H2O)	-17.5 to 7.5"	Varies	Varies	EVAP
FTP Sensor (V)	0-5.0 volts	Varies	Varies	EVAP
Fuel Trim Cell	Cell #0-22	18-21	0-22	EVAP
HO2S-11 (B1 S1)	0-1132 mv	10-1000	10-1000	EVAP
HO2S-12 (B1 S2)	0-1132 mv	10-1000	10-1000	EVAP

1999-2004 Alero 2.4L I4 MFI VIN T (A/T) - EVAP Data List

Parameter Identification (PID)	PID Value Range	PID Value at Hot Idle	PID Value at 30 mph	Data List Type
IAC Position	0-255 counts	5-60	---	EVAP
IAT Sensor (°F)	-40 to 304°F	50-176	50-176	EVAP
Ignition 1 Signal	0.0-25.5v	14.2	14.2	EVAP
Long Term Fuel Trim	-10% to 10%	Varies	Varies	EVAP
Loop Status	Closed Loop / Open Loop	Closed Loop	Closed Loop	EVAP
MAP Sensor (kPa)	10-110 kPa	25-48	Varies	EVAP
MAP Sensor (V)	0-5.0 volts	1.0-3.0	Varies	EVAP
Number of Trouble Codes	Number	0	0	EVAP
Purge Learn Memory	0-255 counts	1.00	Varies	EVAP
Short Term Fuel Trim	-10% to 10%	Varies	Varies	EVAP
Startup ECT Sensor	-40 to 304°F	Varies	Varies	EVAP
TP Angle (%)	0-100%	0	Varies	EVAP
TP Sensor (V)	0-5.0 volts	0.20-0.90	Varies	EVAP
Vehicle Speed	0-155 mph	0	30	EVAP

1999-2004 Alero 2.4L I4 MFI VIN T (A/T) - Fuel Trim Data List

Parameter Identification (PID)	PID Value Range	PID Value at Hot Idle	PID Value at 30 mph	Data List Type
Air flow Calculated	0-512 g/sec	Varies	Varies	Fuel Trim
Air Fuel Ratio	0.0-25:1	14.7:1	14.7:1	Fuel Trim
BARO Sensor (kPa)	10-110 kPa	65-104	65-104	Fuel Trim
Current Gear	1-6	6	3	Fuel Trim
ECT Sensor (°F)	-40 to 304°F	185-239	185-239	Fuel Trim
Engine Load	0-100%	11-21	Varies	Fuel Trim
Engine Run Time	Hr: Min: Sec	00:00:00	00:00:00	Fuel Trim
Engine Speed	0-9999 rpm	± 50 rpm of Desired rpm	---	Fuel Trim
EVAP Purge Solenoid	0-100%	Varies	Varies	Fuel Trim
EVAP Vent Solenoid	Venting/Non	Varies	Varies	Fuel Trim
Fuel Trim Cell	Cell #0-22	18-21	0-22	Fuel Trim
HO2S-11 (B1 S1)	0-1132 mv	10-1000	10-1000	Fuel Trim
HO2S-12 (B1 S2)	0-1132 mv	10-1000	10-1000	Fuel Trim
IAC Position (Counts)	0-255 counts	5-60	---	Fuel Trim
IAT Sensor (°F)	-40 to 304°F	50-176	50-176	Fuel Trim
Ignition 1 Signal	0.0-25.5v	14.2	14.2	Fuel Trim
Injector Pulsewidth	0-985 ms	3-4	Varies	Fuel Trim
Long Term Fuel Trim	-10% to 10%	Varies	Varies	Fuel Trim
Loop Status	Closed Loop / Open Loop	Closed Loop	Closed Loop	Fuel Trim
MAP Sensor (kPa)	10-110 kPa	25-48	Varies	Fuel Trim
Number of Trouble Codes	NUMBER	0	0	Fuel Trim
Short Term Fuel Trim	-10% to 10%	Varies	Varies	Fuel Trim
Spark Advance (°)	-64° to 64°	20	Varies	Fuel Trim
Startup ECT Sensor	-40 to 304°F	Varies	Varies	Fuel Trim
TP Angle (%)	0-100%	0	0-100	Fuel Trim
TR Switch P-A-B-C	Low / High	LO/LO/High/High	---	Fuel Trim
TWC Temp. (Calc.)	32-1409°F	Varies	Varies	Fuel Trim
TWC Test Desired	Yes / No	No	No	Fuel Trim
Vehicle Speed	0-155 mph	0	30	Fuel Trim

1999-2004 Alero 2.4L I4 MFI VIN T (A/T) - HO2S Data List

Parameter Identification (PID)	PID Value Range	PID Value at Hot Idle	PID Value at 30 mph	Data List Type
BARO Sensor (kPa)	10-110 kPa	65-104	65-104	HO2S
Calculated Airflow	0-512 g/sec	Varies	Varies	HO2S
Catalyst Monitor Tests	Number	Varies	Varies	HO2S
Current TWC Test	Yes / No	Yes	Yes	HO2S
Desired Idle Speed	0-3187 rpm	PCM control	---	HO2S
ECT Sensor (°F)	-40 to 304°F	185-239	185-239	HO2S
Engine Speed	0-9999 rpm	±99 Desired	---	HO2S
Fuel Trim Index	-10% to 10%	Varies	Varies	HO2S
HO2S (B1 S1)	0-1132 mv	10-1000	10-1000	HO2S
HO2S (B1 S2)	0-1132 mv	10-1000	10-1000	HO2S
IAT Sensor (°F)	-40 to 304°F	50-176	50-176	HO2S
L-R or R-L Transition	0-255 counts	0	Varies	HO2S
L-R or R-L Avg. Time	0-985 ms	Varies	Varies	HO2S
Long Term Fuel Trim	-10% to 10%	Varies	Varies	HO2S
MAP Sensor (kPa)	10-110 kPa	25-48	Varies	HO2S
Short Term Fuel Trim	-10% to 10%	Varies	Varies	HO2S
TP Sensor (V)	0-5.0 volts	0.20-0.90	Varies	HO2S
TWC Temp. (Calc.)	32-1409°F	Varies	Varies	HO2S
TWC Related DTC set	Yes / No	No	No	HO2S
TWC Stage 1-2 Start	Yes / No	Yes	Yes	HO2S
TWC Test Conditions	Yes / No	Yes	Yes	HO2S
TWC Test Desired	Yes / No	No	No	HO2S
TWC Test Enabled	Yes / No	No	No	HO2S
TWC Warmup Time	Yes / No	Yes	Yes	HO2S

1999-2004 Alero 2.4L I4 MFI VIN T (A/T) - Misfire Data List

Parameter Identification (PID)	PID Value Range	PID Value at Hot Idle	PID Value at 30 mph	Data List Type
A/C Relay Command	On / Off	Off	Off	Misfire
Calculated Airflow	0-512 g/sec	Varies	Varies	Misfire
BARO Sensor (kPa)	10-110 kPa	65-104	65-104	Misfire
Base PWM CYL 1-4	0-985 ms	Varies	Varies	Misfire
Calculated Airflow	0-512 g/sec	Varies	Varies	Misfire
Cycles Of Misfire Data	0-255 counts	0-99	0-99	Misfire
Desired Idle Speed	0-3187 rpm	PCM control	---	Misfire
ECT Sensor (°F)	-40 to 304°F	185-239	185-239	Misfire
Engine Load	0-100%	11-21	Varies	Misfire
Engine Speed	0-9999 rpm	±99 Desired	---	Misfire
IAT Sensor (°F)	-40 to 304°F	50-176	50-176	Misfire
Injector Pulsewidth	0-985 ms	3-4	Varies	Misfire
MAP Sensor (kPa)	10-110 kPa	25-48	Varies	Misfire
Misfire current Cyl 1-4	0-255 counts	0-255 counts	0-255 counts	Misfire
Misfire History Cyl 1-4	0-255 counts	0-255 counts	0-255 counts	Misfire
Power Enrichment	Active / Inactive	Inactive	Inactive	Misfire
Spark Advance (°)	-64° to 64°	20	Varies	Misfire
TCC Enable Solenoid	On / Off	On	On	Misfire
Total Misfire Current	0-255 counts	0-255 counts	0-255 counts	Misfire
TP Angle (%)	0-100%	0	Varies	Misfire

1999-2004 Alero 3.4L V6 MFI VIN E (A/T) - PID Data List

Parameter Identification (PID)	PID Value Range	PID Value at Hot Idle	PID Value at 30 mph	Data List Type
3X Crank Sensor	0-9999 rpm	Varies	Varies	Engine Data 2
24X Crank Sensor	0-1600 rpm	Varies	Varies	Engine Data 2
Abuse Management	Active / Inactive	Inactive	Inactive	Misfire
A/C High Side Pressure	0-5.0v	Varies	Varies	Engine Data 1
A/C Pressure Disable	Yes / No	No	No	Engine Data 2
A/C Pressure Disable	Yes / No	No	No	Engine Data 2
A/C Off for WOT	Yes / No	No	No	Engine Data 2
A/C Pressure Disable	Yes / No	No	No	Engine Data 2
A/C Relay Circuit Status	Okay / Fault	Okay	Okay	Output Driver
A/C Relay Command	On / Off	Off	Off	Engine Data 1-2 EGR, Misfire
A/C Request Signal	Yes / No	No	No	Engine Data 2
Actual EGR Position	0-100%	0	Varies	Engine Data 1, EGR, Misfire
Air Fuel Ratio	0.0-25.5:1	14.7:1	14.7:1	Engine Data 2, Fuel Trim
BARO Sensor (kPa)	10-110 kPa	98	98	Engine Data 1, EGR, EVAP, Fuel Trim
BARO Sensor (V)	0-5.0 volts	3.5-4.5	3.5-4.5	Engine Data 1, EGR, EVAP, Fuel Trim
Brake Switch Signal	Applied / Released	Released	Released	Engine Data 1
CMP Sensor Signal	Yes / No	Yes	Yes	Engine Data 1, Misfire
Crank Request Signal	Yes / No	No	No	Engine Data 2
Cruise Control Active	Yes / No	No	No	Engine Data 1
Cruise Inhibit Reason	Several	Varies	Varies	Engine Data 1
Cruised Inhibit Circuit Status	Okay / Fault/ Invalid State	Okay	Okay	Output Driver
Cruise Inhibit Signal Command	On / Off	On	On	Engine Data 1
Current Gear	1-2-3-4	1	4	Engine Data 1-2 EGR, Fuel Trim
Cycles of Misfire Data	0-99 counts	Varies	Varies	Misfire
Cylinder 1-6 Injector Circuit Status	Okay / Stuck Low / Stuck High / Fault	Okay	Okay	Output Driver
Cylinder 1-6 Injector Circuit History	Okay / Stuck Low / Stuck High / Fault	Okay	Okay	Output Driver
Decel Fuel Cutoff	Active / Inactive	Inactive	Inactive	EGR, Fuel Trim, Misfire
Desired Idle Speed	0-3187 rpm	PCM control	---	Engine Data 2
Desired EGR Position	0-100%	0	Varies	Engine Data 1-2 EGR, EVAP

1999-2004 Alero 3.4L V6 MFI VIN E (A/T) - PID Data List

Parameter Identification (PID)	PID Value Range	PID Value at Hot Idle	PID Value at 30 mph	Data List Type
Driver Module 1 Status	Enabled/ Disabled	Enabled	Enabled	Output Driver
Driver Module 2 Status	Enabled/ Disabled	Enabled	Enabled	Output Driver
Driver Module 3 Status	Enabled/ Disabled	Enabled	Enabled	Output Driver
Driver Module 4 Status	Enabled/ Disabled	Enabled	Enabled	Output Driver
ECT Sensor (°F)	-40 to 304°F	185-220	185-220	Engine Data 1-2 EGR Fuel Trim, EVAP, Misfire
EGR Closed Valve Pintle Position	0-5.1volts	0.14-1.0	Varies	EGR
EGR Solenoid Duty Cycle Command	0-100	0	Varies	Engine Data 1, EGR
EGR Feedback	0-5.1 volts	0.14-1.0	Varies	EGR
EGR Flow Test Count	0-255 counts	0	Varies	EGR
EGR Learned Position	0-5.1 volts	0.68	Varies	EGR
EGR Position Error	0-100%	0	Varies	EGR
EGR Pos. Variance	0-100%	0	Varies	EGR
Engine Load (%)	0-100%	17	Varies	Engine Data 1-2 EGR, Fuel Trim EVAP, Misfire
Engine Oil Level Switch	Okay/Low	Okay	Okay	Engine Data 2
Engine Oil Life Remaining	0-100%	Varies	Varies	Engine Data 2
EOP Switch Signal	Okay/Low/High	Okay	Okay	Engine Data 2
Engine Run Time	Hr: Min: Sec	00:00:00	00:00:00	Engine Data 1-2 EGR, Fuel Trim EVAP, Misfire
Engine Speed	0-9999 rpm	± 50 rpm of Desired Speed	Varies	Engine Data 1-2 EGR, Fuel Trim EVAP, Misfire
EVAP Fault History	No Fault / Excess VAC / Purge Valve Leak / Small Leak / Weak Vacuum	No Fault	No Fault	EVAP
EVAP Purge Solenoid Circuit Status	Okay / Fault	Okay	Okay	Output Driver
EVAP Purge Solenoid Command	0-100%	19	Varies	Engine Data 1, EVAP, HO2S
EVAP Vent Solenoid Control Circuit Status	Okay / Fault/ Invalid state	Okay	Okay	Output Driver
EVAP Vent Solenoid Command	Venting/Not Venting	Venting	Venting	Engine Data 1, EVAP, Fuel Trim

1999-2004 Alero 3.4L V6 MFI VIN E (A/T) - PID Data List

Parameter Identification (PID)	PID Value Range	PID Value at Hot Idle	PID Value at 30 mph	Data List Type
FC Relay 1 Circuit Status	Okay / Fault	Okay	Okay	Output Driver
FC Relay 1 Command	On / Off	Off	Off	Engine Data 2
FC Relay 2 and 3 Circuit Status	Okay / Fault	Okay	Okay	Output Driver
FC Relay 2 and 3 Command	On / Off	Off	Off	Engine Data 2
Fuel Level Sensor	0-5.0 volts	0.8-2.5	Varies	EVAP
Fuel Pump Command	On / Off	On	On	Engine Data 1
Fuel Pump Relay Circuit Status	Okay / Fault	Okay	Okay	Output Driver
Fuel Pump Relay Circuit History	Okay / Fault	Okay	Okay	Output Driver
Fuel Pump Relay Command	On / Off	On	On	Engine Data 1
Fuel Tank Level Remaining	0-100%	Varies	Varies	EVAP
Fuel Tank Pressure Sensor	Inches H2O, mm Hg, volts	-17.5 to +7.5 Inches H2O	-17.5 to +7.5 Inches H2O	Engine Data 1, EVAP
Fuel Tank Pressure Sensor	0-5.0 volts	Varies	Varies	Engine Data 1, EVAP
Fuel Trim Cell	Cell 0-10	Varies	Varies	Engine Data 1, EVAP, Fuel Trim
Fuel Trim Learn	Enabled or Disabled	Enabled	May Toggle	Engine Data 1, EVAP, Fuel Trim
Generator Lamp	On / Off	Off	Off	Engine Data 2
Generator L-Terminal	On / Off	On	On	Engine Data 2
HO2S-11 (B1 S1)	0-1132 mv	10-1000	10-1000	Engine Data 1, EVAP, Fuel Trim
HO2S-12 (B1 S2)	0-1132 mv	10-1000	10-1000	Engine Data 1, EVAP, Fuel Trim
HO2S-11 XCounts	0-255	Varies	Varies	Engine Data 1, HO2S
HO2S-11 Heater Command	On / Off	On	On	Fuel Trim
Hot Mode	On / Off	Off	Off	Misfire
IAC Position	0-255 counts	15	Varies	Engine Data 1, EGR, Fuel Trim
IAT Sensor (°F)	-40 to 304°F	50-194	50-194	Engine Data 1-2 EGR, EVAP, Fuel Trim
Ignition 1 Signal	0.0-25.5v	12.8-14.2	12.8-14.2	Engine Data 1-2 EGR, EVAP, Fuel Trim
Ignition Mode	Bypass/IC	IC	IC	Engine Data 2
Injector Pulsewidth	0-1000 ms	2.90	Varies	Engine Data 2, Fuel Trim, M/Fire
Internal Mode Switch	P/R/N/D4-1	Park	Drive 4	Engine Data 1
Knock Retard (°)	0-25.5°	0	0	Engine Data 1, EGR

1999-2004 Alero 3.4L V6 MFI VIN E (A/T) - PID Data List

Parameter Identification (PID)	PID Value Range	PID Value at Hot Idle	PID Value at 30 mph	Data List Type
Long Term Fuel Trim	-27% to 27%	0 (± 5%)	0 (± 5%)	Engine Data 1-2 EVAP, Fuel Trim
Loop Status	Closed Loop or Open Loop	Closed Loop	Closed Loop	Engine Data 1-2 EGR, EVAP, Fuel Trim
MAF Sensor (g/sec)	0-512 g/sec	4.25	Varies	Engine Data 1-2 EGR, EVAP, Fuel Trim, M/Fire
MAF Sensor (Hz)	0-32,000 Hz	2210	Varies	Engine Data 1-2 EGR, EVAP, Fuel Trim, M/Fire
MAP Sensor (kPa)	10-105 kPa	38	Varies	Engine Data 1-2 EGR, EVAP, Fuel Trim, M/Fire
MAP Sensor (V)	0-5.1 volts	1.45	Varies	Engine Data 2
MIL Circuit Status	Okay / Fault	Okay	Okay	Output Driver
MIL Command	On / Off	Off	Off	Engine Data 2
M/Fire Current Cyl 1-6	0-198 counts	0	0	Misfire
M/Fire History Cyl 1-6	0-65,535	0	0	Misfire
Non-Volatile Memory	Pass / Fail	Pass	Pass	Engine Data 1
Number of DTC(s)	0-255 counts	0	0	Engine Data 1, EVAP, Fuel Trim
HO2S Heater Current	0-10.0 amps	0.54	0.54	Fuel Trim
PCM Reset	Yes / No	No	No	Engine Data 1-2 EGR, EVAP, Fuel Trim, ODD
PCM/VCM in VTD Fuel Enable	Yes / No	No	No	Engine Data 1
Power Enrichment	Active / Inactive	Inactive	Inactive	CAT, EGR, HO2S, M/Fire
Short Term Fuel Trim	-27% to 27%	0 (± 5%)	0 (± 5%)	Engine Data 1-2 EVAP, Fuel Trim
Spark Advance (°)	-64° to 64°	20	Varies	Engine Data 1-2 Fuel Trim, M/Fire
Startup ECT	-40 to 419°F	Varies	Varies	Engine Data 2, EVAP, Fuel Trim
Startup IAT	-40 to 419°F	80°F	80°F	Engine Data 2, EVAP, Fuel Trim
Starter Enable Relay Circuit Status	Okay / Fault/ Invalid State	Okay	Okay	Output Driver
Starter Relay Control	On / Off	Off	Off	Engine Data 2
TCC Brake Pedal Switch	Applied / Released	Released	Released	Engine Data 1-2
TCC Solenoid Circuit Status	Okay / Fault/ Invalid State	Okay	Okay	Output Driver
TCC Solenoid Control	On / Off	Off	On	Engine Data 2, Misfire
TFP Switch Signal	P/R/N/D4-1	Park	Drive 4	Engine Data 2, EGR, Fuel Trim

1999-2004 Alero 3.4L V6 MFI VIN E (A/T) - PID Data List

Parameter Identification (PID)	PID Value Range	PID Value at Hot Idle	PID Value at 30 mph	Data List Type
Torque Delivered Signal	0-100%	86	86	Engine Data 2
Torque Requested Signal	0-100%	100	100	Engine Data 2
Total Misfire Current Count	0-65,535	Varies	Varies	Misfire
TP Angle (%)	0-100%	0	0-100	Engine Data 1-2 EGR, EVAP, Fuel Trim, M/Fire
TP Sensor (V)	0-5.1 volts	0.20-0.74	Varies	Engine Data 1-2 EGR, EVAP, Fuel Trim, M/Fire
Traction Control Status	Active / Inactive	Inactive	Inactive	Engine Data 2
Transmission Range	Park/Neutral, Drive 4-3-2-1	Park/Neutral	Drive 4	Engine Data 1, AIR, EGR, Catalyst
TR Switch P/A/B/C	High / Low	High	Low	Engine Data 1
TWC Protection	Active / Inactive	Inactive	Inactive	Engine Data 2, Misfire
Vehicle Speed	0-155 mph	0	30	Engine Data 1-2 EGR, EVAP, Fuel Trim, M/Fire
VTD Auto Learn Timer	Active / Inactive	Inactive	Inactive	Engine Data 1
VTD Fuel Disable	Active / Inactive	Inactive	Inactive	Engine Data 1
VTD Fuel Disable Until Ignition Off	Yes / No	No	No	Engine Data 1

AURORA PID DATA

2001-02 Aurora 3.5L V6 VIN H (A/T) - PID Data List

Parameter Identification (PID)	PID Value Range	PID Value at Hot Idle	PID Value at 30 mph	Data List Type
24X Crank Sensor	0-1600 rpm	Varies	Varies	Engine Data 2, Misfire
A/T 1-2 Solenoid Circuit Status	Okay / Fault	Okay	Okay	Output Driver
A/T 2-3 Solenoid Circuit Status	Okay / Fault	Okay	Okay	Output Driver
A/C High Side Pressure	0-450 psi	Varies	Varies	Engine Data 2
A/C High Side Pressure	0-5.1 volts	Varies	Varies	Engine Data 2
A/C Off for WOT	Yes / No	No	No	Engine Data 2
A/C Pressure Disable	Yes / No	No	No	Engine Data 2
A/C Relay Control Circuit Status	Okay / Fault	Okay	Okay	Output Driver
A/C Relay Command	On / Off	Off	Off	Engine Data 1-2 EGR, Misfire
A/C Request Signal	Yes / No	No	No	Engine Data 2
Air Fuel Ratio	0.0-25.5:1	14.2-14.7:1	14.2-14.7:1	Engine Data 2, Fuel Trim
Air Pump Relay Circuit Status	Okay / Fault	Okay	Okay	Output Driver
Air Pump Relay Command	On / Off	Off	Off	Engine Data 1, Fuel Trim
Air Pump Solenoid Circuit Status	Okay / Fault / Invalid State	Okay	Okay	Output Driver
Air Pump Solenoid Command	On / Off	Off	Off	Engine Data 1, Fuel Trim
BARO Sensor (kPa)	10-110 kPa	98	98	Engine Data 1, EGR, EVAP, Fuel Trim
CKP Sensor Status	Angle/Time A / Time B	Angle	Angle	Engine Data 2
CMP Sensor	0-9999 rpm	Varies	Varies	Engine Data 2, Misfire
CMP Sensor Signal	Yes / No	Yes	Yes	Engine Data 2, Misfire
Crank Request Signal	Yes / No	No	No	Engine Data 2
Cruise Control Active	Yes / No	No	No	Engine Data 1
Cruise Inhibit Reason	Engine Not Running, Runtime, VSS, Brake	Vehicle Speed	Vehicle Speed	Engine Data 1
Cruise Inhibit Command	On / Off	Off	Off	Engine Data 1
Cruise Inhibit Signal Circuit	Okay / Fault	Okay	Okay	Output Device
Current Gear	1-2-3-4	1	4	Engine Data 1, EGR, Fuel Trim
Cycles of Misfire Data	0-99	Varies	Varies	Misfire
Cylinder 1-6 Ignition Control Circuit Status	Okay / Fault	Okay	Okay	Output Driver

2001-02 Aurora 3.5L V6 VIN H (A/T) - PID Data List

Parameter Identification (PID)	PID Value Range	PID Value at Hot Idle	PID Value at 30 mph	Data List Type
Cylinder 1-6 Injector Circuit Status	Okay / Fault	Okay	Okay	Output Driver
Decel Fuel Cutoff	Active / Inactive	Inactive	Inactive	EGR, Fuel Trim
Desired EGR Position	0-100%	0	Varies	Engine Data 1, EGR, Misfire
Desired EGR Position	0-5.0 volts	0	Varies	EGR
Desired Idle Speed	0-3187 rpm	PCM control	---	Engine Data 1, Data 2, EVAP
Driver Module 1-4 Status	Enabled / Off/ High Volts / High Temperature	Enabled	Enabled	Output Driver
ECT Sensor (°F)	-40 to 304°F	185-220	185-220	Engine Data 1-2 EGR, Fuel Trim, EVAP, Misfire
EGR Flow Test Count	0-255 counts	0-10	Varies	EGR
EGR Learned Minimum Position	0-5.0 volts	Varies	Varies	EGR
EGR Position Sensor	0-100%	0	Varies	Engine Data 1, EGR, Misfire
EGR Position Sensor	0-5.0volts	0.0	Varies	EGR
EGR Position Error	-100 to 100%	Varies	Varies	EGR
EGR Solenoid Circuit Status	Okay / Fault	Okay	Okay	Output Driver
EGR Solenoid Command	0-100%	0	Varies	EGR
Engine Load (%)	0-100%	17	Varies	Engine Data 1-2 EGR, Fuel Trim, EVAP, Misfire
Engine Oil Level Switch	Okay / Not Okay	Okay	Okay	Engine Data 2
Engine Oil Life Left	0-100%	Varies	Varies	Engine Data 2
EOP Sensor	0-5.1 volts	2-3	2-4	Engine Data 2
Engine Run Time	Hr: Min: Sec	00:00:00	00:00:00	Engine Data 1-2 EGR, Fuel Trim, EVAP, Misfire
Engine Speed	0-9999 rpm	± 50 rpm of Desired Speed	Varies	Engine Data 1-2 EGR, Fuel Trim, EVAP, Misfire
EVAP Fault History	No Fault / Excess VAC / Purge Valve Leak / Small Leak / Weak Vacuum	No Fault	No Fault	EVAP
EVAP Purge Solenoid Control Circuit Status	Okay / Fault	Okay	Okay	Output Driver
EVAP Purge Solenoid Command	0-100%	18	Varies	Engine Data 1, EVAP, Fuel Trim

2001-02 Aurora 3.5L V6 VIN H (A/T) - PID Data List

Parameter Identification (PID)	PID Value Range	PID Value at Hot Idle	PID Value at 30 mph	Data List Type
EVAP Test Abort Reason	Not Aborted / Lost Enable / Small Leak / Vehicle Not at Rest	Not Aborted	Not Aborted	EVAP
EVAP Test Result	No Result / Aborted / Fail / Passed	Passed	Passed	EVAP
EVAP Vent Solenoid Circuit Status	Okay / Fault	Okay	Okay	Output Driver
EVAP Vent Solenoid	Venting / Not Venting	Venting	Venting	Engine Data 1, EVAP, Fuel Trim
Extended Travel Brake Pedal Switch	Applied / Released	Released	Released	Engine Data 2
Fan Control Relay 1 Circuit Status	Okay / Fault	Okay	Okay	Output Driver
Fan Control Relay 1 Command	On / Off	Off	Off	Engine Data 2
Fan Control Relay 2-3 Command	On / Off	Off	Off	Engine Data 2
Fan Control Relay 2-3 Circuit Status	Okay / Fault	Okay	Okay	Output Driver
Fuel Pump Relay Circuit Status	Okay / Fault	Okay	Okay	Output Driver
Fuel Pump Relay Command	On / Off	On	On	Engine Data 1
Fuel Level Remaining	0-100%	Varies	Varies	EVAP
Fuel Level Sensor	0-5.0 volts	Varies	Varies	EVAP
Fuel Tank Pressure Sensor	Inches H2O, mm Hg	-17.5 to +7.5 Inches H2O	-17.5 to +7.5 Inches H2O	Engine Data 1, EVAP
Fuel Tank Pressure Sensor	0-5.1 volts	Varies	Varies	Engine Data 1, EVAP
Fuel Trim Cell	Cell 0-9	0	1	Engine Data 1, EVAP, Fuel Trim
Fuel Trim Learn	Enabled or Disabled	Enabled	May Toggle	Engine Data 1, EVAP, Fuel Trim
Generator F-Terminal	0-100%	Varies	Varies	Engine Data 2
Generator L-Terminal	On / Off	On	On	Engine Data 2
HO2S-11 (B1 S1)	0-1132 mv	10-1000	10-1000	Engine Data 1, EVAP, Fuel Trim
HO2S-11 Heater	On / Off	Off	Off	Fuel Trim
HO2S-12 (B1 S2)	0-1132 mv	10-1000	10-1000	Engine Data 1, Fuel Trim
IAC Position	0-255 counts	10-40	10-40	Engine Data 1, EGR, Fuel Trim
IAT Sensor (°F)	-40 to 304°F	50-194	50-194	Engine Data 1-2 EGR, EVAP, Fuel Trim

2001-02 Aurora 3.5L V6 VIN H (A/T) - PID Data List

Parameter Identification (PID)	PID Value Range	PID Value at Hot Idle	PID Value at 30 mph	Data List Type
Ignition 1 Signal	0.0-25.5v	14.2	14.2	Engine Data 1-2 EGR, EVAP, Fuel Trim
Ignition Mode	Bypass/IC	IC	IC	Engine Data 2
Injector Pulsewidth	0-1000 ms	2.0-4.0	Varies	Engine Data 2, Fuel Trim, Misfire
Knock Retard (°)	0-25.5°	0°	0°	Engine Data 1, EGR
Long Term Fuel Trim	-10% to 10%	0 (± 5%)	0 (± 5%)	Engine Data 1-2 EVAP, Fuel Trim
Loop Status	Closed Loop / Open Loop	Closed Loop	Closed Loop	Engine Data 1-2 EGR, EVAP, Fuel Trim
MAF Sensor (g/sec)	0-512 g/sec	3-6	Varies	Engine Data 1-2 EGR, Fuel Trim, EVAP, Misfire
MAF Sensor (Hz)	0-32,000 Hz	1200-3000	Varies	Engine Data 2
MAP Sensor (kPa)	10-105 kPa	20-48	Varies	Engine Data 1-2 EGR, Fuel Trim, EVAP, Misfire
MAP Sensor (V)	0-5.1 volts	0.75-2.0	Varies	Engine Data 1-2
MIL Circuit Status	Okay / Fault	Okay	Okay	Output Driver
MIL Command	On / Off	Off	Off	Engine Data 2
Mileage Since DTC Cleared	Kilometer or Miles	Varies	Varies	Engine Data 2
Misfire Current Cylinder 1-6	0-198 counts	0-4	0-4	Misfire
Misfire History Cylinder 1-6	0-65,535	0	0	Misfire
Number of DTC(s)	Number	0	0	Engine Data 1, EVAP, Fuel Trim
O2S Heater Current	0-10.0 amps	0.50-0.60	0	Fuel Trim
PCM/VCM in VTD Fail Enable	Yes / No	Yes	Yes	Engine Data 1
PCM/VCM in VTD Fail Enable	Yes / No	Yes	Yes	Engine Data 1
Power Enrichment	Active / Inactive	Inactive	Inactive	Engine Data 2, Misfire
Reduced Engine Power	Active / Inactive	Inactive	Inactive	Engine Data 1, EGR
Short Term Fuel Trim	-10% to 10%	0 (± 5%)	0 (± 5%)	Engine Data 1-2 EVAP, Fuel Trim
Spark Advance (°)	-64° to 64°	20-40	Varies	Engine Data 1-2 Fuel Trim, Misfire
Startup ECT	-40 to 419°F	Varies	Varies	EVAP, Fuel Trim
Startup IAT	-40 to 419°F	80°F	80°F	EVAP, Fuel Trim

2001-02 Aurora 3.5L V6 VIN H (A/T) - PID Data List

Parameter Identification (PID)	PID Value Range	PID Value at Hot Idle	PID Value at 30 mph	Data List Type
Starter Enable Relay Control Circuit Status	Okay / Fault	Okay	Okay	Output Driver
Starter Relay Command	On / Off	Off	Off	Engine Data 2
Tachometer Circuit Status	Okay / Fault	Okay	Okay	Output Driver
TCC Brake Pedal Switch	Applied / Released	Released	Released	Engine Data 1-2
TCC Enable Circuit Status	On / Off	Off	On	Output Driver
TCC Enable Solenoid Command	On / Off	Off	On	Engine Data 1, Misfire
TCC PWM Solenoid Command	On / Off	Off	On	Engine Data 2
TFP Switch	Park/ Neutral / Drive 4, 3, 2 and 1	Park	Drive 4	Engine Data 2, EGR, Fuel Trim
Torque Delivered	0-100%	86	Varies	Engine Data 2
Torque Request	0-100%	100%	Varies	Engine Data 2
TP Angle (%)	0-100%	0	Varies	Engine Data 1-2 EGR, Fuel Trim, EVAP, Misfire
TP Sensor (V)	0-5.1 volts	0.5-1.2	Varies	Engine Data 1-2
Traction Control Circuit History	Okay / Fault	Okay	Okay	Output Driver
Traction Control Circuit Status	Okay / Fault	Okay	Okay	Output Driver
Traction Control Circuit Status	Active / Inactive	Inactive	Inactive	Engine Data 2
TWC Temperature Calculated	32-1409°F	Varies	Varies	Fuel Trim
Vehicle Speed	0-155 mph	0	30	Engine Data 1-2 EGR, Fuel Trim, EVAP, Misfire
VTD Auto Learn timer	Active / Inactive	Inactive	Inactive	Engine Data 1
VTD Fuel Disable	Active / Inactive	Inactive	Inactive	Engine Data 1
VTD Fuel Disable Until Key Off	Yes / No	No	No	Engine Data 1

1996-99 Aurora 4.0L V8 MFI VIN C (A/T) - PID Data List

Parameter Identification (PID)	PID Value Range	PID Value at Hot Idle	PID Value at 30 mph	Data List Type
Actual EGR Position	0-100%	0%	0-100%	Engine Data 1
BARO Sensor (kPa)	10-110 kPa	85-103	85-103	Engine Data 1
Commanded EGR	0-100%	0	0-5-	Engine Data 1
Desired Idle Speed	0-2040 rpm	720	---	Engine Data 1
ECT Sensor (°F)	-40-419°F	185-226	185-226	Engine Data 1
EGR Closed Pintle Position	0.0-5.0 volts	0.65-0.80	---	Engine Data 1
EGR Pintle Position	0.0-5.0 volts	0.65-0.80	---	Engine Data 1
Engine Load	0-100%	1-5	---	Engine Data 1
Engine Run Time	Hr: Min: Sec	00:00:00	00:00:00	Engine Data 1
Engine Speed	0-9999 rpm	±50 Desired	---	Engine Data 1
Extended Brake Travel Switch	Applied / Released	Released	Released	Engine Data 1
FC Relay 1, 2, 3	Off or On	Off	Off	Engine Data 1
Fuel Pump Relay	On / Off	On	On	Engine Data 1
Fuel Trim Cell	1-16	16	4	Engine Data 1
Fuel Trim Learn	Disabled / Enabled	Enabled	Enabled	Engine Data 1
Generator PWM	0-100%	0-100	0-100	Engine Data 1
High Electrical Load	Yes / No	No	No	Engine Data 1
HO2S (B1 S1, B1 S2)	0-1132 mv	10-1000	10-1000	Engine Data 1
HO2S-11 (B1 S3)	0-1132 mv	10-1000	10-1000	Engine Data 1
HO2S-21 (B2 S1)	0-1132 mv	10-1000	10-1000	Engine Data 1
IAC Position	0-255 counts	30-80	---	Engine Data 1
IAT Sensor (°F)	-40 to 304°F	50-194	50-194	Engine Data 1
I/C Mode	ICM / PCM	PCM	PCM	Engine Data 1
Ignition 0, Ignition 1	0.0-25.5v	14.2	14.2	Engine Data 1
Injector Pulsewidth	10-1000 ms	3.75	4.2	Engine Data 1
Knock Retard	0° to 45°	0	0	Engine Data 1
Knock Too Long	Yes / No	No	No	Engine Data 1
KS Active Counter	0-255 counts	0	0	Engine Data 1
Loop Status	Closed Loop / Open Loop	Closed Loop	Closed Loop	Engine Data 1
Long Term Fuel Trim	-10% to 10%	Varies	Varies	Engine Data 1
MAF Sensor (g/sec)	0-512 g/sec	5-7	12-20	Engine Data 1
MAP Sensor (kPa)	10-110 kPa	30-50	Varies	Engine Data 1
MIL Command	On / Off	Off	Off	Engine Data 1
P/NP Switch	P/N-R-OD-21	P/N	R-OD-21	Engine Data 1
PSP Switch	High/Normal	Normal	Normal	Engine Data 1
HO2S-11 (B1 S1) R/L	L or R	L-R-L-R	L-R-L-R	Engine Data 1
HO2S-13 (B1 S3) R/L	L or R	L-R-L-R	L-R-L-R	Engine Data 1
HO2S-21 (B1 S1) R/L	L or R	L-R-L-R	L-R-L-R	Engine Data 1
Short Term Fuel Trim	-10% to 10%	Varies	Varies	Engine Data 1
Spark Advance (°)	-90° to 90°	Varies	Varies	Engine Data 1
TCC Brake Switch	Applied / Released	Released	Released	Engine Data 1
TCC Slip Speed rpm	-4096-4096	+15 - +50	---	Engine Data 1
Throttle At Idle	Yes / No	Yes	No	Engine Data 1
TP Normalized	0-100%	0	6	Engine Data 1
TP Normalized	0.0-81.6°	0	6	Engine Data 1
TP Sensor	0-5.0 volts	0.55	Varies	Engine Data 1
Vehicle Speed	0-155 mph	0	30	Engine Data 1

1996-99 Aurora 4.0L V8 MFI VIN C (A/T) - PID Data List

Parameter Identification (PID)	PID Value Range	PID Value at Hot Idle	PID Value at 30 mph	Data List Type
4X Ref. Pulse Counter	0-255 counts	8	8	8
A/C Clutch	Yes / No	No	No	Engine Data 2
Air Fuel Ratio	0.1-25.5:1	14.5-14.9:1	Varies	Engine Data 2
Current Gear	1-2-3-4	1	3	Engine Data 2
Desired Idle Speed	0-2040 rpm	720	---	Engine Data 2
Desired Torque ABS/TCS	0-100%	100%	100%	Engine Data 2
ECT Sensor (°F)	-40-419°F	185-226	185-226	Engine Data 2
ECT At Start Up °F	-40-419°F	Varies	Varies	Engine Data 2
Engine Load	0-100%	2-5	---	Engine Data 2
Engine Run Time	Hr: Min: Sec	00:00:00	00:00:00	Engine Data 2
Engine Speed	0-9999 rpm	±50 Desired	---	Engine Data 2
Engine Oil Level	Okay / Low	Okay	Okay	Engine Data 2
Engine Oil Life	0-100%	Varies	Varies	Engine Data 2
EVAP Canister Purge	0-100%	10-20	Varies	Engine Data 2
HO2S-11 (B1 S1)	0-1132 mv	10-1000	10-1000	Engine Data 2
HO2S-12 (B1 S2)	0-1132 mv	10-1000	10-1000	Engine Data 2
IAC Position	0-255 counts	30 - 80	---	Engine Data 2
I/C Mode	ICM / PCM	PCM	PCM	Engine Data 2
Ignition Cycle Counter	0-255 counts	Varies	Varies	Engine Data 2
Ignition 1 Signal	0.0-25.5v	14.2	14.2	Engine Data 2
Loop Status	Closed Loop / Open Loop	Closed Loop	Closed Loop	Engine Data 2
MAF Sensor (g/sec)	0-512 g/sec	5-7	---	Engine Data 2
MAP Sensor (kPa)	10-110 kPa	30 - 50	---	Engine Data 2
MAP Sensor (V)	0-5.0 volts	1 – 2v	---	Engine Data 2
Non-Driven W/Speed	0-155 mph	0	30	Engine Data 2
Outside Air Temp.	-40-215°	Varies	Varies	Engine Data 2
P/NP Switch	P/N-R-OD-21	P/N	R-OD-21	Engine Data 2
Power Enrichment	Active / Inactive	Inactive	Inactive	Engine Data 2
Reference Volts Low	Volts	0.00v ±0.05	0.00v ±0.05	Engine Data 2
Service Spark Retard	0-4°	0°	Varies	Engine Data 2
Shift Solenoid No. 1	On / Off	Off	Off	Engine Data 2
Shift Solenoid No. 2	On / Off	Off	On	Engine Data 2
TCC Brake Switch	Applied / Released	Released	Released	Engine Data 2
TCC Duty Cycle	0-100%	0%	---	Engine Data 2
TCC Enable	Yes / No	No	Yes	Engine Data 2
TCC Slip Speed rpm	-4096-4096	+15 to +50	---	Engine Data 2
TCM Calibration No.	0-9999 rpm	Varies	Varies	Engine Data 2
TCM Software No.	0-9999 rpm	Varies	Varies	Engine Data 2
TFT Sensor (°F)	-40-419°F	Varies	Varies	Engine Data 2
TFP Switch A/B/C	On / Off	Off/On / Off	---	Engine Data 2
TP Angle (%)	0-100%	0	---	Engine Data 2
TP Sensor	0-5.0 volts	0.55v	Varies	Engine Data 2
Traction Control	Active / Inactive	Inactive	Inactive	Engine Data 2
Traction Control Torque	0-100%	8 - 18	---	Engine Data 2
Trans. Oil Life	0-100%	Varies	Varies	Varies
Vehicle Speed	0-155 mph	0	30	55

1996-99 Aurora 4.0L V8 MFI VIN C (A/T) - PID Data List

Parameter Identification (PID)	PID Value Range	PID Value at Hot Idle	PID Value at 30 mph	Data List Type
A/C Clutch	On / Off	Off	Off	Engine Data 3
A/C Request	Yes / No	No	No	Engine Data 3
A/C Relay Driver	Several	Okay	Okay	Engine Data 3
A/C Low Side Temp.	-40-215°F	Varies	Varies	Engine Data 3
A/C High Side Temp.	9-455°F	Varies	Varies	Engine Data 3
A/C High Side Pressure	Normal/High	Normal	Normal	Engine Data 3
Current Octane Level	87-93	93	93	Engine Data 3
Delivered Torque	Several	Okay	Okay	Engine Data 3
ECL	Okay / Low	Okay	Okay	Engine Data 3
ECT Sensor (°F)	-40 to 304°F	185-226	185-226	Engine Data 3
Engine Speed	0-9999 rpm	±50 Desired	---	Engine Data 3
EVAP Canister Purge	0-100%	10 - 20	Varies	Engine Data 3
EVAP Purge Solenoid	Several	Okay	Okay	Engine Data 3
EVAP Vent Solenoid	Open/Close	Open	Open	Engine Data 3
EVAP Vent Solenoid	Several	Okay	Okay	Engine Data 3
FC Relay 1, 2, 3	Off or On	Off	Off	Engine Data 3
Generator L-Terminal	No output/ok	Okay	Okay	Engine Data 3
Generator PWM	0-100%	0-100	0-100	Engine Data 3
High Electrical Load	Yes / No	No	No	Engine Data 3
HO2S (B1 S1, B1 S2)	0-1132 mv	10-1000	10-1000	Engine Data 3
IAC Position	0-255 counts	30 - 80	---	Engine Data 3
IAT Sensor (°F)	-40 to 304°F	50-194	50-194	Engine Data 3
Ignition 0, Ignition 1	0.0-25.5v	14.2	14.2	Engine Data 3
Injector 1-8 Fault	Yes / No	No	No	Engine Data 3
Loop Status	Closed Loop / Open Loop	Closed Loop	Closed Loop	Engine Data 3
MAF Sensor (g/sec)	0-512 g/sec	5-7	---	Engine Data 3
MAF Frequency (Hz)	0-32000 Hz	2800	3500	Engine Data 3
MAP Sensor (kPa)	10-110 kPa	30-50	49	Engine Data 3
MIL Driver	Okay / Fault 1-3	Okay	Okay	Engine Data 3
Odometer Miles	0-1677722	Number	Number	Engine Data 3
Outside Air Temp.	-40-215°	Varies	Varies	Engine Data 3
Pass Key Fuel	Disabled / Enabled	Enabled	Enabled	Engine Data 3
Pass Key Enabled	Yes / No	Yes	Yes	Engine Data 3
Pass Key Starter	Disabled / Enabled	Disabled	Disabled	Engine Data 3
P/NP Switch	P/N-R-OD-21	P/N	R-OD-21	Engine Data 3
Power Enrichment	Active / Inactive	Inactive	Inactive	Engine Data 3
Quad Driver No. 1	Okay / Fault	Okay	Okay	Engine Data 3
Quad Driver 1-2 GND	Okay / Fault	Okay	Okay	Engine Data 3
Shift Mode Switch	Normal / Performance	Normal	Normal	Engine Data 3
Shift Solenoid 1, 2	Okay / Fault 1-3	Okay	Okay	Engine Data 3
Shift Solenoid 1, 2	On / Off	Off	Off	Engine Data 3
TCC Brake Switch	Applied / Released	Released	Released	Engine Data 3
TCC Duty Cycle	0-100%	0	---	Engine Data 3
TCC Enable	Yes / No	No	Yes	Engine Data 3
TCC Solenoid Driver	Okay / Fault 1-3	Okay	Okay	Engine Data 3
TFP Switch	Park/Neutral	Park/Neutral	Park/Neutral	Engine Data 3
TP Sensor	0-5.0 volts	0.55v	---	Engine Data 3
T/Control Torque	0-100%	8 - 18	---	Engine Data 3

1996-99 Aurora 4.0L V8 MFI VIN C (A/T) - PID Data List

Parameter Identification (PID)	PID Value Range	PID Value at Hot Idle	PID Value at 30 mph	Data List Type
Air Fuel Ratio	0.1-25.5:1	14.7:1	14.7:1	Catalyst
BARO Sensor (kPa)	10-110 kPa	85-103	85-103	Catalyst
Calculated Airflow	0-511.99	Varies	Varies	Catalyst
Calc. TWC Temp.	32-1409°F	Varies	Varies	Catalyst
Catalyst EWMA	0-255 counts	Varies	Varies	Catalyst
Catalyst Test Fail	-25-25.5 sec	0.05	0.05	Catalyst
No. Of Catalyst Tests	0-255 counts	Varies	Varies	Catalyst
Desired Idle Speed	0-2040 rpm	720	---	Catalyst
ECT Sensor (°F)	-40 to 304°F	185-226	185-226	Catalyst
Engine Load	0-100%	1-5	Varies	Catalyst
Engine Run Time	Hr: Min: Sec	00:00:00	00:00:00	Catalyst
Engine Speed	0-9999 rpm	±50 Desired	---	Catalyst
HO2S-11 (B1 S1)	0-1132 mv	10-1000	10-1000	Catalyst
HO2S-12 (B1 S2)	0-1132 mv	10-1000	10-1000	Catalyst
HO2S-13 (B1 S3)	0-1132 mv	10-1000	10-1000	Catalyst
HO2S XCounts B1-B2	0-255 counts	Varies	Varies	Catalyst
IAC Position	0-255 counts	30 - 80	---	Catalyst
IAT Sensor (°F)	-40 to 304°F	50-194	50-194	Catalyst
Injector Pulsewidth	10-1000 ms	3 - 3.75	---	Catalyst
Loop Status	Closed Loop / Open Loop	Closed Loop	Closed Loop	Catalyst
Long Term Fuel Trim	-10% to 10%	0 (± 5%)	0 (± 5%)	Catalyst
MAF Sensor (g/sec)	0-512 g/sec	5-7	---	Catalyst
MAP Sensor (kPa)	65-104 kPa	30-50	---	Catalyst
OSC Test Complete	Yes / No	Varies	No	Catalyst
Short Term Fuel Trim	-10% to 10%	0 (± 5%)	0 (± 5%)	Catalyst
TP Angle (%)	0-100%	0	5	Catalyst
TP Sensor (V)	0-5.0 volts	0.55v	---	Catalyst
TWC Stage 1-2 Start	Yes / No	Varies	No	Catalyst
TWC Test Completed	Yes / No	Varies	No	Catalyst
TWC Test Enabled	Yes / No	Varies	No	Catalyst
TWC Test Started	Yes / No	Varies	No	Catalyst
TWC Test Time Diff.	-25-25.5 sec	---	---	Catalyst

1996-99 Aurora 4.0L V8 MFI VIN C (A/T) - EVAP Data List

Parameter Identification (PID)	PID Value Range	PID Value at Hot Idle	PID Value at 30 mph	Data List Type
BARO Sensor (kPa)	10-110 kPa	85-103	85-103	EVAP
Calculated ECT°F	-40 to 304°F	Varies	Varies	EVAP
Desired Idle Speed	0-2040 rpm	720	---	EVAP
ECT Sensor (°F)	-40 to 304°F	185-226	185-226	EVAP
Engine Load	0-100%	1-5	---	EVAP
Engine Run Time	Hr: Min: Sec	00:00:00	00:00:00	EVAP
Engine Speed	0-9999 rpm	±50 Desired	---	EVAP
EVAP Canister Purge	0-100%	10-20%	Varies	EVAP
EVAP Tank Vacuum Decay Slope	0-2 H2O" per sec	> 0.10	> 0.10	EVAP
EVAP Test Result	Pass / Fail	Pass	Pass	EVAP
EVAP Test State	Wait / Done	Done	Done	EVAP

1996-99 Aurora 4.0L V8 MFI VIN C (A/T) - PID Data List

Parameter Identification (PID)	PID Value Range	PID Value at Hot Idle	PID Value at 30 mph	Data List Type
EVAP Test Abort	Several	Varies	Varies	EVAP
EVAP Vent Solenoid	Open/Close	Open	Open	EVAP
Fuel Control State B1	Normal/Fault	Normal	Normal	EVAP
Fuel Control State B2	Normal/Fault	Normal	Normal	EVAP
FTP Sensor	0-2 H2O"sec	Varies	Varies	EVAP
Fuel Tank Pressure	0-5.1v	Varies	Varies	EVAP
Fuel Tank Vapor (In.)	0-2 H2O"sec	Varies	Varies	EVAP
Fuel Tank Vapor (V)	0-5.1v	Varies	Varies	EVAP
HO2S (B1 S1, B1 S2)	0-1132 mv	10-1000	10-1000	EVAP
IAT Sensor (°F)	-40 to 304°F	50-194	50-194	EVAP
Injector Pulsewidth	10-1000 ms	3.75	4.2	EVAP
Loop Status	Closed Loop / Open Loop	Closed Loop	Closed Loop	EVAP
Long Term Fuel Trim	-10% to 10%	0 (± 5%)	0 (± 5%)	EVAP
MAF Sensor (g/sec)	0-512 g/sec	5-7	---	EVAP
MAP Sensor (kPa)	10-110 kPa	30-50	---	EVAP
MAP Sensor (V)	0-5.0 volts	1 - 2v	---	EVAP
Short Term Fuel Trim	-10% to 10%	0 (± 5%)	0 (± 5%)	EVAP
TP Angle (%)	0-100%	0	5	EVAP
TP Sensor (V)	0-5.0 volts	0.55	0.89	EVAP
Vehicle Speed	0-155 mph	0	30	EVAP

1996-99 Aurora 4.0L V8 MFI VIN C (A/T) - PID Data List

Parameter Identification (PID)	PID Value Range	PID Value at Hot Idle	PID Value at 30 mph	Data List Type
Air Fuel Ratio	0.1-25.5:1	14.5-14.9:1	Varies	Fuel Trim 1
Desired Idle Speed	0-2040 rpm	720	---	Fuel Trim 1
ECT Sensor (°F)	-40 to 304°F	185-226	185-226	Fuel Trim 1
Engine Load	0-100%	1-5	5	Fuel Trim 1
Engine Run Time	Hr: Min: Sec	00:00:00	00:00:00	Fuel Trim 1
Engine Speed	0-9999 rpm	±50 Desired	---	Fuel Trim 1
HO2S (B1 S1, B1 S2)	0-1107 mv	1-1000	1-1000	Fuel Trim 1
HO2S-13 (B1 S3)	0-1107 mv	1-1000	1-1000	Fuel Trim 1
HO2S1 L/R Trans. B1	0-3187.5ms	Varies	Varies	Fuel Trim 1
HO2S2 L/R Trans. B1	0-3187.5ms	Varies	Varies	Fuel Trim 1
HO2S-11 (B1 S1) R/L	RICH/LEAN	L-R-L-R	L-R-L-R	Fuel Trim 1
HO2S-13 (B1 S3) R/L	RICH/LEAN	L-R-L-R	L-R-L-R	Fuel Trim 1
HO2S-21 (B1 S1) R/L	RICH/LEAN	L-R-L-R	L-R-L-R	Fuel Trim 1
HO2S1R/L Trans. B1	0-3187.5ms	Varies	Varies	Fuel Trim 1
HO2S2 R/L Trans. B1	0-3187.5ms	Varies	Varies	Fuel Trim 1
HO2S-1 Warmup B1	0:00-4:15	30-50	30-50	Fuel Trim 1
HO2S-3 Warmup B1	0:00-4:15	60-200	60-200	Fuel Trim 1
HO2S-1 Warmup B2	0:00-4:15	30-50	30-50	Fuel Trim 1
HO2S XCounts B1	0-255 counts	Varies	Varies	Fuel Trim 1
HO2S XCounts B2	0-255 counts	Varies	Varies	Fuel Trim 1
IAC Position	0-255 counts	30 - 80	---	Fuel Trim 1
IAT Sensor (°F)	-40 to 304°F	50-194	50-194	Fuel Trim 1
Injector Pulsewidth	10-1000 ms	3.7	4.3	Fuel Trim 1
Loop Status	Closed Loop / Open Loop	Closed Loop	Closed Loop	Fuel Trim 1

1996-99 Aurora 4.0L V8 MFI VIN C (A/T) - PID Data List

Parameter Identification (PID)	PID Value Range	PID Value at Hot Idle	PID Value at 30 mph	Data List Type
Long Term F/T B1, B2	-10% to 10%	Varies	Varies	Fuel Trim 2
MAF Sensor (g/sec)	0-512 g/sec	5-7	---	Fuel Trim 2
MAP Sensor (kPa)	10-110 kPa	30 - 50	---	Fuel Trim 2
MAP Sensor (V)	0-5.0 volts	1-2	000	Fuel Trim 2
Power Enrichment	Active / Inactive	Inactive	Inactive	Fuel Trim 2
PSP Switch	High/Normal	Normal	Normal	Fuel Trim 2
Short Term Fuel Trim	-10% to 10%	Varies	Varies	Fuel Trim 2
Spark Advance (°)	-90° to 90°	20	Varies	Fuel Trim 2
TFP Switch	Several	Park	Several	Fuel Trim 2
Throttle At Idle	Yes / No	Yes	No	Fuel Trim 2
TP Sensor (V)	0-5.0 volts	0.55	Varies	Fuel Trim 2
Vehicle Speed	0-155 mph	0	30	Fuel Trim 2

1996-99 Aurora 4.0L V8 MFI VIN C (A/T) - PID Data List

Parameter Identification (PID)	PID Value Range	PID Value at Hot Idle	PID Value at 30 mph	Data List Type
Air Fuel Ratio	0.1-25.5:1	14.7:1	14.7:1	Fuel Trim 3
Desired Idle Speed	0-2040 rpm	720	---	Fuel Trim 3
ECT Sensor (°F)	-40 to 304°F	185-226	185-226	Fuel Trim 3
Engine Load	0-100%	2-5	8-10	Fuel Trim 3
Engine Speed	0-9999 rpm	±50 Desired	---	Fuel Trim 3
HO2S-11 (B1 S1)	0-1132 mv	10-1000	10-1000	Fuel Trim 3
HO2S-21 (B2 S1)	0-1132 mv	10-1000	10-1000	Fuel Trim 3
HO2S-13 (B1 S3)	0-1132 mv	10-1000	10-1000	Fuel Trim 3
HO2S-11 R/L Status	RICH/LEAN	L-R-L-R	L-R-L-R	Fuel Trim 3
HO2S-21 R/L Status	RICH/LEAN	L-R-L-R	L-R-L-R	Fuel Trim 3
HO2S-11 L/R Status	LEAN/RICH	L-R-L-R	L-R-L-R	Fuel Trim 3
HO2S-12 L/R Status	LEAN/RICH	L-R-L-R	L-R-L-R	Fuel Trim 3
HO2S-13 L/R Status	LEAN/RICH	L-R-L-R	L-R-L-R	Fuel Trim 3
HO2S-13 R/L Status	RICH/LEAN	L-R-L-R	L-R-L-R	Fuel Trim 3
HO2S-11 R/L Switch	0-65535	< 12	< 12	Fuel Trim 3
HO2S-11 L/R Switch	0-65535	< 12	< 12	Fuel Trim 3
HO2S-21 L/R Switch	0-65535	< 12	< 12	Fuel Trim 3
HO2S-11 R/L Status	RICH/LEAN	L-R-L-R	L-R-L-R	Fuel Trim 3
HO2S-11 Warmup	Min: Sec	00:00	00:00	Fuel Trim 3
HO2S-21 Warmup	0-65535	< 12	< 12	Fuel Trim 3
HO2S XCounts	0-255 counts	Varies	Varies	Fuel Trim 3
IAC Position	0-255 counts	30-80	---	Fuel Trim 3
Ignition 1 Signal	0.0-25.5v	14.2	14.2	Fuel Trim 3
Long Term Fuel Trim	-10% to 10%	0 (± 5%)	0 (± 5%)	Fuel Trim 3
Loop Status	Closed Loop / Open Loop	Closed Loop	Closed Loop	Fuel Trim 3
MAF Sensor (g/sec)	0-512 g/sec	5-7	---	Fuel Trim 3
Short Term Fuel Trim	-10% to 10%	0 (± 5%)	0 (± 5%)	Fuel Trim 3
TFP Switch	Several	Park	Several	Fuel Trim 3
Throttle At Idle	Yes / No	Yes	No	Fuel Trim 3
TP Sensor (V)	0-5.0 volts	0.55	Varies	Fuel Trim 3
TWC Temperature	32-1409°F	Varies	Varies	Fuel Trim 3
Vehicle Speed	0-155 mph	0	30	Fuel Trim 3

1996-99 Aurora 4.0L V8 MFI VIN C (A/T) - PID Data List

Parameter Identification (PID)	PID Value Range	PID Value at Hot Idle	PID Value at 30 mph	Data List Type
Cycles Of Misfire Data	0-100	Varies	Varies	Misfire
Desired Idle Speed	0-2040 rpm	720	---	Misfire
ECT Sensor (°F)	-40 to 304°F	185-226	185-226	Misfire
Engine Load	0-100%	1-5	---	Misfire
Engine Speed	0-9999 rpm	±50 Desired	---	Misfire
Engine Speed At Misfire	0-9999 rpm	0	0	Misfire
HO2S-11 (B1 S1)	0-1132 mv	10-1000	10-1000	Misfire
HO2S-12 (B1 S2)	0-1132 mv	10-1000	10-1000	Misfire
Knock Retard	0° to 45°	0	0	Misfire
Load At Misfire	0-100%	0	0	Misfire
Loop Status	Closed Loop / Open Loop	Closed Loop	Closed Loop	Misfire
MAF Sensor (g/sec)	0-512 g/sec	5-7	---	Misfire
MAP Sensor (kPa)	10-110 kPa	30 - 50	---	Misfire
MAP Sensor (V)	0-5.0 volts	1 - 2v	---	Misfire
Misfire Current Cylinder 1-8	0-255 counts	0	0	Misfire
Misfire History Cylinder 1-8	0-65535	0	0	Misfire
Misfire Delay Counter	0-255 counts	0	0	Misfire
Number Failed Emission M/F Tests Out Of Last 16	0-16	0	0	Misfire
Spark Advance (°)	-90° to 90°	20	25	Misfire
Misfire Failures Since First Failure	0-65535	0	0	Misfire
Misfire Passes Since First Failure	0-65535	0	0	Misfire
TP Sensor	0-5.0 volts	0.55v	Varies	Misfire

2001-03 Aurora 4.0L V8 MFI VIN C (A/T) - PID Data List

Parameter Identification (PID)	PID Value Range	PID Value at Hot Idle	PID Value at 30 mph	Data List Type
5 Volt Reference 1	4.9-5.1 volts	5.0	5.0	Engine Data 1
5 Volt Reference 2	4.9-5.1 volts	5.0	5.0	Engine Data 1
A/C Relay Command	On / Off	Off	Off	Engine Data 1
AIR Relay Command	On / Off	Off	Varies	Engine Data 1
AIR Solenoid Command	On / Off	Off	Off	Engine Data 1
BARO Sensor (kPa)	10-110 kPa	85-103	85-103	Engine Data 1
Current Gear	1-2-3-4	1	3	Engine Data 1
Desired Idle Speed	0-3187 rpm	PCM control	---	Engine Data 1
Engine Load	0-100%	Varies	Varies	Engine Data 1
Engine Run Time	Hr: Min: Sec	00:00:00	00:00:00	Engine Data 1
Engine Speed	0-9999 rpm	±50 Desired	---	Engine Data 1
EVAP Purge Solenoid	On / Off	Off	Off	Engine Data 1
EVAP Vent Solenoid	Open/closed	Open	Open	Engine Data 1
Extended Travel Brake Pedal Switch	Applied / Released	Released	Released	Engine Data 1
F/P Relay Command	On / Off	On	On	Engine Data 1
FTP Sensor	-17.5 to +7.5" In. of H2O	Varies	Varies	Engine Data 1
Fuel Trim Cell (#)	Cell #0-4	0	1-4	Engine Data 1
HO2S-11 (B1 S1)	0-1275 mv	10-1000	10-1000	Engine Data 1
HO2S-12 (B1 S2)	0-1275 mv	10-1000	10-1000	Engine Data 1
HO2S-21 (B2 S1)	0-1275 mv	10-1000	10-1000	Engine Data 1
IAC Position (Counts)	0-255 counts	Varies	Varies	Engine Data 1
IAT Sensor (°F)	-40 to 304°F	50-194	50-194	Engine Data 1
Ignition 1 Signal (V)	0.0-25.5v	14.2	14.2	Engine Data 1
Knock Retard (°)	0.0-25.5°	0	Varies	Engine Data 1
Long Term Fuel Trim Bank 1	-25 to +25%	0 (± 5%)	0 (± 5%)	Engine Data 1
Long Term Fuel Trim Bank 2	-25 to +25%	0 (± 5%)	0 (± 5%)	Engine Data 1
Loop Status	Open Loop or Closed Loop	Closed Loop	Closed Loop	Engine Data 1
MAF Sensor (g/sec)	0-512 g/sec	5-7	Varies	Engine Data 1
MAP Sensor (kPa)	10-110 kPa	30-50	Varies	Engine Data 1
MAP Sensor (V)	0-5.1v	1.5-2.0	Varies	Engine Data 1
MIL Command	On / Off	Off	Off	Engine Data 1
Short Term Fuel Trim Bank 1	-25 to +25%	0 (± 5%)	0 (± 5%)	Engine Data 1
Short Term Fuel Trim Bank 2	-25 to +25%	0 (± 5%)	0 (± 5%)	Engine Data 1
Spark Advance (°)	-64° to 64°	20-40	Varies	Engine Data 1
TCC Enable Solenoid	Enabled/DIS	Disabled	Disabled	Engine Data 1
TP Angle (%)	0-100%	0	Varies	Engine Data 1
TP Sensor (V)	0-5.1v	0.55	Varies	Engine Data 1
Vehicle Speed	0-155 mph	0	30	Engine Data 1
5 Volt Reference 1 (V)	Okay / Fault	Okay	Okay	Engine Data 1
5 Volt Reference 2 (V)	Okay / Fault	Okay	Okay	Engine Data 1
12 Volt Reference (V)	Okay / Fault	Okay	Okay	Engine Data 1

2001-03 Aurora 4.0L V8 MFI VIN C (A/T) - PID Data List

Parameter Identification (PID)	PID Value Range	PID Value at Hot Idle	PID Value at 30 mph	Data List Type
24X Crank Sensor	0-9999 rpm	Varies	Varies	Engine Data 2
A/C High Side Pressure	0-5.1v	Varies	Varies	Engine Data 2
A/C Relay Command	On / Off	Off	Off	Engine Data 2
CMP Sensor Signal	RPM	Varies	Varies	Engine Data 2
Cruise Control Active	Yes / No	No	No	Engine Data 2
Current Gear	1-2-3-4	1	3	Engine Data 2
Desired EGR Position	0-100%	0	0-100	Engine Data 2
Desired Idle Speed	0-3187 rpm	650	---	Engine Data 2
ECT Sensor (°F)	-40 to 304°F	185-239	185-239	Engine Data 2
EGR Position Sensor	0-100%	0	Varies	Engine Data 2
EGR Position Sensor	0-5.1v	0.60	Varies	Engine Data 2
Engine Load	0-100%	2-5	Varies	Engine Data 2
Engine Oil Level Switch	Okay / Low	Okay	Okay	Engine Data 2
Engine Run Time	Hr: Min: Sec	00:00:00	00:00:00	Engine Data 2
Engine Speed	0-9999 rpm	±50 Desired	---	Engine Data 2
Extended Travel Brake Pedal	Applied / Released	Released	Released	Engine Data 2
FC Relay 1 Command	On / Off	Off	Off	Engine Data 2
FC Relay 2 & 3	On / Off	Off	Off	Engine Data 2
Generator F-Terminal	0-100%	0-100%	0-100%	Engine Data 2
Generator L-Terminal	Ok/no output	Okay	Okay	Engine Data 2
IAT Sensor (°F)	-40 to 304°F	50-194	50-194	Engine Data 2
Ignition 1 Signal (V)	0.0-25.5v	14.2	14.2	Engine Data 2
Injector Pulsewidth B1	0-985 ms	3.6	Varies	Engine Data 2
Injector Pulsewidth B2	0-985 ms	3.6	Varies	Engine Data 2
Long Term Fuel Trim	-25 to +25%	0 (± 5%)	0 (± 5%)	Engine Data 2
Loop Status	Open Loop or Closed Loop	Closed Loop	Closed Loop	Engine Data 2
MAF Sensor (g/sec)	0-512 g/sec	5-7	Varies	Engine Data 2
MAP Sensor (kPa)	10-110 kPa	30-50	Varies	Engine Data 2
MAP Sensor (V)	0-5.1v	1.5-2.0	Varies	Engine Data 2
Non-Driven Wheel Speed	0-155 mph	0	30	Engine Data 2
Short Term Fuel Trim	-25 to +25%	0 (± 5%)	0 (± 5%)	Engine Data 2
Spark Advance (°)	-64° to 64°	20-40	Varies	Engine Data 2
Startup ECT Sensor	-40 to 304°F	Varies	Varies	Engine Data 2
Startup IAT Sensor	-40 to 304°F	Varies	Varies	Engine Data 2
TCC Duty Cycle (%)	0-100%	0	Varies	Engine Data 2
TFP Switch Position	Several	P-N	Varies	Engine Data 2
TP Angle (%)	0-100%	0	0-100	Engine Data 2
TP Sensor (V)	0-5.1v	0.55	Varies	Engine Data 2
Torque Delivered Signal	0-100%	Varies	Varies	Engine Data 2
Traction Control Torque	0-100%	Varies	Varies	Engine Data 2
Vehicle Speed (mph)	0-155 mph	0	30	Engine Data 2
VTD Password	Okay/Incorrect	Okay	Okay	Engine Data 2

2001-03 Aurora 4.0L V8 MFI VIN C (A/T) - PID Data List

Parameter Identification (PID)	PID Value Range	PID Value at Hot Idle	PID Value at 30 mph	Data List Type
A/C Relay Command	On / Off	Off	Off	Engine Data 3
Air Fuel Ratio	0.0-25:1	14.7:1	14.7:1	Engine Data 3
Crank Request Signal	Yes / No	No	No	Engine Data 3
Cruise Control Active	Yes / No	No	No	Engine Data 3
Cruise Inhibited Signal	On / Off	On	On	Engine Data 3
Desired Idle Speed	0-3187 rpm	650	---	Engine Data 3
ECT Sensor (°F)	-40 to 304°F	185-239	185-239	Engine Data 3
Engine Oil Level Switch	Okay / Low	Okay	Okay	Engine Data 3
Engine Oil Pressure Switch	Okay / Low	Okay	Okay	Engine Data 3
Engine Run Time	Hr: Min: Sec	00:00:00	00:00:00	Engine Data 3
Engine Speed	0-9999 rpm	±50 Desired	---	Engine Data 3
Extended Travel Brake Pedal	Applied / Released	Released	Released	Engine Data 3
FC Relay 1 Command	On / Off	Off	Off	Engine Data 3
FC Relay 2 & 3	On / Off	Off	Off	Engine Data 3
IAT Sensor (°F)	-40 to 304°F	50-194	50-194	Engine Data 3
Ignition 1 Signal (V)	0.0-25.5v	14.2	14.2	Engine Data 3
Internal Mode Switch	Several	PARK	DRIVE 3	Engine Data 3
Long Term Fuel Trim	-25 to +25%	0 (± 5%)	0 (± 5%)	Engine Data 3
Loop Status	Open Loop or Closed Loop	Closed Loop	Closed Loop	Engine Data 3
MAF Sensor (g/sec)	0-512 g/sec	5-7	Varies	Engine Data 3
MAF Sensor (Hz)	0-31,999 Hz	2800	Varies	Engine Data 3
MAP Sensor (kPa)	10-110 kPa	30-50	Varies	Engine Data 3
Miles Since DTC Cleared	0-1677722	Varies	Varies	Engine Data 3
Odometer Miles	0-1677722	Varies	Varies	Engine Data 3
PCM Induced Chassis	Normal/Firm	Normal	Varies	Engine Data 3
PCM/VCM In VTD Fail Mode	Yes / No	No	No	Engine Data 3
Short Term Fuel Trim	-25 to +25%	0 (± 5%)	0 (± 5%)	Engine Data 3
Spark Advance (°)	-64° to 64°	20-40	Varies	Engine Data 3
Starter Relay Control	On / Off	Off	Off	Engine Data 3
TCC Brake Pedal Switch	Applied / Released	Released	Released	Engine Data 3
TFP Switch Position	Several	P-N	Varies	Engine Data 3
TP Angle (%)	0-100%	0	0-100	Engine Data 3
TP Sensor (V)	0-5.1v	0.55	Varies	Engine Data 3
TFT Sensor (°F)	-40 to 304°F	195	195	Engine Data 3
Vehicle Speed (mph)	0-155 mph	0	30	Engine Data 3
VTD Auto Learn Timer	Active / Inactive	Inactive	Inactive	Engine Data 3
VTD Fuel Disable	Enabled/DIS	Disabled	Disabled	Engine Data 3
Warmups without an Emission Fault	0-255 counts	0	0	Engine Data 3
Warmups without a Non-Emission Fault	0-255 counts	0	0	Engine Data 3

2001-03 Aurora 4.0L V8 MFI VIN C (A/T) - PID Data List

Parameter Identification (PID)	PID Value Range	PID Value at Hot Idle	PID Value at 30 mph	Data List Type
Air Fuel Ratio	0.0-25:1	14.7:1	14.7:1	Fuel Trim
Air Pump Relay	On / Off	Off	Off	Fuel Trim
Air Pump Solenoid	On / Off	Off	Off	Fuel Trim
BARO Sensor (kPa)	10-110 kPa	85-103	85-103	Fuel Trim
Current Gear	1-2-3-4	1	3	Fuel Trim
ECT Sensor (°F)	-40 to 304°F	185-239	185-239	Fuel Trim
Engine Load	0-100%	2-5	Varies	Fuel Trim
Engine Run Time	Hr: Min: Sec	00:00:00	00:00:00	Fuel Trim
Engine Speed	0-9999 rpm	±50 Desired	---	Fuel Trim
EVAP Purge Solenoid	On / Off	Off	Off	Fuel Trim
EVAP Vent Solenoid	Open/Closed	Open	Open	Fuel Trim
Fuel Trim Cell	Cell #0-4	0	1-4	Fuel Trim
HO2S-11 (B1 S1)	0-1275 mv	10-1000	10-1000	Fuel Trim
HO2S-12 (B1 S2)	0-1275 mv	10-1000	10-1000	Fuel Trim
HO2S-21 (B2 S1)	0-1275 mv	10-1000	10-1000	Fuel Trim
HO2S-12 Warmup	Seconds	Varies	Varies	Fuel Trim
IAC Position (Counts)	0-255 counts	Varies	Varies	Fuel Trim
IAT Sensor (°F)	-40 to 304°F	50-194	50-194	Fuel Trim
Ignition 1 Signal (V)	0.0-25.5v	14.2	14.2	Fuel Trim
Injector Pulsewidth B1	0-985 ms	3.6	Varies	Fuel Trim
Injector Pulsewidth B2	0-985 ms	3.6	Varies	Fuel Trim
Long Term Fuel Trim Average	-25 to +25%	0 (± 5%)	0 (± 5%)	Fuel Trim
Long Term Fuel Trim	-25 to +25%	0 (± 5%)	0 (± 5%)	Fuel Trim
Loop Status	Open Loop or Closed Loop	Closed Loop	Closed Loop	Fuel Trim
MAF Sensor (g/sec)	0-512 g/sec	5-7	Varies	Fuel Trim
Short Term F/T Avg.	-25 to +25%	0 (± 5%)	0 (± 5%)	Fuel Trim
Short Term Fuel Trim	-25 to +25%	0 (± 5%)	0 (± 5%)	Fuel Trim
Spark Advance (°)	-64° to 64°	20-40	Varies	Fuel Trim
Startup ECT Sensor	-40 to 304°F	Varies	Varies	Fuel Trim
Startup IAT Sensor	-40 to 304°F	Varies	Varies	Fuel Trim
TFP Switch Position	Several	P-N	Varies	Fuel Trim
TP Angle (%)	0-100%	0	0-100	Fuel Trim
Vehicle Speed (mph)	0-155 mph	0	30	Fuel Trim

2001-03 Aurora 4.0L V8 MFI VIN C (A/T) - PID Data List

Parameter Identification (PID)	PID Value Range	PID Value at Hot Idle	PID Value at 30 mph	Data List Type
Air Fuel Ratio	0.0-25:1	14.2-14.9:1	Varies	EVAP
BARO Sensor (kPa)	10-110 kPa	85-103	85-103	EVAP
Desired Idle Speed	0-3187 rpm	650	---	EVAP
ECT Sensor (°F)	-40 to 304°F	185-239	185-239	EVAP
Engine Load	0-100%	2-5	Varies	EVAP
Engine Run Time	Hr: Min: Sec	00:00:00	00:00:00	EVAP
Engine Speed	0-9999 rpm	±50 Desired	---	EVAP
EVAP Test Result	Not Aborted, Lost Enable, Small Leak, Not at Rest	Vehicle Not at Rest	Vehicle Not at Rest	EVAP

2001-03 Aurora 4.0L V8 MFI VIN C (A/T) - PID Data List

Parameter Identification (PID)	PID Value Range	PID Value at Hot Idle	PID Value at 30 mph	Data List Type
EVAP Test State	Test running, Wait - Purge, test complete	Complete	Complete	EVAP
EVAP Vent Solenoid	Open/closed	Open	Open	EVAP
Fuel Level	Gallons	Varies	Varies	EVAP
Fuel Level Sensor (V)	0-5.1v	Varies	Varies	EVAP
FTP Sensor (V)	0-5.1v	Varies	Varies	EVAP
FTP Sensor (In. H2O)	-17.5 to 7.5"	Varies	Varies	EVAP
Fuel Trim Cell	Cell #0-4	0	1-4	EVAP
HO2S-11 (B1 S1)	0-1275 mv	10-1000	10-1000	EVAP
HO2S-21 (B2 S1)	0-1275 mv	10-1000	10-1000	EVAP
IAT Sensor (°F)	-40 to 304°F	50-194	50-194	EVAP
Ignition 1 Signal (V)	0.0-25.5v	14.2	14.2	EVAP
Long Term Fuel Trim Average	-25 to +25%	0 (± 5%)	0 (± 5%)	EVAP
Long Term Fuel Trim	-25 to +25%	0 (± 5%)	0 (± 5%)	EVAP
Loop Status	Open Loop or Closed Loop	Closed Loop	Closed Loop	EVAP
MAF Sensor (g/sec)	0-512 g/sec	5-7	Varies	EVAP
MAP Sensor (kPa)	10-110 kPa	30-50	Varies	EVAP
MAP Sensor (V)	0-5.1v	1.5-2.0	Varies	EVAP
Short Term F/T Avg.	-25 to +25%	0 (± 5%)	0 (± 5%)	EVAP
Short Term Fuel Trim	-25 to +25%	0 (± 5%)	0 (± 5%)	EVAP
Startup ECT Sensor	-40 to 304°F	Varies	Varies	EVAP
Startup IAT Sensor	-40 to 304°F	Varies	Varies	EVAP
TP Angle (%)	0-100%	0	0-100	EVAP
Vehicle Speed	0-155 mph	0	30	EVAP

2001-03 Aurora 4.0L V8 MFI VIN C (A/T) - PID Data List

Parameter Identification (PID)	PID Value Range	PID Value at Hot Idle	PID Value at 30 mph	Data List Type
24X Crank Sensor	0-9999 rpm	Varies	Varies	Misfire
A/C Relay Command	On / Off	Off	Off	Misfire
CKP Sensor Status	Several	ANGLE	ANGLE	Misfire
CMP Sensor Signal	RPM	Varies	Varies	Misfire
CMP Sensor Signal Present	Yes / No	Yes	Yes	Misfire
Cycles Of Misfire Data	0-255 counts	Varies	Varies	Misfire
ECT Sensor (°F)	-40 to 304°F	185-239	185-239	Misfire
Engine Load	0-100%	2-5	Varies	Misfire
Engine Run Time	Hr: Min: Sec	00:00:00	00:00:00	Misfire
Engine Speed	0-9999 rpm	±50 Desired	---	Misfire
Injector Pulsewidth B1	0-985 ms	3.6	Varies	Misfire
Injector Pulsewidth B2	0-985 ms	3.6	Varies	Misfire
MAF Sensor (g/sec)	0-512 g/sec	5-7	Varies	Misfire
Misfire current Cyl 1-8	0-255 counts	0-10	Varies	Misfire
Misfire History Cyl 1-8	0-255 counts	0	Varies	Misfire
Spark Advance (°)	-64° to 64°	20-40	Varies	Misfire
TP Angle (%)	0-100%	0	0-100	Misfire
Vehicle Speed (mph)	0-155 mph	0	30	Misfire

CUTLASS PID DATA

1997-99 Cutlass 3.1L V6 MFI VIN M (A/T) - PID Data List

Parameter Identification (PID)	PID Value Range	PID Value at Hot Idle	PID Value at 30 mph	Data List Type
3X Crank Sensor	0-9999 rpm	Varies	Varies	Engine Data 1
24X Crank Sensor	0-1280 rpm	Varies	Varies	Engine Data 1
Abuse Management	Active / Inactive	Inactive	Inactive	Engine Data 1
A/C High Side Pressure	0-5.0 volts	Varies	Varies	Engine Data 1
A/C Off for WOT	On / Off	Off	Off	Engine Data 1
A/C Pressure Disable	Yes / No	No	No	Engine Data 1
A/C Request	Yes / No	No	No	Engine Data 1
Actual EGR Position	0-100%	0%	0-100%	Engine Data 1
Air Fuel Ratio	0:0-25:1	14.7:1	14.7:1	Engine Data 1
BARO Sensor (kPa)	10-110 kPa	65-110	65-110	Engine Data 1
Brake Switch	Applied / Released	Released	Released	Engine Data 1
Cam Signal Present	Yes / No	Yes	Yes	Engine Data 1
Commanded A/C	On / Off	Off	Off	Engine Data 1
Cruise Mode	On / Off	Off	On	Engine Data 1
Cruise Inhibited	Yes / No	Yes	Yes	Engine Data 1
Current Gear	1-2-3-4	1	3	Engine Data 1
Decel Fuel Mode	Active / Inactive	Inactive	Inactive	Engine Data 1
Desired EGR Position	0-100%	0%	0-100%	Engine Data 1
Desired Idle Speed	0-9999 rpm	PCM control	---	Engine Data 1
ECT Sensor (°F)	-40 to 304°F	185-220	185-220	Engine Data 1
EGR Closed Valve Pintle Position	0-5.0 volts	0.14-1.00v	---	Engine Data 1
EGR Duty Cycle	0-100%	0%	0-100%	Engine Data 1
EGR Feedback	0-5.0 volts	0.14-1.00v	---	Engine Data 1
EGR Position Error	0-100%	0-9	---	Engine Data 1
EGR Flow Test Count	0-255 counts	0-10	0-10	Engine Data 1
Engine Load	0-100%	2-5	Varies	Engine Data 1
Engine Run Time	Hr: Min: Sec	00:00:00	00:00:00	Engine Data 1
Engine Speed	0-9999 rpm	±50 Desired	---	Engine Data 1
EVAP Canister Purge	0-100%	0-25	0-75	Engine Data 1
EVAP Vent Solenoid	Open/Close	Open	Open	Engine Data 1
FC Relay Low Speed	On / Off	Off	Off	Engine Data 1
FC Relay High Speed	On / Off	Off	Off	Engine Data 1
Fuel Pump Command	On / Off	On	On	Engine Data 1
Fuel Trim Cell	0-21	0	1	Engine Data 1
Fuel Trim Learn	Disabled / Enabled	Enabled	Enabled	Engine Data 1
Generator Command	On / Off	On	On	Engine Data 1
Generator Lamp	On / Off	Off	Off	Engine Data 1
HO2S-11 (B1 S1)	0-1132 mv	10-1000	10-1000	Engine Data 1
HO2S-12 (B1 S2)	0-1132 mv	10-1000	10-1000	Engine Data 1
HO2S-11 (B1 S1)	Not / Ready	Ready	Ready	Engine Data 1
HO2S XCounts	0-255 counts	Varies	Varies	Engine Data 1
Hot Mode	On / Off	Off	---	Engine Data 1
IAC Position	0-255 counts	10-40	---	Engine Data 1
IAT Sensor (°F)	-40 to 304°F	50-194	50-194	Engine Data 1
Ignition 1 Signal	0.0-25.5v	14.2	14.2	Engine Data 1
Ignition Mode	IC/Bypass	IC	IC	Engine Data 1

1997-99 Cutlass 3.1L V6 MFI VIN M (A/T) - PID Data List

Parameter Identification (PID)	PID Value Range	PID Value at Hot Idle	PID Value at 30 mph	Data List Type
Injector Pulsewidth	0-985 ms	1.5-3.5	Varies	Engine Data 1
Injector Status	Okay / Stuck	Okay	Okay	Engine Data 1
Knock Retard	0-25.5°	0	0	Enginé Data 1
Long Term Fuel Trim	-10% to 10%	0 (± 5%)	0 (± 5%)	Engine Data 1
Loop Status	Closed Loop / Open Loop	Closed Loop	Closed Loop	Engine Data 1
Low Oil Lamp	On / Off	Off	Off	Engine Data 1
MAF Sensor (g/sec)	0-512 g/sec	4-6	Varies	Engine Data 1
MAF Sensor (Hz)	0-31,999hz	1200-3000	Varies	Engine Data 1
MAP Sensor (kPa)	10-110 kPa	29-48	Varies	Engine Data 1
MAP Sensor (V)	0-5.0 volts	1-2	Varies	Engine Data 1
MIL Command	On / Off	Off	Off	Engine Data 1
Misfire Current Cylinder 1-6	0-198	0	0	Engine Data 1
Misfire History Cylinder 1-6	0-65535	0	0	Engine Data 1
Non-Volatile Memory	Pass / Fail	Pass	Pass	Engine Data 1
Power Enrichment	Active / Inactive	Active	Active	Engine Data 1
Short Term Fuel Trim	-10% to 10%	0 (± 5%)	0 (± 5%)	Engine Data 1
Spark Advance (°)	-64° to 64°	20	Varies	Engine Data 1
Startup ECT Sensor	-40 to 304°F	Varies	Varies	Engine Data 1
Startup IAT Sensor	-40 to 304°F	Varies	Varies	Engine Data 1
TCC Engaged	Engaged / Disengaged	Disengaged	Disengaged	Engine Data 1
Total Misfire Current Count	0-99 counts	0 – 5	0 - 5	Engine Data 1
Total Misfire Failures	0-65535	0	0	Engine Data 1
Total Misfire Passes Since First Failure	0-65535	0	0	Engine Data 1
TP Angle (%)	0-100%	0	Varies	Engine Data 1
TP Sensor	0-5.0 volts	0.20-0.74	Varies	Engine Data 1
Transmission Range	P-R-N-4-3-2	P	3	Engine Data 1
Transmission Hot Mode Status	On / Off	Off	Off	Engine Data 1
TR Switch P-A-B-C	High / Low	LO/LO/High/High	LO/High/High / Low	Engine Data 1
TWC Protection	Active / Inactive	Inactive	Inactive	Engine Data 1
Vehicle Speed	0-155 mph	0	30	Engine Data 1
VTD Fuel Disable	Active / Inactive	Inactive	Inactive	Engine Data 1

1999 Cutlass 3.1L V6 MFI VIN J (A/T) - PID Data List

Parameter Identification (PID)	PID Value Range	PID Value at Hot Idle	PID Value at 30 mph	Data List Type
3X Crank Sensor	0-9999 rpm	Varies	Varies	Engine Data 2
24X Crank Sensor	0-1600 rpm	Varies	Varies	Engine Data 2
1-2 Solenoid Control	Okay / Fault/ Invalid State	Varies	Varies	Output Driver
2-3 Solenoid Control	Okay / Fault/ Invalid State	Varies	Varies	Output Driver
1-2 Solenoid Control	Okay / Fault/ Invalid State	Varies	Varies	Output Driver
Abuse Management	Active / Inactive	Inactive	Inactive	Misfire
A/C High Side Pressure	0-5.0v	Varies	Varies	Engine Data 1
A/C Relay Circuit Status	Okay / Fault/ Invalid State	Varies	Varies	Output Driver
A/C Request Signal	Yes / No	No	No	Engine Data 2
Actual EGR Position	0-100%	0	Varies	Engine Data 1, EGR, Misfire
Air Active Test Inhibit	Yes / No	No	No	AIR
Air Active Test Injection	Yes / No	No	No	AIR
Air Diagnostic Complete	Yes / No	No	No	AIR
Air Passive Test 1 Pass	Yes / No	No	No	AIR
Air Passive Test 2 Fail	Yes / No	No	No	AIR
Air Passive Test Inhibit	Yes / No	No	No	AIR
Air Pump Relay Circuit Status	Okay / Fault/ Invalid State	Okay	Okay	AIR, Output Driver
Air Pump Relay Command	On / Off	Off	Off	Engine Data 1 Data, AIR, HO2S
Air Solenoid Circuit Status	Okay / Fault/ Invalid State	Okay	Okay	AIR, Output Driver
Air Solenoid Command	On / Off	Off	Off	Engine Data 1 Data, AIR, HO2S
Air Fuel Ratio	0.0-25.5:1	14.7:1	14.7:1	Engine Data 2, Fuel Trim
BARO Sensor (kPa)	10-110 kPa	98	98	Engine Data 1, EGR, EVAP, Catalyst
BARO Sensor (V)	0-5.0 volts	3.5-4.5	3.5-4.5	Engine Data 1, EGR, EVAP, Catalyst
Brake Switch Signal	Applied / Released	Released	Released	Engine Data 1
CMP Sensor Signal	Yes / No	Yes	Yes	Engine Data 1, Misfire
Commanded A/C	On / Off	Off	Off	Engine Data 1, EGR, Misfire, Catalyst
Cruised Inhibited	Okay / Fault/ Invalid State	Okay	Okay	Output Driver

1999 Cutlass 3.1L V6 MFI VIN J (A/T) - PID Data List

Parameter Identification (PID)	PID Value Range	PID Value at Hot Idle	PID Value at 30 mph	Data List Type
Cruise Inhibited	On / Off	On	On	Engine Data 1
Cruise Mode	Yes / No	No	No	Engine Data 1
Current Gear	1-2-3-4	1	4	EGR, Catalyst
Cycles of Misfire Data	0-99 counts	Varies	Varies	Misfire
Cylinder 1-6 Injector Circuit Status	Okay/Stuck Low or High	Okay	Okay	Output Driver
Cylinder 1-6 Injector Circuit History	Okay/Stuck Low or High	Okay	Okay	Output Driver
Decel Fuel Mode	Active / Inactive	Inactive	Inactive	EGR, HO2S, Misfire
Desired Idle Speed	0-3187 rpm	PCM control	----	Engine Data 2
Desired EGR Position	0-100%	0	Varies	Engine Data 1, EGR, Misfire
Desired Idle Speed	0-3187 rpm	± 50 rpm of Actual Speed	Varies	Engine Data 1
Driver Module 1-4 Status	Enabled/Off/ High Temp	Enabled	Enabled	Output Driver
ECT Sensor (ºF)	-40 to 304ºF	185-220	185-220	Engine Data 1, AIR, Catalyst, EGR EVAP, Misfire
EGR Closed Valve Pintle Position	0-5.1volts	0.14-1.0	Varies	EGR
EGR Duty Cycle	0-100	0	Varies	Engine Data 1, EGR
EGR Feedback	0-5.1 volts	0.14-1.0	Varies	EGR
EGR Flow Test Count	0-255 counts	0	Varies	EGR
EGR Position Error	0-100%	0	Varies	EGR
Engine Load (%)	0-100%	17	Varies	Engine Data 1, AIR, Catalyst, EGR EVAP, Misfire
Engine Oil Level Switch	Okay/Low	Okay	Okay	IPC
Engine Oil Life Left	0-100%	Varies	Varies	IPC
EOP Switch Signal	Okay/Low/High	Okay	Okay	IPC
Engine Run Time	Hr: Min: Sec	00:00:00	00:00:00	Engine Data 1, AIR, Catalyst, EGR EVAP, Misfire
Engine Speed	0-9999 rpm	± 50 rpm of Desired Speed	Varies	Engine Data 1, AIR, Catalyst, EGR EVAP, Misfire
EVAP Canister Purge	0-100%	19	Varies	Engine Data 1, AIR, EVAP, HO2S
EVAP Fault History	Several	No Fault	No Fault	EVAP
EVAP Purge Solenoid Control Circuit Status	Okay / Fault/ Invalid state	Okay	Okay	Output Driver
EVAP Vent Solenoid Control Circuit Status	Okay / Fault/ Invalid state	Okay	Okay	Output Driver, EVAP
EVAP Vent Solenoid	Open/Closed	Open	Open	Engine Data 1, EVAP, HO2S
FC Relay High Speed	On / Off	Off	Off	Engine Data 1
FC Relay Low Speed	On / Off	Off	Off	Engine Data 1

1999 Cutlass 3.1L V6 MFI VIN J (A/T) - PID Data List

Parameter Identification (PID)	PID Value Range	PID Value at Hot Idle	PID Value at 30 mph	Data List Type
FC Relay High or Low Speed Circuit Status	Okay / Fault/ Invalid State	Okay	Okay	Output Driver
Fuel Pump Control	On / Off	On	On	Engine Data 1
Fuel Pump Relay Control Circuit Status	Okay / Fault	Okay	Okay	Output Driver
Fuel Pump Relay Control Circuit History	Okay / Fault	Okay	Okay	Output Driver
Fuel Tank Level	0-100%	Varies	Varies	EVAP, IPC
Fuel Tank Pressure (FTP) Sensor	0-5.0 volts	Varies	Varies	EVAP
Fuel Tank Pressure (FTP) Sensor	Inches H2O, mm Hg, volts	-17 to +7.5 Inches H2O	-17 to +7.5 Inches H2O	EVAP
Fuel Trim Cell	Cell 0-10	Varies	Varies	Engine Data 1, AIR, Catalyst
Fuel Trim Learn	Enabled or Disabled	Enabled	May Toggle	Engine Data 1, AIR, Catalyst
Generator Control	On / Off	On	On	IPC
Generator PWM	0-100%	Varies	Varies	IPC
HO2S-11 (B1 S1) Ready	Ready/Not Ready	Ready	Ready	Engine Data 1 HO2S, Catalyst
HO2S-11 (B1 S1)	0-1132 mv	10-1000	10-1000	Engine Data 1, EVAP, Fuel Trim
HO2S-12 (B1 S2)	0-1132 mv	10-1000	10-1000	Engine Data 1, HO2S, Catalyst
HO2S-11 XCounts	0-233	Varies	Varies	Engine Data 1, HO2S
Hot Mode	On / Off	Off	Off	Misfire
Hot Open Loop	Active / Inactive	Inactive	Inactive	Catalyst, EGR
IAC Position	0-255 counts	15	15	Engine Data 1, Catalyst
IAT Sensor (°F)	-40 to 304°F	50-194	50-194	Engine Data 1, AIR, Catalyst, EGR EVAP, Misfire
Ignition 1 Signal	0.0-25.5v	14.2	14.2	Engine Data 1, AIR, Catalyst, EGR EVAP, Misfire
Ignition Mode	Bypass/IC	IC	IC	Engine Data 1
Injector Pulsewidth	0-1000 ms	1.5-3.5	Varies	Engine Data 1, CAT, HO2S, M/Fire
Knock Retard (°)	0-25.5°	0°	0°	Engine Data 1, EGR
Long Term Fuel Trim	-27% to 27%	0 (± 5%)	0 (± 5%)	Engine Data 1, AIR, EVAP, HO2S
Loop Status	Closed Loop / Open Loop	Closed Loop	Closed Loop	Engine Data 1, AIR, Catalyst, EGR EVAP, Misfire
MAF Sensor (g/sec)	0-512 g/sec	3.41	Varies	Engine Data 1, AIR, Catalyst, EGR EVAP, Misfire
MAF Sensor (Hz)	0-32,000 Hz	2000	Varies	Engine Data 1

1999 Cutlass 3.1L V6 MFI VIN J (A/T) - PID Data List

Parameter Identification (PID)	PID Value Range	PID Value at Hot Idle	PID Value at 30 mph	Data List Type
MAP Sensor (kPa)	10-105 kPa	35	Varies	Engine Data 1, EGR, EVAP, Catalyst
MAP Sensor (V)	0-5.1 volts	1.35	Varies	Engine Data 1, EGR, EVAP, Catalyst
MIL Status	Okay / Fault	Okay	Okay	Output Driver
MIL Command	On / Off	Off	Off	IPC
Misfire Current Cylinder 1-6	0-198 counts	0	0	Misfire
Misfire History Cylinder 1-6	0-65,535	0	0	Misfire
Non-Volatile Memory	Pass / Fail	Pass	Pass	Engine Data 1, AIR, Catalyst, EGR EVAP, IPC, Misfire, ODD
Power Enrichment	Active / Inactive	Inactive	Inactive	CAT, EGR, HO2S, M/Fire
Short Term Fuel Trim	-27% to 27%	0 (± 5%)	0 (± 5%)	Engine Data 1, AIR, HO2S
Spark Advance (º)	-64º to 64º	20	Varies	Engine Data 1 Misfire
Startup ECT	-40 to 419ºF	Varies	Varies	AIR, Catalyst, EVAP, HO2S
Startup IAT	-40 to 419ºF	80ºF	80ºF	AIR, Catalyst, EVAP, HO2S
TCC Engaged	Engaged, Disengaged	Disengaged	Disengaged	Engine Data 1, EGR, Misfire
TCC Solenoid Circuit Status	Okay / Fault/ Invalid State	Okay	Okay	Output Driver
Total Misfire Current Count	0-99	Varies	Varies	Misfire
Total Misfire Failures	0-65,535 counts	Varies	Varies	Misfire
Total Misfire Passed	0-65,535 counts	Varies	Varies	Misfire
TP Angle (%)	0-100%	0	0-100	Engine Data 1, AIR, Catalyst, EGR EVAP, M/Fire
TP Sensor (V)	0-5.1 volts	0.20-0.74	Varies	Engine Data 1
Transmission Range	Park/Neutral, Drive 4-3-2-1	Park/Neutral	Drive 4	Engine Data 1, AIR, EGR, Catalyst
TR Switch 'A' Signal	High / Low	Low	High	Engine Data 1
TR Switch 'B' Signal	High / Low	High	Low	Engine Data 1
TR Switch 'C' Signal	High / Low	High	Low	Engine Data 1
TR Switch 'P' Signal	High / Low	Low	High	Engine Data 1
TWC Protection	Active / Inactive	Inactive	Inactive	Catalyst, EGR
Vehicle Speed	0-155 mph	0	30	Engine Data 1, AIR, EGR, Catalyst
VTD Fuel Disable	Active / Inactive	Inactive	Inactive	Engine Data 1

CUTLASS CIERA PID DATA

1996 Cutlass Ciera 2.2L I4 MFI VIN 4 (A/T) - PID Data List

Parameter Identification (PID)	PID Value Range	PID Value at Hot Idle	PID Value at 30 mph	Data List Type
BARO Sensor (V)	0-5.0 volts	3.5-4.5	3.5-4.5	Engine Data 1
Calculated Airflow	0-512	3-7	Varies	Engine Data 1
CKP Active Counter	0-255 counts	Increments	Increments	Engine Data 1
CMP Active Counter	0-255 counts	Increments	Increments	Engine Data 1
CMP Resync Counter	0-255 counts	0	0	Engine Data 1
Desired Idle Speed	0-3187	PCM control	---	Engine Data 1
ECT Sensor (°F)	-40-419°F	185-239	185-239	Engine Data 1
Engine Run Time	Hr: Min: Sec	00:00:00	00:00:00	Engine Data 1
Engine Speed	0-9999 rpm	±50 Desired	---	Engine Data 1
EVAP Purge Solenoid	0-100%	0	0-100	Engine Data 1
EVAP Vent Solenoid	On / Off	Off	Off	Engine Data 1
Fuel Pump Relay	On / Off	On	On	Engine Data 1
Fuel Trim Cell	Cell #	18-21	---	Engine Data 1
Fuel Trim Index	0-255 counts	58-198	Varies	Engine Data 1
Fuel Trim Index	-10% to 10%	Varies	Varies	Engine Data 1
Generator F-Terminal	0-100%	0-100%	0-100%	Engine Data 1
Generator L-Terminal	Active / Inactive	Inactive	Inactive	Engine Data 1
HO2S-12 (B1 S2)	0-1132	10-1000	10-1000	Engine Data 1
IAC Motor	0-255 counts	5-60	---	Engine Data 1
IAT Sensor (°F)	-40 to 304°F	50-176	50-176	Engine Data 1
Ignition Voltage	0.0-25.5v	14.2	14.2	Engine Data 1
Knock Retard (°)	0-128°	0	0	Engine Data 1
Knock Signal	Yes / No	No	No	Engine Data 1
KS Noise Channel	Yes / No	No	No	Engine Data 1
KS Active Counts	0-255 counts	35-50	35-50	Engine Data 1
Engine Load	0-100%	22 - 26	Varies	Engine Data 1
Low Octane Spark Modifier	0-90°	Varies	Varies	Engine Data 1
Long Term Fuel Trim	0-255 counts	58-198	58-198	Engine Data 1
Long Term Fuel Trim	-10% to 10%	Varies	Varies	Engine Data 1
Long Term Fuel Trim Average	0-255 counts	58-198	58-198	Engine Data 1
Loop Status	Closed Loop / Open Loop	Closed Loop	Closed Loop	Engine Data 1
MAP Sensor (kPa)	10-110 kPa	25-48	Varies	Engine Data 1
Medium Resolution Resync Counter	0-255 counts	0	0	Engine Data 1
Medium Resolution Engine Resync	0-255 counts	0	0	Engine Data 1
Number of Current Trouble Codes	0-255 counts	0	0	Engine Data 1
O2S-11 (B1 S1)	0-1132	10-1000	10-1000	Engine Data 1
Power Enrichment	Active / Inactive	Inactive	Inactive	Engine Data 1
Purge Learn Memory	0-100%	0	0-100	Engine Data 1
Short Term Fuel Trim	0-255 counts	58-198	58-198	Engine Data 1
Short Term Fuel Trim	-10% to 10%	Varies	Varies	Varies
Short Term F/T Avg.	-10% to 10%	Varies	Varies	Varies
Spark Advance	-64° to 64°	Varies	Varies	Varies
TCC Brake Switch	On / Off	Off	Off	Off
TP Angle (%)	0-100%	0	Varies	Varies
TP Sensor (V)	0.0-5.0 volts	0.20-0.90	Varies	Varies

1996 Cutlass Ciera 2.2L I4 MFI VIN 4 (A/T) - PID Data List

Parameter Identification (PID)	PID Value Range	PID Value at Hot Idle	PID Value at 30 mph	Data List Type
A/C High Side (psi)	0-459 psi	139-399	139-399	Engine Data 2
A/C High Side (V)	0-5.0 volts	1-4	1-4	Engine Data 2
A/C Relay	On / Off	Off	Off	Engine Data 2
A/C Request	Yes / No	No	Off	Engine Data 2
Air Fuel Ratio	0.0-25.5:1	14.7:1	14.7:1	Engine Data 2
BARO Sensor (V)	0-5.0 volts	3.5-4.5	3.5-4.5	Engine Data 2
Cruise Engaged	Yes / No	No	No	Engine Data 2
Desired Idle Speed	0-3187	PCM control	---	Engine Data 2
ECT Sensor (°F)	-40-419°F	185-239	185-239	Engine Data 2
Engine Oil Level	Low / Okay	Okay	Okay	Engine Data 2
Engine Oil Pressure	Low / Okay	Okay	Okay	Engine Data 2
Engine Speed	0-9999 rpm	±50 Desired	---	Engine Data 2
IAT Sensor (°F)	-40 to 304°F	50-176	50-176	Engine Data 2
INJ Pulsewidth	0-985 ms	1-4	Varies	Engine Data 2
MAP Sensor (V)	0.0-5.0 volts	1.5	Varies	Engine Data 2
MPH / KPH	0-155 mph	0	30	Engine Data 2
Output Driver 1, 2 Open	00000000	00000000	00000000	Engine Data 2
Output Driver 1, 2 Short	00000000	00000000	00000000	Engine Data 2
Stepper Cruise	DIS/ENABL	Disabled	Disabled	Engine Data 2
Total Misfire Counts	0-255 counts	0	0	Engine Data 2
TP Angle (%)	0-100%	0	Varies	Engine Data 2
TP Sensor (V)	0.0-5.0 volts	0.20-0.90	Varies	Engine Data 2
TR Range	P/N-R-O-3-2	P/N	3	Engine Data 2
TR Switch (P)	High / Low	High	Low	Engine Data 2
TR Switch (A)	High / Low	High	Low	Engine Data 2
TR Switch (B)	High / Low	Low	High	Engine Data 2
TR Switch (C)	High / Low	Low	High	Engine Data 2

1996 Cutlass Ciera 2.2L I4 MFI VIN 4 (A/T) - PID Data List

Parameter Identification (PID)	PID Value Range	PID Value at Hot Idle	PID Value at 30 mph	Data List Type
BARO Sensor (kPa)	10-110 kPa	11-105	11-105	Misfire
Calculated Airflow	0-512 g/sec	Varies	Varies	Misfire
ECT Sensor (°F)	-40-419°F	185-239	185-239	Misfire
Desired Idle Speed	0-3187	PCM control	---	Misfire
Engine Run Time	Hr: Min: Sec	00:00:00	00:00:00	Misfire
Engine Speed	0-9999 rpm	±50 Desired	---	Misfire
IAT Sensor (°F)	-40 to 304°F	50-176	50-176	Misfire
INJ Pulsewidth CYL	0-985 ms	1-4	Varies	Misfire
Engine Load	0-100%	22-26	Varies	Misfire
MAP Sensor (kPa)	10-110 kPa	25-48	Varies	Misfire
Misfire current Cyl 1-4	0-255 counts	0	0	Misfire
Misfire History Cyl 1-4	0-255 counts	0	0	Misfire
TP Angle (%)	0-100%	0	Varies	Misfire
TP Sensor (V)	0.0-5.0 volts	0.20-0.90	Varies	Misfire
Total Misfire	0-255 counts	0	0	Misfire

1996 Cutlass Ciera 2.2L I4 MFI VIN 4 (A/T) - Specific EGR & EVAP

Parameter Identification (PID)	PID Value Range	PID Value at Hot Idle	PID Value at 30 mph	Data List Type
Actual EGR Position	0-100%	0	0-100	EGR, EVAP
BARO Sensor (V)	0-5.0 volts	3.5-4.5	3.5-4.5	EGR, EVAP
Decel EWMA	-10 - 10 kPa	-3 to - 4	Varies	EGR, EVAP
Desired Idle Speed	0-3187	PCM control	---	EGR, EVAP
Delta MAP Change	-20 - 20 kPa	0 / - number	0	EGR, EVAP
Desired EGR Position	0-100%	0	0-100	EGR, EVAP
ECT Sensor (°F)	-40-419°F	185-239	185-239	EGR, EVAP
EGR Trip Samples	0-255 counts	Varies	Varies	EGR, EVAP
Engine Run Time	Hr: Min: Sec	00:00:00	00:00:00	EGR, EVAP
Engine Speed	0-9999 rpm	±50 Desired	---	EGR, EVAP
EVAP Solenoid	0-100%	0-25	0-50	EGR, EVAP
EVAP Vent Solenoid	0=off / 1=on	0	0	EGR, EVAP
Fuel Tank Level	0-5.0 volts	Varies	Varies	EGR, EVAP
Fuel Level Sensor %	0-100%	Varies	Varies	EGR, EVAP
Fuel Tank Vapor Pressure Sensor	0-5.0 volts	2.5v: Fuel Cap removed	Varies	EGR, EVAP
Fuel Trim Cell	Cell #	18-21	18-21	EGR, EVAP
HO2S-12 (B1 S2)	0-1132v	10-1000	10-1000	EGR, EVAP
IAC Motor	0-255 counts	5-60	---	EGR, EVAP
IAT Sensor (°F)	-40 to 304°F	50-176	50-176	EGR, EVAP
Long Term Fuel Trim	0-255 counts	58-198	58-198	EGR, EVAP
Long Term Fuel Trim	-10% to 10%	0 (± 5%)	0 (± 5%)	EGR, EVAP
MAP Sensor (kPa)	10-110 kPa	25-48	Varies	EGR, EVAP
Number of Current DTC(s)	Number	0	0	EGR, EVAP
O2S-11 (B1 S1)	0-1132 mv	10-1000	10-1000	EGR, EVAP
Short Term Fuel Trim	0-255 counts	0 (± 5%)	0 (± 5%)	EGR, EVAP
Short Term Fuel Trim	-10% to 10%	Varies	Varies	EGR, EVAP
TP Angle (%)	0-100%	0	Varies	EGR, EVAP
TP Sensor (V)	0.0-5.0 volts	0.20-0.90	Varies	EGR, EVAP

1996 Cutlass Ciera 2.2L I4 MFI VIN 4 (A/T) - PID Data List

Parameter Identification (PID)	PID Value Range	PID Value at Hot Idle	PID Value at 30 mph	Data List Type
BARO Sensor (V)	0-5.0 volts	3.5-4.5	3.5-4.5	Fuel Trim
Calculated Airflow	0-512 g/sec	Varies	Varies	Fuel Trim
Desired Idle Speed	0-3187	PCM control	---	Fuel Trim
ECT Sensor (°F)	-40-419°F	185-239	185-239	Fuel Trim
Engine Run Time	Hr: Min: Sec	00:00:00	00:00:00	Fuel Trim
Engine Speed	0-9999 rpm	±50 Desired	---	Fuel Trim
Fuel Trim Cell	Cell #	18-21	18-21	Fuel Trim
Fuel Trim Index	0-255 counts	58-198	58-198	Fuel Trim
HO2S-12 (B1 S2)	0-1132 mv	10-1000	10-1000	Fuel Trim
IAT Sensor (°F)	-40 to 304°F	50-176	50-176	Fuel Trim
Lean / Rich Average	0-249 ms	0	0	Fuel Trim
Lean / Rich Transition	0-255 counts	0	0	Fuel Trim
HO2S-11 L/R to R/L Ratio	0:1-15.93:1	Varies	Varies	Fuel Trim
Engine Load	0-100%	22 - 26	Varies	Fuel Trim
Long Term Fuel Trim Average	0-255 counts	58-198	58-198	Fuel Trim
Long Term Fuel Trim	0-255 counts	58-198	58-198	Fuel Trim
Long Term Fuel Trim	-10% to 10%	0 (± 5%)	0 (± 5%)	Fuel Trim
Loop Status	Closed Loop / Open Loop	Closed Loop	Closed Loop	Fuel Trim
MAP Sensor (kPa)	10-110 kPa	25-48	Varies	Fuel Trim
O2S-11 (B1 S1)	0-1132 mv	10-1000	10-1000	Fuel Trim
Rich / Lean Average	0-249 ms	0	0	Fuel Trim
Rich / Lean Transition	0-255 counts	0	0	Fuel Trim
Short Term F/T Avg.	0-255 counts	58-198	58-198	Fuel Trim
Short Term Fuel Trim	0-255 counts	58-198	58-198	Fuel Trim
Short Term Fuel Trim	-10% to 10%	0 (± 5%)	0 (± 5%)	Fuel Trim
TP Angle (%)	0-100%	0	Varies	Fuel Trim
TP Sensor (V)	0.0-5.0 volts	0.20-0.90	Varies	Fuel Trim

1996 Cutlass Ciera 2.2L I4 MFI VIN 4 - Freeze Frame Data & Failure Records

Parameter Identification (PID)	PID Value Range	PID Value at Hot Idle	PID Value at 30 mph	Data List Type
BARO Sensor (V)	0-5.0 volts	3.5-4.5	3.5-4.5	F/F, F/R
Calculated Airflow	0-512 g/sec	Varies	Varies	F/F, F/R
Desired Idle Speed	0-3187	PCM control	---	F/F, F/R
ECT Sensor (°F)	-40-419°F	185-239	185-239	F/F, F/R
Engine Speed	0-9999 rpm	±50 Desired	---	F/F, F/R
INJ Pulsewidth	0-985 ms	1-4	Varies	F/F, F/R
Long Term Fuel Trim	0-255 counts	58-198	58-198	F/F, F/R
Long Term Fuel Trim	-10% to 10%	0 (± 5%)	0 (± 5%)	F/F, F/R
Engine Load	0-100%	22-26	Varies	F/F, F/R
Loop Status	Closed Loop / Open Loop	Closed Loop	Closed Loop	F/F, F/R
MAP Sensor (kPa)	10-110 kPa	25-48	Varies	F/F, F/R
MPH / KPH	0-155 mph	0	30	F/F, F/R
Short Term Fuel Trim	0-255 counts	58-198	58-198	F/F, F/R
Short Term Fuel Trim	-10% to 10%	0 (± 5%)	0 (± 5%)	F/F, F/R
TP Angle (%)	0-100%	0	Varies	Varies

1996 Cutlass Ciera 3.1L V6 MFI VIN M (A/T) - PID Data List

Parameter Identification (PID)	PID Value Range	PID Value at Hot Idle	PID Value at 30 mph	Data List Type
3X Crank Sensor	0-9999 rpm	Varies	Varies	Engine Data 1
24X Crank Sensor	0-1280 rpm	Varies	Varies	Engine Data 1
Abuse Management	Active / Inactive	Inactive	Inactive	Engine Data 1
A/C High Side Pressure	0-5.0 volts	Varies	Varies	Engine Data 1
A/C Off for WOT	On / Off	Off	Off	Engine Data 1
A/C Pressure Disable	Yes / No	No	No	Engine Data 1
A/C Request	Yes / No	No	No	Engine Data 1
Actual EGR Position	0-100%	0%	0-100%	Engine Data 1
Air Fuel Ratio	0.0-25.5:1	14.7:1	14.7:1	Engine Data 1
BARO Sensor (kPa)	10-110 kPa	65-110	65-110	Engine Data 1
Brake Switch	Applied / Released	Released	Released	Engine Data 1
Cam Signal Present	Yes / No	Yes	Yes	Engine Data 1
Commanded A/C	On / Off	Off	Off	Engine Data 1
Cruise Mode	On / Off	Off	On	Engine Data 1
Cruise Inhibited	Yes / No	Yes	Yes	Engine Data 1
Current Gear	1-2-3-4	1	3	Engine Data 1
Decel Fuel Mode	Active / Inactive	Inactive	Inactive	Engine Data 1
Desired EGR Position	0-100%	0	0-100	Engine Data 1
Desired Idle Speed	0-9999 rpm	PCM control	---	Engine Data 1
ECT Sensor (°F)	-40 to 304°F	185-220	185-220	Engine Data 1
EGR Closed Valve Pintle Position	0-5.0 volts	0.14-1.00v	Varies	Engine Data 1
EGR Duty Cycle	0-100%	0	0-100	Engine Data 1
EGR Feedback	0-5.0 volts	0.14-1.00v	Varies	Engine Data 1
EGR Position Error	0-100%	0-9	---	Engine Data 1
EGR Flow Test Count	0-255 counts	0-10	0-10	Engine Data 1
Engine Load	0-100%	2-5	Varies	Engine Data 1
Engine Run Time	Hr: Min: Sec	00:00:00	00:00:00	Engine Data 1
Engine Speed	0-9999 rpm	±50 Desired	---	Engine Data 1
EVAP Canister Purge	0-100%	0-25	0-75	Engine Data 1
EVAP Vent Solenoid	Open/Close	Open	Varies	Engine Data 1
FC Relay Low Speed	On / Off	Off	Off	Engine Data 1
FC Relay High Speed	On / Off	Off	Off	Engine Data 1
Fuel Pump Command	On / Off	On	On	Engine Data 1
Fuel Trim Cell	Cell #	0	1	Engine Data 1
Fuel Trim Learn	Disabled / Enabled	Enabled	Enabled	Engine Data 1
Generator Command	On / Off	Off	Off	Engine Data 1
Generator Lamp	On / Off	Off	Off	Engine Data 1
HO2S-11 (B1 S1)	0-1132 mv	10-1000	10-1000	Engine Data 1
HO2S-21 (B2 S1)	0-1132 mv	10-1000	10-1000	Engine Data 1
HO2S-11 (B1 S1) Status	Ready / Not Ready	Ready	Ready	Engine Data 1
HO2S XCounts	0-255 counts	Varies	Varies	Engine Data 1
IAC Position	0-255 counts	10-40	---	Engine Data 1
IAT Sensor (°F)	-40 to 304°F	50-194	50-194	Engine Data 1
Ignition 1 Signal	0.0-25.5v	14.2	14.2	Engine Data 1
Ignition Mode	Bypass / IC	IC	IC	Engine Data 1

1996 Cutlass Ciera 3.1L V6 MFI VIN M (A/T) - Engine Data List

Parameter Identification (PID)	PID Value Range	PID Value at Hot Idle	PID Value at 30 mph	Data List Type
Injector Pulsewidth	0-985 ms	1.5-3.5	Varies	Engine Data 1
Injector Status	Okay/STUCK	Okay	Okay	Engine Data 1
Knock Retard	0-128°	0	0	Engine Data 1
Long Term Fuel Trim	-10% to 10%	0 (± 5%)	0 (± 5%)	Engine Data 1
Loop Status	Closed Loop / Open Loop	Closed Loop	Closed Loop	Engine Data 1
Low Oil Lamp	On / Off	Off	Off	Engine Data 1
MAF Sensor (g/sec)	0-512 g/sec	4-6	Varies	Engine Data 1
MAF Sensor (Hz)	0-31,999hz	1200-3000	Varies	Engine Data 1
MAP Sensor (kPa)	10-110 kPa	29-48	Varies	Engine Data 1
MAP Sensor (V)	0-5.0 volts	1-2	Varies	Engine Data 1
MIL Command	On / Off	Off	Off	Engine Data 1
Misfire Current CYL 1-6	0-198	0	0	Engine Data 1
Misfire History CYL 1-6	0-65535	0	0	Engine Data 1
Non-Volatile Memory	Pass / Fail	Pass	Pass	Engine Data 1
Power Enrichment	Active / Inactive	Inactive	Inactive	Engine Data 1
Rich / Lean	L or R	L-R-L-R	L R L R	Engine Data 1
Short Term Fuel Trim	-10% to 10%	0 (± 5%)	0 (± 5%)	Engine Data 1
Spark Advance (°)	-64° to 64°	20	Varies	Engine Data 1
Startup ECT Sensor	-40 to 304°F	Varies	Varies	Engine Data 1
Startup IAT Sensor	-40 to 304°F	Varies	Varies	Engine Data 1
TCC Engaged	Engaged / Disengaged	Disengaged	Disengaged	Engine Data 1
Total Misfire Current Count	0-99	0	0	Engine Data 1
Total Misfire Failures	0-65535	0	0	Engine Data 1
Total Misfire Passes Since First Failure	0-65535	0	0	Engine Data 1
TP Angle (%)	0-100%	0	Varies	Engine Data 1
TP Sensor	0-5.0 volts	0.20-0.74	Varies	Engine Data 1
Transmission Hot Mode Status	On / Off	Off	Off	Engine Data 1
Transmission Range	P-R-N-4-3-2	P	3	Engine Data 1
TR Switch P, A	High / LO	High	Low	Engine Data 1
TR Switch B, C	High / LO	Low	High	Engine Data 1
TWC Protection	Active / Inactive	Inactive	Inactive	Engine Data 1
Vehicle Speed	0-155 mph	0	30	Engine Data 1
VTD Fuel Disable	Active / Inactive	Inactive	Inactive	Engine Data 1

CUTLASS CIERA, CRUISER PID DATA

1995 Cutlass Ciera, Cruiser 3.1L V6 MFI VIN M - PID Data List

Parameter Identification (PID)	PID Value Range	PID Value at Hot Idle	PID Value at 30 mph	Data List Type
2nd Gear Start Switch	On / Off	On	Off: 2nd	Engine Data 1
2nd, 3rd Gear Switch	Yes / No	No	Yes: 2nd	Engine Data 1
A/C Clutch	On / Off	Off	Off	Engine Data 1
A/C Head Pressure	0-459 psi	139-399	139-399	Engine Data 1
A/C Request	Yes / No	No	No	Engine Data 1
BARO Sensor (V)	0-5.0 volts	3.5-4.5	3.5-4.5	Engine Data 1
CMP Sensor Signal	0-1	0-1	0-1	Engine Data 1
Cooling Fan 1, 2	On / Off	Off	Off	Engine Data 1
DTC Ignition Counter	0-255 counts	Counts Up	Counts Up	Engine Data 1
DTC Stored	0-255 counts	0	0	Engine Data 1
ECT Sensor (°F)	-40-419°F	202	202	Engine Data 1
EGR Solenoid 1, 2, 3	On / Off	Off	On or Off	Engine Data 1
Engine Speed	0-9999 rpm	800 ±50	---	Engine Data 1
Fuel Pump Signal	0.0-25.5v	14.2	14.2	Engine Data 1
Fuel Trim Cell	0-10	0	4	Engine Data 1
Fuel Trim Enabled	Yes / No	Yes	Yes	Engine Data 1
HO2S-11 (B1 S1)	0-1132 mv	10-1000	10-1000	Engine Data 1
IAC Motor	0-255 counts	35	---	Engine Data 1
IAT Sensor (°F)	-40 to 304°F	95	95	Engine Data 1
Ignition Signal (24X)	0-3187 rpm	Varies	Varies	Engine Data 1
Injector Fault	Okay / Fault	Okay	Okay	Engine Data 1
Injector Pulsewidth	0-985 ms	2.8	3.0	Engine Data 1
Knock Retard (°)	0-128°	0	0	Engine Data 1
Knock Signal	Yes / No	No	No	Engine Data 1
Long Term Fuel Trim	0-255 counts	110-150	110-150	Engine Data 1
Loop Status	Closed Loop / Open Loop	Closed Loop	Closed Loop	Engine Data 1
Low Oil Light	On / Off	Off	Off	Engine Data 1
MAP Sensor (kPa)	10-110 kPa	35	45	Engine Data 1
P/NP Switch	P/N / R-DL	P/N	R-DL	Engine Data 1
PRNDL P, A	Several	High	Low	Engine Data 1
PRNDL P, B	Several	Low	High	Engine Data 1
Purge Duty Cycle	0-100%	0	20	Engine Data 1
Shift Solenoid 'A', B	On / Off	On	Off	Engine Data 1
Short Term Fuel Trim	0-255 counts	110-150	110-150	Engine Data 1
SMCC Engaged	Yes / No	No	No	Engine Data 1
SMCC Inhibited	Yes / No	No	No	Engine Data 1
Spark Advance (°)	-90° to 90°	18-24	30-34	Engine Data 1
TCC Apply Solenoid	On / Off	Off	Off	Engine Data 1
TCC Brake Switch	Applied / Released	Released	Released	Engine Data 1
TCC Slip Speed	-4080-4079	720-760	5	Engine Data 1
TCC PWM Solenoid	0-100%	0	0-100	Engine Data 1
Throttle Angle	0-100%	0	4	Engine Data 1
Time From Start	Hr: Min: Sec	00:00:00	00:00:00	Engine Data 1
TCC Solenoid Control	On / Off	Off	Off	Engine Data 1
TP Sensor (V)	0.0-5.0 volts	0.54	0.74	Engine Data 1
TFT Sensor (°F)	-40 to 304°F	180	182	Engine Data 1
Trans. Range Switch	P-A-B-C	Park	DRIVE 2/3	Engine Data 1
Vehicle Speed	0-155 mph	0	30	Engine Data 1

CUTLASS SUPREME PID DATA

1995 Cutlass Supreme 3.1L V6 MFI VIN M (A/T) - PID Data List

Parameter Identification (PID)	PID Value Range	PID Value at Hot Idle	PID Value at 30 mph	Data List Type
2nd Gear Start Switch	On / Off	On	Off: 2nd	Engine Data 1
3rd, 4th Gear Switch	Yes / No	No	Yes: 3rd	Yes: 4th
A/C Clutch	On / Off	Off	Off	Engine Data 1
A/C Head Pressure	0-459 psi	139-399	139-399	Engine Data 1
A/C Request	Yes / No	No	No	Engine Data 1
BARO Sensor (V)	0-5.0 volts	3.5-4.5	3.5-4.5	Engine Data 1
CMP Sensor Signal	0-1	Varies	Varies	Engine Data 1
Cooling Fan 1, 2	On / Off	Off	Off	Engine Data 1
DTC Ignition Counter	0-255 counts	Increments	Increments	Engine Data 1
DTC Stored	0-255 counts	0	0	Engine Data 1
ECT Sensor (°F)	-40-419°F	185-223	185-223	Engine Data 1
EGR Solenoid 1, 2, 3	On / Off	Off	On or Off	Engine Data 1
Engine Speed	0-9999 rpm	800 ±50	---	Engine Data 1
Fuel Pump Signal	0.0-25.5v	14.2	14.2	Engine Data 1
Fuel Trim Enabled	Yes / No	Yes	Yes	Engine Data 1
HO2S-11 (B1 S1)	0-1132ms	10-1000	10-1000	Engine Data 1
IAC Motor	0-255 counts	5-50	---	Engine Data 1
IAT Sensor (°F)	-40 to 304°F	50-184	50-184	Engine Data 1
Ignition 24X Signal	0-3187 rpm	Varies	Varies	Engine Data 1
Injector Fault	Okay / Fault	Okay	Okay	Engine Data 1
Injector Pulsewidth	0-985 ms	3-6	Varies	Engine Data 1
Knock Retard (°)	0-128°	0	0	Engine Data 1
Knock Signal	Yes / No	No	No	Engine Data 1
Long Term Fuel Trim	Cell #0-10	0-1	0-10	Engine Data 1
Long Term Fuel Trim	0-255 counts	110-150	110-150	Engine Data 1
Low Oil Light	On / Off	Off	Off	Engine Data 1
MAP Sensor (kPa)	10-110 kPa	29-48	Varies	Engine Data 1
P/NP Switch	P/N / R-DL	P/N	R-DL	Engine Data 1
PRNDL P, A	High / LO	High	Low	Engine Data 1
PRNDL B, C	High / LO	Low	High	Engine Data 1
Purge Duty Cycle	0-100%	0	20	Engine Data 1
Shift Solenoid 'A'	On / Off	On	Off	Engine Data 1
Shift Solenoid 'B'	On / Off	On	Off	Engine Data 1
Short Term Fuel Trim	0-255 counts	Varies	Varies	Engine Data 1
SMCC Engaged	Yes / No	No	No	Engine Data 1
SMCC Inhibited	Yes / No	No	No	Engine Data 1
Spark Advance (°)	-90° to 90°	18-24	30-34	Engine Data 1
TCC Apply Solenoid	On / Off	Off	Off	Engine Data 1
TCC Brake Switch	Applied / Released	Released	Released	Engine Data 1
TCC Slip Speed	-4080-4079	730-768	2-4	Engine Data 1
TCC PWM Solenoid	0-100%	0	0-100	Engine Data 1
Throttle Angle	0-100%	0	4	Engine Data 1
TP Sensor (V)	0.0-5.0 volts	0.54	0.74	Engine Data 1
Time From Start	Hr: Min: Sec	00:00:00	00:00:00	Engine Data 1
TCC Solenoid Control	On / Off	Off	Off	Engine Data 1
TFT Sensor (°F)	-40 to 304°F	168	169	Engine Data 1
Trans. Range Switch	P-A-B-C	Park	DRIVE 2/3	Engine Data 1
Vehicle Speed	0-155 mph	0	30	Engine Data 1

1996-97 Cutlass Supreme 3.1L V6 MFI VIN M (A/T) - PID Data List

Parameter Identification (PID)	PID Value Range	PID Value at Hot Idle	PID Value at 30 mph	Data List Type
3X Crank Sensor	0-9999 rpm	Varies	Varies	Engine Data 1
24X Crank Sensor	0-1280 rpm	Varies	Varies	Engine Data 1
Abuse Management	Active / Inactive	Inactive	Inactive	Engine Data 1
A/C High Side Pressure	0-5.0 volts	Varies	Varies	Engine Data 1
A/C Off for WOT	On / Off	Off	Off	Engine Data 1
A/C Pressure Disable	Yes / No	No	No	Engine Data 1
A/C Request	Yes / No	No	No	Engine Data 1
Actual EGR Position	0-100%	0	0-100	Engine Data 1
Air Fuel Ratio	0.0-25.5:1	14.7:1	14.7:1	Engine Data 1
BARO Sensor (kPa)	10-110 kPa	65-110	65-110	Engine Data 1
Brake Switch	Applied / Released	Released	Released	Engine Data 1
Cam Signal Present	Yes / No	Yes	Yes	Engine Data 1
Commanded A/C	On / Off	Off	Off	Engine Data 1
Cruise Mode	On / Off	Off	On	Engine Data 1
Cruise Inhibited	Yes / No	Yes	Yes	Engine Data 1
Current Gear	1-2-3-4	1	3	Engine Data 1
Decel Fuel Mode	Active / Inactive	Inactive	Inactive	Engine Data 1
Desired EGR Position	0-100%	0	0-100	Engine Data 1
Desired Idle Speed	0-9999 rpm	PCM control	---	Engine Data 1
ECT Sensor (°F)	-40 to 304°F	185-220	185-220	Engine Data 1
EGR Closed Valve Pintle Position	0-5.0 volts	0.14-1.00v	---	Engine Data 1
EGR Duty Cycle	0-100%	0	0-100	Engine Data 1
EGR Feedback	0-5.0 volts	0.14-1.00v	---	Engine Data 1
EGR Position Error	0-100%	0-9	---	Engine Data 1
EGR Flow Test Count	0-255 counts	0-10	0-10	Engine Data 1
Engine Load	0-100%	2-5	Varies	Engine Data 1
Engine Run Time	Hr: Min: Sec	00:00:00	00:00:00	Engine Data 1
Engine Speed	0-9999 rpm	±50 Desired	---	Engine Data 1
EVAP Canister Purge	0-100%	0	20	Engine Data 1
EVAP Vent Solenoid	Open/Close	Open	Open	Engine Data 1
FC Relay Low Speed	On / Off	Off	Off	Engine Data 1
FC Relay High Speed	On / Off	Off	Off	Engine Data 1
Fuel Pump Control	On / Off	On	On	Engine Data 1
Fuel Trim Cell	Cell #0-4	0	0-4	Engine Data 1
Fuel Trim Learn	Disabled / Enabled	Enabled	Enabled	Engine Data 1
Generator Lamp	On / Off	Off	Off	Engine Data 1
HO2S-11 (B1 S1)	0-1132 mv	10-1000	10-1000	Engine Data 1
HO2S-21 (B2 S1)	0-1132 mv	10-1000	10-1000	Engine Data 1
HO2S-11 Status	Not / Ready	Ready	Ready	Engine Data 1
HO2S-11 XCounts	0-255 counts	Varies	Varies	Engine Data 1
IAC Position	0-255 counts	10-40	---	Engine Data 1
IAT Sensor (°F)	-40 to 304°F	50-194	50-194	Engine Data 1
Ignition 1 Signal	0.0-25.5v	14.2	14.2	Engine Data 1
Ignition Mode	Bypass / IC	IC	IC	Engine Data 1
Injector Pulsewidth	0-985 ms	1.5-3.5	Varies	Engine Data 1
Injector Status	Okay/STUCK	Okay	Okay	Engine Data 1
Knock Retard (°)	0-128°	0	0	Engine Data 1

1996-97 Cutlass Supreme 3.1L V6 MFI VIN M (A/T) - PID Data List

Parameter Identification (PID)	PID Value Range	PID Value at Hot Idle	PID Value at 30 mph	Data List Type
Long Term Fuel Trim	-10% to 10%	0 (± 5%)	0 (± 5%)	Engine Data 1
Loop Status	Closed Loop / Open Loop	Closed Loop	Closed Loop	Engine Data 1
Low Oil Lamp	On / Off	Off	Off	Engine Data 1
MAF Sensor (g/sec)	0-512 g/sec	4-6	Varies	Engine Data 1
MAF Sensor (Hz)	0-31,999hz	1200-3000	Varies	Engine Data 1
MAP Sensor (kPa)	10-110 kPa	29-48	Varies	Engine Data 1
MAP Sensor (V)	0-5.0 volts	1-2	Varies	Engine Data 1
MIL Command	On / Off	Off	Off	Engine Data 1
Misfire Current Cylinder 1-6	0-198	0	0	Engine Data 1
Misfire History Cylinder 1-6	0-65535	0	0	Engine Data 1
Non-Volatile Memory	Pass / Fail	Pass	Pass	Engine Data 1
Power Enrichment	Active / Inactive	Active	Active	Engine Data 1
Rich / Lean Status	Rich or Lean	L-R-L-R	L-R-L-R	Engine Data 1
Short Term Fuel Trim	-10% to 10%	0 (± 5%)	0 (± 5%)	Engine Data 1
Spark Advance (°)	-64° to 64°	20	Varies	Engine Data 1
Startup ECT Sensor	-40 to 304°F	Varies	Varies	Engine Data 1
Startup IAT Sensor	-40 to 304°F	Varies	Varies	Engine Data 1
TCC Engaged	Engaged / Disengaged	Disengaged	Disengaged	Engine Data 1
Total Misfire Current Count	0-99	0	0	Engine Data 1
Total Misfire Failures	0-65535	0	0	Engine Data 1
Total Misfire Passes Since First Failure	0-65535	0	0	Engine Data 1
TP Angle (%)	0-100%	0	Varies	Engine Data 1
TP Sensor (V)	0-5.0 volts	0.54	0.74	Engine Data 1
Transmission Hot Mode	On / Off	Off	Off	Engine Data 1
Transmission Range	P-R-N-4-3-2	P	3	Engine Data 1
TR Switch P, A	High / LO	High	Low	Engine Data 1
TR Switch B, C	High / LO	Low	High	Engine Data 1
TWC Protection	Active / Inactive	Inactive	Inactive	Engine Data 1
Vehicle Speed	0-155 mph	0	30	Engine Data 1
VTD Fuel Disable	Active / Inactive	Inactive	Inactive	Engine Data 1

1995 Cutlass Supreme 3.4L V6 MFI VIN X (A/T) - PID Data List

Parameter Identification (PID)	PID Value Range	PID Value at Hot Idle	PID Value at 30 mph	Data List Type
2nd Gear Start Switch	On / Off	Off	On: 2nd	Engine Data 1
A/C Clutch	On / Off	Off	Off	Engine Data 1
A/C Pressure (psi)	0-459 psi	110-399	110-399	Engine Data 1
A/C Request	Yes / No	No	No	Engine Data 1
Calc. Engine Load	0-100%	70-80%	Varies	Engine Data 1
CMP Sensor Signal	0 or 1	0-1-0-1	0-1-0-1	Engine Data 1
Commanded Gear	1-2-3-4	1st	3rd	Engine Data 1
Cooling Fan 1, 2	On / Off	On: >228°F	Off	Engine Data 1
Current Weak CYL	CYL #	None	None	Engine Data 1
Desired Idle Speed	0-3187	800	---	Engine Data 1
ECT Sensor (°F)	-40-419°F	185-223	185-223	Engine Data 1
EGR Solenoid 1-2-3	On / Off	Off	On	Engine Data 1
Engine Speed	0-9999 rpm	±50 Desired	---	Engine Data 1
Fuel Trim Cell	Cell #	0 or 1	0 or 2	Engine Data 1
HO2S Cross Counts	0-255 counts	0-20	0-20	Engine Data 1
HO2S-11 (B1 S1)	0-1132 mv	10-1000	10-1000	Engine Data 1
IAC Motor	0-255 counts	5-50	---	Engine Data 1
IAT Sensor (°F)	-40-304°F	50-176F	50-176F	Engine Data 1
Ignition Voltage	0.0-25.5v	14.2	14.2	Engine Data 1
INJ Pulsewidth	0-985 ms	1-4	Varies	Engine Data 1
Knock Retard (°)	0-128°	0	0	Engine Data 1
Knock Signal	Yes / No	No	No	Engine Data 1
Long, Short Term F/T	0-255 counts	95-138	95-138	Engine Data 1
Loop Status	Closed Loop / Open Loop	Closed Loop	Closed Loop	Engine Data 1
MAF Sensor (g/sec)	0-512 g/sec	5-7	Varies	Engine Data 1
MAP Sensor (kPa)	10-110 kPa	29-48	Varies	Engine Data 1
MAP Sensor (V)	0-5.0 volts	1-2	Varies	Engine Data 1
PRNDL P/A/B/C	High or Low	High	Low	Engine Data 1
Park Neutral Switch	Yes / No	Yes	No	Engine Data 1
Pass Key ® II Fuel	Disabled / Enabled	Enabled	Enabled	Engine Data 1
P/NP Switch	P/N / R-DL	P/N	R-D-L	Engine Data 1
PROM ID Number	Number	0000	0000	Engine Data 1
PSP Switch	High/Normal	Normal	Normal	Engine Data 1
Purge Duty Cycle	0-100%	0-25	0-50	Engine Data 1
QDM (module) A, B	High / Low	Low	---	Engine Data 1
SMCC Status	On / Off	Off	Off	Engine Data 1
SMCC Inhibited	Yes / No	Yes	---	Engine Data 1
Spark Advance (°)	-90° to 90°	Varies	Varies	Engine Data 1
TCC Apply Solenoid	On / Off	Off	On	Engine Data 1
TCC Brake Switch	Applied / Released	Released	Released	Engine Data 1
TCC Slip Speed	-4080-4079	0	0-4000	Engine Data 1
TCC PWM Solenoid	0-100%	0	0-100	Engine Data 1
Throttle Angle	0-100%	0	Varies	Engine Data 1
Time From Startup	Hr: Min: Sec	00:00:00	00:00:00	Engine Data 1
TP Sensor (V)	0-5.0 volts	0.29-0.98v	---	Engine Data 1
TFT Sensor (°F)	-40 to 304°F	180-200	180-200	Engine Data 1
Trans. Range Switch	On / Off	On	Off	Engine Data 1
Vehicle Speed	0-155 mph	0	30	Engine Data 1

1996 Cutlass Supreme 3.4L V6 MFI VIN X (A/T) - PID Data List

Parameter Identification (PID)	PID Value Range	PID Value at Hot Idle	PID Value at 30 mph	Data List Type
24X Crank Sensor	0-1600 rpm	Varies	Varies	Varies
A/C Clutch	On / Off	Off	Off	Engine Data 1
A/C High Side Pressure	0-459 psi	139-399	139-399	Engine Data 1
A/C Off for WOT	On / Off	Off	Off	Engine Data 1
A/C Pressure Disable	Yes / No	No	No	Engine Data 1
A/C Request	Yes / No	No	No	Engine Data 1
Actual EGR Position	0-100%	0	0-100	Engine Data 1
AIR Pump	On / Off	Off	Off	Engine Data 1
BARO Sensor (kPa)	10-110 kPa	65-110	65-110	Engine Data 1
Cam Signal Present	Yes / No	Yes	Yes	Engine Data 1
Commanded A/C	On / Off	Off	Off	Engine Data 1
Commanded Fan 1, 2	On / Off	Off	On	Engine Data 1
Cruise Engaged	Yes / No	No	No	Engine Data 1
Cruise Inhibited	Yes / No	Yes	Yes	Engine Data 1
Desired EGR Position	0-100%	0	0-100	Engine Data 1
Desired Idle Speed	0-9999 rpm	PCM control	---	Engine Data 1
ECT Sensor (°F)	-40 to 304°F	185-220	105-220	Engine Data 1
EGR Duty Cycle	0-100%	0	0-100	Engine Data 1
Engine Load	0-100%	2-5	Varies	Engine Data 1
Engine Run Time	Hr: Min: Sec	00:00:00	00:00:00	Engine Data 1
Engine Speed	0-9999 rpm	±50 Desired	---	Engine Data 1
EVAP Purge PWM	0-100%	0-25	0-75	Engine Data 1
EVAP Vacuum Switch	Open/closed	Open	Open	Engine Data 1
Fuel Pump	On / Off	On	On	Engine Data 1
Fuel Trim Cell	Cell #0-10	0 or 1	0-10	Engine Data 1
Fuel Trim Learn	Disabled / Enabled	Enabled	Enabled	Engine Data 1
HO2S (B1 S1, B1 S2)	0-1132 mv	1-1000	1-1000	Engine Data 1
HO2S-11 Status	Not / Ready	Ready	Ready	Engine Data 1
HO2S-11 XCounts	0-255 counts	Varies	Varies	Engine Data 1
IAC Position	0-255 counts	10-40	---	Engine Data 1
IAT Sensor (°F)	-40 to 304°F	50-194	50-194	Engine Data 1
Ignition Bypass	Active / Inactive	Inactive	Inactive	Engine Data 1
Ignition IC Mode	Active / Inactive	Active	Active	Engine Data 1
Injector Fault	Yes / No	No	No	Engine Data 1
Injector Pulsewidth	0-985 ms	1.5-3.5	Varies	Engine Data 1
Knock Retard (°)	0-128°	0	0	Engine Data 1
Long, Short Fuel Trim	-23 to 16%	Varies	Varies	Engine Data 1
Loop Status	Closed Loop / Open Loop	Closed Loop	Closed Loop	Engine Data 1
MAF Sensor (g/sec)	0-512 g/sec	3-6	Varies	Engine Data 1
MAF Sensor (Hz)	0-31,999hz	2000-2500	Varies	Engine Data 1
MAP Sensor (kPa)	10-110 kPa	25-34	Varies	Engine Data 1
Spark Advance (°)	-64° to 64°	16	Varies	Engine Data 1
TCC Brake Switch	Applied / Released	Released	Released	Engine Data 1
TCC Engaged	Yes / No	No	No	Engine Data 1
TP Angle (%)	0-100%	0	Varies	Engine Data 1
TP Sensor	0-5.0 volts	0.20-0.74	Varies	Engine Data 1
Transmission Range	X or O	OOXX	Varies	Engine Data 1
VTD Fuel Enable	Disabled / Enabled	Disabled	Disabled	Engine Data 1

1996 Cutlass Supreme 3.4L V6 MFI VIN X (A/T) - PID Data List

Parameter Identification (PID)	PID Value Range	PID Value at Hot Idle	PID Value at 30 mph	Data List Type
A/C Clutch	On / Off	Off	Off	EGR
A/C Request	Yes / No	No	No	EGR
Actual EGR Position	0-100%	0	0-100	EGR
BARO Sensor (kPa)	10-110 kPa	65-110	65-110	EGR
Current Gear	1-2-3-4	1	3	EGR
Decel Fuel Mode	Active / Inactive	Inactive	Inactive	EGR
Desired EGR Position	0-100%	0	0-100	EGR
ECT Sensor (°F)	-40 to 304°F	185-220	185-220	EGR
EGR Closed Valve Pintle Position	0-5.0 volts	0.14-1.00v	----	EGR
EGR Duty Cycle	0-100%	0	0-100	EGR
EGR Position Error	0-100%	0-9	----	EGR
EGR Flow Test Count	0-255 counts	0-10	0-10	EGR
Engine Load	0-100%	2-5	Varies	EGR
Engine Run Time	Hr: Min: Sec	00:00:00	00:00:00	EGR
Engine Speed	0-9999 rpm	±50 Desired	----	EGR
Hot Open Loop	Active / Inactive	Inactive	Inactive	EGR
Ignition 1 Signal	0.0-25.5v	14.2	14.2	EGR
Knock Activity	Yes / No	No	No	EGR
Knock Retard (°)	0-128°	0	0	EGR
Loop Status	Closed Loop / Open Loop	Closed Loop	Closed Loop	EGR
MAF Sensor (g/sec)	0-512 g/sec	3-6	Varies	EGR
MAP Sensor (kPa)	10-110 kPa	25-34	Varies	EGR
MAP Sensor (V)	0-5.0 volts	0.78-1.30	Varies	EGR
MPH / KPH	0-155 mph	0	30	EGR
Power Enrichment	Active / Inactive	Inactive	Inactive	EGR
TCC Engaged	Engaged / Disengaged	Disengaged	Disengaged	EGR
TP Angle (%)	0-100%	0	Varies	EGR
Transmission Range	X or O	OOXX	Varies	EGR
TWC Protection	Active / Inactive	Inactive	Inactive	EGR

1996 Cutlass Supreme 3.4L V6 MFI VIN X (A/T) - Dash Lamps List

Parameter Identification (PID)	PID Value Range	PID Value at Hot Idle	PID Value at 30 mph	Data List Type
Commanded Generator	On / Off	On	On	IPC
Engine Oil Level	Okay / Low	Okay	Okay	IPC
Generator Lamp	On / Off	Off	Off	IPC
Ignition 1 Signal	0.0-25.5v	14.2	14.2	IPC
Low Oil Lamp	On / Off	Off	Off	IPC
MIL Command	On / Off	Off	Off	IPC

1996 Cutlass Supreme 3.4L V6 MFI VIN X (A/T) - PID Data List

Parameter Identification (PID)	PID Value Range	PID Value at Hot Idle	PID Value at 30 mph	Data List Type
Air Fuel Ratio	0.0.0-25.5:1	14.7:1	14.7:1	Fuel Trim
AIR Pump	On / Off	Off	Off	Fuel Trim
Decel Fuel Mode	Active / Inactive	Inactive	Inactive	Fuel Trim
ECT Sensor (°F)	-40 to 304°F	185-220	185-220	Fuel Trim
Engine Load	0-100%	2-5	Varies	Fuel Trim
Engine Run Time	Hr: Min: Sec	00:00:00	00:00:00	Fuel Trim
Engine Speed	0-9999 rpm	±50 Desired	----	Fuel Trim
EVAP Purge PWM	0-100%	0-25%	0-75%	Fuel Trim
HO2S (B1 S1, B1 S2)	0-1132 mv	1-1000	1-1000	Fuel Trim
HO2S-11 Status	Not / Ready	Ready	Ready	Fuel Trim
HO2S-11 XCounts	0-255 counts	Varies	Varies	Fuel Trim
IAT Sensor (°F)	-40 to 304°F	50-194	50-194	Fuel Trim
Injector Pulsewidth	0-985 ms	1.5-3.5	Varies	Fuel Trim
Long, Short Term F/T	-23 to 16%	0 (± 5%)	0 (± 5%)	Fuel Trim
MAF Sensor (g/sec)	0-512 g/sec	3-6	Varies	Fuel Trim
Power Enrichment	Active / Inactive	Inactive	Inactive	Fuel Trim
Startup ECT Sensor	-40 to 304°F	Varies	Varies	Fuel Trim
Startup IAT Sensor	-40 to 304°F	Varies	Varies	Fuel Trim

1996 Cutlass Supreme 3.4L V6 MFI VIN X (A/T) - PID Data List

Parameter Identification (PID)	PID Value Range	PID Value at Hot Idle	PID Value at 30 mph	Data List Type
3X Crank Sensor	0-9999 rpm	Varies	Varies	Misfire
24X Crank Sensor	0-1600 rpm	Varies	Varies	Misfire
Abuse Management	Active / Inactive	Inactive	Inactive	Misfire
A/C Clutch	On / Off	Off	Off	Misfire
A/C Request	Yes / No	No	No	Misfire
Actual EGR Position	0-100%	0	0-100	Misfire
AIR Active Test	Active / Inactive	Inactive	Inactive	Misfire
Cam Signal Present	Yes / No	Yes	Yes	Misfire
Current Gear	1/2/3/4	1	3	Misfire
Decel Fuel Mode	Active / Inactive	Inactive	Inactive	Misfire
Desired EGR Position	0-100%	0	0-100	Misfire
ECT Sensor (°F)	-40 to 304°F	185-220	185-220	Misfire
Engine Load	0-100%	2-5	Varies	Misfire
HO2S-11 (B1 S1)	0-1132 mv	1-1000	1-1000	Misfire
MAF Sensor (g/sec)	0-512 g/sec	3-6	Varies	Misfire
Misfire current Cyl 1-6	0-198	0 - 4	Varies	Misfire
Misfire History Cyl 1-6	0-65535	0	0	Misfire
M/F Fail since 1st Fail	0-65535	0	0	Misfire
M/Fire Pass / 1st Fail	0-65535	0	0	Misfire
Non-Volatile Memory	Pass / Fail	Pass	Pass	Misfire
Power Enrichment	Active / Inactive	Active	Active	Misfire
Spark Advance (°)	-64° to 64°	16	Varies	Misfire
TCC Engaged	Yes / No	No	No	Misfire
Total Misfire Current	0-255 counts	0	0	Misfire
TP Angle (%)	0-100%	0	Varies	Misfire
Tran. Hot Mode	Active / Inactive	Inactive	Inactive	Misfire

EIGHTY EIGHT, NINETY EIGHT PID DATA

1995 Eighty Eight, Ninety Eight 3.8L V6 VIN K, 1 - PID Data List

Parameter Identification (PID)	PID Value Range	PID Value at Hot Idle	PID Value at 30 mph	Data List Type
2nd-3rd-4th Gear Switch	Yes / No	No	Yes / No	Engine Data 1
A/C Clutch	On / Off	Off	Off	Engine Data 1
A/C High Pressure	Not Okay / Okay	Okay	Okay	Engine Data 1
A/C Request	Yes / No	No	No	Engine Data 1
BLM Cell	0-10	0-1	0-10	Engine Data 1
Boost PWM (VIN 1)	0-100%	100	0-100	Engine Data 1
Cam/Crank Error	0-255 counts	0-2	1	Engine Data 1
C/C Brake Switch	Applied / Released	Released	Released	Engine Data 1
Commanded Gear	1-2-3-4	P-N	3	Engine Data 1
Cooling Fan 1, 2	On / Off	Off	Off	Engine Data 1
Current Weak CYL	CYL #/None	None	None	Engine Data 1
Desired Idle Speed	0-3187	---	620	Engine Data 1
Desired EGR Position	0-100%	0%	0%	Engine Data 1
Driver Select Switch	Normal / Sport	Normal	Normal	Engine Data 1
ECT Sensor (°F)	-40-419°F	185-220	210%	Engine Data 1
EGR Actual Position	0-100%	0%	0%	Engine Data 1
EGR Desired Position	0-100%	0%	0%	Engine Data 1
EGR Pintle Position	0-5.0 volts	0-1v	0.76v	Engine Data 1
Engine Load	0-100%	---	38%	Engine Data 1
Engine Speed	0-9999 rpm	±50 Desired	1192	Engine Data 1
EVAP PWM Control	0-100%	0-25%	14%	Engine Data 1
Fuel Trim Cell	0-22	3	2	Engine Data 1
HO2S Cross Counts	0-255 counts	0-20	3	Engine Data 1
HO2S-11 (B1 S1)	0-1132 mv	10-1000	40ms	Engine Data 1
IAC Motor	0-255 counts	16-20	18	Engine Data 1
IAT or MAT Sensor	-38-284°F	50-176	77°F	Engine Data 1
Injector Pulsewidth	0-985 ms	---	2.7ms	Engine Data 1
Knock Retard (°)	0-128°	0	0	Engine Data 1
Knock Signal	Yes / No	No	No	Engine Data 1
Long, Short Term F/T	0-255 counts	95-138	118	Engine Data 1
Loop Status	Closed Loop / Open Loop	Closed Loop	Closed Loop	Engine Data 1
MAF Sensor (g/sec)	0-512 g/sec	4 - 9	Varies	Engine Data 1
Pass-Key Fuel Enable	On / Off	On	On	Engine Data 1
PRNDL P/A/B/C	Several	High	Several	Engine Data 1
QDM 1, 2 & 4	Several	Low	Several	Engine Data 1
Shift Solenoid 'A', 'B'	On / Off	On	Off	Engine Data 1
Spark Advance (°)	-90° to 90°	16°	25°	Engine Data 1
TCC Brake Switch	Applied / Released	Released	Released	Engine Data 1
TCC Duty Cycle	0-100%	0	0-100	Engine Data 1
TCC Mode	On/APP/Off	Off	APPLY	Engine Data 1
TCC Slip Speed	-4080 - 4079	+254	+45	Engine Data 1
Throttle Angle	0-100%	0	2	Engine Data 1
Time From Startup	Hr: Min: Sec	00:00:00	00:00:00	Engine Data 1
TP Sensor (V)	0-5.0 volts	0.20-0.74	0.68v	Engine Data 1
TFT Sensor	40-304°F	180-200°F	199°F	Engine Data 1
Trans. N/V Ratio	0-255 counts	0	39	Engine Data 1
Trans. Range Switch	P-R-N-4-3-2	P	3	Engine Data 1
Vehicle Speed	0-155 mph	0	30	Engine Data 1

1996-99 Eighty Eight, Ninety Eight 3.8L V6 VIN K - PID Data List

Parameter Identification (PID)	PID Value Range	PID Value at Hot Idle	PID Value at 30 mph	Data List Type
Actual EGR Position	0-100%	0	0-100	Engine Data 1
Air Fuel Ratio	0.1-25.5:1	14.7:1	14.7:1	Engine Data 1
BARO Sensor (kPa)	10-110 kPa	65-110	65-110	Engine Data 1
Commanded Generator	On / Off	On	On	Engine Data 1
Cruise Engaged	Yes / No	No	No	Engine Data 1
Cruise Inhibited	Yes / No	Yes	Yes	Engine Data 1
Current Gear	1-2-3-4	1	3	Engine Data 1
Decel Fuel Mode	Active / Inactive	Inactive	Inactive	Engine Data 1
Desired EGR Position	0-100%	0%	0-100%	Engine Data 1
Desired Idle Speed	0-3187	PCM control	---	Engine Data 1
ECT Sensor (°F)	-40 to 304°F	185-220	185-220	Engine Data 1
Engine Load	0-100%	2-5	Varies	Engine Data 1
Engine Run Time	Hr: Min: Sec	00:00:00	00:00:00	Engine Data 1
Engine Speed	0-9999 rpm	±50 Desired	---	Engine Data 1
Fuel Trim Cell	Cell #0-10	0	0-10	Engine Data 1
Fuel Trim Learn	Disabled / Enabled	Enabled	Enabled	Engine Data 1
HO2S-11 (B1 S1)	0-1132 mv	10-1000	10-1000	Engine Data 1
HO2S-12 (B1 S2)	0-1132 mv	10-1000	10-1000	Engine Data 1
HO2S-11 (B1 S1) Status	Not/Ready	Ready	Ready	Engine Data 1
IAC Position	0-255 counts	10-40	---	Engine Data 1
IAT Sensor (°F)	-40 to 304°F	50-194	50-194	Engine Data 1
Ignition 1 Signal	0.0-25.5v	14.2	14.2	Engine Data 1
Injector Pulsewidth	0-985 ms	1.5-3.5	Varies	Engine Data 1
Long Term Fuel Trim Bank 1	-10% to 10%	0 (± 5%)	0 (± 5%)	Engine Data 1
Long Term Fuel Trim Bank 2	-10% to 10%	0 (± 5%)	0 (± 5%)	Engine Data 1
Loop Status	Closed Loop / Open Loop	Closed Loop	Closed Loop	Engine Data 1
MAF Sensor (g/sec)	0-512 g/sec	3-6	Varies	Engine Data 1
MAF Sensor (Hz)	0-31, 999 Hz	1200-3000	Varies	Engine Data 1
MAP Sensor (kPa)	10-110 kPa	29-48	Varies	Engine Data 1
MAP Sensor (V)	0-5.0 volts	1-2	Varies	Engine Data 1
Power Enrichment	Active / Inactive	Active	Active	Engine Data 1
Rich / Lean	L or R	L-R-L-R	L-R-L-R	Engine Data 1
Short Term Fuel Trim Bank 1	-10% to 10%	0 (± 5%)	0 (± 5%)	Engine Data 1
Short Term Fuel Trim Bank 2	-10% to 10%	0 (± 5%)	0 (± 5%)	Engine Data 1
Throttle at Idle	Yes / No	Yes	No	Engine Data 1
TP Angle (%)	0-100%	0	Varies	Engine Data 1
TP Sensor	0-5.0 volts	0.20-0.74	Varies	Engine Data 1
TWC Protection	Active / Inactive	Inactive	Inactive	Engine Data 1
Vehicle Speed	0-155 mph	0	30	Engine Data 1

1996-99 Eighty Eight, Ninety Eight 3.8L V6 VIN K - PID Data List

Parameter Identification (PID)	PID Value Range	PID Value at Hot Idle	PID Value at 30 mph	Data List Type
A/C High Side Pressure	0-5.0 volts	1-4	1-4	Engine Data 2
A/C Pressure Out of Range	Yes / No	No	No	Engine Data 2
A/C Request	Yes / No	No	No	Engine Data 2
Actual EGR Position	0-100%	0	0-100	Engine Data 2
BARO Sensor (kPa)	10-110 kPa	65-110	65-110	Engine Data 2
Boost Duty Cycle Command (VIN 1)	0-100%	100	0-100	Engine Data 2
Commanded A/C	On / Off	Off	Off	Engine Data 2
Commanded Fan 1	On / Off	Off	Off	Engine Data 2
Commanded Fan 2	On / Off	Off	Off	Engine Data 2
Commanded Starter	Disabled / Enabled	Disabled	Disabled	Engine Data 2
Decel Fuel Mode	Active / Inactive	Inactive	Inactive	Engine Data 2
Desired Idle Speed	0-3187 rpm	PCM control	---	Engine Data 2
ECT Sensor (°F)	-40 to 304°F	185-220	185-220	Engine Data 2
Engine Load	0-100%	2 5	Varies	Engine Data 2
Engine Oil Level	Low / Okay	Okay	Okay	Engine Data 2
Engine Oil Life	0-100%	Varies	Varies	Engine Data 2
Engine Oil Pressure	0-5.0 volts	2-3	Varies	Engine Data 2
Engine Run Time	Hr: Min: Sec	00:00:00	00:00:00	Engine Data 2
Engine Speed	0-9999 rpm	±50 Desired	---	Engine Data 2
Fuel Pump	On / Off	On	On	Engine Data 2
Fuel Pump Speed	High/Normal	Normal	Normal	Engine Data 2
Generator PWM	0-100%	0-100	0-100	Engine Data 2
IAC Position	0-255 counts	10-40	---	Engine Data 2
IAT Sensor (°F)	-40 to 304°F	50-194	50-194	Engine Data 2
Ignition 1 Signal	0.0-25.5v	14.2	14.2	Engine Data 2
Ignition Mode	IC/Bypass	IC	IC	Engine Data 2
Knock Retard (°)	0.0-25.5°	0	0	Engine Data 2
MAF Sensor (g/sec)	0-512 g/sec	3-6	Varies	Engine Data 2
MAP Sensor (kPa)	10-110 kPa	29-48	Varies	Engine Data 2
MAP Sensor (V)	0-5.0 volts	1-2	Varies	Engine Data 2
MIL Command	On / Off	Off	Off	Engine Data 2
Power Enrichment	Active / Inactive	Inactive	Inactive	Engine Data 2
Spark Advance	-64° to 64°	20	Varies	Engine Data 2
TP Angle (%)	0-100%	0	Varies	Engine Data 2
Traction Control	Active / Inactive	Inactive	Inactive	Engine Data 2
TCM Desired Torque	0-100%	80-95	0-100	Engine Data 2
TCM Control Torque	0-100%	5-20	0-100	Engine Data 2
Transmission Range	P-R-N-4-3-2	Park	DRIVE 3	Engine Data 2
TR Switch P, A	High / Low	High	Low	Engine Data 2
TR Switch B, C	High / LO	Low	High	Engine Data 2
TWC Protection	Active / Inactive	Inactive	Inactive	Engine Data 2
Vehicle Speed	0-155 mph	0	30	Engine Data 2

1996-99 Eighty Eight, Ninety Eight 3.8L V6 VIN K - PID Data List

Parameter Identification (PID)	PID Value Range	PID Value at Hot Idle	PID Value at 30 mph	Data List Type
A/C Request	Yes / No	No	No	Catalyst
Air Fuel Ratio	0.1-25.5:1	14.7:1	14.7:1	Catalyst
BARO Sensor (kPa)	10-110 kPa	65-110	65-110	Catalyst
Commanded A/C	On / Off	Off	Off	Catalyst
Commanded Fan 1, 2	On / Off	Off	Off	Catalyst
Decel Fuel Mode	Active / Inactive	Inactive	Inactive	Catalyst
Desired Idle Speed	0-3187 rpm	PCM control	---	Catalyst
ECT Sensor (°F)	-40 to 304°F	185-220	185-220	Catalyst
Engine Load	0-100%	2-5	Varies	Catalyst
Engine Run Time	Hr: Min: Sec	00:00:00	00:00:00	Catalyst
Engine Speed	0-9999 rpm	±50 Desired	---	Catalyst
HO2S-11 (B1 S1)	0-1132 mv	10-1000	10-1000	Catalyst
HO2S-12 (B1 S2)	0-1132 mv	10-1000	10-1000	Catalyst
IAC Position	0-255 counts	10-40	---	Catalyst
IAT Sensor (°F)	-40 to 304°F	50-194	50-194	Catalyst
Injector Pulsewidth	0-985 ms	1.5-3.5	Varies	Catalyst
Loop Status	Closed Loop / Open Loop	Closed Loop	Closed Loop	Catalyst
MAF Sensor (g/sec)	0-512 g/sec	3-6	Varies	Catalyst
Power Enrichment	Active / Inactive	Inactive	Inactive	Catalyst
Short Term Fuel Trim	-10% to 10%	0 (± 5%)	0 (± 5%)	Catalyst
TP Angle (%)	0-100%	0	Varies	Catalyst
Transmission Range	P-R-N-4-3-2	Park	D3	Catalyst
TWC Diagnostic	Disabled / Enabled	Enabled	Enabled	Catalyst
TWC Monitor Test	0-49 counts	0	0	Catalyst
TWC Protection	Active / Inactive	Inactive	Inactive	Catalyst
Vehicle Speed	0-155 mph	0	30	Catalyst

1996-99 Eighty Eight, Ninety Eight 3.8L V6 VIN K - PID Data List

Parameter Identification (PID)	PID Value Range	PID Value at Hot Idle	PID Value at 30 mph	Data List Type
Actual EGR Position	0-100%	0	0-100	EGR
BARO Sensor (kPa)	10-110 kPa	65-110	65-110	EGR
Desired EGR Position	0-100%	0	0-100	EGR
ECT Sensor (°F)	-40 to 304°F	185-220	185-220	EGR
EGR Closed Pintle Position	0-5.0 volts	0.14-1.0v	---	EGR
EGR Duty Cycle	0-100%	0	0-100	EGR
EGR Feedback	0-5.0 volts	0.14-1.0v	---	EGR
EGR Flow Test Count	0-255 counts	0-10	0-10	EGR
EGR Position Error	0-100%	0-9	Varies	EGR
Engine Speed	0-9999 rpm	±50 Desired	---	EGR
IAC Position	0-255 counts	10-40	---	EGR
Ignition 1 Signal	0.0-25.5v	14.2	14.2	EGR
MAP Sensor (kPa)	10-110 kPa	29-48	Varies	EGR
MAP Sensor (V)	0-5.0 volts	1-2	Varies	EGR
Power Enrichment	Active / Inactive	Active	Active	EGR
TP Angle (%)	0-100%	0	Varies	EGR
Vehicle Speed	0-155 mph	0	30	EGR

1996-99 Eighty Eight, Ninety Eight 3.8L V6 VIN K - PID Data List

Parameter Identification (PID)	PID Value Range	PID Value at Hot Idle	PID Value at 30 mph	Data List Type
BARO Sensor (kPa)	10-110 kPa	65-110	65-110	EVAP
Desired Idle Speed	0-3187 rpm	PCM control	----	EVAP
ECT Sensor (°F)	-40 to 304°F	185-220	185-220	EVAP
Engine Run Time	Hr: Min: Sec	00:00:00	00:00:00	EVAP
Engine Speed	0-9999 rpm	±50 Desired	----	EVAP
Fuel Level	0-100%	Varies	Varies	EVAP
EVAP Canister Purge	0-100%	0-25	0-50	EVAP
EVAP Fault History	Fault/No	No Fault	No Fault	EVAP
FTP Sensor (In. H2O)	-17.5 to 7.5"	Varies	Varies	EVAP
Fuel Tank Pressure	0-5.0 volts	Varies	Varies	EVAP
IAT Sensor (°F)	-40 to 304°F	50-194	50-194	EVAP
MAP Sensor (kPa)	10-110 kPa	29-48	Varies	EVAP
MAP Sensor (V)	0-5.0 volts	1-2	Varies	EVAP
Startup ECT Sensor	-40 to 304°F	Varies	Varies	EVAP
Startup IAT Sensor	-40 to 304°F	Varies	Varies	EVAP

1996-99 Eighty Eight, Ninety Eight 3.8L V6 VIN K - PID Data List

Parameter Identification (PID)	PID Value Range	PID Value at Hot Idle	PID Value at 30 mph	Data List Type
Air Fuel Ratio	0.1-25.5:1	14.7:1	14.7:1	Fuel Trim
Decel Fuel Mode	Active / Inactive	Inactive	Inactive	Fuel Trim
ECT Sensor (°F)	-40 to 304°F	185-220	185-220	Fuel Trim
Engine Speed	0-9999 rpm	±50 Desired	----	Fuel Trim
EVAP Canister Purge	0-100%	0-25	0-50	Fuel Trim
HO2S-11 (B1 S1)	0-1132 mv	10-1000	10-1000	Fuel Trim
HO2S-12 (B1 S2)	0-1132 mv	10-1000	10-1000	Fuel Trim
HO2S-11 (B1 S1) Status	Not Ready/Ready	Ready	Ready	Fuel Trim
HO2S-11 (B1 S1) warmup	Min: Sec	00:00	00:00	Fuel Trim
HO2S-12 (B1 S2) Warmup	Min: Sec	00:00	00:00	Fuel Trim
HO2S-11 XCounts	0-255 counts	Varies	Varies	Fuel Trim
IAT Sensor (°F)	-40 to 304°F	50-194	50-194	Fuel Trim
Ignition 1 Signal	0.0-25.5v	14.2	14.2	Fuel Trim
Loop Status	Closed Loop / Open Loop	Closed Loop	Closed Loop	Fuel Trim
Power Enrichment	Active / Inactive	Inactive	Inactive	Fuel Trim
Startup ECT Sensor	-40 to 304°F	Varies	Varies	Fuel Trim
Startup IAT Sensor	-40 to 304°F	Varies	Varies	Fuel Trim
TP Angle (%)	0-100%	0	Varies	Fuel Trim
TWC Protection	Active / Inactive	Inactive	Inactive	Fuel Trim

1996-99 Eighty Eight, Ninety Eight 3.8L V6 VIN K - PID Data List

Parameter Identification (PID)	PID Value Range	PID Value at Hot Idle	PID Value at 30 mph	Data List Type
A/C Request	Yes / No	No	No	Misfire
Actual EGR Position	0-100%	0	0-100	Misfire
Commanded A/C	On / Off	Off	Off	Misfire
Decel Fuel Mode	Active / Inactive	Inactive	Inactive	Misfire
Desired EGR Position	0-100%	0	0-100	Misfire
ECT Sensor (°F)	-40 to 304°F	185-220	185-220	Misfire
Engine Load	0-100%	2-5	Varies	Misfire
Engine Speed	0-9999 rpm	±50 Desired	---	Misfire
EVAP Canister Purge	0-100%	0-25	0-50	Misfire
IAC Position	0-255 counts	10-40	---	Misfire
Long Term Fuel Trim	-10% to 10%	0 (± 5%)	0 (± 5%)	Misfire
Loop Status	Closed Loop / Open Loop	Closed Loop	Closed Loop	Misfire
Misfire Current Cylinder 1-6	0-198	0-2	0-2	Misfire
Misfire History Cylinder 1-6	0-65535	0	0	Misfire
Misfiring Cylinder	CYL 1-6	0	0	Misfire
Power Enrichment	Active / Inactive	Inactive	Inactive	Misfire
Short Term Fuel Trim	-10% to 10%	0 (± 5%)	0 (± 5%)	Misfire
Total Misfire Current Count	0-99	0	0	Misfire
Misfire Failures Since First Fail	0-65535	0	0	Misfire
Misfire Passes Since First Fail	0-65535	0	0	Misfire
TP Angle (%)	0-100%	0	Varies	Misfire
TWC Protection	Active / Inactive	Inactive	Inactive	Misfire
Vehicle Speed	0-155 mph	0	30	Misfire

INTRIGUE PID DATA

1998-99 Intrigue 3.8L V6 MFI VIN K - PID Data List

Parameter Identification (PID)	PID Value Range	PID Value at Hot Idle	PID Value at 30 mph	Data List Type
Actual EGR Position	0-100%	0	0-100	Engine Data 1
Air Fuel Ratio	0.1-25.5:1	14.7:1	14.7:1	Engine Data 1
BARO Sensor (kPa)	10-110 kPa	65-110	65-110	Engine Data 1
Commanded Generator	On / Off	On	On	Engine Data 1
Cruise	Engaged / Disengaged	Disengaged	Disengaged	Engine Data 1
Cruise Inhibited	Yes / No	Yes	Yes	Engine Data 1
Current Gear	1-2-3-4	1	3	Engine Data 1
Decel Fuel Mode	Active / Inactive	Inactive	Inactive	Engine Data 1
Desired EGR Position	0-100%	0	0-100	Engine Data 1
Desired Idle Speed	0-3187	PCM control	---	Engine Data 1
ECT Sensor (°F)	-40 to 304°F	185-220	185-220	Engine Data 1
Engine Load	0-100%	2-5	Varies	Engine Data 1
Engine Run Time	Hr: Min: Sec	00:00:00	00:00:00	Engine Data 1
Engine Speed	0-9999 rpm	±50 Desired	---	Engine Data 1
Fuel Trim Cell	Cell #0-10	0	0-10	Engine Data 1
Fuel Trim Learn	Disabled / Enabled	Enabled	Enabled	Engine Data 1
HO2S-11 (B1 S1)	0-1132 mv	10-1000	10-1000	Engine Data 1
HO2S-12 (B1 S2)	0-1132 mv	10-1000	10-1000	Engine Data 1
HO2S-11 (B1 S1) Status	Not/Ready	Ready	Ready	Engine Data 1
IAC Position	0-255 counts	10-40	---	Engine Data 1
IAT Sensor (°F)	-40 to 304°F	50-194	50-194	Engine Data 1
Ignition 1 Signal	0.0-25.5v	14.2	14.2	Engine Data 1
Injector Pulsewidth	0-985 ms	1.5-3.5	Varies	Engine Data 1
Long Term Fuel Trim	-10% to 10%	0 (± 5%)	0 (± 5%)	Engine Data 1
Loop Status	Closed Loop / Open Loop	Closed Loop	Closed Loop	Engine Data 1
MAF Sensor (g/sec)	0-512 g/sec	3-6	Varies	Engine Data 1
MAF Sensor (Hz)	0-32000 Hz	1200-3000	Varies	Engine Data 1
MAP Sensor (kPa)	10-110 kPa	29-48	Varies	Engine Data 1
MAP Sensor (V)	0-5.0 volts	1-2	Varies	Engine Data 1
Power Enrichment	Active / Inactive	Inactive	Inactive	Engine Data 1
Rich / Lean	L or R	L-R-L-R	L-R-L-R	Engine Data 1
Short Term Fuel Trim	-10% to 10%	0 (± 5%)	0 (± 5%)	Engine Data 1
Throttle at Idle	Yes / No	Yes	No	Engine Data 1
TP Angle (%)	0-100%	0	Varies	Engine Data 1
TP Sensor	0-5.0 volts	0.20-0.74	---	Engine Data 1
TWC Protection	Active / Inactive	Inactive	Inactive	Engine Data 1
Vehicle Speed	0-155 mph	0	30	Engine Data 1

1998-99 Intrigue 3.8L V6 MFI VIN K - PID Data List

Parameter Identification (PID)	PID Value Range	PID Value at Hot Idle	PID Value at 30 mph	Data List Type
A/C High Side Pressure	0-5.0 volts	Varies	Varies	Engine Data 2
A/C Press Too Hi/Low	Yes / No	No	No	Engine Data 2
A/C Request	Yes / No	No	No	Engine Data 2
Actual EGR Position	0-100%	0	0-100	Engine Data 2
BARO Sensor (kPa)	10-110 kPa	65-110	65-110	Engine Data 2
Commanded A/C	On / Off	Off	Off	Engine Data 2
Commanded Fan 1, 2	On / Off	Off	Off	Engine Data 2
Commanded Starter	Disabled / Enabled	Disabled	Disabled	Engine Data 2
Desired EGR Position	0-100%	0	0-100	Engine Data 2
Decel Fuel Mode	Active / Inactive	Inactive	Inactive	Engine Data 2
Desired Idle Speed	0-3187 rpm	PCM control	---	Engine Data 2
ECT Sensor (°F)	-40 to 304°F	185-220	185-220	Engine Data 2
Engine Load	0-100%	2-5	Varies	Engine Data 2
Engine Oil Level	Low / Okay	Okay	Okay	Engine Data 2
Engine Oil Pressure	Low / Okay	Okay	Okay	Engine Data 2
Engine Speed	0-9999 rpm	±50 Desired	---	Engine Data 2
Generator PWM	0-100%	0-100	0-100	Engine Data 2
IAC Position	0-255 counts	10-40	---	Engine Data 2
IAT Sensor (°F)	-40 to 304°F	50-194	50-194	Engine Data 2
Knock Retard (°)	0.0-25.5°	0	0	Engine Data 2
MAF Sensor (g/sec)	0.0-512	3-6	Varies	Engine Data 2
MAP Sensor (kPa)	10-110 kPa	29-48	Varies	Engine Data 2
MIL Command	On / Off	Off	Off	Engine Data 2
Power Enrichment	Active / Inactive	Inactive	Inactive	Engine Data 2
Spark Advance (°)	-64° to 64°	20	Varies	Engine Data 2
TP Angle (%)	0-100%	0	Varies	Engine Data 2
Transmission Range	P-R-N-4-3-2	PARK	DRIVE 3	Engine Data 2
TR Switch P/A/B/C	High / Low	High	Several	Engine Data 2
TWC Protection	Active / Inactive	Inactive	Inactive	Engine Data 2

1998-99 Intrigue 3.8L V6 MFI VIN K - PID Data List

Parameter Identification (PID)	PID Value Range	PID Value at Hot Idle	PID Value at 30 mph	Data List Type
Actual EGR Position	0-100%	0	0-100	EGR
BARO Sensor (kPa)	10-110 kPa	110 S/Level	65-110	EGR
Desired EGR Position	0-100%	0	0-100	EGR
ECT Sensor (°F)	-40 to 304°F	185-220	185-220	EGR
EGR Closed Pintle	0-5.0 volts	0.14-1.0v	---	EGR
EGR Duty Cycle	0-100%	0	0-100	EGR
EGR Feedback	0-5.0 volts	0.14-1.0v	---	EGR
EGR Flow Test Count	0-255 counts	0-10	0-10	EGR
EGR Position Error	0-100%	0-9	Varies	EGR
Engine Speed	0-9999 rpm	±50 Desired	---	EGR
IAC Position	0-255 counts	10-40	---	EGR
IAT Sensor (°F)	-40 to 304°F	50-194	50-194	EGR
MAP Sensor (kPa)	10-110 kPa	29-48	Varies	EGR
TP Angle (%)	0-100%	0	0-100	EGR
Vehicle Speed	0-155 mph	0	30	EGR

1998-99 Intrigue 3.8L V6 MFI VIN K (A/T) - PID Data List

Parameter Identification (PID)	PID Value Range	PID Value at Hot Idle	PID Value at 30 mph	Data List Type
Air Fuel Ratio	0.1-25.5:1	14.7:1	14.7:1	Catalyst
BARO Sensor (kPa)	10-110 kPa	65-110	65-110	Catalyst
Commanded A/C	On / Off	Off	Off	Catalyst
Commanded Fan 1, 2	On / Off	Off	Off	Catalyst
Decel Fuel Mode	Active / Inactive	Inactive	Inactive	Catalyst
Desired Idle Speed	0-3187 rpm	PCM control	----	Catalyst
ECT Sensor (°F)	-40 to 304°F	185-220	185-220	Catalyst
Engine Load	0-100%	2-5	Varies	Catalyst
Engine Run Time	Hr: Min: Sec	00:00:00	00:00:00	Catalyst
Engine Speed	0-9999 rpm	±100 rpm	----	Catalyst
Fuel Trim Learn	DIS/ENABL	Enabled	Enabled	Catalyst
HO2S-11 (B1 S1)	0-1132 mv	10-1000	10-1000	Catalyst
HO2S-12 (B1 S2)	0-1132 mv	10-1000	10-1000	Catalyst
IAC Position	0-255 counts	10-40	----	Catalyst
IAT Sensor (°F)	-40 to 304°F	50-194	50-194	Catalyst
Injector Pulsewidth	0-985 ms	1.5-3.5	Varies	Catalyst
Loop Status	Closed Loop / Open Loop	Closed Loop	Closed Loop	Catalyst
MAF Sensor (g/sec)	0-512 g/sec	3-6	Varies	Catalyst
Power Enrichment	Active / Inactive	Active	Active	Catalyst
Short Term Fuel Trim	-10% to 10%	Varies	Varies	Catalyst
TP Angle (%)	0-100%	0	Varies	Catalyst
Transmission Range	P-R-N-4-3-2	PARK	DRIVE 3	Catalyst
TWC Diagnostic	Active / Inactive	Inactive	Inactive	Catalyst
TWC Monitor Test Counter	0-49 counts	0	0	Catalyst
TWC Protection	Active / Inactive	Inactive	Inactive	Catalyst
Vehicle Speed	0-155 mph	0	30	Catalyst

1998-99 Intrigue 3.8L V6 MFI VIN K - PID Data List

Parameter Identification (PID)	PID Value Range	PID Value at Hot Idle	PID Value at 30 mph	Data List Type
BARO Sensor (kPa)	10-110 kPa	65-110	65-110	EVAP
Desired Idle Speed	0-3187 rpm	PCM control	----	EVAP
ECT Sensor (°F)	-40 to 304°F	185-220	185-220	EVAP
Engine Run Time	Hr: Min: Sec	00:00:00	00:00:00	EVAP
Engine Speed	0-9999 rpm	±50 Desired	----	EVAP
Fuel Tank Level	0-100%	Varies	Varies	EVAP
EVAP Canister Purge	0-100%	0-25	0-100	EVAP
EVAP Vent Solenoid	Open/Close	Open	Open	EVAP
FTP Sensor (In. H2O)	-17.5 to 7.5"	Varies	Varies	EVAP
FTP Sensor (V)	0-5.0 volts	Varies	Varies	EVAP
IAT Sensor (°F)	-40 to 304°F	50-194	50-194	EVAP
MAP Sensor (kPa)	10-110 kPa	29-48	Varies	EVAP
MAP Sensor (V)	0-5.0 volts	1-2	Varies	EVAP
Startup ECT Sensor	-40 to 304°F	Varies	Varies	EVAP
Startup IAT Sensor	-40 to 304°F	Varies	Varies	EVAP
Vehicle Speed	0-155 mph	0	30	EVAP

1998-99 Intrigue 3.8L V6 MFI VIN K (A/T) - PID Data List

Parameter Identification (PID)	PID Value Range	PID Value at Hot Idle	PID Value at 30 mph	Data List Type
Air Fuel Ratio	0.1-25.5:1	14.7:1	14.7:1	Fuel Trim
Decel Fuel Mode	Active / Inactive	Inactive	Inactive	Fuel Trim
ECT Sensor (°F)	-40 to 304°F	185-220	185-220	Fuel Trim
Engine Speed	0-9999 rpm	±50 Desired	---	Fuel Trim
EVAP Canister Purge	0-100%	0-25	0-50	Fuel Trim
HO2S-11 (B1 S1)	0-1132 mv	10-1000	10-1000	Fuel Trim
HO2S-12 (B1 S2)	0-1132 mv	10-1000	10-1000	Fuel Trim
HO2S-11 Status	Ready / Not Ready	Ready	Ready	Fuel Trim
HO2S-11 Warmup	00:00-99:99	Varies	Varies	Fuel Trim
HO2S-12 Warmup	00:00-99:99	Varies	Varies	Fuel Trim
HO2S-11 XCounts	0-255 counts	Varies	Varies	Fuel Trim
IAT Sensor (°F)	-40 to 304°F	50-194	50-194	Fuel Trim
Loop Status	Closed Loop / Open Loop	Closed Loop	Closed Loop	Fuel Trim
MAF Sensor (g/sec)	0-512 g/sec	3-6 g/sec	Varies	Fuel Trim
Power Enrichment	Active / Inactive	Inactive	Inactive	Fuel Trim
Startup ECT Sensor	-40 to 304°F	Varies	Varies	Fuel Trim
Startup IAT Sensor	-40 to 304°F	Varies	Varies	Fuel Trim
TP Angle (%)	0-100%	0	Varies	Fuel Trim
TWC Protection	Active / Inactive	Inactive	Inactive	Fuel Trim

1998-99 Intrigue 3.8L V6 MFI VIN K (A/T) - PID Data List

Parameter Identification (PID)	PID Value Range	PID Value at Hot Idle	PID Value at 30 mph	Data List Type
A/C Request	Yes / No	No	No	Misfire
Actual EGR Position	0-100%	0	0-100	Misfire
Commanded A/C	On / Off	Off	Off	Misfire
Decel Fuel Mode	Active / Inactive	Inactive	Inactive	Misfire
Desired EGR Position	0-100%	0	0-100	Misfire
ECT Sensor (°F)	-40 to 304°F	185-220	185-220	Misfire
Engine Load	0-100%	2-5	Varies	Misfire
Engine Speed	0-9999 rpm	±50 Desired	---	Misfire
Fuel Trim Cell	Cell #	0	1	Misfire
IAC Position	0-255 counts	10-40	---	Misfire
IAT Sensor (°F)	-40 to 304°F	50-194	50-194	Misfire
Long Term Fuel Trim	-10% to 10%	0 (± 5%)	0 (± 5%)	Misfire
Loop Status	Closed Loop / Open Loop	Closed Loop	Closed Loop	Misfire
Misfire Cyl PRI/SEC	CYL 1-6	0	0	Misfire
Misfire current Cyl 1-6	0-11998	0-2	0-2	Misfire
Misfire History Cyl 1-6	0-65535	0	0	Misfire
Misfires Since 1st Fail	0-65535	0	0	Misfire
Total Misfire Failures Since 1st Pass	0-65535	0	0	Misfire
Power Enrichment	Active / Inactive	Inactive	Inactive	Misfire
Short Term Fuel Trim	-10% to 10%	0 (± 5%)	0 (± 5%)	Misfire
Total Misfire Current	0-99	0	0	Misfire
TP Angle (%)	0-100%	0	Varies	Misfire
TWC Protection	Active / Inactive	Inactive	Inactive	Misfire
Vehicle Speed	0-155 mph	0	30	Misfire

2000-02 Intrigue 3.5L V6 VIN H (A/T) - PID Data List

Parameter Identification (PID)	PID Value Range	PID Value at Hot Idle	PID Value at 30 mph	Data List Type
24X Crank Sensor	0-1600 rpm	Varies	Varies	Engine Data 2, Misfire
A/T 1-2 Solenoid Circuit Status	Okay / Fault	Okay	Okay	Output Driver
A/T 2-3 Solenoid Circuit Status	Okay / Fault	Okay	Okay	Output Driver
A/C High Side Pressure	0-450 psi	Varies	Varies	Engine Data 2
A/C High Side Pressure	0-5.1 volts	Varies	Varies	Engine Data 2
A/C Off for WOT	Yes / No	No	No	Engine Data 2
A/C Pressure Disable	Yes / No	No	No	Engine Data 2
A/C Relay Control Circuit Status	Okay / Fault	Okay	Okay	Output Driver
A/C Relay Command	On / Off	Off	Off	Engine Data 1-2 EGR, Misfire
A/C Request Signal	Yes / No	No	No	Engine Data 2
Air Fuel Ratio	0.0-25.5:1	14.2-14.7:1	14.2-14.7:1	Engine Data 2
Air Pump Relay Circuit Status	Okay / Fault	Okay	Okay	Output Driver
Air Pump Relay Command	On / Off	Off	Off	Engine Data 1, Fuel Trim
Air Pump Solenoid Circuit Status	Okay / Fault / Invalid State	Okay	Okay	Output Driver
Air Pump Solenoid Command	On / Off	Off	Off	Engine Data 1, Fuel Trim
BARO Sensor (kPa)	10-110 kPa	Varies	Varies	Engine Data 1, EGR, EVAP, Fuel Trim
CKP Sensor Status	Angle/Time A / Time B	Angle	Angle	Engine Data 2, Misfire
CMP Sensor	0-9999 rpm	Varies	Varies	Engine Data 2, Misfire
CMP Sensor Signal	Yes / No	Yes	Yes	Engine Data 2, Misfire
Crank Request Signal	Yes / No	No	No	Engine Data 2
Cruise Control Active	Yes / No	No	No	Engine Data 1
Cruise Inhibit Reason	Engine Not Running, Runtime, VSS, Brake	Vehicle Speed	Vehicle Speed	Engine Data 1
Cruise Inhibit Signal Command	On / Off	Off	Off	Engine Data 1
Cruise Inhibit Signal Circuit	Okay / Fault	Okay	Okay	Output Device
Current Gear	1-2-3-4	1	4	Engine Data 1, EGR, Fuel Trim
Cycles of Misfire Data	0-99 counts	Varies	Varies	Misfire
Cylinder 1-6 Ignition Control Circuit Status	Okay / Fault	Okay	Okay	Output Driver

2000-02 Intrigue 3.5L V6 VIN H (A/T) - PID Data List

Parameter Identification (PID)	PID Value Range	PID Value at Hot Idle	PID Value at 30 mph	Data List Type
Cylinder 1-6 Injector Circuit Status	Okay / Fault	Okay	Okay	Output Driver
Decel Fuel Cutoff	Active / Inactive	Inactive	Inactive	EGR, Fuel Trim
Desired EGR Position	0-100%	0	Varies	Engine Data 1, EGR, Misfire
Desired EGR Position	0-5.0 volts	0	Varies	EGR
Desired Idle Speed	0-3187 rpm	PCM control	---	Engine Data 1, Data 2, EVAP
Driver Module 1-4 Status	Enabled / Off/ High Volts / High Temperature	Enabled	Enabled	Output Driver
ECT Sensor (ºF)	-40 to 304ºF	185-220	185-220	Engine Data 1-2 EGR, Fuel Trim, EVAP, Misfire
EGR Flow Test Count	0-255 counts	0-10	Varies	EGR
EGR Learned Minimum Position	0-5.0 volts	Varies	Varies	EGR
EGR Position Sensor	0-100%	0	Varies	Engine Data 1, EGR, Misfire
EGR Position Sensor	0-5.0volts	0.0	Varies	EGR
EGR Position Error	-100 to 100%	Varies	Varies	EGR
EGR Solenoid Circuit Status	Okay / Fault	Okay	Okay	Output Driver
EGR Solenoid Command	0-100%	0	Varies	EGR
Engine Load (%)	0-100%	17	Varies	Engine Data 1-2 EGR, Fuel Trim, EVAP, Misfire
Engine Oil Level Switch	Okay / Not Okay	Okay	Okay	Engine Data 2
Engine Oil Life Left	0-100%	Varies	Varies	Engine Data 2
EOP Sensor	0-5.1 volts	2-3	2-4	Engine Data 2
Engine Run Time	Hr: Min: Sec	00:00:00	00:00:00	Engine Data 1-2 EGR, Fuel Trim, EVAP, Misfire
Engine Speed	0-9999 rpm	± 50 rpm of Desired Speed	Varies	Engine Data 1-2 EGR, Fuel Trim, EVAP, Misfire
EVAP Fault History	No Fault / Excess VAC / Purge Valve Leak / Small Leak / Weak Vacuum	No Fault	No Fault	EVAP
EVAP Purge Solenoid Control Circuit Status	Okay / Fault	Okay	Okay	Output Driver
EVAP Purge Solenoid Command	0-100%	18	Varies	Engine Data 1, EVAP, Fuel Trim

2000-02 Intrigue 3.5L V6 VIN H (A/T) - PID Data List

Parameter Identification (PID)	PID Value Range	PID Value at Hot Idle	PID Value at 30 mph	Data List Type
EVAP Test Abort Reason	Not Aborted / Lost Enable / Small Leak / Vehicle Not at Rest	Not Aborted	Not Aborted	EVAP
EVAP Test Result	No Result / Aborted / Fail / Passed	Passed	Passed	EVAP
EVAP Vent Solenoid Circuit Status	Okay / Fault	Okay	Okay	Output Driver
EVAP Vent Solenoid	Venting / Not Venting	Venting	Venting	Engine Data 1, EVAP, Fuel Trim
Extended Travel Brake Pedal Switch	Applied / Released	Released	Released	Engine Data 2
Fan Control Relay 1 Circuit Status	Okay / Fault	Okay	Okay	Output Driver
Fan Control Relay 1 Command	On / Off	Off	Off	Engine Data 2
Fan Control Relay 2-3 Command	On / Off	Off	Off	Engine Data 2
Fan Control Relay 2-3 Circuit Status	Okay / Fault	Okay	Okay	Output Driver
Fuel Level Remaining	0-100%	Varies	Varies	EVAP
Fuel Level Sensor	0-5.0 volts	Varies	Varies	EVAP
Fuel Pump Relay Circuit Status	Okay / Fault	Okay	Okay	Output Driver
Fuel Pump Relay Command	On / Off	On	On	Engine Data 1
Fuel Tank Pressure Sensor	Inches H2O, mm Hg	-17.5 to +7.5 Inches H2O	-17.5 to +7.5 Inches H2O	Engine Data 1, EVAP
Fuel Tank Pressure Sensor	0-5.1 volts	Varies	Varies	Engine Data 1, EVAP
Fuel Trim Cell	Cell 0-9	0	1	Engine Data 1, EVAP, Fuel Trim
Fuel Trim Learn	Enabled or Disabled	Enabled	May Toggle	Engine Data 1, EVAP, Fuel Trim
Generator F-Terminal	0-100%	Varies	Varies	Engine Data 2
Generator L-Terminal	On / Off	On	On	Engine Data 2
HO2S-11 (B1 S1)	0-1132 mv	10-1000	10-1000	Engine Data 1, EVAP, Fuel Trim
HO2S-11 Heater	On / Off	Off	Off	Fuel Trim
HO2S-11 Status	Ready / Not Ready	Ready	Ready	Engine Data 1, Fuel Trim
HO2S-12 (B1 S2)	0-1132 mv	10-1000	10-1000	Engine Data 1, Fuel Trim
IAC Position	0-255 counts	10-40	10-40	Engine Data 1, EGR, Fuel Trim

2000-02 Intrigue 3.5L V6 VIN H (A/T) - PID Data List

Parameter Identification (PID)	PID Value Range	PID Value at Hot Idle	PID Value at 30 mph	Data List Type
IAT Sensor (°F)	-40 to 304°F	50-194	50-194	All Lists
Ignition Mode	Bypass/IC	IC	IC	Engine Data 2
Ignition 1 Signal	0.0-25.5v	14.2	14.2	All Lists
Injector Pulsewidth	0-1000 ms	2.0-4.0	Varies	Engine Data 2, Fuel Trim, Misfire
Knock Retard (°)	0-25.5°	0°	0°	Engine Data 1, EGR
Long Term Fuel Trim	-10% to 10%	0 (± 5%)	0 (± 5%)	Engine Data 1-2 EVAP, Fuel Trim
Loop Status	Closed Loop / Open Loop	Closed Loop	Closed Loop	Engine Data 1-2 EGR, EVAP, Fuel Trim
MAF Sensor (g/sec)	0-512 g/sec	3-6	Varies	Engine Data 1-2 EGR, Fuel Trim, EVAP, Misfire
MAF Sensor (Hz)	0-32,000 Hz	1200-3000	Varies	Engine Data 2
MAP Sensor (kPa)	10-105 kPa	20-48	Varies	Engine Data 1-2 EGR, Fuel Trim, EVAP, Misfire
MAP Sensor (V)	0-5.1 volts	0.75-2.0	Varies	Engine Data 1-2
MIL Circuit Status	Okay / Fault	Okay	Okay	Output Driver
MIL Command	On / Off	Off	Off	Engine Data 2
Mileage Since DTC Cleared	Kilometer or Miles	Varies	Varies	Engine Data 2
Misfire Current Cylinder 1-6	0-198 counts	0-4	0-4	Misfire
Misfire History Cylinder 1-6	0-65,535	0	0	Misfire
Number of DTC(s)	Number	0	0	Engine Data 1, EVAP, Fuel Trim
O2S Heater Current	0-10.0 amps	0.50-0.60	0	Fuel Trim
PCM Reset	Yes / No	No	No	All Lists
PCM/VCM in VTD Fail Enable	Yes / No	Yes	Yes	Engine Data 1
Power Enrichment	Active / Inactive	Inactive	Inactive	Engine Data 2, Misfire
Reduced Engine Power	Active / Inactive	Inactive	Inactive	Engine Data 1, EGR
Short Term Fuel Trim	-10% to 10%	0 (± 5%)	0 (± 5%)	Engine Data 1-2 EVAP, Fuel Trim
Spark Advance (°)	-64° to 64°	20-40	Varies	Engine Data 1-2 Fuel Trim, Misfire
Startup ECT	-40 to 419°F	Varies	Varies	Engine Data 2, EVAP, Fuel Trim
Startup IAT	-40 to 419°F	80°F	80°F	Engine Data 2, EVAP, Fuel Trim
Starter Enable Relay Control Circuit Status	Okay / Fault	Okay	Okay	Output Driver
Starter Relay Command	On / Off	Off	Off	Engine Data 2
Tachometer Circuit Status	Okay / Fault	Okay	Okay	Output Driver
TCC Brake Pedal Switch	Applied / Released	Released	Released	Engine Data 1-2
TCC Enable Solenoid Circuit Status	On / Off	Off	On	Output Driver
TCC Enable Solenoid Command	On / Off	Off	On	Engine Data 1, Misfire
TCC PWM Solenoid Command	On / Off	Off	On	Engine Data 2
TFP Switch	Park/ Neutral / Drive 4, 3, 2 and 1	Park	Drive 4	Engine Data 2, EGR, Fuel Trim
Torque Delivered	0-100%	86	Varies	Engine Data 2
Torque Request	0-100%	100%	Varies	Engine Data 2
TP Angle (%)	0-100%	0	Varies	All Lists
TP Sensor (V)	0-5.1 volts	0.5-1.2	Varies	Engine Data 1-2
Traction Control Circuit History	Okay / Fault	Okay	Okay	Output Driver
Traction Control Circuit Status	Okay / Fault	Okay	Okay	Output Driver
Traction Control Circuit Status	Active / Inactive	Inactive	Inactive	Engine Data 2
TWC Temperature Calculated	32-1409°F	Varies	Varies	Fuel Trim
Vehicle Speed	0-155 mph	0	30	All Lists
VTD Auto Learn timer	Active / Inactive	Inactive	Inactive	Engine Data 1
VTD Fuel Disable	Active / Inactive	Inactive	Inactive	Engine Data 1
VTD Fuel Disable Until Key Off	Yes / No	No	No	Engine Data 1

LSS, REGENCY PID DATA

1996-98 LSS, Regency 3.8L V6 MFI VIN K, 1 (A/T) - PID Data List

Parameter Identification (PID)	PID Value Range	PID Value at Hot Idle	PID Value at 30 mph	Data List Type
Actual EGR Position	0-100%	0	0-100	Engine Data 1
Air Fuel Ratio	0.1-25.5:1	14.7:1	14.7:1	Engine Data 1
BARO Sensor (kPa)	10-110 kPa	65-110	65-110	Engine Data 1
Commanded TCC	Engaged / Disengaged	Disengaged	Disengaged	Engine Data 1
Cruise Engaged	Yes / No	No	No	Engine Data 1
Cruise Inhibited	Yes / No	Yes	Yes	Engine Data 1
Current Gear	1-2-3-4	1	3	Engine Data 1
Decel Fuel Mode	Active / Inactive	Inactive	Inactive	Engine Data 1
Desired EGR Position	0-100%	0%	0-100%	Engine Data 1
Desired Idle Speed	0-3187	PCM control	---	Engine Data 1
ECT Sensor (°F)	-40 to 304°F	185-220	185-220	Engine Data 1
Engine Load	0-100%	2-5	Varies	Engine Data 1
Engine Run Time	Hr: Min: Sec	00:00:00	00:00:00	Engine Data 1
Engine Speed	0-9999 rpm	±50 Desired	---	Engine Data 1
Fuel Trim Cell	Cell #0-10	0	0-10	Engine Data 1
Fuel Trim Learn	Disabled / Enabled	Enabled	Enabled	Engine Data 1
Generator Command	On / Off	On	On	Engine Data 1
HO2S-12 (B1 S1)	0-1132 mv	10-1000	10-1000	Engine Data 1
HO2S-12 (B1 S2)	0-1132 mv	10-1000	10-1000	Engine Data 1
HO2S-11 Status	Not/Ready	Ready	Ready	Engine Data 1
IAC Position	0-255 counts	10-40	---	Engine Data 1
IAT Sensor (°F)	-40 to 304°F	50-194	50-194	Engine Data 1
Ignition 1 Signal	0.0-25.5v	14.2	14.2	Engine Data 1
Injector Pulsewidth	0-985 ms	1.5-3.5	Varies	Engine Data 1
Long, Short Term F/T	-10% to 10%	0 (± 5%)	0 (± 5%)	Engine Data 1
Loop Status	Closed Loop / Open Loop	Closed Loop	Closed Loop	Engine Data 1
MAF Sensor (g/sec)	0-512 g/sec	3-6	Varies	Engine Data 1
MAF Sensor (Hz)	0-32000 Hz	1200-3000	Varies	Engine Data 1
MAP Sensor (kPa)	10-110 kPa	29-48	Varies	Engine Data 1
Power Enrichment	Active / Inactive	Inactive	Inactive	Engine Data 1
Rich / Lean	L or R	L-R-L-R	L-R-L-R	Engine Data 1
Throttle at Idle	Yes / No	Yes	No	Engine Data 1
TP Angle (%)	0-100%	0	Varies	Engine Data 1
TP Sensor (V)	0-5.0 volts	0.20-0.74	Varies	Engine Data 1
TWC Protection	Active / Inactive	Inactive	Inactive	Engine Data 1
Vehicle Speed	0-155 mph	0	30	Engine Data 1

1996-98 LSS, Regency 3.8L V6 MFI VIN K, 1 - PID Data List

Parameter Identification (PID)	PID Value Range	PID Value at Hot Idle	PID Value at 30 mph	Data List Type
A/C High Side Pressure	0-5.0 volts	Varies	Varies	Engine Data 2
A/C Press. Too High / Low	Yes / No	No	No	Engine Data 2
A/C Request	Yes / No	No	No	Engine Data 2
Actual EGR Position	0-100%	0%	0-100%	Engine Data 2
BARO Sensor (kPa)	10-110 kPa	110 S/Level	65-110	Engine Data 2
Boost PWM (VIN 1)	0-100%	100	0-100	Engine Data 2
Commanded A/C	On / Off	Off	Off	Engine Data 2
Commanded Fan 1, 2	On / Off	Off	Off	Engine Data 2

1996-98 LSS, Regency 3.8L V6 MFI VIN K, 1 (A/T) - PID Data List

Parameter Identification (PID)	PID Value Range	PID Value at Hot Idle	PID Value at 30 mph	Data List Type
Commanded Starter	Disabled / Enabled	Disabled	Disabled	Engine Data 2
Decel Fuel Mode	Active / Inactive	Inactive	Inactive	Engine Data 2
Desired Idle Speed	0-3187 rpm	PCM control	---	Engine Data 2
ECT Sensor (°F)	-40 to 304°F	185-220	185-220	Engine Data 2
Engine Load	0-100%	2-5	Varies	Engine Data 2
Engine Oil Level	Low / Okay	Okay	Okay	Engine Data 2
Engine Oil Life	0-100%	Varies	Varies	Engine Data 2
EOP Sensor (V)	0-5.0 volts	2-3	Varies	Engine Data 2
Engine Run Time	Hr: Min: Sec	00:00:00	00:00:00	Engine Data 2
Engine Speed	0-9999 rpm	±50 Desired	---	Engine Data 2
Fuel Pump	On / Off	On	On	Engine Data 2
Fuel Pump Speed	High/Normal	Normal	Normal	Engine Data 2
Generator PWM	0-100%	0-100	0-100	Engine Data 2
IAC Position	0-255 counts	10-40	---	Engine Data 2
IAT Sensor (°F)	-40 to 304°F	50-194	50-194	Engine Data 2
Ignition Mode	IC/Bypass	IC	IC	Engine Data 2
Knock Retard (°)	0.0-25.5°	0	0	Engine Data 2
MAF Sensor (g/sec)	0-512 g/sec	3-6	Varies	Engine Data 2
MAP Sensor (kPa)	10-110 kPa	29-48	Varies	Engine Data 2
MIL Command	On / Off	Off	Off	Engine Data 2
Power Enrichment	Active / Inactive	Active	Active	Engine Data 2
Spark Advance (°)	-64° to 64°	20	Varies	Engine Data 2
TCM Desired Torque	0-100%	80-95	0-100	Engine Data 2
TCM Control Torque	0-100%	5-20	0-100	Engine Data 2
TP Angle (%)	0-100%	0	0-100	Engine Data 2
Traction Control	Active / Inactive	Inactive	Inactive	Engine Data 2
Transmission Range	P-R-N-4-3-2	P	D3	Engine Data 2
TR Switch P, A, B & C	High / Low	High	Several	Engine Data 2
TWC Protection	Active / Inactive	Inactive	Inactive	Engine Data 2

1996-98 LSS, Regency 3.8L V6 MFI VIN K, 1 (A/T) - PID Data List

Parameter Identification (PID)	PID Value Range	PID Value at Hot Idle	PID Value at 30 mph	Data List Type
A/C Request	Yes / No	No	No	Catalyst
Air Fuel Ratio	0.1-25.5:1	14.7:1	14.7:1	Catalyst
BARO Sensor (kPa)	10-110 kPa	65-110	65-110	Catalyst
Commanded A/C	On / Off	Off	Off	Catalyst
Commanded Fan 1, 2	On / Off	Off	Off	Catalyst
Decel Fuel Mode	Active / Inactive	Inactive	Inactive	Catalyst
Desired Idle Speed	0-3187 rpm	PCM control	---	Catalyst
ECT Sensor (°F)	-40 to 304°F	185-220	185-220	Catalyst
Engine Load	0-100%	2-5	Varies	Catalyst
Engine Run Time	Hr: Min: Sec	00:00:00	00:00:00	Catalyst
Engine Speed	0-9999 rpm	±50 Desired	---	Catalyst
HO2S (B1 S2, B1 S2)	0-1132 mv	10-1000	10-1000	Catalyst
IAC Position	0-255 counts	10-40	---	Catalyst
IAT Sensor (°F)	-40 to 304°F	50-194	50-194	Catalyst

1996-98 LSS, Regency 3.8L V6 MFI VIN K, 1 (A/T) - PID Data List

Parameter Identification (PID)	PID Value Range	PID Value at Hot Idle	PID Value at 30 mph	Data List Type
Injector Pulsewidth	0-985 ms	1.5-3.5	Varies	Catalyst
Loop Status	Closed Loop / Open Loop	Closed Loop	Closed Loop	Catalyst
MAF Sensor (g/sec)	0-512 g/sec	3-6	Varies	Catalyst
Power Enrichment	Active / Inactive	Inactive	Inactive	Catalyst
Short Term Fuel Trim	-10% to 10%	0%	1%	Catalyst
TP Angle (%)	0-100%	0%	0-100%	Catalyst
Transmission Range	P-R-N-4-3-2	PARK	DRIVE 3	Catalyst
TWC Diagnostic	Disabled / Enabled	Enabled	Enabled	Catalyst
TWC Monitor Test	0-255 counts	0	0	Catalyst
TWC Protection	Active / Inactive	Inactive	Inactive	Catalyst
Vehicle Speed	0-155 mph	0	30	Catalyst

1996-98 LSS, Regency 3.8L V6 MFI VIN K, 1 (A/T) - PID Data List

Parameter Identification (PID)	PID Value Range	PID Value at Hot Idle	PID Value at 30 mph	Data List Type
Actual EGR Position	0-100%	0%	0-100%	EGR
BARO Sensor (kPa)	10-110 kPa	65-110	65-110	EGR
Desired EGR Position	0-100%	0%	0-100%	EGR
ECT Sensor (°F)	-40 to 304°F	185-220	185-220	EGR
EGR Closed Pintle	0-5.0 volts	0.14-1.0v	---	EGR
EGR Duty Cycle	0-100%	0%	0-100%	EGR
EGR Feedback	0-5.0 volts	0.14-1.0v	---	EGR
EGR Flow Test Count	0-255 counts	0-10	0-10	EGR
EGR Position Error	0-100%	0-9	Varies	EGR
Engine Speed	0-9999 rpm	±50 Desired	---	EGR
IAC Position	0-255 counts	10-40	---	EGR
MAP Sensor (kPa)	10-110 kPa	29-48	Varies	EGR
MAP Sensor (V)	0-5.0 volts	1-2	Varies	EGR
Power Enrichment	Active / Inactive	Active	Active	EGR
TP Angle (%)	0-100%	0	Varies	EGR
Trans. Hot Mode	Active / Inactive	Inactive	Inactive	EGR

1996-98 LSS, Regency 3.8L V6 MFI VIN K, 1 (A/T) - PID Data List

Parameter Identification (PID)	PID Value Range	PID Value at Hot Idle	PID Value at 30 mph	Data List Type
BARO Sensor (kPa)	10-110 kPa	65-110	65-110	65-110
Desired Idle Speed	0-3187 rpm	PCM control	---	---
ECT Sensor (°F)	-40 to 304°F	185-220	185-220	185-220
Engine Run Time	Hr: Min: Sec	00:00:00	00:00:00	00:00:00
Engine Speed	0-9999 rpm	±50 rpm	---	---
Fuel Level	0-100%	0-100%	0-100%	0-100%
EVAP Canister Purge	0-100%	0-25%	0-50%	0-100%
EVAP Fault History	Fault/No	No Fault	No Fault	No Fault
FTP Sensor (In. H2O)	-17.5 to 7.5"	Varies	Varies	Varies
IAT Sensor (°F)	-40 to 304°F	50-194	50-194	50-194
MAP Sensor (kPa)	10-110 kPa	29-48	Varies	Varies
Startup ECT Sensor	-40 to 304°F	Varies	Varies	Varies
Startup IAT Sensor	-40 to 304°F	Varies	Varies	Varies

1996-98 LSS, Regency 3.8L V6 MFI VIN K, 1 (A/T) - PID Data List

Parameter Identification (PID)	PID Value Range	PID Value at Hot Idle	PID Value at 30 mph	Data List Type
Air Fuel Ratio	0.1-25.5:1	14.7:1	14.7:1	HO2S
Decel Fuel Mode	Active / Inactive	Inactive	Inactive	HO2S
ECT Sensor (°F)	-40 to 304°F	185-220	185-220	HO2S
Engine Load	0-100%	2-5	Varies	HO2S
Engine Speed	0-9999 rpm	±50 Desired	---	HO2S
EVAP Canister Purge	0-100%	0-25	0-50	HO2S
HO2S (B1 S1, B1 S2)	0-1132 mv	10-1000	10-1000	HO2S
HO2S-11 Status	Not Ready/Ready	Ready	Ready	HO2S
HO2S-11 Warmup	Min: Sec	00:00	00:00	HO2S
HO2S-12 Warmup	Min: Sec	00:00	00:00	HO2S
HO2S-11 XCounts	0-255 counts	Varies	Varies	HO2S
IAT Sensor (°F)	-40 to 304°F	50-194	50-194	HO2S
MAF Sensor (g/sec)	0-512 g/sec	3-6	Varies	HO2S
Power Enrichment	Active / Inactive	Active	Active	HO2S
Startup ECT Sensor	-40 to 304°F	Varies	Varies	HO2S
Startup IAT Sensor	-40 to 304°F	Varies	Varies	HO2S
TP Angle (%)	0-100%	0	0-100	HO2S
TWC Protection	Active / Inactive	Inactive	Inactive	HO2S

1996-98 LSS, Regency 3.8L V6 MFI VIN K, 1 (A/T) - PID Data List

Parameter Identification (PID)	PID Value Range	PID Value at Hot Idle	PID Value at 30 mph	Data List Type
A/C Request	Yes / No	No	No	Misfire
Actual EGR Position	0-100%	0	0-100	Misfire
Commanded A/C	On / Off	Off	Off	Misfire
Current Gear	1-2-3-4	1	3	Misfire
Decel Fuel Mode	Active / Inactive	Inactive	Inactive	Misfire
Desired EGR Position	0-100%	0%	0-100%	Misfire
ECT Sensor (°F)	-40 to 304°F	185-220	185-220	Misfire
Engine Load	0-100%	2-5	Varies	Misfire
Engine Speed	0-9999 rpm	±50 Desired	---	Misfire
EVAP Canister Purge	0-100%	0-25%	0-50%	Misfire
HO2S-11 XCounts	0-255 counts	Varies	Varies	Misfire
IAC Position	0-255 counts	10-40	---	Misfire
Long, Short Term F/T	-10% to 10%	0 (± 5%)	0 (± 5%)	Misfire
Misfire current Cyl 1-6	0-198	0-2	0-2	Misfire
Misfire History Cyl 1-6	0-65535	0	0	Misfire
Misfiring Cylinder	CYL 1-6	0	0	Misfire
M/Fire Since 1st Fail	0-65535	0	0	Misfire
M/Fire Since 1st Pass	0-65535	0	0	Misfire
Misfire Pass Since 1st Fail	0-65535	0	0	Misfire
Power Enrichment	Active / Inactive	Inactive	Inactive	Misfire
Spark Advance	-64° to 64°	20	Varies	Misfire
TCC Enable	On / Off	Off	On or Off	Misfire
Total Misfire Current	0-99	0-5	0-5	Misfire
TP Angle (%)	0-100%	0	Varies	Misfire
TWC Protection	Active / Inactive	Inactive	Inactive	Misfire

SILHOUETTE

1996-2004 Silhouette Light Van 3.4L VIN E - PID Data List

Parameter Identification (PID)	PID Value Range	PID Value at Hot Idle	PID Value at 30 mph	Data List Type
3X Crank Sensor	0-9999 rpm	Varies	Varies	Engine Data 2
24X Crank Sensor	0-1600 rpm	Varies	Varies	Engine Data 2
A/C HI Side Pressure	0-5.0v	Varies	Varies	Engine Data 1
A/C Pressure Disable	Yes/No	No	No	Engine Data 2
A/C Pressure Disable	Yes/No	No	No	Engine Data 2
A/C Off for WOT	Yes/No	No	No	Engine Data 2
A/C Pressure Disable	Yes/No	No	No	Engine Data 2
A/C Relay Circuit Status	Okay/Fault	Okay	Okay	Output Driver
A/C Relay Command	On/Off	Off	Off	Engine Data 1-2, EGR, Misfire
A/C Request Signal	Yes/No	No	No	Engine Data 2
Actual EGR Position	0-100%	0	Varies	Engine Data 1, EGR, Misfire
Air Fuel Ratio	0.0-25.5:1	14.7:1	14.7:1	Engine Data 2, Fuel Trim
BARO Sensor (kPa)	10-110 kPa	98	98	Engine Data 1, EGR, EVAP, Fuel Trim
BARO Sensor (V)	0-5.0 volts	3.5-4.5	3.5-4.5	Engine Data 1, EGR, EVAP, Fuel Trim
Brake Switch Signal	Applied/REL	Released	Released	Engine Data 1
CMP Sensor Signal	Yes/No	Yes	Yes	Engine Data 1, Misfire
Crank Request Signal	Yes/No	No	No	Engine Data 2
Cruise Control Active	Yes/No	No	No	Engine Data 1
Cruise Inhibit Reason	Several	Varies	Varies	Engine Data 1
Cruised Inhibit Signal Circuit Status	Okay/Fault/ Invalid State	Okay	Okay	Output Driver
Cruise Inhibit Signal Command	On/Off	On	On	Engine Data 1
Current Gear	1-2-3-4	1	4	Engine Data 1-2, EGR, Fuel Trim
Cycles of Misfire Data	0-99 counts	Varies	Varies	Misfire
Cylinder 1-6 Injector Circuit Status	Okay/Stuck Low/Stuck High/Fault	Okay	Okay	Output Driver
Cylinder 1-6 Injector Circuit History	Okay/Stuck Low/Stuck High/Fault	Okay	Okay	Output Driver
Deceleration Fuel Cutoff	Active/Inactive	Inactive	Inactive	EGR, Fuel Trim, Misfire
Desired EGR Position	0-100%	0	Varies	Engine Data 1-2, EGR, Misfire
Desired Idle Speed	0-3187 rpm	PCM control	---	Engine Data 1, Data 2, EVAP
Driver Module 1 Status	Enabled/ Disabled	Enabled	Enabled	Output Driver
Driver Module 2 Status	Enabled/ Disabled	Enabled	Enabled	Output Driver
Driver Module 3 Status	Enabled/ Disabled	Enabled	Enabled	Output Driver
Driver Module 4 Status	Enabled/ Disabled	Enabled	Enabled	Output Driver
ECT Sensor (°F)	-40° to 304°F	185-220	185-220	Engine Data 1-2, EGR, Fuel Trim, EVAP, Misfire
EGR Flow Test Count	0-255 counts	0	Varies	EGR
EGR Position Sensor	0-100%	0	Varies	Engine Data 1, EGR, Misfire
EGR Position Sensor	0-5.1 volts	0.14-1.0	Varies	Engine Data 1, EGR, Misfire
EGR Learned Minimum Position	0-5.1 volts	0.71	Varies	EGR
EGR Position Error	0-100%	0	Varies	EGR
EGR Pos. Variance	0-100%	0	Varies	EGR
EGR Solenoid Command	0-100%	0	Varies	Engine Data 1, EGR, Misfire
Engine Load (%)	0-100%	17	Varies	Engine Data 1-2, EGR, Fuel Trim EVAP, Misfire
Engine Oil Level Switch	Okay/Low	Okay	Okay	Engine Data 2
Engine Oil Life Remaining	0-100%	Varies	Varies	Engine Data 2

1996-2004 Silhouette Light Van 3.4L VIN E - PID Data List

Parameter Identification (PID)	PID Value Range	PID Value at Hot Idle	PID Value at 30 mph	Data List Type
EOP Switch Signal	Okay/Low/HI	Okay	Okay	Engine Data 2
Engine Run Time	Hr: Min: Sec	00:00:00	00:00:00	Engine Data 1-2, EGR, Fuel Trim EVAP, Misfire
Engine Speed	0-9999 rpm	± 50 rpm of Desired Speed	Varies	Engine Data 1-2, EGR, Fuel Trim EVAP, Misfire
EVAP Fault History	No Fault / Excess VAC / Purge Valve Leak / Small Leak / Weak Vacuum	No Fault	No Fault	EVAP
EVAP Purge Solenoid Circuit Status	Okay / Fault	Okay	Okay	Output Driver
EVAP Purge Solenoid Command	0-100%	19	Varies	Engine Data 1, EVAP, HO2S
EVAP Vent Solenoid Control Circuit Status	Okay / Fault	Okay	Okay	Output Driver
EVAP Vent Solenoid Command	Venting/Not Venting	Venting	Venting	Engine Data 1, EVAP, Fuel Trim
FC Relay 1 Command	On/Off	Off	Off	Engine Data 2
FC Relay 2 and 3 Circuit Status	Okay/Fault	Okay	Okay	Output Driver
FC Relay 2 and 3 Command	On/Off	Off	Off	Engine Data 2
Fuel Level Sensor	0-5.0 volts	0.8-2.5	Varies	EVAP
Fuel Pump Command	On/Off	On	On	Engine Data 1
Fuel Pump Relay Circuit Status	Okay/Fault	Okay	Okay	Output Driver
Fuel Pump Relay Circuit History	Okay/Fault	Okay	Okay	Output Driver
Fuel Pump Relay Command	On/Off	On	On	Engine Data 1
Fuel Tank Level Remaining	0-100%	Varies	Varies	EVAP
Fuel Tank Pressure Sensor	Inches H2O, mm Hg, volts	-17 to +7.5 Inches H2O	-17 to +7.5 Inches H2O	Engine Data 1, EVAP
Fuel Tank Pressure Sensor	0-5.0 volts	Varies	Varies	Engine Data 1, EVAP
Fuel Trim Cell	Cell 0-10	Varies	Varies	Engine Data 1, EVAP, Fuel Trim
Fuel Trim Learn	Enabled or Disabled	Enabled	May Toggle	Engine Data 1, EVAP, Fuel Trim
FC Relay 1 Circuit Status	Okay / Fault	Okay	Okay	Output Driver
Generator Lamp	On/Off	Off	Off	Engine Data 2
Generator L-Terminal	On/Off	On	On	Engine Data 2
HO2S-11 (B1 S1)	0-1132 mv	10-1000	10-1000	Engine Data 1, EVAP, Fuel Trim
HO2S-12 (B1 S2)	0-1132 mv	10-1000	10-1000	Engine Data 1, EVAP, Fuel Trim
HO2S-11 XCounts	0-255	Varies	Varies	Engine Data 1, HO2S
HO2S-11 Heater Command	On/Off	On	On	Fuel Trim
Hot Mode	On/Off	Off	Off	Misfire
IAC Position	0-255 counts	15	Varies	Engine Data 1, EGR, Fuel Trim
IAT Sensor (°F)	-40° to 304°F	50-194	50-194	Engine Data 1-2, EGR, EVAP, Fuel Trim
Ignition 1 Signal	0.0-25.5v	12.8-14.2	12.8-14.2	Engine Data 1-2, EGR, EVAP, Fuel Trim
Ignition Mode	Bypass/IC	IC	IC	Engine Data 2
IMS	P/R/N/D4-1	Park	Drive 4	Engine Data 1
Injector Pulsewidth	0-1000 ms	2.90	Varies	Engine Data 2, Fuel Trim, Misfire
Knock Retard (°)	0-25.5°	0	0	Engine Data 1, EGR
Long Term Fuel Trim	-27% to 27%	0 (± 5%)	0 (± 5%)	Engine Data 1-2, EVAP, Fuel Trim
Loop Status	Closed Loop / Open Loop	Closed Loop	Closed Loop	Engine Data 1-2, EGR, EVAP, Fuel Trim

1996-2004 Silhouette Light Van 3.4L VIN E - PID Data List

Parameter Identification (PID)	PID Value Range	PID Value at Hot Idle	PID Value at 30 mph	Data List Type
MAF Sensor (g/sec)	0-512 g/sec	3.41	Varies	Engine Data 1-2, EGR, EVAP, Fuel Trim, Misfire
MAF Sensor (Hz)	0-32,000 Hz	2210	Varies	Engine Data 1-2, EGR, EVAP, Fuel Trim, Misfire
MAP Sensor (kPa)	10-105 kPa	38	Varies	Engine Data 1-2, EGR, EVAP, Fuel Trim, Misfire
MAP Sensor (V)	0-5.1 volts	1.45	Varies	Engine Data 2
MIL Circuit Status	Okay/Fault	Okay	Okay	Output Driver
MIL Command	On/Off	Off	Off	Engine Data 2
Misfire Current Cylinder 1-6	0-198 counts	0	0	Misfire
Misfire History Cylinder 1-6	0-65,535	0	0	Misfire
Non-Volatile Memory	PASS/FAIL	PASS	PASS	Engine Data 1
Number of DTC(s)	0-255 counts	0	0	Engine Data 1, EVAP, Fuel Trim
HO2S Heater Current	0-10.0 amps	0.54	0.54	Fuel Trim
PCM Reset	Yes/No	No	No	Engine Data 1-2, EGR, EVAP, Fuel Trim, ODD
PCM/VCM in VTD Fuel Enable	Yes/No	No	No	Engine Data 1
Power Enrichment	Active/Inactive	Inactive	Inactive	Engine Data 2, Misfire
Reduced Engine Power	Active/Inactive	Inactive	Inactive	EGR
Short Term Fuel Trim	-27% to 27%	0 (± 5%)	0 (± 5%)	Engine Data 1-2, EVAP, Fuel Trim
Spark Advance (º)	-64º to 64º	20	Varies	Engine Data 1-2 Fuel Trim, Misfire
Startup ECT	-40 to 419ºF	Varies	Varies	Engine Data 2, EVAP, Fuel Trim
Startup IAT	-40 to 419ºF	80ºF	80ºF	Engine Data 2, EVAP, Fuel Trim
Starter Enable Relay Circuit Status	Okay / Fault/ Invalid State	Okay	Okay	Output Driver
Starter Relay Control	On/Off	Off	Off	Engine Data 2
TCC Brake Pedal Switch	Applied / Released	Released	Released	Engine Data 1-2
TCC PWM Solenoid Command	On/Off	Off	On	Engine Data 2, Misfire
TFP Switch Signal	P/R/N/D4-1	Park/Neutral	Drive 4	Engine Data 2, EGR, Fuel Trim
Torque Delivered Signal	0-100%	86	86	Engine Data 2
Torque Request Signal	0-100%	100	100	Engine Data 2
Total Misfire Current Count	0-65,535	Varies	Varies	Misfire
TP Angle (%)	0-100%	0	0-100	Engine Data 1-2, EGR, EVAP, Fuel Trim, Misfire
TP Sensor (V)	0-5.1 volts	0.20-0.74	Varies	Engine Data 1-2, EGR, EVAP, Fuel Trim, Misfire
Traction Control Status	Active or Inactive	Inactive	Inactive	Engine Data 2
Transmission Range	Park/Neutral, Drive 4-3-2-1	Park/Neutral	Drive 4	Engine Data 1, AIR, EGR, Catalyst
TR Switch P/A/B/C	High/Low	High	Low	Engine Data 1
TWC Protection	Active or Inactive	Inactive	Inactive	Engine Data 2, Misfire
Vehicle Speed	0-155 mph	0	30	Engine Data 1-2, EGR, EVAP, Fuel Trim, Misfire
VTD Auto Learn Timer	Active/Inactive	Inactive	Inactive	Engine Data 1
VTD Fuel Disable	Active/Inactive	Inactive	Inactive	Engine Data 1
VTD Fuel Disable Until Ignition Off	Yes/No	No	No	Engine Data 1

AZTEK

2001-05 Aztek 4D Utility 3.4L VIN E - PID Data List

Parameter Identification (PID)	PID Value Range	PID Value at Hot Idle	PID Value at 30 mph	Data List Type
3X Crank Sensor	0-9999 rpm	Varies	Varies	Engine Data 2
24X Crank Sensor	0-1600 rpm	Varies	Varies	Engine Data 2
A/C HI Side Pressure	0-5.0v	Varies	Varies	Engine Data 1
A/C Pressure Disable	Yes/No	No	No	Engine Data 2
A/C Pressure Disable	Yes/No	No	No	Engine Data 2
A/C Off for WOT	Yes/No	No	No	Engine Data 2
A/C Pressure Disable	Yes/No	No	No	Engine Data 2
A/C Relay Circuit Status	Okay/Fault	Okay	Okay	Output Driver
A/C Relay Command	On/Off	Off	Off	Engine Data 1-2, EGR, Misfire
A/C Request Signal	Yes/No	No	No	Engine Data 2
Actual EGR Position	0-100%	0	Varies	Engine Data 1, EGR, Misfire
Air Fuel Ratio	0.0-25.5:1	14.7:1	14.7:1	Engine Data 2, Fuel Trim
BARO Sensor (kPa)	10-110 kPa	98	98	Engine Data 1, EGR, EVAP, Fuel Trim
BARO Sensor (V)	0-5.0 volts	3.5-4.5	3.5-4.5	Engine Data 1, EGR, EVAP, Fuel Trim
Brake Switch Signal	Applied/REL	Released	Released	Engine Data 1
CMP Sensor Signal	Yes/No	Yes	Yes	Engine Data 1, Misfire
Crank Request Signal	Yes/No	No	No	Engine Data 2
Cruise Control Active	Yes/No	No	No	Engine Data 1
Cruise Inhibit Reason	Several	Varies	Varies	Engine Data 1
Cruised Inhibit Signal Circuit Status	Okay/Fault/ Invalid State	Okay	Okay	Output Driver
Cruise Inhibit Signal Command	On/Off	On	On	Engine Data 1
Current Gear	1-2-3-4	1	4	Engine Data 1-2, EGR, Fuel Trim
Cycles of Misfire Data	0-99 counts	Varies	Varies	Misfire
Cylinder 1-6 Injector Circuit Status	Okay/Stuck Low/Stuck High/Fault	Okay	Okay	Output Driver
Cylinder 1-6 Injector Circuit History	Okay/Stuck Low/Stuck High/Fault	Okay	Okay	Output Driver
Deceleration Fuel Cutoff	Active/Inactive	Inactive	Inactive	EGR, Fuel Trim, Misfire
Desired EGR Position	0-100%	0	Varies	Engine Data 1-2, EGR, Misfire
Desired Idle Speed	0-3187 rpm	PCM control	---	Engine Data 1, Data 2, EVAP
Driver Module 1 Status	Enabled/ Disabled	Enabled	Enabled	Output Driver
Driver Module 2 Status	Enabled/ Disabled	Enabled	Enabled	Output Driver
Driver Module 3 Status	Enabled/ Disabled	Enabled	Enabled	Output Driver
Driver Module 4 Status	Enabled/ Disabled	Enabled	Enabled	Output Driver
ECT Sensor (°F)	-40° to 304°F	185-220	185-220	Engine Data 1-2, EGR, Fuel Trim, EVAP, Misfire
EGR Flow Test Count	0-255 counts	0	Varies	EGR
EGR Position Sensor	0-100%	0	Varies	Engine Data 1, EGR, Misfire
EGR Position Sensor	0-5.1 volts	0.14-1.0	Varies	Engine Data 1, EGR, Misfire
EGR Learned Minimum Position	0-5.1 volts	0.71	Varies	EGR
EGR Position Error	0-100%	0	Varies	EGR
EGR Pos. Variance	0-100%	0	Varies	EGR
EGR Solenoid Command	0-100%	0	Varies	Engine Data 1, EGR, Misfire
Engine Load (%)	0-100%	17	Varies	Engine Data 1-2, EGR, Fuel Trim EVAP, Misfire
Engine Oil Level Switch	Okay/Low	Okay	Okay	Engine Data 2
Engine Oil Life Remaining	0-100%	Varies	Varies	Engine Data 2

2001-05 Aztek 4D Utility 3.4L VIN E - PID Data List

Parameter Identification (PID)	PID Value Range	PID Value at Hot Idle	PID Value at 30 mph	Data List Type
EOP Switch Signal	Okay/Low/HI	Okay	Okay	Engine Data 2
Engine Run Time	Hr: Min: Sec	00:00:00	00:00:00	Engine Data 1-2, EGR, Fuel Trim EVAP, Misfire
Engine Speed	0-9999 rpm	± 50 rpm of Desired Speed	Varies	Engine Data 1-2, EGR, Fuel Trim EVAP, Misfire
EVAP Fault History	No Fault / Excess VAC / Purge Valve Leak / Small Leak / Weak Vacuum	No Fault	No Fault	EVAP
EVAP Purge Solenoid Circuit Status	Okay / Fault	Okay	Okay	Output Driver
EVAP Purge Solenoid Command	0-100%	19	Varies	Engine Data 1, EVAP, HO2S
EVAP Vent Solenoid Control Circuit Status	Okay / Fault	Okay	Okay	Output Driver
EVAP Vent Solenoid Command	Venting/Not Venting	Venting	Venting	Engine Data 1, EVAP, Fuel Trim
FC Relay 1 Command	On/Off	Off	Off	Engine Data 2
FC Relay 2 and 3 Circuit Status	Okay/Fault	Okay	Okay	Output Driver
FC Relay 2 and 3 Command	On/Off	Off	Off	Engine Data 2
Fuel Level Sensor	0-5.0 volts	0.8-2.5	Varies	EVAP
Fuel Pump Command	On/Off	On	On	Engine Data 1
Fuel Pump Relay Circuit Status	Okay/Fault	Okay	Okay	Output Driver
Fuel Pump Relay Circuit History	Okay/Fault	Okay	Okay	Output Driver
Fuel Pump Relay Command	On/Off	On	On	Engine Data 1
Fuel Tank Level Remaining	0-100%	Varies	Varies	EVAP
Fuel Tank Pressure Sensor	Inches H2O, mm Hg, volts	-17 to +7.5 Inches H2O	-17 to +7.5 Inches H2O	Engine Data 1, EVAP
Fuel Tank Pressure Sensor	0-5.0 volts	Varies	Varies	Engine Data 1, EVAP
Fuel Trim Cell	Cell 0-10	Varies	Varies	Engine Data 1, EVAP, Fuel Trim
Fuel Trim Learn	Enabled or Disabled	Enabled	May Toggle	Engine Data 1, EVAP, Fuel Trim
FC Relay 1 Circuit Status	Okay / Fault	Okay	Okay	Output Driver
Generator Lamp	On/Off	Off	Off	Engine Data 2
Generator L-Terminal	On/Off	On	On	Engine Data 2
HO2S-11 (B1 S1)	0-1132 mv	10-1000	10-1000	Engine Data 1, EVAP, Fuel Trim
HO2S-12 (B1 S2)	0-1132 mv	10-1000	10-1000	Engine Data 1, EVAP, Fuel Trim
HO2S-11 XCounts	0-255	Varies	Varies	Engine Data 1, HO2S
HO2S-11 Heater Command	On/Off	On	On	Fuel Trim
Hot Mode	On/Off	Off	Off	Misfire
IAC Position	0-255 counts	15	Varies	Engine Data 1, EGR, Fuel Trim
IAT Sensor (°F)	-40° to 304°F	50-194	50-194	Engine Data 1-2, EGR, EVAP, Fuel Trim
Ignition 1 Signal	0.0-25.5v	12.8-14.2	12.8-14.2	Engine Data 1-2, EGR, EVAP, Fuel Trim
Ignition Mode	Bypass/IC	IC	IC	Engine Data 2
IMS	P/R/N/D4-1	Park	Drive 4	Engine Data 1
Injector Pulsewidth	0-1000 ms	2.90	Varies	Engine Data 2, Fuel Trim, Misfire
Knock Retard (°)	0-25.5°	0	0	Engine Data 1, EGR
Long Term Fuel Trim	-27% to 27%	0 (± 5%)	0 (± 5%)	Engine Data 1-2, EVAP, Fuel Trim
Loop Status	Closed Loop / Open Loop	Closed Loop	Closed Loop	Engine Data 1-2, EGR, EVAP, Fuel Trim

2001-05 Aztek 4D Utility 3.4L VIN E - PID Data List

Parameter Identification (PID)	PID Value Range	PID Value at Hot Idle	PID Value at 30 mph	Data List Type
MAF Sensor (g/sec)	0-512 g/sec	3.41	Varies	Engine Data 1-2, EGR, EVAP, Fuel Trim, Misfire
MAF Sensor (Hz)	0-32,000 Hz	2210	Varies	Engine Data 1-2, EGR, EVAP, Fuel Trim, Misfire
MAP Sensor (kPa)	10-105 kPa	38	Varies	Engine Data 1-2, EGR, EVAP, Fuel Trim, Misfire
MAP Sensor (V)	0-5.1 volts	1.45	Varies	Engine Data 2
MIL Circuit Status	Okay/Fault	Okay	Okay	Output Driver
MIL Command	On/Off	Off	Off	Engine Data 2
Misfire Current Cylinder 1-6	0-198 counts	0	0	Misfire
Misfire History Cylinder 1-6	0-65,535	0	0	Misfire
Non-Volatile Memory	PASS/FAIL	PASS	PASS	Engine Data 1
Number of DTC(s)	0-255 counts	0	0	Engine Data 1, EVAP, Fuel Trim
HO2S Heater Current	0-10.0 amps	0.54	0.54	Fuel Trim
PCM Reset	Yes/No	No	No	Engine Data 1-2, EGR, EVAP, Fuel Trim, ODD
PCM/VCM in VTD Fuel Enable	Yes/No	No	No	Engine Data 1
Power Enrichment	Active/Inactive	Inactive	Inactive	Engine Data 2, Misfire
Reduced Engine Power	Active/Inactive	Inactive	Inactive	EGR
Short Term Fuel Trim	-27% to 27%	0 (± 5%)	0 (± 5%)	Engine Data 1-2, EVAP, Fuel Trim
Spark Advance (°)	-64° to 64°	20	Varies	Engine Data 1-2 Fuel Trim, Misfire
Startup ECT	-40 to 419°F	Varies	Varies	Engine Data 2, EVAP, Fuel Trim
Startup IAT	-40 to 419°F	80°F	80°F	Engine Data 2, EVAP, Fuel Trim
Starter Enable Relay Circuit Status	Okay / Fault/ Invalid State	Okay	Okay	Output Driver
Starter Relay Control	On/Off	Off	Off	Engine Data 2
TCC Brake Pedal Switch	Applied / Released	Released	Released	Engine Data 1-2
TCC PWM Solenoid Command	On/Off	Off	On	Engine Data 2, Misfire
TFP Switch Signal	P/R/N/D4-1	Park/Neutral	Drive 4	Engine Data 2, EGR, Fuel Trim
Torque Delivered Signal	0-100%	86	86	Engine Data 2
Torque Request Signal	0-100%	100	100	Engine Data 2
Total Misfire Current Count	0-65,535	Varies	Varies	Misfire
TP Angle (%)	0-100%	0	0-100	Engine Data 1-2, EGR, EVAP, Fuel Trim, Misfire
TP Sensor (V)	0-5.1 volts	0.20-0.74	Varies	Engine Data 1-2, EGR, EVAP, Fuel Trim, Misfire
Traction Control Status	Active or Inactive	Inactive	Inactive	Engine Data 2
Transmission Range	Park/Neutral, Drive 4-3-2-1	Park/Neutral	Drive 4	Engine Data 1, AIR, EGR, Catalyst
TR Switch P/A/B/C	High/Low	High	Low	Engine Data 1
TWC Protection	Active or Inactive	Inactive	Inactive	Engine Data 2, Misfire
Vehicle Speed	0-155 mph	0	30	Engine Data 1-2, EGR, EVAP, Fuel Trim, Misfire
VTD Auto Learn Timer	Active/Inactive	Inactive	Inactive	Engine Data 1
VTD Fuel Disable	Active/Inactive	Inactive	Inactive	Engine Data 1
VTD Fuel Disable Until Ignition Off	Yes/No	No	No	Engine Data 1

<u>BONNEVILLE PID DATA</u>

1995 Bonneville 3.8L V6 MFI VIN K, VIN 1 (A/T) - PID Data List

Parameter Identification (PID)	PID Value Range	PID Value at Hot Idle	PID Value at 30 mph	Data List Type
1st Gear Switch	On / Off	On	Off	Engine Data 1
3rd Gear Switch	On / Off	Off	On	Engine Data 1
4th Gear Switch	On / Off	Off	On	Engine Data 1
A/C Clutch	On / Off	Off	Off	Engine Data 1
A/C Request	Yes / No	No	No	Engine Data 1
BARO Sensor (V)	0.0-5.0v	3.5-4.5	3.5-4.5	Engine Data 1
Battery Signal (V)	0-25.5v	14.1	14.1	Engine Data 1
Boost PWM (VIN 1)	0-100%	100	100	Engine Data 1
Brake Switch	Applied / Released	Released	Released	Engine Data 1
EVAP Purge Solenoid	On / Off	Off	On	Engine Data 1
Clear Flood	On / Off	Off	Off	Engine Data 1
Coolant Fan(s)	On / Off	Off	Off	Engine Data 1
Crank Signal	0-3187 rpm	0-400 rpm	---	Engine Data 1
Cross Counts	0-255	Varies	Varies	Engine Data 1
Desired Idle	0-3187 rpm	± 25 rpm	---	Engine Data 1
ECT Sensor (°F)	-40 to 419°F	210	208	Engine Data 1
EGR Solenoid	On / Off	Off	On	Engine Data 1
EGR Duty Cycle	0-100%	0	0-50	Engine Data 1
Engine Speed	0-9999 rpm	810	---	Engine Data 1
Fan Relay	On / Off	Off	Off	Engine Data 1
Fan Request	On / Off	Off	Off	Engine Data 1
HO2S-11 (B1 S1)	0-1132 mv	600	560	Engine Data 1
IAC Motor	0-255	16-20	---	Engine Data 1
IAT Sensor (°F)	-40 to 304°F	85	85	Engine Data 1
Ignition / Crank	On / Off	On	On	Engine Data 1
Injector Pulsewidth	0-985	0.8-3.0	Varies	Engine Data 1
Knock Retard	0-255 counts	0	0	Engine Data 1
Knock Signal	Yes / No	No	No	Engine Data 1
Loop Status	Closed Loop / Open Loop	Closed Loop	Closed Loop	Engine Data 1
Long Term Fuel Trim	0-255 counts	128	128	Engine Data 1
MAP Sensor (V)	0-5.0v	1.1v	1.4v	Engine Data 1
Park Neutral Signal	P/N-R-DL	P/N	R-DL	Engine Data 1
Power Steering Switch	Normal / High	Normal	Normal	Engine Data 1
PROM ID Number	Number	0000	0000	Engine Data 1
QDM Signal	High / Low	Low	Low	Engine Data 1
Short Term Fuel Trim	0-255 counts	130	129	Engine Data 1
Spark Advance (°)	0° to 45°	18	Varies	Engine Data 1
TCC Solenoid	On / Off	Off	On	Engine Data 1
TFT Sensor (V)	-40 to 304°F	105	105	Engine Data 1
Throttle Angle (%)	0-100%	0	2	Engine Data 1
TP Sensor (V)	0.0-5.0v	1.0	1.2	Engine Data 1
Throttle Switch	Open/Close	Open	Closed	Engine Data 1
Trouble Codes	Yes / No	No	No	Engine Data 1
Vehicle Speed	0-155 mph	0	30	Engine Data 1

1996-99 Bonneville 3.8L V6 MFI VIN K, VIN 1 (A/T) - PID Data List

Parameter Identification (PID)	PID Value Range	PID Value at Hot Idle	PID Value at 30 mph	Data List Type
A/C Off for WOT	On / Off	Off	Off	Engine Data 1
A/C Refrigerant Press	0-5.0v	Varies	Varies	Engine Data 1
A/C Press. Too HI/LO	Yes / No	No	No	Engine Data 1
A/C Request	Yes / No	No	No	Engine Data 1
A/C Slugging	Active / Inactive	Inactive	Inactive	Engine Data 1
Actual EGR Position	0.0-5.0v	0.22v	Varies	Engine Data 1
Air Fuel Ratio	0.1 to 25:1	14.6:1	14.6:1	Engine Data 1
BARO Sensor (V)	0.0-5.0v	3.5-4.5	3.5-4.5	Engine Data 1
Boost PWM (VIN 1)	0-100%	100%	100%	Engine Data 1
Commanded A/C	On / Off	Off	Off	Engine Data 1
Commanded Fan 1, 2	On / Off	Off	Off	Engine Data 1
Commanded TCC	On / Off	Off	On	Engine Data 1
Cruise Engaged	On / Off	Off	Off	Engine Data 1
Cruise Inhibited	On / Off	Yes	Yes	Engine Data 1
Decel Fuel Mode	Active / Inactive	Inactive	Inactive	Engine Data 1
Delivered Torque (%)	0-100%	5-20	Varies	Engine Data 1
Desired EGR Position	0-100%	0	Varies	Engine Data 1
Desired Idle Speed	0-3187	PCM control	Varies	Engine Data 1
ECT Sensor (°F)	-40 to 419°F	210°F	208°F	Engine Data 1
EGR Closed Valve Position	0-5.0v	0.14-1.0	Varies	Engine Data 1
EGR Duty Cycle (%)	0-100%	0	Varies	Engine Data 1
EGR Feedback (V)	0-5.0v	0.14-1.0	Varies	Engine Data 1
EGR Test Count	0-10	0	Varies	Engine Data 1
Engine Load	0-100%	2-5	Varies	Engine Data 1
Engine Oil Level	Okay / Low	Okay	Okay	Engine Data 1
Engine Oil Life	0-100%	Varies	Varies	Engine Data 1
Engine Run Time	Hr: Min: Sec:	Varies	Varies	Engine Data 1
Engine Speed	0-9999 rpm	± 100 rpm	---	Engine Data 1
EVAP Purge Solenoid	0-100%	25	Varies	Engine Data 1
EVAP Vacuum Switch	Purge / No Purge	No Purge	Purge	Engine Data 1
EVAP Vent Solenoid	Open/Close	Closed	Closed	Engine Data 1
Fuel Level Sensor	0-100%	Varies	Varies	Engine Data 1
Fuel Pump Signal	On / Off	On	On	Engine Data 1
Fuel Pump Speed	Normal / High	Normal	Normal	Engine Data 1
Fuel Trim Cell	Cell #0-4	0	Varies	Engine Data 1
Fuel Trim Learn	Enabled / Disabled	Enabled	Enabled	Engine Data 1
Generator Command	On / Off	On	On	Engine Data 1
Generator F-Terminal	0-100%	Varies	Varies	Engine Data 1
Generator Lamp	On / Off	Off	Off	Engine Data 1
HO2S-11 (B1 S1)	0-1132 mv	Varies	Varies	Engine Data 1
HO2S-12 (B1 S2)	0-1132 mv	Varies	Varies	Engine Data 1
HO2S-11 Status	Not Ready / Ready	Ready	Ready	Engine Data 1
HO2S-11 Warmup	Min: Sec	Varies	Varies	Engine Data 1
HO2S-12 Warmup	Min: Sec	Varies	Varies	Engine Data 1
HO2S Cross Counts	0-255	Varies	Varies	Engine Data 1
IAC Motor	0-255	10-40	---	Engine Data 1
IAT Sensor (°F)	-40 to 304°F	95	95	Engine Data 1

1996-99 Bonneville 3.8L V6 MFI VIN K, VIN 1 (A/T) - PID Data List

Parameter Identification (PID)	PID Value Range	PID Value at Hot Idle	PID Value at 30 mph	Data List Type
Idle Speed Error (rpm)	0-9999 rpm	> 100	---	Engine Data 1
Ignition 1 Signal	0-25.5v	14.1	14.1	Engine Data 1
Ignition Mode	IC/Bypass	IC	IC	Engine Data 1
Injector Pulsewidth	0-985	1.5-3.5	Varies	Engine Data 1
Knock Retard (°)	0° to 45°	0°	Varies	Engine Data 1
KS Active Counter	Yes / No	No	No	Engine Data 1
KS Minimum Learned Noise	0-5.0v	0.1-3.0	Varies	Engine Data 1
KS Noise Channel (V)	0-5.0v	0.1-3.0	Varies	Engine Data 1
Long Term Fuel Trim	-10% to 10%	0 (± 5%)	0 (± 5%)	Engine Data 1
Loop Status	Closed Loop / Open Loop	Closed Loop	Closed Loop	Engine Data 1
Low Oil Lamp	On / Off	Off	Off	Engine Data 1
MPH / KPH	0-159	0	30	Engine Data 1
MAF Sensor (g/sec)	0-512 g/sec	3-6	Varies	Engine Data 1
MAF Sensor (Hz)	0-65535 Hz	1200-3000	Varies	Engine Data 1
MAP Sensor (kPa)	10-110 kPa	29-48	1.4v	Engine Data 1
Misfire current Cyl 1-6	0-255	0-4	0-4	Engine Data 1
Misfire History Cyl 1-6	0-255	0	0	Engine Data 1
Misfire Pass Since 1st	0-255	0	0	Engine Data 1
Oil Level Signal	Okay / Low	Okay	Okay	Engine Data 1
Oil Life Switch	Press / Release	Released	Released	Engine Data 1
Park Neutral Signal	P-N-R-L	P-N	R-DL	Engine Data 1
Power Enrichment	Active / Inactive	Inactive	Inactive	Engine Data 1
Rich / Lean	RICH/LEAN	Varies	Varies	Engine Data 1
Shift Mode Switch	Normal / Performance	Normal	Normal	Engine Data 1
Short Term Fuel Trim	-10% to 10%	0 (± 5%)	0 (± 5%)	Engine Data 1
Spark Advance (°)	0° to 45°	20	Varies	Engine Data 1
Startup ECT Sensor	-40 to 304°F	178	178	Engine Data 1
Startup IAT Sensor	-40 to 304°F	95	95	Engine Data 1
TCC Enable	On / Off	Off	On	Engine Data 1
Throttle at Idle	Yes / No	Yes	No	Engine Data 1
Throttle Angle (%)	0-100%	0	2	Engine Data 1
Total Misfire Count	0-255	0-5	0-5	Engine Data 1
Total Passes Since 1st Misfire	0-255	0	0	Engine Data 1
Total Since 1st Failure	0-255	0	0	Engine Data 1
TP Sensor (V)	0.0-5.0v	0.20-0.74	1.2	Engine Data 1
Traction Control	Active / Inactive	Inactive	Inactive	Engine Data 1
Traction Control (T/C) Torque	0-100%	5-20	Varies	Engine Data 1
T/C Desired Torque	0-100%	80-95	Varies	Engine Data 1
Trans. Hot Mode	Inactive / Active	Inactive	Inactive	Engine Data 1
Transmission Range	P-R-N-4-3-2	P	3	Engine Data 1
TWC Diagnostic	Inactive / Active	Inactive	Inactive	Engine Data 1
TWC Monitor Counter	0-255	0	0	Engine Data 1
TWC Protection	Active / Inactive	Inactive	Inactive	Engine Data 1
Vehicle Speed	0-155 mph	0	30	Engine Data 1
VTD Fuel Disable	Active / Inactive	Inactive	Inactive	Engine Data 1

2000-05 Bonneville 3.8L V6 MFI VIN K, VIN 1 (A/T) - PID Data List

Parameter Identification (PID)	PID Value Range	PID Value at Hot Idle	PID Value at 30 mph	Data List Type
3X Crank Sensor	0-9,999 rpm	Varies	Varies	Engine Data 2
18X Crank Sensor	0-1,600 rpm	Varies	Varies	Engine Data 2
1-2 Shift Solenoid Circuit Status	Okay / Fault / Invalid State	Okay	Okay	Output Driver
2-3 Shift Solenoid Circuit Status	Okay / Fault / Invalid State	Okay	Okay	Output Driver
A/C HI Side Pressure	0-459 kPa	90	90	Engine Data 2
A/C HI Side Pressure	0-5.1 volts	0.98	0.98	Engine Data 2
A/C Off for WOT	Yes / No	No	No	Engine Data 2
A/C Pressure Disable	Yes / No	No	No	Engine Data 2
A/C Relay Circuit Status	Okay / Fault / Invalid State	Okay	Okay	Output Driver
A/C Relay Command	On / Off	Off	Off	Engine Data 1-2, EGR, Misfire
A/C Request Signal	Yes / No	No	No	Engine Data 2
Air Fuel Ratio	0.0-25.5:1	14.2-14.7:1	14.2-14.7:1	Engine Data 2, Fuel Trim
Air Pump Relay Circuit Status	Okay / Fault / Invalid State	Okay	Okay	AIR, Output Driver
Air Pump Relay Command	On / Off	Off	Off	Engine Data 1 Data, AIR, HO2S
Air Solenoid Circuit Status	Okay / Fault / Invalid State	Okay	Okay	Output Driver
Air Solenoid Command	On / Off	Off	Off	Engine Data 1 Data, AIR, HO2S
Air Fuel Ratio	0.0-25.5:1	14.7:1	14.7:1	Engine Data 2, Fuel Trim
BARO Sensor (kPa)	10-105 kPa	Varies w/ALT	Varies	Engine Data 1, EGR, EVAP, Fuel Trim
Boost Solenoid Circuit Status	Okay / Fault / Invalid State	Okay	Okay	Output Driver
Boost Solenoid PWM	0-100%	0	Varies	Engine Data 2
CMP Sensor Present	Yes / No	Yes	Yes	Engine Data 2
Crank Request Signal	Yes / No	No	No	Engine Data 2
Cruise Control Active	Yes / No	No	No	Engine Data 1
Cruise Inhibit Reason	Vehicle Speed/Brake/ Clutch	Vehicle Speed	Vehicle Speed	Engine Data 1
Cruise Inhibit Signal Circuit Status	Okay/Stuck Low or High	Okay	Okay	Output Driver
Cruise Inhibit Signal Command	On / Off	Off	Off	Engine Data 1
Current Gear	1, 2, 3, 4	1	4	Engine Data 1-2, EGR, Fuel Trim
Cycles of Misfire Data	0-99 counts	0	0	Misfire
Cylinder 1-6 Injector Circuit History	Okay / Stuck Low/ Stuck High / Fault	Okay	Okay	Output Driver

2000-05 Bonneville 3.8L V6 MFI VIN K, VIN 1 (A/T) - PID Data List

Parameter Identification (PID)	PID Value Range	PID Value at Hot Idle	PID Value at 30 mph	Data List Type
Cylinder 1-6 Injector Circuit Status	Okay / Stuck Low/ Stuck High / Fault	Okay	Okay	Output Driver
Decel Fuel Cutoff	Active / Inactive	Inactive	Inactive	EGR, Fuel Trim
Desired EGR Position	0-100%	0	Varies	Engine Data 1, EGR, Misfire
Desired EGR Position	0-5.1v	0.0	Varies	EGR
Desired Idle Speed	0-3187 rpm	Commanded by the PCM	Varies	Engine Data 1, Data 2, EVAP
Driver Module 1 Status	Enabled / Off- High Volts / Off-HI TEMP / Invalid State	Enabled	Enabled	Output Driver
Driver Module 2 Status	Enabled / Off- HI Volts/ Off-HI TEMP/ Invalid State	Enabled	Enabled	Output Driver
Driver Module 3 Status	Enabled / Off- HI Volts/ Off-HI TEMP/ Invalid State	Enabled	Enabled	Output Driver
Driver Module 4 Status	Enabled / Off- HI Volts/ Off-HI TEMP/ Invalid State	Enabled	Enabled	Output Driver
ECT Sensor (°F)	-40 to 304°F	190-221°F	190-221°F	Engine Data 1-2, EGR, Fuel Trim, EVAP, Misfire
EGR Flow Test Count	0-255 counts	0-12	Varies	EGR
EGR Learned Minimum Position	0-5.1 volts	0.14-1.0	Varies	EGR
EGR Position Sensor	0-100%	Varies	Varies	Engine Data 1, EGR, Misfire
EGR Position Sensor	0-5.1 volts	Varies	Varies	EGR
EGR Position Variance	0-100%	0-9	0-9	EGR
EGR Solenoid Circuit History	Okay / Fault / Invalid State	Okay	Okay	Output Driver
EGR Solenoid Circuit Status	Okay / Fault / Invalid State	Okay	Okay	Output Driver
EGR Solenoid Command	0-100%	Varies	Varies	EGR
Engine Load	0-100%	1-4	Varies	Engine Data 1-2, EGR, Fuel Trim, EVAP, Misfire
Engine Oil Level Switch	Okay / Low	Okay	Okay	Engine Data 2
Engine Oil Life	0-100%	Varies	Varies	Engine Data 2
EOP Switch Signal	0-5.0 volts	2-3	2-3	Engine Data 2

2000-05 Bonneville 3.8L V6 MFI VIN K, VIN 1 (A/T) - PID Data List

Parameter Identification (PID)	PID Value Range	PID Value at Hot Idle	PID Value at 30 mph	Data List Type
Engine Run Time	Hr: Min: Sec 00:00:00 to 99:99:99	00:00:00	00:00:00	Engine Data 1-2, EGR, Fuel Trim, EVAP, Misfire
Engine Speed	0-9999 rpm	± 50 rpm of actual speed	Varies	Engine Data 1-2, EGR, Fuel Trim, EVAP, Misfire
EVAP Fault History	No Fault / Excess VAC / Purge Valve Leak / Small Leak / Weak Vacuum	No Fault	No Fault	EVAP
EVAP Purge Solenoid Circuit Status	Okay / Fault / Invalid State	Okay	Okay	Output Driver
EVAP Purge Solenoid Command	0-100%	15	Varies	Engine Data 1, EVAP, Fuel Trim
EVAP Test Abort Reason	Not Aborted / Lost Enable / Small Leak / Not at Rest	Not Aborted	Not Aborted	EVAP
EVAP Test Result	No Result / Passed / Fail / P0440 / P0442 / P0446 / P1441	No Test Results	No Test Results	EVAP
EVAP Test State	Waiting for Purge/ Test Running/Test Completed	Test Completed	Test Completed	EVAP
EVAP Vent Solenoid Circuit Status	Okay / Fault / Invalid State	Okay	Okay	Output Driver
EVAP Vent Solenoid Command	Venting / Not Venting	Venting	Venting	Engine Data 1, EVAP, Fuel Trim
Ext. Travel Brake Switch	Applied / Released	Released	Released	Engine Data 2
Fan Control Relay 1 Circuit Status	Okay / Fault / Invalid State	Okay	Okay	Output Driver
Fan Control Relay 1 Command	On / Off	Off	Off	Engine Data 2
Fan Control Relay 2-3 Circuit Status	Okay / Fault	Okay	Okay	Output Driver
Fan Control Relay 2-3 Command	On / Off	Off	Off	Engine Data 2
Fuel Pump Relay Circuit History Status	Okay / Fault	Okay	Okay	Output Driver
Fuel Pump Relay Circuit Status	Okay / Fault / Invalid State	Okay	Okay	Output Driver
Fuel Pump Relay Command	On / Off	On	On	Engine Data 1
Fuel Tank level	0-100%	Varies	Varies	EVAP
Fuel Tank Pressure Sensor	Inches H2O, mm Hg	-17.5 to +7.5 Inches H2O	-17.5 to +7.5 Inches H2O	Engine Data 1, EVAP

2000-05 Bonneville 3.8L V6 MFI VIN K, VIN 1 (A/T) - PID Data List

Parameter Identification (PID)	PID Value Range	PID Value at Hot Idle	PID Value at 30 mph	Data List Type
Fuel Tank Pressure Sensor	0-5.0 volts	Varies	Varies	EVAP
Fuel Trim Cell	Cell 0-9	0-4	Varies	Engine Data 1, EVAP, Fuel Trim
Fuel Trim Learn	Enabled or Disabled	Enabled	May Toggle	Engine Data 1, EVAP, Fuel Trim
Generator F-Terminal	0-100%	Varies	Varies	Engine Data 2
Generator L-Terminal	On / Off	On	On	Engine Data 2
HO2S (B1 S1) Signal	0-1132 mv	10-1000	10-1000	Engine Data 1, EVAP, Fuel Trim
HO2S (B2 S1) Signal	0-1132 mv	10-1000	10-1000	Engine Data 1, EVAP, Fuel Trim
IAC Position	0-255 counts	Varies	Varies	Engine Data 1, EGR, Fuel Trim
IAT Sensor (°F)	-40 to 304°F	91	92	Engine Data 1-2, EGR, EVAP, Fuel Trim
Ignition 1	0.0-25.5v	13.5	13.6	Engine Data 1-2, EGR, EVAP, Fuel Trim
Ignition Mode	Bypass/IC	IC	IC	Engine Data 2
Injector Pulsewidth	0-1000 ms	1.5-3.5	Varies	Engine Data 2, Fuel Trim, M/fire
Internal Mode Switch (IMS) Signal	Park / Reverse / Drive 4-3-2-1	Park	Drive 4	Engine Data 1
Knock Retard (°)	0-25.5°	0°	0°	Engine Data 1, EGR
Long Term Fuel Trim Bank 1 & Bank 2	-10 to +10	0 (± 5%)	0 (± 5%)	Engine Data 1-2, EVAP, Fuel Trim
Loop Status	Open / Closed	Closed	Closed	Engine Data 1-2, EGR, EVAP, Fuel Trim
MAF Sensor (g/sec)	0-512 g/sec	3-6	Varies	Engine Data 1-2, EGR, Fuel Trim, EVAP, Misfire
MAF Sensor (Hz)	0-32,000 Hz	1200-3000	Varies	Engine Data 2
MAP Sensor (kPa)	10-105 kPa	20-48	Varies	Engine Data 1-2, EGR, Fuel Trim, EVAP, Misfire
MAP Sensor (V)	0-5.1 volts	0.75-2.0	Varies	Engine Data 1-2
MIL Circuit Status	Okay / Fault / Invalid State	Okay	Okay	Output Driver
MIL Command	On / Off	Off	Off	Engine Data 2
Misfire Current Cylinder 1-6	0-198 counts	0-4	0	Misfire
Misfire History Cylinder 1-6	0-65,535	0	0	Misfire
Number of DTC(s)	Number	0	No	Engine Data 1, EVAP, Fuel Trim

2000-05 Bonneville 3.8L V6 MFI VIN K, VIN 1 (A/T) - PID Data List

Parameter Identification (PID)	PID Value Range	PID Value at Hot Idle	PID Value at 30 mph	Data List Type
PCM/VCM in VTD Fail Enable	Yes / No	Yes	Yes	Engine Data 1
Power Enrichment	Active / Inactive	Inactive	Inactive	Engine Data 2, Misfire
Short Term Fuel Trim Bank 1 and Bank 2	-10 to +10%	0 (± 5%)	0 (± 5%)	Engine Data 1-2, EVAP, Fuel Trim
Spark Advance	-64° to +64°	-20	Varies	Engine Data 1-2, Fuel Trim, M/fire
Starter Enable Relay Circuit Status	Okay / Fault / Invalid State	Okay	Okay	Output Driver
Starter Relay Control	On / Off	Off	Off	Engine Data 2
Startup ECT	-40 to 419°F	Varies	Varies	EVAP, Fuel Trim
Startup IAT	-40 to 419°F	80°F	80°F	EVAP, Fuel Trim
TCC/Cruise Brake Pedal Switch	Apply/REL	Released	Released	Engine Data 1, Data 2
TCC PWM Solenoid Command	0-100%	0	Varies	Engine Data 2
Traction Control System Circuit Status	Okay/Fault/ Invalid State	Okay	Okay	Output Driver
Traction Control System Circuit History	Okay/Fault/ Invalid State	Okay	Okay	Output Driver
TFP Switch	P/N, Reverse, Drive 4, 3, 2, 1, Invalid	Park/Neutral	Drive 4	Engine Data 2, EGR, Fuel Trim
TP Sensor (%)	0-100%	2-6	Varies	Engine Data 1-2, EGR, Fuel Trim, EVAP, Misfire
TP Sensor (V)	0-5.0 volts	0.2-0.8	Varies	Engine Data 1-2
Traction Control Status	Active / Inactive	Inactive	Inactive	Engine Data 2
Torque Delivered Signal	0-100%	85	85	Engine Data 2
Torque Request Signal	0-100%	100	100	Engine Data 2
Vehicle Speed	Km/h, MPH	0	30	Engine Data 1-2, EGR, Fuel Trim, EVAP, Misfire
VTD Auto Learn Timer	Active / Inactive	Inactive	Inactive	Engine Data 1
VTD Fuel Enable	Active / Inactive	Inactive	Inactive	Engine Data 1
VTD Fuel Disable Until Ignition Off	Yes / No	No	No	Engine Data 1

FIREBIRD PID DATA

1995 Firebird 3.8L V6 MFI VIN K (All) - PID Data List

Parameter Identification (PID)	PID Value Range	PID Value at Hot Idle	PID Value at 30 mph	Data List Type
Actual EGR Position	0-100%	0	0	Engine Data 1
Air Fuel Ratio	0:1-25:5.1	14.6:1	14.6:1	Engine Data 1
BARO Sensor (kPa)	10-110 kPa	65-110	65-110	Engine Data 1
Commanded Generator	On / Off	On	On	Engine Data 1
Cruise Engaged	Yes / No	No	No	Engine Data 1
Cruise Inhibited	Yes / No	Yes	Yes	Engine Data 1
Current Gear	1-2-3-4	1st	3rd	Engine Data 1
Commanded TCC	Disengaged / Engaged	Disengaged	Disengaged	Engine Data 1
Decel Fuel Mode	Active / Inactive	Inactive	Inactive	Engine Data 1
Desired EGR Position	0-100%	0	0	Engine Data 1
Desired Idle Speed	0-3187	PCM control	---	Engine Data 1
ECT Sensor (°F)	-40 to 304°F	198	197	Engine Data 1
Engine Load	0-100%	205	45	Engine Data 1
Engine Run Time	Hr: Min: Sec	00:00:00	00:00:00	Engine Data 1
Engine Speed	0-9999 rpm	Varies	---	Engine Data 1
Fuel Trim Cell	Cell #	0	1	Engine Data 1
Fuel Trim Learn	Enabled / Disabled	Enabled	Enabled	Engine Data 1
Fuel Pump	On / Off	On	On	Engine Data 1
Hot Open Loop	Active / Inactive	Inactive	Inactive	Engine Data 1
HO2S1 Bank 1, 2	Ready / Not Ready	Ready	Ready	Engine Data 1
HO2S-11 (B1 S1)	0-1132 mv	10-1000	10-1000	Engine Data 1
HO2S-12 (B1 S2)	0-1132 mv	10-1000	10-1000	Engine Data 1
HO2S-13 (B1 S3)	0-1132 mv	10-1000	10-1000	Engine Data 1
HO2S-21 (B1 S1)	0-1132 mv	10-1000	10-1000	Engine Data 1
HO2S-11 Rich/Lean	L or R	L-R-L-R	L-R-L-R	Engine Data 1
HO2S-12 Rich/Lean	L or R	L-R-L-R	L-R-L-R	Engine Data 1
IAC Position	0-255	35	---	Engine Data 1
IAT Sensor (°F)	-40 to 304°F	94	93	Engine Data 1
Ignition 1 Volts	0.0-25.5v	14.1	14.1	Engine Data 1
Injector Pulsewidth	0-985 ms	3.0	2.9	Engine Data 1
Long Term Fuel Trim	-10% to 10%	0 (± 5%)	0 (± 5%)	Engine Data 1
Loop Status	Closed Loop / Open Loop	Closed Loop	Closed Loop	Engine Data 1
MAF Sensor (g/sec)	0-512 g/sec	5	9.3	Engine Data 1
MAF Sensor (Hz)	0-32000 Hz	1200-3000	Varies	Engine Data 1
MAP Sensor (kPa)	10-110 kPa	36	46	Engine Data 1
MAP Sensor (V)	0.0-5.0v	1.31	1.12	Engine Data 1
Power Enrichment	Active / Inactive	Active	Active	Engine Data 1
Short Term Fuel Trim	-10% to 10%	0 (± 5%)	0 (± 5%)	Engine Data 1
Throttle at Idle	Yes / No	Yes	No	Engine Data 1
TP Angle (%)	0-100%	0	4	Engine Data 1
TP Sensor (V)	0.0-5.0v	0.56	0.76	Engine Data 1
TWC Protection	Active / Inactive	Inactive	Inactive	Engine Data 1
Vehicle Speed	0-155 mph	0	30	Engine Data 1
VTD Fuel Disable	Active / Inactive	Inactive	Inactive	Engine Data 1

1995 Firebird 3.8L V6 MFI VIN K (All) - PID Data List

Parameter Identification (PID)	PID Value Range	PID Value at Hot Idle	PID Value at 30 mph	Data List Type
18X Crank Sensor	0-1280 rpm	Up to 1280 rpm	---	Engine Data 2
A/C HI Side Pressure	0.0-5.0v	Varies	Varies	Engine Data 2
A/C Pressure Out of Range	Yes / No	No	No	Engine Data 2
A/C Request	Yes / No	No	No	Engine Data 2
Actual EGR Position	0-100%	0	0-100	Engine Data 2
A/C Slugging	Active / Inactive	Inactive	Inactive	Engine Data 2
A/C Off for WOT	Yes / Off	No	No	Engine Data 2
BARO Sensor (kPa)	10-110 kPa	65-110	65-110	Engine Data 2
Cam Signal Present	Yes / No	Yes	Yes	Engine Data 2
Commanded A/C	On / Off	On: A/C on	Off	Engine Data 2
Commanded Fan 1	On / Off	On: Fan on	Off	Engine Data 2
Commanded Fan 2	On / Off	On: Fan on	Off	Engine Data 2
Commanded TCC	Disengaged / Engaged	Disengaged	Disengaged	Engine Data 2
Desired EGR Position	0-100%	0	0-50	Engine Data 2
Desired Idle Speed	0-3187 rpm	PCM control	---	Engine Data 2
ECT Sensor (°F)	-40 to 304°F	198F	198	Engine Data 2
Engine Load	0-100%	4	45	Engine Data 2
Engine Run Time	Hr: Min: Sec	00:00:00	00:00:00	Engine Data 2
Engine Speed	0-9999 rpm	Varies	---	Engine Data 2
EVAP Purge PWM	0-100%	0	14	Engine Data 2
EVAP Vacuum Switch	Purge / No	No Purge	Purge	Engine Data 2
Generator Lamp	On / Off	Off	Off	Engine Data 2
Hot Open Loop	Active / Inactive	Inactive	Inactive	Engine Data 2
IAC Position	0-255	35	---	Engine Data 2
IAT Sensor (°F)	-40 to 304°F	94	94	Engine Data 2
Idle Speed Error	0-12800 rpm	>100	---	Engine Data 2
Ignition Mode	Bypass / IC	IC	IC	Engine Data 2
Knock Retard	0.0-25.5°	0°	0°	Engine Data 2
KS Activity	Yes / No	No	No	Engine Data 2
KS Minimum Learned Noise	0.0-5.0v	0.4	0.2	Engine Data 2
KS Noise Channel	0.0-5.0v	1.6	1.5	Engine Data 2
MAF Sensor (g/sec)	0-512 g/sec	5	9	Engine Data 2
MAP Sensor (kPa)	10-110 kPa	35	45	Engine Data 2
MAP Sensor (V)	0.0-5.0v	1-2v	Varies	Engine Data 2
MIL Status	On / Off	Off	Off	Engine Data 2
Power Enrichment	Active / Inactive	Active	Active	Engine Data 2
Spark Advance (°)	-64° to 64°	20	Varies	Engine Data 2
TP Angle	0-100%	0	4	Engine Data 2
Trans. Hot Mode	Active / Inactive	Inactive	Inactive	Engine Data 2
Transmission Range	P-R-N-2-3-4	PARK	DRIVE 3	Engine Data 2
Trans. Range ABC	X / O	X/O/X/O	---	Engine Data 2
TWC Protection	Active / Inactive	Inactive	Inactive	Engine Data 2
Vehicle Speed	0-155 mph	0	30	Engine Data 2

1995 Firebird 3.8L V6 MFI VIN K (All) - PID Data List

Parameter Identification (PID)	PID Value Range	PID Value at Hot Idle	PID Value at 30 mph	Data List Type
Air Fuel Ratio	0:1-25:5.1	14.6:1	14.6:1	Catalyst
Decel Fuel Mode	Active / Inactive	Inactive	Inactive	Catalyst
ECT Sensor (°F)	-40 to 304°F	198	198	Catalyst
Engine Load	0-100%	2-5	45	Catalyst
Engine Run Time	Hr: Min: Sec	00:00:00	00:00:00	Catalyst
Engine Speed	0-9999 rpm	Varies	---	Catalyst
Fuel Trim Learn	Enabled / Disabled	Enabled	Enabled	Catalyst
Hot Open Loop	Active / Inactive	Inactive	Inactive	Catalyst
HO2S-11 (B1 S1)	0-1132 mv	10-1000	10-1000	Catalyst
HO2S-12 (B1 S2)	0-1132 mv	10-1000	10-1000	Catalyst
HO2S-13 (B1 S3)	0-1132 mv	10-1000	10-1000	Catalyst
IAC Position	0-255	35	---	Catalyst
IAT Sensor (°F)	-40 to 304°F	94	94	Catalyst
Injector Pulsewidth	0-985 ms	3.0	2.9	Catalyst
Loop Status	Closed Loop / Open Loop	Closed Loop	Closed Loop	Catalyst
MIL Status	On / Off	Off	Off	Catalyst
MAF Sensor (g/sec)	0-512 g/sec	5	9	Catalyst
Power Enrichment	Active / Inactive	Inactive	Inactive	Catalyst
Throttle at Idle	Yes / No	Yes	No	Catalyst
TP Angle	0-100%	0	4	Catalyst
TP Sensor	0.0-5.0v	0.56	0.76	Catalyst
TWC Diagnostic	Enabled / Disabled	Enabled	Enabled	Catalyst
TWC Monitor Test Counter	0-255	0	0	Catalyst
TWC Protection	Active / Inactive	Inactive	Inactive	Catalyst
Vehicle Speed	0-155 mph	0	30	Catalyst

1995 Firebird 3.8L V6 MFI VIN K (All) - PID Data List

Parameter Identification (PID)	PID Value Range	PID Value at Hot Idle	PID Value at 30 mph	Data List Type
ECT Sensor (°F)	-40 to 304°F	198	198	HO2S
Engine Speed	0-9999 rpm	Varies	---	HO2S
EVAP Purge PWM	0-100%	0-25	0-100	HO2S
HO2S-11 (B1 S1)	0-1132 mv	10-1000	10-1000	HO2S
HO2S-12 (B1 S2)	0-1132 mv	10-1000	10-1000	HO2S
HO2S-13 (B1 S3)	0-1132 mv	10-1000	10-1000	HO2S
HO2S-21 (B2 S1)	0-1132 mv	10-1000	10-1000	HO2S
HO2S-11 W	00:00-99:99	Varies	Varies	HO2S
HO2S-12 W	00:00-99:99	Varies	Varies	HO2S
HO2S-13 W	00:00-99:99	Varies	Varies	HO2S
HO2S XCount B1	0-255	Varies	Varies	HO2S
HO2S XCount B2	0-255	Varies	Varies	HO2S
Ignition 1 Signal	0.0-25.5v	14.1	14.1	HO2S
MAF Sensor (g/sec)	0-512 g/sec	5	9	HO2S
Startup ECT Sensor	-40 to 304°F	Varies	Varies	HO2S
Startup IAT Sensor	-40 to 304°F	Varies	Varies	HO2S
TP Angle	0-100%	0	4	HO2S
Vehicle Speed	0-155 mph	0	30	Catalyst

1995 Firebird 3.8L V6 MFI VIN K (All) - PID Data List

Parameter Identification (PID)	PID Value Range	PID Value at Hot Idle	PID Value at 30 mph	Data List Type
Actual EGR Position	0-100%	0	0	EGR
BARO Sensor (kPa)	10-110 kPa	65-110	65-110	EGR
Current Gear	1-2-3-4	1st	3rd	EGR
Commanded TCC	Disengaged / Engaged	Disengaged	Disengaged	EGR
Decel Fuel Mode	Active / Inactive	Inactive	Inactive	EGR
Desired EGR Position	0-100%	0	0-20	EGR
Desired Idle Speed	0-3187 rpm	PCM control	---	EGR
ECT Sensor (°F)	-38-304°F	198	198	EGR
EGR Closed Pintle Position	0.0-5.0v	0.14-1.0	---	EGR
EGR Duty Cycle	0-100%	0	0-100	EGR
EGR Feedback	0.0-5.0v	0.76	0.86	EGR
EGR Flow Test Count	0-255	0	0	EGR
EGR Position Error	0-100%	0	Varies	EGR
Engine Load	0-100%	2-5	45	EGR
Engine Speed	0-9999 rpm	Varies	---	EGR
Fuel Trim Cell	0-9	0	1	EGR
Fuel Trim Learn	Enabled / Disabled	Enabled	Enabled	EGR
IAC Position	0-255	35	---	EGR
IAT Sensor (°F)	-40 to 304°F	94	94	EGR
Ignition 1 Signal	0.0-25.5v	14.1	14.1	EGR
Knock Retard (°)	0.0-25.5°	0	0	EGR
KS Activity	Yes / No	No	No	EGR
Loop Status	Closed Loop / Open Loop	Closed Loop	Closed Loop	EGR
MAP Sensor (kPa)	10-110 kPa	35	45	EGR
MAP Sensor (V)	0.0-5.0v	1-2v	Varies	EGR
Power Enrichment	Active / Inactive	Active	Active	EGR
HO2S-11 (B1 S1) R/L	0-1132 mv	10-1000	10-1000	EGR
HO2S-21 (B2 S1) R/L	0-1132 mv	10-1000	10-1000	EGR
Spark Advance (°)	-64° to 64°	20	Varies	EGR
TP Angle (%)	0-100%	0	4	EGR
Trans. Hot Mode	Active / Inactive	Inactive	Inactive	EGR
Transmission Range	P-R-N-4-3-2	PARK	3	EGR
TWC Protection	Active / Inactive	Inactive	Inactive	EGR
Vehicle Speed	0-155 mph	0	30	EGR

1995 Firebird 3.8L V6 MFI VIN K (All) - Misfire Data

Parameter Identification (PID)	PID Value Range	PID Value at Hot Idle	PID Value at 30 mph	Data List Type
A/C Request	Yes / No	No	No	Misfire
Actual EGR Position	0-100%	0	0	Misfire
Commanded A/C	On / Off	Off	Off	Misfire
Current Gear	1-2-3-4	1	3	Misfire
Decel Fuel Mode	Active / Inactive	Inactive	Inactive	Misfire
Desired EGR Position	0-100%	0	0	Misfire
ECT Sensor (°F)	-40 to 304°F	198	198	Misfire
Engine Load	0-100%	2-5	45	Misfire
Engine Speed	0-9999 rpm	Varies	---	Misfire
Fuel Trim Cell	0-9	0	1	Misfire
Fuel Trim Learn	Enabled / Disabled	Enabled	Enabled	Misfire
Hot Open Loop	Active / Inactive	Inactive	Inactive	Misfire
HO2S-11 (B1 S1) Status	Ready / Not Ready	Ready	Ready	Misfire
HO2S-21 (B2 S1) Status	Ready / Not Ready	Ready	Ready	Misfire
HO2S-11 (B1 S1) XCount	0-255	Varies	Varies	Misfire
HO2S-21 (B2 S1) XCount	0-255	Varies	Varies	Misfire
IAC Position	0-255	35	---	Misfire
Ignition Mode	Bypass / IC	IC	IC	Misfire
Knock Retard (°)	0.0-25.5°	9	Varies	Misfire
Long Term Fuel Trim Bank 1 and Bank 2	-10% to 10%	0 (± 5%)	0 (± 5%)	Misfire
Loop Status	Closed Loop / Open Loop	Closed Loop	Closed Loop	Misfire
Misfire Current Cylinder 1-6	0-255	0	0	Misfire
Misfire History Cylinder 1-6	0-65535	0	0	Misfire
Misfiring Cylinder	CYL #1-6	0	0	Misfire
Power Enrichment	Active / Inactive	Inactive	Inactive	Misfire
HO2S-11 Rich/Lean	0-1132 mv	10-1000	10-1000	Misfire
HO2S-21 Rich/Lean	0-1132 mv	10-1000	10-1000	Misfire
Short Term Fuel Trim Bank 1 and Bank 2	-10% to 10%	0 (± 5%)	0 (± 5%)	Misfire
Spark Advance (°)	-64° to 64°	16	26	Misfire
Total Misfire Current Count	0-99	0	0	Misfire
Misfire Failures Since First Fail	0-65535	0	0	Misfire
Misfire Passes Since First Fail	0-65535	0	0	Misfire
TCC Enable	On / Off	Off	Off	Misfire
TWC Protection	Active / Inactive	Inactive	Inactive	Misfire
Vehicle Speed	0-155 mph	0	30	Misfire

1996-99 Firebird 3.8L V6 MFI VIN K (All) - PID Data List

Parameter Identification (PID)	PID Value Range	PID Value at Hot Idle	PID Value at 30 mph	Data List Type
Actual EGR Position	0-100%	0%	Varies	Engine Data 1
Air Fuel Ratio	0:1-25.5:1	14.6:1	14.6:1	Engine Data 1
BARO Sensor (kPa)	10-110 kPa	65-110	65-110	Engine Data 1
BARO Sensor (V)	0.0-5.1	3.5-4.5	3.5-4.5	Engine Data 1
Commanded Generator	On / Off	On	On	Engine Data 1
Commanded TCC	Disengaged / Engaged	Disengaged	Disengaged	Engine Data 1
Cruise Engaged	Enabled / Disabled	Disabled	Disabled	Engine Data 1
Cruise Inhibit	Yes / No	Yes	Yes	Engine Data 1
Current Gear	1-2-3-4	1	3	Engine Data 1
Decel Fuel Mode	Active / Inactive	Inactive	Inactive	Engine Data 1
Desired EGR Position	0-100%	0%	Varies	Engine Data 1
Desired Idle Speed	0-3187 rpm	PCM control	---	Engine Data 1
ECT Sensor (°F)	-40 to 304°F	185	185	Engine Data 1
Engine Load	0-100%	2-5%	Varies	Engine Data 1
Engine Run Time	Hr: Min: Sec	00:00:00	00:00:00	Engine Data 1
Engine Speed	0-9999 rpm	Varies	---	Engine Data 1
Fuel Trim Cell	Cell #0-4	0	1	Engine Data 1
Fuel Trim Learn	Enabled / Disabled	Enabled	Enabled	Engine Data 1
Hot Open Loop	Active / Inactive	Inactive	Inactive	Engine Data 1
HO2S-11 Status	Not Ready / Ready	Ready	Ready	Engine Data 1
HO2S-12 Status	Not Ready / Ready	Ready	Ready	Engine Data 1
HO2S-21 Status	Not Ready / Ready	Ready	Ready	Engine Data 1
HO2S-22 Status	Not Ready / Ready	Ready	Ready	Engine Data 1
HO2S-11 (B1 S1)	0-1132 mv	10-1000	10-1000	Engine Data 1
HO2S-12 (B1 S2)	0-1132 mv	10-1000	10-1000	Engine Data 1
HO2S-13 (B1 S3)	0-1132 mv	10-1000	10-1000	Engine Data 1
HO2S-21 (B2 S1)	0-1132 mv	10-1000	10-1000	Engine Data 1
HO2S-22 (B2 S2)	0-1132 mv	10-1000	10-1000	Engine Data 1
IAC Position	0-255	10-40	---	Engine Data 1
IAT Sensor (°F)	-40 to 304°F	50-194	50-194	Engine Data 1
Ignition 1 Volts	0.0-25.5v	14.2	14.2	Engine Data 1
Injector Pulsewidth	0-985 ms	1.5-3.5	Varies	Engine Data 1
Long Term Fuel Trim	-10% to 10%	0 (± 5%)	0 (± 5%)	Engine Data 1
Loop Status	Closed Loop / Open Loop	Closed Loop	Closed Loop	Engine Data 1
MAF Sensor (g/sec)	0-512 g/sec	3 - 6	Varies	Engine Data 1
MAF Sensor (Hz)	0-31,999	1200-3000	Varies	Engine Data 1
MAP Sensor (kPa)	10-110 kPa	29-48	Varies	Engine Data 1
MAP Sensor (V)	0.0-5.0v	1-2v	Varies	Engine Data 1
Power Enrichment	Active / Inactive	Active	Active	Engine Data 1
Short Term Fuel Trim	-10% to 10%	0 (± 5%)	0 (± 5%)	Engine Data 1
Throttle at Idle	Yes / No	Yes	---	Engine Data 1
TP Angle (%)	0-100%	0	Varies	Engine Data 1
TP Sensor (V)	0.0-5.0v	0.20-0.74	Varies	Engine Data 1
TWC Protection	Active / Inactive	Inactive	Inactive	Engine Data 1
Vehicle Speed	0-155 mph	0	30	Engine Data 1
VTD Fuel Disable	Enabled / Disabled	Disabled	Disabled	Engine Data 1

1996-99 Firebird 3.8L V6 MFI VIN K (All) - PID Data List

Parameter Identification (PID)	PID Value Range	PID Value at Hot Idle	PID Value at 30 mph	Data List Type
A/C HI Side Pressure	0.0-5.0v	Varies	Varies	Engine Data 2
A/C Press. Too HI/LO	Yes / No	No	No	Engine Data 2
A/C Request	Yes / No	No	No	Engine Data 2
A/C Off for WOT	Yes / No	No	No	Engine Data 2
A/C Slugging	Active / Inactive	Inactive	Inactive	Engine Data 2
Actual EGR Position	0-100%	0	0-100	Engine Data 2
BARO Sensor (kPa)	10-110 kPa	65-110	65-110	Engine Data 2
Clutch Pedal Switch	Applied / Released	Applied (clutch in)	Released	Engine Data 2
Commanded A/C	On / Off	On: A/C on	Off	Engine Data 2
Commanded Fan 1, 2	On / Off	On: fan on	Off	Engine Data 2
Commanded TCC	Disengaged / Engaged	Disengaged	Disengaged	Engine Data 2
Cruise Engaged	Disengaged / Engaged	Disabled	Disabled	Engine Data 2
Cruise Inhibit	Yes / No	Yes	Yes	Engine Data 2
Current Gear	1-2-3-4	1	3	Engine Data 2
Decel Fuel Mode	Active / Inactive	Inactive	Inactive	Engine Data 2
Desired EGR Position	0-100%	0	0-100	Engine Data 2
Desired Idle Speed	0-3187 rpm	PCM control	----	Engine Data 2
ECT Sensor (°F)	-40 to 304°F	185-220	185-220	Engine Data 2
Engine Load	0-100%	2-5	Varies	Engine Data 2
Engine Run Time	Hr: Min: Sec	00:00:00	00:00:00	Engine Data 2
Engine Speed	0-9999 rpm	Varies	----	Engine Data 2
EVAP Purge PWM	0-100%	0-25	Varies	Engine Data 2
EVAP Vacuum Switch	Purge/No	No Purge	No Purge	Engine Data 2
Fuel Pump	On / Off	On	On	Engine Data 2
Generator Lamp	On / Off	Off	Off	Engine Data 2
Hot Open Loop	Active / Inactive	Inactive	Inactive	Engine Data 2
IAC Position	0-255	10-40	----	Engine Data 2
IAT Sensor (°F)	-40 to 304°F	50-194	50-194	Engine Data 2
Idle Speed Error	0-12800 rpm	PCM control	----	Engine Data 2
Ignition 1 Volts	0.0-25.5v	14.2	14.2	Engine Data 2
Ignition Mode	Bypass / IC	IC	IC	Engine Data 2
Knock Retard	0.0-25.5%	0%	Varies	Engine Data 2
KS Active Counter	0-255	Varies	Varies	Engine Data 2
KS Minimum Learned Noise	0.0-5.0v	0.1-3.0v	Varies	Engine Data 2
KS Noise Channel	0.0-5.0v	0.1-3.0v	Varies	Engine Data 2
MAF Sensor (g/sec)	0-512 g/sec	3-6	Varies	Engine Data 2
MAP Sensor (kPa)	10-110 kPa	29-48	Varies	Engine Data 2
MIL Status	On / Off	Off	Off	Engine Data 2
Power Enrichment	Active / Inactive	Active	Active	Engine Data 2
Spark Advance (°)	-64° to 64°	20	Varies	Engine Data 2
TP Angle	0-100%	0	Varies	Engine Data 2
Trans. Hot Mode	Active / Inactive	Inactive	Inactive	Engine Data 2
Transmission Range	P/N-R-4-3-2	P/H	3	Engine Data 2
TR Switch P/A/B/C	HI / LO	HI	Low	Engine Data 2
TWC Protection	Active / Inactive	Inactive	Inactive	Engine Data 2
Vehicle Speed	0-155 mph	0	30	Engine Data 2
VTD Fuel Disable	Enabled / Disabled	Disabled	Disabled	Engine Data 2

1996-99 Firebird 3.8L V6 MFI VIN K (All) - PID Data List

Parameter Identification (PID)	PID Value Range	PID Value at Hot Idle	PID Value at 30 mph	Data List Type
Air Fuel Ratio	0:1-25.5:1	14.6:1	14.6:1	Catalyst
Decel Fuel Mode	Active / Inactive	Inactive	Inactive	Catalyst
ECT Sensor (°F)	-40 to 304°F	185-220	185-220	Catalyst
Engine Load	0-100%	2-5	Varies	Catalyst
Engine Run Time	Hr: Min: Sec	00:00:00	00:00:00	Catalyst
Engine Speed	0-9999 rpm	±50 rpm	---	Catalyst
Hot Open Loop	Active / Inactive	Inactive	Inactive	Catalyst
HO2S-13 (B1 S3)	0-1132 mv	10-1000	10-1000	Catalyst
IAC Position	0-255	10-40	---	Catalyst
IAT Sensor (°F)	-40 to 304°F	50-194	50-194	Catalyst
Loop Status	Closed Loop / Open Loop	Closed Loop	Closed Loop	Catalyst
MAF Sensor	0-512 g/sec	5-9	Varies	Catalyst
Power Enrichment	Active / Inactive	Active	Active	Catalyst
TP Angle	0-100%	0	Varies	Catalyst
TWC Diagnostic	Enabled / Disabled	Enabled	Enabled	Catalyst
TWC Monitor Test Counter	0-49	0	0	Catalyst
TWC Protection	Active / Inactive	Inactive	Inactive	Catalyst
Vehicle Speed	0-155 mph	0	30	Catalyst

1996-99 Firebird 3.8L V6 MFI VIN K (All) - PID Data List

Parameter Identification (PID)	PID Value Range	PID Value at Hot Idle	PID Value at 30 mph	Data List Type
Actual EGR Position	0-100%	0	Varies	EGR
BARO Sensor (kPa)	10-110 kPa	65-110	65-110	EGR
BARO Sensor (V)	0.0-5.0v	3.5-4.5	3.5-4.5	EGR
Current Gear	1-2-3-4	1	3	EGR
Decel Fuel Mode	Active / Inactive	Inactive	Inactive	EGR
Desired EGR Position	0-100%	0	0-100	EGR
Desired Idle Speed	0-3187 rpm	PCM control	---	EGR
ECT Sensor (°F)	-38-304°F	185-220	185-220	EGR
EGR Closed Pintle Position	0.0-5.0v	0.14-1.0	---	EGR
EGR Duty Cycle	0-100%	0	0-100	EGR
EGR Feedback	0.0-5.0v	0.14-1.0	---	EGR
Engine Speed	0-9999 rpm	Varies	---	EGR
Hot Open Loop	Active / Inactive	Inactive	Inactive	EGR
IAC Position	0-255	10-40	---	EGR
IAT Sensor (°F)	-40 to 304°F	50-194	50-194	EGR
Ignition 1 Signal	0.0-25.5v	14.2	14.2	EGR
MAP Sensor (kPa)	10-110 kPa	29-48	Varies	EGR
MAP Sensor (V)	0.0-5.0v	1-2v	Varies	EGR
TP Angle (%)	0-100%	0%	Varies	EGR
Vehicle Speed	0-155 mph	0	30	EGR

1996-99 Firebird 3.8L V6 MFI VIN K (All) - PID Data List

Parameter Identification (PID)	PID Value Range	PID Value at Hot Idle	PID Value at 30 mph	Data List Type
Engine Run Time	Hr: Min: Sec	00:00:00	00:00:00	EVAP
Engine Speed	0-9999 rpm	Varies	---	EVAP
EVAP Canister Purge	0-100%	0-25	0-100	EVAP
IAT Sensor (°F)	-40 to 304°F	50-194	50-194	EVAP
MAF Sensor (g/sec)	0-512 g/sec	3-6	Varies	EVAP
MAP Sensor (kPa)	10-110 kPa	29-48	Varies	EVAP
MAP Sensor (V)	0.0-5.0v	1-2	Varies	EVAP
Startup ECT Degrees	-40 to 304°F	185-220	185-220	EVAP
Startup IAT Degrees	-40 to 304°F	50-194	50-194	EVAP

1996-99 Firebird 3.8L V6 MFI VIN K (All) - HO2S Data List

Parameter Identification (PID)	PID Value Range	PID Value at Hot Idle	PID Value at 30 mph	Data List Type
Decel Fuel Mode	Active / Inactive	Inactive	Inactive	HO2S
ECT Sensor (°F)	-40 to 304°F	185-220	185-220	HO2S
Engine Speed	0-9999 rpm	Varies	---	HO2S
Engine Load	0-100%	2-5	Varies	HO2S
Hot Open Loop	Active / Inactive	Inactive	Inactive	HO2S
HO2S-11 (B1 S1)	0-1132 mv	10-1000	10-1000	HO2S
HO2S-12 (B1 S2)	0-1132 mv	10-1000	10-1000	HO2S
HO2S-13 (B1 S3)	0-1132 mv	10-1000	10-1000	HO2S
HO2S-21 (B2 S1)	0-1132 mv	10-1000	10-1000	HO2S
HO2S-11 Status	Not Ready / Ready	Ready	Ready	HO2S
HO2S-21 Status	Not Ready / Ready	Ready	Ready	HO2S
HO2S-11 Warmup	Min: Sec	00:00	00:00	HO2S
HO2S-12 Warmup	Min: Sec	00:00	00:00	HO2S
HO2S-21 Warmup	Min: Sec	00:00	00:00	HO2S
HO2S-11 XCount	0-255	Varies	Varies	HO2S
HO2S-21 XCount	0-255	Varies	Varies	HO2S
IAT Sensor (°F)	-40 to 304°F	50-194	50-194	HO2S
MAF Sensor (g/sec)	0-512 g/sec	5-9	Varies	HO2S
Power Enrichment	Active / Inactive	Active	Active	HO2S
Startup ECT Degrees	-40 to 304°F	Varies	Varies	HO2S
Startup IAT Degrees	-40 to 304°F	Varies	Varies	HO2S
TP Angle (%)	0-100%	0	Varies	HO2S
TWC Protection	Active / Inactive	Inactive	Inactive	HO2S

1996-99 Firebird 3.8L V6 MFI VIN K (All) - TWC Data List

Parameter Identification (PID)	PID Value Range	PID Value at Hot Idle	PID Value at 30 mph	Data List Type
IAC Position	0-255	10-40	---	TWC
IAT Sensor (°F)	-40 to 304°F	50-194	50-194	TWC
Loop Status	Closed Loop / Open Loop	Closed Loop	Closed Loop	TWC
MAF Sensor (g/sec)	0-512 g/sec	5-9	Varies	TWC
TP Angle (%)	0-100%	0	0-100	TWC
TWC Diagnostic	Enabled / Disabled	Enabled	Enabled	TWC
TWC Monitor test counter	0-49	0	0	TWC
TWC Protection	Active / Inactive	Inactive	Inactive	TWC

1996-99 Firebird 3.8L V6 MFI VIN K (All) - PID Data List

Parameter Identification (PID)	PID Value Range	PID Value at Hot Idle	PID Value at 30 mph	Data List Type
Actual EGR Position	0-100%	0	0-100	Misfire
Commanded TCC	Disengaged / Engaged	Disengaged	Disengaged	Misfire
Current Gear	1-2-3-4	1	3	Misfire
Decel Fuel Mode	Active / Inactive	Inactive	Inactive	Misfire
Desired EGR Position	0-100%	0	0-100	Misfire
ECT Sensor (°F)	-40 to 304°F	185-220	185-220	Misfire
Engine Load	0-100%	2-5	Varies	Misfire
Engine Speed	0-9999 rpm	Varies	---	Misfire
Fuel Trim Cell	Cell #	0	1	Misfire
Hot Open Loop	Active / Inactive	Inactive	Inactive	Misfire
IAC Position	0-255	10-40	---	Misfire
Long Term Fuel Trim Bank 1	-10% to 10%	0 (± 5%)	0 (± 5%)	Misfire
Long Term Fuel Trim Bank 2	-10% to 10%	0 (± 5%)	0 (± 5%)	Misfire
Loop Status	Closed Loop / Open Loop	Closed Loop	Closed Loop	Misfire
Misfire Current Cylinder 1-6	0-198	0-4	0-4	Misfire
Misfire History Cylinder 1-6	0-65535	0	0	Misfire
Misfiring Cylinder	CYL 1-6	0	0	Misfire
Power Enrichment	Active / Inactive	Active	Active	Misfire
Short Term Fuel Trim Bank 1	-10% to 10%	0 (± 5%)	0 (± 5%)	Misfire
Short Term Fuel Trim Bank 2	-10% to 10%	0 (± 5%)	0 (± 5%)	Misfire
Total Misfire Current Count	0-99	0-5	0-5	Misfire
Misfires Since 1st Failure	0-65535	0	0	Misfire
Passes Since 1st Failure	0-65535	0	0	Misfire
TP Angle (%)	0-100%	0	Varies	Misfire
TWC Protection	Active / Inactive	Inactive	Inactive	Misfire
Vehicle Speed	0-155 mph	0	30	Misfire

2000-02 Firebird 3.8L V6 MFI VIN K (All) - PID Data List

Parameter Identification (PID)	PID Value Range	PID Value at Hot Idle	PID Value at 30 mph	Data List Type
3X Crank Sensor	0-9,999 rpm	Varies	Varies	Engine Data 2
18X Crank Sensor	0-1,600 rpm	Varies	Varies	Engine Data 2
1-2 Shift Solenoid Circuit Status	Okay / Fault / Invalid State	Okay	Okay	Output Driver
2-3 Shift Solenoid Circuit Status	Okay / Fault / Invalid State	Okay	Okay	Output Driver
A/C HI Side Pressure	0-459 kPa	90	90	Engine Data 2
A/C HI Side Pressure	0-5.1 volts	0.98	0.98	Engine Data 2
A/C Off for WOT	Yes / No	No	No	Engine Data 2
A/C Pressure Disable	Yes / No	No	No	Engine Data 2
A/C Relay Circuit Status	Okay / Fault / Invalid State	Okay	Okay	Output Driver
A/C Relay Command	On / Off	Off	Off	Engine Data 1-2, EGR, Misfire
A/C Request Signal	Yes / No	No	No	Engine Data 2
Air Fuel Ratio	0.0-25.5:1	14.7:1	14.7:1	Engine Data 2, Fuel Trim
APP Average	0-150	5	WOT: 122	TAC
APP Indicated Angle	0-100%	0	WOT: 100%	Engine Data 1-2, EGR, EVAP, Fuel Trim, Cruise, TAC
APP Sensor 1 (%)	0-100%	0	Varies	TAC
APP Sensor 1 (V)	0-5.0 volts	0.8-1.0	Varies	TAC
APP Sensor 2 (%)	0-100%	0	Varies	TAC
APP Sensor 2 (V)	0-5.0 volts	3.9-4.8	Varies	TAC
APP Sensor 3 (%)	0-100%	0	Varies	TAC
APP Sensor 3 (V)	0-5.0 volts	3.2-4.5	Varies	TAC
BARO Sensor (kPa)	10-105 kPa	Varies w/ALT	Varies	Engine Data 1, EGR, EVAP, Fuel Trim
BARO Sensor (V)	0-5.1 volts	Varies w/ALT	Varies	Engine Data 1, EGR, EVAP, Fuel Trim
Clutch Pedal Switch	Applied / Released	Released	Released	EGR, Cruise
CMP Sensor Signal	Yes / No	Yes	Yes	Engine Data 2
Cruise Control Active	Yes / No	No	No	Engine Data 1, Cruise, TAC
Cruise Disengage History 1-6	30 possible causes	Reason	Reason	Cruise
Cruise On / Off Switch	On / Off	Off	Off	Cruise, TAC
Cruise Release Brake Pedal Switch	Applied / Released	Released	Released	Cruise, TAC
Cruise Resume/Accel Switch	On / Off	Off	Off	Cruise, TAC
Cruise Set/Coast Switch	On / Off	Off	Off	Cruise, TAC
Current Gear	0, 1, 2, 3, 4	3	4	Engine Data 1-2, EGR, Fuel Trim

2000-02 Firebird 3.8L V6 MFI VIN K (All) - PID Data List

Parameter Identification (PID)	PID Value Range	PID Value at Hot Idle	PID Value at 30 mph	Data List Type
Cycles of Misfire Data	0-99 counts	0	0	Misfire
Cylinder 1-6 Injector Circuit Status	Okay / Stuck Low/ Stuck High / Fault	Okay	Okay	Output Driver
Cylinder 1-6 Injector Circuit History	Okay / Stuck Low/ Stuck High / Fault	Okay	Okay	Output Driver
Decel Fuel Cutoff	Active / Inactive	Inactive	Inactive	EGR, Fuel Trim
Desired EGR Position	0-100%	0	Varies	Engine Data 1, EGR
Desired EGR Position	0-5.1v	0.0	Varies	EGR
Desired Idle Speed	0-3187 rpm	Commanded by the PCM	Varies	Engine Data 1-2, EVAP, TAC
Driver Module 1 Status	Enabled / Off- High Volts / Off-HI TEMP / Invalid State	Enabled	Enabled	Output Driver
Driver Module 2 Status	Enabled / Off- HI Volts/ Off-HI TEMP/ Invalid State	Enabled	Enabled	Output Driver
Driver Module 3 Status	Enabled / Off- HI Volts/ Off-HI TEMP/ Invalid State	Enabled	Enabled	Output Driver
Driver Module 4 Status	Enabled / Off- HI Volts/ Off-HI TEMP/ Invalid State	Enabled	Enabled	Output Driver
ECT Sensor (°F)	-40 to 304°F	190-221°F	190-221°F	Engine Data 1-2, EGR, Fuel Trim, EVAP, Misfire
EGR Flow Test Count	0-255 counts	0-12	Varies	EGR
EGR Learned Minimum Position	0-5.1 volts	0.16-1.0	Varies	EGR
EGR Position Sensor	0-100%	0	Varies	Engine Data 1, Misfire
EGR Position Sensor	0-5.1 volts	0.58-0.85	Varies	EGR
EGR Position Variance	0-100%	0	0	EGR
EGR Solenoid Circuit Status	Okay / Fault / Invalid State	Okay	Okay	Output Driver
EGR Solenoid Command	0-100%	0	Varies	EGR
Engine Load	0-100%	1-4	Varies	Engine Data 1-2, EGR, Fuel Trim, TAC, Misfire, Cruise, EVAP
Engine Oil Level Switch	Okay / Low	Okay	Okay	Engine Data 2

2000-02 Firebird 3.8L V6 MFI VIN K (All) - PID Data List

Parameter Identification (PID)	PID Value Range	PID Value at Hot Idle	PID Value at 30 mph	Data List Type
Engine Run Time	Hr: Min: Sec 00:00:00 to 99:99:99	00:00:00	00:00:00	Engine Data 1-2, EGR, Fuel Trim, TAC, Misfire, Cruise, EVAP
Engine Speed	0-9999 rpm	± 50 rpm of actual speed	Varies	Engine Data 1-2, EGR, Fuel Trim, TAC, Misfire, Cruise, EVAP
EVAP Fault History	No Fault / Excess VAC / Purge Valve Leak / Small Leak / Weak Vacuum	No Fault	No Fault	EVAP
EVAP Purge Solenoid Circuit Status	Okay / Fault / Invalid State	Okay	Okay	Output Driver
EVAP Purge Solenoid Command	0-100%	15	Varies	Engine Data 1, EVAP, Fuel Trim
EVAP Test Abort Reason	Not Aborted / Lost Enable / Small Leak / Not at Rest	Not Aborted	Not Aborted	EVAP
EVAP Test Result	No Result / Passed / Fail / P0440 / P0442 / P0446 / P1441	Okay	Okay	EVAP
EVAP Test State	Waiting for Purge/ Test Running/Test Completed	Test Completed	Test Completed	EVAP
EVAP Vent Solenoid Circuit Status	Okay / Fault / Invalid State	Okay	Okay	Output Driver
EVAP Vent Solenoid Command	Venting / Not Venting	Not Venting	Not Venting	Engine Data 1, EVAP, Fuel Trim
Fan Control Relay 1 Circuit Status	Okay / Fault / Invalid State	Okay	Okay	Output Driver
Fan Control Relay 1 Command	On / Off	Off	Off	Engine Data 2
Fan Control Relay 2-3 Circuit Status	Okay / Fault	Okay	Okay	Output Driver
Fan Control Relay 2-3 Command	On / Off	Off	Off	Engine Data 2
Fuel Pump Relay Circuit Status	Okay / Fault / Invalid State	Okay	Okay	Output Driver
Fuel Pump Relay Circuit History Status	Okay / Fault	Okay	Okay	Output Driver
Fuel Pump Relay Command	On / Off	On	On	Engine Data 2
Fuel Tank Level Remaining	0-100%	Varies	Varies	EVAP

2000-02 Firebird 3.8L V6 MFI VIN K (All) - PID Data List

Parameter Identification (PID)	PID Value Range	PID Value at Hot Idle	PID Value at 30 mph	Data List Type
Fuel Tank Pressure Sensor	Inches H2O, mm Hg	-17 to +7.5 Inches H2O	-17 to +7.5 Inches H2O	Engine Data 1
Fuel Tank Pressure Sensor	0-5.0 volts	Varies	Varies	EVAP
Fuel Trim Cell	Cell 0-10	Varies	Varies	Engine Data 1, EVAP, Fuel Trim
Fuel Trim Learn	Enabled or Disabled	Enabled	May Toggle	Engine Data 1, EVAP, Fuel Trim
Generator L-Terminal	On / Off	On	On	Engine Data 2
HO2S (B1 S1) Signal	0-1132 mv	10-1000	10-1000	Engine Data 1, EVAP, Fuel Trim
HO2S (B1 S2) Signal	0-1132 mv	10-1000	10-1000	Engine Data 1, Fuel Trim
HO2S (B2 S1) Signal	0-1132 mv	10-1000	10-1000	Engine Data 1, EVAP, Fuel Trim
IAT Sensor (ºF)	-40 to 304ºF	91	92	Engine Data 1-2, EGR, EVAP, Fuel Trim
Ignition 1	0.0-25.5v	13.5	13.6	Engine Data 1-2, EGR, EVAP, Fuel Trim, Cruise, TAC
Ignition Mode	Bypass/IC	IC	IC	Engine Data 2
Injector Pulsewidth	0-1000 ms	1.5-3.5	Varies	Engine Data 2, Fuel Trim, M/fire
Knock Retard (º)	0-25.5º	0º	0º	Engine Data 1, EGR
Long Term Fuel Trim Bank 1	-23 to +16	0 (± 5%)	0 (± 5%)	Engine Data 1-2, EVAP, Fuel Trim
Long Term Fuel Trim Bank 2	-23 to +16	0 (± 5%)	0 (± 5%)	Engine Data 1-2, EVAP, Fuel Trim
Loop Status	Open / Closed	Closed	Closed	Engine Data 1-2, EGR, EVAP, Fuel Trim
MAF Sensor (g/sec)	0-512 g/sec	3-6	Varies	Engine Data 1-2, EGR, EVAP, Fuel Trim, Misfire, TAC
MAF Sensor (Hz)	0-32,000 Hz	2000-3000	Varies	Engine Data 2
MAP Sensor (kPa)	10-105 kPa	20-48	Varies	Engine Data 1-2, EGR, EVAP, Fuel Trim, Misfire, TAC
MAP Sensor (V)	0-5.1 volts	0.75-2.0	Varies	Engine Data 1-2
MIL Circuit Status	Okay / Fault / Invalid State	Okay	Okay	Output Driver
MIL Command	On / Off	Off	Off	Engine Data 2
Misfire Current Cylinder 1-6	0-198 counts	0-4	0	Misfire
Misfire History Cylinder 1-6	0-65,535	0	0	Misfire

2000-02 Firebird 3.8L V6 MFI VIN K (All) - PID Data List

Parameter Identification (PID)	PID Value Range	PID Value at Hot Idle	PID Value at 30 mph	Data List Type
Number of DTC(s)	Number	0	No	Engine Data 1, EVAP, Fuel Trim, TAC
Power Enrichment	Active / Inactive	Inactive	Inactive	Engine Data 2, Misfire
Reduced Engine Power	Active / Inactive	Inactive	Inactive	Engine Data 1, EGR, Cruise, TAC
Short Term Fuel Trim Bank 1 and Bank 2	-10 to +10%	0 (± 5%)	0 (± 5%)	Engine Data 1-2, EVAP, Fuel Trim
Spark Advance	-64° to +64°	-20	Varies	Engine Data 1-2, Fuel Trim, M/fire
Starter Enable Relay Circuit Status	Okay / Fault / Invalid State	Okay	Okay	Output Driver
Starter Relay Control	On / Off	Off	Off	Engine Data 2
Startup ECT	-40 to 419°F	Varies	Varies	EVAP, Fuel Trim
Startup IAT	-40 to 419°F	Varies	Varies	EVAP, Fuel Trim
Stop Lamp Pedal Switch	Applied or Released	Released	Released	Engine Data 2, Cruise, TAC
TAC/PCM Communication signal	Okay / Fault	Okay	Okay	Engine Data 1, Cruise, TAC
TCC/Cruise Brake Pedal Switch	Applied or Released	Released	Released	Engine Data 1-2, Cruise, TAC
TCC Enable Solenoid Circuit Status	Okay / Fault / Invalid State	Okay	Okay	Output Driver
TCC Enable Solenoid Command	On / Off	Off	Off	Engine Data 2, Cruise, Misfire
TCS Circuit History	Okay / Fault	Okay	Okay	Output Driver
TCS Circuit Status	Okay / Fault	Okay	Okay	Output Driver
Torque Delivered	0-100%	70	70	ENG 2, TAC
Torque Requested	0-100%	100	100	ENG 2, TAC
TP Desired Angle	0-100%	2-6	Varies	Engine Data 1-2, EGR, EVAP, Cruise, TAC
TP Indicated Angle	0-100%	2-6	Varies	Engine Data 1-2, EGR, EVAP, Fuel Trim, Misfire Cruise, TAC
TP Sensor 1	0-100%	0-6	Varies	TAC
TP Sensor 2	0-100%	0-6	Varies	TAC
TP Sensor 1-2	Agree/DIS	Agree	Agree	TAC
Traction Control Status	Active / Inactive	Inactive	Inactive	Engine Data 2, Cruise, TAC
Transaxle Range Switch	Park/Neutral/ Reverse	Park	---	ENG 2, EGR, Cruise, Fuel Trim
Vehicle Speed	Km/h, MPH	0	30	Engine Data 1-2, EGR EVAP, Fuel Trim Cruise, M/fire
VTD Fuel Enable	Active / Inactive	Inactive	Inactive	Engine Data 1

1996-97 Firebird 5.7L V8 MFI VIN P (All) - PID Data List

Parameter Identification (PID)	PID Value Range	PID Value at Hot Idle	PID Value at 30 mph	Data List Type
A/C Clutch	On / Off	On: A/C on	Off	Engine Data 1
A/C Evaporative Temperature	19°F to 90°F	Varies	Varies	Engine Data 1
A/C High Side (psi)	-15-452 psi	139-399	139-399	Engine Data 1
A/C Request	Yes / No	No	No	Engine Data 1
A/C Status	On / Off	Off	Off	Engine Data 1
Actual EGR Position	0-100%	0	0-100	Engine Data 1
AIR Pump	Enabled / Disabled	Disabled	Disabled	Engine Data 1
ASR/TSR spark retard	Yes / No	No	No	Engine Data 1
BARO Sensor (kPa)	10-110 kPa	65-104	65-104	Engine Data 1
BARO Sensor (V)	0.0-5.0v	3-5	3-5	Engine Data 1
CKP Engine Speed	0-9999 rpm	Varies	---	Engine Data 1
CKP LO Resolution Angle	-30 to +60°	0 Deviation	Varies	Engine Data 1
Cold Startup	Yes / No	Yes: Cold	No	Engine Data 1
Current Gear	1-2-3-4	1	3	Engine Data 1
Desired Idle Speed	0-3187	PCM control	---	Engine Data 1
DTC Number	0000	Number	Number	Engine Data 1
DTC Set This Ignition	Yes / No	No	No	Engine Data 1
EGR Duty Cycle (%)	0-100%	0	0-100	Engine Data 1
ECT Sensor (°F)	-40 to 304°F	185-221	185-221	Engine Data 1
Engine Run Time	Hr: Min: Sec	00:00:00	00:00:00	Engine Data 1
Engine Speed	0-9999 rpm	Varies	---	Engine Data 1
EVAP Purge (%)	0-100%	0-50	0-100	Engine Data 1
EVAP Vacuum Switch	No/Purge	No Purge	No Purge	Engine Data 1
FC Relay 1	On / Off	Off	Off	Engine Data 1
FC Relay 2, 3	On / Off	Off	Off	Engine Data 1
Fuel Trim Cell	Cell #1-16	16	Varies	Engine Data 1
Fuel Trim Learn	Enabled / Disabled	Enabled	Enabled	Engine Data 1
HI Resolution Signal	Active / Inactive	Active	Active	Engine Data 1
HO2S-11 (B1 S1)	0-1132 mv	10-1000	10-1000	Engine Data 1
HO2S-12 (B1 S2)	0-1132 mv	10-1000	10-1000	Engine Data 1
HO2S-21 (B2 S1)	0-1132 mv	10-1000	10-1000	Engine Data 1
HO2S-22 (B2 S2)	0-1132 mv	10-1000	10-1000	Engine Data 1
IAC Learned	0-255	10-50	---	Engine Data 1
IAC Position	0-255	10-50	---	Engine Data 1
IAT Sensor (°F)	-40 to 304°F	50-194	50-194	Engine Data 1
Ignition System (V)	0.0-25.5v	14.2	14.2	Engine Data 1
Injector Pulsewidth	0-985 ms	1-4	Varies	Engine Data 1
Knock Retard	0-25.5°	0°	0°	Engine Data 1
Long Term Fuel Trim	-13 to 22%	0 (± 5%)	0 (± 5%)	Engine Data 1
Long Term F/T Avg.	-13 to 22%	0 (± 5%)	0 (± 5%)	Engine Data 1
Loop Status	Closed Loop / Open Loop	Closed Loop	Closed Loop	Engine Data 1
LO Resolution Signal	0-332.7ms	Varies	Varies	Engine Data 1
MAF Sensor (g/sec)	0-512 g/sec	5-9	Varies	Engine Data 1
MAP Sensor (V)	0.0-5.0v	1-2	Varies	Engine Data 1
MAP Sensor (kPa)	10-110 kPa	20-48	Varies	Engine Data 1
MIL Status	On / Off	Off	Off	Engine Data 1

1996-97 Firebird 5.7L V8 MFI VIN P (All) - PID Data List

Parameter Identification (PID)	PID Value Range	PID Value at Hot Idle	PID Value at 30 mph	Data List Type
PASS Key® Fuel	Disabled/Okay	Okay	Okay	Engine Data 1
PCM Reset	Yes / No	No	No	Engine Data 1
P/NP Switch	P-R-0-3-2-1	Park	2	Engine Data 1
PSP Switch	HI/Normal	HI: Turning	Normal	Engine Data 1
Performance Mode	Normal / Performance	Normal	Normal	Engine Data 1
Performance Switch	Active / Inactive	Inactive	Inactive	Engine Data 1
Reverse Inhibit Switch	Yes / No	Yes: in R	No	Engine Data 1
Short Term Fuel Trim	-10% to 10%	0 (± 5%)	0 (± 5%)	Engine Data 1
Short Term F/T Avg.	-10% to 10%	0 (± 5%)	0 (± 5%)	Engine Data 1
Skip Shift Solenoid	Enabled / Disabled	Disabled	Disabled	Engine Data 1
Spark Advance (°)	-64° to 64°	Varies	Varies	Engine Data 1
Startup ECT Sensor	-40 to 304°F	Varies	Varies	Engine Data 1
TCC Brake Switch	Applied / Released	Released	Released	Engine Data 1
TCC Duty Cycle	Enable/DIS	Disabled	Enabled	Engine Data 1
TCC Enable	Enable/DIS	Disabled	Enabled	Engine Data 1
TP Angle (%)	0-100%	0	Varies	Engine Data 1
TP Sensor (V)	0.0-5.0v	0.55-0.90	Varies	Engine Data 1
Trans. Range Switch	P/N-D-L-R	P/N	DRIVE	Engine Data 1
Traction Control	Active / Inactive	Inactive	Inactive	Engine Data 1
Vehicle Speed	0-155 mph	0	30	Engine Data 1

1996-97 Firebird 5.7L V8 MFI VIN P (All) - PID Data List

Parameter Identification (PID)	PID Value Range	PID Value at Hot Idle	PID Value at 30 mph	Data List Type
A/C Clutch	On / Off	Off	Off	Engine Data 2
AIR Pump Control	Enabled / Disabled	Disabled	Disabled	Engine Data 2
ASR/TSR spark retard	Yes / No	No	No	Engine Data 2
BARO Sensor (kPa)	10-110 kPa	65-104	65-104	Engine Data 2
CKP Engine Speed	0-9999 rpm	Varies	---	Engine Data 2
CKP LO RES Angle	-30° to +60°	0 Deviation	0 Deviation	Engine Data 2
Cold Startup	Yes / No	Yes: Cold	No	Engine Data 2
Desired Idle Speed	0-3187	PCM control	---	Engine Data 2
DTC Set This Ignition	Yes / No	No	No	Engine Data 2
ECT Sensor (°F)	-40 to 304°F	185-221	185-221	Engine Data 2
EGR Duty Cycle (%)	0-100%	0	0-100	Engine Data 2
Engine Run Time	Hr: Min: Sec	00:00:00	00:00:00	Engine Data 2
Engine Speed	0-9999 rpm	Varies	Varies	Engine Data 2
EVAP Purge	0-100%	0	0-100	Engine Data 2
EVAP Vacuum Switch	No/Purge	No Purge	Purge	Engine Data 2
FC Relay 1	On / Off	On: If hot	Off	Engine Data 2
FC Relay 2, 3	On / Off	On: A/C on	Off	Engine Data 2
Fuel Trim Cell	Cell #1-18	16-18	16-18	Engine Data 2
Fuel Trim Learn	Enabled / Disabled	Enabled	Enabled	Engine Data 2
HI Resolution Signal	Active / Inactive	Active	Active	Engine Data 2
HO2S-11 (B1 S1)	0-1132 mv	10-1000	10-1000	Engine Data 2
HO2S-12 (B1 S2)	0-1132 mv	10-1000	10-1000	Engine Data 2
HO2S-21 (B2 S1)	0-1132 mv	10-1000	10-1000	Engine Data 2
HO2S-22 (B2 S2)	0-1132 mv	10-1000	10-1000	Engine Data 2

1996-97 Firebird 5.7L V8 MFI VIN P (All) - PID Data List

Parameter Identification (PID)	PID Value Range	PID Value at Hot Idle	PID Value at 30 mph	Data List Type
IAC Learned	0-255	10-50	---	Engine Data 2
IAC Position	0-255	10-50	---	Engine Data 2
IAT Sensor (°F)	-40 to 304°F	50-194	50-194	Engine Data 2
Knock Activity	Yes / No	Yes: Knock	No	Engine Data 2
Knock Retard (°)	0-128°	0-16	Varies	Engine Data 2
LO Resolution Signal	0-332.7 ms	Varies	Varies	Engine Data 2
Long Term Fuel Trim B1	-10% to 10%	0 (± 5%)	0 (± 5%)	Engine Data 2
Long Term Fuel Trim B2	-10% to 10%	0 (± 5%)	0 (± 5%)	Engine Data 2
Long Term F/T Avg. B1	-13 to 22%	0 (± 5%)	0 (± 5%)	Engine Data 2
Long Term F/T Avg. B2	-13 to 22%	0 (± 5%)	0 (± 5%)	Engine Data 2
Loop Status	Closed Loop / Open Loop	Closed Loop	Closed Loop	Engine Data 2
MAF Sensor (g/sec)	0-512 g/sec	5-9	Varies	Engine Data 2
MAP Sensor (kPa)	10-110 kPa	20-48	Varies	Engine Data 2
MAP Sensor (V)	0.0-5.0v	1-2	Varies	Engine Data 2
Miles Since DTC Cleared	MPH / KPH	Varies	Varies	Engine Data 2
MPH / KPH	0-155	0	30	Engine Data 2
PASS Key® Fuel	Disabled/Okay	Okay	Okay	Engine Data 2
PCM Reset	Yes / No	No	No	Engine Data 2
Short Term Fuel Trim B1	-10% to 10%	0 (± 5%)	0 (± 5%)	Engine Data 2
Short Term Fuel Trim B2	-10% to 10%	0 (± 5%)	0 (± 5%)	Engine Data 2
Spark Advance (°)	-64° to 64°	Varies	Varies	Engine Data 2
Startup ECT Sensor	-40 to 304°F	Varies	Varies	Engine Data 2
TP Angle (%)	0-100%	0	Varies	Engine Data 2
TP Sensor (V)	0.0-5.0v	0.55-0.90	Varies	Engine Data 2
Traction Control	Active / Inactive	Inactive	Inactive	Engine Data 2
Warmup Cycles without an Emission Fault	0-40	Varies	Varies	Engine Data 2
Warmup Cycles without a Non-Emission Fault	0-40	Varies	Varies	Engine Data 2

1996-97 Firebird 5.7L V8 MFI VIN P (All) - PID Data List

Parameter Identification (PID)	PID Value Range	PID Value at Hot Idle	PID Value at 30 mph	Data List Type
Air Pump Control	On / Off	On: Cold	Off	Fuel Trim
BARO Sensor (V)	0.0-5.0v	3.5-4.5	3.5-4.5	Fuel Trim
BARO Sensor (kPa)	10-110 kPa	65-104	65-104	Fuel Trim
ECT Sensor (°F)	-40 to 304°F	185-221	185-221	Fuel Trim
Engine Speed	0-9999 rpm	Varies	Varies	Fuel Trim
EGR Duty Cycle	0-100%	0-100%	0-100%	Fuel Trim
Engine Run Time	Hr: Min: Sec	00:00:00	00:00:00	Fuel Trim
EVAP Purge	0-100%	0%	0-100%	Fuel Trim
EVAP Vacuum Switch	No/Purge	No Purge	Purge	Fuel Trim
Fuel Trim Cell	Cell #1-18	16-18	16-18	Fuel Trim
Fuel Trim Learn	ENAB/DIS	Enabled	Enabled	Fuel Trim
HO2S-11 (B1 S1)	0-1132 mv	10-1000	10-1000	Fuel Trim
HO2S-12 (B1 S2)	0-1132 mv	10-1000	10-1000	Fuel Trim
HO2S-21 (B2 S1)	0-1132 mv	10-1000	10-1000	Fuel Trim
HO2S-22 (B2 S2)	0-1132 mv	10-1000	10-1000	Fuel Trim

1996-97 Firebird 5.7L V8 MFI VIN P (All) - PID Data List

Parameter Identification (PID)	PID Value Range	PID Value at Hot Idle	PID Value at 30 mph	Data List Type
IAC Learned	0-255	10-50	---	Fuel Trim
IAC Position	0-255	10-50	---	Fuel Trim
IAT Sensor (°F)	-40 to 304°F	50-194	50-194	Fuel Trim
Ignition System (V)	0.0-25.5v	14.2	14.2	Fuel Trim
INJ Pulsewidth B1, B2	0-985 ms	1-4	Varies	Fuel Trim
Long Term Fuel Trim	-10% to 10%	0 (± 5%)	0 (± 5%)	Fuel Trim
Long Term F/T Avg.	-13 to 22%	0 (± 5%)	0 (± 5%)	Fuel Trim
Loop Status	Closed Loop / Open Loop	Closed Loop	Closed Loop	Fuel Trim
MAF Sensor (g/sec)	0-512 g/sec	5-9	Varies	Fuel Trim
MAP Sensor (kPa)	10-110 kPa	20-48	Varies	Fuel Trim
MAP Sensor (V)	0.0-5.0v	1-2v	Varies	Fuel Trim
Miles Per Hour	0-155 mph	0	30	Fuel Trim
Short Term Fuel Trim	-10% to 10%	0 (± 5%)	0 (± 5%)	Fuel Trim
Short Term F/T Avg.	-13 to 22%	0 (± 5%)	0 (± 5%)	Fuel Trim
Spark (°)	-64° to 64°	Varies	Varies	Fuel Trim
TP Angle (%)	0-100%	0	Varies	Fuel Trim

1996-97 Firebird 5.7L V8 VIN P (All) - PID Data List

Parameter Identification (PID)	PID Value Range	PID Value at Hot Idle	PID Value at 30 mph	Data List Type
Desired Idle Speed	0-3187	PCM control	---	F/F, F/R
DTC Number	0000	Number	Number	F/F, F/R
ECT Sensor (°F)	-40 to 304°F	185-221	185-221	F/F, F/R
Emission Failure Counter	0-255	0	0	F/F, F/R
Engine Load	0-100%	2-5	Varies	F/F, F/R
Engine Speed	0-9999 rpm	Varies	Varies	F/F, F/R
Failure Counter	0-255	0	0	F/F, F/R
FC Relay 1	On / Off	Off	Off	F/F, F/R
FC Relay 2, 3	On / Off	Off	Off	F/F, F/R
Fuel Trim Cell	Cell #1-16	16	Varies	F/F, F/R
Fuel Trim Learn	Enabled / Disabled	Enabled	Enabled	F/F, F/R
HO2S-11, HO2S-12	0-1132 mv	10-1000	10-1000	F/F, F/R
HO2S-21, HO2S-22	0-1132 mv	10-1000	10-1000	F/F, F/R
IAT Sensor (°F)	-40 to 304°F	50-194	50-194	F/F, F/R
Long Term F/T B1, B2	-10% to 10%	0 (± 5%)	0 (± 5%)	F/F, F/R
MAF Sensor (g/sec)	0-512 g/sec	5-9	Varies	F/F, F/R
Miles Since DTC Cleared	0-10,000	Varies	Varies	F/F, F/R
Miles Since First Failure	0-10,000	Varies	Varies	F/F, F/R
Miles Since MIL Request	0-10,000	Varies	Varies	F/F, F/R
MPH / KPH	0-155 mph	0	30	F/F, F/R
Not Run Counter	0-255	0	0	F/F, F/R
Pass Counter	0-255	0	0	F/F, F/R
Short Term Fuel Trim	-10% to 10%	0 (± 5%)	0 (± 5%)	F/F, F/R
TP Angle (%)	0-100%	0	Varies	F/F, F/R

1996-97 Firebird 5.7L V8 MFI VIN P (All) - Misfire Data List

Parameter Identification (PID)	PID Value Range	PID Value at Hot Idle	PID Value at 30 mph	Data List Type
A/C Request	Yes / No	No	No	Misfire
A/C Status	On / Off	On: A/C on	Off	Misfire
CKP Engine Speed	0-9999 rpm	Varies	---	Misfire
CKP LO Resolution Angle	-30° to +60°	0 Deviation	0 Deviation	Misfire
Cycles of Misfire Data	0-100	Varies	Varies	Misfire
CYL Mode Misfire Index	0-65535	Varies	Varies	Misfire
Desired Idle Speed	0-3187	PCM control	---	Misfire
ECT Sensor (°F)	-40 to 304°F	185-221	185-221	Misfire
EGR Duty Cycle	0-100%	0%	0-100%	Misfire
Engine Load	0-100%	2 - 5	Varies	Misfire
Engine Speed	0-9999 rpm	Varies	---	Misfire
IAT Sensor (°F)	-40 to 304°F	50-194	50-194	Misfire
MAF Sensor (g/sec)	0-512 g/sec	5-9	Varies	Misfire
Misfire Current Cylinder 1-8	0-65535	Varies	Varies	Misfire
Misfire Failures	0-65535	0	0	Misfire
Misfiring History Cylinder 1-8	0-65535	0	0	Misfire
Misfire Passes	0-65535	0	0	Misfire
Misfire Per Cycle Status	0-255	Varies	Varies	Misfire
Misfire Revolution Status	Accept/REJ	ACCEPT	ACCEPT	Misfire
Revolutions Within Misfire	0-65535	Varies	Varies	Misfire
Total Misfire	0-255	0	0	Misfire
TP Angle	0-100%	0%	0-100%	Misfire
Vehicle Speed	0-155 mph	0	30	Misfire

1998-2002 Firebird 5.7L V8 MFI VIN G (All) - PID Data List

Parameter Identification (PID)	PID Value Range	PID Value at Hot Idle	PID Value at 30 mph	Data List Type
A/C Relay Command	On / Off	Off	Off	Engine Data 1
A/C Request	Yes / No	No	No	Engine Data 1
A/C Status	On / Off	No	Off	Engine Data 1
Actual EGR Position	0-100%	0	Varies	Engine Data 1
AIR Pump Relay	On / Off	Off	Off	Engine Data 1
AIR Solenoid	On / Off	Off	Off	Engine Data 1
BARO Sensor (kPa)	10-110 kPa	65-104	65-104	Engine Data 1
Cam Signal HI to LO	0-65535	Varies	Varies	Engine Data 1
Cam Signal LO to HI	0-65535	Varies	Varies	Engine Data 1
Clutch Pedal Switch	DEP/REL	Released	Released	Engine Data 1
Cold Startup	Yes / No	No	No	Engine Data 1
Commanded EGR	0.0-5.0v	< 0.13	Varies	Engine Data 1
Cruise Control Active	Yes / No	No	No	Engine Data 1
Cruise Control Inhibit	Yes / No	Yes	Yes	Engine Data 1
Cruise Requested	Yes / No	No	No	Engine Data 1
Desired IAC Airflow	0-64 g/sec	Varies	Varies	Engine Data 1
Desired EGR Position	0-100%	0	0-100	Engine Data 1
Desired Idle Speed	0-3187 rpm	PCM control	---	Engine Data 1
DTC Set This Ignition	Yes / No	No	No	Engine Data 1
ECT Sensor (°F)	-38 to 284°F	185-221	185-221	Engine Data 1
EGR Closed Pintle	0.0-5.0v	Varies	Varies	Engine Data 1
EGR Pintle Position	0.0-5.0v	< 0.13	Varies	Engine Data 1
Engine Run Time	Hr: Min: Sec	00:00:00	00:00:00	Engine Data 1
Engine Speed	0-9999 rpm	Varies	Varies	Engine Data 1
EVAP Canister Purge	0-100%	0-25	0-50	Engine Data 1
EVAP Canister Vent	Venting / Not Venting	Venting	Venting	Engine Data 1
FC Relay 1, 2-3	On / Off	Off	Off	Engine Data 1
FTP Sensor (In. H2O)	-17.5 to 7.5"	Varies	Varies	Engine Data 1
Fuel Trim Cell	Cell #1-23	16, 17 & 20	Varies	Engine Data 1
Fuel Trim Learn	Enabled / Disabled	Enabled	Enabled	Engine Data 1
HO2S-11 (B1 S1)	0-1132 mv	10-1000	10-1000	Engine Data 1
HO2S-21 (B1 S2)	0-1132 mv	10-1000	10-1000	Engine Data 1
HO2S-21 (B2 S1)	0-1132 mv	10-1000	10-1000	Engine Data 1
HO2S-22 (B2 S2)	0-1132 mv	10-1000	10-1000	Engine Data 1
IAC Position	0-255 counts	Varies	Varies	Engine Data 1
IAT Sensor (°F)	-38 to 284°F	50-194	50-194	Engine Data 1
Knock Retard (°)	0° to 16°	0	Varies	Engine Data 1
Long Term Fuel Trim	-10% to 10%	0 (± 5%)	0 (± 5%)	Engine Data 1
Loop Status	Closed Loop / Open Loop	Closed Loop	Closed Loop	Engine Data 1
MAF Sensor (g/sec)	0-512 g/sec	5-9	Varies	Engine Data 1
MAP Sensor (kPa)	10-110 kPa	20-48	Varies	Engine Data 1
MIL Status	On / Off	Off	Off	Engine Data 1
PCM Reset	Yes / No	No	No	Engine Data 1
Reverse Inhibit (M/T)	Yes / No	Yes: In Rev	No	Engine Data 1
Short Term Fuel Trim	-10% to 10%	0 (± 5%)	0 (± 5%)	Engine Data 1
Skip Shift Lamp (M/T)	On / Off	Off	Off	Engine Data 1
Skip Shift Solenoid	Enabled / Disabled	Disabled	Disabled	Engine Data 1
Spark Advance (°)	-64° to 64°	Varies	Varies	Engine Data 1

1998-2002 Firebird 5.7L V8 MFI VIN G (All) - PID Data List

Parameter Identification (PID)	PID Value Range	PID Value at Hot Idle	PID Value at 30 mph	Data List Type
Startup ECT Sensor	-40 to 304°F	Varies	Varies	Engine Data 1
TCC Brake Switch	Applied / Released	Released	Released	Engine Data 1
TCC PWM Solenoid	Enabled / Disabled	Disabled	Enabled	Engine Data 1
TCC Enable Solenoid	Enabled / Disabled	Disabled	Enabled	Engine Data 1
TP Angle (%)	0-100%	0	Varies	Engine Data 1
Traction Control	Active / Inactive	Inactive	Inactive	Engine Data 1
Transmission Range	0-4	0	4	Engine Data 1
VTD Fuel Disable	Active / Inactive	Inactive	Inactive	Engine Data 1

1998-2002 Firebird 5.7L V8 MFI VIN G (All) - Engine Data 2 List

Parameter Identification (PID)	PID Value Range	PID Value at Hot Idle	PID Value at 30 mph	Data List Type
A/C Clutch Feedback	On / Off	Off	Off	Engine Data 2
A/C HI Side Pressure	0-450 psi	139-399	139-399	Engine Data 2
A/C High Side (V)	0.0-5.0v	1-4	1-4	Engine Data 2
A/C Request Signal	Yes / No	No	No	Engine Data 2
A/C Status	On / Off	Off	Off	Engine Data 2
AIR Relay Control	On / Off	Off	Off	Engine Data 2
AIR Solenoid Control	On / Off	Off	Off	Engine Data 2
Clutch Pedal Switch	DEP/REL	Released	Released	Engine Data 2
Cold Startup	Yes / No	Yes: If cold	No	Engine Data 2
Cruise Control Inhibit	Yes / No	Yes	Yes	Engine Data 2
Cruise Requested	Yes / No	No	No	Engine Data 2
Current Gear	1-2-3-4	1	3	Engine Data 2
Desired Idle Speed	0-3187 rpm	PCM control	---	Engine Data 2
DTC Set this Ignition	Yes / No	Yes: If set	No	Engine Data 2
ECT Sensor (°F)	-38 to 284°F	185-221	185-221	Engine Data 2
Engine Run Time	Hr: Min: Sec	00:00:00	00:00:00	Engine Data 2
Engine Speed	0-9999 rpm	Varies	Varies	Engine Data 2
EVAP Canister Purge	0-100%	0	0-50	Engine Data 2
EVAP Canister Vent	No/VENT	VENT	VENT	Engine Data 2
FC Relay 1, 2-3	On / Off	Off	Off	Engine Data 2
Fuel Gauge Control	0-100%	Varies	Varies	Engine Data 2
Fuel Level Sensor (V)	0.0-5.0v	Varies	Varies	Engine Data 2
Fuel Tank Level	0-16.6 gal.	Varies	Varies	Engine Data 2
Fuel Tank Level	0-100%	Varies	Varies	Engine Data 2
Fuel Tank Capacity	0-16.6 gal.	Varies	Varies	Engine Data 2
FTP Sensor (In. H2O)	-17.5 to 7.5"	Varies	Varies	Engine Data 2
FTP Sensor (V)	0.0-5.0v	Varies	Varies	Engine Data 2
Fuel Trim Learn	Enabled / Disabled	Enabled	Enabled	Engine Data 2
Generator L-Terminal	Active / Inactive	Active	Active	Engine Data 2
IAT Sensor (°F)	-38 to 284°F	50-194	50-194	Engine Data 2
Injector Pulsewidth	0-985 ms	1-4	Varies	Engine Data 2
Long Term F/T B1, B2	-10% to 10%	0 (± 5%)	0 (± 5%)	Engine Data 2
Loop Status	Closed Loop / Open Loop	Closed Loop	Closed Loop	Engine Data 2
Low Oil Lamp	On / Off	On: If low	Off	Engine Data 2
Low Oil Level	Yes / No	Yes: If low	No	Engine Data 2
MAF Sensor	0-512 g/sec	5-9	Varies	Engine Data 2

1998-2002 Firebird 5.7L V8 MFI VIN G (All) - PID Data List

Parameter Identification (PID)	PID Value Range	PID Value at Hot Idle	PID Value at 30 mph	Data List Type
MAF Sensor (Hz)	0-31,999 Hz	2100-3000	Varies	Engine Data 2
MAP Sensor (kPa)	10-110 kPa	20-48	Varies	Engine Data 2
MAP Sensor (V)	0.0-5.0v	1-2	Varies	Engine Data 2
MIL Status	On / Off	Off	Off	Engine Data 2
Mile Since DTC Clear	Km/Miles	Varies	Varies	Engine Data 2
PCM Reset	Yes / No	No	No	Engine Data 2
Park Neutral Position	P/N-R-D321	P/N	3	Engine Data 2
PCM/VCM in VTD Fail	Enabled / Disabled	Disabled	Disabled	Engine Data 2
Reverse Inhibit (M/T)	Yes / No	Yes: in Rev	No	Engine Data 2
Short Term Fuel Trim	-10% to 10%	0 (± 5%)	0 (± 5%)	Engine Data 2
Skip Shift Lamp (M/T)	On / Off	Off	Off	Engine Data 2
Skip Shift Solenoid	Enabled / Disabled	Disabled	Disabled	Engine Data 2
Spark Advance (°)	-64° to 64°	Varies	Varies	Engine Data 2
TCC Brake Switch	Applied / Released	Released	Released	Engine Data 2
TCC PWM Solenoid	Enabled / Disabled	Disabled	Enabled	Engine Data 2
TCC Enable Solenoid	Enabled / Disabled	Disabled	Enabled	Engine Data 2
TP Angle (%)	0-100%	0	Varies	Engine Data 2
TP Sensor (V)	0.0-5.0v	0.4-0.9	Varies	Engine Data 2
Traction Control	Active / Inactive	Inactive	Inactive	Engine Data 2
Transmission Range	0-4	1	3	Engine Data 2
VTD Auto Learn	Min: Sec	00:00	00:00	Engine Data 2
Warmups w/o E/Fault	0-255	0	0	Engine Data 2
Warmup w/o N/E fault	0-255	0	0	Engine Data 2

1998-2002 Firebird 5.7L V8 MFI VIN G (All) - Enhanced EVAP Data

Parameter Identification (PID)	PID Value Range	PID Value at Hot Idle	PID Value at 30 mph	Data List Type
A/C Relay Command	On / Off	Off	Off	EVAP
A/C Request Signal	Yes / No	No	No	EVAP
A/C Status	On / Off	Off	Off	EVAP
BARO Sensor	10-110 kPa	65-104	65-104	EVAP
Cold Startup	Yes / No	No	No	EVAP
DTC Set This Ignition	Yes / No	No	No	EVAP
ECT Sensor (°F)	-38 to 284°F	185-221	185-221	EVAP
Engine Run Time	Hr: Min: Sec	00:00:00	00:00:00	EVAP
Engine Speed	0-9999 rpm	Varies	Varies	EVAP
EVAP Canister Purge	0-100%	0	0-50	EVAP
EVAP Canister Vent	No/VENT	Venting	Venting	EVAP
FC Relay 1, 2-3	On / Off	On: Fan on	Off	EVAP
Fuel Gauge Control	0-100%	Varies	Varies	EVAP
Fuel Level Sensor (V)	0.0-5.0v	Varies	Varies	EVAP
Fuel Tank Level	0-100%	Varies	Varies	EVAP
FTP Sensor (In. H2O)	-17.5 to 7.5"	Varies	Varies	EVAP
FTP Sensor (V)	0.0-5.0v	2.5 cap off	Varies	EVAP
Fuel Tank Capacity	0-16.5 gal.	Varies	Varies	EVAP
Fuel Trim Learn	Enabled / Disabled	Enabled	Enabled	EVAP
IAT Sensor (°F)	-38 to 284°F	50-194	50-194	EVAP
Injector Pulsewidth	0-512ms	1-4	Varies	EVAP

1998-2002 Firebird 5.7L V8 MFI VIN G (All) - Enhanced EVAP Data

Parameter Identification (PID)	PID Value Range	PID Value at Hot Idle	PID Value at 30 mph	Data List Type
Knock Retard (°)	0° to 16°	0	0	EVAP
Long Term F/T B1, B2	-10% to 10%	0 (± 5%)	0 (± 5%)	EVAP
MAF Sensor (g/sec)	0-512 g/sec	5-9	Varies	EVAP
MAP Sensor (kPa)	10-110 kPa	20-48	Varies	EVAP
MIL Status	On / Off	Off	Off	EVAP
PCM Reset	Yes / No	No	No	EVAP
Short Term Fuel Trim	-10% to 10%	0 (± 5%)	0 (± 5%)	EVAP
Startup ECT Sensor	-40 to 304°F	Varies	Varies	EVAP
TP Angle (%)	0-100%	0	Varies	EVAP
Vehicle Speed	0-155 mph	0	30	EVAP

1998-2002 Firebird 5.7L V8 MFI VIN G (All) - Freeze Frame & Failure Records

Parameter Identification (PID)	PID Value Range	PID Value at Hot Idle	PID Value at 30 mph	Data List Type
Air Fuel Ratio	0.0-25.5:1	14.6:1	14.6:1	F/F, F/R
BARO Sensor (kPa)	10-110 kPa	65-104	65-104	F/F, F/R
Current Gear	1-2-3-4	1	3	F/F, F/R
Desired Idle Speed	0-3187 rpm	PCM control	---	F/F, F/R
ECT Sensor (°F)	-38 to 284°F	185-221	185-221	F/F, F/R
Engine Load	0-100%	2	Varies	F/F, F/R
Engine Run Time	Hr: Min: Sec	00:00:00	00:00:00	F/F, F/R
Engine Speed	0-9999 rpm	Varies	---	F/F, F/R
EVAP Canister Purge	0-100%	0	0-50	F/F, F/R
Failure Counter	0-65535	0	0	F/F, F/R
Faults Since 1st Fault	0-65535	0	0	F/F, F/R
Injector Pulsewidth	0-985 ms	1-4	Varies	F/F, F/R
Long Term F/T B1, B2	-10% to 10%	0 (± 5%)	0 (± 5%)	F/F, F/R
Loop Status	Closed Loop / Open Loop	Closed Loop	Closed Loop	F/F, F/R
MAF Sensor (g/sec)	0-512 g/sec	5-9	Varies	F/F, F/R
MAP Sensor (kPa)	10-110 kPa	20-48	Varies	F/F, F/R
MAP Sensor (V)	0-5.0v	1-2	Varies	F/F, F/R
Miles Since 1st Fail	0-9999 rpm	Varies	Varies	F/F, F/R
Miles Since Last Fail	0-9999 rpm	Varies	Varies	F/F, F/R
Mile Since MIL Req.	0-9999 rpm	Varies	Varies	F/F, F/R
Not Run Counter	0-65535	Varies	Varies	F/F, F/R
Pass Counter	0-65535	Varies	Varies	F/F, F/R
Passes Since 1st Fail	0-65535	Varies	Varies	F/F, F/R
Short Term Fuel Trim	-10% to 10%	0 (± 5%)	0 (± 5%)	F/F, F/R
Startup ECT Sensor	-40 to 304°F	Varies	Varies	F/F, F/R
TCC Brake Switch	Applied / Released	Released	Released	F/F, F/R
TCC PWM Solenoid	Enabled / Disabled	Disabled	Disabled	F/F, F/R
TCC Enable Solenoid	Enabled / Disabled	Disabled	Disabled	F/F, F/R
TP Angle (%)	0-100%	0	Varies	F/F, F/R
Traction Control	Active / Inactive	Inactive	Inactive	F/F, F/R
Traction Output RPM	0-10,000	Varies	Varies	F/F, F/R
Transmission Range	0-4	1	3	F/F, F/R
Vehicle Speed	0-155 mph	0	30	F/F, F/R
VTD Fuel Disable	Active / Inactive	Inactive	Inactive	F/F, F/R

1998-2002 Firebird 5.7L V8 MFI VIN G (All) - PID Data List

Parameter Identification (PID)	PID Value Range	PID Value at Hot Idle	PID Value at 30 mph	Data List Type
AIR Relay Control	On / Off	Off	Off	Fuel Trim
AIR Solenoid Control	On / Off	Off	Off	Fuel Trim
BARO Sensor	10-110 kPa	65-104	65-104	Fuel Trim
Current Gear	1-2-3-4	1	3	Fuel Trim
DTC Set This Ignition	Yes / No	Yes: If set	No	Fuel Trim
ECT Sensor (°F)	-38 to 284°F	185-221	185-221	Fuel Trim
Engine Load	0-100%	0-9%	Varies	Fuel Trim
Engine Speed	0-9999 rpm	Varies	---	Fuel Trim
EVAP Canister Purge	0-100%	0-25	0-50	Fuel Trim
EVAP Canister Vent	No/VENT	Venting	Venting	Fuel Trim
Fuel Tank Level	0-100%	Varies	Varies	Fuel Trim
Fuel Tank Capacity	0-16.6 gal.	Varies	Varies	Fuel Trim
Fuel Trim Cell	Cell #	16-20	16-20	Fuel Trim
Fuel Trim Learn	Enabled / Disabled	Enabled	Enabled	Fuel Trim
F/T Diagnostic Inhibit	Enabled / Disabled	Disabled	Disabled	Fuel Trim
HO2S-11 (B1 S1)	0-1132 mv	10-1000	10-1000	Fuel Trim
HO2S-12 (B1 S2)	0-1132 mv	10-1000	10-1000	Fuel Trim
HO2S-21 (B2 S1)	0-1132 mv	10-1000	10-1000	Fuel Trim
HO2S-22 (B2 S2)	0-1132 mv	10-1000	10-1000	Fuel Trim
IAC Position	0-255 counts	Varies	Varies	Fuel Trim
IAT Sensor (°F)	-40 to 284°F	50-194	50-194	Fuel Trim
Ignition 1 Signal	0.0-25.5v	14.1	14.1	Fuel Trim
Injector Pulsewidth B1	0-985 ms	1-4	Varies	Fuel Trim
Injector Pulsewidth B2	0-985 ms	1-4	Varies	Fuel Trim
Knock Retard (°)	0° to 16°°	0	0	Fuel Trim
Long Term Fuel Trim Bank 1 and Bank 2	-10% to 10%	0 (± 5%)	0 (± 5%)	Fuel Trim
Long Term F/T Avg. Bank 1 and Bank 2	-10% to 10%	0 (± 5%)	0 (± 5%)	Fuel Trim
Loop Status	Closed Loop / Open Loop	Closed Loop	Closed Loop	Fuel Trim
MAF Sensor (g/sec)	0-512 g/sec	5-9	Varies	Fuel Trim
MAP Sensor (kPa)	10-110 kPa	20-48	Varies	Fuel Trim
MAP Sensor (V)	0-5.0v	1-2	Varies	Fuel Trim
PCM Reset	Yes / No	No	No	Fuel Trim
Short Term Fuel Trim Bank 1 and Bank 2	-10% to 10%	0 (± 5%)	0 (± 5%)	Fuel Trim
Short Term F/T Avg. Bank 1 and Bank 2	-10% to 10%	0 (± 5%)	0 (± 5%)	Fuel Trim
Spark Advance (°)	-64° to 64°	Varies	Varies	Fuel Trim
Startup ECT Sensor	-40 to 304°F	Varies	Varies	Fuel Trim
TFT Sensor (°F)	-40 to 304°F	Varies	Varies	Fuel Trim
TP Angle (%)	0-100%	0	Varies	Fuel Trim
Transmission Range	0-4	1	3	Fuel Trim
Vehicle Speed	0-155 mph	0	30	Fuel Trim

1998-2002 Firebird 5.7L V8 MFI VIN G (All) - PID Data List

Parameter Identification (PID)	PID Value Range	PID Value at Hot Idle	PID Value at 30 mph	Data List Type
A/C Relay Command	On / Off	Off	Off	Misfire
A/C Request Signal	Yes / No	No	No	Misfire
A/C Status	On / Off	Off	Off	Misfire
Clutch Pedal Switch	DEP/REL	Released	Released	Misfire
Cycles of Misfire Data	0-100	Varies	Varies	Misfire
Cylinder Mode M/F Index	0-65535	Varies	Varies	Misfire
ECT Sensor (°F)	-38 to 284°F	185-221	185-221	Misfire
Engine Load	0-100%	0-9	Varies	Misfire
Engine Speed	0-9999 rpm	Varies	Varies	Misfire
IC Circuit Cylinder 1-8	Okay / Fault	Okay	Okay	Misfire
Loop Status	Closed Loop / Open Loop	Closed Loop	Closed Loop	Misfire
MAF Sensor (g/sec)	0-512 g/sec	5-9	Varies	Misfire
MAP Sensor (kPa)	10-110 kPa	20-48	Varies	Misfire
MAP Sensor (V)	0-5.0v	1-2	Varies	Misfire
Misfire Current Cylinder 1-8	0-200	0	0	Misfire
Misfire History Cylinder 1-8	0-65535	0	0	Misfire
Misfiring Cylinder 1-8	CYL 1-8	0	0	Misfire
Misfire Revolution Status	Accept/REJ	ACCEPT	ACCEPT	Misfire
PCM Reset	Yes / No	No	No	Misfire
Revolutions with Misfire	0-65535	Varies	Varies	Misfire
Spark Advance (°)	-64° to 64°	Varies	Varies	Misfire
TCC Enable Solenoid	Enabled / Disabled	Disabled	Enabled	Misfire
TCC PWM Solenoid	0-100%	Enabled / Disabled	Enabled	Misfire
Total Misfire Count	0-65535	Varies	Varies	Misfire
Total Misfire Failures Since First Fail	0-65535	Varies	Varies	Misfire
Total Misfire Passes Since First Fail	0-65535	Varies	Varies	Misfire
TP Angle (%)	0-100%	0	Varies	Misfire
TP Sensor (V)	0-5.0v	0.4-0.9	Varies	Misfire
Vehicle Speed	0-155 mph	0	30	Misfire

G6 PID DATA

2005 G6 3.5L VIN 8 (ALL) – PID Data List

Scan Tool Parameter	Data List	Units Displayed	Typical Data Value
24X Crank Sensor	Ignition	RPM	0-6,400
A/C Relay Circuit Status	ODD	OK/Fault/Invalid State	OK
A/C Relay Circuit History	ODD	OK/Fault/Invalid State	OK
A/C Relay Command	Eng, Misfire, EGR, TAC	On/Off	Off
Air Fuel Ratio	Eng, Fuel Trim, HO2S	Ratio	14.2:1-14.7:1
APP Indicated Angle	Eng, EGR, EVAP, HO2S, Ignition, Misfire, TAC, Fuel Trim	Percent	0
APP Sensor 1	TAC	Volts	Varies
APP Sensor 1 Circuit	TAC	OK/Not OK	OK
APP Sensor 2	TAC	Volts	Varies
APP Sensor 2 Circuit	TAC	OK/Not OK	OK
APP Sensor 1	TAC	Percent	Varies
APP Sensor 2	TAC	Percent	Varies
APP Sensor 1 and 2	TAC	Agree/Disagree	Agree
BARO	Eng, EGR, EVAP, Fuel Trim, HO2S, Ignition	kPa	65-110 kPa
Catalytic Converter Protection Active	Fuel Trim, HO2S	Yes/No	No
CMP Sensor	Ignition	RPM	Varies
CMP Sensor Signal Present	Ignition, Misfire	Yes/No	Yes
Cruise Control Active	TAC	Yes/No	No
Cycles of Misfire Data	Misfire	Counts	0-99
Cyl. 1-6 Injector Circuit Status	ODD	OK/Stuck Low (open)/Stuck High	OK
Cylinder 1 and 4 IC Circuit Status	ODD	OK/Not OK	OK
Cylinder 2 and 5 IC Circuit Status	ODD	OK/Not OK	OK
Cylinder 3 and 6 IC Circuit Status	ODD	OK/Not OK	OK
Cylinder 1-6 Injector Circuit History	ODD	OK/Stuck Low (open)/Stuck High	OK
Decel Fuel Cut-off	Eng, EGR, Fuel Trim, HO2S	Active/Inactive	Inactive
Desired EGR Position	EGR, Misfire	Percent	0
Desired Idle Speed	Eng, EVAP, TAC	RPM	Varies
Driver Module 1 Status	ODD	Enabled/Off-High Volts/Off High Temp/Invalid State	Enabled
Driver Module 2 Status	ODD	Enabled/Off-High Volts/Off High Temp/Invalid State	Enabled
Driver Module 3 Status	ODD	Enabled/Off-High Volts/Off High Temp/Invalid State	Enabled
EC Ignition Relay Circuit Status	ODD	OK/Fault/Invalid State	OK
EC Relay Circuit History	ODD	OK/Fault/Invalid State	OK
ECT Sensor	Eng, EGR, EVAP, HO2S, Ignition, Misfire, Fuel Trim, TAC	°C/°F	Varies
EGR Flow Test Count	EGR	Counts	0-12
EGR Learned Minimum Position	EGR	Volts	0.14-1.0
EGR Position Sensor	Eng, EGR, Misfire	Percent	Varies
EGR Position Sensor	EGR	Volts	0
EGR Position Variance	EGR	Percent	0
EGR Solenoid Circuit History	ODD	OK/Open/Fault/Invalid State	OK
EGR Solenoid Circuit Status	ODD	OK/Fault/Invalid State	OK
EGR Solenoid Command	EGR	Percent	0

2005 G6 3.5L VIN 8 (ALL) – PID Data List (CONT)

Scan Tool Parameter	Data List	Units Displayed		Typical Data Value
Engine Load	Eng, Misfire, Fuel Trim, EVAP, EGR, HO2S, Ignition, TAC	Percent		0
Engine Oil Pressure Switch	Eng	Low/OK		OK
Engine Run Time	Eng, EVAP, Misfire, Fuel Trim, EGR, HO2S, Ignition, TAC, ODD	Hr: Min: Sec		Varies
Engine Speed	Eng, EGR, EVAP, Ignition, Misfire, Fuel Trim, TAC	RPM		Varies
EVAP Fault History	EVAP	No Fault/Excess Vacuum/Purge Valve Leak/Small Leak/Weak Vacuum/No Test Result		No Fault
EVAP Purge Solenoid Circuit History	ODD	OK/Open/Short		OK
EVAP Purge Solenoid Circuit Status	ODD	OK/Open/Short		OK
EVAP Purge Solenoid Command	Eng, EVAP, Fuel Trim, HO2S	Percent		Varies
EVAP Test Abort Reason	EVAP	No Aborted/Lost Enable/Small Leak/Veh. Not At Rest		Varies
EVAP Test Result	EVAP	•	No Result/Passed/Aborted OR	Varies
		•	Fail-DTC P0455 OR	
		•	Fail-DTC P0442 OR	
		•	Fail-DTC P0446 OR	
		•	Fail-DTC P0496	
EVAP Test State	EVAP	Waiting for Purge/Test Running/Test Completed		Varies
EVAP Vent Solenoid Circuit History	ODD	OK/Open/Short		OK
EVAP Vent Solenoid Circuit Status	ODD	OK/Open/Short		OK
EVAP Vent Solenoid Command	Eng, EVAP	Venting/Not Venting		Varies
FC Relay 1 Circuit History	ODD	OK/Fault/Invalid State		OK
FC Relay 1 Circuit Status	ODD	OK/Fault/Invalid State		OK
FC Relay 2 and 3 Circuit History	ODD	OK/Fault/Invalid State		OK
FC Relay 2 and 3 Circuit Status	ODD	OK/Fault/Invalid State		OK
Fuel Level Sensor	EVAP	Volts		Varies
Fuel Pump Relay Circuit History	ODD	OK/Fault/Invalid State		OK
Fuel Pump Relay Circuit Status	ODD	OK/Fault/Invalid State		OK
Fuel Pump Relay Command	Eng	On/Off		On
Fuel Tank Level Remaining	EVAP	Percent		0-104
Fuel Tank Pressure Sensor	Eng, EVAP	mm Hg/in H2O		Varies
Fuel Tank Pressure Sensor	EVAP	Volts		Varies
Fuel Trim Cell	EVAP, Eng, Fuel Trim	0-4		0
Fuel Trim Learn	Eng, EVAP, Fuel Trim	Enabled/Disabled		Varies
HO2S Bank 1 Sensor 1	EVAP, Eng, Fuel Trim, HO2S	millivolts		0-1000 and Varying
HO2S Bank 1 Sensor 1 Ready	Fuel Trim, HO2S	Yes/No		Yes
HO2S Bank 1 Sensor 2	Eng, Fuel Trim, HO2S	millivolts		0-1000 and Varying
HO2S Bank 2 Sensor 1 Ready	Fuel Trim, HO2S	Yes/No		Yes
HO2S Bank 2 Sensor 1	Eng, EVAP, Fuel Trim, HO2S	millivolts		0-1000 and Varying
HO2S Bank 2 Sensor 2	Eng, Fuel Trim, HO2S	millivolts		0-1000 and Varying
HO2S Htr. Bn. 1 Sen. 1	HO2S	On/Off		On

2005 G6 3.5L VIN 8 (ALL) – PID Data List (CONT)

Scan Tool Parameter	Data List	Units Displayed	Typical Data Value
HO2S Htr. Bank 1 Sensor 1 Command	HO2S	On/Off	On
HO2S Htr. Bn. 1 Sen. 2	HO2S	On/Off	On
HO2S Htr. Bank 1 Sensor 2 Command	HO2S	On/Off	On
HO2S Heater Bn. 1 Sen. 1	HO2S	Amps	Varies
HO2S Heater Bn. 1 Sen. 2	HO2S	Amps	Varies
HO2S Heater Bn. 2 Sen. 1	HO2S	Amps	Varies
HO2S Heater Bn. 2 Sen. 2	HO2S	Amps	Varies
HO2S Htr. Bn. 1 Sen. 1 Circuit History	ODD	OK/Fault/Invalid State	OK
HO2S Htr. Bn. 1 Sen. 2 Circuit History	ODD	OK/Fault/Invalid State	OK
HO2S Htr. Bn. 2 Sen. 1	HO2S	On/Off	On
HO2S Htr. Bn. 2 Sen. 1 Circuit History	ODD	OK/Fault/Invalid State	OK
HO2S Htr. Bn. 2 Sen. 2	HO2S	On/Off	On
HO2S Htr. Bn. 2 Sen. 2 Circuit History	ODD	OK/Fault/Invalid State	OK
HO2S Htr. Bn. 1 Sen. 1 Circuit Status	ODD	OK/Fault/Invalid State	OK
HO2S Htr. Bn. 1 Sen. 2 Circuit Status	ODD	OK/Fault/Invalid State	OK
HO2S Htr. Bn. 2 Sen. 1 Circuit Status	ODD	OK/Fault/Invalid State	OK
HO2S Htr. Bn. 2 Sen. 2 Circuit Status	ODD	OK/Fault/Invalid State	OK
Hot Open Loop	Fuel Trim, HO2S	Active/Inactive	Inactive
IAT Sensor	Eng, EVAP, EGR, Fuel Trim, HO2S, Ignition, Misfire, TAC	°C/°F	Varies
Ignition 1 Signal	EVAP, Eng, EGR, Fuel Trim, HO2S, Ignition, Misfire, ODD, TAC	Volts	Varies
Ignition Off Time	EVAP, Eng	Minutes/Varies	Varies
Initial Brake Apply Signal	TAC	Released/Applied	Released
Injector PWM	Fuel Trim, Misfire	milliseconds	Varies
Knock Retard	EGR, Ignition	Degrees	0
Long Term FT Bank 1	Eng, EVAP, Fuel Trim, HO2S	Percent	-10% to +10%
Long Term FT Bank 2	Eng, EVAP, Fuel Trim, HO2S	Percent	-10% to +10%
Loop Status	Eng, Fuel Trim, EVAP, HO2S	Open/Closed	Closed
MAF Sensor	Eng, EGR, Misfire, EVAP, Fuel Trim, HO2S, Ignition, TAC	g/s	3-6 (Varies with altitude)
MAF Sensor	EVAP, Fuel Trim, HO2S, Misfire, TAC	Hz	1,200-3,000 (depends on altitude)
MAP Sensor	Eng, EGR, Fuel Trim, HO2S, Misfire, TAC	Volts	0.4-2.0 Volts
MAP Sensor	Eng, EVAP, Fuel Trim, HO2S, Ignition, Misfire, EGR, TAC	kPa	20-48 kPa (Varies with altitude)
MIL Circuit History	ODD	OK/Fault/Invalid State	OK
MIL Command	EVAP, Eng, EGR, Ignition	On/Off	Off
MIL Circuit Status	ODD	OK/Fault/Invalid State	OK
Mileage Since DTC Cleared	Eng	km/h/mph	Varies
Misfire Current Cyl. 1-6	Misfire	Counts	0
Misfire History Cyl. 1-6	Misfire	Counts	0
Moderate Brake Apply Signal	TAC	Applied/Released	Released
Number of DTCs	Eng EVAP, Fuel Trim, TAC	Number	0
PCM Reset	Eng	Yes/No	No

2005 G6 3.5L VIN 8 (ALL) – PID Data List (CONT)

Scan Tool Parameter	Data List	Units Displayed	Typical Data Value
Power Enrichment	Eng, Fuel Trim, HO2S, Misfire	Active/Inactive	Inactive
Reduced Engine Power	TAC	Active/Inactive	Inactive
Short Term FT Bank 1	Eng, EVAP, Fuel Trim, HO2S,	Percent	-10 to +10
Short Term FT Bank 2	Eng, EVAP, Fuel Trim, HO2S,	Percent	-10 to +10
Spark	Eng, Misfire, Fuel Trim, HO2S, Ignition	Degrees	22 (varies)
Start Up ECT	EVAP, Fuel Trim, HO2S	°C/°F	Varies
Start Up IAT	EVAP, Fuel Trim, HO2S	°C/°F	Varies
Starter Relay Circuit Status	ODD	OK/Fault	OK
Starter Relay Circuit History	ODD	OK/Fault/Invalid State	OK
Stop Lamp Pedal Switch	Eng, TAC	Released/Applied	Released
TAC/PCM Communications Signal	TAC	OK/Fault	OK
TAC Stop Lamp Pedal Switch	Eng	Released/Applied	Released
TAC Stop Lamp Pedal Switch	TAC	Applied/Released	Released
TAC Vehicle Speed Signal	TAC	OK/Fault	OK
TCS Delivered Torque Circuit Status	ODD	OK/Fault	OK
Torque Request Signal	TAC	Percent	100%
Torque Delivered Signal	TAC	Percent	75%
Total Misfire Count	Misfire	1-255	0
TP Desired Angle	TAC	Percent	Varies
TP Indicated Angle	Eng, EGR, Fuel Trim, HO2S, Ignition, EVAP, Misfire, TAC	Percent	%
TP Sensor 1	TAC	Percent	Varies
TP Sensor 1	TAC	Volts	Varies
TP Sensors 1 and 2	TAC	Agree/Disagree	Agree

<u>GRAND AM PID DATA</u>

1996-99 Grand Am 2.4L I4 MFI VIN T (All) - PID Data List

Parameter Identification (PID)	PID Value Range	PID Value at Hot Idle	PID Value at 30 mph	Data List Type
BARO Sensor (V)	0.0-5.0v	3.5-4.5	3.5-4.5	Engine Data 1
Calculated Airflow	0-512	3-7	Varies	Engine Data 1
CKP Active Counter	0-255	Counts up	Counts up	Engine Data 1
CMP Active Counter	0-255	Counts up	Counts up	Engine Data 1
CMP Resync Counter	0-255	0	0	Engine Data 1
Desired Idle Speed	0-3187	PCM control	---	Engine Data 1
ECT Sensor (°F)	-40 to 419°F	185-239	185-239	Engine Data 1
Engine Load	0-100%	22-26	Varies	Engine Data 1
Engine Run Time	Hr: Min: Sec	00:00:00	00:00:00	Engine Data 1
Engine Speed	0-9999 rpm	Varies	---	Engine Data 1
EVAP Purge Solenoid	0-100%	0	0-100	Engine Data 1
EVAP Vent Solenoid	On / Off	Off	Off	Engine Data 1
Fan Control Relay	On / Off	Off	Off	Engine Data 1
Fuel Trim Cell	Cell #0-22	18-21	18-21	Engine Data 1
Fuel Trim Index	0-255 counts	58-198	58-198	Engine Data 1
Fuel Trim Index (%)	-10% to 10%	Varies	Varies	Engine Data 1
Generator F-Terminal	0-100%	0-100%	0-100%	Engine Data 1
Generator L-Terminal	Active / Inactive	Inactive	Inactive	Engine Data 1
HO2S-12 (B1 S2)	0-1132 mv	10-1000	10-1000	Engine Data 1
IAC Position	0-255 counts	5-60	---	Engine Data 1
IAT Sensor (°F)	-40 to 304°F	50-176	50-176	Engine Data 1
Ignition Voltage	0.0-25.5v	14.2	14.2	Engine Data 1
Knock Retard (°)	0-128°	0	0	Engine Data 1
KS Noise Channel	Yes / No	No	No	Engine Data 1
KS Active Signal	0-255 counts	35-50	35-50	Engine Data 1
Low Octane Spark Modifier	0-90°	Varies	Varies	Engine Data 1
Long Term Fuel Trim	0-255 counts	58-198	58-198	Engine Data 1
Long Term Fuel Trim	-10% to 10%	0 (± 5%)	0 (± 5%)	Engine Data 1
Long Term F/T Avg.	-10% to 10%	0 (± 5%)	0 (± 5%)	Engine Data 1
Loop Status	Closed Loop / Open Loop	Closed Loop	Closed Loop	Engine Data 1
MAP Sensor (V)	0.0-5.0v	1.5	Varies	Engine Data 1
Medium Resolution Resync Counter	0-255	0	0	Engine Data 1
Medium Resolution Engine Sync	Yes / No	Yes	Yes	Engine Data 1
No. of Current DTC(s)	0-255	0	0	Engine Data 1
O2S-11 (B1 S1)	0-1132 mv	10-1000	10-1000	Engine Data 1
Power Enrichment	Active / Inactive	Inactive	Inactive	Engine Data 1
Purge Learn Memory	0-100%	0	0-50	Engine Data 1
Short Term Fuel Trim	0-255 counts	58-198	58-198	Engine Data 1
Short Term Fuel Trim	-10% to 10%	0 (± 5%)	0 (± 5%)	Engine Data 1
Short Term F/T Avg.	-10% to 10%	0 (± 5%)	0 (± 5%)	Engine Data 1
Spark Advance (°)	-64° to 64°	Varies	Varies	Engine Data 1
TCC Brake Switch	On / Off	Off	Off	Engine Data 1
Throttle Angle (%)	0-100%	0	Varies	Engine Data 1
TP Sensor (V)	0.0-5.0v	0.52	0.74	Engine Data 1
Vehicle Speed	0-155 mph	0	30	Engine Data 1

GRAND AM PID DATA - OBD II

1996-99 Grand Am 2.4L I4 MFI VIN T (All) - PID Data List

Parameter Identification (PID)	PID Value Range	PID Value at Hot Idle	PID Value at 30 mph	Data List Type
A/C High Side (psi)	0-450 psi	139-399	139-399	Engine Data 2
A/C High Side (V)	0.0-5.0v	1-4	1-4	Engine Data 2
A/C Relay	On / Off	Off	Off	Engine Data 2
A/C Request	Yes / No	Off	Off	Engine Data 2
Adaptive Knock Retard	0-128° (degrees)	0	Varies	Engine Data 2
Air Fuel Ratio	0.0-25.5:1	14.6:1	14.6:1	Engine Data 2
BARO Sensor (kPa)	10-110 kPa	65-110	65-110	Engine Data 2
Cruise Engaged	Yes / No	No	No	Engine Data 2
Desired Idle Speed	0-3187	PCM control	----	Engine Data 2
Engine Coolant Level	Okay / Low	Okay	Okay	Engine Data 2
ECT Sensor (°F)	-40 to 419°F	185-239	185-239	Engine Data 2
Engine Oil Level	Okay / Low	Okay	Okay	Engine Data 2
Engine Oil Pressure	Okay / Low	Okay	Okay	Engine Data 2
Engine Speed	0-9999 rpm	Varies	----	Engine Data 2
IAT Sensor (°F)	-40 to 304°F	50-176	50-176	Engine Data 2
Injector Pulsewidth	0-985 ms	1-4	Varies	Engine Data 2
MAP Sensor (V)	0.0-5.0v	1.5	Varies	Engine Data 2
MPH / KPH	0-155 mph	0	30	Engine Data 2
ODD 1, 2 Open	0000000	0000000	0000000	Engine Data 2
ODD 1, 2 Short	0000000	0000000	0000000	Engine Data 2
Stepper Motor Cruise Control	Enabled / Disabled	Disabled	Disabled	Engine Data 2
Total Misfire	0-255 counts	0	0	Engine Data 2
Throttle Angle (%)	0-100%	0	Varies	Engine Data 2
TP Sensor (V)	0.0-5.0v	0.52	0.74	Engine Data 2
TR Range	P/N-R-O-3-2	P/N	3	Engine Data 2
TR Switch P, A	High / Low	Low	HI	Engine Data 2
TR Switch B, C	High / Low	HI	Low	Engine Data 2

1996-99 Grand Am 2.4L I4 MFI VIN T (All) - PID Data List

Parameter Identification (PID)	PID Value Range	PID Value at Hot Idle	PID Value at 30 mph	Data List Type
BARO Sensor (V)	0.0-5.0v	3.5-4.5	3.5-4.5	Misfire
Calculated Airflow	0-512 g/sec	Varies	Varies	Misfire
ECT Sensor (°F)	-40 to 419°F	185-239	185-239	Misfire
Engine Load	0-100%	22-26	Varies	Misfire
Engine Run Time	Hr: Min: Sec	00:00:00	00:00:00	Misfire
IAT Sensor (°F)	-40 to 304°F	50-176	50-176	Misfire
Injector Pulsewidth	0-985 ms	1-4	Varies	Misfire
MAP Sensor (V)	0.0-5.0v	1.5	Varies	Misfire
Mileage Since 1st Fail	0-9999 rpm	Varies	Varies	Misfire
Miles Since Last Fail	0-9999 rpm	Varies	Varies	Misfire
Misfire current Cyl 1-4	0-255	0	0	Misfire
Misfire History Cyl 1-4	0-255	0	0	Misfire
Throttle Angle (%)	0-100%	0	Varies	Misfire
TP Sensor (V)	0.0-5.0v	0.52	0.74	Misfire
Total Misfire	0-255	0	0	Misfire

1996-99 Grand Am 2.4L I4 MFI VIN T - PID Data List

Parameter Identification (PID)	PID Value Range	PID Value at Hot Idle	PID Value at 30 mph	Data List Type
Actual EGR Position	0-100%	0	0-100	EGR, EVAP
BARO Sensor (V)	0.0-5.0v	3.5-4.5	3.5-4.5	EGR, EVAP
Catalyst Converter TT	Not Okay / Okay	Okay	Okay	EGR, EVAP
Decel EWMA	-10 - 10 kPa	-3 to – 4	Varies	EGR, EVAP
Delta MAP Change	-20 - 20 kPa	0 / - number	0	EGR, EVAP
Desired Idle Speed	0-3187	PCM control	---	EGR, EVAP
Desired EGR Position	0-100%	0	0-100	EGR, EVAP
ECT Sensor (°F)	-40 to 419°F	185-239	185-239	EGR, EVAP
EGR Decel Filter	-10 - 10 kPa	-3 to – 4	Varies	EGR, EVAP
EGR Trip Samples	0-255 counts	Varies	Varies	EGR, EVAP
Engine Run Time	Hr: Min: Sec	00:00:00	00:00:00	EGR, EVAP
Engine Speed	0-9999 rpm	Varies	---	EGR, EVAP
EVAP Solenoid	0-100%	0	0-100	EGR, EVAP
EVAP Vent Solenoid	0=off / 1=on	0	0	EGR, EVAP
Fuel Level Sensor (V)	0.0-5.0v	Varies	Varies	EGR, EVAP
Fuel Level Sensor	0-100%	Varies	Varies	EGR, EVAP
FTP Sensor (V)	0.0-5.0v	2.5v: cap off	Varies	EGR, EVAP
Fuel Trim Cell	Cell #0-22	18-21	---	EGR, EVAP
HO2S-12 (B1 S2)	0-1132 mv	10-1000	10-1000	EGR, EVAP
IAC Position	0-255 counts	5-60	---	EGR, EVAP
IAT Sensor (°F)	-40 to 304°F	50-176	50-176	EGR, EVAP
Long, Short Term F/T	-10% to 10%	0 (± 5%)	0 (± 5%)	EGR, EVAP
MAP Sensor (V)	0.0-5.0v	1.5	Varies	EGR, EVAP
No. of Catalyst Tests	Number	Varies	Varies	EGR, EVAP
Number of Current Trouble Codes	0-255	0	0	EGR, EVAP
O2S-11 (B1 S1)	0-1132 mv	10-1000	10-1000	EGR, EVAP
QDM 1, 2 Open, Short	0000000	0000000	0000000	EGR, EVAP
Throttle Angle (%)	0-100%	0	Varies	EGR, EVAP

1996-99 Grand Am 2.4L I4 MFI VIN T (All) - PID Data List

Parameter Identification (PID)	PID Value Range	PID Value at Hot Idle	PID Value at 30 mph	Data List Type
BARO Sensor (V)	0.0-5.0v	3.5-4.5	3.5-4.5	Fuel Trim
Calculated Airflow	0-512 g/sec	Varies	Varies	Fuel Trim
Desired Idle Speed	0-3187	PCM control	---	Fuel Trim
ECT Sensor (°F)	-40 to 419°F	185-239	185-239	Fuel Trim
Engine Load	0-100%	22-26	Varies	Fuel Trim
Engine Speed	0-9999 rpm	Varies	---	Fuel Trim
Fuel Trim Cell	Cell #0-22	18-21	0-21	Fuel Trim
Fuel Trim Index	0-255 counts	Varies	Varies	Fuel Trim
HO2S-11, HO2S-12	0-1132 mv	10-1000	10-1000	Fuel Trim
IAT Sensor (°F)	-40 to 304°F	50-176	50-176	Fuel Trim
L/R & R/L Average	0-249 ms	0	0	Fuel Trim
L/R & R/L Transition	0-255 counts	0	0	Fuel Trim
Long, Short Term F/T	-10% to 10%	0 (± 5%)	0 (± 5%)	Fuel Trim
MAP Sensor (V)	0.0-5.0v	1.5	Varies	Fuel Trim
O2S-11 L-R/R-L Ratio	0:1-15.93:1	Varies	Varies	Fuel Trim
Throttle Angle (%)	0-100%	0	Varies	Fuel Trim

2000-05 Grand Am 2.2L I4 VIN F, 2.4L VIN T (All) - PID Data List

Parameter Identification (PID)	PID Value Range	PID Value at Hot Idle	PID Value at 30 mph	Data List Type
Active Knock Retard	0-90 degrees	Varies	Varies	Engine Data 1
A/C Relay Command	On / Off	Off	Off	Engine Data 1
Airflow Calculated	0-512 g/sec	3-7	Varies	Engine Data 1
BARO Sensor (V)	10-110 kPa	65-104	65-104	Engine Data 1
CKP Active Counter	0-255 counts	Counts up	Counts up	Engine Data 1
CMP Active Counter	0-255 counts	Counts up	Counts up	Engine Data 1
CMP Resync Counter	0-255	0	0	Engine Data 1
Cruise Control Active	Yes / No	No	No	Engine Data 1
Cruise/TCC Switch	Applied/Not	Not Applied	Not Applied	Engine Data 1
Current Gear	1-4	1	3	Engine Data 1
Desired Idle Speed	0-3187	PCM control	---	Engine Data 1
ODD 1, 2 Status	Several	Okay	Okay	Engine Data 1
ECT Sensor (°F)	-40 to 304°F	185-239	185-239	Engine Data 1
Engine Load	0-100%	11-21	Varies	Engine Data 1
Engine Run Time	Hr: Min: Sec	00:00:00	00:00:00	Engine Data 1
Engine Speed	0-9999 rpm	Varies	---	Engine Data 1
EVAP Purge Solenoid	0-100%	0	0-100	Engine Data 1
EVAP Purge Solenoid	On / Off	Off	On	Engine Data 1
EVAP Vent Solenoid	Venting / Not Venting	Not Venting	Not Venting	Engine Data 1
EVAP Vent Solenoid	Several	Several	Several	Engine Data 1
Fuel Level Sensor	0-100%	Varies	Varies	Engine Data 1
Fuel Pump Relay	On / Off	On	On	Engine Data 1
FTP Sensor (In. H2O)	-17.5 to 7.5"	Varies	Varies	Engine Data 1
Fuel Trim Cell	Cell #0-22	18-21	18-21	Engine Data 1
Generator F-Terminal	0-100%	Varies	Varies	Engine Data 1
Generator L-Terminal	Voltage/No	Voltage	Voltage	Engine Data 1
HO2S-12 (B1 S2)	0-1132 mv	10-1000	10-1000	Engine Data 1
IAC Position	0-255 counts	5-60	---	Engine Data 1
IAT Sensor (°F)	-40 to 304°F	50-176	50-176	Engine Data 1
KS Noise Channel	0-5.0v	1.1	Varies	Engine Data 1
Knock Retard (°)	0-90°	0	0	Engine Data 1
Long Term Fuel Trim	-10% to 10%	0 (± 5%)	0 (± 5%)	Engine Data 1
Loop Status	Closed Loop / Open Loop	Closed Loop	Closed Loop	Engine Data 1
MAP Sensor (kPa)	10-110 kPa	25-48	Varies	Engine Data 1
MAP Sensor (V)	0.0-5.0v	1-2	Varies	Engine Data 1
Medium Resolution Resync Counter	0-255 counts	Varies	Varies	Engine Data 1
Medium Resolution Engine Sync	Yes / No	Yes	Yes	Engine Data 1
MIL (lamp) Command	On / Off	Off	Off	Engine Data 1
No. of Current DTC(s)	0-255	0	0	Engine Data 1
O2S-11 (B1 S1)	0-1132 mv	10-1000	10-1000	Engine Data 1
Power Enrichment	Active / Inactive	Inactive	Inactive	Engine Data 1
Short Term Fuel Trim	-10% to 10%	0 (± 5%)	0 (± 5%)	Engine Data 1
Spark Advance (°)	-64° to 64°	Varies	Varies	Engine Data 1
TCC Enable Solenoid	On / Off	On	Off	Engine Data 1
Throttle Angle (%)	0-100%	0	Varies	Engine Data 1
TP Sensor (V)	0.0-5.0v	0.20-0.90	Varies	Engine Data 1

2000-05 Grand Am 2.2L I4 VIN F, 2.4L VIN T (All) - PID Data List

Parameter Identification (PID)	PID Value Range	PID Value at Hot Idle	PID Value at 30 mph	Data List Type
A/C High Side (psi)	0-450 psi	139-399	139-399	Engine Data 2
A/C High Side (V)	0.0-5.0v	1-4	1-4	Engine Data 2
A/C Relay Command	On / Off	Off	Off	Engine Data 2
A/C Request Signal	Yes / No	Off	Off	Engine Data 2
Airflow Calculated	0-512 g/sec	3-7	Varies	Engine Data 2
Air Fuel Ratio	0.0-25.5:1	14.6:1	14.6:1	Engine Data 2
BARO Sensor (kPa)	10-110 kPa	65-104	65-104	Engine Data 2
CKP Active Counter	0-255 counts	Counts up	Counts up	Engine Data 2
CMP Active Counter	0-255 counts	Counts up	Counts up	Engine Data 2
CMP Resync Counter	0-255	0	0	Engine Data 2
Cruise Control Active	Yes / No	No	No	Engine Data 2
Cruise Range	Disengaged / Engaged	Disengaged	Disengaged	Engine Data 2
Cruise Control/TCC Switch	Applied/Not	Not Applied	Not Applied	Engine Data 2
Current Gear	1-4	1	3	Engine Data 2
Desired Idle Speed	0-3187	PCM control	---	Engine Data 2
ECT Sensor (°F)	-40 to 419°F	185-239	185-239	Engine Data 2
Engine Coolant Level	Okay / Low	Okay	Okay	Engine Data 2
Engine Oil Pressure	Okay / Low	Okay	Okay	Engine Data 2
Engine Load	0-100%	4-8	Varies	Engine Data 2
Engine Run Time	Hr: Min: Sec	00:00:00	00:00:00	Engine Data 2
Engine Speed	0-9999 rpm	Varies	---	Engine Data 2
FC Relay 1 Command	On / Off	Off	Off	Engine Data 2
FC Relay 2 Command	On / Off	Off	Off	Engine Data 2
Generator F-Terminal	0-100%	Varies	Varies	Engine Data 2
Generator L-Terminal	Voltage/No	Voltage	Voltage	Engine Data 2
IAT Sensor (°F)	-40 to 304°F	50-176	50-176	Engine Data 2
Ignition 1 Signal	0-25.5v	14.1	14.1	Engine Data 2
Injector Pulsewidth Cylinder 1-4	0-985 ms	3-4	Varies	Engine Data 2
Long Term Fuel Trim	-10% to 10%	0 (± 5%)	0 (± 5%)	Engine Data 2
Loop Status	Closed Loop / Open Loop	Closed Loop	Closed Loop	Engine Data 2
MAP Sensor (kPa)	10-110 kPa	25-48	Varies	Engine Data 2
MAP Sensor (V)	0.0-5.0v	1-2	Varies	Engine Data 2
Medium Resolution Resync Counter	0-255 counts	Varies	Varies	Engine Data 2
Power Enrichment	Active / Inactive	Inactive	Inactive	Engine Data 2
Short Term Fuel Trim	-10% to 10%	0 (± 5%)	0 (± 5%)	Engine Data 2
Spark Advance (°)	-64° to 64°	Varies	Varies	Engine Data 2
Startup ECT Sensor	-40 to 419°F	Varies	Varies	Engine Data 2
TCC Brake Switch	Open/Close	Open	Open	Engine Data 2
TCC Enable Solenoid	On / Off	On	Off	Engine Data 2
Throttle Angle (%)	0-100%	0	Varies	Engine Data 2
TP Sensor (V)	0.0-5.0v	0.20-0.90	Varies	Engine Data 2
TR Range Switch	Several	Several	Several	Engine Data 2
Vacuum Calculated	10-110 kPa	Varies	Varies	Engine Data 2
Vehicle Speed	0-155 mph	0	30	Engine Data 2

2000-05 Grand Am 2.2L I4 VIN F, 2.4L VIN T (All) - PID Data List

Parameter Identification (PID)	PID Value Range	PID Value at Hot Idle	PID Value at 30 mph	Data List Type
A/C Relay Command	On / Off	Off	Off	Engine Data 3
Airflow Calculated	0-512 g/sec	3-7	Varies	Engine Data 3
Air Fuel Ratio	0.0-25.5:1	14.6:1	14.6:1	Engine Data 3
Cruise Control Active	Yes / No	No	No	Engine Data 3
Cruise/TCC Switch	Applied/Not	Not Applied	Not Applied	Engine Data 3
Desired Idle Speed	0-3187	PCM control	---	Engine Data 3
ECT Sensor (°F)	-40 to 419°F	185-239	185-239	Engine Data 3
Engine Load	0-100%	4-8	Varies	Engine Data 3
Engine Oil Level	Okay / Low	Okay	Okay	Engine Data 3
Engine Oil Pressure	Okay / Low	Okay	Okay	Engine Data 3
Engine Run Time	Hr: Min: Sec	00:00:00	00:00:00	Engine Data 3
Engine Speed	0-9999 rpm	Varies	---	Engine Data 3
FC Relay 1 Command	On / Off	Off	Off	Engine Data 3
IAT Sensor (°F)	-40 to 304°F	50-176	50-176	Engine Data 3
Injector 1-4 Command	0-985 ms	3-4	Varies	Engine Data 3
Long Term Fuel Trim	10% to 10%	0 (± 5%)	0 (± 5%)	Engine Data 3
Loop Status	Closed Loop / Open Loop	Closed Loop	Closed Loop	Engine Data 3
MAP Sensor (kPa)	10-110 kPa	25-48	Varies	Engine Data 3
MAP Sensor (V)	0.0-5.0v	1-2	Varies	Engine Data 3
Medium Resolution Resync Counter	0-255 counts	Varies	Varies	Engine Data 3
Short Term Fuel Trim	-10% to 10%	0 (± 5%)	0 (± 5%)	Engine Data 3
Spark Advance (°)	-64° to 64°	Varies	Varies	Engine Data 3
TP Sensor (V)	0.0-5.0v	0.20-0.90	Varies	Engine Data 3
TR Range Switch	Several	Several	Several	Engine Data 3
VTD Auto Learn	Active / Inactive	Inactive	Inactive	Engine Data 3
VTD Fuel Disable	Active / Inactive	Inactive	Inactive	Engine Data 3

2000-05 Grand Am 2.2L I4 VIN F, 2.4L VIN T (All) - PID Data List

Parameter Identification (PID)	PID Value Range	PID Value at Hot Idle	PID Value at 30 mph	Data List Type
A/C Relay Command	On / Off	Off	Off	Misfire
Airflow Calculated	0-512 g/sec	Varies	Varies	Misfire
Cycles of Misfire Data	0-99	Varies	Varies	Misfire
Desired Idle Speed	0-3187	PCM control	---	Misfire
ECT Sensor (°F)	-40 to 419°F	185-239	185-239	Misfire
Engine Load	0-100%	22-26	Varies	Misfire
Engine Run Time	Hr: Min: Sec	00:00:00	00:00:00	Misfire
Engine Speed	0-16384	Varies	---	Misfire
IAT Sensor (°F)	-40 to 304°F	50-176	50-176	Misfire
Injector Pulsewidth	0-985 ms	1-4	Varies	Misfire
MAP Sensor (kPa)	11-104	25-48	Varies	Misfire
Misfire current Cyl 1-4	0-255	0	0	Misfire
Misfire History Cyl 1-4	0-255	0	0	0
No. of Current DTC(s)	0-255	0	0	0
Spark Advance (°)	-64° to 64°	Varies	Varies	Varies
Throttle Angle (%)	0-100%	0	Varies	Varies
TP Sensor (V)	0.0-5.0v	0.20-0.90	Varies	Varies

2000-05 Grand Am 2.2L I4 VIN F, 2.4L VIN T (All) - PID Data List

Parameter Identification (PID)	PID Value Range	PID Value at Hot Idle	PID Value at 30 mph	Data List Type
BARO Sensor (kPa)	11-105	65-104	65-104	EVAP
Desired Idle Speed	0-3187	PCM control	---	EVAP
ECT Sensor (°F)	-40 to 419°F	185-239	185-239	EVAP
Engine Load	0-100%	4-6	Varies	EVAP
Engine Speed	0-9999 rpm	Varies	---	EVAP
EVAP Purge Solenoid	0-100%	Varies	Varies	EVAP
EVAP Test Abort	Abort / Not Aborted	Not Aborted	Not Aborted	EVAP
EVAP Test Results	Passed / Not Passed	Passed	Passed	EVAP
EVAP Test State	Several	Several	Several	EVAP
EVAP Vent Solenoid	Venting/Not	Venting	Venting	EVAP
Fuel Level Sensor	0-100%	Varies	Varies	EVAP
FTP Sensor (In. H2O)	-17.5 to 7.5"	Varies	Varies	EVAP
FTP Sensor (V)	0-5.0v	Varies	Varies	EVAP
Fuel Trim Cell	Cell #0-22	18-21	---	EVAP
IAT Sensor (°F)	-40 to 304°F	50-176	50-176	EVAP
Long Term Fuel Trim	-10% to 10%	0 (± 5%)	0 (± 5%)	EVAP
Loop Status	Closed Loop / Open Loop	Closed Loop	Closed Loop	EVAP
MAP Sensor (kPa)	11-105	65-104	65-104	EVAP
No. of Current DTC(s)	0-255	0	0	EVAP
O2S-11 (B1 S1)	0-1132 mv	10-1000	10-1000	EVAP
Short Term Fuel Trim	-10% to 10%	0 (± 5%)	0 (± 5%)	EVAP
Throttle Angle (%)	0-100%	0	Varies	EVAP
Vehicle Speed	0-159	0	30	EVAP

2000-05 Grand Am 2.2L I4 VIN F, 2.4L VIN T (All) - PID Data List

Parameter Identification (PID)	PID Value Range	PID Value at Hot Idle	PID Value at 30 mph	Data List Type
Another TWC Test	Yes / No	No	No	Fuel Trim
Airflow Calculated	0-512 g/sec	3-7	Varies	Fuel Trim
Air Fuel Ratio	0.0-25.5:1	14.6:1	14.6:1	Fuel Trim
BARO Sensor (kPa)	11-104 kPa	65-104	65-104	Fuel Trim
ECT Sensor (°F)	-40 to 419°F	185-239	185-239	Fuel Trim
Engine Speed	0-9999 rpm	Varies	---	Fuel Trim
EVAP Purge Solenoid	0-100%	Varies	Varies	Fuel Trim
EVAP Vent Solenoid	Venting / Not Venting	Not Venting	Not Venting	Fuel Trim
Fuel Trim Cell	Cell #0-22	18-21	18-21	Fuel Trim
HO2S-12 (B1 S1)	0-1132 mv	10-1000	10-1000	Fuel Trim
IAC Position	0-255 counts	5-60	---	Fuel Trim
IAT Sensor (°F)	-40 to 304°F	50-176	50-176	Fuel Trim
Injector Pulsewidth	0-985 ms	3-4	Varies	Fuel Trim
Long Term Fuel Trim	-10% to 10%	0 (± 5%)	0 (± 5%)	Fuel Trim
Loop Status	Closed Loop / Open Loop	Closed Loop	Closed Loop	Fuel Trim
MAP Sensor (kPa)	11-105	25-48	Varies	Fuel Trim
No. of Current DTC(s)	0-255	0	0	Fuel Trim
O2S-11 (B1 S1)	0-1132 mv	10-1000	10-1000	Fuel Trim
Short Term Fuel Trim	-10% to 10%	0 (± 5%)	0 (± 5%)	Fuel Trim
Spark Advance (°)	-64° to 64°	Varies	Varies	Fuel Trim
Throttle Angle (%)	0-100%	0	Varies	Varies

1996-98 Grand Am 3.1L V6 MFI VIN M (A/T) - PID Data List

Parameter Identification (PID)	PID Value Range	PID Value at Hot Idle	PID Value at 30 mph	Data List Type
3X Crank Sensor	0-9999 rpm	Varies	Varies	Engine Data 1
24X Crank Sensor	0-1600 rpm	Varies	Varies	Engine Data 1
Abuse Management	Active / Inactive	Inactive	Inactive	Engine Data 1
A/C HI Side Pressure	0-450 psi	139-399	139-399	Engine Data 1
A/C HI Side Pressure	0.0-5.0v	1-3	1-3	Engine Data 1
A/C Press. Too HI/LO	Yes / No	No	No	Engine Data 1
A/C Pressure Disable	Yes / No	No	No	Engine Data 1
A/C Clutch	Yes / No	No	No	Engine Data 1
A/C Request	Yes / No	No	No	Engine Data 1
A/C Off for WOT	Yes / No	No	No	Engine Data 1
A/C Slugging	Active / Inactive	Inactive	Inactive	Engine Data 1
Actual EGR Position	0-100%	0	0-100	Engine Data 1
Air Fuel Ratio	0:1-25.5:1	14.6:1	14.6:1	Engine Data 1
AIR Pump Control	On / Off	Off	Off	Engine Data 1
BARO Sensor (kPa)	10-110 kPa	65-110	65-110	Engine Data 1
Cam Signal Present	Yes / No	Yes	Yes	Engine Data 1
Commanded A/C	On / Off	Off	Off	Engine Data 1
Commanded Fan 1	On / Off	Off	Off	Engine Data 1
Commanded Fan 2	On / Off	Off	Off	Engine Data 1
Cruise Engaged	Disengaged / Engaged	Disabled	Disabled	Engine Data 1
Cruise Inhibit	Yes / No	Yes	Yes	Engine Data 1
Cruise Mode	On / Off	Off	On	Engine Data 1
Current Gear	1-2-3-4	1	3	Engine Data 1
Decel Fuel Mode	Active / Inactive	Inactive	Inactive	Engine Data 1
Desired EGR Position	0-100%	0%	0-100%	Engine Data 1
Desired Idle Speed	0-9999 rpm	PCM control	----	Engine Data 1
ECT Sensor (°F)	-40 to 304°F	185-220	185-220	Engine Data 1
Engine Load	0-100%	2-5	Varies	Engine Data 1
Engine Run Time	Hr: Min: Sec	00:00:00	00:00:00	Engine Data 1
Engine Speed	0-9999 rpm	Varies	----	Engine Data 1
EGR Closed Pintle Position	0.0-5.0v	0.14-1.0	----	Engine Data 1
EGR Duty Cycle	0-100%	0	0-100	Engine Data 1
EGR Feedback	0.0-5.0v	0.14-1.0	----	Engine Data 1
EVAP PWM Solenoid	0-100%	0-25	0-100	Engine Data 1
EVAP Vacuum Switch	Purge/No	No Purge	No Purge	Engine Data 1
EVAP Canister Purge	0-100%	0-25	0-100	Engine Data 1
EVAP Vent Solenoid	Open/Close	Open	Open	Engine Data 1
FC Relay Low Speed	On / Off	Off	Off	Engine Data 1
FC Relay High Speed	On / Off	Off	Off	Engine Data 1
Fuel Pump Command	On / Off	On	On	Engine Data 1
Fuel Trim Cell	Cell #0-4	0	1	Engine Data 1
Fuel Trim Learn	Enabled / Disabled	Enabled	Enabled	Engine Data 1
Fuel Pump Status	On / Off	On	On	Engine Data 1
Generator Command	On / Off	On	On	Engine Data 1
Generator Lamp	On / Off	Off	Off	Engine Data 1

1996-98 Grand Am 3.1L V6 MFI VIN M (A/T) - PID Data List

Parameter Identification (PID)	PID Value Range	PID Value at Hot Idle	PID Value at 30 mph	Data List Type
HO2S-11 Status	Not Ready / Ready	Ready	Ready	Engine Data 1
HO2S-11 (B1 S1)	0-1132 mv	10-1000	10-1000	Engine Data 1
HO2S-12 (B1 S2)	0-1132 mv	10-1000	10-1000	Engine Data 1
IAC Position	0-255 counts	5-50	---	Engine Data 1
IAT Sensor (°F)	-40 to 304°F	50-194	50-194	Engine Data 1
IC Mode	Active / Inactive	Active	Active	Engine Data 1
Idle Speed Error	0-12800 rpm	<100	---	Engine Data 1
Ignition 1 Volts	0.0-25.5v	14.2	14.2	Engine Data 1
Ignition Bypass	Active / Inactive	Inactive	Inactive	Engine Data 1
Ignition Mode	Bypass / IC	IC	IC	Engine Data 1
Injector Fault	Yes / No	No	No	Engine Data 1
Injector Pulsewidth	0-985 ms	1.5-3.5	Varies	Engine Data 1
Injector Status	Okay / STUCK	Okay	Okay	Engine Data 1
Knock Retard (°)	0.0-25.5°	0	Varies	Engine Data 1
KS Active Counter	0-255	Varies	Varies	Engine Data 1
KS Noise Channel	0.0-5.0v	0.1-3.0	Varies	Engine Data 1
Long Term Fuel Trim	-10% to 10%	0 (± 5%)	0 (± 5%)	Engine Data 1
Loop Status	Closed Loop / Open Loop	Closed Loop	Closed Loop	Engine Data 1
Low Oil Lamp	On / Off	Off	Off	Engine Data 1
MAF Sensor (g/sec)	0-512 g/sec	3-6	Varies	Engine Data 1
MAF Sensor (Hz)	0-31,999	1200-3000	Varies	Engine Data 1
MAP Sensor (kPa)	10-110 kPa	29-48	Varies	Engine Data 1
MAP Sensor (V)	0.0-5.0v	1-2	Varies	Engine Data 1
MIL Status	On / Off	Off	Off	Engine Data 1
Non-Volatile Memory	PASS/FAIL	PASS	PASS	Engine Data 1
Power Enrichment	Active / Inactive	Active	Active	Engine Data 1
Spark Advance (°)	-64° to 64°	20	Varies	Engine Data 1
TCC Brake Switch	Applied / Released	Released	Released	Engine Data 1
TCC Engaged	Yes / No	No	No	Engine Data 1
TP Angle (%)	0-100%	0	Varies	Engine Data 1
TP Sensor (V)	0.0-5.0v	0.20-0.74	---	Engine Data 1
Trans. Hot Mode	Active / Inactive	Inactive	Inactive	Engine Data 1
Transmission Range	X / O	O/X/X/O	Varies	Engine Data 1
Transmission Range	P/N-R-4-3-2	P	3	Engine Data 1
TR Switch P/A/B/C	HI/LO	LO/LO/HI/HI	---	Engine Data 1
TWC Protection	Active / Inactive	Inactive	Inactive	Engine Data 1
Vehicle Speed	0-155 mph	0	30	Engine Data 1
VTD Fuel Disable	Enabled / Disabled	Disabled	Disabled	Engine Data 1

1996-98 Grand Am 3.1L V6 MFI VIN M (A/T) - PID Data List

Parameter Identification (PID)	PID Value Range	PID Value at Hot Idle	PID Value at 30 mph	Data List Type
Engine Oil Level	Okay / Low	Okay	Okay	IPC
Generator L-Terminal	Okay / Fault	Okay	Okay	IPC
Generator Lamp	On / Off	Off	Off	IPC
Ignition 1 Signal	0.0-25.5v	14.2	14.2	IPC
Low Oil Lamp	On / Off	Off	Off	IPC
MIL Status	On / Off	Off	Off	IPC
PCM Commanded Generator Status	Yes / No	No	No	IPC

1996-98 Grand Am 3.1L V6 MFI VIN M (A/T) - PID Data List

Parameter Identification (PID)	PID Value Range	PID Value at Hot Idle	PID Value at 30 mph	Data List Type
Actual EGR Position	0-100%	0%	0-100%	EGR
BARO Sensor (kPa)	10-110 kPa	65-110	65-110	EGR
BARO Sensor (V)	0.0-5.0v	3.5-4.5	3.5-4.5	EGR
Current Gear	1-2-3-4	1	3	EGR
Decel Fuel Mode	Active / Inactive	Inactive	Inactive	EGR
Desired EGR Position	0-100%	0	0-100	EGR
Desired Idle Speed	0-3187 rpm	PCM control	---	EGR
ECT Sensor (°F)	-38-304°F	185-220	185-220	EGR
EGR Closed Pintle Position	0.0-5.0v	0.14-1.0	---	EGR
EGR Duty Cycle	0-100%	0	0-100	EGR
EGR Feedback	0.0-5.0v	0.14-1.0	---	EGR
EGR Flow Test Count	0-255	0-10	0-10	EGR
EGR Position Error	0-100%	0-9	---	EGR
Engine Load	0-100%	2-5	Varies	EGR
Engine Run Time	Hr: Min: Sec	00:00:00	00:00:00	EGR
Engine Speed	0-9999 rpm	Varies	---	EGR
Hot Open Loop	Active / Inactive	Inactive	Inactive	EGR
IAC Position	0-255	10-40	---	EGR
IAT Sensor (°F)	-40 to 304°F	50-194	50-194	EGR
Ignition 1 Signal	0.0-25.5v	14.2	14.2	EGR
Knock Retard (°)	0.0-25.5°	0	Varies	EGR
KS Active Counter	0-255 counts	Varies	Varies	EGR
Loop Status	Closed Loop / Open Loop	Closed Loop	Closed Loop	EGR
MAF Sensor (g/sec)	0-512 g/sec	3-6	Varies	EGR
MAP Sensor (kPa)	10-110 kPa	29-48	Varies	EGR
MAP Sensor (V)	0.0-5.0v	1-2	Varies	EGR
Power Enrichment	Active / Inactive	Active	Active	EGR
Rich/Lean Status	Lean or Rich	L-R-L-R	L-R-L-R	EGR
TCC Engaged	Yes / No	No	Yes or No	EGR
TP Angle	0-100%	0	Varies	EGR
Transmission Range	X/X/O/O	O/X/X/O	X/O/O/X	EGR
Transmission Range	P/N-R-4-3-2	P	4	EGR
TWC Protection	Active / Inactive	Inactive	Inactive	EGR
Vehicle Speed	0-155 mph	0	30	EGR

1996-98 Grand Am 3.1L V6 MFI VIN M (A/T) - EVAP System Data

Parameter Identification (PID)	PID Value Range	PID Value at Hot Idle	PID Value at 30 mph	Data List Type
Engine Run Time	Hr: Min: Sec	00:00:00	00:00:00	EVAP
Engine Speed	0-9999 rpm	Varies	---	EVAP
EVAP Canister Purge	0-100%	0-25	0-50	EVAP
EVAP Vent Solenoid	Open/Close	Closed	Closed	EVAP
Fuel Tank Pressure Sensor	0.0-5.0v	2.5v: cap off	Varies	EVAP
IAT Sensor (°F)	-40 to 304°F	50-194	50-194	EVAP
MAF Sensor (g/sec)	0-512 g/sec	3-6	Varies	EVAP
MAP Sensor (kPa)	10-110 kPa	29-48	Varies	EVAP
MAP Sensor (V)	0.0-5.0v	1-2v	Varies	EVAP
Startup ECT Sensor	-40 to 304°F	Varies	Varies	EVAP
Startup IAT Degrees	-40 to 304°F	Varies	Varies	EVAP

1996-98 Grand Am 3.1L V6 MFI VIN M (A/T) - PID Data List

Parameter Identification (PID)	PID Value Range	PID Value at Hot Idle	PID Value at 30 mph	Data List Type
Air Fuel Ratio	0:1-25.5:1	14.6:1	14.6:1	Fuel Trim
AIR Pump Control	On / Off	Off	Off	Fuel Trim
AIR Pump Relay	On / Off	Off	Off	Fuel Trim
Decel Fuel Mode	Active / Inactive	Inactive	Inactive	Fuel Trim
ECT Sensor (°F)	-40 to 304°F	185-220	185-220	Fuel Trim
Engine Load	0-100%	2-5	Varies	Fuel Trim
Engine Run Time	Hr: Min: Sec	00:00:00	00:00:00	Fuel Trim
Engine Speed	0-9999 rpm	Varies	---	Fuel Trim
EVAP Purge PWM	0-100%	0	0-50	Fuel Trim
Hot Open Loop	Active / Inactive	Inactive	Inactive	Fuel Trim
HO2S-11 Status	Not Ready / Ready	Ready	Ready	Fuel Trim
HO2S-11 (B1 S1)	0-1132 mv	10-1000	10-1000	Fuel Trim
HO2S-12 (B1 S2)	0-1132 mv	10-1000	10-1000	Fuel Trim
HO2S-11 Warmup	Min: Sec	Varies	Varies	Fuel Trim
HO2S-12 Warmup	Min: Sec	Varies	Varies	Fuel Trim
HO2S-11 XCount	0-255 counts	Varies	Varies	Fuel Trim
IAT Sensor (°F)	-40 to 304°F	50-194	50-194	Fuel Trim
Ignition 1 Signal	0.0-25.5v	14.2	14.2	Fuel Trim
Injector Pulsewidth	0-985 ms	1.5-3.5	Varies	Fuel Trim
Loop Status	Closed Loop / Open Loop	Closed Loop	Closed Loop	Fuel Trim
Long Term Fuel Trim Average Bank 1	-10% to 10%	0 (± 5%)	0 (± 5%)	Fuel Trim
MAF Sensor (g/sec)	0-512 g/sec	5-9	Varies	Fuel Trim
Power Enrichment	Active / Inactive	Inactive	Inactive	Fuel Trim
Short Term Fuel Trim Average Bank 1	-10% to 10%	0 (± 5%)	0 (± 5%)	Fuel Trim
Startup ECT Sensor	-40 to 304°F	Varies	Varies	Fuel Trim
Startup IAT Sensor	-40 to 304°F	Varies	Varies	Fuel Trim
TP Angle (%)	0-100%	0	Varies	Fuel Trim
TWC Protection	Active / Inactive	Inactive	Inactive	Fuel Trim
Vehicle Speed	0-155 mph	0	30	Fuel Trim

1996-98 Grand Am 3.1L V6 MFI VIN M (A/T) - PID Data List

Parameter Identification (PID)	PID Value Range	PID Value at Hot Idle	PID Value at 30 mph	Data List Type
3X Crank Sensor	0-9999 rpm	Varies	Varies	Misfire
24X Crank Sensor	0-1600 rpm	Varies	Varies	Misfire
Abuse Management	Active / Inactive	Inactive	Inactive	Misfire
Actual EGR Position	0-100%	0%	0-100%	Misfire
Air Active Test	Active / Inactive	Inactive	Inactive	Misfire
CMP Sensor Signal Since Last 3X Signal	Yes / No	Yes	Yes	Misfire
Decel Fuel Mode	Active / Inactive	Inactive	Inactive	Misfire
Desired EGR Position	0-100%	0	0-100	Misfire
ECT Sensor (°F)	-40 to 304°F	185-220	185-220	Misfire
Engine Load	0-100%	2-5	Varies	Misfire
Engine Speed	0-9999 rpm	Varies	---	Misfire
HO2S-11 XCount	0-255	Varies	Varies	Misfire
IAC Position	0-255 counts	10-40	---	Misfire
Ignition 1 Signal	0.0-25.5v	14.2	14.2	Misfire
Injector Pulsewidth	0-985 ms	1.5-3.5	Varies	Misfire
Loop Status	Closed Loop / Open Loop	Closed Loop	Closed Loop	Misfire
MAF Sensor (g/sec)	0-512 g/sec	5-9	Varies	Misfire
Misfire Current Cylinder 1-6	0-198	0-4	0-4	Misfire
Misfire History Cylinder 1-6	0-65535	0	0	Misfire
Misfiring Cylinder	CYL 1-6	0	0	Misfire
Non-Volatile Memory	PASS/FAIL	PASS	PASS	Misfire
Power Enrichment	Active / Inactive	Active	Active	Misfire
Total Misfire Current Count	0-99	0-5	0-5	Misfire
Total Misfire Failures Since First Failure	0-65535	0	0	Misfire
Total Misfire Passes Since First Failure	0-65535	0	0	Misfire
Spark Advance (°)	-64° to 64°	20	Varies	Misfire
TCC Engaged	Yes / No	No	No	Misfire
TP Angle (%)	0-100%	0	Varies	Misfire
Trans. Hot Mode	Active / Inactive	Inactive	Inactive	Misfire
TWC Protection	Active / Inactive	Inactive	Inactive	Misfire
Vehicle Speed	0-155 mph	0	30	Misfire

1999-2005 Grand Am 3.4L V6 MFI VIN E (A/T) - PID Data List

Parameter Identification (PID)	PID Value Range	PID Value at Hot Idle	PID Value at 30 mph	Data List Type
3X Crank Sensor	0-9999 rpm	Varies	Varies	Engine Data 1
24X Crank Sensor	0-1600 rpm	Varies	Varies	Engine Data 1
A/C HI Side Pressure	0-450 psi	139-399	139-399	Engine Data 1
A/C HI Side Pressure	0.0-5.0v	1-3	1-3	Engine Data 1
A/C Pressure Disable	Yes / No	No	No	Engine Data 1
A/C Relay Signal	On / Off	Off	Off	Engine Data 1
A/C Request	Yes / No	No	No	Engine Data 1
Actual EGR Position	0-100%	0	0-100	Engine Data 1
Air Fuel Ratio	0:1-25.5:1	14.6:1	14.6:1	Engine Data 1
AIR Relay Control	On / Off	Off	Off	Engine Data 1
AIR Solenoid Control	On / Off	Off	Off	Engine Data 1
BARO Sensor (kPa)	10-110 kPa	65-110	65-110	Engine Data 1
Commanded A/C	On / Off	Off	Off	Engine Data 1
Commanded Fan 1	On / Off	Off	Off	Engine Data 1
Commanded Fan 2	On / Off	Off	Off	Engine Data 1
Cruise Control Active	On / Off	Off	On	Engine Data 1
Cruise Control Inhibit	Yes / No	Yes	Yes	Engine Data 1
Cruise Inhibit Reason	Several	Several	Several	Engine Data 1
Current Gear	1-2-3-4	1	3	Engine Data 1
Desired EGR Position	0-100%	0	0-100	Engine Data 1
Desired Idle Speed	0-9999 rpm	PCM control	---	Engine Data 1
ECT Sensor (°F)	-40 to 304°F	185-220	185-220	Engine Data 1
Engine Load	0-100%	2-5	Varies	Engine Data 1
Engine Run Time	Hr: Min: Sec	00:00:00	00:00:00	Engine Data 1
Engine Speed	0-9999 rpm	Varies	---	Engine Data 1
EGR Position Sensor	0-100%	0	0-100	Engine Data 1
Engine Load	0-100%	2-5	Varies	Engine Data 1
EVAP Purge Solenoid	0-100%	0-25	0-100	Engine Data 1
EVAP Vent Solenoid	Open/Close	Open	Open	Engine Data 1
Fan - Low Speed	On / Off	Off	Off	Engine Data 1
Fan - High Speed	On / Off	Off	Off	Engine Data 1
Fuel Pump Relay	On / Off	On	On	Engine Data 1
Fuel Pump Status	On / Off	On	On	Engine Data 1
Fuel Trim Cell	Cell #0-4	0	1	Engine Data 1
Fuel Trim Learn	Enabled / Disabled	Enabled	Enabled	Engine Data 1
Generator Lamp	On / Off	Off	Off	Engine Data 1
HO2S-11 (B1 S1)	0-1132 mv	10-1000	10-1000	Engine Data 1
HO2S-12 (B1 S2)	0-1132 mv	10-1000	10-1000	Engine Data 1
HO2S-11 Status	Ready / Not Ready	Ready	Ready	Engine Data 1
HO2S-11 XCount	0-255 counts	Varies	Varies	Engine Data 1
IAC Position	0-255 counts	5-50	---	Engine Data 1
IAT Sensor (°F)	-40 to 304°F	50-194	50-194	Engine Data 1
Ignition 1 Signal	0.0-25.5v	14.2	14.2	Engine Data 1
Ignition Mode	Bypass / IC	IC	IC	Engine Data 1
Injector Pulsewidth	0-985 ms	1.5-3.5	Varies	Engine Data 1
Knock Retard (°)	0.0-25.5°	0	0	Engine Data 1
Long Term Fuel Trim	-10% to 10%	0 (± 5%)	0 (± 5%)	Engine Data 1
Loop Status	Closed Loop / Open Loop	Closed Loop	Closed Loop	Engine Data 1

1999-2005 Grand Am 3.4L V6 MFI VIN E (A/T) - PID Data List

Parameter Identification (PID)	PID Value Range	PID Value at Hot Idle	PID Value at 30 mph	Data List Type
MAF Sensor (g/sec)	0-512 g/sec	3-6	Varies	Engine Data 1
MAF Sensor (Hz)	0-31,999	1200-3000	Varies	Engine Data 1
MAP Sensor (kPa)	10-110 kPa	29-48	Varies	Engine Data 1
MAP Sensor (V)	0.0-5.0v	1-2	Varies	Engine Data 1
Non-Volatile Memory	PASS/FAIL	PASS	PASS	Engine Data 1
No. of Current DTC(s)	Number	0	0	Engine Data 1
PCM Reset	Yes / No	No	No	Engine Data 1
PCM/VCM in VTD Fail	Yes / No	No	No	Engine Data 1
Short Term Fuel Trim	-10% to 10%	0 (± 5%)	0 (± 5%)	Engine Data 1
Spark Advance (º)	-64º to 64º	20	Varies	Engine Data 1
TCC Brake Switch	Apply/REL	Released	Released	Engine Data 1
TCC Engaged	Yes / No	No	No	Engine Data 1
TP Angle (%)	0-100%	0	Varies	Engine Data 1
TP Sensor (V)	0.0-5.0v	0.50-0.90	---	Engine Data 1
Traction Control	Active / Inactive	Inactive	Inactive	Engine Data 1
T/C Desired Torque	0-100%	100%	Varies	Engine Data 1
T/C Command Torque	0-100%	80-90%	Varies	Engine Data 1
Transmission Range	Several	PARK	Several	Engine Data 1
TR Switch A, B, C, P	Low/HIGH	Low	Low/HIGH	Engine Data 1
Vehicle Speed	0-155 mph	0	30	Engine Data 1
VTD Auto Learn Timer	Enabled / Disabled	Disabled	Disabled	Engine Data 1
VTD Fuel Disable	Enabled / Disabled	Disabled	Disabled	Engine Data 1
VTD Fuel Disable Until Key Off	Yes / No	No	No	Engine Data 1

1999-2005 Grand Am 3.4L V6 MFI VIN E (A/T) - PID Data List

Parameter Identification (PID)	PID Value Range	PID Value at Hot Idle	PID Value at 30 mph	Data List Type
1-2, 2-3 Solenoids	Several	Okay	Okay	Output Driver
A/C Relay CKT Status	Several	Okay	Okay	Output Driver
AIR Pump Relay CKT	Several	Okay	Okay	Output Driver
AIR Solenoid Status	Several	Okay	Okay	Output Driver
Cruise Inhibit Circuit Status	Several	Okay	Okay	Output Driver
CYL 1-6 CKT History	Several	Okay	Okay	Output Driver
CYL 1-6 Circuit Status	Several	Okay	Okay	Output Driver
ODD 1-4 Status	Okay / Fault	Okay	Okay	Output Driver
EVAP Purge Solenoid	Several	Okay	Okay	Output Driver
EVAP Vent Solenoid	Several	Okay	Okay	Output Driver
FC Relay 1, 2-3 status	Several	Okay	Okay	Output Driver
F/P Circuit History	Several	Okay	Okay	Output Driver
F/P Circuit Status	Several	Okay	Okay	Output Driver
MIL Circuit Status	Several	Okay	Okay	Okay
Non-Volatile Memory	Pass/Fail	PASS	PASS	PASS
PCM Reset	Yes / No	No	No	No
Starter Enable Circuit	Several	Okay	Okay	Okay
TCC Solenoid	Several	Okay	Okay	Okay

1999-2005 Grand Am 3.4L V6 MFI VIN E (A/T) - PID Data List

Parameter Identification (PID)	PID Value Range	PID Value at Hot Idle	PID Value at 30 mph	Data List Type
3X Crank Sensor	0-9999 rpm	Max: 1700	Varies	Engine Data 2
24X Crank Sensor	0-1600 rpm	Varies	Varies	Engine Data 2
A/C HI Side Pressure	0-450 psi	139-399	139-399	Engine Data 2
A/C Off for WOT	Yes / No	No	No	Engine Data 2
A/C Pressure Disable	Yes / No	No	No	Engine Data 2
A/C Request	Yes / No	No	No	Engine Data 2
A/C Relay Command	On / Off	Off	Off	Engine Data 2
Air Fuel Ratio	0:1-25.5:1	14.6:1	14.6:1	Engine Data 2
CMP Sensor Signal	Yes / No	Yes	Yes	Engine Data 2
Crank Request Signal	Yes / No	No	No	Engine Data 2
Current Gear	0-4	1	3	Engine Data 2
Desired Idle Speed	0-9999 rpm	PCM control	---	Engine Data 2
ECT Sensor (°F)	-40 to 304°F	185-220	185-220	Engine Data 2
Engine Load	0-100%	2-5	Varies	Engine Data 2
Engine Oil Level	Okay / Low	Okay	Okay	Engine Data 2
Engine Oil Life	0-100%	Varies	Varies	Engine Data 2
Engine Oil Pressure	Low/Okay/HI	Okay	Okay	Engine Data 2
Engine Run Time	Hr: Min: Sec	00:00:00	00:00:00	Engine Data 2
Engine Speed	0-9999 rpm	Varies	---	Engine Data 2
FC Relay 1, 2-3	On / Off	Off	Off	Engine Data 2
Generator F-Terminal	0-100%	Varies	Varies	Engine Data 2
Generator L-Terminal	On / Off	On	On	Engine Data 2
IAT Sensor (°F)	-40 to 304°F	50-194	50-194	Engine Data 2
Ignition Mode	IC/Bypass	IC	IC	Engine Data 2
Ignition 1 Signal	0.0-25.5v	14.2	14.2	Engine Data 2
Long Term Fuel Trim	-10% to 10%	0 (± 5%)	0 (± 5%)	Engine Data 2
Loop Status	Closed Loop / Open Loop	Closed Loop	Closed Loop	Engine Data 2
MAF Sensor (g/sec)	0-512 g/sec	3-6	Varies	Engine Data 2
MAF Sensor (Hz)	0-31,999	1200-3000	Varies	Engine Data 2
MAP Sensor (kPa)	10-110 kPa	29-48	Varies	Engine Data 2
MIL (lamp) Command	On / Off	Off	Off	Engine Data 2
PCM Reset	Yes / No	No	No	Engine Data 2
PCM/VCM in VTD Fail	Yes / No	No	No	Engine Data 2
Power Enrichment	Active / Inactive	Inactive	Inactive	Engine Data 2
Short Term Fuel Trim	-10% to 10%	0 (± 5%)	0 (± 5%)	Engine Data 2
Spark Advance (°)	-64° to 64°	20	Varies	Engine Data 2
Startup ECT Sensor	-40 to 304°F	Varies	Varies	Engine Data 2
Startup IAT Sensor	-40 to 304°F	Varies	Varies	Engine Data 2
Starter Relay Control	On / Off	Off	Off	Engine Data 2
TCC Brake Pedal Switch	Applied / Released	Released	Released	Engine Data 2
TCC PWM Solenoid	On / Off	Off	On	Engine Data 2
TFP Switch Signal	Several	PARK	D3	Engine Data 2
Torque Delivered Signal	0-100%	86	Varies	Engine Data 2
Torque Request Signal	0-100%	100	Varies	Engine Data 2
TP Angle (%)	0-100%	0	Varies	Engine Data 2
Vehicle Speed (rpm)	0-255	0	30	Engine Data 2

1999-2005 Grand Am 3.4L V6 MFI VIN E (A/T) - PID Data List

Parameter Identification (PID)	PID Value Range	PID Value at Hot Idle	PID Value at 30 mph	Data List Type
A/C Relay Command	On / Off	Off	Off	EGR
A/C Request Signal	Yes / No	No	No	EGR
BARO Sensor (kPa)	10-110 kPa	65-110	65-110	EGR
BARO Sensor (V)	0.0-5.0v	3.5-4.5	3.5-4.5	EGR
Current Gear	1-2-3-4	1	3	EGR
Decel Fuel Mode	Active / Inactive	Inactive	Inactive	EGR
Desired EGR Position	0-100%	0	0-100	EGR
ECT Sensor (°F)	-38-304°F	185-220	185-220	EGR
EGR Closed Pintle Position	0.0-5.0v	0.14-1.0	---	EGR
EGR Duty Cycle	0-100%	0	0-100	EGR
EGR Feedback	0.0-5.0v	0.14-1.0	---	EGR
EGR Flow Test Count	0-255	0-10	0-10	EGR
EGR Learned Min.	0.0-5.0v	0.14-1.0	---	EGR
EGR Position Sensor	0.0-5.0v	0.14-1.0	---	EGR
EGR Pos. Variance	0-100%	0-9	---	EGR
EGR Solenoid Control	0-100%	Varies	Varies	EGR
Engine Load	0-100%	2-5	Varies	EGR
Engine Run Time	Hr: Min: Sec	00:00:00	00:00:00	EGR
Engine Speed	0-9999 rpm	Varies	---	EGR
IAC Position	0-255 counts	15	Varies	EGR
IAT Sensor (°F)	-40 to 304°F	50-194	50-194	EGR
Knock Retard (°)	0.0-25.5°	0	Varies	EGR
Loop Status	Closed Loop / Open Loop	Closed Loop	Closed Loop	EGR
MAF Sensor (g/sec)	0-512 g/sec	3-6	Varies	EGR
MAP Sensor (kPa)	10-110 kPa	29-48	Varies	EGR
MAP Sensor (V)	0.0-5.0v	1-2	Varies	EGR
Non-Volatile Memory	Pass/Fail	PASS	PASS	EGR
PCM Reset	Yes / No	No	No	EGR
TCC Engaged	Yes / No	No	Yes or No	EGR
TFP Switch Signal	Several	PARK	D3	EGR
TP Angle (%)	0-100%	0	Varies	EGR
Transmission Range	P/N-R-4-3-2	P	3	EGR
TWC Protection	Active / Inactive	Inactive	Inactive	EGR

1999-2005 Grand Am 3.4L V6 MFI VIN E (A/T) - PID Data List

Parameter Identification (PID)	PID Value Range	PID Value at Hot Idle	PID Value at 30 mph	Data List Type
Command Generator	On / Off	On	On	IPC
Engine Oil Level	Okay / Low	Okay	Okay	IPC
Engine Oil Life	0-100%	Varies	Varies	IPC
Engine Oil Pressure	Low/Okay/HI	Okay	Okay	IPC
Fuel Level	0-100%	Varies	Varies	IPC
F/P Circuit History	Several	Okay	Okay	IPC
F/P Circuit Status	Several	Okay	Okay	IPC
Generator PWM	0-100%	Varies	Varies	IPC
Ignition 1 Signal	0.0-25.5v	14.2	14.2	IPC
MIL Status	Several	Okay	Okay	IPC

1999-2005 Grand Am 3.4L V6 MFI VIN E (A/T) - PID Data List

Parameter Identification (PID)	PID Value Range	PID Value at Hot Idle	PID Value at 30 mph	Data List Type
BARO Sensor (kPa)	10-110 kPa	65-110	65-110	EVAP
ECT Sensor (°F)	-40 to 304°F	185-220	185-220	EVAP
Engine Load	0-100%	2-5	Varies	EVAP
Engine Speed	0-9999 rpm	Varies	---	EVAP
EVAP Canister Purge	0-100%	0-25	0-50	EVAP
EVAP Fault History	Several	No Fault	No Fault	EVAP
EVAP Vent Solenoid	Open/Close	Closed	Closed	EVAP
Fuel Level (%)	0-100%	Varies	Varies	EVAP
FTP Sensor (In. H2O)	-17.5 to 7.5"	Varies	Varies	EVAP
Fuel Trim Learn	Enabled / Disabled	Enabled	Enabled	EVAP
IAT Sensor (°F)	-40 to 304°F	50-194	50-194	EVAP
Long Term Fuel Trim	-10% to 10%	0 (± 5%)	0 (± 5%)	EVAP
Loop Status	Closed Loop / Open Loop	Closed Loop	Closed Loop	EVAP
MAF Sensor (g/sec)	0-512 g/sec	3-6	Varies	EVAP
MAP Sensor (kPa)	10-110 kPa	29-48	Varies	EVAP
Non-Volatile Memory	Pass/Fail	PASS	PASS	EVAP
Startup ECT Sensor	-40 to 304°F	Varies	Varies	EVAP
Startup IAT Sensor	-40 to 304°F	Varies	Varies	EVAP
TP Angle (%)	0-100%	0	Varies	EVAP

1999-2005 Grand Am 3.4L V6 MFI VIN E (A/T) - PID Data List

Parameter Identification (PID)	PID Value Range	PID Value at Hot Idle	PID Value at 30 mph	Data List Type
Air Fuel Ratio	0:1-25.5:1	14.6:1	14.6:1	Fuel Trim
AIR Relay Control	On / Off	Off	Off	Fuel Trim
AIR Solenoid Control	On / Off	Off	Off	Fuel Trim
BARO Sensor (kPa)	10-110 kPa	65-110	65-110	Fuel Trim
Current Gear	1-2-3-4	1	3	Fuel Trim
Decel Fuel Mode	Active / Inactive	Inactive	Inactive	Fuel Trim
ECT Sensor (°F)	-40 to 304°F	185-220	185-220	Fuel Trim
Engine Speed	0-9999 rpm	Varies	---	Fuel Trim
EVAP Purge Solenoid	0-100%	0-25	0-50	Fuel Trim
EVAP Vent Solenoid	Open/Close	Open	Open	Fuel Trim
Hot Open Loop	Active / Inactive	Inactive	Inactive	Fuel Trim
HO2S-11 Status	Not Ready / Ready	Ready	Ready	Fuel Trim
HO2S (B1 S1, B1 S2)	0-1132 mv	10-1000	10-1000	Fuel Trim
HO2S-11 XCount	0-255 counts	Varies	Varies	Fuel Trim
IAT Sensor (°F)	-40 to 304°F	50-194	50-194	Fuel Trim
Injector Pulsewidth	0-985 ms	1.5-3.5	Varies	Fuel Trim
Long Term Fuel Trim	-10% to 10%	0 (± 5%)	0 (± 5%)	Fuel Trim
Loop Status	Closed Loop / Open Loop	Closed Loop	Closed Loop	Fuel Trim
MAF Sensor (g/sec)	0-512 g/sec	3-6	Varies	Fuel Trim
MAP Sensor (kPa)	10-110 kPa	29-48	Varies	Fuel Trim
Power Enrichment	Active / Inactive	Inactive	Inactive	Fuel Trim
Short Term Fuel Trim	-10% to 10%	0 (± 5%)	0 (± 5%)	Fuel Trim
Startup ECT Sensor	-40 to 304°F	Varies	Varies	Fuel Trim
Startup IAT Sensor	-40 to 304°F	Varies	Varies	Fuel Trim
TP Angle (%)	0-100%	0	Varies	Fuel Trim

1999-2005 Grand Am 3.4L V6 MFI VIN E (A/T) - PID Data List

Parameter Identification (PID)	PID Value Range	PID Value at Hot Idle	PID Value at 30 mph	Data List Type
3X Crank Sensor	0-9999 rpm	Varies	Varies	Misfire
24X Crank Sensor	0-1600 rpm	Varies	Varies	Misfire
Abuse Management	Active / Inactive	Inactive	Inactive	Misfire
Actual EGR Position	0-100%	0%	0-100%	Misfire
A/C Request Control	On / Off	Off	Off	Misfire
Cam Signal Present	Yes / No	Yes	Yes	Misfire
Commanded A/C	On / Off	Off	Off	Misfire
Cycles of Misfire Data	0-99	Varies	Varies	Misfire
Decel Fuel Mode	Active / Inactive	Inactive	Inactive	Misfire
Desired EGR Position	0-100%	0	0-100	Misfire
EGR Position Sensor	0-5.0v	0.14-1.0	Varies	Misfire
ECT Sensor (°F)	-40 to 304°F	185-220	185-220	Misfire
Engine Load	0-100%	2-5	Varies	Misfire
Engine Run Time	Hr: Min: Sec	00:00:00	00:00:00	Misfire
Engine Speed	0-9999 rpm	Varies	---	Misfire
HO2S-11 (B1 S1)	0-1132 mv	10-1000	10-1000	Misfire
Hot Mode	On / Off	Off	Off	Misfire
Ignition 1 Signal	0.0-25.5v	14.2	14.2	Misfire
Injector Pulsewidth	0-985 ms	1.5-3.5	Varies	Misfire
Long Term Fuel Trim	-10% to 10%	0 (± 5%)	0 (± 5%)	Misfire
MAF Sensor (g/sec)	0-512 g/sec	3-6	Varies	Misfire
MAP Sensor (kPa)	10-110 kPa	29-48	Varies	Misfire
MAP Sensor (V)	0.0-5.0v	1-2	Varies	Misfire
Misfire Current Cylinder 1-6	0-198	0-4	0-4	Misfire
Misfire History Cylinder 1-6	0-65535	0	0	Misfire
PCM Reset	Yes / No	No	No	Misfire
Power Enrichment	Active / Inactive	Active	Active	Misfire
Short Term Fuel Trim	-10% to 10%	0 (± 5%)	0 (± 5%)	Misfire
Spark Advance (°)	-64° to 64°	20	Varies	Misfire
TCC Engaged	Yes / No	No	No	Misfire
TCC PWM Solenoid	On / Off	Off	On	Misfire
Total Misfire Current Count	0-99	0-5	0-5	Misfire
Total Misfire Failures Since First Failure	0-65535	0	0	Misfire
Total Misfire Passes Since First Failure	0-65535	0	0	Misfire
TP Angle (%)	0-100%	0	Varies	Misfire
Traction Control	Active / Inactive	Inactive	Inactive	Misfire
Vehicle Speed	0-155 mph	0	30	Misfire

1999-2005 Grand Am 3.4L V6 MFI VIN E (A/T) - PID Data List

Parameter Identification (PID)	PID Value Range	PID Value at Hot Idle	PID Value at 30 mph	Data List Type
Air Fuel Ratio	0:1-25.5:1	14.6:1	14.6:1	AIR
AIR Pump Relay	On / Off	Off	Off	AIR
AIR Solenoid	On / Off	Off	Off	AIR
AIR Active Test Injection	Yes / No	No	No	AIR
AIR Active Test Inhibit	Yes / No	Yes	Yes	AIR
AIR Diagnostic Complete	Yes / No	Yes	Yes	AIR
AIR Passive Test 1 Pass	Yes / No	Yes	Yes	AIR
AIR Passive Test 1 Fail	Yes / No	No	No	AIR
AIR Passive Test Passed	Yes / No	Yes	Yes	AIR
AIR Pass Test in progress	Yes / No	No	No	AIR
AIR Pass Test Inhibited	Yes / No	Yes	Yes	AIR
ECT Sensor (°F)	-40 to 304°F	185-220	185-220	AIR
Engine Load	0-100%	2-5	Varies	AIR
Engine Run Time	Hr: Min: Sec	00:00:00	00:00:00	AIR
Fuel Trim Cell	Cell #0-4	1	3	AIR
HO2S-11 (B1 S1)	0-1132 mv	10-1000	10-1000	AIR
IAT Sensor (°F)	-40 to 304°F	50-194	50-194	AIR
Loop Status	Closed Loop / Open Loop	Closed Loop	Closed Loop	AIR
Long Term Fuel Trim	-10% to 10%	0 (± 5%)	0 (± 5%)	AIR
MAF Sensor (g/sec)	0-512 g/sec	3-6	Varies	AIR
Non-Volatile Memory	PASS/FAIL	PASS	PASS	AIR
Short Term Fuel Trim	-10% to 10%	0 (± 5%)	0 (± 5%)	AIR
Startup ECT Sensor	-40 to 304°F	Varies	Varies	AIR
Startup IAT Sensor	-40 to 304°F	Varies	Varies	AIR
TP Angle (%)	0-100%	0	Varies	AIR

1999-2005 Grand Am 3.4L V6 MFI VIN E (A/T) - PID Data List

Parameter Identification (PID)	PID Value Range	PID Value at Hot Idle	PID Value at 30 mph	Data List Type
Air Fuel Ratio	0:1-25.5:1	14.6:1	14.6:1	Catalyst
Current Gear	1-2-3-4	1	3	Catalyst
Commanded A/C	On / Off	Off	Off	Catalyst
Engine Load	0-100%	2-5	Varies	Catalyst
Engine Run Time	Hr: Min: Sec	00:00:00	00:00:00	Catalyst
Engine Speed	0-9999 rpm	Varies	---	Catalyst
FC Relay Control	On / Off	Off	Off	Catalyst
Fuel Trim Cell	Cell # 0-4	1	3	Catalyst
Fuel Trim Learn	Enabled / Disabled	Enabled	Enabled	Catalyst
HO2S (B1 S1)	0-1132 mv	10-1000	10-1000	Catalyst
HO2S (B1 S2)	0-1132 mv	10-1000	10-1000	Catalyst
HO2S-11 Status	Ready / Not Ready	Ready	Ready	Catalyst
Hot Open Loop	Active / Inactive	Inactive	Inactive	Catalyst
IAT Sensor (°F)	-40 to 304°F	50-194	50-194	Catalyst
Injector Pulsewidth	0-985 ms	1.5-3.5	Varies	Catalyst
MAF Sensor (g/sec)	0-512 g/sec	3-6	Varies	Catalyst
MAP Sensor (kPa)	10-110 kPa	29-48	Varies	Catalyst
Startup ECT Sensor	-40 to 304°F	Varies	Varies	Catalyst
TP Angle (%)	0-100%	0	Varies	Catalyst

GRAND PRIX PID DATA

1995 Grand Prix 3.1L V6 MFI VIN M (A/T) - PID Data List

Parameter Identification (PID)	PID Value Range	PID Value at Hot Idle	PID Value at 30 mph	Data List Type
2nd Gear Start	On / Off	On	Off	Engine Data 1
2nd-3rd-4th Gear Switch	Yes / No	No	Yes / No	Engine Data 1
A/C Clutch	On / Off	Off	Off	Engine Data 1
A/C Pressure	0-450 psi	Varies	Varies	Engine Data 1
A/C Request	Yes / No	No	Off	Engine Data 1
Calc. Engine Load	0-100%	50-60	Varies	Engine Data 1
Camshaft Signal	0 or 1	0-1-0-1	0-1-0-1	Engine Data 1
Commanded Gear	1-2-3-4	1	3	Engine Data 1
Cranking RPM	0-999	N/A	N/A	Engine Data 1
Cruise Control	On / Off	Off	Off	Engine Data 1
Current Weak Cyl	CYL# / None	None	None	Engine Data 1
Desired Idle Speed	0-3187	800	---	Engine Data 1
ECT Sensor (°F)	-40 to 419°F	220	223	Engine Data 1
EGR Solenoid 1, 2	On / Off	Off	On	Engine Data 1
EGR Solenoid 3	On / Off	Off	On	Engine Data 1
Engine Speed	0-9999 rpm	Varies	---	Engine Data 1
Fan Relay1, 2	On / Off	Off	Off	Engine Data 1
Fuel Trim Cell	0-10	0 or 1	0 or 2	Engine Data 1
HO2S Cross Counts	0-255	0-20	0-20	Engine Data 1
HO2S-11 (B1 S1)	0-1132 mv	10-1000	10-1000	Engine Data 1
HO2S-12 (B1 S2)	0-1132 mv	10-1000	10-1000	Engine Data 1
IAC Motor	0-255 counts	5-50	---	Engine Data 1
IAT Sensor (°F)	-40 to 304°F	98	98	Engine Data 1
Injector Pulsewidth	0-985 ms	3.1	3.0	Engine Data 1
Knock Retard (°)	0-128°	0	0	Engine Data 1
Knock Signal	Yes / No	No	No	Engine Data 1
Long Term Fuel Trim	0-255 counts	130	130	Engine Data 1
Loop Status	Closed Loop / Open Loop	Closed Loop	Closed Loop	Engine Data 1
Low Oil Light	On / Off	On: If low	Off	Engine Data 1
MAF Sensor (g/sec)	0-512 g/sec	6	9.8	Engine Data 1
MAP Sensor (kPa)	10-110 kPa	35	45	Engine Data 1
Quad Driver 1	HI or Low	HI	Low	Engine Data 1
Quad Driver 2	HI or Low	HI	Low	Engine Data 1
PSP Switch	HI/Normal	Normal	Normal	Engine Data 1
PRNDL 'A', 'B', 'C', 'P'	HI or Low	HI	Low	Engine Data 1
Purge Duty Cycle	0-100%	0	20	Engine Data 1
Short Term Fuel Trim	0-255 counts	127	125	Engine Data 1
Spark Advance (°)	-90° to 90°	13	30-34	Engine Data 1
TCC Solenoid	On / Off	Off	Off	Engine Data 1
TCC Brake Switch	Applied / Released	Released	Released	Engine Data 1
TCC Slip Speed	-255/+255	730-768	5	Engine Data 1
TCC Duty Cycle	0-100%	0%	0%	Engine Data 1
Throttle Angle (%)	0-100%	0%	4%	Engine Data 1
Time From Startup	Hr: Min: Sec	00:00:00	00:00:00	Engine Data 1
TCC System	On / Off	Off	Off	Engine Data 1
TP Sensor (V)	0-5.0v	0.54	0.74	Engine Data 1
TFT Sensor (°F)	-40 to 304°F	125	125	Engine Data 1
Trans. Range Switch	P, A, B, C	PARK	DRIVE 3	Engine Data 1

1996-99 Grand Prix 3.1L V6 MFI VIN M (A/T) - Engine Data List

Parameter Identification (PID)	PID Value Range	PID Value at Hot Idle	PID Value at 30 mph	Data List Type
3X Crank Sensor	0-9999 rpm	Varies	Varies	Engine Data 1
24X Crank Sensor	0-1600 rpm	Varies	Varies	Engine Data 1
A/C HI Side Pressure	0-450 psi	139-399	139-399	Engine Data 1
A/C HI Side Pressure	0.0-5.0v	1-3	1-3	Engine Data 1
A/C Press. Too HI/LO	Yes / No	No	No	Engine Data 1
A/C Pressure Disable	Yes / No	No	No	Engine Data 1
A/C Clutch	Yes / No	No	No	Engine Data 1
A/C Request	Yes / No	No	No	Engine Data 1
A/C Off for WOT	Yes / No	No	No	Engine Data 1
A/C Slugging	Active / Inactive	Inactive	Inactive	Engine Data 1
Actual EGR Position	0-100%	0	0-100	Engine Data 1
Air Fuel Ratio	0:1-25.5:1	14.6:1	14.6:1	Engine Data 1
AIR Pump Control	On / Off	On: Cold	Off	Engine Data 1
BARO Sensor (kPa)	10-110 kPa	65-110	65-110	Engine Data 1
BARO Sensor (V)	0.0-5.0v	3.5-4.5	3.5-4.5	Engine Data 1
Brake Switch	Applied / Released	Released	Released	Engine Data 1
Cam Signal Present	Yes / No	Yes	Yes	Engine Data 1
Commanded A/C	On / Off	Off	Off	Engine Data 1
Commanded Fan 1, 2	On / Off	Off	Off	Engine Data 1
Cruise Engaged	Disengaged / Engaged	Disabled	Disabled	Engine Data 1
Cruise Inhibit	Yes / No	Yes	Yes	Engine Data 1
Cruise Mode	On / Off	Off	On	Engine Data 1
Current Gear	1-2-3-4	1	3	Engine Data 1
Decel Fuel Mode	Active / Inactive	Inactive	Inactive	Engine Data 1
Desired EGR Position	0-100%	0	0-100	Engine Data 1
Desired Idle Speed	0-3187 rpm	PCM control	---	Engine Data 1
ECT Sensor (°F)	-40 to 304°F	185-220	185-220	Engine Data 1
Engine Load	0-100%	2-5	Varies	Engine Data 1
Engine Run Time	Hr: Min: Sec	00:00:00	00:00:00	Engine Data 1
Engine Speed	0-9999 rpm	Varies	---	Engine Data 1
EGR Closed Pintle Position	0.0-5.0v	0.14-1.0	---	Engine Data 1
EGR Duty Cycle	0-100%	0	0-100	Engine Data 1
EGR Feedback	0.0-5.0v	0.14-1.0	---	Engine Data 1
EVAP Purge PWM	0-100%	0-25%	0-100%	Engine Data 1
EVAP Vacuum Switch	Purge/No	No Purge	Purge	Engine Data 1
EVAP Canister Purge	0-100%	0	0-50	Engine Data 1
EVAP Vent Solenoid	Open/Close	Open	Open	Engine Data 1
Fan - Low Speed	On / Off	Off	Off	Engine Data 1
Fan - High Speed	On / Off	Off	Off	Engine Data 1
Fuel Pump Control	On / Off	On	On	Engine Data 1
Fuel Trim Cell	Cell #	0	1	Engine Data 1
Fuel Trim Learn	Enabled / Disabled	Enabled	Enabled	Engine Data 1
Fuel Pump	On / Off	On	On	Engine Data 1
Generator Control	On / Off	On	On	Engine Data 1
HO2S-11 Status	Not Ready / Ready	Ready	Ready	Engine Data 1
HO2S-11 (B1 S1)	0-1132 mv	10-1000	10-1000	Engine Data 1
HO2S-12 (B1 S2)	0-1132 mv	10-1000	10-1000	Engine Data 1

1996-99 Grand Prix 3.1L V6 MFI VIN M (A/T) - PID Data List

Parameter Identification (PID)	PID Value Range	PID Value at Hot Idle	PID Value at 30 mph	Data List Type
IAC Position	0-255 counts	10-40	---	Engine Data 1
IAT Sensor (°F)	-40 to 304°F	50-194	50-194	Engine Data 1
IC Mode	Active / Inactive	Active	Active	Engine Data 1
Idle Speed Error	0-9999 rpm	PCM control	---	Engine Data 1
Ignition Bypass	Active / Inactive	Inactive	Inactive	Engine Data 1
Ignition Mode	Bypass / IC	IC	IC	Engine Data 1
Injector Fault	Yes / No	No	No	Engine Data 1
Injector Pulsewidth	0-985 ms	1.5-3.5	Varies	Engine Data 1
Injector Status	Okay / STUCK	Okay	Okay	Engine Data 1
Knock Retard (°)	0.0-25.5°	0	0	Engine Data 1
KS Active Counter	0-255	Varies	Varies	Engine Data 1
KS Noise Channel	0.0-5.0v	0.1-3.0	Varies	Engine Data 1
Long Term Fuel Trim	-10% to 10%	0 (± 5%)	0 (± 5%)	Engine Data 1
Loop Status	Closed Loop / Open Loop	Closed Loop	Closed Loop	Engine Data 1
MAF Sensor (g/sec)	0-512 g/sec	3-6	Varies	Engine Data 1
MAF Sensor (Hz)	0-31,999	1200-3000	Varies	Engine Data 1
MAP Sensor (kPa)	10-110 kPa	29-48	Varies	Engine Data 1
MAP Sensor (V)	0.0-5.0v	1-2v	Varies	Engine Data 1
MIL Status	On / Off	Off	Off	Engine Data 1
Non-Volatile Memory	PASS/FAIL	PASS	PASS	Engine Data 1
Power Enrichment	Active / Inactive	Inactive	Inactive	Engine Data 1
Short Term Fuel Trim	-10% to 10%	0 (± 5%)	0 (± 5%)	Engine Data 1
Spark Advance (°)	-64° to 64°	20	Varies	Engine Data 1
TCC Brake Switch	Applied / Released	Released	Released	Engine Data 1
TCC Engaged	Yes / No	No	No	Engine Data 1
TP Angle (%)	0-100%	0	Varies	Engine Data 1
TP Sensor (V)	0.0-5.0v	0.20-0.74	Varies	Engine Data 1
Trans. Hot Mode	Active / Inactive	Inactive	Inactive	Engine Data 1
Transmission Range	X / O	O/X/X/O	Varies	Engine Data 1
Transmission Range	P/N-R-4-3-2	P	3	Engine Data 1
TR Switch P/A/B/C	HI/LO	LO/LO/HI/HI	LO/HI/HI/LO	Engine Data 1
TWC Protection	Active / Inactive	Inactive	Inactive	Engine Data 1
Vehicle Speed	0-155 mph	0	30	Engine Data 1
VTD Fuel Disable	Enabled / Disabled	Disabled	Disabled	Engine Data 1

1996-99 Grand Prix 3.1L V6 MFI VIN M (A/T) - PID Data List

Parameter Identification (PID)	PID Value Range	PID Value at Hot Idle	PID Value at 30 mph	Data List Type
Engine Run Time	Hr: Min: Sec	00:00:00	00:00:00	EVAP
Engine Speed	0-9999 rpm	Varies	---	EVAP
EVAP Canister Purge	0-100%	0-25	0-100	EVAP
EVAP Vent Solenoid	Open/Close	Open	Open	EVAP
Fuel Tank Pressure	0.0-5.0v	2.5v: cap off	Varies	EVAP
IAT Sensor (°F)	-40 to 304°F	50-194	50-194	EVAP
MAF Sensor (g/sec)	0-512 g/sec	3-6	Varies	EVAP
MAP Sensor (kPa)	10-110 kPa	29-48	Varies	EVAP
Startup ECT Sensor	-40 to 304°F	Varies	Varies	EVAP
Startup IAT Sensor	-40 to 304°F	Varies	Varies	EVAP

1996-99 Grand Prix 3.1L V6 MFI VIN M (A/T) - PID Data List

Parameter Identification (PID)	PID Value Range	PID Value at Hot Idle	PID Value at 30 mph	Data List Type
Actual EGR Position	0-100%	0	0-100	EGR
BARO Sensor (kPa)	10-110 kPa	65-110	65-110	EGR
BARO Sensor (V)	0.0-5.0v	3.5-4.5	3.5-4.5	EGR
Current Gear	1-2-3-4	1	3	EGR
Decel Fuel Mode	Active / Inactive	Inactive	Inactive	EGR
Desired EGR Position	0-100%	0%	0-100%	EGR
ECT Sensor (°F)	-38-304°F	185-220	185-220	EGR
EGR Closed Pintle	0.0-5.0v	0.14-1.0	---	EGR
EGR Duty Cycle	0-100%	0	0-100	EGR
EGR Feedback	0.0-5.0v	0.14-1.0	---	EGR
EGR Flow Test Count	0-255	0-10	0-10	EGR
EGR Position Error	0-100%	0-9	---	EGR
Engine Load	0-100%	2-5	Varies	EGR
Engine Run Time	Hr: Min: Sec	00:00:00	00:00:00	EGR
Engine Speed	0-9999 rpm	Varies	---	EGR
Hot Open Loop	Active / Inactive	Inactive	Inactive	EGR
IAC Position	0-255	10-40	---	EGR
IAT Sensor (°F)	-40 to 304°F	50-194	50-194	EGR
Knock Retard	0.0-25.5%	0	Varies	EGR
KS Active Counter	0-255	Varies	Varies	EGR
Loop Status	Closed Loop / Open Loop	Closed Loop	Closed Loop	EGR
MAF Sensor (g/sec)	0-512 g/sec	3-6	Varies	EGR
MAP Sensor (kPa)	10-110 kPa	29-48	Varies	EGR
MAP Sensor (V)	0.0-5.0v	1-2	Varies	EGR
Power Enrichment	Active / Inactive	Inactive	Inactive	EGR
Rich/Lean Status	L or R	L-R-L-R	L-R-L-R	EGR
TCC Engaged	Yes / No	No	Yes or No	EGR
TP Angle (%)	0-100%	0	0-100	EGR
Transmission Range	X / O	O/X/X/O	Varies	EGR
Transmission Range	P/N-R-4-3-2	P	3	EGR
TWC Protection	Active / Inactive	Inactive	Inactive	EGR
Vehicle Speed	0-155 mph	0	30	EGR

1996-99 Grand Prix 3.1L V6 MFI VIN M (A/T) - HO2S Data List

Parameter Identification (PID)	PID Value Range	PID Value at Hot Idle	PID Value at 30 mph	Data List Type
Air Fuel Ratio	0:1-25.5:1	14.6:1	14.6:1	HO2S
AIR Relay Control	On / Off	On: Cold	Off	HO2S
AIR Pump Solenoid	On / Off	On: Cold	Off	HO2S
Decel Fuel Mode	Active / Inactive	Inactive	Inactive	HO2S
ECT Sensor (°F)	-40 to 304°F	185-220	185-220	HO2S
Engine Load	0-100%	2-5	Varies	HO2S
Engine Run Time	Hr: Min: Sec	00:00:00	00:00:00	HO2S
Engine Speed	0-9999 rpm	Varies	---	HO2S
EVAP Purge PWM	0-100%	0-25	0-50	HO2S
Hot Open Loop	Active / Inactive	Inactive	Inactive	HO2S
HO2S-11 Status	Ready / Not Ready	Ready	Ready	HO2S
HO2S -11, HO2S-12	0-1132 mv	10-1000	10-1000	HO2S

1996-99 Grand Prix 3.1L V6 MFI VIN M (A/T) - PID Data List

Parameter Identification (PID)	PID Value Range	PID Value at Hot Idle	PID Value at 30 mph	Data List Type
HO2S-11 Warmup	Min: Sec	00:00	00:00	HO2S
HO2S-12 Warmup	Min: Sec	00:00	00:00	HO2S
HO2S-11 XCount	0-255 counts	Varies	Varies	HO2S
IAT Sensor (°F)	-40 to 304°F	50-194	50-194	HO2S
Injector Pulsewidth	0-985 ms	1.5-3.5	Varies	HO2S
Loop Status	Closed Loop / Open Loop	Closed Loop	Closed Loop	HO2S
Long Term Fuel Trim Average	-10% to 10%	0 (± 5%)	0 (± 5%)	HO2S
MAF Sensor (g/sec)	0-512 g/sec	5-9	Varies	HO2S
Power Enrichment	Active / Inactive	Inactive	Inactive	HO2S
Short Term F/T Avg.	-10% to 10%	0 (± 5%)	0 (± 5%)	HO2S
Startup ECT Sensor	-40 to 304°F	Varies	Varies	HO2S
Startup IAT Sensor	-40 to 304°F	Varies	Varies	HO2S
TP Angle (%)	0-100%	0	Varies	HO2S
TWC Protection	Active / Inactive	Inactive	Inactive	HO2S

1996-99 Grand Prix 3.1L V6 MFI VIN M (A/T) - PID Data List

Parameter Identification (PID)	PID Value Range	PID Value at Hot Idle	PID Value at 30 mph	Data List Type
3X Crank Sensor	1600-9999	Varies	Varies	Misfire
24X Crank Sensor	0-1600 rpm	Varies	Varies	Misfire
Abuse Management	Active / Inactive	Inactive	Inactive	Misfire
Actual EGR Position	0-100%	0%	0-100%	Misfire
Air Active Test	Active / Inactive	Inactive	Inactive	Misfire
Cam Signal Detected	Yes / No	Yes	Yes	Misfire
Decel Fuel Mode	Active / Inactive	Inactive	Inactive	Misfire
Desired EGR Position	0-100%	0%	0-100%	Misfire
ECT Sensor (°F)	-40 to 304°F	185-220	185-220	Misfire
Engine Load	0-100%	2-5	Varies	Misfire
Engine Speed	0-9999 rpm	Varies	---	Misfire
HO2S-11 XCount	0-255 counts	Varies	Varies	Misfire
IAC Position	0-255 counts	10-40	---	Misfire
Ignition 1 Volts	0.0-25.5v	14.2	14.2	Misfire
Injector Pulsewidth	0-985 ms	1.5-3.5	Varies	Misfire
Loop Status	Closed Loop / Open Loop	Closed Loop	Closed Loop	Misfire
MAF Sensor (g/sec)	0-512 g/sec	5-9	Varies	Misfire
Misfire current Cyl 1-6	0-198	0-4	0-4	Misfire
Misfire History Cyl 1-6	0-65535	0	0	Misfire
Misfiring Cylinder	CYL 1-6	0	0	Misfire
Non-Volatile Memory	PASS/FAIL	PASS	PASS	Misfire
Power Enrichment	Active / Inactive	Inactive	Inactive	Misfire
Total Misfire Current	0-99 counts	0-5	0-5	Misfire
Misfire Since 1st Fail	0-65535	0	0	Misfire
Passes Since 1st Fail	0-65535	0	0	Misfire
Spark Advance (°)	-64° to 64°	20	Varies	Misfire
TCC Engaged	Yes / No	No	No	Misfire
TP Angle (%)	0-100%	0	Varies	Misfire
Trans. Hot Mode	Active / Inactive	Inactive	Inactive	Misfire
TWC Protection	Active / Inactive	Inactive	Inactive	Misfire

2000-05 Grand Prix 3.1L V6 MFI VIN J (A/T) - PID Data List

Parameter Identification (PID)	PID Value Range	PID Value at Hot Idle	PID Value at 30 mph	Data List Type
24X Crank Sensor	0-1600 rpm	Varies	Varies	Engine Data 1
A/C HI Side Pressure	0-5v	1-4	1-4	Engine Data 1
A/C Off for WOT	Yes / No	No	No	Engine Data 1
A/C Pressure Disable	Yes / No	No	No	Engine Data 1
A/C Request Signal	Yes / No	No	No	Engine Data 1
Actual EGR Position	0-100%	0	0-100	Engine Data 1
AIR Relay Command	On / Off	Off	Off	Engine Data 1
AIR Solenoid Control	On / Off	Off	Off	Engine Data 1
BARO Sensor (kPa)	10-110 kPa	65-110	65-110	Engine Data 1
Brake Switch	Applied/RE	Released	Released	Engine Data 1
Cam Signal Present	Yes / No	Yes	Yes	Engine Data 1
Commanded A/C	On / Off	Off	Off	Engine Data 1
Commanded Starter	Enabled / Disabled	Disabled	Disabled	Engine Data 1
Cruise Active Mode	On / Off	Off	Off	Engine Data 1
Cruise Inhibit Signal	On / Off	On	On	Engine Data 1
Current Gear	1-2-3-4	1	3	Engine Data 1
Desired EGR Position	0-100%	0	0-100	Engine Data 1
Desired Idle Speed	0-3187 rpm	PCM control	----	Engine Data 1
ECT Sensor (°F)	-40 to 304°F	185-220	185-220	Engine Data 1
EGR Duty Cycle	0-100%	0	0-100	Engine Data 1
EGR Position Sensor	0-100%	0	0-100	Engine Data 1
Engine Load	0-100%	2-10	Varies	Engine Data 1
Engine Run Time	Hr: Min: Sec	00:00:00	00:00:00	Engine Data 1
Engine Speed	0-9999 rpm	Varies	----	Engine Data 1
EVAP Purge Solenoid	0-100%	0-25	0-100	Engine Data 1
EVAP Vent Solenoid	Open/Close	Open	Open	Engine Data 1
FTP Sensor (In. H2O)	-17.5 to 7.5"	Varies	Varies	Engine Data 1
Fuel Pump Command	On / Off	On	On	Engine Data 1
FC Relay Low, High	On / Off	Off	Off	Engine Data 1
F/P Relay Control	On / Off	On	On	Engine Data 1
Fuel Trim Cell	Cell #0-4	0	0-4	Engine Data 1
Fuel Trim Learn	Enabled / Disabled	Enabled	Enabled	Engine Data 1
HO2S-11 (B1 S1)	0-1132 mv	10-1000	10-1000	Engine Data 1
HO2S-12 (B1 S2)	0-1132 mv	10-1000	10-1000	Engine Data 1
HO2S-11 Status	Not / Ready	Ready	Ready	Engine Data 1
HO2S-11 XCount	0-255 counts	Varies	Varies	Engine Data 1
IAC Position	0-255 counts	10-60	----	Engine Data 1
IAT Sensor (°F)	-40 to 304°F	50-194	50-194	Engine Data 1
Ignition 1 Signal	0.0-25.5v	14.2	14.2	Engine Data 1
Ignition Mode	Bypass / IC	IC	IC	Engine Data 1
Injector Pulsewidth	0-985 ms	1.5-3.5	Varies	Engine Data 1
Knock Retard (°)	0-128°	0	0	Engine Data 1
Long Term Fuel Trim	-10% to 10%	0 (± 5%)	0 (± 5%)	Engine Data 1
Loop Status	Closed Loop / Open Loop	Closed Loop	Closed Loop	Engine Data 1
MAF Sensor (g/sec)	0-512 g/sec	3-6	Varies	Varies
MAF Sensor (Hz)	0-31,999hz	1200-3000	Varies	Varies
MAP Sensor (kPa)	10-110 kPa	20-48	Varies	Varies
No. of Current DTC(s)	Number	00000	00000	00000

2000-05 Grand Prix 3.1L V6 MFI VIN J (A/T) - PID Data List

Parameter Identification (PID)	PID Value Range	PID Value at Hot Idle	PID Value at 30 mph	Data List Type
Power Enrichment	Active / Inactive	Inactive	Inactive	Engine Data 1
Short Term Fuel Trim	-10% to 10%	0 (± 5%)	0 (± 5%)	Engine Data 1
Spark Advance (°)	-64° to 64°	20	Varies	Engine Data 1
TCC Engaged	Engage/DIS	Disengaged	Disengaged	Engine Data 1
TP Angle (%)	0-100%	0	Varies	Engine Data 1
Traction Control Mode	Active / Inactive	Inactive	Inactive	Engine Data 1
T/Control Desired	0-100%	100	0-100	Engine Data 1
T/Control Requested	0-100%	80-90	0-100	Engine Data 1
Transmission Range	P-R-N-4-3-2	P	3	Engine Data 1
TR Switch P-A-B-C	Several	Several	Several	Engine Data 1
Vehicle Speed	0-155 mph	0	30	Engine Data 1
VTD Fuel Disable	Active / Inactive	Inactive	Inactive	Engine Data 1

2000-05 Grand Prix 3.1L V6 MFI VIN J (A/T) - PID Data List

Parameter Identification (PID)	PID Value Range	PID Value at Hot Idle	PID Value at 30 mph	Data List Type
3X Crank Sensor	1600-9999	Varies	Varies	Engine Data 2
24X Crank Sensor	0-1600 rpm	Varies	Varies	Engine Data 2
Air Fuel Ratio	0:1-25.5:1	14.6:1	14.6:1	Engine Data 2
A/C HI Side Pressure	0-5v	1-4	1-4	Engine Data 2
A/C Off for WOT	Yes / No	No	No	Engine Data 2
A/C Pressure Disable	Yes / No	No	No	Engine Data 2
A/C Request Signal	Yes / No	No	No	Engine Data 2
Actual EGR Position	0-100%	0	0-100	Engine Data 2
AIR Relay Command	On / Off	Off	Off	Engine Data 2
AIR Solenoid	On / Off	Off	Off	Engine Data 2
Boost PWM (VIN 1)	0-100%	100%	Varies	Engine Data 2
CMP Signal Present	Yes / No	Yes	Yes	Engine Data 2
Current Gear	1-2-3-4	1	3	Engine Data 2
Desired EGR Position	0-100%	0	0-100	Engine Data 2
Desired Idle Speed	0-3187 rpm	PCM control	---	Engine Data 2
Engine Hot Lamp	On / Off	Off	Off	Engine Data 2
Engine Load	0-100%	2-10	Varies	Engine Data 2
Engine Oil Level Switch	Okay / Low	Okay	Okay	Engine Data 2
Engine Oil Life	0-100%	Varies	Varies	Engine Data 2
Engine Run Time	Hr: Min: Sec	00:00:00	00:00:00	Engine Data 2
Engine Speed	0-9999 rpm	Varies	---	Engine Data 2
Fan Control Relay 1	On / Off	Off	Off	Engine Data 2
Fan Control Relay 2-3	On / Off	Off	Off	Engine Data 2
Generator Lamp	On / Off	Off	Off	Engine Data 2
Generator 'L' Signal	On / Off	On	On	Engine Data 2
IAC Position	0-255 counts	10-60	---	Engine Data 2
IAT Sensor (°F)	-40 to 304°F	50-194	50-194	Engine Data 2
Ignition 1 Signal	0.0-25.5v	14.2	14.2	Engine Data 2
Ignition Mode	Bypass / IC	IC	IC	Engine Data 2
Injector Pulsewidth	0-985 ms	1.5-3.5	Varies	Engine Data 2
Knock Retard (°)	0.0-25.5°	0	0	Engine Data 2
Long Term Fuel Trim	-10% to 10%	0 (± 5%)	0 (± 5%)	Engine Data 2

2000-05 Grand Prix 3.1L V6 MFI VIN J (A/T) - PID Data List

Parameter Identification (PID)	PID Value Range	PID Value at Hot Idle	PID Value at 30 mph	Data List Type
Loop Status	Open Loop or Closed Loop	Closed Loop	Closed Loop	Engine Data 2
Low Oil Lamp	On / Off	Off	Off	Engine Data 2
MAF Sensor (g/sec)	0-512 g/sec	3-6	Varies	Engine Data 2
MAF Sensor (Hz)	0-31,999hz	1200-3000	Varies	Engine Data 2
MAP Sensor (kPa)	10-110 kPa	20-48	Varies	Engine Data 2
MIL Command	On / Off	Off	Off	Engine Data 2
Power Enrichment	Active / Inactive	Inactive	Inactive	Engine Data 2
Short Term Fuel Trim	-10% to 10%	0 (± 5%)	0 (± 5%)	Engine Data 2
Spark Advance (°)	-64° to 64°	20	Varies	Engine Data 2
TCC Brake Pedal Switch	Applied / Released	Released	Released	Engine Data 2
TCC PWM Solenoid	On / Off	Off	On	Engine Data 2
TFP Switch	Several	P/N	Several	Engine Data 2
TP Angle (%)	0-100%	0	Varies	Engine Data 2
T/Control Status	Active / Inactive	Inactive	Inactive	Engine Data 2
T/C Request Signal	0-100%	100%	Varies	Engine Data 2
T/C Delivered Signal	75%	100%	Varies	Engine Data 2

2000-05 Grand Prix 3.1L V6 MFI VIN J (A/T) - PID Data List

Parameter Identification (PID)	PID Value Range	PID Value at Hot Idle	PID Value at 30 mph	Data List Type
A/C Request	Yes / No	No	No	EGR
Actual EGR Position	0-100%	0	0-100	EGR
BARO Sensor (kPa)	10-110 kPa	65-110	65-110	EGR
Commanded A/C	On / Off	Off	Off	EGR
Current Gear	1-2-3-4	1	3	EGR
Decel Fuel Mode	Active / Inactive	Inactive	Inactive	EGR
Desired EGR Position	0-100%	0	0-100	EGR
ECT Sensor (°F)	-40 to 304°F	185-220	185-220	EGR
EGR Closed Valve Pintle Position	0-5v	0.14-1.00	Varies	EGR
EGR Duty Cycle	0-100%	0	0-100	EGR
EGR Feedback	0-5v	0.14-1.00	Varies	EGR
EGR Position Error	0-100%	0	0	EGR
EGR Flow Test Count	0-255	0-10	0-10	EGR
Engine Load	0-100%	2-10	Varies	EGR
Engine Run Time	Hr: Min: Sec	00:00:00	00:00:00	EGR
Engine Speed	0-9999 rpm	Varies	---	EGR
Hot Open Loop	Active / Inactive	Inactive	Inactive	EGR
IAT Sensor (°F)	-40 to 304°F	50-194	50-194	EGR
Ignition 1 Signal	0.0-25.5v	14.2	14.2	EGR
Knock Retard (°)	0.0-25.5°	0	0	EGR
Loop Status	Open Loop or Closed Loop	Closed Loop	Closed Loop	EGR
MAF Sensor (g/sec)	0-512 g/sec	3-6	Varies	EGR
MAP Sensor (kPa)	10-110 kPa	20-48	Varies	EGR
Power Enrichment	Active / Inactive	Inactive	Inactive	EGR
TCC Engaged	Engage/DIS	Disengaged	Disengaged	EGR
TP Angle (%)	0-100%	0	Varies	EGR
Transmission Range	P-R-N-4-3-2	P	3	EGR

2000-05 Grand Prix 3.1L V6 MFI VIN J (A/T) - PID Data List

Parameter Identification (PID)	PID Value Range	PID Value at Hot Idle	PID Value at 30 mph	Data List Type
Air Fuel Ratio	0.0-25:1	14.6:1	14.6:1	Fuel Trim
AIR Active Test Inhibit	Yes / No	Yes	Yes	Fuel Trim
AIR Active Test	Yes / No	No	No	Fuel Trim
AIR Diagnostic Done	Yes / No	Yes	Yes	Fuel Trim
AIR Pass. Test 2 Fail	Yes / No	No	No	Fuel Trim
AIR Pass. Test 1 pass	Yes / No	Yes	Yes	Fuel Trim
AIR Pass. Test Pass	Yes / No	Yes	Yes	Fuel Trim
AIR Pass. Test Inhibit	Yes / No	Yes	Yes	Fuel Trim
AIR Pass. Test Active	Yes / No	No	No	Fuel Trim
AIR Pump Relay	On / Off	Off	Off	Fuel Trim
AIR Pump Solenoid	On / Off	Off	Off	Fuel Trim
BARO Sensor (kPa)	10-110 kPa	65-110	65-110	Fuel Trim
ECT Sensor (°F)	-40 to 304°F	185-220	185-220	Fuel Trim
Engine Load	0-100%	2-10	Varies	Fuel Trim
Engine Run Time	Hr: Min: Sec	00:00:00	00:00:00	Fuel Trim
Engine Speed	0-9999 rpm	Varies	---	Fuel Trim
EVAP Purge Solenoid	0-100%	0-25	0-100	Fuel Trim
Fuel Trim Cell	Cell #0-4	0	0-4	Fuel Trim
Fuel Trim Learn	Enabled / Disabled	Enabled	Enabled	Fuel Trim
HO2S (B1 S1, B1 S2)	0-1132 mv	10-1000	10-1000	Fuel Trim
IAT Sensor (°F)	-40 to 304°F	50-194	50-194	Fuel Trim
Long Term Fuel Trim	-10% to 10%	0 (± 5%)	0 (± 5%)	Fuel Trim
Loop Status	Closed Loop / Open Loop	Closed Loop	Closed Loop	Fuel Trim
MAF Sensor (g/sec)	0-512 g/sec	3-6	Varies	Fuel Trim
Number of DTC(s)	Number	00000	00000	Fuel Trim
Power Enrichment	Active / Inactive	Inactive	Inactive	Fuel Trim
Startup ECT Sensor	-40 to 304°F	Varies	Varies	Fuel Trim
Startup IAT Sensor	-40 to 304°F	Varies	Varies	Fuel Trim
TP Angle (%)	0-100%	0	Varies	Fuel Trim

2000-05 Grand Prix 3.1L V6 MFI VIN J (A/T) - PID Data List

Parameter Identification (PID)	PID Value Range	PID Value at Hot Idle	PID Value at 30 mph	Data List Type
1-2 Solenoid Status	Okay / Fault	Okay	Okay	Output Driver
2-3 Solenoid Status	Okay / Fault	Okay	Okay	Output Driver
A/C Relay Status	Okay / Fault	Okay	Okay	Output Driver
AIR Relay Status	Okay / Fault	Okay	Okay	Output Driver
AIR Solenoid Status	Okay / Fault	Okay	Okay	Output Driver
Cruise Inhibit Status	Okay / Fault	Okay	Okay	Output Driver
CYL 1-6 INJ Circuit	Okay / Fault	Okay	Okay	Output Driver
CYL 1-6 INJ History	Okay / Fault	Okay	Okay	Output Driver
ODD 1-4 Status	Okay / Fault	Okay	Okay	Output Driver
EGR Solenoid Status	Okay / Fault	Okay	Okay	Output Driver
EVAP Purge Solenoid	Okay / Fault	Okay	Okay	Output Driver
EVAP Vent Solenoid	Okay / Fault	Okay	Okay	Output Driver
FC Relay 1, 2-3	Okay / Fault	Okay	Okay	Output Driver
MIL Circuit Status	Okay / Fault	Okay	Okay	Output Driver
TCC Solenoid Status	Okay / Fault	Okay	Okay	Output Driver

2000-05 Grand Prix 3.1L V6 MFI VIN J (A/T) - PID Data List

Parameter Identification (PID)	PID Value Range	PID Value at Hot Idle	PID Value at 30 mph	Data List Type
Air Fuel Ratio	0.0-25:1	14.6:1	14.6:1	Catalyst
BARO Sensor (kPa)	10-110 kPa	65-110	65-110	Catalyst
BARO Sensor (V)	0.0-5.0v	3.5-4.5	---	Catalyst
Commanded A/C	On / Off	Off	Off	Catalyst
Current Gear	1-2-3-4	1	3	Catalyst
ECT Sensor (°F)	-40 to 304°F	185-220	185-220	Catalyst
Engine Load	0-100%	2-10	Varies	Catalyst
Engine Run Time	Hr: Min: Sec	00:00:00	00:00:00	Catalyst
Engine Speed	0-9999 rpm	Varies	---	Catalyst
Fan High Speed	On / Off	Off	Off	Catalyst
Fan Low Speed	On / Off	Off	Off	Catalyst
Fuel Trim Cell	Cell #0-4	0	0-4	Catalyst
Fuel Trim Learn	Enabled / Disabled	Enabled	Enabled	Catalyst
HO2S-11 (B1 S1)	0-1132 mv	10-1000	10-1000	Catalyst
HO2S-12 (B1 S2)	0-1132 mv	10-1000	10-1000	Catalyst
HO2S-11 Status	Ready/Not	Ready	Ready	Catalyst
HO2S-11 XCount	0-255 counts	Varies	Varies	Catalyst
Hot Open Loop	Active / Inactive	Inactive	Inactive	Catalyst
IAC Position	0-255 counts	10-40	---	Catalyst
IAT Sensor (°F)	-40 to 302°F	50-194	50-194	Catalyst
Ignition 1 Signal	0.0-25.5v	14.2	14.2	Catalyst
Injector Pulsewidth	0-985 ms	1.5-3.5	Varies	Catalyst
Loop Status	Open Loop or Closed Loop	Closed Loop	Closed Loop	Catalyst
MAF Sensor (g/sec)	0-512 g/sec	4-6 g/sec	Varies	Catalyst
MAP Sensor (kPa)	10-110 kPa	20-48	Varies	Catalyst
MAP Sensor (V)	0-5v	0.75-2.0	Varies	Catalyst
Non-Volatile Memory	PASS/FAIL	PASS	PASS	Catalyst
Power Enrichment	Active / Inactive	Inactive	Inactive	Catalyst
Startup ECT Sensor	-40 to 304°F	Varies	Varies	Catalyst
Startup IAT Sensor	-40 to 304°F	Varies	Varies	Catalyst
TP Angle (%)	0-100%	0	Varies	Catalyst
TP Sensor (V)	0-5v	0.20-0.74	Varies	Catalyst
Transmission Range	P-R-N-4-3-2	P	3	Catalyst
TWC Protection	Active / Inactive	Inactive	Inactive	Catalyst
Vehicle Speed	0-155 mph	0	30	Catalyst

2000-05 Grand Prix 3.1L V6 MFI VIN J (A/T) - PID Data List

Parameter Identification (PID)	PID Value Range	PID Value at Hot Idle	PID Value at 30 mph	Data List Type
Engine Oil Level	Okay / Low	Okay	Okay	IPC
Engine Oil Life	0-100%	0-100%	0-100%	IPC
Engine Oil Pressure	Okay / Low/HI	Okay	Okay	IPC
Fuel Tank Level	0-100%	Varies	Varies	IPC
Generator Command	On / Off	On	On	IPC
Generator PWM	0-100%	0-100	0-100	IPC
Ignition 1 Signal	0.0-25.5v	14.2	14.2	IPC
MIL (lamp)	On / Off	Off	Off	IPC
Non-Volatile Memory	PASS/FAIL	PASS	PASS	IPC

2000-05 Grand Prix 3.1L V6 MFI VIN J (A/T) - EVAP Data List

Parameter Identification (PID)	PID Value Range	PID Value at Hot Idle	PID Value at 30 mph	Data List Type
BARO Sensor (kPa)	10-110 kPa	65-110	65-110	EVAP
ECT Sensor (°F)	-40 to 304°F	185-220	185-220	EVAP
Engine Load	0-100%	2-10	Varies	EVAP
Engine Run Time	Hr: Min: Sec	00:00:00	00:00:00	EVAP
Engine Speed	0-9999 rpm	Varies	---	EVAP
EVAP Canister Purge	0-100%	0-100	0-100	EVAP
EVAP Fault History	Several	No FAULT	No FAULT	EVAP
EVAP Purge Solenoid	Several	Okay	Okay	EVAP
EVAP Vent Solenoid	Open/Close	Open	Open	EVAP
Fuel Tank Level	0-100%	Varies	Varies	EVAP
FTP Sensor (In. H2O)	-17.5 to 7.5"	Varies	Varies	EVAP
Fuel Trim Learn	Enabled / Disabled	Enabled	Enabled	EVAP
IAT Sensor (°F)	-40 to 304°F	50-194	50-194	EVAP
Long Term Fuel Trim	-10% to 10%	0 (± 5%)	0 (± 5%)	EVAP
Loop Status	Closed Loop / Open Loop	Closed Loop	Closed Loop	EVAP
MAF Sensor (g/sec)	0-512 g/sec	3-6	Varies	EVAP
MAP Sensor (kPa)	10-110 kPa	20-48 kPa	Varies	EVAP
Non-Volatile Memory	PASS/FAIL	PASS	PASS	EVAP
Startup ECT Sensor	-40 to 304°F	Varies	Varies	EVAP
Startup IAT Sensor	-40 to 304°F	Varies	Varies	EVAP
TP Angle (%)	0-100%	0	Varies	EVAP

2000-05 Grand Prix 3.1L V6 MFI VIN J (A/T) - Misfire Data List

Parameter Identification (PID)	PID Value Range	PID Value at Hot Idle	PID Value at 30 mph	Data List Type
3X Crank Sensor	1600-9999	Varies	Varies	Misfire
24X Crank Sensor	0-1600 rpm	Varies	Varies	Misfire
Abuse Management	Active / Inactive	Inactive	Inactive	Misfire
A/C Request Signal	Yes / No	No	No	Misfire
Actual EGR Position	0-100%	0	0-100	Misfire
Cam Signal Present	Yes / No	Yes	Yes	Misfire
Commanded A/C	On / Off	Off	Off	Misfire
Cycles of Misfire Data	0-255	0-99	Varies	Misfire
Decel Fuel Mode	Active / Inactive	Inactive	Inactive	Misfire
Desired EGR Position	0-100%	0	0-100	Misfire
ECT Sensor (°F)	-40 to 302°F	185-220	185-220	Misfire
Engine Load	0-100%	2-10	Varies	Misfire
Engine Speed	0-9999 rpm	Varies	---	Misfire
Hot Mode	On / Off	Off	Off	Misfire
HO2S-11 (B1 S1)	0-1132 mv	10-1000	10-1000	Misfire
Injector Pulsewidth	0-985 ms	1.5-3.5	Varies	Misfire
MAF Sensor (g/sec)	0-512 g/sec	4-6 g/sec	Varies	Misfire
Misfire current Cyl 1-6	0-198	Varies	Varies	Misfire
Misfire History Cyl 1-6	0-65535	0	0	Misfire
Non-Volatile Memory	PASS/FAIL	PASS	PASS	Misfire
Total Misfire Current	0-99 counts	Varies	Varies	Misfire
Total Misfire Failures	0-65535	Varies	Varies	Misfire
Total Misfire Passes	0-65535	Varies	Varies	Misfire

1995 Grand Prix 3.4L V6 MFI VIN X (A/T) - PID Data List

Parameter Identification (PID)	PID Value Range	PID Value at Hot Idle	PID Value at 30 mph	Data List Type
1st Gear - Low Switch	On / Off	Off	On: 1st	Engine Data 1
4th Gear - O/D Switch	On / Off	Off	On: 4th	Engine Data 1
A/C Clutch	On / Off	On: A/C on	Off	Engine Data 1
AC Pressure psi	0-499 psi	139-399	139-399	Engine Data 1
A/C Request	Yes / No	No	No	Engine Data 1
AIR Pump Control	On / Off	On: Cold	Off	Engine Data 1
BARO Sensor (V)	0.0-5.0v	3.5-4.5	3.5-4.5	Engine Data 1
BARO Sensor (kPa)	10-110 kPa	65-105	65-105	Engine Data 1
Desired Idle Speed	0-3187	800	---	Engine Data 1
Block Learn Cell	0-4	0-1	0-2	Engine Data 1
Block Learn Counts	0-255	110-156	110-156	Engine Data 1
Cooling Fan 1, 2	On / Off	On: >228ºF	Off	Engine Data 1
Crank RPM	0-900	N/A	N/A	Engine Data 1
ECT Sensor (ºF)	-40 to 419ºF	185-223	185-223	Engine Data 1
EGR Solenoid 1, 2	On / Off	Off	On	Engine Data 1
EGR Solenoid 3	On / Off	Off	On	Engine Data 1
Engine Speed	0-9999 rpm	Varies	---	Engine Data 1
Fuel Pump (V)	0.0-25.5v	14.2	14.2	Engine Data 1
HO2S Signal (V)	0-1132 mv	10-1000	10-1000	Engine Data 1
IAC Motor	0-255 counts	5-50	---	Engine Data 1
IAT Sensor (ºF)	-40 to 304ºF	50-176	50-176	Engine Data 1
Ignition Voltage	0.0-25.5v	14.2	14.2	Engine Data 1
Injector Pulsewidth	0-985 ms	1-4	Varies	Engine Data 1
Fuel Integrator	0-255 counts	118-138	118-138	Engine Data 1
Knock Retard (º)	0-128º	0	0	Engine Data 1
Knock Signal	Yes / No	Yes: Knock	No	Engine Data 1
Loop Status	Closed Loop / Open Loop	Closed Loop	Closed Loop	Engine Data 1
MAP Sensor (kPa)	10-110 kPa	29-48	Varies	Engine Data 1
MAP Sensor (V)	0-5.0v	1-2	Varies	Engine Data 1
P/NP Switch	P/N / R-DL	P/N	R-DL	Engine Data 1
PROM ID Number	Number	0000	0000	Engine Data 1
PSP Switch	HI/Normal	Normal	Normal	Engine Data 1
Purge Duty Cycle	0-100%	0	0-50	Engine Data 1
Shift Solenoid 'A'	On / Off	On	Off	Engine Data 1
Shift Solenoid 'B'	On / Off	On	Off	Engine Data 1
Spark Advance (º)	-90º to 90º	20	Varies	Engine Data 1
TCC Solenoid	On / Off	Off	Off	Engine Data 1
Throttle Angle (%)	0-100%	0	Varies	Engine Data 1
Time From Startup	Hr: Min: Sec	00:00:00	00:00:00	Engine Data 1
TP Sensor (V)	0.0-5.0v	0.29-0.98	Varies	Engine Data 1
Transmission Hot Mode	On / Off	On	Off	Engine Data 1
Vehicle Speed	0-155 mph	0	30	Engine Data 1

1996 Grand Prix 3.4L V6 MFI VIN X (A/T) - PID Data List

Parameter Identification (PID)	PID Value Range	PID Value at Hot Idle	PID Value at 30 mph	Data List Type
24X CKP Sensor	0-1600 rpm	Varies	Varies	Engine Data 1
A/C Clutch	Yes / No	No	No	Engine Data 1
A/C HI Side Pressure	0-450 psi	199-399	199-399	Engine Data 1
A/C Press. Too Hi/LO	Yes / No	No	No	Engine Data 1
A/C Pressure Disable	Yes / No	No	No	Engine Data 1
A/C Request	Yes / No	No	No	Engine Data 1
A/C Off for WOT	Yes / No	No	No	Engine Data 1
Actual EGR Position	0-100%	0	0-100	Engine Data 1
Air Pump	On / Off	On: Cold	Off	Engine Data 1
BARO Sensor (kPa)	10-110 kPa	65-110	65-110	Engine Data 1
BARO Sensor (V)	0.0-5.0v	3.5-4.5	3.5-4.5	Engine Data 1
Cam Signal Present	Yes / No	Yes	Yes	Engine Data 1
Commanded Fan 1, 2	On / Off	Off	Off	Engine Data 1
Cruise Engaged	Disengaged / Engaged	Disabled	Disabled	Engine Data 1
Cruise Inhibit	Yes / No	Yes	Yes	Engine Data 1
Desired EGR Position	0-100%	0	0-100	Engine Data 1
Desired Idle Speed	0-3187 rpm	PCM control	---	Engine Data 1
ECT Sensor (°F)	-40 to 304°F	185-220	185-220	Engine Data 1
Engine Load	0-100%	2-5	Varies	Engine Data 1
Engine Run Time	Hr: Min: Sec	00:00:00	00:00:00	Engine Data 1
Engine Speed	0-9999 rpm	Varies	---	Engine Data 1
EGR Duty Cycle	0-100%	0	0-100	Engine Data 1
EVAP Purge PWM	0-100%	0-25	0-100	Engine Data 1
EVAP Vacuum Switch	Open/Close	Open	Closed	Engine Data 1
Fuel Pump Command	On / Off	On	On	Engine Data 1
Fuel Pump Status	On / Off	On	On	Engine Data 1
Fuel Trim Cell	Cell #0-4	0	0-4	Engine Data 1
Fuel Trim Learn	Enabled / Disabled	Enabled	Enabled	Engine Data 1
HO2S-11 (B1 S1)	0-1132 mv	10-1000	10-1000	Engine Data 1
HO2S-12 (B1 S2)	0-1132 mv	10-1000	10-1000	Engine Data 1
HO2S-11 Status	Ready / Not Ready	Ready	Ready	Engine Data 1
HO2S-11 XCount	0-255 counts	Varies	Varies	Engine Data 1
IAC Position	0-255 counts	5-50	---	Engine Data 1
IAT Sensor (°F)	-40 to 304°F	50-194	50-194	Engine Data 1
IC Mode	Active / Inactive	Active	Active	Engine Data 1
Ignition 1 Signal	0.0-25.5v	14.2	14.2	Engine Data 1
Ignition Bypass	Active / Inactive	Inactive	Inactive	Engine Data 1
Injector Fault	Yes / No	No	No	Engine Data 1
Injector Pulsewidth	0-985 ms	1.5-3.5	Varies	Engine Data 1
Knock Retard (°)	0.0-25.5°	0	0	Engine Data 1
KS Activity	Yes / No	No	No	Engine Data 1
KS Noise Channel	0.0-5.0v	0.1-3.0v	Varies	Engine Data 1
Long Term Fuel Trim	-23-16%	0 (± 5%)	0 (± 5%)	Engine Data 1
Loop Status	Closed Loop / Open Loop	Closed Loop	Closed Loop	Engine Data 1
MAF Sensor (g/sec)	0-512 g/sec	3-6	Varies	Engine Data 1
MAF Sensor (Hz)	0-31,999	1200-3000	Varies	Engine Data 1
MAP Sensor (kPa)	10-110 kPa	29-48	Varies	Varies
MAP Sensor (V)	0.0-5.0v	1-2v	Varies	Varies

1996 Grand Prix 3.4L V6 MFI VIN X (A/T) - PID Data List

Parameter Identification (PID)	PID Value Range	PID Value at Hot Idle	PID Value at 30 mph	Data List Type
MPH / KM/H	0-155 mph	0	30	55
Short Term Fuel Trim	-11-20%	0 (± 5%)	0 (± 5%)	Engine Data 1
Spark Advance	-64° to 64°	-20	Varies	Engine Data 1
TCC Brake Switch	Applied / Released	APPLY: on	Released	Engine Data 1
TCC Engaged	Yes / No	No	No	Engine Data 1
TP Angle (%)	0-100%	0%	Varies	Engine Data 1
Transmission Range	X / O	O/X/X/O	Varies	Engine Data 1
VTD Fuel Disable	Enabled / Disabled	Disabled	Disabled	Engine Data 1

1996 Grand Prix 3.4L V6 MFI VIN X (A/T) - PID Data List

Parameter Identification (PID)	PID Value Range	PID Value at Hot Idle	PID Value at 30 mph	Data List Type
Commanded GEN	On / Off	Off	Off	IPC
Engine Oil Level	Okay / Low	Okay	Okay	IPC
Generator Lamp	On / Off	Off	Off	IPC
Ignition 1 Volts	0.0-25.5v	14.2	14.2	IPC
Low Oil Lamp	On / Off	On: If low	Off	IPC
MIL Status	On / Off	Off	Off	IPC

1996 Grand Prix 3.4L V6 MFI VIN X (A/T) - PID Data List

Parameter Identification (PID)	PID Value Range	PID Value at Hot Idle	PID Value at 30 mph	Data List Type
A/C Clutch	Yes / No	No	No	EGR
A/C Request	Yes / No	No	No	EGR
Actual EGR Position	0-100%	0	0-100	EGR
BARO Sensor (kPa)	10-110 kPa	65-110	65-110	EGR
Current Gear	1-2-3-4	1	3	EGR
Decel Fuel Mode	Active / Inactive	Inactive	Inactive	EGR
Desired EGR Position	0-100%	0	0-100	EGR
ECT Sensor (°F)	-38-304°F	185-220	185-220	EGR
EGR Closed Pintle	0.0-5.0v	0.14-1.0	---	EGR
EGR Duty Cycle	0-100%	0	0-100	EGR
EGR Position Error	0-100%	0-9	0-9	EGR
EGR Flow Test Count	0-255	0-10	0-10	EGR
Engine Run Time	Hr: Min: Sec	00:00:00	00:00:00	EGR
Engine Speed	0-9999 rpm	Varies	---	EGR
Hot Open Loop	Active / Inactive	Inactive	Inactive	EGR
IAT Sensor (°F)	-40 to 304°F	50-194	50-194	EGR
Knock Retard (°)	0.0-25.5°	0	0	EGR
Loop Status	Closed Loop / Open Loop	Closed Loop	Closed Loop	EGR
MAF Sensor (g/sec)	0-512 g/sec	3-6	Varies	EGR
MAP Sensor (kPa)	10-110 kPa	29-48	Varies	EGR
MPH / KM/H	0-155 mph	0	30	EGR
Power Enrichment	Active / Inactive	Inactive	Inactive	EGR
TCC Engaged	Yes / No	No	No	EGR
TP Angle (%)	0-100%	0	Varies	EGR
Transmission Range	X / O	O/X/X/O	Varies	EGR
TWC Protection	Active / Inactive	Inactive	Inactive	EGR

1996 Grand Prix 3.4L V6 MFI VIN X (A/T) - PID Data List

Parameter Identification (PID)	PID Value Range	PID Value at Hot Idle	PID Value at 30 mph	Data List Type
Air Fuel Ratio	0:1-25.5:1	14.6:1	14.6:1	HO2S
AIR Pump	On / Off	Off	Off	HO2S
Decel Fuel Mode	Active / Inactive	Inactive	Inactive	HO2S
ECT Sensor (°F)	-40 to 304°F	185-220	185-220	HO2S
Engine Load	0-100%	2-5	Varies	HO2S
Engine Run Time	Hr: Min: Sec	0:00:00	0:00:00	HO2S
Engine Speed	0-9999 rpm	Varies	---	HO2S
EVAP Purge PWM	0-100%	0-25	0-50	HO2S
HO2S (B1 S1, B1 S2)	0-1132 mv	10-1000	10-1000	HO2S
HO2S-11 Status	Not Ready / Ready	Ready	Ready	HO2S
HO2S B1 Warmup	Min: Sec	00:00	00:00	HO2S
HO2S-11 XCount	0-255 counts	Varies	Varies	HO2S
IAT Sensor (°F)	-40 to 304°F	50-194	50-194	HO2S
Injector Pulsewidth	0-985 ms	1.5-3.5	Varies	HO2S
Long, Short Fuel Trim	-23-16%	0 (± 5%)	0 (± 5%)	HO2S
MAF Sensor	0-512 g/sec	5-9	Varies	HO2S
Power Enrichment	Active / Inactive	Active	Active	HO2S
Startup ECT Sensor	-40 to 304°F	Varies	Varies	HO2S
Startup IAT Sensor	-40 to 304°F	Varies	Varies	HO2S
TP Angle (%)	0-100%	0%	Varies	HO2S

1996 Grand Prix 3.4L V6 MFI VIN X (A/T) - PID Data List

Parameter Identification (PID)	PID Value Range	PID Value at Hot Idle	PID Value at 30 mph	Data List Type
3X Crank Sensor	0-9999 rpm	Varies	Varies	Misfire
24X Crank Sensor	0-1600 rpm	Varies	Varies	Misfire
Abuse Management	Active / Inactive	Inactive	Inactive	Misfire
A/C Clutch	Yes / No	No	No	Misfire
A/C Request	Yes / No	No	No	Misfire
Actual EGR Position	0-100%	0	0-100	Misfire
Air Active Test	Active / Inactive	Inactive	Inactive	Misfire
Cam Signal Present	Yes / No	Yes	Yes	Misfire
Current Gear	1-2-3-4	1	3	Misfire
Desired EGR Position	0-100%	0%	0-100%	Misfire
ECT Sensor (°F)	-40 to 304°F	185-220	185-220	Misfire
Engine Load	0-100%	2-5%	Varies	Misfire
HO2S-11 (B1 S1)	0-1132 mv	10-1000	10-1000	Misfire
Injector Pulsewidth	0-985 ms	1.5-3.5	Varies	Misfire
MAF Sensor (g/sec)	0-512 g/sec	5-9	Varies	Misfire
Misfire current Cyl 1-6	0-198	0-4	0-4	Misfire
Misfire History Cyl 1-6	0-65535	0	0	Misfire
Total Misfire Current	0-99 counts	0-5	0-5	Misfire
Misfires Since 1st Fail	0-65535	0	0	Misfire
Passes Since 1st Fail	0-65535	0	0	Misfire
Spark Advance (°)	-64° to 64°	20	0	Misfire
TCC Engaged	Yes / No	No	No	Misfire
TP Angle (%)	0-100%	0%	Varies	Misfire
Trans. Hot Mode	Active / Inactive	Inactive	Inactive	Misfire

1997-2000 Grand Prix 3.8L V6 MFI VIN K, VIN 1 (A/T) - PID Data List

Parameter Identification (PID)	PID Value Range	PID Value at Hot Idle	PID Value at 30 mph	Data List Type
A/C Relay Command	On / Off	Off	Off	Engine Data 1
Actual EGR Position	0-100%	0	Varies	Engine Data 1
Air Fuel Ratio	0.0-25.5:1	14.6:1	14.6:1	Engine Data 1
APP Indicated Angle	0-100%	0	Varies	Engine Data 1
BARO Sensor (kPa)	10-110 kPa	65-110	65-110	Engine Data 1
Cruise Control Active	Yes / No	No	No	Engine Data 1
Cruise Inhibit Control	Yes / No	Yes	Yes	Engine Data 1
Cruise Inhibit Reason	Several	Several	Several	Engine Data 1
Cruise Inhibit Signal	On / Off	Off	Off	Engine Data 1
Current Gear	1-2-3-4	1	3	Engine Data 1
Decel Fuel Mode	Active / Inactive	Inactive	Inactive	Engine Data 1
Desired EGR Position	0-100%	0	0-100	Engine Data 1
Desired Idle Speed	0-3187 rpm	PCM control	---	Engine Data 1
ECT Sensor (°F)	-40 to 304°F	195	195	Engine Data 1
EGR Position Sensor	0-100%	0	Varies	Engine Data 1
Engine Load	0-100%	2-5	Varies	Engine Data 1
Engine Run Time	Hr: Min: Sec	00.00.00	00:00:00	Engine Data 1
Engine Speed	0-9999 rpm	Varies	Varies	Engine Data 1
EVAP Purge Control	0-100%	0-25	0-100	Engine Data 1
EVAP Vent Command	Open/Close	Open	Open	Engine Data 1
Fuel Trim Cell	Cell #0-9	0	0-9	Engine Data 1
Fuel Trim Learn	Yes / No	Yes	Yes	Engine Data 1
Fuel Pump Relay	On / Off	On	On	Engine Data 1
FTP Sensor (In. H2O)	-17.5 to 7.5"	Varies	Varies	Engine Data 1
Generator Command	On / Off	On	On	Engine Data 1
HO2S-11 (B1 S1)	0-1132 mv	10-1000	10-1000	Engine Data 1
HO2S-12 (B1 S2)	0-1132 mv	10-1000	10-1000	Engine Data 1
IAT Sensor (°F)	-40 to 304°F	50-194	50-194	Engine Data 1
Ignition 1 Signal	0-25.5v	14.2	14.2	Engine Data 1
Injector Pulsewidth	0-985 ms	1.5-3.5	Varies	Engine Data 1
Knock Retard (°)	0.0-25.5°	0	Varies	Engine Data 1
Long Term Fuel Trim	-10% to 10%	0 (± 5%)	0 (± 5%)	Engine Data 1
MAF Sensor (g/sec)	0-512 g/sec	3-6	Varies	Engine Data 1
MAF Sensor (Hz)	0-31,999 Hz	1200-3000	Varies	Engine Data 1
MAP Sensor (kPa)	10-110 kPa	29-48	Varies	Engine Data 1
Number of DTC(s)	Number	0	0	Engine Data 1
PCM Reset	Yes / No	No	No	Engine Data 1
PCM/VCM in VTD Fail	Yes / No	No	No	Engine Data 1
Power Enrichment	Active / Inactive	Inactive	Inactive	Engine Data 1
Reduced Power	Active / Inactive	Inactive	Inactive	Engine Data 1
Short Term Fuel Trim	-10% to 10%	0 (± 5%)	0 (± 5%)	Engine Data 1
Spark Advance (°)	-64° to 64°	19	Varies	Engine Data 1
TCC Brake Pedal Switch	Engaged / Released	Released	Released	Engine Data 1
TP Angle (%)	0-100%	0	Varies	Engine Data 1
TP Indicated Angle	0-100%	2-6	Varies	Engine Data 1
TWC Protection	Active / Inactive	Inactive	Inactive	Engine Data 1
VTD Auto Learn	Active / Inactive	Inactive	Inactive	Engine Data 1
VTD Fuel Disable	Active / Inactive	Inactive	Inactive	Engine Data 1

1997-2000 Grand Prix 3.8L V6 MFI VIN K, VIN 1 (A/T) - PID Data List

Parameter Identification (PID)	PID Value Range	PID Value at Hot Idle	PID Value at 30 mph	Data List Type
3X Crank Sensor	0-9999 rpm	Varies	Varies	Engine Data 2
24X Crank Sensor	0-1600 rpm	Varies	Varies	Engine Data 2
A/C HI Side Pressure	0-5.0v	Varies	Varies	Engine Data 2
A/C Off for WOT	Yes / No	No	No	Engine Data 2
A/C Pressure Disable	Yes / No	No	No	Engine Data 2
A/C Press. Too HI/LO	Yes / No	No	No	Engine Data 2
A/C Relay Command	On / Off	Off	Off	Engine Data 2
A/C Request Signal	Yes / No	No	No	Engine Data 2
Air Fuel Ratio	0.0-25:1	14.6:1	14.6:1	Engine Data 2
BARO Sensor (kPa)	10-110 kPa	65-110	65-110	Engine Data 2
Change Oil Lamp	Okay / Fault	Okay	Okay	Engine Data 2
CMP Sensor Signal	Yes / No	Yes	Yes	Engine Data 2
Commanded A/C	On / Off	On	On	Engine Data 2
Commanded Starter	Enabled / Disabled	Disabled	Disabled	Engine Data 2
Crank Request Signal	Yes / No	No	No	Engine Data 2
Current Gear	1-2-3-4	1	3	Engine Data 2
Decel Fuel Mode	Active / Inactive	Inactive	Inactive	Engine Data 2
Desired EGR Position	0-100%	0	Varies	Engine Data 2
Desired Idle Speed	0-3187 rpm	PCM control	----	Engine Data 2
ECT Sensor (°F)	-40 to 304°F	195	195	Engine Data 2
Engine Load	0-100%	2-5	Varies	Engine Data 2
Engine Hot Lamp	On / Off	Off	Off	Engine Data 2
Engine Oil Level	Okay / Low	Okay	Okay	Engine Data 2
Engine Oil Level Life	0-100%	Varies	Varies	Engine Data 2
Engine Oil Pressure	Okay / Low	Okay	Okay	Engine Data 2
Engine Run Time	Hr: Min: Sec	0:00:00	0:00:00	Engine Data 2
Engine Speed	0-9999 rpm	Varies	----	Engine Data 2
FC Relay 1, 2-3	On / Off	On	Off	Engine Data 2
F/P Speed (VIN 1)	High/Normal	Normal	Normal	Engine Data 2
Generator Lamp	On / Off	Off	Off	Engine Data 2
Generator L-Terminal	On / Off	On	On	Engine Data 2
Generator PWM	0-100%	Varies	Varies	Engine Data 2
IAC Position	0-255 counts	10-60	----	Engine Data 2
IAT Sensor (°F)	-40 to 304°F	85	85	Engine Data 2
Ignition Mode	BYPASS/IC	IC	IC	Engine Data 2
Injector Pulsewidth	0-985 ms	1.5-3.5	Varies	Engine Data 2
Knock Retard (°)	0.0-25.5°	0	Varies	Engine Data 2
Long Term Fuel Trim	-10% to 10%	0 (± 5%)	0 (± 5%)	Engine Data 2
Loop Status	Closed Loop / Open Loop	Closed Loop	Closed Loop	Engine Data 2
Low Oil Lamp	On / Off	Off	Off	Engine Data 2
MAF Sensor (g/sec)	0-512 g/sec	3-6	Varies	Engine Data 2
MAF Sensor (Hz)	0-32000 Hz	1200-3000	Varies	Engine Data 2
MAP Sensor (kPa)	10-110 kPa	20-48	Varies	Engine Data 2
MIL (lamp) Command	Okay / Fault	Okay	Okay	Engine Data 2
PCM Reset	Yes / No	No	No	Engine Data 2
Power Enrichment	Active / Inactive	Inactive	Inactive	Engine Data 2
Short Term Fuel Trim	-10% to 10%	0 (± 5%)	0 (± 5%)	Engine Data 2
Spark Advance (°)	-64° to 64°	19°	Varies	Engine Data 2

1997-2000 Grand Prix 3.8L V6 MFI VIN K, VIN 1 (A/T) - PID Data List

Parameter Identification (PID)	PID Value Range	PID Value at Hot Idle	PID Value at 30 mph	Data List Type
Starter Relay (Inhibit)	On / Off	On	On	Engine Data 2
TCC Brake Pedal Switch	Applied / Released	Released	Released	Engine Data 2
TCC PWM Solenoid	On / Off	Off	Off	Engine Data 2
TFP Switch	Several	Park	Several	Engine Data 2
TP Angle (%)	0-100%	0	Varies	Engine Data 2
Torque Delivered	0-100%	86	Varies	Engine Data 2
Torque Requested	0-100%	100	Varies	Engine Data 2
T/Control Status	Active / Inactive	Inactive	Inactive	Engine Data 2
VTD Fuel Disable	Active / Inactive	Inactive	Inactive	Engine Data 2
Vehicle Speed	0-155 mph	0	30	Engine Data 2

1997-2000 Grand Prix 3.8L V6 MFI VIN K, VIN 1 (A/T) - PID Data List

Parameter Identification (PID)	PID Value Range	PID Value at Hot Idle	PID Value at 30 mph	Data List Type
Air Fuel Ratio	0.0-25.5:1	14.6:1	14.6:1	Fuel Trim
BARO Sensor (kPa)	10-110 kPa	65-110	65-110	Fuel Trim
Decel Fuel Cutoff	Active / Inactive	Inactive	Inactive	Fuel Trim
ECT Sensor (°F)	-40 to 304°F	195	195	Fuel Trim
Engine Load	0-100%	2-5	Varies	Fuel Trim
Engine Run Time	Hr: Min: Sec	00:00:00	00:00:00	Fuel Trim
Engine Speed	0-9999 rpm	Varies	---	Fuel Trim
EVAP Purge Control	0-100%	0-50	50-100	Fuel Trim
EVAP Vent Command	Open/Close	Open	Open	Fuel Trim
Fuel Trim Cell	Cell #0-9	0	0-9	Fuel Trim
Fuel Trim Learn	Enabled / Disabled	Enabled	Enabled	Fuel Trim
HO2S (B1 S1, B1 S2)	0-1132 mv	10-1000	10-1000	Fuel Trim
HO2S-11 Heater	On / Off	On	On	Fuel Trim
HO2S-11 Status	Ready/Not	Ready	Ready	Fuel Trim
HO2S-11 XCount	0-1132 mv	10-1000	10-1000	Fuel Trim
IAC Position	0-255 counts	10-60	---	Fuel Trim
IAT Sensor (°F)	-40 to 304°F	85	85	Fuel Trim
Injector Pulsewidth	0-985 ms	1.5-3.5	Varies	Fuel Trim
Long Term Fuel Trim	-10% to 10%	0 (± 5%)	0 (± 5%)	Fuel Trim
Loop Status	Open Loop or Closed Loop	Closed Loop	Closed Loop	Fuel Trim
MAF Sensor (g/sec)	0-512 g/sec	3-6	Varies	Fuel Trim
MAP Sensor (kPa)	10-110 kPa	20-48	Varies	Fuel Trim
Number of DTC(s)	Number	00000	00000	Fuel Trim
HO2S Heater Current	0-10 amps	0.54	0.54	Fuel Trim
PCM Reset	Yes / No	No	No	Fuel Trim
Power Enrichment	Active / Inactive	Inactive	Inactive	Fuel Trim
Short Term Fuel Trim	-10% to 10%	0 (± 5%)	0 (± 5%)	Fuel Trim
Spark Advance (°)	-64° to 64°	19°	Varies	Fuel Trim
Startup ECT Sensor	-40 to 304°F	Varies	Varies	Fuel Trim
Startup IAT Sensor	-40 to 304°F	Varies	Varies	Fuel Trim
TFP Switch	Several	Park	Several	Fuel Trim
TP Angle (%)	0-100%	1-4	Varies	Fuel Trim
TWC Protection	Active / Inactive	Inactive	Inactive	Fuel Trim
Vehicle Speed	0-150	0	30	Fuel Trim

1997-2000 Grand Prix 3.8L V6 MFI VIN K, VIN 1 (A/T) - PID Data List

Parameter Identification (PID)	PID Value Range	PID Value at Hot Idle	PID Value at 30 mph	Data List Type
Air Fuel Ratio	0.0-25:1	14.6:1	14.6:1	Catalyst
BARO Sensor (kPa)	10-110 kPa	65-110	65-110	Catalyst
Commanded A/C	On / Off	Off	Off	Catalyst
Decel Fuel Mode	Active / Inactive	Inactive	Inactive	Catalyst
ECT Sensor (°F)	-40 to 304°F	195	195	Catalyst
Engine Load	0-100%	2-10	Varies	Catalyst
Engine Run Time	Hr: Min: Sec	00:00:00	00:00:00	Catalyst
Engine Speed	0-9999 rpm	Varies	----	Catalyst
Fan High, Low Speed	On / Off	Off	Off	Catalyst
Fuel Trim Learn	Enabled / Disabled	Enabled	Enabled	Catalyst
HO2S-11 (B1 S1)	0-1132 mv	10-1000	10-1000	Catalyst
HO2S-12 (B1 S2)	0-1132 mv	10-1000	10-1000	Catalyst
IAC Position	0-255 counts	10-60	----	Catalyst
IAT Sensor (°F)	-40 to 304°F	50-194	50-194	Catalyst
Injector Pulsewidth	0-985 ms	1.5-3.5	Varies	Catalyst
Loop Status	Open Loop or Closed Loop	Closed Loop	Closed Loop	Catalyst
MAF Sensor (g/sec)	0-512 g/sec	3-6	Varies	Catalyst
Power Enrichment	Active / Inactive	Inactive	Inactive	Catalyst
Short Term Fuel Trim	-10% to 10%	0 (± 5%)	0 (± 5%)	Catalyst
TP Angle (%)	0-100%	0	Varies	Catalyst
TWC Protection	Active / Inactive	Inactive	Inactive	Catalyst
Vehicle Speed	0-150	0	30	Catalyst

1997-2000 Grand Prix 3.8L V6 MFI VIN K, VIN 1 (A/T) - PID Data List

Parameter Identification (PID)	PID Value Range	PID Value at Hot Idle	PID Value at 30 mph	Data List Type
1-2. 2-3 Solenoid	Several	Okay	Okay	Output Driver
A/C Relay Circuit	Okay / Fault	Okay	Okay	Output Driver
Air Pump Relay	Okay / Fault	Okay	Okay	Output Driver
Air Solenoid Circuit	Okay / Fault	Okay	Okay	Output Driver
Change Oil Lamp	Okay / Fault	Okay	Okay	Output Driver
Cruise Inhibit Status	Okay / Fault	Okay	Okay	Output Driver
Cyl 1-6 INJ Circuit	Okay / Fault	Okay	Okay	Output Driver
Cyl 1-6 History Status	Okay / Fault	Okay	Okay	Output Driver
CYL 1-6 INJ Current	Several	Okay	Okay	Output Driver
CYL 1-6 INJ History	Several	Okay	Okay	Output Driver
ODD 1-4 Status	Several	Enabled	Enabled	Output Driver
Engine Hot Lamp	Okay / Fault	Okay	Okay	Output Driver
Engine Oil Lamp	Okay / Fault	Okay	Okay	Output Driver
EVAP Purge Solenoid	Okay / Fault	Okay	Okay	Output Driver
EVAP Vent Solenoid	Okay / Fault	Okay	Okay	Output Driver
F/C Relay 1, 2-3	Okay / Fault	Okay	Okay	Output Driver
F/P Relay Circuit	Several	Okay	Okay	Output Driver
F/P Relay History	Okay / Fault	Okay	Okay	Output Driver
Generator Lamp	Okay / Fault	Okay	Okay	Output Driver
Low Oil Lamp Status	Okay / Fault	Okay	Okay	Output Driver
MIL Control Circuit	Several	Okay	Okay	Output Driver
TCC Solenoid Circuit	Several	Okay	Okay	Output Driver

1997-2000 Grand Prix 3.8L V6 MFI VIN K, VIN 1 (A/T) - PID Data List

Parameter Identification (PID)	PID Value Range	PID Value at Hot Idle	PID Value at 30 mph	Data List Type
Actual EGR Position	0-100%	0	Varies	EGR
A/C Relay Command	On / Off	Off	Off	EGR
BARO Sensor (kPa)	10-110 kPa	65-110	65-110	EGR
Current Gear	1-2-3-4	1	3	EGR
Decel Fuel Cutoff	Active / Inactive	Inactive	Inactive	EGR
Desired EGR Position	0-100%	0	Varies	EGR
Desired Idle Speed	0-3187 rpm	PCM control	---	EGR
ECT Sensor (°F)	-40 to 304°F	195	195	EGR
EGR Closed Valve	0-5.0v	0.14-1.0	---	EGR
EGR Duty Cycle	0-100%	0	Varies	EGR
EGR Feedback Signal	0-5.0v	0.14-1.0	---	EGR
EGR Flow Test Count	0-255	0-10	0-10	EGR
EGR Learned Min.	0-5.0v	0.14-1.0	---	EGR
EGR Pos. Variance	0-100%	0-9	Varies	EGR
EGR Solenoid	0-100%	0	Varies	EGR
Engine Load	0-100%	2-5	Varies	EGR
Engine Run Time	Hr: Min: Sec	00:00:00	00:00:00	EGR
Engine Speed	0-9999 rpm	Varies	---	EGR
IAC Position	0-255 counts	10-60	---	EGR
IAT Sensor (°F)	-40 to 304°F	85	85	EGR
Ignition 1 Signal	0.0-25.5v	14.1	14.1	EGR
Knock Retard (°)	0.0-25.5°	0	Varies	EGR
Loop Status	Open Loop or Closed Loop	Closed Loop	Closed Loop	EGR
MAF Sensor (g/sec)	0-512 g/sec	3-6	Varies	EGR
MAP Sensor (kPa)	10-110 kPa	20-48	Varies	EGR
PCM Reset	Yes / No	No	No	EGR
TFP Switch	Several	Park	Several	EGR
TP Angle (%)	0-100%	0	Varies	EGR
Vehicle Speed	0-155 mph	0	30	EGR

1997-2000 Grand Prix 3.8L V6 MFI VIN K, VIN 1 (A/T) - PID Data List

Parameter Identification (PID)	PID Value Range	PID Value at Hot Idle	PID Value at 30 mph	Data List Type
BARO Sensor (kPa)	10-110 kPa	65-110	65-110	EVAP
Desired Idle Speed	0-3187 rpm	PCM control	---	EVAP
ECT Sensor (°F)	-40 to 304°F	195	195	EVAP
Engine Load	0-100%	2-5	Varies	EVAP
Engine Run Time	Hr: Min: Sec	00:00:00	00:00:00	EVAP
Engine Speed	0-9999 rpm	Varies	---	EVAP
EVAP Fault History	Several	No Fault	No Fault	EVAP
EVAP Purge PWM	0-100%	Varies	Varies	EVAP
EVAP Test Abort	Several	Not Aborted	Not Aborted	EVAP
EVAP Test Result	Several	Passed	Passed	EVAP
EVAP Test State	Several	Completed	Completed	EVAP
EVAP Vent Command	Open/closed	Open	Open	EVAP
Fuel Tank Level	0-100%	Varies	Varies	EVAP
Fuel Level Remaining	0-100%	Varies	Varies	EVAP
FTP Sensor (In. H2O)	-17.5 to 7.5"	Varies	Varies	EVAP

1997-2000 Grand Prix 3.8L V6 MFI VIN K, VIN 1 (A/T) - PID Data List

Parameter Identification (PID)	PID Value Range	PID Value at Hot Idle	PID Value at 30 mph	Data List Type
FTP Sensor (V)	0-5.0v	2.5: cap off	Varies	EVAP
Fuel Trim Cell	Cell #0-9	0	0-9	EVAP
Fuel Trim Learn	Enabled / Disabled	Enabled	Enabled	EVAP
HO2S-11 (B1 S1)	0-1132 mv	10-1000	10-1000	EVAP
IAT Sensor (°F)	-40 to 304°F	85	85	EVAP
Long, Short Fuel Trim	-10% to 10%	0 (± 5%)	0 (± 5%)	Misfire
MAF Sensor (g/sec)	0-512 g/sec	3-6	Varies	EVAP
MAP Sensor (kPa)	10-110 kPa	20-48	Varies	EVAP
No. of Current DTC(s)	Number	0	0	EVAP
PCM Reset	Yes / No	No	No	EVAP
Startup ECT Sensor	-40 to 304°F	Varies	Varies	EVAP
Startup IAT Sensor	-40 to 304°F	Varies	Varies	EVAP
TP Angle (%)	0-100%	0	Varies	EVAP

1997-2000 Grand Prix 3.8L V6 MFI VIN K, VIN 1 (A/T) - PID Data List

Parameter Identification (PID)	PID Value Range	PID Value at Hot Idle	PID Value at 30 mph	Data List Type
3X Crank Sensor	0-9999 rpm	Varies	Varies	Misfire
24X Crank Sensor	0-1600 rpm	Varies	Varies	Misfire
Actual EGR Position	0-100%	0	Varies	Misfire
Air Relay Command	On / Off	Off	Off	Misfire
Commanded A/C	On / Off	Off	Off	Misfire
Cycles of Misfire Data	0-99	Varies	Varies	Misfire
Decel Fuel Mode	Active / Inactive	Inactive	Inactive	Misfire
Desired EGR Position	0-100%	0	0-100	Misfire
ECT Sensor (°F)	-40 to 304°F	195	195	Misfire
EGR Position Sensor	0-100%	0	Varies	Misfire
Engine Run Time	Hr: Min: Sec	00:00:00	00:00:00	Misfire
Engine Speed	0-9999 rpm	Varies	---	Misfire
Fuel Trim Cell	Cell #0-9	0	0-9	Misfire
IAC Position	0-255 counts	10-60	---	Misfire
Injector Pulsewidth	0-985 ms	1.5-3.5	Varies	Misfire
Long, Short Fuel Trim	-10% to 10%	0 (± 5%)	0 (± 5%)	Misfire
MAF Sensor (g/sec)	0-512 g/sec	3-6	Varies	Misfire
MAP Sensor (kPa)	10-110 kPa	20-48	Varies	Misfire
Misfire current Cyl 1-6	0-198	Varies	Varies	Misfire
Misfire History Cyl 1-6	0-65535	Varies	Varies	Misfire
Misfiring Cylinder	0-6	0	0	Misfire
Power Enrichment	Active / Inactive	Inactive	Inactive	Misfire
Spark Advance (°)	-64° to 64°	19°	Varies	Misfire
TCC PWM Solenoid	On / Off	Off	On	Misfire
Total Fail since 1st fail	0-65535	0	0	Misfire
Total Pass Since 1st Pass	0-65535	0	0	Misfire
Total Misfire Current	0-99 counts	0	0	Misfire
Spark Advance (°)	-64° to 64°	19°	Varies	Misfire
TP Angle (%)	0-100%	0	Varies	Misfire
TWC Protection	Active / Inactive	Inactive	Inactive	Misfire

2001-05 Grand Prix 3.8L V6 VIN K, VIN 1 (A/T) - PID Data List

Parameter Identification (PID)	PID Value Range	PID Value at Hot Idle	PID Value at 30 mph	Data List Type
3X Crank Sensor	0-9999 rpm	Varies	Varies	Engine Data 2
18X Crank Sensor	0-1600 rpm	Varies	Varies	Engine Data 2
A/T 1-2 Solenoid Circuit Status	Okay / Fault / INVALID	Okay	Okay	Output Driver
A/T 2-3 Solenoid Circuit Status	Okay / Fault / INVALID	Okay	Okay	Output Driver
A/C HI Side Pressure	0-5.1 volts	Varies	Varies	Engine Data 2
A/C Off for WOT	Yes / No	No	No	Engine Data 2
A/C Pressure Disable	Yes / No	No	No	Engine Data 2
A/C Relay Control Circuit Status	Okay / Fault / INVALID	Okay	Okay	Output Driver
A/C Relay Command	On / Off	Off	Off	Engine Data 1-2, EGR, Misfire
A/C Request Signal	Yes / No	No	No	Engine Data 2
Air Fuel Ratio	0.0-25.5:1	14.2-14.7:1	14.2-14.7:1	Engine Data 2, Fuel Trim
AIR Active Test Air Injection	Yes / No	No	No	AIR
AIR Active Test Inhibit	Yes / No	No	No	AIR
AIR Active Test Passed	Yes / No	Yes	Yes	AIR
AIR Passive Test 1 Pass	Yes / No	No	No	AIR
AIR Passive Test 2 Failed	Yes / No	No	No	AIR
AIR Passive Test In Progress	Yes / No	No	No	AIR
AIR Passive Test Inhibit	Yes / No	No	No	AIR
AIR Passive Test Passed	Yes / No	No	No	AIR
Air Pump Relay Circuit Status	Okay / Fault / Invalid State	Okay	Okay	Output Driver
Air Pump Relay Command	On / Off	Off	Off	Engine Data 1, Fuel Trim
Air Pump Solenoid Circuit Status	Okay / Fault / Invalid State	Okay	Okay	Output Driver
Air Pump Solenoid Command	On / Off	Off	Off	Engine Data 1, Fuel Trim
BARO Sensor (kPa)	10-110 kPa	98	98	Engine Data 1, EGR, EVAP, Fuel Trim
Boost Solenoid Circuit Status (VIN 1 only)	Okay / Fault / INVALID	Okay	Okay	Output Driver
Boost Solenoid Control (VIN 1 only)	0-100%	0	0	Engine Data 2

2001-05 Grand Prix 3.8L V6 VIN K, VIN 1 (A/T) - PID Data List

Parameter Identification (PID)	PID Value Range	PID Value at Hot Idle	PID Value at 30 mph	Data List Type
CMP Sensor Signal	Yes / No	Yes	Yes	Engine Data 2
Cruise Control Active	Yes / No	No	No	Engine Data 1
Cruise Inhibit Reason	VSS (speed)	0	30	Engine Data 1
Cruise Inhibit Signal Circuit Status	Okay / Stuck Low (open) / Stuck High	Okay	Okay	Engine Data 1
Current Gear	1-2-3-4	1	4	Engine Data 1-2, EGR, Fuel Trim
Cycles of Misfire Data	0-99	Varies	Varies	Misfire
Cylinder 1-6 Injector Circuit History	Okay / Stuck Low / Stuck High / Fault	Okay	Okay	Output Driver
Cylinder 1-6 Injector Circuit Status	Okay / Stuck Low (open) / Stuck High	Okay	Okay	Output Driver
Desired EGR Position	0-100%	0	Varies	Engine Data 1, EGR, Misfire
Decel Fuel Cutoff	Active / Inactive	Inactive	Inactive	EGR, Fuel Trim
Desired Idle Speed	0-3187 rpm	± 50 rpm of Actual Speed	Varies	Engine Data 1, Data 2, EVAP
Driver Module 1-4 Status	Enabled/Off/ High Volts / High TEMP / Invalid State	Enabled	Enabled	Output Driver
ECT Sensor (°F)	-40 to 304°F	185-220	185-220	Engine Data 1-2, EGR, Fuel Trim, EVAP, Misfire
EGR Flow Test Count	0-255 counts	0-10	Varies	EGR
EGR Learned Minimum Position	0-5.1 volts	0.14-1.0	Varies	EGR
EGR Position Sensor	0-100%	0	Varies	Engine Data 1, EGR, Misfire
EGR Position Sensor	0-5.1volts	0.0	Varies	Engine Data 1, EGR, Misfire
EGR Position Variance	0-100%	0-9	Varies	EGR
EGR Solenoid Circuit History	Okay / Fault	Okay	Okay	Output Driver
EGR Solenoid Circuit Status	Enabled/Off/ High Volts / High TEMP / Invalid State	0	Varies	EGR
Engine Load (%)	0-100%	17	Varies	Engine Data 1-2, EGR, Fuel Trim, EVAP, Misfire
Engine Oil Level Switch	Okay / Not Okay	Okay	Okay	Engine Data 2
Engine Oil Life Left	0-100%	Varies	Varies	Engine Data 2
EOP Sensor	0-5.1 volts	2-3	2-4	Engine Data 2

2001-05 Grand Prix 3.8L V6 VIN K, VIN 1 (A/T) - PID Data List

Parameter Identification (PID)	PID Value Range	PID Value at Hot Idle	PID Value at 30 mph	Data List Type
Engine Run Time	Hr: Min: Sec	00:00:00	00:00:00	Engine Data 1-2, EGR, Fuel Trim, EVAP, Misfire
Engine Speed	0-9999 rpm	± 50 rpm of Desired Speed	Varies	Engine Data 1-2, EGR, Fuel Trim, EVAP, Misfire
EVAP Fault History	No Fault / Excess VAC / Purge Valve Leak / Small Leak / Weak Vacuum	No Fault	No Fault	EVAP
EVAP Purge Solenoid Control Circuit Status	Okay / Fault / Invalid State	Okay	Okay	Output Driver
EVAP Purge Solenoid Command	0-100%	18	Varies	Engine Data 1, EVAP, Fuel Trim
EVAP Vent Solenoid Circuit Status	Okay / Fault / Invalid State	Okay	Okay	Output Driver
EVAP Vent Solenoid	Venting / Not Venting	Not Venting	Not Venting	Engine Data 1, EVAP, Fuel Trim
Extended Travel Brake Pedal Switch	Engaged / Released	Released	Released	Engine Data 2
Fan Control Relay 1 Circuit Status	Okay / Fault / Invalid State	Okay	Okay	Output Driver
Fan Control Relay 1 Command	On / Off	Off	Off	Engine Data 2
Fan Control Relay 2-3 Command	On / Off	Off	Off	Engine Data 2
Fan Control Relay 2-3 Circuit Status	Okay / Fault / Invalid State	Okay	Okay	Output Driver
Fuel Pump Relay Circuit Status	Okay / Fault / Invalid State	Okay	Okay	Output Driver
Fuel Pump Relay History Status	Okay / Fault	Okay	Okay	Output Driver
Fuel Pump Relay Command	On / Off	On	On	Engine Data 1
Fuel Tank Level Remaining	0-100%	Varies	Varies	EVAP
Fuel Tank Pressure (FTP) Sensor	Inches H2O, mm Hg	-17 to +7.5 Inches H2O	-17 to +7.5 Inches H2O	Engine Data 1, EVAP
Fuel Tank Pressure (FTP) Sensor	0-5.1 volts	Varies	Varies	Engine Data 1, EVAP
Fuel Trim Cell	Cell 0-9	0	1	Engine Data 1, EVAP, Fuel Trim
Fuel Trim Learn	Enabled or Disabled	Enabled	May Toggle	Engine Data 1, EVAP, Fuel Trim
Generator F-Terminal	0-100%	Varies	Varies	Engine Data 2
Generator L-Terminal	On / Off	On	On	Engine Data 2

2001-05 Grand Prix 3.8L V6 VIN K, VIN 1 (A/T) - PID Data List

Parameter Identification (PID)	PID Value Range	PID Value at Hot Idle	PID Value at 30 mph	Data List Type
HO2S-11 (B1 S1)	0-1132 mv	10-1000	10-1000	Engine Data 1, EVAP, Fuel Trim
HO2S-12 (B1 S2)	0-1132 mv	10-1000	10-1000	Engine Data 1, EVAP, Fuel Trim
IAC Position	0-255 counts	10-40	10-40	Engine Data 1, EGR, Fuel Trim
IAT Sensor (°F)	-40 to 304°F	50-194	50-194	Engine Data 1-2, EGR, EVAP, Fuel Trim
Ignition 1 Signal	0.0-25.5v	14.2	14.2	Engine Data 1-2, EGR, EVAP, Fuel Trim
Ignition Mode	Bypass/IC	IC	IC	Engine Data 2
Injector Pulsewidth	0-1000 ms	1.5-3.5	Varies	Engine Data 2, Fuel Trim, Misfire
Knock Retard (°)	0-25.5°	0°	0°	Engine Data 1, EGR
Long Term Fuel Trim	-27% to 27%	0 (± 5%)	0 (± 5%)	Engine Data 1-2, EVAP, Fuel Trim
Loop Status	Closed Loop / Open Loop	Closed Loop	Closed Loop	Engine Data 1-2, EGR, EVAP, Fuel Trim
MAF Sensor (g/sec)	0-512 g/sec	3-6	Varies	Engine Data 1-2, EGR, Fuel Trim, EVAP, Misfire
MAF Sensor (Hz)	0-32,000 Hz	1200-3000	Varies	Engine Data 2
MAP Sensor (kPa)	10-105 kPa	20-48	Varies	EGR, EVAP, Fuel Trim, Misfire
MAP Sensor (V)	0-5.1 volts	0.75-2.0	Varies	Engine Data 1-2
MIL Status	Okay / Fault / Invalid State	Okay	Okay	Output Driver
MIL Command	On / Off	Off	Off	Engine Data 2
Misfire Current Cyl 1-6	0-198 counts	0-4	0-4	Misfire
Misfire History Cyl 1-6	0-65,535	0	0	Misfire
Number of DTC(s)	Number	0	0	Engine Data 1, EVAP, Fuel Trim
PCM/VCM in VTD Fail Enable	Yes / No	Yes	Yes	Engine Data 1
PCM Reset	Yes / No	No	No	Engine Data 1
Power Enrichment	Active / Inactive	Inactive	Inactive	Engine Data 2, Misfire
Short Term Fuel Trim	-27% to 27%	0 (± 5%)	0 (± 5%)	Engine Data 1-2, EVAP, Fuel Trim
Spark Advance (°)	-64° to 64°	20	Varies	Engine Data 1-2, Fuel Trim, Misfire

2001-05 Grand Prix 3.8L V6 VIN K, VIN 1 (A/T) - PID Data List

Parameter Identification (PID)	PID Value Range	PID Value at Hot Idle	PID Value at 30 mph	Data List Type
Startup ECT	-40 to 419°F	Varies	Varies	EVAP, Fuel Trim
Startup IAT	-40 to 419°F	80°F	80°F	EVAP, Fuel Trim
Starter Enable Relay Control Circuit Status	Okay / Fault / Invalid State	Okay	Okay	Output Driver
Starter Enable Command	On / Off	Off	Off	Engine Data 2
TCC Brake Pedal Switch	Applied / Released	Released	Released	Engine Data 1-2
TFP Switch A/B/C	P, R, N, D, 4, 3, 2, 1	Park	4	Engine Data 2, EGR, Fuel Trim
Torque Delivered	0-100%	86	Varies	Engine Data 2
Torque Request	0-100%	100%	Varies	Engine Data 2
TP Angle (%)	0-100%	0	Varies	Engine Data 1-2, EGR, Fuel Trim, EVAP, Misfire
TP Sensor (V)	0-5.1 volts	0.5-1.2	Varies	Engine Data 1-2
Traction Control Circuit History	Okay / Fault	Okay	Okay	Output Driver
Traction Control Circuit Status	Okay / Fault	Okay	Okay	Output Driver
Traction Control Status	Active / Inactive	Inactive	Inactive	Engine Data 2
TWC Protection	Active / Inactive	Inactive	Inactive	Engine Data 1-2, HO2S, Misfire
Vehicle Speed	0-155 mph	0	30	Engine Data 1-2, EGR, Fuel Trim, EVAP, Misfire
VTD Auto Learn timer	Active / Inactive	Inactive	Inactive	Engine Data 1
VTD Fuel Disable	Active / Inactive	Inactive	Inactive	Engine Data 1
VTD Fuel Disable Until Key Off	Yes / No	No	No	Engine Data 1

<u>GTO PID DATA</u>

2004 GTO 5.7L V8 MFI VIN G (All) - PID Data List

Scan Tool Parameter	Data List	Parameter Range/Units	Typical Data Values
A/C Clutch Feedback Signal	ENG 2	On/Off	Off
A/C High Side Pressure	ENG 2	0-5 volts	Varies
A/C High Side Pressure	ENG 2	kPa/psi	Varies
A/C Relay Command	ENG 1, 2, 3, MF	On/Off	Off
A/C Request Signal	ENG 2	Yes/No	No
Air Fuel Ratio	ENG 2, 3	Ratio	14.6:1
BARO	ENG 1, EE, FT	kPa	65-104 kPa (varies w/altitude)
CMP Sensor High to Low	ENG 2	Counts	Varies
CMP Sensor Low to High	ENG 2	Counts	Varies
Cold Startup	ENG 2, EE	Yes/No	Varies
Column Lock Fuel Disable-ABS DTC	ENG 3, EE	Yes/No	No
Column Lock Fuel Disable-BCM DTC	ENG 3, EE	Yes/No	No
Column Lock-PCM BCM Communication	ENG 3, EE	Fault/OK	OK
Current Gear	ENG 1, 2, FT	0-4	1
Cycles of Misfire Data	MF	0-100 Counts	Varies
DTC Set This Ignition	ENG 1, 2, EE, FT, HO2S	Yes/No	No
Decel Fuel Cutoff	HO2S	Active/Inactive	Inactive
Desired IAC Airflow	ENG 1	0-64 g/s	Varies
Desired Idle Speed	ENG 1, 2, EE	RPM	PCM controlled, 800 RPM typical
ECT Sensor	ALL	-39° to +140°C (-38° to +284°F)	85-113°C (185-235°F)
Engine Load	ALL	0-100%	1-4%
Engine Run Time	ALL	Hrs, Min, Sec	Varies
Engine Speed	ALL	0-10,000 RPM	Varies, 750-850 RPM typical
Engine Oil Life Remaining	ENG 2, 3	%	Varies
Engine Oil Pressure Sensor	ENG 3	kPa/psi	Varies, 215-240 kPa (30-34 psi) typical
Engine Oil Pressure Sensor	ENG 3	Volts	Varies, 1.4-1.6V typical
EVAP Purge Solenoid Command	ENG 1, EE, FT	0-100%	0-25%
EVAP Vent Solenoid Command	ENG 1, EE, FT	Not Venting/Venting	Venting
FC Relay 1 Command	ENG 2, 3	On/Off	Depends on engine coolant temperature and A/C pressure
FC Relay 2 Command	ENG 2, 3	On/Off	Depends on engine coolant temperature and A/C pressure
Fuel Level Sensor	EE	5-0 volts	0.7-2.5 volts
Fuel Tank Level Remaining	EE	0-62.8 L (0-16.6 gal)	Varies
Fuel Tank Level Remaining	EE	0-100%	Varies
Fuel Tank Pressure Sensor	ENG 1, EE	-32.7 to +14.0 mm/Hg (-17.5 to +7.5 in/H2O)	Varies
Fuel Tank Pressure Sensor	EE	0-5.0 volts	Varies
Fuel Tank Rated Capacity	EE	62.8 L (16.6 gal)	62.8 L (16.6 gal)
Fuel Trim Cell	ENG 1, EE, FT	0-23	16, 17, 20
Fuel Trim Learn	ENG 1, EE, FT	Enabled/Disabled	Enabled, may toggle
HO2S Bank 1 Sensor 1	ENG 1, EE, FT	Millivolts	10-1,000 mV and varying
HO2S Bank 1 Sensor 2	ENG 1, FT	Millivolts	10-1,000 mV and varying
HO2S Bank 2 Sensor 1	ENG 1, FT, EE	Millivolts	10-1,000 mV and varying
HO2S Bank 2 Sensor 2	ENG 1, FT	Millivolts	10-1,000 mV and varying
HO2S Heater Bn 1 Sen. 1	HO2S	amps	0.40-0.59 amps
HO2S Heater Bn 1 Sen. 2	HO2S	amps	0.71-0.89 amps
HO2S Heater Bn 2 Sen. 1	HO2S	amps	0.40-0.59 amps
HO2S Heater Bn 2 Sen. 2	HO2S	amps	0.71-0.89 amps
IAC Position	ENG 1, FT	Counts	Varies
IAT	ENG 1, 2, 3, EE, FT	-39 to +140°C (-38 to +284°F)	Varies
Ignition 1 Signal	ENG 1, 2, 3, EE, FT	0-25 volts	11.5-14.5 volts
Injector PWM Bank 1 Average	ENG 2, FT, MF	ms	1-4 ms
Injector PWM Bank 2 Average	ENG 2, FT, MF	ms	1-4 ms

2004 GTO 5.7L V8 MFI VIN G (All) - PID Data List (CONT)

Scan Tool Parameter	Data List	Parameter Range/Units	Typical Data Values
Knock Retard	ENG 1	0.0-16°	0°
Long Term FT Avg. Bank 1	FT	Percentage	Near 0%
Long Term FT Avg. Bank 2	FT	Percentage	Near 0%
Long Term FT Bank 1	ENG 1, 2, EE, FT, HO2S	Percentage	Near 0%
Long Term FT Bank 2	ENG 1, 2, EE, FT, HO2S	Percentage	Near 0%
Loop Status	ENG 1, 2, 3, FT, HO2S	Open/Closed	Closed
MAF Sensor	ALL	Grams per second (g/s)	5-9 g/s (depends on altitude)
MAF Frequency	ENG 2, 3	0-31,999 Hz	Varies
MAP	ALL	kPa	20-48 kPa (varies w/ altitude)
MAP	ENG 1, 2	Volts	1.0-2.0 volts (varies w/ altitude)
Mileage Since DTC Cleared	ENG 2, 3	km/Miles	Varies
Misfire Current Cyl. #1-#8	MF	0-200 Counts	0
Misfire History Cyl. #1-#8	MF	0-65,535 Counts	0
MIL Command	ENG 2	ON/OFF	OFF
PCM Reset	ENG 1, 2, EE, FT	Yes/No	No
PCM/VCM in VTD Fail Enable	ENG 1	Enable/Disable	Disable
PNP	ENG 2	Park/Neutral	Park/Neutral
		In Gear	
Power Enrichment	FT, HO2S	Active/Inactive	Inactive
Reverse Inhibit Solenoid (M/T only)	ENG 1	Yes/No	No
Short Term FT Avg. Bn 1	ENG 3, FT	Percentage	Near 0%
Short Term FT Avg. Bn 2	ENG 3, FT	Percentage	Near 0%
Short Term FT Bank 1	ENG 1, 2, EE, FT, HO2S	Percentage	Near 0%
Short Term FT Bank 2	ENG 1, 2, EE, FT, HO2S	Percentage	Near 0%
Skip Shift Solenoid (M/T only)	ENG 1	Skip/No Skip	No Skip
Spark	ENG 1, 2, 3, FT, MF, HO2S	Degrees	Varies, 10-15 degrees typical
Start Up ECT	ENG 2, EE, FT	C°/F°	Varies
TCC Brake Pedal Switch	ENG 1, 2	Applied/Released	Released
TCC/Cruise Brake Pedal Switch	ENG 3	Applied/Released	Released
TCC Enable Solenoid Command	Eng 1, 2, MF	On/Off	Off
TCC PWM Solenoid Command	ENG 2	On/Off	Off
TFP Sw.	Eng 2, 3, FT	Park/Neutral	Park/Neutral
		Reverse, Drive 4, 3, 2, 1	
TP Sensor	ENG 1, 2, 3, HO2S	0-5.0 volts	0.4-0.9 volts
TP Sensor	ALL	%	0%
Traction Control Status	ENG 2	Active/Inactive	Inactive
TR Sw.	ENG 3, FT	Park, Reverse, Neutral, Drive 4, 3, 2, 1	Park
Vehicle Speed Sensor	ALL	km/h (mph)	0 km/h (mph)
VTD Auto. Learn Timer	ENG 1, 3	Minutes/Seconds	Varies
VTD Fuel Disable	ENG 1, 3	Active/Inactive	Inactive
VTD Fuel Disable Until Ign. Off	ENG 1, 3	Yes/No	No
Warm-Ups w/o Emission Faults	ENG 2, 3	0-255 Counts	Varies
Warm-Ups w/o Non-Emission Faults	ENG 2, 3	0-255 Counts	Varies

2005 GTO 6.0L V8 MFI VIN U (All) - PID Data List

Scan Tool Parameter	Data List	Parameter Range/Units	Typical Data Values
5-Volt Reference 1 Circuit Status	EE, Eng, Ign, TAC	OK/Fault	OK
5-Volt Reference 2 Circuit Status	EE, Eng, Ign, TAC	OK/Fault	OK
5-Volt Reference 1	EE, Eng, Ign, TAC	Volts	4.5 V
5-Volt Reference 2	EE, Eng, Ign, TAC	Volts	4.5 V
A/C Relay Circuit Status	OD	OK, Incomplete, Short B+, Short Gnd/Open	OK/Incomplete
A/C Relay Command	Eng , MF, TAC	On/Off	Varies
Air Flow Calculated	EE, Eng, FT, HO2S, MF, TAC	Grams Per Seconds (g/s)	1-9 g/s @ idle
Air Fuel Ratio	Eng, FT, HO2S	X:1	14.7:1
Ambient Air Temperature	Eng	°C/°F	Varies
APP Indicated Angle	EE, Eng, FT, HO2S, Ign, MF, TAC	0-100%	0
APP Sensor 1	TAC	0-5.0 Volts	0.4-1.0 Volt
APP Sensor 2	TAC	5.0-0 Volts	4.5-4.1 Volts
APP Sensor 1	TAC	0-100%	0%
APP Sensor 2	TAC	0-100%	0%
APP Sensor 1 and 2	TAC	Agree/Disagree	Agree
APP Sensor 1 Indicated Position	TAC	%	0%
APP Sensor 2 Indicated Position	TAC	%	--
APP Sensors	TAC	%	0%
BARO	EE, Eng, FT, HO2S, Ign	kPa	50-104 kPa/ Varies w/Altitude
BPP Circuit Signal	TAC	Applied/Released	--
Calculated TWC Temp Bank 1	FT, MF	Degrees C/F	315-427°C (600-800°F)
Calculated TWC Temp Bank 2	FT, MF	Degrees C/F	315-427°C (600-800°F)
Catalytic Converter Protection Active	FT, HO2S	Yes/No	No
CKP Active Counter	Ign	0-250 Counts	Varies
CKP Resync Counter	Ign	Counts	0
CKP Sensor	Eng, Ign	RPM	500-700 RPM
CMP Active Counter	Ign	0-250 Counts	Varies
CMP Sensor	Eng, Ign, MF	RPM	1,000-1,400 RPM
Cold Start-Up	Eng, EE	Yes/No	Varies
Cruise Control Active	Eng, TAC	Active/Inactive	Inactive
Cycles of Misfire Data	MF	0-100 Counts	Varies
Cylinder 1-8 IC Circuit Status	OD	OK, Incomplete, Short B+, Short Gnd/Open	OK
Cylinder 1-8 Injector Circuit Status	OD	OK, Incomplete, Short B+, Short Gnd/Open	OK
Decel Fuel Cutoff	Eng, FT, HO2S	Active/Inactive	Inactive
Desired Idle Speed	EE, Eng, TAC	RPM	ECM Controlled
Distance Since DTC Cleared	Eng	km/miles	Varies
EC Ignition Relay Circuit Status	OD	OK, Incomplete, Short B+, Short Gnd/Open	OK/Incomplete
EC Ignition Relay Command	Eng, TAC	On/Off	On
EC Ignition Relay Feedback	Eng, TAC	Volts	2.25-2.95 Volts
ECM Reset	EE, Eng, Ign	Yes/No	No
ECT Sensor	All	-39 to +140°C (-38 to +284°F)	88-105°C (190-221°F)
Engine Load	All	0-100%	18% @ Idle 21% @ 2500 RPM
Engine Off EVAP Test Connections Met	EE	Yes/No	Yes
Engine Oil Level Switch	Eng	OK/Low	OK
Engine Oil Pressure Sensor	Eng, MF	PSI	30
Engine Oil Pressure Sensor	EVAP	Volts	1.5

2005 GTO 6.0L V8 MFI VIN U (All) - PID Data List (CONT)

Scan Tool Parameter	Data List	Parameter Range/Units	Typical Data Values
Engine Run Time	All	Hrs, Min, Sec	Varies
Engine Speed	All	0-10,000 RPM	500-700 RPM
Engine Speed Circuit Status	OD	OK, Incomplete, Short B+, Short Gnd/Open	OK
EVAP Purge Solenoid Circuit Status	OD	OK, Incomplete, Short B+, Short Gnd/Open	OK/Incomplete
EVAP Purge Solenoid Command	EE, Eng, FT, HO2S	0-100%	10-25%
EVAP Vent Solenoid Circuit Status	OD	OK, Incomplete, Short B+, Short Gnd/Open	OK/Incomplete
EVAP Vent Solenoid Command	EE, Eng, FT	Not Venting/Venting	Venting
FC Circuit Status	OD	OK, Incomplete, Short B+, Short Gnd/Open	OK/Incomplete
Fuel Level Sensor	EE	0-5 Volts	0.7-2.5 Volts
Fuel Pump Relay Circuit Status	OD	OK, Incomplete, Short B+, Short Gnd/Open	OK/Incomplete
Fuel Pump Relay Command	Eng, FT	On/Off	ON
Fuel Tank Level Remaining	Eng, MF	0-100%	Varies
Fuel Tank Pressure Sensor	EE, Eng	-32.7 to +14.0 mm/Hg (-17.5 to +7.5 in/H2O)	Varies
Fuel Tank Pressure Sensor	EE, Eng	0-5.0 Volts	Varies
Fuel Tank Rated Capacity	EE	98 Liters (25.9 Gallons) or 96.2 Liters (25 Gallons)	Varies with Fuel Tank Option
Fuel Trim Cell	EE, Eng, FT	0-23	19
Fuel Trim Learn	EE, Eng, FT	Enabled/Disabled	Enabled, May Toggle
Fuel Volatility	EE, FT, HO2S	Low/Moderate/High	Varies
HO2S Bank 1 Sensor 1	EE, Eng, FT, HO2S	Millivolts	10-1,000 mV and Varying
HO2S Bank 1 Sensor 2	Eng, FT, HO2S	Millivolts	10-1,000 mV and Varying
HO2S Bank 2 Sensor 1	EE, Eng, FT, HO2S	Millivolts	10-1,000 mV and Varying
HO2S Bank 2 Sensor 2	Eng, FT, HO2S	Millivolts	10-1,000 mV and Varying
HO2S BN 1 Sensor 1 Heater	HO2S	Amps	0.7-0.9 Amps
HO2S BN 1 Sensor 2 Heater	HO2S	Amps	0.4-0.6 Amps
HO2S BN 2 Sensor 1 Heater	HO2S	Amps	0.7-0.9 Amps
HO2S BN 2 Sensor 2 Heater	HO2S	Amps	0.4-0.6 Amps
HO2S BNK 1 Sen 1 Heater Circuit Status	OD	OK, Incomplete, Short B+, Short Gnd/Open	OK/Incomplete
HO2S BNK 1 Sen 2 Heater Circuit Status	OD	OK, Incomplete, Short B+, Short Gnd/Open	OK/Incomplete
HO2S BNK 2 Sen 1 Heater Circuit Status	OD	OK, Incomplete, Short B+, Short Gnd/Open	OK/Incomplete
HO2S BNK 2 Sen 2 Heater Circuit Status	OD	OK, Incomplete, Short B+, Short Gnd/Open	OK/Incomplete
HO2S Bnk 1 Sensor 1 Heater Command	HO2S	On/Off	ON
HO2S Bnk 1 Sensor 1 Heater Command	HO2S	%	100%
HO2S Bnk 1 Sensor 2 Heater Command	HO2S	On/Off	ON
HO2S Bnk 1 Sensor 2 Heater Command	HO2S	%	100%
HO2S Bnk 2 Sensor 1 Heater Command	HO2S	On/Off	ON
HO2S Bnk 2 Sensor 1 Heater Command	HO2S	%	100%
HO2S Bnk 2 Sensor 2 Heater Command	HO2S	On/Off	ON
HO2S Bnk 2 Sensor 2 Heater Command	HO2S	%	100%
Hot Open Loop	FT, HO2S	Active/Inactive	Inactive
IAT Sensor	All	-39 to +140°C (-38 to +284°F)	35°C (91°F) Depends on Ambient Temperature
Ignition Accessory Signal	Eng, TAC	On/Off	ON
Ignition 1 Signal	All	0-25 Volts	11.5-14.5 Volts

2005 GTO 6.0L V8 MFI VIN U (All) - PID Data List (CONT)

Scan Tool Parameter	Data List	Parameter Range/Units	Typical Data Values
Ignition Off Timer	Eng	Seconds, Minutes, Hours	Varies
Inj. PWM Bank 1	Eng, FT, HO2S	Milliseconds	6-Feb
Inj. PWM Bank 2	Eng, FT, HO2S	Milliseconds	6-Feb
Knock Retard	Ign	0.0-16°	0°
KS Active Counter	Ign	Counts	Varies
KS Bank 1 Circuit Status	Ign	OK/Fault/Incomplete	OK/Incomplete
KS Bank 2 Circuit Status	Ign	OK/Fault/Incomplete	OK/Incomplete
KS Module Status	Ign	OK/Fault/Incomplete	OK/Incomplete
Long Term FT Bank 1	EE, Eng, FT, HO2S	Percentage	Near 0%
Long Term FT Bank 2	EE, Eng, FT, HO2S	Percentage	Near 0%
Loop Status	EE, Eng, FT, HO2S	Open/Closed	Closed
MAF Sensor	EE, Eng, FT, HO2S, Ign, MF, TAC,	Grams Per Seconds (g/s)	1-9 g/s @ Idle, Depends on Altitude 15-26 g/s @ 2,500 RPM, Depends on Altitude
MAF Sensor	EE, Eng, FT, HO2S, MF	0-31,999 Hz	1,150-1,350 Hz
MAP Sensor	EE, Eng, FT, HO2S, Ign, MF, TAC,	kPa	20-48 kPa
MAP Sensor	Eng, FT, HO2S, MF, TAC	Volts	1.0-2.0 Volts Varies with Altitude
MIL Circuit Status	OD	OK, Incomplete, Short B+, Short Gnd/Open	OK/Incomplete
MIL Command	EE, Eng, Ign	Off/On	Off
MIL Requested by DTC	EE, Eng, Ign	Yes/No	No
Misfire Current Cyl. 1-8	MF	0-200 Counts	0
Misfire History Cyl. 1-8	MF	0-65,535 Counts	0
PNP Switch	Eng	P/N Yes/No in gear	P/N
Power Enrichment	Eng, FT, HO2S, MF	Active/Inactive	Inactive
Reduced Engine Power	TAC	Active/Inactive	Inactive
Short Term FT Bank 1	EE, Eng, FT, HO2S	Percentage	Near 0%
Short Term FT Bank 2	EE, Eng, FT, HO2S	Percentage	Near 0%
Spark	Eng, FT, HO2S, Ign, MF	Degrees	10-17°
Starter Relay Circuit Status	OD	OK, Incomplete, Short B+, Short Gnd/Open	OK/Incomplete
Start-Up ECT	EE, FT	°C/°F	Varies
Start-Up IAT	EE, HO2S	°C/°F	Varies
TAC Forced Engine Shutdown	TAC	Yes/No	No
TAC Motor	TAC	Enabled/Disabled	Enabled
TAC Motor Command	TAC	0-100%	15-35%
TCC/Cruise Brake Pedal	Eng	Applied/Released/Invalid	Released
TCC/Cruise Brake Pedal Switch	TAC	Applied/Released/Invalid	Released
TCS Circuit Status	OD	OK, Incomplete, Short B+, Short Gnd/Open	OK/Incomplete
TCS Torque Request Signal	TAC	0-100%	100%
Torque Delivered Signal	TAC	0-100%	5-15%
Torque Management Spark Retard	Ign	Degrees	0°
Total Knock Retard	Ign	Degrees	0°
Total Misfire	MF	Counts	Varies
TP Desired Angle	TAC	0-100%	5.50%
TP Indicated Angle	EE, Eng, FT, HO2S, Ign, MF, TAC	0-100%	5.50%

2005 GTO 6.0L V8 MFI VIN U (All) - PID Data List (CONT)

Scan Tool Parameter	Data List	Parameter Range/Units	Typical Data Values
TP Sensor 1	TAC	0-5.0 Volts	4.1-4.95 Volts
TP Sensor 1	TAC	0-100%	Varies near 5%
TP Sensor 1 Learned Minimum	TAC	Volts	0.55
TP Sensor 2	TAC	5.0-0 Volts	0.4-0.85 V
TP Sensor 2	TAC	100-0%	Varies near 5%
TP Sensor 2 Learned minimum	TAC	Volts	0.55
TP Sensors 1 and 2	TAC	Agree/Disagree	Agree
TP Sensor 1 Indicated Position	TAC	%	5%
TP Sensor 2 Indicated Position	TAC	%	5%
TWC Temperature Calculated	FT, MF	Degrees C/F	--570°C (1,075°F)
Vacuum Calculated	EE, Eng, FT, HO2S	kPa/in Hg	59 kPa/16 in Hg
Vehicle Speed Circuit Status	OD	OK, Incomplete, Short B+, Short Gnd/Open	OK/Incomplete
Vehicle Speed Circuit 2 Status	OD	OK, Incomplete, Short B+, Short Gnd/Open	OK/Incomplete
Vehicle Speed Sensor	EE, Eng, FT, HO2S, Ign, MF, TAC	km/h mph	0
Warm-Ups w/o Emission Faults	Eng	0-255 Counts	Varies
Warm-Ups w/o Non-Emission Faults	Eng	0-255 Counts	Varies
Warm-Ups Since DTC Cleared	Eng	Counts	Varies
Wide Open Throttle	TAC	Yes/No	No

MONTANA SV6 PID DATA

2005 Montana SV6 3.5L VIN L (ALL) – PID Data List

Scan Tool Parameter	Data List	Units Displayed	Typical Data Value
24X Crank Sensor	Ignition	RPM	0-6,400
A/C Relay Circuit Status	ODD	OK/Fault/Invalid State	OK
A/C Relay Circuit History	ODD	OK/Fault/Invalid State	OK
A/C Relay Command	Eng, Misfire, EGR, TAC	On/Off	Off
Air Fuel Ratio	Eng, Fuel Trim, HO2S	Ratio	14.2:1-14.7:1
APP Indicated Angle	Eng, EGR, EVAP, HO2S, Ignition, Misfire, TAC, Fuel Trim	Percent	0
APP Sensor 1	TAC	Volts	Varies
APP Sensor 1 Circuit	TAC	OK/Not OK	OK
APP Sensor 2	TAC	Volts	Varies
APP Sensor 2 Circuit	TAC	OK/Not OK	OK
APP Sensor 1	TAC	Percent	Varies
APP Sensor 2	TAC	Percent	Varies
APP Sensor 1 and 2	TAC	Agree/Disagree	Agree
BARO	Eng, EGR, EVAP, Fuel Trim, HO2S, Ignition	kPa	65-110 kPa
Catalytic Converter Protection Active	Fuel Trim, HO2S	Yes/No	No
CMP Sensor	Ignition	RPM	Varies
CMP Sensor Signal Present	Ignition, Misfire	Yes/No	Yes
Cruise Control Active	TAC	Yes/No	No
Cycles of Misfire Data	Misfire	Counts	0-99
Cyl. 1-6 Injector Circuit Status	ODD	OK/Stuck Low (open)/Stuck High	OK
Cylinder 1 and 4 IC Circuit Status	ODD	OK/Not OK	OK
Cylinder 2 and 5 IC Circuit Status	ODD	OK/Not OK	OK
Cylinder 3 and 6 IC Circuit Status	ODD	OK/Not OK	OK
Cylinder 1-6 Injector Circuit History	ODD	OK/Stuck Low (open)/Stuck High	OK
Decel Fuel Cut-off	Eng, EGR, Fuel Trim, HO2S	Active/Inactive	Inactive
Desired EGR Position	EGR, Misfire	Percent	0
Desired Idle Speed	Eng, EVAP, TAC	RPM	Varies
Driver Module 1 Status	ODD	Enabled/Off-High Volts/Off High Temp/Invalid State	Enabled
Driver Module 2 Status	ODD	Enabled/Off-High Volts/Off High Temp/Invalid State	Enabled
Driver Module 3 Status	ODD	Enabled/Off-High Volts/Off High Temp/Invalid State	Enabled
EC Ignition Relay Circuit Status	ODD	OK/Fault/Invalid State	OK
EC Relay Circuit History	ODD	OK/Fault/Invalid State	OK
ECT Sensor	Eng, EGR, EVAP, HO2S, Ignition, Misfire, Fuel Trim, TAC	°C/°F	Varies
EGR Flow Test Count	EGR	Counts	0-12
EGR Learned Minimum Position	EGR	Volts	0.14-1.0
EGR Position Sensor	Eng, EGR, Misfire	Percent	Varies
EGR Position Sensor	EGR	Volts	0
EGR Position Variance	EGR	Percent	0
EGR Solenoid Circuit History	ODD	OK/Open/Fault/Invalid State	OK
EGR Solenoid Circuit Status	ODD	OK/Fault/Invalid State	OK
EGR Solenoid Command	EGR	Percent	0

2005 Montana SV6 3.5L VIN L (ALL) – PID Data List (CONT)

Scan Tool Parameter	Data List	Units Displayed	Typical Data Value
Engine Load	Eng, Misfire, Fuel Trim, EVAP, EGR, HO2S, Ignition, TAC	Percent	0
Engine Oil Pressure Switch	Eng	Low/OK	OK
Engine Run Time	Eng, EVAP, Misfire, Fuel Trim, EGR, HO2S, Ignition, TAC, ODD	Hr: Min: Sec	Varies
Engine Speed	Eng, EGR, EVAP, Ignition, Misfire, Fuel Trim, TAC	RPM	Varies
EVAP Fault History	EVAP	No Fault/Excess Vacuum/Purge Valve Leak/Small Leak/Weak Vacuum/No Test Result	No Fault
EVAP Purge Solenoid Circuit History	ODD	OK/Open/Short	OK
EVAP Purge Solenoid Circuit Status	ODD	OK/Open/Short	OK
EVAP Purge Solenoid Command	Eng, EVAP, Fuel Trim, HO2S	Percent	Varies
EVAP Test Abort Reason	EVAP	No Aborted/Lost Enable/Small Leak/Veh. Not At Rest	Varies
EVAP Test Result	EVAP	• No Result/Passed/Aborted OR • Fail-DTC P0455 OR • Fail-DTC P0442 OR • Fail-DTC P0446 OR • Fail-DTC P0496	Varies
EVAP Test State	EVAP	Waiting for Purge/Test Running/Test Completed	Varies
EVAP Vent Solenoid Circuit History	ODD	OK/Open/Short	OK
EVAP Vent Solenoid Circuit Status	ODD	OK/Open/Short	OK
EVAP Vent Solenoid Command	Eng, EVAP	Venting/Not Venting	Varies
FC Relay 1 Circuit History	ODD	OK/Fault/Invalid State	OK
FC Relay 1 Circuit Status	ODD	OK/Fault/Invalid State	OK
FC Relay 2 and 3 Circuit History	ODD	OK/Fault/Invalid State	OK
FC Relay 2 and 3 Circuit Status	ODD	OK/Fault/Invalid State	OK
Fuel Level Sensor	EVAP	Volts	Varies
Fuel Pump Relay Circuit History	ODD	OK/Fault/Invalid State	OK
Fuel Pump Relay Circuit Status	ODD	OK/Fault/Invalid State	OK
Fuel Pump Relay Command	Eng	On/Off	On
Fuel Tank Level Remaining	EVAP	Percent	0-104
Fuel Tank Pressure Sensor	Eng, EVAP	mm Hg/in H2O	Varies
Fuel Tank Pressure Sensor	EVAP	Volts	Varies
Fuel Trim Cell	EVAP, Eng, Fuel Trim	0-4	0
Fuel Trim Learn	Eng, EVAP, Fuel Trim	Enabled/Disabled	Varies
HO2S Bank 1 Sensor 1	EVAP, Eng, Fuel Trim, HO2S	millivolts	0-1000 and Varying
HO2S Bank 1 Sensor 1 Ready	Fuel Trim, HO2S	Yes/No	Yes
HO2S Bank 1 Sensor 2	Eng, Fuel Trim, HO2S	millivolts	0-1000 and Varying
HO2S Bank 2 Sensor 1 Ready	Fuel Trim, HO2S	Yes/No	Yes
HO2S Bank 2 Sensor 1	Eng, EVAP, Fuel Trim, HO2S	millivolts	0-1000 and Varying
HO2S Bank 2 Sensor 2	Eng, Fuel Trim, HO2S	millivolts	0-1000 and Varying
HO2S Htr. Bn. 1 Sen. 1	HO2S	On/Off	On

2005 Montana SV6 3.5L VIN L (ALL) – PID Data List (CONT)

Scan Tool Parameter	Data List	Units Displayed	Typical Data Value
HO2S Htr. Bank 1 Sensor 1 Command	HO2S	On/Off	On
HO2S Htr. Bn. 1 Sen. 2	HO2S	On/Off	On
HO2S Htr. Bank 1 Sensor 2 Command	HO2S	On/Off	On
HO2S Heater Bn. 1 Sen. 1	HO2S	Amps	Varies
HO2S Heater Bn. 1 Sen. 2	HO2S	Amps	Varies
HO2S Heater Bn. 2 Sen. 1	HO2S	Amps	Varies
HO2S Heater Bn. 2 Sen. 2	HO2S	Amps	Varies
HO2S Htr. Bn. 1 Sen. 1 Circuit History	ODD	OK/Fault/Invalid State	OK
HO2S Htr. Bn. 1 Sen. 2 Circuit History	ODD	OK/Fault/Invalid State	OK
HO2S Htr. Bn. 2 Sen. 1	HO2S	On/Off	On
HO2S Htr. Bn. 2 Sen. 1 Circuit History	ODD	OK/Fault/Invalid State	OK
HO2S Htr. Bn. 2 Sen. 2	HO2S	On/Off	On
HO2S Htr. Bn. 2 Sen. 2 Circuit History	ODD	OK/Fault/Invalid State	OK
HO2S Htr. Bn. 1 Sen. 1 Circuit Status	ODD	OK/Fault/Invalid State	OK
HO2S Htr. Bn. 1 Sen. 2 Circuit Status	ODD	OK/Fault/Invalid State	OK
HO2S Htr. Bn. 2 Sen. 1 Circuit Status	ODD	OK/Fault/Invalid State	OK
HO2S Htr. Bn. 2 Sen. 2 Circuit Status	ODD	OK/Fault/Invalid State	OK
Hot Open Loop	Fuel Trim, HO2S	Active/Inactive	Inactive
IAT Sensor	Eng, EVAP, EGR, Fuel Trim, HO2S, Ignition, Misfire, TAC	°C/°F	Varies
Ignition 1 Signal	EVAP, Eng, EGR, Fuel Trim, HO2S, Ignition, Misfire, ODD, TAC	Volts	Varies
Ignition Off Time	EVAP, Eng	Minutes/Varies	Varies
Initial Brake Apply Signal	TAC	Released/Applied	Released
Injector PWM	Fuel Trim, Misfire	milliseconds	Varies
Knock Retard	EGR, Ignition	Degrees	0
Long Term FT Bank 1	Eng, EVAP, Fuel Trim, HO2S	Percent	-10% to +10%
Long Term FT Bank 2	Eng, EVAP, Fuel Trim, HO2S	Percent	-10% to +10%
Loop Status	Eng, Fuel Trim, EVAP, HO2S	Open/Closed	Closed
MAF Sensor	Eng, EGR, Misfire, EVAP, Fuel Trim, HO2S, Ignition, TAC	g/s	3-6 (Varies with altitude)
MAF Sensor	EVAP, Fuel Trim, HO2S, Misfire, TAC	Hz	1,200-3,000 (depends on altitude)
MAP Sensor	Eng, EGR, Fuel Trim, HO2S, Misfire, TAC	Volts	0.4-2.0 Volts
MAP Sensor	Eng, EVAP, Fuel Trim, HO2S, Ignition, Misfire, EGR, TAC	kPa	20-48 kPa (Varies with altitude)
MIL Circuit History	ODD	OK/Fault/Invalid State	OK
MIL Command	EVAP, Eng, EGR, Ignition	On/Off	Off
MIL Circuit Status	ODD	OK/Fault/Invalid State	OK
Mileage Since DTC Cleared	Eng	km/h/mph	Varies
Misfire Current Cyl. 1-6	Misfire	Counts	0
Misfire History Cyl. 1-6	Misfire	Counts	0
Moderate Brake Apply Signal	TAC	Applied/Released	Released
Number of DTCs	Eng EVAP, Fuel Trim, TAC	Number	0
PCM Reset	Eng	Yes/No	No

2005 Montana SV6 3.5L VIN L (ALL) – PID Data List (CONT)

Scan Tool Parameter	Data List	Units Displayed	Typical Data Value
Power Enrichment	Eng, Fuel Trim, HO2S, Misfire	Active/Inactive	Inactive
Reduced Engine Power	TAC	Active/Inactive	Inactive
Short Term FT Bank 1	Eng, EVAP, Fuel Trim, HO2S,	Percent	-10 to +10
Short Term FT Bank 2	Eng, EVAP, Fuel Trim, HO2S,	Percent	-10 to +10
Spark	Eng, Misfire, Fuel Trim, HO2S, Ignition	Degrees	22 (varies)
Start Up ECT	EVAP, Fuel Trim, HO2S	°C/°F	Varies
Start Up IAT	EVAP, Fuel Trim, HO2S	°C/°F	Varies
Starter Relay Circuit Status	ODD	OK/Fault	OK
Starter Relay Circuit History	ODD	OK/Fault/Invalid State	OK
Stop Lamp Pedal Switch	Eng, TAC	Released/Applied	Released
TAC/PCM Communications Signal	TAC	OK/Fault	OK
TAC Stop Lamp Pedal Switch	Eng	Released/Applied	Released
TAC Stop Lamp Pedal Switch	TAC	Applied/Released	Released
TAC Vehicle Speed Signal	TAC	OK/Fault	OK
TCS Delivered Torque Circuit Status	ODD	OK/Fault	OK
Torque Request Signal	TAC	Percent	100%
Torque Delivered Signal	TAC	Percent	75%
Total Misfire Count	Misfire	1-255	0
TP Desired Angle	TAC	Percent	Varies
TP Indicated Angle	Eng, EGR, Fuel Trim, HO2S, Ignition, EVAP, Misfire, TAC	Percent	%
TP Sensor 1	TAC	Percent	Varies
TP Sensor 1	TAC	Volts	Varies
TP Sensors 1 and 2	TAC	Agree/Disagree	Agree

MONTANA & TRANS SPORT

1996-2005 Montana & Trans Sport Light Van 3.4L VIN E - PID Data List

Parameter Identification (PID)	PID Value Range	PID Value at Hot Idle	PID Value at 30 mph	Data List Type
3X Crank Sensor	0-9999 rpm	Varies	Varies	Engine Data 2
24X Crank Sensor	0-1600 rpm	Varies	Varies	Engine Data 2
A/C HI Side Pressure	0-5.0v	Varies	Varies	Engine Data 1
A/C Pressure Disable	Yes/No	No	No	Engine Data 2
A/C Pressure Disable	Yes/No	No	No	Engine Data 2
A/C Off for WOT	Yes/No	No	No	Engine Data 2
A/C Pressure Disable	Yes/No	No	No	Engine Data 2
A/C Relay Circuit Status	Okay/Fault	Okay	Okay	Output Driver
A/C Relay Command	On/Off	Off	Off	Engine Data 1-2, EGR, Misfire
A/C Request Signal	Yes/No	No	No	Engine Data 2
Actual EGR Position	0-100%	0	Varies	Engine Data 1, EGR, Misfire
Air Fuel Ratio	0.0-25.5:1	14.7:1	14.7:1	Engine Data 2, Fuel Trim
BARO Sensor (kPa)	10-110 kPa	98	98	Engine Data 1, EGR, EVAP, Fuel Trim
BARO Sensor (V)	0-5.0 volts	3.5-4.5	3.5-4.5	Engine Data 1, EGR, EVAP, Fuel Trim
Brake Switch Signal	Applied/REL	Released	Released	Engine Data 1
CMP Sensor Signal	Yes/No	Yes	Yes	Engine Data 1, Misfire
Crank Request Signal	Yes/No	No	No	Engine Data 2
Cruise Control Active	Yes/No	No	No	Engine Data 1
Cruise Inhibit Reason	Several	Varies	Varies	Engine Data 1
Cruised Inhibit Signal Circuit Status	Okay/Fault/ Invalid State	Okay	Okay	Output Driver
Cruise Inhibit Signal Command	On/Off	On	On	Engine Data 1
Current Gear	1-2-3-4	1	4	Engine Data 1-2, EGR, Fuel Trim
Cycles of Misfire Data	0-99 counts	Varies	Varies	Misfire
Cylinder 1-6 Injector Circuit Status	Okay/Stuck Low/Stuck High/Fault	Okay	Okay	Output Driver
Cylinder 1-6 Injector Circuit History	Okay/Stuck Low/Stuck High/Fault	Okay	Okay	Output Driver
Deceleration Fuel Cutoff	Active/Inactive	Inactive	Inactive	EGR, Fuel Trim, Misfire
Desired EGR Position	0-100%	0	Varies	Engine Data 1-2, EGR, Misfire
Desired Idle Speed	0-3187 rpm	PCM control	---	Engine Data 1, Data 2, EVAP
Driver Module 1 Status	Enabled/ Disabled	Enabled	Enabled	Output Driver
Driver Module 2 Status	Enabled/ Disabled	Enabled	Enabled	Output Driver
Driver Module 3 Status	Enabled/ Disabled	Enabled	Enabled	Output Driver
Driver Module 4 Status	Enabled/ Disabled	Enabled	Enabled	Output Driver
ECT Sensor (°F)	-40° to 304°F	185-220	185-220	Engine Data 1-2, EGR, Fuel Trim, EVAP, Misfire
EGR Flow Test Count	0-255 counts	0	Varies	EGR
EGR Position Sensor	0-100%	0	Varies	Engine Data 1, EGR, Misfire
EGR Position Sensor	0-5.1 volts	0.14-1.0	Varies	Engine Data 1, EGR, Misfire
EGR Learned Minimum Position	0-5.1 volts	0.71	Varies	EGR
EGR Position Error	0-100%	0	Varies	EGR
EGR Pos. Variance	0-100%	0	Varies	EGR
EGR Solenoid Command	0-100%	0	Varies	Engine Data 1, EGR, Misfire
Engine Load (%)	0-100%	17	Varies	Engine Data 1-2, EGR, Fuel Trim EVAP, Misfire
Engine Oil Level Switch	Okay/Low	Okay	Okay	Engine Data 2
Engine Oil Life Remaining	0-100%	Varies	Varies	Engine Data 2

1996-2005 Montana & Trans Sport Light Van 3.4L VIN E - PID Data List

Parameter Identification (PID)	PID Value Range	PID Value at Hot Idle	PID Value at 30 mph	Data List Type
EOP Switch Signal	Okay/Low/HI	Okay	Okay	Engine Data 2
Engine Run Time	Hr: Min: Sec	00:00:00	00:00:00	Engine Data 1-2, EGR, Fuel Trim EVAP, Misfire
Engine Speed	0-9999 rpm	± 50 rpm of Desired Speed	Varies	Engine Data 1-2, EGR, Fuel Trim EVAP, Misfire
EVAP Fault History	No Fault / Excess VAC / Purge Valve Leak / Small Leak / Weak Vacuum	No Fault	No Fault	EVAP
EVAP Purge Solenoid Circuit Status	Okay / Fault	Okay	Okay	Output Driver
EVAP Purge Solenoid Command	0-100%	19	Varies	Engine Data 1, EVAP, HO2S
EVAP Vent Solenoid Control Circuit Status	Okay / Fault	Okay	Okay	Output Driver
EVAP Vent Solenoid Command	Venting/Not Venting	Venting	Venting	Engine Data 1, EVAP, Fuel Trim
FC Relay 1 Command	On/Off	Off	Off	Engine Data 2
FC Relay 2 and 3 Circuit Status	Okay/Fault	Okay	Okay	Output Driver
FC Relay 2 and 3 Command	On/Off	Off	Off	Engine Data 2
Fuel Level Sensor	0-5.0 volts	0.8-2.5	Varies	EVAP
Fuel Pump Command	On/Off	On	On	Engine Data 1
Fuel Pump Relay Circuit Status	Okay/Fault	Okay	Okay	Output Driver
Fuel Pump Relay Circuit History	Okay/Fault	Okay	Okay	Output Driver
Fuel Pump Relay Command	On/Off	On	On	Engine Data 1
Fuel Tank Level Remaining	0-100%	Varies	Varies	EVAP
Fuel Tank Pressure Sensor	Inches H2O, mm Hg, volts	-17 to +7.5 Inches H2O	-17 to +7.5 Inches H2O	Engine Data 1, EVAP
Fuel Tank Pressure Sensor	0-5.0 volts	Varies	Varies	Engine Data 1, EVAP
Fuel Trim Cell	Cell 0-10	Varies	Varies	Engine Data 1, EVAP, Fuel Trim
Fuel Trim Learn	Enabled or Disabled	Enabled	May Toggle	Engine Data 1, EVAP, Fuel Trim
FC Relay 1 Circuit Status	Okay / Fault	Okay	Okay	Output Driver
Generator Lamp	On/Off	Off	Off	Engine Data 2
Generator L-Terminal	On/Off	On	On	Engine Data 2
HO2S-11 (B1 S1)	0-1132 mv	10-1000	10-1000	Engine Data 1, EVAP, Fuel Trim
HO2S-12 (B1 S2)	0-1132 mv	10-1000	10-1000	Engine Data 1, EVAP, Fuel Trim
HO2S-11 XCounts	0-255	Varies	Varies	Engine Data 1, HO2S
HO2S-11 Heater Command	On/Off	On	On	Fuel Trim
Hot Mode	On/Off	Off	Off	Misfire
IAC Position	0-255 counts	15	Varies	Engine Data 1, EGR, Fuel Trim
IAT Sensor (°F)	-40° to 304°F	50-194	50-194	Engine Data 1-2, EGR, EVAP, Fuel Trim
Ignition 1 Signal	0.0-25.5v	12.8-14.2	12.8-14.2	Engine Data 1-2, EGR, EVAP, Fuel Trim
Ignition Mode	Bypass/IC	IC	IC	Engine Data 2
IMS	P/R/N/D4-1	Park	Drive 4	Engine Data 1
Injector Pulsewidth	0-1000 ms	2.90	Varies	Engine Data 2, Fuel Trim, Misfire
Knock Retard (°)	0-25.5°	0	0	Engine Data 1, EGR
Long Term Fuel Trim	-27% to 27%	0 (± 5%)	0 (± 5%)	Engine Data 1-2, EVAP, Fuel Trim
Loop Status	Closed Loop / Open Loop	Closed Loop	Closed Loop	Engine Data 1-2, EGR, EVAP, Fuel Trim

1996-2005 Montana & Trans Sport Light Van 3.4L VIN E - PID Data List

Parameter Identification (PID)	PID Value Range	PID Value at Hot Idle	PID Value at 30 mph	Data List Type
MAF Sensor (g/sec)	0-512 g/sec	3.41	Varies	Engine Data 1-2, EGR, EVAP, Fuel Trim, Misfire
MAF Sensor (Hz)	0-32,000 Hz	2210	Varies	Engine Data 1-2, EGR, EVAP, Fuel Trim, Misfire
MAP Sensor (kPa)	10-105 kPa	38	Varies	Engine Data 1-2, EGR, EVAP, Fuel Trim, Misfire
MAP Sensor (V)	0-5.1 volts	1.45	Varies	Engine Data 2
MIL Circuit Status	Okay/Fault	Okay	Okay	Output Driver
MIL Command	On/Off	Off	Off	Engine Data 2
Misfire Current Cylinder 1-6	0-198 counts	0	0	Misfire
Misfire History Cylinder 1-6	0-65,535	0	0	Misfire
Non-Volatile Memory	PASS/FAIL	PASS	PASS	Engine Data 1
Number of DTC(s)	0-255 counts	0	0	Engine Data 1, EVAP, Fuel Trim
HO2S Heater Current	0-10.0 amps	0.54	0.54	Fuel Trim
PCM Reset	Yes/No	No	No	Engine Data 1-2, EGR, EVAP, Fuel Trim, ODD
PCM/VCM in VTD Fuel Enable	Yes/No	No	No	Engine Data 1
Power Enrichment	Active/Inactive	Inactive	Inactive	Engine Data 2, Misfire
Reduced Engine Power	Active/Inactive	Inactive	Inactive	EGR
Short Term Fuel Trim	-27% to 27%	0 (± 5%)	0 (± 5%)	Engine Data 1-2, EVAP, Fuel Trim
Spark Advance (º)	-64º to 64º	20	Varies	Engine Data 1-2 Fuel Trim, Misfire
Startup ECT	-40 to 419ºF	Varies	Varies	Engine Data 2, EVAP, Fuel Trim
Startup IAT	-40 to 419ºF	80ºF	80ºF	Engine Data 2, EVAP, Fuel Trim
Starter Enable Relay Circuit Status	Okay / Fault/ Invalid State	Okay	Okay	Output Driver
Starter Relay Control	On/Off	Off	Off	Engine Data 2
TCC Brake Pedal Switch	Applied / Released	Released	Released	Engine Data 1-2
TCC PWM Solenoid Command	On/Off	Off	On	Engine Data 2, Misfire
TFP Switch Signal	P/R/N/D4-1	Park/Neutral	Drive 4	Engine Data 2, EGR, Fuel Trim
Torque Delivered Signal	0-100%	86	86	Engine Data 2
Torque Request Signal	0-100%	100	100	Engine Data 2
Total Misfire Current Count	0-65,535	Varies	Varies	Misfire
TP Angle (%)	0-100%	0	0-100	Engine Data 1-2, EGR, EVAP, Fuel Trim, Misfire
TP Sensor (V)	0-5.1 volts	0.20-0.74	Varies	Engine Data 1-2, EGR, EVAP, Fuel Trim, Misfire
Traction Control Status	Active or Inactive	Inactive	Inactive	Engine Data 2
Transmission Range	Park/Neutral, Drive 4-3-2-1	Park/Neutral	Drive 4	Engine Data 1, AIR, EGR, Catalyst
TR Switch P/A/B/C	High/Low	High	Low	Engine Data 1
TWC Protection	Active or Inactive	Inactive	Inactive	Engine Data 2, Misfire
Vehicle Speed	0-155 mph	0	30	Engine Data 1-2, EGR, EVAP, Fuel Trim, Misfire
VTD Auto Learn Timer	Active/Inactive	Inactive	Inactive	Engine Data 1
VTD Fuel Disable	Active/Inactive	Inactive	Inactive	Engine Data 1
VTD Fuel Disable Until Ignition Off	Yes/No	No	No	Engine Data 1

SUNFIRE PID DATA

1996-2005 Sunfire 2.2L VIN 4, 2.2L VIN F, 2.4L I4 VIN T - PID List

Parameter Identification (PID)	PID Value Range	PID Value at Hot Idle	PID Value at 30 mph	Data List Type
A/C Relay Command	On / Off	Off	Off	Engine Data 1
Airflow Calculated	0-512 g/sec	3-7	Varies	Engine Data 1
BARO Sensor (V)	10-110 kPa	65-104	65-104	Engine Data 1
Cruise Control Active	Yes / No	No	No	Engine Data 1
Cruise/TCC Switch	Applied/Not	Not Applied	Not Applied	Engine Data 1
Current Gear	1-6	6	3	Engine Data 1
Desired Idle Speed	0-3187	900	---	Engine Data 1
ECT Sensor (°F)	-40 to 304°F	190	190	Engine Data 1
Engine Load	0-100%	11-25	Varies	Engine Data 1
Engine Run Time	Hr: Min: Sec	00:00:00	00:00:00	Engine Data 1
Engine Speed	0-9999 rpm	Varies	---	Engine Data 1
EVAP Purge Solenoid	0-100%	0	0-100	Engine Data 1
EVAP Vent Solenoid	Venting / Not Venting	Venting	Venting	Engine Data 1
Fuel Level Sensor	0-100%	Varies	Varies	Engine Data 1
Fuel Pump Relay	On / Off	On	On	Engine Data 1
FTP Sensor (In. H2O)	-17.5 to 7.5"	Varies	Varies	Engine Data 1
Fuel Trim Cell	Cell #0-22	18-21	18-21	Engine Data 1
Generator F-Terminal	0-100%	Varies	Varies	Engine Data 1
Generator L-Terminal	Voltage/No	Voltage	Voltage	Engine Data 1
HO2S-12 (B1 S2)	0-1132 mv	10-1000	10-1000	Engine Data 1
IAC Position	0-255 counts	42	---	Engine Data 1
IAT Sensor (°F)	-40 to 304°F	50-176	50-176	Engine Data 1
Ignition 1 Signal	0.0-25.5v	13.9	13.9	Engine Data 1
Knock Retard (°)	0-90°	0	0	Engine Data 1
Long Term Fuel Trim	-10% to 10%	0 (± 5%)	0 (± 5%)	Engine Data 1
Loop Status	Closed Loop / Open Loop	Closed Loop	Closed Loop	Engine Data 1
MAP Sensor (kPa)	10-110 kPa	33	Varies	Engine Data 1
MAP Sensor (V)	0.0-5.0v	1-2	Varies	Engine Data 1
MIL (lamp) Command	On / Off	Off	Off	Engine Data 1
No. of Current DTC(s)	0-255	0	0	Engine Data 1
O2S-11 (B1 S1)	0-1132 mv	10-1000	10-1000	Engine Data 1
Short Term Fuel Trim	-10% to 10%	0 (± 5%)	0 (± 5%)	Engine Data 1
Spark Advance (°)	-64° to 64°	18	Varies	Engine Data 1
TCC Enable Solenoid	On / Off	On	Off	Engine Data 1
Throttle Angle (%)	0-100%	0	Varies	Engine Data 1
TP Sensor (V)	0.0-5.0v	0.20-0.90	Varies	Engine Data 1
Vehicle Speed	0-155 mph	0	30	Engine Data 1

1996-2005 Sunfire 2.2L VIN 4, 2.2L VIN F, 2.4L I4 VIN T - PID List

Parameter Identification (PID)	PID Value Range	PID Value at Hot Idle	PID Value at 30 mph	Data List Type
A/C HI Side Pressure	0-450 psi	139-399	139-399	Engine Data 2
A/C HI Side Pressure	0.0-5.0v	1-4	1-4	Engine Data 2
A/C Relay Command	On / Off	Off	Off	Engine Data 2
A/C Request Signal	Yes / No	Off	Off	Engine Data 2
Calculated Airflow	0-512 g/sec	3-7	Varies	Engine Data 2
Air Fuel Ratio	0.0-25.5:1	14.6:1	14.6:1	Engine Data 2
BARO Sensor (kPa)	10-110 kPa	65-104	65-104	Engine Data 2

1996-2005 Sunfire 2.2L VIN 4, 2.2L VIN F, 2.4L I4 VIN T - PID List

Parameter Identification (PID)	PID Value Range	PID Value at Hot Idle	PID Value at 30 mph	Data List Type
CKP Active Counter	0-255 counts	Counts up	Counts up	Engine Data 2
CMP Active Counter	0-255 counts	Counts up	Counts up	Engine Data 2
CMP Resync Counter	0-255	0	0	Engine Data 2
Cruise Control Active	Yes / No	No	No	Engine Data 2
C/Control/TCC Switch	Applied/Not	Not Applied	Not Applied	Engine Data 2
Current Gear	1-4	1	3	Engine Data 2
Desired Idle Speed	0-3187	900	---	Engine Data 2
ECT Sensor (°F)	-40 to 419°F	185-239	185-239	Engine Data 2
Engine Load	0-100%	11-25	Varies	Engine Data 2
Engine Run Time	Hr: Min: Sec	00:00:00	00:00:00	Engine Data 2
Engine Speed	0-9999 rpm	Varies	---	Engine Data 2
FC Relay 1 Command	On / Off	Off	Off	Engine Data 2
Generator F-Terminal	0-100%	37	Varies	Engine Data 2
Generator L-Terminal	Voltage/No	Voltage	Voltage	Engine Data 2
IAT Sensor (°F)	-40 to 304°F	50-176	50-176	Engine Data 2
Injector Pulsewidth	0-985 ms	3-4	Varies	Engine Data 2
Long Term Fuel Trim	-10% to 10%	0 (± 5%)	0 (± 5%)	Engine Data 2
Loop Status	Closed Loop / Open Loop	Closed Loop	Closed Loop	Engine Data 2
MAP Sensor (kPa)	10-110 kPa	33	Varies	Engine Data 2
Medium Resolution Resync Counter	0-255 counts	0	0	Engine Data 2
Power Enrichment	Active / Inactive	Inactive	Inactive	Engine Data 2
Short Term Fuel Trim	-10% to 10%	0 (± 5%)	0 (± 5%)	Engine Data 2
Spark Advance (°)	-64° to 64°	18	Varies	Engine Data 2
Startup ECT Sensor	-40 to 419°F	Varies	Varies	Engine Data 2
TCC Enable Solenoid	On / Off	On	Off	Engine Data 2
Throttle Angle (%)	0-100%	0	Varies	Engine Data 2
TR Range Switch	Several	Several	Several	Engine Data 2
Vacuum Calculated	10-110 kPa	Varies	Varies	Engine Data 2

1996-2005 Sunfire 2.2L VIN 4, 2.2L VIN F, 2.4L I4 VIN T - PID List

Parameter Identification (PID)	PID Value Range	PID Value at Hot Idle	PID Value at 30 mph	Data List Type
A/C Relay Command	On / Off	Off	Off	Misfire
Airflow Calculated	0-512 g/sec	Varies	Varies	Misfire
Cycles of Misfire Data	0-99	Varies	Varies	Misfire
ECT Sensor (°F)	-40 to 419°F	185-239	185-239	Misfire
Engine Load	0-100%	11-25	Varies	Misfire
Engine Run Time	Hr: Min: Sec	00:00:00	00:00:00	Misfire
Engine Speed	0-16384	Varies	---	Misfire
Injector Pulsewidth	0-985 ms	3-4	Varies	Misfire
MAP Sensor (kPa)	11-104	33	Varies	Misfire
Misfire current Cyl 1-4	0-255	0	0	Misfire
Misfire History Cyl 1-4	0-255	0	0	Misfire
Power Enrichment	Active / Inactive	Inactive	Inactive	Misfire
Spark Advance (°)	-64° to 64°	18	Varies	Misfire
Throttle Angle (%)	0-100%	0	Varies	Misfire
Vehicle Speed	0-155 mph	0	30	Misfire

1996-2005 Sunfire 2.2L VIN 4, 2.2L VIN F, 2.4L I4 VIN T - PID List

Parameter Identification (PID)	PID Value Range	PID Value at Hot Idle	PID Value at 30 mph	Data List Type
A/C Relay Command	On / Off	Off	Off	Engine Data 3
Air Fuel Ratio	0.0-25.5:1	14.6:1	14.6:1	Engine Data 3
Cruise Control Active	Yes / No	No	No	Engine Data 3
C/Control/TCC Switch	Applied/Not	Not Applied	Not Applied	Engine Data 3
Desired Idle Speed	0-3187	900	---	Engine Data 3
ECT Sensor (°F)	-40 to 419°F	185-239	185-239	Engine Data 3
Engine Load	0-100%	11-25	Varies	Engine Data 3
Engine Oil Life	0-100%	Varies	Varies	Engine Data 3
Engine Oil Pressure	Okay / Low	Okay	Okay	Engine Data 3
Engine Run Time	Hr: Min: Sec	00:00:00	00:00:00	Engine Data 3
Engine Speed	0-9999 rpm	Varies	Varies	Engine Data 3
FC Relay 1 Command	On / Off	Off	Off	Engine Data 3
IAT Sensor (°F)	-40 to 304°F	50-176	50-176	Engine Data 3
Ignition 1 Signal	0-25.5v	14.1	14.1	Engine Data 3
Injector 1-4 Command	0-985 ms	3-4	Varies	Engine Data 3
Long Term Fuel Trim	-10% to 10%	0 (± 5%)	0 (± 5%)	Engine Data 3
Loop Status	Closed Loop / Open Loop	Closed Loop	Closed Loop	Engine Data 3
MAP Sensor (kPa)	10-110 kPa	33	Varies	Engine Data 3
MAP Sensor (V)	0.0-5.0v	1-2	Varies	Engine Data 3
Short Term Fuel Trim	-10% to 10%	0 (± 5%)	0 (± 5%)	Engine Data 3
Spark Advance (°)	-64° to 64°	18	Varies	Engine Data 3
TP Sensor (V)	0.0-5.0v	0.20-0.90	Varies	Engine Data 3
TR Range Switch	Several	Several	Several	Engine Data 3
Vehicle Speed	0-155 mph	0	30	Engine Data 3
VTD Fuel Disable	Active / Inactive	Inactive	Inactive	Engine Data 3
VTD Auto Learn	Active / Inactive	Inactive	Inactive	Engine Data 3

1996-2005 Sunfire 2.2L VIN 4, 2.2L VIN F, 2.4L I4 VIN T - PID List

Parameter Identification (PID)	PID Value Range	PID Value at Hot Idle	PID Value at 30 mph	Data List Type
BARO Sensor (kPa)	11-105	65-104	65-104	EVAP
Desired Idle Speed	0-3187	900	---	EVAP
ECT Sensor (°F)	-40 to 419°F	185-239	185-239	EVAP
Engine Load	0-100%	11-25	Varies	EVAP
Engine Run Time	Hr: Min: Sec	00:00:00	00:00:00	EVAP
Engine Speed	0-9999 rpm	Varies	Varies	EVAP
EVAP Purge Solenoid	0-100%	Varies	Varies	EVAP
EVAP Test Abort Reason	Abort/Not	Not Aborted	Not Aborted	EVAP
EVAP Test Results	Passed / Not Passed	Passed	Passed	EVAP
EVAP Test State	Several	Several	Several	EVAP
EVAP Vent Solenoid	Venting/Not	Venting	Venting	EVAP
Fuel Level Sensor	0-100%	Varies	Varies	EVAP
FTP Sensor (In. H2O)	-17.5 to 7.5"	Varies	Varies	EVAP
FTP Sensor (V)	0-5.0v	Varies	Varies	EVAP
Fuel Trim Cell	Cell #0-22	18-21	---	EVAP
IAT Sensor (°F)	-40 to 304°F	50-176	50-176	EVAP
Ignition 1 Signal	0-25.5v	14.1	14.1	EVAP

1996-2005 Sunfire 2.2L VIN 4, 2.2L VIN F, 2.4L I4 VIN T - PID List

Parameter Identification (PID)	PID Value Range	PID Value at Hot Idle	PID Value at 30 mph	Data List Type
Long Term Fuel Trim	-10% to 10%	0 (± 5%)	0 (± 5%)	EVAP
Loop Status	Closed Loop / Open Loop	Closed Loop	Closed Loop	EVAP
MAP Sensor (kPa)	11-105	33	Varies	EVAP
No. of Current DTC(s)	0-255	0	0	EVAP
O2S-11 (B1 S1)	0-1132 mv	10-1000	10-1000	EVAP
Short Term Fuel Trim	-10% to 10%	0 (± 5%)	0 (± 5%)	EVAP
Startup ECT Sensor	-40 to 419°F	Varies	Varies	EVAP
Throttle Angle (%)	0-100%	0	Varies	EVAP
Vehicle Speed	0-159	0	30	EVAP

1996-2005 Sunfire 2.2L VIN 4, 2.2L VIN F, 2.4L I4 VIN T - PID List

Parameter Identification (PID)	PID Value Range	PID Value at Hot Idle	PID Value at 30 mph	Data List Type
Airflow Calculated	0-512 g/sec	3-7	Varies	Fuel Trim
Air Fuel Ratio	0.0-25.5:1	14.6:1	14.6:1	Fuel Trim
BARO Sensor (kPa)	11-104 kPa	65-104	65-104	Fuel Trim
Current Gear	1-4	1	3	Fuel Trim
ECT Sensor (°F)	-40 to 419°F	185-239	185-239	Fuel Trim
Engine Load	0-100%	11-25	Varies	Fuel Trim
Engine Run Time	Hr: Min: Sec	00:00:00	00:00:00	Fuel Trim
Engine Speed	0-9999 rpm	Varies	Varies	Fuel Trim
EVAP Purge Solenoid	0-100%	Varies	Varies	Fuel Trim
EVAP Vent Solenoid	Venting/Not	Venting	Venting	Fuel Trim
Fuel Trim Cell	Cell #0-22	18-21	18-21	Fuel Trim
HO2S-12 (B1 S1)	0-1132 mv	10-1000	10-1000	Fuel Trim
IAC Position	0-255 counts	42	---	Fuel Trim
IAT Sensor (°F)	-40 to 304°F	50-176	50-176	Fuel Trim
Injector Pulsewidth	0-985 ms	3-4	Varies	Fuel Trim
Long Term Fuel Trim	-10% to 10%	0 (± 5%)	0 (± 5%)	Fuel Trim
Loop Status	Closed Loop / Open Loop	Closed Loop	Closed Loop	Fuel Trim
MAP Sensor (kPa)	11-105	25-48	Varies	Fuel Trim
Number of Current DTC(s)	0-255	0	0	Fuel Trim
O2S-11 (B1 S1)	0-1132 mv	10-1000	10-1000	Fuel Trim
Short Term Fuel Trim	-10% to 10%	0 (± 5%)	0 (± 5%)	Fuel Trim
Spark Advance (°)	-64° to 64°	18	Varies	Fuel Trim
Throttle Angle (%)	0-100%	0	Varies	Fuel Trim
TR Range Switch	Several	Several	Several	Fuel Trim
TWC Calculated Temperature	32-1409°F	Varies	Varies	Fuel Trim
Vehicle Speed	0-155 mph	0	30	Fuel Trim

L100, L200, LS, LS1, LS2, LW1, LW2, LW200 & VUE PIDS

2000-03 L100-200, LS-LS1, LW1-LW2, LW200, Vue 2.2L I4 VIN F

Parameter Identification (PID)	PID Value Range	PID Value at Hot Idle	PID Value at 30 mph	Data List Type
A/C HI Side Pressure	0-459 psi	199-399	199-399	Engine Data 1
A/C HI Side Pressure	0-5.0 volts	Varies	Varies	Engine Data 1
A/C Relay Circuit Open or Short to GND	Yes / No	No	No	Output Device
A/C Relay Circuit Short to Voltage	Yes / No	No	No	Output Device
A/C Relay Command	Yes / No	No	No	Engine Data 1, Idle Speed
A/C Request	Yes / No	No	No	Engine Data 1, Idle Speed
Airflow (g/sec)	0-512 g/sec	1-3	Varies	PCM Calculation
Air Fuel Ratio	0.1-25.5:1	14.7:1	14.7:1	Engine Data 1
AIR Relay Command	Yes / No	No	No	Engine Data 1, AIR System
AIR Relay Circuit Open or Short to GND	Yes / No	No	No	Output Device
AIR Relay Circuit Short to Voltage	Yes / No	No	No	Output Device
BARO Sensor (V)	0.0-5.0 volts	3.5-4.7	3.5-4.7	Engine Data 1
Calibrated Compression Output	00000110 or 000001001	00000110	000001001	Misfire
Calculated Converter Temperature	32-1409°F	Varies	Varies	PCM Calculation
CKP Active Counter (7X)	0-255 counts	Counts up	Counts up	Engine Data 1, Ignition
CMP Active Counter (CMP Pulses)	0-255 counts	Counts up	Counts up	Engine Data 1, Ignition
CMP Resync Counter (7X Signal is Lost)	0-255 counts	0	0	Engine Data 1, Ignition
Cruise Brake/Clutch Switch	Applied or Released	Released	Released	Engine Data 1, Idle Speed
Cruise Enabled	Yes / No	No	No	Engine Data 1, Idle Speed
Cruise Engaged	Yes / No	No	No	Engine Data 1, Idle Speed
Desired Idle Speed	0-3187	Varies	Varies	Idle Speed
ECT Sensor (°F)	-40 to 419°F	Varies	Varies	All Lists
Engine Load	0-100%	Varies	Varies	All Lists
Engine Oil Life Remaining	0-100%	Varies	Varies	Engine Data 1
Engine Run Time	Hr: Min: Sec	00:00:00	00:00:00	Engine Data 1
Engine Speed	0-9999 rpm	Varies	Varies	All Lists
EVAP Large Leak	Pass / Fail / Aborted / Not Ran	Pass	Pass	EVAP (auxiliary test)

2000-03 L100-200, LS-LS1, LW1-LW2, LW200, Vue 2.2L I4 VIN F

Parameter Identification (PID)	PID Value Range	PID Value at Hot Idle	PID Value at 30 mph	Data List Type
EVAP Purge Circuit Open or Short to GND	Yes / No	No	No	Output Device
EVAP Purge Circuit Short to Voltage	Yes / No	No	No	Output Device
EVAP Purge PWM	0-100%	Varies	Varies	Engine Data 1, EVAP
EVAP Vent Circuit Open or Short to GND	Yes / No	No	No	Output Device
EVAP Vent Circuit Short to Voltage	Yes / No	No	No	Output Device
EVAP Vent Solenoid	Yes / No	No	No	Engine Data 1
FC Relay 1 Circuit Open or Short to GND	Yes / No	No	No	Output Device
FC Relay 1 Circuit Short to Voltage	Yes / No	No	No	Output Device
FC Relay 1 Command	Yes / No	No	No	Engine Data 1
FC Relay 2 Circuit Open or Short to GND	Yes / No	No	No	Output Device
FC Relay 2 Circuit Short to Voltage	Yes / No	No	No	Output Device
FC Relay 2 Command	Yes / No	No	No	Engine Data 1
Fuel Cutoff Command	On / Off	Off	Off	Engine Data 2
Fuel Level Remaining	0-100%	Off	Off	Engine Data 1, EVAP, Fuel & Emissions
Fuel Pump Relay Command	On / Off	On	On	Engine Data 1
Fuel Tank Pressure	0-5.0 volts	Cap off: 2.5v	Varies	Engine Data 1, EVAP
Fuel Trim Cell Number	0-22 cells	Varies	Varies	Fuel & Emissions
Generator F-Terminal	0-100%	Varies	Varies	Engine Data 1
Generator T-Terminal	Enabled or Disabled	Enabled	Enabled	Engine Data 1
HO2S Transition Time Ratio	0-249 ms	Varies	Varies	PCM Calculation
HO2S-11 (B1 S1)	0-1100 mv	Varies	Varies	Engine Data 1, EVAP, Fuel & Emissions
HO2S-12 (B1 S2)	0-1100 mv	Varies	Varies	Engine Data 1, EVAP, Fuel & Emissions
HO2S-11 Heater Command	10-1065 mv	399-499 when cold	10-1065 mv	Fuel Trim
HO2S-12 Heater Command	10-1065 mv	425-460 when cold	500-800	Fuel Trim
IAC Position	0-255 counts	10-60	Varies	Idle Speed
IAT Sensor °F	-40 to 304°F	123	122	All Lists
Ignition Voltage	0.0-25.5v	14.1v	14.1v	All Lists

2000-03 L100-200, LS-LS1, LW1-LW2, LW200, Vue 2.2L I4 VIN F

Parameter Identification (PID)	PID Value Range	PID Value at Hot Idle	PID Value at 30 mph	Data List Type
Injector Pulsewidth	0-1000 ms	2-3	Varies	Engine Data 2
Knock Retard (°)	0-25.5 degrees	0	0	Ignition
Knock Sensor Active Counter	0-255 counts	0	0	Ignition
Knock Sensor Noise	0-5.0 volts	Varies	Varies	Ignition
Lean/Rich Average Time	0-249 ms	Varies	Varies	PCM Calculation
Lean/Rich Transitions	0-255 counts	Varies	Varies	PCM Calculation
Long Term Fuel Trim	-10 to 10%	0 (± 5%)	0 (± 5%)	EVAP, Fuel & Emissions, Ignition
Long Term Fuel Trim (%)	-10 to 10%	0 (± 5%)	0 (± 5%)	Engine Data 1
Loop Status	Closed or Open Loop	Closed	Closed	Engine Data 2
Low Oil Pressure	Yes / No	No	No	Engine Data 1
Low Spark Modifier	0-10	0	0	Ignition
MAP Sensor (kPa)	11-105 kPa	Varies	Varies	All Lists
MAP Sensor (V)	0-5.0 volts	Varies	Varies	Engine Data 1
Misfire Current Cylinder 1-4	0-255 counts	0	0	Misfire
Misfire History Cylinder 1-4	0-255 counts	0	0	Misfire
Module Driver Over-Temperature	Yes / No	No	No	Engine Data 2
Module Driver Over-Voltage	Yes / No	No	No	Engine Data 2
Park/Neutral Switch	P/N/RDL	P/N	RDL	Idle Speed
Rich/Lean Average Time	0-249 ms	Varies	Varies	PCM Calculation
Rich/Lean Transitions	0-255 counts	Varies	Varies	PCM Calculation
Short Term Fuel Trim (%)	-10 to 10%	0 (± 5%)	0 (± 5%)	EVAP, Fuel & Emissions, Ignition
Spark (Cylinder No. 2)	-90 to +90°	0	Varies	Ignition
Startup ECT Sensor	-40 to 304°F	Varies	Varies	Engine Data 2
Torque Requested	0-100%	Varies	Varies	Engine Data 1
Throttle Angle (%)	0-100%	0	6	All Lists
TP Sensor (V)	0.0-5.0 volts	0.35	0.67	Engine Data 1
Vehicle Speed	0-159 mph	0	30	All Lists

<u>L300, LS2, LW300 & VUE PID DATA - OBD II</u>

2000-05 L300, LS2, LW300, Vue 3.0L V6 VIN R (All) - PID Data List

Parameter Identification (PID)	PID Value Range	PID Value at Hot Idle	PID Value at 30 mph	Data List Type
A/C HI Side Pressure	0-459 psi	199-399	199-399	Engine Data 1-2
A/C HI Side Pressure	0-5.0 volts	Varies	Varies	Engine Data 1
A/C Relay Command	Yes / No	No	No	Engine Data 1-2 Idle Speed
APP at Full Position	Yes / No	No	No	Idle Speed
APP at Idle Position	Yes / No	Yes	No	Idle Speed
APP Sensor 1 (V)	0-5.0 volts	0.8-1.0	Varies	Engine Data 1
APP Sensor 2 (V)	0-5.0 volts	3.9-4.8	Varies	Engine Data 1
Air Fuel Ratio	0.1-25.5:1	14.7:1	14.7:1	Engine Data 1
Battery Voltage	0-25.5 volts	12-14v	12-14v	All Lists
Brake Lamp Stop Switch	Applied or Released	Released	Released	Engine Data 1, Idle Speed
CMP Active Counter (CMP Pulses)	0-255 counts	Counts up	Counts up	Engine Data 1, Ignition
Cruise Brake Switch	Applied or Released	Released	Released	Engine Data 1, Idle Speed
Cruise Resume/Accel	On / Off	Off	Off	Engine Data 1, Idle Speed
Cruise Set/Coast	On / Off	Off	Off	Engine Data 1, Idle Speed
Cruise Switch	On / Off	Off	Off	Engine Data 1, Idle Speed
Desired Idle Speed	0-3187	Varies	Varies	Idle Speed
ECT Sensor (°F)	-40 to 419°F	Varies	Varies	All Lists
EGR Solenoid PWM	0-100%	0	Varies	All Lists
EGR Sensor	0-5.0 volts	0	Varies	Engine Data 1, EGR
Engine Load	0-100%	Varies	Varies	All Lists
Engine Oil Life Remaining	0-100%	Varies	Varies	Engine Data 1
Engine Run Time	Hr: Min: Sec	00:00:00	00:00:00	Engine Data 1
Engine Speed	0-9999 rpm	Varies	Varies	All Lists
EVAP Purge Solenoid	On / Off	Off	On	Engine Data 1
EVAP Large Leak	Pass / Fail / Aborted / Not Ran	Pass	Pass	Engine Data 2, EVAP, Fuel & Emissions
EVAP Purge Circuit Open or Short to GND	Yes / No	No	No	Output Device
EVAP Purge Circuit Short to Voltage	Yes / No	No	No	Output Device
EVAP Purge PWM	0-100%	Varies	Varies	Engine Data 1, EVAP
EVAP Vent Solenoid	On / Off	Off	Off	Engine Data 2
FC Relay 1 Command	On / Off	Off	Off	Engine Data 2
FC Relay 2 Command	On / Off	Off	Off	Engine Data 2
Fuel Level Remaining	0-100%	Off	Off	Engine Data 1
Fuel Level Sensor	0-5.0 volts	Varies	Varies	Engine Data 1, Ignition, Fuel & Emissions

2000-05 L300, LS2, LW300, Vue 3.0L V6 VIN R (All) - PID Data List

Parameter Identification (PID)	PID Value Range	PID Value at Hot Idle	PID Value at 30 mph	Data List Type
Fuel Pump Relay Command	On / Off	On	On	Engine Data 2
Fuel Pump Pressure (kPa)	Inches of H2O	-17 to +7.5 Inches H2O	-17 to +7.5 Inches H2O	Engine Data 1, EVAP
Fuel Tank Pressure Sensor (V)	0-5.0 volts	Fuel Cap off: 2.35-2.85v	Varies	Engine Data 1, Fuel & Emissions
Fuel Trim Cell Number	0-22 cells	Varies	Varies	Fuel & Emissions
Generator F-Terminal	0-100%	Varies	Varies	Engine Data 1
Generator T-Terminal	Enabled / Disabled	Enabled	Enabled	Engine Data 1
HO2S-11 (B1 S1) Lambda Value	0-2	1	1	Engine Data 1, HO2S, Fuel & Emissions
HO2S-12 (B1 S2) Value	0-4998 mv	421-479 when cold	400-700 mv	Engine Data 1, HO2S, Fuel & Emissions
HO2S-21 (B2 S1) Lambda Value	0-2	1	30-800 mv	Engine Data 1, HO2S, Fuel & Emissions
HO2S-22 (B2 S2) Value	0-4998 mv	421-479 when cold	400-700 mv	Engine Data 1, HO2S, Fuel & Emissions
HO2S-11 Heater	On / Off	Off	Off	Fuel & Emissions
HO2S-12 Heater	On / Off	Off	Off	Fuel & Emissions
HO2S-21 Heater	On / Off	Off	Off	Fuel & Emissions
HO2S-22 Heater	On / Off	Off	Off	Fuel & Emissions
HO2S-11 Heater (Volts)	0-25.5 volts	Varies	Varies	Engine Data 1, Fuel & Emissions
HO2S-21 Heater (Volts)	0-25.5 volts	Varies	Varies	Engine Data 1, Fuel & Emissions
HO2S-12 Heater (Ohms)	0-1000 ohms	20-40	20-40	Fuel & Emissions
HO2S-22 Heater (Ohms)	0-1000 ohms	20-40	20-40	Fuel & Emissions
IAT Sensor °F	-40 to 304°F	123	122	All Lists
Ignition On	Yes / No	Yes	Yes	Engine Data 1
Injector Pulsewidth	0-1000 ms	1-3	Varies	Fuel & Emissions
Intake Plenum Switchover	On / Off	Off	Off	Engine Data 2
Knock Control Bank 1	Active or Inactive	Inactive	Inactive	Engine Data 2
Knock Control Bank 2	Active or Inactive	Inactive	Inactive	Engine Data 2
Knock Retard Cylinder 1-6 (°)	0-25.5 degrees	0	0	Ignition
Knock Sensor Bank 1	0-5.0 volts	Varies	Varies	Engine Data 1, Ignition
Knock Sensor Bank 2	0-5.0 volts	Varies	Varies	Engine Data 1, Ignition

2000-05 L300, LS2, LW300, Vue 3.0L V6 VIN R (All) - PID Data List

Parameter Identification (PID)	PID Value Range	PID Value at Hot Idle	PID Value at 30 mph	Data List Type
Long Term Fuel Trim Bank 1 Cruise/Accel	-10 to 10%	0 (± 5%)	0 (± 5%)	HO2S, Fuel & Emissions
Long Term Fuel Trim Bank 1 Cruise/Accel	-10 to 10%	0 (± 5%)	0 (± 5%)	HO2S, Fuel & Emissions
Long Term Fuel Trim Bank 1 Idle/Decel	-10 to 10%	0 (± 5%)	0 (± 5%)	HO2S, Fuel & Emissions
Long Term Fuel Trim Bank 1 Idle/Decel	-10 to 10%	0 (± 5%)	0 (± 5%)	HO2S, Fuel & Emissions
Loop Status - HO2S Bank 1 Sensor 1	Closed or Open Loop	Closed	Closed	Fuel & Emissions
Loop Status - HO2S Bank 1 Sensor 2	Closed or Open Loop	Closed	Closed	Fuel & Emissions
Loop Status - HO2S Bank 2 Sensor 1	Closed or Open Loop	Closed	Closed	Fuel & Emissions
Loop Status - HO2S Bank 2 Sensor 2	Closed or Open Loop	Closed	Closed	Fuel & Emissions
Low Oil Pressure	Yes / No	No	No	Engine Data 1
MAF Sensor (g/sec)	0-655 g/sec	Varies	Varies	All Lists
MAF Sensor (V)	0-5.0 volts	Varies	Varies	Engine Data 1
Main Relay Command	On / Off	On	On	Engine Data 2
MAP Sensor (kPa)	11-105 kPa	Varies	Varies	All Lists
MAP Sensor (V)	0-5.0 volts	Varies	Varies	Engine Data 1
Misfire Current Cylinder 1-6	0-255 counts	0	0	Misfire
Misfire History Cylinder 1-6	0-255 counts	0	0	Misfire
Park/Neutral Switch	P/N/RDL	P/N	RDL	Idle Speed
Short Term Fuel Trim Bank 1	-10 to 10%	0 (± 5%)	0 (± 5%)	HO2S, Fuel & Emissions
Short Term Fuel Trim Bank 2	-10 to 10%	0 (± 5%)	0 (± 5%)	HO2S, Fuel & Emissions
Spark (Cylinder No. 1)	-90 to +90º	0	Varies	Ignition
TAC Fuel Shutoff	Yes / No	No	No	PCM Control
TAC Learn Counter	0-9 counts	0-9	0-9	PCM Control
TAC Limp Home (Actuator)	Yes / No	No	No	PCM Control
TAC Limp Home (Pedal)	Yes / No	No	No	PCM Control
Torque Reduction Driver	0-100%	0	0	Engine Data 1
Torque Reduction Request	0-100%	0	0	Engine Data 1
TP Angle (%)	0-100%	Varies	Varies	All Lists
TP Sensor 1 (V)	0-5.0 volts	0.5-0.8	8.1-9.1	Engine Data 1, F & E, Idle Speed
TP Sensor 2 (V)	0-5.0 volts	5.0-0	4.1-4.5	Engine Data 1, F & E, Idle Speed
Vehicle Speed	0-159 mph	0	30	All Lists

RELAY PID DATA

2005 Relay 3.5L VIN L (ALL) – PID Data List

Scan Tool Parameter	Data List	Units Displayed	Typical Data Value
24X Crank Sensor	Ignition	RPM	0-6,400
A/C Relay Circuit Status	ODD	OK/Fault/Invalid State	OK
A/C Relay Circuit History	ODD	OK/Fault/Invalid State	OK
A/C Relay Command	Eng, Misfire, EGR, TAC	On/Off	Off
Air Fuel Ratio	Eng, Fuel Trim, HO2S	Ratio	14.2:1-14.7:1
APP Indicated Angle	Eng, EGR, EVAP, HO2S, Ignition, Misfire, TAC, Fuel Trim	Percent	0
APP Sensor 1	TAC	Volts	Varies
APP Sensor 1 Circuit	TAC	OK/Not OK	OK
APP Sensor 2	TAC	Volts	Varies
APP Sensor 2 Circuit	TAC	OK/Not OK	OK
APP Sensor 1	TAC	Percent	Varies
APP Sensor 2	TAC	Percent	Varies
APP Sensor 1 and 2	TAC	Agree/Disagree	Agree
BARO	Eng, EGR, EVAP, Fuel Trim, HO2S, Ignition	kPa	65-110 kPa
Catalytic Converter Protection Active	Fuel Trim, HO2S	Yes/No	No
CMP Sensor	Ignition	RPM	Varies
CMP Sensor Signal Present	Ignition, Misfire	Yes/No	Yes
Cruise Control Active	TAC	Yes/No	No
Cycles of Misfire Data	Misfire	Counts	0-99
Cyl. 1-6 Injector Circuit Status	ODD	OK/Stuck Low (open)/Stuck High	OK
Cylinder 1 and 4 IC Circuit Status	ODD	OK/Not OK	OK
Cylinder 2 and 5 IC Circuit Status	ODD	OK/Not OK	OK
Cylinder 3 and 6 IC Circuit Status	ODD	OK/Not OK	OK
Cylinder 1-6 Injector Circuit History	ODD	OK/Stuck Low (open)/Stuck High	OK
Decel Fuel Cut-off	Eng, EGR, Fuel Trim, HO2S	Active/Inactive	Inactive
Desired EGR Position	EGR, Misfire	Percent	0
Desired Idle Speed	Eng, EVAP, TAC	RPM	Varies
Driver Module 1 Status	ODD	Enabled/Off-High Volts/Off High Temp/Invalid State	Enabled
Driver Module 2 Status	ODD	Enabled/Off-High Volts/Off High Temp/Invalid State	Enabled
Driver Module 3 Status	ODD	Enabled/Off-High Volts/Off High Temp/Invalid State	Enabled
EC Ignition Relay Circuit Status	ODD	OK/Fault/Invalid State	OK
EC Relay Circuit History	ODD	OK/Fault/Invalid State	OK
ECT Sensor	Eng, EGR, EVAP, HO2S, Ignition, Misfire, Fuel Trim, TAC	°C/°F	Varies
EGR Flow Test Count	EGR	Counts	0-12
EGR Learned Minimum Position	EGR	Volts	0.14-1.0
EGR Position Sensor	Eng, EGR, Misfire	Percent	Varies
EGR Position Sensor	EGR	Volts	0
EGR Position Variance	EGR	Percent	0
EGR Solenoid Circuit History	ODD	OK/Open/Fault/Invalid State	OK
EGR Solenoid Circuit Status	ODD	OK/Fault/Invalid State	OK
EGR Solenoid Command	EGR	Percent	0

2005 Relay 3.5L VIN L (ALL) – PID Data List (CONT)

Scan Tool Parameter	Data List	Units Displayed	Typical Data Value
Engine Load	Eng, Misfire, Fuel Trim, EVAP, EGR, HO2S, Ignition, TAC	Percent	0
Engine Oil Pressure Switch	Eng	Low/OK	OK
Engine Run Time	Eng, EVAP, Misfire, Fuel Trim, EGR, HO2S, Ignition, TAC, ODD	Hr: Min: Sec	Varies
Engine Speed	Eng, EGR, EVAP, Ignition, Misfire, Fuel Trim, TAC	RPM	Varies
EVAP Fault History	EVAP	No Fault/Excess Vacuum/Purge Valve Leak/Small Leak/Weak Vacuum/No Test Result	No Fault
EVAP Purge Solenoid Circuit History	ODD	OK/Open/Short	OK
EVAP Purge Solenoid Circuit Status	ODD	OK/Open/Short	OK
EVAP Purge Solenoid Command	Eng, EVAP, Fuel Trim, HO2S	Percent	Varies
EVAP Test Abort Reason	EVAP	No Aborted/Lost Enable/Small Leak/Veh. Not At Rest	Varies
EVAP Test Result	EVAP	• No Result/Passed/Aborted OR • Fail-DTC P0455 OR • Fail-DTC P0442 OR • Fail-DTC P0446 OR • Fail-DTC P0496	Varies
EVAP Test State	EVAP	Waiting for Purge/Test Running/Test Completed	Varies
EVAP Vent Solenoid Circuit History	ODD	OK/Open/Short	OK
EVAP Vent Solenoid Circuit Status	ODD	OK/Open/Short	OK
EVAP Vent Solenoid Command	Eng, EVAP	Venting/Not Venting	Varies
FC Relay 1 Circuit History	ODD	OK/Fault/Invalid State	OK
FC Relay 1 Circuit Status	ODD	OK/Fault/Invalid State	OK
FC Relay 2 and 3 Circuit History	ODD	OK/Fault/Invalid State	OK
FC Relay 2 and 3 Circuit Status	ODD	OK/Fault/Invalid State	OK
Fuel Level Sensor	EVAP	Volts	Varies
Fuel Pump Relay Circuit History	ODD	OK/Fault/Invalid State	OK
Fuel Pump Relay Circuit Status	ODD	OK/Fault/Invalid State	OK
Fuel Pump Relay Command	Eng	On/Off	On
Fuel Tank Level Remaining	EVAP	Percent	0-104
Fuel Tank Pressure Sensor	Eng, EVAP	mm Hg/in H2O	Varies
Fuel Tank Pressure Sensor	EVAP	Volts	Varies
Fuel Trim Cell	EVAP, Eng, Fuel Trim	0-4	0
Fuel Trim Learn	Eng, EVAP, Fuel Trim	Enabled/Disabled	Varies
HO2S Bank 1 Sensor 1	EVAP, Eng, Fuel Trim, HO2S	millivolts	0-1000 and Varying
HO2S Bank 1 Sensor 1 Ready	Fuel Trim, HO2S	Yes/No	Yes
HO2S Bank 1 Sensor 2	Eng, Fuel Trim, HO2S	millivolts	0-1000 and Varying
HO2S Bank 2 Sensor 1 Ready	Fuel Trim, HO2S	Yes/No	Yes
HO2S Bank 2 Sensor 1	Eng, EVAP, Fuel Trim, HO2S	millivolts	0-1000 and Varying
HO2S Bank 2 Sensor 2	Eng, Fuel Trim, HO2S	millivolts	0-1000 and Varying
HO2S Htr. Bn. 1 Sen. 1	HO2S	On/Off	On

2005 Relay 3.5L VIN L (ALL) – PID Data List (CONT)

Scan Tool Parameter	Data List	Units Displayed	Typical Data Value
HO2S Htr. Bank 1 Sensor 1 Command	HO2S	On/Off	On
HO2S Htr. Bn. 1 Sen. 2	HO2S	On/Off	On
HO2S Htr. Bank 1 Sensor 2 Command	HO2S	On/Off	On
HO2S Heater Bn. 1 Sen. 1	HO2S	Amps	Varies
HO2S Heater Bn. 1 Sen. 2	HO2S	Amps	Varies
HO2S Heater Bn. 2 Sen. 1	HO2S	Amps	Varies
HO2S Heater Bn. 2 Sen. 2	HO2S	Amps	Varies
HO2S Htr. Bn. 1 Sen. 1 Circuit History	ODD	OK/Fault/Invalid State	OK
HO2S Htr. Bn. 1 Sen. 2 Circuit History	ODD	OK/Fault/Invalid State	OK
HO2S Htr. Bn. 2 Sen. 1	HO2S	On/Off	On
HO2S Htr. Bn. 2 Sen. 1 Circuit History	ODD	OK/Fault/Invalid State	OK
HO2S Htr. Bn. 2 Sen. 2	HO2S	On/Off	On
HO2S Htr. Bn. 2 Sen. 2 Circuit History	ODD	OK/Fault/Invalid State	OK
HO2S Htr. Bn. 1 Sen. 1 Circuit Status	ODD	OK/Fault/Invalid State	OK
HO2S Htr. Bn. 1 Sen. 2 Circuit Status	ODD	OK/Fault/Invalid State	OK
HO2S Htr. Bn. 2 Sen. 1 Circuit Status	ODD	OK/Fault/Invalid State	OK
HO2S Htr. Bn. 2 Sen. 2 Circuit Status	ODD	OK/Fault/Invalid State	OK
Hot Open Loop	Fuel Trim, HO2S	Active/Inactive	Inactive
IAT Sensor	Eng, EVAP, EGR, Fuel Trim, HO2S, Ignition, Misfire, TAC	°C/°F	Varies
Ignition 1 Signal	EVAP, Eng, EGR, Fuel Trim, HO2S, Ignition, Misfire, ODD, TAC	Volts	Varies
Ignition Off Time	EVAP, Eng	Minutes/Varies	Varies
Initial Brake Apply Signal	TAC	Released/Applied	Released
Injector PWM	Fuel Trim, Misfire	milliseconds	Varies
Knock Retard	EGR, Ignition	Degrees	0
Long Term FT Bank 1	Eng, EVAP, Fuel Trim, HO2S	Percent	-10% to +10%
Long Term FT Bank 2	Eng, EVAP, Fuel Trim, HO2S	Percent	-10% to +10%
Loop Status	Eng, Fuel Trim, EVAP, HO2S	Open/Closed	Closed
MAF Sensor	Eng, EGR, Misfire, EVAP, Fuel Trim, HO2S, Ignition, TAC	g/s	3-6 (Varies with altitude)
MAF Sensor	EVAP, Fuel Trim, HO2S, Misfire, TAC	Hz	1,200-3,000 (depends on altitude)
MAP Sensor	Eng, EGR, Fuel Trim, HO2S, Misfire, TAC	Volts	0.4-2.0 Volts
MAP Sensor	Eng, EVAP, Fuel Trim, HO2S, Ignition, Misfire, EGR, TAC	kPa	20-48 kPa (Varies with altitude)
MIL Circuit History	ODD	OK/Fault/Invalid State	OK
MIL Command	EVAP, Eng, EGR, Ignition	On/Off	Off
MIL Circuit Status	ODD	OK/Fault/Invalid State	OK
Mileage Since DTC Cleared	Eng	km/h/mph	Varies
Misfire Current Cyl. 1-6	Misfire	Counts	0
Misfire History Cyl. 1-6	Misfire	Counts	0
Moderate Brake Apply Signal	TAC	Applied/Released	Released
Number of DTCs	Eng EVAP, Fuel Trim, TAC	Number	0
PCM Reset	Eng	Yes/No	No

2005 Relay 3.5L VIN L (ALL) – PID Data List (CONT)

Scan Tool Parameter	Data List	Units Displayed	Typical Data Value
Power Enrichment	Eng, Fuel Trim, HO2S, Misfire	Active/Inactive	Inactive
Reduced Engine Power	TAC	Active/Inactive	Inactive
Short Term FT Bank 1	Eng, EVAP, Fuel Trim, HO2S,	Percent	-10 to +10
Short Term FT Bank 2	Eng, EVAP, Fuel Trim, HO2S,	Percent	-10 to +10
Spark	Eng, Misfire, Fuel Trim, HO2S, Ignition	Degrees	22 (varies)
Start Up ECT	EVAP, Fuel Trim, HO2S	°C/°F	Varies
Start Up IAT	EVAP, Fuel Trim, HO2S	°C/°F	Varies
Starter Relay Circuit Status	ODD	OK/Fault	OK
Starter Relay Circuit History	ODD	OK/Fault/Invalid State	OK
Stop Lamp Pedal Switch	Eng, TAC	Released/Applied	Released
TAC/PCM Communications Signal	TAC	OK/Fault	OK
TAC Stop Lamp Pedal Switch	Eng	Released/Applied	Released
TAC Stop Lamp Pedal Switch	TAC	Applied/Released	Released
TAC Vehicle Speed Signal	TAC	OK/Fault	OK
TCS Delivered Torque Circuit Status	ODD	OK/Fault	OK
Torque Request Signal	TAC	Percent	100%
Torque Delivered Signal	TAC	Percent	75%
Total Misfire Count	Misfire	1-255	0
TP Desired Angle	TAC	Percent	Varies
TP Indicated Angle	Eng, EGR, Fuel Trim, HO2S, Ignition, EVAP, Misfire, TAC	Percent	%
TP Sensor 1	TAC	Percent	Varies
TP Sensor 1	TAC	Volts	Varies
TP Sensors 1 and 2	TAC	Agree/Disagree	Agree

SC1, SC2, SL, SL1, SL2, SW1, SW2 PID DATA - OBD II

1996-2002 SC1, SC2, SL1, SL2, SW1, SW2 1.9L I4 MFI VIN 7, VIN 8

Parameter Identification (PID)	PID Value Range	PID Value at Hot Idle	PID Value at 30 mph	Data List Type
A/C Clutch	Yes / No	No	No	Engine Data 2
Brake Switch	On / Off	Off	Off	Engine Data 2
Braking Gear	1-2-3-4	P/N	Varies	Engine Data 2
Clutch Slip Speed	0-255	0	Varies	Engine Data 2
Commanded Gear	1-2-3-4	TCM	Varies	Engine Data 2
Commanded Gear	1-2-3-4	1st Gear	3rd Gear	Engine Data 2
Coolant Temp. Fail	Yes / No	Varies	Varies	Engine Data 2
Engine Speed	0-9999 rpm	856 (M/T)	2416 (M/T)	Engine Data 2
Engine Torque	0-255	Varies	Varies	Engine Data 2
Fuel Pump Relay	On / Off	On	On	Engine Data 2
Grade Gear	2-3-4	P/N	2-3	Engine Data 2
Ignition Voltage	0.0-25.5v	14.1	14.1	Engine Data 2
Line Pressure (kPa)	396-1530	Varies	Varies	Engine Data 2
Drive Mode	Performance / Normal	Normal	Normal	Engine Data 2
Oil Temperature (°F)	-40 to 304°F	Varies	Varies	Engine Data 2
Percent Grade	0-100%	0	0-100	Engine Data 2
Selector Switch	P-R-N-D-3-2	P/N	R-D-3-2	Engine Data 2
Shift Fail Control	0-255	Varies	Varies	Engine Data 2
TCC Delta RPM	0-9999 rpm	Varies	Varies	Engine Data 2
TCC Locked	On / Off	Off	Off	Engine Data 2
TCC Status	On / Off	Off	Off	Engine Data 2
Throttle Angle (%)	0-100%	0	6	Engine Data 2
Traction Allowed at Idle	Yes / No	Yes	---	Engine Data 2
Traction Gear	1-2-3	P/N	2-3	Engine Data 2
Traction Slip In Control	Yes / No	Varies	Varies	Engine Data 2
Traction Spark Is 0	Yes / No	Varies	Varies	Engine Data 2
Turbine Speed RPM	-4080-4079	0	Varies	Engine Data 2
Vehicle Speed	0-159 mph	0	30	Engine Data 2

1996-2002 SC1, SC2, SL1, SL2, SW1, SW2 1.9L I4 MFI VIN 7, VIN 8

Parameter Identification (PID)	PID Value Range	PID Value at Hot Idle	PID Value at 30 mph	Data List Type
Cylinder 1-2-3-4 Misfire	0-255	0	0	Idle Control
IAC Position with A/C	0-255	31 counts	Varies	Idle Control
IAC Position w/o A/C	0-255	21 counts	Varies	Idle Control

1996-2002 SC1, SC2, SL1, SL2, SW1, SW2 1.9L I4 MFI VIN 7, VIN 8

Parameter Identification (PID)	PID Value Range	PID Value at Hot Idle	PID Value at 30 mph	Data List Type
HO2S-12 Time To Activity	0-1132 mv	Varies	Varies	Sensors
HO2S-12 Average Airflow	0-512 g/sec	Varies	Varies	Sensors
KS Active	0-255	Varies	Varies	Sensors
KS Idle Noise	0-255	Varies	Varies	Sensors
CKP Counter (7X)	0-255	14: Normal	14: Normal	Sensors

1996-2002 SC1, SC2, SL1, SL2, SW1, SW2 1.9L I4 MFI VIN 7, VIN 8

Parameter Identification (PID)	PID Value Range	PID Value at Hot Idle	PID Value at 30 mph	Data List Type
Knock Retard (°)	0-128°	0	0	Spark
Knock Too Long	Yes / No	No	No	Spark
KS Active Counter	0-255	38-40	Varies	Spark
KS Idle Noise	0-255	Varies	Varies	Spark
Low/Mid/High Spark Advance	0-9999 rpm	Varies	Varies	Spark
CKP 7X Counter	0-255	14: Normal	14: Normal	Spark

1996-2002 SC1, SC2, SL1, SL2, SW1, SW2 1.9L I4 MFI VIN 7, VIN 8

Parameter Identification (PID)	PID Value Range	PID Value at Hot Idle	PID Value at 30 mph	Data List Type
Actual EGR Position	0-100%	0	30	EGR
EGR Closed Position	0-255	31 counts	---	EGR
EGR Decel Threshold	0-255	Varies	Varies	EGR
EGR Decel Trip	0-255	Varies	Varies	EGR
EGR Decel Value	0-255	Varies	Varies	EGR
EGR Desired Position	0-100%	0	0-100	EGR
EGR Duty Cycle	0-100%	0	0-100	EGR
EGR Feedback	0.0-5.0 volts	0.61	Varies	EGR
EGR Position Error	0-100%	0	0-100	EGR
Engine Load	0-100%	40	55	EGR

1996-2002 SC1, SC2, SL1, SL2, SW1, SW2 1.9L I4 MFI VIN 7, VIN 8

Parameter Identification (PID)	PID Value Range	PID Value at Hot Idle	PID Value at 30 mph	Data List Type
EVAP Canister Purge	0-100%	0-25	0-100	EVAP
EVAP Canister Purge Circuit Open	Yes / No	No	No	EVAP
EVAP Canister Purge Circuit Short	Yes / No	No	No	EVAP
EVAP Canister Purge Feedback (V)	High/Low	High	Low	EVAP
EVAP C/P Intermittent	Yes / No	No	No	EVAP
EVAP C/P Valve Leak	Pass/Fail	Pass	Pass	EVAP
EVAP C/P Vent Blockage	Pass/Fail	Pass	Pass	EVAP
EVAP DTC Pending	Yes / No	No	No	EVAP
EVAP Fuel Tank Pressure	0-5v	2.5: cap off	Varies	EVAP
EVAP Large Leak/Blockage	Pass/Fail	Pass	Pass	EVAP
EVAP Small Leak/Blockage	Pass/Fail	Pass	Pass	EVAP
EVAP Tank Pressure	Mm HG/S	Varies	Varies	EVAP
EVAP Tank Vacuum Slope	Mm HG/S	Varies	Varies	EVAP
Fuel Level I/P %	0-100%	Varies	Varies	EVAP

1996-2002 SC1, SC2, SL1, SL2, SW1, SW2 1.9L I4 MFI VIN 7, VIN 8

Parameter Identification (PID)	PID Value Range	PID Value at Hot Idle	PID Value at 30 mph	Data List Type
Air Fuel Ratio	0.1-25.5:1	14.2-14.7:1	14.2-14.7:1	Fuel Trim
Catalyst Samples	0-50 Counts	29	---	Fuel Trim
Catalyst Stage 1 Fail	Yes / No	No	No	Fuel Trim
Decel Rich Fault	Yes / No	No	No	Fuel Trim
Decel Fuel Mode	Yes / No	No	No	Fuel Trim
Fuel Cutoff P/N	Yes / No	---	---	Fuel Trim
Fuel Cutoff MPH	0-155 mph	---	---	Fuel Trim
Fuel Cutoff Reverse	Yes / No	---	---	Fuel Trim
Fuel Cutoff Drive	Yes / No	---	---	Fuel Trim
Fuel Learned	Yes / No	Yes	Yes	Fuel Trim
HO2S-12 Time To Activity	Min: Sec	55 seconds	---	Fuel Trim
HO2S R/L - L/R Ratio	0:1 - 15.93:1	79 ms	Varies	Fuel Trim
Injectors Turned Off	CYL 1-4	0		Fuel Trim
INJ Pulsewidth CYL 2	0-985 ms	2.40	2.92	Fuel Trim
L/ R & R/L Average	0-249 ms	101	---	Fuel Trim
L/R & R/L Transition	0-255	110	---	Fuel Trim
FT Accel Cell Learned	Yes / No	Yes	Yes	Fuel Trim
FT Cruise Cell Learn	Yes / No	Yes	Yes	Fuel Trim
FT Decel Cell Learn	Yes / No	Yes	Yes	Fuel Trim
FT Idle Cell Learned	Yes / No	Yes	Yes	Fuel Trim
Long Fuel Trim Cell Number	0-3	1	3	Fuel Trim
Long Fuel Trim Count Accel	0-255	---	---	Fuel Trim
Long Fuel Trim Count cruise	0-255	123	---	Fuel Trim
Long Fuel Trim Count Decel	0-255	128	---	Fuel Trim
Long Fuel Trim Count Idle	0-255	114	---	Fuel Trim
Long, Short Term Fuel Trim	-10 to 10%	0 (± 5%)	0 (± 5%)	Fuel Trim
Long Term Fuel Trim Average	-10 to 10%	0 (± 5%)	0 (± 5%)	Fuel Trim
Long Fuel Trim Purge Acceleration	0-255	Varies	Varies	Fuel Trim
Long Fuel Trim Purge cruise	0-255	---	---	Fuel Trim
Long Fuel Trim Purge Decel	0-255	---	---	Fuel Trim
Long Fuel Trim Purge Idle	0-255	---	---	Fuel Trim
MAP Sensor (kPa)	10-105 kPa	38	48	Fuel Trim
Short Term Fuel Trim Average	-10 to 10%	Varies	Varies	Fuel Trim
Spark Advance (°)	-45° to 90°	14	Varies	Fuel Trim
T/Control Active	Yes / No	No	No	Fuel Trim
TWC Average Deviation	0-5v	1v	---	Fuel Trim
Wide Open Throttle	Yes / No	No	No	Fuel Trim
WOT Lean Fault	Yes / No	No	No	Fuel Trim

1996-2002 SC1, SC2, SL1, SL2, SW1, SW2 1.9L I4 MFI VIN 7, VIN 8

Parameter Identification (PID)	PID Value Range	PID Value at Hot Idle	PID Value at 30 mph	Data List Type
Crank Learned	Yes / No	Yes	Yes	Misfire
Misfire current Cyl 1-4	0-255	0	0	Misfire
Misfire Enabled	Yes / No	Yes	Yes	Misfire
Misfire History Cyl 1-4	0-255	0	0	Misfire
Total Misfire Failures	0-255	0	0	Misfire

1996-2002 SC1, SC2, SL1, SL2, SW1, SW2 1.9L I4 MFI VIN 7, VIN 8

Parameter Identification (PID)	PID Value Range	PID Value at Hot Idle	PID Value at 30 mph	Data List Type
Actual EGR Position	0-100%	0	30	Cruise
Brake Switch	On / Off	Off	Off	Cruise
Clutch Disengaged (MT)	Yes / No	Yes	---	Cruise
Cruise Accel	Yes / No	No	---	Cruise
Cruise Clutch	On / Off	Off	---	Cruise
Cruise Delta Speed	0-155 mph	0 mph	---	Cruise
Cruise direction command	Yes / No	No	---	Cruise
Cruise Move Command	Yes / No	No	---	Cruise
Cruise Resume / Acceleration	On / Off	Off	---	Cruise
Cruise Set / Coast	On / Off	Off	---	Cruise
Cruise Set Speed	0-159 mph	0 mph	---	Cruise
Cruise Switch	On / Off	Off	---	Cruise
Cruise Switch Voltage	0-12v	0v	---	Cruise
ECT Sensor (°F)	-40 to 419°F	208	207	Cruise
Engine Speed	0-9999 rpm	Varies	Varies	Cruise
IAT Sensor °F	-40 to 304°F	123	122	Cruise
Ignition Voltage	0.0-25.5v	14.1	14.1	Cruise
Long Term Fuel Trim	-10 to 10%	Varies	Varies	Cruise
MAP Sensor (kPa)	10-110 kPa	38	48	Cruise
MPH From Set	0-155	0 mph	---	Cruise
P/N Switch	PN / R-D-L	PN	R-D-L	Cruise
Resume / Acceleration	Yes / No	No	---	Cruise
Release Cruise	Yes / No	No	---	Cruise
Set / Coast	Yes / No	No	---	Cruise
Short Term Fuel Trim	-10 to 10%	Varies	Varies	Cruise
TCC Allowed	Yes / No	Yes	---	Cruise
TP Angle (%)	0-100%	0	6	Cruise
TP Sensor (V)	0.0-5.0 volts	0.35	0.67	Cruise
Vehicle Speed	0-159 mph	0	30	Cruise

1996-2002 SC1, SC2, SL1, SL2, SW1, SW2 1.9L I4 MFI VIN 7, VIN 8

Parameter Identification (PID)	PID Value Range	PID Value at Hot Idle	PID Value at 30 mph	Data List Type
Manual Force Shift	Yes / No	Varies	---	Transmission
Normal Mode D/shift	-192 to 768	---	---	Transmission
Normal Mode Upshift	-192 to 768	N 1-2	---	Transmission
Overrun Shift	Yes / No	No	---	Transmission
Performance Mode DN/S	-192 to 768	---	---	Transmission
Performance Mode Upshift	-288 to 768	P 1-2 kPa	P 2-3 kPa	Transmission
Sequential Upshift	Yes / No	No	---	Transmission
Shift Cancelled	Yes / No	No	---	Transmission
Slip Monitor 1-2-3-4 Gear	0-600 kPa	Varies	---	Transmission
TCC Adapter	Varies	Varies	---	Transmission
TCC Off With Shift	Yes / No	No	---	Transmission
TCC On With Shift	Yes / No	No	---	Transmission

1996-2002 SC1, SC2, SL1, SL2, SW1, SW2 1.9L I4 MFI VIN 7, VIN 8

Parameter Identification (PID)	PID Value Range	PID Value at Hot Idle	PID Value at 30 mph	Data List Type
A/C Clutch	Yes / No	No	No	Engine Data 1
A/C Request	Yes / No	No	No	Engine Data 1
ABS (Allow) Traction	Yes / No	No	No	Engine Data 1
BARO Sensor (V)	0.0-5.0 volts	3.5-4.7	3.5-4.7	Engine Data 1
Brake Switch	On / Off	Off	Off	Engine Data 1
Current Gear	1-2-3-4	1	3	Engine Data 1
Desired Idle Speed	0-3187	875	---	Engine Data 1
ECT Sensor (°F)	-40 to 419°F	208	207	Engine Data 1
EGR Duty Cycle	0-100%	0	30	Engine Data 1
EGR Pintle Position	0-5v	0.14-1.0v	---	Engine Data 1
EGR Solenoid	On / Off	Off	On	Engine Data 1
Engine At Idle	Yes / No	Yes	No	Engine Data 1
Engine Run Time	Hr: Min: Sec	00:00:00	00:00:00	Engine Data 1
Engine Speed	0-9999 rpm	Varies	Varies	Engine Data 1
EVAP Purge Solenoid	On / Off	Off	On	Engine Data 1
EVO Feedback	0-12v	0 or 12	0 or 12	Engine Data 1
EVO Output PWM	0-100%	0	62	Engine Data 1
Fan Control Relay	On / Off	Off	Off	Engine Data 1
Fuel Pump Command	On / Off	On	On	Engine Data 1
Fuel Pump Feedback	0.0-25.5v	14.1	14.1	Engine Data 1
Hot Light	On / Off	On: >244°F	Off	Engine Data 1
HO2S-12 (B1 S2)	0-1132 mv	10-1000	10-1000	Engine Data 1
IAC Position	0-255	5-60	---	Engine Data 1
IAT Sensor °F	-40 to 304°F	123	122	Engine Data 1
Ignition Voltage	0.0-25.5v	14.1v	14.1v	Engine Data 1
Long Term Fuel Trim	0-255	58-198	58-198	Engine Data 1
Long Term Fuel Trim (%)	-10 to 10%	0 (± 5%)	0 (± 5%)	Engine Data 1
Loop Status	CL or OL	CL	CL	Engine Data 1
Low Coolant	Yes / No	No	No	Engine Data 1
Calculated Airflow	0-512 g/sec	3	Varies	Engine Data 1
MAP Sensor (kPa)	10-110 kPa	38	48	Engine Data 1
MAP Sensor (V)	0.0-5.0 volts	1.02v	0.97v	Engine Data 1
O2S-11 (B1 S1)	0-1132 mv	10-1000	10-1000	Engine Data 1
PRNDL Switch	P/N / R-D-L	P/N	R-D-L	Engine Data 1
RKE Feedback	High/Low	High: >3.25	Low: <2.7	Engine Data 1
Service (Non-E) Light	On / Off	Off	Off	Engine Data 1
SES Lamp Status	On / Off	Off	Off	Engine Data 1
Short Term Fuel Trim	0-255	58-198	58-198	Engine Data 1
Short Term Fuel Trim (%)	-10 to 10%	Varies	Varies	Engine Data 1
Speedo Feedback	High/Low	High	Low	Engine Data 1
Startup ECT Sensor	-40 to 304°F	Varies	Varies	Engine Data 1
TCC Engaged	On / Off	Off	Off	Engine Data 1
Throttle Angle (%)	0-100%	0	6	Engine Data 1
TP Sensor (V)	0.0-5.0 volts	0.35	0.67	Engine Data 1
Trunk Release (Allow)	Yes / No	Yes: 0 mph	No	Engine Data 1
Upshift Light	On / Off	Off	On: Shift	Engine Data 1
Vehicle Speed	0-159 mph	0	30	Engine Data 1

1996-2002 SC1, SC2, SL1, SL2, SW1, SW2 1.9L I4 MFI VIN 7, VIN 8

Parameter Identification (PID)	PID Value Range	PID Value at Hot Idle	PID Value at 30 mph	Data List Type
A/C Clutch	Yes / No	No	No	Engine Data 2
Brake Switch	On / Off	Off	Off	Engine Data 2
Braking Gear	1-2-3-4	P/N	Varies	Engine Data 2
Clutch Slip Speed	0-255	0	Varies	Engine Data 2
Commanded Gear	1-2-3-4	TCM	Varies	Engine Data 2
Commanded Gear	1-2-3-4	1st Gear	3rd Gear	Engine Data 2
Coolant Temp. Fail	Yes / No	Varies	Varies	Engine Data 2
Engine Speed	0-9999 rpm	856 (M/T)	2416 (M/T)	Engine Data 2
Engine Torque	0-255	Varies	Varies	Engine Data 2
Fuel Pump Relay	On / Off	On	On	Engine Data 2
Grade Gear	2-3-4	P/N	2-3	Engine Data 2
Ignition Voltage	0.0-25.5v	14.1	14.1	Engine Data 2
Line Pressure (kPa)	396-1530	Varies	Varies	Engine Data 2
Drive Mode	Performance / Normal	Normal	Normal	Engine Data 2
Oil Temperature (°F)	-40 to 304°F	Varies	Varies	Engine Data 2
Percent Grade	0-100%	0	0-100	Engine Data 2
Selector Switch	P-R-N-D-3-2	P/N	R-D-3-2	Engine Data 2
Shift Fail Control	0-255	Varies	Varies	Engine Data 2
TCC Delta RPM	0-9999 rpm	Varies	Varies	Engine Data 2
TCC Locked	On / Off	Off	Off	Engine Data 2
TCC Status	On / Off	Off	Off	Engine Data 2
Throttle Angle (%)	0-100%	0	6	Engine Data 2
Traction Allowed at Idle	Yes / No	Yes	---	Engine Data 2
Traction Gear	1-2-3	P/N	2-3	Engine Data 2
Traction Slip In Control	Yes / No	Varies	Varies	Engine Data 2
Traction Spark Is 0	Yes / No	Varies	Varies	Engine Data 2
Turbine Speed RPM	-4080-4079	0	Varies	Engine Data 2
Vehicle Speed	0-159 mph	0	30	Engine Data 2

1996-2002 SC1, SC2, SL1, SL2, SW1, SW2 1.9L I4 MFI VIN 7, VIN 8

Parameter Identification (PID)	PID Value Range	PID Value at Hot Idle	PID Value at 30 mph	Data List Type
Cylinder 1-2-3-4 Misfire	0-255	0	0	Idle Control
IAC Position with A/C	0-255	31 counts	Varies	Idle Control
IAC Position w/o A/C	0-255	21 counts	Varies	Idle Control

1996-2002 SC1, SC2, SL1, SL2, SW1, SW2 1.9L I4 MFI VIN 7, VIN 8

Parameter Identification (PID)	PID Value Range	PID Value at Hot Idle	PID Value at 30 mph	Data List Type
HO2S-12 Time To Activity	0-1132 mv	Varies	Varies	Sensors
HO2S-12 Average Airflow	0-512 g/sec	Varies	Varies	Sensors
KS Active	0-255	Varies	Varies	Sensors
KS Idle Noise	0-255	Varies	Varies	Sensors
CKP Counter (7X)	0-255	14: Normal	14: Normal	Sensors

1996-2002 SC1, SC2, SL1, SL2, SW1, SW2 1.9L I4 MFI VIN 7, VIN 8

Parameter Identification (PID)	PID Value Range	PID Value at Hot Idle	PID Value at 30 mph	Data List Type
Knock Retard (°)	0-128°	0	0	Spark
Knock Too Long	Yes / No	No	No	Spark
KS Active Counter	0-255	38-40	Varies	Spark
KS Idle Noise	0-255	Varies	Varies	Spark
Low/Mid/High Spark Advance	0-9999 rpm	Varies	Varies	Spark
CKP 7X Counter	0-255	14: Normal	14: Normal	Spark

1996-2002 SC1, SC2, SL1, SL2, SW1, SW2 1.9L I4 MFI VIN 7, VIN 8

Parameter Identification (PID)	PID Value Range	PID Value at Hot Idle	PID Value at 30 mph	Data List Type
Actual EGR Position	0-100%	0	30	EGR
EGR Closed Position	0-255	31 counts	---	EGR
EGR Decel Threshold	0-255	Varies	Varies	EGR
EGR Decel Trip	0-255	Varies	Varies	EGR
EGR Decel Value	0-255	Varies	Varies	EGR
EGR Desired Position	0-100%	0	0-100	EGR
EGR Duty Cycle	0-100%	0	0-100	EGR
EGR Feedback	0.0-5.0 volts	0.61	Varies	EGR
EGR Position Error	0-100%	0	0-100	EGR
Engine Load	0-100%	40	55	EGR

1996-2002 SC1, SC2, SL1, SL2, SW1, SW2 1.9L I4 MFI VIN 7, VIN 8

Parameter Identification (PID)	PID Value Range	PID Value at Hot Idle	PID Value at 30 mph	Data List Type
EVAP Canister Purge	0-100%	0-25	0-100	EVAP
EVAP Canister Purge Circuit Open	Yes / No	No	No	EVAP
EVAP Canister Purge Circuit Short	Yes / No	No	No	EVAP
EVAP Canister Purge Feedback (V)	High/Low	High	Low	EVAP
EVAP C/P Intermittent	Yes / No	No	No	EVAP
EVAP C/P Valve Leak	Pass/Fail	Pass	Pass	EVAP
EVAP C/P Vent Blockage	Pass/Fail	Pass	Pass	EVAP
EVAP DTC Pending	Yes / No	No	No	EVAP
EVAP Fuel Tank Pressure	0-5v	2.5: cap off	Varies	EVAP
EVAP Large Leak/Blockage	Pass/Fail	Pass	Pass	EVAP
EVAP Small Leak/Blockage	Pass/Fail	Pass	Pass	EVAP
EVAP Tank Pressure	Mm HG/S	Varies	Varies	EVAP
EVAP Tank Vacuum Slope	Mm HG/S	Varies	Varies	EVAP
Fuel Level I/P %	0-100%	Varies	Varies	EVAP

1996-2002 SC1, SC2, SL1, SL2, SW1, SW2 1.9L I4 MFI VIN 7, VIN 8

Parameter Identification (PID)	PID Value Range	PID Value at Hot Idle	PID Value at 30 mph	Data List Type
Air Fuel Ratio	0.1-25.5:1	14.2-14.7:1	14.2-14.7:1	Fuel Trim
Catalyst Samples	0-50 Counts	29	---	Fuel Trim
Catalyst Stage 1 Fail	Yes / No	No	No	Fuel Trim
Decel Rich Fault	Yes / No	No	No	Fuel Trim
Decel Fuel Mode	Yes / No	No	No	Fuel Trim
Fuel Cutoff P/N	Yes / No	---	---	Fuel Trim
Fuel Cutoff MPH	0-155 mph	---	---	Fuel Trim
Fuel Cutoff Reverse	Yes / No	---	---	Fuel Trim
Fuel Cutoff Drive	Yes / No	---	---	Fuel Trim
Fuel Learned	Yes / No	Yes	Yes	Fuel Trim
HO2S-12 Time To Activity	Min: Sec	55 seconds	---	Fuel Trim
HO2S R/L - L/R Ratio	0:1 - 15.93:1	79 ms	Varies	Fuel Trim
Injectors Turned Off	CYL 1-4	0		Fuel Trim
INJ Pulsewidth CYL 2	0-985 ms	2.40	2.92	Fuel Trim
L/ R & R/L Average	0-249 ms	101	---	Fuel Trim
L/R & R/L Transition	0-255	110	---	Fuel Trim
FT Accel Cell Learned	Yes / No	Yes	Yes	Fuel Trim
FT Cruise Cell Learn	Yes / No	Yes	Yes	Fuel Trim
FT Decel Cell Learn	Yes / No	Yes	Yes	Fuel Trim
FT Idle Cell Learned	Yes / No	Yes	Yes	Fuel Trim
Cell Number	0-3	1	3	Fuel Trim
Long Fuel Trim Count Accel	0-255	---	---	Fuel Trim
Long Fuel Trim Count cruise	0-255	123	---	Fuel Trim
Long Fuel Trim Count Decel	0-255	128	---	Fuel Trim
Long Fuel Trim Count Idle	0-255	114	---	Fuel Trim
Long, Short Term Fuel Trim	-10 to 10%	0 (± 5%)	0 (± 5%)	Fuel Trim
Long Term Fuel Trim Average	-10 to 10%	0 (± 5%)	0 (± 5%)	Fuel Trim
Long Fuel Trim Purge Accel	0-255	Varies	Varies	Fuel Trim
Long Fuel Trim Purge cruise	0-255	---	---	Fuel Trim
Long Fuel Trim Purge Decel	0-255	---	---	Fuel Trim
Long Fuel Trim Purge Idle	0-255	---	---	Fuel Trim
MAP Sensor (kPa)	10-105 kPa	38	48	Fuel Trim
Short Term Fuel Trim Average	-10 to 10%	Varies	Varies	Fuel Trim
Spark Advance (º)	-45º to 90º	14	Varies	Fuel Trim
T/Control Active	Yes / No	No	No	Fuel Trim
TWC Average Deviation	0-5v	1v	---	Fuel Trim
Wide Open Throttle	Yes / No	No	No	Fuel Trim
WOT Lean Fault	Yes / No	No	No	Fuel Trim

1996-2002 SC1, SC2, SL1, SL2, SW1, SW2 1.9L I4 MFI VIN 7, VIN 8

Parameter Identification (PID)	PID Value Range	PID Value at Hot Idle	PID Value at 30 mph	Data List Type
Crank Learned	Yes / No	Yes	Yes	Misfire
Misfire current Cyl 1-4	0-255	0	0	Misfire
Misfire Enabled	Yes / No	Yes	Yes	Misfire
Misfire History Cyl 1-4	0-255	0	0	Misfire
Total Misfire Failures	0-255	0	0	Misfire

1996-2002 SC1, SC2, SL1, SL2, SW1, SW2 1.9L I4 MFI VIN 7, VIN 8

Parameter Identification (PID)	PID Value Range	PID Value at Hot Idle	PID Value at 30 mph	Data List Type
Actual EGR Position	0-100%	0	30	Cruise
Brake Switch	On / Off	Off	Off	Cruise
Clutch Disengaged (MT)	Yes / No	Yes	---	Cruise
Cruise Accel	Yes / No	No	---	Cruise
Cruise Clutch	On / Off	Off	---	Cruise
Cruise Delta Speed	0-155 mph	0 mph	---	Cruise
Cruise direction command	Yes / No	No	---	Cruise
Cruise Move Command	Yes / No	No	---	Cruise
Cruise Resume / Acceleration	On / Off	Off	---	Cruise
Cruise Set / Coast	On / Off	Off	---	Cruise
Cruise Set Speed	0-159 mph	0 mph	---	Cruise
Cruise Switch	On / Off	Off	---	Cruise
Cruise Switch Voltage	0-12v	0v	---	Cruise
ECT Sensor (°F)	-40 to 419°F	208	207	Cruise
Engine Speed	0-9999 rpm	Varies	Varies	Cruise
IAT Sensor °F	-40 to 304°F	123	122	Cruise
Ignition Voltage	0.0-25.5v	14.1	14.1	Cruise
Long Term Fuel Trim	-10 to 10%	Varies	Varies	Cruise
MAP Sensor (kPa)	10-110 kPa	38	48	Cruise
MPH From Set	0-155	0 mph	---	Cruise
P/N Switch	PN / R-D-L	PN	R-D-L	Cruise
Resume / Acceleration	Yes / No	No	---	Cruise
Release Cruise	Yes / No	No	---	Cruise
Set / Coast	Yes / No	No	---	Cruise
Short Term Fuel Trim	-10 to 10%	Varies	Varies	Cruise
TCC Allowed	Yes / No	Yes	---	Cruise
TP Angle (%)	0-100%	0	6	Cruise
TP Sensor (V)	0.0-5.0 volts	0.35	0.67	Cruise
Vehicle Speed	0-159 mph	0	30	Cruise

1996-2002 SC1, SC2, SL1, SL2, SW1, SW2 1.9L I4 MFI VIN 7, VIN 8

Parameter Identification (PID)	PID Value Range	PID Value at Hot Idle	PID Value at 30 mph	Data List Type
Manual Force Shift	Yes / No	Varies	---	Transmission
Normal Mode D/shift	-192 to 768	---	---	Transmission
Normal Mode Upshift	-192 to 768	N 1-2	---	Transmission
Overrun Shift	Yes / No	No	---	Transmission
Performance Mode DN/S	-192 to 768	---	---	Transmission
Performance Mode Upshift	-288 to 768	P 1-2 kPa	P 2-3 kPa	Transmission
Sequential Upshift	Yes / No	No	---	Transmission
Shift Cancelled	Yes / No	No	---	Transmission
Slip Monitor 1-2-3-4 Gear	0-600 kPa	Varies	---	Transmission
TCC Adapter	Varies	Varies	---	Transmission
TCC Off With Shift	Yes / No	No	---	Transmission
TCC On With Shift	Yes / No	No	---	Transmission

Domestic Manual ISBN 1-4180-0606-8/Part No. 130606
Import Manual ISBN 1-4180-1537-7/Part No. 131537

Chilton has added so much to its labor guide manual that we've had to put it in two volumes! We've added hundreds of new labor operations—including maintenance services and electronic system diagnosis—to the *Chilton® 2006 Import* and *Domestic Labor Guide Manuals*. All labor times for 1981 through 2006 vehicles consider the real world environment in which technicians work: worn, rusted, or dirty components, being serviced with tools commonly used in the aftermarket. Chilton labor times are accepted by most insurance and extended warranty companies. Vehicle makes and models conform to current Automotive Aftermarket Industry Association standards.

Labor Guide Manual Benefits:

- parts terminology is more standardized across different OEMs to simplify reference
- a total of more than 2,500 pages of updated Chilton labor times appear in these two volumes
- make sure your students have this latest edition because our experts have updated hundreds of labor times for earlier models

Hardcover Manuals are 8 1/2" x 11", ©2006

Labor Guide CD-ROM Benefits:

- easy-to-use software to create and print professional-quality estimates and invoices
- three user-defined levels of labor rates correspond to different types of job scenarios, for "real-world" application
- functions as a database of aftermarket labor times for monitoring warranty and insurance claims
- software keeps track of customers and prior estimates for time-saving recall
- customizable application allows service writers to add labor operations and times, and parts companies to add labor times to existing parts ordering systems

CD ISBN 1-4180-0605-X/Part No. 130605
©2006

Previous Year Editions:
Chilton 2005 Labor Guide Manual, ISBN 1-4018-7412-6/Part No. 27412
Chilton 2005 Labor Guide CD-ROM, ISBN 1-4018-7818-0/Part No. 27818

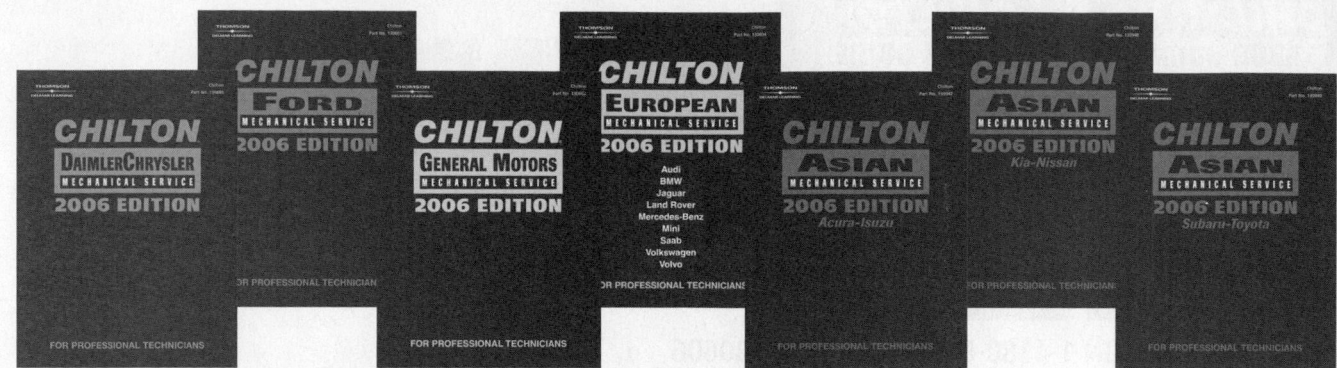

The *Chilton® 2006 Mechanical Service Manuals* provide updated coverage through 2005 models and e
many 2006 models, as made available from original equipment manufacturers (OEMs). Chilton is still y
reliable source for fast, accurate repairs and reassembly and it still provides the lowest-priced professic
repair manuals on the market! These manuals are organized by make, model, and system so informa'
gathering is easier. Now with even more illustrations and a streamlined index, it's no wonder m
automotive professionals turn to Chilton Professional Manuals for their mechanical service and repair informat

Mechanical Service Manual Benefits:

- access up-to-date service and repair information covering model years 2002-2006, all logically arranged by manufact
- follow clear, step-by-step procedures—from drive train to chassis—to yield fast, accurate results
- service more mechanical systems, including brakes, engines, suspensions, steering, and related components
- know what special tools are required for specific jobs, as Chilton editors describe and illustrate them to make
  repair work go more smoothly

2006 Editions

Chilton 2006 DaimlerChrysler Mechanical Service Manual—ISBN 1-4180-0600-9/Part No. 130600
Chilton 2006 Ford Mechanical Service Manual—ISBN 1-4180-0601-7/Part No. 130601
Chilton 2006 General Motors Mechanical Service Manual—ISBN 1-4180-0602-5/Part No. 130602
Chilton 2006 Asian Mechanical Service Manual—Volume I—ISBN 1-4180-0947-4/Part No. 130947
Chilton 2006 Asian Mechanical Service Manual—Volume II—ISBN 1-4180-0948-2/Part No. 130948
Chilton 2006 Asian Mechanical Service Manual—Volume III—ISBN 1-4180-0949-0/Part No. 130949
Chilton 2006 Asian Mechanical Service Manual—3 Volume Set—ISBN 1-4180-0603-3/Part No. 130603
Chilton 2006 European Mechanical Service Manual—ISBN 1-4180-0604-1/Part No. 130604

Asian Manuals Expected Release Date—Summer 2006
European Manuals Expected Release Date—Fall 2006
Manuals are 8 1/2" x 11", ©2006

2005 Editions

Chilton 2005 General Motors Mechanical Service Manual—ISBN 1-4018-7146-1/Part No. 27146
Chilton 2005 Chrysler Mechanical Service Manual—ISBN 1-4018-6718-9/Part No. 26718
Chilton 2005 Ford Mechanical Service Manual—ISBN 1-4018-6719-7/Part No. 26719
Chilton 2005 European Mechanical Service Manual—ISBN 1-4018-6720-0/Part No. 126720
Chilton 2005 Asian Mechanical Service Manual – Volume I—(Acura-Mazda) ISBN 1-4018-6716-2/Part No. 26716
Chilton 2005 Asian Mechanical Service Manual – Volume II—(Mitsubishi-Toyota)
       ISBN 1-4018-6717-0/Part No. 26717
Chilton 2005 Asian Mechanical Service Manual – 2 Volume Set—ISBN 1-4018-7180-1/Part No. 27180

Manuals are 8 1/2" x 11", ©2005

The *Chilton® Perennial Editions* contain repair and maintenance information for popular mechanical systems that may not be available elsewhere. They offer a wide range of repair information on cars, trucks, vans, and SUVs dating back to the early 1960s, and as current as 2002. Information for 1993 and later model years includes scheduled maintenance interval charts.

Benefits:

- covers the most common vehicle models found in the repair aftermarket today
- gain quick understanding of systems using exploded-view illustrations, diagrams, and charts
- simplify tough jobs with easy-to-follow removal and installation instructions for heater core and other components
- obtain complete coverage of repair procedures from drive train to chassis and associated components

Auto Repair Manual, 1998-2002, 1,426 pages
ISBN 0-8019-9362-8/Part No. 9362
Auto Repair Manual, 1993-1997, 2,064 pages
ISBN 0-8019-7919-6/Part No. 7919
Auto Repair Manual, 1988-1992, 1,284 pages
ISBN 0-8019-7906-4/Part No. 7906
Auto Repair Manual, 1980-1987, 1,344 pages
ISBN 0-8019-7670-7/Part No. 7670

Import Car Repair Manual, 1998-2002, 1,792 pps
ISBN 0-8019-9363-6/Part No. 9363
Import Car Repair Manual, 1993-1997, 2,080 pps
ISBN 0-8019-7920-X/Part No. 7920
Import Car Repair Manual, 1988-1992, 1,632 pages
ISBN 0-8019-7907-2/Part No. 7907
Import Car Repair Manual, 1980-1987, 1,488 pages
ISBN 0-8019-7672-3/Part No. 7672

Truck & Van Repair Manual, 1998-2002, 1,408 pages
ISBN 0-8019-9364-4/Part No. 9364
Truck & Van Repair Manual, 1993-1997, 2,096 pages
ISBN 0-8019-7921-8/Part No. 7921
Truck & Van Repair Manual, 1991-1995, 1,664 pages
ISBN 0-8019-7911-0/Part No. 7911
Truck & Van Repair Manual, 1986-1990, 1,536 pages
ISBN 0-8019-7902-1/Part No. 7902
Truck & Van Repair Manual, 1979-1986, 1,440 pages
ISBN 0-8019-7655-3/Part No. 7655

SUV Repair Manual, 1998-2002, 1,292 pages
ISBN 0-8019-9365-2/Part No. 9365

Hardcover manuals are 8 1/2" x 11".

Chilton Collector's Editions—*Reference Manuals for Vintage Vehicles*
Auto Repair Manual, 1964-1971, ISBN 0-8019-5974-8/Part No. 5974,
Truck & Van Repair Manual, 1961-1971, ISBN 0-8019-6198-X/Part No. 6198
Truck & Van Repair Manual, 1971-1978, ISBN 0-8019-7012-1/Part No. 7012